Vegetational Areas of North Central Texas

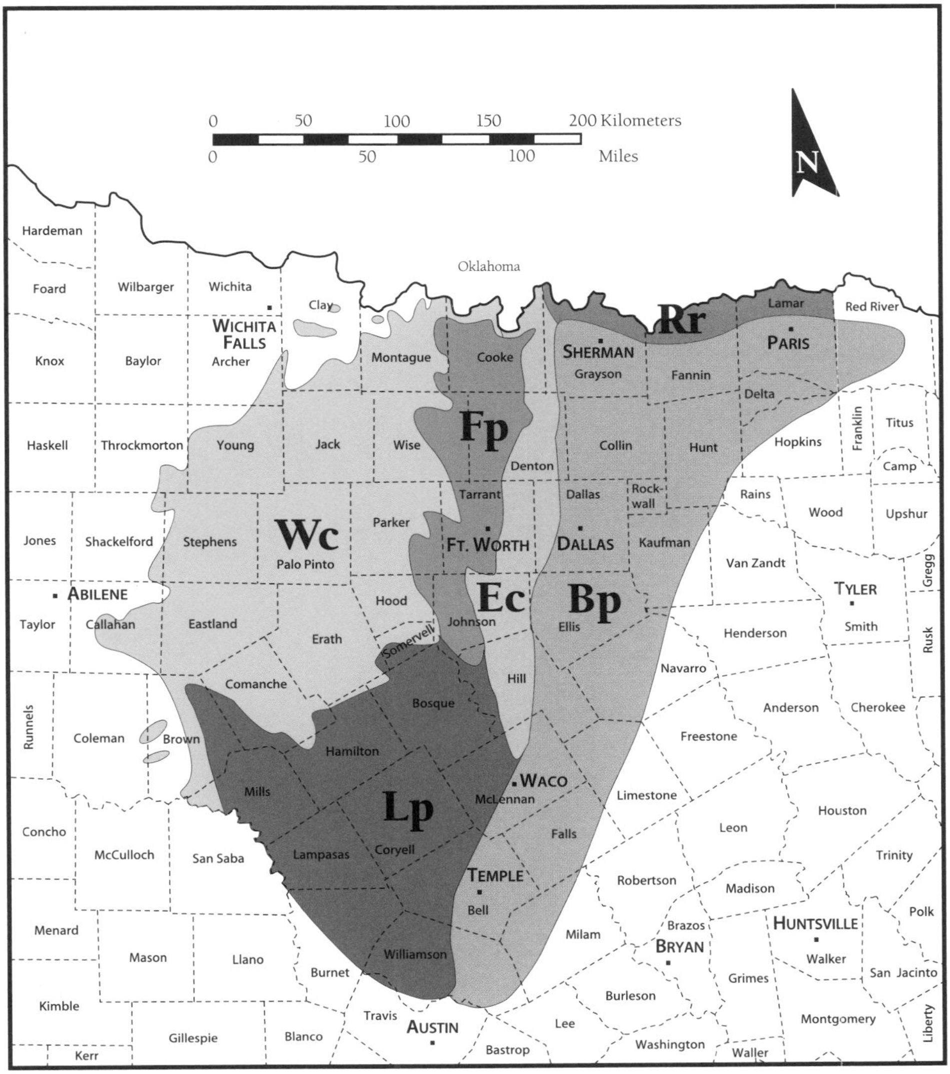

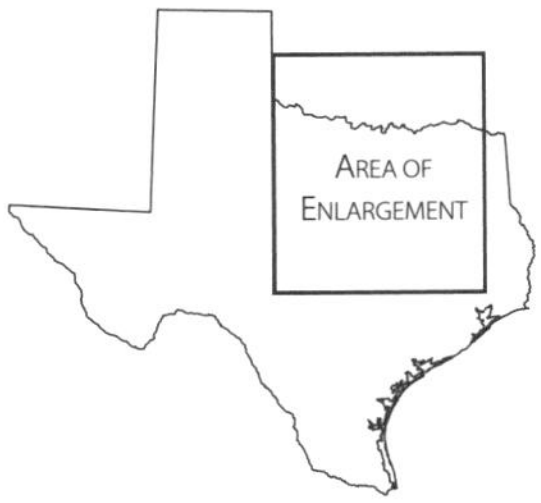

0 50 100 150 200 Kilometers
0 50 100 Miles
N

Oklahoma
Hardeman
Foard
Wilbarger
Wichita
Clay
WICHITA FALLS
Knox
Baylor
Archer
Montague
Cooke
Rr
Lamar
Red River
SHERMAN
Grayson
Fannin
PARIS
Delta
Haskell
Throckmorton
Young
Jack
Wise
Fp
Denton
Collin
Hunt
Hopkins
Franklin
Titus
Camp
Tarrant
Dallas
Rock-wall
Rains
Wood
Upshur
Jones
Shackelford
Stephens
Wc
Parker
FT. WORTH
DALLAS
Kaufman
Palo Pinto
Van Zandt
Gregg
ABILENE
Hood
Ec
Bp
TYLER
Taylor
Callahan
Eastland
Erath
Johnson
Ellis
Henderson
Smith
Rusk
Somervell
Navarro
Comanche
Hill
Bosque
Anderson
Cherokee
Runnels
Coleman
Brown
Freestone
Hamilton
Mills
WACO
Limestone
Houston
Lp
McLennan
Concho
Leon
Falls
McCulloch
San Saba
Lampasas
Coryell
Trinity
TEMPLE
Robertson
Madison
Bell
Polk
Menard
HUNTSVILLE
Brazos
Milam
BRYAN
Walker
Mason
Llano
Williamson
Grimes
San Jacinto
Burnet
Burleson
Kimble
Travis
Lee
Montgomery
AUSTIN
Liberty
Gillespie
Blanco
Washington
Bastrop
Waller
Kerr

AREA OF ENLARGEMENT

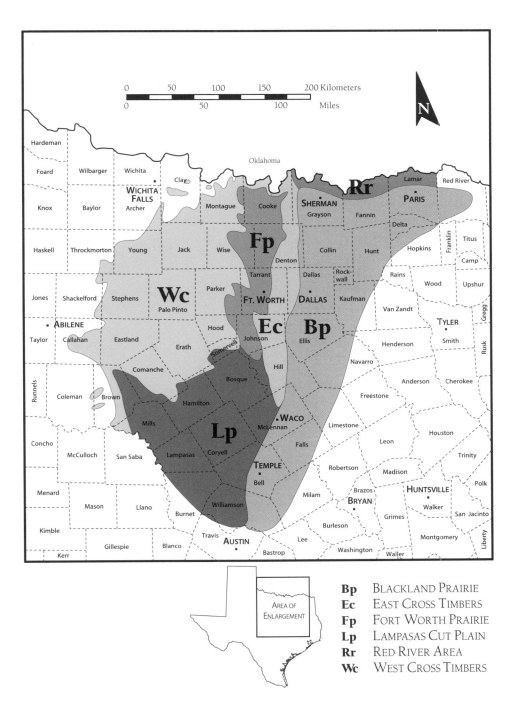

Bp BLACKLAND PRAIRIE
Ec EAST CROSS TIMBERS
Fp FORT WORTH PRAIRIE
Lp LAMPASAS CUT PLAIN
Rr RED RIVER AREA
Wc WEST CROSS TIMBERS

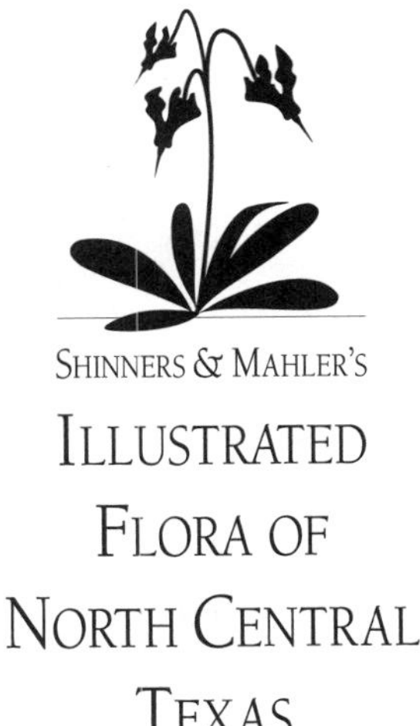

Shinners & Mahler's

Illustrated Flora of North Central Texas

Shinners & Mahler's

Illustrated Flora of North Central Texas

IS PUBLISHED WITH THE SUPPORT OF:

MAJOR BENEFACTORS:

New Dorothea L. Leonhardt Foundation (Andrea C. Harkins)

Bass Foundation

Robert J. O'Kennon

Ruth Andersson May

Mary G. Palko

Amon G. Carter Foundation

Margret M. Rimmer

Mike and Eva Sandlin

INSTITUTIONAL SUPPORT:

Austin College

Botanical Research Institute of Texas

Sid Richardson Career Development Fund of Austin College

OTHER CONTRIBUTORS:

Peg and Ben Keith

Friends of Hagerman National Wildlife Refuge

Summerlee Foundation

John D. and Beth A. Mitchell

Waldo E. Stewart

Dora Sylvester

Founders Garden Club of Dallas

Lorine Gibson

Sue Paschall John

Barbara G. Paschall

Sida, Botanical Miscellany 16
Botanical Research Institute of Texas, Inc.

Shinners & Mahler's

Illustrated Flora of North Central Texas

George M. Diggs, Jr. / Barney L. Lipscomb / Robert J. O'Kennon

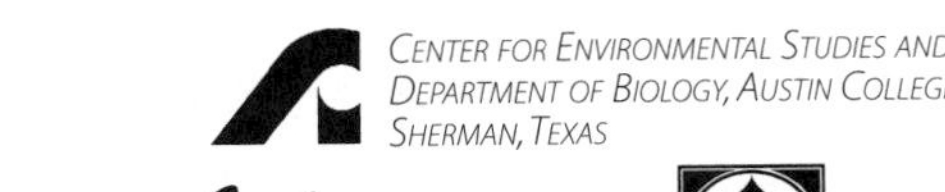

Botanical Research Institute of Texas (BRIT)
Fort Worth, Texas

1999

ISSN 0833-1475
ISBN 1-889878-01-4

SIDA, BOTANICAL MISCELLANY, NO. 16
FOUNDED BY WM. F. MAHLER AND BARNEY L. LIPSCOMB, 1987
EDITOR: BARNEY L. LIPSCOMB
DESIGN CONSULTANT: LINNY HEAGY
BOTANICAL RESEARCH INSTITUTE OF TEXAS
509 PECAN STREET
FORT WORTH, TEXAS 76102-4060 USA

COVER DESIGN/ILLUSTRATION: LINNY HEAGY
DATABASE MANAGEMENT AND ILLUSTRATIONS COORDINATION: SAMUEL BURKETT
PUBLICATION ASSISTANT: AMBERLY ZIJEWSKI

DISTRIBUTION OF COPIES BY:

BOTANICAL RESEARCH INSTITUTE OF TEXAS
509 PECAN STREET
FORT WORTH, TEXAS 76102-4060 USA
TELEPHONE: 817/ 332-4441
FAX: 817/ 332-4112
E-MAIL: sida@brit.org

Shinners & Mahler's
ILLUSTRATED FLORA OF NORTH CENTRAL TEXAS

IS THE FIRST PUBLICATION OF THE

ILLUSTRATED TEXAS FLORAS PROJECT

A COLLABORATIVE PROJECT OF THE

AUSTIN COLLEGE CENTER FOR ENVIRONMENTAL STUDIES

AND THE

BOTANICAL RESEARCH INSTITUTE OF TEXAS

TO PRODUCE ILLUSTRATED FLORISTIC TREATMENTS

DESIGNED TO BE USEFUL TO BOTH BOTANICAL SPECIALISTS

AND A MORE GENERAL AUDIENCE.

"I CAN SIT ON THE PORCH BEFORE MY DOOR AND SEE MILES OF THE MOST BEAUTIFUL PRAIRIE INTERWOVEN WITH GROVES OF TIMBER, SURPASSING, IN MY IDEA, THE BEAUTIES OF THE SEA. THINK OF SEEING A TRACT OF LAND ON A SLIGHT INCLINE COVERED WITH FLOWERS AND RICH MEADOW GRASS FOR 12 TO 20 MILES. . . ."

—JOHN BROOKE, EARLY SETTLER IN GRAYSON CO., TEXAS, 1849

To

Lloyd H. Shinners

and

William F. Mahler

for their contributions to

Texas botany

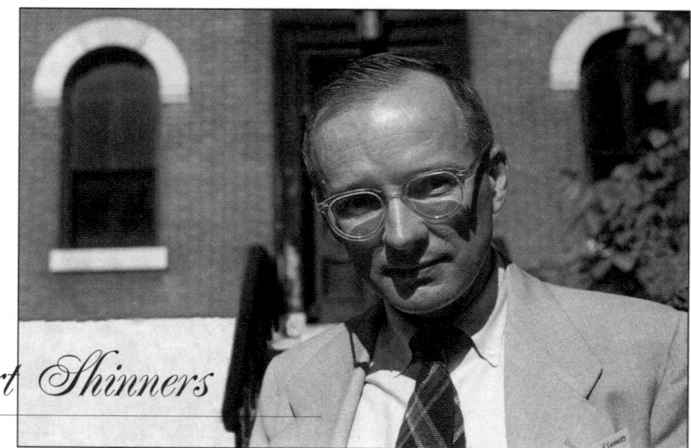

Lloyd Herbert Shinners

LLOYD HERBERT SHINNERS (1918–1971) was born in Bluesky (population 16), near Waterhole in the Peace River country of northwestern Alberta, Canada, on September 22, 1918. He was the child of homesteaders who had come from Wisconsin apparently under the National Policy [of Building Up Canada]. At the age of five, his family returned to Wisconsin where he attended public schools in Milwaukee and graduated from Lincoln High School as valedictorian of his class. He attended the University of Wisconsin-Milwaukee and later transferred to the University of Wisconsin-Madison from which he graduated Phi Beta Kappa in June, 1940. He also received his M.S. (1941) and Ph.D. (1943—Grasses of Wisconsin) degrees from the University of Wisconsin-Madison. Shinners came to Southern Methodist University in Dallas in 1945, became the Director of the Herbarium in 1949, and was on the faculty there until his death in 1971. Not only did he almost single-handedly develop the herbarium which today forms the core of the Botanical Research Institute of Texas (BRIT) collection, but he also created one of the best botanical libraries in the United States, did extensive field work, and published a total of 276 articles and a 514-page flora (Flook 1973). Under his supervision the SMU herbarium grew from ca. 20,000 to 340,000 specimens. His contributions to botanical nomenclature are also particularly impressive, totaling 558 new scientific names and combinations (Flook 1973). Among his most lasting achievements are the *Spring Flora of the Dallas-Fort Worth Area Texas* (Shinners 1958a) and the journal, *Sida, Contributions to Botany*, which he founded in 1962 (Mahler 1973b). His Spring Flora was the first completed, original, technical book on Texas plants prepared by a resident of the state. It was extensively used by high schools, colleges, and universities as a textbook for classes, and is still in use today. Shinners was also one of the organizers in 1953 of the Southwestern Association of Naturalists and was the first editor of its journal, *Southwestern Naturalist*. He was a tireless worker and an individual of varied intellectual pursuits ranging from poetry to linguistics, music, and a proficiency in seven languages. He once wrote "I sometimes feel too that all my passionate desire to be a scientist, compose music and to write philosophy at one and the same time are in some measure owing to the land I live in." His love of America was reflected in his gift to the Fondren Library at SMU of many books on American history. To quote Rowell (1972), he was "...a 'scholar' in the truest sense of the word." For synopses of Shinners' life see Correll (1971), Mahler (1971b), and Rowell (1972); for a guide to his botanical contributions see Flook (1973). Details given here about Shinners' life are from Correll (1971), Mahler (1971), Rowell (1972), and particularly from an extensive unpublished biographical manuscript by Ruth Ginsburg (1998), a BRIT archivist who has organized all of Shinners' correspondence and other papers.

William Fred "Bill" Mahler

WILLIAM F. "BILL" MAHLER grew up in Iowa Park, Texas, where he was born August 30, 1930. Upon graduation from W.F. George High School in 1947, he enrolled at Hardin College in Wichita Falls, Texas. After three years he enlisted in the U.S. Army instead of enrolling his last year in college and served from September 1950 to September 1953. After basic and advanced training in Headquarters Co., 8th Inf. Reg., 4th Inf. Div., he volunteered for airborne and ranger training. He served with the 14th Ranger Infantry Company (Airborne) at Fort Benning, Georgia, and Fort Carson, Colorado, until they were deactivated in 1951 (Black 1989; Taylor n.d.). In the meantime, the 4th Division had been sent to Friedberg, Germany. He returned to his old company and spent nearly two years in Germany. In 1954, he returned to school and received his B.S. degree in 1955 in Agriculture from Midwestern State University (previously Hardin College) with a major in Soil and Plant Science and a minor in Animal Husbandry. Mahler and Lorene Lindesmith, from Addington, Oklahoma, met in his home town and were married in 1955.

In 1958 he went to Oklahoma State University (OSU) in Stillwater to pursue graduate work. Mahler received his M.S. degree in Botany/Plant Taxonomy from OSU in 1960, working under U.T. Waterfall. For the next six years he served as an assistant professor at Hardin-Simmons University (HSU) in Abilene, Texas, teaching botany and establishing the HSU herbarium. Subsequently he continued his graduate studies by attending the University of Tennessee at Knoxville where he received the Ph.D. in Botany/Plant Taxonomy in 1968. Upon graduation he joined the faculty of Southern Methodist University in Dallas, became editor and publisher of *Sida, Contributions to Botany* in 1971 following the death of L.H. Shinners, and assumed leadership of the SMU herbarium in 1973. Mahler was publisher of *Sida, Botanical Miscellany* after he and Barney Lipscomb founded the journal in 1987. Under his guidance, the SMU herbarium grew by 72,000 specimens, eventually reaching about 400,000.

Mahler published *Shinners' Manual of the North Central Texas Flora* (1984, 1988), well known for its clarity and ease of use. The manual, that included the summer and fall flora for North Central Texas, was an expanded version of Shinners' (1958) *Spring Flora of the Dallas-Fort Worth Area Texas*. For his work, Mahler received the Donovan Stewart Correll Memorial Award for scientific writing on the native flora of Texas in 1991 from the Native Plant Society of Texas. Other notable publications included the *Keys to the Plants of Black Gap Wildlife Management Area, Brewster County, Texas* (1971), *Flora of Taylor County, Texas* (1973) and *The Mosses of Texas* (1980). Mahler's specialties include Fabaceae, *Baccharis* (Asteraceae), mosses, floristics, pollen morphology, and the study of endangered plant species. In 1988, Mahler was the first recipient of the Harold Beaty Award for his work with endangered plant species in Texas from the Texas Organization of Endangered Species. The Native Plant Society of Texas again honored Mahler in 1995 with the Charles Leonard Weddle Memorial Award in recognition of a lifetime of service and devotion to Texas native plants.

In 1987 SMU put its herbarium on permanent loan to a newly created organization, The Botanical Research Institute of Texas (BRIT). Mahler received early retirement from SMU (Associate Professor Emeritus) and served as the first Director of BRIT (1987–1992). Along with Andrea McFadden and long-time associate Barney Lipscomb, they were instrumental in its establishment as a free-standing research institution. Currently, Mahler is BRIT Director Emeritus and he and his wife are retired and living in Iowa Park, Texas.

Shinners & Mahler's

Illustrated Flora of North Central Texas

TABLE OF CONTENTS/ PAGE

ABSTRACT/RESUMEN 1
OVERVIEW OF THE BOOK 2
GEOGRAPHIC AREA COVERED 3
INFORMATION HELPFUL IN USING THE FLORA 7
PLANTS TREATED 7
ARRANGEMENT OF TAXA AND GENERAL METHODS 8
DESCRIPTIONS 8
KEYS 9
SOURCES OF INFORMATION 10
NOMENCLATURE 10
GEOGRAPHIC DISTRIBUTIONS 11
INFORMATION ON TOXIC/POISONOUS PLANTS 12
INFORMATION ON ENDANGERED AND THREATENED TAXA 12
INFORMATION ON ILLUSTRATIONS AND PHOTOGRAPHS 13
INFORMATION ON THE GLOSSARY 13
INFORMATION ON REFERENCES AND LITERATURE CITED 13
ABBREVIATIONS AND SYMBOLS 14
SUMMARY DATA ON THE FLORA 16
NEW COMBINATIONS MADE IN THIS BOOK 16
ACKNOWLEDGMENTS 17
AUTHORS' NOTE 19
INTRODUCTION TO NORTH CENTRAL TEXAS 20
OVERVIEW 20
GENERAL GEOLOGY OF NORTH CENTRAL TEXAS 20
SOILS OF NORTH CENTRAL TEXAS 23
CLIMATE OF NORTH CENTRAL TEXAS 28
THE BLACKLAND PRAIRIE 32
OCCURRENCE OF THE BLACKLAND PRAIRIE 32
PRESETTLEMENT AND EARLY SETTLEMENT CONDITIONS ON THE BLACKLAND PRAIRIE 34
GEOLOGY OF THE BLACKLAND PRAIRIE 38
VEGETATION OF THE BLACKLAND PRAIRIE 39
CROSS TIMBERS AND PRAIRIES 42
OCCURRENCE OF THE CROSS TIMBERS AND PRAIRIES 42
PRESETTLEMENT AND EARLY SETTLEMENT CONDITIONS IN THE CROSS TIMBERS 43
GEOLOGY OF THE EAST CROSS TIMBERS 45
GEOLOGY OF THE WEST CROSS TIMBERS 45
VEGETATION OF THE CROSS TIMBERS 46
PRESETTLEMENT AND EARLY SETTLEMENT CONDITIONS IN THE GRAND PRAIRIE (FORT WORTH PRAIRIE AND LAMPASAS CUT PLAIN) 48

CROSS TIMBERS AND PRAIRIES *(CONTINUED)*
GEOLOGY OF THE GRAND PRAIRIE 50
VEGETATION OF THE FORT WORTH PRAIRIE 52
VEGETATION OF THE LAMPASAS CUT PLAIN 53
RED RIVER AREA (AREA ADJACENT TO THE RED RIVER) 54
ORIGIN OF THE NORTH CENTRAL TEXAS FLORA 56
CONSERVATION IN NORTH CENTRAL TEXAS 62
A SKETCH OF THE HISTORY OF BOTANY IN TEXAS WITH EMPHASIS ON NORTH CENTRAL TEXAS 63
COLOR PHOTOGRAPHS 77
GENERAL KEYS 109
TAXONOMIC TREATMENTS 173
LYCOPODIOPHYTA (CLUBMOSSES) 173
EQUISETOPHYTA (HORSETAILS) 176
POLYPODIOPHYTA (FERNS) 178
PINOPHYTA (CONIFERS) 201
GNETOPHYTA (JOINT-FIRS AND RELATIVES) 207
MAGNOLIOPHYTA (FLOWERING PLANTS) 208
CLASS DICOTYLEDONAE (DICOTS) 210
CLASS MONOCOTYLEDONAE (MONOCOTS) 1077
APPENDICES
APPENDIX ONE (PHYLOGENY/CLASSIFICATION OF FAMILIES) 1353
APPENDIX TWO (GRASS PHYLOGENY/CLASSIFICATION) 1357
APPENDIX THREE (LIST OF TEXAS ENDEMIC SPECIES OCCURRING IN NORTH CENTRAL TEXAS) 1358
APPENDIX FOUR (ILLUSTRATION SOURCES) 1360
APPENDIX FIVE (LIST OF SELECTED BOTANICALLY RELATED INTERNET ADDRESSES) 1367
APPENDIX SIX (TAXONOMY, CLASSIFICATION, AND THE DEBATE ABOUT CLADISTICS) 1372
APPENDIX SEVEN (CHANGES IN THE SCIENTIFIC NAMES OF PLANTS) 1381
APPENDIX EIGHT (COLLECTING HERBARIUM SPECIMENS) 1382
APPENDIX NINE (LIST OF CONSERVATION ORGANIZATIONS) 1389
APPENDIX TEN (LARVAL HOST PLANTS OF LEPIDOPTERA OF NORTH CENTRAL TEXAS) 1394
APPENDIX ELEVEN (BOOKS FOR THE STUDY OF TEXAS NATIVE PLANTS) 1404
APPENDIX TWELVE (LIST OF NATIVE PLANTS SUGGESTED FOR USE AS ORNAMENTALS) 1409
APPENDIX THIRTEEN (LIST OF SOURCES FOR NATIVE PLANTS) 1415
APPENDIX FOURTEEN (STATE BOTANICAL SYMBOLS FOR TEXAS AND OKLAHOMA) 1418
APPENDIX FIFTEEN (SPECIAL RECOGNITION—BENNY J. SIMPSON) 1419
GLOSSARY/ILLUSTRATED GLOSSARY 1421
LITERATURE CITED 1457
INDEX 1525
COLOPHON 1623
INFORMATION ON AUTHORS AND PARTICIPATING INSTITUTIONS 1624

©Linny '98
(Dodecatheon meadia)

Abstract

Shinners & Mahler's Illustrated Flora of North Central Texas treats all native and naturalized vascular plant species known to occur in North Central Texas. The flora includes 2,223 species, about 46 percent of the species known for Texas, and 2,376 taxa (species, subspecies, and varieties). An introduction to the vegetation, geology, soils, climate, and presettlement and early settlement conditions is included as well as appendices on topics such as phylogeny and endemic species. The taxonomic treatments include family and generic synopses, keys and descriptions, derivations of scientific names, notes on toxic/poisonous and useful plants, and references to supporting literature. Line drawing illustrations are provided for all species with color photographs for 174. Three new combinations, *Gutierrezia amoena* (Shinners) Diggs, Lipscomb, & O'Kennon, *Manfreda virginica* (L.) Rose subsp. *lata* (Shinners) O'Kennon, Diggs, & Lipscomb, and *Mirabilis latifolia* (A. Gray) Diggs, Lipscomb, & O'Kennon, are made.

Resumen

Shinners & Mahler's Illustrated Flora of North Central Texas (La flora ilustrada del norte central de Texas, de Shinners & Mahler) trata todas las especies de plantas vasculares nativas y naturalizadas en la parte central del norte de Texas. La flora incluye 2,223 especies, aproximadamente el 46 por ciento de las especies conocidas en Texas y 2,376 taxa (especies, subespecies y variedades). Se incluye una introducción a la vegetación, geología, suelos, clima y condiciones de preasentamiento y asentamiento, así como apéndices sobre tópicos tales como filogenia y especies endémicas. Los tratamientos taxonomicos incluyen sinopsis de familias y géneros, claves y descripciones, etimología de los nombres científicos, notas sobre plantas tóxicas y útiles, y referencias bibliográficas. En cuanto a las ilustraciones se ofrecen dibujos de todas las especies y fotografías en color de 174. Se hacen tres nuevas combinaciones: *Gutierrezia amoena* (Shinners) Diggs, Lipscomb & O'Kennon, *Manfreda virginica* (L.) Rose subsp. *lata* (Shinners) O'Kennon, Diggs & Lipscomb y *Mirabilis latifolia* (A. Gray) Diggs, Lipscomb & O'Kennon.

Shinners & Mahler's

Illustrated Flora of North Central Texas

Overview of the Book

Shinners & Mahler's Illustrated Flora of North Central Texas is a floristic treatment of all native and naturalized vascular plant species known to occur in North Central Texas. The flora includes 2,223 species, about 46 percent of the species known for Texas, and a total of 2,376 taxa (species, subspecies, and varieties). Representatives of 168 families and 854 genera are included. It is a continuation in the tradition of Lloyd Shinners' *Spring Flora of the Dallas-Fort Worth Area Texas* (1958a) and William Mahler's *Shinners' Manual of the North Central Texas Flora* (1988). It differs from Mahler's (1988) work in the following ways: the total number of taxa is expanded from about 1,550 to 2,376; in addition to flowering plants, all ferns and similar plants (pteridophytes) and gymnosperms are included; it is fully illustrated; an introduction and appendices are provided; the taxonomic treatments have been expanded, including the addition of references to supporting literature; families, genera, and species are listed in alphabetical order within the major groups of plants; and a literature section with references pertinent to the plants of North Central Texas is provided.

A number of features have been incorporated to make the book more useful to non-specialists. Line drawings are provided for all species, making it the first fully illustrated flora for any region of Texas or the adjacent states. Color photographs are provided for 174 taxa. An introduction, covering general aspects of the vegetation, geology, soils, climate, and presettlement and early settlement conditions, has been included to provide background and context concerning North Central Texas. Further, the taxonomic treatments include brief synopses about each family and genus, derivations of generic names and specific epithets, characters helpful in family recognition in the field, notes on useful and toxic plants (ethnobotanical information), and references to supporting literature. Finally, appendices are provided on phylogeny (evolutionary relationships) at the family level, grass phylogeny, endemics, illustration sources, botanically related internet addresses, cladistics (a current controversy/approach in taxonomy), changes in scientific names, collecting herbarium specimens, conservation organizations, lepidopteran (butterfly and moth) host plant information, books for the study of native plants, suggested native ornamentals, sources for native plants, and state botanical symbols. When possible and practical, we have attempted to conform to the suggestions in Schmid's (1997) article on suggestions to make floras more user friendly.

GEOGRAPHIC AREA COVERED

North Central Texas is an area of roughly 40,000 square miles (103,700 square kilometers) or nearly the size of Kentucky. This 50 county region stretches from the Red River border with Oklahoma on the north, south nearly to Austin, east to Paris, and west nearly to Wichita Falls and Abilene. Vegetational areas included are the Blackland Prairie, the Grand Prairie, the East and West cross timbers, and the Red River Area (Fig. 1).

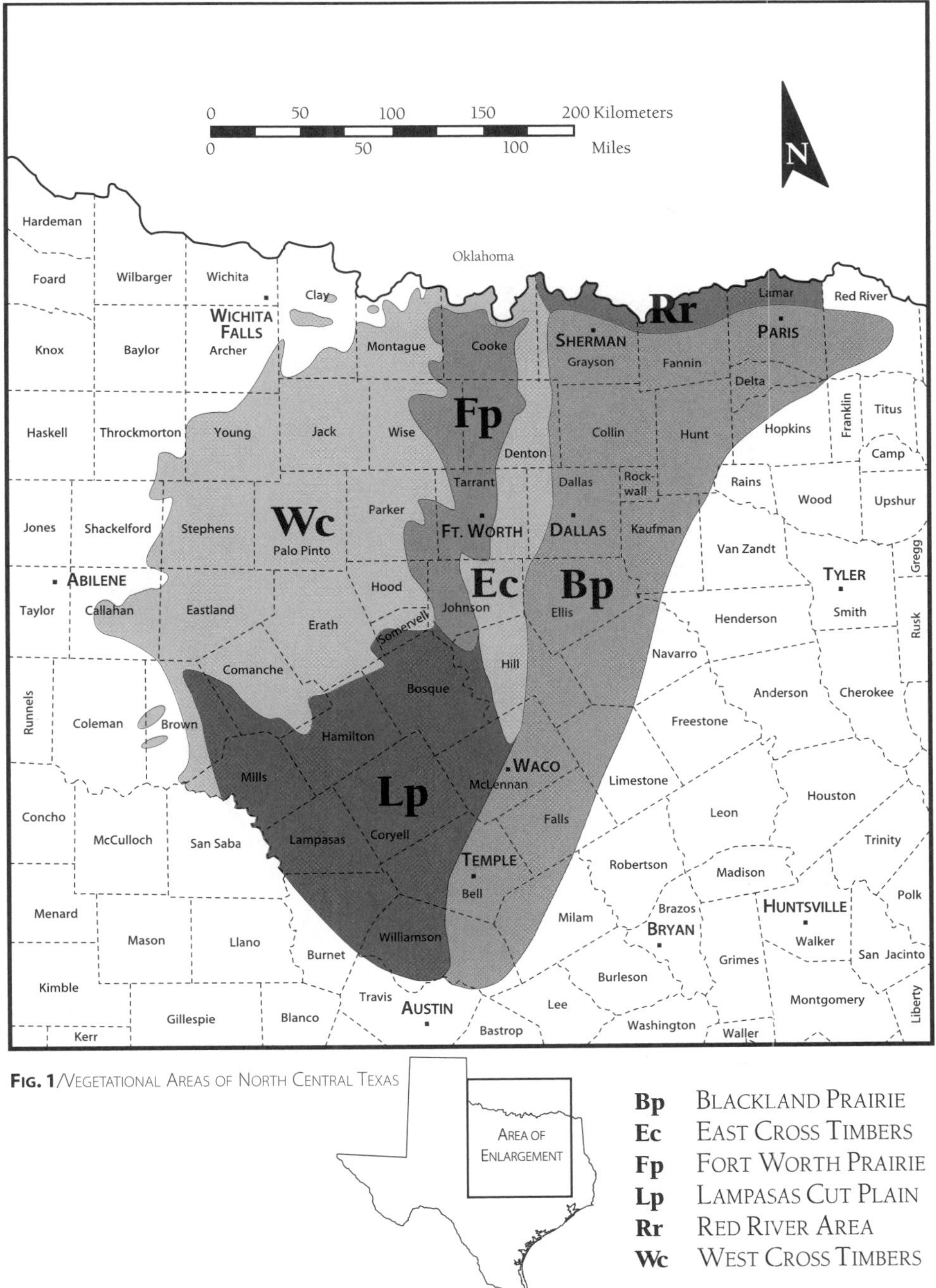

FIG. 1/VEGETATIONAL AREAS OF NORTH CENTRAL TEXAS

How does one define the limits of an area like North Central Texas for a floristic work such as this? On one level, the region can be defined geologically as encompassing all the Texas cross timbers and prairies occurring on soils derived from outcropping Cretaceous rocks. In a different manner, precipitation levels can be used (area of northern Texas with an average precipitation of 24 to 46 inches per year). In still another, the region is basically a broad ecotone between eastern deciduous forest and western grassland. However, for the purpose of this work, North Central Texas corresponds roughly with vegetational areas 4 (Blackland Prairie) and 5 (Cross Timbers and Prairies) of Correll and Johnston (1970) and Hatch et al. (1990) (Fig. 2) and is essentially the same as that treated by Mahler (1988) (Fig. 3). An alphabetical list of the counties wholly or partially included can be found in Fig. 3.

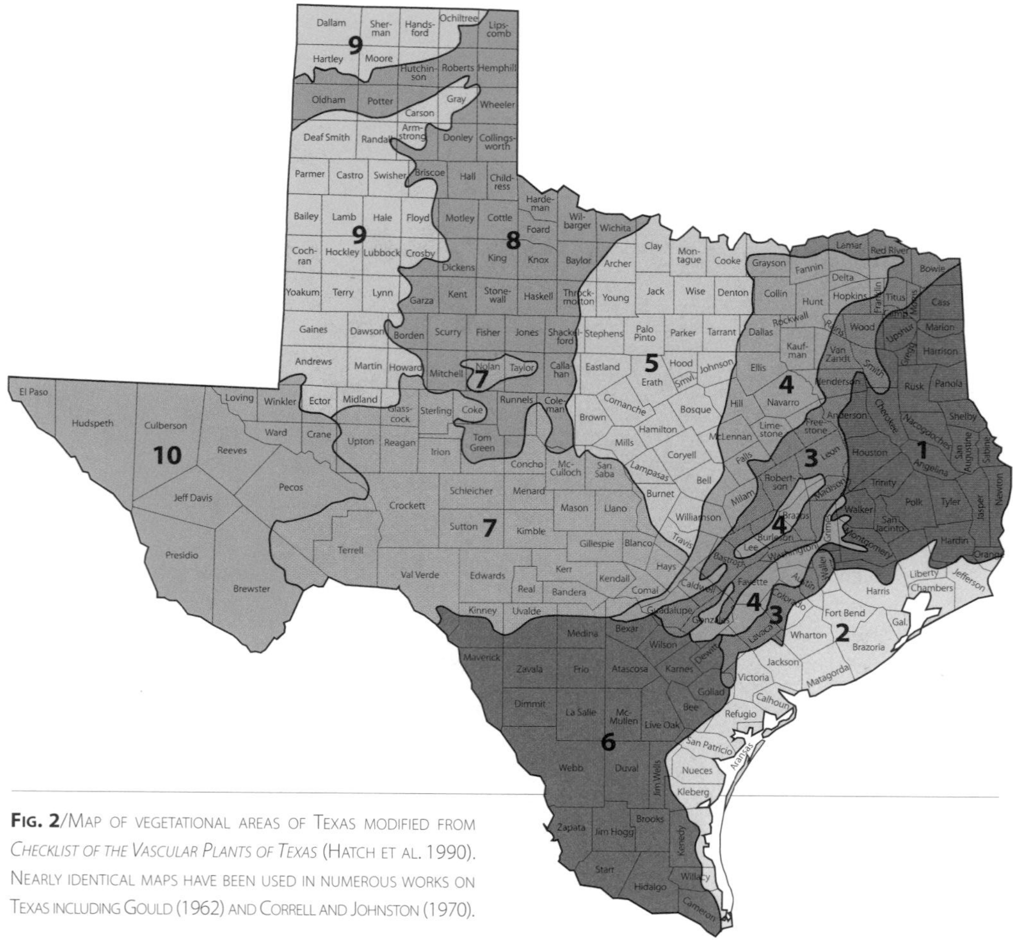

FIG. 2/MAP OF VEGETATIONAL AREAS OF TEXAS MODIFIED FROM *CHECKLIST OF THE VASCULAR PLANTS OF TEXAS* (HATCH ET AL. 1990). NEARLY IDENTICAL MAPS HAVE BEEN USED IN NUMEROUS WORKS ON TEXAS INCLUDING GOULD (1962) AND CORRELL AND JOHNSTON (1970).

1 PINEYWOODS
2 GULF PRAIRIES AND MARSHES
3 POST OAK SAVANNAH
4 BLACKLAND PRAIRIES
5 CROSS TIMBERS AND PRAIRIES
6 SOUTH TEXAS PLAINS
7 EDWARDS PLATEAU
8 ROLLING PLAINS
9 HIGH PLAINS
10 TRANS-PECOS, MOUNTAINS AND BASINS

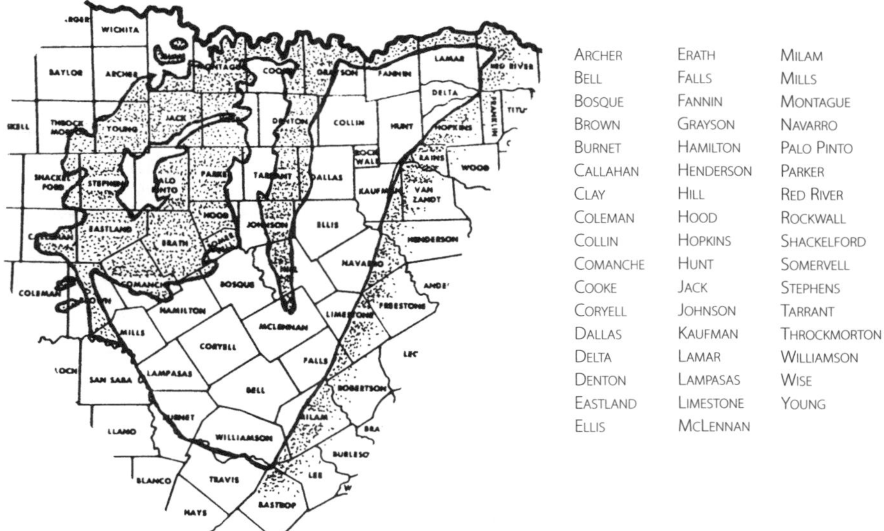

VEGETATIONAL AREAS WITH COUNTIES

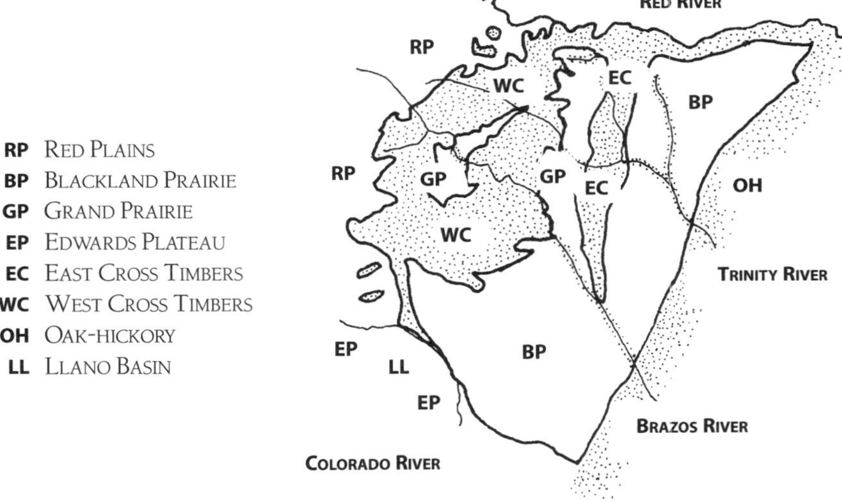

VEGETATIONAL AREAS WITH MAJOR RIVERS

FIG. 3/VEGETATIONAL AREAS OF NORTH CENTRAL TEXAS MODIFIED FROM *SHINNERS' MANUAL OF THE NORTH CENTRAL TEXAS FLORA* (MAHLER 1988) INCLUDING A LIST OF COUNTIES WHOLLY OR PARTIALLY COVERED. THE GEOGRAPHIC AREA COVERED BY THE PRESENT WORK IS ESSENTIALLY THE SAME (EXCEPT THE RED RIVER AREA HAS BEEN ADDED); THE LIST OF COUNTIES TREATED BY MAHLER IS THEREFORE ALSO ACCURATE FOR THIS WORK.

The area includes the Blackland Prairie (excluding the San Antonio and Fayette prairies), the Grand Prairie—here divided into the Fort Worth Prairie and the Lampasas Cut Plain, and the East and West cross timbers (discrete belts of forest surrounded by prairie). Extensions of vegetational area 3 (Post Oak Savannah) and even components of vegetational area 1 (Pineywoods) also enter the area from the east along the major rivers (Fig. 4). In particular, the number of species treated was increased significantly in comparison with Mahler's 1988 work by the inclusion of the Red River Area (Fig. 1). This is a vegetationally different, narrow strip of land near the Red River with sandy soils supporting numerous plants more typically found farther east. Similar range extensions exist farther south where eastern Texas plants extend west along the Trinity and Brazos rivers. Analogous situations also exist in the western and southwestern parts of North Central Texas where western Texas plants extend east to the West Cross Timbers and plants typical of the central Texas Edwards Plateau extend north well into the area, particularly in

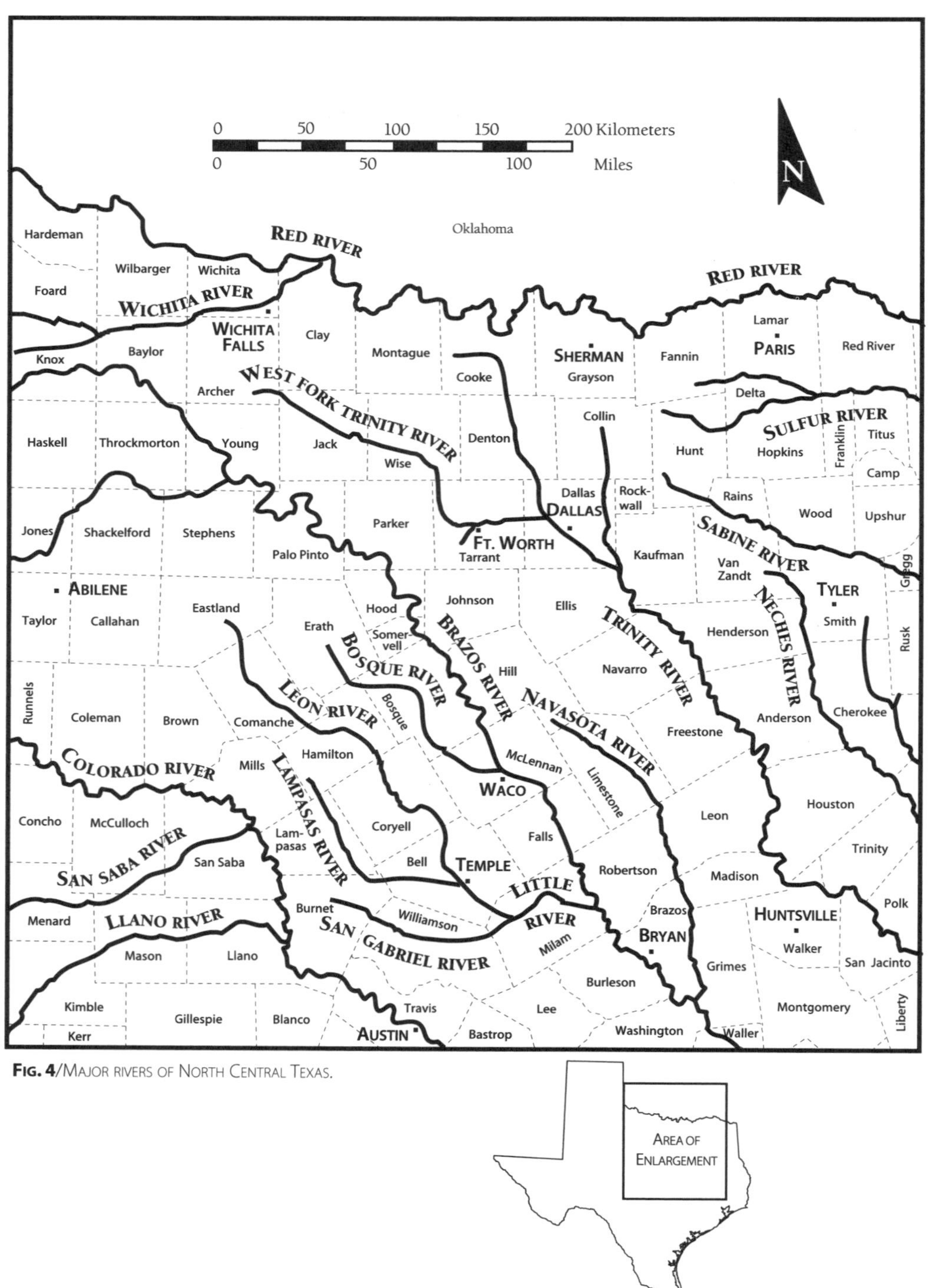

FIG. 4/MAJOR RIVERS OF NORTH CENTRAL TEXAS.

the rocky habitats associated with the Lampasas Cut Plain and the Palo Pinto Country. In virtually any geographic region, the microclimates and migration corridors provided by major rivers or specialized geology allow the occurrence of species not otherwise typical of that particular region. However, while these species may be neither abundant nor widespread, they are important biogeographically. Given their occurrence at the margins of their ranges, they may serve as ecological indicator species, possibly providing information in the future on important issues such as climate change or habitat alteration.

The vegetation map (Fig. 1) is a modification of that used by Mahler (1988) (Fig. 3) and is adapted from Dyksterhuis (1946, 1948) for the Fort Worth Prairie and West Cross Timbers. The eastern edge of the Blackland Prairie is defined by the boundary between the upland clay soils with grassland vegetation and the adjacent sandy soils of the Post Oak Savannah (also known as the Oak-hickory forest). The far northeastern corner of the study area goes east along the Red River as far as Lamar County; only the Blackland Prairie portion of central Red River County is included. To the west, the region extends to the western boundary of the West Cross Timbers; this vegetation type ends rather abruptly where the bedrock changes to Permian-age material underlying the Rolling or Red Plains. The topographically diverse southwestern part of North Central Texas is probably best referred to as the Lampasas Cut Plain. This area is geologically, and to a significant extent botanically, related to the Edwards Plateau and extends south and west to the Edwards Plateau and the Llano Basin. The least biologically meaningful boundary of North Central Texas is the southern one, corresponding roughly to the Williamson-Travis county border. The Blackland Prairie continues south past this line to the vicinity of San Antonio; that area is beyond the scope of the present study. Other sources for the vegetation map include Tharp (1926), Stanford (1971), Renfro et al. (1973), and county soil surveys (Soil Conservation Service). Plants of the adjacent Rolling Plains to the west, the Post Oak Savannah and Pineywoods to the east, and the Edwards Plateau to the southwest are included only if they enter the area under study, typically along river drainages.

INFORMATION HELPFUL IN USING THE FLORA

PLANTS TREATED

All known native and naturalized vascular plant species occurring in North Central Texas (Fig. 1) have been treated taxonomically. For the purposes of this work, a naturalized species is simply a non-native that is reproducing in the area without human assistance. For a species to be included, voucher specimens must have been seen, literature citations found, or in a several cases, plants observed in the field. If a taxon was included based on a literature citation, the citation is given in the text. A number of species were included based on citation of vegetational areas 4 (Blackland Prairie) or 5 (Cross Timbers and Prairies) by Hatch et al. (1990) and are indicated as such. In some instances, plants cited for vegetational area 4 were included only as notes since their distributions were well to the south (e.g., San Antonio Prairie) or east (e.g., Fayette Prairie) of North Central Texas. Such plants, with their scientific names in italics but not in bold, are listed as notes after all the alphabetically arranged species of a genus; they are not illustrated. If such plants are in a genus not treated in the flora, they are included in the family synopsis (e.g., *Ehretia anacua* in the Boraginaceae; *Campanula reverchonii* in the Campanulaceae). In a few instances, species were included based on field observations by individuals. These are listed as such in the treatments as "pers. obs." (personal observation, which denotes observation by one of the authors) or "pers. comm." (personal communication, which indicates an individual's observation communicated to the authors; such individuals are listed in the literature cited with a one or two line biography). A few long-persistent (e.g., *Ficus carica*—common fig), but apparently non-reproducing taxa have been included because of the likelihood of them being encountered. Also, a few taxa in areas immediately adjacent to the boundaries of North Central Texas (e.g., in adjacent parts of a partially treated county) have also been included to avoid confusion, improve clarity, or for general interest.

No attempt has been made to include the hundreds of non-native crop, landscape, and greenhouse plants cultivated in North Central Texas but not naturalized in the area. Information on cultivated plants can be found in such works as Bailey (1949), Shinners (1958), Bailey and Bailey (1976), Huxley et al. (1992), Sperry (1991), Garrett (1994, 1996), and Brickell and Zuk (1997).

The North Central Texas flora has about 46 percent of the 4,834 species of native and naturalized vascular plants recognized as occurring in Texas by Hatch et al. (1990) and over 40 percent of the 5,524 taxa. Since non-naturalized, cultivated plants are not included in our flora, a direct comparison is not possible with the most recent checklist of Texas plants (Jones et al. 1997), which lists 6,871 taxa including cultivated plants.

Arrangement of Taxa and General Methods

Families are listed alphabetically within divisions (Lycopodiophyta, Equisetophyta, Polypodiophyta, Pinophyta, Gnetophyta, and Magnoliophyta). The flowering plants (Magnoliophyta) make up the vast majority of North Central Texas species; within this group, class Dicotyledonae (dicots) is listed before class Monocotyledonae (monocots). For each family a taxonomic description, brief synopsis (indicated by the symbol) including such information as number of genera and species, a short section on family recognition in the field, and references, if appropriate, are given. If the type genus (genus after which the family is named) of a family is not treated in the flora, a brief synopsis of the type genus and the derivation of its name are given at the end of the family synopsis. When only one genus of a family is represented in the flora, the family and generic descriptions are combined. Appendix 1 is a phylogenetic classification of all treated families modified from those of Cronquist (1981, 1988), Lellinger (1985), and Hickman et al. (1993).

Genera are listed alphabetically within families and species within genera. A taxonomic description, brief synopsis (indicated by the symbol), derivation of the generic name, and references if appropriate are given for each genus. When only one species of a genus is represented in the flora, the generic and specific descriptions are combined. References for both families and genera are intended to provide an entry point into the more detailed taxonomic literature and should not be viewed as inclusive. Additional references can be found in Kent (1967), Hatch et al. (1990), Taylor and Taylor (1994), and Jones et al. (1997).

For each taxon treated at the rank of species, subspecies, or variety, all or most of the following are provided: 1) scientific name (**in bold type**) including authority followed by a comma to allow certainty over whether the name of the authority is abbreviated or not; 2) derivation of the specific or infraspecific epithet (in parentheses); 3) common name(s) if available (IN SMALL CAPITAL LETTERS); 4) taxonomic description; 5) habitat; 6) range; 7) phenology (period of flowering); 8) area of origin if not native to North Central Texas; 9) synonyms (in *italics* in brackets, [ ]); 10) notes on toxic/poisonous nature (indicated by the symbol ☠) or other short notes of ethnobotanical or taxonomic interest; and 11) for taxa introduced to the United States, the symbol . A line drawing illustration is provided for each species and in some cases for infraspecific taxa. The illustrations are grouped together on full pages and are as close to the description of a species as possible, typically within a few pages.

Descriptions

Because of space limitations due to the inclusion of illustrations, descriptions are as brief as possible while still allowing accurate identification. Characters useful in identification or helpful in confirming the identity of a plant have been stressed. Information given in the keys is generally not repeated in the descriptions. When only one species of a genus is represented in the flora, the generic and specific descriptions are combined. Therefore, the species descriptions in such cases are generally more ample than for other species. Characters described for a taxon at a higher rank (e.g., family) are not usually repeated for included taxa (e.g., genera). Descriptions were written for North Central Texas taxa and may not apply to taxa from other parts of the world; this is sometimes emphasized in the descriptions by the qualifier "ours" to denote species within the North Central Texas area.

KEYS

Keys are tools or shortcuts by which unknown plants can be identified. They provide a method whereby a choice between alternative statements about plant characteristics can be made, for instance:

1. Petals red; leaf blades pubescent on lower surface.
   2. Petals < 1 cm long; leaf blades entire ____ **Species a**
   2. Petals > 1 cm long; leaf blades toothed ____ **Species b**
1. Petals white; leaf blades glabrous on lower surface.
   3. Plant a shrub; leaf blades with acute apex ____ **Species c**
   3. Plant a tree; leaf blades with obtuse apex ____ **Species d**

The first choice (here lines beginning with the number 1) is followed by another choice indented under it (here lines beginning with the number 2) and so on until the identity of a plant is determined. In other words, after a choice has been made between the two alternatives of a pair (= couplet), the user goes to the more indented next couplet where another choice is presented. The keys provided in this work all have successive choices indented for ease of use and are strictly dichotomous; that is, the user must decide between only two choices at a time.

The keys have also been written to be as parallel as practical. In other words, when a character is given for one choice, it is also given for the other choice. However, in some cases, clarity, practicality, or the avoidance of ambiguity prevented absolute parallelism. Occasionally, a taxon, particularly a highly variable one, is keyed in more than one way to enhance ease of use and clarity. When possible, several characters are used for each choice in the keys; optimally both reproductive and vegetative characters are given. Sometimes, the plants falling under one alternative are variable and exhibit two character states; in order to emphasize this situation, the OR given between these two states is sometimes capitalized, for instance:

1. Leaves usually 30 cm or more long OR if shorter with a hard spiny tip.
1. Leaves 10–30 cm long, without a hard spiny tip.

While not preferred, such characters can still be helpful in identification.

Keys to genera and species were specifically written for the plants of North Central Texas and are not intended to be inclusive of plants occurring in other parts of the world. The General Key to All Families is modified from a key to families generously provided by the Oklahoma Flora Editorial Committee (Tyrl et al. 1994). While numerous couplets have been added to cover plants that occur in North Central Texas but not in Oklahoma, no couplets have been deleted from the Oklahoma family key. Therefore, some families/taxa occurring in Oklahoma are included that do not occur in North Central Texas. This was done so that the family key would be of maximum benefit to Oklahoma users as well as those in Texas. Such families are indicated in the *General Key to All Families* by a note in brackets, e.g., [Family in OK, not in nc TX]. In a number of instances, it is possible to key to the correct family even if a particular, easily confused dichotomy is misinterpreted. For such cases, explanatory notes are given in brackets in the key. The key to genera of Asteraceae is modified from one contributed by Constance Taylor (Taylor 1997).

In addition to the General Key to All Families, through which all families can be reached, several supplemental keys have been added for some groups. These include a key to ferns and similar plants (pteridophytes), a key to gymnosperms, a key to aquatic plants, a key to the families of monocots, and a key to woody vines.

Sources of Information

In addition to original observations and measurements, materials for the keys and descriptions have been obtained from a variety of sources listed in the literature cited. Of particular assistance were the following works: *Manual of the Vascular Plants of Texas* (Correll & Johnston 1970); *Grasses of Texas* (Gould 1975b); *Flora of the Great Plains* (Great Plains Flora Association 1986); *Flora of North America North of Mexico, Vol. 2, Pteridophytes and Gymnosperms* (Flora of North America Editorial Committee 1993); *Flora of North America North of Mexico, Vol. 3, Magnoliophyta: Magnoliidae and Hamamelidae* (Flora of North America Editorial Committee 1993); *Manual of the Vascular Flora of the Carolinas* (Radford et al. 1968); and *Flora of Missouri* (Steyermark 1963). In addition to the references mentioned above, the *Checklist of the Vascular Plants of Texas* (Hatch et al. 1990) was extensively used to determine ranges and as a source of common names. Information for the family synopses was obtained from *The Plant Book* (Mabberley 1987, 1997); *Flowering Plants of the World* (Heywood 1993); *Guide to Flowering Plant Families* (Zomlefer 1994); *Vascular Plant Taxonomy* (Walters & Keil 1995); and *Contemporary Plant Systematics, 2nd ed.* (Woodland 1997); in the interest of space, citations are given only for material from other sources. Material for the brief FAMILY RECOGNITION IN THE FIELD section given for each family was obtained from Smith (1977), Davis and Cullen (1979), Baumgardt (1982), Jones and Luchsinger (1986), and Heywood (1993). Generic synopses were modified from Mabberley (1987, 1997); here also, citations are given only for material from other sources. Derivations of generic names and specific and infraspecific epithets (etymology) were obtained or modified from a variety of sources including *Plant Names Scientific and Popular* (Lyons 1900); *The Standard Cyclopedia of Horticulture* (Bailey 1922); *How Plants Get Their Names* (Bailey 1933); *Gray's Manual of Botany* (Fernald 1950a); *Composition of Scientific Words* (Brown 1956); *Dictionary of Word Roots and Combining Forms* (Borror 1960); *A Gardener's Book of Plant Names* (Smith 1963); *Flora of West Virginia* (Strausbaugh & Core 1978); *Dictionary of Plant Names* (Coombes 1985); *The New Royal Horticultural Society Dictionary of Gardening* (Huxley et al. 1992); *Botanical Latin* (Stern 1992); and *Plants and Their Names* (Hyam & Pankhurst 1995). References of particular importance for the Introduction to North Central Texas included Hill's (1901) classic *Geography and Geology of the Black and Grand Prairies, Texas*, works on the Blackland and Grand prairies by Hayward and Yelderman (1991) and Hayward et al. (1992), a volume on the Blackland Prairie edited by Sharpless and Yelderman (1993), and articles on the Fort Worth Prairie and West Cross Timbers by Dyksterhuis (1946, 1948).

Nomenclature

Nomenclature, including authorities, in general follows *A Synonymized Checklist of the Vascular Flora of the United States, Canada, and Greenland* (Kartesz 1994) unless specifically indicated otherwise. An exception is that nomenclature for ferns, and similar plants, and gymnosperms follows the recent treatments in *Flora of North America* (Flora of North America Editorial Committee 1993). In a number of cases indicated in the treatments, nomenclature follows recent taxonomic works or the recently published *Vascular Plants of Texas* (Jones et al. 1997). While the decision over which source or sources to follow for nomenclature was not an easy one, in our minds the advantages of a standard source outweigh the advantages of other possible choices. Thus, only in instances where more recent works have been followed or where we believe biological reality or clarity is compromised by nomenclature do we differ from Kartesz. Unless other varieties or subspecies are specifically mentioned in the text, the type variety or subspecies is assumed.

Following the rules of the International Code of Botanical Nomenclature (Greuter et al. 1994), the scientific name of each species (or variety or subspecies) is followed by the authority, i.e., the author(s) who originally published that name. If the name is transferred to a different genus or rank, the name of the original author is placed in parentheses, followed by the name of the author(s) who made the transfer. For example, *Erythraea calycosa* Buckley was originally named by Samuel B. Buckley; later Merritt L. Fernald transferred the species to the genus

Centaurium with the correct citation becoming *Centaurium calycosum* (Buckley) Fernald. In some cases, the word "ex" is inserted between the names of authors (e.g., *Hydrolea ovata* Nutt. ex Choisy); this is used when an author such as Choisy publishes a new species (or variety or subspecies) based on a name attributed to but not validly published by another author (in this case Nuttall). Abbreviations for authorities of scientific names follow Brummitt and Powell (1992), which is now widely considered the standard for such abbreviations.

Nomenclatural change is inevitable as more is learned about various plant groups (see Appendix 7). These changes, especially when involving well known species, can be particularly irritating to both professional and amateur botanists as well as others needing to know correct scientific names. In order to avoid confusion regarding name changes, limited synonymy is provided. In particular, no longer recognized names used in Mahler (1988) and many from Correll and Johnston (1970) and Hatch et al. (1990) are listed as synonyms. Other synonyms are given to help clarify nomenclature or for general interest. However, no attempt was made to give complete synonymy. For detailed synonymy of Texas plants see Kartesz (1994) and Jones et al. (1997).

Common names are included in the treatments and in the index, enabling the identification of plants for which little other information is available. These names have been obtained from a variety of literature sources; none has been manufactured for this publication.

GEOGRAPHIC DISTRIBUTIONS

For taxa with limited known geographic distributions within North Central Texas, individual counties are cited. These citations represent specimens in the BRIT/SMU Herbarium (Botanical Research Institute of Texas) or in the private collections of G. Diggs and R. O'Kennon, that are being processed for deposit at BRIT; county records based on literature citations have also been included, as have records supplied by Jack Stanford of Howard Payne University—these are indicated by the herbarium abbreviation HPC, and Stanley Jones of the Botanical Research Center Herbarium—these are indicated by BRCH. A more general distribution within Texas usually follows the counties listed; examples include: e TX w to Blackland Prairie, West Cross Timbers s and w to w TX, and nearly throughout TX. When a taxon is well represented in North Central Texas, only the more general distribution within the state is given. Several taxa collected in the late 1800s and early 1900s by Reverchon, Ruth, and other early collectors have not been reported in the area since; these are mentioned as such. Plants of the eastern and southeastern parts of Texas penetrate into the Blackland Prairie up the Trinity and Brazos rivers, and some are becoming relatively scarce today in these bottomland extensions of their habitats. As mentioned earlier, plants of eastern Texas also enter the region along the northern edge of the Blackland Prairie in the Red River drainage. In both cases, these records have been mentioned.

Very few plants are endemic to North Central Texas; these are indicated by the symbol in front of the scientific name. Many plants listed as endemic to Texas in Correll and Johnston (1970) have since been found in immediately adjacent areas. For information on endemics we are therefore following Bonnie Amos, Paula Hall, and Kelly McCoy (Amos et al. 1998) of Angelo State University who generously contributed their data on Texas endemics; such information is given in the descriptions following a plant's Texas distribution. In order to make Texas endemics easily recognizable in the text, the symbol is placed at the end of such species' taxonomic treatments.

For naturalized plants whose place of origin is outside the continental United States, the symbol is placed at the end of the species' taxonomic treatment; plants for which this symbol is not given are native to the continental United States. A symbol to allow introduced species to be recognizable at a glance seemed a useful inclusion (Schmid 1997) and was an easy decision. However, the question of defining "introduced" was more difficult. For example, all species native outside North Central Texas could have been considered introduced; similarily, introduced species could have been defined as all species not native to Texas. Ultimately, we decided to use symbolic representation only for species native outside the United States. However, all species not native to North Central Texas have their area of origin indicated in the descriptions.

Information on Toxic/Poisonous Plants

Notes on toxic/poisonous properties (indicated by the symbol ☠) are given in the synopses and at the end of the treatments of various taxa. This information has been obtained from a variety of cited sources. However, lack of information about toxicity does not indicate that a plant is safe and no plant material should be eaten unless one is sure of its safety. Indeed, most plants have not been tested for toxicity and all should be considered potentially dangerous unless known otherwise. Technically, a poison is a substance that in suitable quantities has properties harmful or fatal to an organism when it is brought into contact with or absorbed by the organism. Toxin, a more specific term, is any of various poisonous substances that are specific products of the metabolic activities of living organisms (Gove 1993). In referring to such material in plants, the terms have been used synonymously in the text.

In case of toxicity/poisoning by plant material or any other source, the **Texas Poison Center Network** can be reached at **1-800-POISON-1 (1-800-764-7661)** or indirectly via the emergency number **9-1-1**. This is a state-wide 800 service available 24 hours a day.

Information on Endangered and Threatened Taxa

Taxa listed by the Texas Organization for Endangered Species (TOES 1993) are indicated by having (TOES 1993: Roman numeral) at the end of their treatment. The Roman numeral signifies the category as indicated by TOES:

Category I:

Endangered species—legally protected.

Category II:

Threatened species—legally protected.
Likely to become endangered

Category III:

Texas endangered—listed species.
Endangered in Texas portion of range

Category IV:

Texas Threatened—listed species.
Likely to become endangered in Texas portion of range

Category V:

Watch List—listed species.
Either with low population numbers or restricted range in Texas

Such species are also signified by having the symbol ⚠ placed at the end of their taxonomic treatments.

Information on Illustrations and Photographs

The more than 2,300 line-drawing illustrations have been obtained from a variety of sources in the botanical literature dating back to the 1500s (Fuchs 1542). We thank the appropriate individuals or organizations for allowing their use. Three hundred twenty-six illustrations are published here for the first time. These include many drawings of Cyperaceae done by Brenda Mahler and Jessica Procter as part of B. Lipscomb's research on that family. A significant number of the never-before-published drawings were done decades ago by the late Eula Whitehouse, Pat Mueller, and unknown SMU students. These illustrations were made for Lloyd Shinners, whose untimely death prevented publication of a flora for North Central Texas. The drawings were in the archives at BRIT. Finally, Linny Heagy has produced 226 original drawings for all those North Central Texas species either not previously illustrated or for which suitable illustrations could not be found.

Beneath each illustration is the scientific name of the plant represented. The name is followed by a code in parentheses indicating the source of the illustration. A list of illustration sources with codes is given in Appendix 4. Because all species are illustrated, reference to illustrations is not made in the text. Illustrations are as close to the taxonomic descriptions as possible and in general follow the taxonomic descriptions.

The 174 color photographs are arranged alphabetically by genus and are grouped together in plates. Following the common name of each species and the page number of its description, a three letter code in brackets is given to designate the photographer: [JAC] = J. Andrew Crosthwaite, [GMD] = George M. Diggs, Matthew A. Kosnik [MAK], and [RJO] = Robert J. O'Kennon. The symbol ❁/80 at the end of a species description indicates a color photograph is provided on page 80.

Information on the Glossary

The Glossary is modified from those of Shinners (1958a) and Mahler (1988), with additional entries obtained or modified from a variety of sources including Lawrence (1951), Featherly (1954), Correll (1956), Gleason and Cronquist (1963, 1991), Radford et al. (1968), Correll and Johnston (1970), Gould (1975b), Lewis and Elvin-Lewis (1977), Benson (1979), Smutz and Hamilton (1979), Fuller and McClintock (1986), Jones and Luchsinger (1986), Schofield (1986), Gandhi and Thomas (1989), Blackwell (1990), Isely (1990), Harris and Harris (1994), Spjut (1994), and Hickey and King (1997). The glossary is rather extensive and includes terms not otherwise found in the book. This was done so that when using this work in conjunction with other taxonomic treatments, the meaning of obscure terms can be readily found.

Information on References and Literature Cited

The Literature Cited section contains bibliographic citations for all sources cited including those listed immediately following family and generic synopses (e.g., References: Wood 1958; Kral 1997). Originally, we had not intended to include such references. However, during preparation of the taxonomic treatments, we needed to refer repeatedly to the supporting literature; having references readily available in the developing manuscript proved helpful. We hope their inclusion will be useful to users of the treatments. While an attempt was made to be as thorough as possible, the magnitude of the botanical literature makes complete coverage impossible; the references given are intended to provide an entry point into the more detailed taxonomic literature and should not be viewed as inclusive. Abbreviations for periodicals follow *Botanico-Periodicum-Huntianum* (B-P-H) (Lawrence et al. 1968) and *Botanico-Periodicum-Huntianum/Supplementum* (B-P-H/S) (Bridson & Smith 1991). For each individual cited in the text as having personally communicated unpublished information to the authors (indicated by the abbreviation, pers. comm.), a short biographical entry is given in the Literature Cited section.

Abbreviations and Symbols

ABBREVIATIONS/SYMBOLS	MEANING
	endemic to North Central Texas
	endemic to Texas
	family or generic synopsis
	introduced species, subspecies, or variety
	endangered or threatened taxa; a TOES rating is also given for such taxa
	toxic/poisonous plant
	color photograph provided; page number follows symbol
$<$	less than
$\leq$	less than or equal to
$>$	more than
$\geq$	more than or equal to
$\pm$	more or less
+	or more (e.g., small tree 2–5+ m tall)
$\times$	times or to indicate hybridization
auct.	auctorum = author
BAYLU	herbarium abbreviation for Baylor University Herbarium, Waco, TX.
B. P.	before present
BRCH	herbarium abbreviation for Botanical Research Center Herbarium, Bryan, TX
BRIT	herbarium abbreviation for Botanical Research Institute of Texas, Fort Worth
c	central
ca.	circa (about)
cm	centimeter
comb. no	Latin: *combinatio nova*, new combination of name and epithet
diam.	diameter
dm	decimeter
DUR	herbarium abbreviation for Southeastern Oklahoma State University, Durant, OK
e	east
e.g.	Latin: *exempli gratia*, for example
ex.	see preceding section on nomenclature for detailed explanation
f.	Latin: *filius*, son; e.g., L. f. indicates the younger Linnaeus
HPC	herbarium abbreviation for Howard Payne University, Brownwood, TX
i.e.	Latin: *id est*, that is
m	meter

ABBREVIATIONS/SYMBOLS	MEANING
MICH	herbarium abbreviation for the University of Michigan, Ann Arbor.
mm	millimeter
n	north
n =	haploid chromosome number
2*n* =	diploid chromosome number
nom. illeg.	nomen illegitimum (illegitimate name)
of authors, not	used to indicate a name was used in the sense of certain authors, but not in the sense of the author making the combination; technically written as: auct. non
p.p.	pro parte (in part)
pers. comm.	personal communication of information to the authors
per. obs.	personal observation by one of the authors
s	south
sensu	in sense of; used to indicate that a name is used in the sense of one author, not another
sensu lato	in the broad sense, e.g., if a genus is broadly treated to include many species
sensu stricto	in the strict sense, e.g., if a genus is narrowly treated to include few species
SMU	herbarium abbreviation for Southern Methodist University Herbarium, now part of the Botanical Research Institute of Texas (BRIT), Fort Worth
s.n.	sine numero (without number)
spp.	species
subsp.	subspecies
TAES	herbarium abbreviation for S. M. Tracy Herbarium, Department of Rangeland Ecology & Management, Texas A&M University, College Station
TAMU	herbarium abbreviation for Biology Department Herbarium, Texas A&M University, College Station
TEX	herbarium abbreviation for University of Texas at Austin
TOES: (roman numeral)	Texas Organization for Endangered Species (category/status)
Univ.	university
VDB	herbarium abbreviation for Vanderbilt University Herbarium; currently housed at the Botanical Research Institute of Texas, Fort Worth
w	west
var.	variety

Ranges for measurements, e.g., (10–)12–23 mm long, should be interpreted as "typically 12 to 23 mm long, rarely as little as 10 mm long"

States are abbreviated using standard, two letter, United States Postal zip-code abbreviations (e.g., TX = Texas, OK = Oklahoma)

SUMMARY DATA ON THE FLORA

AND COMPARISON WITH OTHER FLORAS

SUMMARY OF THE FLORA OF NORTH CENTRAL TEXAS

	FERNS & SIMILAR PLANTS	GYMNOSPERMS	MONOCOTYLEDONS	DICOTYLEDONS	ANGIOSPERMS	TOTAL
Families	16	3	25	124	149	168
Genera	26	4	176	648	824	854
Species	47	10	567	1599	2166	2223
Additional Infraspecific taxa	2	0	40	111	151	153

COMPARISON WITH OTHER FLORAS

	NORTH CENTRAL TX	TX[1]	OK[2]	KS[3]	AR[4]	TN[5]	WV[6]	NC&SC[7]	CA[8]
Families	168	180	172	139	154	167	143	180	173
Genera	854	1284	850	646	818	850	693	951	1222
Species	2223	4834	2549	1807	2356		2155	3360	5862
Native Species	1829								4739
Introduced Spp.	394								1023
Total Taxa	2376	5524	2844	2226	2469	2745			
Area (in 1000s of square miles)	40	269	70	82	53	42	24	86	164

NORTH CENTRAL TEXAS:

46 % of the species in Texas (in 15 % the land area)
87 % as many species as Oklahoma (in 57 % the land area)
82.3 % native species (17.7 % introduced from outside the United States)
94 Texas endemics and 5 North Central Texas endemics

	Genera	Species
Number of Genera and Species of Asteraceae (Largest North Central Texas family)	103	263
Number of Genera and Species of Poaceae	86	249
Number of Genera and Species of Fabaceae	56	176
Number of Genera and Species of Cyperaceae	15	140
Number of Species of *Carex* (Cyperaceae, largest North Central Texas genus)		56

[1]Hatch et al. 1990; [2]Taylor & Taylor 1994; [3]McGregor 1976; [4]Smith 1988; [5]Wofford & Kral 1993; [6]Strausbaugh & Core 1978; [7]Radford et al. 1968; [8]Hickman 1993

NEW COMBINATIONS MADE IN THIS BOOK

ASTERACEAE

Gutierrezia amoena (Shinners) Diggs, Lipscomb, & O'Kennon
New combination on page 364, illustration on page 363

NYCTAGINACEAE

Mirabilis latifolia (A. Gray) Diggs, Lipscomb, & O'Kennon
New combination on page 840, illustration on page 843

AGAVACEAE

Manfreda virginica (L.) Rose subsp. **lata** (Shinners) O'Kennon, Diggs, & Lipscomb
New combination on page 1079; illustrations on pages 98 and 1081

ACKNOWLEDGMENTS

Contributions to an understanding of the flora of North Central Texas by the late Lloyd Shinners and the very lively William (Bill) Mahler deserve special recognition; this volume is dedicated to them. For a synopsis of Shinners' life see Mahler (1971b); for a guide to his botanical contributions see Flook (1973). This book is published as a volume of *Sida, Botanical Miscellany*, associated with the botanical journal *Sida, Contributions to Botany*, founded by Shinners in 1962 (Mahler 1973b).

Three individuals were indispensible members of the team that produced this work. Special thanks to Linny Heagy, for designing and illustrating the cover and dust jacket, giving creative direction/art direction throughout the whole project, and creating 226 original botanical line drawings; Samuel Burkett, for scanning the more than 2,300 line drawings, laying out the illustration pages and visual glossary pages, coordinating copyright issues, and managing the project databases; and Amberly Zijewski, an Austin College student, for work on derivation of scientific names, authorities of scientific names, literature research, and assistance with layout of illustrations.

Several sections of the manuscript were generously contributed by other authors. Thanks to Stanley Jones of the Botanical Research Center Herbarium for contributing the treatment of the genus *Carex* (Cyperaceae), Connie Taylor of Southeastern Oklahoma State University for the key to genera of Asteraceae, and the Oklahoma Flora Editorial Committee (Ronald J. Tyrl, Susan C. Barber, Paul Buck, James R. Estes, Patricia Folley, Lawrence K. Magrath, Constance E.S. Taylor, and Rahmona A. Thompson—Tyrl et al. 1994) for allowing us to modify their key to families; we are especially indebted to our Oklahoma colleagues for the use of their fine key which represents many years of hard work. Thanks also to Bonnie Amos, Paula Hall, and Kelly McCoy (Amos et al. 1998) of Angelo State University who generously contributed data on plants endemic to Texas.

We are also particularly indebted to three other colleagues for major contributions: Jack Stanford of Howard Payne University reviewed the entire manuscript, made many helpful suggestions, and contributed plant distributional data on the southwestern part of North Central Texas; John Thieret of Northern Kentucky University reviewed large segments of the manuscript and provided valuable constructive criticism; and Ruth Andersson May of Dallas proofread the entire manuscript.

A number of specialists generously reviewed taxonomic treatments. These include: Daniel Austin, John Bacon, Harvey Ballard, Theodore Barkley, Karen Clary, Robert Faden, Paul Fryxell, Gustav Hall, Ronald Hartman, Stephan Hatch, Robert Haynes, William Hess, Gloria Hoggard, Ron Hoggard, Stanley Jones, Carl Keener, Robert Kral, Lawrence Magrath, Mark Mayfield, Edward McWilliams, Nancy Morin, Michael Nee, Guy Nesom, James Phipps, James Pringle, Richard Rabeler, Roger Sanders, Richard Spellenberg, Jack Stanford, Connie Taylor, John Thieret, Warren Wagner, Donna Ware, Sherry Whitmore, Justin Williams, and Lindsay Woodruff.

Several individuals deserve special recognition for their contributions. Special thanks go to Rob Maushardt and Travis Plummer for computer consultation and design of the project database; Rebecca Horn for page layout and production; Matthew Kosnik, an Austin College student, for extensive work on graphics for the introduction and appendices including creation of the color vegetation map; Cathy Stewart for typing portions of the manuscript and providing invaluable assistance at many steps; Yonie Hudson for interlibrary loan coordination; Glenda Ricketson for office assistance; Juliana Lobrecht, an Austin College student, for word processing and assisting with the glossary; Millet the Printer for publication consultation and an excellent printing job; and colleagues and staff at Austin College and the Botanical Research Institute of Texas for support and assistance. Numerous other Austin College students have contributed in various ways; they include Allison Ball, Carrie Beach, Blake Boling, Ricky Boyd, Rhome Hughes, Nichole Knesek, Chris Munns, Matthew Nevitt, Mary Paggi, Kristin Randall, Carla Schwartz, and Laura Wright.

In addition, the following individuals were of particular assistance: Paul Baldon, James Beach, Susan Barber, Bruce Benz, Andy Buckner, Narcadean Buckner, William Carr, Wayne Clark, Karen Clary, Andrew Crosthwaite, Sally Crosthwaite, Carol Daeley, Arnold Davis, Gary Dick, Laurence Dorr, James Eidson, Barbara Ertter, Richard Francaviglia, Kancheepuram Gandhi, Hugh Garnett, Howard Garrett, Jane Gates, Ruth Ginsburg, Steven Goldsmith, Alan Graham, Marcia Hackett, Clyde Hall, Gus Hall, Karl Haller, Neil Harriman, Larry Hartman, Michael Imhoff, Joe Hennen, Jim Hunt, Michael Imhoff, James Johnson, Joann Karges, John Kartesz, Joe Kuban, Jody Lee, Ted Len, Félix Llamas, Peter Loos, Shirley Lusk, William Martin, James Matthews, Howard McCarley, Marion McCarley, Larry McCart, David Montgomery, Jerry Niefoff, Fiona Norris, Richard Norris, Martha O'Kennon, Chetta Owens, John Pipoly, Georgia Prakash, Jeff Quayle, Peggy Redshaw, David Riskind, Laura Sanchez, Rudolf Schmid, Lee Schmitt, Daniel Schores, Peter Schulze, Chuck Sexton, Benny Simpson, Beryl Simpson, Michael Smart, Calvin Smith, Galen Smith, David Stahle, Geoffrey Stanford, John Steele, John Strother, Ronald Stuckey, Dora Sylvester, Eskew Talbot, Lloyd Talbot, John Taylor, Dorothy Thetford, Rahmona Thompson, Dana Tucker, Ron Tyrl, Susan Urshel, Kevin Walker, Sally Wasowski, Grady Webster, Jim Williams, Joseph Wipff, Lindsay Woodruff, and Joe Yelderman.

Thanks go also to all those individuals who have collected specimens in the North Central Texas region for well over one hundred years. Early collections housed in herbaria such as BRIT/SMU, TAES, TAMU, TEX, and represent irreplaceable windows to a time before the vegetation of this area was radically altered. While not all collectors can be recognized here, specimens collected by the following have been of particular importance in allowing the completion of this work: M.D. (Bud) Bryant, William Carr, Donovan Correll, Victor Cory, Sally Crosthwaite, Delzie Demaree, George Diggs, Harold Gentry, Joe Hennen, Barney Lipscomb, Shirley Lusk, Cyrus Lundell, William Mahler, William McCart, Robert O'Kennon, Jeff Quayle, Julien Reverchon, Albert Ruth, Lloyd Shinners, Jack Stanford, Dora Sylvester, Connie Taylor, John Taylor, Billy Turner, Eula Whitehouse, and numerous students at colleges and universities in the area including Austin College, Baylor University, Howard Payne University, Southeastern Oklahoma State University, Texas Christian University, and Southern Methodist University. Geiser (1948) gave historical information on a number of the early collectors.

The frontispiece and dust jacket were created by Linny Heagy. Color photographs are by Andrew Crosthwaite, George Diggs, Matthew Kosnik, and Robert O'Kennon; in the color photographs, following the name of each species, a three letter code in brackets is given to designate the photographer.

An important debt of gratitude is owed to innumerable landowners who kindly allowed access to their property. Thanks also to the Texas Parks and Wildlife Department, the U.S. Department of the Interior, the U.S. Army Corps of Engineers, The Nature Conservancy, Austin College, the Fort Worth Nature Center and Refuge, and the U.S. Army for allowing access to the property under their stewardship.

Additional individuals who contributed to *Shinners' Manual of the North Central Texas Flora* (Mahler 1988), the precursor of the present volume, should also be recognized. These include Geyata Ajilvsgi, Gerald Arp, Barry Comeaux, Charlotte Daugirda, Victor Engel, Kathie Parker, Andrea McFadden, Jane Molpus, Ann Nurre, Tami Sanger, Harriet Schools, Geoffrey Stanford, and numerous students who have used earlier versions of either Shinners' or Mahler's works.

This project is part of the ongoing collaboration between the Austin College Center for Environmental Studies and the Botanical Research Institute of Texas. Without the support of both institutions it would not have been possible. Thanks to Austin College, its president, Oscar Page, Michael Imhoff, Director of the Center for Environmental Studies, and the Austin College Board of Trustees. We are also grateful to the Botanical Research Institute of Texas, its director, S.H. Sohmer, and the BRIT Board of Trustees. We also wish to express our thanks to the founding organizations of BRIT: Dallas Arboretum and Botanical Gardens, Inc.; Fort Worth Botanical Society; Fort Worth Garden Club; Fort Worth Park and Recreation Department; Southern Methodist University; and Texas Garden Club, Inc.

Financial support for this work has come from a number of sources. Major support has been provided by the New Dorothea L. Leonhardt Foundation (Andrea C. Harkins), the Bass Foundation, Robert J. O'Kennon, Ruth Andersson May, Mary G. Palko, the Amon G. Carter Foundation, Margret M. Rimmer, and Mike and Eva Sandlin. Other supporters include Austin College, the Botanical Research Institute of Texas, the Sid Richardson Career Development Fund of Austin College, Peg and Ben Keith, the Friends of Hagerman National Wildlife Refuge, the Summerlee Foundation, John D. and Beth A. Mitchell, Waldo E. Stewart, Dora Sylvester, Founders Garden Club of Dallas, Lorine Gibson, Sue Paschall John, and Barbara G. Paschall.

We would like to thank those individuals who gave us our taxonomic training, including Gary Breckon, Mickey Cooper, Gustav Hall, Hugh Iltis, Marshall Johnston, Robert Kowal, William Mahler, James Phipps, Benny Simpson, Edwin Smith, John Thomson, Donna Ware, Barton Warnock, and Robert Ziegler.

Finally, thanks to our parents, Helen and Minor Diggs, Jack and Christa Lipscomb, and Robert and Elizabeth O'Kennon.

AUTHORS' NOTE

In a work such as this, it is inevitable for omissions and errors, both large and small, to escape attention. Because of the possibility of future editions, we would appreciate corrections, suggestions, or additions from individuals using the book. Also, as part of the Illustrated Texas Floras Project (a collaborative project between BRIT and the Austin College Center for Environmental Studies), we are currently working on a companion volume, to be titled the *Illustrated Flora of East Texas*, projected to be published in the year 2004. Because there is substantial overlap between the plants of North Central and East Texas, corrections and suggestions on the present volume would be very helpful for the next work. Such information can be sent to:

George M. Diggs, Jr.	Barney L. Lipscomb	Robert J. O'Kennon
gdiggs@austinc.edu	barney@brit.org	okennon@brit.org.

Also, we hope that this book will spur additional interest in, and collecting of, plants in the area. Plant specimens, particularly county, regional, or state records would be much appreciated and can be deposited at the:

BOTANICAL RESEARCH INSTITUTE OF TEXAS HERBARIUM (BRIT)
509 PECAN STREET, FORT WORTH, TX 76102
PHONE: 817/332-4441

Such specimens will be important scientific contributions, will have permanent protection, and will be important resources for the future.

Introduction to North Central Texas

Overview

North Central Texas is an area of roughly 40,000 square miles, bound on the north by the Red River and extending south nearly to Austin. While small by Texas standards (260,000 square miles), it is about the size of Kentucky. It includes the Blackland Prairie, the Grand Prairie (Fort Worth Prairie and Lampasas Cut Plain), the East and West cross timbers, and the Red River Area (Fig. 1). The flora includes 2,223 species, about 46 percent of the species known for Texas (Hatch et al. 1990), and a total of 2,376 taxa (species, subspecies, and varieties). This biological diversity is the result of numerous factors, including the region's geologic and climatic variation and its location in the ecotone or transition zone between the eastern deciduous forests and the central North American grasslands. North Central Texas is a mixing ground for plants from the east and west, with different microhabitats even within the same county having radically different plant communities. For the past two centuries, humans have had, and are continuing to have, a tremendous impact on the plants and animals of the region. Presettlement and early settlement conditions were radically different from those found today; the current generation may be the last with the opportunity to preserve even small remnants of the once extensive natural ecosystems.

General Geology of North Central Texas

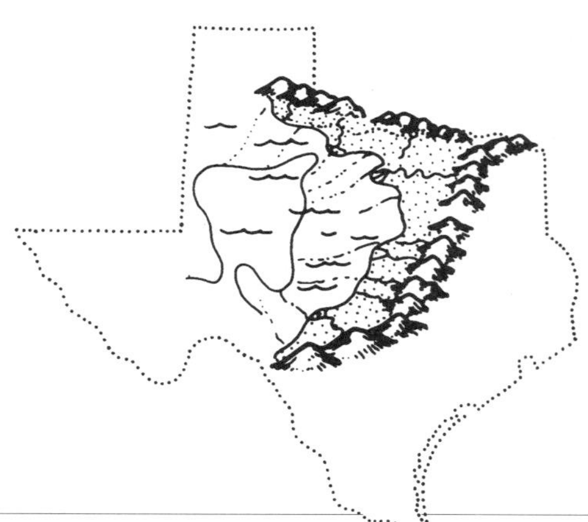

Fig. 5/Ancient Ouachita Mountains (part of the Appalachian-Ouachita-Marathon Mountain system). This belt cut north-south across Texas in a line roughly from Sherman to Dallas, Waco, Austin, and beyond. This range rose approximately 300 million years ago as a result of a plate tectonic collision (adapted from Spearing 1991 with permission of Mountain Press Publishing Co.; ©1991).

The geology of North Central Texas is much more interesting and complex than indicated by the gently rolling topography. If one begins about 300 million years ago, during the Pennsylvanian Period (320–286 million years ago), what is now North Central Texas was on the southern edge of a North American continent shaped very differently than it is today. As a result of plate tectonic movements, North America collided with Africa and South America to become part of the supercontinent Pangaea. The outcome of this collision was the uplift and formation of an extensive mountain system including the ancient Appalachians, Wichitas, and Ouachitas. The Ouachita Mountains formed in a line roughly following the western edge of the current Blackland Prairie and farther south, the Balcones Escarpment. They extended across much of Texas, in a line from near Sherman to Dallas to Austin and beyond (Fig. 5). The ancient Ouachita mountain belt also continued to the northeast; the eroded Ouachitas seen today in southeastern Oklahoma and southwestern Arkansas are remnants of this once much more extensive range. In western Texas, to the west of the ancient Ouachita Mountains, crustal areas sagged and low basins formed. Shallow inland seas invaded these low areas, and during the Pennsylvanian, and later the Permian, western Texas served as a collection basin for the sediments that eroded from the Ouachita Mountains to the east. These deep sediments are world-famous for their oil-bearing layers. They are also the source of the bright red, iron-rich (hematite) Permian layers that easily erode and give the modern Red River its name. Over tens of millions of years the Ouachita Mountains gradually eroded, until today all that is left over most of Texas are their roots, deeply buried under thousands of feet of sediments (Spearing 1991).

During the Triassic (245–208 million years ago) and Jurassic (208–145 million years ago) periods and continuing into the Cretaceous Period (145–65 million years ago), Pangaea eventually split apart into separate continents. The North Central Texas region once again became very active geologically with the zone of weakness where the Ouachitas had originally formed serving as the site of continental rifting or breakup between North and South America. It was here, where the continents pulled apart, that huge shallow seas, eventually retreating to become the present-day Gulf of Mexico, began to form (Fig. 6). Into these seas, thick layers of sediment were deposited during much of the Cretaceous Period, the material coming in part from erosion of the Rocky Mountains rising to the west. The shallow seas repeatedly advanced and retreated over much of Texas. In fact, these seas extended to the Big Bend area and at times even connected all the way from the Gulf of Mexico north to the Arctic Ocean. As a result of the varying water depths and other conditions, a number of different layers of Cretaceous sediments were laid down across the state. Fossil-bearing limestones, so common in North Central Texas, were deposited in the shallow seas. Near the ancient coasts, where muds and sands were laid down, dinosaur tracks were sometimes preserved. Excellent examples of these can be seen in Dinosaur Valley State Park near Glen Rose in Somervell County, and others have been found in the Glen Rose Limestone in Comanche and Hamilton counties (Shuler 1935, 1937; Albritton 1942; Spearing 1991).

FIG. 6/FIGURE SHOWING AN EXAMPLE OF HOW TEXAS WAS COVERED BY AN INLAND SEA AT VARIOUS TIMES DURING THE CRETACEOUS. ONE RESULT IS THAT CRETACEOUS SEDIMENTS FORM MOST OF THE SURFACE ROCKS IN NORTH CENTRAL TEXAS (ADAPTED FROM SPEARING 1991 WITH PERMISSION OF MOUNTAIN PRESS PUBLISHING CO.; ©1991).

As a result of the depositional processes described above and subsequent erosion of overlying material, nearly all surface rocks of modern North Central Texas are Cretaceous in origin (Renfro et al. 1973; McGowen et al. 1991) (Fig. 7). An exception is the more recent, stream-deposited sediment along some of the drainages. As indicated above, all these Cretaceous layers were laid down near the margin of an ocean that can be thought of as a greatly expanded Gulf of Mexico. After the Cretaceous Period, as the Gulf retreated farther to the southeast, sediments continued to be deposited on eastern Texas during much of the Tertiary Period (65–2 million years ago). The youngest of these sediments are therefore found near the present-day Gulf Coast (Bullard 1931). During the Tertiary, the major geologic factor shaping North Central Texas was the removal of material by erosion (Sellards et al. 1932; Baker 1960), revealing layers that were once buried. This process can be clearly seen as one travels west across the region. The eastern Blackland Prairie at the eastern margin of the area is developed on relatively young Upper (= Late) Cretaceous layers. Farther west, at higher elevations subject to greater erosion, more and more Cretaceous material was stripped away and progressively older rocks exposed. The only significant non-Cretaceous rocks found in North Central Texas are the older Pennsylvanian-age rocks uncovered by extensive erosion in the Palo Pinto Country in the extreme western portion of the region (Hill 1901; Sellards et al. 1932) (Figs. 7, 8). Here, all Cretaceous layers have been removed, and the much older, Pennsylvanian-age rocks are exposed at the surface.

As indicated above, North Central Texas was near the edge of the ocean during much of the Cretaceous, and as sea levels rose and fell, shallow seas repeatedly covered and then retreated from much of Texas (Sellards et al. 1932; Spearing 1991). Numerous layers of limestone, marl (chalky or limey clays), shale, and sand (Bullard 1931; Baker 1960) were deposited over the area, the type of layer depending on water depth, distance from shore, and other factors. According to Hill (1901),

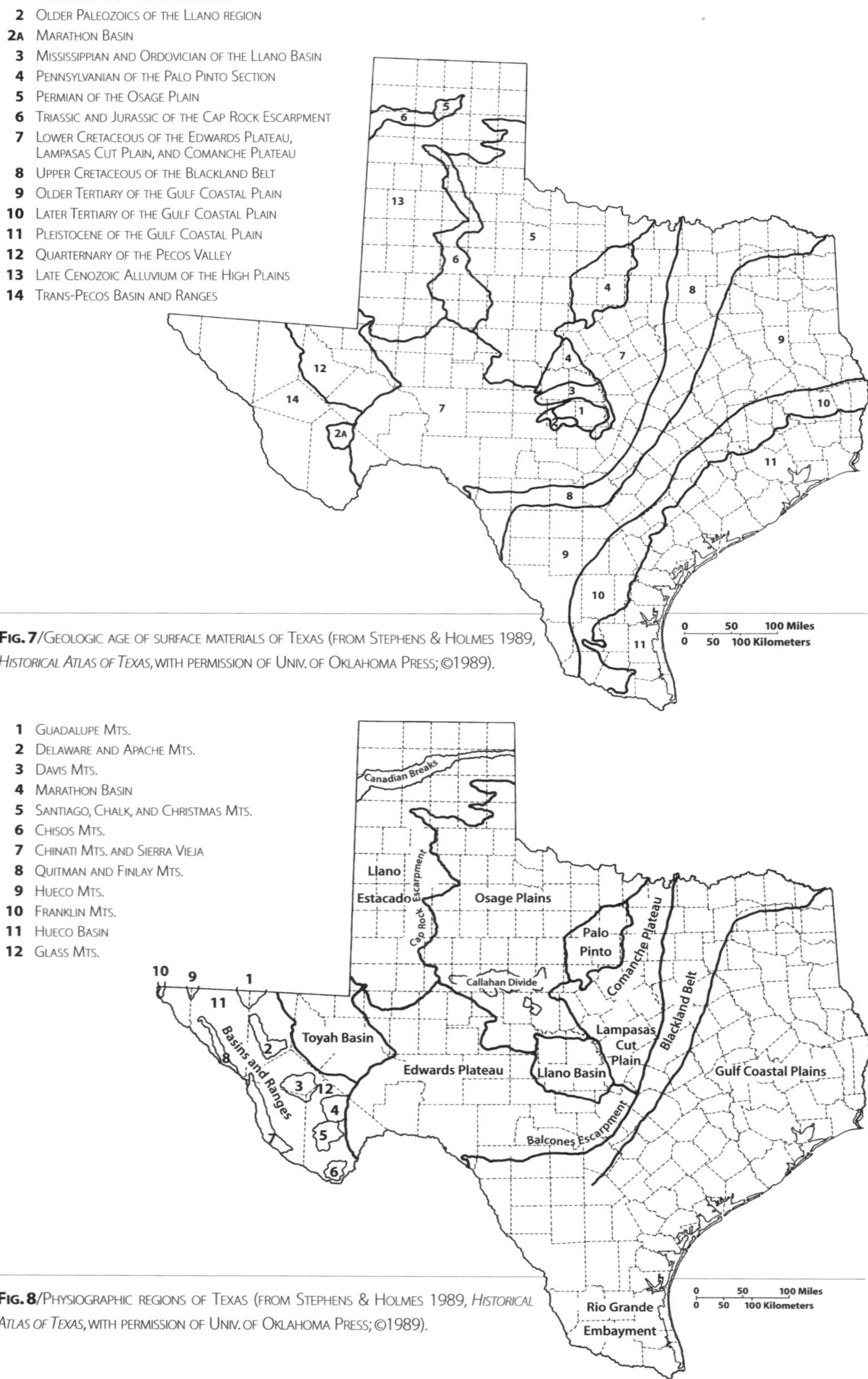

Fig. 7/Geologic age of surface materials of Texas (from Stephens & Holmes 1989, *Historical Atlas of Texas*, with permission of Univ. of Oklahoma Press; ©1989).

Fig. 8/Physiographic regions of Texas (from Stephens & Holmes 1989, *Historical Atlas of Texas*, with permission of Univ. of Oklahoma Press; ©1989).

> In general the sands are near-shore deposits, such as are seen to-day on most ocean beaches. The finer sands were carried a little further seaward than the coarse material. The clays are the lighter débris of the land, which were laid down a little farther from the land border; and so on through the various gradations to the chalky limestones, which largely represent oceanic sediments deposited in relatively purer waters farthest away from the land. The limestones are not all chalky. Some are agglomerates of shells of animals which inhabited the sandy or muddy bottoms; others are old beach wash. The vast numbers of sea shells occurring upon the mountains and prairies of Texas have not been transported, as some people believe. Save that they have been subjected to general regional uplift whereby the sea bottom was converted into land, they are now in the exact locality where they lived and flourished, and the clays and limestones in which they were buried were once the muds of the old ocean bottom.

Because all these layers are ocean sediments, many have excellent fossils of marine organisms. Some of the most obvious include oyster-like bivalve mollusks, sharks teeth, a type of echinoderm known as heart urchins, and ammonites, the large, extinct, coiled-shell relatives of the octopus and squid. In fact, some North Central Texas rock layers are composed almost completely of fossilized animal remains and the area is well known among fossil hunters.

To the southwest of North Central Texas occurs a rugged area which includes granite and other Precambrian outcrops, variously known as the Burnet Country, Central Mineral Region, or Llano Basin; it has been exposed by the extensive erosion of overlying sediments. To the west of the West Cross Timbers, and thus like the Central Mineral Region outside of North Central Texas, lies the vast area known as the Rolling Plains, underlain by the famous Permian Red Beds. This region at least in part is sometimes referred to as the Red Plains. The strikingly colored, iron oxide-rich, erosional products of these Permian layers give the Red River (originating far to the west) its name. The salinity of the Red River (and thus of Lake Texoma) is also the result of erosion from salt-rich Permian-age evaporation flats through which the river passes on its course east from the Texas Panhandle (Spearing 1991).

Soils of North Central Texas

Soils in North Central Texas vary dramatically, ranging from the characteristic black soil of the Blackland Prairie to the easily erodible sands underlying the West Cross Timbers.

The "black waxy" soils of the Blackland Prairie are derived from Upper Cretaceous rocks, which are sometimes strikingly white in color (e.g., Austin Chalk); through the process of weathering there is a dramatic change in color (Fig. 9). In the words of Hill (1901),

> The Black Prairie owes its name to the deep regolith of black calcareous clay soils which cover it. When wet these assume an excessively plastic and tenacious character, which is locally called "black waxy." These soils are the residue of the underlying marls and chalks, or local surficial deposits derived from them, and hence are rich in lime. Complicated chemical changes, probably due to humic acid acting upon vegetable roots, are believed to cause the black color. The region is exceedingly productive, and nearly every foot of its area is susceptible to high cultivation. In fact, the prairies are the richest and largest body of agricultural land in Texas, constituting a practically continuous area of soil extending from Red River to the Comal. . . .

More specifically, the Blackland Prairie (also referred to as the Blacklands) has three dominant soil orders: Vertisols, Mollisols, and Alfisols (Fig. 10). The Vertisols develop mainly on the Eagle Ford shale and rocks of the Taylor Group and are characterized by abundant smectitic (= shrink-swell) clays (Hallmark 1993). Upon wetting and drying, these soils often undergo dramatic changes in volume, which can result in significant soil movements. Swelling and shrinking causes cracks up to 50 centimeters or more deep and as much as 10 centimeters wide at the surface (Hallmark 1993). Stories of golf balls or even baseballs or other objects disappearing in deep cracks are not uncommon from longtime residents of Blackland soil areas. The associated soil movements can have dramatic effects on human activities, resulting in uneven or cracked roadways,

Fig. 9/Photograph showing striking contrast between the extremely dark "black waxy" soil typical of the Blackland Prairie and a piece of the nearly white underlying Austin Chalk bedrock from which it developed. The photograph was taken during trenching on the Austin College campus (photo by GMD).

shifted buildings, and cracked foundations (Hallmark 1993). Only the most elaborately protected houses on many Vertisols are free from at least some cracks or other soil stability problems. These smectitic clay soils are also quite sticky and difficult to manage agriculturally, being easily compacted by farm machinery when wet and forming large clods when plowed dry. Because they can be effectively tilled only within a narrow moisture range, they gained the nickname "nooner soils"—too wet to plow before noon and too dry after noon (Hallmark 1993). The smectitic clays also result in both slickensides and gilgai, two phenomena often seen in Vertisols. Slickensides are planes of weakness in the soil caused by movements associated with shrinkage and swelling. These can result in rather large-scale slippage or failure of soil blocks, which can be problematic in construction (Hallmark 1993). According to Hallmark (1993), slickenside slippage, causing the collapse of the walls of construction trenches, results in Texas workers being crushed to death in trenches almost every year. Gilgai are the microhigh, microlow topography or relief features found on essentially all Vertisols (Diamond & Smeins 1993). On flat areas in the prairie landscape, gilgai typically form circular, almost tub-like depressions, called "hog wallows" by early settlers. These range from about three to six meters across and up to about one-half meter deep (Hayward & Yelderman 1991). On slopes, gilgai take the form of microridges and microvalleys up to about 20 centimeters deep (Miller & Smeins 1988; Diamond & Smeins 1993) (Fig. 11). Both gilgai and the great soil depth of this region are the result of the constant churning and overturn of the shrink-swell, clay-based soils. When these soils shrink during dry weather and large cracks form, loose pieces of soil fall deep into the cracks. Upon wetting, these pieces swell and exert lateral pressure. Material is pushed outward and eventually upward, resulting in depressions rimmed by slightly raised areas (Hayward & Yelderman 1991) (Fig. 12). Gilgai are thus formed and the soil is slowly but constantly churned; the name Vertisol (Latin: *verto*, turn upward, *sol*, soil) is derived from this continuous cycle of overturning of the soil (Steila 1993). On the native Blackland Prairie, soil erosion was low because of the dense tall grass community, and also because of the water-trapping capacity of gilgai. Temporary water storage in gilgai depressions of one-half acre foot of water per acre of flat prairie have been estimated. As much as six inches of rain could be temporarily trapped in these structures before runoff began (Hayward & Yelderman 1991). This would have greatly reduced runoff and allowed significant infiltration, particularly important considering that clay soils are often rather impermeable. In fact, early accounts refer to clear runoff and clear streams on the Blacklands (Hayward & Yelderman 1991), in stark contrast to the current situation. However, because thousands of gilgai covered the prairies and created pools of standing water during wet weather, the prairies were at times virtually impassable (Hayward & Yelderman 1991). Under present agricultural conditions, with no plant cover during much of the year and with the suppression of gilgai formation by plowing, erosion rates in the Blacklands are high. Thompson (1993) noted that the Blacklands have one of the highest rates of soil loss on cropland of any major area in Texas. Estimates run from tens to hundreds of times higher than under the original native prairie vegetation (Hayward & Yelderman 1991). Richardson (1993) cited annual erosion figures of 15 tons per acre (t/a) for a cultivated Blackland area compared with only 0.2 t/a for a native grass meadow, a 70-fold increase. Even though gilgai were one of the most evident surface features on the original Blackland Prairie, because they are destroyed by plowing they are rarely observed today. Excellent examples of these "hog wallows," however, can still be seen at the Nature Conservancy's Clymer Meadow preserve in Hunt County as well as on scattered prairie remnants (Fig. 13).

Mollisols are found on the Fort Worth Prairie and the Lampasas Cut Plain on various limestone layers and on the Blackland Prairie on rocks of the Austin Group. All of these areas have high calcium carbonate levels and consolidated parent rocks. Because bedrock is usually just below the surface, rooting and soil water storage are restricted. Typically Mollisols are less useful for agriculture than are Vertisols and at present they tend to be used as pastures or homesites. Shrink-swell phenomena, while still occurring on Mollisols, are less problematic than on Vertisols (Diamond & Smeins 1993; Hallmark 1993). Laws (1962) and Brawand (1984) have studied the characteristics of soils formed from the Austin Chalk in the Dallas area.

Alfisols, which develop principally on bedrocks which are higher in sand and lower in calcium carbonate, are found mainly on the eastern and northern margins of the Blacklands (Hallmark 1993) (Fig. 10). These soils are less fertile than either of the other two types (Hallmark 1993). Another microtopographical feature, mima mounds (also called pimple mounds or prairie mounds) (Fig. 11), were found on essentially all Alfisols within the Blackland Prairie region and can still be observed on certain unplowed prairie remnants (e.g., northern Grayson County (Fig.14) and Tridens Prairie in Lamar County). These are circular, saucer-shaped hills roughly 1 to 14 meters in diameter and up to more than a meter tall. While numerous hypotheses have been proposed, the structures are of unknown and possibly multiple origins (Collins et al. 1975; Diamond & Smeins 1993). Both gilgai and mima mounds increase microhabitat diversity and thus cause vegetational differences over small distances. The overall biological diversity of the prairie is therefore increased (Miller & Smeins 1988; Diamond & Smeins 1993). Due to the different vegetation associated with the different microhabitats of both gilgai and mima mounds, these features are easily discernible in the field at certain seasons of the year (Figs. 13, 14).

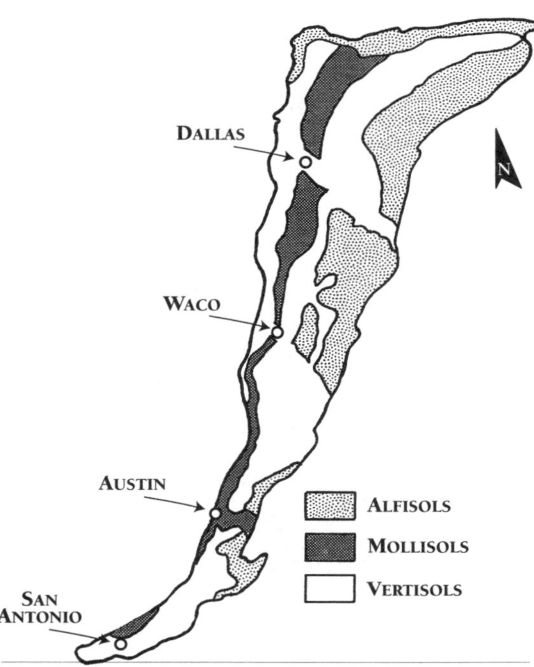

FIG. 10/DOMINANT SOIL ORDERS OF THE MAIN BELT OF BLACKLAND PRAIRIES IN NORTH CENTRAL TEXAS (FROM HALLMARK 1993, *IN* M.R. SHARPLESS AND J.C. YELDERMAN, EDS. THE TEXAS BLACKLAND PRAIRIE, LAND, HISTORY, AND CULTURE; WITH PERMISSION OF BAYLOR UNIV.; ©1993).

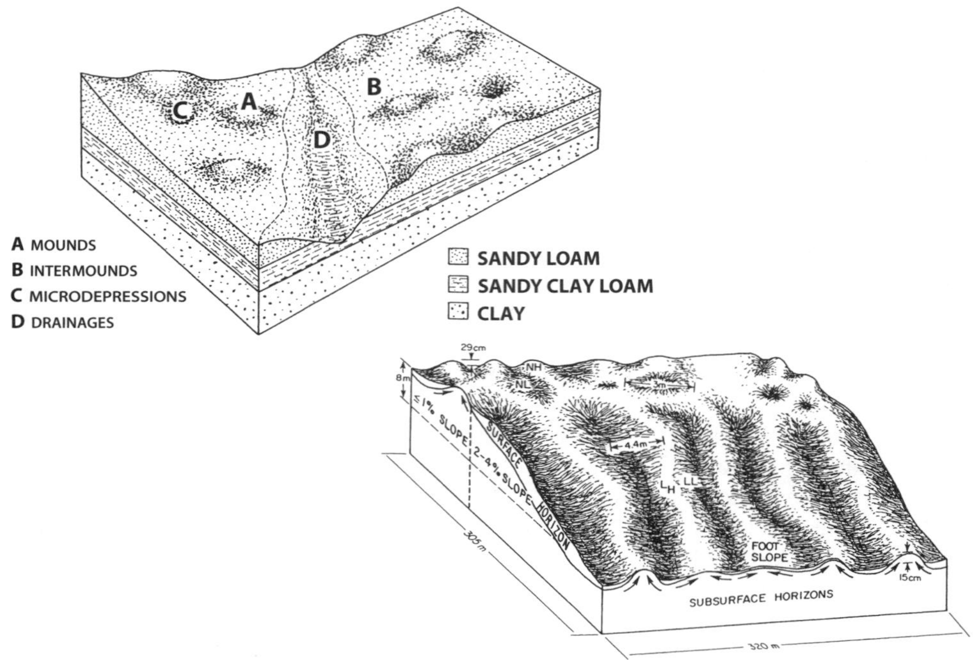

FIG. 11/DIAGRAMS OF MIMA MOUNDS AND GILGAI. **TOP**—MICROHABITAT VARIATION IN TEXAS TALLGRASS PRAIRIE SHOWING TYPICAL MIMA MOUNDS ON ALFISOL SOILS. **BOTTOM**—MICROHABITAT VARIATION IN TEXAS TALLGRASS PRAIRIE SHOWING TYPICAL GILGAI MICRORELIEF ON VERTISOL SOILS. NH=NORMAL HIGH; NL=NORMAL LOW; LH=LATERAL HIGH; LL=LATERAL LOW. (FROM DIAMOND & SMEINS 1993, *IN* M.R. SHARPLESS AND J.C. YELDERMAN, EDS. THE TEXAS BLACKLAND PRAIRIE, LAND, HISTORY, AND CULTURE; WITH PERMISSION OF BAYLOR UNIV.; ©1993).

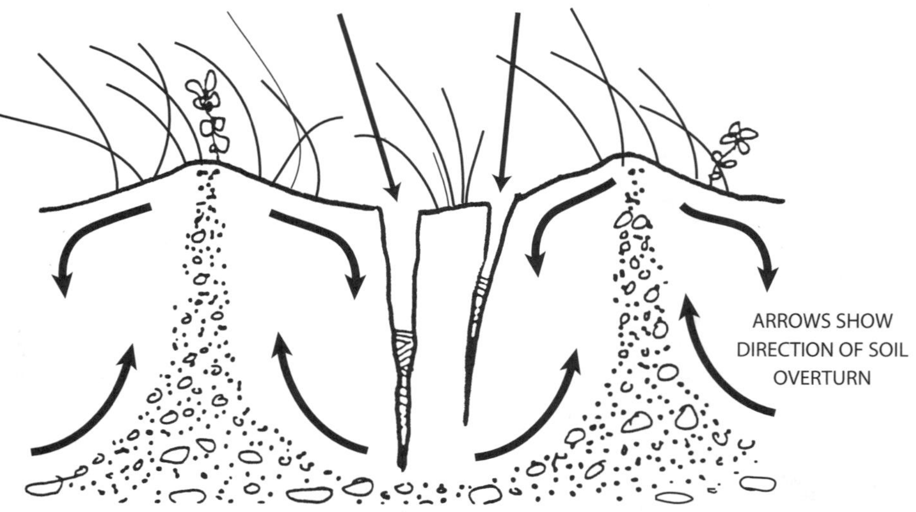

FIG. 12/DIAGRAM SHOWING GILGAI FORMATION (ADAPTED FROM HAYWARD & YELDERMAN 1991). THE TERM "GILGAI" DESCRIBES A PECULIAR FORM OF SURFACE CONFIGURATION, IN WHICH THE LANDSCAPE IS COVERED BY A VAST NUMBER OF SMALL DEPRESSIONS, TEN TO TWENTY FEET ACROSS, AND AS MUCH AS A FOOT-AND-A-HALF DEEP. IN WET SEASONS THESE FILLED WITH WATER, MAKING THE PRAIRIE ALMOST IMPASSABLE. THEY FORMED BECAUSE OF OVERTURN THROUGHOUT THE FULL DEPTH OF THE HIGHLY EXPANSIVE "BLACK WAXY" SOILS OF THE BLACKLANDS.

FIG. 13/PHOTOGRAPH OF GILGAI ON SLOPE (SHOWING MICROVALLEY AND MICRORIDGE EFFECT) ON VERTISOL IN NORTHERN GRAYSON COUNTY, TX (PHOTO BY GMD).

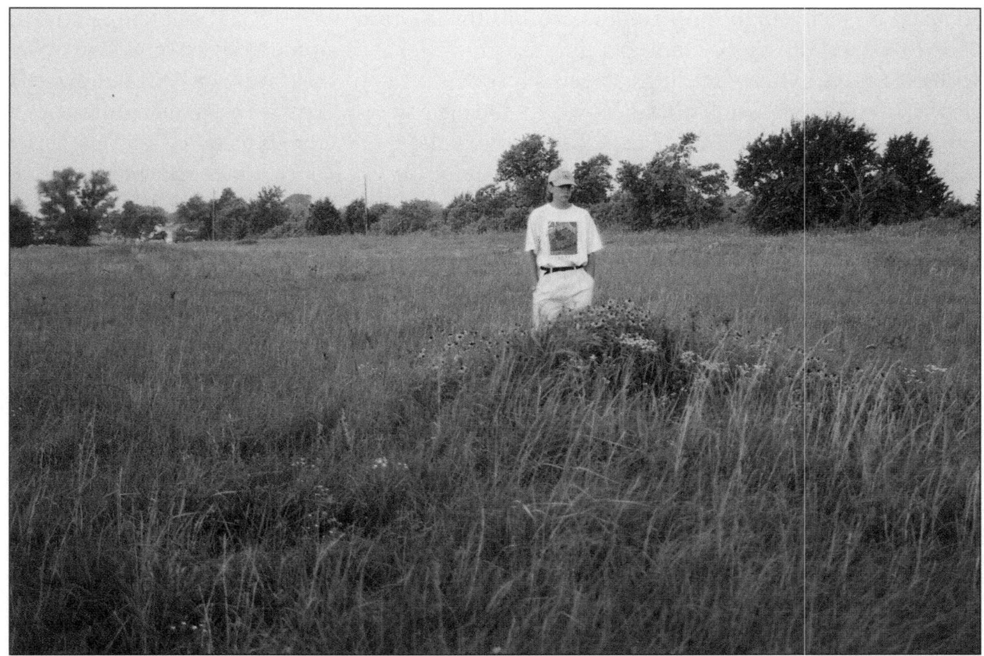

FIG. 14/PHOTOGRAPH OF MIMA MOUND ON ALFISOL IN NORTHERN GRAYSON COUNTY, TX (PHOTO BY GMD).

The soils of the East and West cross timbers are mainly developed on sandy Cretaceous Woodbine and Trinity strata. An exception are the soils in the western portion of the West Cross Timbers which are developed on gravelly and rocky Pennsylvanian strata. While somewhat variable (due to areas of clay in the Woodbine or Trinity Group sands and the complex nature of the Pennsylvanian strata), in general the soils of the cross timbers are rather loose, often deep sands. In fact, the sandy soils in some areas of the West Cross Timbers were so deep that early wagon travel was difficult because wagon wheels would sink in (Hill 1901). The loose nature of these soils also makes them extremely susceptible to erosion. Dyksterhuis (1948) discussed that tremendous gullies formed due to inappropriate cultivation and wind and water erosion, resulting in the subsequent abandonment of fields and even whole farms. However, it is because of these sandy soils, which are much more conducive to the growth of woody vegetation than the heavier clay soils typical of the Blackland and Grand prairies, that the cross timbers vegetation is found in this region.

CLIMATE OF NORTH CENTRAL TEXAS

North Central Texas is located in a zone of dramatic transition between regional climates. The striking vegetational change from the East Texas deciduous forests on the eastern margin of North Central Texas to the grasslands of the Great Plains just to the west is a vivid reflection of this climatic transition (Stahle & Cleaveland 1995). The climate of North Central Texas is often considered subtropical (Yelderman 1993; Norwine et al. 1995; Peterson 1995), but a wide range of extremes can be found. Like the rest of the state, the area can be referred to climatically as a "land of contrasts" (Bomar 1983). There is ample reason why locals say, "If you don't like the weather, just wait a few minutes." "Blue northers," cold fronts swinging down from the north and accompanied by rapid drops in temperature of dozens of degrees, are not uncommon (Bomar 1983). Mean annual temperature varies from about 68° F (20° C) in the south (Williamson County) to about 64° F (18° C) in the north and west (Griffiths & Orton 1968) (Fig. 15), but temperatures of 0° F (-18° C) and 110° F (43° C) are not unknown for winter and summer respectively (Tharp 1926), with even more extreme readings observed on occasion. In the heat wave of 1980, temperatures climbed to 113° F (45° C) in Dallas-Fort Worth, and there were 69 days with a temperature of 100° F (38° C) or above (Bomar 1983). The highest reading, 119° F (48° C), was recorded for Weatherford in Parker County in 1980 (Bomar 1983). In the extreme cold spell of December 1983, temperatures were below freezing in Dallas-Fort Worth for 12 straight days (Bomar 1983). The coldest temperatures recorded in North Central Texas include minus 8° F (-13° C) (Dallas-Fort Worth) and minus 13° F (11° C) (Paris), both in the unusually cold winter of 1899 (Bomar 1983). Native vegetation has evolved with, and is adapted to, such recurrent extremes. A good example of the different effects of the climate on native versus non-native plants was the extensive damage to introduced landscape plants during the winter of 1983, while most native plants were not adversely affected. The mean length of the frost-free period in the area is given in Figure 16.

While native plants are in general adapted to local weather conditions, they can be damaged under exceptional circumstances. An example was the unseasonably late freeze on the night of 11–12 April 1997. Following a period of relatively warm weather, temperatures dropped substantially below freezing over a large part of North Central Texas. For example, a low of 22° F (-6° C) was recorded for a native habitat (Garnett Preserve) in Montague County (H. Garnett, pers. comm.). The result was substantial damage to the young foliage of many native species and in some cases nearly complete defoliation. Some of the species significantly damaged in Grayson County include *Berchemia scandens* (supple-jack), *Cercis canadensis* (redbud), *Diospyros virginiana* (common persimmon), *Fraxinus americana* (white ash), *Gleditsia triacanthos* (common honey-locust), *Morus rubra* (red mulberry), *Platanus occidentalis* (sycamore), *Quercus macrocarpa* (bur oak), *Quercus marilandica* (blackjack oak), *Quercus muhlenbergii* (chestnut oak), *Quercus shumardii* (Shumard's red oak), *Quercus stellata* (post oak), and *Rhus glabra* (smooth sumac). Effects on post and blackjack oaks in areas of Cross Timbers vegetation at Hagerman National Wildlife Area (Grayson County) were serious enough that leaf damage was still obvious at a glance in late May. This event is a good example of a false spring weather anomaly. As discussed by Stahle (1990) and Stahle and Cleaveland (1995), a false spring episode includes late winter warmth followed by the movement of polar or arctic air into southern regions. The resulting intense subfreezing temperatures following a warm spell cause widespread damage to cultivated crops as well as native plant species which were advanced in their development by the unusually mild winter temperatures. Forty-four such false spring episodes have been documented in Texas between A.D. 1650 and 1980 (Stahle & Cleaveland 1995). This detailed information can be obtained because frost-damaged cambial tissues leave a permanent record in the annual growth rings of trees and these can be dated dendrochronologically (tree-ring dating) to the exact year of their formation. While there has been a notable decline in the frequency and intensity of false spring episodes in Texas in the last 100 years, the cause of this decrease is not clear (Stahle & Cleaveland 1995).

In terms of precipitation, there is a steep east-west gradient across North Central Texas. Mean annual precipitation ranges from about 46 inches in the northeastern corner of the area in Lamar County to about 24 inches in the westernmost portion of the West Cross Timbers (Dyksterhuis 1948; Griffiths & Orton 1968) (Fig. 17). In general, mean annual precipitation decreases about one inch for each 15 miles across Texas from east to west (Bomar 1983), in part as a result of the decreasing influence of moist air from the Gulf of Mexico. Thus there is a rainfall difference of more than 20 inches between the eastern and western boundaries of North Central Texas. This region, like much of the state, is prone to drought, while at other times it receives too much rain in a short time (Sharpless & Yelderman 1993). Severe storms and some of the largest rainfalls in the United States have occurred in this area. According to Hayward et al. (1992), all the point rainfall records for North America are held within a belt 50 miles east and west of a line from Dallas through Waco, Austin, and San Antonio. The town of Thrall, in Williamson County on the eastern edge of the Blacklands, had one of the largest recorded United States rainfalls on 9–10 September 1921, receiving 38.2 inches in 24 hours (Yelderman 1993). In 1978, Albany, in Shackelford County on the western edge of North Central Texas, received 29.05 inches in one day as a result of Hurricane Amelia. The yearly precipitation record for Texas, 109.38 inches, is also from this region. It occurred in 1873 at Clarksville in Red River County near the extreme northeastern tip of the Blackland Prairie (Bomar 1983).

In years past, the incredibly sticky "black waxy" soil of the Blackland Prairie was particularly problematic during wet weather. Personal accounts (e.g., Mosely in Yelderman 1993) described that under wet conditions the dirt roads were virtually impassable and families actually went hungry until the ground dried enough for people to get to town to obtain food. Drought, however, has been more of a problem, with the lack of water probably always being a limiting factor for humans in the area. The impermeable clay soils, the lack of dependable shallow water-bearing layers, and the scarcity and transitory nature of surface streams made the early Blacklands a particularly inhospitable environment for humans. This is exemplified by early accounts such as the one by D.P. Smythe (1852) who described a trip across the Blacklands:

> The soil improves now at every step becoming more level, and uniformly of a dark rich color, but the water is very bad and scarce, drying up entirely during the heat of the summer.... During the forenoon of today we must have traveled some twenty miles without passing over a spot of thin soil; being chiefly the black stiff 'hog wallow' prairie, rolling just enough to drain itself, but entirely destitute of water during the summer....

Josiah Gregg, another early explorer, indicated that in addition to droughts, the lack of springs or dependable water was "one of the greatest defects of this country" (Fulton 1941).

The concentration of rainfall in spring and fall, coupled with hot dry summers, makes the water problem even more acute (Yelderman 1993). The severe drought of the mid-1950s exemplifies water difficulties in the area. City water supplies were alarmingly low (e.g., Lake Dallas at 11%

of capacity), cattle and sheep ranchers were desperate, wells declined to record levels, and streams either barely flowed or dried up (Bomar 1983). At the time of settlement, while shallow wells were used in some areas, cisterns were the only source of water in others (Hayward & Yelderman 1991; Yelderman 1993). Currently, access to deep aquifers, such as the Trinity, and surface storage in large reservoirs (e.g., Lake Texoma, Lake Lewisville, Lake Ray Hubbard), provide water in this water-poor environment (Hayward & Yelderman 1991). However, Simpson (1993) has emphasized that, "Texas has been a water-deficit state since the dawn of recorded history" and that, "The problem will only be exaggerated as population growth expands." Currently, many cities and water-supply corporations in Texas are actively seeking access to more water, and some cities, such as San Antonio, have a serious water supply problem (Simpson 1993). The recurrent water difficulties seen locally are a reminder of the overall scarcity of water in the southwestern United States.

Major storms have also long been a problem in the region, as can be seen by Parker's (1856) account of the effects of a tornado in 1854 near Gainesville in Cooke County. He spoke vividly of trees broken off ten feet above the ground, an ox wagon carried a quarter of a mile, and a sheep blown into the top of a high tree. More recent destructive tornadoes (e.g., Waco 1953, Paris 1982—Bomar 1983, Jarrell in Williamson Co. 1997) and hail storms (e.g., grapefruit size hail in Fort Worth in May 1995) are present-day reminders of the ongoing power of extreme weather events.

From a longer term perspective, pollen, plant macrofossils, and other types of evidence demonstrate that the climate of Texas has changed substantially over the past 15,000 years. At 15,000 years BP, there was a more widespread forest mosaic over most of Texas with boreal species such as *Picea glauca* (white spruce) in specialized microhabitats (Stahle & Cleaveland 1995). Certain present-day plant distributions, such as the rare western occurrence of plants normally found predominantly in eastern Texas, may thus be relicts of these past climatic conditions. While long-term climate change is well-documented, attention has focused recently on the possibility

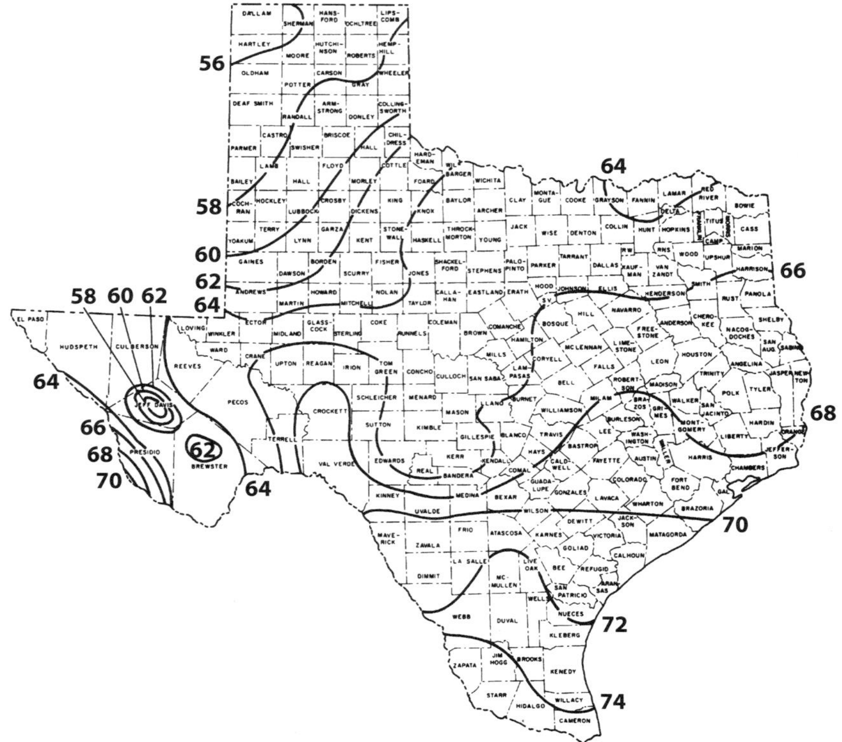

FIG. 15/MEAN ANNUAL TEMPERATURE (°F) FOR TEXAS (ADAPTED FROM GRIFFITHS & ORTON (1968) BY HATCH ET AL. (1990)).

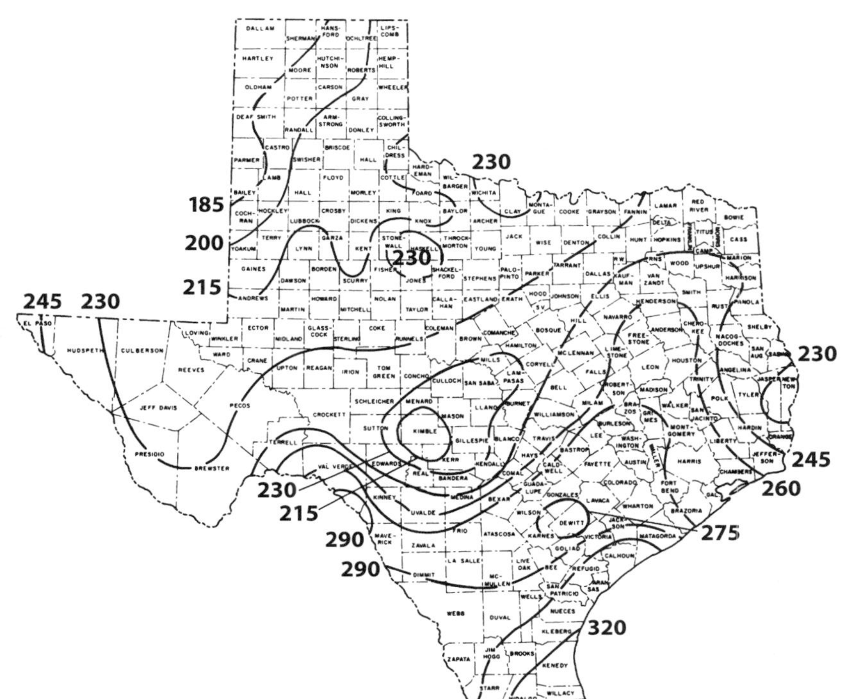

FIG. 16/MEAN LENGTH (IN DAYS) OF FROST-FREE PERIOD FOR TEXAS (ADAPTED FROM GRIFFITHS & ORTON (1968) BY HATCH ET AL. (1990)).

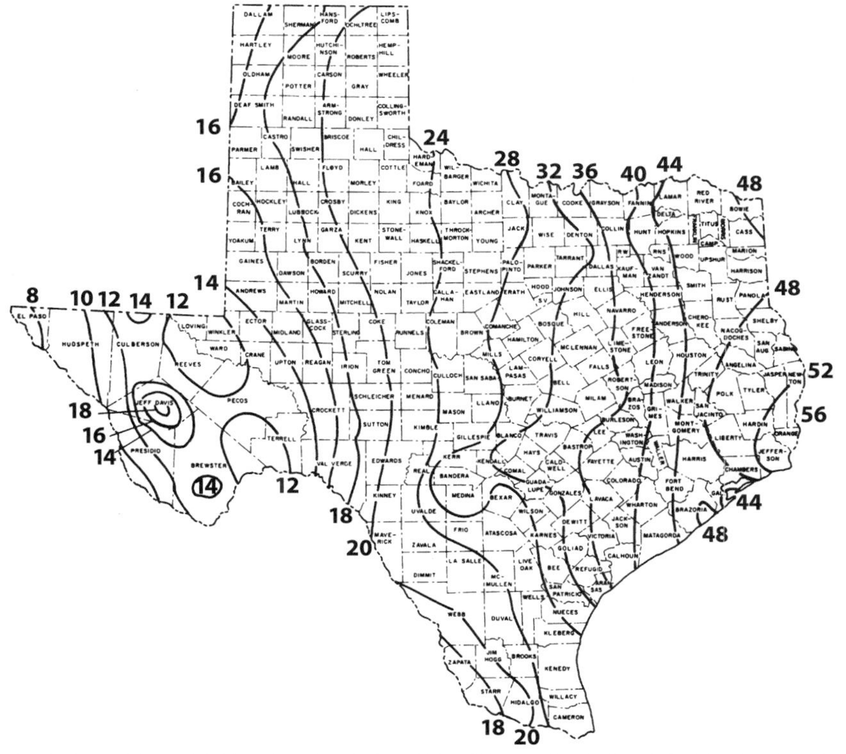

FIG. 17/MEAN ANNUAL TOTAL PRECIPITATION (IN INCHES) FOR TEXAS (ADAPTED FROM GRIFFITHS & ORTON (1968) BY HATCH ET AL. (1990)).

of future climate change in Texas due to human-induced modifications of the atmosphere (e.g., increased CO_2 concentrations) and the resulting increased greenhouse effect and global warming (e.g., Norwine et al. 1995). While considerable controversy exists over details, there is solid evidence that global atmospheric CO_2 concentrations have increased by about 30% since preindustrial times and that this trend can be attributed primarily to human activities (e.g., fossil fuel use, land-use changes, and agriculture) (Houghton et al. 1995). Further, consensus now exists that there is "... a discernible human influence on climate" (Houghton et al. 1995).

Plants can contribute to an understanding of climate change in several ways. First, dendrochronology, the study of tree rings, can provide information on past climate and thus a reference point for present and future studies (Stahle & Cleaveland 1992). Extensive tree-ring chronologies based on remnant old-growth *Taxodium disticum* (bald cypress) stands have provided accurate and well-verified climatic reconstructions for the past 1,000 years for some areas of the southeastern United States, including northwestern Louisiana (Stahle et al. 1988; Stahle & Cleaveland 1992). In North Central Texas, well-documented chronologies based on remnant populations of *Quercus stellata* (post oak) have yielded detailed information on climate for the past 300 years (Stahle & Hehr 1984; Stahle et al. 1985; Stahle & Cleaveland 1988; 1993; 1995).

Changes in phytogeography (plant distributions) can also indicate climate change. An example is the long-term McWilliams (1995) study of the distribution of *Tillandsia recurvata* (ball-moss). This species has expanded its geographical range in Texas over the last 80 years, with much of the expansion occurring since the 1940s. McWilliams demonstrated that even slight changes in temperature or moisture conditions can have significant implications for the survival of plants at the margins of their ranges. The eastward expansion into North Central Texas of species adapted to the drier western Rolling Plains and the northward shift of southerly species would both be expected based on climate models which predict increased temperature (and thus evapotranspiration) and decreased regional precipitation and soil moisture (Houghton et al. 1990; Packard & Cook 1995; Schmandt 1995).

THE BLACKLAND PRAIRIE

OCCURRENCE OF THE BLACKLAND PRAIRIE

The Blackland Prairie of Texas is a well-defined band stretching roughly three hundred miles from the Red River (Oklahoma border) south to near San Antonio (Chambers 1948; Sharpless & Yelderman 1993). It is widest at the north, extending from Grayson County east to near Clarksville in Red River County. It narrows to the south, tapering to a point near San Antonio (Fig. 18). The main belt of the Blackland Prairie totals about 4.3 million hectares or roughly six percent of the total land area of Texas (Collins et al. 1975), and is a region slightly larger than the state of Maryland. It coincides almost exactly with a belt of outcropping Upper Cretaceous marine chalks, marls, and shales (Hayward & Yelderman 1991) that upon weathering forms the characteristic black, calcareous, alkaline, heavy clay, "black waxy" soil. In this work we are including the main body of the Blackland Prairie from the Red River south to the Travis County-Williamson County line just north of Austin, but not the San Antonio Prairie or the outlying Fayette Prairie to the east. Topographically, the Blackland Prairie is a nearly level to gently rolling dissected plain (Hallmark 1993); elevations range from about 300 to 800 feet (92 to 244 meters) (Thomas 1962). Roughly speaking, the Blacklands are bounded on the north by the Red River, on the east by the Post Oak Savannah (also called the Oak-hickory) vegetational area, and on the west by the East Cross Timbers and the Lampasas Cut Plain. North of Sherman in Grayson County, the trend of the Blacklands undergoes a shift in direction, turning from north-south to east-west, before ending near Clarksville in Red River County. In this work we use the terms Blackland Prairie and Blacklands interchangeably.

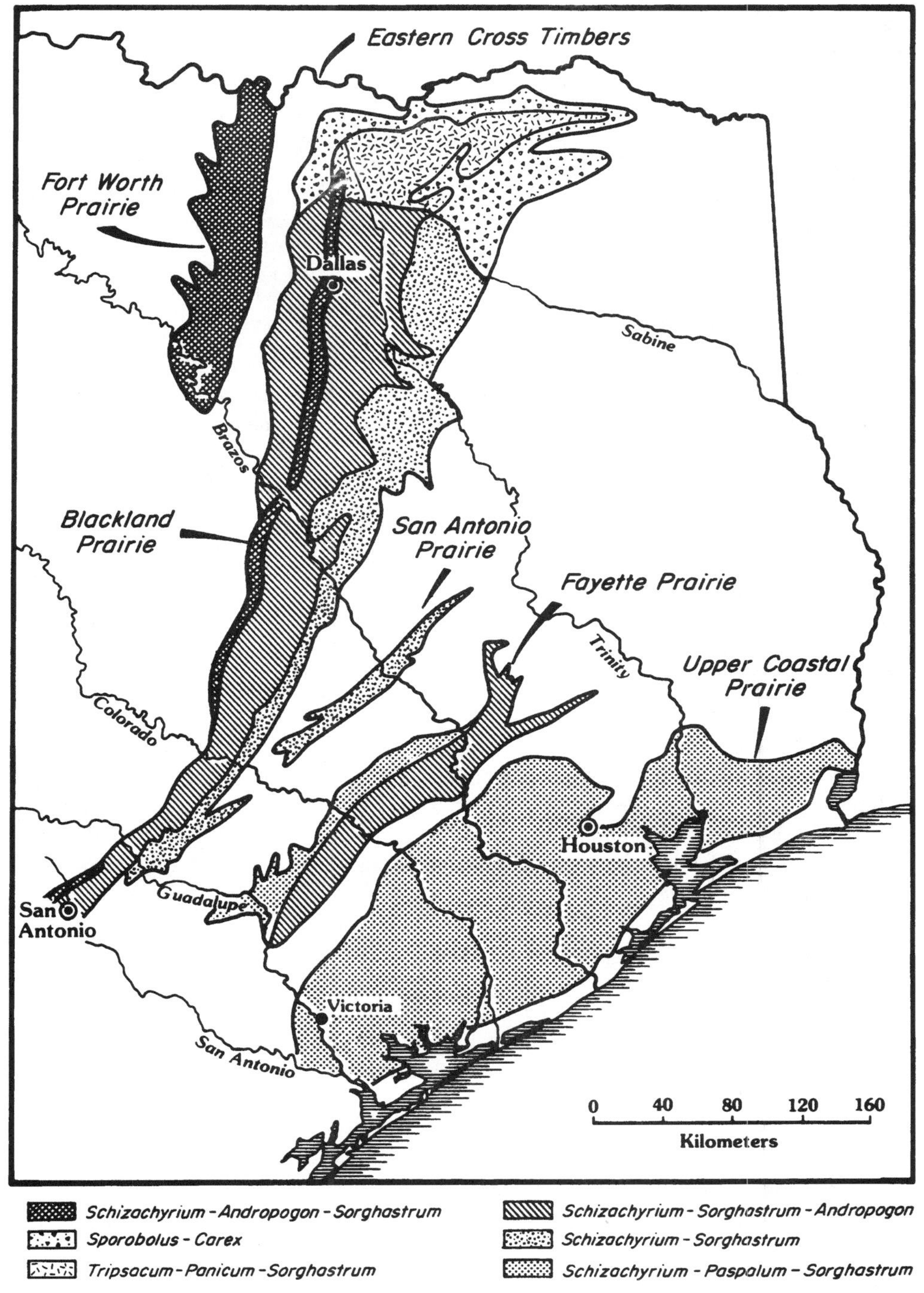

FIG. 18/MAJOR PLANT COMMUNITY TYPES OF THE BLACKLAND PRAIRIE AND RELATED TALLGRASS REGIONS OF TEXAS (FROM DIAMOND & SMEINS 1993, *IN* M.R. SHARPLESS AND J.C. YELDERMAN, EDS. THE TEXAS BLACKLAND PRAIRIE, LAND, HISTORY, AND CULTURE; WITH PERMISSION OF BAYLOR UNIV.; ©1993).

PRESETTLEMENT AND EARLY SETTLEMENT CONDITIONS ON THE BLACKLAND PRAIRIE

Conditions on the presettlement Blackland Prairie were strikingly different from those found today. Probably the most conspicuous difference was the presence of vast expanses of tall grass prairie. In the words of Parker (1856), traveling with the 1854 Marcy expedition, "After leaving Preston [northern Grayson County], we entered upon the vast plains. . . ." Dr. John Brooke, who emigrated from England in 1848, stated on arriving at the edge of the Blackland Prairie, "It was the finest sight I ever saw; immense meadows 2 or 3 feet deep of fine grass & flowers. Such beautiful colours I never saw. . . ." (Brooke 1848). In describing the area where he settled near Dorchester in south central Grayson County, Brooke (1849) said,

> I can sit on the porch before my door and can see miles of the most beautiful Prairie interwoven with groves of timber, surpassing, in my idea, the beauties of the sea. Think of seeing a tract of land on a slight incline covered with flowers and rich meadow grass for 12 to 20 miles. . . .

Hill (1901), speaking of the Blacklands in general, said,

> The surfaces of the prairies are ordinarily clad with thick mantles of grass, liberally sprinkled with many-colored flowers, broken here and there by low growths of mesquite trees, or in exceptional places by 'mottes' or clumps of live oaks on uplands, pecan, bois d'arc, walnut and oaks in the stream bottoms; juniper and sumac where stony slopes exist, and post oak and black-jack in the sandy belts.

Smythe (1852) described the eastern edge of the Blackland Prairie as having

> . . . a view of almost boundless Prairie stretching to the north, as far as the eye could reach. . . .

He further referred to it as

> . . . a boundless plain scarcely broken by a single slope or valley, and nearly destitute of trees; (the mesquite appearing but seldom.) Several times during the forenoon not a single shrub or tree could be seen in any direction. . . . The grazing has reached its climax, it would be impossible for natural pasturage to excell [excel] this.

Kendall (1845) described the southern part of the Blackland Prairie as "beautiful rolling prairies, the land rich, and susceptible of cultivation." Roemer's (1849) descriptions of the same region included "open prairie," "extensive prairies" with mesquite trees and scattered oak groves, "undulating prairie extending. . . an immeasurable distance," and "gently rolling, almost treeless plain." Indeed, on the Blackland Prairie, trees were often rare except as riverine forests along streams or as occasional scattered groves or mottes "such as the one near Kentuckytown that gave Pilot Grove [in southeastern Grayson County] its name, the trees being a major landmark in a featureless terrain." (McLeRoy 1993). The riverine forests along Big Mineral Creek [Grayson County] were described by Parker (1856) as "a rich bottom, thickly grown up with large cotton wood, honey locust, overcup [bur oak], and other heavy timber, besides plenty of the bois d'arc." Roemer (1849) described a trading post he visited in Falls County as "on a hill covered with oak trees, two miles distant from the Brazos, above the broad forested bottom of Tohawacony Creek." He further described the wooded bottomland as having "high, dense trees."

Fire was probably an important factor in maintenance of the original prairie vegetation and had a major impact on community structure (Anderson 1990; Collins & Gibson 1990; Strickland & Fox 1993). Tall grass prairie fires, intensely hot, would have been stopped only by the lack of dry fuel or a change in topography. Even streambank vegetation was susceptible during dry years. The end result was that trees were rare even along some stream banks, and prairie margins probably extended somewhat beyond the limits of the soil types usually associated with prairie (Hayward & Yelderman 1991). Roemer (1849) wrote of a prairie fire as follows:

> . . .we, ourselves, were entertained before going to sleep by the spectacle of a prairie fire. Like a sparkling diamond necklace, the strip of flame, a mile long, raced along over hill and dale, now moving slowly, now faster, now flickering brightly, now growing dim. We could the more enjoy this spectacle undisturbed, since the direction of the wind kept it from approaching us. My companion was of the opinion that Indians had without doubt started the fire, since they do this often to drive the game in a certain

> direction, and also to expedite the growth of the grass by burning off the dry grass.

While lightning was an important source of naturally started fires, Native Americans were long present in the region and their use of fire is considered by some to be equally important in having maintained North American grasslands (Bragg 1995).

In summary, the original Blackland Prairie seems to have been predominantly tall grass prairie with trees along watercourses, sometimes scattered on the prairie or concentrated in certain areas (e.g., Pilot Grove) possibly as the result of locally favorable soil conditions or topography.

It is interesting to note that early (back to the 1830s) surveyor records of mesquite as the most common tree in presettlement upland prairies in Navarro County suggest "...the legendary spread of mesquite into North Texas by longhorn cattle may be an errant concept" (Jurney 1987). Roemer's (1849) mention of "extensive prairies covered with mesquite trees" also points to mesquite as a natural component of the vegetation. However, mesquite has increased in many areas and the observations mentioned above are not so early as to preclude mesquite having already been spread to some extent by land use changes.

While some question the degree to which mesquite was spread by longhorns, animals have had profound impacts on the vegetation since long before settlement. These range from the obvious effects of the bison or beaver to the more subtle but essential roles of pollination and seed dispersal. Present animal life is much different and some species reduced compared to presettlement days. In addition to relatively large present-day species such as the white-tailed deer, coyote, fox, and bobcat, a number of other large or interesting species occurred. According to Brooke (1848) writing of Grayson County, black bears were quite common ("I...have never tasted any meat I like better.") as were deer; panthers [mountain lions] and wolves were also present. In Brooke's (1848) words, "I have been out a-shooting Deer and Turkeys alone, and when going up the branches of the Rivers I often come across either bear or wolf...." Strecker (1926a) (based on early fur-trader records) indicated that next to the skins of deer, "those of the black bear were of the most value to the Indians of McLennan County." Strecker (1926a) also reported that gray wolves occurred as far east as McLennan County. He indicated that they

> ... may never have been very common permanent residents of McLennan County, but in late fall and winter, small packs followed the great herds of buffalo and deer from northwestern Texas and remained here for several months. It was probably only a minority that remained throughout the year. Old settlers refer to packs of from five to eight wolves which they considered small family groups.

Another predator, the ocelot, is thought to have ranged as far north as the Red River (Hall & Kelson 1959). Strecker (1924), for example, reported that ocelot occurred in the bottoms of the Brazos River near Waco in McLennan County. Even jaguar are believed to have ranged north to the Red River; the last jaguar record from North Central Texas was a large male killed in Mills County (Lampasas Cut Plain) in 1903 (Bailey 1905). Mountain lions probably occurred throughout North Central Texas (Schmidly 1983), with Strecker (1926a) indicating they were common in McLennan County in the middle of the 1800s. However, they were rare by the beginning of the 20th century (Bailey 1905) and since that time have been eliminated over most of the region (Schmidly 1983). The collared peccary or javelina, a small wild pig, was also originally present in the southern portion of the area, north to at least the Brazos River valley in McLennan County near Waco (Strecker 1926a; Schmidly 1983; Davis & Schmidly 1994). Other noteworthy large mammals that previously occurred in appropriate habitats of the Blackland Prairie as well as throughout the rest of North Central Texas include river otter, ringtail, and badger (Schmidly 1983; Davis & Schmidly 1994).

The occurrence of bison was documented by Judge John Simpson of Bonham (Fannin County). Simpson, describing a bison hunt in 1833, reported that hunters found "an immense herd" "on the prairie around Whitewright [Grayson County]" (McLeRoy 1993). Parker (1856), in his 1854 journal, stated, "But eight years since, herds roamed around the City of Austin, and were frequently seen in the streets; now there are but few to be found south of Red River." Roemer (1849) described bison on the southern Blackland Prairie as follows:

> When on the following morning at daybreak we entered the prairie on which mesquite trees grew scatteringly, the first object that met our view was a buffalo herd, quietly grazing near us.... The whole prairie was covered with countless buffalo trails, crossing in all directions, reminding one of a European grazing ground.

On a different day, Roemer (1849) indicated,

> They covered the grassy prairie separated into small groups and far distant on the horizon they were visible as black specks. The number of those clearly seen must have been not less than a thousand.

Pronghorn antelope were also native, occurring at least as far east as Fannin County (Hall & Kelson 1959). Smythe (1852) described a small herd on the eastern edge of the Blacklands, Roemer (1849) mentioned sighting pronghorn antelopes near where the Blackland Prairie and Lampasas Cut Plain come together, and Major G.B. Erath, a pioneer of Waco, indicated that antelope were common in what is now McLennan County in the early to middle 1800s (Schmidly 1983). Erath also reported that small herds penetrated as far east as Milam County on the eastern edge of the Blackland Prairie (Strecker 1926a).

While not native, wild horses, descended from those escaped from the Spanish, were by the early 1800s extremely common in Texas and were probably having a significant impact on the vegetation. Ikin (1841), speaking of Texas as a whole, indicated,

> The wild horse which now roams every prairie, sometimes alone, sometimes in herds of more than a thousand, is not native, but the progeny of those which escaped from the early conquerors of Mexico. He is usually a small bony animal about fourteen hands high, with remarkably clean legs, and other signs indicative of good blood. When congregated in bodies of a thousand, these horses form the most imposing spectacle which the prairies present.

Strecker (1926a) also reported the wild horse as abundant throughout the Brazos Valley of McLennan County at the time of arrival of the first American settlers. He further indicated that early settlers sometimes shot the wild horses to prevent interference with their domesticated stock.

The bird, reptile, and fish faunas were also conspicuously different in significant ways from those today. Brooke (1848), writing about early Grayson County, mentioned both turkeys and prairie chickens, and Smythe (1852) spoke of hunting "Prairie Hens" in what is now Limestone County on the eastern edge of the Blackland Prairie. According to Pulich (1988), both greater and lesser prairie chickens were common in North Central Texas until the 1880s; these two species were locally extinct by the early 1900s. Oberholser (1974) mentioned a specimen record for the greater prairie chicken from Dallas County with a number of other North Central Texas records west of the Blackland Prairie in Clay, Cooke, Denton, and Navarro counties. There is a questionable record for the lesser prairie chicken from Dallas and also records for this species from Cooke and Young counties to the west of the Blacklands (Oberholser 1974). The extinct passenger pigeon is also well documented for the Blackland Prairie. These birds, known as "wild pigeons" by early settlers, were recorded from Collin, Fannin, and Henderson counties, with a number of records even farther west in the Grand Prairie, Lampasas Cut Plain, and West Cross Timbers (Oberholser 1974; Pulich 1988). This once very numerous species rapidly became extinct in North Central Texas, with 1896 being the last record in the area (Oberholser 1974; Pulich 1988). The ivory-billed, one of the world's largest woodpecker species, was also present in bottomland forests in the Blacklands. Oberholser (1974) listed records for Cooke, Dallas, Fannin, and Kaufman counties with sightings as late as the early 1900s (Pulich 1988). Another extinct species, the Carolina parakeet, was known from eastern Texas (Greenway 1958) and was probably also present in the riverine forests of the Blackland Prairie (Goodwin 1983), especially along the Red and Trinity rivers. Even more surprising, alligators were abundant in places, with Kendall (1845) describing them along the San Gabriel in the southern Blackland Prairie as "too plentiful for any useful purposes." This large reptile occurred in appropriate habitats throughout most of the Blackland Prairie, west to Grayson, Dallas, McLennan, and Williamson counties

(Brown 1950; Hibbard 1960; Dixon 1987), and are still known to occur in Dallas County. Kendall (1845) further indicated concerning the San Gabriel that "The stream abounds with trout, perch, and catfish, as do nearly all the watercourses in this section of Texas."

During the Pleistocene, an even more extensive megafauna occurred in the area (Smeins 1988), as shown by the excavation of a mammoth from near Flowing Wells (Grayson County) by Dr. Daniel Schores and a student team from Austin College (D. Schores, pers. comm.). Further, fossils of at least three elephant species, including mammoth and mastodon, are known from the Dallas area (Shuler 1934). An even more impressive site containing a large (20+) mammoth herd was found near Waco in a Brazos River terrace dated around 28,000 years BP (Fox et al. 1992; C. Smith, pers. comm.). Several woody plants found in the Blackland Prairie region seem to have adaptations that are difficult to explain based on interactions with the present fauna. Bois d'arc (*Maclura pomifera*), honey-locust (*Gleditsia triacanthos*), and mesquite (*Prosopis glandulosa*) all have fruits that are adapted for dispersal by large animals (megafauna) and seem to fit Janzen and Martin's (1982) hypothesis that large, now extinct animals were involved in the evolution of certain "anachronistic" plant characteristics we see today. Another such possible characteristic is the protective armature displayed by honey-locust. The long, stout, branched thorns, up to a foot or more long, would seem perfectly reasonable in Africa where there are abundant large herbivores, but rather out of place in northern Texas where currently no large native browsers exist.

In general, the animals of the Blacklands have faunal affinities with the eastern woodlands, the Great Plains, and the southwestern United States (Schmidly et al. 1993). A recent, now very abundant, southern addition to the fauna is the nine-banded armadillo. This species is originally native to South America, and as recently as the 1870s to 1880 was found only at the southern tip of Texas (Strecker 1926b; Phelan 1976). Since that time it has spread extensively and is now found hundreds of miles north of Texas (Hall & Kelson 1959). Armadillos were at least sporadic as far north as the Red River by the early 1930s, but did not become common there until the 1950s (H. McCarley, pers. comm.).

The earliest use of the Blackland Prairie by settlers was as grazing areas for herds of cattle or horses. According to Hayward and Yelderman (1991) "... the Blackland Prairie supported some of the earliest of large-scale ranching efforts in Texas, complete with pre-Civil War cattle drives to St. Louis and Chicago." Brooke (1848) stated that, "... the cattle and horses feed on the prairies all winter; no need of laying up winter food." Parker (1856) wrote of a herd of 1,200 wild cattle being driven north across the Red River at Preston (Grayson County).

While limited "sod plowing" occurred quite early (Smythe 1852), it wasn't until the 1870s and 1880s, with the coming of the railroads and the development of special plows and favorable economic conditions, that extensive "breaking of the prairie" and exploitation of its agricultural potential finally occurred (Hayward & Yelderman 1991). Once farming on the Blacklands was possible, widespread cultivation of the rich soils, perhaps as rich as any in the nation (Hayward & Yelderman 1991), was inevitable and farming quickly replaced ranching. Cotton soon became an important crop and thus began the era referred to as the Cotton Kingdom. According to Sharpless and Yelderman (1993), for seventy years more cotton was grown on the Blackland Prairie than any other region of the world. Hill (1901) said, "In fact these calcareous soils ... of the Black Prairies are the most fertile of the whole trans-Mississippi region." Others (e.g., Sharpless and Yelderman 1993) have said the soil is arguably the most fertile west of the Mississippi River. Within a very short time, most of the accessible and desirable land was put into cultivation, and according to Burleson (1993), by 1915 the human population on the Blacklands was greater than on any other United States area of comparable size west of the Mississippi. The result was the virtually complete destruction of native Blackland Prairie communities. With the exception of small or inaccessible areas and a relatively few hay meadows valued for their native grasses, almost nothing remains of the tall grass prairies that were once so abundant. Estimates of the destruction of this ecosystem range from 98% (Hatch et al. 1990) or 99% (Riskind & Collins 1975) to more than 99.9 % (Burleson 1993).

GEOLOGY OF THE BLACKLAND PRAIRIE

The Blackland Prairie is on an erosional landscape developed from easily erodible Cretaceous shales, marls, and limestones that dip gently to the east (Hayward & Yelderman 1991). It originally consisted of four somewhat different parallel north-south bands of vegetation: the Eagle Ford Prairie, the Whiterock Prairie, the Taylor Black Prairie, and the Eastern Transitional Prairie. These all correspond to underlying geologic layers (Hayward & Yelderman 1991).

The westernmost and geologically oldest portion of the Blackland Prairie, known as the Eagle Ford Prairie, is developed on the Eagle Ford Shale, Upper Cretaceous material deposited about 92 to 90 million years ago (Hayward & Yelderman 1991). This layer crops out just east of the Woodbine Sand, on which the East Cross Timbers are found. While variable, the Eagle Ford Shale is principally a dark bluish-gray to nearly black shaly clay (Bullard 1931) that weathers to form black vertisol soils supporting prairie vegetation.

Cropping out to the east of the Eagle Ford Shale is the slightly younger Austin Chalk, deposited about 90 to 85 million years ago. This layer, which supports the Whiterock Prairie, forms the elevated backbone or "axis" of the Blacklands (Hayward & Yelderman 1991). It is a strikingly white, very fine-grained limestone, called chalk, made primarily of millions of calcium carbonate cell walls of tiny marine algae. Similar deposits make up the famous white cliffs of Dover in southern England and are used commercially as writing chalk. The Austin Chalk is a relatively resistant hard layer (Dallas Petroleum Geologists 1941) compared to the surrounding shales, and because of this hardness, it forms a rather conspicuous escarpment from Sherman to Dallas and south to Austin. This topographic feature is sometimes referred to as the "white rock escarpment," "white rock scarp," (Hill 1901) or "white rock cuesta," and although it never exceeds 200 feet in elevational difference from the surrounding terrain (usually much less), it is the most conspicuous topographic feature in the Texas Blacklands (Hill 1901; Montgomery 1993). It typically crops out as a west-facing bluff or escarpment overlooking a prairie formed on the less resistant Eagle Ford Shale (Bullard 1931). In striking contrast to the Grand Prairie with its numerous resistant layers, the Austin Chalk is the only resistant, escarpment-forming layer underlying the entire Blackland Prairie. As a result, most of the Blackland Prairie is gently rolling, in contrast to the sharper, more angular topography of the Grand Prairie (Hill 1901). Surprisingly, the extremely white Austin Chalk weathers to form a sticky black soil, typically thinner than, but similar to that derived from the Eagle Ford Shale (Bullard 1931) (Fig. 9). Where this soil is eroded away, as on stream banks, a distinctive flora can be found on the exposed chalky limestone (see description under vegetation). Despite their biological diversity, these exposed chalky areas are of little commercial value and are thus often destroyed by contouring or other types of "remediation."

The layers that crop out to the east of the Austin Chalk are the Taylor marls and sandy marls, laid down about 79 to 72 million years ago. The Taylor Blacklands, the largest of the four Blackland Prairie belts, occurs on the soils derived from these rocks (Hill 1901; Hayward & Yelderman 1991). In fact, Taylor sediments underlie about two-thirds of the total Blackland Prairie (Hill 1901). The soils developed on Taylor rocks are the classic deep, rich, calcareous, "black waxy" soils that were formerly so valuable for cotton production.

Finally, the easternmost and youngest Cretaceous rocks supporting Blackland Prairie are those of the Navarro group, deposited about 72 to 68 million years ago (Hayward & Yelderman 1991). These deposits crop out from Red River County in the north, through Kaufman and Navarro counties, south to Williamson County on the southeastern margin of North Central Texas. They break down into a soil with a somewhat higher sand content than the Blackland soils farther west, and support the easternmost of the Blackland Prairies, the Eastern Transitional or Marginal Prairie (Hill 1901; Hayward & Yelderman 1991). While easier to till, these soils are poorer in nutrients and thus not as valuable for farming (Hayward & Yelderman 1991). Immediately to the east of the Navarro Group and east of North Central Texas, on younger sandy deposits of Tertiary-age, the Post Oak Savannah begins, marking the western edge of the eastern deciduous forest.

Vegetation of the Blackland Prairie

According to Gould and Shaw (1983), the Blackland Prairie (and in fact all of North Central Texas) is part of the True Prairie grassland association, extending from Texas to southern Manitoba. This is one of the seven grassland associations of North America recognized in the classification system of Gould (1968a) and Gould and Shaw (1983) (Fig. 19). Based on location, climate, and vegetational characteristics, the tall grass prairies of the Texas Blacklands can be considered part of either the True Prairie or Coastal Prairie associations (Collins et al. 1975). They lie at the very southern end of the True Prairie association, but are also connected to the Texas Coastal Prairie. Rainfall values are intermediate and the Blackland Prairies have most of the vegetational dominants of both these areas. According to Collins et al. (1975), adequate data are not currently available for a clear determination. Many authorities, however, recognize the tall grass prairies of the Blacklands as an extension of the True Prairie with little bluestem (*Schizachyrium scoparium*) as a climax dominant (Allred & Mitchell 1955; Thomas 1962; Correll 1972; Gould & Shaw 1983; Simpson & Pease 1995).

Seven different specific grassland communities occurring on three main soil associations are recognized by Collins et al. (1975) as occurring in the Blackland Prairie. Diamond and Smeins (1993), however, recognized five major tall grass communities in the main body of the Blacklands (Fig. 18). Three of these types, the *Schizachyrium-Andropogon-Sorghastrum* (little bluestem-big bluestem-Indian grass), *Schizachyrium-Sorghastrum-Andropogon* (little bluestem-Indian grass-big bluestem) and *Schizachyrium-Sorghastrum* (little bluestem-Indian grass), are relatively similar, have little bluestem as the prevailing dominant, and occur over the majority of the Blacklands (Diamond & Smeins 1993). Associated species include *Bouteloua curtipendula* (side-oats grama), *Carex microdonta* (small-toothed caric sedge), *Sporobolus compositus* (tall dropseed), *Nassella leucotricha* (Texas winter grass), *Acacia angustissima* var. *hirta* (prairie acacia), *Bifora americana* (prairie-bishop), *Hedyotis nigricans* (prairie bluets), and *Hymenopappus scabiosaeus* (old-plainsman) (Diamond & Smeins 1985). The microtopographical features known as "hog wallows" or gilgai are often found on prairies of these types and provide important microhabitat variation based on differences in water, nutrient relations, and frequency of disturbance (Diamond & Smeins 1993). Vegetational differences associated with the microhighs and microlows are easily observed.

The other two Blackland communities are quite different vegetationally and are relatively limited in occurrence. The *Tripsacum-Panicum-Sorghastrum* (eastern gamma grass-switch grass-Indian grass) community is ". . . found over poorly drained Vertisols in uplands of the northern Blackland Prairie and in lowlands throughout the Texas tallgrass prairie region" (Diamond & Smeins 1993). Examples can be found in Grayson and Fannin counties. Additional common species include *Bouteloua curtipendula* (side-oats grama), *Carex microdonta* (small-toothed caric sedge), *Paspalum floridanum* (Florida paspalum), *Sporobolus compositus* (tall dropseed), *Acacia angustissima* var. *hirta* (prairie acacia), *Aster ericoides* (heath aster), *Bifora americana* (prairie-bishop), *Hedyotis nigricans* (prairie bluet), *Rudbeckia hirta* (black-eyed susan), and *Ruellia humilis* (prairie-petunia) (Diamond & Smeins 1985). The *Sporobolus-Carex* (silveanus dropseed-mead sedge) community, dominated by *Sporobolus silveanus* (Silveus' dropseed) and *Carex meadii* (Mead's caric sedge), is found in the northern Blackland Prairie on low pH Alfisols in areas of relatively high precipitation (Diamond & Smeins 1993). An example can be seen on the Nature Conservancy's Tridens Prairie in Lamar County. Other common species found in this community type include *Panicum oligosanthes* (Scribner's rosette grass), *Fimbristylis puberula*, *Coelorachis cylindrica* (Carolina joint-tail), *Panicum virgatum* (switch grass), *Paspalum floridanum* (Florida paspalum), *Sporobolus compositus* (tall dropseed), *Aster pratensis* (silky aster), *Linum medium* (Texas flax), and *Neptunia lutea* (yellow-puff) (Diamond & Smeins 1985). The microtopographical feature known as mima mounds is commonly associated with this community. Like gilgai, mima mounds provide microhabitat variation, increasing the overall biological diversity of the prairie ecosystem (Diamond & Smeins 1985).

Also worth mention is the special assemblage of herbaceous plants often seen on areas of very thin soil and especially on exposed outcrops of the Austin Chalk (Stanford 1995). Species seen in this type of setting in the northern Blackland Prairie (Grayson County) include *Baptisia australis* (wild blue-indigo), *Callirhoe pedata* (finger poppy-mallow), *Eriogonum longifolium* (long-leaf wild buckwheat), *Grindelia lanceolata* (gulf gumweed), *Ipomopsis rubra* (standing-cypress), *Linum pratense* (meadow flax), *Marshallia caespitosa* (Barbara's-buttons), *Oenothera macrocarpa* (Missouri primrose), *Paronychia jamesii* (James' nailwort), and *Thelesperma filifolium* (greenthread). At some seasons, these outcrops have the aspect of barren eroded rock; in the spring, however, they are covered with spectacular displays of color.

As can be seen above, there is considerable variation in the tall grass prairie communities of the Blacklands (Diamond & Smeins 1993), and disagreement about specific community types (Simpson & Pease 1995). However, common dominant grasses of this tall grass prairie ecosystem include *Schizachyrium scoparium* (little bluestem), *Andropogon gerardii* (big bluestem), *Sorgastrum nutans* (Indian grass), *Panicum virgatum* (switch grass), *Tripsacum dactyloides* (eastern gamma grass), *Sporobolus compositus* (tall dropseed), *Eriochloa sericea* (Texas cup grass), *Paspalum floridanum* (*Florida paspalum*), and *Tridens strictus* (long-spike tridens) (Collins et al. 1975). Despite similarities in general aspect and even the occurrence of certain species over broad areas, the particular community present and the dominants observed can vary considerably even over short distances, primarily on the basis of differences in soil. Localized patches of a community type well beyond its main zone of occurrence are common, based on soil or other factors. Therefore most of the Blackland Prairie is a complex mosaic of tall grass communities; an example of this can be seen in northern Grayson County where four of the community types discussed above can be seen within a few miles.

Although prairie predominated, some wooded areas were also natural components of the Blackland Prairie region at the time of settlement. Examples include bottomland forests and wooded ravines along the larger rivers and streams, mottes or clumps in protected areas or on certain soils, scarp woodlands on slopes at the contact zones with the Edwards Plateau and Lampasas Cut Plain, and scattered upland oak woodlands similar to the Cross Timbers (Gehlbach 1988; Nixon et al. 1990; Diamond & Smeins 1993). In areas such as Dallas, where the Austin Chalk forms a conspicuous escarpment or bluff, a characteristic woody vegetation is also found in the varied microhabitats associated with this topographic feature. Kennemer (1987) indicated that *Fraxinus texensis* (Texas ash), *Quercus sinuata* var. *breviloba* (shin oak), and *Ulmus crassifolia* (cedar elm) are dominant. Other noteworthy woody plants of the escarpment include *Cercis canadensis* var. *texensis* (Texas redbud), *Juniperus ashei* (Ashe juniper), *Morus microphylla* (Texas mulberry), and *Ungnadia speciosa* (Texas buckeye). Farther south, in Bell, Hill, and McLennan counties, the Austin Chalk scarp vegetation is similar. Depending on slope and moisture conditions, characteristic species include *Celtis laevigata* (sugarberry), *Diospyros texana* (Texas persimmon), *Forestiera pubescens* (elbow-bush), *Fraxinus texensis*, *Ilex decidua* (deciduous holly), *Juniperus ashei*, *Juniperus virginiana* (eastern red-cedar), *Ptelea trifoliata* (hoptree), *Quercus buckleyi* (Texas red oak), *Quercus fusiformis* (Plateau live oak), *Quercus sinuata* var. *breviloba*, and *Ulmus crassifolia* (Gehlbach 1988).

As indicated earlier, with the exception of preserves, small remnants, or native hay meadows, almost nothing remains of the original Blackland Prairie communities. According to Diamond et al. (1987), all of the tall grass community types of the Blackland Prairie are "...endangered or threatened, primarily due to conversion of these types to row crops." Three specific Blackland communities are considered "threatened natural communities" by the Texas Organization for Endangered Species (TOES 1992). Conversion of the Blackland Prairie for agriculture was the most important cause of the destruction of this ecosystem, with only marginal, often steeply sloped land not rapidly brought under cultivation. High prices for cotton and grains eventually resulted in the cultivation of even these marginal areas, "...with disastrous effects. Blackland soils on steep slopes, stripped of their protective grass, eroded rapidly. Gullying was everywhere,

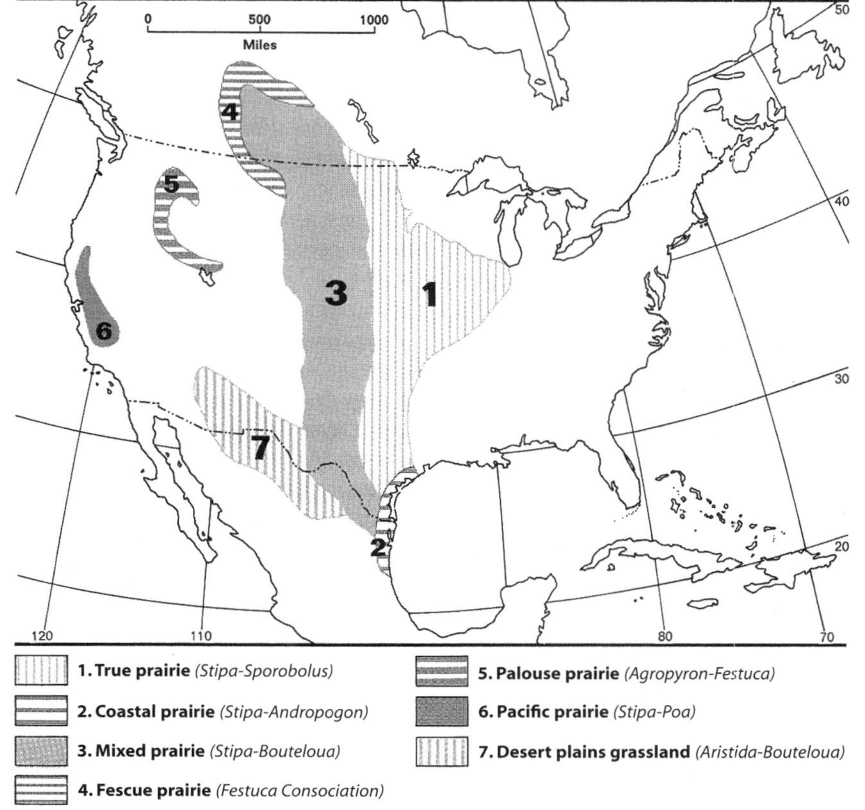

FIG. 19/GRASSLAND ASSOCIATIONS IN NORTH AMERICA (FROM GOULD 1968A; USED WITH PERMISSION OF LUCILE GOULD BRIDGES; ©1968). NOTE THAT ACCORDING TO GOULD (1968A) AND GOULD AND SHAW (1983), ALL OF NORTH CENTRAL TEXAS IS PART OF THE TRUE PRAIRIE ASSOCIATION.

and in a few years, over much of the marginal slope-lands, as much as three feet of soil had been eroded, exposing barren rock where once was prairie soil" (Hayward & Yelderman 1991). Today, extensive eroded areas and large sections that have been contoured to remedy erosion can be seen in many places throughout the Blacklands.

Existing prairie is still being lost due to a variety of causes. An example is the destruction of Stults Meadow in Dallas, studied in detail by Laws (1962) and Correll (1972). In addition to direct destruction of prairie through cultivation or other uses (e.g., urbanization), existing isolated small prairie remnants are currently being lost through invasion by woody vegetation and introduced species. Given the relatively high rainfall over most of the Blacklands, with the suppression of fire by humans, native trees and shrubs (e.g., *Juniperus virginiana*—eastern red-cedar, *Ulmus crassifolia*—cedar elm) as well as introduced species are able to invade and eventually take over areas that were formerly prairie.

Recurrent fire and grazing by bison were natural processes that maintained the Blackland ecosystem; the removal of these processes is a disturbance that causes changes in the vegetation (Smeins 1984; Smeins & Diamond 1986; Diamond & Smeins 1993). In this region, periodic disturbance is essential for the maintenance of prairie. However, even native hay meadows, which are routinely disturbed, are often markedly different from the original vegetation because of the substitution of mowing and particularly past herbicide use in place of fire and grazing. The results include a reduction in broad-leaved plants and an increased abundance of grasses (Diamond & Smeins 1993). While grazing was a natural component of the Blacklands and many other Texas ecosystems, overstocking and thus overgrazing by domesticated animals has also caused a dramatic decline and even near elimination of numerous plants from many areas (Cory 1949). The cumulative effect of all these human-induced changes is that the Blackland Prairie communities have been largely destroyed. Large areas that were once tall grass prairie are now covered by crops or other introduced and now naturalized species such as *Bothriochloa ischaemum* (King Ranch bluestem), *Cynodon dactylon* (Bermuda grass), and *Sorghum halepense* (Johnson grass). Roadsides and pastures are particularly obvious examples; in many cases hardly any native grasses can be found. In these areas there has also been an accompanying dramatic reduction in native forb diversity.

In striking contrast to the terrestrial communities of the Blackland Prairie is the tremendous increase in aquatic habitats. Most native wetlands, including prairie "pothole-like" wetlands, have been lost. However, with the construction of numerous reservoirs, lakes, ponds, and tanks, there is vastly more habitat for aquatic vegetation than in presettlement days. With the exception of oxbow lakes along some of the larger streams, the only permanent surface water prior to human intervention was in rivers, streams, swampy or marshy areas, beaver ponds, and springs. Introduced, as well as native, aquatic plants are now widespread and in some cases so abundant as to be problematic weeds. Many aquatic plants probably have populations several orders of magnitude greater than in the relatively recent past. This same pattern holds not just for the Blackland Prairie, but for all vegetational areas within North Central Texas.

Cross Timbers and Prairies

Occurrence of the Cross Timbers and Prairies

The Cross Timbers and Prairies (vegetational area 5 of Hatch et al. 1990), an area of about 26,000 square miles (about the size of West Virginia), occupies the region south of the Red River between the Blackland Prairie to the east, the Rolling Plains to the west, and the Llano Basin (Central Mineral Region) and Edwards Plateau to the southwest and south. Vegetationally it is quite diverse and includes the East and West cross timbers, the Fort Worth Prairie, and the Lampasas Cut Plain (Fig. 1). Notable physiographic features included are the Comanche Plateau, the Palo Pinto Country (sometimes referred to as the Palo Pinto Mountains), the mesa and butte country of the Lampasas Cut Plain, and the eastern portion of the Callahan Divide (Fig. 8).

The Cross Timbers, which stretch from Texas north through Oklahoma to Kansas (Marriott 1943; Dyksterhuis 1948; Kuchler 1974), are found in Texas from the Red River south for about 150 miles. They are actually two discrete belts of forest divided by the enclosed Grand Prairie (Dyksterhuis 1948). Surrounded by prairie on both sides (Blackland Prairie to the east, Rolling Plains to the west), they represent a final disjunct western extension of woody components of the eastern deciduous forest before the vegetation changes into the vast expanse of central U.S. grasslands known as the Great Plains. The two separate belts are the East Cross Timbers and the West Cross Timbers, sometimes referred to as the Lower Cross Timbers and Upper Cross Timbers respectively. According to Hill (1887), these names developed because the West or Upper Cross Timbers is at a greater altitude and in a more upstream position relative to the flow of rivers in the area. The East Cross Timbers is a narrow strip (roughly along the 97th meridian) extending from the Red River, in eastern Cooke and western Grayson counties, south to near Waco where it merges with the riverine forests of the Brazos River (Hayward & Yelderman 1991). This southernmost portion of the East Cross Timbers, developed on the sandy terraces of the Brazos River, is sometimes referred to as the "false" East Cross Timbers to distinguish it from the upland or "true" East Cross Timbers farther north. The two areas are continuous but can be distinguished by topography—flat on the river terraces and gently rolling in the uplands (Hayward et al. 1992). The somewhat wider West Cross Timbers stretches west from the Grand Prairie to the beginning of the Rolling Plains and includes the rather rugged Palo Pinto Country (in Eastland, Jack, Palo Pinto, Stephens, and Young counties).

The Fort Worth Prairie portion of the Grand Prairie extends as a continuous body of open grasslands, roughly 10 to 30 miles wide, from near the Red River in the north, south about 110 miles to where it ends in the wooded area along the Brazos River near the Johnson County-Hill County line (Dyksterhuis 1946) (Fig. 1).

The Lampasas Cut Plain, the largest portion of the Grand Prairie, is highly dissected butte and mesa country with extensive lowlands, and can in some ways be considered a northern extension of the Texas Hill Country and Edwards Plateau. It has strong geologic and floristic links with the Edwards Plateau as discussed in the sections below on the geology of the Grand Prairie and the vegetation of the Lampasas Cut Plain. It is, in fact, considered a part of the Edwards Plateau by some authorities (e.g., Riskind & Diamond 1988). The Lampasas Cut Plain extends from the Fort Worth Prairie south and west to the Llano Basin (also called the Central Mineral Region or Burnet Country) and the Colorado River (Figs. 1, 8).

Presettlement and Early Settlement Conditions in the Cross Timbers

As has been noted by historical geographer Richard Francaviglia, the natural and cultural histories of the Cross Timbers are inseparable, for human populations have had an impact on the region for thousands of years—first through the Native Americans' use of fire, then in the 19th century through the European Americans' agricultural and grazing practices, and more recently in the form of urbanization, suburbanization, and conservation activities (Francaviglia, forthcoming). Francaviglia notes that the Texas Cross Timbers has become what geographers call a "vernacular region" in the late 20th century, but that the term was used as early as the 1820s to characterize its distinctive vegetation. The exact origin of the term Cross Timbers is not known, but Dyksterhuis (1948) stated that the name

> ... presumably alludes either to the fact that this forest extends north and south across, rather than along, the major streams all of which flow eastward; or to the fact that early westward travelers who had left the main body of the great eastern forest and entered upon open prairie found it necessary to cross yet another body of forest before entering upon the grasslands that extended to the Rocky Mountains.

The abrupt appearance of the East Cross Timbers was quite striking for westbound travelers who had just crossed the extensive open Blackland Prairie. This conspicuous change in the vegetation served as a landmark recognized by almost all early travelers, was a principal marker on early maps (e.g., Holley 1836; Ikin 1941; Gregg 1844; Kendall 1845), and was discussed in many immigrant guides and early explorer accounts (e.g., Marryat 1843; Roemer 1849; Parker 1856). Kennedy (1841), based on accounts of local residents, wrote of the Cross Timbers [not distinguishing East and West],

> This belt of timber varies in width from five to fifty miles. Between the Trinity and Red Rivers it is generally from five to nine miles wide, and is so remarkably straight and regular, that it appears to be a work of art. When viewed from the adjoining prairies on the east or west, it appears in the distance as an immense wall of woods stretching from south to north in a straight line, the extremities of which are lost in the horizon.

Regarding their use as a landmark, he further stated,

> As might naturally be supposed, the Cross Timber forms the great landmark of the western prairies; and the Indians and hunters, when describing their routes across the country, in their various expeditions, refer to the Cross Timber, as the navigators of Europe refer to the meridian of Greenwich. If they wish to furnish a sketch of the route taken in any expedition, they first draw a line representing the Cross Timber, and another representing the route taken, intersecting the former.

Together with the west-east rivers such as the Red and farther south, the Trinity, they formed a kind of navigational grid in an otherwise rather featureless landscape (Phelan 1976). Parker (1856) noted that "the long stretches of prairie, although undulating, present no object so prominent as the belt of timber which bounds them." He further described the East Cross Timbers as

> ... a very singular growth. The one we had now entered is called the Lower Cross Timbers, and is about six miles wide.... The timber is a short, stunted oak, not growing in a continuous forest, but interspersed with open glades, plateaus, and vistas of prairie scenery, which give a very picturesque and pleasing variety.

Marcy (1853, 1866), based on extensive travel in the area, said,

> At six different points where I have passed through it [Cross Timbers], I have found it characterized by the same peculiarities; the trees, consisting principally of post-oak and black-jack, standing at such intervals that wagons can without difficulty pass between them in any direction. The soil is thin, sandy, and poorly watered.

This statement agrees with numerous early settler accounts (Dyksterhuis 1948). Other early references, however, such as that by Kendall (1845) with the Texan Santa Fe Expedition in 1841, referred to the Cross Timbers in places as "almost impenetrable" and "full of deep and almost impassable gullies." Kendall (1845) further stated,

> The ground was covered with a heavy undergrowth of briers and thorn-bushes, impenetrable even by mules, and these, with the black jacks and post oaks which thickly studded the broken surface, had to be cut away, their removal only showing, in bolder relief, the rough and jagged surface of the soil which had given them existence and nourishment.

Some other early travelers also considered the Cross Timbers difficult to cross and an obstacle to travel because of the vegetation and topography (Dyksterhuis 1948). Gregg (1844), for example, wrote of them as follows:

> Most of the timber appears to be kept small by the continual inroads of the 'burning prairies;' for, being killed almost annually, it is constantly replaced by scions of undergrowth; so that it becomes more and more dense every reproduction. In some places, however, the oaks are of considerable size, and able to withstand the conflagrations. The underwood is so matted in many places with grape-vines, green-briars, etc., as to form almost impenetrable 'roughs'....

Another example is Smythe (1852), who referred to the "Lower Cross Timbers" as

> ... chiefly low scrubby Post Oak groves, extremely tangled and thick, with millions of green briers, ... making it truly a difficult task to make your way without serious damage to your skin, and clothes.

Marryat (1843) described the Cross Timbers in similar fashion:

> During two or three days we followed the edge of the wood, every attempt to penetrate into the interior proving quite useless, so thick were the bushes and thorny briers.

The Cross Timbers vegetation at the time of contact by Europeans thus probably exhibited considerable variation. The boundaries in particular were probably variable and at least in places were not completely distinct. Parker (1856) described the area just west of the East Cross Timbers but east of what he referred to as the Grand Prairie (east of Gainesville in Cooke County) as follows:

> ... soon leaving the timber, we entered upon a broken country, consisting of ridges of sand and limestone, interspersed with small prairies and small strips of timber, principally black jack, until we emerged upon and crossed Elm Fork of the Trinity, where, on account of the intense heat, Captain Marcy determined to halt and encamp, thereafter, intending to march by moonlight, until we reached the Grand Prairie.

The variable nature of the Cross Timbers is also reflected in the following description from Kendall (1845):

> The growth of timber is principally small, gnarled, post oaks and black jacks, and in many places the traveller will find an almost impenetrable undergrowth of brier and other thorny bushes. Here and there he will also find a small valley where timber is large and the land rich and fertile, and occasionally a small prairie intervenes; but the general face of the country is broken and hilly, and the soil thin.

In at least some areas the timber was extensive. An example is Jordan's (1973) assertion that Forestburg, a small community in southeastern Montague County, boasted six sawmills by 1895.

The animal life of the Cross Timbers was probably similar to that described earlier for the Blackland Prairie. Because of the lower human population size and more native vegetation, slightly more animal life probably survives in the Cross Timbers. A recent (spring 1996) mountain lion sighting in Lake Mineral Wells State Park is an example.

Geology of the East Cross Timbers

The narrow band of woody vegetation between the Blackland Prairie and the Grand Prairie, known as the East Cross Timbers, occurs largely on sandy soil derived from rocks of the Woodbine formation (Hill 1901), which were deposited about 96 to 92 million years ago (Hayward & Yelderman 1991). According to Hill (1901), the formation is "largely made up of ferruginous [iron containing], argillaceous sands, characterized by intense brownish discoloration in places, which are accompanied by bituminous laminated clays." Like the Trinity Group sands underlying the West Cross Timbers, the Woodbine sands are often unconsolidated and rather loose, but differ in having more iron and other minerals. The post oak-blackjack oak vegetation typical of the East Cross Timbers does well on the deep loose soils developed from these unconsolidated layers. In some instances, however, the iron minerals consolidate the sand layers into dark-brown siliceous iron ore (Hill 1901). These iron deposits are so abundant in places that they cap low, wooded, erosion-resistant hills and isolated knobs known as "iron-ore knobs." These are found from Hill County north through Johnson, Tarrant, Denton, and Cooke counties and east into Grayson County (Hill 1901) where the Woodbine formation swings east along the Preston Anticline (a buckle in the strata exposing deeper older layers). The bed of the Red River follows the outcrop of the Woodbine from Grayson County east along the northern boundary of North Central Texas all the way to Red River County in far northeastern Texas (Hill 1901). In general, the Woodbine formation forms a layer of loose brownish sand cropping out just west of the Eagle Ford Shale. Economically the Woodbine is important because its sands contain significant amounts of water and it serves as one of the main aquifers for some areas (Baker 1960).

Geology of the West Cross Timbers

In general, west of the Fort Worth Prairie and north of the Lampasas Cut Plain lies an area of easily erodible sandy soils developed from the Trinity Group (Paluxy, Antlers, and Twin Mountain-Travis Peak sands), the oldest of the Cretaceous layers in North Central Texas (Hill 1901; Renfro et al. 1973). These sands, which represent shallow-water or near-shore sea deposits, are fine grained and so loose that they are "...readily cut with pick and spade." They are therefore locally known as "pack-sands" (Hill 1901). The West Cross Timbers has developed in part on the sandy soils derived from such strata. These permeable layers are also important as a source of ground water (Baker 1960), which is still used today by many North Central Texas communities. Because of the rather irregular pattern of outcropping and numerous remnant areas of these sands (Hill 1901), the West Cross Timbers is not an easily delineated region. Further, the sedimentary layers from which the soils of the West Cross Timbers formed are not homogeneous; instead these mainly sandy strata have substantial sections of clay and sandy clay. Therefore, numerous glade-like prairies formed on soils derived from these clayey outcrops can be found scattered through this mostly timbered region (Hayward et al. 1992). The area is further complicated by topographic extensions of the Lampasas Cut Plain that extend north into the West Cross Timbers in areas including Erath and Hood counties. Comanche Peak, a noted landmark in Hood County (Hill 1901), is a good example of such an extension.

North of the easternmost part of the Callahan Divide mesa country, and forming the northwestern portion of the West Cross Timbers, is an area known as the Palo Pinto Country (Fig. 8). This rather rugged region is underlain by the oldest rocks exposed in North Central Texas, deposited during the Pennsylvanian Period (Fig. 7) (the Pennsylvanian extends from 320 to 286 million years ago). According to Hill (1901) the Pennsylvanian (Carboniferous in his terminology) is

> ... largely made up of soft, impure shales alternating with harder, coarse, brown sandstone and conglomerates, produces ridge-like mountains and a broken belt of country ... composed of rough-scarped and flat-topped sandstone plains and hills of circumdenudation, surrounded by and overlooking wide clay valleys called 'flats.'

The area is deeply dissected and is essentially a cut plain marked by scarps, mesas, and canyons with flat areas of extensive beds of shale outcrop (Hill 1901). The same strata, and consequently landscapes of similar character, extend south of the Callahan Divide and form the Brownwood Country (parts of Brown, Coleman, and Mills counties) (Hill 1901; Renfro et al. 1973).

VEGETATION OF THE CROSS TIMBERS

The East and West cross timbers, with their woody overstory consisting primarily of post oak (*Quercus stellata*) and blackjack oak (*Quercus marilandica*), owe their existence to the presence of sandy, slightly acidic soils derived from the Cretaceous Woodbine and Trinity strata (and in the westernmost area to gravelly and rocky Pennsylvanian strata). These soils allow more efficient water infiltration, permit easier penetration of tree roots, and provide more moisture to plants than do heavier clay soils (Allred & Mitchell 1955). The result is that the survival of trees is favored in these areas even though they receive less rainfall than the Blackland Prairie farther east. Hill (1887) first pointed out that the Cross Timbers were developed on sandy soils, and contrasted this vegetation with the adjacent treeless prairies growing on the tight calcareous clay soils developed from limestones.

The original vegetation of both the East and West cross timbers, as based on early accounts (discussed in the section on presettlement conditions), was almost certainly variable, ranging from quite open to dense thickets. However, based on these accounts and on an extensive vegetational study of the West Cross Timbers by Dyksterhuis (1948), presettlement vegetation can probably best be described as a savannah with an oak overstory, but dominated by *Schizachyrium scoparius* (little bluestem), with two other grasses, *Andropogon gerardii* (big bluestem), and *Sorghastrum nutans* (Indian grass), as lesser dominants. Weaver and Clements (1938) also regarded the Cross Timbers as ". . .chiefly oak savanna, in which the grasses are climax dominants." Dyksterhuis (1948) concluded that the current vegetation, even where not cleared for cultivation or pasture, is considerably modified from that present before settlement. He considered the existing understory vegetation a disclimax resulting mainly from overgrazing.

At present, in many Cross Timbers localities, younger trees are often branched to the ground, making movement through the vegetation extremely difficult and denying habitat for the originally dominant grasses; dense cedar brakes are particularly problematic in this regard. Fire suppression, apparently the chief cause of such changes, has thus probably been another major factor responsible for differences from the original vegetation.

Currently, where not completely destroyed, the vegetation ranges from open savannah to dense brush (Correll & Johnston 1970). In addition to the characteristic oaks, other woody species commonly found in the Cross Timbers today include *Ulmus crassifolia* (cedar elm), *Celtis* spp. (hackberry), *Carya illinoinensis* (pecan), *Juniperus* spp. (juniper), and *Prosopis glandulosa* (mesquite). Additional common grasses include *Bouteloua hirsuta* (hairy grama), *Bouteloua curtipendula* (side-oats grama), *Sporobolous compositus* (tall dropseed), *Panicum virgatum* (switch grass), *Elymus canadensis* (Canada wild-rye), and *Nassella leucotrica* (Texas winter grass) (Dyksterhuis 1948; Correll & Johnston 1970). Past mismanagement and cultivation have caused many uplands to be covered primarily by scrub oak, mesquite, and juniper with mid- and short-grasses beneath (Hatch et al. 1990).

As the early accounts mentioned above indicate, even in presettlement days the Cross Timbers were probably not continuous unbroken areas of woodland. In the East Cross Timbers some clay is found in the Woodbine formation, and where this clay crops out small prairies are found (Hill 1901). The West Cross Timbers in particular represents a complex pattern of timbered areas interspersed with grasslands or with grassland inclusions (Dyksterhuis 1948).

Dyksterhuis (1948) distinguished two areas of woody vegetation within the West Cross Timbers, the "main belt" (developed on sandy Cretaceous strata), and the "fringe" (developed on rocky or gravelly Pennsylvanian strata in the topographically more rugged Palo Pinto Country). The two areas share the same woody species, but differ in other aspects of their vegetation. For instance, Dyksterhuis (1948) found that *Buchloe dactyloides* (buffalo grass), an important dominant of the drier western plains, was four times as abundant in the fringe as in the main belt. A variety of other plants are more common in one area than the other. Examples include *Bouteloua gracilis* (blue grama) being more frequent in the fringe, with *Bouteloua hirsuta* (hairy grama) more common in the main belt (Dyksterhuis 1948). In the fringe in particular, woody vegetation tends to occupy areas of rugged relief with grasses dominating areas of gentler relief (Dyksterhuis 1948).

Tharp (1926) considered the Cross Timbers to be part of the Oak-hickory Association, and as such, part of the eastern deciduous forest. Allred and Mitchell (1955), however, in their broad classification of Texas vegetation, considered the Cross Timbers and even the eastern Texas Post Oak belt to be Post Oak Savannahs that are part of the True Prairie Association. They supported this contention by pointing out that the grasses of the True Prairie are important components in the vegetation of the savannahs. Dyksterhuis (1948), as pointed out above, indicated that little bluestem was the primary dominant. Barbour and Christensen (1993) stated that in the southern part of the tall grass prairie-deciduous forest boundary [including Texas], the ecotone is an oak savannah 50–100 km wide. Clearly, whether classified as forest or grassland, the Cross Timbers are part of this ecotone. With their rather limited tree diversity and high grass diversity and dominance, they are intermediate vegetationally, as well as geographically, between the eastern deciduous forests and the western grasslands.

One of the most striking features of the Cross Timbers is that this vegetational area contains significant remnants of virgin forests (Stahle & Hehr 1984; Stahle et al. 1985). According to Stahle (1996a), "... literally thousands of ancient post oak-blackjack oak forests still enhance the landscapes and biodiversity of... the Cross Timbers along the eastern margin of the southern Great Plains...." As as result, this is one of the largest relatively unaltered forest vegetation types in the eastern United States (Stahle & Hehr 1984). The small stature and often poor growth form of post and blackjack oaks made these species commercially unattractive and therefore less subject to systematic logging than other more productive forest types. Extensive dendrochronological (tree-ring) data from post oaks in the Cross Timbers indicate that old-growth remnants of post oak-blackjack forest can be found in numerous localities throughout the region. However, while extensive remnants remain, they are often degraded by various human activities such as heavy grazing or selective cutting and their authenticity is rarely noticed or protected (Stahle & Hehr 1984; Stahle 1996a). In comparison with areas in Oklahoma, the Cross Timbers of Texas are more degraded, in part because of the longer history of settlement (D. Stahle, pers. comm.). Examples of old-growth forests in North Central Texas are found in Comanche County (Leon River), Tarrant County (Fort Worth Nature Center), and Throckmorton County (Nichols Ranch) (Stahle et al. 1985; Stahle 1996a). Tree-ring chronologies extending from about 200 to 300 years have been obtained from these North Central Texas sites, with individual trees dating back to 1681. Such data are readily and harmlessly secured by using a Swedish increment borer to obtain a small diameter (<1 mm) core from the bark to the center of a tree; after careful polishing and under magnification, the annual growth rings can be counted (Stahle et al. 1985; 1996b). Because of the low availability of moisture, rocky or infertile soil, and other factors, the trees of these relict forests, while old, have a slow rate of growth and are of relatively small size, the canopy ranging from only about 6 to 15 meters high (Stahle et al. 1985). Such old-growth forests or ancient individual trees can often be located by environmental factors such as steep, rocky, infertile soils or by the appearance of the individual trees (Stahle & Chaney 1994). Twisted stems, dead tops and branches, canopies restricted to a few heavy limbs, branch stubs, fire and lightning scars, leaning stems, exposed roots or root collars, and hollow voids are all hints of significant age (Stahle 1996a, 1996b) (Fig. 20).

The centuries-long tree-ring chronologies obtained from these relict forests are a valuable source of information about past climate and are particularly important at a time when climate change is a topic of national and global concern. These forests also represent an irreplaceable resource and an unparalleled living record about the North Central Texas area prior to the time of European settlement. Further, because they are relatively unaltered, these remnants may represent areas of significant remaining biodiversity in an otherwise highly altered and reduced diversity environment. Finally, they provide a unique conservation opportunity to preserve some of the last remaining virgin North American forests.

Presettlement and Early Settlement Conditions in the Grand Prairie (Fort Worth Prairie and Lampasas Cut Plain)

Like the Blackland Prairie, the presettlement Grand Prairie was largely a vast grassland, with woody vegetation generally limited to areas along the larger watercourses, as scattered mottes on hilltops, or associated with mesas and buttes. Hill (1887) summarized the Grand Prairie as "a prairie region, utterly destitute of timber" and Kendall (1845) described the area as follows:

> To the east, for miles, the prairie gently sloped, hardly presenting a bush to relieve the eye. In the distance, the green skirting of woods, which fringed either border of a large stream, softened down the view. . . .To the west . . . the immediate vicinity was even more desolate, but the fertile bottoms of the Brazos, with their luxuriant growth of timber, were still visible, and the Camanche [Comanche] Peak, rising high above the other hills, gave grandeur and sublimity to a scene which would otherwise have been far from monotonous.

Referring to another Grand Prairie locality, Kendall (1845) said, "As far as the eye could reach . . ., nothing could be seen but a succession of smooth, gently-undulating prairies." In reference to the Brazos River valley where it cuts through the Lampasas Cut Plain, Kendall (1845) said,

> The valley of the Brazos at this place abounded with every species of timber known in Texas; grapes, plums, and other fruit were found in profusion; honey could be obtained in almost every hollow tree; trout and other fish were plentiful in the small creeks in the neighborhood, and the woods and prairies about us not only afforded excellent grazing for our cattle and horses, but teemed with every species of game—elk, deer, bears, wild turkeys, and, at the proper season, buffalo and mustang.

Greer (1935), in describing the Grand Prairie in the period 1850–1890, wrote of an ". . .indescribably beautiful prairie where lush grass swept my mount's sides. . . ." An early settler account (Hattie Richards Sparks in Pool 1964) of Bosque County describes it as follows:

> It was a beautiful prairie country, with sage grass as tall as your head. . . . It was a common occurrence for prairie chickens to fly into the house. I shall never forget when one flew in during our noonday meal and lit in a bowl of soft butter. . . . Wolves were plentiful and I will always remember their howling around our house at night. It was a mournful sound in that open prairie country.

Hill (1887) described this region between the two bands of Cross Timbers as "utterly destitute of timber." However, while Smythe (1852) mentioned "immense fields, with the greatest profusion of delicately painted flowers" and "grassy prairies only bounded by the horizon," he also spoke of "prairie, with an occasional strip of woodland," "beautiful groves of Live Oak . . . crowning every hill . . .," and areas with "densely tangled cedar ravines." The Grand Prairie vegetation thus showed considerable variability.

Based on the similarity of vegetation and climate, and on early traveler and settler reports, the original animal life was probably quite similar to that of the Blackland Prairie described earlier. From Kendall's (1845) observations, bison were apparently particularly common:

> . . . I have stood upon a high roll of the prairie, with neither tree nor bush to obstruct the vision in any direction, and seen these animals grazing upon the plain and darkening it at every point. . . . In the distance, as far as the eye could reach, they were seen quietly feeding upon the short prairie grass. . . .

FIG. 20/AN ANCIENT *QUERCUS STELLATA* FOREST DRAWN BY RICHARD P. GUYETTE; USED WITH PERMISSION. TWISTED STEMS, DEAD TOPS AND BRANCHES, CANOPIES RESTRICTED TO A FEW HEAVY LIMBS, BRANCH STUBS, FIRE AND LIGHTNING SCARS, LEANING STEMS, EXPOSED ROOTS OR ROOT COLLARS, AND HOLLOW VOIDS ARE ALL HINTS OF SIGNIFICANT AGE (STAHLE 1996A, 1996B). WHILE THE DRAWING IS FROM THE OZARK PLATEAU, SUCH TREES, SOME OF WHICH ARE MORE THAN 300 YEARS OLD, CAN BE FOUND IN NORTH CENTRAL TEXAS.

Kendall (1845) also described seeing many pronghorn antelope on the Upper Brazos. There is even a record of the American alligator as far west as Hamilton County in the central part of the Lampasas Cut Plain (Dixon 1987). Other noteworthy species recorded from the Grand Prairie, but not present there for many decades, include jaguar (Bailey 1905), greater prairie chicken, and passenger pigeon (Oberholser 1974).

In some areas animal populations were apparently reduced early on. Kendall (1845), speaking of another part of the Grand Prairie, said,

> Occasionally a deer would jump suddenly from his noonday rest, and scamper off across the prairie, but other than this no game was seen. The few deer we saw were exceedingly wild, from the fact of there being so many Indians in the vicinity; while the buffalo had evidently all been driven to the south.

Native American influence had thus in some areas already modified the fauna, and probably the vegetation as well. Roemer (1849), for example, discussed a Caddo village near the Hill County-Bosque County line as having about one thousand horses and cultivated maize and watermelons.

As with the Blackland Prairie, fire was an important aspect of the presettlement Grand Prairie. Kendall (1845), for example, observed an extensive fire and commented,

> All night the long and bright line of fire, which was sweeping across the prairie to our left, was plainly seen, and the next morning it was climbing the narrow chain of low hills which divided the prairie from the bottoms of the Brazos.

Settlement, basically the conversion of the prairie to ranching, swept across the Grand Prairie in the relatively short time span of 1850–1860. The importance of ranching and cattle to early Texas is also reflected in the fact that the famous Chisholm and Shawnee cattle trails crossed the Grand Prairie in the 1860s to the 1880s. The general availability of barbed wire in the 1870s and 1880s and subsequent droughts resulted in overstocking and severe overgrazing and thus had a significant detrimental effect on the vegetation (Dyksterhuis 1946). The result of these changes can be seen today in the wide variability of range quality in the area of Grand Prairie still devoted to grazing.

Geology of the Grand Prairie (Fort Worth Prairie and Lampasas Cut Plain)

The following excellent description of the Grand Prairie is from Hill (1901):

> Although often confounded with the Black Prairie, the Grand Prairie differs from it in many minor physical features. In general the surfaces are flat rather than undulating, and the valley slopes are angular (scarped or terraced) rather than rounded. The residual soils and regolith are shallow in comparison with those of the Black Prairie belts, and are of chocolate or brown colors instead of black, although in at least one belt (the Del Rio) the latter color prevails. Owing to the more shallow soil and the decreased rainfall many of the upland areas of the western part of the Grand Prairie are not so well adapted to agriculture, other than grazing, as are those of the Black Prairie, but the valley lands are very fertile and are extensively utilized.
>
> The chief difference between the two regions is that the Grand Prairie is established upon firm, persistent bands of limestones, which are harder than the underlying clay substructure of the Black Prairie region, and which under erosion, result in more extensive stratum plains and more angular cliffs and slopes. These limestone sheets of the Grand Prairie belts also alternate with marls and chalky strata of varying degrees of induration and thickness, and at the base of the whole are unconsolidated sands. The rock sheets of the Grand Prairie are so much harder than those underlying the Black Prairie region and are so conspicuous features in the landscape that, in distinction, the Grand Prairie country has been appropriately called 'the hard lime rock region'.
>
> In general the surface of the Grand Prairie, especially north of the Brazos [Fort Worth Prairie], is composed of gently sloping, almost level, and usually treeless dip plains, broken only by the valleys of the transecting drainage. These prairies are more continuous and comparatively void of inequalities of erosion along the eastern portion of the area. In the western half, especially south of the Brazos [Lampasas Cut Plain], their surfaces are broken into cut plains, buttes, mesas, and flat-topped divides and are etched by deeply eroded valleys.

The overall character of the Fort Worth Prairie and Lampasas Cut Plain (Figs. 1, 8) is influenced by such factors as the hardness and slope of the underlying Lower Cretaceous rock layers (Hill 1901). Specific layers cropping out at the surface, from oldest to youngest, include the Twin Mountains-Travis Peak, Antlers, Glen Rose, Paluxy, Walnut, Comanche Peak, Edwards, Kiamichi, Duck Creek, Fort Worth, Denton, Weno, Pawpaw, Mainstreet, and Grayson (Sellards et al. 1932; Renfro et al. 1973). The oldest of these date from the early Cretaceous (which began 145 million years ago), while the youngest, the Grayson, dates from as recently as 98 million years ago (Hayward & Yelderman 1991). Many of these strata include resistant limestone which contributes to the character of what Hill (1901) called the "hard lime rock region." In particular, the Fort Worth Prairie is mostly underlain by layers of firm limestone sloping gently eastward. As one moves from the west to the east, sequentially younger layers are encountered. In general the older western layers disappear under low west-facing escarpments produced by the strata of the next,

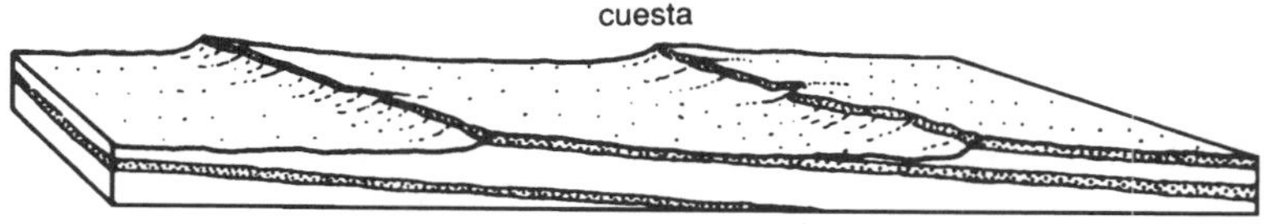

FIG. 21/DIAGRAM OF CUESTA TOPOGRAPHY (MODIFIED FROM SPEARING 1991; USED WITH PERMISSION OF MOUNTAIN PRESS PUBLISHING CO.; ©1991). A CUESTA IS A HILL OR RISE WITH A GENTLE SLOPE ON ONE SIDE AND A STEEP SLOPE, BLUFF, OR ESCARPMENT ON THE OTHER. IT RESULTS FROM AN INCLINED ROCK LAYER RESISTANT TO EROSION OUTCROPPING ADJACENT TO A LESS RESISTANT LAYER. THE STEEP SIDE IS THE END OF THE GENTLY DIPPING OR INCLINED RESISTANT LAYER. THE LESS RESISTANT ADJACENT LAYER ERODES MORE RAPIDLY EXPOSING THE END OF THE HARDER LAYER. IN NORTH CENTRAL TEXAS THE BLUFFS—OF AUSTIN CHALK IN THE BLACKLAND PRAIRIE OR OF VARIOUS LAYERS IN THE GRAND PRAIRIE—FACE WEST.

younger, overlying layer to the east (Hill 1901). It is the exposed ends of the younger layers that form the escarpments and produce the "cuesta" type of topography (Fig. 21) for which the Fort Worth Prairie is known. The Fort Worth Prairie occurs roughly from the Brazos River (northern Hill County) north through Johnson, Tarrant, Denton, and Cooke counties. The area is characterized in large part by gently sloping flat surfaces with thin soil over resistant limestone, extending for miles and ". . . making grass-covered uplands resembling the boundless views of the Great Plains proper" (Hill 1901). Fort Worth itself is built on one of these relatively hard, nearly flat layers, the Fort Worth Limestone.

To the south and southwest these rather flat plains grade into a more rugged, scarped, and dissected area with numerous low buttes and mesas known as the Lampasas Cut Plain (Hill 1901) (Fig. 22). Hill (1901) considered this region to be a modified northern extension of the Edwards Plateau. The term cut plain (or dissected plain) is defined by Hill (1901) as a plain that has been so dissected into remnants by erosion that the level of the original stratum is still recognizable in the summits of the dissected members. The Lampasas Cut Plain is a good example of this type of topography. The eastern portion of the Lampasas Cut Plain, sometimes referred to as the Washita Prairie, is a rolling landscape representing a remnant of the original surface, most of which eventually eroded away to form the highly dissected main portion of the Lampasas Cut Plain to the west (Hayward et al. 1992). This eastern undissected area, while somewhat different topographically, is thus clearly related to the rest of the Lampasas Cut Plain.

The landscape of the main portion of the Lampasas Cut Plain consists of broad grassland valleys that are separated by higher, narrow, often wooded, mesa-like divides (Hayward et al. 1992). These topographic features give the Lampasas Cut Plain a striking and distinctive appearance. The divides have been so eroded that in many places they remain only as flat-topped hills or buttes, often isolated or sometimes in chains. This type of topography can be seen in Bell, Bosque, Comanche, Coryell, Hamilton, Lampasas, Mills, and Williamson counties (Hill 1901); it owes its existence to the hard, white Edwards limestone, the same layer that forms the Edwards Plateau

FIG. 22/DRAWING OF SUMMITS OF THE LAMPASAS CUT PLAIN, LAMPASAS COUNTY (FROM HILL 1901).

(Hill 1901). In the Lampasas Cut Plain, however, much more of this layer has eroded away, the original surface remaining only as caps on the tops of the relatively few buttes and mesas. Figure 23 is a cross-sectional diagram of such an area. In fact, even though these uplands are the defining feature in the overall appearance of the region, less than ten percent of the area referred to as Lampasas Cut Plain is actually represented by these distinctive flat-topped features (Hayward et al. 1992). The majority of the area is made up of broad sloping valleys between the isolated upland fragments (Hill 1901). Most of the valley floors are formed from the Walnut Clay (Hayward et al. 1992) and the soils derived from this layer supported the original grassland vegetation typical of these valleys. On some of the valley floors, primarily in western portions of the Lampasas Cut Plain, the overlying material has been stripped away to expose the underlying Glen Rose Limestone; in these instances a rockier Glen Rose Prairie developed (Hayward et al. 1992). Often the slopes above the valleys are terraced or benched as the result of differential erosion of the various outcropping layers (Hill 1901).

In some places at the western edge of the Lampasas Cut Plain there is a conspicuous escarpment; this represents the edge of the Edwards limestone. To the south, this layer is less dissected and continues as the surface of that famous physiographic feature of central and western Texas, the Edwards Plateau. Also, some isolated fragments of Edwards limestone extend west through the West Cross Timbers into the Rolling Plains. These fragments cap scattered buttes and mesas in the region known as the Callahan Divide in counties including Brown, Callahan, Coleman, Comanche, and Eastland (Hill 1901) (Fig. 8). The erosion-resistant Edwards limestone, with its tendency to cap and protect more erosion-prone layers, has thus had a major effect on the appearance of large areas of Texas.

At the southern edge of the Lampasas Cut Plain near its boundary with the Blackland Prairie in Bell and Williamson counties, the subsurface geology is particularly important economically. The Edwards Aquifer, the main water source for 1.5 million people in central Texas, extends along the Balcones Fault Zone north to this region. The aquifer, which provides habitat for a number of threatened and endangered species, occurs in the porous Cretaceous layers overlying the Glen Rose Limestone. From the southern part of North Central Texas, the aquifer stretches south to the San Antonio area and west across the Edwards Plateau (Longley 1996). The maintenance of this aquifer is critical not only for the continued existence of a number of plants and animals, but also for the economy of central Texas.

VEGETATION OF THE FORT WORTH PRAIRIE

Based on early accounts as described in the section on presettlement conditions, and on a study of relict climax vegetation by Dyksterhuis (1946), a reasonable picture of the original vegetation of the Fort Worth Prairie can be gained. The most striking fact was the absence of trees. In Dyksterhuis' (1946) study of relict areas, *Schizachyrium scoparius* (little bluestem) was the overwhelming dominant, constituting nearly two-thirds of the total plant cover. Second in importance was *Bouteloua curtipendula* (side-oats grama), with other significant species including *Sorghastrum nutans* (Indian grass), *Sporobolus compositus* (tall dropseed), *Bouteloua hirsuta* (hairy grama), and *Andropogon gerardii* (big bluestem). Diamond and Smeins (1993) considered the Fort Worth Prairie to be part of the *Schizachyrium-Andropogon-Sorghastrum* (little bluestem-big bluestem-Indian grass) community, the same tall grass community type found on the shallow soils of the Austin Chalk outcrop in the Blackland Prairie. While the specific underlying strata differ, much of the Fort Worth Prairie, part of Hill's (1901) "hard lime rock region," is also developed on shallow limestone soils. *Nassella leucotricha* (Texas winter grass) and *Bothriochloa laguroides* subsp. *torreyana* (silver bluestem) [referred to by Dyksterhuis as *Andropogon saccharoides*], species that Dyksterhuis (1946) found to be most abundant in his broad study of the present-day grazing disclimax vegetation, were of almost negligible importance in the relatively undisturbed relicts. The major biotic influence on current vegetation is livestock grazing pressure (Dyksterhuis 1946). In comparison with the Blackland Prairie, much more of the original vegetation of the

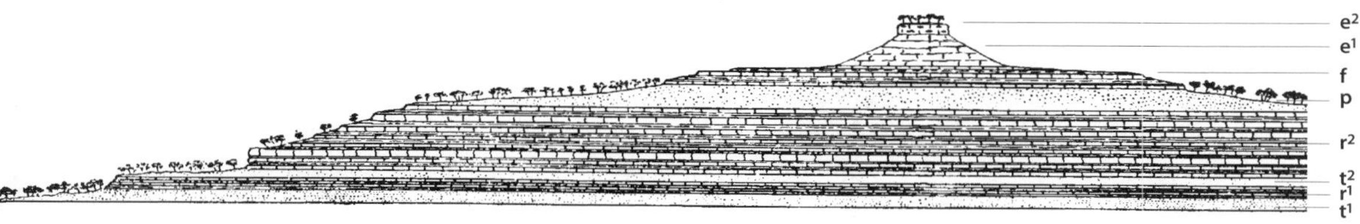

FIG. 23/DIAGRAM OF A DIVIDE OF THE LAMPASAS CUT PLAIN (FROM HILL 1901).

"Diagrammatic representation of a divide of the Lampasas Cut Plain, showing relation of agriculture, forestry, topography, and underground water to the geology. **e2**, Caprina limestone; summit divide of Lampasas Cut Plain; soilless except in large areas; growth of scrub oak, live oak, and post oak. **e1**, Comanche Peak limestone; chalky, soilless slopes; growth of shin oak. **f**, Walnut formation, Walnut Prairie; rich, fertile land in many places. **p**, Paluxy sand; Eastern and Western Cross Timbers, forested with post oak and black-jack; fair cotton and good fruit land; water bearing. **r2**, Barren limestone slopes with growth of juniper, live oak, and sumac; occasional fertile soils from interbedded marly layers; water bearing. **t2**, Trinity sands; Western Cross timbers; same forest growth as p. **r1**, small prairie spots; good soil. **t1**, Trinity sands; Western Cross Timbers; same forest growth as p."

Fort Worth Prairie has survived, in large part due to the extensive areas of shallow untillable soils that are still used primarily for grazing. Fire suppression and the consequent invasion of woody species has also had an important impact on the vegetation of many areas of the Fort Worth Prairie.

VEGETATION OF THE LAMPASAS CUT PLAIN

The vegetation of the Lampasas Cut Plain is more variable than that of the other vegetational areas considered here because of its greater topographic diversity. Depending on the particular conditions present, the vegetation ranges from prairie similar to that found on the Fort Worth Prairie or even the Blackland Prairie, to post oak-blackjack oak woodland similar to the Cross Timbers, to vegetation resembling that of the Edwards Plateau. For example, the easternmost part of the Lampasas Cut Plain, sometimes referred to as the Washita Prairie, is a gently rolling landscape of prairie with scattered oaks on hilltops. This area is characterized in places by rather deep soils with tall grass prairie vegetation similar to the Blackland Prairie, and in other places by shallow soils over hard limestone, with vegetation more closely resembling that of the Fort Worth Prairie (Hayward et al. 1992). To the west, in the more typical butte and mesa country of the main part of the Lampasas Cut Plain, oaks, including *Quercus buckleyi* (Texas red oak), *Quercus fusiformis* (live oak), and *Quercus sinuata* var. *breviloba* (shin oak), may be found on the rocky Edwards limestone summits of the smaller divides. On the larger divides, areas of deeper soil remain and support westward extensions of the Washita Prairie (Hayward et al. 1992). On the chalky nearly soil-less slopes derived from the underlying Comanche Peak limestone, *Quercus sinuata* var. *breviloba* (shin oak), *Rhus* species (sumac), and *Juniperus* species (juniper) may be seen; these dry rocky areas have a distinctly desert-like microclimate (Hayward et al. 1992) and thus support plants with xerophytic adaptations. Below these slopes, on benches in valleys or on the summits of uplands lacking caprock, extensive areas of prairie can be found on the clay soils derived from the Walnut formation where it is exposed. The basal Trinity Group sands (Paluxy, Antlers, Twin Mountains-Travis Peak) underlying the Walnut formation develop Cross Timbers vegetation with *Quercus stellata* (post oak) and *Quercus marilandica* (blackjack oak) (Hill 1901). In parts of the western Lampasas Cut Plain, in areas where the overlying strata have been removed to expose the Glen Rose limestone (occurring between layers of the Trinity sands), short-grass prairie, oak-savannah, and woodlands with abundant junipers are found on the thin-soiled, rough, rocky, stair-stepped landscape (Hill 1901; Hayward et al. 1992).

The topographic diversity and deeply cut streams found in various parts of the Lampasas Cut Plain provide important microhabitat variation. In particular, the diverse microhabitats allow the northward extension of many species otherwise found primarily on the Edwards Plateau to the south and southwest. Plants traditionally considered Edwards Plateau endemics (e.g., Amos & Rowell 1988) but found in the Lampasas Cut Plain include *Acer grandidentatum* Nutt. var. *sinuosum* (Plateau big-tooth maple), *Agalinis edwardsiana* (Plateau gerardia), *Argythamnia aphoroides* (Hill Country wild mercury), *Astragalus wrightii* (Wright's milk-vetch), *Chamaesaracha edwardsiana* (Plateau false nightshade), *Clematis texensis* (scarlet clematis), *Garrya ovata* var. *lindheimeri* (Lindheimer's silktassel), *Matelea edwardsensis*, (Plateau milkvine), *Muhlenbergia lindheimeri* (Lindheimer's muhly), *Nolina lindheimeriana*, (devil's-shoestring), *Onosmodium helleri* (Heller's marbleseed), *Perityle lindheimeri* (Lindheimer's rock daisy), *Prunus serotina* var. *eximia* (escarpment blackcherry), *Styrax platanifolius* (sycamore-leaf styrax), *Pediomelum cyphocalyx* (turnip-root scrufpea), *Tradescantia edwardsiana* (Plateau spiderwort), *Triodanis coloradoensis* (Colorado Venus'-looking-glass), *Verbesina lindheimeri* (Lindheimer's crownbeard), and *Yucca rupicola* (twisted-leaf yucca). When considering vegetation, soils, geologic layers, and general aspects of the landscape, some parts of the Lampasas Cut Plain (e.g., Fort Hood—Bell and Coryell cos.; Meridian State Park—Bosque Co.; bluffs in southern Johnson Co. overlooking the Brazos River) are remarkably similar to the Edwards Plateau; in fact, it could be argued that the Lampasas Cut Plain is simply a northern extension of the Edwards Plateau. A number of plants widely known from the Edwards Plateau also occur not only in the Lampasas Cut Plain, but also in the topographically complex Palo Pinto Country to the north and northwest. Examples include a number of fern species unusual in North Central Texas such as *Astrolepis integerrima* (star-scaled cloak fern), *Cheilanthes eatonii* (Eaton's lip fern), *Cheilanthes feei* (slender lip fern), *Cheilanthes horridula* (rough lip fern), *Pellaea ovata* (cliff-brake), and *Pellaea wrightiana* (Wright's cliff-brake).

Currently in many places in the Lampasas Cut Plain, as well as in the Cross Timbers, junipers (*Juniperus* spp.) are a conspicuous component of the vegetation, often crowding out other native species. Because of the control of fire, overgrazing, and other human-caused changes, juniper has become much more common during the last century (Hayward et al. 1992). In fact, juniper is currently one of the most problematic species invading and eliminating native grassland. Fonteyn et al. (1988) emphasized fire suppression as causing a similar transformation from relatively open savannah to shrubland or woodland (in large part due to invasion by *Juniperus ashei*) on some parts of the Edwards plateau. Mesquite (*Prosopis glandulosa*), historically much less abundant than at present, shows the same pattern. Originally limited by fire, it has increased greatly in abundance as the result of fire suppression, overgrazing, and the plowing and other disturbances associated with agriculture (Hayward et al. 1992). In general, with the suppression of fire, woody vegetation is currently increasing at the expense of grassland throughout North Central Texas.

RED RIVER AREA

AREA ADJACENT TO THE RED RIVER

The narrow band of vegetation found on the primarily sandy soils adjacent to the Red River in the northeastern portion of North Central Texas, specifically in the northern parts of Lamar, Fannin, and Grayson counties, is quite different from the vegetation of the rest of North Central Texas. At least part of this band, which we refer to as the Red River Area (Fig. 1), is often classified as part of vegetational area 3 (Post Oak Savannah) (Correll & Johnston 1970; Hatch et al. 1990). Such a classification is justified because a significant component of the vegetation more typically associated with eastern or southeastern Texas extends west along the Red River in microhabitats with special soil or moisture conditions. Even components of vegetational area 1 (Pineywoods), characteristic of extreme eastern Texas, can be found in this area. In northern Lamar County, the aspect of the vegetation is definitely similar to the eastern deciduous forest. Tall stands of *Quercus falcata* (southern red oak), abundant *Liquidambar styraciflua* (sweetgum), *Pinus taeda* (loblolly pine), *Acer rubrum* (red maple), *Betula nigra* (river birch), *Carpinus caroliniana* (American hornbeam), *Crataegus marshallii* (parsley hawthorn), bottomland brakes of *Arundinaria gigantea* (giant cane), *Calycocarpum lyonii* (cupseed), *Osmunda cinnamomea* (cinnamon fern), *Trachelospermum difforme* (climbing dogbane), and herbs such as *Lysimachia lanceolata* (lance-leaf loosestrife), *Monotropa hypopithys* (American pinesap), *Polygala sanguinea* (blood milkwort), *Porteranthus stipulatus* (Indian-physic), *Saccharum contortum* (bent-awn plume grass), *Sacciolepis striata* (American cupscale), *Saururus cernuus* (lizard's-tail), *Stachys tenuifolia* (slender-leaf betony), and *Veronicastrum virginicum* (culver's-physic), are just a few examples of eastern plants found in Lamar County. Even farther west in Fannin County, there are still isolated pockets of eastern Texas vegetation (e.g., Talbot property). Species reaching their known western limits there include *Quercus falcata* (southern red oak), *Quercus nigra* (water oak), *Quercus phellos* (willow oak), *Nyssa sylvatica* (black-gum), *Sassafras albidum* (sassafras), *Chasmanthium laxum* var. *sessiliflorum* (narrow-leaf wood-oats), *Cirsium horridulum* (bull thistle), *Erechtites hieraciifolia* (American burnweed), *Impatiens capensis* (spotted touch-me-not), *Luzula bulbosa* (bulb woodrush), *Monotropa uniflora* (Indian-pipe), *Pedicularis canadensis* (common lousewort), *Pycnanthemum albescens* (white-leaf mountain-mint), *Sorghastrum elliottii* (slender Indian grass), and *Woodwardia areolata* (narrow-leaved chain fern). Grayson County, the next county to the west, does not have areas dominated by eastern Texas plants as do Lamar and Fannin counties, but there is a significant eastern Texas component to the vegetation. Numerous plant species reach their western limits in Grayson County including *Agrimonia rostellata* (woodland groovebur), *Asimina triloba* (pawpaw), *Desmodium glutinosum* (tick-clover), *Liatris aspera* (tall gayfeather), *Monarda lindheimeri* (Lindheimer's beebalm), *Podophyllum peltatum* (may-apple), *Polygonatum biflorum* (Solomon's-seal), *Quercus velutina* (black oak), *Thalictrum arkansanum* (meadowrue), *Triosteum angustifolium* (yellow-flowered horse-gentian), and *Vaccinium arboreum* (farkleberry). A few typically eastern plants extend even farther west into Cooke and Montague counties and beyond.

The area adjacent to the Red River in Grayson County is further complicated by the presence of the Preston Anticline, a post-Cretaceous (Bradfield 1957) fold in the sedimentary strata that brought deeper layers to the surface (Bullard 1931). In places the river valley is two hundred feet below the surrounding area and creeks have cut deep canyon-like valleys. The overall topography near the Red River is thus very rugged (Bullard 1931). Parker (1856) in an early account described the Texas shore of the Red River as "very bold, presenting a stratification of red clay and white sand, giving a striking and very peculiar appearance in the distance, like chalk cliffs." This different topography and the appearance at the surface of deeper strata otherwise only found far to the west in areas such as the Grand Prairie and West Cross Timbers (e.g., Goodland limestone, Duck Creek limestone, Trinity Group sands) makes the vegetational picture of the area more complex. Many microhabitats, and thus increased biological diversity, result from the cropping out of these deeper strata in the county. For example, in a number of places along the Red River (e.g., Eisenhower State Park, Preston Peninsula, Delaware Bend), the Goodland Limestone forms flat limestone outcrops at the top of rugged cliffs. These areas of very thin soil over flat rock and adjacent slopes and ravines have numerous interesting plant species found nowhere else in Grayson County including *Coryphantha missouriensis* (plains nipple cactus), *Minuartia michauxii* var. *texana* (rock sandwort), *Talinum calycinum* (rock-pink), *Dodecatheon meadia* (common shooting-star), and *Melica nitens* (tall melic).

The proximity of sandy and clayey soils, as well as some intermediate type soils, in the counties adjacent to the Red River, also allows species normally separated ecologically to occur together; this sometimes results in hybridization. An excellent example can be seen in Fannin and Grayson counties where three species of *Baptisia* (wild indigo) and all three possible hybrids are found in close proximity (Kosnik et al. 1996). These occur either in what early settlers locally called "mixed soil" or in the area of the Preston Anticline where radically different soil types are found over quite small distances.

The basic pattern of the Red River Area is thus one of the eastern Texas forests grading gradually into the much less diverse and more xeric woodlands usually referred to as the Cross Timbers. From an even broader perspective, as discussed in the overview, the whole North Central Texas region is in an ecotone or ecological transition zone between two extensive ecosystems, the eastern North American deciduous forest and the central North American grassland or prairie. In virtually any ecotone, significant areas of vegetational interdigitation are seen; rarely is there a clearcut boundary. One type of vegetation extends deep into another along streams, in-pockets are found in protected areas, and special soil conditions often result in a patchwork pattern of vegetation that at the strictly local level seems confusing. The East and West cross timbers, the enclosed Grand Prairie, and the Red River Area are all excellent examples of these phenomena.

ORIGIN OF THE NORTH CENTRAL TEXAS FLORA

The flora of North Central Texas, like that of any relatively large region, has a complex and varied origin. Ultimately, it is the result of the evolutionary and distributional history of each of the component species. However, several influences can be observed which together allow at least a broad understanding of how the present flora originated. North Central Texas contains components of four major floristic provinces as defined by Thorne (1993a): the Appalachian Province, the Atlantic and Gulf Coastal Plain Province, the North American Prairies Province, and the Sonoran Province. There are also considerable numbers of Texas endemics. In addition, the modern flora contains 17.7 percent introduced species, these coming from various parts of the world.

INFLUENCE OF THE EASTERN DECIDUOUS FOREST

Plants from the first two of these floristic provinces, the Appalachian Province and the Atlantic and Gulf Coastal Plain Province, represent the influence of the eastern deciduous forest. This component of the flora is particularly important in the Red River drainage in the northeastern part of North Central Texas, but eastern deciduous forest elements occur across all of North Central Texas, and even make up an important component of the flora of the Edwards Plateau to the south and west of the region studied here (Amos & Rowell 1988). The vast deciduous forest biome of eastern North America is composed of a number of plant communities, and the forests/savannahs of North Central Texas represent *Quercus-Carya* or *Quercus* communities on the relatively dry western fringe of the eastern deciduous forest (Thorne 1993a).

From the phytogeographical standpoint, eastern deciduous forest elements are one of the most fascinating components of the North Central Texas flora. In the geologic past, dispersal between the Eurasian and North American continents was possible, and the combined area is considered a single "Holarctic" biogeographic region. The fossil record shows that many plants had distributions across the Northern Hemisphere—temperate forests, for example, occurred very broadly and reached their maximum extension in the mid-Tertiary (the Tertiary extended from 65 to 5 million years ago). This flora has been referred to as the Arcto-Tertiary flora or the Tertiaro-mesophytic flora. Geohistorical events from the mid-Tertiary to the present have included alterations in the shapes of the northern land masses, fluctuations in sea levels, mountain building, and profound changes in the climate. As a result, there have been great changes in both the composition and the disposition of the flora.

A number of species of the once widespread Arcto-Tertiary flora have survived in one or more of four widely separated Tertiary relict areas—1) eastern Asia; 2) eastern North America; 3) southeastern Europe; and 4) western North America (Li 1952b; Little 1970; Wood 1970; Graham 1972a, 1972b; Boufford & Spongberg 1983; Hamilton 1983; Hsü 1983; Wu 1983; Ying 1983; Cox & Moore 1993; Graham 1993). Examples of North Central Texas genera found in all four of these areas include *Aesculus, Cercis, Erythronium, Juglans, Ostrya, Philadelphus,* and *Platanus* (Wood 1970). Wood (1970) and Thorne (1993a) emphasized the strong floristic relationships between the eastern United States and western North America, and indicated that about 65% of the genera

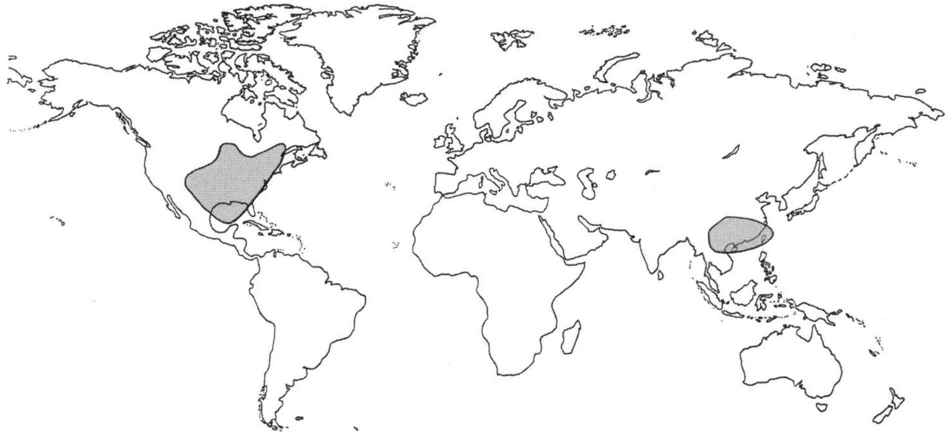

FIG. 24/WORLDWIDE DISTRIBUTION MAP OF *CARYA* (JUGLANDACEAE) SHOWING ITS DISJUNCT OCCURRENCE IN EASTERN ASIA AND EASTERN NORTH AMERICA (FROM WU 1983).

of southern Appalachian seed plants also occur in western North America. Examples of North Texas genera with such western North American connections include *Ceanothus*, *Oxypolis*, *Pycnanthemum*, and *Trichostema* (Wood 1970).

A significant number of species, however, have survived in only two very distant Tertiary relict areas, eastern Asia and eastern North America, and this striking distribution pattern has long been of interest to botanists (e.g., Gray 1846, 1859) and continues to be so today (e.g., Li & Adair 1994, 1997). The genus *Carya* is such an example (Fig. 24); other North Central Texas examples include *Campsis*, *Menispermum*, *Penthorum*, *Phryma*, *Sassafras*, *Saururus*, *Triadenum*, *Triosteum*, and *Veronicastrum* (Li 1952b; Little 1970; Boufford & Spongberg 1983; Hamilton 1983; Hsü 1983; Wu 1983; Ying 1983; Cox & Moore 1993; Graham 1993). In the words of Graham (1993), "It is well known that the broad-leaved deciduous forests of eastern North America and eastern Asia are floristically related.... It results from the maximum extension of the temperate deciduous forest in the mid-Tertiary and its disruption in western North America during the Pliocene and in western Europe during the Quaternary." This is one of the most ancient components of the North Central Texas flora. By at least the early Tertiary period (Eocene epoch—54.9–38 million years ago) deciduous vegetation was present across the middle of the North American continent (familiar genera include *Acer*, *Celtis*, *Liquidambar*, *Populus*, and *Rhus*) (Graham 1993). Dilcher (1998) indicated that different routes between the Old and New worlds have been open at different times in the past and that the shared vegetational elements between Asia and the United States are possibly derived from multiple introductions.

A related floristic relationship is the similarity seen between some forests in the mountains of Mexico and those in the eastern United States. Numerous North Central Texas genera (e.g., *Carpinus*, *Crataegus*, *Cornus*, *Liquidambar*, *Myrica*, *Nyssa*, *Pedicularis*, *Quercus*, *Smilax*, and *Vaccinium*) and even species (e.g., *Liquidambar styraciflua*) occur broadly across the eastern United States as far west as Texas and then reappear in the Mexican highlands and in some cases even in Guatemala (Miranda & Sharp 1950; Martin & Harrell 1957; Thorne 1993a). This relationship represents a middle to late Miocene (Miocene epoch—24.6–5.1 million years ago) extension of deciduous forest and associated fauna (particularly amphibians) into Mexico during a period of climatic cooling. Subsequently, during the Pliocene (Pliocene epoch—5.1–2 million years ago) and later times as the climate warmed and dried, these deciduous forests became disjunct, surviving in Mexico only in isolated pockets of appropriate microclimate in the highlands (Miranda & Sharp 1950; Graham 1993).

Geohistorical events in Tertiary and post-Tertiary times brought tropical elements into the present eastern deciduous forests, including North Central Texas. According to Graham (1993), "In the southeastern U.S., tropical forests reached their maximum northern expansion (to about 50°–60° N) in the Eocene [Eocene epoch—54.9–38 million years ago] with Annonaceae, Lauraceae, and Menispermaceae known from western Kentucky and Tennessee." A few North Central Texas species such as *Asimina triloba* may be a reflection of such an influence.

Influence of the North American Prairies Province

The second major floristic component of North Central Texas is derived from the North American Prairies Province. Grassland vegetation historically covered much of the area currently referred to as North Central Texas, but human activities, particularly the conversion to cropland and the suppression of fire, have greatly reduced the amount of grassland. Allred and Mitchell (1955) viewed virtually all of North Central Texas to be prairie. In their broad classification of Texas vegetation, they considered the Cross Timbers and even the eastern Texas Post Oak belt to be Post Oak Savannahs that are part of the True Prairie Association. They supported this contention by pointing out that the grasses of the True Prairie are important components in the vegetation of the Post Oak Savannahs. Barbour and Christensen (1993) stated that in the southern part of the tall grass prairie-deciduous forest boundary (including Texas), the ecotone is an oak savannah 50–100 km wide. Since there are clearly components of both prairie and deciduous forest, viewing the region as an ecotone seems the most reasonable approach.

Axelrod (1985) argued that North American grasslands are geologically recent and that the rise of extensive grasslands probably dates to the Miocene-Pliocene transition (about 7–5 million years ago), the driest part of the Tertiary. Fossil evidence shows the Great Plains were largely forested from the middle Miocene into the early Pliocene. As the climate dried at the end of the Miocene, forests were more restricted and grasslands were able to spread rapidly. According to Graham (1993), the widespread late Miocene-Pliocene appearance of prairie vegetation in the middle of the North American continent was part of the first major disruption of the vast temperate deciduous forest (the above-noted Arcto-Tertiary flora) that had extended across north temperate latitudes since the late Eocene. During the Pleistocene (the Pleistocene epoch began 2 million years ago), again based on fossil evidence, there was great fluctuation in grassland versus forest vegetation associated with glaciation. From 15,000 to 12,000 years B.P. (before present), areas now covered with grassland vegetation (e.g., Texas panhandle) supported forest. The widespread central North American grasslands (including those of North Central Texas) present at the time of European settlement, probably date to post-glacial times only 12,000 to 10,000 years B.P. (Axelrod 1985). Axelrod (1985) supported his view of the grassland as a young biome with the following evidence: 1) there are few endemic taxa, with most of the grassland species being present in adjacent forests; 2) populations of trees scattered over the region are readily interpreted as relicts of a once more widely distributed forest; and 3) fossil evidence of forests in the recent past occurs over much of the present grasslands. According to Axelrod (1985), "That grasslands spread following the last glacial is apparent from data provided by bogs at Boriack, Gause and Soefje, central Texas (Bryant, 1977). During late glacial time, central Texas was covered with an open deciduous forest with some conifers and an understory of mixed grasses and shrubs. With retreating glaciers, warmer and/or drier climates developed over central Texas. Forests were restricted, leaving parkland vegetation dominated by grasses, shrubs, and herbs, but including trees in protected sites. During post-glacial time many mesic trees disappeared from the pollen record. It was the continual increase in non-arboreal taxa, and especially grasses, throughout post-glacial time that led to the establishment of the present post oak-grassland vegetation of central Texas (Bryant, 1977)." Thorne (1993a) also considered the grasslands of the Prairies Province to be "... mostly recent and adventive. ..."

Influence of the North American Sonoran Province

A third floristic component of North Central Texas is derived from the Sonoran Province (southwestern United States and northwestern Mexico). According to Thorne (1993a), this province is part of the broader Madrean Region which has an exceedingly diverse and distinctive flora that is mostly locally derived and rich in endemics. He further indicated that the xerophytic flora of the Sonoran Province is subtropical and largely Madro-Tertiary in origin. In North Central Texas such Sonoran elements (e.g., *Aloysia* (Verbenaceae), *Colubrina*, *Condalia* (Rhamnaceae), *Garrya* (Garryaceae), *Karwinskia* (Rhamnaceae), and *Nolina* (Agavaceae)) are found mostly in the drier southern and southwestern parts of the region, but others (e.g., some *Acacia* (Fabaceae), *Opuntia* (Cactaceae), and *Yucca* (Agavaceae) species) occur more broadly. Cylindropuntias and *Yucca* species, for example, are the commonest tall plants in some parts of the Sonoran Province (Thorne 1993a) and a strong connection is seen to North Central Texas which has four species of cylindropuntias and seven species of *Yucca* (two of these are endemic to North Central Texas). Thorne indicated that the plants of the Sonoran Province "... seem to have originated as the arid areas of North America expanded through the Tertiary—for the past 65 million years, and especially in the last 15 million years." Families in the North Central Texas flora that he emphasized as examples of this diversification include Agavaceae, Cactaceae, Menispermaceae, Nyctaginaceae, Rafflesiaceae, and Sapotaceae. While some desert species are quite old, this emphasis on the last 15 million years seems to mesh with the prevailing opinion that the modern North American deserts and their floras are relatively recent geologically (Axelrod 1950, 1979; Barbour & Christensen 1993).

Thorne (1993a) indicated that there are other minor components of the North American flora which have very different origins. North Central Texas genera such as *Menodora* (Oleaceae), *Prosopis* (Fabaceae), and *Nicotiana* (Solanaceae) seem to have strong links with South America, while *Thamnosma* (Rutaceae) and *Selinocarpus* (Nyctaginaceae) are related to African taxa.

Endemics

North Central Texas itself has only five endemic taxa (*Yucca necopina*, *Y. pallida* (Agavaceae), *Evonymus atropurpurea* var. *cheatumii* (Celastraceae), *Croton alabamensis* var. *texensis* (Euphorbiaceae), *Dalea reverchonii* (Fabaceae)) presumably because of the lack of geographic and climatic isolation and the ecotonal nature of the area. However, 94 Texas endemics range into North Central Texas (see Appendix 3), with many endemics that were once thought to be restricted to the Edwards Plateau now known from the Lampasas Cut Plain in the southern part of North Central Texas (Amos & Rowell 1988). The explanation for the endemism seen in the Edwards Plateau and adjacent areas, while unclear, may be the result of the climatic history of the last two million years. During the Quaternary period (beginning 2 million years ago) there was significant climatic variability and at least 20 glacial-interglacial cycles. Widespread changes in vegetation were associated with these climatic fluctuations (Delcourt & Delcourt 1993); for example, during the last full-glacial interval (20,000–15,000 years B.P.), across the unglaciated parts of southwestern North America, there was a cool, moist "pluvial" climate (Delcourt & Delcourt 1993) with forest species presumably expanding their ranges. Indeed, Bryant's data (1977) showed an open deciduous forest in central Texas during the last full-glacial interval. The climate moderated from 15,000–10,000 years B.P., with interglacial conditions (i.e., warmer and drier) for the last 10,000 years (Delcourt & Delcourt 1993). The Edwards Plateau and Lampasas Cut Plain endemics are typically found in moist areas such as canyons along wooded streams and have presumably survived in the favorable microclimatic pockets as the overall climate of the area has warmed and/or dried. Many of these species have affinities with eastern taxa and may be relicts of a more widespread flora that became restricted as the result of climatic or geologic changes (Palmer 1920; Amos & Rowell 1988).

INTRODUCED SPECIES

Finally, the 394 species introduced from outside the United States since the time of Columbus make up 17.7 % of the North Central Texas flora (the percentage would be slightly higher if species that have invaded Texas from elsewhere in the United States were included). These introduced taxa are variously referred to as alien, exotic, or foreign species. They are also sometimes called "weeds," but that word can have different meanings (Randall 1997). From the sociological standpoint a weed is a plant growing where it is not wanted or a "plant-out-of-place" (Stuckey & Barkley 1993); if defined in this way, introduced species are indeed often weeds. Biologically, weeds (sometimes termed colonizing or invasive plants) are species that "have the genetic endowment to inhabit and thrive in places of continual disturbance, most especially in areas that are repeatedly affected by the activities of humankind" (Stuckey & Barkley 1993). Again, many introduced plants fall within this definition of weedy species.

While introduced species include some of our most beautiful ornamentals (e.g., *Iris*, *Narcissus*, and *Wisteria* species), some are also extremely invasive taxa capable of becoming serious agricultural pests or of destroying native habitats. Luken and Thieret (1997) examined the assessment and management of plant invasions and gave a selected list of species interfering with resource management goals in North America. Particularly problematic are those that aggressively invade native ecosystems, reproduce extensively, and occupy the habitat of indigenous species. In some cases, single invasive species can come to dominate communities and occur in near monocultures, completely changing the species composition, structure, and aspect of an ecosystem. After habitat destruction, invasion by exotics may be the most serious threat facing native plants in North Central Texas and it is a common but underestimated problem in many ecosystems around the world (Cronk & Fuller 1995; Bryson 1996; Westbrooks & Epler 1996). It is also a potentially lasting and pervasive threat (Coblentz 1990). According to Cronk and Fuller (1995), "It is a lasting threat because when exploitation or pollution stops, ecosystems often begin to recover. However, when the introduction of alien organisms stops the existing aliens do not disappear; in contrast they sometimes continue to spread and consolidate, and so may be called a more pervasive threat." Invasive exotics are an example of the phenomenon of ecological release—an introduced species is released from the ecological constraints of its native area (e.g., diseases, parasites, pests, predators, nutrient deficiencies, etc.) and is consequently able to undergo explosive population growth. There are numerous examples in North Central Texas, some of the most serious including *Bothriochloa ischaemum* var. *songarica*, King Ranch bluestem, *Festuca arundinacea*, tall fescue, *Hydrilla verticillata*, hydrilla, *Lespedeza cuneata*, sericea lespedeza or Chinese bush-clover, *Ligustrum sinense*, Chinese privet, *Lonicera japonica*, Japanese honeysuckle, *Pueraria montana* var. *lobata*, kudzu, and *Sorghum halapense*, Johnston grass. For example, kudzu, an aggressive vine, can completely cover native forests (e.g., in the southeastern United States) and, unfortunately, it is well-established in a number of North Central Texas counties (Grayson, Lamar, and Tarrant). *Festuca arundinacea* is capable of invading intact native tall grass prairies and is considered by some (e.g., Fred Smiens, pers. comm.) to be the most serious invasive threat to tall grass Blackland Prairie remnants (such as the Nature Conservancy's Clymer Meadow in Hunt County).

Some exotic species are currently spreading in North Central Texas. For example, the offensive *Carduus nutans* subsp. *macrocephalus*, musk-thistle or nodding-thistle, is each year becoming more abundant in the northern part of North Central Texas (e.g., Grayson Co.). A possibly even more serious threat, *Scabiosa atropurpurea*, pincushions or sweet scabious, is currently taking over roadsides and adjacent areas in the northern part of North Central Texas (e.g., Collin Co.) and has the potential of becoming one of the most destructive invasive exotics in the area. From the aquatic standpoint, *Hydrilla verticillata* is a serious pest which can completely dominate aquatic habitats eliminating native species, clogging waterways, and severely curtailing recreational use (Steward et al. 1984; Flack & Furlow 1996). It is rapidly spreading at present in North Central Texas (M. Smart, pers. comm.), probably from lake to lake by boats or boat trailers and also intentionally by fishermen (L. Hartman, pers. comm.) to "improve" the habitat. This activity is both illegal and ill-advised since it ultimately degrades the fishery. In fact, because of their potential as problematic invaders, five aquatic species that occur in North Central Texas, *Alternanthera philoxeroides*, alligator-weed (Amaranthaceae), *Eichhornia crassipes*, common water-hyacinth (Pontederiaceae), *Hydrilla verticillata* (Hydrocharitaceae), *Myriophyllum spicatum*, Eurasian water-milfoil (Haloragaceae), and *Pistia stratiotes*, water-lettuce (Araceae), are considered "harmful or potentially harmful exotic plants" and it is illegal to release, import, sell, purchase, propagate, or possess them in the state (Harvey 1998).

These alien taxa are from nearly all parts of the world (e.g., *Bromus catharticus*, rescue grass, from South America; *Chenopodium pumilo*, ridged goosefoot, from Australia; *Eragrostis curvula*, weeping love grass, from Africa; *Bothriochloa ischaemum* var. *songarica*, King Ranch bluestem, from Asia; *Stellaria media*, common chickweed, from Europe) and have gotten to North Central Texas in assorted ways. However, most introduced weeds in eastern North America, including many in North Central Texas, are from central and western Europe. It is thought that many weedy colonizing species evolved in Europe over thousands of years as humans disturbed and modified the environment for agricultural purposes; these same species do well in the disturbed habitats of the eastern United States (Stuckey & Barkley 1993). Numerous such European species entered North America at seaport cities along the Atlantic coast and spread westward across the continent (Stuckey & Barkley 1993). An excellent example of this phenomenon can be seen with *Chaenorrhinum minus*, dwarf snapdragon, which was first observed growing in North America in New Jersey in 1874 (Martindale 1876) and has since spread to over 30 states and nine Canadian provinces (Widrlechner 1983). In some cases, seeds were introduced with soil, sand, or rocks being used as ballast in seagoing ships; Mühlenbach (1979) discussed the role of maritime commerce in dispersal. Other currently problematic taxa were intentionally introduced as ornamentals (e.g., *Ligustrum* species, privets), as windbreaks (e.g., *Tamarix* species, saltcedars), or in misguided attempts at habitat improvement, erosion control, soil stabilization, etc. In yet other cases, exotics are thought to have been accidentally introduced with crop seeds (e.g., *Myagrum perfoliatum*), hay (e.g., *Carduus nutans* subsp. *macrocephalus*), cotton or wool, or are associated with livestock yards (e.g., *Onopordum acanthium*, Scotch-thistle). Still others are transported by trains (e.g., *Chaenorrhinum minus*—Widrlechner 1983); Mühlenbach (1979) discussed the importance of railroads as a means of dispersal. A particularly unusual dispersal mechanism is suspected for *Soliva pterosperma*, lawn burweed or stickers, which is thought to have been introduced into North Central Texas at a soccer field by the spinulose achenes sticking in athletic shoes.

The percentage of exotics in the North Central Texas flora—17.7% as stated above—is approximately what would be expected based on data from other parts of the United States. Elias (1977) estimated the level of exotics at 22% in the northeastern United States and more recently Stuckey and Barkley (1993) indicated that in northeastern states the percentage of foreign species ranged from 20% to over 30%. Their data, compiled from a number of sources, showed that there are higher percentages of foreign species in those states that have been occupied the longest by non-native inhabitants and in those that have been most extensively involved in agriculture. Some northern and western states, with less human influence and disturbance, have figures below 20%. While rather recently colonized by European settlers, North Central Texas, particularly the Blackland Prairie portion, has been extensively cultivated and numerous exotic species have arrived and become naturalized. Comparable percentages of foreign species are seen in the floras of California (17.5%) (Rejmánek & Randall 1994), Colorado (16%), Iowa (22.3%), Kansas (17.4%), and North Dakota (15%) (Stuckey & Barkley 1993).

Several introduced species have only recently been reported in North Central Texas, including *Chaenorrhinum minus*, dwarf snapdragon (Diggs et al. 1997), *Cerastium pumilum*, dwarf mouse-ear chickweed, and *Stellaria pallida*, lesser chickweed (Rabeler & Reznicek 1997). As this book was nearing completion, another European species, *Agrostemma githago*, corn-cockle, was discovered in the area (*O'Kennon s.n.*, Parker Co.) as were two exotics new to Texas, *Cerastium brachypetalum*

(*Rabeler 1333*, Red River Co.), gray chickweed, and *Plantago coronopus* (*O'Kennon 14221*, Tarrant Co.), buck-horn plantago (O'Kennon et al. 1998). Additional exotics can be expected to become part of the North Central Texas flora in the future, many with serious negative consequences to the remnant native flora.

CONSERVATION IN NORTH CENTRAL TEXAS

Human activities have profoundly altered the biological picture of North Central Texas. Only small remnants of the original habitats have survived to the present day. However, numerous conservation efforts are currently underway in the region. Addresses and telephone numbers of the organizations mentioned below are provided in Appendix 9.

Substantial areas of land are controlled by the federal government including Hagerman National Wildlife Refuge (an 11,000-acre tract in Grayson County), Balcones Canyonlands National Wildlife Refuge in Burnet, Travis, and Williamson counties, protected areas in Fort Hood in Bell and Coryell counties, U.S. Army Corps of Engineers land around numerous impoundments, and the Caddo and Lyndon B. Johnson National Grasslands in Fannin and Wise counties. The Texas Department of Parks and Wildlife is also protecting, and in some cases attempting restoration on, numerous tracts in state parks and state wildlife management areas throughout North Central Texas. Examples of state land include Bonham State Park in Fannin County, Cedar Hill State Park in Dallas County, Cleburne State Park in Johnson County, Cooper Lake State Park in Delta and Hopkins counties, Dinosaur Valley State Park in Somervell County, Eisenhower State Park in Grayson County, Lake Brownwood State Park in Brown County, Lake Mineral Wells Park in Parker County, Lake Whitney State Park in Hill County, Meridian State Park in Bosque County, Mother Neff State Park in Coryell County, Possum Kingdom State Park in Palo Pinto County, and Pat Mayse State Wildlife Management Area in Lamar County. A number of far-sighted local governments are also protecting natural habitats. Examples of these include the Dallas Nature Center, the 3,000-acre Fort Worth Nature Center and Refuge, the Gambill Wildlife Refuge in Lamar County, which is maintained by the City of Paris, Harry S. Moss Park in Dallas, Parkhill Prairie Preserve in Collin County, River Legacy Living Science Center in Arlington, and Tandy Hills Park in Fort Worth.

Non-governmental organizations such as the Nature Conservancy and the Natural Area Preservation Association protect particularly critical pieces of habitat. Two well known examples are the Nature Conservancy's Clymer Meadow in Hunt County and Tridens Prairie in Lamar County. The Heard Natural Science Museum and Wildlife Sanctuary, a 287-acre protected area in Collin County, has numerous conservation activities including a raptor rehabilitation program and a tall grass prairie restoration project (e.g., Steigman & Ovenden 1988). Austin College and its Center for Environmental Studies protects three field laboratories and preserves totaling nearly 300 acres in Grayson County. Other organizations, such as the Native Plant Society of Texas, the Native Prairies Association of Texas, the Texas Committee on Natural Resources, and the Thompson Foundation are actively engaged in educating the public and promoting the importance of plants, natural areas, and conservation. The Lady Bird Johnson Wildflower Center, located in Travis County just south of North Central Texas, is dedicated to the study, preservation, and reestablishment of North American native plants in planned landscapes; it has had an important impact throughout Texas and beyond. The Texas Organization for Endangered Species (TOES) monitors and regularly publishes information about endangered and threatened species and natural communities in North Central Texas as well as throughout the state. The Botanical Research Institute of Texas, in addition to its research activities, has an environmental education program, providing appropriate publications and educational opportunities for school children in the North Central Texas area.

Finally, many individual landowners are also making significant contributions by managing their properties in ways that preserve the natural diversity of the area. Enlightened grazing regimens, setting aside particularly fragile or erosion-prone parcels, or simply purchasing areas to protect are some of the strategies being undertaken by landowners throughout the North Central

Texas region. Other individuals such as Rosa Finsley, Howard Garrett, the late Lynn Lowrey, the late Benny Simpson, and Sally Wasowski have also made large contributions by bringing attention to the superiority of native plants in landscapes and other environmentally sensitive strategies.

All of these efforts are critical because given the rate at which remaining areas of natural habitat are disappearing, unless action is taken by those living today, the opportunity to provide future generations with the chance to experience natural areas in North Central Texas will soon be lost.

A Sketch of the History of Botany in Texas with Emphasis on North Central Texas

Botany in Texas

Prior to the Republic of Texas / before 1836

Fig. 25/ Thomas Drummond, 1780–1835. Redrawn from a portrait in Popular Science Monthly 74:49. 1909.

Much of the earliest natural history work in Texas was botanical in nature (McCarley 1986). According to Winkler (1915), "The study of Texas plants . . . is as old as the state itself. Prior to her annexation to the Union, and even before the period of the Republic of Texas, Texas had become an interesting field of observation and research for botanists and naturalists." The first known collection of plants from what is now the state was made by Edwin James in August 1820 in the Texas Panhandle as part of Major S.H. Long's expedition to the Rocky Mountains (Shinners 1949h). Details of the expedition's route are provided by Goodman and Lawson (1995). However, the first person to make more extensive collections in the area that would become Texas was Jean Louis Berlandier (1805–1851), a French (or Swiss if today's borders are accepted) botanist. Berlandier collected in Texas during the years 1828 to 1834 with the earliest of his collections being made in 1828 between Laredo and San Antonio while on a Mexican Boundary Commission expedition to explore the area along the proposed United States-Mexico border (Winkler 1915; Geiser 1948; Berlandier 1980). His name is recognized in many scientific binomials including the genus *Berlandiera*, greeneyes, a composite group of four species native to the southern United States and Mexico. Berlandier apparently made the first collection of *Lupinus texensis*, one of the six *Lupinus* species which are the state flowers of Texas (Andrews 1986; Turner & Andrews 1986). A two volume translation of his journal has been published (Berlandier 1980).

Another early plant collector was Thomas Drummond (1780–1835), a Scottish botanist and naturalist who came to Texas in 1833 (Fig. 25). While in the area for only a brief period (1833–1834), he made important collections in southeast Texas and stimulated such later collectors as Lindheimer and Wright (discussed below). Drummond's were the first Texas collections ". . . that were extensively distributed among the museums and scientific institutions of the world" (Geiser 1948). While many Texas plants are named for him, perhaps none is better known than *Phlox drummondii*, commonly known as Drummond's phlox or pride-of-Texas. Also of note is that it was from several of Drummond's collections that Sir William Jackson Hooker described both *Lupinus subcarnosus* and *Lupinus texensis* (Hooker 1836; Turner & Andrews 1986).

DURING REPUBLIC OF TEXAS TIMES AND EARLY STATEHOOD / 1836–1865

While not chronologically the first collector in the state, Ferdinand Jacob Lindheimer (1801–1879), a German-born collector, is often referred to as the "father of Texas botany" because of his important botanical contributions, particularly on the central Texas flora (Fig. 26). Lindheimer's botanical work in the state, supported in part by George Engelmann and Asa Gray (the pre-eminent Harvard botanist), stretched from 1836 to 1879 (Geiser 1948). Lindheimer's collections were widely distributed by Englemann and Gray under the title "Flora Texana Exsiccata" (Blankinship 1907) and numerous new species were described in the well known *Plantae Lindheimerianae* (Engelmann & Gray 1845). Many Texas plants including *Lindheimera texana*, Texas-star, yellow Texas-star, or Lindheimer's daisy, and *Gaura lindheimeri*, white gaura, are named after him. Details about his life and botanical contributions can be found in Blankinship (1907) and Geiser (1948). Lindheimer's letters to Engelmann have been edited, translated, and discussed by Goyne (1991).

FIG. 26/ FERDINAND JACOB LINDHEIMER, 1801–1879. USED WITH PERMISSION OF SOPHIENBURG MUSEUM & ARCHIVES, NEW BRAUNFELS, TX.

A friend and sometimes collecting companion of Lindheimer was another German, Ferdinand Roemer (1818–1891), who spent the years 1845 to 1847 in Texas (Geiser 1948). While a geologist, sometimes referred to as the "father of the geology of Texas," he is probably best known for his book, *Texas with Particular Reference to German Immigration and the Physical Appearance of the Country* (Roemer 1849). Roemer, however, also collected plants (Winkler 1915) and his botanical contributions are recognized in such names as *Phlox roemeriana*, gold-eye phlox, and *Salvia roemeriana*, cedar sage.

A further early Texas collector was Charles Wright (1811–1885), whose collections for Asa Gray spanned the years 1837 to 1852 (Geiser 1948). Much of his collecting in western Texas was conducted while accompanying troops to that part of the state, an example being his 1849 expedition across the unexplored region between San Antonio and El Paso. This expedition is of special interest because the Smithsonian's $150 contribution to defray Wright's expenses was, according to some, one of the early steps taken by that institution toward the formation of a national herbarium (Winkler 1915). Wright is commemorated by such plants as *Datura wrightii*, angel-trumpet, and *Ipomoea wrightii*, Wright's morning-glory. Further information on Wright's Texas travels can be found in Shaw (1987).

Another German-born naturalist was Louis Cachand Ervendberg (1809–1863), active in Texas from 1839 to 1855. He corresponded with and collected plants for Asa Gray in Comal County and later in Veracruz, Mexico (Geiser 1948).

John Leonard Riddell (1807–1865), a botanist and geologist, visited Texas briefly in 1839 and contributed to early knowledge about the plants of the state. His name can be seen in *Aphanostephus riddellii*, Riddell's lazy daisy. Detailed information about his travels in Texas are given in Breeden (1994).

Another student of Texas natural history was Gideon Lincecum (1793-1874), a Georgia-born frontier naturalist and pioneer physician who lived and worked in Texas and later Mexico from 1848 to 1874 (Fig. 27). During his career he corresponded with such eminent scientists as Charles Darwin, Spencer Baird, and Joseph Henry. Though self-taught, he published at least two dozen scientific articles and was elected a corresponding member of the Philadelphia Academy of Natural Sciences. Lincecum sent botanical specimens to such prestigious museums as the Academy of Natural Sciences of Philadelphia, the British Museum, and the Smithsonian Institution. Not only did Lincecum make plant collections but he also became an authority on Texas grasses. Additionally, he made extensive observations of the Texas agricultural (harvester) ant. His work with ants was eventually read by Darwin before the Linnaean Society in London and published in the Society's journal in 1862 (Lincecum 1861, 1862; Geiser 1948; Burkhalter 1965; Lincecum & Phillips 1994; Lincecum et al. 1997). His name is remembered in *Vitis aestivalis* var. *lincecumii*, the pinewoods grape, of East Texas. Detailed information and much of his correspondence can be found in Lincecum and Phillips (1994) and Lincecum et al. (1997).

FIG. 27/ GIDEON LINCECUM, 1793–1874.
USED WITH PERMISSION OF THE CENTER FOR AMERICAN HISTORY, THE UNIVERSITY OF TEXAS AT AUSTIN.

Important Texas collections were also made in 1849–1850 by the French botanist Auguste Adolph Lucien Trécul (1818–1896). According to Geiser (1948) he "... visited Texas on his scientific mission to North America to study and collect farinaceous-rooted plants used for food by the Indians." *Stillingia treculeana*, Trecul's stillingia, and *Yucca treculeana*, Trecul's yucca or Spanish-dagger, are both named in his honor. McKelvey (1955, 1991) gave detailed information about Trécul's travels in southern and central Texas including an outline of his route and some collection numbers. Further information on Trécul can be found in Jovet and Willmann (1957).

In 1852, Captain R.B. Marcy's expedition to explore the Red River to its source (Marcy 1853) resulted in the collection by George G. Shumard (1825–1867), surgeon of the expedition, of 200 plant species (Winkler 1915). This expedition also yielded a published list of species by Torrey (1853) with 20 excellent illustrations, some of which are reprinted in the present volume.

Another interesting early contributor to Texas botany was Samuel Botsford Buckley (1809–1884). He first came to Texas in 1859, twice served as State Geologist of Texas, and collected plants in various parts of the state. According to L. Dorr (pers. comm.), "... it should be noted that Buckley was the first botanist to collect in Texas who then described new taxa from his own collections. Asa Gray took great exception to this infringement upon his virtual monopoly on publishing on Texas plants and Gray published several scathing reviews of Buckley's work. Buckley published in excess of 100 taxa of Texas plants, a number of which are recognized today." Among his scientific papers, several were published in the Proceedings of the Academy of Natural Sciences of Philadelphia (e.g., Buckley 1861 [1862]) including his rebuttal to Gray's criticisms (Buckley 1870). One of the best known species described by Buckley is *Quercus shumardii*, Shumard's red oak, which he named for B.F. Shumard, a geologist under whose direction he at one time worked. Buckley's name is remembered in *Quercus buckleyi*, Texas red oak (Dorr & Nixon 1985). Detailed information about Buckley's life and work can be found in Dorr and Nixon (1985) and Dorr (1997).

After the Civil War to the End of World War II / 1865–1945

The first woman botanist in Texas, Mrs. Maude Jeannie Young (1826–1882) taught botany in Houston, collected plants, and in 1873 published *Familiar Lessons in Botany with Flora of Texas.* This extensive work (646 pages) is reported to be the first scientific text for the state (Studhalter 1931; Todzia 1998). According to Dorr and Nixon (1985), "It is a curious book. The major portion of the Flora was copied verbatim from Chapman's *Flora of the Southern United States* (1860), Mrs. Young's editorial contribution consisting of the deletion of taxa not present or expected to be present in Texas, occasional notes on the distribution of species within Texas and the description of one new species of plant." Another early Texas female botanist was Mary S. Young (1872–1919) (apparently unrelated to M.J. Young), one of the first botanists at the University of Texas (Fig. 28). She made important plant collections in various parts of the state including the Panhandle and Trans-Pecos and expanded the herbarium of the University of Texas by doubling the number of specimens (Young 1920; Tharp & Kielman 1962; Bonata 1995; Todzia 1998). Her publications included *A Key to the Families and Genera of the Wild Plants of Austin Texas* (Young 1917) and *The Seed Plants, Ferns, and Fern Allies of the Austin Region* (Young 1920).

Other relatively early (pre-1940) contributions to the understanding of Texas botany were those by E. Hall (1873) *Plantae Texanae: A List of the Plants Collected in Eastern Texas*; T.V. Munson (1883) *Forests and Forest Trees of Texas*; V. Havard (1885) *Report on the Flora of Western and Southern Texas*; J.M. Coulter (1891–1894) *Botany of Western Texas*; various works by W.L. Bray among them the *Ecological Relations of the Vegetation of Western Texas* (Bray 1901); J.W. Blankinship (1907) *Plantae Lindheimerianae, Part III*; I.M. Lewis (1915) *The Trees of Texas*; C.H. Winkler (1915) *The Botany of Texas*; B.C. Tharp (1926) *Structure of Texas Vegetation East of the 98th Meridian*; E.D. Schulz (1922) *500 Wild Flowers of San Antonio and Vicinity* and (1928) *Texas Wild Flowers*; M.C. Metz (1934) *A Flora of Bexar County, Texas*; H.B. Parks and V.L. Cory (1936) *The Fauna and Flora of the Big Thicket Area*; E. Whitehouse (1936) *Texas Flowers in Natural Colors*; and V.L. Cory and H.B. Parks (1937) *Catalogue of the Flora of the state of Texas.* This latter work was the earliest attempt to compile a complete list of the vascular plants of Texas.

Fig. 28/ Mary Sophie Young, 1872–1919, (and Nebuchadnezzar). Used with permission of the Texas State Historical Association, Austin.

POST WORLD WAR II TO THE PRESENT / 1945–1998

Following Cory and Parks' (1937) first list, a number of subsequent checklists have been produced by botanists associated with or trained at Texas A&M University, one of the centers of research on Texas botany. These include *Texas Plants—A Checklist and Ecological Summary* (Gould 1962, 1969, 1975a), *Checklist of the Vascular Plants of Texas* (Hatch et al. 1990), and *Vascular Plants of Texas: A Comprehensive Checklist including Synonymy, Bibliography, and Index* (Jones et al. 1997). Other large scale taxonomic works covering the entire state are *Ferns and Fern Allies of Texas* (Correll 1956), *The Legumes of Texas* (Turner 1959), and *The Grasses of Texas* (Gould 1975b). The second of these was one of the numerous contributions by Billie Lee Turner of the University of Texas at Austin, who has published extensively on the plants of Texas with particular emphasis on the Asteraceae. Turner was one of the individuals responsible for developing the Botany Department at the University of Texas at Austin into one of the best known and most respected departments in the United States. The extensive work by Frank Gould on grasses (e.g., 1968a, 1975b) at Texas A&M University received national and even international recognition, and his book on Texas grasses is one of the best treatments in the country for a large taxonomic group at the state level.

FIG. 29/ DONOVAN STEWART CORRELL, 1908–1983. USED WITH PERMISSION OF THE NEW YORK BOTANICAL GARDEN, BRONX.

The first attempt at a comprehensive state-wide flora was the three volume *Flora of Texas* (3 vols.) by C.L. Lundell (1961, 1966, 1969). While never completed, this project of the Texas Research Foundation at Renner (near Dallas) was a valuable contribution to the knowledge of Texas plants. The Texas Research Foundation subsequently published the *Manual of the Vascular Plants of Texas* (Correll & Johnston 1970), which after nearly four decades is still the only comprehensive source of information about the flora of the entire state. This work was authored by Donovan Stewart Correll (1908–1983) (Fig. 29) and Marshall Conring Johnston (1930–). After service at Harvard University and the United States Department of Agriculture, Correll, who was born in North Carolina and trained at Duke, in 1956 came to the Texas Research Foundation where he directed the *Manual* project. His research specialties included potatoes (*Solanum*), ferns, the Orchidaceae, and economic botany (Schubert 1984). With his wife, Helen B. Correll, he authored the influential and still widely used *Aquatic and Wetland Plants of Southwestern United States* (1972). Marshall Johnston, the second author of the *Manual* and a native Texan reared in the brush country of the Rio Grande delta, spent his career in the Botany Department at the University of Texas at Austin. His research specialties include the Euphorbiaceae, Rhamnaceae, and floristics of Texas and Mexico. Subsequent to the publication of the *Manual*, Johnston published two lists updating that work (Johnston 1988, 1990).

Floras are also available for various regions of the state including South Central Texas (Reeves & Bain 1947), the Big Bend (McDougall & Sperry 1951), North Central Texas (Shinners 1958a; Mahler 1984, 1988), the Texas Coastal Bend (Jones et al. 1961; Jones 1975, 1977, 1982), Central Texas (Reeves 1972, 1977), and the Edwards Plateau (Stanford 1976). More specialized works (e.g., treatments of trees and shrubs or grasses) are available for some regions of the state (e.g., Austin and the Hill Country—Lynch 1981; East Texas—Nixon 1985; Trans-Pecos—Powell 1988, 1994, 1998).

At present there are several long-term flora projects ongoing in Texas. These are a revision of the *Manual* being undertaken by David Lemke of Southwest Texas State University in San Marcos, and the new Flora of Texas Project, conceived by the Botanical Research Institute of Texas (BRIT), with founding members including BRIT, Southwest Texas State University, Texas A&M University, and the University of Texas at Austin. The goal of this latter project is to create an electronic database of information about the approximately 6,000 taxa of native and naturalized vascular plants of Texas, to make these data accessible via the internet, and to use the information to support botanical studies including the production of floras. At a more local scale, the Illustrated Texas Floras Project, a collaboration between BRIT and the Austin College Center for Environmental Studies, is attempting to produce illustrated floras for various parts of the state. This volume is the first in that series and is the first fully illustrated flora to be published for any region of Texas or surrounding states. Currently BRIT is an active center of plant research with one of the largest concentrations of professional taxonomic botanists in the southwestern United States. Five nationally prominent scientists have located at BRIT to continue their research. These are Theodore Barkley (formerly of Kansas State University), Robert Kral (formerly of Vanderbilt University), Joe Hennen (formerly of Purdue University), Henri Alain Liogier (formerly of the Botany Garden of the University of Puerto Rico-San Juan), and Richard Norris (formerly of the University of Washington and the University of California-Berkeley). Other professional biologists or research associates in residence at BRIT are Bruce Benz, Charlotte Bryant, George Diggs, Harold Keller, Barney Lipscomb, Fiona Norris, Robert O'Kennon, John Pipoly, Roger Sanders, S.H. Sohmer, Dora Sylvester, and Lindsay Woodruff.

While a great deal of work was conducted in the 1800s on Texas plants, most of the research was accomplished by non-residents or was funded by outside sources. The result was that few of the early collections remained in the state. According to Shinners (1949h),

> Pioneer collectors [in Texas] were either sent from Europe, or were patronized by botanists in the older parts of the United States. Not until the late 1890s did a Texas institution begin serious study of the flora of the state. Just fifty years ago [now about 100 years ago], W.L. Bray made collections more or less incidentally to ecological studies of the vegetation. These were the earliest collections to remain permanently in Texas and were the beginning of what is now the largest herbarium in the state, that of the University of Texas [at Austin].

Over the past century this situation has changed greatly. As a result of various state and local floristic projects and the collecting efforts of numerous individuals, currently well over two million herbarium specimens are kept in Texas. About 27 herbaria are active in the state, the three largest are the Plant Resources Center at the University of Texas at Austin (about 1,100,000 specimens including the University of Texas and Lundell Herbaria), the Botanical Research Institute of Texas in Fort Worth (860,000 specimens including the Southern Methodist University and Vanderbilt University collections), and the S.M. Tracy Herbarium of the Range Science Department of Texas A&M University (over 217,000 specimens) (Simpson 1996). A substantial number of very early Texas collections have returned to the state through the efforts of Lloyd Shinners and exchanges with the Milwaukee Public Museum and the Missouri Botanical Garden. For example, slightly less than 1,400 early Texas specimens (dating back to 1839) collected by Ferdinand Lindheimer, Julien Reverchon, Charles Wright, and others are now in the collection at the Botanical Research Institute of Texas (Shinners 1949h).

Further information on the history of botany in Texas can be obtained from Winkler (1915), Geiser (1945, 1948), (Shinners 1949h, 1958a), and McKelvey (1955, 1991).

BOTANY IN NORTH CENTRAL TEXAS

EARLY CONTRIBUTIONS / PRIOR TO 1970

While botanical exploration, observation, and collecting occurred early in North Central Texas (e.g., Smythe 1852; Parker 1856; Buckley—See Dorr & Nixon 1985; Munson 1883, 1909), the first botanist to extensively collect in the north central part of the state was Julien Reverchon (Fig. 30). By the time of his death in 1905, Reverchon's collection numbered about 20,000 specimens of more than 2,600 Texas species. It was the best collection of the state's flora then in existence (Geiser 1948). Reverchon corresponded extensively with Asa Gray, one of the leading American botanists of the nineteenth-century, and was even visited by Gray. In addition to his collecting, Reverchon was a member of the Torrey Botanical Club, published a number of scientific papers (e.g., Reverchon 1879, 1880, 1903), and during the last decade of his life served as Professor of Botany in the Baylor University College of Medicine and Pharmacy at Dallas (Geiser 1948). Gray eventually named the monotypic genus *Reverchonia* (Euphorbiaceae) in his honor (Geiser 1948) as well as the Texas endemic *Campanula reverchonii*, basin bellflower. According to Shinners (1958a),

FIG. 30/ JULIEN REVERCHON
1837–1905

> Born at Lyons, France, in 1837, he [Reverchon] came with his father to La Reunion (now part of the city of Dallas) in 1856. Though early interested in plants, he did not begin serious collecting until 1876. His early specimens went to Asa Gray at Harvard University, who encouraged him to make extended trips west of our area. Later he collected much more prolifically for William Trelease of the Missouri Botanical Garden, aided by the grant of passes on railroads. After his death in 1905 his entire personal herbarium went to the Garden, in St. Louis. Through the good offices of Dr. Robert E. Woodson [and the work of Lloyd Shinners] over a thousand duplicates of Reverchon's specimens came back to Dallas, starting in 1949, and are now incorporated in the Herbarium of Southern Methodist University [now at BRIT]. The specimens and the manuscript field-notes which often accompany them show Reverchon to have been a keen and discerning collector, but more than this, a perceptive naturalist, recording a wealth of information about the plants and their habitats. Whether from diffidence or under terms of agreement with the eminent botanists who were purchasing his specimens, he unfortunately published almost nothing. The Missouri store-keeper, Benjamin Franklin Bush, who was for a time herbarium assistant to Trelease, visited Reverchon and also collected in our area....The growth of a large city [Dallas] and its suburbs has eliminated many species which he found here, and our knowledge of Texas plants would be faulty indeed were it not for his work. Undoubtedly still more species should appear in this book than are cited. But his collections are dispersed among the 1,500,000 at St. Louis, and though I have spent periods of a few days to several months there in each of some eight years, I have by no means checked them all. In the city where he worked, Reverchon is remembered now only by the small Reverchon Park, but hardly anyone knows for whom it was named.

Regarding the history of North Central Texas botany, Shinners (1958a) went on to say,

> Fort Worth can boast our next resident botanist, Albert Ruth, born in 1844 [Fig. 31]. Forced into unwilling retirement from his position as superintendent of schools in Knoxville, Tennessee, he turned his back on the state where he had been an active amateur botanist and spent his last twenty years (1912–1932) collecting in Texas. He was almost as prolific as Reverchon.... Ruth's specimens (unlike Reverchon's) were very widely distributed, and he apparently did not attempt to preserve a complete collection himself. His quite incomplete personal herbarium was purchased by the Fort Worth Park Board after his death.... It is now on deposit at Texas Christian University [currently at Fort Worth Museum of Science and History]. Though Ruth did some collecting as far away as Bexar and Garza counties, most of his activity was confined to our local area, chiefly Tarrant County, and to a much smaller extent Dallas and Denton counties. In 1929 a set of 300 specimens was collected for Dr. W.M. Longnecker of Southern Methodist University, primarily for class use. These are the first specimens now part of the [SMU, now BRIT] Herbarium there to have been acquired. About 500 additional Ruth collections, obtained from several sources, have augmented this original set. [BRIT has in its library a lengthy unpublished typescript by Ruth of a *Manual of Texas Flora*.]
>
> Fort Worth can also claim our third resident botanist, William Larrey McCart, whose interests in science began while he was a student at Central High School there, and were continued, with special attention to plants, by Mrs. Hortense Winton under whom he took freshman work at Texas Christian University. He attended The University of Texas briefly, to take a course in plant taxonomy under Dr. B.C. Tharp, subsequently going to North Texas State College in Denton, where in due course he took his master's degree. He then returned to Fort Worth, where family affairs made it necessary for him to take employment, and severely curtailed botanical activities for more than a decade. A little work was done at the Botanical Garden, where his own herbarium shared storage space with that of Albert Ruth. From 1954 to 1957 he took additional work at The University of Texas, resumed collecting, ordered up the W.A. Silveus Grass Collection which had been bequeathed to the University, and then returned to Denton as a member of the staff of North Texas State College. His chief early period of activity was from 1937 to 1940, during which time he set out to collect systematically, county by county, to establish distributions of species — the first time work of this kind had been done in our area. He also made great efforts to send off collections to specialists for accurate determinations. During that time, the best organised [organized] and most thorough work on the state's flora being carried out was done by him. His personal herbarium of some 4,000 specimens is now incorporated in that of Southern Methodist University [now BRIT], and has been an invaluable help in the completion of this book.

FIG. 31/ ALBERT RUTH (1844–1932).
FROM THE COLLECTION OF THE FORT WORTH PUBLIC LIBRARY.

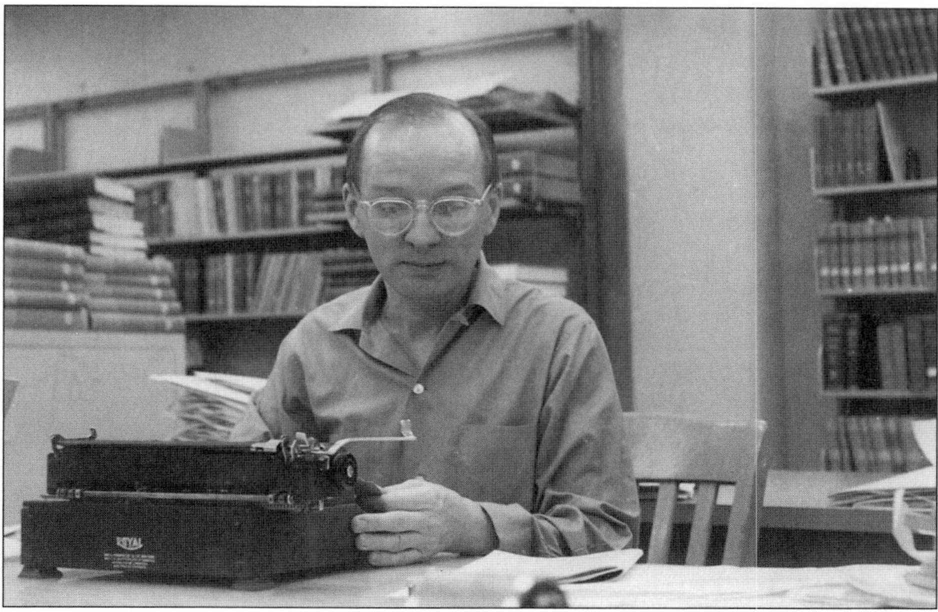

FIG. 32/ LLOYD HERBERT SHINNERS, 1918–1971.

In 1939 Mrs. Norma Stillwell published her "Key and Guide to the Woody Plants of Dallas County," a pamphlet treating some 90 species, the first independent publication dealing with our local flora (and only the third such publication of any kind, its only predecessors being Reverchon's brief note "Notes on some introduced plants in Dallas County, Texas," in the Botanical Gazette vol. 5 p. 10, 1880; and the "Directions for Plant Collections" and "Flowering Plants" section in "Natural History Manual of T.C.U. Vicinity," by Hortense Winton and Sadie Mahon, which reached a fifth edition in 1929). Mrs. Stillwell was an enthusiastic and talented amateur who attempted with the aid of garden-club friends to compile a popular local flora, portions of which were mimeographed. Unfortunately only a few permanent specimens were made which were adequate for preservation in a herbarium. Some, however, are of special interest - for example, the type specimen of *Rosa ignota*. Mrs. Stillwell's manuscript notes and specimens were turned over to S.M.U. when she moved from the city.

Beginning in 1940, Dr. C. L. Lundell, an alumnus of S.M.U., then at the University of Michigan, began systematic collecting toward an ambitious *Flora of Texas*. His work paralleled that of McCart — organised [organized] geographic exploration, getting identifications from specialists — but this time it was done by an experienced, professional botanist, with ample financial backing. The S.M.U. Herbarium was formally organised by him in 1944. But promotional and public relations work with what was to become the Texas Research Foundation made increasing inroads on his time, almost completely halting his botanical field work. From 1945 on, this was continued by three additions to the staff of the new Herbarium: myself [L.H. Shinners] in February, Dr. Eula Whitehouse [see Flook 1974 for more information] in June, and Mr. V.L. Cory in September. During several years we were assisted by visiting botanists: Drs. Donovan S. Correll, C.H. Muller, and Rogers McVaugh. Collecting was done in all parts of the state, that of the visiting botanists especially being carried on chiefly away from our local area. But since it would hardly have been possible to reach satisfactory conclusions without examining collections from elsewhere, all this work contributed to the preparation of a local flora.

When I [L.H. Shinners] assumed charge in 1948, the Herbarium had reached a total of slightly under 21,000 specimens. Now, ten years later, the higher plants total 150,000 [from this beginning BRIT/SMU, including VDB, has grown to about 860,000 specimens as of 1998]. My own collection numbers since coming to Texas amount to almost 20,000 (which with duplicates means perhaps 100,000 specimens), of which about 40% are from the local area. This is really not very much compared with the amount of work that has been done in New England, the Philadelphia area, the central California Coast, or many parts of Europe; it seems like even less when one considers the richness of the flora and the size of the area being covered. I have often remarked that the first edition of this book [*Spring flora of the Dallas-Fort Worth area Texas*] will be a Flora of the Main Highways, the second will include the back roads, and perhaps the third will begin to cover the country.

My [L.H. Shinners] earliest studies depended chiefly on the specimens collected by Dr. and Mrs. Lundell, Miss Whitehouse, and Mr. Cory, and on loans from the Missouri Botanical Garden (these often extended for periods of as much as several years). These were gradually augmented by my own collections and those of Reverchon, Ruth and McCart, as already noted. But nothing of course can take the place of seeing the plants live, again and again, year after year.

FIG. 33/ EULA WHITEHOUSE, 1892–1974.

Without a doubt, Lloyd Herbert Shinners (1918–1971), a native Canadian who received his botanical training at the University of Wisconsin-Madison, is the most important twentieth-century North Central Texas botanist (Fig. 32). He came to Southern Methodist University in Dallas in 1945, became the Director of the Herbarium in 1949, and was on the faculty there until his death (Mahler 1971b). Not only did he almost single-handedly develop the herbarium which today forms the core of the collection at BRIT, but he also created one of the best botanical libraries in the United States, did extensive field work, and published a total of 276 articles and a 514-page flora (Flook 1973). His contributions to botanical nomenclature are particularly impressive, totaling 558 new scientific names and combinations (Flook 1973). Among his most lasting achievements are the *Spring Flora of the Dallas-Fort Worth Area Texas* (Shinners 1958a) and the journal, *Sida, Contributions to Botany*, which he founded in 1962 (Mahler 1973b). Shinners' *Spring Flora* was the first completed, original, technical book on Texas plants prepared by a resident of the state. It was extensively used by high schools, colleges, and universities as a textbook for classes, and is still in use today. For a synopsis of Shinners' life see Mahler (1971b); for a guide to his botanical contributions see Flook (1973).

Eula Whitehouse (1892–1974) (Fig. 33), mentioned above, is best known for her *Texas Flowers in Natural Colors* (1936), the first color-illustrated guide to Texas wildflowers (Flook 1974). Her career was at the Houston Municipal Hospital, the Texas Memorial Museum in Austin, the University of Texas College of Mines, and Southern Methodist University. While at SMU she studied bryophytes (Whitehouse & McAllister 1954), published taxonomic revisions (e.g., Whitehouse 1945, 1949), and did extensive art work. Some of her illustrations were used in Shinners' *Spring Flora* and are reproduced in this volume.

Another important North Central Texas botanist was Cyrus Longworth Lundell (1907–1994), mentioned above (Fig. 34). Lundell is best known as founder of the Texas Research Foundation, author (with collaborators) of the *Flora of Texas*, and as a specialist on the Myrsinaceae. His institute was instrumental in establishing Texas as an important center of taxonomic botany.

Recent Contributions / 1970–1998

More recently, Wm. F. "Bill" Mahler (1930–) (Fig. 35), Director Emeritus of BRIT, had an extensive role in the botany of the north central part of the state. After receiving his Ph.D. from the University of Tennessee at Knoxville, he joined the faculty of Southern Methodist University in 1968, became editor of *Sida* in 1971, and assumed leadership of the herbarium in 1973. Mahler is probably best known for his *Shinners' Manual of the North Central Texas Flora* (1984, 1988), well known for its clarity and ease of use. This manual was an expanded version of Shinners' (1958) *Spring Flora of the Dallas-Fort Worth Area Texas* that also included the summer and fall flora for North Central Texas. Other notable publications by Mahler were the *Keys to the Plants of Black Gap Wildlife Management Area, Brewster County, Texas* (1971a), *Flora of Taylor County, Texas* (1973a) and *The Mosses of Texas* (1980), an elaboration upon Eula Whitehouse's research on the mosses of Texas. Mahler's specialties included Fabaceae, *Baccharis* (Asteraceae), mosses, floristics, and the study of endangered species. He served as the first Director of the Botanical Research Institute of Texas (1987–1992) and along with Barney Lipscomb and Andrea McFadden, was instrumental in its establishment as a free-standing research institution.

Jack Stanford (1935–), of Howard Payne University in Brownwood on the very southwest margin of North Central Texas, also made an important contribution to the knowledge of Texas botany with his publication in 1976 of *Keys to the Vascular Plants of the Texas Edwards Plateau and Adjacent Areas*. This work covered portions of the Lampasas Cut Plain, which is included in the current delineation of North Central Texas. Stanford has also done extensive collecting in the Lampasas Cut Plain and Edwards Plateau and has found many important distributional records (e.g., Stanford & Diggs 1998).

Another important figure in the history of botany in North Central Texas and the state as a whole is Benny Simpson (1928–1996) (see Appendix 15). Serving for many years with the Texas Research Foundation and later with the Texas A&M Research and Extension Center at Dallas,

Fig. 34/ Cyrus Longworth Lundell, 1907–1994. Working in front of his thatched hut headquarters at Tikal, Guatemala.

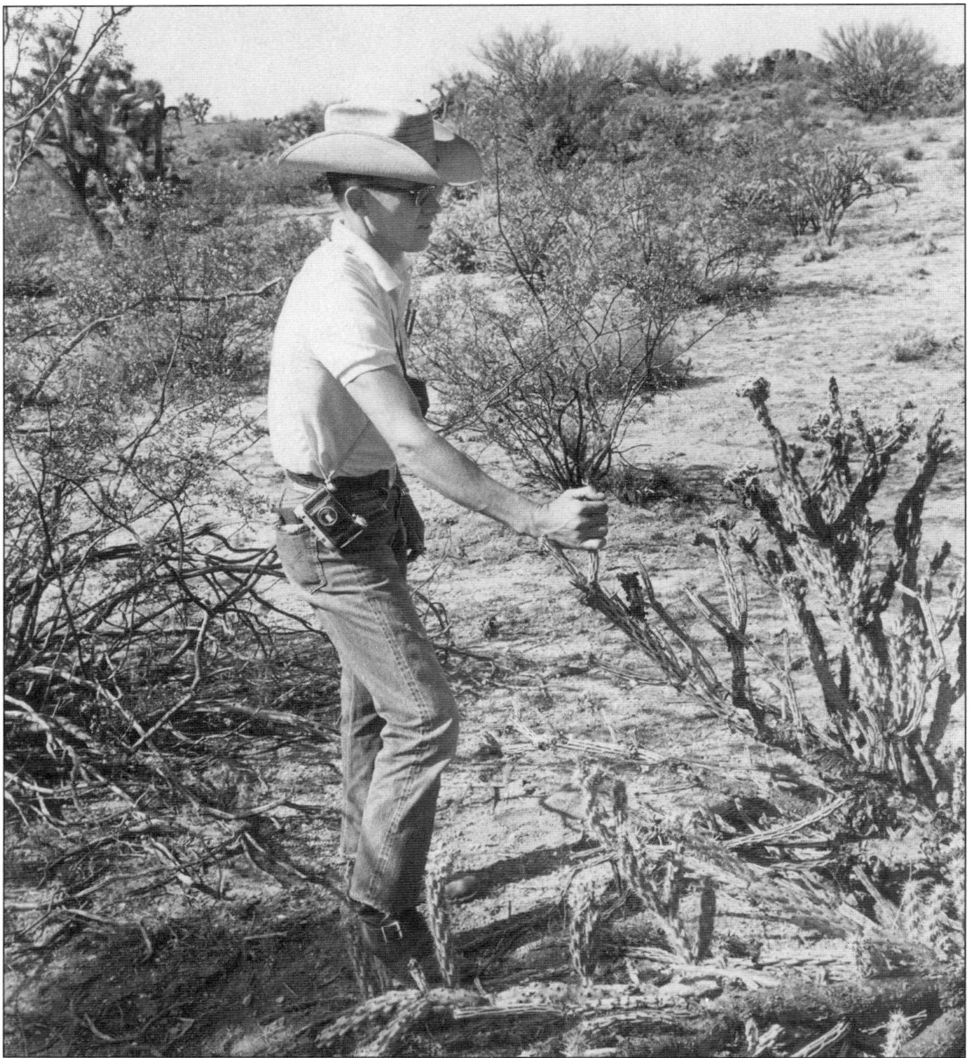

FIG. 35/ WILLIAM FRED "BILL" MAHLER

Simpson is possibly best known as the author of *A Field Guide to Texas Trees* (Simpson 1988). For a list of his publications see Davis (1997). However, among botanists and native plant enthusiasts he is correctly best remembered as the "Pioneer of the Native Plant Movement" in Texas (Nokes 1997). Simpson understood that the scarcity of water is one of the biggest challenges facing Texas' future and that native plants, well-adapted to the state's climate, are an important resource (e.g., Simpson & Hipp 1984; Simpson 1993). Through his research, nine superior selections of native plants were released to the nursery industry including three forms of *Leucophyllum* (Scrophulariaceae), widely known as Texas purple-sage (Nokes 1997; Kiphart 1997). In addition to his other contributions, Simpson was one of the founding members and a former president of the Native Plant Society of Texas and was active in that organization until his death (Nokes 1997; Pickens 1997).

Other notable contributors to the botany of North Central Texas include Robert Adams (Baylor University), Geyata Ajilvsgi (Austin), John Bacon (University of Texas at Arlington), Lewis Bragg (University of Texas at Arlington), M.D. "Bud" Bryant (Austin College), William Carr (The Nature Conservancy of Texas), Wayne Clark (Fort Worth Nature Center), Sally Crosthwaite (Austin College), Arnold Davis (Native Prairies Association of Texas), Charles Finsley (Dallas Museum of Natural History), Hugh Garnett (Austin College), Harold Gentry (Grayson County), Glenn Kroh (Texas Christian University), Fred Gelbach (Baylor University), Joe Hennen (BRIT), George High (Austin), Walter Holmes (Baylor University), Harold Keller (Central Missouri State University), Joe Kuban (Nolan High School, Fort Worth), Shirley Lusk (Gainesville), David Montgomery (Paris Junior College), Jeff Quayle (Fort Worth), Elray Nixon (Las Vegas, Nevada), Donald Smith (University of North Texas), John Steele (BRIT), Connie and John Taylor (Southeastern Oklahoma State University), Dora Sylvester (Fort Worth Nature Center & BRIT), Geoffrey Stanford (Dallas Nature Center), Jerry Vertrees (Texas Wesleyan University), and Sally Wasowski (Taos, New Mexico).

As noted above, with Lundell's (1961, 1966, 1969) unfinished but important *Flora of Texas* and the *Manual of the Vascular Plants of Texas* (Correll & Johnston 1970), both published in the Dallas-Fort Worth Metroplex, North Central Texas has been one of the centers of research on the state's flora. The publications by Lloyd Shinners (1958a) *Spring Flora of the Dallas-Fort Worth Area Texas*, Jack Stanford (1976) *Keys to the Vascular Plants of the Texas Edwards Plateau and Adjacent Areas*, and William Mahler (1984, 1988) *Shinners' Manual of the North Central Texas Flora* have been extremely valuable and useful guides to the region's flora. In addition to such books, a number of scientific journals originated in North Central Texas including *Field & Laboratory*; *Wrightia*; *Sida, Contributions to Botany*; and *Sida, Botanical Miscellany*. *The Southwest Naturalist*, a prominent, regional, natural history journal also has close ties to North Central Texas, with Lloyd Shinners serving as its first editor. Most of the botanical work in North Central Texas has been completed at private institutions, a tradition which continues today. Until the 1970s and 1980s respectively, the Texas Research Foundation and Southern Methodist University were leaders in the field. In recent years, Austin College, Baylor University, the Botanical Research Institute of Texas, and Howard Payne University have all been actively engaged in botanical research. A number of public colleges and universities in the area also have taxonomic botanists. Among these are Paris Junior College, Tarleton State University, Texas Christian University, Texas Wesleyan University, the University of North Texas, and the University of Texas at Arlington.

ABRONIA AMELIAE / AMELIA'S SAND-VERBENA / P. 836 / [RJO]

AGALINIS HOMALANTHA / FLAT-FLOWER GERARDIA / P. 993 / [GMD]

ALETRIS AUREA / YELLOW STAR-GRASS / P. 1194 / [RJO]

AESCULUS PAVIA VAR. **PAVIA** / RED BUCKEYE / P. 738 / [RJO]

ALOPHIA DRUMMONDII / PURPLE PLEAT-LEAF / P. 1173 / [RJO]

ANEMONE BERLANDIERI / TEN-PETAL ANEMONE / P. 918 / [GMD]

ANDROPOGON GERARDII SUBSP. **GERARDII** / BIG BLUESTEM / P. 1238 / [GMD]

◄ **APIOS AMERICANA** / GROUNDNUT / P. 628 / [GMD]

AQUILEGIA CANADENSIS / COMMON COLUMBINE / P. 918 / [RJO]

ARISAEMA DRACONTIUM / GREEN-DRAGON / P. 1092 / [RJO]

APHANOSTEPHUS SKIRRHOBASIS / ARKANSAS LAZY DAISY / P. 314 / [RJO] ►

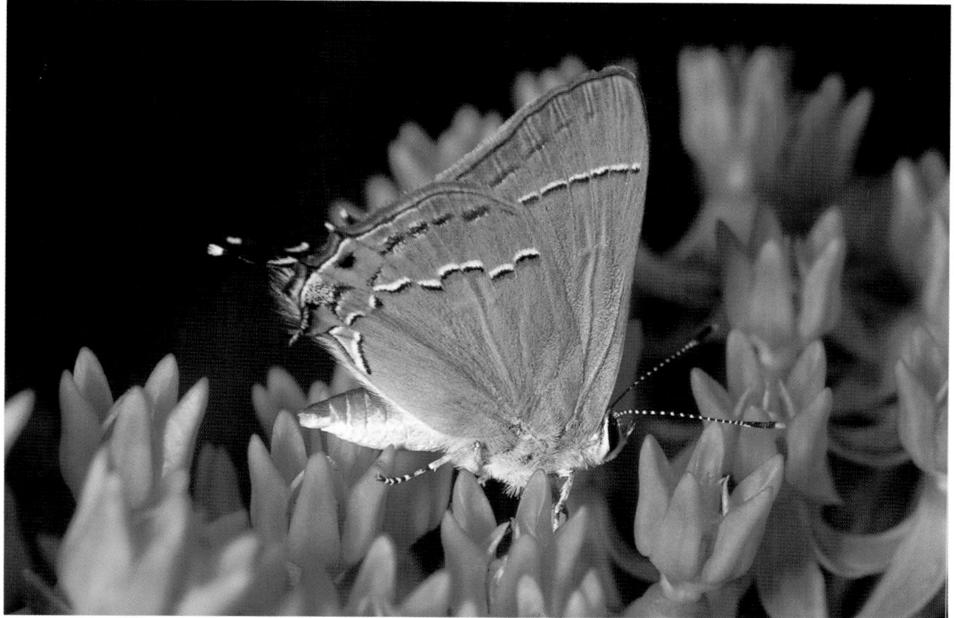

Asclepias tuberosa subsp. **interior** / butterfly-weed with gray hairstreak butterfly / p. 280 / [JAC]

Asclepias asperula subsp. **capricornu** / antelope-horns / p. 278 / [RJO]

Asclepias viridiflora / green-flower milkweed / p. 282 / [RJO]

Asclepias variegata / white-flower milkweed / p. 280 / [JAC]

Aureolaria grandiflora var. **serrata** / downy oakleech / p. 994 / [RJO]

BAPTISIA AUSTRALIS VAR. **MINOR**
BLUE WILD INDIGO / P. 636 / [MAK]

BAPTISIA ×BICOLOR / TWO-COLOR WILD INDIGO / P. 636 / [MAK]

BAPTISIA SPHAEROCARPA / GREEN WILD INDIGO / P. 638 / [GMD]

BAPTISIA ×VARIICOLOR
VARICOLORED WILD INDIGO / P. 638 / [GMD]

BAPTISIA BRACTEATA VAR. **LEUCOPHAEA**
PLAINS WILD INDIGO
P. 638 / [MAK]

BAPTISIA ×BUSHII / BUSH'S WILD INDIGO / P. 638 / [MAK]

BAPTISIA ×VARIICOLOR
VARICOLORED WILD INDIGO
P. 638 / [GMD]

Callicarpa americana / AMERICAN BEAUTY-BERRY / P. 1049 / [GMD]

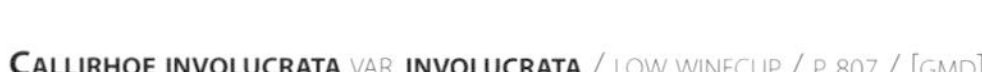

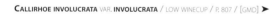

▲ **Bouteloua pectinata** / TALL GRAMA / P. 1248 / [GMD]

Callirhoe involucrata var. **involucrata** / LOW WINECUP / P. 807 / [GMD] ➤

Camassia scilloides / WILD-HYACINTH / P. 1200 / [GMD]

Campsis radicans / COMMON TRUMPET-CREEPER / P. 442 / [RJO]

CASTILLEJA PURPUREA VAR. **PURPUREA** / PURPLE PAINTBRUSH / P. 998 / [GMD]

CASTILLEJA INDIVISA / TEXAS PAINTBRUSH / P. 996 / [RJO]

CASTILLEJA PURPUREA VAR. **CITRINA** / YELLOW PAINTBRUSH / P. 998 / [RJO]

CEPHALANTHUS OCCIDENTALIS
COMMON BUTTONBUSH / P. 962 / [GMD]

CENTAUREA AMERICANA / AMERICAN BASKET-FLOWER / P. 332 / [RJO]

CENTAURIUM BEYRICHII / ROCK CENTAURY, MOUNTAIN-PINK / P. 724 / [RJO]

CERCIS CANADENSIS VAR. **CANADENSIS** / EASTERN REDBUD WITH FEMALE CARDINAL P. 640 / [JAC]

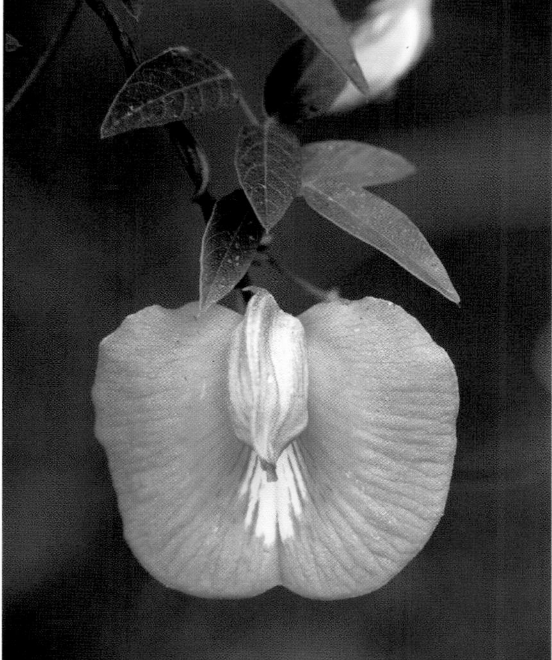

CENTROSEMA VIRGINIANUM / BUTTERFLY-PEA / P. 639 / [RJO]

CHAMAECRISTA FASCICULATA / PARTRIDGE-PEA / P. 642 / [RJO]

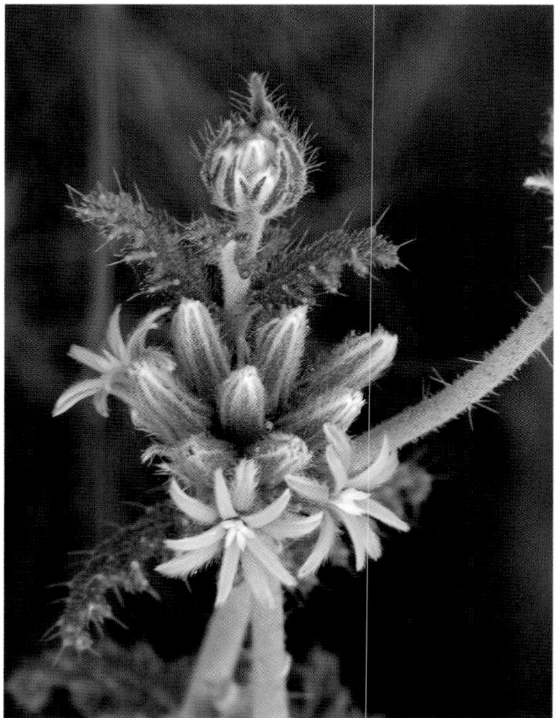

CEVALLIA SINUATA / STINGING CEVALLIA / P. 794 / [RJO]

Cirsium horridulum / BULL THISTLE / P. 340 / [RJO]

Cirsium texanum / TEXAS THISTLE / P. 340 / [RJO]

Cirsium undulatum / WAVY-LEAF THISTLE / P. 340 / [RJO]

Claytonia virginica / VIRGINIA SPRING-BEAUTY
P. 908 / [RJO]

Clematis texensis / SCARLET CLEMATIS
P. 922 / [RJO]

CNIDOSCOLUS TEXANUS / TEXAS BULL-NETTLE
P. 596 / [GMD]

▲ **COREOPSIS TINCTORIA** / PLAINS COREOPSIS / P. 342 / [GMD]

COMMELINA ERECTA VAR. **ERECTA** / ERECT DAYFLOWER / P. 1100 / [RJO]

COOPERIA PEDUNCULATA / GIANT RAIN-LILY / P. 1200 / [RJO]

CORYPHANTHA SULCATA / PINEAPPLE CACTUS / P. 485 / [RJO]

DALEA AUREA / GOLDEN DALEA / P. 646 / [RJO]

DALEA LASIATHERA / PURPLE DALEA / P. 650 / [RJO]

CUCURBITA FOETIDISSIMA / BUFFALO GOURD / P. 568 / [RJO]

CUCURBITA TEXANA / TEXAS GOURD / P. 568 / [JAC]

DASISTOMA MACROPHYLLA / MULLEIN SEYMERIA / P. 999 / [GMD]

▲ **DATURA WRIGHTII**
INDIAN-APPLE
P. 1020 / [RJO]

DODECATHEON MEADIA / COMMON SHOOTING-STAR / P. 913 / [GMD]

◀ **DELPHINIUM CAROLINIANUM** SUBSP. **VIRESCENS** / PRAIRIE LARKSPUR
P. 924 / [RJO]

DICLIPTERA BRACHIATA / FALSE MINT
P. 212 / [RJO]

DRACOPIS AMPLEXICAULIS
▼ CLASPING CONEFLOWER / P. 346 / [RJO]

ECHINOCACTUS TEXENSIS / HORSECRIPPLER
P. 486 / [RJO]

DYSCHORISTE LINEARIS / NARROW-LEAF SNAKEHERB
P. 213 / [RJO]

ERODIUM TEXANUM / STORK'S-BILL / P. 730 / [JAC]

ECHINOCEREUS COCCINEUS VAR. **PAUCISPINUS** / CLARET-CUP CACTUS / P. 486 / [RJO]

ECHINOCEREUS REICHENBACHII / LACE CACTUS / P. 486 / [RJO]

Eryngium leavenworthii
LEAVENWORTH'S ERYNGO / P. 252 / [RJO]

Erythrina herbacea / CORAL-BEAN / P. 658 / [RJO]

Eryngium yuccifolium
RATTLESNAKE-MASTER / P. 252 / [RJO]

Erysimum asperum
WESTERN WALLFLOWER / P. 468 / [RJO]

Erythronium mesochoreum
DOG-TOOTH-VIOLET / P. 1201 / [RJO]

Escobaria missouriensis VAR. **similis**
PLAINS NIPPLE CACTUS / P. 488 / [GMD]

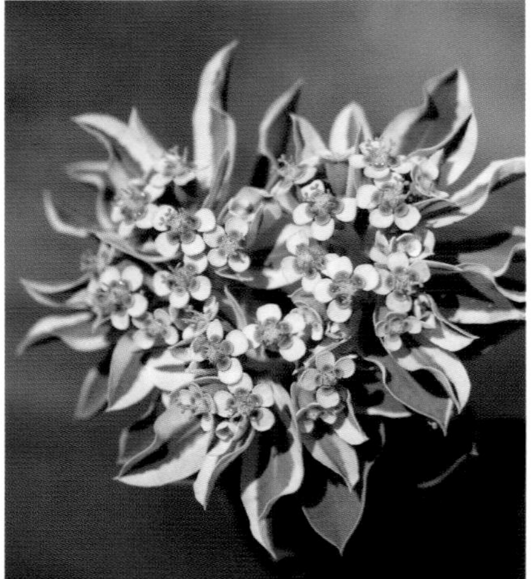

EUPHORBIA MARGINATA / SNOW-ON-THE-MOUNTAIN / P. 608 / [RJO]

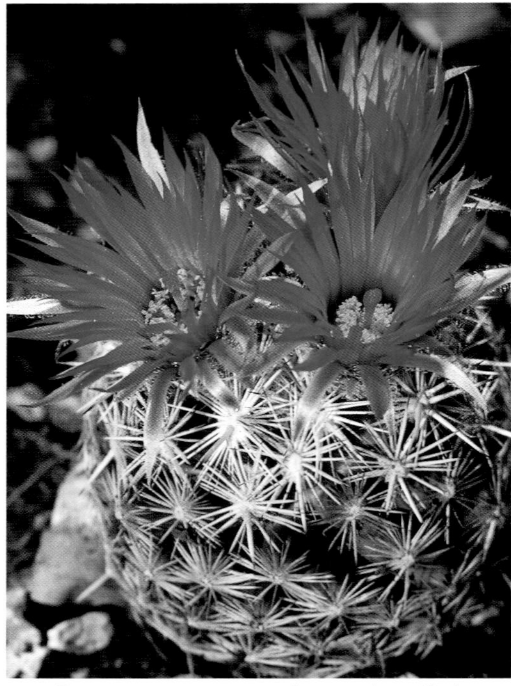

ESCOBARIA VIVIPARA VAR. **RADIOSA** / SPINY-STAR / P. 488 / [RJO]

EUSTOMA RUSSELLIANUM / BLUEBELL GENTIAN / P. 727 / [GMD]

FUNASTRUM CYNANCHOIDES / CLIMBING-MILKWEED / P. 283 / [GMD]

FUNASTRUM CRISPUM / WAVY-LEAF TWINEVINE / P. 283 / [JAC]

GAILLARDIA PULCHELLA / INDIAN-BLANKET WITH GAILLARDIA MOTH / P. 358 / [RJO]

Glandularia bipinnatifida / Dakota vervain / p. 1050 / [RJO]

Helianthus maximiliani / Maximilian sunflower with Eryngium leavenworthii / Leavenworth's eryngo / p. 370 / [JAC]

▲ **Grindelia papposa** / saw-leaf daisy p. 362 / [RJO]

Helianthemum georgianum georgia sun-rose / p. 543 / [RJO]

Helianthus annuus / common sunflower / p. 369 / [GMD]

Herbertia lahue subsp. **caerulea**
Herbertia / p. 1173 / [RJO]

Hesperaloe parviflora / red-flowered-yucca / p. 1079 / [RJO]

Hexalectris nitida / shining hexalectris
p. 1216 / [RJO]

Hexalectris warnockii / Texas purple-spike
p. 1218 / [RJO]

Hibiscus laevis / halberd-leaf rose-mallow
p. 810 / [RJO]

Hibiscus trionum / flower-of-an-hour
p. 812 / [RJO]

HYMENOCALLIS LIRIOSME / WESTERN SPIDER-LILY / P. 1204 / [JAC]

HOFFMANSEGGIA GLAUCA / SICKLE-POD RUSH-PEA / P. 663 / [RJO]

HYDROLEA OVATA / HAIRY HYDROLEA / P. 740 / [JAC]

IBERVILLEA LINDHEIMERI / BALSAM GOURD / P. 569 / [GMD]

IPOMOPSIS RUBRA / STANDING-CYPRESS / P. 890 / [GMD]

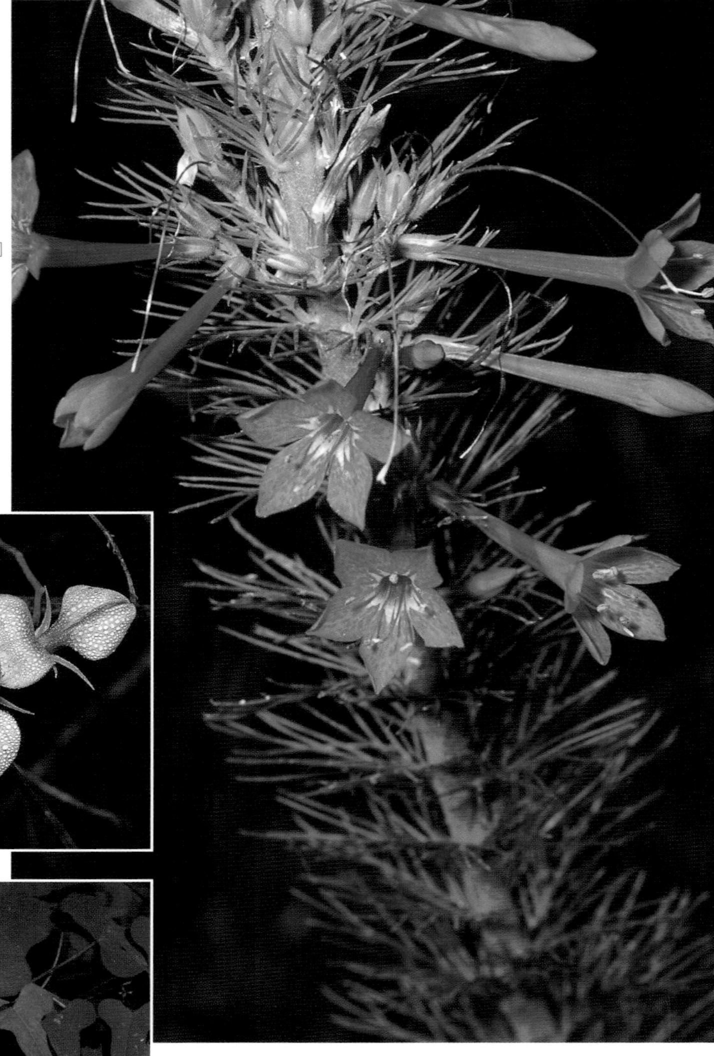

INDIGOFERA MINIATA VAR. **LEPTOSEPALA**
WESTERN SCARLET-PEA / P. 664 / [GMD]

➤ **IPOMOEA PANDURATA** / BIG-ROOT MORNING-GLORY / P. 556 / [RJO]

JUSTICIA AMERICANA / AMERICAN WATER-WILLOW / P. 213 / [RJO]

➤ **IPOMOEA CORDATOTRILOBA** VAR. **CORDATOTRILOBA** / SHARP-POD MORNING-GLORY
P. 556 / [RJO]

LIATRIS ASPERA / TALL GAYFEATHER / P. 384 / [GMD]

LIATRIS SQUARROSA VAR. **GLABRATA**
SMOOTH GAYFEATHER / P. 386 / [GMD]

LINUM RIGIDUM VAR. **BERLANDIERI**
BERLANDIER'S FLAX / P. 792 / [JAC]

KRAMERIA LANCEOLATA / TRAILING RATANY / P. 750 / [RJO]

▼ **LANTANA URTICOIDES** / COMMON LANTANA / P. 1053 / [RJO]

LITHOSPERMUM INCISUM / NARROW-LEAF GROMWELL / P. 452 / [RJO]

LOBELIA CARDINALIS / CARDINAL-FLOWER / P. 498 / [RJO]

LONICERA SEMPERVIRENS / CORAL HONEYSUCKLE / P. 510 / [RJO]

LOBELIA SIPHILITICA VAR. **LUDOVICIANA**
BIG BLUE LOBELIA / P. 498 / [JAC]

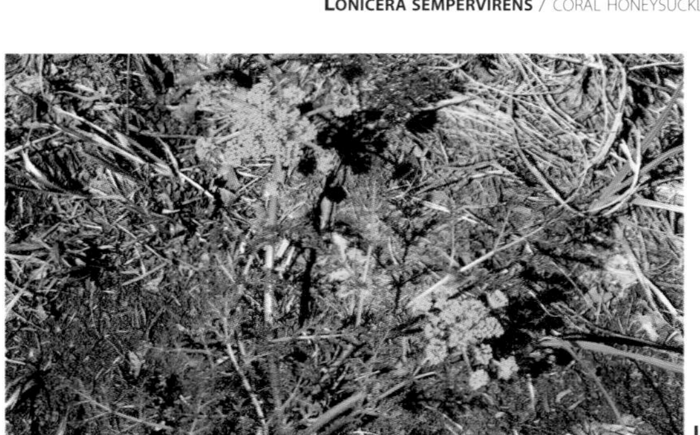

LOMATIUM FOENICULACEUM SUBSP. **DAUCIFOLIUM** / CARROT-LEAF LOMATIUM / P. 255 / [RJO]

LUDWIGIA PEPLOIDES / WATER-PRIMROSE / P. 860 / [GMD] ▲

LUPINUS TEXENSIS / TEXAS BLUEBONNET / P. 672 / [GMD]

LUPINUS TEXENSIS / TEXAS BLUEBONNET
WITH OENOTHERA SPECIOSA / SHOWY PRIMROSE / P. 672 / [JAC]

MAMMILLARIA HEYDERI / LITTLE-CHILIS / P. 489 / [RJO]

LYGODESMIA TEXANA / TEXAS SKELETON-PLANT / P. 386 / [RJO]

Manfreda virginica subsp. **lata**
wide-leaf false aloe / p. 1079 / [RJO]

Marshallia caespitosa var. **signata** / Barbara's-buttons / p. 387 / [GMD]

Matelea edwardsensis
plateau milkvine / p. 284 / [RJO]

Matelea reticulata / net-vein milkvine / p. 284 / [RJO]

Matelea biflora / two-flower milkvine / p. 284 / [RJO]

◄**MAURANDYA ANTIRRHINIFLORA** / SNAPDRAGON-VINE / P. 1002 / [RJO]

MIMOSA ROEMERIANA / ROEMER'S SENSITIVE-BRIAR / P. 678 / [RJO]

MIRABILIS NYCTAGINEA / WILD FOUR-O'CLOCK
P. 840 / [JAC]

◄**MONARDA FISTULOSA** VAR. **MOLLIS**
WILD BERGAMOT / P. 765 / [GMD]

MONARDA CITRIODORA / LEMON BEEBALM
P. 765 / [JAC]

MONARDA PUNCTATA VAR. **INTERMEDIA**
SPOTTED BEEBALM / P. 766 / [GMD]

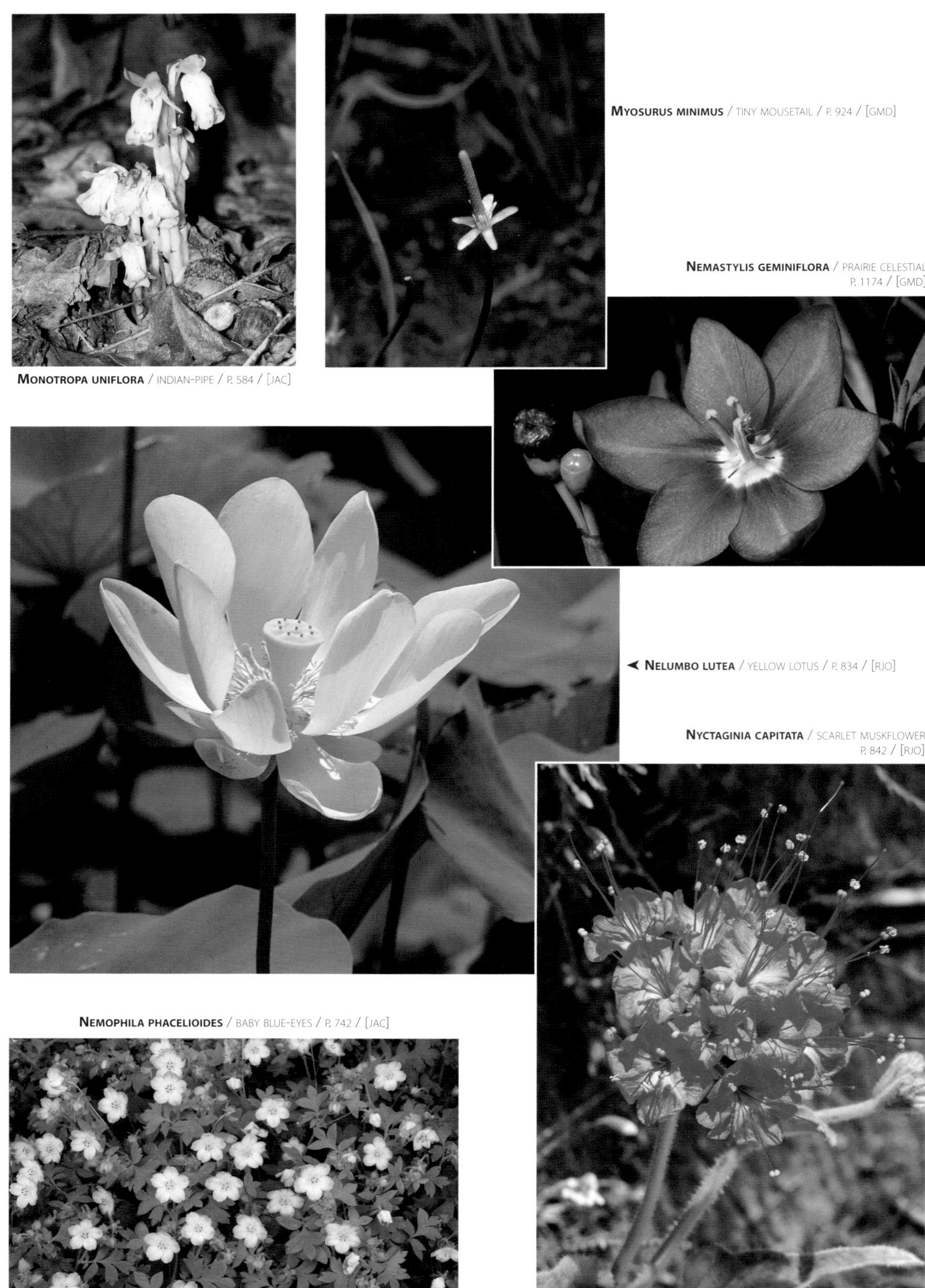

Monotropa uniflora / INDIAN-PIPE / P. 584 / [JAC]

Myosurus minimus / TINY MOUSETAIL / P. 924 / [GMD]

Nemastylis geminiflora / PRAIRIE CELESTIAL P. 1174 / [GMD]

➤ **Nelumbo lutea** / YELLOW LOTUS / P. 834 / [RJO]

Nyctaginia capitata / SCARLET MUSKFLOWER P. 842 / [RJO]

Nemophila phacelioides / BABY BLUE-EYES / P. 742 / [JAC]

NYMPHAEA ODORATA / WHITE WATER-LILY / P. 845 / [RJO]

OENOTHERA MACROCARPA SUBSP. **MACROCARPA** / FLUTTER-MILL / P. 864 / [RJO]

OENOTHERA SPECIOSA / SHOWY-PRIMROSE / P. 866 / [JAC]

OPUNTIA TUNICATA VAR. **DAVISII**
GREEN-FLOWER CHOLLA / P. 492 / [RJO]

OXYTROPIS LAMBERTII / LOCOWEED / P. 682 / [RJO]

OPUNTIA ENGELMANNII VAR. **LINDHEIMERI** / TEXAS PRICKLY-PEAR / P. 490 / [RJO]

Passiflora incarnata / maypop passion-flower / p. 878 / [GMD]

Passiflora affinis / bracted passion-flower
p. 878 / [RJO]

Passiflora tenuiloba
spread-lobe passion-flower / p. 878 / [RJO]

Pavonia lasiopetala / Wright's pavonia
p. 816 / [RJO] ➤

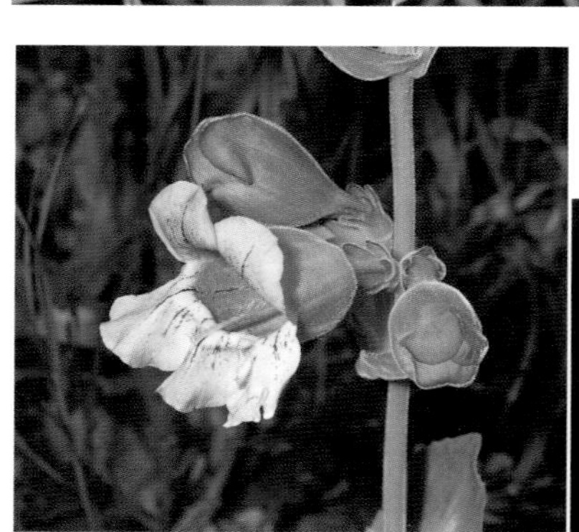

Penstemon cobea / wild foxglove
p. 1007 / [GMD]

Penstemon digitalis / smooth penstemon / p. 1007 / [GMD]

PHLOX DRUMMONDII SUBSP. **DRUMMONDII** / DRUMMOND'S PHLOX / P. 892 / [JAC]

PHLOX DRUMMONDII SUBSP. **WILCOXIANA** / DRUMMOND'S PHLOX / P. 892 / [RJO]

PHLOX ROEMERIANA / GOLD-EYE PHLOX / P. 892 / [JAC]

PHYSOSTEGIA PULCHELLA / BEAUTIFUL FALSE DRAGON'S-HEAD / P. 770 / [RJO] ➤

➤ **PLATANTHERA CILIARIS** / YELLOW FRINGED ORCHID / P. 1218 / [GMD]

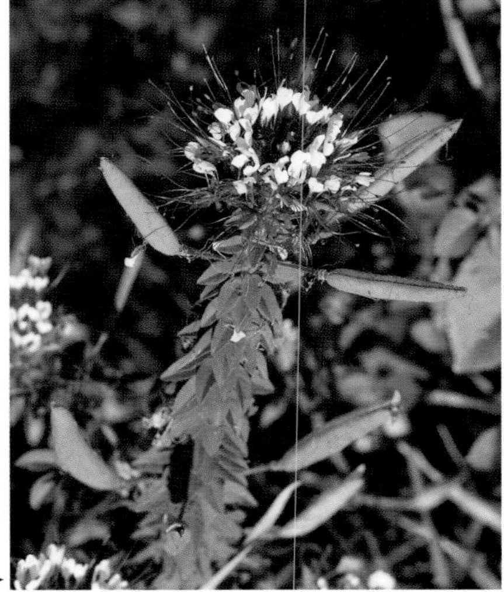

POLANISIA DODECANDRA SUBSP. **TRACHYSPERMA** / CLAMMYWEED / P. 506 / [RJO] ➤

Proboscidea louisianica / COMMON DEVIL'S-CLAW / P. 880 / [RJO]

Quincula lobata / PURPLE GROUND-CHERRY / P. 1026 / [RJO]

GALLS ON **Quercus falcata** / GALLS ON SOUTHERN RED OAK / P. 714 / [GMD] ➤

Rhynchosida physocalyx / SPEAR-LEAF SIDA / P. 816 / [RJO]

Ratibida columnifera / MEXICAN-HAT / P. 400 / [GMD] ➤

Sabatia campestris / PRAIRIE ROSE GENTIAN / P. 728 / [GMD]

Scutellaria wrightii / WRIGHT'S SKULLCAP / P. 780 / [JAC]

Silphium albiflorum / WHITE ROSINWEED / P. 404 / [RJO]

Salvia farinacea / MEALY SAGE / P. 776 / [RJO]

Salvia azurea var. **grandiflora** / BLUE SAGE P. 774 / [GMD] ➤

Silphium radula / ROUGH-STEM ROSINWEED / P. 406 / [RJO] ➤

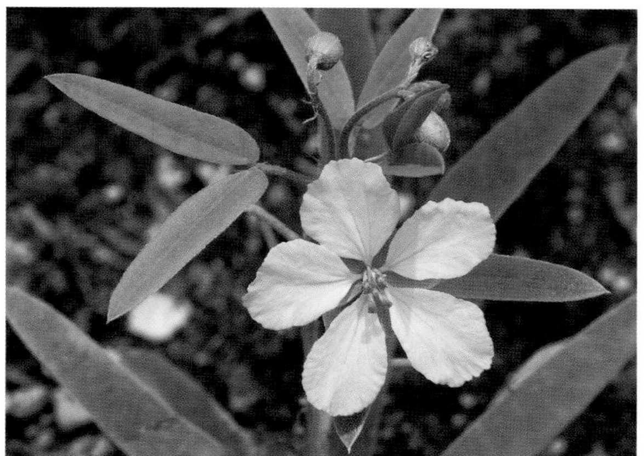

Senna roemeriana / TWO-LEAF SENNA / P. 696 / [RJO]

Silphium laciniatum / COMPASSPLANT / P. 404 / [RJO] ➤

Sophora affinis / EVE'S-NECKLACE / P. 697 / [GMD]

Sphaeralcea angustifolia subsp. **cuspidata**
NARROW-LEAF GLOBE-MALLOW / P. 820 / [RJO]

Sisyrinchium pruinosum / DOTTED BLUE-EYED-GRASS / P. 1178 / [GMD]

Solanum citrullifolium / MELON-LEAF NIGHTSHADE / P. 1028 / [RJO]

Solanum dimidiatum / WESTERN HORSE-NETTLE / P. 1028 / [GMD]

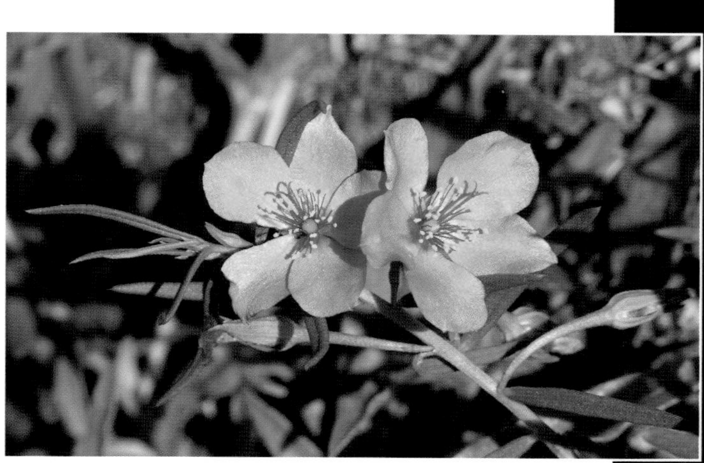

TALINUM AURANTIACUM / ORANGE FLAMEFLOWER / P. 911 / [RJO]

SPIRANTHES LACERA VAR. **GRACILIS** / SLENDER LADIES'-TRESSES / P. 1220 / [GMD] ➤

TALINUM CALYCINUM / ROCK-PINK / P. 911 / [GMD]

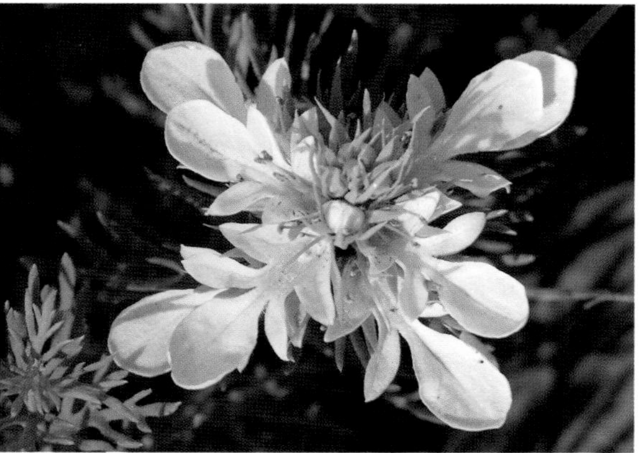

TEUCRIUM LACINIATUM / CUT-LEAF GERMANDER / P. 783 / [RJO]

TINANTIA ANOMALA / FALSE DAYFLOWER / P. 1101 / [RJO]

ZINNIA GRANDIFLORA / PLAINS ZINNIA / P. 433 / [RJO]

YUCCA NECOPINA / GLEN ROSE YUCCA / P. 1083 / [RJO] ➤

TRADESCANTIA OHIENSIS / OHIO SPIDERWORT
P. 1104 / [GMD]

UNGNADIA SPECIOSA / MEXICAN-BUCKEYE
P. 982 / [RJO]

VERBENA HALEI ➤
SLENDER VERVAIN / P. 1056 / [GMD]

➤ **UTRICULARIA GIBBA** / CONE-SPUR BLADDERWORT / P. 787 / [GMD]

Keys To The Vascular Plants Of North Central Texas

Including General Key To All Families[1] on Pp. 126

KEY TO MAJOR VASCULAR PLANT GROUPS

1. Plants without seeds or flowers, reproducing by microscopic spores borne in sporangia (= spore cases), these either (usually) on the surface of leaves or leaf-like structures (Polypodiophyta—Ferns) OR at the base of quill-like leaves (*Isoetes*) OR in small usually more or less cone-like structures (*Equisetum, Lycopodium,* and *Selaginella*); plants fern-like, moss-like, with quill-like leaves, or leaves reduced and stems green and hollow ____ **Lycopodiophyta, Equisetophyta, and Polypodiophyta**
(Pteridophytes = Ferns & Similar Plants)
—see Key on pp. 110 OR Group K on pp. 154

1. Plants reproducing by seeds, these developing either from flowers or on the surface of thin or thick, sometimes woody cone scales; plants usually not with growth forms as above (Spermatophytes = Seed Plants).
   2. Plants without flowers, the seeds on the surface of thick or thin, sometimes woody cone scales (cone scales fleshy in *Juniperus* with berry-like cones or thin, becoming fleshy in *Ephedra*); leaves needle-like (*Pinus* and *Taxodium*) OR very small, scale-like, and closely appressed to the stem (*Juniperus*) OR reduced to non-leaf-like scales in whorls at the joints of the stem (*Ephedra*) ____ **Pinophyta and Gnetophyta**
   (Gymnosperms ("naked seeds"))
   —see Key on pp. 113 OR General Key to all Families on pp. 126
   2. Plants with flowers, the seeds developing inside a closed carpel, the base of which (= ovary) becomes the fruit; leaves usually broader than needles or scales, rarely needle-like or scale-like—Magnoliophyta (Angiosperms ("vessel seeds") = Flowering Plants).
      3. Plants with 2 or more of the following characters: leaves parallel-veined; cotyledon (= seed leaf) 1; floral parts in 3s or 6s; mostly herbaceous plants with vascular bundles of stem usually scattered throughout the pith; cambium usually absent ____ **Monocotyledonae**
      (Monocots)
      —see Key on pp. 121 OR Key to Keys on pp. 110
      3. Plants with 2 or more of the following characters: leaves net-veined; cotyledons 2; floral parts usually not in 3s or 6s (usually in 2s, 4s, or 5s); herbaceous and woody plants with vascular bundles of stem in a ring around the pith; cambium usually present except in some annuals ____ **Dicotyledonae**
      (Dicots)
      —see Key to Keys on pp. 110

[1]The *General Key to All Families* was modified from *Key and Descriptions for the Vascular Plant Families of Oklahoma* contributed by Oklahoma Flora Editorial Committee (Tyrl et al. 1994).

KEY TO KEYS

1. Plants aquatic (either free-floating on or in water OR entirely submersed OR rooted in bottom and floating OR basal part in water and upper part emergent) ____ **Key to Aquatic Plants on pp. 113 OR Group G on pp. 141**
1. Plants terrestrial.
   2. Plants ferns OR fern-like plants OR gymnosperms.
      3. Plants ferns OR fern-like plants ____ **Key to Ferns and Similar Plants on pp. 110 OR Group K on pp. 154**
      3. Plants gymnosperms ____ **Key to Gymnosperms on pp. 113 or General Key to all families on pp. 126**
   2. Plants angiosperms (= flowering plants).
      4. Plants woody vines ____ **Key to Woody Vines on pp. 119 OR Group A on pp. 127**
      4. Plants not woody vines.
         5. Plants monocots ____ **Key to Monocots on pp. 121 OR General Key to all Families on pp. 126**
         5. Plants dicots ____ **General Key to all Families on pp. 126**

Key to Ferns and Similar Plants (Pteridophytes)

1. Plants small floating aquatics ____ (*Azolla*) **Azollaceae**
1. Plants not small floating aquatics, either terrestrial (rooted in soil or mud) OR aquatic (rooted on bottom) OR growing on rocks or tree trunks.
   2. Stems conspicuously jointed, green and hollow, the segments separating easily at the joints (= nodes), unbranched or branched at the nodes; leaves reduced to small essentially non-photosynthetic (non-green) scales in whorls at the nodes; sporangia (= spore cases) in terminal strobili (= cone-like structures) on stems without green leaves ____ (*Equisetum*) **Equisetaceae**
   2. Stems not jointed, not green and hollow; leaves usually green; sporangia either on the surface of leaves or leaf-like structures OR in terminal strobili on leafy stems OR in short-stalked sporocarps at leaf bases.
      3. Leaves simple, linear, grass-like or thread-like, the blades not expanded; spore-bearing structures embedded in leaf bases or on very short stalks (1–2 mm long) at leaf bases; plants often rooted in mud or in temporary pools.
         4. Leaves tightly clustered together (plants superficially resembling a green onion), arising from a corm-like rootstock, quill-like (larger at base); sporangia embedded in the leaf bases, borne one per leaf ____ (*Isoetes*) **Isoetaceae**
         4. Leaves scattered along a creeping rhizome, separate, not clustered; sporangia in stalked (stalks 1–2 mm long), globose, hairy sporocarps (= nut-like or hard bean- or pea-like sporangia-bearing cases) arising at the base of the leaves, numerous per sporocarp ____ (*Pilularia*) **Marsileaceae**
      3. Leaves compound or simple, with expanded blades OR leaves needle-like or scale-like, neither grass-like nor thread-like; spore-bearing structures neither embedded in leaf bases nor on very short stalks (1–2 mm long) at leaf bases; plants rooted in various substrates including mud or soil or on rocks or tree trunks.
         5. Leaf blades deeply 4-parted (resembling a 4-leaf clover) on petioles usually much longer than the blades; sporangia in sporocarps borne near the base of the plant ____ (*Marsilea*) **Marsileaceae**
         5. Leaf blades variously compound or simple but not 4-parted; petioles usually much shorter than leaf blades to absent; sporangia not in sporocarps near the base of the plant.
            6. Leaves 8 mm or less long; plants with numerous, small, usually overlapping or divergent, scale-like or needle-like leaves with a single unbranched vein (this type of leaf is

a microphyll); stems well-exposed above the ground surface, covered with the numerous small leaves; sporangia in the axils of the microphylls, these often aggregated into cone-like strobili.

7. Sporangia in cylindrical strobili at the tips of elongate, distinctly erect, leafy, fertile stems; leaves 4–8 mm long; plants homosporous (= with 1 spore type of a single size) ________ (*Lycopodiella*) **Lycopodiaceae**

7. Sporangia in ± 4-angled strobili at the tips of leafy stems; fertile stems ascending or spreading, not distinctly erect; leaves 1–3 mm long; plants heterosporous (= with 2 spore types which are of different sizes) ________ (*Selaginella*) **Selaginellaceae**

6. Leaves usually much more than 10 mm long; plants with relatively few large leaves with numerous branched veins (this type of leaf is a megaphyll); stems underground rhizomes or short crowns or caudices, not well-exposed above the ground surface and relatively inconspicuous; sporangia in clusters (= sori) on the surface of the leaf blades (the blades can sometimes be considerably modified).

8. Plants with 1(–2) leaves per stem; leaves with 2 distinct parts, the sterile portion either entire or ternately (= in 3 parts) to pinnately compound to dissected, the fertile portion being an elongate stalk with a spike-like or panicle-like sporangia-bearing terminal part ________ **Ophioglossaceae**

8. Plants usually with numerous leaves per stem; leaves not as described above.

9. Lowermost 2 pinnae (= primary divisions of a leaf, here one on each side of the leaf) of the fertile leaf greatly elongated and bearing the sporangia near their tips ________ (*Anemia*) **Anemiaceae**

9. Lowermost 2 pinnae of the fertile leaf neither greatly elongated nor bearing the sporangia near their tips.

10. Sori linear-oblong, in one row on each side of, immediately adjacent to, and parallel with the costae (= midveins of the pinnae) or costules (= midveins of the pinnules), chain-like in arrangement ________ (*Woodwardia*) **Blechnaceae**

10. Sori various, but arrangement not chain-like in one row on each side of, immediately adjacent to, and parallel with the costae or costules.

11. Fertile and sterile leaves either completely different OR fertile portion of fertile leaves essentially without any blade tissue (leaves extremely dimorphic).

12. Fertile leaves completely different from sterile leaves, essentially without photosynthetic tissue, solely sporangia-bearing, glabrous; sterile leaves 1-pinnatifid (= deeply divided but not completely pinnate), the rachis (= central axis of a fern frond/leaf) with a conspicuous flange or wing of photosynthetic tissue ________ (*Onoclea*) **Dryopteridaceae**

12. Fertile leaves either with numerous ± normal photosynthetic pinnae OR fertile leaves with conspicuous pubescence; sterile leaves pinnate (= pinnae narrowed to petiole-like attachment to rachis, the rachis without a flange or wing of photosynthetic tissue except possibly at very tip of blade) ________ (*Osmunda*) **Osmundaceae**

11. Fertile and sterile leaves or portions of leaves similar or somewhat modified, the fertile portion never so different as to be without blade tissue (leaves not extremely dimorphic).

13. Sori marginal or submarginal (= located at or near the edges of the leaves) with leaf margins recurved over the sori, protecting them and forming a false indusium (= thin scale-like outgrowth covering the clusters of sporangia) (except not recurved in *Astrolepis* which has stellate or coarsely ciliate scales on the adaxial leaf surfaces).

14. Plants stout, to 1 m or more tall; leaf blades triangular in shape or nearly so, with 3 main divisions (each division usually bipinnate); petioles grooved, greenish or ± straw-colored; stems and petiole bases with slender hairs but without scales ___ (*Pteridium*) **Dennstaedtiaceae**

14. Plants slender, usually 0.1–0.5 m tall; leaf blades neither triangular nor with 3 main divisions; petioles rounded, often dark brown or black; stems and petiole bases generally with scales ___ **Pteridaceae**

13. Leaf margins never recurved to form a false indusium; sori variously located on the abaxial (= beneath) leaf surfaces, often near veins, occasionally near the margins of the leaves; adaxial leaf surfaces without stellate or coarsely ciliate scales.

15. Sori elongate along the veins; leaf blades 1-pinnate ___ (*Asplenium*) **Aspleniaceae**

15. Sori round or variously shaped, not elongate (in our species, except elongate in *Athyrium* with leaf blades 2-pinnate-pinnatifid); leaf blades in our species 1-pinnatifid, 1-pinnate, or more than 1-pinnate.

16. Leaf blades in our species only 1-pinnate or 1-pinnatifid (pinnae not further subdivided into pinnules).

17. Leaf blades only pinnatifid (in our species), divided nearly but not all the way to the rachis (= midrib); pinnae essentially as wide at base as towards their tips; margins of pinnae (in our species) without any teeth or basal auricles ___ (*Pleopeltis*) **Polypodiaceae**

17. Leaf blades pinnate, divided all the way to the rachis; pinnae narrowed basally to a very narrow petiole-like attachment to the rachis; pinnae with small teeth on the margins, sometimes with basal auricles ___ **Dryopteridaceae** (in part)

16. Leaf blades more than 1-pinnate, the pinnae themselves either pinnate or pinnatifid.

18. Ultimate leaf segments variously incised, serrate, dentate, crennate, or lobulate, not entire; sori round or elongate; indusia elongate or splitting into spreading lobes ___ **Dryopteridaceae** (in part)

18. Ultimate leaf segments entire; sori round; indusia round to kidney-shaped ___ (*Thelypteris*) **Thelypteridaceae**

Key to Gymnosperms

1. Shrubs 0.25–1 m tall OR plants with clambering vine-like habit; leaves inconspicuous, the main photosynthetic structures being the green to gray-green or yellow-green stems; stems ± jointed; seed-producing cones 6–12 mm long, the scales thin, the inner scales becoming fleshy and red; longest internodes 1–8 cm long; plants neither resinous nor fragrant (Gnetophyta) ________ **Ephedraceae**
1. Trees or shrubs much more than 1 m tall; leaves conspicuous (though often small) and serving as the primary photosynthetic structures; stems not jointed; seed-producing cones either large (15 mm or more long) and woody or small (to 10 mm long), berry-like, and blue to bluish black, bluish purple, reddish, or copper-colored; longest internodes usually 0–1 cm long; plants usually resinous and fragrant (Pinophyta).
   2. Adult foliage leaves needle-like, not flattened, 50–450 mm long, in fascicles of 2–5 surrounded at the base by a membranous sheath; seed-producing cones 40 mm or more long (often much longer) ________ **Pinaceae**
   2. Adult foliage leaves scale-like OR flat and linear, ca. 15 mm or less long, not in fascicles; seed-producing cones 5–25(–40) mm long ________ **Cupressaceae**

Key to Aquatic Plants

1. Entire plants (single unit or small chain-like cluster) small, usually less than 2 cm long, lacking leaves or stems OR with minute leaves 1 mm or less in diam.; plants floating-free on the surface or just beneath.
   2. Plants with numerous very small leaves; minute branching stems present ________ (*Azolla*) **Azollaceae**
   2. Plants of 1–several joints or thalli, not differentiated into leaves or stems ________ **Lemnaceae**
1. Entire plants more than 2 cm long; leaves and/or stems present; plants free-floating or bottom-rooted.
   3. Plants reproducing by spores produced in basal sporangia or sporocarps, without flowers, fruits, or seeds; leaves either linear and grass-like or narrowly filiform (= thread-like) OR with 4 leaflets (resembling a 4-leaf clover) (Ferns and Similar Plants).
      4. Leaves simple, linear, grass-like or thread-like, the blades not expanded.
         5. Leaves tightly clustered together (plants superficially resembling a green onion), arising from a corm-like rootstock, quill-like (larger at base); sporangia embedded in the leaf bases, borne one per leaf ________ (*Isoetes*) **Isoetaceae**
         5. Leaves scattered along a creeping rhizome, separate, not clustered; sporangia in stalked (stalks 1–2 mm long) globose, hairy sporocarps arising at the base of the leaves, numerous per sporocarp ________ (*Pilularia*) **Marsileaceae**
      4. Leaf blades expanded, deeply 4-parted (resembling a 4-leaf clover), on petioles usually much longer than the blades ________ (*Marsilea*) **Marsileaceae**
   3. Plants reproducing by flowers, fruits, and seeds; leaves various (Flowering Plants).
      6. Leaves (or stems if plants leafless) all attached at base of the plant.
         7. Plants with leaves (possibly leaf-like branches) or branches either thread-like or divided into thread-like segments, with numerous, small, bladder-like traps for capturing small organisms; corollas bilaterally symmetrical, spurred, yellow ________ (*Utricularia*) **Lentibulariaceae**
         7. Plants with linear to lanceolate, elliptic, ovate, or orbicular leaves OR plants leafless with unbranched, cylindrical or thread-like, green stems, without bladder-like traps; corollas not as above.
            8. Leaves modified into hollow, tubular, trumpet-shaped pitchers ________ (*Sarracenia*) **Sarraceniaceae**
            8. Leaves not modified into pitchers.
               9. Leaf blades linear to lanceolate, entire OR plants leafless with unbranched, cylindrical or thread-like, green stems about as thick as wide (these sometimes flattened).

10. Plants completely submersed aquatics; male flowers numerous, tiny, breaking from a spathe and free-floating at maturity; female flowers solitary in pedunculate spathes at the water surface at flowering time _________ (*Vallisneria*) **Hydrocharitaceae**

10. Plants partly emergent; flowers borne above the water surface.

11. Plants without obvious leaves (only bladeless sheaths present), with unbranched, cylindrical or thread-like, green stems about as thick as wide (these sometimes flattened); perianths absent or of inconspicuous bristles or small scales _________ **Cyperaceae**

11. Plants with obvious leaves (rarely reduced to bracts); perianths conspicuous, white or yellow OR inconspicuous, but with conspicuous whitish hairs.

12. Leaves without distinct petioles; flowers in a single, small (up to 30 mm long), dense head or spike terminating an elongate naked scape, the head or spike either cone-like with numerous, overlapping, brownish, thin, ± woody bracts (subtending yellow flowers) OR head whitish or grayish due to numerous hairs on the subtending involucral bracts and flower parts.

13. Inflorescences cone-like, with spirally imbricated, brownish, thin, ± woody bracts; flowering inflorescences usually with a single yellow flower exposed _________ (*Xyris*) **Xyridaceae**

13. Inflorescences small whitish or grayish heads, not cone-like, lacking brownish woody bracts; flowering inflorescences without yellow flowers _________ (*Eriocaulon*) **Eriocaulaceae**

12. Leaves with distinct spongy petioles; flowers in whorls on an elongate inflorescence _________ **Alismataceae**

9. Leaf blades elliptic to ovate to orbicular, entire, shallowly toothed, or lobed.

14. Plants free-floating, with abundant and conspicuous roots in water.

15. Leaves distinctly petiolate, the petioles swollen, ca. as long as the blade or longer, the blades glabrous; perianth 4–6 cm long, bluish lavender, the upper segment with a yellow spot, very showy _________ (*Eichhornia*) **Pontederiaceae**

15. Leaves sessile, velvety-hairy; perianth absent _________ (*Pistia*) **Araceae**

14. Plants rooted in bottom (broken off stem sections can sometimes be found free-floating but these without abundant conspicuous roots).

16. Leaf blades with wide, rounded teeth or shallow lobes; largest leaf blades 8 cm or less broad, often nearly orbicular _________ **Apiaceae**

16. Leaf blades entire or nearly so (but basal notch often present); largest leaf blades 5–90 cm or more long or broad, variously shaped.

17. Plants emergent, 1–2 m tall; flowers 1 cm long or less, purplish; inflorescence a panicle with zigzag branches and a striking white-powdery appearance _________ (*Thalia*) **Marantaceae**

17. Plants without the above combination.

18. Inflorescence a fleshy spike (= spadix) with a yellow or creamy white leafy bract (= spathe) subtending or enclosing it; flowers very small, numerous and crowded on the spike, individually inconspicuous; perianth absent or minute _________ **Araceae**

18. Inflorescence not a fleshy spike; individual flowers large and conspicuous, individually easily seen even at a glance; perianth obvious.

19. Petals 3 (or apparently 6 due to 3 colored, petal-like sepals in some species); leaf blades not lying flat on water surface, variously shaped (elliptic, ovate, often sagittate), but never peltate

and usually without a single, more or less parallel-sided, basal notch (except in *Pontederia*), usually longer than wide.

20. Perianth with a well-developed tube, white to purplish blue, with one petal having a pair of light yellow dots at base OR blue with yellow markings; pistils solitary per flower, made up of a single carpel or of several carpels fused together; fruits various, 1–many-seeded; stamens 3 or 6 ________ **Pontederiaceae**

20. Perianth without a tube, white or rarely pink; pistils numerous per flower, free from each other or nearly so, each pistil developing into a 1-seeded indehiscent fruit (achene); stamens 6–numerous ________ **Alismataceae**

19. Petals numerous; leaf blades often lying flat on water surface (under dry conditions sometimes above the water), either peltate OR with a single more or less parallel-sided basal notch (never sagittate), usually nearly as wide as long.

21. Leaf blades peltate, not notched; pistils (and later fruits) sunken into the greatly enlarged, inverted-conical receptacle; fruiting stalks held well above the water surface; fruits nut-like ________ (*Nelumbo*) **Nelumbonaeaceae**

21. Leaf blades not peltate, the petiole attached at base of deep notch in blade; pistils not sunken into the receptacle; fruits maturing underwater; fruits fleshy ________ **Nymphaeaceae**

6. Leaves not all attached at base of plant, rather borne along the stems.

22. Leaves pinnately divided.

23. Leaf divisions broad, flat, blunt; fruits many-seeded, 2-valved, dehiscent, linear capsules; stamens 6 per flower (2 short, 4 long) ________ (*Rorippa*) **Brassicaceae**

23. Leaf divisions linear or thread-like, pointed; fruits indehiscent, either nutlets OR 4-lobed and eventually splitting into 4 nutlets; stamens 3, 4, or 8 per flower ________ **Haloragaceae**

22. Leaves not pinnately divided.

24. Leaves all opposite or whorled.

25. Leaves in distinct whorls of 3–8, giving the stem a "bottle brush" appearance; plants completely submersed; flowers (male and/or female) borne to the water surface on a thread-like stalk 3–6 cm long; perianths 3–10 mm long, white or translucent, visible with the naked eye ________ **Hydrocharitaceae**

25. Leaves opposite, or if whorled, the stem not appearing like a "bottle brush"; plants completely submersed OR partly floating OR partly emergent; flowers and perianths various.

26. Leaves deeply palmately divided or dichotomously forked, the ultimate segments linear or thread-like.

27. Leaves usually opposite (rarely whorled); perianth usually whitish (rarely purplish or yellowish), 4–12 mm long; small (blades ca. 1–3 cm long), alternate, peltate, entire, floating leaves usually present in addition to deeply palmately divided leaves ________ (*Cabomba*) **Cabombaceae**

27. Leaves whorled; perianth absent; floating leaves absent (algae in the Characeae with whorled "branches" can superficially resemble *Ceratophyllum*; however, *Ceratophyllum* can be recognized by the dichotomously forked leaves) ________ (*Ceratophyllum*) **Ceratophyllaceae**

26. Leaves entire or with small teeth to undulate-dentate or coarsely so, neither palmately divided nor dichotomously forked.

28. Leaves (2–)3–60 mm wide, linear to lanceolate to spatulate to suborbicular.
    29. Inflorescences silvery whitish pedunculate heads lacking corollas (silvery whitish color due to numerous tepals and bracts) ______ (*Alternanthera*) **Amaranthaceae**
    29. Inflorescences various, but not silvery whitish pedunculate heads with numerous bracts and tepals; corollas present OR absent.
        30. Corollas white to cream, violet, lavender, pale blue, pink, rose-purple, or red-purple, sometimes with darker markings, sometimes quickly deciduous.
            31. Corollas sympetalous, slightly to strongly bilaterally symmetrical, usually 2-lipped, 6–28 mm long; calyces 5-merous; flowers axillary OR in pedunculate heads or spikes.
                32. Flowers in pedunculate heads or spikes; seeds 2–4 per fruit ______ (*Justicia*) **Acanthaceae**
                32. Flowers axillary; seeds 12 or more per fruit ______ **Scrophulariaceae**
            31. Corollas of separate petals, radially symmetrical, 2.5 mm or less long; calyces 4-merous; flowers axillary ______ **Lythraceae**
        30. Corollas yellow OR absent.
            33. Corollas slightly to strongly bilaterally symmetrical, often 2-lipped, with a definite tube ______ **Scrophulariaceae**
            33. Corollas if present radially symmetrical, without a definite tube, either of separate petals OR rotate with petals united only at base.
                34. Flowers in umbels; petals united at base; leaves suborbicular; ovaries superior ______ (*Nymphoides*) **Menyanthaceae**
                34. Flowers solitary in the leaf axils; petals separate or absent; leaves lanceolate to spatulate to suborbicular; ovaries inferior ______ (*Ludwigia*) **Onagraceae**
28. Leaves 3 mm or less wide, variously shaped, often linear.
    35. Leaves obviously toothed to the naked eye ______ (*Najas*) **Hydrocharitaceae**
    35. Leaves not obviously toothed to the naked eye.
        36. Leaf blades linear or thread-like, mostly < than 1 mm wide; a rosette of floating leaves never present.
            37. Leaf blades usually very minutely denticulate (under a scope); fruits not stipitate, without a beak _ (*Najas*) **Hydrocharitaceae**
            37. Leaf blades entire; fruits short stipitate (= stalked), also with a beak to 1.5 mm long ______ (*Zannichellia*) **Zannichelliaceae**
        36. Leaf blades linear to obovate, at least the uppermost usually 1–3 mm wide; a rosette of floating leaves sometimes present.
            38. Stipules present; flowers perfect; fruits subglobose cap sules; leaves not forming rosettes at the stem tips ______ (*Elatine*) **Elatinaceae**
            38. Stipules absent; flowers imperfect, the plants monoecious; fruits somewhat flattened laterally, often slightly heart-shaped and appearing to have 2 lobes, eventually splitting into 4 achene-like mericarps; leaves sometimes forming rosettes at the stem tips ______ (*Callitriche*) **Callitrichaceae**

24. Leaves alternate, at least on lower part of stem.

39. Leaf blades peltate, floating (submersed dissected leaves can also be present) ____ **Cabombaceae**
39. Leaf blades not peltate, either floating OR not floating.
    40. Leaves of 2 kinds, the emersed leaves toothed, the submersed leaves deeply pinnatifid or pectinate ____ (*Proserpinaca*) **Haloragaceae**
    40. Leaves of 1 or 2 kinds, but none either deeply pinnatifid or pectinate.
        41. Leaves palmately compound or palmately divided (segments not thread-like) ____ (*Ranunculus*) **Ranunculaceae**
        41. Leaves entire or finely toothed OR leaves (possibly leaf-like branches) irregularly or palmately divided into thread-like segments.
            42. Leaves (possibly leaf-like branches) or branches irregularly or palmately divided into linear, thread-like segments; plants with numerous, small, bladder-like traps for capturing small aquatic organisms ____ (*Utricularia*) **Lentibulariaceae**
            42. Leaves entire or finely toothed; plants without bladder-like traps.
                43. Perianth parts greenish, unequal, one of them differentiated into a lip divided into three narrow lobes and extended at base into a spur 9–14 mm long ____ (*Habenaria*) **Orchidaceae**
                43. Perianth parts equal, white, green, or yellow OR absent OR reduced to bristles or scales, none differentiated into a lip; spur absent.
                    44. Plants large (1–3 m tall) emergents with linear leaves and an extremely dense, large (12–40 cm long), cylindrical, brownish spike with thousands of very tiny flowers ____ (Typha) **Typhaceae**
                    44. Plants not as above.
                        45. Leaves ovate, cordate basally; inflorescences spike-like, 10–20(–30) cm long, 10–15 mm in diam., erect, the tip often drooping, of numerous (to 300 per inflorescence), crowded, whitish flowers ____ (*Saururus*) **Saururaceae**
                        45. Leaves usually much narrower than ovate, not cordate basally; inflorescences not as above.
                            46. Leaves differentiated into a basal tubular sheath surrounding the stem and a terminal, usually elongate, linear, parallel-sided blade; petioles not present.
                                47. Perianth present, of 6 small scaly segments; fruits 3–many-seeded; flowers not subtended by chaffy, scale-like bracts ____ **Juncaceae**
                                47. Perianth absent or reduced to bristles or rarely of 3 small scales; fruits 1-seeded; each flower usually subtended by 1–2 chaffy, scale-like bracts.
                                    48. Leaves usually 2-ranked with sheath split down one side; stems with hollow internodes, round, typically jointed (nodes obvious); flowers usually each subtended by 2 bracts (lemma and palea), additional bracts (glumes, sterile lemmas) also sometimes present ____ **Poaceae**

48. Leaves usually 3-ranked with sheath closed; stems with solid internodes, rounded or often triangular, typically not jointed; flowers usually each subtended by 1 bract ______ **Cyperaceae**

46. Leaves not differentiated into a basal sheath and a terminal blade (but may have sheathing stipules); petioles usually present and clearly differentiated from blades.

49. Leaves either all submersed or some submersed and some floating; perianth absent OR pale yellow with an elongated filiform tube; leaf blades usually parallel-veined or with concentrically curved veins about equally prominent from base to tip.

50. Flowers in pedunculate, often dense spikes; perianth absent; leaves all submersed or some floating ______ **Potamogetonaceae**

50. Flowers solitary; perianth present, pale yellow with an elongated filiform tube; leaves all submersed ______ (*Heteranthera*) **Pontederiaceae**

49. Leaves often borne on stems emergent from the water; colorful perianth often present; leaf blades 1-veined OR with branched or irregular veins.

51. Plants armed with 1–2 spines per node; corollas blue (rarely white), showy, 11–17 mm long ______ (*Hydrolea*) **Hydrophyllaceae**

51. Plants unarmed; corollas if present usually much smaller, never blue.

52. Corollas sympetalous; ovaries inferior; stamens inserted near middle of corolla tube ______ (*Sphenoclea*) **Sphenocleaceae**

52. Corollas (or perianth) of separate parts; ovaries superior OR inferior; stamens not attached to perianth.

53. Sheathing stipules (= ocrea) present at base of petiole; fruit a 2- or 3-sided, 1-seeded nutlet; perianth usually white to pink; ovaries superior ______ (*Polygonum*) **Polygonaceae**

53. Sheathing stipules absent at base of petiole; fruit a many-seeded capsule; perianth yellow or absent; ovaries inferior ______ (*Ludwigia*) **Onagraceae**

Key to Woody Vines

MODIFIED FROM

WOODY VINES OF THE SOUTHEASTERN STATES

DUNCAN (1967)

1. Leaves compound (in *Cissus* some leaves, but not all, only deeply 3-parted).
  2. Leaves opposite.
    3. Plants climbing by aerial roots; corollas tubular, red-orange; stamens 4; each flower producing a single capsule ____ **Campsis**
    3. Plants climbing by twisting leaf stalks; corollas absent, but the 4(–6) sepals petal-like and separate, white to lavender to blue-purple or red; stamens numerous; each flower producing numerous achenes (these often with long plumose beaks) ____ **Clematis**
  2. Leaves alternate.
    4. Plants armed, the stems with prickles; pistils 12 or more per flower; fruits aggregates of druplets or achenes.
      5. Hypanthium globose to urn-shaped, with a constricted opening, the achenes concealed inside (the hypanthium is termed a hip, is smooth in outline and typically red or reddish orange) ____ **Rosa**
      5. Hypanthium flat or hemispheric, the ovules and drupelets conspicuously exposed (the cluster of druplets is commonly termed a blackberry or dewberry and is lumpy in outline and red to dark purple or black) ____ **Rubus**
    4. Plants unarmed, the stems without prickles; pistils 1 per flower; fruits drupaceous, legumes, or berries.
      6. Plants climbing by aerial roots; leaves with 3 (or rarely 5) leaflets, pinnate; fruits drupaceous ____ **Toxicodendron**
      6. Plants climbing by twining or by tendrils (the tendrils are sometimes similar to roots; they sometimes have adhesive discs at their tips); leaves with 3–many leaflets, palmate or pinnate; fruits legumes or 1–4-seeded berries.
        7. Stems twining, the plants without tendrils; fruits legumes.
          8. Leaflets 3, the largest 10 cm or more long ____ **Pueraria**
          8. Leaflets 7–19, the largest less than 10 cm long ____ **Wisteria**
        7. Stems not twining, the plants with tendrils (the tendrils sometimes have adhesive discs at their tips); fruits 1–4-seeded berries.
          9. Leaves 2-pinnately or (partially 3-pinnately) compound, the leaflets many ____ **Ampelopsis**
          9. Leaves once palmately compound (or apparently so), the leaflets 3–7.
            10. Leaves with 3 leaflets, conspicuously fleshy; inflorescences resembling compound umbels; flowers 4-merous; leaflets and petioles falling apart when pressed and dried ____ **Cissus**
            10. Leaves with (3–)5–7 leaflets, usually not fleshy; inflorescences paniculate, racemose, or cymose; flowers 5-merous; leaflets and petioles usually not falling apart when pressed and dried ____ **Parthenocissus**
1. Leaves simple (some deeply lobed).
  11. Leaves opposite or rarely whorled.
    12. Sap milky; leaf blades often (but not always) cordate basally; corollas with a corona.
      13. Plants woody nearly throughout; leaf blades acute to broadly rounded basally; corollas brown-purple; introduced species ____ **Periploca**
      13. Plants woody only in lower half; leaf blades cordate (in nc TX species) basally; corollas white to cream or greenish white; native species ____ **Cynanchum**
    12. Sap not milky (except in *Trachelospermum*); leaf blades not cordate basally; corollas without a corona.

14. Corollas often bilaterally symmetrical (sometimes nearly radially symmetrical); uppermost
leaves united around stem OR not so; fruits fleshy berries ______ **Lonicera**
14. Corollas radially symmetrical; uppermost leaves not united around stem; fruits dry and dehiscent at maturity.
15. Leaves evergreen, leathery; corollas conspicuously yellow, with tube > 15 mm long; fruits 2-celled capsules, flattened contrary to the partition; seeds without hairy tufts at ends ______ **Gelsemium**
15. Leaves evergreen or deciduous, leathery OR not so; corollas creamy white to pale yellow, with tube 10 mm or less long; fruits of twin follicles; seeds with hairy tufts at ends ______ **Trachelospermum**
11. Leaves alternate.
16. Pith lacking, the stems solid except sometimes for scattered pores, with scattered vascular strands; tendrils arising in pairs from the petioles of leaves; plants often armed, with prickles often present on stems ______ **Smilax**
16. Pith present or stems rarely hollow; tendrils absent or if present not arising from the petioles of leaves; plants unarmed, prickles absent.
17. Plants climbing by tendrils or aerial roots.
18. Plants climbing by aerial roots; leaves evergreen; inflorescences umbels (solitary or racemosely arranged) OR flowers hidden from view inside a hollow receptacle; introduced species.
19. Leaf blades usually 3–5 lobed; sap not milky; flowers and fruits (small 3–5-seeded berries) in solitary or racemosely arranged umbels ______ **Hedera**
19. Leaf blades unlobed; sap milky; flowers and fruits hidden from view inside a hollow receptacle ______ **Ficus**
18. Plants climbing by tendrils; leaves deciduous; inflorescences various, but flowers neither in umbels nor hidden inside a hollow receptacle; mostly native species.
20. Leaf blades entire, ovate, not lobed; petioles dilated at base and extending into a minutely pubescent ring (stipular) surrounding the stem; tendrils limited to the ends of the branches; stems grooved ______ **Brunnichia**
20. Leaf blades toothed or lobed, or if unlobed or entire the stems not grooved (sometimes with soft corky ridges); petioles not as above; tendrils opposite leaves and thus apparently lateral to stems although basically terminal.
21. Petioles bearing two stalked glands between the middle of the petiole and the blade; stems, except youngest, with tight almost white corky longitudinal strips or sometimes covered with the cork; flowers and fruits one per pedicel, solitary to two in leaf axils ______ **Passiflora**
21. Petioles with glands absent; stems lacking the whitish cork, the older stems sometimes with rough brownish bark or the brownish bark sometimes shredding; flowers and fruits several to many in clusters attached opposite leaves.
22. Tendrils with slender, pointed, curling tips; native species.
23. Inflorescences cymes, wider than long, dichotomously forking; leaf blades truncate to cordate at base; plants essentially glabrous; year-old stems having white continuous pith; petals spreading at flowering time and later dropping singly; mature fruits a turquoise blue, not edible; bark of stems tight ______ **Ampelopsis**
23. Inflorescences panicles, longer than wide, not dichotomously forking; leaf blades cordate at base; plants densely pubescent to nearly glabrate; year-old stems having brown pith with cross

partitions at the nodes, except in *V. rotundifolia*; petals separating only at their bases and falling as a unit; mature fruits black or purple, edible although sometimes sour or bitter; bark of stems of most species loosening into elongated flakes or shreds ________ **Vitis**

22. Tendrils with small, disk-like tips; introduced ornamentals ________ **Parthenocissus**

17. Plants climbing by twining.
24. Leaf blades palmately veined.
25. Petioles attached inside the edge of the leaf blade on the underside (occasionally on some leaves by as little as 1 mm) ________ **Menispermum**
25. Petioles joining the edge of the leaf blade at blade base.
26. Leaf blades not lobed, cordate to broadly ovate; fruits capsules; calyces curved, pipe-like in shape ________ **Aristolochia**
26. Leaf blades usually slightly to deeply lobed; fruits drupes; calyces not curved.
27. Lower surface of leaf blades glabrous beneath except for sparse hairs on the larger veins; drupes black, 15–25 mm long and flattened only on one side; leaf blades deeply 3–5-lobed, the middle lobe narrower at the base than in the middle, the tips of lobes sharply pointed but not mucronate; at least seven veins arising from blade base, the lowermost ones often obscure; bud area neither vertically elongate nor hairy ________ **Calycocarpum**
27. Lower surface of leaf blades silky pubescent; drupes red, 5–8 mm long and flattened on both sides; leaf blades usually only slightly lobed (but variable, ranging from unlobed to 3–5-lobed), the tips of the blades mucronate; not more than 5 (rarely 7) veins arising from the blade base, the lowermost ones often obscure; bud area vertically elongate and densely hairy ________ **Cocculus**
24. Leaf blades pinnately veined.
28. Pinnate veins of leaf blades nearly straight, evenly spaced, and parallel; margins of leaf blades entire or obscurely crenate ________ **Berchemia**
28. Pinnate veins of leaf blades neither straight, evenly spaced, nor parallel; margins of leaf blades with distinct and rather uniformly distributed serrations ________ **Celastrus**

Key to Families of Monocots

1. Leaf blades palmately divided, fan-like, up to 1 m or more wide; plants palm-like (Palmae) ________ **Arecaceae**
1. Leaf blades simple or pinnatifid, usually much narrower; plants not palm-like.
2. Plants epiphytic (growing on branches of other plants, without roots in the ground) ________ (*Tillandsia*) **Bromeliaceae**
2. Plants terrestrial or aquatic.
3. Plants small (of 1–several fronds or thalli each ca. 1 cm or less long), floating aquatics, without definite stems or leaves ________ **Lemnaceae**
3. Plants not as above, usually much larger, terrestrial OR aquatic and rooted in substrate OR floating; stems or leaves distinguishable.
4. Stems woody.
5. Leaves many, clustered close together, either all basal or in a crown, long and sword-

like (usually 0.2 m to > 0.5 m long); inflorescences large terminal racemes or panicles with conspicuous flowers ______ **Agavaceae**

5. Leaves conspicuously scattered all along the elongate stem, not long and sword-like (0.25 m or less long); inflorescences not as above; flowers inconspicuous.
   6. Plants climbing or trailing vines with prickles and/or tendrils; fruits black or blue-black berries; leaf sheaths absent ______ (*Smilax*) **Smilacaceae**
   6. Plants erect, without prickles or tendrils; fruits caryopses; leaf sheaths present ______ **Poaceae**

4. Stems herbaceous (not woody).
   7. Plants aquatics growing completely submersed; leaves opposite or whorled.
      8. Leaves opposite (some can occasionally appear whorled where branches arise); flowers sessile or subsessile, borne underwater; perianth absent or minute, clearish or greenish, virtually indistinguishable without a lens.
         9. Leaves obviously toothed to the naked eye ______ (Najas) **Hydrocharitaceae**
         9. Leaves not obviously toothed to the naked eye.
            10. Leaf blades usually very minutely denticulate (under a scope), sheathing basally; fruits not curved, not short stipitate, without a beak; flowers with a single carpel; sheathing stipules not present ______ (*Najas*) **Hydrocharitaceae**
            10. Leaf blades entire, not sheathing basally; fruits curved, short stipitate (= stalked), also with a beak to 1.5 mm long; flowers with 2–8 separate carpels; sheathing stipules present ______ (*Zannichellia*) **Zannichelliaceae**
      8. Leaves in distinct whorls of 3–8; flowers (staminate and/or pistillate) borne to the water surface on a thread-like stalk 3–6 cm long; perianth (staminate and/or pistillate) 3–10 mm long, white or translucent, visible with the naked eye ______ **Hydrocharitaceae**
   7. Plants terrestrial or aquatic; if leaves completely submersed then alternate or basal.
      11. Plants free-floating aquatics with leaves in rosettes.
         12. Leaves distinctly petiolate, the petioles swollen, ca. as long as the blades, or longer, the blades glabrous; perianth 4–6 cm long, bluish lavender, the upper segment with a yellow spot, very showy ______ (*Eichhornia*) **Pontederiaceae**
         12. Leaves sessile, velvety-hairy; perianth absent ______ (*Pistia*) **Araceae**
      11. Plants not free-floating, either terrestrial or aquatic, but rooted in substrate; leaves variously arranged.
         13. Plants completely submersed rooted aquatics with elongate, linear, basal leaves; flowers at the water surface, the inflorescences never extending above the water ______ (*Vallisneria*) **Hydrocharitaceae**
         13. Plants either terrestrial or aquatic, with leaves various; if aquatic then flowers held above the water surface.
            14. Plants without obvious leaves (only bladeless sheaths present), with unbranched, cylindrical or thread-like, green stems about as thick as wide (these sometimes flattened); perianth absent or of inconspicuous bristles or small scales ______ **Cyperaceae**
            14. Plants usually with obvious leaves (rarely reduced to bracts); perianth various, ranging from conspicuous to absent.
               15. Flowers in a single, small (up to 30 mm long), dense head or spike terminating an elongate naked scape, the head or spike either cone-like with numerous, overlapping, brownish, thin, ± woody bracts (subtending yellow flowers) OR head whitish or grayish due to numerous hairs on the subtending involucral bracts and flower parts.
                  16. Inflorescence cone-like, with spirally imbricated, brownish, thin, ± woody bracts, usually with a single yellow flower exposed per inflorescence ______ (*Xyris*) **Xyridaceae**

16. Inflorescence a small whitish or grayish head, not cone-like, lacking brownish woody bracts, without yellow flowers _ (*Eriocaulon*) **Eriocaulaceae**

15. Flowers not in a single, small, dense head or spike terminating an elongate naked scape; head or spike neither cone-like with numerous overlapping brownish bracts nor whitish nor grayish due to numerous hairs on the bracts and flower parts.

17. Flowers and fruits in the axils of imbricate (= overlapping) or distichous (= 2-ranked) scales, concealed by the scales at least when young; fruits 1-seeded; perianth absent or represented by bristles or small scales (GRASSES and SEDGES).

18. Stems typically round or flat in cross-section but never triangular, typically jointed (nodes obvious), with hollow internodes; leaves usually 2-ranked, with sheaths normally split lengthwise on the side opposite the blade; each flower usually subtended by 2 scales ______ **Poaceae**

18. Stems round or often triangular, typically not jointed, with solid internodes; leaves usually 3-ranked, with sheaths continuous around the stem or splitting only in age or leaves reduced to sheaths only; each flower usually subtended by 1 scale ______ **Cyperaceae**

17. Flowers and fruits not in the axils of imbricate or distichous scales, not concealed by scales, or if so, fruits more than 1-seeded; perianth absent or present, sometimes petal-like or with conspicuous petals.

19. Inflorescence a fleshy spike (= spadix) of numerous, small, imperfect flowers, the inflorescence enclosed in a specially modified bract (= spathe) or diverging at an angle from the side of a spathe-like structure.

20. Plants with elongate, linear, sword-like, parallel-veined leaves; spadix diverging from the side of, but not enclosed in, an elongate linear spathe-like structure ______ (*Acorus*) **Acoraceae**

20. Plants without elongate, linear, sword-like, parallel-veined leaves; spadix enclosed in a spathe ______ **Araceae**

19. Inflorescence not a fleshy spike; flowers usually perfect; inflorescence neither enclosed in a spathe nor diverging at an angle from the side of a spathe-like structure.

21. Plants large (1–3 m tall) emergents with an extremely dense, large (12–40 cm long), brownish spike with thousands of very tiny flowers ______ (*Typha*) **Typhaceae**

21. Plants not as above.

22. Corollas absent; plants aquatic with submersed or floating leaves; fruits drupe-like, 1-seeded ______ (*Potamogeton*) **Potamogetonaceae**

22. Corollas present; plants terrestrial or aquatic; fruits capsules, berries, or achenes (if achenes, these usually winged).

23. Plants climbing vines.

24. Plants with tendrils; leaves alternate; flowers in pedunculate or sessile axillary umbels; ovary superior; fruits berries ______ (*Smilax*) **Smilacaceae**

24. Plants without tendrils, climbing by twining;

leaves (at least of lower nodes) opposite or whorled; flowers in paniculate or spike-like inflorescences; ovary inferior; fruits capsules ______ (*Dioscorea*) **Dioscoreaceae**

23. Plants not climbing vines.
    25. Plants 1–2 m tall; flowers 1 cm long or less, purplish; inflorescence a panicle with zig-zag branches and a striking white-powdery appearance ______ (*Thalia*) **Marantaceae**
    25. Plants without the above combination.
        26. Ovary inferior.
            27. Plants very small, 5–20 cm tall; stems delicately thread-like; leaves scale-like (5 mm or less long); flowers small (to 5 mm long), greenish white or cream, sometimes tinged with blue ______ (*Burmannia*) **Burmanniaceae**
            27. Plants usually > 20 cm tall; stems not thread-like; leaves not scale-like (except in saprophytic species); flowers small OR often large, variously colored.
                28. Stamens 6; flowers radially symmetrical ______ **Liliaceae**
                28. Stamens 3 or less; flowers radially symmetrical or bilaterally symmetrical.
                    29. Flowers radially symmetrical; stamens 3 per flower; filaments present, separate or united; column absent; leaves equitant (= 2-ranked with closely overlapping bases) ______ **Iridaceae**
                    29. Flowers bilaterally symmetrical; stamens 1 or 2 per flower; filaments absent; male and female parts united into a column; leaves not equitant ______ **Orchidaceae**
        26. Ovary superior.
            30. Pistils numerous per flower, free from each other or nearly so, each pistil developing into a 1-seeded indehiscent fruit (achene) ______ **Alismataceae**
            30. Pistils 1 per flower, made up of a single carpel or of several carpels fused together; fruits various but often a many-seeded capsule.
                31. Perianth (sepals and petals) of 6 small, dry, bract-like segments, persistent; plants rush-like ______ **Juncaceae**

31. Perianth not bract-like, at least some of the segments usually petaloid, at least the corolla usually not persistent; plants not rush-like.
    32. Plants with large woody bases or a thick, fibrous-rooted crown; inflorescence a large many-flowered raceme or panicle ________ **Agavaceae**
    32. Plants neither woody-based nor with a thick, fibrous-rooted crown; inflorescences various.
        33. Perianth united in lower part forming a slender tube; flowers solitary; plants aquatic or growing in wet areas ________ **Pontederiaceae**
        33. Perianth of distinct segments; flowers solitary or otherwise; plants of various habitats, often terrestrial.
            34. Perianth segments dissimilar, of more than one type (some petaloid, some sepaloid); leaf bases usually sheathing; flowers in 1- or 2-bracted leaf-like spathes _ **Commelinaceae**
            34. Perianth segments all similar (all petaloid); leaf bases usually not sheathing; flowers not in leaf-like spathes ________ **Liliaceae**

GENERAL KEY TO ALL FAMILIES[2]

MODIFIED FROM

KEY AND DESCRIPTIONS FOR THE VASCULAR PLANT FAMILIES OF OKLAHOMA

CONTRIBUTED BY OKLAHOMA FLORA EDITORIAL COMMITTEE (TYRL ET AL. 1994).

KEY TO GROUPS

1. Plants trees or shrubs or woody vines or woody aerial hemiparasites (growing on trees or shrubs—mistletoes).
   2.Plants woody vines or woody aerial hemiparasites ________ **Group A**
   2. Plants trees or shrubs.
      3. Stems succulent, bearing spines in clusters; flowers showy; ovaries inferior; perianth parts 25 or more; stamens 25 or more ________ **Cactaceae**
      3. Stems not succulent, not bearing spines in clusters; flowers showy or not showy; ovaries superior or inferior; perianth parts of various numbers; stamens of various numbers.
         4. Plants producing flowers or cones before leaves ________ **Group B**
         4. Plants producing flowers or cones simultaneously with leaves OR producing flowers or cones after leaves are formed.
            5. Leaves opposite or whorled or fascicled or in rosettes ________ **Group C**
            5. Leaves alternate.
               6. Leaves compound ________ **Group D**
               6. Leaves simple.
                  7. Leaf margins entire ________ **Group E**
                  7. Leaf margins lobed or toothed ________ **Group F**

1. Plants herbs, some woody at the base.
   8. Plants aquatic (plants floating or submersed in or emergent from water) ________ **Group G**
   8. Plants terrestrial OR growing on other plants (epiphytes and hemiparasites).
      9. Plants vines or epiphytes or aerial hemiparasites (mistletoes) ________ **Group H**
      9. Plants neither vines nor epiphytes nor aerial hemiparasites.
         10. Plants parasitic or saprophytic; chlorophyll absent ________ **Group I**
         10. Plants autophytic; chlorophyll present.
            11. Stems bearing spines and/or glochids in areoles, succulent; foliage leaves absent; ovaries inferior; perianth parts 25 or more; stamens 25 or more ________ **Cactaceae**
            11. Stems not bearing spines or glochids in areoles, succulent or not succulent; foliage leaves present or absent; ovaries superior or inferior; perianth parts of various numbers; stamens of various numbers.
               12. Plants acaulescent, the aerial stems not apparent and leaves not cauline.
                  13. Plants producing flowers and seeds; spores produced in anthers or ovaries ____ **Group J**
                  13. Plants not producing flowers and seeds; spores produced in sori or sporocarps or in aggregations of sporangia at ends of elongated stalks ________ **Group K**

[2]While numerous couplets have been added to cover plants which occur in North Central Texas but not in Oklahoma, no couplets have been deleted from the Oklahoma family key. Therefore, some families/taxa occurring in Oklahoma are included that do not occur in North Central Texas. This was done so that the family key would be of maximum benefit to Oklahoma users as well as those in Texas. Such families are indicated in the *General Key to All Families* by a note in brackets, e.g., [Family in OK, not in nc TX]. In a number of instances, it is possible to key to the correct family even if a particular, easily confused dicotomy is misinterpreted. For such cases, explanatory notes are given in brackets in the key.

12. Plants caulescent, the aerial stems apparent and leaves cauline.
    14. Plants not producing flowers or seeds; spores produced in strobili or sori on abaxial surfaces of leaves or in aggregations of sporangia at ends of elongated stalks ______ **Group K**
    14. Plants producing flowers and seeds; spores produced in anthers or ovaries.
        15. Perianth parts absent ______ **Group L**
        15. Perianth parts present.
            16. Perianth parts in 1 series or parts all similar.
                17. Perianth parts 3 or in multiples of 3 ______ **Group M**
                17. Perianth parts 1 or 2 or 4 or 5 or in multiples of 4 or 5 or many ______ **Group N**
            16. Perianth parts in 2 series.
                18. Petals 3 or in multiples of 3 ______ **Group O**
                18. Petals 1 or 2 or 4 or 5 or in multiples of 4 or 5 or many.
                    19. Corollas bilaterally symmetrical.
                        20. Petals free ______ **Group P**
                        20. Petals fused at least at the base.
                            21. Ovaries inferior, wholly or partially ______ **Group Q**
                            21. Ovaries superior ______ **Group R**
                    19. Corollas radially symmetrical or asymmetrical.
                        22. Petals free.
                            23. Ovaries inferior, wholly or partially ______ **Group S**
                            23. Ovaries superior.
                                24. Pistils or fruits 1 per flower ______ **Group T**
                                24. Pistils or fruits 2 or more per flower ______ **Group U**
                        22. Petals fused at least at the base.
                            25. Ovaries inferior, wholly or partially ______ **Group V**
                            25. Ovaries superior ______ **Group W**

GROUP A

Plants woody vines OR woody aerial hemiparasites.

1. Plants aerial hemiparasites (mistletoes). ______ **Viscaceae**
1. Plants autophytic vines.
    2. Leaves opposite or whorled.
        3. Plants climbing by tendrils or aerial rootlets or prehensile petioles.
            4. Plants climbing by prehensile petioles; flowers radially symmetrical; corollas absent, but the sepals petal-like and separate; fruits achenes ______ **Ranunculaceae**
            4. Plants climbing by aerial rootlets or tendrils from leaf rachises (may be absent on scrambling-bushy forms); flowers bilaterally symmetrical; corollas 5-merous, of fused petals; fruits capsules ______ **Bignoniaceae**
        3. Plants climbing by twining stems.
            5. Leaf margins crenate or serrate ______ **Bignoniaceae**
            5. Leaf margins entire.
                6. Corollas yellow to orange, cream, or white, with a dark purple center; petioles conspicuously winged; stamens 4 [Stems actually herbaceous] ______ (*Thunbergia*—Acanthaceae) **Group H**
                6. Corollas variously colored but not light with a dark center; petioles not winged; stamens 5.
                    7. Sap milky; leaf blades often (but not always) cordate basally; corollas with a corona ______ **Asclepiadaceae**

7. Sap not milky (except in *Trachelospermum*); leaf blades not cordate basally; corollas without a corona.
 8. Corollas bilaterally symmetrical (sometimes nearly radially symmetrical); uppermost leaves united around stem OR not so; fruits fleshy berries ______ **Caprifoliaceae**
 8. Corollas radially symmetrical; uppermost leaves not united around stem; fruits dry and dehiscent at maturity.
  9. Leaves evergreen, leathery; corollas conspicuously yellow, with tube > 15 mm long; fruits 2-celled capsules, flattened contrary to the partition; seeds without hairy tufts at ends ______ **Loganiaceae**
  9. Leaves evergreen or deciduous, leathery OR not so; corollas creamy white to pale yellow, with tube 10 mm or less long; fruits of twin follicles; seeds with hairy tufts at ends ______ **Apocynaceae**

2. Leaves alternate.
 10. Plants climbing by tendrils or aerial rootlets.
  11. Venation parallel-convergent; tendrils paired; inflorescences umbels; pith absent ______ **Smilacaceae**
  11. Venation palmate or pinnate or pinnipalmate; tendrils solitary or absent; inflorescences various, including racemes, panicles, cymes, umbels, or flowers hidden from view inside a hollow receptacle; pith present.
   12. Leaves compound.
    13. Inflorescences cymes; pistils 2-carpellate; fruits berries, dark blue to black ______ **Vitaceae**
    13. Inflorescences panicles; pistils 3-carpellate; fruits drupes, white (*Toxicodendron*—poison ivy) ______ **Anacardiaceae**
   12. Leaves simple.
    14. Leaves evergreen, thickish; inflorescences umbels (solitary or racemosely arranged) OR flowers hidden from view inside a hollow receptacle; introduced species spreading from cultivation, not expected in native habitats.
     15. Leaves usually 3–5-lobed; sap not milky; flowers and fruits (small 3–5-seeded berries) in solitary or racemosely arranged umbels ______ **Araliaceae**
     15. Leaves unlobed; sap milky; flowers and fruits hidden from view inside a hollow receptacle ______ **Moraceae**
    14. Leaves deciduous, not noticeably thickened; inflorescences racemes, panicles, or cymes; widespread native species.
     16. Leaves ovate or oblong-ovate, the margins entire; inflorescences racemose panicles; flowers perfect; calyces deeply 5-parted; corollas absent; fruits achenes, subtended by persistent sepals ______ **Polygonaceae**
     16. Leaves cordate or rotund to broadly ovate, the margins toothed; inflorescences cymes; flowers functionally imperfect; calyces slightly to shallowly 4-lobed; corollas present, may be caducous; fruits berries ______ **Vitaceae**
 10. Plants climbing by twining stems.
  17. Stipules absent; axillary buds 3, 2 may be obscured by leaf scars.
   18. Stems and leaves glabrous or puberulent; leaf scars U-shaped; flowers imperfect, the plants dioecious; perianths radially symmetrical; fruits drupes ______ **Menispermaceae**
   18. Stems and leaves tomentose; leaf scars elliptic; flowers perfect; perianths bilaterally symmetrical; fruits capsules ______ **Aristolochiaceae**
  17. Stipules or stipular scars present; axillary buds 1.
   19. Leaves compound.
    20. Stems bearing prickles; pistils 12 or more per flower; fruits aggregates of drupelets or achenes ______ **Rosaceae**
    20. Stems not bearing prickles; pistils 1 per flower; fruits berries or legumes.

21. Leaves 1–3 times compound; flowers radially symmetrical; corollas greenish, 1–3 mm long; fruits berries 10–15 mm in diam, not conspicuously hairy ______ **Vitaceae**
21. Leaves once compound; flowers bilaterally symmetrical; corollas purplish or lilac or blue, 15–27 mm long; fruits legumes, much > 15 mm long, conspicuously hairy ______ (Papilionoideae) **Fabaceae**

19. Leaves simple.
22. Inflorescences cymes; fruits berries; vascular bundle scars 12; pith dividing into thin plates at periphery ______ **Vitaceae**
22. Inflorescences panicles; fruits drupes or capsules; vascular bundle scars 1; pith continuous.
23. Axillary buds subglobose, the exposed scales 6; fruits capsules, orange; seeds covered by bright red arils ______ **Celastraceae**
23. Axillary buds triangular and elongated, the exposed scales 1–3; fruits drupes, bluish-black; seeds not covered by bright red arils ______ **Rhamnaceae**

GROUP B

Plants trees or shrubs; flowers or cones appearing before leaves.

1. Plants producing cones, not producing flowers; trunks often with buttresses; plants producing knees (= erect woody projections from the roots) when in standing water ______ **Cupressaceae**
1. Plants not producing cones, producing flowers; trunks without buttresses; plants without knees.
2. Leaf scars opposite.
3. Corollas yellow, showy, 20–30 mm across; stamens 2 ______ **Oleaceae**
3. Corollas absent or small (but conspicuous white bracts present in one species); petals if present 5.5 mm or less long, creamy white to yellowish green, greenish, or red; stamens 2–12.
4. Bracts 20–50 mm long, white; ovaries inferior ______ **Cornaceae**
4. Bracts 5 mm or less long or absent, purple or green or yellow; ovaries superior.
5. Staminate flowers with 2 or 4 stamens; styles of pistillate flowers 1; immature ovaries not winged; vascular bundle scars 1 or numerous ______ **Oleaceae**
5. Staminate flowers with 5–12 stamens; styles of pistillate flowers 2; immature ovaries 2-winged; vascular bundle scars 3 ______ **Aceraceae**
2. Leaf scars alternate.
6. Inflorescences catkins.
7. Plants dioecious; ovaries superior; axillary bud scales 1 ______ **Salicaceae**
7. Plants monoecious; ovaries inferior; axillary bud scales 2–numerous.
8. Terminal buds present; pith 5-starred in cross-section ______ **Fagaceae**
8. Terminal buds absent; pith 3-sided to round in cross-section ______ **Betulaceae**
6. Inflorescences of various types, but not catkins.
9. Perianth parts in 1 series.
10. Inflorescences solitary flowers or clusters of 2–3 flowers; pistils subtended by spiny or muricateor involucral cupules; ovaries inferior; terminal buds present ______ **Fagaceae**
10. Inflorescences umbels or fascicles or dense clusters of flowers; pistils not subtended by spiny or muricate or involucral cupules; ovaries superior; terminal buds absent.
11. Inflorescences umbel-like clusters; stamen number greater than number of perianth parts; branchlets aromatic when fresh ______ **Lauraceae**
11. Inflorescences spherical clusters or fascicles or cymes or racemes; stamen number equal to number of perianth parts; branchlets not aromatic.
12. Sap viscous, white; thorns present; flowers imperfect ______ **Moraceae**

12. Sap thin, colorless; thorns absent; flowers perfect or both perfect and imperfect intermixed ________ **Ulmaceae**

9. Perianth parts in 2 series.
13. Corollas bilaterally symmetrical ________ (Caesalpinioideae) **Fabaceae**
13. Corollas radially symmetrical.
14. Flowers 3- or 4-merous; fruits berries or drupes or capsules.
15. Flowers 3-merous, solitary; petals dull purple; fruits large berries to 12 cm long ________ **Annonaceae**
15. Flowers 4-merous, solitary OR in axillary clusters; petals yellow to reddish yellow, yellowish green, pink, or purplish pink; fruits capsules or drupes.
16. Petals 0.5–1.3 mm long, yellowish green; fruits drupes ________ **Rhamnaceae**
16. Petals much > 1.3 mm long, yellow to reddish yellow, pink, or purplish pink; fruits capsules.
17. Leaves simple; petals linear, yellow to reddish yellow; stamens 4 ________ **Hamamelidaceae**
17. Leaves compound; petals obovate, pink to purplish pink; stamens 7–10 ________ **Sapindaceae**
14. Flowers 5-merous; fruits pomes or drupes or legumes or capsules.
18. Petals pink to purplish pink, obovate, with a pilose claw; stamens 7–10, unequal, conspicuously exserted beyond perianth; fruit a 3-lobed, somewhat woody capsule ________ **Sapindaceae**
18. Petals variously colored, not obovate with a pilose claw; stamens variable in number, exserted or included within perianth; fruit a legume, pome, or drupe.
19. Petals fused; stamens conspicuously exserted beyond perianth, radiating; fruits legumes ________ (Mimosoideae) **Fabaceae**
19. Petals free; stamens included within perianth, not radiating; fruits pomes or drupes.
20. Petals white or pink, 10–25 mm long ________ **Rosaceae**
20. Petals yellow-green, 1–2 mm long ________ **Anacardiaceae**

GROUP C

Plants trees or shrubs; leaves opposite or whorled or fascicled or in rosettes.

1. Leaves fascicled or borne in rosettes at ends of stems.
2. Leaves borne in rosettes at end of stems, the venation parallel.
3. Leaves flabellate (= fan-shaped), longitudinally pleated toward base, 100–150 cm wide; perianth parts 3–10 mm long; fruits drupes, spherical, 8–13 mm in diam. ________ **Arecaceae**
3. Leaves lanceolate or ensiform, 0.5–2.5 cm wide, not pleated; perianth parts 30–50 mm long; fruits capsules, oblong, 25–40 mm in diam. ________ **Agavaceae**
2. Leaves fascicled, the venation pinnate or palmate or not apparent.
4. Leaves even pinnately compound with 8–16 leaflets ________ **Zygophyllaceae**
4. Leaves simple.
5. Leaves needle-like; cut surfaces of stems or leaves exuding sticky resin; flowers absent; cones present, woody ________ **Pinaceae**
5. Leaves terete or flat, spatulate or ovate or cordate-orbicular; cut surfaces of stems or leaves without resin; flowers present; cones absent.
6. Leaves terete, spatulate to ovate, the margins entire, the venation pinnate; petals fused; ovaries superior, with locules 2; berries subtended by persisent calyces ________ **Solanaceae**
6. Leaves flat, cordate-orbicular, the margins serrate or crenate, the venation palmate; petals free; ovaries inferior, with locules 1; berries crowned by shriveled hypanthia ________ **Grossulariaceae**

1. Leaves opposite or whorled.
   7. Leaves scale- or awl-like or reduced to membranous sheaths fused at bases and surrounding stems; flowers absent; cones present.
      8. Trees or shrubs more than 3 m tall; leaves imbricate, scale- or awl-like; ovulate cones globose, fleshy, blue to bluish black, bluish purple, reddish, or copper-colored ______ **Cupressaceae**
      8. Shrubs less than 1 m tall OR plant with clambering, vine-like habit; leaves not imbricate, reduced to membranous sheaths fused at bases and surrounding stems; ovulate cones elliptic, the scales thin, stramineous, the inner becoming fleshy and red ______ **Ephedraceae**
   7. Leaves elongated, terete or flattened, neither scale- nor awl-like nor reduced to sheaths; flowers present; cones absent.
      9. Leaves terete, fleshy, 0.1–0.2 cm wide, 1–2 cm long; plants 20–40 cm tall; [*Pseudoclappia* in OK and w TX, not in nc TX] ______ (*Pseudoclappia*) **Asteraceae**
      9. Leaves flat, not fleshy, more than 1 cm wide, more than 2 cm long; plants more than 40 cm tall.
         10. Leaves compound.
            11. Leaflets 3.
               12. Shrubs 5 m or less tall; twigs with longitudinal stripes; pistils 3-carpellate; fruits capsules, inflated; [Family in OK, not in nc TX] ______ **Staphyleaceae**
               12. Trees to 20 m tall; twigs without stripes; pistils 2-carpellate; fruits samaras ______ **Aceraceae**
            11. Leaflets 5–16.
               13. Leaves pinnately compound.
                  14. Leaves evergreen, even pinnately compound, the leaflets 15 mm or less long; fruits flattened, heart-shaped capsules with an apiculate apex ______ **Zygophyllaceae**
                  14. Leaves deciduous, odd pinnately compound, the leaflets usually much > 15 mm long; fruits drupes OR samaras without an apiculate apex.
                     15. Twigs thick but weak, the pith 1/2–3/4 of twigs in cross-section; ovaries inferior; fruits drupes (berry-like) ______ **Caprifoliaceae**
                     15. Twigs slender and strong, the pith < 1/4 of twigs in cross-section; ovaries superior; fruits samaras.
                        16. Axillary buds solitary; leaflet margins coarsely toothed; stamens 3–12; samaras 2-seeded ______ **Aceraceae**
                        16. Axillary buds superposed, the lower small; leaflet margins entire or shallowly toothed; stamens 2; samaras 1-seeded ______ **Oleaceae**
               13. Leaves palmately compound.
                  17. Leaflets sometimes peppery aromatic, lanceolate to elliptic, entire to conspicuously toothed to deeply palmately divided; petals fused; fruits drupes, 3 mm in diam. ______ **Verbenaceae**
                  17. Leaflets without odor, oblanceolate to obovate, toothed; petals free; fruits capsules, 1- or 3-seeded, 30–50 mm in diam. ______ **Hippocastanaceae**
         10. Leaves simple (but blades may be dissected).
            18. Leaf margins palmatifid, the lobes 3–5-parted 1/2–2/3 to midribs; fruits samaras ______ **Aceraceae**
            18. Leaf margins dentate or serrate or entire; fruits capsules or berries or drupes or schizocarps or multiple syncarps of achenes covered by fleshy calyces.
               19. Petals absent.
                  20. Sap viscous, white; fruits multiple syncarps of achenes covered by fleshy calyces ______ **Moraceae**
                  20. Sap thin, colorless; fruits drupes; calyces absent or minute, not fleshy.
                     21. Flowers in pendulous, catkin-like, fascicled racemes; leaves evergreen; lower surface of leaf blades densely pubescent; ovaries inferior ______ **Garryaceae**

21. Flowers in lateral fascicles or axillary glomerules; leaves deciduous; lower surface of leaf blades glabrous or pubescent; ovaries superior ____ **Oleaceae**

19. Petals present.

22. Petals free.

23. Leaves with minute translucent dots when held-up to light; flowers bright yellow; stamens in 3 bundles ____ **Clusiaceae**

23. Leaves without translucent dots; flowers of various colors, may be pale yellowish white; stamens in whorls.

24. Leaves conspicuously pinnately veined, the veins strikingly parallel and when viewed on lower leaf surface with an alternating pattern of light and dark areas; fruits small, globose, black drupes ____ **Rhamnaceae**

24. Leaves without either strikingly parallel veins or an alternating pattern of light and dark areas; fruits capsules OR small red or white drupes.

25. Leaf margins evenly, finely serrate; ovaries superior; seeds with bright red arils ____ **Celastraceae**

25. Leaf margins irregularly serrate or entire; ovaries inferior; seeds without arils.

26. Axillary buds with scales; leaf margins toothed; fruits capsules ____ **Hydrangeaceae**

26. Axillary buds without scales; leaf margins entire; fruits drupes ____ **Cornaceae**

22. Petals fused.

27. Corollas radially symmetrical.

28. Stipules or stipular scars present; inflorescences heads; fruits dry, schizocarps, separating into 2 one-seeded segments ____ **Rubiaceae**

28. Stipules absent; inflorescences cymes or panicles; fruits fleshy, drupes or berries, not separating into 2 one-seeded segments.

29. Stamens 5; ovaries inferior ____ **Caprifoliaceae**

29. Stamens 2 or 4; ovaries superior.

30. Branchlets and leaves stellate-scurfy; inflorescences cymes, axillary, forming verticels; stamens 4 ____ **Verbenaceae**

30. Branchlets and leaves glabrous or variously indumented, but not stellate-scurfy; inflorescences panicles, terminal, not forming verticels; stamens 2 ____ **Oleaceae**

27. Corollas bilaterally symmetrical.

31. Ovaries inferior ____ **Caprifoliaceae**

31. Ovaries superior.

32. Corollas reddish, 25–40 mm long.

33. Stems not square; calyces ca. 5 mm long; corollas 3–4 cm long; fruits 2-seeded capsules; leaves not gland-dotted ____ (*Anisacanthus*) **Acanthaceae**

33. Stems square; calyces 10–15 mm long; corollas 2.5–3 cm long; fruits of 4 one-seeded nutlets; leaves gland-dotted (use lens) ____ (*Salvia*) **Lamiaceae**

32. Corollas not as above, either not reddish OR if reddish then much smaller (4–7 mm long).

34. Plants much-branched shrubs; leaf blades 3–27 mm long ____ (*Aloysia*) **Verbenaceae**

34. Plants shrubs or trees; leaf blades much greater than 27 mm long.
    35. Branchlets and abaxial leaf surfaces densely stellate-scurfy; inflorescences spikes or cymes, axillary, many-flowered; flowers small, 4–7 mm long; fruits drupes ____ **Verbenaceae**
    35. Branchlets and abaxial leaf surfaces not stellate-scurfy; inflorescences panicles, terminal; flowers large, 20–70 mm long; fruits capsules ____ **Bignoniaceae**

Group D

Plants trees or shrubs; leaves alternate, compound.

1. Leaves simple, linear, borne on deciduous branchlets [falsely appearing pinnately compound]; flowers absent; seeds borne in fleshy cones; trunks often with buttresses; plants producing knees (= erect woody projections from the roots) when in standing water ____ (Cupressaceae) **Group E**
1. Leaves compound, of various shapes, deciduous at petioles, not borne on deciduous branchlets; flowers present; seeds borne in fruits; trunks without buttresses; plants without knees.
   2. Leaves 2–3-compound.
      3. Leaves 2–3-compound, evergreen; fruits red berries 6–9 mm in diam.; flowers 3-merous; stamens 6 ____ **Berberidaceae**
      3. Leaves 2-compound, deciduous; fruits legumes, not red; flowers 5-merous; stamens 5–many.
         4. Inflorescences dense heads or spikes; stamen filaments 2–4 times longer than sepals and petals; flowers small and individually inconspicuous, the corollas so small as to be ± inevident; corollas radially symmetrical; stamens 5–many ____ (Mimosoideae) **Fabaceae**
         4. Inflorescences racemes or panicles; stamen filaments as long as or shorter than sepals and petals (except longer in *Caesalpinia*); flowers whether small or large usually individually conspicuous, the corollas usually easily seen; corollas weakly bilaterally symmetrical; stamens 10 or less ____ (Caesalpinioideae) **Fabaceae**
   2. Leaves 1-compound or both 1- and 2-compound.
      5. Fruits nuts, enclosed in involucral husks; flowers imperfect, the plants monoecious; staminate flowers borne in elongated catkins; pistillate flowers solitary or borne in clusters of 2–3 ____ **Juglandaceae**
      5. Fruits of various types but not nuts enclosed in involucral husks; flowers perfect; inflorescences of various types, but not catkins.
         6. Inflorescences racemes or globose spikes; fruits legumes or red berries.
            7. Petals 6, equal; stamens 6; fruits red berries 8–10 mm in diam.; leaves trifoliate, the leaflets with spiny lobe-like teeth ____ **Berberidaceae**
            7. Petals 5, unequal; stamens 5–10; fruits legumes; leaves not as above.
               8. Flowers strongly bilaterally symmetrical; corollas papilionaceous, the upper (= adaxial) petal enclosing other petals in bud ____ (Papilionoideae) **Fabaceae**
               8. Flowers weakly bilaterally symmetrical; corollas not papilionaceous, the upper (= adaxial) petal enclosed by other petals in bud ____ (Caesalpinioideae) **Fabaceae**
         6. Inflorescences corymbs or panicles or fascicles or solitary flowers; fruits achenes or drupes or drupelets or follicles or berries or samara-like schizocarps or capsules or hesperidia.
            9. Pistils 2 or more per flower; fruits achenes or druplets or follicles.
               10. Stipules absent; abaxial surfaces of leaves glandular punctate; fruits follicles ____ **Rutaceae**
               10. Stipules present; abaxial surfaces of leaves not glandular punctate; fruits achenes or druplets ____ **Rosaceae**

9. Pistils 1 per flower; fruits drupes or berries or samaras or samara-like schizocarps or capsules or hesperidia.
   11. Leaves 2- or 3-compound.
      12. Plants shrubs, unarmed; stipules or stipular scars present ________ **Vitaceae**
      12. Plants trees, unarmed OR armed; stipules absent.
         13. Stems and leaves armed with stout prickles; petals 1.5–2 mm long; ovaries inferior; fruits 5-seeded, black, 4–6 mm in diam.; trees sparingly branched ________ **Araliaceae**
         13. Stems and leaves not armed with prickles; petals 9–11 mm long; ovaries superior; fruits 1-seeded, yellow, 12–15 mm in diam.; trees many branched ________ **Meliaceae**
   11. Leaves 1-compound.
      14. Leaflets 3.
         15. Leaflets not gland-dotted; ovaries 1-locular; fruits drupes, red or reddish brown or white to yellowish gray (poisonous species with white to yellowish gray fruits), 5–8 mm in diam. ________ **Anacardiaceae**
         15. Leaflets gland-dotted; ovaries 2–5-locular; fruits samaras OR hesperidia, yellow-brown, ca. 20–50 mm in diam. ________ **Rutaceae**
      14. Leaflets 4–25.
         16. Leaflets gland-dotted or bearing 1–5 dark green glands near bases on lower surfaces; fruits follicles or samara-like schizocarps or samaras.
            17. Branchlets armed with stout prickles; fruits follicles, 5–6 mm long; pith white, occupying less than 1/2 of stem in cross-section; vascular bundle scars 3 ________ **Rutaceae**
            17. Branchlets not armed with prickles; fruits schizocarps, splitting into samaras, 30–50 mm long; pith brown, occupying about 3/4 of stem in cross-section; vascular bundle scars 9 ________ **Simaroubaceae**
         16. Leaflets neither gland-dotted nor bearing 1–5 dark green glands near bases on lower surfaces; fruits drupes or capsules.
            18. Flowers pink to purplish pink; fruits 3-lobed, somewhat woody, stipitate capsules ________ **Sapindaceae**
            18. Flowers white to yellowish or greenish; fruits drupes.
               19. Drupes red or reddish brown, opaque at maturity, 5–8 mm in diam.; sap viscous, white or brown; plants typically thicket-forming shrubs ________ **Anacardiaceae**
               19. Drupes amber or yellow, translucent at maturity, 10–13 mm in diam.; sap thin, colorless; plants typically trees, occasionally forming thickets ________ **Sapindaceae**

GROUP E

Plants trees or shrubs; leaves alternate, simple, the margins entire.

1. Venation parallel.
   2. Stems jointed; branches fascicled at nodes; internodes hollow; leaves with sheaths; flowers borne in spikelets ________ (*Arundinaria*) **Poaceae**
   2. Stems not jointed; branches absent; internodes solid; leaves without sheaths; flowers borne in panicles.
      3. Leaves flabellate (= fan-shaped), longitudinally pleated toward base, 100–150 cm wide; perianth parts 3–10 mm long; fruits drupes, spherical, 8–13 mm in diam. [Leaves large, divided into segments, but the segments mostly entire] ________ **Arecaceae**

3. Leaves lanceolate or ensiform, not pleated, 0.5–8 cm wide; perianth parts 30 mm or more long; fruits capsules, oblong, 25 mm or more in diam. ____ **Agavaceae**

1. Venation pinnate or palmate or not apparent.
   4. Plants subshrubs or shrubs, less than 2 m tall.
      5. Flowers imperfect, the plants monoecious or dioecious.
         6. Inflorescences heads or catkins.
            7. Inflorescences heads; pappus present, of capillary bristles; fruits achenes ____ **Asteraceae**
            7. Inflorescences catkins; pappus absent; fruits drupes or capsules.
               8. Leaves evergreen, resin-dots present, fragrant; fruits drupes, white, waxy; seeds not comose ____ **Myricaceae**
               8. Leaves deciduous, resin-dots absent, not fragrant; fruits capsules; seeds comose ____ **Salicaceae**
         6. Inflorescences racemes or cymes or solitary flowers in leaf axils.
            9. Stipules present, 1–2 mm long (sometimes falling early); pistils 3-lobed, 3-locular, with 3 or more ovules; fruits capsules ____ **Euphorbiaceae**
            9. Stipules absent; pistils not lobed, 1-locular, with 1 ovule; fruits utricles or drupes.
               10. Leaf surfaces scurfy or farinaceous; fruits utricles; bark not spicy aromatic; plants of saline or alkaline sites ____ **Chenopodiaceae**
               10. Leaf surfaces neither scurfy nor farinaceous; fruits drupes; bark spicy aromatic; plants of moist sites ____ **Lauraceae**
      5. Flowers perfect.
         11. Leaves less than 3 mm long, imbricate; branchlets deciduous ____ **Tamaricaceae**
         11. Leaves more than 5 mm long, not imbricate; branchlets not deciduous.
            12. Inflorescences heads, 100–300 per plant, in paniculate arrangement; anthers fused in ring around style ____ **Asteraceae**
            12. Inflorescences of various types, but not heads; anthers not fused in ring around style.
               13. Leaves and stems with silvery peltate scales.
                  14. Plants usually spiny; fruits drupe-like; flowers usually 1–3 in the leaf axils; stamens 4 ____ **Elaeagnaceae**
                  14. Plants not spiny; fruits capsules; flowers in terminal 6–14-flowered racemes; stamens (11–)14–18(–21) ____ **Euphorbiaceae**
               13. Leaves and stems without silvery peltate scales.
                  15. Flowers 5–6 mm across, yellow-green, appearing glomerate on short, twig-like, condensed spur shoots (some leaves also crowded with flowers on spur shoots) ____ **Rhamnaceae**
                  15. Flowers variously colored, not arranged as above.
                     16. Plants armed (branches ending in stout thorns); petals absent ____ **Rhamnaceae**
                     16. Plants unarmed OR if armed, not from the ends of branches; petals present or absent.
                        17. Leaves broadly obovate or broadly elliptic; leaf scars annular, nearly encircling bud; fruits drupes, red; [Family in OK, not in TX] ____ **Thymeliaceae**
                        17. Leaves of various shapes, but neither broadly obovate nor elliptic; leaf scars not annular; fruits berries or capsules or achenes or follicles, of various colors.
                           18. Petals fused; fruits berries or capsules.
                              19. Branchlets armed; axillary buds multiple; anthers opening along longitudinal sutures; pistils 2-carpellate ____ **Solanaceae**
                              19. Branchlets not armed; axillary buds solitary; anthers opening by apical pores; pistils 5-carpellate ____ **Ericaceae**

18. Petals free or absent; fruits achenes or follicles.
    20. Stipules present as ocrea; perianth parts in 1 series; fruits achenes _______________ **Polygonaceae**
    20. Stipules absent; perianth parts in 2 series; fruits follicles; [Family in OK and w TX, not in nc TX] _______________ **Crossosomataceae**

4. Plants large shrubs or trees, more than 2 m tall.
   21. Trunks typically with buttresses; plants producing knees (= erect woody projections from the roots) when in standing water; branchlets deciduous and bearing linear leaves; flowers absent; seeds borne in fleshy cones _______________ **Cupressaceae**
   21. Trunks without buttresses; plants without knees; branchlets not deciduous and bearing linear leaves; flowers present; seeds borne in fruits.
       22. Plants armed and/or with spur branches.
           23. Sap viscous, white; flowers imperfect, the plants dioecious; fruits multiple syncarps of achenes covered by fleshy calyces, 10–15 cm in diam., globose, yellow-green _____ **Moraceae**
           23. Sap thin; colorless; flowers perfect; fruits berries or achenes or drupe-like or pomes, less than 5 cm in diam., of various colors and textures.
               24. Fruits small red pomes 6–8 mm in diam.; inflorescences small corymbs; stamens 20 _______________ **Rosaceae**
               24. Fruits berries, drupe-like, or achenes, not red; inflorescences various; stamens 4, 5, 12, or more.
                   25. Perianth parts in 2 series; stamens 5; fruits berries, black, drupe-like, 1-seeded _______________ **Sapotaceae**
                   25. Perianth parts in 1 series; stamens 4 or 12 or more; fruits achenes, plumose or enclosed by fleshy perianths.
                       26. Stems and leaves with silvery peltate scales; spines present; spur branches absent; stamens 4; achenes enclosed by fleshy perianths __ **Elaeagnaceae**
                       26. Stems and leaves without silvery peltate scales; spines absent; spur branches present; stamens 12 or more; achenes with plumose tails _____ **Rosaceae**
       22. Plants not armed; spur branches absent.
           27. Leaves less than 3 mm long, imbricate; branchlets deciduous _______________ **Tamaricaceae**
           27. Leaves more than 10 mm long, not imbricate; branchlets not deciduous.
               28. Flowers solitary.
                   29. Flowers small, 0.2–0.3 cm in diam., imperfect; stamens 5–12, arrangement whorled; pistils 1 per flower; fruits drupes.
                       30. Flowers imperfect; bark becoming warty; leaf blades 3-veined at base; ovaries superior _______________ **Ulmaceae**
                       30. Flowers perfect; bark not becoming warty; leaf blades 1-veined at base; ovaries inferior _______________ **Nyssaceae**
                   29. Flowers large, 3–25 cm in diam.; perfect; stamens 13 or more, arrangement spiraled; pistils 3 or more per flower; fruits follicles or berries.
                       31. Flowers yellow or white, 10–25 cm in diam.; leaves coriaceous; stipules present, but caducous; fruits follicles; [Family in OK and se and e TX, not in nc TX] _______________ **Magnoliaceae**
                       31. Flowers dull purple, 3–4 cm in diam.; leaves not coriaceous; stipules absent; fruits berries _______________ **Annonaceae**
               28. Flowers borne in clusters.
                   32. Stems and leaves with silvery peltate scales _______________ **Elaeagnaceae**
                   32. Stems and leaves without silvery peltate scales.
                       33. Leaves evergreen.

34. Flowers in panicles; corollas conspicuous, white, ca. 7 mm long; fruits red to yellowish red, berries ________ **Ericaceae**
34. Flowers solitary, in clusters of 2–3, or in catkins; corollas absent; fruits white drupes or brown or green nuts subtended by an involucral cupule (acorn).
    35. Terminal buds absent; leaves with resin-dots, fragrant; pistillate flowers in catkins; fruits drupes, white ________ **Myricaceae**
    35. Terminal buds multiple; leaves without resin-dots, not fragrant; pistillate flowers solitary or in clusters of 2–3; fruits nuts subtended by an involucral cupule (acorn), brown or green ________ **Fagaceae**

33. Leaves deciduous.
    36. Flowers imperfect, the plants monoecious or dioecious.
        37. Inflorescences catkins.
            38. Terminal buds multiple; pith 5-starred in cross-section; plants monoecious; fruits nuts, solitary or in clusters of 2–3, subtended by an involucral cupule (acorn) ________ **Fagaceae**
            38. Terminal buds absent; pith terete in cross-section; plants dioecious; fruits capsules or multiple syncarps of achenes covered by fleshy calyces.
                39. Leaves ovate or lanceolate; sap viscous, white; fruits multiple syncarps of achenes covered by fleshy calyces; seeds not comose ________ **Moraceae**
                39. Leaves obovate or oblanceolate; sap thin, colorless; fruits capsules; seeds comose ________ **Salicaceae**
        37. Inflorescences of various types, but not catkins.
            40. Terminal buds multiple; pith 5-starred in cross-section; plants monoecious; fruits nuts, solitary or in clusters of 2–3, subtended by an involucral cupule (acorn) ________ **Fagaceae**
            40. Terminal buds solitary or absent; pith terete in cross-section; plants dioecious, monoecious, or polygamo-monoecious; fruits berries or drupes or capsules.
                41. Fruits capsules; plants with milky sap ________ **Euphorbiaceae**
                41. Fruits berries or drupes; plants without milky sap.
                    42. Leaf blades 3-veined at base; bark becoming warty; branchlets slender, the growth zigzagged ________ **Ulmaceae**
                    42. Leaf blades 1-veined at base; bark not becoming warty; branchlets stout, the growth not zigzagged.
                        43. Leaf scars with 1 vascular bundle scar; petals fused; fruits berries, 2–5 cm in diam., yellowish orange or black ________ **Ebenaceae**
                        43. Leaf scars with 3 vascular bundle scars; petals absent or free; fruits drupes, 0.5–1 cm in diam., red or blue-black.
                            44. Accessory buds present; young twigs aromatic; perianth parts yellow or yellow-white; ovaries superior ________ **Lauraceae**
                            44. Accessory buds absent; young twigs not aromatic; perianth parts greenish; ovaries inferior ________ **Nyssaceae**

36. Flowers perfect.
    45. Flowers bilaterally symmetrical.
        46. Stamens 36–42; petals separate, conspicuously slender clawed with orbicular-cordate blades, large and conspicuous, to 20 mm long, white to pink or purple _______ **Lythraceae**
        46. Stamens 2–10; petals not as above (not slender clawed, sometimes small and inconspicuous, sometimes fused, sometimes yellow).
            47. Corollas papilionaceous; petals free; stamens 10; pistils 1-carpellate; fruits legumes, flattened; seeds not winged _______ (Caesalpinioideae) **Fabaceae**
            47. Corollas campanulate or funnelform; petals fused; stamens 2 or 4 or 5; pistils 2-carpellate; fruits capsules; seeds winged _______ **Bignoniaceae**
    45. Flowers radially symmetrical.
        48. Leaf blades 3-veined at base; bark typically becoming warty; branchlets slender, the growth zigzagged _______ **Ulmaceae**
        48. Leaf blades 1-veined at base; bark not becoming warty; branchlets stout, the growth not zigzagged.
            49. Leaf scars with 1 vascular bundle scar; flowers borne in dense, sessile clusters along sides of branches; petals fused; [Family in OK and se and e TX, not in nc TX] ___ **Symplocaceae**
            49. Leaf scars with 3 vascular bundle scars; flowers borne in peduncled cymes or panicles; petals free or absent.
                50. Branchlets aromatic; wood yellow; inflorescences panicles, terminal; petals persistent; ovaries superior _______ **Anacardiaceae**
                50. Branchlets not aromatic; wood white; inflorescences cymes, axillary; petals caducous; ovaries inferior _______ **Nyssaceae**

GROUP F

Plants trees or shrubs; leaves alternate, simple, the margins lobed or toothed.

1. Venation palmate.
    2. Leaf blades peltate [*Ricinus*, a large herb, can appear ± like a small tree] _______ **Euphorbiaceae**
    2. Leaf blades not peltate.
        3. Flowers perfect; petals present; fruits berries or capsules or follicles or nut-like or drupe-like.
            4. Leaves flabellate (= fan-shaped), longitudinally pleated toward base, 100–150 cm wide; plants palm-like _______ **Arecaceae**
            4. Leaves neither flabellate nor pleated, much < than 100 cm wide; plants not palm-like.
                5. Stipules absent; stamens 5; ovaries inferior _______ **Grossulariaceae**
                5. Stipules present; stamens 10–50; ovaries superior.
                    6. Filaments fused, forming a tube around the style _______ **Malvaceae**
                    6. Filaments separate.
                        7. Plants trees; peduncles arising from midribs of strap-shaped bracts; pistils 1; fruits nut-like or drupe-like _______ **Tiliaceae**
                        7. Plants shrubs; peduncles not arising from midribs of strap-shaped bracts; pistils 2 or 3; fruits follicles _______ **Rosaceae**
        3. Flowers imperfect; petals absent; fruits syncarps composed of numerous achenes or cap-

sules OR fruits hidden from view inside a fleshy receptacle OR fruits capsules, the carpels separating into 5 stalked follicle-like structures.

8. Leaves noticeably thick, obtuse apically; flowers and fruits hidden from view inside a fleshy hollow receptacle; terminal vegetative bud surrounded by a pair of stipules _____ **Moraceae**
8. Leaves not noticeably thick, acute to acuminate apically (rarely subobtuse); flowers and fruits not hidden inside a receptacle; terminal bud scaly, not surrounded by a pair of stipules.
    9. Flowers in panicles; leaves 3–5 lobed, the lobes entire; fruits capsules, the carpels separating into 5 stalked follicle-like structures which spread open and become leaf-like and bear seeds on their margins _____ **Sterculiaceae**
    9. Flowers in heads or catkins or catkin-like structures; leaves either unlobed or lobed, but if lobed then the lobes with teeth; fruits syncarps, composed of numerous capsules or achenes (these sometimes covered by fleshy perianths and the whole structure berry-like).
        10. Sap viscous, white; plants dioecious; staminate catkins pendulous; fruits covered by fleshy calyces _____ **Moraceae**
        10. Sap thin, colorless; plants monoecious; staminate catkins erect or ascending; fruits not covered by fleshy calyces.
            11. Bark light, gray-green, exfoliating in strips; axillary buds enclosed by petioles; older branches not winged; fruits syncarps of achenes, not spiny _____ **Platanaceae**
            11. Bark dark, brown, not exfoliating in strips; axillary buds not enclosed by petioles; older branches winged; fruits syncarps of capsules, spiny _____ **Hamamelidaceae**

1. Venation pinnate.
    12. Flowers imperfect.
        13. Plants dioecious or polygamo-dioecious.
            14. Inflorescences of various types, but not catkins.
                15. Perianth parts in 1 series _____ **Moraceae**
                15. Perianth parts in 2 series (counting pappus in Asteraceae).
                    16. Inflorescences heads; pappus present, of capillary bristles; fruits achenes _____ **Asteraceae**
                    16. Inflorescences of various types, but not heads; pappus absent; fruits drupes or berry-like.
                        17. Bark spicy-aromatic; inflorescences racemes or umbels _____ **Lauraceae**
                        17. Bark not spicy-aromatic; inflorescences cymes or fascicles or solitary flowers.
                            18. Stipules absent; ovaries inferior; locules 1; fruits 1-seeded _____ **Nyssaceae**
                            18. Stipules or stipular scars present; ovaries superior; locules 2–8; fruits 2–5-seeded.
                                19. Petals clawed, cucullate; stamens opposite the petals; drupes black _____ **Rhamnaceae**
                                19. Petals neither clawed nor cucullate; stamens alternate with the petals; drupes red to orange _____ **Aquifoliaceae**
            14. Inflorescences catkins.
                20. Perianth parts in 1 series; fruits multiple syncarps of achenes _____ **Moraceae**
                20. Perianth parts absent or vestigial; fruits capsules or drupes.
                    21. Leaves oblanceolate, resin-dots present, aromatic; fruits drupes; seeds not comose _____ **Myricaceae**
                    21. Leaves linear to deltoid, resin-dots absent, not aromatic; fruits capsules; seeds comose _____ **Salicaceae**
        13. Plants monoecious or polygamo-monoecious.

22. Perianth parts in 2 series; staminate flowers (1–)2–3 per leaf axil; ovaries superior; fruits drupes ____ **Rhamnaceae**
22. Perianth parts in 1 series or absent; staminate flowers borne in fascicles at bases of branchlets or in pendulous catkins; ovaries superior or inferior; fruits drupes OR nuts subtended by bracts or cap-like involucral cupule.
    23. Leaf bases oblique; plants polygamo-monoecious; staminate flowers borne in fascicles at bases of branchlets; perfect flowers present, borne in axils of leaves; ovaries superior ____ **Ulmaceae**
    23. Leaf bases not oblique; plants monoecious; staminate flowers borne in pendulous catkins; perfect flowers absent; ovaries inferior.
        24. Pistillate flowers in catkins; nuts individually subtended by woody or foliaceous bracts, but not by cupule ____ **Betulaceae**
        24. Pistillate flowers solitary or in clusters of 2–3, but not in catkins; nuts (individually or in clusters of 3) subtended by a spiny or muricate or cap-like involucral cupule; bracts neither woody nor foliaceous ____ **Fagaceae**

12. Flowers perfect.
    25. Ovaries inferior.
        26. Petals fused at least at base.
            27. Leaf blades 0.1–3.5 cm wide; inflorescences heads; fruits achenes ____ **Asteraceae**
            27. Leaf blades 4–10 cm wide; inflorescences clusters of flowers; fruits capsules ____ **Styracaceae**
        26. Petals free or absent.
            28. Plants shrubs; sepals 4; petals 4, yellow ____ **Hamamelidaceae**
            28. Plants trees; sepals 5; petals 5 or 0, white or greenish white.
                29. Leaf margins entire or with 1 or 2 coarse teeth; stipules absent; stamens 5–12; fruits drupes with thin mesocarp and ridged or winged endocarp ____ **Nyssaceae**
                29. Leaf margins serrate or crenate or irregularly lobed; stipules or stipular scars present; stamens 15 or more; fruits pomes or drupes with thick fleshy mesocarp and smooth non-winged endocarp ____ **Rosaceae**
    25. Ovaries superior.
        30. Perianth parts in 1 series, in 1 or 2 whorls.
            31. Leaf margins pinnately lobed or pinnately toothed; leaf bases oblique; bark not spicy-aromatic; nectaries absent ____ **Ulmaceae**
            31. Leaf margins palmately lobed; leaf bases cuneate; bark spicy-aromatic; nectaries present [sometimes resembling stamens hence flowers falsely appearing perfect] ____ **Lauraceae**
        30. Perianth parts in 2 series.
            32. Petals fused.
                33. Petals fused more than half of their length; anthers opening by apical pores; styles present, long; stigmas not subsessile ____ **Ericaceae**
                33. Petals fused only at base; anthers opening along longitudinal sutures; styles absent or short; stigmas subsessile ____ **Aquifoliaceae**
            32. Petals free.
                34. Stamens 15 or more; fruits pomes or follicles or drupes ____ **Rosaceae**
                34. Stamens 4–6; fruits drupes or capsules.
                    35. Branches terminating in straight spiny tips OR with axillary spines ____ **Rhamnaceae**
                    35. Branches unarmed (but leaves can be spiny in some species).
                        36. Plants ± herbaceous shrubs; petals pink or violet, with yellowish base; fruits capsules ____ **Sterculiaceae**
                        36. Plants shrubs or small trees; petals white, yellowish, greenish, or rarely

pinkish; fruits usually drupes (capsules in 1 species in se and e TX and OK).

37. Petals clawed, cucullate; stamens opposite petals; nectary disks present ______ **Rhamnaceae**
37. Petals not clawed, not cucullate; stamens alternate with petals; nectary disks absent.
 38. Inflorescences racemes, terminal; fruits capsules ______ **Grossulariaceae**
 38. Inflorescences solitary flowers or cymose clusters, axillary; fruits drupes with 4 or 5 stones [falsely resembling berries] ______ **Aquifoliaceae**

GROUP G

Plants aquatic herbs, floating on or submersed in or emergent from water.

1. Plants free-floating in water column or on surface of water.
 2. Plants floating on surface.
  3. Leaves 4–15 cm long.
   4. Plants with gray-green, velvety-hairy leaves in rosettes and conspicuously feathery roots ______ **Araceae**
   4. Plants not as above.
    5. Petioles or stems not inflated; abaxial surfaces of blades spongy; flowers imperfect; ovaries inferior; fruits berries ______ **Hydrocharitaceae**
    5. Petioles or stems inflated; abaxial surfaces of blades not spongy; flowers perfect; ovaries superior; fruits capsules.
     6. Leaves simple; leaf blades suborbicular to broadly elliptic; flowers 4–6 cm long; stamens 3 ______ **Pontederiaceae**
     6. Leaves pinnately compound; blades of leaflets filiform; flowers 0.4–0.5 cm long; stamens 5 ______ **Primulaceae**
  3. Leaves 1.5 cm or less long or absent.
   7. Stems 3–7 cm long, inflated, radiating and forming conspicuous floating whorls at surfaces, bearing finely dissected branches with numerous sac-like bladders; flowers conspicuous, borne on scapes 10–15 cm long above water surface; corollas yellow, bilaterally symmetrical ______ **Lentibulariaceae**
   7. Stems less than 0.5 cm long or absent; dissected branches absent; bladders absent; flowers inconspicuous or not produced; scapes absent; corollas absent.
    8. Plants thalloid (= consisting of a flat or solid body a few mm or less across, not differentiated into stems and leaves); flowers and fruits present (but minute and inconspicuous); spores produced in anthers and ovaries ______ **Lemnaceae**
    8. Plants differentiated into stems and fronds (= leaves); flowers and fruits absent; spores produced in soft, thin-walled sporocarps.
     9. Fronds less than 1 mm long, imbricate, dull reddish green, the adaxial surface glabrous ______ **Azollaceae**
     9. Fronds 5–15 mm long, not imbricate, bright green, the adaxial surface with short, branched, multicellular hairs; [Salviniaceae sensu stricto in OK, not in TX] ______ **Salviniaceae**
 2. Plants floating submersed in water column.
  10. Plants thalloid (= consisting of a flat or solid body, not differentiated into stems and leaves); entire plant small, usually < 2 cm long ______ **Lemnaceae**
  10. Plants not thalloid, with stems and often leaves; entire plant much larger than 2 cm long.
   11. Plants without leafy stems.
    12. Branches whorled, not dissected, without sac-like bladders, consisting of 1–5 mac-

roscopic cells; joints of stem consisting of single macroscopic cells; flowers and fruits absent; oogonia and antheridia present [This is a macroscopic non-vascular family of algae occasionally collected in ponds and lakes] ________ **Characeae**

12. Branches alternate, finely dissected with numerous sac-like bladders, consisting of many microscopic cells; joints of stem consisting of many microscopic cells; flowers and fruits present; oogonia and antheridia absent; flowers borne on scapes 10–15 cm above water surface; corollas yellow, bilaterally symmetrical ________ **Lentibulariaceae**

11. Plants with leafy stems.

13. Leaves whorled.

14. Leaves simple, elliptic to linear-lanceolate; petals 3 ________ **Hydrocharitaceae**

14. Leaves compound, linear or filiform; petals 0 or 4.

15. Leaves dichotomously 1–4-compound; flowers submersed ________ **Ceratophyllaceae**

15. Leaves pinnately 1-compound; flowers borne at water surface or above ________ **Haloragaceae**

13. Leaves alternate or opposite.

16. Leaves alternate.

17. Stipules present; flowers borne in terminal spikes above water surface; perianth parts present; fruits globose, not beaked, not curved ________ **Potamogetonaceae**

17. Stipules absent; flowers borne in axils of leaves below water surface; perianths parts absent; fruits flattened, beaked, curved ________ **Zannichelliaceae**

16. Leaves opposite.

18. Leaves elliptic to linear-lanceolate, the bases not sheathing stems; flowers borne at water surface or just above; petals 3, white ________ **Hydrocharitaceae**

18. Leaves filiform, the bases sheathing stems; flowers submersed; petals 0.

19. Leaf blades usually minutely denticulate under a scope OR obviously toothed to the naked eye; pistils 1 per flower; fruits terete, not beaked, not curved, not stipitate ________ (*Najas*) **Hydrocharitaceae**

19. Leaf blades entire; pistils 2–8 per flower; fruits flattened, beaked (the beak to 1.5 mm long), curved, short stipitate (= stalked) ________ **Zannichelliaceae**

1. Plants rooted in substrate; stems and leaves submersed in or floating on or emergent from water.

20. Leaves compound or dissected into filiform or linear segments.

21. Plants attached to rocks by fleshy disks and forming mats or crusts on them; [Family in OK, not in TX]. ________ **Podostemaceae**

21. Plants attached to substrate by roots, not forming mats or crusts.

22. Leaves pinnately compound or pinnately dissected.

23. Leaflets ovate or oval; terminal leaflets larger than lateral ones, somewhat fleshy ________ **Brassicaceae**

23. Leaflets or leaf segments linear or filiform; terminal leaflets if present not larger than laterals, not fleshy.

24. Stems and peduncles inflated; stamens 5; fruits capsules [*Hottonia*—in OK and se and e TX, not in nc TX] ________ **Primulaceae**

24. Stems and peduncles not inflated; stamens 4 or 6 or 8; fruits siliques or silicles or nut-like.

25. Leaves all alike, emergent ones dissected ________ **Haloragaceae**

25. Leaves of 2 forms, emergent ones not dissected.

26. Inflorescences racemes, terminal; petals 4; stamens 6; pistils 2-carpellate; fruits siliques or silicles ________ **Brassicaceae**

26. Inflorescences solitary flowers, axillary; petals 0; stamens 3; pistils 3-carpellate; fruits nut-like ________ **Haloragaceae**

22. Leaves palmately compound or palmately dissected or dichotomously compound.

27. Leaflets 4, obdeltoid or flabellate, the venation dichotomous; flowers absent; spores produced in sori borne in hard sporocarps in axils of leaves ____________ **Marsileaceae**
27. Leaflets of various numbers, filiform or linear, the venation comprising a single vein; flowers present; spores produced in anthers and ovaries.
    28. Leaves alternate.
        29. Leaves dichotomously compound; plants attached to rocks by fleshy disks and forming mats or crusts on them; [Family in OK, not in TX]. ____________ **Podostemaceae**
        29. Leaves palmately compound or dissected; plants attached to substrate by roots, not forming mats or crusts on rocks ____________ **Ranunculaceae**
    28. Leaves opposite or whorled.
        30. Leaves dichotomously 1–4-compound; flowers submersed, inconspicuous; plants may be embedded in substrate, but without roots ____ **Ceratophyllaceae**
        30. Leaves palmately 1-compound; flowers borne at water surface, showy; plants rooted in substrate ____________ **Cabombaceae**

20. Leaves simple, not dissected into filiform or linear segments.
    31. Plants submersed or floating.
        32. Leaves floating on surface.
            33. Leaves orbicular, peltate or cordate, arising from rhizomes.
                34. Pistils 4 or more per flower, simple.
                    35. Perianth parts 6–8; plants covered with mucilage ____________ **Cabombaceae**
                    35. Perianth parts 12 or more; plants not covered with mucilage _____ **Nelumbonaceae**
                34. Pistils 1 per flower, compound.
                    36. Perianth less than 1 cm across; ovaries inferior; styles 2; stylopodia present; fruits schizocarps ____________ (*Hydrocotyle*) **Apiaceae**
                    36. Perianth 2 cm or more across; ovaries superior; styles 0 or 1 or 12 or more; stylopodia absent; fruits capsules or berries.
                        37. Petals 5, fused, valvate in bud, the margins fringed; styles 12 or more; fruits capsules, beaked ____________ **Menyanthaceae**
                        37. Petals 12 or more, free, imbricate in bud, the margins entire; styles 0 or 1; fruits berries, not beaked ____________ **Nymphaeaceae**
            33. Leaves of various shapes, but neither peltate nor cordate, either cauline or basal, but not arising from rhizomes.
                38. Petals 3; stamens 12 or more; pistils 12 or more; venation parallel convergent ____________ **Alismataceae**
                38. Petals 4 or 5 or 0; stamens 1 or 4 or 8 or 10; pistils 1; venation parallel or pinnate or palmate.
                    39. Inflorescences whitish pedunculate heads; leaves opposite ____________ (*Alteranthera*) **Amaranthaceae**
                    39. Infloresences not whitish pedunculate heads; leaves opposite OR alternate.
                        40. Flowers in pedunculate, often dense spikes; stipules present; venation parallel; fruits achenes; corollas absent ____________ **Potamogetonaceae**
                        40. Flowers not in pedunculate spikes; stipules absent; venation pinnate or palmate; fruits capsules OR fruits appearing to have 2 lobes and eventually splitting into 4 achene-like mericarps; corollas present OR absent.
                            41. Leaves less than 15 mm long; flowers imperfect, the plants monoecious; fruits appearing to have 2 lobes and eventually splitting into 4 achene-like mericarps ____________ **Callitrichaceae**

41. Leaves more than 15 mm long; flowers perfect; fruits capsules.
42. Venation pinnate; corollas radially symmetrical or absent; petals free or absent; ovaries inferior. ____ **Onagraceae**
42. Venation palmate; corollas bilaterally symmetrical; petals fused; ovaries superior ____ **Scrophulariaceae**

32. Leaves submersed.
43. Leaves obovate or oblanceolate or ovate or lanceolate or elliptic or linear-lanceolate.
44. Leaves alternate or in basal rosettes; petioles conspicuous, 5–20 cm long; perianth salverform, the segments united below into a distinct tube ____ **Pontederiaceae**
44. Leaves whorled or opposite; petioles much shorter than 5 cm long; perianth of separate segments or absent.
45. Leaves whorled, elliptic or linear-lanceolate ____ **Hydrocharitaceae**
45. Leaves opposite, obovate or oblanceolate.
46. Stipules present; flowers perfect; fruits subglobose capsules; leaves not forming rosettes at the stem tips ____ (*Elatine*) **Elatinaceae**
46. Stipules absent; flowers imperfect, the plants monoecious; fruits somewhat flattened laterally, often slightly heart-shaped and appearing to have 2 lobes, eventually splitting into 4 achene-like mericarps; leaves sometimes forming rosettes at the stem tips __ (*Callitriche*) **Callitrichaceae**

43. Leaves linear or filiform.
47. Plants cespitose, attached to rocks by fleshy disks and forming mats or crusts on them; [Family in OK, not in TX] ____ **Podostemaceae**
47. Plants rhizomatous or with stems rooting at nodes, not attached to rocks by fleshy disks.
48. Leaves alternate or basal.
49. Leaves basal; flowers absent OR present.
50. Leaves thread-like, terete, 1.6–10.2 cm long; flowers absent; spores produced in sori borne in hard sporocarps in axils of leaves ____ **Marsileaceae**
50. Leaves ribbon-like, the flattened blades to 20 mm wide, to 60 cm long; flowers present; spores produced in anthers and ovaries ____ (*Vallisneria*) **Hydrocharitaceae**
49. Leaves alternate; flowers present.
51. Perianth pale yellow, with an elongate tube and a 6-parted limb; stamens 3 ____ **Pontederiaceae**
51. Perianth of 4 inconspicuous greenish segments or absent; stamens 2 or 4.
52. Flowers borne in 2–5 whorls on peduncles elongated above water surface; perianth parts present; stamens 4 ___ **Potamogetonaceae**
52. Flowers borne in 1 whorl on peduncle below water surface; perianth parts absent; stamens 2; [Family in OK and se and s TX, not in nc TX] ____ **Ruppiaceae**

48. Leaves opposite or appearing whorled.
53. Leaf bases not sheathing stems; apices of leaf blades obtuse, notched.
54. Fruits capsules; perianth parts present ____ **Lythraceae**
54. Fruits appearing to have 2 lobes and eventually splitting into 4

achene-like mericarps; perianth parts absent ______ **Callitrichaceae**

53. Leaf bases sheathing stems; apices of leaf blades acute, not notched.
   55. Leaf sheaths conspicuously inflated and elongated, 6–10 mm long; flowers borne on elongated peduncles; [Family in OK and se and s TX, not in nc TX] ______ **Ruppiaceae**
   55. Leaf sheaths neither conspicuously inflated nor elongated, 0.2–4 mm long; flowers borne in axils of leaves.
      56. Leaf blades usually minutely denticulate under a scope OR obviously toothed to the naked eye; pistils 1 per flower; fruits terete, not beaked, not curved, not stipitate ___ (*Najas*) **Hydrocharitaceae**
      56. Leaf blades entire; pistils 2–8 per flower; fruits flattened, beaked (the beak to 1.5 mm long), curved, short stipitate (= stalked) ______ **Zannichelliaceae**

31. Plants emergent from water.
   57. Leaves modified into hollow, tubular, trumpet-shaped pitchers; flowers solitary at the end of a long naked scape ______ **Sarraceniaceae**
   57. Leaves not modified into pitchers; flowers variously arranged.
      58. Venation pinnate or palmate.
         59. Plants acaulescent; leaves basal.
            60. Flowers 5-merous; fruits schizocarps ______ (*Hydrocotyle*) **Apiaceae**
            60. Flowers 3-merous; fruits capsules or berries or achenes.
               61. Corollas bilaterally symmetrical, purple; ovaries inferior; fruits capsules ______ **Marantaceae**
               61. Corollas radially symmetrical or absent, white; ovaries superior; fruits berries or achenes.
                  62. Inflorescences spadices; spathes present; fruits berries ______ **Araceae**
                  62. Inflorescences racemes, the flowers borne in whorls of 3; spathes absent; fruits achenes ______ **Alismataceae**
         59. Plants caulescent; leaves cauline.
            63. Corollas bilaterally symmetrical.
               64. Seeds 2–4; anther apices recurved; anthers borne at 45 degree angle to filaments ______ **Acanthaceae**
               64. Seeds 12 or more; anther apices not recurved; anthers borne vertically or at less than 45 degree angle to filaments ______ **Scrophulariaceae**
            63. Corollas radially symmetrical or absent.
               65. Ovaries inferior.
                  66. Flowers in terminal spikes; capsules with circumscissile dehiscence; corollas sympetalous ______ **Sphenocleaceae**
                  66. Flowers in axils of upper leaves; capsules without circumscissile dehiscence; corollas of separate petals or absent ______ **Onagraceae**
               65. Ovaries superior.
                  67. Leaves opposite.
                     68. Flowers in pedunculate heads; petals absent (tepals silvery white) ______ (*Alternanthera*) **Amaranthaceae**
                     68. Flowers borne in all axils of stem leaves; petals present, lavender to pink to purple-red or rose-purple ______ **Lythraceae**
                  67. Leaves alternate.
                     69. Inflorescences spadices; spathes present; larger leaf blades to 90 cm long, sagittate at base ______ **Araceae**

69. Inflorescences not spadices; spathes not present; leaf blades of various sizes, typically much smaller, usually not sagittate at base.
    70. Plants armed with 1–2 spines per node; corollas blue (rarely white), showy, 11–17 mm long ____________ **Hydrophyllaceae**
    70. Plants unarmed; corollas if present much smaller, never blue.
        71. Leaves of 2 forms, the submersed ones pinnately compound or pinnately dissected, the emergent ones simple; inflorescences racemes or solitary flowers.
            72. Inflorescences racemes, terminal; petals 4; pistils 2-carpellate; fruits siliques or silicles __________ **Brassicaceae**
            72. Inflorescences solitary flowers, axillary; petals 0; pistils 3-carpellate; fruits nut-like ____________ **Haloragaceae**
        71. Leaves all alike; inflorescences spikes or spicate racemes.
            73. Stipules present as ocrea; perianth parts present; pistils 1 per flower; fruits achenes ___________ **Polygonaceae**
            73. Stipules absent; perianth parts absent; pistils 3–4 per flower, fused at base; fruits capsules ________ **Saururaceae**

58. Venation parallel or parallel-convergent.
    74. Leaf blades sagittate or cordate or ovate or elliptic, the venation parallel-convergent.
        75. Plants caulescent; leaves cauline; perianth parts absent ______________ **Saururaceae**
        75. Plants acaulescent; leaves forming a rosette; perianth parts present.
            76. Pistils 12 or more per flower; perianth parts in 2 series, the parts free; fruits achenes ________________________________ **Alismataceae**
            76. Pistils 1 per flower; perianth parts in 1 series, the parts fused; fruits capsules or utricles ____________________________ **Pontederiaceae**
    74. Leaf blades linear or linear-lanceolate, elongated, the venation parallel.
        77. Leaves minute, less than 1 cm long, arising from filiform subterranean stems (leaves are possibly leaf-like branches); sac-like bladders borne laterally on stems; flowers borne on filiform scapes 10–20 cm long, bilabiate ______________________________ **Lentibulariaceae**
        77. Leaves more than 1 cm long, arising from well-developed aerial or subterranean stems; sac-like bladders absent; flowers not borne on filiform scapes, not bilabiate.
            78. Plants caulescent; leaves cauline.
                79. Perianth parts petaloid or sepaloid.
                    80. Inflorescences racemes; ovaries inferior; perianth parts very unequal, one a lip divided into three narrow lobes and extended at base into a spur 9–14 mm long ______________ **Orchidaceae**
                    80. Inflorescences spadices or panicles or glomerules or head-like clusters; ovaries superior; perianth parts equal, none differentiated into a lip; spur absent.
                        81. Inflorescences spadices; spadices diverging from the side of elongate, linear, spathe-like scapes; peduncles 3-angled; fruits berries ______________________ **Acoraceae**
                        81. Inflorescences panicles or glomerules or head-like clus-

ters; spathe-like scapes absent; peduncles terete or flattened; fruits capsules ______ **Juncaceae**

79. Perianth parts absent or perianth of bristles or scales.
    82. Stems jointed, the nodes and internodes distinct; each flower subtended by 2–5 bracts; stigmas feathery ______ **Poaceae**
    82. Stems not jointed, the nodes and internodes not distinct; each flower subtended by 1 bract or bracts absent; stigmas barbellate or smooth.
        83. Leaves 3-ranked; margins of leaf sheaths fused to form tubes ______ **Cyperaceae**
        83. Leaves 2-ranked; margins of leaf sheaths overlapping, not fused.
            84. Inflorescences cylindrical; achenes long stipitate, subtended by bristles ______ **Typhaceae**
            84. Inflorescences globose; achenes sessile or subsessile, not subtended by bristles; [Family in OK and se and s TX, not in nc TX] ______ **Sparganiaceae**

78. Plants acaulescent; leaves basal.
    85. Flowers absent; spores produced in sporangia at bases of leaves or in subterranean sporocarps.
        86. Plants cespitose with corms 2–5 lobed; leaves 5–60 cm long, divided into 4 longitudinal cavities, the bases enlarged; sporangia embedded in leaf bases ______ **Isoetaceae**
        86. Plants rhizomatous; leaves 1.6–10.2 cm long, not divided into 4 longitudinal cavities, the bases not enlarged; sporangia borne in subterranean sporocarps ______ **Marsileaceae**
    85. Flowers present; spores produced in anthers or ovaries.
        87. Perianth parts absent or perianth of 6 inconspicuous bristles or 6 scales.
            88. Leaves inconspicuous, reduced to scales or bladeless sheaths at stem bases; plants appearing to consist only of green leafless stems ______ **Cyperaceae**
            88. Leaves conspicuous, with well-developed blades and petioles or sheaths; plants not appearing to consist only of green leafless stems.
                89. Leaves 2-ranked; margins of leaf sheaths overlapping, not fused; inflorescences heads, globose, 6 or more per peduncle, with multiple staminate heads above multiple pistillate heads; [Family in OK and se and s TX, not in nc TX] ______ **Sparganiaceae**
                89. Leaves 3-ranked; margins of leaf sheaths fused; inflorescences of various types, 1–4 per peduncle, multiple staminate heads not borne above multiple pistillate heads ______ **Cyperaceae**
        87. Perianth parts present, petaloid or sepaloid.
            90. Flowers imperfect, the plants monoecious; pistils 12 or more per flower; stamens 12 or more; inflorescences racemes or multiple heads; fruits achenes.
                91. Inflorescences racemes, the flowers borne in whorls

of 3; perianth parts in 2 series; petals white; achenes beakless ______ **Alismataceae**

91. Inflorescences heads, the flowers numerous; perianth parts in 1 series; petals absent; achenes beaked; [Family in OK and se and s TX, not in nc TX] ______ **Sparganiaceae**

90. Flowers perfect; pistils 1 per flower; stamens 3 or 4 or 6; inflorescences solitary spikes or solitary heads; fruits capsules.

92. Perianth parts yellow, glabrous; stamens 3; anthers yellow ______ **Xyridaceae**

92. Perianth parts gray-black, bearing fleshy trichomes at apices; stamens 4 or 6; anthers black ______ **Eriocaulaceae**

GROUP H

Plants herbaceous vines or epiphytes or aerial hemiparasites.

1. Plants entirely parasitic; chlorophyll absent; stems filamentous, typically forming tangled masses on host plants, or embedded entirely in tissues of host plants; leaves absent or reduced to scales.

2. Stems apparent, filamentous, typically forming tangled masses on host plants, white or yellow or orange; flowers perfect; perianth parts in 2 series; ovaries superior ______ **Cuscutaceae**

2. Stems not apparent, embedded entirely in tissues of *Dalea* spp., only flowers and subtending bracts visible; flowers imperfect; perianth parts in 1 series; ovaries inferior ______ **Rafflesiaceae**

1. Plants autophytic or hemiparasitic (at least partially autophytic); chlorophyll present; stems neither filamentous nor imbedded in host tissues; leaves present.

3. Plants epiphytes or hemiparasites, the plants growing on other plants, without roots in the ground; stems arising from bark of woody hosts.

4. Leaf margins entire; flowers present; spores produced in anthers or ovaries; fruits present.

5. Plants truly epiphytic, growing on branches of other plants but not penetrating the tissues of the host plant; fruits capsules; leaves very narrow to thread-like, 2 mm or less wide ______ **Bromeliaceae**

5. Plants hemiparasitic, penetrating the tissues of the host plant; fruits drupes; leaves elliptical-ovate to orbicular, much > 2 mm wide [Stems woody at base, but falsely appearing herbaceous] ______ (Viscaceae) **Group A**

4. Leaf (frond) margins pinnately lobed; flowers absent; spores produced in sori on abaxial surfaces of fronds; fruits absent ______ **Polypodiaceae**

3. Plants vines; stems arising from soil and climbing or twining among other plants for support.

6. Stems climbing by tendrils.

7. Leaves simple.

8. Leaf margins entire or finely denticulate; leaf venation pinnate or parallel-convergent.

9. Venation pinnate; inflorescences racemes (these can be panicled); perianth parts 5; fruits achenes ______ **Polygonaceae**

9. Venation parallel-convergent; inflorescences umbels; perianth parts 6; fruits berries _ **Smilacaceae**

8. Leaf margins lobed or serrate; leaf venation palmate.

10. Flowers with a conspicuous fringed corona attached to hypanthial cup; petals free; styles 3; ovaries superior ______ **Passifloraceae**

10. Flowers without a fringed corona; petals fused; styles 1; ovaries inferior ______ **Cucurbitaceae**

7. Leaves compound.

11. Leaves opposite; perianth parts in 1 series; fruits achenes with plumose tails ______ **Ranunculaceae**

11. Leaves alternate; perianth parts in 2 series; fruits legumes or capsules.
    12. Leaves 1-compound; tendrils borne on leaves, formed from ultimate leaflets; flowers papilionaceous; fruits legumes ________ (Papilionoideae) **Fabaceae**
    12. Leaves 2- or 3-compound; tendrils borne on peduncles of inflorescences; flowers funnelform; fruits capsules, inflated, 3-loculed, with 3 round black seeds ________ **Sapindaceae**

6. Stems climbing by twining; tendrils absent.
   13. Plants not producing flowers and seeds; spores produced in sporangia borne in 2-rowed aggregations at ends of oblong marginal lobes of pinnules; [Family in OK and se and s TX, not in nc TX] ________ **Lygodiaceae**
   13. Plants producing flowers and seeds; spores produced in sporangia borne in anthers or ovaries.
       14. Leaves alternate.
           15. Leaves compound, at least on upper stems; flowers papilionaceous; fruits legumes ________ (Papilionoideae) **Fabaceae**
           15. Leaves simple; flowers of various forms, but not papilionaceous; fruits capsules or achenes or drupes.
               16. Perianth parts in 2 series.
                   17. Corollas bilabiate (= 2-lipped); stamens 4 ________ (*Maurandya*) **Scrophulariaceae**
                   17. Corollas not bilabiate; stamens 5–12.
                       18. Petals 5, fused; corollas salverform; fruits capsules; seeds 1–4, wedge-shaped ________ **Convolvulaceae**
                       18. Petals 3 or 6, free; corollas bowl-shaped; fruits drupes, red at maturity; seeds 1, the stone curved into a closed spiral [Plants woody, but distal portion of stems falsely appearing herbaceous] ________ (Menispermaceae) **Group A**
               16. Perianth parts in 1 series or absent.
                   19. Leaves thin-fleshy; stipules absent; perianth parts 5 ________ **Basellaceae**
                   19. Leaves not fleshy; stipules absent OR present as ocrea sheathing stems; perianth parts 3 or 6.
                       20. Stipules present as ocrea sheathing stems; perianth parts 3; fruits achenes, trigonous, not winged, black at maturity ________ **Polygonaceae**
                       20. Stipules absent; perianth parts 6; fruits capsules; seeds 1 or 2, flat, winged, golden-brown at maturity ________ **Dioscoreaceae**
       14. Leaves opposite or whorled.
           21. Leaves whorled, becoming opposite or alternate above ________ **Dioscoreaceae**
           21. Leaves opposite at all nodes.
               22. Petals absent.
                   23. Leaves simple; plants dioecious; perianth parts sepaloid; inflorescences dissimilar, the pistillate flowers in drooping clustered spikes, the staminate flowers borne in drooping panicles ________ **Cannabaceae**
                   23. Leaves compound; plants dioecious or polygamous; perianth parts petaloid; pistillate and staminate inflorescences similar, panicles ________ **Ranunculaceae**
               22. Petals present.
                   24. Inflorescences heads, in cymose-paniculate arrangement, the individual heads 4-flowered; ovaries inferior; pappus present, of numerous capillary bristles; fruits achenes ________ (*Mikania*) **Asteraceae**
                   24. Inflorescences umbels or cymes or racemes or flowers solitary; ovaries superior; pappus absent; fruits follicles or capsules, with seeds usually 12 or more.

25. Corollas yellow to orange, cream, or white, with a dark purple center; stamens 4; petioles conspicuously winged ______ (*Thunbergia*) **Acanthaceae**
25. Corollas variously colored but not light with a dark purple center; stamens 5; petioles not winged.
    26. Pistils 1 per flower; fruits 2-valved capsules flattened contrary to the septum; corollas showy, yellow, funnelform, 25–35 mm long [Plants woody, but distal portion of stems falsely appearing herbaceous] ______ (*Gelsemium*—Loganiaceae) **Group A**
    26. Pistils 2 per flower, united at stigmas; fruits follicles; corollas not as above.
        27. Coronas present; pollen aggregated in pollinia; anthers fused to stigmas to form gynostegia; follicles 1 at maturity ______ **Asclepiadaceae**
        27. Coronas absent; pollen not aggregated in pollinia; anthers united but not fused to stigmas; follicles 2 at maturity [Plants woody, but distal portion of stems falsely appearing herbaceous] ______ (Apocynaceae) **Group A**

GROUP I

Plants parasitic or saprophytic; chlorophyll absent.

1. Stems filamentous, typically forming tangled masses on host plants OR embedded entirely in tissues of host plants; leaves absent or reduced to scales.
    2. Stems apparent, filamentous, typically forming tangled masses on host plants, white or yellow or orange; flowers perfect; perianth parts in 2 series; ovaries superior ______ **Cuscutaceae**
    2. Stems not apparent, embedded entirely within tissues of *Dalea* spp., only flowers and subtending bracts visible; flowers imperfect; perianth parts in 1 series; ovaries inferior ______ **Rafflesiaceae**
1. Stems neither filamentous nor imbedded in host tissues; leaves present (but can be reduced and bract-like).
    3. Ovaries inferior; perianth parts 3 or in multiples of 3 ______ **Orchidaceae**
    3. Ovaries superior; perianth parts 4 or 5.
        4. Corollas radially symmetrical; leaves cauline; fertile stamens 8 or 10; sepals deciduous ______ (previously Monotropaceae) **Ericaceae**
        4. Corollas bilaterally symmetrical; leaves basal; fertile stamens 4; sepals persistent ______ **Orobanchaceae**

GROUP J

Plants acaulescent herbs; plants producing flowers and seeds.

1. Leaves inconspicuous, reduced to scales or bladeless sheaths at stem bases; plants appearing to consist only of green leafless stems ______ **Cyperaceae**
1. Leaves conspicuous, with well-developed blades, and petioles or sheaths; plants consisting of scapes and leaves.
    2. Leaves modified into hollow, tubular, trumpet-shaped pitchers ______ **Sarraceniaceae**
    2. Leaves not modified into pitchers.
        3. Leaves emerging from ground singly or in 2s or in 3s, neither forming conspicuous rosettes nor tufts; flowers borne on scapes that emerge from ground separately from leaves.
            4. Leaves compound.
                5. Inflorescences spadices; spathes present; flowers imperfect, borne in same inflorescence, the staminate above the pistillate; fruits berries ______ **Araceae**
                5. Inflorescences racemes or cymes or solitary flowers; spathes absent; flowers perfect; fruits capsules or achenes.

6. Leaves 2-compound; inflorescences racemes; corollas bilaterally symmetrical ______ **Fumariaceae**
6. Leaves 1-compound; inflorescences cymes or solitary flowers; corollas radially symmetrical.
7. Pistils 1; leaflets usually obcordate ______ **Oxalidaceae**
7. Pistils 12 or more; leaflets of various shapes, but not obcordate ______ **Ranunculaceae**
4. Leaves simple.
8. Leaf margins entire or weakly undulate.
9. Inflorescences spadices or heads.
10. Leaves with sheaths; inflorescences spadices; spathes present; perianth parts 6 or 0; fruits berries ______ **Araceae**
10. Leaves without sheaths; inflorescences heads; spathes absent; perianth parts 5; fruits achenes ______ **Asteraceae**
9. Inflorescences racemes or spikes or panicles or umbels or solitary flowers.
11. Corollas bilaterally symmetrical; stamens 1 or 2, united with style to form a column ______ **Orchidaceae**
11. Corollas (or corolla-like calyces) radially symmetrical; stamens 6 or 12, free, not united with a style.
12. Perianth parts 3; stamens 12; leaves cordate-reniform ______ **Aristolochiaceae**
12. Perianth parts 6; stamens 6; leaves linear or lanceolate or ovate ______ **Liliaceae**
8. Leaf margins crenate or toothed or lobed or cleft.
13. Flowers bilaterally symmetrical, with one of the 5 petals with a short basal spur; fruits 3-valved, unarmed capsules ______ **Violaceae**
13. Flowers radially symmetrical, without a spurred petal; fruits schizocarps or achenes or follicles or capsules (if capsules either prickly or not 3-valved).
14. Leaf blades orbicular, peltate or nearly so, the margins crenate.
15. Flowers borne in open or spicate umbels; styles 2; ovaries inferior ______ (*Hydrocotyle*) **Apiaceae**
15. Flowers borne in 2s in axils of leaves; styles 5; ovaries superior ______ **Geraniaceae**
14. Leaf blades flabellate (= fan-shaped) or reniform, not peltate, the margins palmately lobed or cleft.
16. Stamens 5 or 10; fruits schizocarps, dehiscing into 5, one-seeded, beaked mericarps ______ **Geraniaceae**
16. Stamens 20 or more; fruits capsules or achenes.
17. Pistils 1; fruits capsules; sap of rhizomes red-orange ______ **Papaveraceae**
17. Pistils 20 or more; fruits achenes; sap of tubers colorless ______ **Ranunculaceae**
3. Leaves forming rosettes or tufts; flowers borne on scapes that emerge from centers of rosettes or tufts.
18. Leaves compound.
19. Leaves 2- or 3-compound; ovaries inferior ______ **Apiaceae**
19. Leaves 1-compound; ovaries superior.
20. Inflorescences umbels; stamens 5; fruits schizocarps, dehiscing into 5, one-seeded, beaked mericarps ______ **Geraniaceae**
20. Inflorescences of various types, but not umbels; stamens 6 or more; fruits achenes or drupes or berries.
21. Leaflets 11–25, 30–45 cm long, the arrangement conspicuously flabellate (= fan-shaped); stamens 6 ______ **Arecaceae**
21. Leaflets 3–7, 3–5 cm long, the arrangement not flabellate; stamens 10 or more.
22. Stipules present; perianth parts in 2 series ______ **Rosaceae**
22. Stipules absent; perianth parts in 1 series ______ **Ranunculaceae**

18. Leaves simple.
  23. Leaves spatulate or clavate, covered with long glandular hairs that exude a clear, glistening, sticky secretion; plants insectivorous ____ **Droseraceae**
  23. Leaves of various shapes and with various indumentation, but not covered with long glandular hairs; plants not insectivorous.
    24. Perianth parts absent; flowers enclosed by spathes or chaffy bracts.
      25. Leaves sagittate; flowers enclosed by spathes; inflorescences spadices ____ **Araceae**
      25. Leaves linear or linear-lanceolate; flowers enclosed by chaffy bracts; inflorescences spikes ____ **Cyperaceae**
    24. Perianth parts present; flowers not enclosed by either spathes or chaffy bracts.
      26. Perianth parts in 2 series.
        27. Perianth parts in 3s.
          28. Perianth parts (tepals) variously bluish to violet or purple ____ **Iridaceae**
          28. Perianth parts (petals) yellow to white or pink.
            29. Petals yellow; inflorescences cone-like, with spirally imbricated, brownish, thin, ± woody bracts ____ **Xyridaceae**
            29. Petals white or pink; inflorescences not cone-like.
              30. Flowers borne in fascicles at ends of inflorescence branches; pistils 1 per flower ____ **Polygonaceae**
              30. Flowers borne in whorls of 3 along a rachis; pistils 25 or more per flower ____ **Alismataceae**
        27. Perianth parts in 4s or 5s.
          31. Perianth parts in 4s.
            32. Flowers borne at base of plant; ovaries inferior; stamens 8 ____ **Onagraceae**
            32. Flowers (actually inflorescences) borne at ends of elongated peduncles; ovaries superior; stamens 2 or 4 or 6.
              33. Inflorescences racemes; stamens 6; petals free, yellow or white, membranous; fruits siliques or silicles ____ **Brassicaceae**
              33. Inflorescences spikes, dense; stamens 2 or 4; petals fused, chartaceous, hyaline; fruits capsules, circumscissile ____ **Plantaginaceae**
          31. Perianth parts in 5s.
            34. Inflorescences heads, 1–10 per plant, the arrangement solitary or racemose or spicate; pappus of bristles or scales ____ **Asteraceae**
            34. Inflorescences panicles or umbels or cymes or solitary flowers, heads not present; pappus not present.
              35. Inflorescences panicles, large, dichotomously branched, with numerous flowers; sepal apices white; fruits utricles; [Family in OK and se and s TX, not in nc TX] ____ **Plumbaginaceae**
              35. Inflorescences umbels or cymes or solitary flowers; sepal apices green; fruits capsules or achenes.
                36. Pistils 12 or more per flower; sepals spurred at base; fruits achenes ____ **Ranunculaceae**
                36. Pistils 1 per flower; sepals not spurred at base; fruits capsules.
                  37. Corollas bilaterally symmetrical; petals spurred or gibbous.
                    38. Sepals fused; petals fused; leaves soft-fleshy, greasy to the touch ____ **Lentibulariaceae**

38. Sepals free; petals free; leaves not soft-fleshy, not greasy to the touch __________ **Violaceae**

37. Corollas radially symmetrical; petals neither spurred nor gibbous.

39. Petals free; stigmas 2–4 __________ **Saxifragaceae**

39. Petals fused; stigmas 1 __________ **Primulaceae**

26. Perianth parts in 1 series or parts all similar.

40. Inflorescences heads or spadices.

41. Inflorescences heads; perianth parts and stamens 5; ovaries inferior __________ **Asteraceae**

41. Inflorescences spadices; perianth parts and stamens 6; ovaries superior __________ **Araceae**

40. Inflorescences of various types, but neither heads nor spadices.

42. Pistils 12 or more per flower.

43. Flowers imperfect, borne in whorls of 3; perianth parts 3 __________ **Alismataceae**

43. Flowers perfect, borne singly; perianth parts 5 or more __________ **Ranunculaceae**

42. Pistils 1 per flower.

44. Ovaries inferior.

45. Corollas bilaterally symmetrical; stamens 1 or 2, united with style to form a column __________ **Orchidaceae**

45. Corollas radially symmetrical; stamens 3 or 6, free, not united with style.

46. Inflorescences spikes, elongated, 25–45 cm long; leaves conspicuously stiff and succulent; leaf apices spine-tipped; leaf margins minutely spinose __________ **Agavaceae**

46. Inflorescences of various types, but not elongated spikes; leaves flexible and non-succulent; leaf apices not spine-tipped; leaf margins entire.

47. Leaves equitant; stamens 3 __________ **Iridaceae**

47. Leaves not equitant; stamens 6 __________ **Liliaceae**

44. Ovaries superior.

48. Venation pinnate; flowers borne in umbellate fascicles, subtended by whorls of foliaceous bracts; stamens 9 __________ **Polygonaceae**

48. Venation parallel; flowers not borne in umbellate fascicles; bracts if present neither foliaceous nor in whorls; stamens 3 or 6.

49. Leaves conspicuously 3-ranked; each flower enclosed by 1 chaffy bract; fruits achenes __________ **Cyperaceae**

49. Leaves without conspicuous ranking; flowers not enclosed by chaffy bracts; fruits capsules or berries.

50. Leaves conspicuously stiff and succulent or not so, arising from woody caudices or thick, fibrous-rooted crowns; leaf apices spine-tipped or not so; inflorescences many-flowered racemes or panicles __________ **Agavaceae**

50. Leaves flexible and non-succulent, arising from

fibrous roots or bulbs or corms or rhizomes; leaf apices not spine-tipped; inflorescences various.

51. Inflorescences solitary spikes or solitary heads.
    52. Perianth parts yellow, glabrous; stamens 3; anthers yellow ______ **Xyridaceae**
    52. Perianth parts gray-black, bearing fleshy trichomes at apices; stamens 4 or 6; anthers black ______ **Eriocaulaceae**
51. Inflorescences panicles or racemes or umbels.
    53. Perianth parts green or brown, scarious, persistent at fruit maturity ______ **Juncaceae**
    53. Perianth parts of various bright colors, moist, withering by fruit maturity ______ **Liliaceae**

GROUP K

Plants acaulescent or caulescent herbs; spores produced in sori or sporocarps or in aggregations of sporangia at ends of elongated stalks.

1. Leaves (microphylls) scale-like, less than 1 cm long, the veins 1, unbranched; aerial stems present; strobili present, terminal.
    2. Stems jointed, fluted, the internodes hollow; leaves (very reduced) whorled and forming sheaths around stems ______ **Equisetaceae**
    2. Stems not jointed, not fluted, the internodes solid; leaves spiraled and imbricate.
        3. Leaves 1–3 mm long; strobili 4-angled; spores of 2 sizes ______ **Selaginellaceae**
        3. Leaves 6–7 mm long; strobili cylindrical; spores of 1 size ______ **Lycopodiaceae**
1. Leaves (megaphylls) not scale-like, more than 1 cm long, the veins 2 or more, branched; aerial stems absent; strobili absent.
    4. Leaves linear or filiform.
        5. Plants bearing both simple and dichotomously compound leaves; sporangia produced in sori on abaxial surfaces of pinnae; pinnae present, the margins bearing 1–3 teeth [*Asplenium* spp. with linear or filiform leaves in OK, not in TX] ______ **Aspleniaceae**
        5. Plants bearing only simple leaves; sporangia produced in cavities at bases of leaves or in subterranean sporocarps; pinnae absent.
            6. Plants cespitose with corms 2–5 lobed; leaves 5–60 cm long, divided into 4 longitudinal cavities, the leaf bases enlarged; sporangia embedded in leaf bases ______ **Isoetaceae**
            6. Plants rhizomatous; leaves 1.6–10.2 cm long, not divided into 4 longitudinal cavities, the leaf bases not enlarged; sporangia borne in subterranean sporocarps ______ **Marsileaceae**
    4. Leaves of various shapes, but neither linear nor filiform.
        7. Plants climbing; leaves twining; sporangia clustered in 2-rowed aggregations at ends of oblong marginal lobes of pinnules; [Family in in OK and se and s TX, not in nc TX] ______ **Lygodiaceae**
        7. Plants not climbing; leaves not twining; sporangia clustered in sori or in aggregations at ends of elongated stalks.
            8. Leaves palmately compound; leaflets 4, obdeltoid or flabellate; spores produced in sori borne in hard sporocarps in axils of leaves ______ **Marsileaceae**
            8. Leaves simple or pinnately compound; leaflets when present usually neither obdeltoid nor flabellate; spores produced in aggregations of sporangia at ends of stalks or in sori on abaxial surfaces of fronds; sporocarps absent.
                9. Spores produced in aggregations of sporangia at ends of elongated stalks.
                    10. Fronds (leaves) simple.

11. Fronds ovate or elliptic, the margins entire ______ **Ophioglossaceae**
11. Fronds deltoid, the margins pinnatifid ______ **Dryopteridaceae**

10. Fronds (leaves) compound.
12. Lowermost 2 pinnae (= primary divisions of a leaf, here one on each side of the leaf) of the fertile leaf long-stalked and thus greatly elongated (usually longer than the sterile portion of the leaf), very different from the other pinnae, and bearing sporangia near the apex ______ **Anemiaceae**
12. Lowermost 2 pinnae of the fertile leaf not as above, either fronds of 2 different types—sterile and fertile OR fronds differentiated into basal sterile and apical fertile portions.
13. Fronds of 2 types, the sterile fronds foliaceous, the fertile fronds stalk-like and bearing aggregations of sporangia at ends.
14. Pairs of pinnae 15–25; bases of pinnae with tufts of reddish brown hairs ______ **Osmundaceae**
14. Pairs of pinnae 1–12; bases of pinnae without tufts of reddish brown hairs.
15. Blades of vegetative fronds 17–35 cm long; rhizomes present; roots not fleshy ______ **Dryopteridaceae**
15. Blades of vegetative fronds 3–15 cm long; rhizomes absent; roots fleshy [Fronds of 1 type, but divided near base, hence falsely appearing as 2 types of fronds] ______ **Ophioglossaceae**
13. Fronds of 1 type, differentiated into basal sterile and apical fertile portions, the sterile portions foliaceous, the fertile portions bearing paniculate aggregations of sporangia.
16. Blades of sterile portions of fronds 3–15 cm long; rhizomes absent; roots fleshy; reproductive portion of frond arising from base of vegetative portion; sporangia fused to form 2 rows ______ **Ophioglossaceae**
16. Blades of sterile fronds 20–50 cm long; rhizomes present; roots not fleshy; reproductive portion of fronds arising at apex of vegetative portion; sporangia free ______ **Osmundaceae**

9. Spores produced in sori on abaxial surfaces of fronds.
17. Fronds simple.
18. Frond margins pinnatifid; frond bases truncate or acute, the apices acute, neither rooting nor forming new plants; sori orbicular; indusia absent ______ **Polypodiaceae**
18. Frond margins entire; frond bases cordate, the apices attenuate, rooting and forming new plants; sori elongate; indusia present [*Asplenium rhizophyllum* in OK, not in TX] ______ **Aspleniaceae**
17. Fronds 1- or 2- or 3-compound.
19. Fronds of 2 conspicuously different types, sterile and fertile, 1-compound; veins of fronds partly anastomosing ______ **Blechnaceae**
19. Fronds of 1 type, not differentiated into conspicuously different sterile and fertile, 1- or 2- or 3-compound; veins of fronds free OR partly anastomosing.
20. Sori linear-oblong, end to end in one row on each side of, immediately adjacent to, and parallel with the costules (= midveins of the pinnules), chain-like in arrangement; veins of fronds partly anastomosing (veins anastomosing to form a single row of areoles near midvein) ______ **Blechnaceae**
20. Sori various, but not as above; veins of fronds free.
21. Sori located at margins of pinnae or pinnules, completely or partially covered by revolute margins.

22. Blades broadly triangular; sori covered by both margin of pinnule and hyaline indusium; rhizome scales absent ______ **Dennstaedtiaceae**
22. Blades lanceolate or elliptic or rhomboidal or reniform; sori covered only by margin of pinna or pinnule; indusium absent; rhizome scales present ______ **Pteridaceae**

21. Sori not located at margins of pinnae or pinnules, not covered by revolute margins.
23. Indusia absent or seemingly so.
24. Fronds separated; distal portions of rachises winged; veins reaching margins of pinnules ______ **Thelypteridaceae**
24. Fronds clustered together; distal portions of rachises not winged; veins not reaching margins of pinnules ______ **Aspleniaceae**
23. Indusia present, conspicuous.
25. Indusia orbicular or reniform, attached at sinus or in center or at base ______ **Dryopteridaceae**
25. Indusia linear or oblong, attached along edge.
26. Stipes stramineous, angular or flattened; fronds annual, deciduous; indusia crossing veins ______ **Dryopteridaceae**
26. Stipes black or brown or green, terete, neither angular nor flattened; fronds perennial, evergreen; indusia not crossing veins ______ **Aspleniaceae**

GROUP L

Plants caulescent herbs; perianth parts absent.

1. Venation parallel or a single vein.
2. Flowers borne in cyathia; ovaries 3-lobed, the lobes round; fruits capsular-schizocarps, 3-seeded ______ **Euphorbiaceae**
2. Flowers borne in spikelets or spikes or heads or solitary; ovaries not 3-lobed; fruits achenes or caryopses or achene-like mericarps.
3. Flowers subtended by 1–5 chaffy bracts.
4. Leaves 2-ranked; stems rounded, jointed, the nodes and internodes apparent; each flower subtended by 2–5 bracts; stigmas feathery ______ **Poaceae**
4. Leaves 3-ranked; stems rounded or often triangular, not jointed, the nodes and internodes not apparent; each flower subtended by 1 bract; stigmas barbellate or smooth ______ **Cyperaceae**
3. Flowers not subtended by bracts.
5. Flowers solitary, axillary; fruits appearing to have 2 lobes and eventually splitting into 4 achene-like mericarps; stamens 1 ______ **Callitrichaceae**
5. Flowers many, terminal; fruits achenes, 1 per flower; stamens 3.
6. Inflorescences spikes, dense, elongated, cylindrical; achenes long stipitate, subtended by hairs. ______ **Typhaceae**
6. Inflorescences heads, spherical; achenes sessile or subsessile, not subtended by hairs; [Family in OK and se and e TX, not nc TX] ______ **Sparganiaceae**

1. Venation pinnate or palmate.
7. Leaves opposite.
8. Leaves spatulate or obovate or oblanceolate; stems flaccid; flowers solitary, borne in leaf axils; fruits appearing to have 2 lobes and eventually splitting into 4 achene-like mericarps ______ **Callitrichaceae**

8. Leaves ovate or lanceolate or linear; stems rigid or flexible, but not flaccid; flowers borne in heads or cyathia; fruits achenes or capsular-schizocarps, 1 per flower.
  9. Flowers borne in small heads; heads borne in elongated terminal racemes or in axils of leaves; fruits achenes, enclosed in involucre to form a bur ______ (*Ambrosia*) **Asteraceae**
  9. Flowers borne in cyathia; fruits capsular-schizocarps, 3-lobed ______ **Euphorbiaceae**

7. Leaves alternate.
  10. Inflorescences spadices or heads or spiny burs or cyathia.
    11. Root systems fibrous; leaves with sheaths; inflorescences spadices; spathes present; fruits berries ______ **Araceae**
    11. Root systems with a central taproot; leaves without sheaths; inflorescences heads or spiny burs or cyathia; spathes absent; fruits achenes or capsular-schizocarps.
      12. Inflorescences heads or spiny burs; ovaries not lobed; fruits achenes ______ **Asteraceae**
      12. Inflorescences cyathia; ovaries 3-lobed; fruits capsular-schizocarps ______ **Euphorbiaceae**
  10. Inflorescences solitary flowers or panicles or spikes or racemes or glomerules.
    13. Plants dioecious; flowers subtended by 2 or 3 spine-tipped bracts ______ **Amaranthaceae**
    13. Plants monoecious or bearing only perfect flowers or polygamous; flowers not subtended by 2 or 3 spine-tipped bracts.
      14. Plants rhizomatous or stoloniferous; stipules present, fused to petioles; stamens 6–8; seeds 2 or more; leaf blades truncate or cordate basally ______ **Saururaceae**
      14. Plants from taproots, neither rhizomatous nor stoloniferous; stipules absent; stamens 1–5; seeds 1; leaf blades various basally ______ **Chenopodiaceae**

GROUP M

Plants caulescent herbs; perianth parts in 1 series or parts all similar; perianth parts 3 or in multiples of 3.

1. Venation pinnate or palmate or a single vein.
  2. Leaves [branches] fascicled, needle-like or filiform [Leaves reduced to inconspicuous, dry scales; stems cladophylls, hence foliage falsely appearing to comprise fascicled leaves] ______ (*Asparagus*) **Liliaceae**
  2. Leaves alternate or opposite, of various shapes, but neither needle-like nor filiform.
    3. Leaves opposite.
      4. Leaves peltate, the margins palmately lobed; flowers solitary in leaf axils [Sepals 6, but falling off early, and perianth parts thus falsely appearing in 1 series] ______ (Berberidaceae) **Group O**
      4. Leaves not peltate, the margins serrate; flowers 3–12 in axils of leaves ______ **Urticaceae**
    3. Leaves alternate.
      5. Inflorescences umbels; fruits berries, purple-black; tendrils present ______ **Smilacaceae**
      5. Inflorescences spikes or flowers solitary or in clusters of 1–5; fruits capsules or capsular-schizocarps or achenes or utricles, of various colors; tendrils absent.
        6. Perianths tubular, conspicuously curved or S-shaped, the parts fused ______ **Aristolochiaceae**
        6. Perianths bowl-shaped, neither curved nor S-shaped, the parts free.
          7. Flowers imperfect, the plants monoecious.
            8. Pistils 3-lobed; styles 3 (may be divided); fruits capsular-schizocarps, 3- or 6-seeded ______ **Euphorbiaceae**
            8. Pistils not lobed; styles 2; fruits utricles, 1-seeded.
              9. Staminate flowers ebracteate; pistillate flowers without perianth parts ______ **Chenopodiaceae**
              9. Staminate flowers bracteate; pistillate flowers with perianth parts ______ **Amaranthaceae**
          7. Flowers perfect.
            10. Ovaries inferior; seeds 3 ______ **Haloragaceae**
            10. Ovaries superior; seeds 1 or numerous.
              11. Stamens 12 or more; fruits capsules; sap viscous, yellow or white ______ **Papaveraceae**

11. Stamens 3 or 5–9; fruits achenes or utricles; sap thin, colorless.
12. Perianth parts 6; stamens 6–9; fruits achenes, trigonous or lenticular, not winged _____ **Polygonaceae**
12. Perianth parts 3; stamens 3 or 5; fruits utricles, elliptic to orbicular; winged _____ **Chenopodiaceae**

1. Venation parallel or parallel-convergent.
13. Ovaries inferior.
14. Perianth parts bilaterally symmetrical; stamens 1 or 2, fused with style to form a column _____ **Orchidaceae**
14. Perianth parts radially symmetrical; stamens 3 or 6, free or fused to perianth parts.
15. Stamens 3.
16. Leaves more than 2 cm long, equitant; inflorescences racemes or panicles _____ **Iridaceae**
16. Leaves less than 0.5 cm long, not equitant; inflorescences heads, solitary _____ **Burmanniaceae**
15. Stamens 6.
17. Leaves conspicuously stiff, succulent, the apices spine-tipped, the margins minutely spinose or filiferous _____ **Agavaceae**
17. Leaves flexible, not succulent, the apices not spine-tipped, the margins entire _____ **Liliaceae**
13. Ovaries superior.
18. Flowers subtended by 1–5 chaffy bracts.
19. Fruits capsules; seeds 3–many per fruit _____ **Juncaceae**
19. Fruits caryopses or achenes; seeds 1 per fruit _____ **Cyperaceae**
20. Leaves 2-ranked; margins of leaf sheaths overlapping, rarely fused to form tubes; stems rounded, jointed, the nodes and internodes apparent; inflorescences spikelets; each flower subtended by 2–5 bracts; stigmas feathery _____ **Poaceae**
20. Leaves 3-ranked; margins of leaf sheaths fused to form tubes; stems rounded or often triangular, not jointed, the nodes and internodes not apparent; inflorescences spikes; each flower subtended by 1 bract; stigmas barbellate or smooth _____ **Cyperaceae**
18. Flowers not subtended by chaffy bracts.
21. Flowers imperfect, the plants monoecious or dioecious.
22. Tendrils present; inflorescences umbels, axillary; fruits berries; plants dioecious _____ **Smilacaceae**
22. Tendrils absent; inflorescences heads, terminal; fruits achenes; plants monoecious, the staminate inflorescences above pistillate; [Family in OK and se and e TX, not in nc TX] _____ **Sparganiaceae**
21. Flowers perfect.
23. Perianths bilaterally symmetrical _____ **Pontederiaceae**
23. Perianths radially symmetrical.
24. Inflorescences spadices _____ **Araceae**
24. Inflorescences of various types, but not spadices.
25. Perianth parts green or brown or stramineous or black _____ **Juncaceae**
25. Perianth parts white or greenish white or other colors, but neither green nor brown nor stramineous nor black.
26. Leaves spatulate; basal leaf sheaths present; spathes present; stamens 3 _____ **Pontederiaceae**
26. Leaves of various shapes, but not spatulate; basal leaf sheaths absent; spathes absent; stamens 6.
27. Leaves conspicuously stiff, succulent, the apices spine-tipped, the margins minutely spinose or filiferous _____ **Agavaceae**
27. Leaves flexible, not succulent, the apices not spine-tipped, the margins entire _____ **Liliaceae**

Group N

Plants caulescent herbs; perianth parts in 1 series or parts all similar; perianth parts 1 or 2 or 4 or 5 or in multiples of 4 or 5 or many.

1. Inflorescences spikelets or heads with flowers subtended by bracts.
  2. Inflorescences spikelets; leaves with basal sheaths; stamens 3 or 6 or 1; perianth parts 2 ________ **Poaceae**
  2. Inflorescences heads; leaves without basal sheaths; stamens 4 or 5; perianth parts 4 or 5.
    3. Stems and leaves prickly; heads subtended by stiff prickly bracts; perianth parts 4; stamens 4, free ________ **Dipsacaceae**
    3. Stems and leaves not prickly; heads not subtended by stiff prickly bracts; perianth parts 5; stamens 5, either anthers or filaments united.
      4. Ovaries superior; fruits urticles; anthers free; filaments united into a slender tube ____ **Amaranthaceae**
      4. Ovaries inferior; fruits achenes; anthers fused into a ring around style; filaments free ______ **Asteraceae**
1. Inflorescences of various types, but neither spikelets nor heads with flowers subtended by bracts.
  5. Perianths bilaterally symmetrical.
    6. Perianths spurred or saccate.
      7. Stamens 12 or more; pistils 3 or 5 per flower, free or fused slightly at base; fruits follicles ________ **Ranunculaceae**
      7. Stamens 3 or 6; pistils 1 per flower; fruits capsules.
        8. Leaves alternate, pinnately dissected; perianth parts 4; stamens 6; ovaries superior [Sepals 2, but falling off early, and perianth parts thus falsely appearing to be in 1 series] ________ (Fumariaceae) **Groups P or R**
        8. Leaves opposite, not pinnately dissected; perianth parts 5; stamens 3; ovaries inferior ________ **Valerianaceae**
    6. Perianths neither spurred nor saccate.
      9. Ovaries inferior; perianth parts petaloid.
        10. Leaves and peduncles viscid-villous to glandular-puberulent [calyces tightly constricted above ovaries which falsely appear inferior] ________ **Chenopodiaceae**
        10. Leaves and peduncles glabrous or variously pubescent but not viscid-villous to glandular-puberulent.
          11. Leaves alternate, compound; perianth parts free; inflorescences umbels; fruits schizocarps ________ **Apiaceae**
          11. Leaves opposite, simple; perianth parts fused; inflorescences cymes; fruits achene-like ________ **Valerianaceae**
      9. Ovaries superior; perianth parts sepaloid.
        12. Plants annual; perianth parts 1; fruits utricles ________ **Chenopodiaceae**
        12. Plants perennial; perianth parts 4 or 5; fruits capsules or achenes.
          13. Leaves ovate, the margins serrate; flowers perfect; fruits capsules ________ **Cistaceae**
          13. Leaves linear or lanceolate, the margins entire; flowers imperfect; fruits achenes ________ **Urticaceae**
  5. Perianths radially symmetrical or asymmetrical.
    14. Leaves opposite or whorled.
      15. Leaves whorled.
        16. Pistils 4 or more per flower; stamens 12 or more; fruits achenes ________ **Ranunculaceae**
        16. Pistils 1 per flower; stamens 3–10; fruits capsules or schizocarps.
          17. Ovaries superior; pistils 3- or 5- carpellate; fruits capsules ________ **Molluginaceae**
          17. Ovaries inferior; pistils 2-carpellate; fruits schizocarps.
            18. Perianth parts 3 or 4, fused; inflorescences cymes; leaves and foliaceous stipules in numerous whorls ________ **Rubiaceae**
            18. Perianth parts 5, free; inflorescences umbels; leaves in 1 whorl; stipules absent ________ **Araliaceae**

15. Leaves opposite.
    19. Perianth parts bearing long woolly or silky hairs and hidden by them ____________ **Amaranthaceae**
    19. Perianth parts glabrous or variously indumented, but neither bearing long woolly or silky hairs nor hidden by them.
        20. Perianth parts fused.
            21. Ovaries inferior, wholly or partially.
                22. Leaves and peduncles viscid-villous to glandular-puberulent [calyces tightly constricted above ovaries which falsely appear inferior] ________ **Nyctaginaceae**
                22. Leaves and peduncles glabrous or variously pubescent but not viscid-villous to glandular-puberulent.
                    23. Leaves ovate or elliptic; stamens 4 ________________________________ **Onagraceae**
                    23. Leaves obovate or oblanceolate or spatulate; stamens 2 or 3 or 12 or more.
                        24. Inflorescences solitary flowers, axillary; stamens 12 or more; fruits capsules, circumscissile ________________________________ **Aizoaceae**
                        24. Inflorescences cymes, terminal, in dense clusters; stamens 2 or 3; fruits achene-like ________________________________ **Valerianaceae**
            21. Ovaries superior.
                25. Ovaries 3-lobed; flowers borne in cyathia; sap viscous, white ___________ **Euphorbiaceae**
                25. Ovaries not 3-lobed; flowers borne in various inflorescences, but not cyathia; sap thin, colorless.
                    26. Stipules present, conspicuous, scarious; fruits utricles ____________ **Caryophyllaceae**
                    26. Stipules absent; fruits achenes or capsules.
                        27. Flowers subtended by bracts; hypanthia absent; fruits achenes, 5–10 angled or ribbed (actually anthocarps = indehiscent achenes tightly enclosed in persistent base of perianth tube) ___________ **Nyctaginaceae**
                        27. Flowers not subtended by bracts; hypanthia present; fruits capsules.
                            28. Perianth parts 5; capsules circumscissile ______________________ **Aizoaceae**
                            28. Perianth parts 4; capsules not circumscissile ____________________ **Lythraceae**
        20. Perianth parts free.
            29. Leaves compound; pistils 4–15 per flower ____________________________ **Ranunculaceae**
            29. Leaves simple; pistils 1 per flower.
                30. Leaves 1 or 2 per stem, palmately lobed; fruits berries [Sepals 6, but falling off early and perianth parts thus falsely appearing to be in 1 series] ________________________________ (Berberidaceae) **Group O**
                30. Leaves more than 2 per stem, not palmately lobed; fruits utricles or capsules or achenes.
                    31. Flowers subtended by bracts; bracts scarious; perianth parts scarious or lanate; fruits utricles ________________________________ **Amaranthaceae**
                    31. Flowers not subtended by bracts; perianth parts petaloid or sepaloid; fruits capsules or achenes.
                        32 Flowers imperfect, the plants monoecious or dioecious; perianth parts 2 or 4; pistils 1-carpellate; fruits achenes ____________________ **Urticaceae**
                        32. Flowers perfect; perianth parts 5; pistils 2–5 carpellate; fruits capsules.
                            33. Flowers pedicelled in terminal cymes; styles 2–5; locules 1 ________________________________ **Caryophyllaceae**
                            33. Flowers sessile in dense axillary glomerules; styles 1; locules 2–5 ________________________________ **Molluginaceae**

14. Leaves alternate.
    34. Ovaries inferior, wholly or partially.

35. Leaves compound or both compound and simple leaves present; styles 2 __________ **Apiaceae**
35. Leaves simple; styles 1.
  36. Leaves peltate; fruits schizocarps __________ **Apiaceae**
  36. Leaves not peltate; fruits capsules or dry drupes.
    37. Inflorescences panicles, terminal; stamens 5; fruits dry drupes; seeds 1 __________ **Santalaceae**
    37. Inflorescences solitary flowers, axillary; stamens 4; fruits capsules; seeds 12 or more.
      38. Capsules 4-loculed, dehiscent longitudinally or by terminal pore __________ **Onagraceae**
      38. Capsules 1–3 loculed, dehiscent by lateral pore __________ **Campanulaceae**

34. Ovaries superior.
  39. Plants bearing only imperfect flowers.
    40. Leaves palmately compound __________ **Cannabaceae**
    40. Leaves simple.
      41. Pistils 3-loculed; fruits capsular-schizocarps; seeds 3 or more __________ **Euphorbiaceae**
      41. Pistils 1-loculed; fruits achenes or utricles; seeds 1.
        42. Leaf margins serrate or crenate.
          43. Plants with stinging hairs; inflorescences panicles __________ **Urticaceae**
          43. Plants without stinging hairs; inflorescences glomerules __________ **Moraceae**
        42. Leaf margins entire or sinuate or irregularly toothed or lobed.
          44. Flowers subtended by 2 or 3 imbricate, unfused, spine-tipped bracts; stamen filaments fused and forming a short tube; perianth scarious __________ **Amaranthaceae**
          44. Flowers not subtended by 2 or 3 spine-tipped bracts or if subtended by 2 spine-tipped bracts (in 1 species) these fused for 1/2 or more their length; stamen filaments free, not forming a tube; perianth greenish or absent.
            45. Perianth parts 5 __________ **Chenopodiaceae**
            45. Perianth parts 2 or 4.
              46. Styles 2 or 3; fruits utricles __________ **Chenopodiaceae**
              46. Styles 1; fruits achenes __________ **Urticaceae**
  39. Plants bearing only perfect flowers or plants bearing both perfect and imperfect flowers.
    47. Leaves compound.
      48. Perianth parts 4; stamens 4; hypanthia present __________ **Rosaceae**
      48. Perianth parts 5 or more; stamens 12 or more; hypanthia absent __________ **Ranunculaceae**
    47. Leaves simple.
      49. Stamens 12 or more.
        50. Pistils 4–7; fruits follicles __________ **Ranunculaceae**
        50. Pistils 1; fruits capsules.
          51. Inflorescences solitary flowers; sap viscous, yellow or white [Sepals 2 or 3, but falling off early and perianth parts thus falsely appearing to be in 1 series] __________ (Papaveraceae) **Group T**
          51. Inflorescences cymes; sap thin, colorless [Sepals 2, but falling off early and perianth parts thus falsely appearing to be in 1 series] __________ (Portulacaceae) **Group T**
      49. Stamens 1–10.
        52. Perianth parts 4.
          53. Leaf margins palmately lobed; stipules present; hypanthia present __________ **Rosaceae**
          53. Leaf margins pinnately lobed or entire; stipules absent; hypanthia absent.

54. Inflorescences cymes; fruits achenes ______ **Urticaceae**
54. Inflorescences racemes; fruits berries or siliques or silicles.
55. Fruits berries ______ **Phytolaccaceae**
55. Fruits siliques or silicles ______ **Brassicaceae**

52. Perianth parts 5 or more.
56. Plants less than 3 cm in diam. or height [Petals minute and easily overlooked hence perianth parts falsely appearing to be in 1 series] ______ (Saxifragaceae) **Group S**
56. Plants greater than 3 cm in diam. or height.
57. Stipules present as ocrea; fruits achenes ______ **Polygonaceae**
57. Stipules absent; fruits berries or utricles or capsules.
58. Inflorescences racemes or scorpioid cymes.
59. Pistils 1, terete, not horned; fruits berries ______ **Phytolaccaceae**
59. Pistils 5–7, angular, horned, united at bases; fruits follicles ______ **Crassulaceae**
58. Inflorescences solitary flowers or cymes or spikes or glomerules.
60. Perianth parts sepaloid; ovaries superior; fruits utricles ______ **Chenopodiaceae**
60. Perianth parts petaloid; ovaries inferior, wholly or partially; fruits capsules, circumscissile [Sepals 2, but falling off early and perianth parts thus falsely appearing to be in 1 series] ______ (Portulacaceae) **Group T**

GROUP O

Plants caulescent herbs; perianth parts in 2 series; petals 3 or in multiples of 3.

1. Venation pinnate or palmate or a single vein.
2. Petals 6 or 9.
3. Corollas 5–6 mm long, white, sympetalous, 6-lobed ______ **Rubiaceae**
3. Corollas without the above combination.
4. Leaves 1 or 2 per stem; fruits berries ______ **Berberidaceae**
4. Leaves 4 or more per stem; fruits capsules.
5. Petals fused, the sympetalous corollas yellow within and ± red without; plants 25 cm or less tall ______ **Oleaceae**
5. Petals separate, the corollas not as above; plants usually > 25 cm tall.
6. Stems and leaves with prickly bristles; stamens 20–150 or more; sap viscous, yellow or orange-red; hypanthium absent; sepals 2 or 3, falling off early ______ **Papaveraceae**
6. Stems and leaves without prickly bristles; stamens 4–12; sap thin, colorless; hypanthium present; sepals 4–6, persistent ______ **Lythraceae**
2. Petals 3.
7. Ovaries inferior.
8. Petals united into a funnelform corolla 2–4 mm long; stipular bristles present; stamens 4; leaves opposite ______ **Rubiaceae**
8. Petals separate or united; stipules absent or minute; stamen number various; leaves alternate or opposite.
9. Inflorescences spikes; flowers not subtended by an involucre ______ **Onagraceae**
9. Inflorescences heads; flowers subtended by an involucre [Petals 5 but fused and conspicuously 3-lobed hence flowers appearing appearing to have 3 petals] ______ (Asteraceae) **Group Q**
7. Ovaries superior.
10. Sepals 3.

11. Pistils 3 per flower; fruits follicles ________ **Crassulaceae**
11. Pistils 1 per flower; fruits capsules or achenes.
12. Perianths with a spur; fruits capsules [Petals 5, but 4 fused into 2 lateral hence flowers falsely appearing to have 3 petals] ________ (Balsaminaceae) **Group P**
12. Perianths without a spur; fruits achenes [Sepals of 2 sizes, the inner larger and can be mistaken for petals] ________ (Polygonaceae) **Group M**
10. Sepals 5.
13. Corollas radially symmetrical; pistils 3-carpellate; styles 0; stigmas 3; seeds 6 ________ **Cistaceae**
13. Corollas bilaterally symmetrical; pistils 2-carpellate; styles 1; stigmas 1, 2-lobed; seeds 2 ________ **Polygalaceae**
1. Venation parallel or parallel-convergent.
14. Corollas bilaterally symmetrical.
15. Ovaries superior; leaves and stems mucilaginous when crushed ________ **Commelinaceae**
15. Ovaries inferior; leaves and stems not mucilaginous when crushed.
16. Plants terrestrial, less than 1 m tall; stamens united with style to form a column; seeds 12 or more ________ **Orchidaceae**
16. Plants emergent aquatics, more than 1 m tall; stamens not united with style to form a column; seeds 1–3 ________ **Marantaceae**
14. Corollas radially symmetrical.
17. Pistils 12 or more per flower; fruits achenes [Plants acaulescent, but can falsely appear caulescent] ________ (Alismataceae) **Group J**
17. Pistils 1 per flower; fruits capsules.
18. Inflorescences solitary spikes or solitary heads.
19. Perianth parts yellow, glabrous; stamens 3; anthers yellow [Plants acaulescent, but can appear caulescent] ________ (Xyridaceae) **Group J**
19. Perianth parts gray-black, bearing fleshy trichomes at apices; stamens 4 or 6; anthers black ________ **Eriocaulaceae**
18. Inflorescences racemes or cymes or solitary flowers.
20. Leaves equitant; inflorescences racemes; stamens 3 ________ **Iridaceae**
20. Leaves alternate or whorled, not equitant; inflorescences cymes or solitary flowers; stamens 6.
21. Leaves alternate; inflorescences cymes; spathes present; stamen filaments pilose ________ **Commelinaceae**
21. Leaves whorled; inflorescences solitary flowers; spathes absent; stamen filaments glabrous ________ **Liliaceae**

Group P

Plants caulescent herbs; perianth parts in 2 series; petals 1 or 2 or 4 or 5; corollas bilaterally symmetrical; petals free.

1. Perianth parts spurred or cucullate.
2. Stamens 12 or more; pistils simple, free or fused slightly at base; fruits follicles ________ **Ranunculaceae**
2. Stamens 5 or 10; pistils compound; fruits capsules or schizocarps.
3. Spurs or hoods formed from sepals.
4. Venation palmate; sepals 5; perianths slightly bilaterally symmetrical ________ **Geraniaceae**
4. Venation pinnate; sepals 3; perianths strongly bilaterally symmetrical ________ **Balsaminaceae**
3. Spurs or hoods formed from petals.
5. Petals 5; sepals 5 ________ **Violaceae**
5. Petals 4; sepals 2 ________ **Fumariaceae**

1. Perianth parts neither spurred nor cucullate.
   6. Sepals 4.
      7. Hypanthia present; ovaries inferior ____ **Onagraceae**
      7. Hypanthia absent; ovaries superior.
         8. Leaves simple; stamens in 2 whorls ____ **Brassicaceae**
         8. Leaves palmately compound; stamens in 1 whorl.
            9. Stipules absent or minute; petals 4; stamens exserted beyond perianth; fruits capsules ____ **Capparaceae**
            9. Stipules present, large; petals 5; stamens included within perianth; fruits legumes ____ (Papilionoideae) **Fabaceae**
   6. Sepals 5.
      10. Ovaries inferior; fruits schizocarps ____ **Apiaceae**
      10. Ovaries superior; fruits capsules or achenes or legumes.
         11. Leaves simple.
            12. Stipules present; stamens 10; fruits legumes, inflated ____ (Papilionoideae) **Fabaceae**
            12. Stipules absent; stamens 4 or 6 or 8; fruits achenes or capsules.
               13. Stems trailing or prostrate; inflorescences solitary flowers; stamens 4 or 5; fruits indehiscent, 1-seeded, lanate-tomentose, spiny [Petals appearing free, but slightly fused at base] ____ (Krameriaceae) **Group R**
               13. Stems erect or ascending; inflorescences racemes or spikes; stamens 6 or 8; fruits capsules, usually 2-seeded, glabrous, not spiny [Inner sepals petaloid and can be mistaken for petals] ____ (Polygalaceae) **Group O**
         11. Leaves compound.
            14. Petals 1 ____ (*Amorpha*) **Fabaceae**
            14. Petals 5.
               15. Flowers strongly bilaterally symmetrical; corollas papilionaceous; upper (adaxial) petal enclosing other petals in bud ____ (Papilionoideae) **Fabaceae**
               15. Flowers weakly bilaterally symmetrical; corollas not papilionaceous; upper (adaxial) petal enclosed by other petals in bud.
                  16. Inflorescences spikes; bracts present; fruits 1- or 2-seeded ____ (Papilionoideae) **Fabaceae**
                  16. Inflorescences racemes or panicles or umbels; bracts absent; fruits 5- or more-seeded ____ (Caesalpinioideae) **Fabaceae**

GROUP Q

Plants caulescent herbs; perianth parts in 2 series; petals 4 or 5; corollas bilaterally symmetrical; petals fused at least at the base; ovaries wholly or partially inferior.

1. Inflorescences heads.
   2. Stamens 4; anthers free; styles not branched ____ **Dipsacaceae**
   2. Stamens 5; anthers fused into a ring around style; styles 2-branched ____ **Asteraceae**
1. Inflorescences solitary flowers or cymes or thyrses or racemes.
   3. Leaves alternate ____ **Campanulaceae**
   3. Leaves opposite or whorled or appearing whorled due to the presence of stipules.
      4. Corolla lobes 4 ____ **Rubiaceae**
      4. Corolla lobes 5.
         5. Petals yellow or orange to red; stamens 5; fruits berries, 3-seeded ____ **Caprifoliaceae**
         5. Petals white to bluish white; stamens 3; fruits achene-like, 1-seeded ____ **Valerianaceae**

Group R

Plants caulescent herbs; perianth parts in 2 series; petals 2 or 4 or 5; corollas bilaterally symmetrical; petals fused at least at the base; ovaries superior.

1. Plants with slender leafless stems bearing finely dissected branches with numerous sac-like bladders; plants free-floating aquatics, but often stranded in wet areas; corollas yellow ________ **Lentibulariaceae**
1. Plants with stems and foliage leaves; sac-like bladders absent; plants terrestrial; corollas variously colored.
  2. Lower cauline leaves alternate.
    3. Leaves compound.
      4. Petals 5; perianth without spurs; fruits legumes (sometimes reduced to 1-seeded and indeshiscent) [Keel petals distally fused and basally free] ________ (Papilionoideae) **Fabaceae**
      4. Petals 2 or 4; perianth spurred; fruits capsules or follicles.
        5. Stamens 6; fruits capsules ________ **Fumariaceae**
        5. Stamens 10–15; fruits follicles ________ **Ranunculaceae**
    3. Leaves simple.
      6. Sepals of 2 forms, stamens 5, 6, or 8.
        7. Perianth with a spur; stamens 5; flowers solitary or in few-flowered cymes ________ **Balsaminaceae**
        7. Perianth without a spur; stamens 6 or 8; flowers in spike-like or head-like racemes ____ **Polygalaceae**
      6. Sepals of 1 form, all alike; stamens 4 or 5.
        8. Petals clawed; fruits indehiscent, 1-seeded, lanate-tomentose ________ **Krameriaceae**
        8. Petals not clawed; fruits capsules or berries, glabrous or variously indumented, but not lanate-tomentose.
          9. Inflorescences spikes or racemes; fruits capsules ________ **Scrophulariaceae**
          9. Inflorescences cymes; fruits berries ________ **Solanaceae**
  2. Lower cauline leaves opposite or whorled.
    10. Fruits nutlets or achenes (each with a single seed).
      11. Fruits achenes, 1 per flower; flowers paired, oriented at right angles to rachises at anthesis; pedicels conspicuously reflexed and flowers appressed against rachises in fruit ________ **Phrymaceae**
      11. Fruits nutlets, 2–4 per flower; flowers solitary or paired or whorled, but not oriented at right angles to rachises at anthesis; pedicels not reflexed and flowers not appressed against rachises in fruit.
        12. Corollas bilabiate or unilabiate; stigmas distinctly bifid; styles gynobasic ________ **Lamiaceae**
        12. Corollas salverform; stigmas not bifid; styles apical. ________ **Verbenaceae**
    10. Fruits capsules, 1 per flower (seed number various).
      13. Plants with fetid odor; surfaces clammy with glandular hairs; fruits with incurved beak that splits at maturity to form 2 horns ________ **Pedaliaceae**
      13. Plants without fetid odor; surfaces not clammy, with or without hairs; fruits not developing 2 horns.
        14. Stamens 2.
          15. Corollas conspicuously bilaterally symmetrical, bilabiate ________ **Acanthaceae**
          15. Corollas inconspicuously bilaterally symmetrical, only 1 lobe slightly larger or smaller, not bilabiate ________ **Scrophulariaceae**
        14. Stamens 4 or 5.
          16. Petals 4, scarious; capsules circumscissile ________ **Plantaginaceae**
          16. Petals 5, not scarious; capsules septicidal or loculicidal.
            17. Seeds 2–4; anther apices recurved; anthers borne at 45 degree angle to filaments ________ **Acanthaceae**
            17. Seeds 12 or more; anther apices not recurved; anthers borne vertically or at less than 45 degree angle to filaments ________ **Scrophulariaceae**

Group S

Plants caulescent herbs; perianth parts in 2 series; petals 4 or 5 or in multiples of 4 or 5 or many; corollas radially symmetrical or asymmetrical; petals free; ovaries wholly or partially inferior.

1. Stamens 5.
   2. Plants less than 3 cm in diam. or height; inflorescences solitary flowers; seeds 12 or more per fruit. ____ **Saxifragaceae**
   2. Plants greater than 3 cm in diam. or height; inflorescences heads or umbels; seeds 1 or 2 per fruit.
      3. Petals plumose on adaxial surfaces, erect, linear; leaves bearing stinging hairs, sessile or subsessile; inflorescences heads; fruits achenes; sepals obvious ____ **Loasaceae**
      3. Petals not plumose, spreading, not linear; leaves indumented or glabrous, but without stinging hairs, petiolate; inflorescences umbels; fruits schizocarps or drupes; sepals inconspicuous, may be minute.
         4. Leaves whorled, palmately compound; fruits berry-like drupes ____ **Araliaceae**
         4. Leaves alternate, pinnately compound or simple; fruits schizocarps ____ **Apiaceae**
1. Stamens 8 or more.
   5. Sepals 2; styles 3–9; capsules circumscissile; placentation free-central; ovaries partially inferior, the distal 1/2 free from sepals and petals ____ **Portulacaceae**
   5. Sepals 3 or 4 or 5; styles 1; capsules loculicidal or poricidal; placentation axile or parietal; ovaries wholly inferior, the distal portion not free from sepals and petals.
      6. Petals 5 or apparently more with outer stamens sometimes petaloid; stamens 10 or more.
         7. Stamens 15–60; capsules 1-locular, poricidal; herbage with glochidiate, variously ornamented hairs, rough to the touch ____ **Loasaceae**
         7. Stamens 10; capsules 5-locular, loculicidal; herbage indumented or glabrous, but not rough to the touch ____ **Onagraceae**
      6. Petals 4; stamens 8.
         8. Leaves with 3 primary veins; hypanthia urceolate; anthers opening by terminal pores; inflorescences cymes or solitary flowers ____ **Melastomataceae**
         8. Leaves with 1 primary vein; hypanthia tubular; anthers opening by longitudinal slits; inflorescences panicles or spikes or flowers borne in leaf axils ____ **Onagraceae**

Group T

Plants caulescent herbs; perianth parts in 2 series; petals 2 or 4 or 5 or more; corollas radially symmetrical or asymmetrical; petals free; ovaries superior; pistils 1 per flower.

1. Petals 2, gray-black, bearing fleshy trichomes at apices ____ **Eriocaulaceae**
1. Petals 4 or 5 or more, of various colors, but not gray-black, not bearing fleshy trichomes.
   2. Flowers imperfect, the plants monoecious ____ **Euphorbiaceae**
   2. Flowers perfect.
      3. Sepals 2.
         4. Leaves fleshy, entire; sap thin, colorless; placentation basal or free-central ____ **Portulacaceae**
         4. Leaves not fleshy, variously toothed or divided; sap viscous, white or yellow or orange-red; placentation parietal ____ **Papaveraceae**
      3. Sepals 3 or more
         5. Petals 4.
            6. Sepals and petals inserted on a hypanthium.
               7. Anthers basifixed, curved; venation parallel-convergent, the veins 3, conspicuous [Ovaries falsely appearing superior because of their separation from hypanthia at maturity] ____ **Melastomataceae**

7. Anthers dorsifixed, straight; venation pinnate or a single vein ______ **Lythraceae**

6. Sepals and petals inserted on receptacle.

8. Leaves simple, entire or toothed or lobed or pinnatifid, but not compound.

9. Open flowers 7–10 cm in diam.; sepals with prickles; fruits with prickles; sap viscous, yellow or orange-red ______ **Papaveraceae**

9. Open flowers 0.3–5 cm in diam.; sepals without prickles; fruits without prickles; sap thin, colorless.

10. Stamens 12 or more ______ **Clusiaceae**

10. Stamens 2–10.

11. Leaves strongly gland-dotted and aromatic with a citrus-like odor; fruits 2-lobed capsules 3–7 mm long, the upward pointing lobes resembling the inflated legs of a dutchman's breeches ______ (*Thamnosma*) **Rutaceae**

11. Leaves neither gland-dotted nor aromatic; fruits various, but not as above.

12. Stamens equal in length; pistils 4-carpellate; fruits capsules; placentation free-central ______ **Caryophyllaceae**

12. Stamens didynamous or tetradynamous; pistils 2-carpellate; fruits siliques or silicles; placentation parietal. ______ **Brassicaceae**

8. Leaves compound.

13. Leaves palmately compound.

14. Stamens tetradynamous; ovaries 2-locular; fruits siliques ______ **Brassicaceae**

14. Stamens equal in length; ovaries 1-locular; fruits capsules ______ **Capparaceae**

13. Leaves pinnately compound.

15. Leaves 1-pinnately compound; stamens 2 or 4 or 6; fruits siliques or silicles ______ **Brassicaceae**

15. Leaves 2- or 3-pinnately compound; stamens 5 or 10; fruits berries or legumes.

16. Leaflets ovate or lanceolate; inflorescences racemes; fruits berries ___ **Ranunculaceae**

16. Leaflets linear or oblong; inflorescences heads; fruits legumes ______ (Mimosoideae) **Fabaceae**

5. Petals 5 or more.

17. Stamens 12 or more.

18. Filaments fused, forming a tube surrounding styles; stigmas peltate ______ **Malvaceae**

18. Filaments free or fused only at base, not forming a tube surrounding styles; stigmas not peltate.

19. Leaves 2- or 3-pinnately compound; fruits legumes or berries.

20. Leaflets ovate or lanceolate; inflorescences racemes; fruits berries ___ **Ranunculaceae**

20. Leaflets linear or oblong; inflorescences heads; fruits legumes ______ (Mimosoideae) **Fabaceae**

19. Leaves simple; fruits capsules.

21. Leaf margins conspicuously spinose; sap viscous, yellow or orange-red; sepals 3; capsules spiny ______ **Papaveraceae**

21. Leaf margins not spinose; sap thin, colorless; sepals 4 or 5; capsules not spiny.

22. Sepals in 2 whorls, the outer whorl of 2 smaller than inner whorl of 3; styles 1 ______ **Cistaceae**

22. Sepals in 1 whorl, all the same size; styles 2 ______ **Clusiaceae**

17. Stamens 1–11.

23. Stamens 1–5.

24. Leaves compound.

25. Inflorescences cymes; styles 5; fruits schizocarps ______ **Geraniaceae**

25. Inflorescences spikes; styles 1; fruits legumes (can be 1-seeded and indehiscent) ____ (Papilionoideae) **Fabaceae**

24. Leaves simple.
26. Leaves palmately lobed or crenate; pistils 2-carpellate ____ **Saxifragaceae**
26. Leaves entire or toothed or pinnately lobed, not crenate; pistils 3- or 4- or 5-carpellate.
27. Styles 3–5.
28. Leaves alternate; fruits 5-winged, bladdery capsules; petals pink or violet, with yellowish base; flowers axillary, solitary or in small cymes ____ **Sterculiaceae**
28. Leaves opposite or alternate; fruits unwinged capsules; petals pink, white, blue, yellow, yellow-orange, or red; flowers variously arranged.
29. Upper cauline leaves opposite; petals pink or white ____ **Caryophyllaceae**
29. Upper cauline leaves alternate; petals blue or yellow or yellow-orange or red ____ **Linaceae**
27. Styles 1.
30. Leaves lobed; inflorescences cymes; pistils 5-carpellate; fruits schizocarps ____ **Geraniaceae**
30. Leaves entire or toothed; inflorescences solitary flowers; pistils 3- or 4-carpellate; fruits capsules ____ **Saxifragaceae**

23. Stamens 6–11.
31. Leaves compound.
32. Leaves opposite.
33. Leaves pinnately compound; petals yellow ____ **Zygophyllaceae**
33. Leaves palmately compound; petals pink or purple or white ____ **Geraniaceae**
32. Leaves alternate.
34. Leaves palmately compound; styles 5; fruits capsules ____ **Oxalidaceae**
34. Leaves pinnately compound; styles 1; fruits legumes (can be 1-seeded and indehiscent.
35. Leaves 1-pinnately compound ____ (Papilionoideae) **Fabaceae**
35. Leaves 2-pinnately compound ____ (Mimosoideae) **Fabaceae**
31. Leaves simple.
36. Leaves alternate.
37. Petals and stamens arising from a hypanthium; stipules absent ____ **Saxifragaceae**
37. Petals and stamens arising from receptacle; stipules present.
38. Stamens free; fruits beaked ____ **Geraniaceae**
38. Stamens fused, forming a tube surrounding styles; fruits usually not beaked ____ **Malvaceae**
36. Leaves opposite.
39. Leaf margins palmately lobed or palmately parted; fruits schizocarps ____ **Geraniaceae**
39. Leaves margins entire or toothed; fruits capsules.
40. Styles 1; sepals in 2 whorls, the outer whorl of 2 smaller than inner whorl of 3 ____ **Cistaceae**
40. Styles 2–5; sepals in 1 whorl.
41. Stamens 9, in 3 fascicles ____ **Clusiaceae**
41. Stamens 5–10, separate, not in fascicles.
42. Placentation free-central ____ **Caryophyllaceae**
42. Placentation axile ____ **Elatinaceae**

Group U

Plants caulescent herbs; perianth parts in 2 series; petals 4 or 5 or in multiples of 4 or 5 or many; corollas radially symmetrical or asymmetrical; petals free; ovaries superior; pistils 2 or more per flower.

1. Leaves opposite or whorled ________ **Crassulaceae**
1. Leaves alternate or basal.
  2. Hypanthia absent; perianth and stamens inserted on receptacle.
    3. Leaves succulent, terete; stamens 8 or 10 ________ **Crassulaceae**
    3. Leaves neither succulent nor terete; stamens 12 or more.
      4. Filaments free, not forming a tube around styles; stamens spiraled; ovaries free throughout development ________ **Ranunculaceae**
      4. Filaments fused, forming a tube around styles; stamens whorled; ovaries fused until the fruits mature, then separating [hence falsely appearing polycarpous] ________ (Malvaceae) **Group T**
  2. Hypanthia present as a disk or cup or tube; perianth and stamens inserted on hypanthium.
    5. Pistils 5 or more per flower ________ **Rosaceae**
    5. Pistils 2 or 3 per flower.
      6. Leaves compound; stipules present ________ **Rosaceae**
      6. Leaves simple; stipules absent ________ **Saxifragaceae**

Group V

Plants caulescent herbs; perianth parts in 2 series; petals 2 or 4 or 5; corollas radially symmetrical or asymmetrical; petals fused at least at the base; ovaries wholly or partially inferior.

1. Stems trailing or prostrate.
  2. Tendrils present; leaves alternate; flowers imperfect; fruits pepos ________ **Cucurbitaceae**
  2. Tendrils absent; leaves opposite or whorled; flowers perfect; fruits drupes or schizocarps ________ **Rubiaceae**
1. Stems erect or ascending.
  3. Flowers with hypanthium-tube elongated beyond ovary [thus falsely giving the appearance of fused petals] ________ (Onagraceae) **Group S**
  3. Flowers without an elongated hypanthium-tube.
    4. Anthers connivent or fused.
      5. Inflorescences racemes or cymes or mixed; fruits capsules; sepals present, not modified into a pappus ________ **Campanulaceae**
      5. Inflorescences heads; fruits achenes; sepals absent or modified into a pappus ________ **Asteraceae**
    4. Anthers free.
      6. Ovaries partially inferior, the distal 1/3–1/2 free from sepals and petals.
        7. Petals 5; ovaries 5-carpellate, 1-locular; placentation free-central ________ **Primulaceae**
        7. Petals 4; ovaries 2-carpellate, 2-locular; placentation axile ________ **Rubiaceae**
      6. Ovaries wholly inferior, the distal portion not free from sepals and petals.
        8. Leaves alternate.
          9. Corollas 5–10 mm long; rachises of inflorescences visible; stamens attached at middle of corolla tubes; capsules circumscissile ________ **Sphenocleaceae**
          9. Corollas 2.3–2.7 mm long; rachises of inflorescences not visible; stamens attached at bases of corolla tubes; capsules poricidal or loculicidal ________ **Campanulaceae**
        8. Leaves opposite or whorled.
          10. Flowers numerous, borne in dense flat-topped inflorescences; branches conspicuously dichotomous; locules 3, 2 small and empty, 1 large and containing 1 seed ________ **Valerianaceae**

10. Flowers solitary or borne in few-flowered inflorescences that are not flat-topped or in terminal heads; branches not conspicuously dichotomous; locules 1 or 2 or 3 or 5.
11. Sepals 8–10 mm long; corollas gibbous; stipules absent ________ **Caprifoliaceae**
11. Sepals 0.5–5 mm long; corollas not gibbous; stipules present (sometimes leaf-like and the leaves thus appearing whorled) ________ **Rubiaceae**

Group W

Plants caulescent herbs; perianth parts in 2 series; petals 2 or 4 or 5; corollas radially symmetrical or asymmetrical; petals fused at least at the base; ovaries superior.

1. Pistils or fruits 2 or 4 or 5 per flower.
2. Fruits follicles or capsules, multi-seeded.
3. Plants succulent; petals fused only at base and not forming a tube and limb; fruits 5 per flower ________ **Crassulaceae**
3. Plants not succulent; petals fused forming a tube and limb; fruits 2 or 4 per flower.
4. Plants prostrate or decumbent; sap thin; colorless; stigmas not massive; fruits capsules; seeds 2–4; leaves alternate [two ovary lobes united only at base by gynobasic style, and thus falsely appearing separate] ________ *(Dichondra)* **Convolvulaceae**
4. Plants usually erect or ascending; sap typically viscous; white; stigmas massive; fruits follicles; seeds 12 or more; leaves opposite or alternate.
5. Coronas present; stigmas fused to anther and/or corolla tissues; pollinia present; styles 2 ________ **Asclepiadaceae**
5. Coronas absent; sigmas not fused to anther and/or corolla tissues; pollinia absent; styles 1 ________ **Apocynaceae**
2. Fruits nutlets, each 1-seeded.
6. Stamens 5; leaves alternate ________ **Boraginaceae**
6. Stamens 2 or 4; leaves opposite or whorled.
7. Styles gynobasic; stigmas 2; nutlet scars basal ________ **Lamiaceae**
7. Styles apical; stigmas 1; nutlet scars covering the entire inner surface ________ **Verbenaceae**
1. Pistils or fruits 1 per flower.
8. Pistils with 2 separate ovaries, 1 or 2 styles, but only 1 stigma due to fusion; stigmas massive; fruits follicles.
9. Coronas present; stigmas fused to anther and/or corolla tissues; pollinia present; styles 2 ________ **Asclepiadaceae**
9. Coronas absent; stigmas not fused to anther and/or corolla tissues; pollinia absent; styles 1 ________ **Apocynaceae**
8. Pistils with only 1 ovary, 1 or more styles, and 1 or more stigmas; stigmas not massive; fruits capsules or nutlets or anthocarps or legumes or schizocarps or berries.
10. Fruits nutlets OR anthocarps (= indehiscent achene and persistent base of perianth tube), 1–4 per flower.
11. Perianths 35–170 mm long [Petals absent, sepals petaloid, and involucre resembling calyx, hence perianths falsely appearing to be in 2 series] ________ (Nyctaginaceae) **Group N**
11. Perianths 1.2–35 mm long.
12. Stems usually with at least lower nodes swollen; the two leaves at a node often unequal; ovaries apparently inferior (tightly enclosed by base of perianth); fruits anthocarps, 1 per flower [Petals absent, sepals petaloid, and involucre resembling calyx, hence perianths falsely appearing to be in 2 series] ________ (Nyctaginaceae) **Group N**
12. Stems usually without nodes swollen; the two leaves at a node usually equal; ovaries superior; fruits nutlets, 1–4 per flower.

13. Stamens 5; leaves alternate ________ **Boraginaceae**
13. Stamens 2 or 4; leaves opposite or whorled.
14. Styles gynobasic; stigmas 2; nutlet scars basal ________ **Lamiaceae**
14. Styles apical; stigmas 1; nutlet scars covering the entire inner surface ________ **Verbenaceae**

10. Fruits capsules or berries or schizocarps or legumes.
15. Leaves opposite or whorled.
16. Stamens opposite the corolla lobes; pistils 5-carpellate; placentation free-central ________ **Primulaceae**
16. Stamens alternate with the corolla lobes; pistils 2- or 3- carpellate; placentation parietal or axile.
17. Pistils 3-carpellate; stigmas 3 ________ **Polemoniaceae**
17. Pistils 2-carpellate; stigmas 1 or 2.
18. Inflorescences scorpioid cymes ________ **Hydrophyllaceae**
18. Inflorescences of various types, but not scorpioid cymes.
19. Stamen number less than corolla lobe number.
20. Corollas variously colored but not yellow inside and not red outside; capsules not circumscissile; plants of various sizes ________ **Acanthaceae**
20. Corollas yellow inside and ± red outside; capsules circumscissile; plants 25 cm or less tall ________ (*Menodora*) **Oleaceae**
19. Stamen number same as corolla lobe number.
21. Corollas white OR white suffused or lined with pink OR light blue.
22. Leaf margins pinnatifid ________ **Hydrophyllaceae**
22. Leaf margins entire or serrate.
23. Leaf bases connected around the stem by united short stipules or a stipular ridge; corrolla throats indumented OR glabrous; locules 2; placentation axile.
24. Leaves lanceolate or broader, usually 10 mm or more wide; flowers 5-merous ________ **Loganiaceae**
24. Leaves narrowly linear, usually 2 mm or less wide; flowers 4-merous ________ **Buddlejaceae**
23. Leaf bases without a trace of stipules; corolla throats glabrous; locules 1; placentation parietal ________ **Gentianaceae**
21. Corollas of various colors, but not white or light blue.
25. Corollas red and yellow; placentation axile ________ **Loganiaceae**
25. Corollas green or blue-purple or pink; placentation parietal ________ **Gentianaceae**

15. Leaves alternate and/or basal.
26. Corolla lobes 2, gray-black, bearing fleshy trichomes at apices; anthers black ________ **Eriocaulaceae**
26. Corolla lobes 4 or 5, of various colors, but not gray-black, not bearing fleshy trichomes; anthers of various colors, but not black.
27. Pistils 5-many carpellate.
28. Stamen filaments fused, forming a tube surrounding styles.
29. Stamens 5–10 [Petals coherent, and thus falsely appearing fused] ________ (Oxalidaceae) **Group T**
29. Stamens 12–many [Petals fused basally to staminal tube, and thus falsely appearing fused] ________ (Malvaceae) **Group T**
28. Stamen filaments free from each other.

30. Seeds 1; styles 3 or 5; petals fused only at base; [Family in OK and s TX, not in nc TX] ____ **Plumbaginaceae**
30. Seeds 5 or more; styles 1; petals fused more than 1/2 length ____ **Primulaceae**

27. Pistils 1–3-carpellate.
31. Petals 4; fruits circumscissile or septicidal capsules.
32. Inflorescences panicles or racemes, terminal; capsules septicidal ____ **Gentianaceae**
32. Inflorescences terminal spikes or solitary flowers borne in axils of leaves; capsules circumscissile.
33. Inflorescences spikes, terminal; petals scarious, colorless or tan ____ **Plantaginaceae**
33. Inflorescences solitary flowers, axillary; petals not scarious, pink [5-carpellate but falsely appearing 1-carpellate] ____ **Primulaceae**
31. Petals 5; fruits berries or loculicidal capsules or legumes.
34. Stamens 5–12 or more; filaments exserted beyond perianth; inflorescences heads; leaves 2-compound; fruits legumes __ (Mimosoideae) **Fabaceae**
34. Stamens 5 or fewer; filaments not prominently exserted beyond perianth; inflorescences of various types, but not heads; leaves simple, but may be deeply dissected; fruits capsules or berries.
35. Ovaries 3-locular; stigmas 3 ____ **Polemoniaceae**
35. Ovaries 1- or 2- or 4-locular; stigmas 1 or 2.
36. Stamens opposite the corolla lobes; placentation free-central. ____ **Primulaceae**
36. Stamens alternate with the corolla lobes; placentation parietal or axile.
37. Inflorescences helicoid cymes ____ **Hydrophyllaceae**
37. Inflorescences of various types, but not helicoid cymes.
38. Leaves pinnatifid.
39. Petals longer than sepals; fruits berries; seeds 12 or more; placentation axile ____ **Solanaceae**
39. Petals equal to or shorter than sepals; fruits capsules; seeds 4; placentation parietal ____ **Hydrophyllaceae**
38. Leaves entire or variously lobed, but not pinnatifid.
40. Sepals fused.
41. Styles 2; seeds 1–4 ____ **Convolvulaceae**
41. Styles 1; seeds 12 or more ____ **Solanaceae**
40. Sepals free.
42. Corollas 5–9 cm long; styles not divided; seeds 1–4 ____ **Convolvulaceae**
42. Corollas 0.5–2 cm long; styles divided; seeds 12 or more ____ **Hydrophyllaceae**

FERNS AND SIMILAR PLANTS (PTERIDOPHYTES)

Seedless vascular plants (reproducing by spores) formerly lumped together as the Division Pteridophyta, the ferns and similar plants are currently segregated into three separate divisions (Lycopodiophyta, Equisetophyta, and Polypodiophyta) to reflect the great diversity between these ancient plant groups; the group Pteridophyta is thus no longer formally recognized. Together the three divisions have nearly 10,000 species (Wagner & Smith 1993). For a Key to Ferns and Similar Plants see page 110 or Key K on page 154.

REFERENCE: Wagner & Smith 1993.

DIVISION **LYCOPODIOPHYTA**
CLUBMOSSES, SPIKE-MOSSES, AND QUILLWORTS

A group of 1,200–1,250 species in 12–17 genera arranged in 3 families (Flora of North America Editorial Committee 1993). Extinct members of this ancient division (e.g., Lepidodendrales—scale trees to 30 m tall) were dominants of the Carboniferous forests that formed present-day coal deposits; it is one of the oldest plant groups, dating to the Lower Devonian period (408–360 million years ago) (Benson 1979; Bell & Woodcock 1983; Jones & Luchsinger 1986; Raven et al. 1986). The Lycopodiophyta are characterized by microphylls (= leaves with a single vein). There are three extant families, Isoetaceae, Lycopodiaceae, and Selaginellaceae, all with representatives in nc TX. The group is sometimes referred to as the Microphyllophyta (Woodland 1997).

REFERENCES: Benson 1979; Bell & Woodcock 1983; Jones & Luchsinger 1986; Raven et al. 1986; Bold et al. 1987; DiMichele & Skog 1992; Wagner & Smith 1993; Woodland 1997.

LYCOPODIACEAE CLUBMOSS FAMILY

A diverse ancient family with a long fossil history; it is cosmopolitan and contains 10–15 genera and ca. 350–400+ species of terrestrial or epiphytic, evergreen, coarsely moss-like, vascular plants with scale- or needle-like leaves containing a single vein; ligules (= minute, tongue-like, basal protuberance on a leaf) are absent and spores are all of one type. Many species were previously treated in the large genus *Lycopodium*, which is now usually divided into a number of segregate genera; some of these segregates are known to hybridize. Certain species were in the past gathered for making Christmas wreaths; in some areas (e.g., Appalachian Mts.) this resulted in populations being greatly reduced; the very flammable, dust-like, dry spores of some were formerly used in fireworks, for stage-lighting, and in photography as flash powder (Jones & Luchsinger 1986).

FAMILY RECOGNITION IN THE FIELD: evergreen, superficially somewhat moss-like herbs with stems covered by numerous, small, linear-lanceolate to lanceolate, *1-veined leaves*; stems lying flat on the ground with upright shoots terminating in *cylindrical, spore-producing cones*.

REFERENCES: Correll 1949, 1956, 1966a; Wagner & Beitel 1992, 1993.

LYCOPODIELLA BOG CLUBMOSS

Lycopodiella , distinguished by its prostrate stems, has often been treated in a more broadly defined *Lycopodium*. As treated here, *Lycopodiella* is a genus of 8–10 species of the n temperate region and tropical America; a number of the species readily hybridize. (Name derived from the genus *Lycopodium* (Greek: *lykos*, wolf, and *pous* or *podium*, foot; in reference to the resemblance of the branch tips to a wolf's paw), plus the Latin diminutive, *-ella*)

Lycopodiella appressa (Chapm.) Cranfill, (appressed or lying close), CHAPMAN CLUBMOSS, SOUTHERN CLUBMOSS, APPRESSED BOG CLUBMOSS. Plant perennial; horizontal stems flat on ground; upright, usually unbranched leafy shoots (serving as peduncles) scattered along stems; leaves numerous, small, linear-lanceolate to lanceolate, 6–7 mm long, incurved, appressed, 1-nerved; strobili solitary, terminating peduncles, slender, 0–2 mm thicker than the supporting shoot, ca. 25–70 mm long, 3–4 mm wide; sporophylls (= spore-bearing leaves) incurved, appressed, similar to other leaves; sporangia subglobose, solitary at base of sporophylls. Depressions and moist areas; Henderson Co. (Correll 1956), also Carr (1994) listed an unidentified *Lycopodium* (probably *L. appressa)* for Lamar Co.; se and e TX w to e margin of nc TX. Sporulating Jun-Oct. [*Lycopodium appressum* (Chapm.) F.E. Lloyd & Underw.]

Lycopodiella prostrata (R.M. Harper) Cranfill [*Lycopodium prostratum* R.M. Harper], (prostrate), CREEPING CLUBMOSS, PROSTRATE BOG CLUBMOSS, distinguished from *L. appressa* by having its sporophylls ± spreading and the stroboli stout (12–20 mm wide), 3–6 mm wider than the supporting shoot, is known from one TX site just s of nc TX in Travis Co. (Correll 1956).

SELAGINELLACEAE SPIKE-MOSS FAMILY

A cosmopolitan, but primarily tropical and subtropical family currently treated as a single genus with > 700 species of usually terrestrial or epiphytic, superficially moss-like vascular plants bearing spores differentiated into microspores and megaspores; leaves usually have a single vein and ligules (= minute, tongue-like basal protuberance on a leaf; the function is uncertain) are present. This family is apparently only distantly related to the Lycopodiaceae and Isoetaceae. FAMILY RECOGNITION IN THE FIELD: superficially somewhat moss-like, small herbs with numer ous, scale-like, *1-veined leaves*; stems terminating in ± *4-angled, spore-producing cones*.
REFERENCES: Correll 1956, 1966a; Valdespino 1993.

SELAGINELLA SPIKE-MOSS

Ours small terrestrial or lithophytic (= growing on rocks) plants; stems leafy; vegetative leaves small, with ligule on adaxial side near base, all alike or of 2 kinds; sporophylls (= fertile leaves) modified, in strobili (= cones) at branch tips; sporangia solitary in axils of sporophylls, of 2 kinds (plants heterosporous).

Selaginella is the only extant genus in the family; it has an extremely long history in the fossil record; it is currently most diverse in the tropics. Some are well known as "resurrection" plants, capable of reviving after long periods of dessication. (From *Selago*, an ancient name for *Lycopodium*, a genus resembling *Selaginella*, and the Latin diminutive suffix, *-ella)*
REFERENCES: Clausen 1946; Tryon 1955.

1. Plants of moist habitats, delicately thin-herbaceous; stem leaves not overlapping or only slightly so, in 4 ranks, 2 lateral and spreading, 2 smaller and appressed-ascending along the adaxial (= above) surface of the stem; abaxial (= beneath) surface of the stem easily visible; plants annual ____ **S. apoda**
1. Plants of xerophytic habitats, rather rigid; stem leaves crowded, conspicuously overlapping, appressed to stem, not in 4 distinct ranks; abaxial surface of the stem not visible (concealed by leaves completely surrounding the stem); plants perennial.
    2. Vegetative part of plant erect to ascending; leaves not curving upward, the leaf-covered stems therefore appearing radially symmetrical ____ **S. arenicola**
    2. Vegetative part of plant ± completely prostrate; leaves curving upward making the adaxial and abaxial views of the leaf-covered stems distinctly different ____ **S. peruviana**

Selaginella apoda (L.) Spring, (footless), MEADOW SPIKE-MOSS, BASKET SELAGINELLA. Plant prostrate-creeping or ascending, often forming mats; leaves of 2 distinct kinds; lateral leaves ovate to ovate-elliptic, asymmetrical, ca. 1.35–2.25 mm long, 0.75–1.35 mm wide; appressed-ascending leaves smaller, to ca. 1.2(–1.6) mm long; strobili solitary or paired, obscurely quadrangular (= 4-sided)-flattened, 0.5–2 cm long; 2–4 mm in diam.; sporophylls apically acute to acuminate. Moist areas, low fields and woods; Burnet Co., also Ellis (Correll 1956), and Lamar (Carr 1994) cos.; mainly e TX and in several localities in se TX and Edwards Plateau. Sporulating May–Dec.

Selaginella arenicola Underw. subsp. **riddellii** (Van Eselt.) R.M. Tryon, (sp.: growing in sandy places; subsp.: for J.L. Riddell, 1807–1865, botanist), RIDDELL'S SELAGINELLA, RIDDELL'S SPIKE-MOSS. Vegetative part of plant erect to ascending, forming clumps, to ca. 12 cm tall, usually smaller; leaves narrowly triangular-lanceolate to linear-lanceolate, ca. 1.2–3 mm long, 0.4–0.5 mm wide, marginally ciliate, apically with whitish bristle; stroboli solitary, sometimes with apical vegetative growth, quadrangular, ascending, (0.5–)1–3(–3.5) cm long and ca. 1.2 mm in diam.; sporophylls often with a bristle. Rocky areas, sandy or gravelly soils; Bell Co., also Burnet Co. (Correll 1956); e 1/3 of TX w to e Edwards Plateau. Sporulating throughout the year. [*S. riddellii* Van Eselt.]

Selaginella peruviana (J. Milde) Hieronymus, (of Peru, the species ranging to South America), PERUVIAN SPIKE-MOSS. Vegetative part of plant ± completely prostrate, forming loose mats; main stems to ca. 12 cm long; leaves linear-lanceolate to falcate, 1.6–4 mm long, 0.4–0.5 mm wide, marginally ciliate, apically with whitish bristle 0.3–0.8 mm long; strobili solitary, quadrangular, ascending, 0.5–2 cm long, 1–1.5 mm in diam.; sporophylls usually bristle-tipped. On rocks or ground; Comanche Co. (Stanford 1971), also Burnet Co. (Correll 1956); sw part of nc TX through Edwards Plateau to Trans-Pecos. Sporulating Jun–Oct. [*S. sheldonii* Maxon]

ISOETACEAE QUILLWORT FAMILY

A monogeneric, nearly cosmopolitan family of ca. 150 species of superficially grass- or sedge-like plants ranging from perennial evergreen aquatics to ephemeral terrestrials; they are superficially unlike other Lycopodiophyta, but as in other members of the division, the leaves have a single vein; ligules are present as in the Selaginellaceae; spores are differentiated into microspores and megaspores. The long linear leaves have a resemblance to the quills of feathers formerly used as writing implements.

FAMILY RECOGNITION IN THE FIELD: the single nc TX species is a tufted, wet area plant with superficially grass-like or sedge-like leaves and a corm-like rootstock giving it a green onion-like appearance; *sporangia are in the leaf bases.*

REFERENCES: Pfeiffer 1922; Correll 1949, 1956, 1966a; Taylor et al. 1993.

ISOETES QUILLWORT

Interspecific hybrids are frequently seen; the spores are reported to be dispersed in the excreta of earthworms; species are often difficult to identify, sometimes requiring microscopic examination of spores. (Greek: *isos*, equal, and *etos*, year, referring to the evergreen habit of some species)

REFERENCES: Taylor et al. 1975; Boom 1982; Taylor & Hickey 1992.

Isoetes melanopoda J. Gay & Durieu ex Durieu, (black-footed), BLACK-FOOTED QUILLWORT. Plant tufted, with leaves tightly clustered together and superficially resembling a green onion, usually terrestrial or becoming so; rootstock corm-like, globose, 2-lobed; leaves superficially grass-like or sedge-like, to 40 cm long, blackish towards very base; sporangia solitary, embedded in basal cavity of leaf with ligule inserted above, often partly covered by a velum (= thin flap of tissue); spores of 2 types (plant heterosporous), the megaspores whitish, usually with prominent ridges. Seasonally saturated soils, temporary pools, shallow pools; Dallas Co., also Burnet and Tarrant cos. (Correll 1956); se and e TX w to nc TX and Edwards Plateau. Sporulating Mar–Oct.

DIVISION **EQUISETOPHYTA**
HORSETAILS

This a very ancient group consisting of a single extant family; fossil forms date to the Devonian period (408–360 million years ago) and the division reached its maximum diversity and abundance in the Paleozoic era; they were components of the Carboniferous swamp forests that formed present-day coal deposits; some reached the proportions of trees (to 18 m tall) and were probably competitors of the tree Lycopodiophyta. The largest living species is the tropical *Equisetum giganteum* L., which may exceed 5 m in height (Bell & Woodcock 1983; Raven et al. 1986; Bold et al. 1987). The division is sometimes referred to as the Arthrophyta (Woodland 1997) or the Sphenophyta (Raven et al. 1986). The Equisetophyta are characterized by microphylls (= leaves with a single vein). Some species have numerous small branches and bear a slight resemblance to a horse's tail.
REFERENCES: Bell & Woodcock 1983; Raven et al. 1986; Bold et al. 1987; Wagner & Smith 1993; Woodland 1997.

EQUISETACEAE HORSETAIL FAMILY

The family is represented only by the distinctive genus *Equisetum* which is also the only extant genus in the division; it has a long fossil history. *Equisetum* is nearly cosmopolitan and contains ca. 15 species.
FAMILY RECOGNITION IN THE FIELD: plant body consisting primarily of *hollow, jointed, green* stems; leaves inconspicuous, scale-like, in whorls at the very distinct nodes; sporangia in *small, terminal cones.*
REFERENCES: Correll 1949, 1956, 1966a; Hauke 1993.

EQUISETUM HORSETAIL, SCOURING-RUSH

Plants perennial, rhizomatous; stems hollow in center, jointed with very distinct nodes, ridged, green and photosynthetic; leaves small, inconspicuous, whorled, scale-like, fused into sheaths but with tips free and tooth-like; sporangia on the undersurface of pelate sporophylls arranged in discrete terminal stroboli (= cones); spores of 1 kind (plant homosporous).

The coarse stems contain silica and were used by early settlers to scour pots and pans (Woodland 1997); ☠ some species contain alkaloids or other toxins such as thiaminase, an enzyme that destroys thiamine and causes Vitamin B_1 deficiency; they can be poisonous to livestock when included in hay (Kingsbury 1964; Burlage 1968; Fuller & McClintock 1986); hybridization between species is frequent. (Latin: *equis*, horse, and *seta*, bristle, referring to the coarse black roots of *E. fluviatile* L.)

1. Sheaths (= fused leaves) dark girdled at most nodes of stem (in addition to thin dark line at sheath apex where teeth are shed), ashy-gray to brownish above girdle; aerial stems usually persisting more than one year; cone apex pointed; teeth of sheaths promptly shed or persistent ____ **E. hyemale**
1. Most sheaths green (but with a thin dark line at sheath apex where teeth are shed), only some near stem base dark girdled; aerial stems lasting less than a year, occasionally overwintering; cone apex rounded to pointed; teeth of sheaths promptly shed ____ **E. laevigatum**

Equisetum hyemale L. subsp. **affine** (Engelm.) Calder & R.L. Taylor, (sp.: of winter; subsp.: related), TALL SCOURING-RUSH, AMERICAN SCOURING-RUSH, COMMON SCOURING-RUSH, GREAT SCOURING-RUSH, CAÑUELA. Stems 18–220 cm tall; leaves 14–50 per node (number evident as teeth of sheaths). Parker and Tarrant cos., also Erath and Grayson (Correll 1956); throughout TX. Sporu-

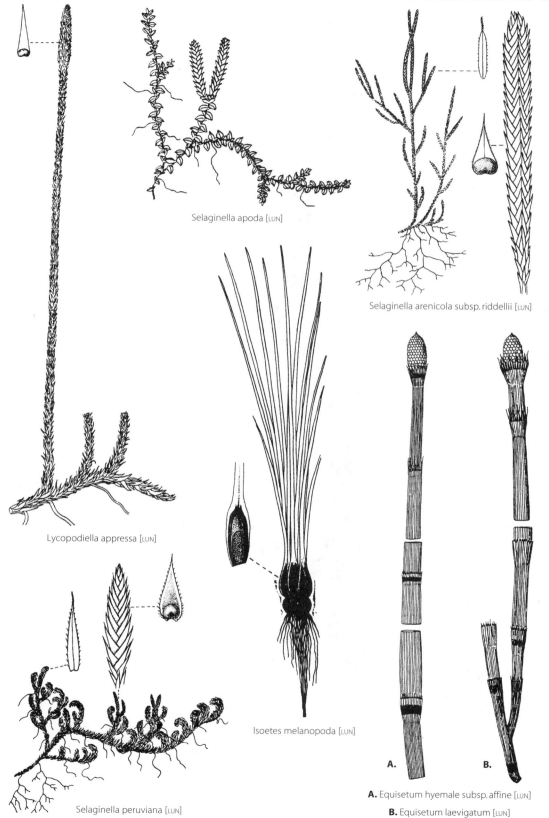

Selaginella apoda [LUN]

Selaginella arenicola subsp. riddellii [LUN]

Lycopodiella appressa [LUN]

Isoetes melanopoda [LUN]

Selaginella peruviana [LUN]

A. Equisetum hyemale subsp. affine [LUN]

B. Equisetum laevigatum [LUN]

lating Mar–late fall. [*E. hyemale* L. var. *affine* (Engelm.) A.A. Eaton, *E. prealtum* Raf.] Poisonous (Burlage 1968). 9

Equisetum laevigatum A. Braun, (smooth), SMOOTH HORSETAIL, SMOOTH SCOURING-RUSH, BRAUN'S SCOURING-RUSH, KANSAS HORSETAIL, KANSAS SCOURING-RUSH, SUMMER SCOURING-RUSH, COLA DE CABALLO, CAÑUELA. Stems 20–150 cm tall; leaves 10–32 per node. Dallas and Somervell cos., also Erath Co. (Correll 1956); throughout much of TX. Sporulating May–Jul. [*E. kansanum* J.F. Schaffn.] These two species are often very difficult to distinguish in nc TX and seem to intergrade. According to Hauke (1993), we are within the range of *E. ×ferrissii* Clute, a hybrid between *E. hyemale* and *E. laevigatum*. Hauke (1993) distinguished *E. ×ferrissii* from the two parental species (with greenish spherical spores) by its white misshapen spores. Poisonous (Burlage 1968).

DIVISION **POLYPODIOPHYTA**
FERNS

A group of 8,550 species in 223 genera arranged in 33 families (Mabberley 1997). The fossil record of ferns dates to the Carboniferous period (360–286 million years ago) and related groups occurred back to the Devonian period. The leaves are megaphylls (with branched veins) which apparently are derived from modified branch systems; spores are of one or two types. Modern species range from tree ferns (to 24 m tall) to free-floating aquatics, but are mostly rhizomatous perennial herbs. The group is sometimes referred to as the Filicophyta or the Pterophyta (Bell & Woodcock 1983; Raven et al. 1986). For a Key to Ferns and Similar Plants see page 110 or Key K on page 154.

REFERENCES: Bush 1903; Reverchon 1903; Small 1938; Correll 1949, 1956, 1966a; Thieret 1980; Tryon & Tryon 1982; Taylor 1984; Lellinger 1985; Bell & Woodcock 1983; Raven et al. 1986; Bold et al. 1987; Flora of North America Editorial Committee 1993; Wagner & Smith 1993.

GENERAL CHARACTERISTICS OF THE FERNS [JEP]

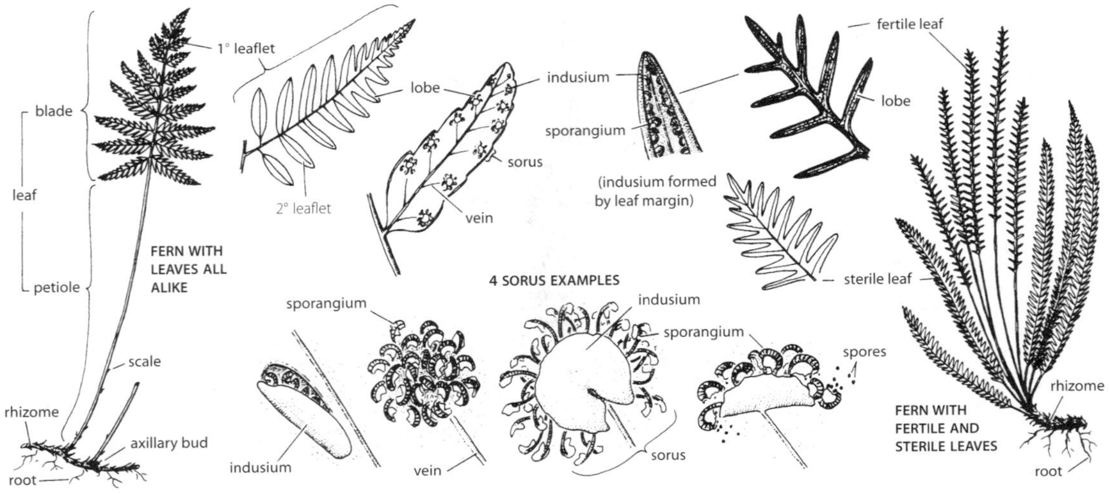

ANEMIACEAE ANEMIA FAMILY

A family of 2 genera and ca. 119 species widespread in the tropics and subtropics. It is sometimes lumped with the Schizaeaceae.

FAMILY RECOGNITION IN THE FIELD: the single local species has 1-pinnate leaves with 2 conspicuously different types of pinnae: 4–6 pairs of sterile pinnae and below these a pair of *very long stalked*, bipinnate, fertile pinnae.

REFERENCES: Mickel 1981, 1993.

ANEMIA

A genus of 117 species of tropical and subtropical regions of the world, especially Brazil and Mexico. *Anemia* is sometimes placed in the Schizaeaceae (Kartesz 1994); however, we are following Mickel (1993) in placing it in the Anemiaceae. (Greek: *aneimon*, without clothing, referring to the absence of blade protection for the sporangia)

REFERENCES: Correll 1956, 1966a.

Anemia mexicana Klotzsch, (Mexican), MEXICAN FERN. This species, found primarily on limestone outcrops on the Edwards Plateau (n to Travis Co. just to the s of nc TX), is also disjunct to Austin Co. to the se of nc TX. It is a small fern (to ca. 50 cm tall) with leaves 1-pinnate, with 4–6 pairs of sterile pinnae and with the lowermost pair of pinnae fertile, very long stalked, bipinnate, highly modified, to 30 cm long, and usually exceeding the sterile portion of the leaf in length. It is included here to alert collectors because reasonable habitat exists in the s portion of nc TX.

ASPLENIACEAE SPLEENWORT FAMILY

A cosmopolitan monogeneric family of ca. 700 species; all species are currently treated as members of a diverse genus *Asplenium*.

FAMILY RECOGNITION IN THE FIELD: leaves 1-pinnate, all alike or the fertile slightly smaller; *sori elongate* along the veins; indusia attached along one side of the sori.

REFERENCE: Wagner et al. 1993.

ASPLENIUM SPLEENWORT

Ours terrestrial (in soil) or on rocks; stems (rhizomes) short-creeping to erect; leaves clustered, 1-pinnate, mostly evergeen; sori elongate along veins; indusia attached along the edge of the sori.

The genus is well known for its intraspecific hybridization and complex polyploid series with numerous allopolyploids; ploidy levels range from diploid to hexaploid; three-fifths of the species are thought to be of hybrid, allopolyploid origin; a number of species are cultivated as ornamentals (e.g., *A. nidus* L.—BIRD'S-NEST FERN). (Greek: *splen*, spleen; thought by Dioscorides to be useful for treating spleen diseases)

REFERENCES: Wagner 1954; Correll 1956, 1966a.

1. Pinnae (leaflets) usually alternate, with their basal auricles overlapping the rachis, their margins subentire to deeply serrate or incised; plants terrestrial or growing on rocks; leaves slightly dimorphic, the fertile erect, the sterile smaller and spreading ____ **A. platyneuron**
1. Pinnae opposite, usually not overlapping the rachis, their margins subentire to crenulate; plants usually growing on rocks; leaves monomorphic, all fertile, erect or ascending ____ **A. resiliens**

Asplenium platyneuron (L.) Britton, Sterns, & Poggenb., (broad-nerved), EBONY SPLEENWORT. Leaves to 50 cm tall; leaf blades linear-lanceolate to narrowly elliptic-lanceolate in outline; petiole and rachis usually reddish brown to dark brown (rarely nearly black), shining. Sandy, moist, wooded banks and slopes, or on rocks; Cooke Co. (Correll 1956), Fannin, Grayson, Tarrant, and Parker cos., also Palo Pinto Co. (R. O'Kennon pers. obs.); se and e TX w to West Cross Timbers. Sporulating Apr–Dec.

Asplenium resiliens Kunze, (recoiling), LITTLE EBONY SPLEENWORT, BLACK-STEM SPLEENWORT. Leaves to ca. 35 cm tall, the blades linear-oblong to linear-lanceolate, usually more coriaceous than in *A. platyneuron*; petiole and rachis black, shining. Usually growing on rocks; Bell, Burnet, Grayson, and Palo Pinto cos.; also Brown and Erath cos. (Correll 1956) and Coryell Co. (Fort Hood—Sanchez 1997); widely scattered in TX. Sporulating Apr–Nov.

AZOLLACEAE
AZOLLA, MOSQUITO FERN, OR WATER FERN FAMILY

A cosmopolitan family of a single genus and only ca. 7 species of floating aquatics (sometimes stranded on mud); it is often included in the Salviniaceae, but according to Lumpkin (1993), the relationship is not close.
FAMILY RECOGNITION IN THE FIELD: tiny, liverwort-like, free-floating or mat-forming plants that sometimes form conspicuous velvet-like, green to red mats on the surface of quiet waters.
REFERENCE: Lumpkin 1993.

AZOLLA WATER FERN, MOSQUITO FERN

The upper leaf lobes (out of the water) of *Azolla* are hollow and inhabited by a symbiotic nitrogen-fixing cyanobacterium (= blue-green bacterium), *Anabaena azollae* Strasb. Because of the resulting nitrogen content, *Azolla* species have been widely used agriculturally as a fertilizer. (Greek: *azo*, to dry, and *ollyo*, to kill, alluding to death from drought)
REFERENCES: Svenson 1944, Correll 1956, 1966a.

Azolla caroliniana Willd., (of Carolina), MOSQUITO FERN, WATER FERN. Plant small, free-floating or mat-forming, superficially resembling some liverworts; stems prostrate, to ca. 1 cm long; leaves minute, deeply bilobed, imbricate, deep green to reddish (under stress); infrequently fertile; sporocarps of two kinds, in the leaf axils, the megasporocarps with 1 megasporangium producing 1 megaspore, the microsporocarps with numerous microsporangia containing numerous microspores. Still water of ponds, lakes, or slow-moving streams or stranded on mud; Grayson, Fannin, Lamar, and Tarrant cos., also Dallas Co. (Reverchon 1903; J. Stanford, pers. comm.); sporadically but widely distributed in TX. Where found, this species is often abundant and huge numbers of individuals can at certain times of the summer turn the surface of ponds a striking red color. Sporulating summer–fall.

BLECHNACEAE CHAIN FERN OR DEER FERN FAMILY

A family of ca. 10 genera and ca. 250 species; it is mostly tropical and s temperate except for the n temperate *Woodwardia*. Family name from *Blechnum*, DEER FERN, a mostly tropical, especially s hemisphere genus of ca. 220 species. (Greek: *blechnon*, classical name for ferns in general)
FAMILY RECOGNITION IN THE FIELD: sori *discrete, linear-oblong*, in a *chain-like row* along each side of the midvein of a pinna or pinnule; indusia attached by their outer margin, opening towards midvein.
REFERENCE: Cranfill 1993a.

WOODWARDIA CHAIN FERN

Terrestrial; stems (rhizomes) in ours long-creeping with leaves scattered along the stems; leaves monomorphic or dimorphic, deciduous, the blades 1-pinnatifid or 1-pinnate; sori discrete, linear-oblong, in a single chain-like row along each side of the midvein; indusia attached by their outer margin, opening on side next to midvein, often obscured by dehisced (= opened) sporangia.

A genus of 14 species of North America, Central America, Mediterranean Europe, and e Asia. (Named for Thomas Jenkinson Woodward, 1745–1820, English botanist)
REFERENCES: Correll 1956, 1966a.

1. Leaves conspicuously dimorphic (pinnae of fertile leaves contracted, linear); sterile blades 1-pinnatifid, with a wing of blade tissue several mm wide along much (at least upper half) of the rachis; pinnae (subdivisions of leaves) themselves not pinnatifid, sometimes sinuate, the margins serrulate **W. areolata**
1. Leaves monomorphic or nearly so; blades 1-pinnate, with no leaf tissue along the rachis; pinnae deeply pinnatifid with entire margins **W. virginica**

Woodwardia areolata (L.) T. Moore, (pitted), CHAIN FERN, NARROW-LEAF CHAIN FERN. Sterile leaves few, 40–58 cm long; pinnae in 7–12 alternate pairs, 1–2.5 cm wide, the veins anastomosing into 2 or more rows of areoles between midvein (= costa) and margin; sori nearly completely covering surface of blade. Low, wet, usually sandy areas; Fannin Co. in Red River drainage; se and e TX w to ne part of nc TX. Sporulating Mar–Nov. This species has sometimes been segregated into the genus *Lorinseria* [as *L. areolata* (L.) C. Presl]. The sterile leaves resemble those of *Onoclea* (subopposite pinnae with entire margins) except *W. areolata* usually has alternate pinnae with minutely serrulate margins.

Woodwardia virginica (L.) Small, (of Virginia), VIRGINIA CHAIN FERN. Leaves numerous, 50–100 cm long; pinnae in 12–23 pairs, the middle pinnae 1–3.5 cm wide, the veins anastomosing to form a single row of areoles near midvein; sori covering only a small part of the blade surface. Low areas; Milam Co. (Correll 1956) on e edge of nc TX; mainly se and e TX. Sporulating Apr–Dec.

DENNSTAEDTIACEAE BRACKEN FAMILY

As currently recognized, the Dennstaedtiaceae is a cosmopolitan, but mostly tropical family of ca. 20 genera and ca. 400 species; it has been variously circumscribed to include as few as 8 genera or in other cases nearly half the genera of higher ferns. Family name from *Dennstaedia*, a cosmopolitan but mostly tropical genus of ca. 70 species. (Named for August Wilhelm Dennstaedt, 1776–1826, German botanist and physician)
FAMILY RECOGNITION IN THE FIELD: the single nc TX species is a terrestrial plant with large leaves with 3 main divisions, each of these being 2-pinnate-pinnatifid; sori linear, *along margins* of the ultimate leaf segments with the leaf *margins recurved* over sori to form a false indusium.
REFERENCE: Cranfill 1993b.

PTERIDIUM BRACKEN FERN

A monotypic, cosmopolitan genus sometimes placed in the Pteridaceae. (Greek: *pteridon*, a small fern, from *pteron*, feather or wing, due to the shape of the leaves)
REFERENCES: Correll 1956, 1966a; Tryon 1941; Page 1976.

Pteridium aquilinum (L.) Kuhn var. **pseudocaudatum** (Clute) A. Heller, (sp.: eagle-like; var.: false-tailed), WESTERN BRACKEN FERN, PASTURE BRAKE, BRACKEN FERN. Terrestrial; stems (rhizomes) deeply underground, long-creeping; leaves monomorphic, deciduous, scattered along the stems, to 1 m or more tall; leaf blades glabrous or nearly so, broadly triangular to triangular-lanceolate in outline, usually of 3 main divisions, each division 2-pinnate-pinnatifid, the pinnae rigidly herbaceous to subcoriaceous; sori marginal, linear, continuous, covered by a false indusium formed by the recurved margin of the ultimate leaf segments and an obscure inner, delicate, true indusium. Open woods, pastures, thickets, often in sandy soils; Grayson Co. (S. Crosthwaite, pers. comm.) in Red River drainage, also Henderson, Milam, and Red River cos. on the e margin of nc TX; mainly e TX. Sporulating Jun–Nov. This variable species, with numerous infraspecific taxa, is virtually worldwide in distribution, is the most widely distributed fern, and is considered by some to be the most widespread of all vascular plants (with the exception of a few annual weeds) (Page 1976). Its tenacity is shown by regeneration through several meters of volcanic ash on Mt. St. Helens in Washington within 1–2 years of the volcanic eruption (Woodland 1997). In some areas (e.g., British Isles) BRACKEN FERN is a problematic weed and the cause of "bracken poisoning," a potentially fatal condition in livestock. Toxins include a cyanide-producing glycoside (prunasin); an enzyme, thiaminase, which can cause fatal thiamine (Vitamin B_1) deficiency in livestock; and at least two carcinogens. Human consumption of the fiddleheads has been suggested as a cause of stomach cancer in some parts of the world. It is also known to be allelopathic, with toxins leaching from the tissues adversely affecting surrounding plants (Mabberley 1987; Turner & Szczawinski 1991). ☠

DRYOPTERIDACEAE WOOD FERN FAMILY

Ours usually terrestrial or on rocks or epiphytic; leaves monomorphic or dimorphic; leaf blades 1-pinnatifid to 1-more-pinnate or pinnate-pinnatifid; sori on abaxial leaf surfaces, on veins or vein tips, usually not marginal, or in berry-like or bead-like structures on fertile leaves conspicuously different from sterile (*Onoclea*).

The family as broadly described here follows Smith (1993b) and includes genera (*Athyrium, Nephrolepis, Onoclea, Woodsia*) at times segregated into other families; it is cosmopolitan and has ca. 60 genera and ca. 3,000 species. The family has sometimes been treated as the Aspidiaceae (an illegitimate name). Family name from *Dryopteris*, WOOD FERN or SHIELD FERN, a mostly temperate (especially Asian) genus of ca. 250 species. (Greek: *drys*, oak or tree, and *pteris*, fern; several species are associated with oak woodlands)

FAMILY RECOGNITION IN THE FIELD: sori in most species on veins or vein tips (usually not marginal), or in *Onoclea* in berry-like or bead-like structures on fertile leaves conspicuously different from the sterile leaves.

REFERENCES: Correll 1956, 1966a; Smith 1993b.

1. Fertile and sterile leaves completely different (leaves extremely dimorphic); fertile leaves without typical blade tissue; sterile leaf 1-pinnatifid (deeply divided but not completely pinnate); rachis with a conspicuous flange of photosynthetic tissue ____ **Onoclea**
1. Fertile and sterile leaves or portions of leaves similar, the fertile portion never so different as to be without blade tissue; leaves at least completely 1-pinnate, often more divided; rachis without a flange of photosynthetic tissue.
   2. Leaf blades only 1-pinnate, the pinnae themselves not further divided, neither pinnate nor pinnatifed (but basal auricles sometimes present).
      3. Sori only on the uppermost somewhat reduced fertile pinnae; indusia orbicular, not at all kidney-shaped; pinnae with bristly teeth on the margins ____ **Polystichum**
      3. Sori not restricted to the uppermost pinnae, the fertile pinnae not reduced; indusia orbicular-kidney-shaped; pinnae without bristly teeth on the margins (but small non-bristly teeth can be present) ____ **Nephrolepis**
   2. Leaf blades more than 1-pinnate, the pinnae themselves further divided, either pinnate or pinnatifid.
      4. Sori elongate; indusia attached to blade along one side of sorus only; basal pinnules often with small auricles; plants to 120 cm tall ____ **Athyrium**
      4. Sori round or nearly so; indusia of lobes or flaps attached at several spots around the sorus; basal pinnules without auricles; plants 60 cm or less tall (often only ca. 30) ____ **Woodsia**

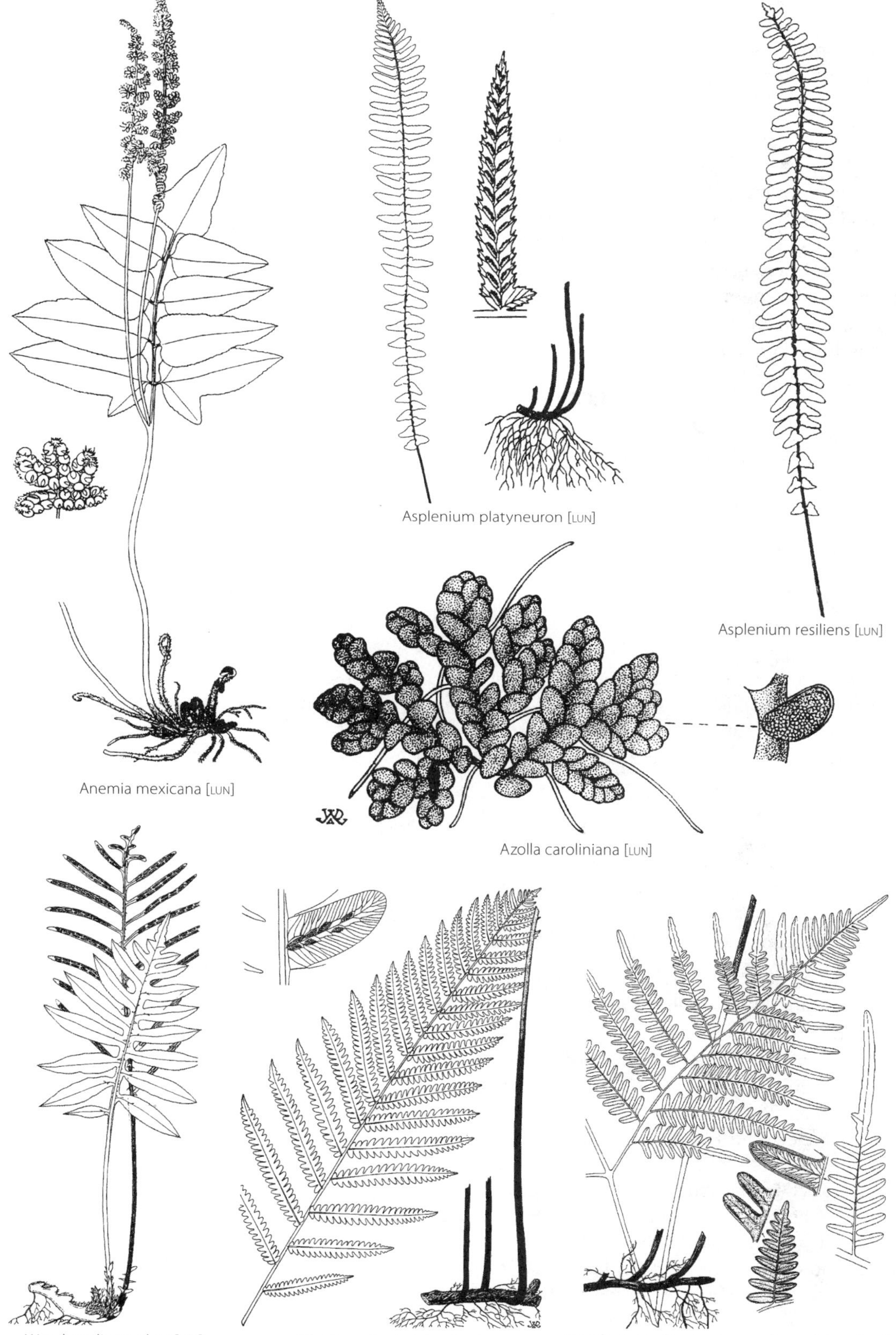

Asplenium platyneuron [LUN]

Asplenium resiliens [LUN]

Anemia mexicana [LUN]

Azolla caroliniana [LUN]

Woodwardia areolata [LUN]

Woodwardia virginica [LUN]

Pteridium aquilinum var. pseudocaudatum [LUN]

ATHYRIUM LADY FERN

A cosmopolitan genus of ca. 180 species. (Greek: *athyros*, doorless; the sporangia only tardily push back the outer edge of the indusium)
REFERENCE: Kato 1993.

Athyrium filix-femina (L.) Roth subsp. **asplenioides** (Michx.) Hultén, (sp.: lady fern; subsp.: resembling *Asplenium*—spleenwort), SOUTHERN LADY FERN, LOWLAND LADY FERN. Stems (rhizomes) short-creeping; leaves monomorphic, deciduous, clustered, to 120 cm tall, 2-pinnate-pinnatifid (rarely sub-3-pinnate), the pinnae usually short stalked; sori elongate, straight to hooked or curved, somewhat resembling those of *Asplenium*, in a single row on each side of the midrib, ca. midway between midrib and margin of ultimate leaf segment; indusia membranous, opening facing midrib. Moist woods, thickets, swamps, stream banks; Williamson Co. (Correll 1956); mainly e TX nw to Red River Co. Sporulating May–Nov. [*A. asplenioides* (Michx.) A.A. Eaton] This species is sometimes cultivated as an ornamental.

NEPHROLEPIS BOSTON FERN

A genus of 25–30 species widespread in tropical areas. *Nephrolepis* is sometimes placed in the Davalliaceae or Nephrolepidaceae. (Greek: *nephros*, kidney, and *lepis*, scale, in reference to the shape of the indusium)
REFERENCE: Nauman 1993.

Nephrolepis exaltata (L.) Schott, (very tall), SWORD FERN, WILD BOSTON FERN. Stems (rhizomes) short, ± erect, with wiry, widely creeping stolons; leaves monomorphic, evergreen, clustered, 1-pinnate, 0.4–1.5(–2) m or more long, the blades linear-lanceolate; sori roundish, somewhat closer to margin than to midvein of pinnae, the indusia ± orbicular-reniform. Escaped, persisting and spreading in yard in Highland Park, Dallas (R. O'Kennon, pers. obs.); apparently naturalized in several sites in e TX and the Edwards Plateau; native to Florida, the West Indies, and scattered Pacific Islands; terrestrial or most often epiphytic in its native habitat. This is a commonly cultivated and commercially important fern with many cultivars including cv. 'Bostoniensis' (BOSTON FERN) and the locally developed DALLAS JEWEL FERN,™ commonly known as the DALLAS FERN.

ONOCLEA SENSITIVE FERN

A monotypic genus of n temperate areas; sometimes cultivated as an ornamental. (Greek: *onos*, vessel, and *cleisto*, closed, in reference to the sori, which are enclosed by the revolute fertile leaf margins)
REFERENCE: Johnson 1993b.

Onoclea sensibilis L., (sensitive), SENSITIVE FERN. Stems (rhizomes) creeping; leaves conspicuously dimorphic, of 2 very different types, scattered along the rhizome, erect, glabrous; sterile leaves to ca. 1(–1.3) m tall, thin herbaceous, deciduous, broadly triangular to ovate in outline, deeply pinnatifid with the pinnae few, the pinnae subopposite (especially the lowermost), undulate to irregularly deeply lobed, with margins entire, the rachis winged; fertile leaves persistent over winter, 2-pinnate, the blades greatly reduced, the ultimate segments rolled into globular, berry-like or bead-like structures concealing the sori, the whole fertile leaf superficially resembling a narrow panicle of small round fruits. Swamps, low woods, and wet areas; Milam Co., also Burnet Co. (Correll 1956) on the s edge of nc TX; mainly se and e TX , the Edwards Plateau, and in the Rio Grande Plains. Sporulating Apr–Dec. The common name is in reference to the sensitivity of the leaves to even a light frost (Johnson 1993b). The sterile leaves superficially resemble those of *Woodwardia areolata*. Reported to be poisonous; horses are said to become unsteady and collapse upon ingesting the plant (Burlage 1968; Turner & Szczawinski 1991). ☠

POLYSTICHUM CHRISTMAS FERN, SWORD FERN, HOLLY FERN

A cosmopolitan genus of ca. 180 species. (Greek: *poly*, many, and *stichos*, row, presumably in reference to the rows of sori on each pinna)
REFERENCE: Wagner 1993.

Polystichum acrostichoides (Michx.) Schott, (resembling *Acrostichum*—another genus of ferns), CHRISTMAS FERN, DAGGER FERN. Stems (rhizomes) erect; leaves essentially evergreen, clustered, to 70 cm long, the blades elliptic-lanceolate to lanceolate in outline, 1-pinnate; pinnae mostly alternate, auricled basally, the margins bristle-toothed; petioles densely scaly; leaf blades partially dimorphic, the proximal pinnae (those near blade base) sterile, the distal pinnae (those near blade tip) of some blades fertile and conspicuously contracted (but blade tissue still evident); sori round, crowded in 2–4 rows, medial, often confluent at maturity; indusia peltate, entire, persistent. Rich wooded slopes, moist areas; included based on citation of vegetational area 4 (Fig. 2) by Hatch et al. (1990); it has been collected a few miles e of the e margin of nc TX in w Red River Co.; mainly e TX. Sporulating May–Nov.

WOODSIA CLIFF FERN

A genus of ca. 30 species found mainly in n temperate regions and at high elevations in the tropics. (Named for Joseph Woods, 1776–1864, English botanist)
REFERENCES: Windham 1987a, 1993d.

Woodsia obtusa (Spreng.) Torr., (obtuse, blunt), COMMON WOODSIA, BLUNT-LOBED WOODSIA, LARGE WOODSIA. Stems (rhizomes) short; leaves monomorphic, semi-evergreen, clustered, erect-ascending, to 40(–60) cm tall, often smaller, the blades elliptic-lanceolate to broadly lanceolate, 2-pinnate or 2-pinnate-pinnatifid; sori round, between midrib and lateral margins of ultimate leaf segments; indusia rather large, at first enclosing the sporangia and later splitting into several spreading, irregular lobes. Rocky areas, outcrops, well-drained often sandy areas; Lamar (Carr 1994) and Kaufman cos. w to Montague and Palo Pinto cos.; mainly e, nc, and c TX. Two subspecies of *W. obtusa*, differing in chromosome number, are recognized by Windham (1993d) as occurring in nc TX and separated and described by him as follows. We, however, have been unable to clearly and consistently separate the specimens from nc TX into the 2 subspecies. Windham (1993d) further indicated that the 2 subspecies hybridize in the area of sympatry and form sterile triploids with malformed spores.

1. Spores averaging 42–47 μm; proximal pinnules of lower pinnae usually shallowly lobed or merely dentate; blades coarsely cut and evidently 2-pinnate; stems compact to short-creeping, individual branches usually 5–10 mm diam. ____ subsp. **obtusa**
1. Spores averaging 35–42 μm; proximal pinnules of lower pinnae usually deeply lobed or pinnatifid; blades finely cut, 2-pinnate-pinnatifid; stems short- to long-creeping, individual branches 3–5 mm diam. ____ subsp. **occidentalis**

subsp. **obtusa**. Cliffs and rocky slopes, also terrestrial. $2n = 152$. E U.S. w to e 1/3 of TX.

subsp. **occidentalis** Windham, (western). Cliffs and rocky slopes, also terrestrial. $2n = 76$. C U.S. including nc TX to c TX.

MARSILEACEAE
WATER-CLOVER OR PEPPERWORT FAMILY

Plants aquatic or of very wet habitats; stems (rhizomes) long-creeping; leaves scattered along the stems long-petioled, palmately divided into 4 pinnae or filiform and lacking expanded blades; sori contained in sporocarps (= hard bean- or pea-like structures which are apparently highly modified pinnae) on stalks from near base of petiole; sporangia of 2 kinds within the same sorus, the megasporangia with 1 megaspore, the microsporangia with numerous microspores.

A nearly cosmopolitan family of 3 genera and ca. 50 species.
FAMILY RECOGNITION IN THE FIELD: plants of wet areas with leaves resembling a *4-leaf clover* (in 1 species apparently rare in nc TX the leaves are thread-like and ± resemble those of a grass); sori in hard, *bean- or pea-like structures* (= sporocarps) near the base of the petioles.
REFERENCES: Correll 1956, 1966a; Johnson 1993a.

1. Leaf blades palmately divided into 4 narrowly to broadly cuneate (= wedge-shaped) pinnae (resembling a 4-leaf clover) ____ **Marsilea**
1. Leaves filiform, very narrow, inconspicuously grass-like in appearance, without expanded blades ____ **Pilularia**

MARSILEA WATER-CLOVER, PEPPERWORT

Small plants, aquatic or of wet habitats, often forming dense colonies; leaves long petiolate with blades palmately divided into 4 pinnae; sporocarps on stalks, the tip of stalk often protruding as a bump or tooth (proximal tooth), a second tooth (distal tooth) sometimes present on sporocarps beyond the attachment point of the stalk.

A nearly cosmopolitan genus of 45 species. The leaves have a superficial resemblance to those of CLOVER; young plants can have unlobed leaves like *Pilularia*. (Named for Count Luigi Marsigli, 1656–1730, Italian mycologist at Bologna)
REFERENCES: Gupta 1957; Thieret 1977b; Johnson 1986, 1988.

1. Pinnae 9–35 mm long, 8–39 mm wide; sporocarps densely villous with long spreading hairs; distal tooth of sporocarps absent or to 0.5 mm long, blunt; sporocarp stalks usually branched, several sporocarps per stalk ____ **M. macropoda**
1. Pinnae 4–19 mm long, 4–16 mm wide; sporocarps pubescent with appressed hairs, often glabrate; distal tooth of sporocarps 0.4–1.2 mm long, acute; sporocarp stalks unbranched, 1 sporocarp per stalk ____ **M. vestita**

Marsilea macropoda Engelm. ex A. Braun, (large-footed), LARGE-FOOT PEPPERWORT, WATER-CLOVER. Petioles 5–39 cm long. Typically in mud, also shallow water; Brown Co., also Travis Co. (Blackland Prairie (Correll 1956)) just s of nc TX; mainly c to s TX. Sporocarps produced nearly year round. An attractive plant that is cultivated as an ornamental.

Marsilea vestita Hook. & Grev., (covered). Petioles 2–20 mm long. Ponds, wet depressions, along streams and rivers. Sporocarps produced Mar–Oct.

1. Pinnae narrow in appearance, 3–7.5 times as long as wide, narrowly and obliquely cuneate (= wedge-shaped), irregularly toothed or crenulate at apex ____ subsp. **tenuifolia**
1. Pinnae broad in appearance, usually 1–2 times as long as wide, fan-shaped or broadly cuneate, with entire or undulate-crenulate apex ____ subsp. **vestita**

subsp. **tenuifolia** (Engelm. ex A. Braun) D.M. Johnson, (slender-leaved), NARROW-LEAF PEPPERWORT. This rare taxon has been variously treated as a separate species (Correll & Johnston 1970), as a subspecies of *M. vestita* (Johnson 1986; Kartesz 1994), or lumped with *M. vestita* (Johnson 1993a; Jones et al. 1997). Because it can usually be easily distinguished in the field (see key above), we are treating it as a subspecies of *M. vestita*. Included based on citation by Hatch et al. (1990) for vegetational area 5 (Fig. 2); "Burnet (or Llano)" and Travis cos. (Correll 1956) at the s margin of nc TX; mainly on the Edwards Plateau. [*M. tenuifolia* Engelm. ex A. Braun]

subsp. **vestita**, HOOKED PEPPERWORT, WATER-CLOVER, HAIRY PEPPERWORT. Coryell (Fort Hood—

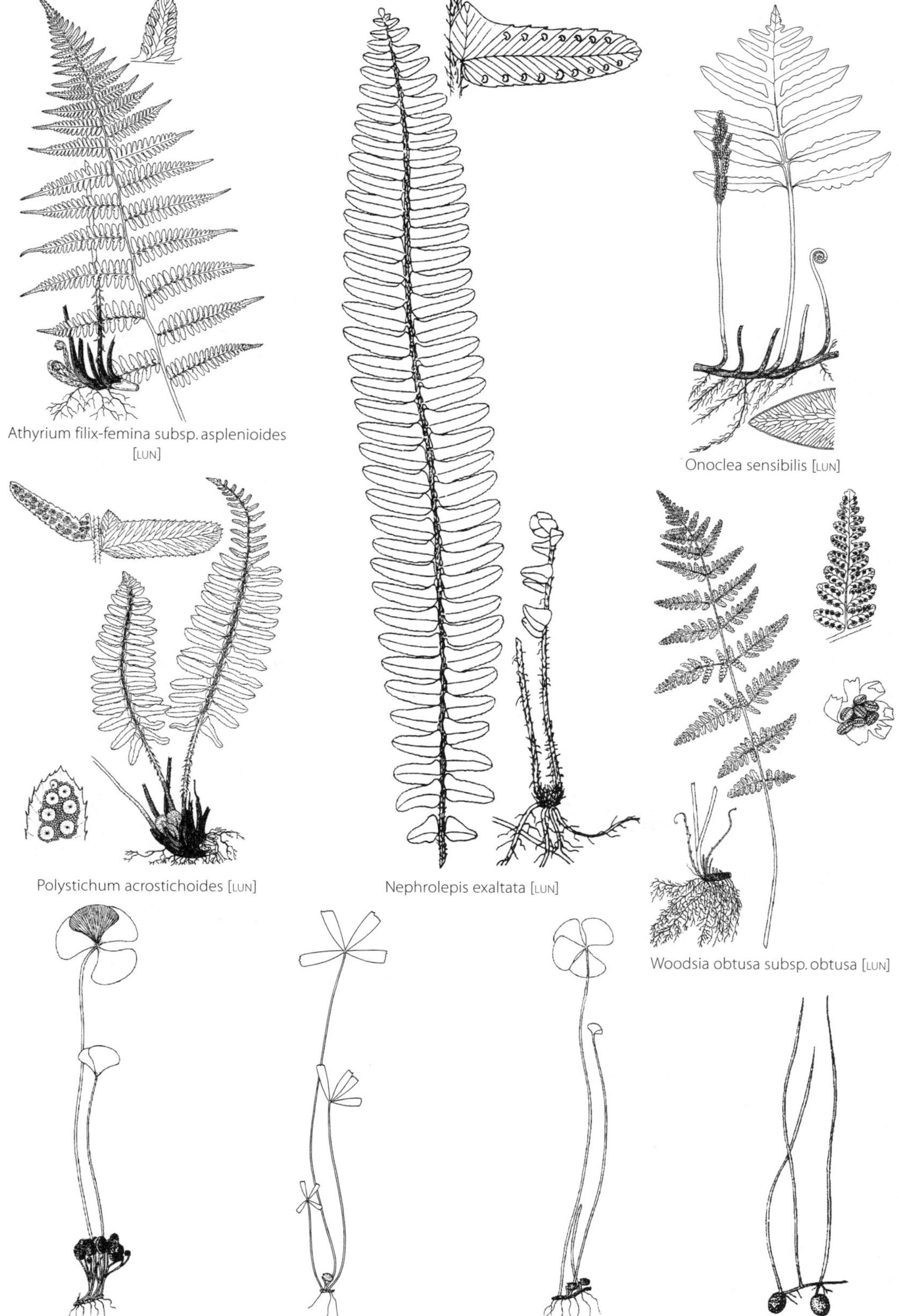

Athyrium filix-femina subsp. asplenioides [LUN]

Onoclea sensibilis [LUN]

Polystichum acrostichoides [LUN]

Nephrolepis exaltata [LUN]

Woodsia obtusa subsp. obtusa [LUN]

Marsilea macropoda [LUN]

Marsilea vestita subsp. tenuifolia [LUN]

Marsilea vestita subsp. vestita [LUN]

Pilularia americana [LUN]

Sanchez 1997), Dallas, Ellis, Tarrant, and Williamson (Correll 1956) cos.; Blackland Prairie s and w to w TX. [*M. mucronata* A. Braun]

PILULARIA PILLWORT

A genus of 6 species of North America, South America, Europe, Pacific Islands, Australia, and New Zealand; sometimes placed in its own family. (Latin: *pilula*, a little ball, in reference to the spheric sporocarps)
REFERENCES: LaMotte 1940; Hill 1980a; Dennis & Webb 1981.

Pilularia americana A. Braun, (of America), AMERICAN PILLWORT, WATER-PEPPER. Small inconspicuous aquatic, underwater or infrequently persisting on bare mud; leaves filiform, 1.6–10.2 cm long, lacking expanded blades; sporocarps produced just below ground surface, globose, 2–6(–10) mm long, 2–3 mm in diam. Temporary pools, ponds, reservoir margins. According to the range map in Johnson (1993a), *P. americana* occurs widely in nc TX and it is included here on that basis; the only known nearby collection we have seen is from Burnet Co. (Granite Mt., just s of nc TX). The species is so inconspicuous that it is rarely recognized or collected.

OPHIOGLOSSACEAE ADDER'S-TONGUE FAMILY

Ours terrestrial; stems (± subterranean) simple, unbranched, upright; leaves 1 or less commonly 2 per stem, with common stalk divided into a blade portion (= trophophore) and a fertile sporangia-bearing portion (= sporophore); blade portion simple, divided, or compound; fertile portion (lacking blade-like tissue) typically consisting of a long stalk with a terminal, branched or unbranched, sporangia-bearing area; sporangia large, spherical, thick-walled, borne in 2 rows on the branches or on the unbranched sporangia-bearing area.

A nearly cosmopolitan family of 5 genera and ca. 70–80 species thought by some to be only distantly related to other ferns; they are apparently relicts of an ancient lineage (Bell & Woodcock 1983). The family is made up of 2 clearly defined subfamilies, Botrychioideae and Ophioglossoideae, sometimes recognized as distinct families. The following treatment draws heavily on Wagner and Wagner (1993).
FAMILY RECOGNITION IN THE FIELD: often small plants with only 1 or sometimes 2 leaves; leaves with a blade portion (simple to compound) and an erect, spike-like, fertile portion consisting of an elongate stalk and a terminal, fertile, sporangia-bearing area.
REFERENCES: Clausen 1938; Correll 1956, 1966a; Thomas 1972; Wagner & Wagner 1993.

1. Leaf blades ternately-pinnately compound, divided, or lobed, the margins usually denticulate to serrate or lacerate; veins of leaf blades dichotomous (= equally 2-forked) and free; sporangia in pinnately branched, panicle-like arrangement **Botrychium**
1. Leaf blades simple, the margins entire; veins of leaf blades reticulate (= in a net-like pattern); sporangia in unbranched, linear, spike-like arrangement **Ophioglossum**

BOTRYCHIUM GRAPE FERN, MOONWORT

Blade portion of leaf compound, divided or lobed, ovate to triangular or broadly triangular in outline; fertile portion of leaf consisting of an elongate stalk terminated by a 1–2-pinnate, panicle-like sporangia-bearing region.

A nearly cosmopolitan genus of 50–60 species with greatest diversity at high latitudes and high elevations; most species are quite variable vegetatively. (Latin: *botry*, bunch (of grapes), and *-oides*, like, in reference to the sporangial clusters)
REFERENCE: Holmes et al. 1996.

1. Plants small, to only ca. 12 cm tall; blade portion of leaf prostrate on ground, small, only 3–8 cm long, short-stalked (petiole-like stalk 1.5–3 cm long); ultimate leaf segments fan-shaped, their tips broadly rounded; leaves commonly 2 per plant _____ **B. lunarioides**
1. Plants usually larger, 8–75 cm tall; blade portion of leaf raised above the ground, not prostrate, usually larger, 4–30 cm long, either sessile (petiole-like stalk absent) or long-stalked (petiole-like stalk 3–20 cm long); ultimate leaf segments not fan-shaped, their tips usually pointed; leaves usually 1 per plant.
   2. Blade portion of leaf appearing to have a long petiole (blade portion well-separated from origin of fertile stalk); blade coarsely divided, the relatively few large ultimate segments with finely denticulate margins; leaves present in winter _____ **B. biternatum**
   2. Blade portion of leaf sessile (fertile stalk originating at very base of blade portion); blade finely divided, the numerous small ultimate segments with coarsely serrate to lacerate (= irregularly cut) margins; leaves absent in winter _____ **B. virginianum**

Botrychium biternatum (Savigny) Underw., (twice-ternate), SOUTHERN GRAPE FERN, SPARSE-LOBED GRAPE FERN. Plant ca. 10–35 cm tall; roots usually 10 or less, blackish; leaves present over winter, rarely bronze in winter if exposed; new leaves appearing in late spring to early summer; sterile blade portion green to dark green, long-stalked (stalk 3–20 cm long), herbaceous, to 18 cm long and 28 cm wide, usually smaller, 2–3-pinnate; pinnules elongate, obliquely lanceolate to narrowly lanceolate, the margins nearly parallel, finely denticulate, the apices short-acuminate. The leaves are much less finely divided than in *B. virginianum*, the 2 species immediately distinguishable in the field, herbarium or illustrations. Low woods; included based on map in Wagner and Wagner (1993) and citation of *B. dissectum* Spreng. for vegetational area 4 (Fig. 2) by Hatch et al. (1990); we have seen no nc TX specimens. Sporulating Apr–Dec. While *B. biternatum* is cited only for vegetational area 1 (Fig. 2) by Hatch et al. (1990), all TX material seen by W.H. Wagner, Jr. (pers. comm.) going under the name of *B. dissectum* is actually *B. biternatum* (with the possible exception of material from very close to the LA border). The map in Wagner and Wagner (1993) clearly shows *B. biternatum* in e TX while *B. dissectum* occurs in the se U.S. w to approximately the LA-TX border. The vegetational area 4 (Fig. 2) citation for *B. dissectum* by Hatch et al. (1990) is therefore assumed to be *B. biternatum*. While sometimes resembling *B. biternatum*, according to W.H. Wagner Jr. (pers. comm.) and Wagner and Wagner (1993), *B. dissectum* has leaves that are more dissected and the pinnules trowel-shaped or linear, apically more pointed, and with the margins more lacerate. [*B. tenuifolium* Underw., *B. dissectum* Spreng. var. *tenuifolium* (Underw.) Farw.]

Botrychium lunarioides (Michx.) Sw., (resembling *Botrychium lunaria*), WINTER GRAPE FERN, PROSTRATE GRAPE FERN. Roots 20–30, yellow to brown; leaves appearing in late fall, overwintering and then dying in early spring; sterile blade portion usually pale green, short-stalked, fleshy, to 12 cm wide, 2–3-pinnate-pinnatifid; ultimate leaf segments fan-shaped, with midrib absent, denticulate, rounded at apex. Open grassy areas; Falls, Hunt, Hopkins, Kaufman, Milam, and Navarro cos. on e edge of Blackland Prairie (Holmes et al. 1996); mainly e TX. According to Wagner and Wagner (1993), a "peculiarity of this species is the tendency for the sporophores to remain curled in late fall and early winter and to become erect in February." This taxon was only recently reported from the Blackland prairie (Holmes et al. 1996).

Botrychium virginianum (L.) Sw., (of Virginia), RATTLESNAKE FERN, VIRGINIA GRAPE FERN, COMMON GRAPE FERN. Plant erect, 8–75 cm tall; roots 15 or fewer, yellow to brown; leaves seasonal, appearing in early spring and dying in summer; sterile blade portion pale green, sessile, thin, herbaceous, 4–30 cm long and wide, 3–5-pinnate-pinnatifid; ultimate leaf segments linear, with midrib present, serrate to lacerate, pointed at apex. Moist, rich woods and thickets; Grayson, Lamar, and Tarrant cos.; also Bell, Burnet, and Dallas cos. (Correll 1956); mainly e TX w to nc TX, also Edwards Plateau. Sporulating Mar–Sep.

OPHIOGLOSSUM ADDER'S-TONGUE

Plant small, ours to ca. 25 cm tall; blade portion of leaf simple; fertile portion of leaf consisting of an elongate stalk terminated by an unbranched, linear, spike-like, sporangia-bearing region.

A nearly cosmopolitan, but mainly tropical and subtropical genus of 25–30 species. *Ophioglossum* species have the highest chromosome numbers known for vascular plants, with numbers as high as $2n$ = 1,200+ being reported. (Greek: *ophis*, snake, and *glossa*, tongue, in reference to the tip of the sporangia-bearing structure)

1. Stems (± subterranean and sometimes called rootstocks) globose-bulbous, 3–12 mm diam.; leaves emerging from cavity in top of stem, the blade portion usually near ground surface, appearing spreading or nearly flat on ground, usually roughly triangular to orbicular-ovate or cordate, to only 35 mm long; sporangial clusters < 1 cm long; common stalk (to where blade and fertile stalk separate) usually < 3 cm long ____ **O. crotalophoroides**
1. Stems cylindric upright, to ca. 4 mm diam.; leaves developing at top of stem, the blade portion well above ground, erect to spreading, usually ovate to lanceolate, to 120 mm long; sporangial clusters 2–4 cm long; common stalk to 10 cm long.
   2. Blade portion of leaf with distinct and prominent apiculate tip, commonly ± folded when alive; principal veins of blade forming large primary areoles (= vein enclosed areas) in which are included numerous veinlets forming secondary areoles ____ **O. engelmannii**
   2. Blade portion of leaf without apiculate tip, usually rounded to acute at apex, commonly plane when alive; principal veins of blade forming areoles but these including only free veinlets ____ **O. vulgatum**

Ophioglossum crotalophoroides Walter, (from Greek: *krotalon*, a rattle, and *-oides*, like or resembling, due to the resemblance of the sporongial clusters to rattles or castanets), BULBOUS ADDER'S-TONGUE. Plant usually to only 15 cm tall; blade portion of leaf to 35 mm long and 25 mm wide, usually smaller; fertile stalk 1–5 times as long as blade portion; sporangia 4–8(–12) on each side of fertile stalk. Usually in moist sand; Fannin, Hopkins, Hunt, Lamar, Limestone, and Red River cos.; se and e TX w to n part of nc TX, also e Edwards Plateau. Sporulating Mar–May.

Ophioglossum engelmannii Prantl, (for George Engelmann, 1809–1884, German-born American botanist), ENGELMANN'S ADDER'S-TONGUE, LIMESTONE ADDER'S-TONGUE. Plant to 25 cm tall; blade portion of leaf to 100 mm long and 45 mm wide, commonly folded when alive, when dried uniformly green without pale central band; fertile stalk 1.3–2.5 times as long as blade portion; sporangia 20–40 on each side of fertile stalk. Usually in thin black soils on limestone, wooded rocky slopes; Dallas, Denton, Grayson, Kaufman, Limestone, Montague, and Tarrant cos.; also Bell, Brown, and McLennan cos. (Correll 1956); se and e TX w to West Cross Timbers, also Edwards Plateau and Deaf Smith Co. in the Panhandle (Floyd Waller collection—J. Stanford, pers. comm.). Sporulating Dec–Jun.

Ophioglossum vulgatum L., (common), ADDER'S-TONGUE, SOUTHERN ADDER'S-TONGUE. Similar to *O. engelmannii*; leaves 1 per stem; blade portion of leaf to 120 mm long and 50 mm wide, dark green, somewhat shiny, rounded at apex; fertile stalk 2–4 times as long as blade portion; sporangia 10–35 on each side of fertile stalk. Moist woods, meadows, swamps, usually in sandy soils; Fannin and Lamar cos. in Red River drainage, also Denton Co. (Clausen in Correll 1956); mainly se and e TX w to n part of nc TX. Sporulating Mar–Jun. [*O. pycnostichum* (Fernald) A. Löve & D. Löve, *O. vulgatum* var. *pycnostichum* Fernald]

According to W.H. Wagner Jr. (pers. comm.), two other species, *O. nudicaule* L., (naked stem), and *O. petiolatum* Hook., (with a petiole or leaf stalk), occur just to the east and may yet be found in nc TX. Both are found in disturbed places, commonly in cemeteries and mowed areas around motels. In the key above, *O. nudicaule* would key to *O. engelmannii*; *O. nudicaule* can be

Botrychium lunarioides [TAY]

Botrychium virginianum [TAY]

Botrychium biternatum [TAY]

Ophioglossum crotalophoroides [LUN]

Osmunda cinnamomea [LUN]

Osmunda regalis var. spectabilis [LUN]

Ophioglossum engelmannii [LUN]

Ophioglossum vulgatum [LUN]

distinguished by the following: blade portion of leaf to only 45 mm long and 17 mm wide, plane when alive, when dried commonly with a pale central band; fertile stalk 2–6 times as long as blade portion. In the key above, *O. petiolatum* would key to *O. vulgatum*. *Ophioglossum petiolatum* can be distinguished by: leaves (= blade portion and fertile portion combined) commonly 2–3 per stem; blade portion of leaf acute at apex, to 60 mm long and 30 mm wide, gray-green, dull.

OSMUNDACEAE CINNAMON FERN FAMILY

A nearly cosmopolitan family with 3 genera and up to ca. 36 species; some are cultivated as ornamentals.

FAMILY RECOGNITION IN THE FIELD: leaves usually large, wholly or partly *dimorphic* (fertile leaves or pinnae conspicuously different from sterile); sporangia not in discrete sori.

REFERENCES: Correll 1956, 1966a; Hewitson 1962; Whetstone & Atkinson 1993.

OSMUNDA

Terrestrial; leaves erect to spreading, in a large crown from a stout woody creeping to erect stem (rhizome), wholly or partly dimorphic; sori absent; sporangia clustered; indusia absent.

A nearly cosmopolitan genus of 10 species. (Saxon: *Osmunder*, name for Thor, god of war)

1. Fertile leaves completely different in appearance from sterile leaves; ultimate leaf segments of sterile leaves not narrowed at base, the area of attachment as broad as segment ____ **O. cinnamomea**
1. Fertile leaves similar in appearance to sterile leaves except with greatly reduced sporangia-bearing pinnae at tip; ultimate leaf segments greatly narrowed at very base, attached at one stalk-like point only ____ **O. regalis**

Osmunda cinnamomea L., (cinnamon-brown), CINNAMON FERN, BUCKHORN FERN, BUCKHORN BRAKE, FLOWERING FERN. Sterile leaves 1-pinnate-pinnatifid, ca. 0.3–1.5 m long, the ultimate segments with margins entire and apically usually mucronate; pinnae with a persistent tuft of tomentum at base; fertile leaves with no expanded pinnae, densely tomentose, much narrower and shorter than sterile leaves; sporangia cinnamon-colored. Wet areas; Lamar Co. in Red River drainage, also Milam Co. (Correll 1956); mainly se and e TX. Sporulating Mar–Jul or later.

Osmunda regalis L. var. **spectabilis** (Willd.) A. Gray, (sp.: royal; var. spectacular), ROYAL FERN, FLOWERING FERN. Leaves 2-pinnate; sterile leaves ca. 0.75–1 m long; pinnules lanceolate, the margins subentire to remotely dentate, apically acute to rounded; pinnae without a persistent tuft of tomentum at base, essentially glabrous; sporangia brown at maturity. Wet areas; Lamar Co. (Carr 1994) in Red River drainage; se and e TX w to ne corner of nc TX and Travis Co. (Correll 1956) just s of nc TX. Sporulating Mar–Jul.

POLYPODIACEAE POLYPODY FAMILY

A cosmopolitan family today treated as composed of ca. 40 genera and ca. 500 species. As previously circumscribed the Polypodiaceae encompassed ca. 7,000 species or nearly two-thirds of the living ferns. Family name from *Polypodium*, POLYPODY, a cosmopolitan genus of ca. 100 species; the genus is currently more narrowly defined than previously. (Greek: *poly*, many, and *pous* or *podiun*, foot, referring to the branched rhizomes)

FAMILY RECOGNITION IN THE FIELD: the single nc TX species is typically epiphytic or found growing on rocks; the *discrete round sori* (without indusia) are found in single rows on each side of the midrib of the lobes of the *deeply pinnatifid* leaves.

REFERENCES: Correll 1956, 1966a; Smith 1993c.

PLEOPELTIS SHIELD-SORUS FERN

A widespread, but primarily neotropical genus of ca. 50 species of mostly epiphytic ferns; some of the species now treated in *Pleopeltis* were formerly included in *Polypodium*. (Greek: *pleos*, many, and *pelte*, shield, in reference to the peltate scales covering immature sori)
REFERENCE: Andrews & Windham 1993.

Pleopeltis polypodioides (L.) E.B. Andrews & Windham subsp. **michauxiana** (Weath.) E.B. Andrews & Windham, (sp: resembling *Polypodium*; subsp.: for André Michaux, 1746–1803, French botanist and explorer of North America), RESURRECTION FERN, GRAY POLYPODY. Usually epiphytic or sometimes growing on rocks; rhizomes slender, widely creeping, densely scaly; leaves monomorphic, evergreen, widely spaced; leaf blades oblong to triangular-oblong in outline, deeply pinnatifid, to 15 cm long and 5 cm wide, thick, opaque, hygroscopic, involute upon drying, glabrous above except for a few scales along midrib, densely covered with peltate scales below, the margins mostly entire; sori in single rows on each side of the midrib of the lobes near the margins, round, discrete, forming conspicuous bumps on the undersurface of leaves; indusia absent. Usually growing on various species of trees, especially oaks, sometimes on rocks, usually in shady damp situations; Dallas, Grayson, and Fannin cos., also Parker Co. (Correll 1956); se and e TX w to nc TX and Edwards Plateau. Previously lumped into the genus *Polypodium* [as *P. polypodioides* (L.). Watt var. *michauxianum* Weath.].

PTERIDACEAE MAIDENHAIR FERN OR BRAKE FAMILY

Ours mostly on rocks, sometimes terrestrial; leaves monomorphic (rarely somewhat dimorphic); leaf blades 1–4(–5) pinnate; sporangia abaxial on the blades, marginal or submarginal; margins of ultimate segments recurved in ours to form false indusia (except in *Astrolepis*).

The taxa included here in the Pteridaceae have been variously treated at the family level. We follow Windham's (1993a) treatment and recognize 5 genera in nc TX; the newer name Adiantaceae has sometimes been applied to the family. The Pteridaceae is a cosmopolitan family of ca. 40 genera and ca. 1,000 species. Family name from *Pteris*, BRAKE FERN, a cosmopolitan, but generally warm and tropical area genus of ca. 300 species. (Greek: *pteris*, fern, from *pteron*, feather or wing, due to the closely spaced pinnae which give the leaves somewhat of a resemblance to feathers)
FAMILY RECOGNITION IN THE FIELD: plants typically growing *on rocks*; *sporangia at or near margins* of the ultimate leaf segments with the leaf *margins usually recurved* over sporangia to form false indusia (except in *Astrolepis*).
REFERENCES: Correll 1956, 1966a; Windham 1993a.

1. Only the apical margin of the ultimate leaf segments recurved; sporangia borne directly on recurved apical margins of ultimate leaf segments; veins of ultimate leaf segments prominent, dichotomously branched (= equally 2-forked), essentially parallel distally (= near their tips) ____ **Adiantum**
1. Apical and lateral margins of ultimate leaf segments usually recurved over sporangia (except margins not recurved in *Astrolepis*); sporangia borne on abaxial (= beneath) leaf surface (and covered by the recurved margins); veins of ultimate leaf segments obscure or, if prominent, pinnately branched and more divergent distally.
   2. Leaf blades 1-pinnate to 1-pinnate-pinnatifid throughout; abaxial leaf surfaces densely covered with coarsely ciliate or stellate scales; adaxial leaf surfaces with coarsely ciliate or stellate scales; margins of ultimate leaf segments not recurved to form false indusia ____ **Astrolepis**
   2. Leaf blades 2–5 pinnate at least at base; abaxial leaf surfaces scaly, pubescent or glabrous; adaxial leaf surfaces without coarsely ciliate or stellate scales; margins of ultimate leaf segments recurved to form false indusia.

3. Leaf blades glabrous abaxially or nearly so; stem scales strongly bicolored (dark central stripe and much lighter margins), or if uniformly colored, then largest ultimate leaf segments more than 4 mm wide ____ **Pellaea**
3. Leaf blades usually tomentose abaxially (except sparsely pubescent to nearly glabrous in *Cheilanthes alabamensis*) OR covered with conspicuous whitish powdery material; stem scales uniformly colored or weakly bicolored; ultimate leaf segments < 4 mm wide.
   4. Leaf blades with conspicuous whitish powdery material and without pubescence abaxially ____ **Argyrochosma**
   4. Leaf blades lacking conspicuous whitish powdery material, usually tomentose abaxially (except sparsely pubescent to nearly glabrous in *C. alabamensis*) ____ **Cheilanthes**

ADIANTUM MAIDENHAIR FERN

A genus of 150–200 species, nearly worldwide in distribution except at higher latitudes (> 60°); sometimes placed in the Adiantaceae. Some are used medicinally and a number are cultivated as ornamentals for their delicate, beautiful foliage. The position of the sporangia is definitive for identification. (Greek: *adiantos*, unwetted, for the glabrous leaves, which shed raindrops)
REFERENCES: Fernald 1950b; Paris 1993.

Adiantum capillus-veneris L., (Venus' hair), VENUS'-HAIR FERN, SOUTHERN MAIDENHAIR, CULANTRILLO. Terrestrial or on rocks; stems (rhizomes) short-creeping; leaves ± monomorphic, weakly deciduous, closely spaced, numerous, lax-arching or pendulous, 15–75 cm tall; leaf blades 2-(-more) pinnate, membranous to thin-herbaceous, bright green, the ultimate segments usually wedge or fan-shaped to irregularly rhombic (4-sided, diamond-shaped), ca. as long as broad, stalked; apical leaf margins recurved to form false indusia; sporangia submarginal, borne on the abaxial (= beneath) surface of the false indusia. Continuously moist calcareous areas, particularly limestone bluffs, rocks and ledges along streams. Bell, Brown, Burnet, Cooke, Somervell, and Tarrant cos.; also Dallas, Kaufman, McLennan (Correll 1956), and Johnson (R. O'Kennon, pers. obs.) cos.; scattered nearly throughout TX, common in some areas such as the Edwards Plateau. Sporulating May–Jan. The species has long been used medicinally for conditions of the skin, scalp, and internal organs (Cheatham & Johnston 1995).

ARGYROCHOSMA

A New World genus of ca. 20 species traditionally recognized in either *Notholaena* or *Pellaea*. (Greek: *argyros*, silver, and *chosma*, powder, referring to whitish farina (= mealy powder) covering the abaxial surface of leaf blades in most species)
REFERENCES: Tryon 1956; Windham 1987b, 1993b.

Argyrochosma dealbata (Pursh) Windham, (white-washed), POWDERY CLOAK FERN, FALSE CLOAK FERN. Usually on rocks; stems (rhizomes) short, ascending; plants small; leaves to only ca. 15 cm long, monomorphic, evergreen, clustered; leaf blades 3–4(–5)-pinnate, less distally, adaxial (= above) surface bluish green, glabrous, abaxial (= beneath) surface with very conspicuous whitish powdery material; pinnae and most pinnules distinctly stalked; sporangia on the abaxial surface of the blades, submarginal, protected by the recurved margins of the ultimate segments. Crevices of limestone and other calcareous rocks; Burnet, Coleman, Hood, Johnson, Parker, and Palo Pinto cos.; also Bell, Bosque, Ellis, Erath, Stephens (Correll 1956), and Brown (Carr 1995; HPC) cos.; nc TX, Edwards Plateau, and Trans-Pecos. While previously placed in a variety of genera, Windham (1987b) segregated *A. dealbata* and related species into the genus *Argyrochosma*. [*Cheilanthes dealbata* Pursh, *Notholaena dealbata* (Pursh) Kunze, *Pellaea dealbata* (Pursh) Prantl]

Argyrochosma microphylla (Mett. ex Kuhn) Windham, (small-leaved), mainly occurring in w

Texas and the Edwards Plateau, is disjunct to the e of nc TX in Brazos Co. (Correll 1956). It is easily distinguished from *A. dealbata* by the lack of whitish powdery material on the abaxial leaf surfaces.

ASTROLEPIS STAR-SCALED CLOAK FERN

Usually on rocks; stems (rhizomes) compact to short-creeping; leaves monomorphic, evergreen, clustered, 1-pinnate to 1-pinnate-pinnatifid, the abaxial (= beneath) leaf surfaces with ciliate scales and usually underlying layer of stellate scales concealing the surface, the adaxial surfaces sparsely to densely covered with stellate or coarsely ciliate scales to glabrescent with age; sporangia marginal or nearly so, forming a ± continuous band; false indusium absent.

A New World genus of ca. 8 species. The taxa treated here as *Astrolepis* have been previously lumped into various genera including *Notholaena* or *Cheilanthes*. Benham and Windham (1992) indicated these and several related species are a monophyletic group worthy of recognition as the genus *Astrolepis*. (Greek: *astro*, star, and *lepis*, scale, in reference to the star-like scales on the adaxial surfaces of the leaf blades)

REFERENCES: Tryon 1956; Benham & Windham 1992, 1993.

1. Adaxial leaf surfaces (= above) densely scaly, particularly near margins, the scales usually persistent; largest pinnae entire or slightly lobed; body of adaxial scales 5–7 cells wide ______ **A. integerrima**
1. Adaxial leaf surfaces only sparsely scaly to glabrescent, most scales deciduous with age; largest pinnae often conspicously lobed; body of adaxial scales 1–2 cells wide ______ **A. sinuata**

Astrolepis integerrima (Hook.) D.M. Benham & Windham, (very entire). Leaves 8–45 cm long; largest pinnae usually 7–15 mm long, symmetrically 6–14 lobed. Rocky slopes, outcrops, or cliffs, usually limestone or other calcareous substrates; Burnet and Palo Pinto cos. (Correll 1956), also Brown Co. (Carr 1995); w and sw parts of nc TX s and w to w TX. Sporulating summer–fall. [*Cheilanthes integerrima* (Hook.) Mickel, *Notholaena integerrima* (Hook.) Hevly, *Notholaena sinuata* (Lag. ex Sw.) Kaulf. var. *integerrima* Hook.]

Astrolepis sinuata (Lag. ex Sw.) D.M. Benham & Windham, (wavy-margined), BULB LIP FERN, WAVY CLOAK FERN, LONG CLOAK FERN. Leaves 11–130 cm long; longest pinnae 7–35 mm long, entire or asymmetrically and shallowly lobed. Rocky slopes, outcrops, or cliffs, calcareous or other substrates; Coleman Co. (Correll 1956); mainly c to w TX.; Hatch et al. (1990) also cited vegetational area 4 (Fig. 2), probably based on a record from Anderson Co. (Correll 1956) near the boundary of the Blackland Prairie and Post Oak Savannah vegetation areas. Sporulating Mar–Nov. [*Acrostichum sinuatum* Lag. ex Sw., *Cheilanthes sinuata* (Lag. ex Sw.) Domin, *Notholaena sinuata* (Lag. ex Sw.) Kaulf.] Burlage (1968) reported this species as toxic to livestock.

CHEILANTHES LIP FERN

Xeric-adapted, usually growing on rocks; stems (rhizomes) compact to long-creeping; leaves monomorphic, evergreen, clustered or scattered along the rhizomes; leaf blades 2-more-pinnate-pinnatifid, usually conspicuously tomentose beneath; petioles dark brown to black; sporangia marginal on the abaxial (= beneath) leaf surfaces; margins of ultimate leaf segments recurved to form false indusia; veins of ultimate segments free or rarely anastomosing, obscure.

A genus of ca. 150 species found primarily in the New World with a few in Europe, Asia, Africa, Pacific Islands, and Australia. According to Windham and Rabe (1993), *Cheilanthes* is the largest and most diverse genus of xeric-adapted ferns. Even after the removal of segregates including *Argyrochosma* and *Astrolepis*, it is still a heterogeneous and possibly polyphyletic genus. (Greek: *cheilos*, margin, and *anthus*, flower, referring to the marginal sporangia)

REFERENCES: Mickel 1979; Windham & Rabe 1993.

1. Midrib of leaf segments (= pinnae) and/or rachis with scales (hairs can also be present) beneath (= abaxially).
   2. Ultimate leaf segments scabrous (= rough to the touch) on adaxial (= above) surface, covered with stiff hairs ____ **C. horridula**
   2. Ultimate leaf segments smooth to the touch, lacking stiff hairs.
      3. Scales linear, inconspicuous, only slightly wider than hairs, the largest 0.1–0.4 mm wide ____ **C. tomentosa**
      3. Scales linear to lanceolate to ovate, conspicuous, obviously much wider than hairs, the largest 0.4–1.0 mm wide.
         4. Scales ovate to lanceolate, long ciliate, the cilia sometimes forming an entangled mass; rhizome slender, widely creeping, with leaves scattered along the rhizome ____ **C. lindheimeri**
         4. Scales linear to lanceolate, not ciliate, rarely with 1–2 cilia at base; rhizome stout, short, with leaves in a dense clump ____ **C. eatonii**
1. Midrib of leaf segments and rachis lacking scales beneath or with extremely narrow inconspicuous hair-like scales (but can be strikingly pubescent to glabrous).
   5. Leaves essentially glabrous to sparsely pubescent beneath; ultimate leaf segments narrowly elliptic to elongate-deltate, not at all sub-orbicular to bead-like ____ **C. alabamensis**
   5. Leaves densely pubescent beneath; ultimate leaf segments sub-orbicular to bead-like OR not so.
      6. Ultimate leaf segments scabrous (= rough to the touch) on adaxial (= above) surface, covered with stiff hairs ____ **C. horridula**
      6. Ultimate leaf segments smooth to the touch, lacking stiff hairs.
         7. Stipe and rachis not densely tomentose, instead very sparsely to densely hispidulose, the hairs noticeably jointed (under strong lens or dissecting scope).
            8. Leaf blades 3-pinnate near base, the fertile ultimate segments nearly round, bead-like ____ **C. feei**
            8. Leaf blades 2-pinnate-pinnatifid near base, the fertile ultimate segments elongate, not bead-like ____ **C. lanosa**
         7. Stipe and rachis densely tomentose, particularly when young, the hairs not noticeably jointed ____ **C. tomentosa**

Cheilanthes alabamensis (Buckley) Kunze, (of Alabama), ALABAMA LIP FERN, SMOOTH LIP FERN. Leaves clustered, 6–50 cm long; leaf blades lanceolate to oblong, 1–7 cm wide, the largest ultimate segments 3–7 mm long; this is the most glabrous of our *Cheilanthes* species. Limestone hillsides, crevices of limestone ledges and cliffs; Coryell, Palo Pinto, and Tarrant cos.; also Bell, Brown, Hamilton (HPC), Somervell, Williamson (Correll 1956), and Parker (B. Carr, pers. comm.) cos.; widely distributed across TX. Sporulating nearly throughout the year, especially Mar–Nov. *Cheilanthes aemula* Maxon, known se of nc TX in Austin Co. (Correll 1956), differs from the similiar *C. alabamensis* in having broadly triangular to ovate leaf blades 5–15 cm wide.

Cheilanthes eatonii Baker, (for its discoverer, A.A. Eaton, 1865–1908), EATON'S LIP FERN. Leaves clustered, 6–35 cm long; leaf blades 1.5–5 cm wide, the ultimate segments oval to round, bead-like, the largest 1–3 mm long; scales conspicuous. Rocky slopes and ledges; Brown Co. (Correll 1956; HPC); mainly Edwards Plateau and Trans-Pecos. Sporulating Mar–Nov. [*C. castanea* Maxon]

Cheilanthes feei T. Moore, (for A.L.A. Fée, 1789–1874, French botanist), SLENDER LIP FERN, WOOLLY LIP FERN, FEE'S LIP FERN. Leaves clustered, 4–20 cm long; leaf blades 1–3 cm wide, the ultimate segments 1–3 mm long; similar to *C. tomentosa* but with jointed hairs and without tomentum on the stipe and rachis. Limestone or calcareous, dry rocky slopes and crevices; Hamilton and Palo Pinto cos. (Correll 1956); w part of nc TX s and w to w TX. Sporulating Mar–Nov.

Cheilanthes horridula Maxon, (prickly), ROUGH LIP FERN. Leaves clustered, 5–30 cm long; leaf

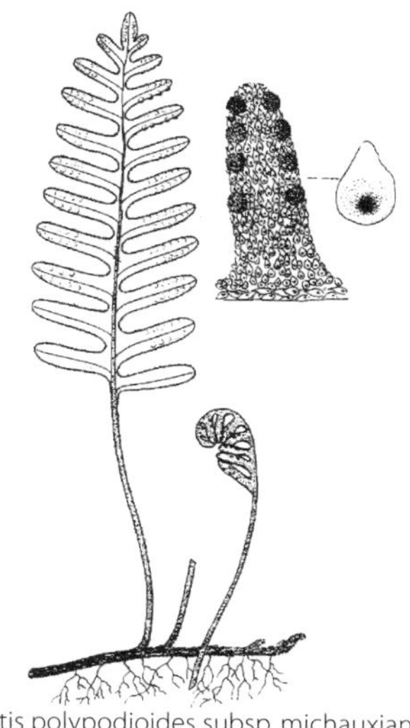
Pleopeltis polypodioides subsp. michauxiana [LUN]

Adiantum capillus-veneris [LUN]

Argyrochosma dealbata [LUN]

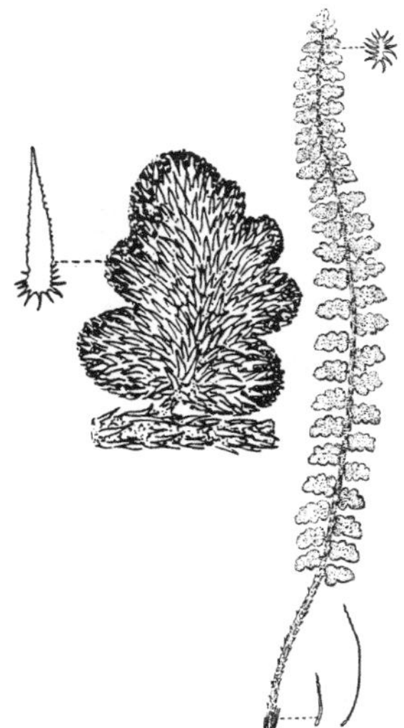
Astrolepis integerrima [LUN]

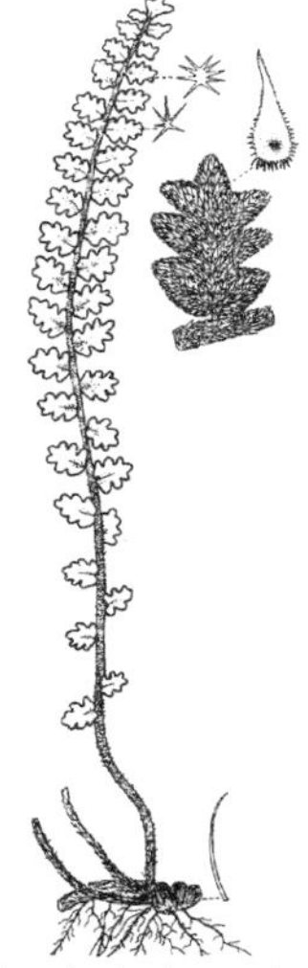
Astrolepis sinuata [LUN]

Cheilanthes alabamensis [LUN]

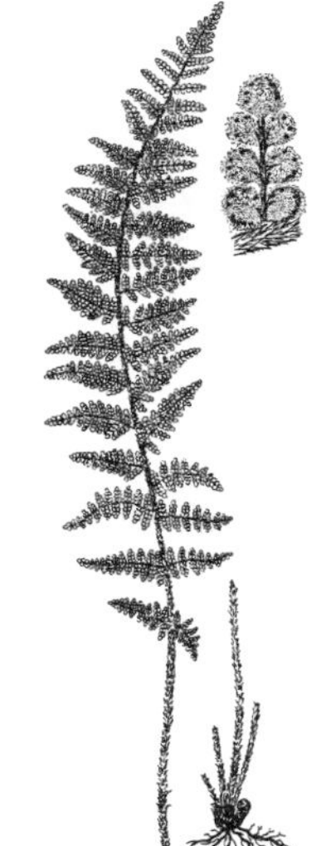
Cheilanthes eatonii [LUN]

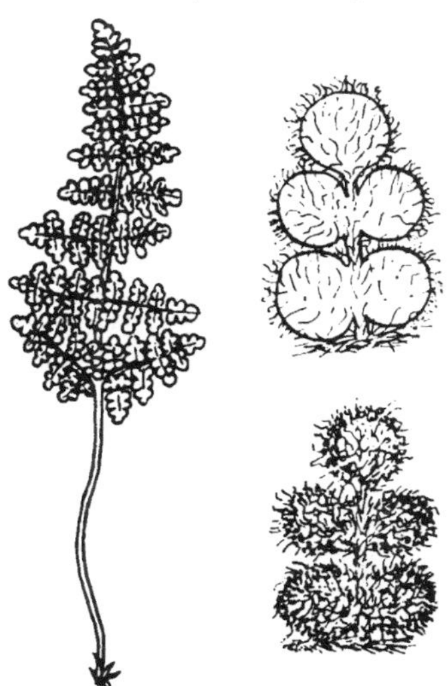
Cheilanthes feei [LUN]

Cheilanthes horridula [LUN]

blades 1–4 cm wide, the ultimate segments narrowly elliptic to elongate-deltate, not bead-like, the largest 3–5 mm long; the distinctive stiff hairs giving the leaf surfaces their scabrous nature are often inflated basally. Rock crevices; Brown, Burnet (HPC), Coleman, and Palo Pinto (Correll 1956) cos.; mainly w 2/3 of TX. Sporulating mainly May–Nov.

Cheilanthes lanosa (Michx.) D.C. Eaton, (woolly), HAIRY LIP FERN, WOOLLY LIP FERN. Leaves clustered, 7–50 cm long; leaf blades 1.5–5 cm wide, the ultimate segments oblong to lanceolate, not bead-like, the largest 3–5 mm long; similar in some respects to *C. tomentosa* but with hispidulous jointed hairs instead of tomentum on the stipe and rachis. Dry rocky slopes and sandstone ledges; known in TX only in McLennan Co. (Correll 1956: *Wherry s.n.*, BAYLU). Sporulating Apr–Oct. Jack Stanford (pers. comm.), who studied the Wherry collection, questioned whether it is actually *C. lanosa*.

Cheilanthes lindheimeri Hook., (for F.J. Lindheimer, 1801–1879, German-born Texas collector), LINDHEIMER'S LIP FERN, FAIRY-SWORDS. The slender creeping rhizomes distinguish this species from other nc TX *Cheilanthes*; scales conspicuous; leaves scattered along the rhizomes, 7–30 cm long; leaf blades 2–5 cm wide, the ultimate segments round to slightly oblong, bead-like, the largest 0.7–1 mm long. Rocky slopes and ledges; Palo Pinto Co. (Correll 1956), also Brown (Carr 1995) and Parker (B. Carr, pers. comm.) cos.; mainly Edwards Plateau and Trans-Pecos. Sporulating Mar–Nov. Jack Stanford (pers. comm.) indicated that this species is found primarily on granite.

Cheilanthes tomentosa Link, (tomentose, densely woolly), WOOLLY LIP FERN. Leaves clustered, 8–45 cm long; leaf blades 1.5–8 cm wide, the ultimate segments oval (rarely oblong), bead-like, the largest 1–2 mm long; scales inconspicuous. Rocky slopes and ledges; Grayson, Denton, Palo Pinto, and Parker cos., also Brown, Comanche (HPC), Milam, and Young (Correll 1956) cos.; widely distributed in TX. Sporulating mainly May–Oct.

PELLAEA CLIFF-BRAKE

Xeric-adapted, usually on rocks; stems (rhizomes) compact to creeping; leaves monomorphic or somewhat dimorphic, evergreen, clustered to scattered, 1–3 pinnate, in ours glabrous or nearly so, thick-herbaceous to coriaceous; sporangia near margins of leaf segments on the abaxial (= beneath) leaf surfaces; margins of ultimate leaf segments recurved to form false indusia.

A genus of ca. 40 species distributed mainly in the New World with a few in Asia, Africa, the Pacific Islands, and Australia. The genus has often been circumscribed more broadly, but as such is probably polyphyletic. Some species previously placed in *Pellaea* are now recognized in *Argyrochosma*. (Greek: *pellos*, dark, possibly referring to bluish gray leaves)
REFERENCES: Tryon 1957; Knobloch & Britton 1963; Windham 1993c.

1. Petiole and rachis straw-colored or tan, not shiny, usually glabrous; rachis uniformly zigzag throughout ____ **P. ovata**
1. Petiole and rachis reddish purple to dark brown or blackish, shiny, glabrous or pubescent adaxially (= above) with curly hairs; rachis not uniformly zigzag, at most slightly flexuous.
    2. Pinnules mucronate (= with a small tip); some scales of the stem (look near attachment of petioles) bicolored with a dark, blackish, linear central region and a lighter brown margin; rachis usually glabrous ____ **P. wrightiana**
    2. Pinnules not mucronate; stem scales uniformly reddish brown or tan; rachis pubescent adaxially ____ **P. atropurpurea**

Pellaea atropurpurea (L.) Link, (dark purple), PURPLE CLIFF-BRAKE, CLIFF-BRAKE, BLUE FERN. Plants to 45 cm tall; leaf blades 1-pinnate or 2-pinnate below, 10–30 cm long, 5–20 cm wide. Rocky slopes and woods, cliffs, usually limestone or calcareous rocks; Bell, Burnet, Coleman,

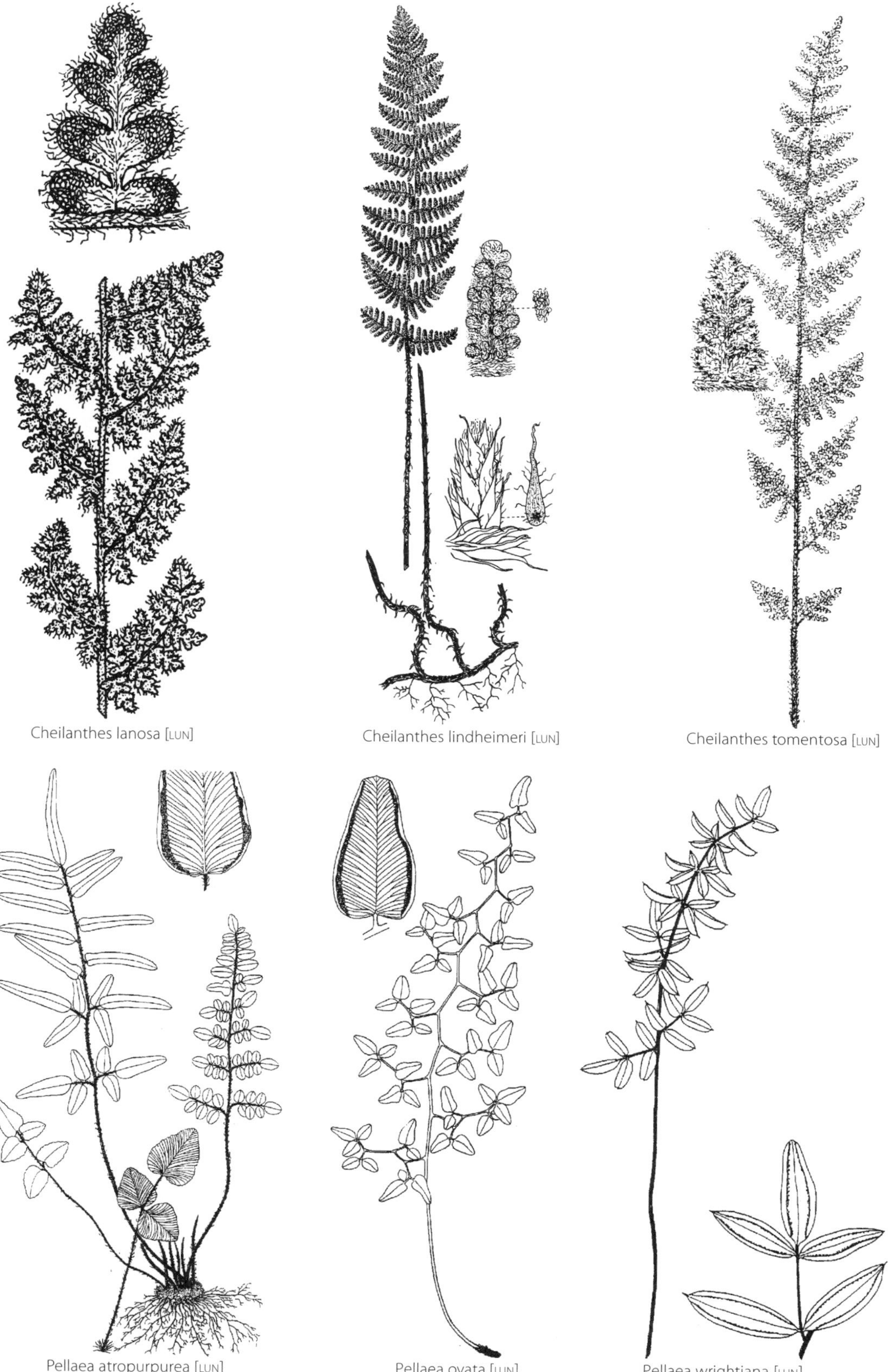

Cheilanthes lanosa [LUN]

Cheilanthes lindheimeri [LUN]

Cheilanthes tomentosa [LUN]

Pellaea atropurpurea [LUN]

Pellaea ovata [LUN]

Pellaea wrightiana [LUN]

Cooke, Denton, Grayson, Hood, Jack, Johnson, Palo Pinto, Tarrant, and Young cos.; also Dallas (Reverchon 1903), Hamilton (HPC), McLennan, and Williamson (Correll 1956) cos.; nearly throughout TX. Sporulating Mar–Nov.

Pellaea ovata (Desv.) Weath., (ovate). Plants usually large, to 1 m or more tall; leaf blades 2-3-pinnate, 15–70 cm long, 5–25 cm wide. Rocky slopes and ledges, including limestone; Burnet Co.; also Brown (J. Stanford, pers. comm.) and Palo Pinto (Correll 1956) cos.; mainly s TX to Edwards Plateau and Trans-Pecos. Sporulating Mar–Nov.

Pellaea wrightiana Hook., (for Charles Wright, 1811-1885, Texas collector), WRIGHT'S CLIFF-BRAKE. Plants 15–30(–50) cm tall; leaf blades 1-pinnate-pinnatifid to 2-pinnate below, usually 8–25 cm long, 1–5 cm wide. Burnet Co.; also Comanche (HPC) and Palo Pinto (Correll 1956) cos.; mainly w 1/2 of TX. Sporulating Mar–Nov. [*P. ternifolia* (Cav.) Link var. *wrightiana* (Hook.) A.F. Tryon]

THELYPTERIDACEAE MARSH FERN FAMILY

A mostly tropical family of ca. 900 species; depending on circumscription, the number of genera can vary from 1 to ca. 30. Many have been historically associated with the Dryopteridaceae, but are not closely related to that family.

FAMILY RECOGNITION IN THE FIELD: leaves all alike, 1-pinnate-pinnatifid with the ultimate segments entire; sori round, located medially to submarginally on the leaf segments; indusia round to kidney-shaped.

REFERENCE: Smith 1993a.

THELYPTERIS FEMALE FERN

Terrestrial; stems (rhizomes) horizontal, short- or long-creeping; leaves monomorphic, spaced (0.5–)1–4 cm apart along the stems; leaf blades 1-pinnate-pinnatifid; ultimate leaf segments entire; petioles about equal to blade in length, straw-colored; sori round, in medial to submarginal position on the leaf segments on the abaxial (= beneath) surfaces; indusia round to kidney-shaped.

A nearly cosmopolitan genus of ca. 875 species; often subdivided into segregates. (Greek: *thelys*, female, and *pteris*, fern)

REFERENCE: Smith 1971.

1. Leaf blades with midveins of pinnae on adaxial (= upper) surface with conspicuous (use lens) hairs usually longer than width of the veins; scales absent on abaxial surfaces of rachises and costae of mature leaves; sori medial to supramedial; n part of nc TX ____ **T. kunthii**
1. Leaf blades with midveins of pinnae on adaxial (= upper) surface glabrous or with a few minute hairs; a few scales often persistent on abaxial surfaces of rachises and costae of mature leaves; sori supramedial to submarginal (sori typically closer to leaf margins than in *T. kunthii*); s part of nc TX ____ **T. ovata**

Thelypteris kunthii (Desv.) C.V. Morton, (for Karl Sigismund Kunth, 1788–1850, German botanist), WIDESPREAD MAIDEN FERN, SOUTHERN SHIELD FERN. Stems short- to long-creeping; leaves up to 2(–3) cm apart along the stems, (15–)50–160 cm long; leaf blades relatively large (pinnae (2–)8–15(–20) cm long), the pinnae cut 3/5–4/5 of width; abaxial (= lower) surface with indument of short hairs on costae, veins, and blade tissue; petioles (5–)20–80 cm long. Moist areas, seeps at base of bluffs; Parker Co. (*Jeff Quayle, s.n.*, 1997, BRIT), also a Dallas Co. specimen of *Dryopteris normalis* cited by Correll (1956) is probably this species; mainly e TX. [*Dryopteris normalis* C. Chr., *T. normalis* (C. Chr.) Moxley]

Thelypteris ovata R.P. St. John var. **lindheimeri** (C. Chr.) A.R. Sm., (sp.: ovate; var.: for F.J. Lindheimer, 1801–1879, German-born Texas collector), LINDHEIMER'S MAIDEN FERN. Stems usually long-creeping; leaves (0.5–)1–4 cm apart along the stems, (30–)55–135(–165) cm long, erect or ascending; leaf blades relatively large (pinnae (5–)10–15(–25) cm long), the basal pinnae usually only slightly shorter than ones just above, the pinnae cut ca. 3/4–4/5 of their width; abaxial (= lower) surface pubescent, the hairs on the abaxial midveins of the pinnae shorter than the width of the midvein; petioles 15–80 cm long; indusia orbicular-reniform, persistent. Low, moist areas, wet bluffs and ledges, including limestone; Bell and Burnet cos., also a Williamson Co. specimen cited by Correll (1956) as *Dryopteris normalis* is probably *T. ovata* var. *lindheimeri*, also a recent Tarrant Co. collection (*Jeff Quayle, s.n.*, 1997, BRIT) from a ditch in the Fort Worth Nature Center may be an escape from cultivation; nc TX w to Edwards Plateau and Trans-Pecos. Sporulating May–Nov. [*Dryopteris normalis* C. Chr. var. *lindheimeri* C. Chr.] This species has often been confused and lumped (Correll 1956, 1966a, Correll & Johnston 1970, Hatch et al. 1990) with *Thelypteris kunthii* (either as *T. kunthii* or under the name *Dryopteris normalis*); while strikingly similar in overall aspect, the two can be readily distinguished by the characters in the key.

GYMNOSPERMS

The term gymnosperm (literally naked seed), referring to those plants with ovules, and subsequently seeds, borne on the surface of an open scale, is not recognized here as a formal taxonomic category (it was formerly treated as the Gymnospermae). The evolution of the seed in the various gymnosperm groups probably occurred independently from non-seed ancestors. The group would thus be polyphyletic and not worthy of formal recognition. The four living gymnosperm groups (surviving remnants of ancient and much more diverse lineages; currently totaling 840 species in 86 genera arranged in 17 families worldwide) are therefore treated as separate divisions (Cycadophyta, Ginkgophyta, Gnetophyta, and Pinophyta); only two of these are native to nc TX.

REFERENCES: Hardin 1971; Eckenwalder 1993.

DIVISION **PINOPHYTA**
CONIFERS

This is the gymnosperm division with the largest number of living representatives (70 genera and 598 species arranged in 8 families—Mabberley 1997); the seeds are typically borne in cones (thus the common name from *conium*, cone, and *-feros*, bearing). The fossil history of the group extends to late in the Carboniferous period (360–286 million years ago). Vast forests of Pinophyta (PINE, SPRUCE, FIR, DOUGLAS-FIR, CEDAR, etc.) are present across the northern part of the world between areas of tundra and deciduous forest; they dominate the biome known as taiga. These mostly evergreen species have xerophytically adapted, desiccation resistant foliage that allows them to maintain their photosynthetic surface through the long winter and make immediate and maximal use of the short growing season available in the taiga. Having evergreen leaves that last for several years also means that the high nutrient demand associated with making a new set of leaves each spring is not required—this is considered a significant advantage on the generally nutrient-poor soils of the taiga (Pielou 1988). The result is that this is one of the few gymnosperm groups that has maintained dominance over flowering plants across vast areas. The small family Taxaceae (YEWS) is important because the bark of *Taxus brevifolia* Nutt. (PACIFIC YEW, CALIFORNIA YEW) is the source of the terpenoid taxol, a promising anti-cancer drug used in the treatment of ovarian and other types of cancer; as a result, PACIFIC YEW populations in some areas have been greatly reduced. While not important as a direct source of taxol, the leaves of the European and Mediterranean *Taxus baccata* L. (EUROPEAN YEW, ENGLISH YEW) contain a compound that is now being used in taxol synthesis. It is interesting to

note that like many medically valuable plants "discovered" by modern medicine, the genus has a long history of medicinal use; e.g., early Europeans used it in treating hydrophobia and heart ailments and Native Americans used it against such conditions as rheumatism, bronchitis, fever, scurvy, and skin cancer. ☠ Also like many medicinal plants, YEWS are poisonous; the species have long been used variously as arrow poisons, to kill fish, and in murder and suicide, and are known to be fatally poisonous to animals and humans. Death from YEW can be sudden with animals sometimes being found close to the plant with foliage still in their mouths (Kingsbury 1964; Hartzell 1991, 1995; USDA Forest Service 1993; Cragg et al. 1995; Suffness & Wall 1995). The Pinophyta is sometimes referred to as the Coniferophyta (Raven et al. 1986).
REFERENCES: Hardin 1971; Bell & Woodcock 1983; Raven et al. 1986; Bold et al. 1987; Eckenwalder 1993; Woodland 1997.

CUPRESSACEAE CYPRESS OR REDWOOD FAMILY

Evergreen or deciduous trees or shrubs; monoecious or in *Juniperus* usually dioecious; leaves alternate and spirally arranged, sometimes appearing 2-ranked due to twisting, sometimes dimorphic, often with an abaxial resin gland; pollen cones usually solitary, terminal; pollen not winged; seed cones with scales fleshy or woody.

This family has often been divided between Cupressaceae (in the strict sense), for those genera having opposite or whorled leaves (including *Juniperus*), and Taxodiaceae, or REDWOOD FAMILY, for those genera having leaves mostly alternate. We follow Eckenwalder (1976), Hart and Price (1990), and Watson and Eckenwalder (1993) in treating them as a single family. Recent molecular evidence (Brunsfeld et al. 1994) shows Cupressaceae (in the strict sense) derived from within Taxodiaceae, supporting the single family treatment. The family is widespread in temperate areas and has ca. 25–30 genera and ca. 110–130 species; it includes many interesting or important genera including *Metasequoia*, *Sequoia*, *Sequoiadendron* (GIANT REDWOOD), and *Thuja* (ARBORVITAE). *Metasequoia glyptostroboides* Hu & W.C. Cheng (DAWN REDWOOD), known from only one remote area of China, was discovered in 1945; it has an extensive fossil record—it was the most abundant conifer in w and arctic North America from the late Cretaceous to the Miocene—and is thus often referred to as a living fossil. *Sequoia sempervirens* (D. Don) Endl. (COAST REDWOOD), of the Pacific coast of the U.S., is the world's tallest tree, reaching heights of over 117 m (Raven et al. 1986); it has been greatly overexploited and is now restricted to a few reserves. Family name from *Cupressus*, CYPRESS, a genus of 10–26 species of warm north temperate areas. (Latin name for the Italian cypress, *C. sempervirens* L.)
FAMILY RECOGNITION IN THE FIELD: EITHER evergreen trees or shrubs of dry habitats with opposite or whorled, *scale-like* leaves and *small, berry-like* cones OR trees of wet habitats with alternate, linear to linear-lanceolate, *flat and feathery, deciduous* leaves, nearly *globose, plum-sized* cones, and often with "*knees*" (erect woody projections) from the roots.
REFERENCES: Dallimore & Jackson 1931; Correll 1966b; Eckenwalder 1976; Price & Lowenstein 1989; Hart & Price 1990; Watson & Eckenwalder 1993; Brunsfeld et al. 1994.

1. Leaves (adult) scale-like, closely appressed to stem, to 2.5 mm long, opposite or whorled, evergreen; cones globose to ovoid, to ca. 10 mm long, berry-like; plants typically of dry habitats ______ **Juniperus**
1. Leaves linear to linear-lanceolate, conspicuously flat and feathery, not appressed, 10–15 mm long, alternate, deciduous; cones usually nearly globose, 15–25(–40) mm in diam., woody; plants of wet habitats ______ **Taxodium**

JUNIPERUS JUNIPER

Dioecious (pollen cones and seed cones on separate trees) or rarely monoecious (pollen cones and seed cones on same tree), evergreen, aromatic, resinous trees or shrubs; bark (in our species) reddish brown to brown or ashy gray, with long, thin, shreddy scales; adult leaves usually scale-like, opposite or in whorls; juvenile leaves needle-like; staminate cones small, cylindric; mature ovulate cones fleshy, berry-like, variously colored, often glaucous, globose to ovoid, to ca. 10 mm long; seeds (in our species) 1–several, wingless.

A genus of ca. 60 species, primarily n hemisphere in distribution with 1 species in e Africa. The decay resistent wood of *Juniperus* species is often used for fence posts; the cones are an important food for birds; also, gin is flavored by the cones of *Juniperus communis* L., of n North America. Numerous cultivars are used in landscaping, particularly those with unusual habits or foliage. The wind borne pollen is one of the most serious allergens in nc TX. JUNIPERS are problematic near apple trees and native hawthorns (*Crataegus* species) since they serve as an alternate host for cedar apple rusts (*Gymnosporangium* spp.). (Latin: *juniperus*, name for JUNIPER)
REFERENCES: Hall 1952; Adams 1972, 1975, 1986, 1993; Flake et al. 1978.

1. Mature ovulate cones (seed cones) reddish or copper-colored; leaf gland often with white crystalline exudate; hilum (= attachment scar) covering seed ca. 1/2 its length ____ **J. pinchotii**
1. Mature ovulate cones blue to bluish black or bluish purple; leaf gland without exudate; hilum covering seed ca. 1/3 or less it length.
   2. Plant usually with one main trunk from base; abaxial (= on side away from twig) leaf glands usually elliptic to elongate, usually not conspicuously raised (10X lens); leaf margins entire, smooth (under a dissecting scope) ____ **J. virginiana**
   2. Plant usually with several trunks from near base; abaxial leaf glands usually roundish in outline, often conspicously raised (10X lens); leaf margins irregularly very minutely cellular-serrulate or cellular-denticulate, not smooth (under a dissecting scope) ____ **J. ashei**

Juniperus ashei J. Buchholz, (for its discoverer, William Willard Ashe, 1872–1932), MOUNTAIN-CEDAR, ROCK-CEDAR, POST-CEDAR, MEXICAN JUNIPER, ASHE'S JUNIPER. Large shrub or small tree to ca. 6 m tall, usually with several trunks from near base; does not resprout after cutting or burning; bark ashy-gray to brown; ovulate cones mostly 7–8.5 mm long when mature, dark blue, glaucous, sweet, resinous; seeds 1(–3), covered by hilum for 1/3 their length. Rocky soils; often forming thickets or "cedar brakes"; Dallas and Cooke cos. s and w; nc TX and Edwards Plateau s and w to w TX. Due to fire supression, this species currently covers much more area than previously (Hall 1952); this has significant negative impacts on other native plants and is problematic for ranchers. *Juniperus ashei* is sometimes distinguished with difficulty from *J. virginiana*; in addition to the characters in the key, *J. ashei* usually has stiffer twigs and more odoriferous herbage; hybridization and introgression are known where the 2 occur together (Correll 1966b, Hall 1952). Hall (1952) noted that *J. ashei* can also hybridize with *J. pinchotii*.

Juniperus pinchotii Sudw., (for botanist Giffard Pinchot, 1865–1946), RED-BERRY JUNIPER, PINCHOT'S JUNIPER. Large shrub or shrub-like small tree to ca. 6 m tall, usually with several trunks from near base; resprouts after cutting or burning; bark ashy-gray to brown; ovulate cones 6–10 mm long, usually not glaucous or only slightly so, sweet, not resinous; seeds 1–2, covered by hilum for ca. 1/2 their length. Gravelly or rocky soils, commonly limestone or gypsum; Montague and Johnson cos., also Little (1971) mapped numerous other counties in the West Cross Timbers and Lampasas Cut Plain; w part of nc TX s and w to w TX. According to Correll (1966b), the branchlets of *J. pinchotii* tend to be more slender and erect than the usually stiffish, recurved branchlets of *J. ashei*.

Juniperus virginiana L., (of Virginia), EASTERN RED-CEDAR, RED-CEDAR, VIRGINIA RED-CEDAR, RED SAVIN, PENCIL-CEDAR, RED JUNIPER. Medium to large tree to 30 m tall, typically much smaller, usually with one main trunk; does not resprout after cutting or burning; bark reddish brown; ovulate cones 5–8 mm long, blue to bluish black or bluish purple, glaucous, resinous; seeds 1–2(–3), the hilum small, inconspicuous. Dry sandy and rocky soils, old fields, fencerows, forest margins; se and e TX w to West Cross Timbers and Edwards Plateau; Little (1971) mapped the species as far west as Wichita Co. in the Rolling Plains. This is a problematic invader of native prairies under conditions of fire suppression. The aromatic, moth-repelling heartwood is used for cedar chests and closets. RED-CEDAR symbolized the tree of life for a number of Native American tribes and was burned in sweat lodges and in purification rituals (Kindscher 1992).

TAXODIUM BALD CYPRESS

A genus of a single species (sometimes divided into 3) ranging from the United States through Mexico to Guatemala; this is one of only 11 tree genera endemic to e North America (and adjacent tropical areas); (only three of these, *Asimina*, *Maclura*, and *Taxodium*, occur in nc TX) (Little 1983). It is frequently segregated with related taxa into the Taxodiaceae. (*Taxus*, generic name of yew, and Greek *-oides*, like)
REFERENCE: Watson 1985.

Taxodium distichum (L.) Rich. var. **distichum**, (in two ranks), BALD CYPRESS, SOUTHERN CYPRESS. Monoecious (pollen cones and seed cones on the same tree), deciduous trees to 50 m tall with a swollen, often buttressed base; in frequently flooded areas often with "knees" (erect woody projections) from the roots; slender leafy twigs deciduous with the leaves in fall; leaves 2-ranked, feathery, linear, flat, 1–1.5 cm long; staminate (pollen) cones ca. 2 mm in diam., in drooping panicles 10–12 cm long; ovulate (seed) cones usually nearly globose, to ca. 25 mm in diam., the scales somewhat peltate. Swamps and along water courses. Pollen shed in spring; seeds in fall. While BALD CYPRESS does not occur naturally in nc TX (native to Edwards Plateau and e TX as far w as Upshur and Red River cos.), it is now extensively planted and does well even in upland situations; trees planted in a swamp in Fannin Co. appear almost native and a volunteer seedling has been found (Talbot property). It is included because given the frequency of cultivation and the often excellent cone production, more extensive reproduction from seeds along water courses is a strong possibility. BALD CYPRESS is an important timber tree known for its decay-resistant wood, even when in contact with soil; the heartwood is so durable that it has been referred to as "the wood eternal" (Hart & Price 1990).

Taxodium distichum var. *mexicanum* Gordon, (of Mexico), [*T. mucronatum* Ten.], the related MEXICAN OR MONTEZUMA BALD CYPRESS, is famous for the "Tule Tree" of Oaxaca, one of the world's largest trees (Hall et al. 1990; Dorado et al. 1996); this ± evergreen variety extends as far n as s TX.

PINACEAE PINE FAMILY

A primarily n hemisphere family of 10 genera and ca. 200 species; it is of great economic importance as a source of softwood timber, pulpwood, naval stores (e.g., turpentine), Christmas trees, and ornamentals. Other important genera include *Abies* (FIRS), *Picea* (SPRUCE), *Pseudotsuga* , and *Tsuga* (HEMLOCK). *Pseudotsuga menziesii* (Mirbel) Franco (DOUGLAS FIR), of w North America, with trunks 3–4 m in diam. and over 90 m tall, is one of the most important lumber trees in the world (Lipscomb 1993; Woodland 1997); it is frequently sold as a Christmas tree in nc TX and can be recognzied by the pointed buds; an individual 133 m tall was reported to have been felled in British Columbia in 1895 (Mabberley 1987).
FAMILY RECOGNITION IN THE FIELD: trees with long, *needle-like leaves in bundles* of 2 or 3 (our species) and large woody *pine cones*; tissues resinous and aromatic.
REFERENCES: Dallimore & Jackson 1931; Correll 1966b; Little 1971; Price 1989; Thieret 1993.

PINUS PINE

Monoecious (pollen cones and seed cones on the same tree), evergreen, resinous, aromatic trees to 30 m or more tall; leaves of 2 kinds; scale-like leaves subtending minute branchlets; each

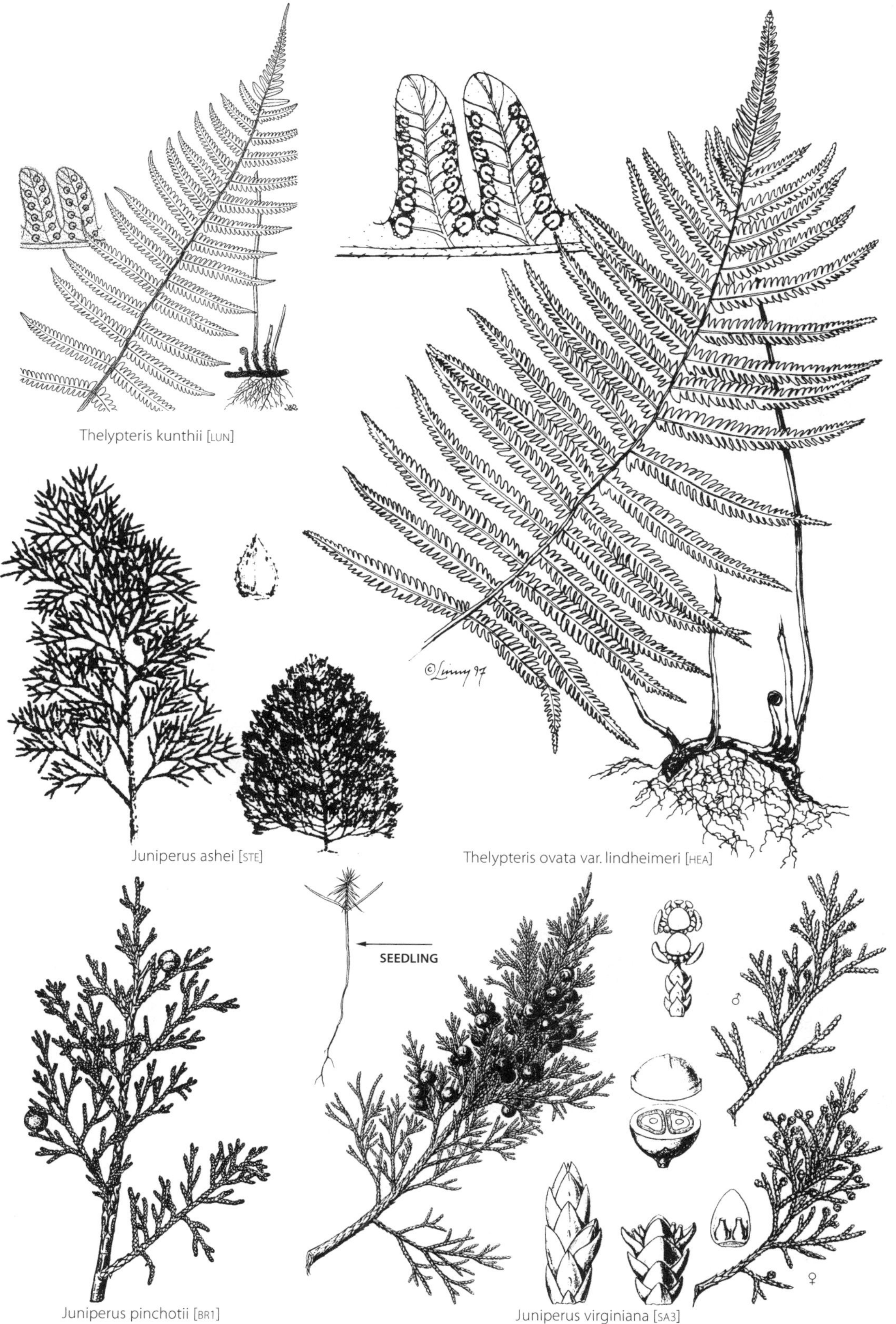

Thelypteris kunthii [LUN]

Juniperus ashei [STE]

Thelypteris ovata var. lindheimeri [HEA]

Juniperus pinchotii [BR1]

Juniperus virginiana [SA3]

branchlet bearing a fascicle of 2–3 (in our species) elongate, needle-like foliage leaves (= needles) surrounded at the base by a membranous sheath; staminate (pollen) cones small, in clusters at the base of the current years growth; pollen winged; ovulate (seed) cones becoming large and woody; each scale of seed cones with a thickened, exposed, apical portion (= apophysis) terminated by a protuberance (= umbo); seeds winged (in our species).

A genus of ca. 100 species widely distributed in the n temperate zone and in mountainous areas of the n tropics; many are cultivated for timber, pulp, and resinous products; others are used for their edible seeds (pignons, pignolia or pine nuts) or as ornamentals. According to Millar (1993), "*Pinus* contains more species than any other group of conifers ..." *Pinus longaeva* D.K. Bailey (BRISTLE-CONE PINE of far w North America) is among the oldest living trees, with individuals approaching 5,000 years old; this species has been important in the development of dendrochronology (= tree-ring dating); when dead specimens (which can last thousands of years before decaying) are used, a tree ring record of 8,200 years is available. The genus is economically important and widely cultivated in e TX as a source of wood products. PINES are native as far w as Lamar Co. (Fannin Co.[?] (Correll & Johnston 1970)) in the extreme ne part of nc TX where they occur on sandy, more acidic alluvium associated with the Red River. However, the calcium-rich, basic soils of much of nc TX are not well-suited for pines. The following treatment relies heavily on Kral (1993). (Latin: *pinus*, name for pine)
REFERENCES: Kral 1993; Millar 1993.

1. Needles (20–)25–45 cm long, 3 per bundle; terminal buds silvery white, 3–4 cm long; bundle sheaths of new needles on young twigs 25 mm or more long; seeds with body ca. 10 mm long and wing 30–40 mm long ____ **P. palustris**
1. Needles 5–23(–29) cm long, 2–3 per bundle; terminal buds brownish, 0.5–2 cm long; bundle sheaths of new needles on young twigs 20 mm or less long; seeds with body 5–7 mm long and wing 12–20 mm long.
   2. Needles (5–)7–11(–12) cm long, usually 2(–3) per bundle; bundle sheaths 5–10(–15) mm long; terminal buds 0.5–0.7(–1) cm long; mature seed cones 4–7 cm long; pollen cones 15–20 mm long at time of pollen release; bark with evident resin pockets ____ **P. echinata**
   2. Needles 12–23(–29) cm long, 2–3 per bundle; bundle sheaths (10–)12–20 mm long; terminal buds 1–2 cm long; mature seed cones 6–18(–20) cm long; pollen cones 20–40 mm long at time of pollen release; bark without resin pockets.
      3. Needles almost always 3 per bundle (very rarely 2), yellowish green to grayish green, not glossy; seed cones sessile or nearly so, mostly dull yellow-brown; surface of the exposed, thickened, apical portion of each seed cone scale (= apophysis) dull; pollen cones yellow to yellow-brown; terminal buds 1–1.2(–2) cm long ____ **P. taeda**
      3. Needles 2–3 per bundle, at least some 2, usually dark green, glossy; seed cones short-stalked, light chocolate brown; surface of exposed, thickened, apical portion of each seed cone scale lustrous as if varnished; pollen cones purplish; terminal buds 1.5–2 cm long ____ **P. elliottii**

Pinus echinata Mill., (spiny), SHORTLEAF PINE, SHORTLEAF YELLOW PINE, LONGTAG PINE. Bark on older stems red-brown and separated into irregular, flat, scaly plates, with evident resin pockets; twigs greenish brown to red-brown, red-brown to gray with age, slender (ca. 5 mm or less thick); terminal buds 0.5–0.7(–1) cm long; pollen cones 15–20 mm long at time of pollen release, yellow- to pale purple-green; seed cones 4–6(–7) cm long, red-brown, aging gray, the scales with an elongate to short, stout, sharp prickle. Uplands, dry forests; native to e TX as far w as Henderson (Correll 1966b), Red River (Little 1971), and possibly Lamar (Simpson 1988) cos.; spreading from cultivation in Fannin Co. in Red River drainage.

Pinus elliottii Engelm., (for Stephen Elliott, 1771–1831, American botanist), SLASH PINE, PITCH PINE, YELLOW SLASH PINE. Bark on older stems orange- to purple-brown, broken up into rather large flat flakes, without resin pockets; twigs orange-brown, darker brown with age, relatively slender (to 10 mm thick); terminal buds 1.5–2 cm long; pollen cones 30–40 mm long at time of pollen release, purplish; seed cones (7–)9–18(–20) cm long, light chocolate brown, the scales with a short stout prickle. Cultivated and used in reforestation; spreading from cultivation on sandy soils in Hood Co. in West Cross Timbers, also spreading on sandy soils in Denton and Tarrant cos. (R. O'Kennon, pers. obs.); mainly se and e TX; native as far w as Louisiana.

Pinus palustris Mill., (of marshes), LONGLEAF PINE, LONGLEAF YELLOW PINE. Bark on older stems orange-brown, of thin papery scales, usually plated on large trees, without resin pockets; twigs orange-brown, darker with age, stout (to 20 mm thick); pollen cones 30–80 mm long at time of pollen release, purplish; seed cones 15–25 cm long, dull brown, the scales with a short reflexed prickle. Sandy soils; se and e TX; cultivated and used in reforestation. Included because it could possibly be found persisting or escaping in the extreme ne part of nc TX.

Pinus taeda L., (ancient name for resinous pines), LOBLOLLY PINE, OLD-FIELD PINE. Bark on older stems dark red-brown and divided into irregular scaly blocks, without resin pockets; twigs orangish to yellow-brown, darker brown with age, relatively slender (to 10 mm thick); terminal buds 1–1.2(–2) cm long; pollen cones 20–40 mm long at time of pollen release, yellow to yellow-brown; seed cones 6–12 cm long, mostly dull yellow-brown, the scales with a stout-based, sharp prickle. Lowlands to dry uplands; native to e TX as far w as Lamar Co. in Red River drainage (Little 1971) and common there; cultivated and escapes further w on sandy soils in Fannin (Lake Fannin) and Grayson (Buckner Preserve and Preston Peninsula) cos.

DIVISION **GNETOPHYTA**
JOINT-FIRS AND RELATIVES

A small group of 3 distinctive families: Ephedraceae, Gnetaceae (1 genus, 28 species), and Welwitschiaceae (monotypic). The division is unusual among the gymnosperms in having double fertilization and xylem with vessels. Recent molecular studies link the three families (i.e., suggest the Gnetophyta is monophyletic) and indicate the Gnetophyta is the sister group of the flowering plants (more closely related to the flowering plants than to any other living gymnosperm group) (Hambry & Zimmer 1992; Chase et al. 1993; Qui et al. 1993; Doyle et al. 1994; Price 1996). Extensive information on the evolution, relationships, and morphology of the Gnetophyta can be found in Friedman (1996).
REFERENCES: Arber & Parkin 1908; Bell & Woodcock 1983; Bold et al. 1987; Doyle 1996; Friedman 1996; Price 1996.

EPHEDRACEAE MORMON-TEA OR JOINT-FIR FAMILY

A monogeneric family of ca. 60 xeric adapted species found mainly in the n hemisphere and South America.
FAMILY RECOGNITION IN THE FIELD: Plants shrubby with *jointed photosynthetic* stems and leaves *reduced to minute scales*; seeds borne in *small cones* at the nodes.
REFERENCES: Correll 1966b; Stevenson 1993.

EPHEDRA MORMON-TEA, JOINT-FIR, MEXICAN-TEA

Erect to vine-like shrubs, dioecious (pollen- and seed-producing cones on separate plants); bark gray; branches jointed, alternate to whorled; twigs green to gray-green or yellow-green, photosynthetic; leaves opposite, scale-like, minute, 1–3 mm long, connate 2/3–7/8 their length, mostly not photosynthetic; cones in ours 1–2 per node on the young branches; pollen-producing (= staminate) cones compound, of 5–12 pairs of membranous bracts, the proximal bracts empty, the distal bracts each subtending a small cone composed of 2 basally fused bracteoles and a stalk-like sporangiophore; sporangiophores 3–5 mm long, exserted to 1/2 their length, bearing 4–6 pollen-producing microsporangia; microsporangia sessile or on stalks to 2 mm long; seed-producing (= ovulate) cones compound, of 3–6 pairs of bracts; inner bracts becoming fleshy and red, the cones thus fruit-like; seeds 1–2 per compound cone.

A number of species have been used medicinally. Ephedrine, an alkaloid commonly used as an antihistamine and in the treatment of asthma and sinusitis, is derived from Asian species; it has been used in China for 5,000 years. The common name MORMON-TEA comes from the use of various sw U.S. species as a beverage by early Mormon settlers (Woodland 1997). (Greek: *ep-*, upon, and *hédra*, seat or sitting upon a place; from the ancient name used by Pliny for *Equisetum*; the stems resemble the jointed stems of *Equisetum*, the segments of which appear to sit one upon the other)
REFERENCES: Cutler 1939; Steeves & Barghoorn 1959.

1. Plant erect to spreading, to ca. 1 m tall; seed-producing cones with 1 (rarely 2) seeds, sessile or nearly so; microsporangia sessile or on stalks < 1 mm long ____ **E. antisyphilitica**
1. Plant with clambering vine-like habit, to ca. 7 m long; seed-producing cones 2-seeded, on short to long peduncles; microsporangia on stalks 1–2 mm long ____ **E. pedunculata**

Ephedra antisyphilitica Berland. ex C.A. Mey., (against syphilis), JOINT-FIR, CLAPWEED, POPOTE, TEPOPOTE, CAÑATILLA. Plant erect to spreading, to ca. 1 m tall; branches, stiff, to ca. 4 mm thick; internodes ca. 2–5 cm long; pollen-producing (= staminate) cones lance-ellipsoid, 5–8 mm long, of 5–8 pairs of bracts; seed-producing (= ovulate) cones ellipsoid, 6–12 mm long, of 4–6 pairs of bracts; seeds 6–9 mm long, 2–4 mm wide. Gravelly or rocky soils; Archer, Brown, Callahan, Palo Pinto, Shackelford, and Young cos.; West Cross Timbers s and w across w 2/3 of TX. With cones late winter–early spring. According to Correll (1966b), this taxon can be distinguished from all other TX *Ephedra* species by the very narrow, pale orange-yellow or tannish band that encircles the stem at the very base of the connate leaves.

Ephedra pedunculata Engelm. ex S. Watson, (stalked), VINE JOINT-FIR, COMIDA DE VÍBORA, CLAPWEED. Plant trailing or clambering, to 7 m long; branches lax, to ca. 3 mm thick; internodes 1–8 cm long; pollen-producing cones lanceoloid, 4–8 mm long, of 6–12 pairs of bracts; seed-producing cones ovoid, 6–10 mm long, of 3–6 pairs of bracts; seeds 4–10 mm long, 2–4 mm wide. Dry, sandy to rocky areas; Brown Co. near w margin of nc TX (Cutler 1939; Vines 1960); w margin of nc TX w to w Edwards Plateau and s to s TX. With cones midwinter–early spring.

DIVISION **MAGNOLIOPHYTA**
ANGIOSPERMS OR FLOWERING PLANTS

Worldwide, the Magnoliophyta is composed of ca. 249,500 species in 13,185 genera arranged into 405 families (Mabberley 1997); 149 of these families occur in nc Texas. Depending on a variety of factors, including taxonomic philosophy (lumping versus splitting), the number of flowering plant families recognized ranges from 387 to 685; these rather different numbers mainly reflect differences in the rank at which groups are recognized (e.g., family versus subfamily) rather than differing views of evolutionary relationships (Cronquist 1988; Reveal 1993a, 1993b). The Magnoliophyta is the dominant and most diverse group of plants on a worldwide basis; it is also the primary group upon which human civilization relies. The angiosperms are seed plants with flowers, seeds developing inside closed carpels, and double fertilization, a process by which cells in addition to the egg unite during fertilization to form a triploid endosperm (Mabberley 1997). Recent, large scale molecular analyses have indicated that the an-

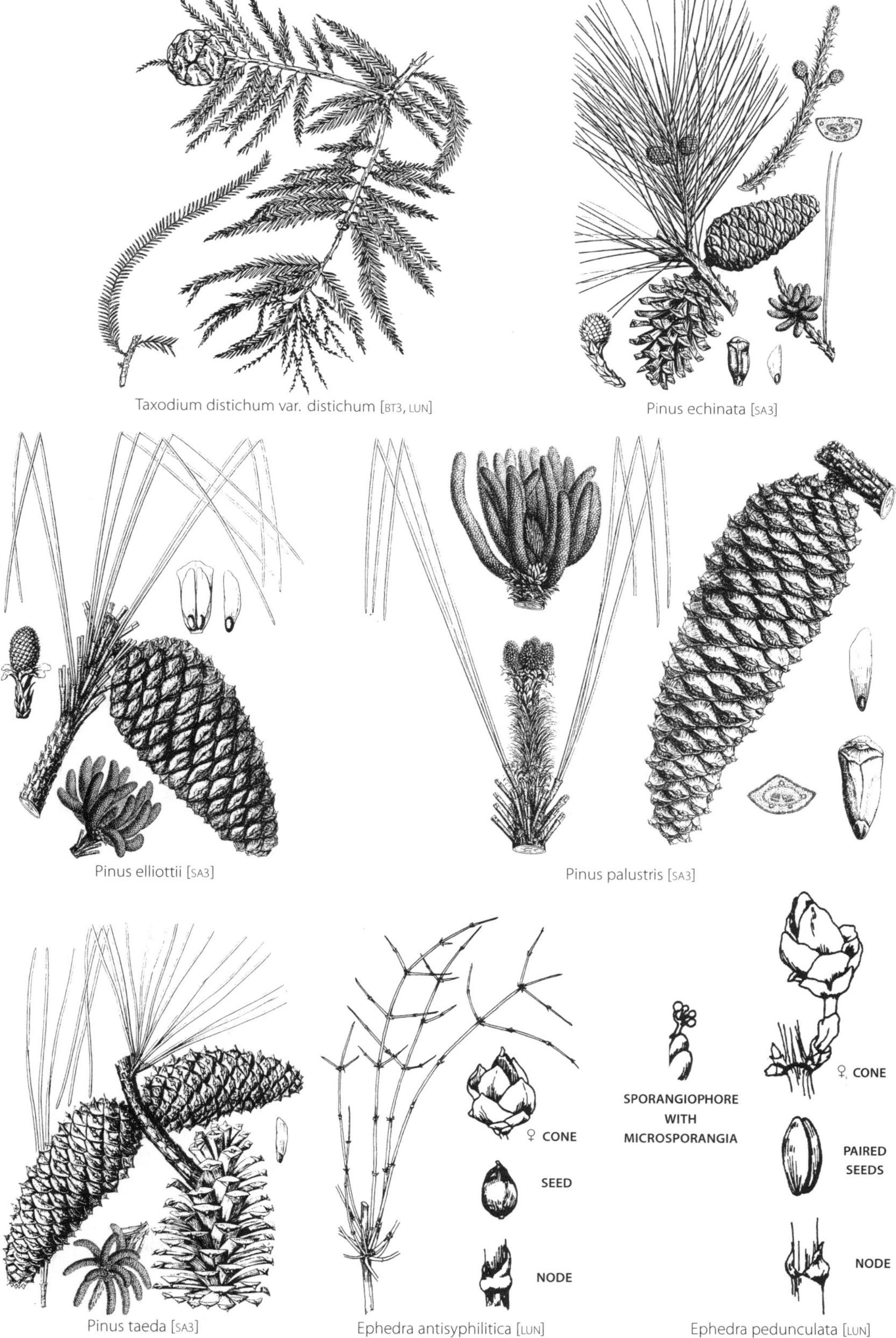

Taxodium distichum var. distichum [BT3, LUN]

Pinus echinata [SA3]

Pinus elliottii [SA3]

Pinus palustris [SA3]

Pinus taeda [SA3]

Ephedra antisyphilitica [LUN]

Ephedra pedunculata [LUN]

giosperms are a monophyletic group (Hamby & Zimmer 1992; Chase et al. 1993; Doyle et al. 1994) and that among living plants the Gnetophyta are the sister group of the angiosperms. In other words, the Gnetophyta are more closely related to the angiosperms than to any other living gymnosperm group (Chase et al. 1993; Doyle et al. 1994); however, this does not necessarily mean that the Gnetophyta is the gymnosperm group that gave rise to the angiosperms. While there is general concensus that the angiosperms evolved from gymnosperms, the exact gymnosperm group is as yet unknown with certainty.

The historical division of flowering plants into Class Monocotyledonae (monocots) and Class Dicotyledonae (dicots) is not supported by molecular data. Rather, the monocots appear to be a well-supported monophyletic group derived from within the Magnoliidae group of dicots (Chase et al. 1993; Duvall et al. 1993; Qiu et al. 1993). The dicots are therefore paraphyletic and thus, from the cladistic stand point, inappropriate for formal recognition (see explantion and Fig. 41 in Appendix 6). However, for practical reasons we are continuing to recognize these two traditional classes. The monocots are listed after the dicots to indicate their derivation from within the dicots.
REFERENCES: Cronquist 1981, 1988, 1993; Hamby & Zimmer 1992; Thorne 1992; Chase et al. 1993; Duvall et al. 1993; Qiu et al. 1993; Reveal 1993a, 1993b; Doyle et al. 1994; Takhtajan 1997.

CLASS **DICOTYLEDONAE**

Plants herbaceous or woody, often with secondary tissues derived from a vascular cambium; seedlings usually with 2 seed leaves or cotyledons; stems or branches elongating by apical growth; leaves not elongating when once expanded; new leaves and branches developing from terminal or axillary buds; leaves when present alternate, opposite, whorled, or basal, commonly with a petiole and expanded blade; leaf blades usually net-veined, with midrib and less prominent, spreading branch-veins (these sometimes nearly parallel in very narrow leaves), but veins sometimes buried in thick or fleshy leaves; perianth herbaceous, with dissimilar inner and outer whorls (= petals and sepals), or all parts about alike (= tepals), the perianth parts separate or united, commonly in 2s or 5s or numerous, occasionally in 3s or perianth absent.

Worldwide, the Dicotyledonae is a group composed of ca. 193,700 species in 10,534 genera divided into 321 families (Mabberley 1997); 124 of these families occur in nc TX. Based on molecular analyses (Chase et al. 1993), the dicots are a paraphyletic group with the monocots derived from within the dicot subclass Magnoliidae (see explanation and Fig. 41 in Appendix 6). From a cladistic standpoint, the dicots are thus not appropriately recognized in a formal taxonomic sense. One traditional dicot family, Ceratophyllaceae, is apparently the sister taxon to, and highly divergent from, all the rest of the flowering plants. The major division within the remaining flowering plants (excluding Ceratophyllaceae) is not between monocots and dicots, but rather between two groups corresponding to the two major angiosperm pollen types: uniaperturate and triaperturate. The uniaperturate (= 1-pored) group includes the "woody magnoliids" (e.g., Magnoliales, Laurales), the "paleoherbs" (e.g., Aristochiales, Piperales, and Nymphaeales), and the monocots. The triaperturate (= 3-pored) group includes all remaining dicots (Chase et al. 1993). This triaperturate group is referred to as the "eudicots" and appears to be monophyletic (Raven et al. 1999).
REFERENCES: Cronquist 1981, 1988, 1993; Downie & Palmer 1992; Hufford 1992; Olmstead et al. 1992; Thorne 1992; Wagenitz 1992; Chase et al. 1993; Kubitzki et al. 1993; Qiu et al. 1993; Reveal 1993a, 1993b; Takhtajan 1997; Raven et al. 1999.

ACANTHACEAE ACANTHUS OR WILD PETUNIA FAMILY

Ours perennial herbs (some woody basally and 1 shrub) often with squarish stems; foliage often with minute cystoliths (= mineral concretions) on surface; leaves opposite, sessile to petioled, simple; leaf blades entire or indistinctly toothed; stipules absent; flowers axillary or terminal, solitary or in peduncled spikes, cymes, or panicles; sepals 5, lanceolate or linear-lanceolate, united at base or up to 1/3 their length; corollas 2-lipped, or nearly radially symmetrical and 4-to 5-lobed, often with a color pattern; stamens 2 or 4; pistil 1; pistil of 2 carpels; ovary superior; style 1; fruit a capsule usually with 2 or 4 seeds; each seed typically supported by a small hook-like outgrowth (called a retinaculum or a jaculator)

A large (3,450 species in 229 genera), cosmopolitan, but mainly tropical family of mostly shrubs with some herbs, twiners, trees, and even aquatics; a number are used ornamentally including *Acanthus*, *Justicia brandegeeana* Wassh. & L.B. Sm. (SHRIMP-PLANT), and *Thunbergia* (CLOCKVINE). The family is closely allied to the Scrophulariaceae and some genera are intermediate between the two. Family name from *Acanthus*, BEAR'S-BREECHES, a genus of 30 mostly spiny xerophytic species of Old World tropical and warm areas; the leaf motif in the capitals of Corinthian columns supposedly originated from *A. spinosus* L. (Greek: *acanthos*, thorn, in reference to the spiny leaves and bracts) (subclass Asteridae)

FAMILY RECOGNITION IN THE FIELD: herbs (and 1 shrub) with opposite, simple, entire or indistinctly toothed leaves; flowers with sympetalous, 2-lipped or sometimes nearly radially symmetrical flowers in often *bracteate* inflorescences; stamens 2 or 4; fruit a capsule with seeds on small hook-like structures. In the somewhat similar Lamiaceae the fruits are of 4 one-seeded nutlets.

REFERENCES: Wasshausen 1966; Long 1970; Daniel 1984.

1. Plant a trailing or climbing vine; corollas yellow to orange, cream, or white, with a conspicuous dark purple center; seeds not supported by a small hook-like retinaculum ____ **Thunbergia**
1. Plant not a vine, either a ± erect herb or small shrub; corollas various, but not as above; seeds each supported by a small hook-like retinaculum.
   2. Corollas not deeply 2-lipped, nearly radially symmetrical to slightly bilabiate, the lobes ± equal; stamens 4 (except 2 in the rare *Carlowrightia*).
      3. Corollas 15–65 mm long, lavender to bluish purple or white, with 5 lobes; stamens 4; widespread in nc TX.
         4. Calyces about 3/4 as long as corollas; anther sacs basally awned or mucronate; leaf blades linear to oblanceolate; corolla tube 7 mm or less long ____ **Dyschoriste**
         4. Calyces less than half as long as corollas; anther sacs basally rounded; leaf blades usually much broader than linear to oblanceolate; corolla tube usually much > 7 mm long ____ **Ruellia**
      3. Corollas 6–7 mm long, white with maroon nerves on lobes, with 4 lobes; stamens 2; rare in nc TX ____ **Carlowrightia**
   2. Corollas deeply 2-lipped and/or lobes clearly unequal; stamens 2.
      5. Corollas reddish, 30–40 mm long; plants definitely woody, small shrubs ____ **Anisacanthus**
      5. Corollas white to violet, lavender, purple, rose, or flesh-colored, sometimes with darker dots, 10–28 mm long; plants herbaceous or woody only at base.
         6. Calyces 4-lobed; corolla tube long, slender, and cylindric, very conspicuous, longer than the corolla lobes; plants woody at base; plants of upland habitats; in w part of nc TX ____ **Siphonoglossa**
         6. Calyces 5-lobed; corolla tube inconspicuous, ca. as long as corolla lobes or shorter; plants not woody at base; plants of wet or moist areas; widespread in nc TX.
            7. Leaves usually long-petioled (petioles (0.5–)2–7 cm or more long); flowers in small clusters, subtended and covered by 2 to 4 obovate bracts; peduncles < 4 cm long, usually shorter than or equal to the petioles of subtending leaves; corollas with lower lip unlobed ____ **Dicliptera**
            7. Leaves usually sessile or with shorter petioles; flowers in a slender, 1-sided spike or in a dense head-like spike, subtended by small, linear to subulate bracts; peduncles elon-

gate, usually > 4 cm long, much longer than the petioles of subtending leaves; corollas with lower lip 3-lobed ________ **Justicia**

ANISACANTHUS ANISACANTH

A genus of 8 species native to the sw United States and Mexico; some are cultivated as ornamentals. (Greek: *anisa*, unequal, and *Acanthus*, from *akanthos*, a thorn)
REFERENCES: Hagen 1941; Henrickson 1986.

Anisacanthus quadrifidus (Vahl) Nees var. **wrightii** (Torr.) Henr., (sp.: four-cut; var.: for Charles Wright, 1811–1885, TX collector), WRIGHT'S ANISACANTHUS, HUMMINGBIRD-BUSH, FLAME-ACANTHUS. Shrub to ca. 1.5 m tall; leaf blades to 5 cm long and 2 cm wide; petioles to 1 cm long; flowers single or in pairs in terminal, 1-sided, spike-like inflorescences; calyces ca. 5 mm long; corollas reddish, 3–4 cm long, 2-lipped, the upper lip entire or nearly so, the lower lip 3-lobed; corolla tube longer than lobes, ± straight, slender, scarcely dilated at throat; stamens 2; capsules ca. 15 mm long; seeds 2(–4). Used in landscapes and escapes; adventive in Dallas Co. but not well-established (E. McWilliams, pers. comm.); native to Edwards Plateau. Jun–Aug. [*A. wrightii* (Torr.) A. Gray]

CARLOWRIGHTIA

A genus of 23 species of warm to arid areas from the sw U.S. to Costa Rica. (Named for Charles Wright, 1811–1885, botanical collector mainly in TX, Cuba, and his native Connecticut; correspondent of Asa Gray)
REFERENCES: Henrickson & Daniel 1979; Daniel 1980, 1983.

Carlowrightia texana Henr. & T.F. Daniel, (of Texas), TEXAS CARLOWRIGHTIA. Strigulose perennial (5–)10–30 cm tall from a much branched twiggy base; leaves petiolate; leaf blades (2.5–)6–16(–42) mm long, oblong-ovate, broadly ovate to orbicular, entire; flowers 1–3 in middle and upper leaf axils or occasionally in terminal spike-like inflorescences with reduced leaves/bracts; calyces 3–5 mm long; corollas 6–7 mm long, white with maroon nerves on lobes; capsules 7.5–11 mm long, glabrous, 4-seeded. Rocky slopes, often disturbed areas; described as a new species by Henrickson and Daniel in 1979; their distribution map showed 2 nc TX localities (no counties given); Daniel (1983) cited Callahan and Tarrant cos.; mainly s TX w across Edwards Plateau to Trans-Pecos. Apr–Nov.

DICLIPTERA

A genus of ca. 150 species of tropical and warm areas; some are cultivated as ornamentals. (Greek: *diklis*, double-folded, as of doors, etc., and *pteron*, a wing, alluding either to the 2-winged fruit or possibly the involucre)

Dicliptera brachiata (Pursh) Spreng., (branched at right angles), FALSE MINT. Perennial herb to 0.7(–1) m tall; leaf blades to 12(–15) cm long, (1.5–)2–7 cm wide; petioles (0.5–)2–7 cm or more long; flowers in axillary, short-peduncled to subsessile clusters, the clusters subtended by an involucre of 2–4 floral bracts; floral bracts obovate, to 7 mm long, 2–5 mm wide; corollas purplish or pinkish (rarely white), 15–20 mm long including tube, the upper lip entire or shallowly 2-lobed, the lower lip usually entire; capsules ovoid, 5–6 mm long, stipitate, emarginate apically, with 2–4 seeds. Moist wooded stream bottoms, wet habitats; se and e TX w to East Cross Timbers. Jun–Nov. ⊠/87

DYSCHORISTE

A genus of ca. 65 tropical and warm area species. (Greek: *dys*, hard, and *chorist*, separate or asunder, referring to the coherent, hard to separate capsule valves)

REFERENCE: Kobuski 1928.

Dyschoriste linearis (Torr. & A. Gray) Kuntze, (linear, narrow with sides nearly parallel), NARROW-LEAF SNAKEHERB, NARROW-LEAF DYSCHORISTE, SNAKEHERB. Rhizomatous, ± pubescent herbaceous perennial; stems erect or partly decumbent, usually 7–30 cm long; leaves sessile, linear to oblanceolate, obtuse; flowers axillary, sessile, shorter than the leaves, with foliaceous bracts; sepals long acuminate, united about 1/4, the lobes 9–13 mm long; corollas lavender, with dark dots in throat, 15–27 mm long; capsules with 2–4 seeds. Rocky or sandy ground; throughout most of TX, especially w 2/3. Apr–Jul, sporadically to Oct. ▣/88

JUSTICIA WATER-WILLOW

Ours perennial herbs; leaves simple, entire; flowers in axillary pedunculate heads or spikes; calyx lobes 5; corollas 2-lipped, the lower lip 3-lobed; stamens 2; capsules 4-seeded, the 2 valves reflexed at maturity.

A genus of ca. 600 species of herbs and shrubs of tropical and warm areas of the world and temperate North America; some such as *J. brandegeeana* Wassh. & L.B. Sm., SHRIMP-PLANT, (formerly treated in *Beloperone*), are cultivated as ornamentals. (Named for James Justice, 18th century Scottish botanist)

1. Flowers closely crowded in dense head-like spikes that become oblong at maturity; lower lip of corolla arched-recurving; widespread throughout nc TX ____ **J. americana**
1. Flowers in loosely-flowered, elongate spikes, usually spaced along 1 side of the axis of the inflorescence; lower lip of corolla ± flat; limited to extreme ne part of nc TX ____ **J. ovata**

Justicia americana (L.) Vahl, (of America), AMERICAN WATER-WILLOW. Plant glabrous, colonial, rhizomatous, producing long leafy shoots late in summer; stems erect, to ca. 100 cm tall; leaves sessile or nearly so; leaf blades narrowly oblong-lanceolate to elliptic-lanceolate; spikes to 3 cm long, peduncles to 15 cm long; sepals separate nearly to base, linear-lanceolate, ca. 7 mm long; corollas 10–12 mm long, white or violet with red-purple dots; capsules ca. 12 mm long. Wet ground, margins of streams and ponds; se and e TX w to West Cross Timbers and Edwards Plateau. May–Jun. ▣/94

Justicia ovata (Walter) Lindau var. **lanceolata** (Chapm.) R.W. Long, (sp.: ovate, egg-shaped; var.: lanceolate, lance-shaped), LANCE-LEAF WATER-WILLOW. Stems erect or spreading, 10–30 cm tall; leaves sessile or nearly so; leaf blades linear to elliptic-lanceolate, to 10 cm long and 3 cm wide, minutely puberulent; spikes 3–10 cm long; corollas ca. 9–10 mm long, lavender with darker markings; capsules 10–15 mm long. Wet areas; Lamar Co. in Red River drainage, also Callahan Co. (Stanford 1976); mainly se and e TX, also Edwards Plateau. Mar–Jun. [*J. lanceolata* (Chapm.) Small]

RUELLIA WILD PETUNIA

Ours glabrous or pubescent perennial herbs, the base sometimes woody; leaves sessile or short-petioled; leaf blades narrowly lanceolate to ovate-lanceolate or oblong-elliptic; sepals united a short distance at base, acuminate; ours with corollas usually lavender to bluish purple, often with reddish or purple spots down one side, or white (in *R. metziae*), large and showy, opening in the morning, falling by late afternoon (some species also produce cleistogamous flowers); capsules usually explosively dehiscent.

A genus of ca. 150 species found mainly in tropical areas of the world with a number in temperate North America; some are cultivated as ornamentals. (Named for Jean Ruelle, 1474–1537, early French herbalist)

REFERENCES: Tharp & Barkley 1949; Long & Uttal 1962; Long 1961, 1966, 1971, 1973, 1974, 1975; Turner 1991a.

1. Flowers terminating the main stem in panicle-like inflorescence on long peduncle (lateral few-flowered inflorescences of cleistogamous flowers also produced early); corollas white or lavender to bluish purple.
   2. Corollas ca. 4–5(–5.5) cm long, lavender to bluish purple, conspicuously curved; calyx lobes 10–15 mm long in flower ____ **R. nudiflora**
   2. Corollas (5–)5.5–6.5 cm long, usually white (sometimes pale bluish purple), very slightly curved; calyx lobes 14–20 mm long in flower ____ **R. metziae**
1. Flowers axillary, sessile or in ± peduncled glomerules or cymes OR at tips of side branches; corollas lavender to bluish purple.
   3. Leaf blades linear to linear-lanceolate < 2 cm wide, ≥10 times as long as wide ____ **R. brittoniana**
   3. Leaf blades wider, much < 10 times as long as wide.
      4. Flowers at leafy-bracted tips of branches OR on peduncles from axils of the main axis OR the axillary cymose inflorescences branched.
         5. Bracts linear or linear-lanceolate; leaf blades usually lanceolate; flowers on once or twice dichotomously branched peduncles ____ **R. malacosperma**
         5. Bracts wider; leaf blades wider, lanceolate, oblong, elliptic, or ovate; flowers solitary–3 (or rarely loosely cymose) at ends of usually unbranched peduncles.
            6. Largest leaf blades usually < 30 mm wide; ovary pubescent; calyx lobes narrowly linear, 1–1.2 mm wide, with prolonged linear-acicular tips ____ **R. pedunculata**
            6. Largest leaf blades usually > 30 cm wide; ovary glabrous; calyx lobes lanceolate to linear-lanceolate, 2–4 mm wide, flat to the tips ____ **R. strepens**
      4. Flowers in sessile or short-peduncled glomerules from axils of main stem or main leafy branches.
         7. Calyx lobes linear-lanceolate to lanceolate, (1.5–)2–4 mm wide, flat to the tips; flowers mainly cleistogamous ____ **R. strepens**
         7. Calyx lobes narrowly linear, the prolonged tips very slender to almost bristle-like; flowers usually chasmogamous, rarely cleistogamous.
            8. Leaves sessile or subsessile (petioles to 3 mm long), erect-ascending (= staying close to the stem); common in nc TX ____ **R. humilis**
            8. Leaves subsessile to distinctly petioled, those on main stem with petioles 2–30 mm long, mostly spreading away from the stem; rare or of limited range in nc TX, mainly e and sc TX.
               9. Glomerules of flowers few, restricted to the upper nodes; leaf blades linear-lanceolate to lanceolate, usually 20 mm or less wide ____ **R. caroliniensis**
               9. Glomerules of flowers several–many, extending well down on the stem and branches; leaf blades linear-lanceolate to broadly elliptic or ovate, up to 90 mm wide.
                  10. Leaf blades relatively narrow, linear to lanceolate; fruits often glabrous or with long hairs, to ca. 17 mm long; mainly e TX, no recent collections from nc TX ____ **R. caroliniensis**
                  10. Leaf blades relatively wide (to 90 mm), broadly elliptic to ovate; fruits minutely canescent, ca. 10 mm long; sc TX n to nc TX ____ **R. drummondiana**

Ruellia brittoniana Leonard, (Nathanial Lord Britton, 1859–1934, botanist at NY Botanical Garden). Plant to 1 m tall; foliage glabrous except for cystoliths (= mineral concretions); bracts linear to linear-lanceolate; cymes axillary, loose, elongate. Common in nursery trade, readily escapes cultivation; Tarrant Co., also Dallas Co. (Wasshausen 1966; Turner 1991a); se and e TX w to nc TX and Edwards Plateau. Jun–Nov. Native to e Mexico.

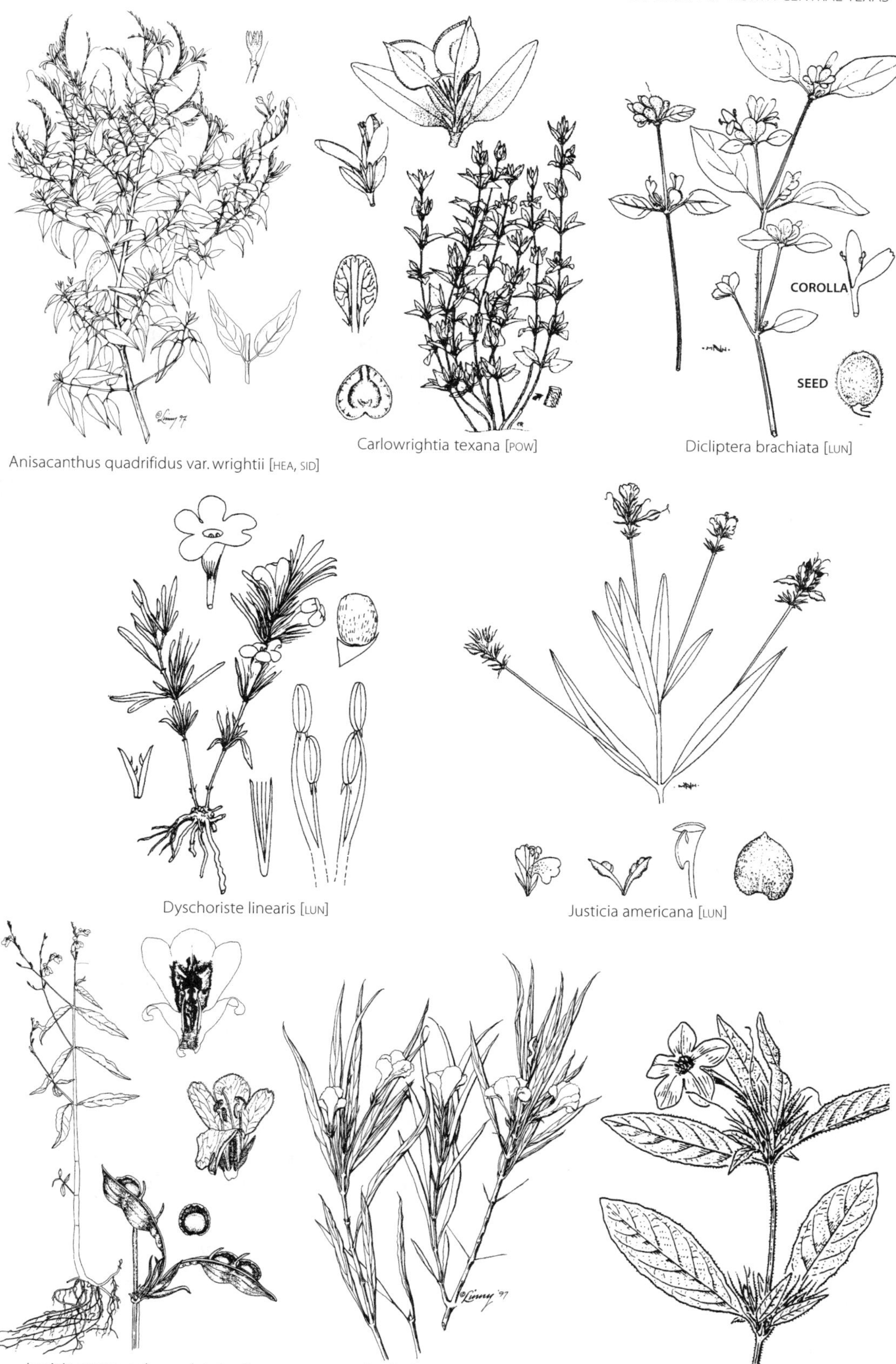

Anisacanthus quadrifidus var. wrightii [HEA, SID]

Carlowrightia texana [POW]

Dicliptera brachiata [LUN]

Dyschoriste linearis [LUN]

Justicia americana [LUN]

Justicia ovata var. lanceolata [CO1]

Ruellia brittoniana [HEA]

Ruellia caroliniensis [ST1]

Ruellia caroliniensis (J.F. Gmel.) Steud., (of Carolina), SMALL-FLOWER RUELLIA. Plant to ca. 0.9 m tall, inconspicuously and sparsely pubescent; flowers in sessile or very short peduncled glomerules mostly at the upper 1–4 nodes. Sandy open woods and open ground; collected by Reverchon in Dallas Co. (Mahler 1988), also Henderson Co.; mainly se and e TX. May–Sep. [*R. caroliniensis* (J.F. Gmel.) Steud. var. *salicina* Fernald, *R. caroliniensis* (J.F. Gmel.) Steud. var. *semicalva* Fernald]

Ruellia drummondiana (Nees) A. Gray, (for its discoverer, Thomas Drummond, 1780–1835, Scottish botanist and collector in North America), DRUMMOND'S RUELLIA. Plant to ca. 0.75 m tall, densely pubescent with short, spreading to appressed hairs. Wooded, riparian, often rocky areas; Bell, Bosque, and Williamson cos., also Hamilton (HPC), Dallas, and McLennan (Turner 1991a) cos.; sc TX n to nc TX; endemic to TX. Aug–Sep.

Ruellia humilis Nutt., (dwarf, low-growing), PRAIRIE-PETUNIA, LOW RUELLIA, FRINGE-LEAF RUELLIA. Plant to ca. 0.8 m tall, usually rather densely short-pilose; leaf blades lanceolate to ovate or broadly elliptic. Prairies and open woods; se and e TX w to Panhandle and Edwards Plateau. May–Sep. [*R. humilis* var. *depauperata* Tharp & F.A. Barkley, *R. humilis* var. *expansa* Fernald, *R. humilis* var. *longiflora* (A. Gray) Fernald] Jones et al. (1997) recognized var. *depauperata*.

Ruellia malacosperma Greenm., (soft-seeded), SOFT-SEED RUELLIA. Plant to ca. 0.8 m tall; cystoliths numerous; leaf blades glabrous or with a few spreading hairs; loose cymes conspicuous; peduncles 1–2 times branched, to 8 cm long; capsules 20–25 mm long. Common in nursery trade, escapes from cultivation; Dallas and Tarrant cos.; probably native to s TX and Mexico. May–Nov. Turner (1991a) suggested that *R. brittoniana* and *R. malacosperma* are "... probably no more than regional populational leaf variants of the same species."

Ruellia metziae Tharp, (for Sister Mary Clare Metz, 1907–?, nun and professor of botany at Our Lady of the Lake College, San Antonio, TX). Plant to ca. 0.6 m tall; leaves petioled; leaf blades oblong-lanceolate to narrowly ovate-oblong, 3–12 cm long, undulate to toothed; corollas usually white (sometimes pale bluish purple); cleistogamous flowers sometimes produced early from axillary peduncles only; this is the only nc TX species that usually has white corollas. Limestone outcrops, gravel, thickets, and open woods; Bosque, McLennan, and Mills cos., also Brown, Grayson, (Turner 1991a), Hamilton (HPC), Burnet, and Williamson (Tharp & Barkley 1949) cos.; mainly Lampasas Cut Plain s and sw to c TX and Edwards Plateau. Apr–Jul.

Ruellia nudiflora (Engelm. & A. Gray) Urb. var. **nudiflora**, (naked-flowered), VIOLET RUELLIA. Stems spreading-pilose, becoming glabrate, to 0.7 m tall; leaves petioled; flowers opening near sunrise, falling during afternoon; sepals with gland-tipped hairs or hispid-pubescent to slightly scabrous, without gland-tipped hairs. Sandy open woods; Bell, Bosque, Dallas, Grayson (Buckner Preserve), Hill, Johnston, McLennan, and Navarro cos.; also Burnet, Williamson (Turner 1991a), and Brown (HPC) cos.; se and c TX n to nc TX. Apr–May. [*R. nudiflora* (Engelm. & A. Gray) Urb. var. *hispidula* Shinners]

Ruellia nudiflora var. *runyonii* (Tharp. & F.A. Barkley) B.L. Turner, (for Robert Runyon of Brownsville, a friend of Tharp and Barkley and student of the vegetation of the lower Rio Grande Valley), is cited (as *R. runyonii*) by Hatch et al. (1990) for vegetational area 4 (Fig. 2), but apparently occurs only to the s of nc TX. It can be distinguished by its small corollas (< 4 cm long).

Ruellia pedunculata Torr. ex A. Gray, (with a flower stalk), STALKED RUELLIA. Plant to ca. 0.7 m tall; leaves short-petioled; peduncles to ca. 7 cm long; calyces pilose with slender pointed hairs. Open woods and along streams; Fannin, Hopkins, and Lamar cos.; se and e TX w to ne part of nc TX. May–Sep.

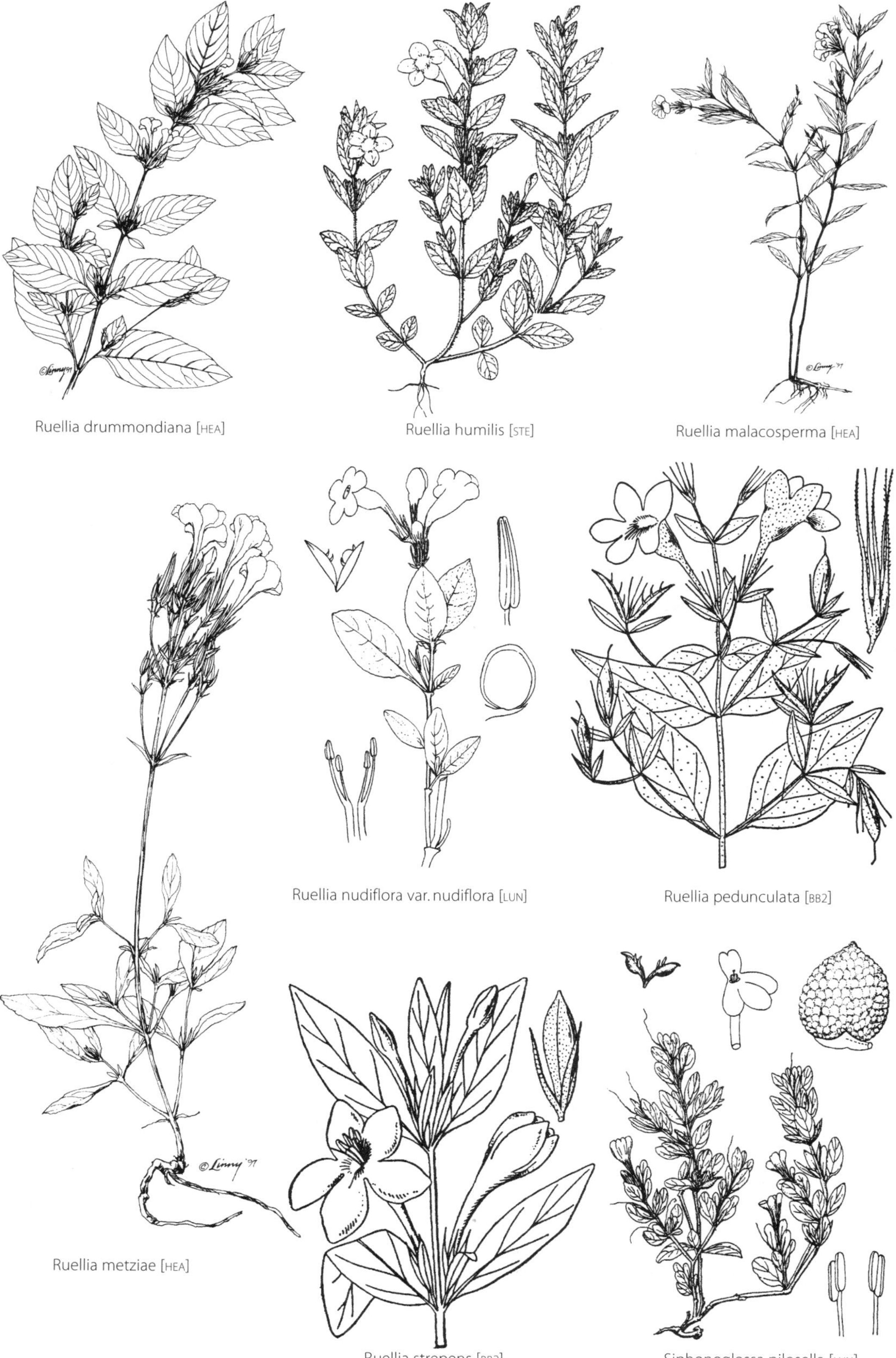

Ruellia drummondiana [HEA]

Ruellia humilis [STE]

Ruellia malacosperma [HEA]

Ruellia nudiflora var. nudiflora [LUN]

Ruellia pedunculata [BB2]

Ruellia metziae [HEA]

Ruellia strepens [BB2]

Siphonoglossa pilosella [LUN]

Ruellia strepens L., (rustling or rattling), LIMESTONE RUELLIA, SMOOTH RUELLIA. Stems to ca. 1 m tall, glabrous or pubescent in lines; leaves petioled; leaf blades ovate- or elliptic-lanceolate, acute; peduncles 1–3-flowered, leafy-bracted, the bracts lanceolate to elliptic-ovate. Stream and hillside woods; se and e TX w to East Cross Timbers, also Hamilton (HPC) and Brown (Stanford 1976) cos., also Edwards Plateau. Apr–Jun, with solitary, open flowers, also Sep–Oct, with clustered, often smaller, open or usually cleistogamous flowers. [*R. strepens* var. *cleistantha* A. Gray]

Ruellia davisiorum Tharp & F.A. Barkley, (for L. Irby Davis and Anna Tarrance Davis of Harlingen, TX, friends of Tharp and Barkley and donors of collections), cited by Hatch et al. (1990) for vegetational area 4 (Fig. 2), apparently occurs only to the s of nc TX. It is similar to *R. metziae*, but differs in having bluish lavender corollas and short (3–6 cm long), ovate to broadly ovate leaf blades.

SIPHONOGLOSSA TUBETONGUE

A genus of 7 species of mostly tropical America. (Greek: *siphono*, a tube or pipe, and *glossa*, tongue, possibly in reference to the stamens of some species being exserted from the corolla tube)

Siphonoglossa pilosella (Nees) Torr., (little pilose—covered with long soft hairs), HAIRY TUBETONGUE, TUBETONGUE, FALSE HONEYSUCKLE. Low suffrutescent perennial; leaves subsessile; leaf blades ovate to oval, to 4 cm long; flowers solitary in axils or on short axillary peduncles 4 mm or less long; floral bracts none; bractlets subtending the calyces small, to 2.5 mm long; corollas lavender to rose or white, to 28 mm long, the tube slender, cylindric, elongate, the upper lip entire or 2-lobed, the lower lip 3-lobed, much larger, typically with a white spot and purple dots at base; stamens inserted near mouth of corolla tube; capsules 8–9 mm long. Rocky, sandy or grassy areas; Brown, Callahan, Mills, and Stephens cos.; mostly s 2/3 of TX. Apr–Oct. Jones et al. (1997) treated this species in the genus *Justicia* [as *J. pilosella* (Nees) Hilsenb. [ined.]].

THUNBERGIA CLOCKVINE

An Old World tropical genus of 90 species of climbing or erect herbs and shrubs with bracts enclosing the calyces; a number are cultivated as ornamentals. (Named for Carl Peter Thunberg, 1743–1828, Swedish student and successor of Linnaeus and traveler to Asia and South Africa)

Thunbergia alata Bojer ex Sims, (winged), BLACK-EYED-SUSAN, BLACK-EYED CLOCKVINE. Perennial, herbaceous, trailing or climbing vine often grown as an annual; stems usually ca. 1 m long; tendrils absent; leaf blades 2.5–8 cm long, ovate to triangular ovate, apically acute, basally cordate to ± hastate, entire to few-toothed, pubescent; petioles wing-margined, often nearly as long as blades; flowers usually axillary, solitary, pedunculate, subtended by 2 conspicuous foliaceous bracts ca. 15 mm long; calyces small, hidden by the bracts; corollas large and showy, 2.5–4 cm long, yellow to orange, cream, or white, with a dark purple center, with a slender tube and a spreading 5-lobed limb; lobes of limb rounded distally, not differing greatly in size or shape; stamens 4; capsules 8–10 mm in diam., with a stout beak ca. 10 mm long. Long cultivated in nc TX (an 1885 Reverchon collection is known from Dallas—Wasshausen 1966) and known to escape in se, e, and s TX and the Edwards Plateau; included because of possibily of escape in nc TX. Mar–Apr. Native of Africa.

ACERACEAE MAPLE FAMILY

A small (113 species in 2 genera) of trees and shrubs of n temperate areas and tropical mountains; the genus *Dipteronia* with 2 species of c and s China has fruits winged all the way around. Aceraceae are closely related to Sapindaceae and appear to represent a clade within

that family. From a cladistic standpoint they should be lumped to form a more inclusive, monophyletic Sapindaceae (Judd et al. 1994). (subclass Rosidae)
FAMILY RECOGNITION IN THE FIELD: trees with leaves opposite, simple and palmately lobed and veined or in 1 species pinnately compound; fruits distinctive—of 2 one-seeded, *winged* samaras.
REFERENCES: Brizicky 1963b; Murray 1970; Judd et al. 1994.

ACER MAPLE

Polygamo-dioecious or dioecious, deciduous trees with watery often sweet sap; leaves opposite, simple and palmately lobed or pinnately compound; flowers pedicellate, in racemes, panicles, or corymbose or in umbellate fascicles, radially symmetrical, small, mostly 5-merous, usually completely or functionally unisexual; ovary superior; fruit a schizocarp of 2 one-seeded, winged samaras which eventually separate and function as wind-dispersed small "helicopters."

A genus of 111 species of trees and shrubs of the n temperate zone and tropical mountains. Some species are important sources of maple syrup (e.g., *A. saccharum* Marsh.—SUGAR MAPLE) or as ornamentals, these often with brightly colored fall foliage. Others are valued for their timber which yields a hard, usually white wood used in furniture, musical instruments, flooring, gunstocks, etc. (Latin name of the maple; also meaning sharp, referring to the hard wood)
REFERENCES: Desmaris 1952; Gehlbach & Gardner 1983.

1. Leaves pinnately compound with 3–5(–9) distinct leaflets __________ **A. negundo**
1. Leaves simple (but lobed).
   2. Central lobe of leaf narrowed basally where it joins the other lobes; leaves silvery-white beneath __________ **A. saccharinum**
   2. Central lobe of leaf not narrowed basally; leaves light green to somewhat silvery-white beneath.
      3. Spaces (= sinuses) between the main lobes of the leaf rounded or U-shaped; margins of leaf lobes entire or somewhat lobulate; plants of areas near the Edwards Plateau __________ **A. grandidentatum**
      3. Spaces between the main lobes of the leaf ± sharply angled or V-shaped; margins of leaf lobes coarsely toothed; plants mainly of e TX __________ **A. rubrum**

Acer grandidentatum Nutt. var. **sinuosum** (Rehder) Little, (sp.: large-toothed; var.: wavy-margined), PLATEAU BIG-TOOTH MAPLE, UVALDE BIG-TOOTH MAPLE, LIMEROCK MAPLE. Flowers appearing as the leaves are expanding or after; petals absent; sinus between the two samaras U-shaped; samaras to ca. 31 mm long, typically much shorter. Limestone canyons; Bell Co. (HPC), also Coryell Co. (Gehlbach & Gardner 1983); Edwards Plateau n to Lampasas Cut Plain; other varieties are found in far w TX. While we are following Hatch et al. (1990), Kartesz (1994), and Jones et al. (1997) in terms of nomenclature, the most appropriate name for this taxon is not completely clear. It is closely related to and is sometimes put into *A. saccharum* Marshall (SUGAR MAPLE) [as *A. saccharum* var. *sinuosum* (Rehder) Sarg.]. Gehlbach and Gardner (1983) argued that all the relict maple populations in nc, c, and w TX should be treated as *A. saccharum* var. *floridanum* (Chapm.) Small & A. Heller (SOUTHERN SUGAR MAPLE).

Acer negundo L., (from the native name of *Vitex negundo*, because of supposed similarity of leaves), BOX-ELDER, ASH-LEAF MAPLE, ARCE, FRESNO DE GUAJUCO. Leaflets pinnately veined; terminal leaflet elliptic to obovate; lateral leaflets narrower and coarsely few-toothed to entire; flowers appearing just before the leaves, greenish; petals absent; samaras ca. 25–35 mm long. Stream banks, low woods. (Feb–)late Mar–Apr. Leaves with 3 leaflets can sometimes resemble and be confused with those of POISON-IVY. However, BOX-ELDER has opposite leaves in contrast to the alternate leaves of POISON-IVY and most other Anacardiaceae. This is by far the most common *Acer* in nc TX.

1. Younger twigs glabrous ____ var. **negundo**
1. Younger twigs coated with short velvety pubescence ____ var. **texanum**

var. **negundo**. Mainly in the e 1/2 of TX.

var. **texanum** Pax, (of Texas), TEXAS BOX-ELDER. Se and e TX w to Blackland Prairie.

Acer rubrum L., (red), RED MAPLE, SCARLET MAPLE. Branchlets and petioles usually reddish; flowers appearing before the leaves develop, reddish; petals ca. 2 mm long; sinus between the 2 samaras V-shaped; samaras usually 15–25 mm long. Low woods; Lamar Co. in Red River drainage, also Fannin Co. (Little 1971); mainly se and e TX. Feb–Apr. The leaves become bright crimson in autumn. Varieties are often recognized (Kartesz 1994; Jones et al. 1997); in var. *rubrum* the lower surface of the leaf blades are glabrous or with hairs only along the veins while in var. *trilobum* Torr. & A. Gray ex K. Koch and var. *drummondii* (Hook. & Arn. ex Nutt.) Sarg. the lower surface of the leaf blades are densely and usually permanently hairy; var. *trilobum* is differentiated by having the leaf blades with only 3 lobes, the smaller lateral basal lobes suppressed, while var. *rubrum* and var. *drummondii* have leaf blades usually with 5 lobes (the lateral basal lobes small). The leaves and bark are poisonous and have caused the death of livestock (Fuller & McClintock 1986). ☠

Acer saccharinum L., (sugary, from the sap), SILVER MAPLE, SOFT MAPLE. Leaves deeply 5-lobed, spaces between the main lobes of the leaf ± sharply angled or V-shaped; margins of leaf lobes coarsely toothed; flowers appearing before the leaves develop or immediately with them, greenish or reddish; petals absent; sinus between the two samaras V-shaped, samaras to 50(–60) mm long. Naturalized in sandy low woods, wooded lake shore, and stream banks; Grayson and Lamar cos.; introduced to TX; native further e in North America. Feb(?)–Mar.

AIZOACEAE ICEPLANT OR FIG-MARIGOLD FAMILY

A medium-large (1,850 species in 128 genera), tropical and subtropical family of betalain-containing, succulent herbs or small shrubs centered in s Africa; most have the CAM type of photosynthesis, a physiological adaptation for hot dry conditions; some are cultivated as ornamentals including the PEBBLE PLANTS or LIVING STONES (*Lithops*) that mimic rocks and *Mesembryanthemum* (ICEPLANT), a huge genus now split into over 100 segregate genera. *Mollugo* and *Glinus*, treated here in the Molluginaceae, were formerly included in the Aizoaceae. Family name from *Aizoon*, a genus of 25 species of the Mediterranean, s Africa, and Australia. (Latin name for an evergreen plant) (subclass Caryophyllidae)
FAMILY RECOGNITION IN THE FIELD: the only nc TX species is a *succulent*, decumbent herb with opposite or subopposite leaves, inconspicuous axillary flowers, and circumscissile capsules.
REFERENCES: Bogle 1970; Hartmann 1993; Behnke & Mabry 1994.

TRIANTHEMA HORSE-PURSLANE, SEA-PURSLANE

A genus of 17 species of warm areas of the world, especially Australia; some contain alkaloids and are used medicinally. (Greek: *tri*, three, and *anthemon*, flower)

Trianthema portulacastrum L., (with flowers like *Portulaca*—purslane), DESERT HORSE-PURSLANE, SEA-PURSLANE, VERDOLAGA BLANCA. Glabrous, succulent, annual herb branching from the base; branches usually decumbent; leaves subopposite to opposite, in unequal pairs, petiolate; leaf blades broadly obovate to nearly orbicular, to ca. 4 cm long; flowers solitary, axillary, sessile; calyx lobes 5, ca. 2.5 mm long, pinkish purple within, green outside; petals absent; stamens 5–10; ovary superior; fruit a 1-several-seeded capsule ca. 4 mm long with prominent winged appendages at apex, circumscissile near the middle. Sandy soils and waste ground; Dallas, Grayson, and Tarrant cos.; mainly s and w TX. May–Oct.

AMARANTHACEAE AMARANTH OR PIGWEED FAMILY

Ours annual or perennial, often weedy herbs; leaves opposite in our species (except *Amaranthus*), simple, entire or nearly so; stipules absent; inflorescences axillary or terminal, of small clusters, head-like, or in spikes or spike-like racemes or compact panicles; flowers very small, each subtended or enclosed by 3 scarious or prominently scarious-margined bracts/ bracteoles, perfect or unisexual; tepals 0–5 in 1 series; stamens usually 5; pistil 1; ovary superior; fruit a dehiscent or indehiscent urticle.

A medium-sized (750 species in 71 genera) family of mostly herbs or shrubs, climbers, or rarely trees of tropical and warm areas of the world with a few in temperate regions; the family includes a number of ornamentals (e.g., *Celosia*—COCKSCOMB) and weeds as well as grain crops (e.g., *Amaranthus*); pigments present are of the nitrogen-containing type known as betalains (Cronquist & Thorne 1994). Molecular evidence indicates the family is most closely related to Chenopodiaceae (Downie & Palmer 1994). (subclass Caryophyllidae)

FAMILY RECOGNITION IN THE FIELD: herbs with simple leaves entire or nearly so and very small apetalous flowers in usually dense inflorescences; flowers subtended by *scarious often colored bracts*; filaments united. Chenopodiaceae are similar in flower structure (e.g., very small, lacking petals), but lack scarious bracts and have separate filaments.

REFERENCES: Standley 1917; Reed 1969b; Robertson 1981; Townsend 1993; Downie & Palmer 1994; Behnke & Mabry 1994.

1. Leaves alternate ____ **Amaranthus**
1. Leaves opposite.
  2. Stems and leaves densely stellate-pubescent (at least young portions); plants with gray-green foliage and inconspicuous flowers in small axillary clusters lacking white or colored bracts; in extreme w part of nc TX ____ **Tidestromia**
  2. Stems and leaves without stellate hairs; plants not as above; widespread in nc TX.
    3. Inflorescence a panicle up to 20 cm broad; flowers unisexual ____ **Iresine**
    3. Inflorescence a h ead or narrow spike (a number of these can be grouped together); flowers perfect.
      4. Flowers in large, globose to short cylindric heads 2–2.8 cm in diam., often variously colored and showy ____ **Gomphrena**
      4. Flowers in much smaller heads, short thick spikes, or elongate inflorescences, neither colored nor showy (but sometimes strikingly whitish).
        5. Flowers in narrow, elongating, single or grouped spikes; tepals united ____ **Froelichia**
        5. Flowers in heads or short thick spikes; tepals separate.
          6. Floral bracts slightly shorter than or longer than the tepals; tepals acuminate or awn-tipped ____ **Alternanthera**
          6. Floral bracts about half as long as the tepals; tepals subacute or acute ____ **Gossypianthus**

ALTERNANTHERA CHAFF-FLOWER

Ours perennials; leaves opposite, entire; inflorescences in head-like spikes, flowers perfect; tepals usually 5, separate; fruit an utricle.

A genus of 100 species of tropical and warm areas, especially the Americas; some are edible while others, because of their colorful leaves, are used as bedding plants. (Latin: *alternans*, alternating, and *anthera*, an anther, referring to the sterile alternate anthers)

REFERENCES: Melville 1958; Buckingham 1996.

1. Inflorescences sessile or nearly so, the peduncles at most 1 cm long; plants terrestrial in sandy disturbed areas and waste places; leaf blades spatulate to suborbicular, 8–20 mm long ____ **A. caracasana**

1. Inflorescences pedunculate, the peduncles ca. 2–7 cm long; plants aquatic or semi-terrestrial; leaf blades linear to linear-lanceolate or obovate, 20–110 mm long _______________ **A. philoxeroides**

Alternanthera caracasana Kunth, (of Caracas, Venezuela), MAT CHAFF-FLOWER, VERDOLAGA DE PUERCO. Stems prostrate, branched, sometimes hirsute; leaves clustered; leaf blades glabrous or sparsely pubescent (chiefly beneath and on margins), dark green and glossy above; floral bracts and tepals creamy white; tepals awn-tipped, unequal, at least the shorter ones pubescent with numerous barbed hairs. Disturbed areas and waste places; w Blackland Prairie s and w to w TX. Jun–Oct. There are questions concerning the native range of this species; Henrickson (1993) indicated it is native to Central and South America; however, Robertson (1981), in his treatment of the family for the se U.S, considered it native to the s U.S.; Hatch et al. (1991) treated it as native to TX. [*A. peploides* (Humb. & Bonpl. ex Schult.) Urb.]

Alternanthera philoxeroides (Mart.) Griseb., (like *Philoxerus*, an Australian genus of Amaranthaceae), ALLIGATOR-WEED. Trailing and mat-forming with erect flowering stems; leaves well-spaced on flowering stems, thick and fleshy, glabrous; floral bracts 1/4 as long as tepals; tepals nearly equal, glabrous, silvery white. Aquatic or in very wet areas; Dallas and Tarrant cos. (both along the Trinity River); mainly se TX. Mar–Aug. Native of South America. [*Achyranthes philoxeroides* (Mart.) Standl.] This species can be a problematic weed contributing to clogging of aquatic habitats (Buckingham 1996). In Texas, ALLIGATOR-WEED is considered a "harmful or potentially harmful exotic plant" and it is illegal to release, import, sell, purchase, propagate, or possess this species in the state (Harvey 1998). The species was first reported in the se U.S. in Florida in 1894; it subsequently spread widely and in some areas formed nearly pure stands, outcompeting native vegetation; biological controls (various insects) have been used; the species apparently does not produce viable seeds in the U.S.—reproduction occurs by plant fragments (Robertson 1981).

AMARANTHUS GREEN PIGWEED, AMARANTH

Ours annuals with taproots; leaves alternate; inflorescence of dense, terminal or axillary spikes or panicles; flowers perfect or unisexual; fruit a 1-seeded, circumcissile, irregularly splitting or indehiscent utricle; seeds lenticular.

A genus of 60 species of tropical and temperate areas of the world. As treated here, *Amaranthus* includes the segregates *Acnida* and *Acantochiton*. The seeds of *Amaranthus* species were used as a major "grain" crop by Native Americans in pre-Colombian times (e.g., *A. caudatus* L.—INCA-WHEAT, *A. hybridus*—especially in Mexico). The protein-rich leaves have been used for millenia and are one of the most common vegetables in the tropics; some species, however, can accumulate toxic levels of nitrates (Cheatham & Johnston 1995); some are cultivated as ornamentals; others are widespread weeds considered by some to be among the worst annual garden weeds—they grow rapidly and can produce hundreds of thousands of seeds which can survive dormant in the soil for decades (Baumgardt 1982). Hybridization is common and widespread in the genus (Sauer 1955). (Greek: *amaranthos*, unfading, because the dry calyx and bracts are persistent)
REFERENCES: Sauer 1955, 1967.

1. Floral bracts subtending individual flowers much enlarged, enfolding and concealing the flowers, and broadly cordate in fruit; rare in nc TX _______________ **A. acanthochiton**
1. Floral bracts subtending individual flowers not much enlarged, neither enfolding and concealing the flower nor cordate in fruit; including species widespread in nc TX.
   2. Flowers in small axillary clusters, main stem and branches leafy to tip.
      3. Tepals with tips broader than base, somewhat spatulate, 3-nerved, apiculate; stems erect; s part of nc TX _______________ **A. polygonoides**

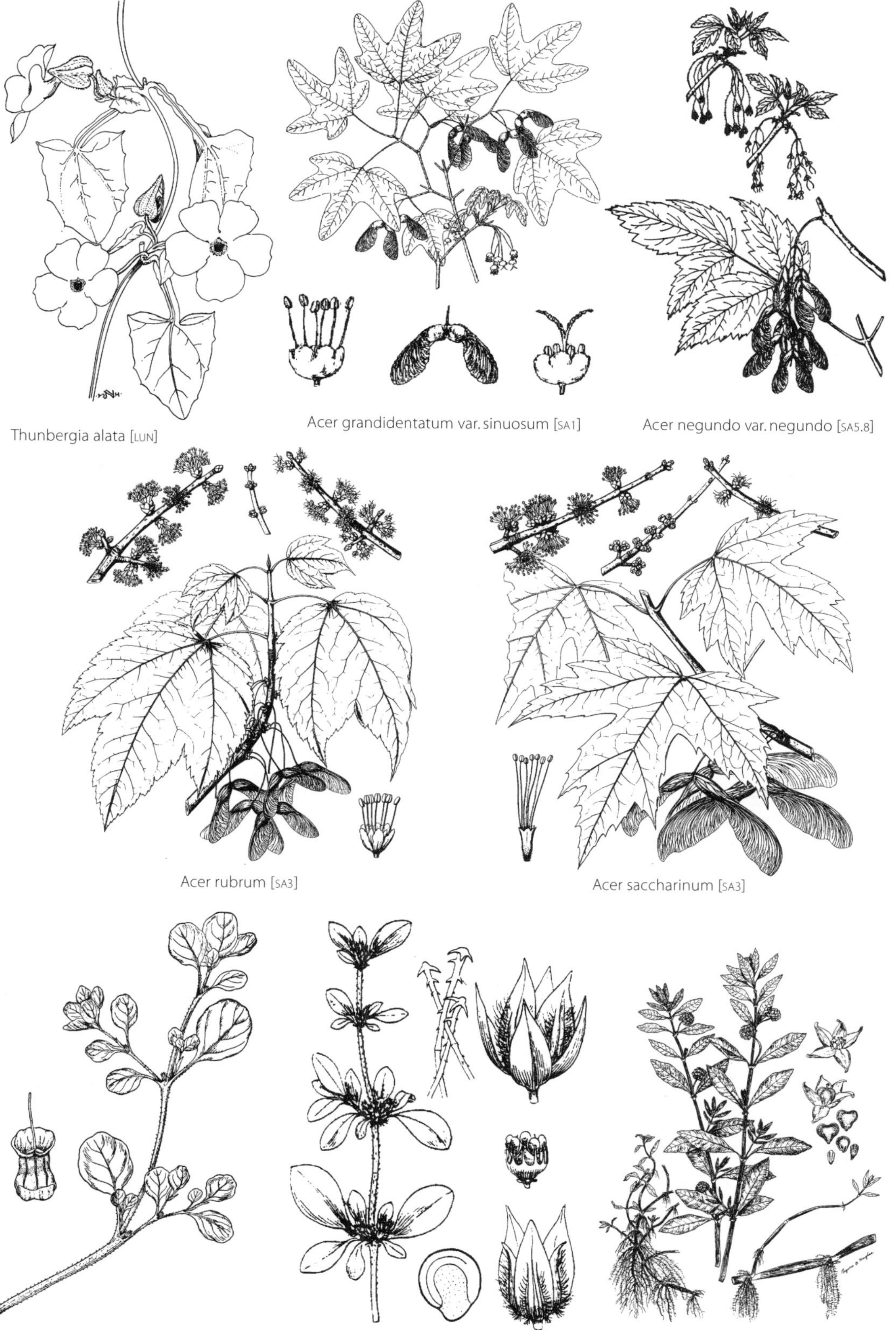

Thunbergia alata [LUN]

Acer grandidentatum var. sinuosum [SA1]

Acer negundo var. negundo [SA5.8]

Acer rubrum [SA3]

Acer saccharinum [SA3]

Trianthema portulacastrum [GLE]

Alternanthera caracasana [FAW]

Alternanthera philoxeroides [REE]

3. Tepals oblong or lanceolate, narrowed at tip, 1-nerved, acute to acuminate; stems prostrate or erect; widespread in nc TX.
    4. Plants prostrate; tepals 4 or 5 ______ **A. blitoides**
    4. Plants stiffly erect; tepals 3 ______ **A. albus**
2. Flowers in terminal, spike-like panicles (often some of the lower ones also axillary); main stem and branches with reduced leaves or none apically.
    5. Upper leaf axils and lower flower clusters with long spines (5–10 mm long) ______ **A. spinosus**
    5. Upper leaf axils and inflorescence without spines (but bracts can be prickly).
        6. Bracts at base of flowers 1.5—2.5 mm long; pistillate tepals 1 or 2 with 1 rudimentary (or 5 in the rare *A. arenicola*); staminate and pistillate flowers on separate plants (= dioecious); inflorescences not noticably prickly to the touch.
            7. Pistillate tepals 0, 1 or 2; bracts with midveins clearly excurrent or not so; outer staminate tepals acuminate; including a species very abundant in nc TX.
                8. Plants pistillate.
                    9. Tepals 2, 1 of these well-developed, ca. 2 mm long; fruits with circumscissile dehiscence; midrib of bracts clearly excurrent; very abundant in nc TX ______ **A. rudis**
                    9. Tepals completely absent or if present rudimentary (< 1 mm long); fruits indehiscent or irregularly dehiscent; midrib of bracts not conspicuously excurrent; rare in nc TX ______ **A. australis**
                8. Plants staminate.
                    10. Bracts ca. 2 mm or slightly more in length, with heavy midribs; outer tepals definitely longer than the inner; very abundant in nc TX ______ **A. rudis**
                    10. Bracts 1.5–1.8 mm long, usually with slender midribs; outer tepals not appreciably longer than the inner; rare in nc TX ______ **A. australis**
            7. Pistillate tepals 5; bracts with midveins barely if at all excurrent; staminate tepals obtuse or retuse (= slightly notched at apex); rare in nc TX, reported only from Bell Co. ______ **A. arenicola**
        6. Bracts at base of flower 3–6 mm long; pistillate tepals usually 3–5; plant dioecious OR staminate and pistillate flowers on the same plant (= monoecious) OR flowers perfect; inflorescences prickly to the touch OR not so.
            11. Plants dioecious; bracts 4–6 mm long; tepals 2.5–4.5 mm long with conspicuous extension of the midvien; terminal spikes long, 15–70(–120) cm, often lax or drooping to erect, usually prickly to the touch ______ **A. palmeri**
            11. Plants monoecious or flowers perfect; bracts 3–5 mm long; tepals 1.5–3.2 mm long, with less obvious extension of the midrib; terminal spikes to ca. 20 cm long, usually erect, usually not prickly to the touch.
                12. Spikes slender, 6–10(–12) mm broad; tepals acute, 1.5–2(–2.5) mm long ______ **A. hybridus**
                12. Spikes thick, 10–20 mm broad; tepals rounded to truncate, usually notched apically, often mucronate, 2.5–3.2 mm long ______ **A. retroflexus**

Amaranthus acanthochiton J.D. Sauer, (named for the genus *Acanthochiton*, that word derived from roots meaning thorny hard outer covering), GREENSTRIPE. Stems erect, to 0.8 m tall; seeds 1–1.25 mm in diam., dark reddish brown. Sandy or rocky areas; included based on citation of vegetational areas 4 and 5 (Fig. 2) by Hatch et al. (1990); se TX nw to nc TX and w to Trans-Pecos. Jul–Oct. [*Acanthochiton wrightii* Torr.]

Amaranthus albus L., (white), TUMBLEWEED, TUMBLEWEED AMARANTH, WHITE AMARANTH. Stems erect, stiff, whitish or pale green, 0.2–1.2 m tall; branches spreading; seeds 0.6–0.8 mm in diam., reddish brown, shiny. Disturbed sites; widespread in TX. Aug–Dec.

Amaranthus arenicola I.M. Johnst., (growing in sandy places), TORREY'S AMARANTH, SANDHILL AMARANTH. Plant erect, to 2(–3) m tall; seeds 1–1.3 mm in diam., dark reddish brown, shiny.

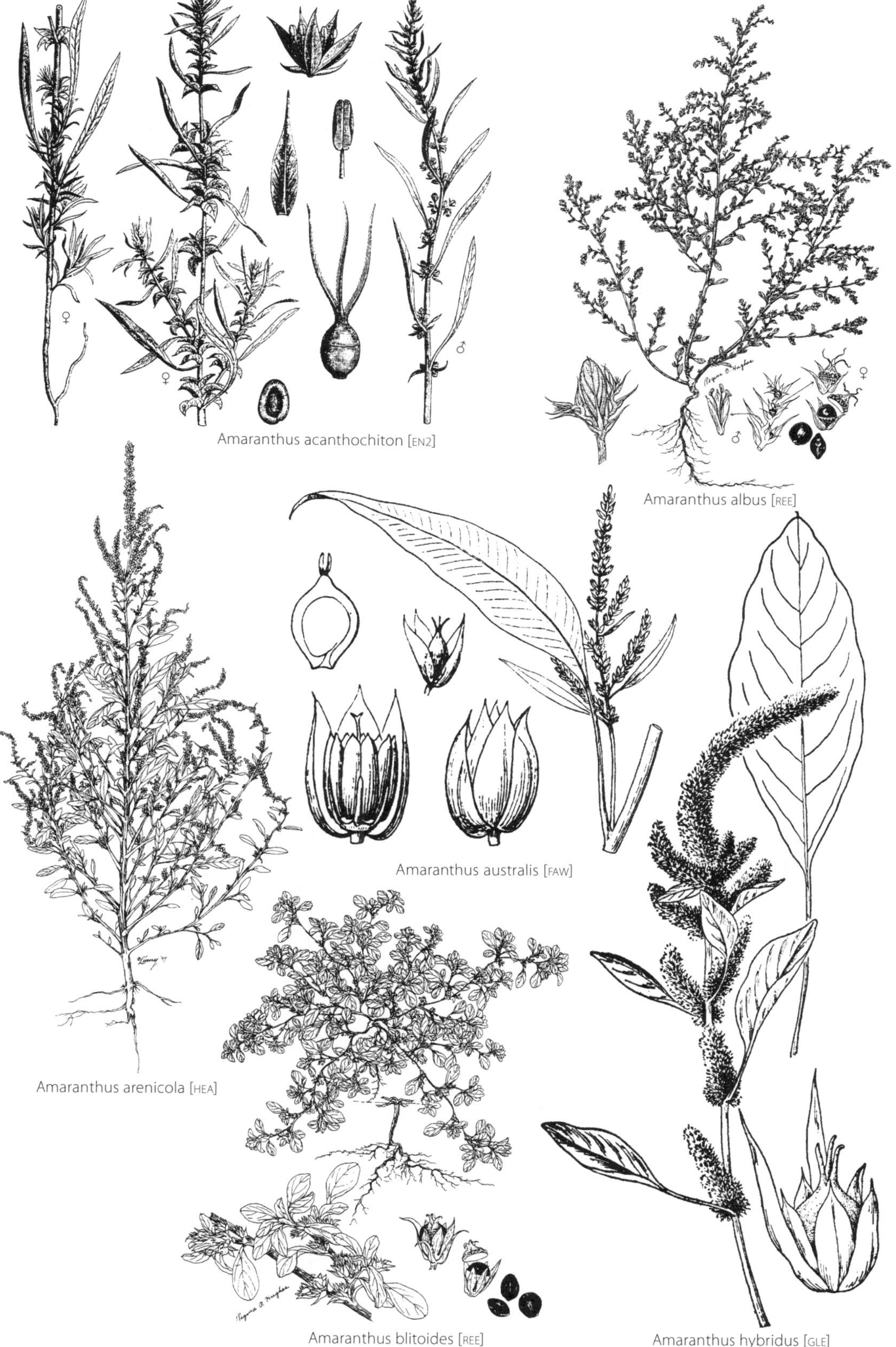

Amaranthus acanthochiton [EN2]

Amaranthus albus [REE]

Amaranthus arenicola [HEA]

Amaranthus australis [FAW]

Amaranthus blitoides [REE]

Amaranthus hybridus [GLE]

Sandy areas; Bell Co. (Reed 1969b); widespread in TX. [*A. torreyi* of authors, not. (A. Gray) Benth. ex S. Watson]

Amaranthus australis (A. Gray) J.D. Sauer, (southern), SOUTHERN WATER-HEMP, SOUTHERN AMARANTH. Plant erect, 2–3(–9!) m tall; seeds 1–1.25 mm in diam., dark reddish brown to black, shiny. Marshy places; Dallas, Denton, and Ellis cos.; central TX nw to nc TX. May–Aug. [*Acnida australis* A. Gray, *Acnida cuspidata* Bertero ex Spreng.]

Amaranthus blitoides S. Watson, (resembling blithe—a common name for *Chenopodium* species), PROSTRATE PIGWEED, QUELITE MANCHADO. Stems prostrate to rarely ascending, trailing and forming mats, to 1 m long, glabrous or nearly so; leaf blades oblong-oblanceolate; seeds 1.3–1.6 mm in diam., black, dull. Disturbed sites, weedy areas; widespread in TX. Jun–Oct. [*Amaranthus graecizans* of authors, not L.]

Amaranthus hybridus L., (hybrid), GREEN AMARANTH, SLENDER PIGWEED, SLIM AMARANTH, SPLEEN AMARANTH, QUELITE DE COCHINO, QUELITE MORADO. Stems erect, stout, 0.5–1.5(–2.5) m tall; spikes slender, 6–10(–12) mm broad, terminal and axillary; fruits dehiscent; seeds round, 1.1–1.3 mm in diam., black, shiny. Pioneer weed, disturbed areas; Brown Co., also Bell Co. (Reed 1969b); widespread but more common in w part of state. May–Oct. Native to e North America, also Mexico to n South America; naturalized weed in Mediterranean region, South Africa, Australia, and e Asia (Correll & Johnston 1970).

Amaranthus palmeri S. Watson, (for its discoverer, Edward Palmer, 1831–1911, English-American surgeon and botanist, collected in w U.S. and Mexico), CARELESSWEED, PALMER'S PIGWEED, PALMER'S AMARANTH, REDROOT. Plant erect, 1–2(–3) m tall; branches of inflorescence often very long, densely flowered, cylindrical, half-drooping; seeds 1–1.3 mm in diam., dark reddish brown. Disturbed sites and weedy areas; widespread in TX, often abundant. Jun–Oct. Implicated in livestock poisoning through the accumulation of nitrates (Kingsbury 1964; Burlage 1968). ☠

Amaranthus polygonoides L., (resembling *Polygonum*—knotweed), TROPICAL AMARANTH, BERLANDIER'S AMARANTH. Stems erect, stiff, 0.15–0.3 m tall, branching from the base; branches spreading, prostrate to ascending; seeds ca. 1 mm in diam., black, shiny. Disturbed sites; McLennan and Brown cos.; s TX n to s part of nc TX and w to Trans-Pecos. May–Oct. [*Amaranthus berlandieri* (Moq.) Uline & W.L. Bray]

Amaranthus retroflexus L., (reflexed or twisted back), RED-ROOT PIGWEED, ROUGH PIGWEED, GREEN AMARANTH, QUELITE. Plants erect, 0.3–3 m tall; fruits deshicent; seeds 1 mm in diam., dark reddish brown, shiny. Disturbed sites; Collin, Dallas, Grayson, Jack, and Tarrant cos., also Bell Co. (Reed 1969b); nearly throughout TX. May–Oct. Apparently introduced in Texas; first found in the state in 1894 (Mahler 1988). A cosmopolitan weed possibly native to the e U.S. Animals may be poisoned from eating the plants due to the accumulation of large amounts of nitrates (Burlage 1968; Stephens 1980). ☠

Amaranthus rudis J.D. Sauer, (wild), NUTTALL'S WATER-HEMP, WATER-HEMP. Plant highly variable, glabrous, 0.15–2.5 m tall; seeds ca. 1 mm in diam., dark reddish brown. Low, moist, disturbed sites, fields; invader species; nearly throughout TX. Late Apr–Oct. [*Acnida tamariscina* of authors, not (Nutt.) A.W. Wood, *Amaranthus tamariscinus* of authors, not Nutt.] One of the most abundant native weeds of late summer and fall.

Amaranthus spinosus L., (full of spines), SPINY PIGWEED, THORNY AMARANTH, QUELITE ESPINSO. Plant erect, 0.2–1.2 m tall; the only species in nc TX with distinct spines; leaf blades ovate to rhombic-lanceolate; seeds 0.7–1 mm in diam., black, shiny. Disturbed sites; Bell, Cooke, Dallas, Grayson, and Tarrant cos., also Denton and Williamson cos. (Reed 1969b); se and e TX w to East

Cross Timbers and Edwards Plateau. May–Oct. Pantropical weed. Reported to cause internal mechanical injury and bloat in livestock if ingested (Burlage 1968).

Amaranthus blitum L., (old name for *Chenopodium capitatum* (L.) Asch.—strawberry-blite), (GREEN AMARANTH), is cited by Hatch et al. (1990) for vegetational area 4 (Fig. 2) but apparently occurs only to the e and s of nc TX. It is monoecious and somewhat similar to *A. retroflexus* and *A. hybridus*, but erect or prostrate, 1 m or less tall, with lateral spikes not much shorter than the terminal, and with indehiscent fruits. Native of Europe. [*A. ascendens* Loisel., *A. lividus* L.]

FROELICHIA SNAKE-COTTON, COTTONWEED

Ours annuals, gray with ± matted pubescence, simple or with long, erect or partly decumbent branches from basal part of plant; leaves few, mostly in basal part of plant; leaf blades oblong or oblanceolate; inflorescences leafless; floral bracts white or yellowish, unequal, shorter than the perianth; flowers perfect; perianth densely and obviously woolly outside, becoming hardened at maturity; petals absent; stamens with united filaments forming a scarious tube, resembling an additional inner perianth; fruit an utricle surrounded by the hardened perianth.

A genus of 18 species of warm areas of the Americas and the Galapagos Islands. Hybridization between species is not uncommon. (Named for Joseph Aloys Froelich, 1766–1841, a German botanist)

1. Fruiting spikes 1–10 cm long, 10–13 mm thick, appearing whitish when fresh; fruiting perianth 4–6 mm long, usually with entire or deeply dentate to lacerate wings or crests; main stem usually erect, 2.5–7 mm thick near base, with few branches; plants 40–130 cm tall ____ **F. floridana**
1. Fruiting spikes 1–3 cm long, 7–8 mm thick, appearing grayish when fresh; fruiting perianth usually 2–3.5 mm long, usually distinctly spiny; main stem bent at base or with several near-basal branches with lower internodes widely spreading or decumbent, 1–3 mm thick; plants 10–50 cm tall ____ **F. gracilis**

Froelichia floridana (Nutt.) Moq., (of Florida), FIELD SNAKE-COTTON. Fruiting perianth with narrow to broad, nearly entire to deeply dentate or lacerate wings or crests down the keels, and with one or few teeth or short spines on the faces at base or occasionally up to the middle. Loose sandy soils, often abundant on roadsides and disturbed sites; throughout TX. Jun–Oct. Virtually all of our plants would seem referable to var. *campestris* (Small) Fernald. However, while this variety is often recognized (e.g., Kartesz 1994; Jones et al. 1997), because of the lack of consistently recognizable differences, we are not formally recognizing varieties. [*F. campestris* Small, *F. floridana* var. *campestris* Small] Wind-pollinated and considered a cause of hay fever (Ajilvsgi 1984).

Froelichia gracilis (Hook.) Moq., (graceful), SLENDER SNAKE-COTTON. Fruiting perianth spinier than in *F. floridana*, with rows of distinct and rather sharp, separate or united, spiny teeth on keels and a tooth at base of each face. Sandy or gravelly ground; throughout much of TX except far e part. Jun–Oct. According to McGregor (1986), *F. floridana* and *F. gracilis* often grow together and "... putative hybrids between the two are all too frequent."

Froelichia drummondii Moq., (for its discoverer, Thomas Drummond, 1780–1835, Scottish botanist and collector in North America), DRUMMOND'S SNAKE-COTTON, is cited by Reed (1969b) for Tarrant Co. based on a 1925 Ruth collection; this taxon, which is extremely similar to if not conspecific with *F. floridana*, is apparently otherwise unknown from nc TX. Reed (1969b) distinguished it from *F. floridana* (calyx crests deeply dentate) by its merely erose, crenulate or entire calyx crests. Waterfall (1972) lumped *F. drummondii* with *F. floridana*.

GOMPHRENA GLOBE-AMARANTH

A genus of ca. 120 species of tropical and warm areas of the New World and Australia; a number are cultivated as ornamentals. (Modification of Pliny's *Gromphaena*, a species of amaranth)
REFERENCE: Mears 1980.

Gomphrena globosa L., (globose, spherical), COMMON GLOBE-AMARANTH. Annual; stems stout, usually much branched, sometimes simple, to 1 m tall; leaves opposite; leaf blades mostly oblong to ovate, entire; flowers perfect, in showy, pedunculate, subglobose heads 2–2.8 cm in diam.; heads variously colored, purple, orange, rose, white, or varigated; perianth usually woolly; filaments united into a slender tube. Widely cultivated ornamental that sometimes escapes; Cooke and Tarrant cos.; also s TX and Edwards Plateau. Jul–Sep. Native of s Asia. The flowering heads superficially resemble those of CLOVER species.

Gomphrena haageana Klotzsch, (for J.N. Haage, 1826–1878, a seed grower of Erfurt, Germany), HAAGE'S GLOBE-AMARANTH, a perennial with leaf blades oblanceolate to oblong-linear and heads light red with yellow florets, is sometimes cultivated in nc TX and might be expected to escape; it is native to the w Edwards Plateau.

GOSSYPIANTHUS COTTON-FLOWER

A genus of 2 species, 1 from OK through TX to adjacent Mexico and in Hispanola, and the other restricted to Cuba (Henrickson 1987); it has often been included in *Guilleminea*. (Latin: *gossypion*, cotton, and Greek: *anthemon*, flower, presumably in reference to the densely lanate tepals)
REFERENCES: Mears 1967; Henrickson 1987.

Gossypianthus lanuginosus (Poir.) Moq., (woolly). Perennial herb; stems often procumbent, sometimes ascending; leaves opposite; rosette leaves with wide-winged petiole without chlorophyll; cauline leaves short-petioled, oval to obovate; inflorescence a spike of 6–12 axillary flowers; flowers sessile, 5-merous, perfect; tepals 3-nerved, green between nerves, densely lanate with silky hairs; stamens 5; style 1; stigmas capitate, bilobed; fruit a membranous, indehiscent utricle; seeds 1 mm long, brown, shiny. Dry soils, sandy, rocky, or disturbed areas. A highly variable species sometimes divided into several varieties; 2 are recognized here.

1. Leaves densely pilose or tomentose, especially beneath; rosette leaves lanceolate or oblanceolate to spatulate, 4–12 mm wide, obtuse to acute ____ var. **lanuginosus**
1. Leaves nearly glabrous; rosette leaves usually linear to lanceolate, 3–6 mm wide, acute ____ var. **tenuiflorus**

var. **lanuginosus**, WOOLLY COTTON-FLOWER. Throughout most of TX. [*Guilleminea lanuginosa* (Poir.) Hook. f., *Guilleminea lanuginosa* (Poir.) Hook. f. var. *rigidiflora* (Hook.) Mears, *Guilleminea lanuginosa* (Poir.) Hook. f. var. *sheldonii* (Uline & W.L. Bray) Mears]

var. **tenuiflorus** (Hook.) Mears ex Henr., (slender-flowered), LANCE-LEAF COTTON-FLOWER. Mainly parts of e 1/3 of TX. [*Guilleminea lanuginosa* (Poir.) Hook. f. var. *tenuiflora* (Hook.) Mears]

IRESINE BLOODLEAF

Ours monoecious or dioecious herbs; leaves opposite, petiolate; leaf blades thin, entire or serrulate; inflorescence a panicle of numerous, small flowers, often whitish to straw-colored due to bracts and perianths; perianth 5-parted; pistillate flowers subtended by conspicuous white wool; stamens 5; fruit an indehiscent utricle; seeds dark red, shiny.

A genus of ca. 80 species of tropical and temperate areas, especially the Americas and Australia; some are used medicinally while others are cultivated for their ornamental foliage. (Greek: *eiresione*, a staff or branch wound with wool, from the calyx, often bearing long hairs)

Amaranthus palmeri [CO1]

Amaranthus polygonoides [HEA]

Amaranthus retroflexus [REE]

Amaranthus rudis [CO1]

Amaranthus spinosus [GLE]

Froelichia gracilis [GLE, IPL]

Froelichia floridana [GLE, RAD]

Gomphrena globosa [NIC]

REFERENCE: Henrickson & Sundberg 1986.

1. Plants annual or perennial from a vertical root; stems usually much-branched; tepals of pistillate flowers 3-nerved, acute or obtuse, longer than the utricle; plants monoecious; rare in nc TX **I. diffusa**
1. Plants perennial with horizontal rhizomes; stems usually unbranched up to the inflorescence; tepals of pistillate flowers (look inside of small tepal-like bracts) faintly 1-nerved, acute, equaling or usually shorter than the utricle; plants dioecious; widespread in nc TX **I. rhizomatosa**

Iresine diffusa Humb. & Bonpl. ex Willd., (diffuse, spreading), JUBA'S-BUSH. Stems erect or spreading, sometimes clambering over adjacent plants, 0.4–3 m long; male and female flowers on the same plant; inflorescences white to straw-colored; tepals 1–1.5 mm long; seed broadly obovoid or suborbicular, ca. 1.5 mm in diam. Often in sandy moist areas; Tarrant and Williamson cos. (Reed 1969b); scattered in TX. Jul–Oct. [*I. celosia* L.]

Iresine rhizomatosa Standl., (very rooted), ROOTSTOCK BLOODLEAF, BLOODLEAF. Stems erect, 0.5–1.5 m tall; male and female flowers on separate plants, bracts and tepals silvery white; tepals 1.2–1.5 mm long; urticle 2–2.5 mm long; seed suborbicular, ca. 0.5 mm in diam., dark red, lustrous. Sandy soils, often moist areas; se and e TX w to East Cross Timbers, also Edwards Plateau. Aug–Oct.

TIDESTROMIA

A genus of 3 species of sw North America. (Named for Ivar T. Tidestrom, 1864–1956, Swedish-born botanist of sw U.S.)

Tidestromia lanuginosa (Nutt.) Standl., (woolly), WOOLLY TIDESTROMIA, ESPANTA VAQUEROS. Annual herb to 15 cm tall and 1 m across; stems usually prostrate to rarely ascending; stems and leaves gray-green to ashy-white, densely stellate-pubescent, glabrate with age; leaves opposite, obovate to rhombic-ovate, entire, 5–20(–30) mm long; petioles as long as blades or shorter; flowers in small axillary clusters, inconspicuous, yellowish, perfect; sepals 5, unequal, 1–3 mm long; stamens 5; utricle subglobose. Dry sandy, gravelly, or rock areas; Archer, Callahan, and Shackelford cos.; coastal, also extreme w part of nc TX w through w 1/2 of TX. Mar–Oct.

ANACARDIACEAE SUMAC OR CASHEW FAMILY

Ours subshrubs, shrubs, small trees, or woody vines, often with somewhat milky juice; leaves alternate, with 3 leaflets or once pinnately compound (simple in 1 species possibly present in nc TX); leaflets entire, toothed, or lobed; flowers small, in terminal or lateral, head-like to loose open panicles, perfect or unisexual, 5-merous; petals white to cream, yellow, or green; stamens 5; pistil 1; fruit a drupe.

A medium-sized (875 species in 70 genera) family of the tropics and subtropics with a few temperate species. Economically important taxa include food plants such as *Anacardium occidentale* L. (CASHEW), *Mangifera indica* L. (MANGO), and *Pistacia vera* L. (PISTACHIO), and ornamentals including *Cotinus* (SMOKETREE), *Rhus* (SUMACS), and *Schinus* (PEPPERTREE). ☠ A number of species cause contact dermatitis. *Pistacia chinensis* Bunge, (genus: Greek *pistake*, pistachio; sp.: of China), CHINESE PISTACHIO, native from Afghanistan to China and the Philippines, is widely planted in nc TX as an ornamental. It is a dioecious deciduous tree to 15(–25) m tall with even-pinnately compound leaves (leaflets 10–20, 1.2–2 cm wide, oblique, acuminate, entire), apetalous flowers, and small, reddish to purple, dry drupes 5–6 mm long in much-branched panicles. In the spring of 1998, a number of seedlings were observed in a woods in Fort Worth (R. O'Kennon, pers. obs.). It is not clear whether they will survive and whether this species will become naturalized. This small genus of 9 species includes the Old World *P. vera* L.,

PISTACHIO, cultivated for the edible nuts; one species, *P. texana* Swingle, is native to sw TX. Family name from *Anacardium*, a genus of 8 species of the New World tropics. (Greek: *ana*, up, and *cardia*, heart, possibly in reference to the large, swollen, fleshy, bright red or yellow, pear-like receptacle above the nut; the receptacle is edible and referred to as the cashew apple) (subclass Rosidae)
FAMILY RECOGNITION IN THE FIELD: subshrubs, shrubs, small trees, or woody vines with alternate leaves that either have 3 leaflets or are pinnately compound (simple in 1 species possibly present in nc TX); flowers small, inconspicuous; fruit a small, red or white to yellowish gray drupe (light brown in 1 species possibly present in nc TX).
REFERENCES: Barkley 1957, 1961; Brizicky 1962d; Baer 1979.

1. Leaves simple; fruiting panicles with plumose sterile pedicels; styles lateral; fruits light brown; rare species of Edwards Plateau questionably present in nc TX ______ **Cotinus**
1. Leaves compound with 3–numerous leaflets; fruiting panicles without plumose sterile pedicels; styles terminal; fruits red or white to yellowish gray; including species widespread in nc TX.
   2. Fruits red, noticeably pubescent with glandular and eglandular hairs; leaves pinnately compound with > 3 leaflets OR palmately 3-foliate (= terminal leaflet sessile); inflorescences terminal (or in *R. aromatica* and *R. trilobata* sometimes appearing axillary); plants without contact poisons ______ **Rhus**
   2. Fruits white to yellowish gray, glabrous or sparingly pubescent with eglandular hairs; leaves usually pinnately 3-foliate (= terminal leaflet stalked), rarely 5-foliate OR in 1 species rare in nc TX the leaves pinnately compound with 5–17 leaflets; inflorescences axillary; plants with contact poisons ______ **Toxicodendron**

COTINUS SMOKETREE, SMOKEWOOD

A genus of 3 species, 1 ranging from s Europe to China, 1 in sw China, and 1 in the se U.S. The widespread Old World species, *C. coggygria* Scop. (WIGTREE, EUROPEAN SMOKETREE), is widely cultivated and the wood yields a yellow dye; it can be distinguished from the native species by its smaller (ca. 4–9 cm long), glabrous, often ovate leaf blades. The smoke-like effect of the plumose fruiting pedicels gave rise to the common name. (Greek: *cotinus*, ancient name of the wild olive, used by Pliny for an unidentified shrub of the Apennines, but applied by some pre-Linnean botanists, such as Tournefort, to *C. coggygria*—Brizicky 1962d)

Cotinus obovatus Raf., (obovate, inversely ovate), AMERICAN SMOKETREE, SMOKEBUSH, CHITTAMWOOD. Usually dioecious, shrub or small tree to ca. 12 m tall with strong-smelling sap; wood yellow; leaves simple; leaf blades obovate to elliptic-obovate, 6–17 cm long, 5–9 cm wide, strikingly yellow to orange, red, or scarlet in fall, apically usually obtuse, with lower surface pubescent; petioles 1.5–6 cm long, often purple or reddish; flowers small, numerous, yellowish white to greenish yellow, in terminal panicles ca. 10 cm long which can enlarge to ca. 30 cm in fruit; sterile pedicels plumose with pale purplish or brownish hairs (panicles thus with a pink "smoky" appearance but not as showy as in the introduced *C. coggygria*); fruits flattened, kidney-shaped or oblique-oblong, ca. 4 mm long. Rocky woods, limestone outcrops; Cox and Leslie (1991) reported and mapped this species for nc TX; however, we have found no nc TX specimens or other literature reports and its presence in nc TX is questionable; mainly Edwards Plateau (Bandera, Kendall, Kerr, and Uvalde cos.—Stanford 1976; Simpson 1988). Apr–May. [*C. americanus* Nutt., *Rhus cotinoides* Nutt.] According to Simpson (1988), the TX occurrence of this tree is relictual with the TX populations more than 500 miles from the nearest locality in OK; the species has an extremely limited and scattered distribution in the e U.S. (Little 1970) and its distribution is considered relictual (Little 1983). The wood yields a yellow dye which was extensively used during the Civil War (Cox & Leslie 1991). The striking fall foliage suggests more widespread use as an ornamental.

RHUS SUMAC

Shrubs or rarely small trees often forming thickets; leaves palmately 3-foliate or pinnately compound, often turning bright red in fall.

A genus of ca. 200 species of temperate and warm areas; *Toxicodendron*, recognized here as a separate genus, is sometimes lumped into *Rhus* as a section or subgenus. According to J. Hennen (pers. comm.), there are rust fungi that attack only *Toxicodendron* (*Pileolaria brevipes* Berk. & Ravenel) and others that attack only *Rhus* (*Pileolaria patzcuarensis* (Holw.) Arthur). *Rhus* species often display very early fall foliage color (often strikingly red); this is considered to serve as a "foliar fruit flag" which attracts birds that act as dispersal agents for the fall-ripening fruits (Stiles 1984). In contrast to *Rhus* species, which are all non-poisonous, *Toxicodendron* species are poisonous and can cause contact dermatitis. Some species are a source of tannin. (Ancient Greek and Latin name for SUMAC)
REFERENCES: Barkley 1937; Brizicky 1963a.

1. Leaves palmately compound; leaflets 3.
   2. Terminal leaflet 25–60 mm long, ± narrowed and often pointed at apex; leaflets usually pubescent below, the margins usually ciliate ____ **R. aromatica**
   2. Terminal leaflet 15–33 mm long (rarely larger), abruptly narrowed to truncate at apex; leaflets usually glabrous below (or minutely puberulent when young), the margins not usually ciliate ____ **R. trilobata**
1. Leaves pinnately compound; leaflets 5 or more.
   3. Leaves with rachis completely unwinged and leaflets entire, 4–12 cm long ____ see **Toxicodendron vernix**
   3. Leaves with rachis winged OR leaflets toothed OR leaflets 4 cm or less long.
      4. Leaflets (3–)5–9, very small, < 2 cm long and < 6 mm wide; branches often ± spinescent ____ **R. microphylla**
      4. Leaflets 5–31, much > 2 cm long and usually much > 6 mm wide; branches not spinescent.
         5. Leaf rachis unwinged; leaflets sharply toothed or entire.
            6. Leaves deciduous, not coriaceous; leaflets mostly sharply and conspicuously toothed, glabrous and glaucous beneath, to 12 cm long ____ **R. glabra**
            6. Leaves evergreen, coriaceous; leaflets entire, pubescent beneath, soft to the touch, 4 cm or less long ____ **R. virens**
         5. Leaf rachis winged; leaflets mostly entire.
            7. Rachis and its wings on some leaves over 4 mm wide (total distance across rachis and its wings); leaflets 7–17, ovate-lanceolate, 2–4 times as long as wide, scarcely falcate ____ **R. copallinum**
            7. Rachis and its wings usually less than 3.5 mm wide; leaflets 13–19, linear-lanceolate, 4–9 times as long as wide, usually strongly falcate ____ **R. lanceolata**

Rhus aromatica Aiton var. **serotina** (Greene) Rehder, (sp.: fragrant; var.: late-flowering or -ripening), FRAGRANT SUMAC. Shrub to ca. 2 m tall with creeping stem-bases, forming loose clumps; foliage fragrant when bruised; flowers in terminal, head-like or irregular, dense clusters; petals yellow; flowering before leaves appear or with unfolding leaves. Sandy woods and ravines; se and e TX w to Dallas and Grayson cos., also Edwards Plateau. Mar.

Rhus copallinum L. var. **latifolia** Engl., (sp.: gummy, resinous; var.: broad-leaved), WING-RIB SUMAC, FLAME-LEAF SUMAC, DWARF SUMAC, SHINING SUMAC. Shrub or small tree to 1.5–3(–10) m tall; leaflets soft-pubescent to nearly glabrous beneath; flowers in dense, subsessile, pyramidal panicles. Sandy woods, hills, and open areas; se and e TX w to East Cross Timbers, also Coryell Co. in Lampasas Cut Plain (Little 1976 [1977]). Jun–Jul. We are following Jones et al. (1997) and J. Kartesz (pers. comm. 1997) in treating all TX examples of this species as var. *latifolia*. According

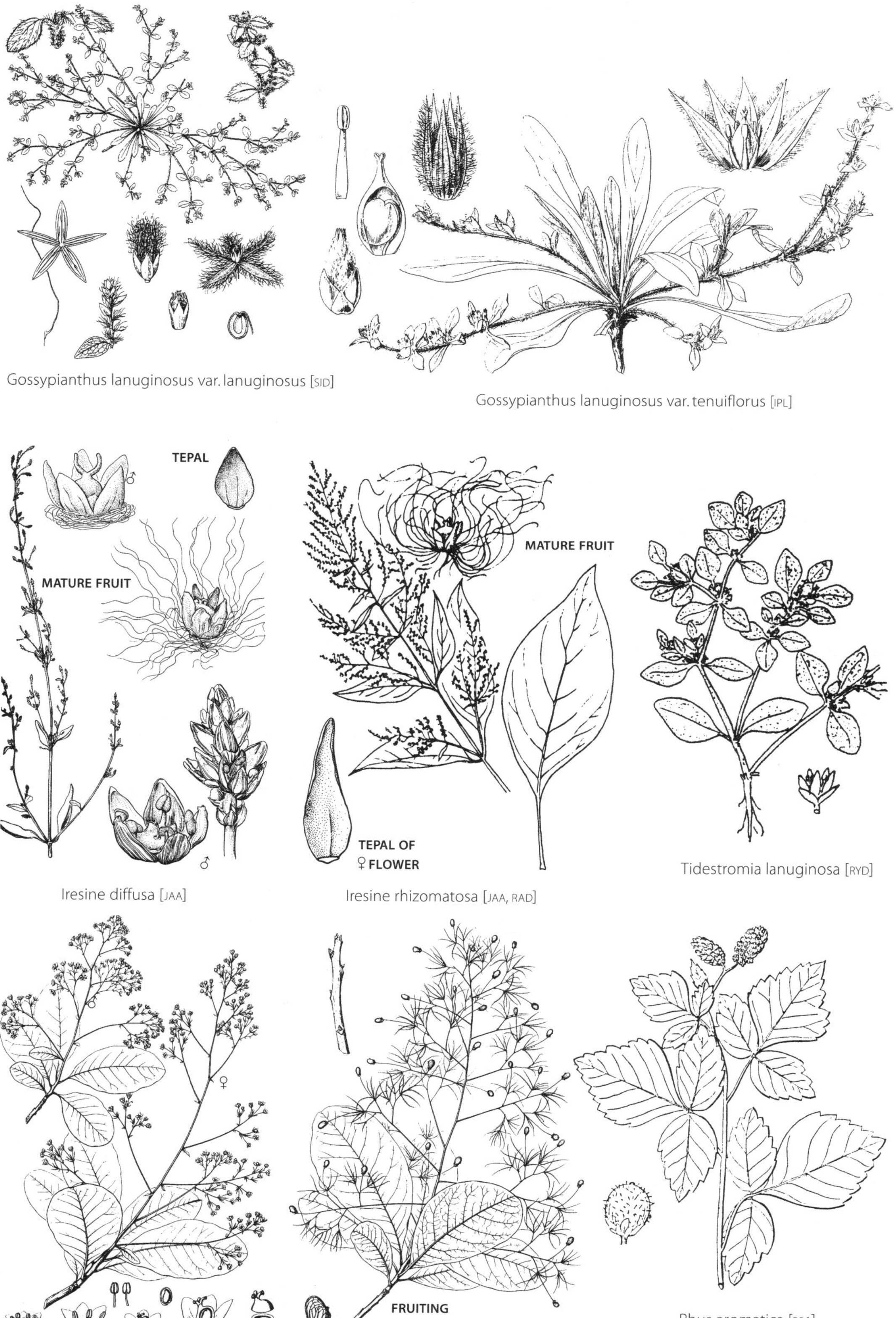

Gossypianthus lanuginosus var. lanuginosus [SID]

Gossypianthus lanuginosus var. tenuiflorus [IPL]

Iresine diffusa [JAA]

Iresine rhizomatosa [JAA, RAD]

Tidestromia lanuginosa [RYD]

Cotinus obovatus [SA3]

Rhus aromatica [BB1]

to Crosswhite (1980), before lemons were readily available, the small fruits of *R. copallinum* were crushed in water to make a tart drink called sumac-ade.

Rhus glabra L., (glabrous, smooth, hairless), SMOOTH SUMAC, SCARLET SUMAC. Tree-shaped shrub (rarely a small tree) to ca. 3 m tall, the leaves and few branches crowded toward summit of stem; leaflets 9–23, short-oblong to narrowly lanceolate, glabrous, glaucous beneath; flowers in dense, subsessile, pyramidal panicles. Sandy or rocky soils, hillside woods, stream banks, and fencerows; e TX w to Rolling Plains and Edwards Plateau. May.

Rhus lanceolata (A. Gray) Britton, (lanceolate, lance-shaped), PRAIRIE SUMAC, PRAIRIE FLAME-LEAF SUMAC. Large shrub or small tree to 10 m tall; leaflets usually glabrous, narrow; flowers in dense, subsessile, pyramidal panicles. On limestone; Hood and McLennan cos., also Brown, Coryell, Hamilton, and Mills cos. (HPC), also Dallas, Palo Pinto, Parker, and Tarrant cos. s to the Edwards Plateau (Little 1976); mainly c and w parts of nc TX and Edwards Plateau; scattered elsewhere in the state. Jul–Aug. [*R. copallinum* L. var. *lanceolata* A. Gray]

Rhus microphylla Engelm. ex A. Gray, (small-leaved), LITTLE-LEAF SUMAC, DESERT SUMAC, SCRUB SUMAC, SMALL-LEAF SUMAC, CORREOSA SODA-POP-BUSH. Much-branched, shrub (rarely a small tree) to ca. 5 m tall (usually much smaller); leaflets sessile, pilose; rachis winged; flowers appearing before (rarely with) the leaves. Dry open areas; Brown and Clay cos. (Little 1976); w part of nc TX s and w to w TX. Apr(–May). According to Crosswhite (1980), before lemons were readily available, the small fruits were crushed in water to make a tart drink called sumac-ade.

Rhus trilobata Nutt., (three-lobed), SKUNKBUSH. Similar to *R. aromatica*; usually flowering before leaves appear or with unfolding leaves. Limestone outcrops, rocky slopes, prairies, fencerows, woods margins, and sandy woods; Blackland Prairie and Edwards Plateau w to w TX. This taxon is distinguished in some instances with difficulty from *R. aromatica* and is possibly only a variety of that species. [*R. aromatica* Aiton var. *flabelliformis* Shinners]

Rhus virens Lindh. ex A. Gray, (green), EVERGREEN SUMAC, TOBACCO SUMAC, LENTISCO. Shrub or small tree to 3 m or more tall; leaflets 4 cm or less long, 2 cm or less wide. Limestone bluffs overlooking the Brazos River, Johnson Co., otherwise known in nc TX only from Bell, Coryell (J. Stanford, pers. comm. and Sanchez 1997), and Brown (Stanford 1976) cos.; mainly Edwards Plateau w to Trans-Pecos. Aug–Oct. The Johnson Co. record is well to the n of any other known localities of this species.

TOXICODENDRON POISON-OAK, POISON-IVY

Poisonous woody vines, shrubs, subshrubs, or small trees; leaves pinnately 3-foliate or pinnately compound, often turning bright red in fall.

A genus of 15+ species of North and South America and e Asia; often included in the genus *Rhus* but rather clearly distinguished (Barkley 1937; Gillis 1971). ☠ These species are **toxic**, causing severe allergic contact dermatitis in some individuals. Physical contact with any part of the plant, exposure to fumes/smoke from burning plants, or contact with pets having touched the plants are common means of exposure. Following a latent period of 12–24+ hours after exposure, there is reddening of the skin, sometimes accompanied by edematous swelling. This can be followed by the formation of fluid-filled blisters; this fluid cannot spread the dermatitis. Instead, additional spread is from unwashed skin, clothing, or other objects (Lampe 1986). Eating the leaves can result in an internal reaction occasionally known to be fatal (Gillis 1975). According to Gillis (1975), "The poisons may be effective for an indefinite period of time in causing dermatitis. Several hundred year-old herbarium specimens have been known to affect a sensitive person who has handled them!" These reactions are caused by resinous phenolic compounds commonly known as urushiols (chemically pentadecylcatechols). Because the

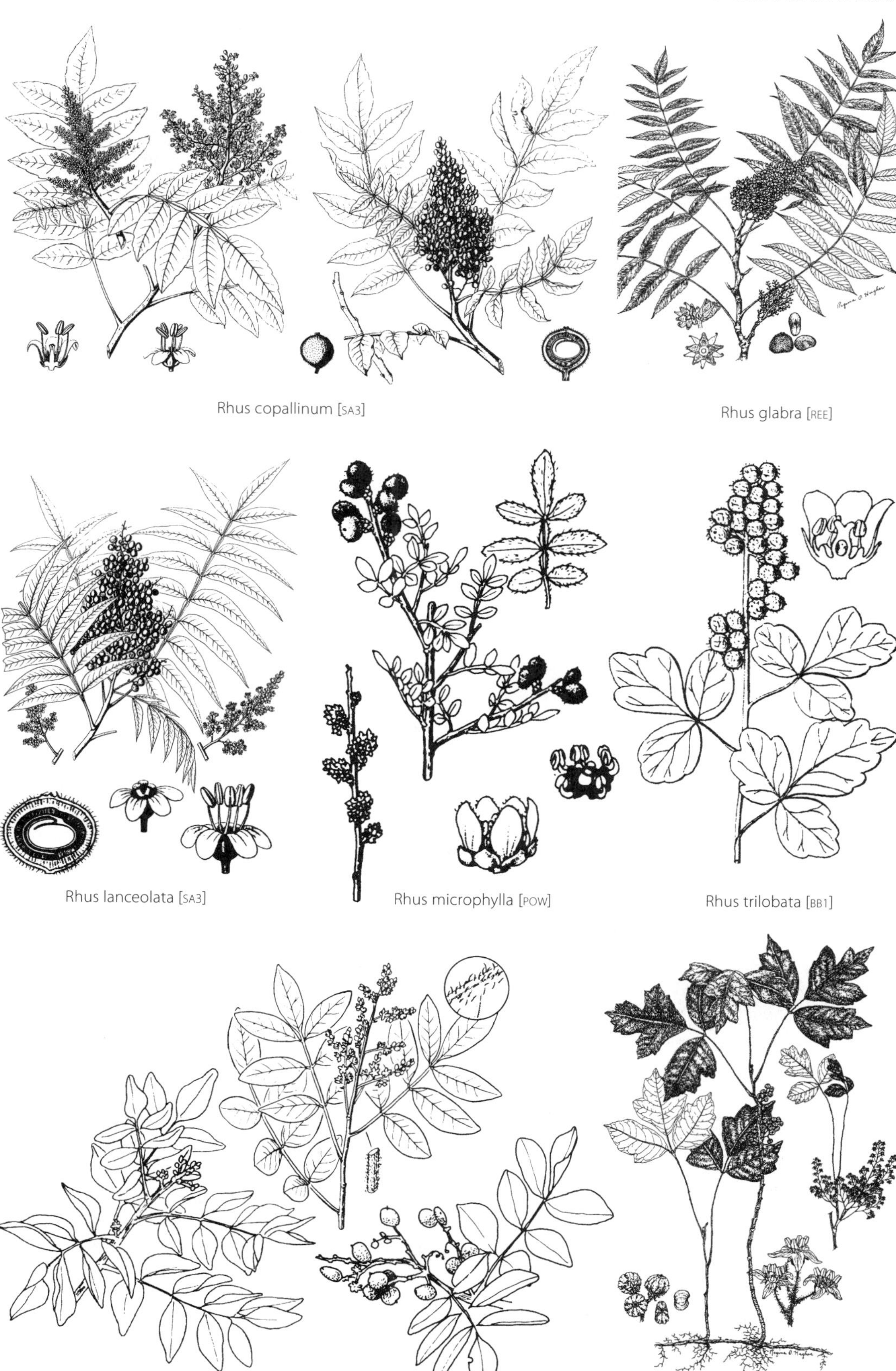

Rhus copallinum [SA3]

Rhus glabra [REE]

Rhus lanceolata [SA3]

Rhus microphylla [POW]

Rhus trilobata [BB1]

Rhus virens [LYN, POW]

Toxicodendron pubescens [REE]

compounds are insoluble in water, washing with a strong soap as soon as possible after contact is recommended. Gillis (1975) and Frankel (1991) discussed *Toxicodendron* dermatitis. Gillis (1975) also gave several citations documenting the use of POISON-IVY in Native American arrow poisons. The sap of an Asian species, *T. vernicifluum* (Stokes) F.A. Barkley (ORIENTAL LACQUER TREE), is a major source of lacquer; furniture treated with the lacquer can cause dermatitis in sensitive individuals. *Toxicodendron* species, like those in the genus *Rhus*, often display very early fall foliage color (often strikingly red); this is considered to serve as a "foliar fruit flag" which attracts birds that act as dispersal agents for the fall-ripening fruits (Stiles 1984). (Latin: *toxicum*, poison, and Greek: *dendron*, tree)
REFERENCES: McNair 1925; Barkley 1937; Brizicky 1963a; Gillis 1971, 1975; Brooks 1977; Baer 1979; Reveal 1990b.

1. Leaflets 5–17; large shrub or small tree; usually in swamps or wet areas; mainly e TX ____ **T. vernix**
1. Leaflets usually 3(–5); vines, subshrubs, or shrubs; usually in habitats of moderate to dry moisture conditions; extremely abundant in nc TX.
   2. Leaflets often with broad, blunt apex, sometimes with narrowed apex; leaflets entire or with usually shallow rounded lobes or broad blunt teeth (can be deeply lobed), pubescent underneath; fruits pubescent or papillose; plants not climbing ____ **T. pubescens**
   2. Leaflets with narrowed apex; usually some leaflets deeply lobed or sharply toothed, glabrous or pubescent underneath; fruits glabrous (or with occasional hairs); plants varying from not climbing to high-climbing ____ **T. radicans**

Toxicodendron pubescens Mill., (pubescent, downy), EASTERN POISON-OAK. Low, creeping shrub or subshrub; leaflet lobes usually shallow but sometimes ± deep. Sandy woods; Montague Co., also Dallas, Grayson, Lamar, and Tarrant cos. (Gillis 1971); widespread in TX. [*Rhus toxicarium* Salisb., *Rhus toxicodendron* L.] Apr–May. Jones et al. (1997) treated this taxon as [*T. diversilobum* (Torr. & A. Gray) Greene var. *pubescens* Mill.]. Causes contact dermatitis. ☠

Toxicodendron radicans (L.) Kuntze, (rooting), POISON-OAK, POISON-IVY, HIEDRA. Low, creeping shrub with aerial roots to erect shrub or low- to high-climbing vine. In a variety of habitats from low woods to forest margins and disturbed areas. Mid-Apr–May(–later). The subspecies of *T. radicans* seem to overlap morphologically and are often difficult to distinguish. In addition to the subspecies below, Hatch et al. (1990) cited subsp. *eximum* (Greene) Gillis for vegetational area 5 (Fig. 2) and subsp. *radicans* for vegetational areas 4 and 5. We have seen no material for nc TX and according to Gillis (1971), these subspecies do not occur in nc TX. Reveal (1990b) believed the variation within this species is better treated at the varietal rank than the subspecific. Causes contact dermatitis. ☠

1. Leaflets usually clearly and sharply lobed or with deeply cut margins ____ subsp. **verrucosum**
1. Leaflets with entire, undulate, notched, or serrate margins.
   2. Leaflets usually glabrous or with sparse pubescence both below and above, the pubescence, if any, appressed; terminal leaflet usually ovate to elliptic ____ subsp. **negundo**
   2. Leaflets with pubescence on lower leaf surfaces, often pilose, velvety, the pubescence erect; upper leaf surface usually pubescent; terminal leaflet broadly ovate ____ subsp. **pubens**

subsp. **negundo** (Greene) Gillis, (the native name of *Vitex negundo*), POISON-IVY. Hunt Co., also Cooke, Dallas, and Kaufman cos. (Gillis 1971); se and e TX w to Rolling Plains and e Edwards Plateau. [*T. radicans* var. *negundo* (Greene) Reveal]

subsp. **pubens** (Engelm. ex S. Watson) Gillis, (downy), POISON-IVY. Cooke and Dallas cos.; se and e TX w to nc TX and e Edwards Plateau.

subsp. **verrucosum** (Scheele) Gillis, (verrucose, warted), POISON-IVY. Bosque, Dallas, Hill, Johnson,

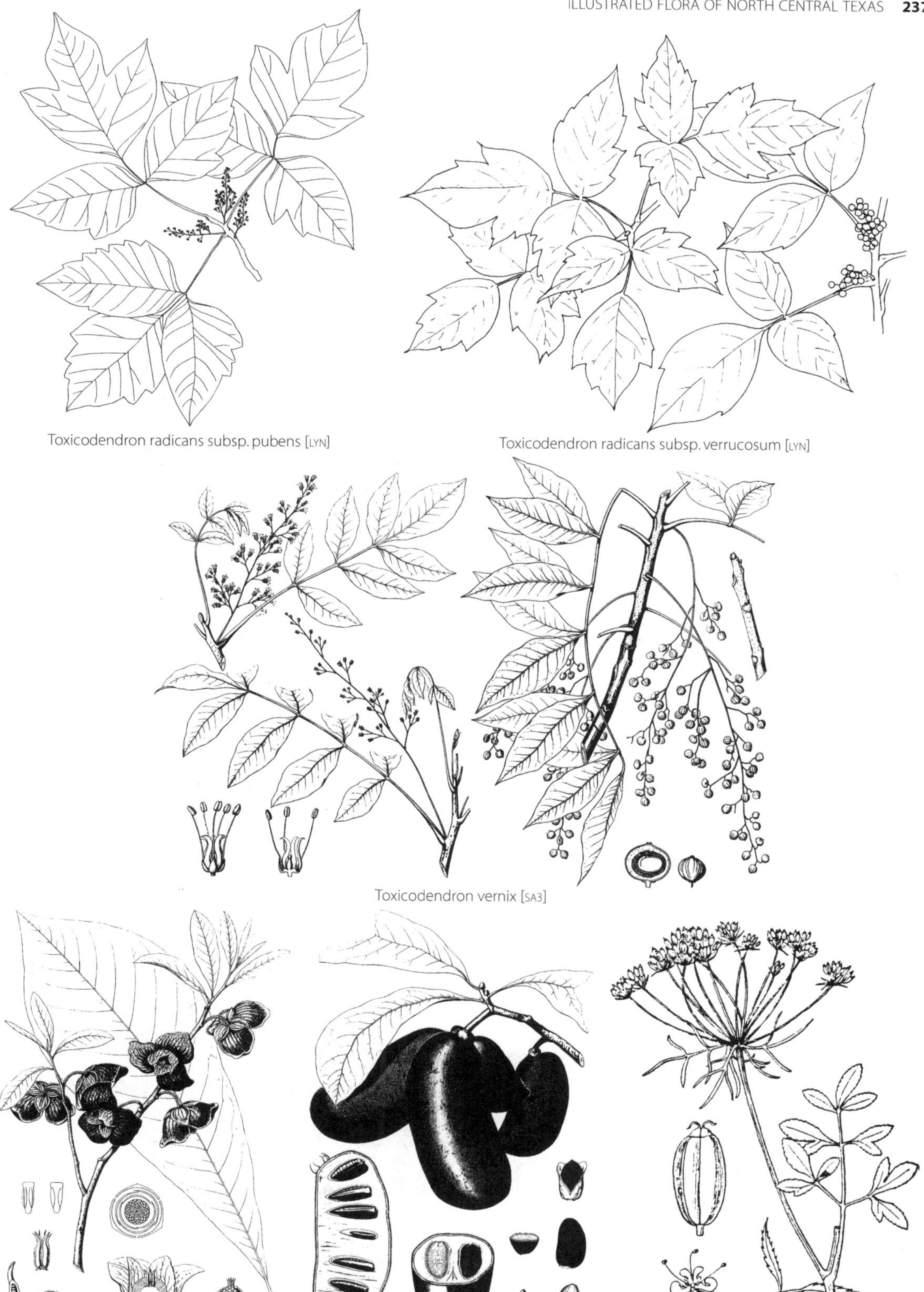

Toxicodendron radicans subsp. pubens [LYN]

Toxicodendron radicans subsp. verrucosum [LYN]

Toxicodendron vernix [SA3]

Asimina triloba [SA3]

Ammi majus [BR2]

McLennan, Palo Pinto, and Tarrant cos., also Grayson Co. (Gillis 1971); se and e TX w to nc TX and Edwards Plateau.

Toxicodendron vernix (L.) Kuntze, (varnish), POISON SUMAC, POISON-ELDER, POISON-DOGWOOD. Branchlets glabrous; leaflets 4–12 cm long, 2–5 cm wide, entire; flowers in axillary panicles, greenish. Swamps or wet areas; included based on citation of vegetational area 4 (Fig. 2) by Hatch et al. (1990); mainly e TX. Apr–Jun. [*Rhus vernix* L.] Vegetatively this species resembles many non-toxic *Rhus* species. It can, however, be easily distinguished by its leaves having entire leaflets and lacking any wing along the rachis; all nc TX *Rhus* species with pinnately compound leaves have either toothed leaflets or a winged rachis or in the case of the locally rare *R. virens*, small leaflets 4 cm or less long. This species causes contact dermatitis. ☠

ANNONACEAE CUSTARD-APPLE OR ANNONA FAMILY

The Annonaceae is a large family of ca. 2,150 species in 112 genera (Kral 1997); it is mainly in tropical areas with *Asimina* extending n to Michigan. The family consists of aromatic (due to ethereal oils) trees, shrubs, and lianas, usually with alkaloids, and with 3-merous flowers (unusual among dicots); they share with the related Magnoliaceae a number of unspecialized flower characters such as numerous free stamens and carpels, laminar stamens, and monocolpate pollen. Many are important economically with tropical species variously used medicinally or for perfumes or cosmetics (Baumgardt 1982); a number are valued for their edible fruits including *Annona* species (e.g., CUSTARD-APPLE, CHERIMOYA, SOURSOP, and SWEETSOP). Family name from *Annona*, a genus of ca. 100 species of tropical America and Africa. (Name either from Haitian, *anon*, the native name for the widely cultivated *A. reticulata* L.—custard-apple, or from Latin: *annona*, yearly produce, crop, grain, food) (subclass Magnoliidae)

FAMILY RECOGNITION IN THE FIELD: the only nc TX species is a large shrub or tree with large, alternate, entire leaves, hairy *naked buds*, *3-merous* flowers, and large banana-like fruits.

REFERENCES: Wood 1958; Kessler 1993a; Kral 1997.

ASIMINA AMERICAN PAWPAW

An e U.S. genus of 8 species; this is one of only 11 genera of trees endemic to the e U.S. (only three of these, *Asimina*, *Maclura*, and *Taxodium*, occur in nc TX) (Little 1983). It is also the northernmost and only extra-tropical genus of a large and otherwise tropical family (Little 1970). (Name through *asiminier* of the French colonists, from the Native American name, *assimin*)

REFERENCES: Nash 1896; Kral 1960a; Wilbur 1970.

Asimina triloba (L.) Dunal, (three-lobed), PAWPAW, COMMON PAWPAW. Large shrub or tree to 12 m tall; young twigs rusty pubescent; buds naked (= without scales) but protected by pubescence; leaves alternate; leaf blades obovate to elliptic-ovate, to 25(–30) cm long and 7(–15) cm wide, apically abruptly acuminate, basally cuneate, marginally entire; petioles 5–10 mm long; sepals 3; petals dull purple, 6, in 2 series, the inner and outer series very unequal; stamens numerous in a mass; carpels 3–15, separate; ovaries superior; fruits to 12 cm long, 3–4 cm thick, green to dark brown when ripe (in fall), the flesh sweet and edible; seeds somewhat flattened, 1.5–2 cm long. Rich woods and banks of streams; Grayson Co. along the Red River near the mouth of Pawpaw Creek; with the exception of the Grayson collection, this species is known in the state only from extreme ne TX. Apr–May. The fruits are edible (and delicious) but can cause gastrointestinal problems for some; the seeds are reported to contain a toxic alkaloid; the plant causes contact dermatitis in a small percentage of people (Peattie 1948; Kingsbury 1964). ☠

APIACEAE (UMBELLIFERAE)
CARROT OR PARSLEY FAMILY

Annual or perennial herbs; leaves basal, alternate, or (in *Bowlesia*) opposite; petioles in nearly all species with enlarged, clasping base; leaf blades simple or compound, entire, toothed, lobed, or finely dissected; flowers axillary or terminal, small, in umbels (varying from loose and open to compact and head-like) or whorls; calyx tube cylindrical to shallowly cup-like, with entire or 5-toothed summit; petals 5, attached at summit of calyx tube, equal or occasionally unequal; stamens 5, attached around a fleshy disk; pistil 2-carpellate; ovary inferior; styles and stigmas 2; fruit a schizocarp that splits into 2 one-seeded segments.

A large (3,540 species in 446 genera), nearly cosmopolitan but more commonly n temperate and tropical mountain family of mainly herbs or less commonly shrubs or even trees. The family yields many important spices and foods including *Anethum* (DILL), *Apium* (CELERY), *Carum* (CARAWAY), *Coriandrum* (CORIANDER, CILANTRO), *Cuminum* (CUMIN), *Daucus* (CARROT), *Foeniculum* (FENNEL), *Pastinaca* (PARSNIP), *Petroselinum* (PARSLEY), and *Pimpinella* (ANISE); ☠ it also includes some extremely poisonous plants. To avoid possibly **fatal** poisoning, no wild member of the family should be eaten without absolute certainty of identification. Many species of Apiaceae are used as food by caterpillars of the black swallowtail butterfly (*Papilio polyxenes*); numerous native species (*Berula erecta, Cicuta maculata, Cryptotaenia canadensis, Osmorhiza longistylis, Polytaenia nuttallii, Ptilimnium capillaceum, Sium suave, Spermolepis divaricata, and Zizia aurea*) are utilized as well as introduced taxa (DILL, FENNEL, PARSLEY, QUEEN-ANNE'S-LACE, POISON-HEMLOCK). Apiaceae typically produce toxic psoralins (furanocoumarins) to protect themselves from herbivore damage; however, black swallowtail larvae are resistant and actually grow faster in the presence of the chemicals (Scott 1986; Tveten & Tveten 1993). The Apiaceae are closely related to the Araliaceae and appear to be a polyphyletic group derived from within a paraphyletic Araliaceae. From a cladistic standpoint the two families should be lumped to form a more inclusive monophyletic family, which based on nomenclatural rules should be called Apiaceae (Judd et al. 1994). *Tauschia texana* A. Gray, an acaulescent perennial with yellow flowers, oval fruits 3–4 mm long, and the seed face deeply sulcate, is cited by Hatch et al. (1990) for vegetational area 4 (Fig. 2); it apparently occurs well to the s of nc TX. Family name from *Apium*, a temperate genus of 20 species including *A. graveolens* L., CELERY, valued for its edible petioles and fruits (celery "seed"). (Classical Latin name for celery and parsnip) (Rosidae)

FAMILIY RECOGNITION IN THE FIELD: often aromatic herbs with small, frequently white to green, yellow, or pink flowers usually in *umbels*; leaves alternate and often pinnately compound, sometimes finely dissected, with sheathing petioles; internodes hollow; fruits small, splitting into 2 one-seeded parts.

REFERENCES: Coulter & Rose 1900; Mathias & Constance 1944–1945, 1961; Mathias 1965; Hiroe 1979; Pimenov & Leonov 1993; Judd et al. 1994; Plunkett et al. 1996 [1997].

1. Leaves simple or apparently so; leaf blades round to reniform, ovate, lanceolate, or linear, either unlobed OR palmately divided 2/3 to base or less, with few, wide lobes.
   2. Leaves reduced to elongate, filiform or linear, entire, hollow, septate phyllodes, the upper bract-like, normal blade tissue absent (leaves rush-like in appearance) ____ **Oxypolis**
   2. Leaves with normal blade tissue.
      3. Leaves perfoliate (= completely surrounding the stem) ____ **Bupleurum**
      3. Leaves not perfoliate.
         4. Leaf blades (at least upper) linear to oblong-lanceolate; flowers sessile, numerous, in very dense, ovoid to globose, whitish to bluish heads 5–25 mm in diam. ____ **Eryngium**
         4. Leaf blades round to reniform, broadly ovate, or ovate; flowers not arranged as above.

5. Inflorescences usually long-peduncled; plants glabrous or with simple pubescence; stipules absent; leaf blades peltate OR not so; leaves basal or solitary at the nodes; creeping perennials, rhizomatous or stems rooting at the nodes.
6. Involucre subtending inflorescence much reduced or absent; leaf blades round to reniform, either peltate OR lobed to about the middle; petioles not sheathing; wide-spread in nc TX ____ **Hydrocotyle**
6. Involucre of 2 conspicuous ovate to suborbicular bracts 2–4 mm long; leaf blades ovate to broadly ovate, neither peltate nor lobed; petioles sheathing; rare in nc TX ____ **Centella**
5. Inflorescences sessile or nearly so; plants with ± gray, stellate pubescence; stipules present, scarious, lacerate; leaf blades not peltate; leaves opposite; prostrate to suberect annuals from slender taproot ____ **Bowlesia**
1. Leaves compound or at least the largest deeply lobed; leaf blades triangular-ovate to lanceolate or linear in outline.
7. Inflorescence a dense "head," globose to elongated and usually with spiny leaves exserted apically ____ **Eryngium**
7. Inflorescence open, loose, not dense, without spiny leaves exserted apically.
8. Ovaries and fruits bristly or prickly.
9. Leaves palmately compound with 3–5 leaflets, the leaflets toothed and lanceolate, the largest > 1 cm wide, usually much wider ____ **Sanicula**
9. Leaves pinnately compound or decompound, with numerous segments; ultimate leaf segments < 1 cm wide, often much less.
10. Umbels with simple, entire basal bracts or bracts absent.
11. Segments of upper leaves narrowly linear or thread-like; plants glabrous; mature fruits 1.5–2 mm long ____ **Spermolepis**
11. Segments of upper leaves lanceolate; plants hispid; mature fruits 3–5 mm long ____ **Torilis**
10. Umbels with toothed or lobed basal bracts ____ **Daucus**
8. Ovaries and fruits glabrous or pubescent, but neither bristly nor prickly.
12. Petals yellow or green.
13. Leaves all basal ____ **Lomatium**
13. Leaves distributed up the stem.
14. Ultimate leaf segments long and thread-like (< 0.5 mm wide); bracts absent below both umbels and umbellets (= secondary or ultimate umbels); plants with strong anise or dill odor.
15. Plants with strong anise odor; mature fruits with the lateral ribs wingless, all the ribs ± the same in appearance ____ **Anethum**
15. Plants with strong dill odor; mature fruits with lateral ribs winged, distinctly different from other ribs ____ **Foeniculum**
14. Ultimate leaf segments not long and thread-like (> 1 mm wide, often much greater); bracts absent below umbels but present below umbellets (sometimes inconspicuous); plants without strong anise or dill odor.
16. Ultimate segments of upper leaves sharply toothed around the margins, not lobed; fruits 2–4 mm long ____ **Zizia**
16. Ultimate segments of upper leaves toothed at apex or entire, conspicuously lobed; fruits 2–11 mm long.
17. Leaves divided into numerous segments, but not parsley-like in appearance (the segments too large), the ultimate segments or their lobes often > 5 mm wide; small calyx teeth present; fruits 5–11 mm long, 4–7 mm wide; widespread native species ____ **Polytaenia**

17. Leaves parsley-like in appearance, the ultimate segments or their lobes usually < 5 mm wide; calyx teeth absent; fruits 2–4 mm long, 1.5–3 mm wide; rarely escaped introduced cultivar ____ **Petroselinum**

12. Petals white, pink, or lavender.

18. Fruits linear with long differentiated beak 20–70 mm long; ovary pubescent ____ **Scandix**

18. Fruits beakless or nearly so; ovary pubescent OR glabrous.

19. Pedicels 0.3–2 times as long as the ovary when petals have expanded (up to 3 times in fruit).

20. Umbels small, simple or compound with 2–5 rays (= main branches of umbel).

21. Small bracts below umbellets linear or narrowly lanceolate, acute OR small bracts absent.

22. Segments of middle and upper leaves obtuse or subacute; plants 2–30 cm tall, branching from base; fruits 2.5–6 mm long ____ **Ammoselinum**

22. Segments of middle and upper leaves sharply acute; plants 20–90 cm tall, branching in upper part; fruits 8–10 mm long ____ **Trepocarpus**

21. Small bracts below umbellets oblong-elliptic, obtuse or subacute ____ **Chaerophyllum**

20. Umbels large, compound with 14–17 rays ____ **Daucosma**

19. Pedicels 2–many times as long as the ovary when petals have expanded.

23. Umbels mostly lateral or axillary and short-peduncled; fruits ovoid, 1–2(–3) mm long ____ **Cyclospermum**

23. Umbels mostly or all terminal, short- or long-peduncled; fruits various.

24. Small bracts below umbellets oblong-elliptic, obtuse, scarious; flowers pinkish; fruits broadly winged; leaves mostly near base of plant ____ **Cymopterus**

24. Small bracts below umbellets absent OR slender, acute, not scarious; flowers various; fruits winged OR not winged; leaves along stem and/or basal.

25. Upper leaves with lanceolate or oblanceolate, usually toothed segments or leaflets.

26. Umbels with 2–8 primary rays; ovary elongating, the fruit elongate subcylindrical.

27. Leaves mostly once compound; small bracts below umbellets linear-lan ceolate, erect, inconspicuous; pedicels very unequal ____ **Cryptotaenia**

27. Leaves 2–3 times compound; small bracts below umbellets lanceolate, reflexed, prominent; pedicels about equal ____ **Osmorhiza**

26. Umbels with many (6–30+) primary rays; ovary remaining short, the fruit subglobose.

28. Leaf blades much-dissected, almost parsley-like in appearance; small bracts below umbellets ovate-lanceolate with wide clasping base, abruptly tapered to ± narrow tips ____ **Conium**

28. Leaf blades 1–3 times pinnately compound but not much-dissected, not at all parsley-like in appearance; small bracts below umbellets linear to lanceolate, scarcely if at all clasping, not abruptly tapered.

29. Leaf blades 1–3 times pinnately compound, the lateral leaflets or lateral leaf segments with distinct stalks; widespread in nc TX.

30. Involucres subtending main umbels absent OR of 1–4 entire, inconspicuous bracts; ribs of fruit prominent, corky, obtuse, covering half or more of the fruitsurface ____ **Cicuta**

30. Involucres subtending main umbels of numerous usually pinnately divided bracts with filiform segments; ribs of fruit narrow, acute, covering much less than half of the fruit surface ____ **Ammi**

29. Leaf blades once pinnately compound (but leaflets can be deeply lobed), the lateral leaflets sessile or nearly so; rare in nc TX.

31. Leaflets not only toothed but also some often deeply lobed; fruits 1.5–2 mm long, with ribs obscure ____ **Berula**

31. Leaflets finely to coarsely toothed (rarely entire), but not lobed; fruits 2–7 mm long, with ribs prominent.

32. Leaflets often coarsely toothed (teeth to 4 mm or more tall); fruits 4–8 mm long; stems not conspicuously corrugated ____ **Oxypolis**

32. Leaflets finely toothed (teeth < 1 mm tall); fruits2–3 mm long; stems conspicuously corrugated ____ **Sium**

25. Upper leaves with thread-like to linear-lanceolate, mostly entire segments OR some leaves undivided, slender, entire.

33. Leaves once compound (some can be simple), with linear-lanceolate, entire lobes or leaflets.

34. Stem leaves palmately compound; fruits with a small beak ____ **Cynosciadium**

34. Stem leaves pinnately compound or pinnatifid OR some simple and entire; fruits beakless ____ **Limnosciadium**

33. Leaves 1–3 times compound, the upper with linear-lanceolate to thread-like segments.

35. Petals obtuse or subacute (a few sometimes bilobed), oblong or elliptic, longer than wide; fruits neither constricted into 2 parts nor conspicuously flattened.

36. Petals 0.5–1.3 mm long, white; fruits minutely tuberculate to smooth, 1.5–2 mm in diam.; native species wide spread in nc TX ____ **Spermolepis**

36. Petals (of outer flowers in umbels) 2.2–4 mm long, white, often tinged pink or lavender; fruits smooth, (1.5)2.5–3(–5) mm in diam; introduced cultivated species rarely escaped in nc TX ____ **Coriandrum**

35. Petals with indented or notched apex, obovate or suborbicular, shorter to slightly longer than wide; fruits constricted into 2 parts OR conspicuously flattened OR neither.

37. Segments of middle and lower leaves linear-lanceolate to thread-like, entire or nearly so; fruits not conspicuously flattened.

38. Petals (of outer flowers in umbels) 0.5–1 mm long; fruits not constricted into 2 parts; calyx tube of

outer flowers cup-shaped, slightly shorter to much longer than its lobes or teeth ____ **Ptilimnium**

38. Petals (of outer flowers in umbels) 1.6–2.1 mm long; fruits conspicuously constricted into 2 parts; calyx tube of outer flowers sub-rotate, much shorter than its lobes ____ **Bifora**

37. Segments of middle and lower leaves lanceolate, sharply toothed; fruits conspicuously flattened ____ **Eurytaenia**

AMMI BISHOP'S-WEED

An Old World genus of 3–4 species; *A. visnaga* (L.) Lam. has been cultivated since the time of the Assyrians for medical uses, especially in treating angina and asthma. (ancient Greek name)

Ammi majus L., (bigger, larger), GREATER AMMI, BISHOP'S-WEED. Erect, glabrous, branching annual 0.2–0.8 m tall; superficially resembling *Cicuta*; leaves ternately or pinnately compound or decompound, the ultimate segments narrowly lanceolate, serrate; inflorescences of compound umbels; involucral bracts numerous, usually pinnately divided, the segments filiform; rays 50–60; flowers white; fruits oblong, 1.5–2 mm long, 1 mm or less wide, the ribs narrow, acute, covering much less than half of the fruit surface. Waste areas, roadsides; Limestone and Tarrant cos., also Hamilton Co. (Stanford 1971); se and e TX w to nc TX and Edwards Plateau. Mar–Jun. Native of the Mediterranean. Cultivated for the cut-flower trade (Mabberley 1987).

AMMOSELINUM SAND-PARSLEY

Glabrous small annuals branching from base; leaves ternately or ternate-pinnately decompound, the ultimate segments linear to spatulate; flowers inconspicuous, white, in small compound umbels; fruit with prominent ribs.

A genus of 3 species of the North America and temperate South America. (Greek: *ammo*, sand, and *selinum*, parsley, presumably from the habitat)

1. Umbels in the axils of stem leaves, ± sessile; fruits 2.5–3 mm long, 1–1.5 mm wide; plants 4–5 (–12) cm tall ____ **A. butleri**

1. Umbels axillary or seemingly terminal, on peduncles (0.5–)1.5–4 cm long; fruits 3–6 mm long, to ca. 3 mm wide; plants 10–35 cm tall ____ **A. popei**

Ammoselinum butleri (Engelm. ex S. Watson) J.M. Coult. & Rose, (for George Dexter Butler, 1850–1910, lawyer, teacher, botanist, correspondent of George Engelmann, and one of the early collectors of this species), BUTLER'S SAND-PARSLEY. Fruits ovoid. Sandy or disturbed ground, fairly common but inconspicuous and overlooked; se and e TX w to e Rolling Plains and Edwards Plateau. Mar–Apr.

Ammoselinum popei Torr. & A. Gray, (for Clara Maria Pope, active 1760s–1838, British flower painter), PLAINS SAND-PARSLEY. Fruits oblong-ovoid. Sandy or gravelly ground; Grand Prairie s and w to w TX. Mar–May.

ANETHUM DILL

A monotypic genus of sw Asia. (From *Anethon*, ancient Greek name of dill, thought to come from *aithein*, blaze, in allusion to the pungent seeds)

Anethum graveolens L., (heavy-scented), DILL. Glabrous, strong-scented annual, 0.5–1.5 m tall; leaves pinnately decompound; rays 25 or more; petals yellow; fruits 3–4 mm long; similar to *Foeniculum* and reported to hybridize with it. Widely cultivated in TX and reseeds in open ground; Grayson Co. May–Summer. Native of sw Asia. This species has been cultivated since at least 400 BC.; the leaves and fruits are widely used in pickling and as a flavoring (Mabberley 1987). Duke (1985) indicated that insects exposed to the insecticide parathion alone suffered only 8% mortality; those exposed to the same doses of parathion plus d-carvone or other DILL components showed 99% mortality; he raised the question of whether certain naturally occurring plant products can synergistically react with pesticides to produce harmful effects in humans.

BERULA WATER-PARSNIP

A monotypic genus of wet areas of the n temperate zone and e and s Africa. (Latin name of some aquatic plant)

Berula erecta (Huds.) Coville, (erect, upright), STALKY BERULA, WATER-PARSNIP. Erect or reclining, glabrous perennial; stems 0.2–1 m tall; leaves once pinnately compound; leaflets toothed to deeply lobed; small bracts below umbellets linear to lanceolate; flowers white; calyx teeth minute, subulate; fruits oval to orbicular, 1.5–2 mm long. Wet areas; included based on citation of vegetational area 4 (Fig. 2) by Hatch et al. (1990); also Edwards Plateau, Plains Country, and Trans-Pecos. May–Nov. Reported to be extremely poisonous and capable of causing death in cattle (Lewis & Elvin-Lewis 1977).

BIFORA

A genus of 3 species of the Mediterranean to c Asia and North America; resembles *Ptilimnium*. (Possibly from Latin: *bi*, two, in reference to the two-parted fruit)

Bifora americana Benth. & Hook. f. ex S. Watson, (of America), PRAIRIE-BISHOP. Low, glabrous annual 25–75 cm tall; leaves ternate-pinnately decompound; flowers white; fruits subglobose and didymous, 2–3 mm long, 4–5 mm wide, the two nearly spherical segments separating from one another at maturity. Prairies, rocky slopes, and roadsides, limestone areas; se and e TX w to e Rolling Plains and Edwards Plateau. May–Jun.

BOWLESIA

A mainly South American genus of 15 species. (Named for William Bowles, 1705–1780, Irish writer on Spanish natural history)
REFERENCE: Mathias & Constance 1965.

Bowlesia incana Ruiz & Pav., (hoary, quite gray), HOARY BOWLESIA. Low, prostrate to suberect annual, ± gray with stellate pubescence; leaves opposite; leaf blades suborbicular in outline, 5–7-lobed, to 3 cm long and 4.5 cm wide, usually much smaller; petioles to 7 cm long; flowers minute, white or purplish; fruits sessile to subsessile, ovate to round, 1–2 mm long. Lawn weed, rarely collected but spreading rapidly in nc TX; Bell, Dallas, Denton, Johnson, Parker, and Tarrant cos., also Brown Co. (HPC); s and se TX n to nc TX and w to Edwards Plateau. Feb–Jun. Native of South America and possibly also s North America; the geographic origin of this species is problematic; it is found in South America and also in the sw U.S. and adjacent Mexico. It was collected in TX in 1828 by Berlandier, but Mathias and Constance (1965) concluded that its North American distribution is probably the result of naturalization from South America; however, Constance (1993) considered it native to California and the sw US.

BUPLEURUM THOROUGHWAX

A genus of ca. 180–190 species of Eurasia, n Africa, Canary Islands, arctic North America, and s Africa. (Greek: *bous*, an ox, and *pleuron*, a rib, referring to another plant)

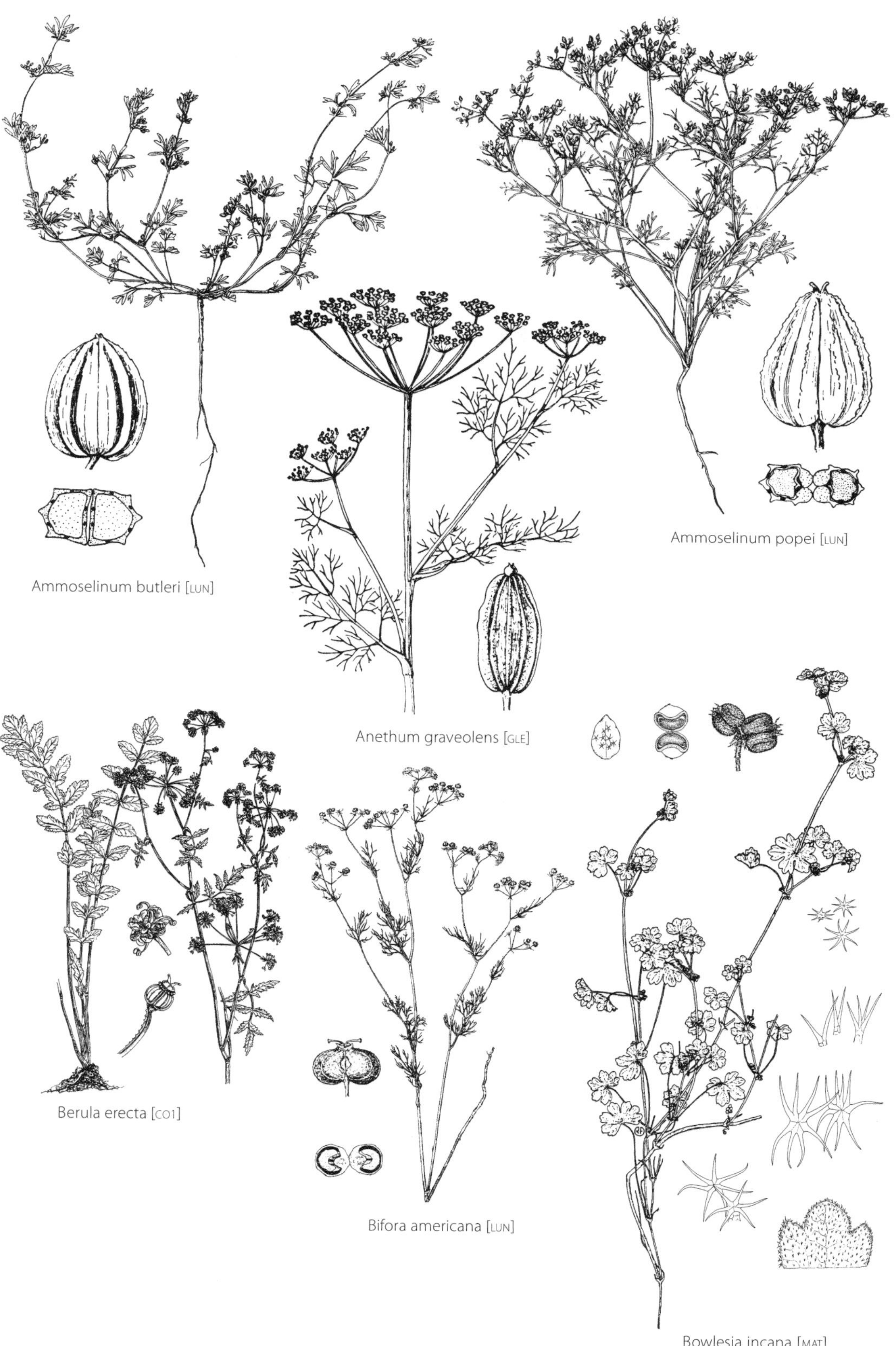

Ammoselinum butleri [LUN]

Ammoselinum popei [LUN]

Anethum graveolens [GLE]

Berula erecta [CO1]

Bifora americana [LUN]

Bowlesia incana [MAT]

Bupleurum rotundifolium L., (round-leaved), ROUND-LEAF THOROUGHWAX. Annual to 60 cm tall; leaves simple, entire; basal and lower leaves oblong- to obovate-lanceolate, subpetiolate to perfoliate basally; upper leaves ovate, perfoliate; rays of inflorescence 4–10; involucel (bractlets below umbellets) of 5–6 conspicuous, broadly ovate to obovate, acuminate bractlets; flowers yellow; fruits oblong-oval, 2.5–3 mm long, purplish brown, smooth. Open, weedy areas; Dallas, Erath, Grayson, Parker, and Tarrant cos.; se and e TX w to nc TX. Apr–Jun. Native of the Mediterranean region.

Bupleurum lancifolium Hornem., (lance-leaved), cited by Hatch et al. (1990) for vegetational area 4 (Fig. 2), apparently occurs only to the se of nc TX. It can be distinguished by its ovoid-globose, tuberculate to rugose fruits and by having only 2–5 rays. A Mediterranean species, possibly introduced in bird-seed (Correll & Johnston 1970).

CENTELLA

A mainly s African genus of 40 species with 1 species widespread. (Possibly diminutive of *centem*, a hundred, from the rounded, coin-like leaf blades of some species)
REFERENCES: Fernald 1940a; Schubert & van Wyk 1995.

Centella erecta (L.f.) Fernald, (upright, erect), SPADELEAF. Acaulescent perennial with creeping rhizomes; leaves simple; leaf blades ovate to broadly ovate, basally cordate to truncate, petiolate; umbels simple; flowers white or rose-tinged; fruits with 3 primary and 2 secondary ribs, reticulate. Edges of streams and other wet places; Tarrant Co.; in parts of e 1/2 of TX. May–Sep. [*C. repanda* (pers.) Small] In Texas this species has long gone under the name *C. asiatica* (L.) Urb. (e.g., Correll & Johnston 1970; Hatch et al. 1990; Jones et al. 1997). However, according to J. Kartesz (pers. comm. 1998), *C. asiatica* is a similar but distinct species which is introduced in the U.S. only in the Pacific Northwest and Hawaii. While *C. asiatica* is sometimes eaten or taken as a tea, it is also said to be used as a fish poison and to contain potentially carcinogenic glucosides and an alkaloid, hydrocotylin; it has been used in folk medicine to treat leprosy and apparently shows activity against the bacterium responsible for tuberculosis (Duke 1985; Mabberley 1997); because *C. erecta* is similar enough to have been lumped with *C. asiatica* by many authorities, *C. erecta* should also be assumed to be toxic.

CHAEROPHYLLUM CHERVIL, WILD CHERVIL

A n temperate genus of ca. 35 species. (Greek: *chaero*, delight or rejoice, and *phyllon*, a leaf, alluding to the agreeable odor of the foliage)

Chaerophyllum tainturieri Hook., (for L.F. Tainturier des Essarts, who sent plants from Louisiana to Sir William Hooker from 1824–1836). Moderately to densely pubescent, erect annual 15–90 cm tall; leaves ternate-pinnately dissected; flowers white; fruits narrowly oblong, 4.5–7 mm long, slightly beaked or narrowed toward the apex. Stream bottoms, woods, roadsides, and waste areas. Mid-Mar–Apr.

1. Ovary and fruit pubescent ____ var. **dasycarpum**
1. Ovary and fruit glabrous or nearly so ____ var. **tainturieri**

var. **dasycarpum** Hook. ex S. Watson, (thick-fruited), HAIRY-FRUIT CHERVIL. Widespread in TX but more commonly in e 1/2.

var. **tainturieri**. Widespread in TX but more commonly in e 1/2.

CICUTA WATER-HEMLOCK

A n temperate genus of 8 species; all are extremely toxic. (Latin: *cicuta*, name of *Conium*

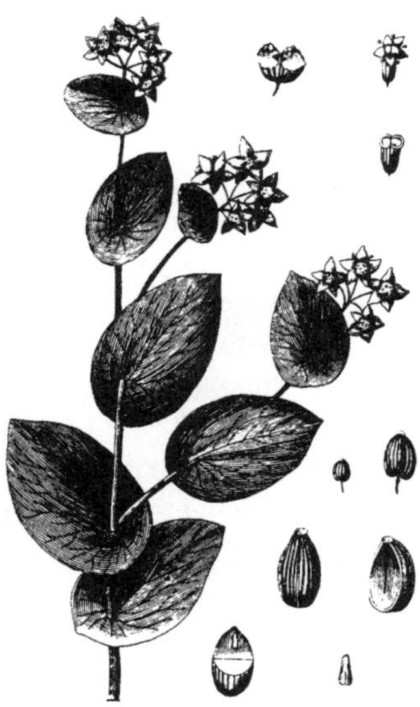

VAR. DASYCARPUM [USH]

VAR. TAINTURIERI [BB2]

Bupleurum rotundifolium [LAM]

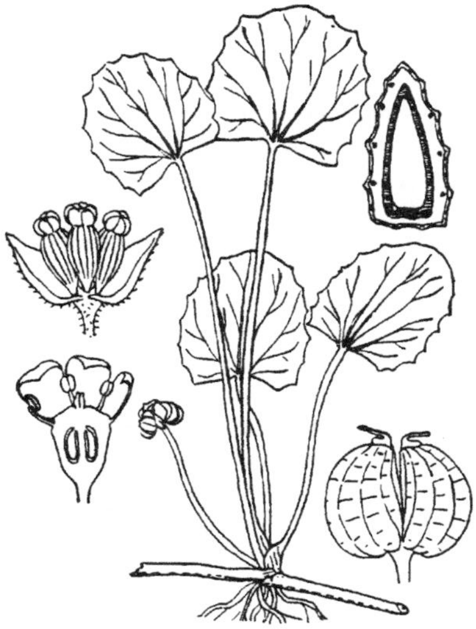

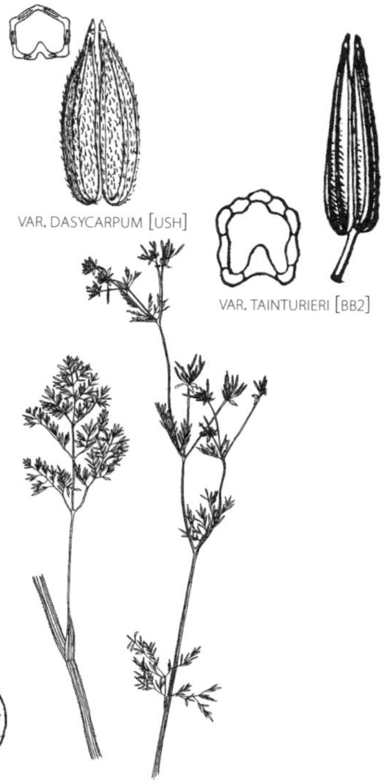

Chaerophyllum tainturieri [BT3]

Centella erecta [BR2]

Cicuta maculata [REE]

Conium maculatum [REE]

maculatum—poison-hemlock, a deadly herb native to the Old World; the name was transferred to this genus)
REFERENCES: Mathias & Constance 1942; Mulligan 1980.

Cicuta maculata L., (spotted), COMMON WATER-HEMLOCK, SPOTTED COWBANE, COWBANE, BEAVER-POISON, SPOTTED WATER-HEMLOCK, MUSQUATROOT, MUSKRAT-WEED. Glabrous and glaucous, usually tall perennial to 2 m tall (occasionally flowering when much smaller); roots fleshy and fascicled; leaves 2–3 times pinnately compound; flowers white; fruits oval to orbicular, 2–4 mm long, with ribs prominent. Low or wet ground; Bell, Dallas, Henderson, and Lamar cos., also Collin Co. (G. Diggs, pers. obs.); se and e TX w to Rolling Plains and Edwards Plateau. Late May–Jul. [*C. mexicana* J.M. Coult. & Rose] **Virulently poisonous** due to cicutoxin, a resinoid which affects the nervous system; a single bite is reported to be sufficient to kill a human; children have even been poisoned by whistles made from the stems; in extreme cases death may occur in less than an hour; some authorities consider this to be the most poisonous plant of the North Temperate zone; fatalities are also known in browsing livestock (Stephens 1980; Turner & Szczawinski 1991; Hardin & Brownie 1993). ☠

CONIUM POISON-HEMLOCK

A temperate Eurasian and s African genus of 6 species; ☠ some are very poisonous. (Greek: *coneion*, the hemlock, by which Socrates and various criminals were put to death at Athens)

Conium maculatum L., POISON-HEMLOCK, POISON-PARSLEY. Glabrous biennial to 3 m tall with distinctive unpleasant odor; root a carrot- or parsnip-like taproot; stems hollow, usually purplish-spotted; leaves pinnately decompound; umbels compound, many-flowered; flowers white; fruits broadly ovoid, 2–4 mm long. Stream bottoms; included based on citation of vegetational area 4 (Fig. 2) by Hatch et al. (1990), apparently from Travis Co. just beyond s margin of nc TX s and sw through the s 1/2 of TX. May–Jun. Native of Europe and Asia. All parts of the plant are **extremely poisonous** to humans and livestock due to coniine and other toxic alkaloids; often fatal if eaten due to paralysis of respiratory muscles. The leaves can be mistaken for parsley, the roots for parsnips, or the seeds for anise. It has been recognized as poisonous since ancient times and is supposedly the source of the poison used to kill the Greek philosopher Socrates in 399 B.C.; it was apparently widely used as a means of execution by the early Greeks (Muenscher 1951; Kingsbury 1964, 1965; Hardin & Arena 1974; Mabberley 1987; Schmutz & Hamilton 1979; Turner & Szczawinski 1991). Significant skin contact with the foliage can result in symptoms of toxicity such as nausea or blurred vision. ☠

CORIANDRUM CORIANDER

A genus of 3 species native to sw Asia. (From the Greek name, *koriandron*, from *koris*, bug, alluding to the aromatic leaves)

Coriandrum sativum L., (cultivated or sown), CILANTRO, CORIANDER, CHINESE-PARSLEY. Low glabrous annual 20–70 cm tall; basal leaves simple and lobed or pinnate, much less finely divided than stem leaves; stem leaves pinnately dissected; outer flowers with rather large petals, some bilobed; inner flowers with smaller petals; petals white, often tinged pink; fruits orbicular, (1.5–)2.5–3(–5) mm in diam. Escapes cultivation, waste places; Dallas and McLennan cos., also Grayson Co. (G. Diggs, pers. obs.); scattered in TX. Apr–May. Native of Mediterranean region. Long cultivated for the fruits (e.g., seeds were found in the tomb of Tutankhamun—Hepper 1990); the foliage of this plant provides the flavoring so widely used in Mexican salsas and other Latin American cuisines.

CRYPTOTAENIA WILD CHERVIL, HONEWORT

A genus 6 species of the n temperate region and tropical African mountains. (Greek: *crypto*, hidden, and *taenia*, band or ribbon, referring to the concealed oil tubes)

Cryptotaenia canadensis (L.) DC., (of Canada). Largely glabrous perennial to 1 m tall; leaflets 3 or 5, thin, broadly lanceolate, sharply double-toothed, sometimes lobed; rays of umbel few; pedicels very unequal; flowers white; fruits 3.5–8 mm broad. Woods; collected by Reverchon in Dallas (Mahler 1988); recent collections in Dallas, Grayson, and Lamar cos.; also e TX. May.

CYCLOSPERMUM SLIM-LOBE CELERY

A monotypic genus native to warm areas of the Americas; related to the genus *Apium* which includes *A. graveolens* L. (CELERY). (Greek: *cyclo*, circle, and *spermum*, seed)

Cyclospermum leptophyllum (Pers.) Sprague ex Britton & P. Wilson, (slender-leaved), SLIM-LOBE CELERY. Low glabrous annual 5–60 cm tall; leaves highly variable, the lowest with wide segments, the uppermost with almost thread-like segments; umbels axillary and terminal, sessile or short-peduncled, typically with 3–5 rays (= main branches of umbel); flowers minute, white; fruits ovoid, 1–2(–3) mm long. Ditches or low areas, disturbed sites; s TX nw to nc TX and w to Trans-Pecos. Apr–Jun, sporadically later. [*Apium leptophyllum* (Pers.) F. Müll. ex Benth.] The name of this genus is sometimes spelled *Ciclospermum*; however, the accepted spelling is *Cyclospermum* (Greuter et al. 1993). Native from s US to South America. Reported to be poisonous (Burlage 1968). ☠

CYMOPTERUS WAVEWING

A w North American genus of 32 species including several eaten by Native Americans. (Greek: *cyma*, a wave, and *pteron*, a wing, referring to the often undulate wings)
REFERENCE: Mathias 1930.

Cymopterus macrorhizus Buckley, (large-rooted), BIG-ROOT WAVEWING. Glabrous, glaucous, dwarf perennial to 35 cm tall; from a thick, soft-woody, fusiform to subglobose root; leaves few, crowded at base of plant, pinnate or bipinnate; flowers pink or nearly white; anthers purple-black, prominent; fruits ovoid to ovoid-oblong, 4–9 mm long, 3–8 mm wide, the wings obvious. Rocky limestone prairies; Blackland Prairie and Edwards Plateau w through Plains country; endemic to TX and OK. Mar–Apr.

CYNOSCIADIUM

A North American genus of 2 species. (Greek: *kyon* or *cyno*, dog, and *skiados*, an umbel)

Cynosciadium digitatum DC., (finger- or hand-like), FINGER DOGSHADE. Glabrous annual; stems to 65 cm tall; basal leaves linear-lanceolate, entire; stem leaves palmately compound or parted, with 3–5 long narrow segments; flowers rather few, not conspicuous, white; fruits ovoid, 2–3 mm long. Wet places; Hunt and Lamar cos.; also Coastal Prairies. May.

DAUCOSMA

A monotypic genus endemic to TX. (Possibly from *Daucus*, the genus including carrots, and Greek: *osma*, smell or odor)

Daucosma laciniatum Engelm. & A. Gray, (laciniate, torn), MEADOW DAUCOSMA. Annual to 1.2 m tall; crushed leaves very aromatic; umbels compound, the rays 2–5 cm long; peduncles to 10 cm long; involucres and involucels equaling or longer than rays and pedicels; flowers white; fruits ovoid, 3–4 mm long, glabrous. Low disturbed areas; Bell Co. (Mathias & Constance 1961); mainly s TX and Edwards Plateau; endemic to TX. May–Aug.

DAUCUS CARROT

Hispid-pubescent annuals or biennials; leaf blades finely cut, the ultimate divisions narrowly lanceolate; umbels compound, the outer rays becoming longer in age, the umbels then hollow in center; fruits bristly.

A genus of ca. 22 species of Europe, the Mediterranean, sw and c Asia, tropical Africa, Australia, New Zealand, and the Americas. (From *daukos*, the classical Greek name)

1. Petals white to yellowish, except those of central flower in umbel usually pink or purple; rays 15–50 mm long in flower, up to 75 mm in fruit; involucral bracts shorter than umbel in flower, divided into filiform divisions; bristles of fruit without prominent apical barbs ____ **D. carota**
1. Petals white (withering yellowish), including those of central flower in umbel; rays 3–30 mm long in flower, up to 40 mm in fruit; involucral bracts ± equal to or exceeding the umbel in flower, divided into linear or lanceolate divisions; bristles of fruit with apical barbs ____ **D. pusillus**

Daucus carota L., (carrot), WILD CARROT, QUEEN-ANNE'S-LACE. Biennial 0.4–1.5(–2) m tall; involucral bracts usually reflexed; fruits 3–4 mm long. Roadsides or railroads, disturbed areas; Ellis, Grayson, and Tarrant cos., also Lamar (Carr 1994) and Dallas (Moss Park—Ruth May, pers. comm.) cos.; still spreading rapidly in e part of TX w to nc TX and Edwards Plateau. First observed in Tarrant Co. (in Keller) in 1996. May–Jul. Native of Europe. This species is an aggressive invader capable of crowding out native vegetation. Handling the wet foliage of cultivated CARROT (*D. carota* subsp. *sativus* (Hoffm.) Arcang.) can apparently cause skin irritation in sensitive individuals; photosensitization may be involved (Tampion 1977; Duke 1985); care should thus be taken with both cultivated CARROT and WILD CARROT.

Daucus pusillus Michx., (very small), RATTLESNAKE-WEED, SOUTHWESTERN CARROT, SEEDTICKS. Annual to ca. 0.9 m tall; fruits 3–5 mm long. Stream banks, roadsides, and waste areas; throughout TX. Apr–Jun.

ERYNGIUM ERYNGO

Plants creeping to erect; essentially glabrous; leaves entire to pinnately or palmately lobed to divided, often spinose; inflorescences conspicuously head-like, globose to cylindrical, sometimes with a coma of bracts from apex of head; flowers white to purple; fruits variously covered with scales or tubercles.

A genus of 230–250 species of tropical and temperate areas of the world; a number are edible or cultivated as ornamentals. (From *eyringion*, the Greek name for *E. campestre* L.)
REFERENCES: Bell 1963; Mathias & Constance 1941b.

1. Leaves parallel-veined, broadly linear, completely unlobed and undivided, the margin entire except for remote bristles, to 1 m long; plants monocot-like in appearance ____ **E. yuccifolium**
1. Leaves reticulate-veined, not linear, broader in shape, unlobed to usually palmately lobed or divided, to only 9 cm long, often much shorter; plants not monocot-like.
   2. Stem leaves conspicuously sharp-spinulose, palmately lobed or divided; bracts (including teeth) subtending heads > 2 mm wide (often much greater), usually spinulose; throughout nc TX.
      3. Heads 20–35 mm long, purplish to reddish; coma of 4–8 spinescent bracts 1–2 cm long projecting from apex of heads ____ **E. leavenworthii**
      3. Heads 8–15 mm long, bluish to purplish; coma bracts absent or inconspicuous, nearly entire.
         4. Plants erect, the stems solitary, branched above; basal leaves serrate or dentate, petioled; fruits 1–2 mm long; plants of limestone soils ____ **E. hookeri**

Coriandrum sativum [BL1]

Cryptotaenia canadensis [BB1]

Cyclospermum leptophyllum [CO1]

Cymopterus macrorhizus [LUN]

Cynosciadium digitatum [LUN]

Daucosma laciniatum [LUN]

Daucus carota [BL1]

Daucus pusillus [BT3, USH]

4. Plants low, much-branched; basal leaves deeply palmately parted, sessile or subsessile; fruits 2.5–3 mm long; plants of sandy soils ____ **E. diffusum**

2. Stem leaves not sharp-spinulose or only slightly so, unlobed to palmately lobed or divided; bracts (including teeth if present) subtending heads very narrow, usually < 2 mm wide, entire or with 3–5 spinulose teeth; extreme e portion of nc TX.

5. Plants erect; bracts subtending heads usually with 3–5 teeth (rarely entire); stem leaves not palmately lobed, usually serrate to laciniate or subentire; heads 5–15 mm wide, few–numerous in a cymose inflorescence terminating the stem ____ **E. integrifolium**

5. Plants prostrate or weakly ascending; bracts subtending heads without teeth or nearly so; stem leaves entire to irregularly toothed, at least some palmately lobed; heads 2–4 mm wide, usually solitary in the leaf axils ____ **E. prostratum**

Eryngium diffusum Torr., (diffuse, spreading), BUSHY ERYNGO, DIFFUSE ERYNGO. Annual or biennial 10–40 cm tall; prostrate to erect; stem leaves deeply palmately parted, the divisions spinulose-dentate or lobed; heads bluish; coma of bracts absent from apex of heads; fruits 2.5–3 mm long. Sandy soils; Montague and Shackelford cos.; nearly throughout TX. May–Aug.

Eryngium hookeri Walp., (for William Jackson Hooker, 1785–1865, director of Kew Gardens), HOOKER'S ERYNGO. Erect annual 30–60 cm tall; lower stem leaves nearly sessile, lanceolate, laciniately toothed, spinulose; upper stem leaves ovate, palmately divided into 5–7 oblong, laciniate or pinnatifid lobes; heads purplish; coma of a few bracts at apex of heads or coma absent; fruits 1–2 mm long. Moist limestone soils; Bell, Dallas, Denton, Ellis, Grayson, and McLennan cos.; Blackland and Coastal prairies. Jul–Sep.

Eryngium integrifolium Walter, (entire-leaved), SIMPLE-LEAF ERYNGO. Erect perennial 30–80 cm tall; stem leaves simple; heads ovoid to globose, 5–15 mm in diam., bluish, subtended by 6–10 linear bracts, the bracts 10–20 mm long; coma of bracts absent from apex of heads. Moist wooded areas; included based on citation of vegetational area 4 (Fig. 2) by Hatch et al. (1990); se and e TX w to at least Henderson Co. Aug–Oct.

Eryngium leavenworthii Torr. & A. Gray, (for its discoverer, Melines Conklin Leavenworth, 1796–1862, s U.S. botanist, explorer, and army surgeon), LEAVENWORTH'S ERYNGO. Erect annual to 100 cm tall; vegetative tissues often purplish; upper stem leaves sessile, orbicular, deeply palmately parted, the divisions pinnatifid, spiny; heads purplish to reddish; coma of 4–8 conspicuous spinescent bracts projecting from apex of heads; sepals spinescent; stamens bluish; fruits 2–4 mm long, conspicuously covered with white scales. Prairies and weedy areas; nearly throughout TX. Jul–Sep. This strikingly colored plant is often common and conspicuous. ▣/89

Eryngium prostratum Nutt. ex DC., (prostrate, flat to the ground). Creeping, prostrate, mat-forming, or ascending perennial; stems 10–70 cm long, slender, 0.5–1 mm in diam.; basal leaves 2–4(–5.5) cm long, entire to irregularly toothed, often palmately lobed; stem leaves reduced; heads very small, to 5–9 mm long, short cylindric, often pale to dark blue or purple-blue, subtended by 5–10 linear bracts to 12 mm long; coma of bracts absent from apex of heads. Lake margins and other moist areas; Henderson, Limestone, and Milam cos. on e margin of nc TX, also Lamar Co. in Red River drainage; mostly se and e TX. May–Sep.

Eryngium yuccifolium Michx., (with leaves like *Yucca*), RATTLESNAKE-MASTER, BUTTON SNAKE-ROOT, BRISTLE-LEAF ERYNGO. Plant glabrous; perennial from tuberous root; stems 30–180 cm tall; basal leaves broadly linear, to 1 m long; stem leaves like the basal but reduced upward; heads globose-ovoid, 10–25 mm in diam.; coma of bracts absent from apex of heads. Prairies; Dallas, Denton, and Grayson cos.; also Hunt, Kaufman, and Lamar cos. (R. O'Kennon, pers. obs.); se and e TX w to nc TX. May–Aug. We are following McGregor (1986) and Jones et al. (1997) in lumping var. *synchaetum*. [*E. yuccifolium* var. *synchaetum* A. Gray ex J.M. Coult. & Rose, *E. synchaetum*

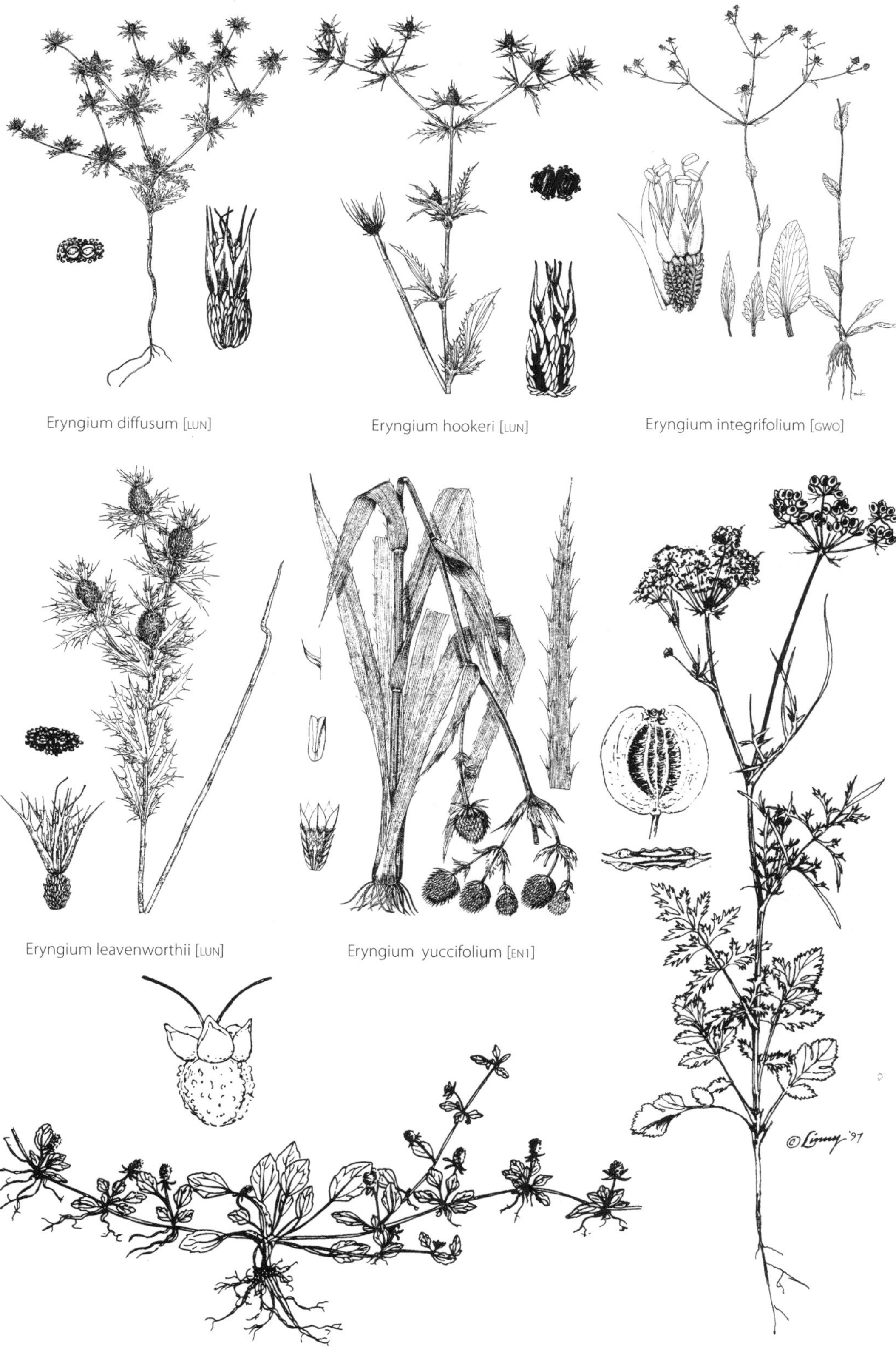

Eryngium diffusum [LUN]

Eryngium hookeri [LUN]

Eryngium integrifolium [GWO]

Eryngium leavenworthii [LUN]

Eryngium yuccifolium [EN1]

Eryngium prostratum [BB2, CO1]

Eurytaenia texana [HEA, LUN]

(A. Gray ex J.M. Coult. & Rose) J.M. Coult. & Rose] The common names refer to Native American and early settler use of poultices from the roots to treat snakebite (Tveten & Tveten 1993). This species is an indicator of native prairie. ▣/89

EURYTAENIA TEXAS SPREADWING

A genus of 2 species native to TX and OK. (Greek: *eury*, wide, and *taeni*, ribbon or band)

Eurytaenia texana Torr. & A. Gray, (of Texas), TEXAS SPREADWING. Largely glabrous annual 30–120 cm tall; basal leaves lobed or pinnatifid; stem leaves pinnately or ternate-pinnately dissected; inflorescence of loose compound umbels; rays 8–16; involucres usually of ca. 5 three-cleft bracts 5–10 mm long; involucel bractlets similar to bracts or entire; pedicels 5–8 mm long; petals white; fruits 4–6 mm long and broad. Rocky or sandy ground; Montague and Tarrant cos.; widespread in TX except far w part; endemic to TX and OK. May–Jun.

FOENICULUM FENNEL

A mainly Asian genus of 4–5 species, widely naturalized; some authorities consider the genus monotypic (e.g., Mabberley 1987). (Latin: *foenum*, hay, presumably from the abundant straw-like leaf segments; classical name for fennel)

Foeniculum vulgare Mill., (common), COMMON FENNEL. Glabrous strong-scented biennial or perennial; stems 0.9–2+ m tall; leaves pinnately decompound; rays 15–40; petals yellow; fruits 3–4 mm long. Widely cultivated and escapes, low areas; included based on citation of vegetational area 4 (Fig. 2) by Hatch et al. (1990); also s TX and Edwards Plateau. Spring–summer. Native of the Mediterranean region. Cultivated for its fruits and essential oils used as flavoring and medicine since the 13th century BC. (Mabberley 1987); however, distilled oil of fennel, even in small quantities can be toxic and can cause symptoms including pulmonary edema, respiratory problems, and seizures (Duke 1985).

HYDROCOTYLE WATER-PENNYWORT, NICKELS-AND-DIMES

Ours glabrous creeping perennials of damp ground or shallow water, rooting at the nodes; leaves peltate or not so; leaf blades shiny above, sometimes ± round; umbels simple, branched or verticillate; flowers inconspicuous, yellow-green or white; fruits without secondary ribs, not reticulate.

A cosmopolitan genus of ca. 130 species often with peltate leaves; some are cultivated as ornamental ground covers. Some workers have suggested that if Araliaceae and Apiaceae are recognized as separate families (as done here), then the Hydrocotylaceae should be recognized as well (e.g., Thorne 1992); however, the traditional approach of including *Hydrocotyle* in the Apiaceae is followed here. (Greek: *hydor*, water, and *kotyle*, cup, alluding to the hollows in the center of the leaves of some species)

1. Leaves peltate (= petiole attached to middle of lower surface of leaf blade); leaf blades shallowly lobed or toothed.
   2. Flowers in a simple umbel (= all attached at the same point) __________ **H. umbellata**
   2. Flowers in whorls (= verticils) forming an interrupted spike or spike-like raceme or rarely a branched spike __________ **H. verticillata**
1. Leaves not peltate, the petiole attached at base of a deep, narrow notch; leaf blades 5–6-lobed to about the middle __________ **H. ranunculoides**

Hydrocotyle ranunculoides L.f., (resembling *Ranunculus*—buttercup), FLOATING WATER-PENNYWORT. Leaf blades roundish reniform, to 8 cm long and wide; petioles to ca. 35 cm long; umbels simple, 5–10-flowered. Wet places; Dallas and Henderson cos.; e TX w to nc TX. Apr–Oct.

Hydrocotyle umbellata L., (with umbels), UMBRELLA WATER-PENNYWORT, OMBILIGO DE VENUS. Leaf blades to 7.5 cm in diam.; petioles to ca. 40 cm long; umbels many-flowered. Wet places; Burnet, Denton and Milam cos.; much of e 1/2 of TX. May–Oct.

Hydrocotyle verticillata Thunb., (verticillate, whorled). Leaf blades to ca. 6 cm in diam., with 8–14 shallow lobes; petioles to 35 cm long; inflorescences interrupted, with 2–15-flowered whorls. Wet places. May–Oct.

1. Fruits sessile or subsessile; inflorescences often bifurcate (= 2-branched) ____ var. **verticillata**
1. Fruits pedicellate, the pedicels 1–10 mm long; inflorescences rarely bifurcate ____ var. **triradiata**

var. **triradiata** (A. Rich.) Fernald, (three-rayed). Inflorescences with few 4–15-flowered verticils. Dallas, Grayson, Parker, and Tarrant cos.; se and e TX w to nc TX and Edwards Plateau, also Trans-Pecos.

var. **verticillata**. WHORLED WATER-PENNYWORT. Inflorescences with 2–7 few-flowered verticils. Bell, Dallas, and Tarrant cos.; se and e TX w to nc TX and Edwards Plateau.

LIMNOSCIADIUM

A genus of 2 species of the sc United States; sometimes treated as part of the genus *Cynosciadium*. (Greek: *limne*, a marsh, and *skiados*, an umbel)
REFERENCE: Mathias & Constance 1941a.

Limnosciadium pinnatum (DC.) Mathias & Constance, (pinnate, feather-like), ARKANSAS DOGSHADE. Glabrous erect or ascending annual 10–80 cm tall; stem leaves pinnately compound with 2–9 linear to linear-lanceolate divisions 3–10 cm long or the lowest and uppermost leaves entire; flowers small but numerous and rather showy; petals white; calyx teeth in fruit shorter than the stylopodium, 0.5 mm or less long; fruits oblong-oval, 2–4 mm long, 1–2 mm long. Ditches or low ground; se and e TX w to East Cross Timbers, also Edwards Plateau. May–Jun. [*Cynosciadium pinnatum* DC.]

Limnosciadium pumilum (DC.) Mathias & Constance, (dwarf), is cited by Hatch et al. (1990) for vegetational area 4 (Fig. 2) but apparently occurs only to the s of nc TX. It can be distinguished by its low and diffuse habit (plant 5–40 cm high or long), the oval to orbicular fruits 2–3 mm long and 2 mm wide, and its calyx teeth in fruit equal in length to the stylopodium (to 1.5 mm long).

LOMATIUM

A w North American genus of 74 species. (Greek, *lomation*, a little border, in reference to the winged fruit)
REFERENCE: Schlessman 1984.

Lomatium foeniculaceum (Nutt.) J.M. Coult. & Rose subsp. **daucifolium** (Torr. & A. Gray) Theobald, (sp.: like *Foeniculum*—fennel; subsp.: leaves like *Daucus*, carrot), CARROT-LEAF LOMATIUM. Pubescent, dwarf, acaulescent perennial to 45 cm tall, from a thick taproot; leaf segments linear or linear-lanceolate; petals yellow; fruits ovate-oblong, 6–9 mm long, 3–6 mm broad. Prairies, in calcareous clay; Collin, Dallas, Ellis, Grayson, and Montague cos.; also Nolan Co. to the w of nc TX. Mar–Apr. [*Lomatium daucifolium* (Nutt. ex Torr. & A. Gray) J.M. Coult. & Rose] ▣/96

OSMORHIZA SWEET-CICELY, ANISEROOT

A genus of 10 species native to the Americas and e Asia. (Greek: *osme*, a scent, and *rhiza*, a root, in reference to the odor of the root)
REFERENCES: Constance & Shan 1948; Lowry & Jones 1984.

Osmorhiza longistylis (Torr.) DC., (long-styled), ANISEROOT, LONG-STYLE SWEETROOT. Pubescent perennial 0.6–1 m tall; leaflets thin, sharply toothed; rays of umbel few; petals white; fruits 15–20 mm long, with some hairs on the ribs, terminated by appendages 2–3(–6) mm long. Woods, along streams; Dallas, Grayson, and Tarrant cos., also Collin Co. (R. O'Kennon, pers. obs.); apparently in TX only in the nc part of the state.

OXYPOLIS HOG-FENNEL, COWBANE

Erect perennials; leaves reduced to phyllodes or once pinnately compound; involucre of linear to lanceolate bracts; petals white; calyx teeth present; fruits with lateral wings.

A North American genus of 7 species. (Greek: *oxys*, sharp, and *polios*, white)
REFERENCE: Tucker et al. 1983.

1. Leaves reduced to elongate, linear, entire, hollow, septate phyllodes; leaflets absent (leaves rush-like in appearance); principal rays of umbels 6–14(–17) ____ **O. filiformis**
1. Leaves once pinnately compound, with distinct leaflets; principal rays of umbels 15–45 ____ **O. rigidor**

Oxypolis filiformis (Walter) Britton, (thread-like), LEAF-LESS COWBANE. Stems to 1.4 m tall; the reduced leaves (phyllodes) 20–60 cm long are unique among nc TX Apiaceae; fruits 5–8 mm long, 3–5 mm wide. Wet areas; included based on citation of vegetational area 4 (Fig. 2) by Hatch et al. (1990); mainly e TX. Jul–Sep.

Oxypolis rigidor (L.) Raf., (rigid, stiff), COWBANE, WATER DROPWORT, WATER DROPWORT COWBANE. Stems 0.6–1.5 m tall; leaflets 5–9, lanceolate to linear, toothed or entire; fruits 4–7 mm long, 2.5–4 mm wide. Along streams, wet areas; included based on citation of vegetational area 4 (Fig. 2) by Hatch et al. (1990); mainly e TX. Aug–Oct. Reported to be poisonous to cattle (Burlage 1968).

PETROSELINUM PARSLEY

A genus of 2 species of Europe and the Mediterranean region. (Greek: *petros*, rock, and genus name *Selinum* from *selinon*, celery)

Petroselinum crispum (Mill.) Nyman ex A.W. Hill, (crisped, curled), PARSLEY, GARDEN PARSLEY, CURLY PARSLEY. Glabrous, erect, pungent-scented biennial; stems 0.2–1.3 m tall; leaves ternately-pinnately or pinnately decompound, 3–25 cm long, the ultimate segments 2–5 mm wide or wider and divided into narrow lobes (usually < 5 mm wide), flat or curled and crisped; flowers yellow or greenish yellow. Cultivated and possibly escapes in weedy areas; included based on citation of vegetational area 4 (Fig. 2) by Hatch et al. (1990). Summer. Native of Europe and w Asia. Extensive contact has been reported to cause skin inflammation (Duke 1985).

POLYTAENIA PRAIRIE-PARSLEY, PRAIRIE-PARSNIP

A North American genus of 1–2 species. (Greek: *poly*, many, and *taenia*, a ribbon or band, which refers to the numerous oil tubes in the fruits)

Polytaenia nuttallii DC., (for Sir Thomas Nuttall, 1786–1859, English-American botanist), PRAIRIE-PARSLEY, TEXAS-PARSLEY, WILD DILL. Stout glabrous biennial to 1 m tall; leaves bipinnately or ternate-pinnately compound; leaflets large, crenate to incised or lobed, ovate to oblong; flowers yellow; fruits 5–11 mm long, 4–7 mm wide, with wings. Common and conspicuous in prairies and open woods, various soils; se and e TX w to nc TX, also Edwards Plateau. Apr–May. While a number of authors recognize two species in this genus (e.g., Kartesz 1994; Jones et al. 1997), because of apparent overlap and inconsistancy in fruit characters and because no means of distinguishing flowering material exist, we are following Shinners (1958a) and Mahler (1988) in

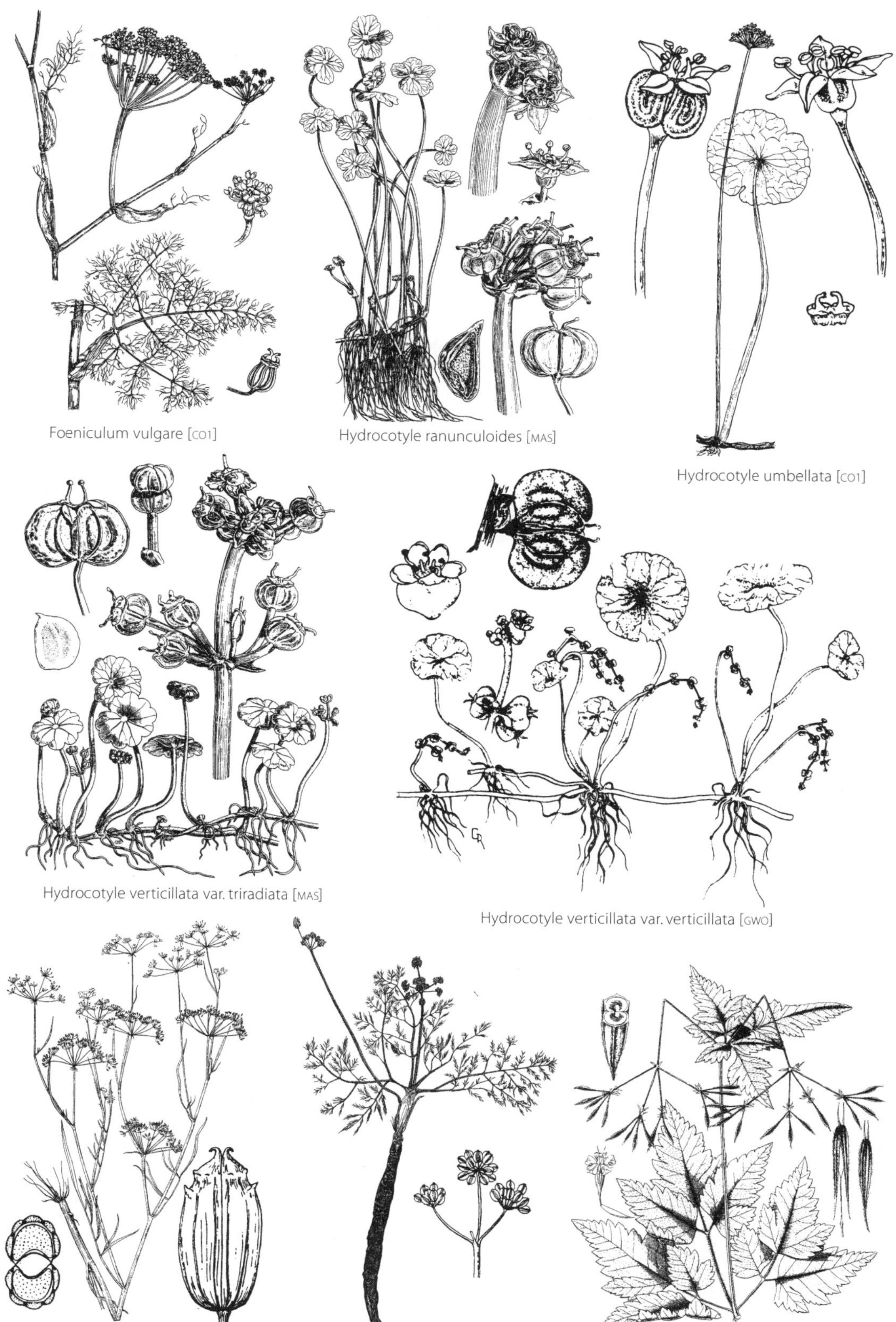

Foeniculum vulgare [CO1]

Hydrocotyle ranunculoides [MAS]

Hydrocotyle umbellata [CO1]

Hydrocotyle verticillata var. triradiata [MAS]

Hydrocotyle verticillata var. verticillata [GWO]

Limnosciadium pinnatum [LUN]

Lomatium foeniculaceum subsp. daucifolium [BB2, BT3]

Osmorhiza longistylis [HO1]

including [*P. texana* (J.M. Coult. & Rose) Mathias & Constance, *P. nuttallii* var. *texana* J.M. Coult. & Rose] in *P. nuttallii*. Mathias and Constance (1970) separated the 2 as follows:

1. Fruits 5–11 mm long, 4–7 mm broad, the lateral wings narrower and thicker than the body; oil several in the tubes indistinct, intervals _______ *P. nuttallii*
1. Fruits 9–11 mm long, 6–7 mm broad, the lateral wings broader and thinner than the body; oil tubes distinct, solitary in the intervals _______ *P. texana*

PTILIMNIUM MOCK BISHOP'S-WEED, MOCK BISHOP

Rather low, glabrous annuals of sandy or boggy ground; leaves pinnately decompound with the ultimate divisions filiform; petals white, rarely pink.

A genus of 5 species endemic to e North America. (Greek: *ptilon*, feather or down, and *limne*, mud, in allusion to the finely divided leaves and the habitat)
REFERENCE: Easterly 1957.

1. Upper stem leaves with petioles 1–7 mm long (measure to attachment of 1st leaf segment), the blades 7–15 times as long as the petioles; leaf segments not crowded; plants smaller, to 85 cm tall; styles 0.2–1.5 mm long.
   2. Upper leaves with 5–9 primary segments, these mostly again divided, the terminal primary segment 0.5–2 cm long, shorter or slightly longer than the rachis, the primary segments usually 3 per node on the rachis; involucral bracts usually branched; styles not strongly recurved, 0.2–0.5 mm long _______ **P. capillaceum**
   2. Upper leaves with 3–5 or rarely 7 primary segments, these all or mostly undivided, the terminal one 1–5 cm long, much longer than the rachis, the primary segments usually 2 per node on the rachis; involucral bracts usually entire; styles strongly recurved, 0.5–1.5 mm long _______ **P. nuttallii**
1. Upper stem leaves with petioles 6–25 mm long, the blades less than 4 times as long as the petioles; leaf segments crowded; plants robust, 80–160 cm tall; styles ca. 1–3 mm long _______ **P. costatum**

Ptilimnium capillaceum (Michx.) Raf., (with fine hairs), THREAD-LEAF MOCK BISHOP'S-WEED. Plant 10–85 cm tall; flowers relatively few and small, not very showy; fruits broadly ovoid, 1.5–3 mm long. Wet places; Bell, Henderson, and Milam cos., once collected in Tarrant Co. (Mahler 1988); se and e TX w to nc TX, also Edwards Plateau. Late May–Aug.

Ptilimnium costatum (Elliott) Raf., (ribbed). Plant with few, widely spaced leaves; involucral bracts usually entire; flowers numerous and prominent; fruits ovoid, 2–4 mm long. Wet places; Grayson and Henderson cos.; e TX w to e part of nc TX; also Edwards Plateau. Jun–Oct.

Ptilimnium nuttallii (DC.) Britton, (for its discoverer, Sir Thomas Nuttall, 1786–1859, English-American botanist), NUTTALL'S MOCK BISHOP'S-WEED. Resembling *Bifora americana* but usually taller, with longer leaf segments, and smaller but rather showy flowers; fruits ovoid, 1–1.5 mm long. Moist prairies, sandy or silty open ground, often abundant on roadsides; Denton, Hopkins, Hunt, Kaufman, Limestone, and Tarrant cos.; se and e TX w to East Cross Timbers, also Edwards Plateau. Apr–Jul. We are following Easterly (1957) in treating *Ptilimnium* ×*texense* J.M. Coult. & Rose [*P. capillaceum* × *P. nuttallii*, recognized by some authorities as a distinct species, as a hybrid.

SANICULA BLACK SNAKEROOT, SANICLE

Largely glabrous biennials or perennials; leaves petiolate below to subsessile above, palmately compound with 3–5 leaflets, the leaflets toothed and lanceolate, the largest > 1 cm wide, usually much wider; inflorescences at first short, barely exceeding the leaves, later elongating;

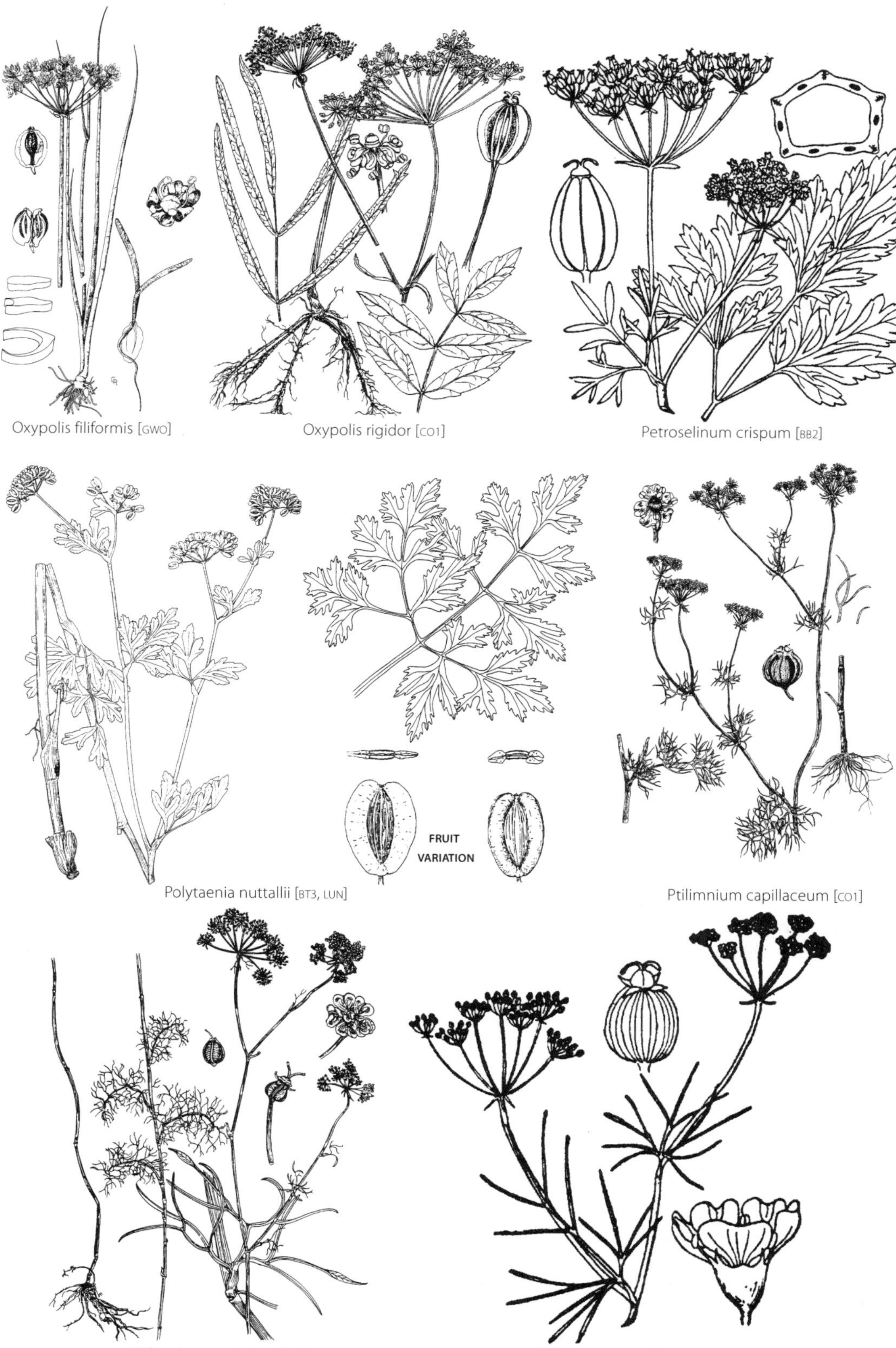

Oxypolis filiformis [GWO]

Oxypolis rigidor [CO1]

Petroselinum crispum [BB2]

Polytaenia nuttallii [BT3, LUN]

Ptilimnium capillaceum [CO1]

Ptilimnium costatum [CO1]

Ptilimnium nuttallii [BB2]

flowers relatively few per ultimate umbel (= umbellet); perfect flowers subsessile, usually three per umbellet; staminate flowers pedicellate, (0–)2–15 or more per umbellet; petals inconspicuous, white to yellow-green; fruits bur-like, with numerous bristles.

A nearly cosmopolitan genus of 39 species. (Latin: *sanare*, to heal, in reference to former medicinal use of some species)
REFERENCES: Bicknell 1895; Shan & Constance 1951.

1. Petals white, shorter than or equaling the narrow, stiff, spine-tipped calyx lobes; styles not exserted, shorter than bristles of fruit; widespread in nc TX ____ **S. canadensis**
1. Petals yellow-green, exceeding the ovate-lanceolate, herbaceous calyx lobes; styles long-exserted in fruit, longer than bristles of fruit; rare in nc TX ____ **S. odorata**

Sanicula canadensis L., (of Canada), CANADA SANICLE. Ultimate umbels with (0–)2–4 staminate flowers; fruit bristles swollen at base. Woods and thickets; se and e TX w to e Rolling Plains (to Taylor Co.—Mahler 1988), also Edwards Plateau. Apr–Jun.

Sanicula odorata (Raf.) Pryer & Phillippe, (fragrant), CLUSTER SANICLE. Ultimate umbels with up to 15 or more staminate flowers; fruit bristles not bulbous at base. Woods, thickets; Dallas and Grayson cos.; e TX w to nc TX, also Edwards Plateau. Apr–May. [*S. gregaria* E.P. Bicknell]

SCANDIX

A genus of 15–20 species of Europe and the Mediterranean region. (Greek name for chervil, another member of the carrot family)

Scandix pecten-veneris L., (comb of Venus), VENUS'-COMB, CROW-NEEDLES, SHEPHERD'S-NEEDLE, LADY'S-COMB. Hispid annual; leaves pinnately decompound; ultimate leaf divisions short and linear; petals white; fruits linear or narrowly oblong, 6–15 mm long, 1–2 mm broad, ciliate; beak of fruit linear, 20–70 mm long, compressed laterally. Low moist areas; Dallas (Kidd Spring Park—Mahler 1988), also Coastal Prairie. Mar–Apr. Native of Eurasia.

SIUM WATER-PARSNIP

A n hemisphere and African genus of 14 species. (Greek: *sion*, the name of some wet area plant)

Sium suave Walter, (sweet), WATER-PARSNIP, HEMLOCK. Perennial; stems erect, 0.5–1.2 m tall; superficially very similar to *Cicuta*; leaves only once pinnate, rarely simple (submerged leaves can be decompound); leaflets lanceolate to linear, 10–40(–60) mm long, serrate or incised; petals white; fruits oval to orbicular, 2–3 mm long, the ribs prominent. Swamps, wet areas; included based on citation of vegetational areas 4 and 5 (Fig. 2) by Hatch et al. (1990); mainly sc TX. May–Sep. [*S. cicutifolium* Schrank] There are reports of this species causing livestock poisoning (Kingsbury 1964).

SPERMOLEPIS SCALESEED

Low, glabrous annuals; leaves ternately or ternate-pinnately decompound, the ultimate divisions linear to filiform; inflorescence of loose compound umbels; petals white; fruits 1.5–2 mm long, with or without bristles.

A genus of 5 species in North America, Argentina, and Hawaii. (Greek: *sperma*, seed, and *lepis*, scale, alluding to the scruffy or bristly fruits of some species)

1. Ovaries and fruits with hooked short bristles ____ **S. echinata**

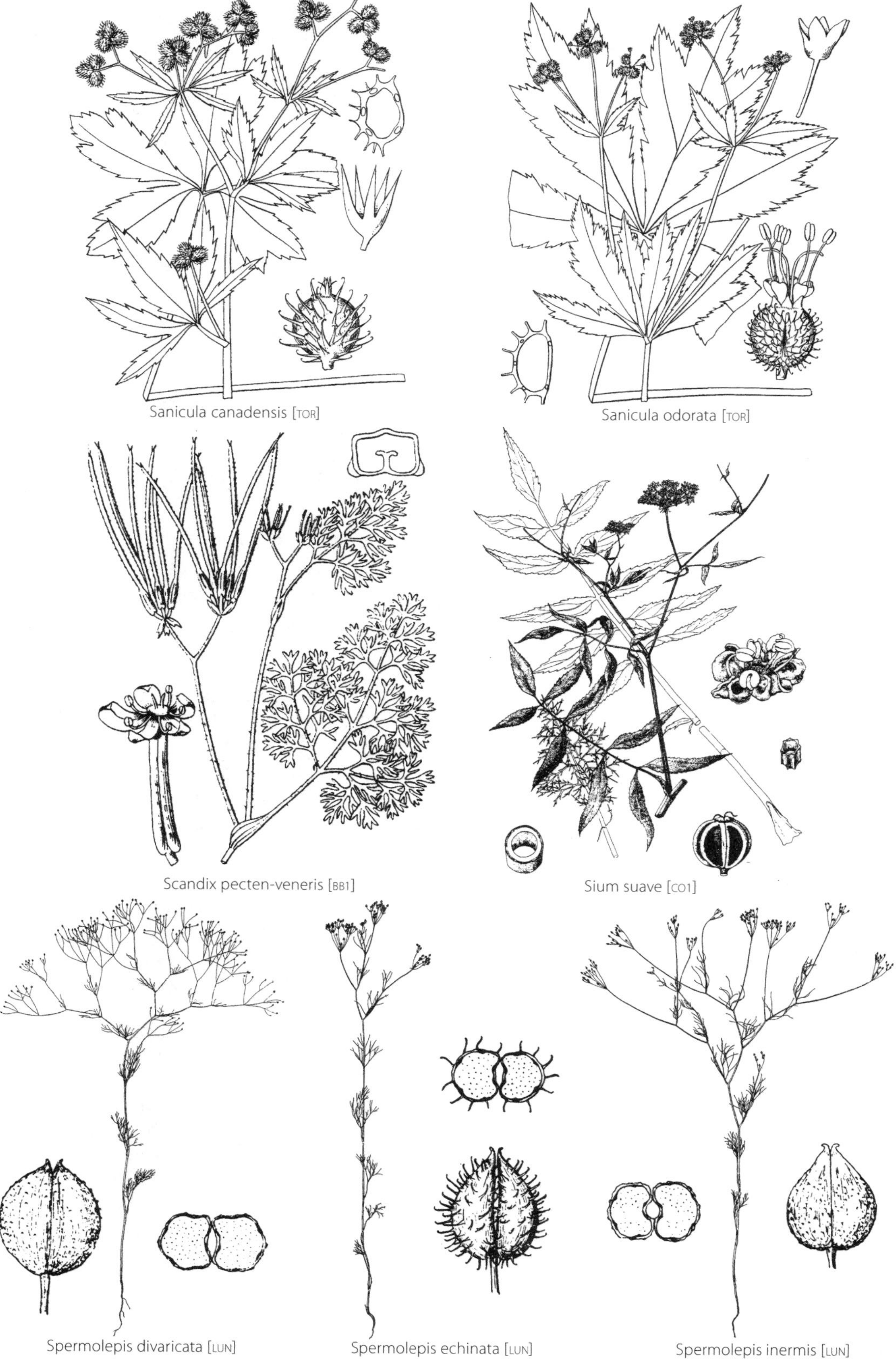

Sanicula canadensis [TOR]

Sanicula odorata [TOR]

Scandix pecten-veneris [BB1]

Sium suave [CO1]

Spermolepis divaricata [LUN]

Spermolepis echinata [LUN]

Spermolepis inermis [LUN]

1. Ovaries and fruits smooth or roughened with minute tubercles, but without bristles.
   2. Umbels with (2–)5–9(–11) rays; rays very unequal, 2–5 mm long in flower, up to 20 mm in age; fruiting pedicels very unequal, 6 mm long or less ________ **S. inermis**
   2. Umbels with 2–4(–7) rays; rays nearly equal, 5–20 mm long in flower, up to 35 mm in age; fruiting pedicels about equal, usually much longer than 6 mm ________ **S. divaricata**

Spermolepis divaricata (Walter) Raf. ex Ser., (spreading, widely divergent), FORKED SCALESEED. Plant erect, 10–70 cm tall. Open sandy areas, open woodlands; Bosque, Callahan, Grayson, Henderson, Denton, Parker, and Tarrant cos.; se and e TX w to West Cross Timbers, also Edwards Plateau. Apr–Jun.

Spermolepis echinata (Nutt. ex DC.) A. Heller, (prickly), BEGGAR'S-LICE, BRISTLY SCALESEED. Plant erect, often spreading, to 40 cm tall. Sandy soils; widespread in TX, but mainly East Cross Timbers w through w part of state. Mar–May.

Spermolepis inermis (Nutt. ex DC.) Mathias & Constance, (unarmed), SPREADING SCALESEED. Plant erect, 10–60 cm tall. Sandy or gravelly soils; Blackland Prairie to s and w TX. Apr–May.

TORILIS HEDGE-PARSLEY

Hispid annuals; leaves once pinnate or pinnately decompound; petals white; fruits in ours 3–5 mm long, bristly.

A genus of 15 species native to the Canary Islands, the Mediterranean region to e Asia, and tropical and s Africa. (Name used by Adanson in 1763; possibly from Latin: *torus*, cushion)

1. Flowers pedicelled, in terminal, peduncled, open umbels; plants erect ________ **T. arvensis**
1. Flowers subsessile, in axillary, sessile or short-peduncled, head-like umbels; plants reclining to suberect ________ **T. nodosa**

Torilis arvensis (Huds.) Link, (pertaining to cultivated fields), HEDGE-PARSLEY. Plant 30–100 cm tall; fruit bristles with microscopic barbs. Waste or disturbed ground, especially low or shady places, chiefly in limestone areas; se and e TX w to West Cross Timbers and Lampasas Cut Plain, also Edwards Plateau. May–Jun. Native of Mediterranean region. A problematic weed in some areas. The mericarps of this species were used in the successful prosecution of a kidnapping/child molestation case in Fort Worth in 1995. A 2 year old girl was pulled from the window of her apartment, kidnapped, sexually molested, and then left in a weedy area ca. 100 meters from where she was abducted. Law enforcement officials found plant fragments on the suspect's shoes and also collected assorted plant material from the crime scene. Botanists at BRIT identified the material from both the shoes and the crime scene as mericarps of *Torilis arvensis*, thus linking the suspect with the scene of the crime. This forensic botanical evidence, in combination with various other types of evidence, led to a conviction and a 99 year jail term (Lipscomb & Diggs 1998).

Torilis nodosa (L.) Gaertn., (with nodes), KNOTTED HEDGE-PARSLEY. Plant to 60(–100) cm tall; ultimate leaf divisions filiform. Waste or disturbed ground, especially low or shady places, chiefly in limestone areas, can become a pest in lawns; se and e TX w to West Cross Timbers, also Edwards Plateau. Apr–Jun. Native of Mediterranean region.

TREPOCARPUS

A monotypic genus of the s United States. (Greek: *trep*, turn, and *karpos*, fruit)

Trepocarpus aethusae Nutt. ex DC., (presumably referring to resemblance to *Aethusa*—fool's parsley, another genus of Apiaceae). Glabrous erect annual 35–50 cm tall; leaves pinnately de-

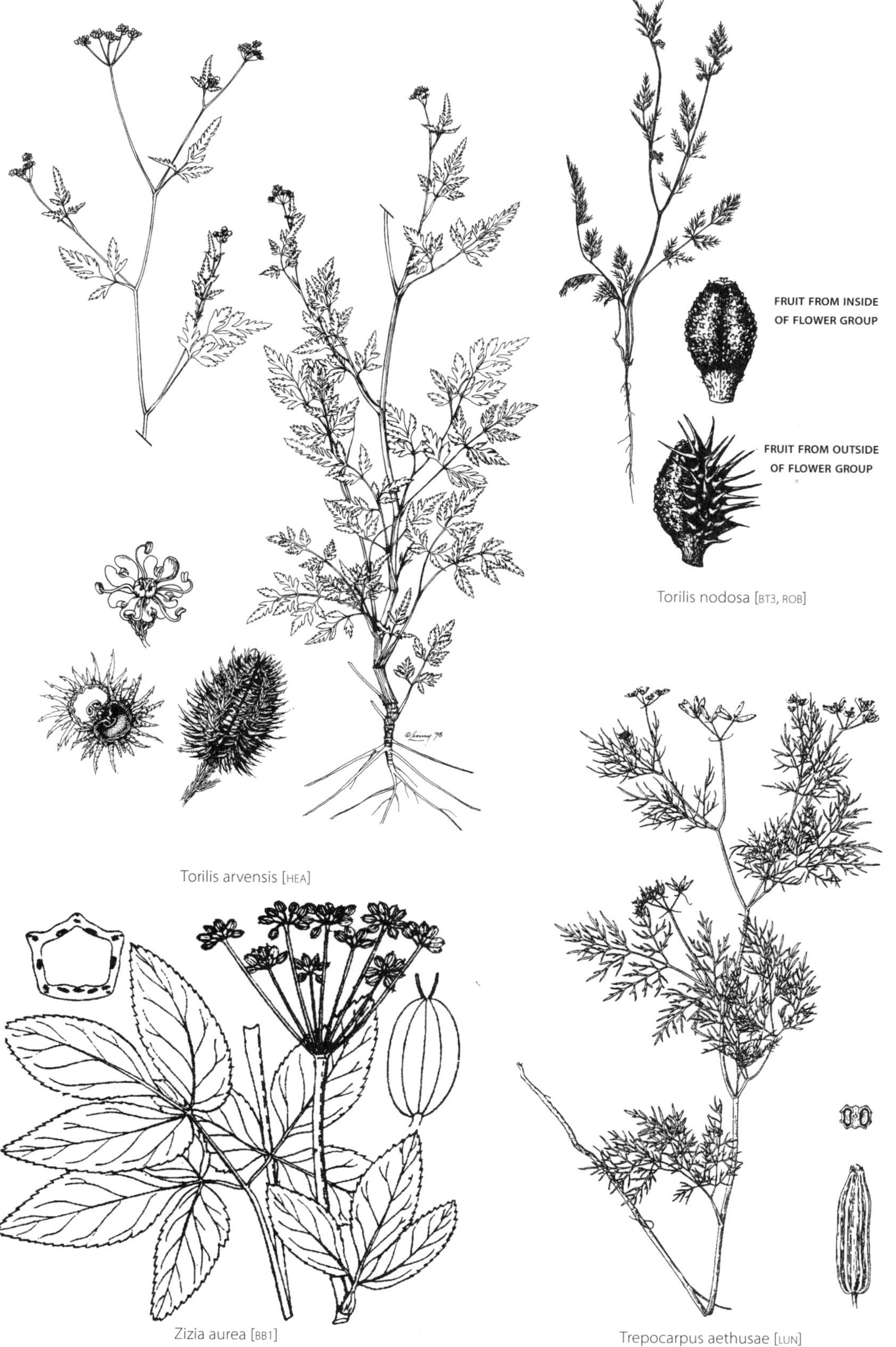

Torilis nodosa [BT3, ROB]

Torilis arvensis [HEA]

Zizia aurea [BB1]

Trepocarpus aethusae [LUN]

compound, the ultimate divisions linear; petals white; fruits oblong or broadly oblong-linear, 8–10 mm long. Woods and thickets, especially in low ground; Collin, Dallas, Grayson, and Hunt cos., also Lamar (Carr 1994) and Tarrant (R. O'Kennon, pers. obs.) cos.; se and e TX w to East Cross Timbers. May–Jun.

ZIZIA

A North American genus of 4 species. (Named for Johann Baptist Ziz, 1779–1829, a Rhenish (= of the Rhine region) botanist)

Zizia aurea (L.) W.D.J. Koch, (golden), GOLDEN-ALEXANDERS, GOLDEN ZIZIA. Glabrous perennial 40–100 cm tall; leaves once or twice compound; ultimate leaflets ovate- or oblong-lanceolate, sharply toothed; petals yellow; fruits 2–4 mm long. Low woods; Grayson and Lamar cos., also Dallas Co. (Mahler 1988); mainly e TX. Apr–May.

APOCYNACEAE OLEANDER OR DOGBANE FAMILY

Ours perennial herbs or vines with milky juice; leaves opposite or whorled (except alternate in *Amsonia*), sessile or very short-petioled, simple, entire, deciduous or evergreen; flowers axillary or terminal, solitary or in cymes; sepals 5, barely united at base; corollas salverform, funnelform, campanulate, cylindric, or urceolate, with 5 lobes, the throat or summit of tube usually with a constriction, crown, or dense zone of hairs; stamens 5, separate; pollen not in pollinia; pistils 2, united by styles and/or stigmas only (superficially flowers can appear to have 2 separate pistils); ovaries superior; fruit of 2 (or 1 by abortion) follicles (or in other areas berries or drupes).

A medium-large (1,900 species in 165 genera), pantropical family with some temperate species; vegetatively it includes mainly lianas, trees, shrubs, and a few temperate herbs; they have ubiquitous laticifer systems, glycosides, and alkaloids. The family contains many ornamental, poisonous, and medicinal species including the beautiful *Plumeria rubra* L. (FRANGIPANI), the widely planted but very poisonous *Nerium oleander* L. (OLEANDER), *Rauvolfia serpentina* (L.) Kurz, the source of the medicinal alkaloid reserpine, *Acokanthera* species, from which deadly arrow poisons (cardiac glycosides including ouabain with an effect like digitalin) are obtained in Africa, *Strophanthus*, an Old World genus yielding arrow poisons and the cardiac drug strophanthin, and *Catharanthus*, the source of alkaloids used in cancer treatments (see below). ☠ OLEANDER, a commonly used ornamental, contains extremely poisonous cardiac glycosides (e.g., neriin and oleandrin) resembling digitalin in action; a single leaf can kill an adult; children have been poisoned by sucking nectar or chewing leaves; poisoning may result from using twigs as skewers for food or even inhaling the smoke; even water in which flowers have been placed is toxic (Schmutz & Hamilton 1979; Morton 1982; Powell 1988). The sap of all species should be avoided. Apocynaceae are closely related to Asclepiadaceae (MILKWEED family) and considered paraphyletic when treated separately (Judd et al. 1994); Liede (1997), for example, treated the asclepiads as subfamily Asclepiadoideae in the Apocynaceae. From a cladistic standpoint the two should be lumped to form a more inclusive monophyletic Apocynaceae (Judd et al. 1994). (subclass Asteridae)

FAMILY RECOGNITION IN THE FIELD: herbs or vines with simple, entire, often opposite leaves, *milky sap*, sympetalous corollas usually with a distinct *tube*, and 2 pistils united only by their styles and/or stigmas. Similar to Asclepiadaceae (e.g., herbs or vines with milky sap) but that family has flowers with a distinctive corona and a gynostegium (= combined structure composed of stamens and pistils).

REFERENCES: Woodson 1938; Rosatti 1989; Judd et al. 1994.

1. Plant a trailing and climbing vine; corollas cream or yellowish white to pale yellow ______ **Trachelospermum**

1. Plant erect or a trailing vine with ascending branches, not climbing; corollas blue to lavender-blue, purple-red, white, green, or yellow; native and introduced species.
   2. Leaves alternate; flowers blue to lavender-blue (rarely white), in many-flowered cymes usually terminating the stems ____ **Amsonia**
   2. Leaves opposite or whorled; flowers either 1–2 in leaf axils OR if in cymes then white to green or yellow.
      3. Corollas salverform to funnelform, blue to lavender-blue, lavender, purple-red, or white, large, > 15 mm broad; seeds without a coma of hairs.
         4. Perennial trailing vines with ascending branches; corollas blue to lavender-blue (rarely white); peduncles (=flower stalks) 10–40 mm long ____ **Vinca**
         4. Erect, tender perennial grown as an annual; corollas purple-red to lavender or white, often with a dark eye; peduncles 2–4 mm long ____ **Catharanthus**
      3. Corollas campanulate to cylindric or urceolate, white to green or yellow, small, 2–4 mm broad; seeds with a coma of hairs ____ **Apocynum**

AMSONIA BLUESTAR, SLIMPOD

Largely glabrous perennials from woody rootstocks; leaves numerous, alternate, filiform to lanceolate or elliptic; flowers in compact terminal cymes; corollas salverform, light to deep blue or lavender, rarely white, with narrow lobes; follicles erect to pendulous, slender; seeds numerous, without a coma.

A genus of 20 species native to North America and Japan; a number have alkaloids and some are cultivated as ornamentals. (Named for Charles Amson, 18th century Gloucester, Virginia physician and friend of John Clayton)
REFERENCES: Woodson 1928, 1929; McLaughlin 1982.

1. Corollas sharply constricted at orifice, with tube 25–40 mm long; leaf blades filiform to linear-lanceolate, to only ca. 4 mm wide; sw part of nc TX ____ **A. longiflora**
1. Corollas not constricted at orifice, with tube 6–10 mm long; leaf blades filiform to lanceolate or elliptic, to 30 mm wide; widespread in nc TX.
   2. Corollas glabrous outside; leaf blades 0.5–17 mm wide, usually lanceolate to linear or filiform (rarely elliptic), usually 5(–8) cm or less long ____ **A. ciliata**
   2. Corollas pubescent outside, at least in bud; leaf blades 8–30 mm wide, narrowly lanceolate to elliptic, often > 5 cm long ____ **A. tabernaemontana**

Amsonia ciliata Walter var. **texana** (A. Gray) J.M. Coult., (sp.: ciliate, fringed; var.: of Texas), TEXAS SLIMPOD, TEXAS AMSONIA. Plant usually 15–70 cm tall, forming loose clumps or patches from creeping rootstocks; corolla tube 6–10 mm long, ca. 1 mm in diam. at base; corolla lobes 3.5–11 mm long; follicles 6–11 cm long; seeds 5–11 mm long. Chiefly gravelly limestone soils, less commonly sandy ground; Blackland Prairie w to Rolling Plains, also Edwards Plateau. Apr–early May. [*A. texana* (A. Gray) A. Heller]

Amsonia ciliata var. *tenuifolia* (Raf.) Woodson, (slender-leaved), [*A. ciliata* Walter var. *filifolia* Woodson], is cited by Hatch et al. (1990) for vegetational areas 4 and 5 (Fig. 2). According to Justin Williams (pers. comm.), who is doing revisionary work on the genus, this is a taxon of the e U.S. that does not reach TX.

Amsonia longiflora Torr. var. **salpignantha** (Woodson) S.P. McLaughlin, (sp.: long-flowered; var.: trumpet flower), TRUMPET SLIMPOD. Plant 20–35 cm tall; stems usually clustered from base; leaves 2–5 cm long; corolla tube ca. 1.5 mm in diam. at base, inflated at insertion of stamens and then constricted at orifice; corolla lobes 5–10 mm long; follicles 7–9 cm long; seeds 5–12 mm long. Limestone hills and rocky prairies; Hamilton Co. (the type locality is on Cowhouse

Creek—Woodson 1928), also Bell Co. (further downstream on Cowhouse Creek on Fort Hood—Sanchez 1997); sw part of nc TX through the Edwards Plateau to the Trans-Pecos. Mar–May. [*A. salpignantha* Woodson]

Amsonia tabernaemontana Walter, (for Jacob Theodore von Bergzabern, (d. 1590), Heidelberg botany professor who Latinized his name as Tabernaemontanus), WILLOW SLIMPOD, CREEPING SLIMPOD. Plant to 120 cm tall; leaves to 12 cm long; corolla tube 6–8 mm long, ca. 1 mm in diam. at base; corolla lobes 5–10 mm long; follicles 8–14 cm long. Stream bottom thickets; se and e TX w to East Cross Timbers. April. Texas material of this species has traditionally gone under the names *A. illustris* Woodson and *A. repens* Shinners. According to J. Williams (pers. comm.), *A. illustris*, which occurs only to the n of TX, should be put into synonymy with *A. tabernaemontana* var. *gattingeri* Woodson. A manuscript in preparation by Williams will name a new variety of *A. tabernaemontana* for the material of this species that occurs in TX; in the interim we are simply recognizing the TX material as *A. tabernaemontana*. [*A. repens* Shinners]

APOCYNUM DOGBANE, INDIAN-HEMP

A genus of ca. 12 species of s Russia to China and temperate America. (Greek: *apo*, far from, and *kyon or cyon*, a dog, from ancient use as dog poison and ancient name of the Old World dogbane)

REFERENCES: Woodson 1930; McGregor 1984b.

Apocynum cannabinum L., (resembling *Cannabis*-hemp), INDIAN-HEMP, HEMP DOGBANE, PRAIRIE DOGBANE, HAIRY DOGBANE, SMOOTH DOGBANE, WILLOW DOGBANE, CHOCTAWROOT. Herbaceous perennial with deep root, glabrous to pubescent; stems erect to ascending, to 1 m tall; leaves usually opposite, the blades variable in shape, linear-lanceolate to oval, sessile or short petiolate, narrowed to rounded or cordate and clasping basally; flowers in small usually terminal cymes, 5-merous; corollas white, yellow, or greenish, 3–6 mm long; follicles 4–22 cm long, pendulous at maturity, glabrous; seeds numerous, each terminated by a coma of long, silky hairs. Open or disturbed, often moist ground, sandy or less often gravelly or eroding clayey soils; nearly throughout TX. Apr–Jul. [*A. cannabinum* L. var. *glaberrimum* A. DC., *A. cannabinum* L. var. *pubescens* (J. Mitch. ex R. Br.) Woodson, *A. sibiricum* Jacq.] The bark has been used as a source of fiber and the root as an emetic and cardiac stimulant (Mabberley 1987); however, it is poisonous to animals such as cats, dogs, and livestock due to the presence of the cardiac glycoside apocynamarin, resins, and other toxins; sickness and death have been reported in humans from medicinal use (Kingsbury 1964; Schmutz & Hamilton 1979; Fuller & McClintock 1986; Mulligan & Munro 1990). We are following Kartesz (1994) and J. Kartesz (pers. comm. 1997) in treating this as a single variable species; several previously recognized taxa (including *A. sibiricum*) that intergrade with *A. cannabinum* have been lumped into this species; Jones et al. (1997) recognized *A. sibiricum* at the specific level. Mahler (1988) separated the two as follows: ☠

1. Leaves narrowed to broadly rounded basally, petiolate; floral bracts scarious, aristate ____ *A. cannabinum*
1. Leaves cordate, sessile, usually some clasping; floral bracts foliaceous ____ *A. sibiricum*

CATHARANTHUS

A tropical genus of 6 species, 5 endemic to Madagascar, 1 India and Sri Lanka; previously recognized in *Vinca*. (Greek: *katharos*, pure, and *anthos*, flower)

Catharanthus roseus (L.) G. Don, (rose-colored), MADAGASCAR PERIWINKLE, ROSE PERIWINKLE, OLD-MAID, CAYENNE-JASMINE, CAPE PERIWINKLE. Erect herb 20–60 cm tall; leaves obovate or oblanceolate, apiculate, 3–7 cm long; flowers 1–2 in upper leaf axils; peduncles 2–4 mm long; corollas showy, salverform, the tube 20–25 mm long; follicles erect, puberulent, striately ridged.

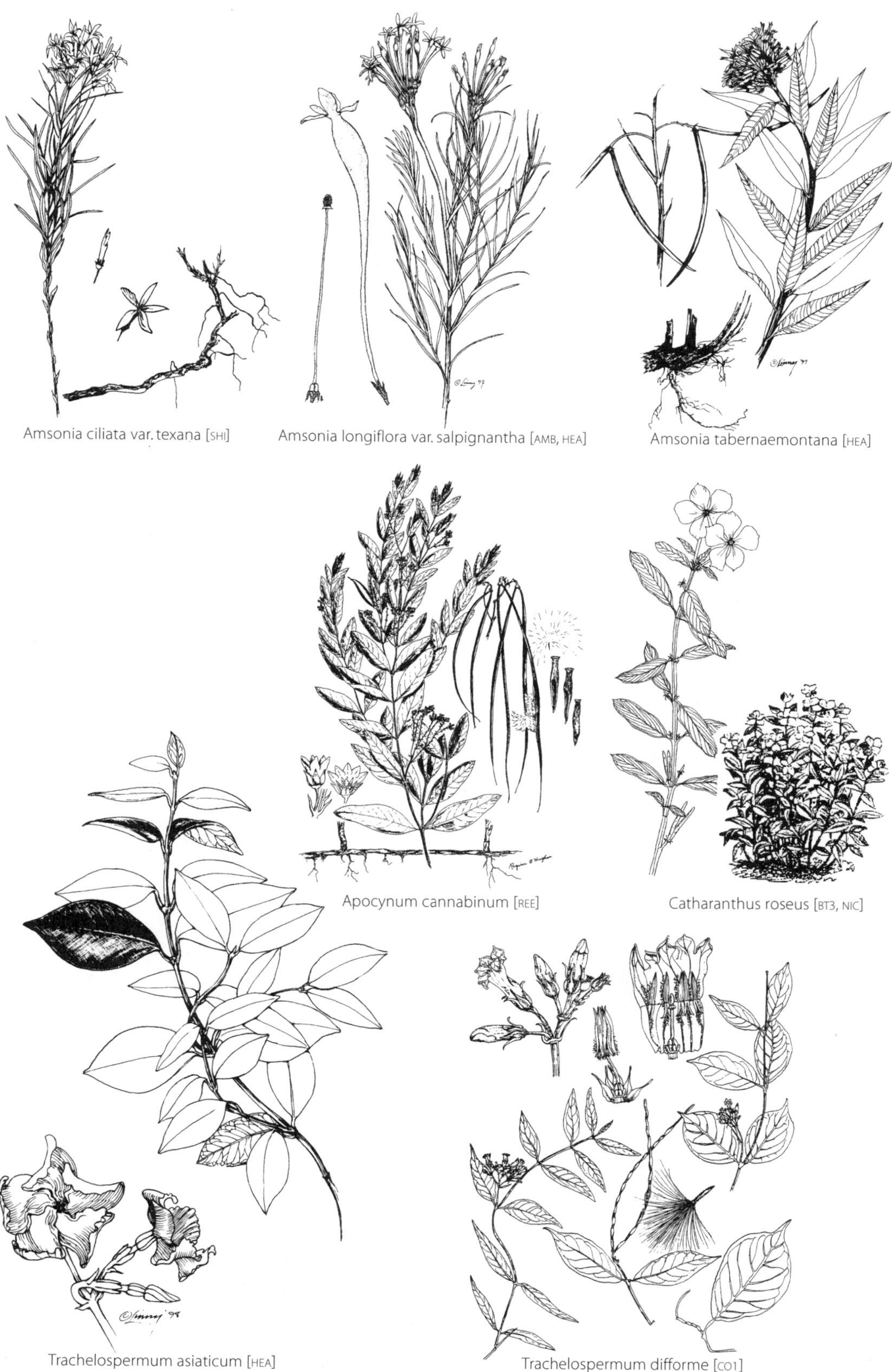

Amsonia ciliata var. texana [SHI]

Amsonia longiflora var. salpignantha [AMB, HEA]

Amsonia tabernaemontana [HEA]

Apocynum cannabinum [REE]

Catharanthus roseus [BT3, NIC]

Trachelospermum asiaticum [HEA]

Trachelospermum difforme [CO1]

Commonly grown in flower beds in summer, and sometimes appearing spontaneously in landscapes; not frost-resistant; Tarrant Co., also included for nc TX by Mahler (1988). Jun–Oct. Native of Madagascar, now pantropical. [*Vinca rosea* L.] Widely cultivated as an ornamental and also produces at least 80 alkaloids including vinblastine, vincristine, and others useful in the treatment of certain cancers including Hodgkin's disease and leukemia; this species is considered the most important plant to date in terms of cancer treatment; if misused the alkaloids are potentially toxic; they can act as an abortifacient (Duke 1985; Mabberley 1987; McGuffin et al. 1997). ☠ ☜

TRACHELOSPERMUM CLIMBING-DOGBANE

Vines; leaves opposite, entire, short-petioled; flowers in loose cymes; corollas salverform or subfunnelform with cylindrical tube and oblong lobes overlapping to the right; corolla tube slightly inflated above insertion of stamens; stamens inserted near middle to upper part of corolla tube; fruit of 2 elongate, slender, terete follicles.

A genus of 20 species of climbing or clambering vines with milky sap native from India to Japan with 1 in the se United States; a number are cultivated as ornamentals. (Greek: *trachelos*, a neck, and *sperma*, seed, referring to the narrow seeds)

1. Leaves thinly membraneous, deciduous, 4–14 cm long; native species in moist and weedy areas in ne part of nc TX ____ **T. difforme**
1. Leaves coriaceous, evergreen, 2–8(–10) cm long; introduced species spreading or persisting from cultivation.
   2. Calyx lobes erect, usually shorter than or ca. equal to narrow part of corolla tube; anthers slightly exserted ____ **T. asiaticum**
   2. Calyx lobes strongly recurved, longer than narrow part of corolla tube; anthers included ____ **T. jasminoides**

Trachelospermum asiaticum (Siebold & Zucc.) Nakai, (of Asia), ASIAN-JASMINE, JAPANESE STAR-JASMINE. Plant trailing and climbing; leaves evergreen, coriaceous, deep green in color, elliptic to ovate or elliptic-oblong, 2–6.5 cm long (usually towards lower end of this range); flowers fragrant; corollas yellowish white. Generally not flowering in nc TX. One of the most widely planted ground covers in nc TX, including in full sun; persists and spreads from cultivation; Tarrant Co. (R. O'Kennon, pers. obs). Early summer. Native of Korea and Japan. ☜

Trachelospermum difforme (Walter) A. Gray, (of unusual or differing forms), AMERICAN STAR-JASMINE, CLIMBING-DOGBANE. Twining vine climbing on trees; leaves variable in shape, lanceolate to elliptic, ovate, or even suborbicular, to 14 cm long and 8 cm wide but mostly to only slightly more than half that size; corollas creamy white to pale yellow; anthers not exserted. Along streams, forest margins, weedy areas; Lamar Co. in Red River drainage (*Correll & Correll 35922*, TEX); se and e TX w to ne corner of nc TX. Apr–Jun.

Trachelospermum jasminoides (Lindl.) Lem., (resembling *Jasminum*—jasmine), STAR-JASMINE, CONFEDERATE-JASMINE. Plant climbing and trailing; leaves evergreen, coriaceous, deep green in color, elliptic to ovate or obovate, 4–8(–10) cm long; flowers fragrant; corollas white, or with yellowish tinge near tube. Cultivated and spreading, climbing up sides of buildings; Grayson Co. (Austin College campus). May–early Jun. Native of China. ☜

VINCA PERIWINKLE

Ours creeping or trailing perennials; leaves opposite, evergreen; flowers axillary, usually solitary; corollas salverform to funnelform with cylindrical tube and 5 equal lobes widely spreading, twisted to the left, thickened or hairy at throat, blue to lavender-blue, rarely white; paired follicles erect, glabrous, smooth; seeds without a coma.

A genus of 7 species native from Europe to n Africa and c Asia including cultivated ornamentals; some species contain alkaloids. (Name abbreviated from Latin: *vinca pervinca*, from *vincio*, to bind, alluding to the use of the foliage in wreaths; the ancient name reflected in the colloquial Italian *pervinca*, French *pervenche*, Middle English *per wynke*, and English *periwinkle*)

1. Leaves short petiolate (1–2 mm long) or sessile; leaf blades usually not narrowed at base, rounded to truncate or subcordate; calyx lobes ca. 10 mm long; corolla tube ca. 15 mm long _____ **V. major**
1. Leaves petiolate, the petioles usually much more than 2 mm long; leaf blades usually narrowed at base; calyx lobes to 3 mm long; corolla tube 3–6 mm long _____ **V. minor**

Vinca major L., (large), BIG-LEAF PERIWINKLE. Leaves to 7 cm long, to 5 cm wide. Cultivated, especially in limestone areas; readily spreads in woods and along streams and can overcome native herbaceous vegetation; Bell, Dallas, Grayson, and Williamson cos., also Tarrant Co. (R. O'Kennon, pers. obs.); scattered in e 1/2 of TX. Mainly Mar–Apr (but year round; observed in cultivation in Sherman flowering virtually every month of the year). Native of se Europe and sw Asia.

Vinca minor L., (little), COMMON PERIWINKLE. Leaves elliptic, 2–4.5(–6) cm long, 1–1.5(–2.5) cm wide. Cultivated, wooded areas; included based on citation of vegetational areas 4 and 5 (Fig. 2) by Hatch et al. (1990); however, we have seen no nc TX material; supposedly naturalized throughout TX (vegetational areas 1–10—Hatch et al. 1990). Native of s Europe. Mar–May.

AQUIFOLIACEAE HOLLY FAMILY

A medium-sized (420 species in 4 genera), almost cosmopolitan family of usually evergreen trees and shrubs. *Ilex* is economically important for its hard white wood, as an ornamental, and as a source of Christmas decorations. Family name conserved from *Aquifolium*, a genus now treated as *Ilex* (the name *Ilex* was published earlier and thus has priority in terms of nomenclature). (Classical Latin name for holly) (subclass Rosidae)
FAMILY RECOGNITION IN THE FIELD: shrubs or small trees with alternate simple leaves and small, inconspicuous, axillary, often 4-merous, frequently unisexual flowers; fruit a small, drupe-like, red or orange berry.
REFERENCES: Lundell 1961; Brizicky 1964a.

ILEX HOLLY

Ours shrubs or small trees with mostly smooth, light gray bark; leaves alternate, short-petioled, simple; leaf blades nearly entire to crenulate or serrulate (spinose-dentate in species outside nc TX), evergreen and leathery or deciduous; stipules minute, falling early; flowers small, axillary, solitary or in sessile or peduncled, umbel-like or cymose clusters, usually unisexual (plants dioecious, often with some perfect flowers); calyces rotate or shallowly funnelform, 4–5-lobed; petals 4–5, oblong-elliptic, white, exceeding the calyces; stamens (or staminodes) 4–5; pistil 1; ovary superior; fruit a small drupaceous berry, globose or nearly so, usually with 4 stones.

A genus of ca. 400 species, cosmopolitan, especially tropical and temperate Asia and America; various species are used as ornamentals, for timber, or as a source of stimulants (contain alkaloids including caffeine and theobromine). *Ilex paraguariensis* A. St.-Hil. (YERBA MATÉ) is used as a tea in South America. Apparently the Roman use of holly for Saturnalia was taken over by Christians for Christmas. The fruits of *Ilex* species, while an important source of food for native birds, are considered somewhat poisonous due to the presence of saponins; they cause vomiting, diarrhea, and stupor if eaten in quantity (Hardin & Arena 1974; Lampe & McCann 1985). (Ancient Latin name of the holly-oak, rather than of the holly)

1. Larger leaf blades on flowering branches spatulate, oblanceolate, or obovate, with long-tapering base, pubescent beneath along midrib; leaves deciduous ____ **I. decidua**
1. Larger leaf blades on flowering branches oblong-lanceolate or oblong-elliptic, abruptly narrowed or rounded-truncate basally, glabrous beneath; leaves evergreen ____ **I. vomitoria**

Ilex decidua Walter, (deciduous), DECIDUOUS HOLLY, POSSUMHAW, WINTERBERRY, BEARBERRY, HOLLY, MEADOW HOLLY, PRAIRIE HOLLY, WELK HOLLY. Petioles 2–11 mm long, leaf blades to 80 mm long and 45 mm wide; fruits bright red or orange, to ca. 7.5 mm in diam. Rock outcrops, ravines, and disturbed areas; se and e TX w to Bell, Dallas (Lundell 1961), and Grayson cos., also Comanche (Little 1976 [1977]), Parker (R. O'Kennon, pers. obs.), and Tarrant (Lundell 1961) cos.; also to Edwards Plateau. Mar–May. The persistent, red or orange fruits are often conspicuous during the winter on the otherwise bare branches. Fruits poisonous (Lampe & McCann 1985). ☠

Ilex vomitoria Sol. in Aiton, (emetic), YAUPON, YAUPON HOLLY, CASSINE, CASSENA, CASSIO-BERRY BUSH, CHOCOLATE DEL INDIO, EMETIC HOLLY, EVERGREEN CASSENA, EVERGREEN HOLLY, INDIAN BLACKDRINK, SOUTH-SEA-TEA. Petioles 1–3 mm long, leaf blades to 55 mm long and 28 mm wide; fruits bright red, to ca. 6.5 mm in diam. Commonly cultivated, low woods, sandy areas; Fannin Co. (Talbot property) in Red River drainage and Limestone Co., also Lamar Co. (Carr 1994), also naturalized in Dallas and Tarrant cos. (R. O'Kennon, pers. obs.); mainly se, e, and sc Texas. Mid–late Apr. This species is one of the most widely used evergreen landscape shrubs in nc TX. The leaves contain caffeine and were used by Native Americans to make a ceremonial drink (known as "black drink"); during religious festivals they drank large amounts of a strong preparation which caused vomiting—thus the specific epithet (Havard 1896; Alston & Schultes 1951; Correll & Johnston 1970; Cox & Leslie 1991). The authority for this species is often given as Aiton (Kartesz 1994; Jones et al. 1997). However, Hatch et al. (1990) gave Solander in Aiton. Stafleu and Cowen (1976) indicated the botanical descriptions in *Hortus Kewensis* were not made by the Aitons but instead by Solander, Dryander, and Robert Brown based on material from Kew (Stafleu & Cowen 1976). On this basis, authorities for a number of taxa may have to be reexamined. Fruits poisonous (Lampe & McCann 1985). ☠

Ilex opaca Aiton, (opaque, shaded), (AMERICAN HOLLY, WHITE HOLLY), occurs in se and e TX just to the e of nc TX. It has large (to 12 cm long and 6 cm wide), evergreen, coriaceous leaves that are usually conspicuously spinose-dentate (sometimes entire or nearly so) and painful to the touch. The species is valued for its strikingly white wood, used for veneer and decorative inlays (Tyrl et al. 1994). Fruits poisonous (Lampe & McCann 1985). ☠

ARALIACEAE GINSENG OR ARALIA FAMILY

A medium-sized (1,325 species in 47 genera), mainly tropical family with a few in temperate regions. It contains trees, shrubs, lianas, woody epiphytes, or rarely herbs; leaves are often large and compound. The family includes a number of ornamentals (e.g., *Hedera*, *Schefflera*) as well as the economically important *Panax quinquefolius* L. (GINSENG), which supposedly acts as a stimulant and aphrodisiac. The Araliaceae are closely related to the Apiaceae and probably paraphyletic when treated separately. From a cladistic standpoint they should be lumped to form a more inclusive monophyletic family, which based on nomenclatural rules should be called Apiaceae (Judd et al. 1994). Family name from *Aralia*, a genus of 36+ species of North America, e Asia, and Malaysia; some are edible or used medicinally. *Aralia spinosa* L., (spiny), HERCULES'-CLUB, DEVIL'S-WALKINGSTICK, occurs in se and e TX just e of nc TX and would not be unexpected on the e margin of nc TX. It is a shrub or tree to 12 m or more tall, with alternate

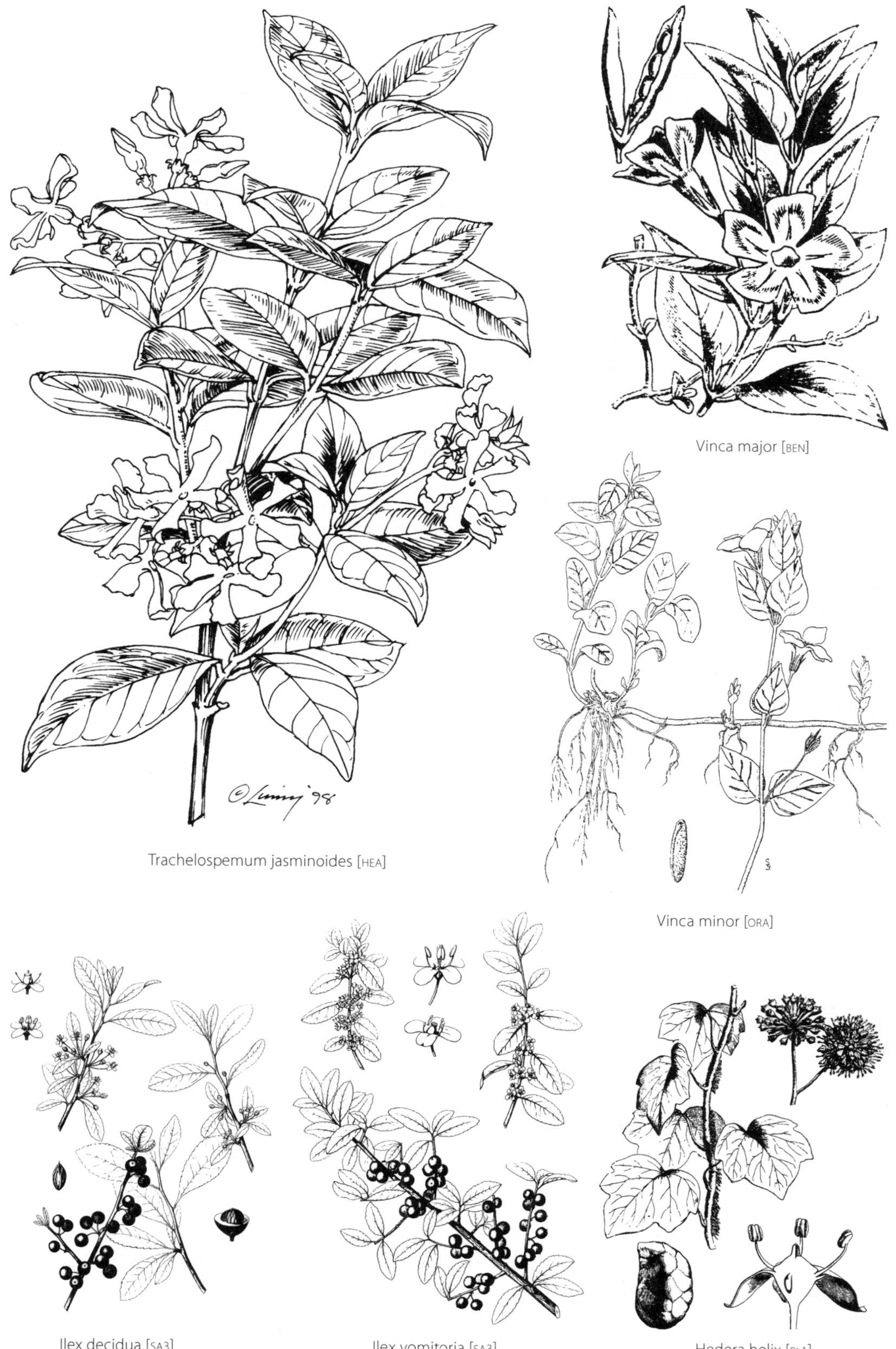

Trachelospemum jasminoides [HEA]

Vinca major [BEN]

Vinca minor [ORA]

Ilex decidua [SA3]

Ilex vomitoria [SA3]

Hedera helix [BL1]

decompound leaves to 1(+) m long, coarsely prickly stems and petioles, umbels in a large, terminal, compound panicle, white petals, 2–3 mm long, and black berries 4–6 mm in diam. The leaves and fruits are reported as potentially poisonous (Hardin & Brownie 1993). ☠ (From the French-Canadian name, *aralie*) (subclass Rosidae)

FAMILY RECOGNITION IN THE FIELD: the only nc TX species is an introduced, woody, evergreen vine with alternate leaves exhibiting leaf dimorphism—at least some leaves are 3–5 palmately lobed; flowers small in solitary or racemose umbels; fruit a small drupe. The similar Apiaceae (e.g., small flowers in umbels; inferior ovary) are herbaceous and have schizocarp fruits while Araliaceae are usually woody with drupes or berries.

REFERENCES: Smith 1944; Graham 1966; Eyde & Tseng 1971; Judd et al. 1994.

HEDERA IVY

A genus of 4–11 species of vines native to Europe, the Mediterranean region, and e Asia; thought to be related to *Schefflera*, a widely used ornamental. (The ancient Latin name)

Hedera helix L., (Greek for a twining plant), ENGLISH-IVY. Woody, evergreen, trailing or high-climbing vine with 4–10-rayed stellate pubescence; leaves alternate, dimorphic, the adult and juvenile differently shaped, 3–5 palmately lobed or ovate to lanceolate, dark green, often with paler markings; flowers small, in terminal, solitary or racemose umbels; ovary inferior; fruit a small, black, 3–5-seeded berry 7–8 mm in diam. Widely cultivated ornamental that long persists around old homesites and occasionally spreads to adjacent vegetation, wooded areas; Grayson Co., also Tarrant Co. (R. O'Kennon, pers. obs.). Native of Europe, w Asia, and n Africa. This species can potentially damage trees by overtopping the foliage and stealing light, thus reducing the tree's photosynthetic output. There are numerous cultivars including variegated forms. The juvenile stems bear lobed leaves in a single plane while the adult flowering stems have unlobed leaves circling the stem (Graham 1966). The leaves and fruits contain the saponic glycoside, hederagenin, which if ingested can cause breathing difficulties and coma; the sap can cause dermatitis with blistering and inflammation—this is apparently due to polyacetylene compounds which are also present; gloves should be used when trimming the plant (Schmutz & Hamilton 1979; Spoerke & Smolinske 1990; Turner & Szczawinski 1991). ☠

ARISTOLOCHIACEAE PIPEVINE OR BIRTHWORT FAMILY

A medium-sized family of ca. 600 species in 5 genera (Barringer & Whittemore 1997); it consists of herbs, shrubs, and lianas (= woody vines) and extends from the tropics to temperate areas, especially in the Americas; they are usually aromatic and contain alkaloids or aristolochic acid; a number are cultivated for their unusual flower shapes. Some taxonomists refer to the Aristolochiaceae as "paleoherbs" (a group including Aristolochiales, Piperales, and Nymphaeales) and believe them to be an early branch off the evolutionary line leading to monocots; this view is supported by characters such as the 3-merous flowers (Zomlefer 1994) and molecular data which place the paleoherbs as the immediate sister group of the monocots (Chase et al. (1993) (see Fig. 41 in Appendix 6). (subclass Magnolidae)

FAMILY RECOGNITION IN THE FIELD: herbs or woody vines with alternate, simple, palmately veined leaves often basally cordate to hastate or sagittate and apetalous flowers with a curved, *pipe-like, 3-merous* calyx; fruit a capsule.

REFERENCES: Gregory 1956; Huber 1993; Barringer & Whittemore 1997.

ARISTOLOCHIA PIPEVINE, BIRTHWORT, TACOPATE

Herbs or woody vines; leaves alternate, simple; leaf blades entire, palmately veined, often cordate to hastate or sagittate at base; flowers in ours solitary in the leaf axils or in racemose inflorescences from near the base of the plant, bisexual; calyces petaloid, curved, strikingly pipe-like in shape; corollas absent; stamens in ours 6; fruit a capsule.

A genus of ca. 300 (Barringer 1997) species of vines, scramblers, or herbs of tropical and warm areas, especially the Americas; some flowers trap insects which then pollinate them; all species contain alkaloids or aristolochic acid (carcinogenic and nephrotoxic—McGuffin et al. 1997); some are used medicinally, in treating snakebites, or as an arrow poison. According to Crosswhite and Crosswhite (1985), "*Aristolochia* was the source of the active ingredient in 'snakeroot oil' sold by itinerant snakeroot doctors who staged medicine shows in the western United States in the Nineteenth Century." The leaves of a number of species are eaten by larvae of pipe-vine swallowtail butterflies such as *Battus philenor*; the aristolochic acid thus obtained is sequestered and presumably makes them unpalatable to birds (Howe 1975; Barringer & Whittemore 1997). (Greek, *aristos*, best, and *lochia*, delivery, from supposed value in aiding childbirth; apparently derived through the ancient Doctrine of Signatures from the fancied resemblance of the flower bud to a fetus; the plants were used as a source of medicine to alleviate the pain of childbirth—Pfeifer 1966)

REFERENCES: Pfeifer 1966; Barringer 1997.

1. High-climbing, twining, woody vines to 25 m long; flowers from leaf axils on upper part of the stem; leaves large, 8–15(–20) cm wide; seeds ca. 10 mm long ____ **A. tomentosa**
1. Erect to sprawling herbs to 60 cm tall; flowers on peduncles borne mostly at the base of the stems; leaf blades ca. 7 cm or less wide; seeds 5 mm or less long.
   2. Leaves subsessile, conspicuously clasping; leaf blades subcoriaceous, prominently reticulate veined beneath, obtuse to subacute at apex; stems with conspicuous spreading hairs ____ **A. reticulata**
   2. Leaves with distinct petioles, not clasping; leaf blades thin, delicate, not prominently reticulate veined beneath, apically acuminate to acute; stems glabrous to with inconspicuous pubescence ____ **A. serpentaria**

Aristolochia reticulata Jacq., (netted, in reference to the leaf venation), TEXAS DUTCHMAN'S-PIPE. Perennial herb; stems to ca. 40 cm tall, ± zigzag; leaves few, subsessile, the petioles only 0.1–0.8 cm long; leaf blades oblong to broadly ovate, 7–12 cm long and 3–7 cm wide; inflorescences racemose, several-flowered, arising from near base of plant (often covered by fallen leaves); perianth brownish purple; capsules subglobose, ca. 10–30 mm in diam. Wooded areas, sandy soils; Young Co. (Pfeifer 1966) on w edge of nc TX; mainly e TX. May–Jul. The dried rhizome has been sold as a serpentary (= treatment for snakebite); it is also used as a tonic; however, the active ingredient is aristolochic acid, a gastric irritant that in large doses can cause respiratory paralysis (Barringer 1997).

Aristolochia serpentaria L., (referring to medicinal use for snakebite), VIRGINIA DUTCHMAN'S-PIPE, VIRGINIA SNAKEROOT, NASH DUTCHMAN'S-PIPE. Perennial herb; stems to 60 cm tall; leaf blades quite variable, ovate to narrowly lanceolate, cordate to hastate, sagittate, or truncate at base, 5–15 cm long, 1–5 cm wide; petioles 0.5–3.5 cm long; flowers solitary, at the ends of scaly branches arising from the lowest nodes of the stem; perianth purplish brown; capsules subglobose, 8–20 mm in diam. Moist wooded areas; Fort Hood (Bell or Coryell cos.—Sanchez 1997), also mapped for e part of nc TX by Cheatham and Johnston (1995); also e TX and Edwards Plateau, and in s OK just n of the Red River. [*A. serpentaria* var. *hastata* (Nutt.) Duch.] Even though it has distinctive leaves, this is a small, easily overlooked plant of the forest floor. According to Barringer (1997), closed, apparently cleistogamous flowers are often present. Also, this species exhibits a great deal of variation in leaf shape; this variability is apparently related to the fact that eastern pipe-vine swallowtail butterflies use leaf shape as a search image when looking for *Aristolochia* leaves on which to lay their eggs (Barringer 1997). This species, which contains a number of physiologically active components including aristolochine and aristolochic acid, was long used by Native Americans and early settlers for a variety of medicinal applications including snakebite; it has been rather popular recently as an herbal medicine; however, it can irritate the gastrointestinal tract and kidneys, and potentially cause coma and death due to respiratory paralysis (Duke 1985; Cheatham & Johnston 1995); aristolochic acid is carcinogenic and nephrotoxic (McGuffin et al. 1997). ☠

Aristolochia tomentosa Sims, (densely woolly, with matted hairs), WOOLLY DUTCHMAN'S-PIPE, PIPEVINE. Vine to 25 m long, densely pubescent except on old growth; leaf blades cordate, 3-ribbed from base; petioles 1–5.5(–9) cm long; flowers solitary (rarely paired), axillary; perianth green or yellowish green with brown or purplish stripes inside, with curved, funnelform tube abruptly inflated above the ovary, and flaring limb; stamens 6, united with style; anthers sessile; pistil 1; ovary appearing inferior (tightly enclosed by base of perianth tube); fruit a cylindric capsule to 8 cm long and 3 cm in diam.; seeds flat, triangular. Thickets and woods, sandy or silty soils, frequent, but not often found in flower; Denton, Grayson, Hood, Somervell, Tarrant, and Wise cos., also Dallas Co. (Pfeifer 1966); se and e TX w to West Cross Timbers Apr. Presumably contains aristolochic acid. ☠

ASCLEPIADACEAE MILKWEED FAMILY

Ours perennial herbs or twining vines, usually with milky juice; leaves usually opposite or whorled or alternate, simple, entire or undulate; flowers axillary or terminal, solitary, in umbels, or umbel-like racemes; calyces deeply 5-lobed; corollas deeply 5-lobed; crown (= corona) of various sorts (e.g., disk-like, inflated, flat, stamen-like—often called hoods; sometimes very conspicuous) present around base of the short stamen-tube; stamens 5, the filaments united (except in *Periploca*), the anthers large, united to form an anther head enclosing the two pistils (the combined column-like or disk-like structure made up of the stamens and pistils is called the gynostegium); anthers usually with a terminal outgrowth of the connective called an anther appendage and with lateral wing-like margins (= anther wings); anther wings from adjacent anthers are positioned close together and thus form a slit-like opening or groove between them (through which the legs or other appendages of pollinators can fit); pollen grains coherent in a mass (= pollinium) (except in *Periploca*); the 2 pollinia from adjacent anthers connected by a wishbone-shaped structure (= translator); clip-like portion (= corpusculum) of translator attaches the pair of pollinia to the pollinator's appendage as it is pulled out of the groove between adjacent anthers (pollinia from other flowers can enter the groove and thus be brought into contact with the enclosed stigmatic surfaces); pistils 2, united only by styles and/or stigmas; fruit of follicles; seeds numerous, usually comose.

A large (2,900 species in 315 genera), mainly tropical and warm area family with a few in temperate regions; it consists of herbs to shrubs, trees, vines, and cactus-like succulents; the family contains a number of ornamentals including *Hoya* (WAX PLANT) and *Stapelia* (CARRION-FLOWER—with flesh-like, foul-smelling flowers that attract carrion flies which serve as pollinators). ☠ There are laticifers throughout and the milky sap often contains alkaloids or other toxins such as glycosides; as a result they are avoided by most animals. However, members of the Danaidae (milkweed butterflies such as the monarch) as larvae feed primarily on members of the Asclepiadaceae and as a result are distasteful to predators (Howe 1975) due to the sequestered toxins. Asclepiadaceae typically have a specialized insect pollination mechanism in which the pollinia become attached to the legs of insects which have been guided into position by grooves or slits in the gynostegium (Macior 1965; Bookman 1981); some species have extremely unpleasant odors (carrion smells) and are pollinated by flesh-flies. The MILKWEEDS are closely related to the Apocynaceae and appear to represent a clade within that family (the relationship can be seen in the shared milky sap and pistils united only by styles and/or stigmas). From a cladistic standpoint the two families should be lumped to form a more inclusive mono-

Aristolochia reticulata [AMB]

Aristolochia serpentaria [AMB, ST2]

Aristolochia tomentosa [AMB, BB2]

Asclepias amplexicaulis [AMB, BT3]

Asclepias arenaria [AMB, BB2]

Asclepias asperula subsp. capricornu [AMB, BT3]

Asclepias engelmanniana [AMB, SHI]

phyletic Apocynaceae (Judd et al. 1994); Liede (1997), for example, treated the family as subfamily Asclepiadoideae in the Apocynaceae. (subclass Asteridae)
FAMILY RECOGNITION IN THE FIELD: herbs or vines usually with *milky sap*, frequently opposite leaves, flowers with a distinctive *corona* and a *gynostegium* (= combined structure composed of stamens and pistil), and wind-dispersed seeds with a tuft of long silky hairs. Similar to Apocynaceae (e.g., herbs or vines with milky sap and often opposite leaves) but that family has flowers without corona or gynostegium.
REFERENCES: Woodson 1941; Shinners 1964b; Bookman 1981; Hartman 1986; Rosatti 1989; Judd et al. 1994; Liede & Albers 1994; Liede 1997.

1. Central column of flower surrounded by 5 separate, fleshy-inflated or fleshy-thickened, erect or spreading stamen-like appendages (= hoods).
   2. Stems not twining, the plants erect to prostrate herbs; corollas without a fleshy disk at base under the separate appendages; corolla appendages usually with an internal horn or crest (horn or crest absent in a few species) ____ **Asclepias**
   2. Stems twining, at least towards the tip, the plants vines; corollas with a fleshy disk at base under the appendages; corolla appendages without internal horn or crest ____ **Funastrum**
1. Central column of flower or its base with 1 or 2 rows of flat thin appendages OR with a single, entire or lobed, fleshy disk or cup.
   3. Plants woody nearly throughout; leaf bases acute to broadly rounded; introduced species ____ **Periploca**
   3. Plants herbaceous or woody only in lower half; leaf bases cordate (in nc TX species); native species.
      4. Flowers brownish to purple to green; corolla lobes spreading or ascending, not erect; open flowers ca. 8–15 mm across; corolla appendages thin and flat, in 2 rows OR flower with a single, entire or lobed, fleshy disk or cup ____ **Matelea**
      4. Flowers white to cream or greenish white; corolla lobes erect, the tips spreading; open flowers ca. 6–8 mm across; corolla appendages thin and flat, in 1 row ____ **Cynanchum**

ASCLEPIAS MILKWEED, SILKWEED

Perennial herbs; stems erect to reclining; leaves alternate, opposite, or whorled, sessile or petioled; flowers in terminal or lateral umbels; corollas becoming reflexed or shallowly campanulate; hoods with or without an exserted appendage (= horn).

A genus of 100 species of the Americas, especially the United States; some are cultivated as ornamentals; others have been used medicinally since early times. Many if not most species are poisonous; however, all species are distasteful to livestock and severe losses usually occur only when animals are forced to eat the plants. Abundant MILKWEEDS in a pasture are often a sign of severe overgrazing. The poisonous principles include resinoids, cardiac glycosides, and alkaloids; the milky latex can cause dermatitis in susceptible individuals (Kingsbury 1964; Lewis & Elvin-Lewis 1977; Turner & Szczawinski 1991). Monarch butterflies (*Danaus plexippus*) feed as larvae on *Asclepias* species and obtain cardiac glycosides which provide them with protection from bird predators; the butterflies are poisonous only if they fed on poisonous plants as larvae (Scott 1986). The long silky hairs on the seeds were formerly used in making candle wicks (Ajilvsgi 1984). (Named from Greek Asklepios, god of medicine, alluding to its medicinal properties)
REFERENCES: Woodson 1947, 1953, 1954, 1962; Edwards & Wyatt 1994.

1. Leaf blades linear or narrowly linear-lanceolate, 1–5 mm wide.
   2. Leaves mostly in whorls of 3 or 4 ____ **A. verticillata**
   2. Leaves opposite or alternate.
      3. Leaves drooping, usually 10–18 cm long; plants 0.6–1.2(–1.4) m tall; hoods without horns ____ **A. engelmanniana**

3. Leaves not drooping, 3–14 cm long; plants to 0.8 m tall; hoods with horns, the horns free for much of their length and longer than the hood bodies or adnate with hood bodies and visible as the middle lobe of the hoods.
    4. Leaves opposite; rootstock of numerous thin segments ca. 1–2 mm thick; leaf blades 3–6(–9) cm long; horns free for much of their length and longer than the hood bodies **A. linearis**
    4. Leaves alternate; rootstock thickly tuberous (to ca. 15 mm thick); leaf blades usually 6–14(–18) cm long; horns adnate and visible as the middle lobe of the hoods **A. stenophylla**

1. Leaf blades lanceolate to oblong-orbicular, mostly over 8 mm wide.
    5. Leaves opposite.
        6. Umbels terminating the stems or some in upper leaf axils, held erect.
            7. Corollas greenish, tinged with rose or purple; umbels long-peduncled, the peduncle greatly exceeding the upper leaves; leaf apex obtuse to rounded, mucronate **A. amplexicaulis**
            7. Corollas white, white tinged with pink, or bright pink; peduncles various; leaf apex acuminate to broadly acute (rarely rounded).
                8. Leaf blades 4–9 cm wide, broadly oval; corollas white, the lobes 4.5–8 mm long **A. variegata**
                8. Leaf blades 4 cm or less wide, oval, ovate-elliptic, narrowly oblong to linear-lanceolate; corollas white to bright pink, the lobes 3–4 mm long.
                    9. Corollas usually bright pink (rarely white); larger leaf blades often abruptly narrowed or truncate basally; stems stout, 0.4–1.5(–2.5) m tall; inflorescences usually paired at upper nodes **A. incarnata**
                    9. Corollas white, sometimes tinted with pale pink; larger leaf blades gradually or abruptly narrowed basally; stems slender, 0.3–0.6 m tall; inflorescences solitary at upper nodes.
                        10. Follicles pendulous; seeds naked, without a coma; leaf blades narrowly lanceolate to lanceolate, to 15(–20) mm wide; mostly in low swampy areas in se and e TX, in nc TX known only from Dallas Co. **A. perennis**
                        10. Follicles erect on erect pedicels; seeds with a coma of hairs to ca. 20 mm long; leaf blades lanceolate to broadly ovate, to 35 mm wide; mostly in rocky shaded areas in c TX; in nc TX known only from s margin of area **A. texana**
        6. Umbels lateral, in leaf axils (but often in upper axils), held in a horizontal or oblique manner.
            11. Calyx lobes, pedicels, and sometimes even stems and lower surface of leaf blades conspicuously and ± densely white tomentulose **A. arenaria**
            11. Calyx lobes, pedicels, and sometimes even stems and lower surface of leaf blades glabrate to puberulent, but not white tomentulose.
                12. Pedicels 15–25 mm long; leaves petiolate, the petioles 5–25 mm long; corolla lobes 8–14 mm long; hoods with horns **A. oenotheroides**
                12. Pedicels 5–15(–20) mm long; leaves sessile or petioles to about 5(–7) mm long; corolla lobes 6–7 mm long; hoods without horns **A. viridiflora**
    5. Leaves, except lowest, alternate.
        13. Stems pilose; corollas yellow, orange, or red; hoods with horns; plants without milky latex **A. tuberosa**
        13. Stems glabrous or short pubescent; corollas green to white; hoods without horns; plants with milky latex; (previously segregated as genus *Asclepiodora*).
            14. Corolla lobes 7–12 mm long; leaf blades 1–3 cm wide, narrowly oblong- or ovate-lanceolate, the apex narrowly acuminate; anther head much wider than tall **A. asperula**
            14. Corolla lobes 13–17 mm long; leaf blades 1.5–6 cm wide, broadly oblong to ovate, the apex usually obtuse to acute; anther head ca. as wide as tall **A. viridis**

Asclepias amplexicaulis Sm., (clasping the stem), BLUNT-LEAF MILKWEED. Plant glabrous; stems erect, (20–)40–80(–100) cm tall; upper leaves sessile, clasping; leaf blades ovate to elliptic-oblong, 4–12 cm long, 1.8–8 cm wide, apically obtuse to rounded and often mucronate; corolla lobes greenish, tinged with rose or purple; gynostegium pale purple or rose; hoods with horns. Sandy open woods, roadsides, and old fields; Dallas, Grayson, and Tarrant cos., also Lamar Co. (Carr 1994); se and e TX w to East Cross Timbers. May.

Asclepias arenaria Torr., (of sand), SAND MILKWEED. Stems solitary, 20–50(–80+) cm tall, pubescent; leaf blades ca. 5–10 cm long, 3–8 cm wide; petioles usually 10–18 mm long; corolla lobes pale green, 7.5–11 mm long; gynostegium white to cream; hoods with horns. Sandy soils, roadsides; included based on citation of vegetational area 4 (Fig. 2) by Hatch et al. (1990); mainly w 1/2 of TX and s part of the state. Jun–Sep.

Asclepias asperula (Decne.) Woodson subsp. **capricornu** (Woodson) Woodson, (sp.: rough; subsp.: like a goat's horn), ANTELOPE-HORNS, TRAILING MILKWEED. Stems partly decumbent or low-spreading, 10–60 cm long, minutely pubescent or largely glabrous; petioles 2–7 mm long; corolla lobes pale yellowish green, sometimes purple-tinged; gynostegium greenish cream to dark purple; hoods widespreading, without horns. Rocky or sandy prairies; Blackland Prairie w to Rolling Plains and Edwards Plateau. Apr–Jun, sporadically through summer, fall. [*A. asperula* (Decne.) Woodson var. *decumbens* (Nutt.) Shinners, *A. capricornu* Woodson, *Asclepiodora decumbens* (Nutt.) A. Gray] The common name ANTELOPE-HORNS is said to come from the curved fruits (Kirkpatrick 1992). Reportedly poisonous to livestock (Hartman 1986). ☠ ▣/79

Asclepias engelmanniana Woodson, (for George Engelmann, 1809–1884, German-born botanist and physician of St. Louis), ENGELMANN'S MILKWEED. Stems (30–)60–120(–140) cm tall, glabrous or nearly so; leaf blades sessile, linear, usually 10–20 cm long, 1–3(–5) mm wide; corolla lobes pale green, purple-tinged; gynostegium yellowish; hoods without horns. Sandy or rocky ground; Brown, Collin, Cooke, Jack, and Tarrant cos., also Hamilton (HPC), Palo Pinto, Somervell (Woodson 1954) cos.; w 2/3 of TX. May–Sep.

Asclepias incarnata L., (flesh-colored), SWAMP MILKWEED. Stems erect, usually (40–)70–150(–250) cm tall; leaf blades linear-lanceolate to lanceolate or ovate-elliptic, (3–)5–15 cm long, to ca. 4 cm wide; petioles 3–10(–17) mm long; corolla lobes bright pink (rarely white); gynostegium pale pink (rarely white); hoods with horns. Wet ground. May–Oct.

1. Plants glabrous or nearly so ____ subsp. **incarnata**
1. Plants pubescent ____ subsp. **pulchra**

subsp. **incarnata**. Dallas Co. (found by Reverchon at Buzzards Spring, not seen recently), also Hood Co. (Woodson 1954); nc TX w through much of w 1/2 of TX.

subsp. **pulchra** (Ehrh. ex Willd.) Woodson, (handsome). Indicated by Woodson (1954) to be introduced in nc TX (mapped apparently as Tarrant Co.). [*A. incarnata* L. var. *pulchra* (Ehrh. ex Willd.) Pers.]

Asclepias linearis Scheele, (narrow, with sides nearly paralled), SLIM MILKWEED. Stems erect, usually 20–50 cm tall, glabrous or inconspicuously pubescent; leaves sessile or subsessile; leaf blades linear, 1–4 mm wide; umbels axillary on upper part of stem; corolla lobes greenish white (rarely brownish or purplish); gynostegium white; hoods with horns. Dry prairies; collected at Buzzards Spring, Dallas, in October 1902 by Reverchon (Mahler 1988), also Milam Co. (Woodson 1954); endemic to TX from coast n to Dallas Co. May–Oct.

Asclepias oenotheroides Cham. & Schltdl., (resembling *Oenothera*—evening primrose), SIDE-CLUSTER MILKWEED, HIERBA DE ZIZOTES. Stems erect or spreading, 10–45 cm long, minutely scabrous-pubescent; leaf blades oblong-lanceolate to elliptic, 2.5–12 cm long, 1.5–6 cm wide; petioles 5–25 mm long; corolla lobes greenish white to yellow, gynostegium pale greenish cream; hoods with horns. Sandy or gravelly ground, especially disturbed places; Clay, McLennan,

Asclepias incarnata subsp. incarnata [AMB, COR]

Asclepias incarnata subsp. pulchra [BB2]

Asclepias linearis [AMB, HEA]

Asclepias oenotheroides [AMB, HEA]

Asclepias perennis [AMB, CO1]

Asclepias stenophylla [AMB, BB2]

Asclepias texana [FMC]

Asclepias tuberosa subsp. interior [AMB, BT3]

Somervell, Tarrant, and Wise cos., also Brown, Comanche, Hamilton (HPC), Bell, and Dallas (Woodson 1954) cos.; mainly the w 2/3 of TX. May–Oct.

Asclepias perennis Walter, (perennial), SHORE MILKWEED, THIN-LEAF MILKWEED. Stems erect, 30–50 cm tall; plant glabrous except for upper stem, peduncles, and pedicels; leaf blades narrowly to broadly lanceolate, 5–14 cm long, to 15(–20) mm wide; petioles to 15 mm long; corolla lobes white, usually suffused with pale pink or lavender; gynostegium white; hoods with horns; seeds ca. 15 mm long. Damp woods, swampy areas; collected by Reverchon at Dallas, not found there recently (Mahler 1988); mainly se and e TX. Jun–Jul. According to Woodson (1954), the pendulous follicles and large naked seeds may be adaptations to dispersal by water. Woodson (1954) indicated that he could not distinguish this species from *A. texana* except by the strikingly different fruit and seed, and the geographical distribution.

Asclepias stenophylla A. Gray, (narrow-leaved), SLIM-LEAF MILKWEED, NARROW-LEAF MILKWEED. Plant superficially resembling *A. verticillata*, densely pubescent above, with larger corollas; stems erect, 20–80(–100) cm tall, minutely puberulent to glabrate; leaves sessile; leaf blades usually 6–14(–18) cm long; corolla lobes pale greenish cream or yellow, ca. 6 mm long; gynostegium pale greenish cream or white; hoods 3–4 mm long, with horns. Prairies or open ground; Cooke, Grayson, and McLennan cos., also Henderson and Tarrant cos. (Woodson 1954); se and e TX w to nc TX, also Edwards Plateau. Jun–Aug.

Asclepias texana A. Heller, (of Texas), TEXAS MILKWEED. Stems erect, to ca. 50 cm tall, with inconspicuous pubescence in lines; leaf blades lanceolate to broadly ovate, 20–70 cm long, to 35 mm wide; inflorescences solitary at the uppermost nodes; flowers showy; corolla lobes white, ca. 3 mm long; gynostegium white; hoods with horns; follicles erect, 9–12 cm long; seeds ca. 8 mm long, with a white coma to ca. 20 mm long. Rocky shaded areas; Burnet Co. on the very s margin of nc TX (Balcones Canyonlands Nat. Wildlife Refuge, C. Sexton, pers. comm.), also Travis Co. just s of nc TX; mainly Edwards Plateau and Trans-Pecos. Jun–Aug.

Asclepias tuberosa L. subsp. **interior** Woodson, (sp.: tuberous; subsp.: presumably referring to its range in the interior of the North American continent), BUTTERFLY-WEED, BUTTERFLY MILKWEED, ORANGE MILKWEED, CHIGGER-FLOWER, CHIGGERWEED, PLEURISY-ROOT. Plant without milky latex; stems spreading to erect, to 90 cm long, pilose; leaves many, crowded, oblong-lanceolate, usually 3–12 cm long, to 23(–30) mm wide, ± sessile and clasping or petioles to 5 mm long; corolla lobes usually orange (rarely yellowish or reddish); gynostegium usually orange, occasionally yellowish; hoods with horns. Sandy open woods or occasionally silty clay prairies; throughout TX, more in e 2/3 of the state. May–Jul. [*A. tuberosa* L. subsp. *terminalis* Woodson] Reportedly poisonsous to livestock (Hartman 1986). ☠ ▣/79

Asclepias variegata L., (variegated, irregularly colored), WHITE-FLOWER MILKWEED. Plant glabrous except for upper stem, peduncles, and pedicels; stems erect, (20–)30–120 cm tall; leaf blades oblong-lanceolate to suborbicular, 5–15 cm long, 4–9 cm wide; petioles 10–20 mm long; corolla lobes white; gynostegium white except for purplish column; hoods with short horns. Sandy woods; found by Reverchon at Dallas, not seen there recently (Mahler 1988); mainly e TX. Apr–Jun. ▣/79

Asclepias verticillata L., (whorled), WHORLED MILKWEED, EASTERN WHORLED MILKWEED. Stems erect, usually 30–90 cm tall, puberulent in lines; leaves mostly in whorls; leaves sessile or nearly so; leaf blades filiform to linear, (1.5–)3–8 cm long, 0.5–2(–3) mm wide, often revolute; corolla lobes greenish white, sometimes with a hint of purple, ca. 3.5 mm long; gynostegium greenish white; hoods ca. 1.5 mm long, with horns. Sandy open woods, rocky slopes, or prairies; se and e TX w to West Cross Timbers and Edwards Plateau, also Trans-Pecos. May–Jul, sporadically later. Reported to be poisonous to livestock (Hartman 1986). ☠

Asclepias variegata [AMB, BTE]

Asclepias verticillata [REE]

Asclepias viridiflora [AMB, BT3]

Asclepias viridis [AMB, STE]

Cynanchum barbigerum [HEA]

Cynanchum laeve [BB2]

Cynanchum racemosum var. unifarium [JAC, TOR]

Asclepias viridiflora Raf., (green-flowered), GREEN-FLOWER MILKWEED, GREEN ANTELOPE-HORNS, WAND MILKWEED. Stems erect, usually 20–80(–100) cm tall, glabrous or densely and finely pubescent; leaves sessile or petioles to about 5(–7) mm long; leaf blades quite variable, from narrowly lanceolate to oblong-obovate or suborbicular, 4–13(–14) cm long, 1–5(–6) cm wide; corolla lobes and gynostegium pale green; hoods without horns. Sandy or calcareous areas, gravelly open ground, or prairies; throughout TX. Jun–Aug. [*A. viridiflora* Raf. var. *lanceolata* Torr.] ▣/79

Asclepias viridis Walter, (green), GREEN MILKWEED, ANTELOPE-HORNS. Plant glabrous or upper stem and young leaves minutely pubescent; stems low-spreading to suberect, usually 25–60 cm long; leaf blades broadly oblong to ovate, 4–13 cm long, 1.5–6 cm wide; petioles 3–10 mm long; corolla lobes pale green; gynostegium pale purplish rose; hoods wide-spreading, without horns. Prairies, ditch banks, pastures, and disturbed ground; se and e TX w to West Cross Timbers and Edwards Plateau. Apr–Jun, sporadically in summer and fall. [*Asclepiodora viridis* (Walter) A. Gray] Probably the most common milkweed in nc TX; sometimes extremely abundant in overgrazed pastures.

CYNANCHUM SWALLOW-WORT

Twining vines to 3 m or more long; leaves opposite, petioled, basally cordate; inflorescences axillary; follicles lanceolate in outline, smooth.

A genus of 200 species of tropical and warm areas of the world. (Greek: *cyon*, dog, and *anchein*, to strangle; an ancient name for some plant supposedly poisonous to dogs)
REFERENCE: Sundell 1981.

1. Leaf blades < 10(–12) mm wide, with broadly rounded base, not cordate; petioles usually 1–4(–8) mm long; rare on s edge of nc TX ____ **C. barbigerum**
1. Leaf blades 10–70 mm wide, cordate at base; petioles to 50(–90) mm long; widespread in nc TX.
   2. Petals ca. 6–7 mm long; peduncle mostly shorter than petiole of subtending leaf; corolla appendages nearly as long as the corolla, deeply divided into linear segments ____ **C. laeve**
   2. Petals ca. 3.5 mm long; peduncle about as long as or longer than petiole of subtending leaf; corolla appendages < 2/3 as long as the corolla, broadly oblong with toothed or lobed summit ____ **C. racemosum**

Cynanchum barbigerum (Scheele) Shinners, (bearded, presumably in reference to the densely long pilose inner surfaces of the corolla lobes), BEARDED SWALLOW-WORT, THICKET THREADVINE. Leaf blades linear to linear-oblong or elliptic-lanceolate, to 5 cm long, usually shorter; flowers white or creamy white, in umbellate clusters of 5 or fewer; corollas 3.6–5.2 mm long; corolla lobes strongly recurved above middle; follicles to 5 cm long. In open rocky areas and on herbs and shrubs; Burnet Co. (Balcones Canyonlands Nat. Wildlife Refuge, C. Sexton, pers. comm.) on s margin of nc TX s to s TX and w to Trans-Pecos. May–Jul. [*Metastelma barbigerum* Scheele]

Cynanchum laeve (Michx.) Pers., (smooth), BLUEVINE, SANDVINE, SMOOTH SWALLOW-WORT, SMOOTH ANGLEPOD. Leaf blades triangular-lanceolate to deltoid; petioles to 45(–90) mm long; flowers white, in peduncled umbellate clusters or abbreviated racemes; follicles to 15 cm long. Low moist woods or fields; Cooke, Collin, Dallas, Grayson, Hill, Hood, Palo Pinto, and Tarrant cos., also Brown Co. (HPC); in much of the e 1/2 of TX. Aug–Sep.

Cynanchum racemosum (Jacq.) Jacq. var. **unifarium** (Scheele) Sundell, (sp.: with flowers in racemes; var.: of a single rank or series), TALAYOTE. Leaf blades broadly ovate to ovate-lanceolate or triangular-lanceolate with stipule-like small leaves from leaf axils; petioles to 50 mm long; flowers cream or greenish white, in racemes; follicles to 10 cm long. Sandy forests and thickets; Palo Pinto and Tarrant (Fort Worth Nature Center) cos.; also Parker Co. (R. O'Kennon, pers. obs.); also Edwards Plateau, s TX, and Trans-Pecos. May–Oct. [*C. unifarium* (Scheele) Woodson]

FUNASTRUM TWINEVINE

Twining vines; leaves opposite, petioled, with cordate bases; inflorescences of peduncled, umbellate clusters in leaf axils; follicles smooth, fusiform, not angled.

The species here treated in *Funastrum* have traditionally been placed in *Sarcostemma* (e.g., Correll & Johnston 1970; Kartesz 1994; Jones et al. 1997), a genus in the broad sense of ca. 40 species of leafy or leafless vines with photosynthetically active stems (Liede 1996); it is native to tropical and warm areas of the world. Based on a cladistic analysis, Liede (1996) argued convincingly that *Sarcostemma* be restricted to Old World taxa, while the ca. 15 species in the New World be recognized in *Funastrum*; the similar floral structures of the 2 groups are interpreted as parallel evolution. We are following Liede (1996) and J. Kartesz (pers. comm. 1997) in recognizing *Funastrum*. (Greek: *funis*, rope, presumably in reference to the vine habit)
REFERENCES: Holm 1950; Jones & Jones 1991; Liede 1996.

1. Leaf margins distinctly crisped or ruffled; corollas brownish green or bronze; calyx lobes mostly > 3 times as long as wide ____ **F. crispum**
1. Leaf margins neither crisped nor ruffled; corollas white to pale lavender or pinkish; calyx lobes mostly < 3 times as long as wide ____ **F. cynanchoides**

Funastrum crispum (Benth.) Schltr., (crisped, curled), WAVY-LEAF TWINEVINE, WAVY-LEAF MILKWEEDVINE. Glabrous, low-climbing or tangle-forming vine; leaf blades narrowly triangular and elongate to linear, acuminate, blue-green, with ruffled, occasionally purple-tinged margin; calyx lobes 3–6 mm long, narrowly lanceolate; follicles to 12 cm long. In gravelly or silty limestone soils; Dallas (*Reverchon* in "light soil, West Dallas"), McLennan, and Palo Pinto cos., also Callahan, Dallas (Holm 1950), Johnson, Parker, and Tarrant (R. O'Kennon, pers. obs.) cos.; nc TX s and w to Trans-Pecos. May–Aug. [*Sarcostemma crispum* Benth.] /90

Funastrum cynanchoides (Decne.) Schltr., (resembling *Cynanchum*—swallow-wort), CLIMBING-MILKWEED, TWINEVINE, WHITE TWINEVINE. Glabrous to sparsely puberulent twining to trailing vine; leaves broadly to narrowly ovate-lanceolate to triangular-lanceolate, acute to acuminate, foliage with noticably very unpleasant odor; calyx lobes 2–3 mm long, ovate to ovate-linear; follicles to 7 cm long. Sandy or rocky soils, waste places; Grayson and Lamar cos., also Archer, Brown, Young (Holm 1950), and Tarrant (R. O'Kennon, pers. obs.) cos.; mainly w 1/2 of TX. Apr–Sep. [*Sarcostemma cynanchoides* Decne.] /90

MATELEA MILKVINE, ANGLEPOD

Plants variously pubescent or pilose, often with burned-rubber odor; stems trailing to suberect, or twining and climbing; leaves opposite, short- to rather long-petioled; leaf blades ovate to oblong-ovate, with cordate base; flowers axillary; corollas rotate; follicles smooth or with spine-like tubercules, angled or not.

A genus of 180 species native to tropical and temperate regions of the New World. (Derivation of generic name not explained by original author)
REFERENCES: Shinners 1950b; Correll 1965.

1. Flowers in peduncled, axillary, several- or many-flowered umbels or umbel-like racemes; stems twining or climbing; leaf blades to 17 cm long, typically > 6 cm.
   2. Corollas greenish brown to brownish purple, without a network of fine veins on the upper surface, the lobes linear-oblong or linear-lanceolate; fruits smooth OR with sharp, spike-like tubercles.
      3. Plants without glandular puberulence; sepals glabrous or ciliate only at apex; peduncle of inflorescence ca. the same length as pedicels; fruits smooth; widespread in nc TX ____ **M. gonocarpos**

3. Plants minutely glandular-puberulent; sepals pubescent; peduncle of inflorescence much longer than pedicels; fruits with sharp, spine-like tubercles; e margin of nc TX ____ **M. decipiens**

2. Corollas green, with a network of fine veins on at least part of the upper surface, the lobes ovate-lanceolate or elliptic; fruits with sharp spine-like tubercules.

4. Upper surface of corollas densely puberulent; only distal part of upper surface of corolla lobes with network of fine veins (with dark green parallel veins below middle); stems and leaves with only sparse short pubescence of curved or appressed hairs; leaves without a strong bad odor; crown (= corona of flower) with 5 short, distinct, rounded, spreading appendages; peduncles to 12 mm long, shorter than the subtending petiole; in nc TX known only from Dallas Co. ____ **M. edwardsensis**

4. Upper surface of corollas glabrous; entire upper surface of corollas with network of fine veins; stems and leaves with both long spreading hairs and short glandular hairs; leaves with a strong bad odor; crown without 5 distinct appendages; peduncles usually as long as or longer than the subtending petiole; scattered in nc TX ____ **M. reticulata**

1. Flowers mostly in 2s, with pedicels attached directly in leaf axils; stems trailing to suberect, not twining; leaf blades 5(–6.5) cm or less long, often much less.

5. Pedicels shorter than adjacent petioles; corolla lobes 4–7 mm long, dark red-brown or purple brown ____ **M. biflora**

5. Pedicels when fully expanded longer than adjacent petioles; corolla lobes 2–4 mm long, dark greenish brown ____ **M. cynanchoides**

Matelea biflora (Raf.) Woodson, (two-flowered), TWO-FLOWER MILKVINE. Plant usually trailing; stems 10–40 cm long. Praires or open ground, usually limestone areas; Blackland Prairie w to Rolling Plains, also Edwards Plateau. Apr–Jun, occasionally in fall. ▣/98

Matelea cynanchoides (Engelm.) Woodson, (resembling *Cynanchum*—swallow-wort), MILKVINE. Plant with odor like that of burned rubber, usually ascending to suberect; stems 15–50 cm long. Loose sandy soils, open oak woods; Montague, Navarro, Parker, Tarrant, and Young cos.; se and e TX w to Edwards Plateau and Rolling Plains. Apr–Aug.

Matelea decipiens (Alexander) Woodson, (deceptive, not obvious), CLIMBING-MILKWEED. Plant climbing on other plants, with spreading pubescence (hairs 1–2 mm long) and minute glandular puberulence; petioles ca. equal in length to inflorescences; flowers 5–25 per inflorescence; calyx lobes with short pubescence; corollas brownish purple; corolla lobes to 15 mm long; fruits somewhat angled. Sandy soils, wooded areas; Grayson and Henderson cos., also Lamar Co. (Carr 1994); mainly se and e TX. Apr–Jun. [*Gonolobus decipiens* (Alexander) L.M. Perry] A revision currently in progress may separate this species into the genus *Gonolobus*.

Matelea edwardsensis Correll, (of the Edwards Plateau), PLATEAU MILKVINE. Plant climbing on other plants; leaf blades to 7.5 cm long and 7 cm wide; petioles to 6 cm long; pedicels to ca. 10 mm long; calyx pubescence short; calyx lobes ca. 3 mm long; corolla lobes ca. 8 mm long and 4 mm wide; fruits ca. 9–10 cm long. On limestone, open woods; Dallas Co. (Cedar Hill State Park—collected by Paul Baldon), also Fort Hood (Bell or Coryell cos.—Sanchez 1997); otherwise known only from a few counties on the Edwards Plateau; endemic to TX. Apr–May. ▣/98

Matelea gonocarpos (Walter) Shinners, (angled-fruit), ANGLEPOD. Plant high-climbing, with spreading pubescence (hairs ca. 1 mm long); leaf blades large, thin; petioles longer than inflorescences; flowers 2–12 per inflorescence; corollas greenish brown to brownish purple; corolla lobes to ca. 14 mm long; fruits smooth, sharply angled. Stream bottom woods; se and e TX w to East Cross Timbers and Lampasas Cut Plain. May–Jun. [*Gonolobus gonocarpus* (Walter) L.M. Perry] A revision currently in progress may separate this species into the genus *Gonolobus*.

Matelea reticulata (Engelm. ex A. Gray) Woodson, (netted) NET-VEIN MILKVINE, GREEN MILK-

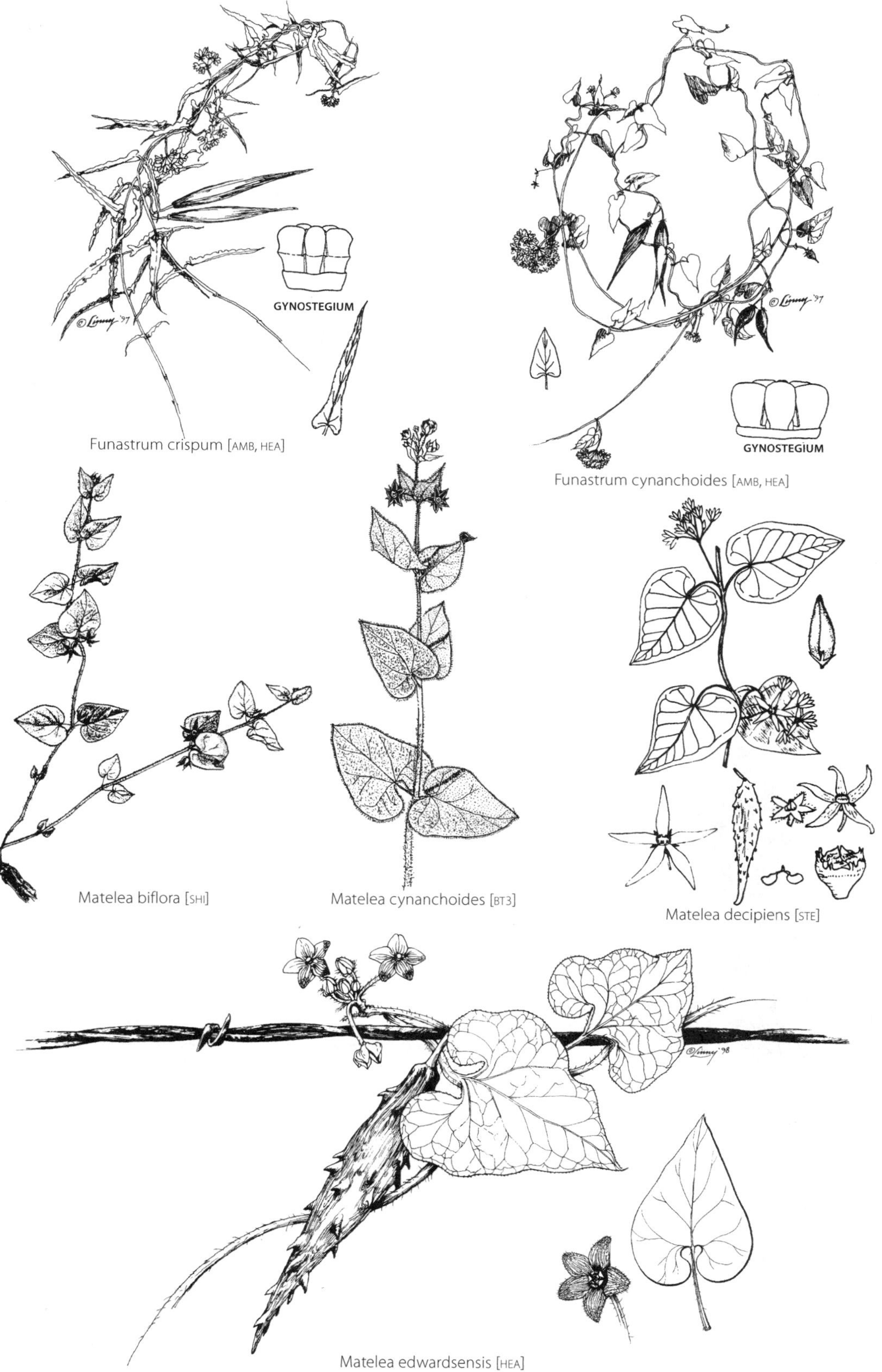

Funastrum crispum [AMB, HEA]

Funastrum cynanchoides [AMB, HEA]

Matelea biflora [SHI]

Matelea cynanchoides [BT3]

Matelea decipiens [STE]

Matelea edwardsensis [HEA]

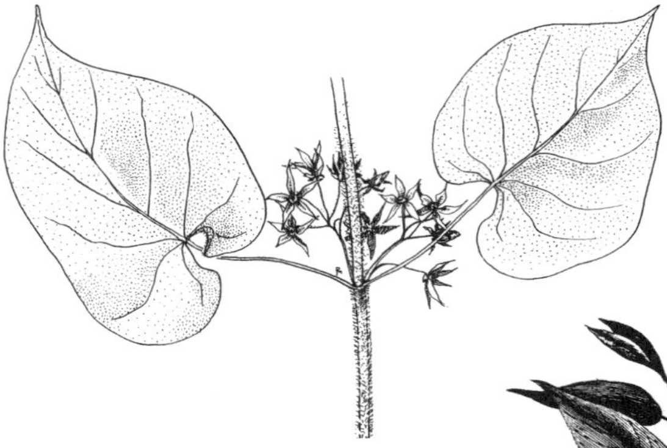

Matelea gonocarpa [BT3]

Matelea reticulata [BT3]

Periploca graeca [LAM]

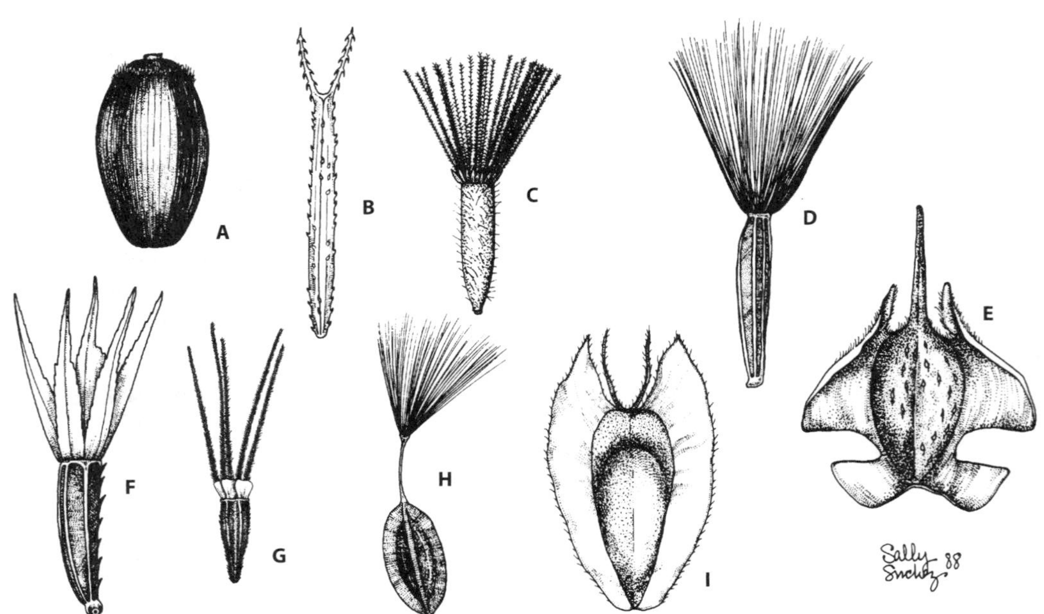

Asteraceae fruit types. A: Without pappus (e.g., *Helianthus* sp.); **B**: With pappus of 2, barbed awns (e.g., *Bidens* sp.); **C**: With dimorphic pappus: Inner long and outer short bristles (as in *Heterotheca* sp.); **D**: Ribbed with nearly uniform pappus bristles (e.g., *Eupatorium* sp.); **E**: Winged, with spinescent style (e.g., *Soliva* sp.); **F**: Ribbed, with pappus of 5 awn-like scales (e.g., *Ageratum* sp.); **G**: With dimorphic pappus: bristles alternating with scales (e.g. *Krigia* sp.); **H**: With pappus bristles at the tip of the beak (e.g., *Lactuca*); **I**: Winged with pappus of 2 awns (e.g., *Verbesina* sp.). [GAN]

WEEDVINE. Plant climbing on other plants, with long, spreading pubescence (hairs 1–2 mm long) and minute glandular pubescence; petioles to 5.5 cm long; peduncle ca. as long as or exceeding subtending petiole; inflorescences few-flowered; calyx pubescence long, the hairs ca. as long as calyx lobes are wide; calyx lobes 3–4 mm long; corollas 10–15 mm across; fruits to 15 cm long. Thickets on rocky hillsides; Bell, Burnet, Palo Pinto, and Parker cos., also Brown, Comanche, Eastland (HPC), and Johnson (R. O'Kennon, pers. obs.) cos.; also c, s, and w TX. May. ▣/98

PERIPLOCA SILKVINE

A genus of 11 species native to the Mediterranean region, e Asia, and tropical Africa; some are cultivated as ornamentals; others are used medicinally. *Periploca* is a member of the Periplocoideae, a less specialized subfamily, in contrast to the rest of the members of the family in North Central Texas which are in the Asclepiadoideae. (Greek: *peri*, around and *ploke*, twining, alluding to the climbing habit of some species)
REFERENCE: Venter & Verhoeven 1997.

Periploca graeca L., (of Greece), SILKVINE. Twining woody vine to 5 m long; leaves opposite, entire, petiolate; flowers in long pedunculate terminal cymes; corollas brown-purple, rotate, deeply lobed, densely villous in lines; corona a ring of 5 broad lobes, each alternating with a filiform apically cleft lobe 5–10 mm tall; filaments distinct; follicles linear, smooth, not angled. Cultivated and escapes; Dallas Co.; in TX apparently only known from the Blackland Prairie. May. Introduced from se Europe and w Asia.

ASTERACEAE (COMPOSITAE) SUNFLOWER OR DAISY FAMILY

Annual or perennial herbs or more rarely shrubs; leaves basal, alternate, opposite, or whorled, simple or compound, entire, toothed, or lobed, not stipulate (but small basal lobes sometimes resemble stipules); inflorescence a single involucrate head, or several or many involucrate heads in corymbs, racemes, or panicles, each head simulating a single flower; involucre of one or more rows of separate or united bracts (termed phyllaries to avoid confusion with bracts on peduncles) imitating sepals; flowers (= florets) without typical calyx, but commonly with modified calyx of hairs, scales, or teeth (= pappus) on summit of the inferior ovary; corollas of two basic types: disk, mainly tubular (varying from thread-like to tubular-funnelform or campanulate with cylindrical basal tube), with 4 or 5 equal or unequal teeth or lobes; and ray, with small basal tube and broad, strap-shaped main portion (= ligule); stamens none or 5, attached inside the corolla tube; anthers separate or united into a ring; pistil 1; style 1, commonly with 2 branches; ovary inferior; fruit an achene (see various types on facing page).

The Asteraceae is one of the two largest families of flowering plants (Orchidaceae is the other), containing ca. 21,000–25,000 species (22,750 species in 1,528 genera—Mabberley 1997). It is a cosmopolitan family of mainly herbs to shrubs and is of significant economic importance as a source of food plants (e.g., *Lactuca*—LETTUCE, *Cynara scolymus* L.—ARTICHOKE), oil (e.g., *Helianthus*—SUNFLOWER and *Carthamus*—SAFFLOWER), and numerous ornamentals (e.g., *Aster*, *Bidens*, *Cosmos*, *Dahlia*, *Helianthus*, *Tagetes*—MARIGOLD). Many are weeds including *Taraxacum* species (DANDELIONS) and *Cirsium* species (THISTLES); ☠ in some parts of the world poisonous species of *Senecio* are a major cause of livestock poisoning; wind-pollinated genera such as *Ambrosia* (RAGWEED), *Iva* (SUMPWEED), and *Artemisia* (SAGEBRUSH), which produce large quantities of allergenic pollen, are important causes of hay fever (Lewis et al. 1983). Cronquist (1981) suggested that the evolutionary success of the family may be in part due to a diversified chemical defense system, including polyacetylenes and sesquiterpene lactones. The 351 species of Asteraceae found in OK make it the largest family in that state (Taylor & Taylor 1994) and the 620 species (almost 13% of the TX flora) in TX likewise make it the largest TX family (Hatch et al. 1990). The Asteraceae is also the largest family in the nc TX flora; the 263 species represent nearly 12% of the 2,223 species known in nc TX. (subclass Asteridae)

FAMILY RECOGNITION IN THE FIELD: usually herbs or rarely shurbs with a characteristic inflorescence: flowers in a compact *head* subtended by bracts (= phyllaries)—the inflorescence resembling a single flower (the heads are often grouped together to form compound inflorescences); corollas sympetalous, 5-merous; stamens united by their anthers; fruit a 1-seeded achene often topped by a *pappus* of hairs, scales, or teeth.

REFERENCES: Rydberg 1914–1927; Sharp 1935; Solbrig 1963; Vuilleumier 1969a, 1973; Moore & Frankton 1974; Carlquist 1976; Cronquist 1977, 1980, 1985; Gandhi & Thomas 1989; Scott 1990; Jansen et al. 1991, 1992; Bremer 1994; Nesom 1994a; Taylor 1997; Arriagada & Miller 1997.

KEYS TO GENERA OF ASTERACEAE

Modified from *Keys to the Asteraceae of Oklahoma* (Taylor 1997) contributed by

CONSTANCE E.S. TAYLOR (DUR)

ARTIFICIAL KEY TO THE GENERA OF ASTERACEAE

(For Technical Key to Tribes of Asteraceae see page 298)

1. Heads with only ray flowers (with strap-shaped ligules), these all perfect; plant sap usually milky ____ Use Key to **Lactuceae Tribe** (page 306)
1. Heads with disk flowers (with tubular eligulate corollas); ray flowers present or absent at the margins of the heads, these pistillate or neutral; plant sap usually watery.
   2. Heads with ray flowers, the ligules evident.
      3. Rays yellow to orange, sometimes with reddish-brown, rusty-brown, or maroon at base.
         4. Receptacle chaffy, a bract associated with each flower, or with bristles. ____ **Key 1**
         4. Receptacle naked ____ **Key 2**
      3. Rays white, or pink, or lavender, or blue, or reddish purple.
         5. Pappus of capillary bristles ____ **Key 3**
         5. Pappus of awns, scales, scaly bristles, a minute crown, or absent. ____ **Key 4**
   2. Heads with only disk flowers or apparently so (ray flowers sometimes present with corollas inconspicuous or reduced to a ring or absent).
      6. Phyllaries with either hooked or straight spines or tubercles OR large teeth at edges, sometimes united into a bur; leaves frequently also spiny; receptacle chaffy or bristly (COCKLEBURS, RAGWEEDS, THISTLES) ____ **Key 5**
      6. Phyllaries without hooked or straight spines or tubercles or large teeth, not united into a bur; leaves not spiny; receptacle various.
         7. Disk flowers pink, blue, purple, green, white, or yellowish cream ____ **Key 6**
         7. Disk flowers yellow, reddish brown, rusty-brown, or maroon ____ **Key 7**

Key 1. Ray and disk flowers both present; ray flowers yellow to orange, sometimes with reddish brown or rusty-brown or maroon marks; receptacle chaffy.

1. Receptacle chaff bristly; pappus of awned scales, the awns ca. as long as the scale bodies ____ **Gaillardia**
1. Receptacle chaff of bracts, associated with each flower; pappus various.
   2. Disk flowers sterile; ray flowers only fertile and maturing achenes, these much larger than those of the sterile disk flowers.
      3. Ray achenes thick, rounded or weakly compressed, without wings; leaf blades usually ± palmately 3–5-lobed ____ **Smallanthus**

General Characteristics of the Asteraceae (Compositae) Family [JEP]

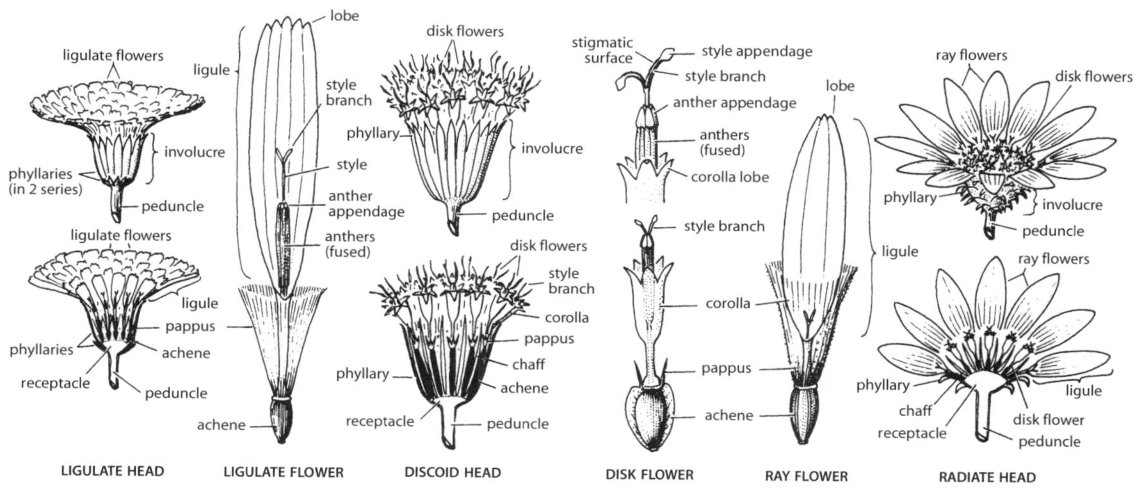

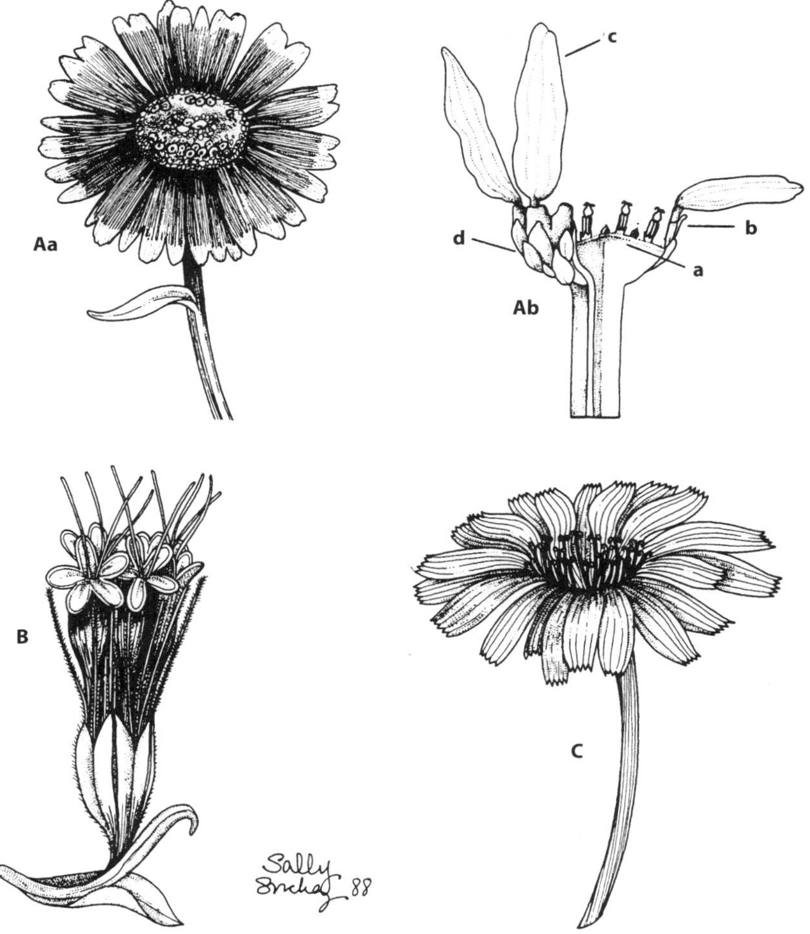

Asteraceae head types. Aa: Radiate head (e.g., *Gallardia*); **Ab**: Vertical section through radiate head; **a**: Receptacle; **b**: Disk flowers; **c**: Ray flowers; **d**: Phyllaries; **B**: Discoid head (e.g., *Eupatorium*); **C**: Ligulate head (e.g., *Cichorium*). [GAN]

3. Ray achenes strongly flattened, with or without wings; leaf blades pinnately lobed or unlobed.
 4. Achenes broadly winged.
  5. Ray flowers usually 5, in 1 row; phyllaries 8 to 10 in 2 rows, the outer row distinctly narrower; disks less than 1.5 cm wide; plants taprooted annuals up to 0.3(–0.65) m tall ____ **Lindheimera**
  5. Ray flowers numerous, in 2 to 3 rows; phyllaries numerous, in several rows, ± similar except for size; disks (1.5–)2–3 cm wide; plants coarse perennials up to 2(–3) m tall ____ **Silphium**
 4. Achenes not winged.
  6. Ray flowers with back of corolla having green, red, or maroon veins; leaves crenate, serrate, dentate, or lyrate-pinnatifid (in one species rare in nc TX), variously pubescent but not hispid; pappus none ____ **Berlandiera**
  6. Ray flowers with back of corolla entirely yellow, without colored veins; leaves deeply pinnatifid, densely hispid; pappus of a few scales ____ **Engelmannia**

2. Disk flowers fertile; ray flowers fertile or neutral.
 7. Receptacle columnar.
  8. Leaves opposite at base, may be alternate or opposite on upper stem.
   9. Plants creeping and rooting at the nodes, the flowering branches ascending; leaves neither connate-perfoliate nor pinnatifid; phyllaries linear, subequal; disks 4–9 mm wide ____ **Acmella**
   9. Plants ± erect or ascending, not creeping, not rooting at the nodes; leaves either connate-perfoliate or pinnatifid; outer phyllaries ovate, foliaceous, much larger than inner phyllaries; disks > 10 mm wide ____ **Tetragonotheca**
  8. Leaves all alternate.
   10. Leaves simple, the blades entire or toothed.
    11. Leaves clasping; achenes terete ____ **Dracopis**
    11. Leaves not clasping; achenes 4-angled ____ **Rudbeckia**
   10. Leaves pinnately compound or deeply lobed ____ **Ratibida**
 7. Receptacle flat, or convex, or very low conic.
  12. Leaf bases decurrent; middle and lower stems winged ____ **Verbesina**
  12. Leaf bases not decurrent; stems not winged.
   13. Phyllaries dimorphic, the inner longer and enclosing the head; the outer smaller, less conspicuous; achenes flattened parallel to the phyllaries (or rarely 4-angled in *Bidens*).
    14. Inner phyllaries united for up to 1/2 their length; disk corolla lobes often linear ____ **Thelesperma**
    14. Phyllaries all separate or nearly so; disk corolla lobes triangular to ovate.
     15. Pappus of 2(–4) rigid, retrorsely barbed awns ____ **Bidens**
     15. Pappus of 2(–4) awns or teeth which are inconspicuously barbed or not barbed OR pappus absent.
      16. Inner phyllaries spreading at anthesis; achenes not winged and disk corollas yellow ____ **Bidens**
      16. Inner phyllaries tightly enclosing heads; achenes winged and all corollas yellow OR if achenes not winged, then disk corollas reddish brown ____ **Coreopsis**
   13. Phyllaries all similar, imbricate, or slightly unequal, but not dimorphic in 2 rows; achenes various.
    17. Leaves, including upper ones, mostly opposite.
     18. Leaves linear, 10–30 mm long, to only ca. 2 mm wide; plants erect, 8–22 cm tall ____ **Zinnia**

18. Leaves larger, usually much so; plants either decumbent OR much greater than 22 cm tall.
   19. Plants decumbent, slender, 5–30 cm tall; ray flowers 3–8, with corollas 4 mm or less long; mainly a lawn weed ______ **Calyptocarpus**
   19. Plants erect, coarse; much taller than 30 cm; ray flowers usually 8–numerous, with corollas more than 4 mm long, usually much more; widespread in native habitats.
      20. Achenes not winged; ray flowers neutral, without styles and stigmas; widespread in nc TX ______ **Helianthus**
      20. Achenes winged or with wing-like margins; ray flowers pistillate (fertile or infertile), with styles present; s and w parts of nc TX.
         21. Leaves with distinct petioles; petioles with conspicuous auricles basally; pappus of 2 short awns, usually absent on mature achenes plants; plants perennial herbs ______ **Simsia**
         21. Leaves sessile; auricles lacking; pappus of 2 or 3 short awns subtended by short, hyaline scales; plants small shrubs ______ **Wedelia**

17. Upper leaves alternate.
   22. Achenes winged; petioles auriculate dilated at base OR not so ______ **Verbesina**
   22. Achenes not winged; petioles not auriculate.
      23. Ray flowers neutral, without styles and stigmas; pappus of 2 awns, these stout and early deciduous; squamellae lacking; widespread in nc TX ______ **Helianthus**
      23. Ray flowers pistillate (but infertile), with styles present; pappus of 2 slender and persistent awns and 4 squamellae; s half of nc TX ______ **Viguiera**

Key 2. Ray and disk flowers both present; ray flowers yellow, sometimes with reddish brown, rusty-brown or maroon; receptacle naked.

1. Pappus of disk flowers of capillary bristles.
   2. Phyllaries in 1 series, equal in length, with very minute outer bracts sometimes present at the base.
      3. Stem leaves deeply toothed or pinnately lobed or compound, the upper not clasping or but weakly so; plants annuals or perennials ______ **Packera**
      3. Stem leaves shallowly toothed to subentire, conspicuously clasping; plants annuals ______ **Senecio**
   2. Phyllaries of unequal lengths; imbricate in 2 or more rows.
      4. Pappus of disk flowers double, the inner of capillary bristles, the outer of shorter (0.3–1.0 mm) stout bristles or scales.
         5. Receptacle with a chaff of long subulate, persistent scales; phyllaries spiny-tipped, white margined; lower and basal leaves often pinnately lobed, sometimes deeply so ______ **Xanthisma**
         5. Receptacle naked; phyllaries not as above; leaves entire to toothed to shallowly lobed.
            6. Lower stem leaves often grass-like in appearance, with "parallel" veins; basal leaves usually much longer than stem leaves; stems and leaves with long, silvery white, silky pubescence; in nc TX known only from Tarrant Co., mainly se and e TX ______ **Pityopsis**
            6. Lower stem leaves not as above; basal leaves various; stem and leaf pubescence various; widespread in nc TX.
               7. Upper and middle stem leaves little or not narrowed at base, slightly to strongly clasping; annual or perennial from taproot.
                  8. Upper leaves clasping, ovate; leaf blades scabrous above, hispid-pilose beneath; achenes of ray flowers without a pappus ______ **Heterotheca**
                  8. Upper leaves slightly clasping, oblanceolate; leaf blades hairy to nearly smooth

on both sides; achenes of ray flowers with a pappus of bristles and sometimes small scales _______ **Chrysopsis**

7. Upper and middle stem leaves tapered to slender petiole-like bases; perennial from woody crown _______ **Heterotheca**

4. Pappus of all flowers a single row (but sometimes some can be unequal in length).

9. Disk flowers with pappus bristles 1 or 2; leaves linear, entire _______ **Chrysopsis**

9. Disk flowers with pappus bristles several to numerous; leaves various.

10. Leaves pinnatifid or bipinnatifid _______ **Machaeranthera**

10. Leaves simple, nearly entire to prominently toothed.

11. Heads (excluding ligules of ray flowers) 2 cm or more broad; leaves spiny-toothed _______ **Grindelia**

11. Heads (excluding ligules of ray flowers) < 2 cm broad; leaves spiny-toothed OR in most species not spiny-toothed.

12. Heads relatively large, hemispheric or broadly conic, wider than tall; stems glandular-pubescent, from a taproot.

13. Upper leaves conspicuously toothed; phyllaries without scarious margins _______ **Rayjacksonia**

13. Upper leaves entire or nearly so (but can be bristly-ciliate); phyllaries with a darker line in the middle and scarious margins _______ **Croptilon**

12. Heads small, cylindrical, taller than wide; stems without glandular hairs, from a fibrous root system.

14. Plants with glandular-punctate dots, these especially obvious on the leaves; inflorescences corymbiform (= flat-topped) _______ **Euthamia**

14. Plants without glandular-punctate dots (except in 1 species with a panicle); inflorescences paniculate, or axillary, or if corymbiform then not gland-dotted _______ **Solidago**

1. Pappus of disk flowers of awns, scales, scaly bristles, a crown, or absent or nearly so.

15. Phyllaries and generally also the leaves conspicuously gland-dotted.

16. Leaves linear, 1–2 mm wide, glandular-punctate marginally; pappus of 0–4 short scales forming a low crown 0.5–1(–2) mm long; phyllaries in 1 series _______ **Pectis**

16. Leaves linear, 2–6 mm wide, with coarse teeth marginally OR leaves pinnately divided into 5–15 linear lobes; pappus of 10–12 awned or awnless scales, often longer than 2 mm; phyllaries in 2 series.

17. Leaves with coarse teeth, but not pinnately divided into linear lobes _______ **Dysodiopsis**

17. Leaves pinnately divided into 5–15 linear lobes _______ **Thymophylla**

15. Phyllaries and leaves without conspicuous glandular dots (but may be minutely punctate-dotted).

18. Phyllaries reflexed at anthesis or curled back; stems winged OR not so.

19. Leaf blades decurrent; stems winged _______ **Helenium**

19. Leaf blades not decurrent on stem; stems not winged.

20. Leaves toothed, broader than linear, much > 2 mm wide; phyllaries curled back even in bud, gummy _______ **Grindelia**

20. Leaves or leaf segments linear to thread-like, usually < 2 mm wide, entire; phyllaries reflexed at anthesis, not gummy _______ **Helenium**

18. Phyllaries erect at anthesis; stems not winged.

21. Stem leaves present, oblanceolate to ovate, much broader than filiform to linear.

22. Leaves sessile, entire; stems and leaf margins with long, wide-spreading, cottony hairs; growing on sandy or rocky open ground _______ **Amblyolepis**

22. Leaves petiolate, with large serrations/small lobes; stems and leaf margins gla-

brous or with minute pubescence; growing in crevices on limestone bluffs ____ **Perityle**

21. Stem leaves or their segments filiform to linear OR stem leaves absent.
    23. Leaves simple and entire, essentially all cauline; phyllaries resinous shiny, glabrous, in several series ____ **Gutierrezia**
    23. Leaves simple and entire OR with 1–5 short lobes OR with 3–15 long, linear-filiform divisions, at least some of the leaves frequently forming basal rosettes; phyllaries dull, typically pubescent, usually in 1 series.
        24. Leaves pinnatifid with 3–15 long, linear-filiform divisions; outer phyllaries united near their noticeably thickened bases ____ **Hymenoxys**
        24. Leaves simple, entire, or with 1–5 short lobes; phyllaries all separate, not thickened ____ **Tetraneuris**

Key 3. Ray and disk flowers both present; ray flowers white, pink, lavender, blue, or reddish purple; pappus of capillary bristles.

1. Ray flowers inconspicuous, barely exceeding the phyllaries ____ **Conyza**
1. Ray flowers conspicuous, twice or more length of phyllaries.
   2. Leaves basal; heads solitary at end of a scape; corollas of disk flowers bilaterally symmetrical, cream-colored, maturing to crimson ____ **Chaptalia**
   2. Leaves cauline; heads numerous at ends of branches; corollas of disk flowers radially symmetrical, yellow, or sometimes becoming pink with age.
      3. Ray flowers more than 50, with corollas less than 1 mm wide; pappus of ray flowers sparse or absent ____ **Erigeron**
      3. Ray flowers less than 50, with corollas more than 1 mm wide; pappus of ray flowers of capillary bristles.
         4. Capillary bristles barbellate; plants scapose; flowers solitary ____ **Townsendia**
         4. Capillary bristles smooth; plants caulescent; flowers various.
            5. Leaves, particularly those in upper half of plant, scale-like; plants less than 15(–20) cm tall ____ **Chaetopappa**
            5. Leaves larger, not scale-like, even the upper reduced leaves not scale-like OR if leaves reduced and the plant almost leafless (in *Chloracantha*), then plants much greater than 20 cm tall; plants more than 20 cm tall.
               6. Plants almost leafless, rush-like, the leaves if present usually minute; thorns often present in upper leaf axils; stems green and glaucous, photosynthetic, alive for up to ca. 4 growing seasons ____ **Chloracantha**
               6. Plants with leaves small to well-developed; thorns absent; stems not as above ____ **Aster**

Key 4. Ray and disk flowers both present; ray flowers white, pink, lavender, or reddish purple; pappus of awns, scales, scaly bristles, a minute crown, or absent.

1. Receptacle chaffy, the bracts associated with each flower.
   2. Only ray flowers maturing achenes, these much larger than those of the sterile disk flowers.
      3. Corollas of ray flowers white or yellow, the ligules 7–20(–30) mm long; ray achenes thick, rounded or weakly compressed; leaves opposite or alternate.
         4. Corollas of ray flowers white; leaf blades entire to pinnately lobed; phyllaries in several series; the inner completely clasping the achenes ____ **Melampodium**
         4. Corollas of ray flowers yellow; leaf blades usually ± palmately 3–5-lobed; phyllaries in 1–2 series, subtending, not clasping the achenes ____ **Smallanthus**
      3. Corollas of ray flowers white, the ligules minute, < 1 mm long; ray achenes strongly flattened, without wings, persistent on mature achenes; leaves alternate ____ **Parthenium**
   2. Ray and disk flowers both maturing achenes.

5. Ray flowers with inconspicuous corollas 3 mm or less long; leaves opposite ____ **Eclipta**
5. Ray flowers with corollas 5 mm or more long, often much longer, these usually very conspicuous; leaves alternate.
6. Stems not winged; receptacle chaff stiff, sharp, longer than disk flowers; leaves mostly basal, the stem leaves much reduced upward; heads borne on long, naked, unbranched peduncles ____ **Echinacea**
6. Stems winged; receptacle chaff soft, hidden by disk flowers; leaves mostly cauline; little reduced upward; heads on short leafy peduncles ____ **Verbesina**

1. Receptacle naked.
7. Pappus a short crown or none.
8. Ray flowers with ligules ca. 2 mm or less long; disk flowers white to pinkish ____ **Achillea**
8. Ray flowers with ligules much longer than 2 mm; disk flowers various colors.
9. Leaves bipinnately dissected ____ **Anthemis**
9. Leaves entire or toothed or lobed.
10. Leaves toothed to lobed; plants rhizomatous perennials 20–100 cm tall; disk 1–2.5+ cm wide; ligules 10–nearly 30 mm long, white ____ **Leucanthemum**
10. Leaves entire; plants taprooted annuals 10–40 cm tall; disk 0.6–1 cm wide; ligules 5–12 mm long, white to bluish- or pinkish-tinged ____ **Astranthium**
7. Pappus of awns, scales, hairs, short bristles, or a cup-like crown.
11. Pappus of 5 scales alternating with 5 slender awns; plants low; dainty; plants taprooted annuals, 5–15(–25) cm tall; ray corollas white to pinkish ____ **Chaetopappa**
11. Pappus not 5 scales and 5 awns, various; plants various heights; plants annuals or perennials; ray corollas various colors.
12. Disk flowers violet ____ **Palafoxia**
12. Disk flowers yellow.
13. Achenes 4-angled to round, not winged; leaves varying from entire to deeply pinnatifid, but at least some usually toothed to pinnatifid, sometimes linear, but usually broader; phyllaries > 0.5 mm wide; annuals or perennials ____ **Aphanostephus**
13. Achenes with winged margins (2 or 3 wings); leaves entire, mostly linear (lower leaves sometimes linear-lanceolate); phyllaries < 0.5 mm wide; perennials ____ **Boltonia**

Key 5. Disk flowers only; phyllaries with hooked or straight spines, or tubercles, sometimes united into a bur, OR with large teeth at edges; plants frequently spiny.

1. All flowers imperfect; pistillate and staminate flowers in different heads (plants monoecious); pistillate flowers apetalous or corolla reduced to a ring; staminate flowers with anthers not joined at the edges; pappus none.
2. Staminate heads in racemes or spikes; bur ca. 5 mm or less long, with straight or hooked spine-like structures OR tubercles, 1-flowered; lower leaves usually opposite ____ **Ambrosia**
2. Staminate heads in terminal clusters; bur (cocklebur) 10–30 mm long, with hooked spine-like structures, 2-flowered; lower leaves alternate ____ **Xanthium**
1. All flowers perfect; corollas large, showy, with lobes at least 1/3 or more of corolla length, variously colored; anthers joined at the edges; pappus of capillary or plumose bristles (THISTLES).
3. Leaves without spines; phyllaries without spines at tip OR spiny-tipped; corollas of outer row of flowers often enlarged, simulating ray flowers ____ **Centaurea**
3. Leaves with spines (can be entire in 1 species of *Carthamus*); phyllaries spiny-tipped; corollas of all flowers similar.
4. Corollas yellow to orange (rarely whitish); pappus absent or of ± rigid, flat, bristle-like scales to ca. 1 cm long ____ **Carthamus**
4. Corollas white to pink, lavender, or purple; pappus of plumose or barbellate, hair-like bristles, these of variable length but often much > 1 cm long.

5. Pappus bristles plumose (= feather-like, with long side branches) at least basally; phyllaries all ± appressed, linear; widespread native and introduced species ______ **Cirsium**
5. Pappus bristles barbellate (with short side branches) or barbless; phyllaries variable (but the outer ones sometimes spreading or reflexed, flat, broad, 2–8 mm wide); introduced species with rather limited distributions in nc TX.
   6. Stems conspicuously prickly winged (by decurrent leaf bases); leaves usually uniformly colored above; filaments separate.
      7. Phyllaries 2–9 mm wide (at least near base); leaves ± glabrous or with cottony pubescence below; heads 10–70 mm in diam.; receptacle with numerous bristles ca. 2 mm long; pappus 11–20 mm long ______ **Carduus**
      7. Phyllaries < 2 mm wide; leaves with cottony pubescence below; heads 25–50 mm in diam.; receptacle bristles absent or very short; pappus 7–9 mm long ______ **Onopordum**
   6. Stems not winged; leaves ± mottled green and white above; filaments united into a tube above their attachment to the corolla, free below anthers ______ **Silybum**

Key 6. Disk flowers only; phyllaries mostly entire; flowers pink, blue, purple, green, white, or yellowish cream.

1. Shrubs.
   2. Pappus of capillary bristles; leaf blades entire to toothed, never lobed or dissected.
      3. Flowers imperfect, the plans dioecious (male and female plants can have quite different appearances); leaves linear to lanceolate, elliptic, rhomboid, or obovate, sessile; plants very variable in size, from 0.25 m to 6 m tall (but if < 1 m tall, then leaves ± linear) ______ **Baccharis**
      3. Flowers perfect; leaves ovate to lanceolate, with petioles to 10 mm long; plants to 1.2 m tall ______ **Brickellia**
   2. Pappus absent; leaf blades entire to more often lobed or dissected ______ **Artemisia**
1. Herbs.
   4. Plants white-woolly, at least on lower leaf surfaces; basal leaves or inflorescence leaves largest; stem leaves smaller or greatly reduced except in one species of *Gnaphalium*.
      5. Plants typically small, usually < 10(–15) cm tall; heads ± completely imbedded in woolly hairs; pappus lacking; disk corollas (central) 4-toothed; receptacle chaffy ______ **Evax**
      5. Plants over 10 cm tall; heads often with woolly hairs, but not completely embedded in the hairs; pappus of hair-like bristles present; disk corollas 3- or 5-toothed; receptacle ± naked.
         6. Pappus of plumose bristles (use lens to see side branches) 10–11 mm long; leaves 1.5–4 mm wide; apparently rare in nc TX ______ **Facelis**
         6. Pappus of smooth or barbellate bristles, the bristles much < 10 mm long except in *Antennaria* which has basal leaves 15–50 mm wide; leaves of various widths, often much wider than 4 mm; widespread in nc TX.
            7. Basal and stem leaves present, the stem leaves smaller in most species; basal leaves 22 mm or less wide (usually much less); phyllaries ± completely scarious (except midrib can be greenish); heads all alike, bisexual, usually numerous in elongate or terminal inflorescences, the individual heads usually ± sessile in glomerules.
               8. Pappus bristles not united basally, separating individually or in clusters; inflorescences often spreading, somewhat flat-topped, sometimes elongate; phyllaries ± obtuse to acutish; flowering (Jul–)Sep–Nov ______ **Pseudognaphalium**
               8. Pappus bristles united basally, deciduous as a ring of hairs; inflorescences usually ± elongate, spike-like; phyllaries acute to acuminate; flowering Mar–Jun ______ **Gamochaeta**
            7. Leaves nearly all basal, the stem leaves very reduced; basal leaves 15–50 mm wide; phyllaries herbaceous at least in lower half; heads of 2 kinds, the sexes on different plants, typically few at end of stem, the individual heads short pedunculate or solitary ______ **Antennaria**
   4. Plants not white-woolly; pubescence various or plants glabrous; leaves various, but in many species stem leaves well-developed and prominent.

9. Pappus of scales or awns, or absent.
  10. Pappus absent; heads small, white or greenish, 8 mm or less wide.
    11. Plants very small, ca. 15 cm or less tall; achenes flattened, with wings or ± wing-like lateral appendages, tipped by the persistent spine-like style (painful in a manner similar to sandburs) ____ **Soliva**
    11. Plants 20–200 cm tall; achenes not flattened or not conspicuously so, without either wing-like lateral appendages or spine-like style.
      12. Leaf blades entire to toothed, neither lobed nor dissected, not gray- or white-woolly; stems and leaves usually resinous-glandular; receptacles chaffy; flowers pistillate or staminate ____ **Iva**
      12. Leaf blades (at least some) usually lobed or dissected but varying to entire, if not lobed or dissected then the leaf blades gray- or white-woolly at least on lower surfaces; stems and leaves either gray to white woolly or nearly glabrous, not resinous glandular; receptacles naked; flowers perfect or pistillate or staminate ____ **Artemisia**
  10. Pappus of awns or scales (sometimes < 1 mm long); heads various.
    13. Corollas lavender or pink; leaves usually entire (can be lobed to pinnately dissected in the one species of *Hymenopappus* that keys here).
      14. Pappus of 14–18 scales (< 1 mm long); leaves variously lobed or pinnately dissected, or rarely entire ____ **Hymenopappus**
      14. Pappus of 5–10 scales; leaves entire.
        15. Plants with 1–few heads; pappus of 5 scarious translucent scales; receptacle chaffy; involucres > 10 mm across ____ **Marshallia**
        15. Plants usually with numerous heads; pappus of 7–10 scarious white scales; receptacle naked; involucres ca. 10 mm or less across ____ **Palafoxia**
    13. Corollas white or greenish; leaves variously lobed or pinnatifid OR entire.
      16. Leaves entire; heads (including corollas) commonly 25–35 mm wide at flowering time ____ **Marshallia**
      16. Leaves dentate to lobed or pinnatifid; heads much < 25 mm wide at flowering time.
        17. All leaves usually pinnatifid; phyllaries petaloid at tips; receptacle naked; ray flowers absent ____ **Hymenopappus**
        17. Upper leaves merely dentate to lobed; phyllaries not petaloid at tips; receptacle chaffy; usually with 5 very short, often overlooked ray flowers ____ **Parthenium**
9. Pappus of bristles.
  18. Leaves opposite or rarely whorled.
    19. Vines; phyllaries 4 per head; flowers usually 4 per head ____ **Mikania**
    19. Not vines; phyllaries more than 4 per head; flowers more than 4 per head ____ **Eupatorium**
  18. Leaves alternate.
    20. Corollas and sometimes phyllaries pink or lavender to purple.
      21. Pappus a double row of bristles, the inner row of long bristles, the outer of very short bristles ____ **Vernonia**
      21. Pappus a single row of bristles.
        22. Several heads clustered and subtended by a whorl of ca. 3 ovate leaf-like bracts ca. 1 cm long; pappus of 5 bristles ____ **Elephantopus**
        22. Heads not in clusters subtended by leafy bracts; pappus of numerous capillary bristles.
          23. Inflorescence an unbranched spike or raceme; leaves essentially linear, entire, punctate; plants without camphor odor ____ **Liatris**

23. Inflorescence much branched; leaves ovate to elliptic, serrate to serrate-dentate or crenate, more or less glandular pubescent; plants aromatic with distinctive camphor-like odor ____ **Pluchea**

20. Corollas white or cream.

24. Leaves pinnately dissected into linear segments less than 2 mm wide ____ **Eupatorium**

24. Leaves not pinnately dissected.

25. Heads with 5 equal phyllaries and 5 flowers; leaves mostly basal, broadly ovate, up to 15+ cm long, 2–8 cm wide, entire or rarely toothed, with prominent, ± parallel, longitudinal nerves (often 7–9) eventually converging toward the tip ____ **Arnoglossum**

25. Heads with numerous phyllaries and more than 5 flowers; leaves basal or cauline, variously shaped, entire or often distinctly toothed, usually smaller, with venation not as above.

26. Leaves mostly basal; several heads clustered and subtended by a whorl of ca. 3 ovate leaf-like bracts ca. 1 cm long; pappus a single whorl of 5 bristles ____ **Elephantopus**

26. Leaves mostly cauline; heads not in clusters subtended by bracts; pappus not as above.

27. Plants perennial from a woody base or woody taproot; pappus plumose OR not so; heads with all flowers perfect ____ **Brickellia**

27. Plants annual; pappus never plumose; heads with pistillate and perfect flowers.

28. Leaves sharply and conspicuously toothed and sometimes irregularly lobed; inflorescences corymbiform cymes of few to many heads; phyllaries in a single series, 9–17 mm long (sometimes with a few minute bracteoles at base); ray flowers not present ____ **Erechtites**

28. Leaves entire or subentire; inflorescences panicles of numerous small heads; phyllaries imbricate, 3–5 mm long; small inconspicuous ray flowers usually present ____ **Conyza**

Key 7. Disk flowers only or apparently so; phyllaries entire or minutely serrated; flowers yellow to reddish brown, rusty-brown, or maroon.

1. Pappus of capillary bristles or stramineous stiff bristles.

2. Leaves opposite ____ **Eclipta**

2. Leaves alternate.

3. Phyllaries in a single series (subtended by a few tiny bractlets); middle and upper leaves undulate to pinnately lobed, auriculate, ± clasping ____ **Senecio**

3. Phyllaries in several series, imbricate; middle and upper leaves entire or toothed, neither auriculate nor clasping.

4. Flowers 4–6 per head; all flowers perfect and tubular, without ligule; phyllaries resinous ____ **Bigelowia**

4. Flowers ca. 20 or more per head; marginal flowers pistillate, with inconspicuous ligule scarcely if at all surpassing the disk; phyllaries not resinous ____ **Conyza**

1. Pappus of scales or awns or short crown or absent.

5. Receptacle chaffy or with bristly setae.

6. Pappus of 2 retrorsely or antrorsely barbed awns; leaves opposite.

7. Leaves decurrent on stems forming wings ____ **Verbesina**

7. Leaves not decurrent, not forming wings on stem.

8. Phyllaries joined at base 1/4 to 1/2 length; corollas yellowish with reddish brown veins ____ **Thelesperma**
8. Phyllaries all separate; corollas all yellow ____ **Bidens**
6. Pappus of scales or smooth, nonbarbed awns; leaves basal or alternate.
9. Corollas reddish purple or brownish red; heads solitary at end of each peduncle; leaves entire to deeply lobed ____ **Gaillardia**
9. Corollas yellowish; heads solitary at end of each peduncle OR heads numerous in corymbose inflorescences; leaves various, entire to tripinnatifid.
10. Heads solitary at end of each peduncle; ray flowers absent; lower and middle leaves entire to toothed or pinnatitifid ____ **Gaillardia**
10. Heads numerous in corymbose inflorescences; ray flowers usually present with very small corollas; lower and middle leaves bipinnatifid to tripinnatifid ____ **Parthenium**
5. Receptacle naked.
11. Phyllaries and leaves with large (up to 1 mm) yellowish brown or orange oil glands ____ **Dysodiopsis**
11. Phyllaries and leaves without conspicuous oil glands (but may be minutely punctate-dotted or have glandular hairs).
12. Stems winged; phyllaries reflexed at anthesis, not gummy ____ **Helenium**
12. Stems not winged; phyllaries curled back even in bud, gummy ____ **Grindelia**

TECHNICAL KEY TO THE TRIBES

1. All flowers ligulate, perfect, the ligule strap-shaped; plant sap usually milky ____ **Lactuceae (Cichorieae)**
1. Disk flowers with tubular eligulate corollas present; ray flowers with strap-shaped ligules present or absent, if present, pistillate or neutral; plant sap usually watery.
2. Ray flowers present (these can rarely lack corollas), these pistillate or neutral.
3. Receptacle chaffy, a bract subtending each flower; lower leaves in most genera opposite; ligules tending to be rather broad, often larger than typical for other tribes ____ **Heliantheae**
3. Receptacle naked or rarely with bristles that are not associated with individual flowers; leaves variable; ligules variable.
4. Phyllaries dry, membranous at margins and tips, usually whitish; ray corollas white or absent; pappus absent or a mere border or crown; leaves generally ± dissected; plants often with a characteristic odor ____ **Anthemideae**
4. Phyllaries herbaceous; with or without membranous margins; ray corollas usually yellow, occasionally pink, lavender, or white; pappus various or absent; leaves generally entire or toothed; plants usually without a characteristic odor.
5. Phyllaries imbricate (= overlapping) in 2–many series ____ **Astereae**
5. Phyllaries not overlapping (the margins of one phyllary can, however, abut the next), in 1–2 series, the second series often minute at base of first.
6. Pappus of capillary bristles; leaves without punctate dots or glandular dots ____ **Senecioneae**
6. Pappus of chaffy scales, a mere border or crown, or absent or nearly so; leaves frequently glandular or with punctate dots ____ **Helenieae**
2. Ray flowers absent (but ray-like marginal flowers sometimes present in Cardueae and Mutisieae).
7. Corollas of central flowers bilaterally symmetrical; leaves in a basal rosette only; a single genus rare in nc TX ____ **Mutisieae**
7. Corollas of central flowers radially symmetrical; leaves basal or cauline; numerous widespread genera in nc TX.
8. Corolla tubes elongate, the lobes at least 1/4 or usually more of the total corolla length; receptacle densely bristly, the bristles not associated with individual flowers; anthers long-tailed at base; phyllaries spiny or their margins with large teeth; leaves usually spiny; (THISTLES) ____ **Cardueae (Cynareae)**

8. Corolla tubes not elongate, the lobes small teeth less than 1/4 the total corolla length; receptacle naked or with chaff associated with each individual flower, or rarely bristly; anthers tail-less or short-tailed (in Inuleae); phyllaries entire or with fine teeth, not spine-tipped; leaves not spiny.
  9. Plants white-woolly pubescent at least on the lower leaf surfaces; phyllaries whitish, with scarious margins; anthers short-tailed ____ **Inuleae**
  9. Plants glabrous or pubescent, not white-woolly; phyllaries various; with or without scarious margins; anthers tail-less or short-tailed (in Inuleae).
    10. Phyllaries of pistillate heads joined and either with hooked or straight spine-like structures or with tubercules, the heads thus bur-like; all flowers imperfect, pistillate and staminate flowers in same or different heads; stamens usually not united by anthers (COCKLEBURS, RAGWEEDS) ____ **Heliantheae**
    10. Phyllaries without either spine-like structures or tubercules, the heads not bur-like; flowers perfect or imperfect; anthers united into a ring around style.
      11. Pappus of scales or awns or none.
        12. Receptacle chaffy, bracts associated with each flower; phyllaries variable but tending to be in several series ____ **Heliantheae**
        12. Receptacle naked or rarely with bristles; phyllaries appearing to be in one series ____ **Heleniae**
      11. Pappus of capillary bristles.
        13. Phyllaries scarious margined or papery, often white or pink or lavender.
          14. Phyllaries imbricate (= overlapping like shingles) in 2–many series; anthers short-tailed at base ____ **Inuleae**
          14. Phyllaries not overlapping (the margins of one phyllary abutting the next), in 1–2 series, the second series often minute at base of first; anthers not tailed at base ____ **Senecioneae**
        13. Phyllaries herbaceous, sometimes pigmented.
          15. Either phyllaries and corollas usually purplish and heads in a corymbose arrangement OR small clusters of several heads subtended by 3 foliaceous ovate bracts; style branches slender to filiform; leaves alternate ____ **Vernonieae**
          15. Phyllaries not purplish (or if so, heads in a spike-like or raceme-like arrangement; clusters of heads not subtended by 3 foliaceous bracts; style branches either slender but wider apically (= ± clavate) OR broad and flattened; leaves alternate OR opposite.
            16. Corollas variously colored, but not distinctly yellow; style branches slender, wider apically ____ **Eupatorieae**
            16. Corollas often yellow, but varying to other colors; style branches usually broad and flattened ____ **Astereae**

ANTHEMIDEAE TRIBE

1. Pappus of short but discrete scales; phyllaries nearly equal, not imbricate, petaloid at the tips (*Hymenopappus* can be reached here and in Helenieae key) ____ **Hymenopappus**
1. Pappus a short crown or absent; phyllaries various, not petaloid.
  2. Heads with disk flowers only (or apparently so—corolla-less ray flowers present in *Solvia*).
    3. Plants very small, 15 cm or less tall; achenes flattened, tipped by the persistent spine-like style (painful in a manner similar to sandburs), with wings or ± wing-like lateral appendages ____ **Soliva**
    3. Plants 20–150 cm tall; achenes not noticeably flattened, without either spine-like style or wing-like lateral appendages ____ **Artemisia**

2. Heads with both evident ray flowers and disk flowers, the ray flowers with evident ligules.
   4. Ray corollas with ligules ca. 2 mm long or less ____ **Achillea**
   4. Ray corollas with ligules 7 mm long or longer.
      5. Leaves toothed or lobed, but not finely dissected; ray flowers 15–35 per head; ray corollas with ligules 10–nearly 30 mm long; heads usually solitary at ends of main stems and long branches; plants rhizomatous perennials ____ **Leucanthemum**
      5. Leaves finely bipinnately dissected; ray flowers ca. 10–14 per head; ray corollas with ligules 7–10(–12) mm long mm long; heads usually several per main stem; plants taprooted annuals ____ **Anthemis**

ASTEREAE TRIBE

1. Heads with only disk flowers.
   2. Plants shrubs, very variable in size, from 0.25–6 m tall; flowers imperfect, the plants dioecious (male and female plants can have quite different appearances) ____ **Baccharis**
   2. Plants herbaceous, 0.1–ca. 2 m tall; at least some flowers perfect, the plants not dioecious.
      3. Phyllaries resinous; corollas yellow.
         4. Flowers numerous per head; leaves serrate to spiny toothed, oblong to lanceolate or deltoid, over 5 mm wide ____ **Grindelia**
         4. Flowers 4–6 per head; leaves entire, linear, ca. 2 mm wide ____ **Bigelowia**
      3. Phyllaries not resinous; corollas white.
         5. Leaves opposite ____ **Eclipta**
         5. Leaves alternate ____ **Conyza**
1. Heads with ray and disk flowers, the ray flowers usually conspicuous but rarely concealed by phyllaries.
   6. Ray flowers yellow.
      7. Pappus of scales or awns or a few stout bristles or none.
         8. Leaves linear, entire; pappus of scales or none; disks less than 0.5 cm wide ____ **Gutierrezia**
         8. Leaves linear-lanceolate or wider, serrate to spiny-toothed; pappus of awns or a few stout bristles; disks 0.7–3 cm wide ____ **Grindelia**
      7. Pappus with numerous capillary bristles, with or without an additional row of scales or short bristles.
         9. Pappus of disk flowers double, the inner of capillary bristles, the outer of shorter stout bristles or scales.
            10. Receptacle with a chaff of long subulate, persistent scales; phyllaries spiny-tipped, white-margined; lower and basal leaves often pinnately lobed, sometimes deeply so ____ **Xanthisma**
            10. Receptacle naked; phyllaries not as above; leaves entire to toothed to shallowly lobed.
                11. Lower stem leaves often grass-like in appearance, with "parallel" veins; basal leaves usually much longer than stem leaves; stems and leaves with long, silvery white, silky pubescence; in nc TX known only from Tarrant Co., mainly se and e TX ____ **Pityopsis**
                11. Lower stem leaves not as above; basal leaves various; stem and leaf pubescence various; widespread in nc TX.
                    12. Upper and middle stem leaves little or not narrowed at base, slightly to strongly clasping; plants annuals or perennials from taproot.
                        13. Upper leaves clasping, ovate; leaf blades scabrous above, hispid-pilose beneath; achenes of ray flowers without a pappus ____ **Heterotheca**
                        13. Upper leaves slightly clasping, oblanceolate; leaf blades hairy to sparsely

12. Upper and middle stem leaves tapered to slender petiole-like bases; plants perennials from woody crown **Heterotheca**
9. Pappus of all flowers a single row of capillary bristles.
14. Disk flowers with pappus bristles 1 or 2; leaves linear, entire **Chrysopsis**
14. Disk flowers with pappus bristles several to numerous; leaves various.
15. Leaves pinnatifid or bipinnatifid **Machaeranthera**
15. Leaves simple, nearly entire to prominently toothed.
16. Heads (excluding ligules of ray flowers) 2 cm or more broad; leaves spiny toothed **Grindelia**
16. Heads (excluding ligules of ray flowers) < 2 cm broad; leaves spiny-toothed OR in most species not spiny-toothed.
17. Heads relatively large, hemispheric or broadly conic, wider than tall; stems and phyllaries glandular pubescent; taprooted annuals OR perennials.
18. Upper leaves conspicuously toothed; phyllaries without scarious margins **Rayjacksonia**
18. Upper leaves entire or nearly so (but can be bristly-ciliate); phyllaries with a darker line in the middle and scarious margins **Croptilon**
17. Heads small, cylindrical, taller than wide; stems without glandular hairs; perennials from a fibrous root system.
19. Inflorescences corymbiform (= flat-topped).
20. Leaves with glandular-punctate dots **Euthamia**
20. Leaves without glandular-punctate dots **Solidago**
19. Inflorescences panicles, or axillary **Solidago**
6. Ray flowers white, or pink, or lavender, or purple.
21. Pappus absent or nearly so (minute ring of tissue can be present) **Astranthium**
21. Pappus present of capillary bristles, awns, scales, hairs, or a cup-like crown.
22. Pappus of disk flowers of awns, scales, hairs, or a cup-like crown.
23. Pappus of 5 scales alternating with 5 slender awns; plants low, dainty, taprooted annuals 5–15(–25) cm tall; ray corollas white to pinkish **Chaetopappa**
23. Pappus not of 5 scales and 5 awns, various; plants annuals or perennials of various heights, often much larger than 25 cm tall; ray corollas various colors.
24. Achenes 4-angled to round, not winged; leaves varying from entire to deeply pinnatifid, but at least some usually toothed to pinnatifid, sometimes linear, but usually broader; phyllaries > 0.5 mm wide; annuals or perennials **Aphanostephus**
24. Achenes with winged margins (2 or 3 wings); leaves entire, mostly linear (lower leaves sometimes linear-lanceolate); phyllaries < 0.5 mm wide; perennials **Boltonia**
22. Pappus of disk flowers of numerous capillary bristles.
25. Ray flowers inconspicuous, barely exceeding the phyllaries **Conyza**
25. Ray flowers conspicuous, twice or more length of phyllaries.
26. Ray flowers more than 50, with corollas less than 1 mm wide; pappus of ray flowers sparse or absent **Erigeron**
26. Ray flowers less than 50, with corollas 1 mm wide or wider; pappus of ray flowers of capillary bristles.
27. Capillary bristles barbellate; plants scapose; flowers solitary **Townsendia**
27. Capillary bristles smooth; plants caulescent; flowers various.
28. Leaves, particularly those in upper half of plant, scale-like; plants less than 15(–20) cm tall **Chaetopappa**
28. Leaves larger, not scale-like, even the upper reduced leaves not scale-like OR if leaves reduced and the plant almost leafless (in *Chloracantha*),

then plants much greater than 20 cm tall; plants more than 20 cm tall.

29. Plants almost leafless, rush-like, the leaves if present usually minute; thorns often present in upper leaf axils; stems green and glaucous, photosynthetic, alive for up to ca. 4 growing seasons ______ **Chloracantha**
29. Plants with leaves small to well-developed; thorns absent; stems not as above ______ **Aster**

Cardueae (Cynareae) Tribe

1. Leaves without spines; phyllaries without spines at tip OR spiny-tipped; corollas of outer row of flowers often enlarged, simulating ray flowers ______ **Centaurea**
1. Leaves with spines (can be entire in 1 species of *Carthamus*); phyllaries spiny-tipped; corollas of all flowers similar.
   2. Corollas yellow to orange (rarely whitish); pappus absent OR of ± rigid, flat, bristle-like, ciliate scales to 1 cm long as well as some outer short scales ______ **Carthamus**
   2. Corollas white to pink, lavender, or purple; pappus of plumose or barbellate, hair-like bristles, these of variable length but often much > 1 cm long.
      3. Pappus bristles plumose (= feather-like, with long side branches) at least basally; phyllaries all ± appressed, linear; widespread native and introduced species ______ **Cirsium**
      3. Pappus bristles barbellate (= with short side branches) or barbless; phyllaries variable (but the outer ones sometimes spreading or reflexed, flat, broad, 2–8 mm wide); introduced species with rather limited distributions in nc TX.
         4. Stems conspicuously prickly winged (by decurrent leaf bases); leaves usually uniformly colored above; filaments separate.
            5. Phyllaries 2–9 mm wide (at least near base); leaves ± glabrous or with cottony pubescence below; heads 10–70 mm in diam.; receptacle with numerous bristles ca. 2 mm long; pappus 11–20 mm long ______ **Carduus**
            5. Phyllaries < 2 mm wide; leaves with cottony pubescence below; heads 25–50 mm in diam.; receptacle bristles absent or very short; pappus 7–9 mm long ______ **Onopordum**
         4. Stems not winged; leaves ± mottled green and white above; filaments united into a tube above their attachment to the corolla, free below anthers ______ **Silybum**

Eupatorieae Tribe

1. Perennial twining vine; leaves opposite; phyllaries 4 per head; flowers usually 4 per head ______ **Mikania**
1. Perennial or annual, not a twining vine; leaves alternate or opposite; phyllaries more than 4 per head; flowers 3–numerous per head.
   2. Leaves alternate, not finely dissected; achenes 10-ribbed.
      3. Leaves essentially linear, entire, sessile; corollas purple (rarely white); inflorescence an unbranched spike or raceme ______ **Liatris**
      3. Leaves lanceolate to ovate, usually toothed (rarely entire), with petioles to 1 cm long; corollas cream to yellowish cream; inflorescence usually branched, ± hemispherical to flat-topped or paniculate or racemose ______ **Brickellia**
   2. Leaves either opposite and not finely dissected OR alternate and finely dissected; achenes 5-ribbed ______ **Eupatorium**

Helenieae Tribe

1. Phyllaries and generally also the leaves conspicuously gland-dotted.
   2. Leaves linear, 1–2 mm wide, glandular-punctate marginally; pappus of 0–4 short scales forming a low crown 0.5–1(–2) mm long; phyllaries in 1 series ______ **Pectis**

2. Leaves linear, 2–6 mm wide, with coarse teeth marginally OR leaves pinnately divided into 5–15 linear lobes; pappus of 10–12 awned or awnless scales, often longer than 2 mm; phyllaries in 2 series.
3. Leaves with coarse teeth, but not pinnately divided into linear lobes ________ **Dysodiopsis**
3. Leaves pinnately divided into 5–15 linear lobes ________ **Thymophylla**

1. Phyllaries and leaves without conspicuous glandular dots (but may be minutely punctate-dotted).
4. Receptacle bristly with spine-like setae usually exceeding the achenes ________ **Gaillardia**
4. Receptacle usually naked, rarely with poorly developed bristles.
5. Flowers white or pink or lavender.
6. Leaves pinnatifid, with conspicuous lobes; flowers usually white (reddish-tinged in 1 species) ________ **Hymenopappus**
6. Leaves entire; flowers pink or lavender. ________ **Palafoxia**
5. Flowers yellow, dark red-maroon, rusty-brown, maroon, or a combination of these colors.
7. Leaf blades decurrent on stem, forming wings ________ **Helenium**
7. Leaf blades not decurrent on stem, the stems thus not winged.
8. Leaves or their segments filiform to linear (all very narrow).
9. Heads (excluding ray flowers) globose; phyllaries and ray flowers reflexed ________ **Helenium**
9. Heads (excluding ray flowers) hemispherical to cylindrical; phyllaries and ray flowers spreading to erect.
10. Leaves pinnatifid with 3–15 long, linear-filiform divisions; outer phyllaries united near their noticeably thickened bases ________ **Hymenoxys**
10. Leaves simple, entire, or with 1–5 short lobes; phyllaries all separate, not thickened ________ **Tetraneuris**
8. Leaves oblanceolate to ovate, much broader than filiform to linear.
11. Leaves sessile, entire; stems and leaf margins with long, wide-spreading, cottony hairs; growing on sandy or rocky open ground ________ **Amblyolepis**
11. Leaves petiolate, with large serrations/small lobes; stems and leaf margins glabrous or with minute pubescence; growing in crevices on limestone bluffs ________ **Perityle**

HELIANTHEAE TRIBE

1. Plants generally wind-pollinated or self-pollinated, the heads small and not at all showy nor attractive to pollinators; all flowers imperfect; pistillate and staminate flowers in same or different heads; ray flowers absent; phyllaries either joined forming a bur-like structure with hooked or straight spine-like structures or tubercles OR not joined and not bur-like; pappus none.
2. Staminate and pistillate flowers in same heads; phyllaries without either spine-like structures or tubercules, the heads not bur-like ________ **Iva**
2. Staminate and pistillate flowers in separate heads, the staminate usually uppermost; phyllaries of pistillate heads joined and either with hooked or straight spine-like structures or with tubercules, the heads thus bur-like.
3. Staminate heads in racemes or spikes; bur ca. 5 mm or less long, with straight or hooked spine-like-structures OR tubercles, 1-flowered; lower leaves usually opposite ________ **Ambrosia**
3. Staminate heads in terminal clusters; bur (cocklebur) 10–30 mm long, with hooked spine-like structures, 2-flowered; lower leaves alternate ________ **Xanthium**

1. Plants generally adapted for attracting pollinating insects, the heads colorful and otherwise attractive; some or all flowers perfect; ray flowers usually present, with strap-shaped ligule; phyllaries separate, not forming a bur-like structure; pappus various.
4. Only ray flowers fertile (= maturing achenes), these achenes much larger than those of the sterile disk flowers.
5. Ray achenes thick, rounded or weakly compressed, without wings.

6. Corollas of ray flowers white; leaf blades entire to pinnately lobed; phyllaries in several series, the inner completely clasping the achenes **Melampodium**
6. Corollas of ray flowers yellow; leaf blades usually ± palmately 3–5-lobed; phyllaries in 1–2 series, subtending, not clasping the achenes **Smallanthus**

5. Ray achenes strongly flattened, with or without wings.
7. Achenes broadly winged.
8. Ray flowers usually 5, in 1 row; phyllaries 8–10 in 2 rows, the outer row distinctly narrower; disks less than 1.5 cm wide; plants taprooted annuals up to 0.3(–0.65) m tall **Lindheimera**
8. Ray flowers numerous, in 2–3 rows; phyllaries numerous, in several rows, ± similar except for size; disks (1.5–)2–3 cm wide; plants coarse perennials up to 2(–3) m tall **Silphium**
7. Achenes not winged.
9. Ray flowers with corollas white, ca. 1 mm long, persistent on mature achenes **Parthenium**
9. Ray flowers with corollas yellow, 10 mm or longer, deciduous from mature achenes.
10. Ray flowers with back of corolla having green, red, or maroon veins; leaves crenate, serrate, dentate, or lyrate-pinnatifid (in one species rare in nc TX), variously pubescent but not hispid; pappus none **Berlandiera**
10. Ray flowers with back of corolla entirely yellow, without colored veins; leaves deeply pinnatifid, densely hispid; pappus of a few scales **Engelmannia**

4. Disk flowers fertile; ray flowers present and fertile OR sterile OR absent.
11. Ray flowers white, pink, or lavender OR if absent, then disk flowers white, pink, or lavender.
12. Ray flowers absent; disk flowers with corollas elongate, hairy, conspicuously lobed **Marshallia**
12. Ray flowers present; disk flowers not elongate, glabrous or pubescent, the corolla lobes only small teeth.
13. Ray flowers with inconspicuous corollas 3 mm or less long; leaves opposite **Eclipta**
13. Ray flowers with corollas 5 mm or more long, sometimes much longer, these usually conspicuous; leaves alternate.
14. Stems not winged; receptacle chaff stiff, sharp, longer than disk flowers; leaves mostly basal, the stem leaves much reduced upward; heads borne on long, naked, unbranched peduncles; ligules of ray corollas 2–9 cm long **Echinacea**
14. Stems winged; receptacle chaff soft, hidden by disk flowers; leaves mostly cauline, little reduced upward; heads on short leafy peduncles; ligules of ray corollas < 1 cm long **Verbesina**
11. Ray flowers yellow, sometimes with maroon, deep red, or rusty-brown toward the base.
15. Receptacle columnar.
16. Leaves opposite at base, may be alternate or opposite on upper stems; plants creeping and rooting at the nodes, the flowering branches ascending **Acmella**
16. Leaves all alternate; plants erect.
17. Leaves simple, the blades entire or toothed.
18. Leaves clasping; achenes terete **Dracopis**
18. Leaves not clasping; achenes 4-angled **Rudbeckia**
17. Leaves pinnately compound or deeply lobed **Ratibida**
15. Receptacle flat, or convex, or very low conic.
19. Leaf bases decurrent, the stems thus winged **Verbesina**
19. Leaf bases not decurrent, the stems not winged.
20. Phyllaries dimorphic, the outer row much smaller than the inner; achenes flattened parallel to the phyllaries (or rarely 4 angled in *Bidens*).
21. Inner phyllaries united for up to 1/3 their length; disk corolla lobes often linear **Thelesperma**
21. Phyllaries all separate or nearly so; disk corolla lobes triangular to ovate.
22. Pappus of 2(–4) rigid, retrorsely barbed awns **Bidens**

22. Pappus of 2(–4) awns or teeth which are inconspicuously barbed or not barbed OR pappus absent.
    23. Inner phyllaries spreading at anthesis; achenes not winged and disk corollas yellow ______ **Bidens**
    23. Inner phyllaries tightly enclosing heads; achenes winged and all corollas yellow OR if achenes not winged, then disk corollas reddish brown ______ **Coreopsis**

20. Phyllaries all similar, imbricate, or slightly unequal, not dimorphic in 2 rows; achenes various.
    24. Leaves, including upper ones, mostly opposite.
        25. Leaves linear, 10–30 mm long, to only ca. 2 mm wide; plants erect, 8–22 cm tall ______ **Zinnia**
        25. Leaves larger, usually much so; plants either decumbent OR much greater than 22 cm tall.
            26. Plants decumbent, slender, 5–30 cm tall; ray flowers 3–8, with corollas 4 mm or less long; mainly a lawn weed ______ **Calyptocarpus**
            26. Plants erect, coarse; much taller than 30 cm; ray flowers usually 8–numerous, with corollas more than 4 mm long, usually much more; widespread in native habitats.
                27. Achenes not winged; ray flowers neutral, styles and stigmas absent; widespread in nc TX ______ **Helianthus**
                27. Achenes winged or with wing-like margins; ray flowers pistillate (fertile or infertile), styles present; s and w parts of nc TX.
                    28. Leaves with distinct petioles; petioles with conspicuous auricles basally; pappus of 2 short awns, usually absent on mature achenes; plants perennial herbs ______ **Simsia**
                    28. Leaves sessile; auricles lacking; pappus of 2 or 3 short awns subtended by short, hyaline scales; plants small shrubs ______ **Wedelia**
    24. Upper leaves alternate.
        29. Achenes winged; petioles auriculate dilated at base OR not so ______ **Verbesina**
        29. Achenes not winged; petioles not auriculate.
            30. Ray flowers neutral, styles and stigmas absent; pappus of 2 awns, these stout and early deciduous; squamellae lacking; widespread in nc TX ______ **Helianthus**
            30. Ray flowers pistillate, styles present (but infertile); pappus of 2 slender and persistent awns and 4 squamellae; s half of nc TX ______ **Viguiera**

INULEAE TRIBE (INCLUDING PLUCHEEAE AND GNAPHALIEAE)

1. Plants typically small, usually < 10(–15) cm tall; heads ± completely imbedded in woolly hairs; pappus lacking; disk corollas (central) 4-toothed; receptacle chaffy ______ **Evax**
1. Plants over 10 cm tall; heads often with woolly hairs, but not completely embedded in the hairs; pappus of hair-like bristles present; disk corollas 3- or 5-toothed; receptacle ± naked.
    2. Plants without conspicuous white or gray woolly pubescence; middle and upper stem leaves broad, 10–70 mm wide; phyllaries ± herbaceous (= greenish, but often tinged purplish), not scarious, sometimes puberulent or with minute resin-globules; plants aromatic with distinctive camphor-like odor ______ **Pluchea**
    2. Plants with conspicuous white or gray woolly pubescence; middle and upper stem leaves narrow, 1–15(–20) mm wide; phyllaries ± scarious (= dry, membranous), glabrous above the often woolly base; plants without camphor-like odor.

3. Pappus of plumose bristles (use lens to see side branches) 10–11 mm long; leaves 1.5–4 mm wide; apparently rare in nc TX ____ **Facelis**
3. Pappus of smooth or barbellate bristles, the bristles much < 10 mm long except in *Antennaria* which has basal leaves 15–50 mm wide; leaves of various widths, often much wider than 4 mm; widespread in nc TX.
    4. Basal and stem leaves present, the stem leaves smaller in most species; basal leaves 22 mm or less wide (usually much less); phyllaries ± completely scarious (except midrib can be greenish); heads all alike, bisexual, usually numerous in elongate or terminal inflorescences, the individual heads usually ± sessile in glomerules.
        5. Pappus bristles not united basally, separating individually or in clusters; inflorescences often spreading, somewhat flat-topped, sometimes elongate; phyllaries ± obtuse to acutish; flowering (Jul–)Sep–Nov ____ **Pseudognaphalium**
        5. Pappus bristles united basally, deciduous as a ring of hairs; inflorescences usually ± elongate, spike-like; phyllaries acute to acuminate; flowering Mar–Jun ____ **Gamochaeta**
    4. Leaves nearly all basal, the stem leaves very reduced; basal leaves 15–50 mm wide; phyllaries herbaceous at least in lower half; heads of 2 kinds, the sexes on different plants, typically few at end of stem, the individual heads short pedunculate or solitary ____ **Antennaria**

LACTUCEAE (CICHORIEAE) TRIBE

1. Pappus of scales OR scales and bristles OR pappus lacking.
    2. Corollas blue, pink, or white; pappus of numerous minute scales ca. 0.2 mm long ____ **Cichorium**
    2. Corollas yellow or yellow-orange or in one species yellow with purple teeth; pappus of scales and bristles, or only scales, or pappus absent.
        3. Corollas completely yellow or yellow-orange; plants ± glabrous; pappus of all flowers the same; achenes < 3 mm long ____ **Krigia**
        3. Corollas yellow with purple teeth; plants with conspicuous pubescence; pappus of outer flowers a fringed crown, ca. 1 mm long, that of inner flowers of bristle-like scales ca. 5 mm long; achenes 5–7.5 mm long ____ **Hedypnois**
1. Pappus of bristles or hairs.
    4. Pappus plumose (= feather-like, the bristles with long side hairs—use lens).
        5. Stems essentially leafless or with a few leaves toward base; basal leaves oblanceolate, toothed or pinnatifid; achene body 4–5 mm long, the beak 5 mm long or less; involucre ca. 0.8–1 cm high at flowering time ____ **Hypochaeris**
        5. Stems with leaves; leaves linear, grass-like, entire; achene body 10–25 mm long, the beak 10–25 mm long; involucre ca. 2.4–4 cm high at flowering time ____ **Tragopogon**
    4. Pappus not plumose.
        6. Corollas blue, red, or white.
            7. Phyllaries imbricated, in several series; plants 10–30 cm tall ____ **Pinaropappus**
            7. Phyllaries in one series or with basal row of short ones; plants usually > 30 cm tall.
                8. Achenes flattened; stems with well-developed leaves ____ **Lactuca**
                8. Achenes terete; most leaves basal, stems only with a few reduced leaves or bracts ____ **Lygodesmia**
        6. Corollas yellow, or in one species yellow with purple teeth.
            9. Achenes flattened.
                10. Achenes beaked (with a thin extension between the body of the achene and the attachment point of the pappus), the beak usually very thin; heads with relatively few flowers (ca. 8–56) ____ **Lactuca**
                10. Achenes beakless; heads with many flowers (ca. 80–250) ____ **Sonchus**
            9. Achenes rounded in cross-section (terete, cylindrical or fusiform).
                11. Achenes at maturity beakless or essentially so, the pappus attached at the end of the achene body.

12. Leaves entire, with conspicuous, long, spreading pubescence ______ **Hieracium**

12. Leaves toothed or lobed, with pubescence various.

13. Lower stems glabrous or with sparse pubescence; flowering involucres < 6 mm long; achenes 1.5–2.5 mm long ______ **Youngia**

13. Lower stems conspicuously pubescent to the naked eye; flowering involucres 7–12 mm long; achenes ca. 4–7.5 mm long.

14. Corollas with purple teeth; pappus of outer flowers a fringed crown, ca. 1 mm long, that of inner flowers of a few bristle-like scales ca. 5 mm long ______ **Hedypnois**

14. Corollas completely yellow; pappus of all flowers the same, of numerous hair-like bristles 4–5 mm long ______ **Crepis**

11. Achenes at maturity with long, slender, conspicuous beaks, the pappus well-separated from the body of the achene (beak not visible in young flowers).

15. Plants stemless, leaves all basal (in a rosette); beak without a ring of hairs just beneath the pappus; stems unbranched, each with only a single head ______ **Taraxacum**

15. Plants usually with leafy stems (often 1–few leaves low on the stem); beak with a ring of microscopic, reflexed hairs just beneath where the pappus attaches (use lens); stems sometimes branched, each with 1–several heads ______ **Pyrrhopappus**

MUTISEAE TRIBE

Only one genus ______ **Chaptalia**

SENECIONEAE TRIBE

1. Corollas yellow; heads with both ray and disk flowers (ray flowers absent in 1 species that is rare in nc TX).

2. Stem leaves deeply toothed or pinnately lobed or compound, the upper not clasping or but weakly so; phyllaries not black-tipped; plants annuals or perennials; ray flowers present ______ **Packera**

2. Stem leaves shallowly toothed to pinnately lobed (if pinnately lobed, then phyllaries with black tips and rays flowers absent), conspicuously clasping; phyllaries black-tipped or not so; plants annuals; ray flowers absent or present ______ **Senecio**

1. Corollas white, cream, or yellowish-white; heads with only disk or disk-like flowers.

3. Leaves entire or rarely toothed, with prominent, ± parallel longitudinal nerves (often 7–9) eventually converging toward the tip; stem leaves few; heads with all florets bisexual; plants perennials; prairies and open woods, widespread in nc TX ______ **Arnoglossum**

3. Leaves sharply double serrate and sometimes irregularly lobed, with but a single prominent midnerve (prominent, parallel, longitudinal nerves lacking); leaves well-distributed up the stem; heads with marginal florets pistillate; plants annuals; wet areas in extreme ne part of nc TX ______ **Erechtites**

VERNONIEAE TRIBE

1. Heads in dense clusters subtended by whorls of ca. 3 leaf-like ovate bracts ca. 1 cm long; leaves mostly basal, much larger than the reduced stem leaves; pappus of 5 bristles; phyllaries not purplish, often pale yellowish ______ **Elephantopus**

1. Heads without a whorl of leaf-like bracts; leaves ± uniform in size and well-distributed up the stem; pappus a double row of numerous bristles, the outer much shorter than the inner; phyllaries often purplish ______ **Vernonia**

ACHILLEA YARROW

A mainly n temperate genus of 115 species; many are used medicinally or cultivated as ornamentals. (So named because its healing powers were said to have been discovered by Achilles of

Greek mythology; he supposedly stopped the flow of blood from the wounds of his Myrmidon warriors, perhaps with this plant—Arriagada & Miller 1997) (tribe Anthemideae)
REFERENCES: Mulligan & Bassett 1959; Arriagada & Miller 1997.

Achillea millefolium L., (thousand-leaved), MILFOIL, WESTERN YARROW, COMMON YARROW. Aromatic, woolly-pubescent, rhizomatous, perennial herb 35–100 cm tall, usually unbranched up to the inflorescence; leaves alternate, sessile, lanceolate, 2–3 times pinnately compound, fern-like in appearance and often confused with ferns by non-botanists, the crowded, narrow leaflets finely lobed or toothed; heads small, in a dense, terminal corymbose arrangement; ray flowers 5–12, white, rarely pinkish, the ligules ca. 2 mm long; disk corollas white or creamy white to pinkish; pappus absent. Roadsides, disturbed sites; widespread in TX. Apr–May. [*A. lanulosa* Nutt., *A. millefolium* var. *lanulosa*, *A. millefolium* subsp. *lanulosa* (Nutt.) Piper, *A. millefolium* var. *occidentalis* DC., *A. occidentalis* (DC.) Raf. ex Rydb.] While this species is often split into subspecific taxa (e.g., Kartesz 1994; Jones et al. 1997), Cronquist (1980) indicated that it is a highly variable polyploid complex with both native and introduced forms not yet satisfactorily sorted into infraspecific taxa. We are thus not recognizing subspecies or varieties. When the foliage is crushed, it releases a spicy odor. MILFOIL was attributed to have extensive curative powers by herbals of the Middle Ages, supposedly being useful in treating such conditions as influenza, gout, and ailments of the kidneys and liver (Wills & Irwin 1961). It was also used medicinally by Native Americans and is still used in folk remedies; however, ingestion was reported to cause irritation of mucous membranes and gastric and abdominal pain (Burlage 1968). ☠

ACMELLA

A mainly tropical genus of. 30 species. Previously included in and closely related to *Spilanthes*. (Derivation of generic name unknown) (tribe Heliantheae)
REFERENCES: Jansen 1981, 1985a, 1985b.

Acmella oppositifolia (Lam.) R.K. Jansen var. **repens** (Walter) R.K. Jansen, (sp.: opposite-leaved; var.: creeping), CREEPING SPOTFLOWER. Herbaceous rhizomatous perennial, creeping, rooting at nodes; flowering branches ascending; leaves opposite, petioled; leaf blades ovate, rhombic to deltoid, 20–40 mm long, serrate; heads few, solitary, on axillary peduncles 5–15 cm long; phyllaries linear, in 2 series; ray flowers few, pistillate, infertile, the ligules ca. 5 mm long, yellow; disk flowers perfect, fertile, yellow; pales (= bracts on receptacle) about equaling disk flowers; receptacles elongate, conical, 8–10 mm long in fruit; achenes black; pappus absent or rarely with 1–2 awns. Low moist areas, ditches, ponds; Grayson, Red River, and Tarrant cos., also Parker Co. (nursery escape), also Ellis Co. (Jansen 1985a); mainly se and e TX, also Edwards Plateau. May–Nov. [*A. repens* (Walter) Rich., *Spilanthes americana* (Mutis) Hieron. var. *repens* (Walter) A.H. Moore]

AMBLYOLEPIS HUISACHE DAISY

A monotypic genus endemic to the sw United States and Mexico; previously treated in *Helenium*. (Greek: *ambly*, blunt, and *lepis*, scale) (tribe Helenieae, sometimes lumped into Heliantheae)
REFERENCE: Bierner 1990.

Amblyolepis setigera DC., (bristle-bearing), HUISACHE DAISY. Small annual 10–50 cm tall with odor resembling recently cut *Medicago* hay or *Melilotus*; stems and leaf margins with long, wide-spreading, cottony hairs; leaves sessile, entire, the lowest oblanceolate, the upper ovate and clasping; heads long-peduncled and large, the involucres 7–11 mm high; phyllaries in about 2 rows, slightly unequal, lanceolate, subappressed; ray and disk corollas yellow; disk hemispherical-ovoid; achenes with a cage-like exterior of stout, separate ribs. Sandy or rocky open ground; Brown, Mills, and Shackelford cos., also Comanche and Hamilton cos. (HPC); se TX and w part of nc TX w to w TX. Mar–Jul. [*Helenium setigerum* (DC.) Britton & Rusby]

AMBROSIA RAGWEED

Annual or perennial herbs, 0.3–3+ m tall; leaves alternate, alternate above and opposite below, or opposite, nearly sessile to petioled, palmately lobed or pinnatifid, sometimes aromatic; flowers unisexual, in separate heads, these on the same plant; staminate heads in spike-like or raceme-like inflorescences; phyllaries united and cup-shaped; anthers distinct; pistillate heads 1 or few-flowered, axillary, below the staminate; phyllaries united, enclosing the achenes at maturity, forming a hard indehiscent "bur" or "fruit," the bur often with the tips of the phyllaries projecting as spines or tubercles; corollas and pappus absent.

A cosmopolitan, but especially American genus of 43 species. The flowers of *Ambrosia* species are small and wind-pollinated; the abundant air-borne pollen is a problematic cause of allergic reactions during the fall and is considered the leading cause of hay fever in the U.S. (Lewis & Elvin-Lewis 1977). The allergic response is initiated when pollen grain proteins (antigens) attach to receptors on antibodies (immunoglobulin E—IgE) linked to immune system cells. This results in the immune cells releasing histamines which are the molecules directly responsible for the symptoms known as hay fever (Kuby 1997; Lim 1998). A cladistic study by Karis (1995) suggested that *Xanthium* is a monophyletic clade in a paraphyletic *Ambrosia*. (Early Greek name for aromatic plants; the mythic food of the gods) (tribe Heliantheae)
REFERENCES: Rydberg 1922; Payne 1964; Lee & Dickinson 1980; Lee 1981; Karis 1995.

1. Leaves 3–5-lobed or sometimes unlobed.
    2. Side lobes of leaves basically a large tooth on each side, these a small fraction as large as the middle lobe; lobes or whole leaf (if unlobed) 12 mm wide or less; plants < 1 m tall; sap not red; staminate heads sessile ____ **A. bidentata**
    2. Side lobes of leaves not much smaller than the middle lobe; lobes or whole leaf > 12 mm wide (often much more); plants 1–3(–5) m tall; sap often red; staminate heads on short stalks ____ **A. trifida**
1. Leaves pinnatifid, divided into numerous narrow segments.
    3. Leaves 1–4 times pinnatifid, parsley-like in appearance, with petioles to 15 cm long (longest on lower leaves); staminate involucres to 10 mm across; spines of the bur (0–)10–20, in several series on the upper 2/3 of the body of the bur; rare in nc TX ____ **A. confertifolia**
    3. Leaves 1–3 times pinnatifid, but not parsley-like, subsessile or with petioles to only 3 cm long; staminate involucres ca. 2.5–3 mm across; spines or tubercles of the bur 5–7, in a single series near the middle to the apex (= beak) of the bur; widespread throughout nc TX.
        4. Stem leaves petioled; plants annuals with a taproot; female involucres with 5–7 short spines ____ **A. artemisiifolia**
        4. Stem leaves essentially sessile; plants perennials from creeping rootstocks; female involucres with 4–6 short, stout tubercules, these sometimes obscure ____ **A. psilostachya**

Ambrosia artemisiifolia L., (with leaves like *Artemisia*—sagebrush, wormwood), COMMON RAGWEED, ROMAN WORMWOOD, SHORT RAGWEED, ALTAMISA, HOGBRAKE. Annual; stems 0.3–1 m tall; leaves opposite or uppermost sometimes alternate, pinnatifid to bipinnatifid, with petioles usually 10–30 mm long; staminate heads short-stalked; "fruits" ca. 3 mm long, obovoid in outline, the spines subulate. Disturbed areas; se and e TX w to nc TX and Edwards Plateau. Aug–Nov.

Ambrosia bidentata Michx., (two-toothed), SOUTHERN RAGWEED, LANCE-LEAF RAGWEED. Annual; stems 0.3–1 m tall; leaves alternate above, opposite towards base, sessile, narrowly ovate, 3–5 cm long, usually with a broad tooth or small lobe on each side at the blade base; staminate heads sessile with oblique involucre; "fruits" ca. 5 mm long, 4-angled, the spines 4. Prairies; Grayson and Lamar cos. in Red River drainage and Bowie Co. and Sabine Co. (G. Nesom, pers. comm.) in extreme e TX are apparently the only records for the state. Aug–Sep.

Ambrosia confertiflora DC., (with flowers crowded together), FIELD RAGWEED, BUR-SAGE. Perennial, forming colonies; stems 0.3–0.6(–1.8) m tall; leaves alternate, 1–4 times pinnatifid, often with 1-several pair of small lobes on the petiolar bases below the main blade; staminate heads short-stalked; "fruits" ca. 5 mm long, the spines usually (0–)10–20, ca. 1–2 mm long, with hooked tips. Disturbed areas; mainly w 1/2 of TX e to Brown and Mills cos. on w margin of nc TX, disjunct e to Dallas Co. Mostly Aug–Nov.

Ambrosia psilostachya DC., (naked spike), WESTERN RAGWEED, PERENNIAL RAGWEED. Perennial, forming colonies; stems 0.3–0.6(–1+) m tall; leaves pinnatifid, sessile; staminate heads short-stalked; "fruits" ca. 2.5 mm long, obovoid in outline, the tubercles acute or blunt. Disturbed areas; nearly throughout TX. Aug–Nov. [*A. cumanensis* of authors, not Kunth] Used as a medicinal tea by Native Americans.; also apparently inhibits the growth of some other organisms through allelopathy (Cheatham & Johnston 1995).

Ambrosia trifida L. var. **texana** Scheele, (sp.: three-parted; var.: of Texas), BLOOD RAGWEED, GIANT RAGWEED, BUFFALOWEED. Annual; sap blood red; stems often very tall, 1–3(–5) m tall; leaves usually opposite, palmately 3(–5)-lobed or uppermost rarely unlobed, scabrous on both sides, extremely so on upper surface, petioled; staminate heads stalked; "fruits" ca. 4 mm long, obovoid in outline, the tubercles 4–8, small or obsolete. Disturbed areas, often extremely abundant; nearly throughout the state except s TX and Trans-Pecos. Aug–Nov. The sap stains the hands red if the tissues are damaged.

ANTENNARIA PUSSY-TOES, EVERLASTING, LADIES'-TOBACCO

A temperate and warm area genus of 71 species of small dioecious herbs; some are cultivated as rock garden subjects. (Latin: *antenna*; the pappus of the male flowers have swollen tips resembling a butterfly's antennae) (tribe Inuleae)
REFERENCES: Bayer & Stebbins 1982; Bayer 1984, 1985; Anderberg 1991.

Antennaria parlinii Fernald subsp. **fallax** R.J. Bayer & Stebbins, (sp.: for its discoverer, John Crawford Parlin, 1863–1948, of Maine; subsp.: deceptive), LARGE-LEAF PUSSY-TOES, PLAIN-LEAF PUSSY-TOES. Small stoloniferous perennial to 25(–35) cm tall, gray-pubescent with matted, woolly hairs; leaves simple, entire, prominently 3- to 5-nerved, with conspicuously different upper and lower surfaces, olive green and glabrate above, gray woolly below; heads terminal, few, in a cymose cluster, unisexual; phyllaries with scarious or white tips; ray flowers absent; staminate flowers with prominent black-brown anthers; pistillate flowers with prominent, white, hair-like pappus; achenes < 1 mm long. Sandy woods; e TX w to West Cross Timbers. Mar–Apr. [*A. fallax* Greene] Thought to be a polyploid of hybrid origin (Bayer & Stebbins 1982; Bayer 1985).

ANTHEMIS MAYWEED, DOG-FENNEL, CHAMOMILE

An Old World genus of ca. 210 species ranging from Europe through the Mediterranean region to Iran and e Africa; some are medicinal; others are cultivated as ornamentals. (Greek name for the related *Chamaemelum nobile* (L.) All., camomile; probably derived from *anthemon*, flower) (tribe Anthemideae)
REFERENCE: Arriagada & Miller 1997.

Anthemis cotula L., (small cup), MAYWEED CHAMOMILE, MAYWEED, DOG-FENNEL, STINKING CAMOMILE, STINKWEED. Rankly aromatic, slightly and inconspicuously pubescent, annual to 60 cm

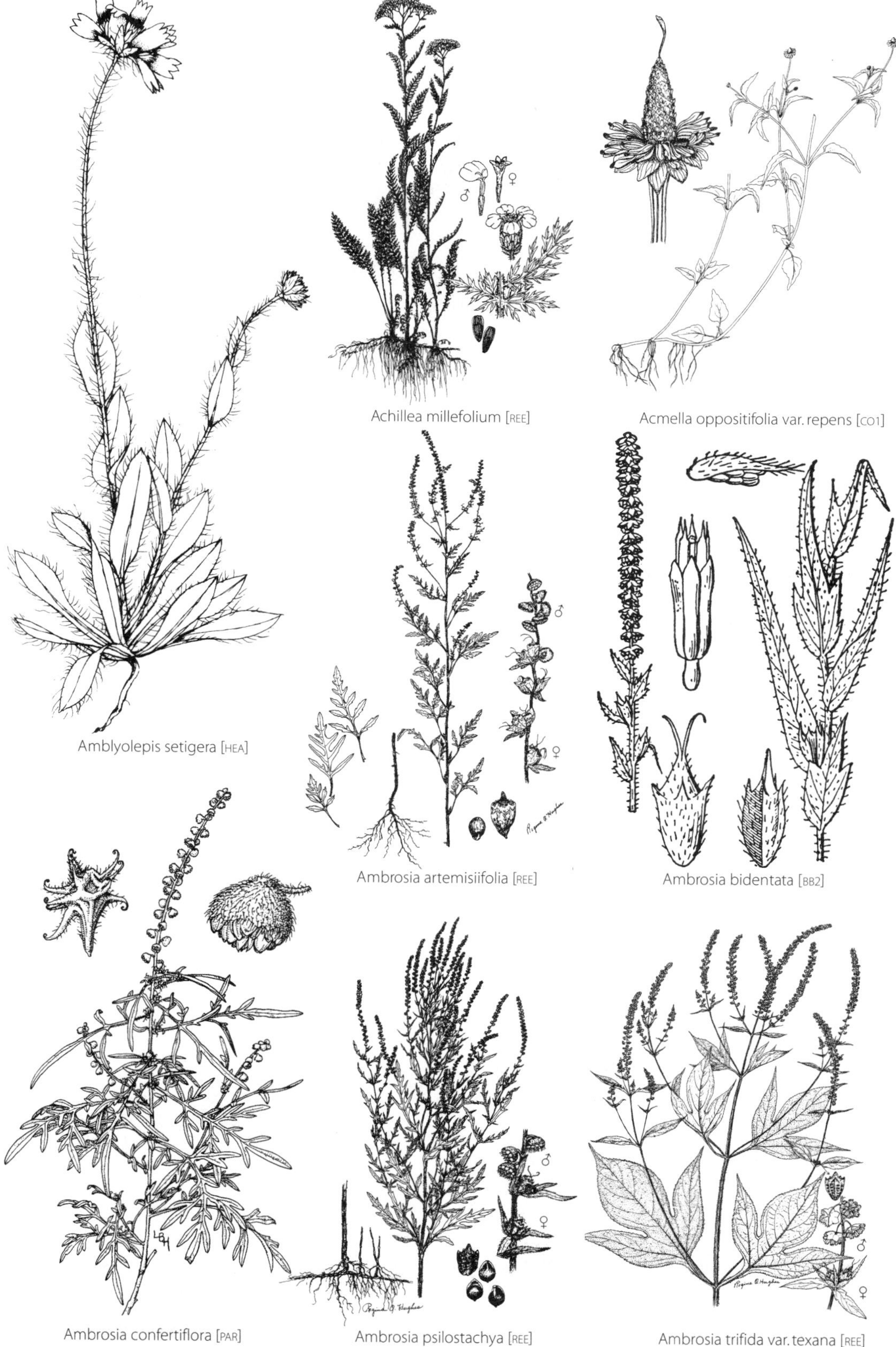

Achillea millefolium [REE]

Acmella oppositifolia var. repens [CO1]

Amblyolepis setigera [HEA]

Ambrosia artemisiifolia [REE]

Ambrosia bidentata [BB2]

Ambrosia confertiflora [PAR]

Ambrosia psilostachya [REE]

Ambrosia trifida var. texana [REE]

tall; leaves alternate, finely and deeply cut, twice compound or pinnately lobed; heads corymbose; phyllaries appressed, scarious-margined; ray flowers 10–14 per head, sterile, persistent and reflexed in age, the ligules white, ca. 7–12 mm long; disk ovoid-conical; disk corollas greenish yellow; achenes 1–1.5 mm long; pappus absent. Sandy roadsides, pastures, disturbed areas; cultivated and escaped; Dallas, Denton, Fannin, Grayson, Hopkins, Kaufman, and Palo Pinto cos.; se and e TX w to nc TX. Apr–May. Native of Europe. Reverchon (1880) stated that its introduction in Dallas Co. dates from 1875. Source of an insecticide and reported to taint cow's milk (Mabberley 1987); related to *Chamaemelum nobile*, which is the source of the stomach remedy camomile.

APHANOSTEPHUS LAZY DAISY, DOZE DAISY

Annual or perennial herbs; leaves alternate, with narrowed, petiolar base; blades linear to oblong-oblanceolate, varying from entire to pinnately lobed; heads solitary or corymbose; ray flowers with ligules white to pink, often rose-red or rose-purple beneath, spreading from midmorning to late afternoon, erect at night; disk corollas yellow; pappus minute to small, 1 mm or less long.

A genus of 4 species endemic to the United States and Mexico. The common names refer to the flowers remaining closed until mid-morning or later (Ajilvsgi 1984); the often rose-red or rose-purple undersides of the unopened ray flowers are conspicuous (Kirkpatrick 1992). (Presumably from Greek: *aphanes*, unseen or hidden, and *stephanus*, crown, presumably from the inconspicuous pappus) (tribe Astereae)
REFERENCES: Shinners 1946c; Turner 1984, Elisens et al. 1992.

1. Plants soft-pubescent to hispid-pubescent with spreading to reflexed hairs (usually < 1 mm long); heads not crowded, on long peduncles naked for 15–100 mm below the heads (to 150 mm in age); ray flowers 16–85 per head.
   2. Ray flowers with ligules 5–10 mm long; pappus a ring of minute subequal hairs 0.25 mm or less long; bases of disk corollas only slightly enlarged at maturity; plants taprooted annuals or clump-forming perennials; w part of nc TX.
      3. Ray flowers usually 16–32 per head; plants annuals from slender taproot ____ **A. ramosissimus**
      3. Ray flowers 40–85 (as few as 25 in second flowering of summer and fall) per head; plants perennials forming clumps from branched, woody root ____ **A. riddellii**
   2. Ray flowers with ligules 8–15 mm long; pappus an uneven crown of ca. 5 small acute scales 0.25–1 mm long; bases of disk corollas becoming swollen (to 2 times or more their original diam.) and hardened at maturity; plants taprooted annuals; widespread in nc TX ____ **A. skirrhobasis**
1. Plants conspicuously hispid with long (0.7–2.2 mm long), coarse, jointed, translucent hairs spreading at right angles; heads ± crowded, on short peduncles naked for only 3–12 mm below the heads (to 25 mm in age); ray flowers 12–18 ____ **A. pilosus**

Aphanostephus pilosus Buckley, (pilose, with long soft hairs), HAIRY LAZY DAISY. Small annual; stems 6–33 cm tall; involucres 4.5–5.5 mm high, 4–6 mm wide; ligule 5–7 mm long, white to rosy-lavender, especially upon withering; pappus a cup-like, small, scaly crown 0.3–0.4 mm tall. Sandy or silty soils; low prairies, draws, and ditches; Archer, Clay, and Jack cos. in w part of nc TX; Rolling Plains and West Cross Timbers; also disjunct e to Red River Co. May–Jun.

Aphanostephus ramosissimus DC., (much-branched), PLAINS LAZY DAISY, ARIZONA LAZY DAISY. More or less soft pubescent annual, often widely branched and partly decumbent; stems 5–45 cm tall; involucres 3.3–5 mm high, 5–9 mm wide; ligule 5.5–7.5 mm long, white or becoming rose to purple, especially on lower surface; pappus minute, to 0.25 mm long. Sandy open woods, fields, prairies, and roadsides; Archer and Erath cos., also disjunct e to Tarrant Co., also Callahan Co. (Shinners 1946c); w part of nc TX s and w to w TX; Apr–Jun(–Aug).

Anthemis cotula [REE]

Antennaria parlinii subsp. fallax [HEA]

Aphanostephus pilosus [BT3]

Aphanostephus ramosissimus [MAR]

Aphanostephus riddellii [JON]

Aphanostephus skirrhobasis [BB2]

Aphanostephus riddellii Torr. & A. Gray, (for John Leonard Riddell, 1807–1865, botanist), RIDDELL'S LAZY DAISY. Hispid-pubescent perennial; stems 10–50 cm tall; heads usually numerous, rather large and long-peduncled; involucres 4.5–6.2 mm high, 9–14 mm wide; ligule 7–10 mm long, white; pappus minute, to 0.2 mm long. Rocky or sandy ground, chiefly on limestone; Burnet, Coleman, and Shackleford cos. on w margin of nc TX s and w to w TX. Apr–Jun(–Jul–Oct).

Aphanostephus skirrhobasis (DC.) Trel., (possibly white parasol rosy at base, from the ligules colored on the lower surface), ARKANSAS LAZY DAISY. Soft, gray-pubescent annual from a taproot; stems to 50 cm tall; heads rather large; involucres 6–8 mm high, 7–13 mm wide; ray flowers 20–44 with ligules 8–15 mm long, white or reddish to rosy on the lower surface, the color often streaked; pappus minute, 0.25–1 mm long. Sandy open woods, fields, prairies, and roadsides; widespread in TX. Apr–Jun, sporadically later. Sometimes cultivated as an ornamental. ▣/78

ARNOGLOSSUM INDIAN-PLANTAIN

A genus of 7 species native to the e United States; previously recognized in *Cacalia*. (Presumably from Greek: *arnos*, lamb, and *glossa*, tongue) (tribe Senecioneae)
REFERENCES: Shinners 1950c; Robinson 1974; Pippen 1978.

Arnoglossum plantagineum Raf., (resembling *Plantago*—plantain), PRAIRIE-PLANTAIN, GROOVE-STEM INDIAN-PLANTAIN, TUBEROUS INDIAN-PLANTAIN. Glabrous, coarse perennial from fleshy-fibrous roots, 0.5–1(–1.5) m tall; stems striate-angled, often dark purple; basal leaves conspicuous, 5–15+ cm long, 2–8 cm wide, long-petioled, the blades elliptic or ovate, entire or rarely toothed, somewhat thick and fleshy, with prominent ± parallel, longitudinal nerves (often 7–9) eventually converging toward the tip; stem leaves relatively few, alternate, much smaller, the upper subsessile; heads numerous in a broad corymbose arrangement at apex of plant; phyllaries 5, 7–10 mm long, with narrow, sharp, prominent keel; ray flowers absent; disk corollas 5, creamy white; achenes 4–5 mm long; pappus of numerous slender bristles. Prairies, fields, and open woods; se and e TX w to Denton, Grayson, McLennan, Tarrant, and Williamson cos., also Hamilton (HPC) and Somervell (Fossil Rim Wildlife Center—R. O'Kennon, pers. obs.) cos.; also Edwards Plateau. Apr–Jun, rarely late Sep–Nov. [*Cacalia plantaginea* (Raf.) Shinners]. The striking leaves and erect flowering stalks make this species a conspicuous spring component of many nc TX prairies and fields.

The only other *Arnoglossum* species in TX, *A. ovatum* (Walter) H. Rob. (LANCE-LEAF INDIAN-PLANTAIN), occurs in se and e TX. It has lanceolate leaves, phyllaries without keels, and flowers from Jul–Oct.

ARTEMISIA SAGEBRUSH, WORMWOOD, SAGEWORT, SAGE, MUGWORT

Ours perennial or biennial herbs or slightly subshrubby below; white-pubescent or nearly glabrous, often aromatic; leaves alternate, entire or deeply lobed, the segments linear to oblong, elliptical, or ovate; heads small, in numerous panicles; involucres white-woolly or nearly glabrous; ray flowers fertile or infertile, 1–few; corollas yellowish white; disk flowers fertile or infertile, corollas minute, yellowish white; achenes ca. 1 mm or less long; pappus absent.

A genus of ca. 350 species of the n temperate area, w South America, and s Africa; usually in dry areas. *Artemisia* species are wind-pollinated and cause allergies; they are typically aromatic shrubs and herbs; *Artemisia tridentata* Nutt. is the famous SAGEBRUSH widespread in the w 1/2 of the U.S.; TARRAGON, a culinary herb, is obtained from the Eurasian *A. dracunculus* L.; the WORMWOOD of the Bible is *A. herba-alba* Asso; the harmful green liqueur, absinthe, is made from the European *A. absinthium* L. (Ancient name of *Artemisia vulgaris* L., MUGWORT, from Greek goddess of the hunt, Artemis, or possibly from Artemisia, wife of Mausolus, king of Caria) (tribe Anthemideae)

REFERENCES: Hall & Clements 1923; McArthur & Welch 1984; Arriagada & Miller 1997.

1. Leaf blades nearly glabrous, usually very deeply and finely dissected, the divisions linear to thread-like, 2 mm wide or less; foliage not aromatic; disk flowers sterile; on extreme w margin of nc TX ______ **A. campestris**
1. Leaf blades conspicuously grayish or whitish woolly on both surfaces or at least on lower surfaces, entire to lobed or dissected, the main axis and lobes usually 3–10+ mm wide; foliage aromatic; disk flowers all fertile; throughout nc TX ______ **A. ludoviciana**

Artemisia campestris L. subsp. **caudata** (Michx.) H.M. Hall & Clem., (sp.: of the fields or plains; subsp.: tailed), WESTERN SAGEWORT, THREAD-LEAF SAGEWORT. Biennial or weakly perennial herb 30–90 cm tall; involucres essentially glabrous, 2.5–4 mm high. Roadsides, prairies, open areas; Callahan Co. on extreme w margin of nc TX; mainly Plains Country and Trans-Pecos. Sep–Oct. [*A. caudata* Michx.]

Artemisia ludoviciana Nutt. subsp. **mexicana** (Willd. ex Spreng.) D.D. Keck, (sp.: of Louisiana; subsp.: of Mexico) MEXICAN SAGEBRUSH. Perennial herb, rarely slightly subshrubby, 20–90(–150) cm tall; leaves very variable in shape, usually some lobed or dissected but varying to entire, also variable in terms of pubescence, the young leaves white woolly above and below to glabrate and dark green above and white woolly below; inflorescences compact to open diffuse; involucres grayish or white woolly, 2.5–4+ mm high. Prairies, disturbed habitats; widespread throughout most of TX. Oct–Nov. [*A. ludoviciana* var. *mexicana* Willd. ex Spreng.] According to Jones et al. (1997) and J. Kartesz (pers. comm. 1997), subsp. *ludoviciana* does not occur in TX.

ASTER

Ours herbaceous perennials (1 species annual); leaves alternate, simple, entire to toothed; heads usually numerous; ray flowers fertile, the ligule often conspicuous, white to pink, blue, violet, or purplish, never yellow; disk flowers perfect, the corollas yellow to rose, purple, or blue; phyllaries imbricated, the tips usually green; receptacles naked; achenes somewhat flattened; pappus of hair-like bristles, white.

Mabberley (1997) viewed *Aster* as a large genus of ca. 250 species found in America, Eurasia, and Africa; a number are cultivated; many species hybridize and intergrade; definite identification is often difficult. Nesom (1994b) treated *Aster* as an essentially Old World genus of ca. 180 species; he divided the New World taxa traditionally recognized in *Aster* into 13 genera (number of species in parentheses): *Ampelaster* (1), *Almutaster* (1), *Chloracantha* (1), *Canadanthus* (1), *Doellingeria* (11 total, 3 from North America, 8 from Asia), *Eucephalus* (11), *Eurybia* (28), *Ionactis* (5), *Oclemena* (3), *Oreostemma* (3), *Psilactis* (6), *Sericocarpus* (5), and *Symphyotrichum* (97). All nc TX *Aster* species fall into *Symphyotrichum*. Until consensus is reached on generic circumscription in the Asteraceae, we are following the conservative approach in maintaining all nc TX species (except *Chloracantha*) in the genus *Aster*; however, appropriate synonymy is provided. Semple et al. (1996) also retained *Symphyotrichum* in *Aster*. (Greek, *aster*, a star, from the radiate heads) (tribe Astereae)
REFERENCES: Shinners 1953e; Jones, A. 1978a, 1978b, 1980, 1983, 1984, 1987, 1992; Semple & Brouillet 1980a, 1980b; Jones, A. & Young 1983; Jones, R. 1983; Semple & Chmielewski 1987; Nesom 1994b, 1997a; Semple et al. 1996.

1. Uppermost leaves crowding heads, not much reduced in size, sometimes canescent, longer and broader than phyllaries, partly hiding phyllaries or grading into them; head-bearing branchlets few, usually only 10–30 per plant; heads 20–30 mm in diam. when ligules fully extended.
   2. Outer phyllaries glabrous to slightly pubescent, 2–3 mm wide, marginally with a fringe of minute cilia; involucres 10–20 mm wide; widespread in nc TX ______ **A. pratensis**
   2. Outer phyllaries with antrorse-appressed whitish pubescence, 1.3–2 mm wide, not distinctly fringed; involucres 7–12 mm wide; rare in nc TX ______ **A. sericeus**

1. Uppermost leaves not crowding or hiding phyllaries, usually reduced in size, not canescent, smaller than phyllaries; head-bearing branchlets usually more numerous than 30 per plant; heads often smaller than 20 mm in diam.
  2. Phyllaries with greenish, linear (very narrow) midrib; leaves on the branches usually subulate, tapered from base to apex, usually ± entire; head-bearing branchlets usually not very leafy (reduced bracts present).
    3. Ligules of ray flowers usually white; leaves small or nearly absent; plants superficially reed-like, usually without well-developed leaves; usually some branchlets developed into thorns; plants perennials forming colonies from creeping rhizomes ______ see **Chloracantha**
    3. Ligules of ray flowers white, violet, purplish, blue, or pink; leaves well-developed, at least basally; plants not reed-like; thorns absent; plants annuals from a taproot ______ **A. subulatus**
  2. Phyllaries with greenish central midrib narrowly to broadly rhomboid (= diamond-shaped) or obovate; leaves on the branches often ovate to lanceolate or linear (rarely subulate), often serrate; head bearing branchlets leafy or not so.
    4. Stem leaves sessile and conspicuously auriculate-clasping ______ **A. patens**
    4. Stem leaves sessile or petiolate, not auriculate-clasping.
      5. Phyllaries, peduncles, and often also the leaves and upper stems glandular (use lens); basal and lower cauline leaves sessile ______ **A. oblongifolius**
      5. Phyllaries, peduncles, leaves, and stems varying from glabrous to hairy, but not glandular; basal and lower cauline leaves sessile to petiolate.
        6. Longest phyllaries usually only ca. 3–4 mm long, obtuse, with a fringe of a few short cilia; ligules of ray flowers white ______ **A. ericoides**
        6. Longest phyllaries usually longer than 4 mm OR ciliate fringe absent; ligules of ray flowers white OR variously colored.
          7. Basal and lower stem leaves with long, winged petiolar bases quite distinct from the much wider well-delimited blade portion.
            8. Stem leaves usually cordate (sometimes truncate) where narrowed to petiolar base; leaves of inflorescence obovate to elliptic or lanceolate, not linear; middle and sometimes even upper stem leaves sharply serrate ______ **A. drummondii**
            8. Stem leaves usually rounded or truncate or even attenuate where narrowed to petiolar base; leaves of inflorescence linear; middle and upper stem leaves usually entire ______ **A. oolentangiensis**
          7. Basal and lower stem leaves sessile or gradually narrowed to base, not distinctly differentiated into petiolar and blade portions.
            9. Mid-stem leaves linear, only 10–25(–34) mm long, 1–3(–5) mm wide; leaves of the head-bearing branchlets numerous, linear-subulate ______ **A. dumosus**
            9. Mid-stem leaves usually wider than linear, usually averaging longer than 20 mm, 3–20 mm wide; leaves of head-bearing branchlets variable but often wider than linear-subulate.
              10. Leaves on the ultimate branchlets just below heads < 1 mm wide, scale-like ______ **A. oolentangiensis**
              10. Leaves on the ultimate branchlets just below heads 2 mm or more wide, not scale-like.
                11. Phyllaries usually not greatly differentiated in size, the outer smallest ones (1/3–)1/2 or more as long as the inner; disk corolla limb (= flaring portion above tube) not deeply divided, the lobes < 1/2 the length of limb; heads 10–25 mm in diam. when ligules fully extended; ray flowers 15–40, the ligules bluish white, lavender, or white.
                  12. Reticulate brownish veins usually conspicuous on lower leaf surface (use lens), the enclosed areolae ± isodiametric; ray flowers 15–25, the

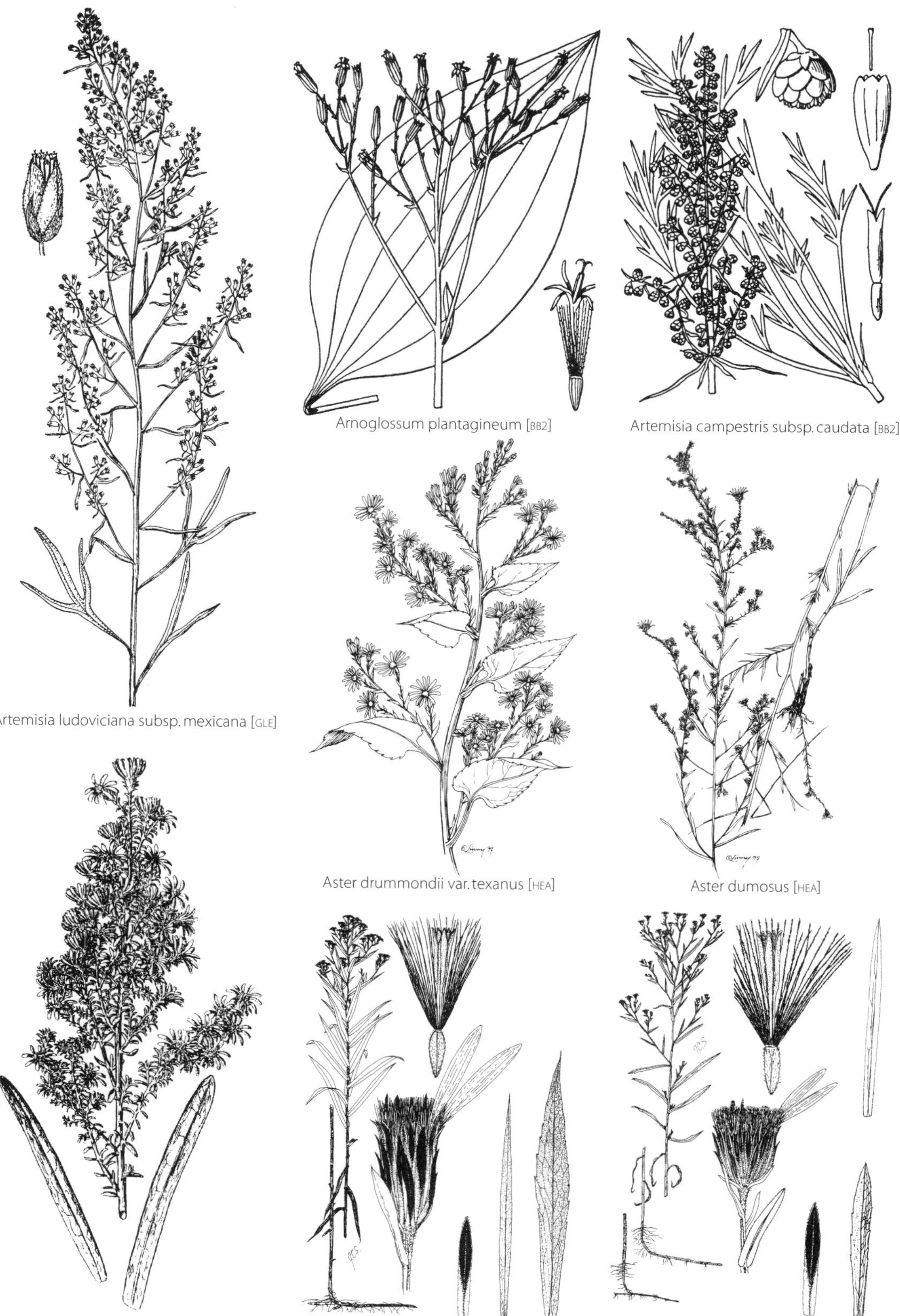

Artemisia ludoviciana subsp. mexicana [GLE]

Arnoglossum plantagineum [BB2]

Artemisia campestris subsp. caudata [BB2]

Aster drummondii var. texanus [HEA]

Aster dumosus [HEA]

Aster ericoides [DIL]

Aster lanceolatus subsp. hesperius [UWA]

Aster lanceolatus subsp. lanceolatus [UWA]

ligules usually bluish white or lavender (rarely white); achenes brown or purple at maturity; widespread in nc TX ____ **A. praealtus**

12. Reticulate brownish veins not conspicuous on the lower leaf surface, the inconspicuous areolae not isodiametric; ray flowers 20–40, the ligules usually white; achenes gray at maturity; apparently rare in nc TX ____ **A. lanceolatus**

11. Phyllaries more differentiated in size, the outer smaller ones 1/3 as long as the inner; disk corolla limb deeply divided, the lobes ca. 1/2–3/4 the length of the limb; heads small, 8–13 mm in diam. when ligules fully extended; ray flowers 9–15(–20), the ligules white ____ **A. lateriflorus**

Aster drummondii Lindl. var. **texanus** (E.S. Burgess) A.G. Jones, (sp.: for its discoverer, Thomas Drummond, 1780–1835, Scottish botanist and collector in North America; var.: of Texas), TEXAS ASTER. Perennial with short rhizomes; stem leaves ovate to cordate, large, 6–12(–15) cm long, 3–5(–6) cm wide, persistent, with broadly winged petioles; uppermost leaves reduced; involucres 5–6 mm high; ray flowers 10–20, the ligules usually deep purple to blue-lavender (rarely white); disk corollas cream to yellow, turning purple with age. Open woods and prairies; se and e TX w to West Cross Timbers and e edge of Edwards Plateau. Mostly Oct–Nov, rarely Mar–Apr. [*A. texanus* E.S. Burgess, *A. texanus* subsp. *texanus* (E.S. Burgess) A.G. Jones, *Symphyotrichum drummondii* (Lindl.) G.L. Nesom var. *texanum* (E.S. Burgess) G.L. Nesom]

Aster dumosus L., (bushy), BUSHY ASTER, RICE-BUTTON ASTER. Perennial, rhizomatous; stem leaves linear, 10–25(–34) mm long, 1–3(–5) mm wide, often falling early; leaves of the head-bearing branches numerous, minute, bract-like, linear-oblong; heads 8–15 mm in diam. when ligules are fully extended; involucres 3–5 mm high; ray flowers 4–7(–8) mm long, pink, lavender, bluish white, or sometimes white. Low areas, roadsides, sandy or clay soils; Lamar Co. in Red River drainage; mainly se and e TX. Aug–Oct. [*A. dumosus* var. *cordifolius* (Michx.) Torr. & A. Gray, *A. dumosus* var. *subulifolius* Torr. & A. Gray, *Symphyotrichum dumosum* (L.) G.L. Nesom] We are following Hatch et al. (1990) and Jones (1992) in not distinguishing varieties.

Aster ericoides L., (resembling *Erica*—heath), HEATH ASTER, WHITE PRAIRIE ASTER, WREATH ASTER. Perennial, rhizomatous, usually much branched; stem leaves linear or oblong, sessile, entire, usually lost before flowering time; leaves of branchlets 2–3 mm long and ca. 1 mm wide, persistent; heads crowded; involucres 4–7 mm broad; ray flowers with ligules white, slightly exceeding the pappus. Disturbed or open areas; widespread in TX. Sep–Oct(–Nov). [*Symphyotrichum ericoides* (L.) G.L. Nesom]

Aster lanceolatus Willd., (lanceolate, lance-shaped), PANICLED ASTER, TALL WHITE ASTER. Rhizomatous perennial; basal and larger stem leaves early deciduous; main stem leaves sessile or subsessile, elliptic or oblanceolate to linear-lanceolate (4–)6–15 cm long, (0.5–)1–2(–3.5) cm wide; ray flowers with ligules white (rarely pink). Low areas. Aug–Nov. Similar to and apparently hybridizes and intergrades with *A. praealtus* (Jones 1984; Jones 1992). Semple and Chmielewski (1987) separated the 2 nc TX subspecies as follows:

1. Outer phyllaries 2/3 length or more of inner ones; heads usually subtended by large leafy bracts ____ subsp. **hesperius**
1. Outer phyllaries 1/3–2/3 length of inner ones; heads not usually subtended by large leafy bracts ____ subsp. **lanceolatus**

subsp. **hesperius** (A. Gray) Semple & Chmiel., (of the west), SISKIYOU ASTER. Included based on citation of vegetational area 4 (Fig. 2) by Hatch et al. (1990) and range map in Semple and Chmielewski (1987); Semple and Chmielewski (1987) mapped this subspecies as occurring in the ne 1/4 of TX. [*A. hesperius* A. Gray, *Symphyotrichum lanceolatum* (Willd.) G.L. Nesom subsp. *hesperium* (A. Gray) G.L. Nesom]

subsp. **lanceolatus**. Dallas Co., also Tarrant Co. (BAYLU); Semple & Chmielewski (1987) map this subspecies as occurring over the w 2/3 of TX. [*A. simplex* Willd., *A. lanceolatus* var. *simplex* (Willd) A.G. Jones, *Symphyotrichum lanceolatum* (Willd.) G.L. Nesom]

Aster lateriflorus (L.) Britton, (with one-sided flower clusters), CALIFORNIA ASTER, CALICO ASTER, SIDE-FLOWER ASTER, STARVED ASTER, WHITE WOODLAND ASTER. Rhizomatous perennial; larger stem leaves persistent at least in part, sessile or subsessile, elliptic-oblanceolate or linear (3–)5–10(–15) cm long, (0.2–)1–2(–3.5) cm wide; heads small, 0.8–1.3 cm in diam. when ligules are fully extended; ray flowers with ligules ca. 3–4 mm long, white. Well-drained uplands; Fannin, Henderson, Hopkins, Kaufman, and Limestone cos.; se and e TX w to e part of nc TX. Sep–Oct. [*A. lateriflorus* var. *flagellaris*, *A. lateriflorus* var. *indutus* Shinners, *Symphyotrichum lateriflorum* (L.) Á. Löve and D. Löve] Because of intergradation, we are following Correll and Johnston (1970), Hatch et al. (1990), and Jones (1992) in not distinguishing varieties in this species.

Aster oblongifolius Nutt., (oblong-leaved), AROMATIC ASTER, OBLONG-LEAF ASTER. Rhizomatous perennial; leaves numerous; stem leaves oblong to linear-lanceolate, often subclasping, some usually persistent throughout flowering; leaves of the branchlets 3–10 mm long, 1–2 mm wide; heads not crowded; involucres 8–12 mm broad; phyllaries glandular-pubescent apically; ray flowers with ligules usually blue or purple. Calcareous soils, prairies; Denton and Grayson cos. in East Cross Timbers w to Rolling Plains and s to Edwards Plateau. Sep–Nov. [*Symphyotrichum oblongifolium* (Nutt.) G.L. Nesom]

Aster oolentangiensis Riddell, (of Oolentangy [Olentangy] River in Ohio), AZURE ASTER, SKY-BLUE ASTER. Perennial; rhizomes short; basal and lower stem leaves truncate or rounded (rarely cordate), persistent; petioles winged; head-bearing branchlets with minute leaves; heads 10–25 mm in diam. when ligules fully expanded; involucres 5–8 mm high; ray flowers 10–25, with ligules blue or violet-purple (rarely white); disk flowers yellow, turning purple with age. Dry, sandy areas; Grayson and Henderson cos.; e TX w rarely to nc TX. Sep–Nov. [*A. azureus* Lindl., *Symphyotrichum oolentangiense* (Riddell) G.L. Nesom]

Aster patens Aiton, (spreading), SPREADING ASTER, LATE PURPLE ASTER, SKY-DROP ASTER. Rhizomatous perennial; stem leaves obovate or spoon-shaped, conspicuously clasping, grayish green; leaves of head-bearing branchlets different, small, appressed; flowering heads showy, 20–35 mm in diam. when ray flowers fully extended; involucres 7–15 mm broad; phyllary tips often glandular-pubescent; ray flowers 10–15(–17) mm long, the ligules usually blue or purple (rarely pink or white). Disturbed or open, often sandy areas. Aug–Nov. Three intergrading and often difficult to distinguish varieties are reported from nc TX and separated by Jones (1992) as follows:

1. Involucres 8–10 mm high, broadly turbinate; median phyllaries ovate-lanceolate, 1.2–1.5 mm in width, obtuse, not squarrose and not obviously glandular, although sometimes with minute glands on the abaxial surface that are obscured by a densely canescent-strigillose indument ______ var. **patentissimus**
1. Involucres usually < 8 mm high, campanulate or slenderly turbinate; median phyllaries linear-lanceolate, usually < 1.2 mm in width, acute or acuminate, often at least somewhat squarrose, and distinctly glandular.
    2. Plants robust; principal cauline leaves 3–6(–8) cm long and 1–2 cm wide; median phyllaries 1–1.2 mm in width, sparsely strigillose on the abaxial surface mainly along the midrib, the glandularity usually very pronounced ______ var. **patens**
    2. Plants slender-stemmed; principal cauline leaves 1–3(–4) cm long and 0.5–1(–1.5) cm wide; median phyllaries slender, usually < 1 mm in width, densely cinereous-puberulent on the abaxial surface, the glandularity less pronounced ______ var. **gracilis**

var. **gracilis** Hook., (graceful). Predominantly diploid ($2n = 10$). Se and e TX w to West Cross Timbers; also Edwards Plateau; by far the most common of the 3 varieties in nc TX. [*Symphyotrichum patens* (Aiton) G.L. Nesom var. *gracile* (Hook.) G.L. Nesom]

var. **patens**. SKY-DROP ASTER. Mostly tetraploid ($2n = 20$). Included based on citation of vegetational areas 4 and 5 (Fig. 2) by Hatch et al. (1990); se and e TX w to nc TX. [*Symphyotrichum patens* var. *patens*]

var. **patentissimus** (Lindl. ex DC.) Torr. & A. Gray, (much-spreading). Mostly tetraploid ($2n = 20$). Included based on citation of vegetational area 4 (Fig. 2) by Hatch et al. (1990); mainly se and e TX. [*Symphyotrichum patens* var. *patentissimum* (Lindl. ex DC.) G.L. Nesom]

Aster praealtus Poir., (very tall), WILLOW-LEAF ASTER, WILL ASTER, TALL ASTER. Perennial, rhizomatous; stem leaves sessile, linear to narrowly elliptic, 4–12(–15) cm long, 0.3–1.5 cm wide; lower ones falling early; leaves of the head-bearing branchlets very small and subulate or linear subulate; flowering heads 15–20(–25) mm in diam. when the ray flowers are fully extended; involucres ca. 4–8 mm high; ray flowers with ligules bluish white or lavender (rarely white). Low disturbed areas; se and e TX w to Rolling Plains and Edwards Plateau. Oct–Nov. While intergrading varieties are sometimes recognized in this species (e.g., Hatch et al. 1990; Kartesz 1994), we have been unable to find clear differences in the available material. We are thus lumping all the nc TX taxa. [*A. coerulescens* DC., *A. praealtus* var. *coerulescens* (DC.) A.G. Jones. *A. praealtus* var. *texicola* Wiegand, *A. salicifolius* Aiton, *Symphyotrichum praealtum* (Poir.) G.L. Nesom]

Aster pratensis Raf., (of the meadows), SILKY ASTER. Rhizomatous perennial similar to *A. sericeus* and often treated as a variety of that species; ray flowers with ligules purple to violet (rarely pinkish or white). Open woods, open areas, usually sandy soils; se and e TX w to East Cross Timbers, also Edwards Plateau. Sep–Nov. [*A. sericeus* var. *microphyllus* DC., *Symphyotrichum pratense* (Raf.) G.L. Nesom] While this taxon is often recognized as a var. of *A. sericeus* (e.g., Kartesz 1994; J. Kartesz, pers. comm. 1997), we are following Nesom (1994b) in recognizing it at the specific level.

Aster sericeus Vent., (silky), SILKY ASTER. Perennial, rhizomatous; stem leaves mostly falling before flowering; leaves of the head-bearing branchlets persistent, elliptic, not reduced below heads; ray flowers with ligules purple to violet (rarely pinkish or white). Open areas, rocky calcareous soils; Coryell Co. in the Lampasas Cut Plain; also Edwards Plateau. Aug–Oct. [*Symphyotrichum sericeum* (Vent.) G.L. Nesom]

Aster subulatus Michx. var. **ligulatus** Shinners, (sp.: awl-shaped; var.: ligulate, strap-shaped), WIREWEED, BLACKWEED, SALTMARSH ASTER, SLIM ASTER. Annual, glabrous, much-branched; leaves linear to subulate, 1–10(–20) cm long, 2–4(–7) mm wide, much reduced up the stem, soon deciduous except those of the head-bearing branchlets; flowering heads 10–12(–15) mm wide when ray flowers are fully extended; involucres 3–5(–7) mm high; ligules of ray flowers blue, purple, violet, pink or white. Weed in lawns, ditches, wet areas; throughout TX. Aug–Nov. [*A. divaricatus* (Nutt.) Torr. & A. Gray [nom. illeg.], *A. exilis* Elliott, *Symphyotrichum divaricatum* (Nutt.) G.L. Nesom] This is the most abundant *Aster* in the state.

Aster ×*eulae* Shinners, (for Eula Whitehouse, 1892–1974, of Southern Methodist Univ., TX collector, artist, and author of *Texas Flowers in Natural Color*), is cited by Kartesz (1994) as a hybrid, *Aster* ? ×*eulae* Shinners [*lanceolatus* × *praealtus*]. Most specimens at BRIT annotated as *A. eulae* seem very similar to *A. praealtus*. However, this entire complex of related asters, including *A.* ×*eulae*, *A. lanceolatus*, *A. lateriflorus*, and *A. praealtus*, is extremely difficult taxonomically. More work is needed to clarify these taxa. Jones et al. (1997) recognized *A. eulae* as a distinct species. Guy Nesom (1997 and pers. comm.) indicated that he believes *A. eulae* is a good

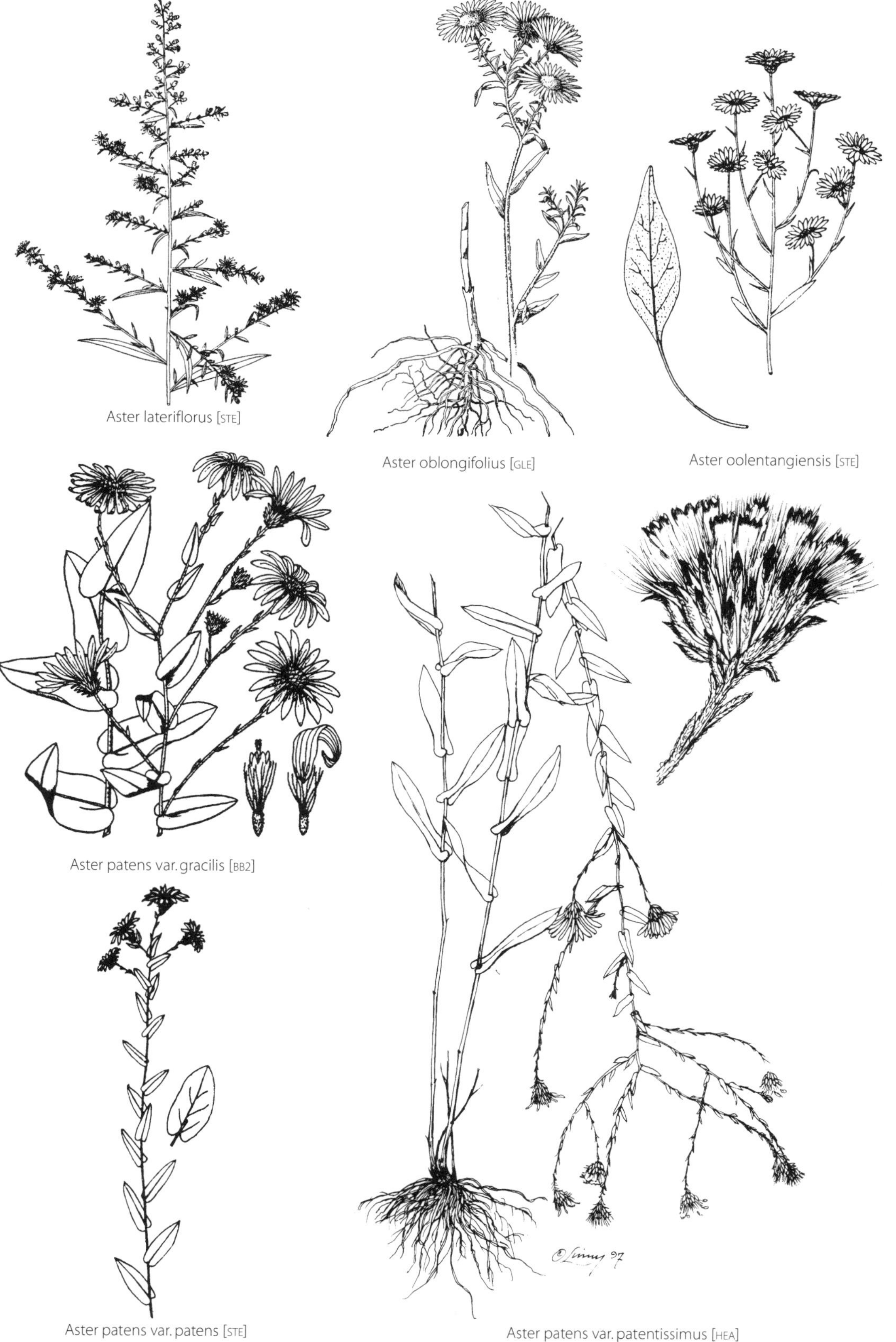

Aster lateriflorus [STE]

Aster oblongifolius [GLE]

Aster oolentangiensis [STE]

Aster patens var. gracilis [BB2]

Aster patens var. patens [STE]

Aster patens var. patentissimus [HEA]

species and that a study of this taxon is in progress. [*Symphyotrichum eulae* (Shinners) G.L. Nesom] Mahler (1988) separated *A. praealtus* and *A. eulae* as follows:

1. Stem leaves narrowly elliptical; outer phyllaries half as long as inner ones; involucres ca. 6 mm high ______ *A. praealtus*
1. Stem leaves elliptic to oblanceolate; outer phyllaries 1/3–1/4 as long as inner ones; involucres 4–5 mm high ______ *A. eulae*

ASTRANTHIUM WESTERN DAISY

A genus of 11 species endemic to the s United States and Mexico. (Greek, *astron*, star, and *anthos*, flower) (tribe Astereae)
REFERENCES: Larsen 1933; DeJong 1965.

Astranthium integrifolium (Michx.) Nutt. subsp. **ciliatum** (Raf.) DeJong, (sp.: entire-leaved; subsp.: ciliate, fringed), WESTERN DAISY, BLUE DAISY. Low pubescent annual from a taproot; stems 10–40 cm tall; leaves alternate, the lowest with narrow, petiolar bases, the upper sessile; blades oblanceolate, entire; heads solitary at the ends of long peduncles, usually several per main stem; ray flowers 8–35, perfect, fertile, the ligules white to bluish white or pinkish white, not curling, 5–12 mm long; disk flowers perfect, fertile, yellow; disk rounded-conical; pappus absent or a minute ring or crown. Open woods, prairies, and roadsides, in sandy or silty clay soils; se and e TX w to Rolling Plains and Edwards Plateau. Mar–May. [*A. integrifolium* var. *ciliatum* (Raf.) Larsen, *A. integrifolium* var. *triflorum* (Raf.) Shinners] Barkley (1986) questioned the recognition of subsp. *ciliatum* by DeJong (1965).

BACCHARIS GROUNDSELTREE

Shrubs or subshrubs, dioecious, pistillate and staminate plants with rather different appearance; leaves alternate, subulate to obovate, entire, serrate, or dentate, 1- or 3-nerved, glandular or punctate glandular; inflorescences somewhat paniculate or corymbose; involucres hemispheric to narrowly cylindrical; pistillate heads: ray flowers absent; disk flowers fertile; corollas filiform, yellowish white to brown; achenes 5–10-ribbed, yellow to reddish; pappus of numerous bristles; staminate heads: ray flowers absent; corollas funnelform, white to yellowish brown; ovary abortive; pappus of numerous bristles.

An American genus of ca. 400 species of dioecious shrubs; some are used medicinally, while others are cultivated as ornamentals. Pistillate plants are necessary for definitive identification. ☠ The leaves and flowers of a number of species contain cardioactive glycosides and are considered dangerous, even fatal, to livestock (Hardin & Brownie 1993) (Name derived from Bacchus, god of wine) (tribe Astereae)
REFERENCES: Mahler 1955; Mahler & Waterfall 1964.

1. Leaf blades linear, ca. 1–2 mm wide; achenes 5–6-ribbed, 3–4.5 mm long, slightly to prominently glandular-scabrous; pistillate pappus of many series, light reddish brown; plants less than 1 m tall ______ **B. texana**
1. Leaf blades linear to oblanceolate, elliptic, rhomboid or obovate, usually > 2 mm wide (often much more); achenes 8–10-ribbed, 1–2 mm long, glabrous; pistillate pappus in 1–2 series, whitish or dull; plants usually 1–3(–6) m tall (but can flower at 0.5 m tall).
    2. Leaf blades elliptic to rhomboid or obovate, the larger ones 22–37 mm wide, the upper leaves gradually reduced, narrower; pistillate involucres 4–6 mm long. ______ **B. halimifolia**
    2. Leaf blades very narrowly elliptic to linear or oblanceolate, 2–8(–15) mm wide; pistillate involucres 4–8 mm long.
        3. Pistillate involucres 5 mm or less long; leaves linear to very narrowly elliptic, usually 2–4(–5) mm wide; widespread in nc TX ______ **B. neglecta**

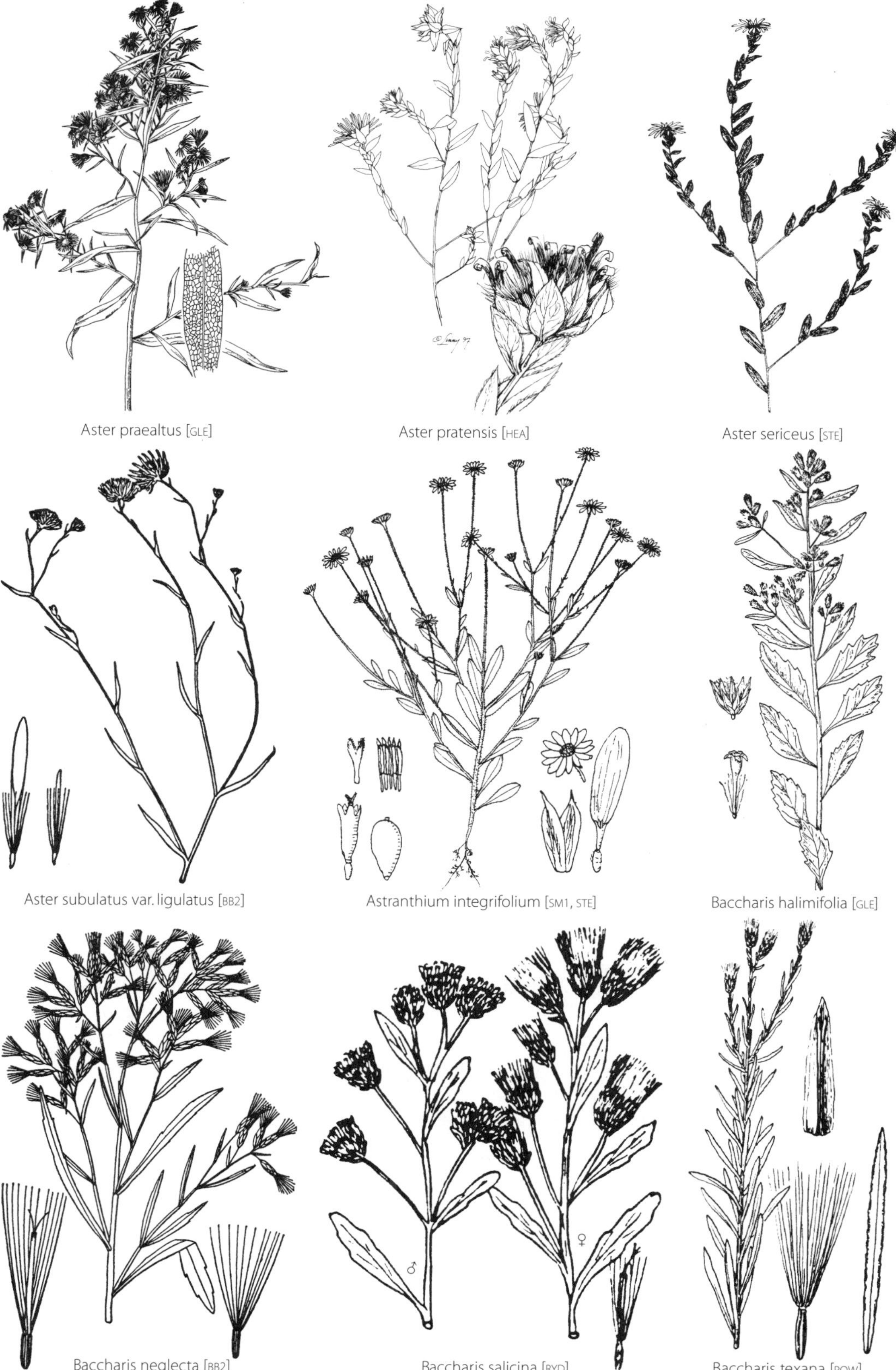

Aster praealtus [GLE]

Aster pratensis [HEA]

Aster sericeus [STE]

Aster subulatus var. ligulatus [BB2]

Astranthium integrifolium [SM1, STE]

Baccharis halimifolia [GLE]

Baccharis neglecta [BB2]

Baccharis salicina [RYD]

Baccharis texana [POW]

3. Pistillate involucres 6–8 mm long; leaves oblanceolate, usually 4–8(–15) mm wide; w TX possibly e to w margin of nc TX ____ **B. salicina**

Baccharis halimifolia L., (with leaves like *Halimium* of the Cistaceae), EASTERN BACCHARIS, SEA-MYRTLE, CONSUMPTION-WEED, TREE-GROUNDSEL, GROUNDSELTREE, MANGLIER, SALTBUSH, SILVERLING. Shrub 1–6 m tall; leaves alternate; leaf blades punctate, entire in lower half, the upper half entire or with few to several teeth, with 1 prominent nerve and 2 lateral nerves, petiolate; achenes 1.0–1.7 mm long; pistillate pappus 9–14 mm long. Wet fields, edges of swamps and marshes, along creeks in woods; Dallas, Fannin, Henderson, and Lamar cos., also Hopkins Co. (Mahler & Waterfall 1964); mainly se and e TX. Oct–Nov. Salt tolerant. This species is a serious invasive pest in Tridens Prairie and similar areas in Lamar Co. Potentially toxic to livestock (Duncan et al. 1957; Burlage 1968). ☠

Baccharis neglecta Britton, (neglected, overlooked), ROOSEVELT-WEED, NEW DEAL WEED, JARA DULCE. Shrub 1–3 m tall; leaves sessile or short petiolate; leaf blades punctate, the upper entire, the lower entire to serrate, 1-nerved with lateral nerves obscure; achenes ca. 1.2 mm long; pistillate pappus 7–12 mm long. Calcareous soils; widely distributed in TX except e TX and Plains Country. Sep–Nov. This is the common *Baccharis* in most of nc TX.

Baccharis salicina Torr. & A. Gray, (resembling *Salix*—willow), SEEP-WILLOW, WATER-WILLOW, WATER-WALLY, JARA, WILLOW BACCHARIS. Shrub 1–3 m tall; leaves nearly sessile; leaf blades serrate, the wider ones 3-nerved; achenes 1.2–2 mm long; pistillate pappus to 12 mm long. Alluvial, often sandy or saline soils; included based on citation of vegetational area 5 (Fig. 2) by Hatch et al. (1990); possibly no futher e than Taylor and Wichita cos.; mainly Trans-Pecos and Plains Country. Sep–Nov.

Baccharis texana (Torr. & A. Gray) A. Gray, (of Texas), PRAIRIE BACCHARIS. Rhizomatous subshrub or shrub 0.25–0.6 m tall; leaves sessile, punctate, essentially linear, minutely undulate, 1-nerved; pistillate involucres 7–10 mm high; pistillate pappus 11–13 mm long. Calcareous soils, prairies; Bell, Callahan, Jack, Palo Pinto, and Tarrant cos., also Dallas Co., but not collected there since Bush in 1900 where he stated it was common (Mahler 1988); widespread in TX except e TX. Jul–Oct.

BERLANDIERA GREENEYES

Pubescent, herbaceous to suffrutescent perennials; leaves alternate, the margins toothed to lyrate-pinnatifid; heads 1-several in corymbose clusters; ray flowers with ligules yellow to orangy with conspicuous green or red to maroon veins on lower surface; disk flowers in ours reddish to maroon; pappus absent or inconspicuous.

A genus of 4 species native to the s United States and Mexico. The disk flowers are subtended by green receptacular bracts giving the disk a green appearance—hence the name common name GREENEYES. (Named for Jean Louis Berlandier, 1805–1851, French botanist who explored TX, NM, and Mexico, and one of the first botanists to make extensive collections in TX). (tribe Heliantheae)

REFERENCES: Turner & Johnston 1956; Pinkava 1967; Nesom & Turner 1998.

1. Ligules of ray flowers with veins on lower (= abaxial) surface red to maroon; at least some leaves lyrate-pinnatifid; peduncles scabrous to subscabrous; stems arising from a persistent basal rosette; known in nc TX only from Brown and Callahan cos. on extreme w margin of area ____ **B. lyrata**
1. Ligules of ray flowers with veins on lower surface green; leaves variously toothed, not lyrate-pinnatifid; peduncles with spreading or matted hairs; stems not arising from a persistent basal rosette; scattered in n and e parts of nc TX.

2. Middle stem leaves usually with evident petioles, the blades ovate; peduncles with matted hairs; stems not densely leafy, the internodes often 3 cm or more long; e margin of nc TX _____ **B. pumila**
2. Middle stem leaves sessile or with inconspicuous short petioles, the blades ovate to triangular; peduncles with ± spreading hairs; stems very densely leafy, with most internodes 3 cm or less long; in nc TX from Dallas Co. n and w _____ **B. betonicifolia**

Berlandiera betonicifolia (Hook.) Small, (with leaves like *Betonica*—betony, now = *Stachys* in Lamiaceae), TEXAS GREENEYES. Plant weakly suffrutescent; stems to 1 m tall, producing a new crop of flowering branches from summit after spring blooming period; leaf blades 4–15 cm long, 2–6 cm wide; leaves evenly distributed or crowded near summit; leaf blades stiffly hirsute to subscabrous or loosely hairy but not velvety, the margins toothed; central stem leaves sessile or short-petioled; ray flowers with ligules 10.5–17 mm long; achenes 4.5–6 mm long, 3–5 mm wide. Sandy soils, post oak woods, ± open areas; Cooke, Dallas, Montague, and Wise cos.; also a single collection apparently as an introduced weed in Grayson Co.; also se and c TX and in the Panhandle. Jun–Sep. [*B. texana* DC.] We are following Nesom and Turner (1998) for nomenclature of this species. The name *B. betonicifolia* must be used since the type collection of *B. betonicifolia* (named in 1835) is of the same species as *B. texana* (named in 1836) (Nesom & Turner 1998). Plants from se TX which had previously gone under the name *B.* ×*betonicifolia* (Hook.) Small are now being recognized as a distinct variety of *B. pumila*.

Berlandiera lyrata Benth., (lyre-shaped), LYRE-LEAF GREENEYES; BERLANDIER'S DAISY, CHOCOLATE DAISY, GREEN-EYED LYRE-LEAF. Perennial herb; stems to 1.2 m tall; leaves crowded toward base of plant; leaf blades velvety; ray flowers with ligules 10–14 mm long, sometimes the entire lower surface red to maroon; achenes 4.5–6 mm long, 2.7–3.7 mm wide. Dry rocky limestone areas, roadsides; Callahan Co., also Brown Co. (Stanford 1971); w margin of nc TX s and w to w TX. Nearly throughout the growing season. The flowers have a chocolate-like aroma (Kirkpatrick 1992).

Berlandiera pumila (Michx.) Nutt., (dwarf), SOFT GREENEYES. Plant herbaceous to suffrutescent; stems to 0.7 m tall; leaf blades velvety; central stem leaves often with long petioles; ray flowers with ligules 12–20 mm long; achenes 4.5–6 mm long, 3–4 mm wide. Roadsides, open wooded areas, often in sandy soils; Henderson Co. near extreme e margin of nc TX, also Hunt, Navarro, and Williamson cos. (Nesom & Turner 1998); mainly se and e TX. Spring–summer(–fall). [*B. dealbata* (Torr. & A. Gray) Small]

BIDENS BEGGAR-TICKS, BUR-MARIGOLD, TICKSEED-SUNFLOWER

Annual, biennial, or perennial herbs; leaves opposite, simple, divided, or compound; phyllaries in 2 series, the outer phyllaries herbaceous, the inner series hyaline- or yellow-margined; ray flowers present or absent, neuter, or pistillate and infertile; corollas in ours yellow or whitish; disk flowers perfect, fertile, yellow or orange-yellow; pappus of 1–4 usually barbed awns or sometimes reduced or absent; those with retrorse barbs are dispersed by attaching to hair or clothing.

A genus of ca. 240 species, cosmopolitan, but especially in Mexico; a few are cultivated ornamentals while a number are considered weeds. (Latin: *bis*, twice, and *dens*, a tooth, in allusion to the two awns on the achenes of some species) (tribe Heliantheae)
REFERENCES: Sherff 1937; Sherff & Alexander 1955; Hall 1967; Lipscomb & Smith 1977; Mesfin et al. 1995a, 1995b.

1. Leaves simple, unlobed, serrate or nearly entire; ray flowers large and conspicuous, the ligules 15–30 mm long (rarely absent) _____ **B. laevis**
1. Leaves incised, lobed, or pinnatifid; ray flowers absent or small and inconspicuous or ligules large, 10–25 mm long.

2. Ray flowers well-developed, the ligules 10–25 mm long; pappus awns usually absent or reduced to a crown, if present, not retrorsely barbed **B. aristosa**
2. Ray flowers absent or very small, the ligules 2–5 mm long; pappus awns present, retrorsely barbed.
    3. Achenes linear, shaped like a fat needle, quadrangular, longer than inner phyllaries, usually 10–18 mm long; leaves 2–3 times pinnately dissected or compound with numerous segments; ray flowers with ligules yellowish white, inconspicuous and not exceeding the disk flowers; pappus of (2–)3–4 awns; outer phyllaries 3–7 mm long **B. bipinnata**
    3. Achenes narrowly wedge-shaped, flattened, equaling or shorter than phyllaries, 5–8(–10) mm long; leaves usually once pinnately or ternately divided or compound with 3–5 segments; ray flowers absent or present, if present ligules golden yellow, 2–3.5 mm long; pappus of 2 awns; outer phyllaries often 30–50 mm long **B. frondosa**

Bidens aristosa (Michx.) Britton, (bearded), BEARDED BEGGAR-TICKS, AWNLESS BEGGAR-TICKS, TICKSEED-SUNFLOWER. Annual or biennial, 0.3–1(–1.5) m tall; leaves 1-2-pinnate, the segments linear to lanceolate or narrowly ovate; petioles to 25 mm long; outer phyllaries 12–20, linear, 7–25 mm long, conspicuously hispid-ciliate; ray flowers ca. 8, golden yellow; achenes flat, 5.5–7.5 mm long, 3–3.5 mm wide; pappus awns absent or slightly developed and with erect-hispid teeth. Low moist areas; Fannin and Lamar cos. in Red River drainage; mainly se and e TX. Apr–Oct. [*B. polylepis* S.F. Blake]

Bidens bipinnata L., (twice-pinnate), SPANISH-NEEDLES. Annual, 0.3–1.5 m tall; stems ± square; leaves 2–3 times pinnately dissected or compound with numerous segments, the segments deltoid-lanceolate; petioles 20–50 mm long; outer phyllaries 7–10, linear, 3–7 mm long; inner phyllaries 5–9 mm long; achenes linear, quadrangular, usually 10–18 mm long, 0.6–1 mm in diam.; pappus awns 2–4 mm long. In moist soils; Bell, Dallas, Grayson, and Hopkins cos., also Tarrant Co. (R. O'Kennon, pers. obs.); se and e TX w to nc TX and Edwards Plateau; rare in Trans-Pecos. Aug–Oct. [*B. bipinnata* var. *biternatoides* Sherff] Jones et al. (1997) recognized TX material of this species as variety *biternatoides* Sherff.

Bidens frondosa L., (leafy), BEGGAR-TICKS, STICKTIGHTS, DEVIL'S BEGGAR-TICKS. Annual 0.2–1.2 m tall; leaves once pinnately or ternately divided or compound, the 3–5 segments lanceolate, serrate; petioles 10–60 mm long; outer phyllaries 5–10, linear-spatulate, conspicuously ciliate, often very long (to 30–50 mm); inner phyllaries shorter, to 5–7 mm long; disk flowers with corollas orange-yellow; achenes 1-nerved on each face, flattened, narrowly wedge-shaped, 5–8(–10) mm long, 2.5–4 mm wide; pappus awns 3–4.5 mm long. Moist areas; se and e TX w to Rolling Plains and Edwards Plateau. Sep–Oct.

Bidens laevis (L.) Britton, Sterns, & Poggenb., (smooth), SMOOTH BEGGAR'S-TICKS, WILD GOLDENGLOW. Annual or perennial, 0.3–1 m tall; leaves sessile, simple, linear to lanceolate or rarely narrowly ovate, serrate or nearly entire; outer phyllaries 6–8, linear-lanceolate, usually not longer than the head; ray flowers 7–8 (rarely absent), golden yellow, sometimes red-tinged, 15–30 mm long; achenes wedge-shaped, 6–9 mm long, to ca. 2 mm wide; pappus awns 2–4, 3–5 mm long, retrorsely barbed. In wet areas, ponds, streams; Bell and Dallas cos.; e TX w to nc TX and Edwards Plateau. Jun–Oct.

BIGELOWIA

A genus of 2 species native to se United States. (Named for Dr. Jacob Bigelow, 1787–1879, Boston physician who assisted George Engelmann on the U.S.-Mexican boundary mission and collected regularly while visiting his patients in the Boston area on horseback) (tribe Astereae)
REFERENCES: Anderson 1970, 1977.

Berlandiera betonicifolia [DEL]

Berlandiera lyrata [BB2]

Berlandiera pumila [SM1]

Bidens aristosa [BB2]

Bidens bipinnata [REE]

Bidens frondosa [REE]

Bidens laevis [MAS]

Bigelowia nuttallii [HEA]

Boltonia diffusa [BB2]

Bigelowia nuttallii L.C. Anderson, (for Sir Thomas Nuttall, 1786–1859, English-American botanist), SLENDER BIGELOWIA, RAYLESS-GOLDENROD. Perennial, rhizomatous, growing in colonies or clumps; stems rigidly erect, 0.5–1 m tall, unbranched except in head-bearing region; leaves alternate, sessile, linear, entire, often resinous; heads crowded in a corymb-like arrangement; involucres 6–8 mm long, to 2.2 mm wide; phyllaries in several series, imbricate; ray flowers absent; disk flowers 4–6 per head; corollas ca. 4.5 mm long, yellow, with 5 lobes; achenes with stiff antrorse hairs; pappus of numerous bristles, ca. 3.8 mm long. Open, sandy or rocky areas; included based on citation of vegetational area 4 (Fig. 2) by Hatch et al. (1990), this possibly based on a Fayette Co. record to the se of nc TX; mainly se and e TX. Aug–Nov. [*B. virgata* of authors, not (Nutt.) DC.]

BOLTONIA

A genus of ca. 5 species of c and e North America and e Asia; some are cultivated as ornamentals. (Named for James Bolton, 1758–99, English botanist) (tribe Astereae)
REFERENCES: Fernald 1940b; Morgan 1966.

Boltonia diffusa Elliott, (diffuse, spreading), SMALL-HEAD BOLTONIA, DOLL'S DAISY. Perennial, ± glabrous herb 0.2–1(–1.5) m tall; leaves alternate, linear, entire, sessile; inflorescences diffusely branched; heads small, solitary at ends of branches; involucres 2–4 mm high with phyllaries in 3 series; receptacles 3–6 mm wide, naked; ray flowers 15–20, pistillate, fertile, the corollas 5–8 mm long, white (can be pink upon drying) or lilac; disk flowers perfect, fertile, the corollas 1.5–2 mm long, yellow; achenes ca. 1.5 mm long; laterally flattened, with 2 wings or in ray achenes with 3 wings; pappus awns < 1 mm long or absent. Disturbed areas, pond margins, moist sand; Limestone and Milam cos. near e margin of nc TX; mainly se and e TX. Jul–Oct.

A similar species, *B. asteroides* (L.) L'Hér., (WHITE BOLTONIA, LARGE-FLOWER DOLL'S DAISY), occurs in se and e TX just e of nc TX. It differs in having ray corollas usually 8–15 mm long, receptacles 6–10 mm wide, and pappus awns ca. 1 mm long.

BRICKELLIA (INCLUDING *KUHNIA*) BRICKELLBUSH, FALSE BONESET

Perennial herbs or shrubs; stems solitary to several; leaves alternate or opposite; leaf blades in ours resin-dotted and usually reticulate-veined beneath; phyllaries in several series; ray flowers absent; achenes 10-ribbed, cylindrical; pappus of bristles.

A New World genus of 100 species native from w United States, Mexico, and Central America to Argentina; some are cultivated as ornamentals. Several taxa were previously treated in *Kuhnia*. (Named for Dr. John Brickell, 1749–1809, of Savanna, GA, amateur botanist and helpful correspondent of Muhlenberg, Fraser, and others) (tribe Eupatorieae)
REFERENCES: Robinson 1917; Shinners 1946d, 1971; Turner 1989.

1. Pappus bristles scabrous, not feathery, the side branches ca. as long as the width of the central bristle axis (use lens); lowest phyllaries lanceolate to ovate, usually only 1–3 times as long as wide; heads usually in a ± elongated racemose or paniculate arrangement ______ **B. cylindracea**
1. Pappus bristles feathery (= plumose), with conspicuous (using lens) side branches many times longer than the width of the central bristle axis; lowest phyllaries linear to subulate, many times as long as wide; heads in ± hemispherical or flat-topped, corymbose groups ______ **B. eupatorioides**

Brickellia cylindracea A. Gray & Engelm., (cylindrical, long and round), GRAVEL-BAR BRICKELLBUSH. Perennial herb or small shrub from a woody base, to 1.2 m tall; leaves sessile or with petioles 1–3(–5) mm long; leaf blades lanceolate to ovate, usually coriaceous, the margins crenate to serrate; inflorescences usually with leafy bracts; involucres usually reddish-tinged, the outer phyllaries much shorter than inner; disk corollas yellowish cream; pappus of ca. 30–35 bristles, usually white or nearly so. Various habitats, but often on rocky limestone soils; Bell and Somervell (Fossil Rim Wildlife Center) cos.; c part of nc TX s and w to Trans-Pecos. Aug–Nov. According to Correll and Johnston (1970), this is a very variable species.

Brickellia eupatorioides (L.) Shinners, (resembling *Eupatorium*—boneset). Herbaceous perennial 0.4–1.3 m tall, from woody taproots; leaf blades lanceolate to broadly rhombic-lanceolate, the margins entire to serrate; petioles to 1 cm long; disk flowers cream to yellowish cream; pappus of 10–20 brownish to white bristles. This species is quite variable vegetatively. Aug–Nov. [*Kuhnia eupatorioides* L.]

1. Middle and outer phyllaries mostly < 3/4 as long as the inner ones, acute or acuminate, appressed.
   2. Involucres 8.7–15 mm high; heads with 14–35 flowers; widespread in nc TX ____ var. **corymbulosa**
   2. Involucres 7–11 mm high; heads with 6–15 flowers; e margin of nc TX ____ var. **eupatorioides**
1. Middle and outer phyllaries elongate, at least some of them 3/4 to fully as long as the inner ones, with conspicuous sickle-shaped or twisted, filiform tips ____ var. **texana**

var. **corymbulosa** (Torr. & A. Gray) Shinners, (with flowers in small corymbs), PLAINS KUHNIA. Sandy soils, post oak woodland; Archer, Dallas, Denton, Fannin, Kaufman, Milam, and Montague cos.; nc TX s to Edwards Plateau and w to Panhandle.

var. **eupatorioides**. FALSE BONESET. Sandy soils, open oak woods; Henderson, Lamar, and Milam cos.; mainly e, se and s TX.

var. **texana** (Shinners) Shinners, (of Texas), PRAIRIE KUHNIA. Limestone soils, prairies; Blackland and Grand prairies s to Edwards Plateau and w to Panhandle.

CALYPTOCARPUS

A genus of 3 species native from TX to Guatemala. (Presumably from Greek: *calypto*, covered, and *carpos*, fruit, referring to the "corticate" achene) (tribe Heliantheae)

Calyptocarpus vialis Less., (of the wayside), PROSTRATE LAWNFLOWER, HIERBA DEL CABALLO, HORSE HERB. Small, trailing, pubescent, perennial herb 5–30 cm tall; leaves opposite, petioled; leaf blades ovate to ovate-lanceolate or deltoid, 1–3(–4) cm long, toothed; heads short-peduncled, solitary in leaf axils, small; ray flowers 3–8, the corollas 1–4 mm long; phyllaries 4–5, subequal in length, obtuse, overlapping laterally; ray and disk corollas yellow or yellow-orange; pappus of 2 awns and sometimes also 1–3 rudimentary awns. Lawn weed; Bell, Brown, Dallas, and Grayson cos., also Hood, Somervell, and Tarrant cos. (R. O'Kennon, pers. obs.); e 1/2 of TX, more common to the s. Apr–Jul. This sometimes troublesome lawn weed is occasionally purposely planted in shady situations to provide ground cover.

CARDUUS THISTLE, PLUMELESS-THISTLE

Ours annual or biennial herbs; leaves alternate, pinnatifid with prickly margins, decurrent as spiny wings along stem; phyllaries many, imbricated, the tips prickly; ray flowers absent; disk corollas pink to deep pink (rarely purple); pappus of numerous hair-like bristles, these not plumose, sometimes barbellate with short projections to ± as long as width of bristle axis, the bristles joined basally and deciduous as a ring.

A genus of 91 species native to Eurasia, the Mediterranean, and e Africa mountains. According to some authors (Correll & Johnston 1970), *Carduus* should probably be enlarged to include *Cirsium* which is similar except for the pappus. (The ancient Latin name) (tribe Cardueae)
REFERENCES: Desrochers et al. 1988; McGregor 1985b.

1. Heads usually ± solitary at the ends of branches, 30–50(–70) mm in diam.; some outer phyllaries spreading or reflexed **C. nutans**
1. Heads usually in close clusters of 3–10 at the ends of branches, 5–15(–20) mm in diam.; phyllaries usually ascending **C. tenuiflorus**

Carduus nutans L. subsp. **macrocephalus** (Desf.) Nyman, (sp.: nodding; subsp.: large-headed), MUSK-THISTLE, NODDING-THISTLE. Extremely prickly, stout winter annual or biennial 0.3–2(–3) m tall; larger leaves 15–40 cm long; leaves ± glabrous or with long hairs mainly along the veins below, deeply lobed, coarsely spiny-toothed, sessile, decurrent as spiny wings 0.5–2 cm wide; heads nodding, large; flowers numerous; corollas slender, to 3 cm long; achenes 3–5 mm long; pappus to 20 mm long. Roadsides, weedy areas; a noxious weed now spreading in nc TX; Collin and Grayson cos.; also Edwards Plateau. Jun–Sep. Native of Mediterranean region. [*C. macrocephalus* Desf.] This species was apparently first collected in TX in Sutton Co. on the Edwards Plateau in 1939; it was supposedly introduced through a shipment of hay from California (Cory 1940). McGregor (1986) questioned the validity of distinguishing subspecies. McGregor (1985b) indicated that this species is an official noxious weed in KS; he gave a detailed account of its invasion.

Carduus tenuiflorus Curtis, (slender-flowered), SLENDER BRISTLE-THISTLE. Prickly annual or biennial 0.3–1 m tall; stems with wings to 1 cm wide; corollas 10–14 mm long; achenes 4–5 mm long; pappus 11–13 mm long. Roadsides, fields, weedy areas; Parker Co., also Bell, Brown, Comanche, Coryell, Hamilton, Mills (J. Stanford, pers. comm.), Somervell, and Tarrant (R. O'Kennon, pers. obs.) cos.; also Edwards Plateau. Apr–May. Native of s Europe. Unknown in nc TX until ca. 1990; this species has the potential of becoming a problematic pernicious weed much like *C. nutans*.

CARTHAMUS SAFFLOWER

Annual herbs; leaves usually alternate, spiny, somewhat clasping, sessile; heads terminal, solitary or corymbose; at least inner phyllaries spiny or spine-tipped, outer phyllaries spreading, conspicuously leafy; receptacle scaly; ray flowers absent; disk flowers tubular, in ours yellow to orange (rarely whitish); pappus of scales or absent.

A genus of 17 species native from the Mediterranean area to c Asia. (From Arabic or Hebrew: *quarthami*, to paint, alluding to the dye obtained from the flowers of *C. tinctorius*) (tribe Cardueae)
REFERENCES: Shinners 1958; Kessler 1987.

1. Leaves ± pinnatifid, coarsely and conspicuously spine-toothed; filaments bearded; achenes (at least outer) rugose; pappus present on inner achenes **C. lanatus**
1. Leaves usually simple (rarely pinnatifid), entire or inconspicuously spine-toothed; filaments glabrous; achenes ± smooth; pappus usually absent (rarely present) **C. tinctorius**

Carthamus lanatus L., (woolly), DISTAFF-THISTLE, SAFFRON-THISTLE. Plant glandular, and sometimes with woolly, spider-web-like pubescence; stems 0.4–1.2 m tall; leaves alternate; corollas yellow with red veins (rarely whitish); achenes straw-colored, 5–6 mm long; pappus of ± rigid, flat, bristle-like, ciliate scales to 1 cm long as well as some outer short scales. Disturbed roadsides; Coryell Co. (Fort Hood—Sanchez 1997); nc TX and Edwards Plateau. Jun. Native of Mediterranean region. Kessler (1987), who documented its first occurrence in OK, indicated it is a potentially serious pest plant.

Carthamus tinctorius L., (belonging to the dyers, of dyes), SAFFLOWER, FALSE SAFFRON, BASTARD SAFFRON. Plant a ± glabrous taprooted herb; stems 0.3–1 m tall; leaves alternate to subopposite above, elliptic to ovate or oblong; heads in a cymose arrangement; corollas bright orange to yel-

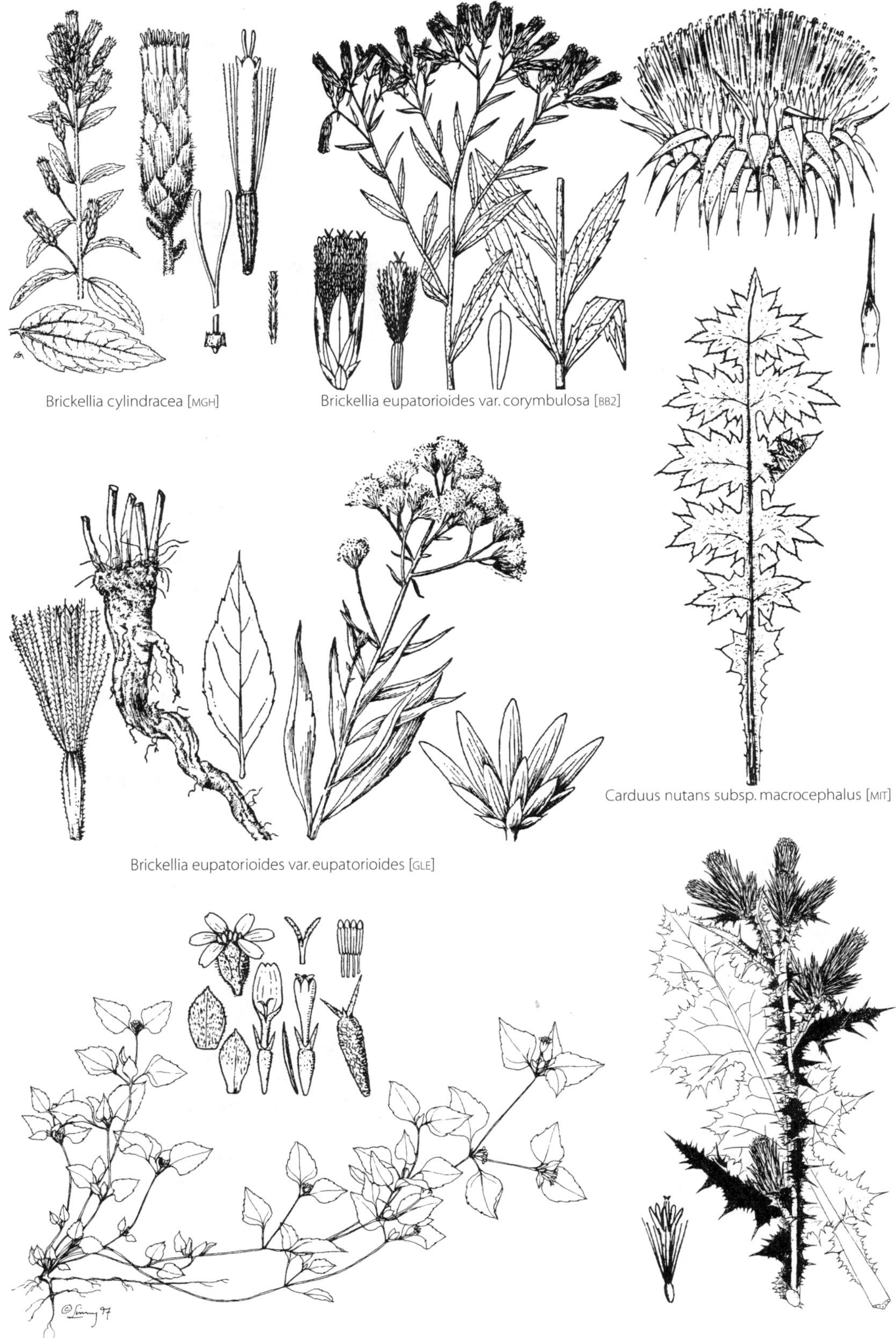

Brickellia cylindracea [MGH]

Brickellia eupatorioides var. corymbulosa [BB2]

Carduus nutans subsp. macrocephalus [MIT]

Brickellia eupatorioides var. eupatorioides [GLE]

Calyptocarpus vialis [HEA, SM1]

Carduus tenuiflorus [SMI]

low-orange; achenes glabrous, ivory white, ca. 6 mm long and 4 mm wide; pappus usually absent (rarely with rudimentary or well-developed bristle-like scales). Yard weed, possibly from seeds in bird feeder; collected in 1996 in Tarrant Co. (Fort Worth; *Sylvester s.n.*); this is apparently the first report from TX. Summer. Native of Old World, possibly from the Near East. Cultivated since ancient times as a dye plant (orange from the flowers used in dyeing food and for rouge) and now for its edible oil (from the seeds); seeds were found in the tomb of Tutankhamun and mummy wrappings were sometimes dyed with this plant (Hepper 1990).

CENTAUREA BASKET-FLOWER, STAR-THISTLE, KNAPWEED

Annuals (rarely biennials); leaves alternate, simple, entire, toothed or pinnatifid, not prickly; heads solitary or corymbose-paniculate; ray flowers absent; outer disk corollas enlarged, simulating rays; corollas deeply (4–)5-lobed, the lobes linear; anthers with elongated appendages; achenes obliquely attached; pappus of bristles.

A large genus (ca. 500 species) of herbs and subshrubs native to the Mediterranean, Turkey, the Near East, n Eurasia, tropical Africa, Australia, and North America; some are cultivated as ornamentals. *Centaurea maculosa* Lam. (SPOTTED KNAPWEED), a European native now naturalized and invasive in parts of the U.S., is suspected of being carcinogenic; a field worker in Idaho developed tumors in his fingers after pulling plants of this species and apparently getting sap in breaks in his skin (Jerry Niefoff, pers. comm.); as a precaution, gloves should be worn when handling *Centaurea* species; some species are known to cause toxic effects in horses (Kingsbury 1964; James & Welsh 1992). The stamens of many *Centaurea* species are touch-sensitive; when touched by visiting insects, they contract suddenly resulting in pollen being forced out the tube formed from the fusion of the anthers (Wills & Irwin 1961). (Greek: *kentaur*, centaur; the centaur Chiron in Greek mythology was said to know the medicinal value of plants) (tribe Cardueae)

REFERENCE: Moore 1972.

1. Heads large, the involucres of well-developed heads 3–6+ cm wide; pappus bristles (5–)6–14 mm long, barbed; phyllaries without a sharp spine at tip (but margins pectinately dissected); native species ____ **C. americana**
1. Heads small, the involucres 2.5 cm or less wide; pappus bristles 2–6 mm long, barbless; phyllaries without OR with a sharp spine at tip; introduced species.
   2. Corollas blue, purple, pink, or white; involucres of well-developed heads 1.5–2.5 cm wide; phyllaries without a sharp spine at tip (margins with teeth or narrow lobes) ____ **C. cyanus**
   2. Corollas yellow; involucres ca. 1 cm wide (not including spines); phyllaries with a conspicuous sharp spine at tip.
      3. Spiny tip of phyllaries 3–9 mm long, often darker than phyllary body ____ **C. melitensis**
      3. Spiny tip of phyllaries 10–17 mm long, straw-colored, usually lighter than phyllary body ____ **C. solstitialis**

Centaurea americana Nutt., (of America), BASKET-FLOWER, AMERICAN BASKET-FLOWER, POWDERPUFF THISTLE, THORNLESS-THISTLE, CARDO DEL VALLE, AMERICAN KNAPWEED, STAR-THISTLE. Annual; stems 0.3–2 m tall; leaves sessile, ± entire, glabrous or sparsely scabrous and gland-dotted; heads solitary at the ends of branches, large, (3–)4–8(–10) cm across including the corollas, quite showy, 1–few per plant; phyllaries appressed, imbricate, of 2 distinct parts, the lower part light green, entire, the upper part straw-colored and pectinately dissected into narrow lobes; corollas pink to occasionally deep purple-red (rarely white); marginal corollas larger, to 5 cm long; central corollas to 3.2 cm long. Disturbed areas; widespread in TX. May–Jul. Sometimes cultivated. The basket-like appearance of the overlapping dissected phyllaries gives this species its common name. The stamens are reported to be sensitive to touch; when touched by insects they suddenly contract and push pollen out onto the pollinator (Kirkpatrick 1992). /82

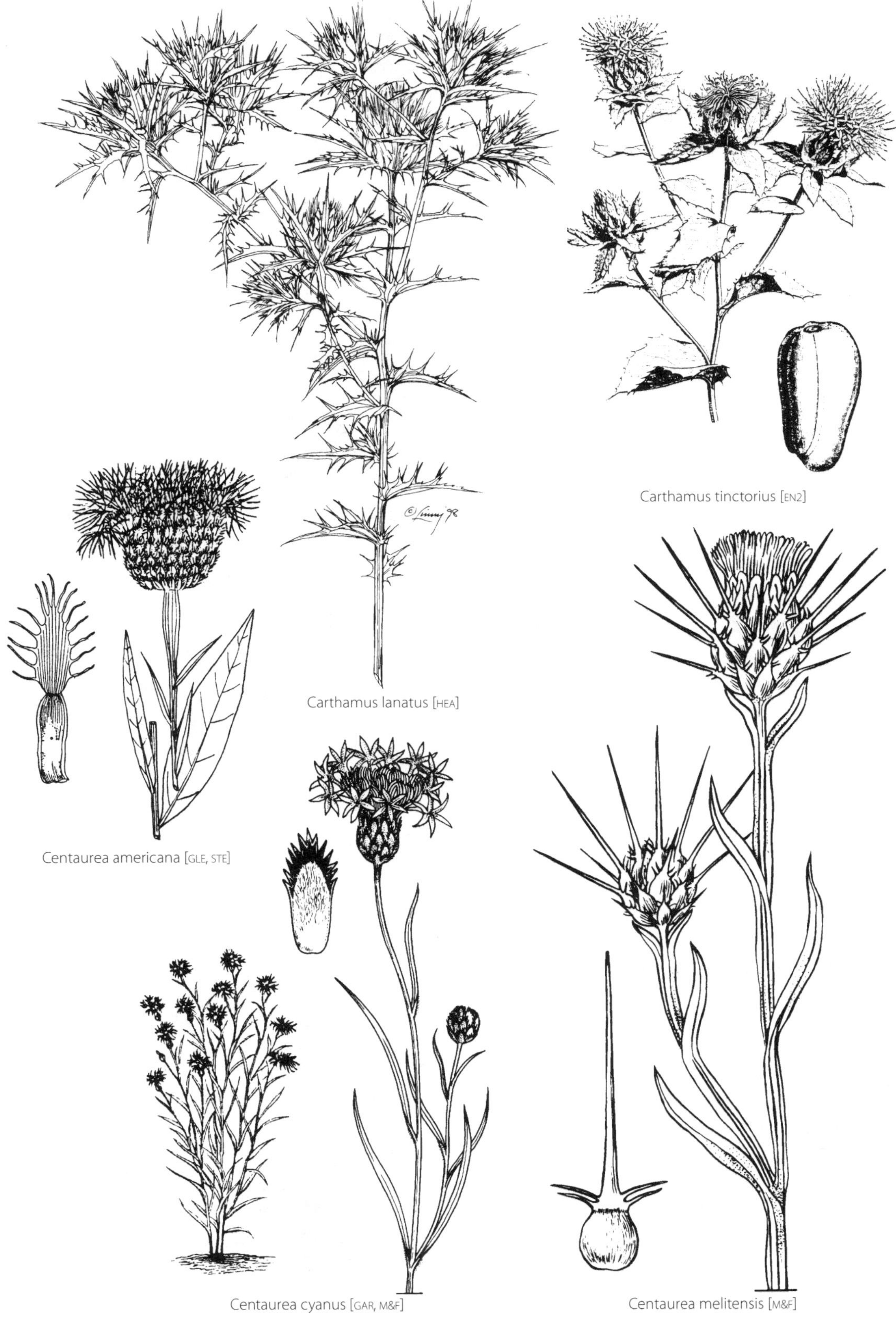

Carthamus tinctorius [EN2]

Carthamus lanatus [HEA]

Centaurea americana [GLE, STE]

Centaurea cyanus [GAR, M&F]

Centaurea melitensis [M&F]

Centaurea cyanus L., (blue), BACHELOR-BUTTON, CORNFLOWER, BLUEBOTTLE. Annual; stems 0.2–1 m tall; basal and lower leaves oblanceolate, pinnately lobed or toothed; upper stem leaves narrowly oblanceolate, entire to toothed, white woolly below, decurrent; phyllaries with numerous small teeth, the margins often darker; marginal corollas enlarged; pappus 0.2–3.5 mm long. Cultivated, naturalized in disturbed areas, roadsides; Clay, Dallas, Grayson, and Kaufman cos.; also Rolling Plains, Edwards Plateau, and Post Oak Savannah. Apr–Jun. Native of Mediterranean region.

Centaurea melitensis L., (of Malta), STAR-THISTLE, MALTA CENTAUREA, MALTA STAR-THISTLE, TOCALOTE. Annual or rarely biennial; stems 0.1–0.8 m tall; basal and lower leaves pinnately lobed, the lobes rounded; upper stem leaves linear, decurrent; pappus 2–6 mm long. Disturbed calcareous soils; Lampasas Co., also Fort Hood (Bell or Coryell cos.—Sanchez 1997); widespread in TX but especially on Edwards Plateau. Native of Europe. Cory (1940) reported that this species was first collected in TX in Bexar Co. in 1934.

Centaurea solstitialis L., (the summer solstice), YELLOW STAR-THISTLE, BARNABY'S STAR-THISTLE. Annual; stems 0.3–0.7 m tall; basal leaves pinnatifid; upper stem leaves linear, decurrent; pappus 2–4 mm long. Disturbed areas, roadsides; Dallas (*Flyr*, 1962) and Tarrant cos.; also Edwards Plateau. Jul–Oct. Native of Eurasia. Reportedly causes brain lesions and a nervous syndrome, "chewing disease," in horses; the toxic principle is not known (Kingsbury 1964). First noted in Tarrant Co. in 1993 and now rapidly spreading on levees along the Trinity River and along highway embankments; it has the potential of becoming a problematic weed.

CHAETOPAPPA LEAST DAISY

Small annuals or perennials; leaves alternate; heads solitary at the ends of the upper branches; ray flowers pistillate and fertile, usually white; disk flowers perfect and fertile, yellow; achenes 5-nerved.

A genus of 10 species endemic to sw North America; including the previously recognized monotypic genus *Leucelene*. (Greek: *chaite*, a bristle, and *pappos*, down, fuzz, pappus) (tribe Astereae)

REFERENCES: Shinners 1946a, 1946b; Van Horn 1973; Nesom 1988.

1. Plants taprooted annuals; leaves usually neither crowded nor conspicously overlapping; ray flowers 5–13; pappus of 5 short scales alternating with as many awns 1–3 mm long or the awns absent; involucres 2–3 mm wide ______ **C. asteroides**
1. Plants perennials from creeping subligneous roots; leaves crowded and conspicuously overlapping; ray flowers 12–24; pappus of 20–30 capillary bristles 4.5–5.5 mm long; involucres 5–10 mm wide ______ **C. ericoides**

Chaetopappa asteroides Nutt. ex DC., (resembling *Aster*), COMMON LEAST DAISY. Plant very small, 5–15(–25) cm tall; stem pubescence appressed to spreading; basal rosette leaves narrowly obovate to orbicular; stem leaves narrower than basal, usually 5–10 mm long; floral bracts linear to subulate; involucres 3.5–4.5 mm high, cylindrical; phyllaries imbricated; ray flowers with ligules 2–4 mm long, white, turning bluish, violet, or pinkish, curling under at night or in age; disk flowers yellow; achenes 1.6–2 mm long, pubescent. Sandy open areas; se and e TX w to Rolling Plains and Edwards Plateau. Apr–Jun, sporadically later.

Chaetopappa ericoides (Torr.) G.L. Nesom, (resembling *Erica*—heath), BABY WHITE ASTER, WHITE ASTER, ROSE-HEATH. Plant very small, 10–15(–20) cm tall, tufted, forming patches from creeping roots; leaves crowded, alternate, narrowly oblanceolate, 2–12(–15) mm long, 2 mm or less wide, entire; heads terminating the numerous erect branches; involucres 5–7 mm high; ray flowers with ligules 2–3 mm long, white or withering rosy, rather showy, curling under at night or in age; disk flowers yellow; achenes 2–3 mm long. Sandy or gravelly prairies or rock outcrops; Palo Pinto Co.; Panhandle to Trans-Pecos e locally to West Cross Timbers. Apr–Sep. [*Aster arenosus* (A. Heller) S.F. Blake, *Leucelene ericoides* (Torr.) Greene]

CHAPTALIA SUNBONNETS

A genus of ca. 60 species of warm areas of the Americas. (Named for J.A.C. Chaptal, 1756–1831, agricultural chemist, who invented the wine-making process of chaptalization, which makes the wine more alcoholic by adding sugar at the same time as squeezing the grape) (tribe Mutisieae)
REFERENCES: Vuilleumier 1969b; Simpson 1978; Nesom 1995.

Chaptalia texana Greene, (of Texas), SILVERPUFF. Acaulescent perennial herb; leaves in a rosette, 6–15 cm long, 2.4–4 cm wide, oblanceolate- or obovate-lyrate, narrowed to a subpetiolar base; lower surface of leaf blades conspicuously whitish tomentose, in striking contrast to green upper surface; peduncles 10–60 cm long, naked or with 1–2 minute bracts; heads solitary, with ca. 150 flowers, nodding in flower, erect in fruit; phyllaries whitish tomentose except often glabrous and darker (sometimes purplish) at apex; peripheral flowers pistillate, ray-like, the upper lip absent, creamy, maturing to crimson, 0.2–0.8 mm wide; inner pistillate flowers with filiform corollas; central flowers perfect; achenes 0.7 mm thick, 9–16 mm long, 2/3 of which is the filiform beak; pappus conspicuous, of numerous buffy-white, hair-like bristles ca. 13–14 mm long. Calcareous areas; Burnet Co. (C. Sexton, pers. comm.); also citation of vegetational area 5 (Fig. 2) by Hatch et al. (1990); also Travis Co. just s of nc TX; se TX w through Edwards Plateau to Trans-Pecos; endemic to TX. Sep–Nov. [*C. nutans* (L.) Polák. var. *texana* (Greene) Burkart]

CHLORACANTHA MEXICAN DEVILWEED, SPINY-ASTER

The genus is composed of a single species occuring from the sw U.S. to Central America (Nesom et al. 1991). It is distinctive, but of uncertain phylogeny, and has been placed in a variety of genera including *Aster*, *Erigeron*, and *Leucosyris*; based on DNA evidence, *C. spinosa* seems more closely related to *Boltonia* and *Heterotheca* than to either *Aster* or *Erigeron*; it is thus recognized in its own genus (Nesom et al. 1991). (Greek: *chloro*, green, and *acanthos*, spine or thorn, in reference to the evergreen thorny stems) (tribe Astereae)
REFERENCES: Nesom et al. 1991; Sundberg 1991; Nesom 1994b.

Chloracantha spinosa (Benth.) G.L. Nesom, (spiny), MEXICAN DEVILWEED, SPINY-ASTER, DEVILWEED-ASTER. Coarse, nearly leafless rhizomatous perennial, the plant superficially rush-like in appearance, sometimes in dense clumps; stems glabrous or glabrate, green and glaucous, photosynthetic, strictly erect, 0.5–1.5(–2.5) m tall, freely branched, alive for up to ca. 4 growing seasons; branchlets sometimes modified into thorns 10–20 mm long; usually ± leafless or leaves early deciduous or with a few small or minute subulate leaves; heads numerous; involucres 3.5–7.5 mm high; ray flowers ca. 10–33; ligules to ca. 5.5 mm long, white, sometimes bluish-tinged. Low disturbed or weedy areas, roadsides; Milam and Limestone cos. on e margin of nc TX w to West Cross Timbers; widespread in TX but more common in s and w parts of the state. Jun–Oct. [*Aster spinosus* Benth., *Leucosyris spinosa* (Benth.) Greene]

CHRYSOPSIS GOLDEN-ASTER

Taprooted annual herbs; leaves alternate; ray flowers pistillate, fertile, yellow; disk flowers yellow; pappus of bristles and sometimes small scales.

A genus of 10 species native from the se United States (especially Florida) to Mexico and the Bahamas; including the previously recognized monotypic genus *Bradburia*. *Chrysopsis* has sometimes been treated in *Heterotheca*. Generic boundaries for the GOLDEN-ASTERS has long been problematic. Semple (1977, 1981) and Nesom (1991a, 1991b) gave convincing justification

for the delineation followed here; more recently Semple (1996) transferred *C. pilosa* to *Bradburia*, making it a genus of 2 species of annuals; he indicated (pers. comm.) that while "... the two genera are most likely more closely related than any of the other GOLDEN-ASTER genera, it remains to be seen whether the (x = 3,4) *pilosa-hirtella* lineage is derived from within the *Chrysopsis* lineage (x = 4,5) or whether the two share an immediate common ancestor (x = 5)." Until these relationships are clarified, we are following Nesom (1991a, 1997b) and J. Kartesz (pers. comm. 1997) in treating the 2 nc TX species in *Chrysopsis*. (Greek: *chrysos*, gold, and *opsis*, aspect, from the golden inflorescences) (tribe Astereae)
REFERENCES: Shinners 1951b; Harms 1968, 1974; Semple 1977, 1981, 1996; Semple et al. 1980; Semple & Chinnappa 1984; Nesom 1991a, 1991b, 1997b.

1. Phyllaries with conspicuous pilose pubescence and sometimes glandular; ray flowers with an inner pappus of bristles and an outer pappus of small scales; disk flowers fertile, with a pappus of numerous bristles, similar to pappus of the ray flowers; widespread in nc TX ____ **C. pilosa**
1. Phyllaries glabrous or nearly so; ray flowers with a pappus of bristles, without an outer pappus of small scales; disk flowers infertile, with a pappus of 1–2 slender bristles or awns, very different from pappus of the ray flowers; possibly present on s margin of nc TX ____ **C. texana**

Chrysopsis pilosa Nutt., (pilose, with long soft hairs), SOFT GOLDEN-ASTER. Plant with pubescence of sparse to dense soft pilose hairs and short glandular hairs; stems 15–90(–160) cm tall; leaves entire or the larger coarsely toothed or shallowly lobed, the lowermost sessile or short petiolate; heads pedunculate, terminating the branches, the central one often overtopped by the laterals; ray flowers ca. 13–21, the ligule to ca. 10 mm long, remaining straight, becoming erect at night; disk flowers perfect, fertile. Sandy woods, old fields, and roadsides; se and e TX w to East Cross Timbers, also Montague Co. May–Oct. n = 4. [*Bradburia pilosa* (Nutt.) Semple, *Heterotheca pilosa* (Nutt.) Shinners]

Chrysopsis texana G.L. Nesom, (of Texas), MAUCHIA. Plant pilose-hispid; stems 10–70 cm tall; basal leaves toothed or short lobed, usually withered before flowering; stem leaves sessile, linear to narrowly oblanceolate, entire; heads pedunculate, solitary at ends of branches; disk flowers perfect but infertile. Open, disturbed areas, usually on sand or gravel; included based on citation of vegetational area 5 (Fig. 2) by Hatch et al. (1990); se and s TX n to at least Travis Co. just s of nc TX (Semple & Chinnappa 1984); endemic to TX (waifs present in Louisiana). Sometimes recognized in the genus *Bradburia* [as *B. hirtella* Torr. & A. Gray]. [not *C. hirtella* DC.]

CICHORIUM CHICORY

A genus of 7 species native to Europe, the Mediterranean, and Ethiopia. The Old World *C. endivia* L., ENDIVE, has long been cultivated for its edible leaves. (Name of disputed origin; according to Pliny from Egyptian and to Forsskål from Arabic; according to Theophrastes and Dioscorides from Greek: *kichorion*, of the fields (Vuilleumier 1973)) (tribe Lactuceae)
REFERENCE: Vuilleumier 1973.

Cichorium intybus L., (old generic name, from Latin for endive or succory), COMMON CHICORY, BLUE-SAILORS, SUCCORY, WITLOOF, RAGGED-SAILORS. Glabrous perennial herb with deep taproot; sap milky; stems 0.3–1(–1.7) m tall; basal leaves toothed or pinnately lobed, oblong-lanceolate; stem leaves alternate, much reduced; heads sessile and axillary, or terminating naked branchlets, the interrupted inflorescences raceme-like; phyllaries in 2 series, the outer much shorter, all with hardened, whitish bases and loose, green tips; flowers all ligulate; corollas lavender-blue, rarely white or pink, open during the morning, later on cloudy days; ligules 15–25 mm long; achenes 2–3 mm long; pappus a minute crown of scales ca. 0.2 mm long. A cool-climate species, rare in TX; roadsides; Grayson and Rockwall cos.; a few scattered records elsewhere in the Panhandle, Trans-Pecos, and Edwards Plateau. Jun–Oct. Native of Eurasia. The

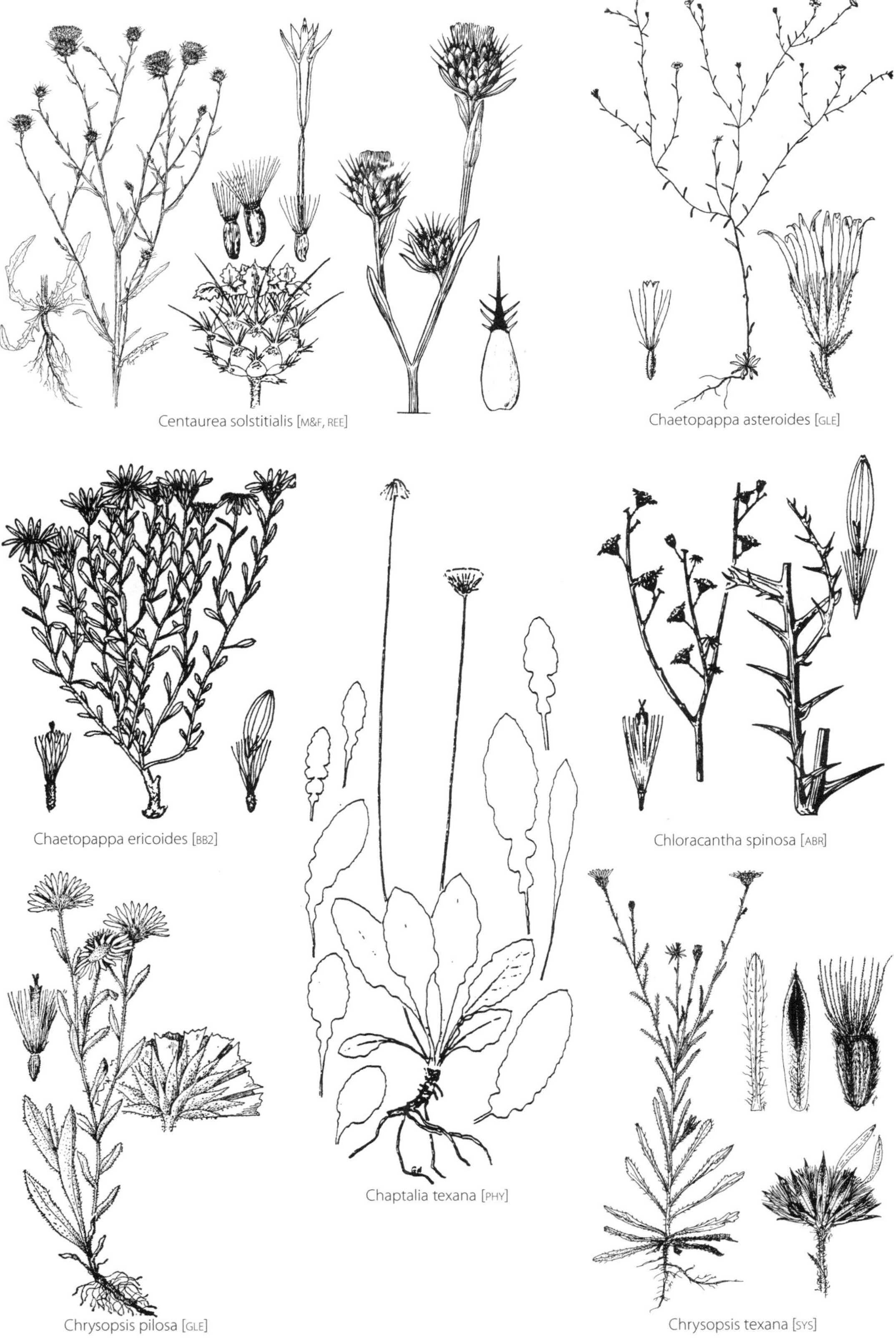

Centaurea solstitialis [M&F, REE]

Chaetopappa asteroides [GLE]

Chaetopappa ericoides [BB2]

Chloracantha spinosa [ABR]

Chaptalia texana [PHY]

Chrysopsis pilosa [GLE]

Chrysopsis texana [SYS]

roots, when ground and roasted, are the chicory often mixed with coffee; the blanched leaves are reported to be edible (Mabberley 1987).

CIRSIUM THISTLE, PLUMED THISTLE, TRUE THISTLE

Erect biennial or perennial herbs; leaves alternate, prickly-toothed or -lobed, glabrate to often strikingly white or gray tomentose on one or both sides; phyllaries many, imbricated, the outer successively shorter, the tips prickly; ray flowers absent; disk flowers perfect; corollas purplish, lavender, white, or rarely yellow, long tubular, the lobes linear; pappus of numerous plumose bristles united at base and deciduous as a unit.

A n temperate genus of ca. 250 species; some are problematic weeds; similar to *Carduus* except for the pappus. (Greek *cirsium*, thistle, from *cirsos*, a swollen vein, for which thistle was a reputed remedy) (tribe Cardueae)
REFERENCES: Howell 1959; Moore & Frankton 1969; O'Kennon & Nesom 1988.

1. Upper surface of leaf blades with numerous short, yellowish, appressed prickles; leaf blades decurrent along stem; rare introduced species ____ **C. vulgare**
1. Upper surface of leaf blades glabrous to variously pubescent but without prickles; leaf blades decurrent along stem OR not so; common native species.
   2. Involucres subtended by a false involucre (= whorl of spiny-pinnatifid, leafy bracts equaling or exceeding the phyllaries present just below the phyllaries); leaves green, glabrous or thinly pubescent beneath ____ **C. horridulum**
   2. Involucres without basal whorl of spiny leafy bracts (= false involucre) just below the phyllaries, the heads thus ± naked at base or with a few linear or narrow bracts shorter than the phyllaries; leaves gray or white with woolly pubescence (= tomentose), at least beneath.
      3. Involucres 2.5–4.5 cm high, oblong or urceolate-oblong, higher than broad in flower (broader in age).
         4. Upper stem leaves with broad, clasping or decurrent bases; perennial from creeping rootstocks; upper leaf surface woolly-pubescent, gray.
            5. Leaf blades mostly shallowly pinnatifid, not decurrent or decurrent for 1 cm or less; prickle tips of middle and outer phyllaries 1–5 mm long; widespread in nc TX ____ **C. undulatum**
            5. Leaf blades uniformly deeply pinnatifid, middle and upper decurrent as wings on stem for > 1 cm; prickle tips of middle and outer phyllaries (5–)7–15 mm long; on extreme w edge of nc TX ____ **C. ochrocentrum**
         4. Upper stem leaves narrowed at base and short-petioled; winter annual or biennial, without creeping rootstocks; upper leaf surface dark green, often nearly glabrous.
            6. Stem leaves entire to shallowly pinnatifid (divided half way to midrib or less); roots coarsely fibrous, not tuberous; flowering mainly mid-Jun–Sep ____ **C. altissimum**
            6. Stem leaves (except uppermost) deeply pinnatifid, divided 1/4–3/4 to midrib; usually some roots with tuberous enlargement; flowering May–Jul ____ **C. engelmannii**
      3. Involucres 1.4–2.2 cm high, nearly as broad as high or broader in flower, much broader in age ____ **C. texanum**

Cirsium altissimum (L.) Hill, (very tall), IOWA THISTLE, TALL THISTLE, ROADSIDE THISTLE. Winter annual, biennial, or perennial 0.5–1.5+ m tall; leaf blades varying from unlobed to deeply lobed (on lower leaves), strikingly white-tomentose beneath, glabrescent above; corollas lavender, to 32 mm long; pappus bristles to 25 mm long. Open stream bottom thickets and pastures, often calcareous soils; Post Oak Savannah w to Tarrant and Wise cos. Jun–Sep. [*Cirsium iowense* sensu Mahler, not (Pammel) Fernald]

Cirsium engelmannii Rydb., (for George Engelmann, 1809–1884, German-born botanist and physician of St. Louis), BLACKLAND THISTLE. Biennial or weak perennial 0.5–1(–1.5) m tall; leaf

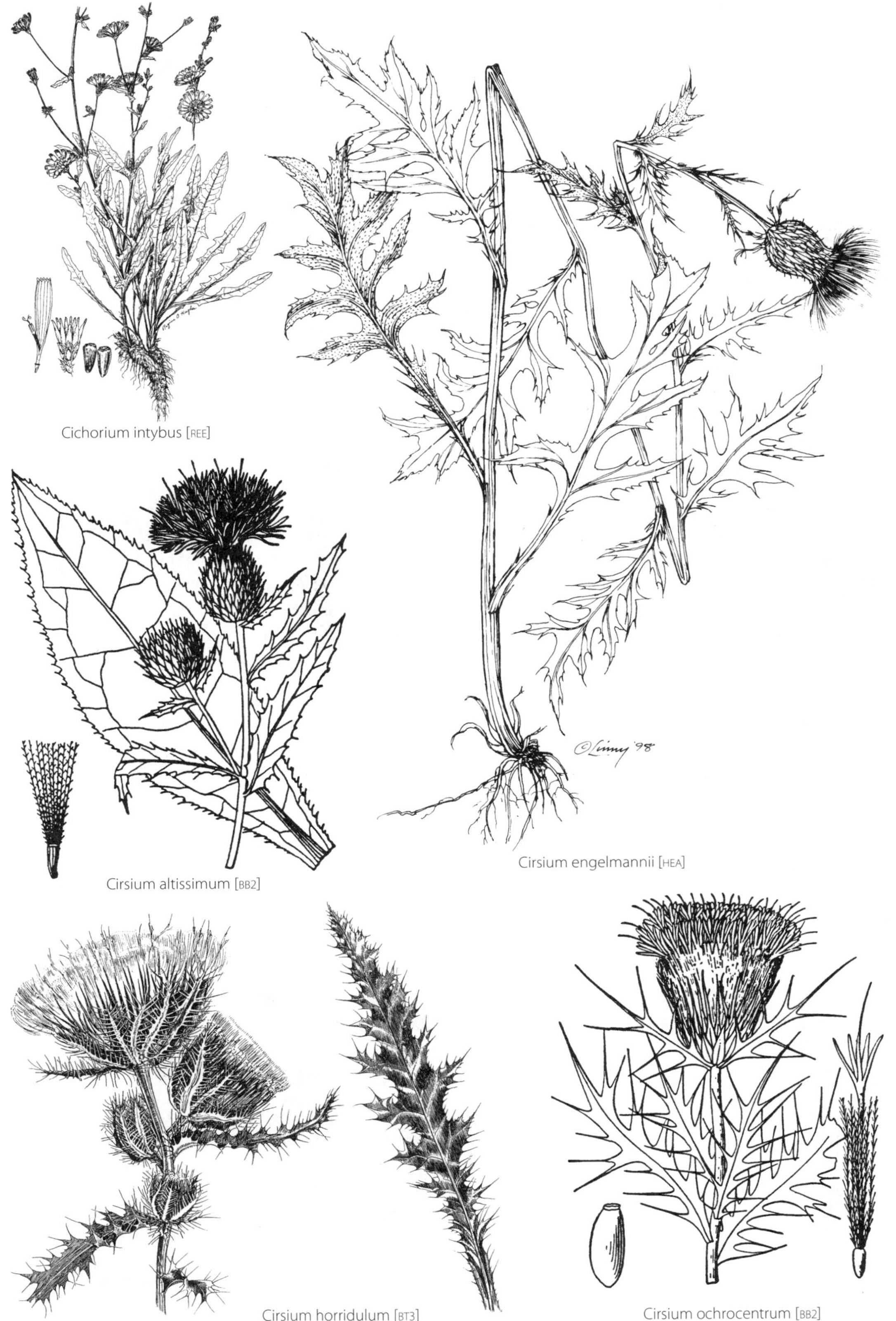

Cichorium intybus [REE]

Cirsium altissimum [BB2]

Cirsium engelmannii [HEA]

Cirsium horridulum [BT3]

Cirsium ochrocentrum [BB2]

blades nearly all deeply lobed, strikingly white-tomentose beneath, green and moderately hairy to often glabrescent above; corollas deep rosy lavender, to 36 mm long; pappus bristles to 25 mm long. Prairies, in calcareous clay, rarely sandy soils; Blackland Prairie w to Grand Prairie and s to Edwards Plateau; endemic to TX and adjacent LA and OK. May–Jul. [*Cirsium terrae-nigrae* Shinners]

Cirsium horridulum Michx., (prickly, horribly armed), BULL THISTLE, YELLOW THISTLE, HORRID THISTLE. Coarse, thinly pubescent or glabrous, winter annual or biennial, beginning to flower when as low as 25 cm, becoming as much as 2 m tall; leaf blades green on both surfaces; flowering stems usually with a solitary terminal head, rarely branched above; involucres wider than long, 4–8 cm wide, immediately subtended by a conspicuous false involucre of ca. 10–12 spiny bracts; corollas rosy lavender, rarely yellow (in se TX). Damp open woods, low areas; Fannin Co. (Talbot property) in Red River drainage, also Denton Co. (G. Diggs, pers. obs.); mainly se and e TX. Mar–May. Jones et al. (1997) recognized TX material of this species as var. *elliottii* Torr. & A. Gray. ▣/84

Cirsium ochrocentrum A. Gray, (yellow-centered), YELLOW-SPINE THISTLE. Biennial or perennial 0.2–1(–1.5) m tall, extremely spiny; leaf blades strikingly white-tomentose beneath, with some tomentum above, the upper surface grayish, often much darker in appearance than lower; corollas rosy lavender or purplish, rarely white. Sandy or rocky prairies and roadsides; Clay and Callahan cos. at w edge of nc TX, s and w to w TX. May–Oct.

Cirsium texanum Buckley, (of Texas), TEXAS THISTLE, SOUTHERN THISTLE. Biennial or perennial 0.5–2 m tall, with deep taproot; leaf blades green and glabrous or thinly tomentose above, thinly grayish or whitish tomentose beneath; upper leaves reduced, sessile, slightly clasping; branches long, erect or ascending; involucres small (1.4–2.2 cm high) in comparison with other nc TX species; corollas pink to deep rosy lavender. Prairies and roadsides; nearly throughout the state except extreme e TX. May–Jun. Goldfinches are reported to use the plumose pappus bristles to line their nests (Ajilvsgi 1984). ▣/84

Cirsium undulatum (Nutt.) Spreng., (undulated, wavy), WAVY-LEAF THISTLE, PASTURE THISTLE. Rhizomatous low perennial 0.2–1(–2) m tall; leaf blades sometimes nearly as white-tomentose above as below, sometimes undulate; corollas purplish, lavender or white. Prairies, pastures, and roadsides; Blackland Prairie s and w to w TX. May–Jul. ▣/84

Cirsium vulgare (Savi) Ten., (common), BULL THISTLE. Biennial; stems 0.5–2 m tall; leaf blades green above, green or with grayish pubescence below; involucres 2–3 cm high; corollas dark purple, 27–35 mm long. Roadsides, pastures, waste areas; Tarrant Co. (R. O'Kennon, pers. obs.); also n Red River Co. (R. O'Kennon, pers. obs.); also Edwards Plateau; first collected in TX in Gillespie Co. in 1987 (O'Kennon & Nesom 1988). Jul–Sep. Native of Eurasia.

CONYZA

Annual herbs; stems glabrous to pubescent; leaves alternate, entire or toothed; phyllaries in 4 series, imbricate, the inner ones larger; ray flowers pistillate, fertile; corollas whitish, cream or pinkish, the ligules inconspicuous, shorter than the tubes and scarcely if at all surpassing the pappus, the tubes filiform; disk flowers perfect, fertile; corollas cream-colored to yellowish or pinkish; achenes 2-ribbed, ± hirsute; pappus of hair-like bristles.

A genus of ca. 60 species of temperate and warm areas; similar to *Erigeron* and previously lumped with that genus by some authorities; not sharply distinct. (From Greek name for safflower) (tribe Astereae)

REFERENCES: Cronquist 1943; Shinners 1949e; Nesom 1978, 1990b.

1. Plants unbranched or branching well above base, erect, (30–)100–200+ cm tall; larger leaves 4–13 cm long, > 2 mm wide; stems glabrous or with widely spreading hairs ____ **C. canadensis**

1. Plants branching from near base, not obviously erect, low, 10–30(–40) cm tall; larger leaves 1–2(–4) cm long, 2 mm or less wide; stem pubescence appressed ____ **C. ramosissima**

Conyza canadensis (L.) Cronquist, (of Canada), HORSEWEED. Leaves many, subsessile, narrowly lanceolate or oblanceolate, to 14 mm wide, entire or toothed; heads many, small, in a narrow terminal panicle-like inflorescence; involucres 3–4 mm high; ray flowers numerous (20–40). Eroding or disturbed ground; various soils. May–Oct.

1. Stems with widely spreading hairs ____ var. **canadensis**
1. Stems glabrous ____ var. **glabrata**

var. **canadensis**. HORSE-TAIL CONYZA, CANADA FLEABANE. Leaf blades prominently pilose on margins. Clay, Comanche, and Dallas cos., also Tarrant and Somervell cos. (R. O'Kennon, pers. obs.); nc TX and Plains Country. [*Erigeron canadensis* L.]

var. **glabrata** (A. Gray) Cronquist, (somewhat smooth or hairless, becoming hairless). Leaf blades pilose or nearly glabrous on margins. Common and widespread in nc TX; throughout TX. [*Erigeron canadensis* var. *glabratus* A. Gray]

Conyza ramosissima Cronquist, (much-branched), LOW FLEABANE, LOW CONYZA, SPREADING FLEABANE. Stem hairs antrorsely appressed; leaves linear, entire; heads numerous, similar to those of *C. canadensis*; involucres 3–4(–5) mm high. Clay soils, disturbed habitats. lawns; Collin, Dallas, Erath, Grayson, and Tarrant cos.; nc TX and Plains Country. Jun–Oct. [*Erigeron divaricatus* Michx.]

COREOPSIS GOLDEN-WAVE, TICKSEED, COREOPSIS

Herbaceous annuals or perennials, glabrous or minutely pubescent; leaves usually opposite, simple or pinnately compound; heads pedunculate, solitary or corymbose; inner phyllaries united at base into a cup with outer phyllaries distinct and different in appearance; ray flowers ca. 8, sterile, the ligules yellow, sometimes with a reddish brown basal spot; disk flowers yellow or reddish brown; achenes flattened, winged or wingless; pappus of 2 awns, sometimes obsolete.

An American genus of 50 species; some are cultivated as ornamentals; closely related to *Bidens*. (Greek: *coris*, a bug, and *opsis*, appearance, from the form of the achene) (tribe Heliantheae)

REFERENCES: Sherff 1936; Sherff & Alexander 1955; Smith & Parker 1971; Smith 1974, 1976; Crawford & Smith 1982, 1984; Jansen et al. 1987; Crawford et al. 1992; Mesfin et al. 1995a, 1995b.

1. Disk corollas reddish brown (at least apically); ray flowers with ligules yellow but typically with a reddish brown spot at base.
   2. Outer phyllaries equal to or longer than the inner, ca. 4–11 mm long, linear; disk corollas apically 5-lobed; anthers 5; ray flowers with ligules 4–5-toothed ____ **C. wrightii**
   2. Outer phyllaries very small, much shorter than the inner, ca. 2 mm long, linear-oblong to triangular; disk corollas apically 4-lobed; anthers 4; ray flowers with ligules bluntly 3-toothed (middle tooth sometimes notched) ____ **C. tinctoria**
1. Disk corollas yellow; ray flowers with ligules completely yellow.
   3. Stems with leaves crowded near base (most on lower 1/2 of stem); leaves much reduced upwards; flowering peduncles usually > 15 cm long; most leaves undivided or with 1 or 2 short side lobes (rarely more divided), the blades or divisions 5–30 mm wide ____ **C. lanceolata**
   3. Stems with well-developed leaves nearly to the top; leaves only slightly reduced upwards; flowering peduncles usually < 15 cm long (rarely longer); most leaves 3–5 parted or divided to the midrib, the divisions 1–10 mm wide ____ **C. grandiflora**

Coreopsis grandiflora T. Hogg ex Sweet var. **longipes** (Hook.) Torr. & A. Gray, (sp.: large-flowered; var.: long-stalked), Perennial or annual; stems 0.3–1 m tall; leaves pinnately compound, the lower terminal leaflets linear, becoming filiform apically; outer phyllaries ± lanceolate, shorter to equaling inner phyllaries, to ca. 10 mm long; disk corollas apically 5-lobed; anthers 5; ray flowers with ligules 1.3–2.5 cm long, 4–5-toothed; achenes winged, the wings entire. Sandy woods, also cultivated as an ornamental and spreading; se and e TX w to West Cross Timbers. May.

Coreopsis lanceolata L., (lanceolate, lance-shaped), LANCE COREOPSIS. Perennial; stems 0.3–0.7 m tall; outer phyllaries ± lanceolate, shorter to equaling inner phyllaries, 5–10 mm long; disk corollas apically 5-lobed; anthers 5; ray flowers with ligules 1.5–3 cm long, 4–5 toothed; achenes with thin flat wings. Apparently planted and spreading along highways; Cooke, Dallas, Grayson, Tarrant, and Williamson cos., also Somervell Co. (R. O'Kennon, pers. obs.); mainly se and e TX. Apr–May.

Coreopsis tinctoria Nutt., (of dyes), PLAINS COREOPSIS, CARDAMINE COREOPSIS, MANZANILLA SILVESTRE, CALLIOPSIS. Annual; stems 0.6–1.2 m tall; leaves pinnately compound, the lower terminal leaflets linear, becoming filiform apically; outer phyllaries very small; ray flowers with ligules 1–1.5 cm long; achenes winged or wingless. Low moist areas, often sandy soils; widely cultivated for its very showy flowers; nearly throughout TX, more abundant in e 1/2. May–Sep. [*C. cardaminaefolia* (DC.) Torr. & A. Gray] This species has been used as a source of a range of red and yellow dyes—hence the specific epithet (Tveten & Tveten 1993). ■/85

Coreopsis wrightii (A. Gray) H.M. Parker, (for Charles Wright, 1811–1885, TX collector), ROCK COREOPSIS. Annual; leaves pinnately compound; middle and upper leaves with narrow segments, the terminal leaflet of middle leaves > 3 times as long as wide, 1–3 mm wide; achenes wingless. Calcareous soils; prairies; Grand Prairie to West Cross Timbers, s to Edwards Plateau; endemic to c and nc TX and s OK. May–Jun. [*C. basalis* (A. Dietr.) S.F. Blake var. *wrightii* (A. Gray) S.F. Blake]

Coreopsis basalis (A. Dietr.) S.F. Blake, (basal), GOLDEN-MANE COREOPSIS, cited by Hatch et al. (1990) for vegetational area 4 (Fig. 2), occurs just to the e (Freestone and Van Zandt cos.) and s of nc TX (Smith 1976). It differs from the similar *C. wrightii* in having the middle and upper leaves with broad segments, the terminal leaflet of the middle leaves usually < 3 times as long as wide (ca. 5–15 mm wide).

Coreopsis nuecensis A. Heller, (presumably of the Nueces River area of TX), with disk corollas yellow, ray flowers with ligules yellow, commonly with several reddish brown flecks a little above the base, inner phyllaries dorsally pubescent, and median and lower stems commonly glabrous or sometimes sparsely pubescent, was cited by Hatch et al. (1990) for vegetational area 4 (Fig. 2). This TX endemic apparently occurs only to the e and s of nc TX (Smith 1976).

Coreopsis nuecensoides E.B. Sm., (resembling *C. nuecensis*), CROWN COREOPSIS, with disk corollas yellow, ray flowers with ligules yellow, commonly with several reddish brown flecks a little above the base, inner phyllaries dorsally pubescent, and median and lower stems commonly densely pubescent or sometimes sparsely pubescent, is cited by Hatch et al. (1990) for vegetational area 4 (Fig. 2). It apparently occurs only to the e and s of nc TX (Smith 1976).

CREPIS HAWK'S-BEARD

A genus of ca. 200 species of the n hemisphere, s Africa, and South America; a number of weeds and a few are cultivated as ornamentals. (Greek: *crepis*, a classical Greek name of some plant that also meant sandal or boot, possibly alluding to the shape of the achene) (tribe Lactuceae)

Cirsium texanum [HEA]

Cirsium undulatum [GLE]

Cirsium vulgare [REE]

Conyza canadensis var. canadensis [REE]

Conyza ramosissima [BB2]

Coreopsis grandiflora var. longipes [CUR]

Coreopsis lanceolata [SID]

Coreopsis tinctoria [SID]

Coreopsis wrightii [SID]

REFERENCES: Babcock 1947; Clausen 1949; Stebbins 1949; Vuilleumier 1973.

Crepis pulchra L., (handsome), SHOWY HAWK'S-BEARD. Taprooted annual 0.3–1 m tall; stem pubescence glandular; sap milky; leaves more basally, alternate and reduced up the stem, with glandular pubescence like the stems, oblanceolate or spatulate, toothed or pinnatifid; heads numerous; involucres 8–12 mm high, glabrous; flowers all ligulate, yellow; achenes 4–6 mm long, narrowly columnar, 10–12 ribbed; pappus of numerous hair-like bristles. Disturbed and weedy areas; Dallas, Grayson, Hunt, and Tarrant cos.; mainly e TX. Apr–May. Native of Eurasia.

CROPTILON SCRATCH DAISY

Ours annual, usually glandular-pubescent herbs; stems not leafy to the top, the leaves decreasing in size up the stem to small bracts near the top (but leafy to top in species outside nc TX); leaves alternate, sessile, narrowly oblanceolate to linear, toothed to entire; heads in an open panicle; ray flowers pistillate, fertile, with yellow ligules; disk flowers perfect, fertile, the corollas yellow; achenes unribbed, 2–3 mm long; pappus of persistent bristles.

A genus of 3 species of the se United States; previously treated by some in *Haplopappus*. (According to Rafinesque, author of the genus, the name means "col. feather," possibly colored feather, from the off white or rusty white pappus) (tribe Astereae)
REFERENCES: Hall 1928; Smith 1965, 1981; Nesom 1991b.

1. Ray flowers 5–11; ligules 3.9–6 mm long; receptacles 2.5–4 mm in diam., usually ca. 3 mm; e TX w to e edge of nc TX ____ **C. divaricatum**
1. Ray flowers 13–21; ligules 6–12 mm long; receptacles 3.5–7 mm in diam., usually 4.7–5.4 mm; e Blackland Prairie w through nc TX to e Rolling Plains ____ **C. hookerianum**

Croptilon divaricatum (Nutt.) Raf., (spreading, widely divergent), SLENDER GOLDENWEED, SCRATCH DAISY, SPREADING GOLDEN-ASTER. Stems erect, 20–70(–115) cm tall; basal leaves 7–14 cm long, 1.5–2 cm wide, the upper leaves smaller; peduncles with stalked glandular hairs. Disturbed areas, sandy soils; Henderson and Limestone cos.; also Fannin, Lamar, and Milam cos. (Smith 1965); e TX w to e edge of nc TX. Aug–Oct. [*Haplopappus divaricatus* (Nutt.) A. Gray]

Croptilon hookerianum (Torr. & A. Gray) House, (for William Jackson Hooker, 1785–1865, director of Kew Gardens). Stems erect, 30–75 cm tall; basal leaves 6–10 cm long, 1.2–3 cm wide, the upper leaves smaller. Disturbed areas, sandy soils. Jul–Oct.

1. Ligules (2.4–)2.6–3.3 mm wide; peduncles never hispid, with short-stalked glandular pubescence (longest hairs ca. 0.2–0.3 mm long, including gland); probably only to the s of nc TX ____ var. **hookerianum**
1. Ligules 1.8–2.8(–3.1) mm wide; peduncles often slightly hispid just below heads, with long-stalked glandular pubescence (longest hairs ca. 0.4–0.5 mm long including gland); Blackland Prairie w to e Rolling Plains ____ var. **validum**

var. **hookerianum**, cited by Hatch et al. (1990) for vegetational area 4 (Fig. 2), is a coastal plain taxon that apparently occurs just s of nc TX (Smith 1965). [*C. divaricatum* var. *hookerianum* (Torr. & A. Gray) Shinners, *Haplopappus divaricatus* (Nutt.) A. Gray var. *hookerianus* (Torr. & A. Gray) Waterfall]

var. **validum** (Rydb.) E.B. Sm., (strong). Blackland Prairie w to e Rolling Plains. [*Haplopappus validus* (Rydb.) Cory, *Haplopappus validus* subsp. *torreyi* E.B. Sm.]

DRACOPIS CLASPING CONEFLOWER

A monotypic North American genus; sometimes cultivated as an ornamental. (Greek: *drakon*, dragon, in allusion to appendages on the style) (tribe Heliantheae)
REFERENCES: Cox & Urbatsch 1990, Urbatsch & Jansen 1995.

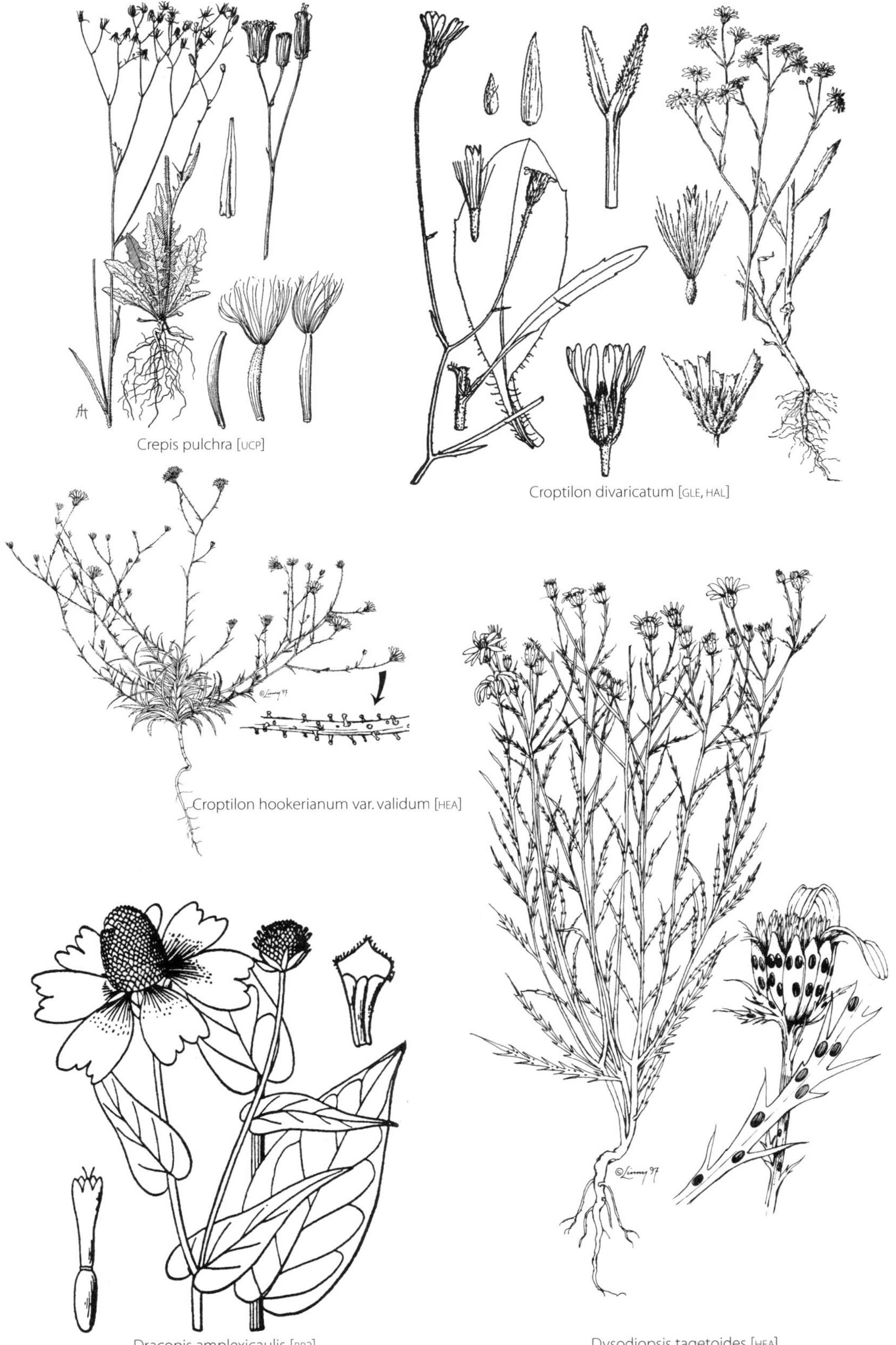

Crepis pulchra [UCP]

Croptilon divaricatum [GLE, HAL]

Croptilon hookerianum var. validum [HEA]

Dracopis amplexicaulis [BB2]

Dysodiopsis tagetoides [HEA]

Dracopis amplexicaulis (Vahl) Cass., (stem-clasping), CLASPING CONEFLOWER, CLASPING-LEAF CONEFLOWER, BLACK-EYED-SUSAN, CONEFLOWER. Annual herb 0.3–0.7 m tall, essentially glabrous; leaves alternate, oblong to ovate, cordate-clasping basally, toothed or entire, acute or acuminate; heads peduncled, short cylindric; receptacles elongated, chaffy, the scales green-tipped; ray flowers 5–10, neuter, infertile; ligules all yellow or yellow with reddish brown or brownish purple blotch basally, ca. 10–25 mm long; disk flowers perfect, fertile, the corollas purplish or brownish; achenes cylindrical to clavate; pappus absent. Disturbed habitats, particularly moist areas; e 2/3 of TX. May. [*Rudbeckia amplexicaulis* Vahl] ▣/87

DYSODIOPSIS DOGWEED, FOETID-MARIGOLD

A monotypic genus native to the s United States (Karis & Ryding 1994); previously recognized in *Dyssodia*; we are following Karis and Ryding (1994), Kartesz (1994), and J. Kartesz (pers. comm. 1997) in treating it separately. (Presumably from related genus *Dyssodia*, Greek: *dysodia*, an ill smell, and *opsis*, aspect or appearance) (tribe Helenieae, sometimes lumped into Heliantheae)

REFERENCES: Johnston 1956; Johnston & Turner 1962; Strother 1969, 1986; Karis & Ryding 1994.

Dysodiopsis tagetoides (Torr. & A. Gray) Rydb., (resembling *Tagetes*—marigold), MARIGOLD DOGWEED. Annual or short-lived perennial 0.4–0.9 m tall, glabrous; leaves mostly alternate, linear, 2–6 mm wide, with conspicuous, yellowish-brown or orange, dot-like glands, aromatic, the margins with coarse, conspicuous teeth; involucres 9–12 mm high; ray flowers 7–12, with ligules 10–15 mm long, bright yellow; disk corollas brownish yellow; achenes 3–3.5 mm long; pappus of 10–12 scales with 1 to 3 awns per scale; $x = 13$. Calcareous soils; Bell, Dallas, and Grayson cos. w to West Cross Timbers and s to c TX. Jun–Aug. [*Dyssodia tagetoides* Torr. & A. Gray, *Hymenatherum tagetoides* (Torr. & A. Gray) A. Gray]

ECHINACEA CONEFLOWER, PURPLE CONEFLOWER

Perennial herbs; leaves alternate, the lower long-petioled; leaf blades simple, entire or subentire, narrowly lanceolate to oblong or ovate, often with 3–5 prominent veins; heads large, solitary, terminal, long-peduncled; ray flowers sterile, with ligules long, narrow, spreading to reflexed, purple to pink or white; disk flowers fertile; disk with stiff, sharp-pointed, brownish to dark reddish purple scales; pappus a short crown.

A genus of 9 species of the e United States; according to McGregor (1968b), considerable natural hybridization occurs. *Echinacea* species were widely used medicinally by Native Americans for a variety of purposes and are still valued today as herbal medicines; they apparently act as immune system stimulants; unfortunately wild populations are under pressure from over-collecting (Foster 1991). (Greek: *echinos*, hedgehog, from the prickly scales of the disk which protrude beyond the disk flowers) (tribe Heliantheae)

REFERENCES: McGregor 1968a, 1968b; Cox & Urbatsch 1990; Baskin et al. 1993; Urbatsch & Jansen 1995.

1. Leaf blades abruptly contracted to the petiole, often rounded at base, broadly to narrowly ovate, the basal and lower ones 2–3(+) times as long as wide, 1–15 cm wide, usually toothed; cultivated species native to the e of nc TX ____ **E. purpurea**
1. Leaf blades gradually tapering to base, lanceolate to lanceolate-linear or narrowly elliptic, the basal and lower ones usually > 5 times as long as wide, 1–4 (+) cm wide, entire; native species.
   2. Ligules dark purple (rarely lighter), strongly recurved-reflexed so their tips nearly touch the peduncle, 2–3.5 cm long ____ **E. atrorubens**
   2. Ligules light purple to pink or white, spreading to drooping but not recurved-reflexed, 2–9 cm long.

3. Ligules when fully expanded (3–)4–9 cm long, clearly drooping, 1.5–2.5 times as long as breadth of disk; plants 0.5–1 m tall; pollen white _______ **E. pallida**
3. Ligules when fully expanded 2–4 cm long, spreading, 1–1.5 times as long as breadth of disk; plants 0.2–0.7 m tall; pollen yellow _______ **E. angustifolia**

Echinacea angustifolia DC., (narrow-leaved), BLACKSAMSON. Ligules light pink to light purplish or white. Gravelly or rocky limestone prairies (on mixed sand and gravel farther w); Blackland Prairie w to Plains Country and s to Edwards Plateau. May–Jun. [*E. pallida* var. *angustifolia* (DC.) Cronquist] Barkley (1986) indicated that this species grades into *E. pallida* to the e; Cronquist (1980) and Gandhi and Thomas (1989) treated it as a variety of *E. pallida*. We are following McGregor (1968a) and Baskin et al. (1993) in recognizing it at the specific level. It is sometimes split into 2 varieties based on pubescence characters (e.g., McGregor 1968a; Kartesz 1994; Jones et al. 1997). Variety *strigosa* McGregor is said to differ in having consistently strigose or strigose-hirsute leaves and upper stems. While numerous individuals annotated as both variety *angustifolia* and variety *strigosa* are known from nc TX, virtually all are ± strigose or strigose-hirsute. We are thus not distinguishing varieties.

Echinacea atrorubens Nutt., (dark red). Stems 0.3–0.9 m tall; pollen yellow. Limestone hillsides; Cooke, Denton, Red River, and Tarrant cos.; endemic to a narrow band from se TX n through nc TX to Ardmore, OK and Topeka, KN. May–Jul.

Echinacea pallida (Nutt.) Nutt., (pale), PALE ECHINACEA. Ligules purplish pink to whitish. Sandy open woods and prairies; Grayson, Lamar, and Tarrant cos., also Fannin Co. (McGregor 1968a); e TX w to East Cross Timbers. May–Jun.

Echinacea purpurea (L.) Moench, (purple). Stems 0.6–1.8 m tall; ligules 3–8 cm long, drooping, reddish purple (rarely pink or white); pollen yellow. Cultivated in nc TX and persists or spreads; Tarrant Co. (R. O'Kennon, pers. obs.); e TX. May–Sep.

Echinacea sanguinea Nutt., (blood red), of se and e TX, occurs just e of nc TX (c Henderson Co.). It can be distinguished from *E. angustifolia* and *E. pallida* using the following characters: it has nearly hemispheric heads (10–20 mm high) vs. conical (15–40 mm high), smaller achenes (ca. 3 mm long vs. 3.5–5 mm long), ligules 4–7 cm long, and pollen yellow. [*E. pallida* var. *sanguinea* Nutt.) Gandhi & R.D. Thomas]

ECLIPTA

A genus of 4 species found in warm areas of the world. (Greek: *ecleipo*, to be deficient or lack, alluding to the absence of pappus) (tribe Heliantheae, can also be keyed using Astereae key)

Eclipta prostrata (L.) L., (prostrate, flat to the ground), PIEPLANT, YERBA DE TAGO. Low, prostrate to erect, appressed-scabrous-pubescent annual, usually freely branched; leaves opposite, sessile or short petiolate, lanceolate to narrowly elliptic, entire or slightly toothed, dark green; heads small, short-peduncled, terminal and in upper leaf axils, rather inconspicuous; ray flowers numerous, fertile, with ligules linear, very short (ca. 1 mm long), white; disk flowers perfect; disk corollas minute, whitish; achenes 2–2.5 mm long; pappus none or very obscure. Ditches, shorelines, stream banks; throughout TX. Jun–Nov. [*E. alba* (L.) Hassk.] Gandhi and Thomas (1989) explained why the correct name is *E. prostrata*. This species has been introduced into the Old World and in India is used as a blackish dye for hair and tattooing (Mabberley 1997).

ELEPHANTOPUS ELEPHANT'S-FOOT

A mainly tropical and warm area genus of ca. 30 species. (Greek: *elephas*, elephant, and *pous*, foot; translation of aboriginal name) (tribe Vernonieae)
REFERENCES: Gleason 1906, 1922; Clonts & McDaniel 1978; Jones 1982.

Elephantopus carolinianus Raeusch., (of Carolina), LEAFY ELEPHANTOPUS. Pubescent perennial herb, 30–90 cm tall; basal leaves absent at flowering time; leaves alternate, oblanceolate-elliptic, 5–18 cm long, to 9 cm wide; 3 foliaceous bracts subtending glomerules of heads ovate, acute; involucres subcylindric, the phyllaries stramineous; ray flowers absent; disk flowers 3 or 4 per head, the corollas pinkish white to lavender, 5-lobed; achenes cylindrical, antrorsely pubescent; pappus bristles persistent, basally dilated and indurated. Low wooded areas; se and e TX w to Grand Prairie, also Edwards Plateau. Aug–Oct.

ENGELMANNIA ENGELMANN'S DAISY, CUT-LEAF DAISY

A monotypic genus native to North America; sometimes cultivated as an ornamental. (Named for Dr. George Engelmann, 1809-1884, German-born botanist and physician of St. Louis) (tribe Heliantheae)

REFERENCES: Turner & Johnston 1956; Goodman & Lawson 1992.

Engelmannia peristenia (Raf.) Goodman & C.A. Lawson, (possibly from Greek: *perisso*, beyond the regular number or size, odd in number, and *tenia*, band, ribbon), ENGELMANN'S DAISY, CUT-LEAF DAISY. Densely hispid-pubescent perennial with woody taproot; stems spreading to erect, 0.2–1 m tall; leaf blades 8–30 cm long, pinnately divided nearly to midrib, the lobes again lobed or toothed; inflorescences of several long-peduncled heads; involucres 6–10 mm high; ray flowers with ligules ca. 1 cm long, yellow, curling under during daytime, expanded from late afternoon to mid-morning; disk corollas numerous, dark yellow; pappus of a few scales. Clayey or occasionally sandy prairies and limestone outcrops; nearly throughout TX except forested sandy areas of e part of state. Apr–Jul (rarely later). [*E. pinnatifida* Torr. & A. Gray] Goodman and Lawson (1992) documented that *E. peristenia* is an older name (published 1832) than the long established *E. pinnatifida* (published 1840). Farmers and ranchers refer to this as an "ice cream" plant because it is preferred by livestock; it is therefore now rarely seen except along roadsides or other areas inaccessible to grazing animals (J. Stanford, pers. comm.).

ERECHTITES BURNWEED, FIREWEED

A genus of 15 species native to the Americas; some contain alkaloids. (An ancient Greek name, *erechthites*, used by Dioscorides for somes species of *Senecio*—groundsel) (tribe Senecioneae)

REFERENCES: Belcher 1956; Vuilleumier 1969a; Barkley & Cronquist 1978.

Erechtites hieraciifolia (L.) Raf. ex DC., (with leaves like *Hieracium*—hawkweed), AMERICAN BURNWEED, FIREWEED. Annual herb, glabrescent or with sparse, jointed hairs; stems 0.1–2.5 m tall; leaves alternate, numerous, elliptic to lanceolate or oblanceolate, 3–20 cm long, sharply doubly serrate and sometimes irregularly lobed, sometimes clasping, well-distributed up stem, the uppermost reduced; inflorescence an elongate to corymbose cyme of few-many heads; involucres cylindrical; main phyllaries ca. 12–14, in 1 series, ± equal, 9–17 mm long, with a few, very small bracts at base; heads of disk-like flowers only; ligulate ray flowers absent; flowers whitish or cream-white; peripheral flowers pistillate, with tubular-filiform corollas; central flowers perfect; achenes all alike, ribbed, columnar, ca. 2–3 mm long; pappus of numerous capillary bristles, deciduous. Wet areas; Fannin (Talbot property) and Lamar cos. in Red River drainage, also Fort Hood (Bell or Coryell cos.—Sanchez 1997); mainly se and e TX. Aug–Nov. [*E. hieraciifolia* var. *intermedia* Fernald, *Senecio hieraciifolia* L.]

ERIGERON FLEABANE, DAISY FLEABANE, DAISY

Annual to perennial herbs; leaves alternate, sessile or the lower petioled, oblong or oblanceolate to linear, entire or toothed; heads solitary or corymbose; ray flowers numerous, often 30–150,

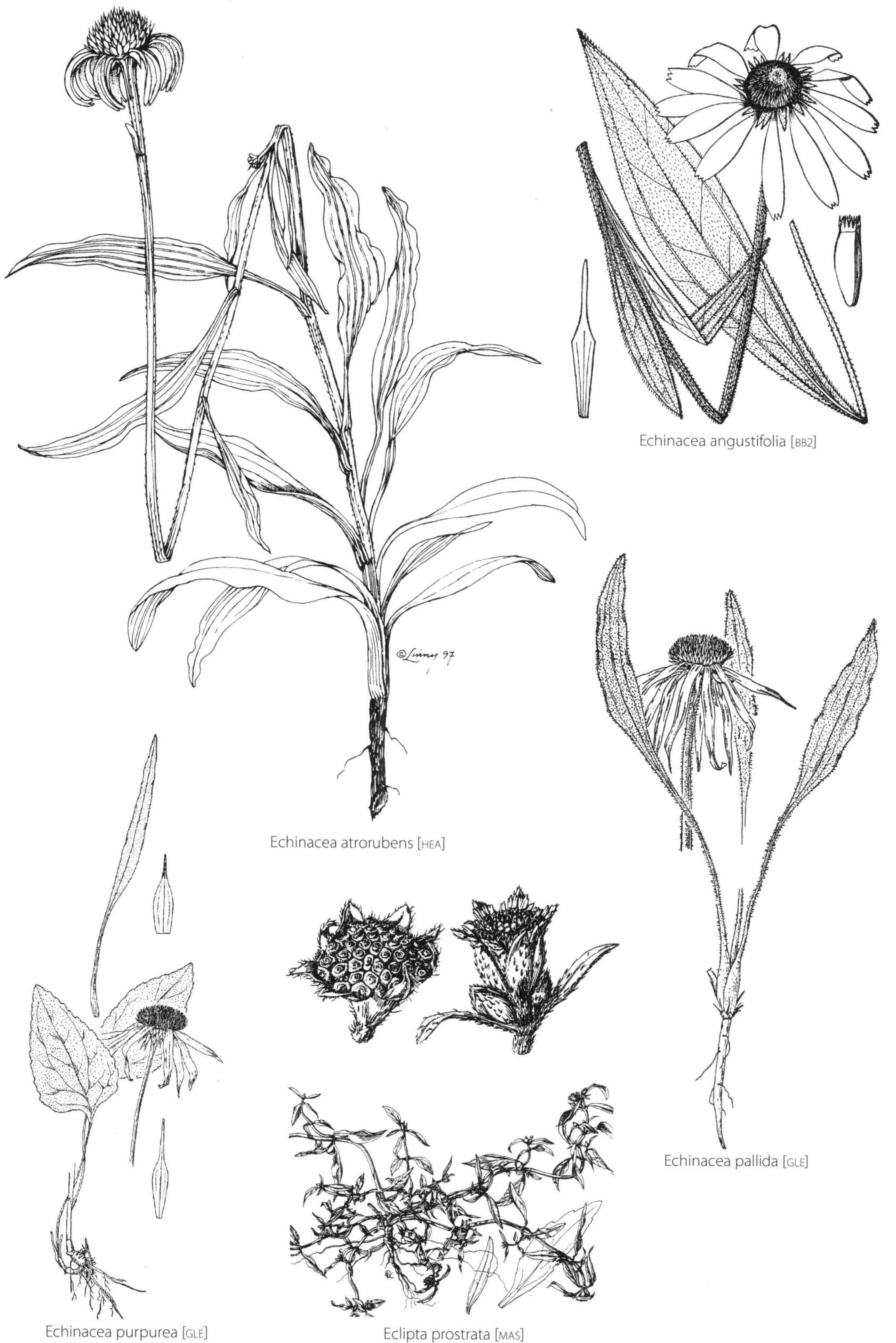

Echinacea angustifolia [BB2]

Echinacea atrorubens [HEA]

Echinacea pallida [GLE]

Echinacea purpurea [GLE]

Eclipta prostrata [MAS]

pistillate, fertile, the ligules narrow (to ca. 1.2 mm wide), white to blue, lavender or pink, never yellow, curling under at night or in age; disk flowers usually perfect and fertile, the corollas yellow; achenes flattened, usually strongly 2-ribbed; pappus various, of hair-like bristles or of 2 whorls (bristles and small bristles or scales).

A cosmopolitan genus (especially North America) of ca. 413 species (G. Nesom, pers. comm.) including a number of cultivated ornamentals. Closely related to *Conyza* (Cronquist 1947) and not sharply distinguished from it. *Erigeron* species are sometimes difficult to distinguish. Barkley (1986) indicated, "As with many widespread genera in the Asteraceae, the species boundaries are not always sharp, and morphological intermediates are to be expected." Guy Nesom (pers. comm.) indicated, however, that very few *Erigeron* species hybridize. (Greek: *eri*, early, and *geron*, old man; the ancient name of an early-flowering plant with fluffy white seed heads) (tribe Astereae)

REFERENCES: Cronquist 1943, 1947; Shinners 1947a; Van Vleet 1951; Nesom 1978, 1989a.

1. Plants small, 10–30 cm tall; with distinct taproot (obviously longer than lateral roots).
   2. Plants annuals; spring form with peduncles less than 4 cm long; usually in sandy habitats; taproot neither noticeably enlarged nor woody ____ **E. geiseri**
   2. Plants perennials; spring form with peduncles 4–12 cm long; usually in calcareous habitats; taproot typically enlarged and woody ____ **E. modestus**
1. Plants of various heights, but often much > 30 cm tall; with shallow, stubby crown and fibrous roots.
   3. Upper and middle stem leaves widest basally or nearly so, clasping, 4–20 mm wide, 2–4 times as long as wide; lower leaves on broadly margined or winged petioles; pappus of disk flowers of a single whorl of ca. 20–30 bristles ____ **E. philadelphicus**
   3. Upper and middle stem leaves narrowed basally, slightly or not clasping, usually 1–10 mm wide, 4–10 times as long as wide; lower leaves on narrowly winged or wingless petioles; pappus of disk flowers of 2 whorls, the inner of 10–15 bristles, the outer of small scales.
      4. Ray flowers with ligules purplish or bluish or white above and colored beneath; ray flowers with pappus of long hairs similar to disk flowers; stems 10–40 cm tall, with 7–15 leaves below inflorescence on well-developed plants; basal leaves usually 15 mm wide or less ____ **E. tenuis**
      4. Ray flowers with ligules white, rarely bluish; ray flowers with minute pappus, the disk flowers with pappus of long hairs; stems 30–150 cm tall, with 17–25 leaves below inflorescence on well-developed plants (some of the leaves quite reduced); basal leaves to 70 mm wide.
         5. Upper stem pubescence of mostly incurved or appressed hairs; stems 30–70 cm tall; stem leaves typically well-developed, linear to lanceolate, entire or the lower or middle slightly toothed, the teeth usually with rounded sides; widespread in nc TX ____ **E. strigosus**
         5. Upper stem pubescence, at least some, long and spreading; stems 60–150 cm tall; stem leaves often rather reduced, lanceolate or wider, all except the upper toothed, often conspicuously so, the teeth straight-sided; rare in nc TX (known only in Red River drainage) ____ **E. annuus**

Erigeron annuus (L.) Pers., (annual), ANNUAL FLEABANE. Annual; involucres 3–5 mm high; disks 6–10 mm across; ray flowers with ligules ca. 5–10 mm long. Prairies or open ground, various soils; in nc TX known only from Grayson Co.; also scattered in e TX and Edwards Plateau. Apr–Jun.

Erigeron geiseri Shinners, (for Samuel Wood Geiser, 1890–1983, professor at Southern Methodist Univ.), BASIN FLEABANE. Stems erect, 7–30 cm tall; ligules white to rosy or lavender; pappus of 2 whorls—10 scabrous bristles and as many small scales. Sandy prairies; Archer, Clay, Dallas, Falls, Palo Pinto, and Young cos.; se TX w to Edwards Plateau and n to nc TX and Rolling Plains. Late Mar–Jun.

Erigeron modestus A. Gray, (modest), PLAINS FLEABANE. Perennial, often grayish pubescent; stems 10–20(–30) cm tall; producing two crops of stems: first one beginning to flower in late

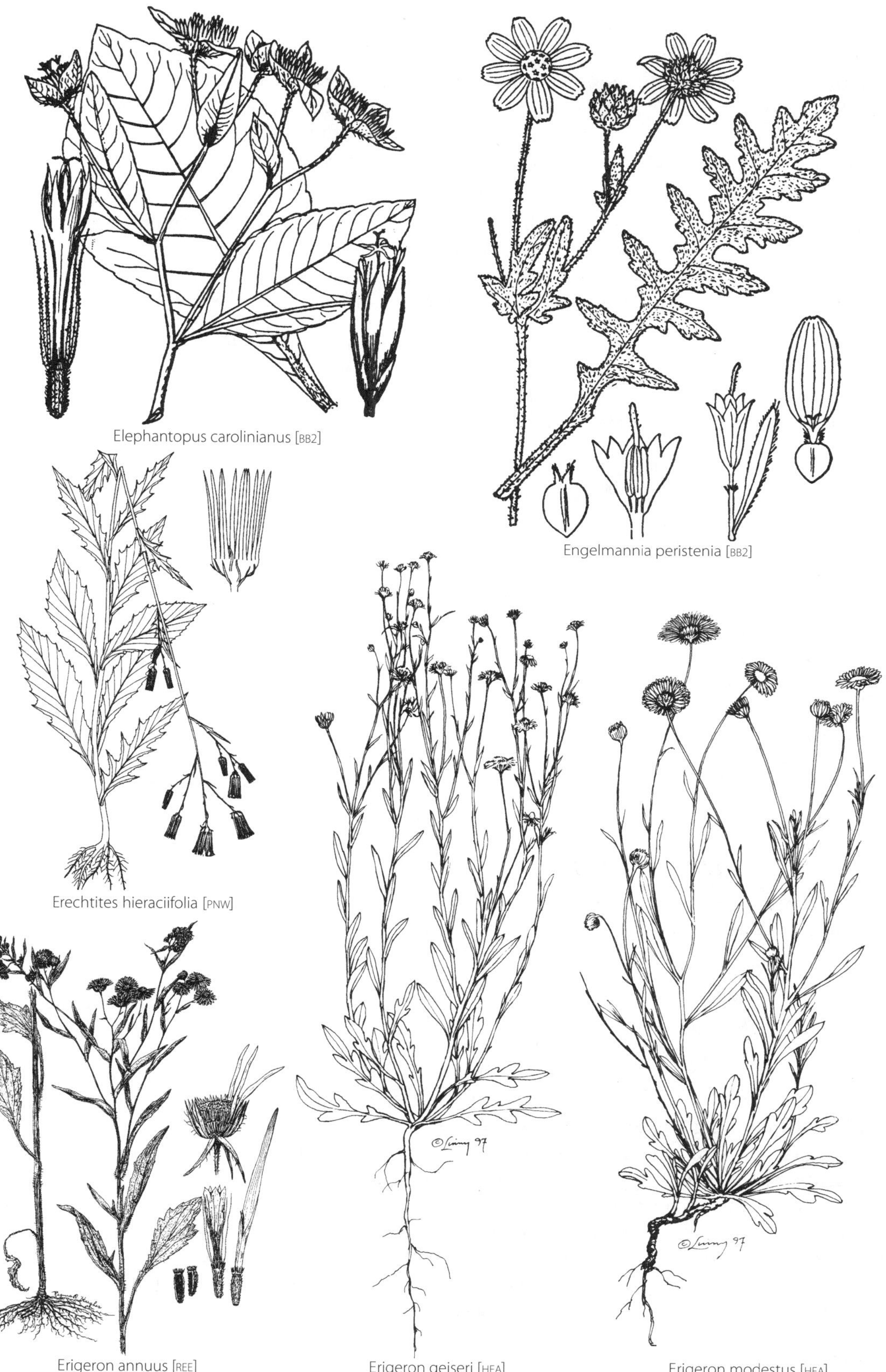

Elephantopus carolinianus [BB2]

Engelmannia peristenia [BB2]

Erechtites hieraciifolia [PNW]

Erigeron annuus [REE]

Erigeron geiseri [HEA]

Erigeron modestus [HEA]

Mar–Apr, with crowded, deeply lobed basal leaves and long, naked peduncles; from Apr–Jun developing into a much-branched plant with narrowly oblanceolate or linear-lanceolate leaves (the basal ones withered); by fall (Sep–Nov) producing a bushy-branched mass of often sprawling or decumbent stems with narrow leaves only, and shorter-peduncled heads; ray flowers with ligules ca. 4–8 mm long; pappus with 2 whorls—bristles and much shorter bristles or scales. Gravelly or rocky limestone; Cooke to Burnet cos. s and w to w TX (ca. w 1/2 of nc TX). Apr–Oct.

Erigeron philadelphicus L., (of the Philadelphia region), PHILADELPHIA FLEABANE. Perennial without rhizomes; stems 15–70 cm tall; involucres 4–6 mm high; disk 6–15 mm across; ray flowers with ligules ca. 5–10 mm long, usually white to light pink. Low prairies and stream banks; calcareous clay; se and e TX w to West Cross Timbers and Edwards Plateau. Mar–May.

Erigeron strigosus Muhl. ex Willd., (strigose, with stiff bristles). Erratic as to duration: commonly biennial, sometimes annual, occasionally perennial; ray flowers with ligules to ca. 6 mm long. Prairies, open woods, pastures, roadsides; se and e TX w to West Cross Timbers; in Red River drainage to e Rolling Plains (Wichita Co.), s to Llano Co. in c TX. Apr–Jun, sporadically to Oct. Sometimes 2 varieties are recognized; Cronquist (1980) considered them ill-defined.

1. Inflorescences diffuse (= widely spreading), subnaked; heads tiny, the involucres 2–3 mm tall ______ var. **beyrichii**

1. Inflorescences not diffuse, or if so, then somewhat leafy; heads usually larger, the involucres (2–) 2.5–5 mm tall ______ var. **strigosus**

var. **beyrichii** (Fisch. & C.A. Mey.) Torr. & A. Gray ex A. Gray, (for Heinrich Karl Beyrich, 1796–1834, Prussian botanist who collected in GA, SC, and TX). Similar to variety *strigosus* and sometimes lumped with that variety (Correll & Johnston 1970).

var. **strigosus**. PRAIRIE FLEABANE, WHITETOP.

Erigeron tenuis Torr. & A. Gray, (slender, thin), SLENDER FLEABANE. Annual or biennial (rarely perennial?); ray flowers with ligules ca. 2.5–5 mm long. Sandy open woods and roadsides; se and e TX w to West Cross Timbers and e Edwards Plateau. Mar–May. There is evidence of introgressive hybridization between this species and *E. strigosus* (Cronquist 1947).

EUPATORIUM BONESET, THOROUGHWORT

Ours herbaceous perennials (1 species a shrub); leaves mostly opposite or in *E. capillifolium* mostly alternate; heads corymbose; ray flowers absent; disk flowers white, blue, pink, or purplish; achenes blackish, 5-ribbed; pappus of persistent bristles.

The genus is mainly American with some Old World species. It is variously treated as including up to 1,200 species or segregated into a number of smaller genera (*Eupatorium* sensu stricto 45 species); if split according to some authorities (e.g., Robinson & King 1985; King & Robinson 1987; Bremer et al. 1994; Jones et al. 1997), the 6 nc TX species go in 4 different genera, *Ageratina*, *Conoclinium*, *Eupatorium* in a strict sense, and *Fleischmannia*. It is difficult to know how to treat this group; for practical reasons, we are following Cronquist (1980), McVaugh (1982), and Gandhi and Thomas (1989) in treating the genus in the traditional broad sense; however, appropriate synonymy is given below. In referring to this group, Stebbins (1993) argued that, "While microscopic characters should by no means be overlooked, their use to establish new genera that are not distinct on the basis of any other characteristics should be discouraged." However, based on the cladistic analysis of the Eupatorieae by Bremer et al. (1994), *Eupatorium* sensu lato is apparently polyphyletic. The common name BONESET is apparently derived from the historic use in treating dengue fever, also known as break-bone fever (Tveten &

Tveten 1993). (Named for Mithridates Eupator, 132–63 B.C., King of Pontus, who is said to have used a species of the genus in medicine or as an antidote for poison) (tribe Eupatorieae)
REFERENCES: Grant 1953; King & Robinson 1970a, 1970b, 1970c, 1970d, 1987; Clewell & Wooten 1971; Wooten & Clewell 1971; Sullivan 1975; Cronquist 1980; McVaugh 1982; Robinson & King 1985; Warnock 1987b; Gandhi & Thomas 1989; Bremer et al. 1994.

1. Leaves connate-perfoliate (= united in pairs around the stem); in nc TX known only from Red River drainage ____ **E. perfoliatum**
1. Leaves distinct, not united; including species widespread in nc TX.
  2. Leaves or their segments finely dissected into filiform (= thread-like) segments < 1 mm wide ____ **E. capillifolium**
  2. Leaves not finely dissected into filiform segments.
    3. Plant a rounded, much-branched shrub with deltoid leaves and pinkish-white to white flowers; mainly rocky limstone areas on Edwards Plateau, in nc TX known only from s part of Lampasas Cut Plain (segregate genus *Ageratina*) ____ **E. havanense**
    3. Plant herbaceous (but stems can be coarse in some species); leaves narrowly elliptic to ovate or deltoid; flowers white to pink, blue, or purplish blue; widespread in nc TX.
      4. Leaves narrowly elliptic, prominently 3-nerved, sessile or with petioles 1–2 mm long ____ **E. altissimum**
      4. Leaves ovate to deltoid, if 3-nerved, at base only, distinctly petiolate.
        5. Corollas blue or purplish blue; receptacles conical (segregate genus *Conoclinium*) ____ **E. coelestinum**
        5. Corollas white or pink; receptacles flat.
          6. Corollas pink or whitish with the distal portion pink (distal portion rarely lavender-pink or white); stems weak, sprawling or scandent; involucres 5–6 mm long (segregate genus *Fleischmannia*) ____ **E. incarnatum**
          6. Corollas white; stems usually erect or ascending; involucres 3–5 mm long.
            7. Phyllaries clearly varying in length, obviously pubescent, resin-dotted (under magnification) ____ **E. serotinum**
            7. Phyllaries all about the same length, nearly glabrous or with pubescent tips, not resin-dotted (segregate genus *Ageratina*) ____ **E. rugosum**

Eupatorium altissimum L., (very tall), TALL EUPATORIUM, TALL THOROUGHWORT. Rhizomatous perennial; leaves mostly opposite; leaf blades narrowly elliptic; petioles 0–2 mm long; involucres 3.8–4.3 mm high; phyllaries rounded apically; corollas white. Prairies; Collin, Denton, Fannin, Grayson, Lamar, and Tarrant cos.; also e TX. Aug–Oct.

Eupatorium capillifolium (Lam.) Small, (hair-leaved), DOGFENNEL, DOGFENNEL EUPATORIUM, CYPRESSWEED. Erect perennial to 3 m tall; leaves mostly alternate, dissected; involucres 2–3 mm high; corollas white. Disturbed areas, sandy soils; Dallas, Lamar, and Tarrant cos.; mainly se and e TX. Sep–Nov.

Eupatorium coelestinum L., (sky-blue), MISTFLOWER, BLUE BONESET. Rhizomatous perennial; leaves opposite; leaf blades deltoid; petioles 3–20 mm long; involucres 3–5 mm high; phyllaries linear-subulate; corollas blue or purplish blue. Moist situations, often wooded areas, sandy or calcareous soils; East and West cross timbers; se and e TX w to nc TX and Edwards Plateau. Aug–Nov. Sometimes (e.g., Jones et al. 1997) recognized in the genus *Conoclinium* [as *C. coelestinum* (L.) DC.]. If segregated, *Conoclinium* is a genus of 3 species of the e United States and Mexico (Mabberley 1997).

Eupatorium havanense Kunth, (of Havana), SHRUBBY BONESET. Rounded, much-branched shrub (0.3–)0.6–1.5(–2) m tall; leaves opposite; leaf blades deltoid, (2–)3–5(–7) cm long; petioles 3–10 (–15) mm long; involucres 4–6 mm high; phyllaries linear; corollas pinkish-white to white. Rocky limestone areas; in nc TX known only from Bell (Fort Hood—Sanchez 1997) and Burnet

(C. Sexton, pers. comm.) cos. in s part of Lampasas Cut Plain; mainly Edwards Plateau. (Sep–) Oct–Nov. Despite the epithet, this species is native to TX; it also occurs in Mexico and on Caribbean islands. [*E. ageratifolium* DC. var. *texense* Torr. & A. Gray, *E. texense* (Torr. & A. Gray) Rydb.] Now sometimes (e.g., Kartesz 1994; Jones et al. 1997) recognized in the segregate genus *Ageratina* [as *A. havanensis* (Kunth) R.M. King & H. Rob.]. If segregated, *Ageratina* is a genus of ca. 290 species of the e United States and c and w South America (Mabberley 1997). *Eupatorium wrightii* A. Gray, a somewhat similar rounded shrub of the Trans-Pecos, has been implicated in the sudden death of cattle (appearing as though "shot and instantly killed"); feeding experiments also resulted in death (Kingsbury 1964). The smaller leaf blades (1–2 cm long) and shorter petioles (2–5 mm long) distinguish *E. wrightii* from *E. havanense.* ☠

Eupatorium incarnatum Walter, (flesh-colored), PINK EUPATORIUM, PINK BONESET. Perennial from fibrous-rooted crown, scandent, to 2 m tall; leaves opposite; leaf blades deltoid; petioles 2 mm or more long; involucres 5–6 mm high; phyllaries lance-subulate; corollas whitish, with pink or lilac apically. Bottomlands, woodlands; Dallas, Fannin, Grayson, Hunt, Kaufman, Limestone, Rockwall, and Tarrant cos.; se and e TX w to nc TX. Oct–Nov. Sometimes (e.g., Jones et al. 1997) recognized in the genus *Fleischmannia* [as *F. incarnata* (Walter) R.M. King & H. Rob.]. If segregated, *Fleischmannia* is a genus of 80 species of North America and w South America (Mabberley 1997).

Eupatorium perfoliatum L., (perfoliate, with leaf surrounding the stem), BONESET, THOROUGHWORT, AGUEWEED. Rhizomatous perennial; leaves opposite, lanceolate, perfoliate; involucres 3–5 mm high; corollas whitish. Sandy soils; Grayson and Lamar cos. in Red River drainage; mainly se and e TX. Aug–Nov. This species contains the glucoside, eupatorin, and has been variously used medicinally (Duke 1985).

Eupatorium rugosum Houtt., (rugose, wrinkled), WHITE SNAKEROOT, SNAKEROOT, RICHWEED, FALLPOISON. Perennial from a fibrous crown; leaves opposite; leaf blades deltoid-ovate to lance-ovate; petioles about 1/3 as long as blades; involucres 4–5 mm high; corolla lobes pubescent dorsally. Low woods; Grayson Co. in Red River drainage (Hagerman N.W.R.); mainly se and e TX; also Edwards Plateau. Sep–Nov. Now sometimes (e.g., Kartesz 1994; Jones et al. 1997) recognized in the segregate genus *Ageratina* [as *A. altissima* (L.) R.M. King & H. Rob.]. All parts of WHITE SNAKEROOT contain tremetol, a complex alcohol, and glycosides; it is poisonous to livestock. During colonial times cows eating SNAKEROOT passed the poison through their milk causing "milk sickness" in humans (Hardin & Arena 1974); numerous deaths resulted, in some areas the human population was reduced to less than 1/2 the original number, and whole villages were abandoned (Kingsbury 1964); in livestock the condition is known as "trembles" (Stephens 1980). Other species are not known to contain the toxin. ☠

Eupatorium serotinum Michx., (late-flowering), LATE EUPATORIUM, LATE-FLOWERING THOROUGHWORT, FALL BONESET. Rhizomatous; leaves mostly opposite; leaf blades narrowly ovate; petioles 10–30 mm long; involucres 3–5 mm high; phyllaries linear; flowers whitish. Open areas, disturbed sites; se and e TX w to Rolling Plains and Edwards Plateau. Aug–Nov.

EUTHAMIA FLAT-TOPPED-GOLDENROD

A North American genus of 8 species; similar to and often lumped into *Solidago* as a section or subgenus (Sieren 1981); some authors (e.g., Cronquist 1980) suggest the genus may be more closely related to *Gutierrezia*. (Possibly either from Greek: *eu*, well, and [?], crowded, from dense inflorescences (Semple 1993), or Greek: *euthemon*, neat or pretty) (tribe Astereae)
REFERENCES: Cronquist 1980; Sieren 1981; Taylor & Taylor 1983; Semple 1992.

Euthamia gymnospermoides Greene, (gymnosperm-like), VISCID EUTHAMIA, FLAT-TOPPED-GOLDENROD. Rhizomatous, ± glabrous perennial; stems to 1 m tall; basal and lower stem leaves early

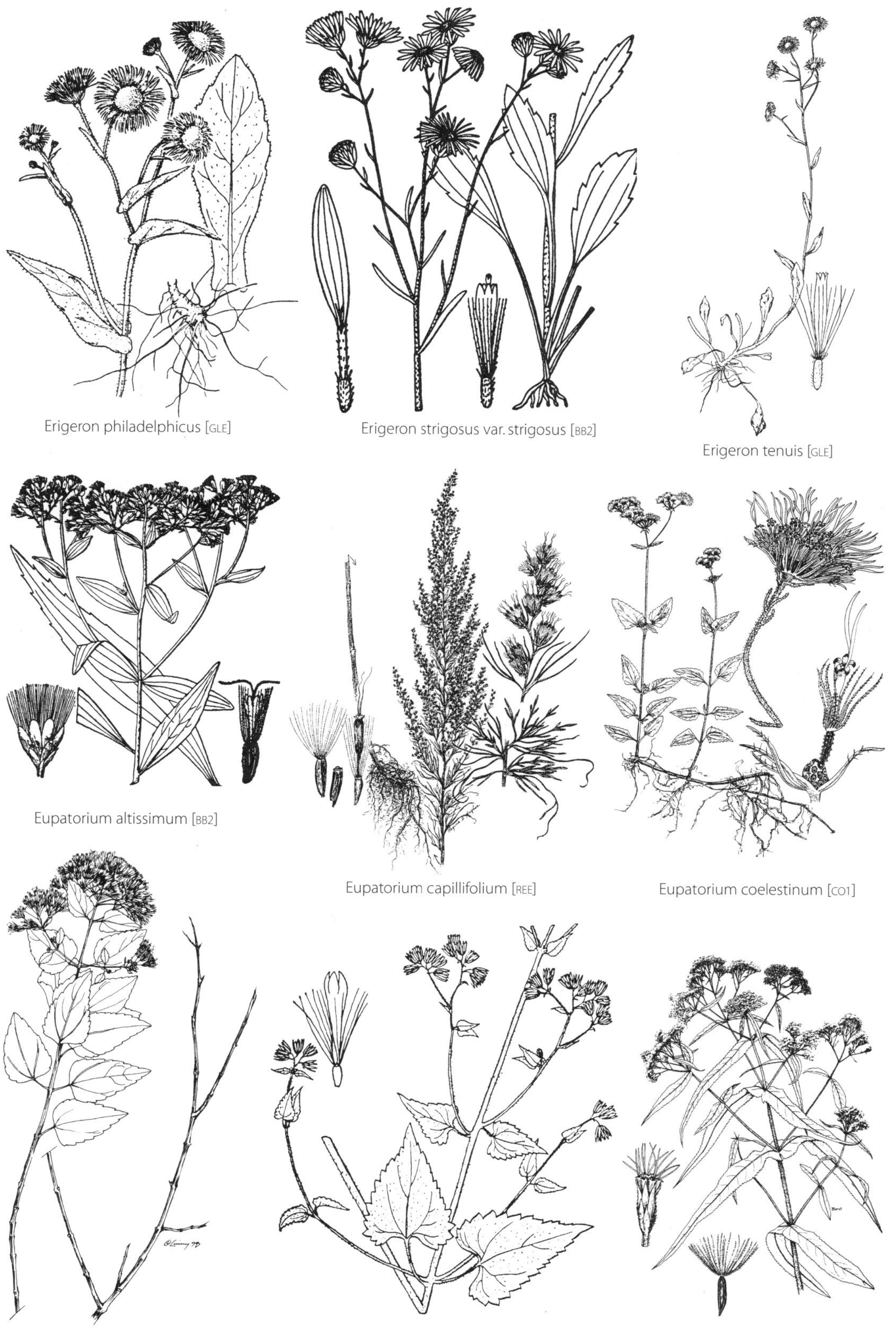

Erigeron philadelphicus [GLE]

Erigeron strigosus var. strigosus [BB2]

Erigeron tenuis [GLE]

Eupatorium altissimum [BB2]

Eupatorium capillifolium [REE]

Eupatorium coelestinum [CO1]

Eupatorium havanense [HEA]

Eupatorium incarnatum [BB2]

Eupatorium perfoliatum [GWO]

deciduous; persistent stem leaves numerous, alternate, linear or linear-lanceolate, 4–9 cm long, 1.5–4(–8) mm wide, entire, strongly glandular-punctate, sessile; inflorescences broad, cymose, with numerous, ± sessile, small heads; involucres 4–6.5 mm high; flowers with sweet fragrance; ray flowers pistillate and fertile; ligules yellow, 1–3 mm long; disk flowers fewer than ray flowers, perfect and fertile, the corollas yellow; achenes hairy; pappus of numerous whitish, hair-like bristles. Open woods, prairies; Lamar and Montague cos; also se and e TX and Plains Country. Oct–Nov. [*E. camporum* Greene, *E. pulverulenta* Greene, *Solidago gymnospermoides* (Greene) Fernald, *Solidago graminifolia* (L.) Nutt. var. *gymnospermoides* (Greene) Croat]

EVAX RABBIT'S-TOBACCO, COTTON-ROSE

Very small annuals with taproots, usually less than 10(–15) cm tall, densely and conspicuously white- to gray-pubescent with woolly hairs; stems usually branched; leaves alternate or uppermost crowded and apparently whorled, small (ca. 2–15 mm long), linear or oblanceolate, entire; heads sessile, in dense clusters, axillary or terminal, small and inconspicuous, woolly-pubescent; true involucres absent; ray flowers absent; pappus absent.

A genus of ca. 15–20 species (Mabberley 1987) native from the Mediterranean to c Asia, also North America; sometimes lumped into *Filago* (e.g., Anderberg 1991; Mabberley 1997). We are following Jones et al. (1997) and J. Kartesz (pers. comm. 1997) in maintaining it as a separate genus. (Derivation of generic name not explained by original author) (tribe Inuleae)
REFERENCES: Shinners 1951e; Anderberg 1991; Morefield 1992.

1. Clusters of heads axillary, distributed along the elongate stems; central flowers of heads usually fertile ____ **E. candida**
1. Clusters of heads nearly all terminal at the branch tips; central flowers of head usually infertile.
   2. Bracts among the heads shorter than or equal to the heads in length (bracts thus largely concealed by dense, woolly pubescence associated with the heads); ultimate clusters of heads subtended by leaves 3–10 mm long that simulate an involucre ____ **E. verna**
   2. Bracts among the heads exceeding the heads in length (bracts thus conspicuously visible protruding from between the crowded heads); ultimate clusters of heads subtended by leaves 6–12 mm long that simulate an involucre ____ **E. prolifera**

Evax candida (Torr. & A. Gray) A. Gray, (pure white, shining), SILVER EVAX. Clusters of heads mostly sessile in the leaf axils. Sandy open woods, prairies, fields, and roadsides; Bell, Burnet, Dallas, Henderson, Limestone, and Parker cos.; also Hood, Somervell, and Tarrant cos. (R. O'Kennon, pers. obs.); se and e TX rarely w to nc TX and Edwards Plateau. Apr–May. [*Filago candida* (Torr. & A. Gray) Shinners]

Evax prolifera Nutt. ex DC., (producing offshoots), BIG-HEAD EVAX. Clayey or rocky ground on limestone; widespread in TX from Blackland Prairie w. Apr–Jun. [*Filago prolifera* (Nutt. ex DC.) Britton]

Evax verna Raf., (of Spring), MANY-STEM EVAX. Sandy or eroding silty or rocky ground; nearly throughout TX. Apr–Jun. [*E. multicaulis* DC., *Filago verna* (Raf.) Shinners]

FACELIS

A South American genus of 3 species. (Derivation of generic name unknown) (tribe Inuleae)
REFERENCE: Anderberg 1991.

Facelis retusa (Lam.) Sch.Bip., (retuse, notched slightly at a rounded apex). Annual herb from a taproot; stems 6–20(–30) cm tall, erect or with decumbent branches, white-gray woolly; leaves alternate, simple, entire, linear-spatulate, 7–20(–30) mm long, 1.5–4 mm wide, ± sessile; heads in a terminal cluster; involucres 8–11 mm high; phyllaries mostly scarious or partly greenish, the

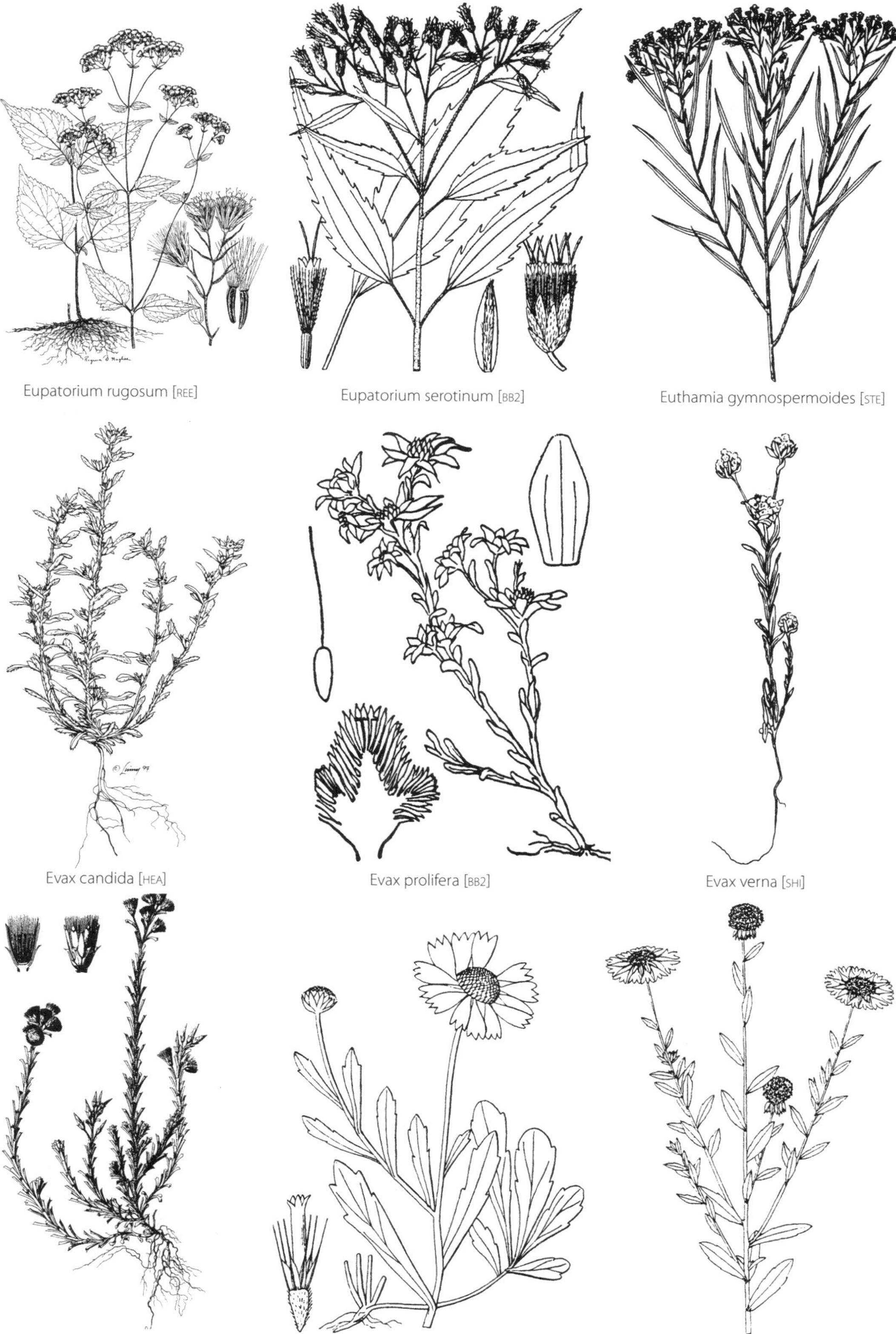

Eupatorium rugosum [REE]

Eupatorium serotinum [BB2]

Euthamia gymnospermoides [STE]

Evax candida [HEA]

Evax prolifera [BB2]

Evax verna [SHI]

Facelis retusa [MAR]

Gaillardia aestivalis var. aestivalis [BB1]

Gaillardia aestivalis var. flavovirens [STE]

inner often pigmented near tip; ray flowers absent; disk flowers of 2 types: the central ca. 3–5 perfect and fertile, with corollas 5-toothed, white; the peripheral 10–25 pistillate, with corollas filiform, truncate or obscurely toothed, white; achenes white hairy; pappus of numerous whitish, strongly plumose, hair-like bristles 10–11 mm long, surpassing the corollas. Lawn weed, roadsides, disturbed areas; Dallas and Tarrant cos.; mainly e TX. Apr–May. Native of South America. [*Gnaphalium retusum* Lam.]

GAILLARDIA INDIAN-BLANKET, BLANKET-FLOWER

Low, pubescent annuals or perennials; leaves basal or alternate, entire, toothed, or lobed, petioled to ± clasping; heads large, solitary or corymbose; phyllaries herbaceous or with hardened base, narrow, loose; ray flowers infertile; ligules widened and lobed apically, yellow to dark orange or red, or ligules absent; disks rounded-conical to subglobose, the perfect disk flowers usually intermixed with bristles or slender toothed scales; disk corollas yellow to reddish purple or reddish brown; pappus of awned scales, the awns ca. as long as the scale body.

A genus of 28 species native to North America and temperate South America; includes cultivated ornamentals. (Named for M. Gaillard de Charentoneau, 18th century French magistrate and patron of botany) (tribe Helenieae, sometimes lumped into Heliantheae)
REFERENCES: Biddulph 1944; Turner & Whalen 1975; Heywood & Levin 1984.

1. Leaves ± all basal; ray flowers with ligules very short and inconspicuous or absent ____ **G. suavis**
1. Leaves conspicuously extending up the stem, alternate; ray flowers with ligules usually present (sometimes absent), the ligules short to long, often showy.
   2. Ligules of ray flowers usually reddish tipped with orangish yellow; lobes of disk corollas narrowly triangular-acuminate, short, much < 1/2 as long as the corolla tube (note: the margins of the disk corolla lobes have long, moniliform, often purplish red hairs which can obscure the lobe shape); receptacles (look between the disk flowers) with well-developed bristles ca. as long as the achenes ____ **G. pulchella**
   2. Ligules of ray flowers usually yellow or yellow with red at base; lobes of disk corollas narrowed to linear tip, very long, the lobes ca. 1/2 as long as the corolla tube or longer (margins of lobes can have hairs as in *G. pulchella*); receptacles naked or nearly so ____ **G. aestivalis**

Gaillardia aestivalis (Walter) H. Rock, (summer), PRAIRIE GAILLARDIA. Annual or short-lived perennial 12–65 cm tall, as broad as tall or usually narrower, with slender to stout taproot, usually single-stemmed; leaves mostly sessile; disks 15–25 mm across; ray flowers usually with ligules (8–)10–15 mm long or ligules sometimes absent; achenes ca. 2 mm long; pappus scales 5–6 mm long, with a slightly longer awn. Sandy open woods, sandy prairies, disturbed areas. May–Oct.

1. Disks purple-brown; ray flowers with ligules partly red or wholly yellow or absent ____ var. **aestivalis**
1. Disks yellow; ray flowers with ligules yellow ____ var. **flavovirens**

var. **aestivalis**. LANCE-LEAF GAILLARDIA, PRAIRIE GAILLARDIA, YELLOW INDIAN-BLANKET. Se and e TX w to East and West cross timbers and Edwards Plateau. [*G. fastigiata* Greene, *G. lanceolata* Michx.]

var. **flavovirens** (C. Mohr) Cronquist, (yellow-green), YELLOW INDIAN BLANKET. Included based on citation of vegetational areas 4 and 5 (Fig. 2) by Hatch et al. (1990); mainly se and e TX. [*G. lanceolata* Michx. var. *flavovirens* C. Mohr]

Gaillardia pulchella Foug., (handsome), INDIAN-BLANKET, FIRE-WHEELS, ROSE-RING GAILLARDIA. **State wildflower of Oklahoma** (Tyrl et al. 1994). Low annual 10–60 cm tall, pubescent to hispid-pilose, developing wide-spreading branches; leaves oblong to ovate-lanceolate, the upper clasping, entire or the lower toothed or lobed; ray corollas with ligules 10–20 mm long, reddish with narrow to wide orangish yellow zone at tip, or rarely all yellow; disk corollas dark red-brown, rarely yellow; achenes ca. 2 mm long; pappus scales 5–6 mm long, with an awn of ca. equal length. Prairies, disturbed areas; nearly throughout TX. May–Jun, sporadically later. Very showy and often abundant on roadsides; one of the most common wildflowers in nc TX. ▣/90

Gaillardia suavis (A. Gray & Engelm.) Britton & Rusby, (sweet), RAYLESS GAILLARDIA, FRAGRANT GAILLARDIA, GLOBE-FLOWER, PERFUME-BALL. Perennial 30–80 cm tall forming beds from obliquely branching roots; leaves slightly fleshy, oblanceolate or obovate, varying from entire to deeply lobed; disk flowers with heavy, sweet scent; ray flowers with ligules very short or absent, if present orange brownish, inconspicuous; achenes ca. 2.5 mm long; pappus scales ca. 4–5 mm long, with an awn of ca. equal length. Sandy or rocky prairies and roadsides; Blackland Prairie w through 2/3 of TX. Apr–May, sporadically later.

GAMOCHAETA CUDWEED, EVERLASTING

Annual or biennial herbs, whitish to grayish with felty or woolly pubescence; leaves alternate, linear to narrowly oblong, oblanceolate, spatulate or obovate, sometimes ± glabrous above, sessile or with narrow petiolar base; heads disciform, nearly sessile in the upper axils, together spike-like in appearance; phyllaries scarious, sometimes pink-tinged; flowers inconspicuous; pappus bristles deciduous as a ring.

A genus of 52 species; mainly South American, with 5–6 in North America, and a few in other parts of the world (Nesom 1990a; Anderberg 1991). *Gamochaeta* was previously included in *Gnaphalium*, and while distinct from that genus, a more narrowly defined *Gamochaeta* is still possibly paraphyletic (Anderberg 1991). (Greek: *gamo*, marriage, and *chaeto*, bristle, in allusion to the pappus bristles falling together) (tribe Inuleae)
REFERENCES: Nesom 1990a; Anderberg 1991.

1. Leaves obovate or spatulate to oblanceolate, 5–20 mm wide, less densely pubescent on upper surface, strongly to weakly bicolored (= the 2 sides different in color); basal leaves present or absent at flowering time.
   2. Leaves whitened or silvery beneath with tight, felty-woolly pubescence, green or brownish and sparsely pubescent above; involucres densely woolly at base, otherwise nearly glabrous ______ **G. purpurea**
   2. Leaves gray-green on both surfaces, loosely woolly beneath, greener and less pubescent above; involucres abundantly loose-woolly, often buried in tangle of hairs, the hairs not limited to the base ______ **G. pensylvanica**
1. Leaves linear to narrowly oblanceolate, 1–6(–9) mm wide, ± equally gray-green pubescent on both surfaces, not bicolored; basal leaves absent at flowering time ______ **G. falcata**

Gamochaeta falcata (Lam.) Cabrera, (falcate, sickle-shaped). Annual similar to *G. purpurea*; involucres 4–6 mm high. Sandy or rocky ground; Burnet, Denton, Fannin, and Palo Pinto cos.; se and e TX w to nc TX and Edwards Plateau. Apr–Jun. Native of se South America. [*Gnaphalium falcatum* Lam.]

Gamochaeta pensylvanica (Willd.) Cabrera, (of Pennsylvania), CUDWEED. Annual or biennial; stems 10–40 cm tall; similar to *G. purpurea*; leaves weakly bicolored; inflorescences usually interrupted; involucres 3–5 mm high. Sandy roadsides, fields, and woods; se and e TX w to West Cross Timbers and Edwards Plateau. Mar–Jun. [*Gnaphalium pensylvanicum* Willd., *Gnaphalium peregrinum* Fernald]

Gamochaeta purpurea (L.) Cabrera, (purple), PURPLE CUDWEED. Annual; stems 10–30 cm tall; leaves strongly bicolored; inflorescences usually continuous; involucres 4–6 mm high. Sandy woods and roadsides; se and e TX w to East Cross Timbers and Edwards Plateau Apr–May. [*Gnaphalium purpureum* L.]

GRINDELIA GUMWEED, TARWEED, ROSINWEED, GUMPLANT

Annual to perennial herbs, usually glutinous; leaves alternate, simple, sessile, serrate to spiny-toothed; receptacles naked; phyllaries linear to linear-subulate; ray flowers 0–45, pistillate, fertile, the corollas yellow; disk flowers perfect, the central sometimes sterile, the peripheral fertile, the corollas yellow; achenes glabrous; pappus similar, of 2-numerous bristles or awns.

A genus of ca. 55 species of w North America and South America; some are used as cultivated ornamentals and others as medicinal herbs. We are following Nesom et al. (1993), Jones et al. (1997), and J. Kartesz (pers. comm. 1997) in including the monotypic genus *Prionopsis*; according to Nesom et al. (1993), the only morphological character distinguishing the genera is a difference in the pappus; molecular data (Suh 1989) indicated that *Grindelia* is paraphyletic when *Prionopsis* is excluded. However, while noting that *Grindelia* and *Prionopsis* are sister groups in all of their analyses including molecular data, Lane and Hartman (1996) indicated there are other morphological characters separating the two and that other genera should possibly be included in the clade. They stated, "... we believe the distinct status of these genera should be maintained until data are obtained that unequivocally support one or another of the possible mergers." (Named for professor David Hieronymus Grindel, 1776–1836, a European botanist from Riga) (tribe Astereae)
REFERENCES: Hall 1928; Steyermark 1934; Suh 1989; Nesom 1990c, 1992b; Nesom et al. 1993.

1. Ray flowers absent ____ **G. nuda**
1. Ray flowers present.
  2. Pappus bristles numerous per achene, united in a ring at base, ± persistent but upon full maturity falling as a unit or in groups ____ **G. papposa**
  2. Pappus bristles or awns only 2–8 per achene, not united at base, falling off at the slightest touch.
    3. Middle stem leaves oblong, with 13–30 teeth per side; upper part of phyllaries usually spreading to squarrose (= spreading rigidly at right angle or more) to reflexed; rare in nc TX ____ **G. squarrosa**
    3. Middle stem leaves lanceolate to deltoid, with 7–15 teeth per side OR oblong, with 25–60 small teeth per side; upper part of phyllaries ascending to spreading, but neither squarrose nor reflexed; wide-spread in nc TX.
      4. Middle stem leaves lanceolate to deltoid, with 7–15 teeth per side; plants perennial ____ **G. lanceolata**
      4. Middle stem leaves oblong, with 25–60 teeth per side; plants annual ____ **G. adenodonta**

Grindelia adenodonta (Steyerm.) G.L. Nesom, (sticky tooth), LITTLE-HEAD GUMWEED, GLAND-TOOTH GUMWEED. Taprooted annual 1–1.5 m tall, glandular pubescence usually absent; leaves not much reduced towards the heads; mature achenes ca. as broad as long. Clay soils; se and s TX n to nc TX and w to Edwards Plateau. Jun–Nov. While often recognized as [*G. microcephala* DC. var. *adenodonta* Steyerm.] (e.g., Kartesz 1994), we are following Nesom (1992b), Jones et al. (1997) and J. Kartesz (pers. comm. 1997) in recognizing it at the specific level.

Grindelia lanceolata Nutt., (lanceolate, lance-shaped), GULF GUMWEED. Biennial or perennial from woody crown; stems 0.3–1.5 m tall; leaves 4–10 cm long, 1–3 cm wide, tapered to apex, gradually reduced towards the heads; ray flowers 20–30, the ligules ca. 10–16 mm long; achenes longer than broad, 2–5 mm long. Calcareous soils; se and e TX w to Rolling Plains and Edwards Plateau. Summer(–fall).

Grindelia nuda A.W. Wood, (nude), RAYLESS GUMWEED. Similar to and sometimes treated as a variety of *G. squarrosa*; taprooted annual; leaves gradually reduced towards the heads; middle

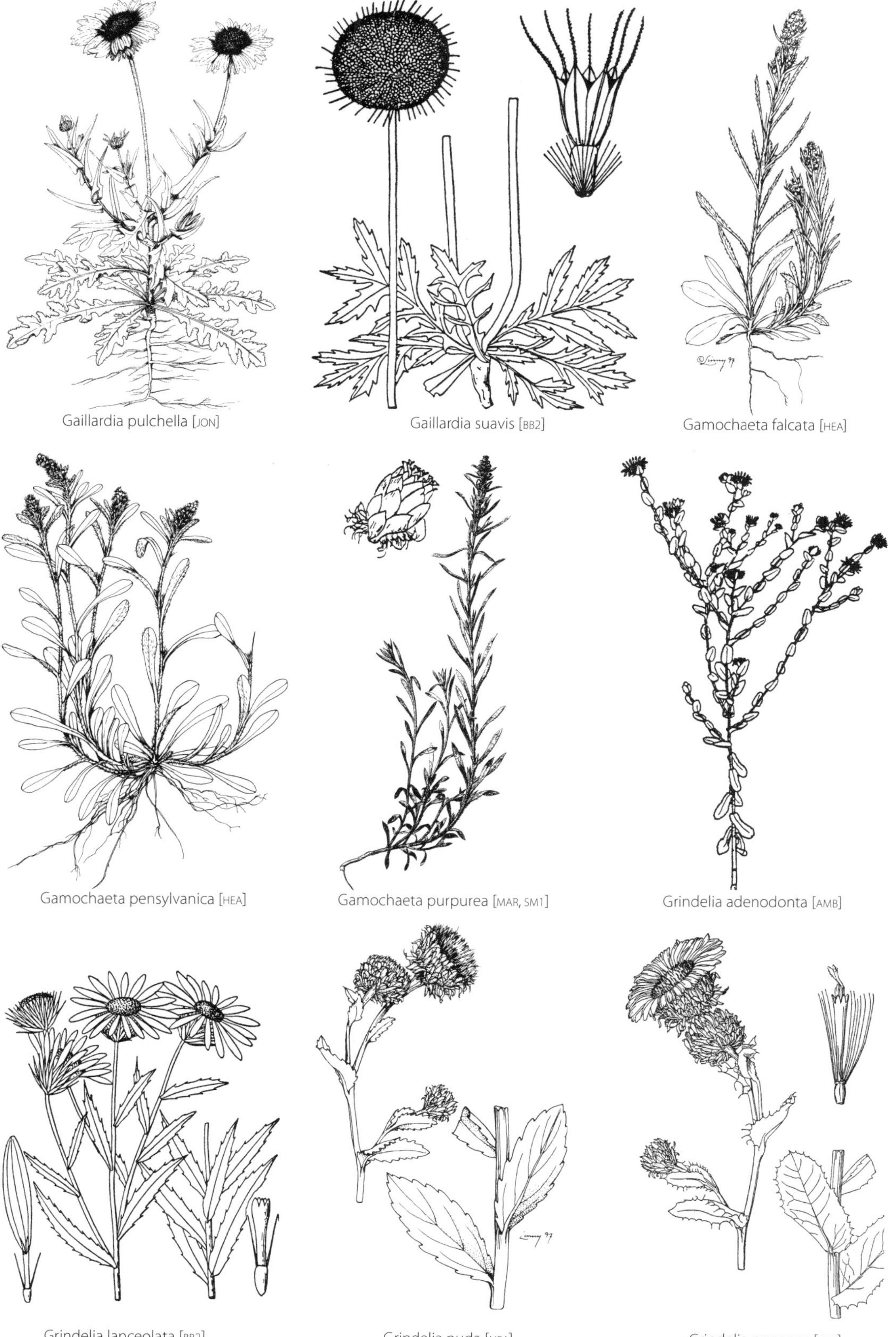

Gaillardia pulchella [JON]

Gaillardia suavis [BB2]

Gamochaeta falcata [HEA]

Gamochaeta pensylvanica [HEA]

Gamochaeta purpurea [MAR, SM1]

Grindelia adenodonta [AMB]

Grindelia lanceolata [BB2]

Grindelia nuda [HEA]

Grindelia papposa [HAL]

stem leaves oblong, with 13–30 teeth per side; phyllaries usually spreading to squarrose or reflexed; achenes oblong, longer than broad, 2.3–3 mm long. Disturbed areas; widespread in n and w parts of TX, most frequent in Plains Country. Summer–fall. [*G. squarrosa* (Pursh) Dunal var. *nuda* (A.W. Wood) A. Gray]

Grindelia papposa G.L. Nesom & Y.B. Suh, (named for its conspicuous pappus), SAW-LEAF DAISY, GOLDENWEED. Stout annual, glabrous, resinous-glandular; stems 0.5–1.5+ m tall, usually unbranched except near top; leaves oblong, to 5 cm long and 1 cm wide, coarsely spiny-toothed; heads few, terminating the branches; receptacles 10–30 mm wide; involucres 10–15 mm high; ray flowers ca. 45, the ligules 10–13 mm long. Disturbed areas; widespread in TX. Aug–Sep. Previously recognized in the monotypic genus *Prionopsis* [as *P. ciliata* (Nutt.) Nutt.]. [*Haplopappus ciliatus* (Nutt.) DC., not *Grindelia ciliata* Spreng.] Avoided by grazing livestock. ▣/91

Grindelia squarrosa (Pursh) Dunal, (with recurved tips), CURLY-CUP GUMWEED. Taprooted annual or biennial 0.1–1 m tall; leaves gradually reduced towards the heads; achenes longer than broad, 2.3–3 mm long. Open or disturbed areas; Clay, Montague, Tarrant, and Wise cos., also Brown, Hamilton, and Mills cos. (HPC); also Plains Country. Summer–fall. Used medicinally by Native Americans and in folk remedies; however, it contains the carcinogen, safrole (Duke 1985). ☠

GUTIERREZIA BROOMWEED, SNAKEWEED

Annuals or perennials, herbaceous or woody, often glutinous (with a gluey or sticky exudate); leaves alternate, essentially sessile, linear to narrowly lanceolate or narrowly elliptical, entire; inflorescences paniculate or corymbose; phyllaries in several series, usually green-tipped; ray flowers pistillate, fertile or sterile, the corollas yellow; disk flowers perfect, fertile or sterile, the corollas yellow; pappus absent, of scales, or a low crown.

A genus of 27 species (number depending on circumscription) of w North America and warm South America. The species treated here have been variously placed in *Amphiachyris*, *Gutierrezia*, and *Xanthocephalum*. Shinners (1951c), Correll and Johnston (1970), and Mahler (1988) placed all in *Xanthocephalum*. Based on chromosome number and morphological data, *Gutierrezia* (including *Amphiachyris*) ($x = 4$ or 5) appears more closely related to a number of other genera than to *Xanthocephalum* sensu stricto ($x = 6$) (Lane 1982). Molecular data (Suh & Simpson 1990) also showed that *Gutierrezia* is not closely related to *Xanthocephalum* (which is more closely related to *Grindelia*). Lane (1979, 1982) argued that *G. amoena* and *G. dracunculoides* should be segregated into the genus *Amphiachyris*. We, however, follow Barkley (1986) and Taylor (1997) in not segregating *Amphiachyris* from *Gutierrezia*. The molecular analysis of Suh and Simpson (1990) indicated that the 2 species sometimes segregated as *Amphiachyris* form a sister group (with 1 other species–*G. triflora* which is sometimes put in the monotypic genus *Thurovia*) to the rest of *Gutierrezia*. Therefore, *Amphiachyris* (and *Thurovia*) can be either lumped with *Gutierrezia* or split depending on other considerations. Because of striking morphological similarity and because *Gutierrezia* is still monophyletic with the inclusion of these three species, the recognition of one monotypic genus (*Thurovia*) and a second genus with only 2 species (*Amphiachyris*) does not seem justified. ☠ Several species are toxic to grazing livestock due to saponins or unknown toxins (Kingsbury 1964; James & Welsh 1992); they increase under overgrazing. (Named in 1816 for Pedro Gutierrez, correspondent of the botanical garden of Madrid) (tribe Astereae)

REFERENCES: Shinners 1951c; Solbrig 1960, 1961; Ruffin 1974; Lane 1979, 1980, 1982, 1985; Barkley 1986; Suh & Simpson 1990; Nesom 1991c.

1. Heads campanulate, about as long as broad, with 7–15 ray flowers; annuals with stems typically branching from ca. the middle upward; widespread in nc TX.
   2. Disk flowers sterile (staminate), not producing achenes; disk pappus usually of 5 long, narrowly

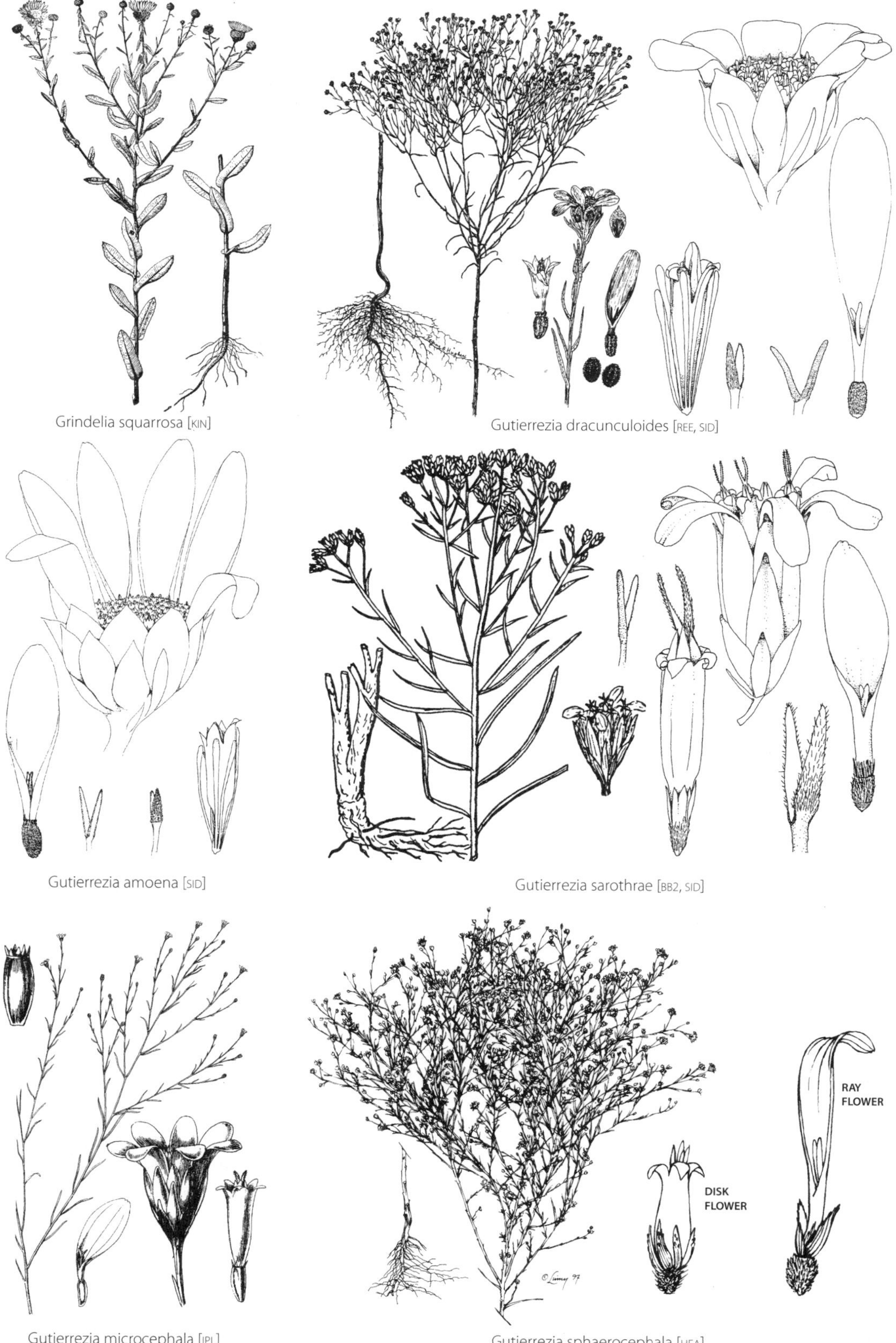

Grindelia squarrosa [KIN]

Gutierrezia dracunculoides [REE, SID]

Gutierrezia amoena [SID]

Gutierrezia sarothrae [BB2, SID]

Gutierrezia microcephala [IPL]

Gutierrezia sphaerocephala [HEA]

spatulate scales as long as or longer than the corollas, united from base to ca. 1/4 their length; ray pappus of several acute scales; phyllaries appear ± without nerves, varnished, shining.

3. Plants 30–40(–60) cm tall; heads scattered in open panicles; achenes long setulose; stamens included; leaves 0.2–1(–2) mm wide; receptacles with hooked, swollen-based hairs; mainly s 1/2 of nc TX ____ **G. amoena**
3. Plants 30–100 cm or more tall; heads many to very numerous in crowded corymbs; achenes short setulose; stamens exerted; leaves 0.5–6 mm wide; receptacles glabrous; throughout nc TX ____ **G. dracunculoides**

2. Disk flowers fertile, producing achenes; disk pappus of several short pointed scales usually < 1/2 as long as the corollas; ray pappus greatly reduced or nearly absent; phyllaries appear nervate (1 or 3 nerves), neither varnished nor shining.

4. Pappus of ray flowers inconspicuous to absent; disk flowers 7–13 per head; achenes puberulent by short purplish hairs, the achene surface not concealed; widespread in nc TX ____ **G. texana**
4. Pappus of ray flowers of small scales; disk flowers 18–37 per head; achenes densely long pubescent by white hairs, the achene surface concealed; a w TX species possibly occurring as far e as w margin of nc TX ____ **G. sphaerocephala**

1. Heads ± cylindrical, longer than broad, with 1–7(–8) ray flowers; perennials with stems usually branching from the base upwards; extreme w part of nc TX.

5. Ray flowers (2–)3–5(–8) per head; disk flowers (2–)3–5(–9) per head, producing achenes; involucres 3–6 mm high; phyllaries 8–14(–21) ____ **G. sarothrae**
5. Ray flowers 1 (very rarely 2) per head; disk flowers 1 (very rarely 2) per head, not producing achenes; involucres 2–4 mm high; phyllaries 4–6(–8) ____ **G. microcephala**

Gutierrezia amoena (Shinners) Diggs, Lipscomb, & O'Kennon, comb. nov. BASIONYM: *Xanthocephalum amoenum* Shinners, Field & Lab. 19:77. 1951. TYPE: TEXAS. Comal Co.: rocky prairies of the Guadaloupe, north of New Braunfels, 1846, *Lindheimer 422*, (HOLOTYPE: BRIT/SMU), (charming), ANNUAL BROOMWEED, BROOMWEED. Main stem usually 5 mm or less in diam.; leaves linear (rarely linear-lanceolate), 0.5–2.5(–3.5) cm long, 0.2–1(–2) mm wide; phyllaries to 3–4 mm long and 1–2 mm wide; achenes 2–3 mm long, purple-black with 4–6 greenish stripes. Calcareous soils on or near limestone outcrops; c to nc TX and the w part of e TX. Late Sep–Oct. [*Amphiachyris amoena* (Shinners) Solbrig, *Xanthocephalum amoenum* Shinners, *X. amoenum* var. *intermedium* Shinners]

Gutierrezia dracunculoides (DC.) S.F. Blake, (resembling *Artemisia dracunculus* L.—taragon), COMMON BROOMWEED. Main stem 0.3–1(–2) cm in diam.; leaves narrowly to broadly lanceolate, 0.5–6 cm long, 0.5–6 mm wide; phyllaries to 2–3 mm long and 1–2 mm wide; achenes 1.2–2.2 mm long, purple-black with 7–9 greenish stripes. Calcareous, clay or sandy soils, disturbed habitats, often in large populations particularly in overgrazed areas; nearly throughout TX. Jul–Nov. [*Amphiachyris dracunculoides* (DC.) Nutt., *Xanthocephalum dracunculoides* (DC.) Shinners] This species can cause contact dermatitis in humans, eye inflamation in humans and livestock, and gastrointestinal upsets in cattle (Gates 1945; Ajilvsgi 1984). The plants were tied to sticks by early settlers and used as brooms (Ajilvsgi 1984). ☠

Gutierrezia microcephala (DC.) A. Gray, (small-headed), THREAD-LEAF SNAKEWEED. Much branched subshrub 0.3–1 m tall; leaves linear, 1–4 mm wide; pappus of disk flowers of several acute scales; pappus of ray flowers similar but shorter. Open sandy or limestone soils; included based on citation of vegetational area 5 (Fig. 2) by Hatch et al. (1990); mainly w 1/2 of TX, possibly e as far as w margin of nc TX (Lane 1985). Jul–Oct. [*Xanthocephalum microcephalum* (DC.) Shinners] This species causes poisoning and abortion in sheep and cattle apparently due to the presence of saponins; this has resulted in significant economic losses in w TX

(Kingsbury 1964); symptoms include loss of appetite, dropping of the head, and hematuria (Burlage 1968); in w TX this species is an indicator of overgrazed range conditions (Powell 1988). ☠

Gutierrezia sarothrae (Pursh) Britton & Rusby, (thought by Pursh to resemble *Hypericum sarothra* Michx.), BROOM SNAKEWEED, PERENNIAL BROOMWEED, MATCHWEED, MATCHBRUSH, KINDLING-WEED, YERBA DE VÍBORA, TURPENTINE-WEED, ESCOBA DE LA VÍBORA. Woody-based subshrub 0.2–1 m tall; leaves linear, entire; involucres 1–3 mm broad; ray flowers with pappus of linear, obtuse scales; disk flowers with pappus of acute scales about twice the length of the ray pappus scales. Open or disturbed sites, often calcareous soils; Brown, Callahan, Erath, Montague, and Young cos.; w part of nc TX s and w to w TX, also se TX. Jul–Nov. [*Xanthocephalum sarothrae* (Pursh) Shinners] This species is an indicator of overgrazing; a number of common names refer to the highly flammable nature of the dried stems which were used historically in starting fires (Ajilvsgi 1984). This species causes toxicity to livestock, as in *G. microcephala*, due to saponins or unknown toxins (Kingsbury 1964; James & Welsh 1992). It is reported to contain a protein with antitumor activity (Mabberley 1997). ☠

Gutierrezia sphaerocephala A. Gray, (spherical-headed), ROUND-HEAD BROOMWEED, ROUNDLEAF. Stems to 0.6 m tall, much branched above; leaves linear, 2–3 mm wide; involucres about 3–5 mm long; achenes densely white pubescent on ribs; pappus of disk flowers of several scales; pappus of ray flowers similar but shorter. Disturbed habitats; included by Mahler (1988) for nc TX; we can find no specimens or other citations for nc TX and according to Lane (1985) this species occurs only far to the w of nc TX; w 1/2 of TX. Jul–Nov. [*Xanthocephalum sphaerocephalum* (A. Gray) Shinners]

Guterrezia texana (DC.) Torr. & A. Gray, (of Texas), TEXAS BROOMWEED, KINDLING-WEED. Stems 0.1–0.8 m high; much branched above; leaves linear, 0.6–2(–5) mm wide; involucres ca. 2.5–4 mm long; ray flowers 10–15; pappus of disk flowers of acute scales. Disturbed habitats; se and e TX w to Rolling Plains and Edwards Plateau. Jul–Nov. [*Xanthocephalum texanum* (DC.) Shinners]

HEDYPNOIS

A genus of 2 species native from Macaronesia to Iran. (Greek name used by Pliny for a kind of wild endive) (tribe Latuceae)
REFERENCE: Sell 1976.

Hedypnois cretica (L.) Dum.Cours., (of Crete). Taprooted annual, ± hairy; stems spreading and ascending, 10–30 cm long, sparingly branched; leaves alternate; basal leaves pinnately lobed, usually with winged petioles; stem leaves few, reduced, sessile; heads solitary at the branch tips; involucres ca. 10 mm high; inner phyllaries scabrous-hispid, after flowering becoming conspicuously curved and boat-shaped (with a very rounded keel); outer phyllaries < 1/2 as long as inner; inner phyllaries strongly incurved in fruit; flowers perfect, mostly fertile; corollas all ligulate, yellow, with 5 purple teeth; achenes ca. 5–7.5 mm long, ± cylindrical, beakless, the outer achenes incurved; pappus of outer achenes a crown ca. 1 mm long, that of inner achenes of several bristle-like scales ca. 5 mm long. Sandy or rocky soils, roadsides; included based on citation of vegetational area 4 (Fig. 2) by Hatch et al. (1990), probably only to the se of nc TX. Mar–Apr. Native of Mediterranean region, Canary Islands. (Cichorieae)

HELENIUM SNEEZEWEED, BITTERWEED

Annual or perennial herbs, glabrous or pubescent; leaves alternate, entire to deeply and finely lobed, resin-dotted; heads terminal, solitary or corymbose; phyllaries few, herbaceous, loosely spreading or reflexed; heads usually radiate; ray flowers with ligules widened and 3-toothed apically, yellow to brown-red, often reflexed; disks ovoid or globose; disk corollas yellow to brown-red; pappus of 5–7 scales elongated into an awn-like tip.

An American genus of 40 species including some cultivated ornamentals. Most species are poisonous and unpalatable to grazing animals (Barkley 1986); some contain sesquiterpene lactones (e.g., helenalin) or a glycoside (Blackwell 1990; Hardin & Brownie 1993). (Greek: *helenion*, the name of another plant said by Linnaeus to be named after Helen of Troy, wife of King Menelaus of Sparta) (tribe Helenieae, sometimes lumped into Heliantheae)
REFERENCES: Rock 1957; Bierner 1972, 1989; Stanford & Turner 1988.

1. Leaves linear to lanceolate, elliptic, or oblong, usually > 2 mm wide, decurrent (= leaf tissue extending down the stem), the stems thus winged.
   2. Perennials with fibrous roots from subrhizomatous base; disk corollas completely yellow; ray flowers with ligules completely yellow ____ **H. autumnale**
   2. Annuals with taproot; disk corollas with lobes reddish brown or yellowish brown; ray flowers with ligules completely yellow or yellow with reddish brown basal blotch.
      3. Ray flowers with ligules 3–11.5 mm long, yellow, often with reddish brown basal blotch; upper stem leaves linear, reduced, entire; heads peduncled, well above the foliage ____ **H. elegans**
      3. Ray flowers with ligules 2–3(–5) mm long, completely yellow; upper stem leaves ovate to oblong, little reduced, often with at least some teeth; heads short-peduncled, barely above the foliage ____ **H. microcephalum**
1. Leaves or leaf segments linear to thread-like, usually < 2 mm wide, not decurrent, the stems thus unwinged ____ **H. amarum**

Helenium amarum (Raf.) H. Rock, (bitter). Annual; plant 10–85 cm tall; leaves all linear or filiform except for lowest which are sometimes pinnatifid (these sometimes withered before flowering time); ray flowers ca. 10, with ligules yellow, 5–10 mm long. Open woods, fields, pastures, and disturbed areas. The following two varieties have often been recognized as species; Stanford and Turner (1988) discussed their relationship.

1. Disk corollas yellow with lobes yellow to yellow-brown; lower and basal leaves typically withered at flowering time; basal leaves entire to toothed to occasionally pinnatifid ____ var. **amarum**
1. Disk corollas yellow with lobes reddish brown or purple; lower and basal leaves sometimes withered at flowering time or often present; basal leaves pinnatifid ____ var. **badium**

var. **amarum**, BITTERWEED. Se and e TX w to Rolling Plains and Edwards Plateau, mostly e of 100th meridian. May–Nov. The foliage is extremely bitter, largely avoided by livestock, and thus greatly increasing under conditions of overgrazing; conspicuously dominant on abused pastures. May–Nov. [*H. tenuifolium* Nutt.] All parts of the plant contain the glycoside dugaldin and cause toxic symptons in animals; cows that have eaten even small amounts of the plant produce extremely bitter, distasteful milk; honey made from the flowers is also reportedly bitter (Correll & Johnston 1970; Stephens 1980; Ajilvsgi 1984).

var. **badium** (A. Gray ex S. Watson) Waterfall, (reddish brown), BASIN SNEEZEWEED. Similar to variety *amarum* except for the characters in the key. Dallas and Grayson cos. w and sw to w TX. Apr–Jul, sporadically to Oct. [*H. badium* (A. Gray) Greene]

Helenium autumnale L., (of autumn), COMMON SNEEZEWEED, TALL SNEEZEWEED, STAGGERWORT, SWAMP-SUNFLOWER. Stems to 1(–2) m tall; leaves linear-elliptic; ray flowers ca. 20, with corollas to 15 mm long. Low, moist, calcareous soils; Tarrant County; also e TX, Rolling Plains, and Edwards Plateau. Aug–Oct. Poisonous to cattle and sheep; it is reported to cause "spewing sickness" with symptoms including weakness, depression, vomiting, excessive salivation, frothing, irregular pulse, and kidney and liver damage (Burlage 1968).

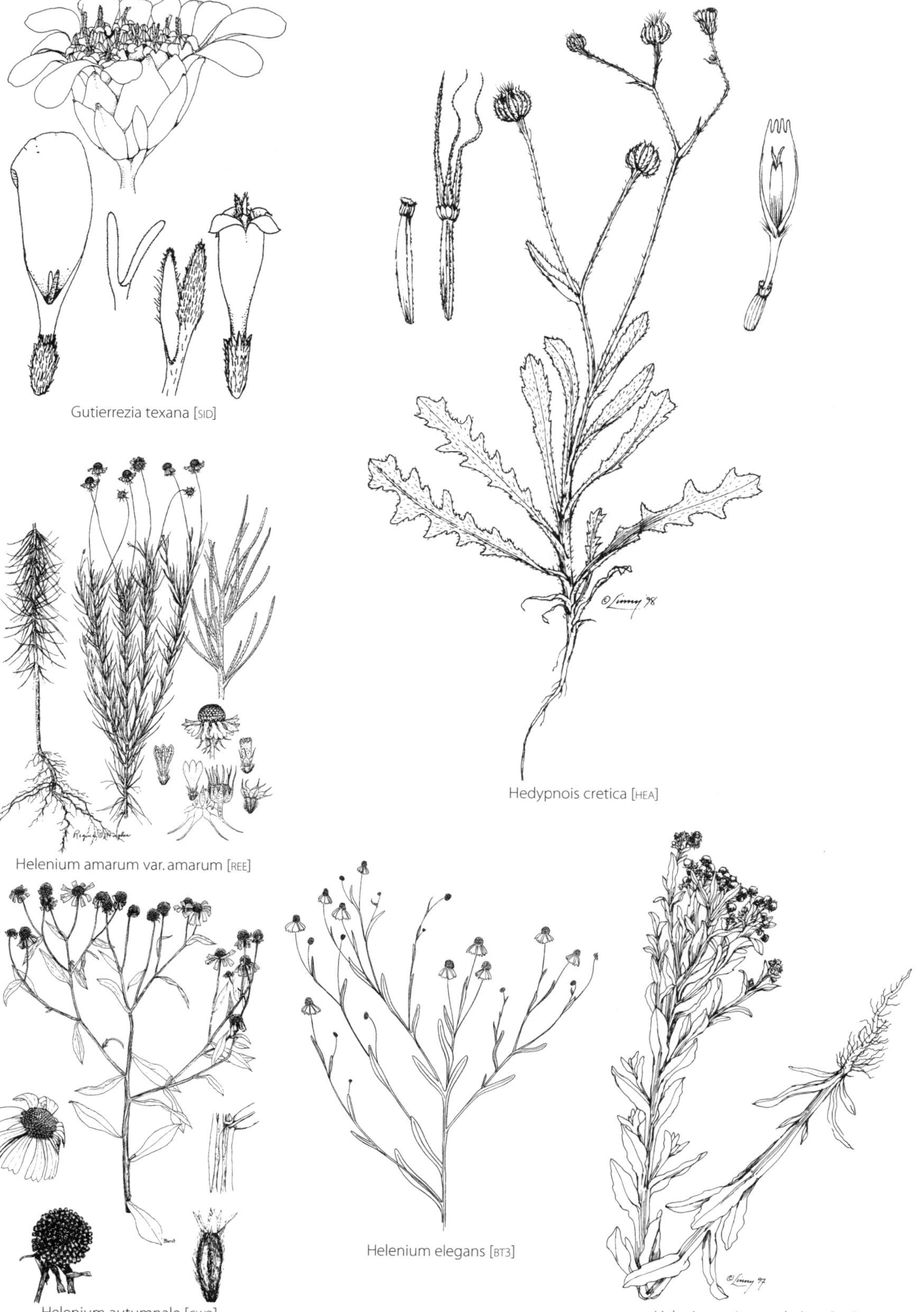

Gutierrezia texana [SID]

Hedypnois cretica [HEA]

Helenium amarum var. amarum [REE]

Helenium autumnale [GWO]

Helenium elegans [BT3]

Helenium microcephalum [HEA]

Helenium elegans DC., (elegant). Plant 20–120 cm tall; leaves narrowly elliptic to lanceolate or linear. Gravelly stream bottoms or seepy slopes on limestone; Dallas and Grayson cos. s and sw to Edwards Plateau. May–Jul.

Helenium microcephalum DC., (small-headed), SMALL-HEAD SNEEZEWEED, SMALL SNEEZEWEED, SNEEZEWEED. Plant 10–120 cm tall; leaves narrowly elliptic to narrowly oblong-elliptic. Sandy or silty ground, dried-up ponds, and fields; West Cross Timbers and Lampasas Cut Plain s and w to w TX. Jun–Jul.

HELIANTHUS SUNFLOWER

Annuals or perennials, herbaceous; leaves sessile or petioled, alternate, opposite, or opposite below and alternate above; heads sessile or peduncled; phyllaries in several series; ray flowers pistillate, infertile (not producing mature achenes), the corollas yellow; disk flowers perfect, fertile, the corollas reddish purple or yellow, producing mature achenes that are compressed laterally but not thin-edged; pappus of 2 awns, readily falling.

A North American genus of 50 species including the widely cultivated SUNFLOWER, *H. annuus*. Species of this genus are often confused with those of *Silphium*; however, in *Silphium* the ray flowers produce mature achenes (disk flowers do not), while in *Helianthus* the disk flowers produce mature achenes (ray flowers do not). (Greek: *helios*, the sun, and *anthos*, flower, presumably in reference to the turning of the inflorescences towards the sun) (tribe Heliantheae)
REFERENCES: Heiser 1951, 1954; Jackson 1963; Heiser et al. 1969; Rogers et al. 1982; Sims & Price 1985; Chandler et al. 1986; Rieseberg et al. 1991.

1. Foliage blue-green glaucous; leaf margins and phyllaries ciliate; perennial with long, slender rhizomes; leaves almost all opposite; plants usually 0.7 m or less tall ____ **H. ciliaris**
1. Foliage not blue-green glaucous; plants without the above combination.
    2. Leaf blades narrowed to a distinct petiole (sometimes as short as 5 mm, often much longer), entire to conspicuously toothed.
        3. Annuals with taproot.
            4. Larger leaves 10–30 cm wide; plants often 1.5–4 m tall; phyllaries conspicuously long ciliate marginally, 4–7(–10) mm wide, ± abruptly contracted near apex; peduncles with wide-spreading hairs ____ **H. annuus**
            4. Larger leaves 2–10 cm wide; plants 0.4–2 m tall; phyllaries scabrous, not marginally ciliate, 4 mm or less wide, ± gradually tapered to apex; peduncles with appressed or ascending hairs.
                5. Leaf blades usually entire, sometimes serrate, cuneate to truncate at base; pales (= bracts on receptacle) in center of head with tips densely white-bearded; w Blackland Prairie ____ **H. petiolaris**
                5. Leaf blades usually serrate, rarely entire, cuneate to truncate or often cordate at base; pales in center of head with tips hispid or slightly villous; extreme e margin of nc TX ____ **H. debilis**
        3. Perennials with erect crowns and/or tough creeping rootstocks or rhizomes.
            6. Stems glabrous, glaucous; leaf blades not or only slightly scabrous above ____ **H. grosseserratus**
            6. Stems pubescent; leaf blades usually conspicuously scabrous above.
                7. Phyllaries with tips ± tightly appressed and clearly imbricate in several series of very different lengths; disk corollas with reddish brown or reddish purple lobes ____ **H. pauciflorus**
                7. Phyllaries with the tips loose-spreading and all about the same length; disk corollas totally yellow.
                    8. Upper leaves alternate; petioles winged, 4–9 cm long; leaf blades 60–150 mm wide ____ **H. tuberosus**

8. Leaves mostly opposite; petioles unwinged, 0.5–2 cm long; leaf blades 7–70 mm wide ________ **H. hirsutus**

2. Leaf blades sessile, without a distinct petiole, entire or with small inconspicuous teeth.
   9. Leaf blades very narrow, linear to linear-lanceolate, 2–10(–12) mm wide; disk corollas with reddish brown or reddish purple lobes ________ **H. salicifolius**
   9. Leaf blades usually much broader, 20–60(–90) mm wide; disk corollas with yellow OR reddish brown or reddish purple lobes.
      10. Phyllaries with tips ± tightly appressed and clearly imbricate in several series of very different lengths; disk corollas with reddish brown or reddish purple lobes; leaves mostly opposite, not clasping ________ **H. pauciflorus**
      10. Phyllaries with the tips loose-spreading and all about the same length; disk corollas totally yellow; leaves either alternate or if opposite then cordate clasping.
         11. Leaves alternate, not clasping, the blades narrowly elliptic, some or all folded lengthwise; flowering mostly Sep–Oct ________ **H. maximiliani**
         11. Leaves mostly opposite, mostly cordate-clasping, the blades ovate to ovate-lanceolate, all flat; flowering Jun–Sep ________ **H. mollis**

Helianthus annuus L., (annual), COMMON SUNFLOWER, MIRASOL. Extremely variable in size, flowering when as little as 10 cm tall or as much as 4 m; leaf blades triangular-ovate to ovate-lanceolate, mostly alternate but lowermost opposite, truncate to cordate at base; ray corollas deep yellow, disk corollas brown-red (rarely dark yellow); achenes glabrous or pubescent. Stream bottoms, roadsides, disturbed areas; throughout TX. May–Oct. [*H. annuus* subsp. *lenticularis* (Douglas ex Lindl.) Cockerell, *H. annuus* subsp. *texanus* Heiser] The cultivated SUNFLOWER, widely grown for the seeds, for oil, and as an ornamental, is derived from *H. annuus*; it is one of the world's most important oil crops (Mabberley 1987; Rieseberg & Seiler 1990). Achenes of *H. annuus* were gathered as food by Native Americans of the w U.S. (Heiser 1951) and the species was being cultivated in parts of North America when Europeans arrived (Heiser 1993); it is one of the few agricultural plants to originate in North America north of Mexico (Nabhan 1979). ▣/91

Helianthus ciliaris DC., (ciliate, fringed with hairs), BLUE-WEED SUNFLOWER, BLUEWEED, TEXAS BLUEWEED. Perennial from slender rhizomes; foliage blue-green glaucous; stems 0.3–0.7 m tall, glabrous or nearly so; leaves mostly opposite, sessile or short petiolate; leaf blades narrowly to broadly lanceolate, usually 0.5–2(+) cm wide; ray flowers 10–18, the ligules ca. 10 mm long or sometimes absent; disk flowers reddish or with reddish lobes; achenes ca. 3 mm long. Sandy or sandy clay soils; Denton, Tarrant, and Young cos., also Brown Co. (HPC); w 2/3 of TX, mainly w of nc TX. Jun–Sep.

Helianthus debilis Nutt. subsp. **cucumerifolius** (Torr. & A. Gray) Heiser, (sp.: weak, frail; subsp.: with leaves like *Cucumis*—cucumber), CUCUMBER-LEAF SUNFLOWER. Stems usually 0.55–0.65 m tall; leaves ± all alternate; leaf blades deltoid-ovate to deltoid-lanceolate, cuneate to truncate or often cordate at base, 3–8 cm wide, usually regularly serrate, rarely entire; petioles ca. as long as blades; ray flowers with ligules ca. 20 mm long; disk corollas with deep red-purple lobes. Sandy soils; included based on citation of vegetational area 4 (Fig. 2) by Hatch et al. (1990); mainly se TX, but at least as far n as Travis Co. (Correll & Johnston 1970) just s of nc TX. Late summer-fall. [*H. cucumerifolius* Torr. & A. Gray]

Helianthus grosseserratus M. Martens, (large-toothed), SAW-TOOTH SUNFLOWER. Stems 1–3(–5) m tall, glabrous, glaucous; leaves mostly alternate; leaf blades narrowly ovate, prominently serrate; petioles about 8 cm long mid-stem; ray flowers with ligules ca. 25–40 mm long, yellow; disk corollas yellow; achenes glabrous, about 5 mm long. Wooded stream bottoms; Cooke, Dallas, Fannin, and Lamar cos., also Collin Co. (R. O'Kennon, pers. obs.); in TX mainly nc part of state, also rare in Edwards Plateau (Gillespie Co.) and sc TX (Gonzales Co.). Oct.

Helianthus hirsutus Raf., (hairy), HAIRY SUNFLOWER, STIFF-HAIR SUNFLOWER. Stems 0.5–1.7 m tall; leaves opposite; leaf blades lanceolate to narrowly ovate or ovate, serrate to rarely entire; petioles 5–20 mm long; ray and disk corollas completely yellow; ray flowers with ligules ca. 15–25 mm long; achenes glabrous. Sandy open woods, forest margins; se and e TX w to West Cross Timbers and Edwards Plateau. Jun–Aug.

Helianthus maximiliani Schrad., (for its discoverer, Prince Maximilian von Wied-Neuwied, 1782–1867, German botanist and traveler in Brazil and North America), MAXIMILIAN SUNFLOWER, MICHAELMAS DAISY. Perennial, sometimes forming extensive colonies; stems 0.3–3 m tall; leaves alternate, often conspicuously folded and sickle-shaped; heads few to numerous in a spike-like or raceme-like arrangement; ray flowers with ligules ca. (15–)25–35(–40) mm long, yellow; disk corollas yellow; achenes 3–4 mm long. Low moist areas, roadsides, prairies; one of the most abundant sunflowers in nc TX; widespread in TX. Mostly Sep–Oct. ▣/91

Helianthus mollis Lam., (soft hairy), ASHY SUNFLOWER, HAIRY SUNFLOWER, DOWNY SUNFLOWER. Perennial; stems 0.5–1 m tall, hirsute to villous; leaves mostly opposite, ovate-lanceolate to broadly ovate, sessile, usually cordate-clasping, grayish green in appearance; heads solitary or in spike-like racemes; ray and disk corollas completely yellow; ray flowers with ligules to ca. 30 mm long; achenes ca. 4 mm long. Sandy, wooded or open areas; Cooke, Grayson, and Tarrant cos; se and e TX w to nc TX. Jun–Sep.

Helianthus pauciflorus Nutt., (few-flowered), STIFF SUNFLOWER. Perennial; stems 0.8–2 m tall; leaves nearly all opposite; leaf blades oblong-lanceolate to lance-ovate, acuminate, subsessile or with winged petiolar base; ray flowers with ligules yellow, sometimes with reddish tinge on back near apex, ca. 20–35 mm long. Prairies or open woods, often on sand; mainly se and e TX disjunct w to Cooke, Montague, and Wise cos., also a report from Hunt Co. (Clymer Meadow—J. Eidson, pers. comm.). Aug–Oct. [*H. laetiflorus* Pers. var. *rigidus* (Cass.) Fernald; *H. rigidus* (Cass.) Desf.]

Helianthus petiolaris Nutt., (with a petiole or leaf stalk), PLAINS SUNFLOWER, PRAIRIE SUNFLOWER. Stems 0.5–2 m tall; leaves alternate; leaf blades triangular-ovate to narrowly ovate; petioles 1–15 cm long, nearly as long as the blades; heads paniculate, pedunculate; ray flowers with ligules ca. 20 mm long; disk corollas usually with red-purple lobes; achenes 3.5–4.5 mm long, villous. Sandy or rocky prairies; Dallas and McLennan cos. s and w to w TX. May–Oct. Known to hybridize with *H. annuus*.

Helianthus salicifolius A. Dietr., (with leaves like *Salix*—willow), WILLOW-LEAF SUNFLOWER. Perennial; stems 0.4–2 m tall, glabrous; leaves alternate, crowded, usually conspicuously linear to linear-lanceolate; disk corollas with red-purple or reddish brown lobes; ray flowers with ligules ca. 10–20(–30) mm long; achenes 4–6 mm long, glabrous. Calcareous soils, prairies; Cooke, Dallas, Grayson, Hill, and Montague cos.; nc TX w to e Rolling Plains and s to Edwards Plateau. Jun–Oct.

Helianthus tuberosus L., (tuberous), JERUSALEM-ARTICHOKE, GIRASOLE. Rhizomes slender with tuberous thickenings at ends; stems 1–3 m tall; leaves opposite below, alternate above; leaf blades ovate, 10–25 cm long; petioles winged, 4–9 cm long; ray and disk corollas yellow; ray flowers with ligules ca. 25–40 mm long; achenes 5–7 mm long, glabrous. Wooded areas; Dallas and Grayson cos.; in TX only in the nc part of the state. Aug–Oct. Cultivated for the edible tubers which are sweet (partly due to the presence of fructose and thus reported to be a useful carbohydrate source for diabetics since fructose is sweeter on a molar basis than glucose); used as a food by Native Americans (Mabberley 1997).

Helianthus simulans E. Watson, (resembling), a rhizomatous perennial with rhizomes 5–10 cm long, to 1 cm thick, alternate (± opposite in small individuals), ± sessile, non-folded leaves, and yellow or purple-tipped disk corollas, has been cited by Hatch et al. (1990) for vegetational area

Helianthus ciliaris [CO1]

Helianthus annuus [BT3]

Helianthus debilis subsp. cucumerifolius [MTB]

Helianthus grosseserratus [GLE, MTB]

Helianthus hirsutus [BB2]

Helianthus maximiliani [STE]

4 (Fig. 2) but apparently occurs only to the s and e of nc TX (Heiser et al. 1969). It is possibly of hybrid origin (Heiser et al. 1969).

HETEROTHECA GOLDEN-ASTER, GOLD-ASTER, CAMPHORWEED

Pubescent or pilose annual or perennial herbs, often aromatic; leaves alternate; heads solitary or corymbose; phyllaries in several rows of unequal lengths, slender; ray flowers with ligules yellow, curling under at night or in age; disk corollas yellow; pappus of long, scabrous hairs, and an outer row of inconspicuous, short, narrow scales, or pappus absent from ray flowers.

A s North American genus of 25 species. *Chrysopsis* is sometimes lumped into *Heterotheca* (Mahler 1988). (Greek: *heteros*, different, and *theca*, case, alluding to the dissimilar ray and disk achenes) (tribe Astereae)

REFERENCES: Shinners 1951b; Wagenknecht 1960; Harms 1965, 1968; Vernon 1965; Semple 1977, 1996; Semple et al. 1980; Nesom 1991b.

1. Blades of upper and middle stem leaves little or not narrowed at base, sessile and slightly to strongly clasping, ovate; plants annual or biennial from taproot; commonly with a single stem from base; pappus of ray flowers absent ____ **H. subaxillaris**
1. Blades of upper and middle stem leaves tapered to slender, petiole-like bases, not clasping, linear to oblanceolate; plants perennial from tough woody base, often rhizomatous; stems usually clustered; pappus of ray flowers present.
   2. Leaf blades conspicuously gray in appearance, softly and densely pubescent, the hairs usually without swollen bases; leaf blades of flowering branches oblanceolate, 3–5 times as long as wide; foliage lacking resin glands or nearly so; widespread in w 1/2 of nc TX ____ **H. canescens**
   2. Leaf blades distinctly green, bristly pubescent, the hairs usually with swollen bases (use lens); leaf blades of flowering branches usually ± linear, sometimes to linear-oblanceolate, mostly 8–15 times as long as wide; foliage with abundant resin glands (use lens); on extreme w margin of nc TX ____ **H. stenophylla**

Heterotheca canescens (DC.) Shinners, (grayish white), GRAY GOLD-ASTER. Perennial 15–60 cm tall, older plants many-stemmed; leaves rather numerous and crowded, entire; ray flowers with ligules 6–8 mm long. Sandy or gravelly prairies, rock outcrops; Panhandle to Trans-Pecos, e to West Cross Timbers and Lampasas Cut Plain; also Tarrant Co. (Semple 1996) and collected at Dallas by Reverchon. Jun–Oct. [*Chrysopsis canescens* (DC.) Torr. & A. Gray]

Heterotheca stenophylla (A. Gray) Shinners, (narrow-leaved), NARROW-LEAF GOLD-ASTER. Perennial 10–40 cm tall; ray flowers with ligules ca. 10 mm long. Sandy soils; Archer and Clay cos., also Burnet, Eastland (Semple 1996) and Erath (Harms 1968) cos.; w TX e to w boundary of nc TX. Jun–Oct. [*Chrysopsis villosa* (Pursh) Nutt. ex DC. var. *stenophylla* (A. Gray) A. Gray]

Heterotheca subaxillaris (Lam.) Britton & Rusby, (subaxillary), CAMPHORWEED, GOLDEN-ASTER, CAMPHOR DAISY. Pilose and glandular-pubescent aromatic annual or biennial; stems 20–200 cm tall. Spring form: dwarf, with long-petioled basal leaves, the blades coarsely toothed; stem leaves short-petioled or sessile, the blades lanceolate, toothed, or entire. Summer form: taller, to 2 m by fall; lower leaves petioled, the petioles with enlarged, winged, clasping base; upper leaves similar or varying to sessile, the blades auricled-clasping, oblong or ovate, slightly or coarsely toothed; heads rather numerous; ray flowers with ligules ca. 5 mm long. Sandy or rocky ground, open woods, fields, roadsides; throughout TX. Apr–Oct. [*H. latifolia* Buckley, *H. subaxillaris* var. *latifolia* (Buckley) Gandhi & R.D. Thomas] The foliage has a strong camphor-like odor and the plants are avoided by livestock (Ajilvsgi 1984).

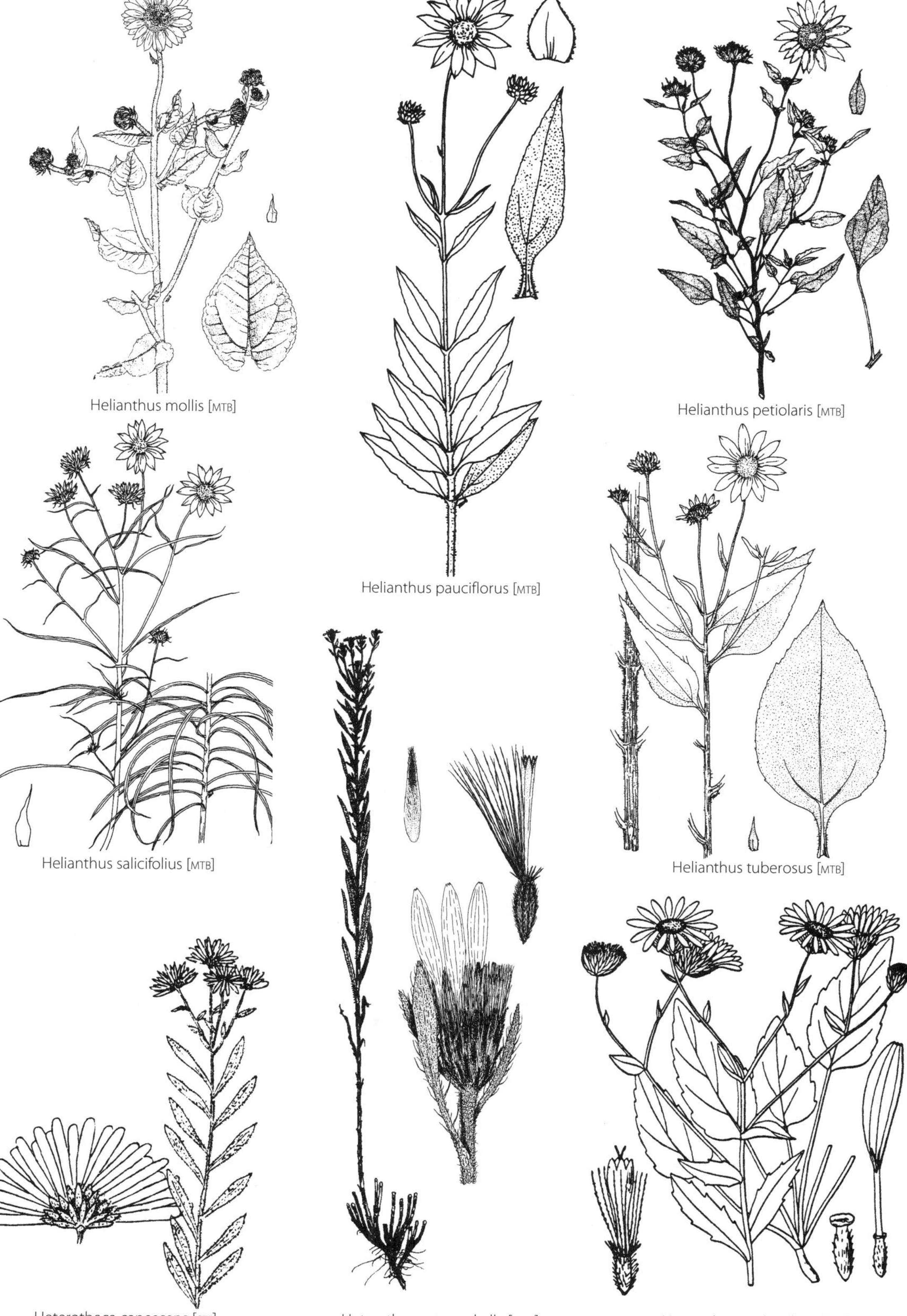

Helianthus mollis [MTB]

Helianthus pauciflorus [MTB]

Helianthus petiolaris [MTB]

Helianthus salicifolius [MTB]

Helianthus tuberosus [MTB]

Heterotheca canescens [STE]

Heterotheca stenophylla [UWA]

Heterotheca subaxillaris [BB2]

HIERACIUM HAWKWEED

A taxonomically complicated temperate and tropical montane genus of ca. 90 species (ca. 1,000 apomictic microspecies); the genus is related to *Crepis*. (Greek: *herax*, a hawk; the ancients, as recorded by Pliny and others, supposing that hawks ate these or similar plants to improve their eyesight) (tribe Lactuceae)
REFERENCE: Vuilleumier 1973.

Hieracium gronovii L., (for Jan Friedriech Gronovius 1690–1762, Dutch botanist at Leiden), GRONOVIUS' HAWKWEED. Perennial herb; sap milky; stems 0.4–1.5 m tall, naked for a considerable distance below the heads; leaves mostly in lower part of stem; leaf blades oblanceolate or broadly elliptic, pilose with long brownish hairs (strikingly pilose in the field); inflorescences leafless, corymbose or narrowly paniclulate, ± glandular-pubescent; heads small, 7–10 mm long; flowers all ligulate; corollas yellow; achenes 2–4 mm long; pappus of bristles. Sandy woods; Denton, Grayson, and Lamar cos.; se and e TX w to East Cross Timbers. May–Jun(–fall).

HYMENOPAPPUS WOOLLY-WHITE

Ours biennial herbs; leaves basal (rosette) and cauline, alternate, the blades usually pinnatifid, glandular-punctate, often conspicuously white hairy on lower surface; heads in corymb-like arrangement, discoid; phyllaries in 2 or 3 series, petaloid apically, herbaceous basally; receptacles naked; ray flowers (in our species) absent; disk flowers perfect, fertile; corollas white or with reddish tinge (yellow in a species to the w of nc TX), the lobes becoming reflexed; anther column long-exserted; achenes obconic, 4-angled, pubescent; pappus of 14–18 ovate to oblong hyaline scales to 1.5(–2) mm long.

A s North American genus of 14 species; some were used medicinally by Native Americans. (Greek: *hymen*, membrane, and *pappus*, down, fuzz, pappus) (tribe Helenieae, sometimes lumped into Heliantheae; *Hymenopappus* can also be keyed using the Anthemidae key)
REFERENCES: Turner 1956; Rieseberg 1991.

1. Basal leaves with ultimate segments to 1.5 mm wide, elongate linear to filiform; corollas 2–3 mm long (to tips of corolla lobes), the corolla tubes 1.5–2(–2.2) mm long ____ **H. tenuifolius**
1. Basal leaves with ultimate segments 2–20(–30) mm wide, variously shaped but wider than linear to filiform; corollas 3–5 mm long (to tips of corolla lobes), the corolla tubes 2–3 mm long.
   2. Flowers white or creamy white; basal leaves finely once-pinnately to bipinnately parted, the ultimate segments linear, 2–8 mm wide; usually on clay or limestone soils ____ **H. scabiosaeus**
   2. Flowers reddish-tinged (rarely completely white); basal leaves simple to coarsely once-pinnately parted (rarely bipinnately parted), the ultimate segments broad, 6–30 mm wide; usually on sandy soils ____ **H. artemisiifolius**

Hymenopappus artemisiifolius DC., (with leaves like *Artemisia*—sagebrush or wormwood), RAGWEED WOOLLY-WHITE, WOOLLY-WHITE, WILD CAULIFLOWER. Plant 0.4–1.3 m tall; corollas reddish-tinged (rarely completely white), 3.5–5 mm long; phyllaries 6–12 mm long, 3–7 mm wide, white or tinged with red. Open woods, fields, disturbed areas; Denton, Henderson, Hill, Milam, Tarrant, and Wise cos.; scattered in nc TX; mainly on sandy soils of se and e TX. Apr–Jun. [*H. scabiosaeus* L'Her. var. *artemisiifolius* (DC.) Gandhi & R.D. Thomas]

Hymenopappus scabiosaeus L'Her. var. **corymbosus** (Torr. & A. Gray) B.L. Turner, (sp.: from Latin: *scabies*, itch; var.: with flowers in corymbs), OLD-PLAINSMAN. Plant 0.4–1 m tall; phyllaries 5–9 mm long, 2–4 mm wide, with white or yellowish white tips 2-4 mm long; corollas white or creamy white, 3–4 mm long. Prairies, open woods, roadsides; widespread in TX except for e TX and Panhandle. Apr–May.

Hymenopappus tenuifolius Pursh, (slender-leaved), CHALKHILL WOOLLY-WHITE, WOOLLY-WHITE, OLD-PLAINSMAN. Plant 0.4–1.5 m tall; phyllaries 5–8 mm long, 2–4 mm wide, yellowish for 1–2 mm from tip; corollas white. Gravelly prairies, limestone outcrops or sandy soils; Bell and Dallas cos. w to Panhandle and s to sc TX. May–Jul.

Hymenopappus flavescens A. Gray, (yellowish), with bright yellow corollas, occurs from w TX e as far as Taylor Co., just w of nc TX.

HYMENOXYS BITTERWEED

A genus of 17–28 species depending on circumscription; they are native from s North America to Argentina; sometimes treated broadly to include *Tetraneuris*. A molecular study by Bierner and Jansen (1998) supported the recognition of *Tetraneuris* as a separate genus. ☠ Some contain sesquiterpene lactones toxic to sheep and goats; other species were formerly used as chewing gum by Native Americans. (Greek: *hymeno*, membrane, and *oxys*, sharp, alluding to the pointed pappus scales) (tribe Helenieae, sometimes lumped into Heliantheae)
REFERENCES: Strother 1966; Bierner & Jansen 1998.

Hymenoxys odorata DC., (odorous), POISON BITTERWEED, WESTERN BITTERWEED, BITTERWEED, BITTER RUBBERWEED. Inconspicuously pubescent, taprooted annual, 7–60 cm tall, becoming bushy-branched when not crowded, forming rounded mounds; foliage aromatic when crushed; leaves alternate, pinnately divided into 3–15 linear-filiform segments, minutely resin-dotted; heads terminal, solitary or corymbose; ray flowers 6–13, the ligules deep yellow, widened and toothed apically, ca. 5–11 mm long; disk corollas orange; pappus of 5(–6) acuminate or awn-tipped scales 1–2.3 mm long. Disturbed sites, becoming abundant in overgrazed areas; Archer Co., also Brown Co. (Stanford 1971); w margin of nc TX s and w to w TX. Apr–Jun. Poisonous to livestock, especially to sheep under starvation conditions; the sesquiterpenes cause gastrointestinal irritation and result in a wasting condition even after short periods of grazing; POISON BITTERWEED causes considerable loss of livestock on ranges in the sw U.S. including the Edwards Plateau and has been one of the main causes of the decline of sheep ranching in the sw U.S. (Kingsbury 1964; James & Welsh 1992). ☠

HYPOCHAERIS CAT'S-EAR

Annual or perennial herbs with leaves basal or flowering stems with a few more or less well-developed leaves, at least toward the base; leaf blades oblanceolate, toothed or pinnatifid; flowers all ligulate; corollas yellow or in one species white; pappus, at least in part, of plumose bristles; achenes all alike and beaked or of two kinds, the outermost beakless and the inner beaked.

A genus of ca. 60 species native to Europe, Asia, n Africa, and especially South America; including some weeds and cultivated ornamentals. The generic name has sometimes been spelled *Hypochoeris* (e.g., Shinners 1966b; Tomb 1974; DeFilipps 1976), probably based on the spelling by Linnaeus (1754) in *Genera Plantarum*. However, the accepted spelling is *Hypochaeris* (Greuter et al. 1993), based on Linnaeus' (1753) original spelling in *Species Plantarum*. While not presently known from nc TX, 3 other species naturalized in the e and se parts of TX are included below. They have been treated to alert collectors, and because of the possibility that they will become part of the nc TX flora; their scientific names are in italics rather than bold to indicate they are not currently known from nc TX. The following key is modified from Shinners (1966b) and Cronquist (1980). (According to Vuilleumier (1973), Greek: *hypochoiris*, used by Dioscorides for some plant, and also mentioned by Pliny as *hypochoeris*, perhaps derived from *hypo*, for [or beneath], and *choiros*, pig, the animals being fond of its roots) (tribe Lactuceae)

REFERENCES: Shinners 1966b; Vuilleumier 1973; Tomb 1974; Cronquist 1980; Diggs et al. 1997.

1. Pappus bristles of 2 types: an outer series of short, merely barbellate bristles and an inner series of much longer, plumose bristles; flowering stems leafless or nearly so (at most with reduced leafy bracts abruptly much smaller than the numerous basal rosette leaves).
   2. Plants mostly annual, essentially glabrous; heads not very showy, the ligules about equaling the involucre and only about twice as long as broad; outermost achenes usually beakless ________ **H. glabra**
   2. Plants perennial, with evidently hispid leaves; heads showy, the ligules evidently surpassing the involucre and about 4 times as long as wide; outermost achenes usually with well-developed beak (like the inner achenes) ________ *H. radicata*
1. Pappus bristles all alike, of long plumose bristles; flowering stems with a few more or less well-developed leaves, at least toward the base.
   3. Flowers yellow; middle and outer involucral bracts hispid; heads relatively broad, the involucres campanulate, nearly as wide as high or wider, mostly 5–8 mm wide at the middle at flowering time ________ *H. brasiliensis*
   3. Flowers white; involucral bracts glabrous or inconspicuously tomentose-puberulent; heads narrow, the involucres cylindric, ca. half as wide as high or less, mostly 2–4 mm wide at the middle at flowering time ________ *H. microcephala*

Hypochaeris glabra L., (smooth, hair-less). Annual or winter annual, ± glabrous; stems 10–40 cm tall, leafless or minutely bracteate; leaves basal; leaf blades oblanceolate, toothed or pinnatifid; heads 1–several; involucres ca. 8–10 mm high at flowering, to 17 mm in fruit; flowers all ligulate; corollas yellow, the ligules ± equal to involucre in length; body of achenes 4–5 mm long, the inner ones with well-developed beaks, the outer beakless; pappus bristles of 2 types, the inner plumose, the outermost shorter and merely barbellate. Frequently mowed hay field; Williamson Co. (*T.J. Watson 1620*, 1993, BRIT, LL/TEX) is the only collection seen from Texas (Diggs et al. 1997). Feb–? Native of Europe.

Hypochaeris brasiliensis (Less.) Griseb. var. *tweedyi* (Hook. & Arn.) Baker, (sp.: of Brazil; var.: for John Tweedie, 1775–1862, Scottish gardener at the Royal Botanic Garden, Edinburgh and collector in South America), a yellow-flowered South American species related to *H. microcephala*, was first reported for TX by Tomb (1974). It is known from vegetational area 1 (Hatch et al. 1990). [*H. tweedyi* Hook. & Arn.]

Hypochaeris microcephala (Sch.Bip.) Cabrera var. *albiflora* (Kuntze) Cabrera, (small-headed; var.: white-flowered), a white-flowered South American taxon first reported for North America (southeastern TX) by Shinners (1966b) is known from both e and se TX (Hatch et al. 1990). At present this weedy species is rapidly spreading in se TX (G. Nesom, pers. comm., 1998) and is likely to soon become a member of the nc TX flora.

Hypochaeris radicata L., (having conspicuous roots), a yellow-flowered perennial species similar to *H. glabra*, is also known from e TX. This European species was apparently first collected in the state in 1970 (Tomb 1974).

IVA SUMPWEED, MARSH-ELDER

Ours annual herbs; leaves opposite or sometimes alternate above, entire to toothed, resinous-glandular, petiolate; inflorescences spike-like or raceme-like in paniculate arrangements; heads small, inconspicuous (plants wind pollinated), sessile or subsessile in the axils of bracts, the peripheral flowers pistillate, the central flowers staminate; phyllaries distinct or united; staminate corollas funnelform; pistillate corollas tubular; pappus none.

A genus of 15 species native from North America to the West Indies; pollen from some spe-

Hieracium gronovii [BB2]

Hymenopappus artemisiifolius [HEA]

Hymenopappus scabiosaeus var. corymbosus [BB2]

Hymenopappus tenuifolius [RYD]

Hymenoxys odorata [REE]

Hypochaeris glabra [BEN]

cies causes hay fever. (Latin name from the mint *Ajuga iva* Schreb., with similar odor) (tribe Heliantheae)
REFERENCES: Rydberg 1922; Jackson 1960; Bolick 1985.

1. Phyllaries united into a cup-like involucre with toothed margin; leaves linear to lanceolate, 2–5 cm long, 2–12 mm wide **I. angustifolia**
1. Phyllaries 3 or 4, distinct; leaves ovate to lanceolate, 3–15 cm long, ca. 20–70 mm wide **I. annua**

Iva angustifolia Nutt. ex DC., (narrow-leaved), NARROW-LEAF SUMPWEED, MARSH-ELDER. Stems 0.5–1.4 m tall; leaf blades entire to serrate, the larger ones conspicuously 3-nerved; heads subsessile in axils of usually linear or filiform bracts; involucres 2–3 mm broad; corollas 1.5–2 mm long; mature achenes black, 2–2.8 mm long. Low moist areas, prairies, and post oak woodlands; widespread in TX. Sep–Nov.

Iva annua L., (annual), MARSH-ELDER, SEA-COAST SUMPWEED, SHARP-BRACT SUMPWEED, PELOCOTE. Stems 0.4–2 m tall; leaf blades serrate, 3-nerved, reduced upward to bracts, aromatic; heads sessile in axils of linear to ovate bracts; involucres 4–5 mm broad; corollas 1.5–2.5 mm long; mature achenes brown, 2–4.5 mm long. Low moist areas, calcareous or sandy soils, sometimes forming extensive stands; widespread, but mainly e 1/2 of TX. Sep–Nov. The wind-borne pollen is a cause of hay fever (Jackson 1960); the achenes were formerly eaten by Native Americans who domesticated this species (Heiser 1990b; Mabberley 1997).

KRIGIA DWARF-DANDELION

Small, ± glabrous annual or perennial herbs; sap milky; leaves basal or alternate; leaf blades oblanceolate or oblong-lanceolate to linear, entire or unevenly toothed or lobed; heads generally small, usually solitary and terminal or few together; phyllaries nearly equal; flowers all ligulate; corollas yellow or yellow-orange, open during morning in sunny weather; pappus none or single or double, of scales or an inner ring of bristles and an outer ring of scales.

A North American genus of 7 species. The 5 nc TX species all differ in chromosome number (Kim & Turner 1992). (Named for David Krig or Krieg, died 1713, a German or Hungarian physician, who was among the first to collect plants in Maryland) (tribe Lactuceae)
REFERENCES: Shinners 1947b; Vuilleumier 1973; Kim & Mabry 1991; Kim & Turner 1992; Kim et al. 1992a, 1992b.

1. Phyllaries 4–8 times as long as wide, linear-lanceolate to oblong-lanceolate, numerous (8–16), becoming shrivelled or reflexed in age, never keeled; inner pappus of bristles 4–8 mm long, outer of as many or fewer short scales less than 1/4 as long; stems leafless.
   2. Plants perennials with a small tuber a few cm below ground, developing slender, whitish rhizomes; involucres 10–15 mm high; corollas 12–17 mm long; pappus with numerous bristles and numerous short scales **K. dandelion**
   2. Plants annuals with fibrous roots; involucres 4–6.5 mm high (to 9 mm in age); corollas 4–7 mm long; pappus with 5 bristles and 5 scales **K. virginica**
1. Phyllaries 1.5–3 times as long as wide, lanceolate to ovate, few (4–10), remaining erect and expanded in age, with midvein which in age becomes a prominent keel; pappus absent OR of short scales or bristles up to 2 mm long; stems leafless or leafy.
   3. Stems leafless, unbranched (but stems buried by over-washed soil or robust plants may develop short, basal internodes); pappus of 5 well-developed short scales or both 5 well-developed short scales and short bristles **K. occidentalis**
   3. Stems leafy, branched (but beginning to flower with leaves crowded at base and branches scarcely developed); pappus absent OR of much reduced scales and bristles.
      4. Pappus of much reduced scales and bristles (use hand lens or scope) **K. wrightii**
      4. Pappus absent **K. cespitosa**

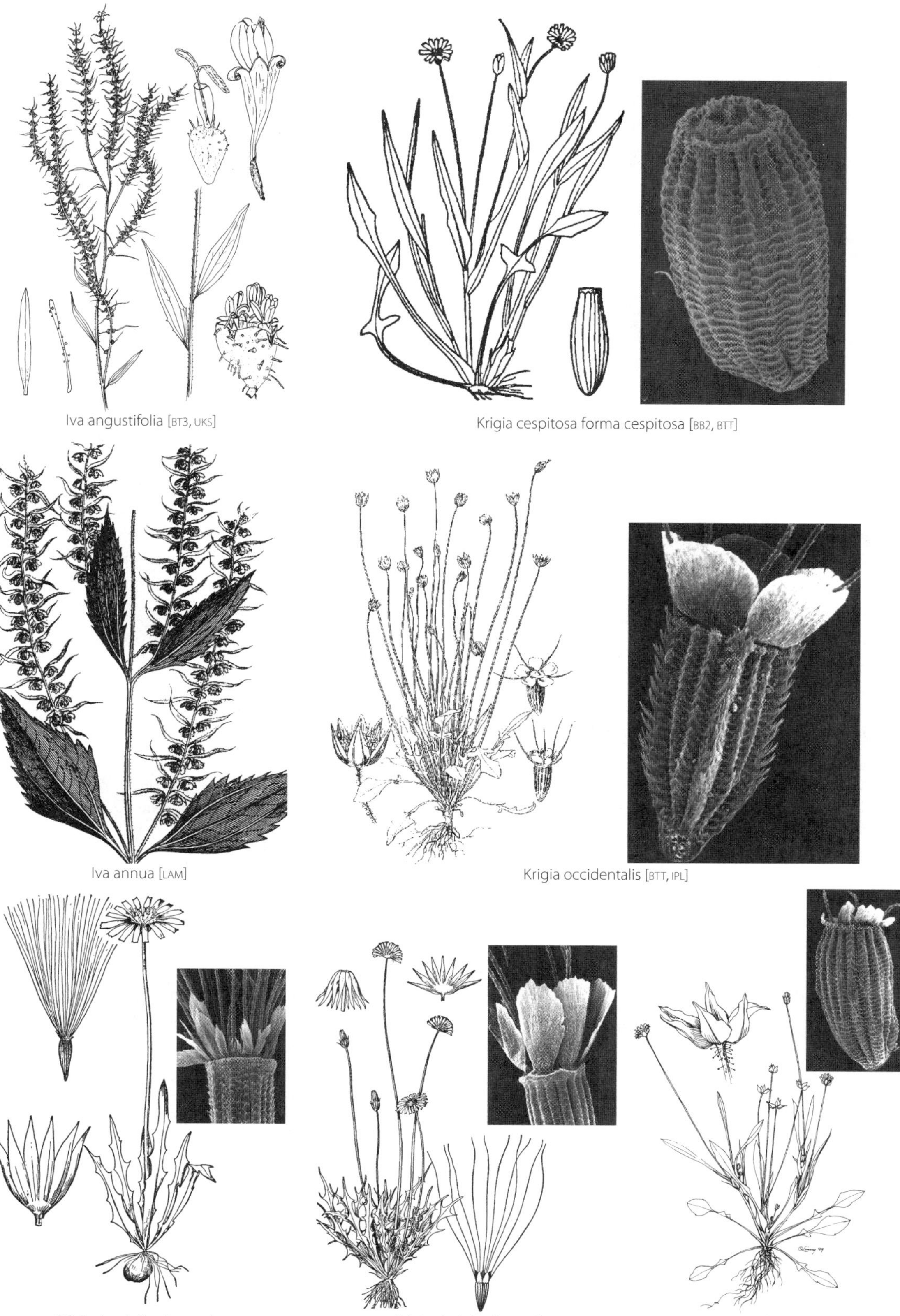

Iva angustifolia [BT3, UKS]

Krigia cespitosa forma cespitosa [BB2, BTT]

Iva annua [LAM]

Krigia occidentalis [BTT, IPL]

Krigia dandelion [BTT, GLE]

Krigia virginica [BTT, GLE]

Krigia wrightii [BTT, HEA]

Krigia cespitosa (Raf.) K.L. Chambers, (growing in tufts). Variable annual; stems leafy, branched. Shinners (1947b) treated the following 2 forms as species. Kim and Mabry (1991), Kim and Turner (1992), and Kim et al. (1992b) have observed intermediate populations, have concluded that the characters Shinners used to separate them are often correlated with environmental differences, and that molecular data could not separate the 2 taxa; *n* = 4.

1. Corollas 2–4 mm long; involucres 3–4.3 mm high in flower, 3–5.3 mm high in age; heads 3–6 mm in diam. ____ forma **cespitosa**
1. Corollas 5–10 mm long; involucres 5.3–8.3 mm high in flower, to 8.5 mm high in age; heads 6–9 mm in diam. ____ forma **gracilis**

forma **cespitosa**, WEEDY DWARF-DANDELION. Plant 35–40 cm tall; phyllaries 5–8. Stream banks, damp woods, roadsides, and disturbed areas, especially on sandy soils; se and e TX w to East Cross Timbers and Edwards Plateau. Mar–May. [*K. oppositifolia* Raf., *Serenia cespitosa* (Raf.) Kuntze]

forma **gracilis** (DC.) K.J. Kim, (graceful). Plant 8–40 cm tall; phyllaries 8–10. Low spots in prairies, or borders of woods, disturbed areas, chiefly calcareous clay, less often sandy soils; Post Oak Savannah and se TX w to West Cross Timbers and Edwards Plateau. Apr–May. [*K. gracilis* (DC.) Shinners]

Krigia dandelion (L.) Nutt., (from its resemblance to *Taraxacum*—dandelion), TUBER DWARF-DANDELION, POTATO-DANDELION. Scapose tuberous perennial 10–45 cm tall; heads large and showy; phyllaries ca. 15; pappus of numerous bristles 5–8 mm long and 5 small scales; *n* = 30. Sandy open woods, roadsides, and lawns; Grayson, Henderson, Hopkins, and Hunt cos.; also Lamar and Parker cos. (Shinners 1947b); se and e TX w to nc TX. Apr–May.

Krigia occidentalis Nutt., (western), WESTERN DWARF-DANDELION. Scapose annual 4–16 cm tall; phyllaries 4–7; corollas 3.5–3.8 mm long; pappus usually of 5 bristles 0.8–2 mm long and 5 small scales; *n* = 6. Sandy open woods, fields, and roadsides; Falls, Hopkins, Hunt, Kaufman, Milam, and Navarro cos.; also Bell, Eastland, and Tarrant cos. (Shinners 1947b); se and e TX w to West Cross Timbers and Edwards Plateau. Mar–May.

Krigia virginica (L.) Willd., (of Virginia), CAROLINA DWARF-DANDELION. Scapose annual 4–35 cm tall; phyllaries ca. 10; pappus of 5 bristles 4–7 mm long and 5 small scales; *n* = 5 or 10. Sandy open woods, fields, and roadsides; Burnet, Denton, Henderson, and Tarrant cos., also Comanche Co.; se and e TX w to Grand Prairie and Edwards Plateau. Mar–Apr, rarely repeating as late as Oct.

Krigia wrightii (A. Gray) K.L. Chambers ex K.J. Kim, (for Charles Wright, 1811–1885, TX collector). Annual extremely similar to the dwarf form of *K. cespitosa* and the taller individuals of *K. occidentalis*; the chromosome number of *n* = 9 is unique in the genus (Kim & Turner 1992); chromosomal and molecular data (Kim & Turner 1992) indicated the species is more distinctive than morphology would suggest. Roadsides, weedy areas, low spots, often in sandy soils; Burnet, Dallas, Denton, Falls, Hunt, Kaufman, Navarro, and Tarrant cos.; according to Kim and Turner (1992), *K. wrightii* occurs throughout most of nc TX; se and e TX w to nc TX and e Edwards Plateau. [*Apogon wrightii* A. Gray]

LACTUCA LETTUCE

Coarse annual or biennial herbs; sap milky or brownish; leaves alternate; leaf blades oblong-lanceolate, coarsely toothed or deeply lobed; heads small, in terminal panicle-like inflorescences (these at first dense and spike-like); involucres narrow; phyllaries in several rows, unequal; flowers all ligulate; corollas yellow in most species to blue or white in a summer- and fall-blooming species, open during the morning, withering greenish or bluish; pappus of numerous hair-like bristles.

A genus of ca. 75 species; cosmopolitan but especially n temperate. *Lactuca sativa* L., the cultivated LETTUCE or GARDEN LETTUCE, lacks the bitter sesquiterpene lactones present in wild species. (The ancient name of the lettuce, *L. sativa*; from Latin: *lac*, milk, in allusion to the white sap) (tribe Lactuceae)

REFERENCES: Stebbins 1937; Vuilleumier 1973.

1. Corollas blue or white; body of achenes 4–5 mm long; beak of achenes stout, 0.5–1(–2) mm long or absent **L. floridana**
1. Corollas usually yellow; body of achenes of various lengths; beak of achenes thread-like, 2–10 mm long.
   2. Plants annuals with taproot; achenes ca. 1 mm wide, 5- to 7-nerved on each face; beak equal to or conspicuously longer than body; latex white; midrib prickly on lower surface of leaf **L. serriola**
   2. Plants biennials with thick crown; achenes at least 2 mm wide, 1- to 3-nerved on each face; beak equal to or shorter than body; latex brownish; midrib on lower surface of leaf prickly or not so.
      3. Body of achenes 4.5–5 mm long, the achenes including beak 7–10 mm long; pappus bristles 7–10 mm long; involucres 13–22 mm long in fruit; upper leaf blades pinnately lobed, the margins conspicuously prickly toothed **L. ludoviciana**
      3. Body of achenes 3.5–4.5 mm long, the achenes including beak 4.5–6.5 mm long; pappus bristles 5–7 mm long; involucres 10–15 mm long in fruit; upper leaf blades usually not lobed, the margins entire to toothed but usually not prickly-toothed **L. canadensis**

Lactuca canadensis L., (of Canada), WILD LETTUCE. Biennial; stems 0.5–2.5(–3) m tall; leaf blades pinnatifid (the upper nearly lobeless), oblanceolate, narrowed to nonclasping base; achenes flat and somewhat winged, the beak filiform. Typically in sandy soils; Denton, Fannin, Grayson, and Tarrant cos.; se and e TX w to nc TX and Edwards Plateau. May–Aug. In some cases *L. canadensis* and *L. ludoviciana* are difficult to distinguish; according to Correll and Johnston (1970), in TX the two intergrade and introgressive hybridization has possibly occurred.

Lactuca floridana (L.) Gaertn., (of Florida), WOODLAND LETTUCE, FLORIDA LETTUCE. Annual 0.5–2.5(–3) m tall; leaf blades deeply runcinate-pinnatifid, the terminal portion large, triangular, the lateral lobes large, 1 or 2 per side; involucres 9–10 mm high; achenes flattened, unwinged, mottled black and gray, the beak tapered, 0.5–1(–2) mm long or absent. Disturbed areas; se and e TX w to West Cross Timbers and Edwards Plateau. Aug–Nov.

Lactuca ludoviciana (Nutt.) Riddell, (of Louisiana), WESTERN WILD LETTUCE. Biennial 1–2 m tall; leaves pinnately lobed becoming deltoid and clasping in upper part, the midrib beneath sometimes prickly; corollas yellow or according to Barkley (1986), possibly blue (this color not observed in nc TX); achenes flattened, somewhat winged, the beak filiform. Usually in calcareous soils; widespread in TX. May–Jul.

Lactuca serriola L., (possibly from Latin: *serrula*, a small saw, referring to the toothed leaves, or a corruption of scariola, an old name for wild lettuce), PRICKLY LETTUCE. Annual 0.5–2 m tall; latex white; leaf blades pinnatifid, the uppermost unlobed, clasping, the margins prickly-toothed, the midrib prickly beneath, sometimes conspicuously so; corollas yellow, drying bluish; achenes 2.5–3 mm long, halfway flattened (lenticular), unwinged, the beak filiform, 3–4 mm long. Disturbed sites; mainly nc TX and Edwards Plateau s and w to w TX. Apr–Aug. Native of the Old World; name often but not originally written *scariola*. The widely cultivated GARDEN LETTUCE, *L. sativa*, is thought to be derived from *L. serriola* (Moore et al. 1976). Cattle have been poisoned by eating large quantities of the young growth (Kingsbury 1964). ☠

Lactuca saligna L., (resembling *Salix*—Willow), WILLOW-LEAF LETTUCE. Just before this book went to press, a large individual (ca. 2 m tall) of this species was discovered in a recently planted landscape in Tarrant Co. (*O'Kennon 14252*, Fort Worth, 21 Aug 1998). While known from OK (Taylor & Taylor 1994), this European native has not previously been reported from TX (O'Kennon et al. 1998). It is not clear whether it will become an established member of the nc TX flora. It is a glabrous winter annual with milky latex. While the Tarrant Co. individual has pinnatifid leaves, leaves in this species can vary from pinnatifid to entire. The corollas are yellow with bluish or purplish on the abaxial side. In the above key it would come out with *L. serriola* except that the midrib is usually not prickly on the lower surface of the leaf. Also, the achenes lack bristles at the base of the beak (in contrast to *L. serriola*, whose achenes have conspicuous bristles at the base of the beak), the leaves are linear to linear-lanceolate, 0.3–5 cm wide, with narrow lobes and entire or few-toothed margins (in contrast to *L. serriola*, whose leaves are lanceolate to ovate, oblong-elliptic, or obovate, 1–10(–15+) cm wide, with broad lobes and prickly-toothed margins), and there are usually 8–15 flowers per head (versus 14–25 in *L. serriola*).

LEUCANTHEMUM

A genus of 33 species of Europe and n Asia including a number of cultivated ornamentals; previously treated in *Chrysanthemum*. The widely cultivated CHRYSANTHEMUMS or MUMS are now treated in the segregate genus *Dendranthema*. Another segregate of *Chrysanthemum*, *Tanacetum*, includes the European *T. cinerariifolium* (Trevir.) Sch.Bip. [*Pyrethrum cinerariifolium* Trevir.] (PYRETHRUM, DALMATIAN INSECT-FLOWER), the flower heads of which yield the monoterpene, pyrethrum, used since ancient times as an insecticide and now widely used by organic gardeners (Fuller & McClintock 1986). ☠ The leaves and flowers of some species can cause contact dermatitis in sensitive individuals (Spoerke & Smolinske 1990). (Greek: *leucos*, white, and *anthemon*, flower) (tribe Anthemideae)
REFERENCE: Arriagada & Miller 1997.

Leucanthemum vulgare Lam., (common), OX-EYE DAISY, WHITE DAISY, WHITEWEED, MOON DAISY, DOG DAISY, MARGUERITE. Perennial; stems glabrous, 0.3–1 m tall; leaves mostly basal, the blades spatulate-obovate, on long slender petioles; stem leaves alternate, the blades serrate to pectinate to pinnately lobed, attenuate basally; heads solitary on long peduncles; ray flowers pistillate, the ligules 3-toothed at apex, white, 10–nearly 30 mm long, 2–10 mm wide; disk flowers perfect, the corollas yellow; receptacles naked; achenes 1.5–2 mm long; pappus a very short crown. Open grassy areas, spreading in native prairie from "wildflower" planting; Grayson and Tarrant cos.; otherwise not reported from TX; escaping more commonly in the e U.S. May–Jul(–fall). Native of Europe and Asia. While formerly included in *Chrysanthemum* and still treated there by some authors (e.g., Jones et al. 1997; Taylor 1997), we are following Bremer (1994), Kartesz (1994), and Arriagada and Miller (1997) in recognizing it in *Leucanthemum*. [*Chrysanthemum leucanthemum* L., *C. leucanthemum* var. *pinnatifidum* Lecoq & Lemotte, *Leucanthemum leucanthemum* Rydb.]

LIATRIS BLAZINGSTAR, GAYFEATHER, BUTTON-SNAKEROOT

Herbaceous perennials from swollen, underground, corm-like structures; stem leaves alternate, simple, ± sessile, the blades entire, essentially linear, usually punctate with resinous dots; inflorescences spike-like or raceme-like, showy; phyllaries sometimes resinous-dotted, greenish to purplish, pink, or white; ray flowers absent; disk flowers fertile; corollas purplish or rarely white; pappus of bristles, these barbellate or plumose; achenes ± cylindrical, usually ca. 10-ribbed.

An e North American (including ne Mexico) genus of 43 species including some cultivated as ornamentals. The common name BUTTON SNAKEROOT is derived from the use of the corm-like structures in treating snake bites (Ajilvsgi 1984). (Derivation of generic name unknown) (tribe Eupatorieae)
REFERENCE: Gaiser 1946.

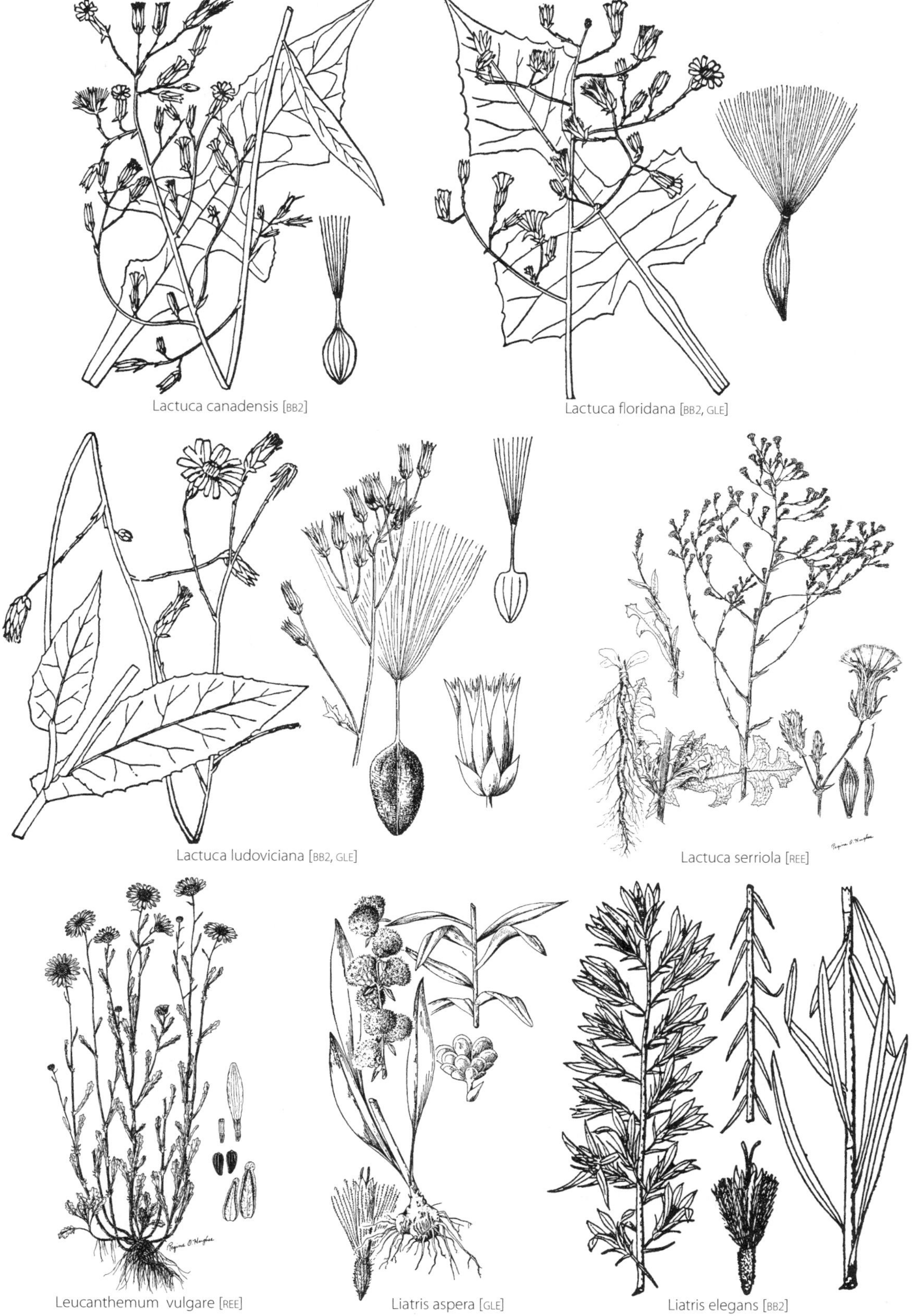

Lactuca canadensis [BB2]

Lactuca floridana [BB2, GLE]

Lactuca ludoviciana [BB2, GLE]

Lactuca serriola [REE]

Leucanthemum vulgare [REE]

Liatris aspera [GLE]

Liatris elegans [BB2]

1. Phyllary tips usually ± rounded; phyllaries glabrous; heads globose ____ **L. aspera**
1. Phyllary tips usually acute to acuminate; phyllaries glabrous or pubescent; heads longer than broad.
   2. Inner phyllaries with prolonged petal-like tips conspicuously longer than the corollas or pappus, the tips white or pink in color; corollas not pilose within the tube ____ **L. elegans**
   2. Inner phyllaries without prolonged petal-like tips, the phyllaries green or purplish and not exceeding the corollas or pappus; corollas pilose within the tube (except in *L. pycnostachya*).
      3. Heads densely crowded in the racemes, the inflorescence axis thus not easily visible; bracts subtending the individual heads usually not or only slightly longer than the heads; heads with 3–12 flowers each, cylindrical, 8–15 mm long to end of pappus; inner phyllaries usually < 12 mm long (often much less).
         4. Pappus bristles merely barbellate (the length of the side hairs on the bristles only 3–6 times the thickness of the central bristle axis), the side hairs inconspicuous even with a lens; heads with 5–12 flowers; plants 0.6–1.5 m tall; n part of nc TX ____ **L. pycnostachya**
         4. Pappus bristles feathery (the length of the side hairs on the bristles 15 times or more the thickness of the central bristle axis), the side hairs conspicuous with a lens and often even with the naked eye; heads with 3–6 flowers; plants 0.8 m or less tall; widespread in Blackland and Grand prairies ____ **L. mucronata**
      3. Heads not densely crowded in the racemes, the inflorescence axis thus easily visible; bracts subtending the individual heads often conspicuously longer than the heads; heads with 20–40 or more flowers each, broadly cylindrical, the well-developed ones (13–)15–20+ mm long to end of pappus; inner phyllaries usually 10–20 mm long ____ **L. squarrosa**

Liatris aspera Michx., (rough), TALL GAYFEATHER, ROUGH GAYFEATHER. Stems 0.4–1.1(–1.5) m tall; inflorescences with heads not crowded, the central axis usually easily visible; heads 15–25 mm broad, with 25–40 flowers; phyllaries ± puckered; corollas purple; pappus bristles merely barbellate, 7–8 mm long. Sandy soils; Fannin, Grayson, Hunt, and Lamar cos., also Denton Co. (J. Quayle, pers. comm.); se and e TX w to n part of nc TX. Jul–Oct. ▣/95

Liatris elegans (Walter) Michx., (elegant), PINK-SCALE GAYFEATHER, HANDSOME BLAZINGSTAR. Stems 0.3–1 m tall; inflorescences spike-like, with heads ± crowded; heads usually 5-flowered; prolonged inner phyllary tips serrulate; corollas white or light purple; pappus plumose, 9–11 mm long. Sandy soils, often in open woods; Cooke, Dallas, Denton, Grayson, Henderson, Lamar, Limestone, and Milam cos; se and e TX w to East Cross Timbers. Aug–Oct.

Liatris mucronata DC., (mucronate, pointed), NARROW-LEAF GAYFEATHER. Stems 0.3–0.8 m tall; inflorescences 8–60 cm long, spike-like, the heads densely crowded; corollas purple, 9–10 mm long; pappus 6–7 mm long. Calcareous upland soils; Blackland and Grand prairies in nc TX; se and e TX w to Edwards Plateau and Rolling Plains. (Jun–)Aug–Nov.

Liatris pycnostachya Michx., (thick-spiked), KANSAS GAYFEATHER, HAIRY BUTTON-SNAKEROOT. Stems 0.6–1.5 m tall; inflorescences 15–30 cm long, spike-like, the heads densely crowded; phyllary tips reflexed or spreading; corollas purple (rarely white); pappus 6–7 mm long. Sandy open areas, prairies, in or around boggy places; Denton (wet area near Lake Ray Roberts) and Lamar (Tridens Prairie) cos., also a plant growing on a roadside in Cooke Co. was observed in 1997 (S. Lusk and J. Quayle, pers. comm.); mainly se and e TX. Jun–Oct. At one time, a poultice of the roots was used in treating snakebite, leading to the common name (Tveten & Tveten 1993).

Liatris squarrosa (L.) Michx., (with parts spreading or recurved at the ends). Stems 0.3–0.6 m tall; inflorescences raceme-like or paniculate, the heads not crowded; heads sessile or on short peduncles, the terminal head often larger; corollas purple; pappus plumose. Sandy soils.

Liatris mucronata [ROE]

Liatris pycnostachya [CO1, GLE]

Liatris squarrosa var. glabrata [RYD]

Lindheimera texana [HEA, KEM]

Lygodesmia texana [SBM]

1. Stems, leaves, and often outer phyllaries glabrous OR pubescent; pappus 10 mm long; outer phyllaries 5–8 mm long, usually ciliate-margined; se and e TX w to e margin of nc TX ____ var. **alabamensis**
1. Plants entirely glabrous; pappus 7–8 mm long; outer phyllaries 7–10 mm long, not ciliate-margined; widespread, East and West cross timbers and along Red River ____ var. **glabrata**

var. **alabamensis** (Alexander) Gaiser., (of Alabama), Heads 10–15 mm wide; corollas 10–12 mm long. Limestone Co. near e margin of nc TX; mainly se and e TX. Aug–Nov. [*L. glabrata* Rydb. var. *alabamensis* (Alexander) Shinners]

var. **glabrata** (Rydb.) Gaiser, (somewhat glabrous, becoming glabrous), SMOOTH GAYFEATHER. Heads averaging 10 mm wide; corollas ca. 10 mm long. Cooke, Denton, Grayson, Lamar, and Tarrant cos.; East and West cross timbers, and along Red River; also e TX. Jul–Oct. [*Liatris glabrata* Rydb.] ▣/95

Liatris acidota (Engelm. & A. Gray) Kuntze, (possibly from Latin: *acidus*, sour, sharp), which would key to *L. pycnostachya* or *L. mucronata* in the key, is cited by Hatch et al. (1990) for vegetational area 4 (Fig. 2). This species can be distinguished by the merely barbellate pappus bristles, flowers 3–5 per head, and the phyllary tips appressed (not reflexed or spreading). It apparently occurs only to the se of nc TX.

LINDHEIMERA TEXAS-STAR, STAR DAISY

A monotypic genus native to the sw United States and n Mexico. (Named for Ferdinand Jacob Lindheimer, 1801–1879, German-born collector of Texas plants and correspondent of Asa Gray and George Engelmann, resided in New Braunfels—Geiser 1948) (tribe Heliantheae)
REFERENCES: Turner & Johnston 1956; Turner & Woodruff 1993.

Lindheimera texana Engelm. & A. Gray, (of Texas), TEXAS-STAR, YELLOW TEXAS-STAR, LINDHEIMER'S DAISY. Hispid-pubescent taprooted annual 10–30(–65) cm tall; leaves alternate and subopposite or the uppermost opposite; leaf blades oblanceolate to rhombic- or ovate-lanceolate, entire or coarsely toothed, sessile or short-petioled; heads usually solitary or few and corymbose, on peduncles 1–4 cm long; involucres 10–15 mm tall, the phyllaries in 2 series, the outer series narrower, longer, and more acute than the conspicuous inner; ray flowers usually 5, pistillate and fertile, the ligules ca. 1 cm long, yellow or orange-yellow, 2-toothed at apex; disk flowers few, infertile; achenes flattened, large, nearly as large as the inner phyllaries; pappus of 2 awns. Clayey or occasionally sandy prairies, roadsides; Post Oak Savannah and se TX w to Edwards Plateau and Rolling Plains. Mar–May.

LYGODESMIA
SKELETON-WEED, SKELETON-PLANT, RUSH-PINK, ROSE-RUSH, RUSHWEED

A genus of 7 species of the Americas, especially w North America. (Greek: *lygos*, a plant twig, and *desme*, a bundle, from clustered stick-like or twiggy stems) (tribe Lactuceae)
REFERENCES: Vuilleumier 1973; Tomb 1980.

Lygodesmia texana (Torr. & A. Gray) Greene, (of Texas), TEXAS SKELETON-PLANT, PURPLE-DANDELION, FLOWERING-STRAW, MILK-PINK. Glabrous and glaucous, low perennial with deep, thick, woody root producing in age numerous vertical to oblique rhizomes, terminating in above-ground stems, forming clumps; sap milky; stems 20–65 cm tall, with few reduced leaves or bracts, openly and stiffly branched toward summit; leaf blades mostly basal, linear to lanceolate, entire or usually deeply pinnately lobed; heads large; involucres subcylindrical, 10–25 mm high; flowers all ligulate, with heavy sweet scent; corollas purple to pale lavender (rarely white), open during morning; achenes 13–17 mm long; pappus of hair-like bristles, 10–15 mm long. Limestone outcrops and rocky calcareous soils; in nc TX mainly Grand Prairie and w, also Dallas Co. to the e (Tomb 1980); widespread in TX except the extreme ne part of the state. May–Sep. ▣/97

MACHAERANTHERA

A w North American genus of 36 species including some used as indicators of selenium and uranium. The genus is sometimes included in *Haplopappus*; some taxa previously treated in *Machaeranthera* are now recognized in the recently named genus *Rayjacksonia* (Lane & Hartman 1996). (Greek: *macha*, sword, and *anthera*, anther) (tribe Astereae)

REFERENCES: Hall 1928; Cronquist & Keck 1957; Turner & Hartman 1976; Hartman 1990; Morgan & Simpson 1992; Morgan 1997 [1998].

Machaeranthera pinnatifida (Hook.) Shinners, (pinnately cut), SPINY IRONPLANT, CUT-LEAF IRONPLANT. Herbaceous perennial from a woody, branching, underground structure, nearly glabrous to variously pubescent, often minutely glandular-pubescent; stems 0.2–0.8 m tall, branched only in uppermost part; leaves alternate, ± sessile, often not much reduced upwards; leaf blades 1–3 cm long, usually pinnately (rarely bipinnately) lobed, the sinuses usually extending 1/2 way to the midrib or more; heads solitary at the ends of the branches; involucres 5–8 mm high; phyllaries in several series, often acute and minutely bristle-tipped, eglandular or with scattered glands, straw-colored but the upper part with a dark often green area; ray and disk flowers yellow; ray flowers pistillate, fertile; ligules 8–10 mm long; disk flowers perfect, fertile; pappus of ray and disk flowers similar, of hair-like bristles; achenes ca. 2–2.5 mm long, pubescent. Grasslands; Clay, Jack, Montague, and Young cos.; mainly w 1/2 of TX rare e to w part of nc TX. Late May–Sep. [*Haplopappus spinulosus* (Pursh) DC.]

MARSHALLIA BARBARA'S-BUTTONS

A genus of 7 species native to the United States. Watson and Estes (1990) indicated that morphological variation is continuous among the species and that there are few abrupt boundaries. (Named at the request of Muhlenberg for Dr. Moses Marshall, 1785–1813, nephew of the more distinguished Humphrey Marshall) (tribe Heliantheae)

REFERENCES: Channell 1957; Watson & Estes 1990.

Marshallia caespitosa Nutt. ex DC., (growing in tufts). Small perennial; stems glabrous or minutely pubescent above; leaves sessile or basal ones petioled; leaf blades oblanceolate to linear, 5–15 cm long, to 1(–1.5) cm wide, entire; heads terminal, solitary or few, 2.5–3.5 cm across, long-peduncled; phyllaries grading into bracts of disk, narrow, green, herbaceous; ray flowers absent; disk flowers sweet-scented; corollas white or occasionally pink-tinged; pappus of 5 scales 2–3 (–4) mm long, ca. as long as the achene.

1. Leaves crowded near base of plant, with no or few leaves up the stems; heads usually solitary per stem ______ var. **caespitosa**
1. Leaves gradually reduced upwards, the stems leafy except for peduncles, the leafy part of the stems usually as long or longer than the peduncles; heads solitary–several per stem ______ var. **signata**

var. **caespitosa**, SEEP MARSHALLIA, BARBARA'S-BUTTONS. Clayey prairies and limestone outcrops; Cooke, Dallas, and Denton cos.; se and e TX w to nc TX and Edwards Plateau. Apr–May.

var. **signata** Beadle & F.E. Boynton, (marked, designated), BARBARA'S BRUTTONS. Limestone outcrops; nc TX s to s TX and w to e Rolling Plains; more common in nc TX than var. *caespitosa*. Apr–May. ▣/98

MELAMPODIUM BLACKFOOT

A genus of 37 species native to tropical and warm areas of the Americas, especially Mexico.

(Greek: *melas*, black, and *podos*, foot, alluding to the fact that each ray flower is subtended by a small, foot-shaped bract that turns black upon maturity) (tribe Heliantheae)
REFERENCES: Turner & King 1962; Stuessy 1971, 1972, 1979; Stuessy & Crisci 1984.

Melampodium leucanthum Torr. & A. Gray, (white-flowered), ROCK DAISY, PLAINS BLACKFOOT, BLACK-FOOT DAISY. Herbaceous or weakly subshrubby perennial; stems 15–50 cm tall; leaves opposite; leaf blades linear to narrowly oblong, 2–4.5 cm long, to 1 cm wide, entire to pinnately lobed, sessile; heads solitary, on peduncles 3-7 cm long; outer phyllaries united more than half way; inner phyllaries each surrounding the achene of a ray flower, expanded hood-like above the achene; ray flowers 8-13, pistillate, fertile; ligules 7–13 mm long, white; disk flowers staminate with abortive ovaries, the corollas yellow; achenes 1.5–2.5 mm long; pappus none. Calcareous soils; Bell, Eastland, Erath, Hamilton, Lampasas, and Mills cos.; Lampasas Cut Plain and East Cross Timbers s and w to w TX. Mar–Nov.

MIKANIA HEMPWEED, CLIMBING HEMPWEED

A mainly tropical genus of ca. 430 species of climbers; the single nc TX representative of this genus is the only Asteraceae in the nc TX flora that is a twining vine. (Named for Joseph Gottfried Mikan, 1743–1814, professor in the University of Prague) (tribe Eupatorieae)
REFERENCES: Holmes 1981, 1990.

Mikania scandens (L.) Willd., (scandent, climbing), CLIMBING HEMPWEED, HEMPVINE, CLIMBING-BONESET. Perennial twining vine; leaves opposite, petioled, the blades ovate to deltoid, subentire or undulate, generally acuminate; heads corymbose; phyllaries 4, equal, 4–5 mm long, acuminate; ray flowers absent; disk flowers 4 per head; corollas whitish, sometimes pink-tinged; achenes 5-ribbed; pappus persistent, of numerous bristles. Woodlands, low areas; Bell, Dallas, Grayson, Hopkins, and Rockwall cos.; se and e TX w to East Cross Timbers and Edwards Plateau. Aug–Nov.

ONOPORDUM COTTON-THISTLE, SCOTCH-THISTLE

A genus of ca. 60 species of prickly herbs native to Europe, the Mediterranean region, and w Asia. (Ancient Greek name of the plant, from *onos*, donkey, and *porde*, flatulence, Pliny stating that it produced flatulence in donkeys) (tribe Cardueae)

Onopordum acanthium L., (prickly), SCOTCH-THISTLE. Coarse biennial 0.5–3 m tall, often gray from sparse to dense cottony pubescence; stems with wings 5-15(–20) mm wide, these with spine-tipped lobes; leaves alternate, lobed or toothed, conspicuously spiny-margined, sessile, decurrent; heads terminal, solitary or 2–5 in ± loose clusters, large (2.5–5 cm in diam.); phyllaries linear-subulate, ascending-spreading, tapering from base to spine-tip; ray flowers absent; disk corollas purplish or pinkish white, slender, 20–25 mm long; pappus of numerous barbellate bristles 7–9 mm long. Roadsides and disturbed areas; first seen in TX at the Fort Worth stockyards (Tarrant Co.) in 1938, also Erath, Johnson, and Parker cos.; otherwise known in TX only from 2 counties in the Edwards Plateau (Gillespie and Kerr), but given its local spread, this species is potentially a problematic weed for the state. May–Jul. Native of Eurasia.

PACKERA RAGWORT, GROUNDSEL, BUTTERWEED

Ours annual or perennial herbs, glabrous to woolly pubescent; leaves alternate, toothed to pinnately compound; inflorescences terminal, the heads cymosely arranged; principal phyllaries in a single series, linear, with hyaline margins, subtended by an outer series of short, bract-like phyllaries; ray flowers in ours ca. 8 or 13, pistillate, fertile, with yellow ligules; disk flowers perfect, fertile, the corollas yellow; pappus of numerous, white, capillary hairs; achenes cylindrical.

A genus of ca. 60–65 species centered in temperate North America but with 16 species in

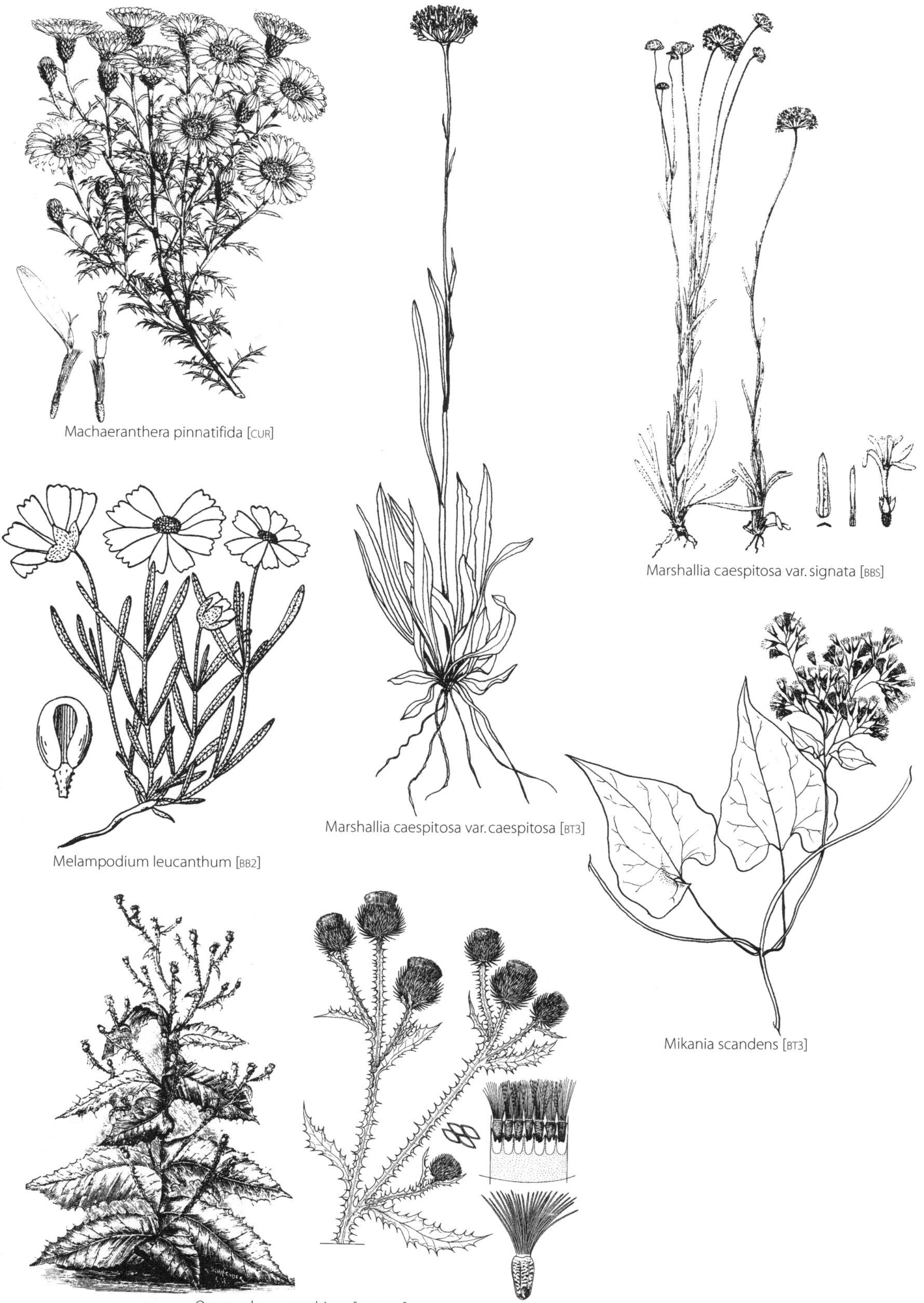

Machaeranthera pinnatifida [CUR]

Marshallia caespitosa var. signata [BBS]

Melampodium leucanthum [BB2]

Marshallia caespitosa var. caespitosa [BT3]

Mikania scandens [BT3]

Onopordum acanthium [GAR, M&F]

Mexico and some in Siberia; it is well known for imprecise and intergrading species (Barkley et al. 1996). While sometimes still treated in the genus *Senecio* in the broad sense (Kartesz 1994; Jones et al. 1997), we are following Freeman and Barkley (1995), Barkley et al. (1996), and T. Barkley (pers. comm.) in treating the aureoid group of Senecios as *Packera*. In addition to morphological differences, the genera differ in chromosome number; in *Packera* $n = 23$ (more rarely 22) or a derivative, while in *Senecio* $n = 20$ (more rarely 10) or a derivative. ☠ As with *Senecio*, some contain pyrrolizidine alkaloids (e.g., senecionine) which can cause severe liver damage and even death in livestock (Morton 1982); because similar effects are to be expected in people, the plants should not be used as herbal teas or ingested in other ways. (Named for John G. Packer, botanist and friend of Á. and D. Löve, who named the genus) (tribe Senecioneae)
REFERENCES: Greenman 1915–1918; Barkley 1962, 1978, 1981, 1985a, 1985b, 1986, 1988; Vuilleumier 1969a; Löve & Löve 1975 [1976]; Freeman & Barkley 1995; Barkley et al. 1996.

1. Basal leaves pinnately compound or very deeply lobed; basal leaves and stem leaves usually similar; leaves ± well-distributed along the stem, only gradually reduced upwards; annuals with delicate taproot or fibrous roots, without rhizomes or stolons.
   2. Plants usually with a delicate, distinct, main taproot; lateral lobes of lower and middle stem leaves often contracted to very narrow linear basal portions that attaches to the midrib (this character, while frequently used to separate these 2 species, does not appear to always be dependable in nc TX); widespread weed in open disturbed sites in nc TX ____ **P. tampicana**
   2. Plants with a cluster of fibrous roots, a taproot lacking; lateral lobes of lower and middle stem leaves tapering to base, but usually with a broad attachment to the midrib; disturbed damp sites, often in partial shade; rare in nc TX, mainly se and e TX ____ **P. glabella**
1. Basal leaves shallowly toothed or deeply lobed; basal leaves and stem leaves quite different in appearance; largest leaves crowded near base of plant, progressively reduced up the stem; perennials from an abruptly foreshortened rhizome or an elongated rhizome, sometimes with stolons.
   3. Plants glabrous except in leaf axils; inflorescences without woolly pubescence; blades of the basal leaves ovate to orbicular, < 1.5 times as long as wide; in wooded bottomlands ____ **P. obovata**
   3. Plants unevenly woolly-pubescent when young, becoming partly glabrous but with at least some woolly pubescence at the nodes of the inflorescences; blades of the basal leaves oblong-elliptic or lanceolate, > 1.5 times as long as wide; in prairies and upland woods ____ **P. plattensis**

Packera glabella (Poir.) C. Jeffrey, (somewhat smooth, hairless), BUTTERWEED, YELLOWTOP. Annual 10–80(–100) cm tall with a tuft of fibrous roots and no persistent taproot; terminal leaf segments variously sub-orbicular to ovate or obovate, usually undulate; phyllaries ca. 5 mm long; ligules ca. 4.5 mm long; achenes with some fine hairs (rarely glabrous), ca. 1.7 mm long. Moist, disturbed, often sandy soils, usually with at least partial shade; se and e TX w to n Red River Co. just e of nc TX, also reported for Fort Hood (Bell or Coryell cos.—Sanchez 1997). This species is superficially similar to *P. tampicana*. Mar–May. [*Senecio glabellus* Poir.] Liver damage and death have been reported in livestock; symptoms may not appear for 1 or 2 months after grazing (Morton 1982). ☠

Packera obovata (Muhl. ex Willd.) W.A. Weber & Á. Löve, (obovate), GOLDEN GROUNDSEL, ROUND-LEAF GROUNDSEL, OVATE-LEAF RAGWORT. Perennial 20–50(–70) cm tall, fibrous-rooted and rhizomatous, usually stoloniferous; basal leaves often purplish beneath; phyllaries 3–7 mm long; ligules 5–10 mm long; achenes usually glabrous, ca. 2.3 mm long; pappus bristles to ca. 6 mm long. Stream bottom woods; e TX w to nc TX and Edwards Plateau; also Trans-Pecos. Feb–Apr. [*Senecio obovatus* Muhl. ex Willd.] This is one of the earliest flowering native wildflowers in nc TX.

Packera plattensis (Nutt.) Á. Löve & D. Löve, (of the Platte River region), PRAIRIE GROUNDSEL,

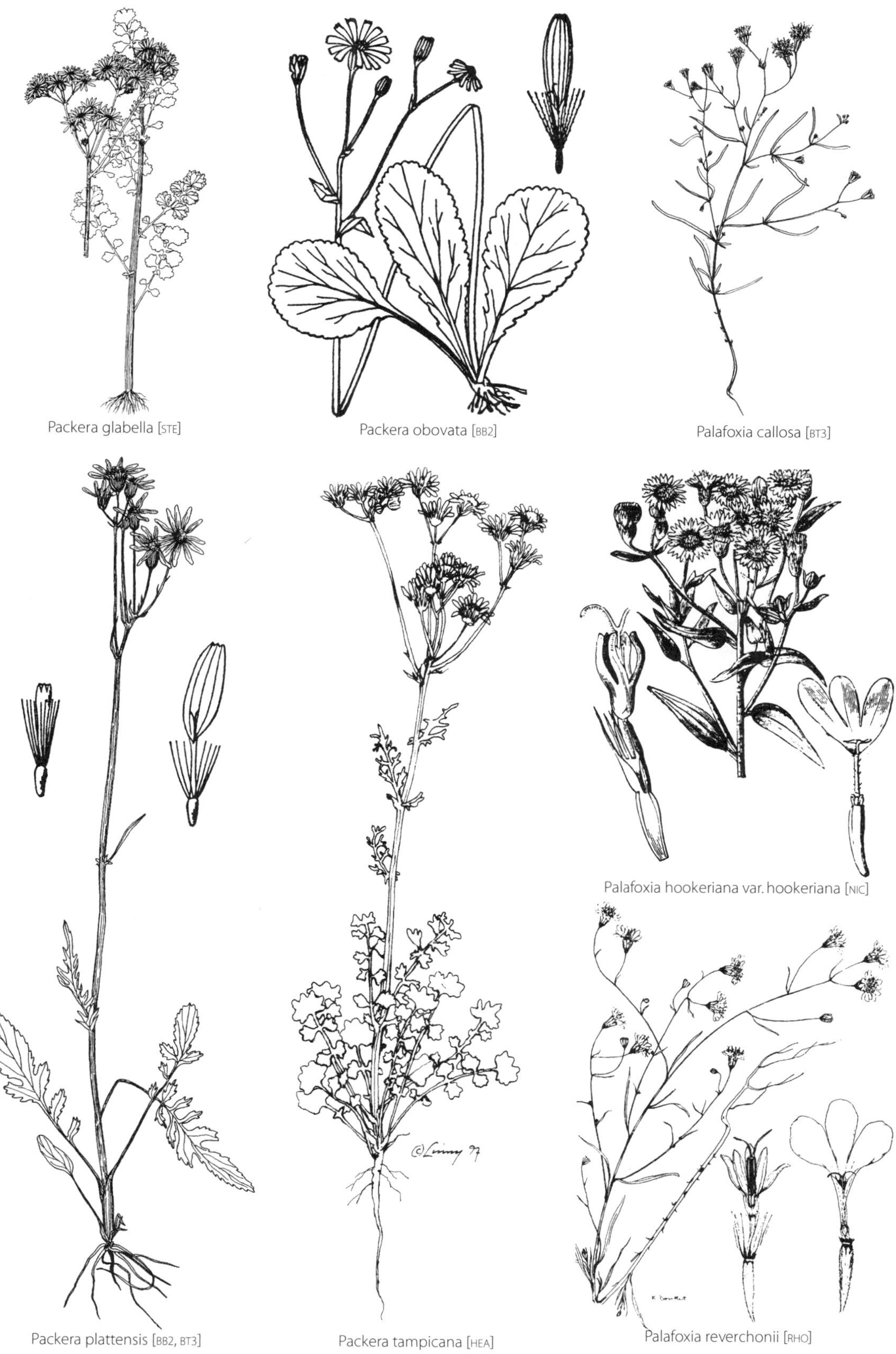

Packera glabella [STE]

Packera obovata [BB2]

Palafoxia callosa [BT3]

Packera plattensis [BB2, BT3]

Packera tampicana [HEA]

Palafoxia hookeriana var. hookeriana [NIC]

Palafoxia reverchonii [RHO]

PRAIRIE RAGWORT. Biennial or perennial, fibrous-rooted, short rhizomatous, sometimes stoloniferous; stems 10–50(–70) cm tall; leaves sometimes purplish-tinged; phyllaries 5–6(–7) mm long; ligules 4–6 mm long; achenes with fine hairs, ca. 2.5 mm long. Prairies and open woods, calcareous soils; nc TX w to Rolling Plains and s to Edwards Plateau; rare in e TX. Mar–May. [*Senecio plattensis* Nutt.] Reported to be poisonous (Kingsbury 1964). ☠

Packera tampicana (DC.) C. Jeffrey, (of Tampico, Mexico), YELLOWTOP. Glabrous, low annual 10–40(–80) cm tall; similar to *P. glabella*; taprooted, but taproot sometimes poorly developed and masked by lateral roots; lower herbage often variously anthocyanic; terminal leaf segments often subreniform and dentate; phyllaries 3–7 mm long. Weedy in low prairies, disturbed areas, roadsides, typically in open places; populations in some areas (e.g., Hagerman N.W.R. in Grayson Co.) can number in the tens of thousands; widespread in TX, more abundant in e 1/2. Apr–May. [*Senecio imparipinnatus* Klatt., *S. greggii* Rydb., *S. tampicanus* DC.]

PALAFOXIA

Ours annual herbs, often glandular; leaves alternate or first ones opposite; leaf blades linear to narrowly elliptical, lanceolate, or narrowly ovate; heads peduncled, irregularly corymbose; phyllaries in 2 series, narrowly obovate, scarious-margined or -tipped; receptacles naked; ray flowers absent or present, if present fertile, with pink to violet ligules; disk flowers radially symmetrical or irregularly lobed, perfect, fertile; corollas pink to purplish, the limb divided nearly to base, the lobes linear to narrowly oblong, the tube slender; achenes narrowly obconic, 4-angled, pubescent; pappus of 7–10 scales, minute to long acuminate.

A genus of 12 species native to the s U.S. and Mexico; some are cultivated as ornamentals. (Named either for José de Palafox y Melzi, 1780–1847, a Spanish general, or for Juan de Palafox y Mendoza, 1600–1659, a prelate—Wagner et al. 1990) (tribe Helenieae, sometimes lumped into Heliantheae)

REFERENCES: Ammerman (or as Baltzer) 1944; Cory 1946; Shinners 1952; Turner & Morris 1976.

1. Heads with disk flowers only; all achenes with similar pappus.
   2. Phyllaries 0.6–1.2 mm wide, narrowly linear, usually 3–5 mm long; pappus scales 0.5–2 mm long; mainly on limestone derived soils; widespread in nc TX ____ **P. callosa**
   2. Phyllaries 1.2–2.5 mm wide, linear to obovate, 5–10 mm long; pappus scales 1–8 mm long; mainly on sandy soils; rare in nc TX ____ **P. rosea**
1. Heads with conspicuous ray flowers; ray and disk achenes with pappus greatly different in size.
   3. Leaf blades linear to linear-lanceolate, 2–4(–6) mm wide; stems (except for inflorescence) not glandular; phyllaries 6–8 mm long; achenes 5–6 mm long ____ **P. reverchonii**
   3. Leaf blades narrowly to broadly lanceolate, 3–25 mm wide; stems usually glandular for some distance below the inflorescence; phyllaries 7–20(–25) mm long; achenes 6–9 mm long.
      4. Phyllaries 2–2.5(–3) mm wide; ligules of ray flowers mostly 10 mm long or less; stems usually branched at or below middle; extreme n part of nc TX ____ **P. sphacelata**
      4. Phyllaries 2–5 mm wide; ligules of ray flowers mostly 10–12 mm long; stems usually branched at or above middle; extreme se part of nc TX ____ **P. hookeriana**

Palafoxia callosa (Nutt.) Torr. & A. Gray, (thick-skinned), SMALL PALAFOXIA. Plant scabrous, glandular; stems 20–60 cm tall; leaf blades linear, 1–4 mm wide, with glandular-based hairs; peduncles densely stipitate-glandular; anthers brownish to reddish purple; achenes 3–5 mm long. Calcareous soils, disturbed habitats; e TX w to Rolling Plains and Edwards Plateau. Aug–Nov.

Palafoxia hookeriana Torr. & A. Gray var. **hookeriana**, (for William Jackson Hooker, 1785–1865, director of Kew Gardens), SHOWY PALAFOXIA. Stems 25–180 cm tall, densely viscid-glandular throughout; leaf blades narrowly to broadly lanceolate, 5–10 cm long, 4–25 mm wide; phyllaries (7–)10–15 mm long, with glandular and eglandular hairs; ray flowers with ligules pink, showy, 3-lobed; disk corollas deep pink, 10–12 mm long; achenes 6–9 mm long; pappus of ray flowers < 1 mm long; pappus of disk flowers 5–8 mm long; $n = 12$. Sandy soils, usually in forested areas; Milam Co. (Shinners 1952) on e margin of nc TX; mainly se to sc TX; endemic to TX. Sep–Oct.

var. *minor* Shinners, (smaller), with the lower half of the stem without glandular hairs, is found just to the e of nc TX in e Milam and Freestone cos.; endemic to TX (Turner & Morris 1976).

Palafoxia reverchonii (Bush) Cory, (for Julien Reverchon, 1837-1905, a French-American immigrant to Dallas and important botanical collector of early TX), REVERCHON'S PALAFOXIA. Stems 10–90 cm tall; ray flowers with ligules pale to dark violet, with 3 lobes, the lobes 4–5 mm long; disk corollas violet; pappus of ray flowers ca. 0.5 mm long; pappus of disk flowers 3–6 mm long. Sandy soils, wooded areas; Limestone Co. at extreme e margin of nc TX, also Henderson Co. (Turner & Morris 1976); mainly e TX; endemic to TX. Sep–Oct. Closely related to *P. hookeriana* (Turner & Morris 1976).

Palafoxia rosea (Bush) Cory, (rose-colored), ROSE PALAFOXIA. Plants scabrous, glandular; stems 10–50 cm tall; leaf blades with glandular-based hairs; peduncles stipitate-glandular; inflorescences corymbose, with ca. 3–10 heads; heads with 10–30 flowers; corollas pale violet, 7–10 mm long; anthers brownish to reddish purple. Disturbed habitats. Jun–Nov.

1. Pappus scales 3–8 mm long, acute to long acuminate; phyllaries 7–10 mm long ______ var. **macrolepis**
1. Pappus scales 1–3 mm long, obtuse to acute; phyllaries 5–7 mm long ______ var. **rosea**

var. **macrolepis** (Rydb.) B.L. Turner & M.I. Morris, (large-scaled). Sandy soils, open areas; Eastland Co. (Turner & Morris 1976) on w edge of nc TX; sc TX nw to Panhandle.

var. **rosea**. ROSE PALAFOXIA. Forested areas; Dallas Co. (Buzzards Spring, *Reverchon*, 1902); mainly se and e TX, rarely further w; endemic to TX and OK (Turner & Morris 1976).

Palafoxia sphacelata (Nutt. ex Torr.) Cory, (dead, withered), RAYED PALAFOXIA. Similar to *P. hookeriana*; stems 10–90 cm tall, densely viscid-glandular; leaf blades narrowly to broadly lanceolate, 3–9 cm long, 3–20 mm wide; ray flowers with ligules pale to dark violet, with 3 narrow lobes, the lobes 4–8 mm long; disk corollas pale violet, 10–14 mm long; achenes 6–9 mm long; pappus of ray flowers < 1 mm long; pappus of disk flowers of ca. 8 scales 7–9 mm long; $n = 12$. Sandy soils, grasslands; disjunct e to Grayson Co. in Red River drainage (Turner & Morris 1976); mainly Trans-Pecos and Panhandle e to Rolling Plains. May–Nov.

PARTHENIUM

A genus of 16 species native to North America and the West Indies. The genus also contains the GUAYULE RUBBER PLANT (*P. argentatum* A. Gray), a source of natural rubber (Rollins 1950), native to w TX and Mexico; during World War II, because of rubber shortages, a program to grow the GUAYULE RUBBER PLANT was carried out by the U.S. Forest Service (McGinnies in Foster et al. 1983). (An ancient name of some plant from the Greek: *parthenos*, virgin; with only the pistillate ray flowers fertile, i.e., producing achenes) (tribe Heliantheae)
REFERENCES: Rollins 1950; Mears 1975.

Parthenium hysterophorus L., (old generic name), FALSE RAGWEED, RAGWEED PARTHENIUM, SANTA-MARIA, FEVERFEW, CICUTILLA. Pubescent, somewhat glandular annual, 0.3–1 m tall; leaf blades, except uppermost, deeply once or twice pinnatifid; heads small (3–4 mm across), numerous, in an open inflorescence; ray flowers pistillate, fertile, with ligules minute (< 1 mm long), white; disk flowers seemingly perfect but functionally staminate, not maturing achenes, the corollas white; achenes black, 2–2.5(–3.5) mm long; pappus of 2 petaloid scales. In and near towns and farms, disturbed areas; Dallas, Grayson, and Tarrant cos.; widespread in TX. Summer–fall. Hairs and sessile capitate glands on the leaf surfaces contain sesquiterpene lactones (parthenin and ambrosin) which can cause serious dermatitis in some individuals; this New World native was introduced in India (first noted in 1956) presumably as a contaminant with cereal grains; there it has become a problematic alien invader, crowding out the native flora and causing dermatitis in agricultural workers (Lampe 1986; Mabberley 1987).

PECTIS

A genus of ca. 100 species of tropical and warm areas of the New World and the Galápagos Islands. Like many members of the genus, *Pectis papposa* A. Gray, of the sw U.S. and Mexico, is strong-scented; it was used by Native Americans for flavoring foods and as a perfume (Bradley & Haagen-Smit 1949). (Greek: *pectis*, to comb, when the marginal leaf glands become broken, the leaf has a somewhat serrated effect marginally—Bradley & Haagen-Smit 1949) (tribe Heliantheae)
REFERENCES: Fernald 1897; Bradley & Haagen-Smit 1949; Keil 1977.

Pectis angustifolia Torr. var. **fastigiata** (A. Gray) D.J. Keil, (sp.: narrow-leaved; var.: having branches close together and erect). Strong-scented, fibrous-rooted, much-branched perennial or some individuals annual; stems 5–15 cm long, sometimes woody at base; leaves linear, 10–40 mm long, 1–2 mm wide, glabrous, ciliate basally with bristles 1–2 mm long, often revolute, marginally glandular-punctate; heads congested at ends of branches; peduncles 3–30 mm long; phyllaries in one row, subequal, 2.5–4.5 mm long, keeled, broadest near apex, with a conspicuous subterminal oil gland 0.5–1 mm long and 1 or 2 pairs of smaller submarginal glands; ray flowers 8, pistillate, fertile; ray corollas 4–6 mm long, yellow; disk flowers 8–21, perfect, fertile; disk corollas 2.7–4 mm long, yellow; achenes 2.5–3.5 mm long; pappus 0.5–1(–2) mm long, of 0–4 short scales forming a low crown. Usually on limestone, dry uplands; Bell and Bosque cos., also Coryell Co. (Fort Hoot—Sanchez 1997); also Travis Co. just s of nc TX; Lampasas Cut Plain and Edwards Plateau; endemic to TX. Sep–Nov. Two other varieties of this species occur in s and w TX and Mexico. [*P. fastigiata* A. Gray, *P. texana* Cory]

PERITYLE ROCK DAISY

A genus of ca. 63 species native to sw North America with 1 in Chile and Peru. (Greek: *peri*, around, and *tyle*, knot, knob, or callus, from the thickened margin around the fruit) (tribe Helenieae)
REFERENCE: Powell 1973.

Perityle lindheimeri (A. Gray) Shinners, (for Ferdinand Jacob Lindheimer, 1801–1879, German-born TX collector), LINDHEIMER'S ROCK DAISY. Perennial from a woody base; plants (10–)18–45(–60) cm tall; leaves opposite below, alternate above; leaf blades broadly ovate to ovate-lanceolate, 2–5 cm long, minutely punctate, essentially glabrous, with large serrations/small lobes; petioles 4–10 mm long; inflorescences cymose; involucres ca. 4–5 mm high; ray flowers 3–5, pistillate and fertile, the ligules 2.5–3 mm long, 1.5–2 mm wide, yellow; disk corollas 3–3.5 mm long, yellow; achenes 2–2.8 mm long, linear to narrowly obconical; pappus usually of a single bristle 0.5–1.8 mm long and vestigial squamellae. Crevices in limestone bluffs; Burnet Co. (Balcones Canyonlands Nat. Wildlife Refuge, C. Sexton, pers. comm.); s margin of nc TX and Edwards Plateau; endemic to TX and NM. Spring–fall.

PINAROPAPPUS PINK-DANDELION, ROCK-LETTUCE

A genus of 6 species native to s North America. (Greek: *pinaro*, dirty, and *pappus*, down, fuzz, pappus) (tribe Lactuceae)

Pinaropappus roseus (Less.) Less., (rose-colored), SMALL ROCK-LETTUCE. Glabrous, small (10–30

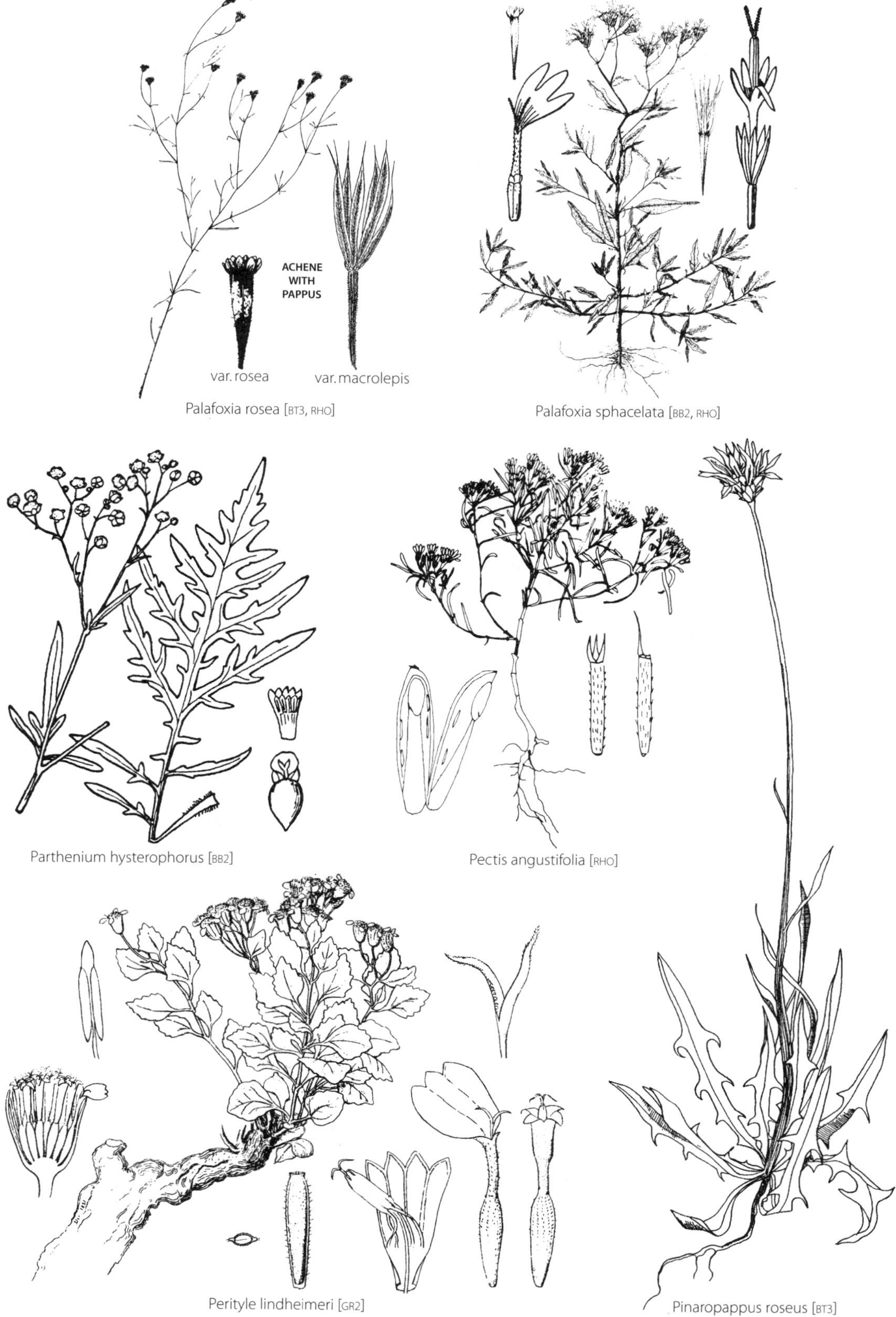

Palafoxia rosea [BT3, RHO]

Palafoxia sphacelata [BB2, RHO]

Parthenium hysterophorus [BB2]

Pectis angustifolia [RHO]

Perityle lindheimeri [GR2]

Pinaropappus roseus [BT3]

cm tall) perennial with slender to thick, woody, branching root; leaves mostly basal; leaf blades oblong-lanceolate to linear, entire or coarsely toothed or lobed; heads large (15–25 mm high including corollas), terminal, solitary; phyllaries semi-scarious, in several overlapping rows, usually with prominent dark tips; flowers all ligulate; corollas pink to deep rose-lavender beneath, paler to white above, the innermost ones yellowish; pappus of numerous hair-like bristles. Limestone outcrops; Bell, Brown, Burnet, Callahan, Coleman, Shackelford, and Throckmorton cos.; w part of nc TX s and w to Trans-Pecos. Apr–May.

PITYOPSIS GRASS-LEAF GOLDEN-ASTER

A genus of 8 species of the e U.S. and Central America. Previously often included in *Chrysopsis* or *Heterotheca*. (Greek: *pitys*, pine, fir, and *opsis*, appearance or resembling; possibly for pine-like appearance resulting from the narrow leaves) (tribe Astereae)
REFERENCES: Shinners 1951b; Semple 1977; Semple et al. 1980; Semple & Bowers 1985.

Pityopsis graminifolia (Michx.) Nutt. var. **latifolia** (Fernald) J.C. Semple & F.D. Bowers, (sp.: grass-leaved; var.: broad-leaved), SILK-GRASS. Rhizomatous perennial; stems erect, 20–80 cm tall, usually 1–5(–more) per plant, with long, appressed, silvery white, silky pubescence; leaves linear to narrowly lanceolate, ± grass-like in appearance, 8–25(–40) cm long, 2–20 mm wide, with pubescence like the stems; basal leaves usually much longer than stem leaves; upper leaves reduced; inflorescences cymose-paniculate, with several to many pedunculate heads; peduncles 1–10 cm long; involucres at anthesis 8–12 mm high; phyllaries imbricate in 4–6 series; ray flowers 9–13, with ligules 4–14 mm long, yellow; disk flowers 30–50, the corollas yellow; achenes fusiform, 2.5–4.5 mm long, ribbed, strigose; pappus double, the outer whorl of setiform squamellae 0.4–0.9 mm long, the inner whorl of capillary bristles 5–9 mm long. Wooded areas, roadsides, waste areas, often in sandy soils; Tarrant Co. (Semple & Bowers 1985 cited *Ruth s.n.* at TEX); mainly se and e TX. Aug–Nov. [*Chrysopsis graminifolia* var. *latifolia* Fernald] Jones et al. (1997) did not cite this variety for TX.

PLUCHEA MARSH-FLEABANE, STINKWEED, FLEABANE

Annual or perennial aromatic herbs; stems erect, 0.5–2 m tall, glabrate below, pubescent above; leaves alternate; leaf blades simple, ovate to elliptic, glabrous to puberulent or tomentose, sometimes with resin-globules, usually serrate to serrate-dentate or crenate; heads numerous; phyllaries imbricated, sometimes pink or purplish; ray flowers absent; disk flowers of two types: a few central flowers perfect, infertile, the corollas 5-lobed; numerous outer flowers pistillate and fertile, the corollas 3-lobed; corollas rose, rose-purple, or creamy white or yellowish; achenes cylindrical, < 1 mm long; pappus of fine barbellate bristles in a single series.

A genus of ca. 40 species of warm areas of the world. The key is adapted from Godfrey (1952). The foliage is very aromatic with a strong camphor-like smell. (Named for Noel Ant. Pluche, 1688–1761, French naturalist) (tribe Inuleae)
REFERENCES: Godfrey 1952; Robinson & Cuatrecasas 1973; Gillis 1977; Nesom 1989b; Keeley & Jansen 1991.

1. Middle and upper leaves sessile, not narrowed at base, usually ± auriculate-clasping; corollas cream-white or yellowish; near extreme e edge of nc TX ____ **P. foetida**
1. Middle and upper leaves sessile or petiolate, tapering to a narrow base, not auriculate-clasping; corollas rose to rose-purple; widespread in nc TX.
   2. Phyllaries granular from minute, sessile, golden resin-globules; only the outermost phyllaries sparsely puberulent and ciliate, the median and inner phyllaries ± glabrous except for resin-globules; inflorescence characteristically an elongate panicle, the branches numerous and

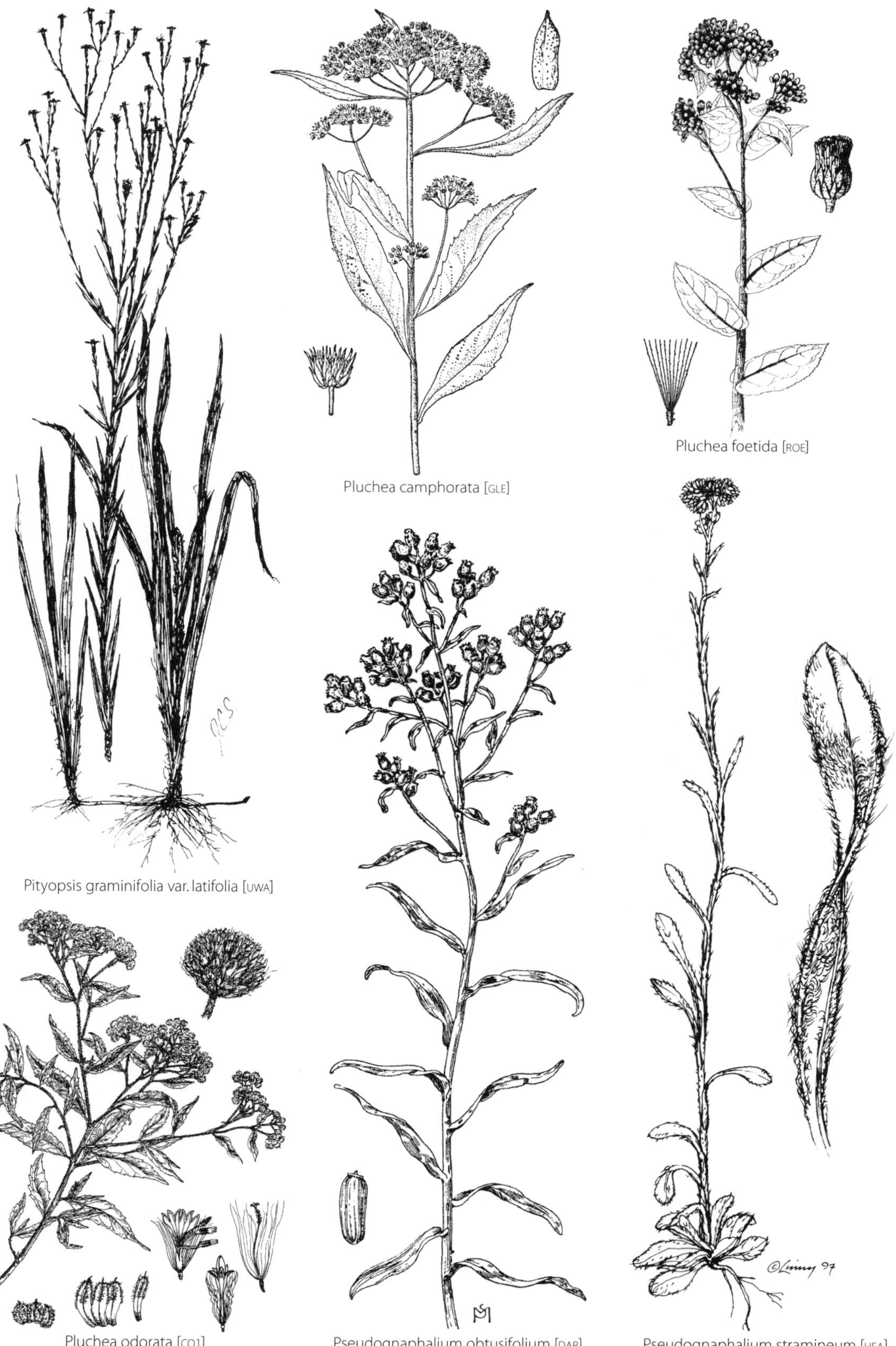

Pluchea camphorata [GLE]

Pluchea foetida [ROE]

Pityopsis graminifolia var. latifolia [UWA]

Pluchea odorata [CO1]

Pseudognaphalium obtusifolium [DAR]

Pseudognaphalium stramineum [HEA]

terminating in convex, panicled cymes, the central uppermost axis maturing first, but the lateral ones never equaling or exceeding it; leaves petioled _______ **P. camphorata**

2. Phyllaries with resin globules or not; outermost and median phyllaries copiously puberulent and ciliate, the inner sparsely puberulent on their summits; inflorescence characteristically more truly cymose, the younger lateral branches elongating and exceeding the more central ones, thus producing a flat-topped or layered inflorescence; leaves sessile or petioled _______ **P. odorata**

Pluchea camphorata (L.) DC., (camphor), CAMPHORWEED, CAMPHOR PLUCHEA. Aromatic annual or perennial to 2 m tall; leaf blades elliptic to oblong-elliptic, entire to serrate, the lower surface puberulent, both surfaces sparsely dotted with resin-globules. Low drainage areas; Collin, Dallas, Grayson, and Lamar cos.; se and e TX w to n part of nc TX, also Edwards Plateau. Summer–fall.

Pluchea foetida (L.) DC., (fetid, bad-smelling), STINKING PLUCHEA, STINKING-FLEABANE. Aromatic perennial 0.4–1 m tall; leaves usually auriculate-clasping but sometimes with cuneate bases; inflorescences often flat-topped. Wet areas; Henderson Co. near extreme e margin of nc TX; mainly se and e TX. Summer–fall.

Pluchea odorata (L.) Cass., (fragrant), CANELA, PURPLE PLUCHEA. Aromatic annual to 1.5 m tall; leaf blades mostly ovate to lanceolate or elliptic, entire to serrate, essentially glabrous to tomentose, resin-globules often present. Low drainage areas; Brown, Comanche, Cooke, Coryell, Dallas, and Denton cos.; nearly throughout TX. Summer–fall. [*Conyza odorata* L., *P. purpurascens* (Sw.) DC.] Jones et al. (1997) used the name *P. purpurascens* for this taxon.

PSEUDOGNAPHALIUM CUDWEED, EVERLASTING

Annual or biennial herbs, whitish to grayish with felty or woolly pubescence at least when young; stems erect or ascending; leaves alternate, sessile; inflorescences usually with numerous disciform heads densely clustered, somewhat flat-topped; involucres woolly near base; phyllaries scarious; receptacles naked; ray flowers absent; disk flowers all fertile; achenes glabrous; pappus of numerous bristles, these separating individually or in clusters, not deciduous as a ring.

A genus of ca. 80 species in the New World, Africa, and Asia; often lumped into *Gnaphalium*. While often treated in *Gnaphalium* (e.g., Mahler 1988; Kartesz 1994), we are following Anderberg (1991), Jones et al. (1997), and J. Kartesz (pers. comm. 1997) in recognizing the following species in *Pseudognaphalium*. (Greek: *pseudo*, false, and the genus name *Gnaphalium*, from an ancient Greek name of some downy plant, derived from *gnaphalon*, lock of wool) (tribe Inuleae)
REFERENCES: Hillard & Burtt 1981; Anderberg 1991.

1. Leaves at maturity usually bright green and somewhat glandular-pubescent on the upper surface, not decurrent; widespread in nc TX _______ **P. obtusifolium**
1. Leaves at maturity gray- or white-woolly on both surfaces, decurrent; rare in nc TX, known only from Brown Co. on extreme w margin of nc TX _______ **P. stramineum**

Pseudognaphalium obtusifolium (L.) Hilliard & Burtt, (blunt-leaved), FRAGRANT CUDWEED, CAT'S-FOOT, FRAGRANT EVERLASTING. Aromatic annual; stems usually erect, 0.1–1 m tall, usually white-woolly to subglabrate; rosette leaves with blades oblanceolate to spatulate; stem leaves with blades linear to narrowly lanceolate or narrowly oblanceolate, 3–10 cm long, 2–10 mm wide, white woolly below, green, ± glabrous or sparsely woolly or glandular above; involucres 5–7 mm tall, whitish, developing a slight rusty tinge; achenes 0.8–1.2 mm long. Prairies, open woods, roadsides, often in sandy soils; se and e TX w to East Cross Timbers, also Edwards Plateau. (Jul–)Sep–Nov. [*Gnaphalium obtusifolium* L.]

Pseudognaphalium stramineum (Kunth) W.A. Weber, (straw-colored), COTTON-BATTING CUD-

WEED. Annual or biennial; stems and leaves gray- or white-woolly, not glandular; stems erect or ascending, 0.2–0.8 m tall; rosette leaves oblanceolate to spatulate; leaves mainly cauline, linear to narrowly lanceolate or narrowly oblanceolate, 2–6 cm long, 2–5 mm wide, adnate-auriculate; involucres 4–6 mm tall, shiny yellowish white; achenes 0.6–0.8 mm long. Rocky prairies and fields; Brown Co. (Stanford 1971) on extreme w margin of nc TX; Rolling Plains and Edwards Plateau w to Trans-Pecos. May–Oct. [*Gnaphalium chilense* Spreng.; *Gnaphalium stramineum* Kunth]

PYRRHOPAPPUS FALSE DANDELION, NATIVE-DANDELION

Annuals or perennials; sap milky; leaves basal and alternate, reduced upwards; leaf blades oblanceolate or oblong- to elliptic-lanceolate, entire or usually toothed or lobed; heads solitary or few, large and showy; flowers all ligulate; corollas yellow, open during morning; achenes beaked; pappus of abundant hair-like bristles 7–12 mm long.

A North American genus of 3 species; the roots of some were eaten by Native Americans. (Greek: *pyrros*, flame-colored or red, and *pappus*, down, fuzz, pappus) (tribe Lactuceae)
REFERENCES: Northington 1971, 1974; Vuilleumier 1973; Barber & Estes 1978; Petersen et al. 1990; Turner & Kim 1990.

1. Perennials with small tuberous thicking 5–20 mm thick located 2–15 cm below ground at end of roots; stems with 0–2 leaves or leafy bracts, usually pubescent or short-pilose; stems 10–30 (–40) cm tall ____ **P. grandiflorus**
1. Annuals with tapering taproot; stems with 1–9 leaves, pubescent or glabrous; stems 20–100 cm tall.
   2. Stems pubescent with curly hairs (rarely glabrous), mostly branched from the base; stem leaves 1–5, the uppermost with blades deeply several-lobed; outer phyllaries usually 1/3 as long as inner or less; plants (5–)20–50(–60) cm tall ____ **P. pauciflorus**
   2. Stems glabrous, mostly unbranched from the base; stem leaves 3–9, the uppermost with blades entire or merely toothed or with one pair of basal lobes; outer phyllaries mostly 1/3–2/3 as long as inner; plants (5–)30–100 cm tall ____ **P. carolinianus**

Pyrrhopappus carolinianus (Walter) DC., (of Carolina), CAROLINA FALSE DANDELION. Annual or biennial from a taproot; stems 30–100 cm tall; basal leaves in a rosette, often early deciduous; stem leaves reduced upwards; principal phyllaries 15–25 mm long; achenes 14–17 mm long (including beak). Sandy soils; woodlands and fields; in nc TX in East and West cross timbers, also along Red River; se and e TX w to West Cross Timbers. Apr–Jul, sporadically to Sep. [*P. carolinianus* var. *georgianus* (Shinners) Ahles, *P. georgianus* Shinners]

Pyrrhopappus grandiflorus (Nutt.) Nutt., (large-flowered), TUBER FALSE DANDELION. Perennial with a tuber giving rise to 1 or several, simple or branched, erect or suberect rhizomes, these developing near ground level; leaf rosettes and eventually flowering stems 10–30(–40) cm tall; stems usually rather densely; pubescent with curly hairs, rarely glabrous; principal phyllaries 15–20 mm long; achenes 10–13 mm long. Rocky, clayey, or sandy prairies; Grand Prairie s and w to w TX. Apr–May.

Pyrrhopappus pauciflorus (D. Don) DC., (few-flowered), MANY-STEM FALSE DANDELION, TEXAS DANDELION, PATA DE LEON. Annual 20–60 cm tall; basal leaves in a rosette; principal phyllaries 16–20 mm long; achenes 12–14 mm long. Clay soils, prairies, roadsides; Blackland Prairie and Grand Prairie; se and e TX w to nc TX and Edwards Plateau. Mar–Jun. Hybridizes with *P. carolinianus* along highways where sand and limestone gravel have been brought in for fill. Petersen et al. (1990) indicated that infraspecific taxa are not warranted. [*P. geiseri* Shinners, *P. multicaulis* DC., *P. multicaulis* var. *geiseri* (Shinners) North.]

RATIBIDA MEXICAN-HAT

A North American genus of 7 species; some are cultivated as ornamentals. (According to Fernald (1950a), "... meaning, like most work of its author [Rafinesque], not clear") (tribe Heliantheae)

REFERENCES: Jackson 1963; Richards 1968; Cox & Urbatsch 1990; Urbatsch & Jansen 1995.

Ratibida columnifera (Nutt.) Wooton & Standl., (bearing columns), MEXICAN-HAT, THIMBLE-FLOWER, PRAIRIE CONEFLOWER, UPRIGHT PRAIRIE CONEFLOWER. Minutely scabrous-pubescent perennial, flowering the first year; stems 0.2–1.2 m tall; leaves alternate, compound or deeply lobed, the ultimate segments linear or lanceolate; heads long-peduncled; ray flowers 3–7, infertile, with ligules yellow or orange-yellow, usually but not always with dark reddish brown base, sometimes with only the tip yellow, 10–20(–35) mm long, spreading or often reflexed; disk flowers perfect, fertile, with corollas yellow-brown; receptacles columnar, conspicuously elongated, 1.5–4.5 cm long; scales gray-pubescent; pappus of a minute awn ca. 0.6 mm long and often also a second shorter awn. Sandy, silty, or rocky open ground; throughout most of TX. May–Oct. [*R. columnaris* (Sims) D. Don] ▣/104

RAYJACKSONIA

A North American genus of 3 species only recently named (Lane & Hartman 1996); it was previously included in a broadly defined and polyphyletic *Haplopappus*; the taxa have been treated in a variety of genera including *Machaeranthera* (e.g., Kartesz 1994; Jones et al. 1997). We are following Lane and Hartman (1996) for nomenclature of the genus. Hartman (1990), referring to what is now *Rayjacksonia* as the "Phyllocephalus" group of *Haplopappus*, indicated it is part of a $x = 6$ chromosome line including such genera as *Grindelia* and *Xanthocephalum*. (Named for Dr. Raymond C. Jackson, student of *Haplopappus* sensu lato who first reported the correlation in the Astereae between a base chromosome number of $x = 6$ and a particular disk floret corolla shape known as "goblet-shaped") (tribe Astereae)

REFERENCES: Hall 1928; Hartman 1990; Lane & Hartman 1996; Morgan 1997 [1998].

Rayjacksonia annua (Rydb.) R.L. Hartman & M.A. Lane, (annual). Annual or weakly biennial herb from a taproot, sparsely to densely glandular; stems erect, 0.3–0.7 m tall, branched above; leaves alternate, ± sessile, often much reduced upwards; leaf blades oblanceolate to narrowly obovate, 2–5+ cm long, 3–15 mm wide, with ca. 7–10 prominent marginal teeth on each side, these mucronate or with a short bristle; heads solitary at the ends of the branches or ± corymbosely arranged; involucres 6–8 mm high; phyllaries in several series, acute to acuminate or tapered to a minute bristle, densely glandular-pubescent (use lens), mostly straw-colored basally, with darker (usually green) tip; ray and disk flowers yellow; ray flowers pistillate, fertile; ligules 7–10 mm long; disk flowers perfect, fertile; pappus of ray and disk flowers similar, of hair-like bristles; achenes ca. 2 mm long, pubescent. Open areas; included based on citation of vegetational area 4 (Fig. 2) by Hatch et al. (1990); also s part of e TX, Plains Country, and Trans-Pecos. Aug–Oct. [*Haplopappus annuus* (Rydb.) Cory, *Haplopappus phyllocephalus* (DC.) Shinners subsp. *annuus* (Rydb.) H.M. Hall, *Machaeranthera annua* (Rydb.) Shinners, *Sideranthus annuus* Rydb.]

RUDBECKIA CONEFLOWER, BROWN-EYED-SUSAN

Annual or perennial herbs; leaves alternate, the blades usually unlobed or lobed to pinnatifid in 1 species; heads terminal, large, solitary or corymbose; phyllaries few, herbaceous, loose; ray flowers infertile, with ligules deep yellow or orangish, sometimes partly reddish brown; disk conical or ovoid to subcylindrical, often elongating after flowering; disk flowers perfect, fertile, with corollas yellow-brown to blackish red-brown or dark purple.

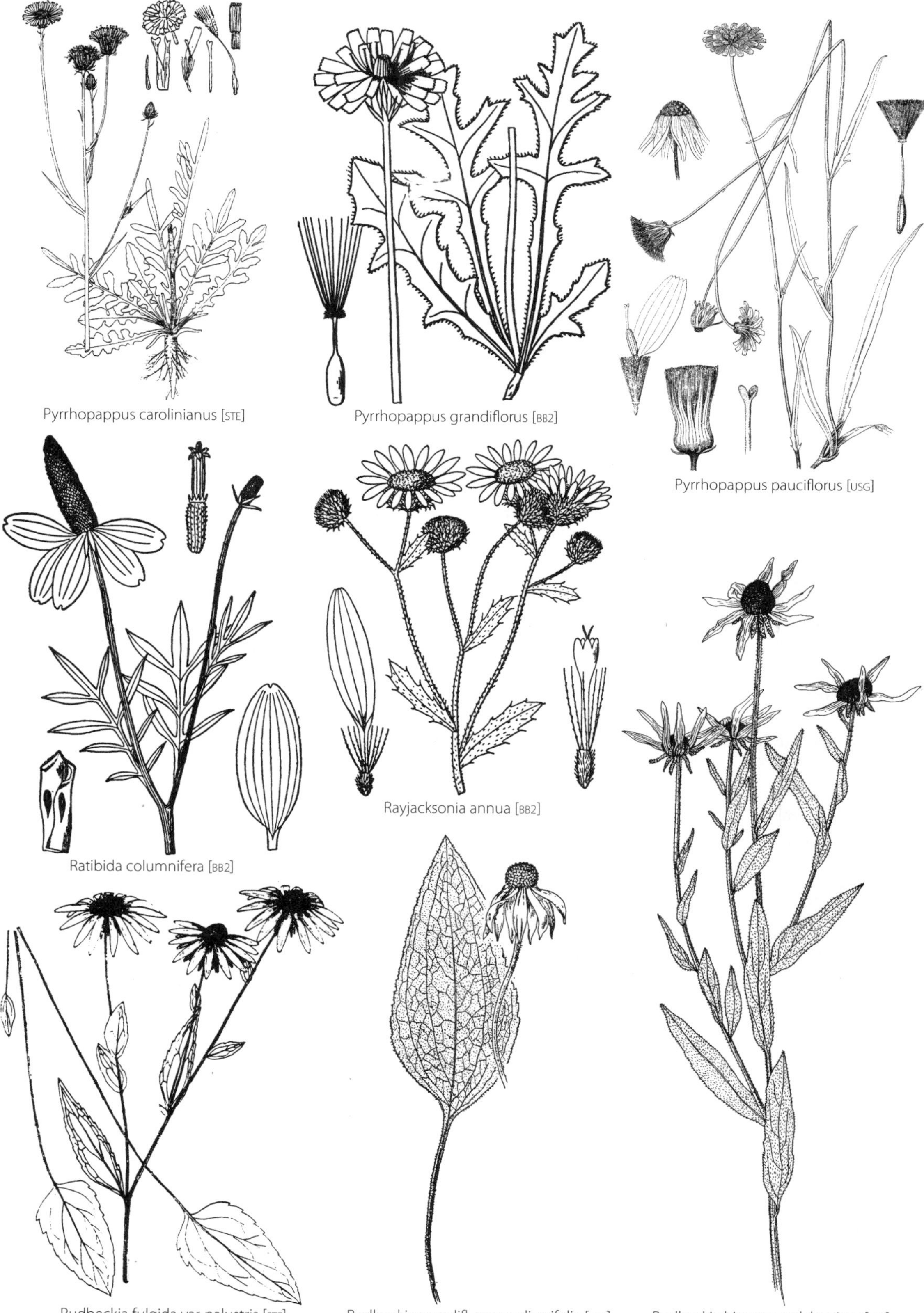

Pyrrhopappus carolinianus [STE]

Pyrrhopappus grandiflorus [BB2]

Pyrrhopappus pauciflorus [USG]

Ratibida columnifera [BB2]

Rayjacksonia annua [BB2]

Rudbeckia fulgida var. palustris [STE]

Rudbeckia grandiflora var. alismifolia [GLE]

Rudbeckia hirta var. pulcherrima [BT3]

A North American genus of 15 species; some are cultivated as ornamentals. (Named for Swedish professors Rudbeck: Olaf, 1630–1702, the father, and Olaf, 1660–1740, the son, predecessor of Linnaeus at Uppsala) (tribe Heliantheae)
REFERENCES: Perdue 1957; Cox & Urbatsch 1990, 1994; Urbatsch & Jansen 1995.

1. Some leaf blades usually tri-lobed or pinnatifid; pales (= bracts on receptacle) subtending disk flowers with glabrous, long smooth points; rare in nc TX ____ **R. triloba**
1. Leaf blades usually merely toothed or subentire; pales subtending disk flowers ciliate margined or pubescent- or bristle-tipped; includes species common in nc TX.
   2. Upper stem leaves auricled-clasping; leaf blades glabrous, glaucous; plants often very large, 1–3 m tall ____ **R. maxima**
   2. Upper stem leaves slightly or not clasping; leaf blades scabrous-pubescent or hispid-pilose; plants 0.3–1.3 m tall.
      3. Style branches of disk flowers short and blunt; pappus present (often minute); foliage pubescent but not coarsely hirsute; pales without a bristle at tip.
         4. Ligules 1–3 cm long; plants rhizomatous; pales glabrous except for ciliate margin; leaves without microscopic resin globules ____ **R. fulgida**
         4. Ligules 3–5(–7) cm long; plants with thick woody root or crown; pales canescent near tip with viscid hairs; leaves with microscopic resin globules (appear golden in light with lens) ____ **R. grandiflora**
      3. Style branches of disk flowers elongate, awl-like; pappus absent; foliage coarsely hirsute; pales with a bristle (or bristles) at tip ____ **R. hirta**

Rudbeckia fulgida Aiton var. **palustris** (Eggert ex C.L. Boynton & Beadle) Perdue, (sp.: fulgid, shining; var.: marsh-loving), MARSH CONEFLOWER. Nearly glabrous to strigose or hirsute perennial; basal leaf blades elliptic to orbicular; stem leaves gradually reduced upward, with blades ovate-lanceolate to lanceolate; ligules yellow to orangish. Disturbed areas; included based on citation of vegetational areas 4 and 5 (Fig. 2) by Hatch et al. (1990); e and nc TX and Edwards Plateau. May–Jul. [*R. coryi* Shinners]

Rudbeckia grandiflora (D. Don) J.F. Gmel. ex DC. var. **alismifolia** (Torr. & A. Gray) Cronquist, (sp.: large-flowered; var.: with leaves like *Alisma*—water-plantain), ROUGH CONEFLOWER. Perennial 0.5–1.0 m tall; stems glabrous to scabrous-pubescent below; basal leaves persistent, long-petioled, with broadly lanceolate or elliptic blades; ligules orange-yellow. Sandy open woods, chiefly low ground; Tarrant Co. (Mahler 1988); mainly se and e TX. Jun–Jul.

Rudbeckia hirta L. var. **pulcherrima** Farw., (sp.: hairy; var.: very handsome), BLACK-EYED-SUSAN. Coarsely pubescent annual or short-lived perennial; leaves narrowed basally, subpetiolate to sessile; heads peduncled; ligules yellow to orangish, sometimes reddish brown or purplish basally. Disturbed areas; se and e TX w to West Cross Timbers and Edwards Plateau. May–Jul, sporadically to Sep. *Rudbeckia hirta* is reported to cause poisoning in cattle, sheep, and hogs (Pammel 1911; Kingsbury 1964). ☠

Rudbeckia maxima Nutt., (largest), GIANT CONEFLOWER, GREAT CONEFLOWER, LARGEST BROWN-EYED-SUSAN. Glabrous, glaucous perennial, large and conspicuous; leaf blades ovate to oblong; upper leaves sessile, partly clasping; ligules 3–8 cm long, to 12 mm wide, yellow or orangish yellow; pales pubescent apically; disks conspicuously elongating in fruit, 4–8 cm long. Forming beds in low ground, sandy or silty soils; e TX w to East Cross Timbers, also Tarrant Co. (Fort Worth Nature Center). May–Jun.

Rudbeckia triloba L., (three-lobed), YELLOW DAISY, BROWN-EYED-SUSAN. Short-lived perennial, hispid to glabrous; leaves simple and cordate when young, often becoming 3-lobed to pinnately 5-7-parted; upper leaves ovate to narrowly ovate; ligules yellow or orange toward base. Creek bottoms, roadsides; Lamar Co. in Red River drainage; otherwise apparently only known in TX from Nacogdoches Co. in e TX; first reported for TX by Taylor and Taylor (1981). May–Oct.

SENECIO RAGWORT, SQUAW-WEED, GROUNDSEL

Ours annual herbs, glabrate to usually pubescent; leaves alternate, the blades toothed to pinnately lobed; inflorescences terminal, the heads cymosely arranged; principal phyllaries in a single series, linear, with hyaline margins, subtended by an outer series of short, bract-like phyllaries; ray flowers usually pistillate, fertile, with yellow ligules, or ray flowers absent; disk flowers perfect, fertile, the corollas yellow; achenes cylindrical; pappus of numerous, capillary, white hairs.

When broadly considered, *Senecio* is a huge, cosmopolitan genus of ca. 1,250–3,000 species depending on circumscription. Species vary in growth form from trees to shrubs, vines, herbs, or desert succulents, and include the GIANT GROUNDSELS on the mountains of e Africa. According to Barkley et al. (1996), new information and new concepts of taxa indicate that the genus should be divided into various segregates. We are following Freeman and Barkley 1995, Barkley et al. 1996, and T. Barkley (pers. comm.) in treating the aureoid group of Senecios as *Packera*. Even when this and numerous other segregates are removed, *Senecio* is still one of the largest genera of seed plants. A number are toxic and can be problematic if eaten by livestock, especially horses; acute illness and death can result from hepatotoxic pyrrolizidine alkaloids which cause liver damage. Milk from grazing animals and honey from *Senecio* nectar reportedly contain the alkaloids (Lewis & Elvin-Lewis 1977; Fuller & McClintock 1986; Blackwell 1990). Human deaths have been reported from ingestion of *Senecio* herbal teas as the result of liver damage—there is no known cure (Lampe & McCann 1985; Tveten & Tveten 1993). Some *Senecio* species were in the past used in poultices for wounds and abscesses; the common name GROUNDSEL apparently is derived through a series of changes from the Anglo-Saxon: *grundeswelge*, pus-absorber (Tveten & Tveten 1993). (Latin name of a plant, from *senex*, an old man, alluding to the hoariness of many species, or to the white hairs of the pappus) (tribe Senecioneae).

REFERENCES: Greenman 1915–1918; Vuilleumier 1969a; Barkley 1978, 1985a, 1985b, 1986; Barkley et al. 1996.

1. Ray flowers yellow, prominent; main phyllaries and short, bract-like outer phyllaries below main phyllaries green-tipped; leaves shallowly toothed to subentire; native species ____ **S. ampullaceus**
1. Ray flowers usually absent, but if present the ligules greatly reduced and inconspicuous; main phyllaries and short, bract-like outer phyllaries prominently black-tipped; leaves usually shallowly pinnately lobed; introduced weedy species ____ **S. vulgaris**

Senecio ampullaceus Hook., (flask-shaped), TEXAS GROUNDSEL. Loosely woolly-pubescent to nearly glabrate annual 30–80 cm tall; lowermost leaves narrowed to a subpetiolar base, auriculate-clasping at very base; middle and upper leaves often truncate-clasping; involucres 7–11 mm high. Sandy open woods, fields, roadsides; Burnet, Callahan, Comanche, Dallas, and Denton cos.; se and e TX w to West Cross Timbers and Edwards Plateau; endemic to TX. Mar–May.

Senecio vulgaris L., (common), COMMON GROUNDSEL. Annual 10–30(–60) cm tall, woolly-pubescent to glabrate; lower leaves petiolate; middle and upper leaves sessile, auriculate, ± clasping, undulate to pinnately lobed, 2–10 cm long, 0.5–2(–4.5) cm wide; involucres 6–10 mm high; achenes with fine hairs, 2.2–2.5 mm long; pappus 5–6 mm long. Disturbed areas; Dallas, Eastland, Grayson (first noted Mar 1998), and Tarrant (first noted Mar 1998) cos., also Brown Co. (first noted 1987; HPC); also Wichita Co. in Rolling Plains and Harris Co. in se TX; also Dimmit Co. (Cory 1948a); apparently first collected in TX in Dimmit Co. in 1944 (Cory 1947, 1948a). Mar–May. Native of

Europe. This species is among a number of Senecios that contain toxic pyrrolizidine alkaloids; animal deaths have been reported; the toxins are not lost upon drying and are still present in hay (Burlage 1968; Lampe & McCann 1985; Fuller & McClintock 1986).

SILPHIUM ROSINWEED

Herbaceous, coarse, sunflower-like perennials with woody crown or taproot; stems erect; leaves alternate to opposite, simple, the blades toothed or deeply lobed, entire, stiff, leathery, often conspicuously scabrous; heads large and showy, terminal, solitary, corymbose, or spicate-racemose; ray flowers pistillate, fertile, with ligules yellow or in one species white, producing mature achenes that are flattened, thin-edged, and winged; disk flat or nearly so; disk flowers functionally staminate, not maturing achenes; pappus absent or of 2 awns.

An e North American genus of 23 species. The common name refers to the sticky secretions on the stems and leaves of some species (Ajilvsgi 1984). Species of this genus are often confused with those of *Helianthus*; however, in *Helianthus* the disk flowers produce mature achenes (ray flowers do not), while in *Silphium* the ray flowers produce mature achenes (disk flowers do not). (Greek: *silphion*, the ancient name of some resinous plant, transferred by Linnaeus to this genus) (tribe Heliantheae)
REFERENCES: Perry 1937; Settle & Fisher 1970.

1. Leaves deeply pinnatifid.
   2. Ray flowers with ligules white, ca. 25 mm long; stems 1 m or less tall; achenes puberulent, with wings prolonged beyond the apex of the body of the achenes forming a V-shaped notch ca. 3–5 mm deep ____ **S. albiflorum**
   2. Ray flowers with ligules yellow, 25–40 mm long; stems 1–3 m tall; achenes essentially glabrous, with wings prolonged beyond the body of the achenes forming a notch ca. 1–3 mm deep ____ **S. laciniatum**
1. Leaves entire to toothed.
   3. Leaves alternate or opposite above, with some subtending heads (leaves only very gradually reduced up the stem so that well-developed leaves are usually very near the heads); peduncles usually short ____ **S. radula**
   3. Leaves alternate above, reduced, not subtending heads (no well-developed leaf close to a head); peduncles typically long and ± naked ____ **S. gracile**

Silphium albiflorum A. Gray, (white-flowered), WHITE ROSINWEED, WHITE COMPASSPLANT. Stems (0.2–)0.4–1 m tall; leaves alternate, rigid, very scabrous, vertically oriented, ca. as broad as long; heads in a raceme-like arrangement; ray flowers with ligules white. Calcareous soils, open areas and prairies; Blackland Prairie and Grand Prairie; nc TX s to Edwards Plateau and w to Rolling Plains; endemic to TX. May–Jul. /105

Silphium gracile A. Gray, (graceful), SIMPSON ROSINWEED, SLENDER ROSINWEED. Plant subscapose; stems (0.2–)0.3–0.6 m tall; leaves opposite below, alternate above and reduced in size and number; heads usually solitary on the peduncles; ray flowers with ligules yellow. Calcareous or sandy soils; Bell, Hunt, and Milam cos.; se and e TX w to e part of nc TX. Jun–Jul. [*S. reverchonii* Bush, *Silphium simpsonii* Greene var. *wrightii* L.M. Perry] The common name refers to the sticky resinous material secreted along the stems and leaves; it was used medicinally by Native Americans and also as a chewing gum (Ajilvsgi 1984).

Silphium laciniatum L., (laciniate, torn), COMPASSPLANT. Stems 1–3 m tall; leaves alternate, rigid, vertically oriented, usually longer than wide; heads in a raceme-like arrangement; ray flowers with ligules yellow. Calcareous or sandy soils, prairies; Cooke, Dallas, Denton, Grayson, Kaufman, Tarrant, and Wise cos.; also se and e TX. Jun–Aug. The common name refers to the vertical, supposedly north-south orientation of the leaves. /105

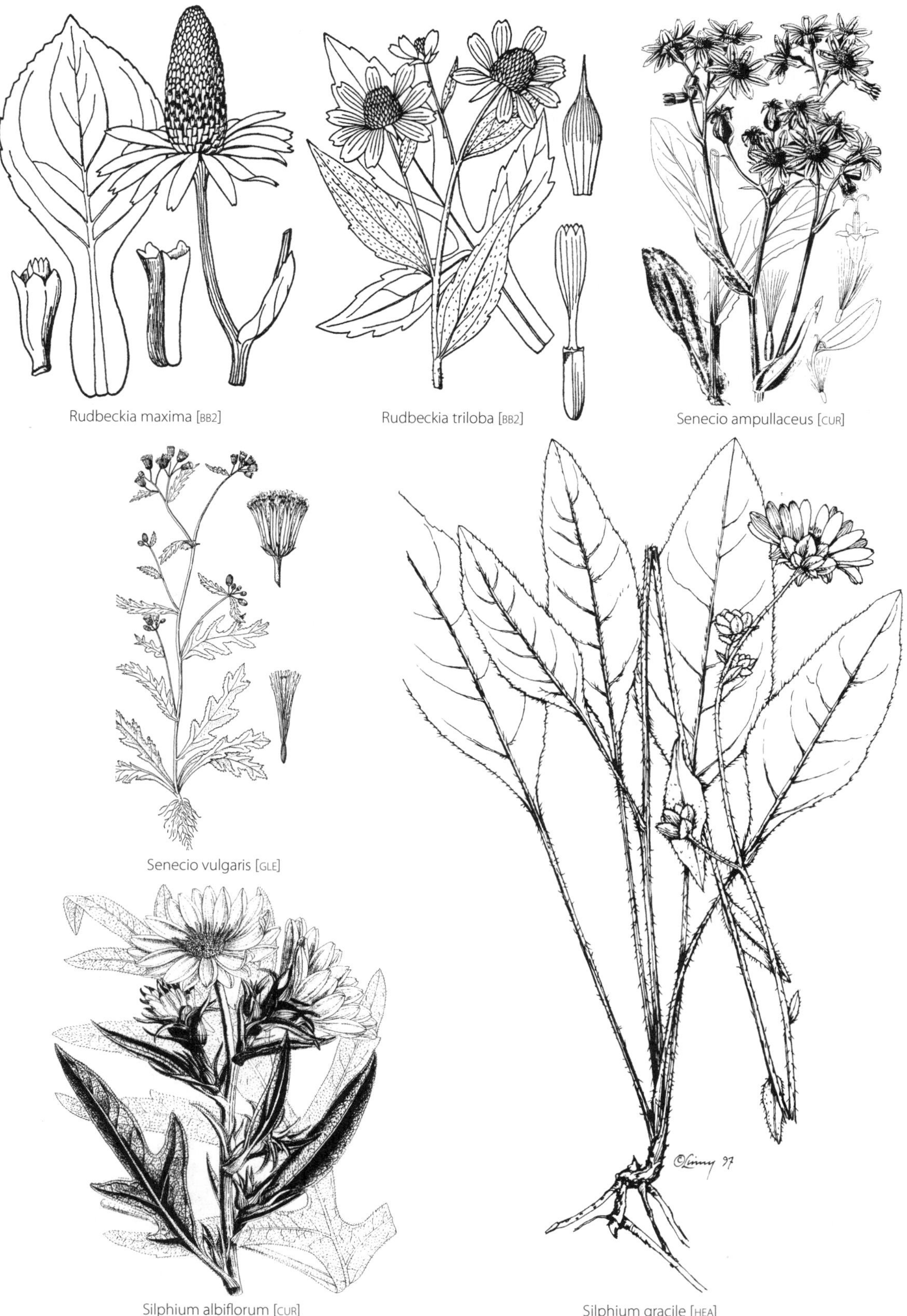

Rudbeckia maxima [BB2]

Rudbeckia triloba [BB2]

Senecio ampullaceus [CUR]

Senecio vulgaris [GLE]

Silphium albiflorum [CUR]

Silphium gracile [HEA]

Silphium radula Nutt., (rough), ROUGH-STEM ROSINWEED. Stems 0.3–1.2+ m tall; leaves alternate to opposite below and above (varying in same population); median and upper leaves sessile and entire or slightly toothed; heads solitary or a few in a corymbose arrangement; ray flowers with ligules yellow. Calcareous or sandy soils; se and e TX w to West Cross Timbers and e Edwards Plateau. Jun–Jul. [*Silphium asperrimum* Hook.] According to Correll and Johnston (1970), *S. radula* apparently intergrades with *S. gracile* (as *S. simpsonii* var. *wrightii*). *Silphium radula* seems very similar to *S. integrifolium* Michx., which occurs to the n and in TX only in the extreme e part of the state (vegetational area 1) (Settle & Fisher 1970; Hatch et al. 1990). According to Cronquist (1980), *S. radula* has longer, coarser stem pubescence with many or all of the hairs ca. 1 mm or more long and leaves often alternate. *Silphium integrifolium*, on the other hand, has most stem hairs ca. 0.5 mm or less long, varying to largely or wholly glabrous and leaves mostly opposite. Our plants seem to have mostly shorter hairs with individuals varying from having alternate to opposite leaves. While we are following Hatch et al. (1990) in treating the nc TX plants as *S. radula*, more work needs to be done to determine the actual patterns of variation of these very similar entities. /105

SILYBUM MILK-THISTLE

A genus of 2 species native to the Mediterranean region. (Greek: *silybon*, name for similar thistle-like plants) (tribe Cardueae)

Silybum marianum (L.) Gaertn., (with white-mottled leaves, the spots were supposed to have resulted from drops of Mary's milk falling on the leaves), BLESSED MILK-THISTLE, OUR-LADY'S-THISTLE, HOLY-THISTLE. Coarse annual or biennial (rarely perennial?) 0.6–1.8+ m tall; stems loosely woolly-pubescent, becoming glabrous; leaves alternate, mottled green and white, the lower pinnately lobed, the middle and upper toothed; stem leaves auricled-clasping, all prickly or spiny-margined, not decurrent; heads large (2.5–6 cm in diam.), solitary, terminal; phyllaries to 5 cm long, with spinescent margins, constricted below middle, with broad bases and green, sub-leafy apical portions tapering to a spiny tip to 8 mm long; ray flowers absent; disk corollas rosy purple; pappus of numerous white, non-plumose bristles 15–20 mm long, deciduous as a ring. Pastures and roadsides; Dallas, Johnson, and Navarro cos., also Hamilton Co. (HPC); also Edwards Plateau; first collected in TX in Sutton Co. on the Edwards Plateau in 1938 (supposedly introduced through a shipment of hay from California) and in nc TX in Navarro Co. in 1949 (Cory 1940, 1950). May–Jun. Native of Mediterranean region. The fruit is used medicinally and considered since the time of Dioscorides to protect the liver; it contains flavonoids effective as an antidote for mushroom (*Amanita*) poisoning—the mode of action is to displace the toxin from cell membrane receptors (Mabberley 1987); its clinical use in Europe is widespread (Leung & Foster 1996).

SIMSIA

A New World genus of 18 species extending from the sw U.S. to Argentina with 1 taxon endemic to Jamaica (Spooner 1987; Schilling & Spooner 1988). (Named for John Sims, 1749–1831, of Dorking, editor of *Curtis's Botanical Magazine*) (tribe Heliantheae)
REFERENCES: Spooner 1987; Schilling & Spooner 1988.

Simsia calva (Engelm. & A. Gray) A. Gray, (naked), AWNLESS BUSH-SUNFLOWER. Perennial, herbaceous from woody base, coarsely pubescent; roots fusiform or woody; herbage dark green; leaves opposite, petioled, the petioles sometimes auricled; leaf blades deltoid, often lobed, serrate; stipules toothed or serrate, often united; heads solitary at the end of long peduncles; phyllaries in several series; ray flowers 15–30, pistillate, infertile, the ligules yellow or orange-yellow, sometimes spotted or striped beneath with purple or red; pappus absent; disk flowers perfect, fertile; corollas yellow or orange-yellow, sometimes with purplish lines, gibbous basally; achenes flat-

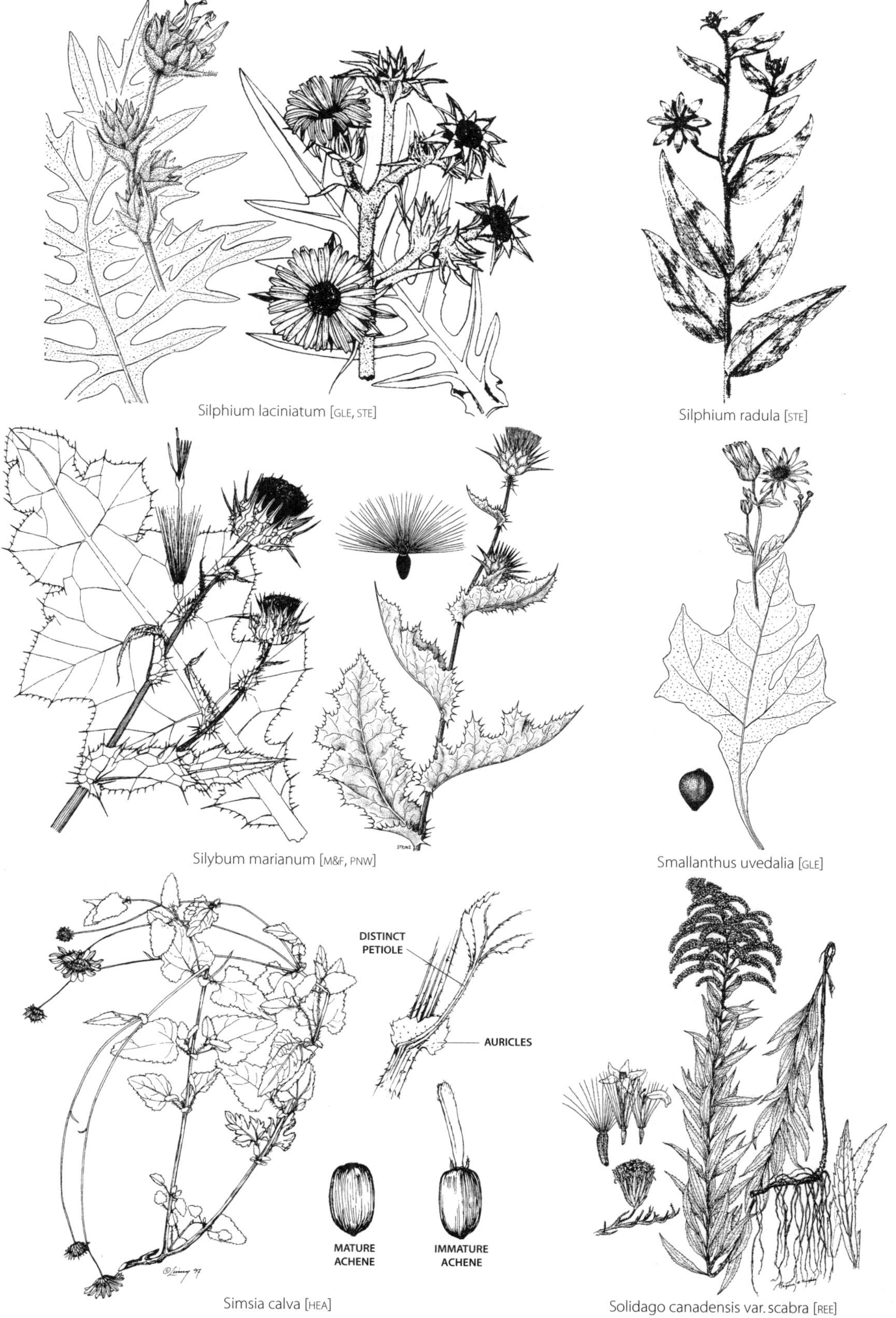

Silphium laciniatum [GLE, STE]

Silphium radula [STE]

Silybum marianum [M&F, PNW]

Smallanthus uvedalia [GLE]

Simsia calva [HEA]

Solidago canadensis var. scabra [REE]

tened, emarginate, appearing winged; pappus of 2 short awns, usually absent on mature achenes. Calcareous soils; Coryell, Hamilton, Mills, and Shackelford cos., also Brown and Comanche cos. (Mahler 1988); w and sw parts of nc TX through Edwards Plateau to Trans-Pecos, also se and s TX. May–Nov.

SMALLANTHUS LEAFCUP

A genus of ca. 19 species of tropical and warm areas of the Americas; previously recognized in the genus *Polymnia*. (Named for John Kunkel Small, 1869–1938, American botanist and author of numerous works including *Manual of the Southeastern Flora*) (tribe Heliantheae)
REFERENCES: Wells 1965; Robinson 1978; Turner 1988b.

Smallanthus uvedalia (L.) Mack. ex Small, (derivation unknown, not indicated by Linnaeus), BEAR'S-FOOT, HAIRY LEAFCUP. Herbaceous perennial 1–3 m tall; stems erect, purple-spotted; leaves at least lower, opposite, often huge, to ca. 70 cm long and 40 cm wide, ovate to deltoid, palmately 3- to 5-lobed and veined, sessile or with broad conspicuous wings to base of petiole; heads clustered in loose leafy cymes; phyllaries 4–6, 10–20 mm long, to ca. 10–12 mm wide, ovate or ovate lanceolate; ray flowers 7–13, pistillate, fertile, the ligules yellow, 10–20(–30) mm long; disk flowers numerous, infertile, staminate, yellow; achenes ca. 6 mm long and 4 mm wide; pappus absent. Wooded bottomlands; Bell, Dallas, Fannin, Grayson (Hagerman N.W.R.), and Tarrant cos.; mainly se and e TX, w to nc TX and e edge of Edwards Plateau. Jul–Sep. [*Polymnia uvedalia* (L.) L., *Polymnia uvedalia* (L.) L. var. *densipilis* S.F. Blake]

SOLIDAGO GOLDENROD

Herbaceous perennials; leaves alternate, sessile or short-petioled; inflorescence terminal, branched, of numerous, relatively small, often secund heads; phyllaries imbricated, in several series, inconspicuously green-tipped; ray flowers pistillate, fertile, few, usually 14(–20) or less, the ligules linear, yellow in ours; disc flowers perfect, fertile, the corollas cylindrical to funnelform, yellow in ours; achenes cylindrical, ribbed; pappus of numerous, white, hair-like bristles.

A mainly North American genus of ca. 80 species with a few taxa in South America, Macaronesia, and Eurasia. Because of marked phenotypic plasticity, polyploidy, and hybridization, *Solidago* is a taxonomically difficult genus. Complete specimens including mature inflorescences and lower leaves are often needed for definitive identification. Despite their bad reputation, GOLDENRODS are insect-pollinated and thus do not produce significant windblown pollen which could cause hay fever; most fall allergies are probably due to wind-pollinated species of *Ambrosia* (RAGWEED) that flower at the same time as GOLDENRODS. The related genus *Euthamia*, sometimes included in *Solidago*, is here treated separately. The key to *Solidago* species is adapted from Taylor and Taylor (1984). Nesom (1993) split the genus *Oligoneuron* from *Solidago*; it is a segregate of 6 species including 2 in nc TX, *S. nitida* and *S. rigida*. Until consensus is reached, we are following Taylor and Taylor (1984) and Semple (1992) in treating it traditionally and keeping all species in *Solidago*; appropriate synonymy is given below; nomenclature in general follows Taylor and Taylor (1984). (Latin, *solido*, to make whole or strengthen, alluding to medicinal properties) (tribe Astereae)
REFERENCES: Kapoor & Beaudry 1966; Croat 1972; Anderson & Creech 1975; Taylor & Taylor 1983, 1984; Heard & Semple 1988; Gandhi & Thomas 1989; Nesom 1990d, 1991c, 1993; Semple 1992.

1. Inflorescence flat-topped or nearly so (= corymbose) (segregate genus *Oligoneuron*).
   2. Leaf blades usually densely pubescent, ovate, ca. 2–3.5 times as long as wide; ray flowers 7–14 per head; usually flowering Sep–Oct(–Nov) **S. rigida**
   2. Leaf blades glabrous, linear to narrowly lanceolate, ca. 5–10 times as long as wide; ray flowers 1–4 per head; usually flowering Jun–Sep **S. nitida**

1. Inflorescence elongate, not flat-topped, either racemose, paniculate, cylindric, or pyramid-shaped and broadest near base or midsection (*Solidago* in a narrower sense).
   3. Inflorescence racemose or very narrowly paniculate, less than 4 cm wide, ± cylindric in shape.
      4. Primary stems glabrous; leaf blades either usually glabrous OR scabrous on upper surface; basal leaves elliptic to ovate or ovate; upper stem leaves small, appressed; atypical forms (these species usually have broadly paniculate inflorescences).
         5. Stems subcylindrical; leaf blades with upper surface usually glabrous ____ **S. ludoviciana**
         5. Stems, at least towards the base, bluntly 4-angled; leaf blades with upper surface strongly scabrous ____ **S. patula**
      4. Primary stems pubescent (pubescence often short); leaf blades usually pubescent on upper surface; basal leaves absent at flowering time; upper stem leaves spreading to widely spreading ____ **S. petiolaris**
   3. Inflorescence paniculate, broadest at or near base, wider than 4 cm, not cylindric in shape.
      6. Primary stems essentially glabrous, rarely with sparse hairs; leaves glabrous.
         7. Inflorescence essentially glabrous.
            8. Leaf blades ovate, toothed along most of margin; in wooded habitats ____ **S. ulmifolia**
            8. Leaf blades oblanceolate, toothed only towards tip; in grasslands ____ **S. missouriensis**
         7. Inflorescence pubescent, usually densely so.
            9. Inflorescence usually narrowly paniculate, with ends of branches erect; heads not secund ____ **S. speciosa**
            9. Inflorescence broadly paniculate, triangular in outline, with ends of branches recurved; heads secund.
               10. Basal leaves absent; lower stem leaves absent at flowering time; stem leaves uniform in size.
                   11. Leaf blades lanceolate, usually with 3 veins (central and 2 somewhat weaker laterals) more prominent than the others ____ **S. gigantea**
                   11. Leaf blades ovate, usually with only 1 prominent vein (central) ____ **S. ulmifolia**
               10. Basal leaves present; lower stem leaves 10–30 cm long; leaves reduced upwards.
                   12. Stems subcylindrical; leaf blades with upper surface usually glabrous ____ **S. ludoviciana**
                   12. Stems, at least towards the base, bluntly 4-angled; leaf blades with upper surface strongly scabrous ____ **S. patula**
      6. Primary stems pubescent (pubescence often short but dense—use lens); leaves usually pubescent.
         13. Leaf blades linear, minutely gland-dotted (view with back lighting); on extreme e margin of nc TX ____ **S. odora**
         13. Leaf blades lanceolate, ovate, or elliptic, not gland-dotted; widespread in nc TX.
             14. Inflorescence branches erect, not recurved apically; heads not secund or only slightly so; panicle narrow ____ **S. speciosa**
             14. Inflorescence branches recurved; heads secund (= on 1 side of branches); panicle triangular in outline.
                 15. Leaf blades serrate.
                     16. Leaf blades ovate (rarely lanceolate), the venation often prominently reticulate beneath, with 1 main vein (central), the smaller laterals all ± equally prominent, with long hairs on the veins (hairs much longer than vein width) ____ **S. rugosa**
                     16. Leaf blades lanceolate, the venation inconspicuously reticulate beneath, with 3 main veins (central and 2 laterals) more prominent than the others, with short hairs on the veins (ca. as long as vein width or slightly longer) ____ **S. canadensis**
                 15. Leaf blades essentially entire except for apical teeth on larger leaves of some taxa.

17. Leaf blades usually more than 5 times as long as wide, lanceolate or linear to spatulate.
    18. Leaf blades 1-nerved, spatulate, reduced up the stem; inflorescences mostly longer than broad ____ **S. nemoralis**
    18. Leaf blades 3-nerved, lanceolate, ± uniform up the stem; inflorescences pyramidal, often nearly as broad as long ____ **S. canadensis**
17. Leaf blades 2–4 times as long as wide, lanceolate or broadly oblanceolate to elliptic or ovate ____ **S. radula**

Solidago canadensis L., (of Canada), COMMON GOLDENROD, CANADA GOLDENROD. Plant to 1.5(–2) m tall; stems pubescent; leaves chiefly cauline; leaf blades 3-nerved, lanceolate, pubescent on both surfaces; inflorescence triangular in outline; heads secund; involucres 2–5 mm high; achenes short-hairy. (Jun–)Aug–Nov. According to Taylor and Taylor (1984), the differences between the 2 varieties are not clear and many intermediates occur.

1. Plants usually < 0.7 m tall; leaf blades < 7 cm long, ± thick, gray-green, with pubescence nearly the same on both sides; prairies and floodplains ____ var. **gilvocanescens**
1. Plants usually > 0.8 m tall; leaf blades usually > 7 cm long, ± thin, green, with pubescence on upper surface shorter and less dense than below; open areas in woods ____ var. **scabra**

var. **gilvocanescens** Rydb., (yellowish gray). E TX w to West Cross Timbers and Edwards Plateau. [*S. gilvocanescens* (Rydb.) Smyth] This is the common variety on the prairies of nc TX.

var. **scabra** Torr. & A. Gray, (rough). Hopkins, Kaufman, Lamar, and Montague cos., also Dallas Co. (Taylor & Taylor 1984); se and e TX w to nc TX. [*S. altissima* L., *S. altissima* var. *pluricephala* M.C. Johnst.]

Solidago gigantea Aiton, (gigantic), GIANT GOLDENROD, LATE GOLDENROD. Plant large, to 1.5 m or more tall; stems glabrous, usually red; leaf blades lanceolate, toothed along most of the edge, glabrous; inflorescence broadly paniculate, triangular in outline; heads secund; involucres ca. 2.5–4 mm high; achenes short-hairy. Low wet areas, roadsides, margins of ponds and streams, generally mesic areas; widespread in TX. Mid-summer–late fall. [*S. gigantea* var. *serotina* (Kuntze) Cronquist]

Solidago ludoviciana (A. Gray) Small, (of Louisiana), WILLOW GOLDENROD. Leaves basal and cauline; basal leaves with blades elliptic to ovate, abruptly contracted into petioles; mid-stem leaves with blades narrowly elliptic to ovate, 10–35 mm wide, serrulate, glabrous or with sparse pubescence, short petiolate to subsessile; inflorescences usually broadly paniculate; heads often secund; involucres ca. 4.5 mm high; achenes short-hairy. Disturbed areas, usually sandy soils; Taylor and Taylor (1984) lumped this species with *S. patula* var. *strictula* (as *S. salicina*) and mapped a number of nc TX counties (Dallas, Henderson, Hopkins, Johnson, and Navarro); they indicated that most TX plants are glabrous (and thus *S. ludoviciana*); Hatch et al. (1990) also cited vegetational area 4 (Fig. 2); se and e TX w to nc TX. Late summer–fall. [*S. boothii* Hook. var. *ludoviciana* A. Gray]

Solidago missouriensis Nutt. var. **fasciculata** Holz., (sp.: of Missouri; var.: fascicled, clustered), MISSOURI BASIN GOLDENROD. Plant 0.4–0.9 + m tall; stems glabrous; leaves frequently fastigiate; leaf blades oblanceolate, toothed in upper half; inflorescence essentially glabrous, with sparse pubescence, paniculate, broadest basally; achenes glabrous or with sparse hairs. Upland prairies; Clay, Lamar, Montague, Tarrant, and Wise cos. in n part of nc TX; otherwise known in TX only from Bowie and Wichita cos. (Taylor & Taylor 1984). Jul–Oct. [*S. glaberrima* M. Martens]

Solidago nemoralis Aiton, (growing in shady places), OLD-FIELD GOLDENROD. Plant to 1(–1.3) m tall; stems pubescent, with basal rosette and persistent lower leaves; leaves reduced upward; leaf

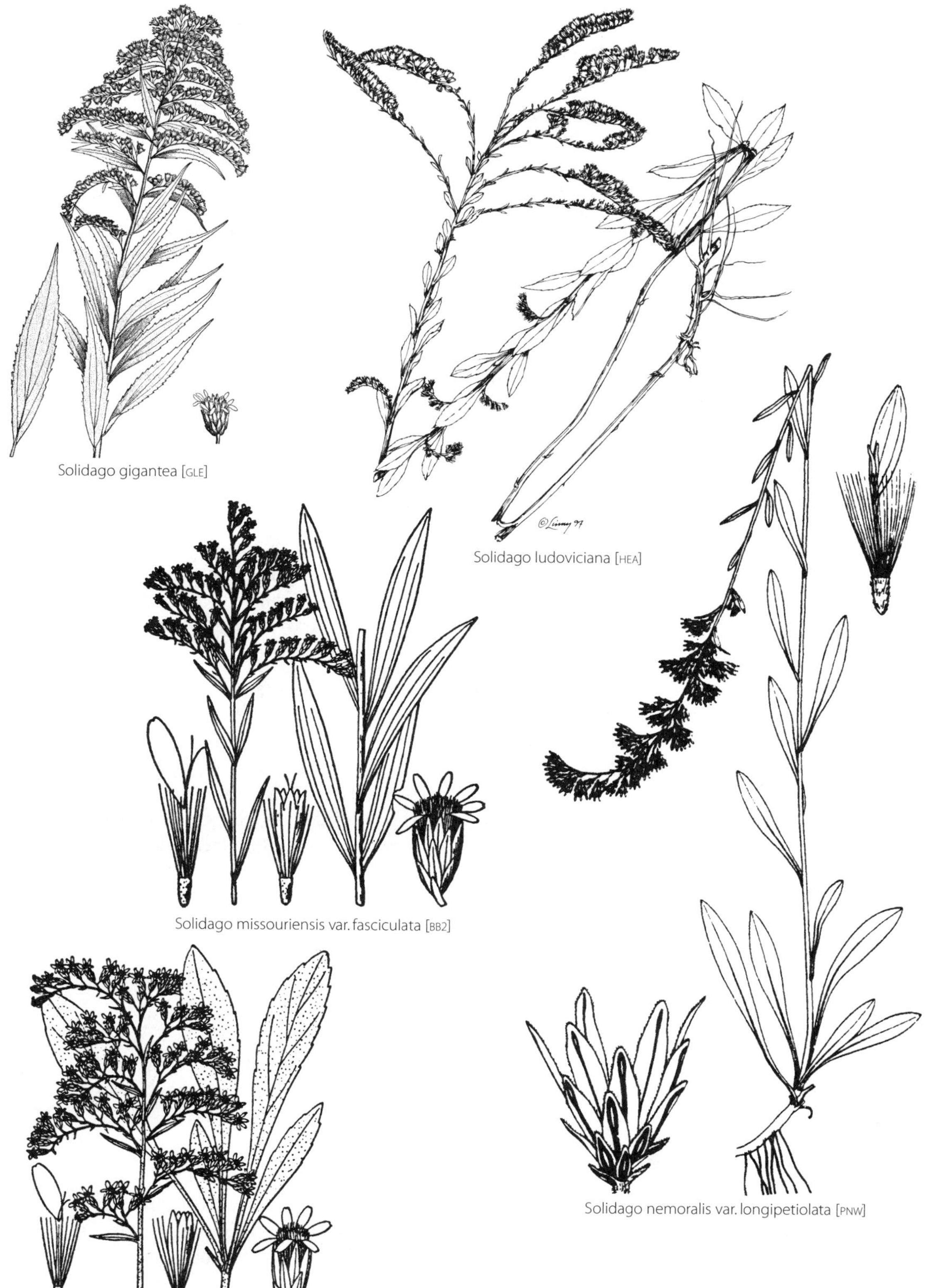

Solidago gigantea [GLE]

Solidago ludoviciana [HEA]

Solidago missouriensis var. fasciculata [BB2]

Solidago nemoralis var. longipetiolata [PNW]

Solidago nemoralis var. nemoralis [BB2]

blades spatulate, 1-nerved (only the midvein apparent), mostly entire, pubescent; inflorescence paniculate, longer than broad; heads secund; achenes short-hairy. Dry upland sites. Aug–Nov.

1. Main stem leaves with blades linear-oblanceolate to linear, the upper sometimes much reduced in size, the lower ones 7–10 times as long as wide; involucres 4.5–6.5 mm high; widespread in nc TX _______ var. **longipetiolata**
1. Main stem leaves with blades oblanceolate to obovate, the upper gradually reduced in size, the lower ones 3–6 times as long as wide; involucres 3–4.5 mm high; mainly e TX _______ var. **nemoralis**

var. **longipetiolata** (Mack. & Bush) E.J. Palmer & Steyerm., (long-petioled). Bell and Grayson cos. w to e Rolling Plains and Edwards Plateau. [*S. decemflora* DC., *S. nemoralis* var. *decemflora* Fernald]

var. **nemoralis**. OLD-FIELD GOLDENROD. A BRIT sheet (*Taylor 27009*) from Lamar Co. is possibly var. *nemoralis*; mainly e TX.

Solidago nitida Torr. & A. Gray, (shining), SHINY GOLDENROD, FLAT-TOPPED GOLDENROD. Plant 0.5–1.2 m tall; stems essentially glabrous; leaf blades linear, almost grass-like, glabrous; inflorescence corymbose, flat-topped; involucres ca. 5 mm high, the phyllaries multiveined; achenes glabrous. Prairies and disturbed areas; se and e TX w to West Cross Timbers. Jun–Sep, a primarily summer-blooming species. Sometimes placed in the segregate genus *Oligoneuron* [as *O. nitidum* (Torr. & A. Gray) Small].

Solidago odora Aiton, (odorous, fragrant), FRAGRANT GOLDENROD, SWEET GOLDENROD. Plant to 1.2 m tall; stems pubescent; leaf blades linear, essentially glabrous except for margin, minutely gland-dotted (when viewed with back lighting), often with an anise odor when crushed; involucres 4–5 mm high; achenes glabrate or short-hairy. Roadsides, woods; Henderson Co., also Lamar Co. (Taylor & Taylor 1984); mainly se and e TX. Sep–Nov.

Solidago patula Muhl. var. **strictula** Torr. & A. Gray, (sp.: spreading; var.: somewhat erect, upright), WILLOW GOLDENROD. Stems glabrous below the rough-puberulent inflorescence, bluntly 4-angled at least below; leaves basal and cauline; basal rosettes of large, long-petiolate leaves, the blades broadly elliptic to obovate; mid-stem leaves subsessile or short-petioled, the blades elliptic, strongly scabrous on upper surface; inflorescences usually broadly paniculate; heads often secund; involucres ca. 4.7 mm high; achenes glabrate. Disturbed areas; possibly rare in e part of nc TX; Taylor and Taylor (1984) lumped this taxon with *S. ludoviciana*, recognized it under the name *S. salicina*, and cited a number of nc TX counties; however, they indicated that most TX plants have glabrous leaves (and are thus *S. ludoviciana*); they stated of the two taxa that "Our own field and herbarium studies indicate they are the same entity. . ." ; mainly e TX possibly w to nc TX. Late Aug–Oct. [*S. salicina* Elliott]

Solidago petiolaris Aiton, (with a petiole or leaf stalk). Plant to 1.5 m tall, growing in dense bunches with many stems; stems pubescent, the hairs short; leaf blades lanceolate to occasionally ovate, pubescent, at least on upper surface; inflorescence racemose to very narrowly paniculate; heads 6–11 mm high (including pappus)–among the largest heads of any nc TX GOLDENROD; involucres to 6.3 mm high; achenes glabrous. Prairies; Post Oak Savannah w to Panhandle and Edwards Plateau. Sep–Oct. [*S. petiolaris* var. *angustata* (Torr. & A. Gray) A. Gray]

var. *angustata* (Torr. & A. Gray) A. Gray, (narrow), was recognized by Taylor and Taylor (1984); they indicated it is more common in e TX but occasional further w; while we have seen no nc TX material of this variety, Hatch et al. (1990) cited vegetational areas 4 and 5 (Fig. 2) and Nesom (1990d) cited Cooke, Grayson, Palo Pinto, and Parker cos.; we are following Nesom (1990d), Kartesz (1994), and Jones et al. (1997) in lumping var. *angustata* with the more common var. *petiolaris*. Taylor and Taylor (1984) separated the 2 varieties as follows:

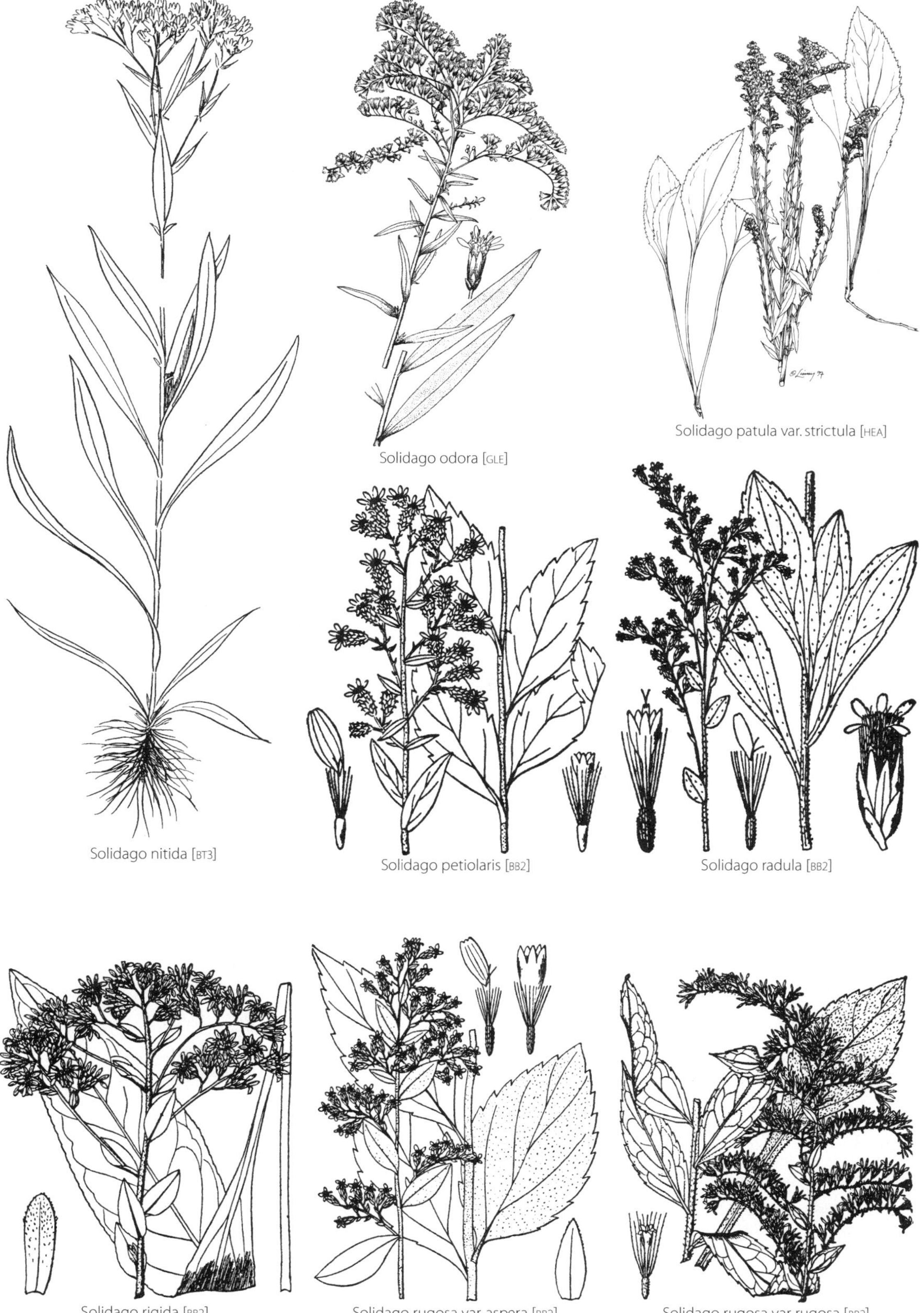

Solidago odora [GLE]

Solidago patula var. strictula [HEA]

Solidago nitida [BT3]

Solidago petiolaris [BB2]

Solidago radula [BB2]

Solidago rigida [BB2]

Solidago rugosa var. aspera [BB2]

Solidago rugosa var. rugosa [BB2]

1. Leaf blades prominently sticky or resinous, lower surface glabrous; phyllaries glandular-dotted, essentially glabrous ______ var. *angustata*
1. Leaf blades scarcely or not at all sticky or resinous, lower surface and main veins ± hairy with spreading hairs; phyllaries minutely hairy with hairs often sticky, but sometimes glabrous ______ var. *petiolaris*

Solidago radula Nutt., (rough), ROUGH GOLDENROD. Plant 0.4–1.2 m tall; stems pubescent; inflorescence paniculate, dense to open, with long floriferous branches with small (< 1 cm long) elliptic leaves on their lower portions; heads not secund in the field, but often appearing so when pressed; involucres to 5.5 mm high; achenes short-hairy. Dry bluffs, upland xeric sites; e TX w to Rolling Plains and Edwards Plateau. Aug–Nov.

Solidago rigida L., (rigid, stiff), STIFF GOLDENROD, RIGID GOLDENROD. Plant variable, to 1.6 m tall; basal leaves with long conspicuous petioles; middle stem leaves sessile or nearly so; leaf blades narrowly ovate to ovate, usually densely pubescent; inflorescence flat-topped, dense; heads large, the involucres 5–9 mm high, the phyllaries multinerved; achenes glabrous or nearly so, ribbed. Prairies; se and e TX w to West Cross Timbers. Sep–Oct(–Nov). Sometimes put in the segregate genus *Oligoneuron* [as *O. rigidum* (L.) Small]. [*S. rigida* var. *laevicaulis* Shinners, *S. rigida* var. *glabrata* E.L. Braun, *S. rigida* subsp. *glabrata* (E.L. Braun) S.B. Heard & J.C. Semple, *S. rigida* subsp. *humilis* (Porter) S.B. Heard & J.C. Semple] We are following Jones et al. (1997) in not recognizing infraspecific taxa in this species. Heard and Semple (1988) divided this species into three subspecies which, according to their range map, all occur in nc TX. Their key divided the subspecies as follows:

1. Outer series phyllaries glabrous; leaves and stems glabrous to somewhat hispid ______ subsp. *glabrata*
1. Outer series phyllaries pubescent; leaves and stems hispid to densely so.
   2. Inner series phyllaries conspicuously pubescent, often linear; plants usually short (0.3–0.7 m) but may be taller; capitulescence compact; pubescence of leaves and stems fine and very dense (> 50 hairs/mm^3) ______ subsp. *humilis*
   2. Inner series phyllaries glabrate to very sparsely pubescent, oblong and bluntly rounded; plants more robust (0.6–1.4 m) with loose, open capitulescence; pubescence coarsely hispid (< 50 hairs/mm^3) ______ subsp. *rigida*

Solidago rugosa Mill., (rugose, wrinkled), ROUGH-LEAF GOLDENROD, WRINKLED GOLDENROD. Plant usually < 1 m tall; stems pubescent; leaf blades serrate, the lower surface with long hairs; panicle triangular in outline, the branches recurved; heads secund; involucres ca. 4 mm high; achenes short-hairy. Roadside ditches, along streams, moist sites in oak-hickory forests. Sep–Nov. We are following Taylor and Taylor (1984), Gandhi and Thomas (1989), and Jones et al. (1997) in treating the following infraspecific taxa as varieties. Kartesz (1994) recognized them as subspecies.

1. Leaf blades broadly lanceolate to ovate, with pronounced rugose-venation on lower surface ______ var. **aspera**
1. Leaf blades lanceolate, without pronounced rugose-venation on lower surface ______ var. **rugosa**

var. **aspera** (Aiton) Fernald, (rough), CELTIS-LEAF GOLDENROD. Included based on citation of vegetational area 4 (Fig. 2) by Hatch et al. (1990); mainly se and e TX. [*S. aspera* Aiton, *S. celtidifolia* Small, *S. rugosa* var. *celtidifolia* (Small) Fernald, *S. rugosa* subsp. *aspera* (Aiton) Cronquist]

var. **rugosa**. HARSH GOLDENROD, ROUGH-LEAF GOLDENROD. Fannin, Henderson, Hopkins, and Lamar cos. on e edge of nc TX; mainly se and e TX.

Solidago speciosa Nutt. var. **rigidiuscula** Torr. & A. Gray, (sp.: showy; var.: somewhat rigid), PRAIRIE GOLDENROD, NOBLE GOLDENROD. Stems usually pubescent at least above; leaf blades narrowly elliptic to elliptic, entire or nearly so, glabrate; inflorescence a narrow dense panicle; involucres ca. 6 mm high; achenes glabrous. Sandy prairies; Henderson, Kaufman, Jack, and

Montague, cos., also Navarro and Palo Pinto cos. (Mahler 1988); e TX w to West Cross Timbers. (Jun–) Sep–Oct. [*S. rigidiuscula* (Torr. & A. Gray) Porter]

Solidago ulmifolia Muhl. ex Willd. var. **microphylla** A. Gray, (sp.: with leaves like *Ulmus*—elm; var.: small-leaved), ELM-LEAF GOLDENROD. Plant 0.4–1.5 m tall; stems glabrous; basal rosette absent; leaf blades ovate, toothed, essentially glabrous; lower inflorescence branches long, floriferous apically; heads secund; involucres 2.5–5 mm high; achenes short-hairy, ribbed. Sandy soils, roadsides, openings in woodlands, drier upland sites; Grayson and Lamar cos., also Dallas, Denton, Fannin, Hunt, Kaufman, Montague, Tarrant, and Wise cos. (Taylor & Taylor 1984); e TX w to West Cross Timbers and sc TX. (Jul–)Sep–Oct(–Nov), Taylor and Taylor (1984) indicated *S. ulmifolia* begins blooming in summer and is often finished flowering before some of the later flowering species begin. [*S. delicatula* Small, *S. microphylla* (A. Gray) Engelm. ex Small]

Solidago altiplanities C.E.S. Taylor & J. Taylor, (high plains), HIGH PLAINS GOLDENROD, occurs in OK just across the Red River from Clay Co.; it also occurs in the Plains Country of TX. It would key to *S. odora* in the above key but can be distinguished by its erect branches (vs. recurved apically), conical inflorescences (vs. pyramidal), and larger leaves to 9 cm long (vs. 5 cm or less) and not punctate.

Solidago sempervirens L., (evergreen), SEASIDE GOLDENROD. Two varieties, var. *mexicana* (L.) Fernald and var. *sempervirens*, are cited by Hatch et al. (1990) for vegetational area 4 (Fig. 2), but according to Taylor and Taylor (1984) occur only in se TX to the s of nc TX; C. Taylor (pers. comm.) indicated that they do not occur close to nc TX. The species can be distinguished by the primary stems and leaves usually glabrous; the inflorescence racemose or paniculate, pubescent; a rosette of linear-lanceolate leaves; and linear to narrowly lanceolate, entire stem leaves.

SOLIVA BURWEED, PIQUANTE, STICKERWEED, STICKERS

Low growing annuals with short stems (to only ca. 15 cm tall) or nearly stemless; leaves alternate, often crowded; leaf blades 1–3-pinnatifid; heads sessile in clusters of leaves, ca. 3–8 mm wide, apparently only of disk flowers; outer flowers pistillate, without corollas; inner disk flowers functionally staminate with minute 4-toothed whitish or green-translucent corollas; achenes flattened, with wings or ± wing-like lateral appendages, tipped by the persistent spine-like style, conspicuously hairy to glabrous apically; pappus absent.

A South American genus of 8 species. Easily recognized by the spine-like styles which make *Soliva* species painfully noxious weeds, similar in effect to *Cenchrus* (SANDBURS). The segregate genus *Gymnostyles* has been recognized by some (Tutin 1976; Kartesz 1994; Jones et al. 1997). However, the differences used to distinguish it from *Soliva* seem minute and we are following Cronquist (1980), Gandhi and Thomas (1989), Bremer (1994), and Arriagada and Miller (1997) in retaining all the species in *Soliva*. (Named for Dr. Salvador Soliva, 18th century physician to the Spanish court and botanist at the Royal Botanic Garden, Madrid, where he studied medicinally useful plants—Arriagada & Miller 1997) (tribe Anthemideae)
REFERENCES: Cabrera 1949; Tutin 1976; Cronquist 1980; Ray 1987; Gandhi & Thomas 1989; Arriagada & Miller 1997.

1. Achenes 2.5–4 mm wide, with only short hairs; wings (lateral appendages) of achenes broad, thin, smooth, without raised lines, conspicuously indented toward lower half forming 2 basal lobes; lateral pinnae of leaves palmately divided; heads scattered, not clustered at base of plant; plants caulescent ______ **S. pterosperma**
1. Achenes 1–2 mm wide, conspicuously long-hairy at tip; wings (lateral appendages) of achenes thick, only slightly wing-like, with conspicuous raised lines running across the wings, not indented; lateral pinnae of leaves not palmately divided (either pinnately divided or not further divided); heads mostly clustered near base of plant; plants ± acaulescent.

2. Lateral pinnae of leaves usually not further divided; lateral projections of achenes with raised lines nearly to apex, with 2 divergent acute awns, projections, or shoulders at apex of achene, 1 on each side of the spine-like style ____ **S. stolonifera**
2. Lateral pinnae of leaves pinnately divided; lateral projections of achenes ± smooth in distal 1/3, without any apical awns, projections, or shoulders near the spine-like style ____ **S. mutisii**

Soliva mutisii Kunth, (for Jose Celastino Bruno Mutis y Bosio, 1732–1808, Spanish physician and botanist of Madrid), BUTTON BURWEED, SMOOTH-STICKERS. Heads clustered in basal rosette. This species, not known from nc TX, is included in the key. It is widespread in se and e TX and is to be expected in nc TX. Mar–Apr. Native of South America.

Soliva pterosperma (Juss.) Less., (winged seed), LAWN BURWEED, STICKERS, JO-JO WEED). Much branched, small, mat-forming or ascending annual herb to 15 cm tall; leaves small (1–3.5 cm long), usually tri-pinnatifid; heads scattered, not clustered at very base; corollas of central flowers ca. 1.6 mm long; achenes ca. 3.2 mm long, 2.5–4 mm wide, with broad lateral wings indented toward lower half forming basal lobes, glabrous apically, with a sharp, spinose, 1.5–2 mm long, persistent style and in addition a pair of spinulose projections ca. 1 mm long. Collected from soccer field near Arlington, Tarrant Co. (1995), possibly spread by atheletic shoes; also a lawn weed in se and e TX. Apr–May. Native of South America. Ray (1987) and Kartesz (1994) treated *S. pterosperma* as a synonym of *S. sessilis* Ruiz & Pav. We are following Cronquist (1980), Gandhi and Thomas (1989), and Jones et al. (1997) in recognizing it as a separate species.

Soliva stolonifera (Brot.) Loudon, (bearing stolons), TRAILING-STICKERS. Stoloniferous, ± acaulescent; leaves to 3 cm long and 1 cm wide; heads solitary per node; corollas of central flowers 1.8 mm long; styles of pistillate flowers to ca. 2 mm long; achenes ca. 2 mm long and 1.7 mm wide. Lawn weed; included based on citation of vegetational area 4 (Fig. 2) by Hatch et al. (1990); mainly se and e TX. Mar–Apr. Native of South America. [*Gymnostyles stolonifera* (Brot.) Tutin]

SONCHUS SOW-THISTLE

Ours annuals with taproot, glabrous or with glandular-pubescent peduncles and upper stems; sap milky; stem leaves auricled-clasping, the blades obovate to oblong or ovate-lanceolate, coarsely or spiny-toothed or lobed; heads small, corymbose; phyllaries narrow, green, in several rows, the outer successively shorter; flowers all ligulate; corollas yellow; achenes flattened, 2–3 mm long; pappus of white, hair-like bristles 5–9 mm long.

A genus of 62 species native from Eurasia to Australasia and tropical Africa. (Greek: *sonchos*, ancient name used for a prickly plant) (tribe Lactuceae)
REFERENCES: Boulos 1972–74, 1976; Vuilleumier 1973.

1. Leaves stiff and painfully prickly when grasped; leaf auricles (lobe-like extensions on both sides at base) rounded, sometimes made irregularly so by prominent, spiny teeth; mature achenes with several nerves but not wrinkled-roughened ____ **S. asper**
1. Leaves softer, not painfully prickly when grasped; leaf auricles triangular or triangular-lanceolate, pointed, as well as sharp-toothed; mature achenes with several nerves and cross wrinkled-roughened ____ **S. oleraceus**

Sonchus asper (L.) Hill, (rough), PRICKLY SOW-THISTLE, ACHICORIA DULCE. Stems 0.1–1(–1.5) m tall; corolla tubes longer than the ligules; nerves of achenes evident. Gardens, roadsides, disturbed areas; nearly throughout TX. Mar–Jun, sporadically all year. Native of Eurasia.

Sonchus oleraceus L., (of the vegetable garden, a potherb used in cooking), COMMON SOW-THISTLE. Similar to *S. asper*, but less prickly and often more glaucous; corolla tubes ca. equaling the ligules; nerves of achenes evident to obscure. Same habitats and flowering period as *S. asper*; less common than *S. asper*, but widespread in TX. Native of Eurasia.

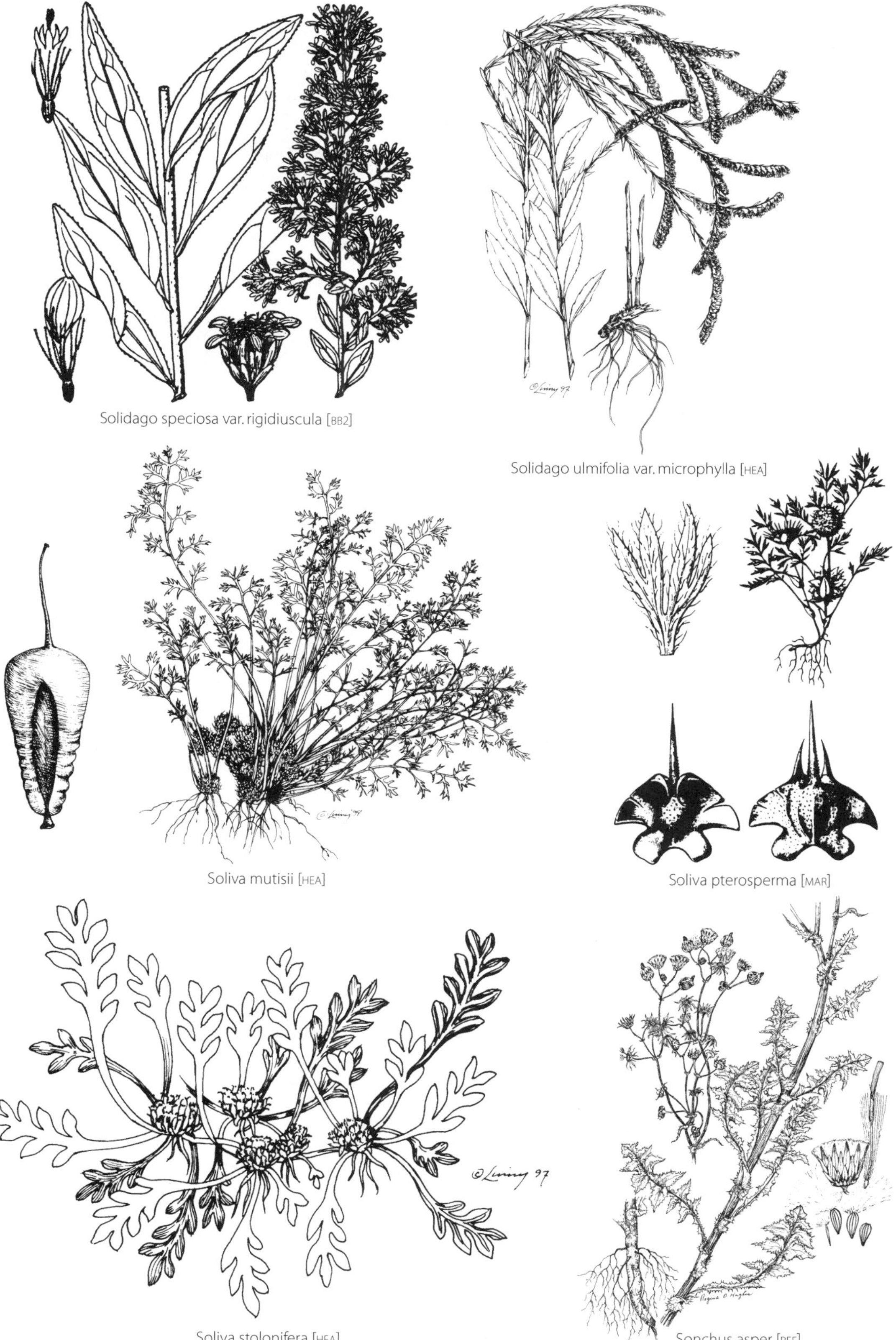

Solidago speciosa var. rigidiuscula [BB2]

Solidago ulmifolia var. microphylla [HEA]

Soliva mutisii [HEA]

Soliva pterosperma [MAR]

Soliva stolonifera [HEA]

Sonchus asper [REE]

TARAXACUM DANDELION

Small perennials, commonly behaving as winter annuals, with coarse taproot; sap milky; leaves all in a basal rosette, the blades oblong-lanceolate, coarsely toothed to deeply pinnately lobed; scapes hollow, ± pubescent, 4–30(–50) cm tall; heads solitary, terminal, large; phyllaries green, the outer short and ± curved-spreading or reflexed, the inner long, erect; flowers all ligulate; corollas yellow; achenes with beak 0.5–4 times as long as the body; beak tipped with parachute-like pappus of numerous capillary bristles; achenes with their pappuses in a conspicuous, easily disrupted, ball-like arrangement.

A genus of 60 species of the n temperate zone and temperate South America including some cosmopolitan weeds. Apomixis and polyploidy have made the taxonomy of this group exceedingly complex. The name DANDELION comes from the French: *dent de lion*, tooth of the lion, apparently in reference to the leaf shape (Tveten & Tveten 1993). Some authors (e.g., Hatch et al. 1990) lump the following 2 very similar species. (Persian: *talkh chakok*, bitter herb, through medieval Latin *tarasacon*) (tribe Lactuceae)
REFERENCES: Shinners 1949a; Vuilleumier 1973.

1. Mature achenes red or purplish red or brownish red; tips of some of the inner phyllaries with a low knob or blunt ridge on back (visible only on living plants, sometimes quite obscure); leaf blades typically deeply lobed for their whole length, the terminal lobe usually not larger than the lateral lobes ____ **T. laevigatum**
1. Mature achenes greenish tan or brownish or straw-colored; tips of inner phyllaries all flat; leaf blades usually less deeply lobed, the terminal lobe usually larger than the lateral lobes ____ **T. officinale**

Taraxacum laevigatum (Willd.) DC., (smooth), RED-SEED DANDELION. Terminal lobe of leaf blades usually triangular with straight or concave, rarely convex, sides. In lawns, disturbed areas, and under trees in towns; Bell, Dallas, Denton, Hopkins, Kaufman, and McLennan cos., also Brown Co. (HPC); widespread in TX. Dec–Jun, sporadically to Nov, often drying up and disappearing during summer heat. Native of Europe and w Asia. [*T. erythrospermum* Andrz. ex Besser]

Taraxacum officinale F.H. Wigg., (medicinal), COMMON DANDELION. Terminal lobe of leaf blades rounded-triangular with convex sides. Same habitats, flowering, and growth habits as *T. laevigatum*; throughout TX. Native of Eurasia. A cosmopolitan weed nearly throughout the temperate zone.

TETRAGONOTHECA NERVERAY, SQUAREHEAD

Herbaceous perennials; leaves opposite, distinct or connate-perfoliate, the blades deltoid to oblong; heads large, 2–4 cm broad (including ligules); phyllaries apparently in 2 series; outer series 4(–5) foliaceous, conspicuous, to 14 mm long and 9 mm wide; inner series (actually bracts of receptacle) 6–15, small, each subtending a ray flower; ray flowers pistillate, fertile, with ligules conspicuous, yellow, often with some reddish or brown venation; disk flowers numerous, perfect, fertile, the corollas yellow, sometimes with dark stripes; achenes 4-angled; pappus of scales or absent.

A genus of 4 species of the se United States. (Greek: *tetragonos*, four-angled, and *thece*, a case, from the shape of the involucre) (tribe Heliantheae)
REFERENCE: Turner & Dawson 1980.

1. Leaf blades toothed, usually very broad, the upper leaves often broadest, or at least very broad at base, very conspicuously connate-perfoliate, to 11 cm wide; stems (0.3–)0.6–1.2 m tall; in sandy habitats in n part of nc TX ____ **T. ludoviciana**
1. Leaf blades pinnatifid, usually narrow, the upper leaves usually not broadest near base, incon-

Sonchus oleraceus [REE]

Taraxacum laevigatum [BB1]

Taraxacum officinale [REE]

Tetragonotheca ludoviciana [HEA]

Tetragonotheca texana [HEA]

Tetraneuris linearifolia [IPL]

spicuously connate-perfoliate, to 4.5 cm wide, usually much less; stems 0.2–0.5(–0.7) m tall; in calcareous habitats in s part of nc TX ______ **T. texana**

Tetragonotheca ludoviciana (Torr. & A. Gray) A. Gray ex Hall, (of Louisiana), SAWTOOTH NERVERAY. Leaf blades ovate to oblong, 5–7(–20) cm long, prominently toothed; ray flowers ca. 10, the ligules to 14 mm long; achenes ca. 4 mm long. Sandy woodlands; Denton, Parker, and Tarrant cos.; mainly se and e TX. May–Aug.

Tetragonotheca texana Engelm. & A. Gray ex A. Gray, (of Texas), PLATEAU NERVERAY, SQUARE-BUD DAISY. Leaf blades pinnatifid, narrowly ovate to elliptic or oblong in outline, 3–10 cm long. Calcareous soils, uplands; s Blackland Prairie (Mahler 1988) s to Edwards Plateau and s TX. Apr–Sep.

TETRANEURIS

Small annuals or perennials, ± soft-pilose or woolly-pubescent to glabrate; leaves basal or cauline, alternate, or subopposite, simple, the blades entire or few-lobed, linear to oblanceolate with petiolar base, minutely resin-dotted; heads terminal, rather large, long-peduncled, solitary; ray flowers pistillate, fertile; ligules deep yellow or orangish yellow with brown-red lines beneath, toothed at tip, reflexed and persistent in age and becoming almost whitish; disk corollas perfect, fertile, orangish yellow; pappus scales awn-tipped.

A New World genus of ca. 11–15 species; it is a segregate of *Hymenoxys* and has sometimes been treated in that genus (e.g., Karis & Ryding 1994; Mabberley 1997). We are following Kartesz (1994), Jones et al. (1997), J. Kartesz (pers. comm. 1997), and Bierner and Jansen (1998) in recognizing the genus. A molecular study by Bierner and Jansen (1998) showed *Tetraneuris* to be a monophyletic clade and supported its recognition as a separate genus. (Greek: *tetra*, four, and *neura*, cord or nerve) (tribe Helenieae, sometimes lumped into Heliantheae)
REFERENCES: Seeligmann & Alston 1967; Bierner & Jansen 1998.

1. Annuals with branched, conspicuously leafy stems; ray flowers with ligules 5–10 mm long; phyllaries 2.5–4(–4.5) mm long ______ **T. linearifolia**
1. Perennials with basal or nearly basal leaves only, the flowering stalks (= peduncles) apparently unbranched, very scapose in appearance; ray flowers with ligules 8–22 mm long; phyllaries 4.5–6 mm long ______ **T. scaposa**

Tetraneuris linearifolia (Hook.) Greene, (linear-leaved). Plant to ca. 40(–60) cm tall; stems ranging from heavily pilose to glabrate; rosette leaves soon deciduous, with blades oblanceolate to linear-lanceolate, entire to few-lobed; cauline leaves persistent, with blades narrowly linear, 3–6(–8) cm long, 3–10 mm wide; heads on naked peduncles above the leaves; ray flowers variable in number, 6–22; achenes ca. 2 mm long; pappus scales ca. 2 mm long. Prairies, in calcareous clay or rocky soils; nearly throughout TX. Mar–May, re-branching and producing additional smaller heads to Jul. [*Hymenoxys linearifolia* Hook.]

Tetraneuris scaposa (DC.) Greene, (with scapes), PLAINS YELLOW DAISY, YELLOW PAPER-FLOWER, FOUR-NERVE DAISY. Plant 10–30+ cm tall; stems densely to slightly pilose; leaf blades linear to linear-lanceolate, entire to with a few short lobes, 2–10 cm long, 1.5–9(–11) mm wide; petioles 0.5–1.5 mm wide, with clasping base; axils densely long hairy; heads solitary on naked scapes; ray flowers 12–31; achenes 2–3.5 mm long; pappus scales 1–ca. 3 mm long. Gravelly or rocky prairies on limestone; in nc TX mainly Grand Prairie, also Dallas Co. (Correll & Johnston 1970); nc TX s and w to s TX and Trans-Pecos. Mar–May, sporadically through summer, often repeating in Sep–Oct. [*Hymenoxys scaposa* (DC.) K.L. Parker] The leaves are dotted with thin granules of a resin-like substance which gives them a bitter taste (Ajilvsgi 1984). The common name YELLOW PAPER-FLOWER is probably derived from the tendency of the rays to remain on the heads long after maturity, eventually turning almost white.

Tetraneuris turneri (K.L. Parker) K.L. Parker, (for Billie Lee Turner, 1925–, botanist at Univ. of TX), cited for vegetational area 4 (Fig. 2) by Hatch et al. (1990), apparently occurs only to the s of nc TX. It is a perennial with leaves mostly basal, phyllaries 5.5–8.5 mm long, and petioles 2–6 mm wide. [*Hymenoxys turneri* K.L. Parker]

THELESPERMA GREENTHREAD

Perennial (rarely annual) herbs; leaves opposite, compound or deeply lobed, the segments linear-lanceolate to filiform; heads terminal on naked peduncles, solitary or corymbose, large; phyllaries in 2 series, the outer phyllaries herbaceous, the inner hyaline- or yellow-margined, basally fused for 1/4–1/2 their length; ray flowers with ligules yellow or golden yellow or ray flowers absent; disk corollas unequally lobed, yellow with reddish brown veins to dark reddish brown; achenes linear to linear-oblong, 2–7 mm long; pappus of 2(–3) awns or minute or absent.

A genus of 15 species native to w North America and s South America; similar in appearance to and sometimes confused with *Coreopsis*; however, *Thelesperma* has the inner phyllaries basally united for 1/4 to 1/2 their length and the disk corolla lobes often linear, while *Coreopsis* has the phyllaries all separate or nearly so and the disk corolla lobes triangular to ovate. (Greek: *thele*, nipple, and *sperma*, seed, in reference to the papillose achenes) (tribe Heliantheae)
REFERENCES: Shinners 1950a; Alexander 1955; Greer 1997.

1. Ray flowers present.
  2. Corollas of disk flowers dark reddish brown.
    3. Outer phyllaries 3–12 mm long, usually at least 1/2 as long as inner phyllaries; leaves ± evenly distributed on stem ____ **T. filifolium**
    3. Outer phyllaries very small, usually 1–2 mm long, < 1/4 as long as the inner phyllaries; leaves mostly in a rosette and on basal 1/3 of stem ____ **T. ambiguum**
  2. Corollas of disk flowers yellow (veins can be reddish brown).
    4. Leaves on upper 1/3 of stem (and sometimes most leaves) with only 1–3 narrow divisions; pappus absent or of minute tooth-like awns; disk corollas with shallowly lobed limb (to ca. 1/5 limb length) ____ **T. simplicifolium**
    4. All leaves much divided with many narrow divisions; pappus of well-developed retrorsely barbed awns 0.5–2 mm long; disk corollas with deeply lobed limb (1/2–3/4 limb length) ____ **T. filifolium**
1. Ray flowers absent.
  5. Corollas of disk flowers reddish brown; leaves mostly on basal 1/3 of stem ____ **T. ambiguum**
  5. Corollas of disk flowers yellow (veins can be reddish brown); leaves often ± evenly distributed on stem or sometimes mostly on basal 1/2 ____ **T. megapotamicum**

Thelesperma ambiguum A. Gray, (ambiguous). Perennial 22–50 cm tall; ray flowers usually present, yellow, occasionally absent; disk corollas deeply lobed. Open limestone areas; mainly s 1/2 of TX, possibly to sw part of nc TX. Summer–fall. [*T. megapotamicum* var. *ambiguum* (A. Gray) Shinners]

Thelesperma filifolium (Hook.) A. Gray, (thread-leaved), GREENTHREAD, THREAD-LEAF THELESPERMA. Annual or short-lived perennial blooming the first year, 20–70 cm tall; ray corollas with ligules yellow or golden yellow, sometimes with reddish tinge near base, ca. 9–22 mm long, 9–17 mm wide, 3-lobed at apex; disk corollas yellow or reddish brown; achenes 3.5–6.5 mm long. Prairies, roadsides, and disturbed areas. Apr–Jun, sporadically to Sep.

1. Plants 25–70 cm tall; ray corollas with ligules golden yellow; outer phyllaries usually > 1/2 as long as inner; widespread in nc TX ____ var. **filifolium**
1. Plants 20–30(–40) cm tall; ray corollas with ligules yellow; outer phyllaries ca. 1/2 as long as inner; sw margin of nc TX; mainly w part of TX ____ var. **intermedium**

var. **filifolium**. Blackland Prairie and Grand Prairie; widespread in TX. [*T. trifidum* (Poir.) Britton]

var. **intermedium** (Rydb.) Shinners, (intermediate). Usually smaller than var. *filifolium*. Burnet Co. (Shinners 1950a) near sw margin of nc TX; mainly Rolling Plains, also Edwards Plateau and Trans-Pecos. [*T. intermedium* Rydb.]

Thelesperma megapotamicum (Spreng.) Kuntze, (of the big river), COTA, INDIAN TEA, NAVAJO TEA, COLORADO GREENTHREAD, RAYLESS THELESPERMA. Perennial 30–75 cm tall, spreading by creeping rootstocks; outer phyllaries 1–2(–3) mm long; ray flowers very rarely present; if so this species would key to *T. simplicifolium* from which it can be distinguished by pappus and disk corolla differences; disk corollas deep or brownish yellow, with deeply lobed limb (lobed 1/2–5/6 limb length); pappus well-developed, of 2 triangular, retrorsely barbed awns 1.5–3 mm long. Sandy or rocky prairies and roadsides; Clay and Young cos., also Archer Co. (Shinners 1950a); w margin of nc TX s and w to w TX. Apr–Oct. Reportedly used as a tea by the Pueblo peoples of Arizona and New Mexico (Kirkpatrick 1992).

Thelesperma simplicifolium A. Gray, (simple-leaved), SLENDER GREENTHREAD. Perennial, glabrous and usually glaucous; stems 30–70 cm tall; leaves often few, the plant thus often with a naked appearance; disk corollas with shallowly lobed limb (to ca. 1/5 its length); pappus of 2 minute tooth-like awns. Limestone outcrops; West Cross Timbers and Lampasas Cut Plain s and w to w TX. May–Sep. Greer (1997) indicated that *T. curvicarpum* Melchert, originally described as a TX endemic from Burnet and Coleman cos. (Melchert 1963), is "... only a rare achene form that occurs in populatons of both *T. simplicifolium* and *T. filifolium*." He further suggested, "I therefore feel that *T. curvicarpum* should no longer be considered a separate species deserving nomenclatural recognition."

THYMOPHYLLA DOGWEED, FOETID-MARIGOLD

Annual or perennial, aromatic herbs; plants 10–30 cm tall; leaves pinnately divided, opposite or alternate, often with scattered oil glands; involucres 3–7 mm high; ray flowers with corollas yellow to yellowish orange; disk corollas yellow to yellowish orange; $x = 8$.

A genus of 17 species native to the sw U.S. and Mexico. It is sometimes placed in *Dyssodia* (Jones et al. 1997); we are following Karis and Ryding (1994), Kartesz (1994), and J. Kartesz (pers. comm. 1997) in treating it separately. (Greek: *thymos*, thyme, and *phyllum*, leaf) (tribe Helenieae, sometimes lumped into Heliantheae)

REFERENCES: Johnston 1956; Johnston & Turner 1962; Strother 1969, 1986; Gandhi & Thomas 1984.

1. Phyllaries united up to ca. 1/2 their length, each free phyllary apex with 1 or more linear to oblong glands; leaves opposite ____ **T. pentachaeta**
1. Phyllaries united nearly to apex (only short, ± triangular teeth free); top of the united portion of the phyllaries with a ring of oblong to round glands; leaves mostly alternate ____ **T. tenuiloba**

Thymophylla pentachaeta (DC.) Small, (five-bristled), COMMON DOGWEED, PARRALENA. Short-lived perennial 10–20 cm tall; leaf lobes 5–7, linear, sparsely gland-dotted; peduncles 5–10 cm long; involucres 4–6 mm high; ray flowers 8–13, the ligules 2–8 mm long, yellow to yellow-orange; disk corollas dull yellow; achenes 2–3 mm long; pappus of ca. 10 awnless and awned scales, often alternating. Sandy, often calcareous soils; Brown and Coleman cos. on the w margin of nc TX s and w to w TX. Mar–Jul, Sep–Nov. [*Dyssodia pentachaeta* (DC.) B.L. Rob.] The foliage has an unpleasant scent if handled (Ajilvsgi 1984).

Thymophylla tenuiloba (DC.) Small var. **tenuiloba**, (slender-lobed), BRISTLE-LEAF DYSSODIA, TINY-TIM. Annual or short-lived perennial 10–30 cm tall, sometimes forming dense clumps; leaves dissected into 7–15 linear lobes; peduncles 3–8 cm long; involucres 5–7 mm high; ray flowers

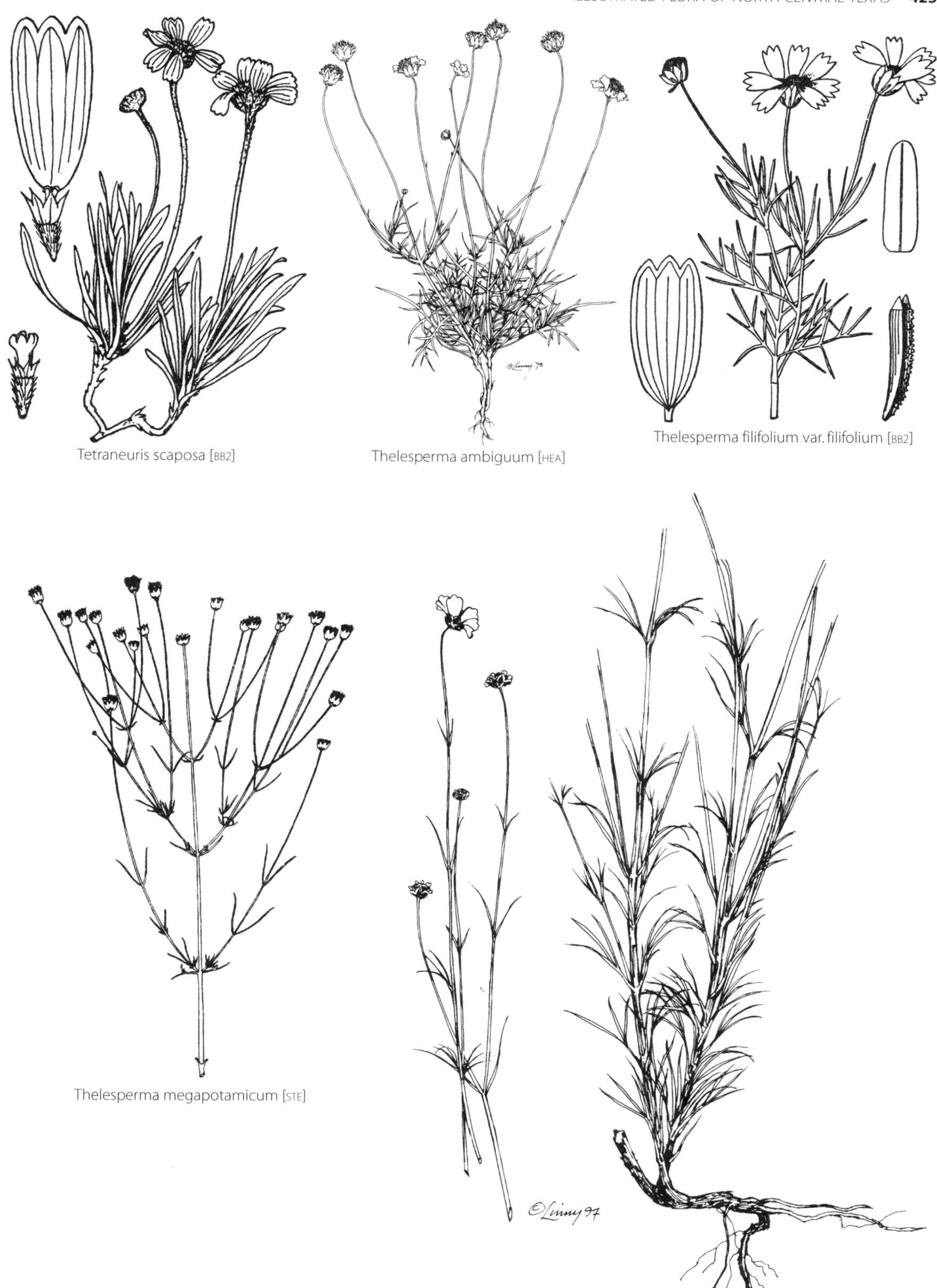

Tetraneuris scaposa [BB2]

Thelesperma ambiguum [HEA]

Thelesperma filifolium var. filifolium [BB2]

Thelesperma megapotamicum [STE]

Thelesperma simplicifolium [HEA]

with ligules golden yellow to yellow-orange, 4–10 mm long; disk corollas yellow-orange; achenes 2–3.2 mm long; pappus of 10–12 similar small scales 2–3.4 mm long, each bearing 3–5 awns. Sandy, often calcareous soils; Brown and Burnet cos., also Hamilton Co. (Mahler 1988); se and e TX w to nc and c TX. May–Oct. [*Dyssodia tenuiloba* (DC.) B.L. Rob.]

Thymophylla tenuiloba (DC.) Small var. *wrightii* (A. Gray) Strother, (for Charles Wright, 1811–1885, TX collector), cited by Hatch et al. (1990) for vegetational area 4 (Fig. 2), apparently occurs only to the s of nc TX, mostly on sandy soils of the coastal plain. This variety can be distinguished by its mostly spatulate, entire leaves (sometimes with a few lateral lobes).

TOWNSENDIA EASTER DAISY

A genus of 25 species native to w North America; some are cultivated as ornamentals. (Named for D. Townsend, 1787–1858, amateur U.S. botanist from Pennsylvania) (tribe Astereae)
REFERENCES: Larsen 1927; Beaman 1957.

Townsendia exscapa (Richardson) Porter, (without scapes), STEMLESS TOWNSENDIA, EASTER DAISY. Tufted or matted, appressed-pubescent, dwarf perennial, the woody stems barely out of the ground; leaves ± in rosettes; leaf blades linear-oblanceolate, 1–5(–8) cm long, 2–6 mm wide, entire; heads large in relation to plant size, 3–4 cm across, sessile or short-pedunculate; ray flowers 20–40, with ligules white to pinkish or rosy-lavender, often with a darker stripe below, showy, 12–22 mm long, curling under at night or in age; disk corollas yellow, often pink- or purple-tipped or -tinged; pappus of barbellate bristles 6–13 mm long. Eroding limestone slopes; Hamilton, Hood, and Johnson cos., also Bell (Beaman 1957) and Brown (HPC) cos.; Grand Prairie s and w to w TX. Jan–Apr. This is one of the earliest blooming native wildflowers in nc TX.

TRAGOPOGON SALSIFY, GOAT'S-BEARD

Biennial herbs with stout taproot; sap milky; stems 20–80 cm tall; leaves alternate, sessile and clasping, long and narrow (monocot-like), to 30 cm long, tapered from base to a long, narrow tip, entire; heads solitary, terminal, large, on swollen peduncles, opening mainly in morning; flowers all ligulate; phyllaries equaling or longer than corollas, ca. 2.4–4 cm long in flower; achenes 2–5 cm long including elongate beak; pappus of very conspicuous plumose (= feather-branched) bristles; achenes with their pappuses in a ball-like arrangement resembling the infructescence of a giant dandelion.

A temperate Eurasian and Mediterranean genus of ca. 110 species of taprooted herbs with monocot-like leaves. While both species occurring in nc TX are diploids, hybridization and polyploidy are well-documented in the genus; *T. miruus* Ownbey, an allotetraploid resulting from hybridization between *T. dubius* and *T. porrifolius*, is known from the Pacific northwest (Ownbey 1950a; Roose & Gottlieb 1976; Soltis & Soltis 1989, 1991; Novak et al. 1991; Soltis et al. 1995). (Greek: *tragos*, goat, and *pogon*, beard, in reference to the conspicuous pappus) (tribe Lactuceae)
REFERENCES: Shinners 1949a; Ownbey 1950a; Vuilleumier 1973; Roose & Gottlieb 1976; Soltis & Soltis 1989, 1991; Novak et al. 1991; Soltis et al. 1995.

1. Corollas lemon-yellow; phyllaries typically 13 per head (sometimes less on depauperate plants) ____ **T. dubius**
1. Corollas purple; phyllaries ca. 8 per head ____ **T. porrifolius**

Tragopogon dubius Scop., (doubtful), GOAT'S-BEARD, WESTERN SALSIFY, NOON-FLOWER. Plant 0.3–1 m tall. Becoming rather common in West Cross Timbers (Mahler 1988); migrating from the Panhandle and South Plains e to at least Grayson Co. Apr–Jun. Native of Eurasia. [*T. major* Jacq.]

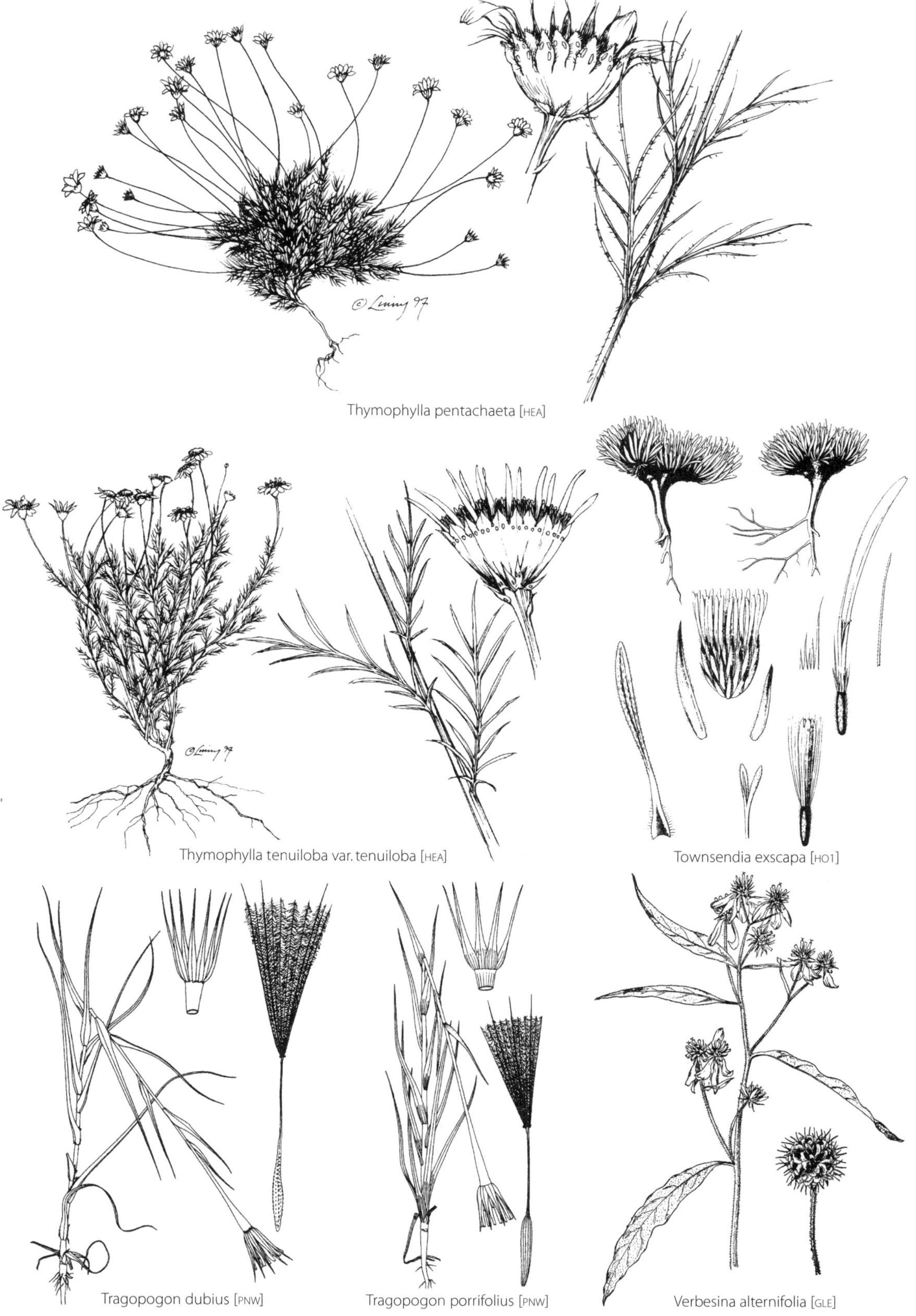

Thymophylla pentachaeta [HEA]

Thymophylla tenuiloba var. tenuiloba [HEA]

Townsendia exscapa [HO1]

Tragopogon dubius [PNW]

Tragopogon porrifolius [PNW]

Verbesina alternifolia [GLE]

Tragopogon porrifolius L., (with leaves like leek, in the Porrum group of genus *Allium*), SALSIFY, VEGETABLE-OYSTER SALSIFY, OYSTERPLANT. Plant 0.4–1 m tall. Reported by Reverchon as "naturalized in gardens" at Dallas in 1903 (Mahler 1988); a single plant found in vacant lot in University Park, Dallas, in 1946; 1977 collection from freeway waste area in downtown Dallas; also observed in Harry S. Moss Park, Dallas (R. May, pers. comm.); also w TX. Apr–Jun. Native of Eurasia. Cultivated for its edible root (Mabberley 1987).

VERBESINA CROWN-BEARD, FLAT-SEED-SUNFLOWER, WINGSTEM

Ours herbaceous annuals or perennials, pubescent; leaves usually alternate, sometimes opposite; leaf blades ovate to deltoid, usually serrate; petioles usually winged, the wings in some species extending along the stem; phyllaries in 2 to several series, subequal; ray flowers pistillate, fertile or infertile, the ligules yellow or in one species white; disk flowers perfect, fertile, the corollas yellow or in one species white; achenes flattened, often winged; pappus of 2 deciduous or persistent awns or absent.

A genus of ca. 300 species of trees, shrubs, and herbs of warm areas of the Americas; some are cultivated as ornamentals. (Name said to be modified from *Verbena*, reason unknown) (tribe Heliantheae)

REFERENCES: Robinson & Greenman 1899; Coleman 1966, 1968, 1977; Olsen 1979, 1985.

1. Stems not winged; ligules of ray flowers prominently 3-lobed apically, the lobes 1.5–4 mm long; plants annuals with taproot; leaf blades usually gray-canescent beneath ____ **V. encelioides**
1. Stems usually winged, a thin wing of leaf-like tissue extending from the leaves along the stem (but not winged in the rare *V. lindheimeri* and rarely not winged in other species); ligules of ray flowers only slightly 3-toothed apically; plants perennials with fibrous or fleshy fibrous roots; leaf blades not gray-canescent beneath.
   2. Ray flowers 1–5, the ligules white, ca. 5–10 mm long ____ **V. virginica**
   2. Ray flowers 2–15+, the ligules yellow, ca. 10–30 mm long.
      3. Heads usually many, (10–)20–100 per plant; phyllaries few, soon deflexed (= bent downward); ray flowers 2–10, the ligules drooping; achenes spreading in all directions forming globose heads, some achenes deflexed; known in nc TX only from Dallas and Grayson cos. ____ **V. alternifolia**
      3. Heads solitary–few, usually 1–10(–20) per plant; phyllaries numerous, imbricated, not deflexed; ray flowers 8–15+, the ligules spreading ± horizontally; achenes only slightly spreading, none deflexed; widespread in nc TX.
         4. Stems winged; leaves mostly alternate, not harshly scabrous or only somewhat so; phyllaries lanceolate to linear-oblong, acute to acuminate apically; widespread in nc TX w to East Cross Timbers and in Red River drainage to Cooke Co. ____ **V. helianthoides**
         4. Stems not winged; leaves mostly opposite (but alternate in head-bearing region), mostly harshly scabrous; phyllaries broadly oblong to ovate or obovate, rounded to subacute apically; rare in nc TX, found locally only in Lampasas Cut Plain ____ **V. lindheimeri**

Verbesina alternifolia (L.) Britton ex Kearney, (alternate-leaved), WINGSTEM. Stems 1–2(–3) m tall, usually winged; leaves alternate, the blades lanceolate to elliptic or ovate, coarsely serrate to subentire; heads numerous in a ± open arrangement; achenes winged. Woods of tributaries; Dallas and Grayson cos. are the only records located for TX. Sep.

Verbesina encelioides (Cav.) Benth. & Hook.f. ex A. Gray, (resembling *Encelia*, another genus of Asteraceae), COWPEN DAISY, GOLDEN CROWNBEARD, FEVERWEED, BUTTER DAISY. Stems 0.1–0.9 m tall; leaves opposite towards base, tending to alternate upward; leaf blades deltoid to ovate, toothed; petioles usually winged and auriculate at base; heads 1(–3) at ends of branches, large and conspicuous; ray flowers 10–15, the ligules ca. 1–2.5 cm long, yellow, showy; achenes usually winged. Disturbed areas, often in sandy soils; nearly throughout TX. May–Nov. [*Ximenesia encelioides* Cav.] The foliage has an unpleasant odor if crushed or touched; Native Americans reportedly used the plant in treating skin diseases (Ajilvsgi 1984).

Verbesina helianthoides Michx., (resembling *Helianthus*—sunflower), GRAVELWEED CROWNBEARD. Stems 0.5–1.2 m tall; leaves alternate, the blades lanceolate to narrowly ovate, serrate; heads usually few in a compact arrangement; achenes winged. Disturbed areas; sandy woods; e TX w to East Cross Timbers, in Red River drainage to Cooke Co. May–Jun.

Verbesina lindheimeri B.L. Rob. & Greenm., (for Ferdinand Jacob Lindheimer, 1809–1879, German-born TX collector), LINDHEIMER'S CROWNBEARD. Stems (0.2–)0.4–0.6(–1) m tall, not winged; leaf blades ovate-deltoid, cuneate basally; heads few; achenes winged. Juniper-oak woodlands, shaded slopes, on limestone; Coryell Co., also Bell (Fort Hood—Sanchez 1997) and Burnet (Balcones Canyonlands Nat. Wildlife Refuge, C. Sexton, pers. comm.) cos.; endemic to Lampasas Cut Plain and e part of Edwards Plateau. Late summer–fall.

Verbesina virginica L., (of Virginia), FROSTWEED, ICEPLANT, WHITE CROWNBEARD, VIRGINIA CROWNBEARD, RICHWEED, SQUAW-WEED, INDIAN-TOBACCO. Stems 4–5-winged, coarse, to 2 m tall; leaves alternate; heads numerous in cymose arrangement; ray and disk corollas white; ligules of ray corollas to 10 mm long. Disturbed woody areas; se and e TX w to West Cross Timbers and Edwards Plateau. Aug–Oct. When exposed to the first freezing temperatures of the year, the stems split and exude a sap with freezes into conspicuous ice formations—hence several of the common names (Ajilvsgi 1984). The roots were used by Native Americans to relieve cramps, chills, and fevers (Burlage 1968).

VERNONIA IRONWEED

Ours clump-forming or rhizomatous perennial herbs; leaves cauline, evenly distributed along the stem, alternate, simple, toothed; heads many, rather small, corymbose; phyllaries herbaceous but rather stiff or hard, sometimes purple-tinged, in several rows, strongly imbricated, the outer successively shorter; ray flowers absent; disk corollas usually purple (rarely white), deeply lobed, usually more than 12 per head; pappus of many, long, scabrous, white to brownish or purplish hairs, plus a few very small outer scales.

A genus of ca. 500 species of the Old World tropics and tropical and warm areas of North America; timber trees to herbs; some medicinally important. Interspecific hybridization is well-known in *Vernonia* (Urbatsch 1972; Jones & Faust 1978). The rich purple corollas make *Vernonia* species a showy component of the summer flora of ne TX. (Named for William Vernon, 16-?–1711, English botanist who collected in Maryland) (tribe Vernonieae)
REFERENCES: Gleason 1906, 1922, 1923; Jones 1970, 1982; Faust 1972; Urbatsch 1972; Faust & Jones 1973; King & Jones 1975; Chapmam & Jones 1978; Jones & Faust 1978; Keeley & Jones 1979.

1. Phyllaries appressed-pubescent over the back (= abaxially), white or grayish; leaf blades narrowly linear, densely white-woolly beneath; stems densely gray-woolly ____ **V. lindheimeri**
1. Phyllaries glabrous or spreading-pubescent, not whitish or grayish; leaf blades narrow to wide, not densely white-woolly beneath; stems glabrous to pubescent, but not gray-woolly.
   2. Leaf blades scabrous or pubescent to tomentose on the lower surface, the larger ones linear to lanceolate or ovate, not pitted below OR pitted (only in *V. texana*); upper stem leaves with blades of various widths, often wider than 10 mm; widespread in nc TX.
      3. Stems usually with curly, spreading hairs (sometimes short, use lens); leaf blades slightly pubescent to tomentose on the lower surface, not pitted, those at mid-stem lanceolate to ovate-lanceolate to elliptic, 10–45 mm wide.
         4. Involucres in flower 2–6 mm in diam. at equator; flowers 9–34 per head; leaves almost glabrous to with various pubescence beneath; including species widespread in nc TX.

5. Phyllaries with shiny resiniferous glands; middle phyllaries with tips acute to acuminate, usually recurved or spreading; hairs on lower leaf surfaces curly; widespread in nc TX ____ **V. baldwinii**
5. Phyllaries without resiniferous glands; middle phyllaries with tips subacute to obtuse or mucronate, appressed, not recurved or spreading; hairs on lower leaf surfaces appressed; rare in nc TX, apparently known from a single Blackland Prairie location ____ **V. gigantea**
4. Involucres in flower 5.5–10 mm in diam. at equator; flowers 32–55(–60) per head; leaves densely tomentose beneath; on e margin of nc ____ **V. missurica**
3. Stems glabrous except at summit; leaf blades sparsely scabrous-pubescent on the lower surface, pitted, those at mid-stem linear to lanceolate, 1–14 mm wide ____ **V. texana**
2. Leaf blades glabrous, the larger ones linear, with numerous small but conspicuous pits on the lower surface (can appear as black dots under low magnification); upper stem leaves with blades 3–10 mm wide; in nc TX known only from West Cross Timbers ____ **V. marginata**

Vernonia baldwinii Torr., (for its discoverer, William Baldwin, 1779–1819, botanist and physician of Pennsylvania), BALDWIN'S IRONWEED, WESTERN IRONWEED. Plant 40–150 cm tall; middle stem leaves with blades lanceolate to narrowly ovate, 3.7–17 cm long, 2–6 cm wide. Open woods, low ground; widespread in TX. Jul–Sep. The bitter foliage apparently prevents herbivory by cattle and the plant is thus sometimes common in overgrazed pastures (Ajilvsgi 1984).

Vernonia gigantea (Walter) Trel. ex Branner & Coville, (gigantic), TALL IRONWEED. Plant 100–150(–350) cm tall; middle stem leaves with blades linear-lanceolate to lanceolate or oblanceolate, 10–75 mm wide. Open weedy areas; included based on citation of vegetational area 4 (Fig. 2) by Hatch et al. (1990) and locality mapped apparently in e part of nc TX by Urbatsch (1972); mainly e U.S. to the e of TX. Aug–Nov. [*V. altissima* Nutt.] Urbatsch indicated that this species hybridizes with *V. missurica* and *V. baldwinii*.

Vernonia lindheimeri A. Gray & Engelm., (for Ferdinand Jacob Lindheimer, 1809–1879, German-born TX collector), WOOLLY IRONWEED. Plant 25–75 cm tall, clump-forming. Limestone outcrops; Dallas and Tarrant cos. s and w to Edwards Plateau and Trans-Pecos, also se TX. Summer.

Vernonia marginata (Torr.) Raf., (margined), PLAINS IRONWEED, NARROW-LEAF IRONWEED. Plant 30–100 cm tall; middle stem leaves with blades broadly linear; upper stem leaves sessile, 3–10 mm wide. Low areas; Palo Pinto and Shackelford cos.; West Cross Timbers s and w to w TX. Jul–Aug. Avoided by grazing livestock because of its bitter taste (Kirkpatrick 1992).

Vernonia missurica Raf., (of Missouri), MISSOURI IRONWEED. Plant to 150 cm tall; middle stem leaves with blades lanceolate to narrowly ovate. Open woods, low ground; Henderson and Lamar cos. on e margin of nc TX; mainly se and e TX. Jul–Oct.

Vernonia texana (A. Gray) Small, (of Texas), TEXAS IRONWEED. Plant 40–150 cm tall; upper leaves markedly reduced. Sandy woods; Bell, Fannin, Henderson, Lamar, Limestone, and Milam cos., also collected at Dallas by Reverchon; mainly se and e TX. Summer.

Two hybrids are also known from nc TX:

Vernonia ×guadalupensis A. Heller [*baldwinii* × *lindheimeri*], (for the Guadalupe River area of the Edwards Plateau). Phyllaries appressed pubescent over the back as in *V. lindheimeri*, but differing in having leaf blades gray-green beneath with thin pubescence to nearly glabrous (versus densely white-woolly beneath in *V. lindheimeri*). This hybrid is known from several Dallas Co. specimens, also Bell Co. (Fort Hood—Sanchez 1997).

Vernonia ×vultrina Shinners [*baldwinii* × *marginata*], (for Buzzards Spring in Dallas, TX). Leaves glabrous as in *V. marginata* but differing in the upper stem leaves with petioles 2–3

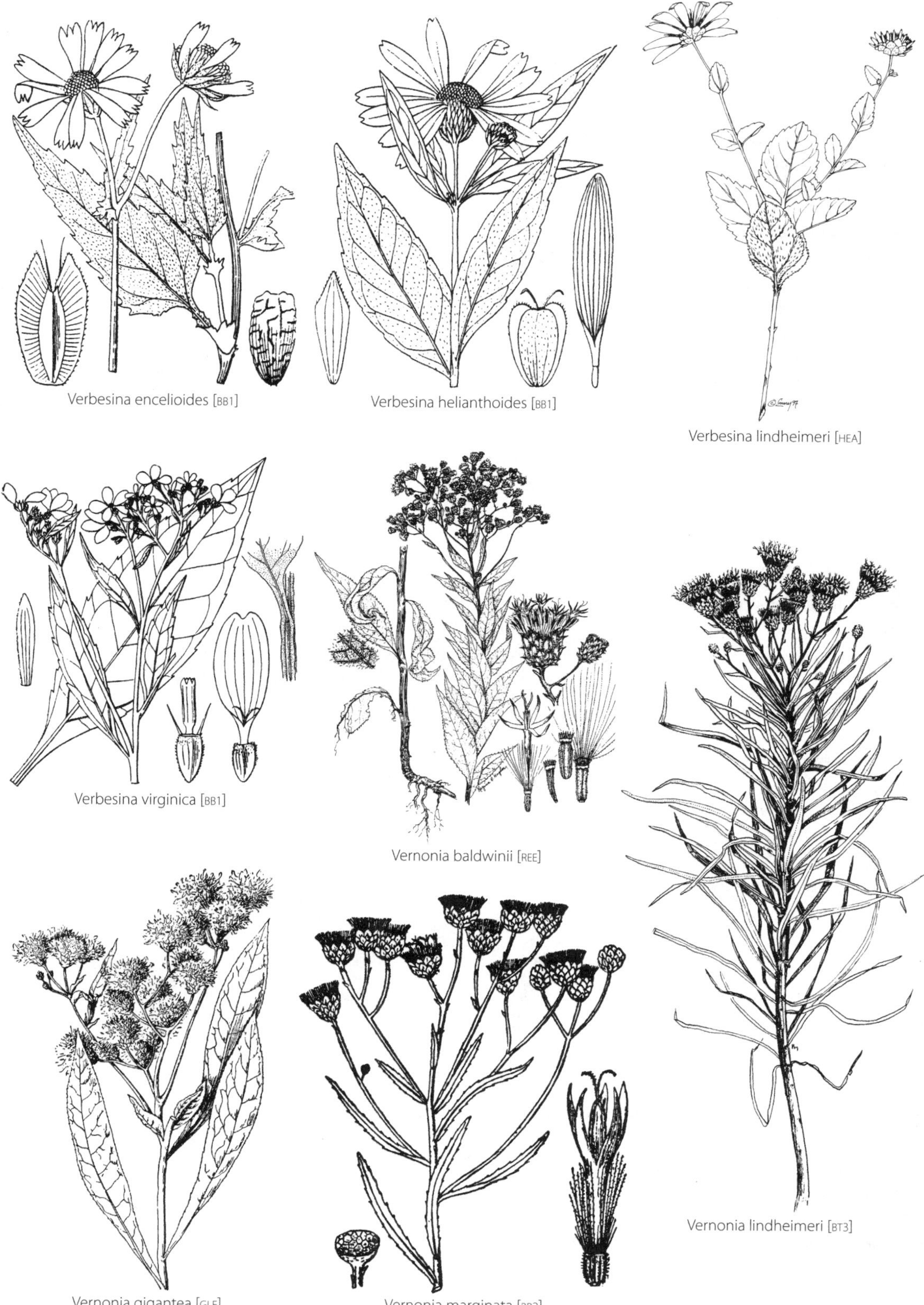

Verbesina encelioides [BB1]

Verbesina helianthoides [BB1]

Verbesina lindheimeri [HEA]

Verbesina virginica [BB1]

Vernonia baldwinii [REE]

Vernonia lindheimeri [BT3]

Vernonia gigantea [GLE]

Vernonia marginata [BB2]

mm long and blades 15–25 mm wide. This hybrid is known only from a top of an apparently large plant, collected by Reverchon at Buzzards Spring, Dallas (now a residential area).

VIGUIERA GOLDEN-EYE

A genus of 180 species native to warm and tropical areas of the Americas; a few are cultivated as ornamentals. (Named for L.G.A. Viguier, 1790–1867, French physician and botanist) (tribe Heliantheae)
REFERENES: Blake 1918; Robinson 1977.

Viguiera dentata (Cav.) Spreng., (toothed), SUNFLOWER GOLDEN-EYE. Herbaceous perennial; leaves opposite below, alternate above; leaf blades ovate or rhombic-ovate, cuneate to truncate, serrate; phyllaries in 3 series, imbricated, ovate basally, indurate with herbaceous linear tips; ray flowers 10–12, pistillate, infertile, the ligules yellow, ca. 7–15 mm long, 3–7.5 mm wide; disk flowers perfect, fertile, the corollas yellow; achenes black or mottled, appressed-pubescent, 3.5–4 mm long; pappus awns 2, 2.2–2.8 mm long, squamellae (additional small pappus scales) 4, squarish, fimbriate. Calcareous soils; Bell, Hill, McLennan, and Williamson cos., also Coryell Co. (Fort Hood—Sanchez 1997); s part of nc TX through Edwards Plateau to Trans-Pecos. Oct–Nov.

WEDELIA

A genus of 100 species of tropical and warm areas of the world; 4 species are found in the United States; a number of taxa have sometimes been treated in *Zexmenia*. (Named for Georg Wolfgang Wendel, 1645–1721, professor of botany at Jena, Germany) (tribe Heliantheae)
REFERENCES: Turner 1988a, 1992; Strother 1991.

Wedelia texana (A. Gray) B.L. Turner, (of Texas) ORANGE ZEXMENIA, HAIRY WEDELIA, ORANGE DAISY. Small shrub 0.5–1 m tall; herbage strigose-hispid; leaves opposite, sessile; leaf blades narrowly ovate, usually 5 cm or more long, cuneate basally, coarsely toothed to slightly lobed; heads long-peduncled, solitary or in a cyme of 3; phyllaries in 2 series; ray flowers pistillate, fertile, the ligules conspicuously yellowish orange; disk flowers perfect, fertile, the corollas similar in color to ligules of ray flowers; achenes of ray and disk flowers usually 2- or 3-angled, usually winged; pappus of 2 or 3 short awns subtended by short, hyaline scales. Calcareous soils; Burnet Co., also Brown, Comanche (Mahler 1988), and Somervell (R. O'Kennon, pers. obs.) cos.; sw part of nc TX s through Edwards Plateau to s TX and w to Trans-Pecos. Apr–Nov. [*W. acapulcensis* Kunth var. *hispida* of authors, not (Kunth) Strother, *W. hispida* of authors, not Kunth, *Zexmenia hispida* of authors, not (Kunth) A. Gray ex Small] We are following Jones et al. (1997) and J. Kartesz (pers. comm. 1997) for nomenclature of this taxon which has long been associated with the epithet *hispida*.

XANTHISMA SLEEPY DAISY

A monotypic genus native to TX, NM, and OK. The common name probably refers to the late morning expansion of the ray flowers. (Greek: *xanthisma*, that which dyes yellow) (tribe Astereae)
REFERENCE: Semple 1974.

Xanthisma texanum DC. subsp. **drummondii** (Torr. & A. Gray) Semple, (sp.: of Texas; subsp.: for its discoverer, Thomas Drummond, 1780–1835, Scottish botanist and collector in North America), TEXAS SLEEPY DAISY. Glabrous annual 10–75 cm tall; leaves sessile or subsessile, the blades lanceolate, entire, toothed, or the lower often pinnately lobed; heads rather large, solitary and terminal on main stem and branches; receptacles with chaffy scales; ray flowers yellow, expanded from late morning to late afternoon, becoming erect and with tips nearly touching, the head closed during the night. Sandy open woods, roadsides, and pastures; Hill, Montague,

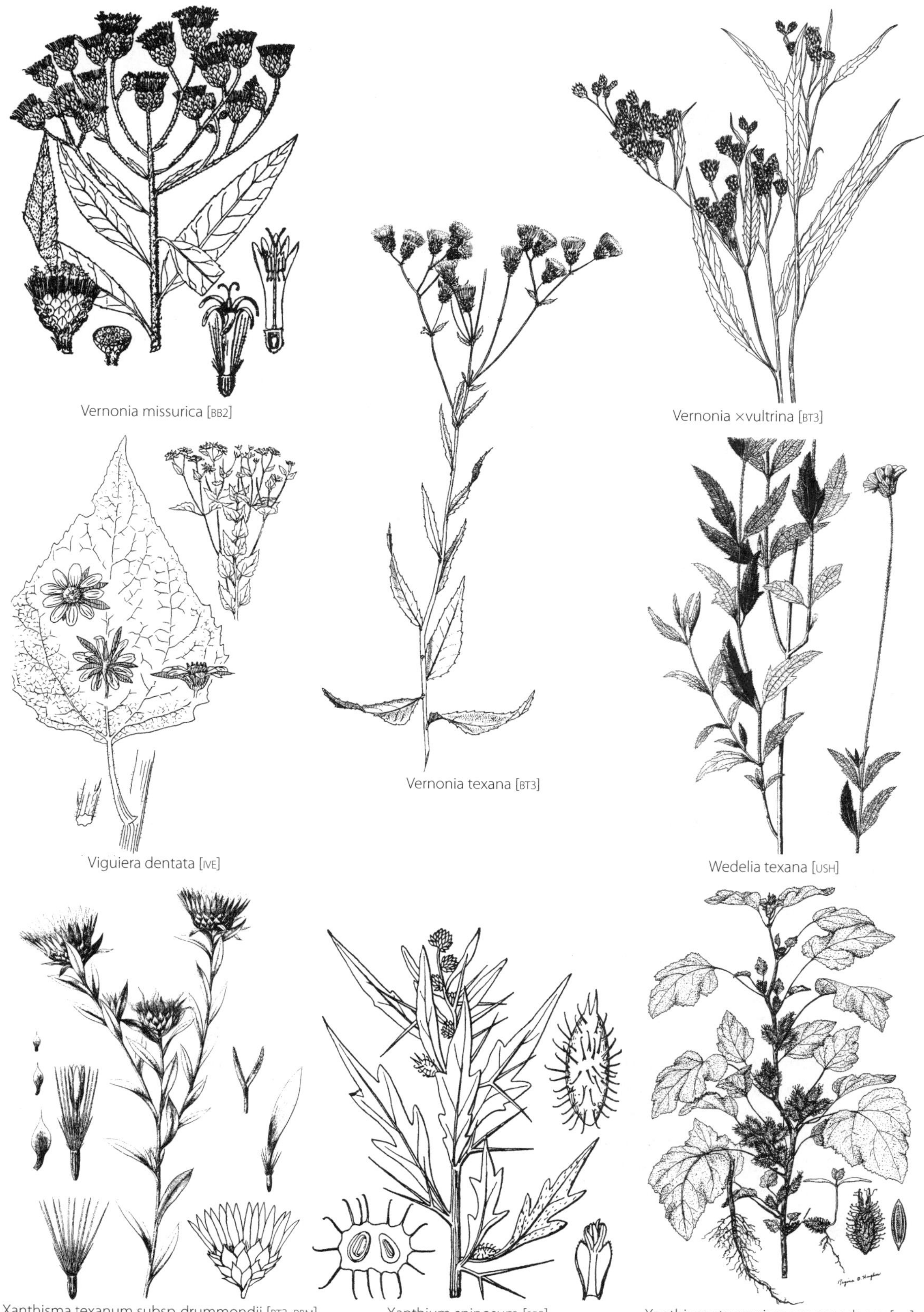

Vernonia missurica [BB2]

Vernonia ×vultrina [BT3]

Vernonia texana [BT3]

Viguiera dentata [IVE]

Wedelia texana [USH]

Xanthisma texanum subsp. drummondii [BT3, RBM]

Xanthium spinosum [BB2]

Xanthium strumarium var. canadense [REE]

and Tarrant cos. w through West Cross Timbers to Panhandle; also Edwards Plateau. May–Oct. [*Xanthisma texanum* DC. var. *drummondii* (Torr. & A. Gray) A. Gray] Burlage (1968) reported the plant to contain a saponin and to be poisonous. ☠

XANTHIUM COCKLEBUR, BURWEED

Coarse, monoecious annuals with taproots; leaves alternate; heads small, axillary, nearly sessile, unisexual; ray flowers absent; staminate heads with 1–3 series of separate phyllaries; disk flowers with minute corollas; anthers 5, separate; pistillate heads with phyllaries united into a prickly bur or "fruit"; bur completely enclosing 2 flowers which lack corolla and pappus; achenes 2 per bur.

A cosmopolitan (now) genus of ca. 3 species with conspicuous, accrescent involucres covered with hooked prickles used for animal dispersal. A cladistic study by Karis (1995) suggested that *Xanthium* is a monophyletic ingroup in a paraphyletic *Ambrosia*. According to the VELCRO® Industries homepage (www.velcro.com), in the early 1940s, a Swiss inventor, George de Mestral, after a walk noticed "cockleburrs" [presumably *Xanthium*] on his dog and his pants. He examined the hooked prickles under a microscope and derived the idea for the well-known two-sided fastener—one side with stiff, cocklebur-like "hooks" and the other side with soft "loops" like the cloth of his pants. The word velcro comes from the French words *velours*, velvet, and *croché*, hooked. (Greek name of some plant used to dye the hair, from *xanthos*, yellow) (tribe Heliantheae)

REFERENCES: Millspaugh & Sherff 1919; Rydberg 1922; Löve & Dansereau 1959; Karis 1995.

1. Nodes with a conspicuous, 3-pronged, yellowish spine; bur ca. 1 cm long; leaf blades tapering or wedge-shaped at base; rare in nc TX ____ **X. spinosum**
1. Nodes without spines (the only spiny or prickly part of the plant is the bur or "fruit"); bur usually (1–)2–3 cm long; leaf blades truncate or cordate at base; widespread in nc TX ____ **X. strumarium**

Xanthium spinosum L., (spiny), SPINY COCKLEBUR, CLOTBUR. Stems 0.3–1(+) m tall; leaves entire or with a few teeth or lobes, slightly pubescent or glabrate on upper surface, densely silvery-pubescent on lower surface, tapering to a short petiole; bur finely pubescent, with numerous, hooked prickles ca. 2 mm long. Disturbed and waste places; included based on citation of vegetational area 5 (Fig. 2) by Hatch et al. (1990), also Brown Co. (J. Stanford, pers. comm.); mainly w 1/2 of TX. Jul–Oct. Introduced, region of origin unclear, probably the Americas. The plant can cause mechanical injuries; ingestion by pigs can result in toxic symptoms including prostration and convulsions (Burlage 1968). ☠

Xanthium strumarium L. var. **canadense** (Mill.) Torr. & A. Gray, (sp.: of swellings or tumors; var.: of Canada), COCKLEBUR, COMMON COCKLEBUR, ABROJO. Stems usually 0.4–2 m tall; leaf blades to 15 cm long, ovate to deltoid, suborbicular, or reniform, serrate or with shallow lobes, scabrous; petioles elongate, to 15 cm long; bur (1–)2–3+ cm long, ca. 1.5 cm in diam., terminated by 2 prominent spines, covered with stiff, hooked prickles ca. 5 mm long; bur and bases of spines densely pubescent. Disturbed moist areas or sandy soils; throughout TX. Fruiting Jul–Nov. [*Xanthium italicum* Moretti] *Xanthium strumarium* is a very variable taxon questionably divisible into infraspecific taxa. Our plants seem to best fit var. *canadense*. Poisonous and potentially fatal to pigs and other livestock; the poisonous principle is hydroquinone or a diterpenoid glycoside and occurs in the seeds and seedlings (Sperry et al. 1955; Kingsbury 1964, 1965; Stephens 1980; Hardin & Brownie 1993). The burs can cause mechanical injury when eaten by livestock and are sometimes referred to as "porcupine eggs" (Barkley 1986). ☠

YOUNGIA

An Asian genus of ca. 40 species. (Named for William Young, 1742–1785, German-born American botanist, nurseryman, and gardener) (tribe Lactuceae)
REFERENCES: Babcock & Stebbins 1939; Vuilleumier 1973.

Youngia japonica (L.) DC., (of Japan), JAPANESE-HAWKWEED. Annual herb from taproot; sap milky; stems 0.1–0.9 m tall; basal leaves pinnatifid, crowded; stem leaves few and reduced; heads numerous, small, 6–7 mm long including pappus; involucres 3.5–5.7 mm high, glabrous; corollas all ligulate, yellow or yellow-orange; achenes 1.5–2.5 mm long, strongly compressed; pappus of numerous hair-like bristles 2.5–3.5 mm long. Flowerbeds, gardens, weedy areas; Dallas and Tarrant cos.; also scattered in se and e TX. Oct–Nov. Native of Asia.

ZINNIA

A genus of 11 species from the U.S. to Argentina, especially Mexico; herbs and low shrubs with opposite or whorled leaves and alkaloids; some are cultivated as ornamentals; *Z. elegans* Jacq., COMMON ZINNIA, YOUTH-AND-OLD-AGE, was cultivated by the Aztecs. (for Johann Gottfried Zinn, 1727–1759, German professor of botany at Göttingen known for his botanical work in Mexico) (tribe Heliantheae)
REFERENCE: Torres 1963.

Zinnia grandiflora (L.) DC., (large-flowered), PLAINS ZINNIA, ROCKY MT. ZINNIA. Low much-branched perennial 8–22 cm tall; stems from a woody base; leaves opposite, linear, 10–30 mm long, to ca. 2 mm wide, strigose; heads terminating stems, not much raised above the leaves; involucres 5–8 mm high; phyllaries broadly obtuse and often red-tipped apically; receptacles chaffy; ray flowers 3–6, pistillate, fertile; ligules to 18 mm long, ovate to orbicular, yellow; disk flowers perfect, fertile, the corollas red or green, 5-toothed with 1 tooth often larger; chaffy bracts membranous, enclosing the achenes; ray achenes 3-angled, with lateral awns ± adnate to ligule or only the tip free, the awn of inner angle minute or absent; disk achenes oblanceolate, 4–5 mm long, with (1–)2(–4) awns or awns absent. Dry calcareous areas; Brown Co.; extreme w margin of nc TX w to w TX. Summer–fall. /108

BALSAMINACEAE TOUCH-ME-NOT FAMILY

A medium-sized (850 species in 2 genera), mainly Old World tropical family with a few in temperate regions; they are generally herbs or rarely subshrubs. *Hydrocera*, native to Indomalesia, has only a single species. Family name conserved from *Balsamina*, a genus now treated as *Impatiens* (the name *Impatiens* was published earlier and thus has priority in terms of nomenclature). (Possibly from Greek: *balsamon*, a fragrant gum, or from ancient Arabic: *balassam*, for some species of *Impatiens*) (subclass Rosidae)
FAMILY RECOGNITION IN THE FIELD: herbs with translucent watery stems, alternate simple leaves, showy bilaterally symmetrical flowers with a conspicuous *spur*, and *explosively dehiscing* capsules.
REFERENCES: Rydberg 1910; Wood 1975.

IMPATIENS TOUCH-ME-NOT, JEWELWEED, BALSAM, SNAPWEED

Ours erect annual herbs; leaves alternate, simple; sepals 3, 2 small, 1 large and conspicuous, petal-like, saccate, and with a long nectar spur; petals apparently 3, the 2 laterals 2-lobed; anthers 5, united around stigma; pistil 1; ovary superior; fruit an explosively dehiscent, 5-valved capsule.

A genus of 849 species of tropical and n temperate areas, especially India. Many *Impatiens* species are used as ornamentals (e.g., *I. walleriana* Hook.f.—BUSY-LIZZIE, SULTAN'S-FLOWER). Several

Middle Eastern and African species were used as a source of dyes (red, yellow, or black) (Baumgardt 1982). (Latin: *impatiens*, impatient, alluding to the explosive release of the seeds when the capsule is touched)

REFERENCE: Rust 1977.

1. Flowers of various colors, with extremely narrow opening; capsules densely pubescent; stems usually pubescent; petioles 0.5–2 cm long, often with conspicuous glands ______ **I. balsamina**
1. Flowers orange with crimson spots, with obvious, open, funnel-like shape; capsules glabrous; stems glabrous; petioles 0.5–10 cm long, without glands ______ **I. capensis**

Impatiens balsamina L., (balsamic, similar to balsam), GARDEN BALSAM, ROSE BALSAM. Stems to ca. 0.75 m tall; leaf blades elliptic-lanceolate to lanceolate, 5–10(–15) cm long, 1–3 cm wide, serrate; petioles 1–2 cm long; flowers axillary, solitary or in groups of 2–3, large and showy, ca. 25–50 mm across, of various colors including white to pink, red, or yellow, often spotted, often double-flowered; capsules 1–3 cm long. Cultivated; escaped to a creek bottom in Dallas Co. ?–Nov. Native of Asia. The flowers have been used to dye fingernails (Mabberley 1997).

Impatiens capensis Meerb., (of the Cape of Good Hope), SPOTTED TOUCH-ME-NOT, JEWELWEED, LADY'S-EARRINGS. Plant glabrous; stems to 1.5 m tall, succulent; leaves ovate to ovate-elliptic or elliptic, to 10(–13) cm long and to 8 cm wide (often much smaller), crenate, mucronate; petioles 0.5–10 cm long, those on flowering branches often rather short; flowers usually 2–4 on axillary inflorescences, pendulous on pedicels to ca. 20 mm long; flowers of 2 kinds, some large and showy, others small and cleistogamous; larger flowers 20–30 mm long; large sepal spurred, the spur ca. 6–25 mm long and bent backward parallel to the body of the sepal; stamens 5, pistil 1, ovary superior; capsules ca. 2 cm long. Stream bottom woods; Fannin and Lamar cos. in Red River drainage; mainly e TX. May–Jul. [*I. biflora* Walter] The common name is derived from the fruits being explosively dehiscent when touched or disturbed. The sap is reported by some to prevent POISON-IVY dermatitis when rubbed on the skin (McGregor 1986). Despite the specific epithet, this species is native to North America including e TX. Austin (1975) considered this species adapted for pollination by ruby-throated hummingbirds (*Archilochus colubris*).

BASELLACEAE MADEIRA-VINE OR BASELLA FAMILY

A very small (20 species in 4 genera), mainly tropical and warm area, especially American family of climbing vines containing betalain pigments; some are cultivated as ornamentals. The family is thought to be related to the Portulacaceae. Family name from *Basella*, MALABAR-NIGHTSHADE, a genus of 5 species of Madagascar, e Africa, and one pantropical. (Latinized version of the Malabar name for the plant) (subclass Caryophyllidae)

FAMILY RECOGNITION IN THE FIELD: the only nc TX species is an introduced, unarmed, non-tendriled, herbaceous vine with alternate, simple, entire, broadly ovate leaves and small white flowers on axillary spike-like racemes or panicles.

REFERENCES: Wilson 1932; Bogle 1969; Sperling & Bittrich 1993; Behnke & Mabry 1994; Nowicke 1996.

ANREDERA MADEIRA-VINE

A genus of 10–15 species of warm areas of the Americas; a number are cultivated as ornamentals. (Derivation of generic name unknown, possibly a personal name)

Anredera cordifolia (Ten.) Steenis, (with heart-shaped leaves), MADEIRA-VINE, MIGNONETTE-VINE. Twining, glabrous, herbaceous vine to ca. 6 m long, without tendrils or armature; perennial from a tuber-like root, hardy in nc TX; leaves alternate, simple, entire, broadly ovate, subcordate to cordate, to ca. 11 cm long, thin-fleshy, petiolate; inflorescence an axillary spike-like raceme or

Youngia japonica [BR2]

Zinnia grandiflora [BB2]

Impatiens balsamina [BL1]

Berberis trifoliolata [POW]

Anredera cordifolia [HEA]

Impatiens capensis [CO1]

Nandina domestica [LAM]

panicle with 2–4 branches, with numerous, small, usually perfect flowers, 4–30 cm long; pedicels 1.5–2.5 mm long, with 2 minute bracteoles persistent at tip; 2 larger unwinged bracteoles slightly shorter than the perianth lobes present immediately below the 5-parted perianth; perianth to ca. 6 mm wide, white; stamens 5; pistil 1; style 3-divided ca. 1/4–3/4 of way to base; ovary superior; fruit not seen. Cultivated as an ornamental and escaping; Tarrant and Williamson cos.; we are not aware of other TX localities. Aug–Oct. Native of tropical South America.

This species is often confused with the similar *A. baselloides* (Kunth) Baill., (resembling *Basella*—Malabar nightshade), which differs in having an undivided style and upper bracteoles with narrowly winged keels and slightly exceeding the perianth in length.

BERBERIDACEAE BARBERRY FAMILY

Ours perennial herbs or shrubs; leaves alternate or apparently opposite, simple or compound, entire, toothed, or lobed; flowers solitary or in racemes or panicles; perianth parts in 3s, all nearly alike or distinctly differentiated into sepals and petals (1 to several rows of each), the outer usually falling as or soon after the flowers open, the innermost sometimes with glands or modified into prominent staminode-like nectaries; perianth sometimes with basal bracts; stamens as many as the petals or twice as many; anthers opening by apical valves or by a longitudinal slit; pistil 1; ovary superior; fruit in ours a berry.

A medium-sized family of ca. 650 species in 15 genera (Whetstone et al. 1997a); a widespread, but especially n temperate family of trees, shrubs, or perennial herbs; often with alkaloids; tissues are often colored yellow due to the isoquinoline berberine; many are valued as ornamentals including *Berberis, Mahonia* (if segregated), and *Nandina*. Although the family has sometimes been split with *Podophyllum* and related genera recognized as the Podophyllaceae, a molecular study by Kim and Jansen (1996 [1997]) suggested the Berberidaceae sensu lato is a monophyletic group related to (sister group of) the Ranunculaceae. (subclass Magnoliidae)
FAMILY RECOGNITION IN THE FIELD: evergreen shrubs or perennial herbs with compound or deeply palmately lobed leaves; *flower parts in 3s* is a good field character for this otherwise obviously dicot family.
REFERENCES: Ernst 1964a; Norwicke & Skvarla 1981; Loconte & Estes 1989; Loconte 1993; Kim & Jansen 1996 [1997]; Whetstone et al. 1997a.

1. Evergreen shrubs; leaves numerous, with 3 leaflets or 2–3 times pinnately compound, not peltate; flowers numerous.
    2. Leaves with 3 leaflets; leaflets with spiny lobe-like teeth ____ **Berberis**
    2. Leaves 2–3 times pinnately compound; leaflets without spiny teeth ____ **Nandina**
1. Herbs; leaves 1 or 2 per plant, deeply palmately lobed, peltate; flower 1 per plant ____ **Podophyllum**

BERBERIS BARBERRY

A nearly worldwide genus of ca. 500 species (Whittemore 1997d) of usually spiny shrubs with alkaloids. Some serve as alternate hosts for cereal rusts; as a result, the sale or transport of certain species is illegal in the U.S. and Canada; a number are cultivated as ornamentals (e.g., *B. aquifolium* Pursh—OREGON GRAPE) and the berries of some species are edible. Some species were previously segregated as the genus *Mahonia*. However, species intermediate between *Mahonia* and *Berberis* (in the narrow sense) are known and recognizing segregate genera does not seem warranted (Whittemore 1997d). (Latinized from of *berberys*, Arabic name of the fruit)
REFERENCES: Ahrendt 1961; McCain & Hennen 1982; Whittemore 1997d.

Berberis trifoliolata Moric., (with three leaflets), ALGERITAS, AGARITO, CURRANT-OF-TEXAS.

Densely leafy evergreen shrub to 2 m tall; wood yellow; leaves with 3 leaflets; leaflets narrowly oblong-lanceolate, blue-gray above, green beneath, with 1–3 pairs of spiny lobe-like teeth resembling those of HOLLY, glabrous; inflorescences of compact, axillary, umbel-like racemes; flowers small, yellow; perianth of 6–9 ± petal-like sepals, and 6 narrower, longitudinally cupped petals; anthers opening by apical valves; fruits globose, 8–10 mm in diam., red. Rocky limestone soils; Bell, Brown, Johnson, Palo Pinto, and Somervell cos., also McLennan and Parker cos. (Mahler 1988); s and w parts of nc TX s to s TX and w to Trans-Pecos. Late Feb–Mar. This species is sometimes recognized in the genus *Mahonia* [as *M. trifoliolata* (Moric.) Fedde] (e.g., Kartesz 1994). However, we are following Whittemore (1997d) and Jones et al. (1997) in treating it in *Berberis*. [*B. trifoliolata* var. *glauca* (I.M. Johnst.) M.C. Johnst.] Jones et al. (1997) spelled the epithet *trifoliata*. According to Correll and Johnston (1970), an excellent jelly can be made from the fruits. The wood and roots were used as a source of yellow dye by early settlers (Kirkpatrick 1992). The stamens are reported to be touch sensitve; when touched by a pollinating insect, they spring out throwing pollen onto the pollinator (Wills & Irwin 1961; Kirkpatrick 1992). This species was reported to be susceptible to infection by the fungus, *Puccinia graminis* Pers., black stem rust of wheat (Whittemore 1997d); however, according to J. Hennen (pers. comm.), it is not susceptible; the species of *Berberis* in section *Mahonia* (sometimes treated as a separate genus) are not infected by *P. graminis*; rather they are susceptible to *Cumminsiella texana* (Holw. & Long.) Arthur, an autoecious (= requires only 1 host) rust (J. Hennen, pers. comm.). The roots contain berberine and associated alkaloids and were used in folk medicine; however, in high concentrations they can cause poisoning (Powell 1988). ☠

NANDINA SACRED-BAMBOO, HEAVENLY-BAMBOO

A monotypic genus native from India to Japan; the single species is a much cultivated ornamental shrub, especially in Japan; sometimes segregated into the Nandianceae. (Chinese name meaning plant from the south or from the Japanese name, *nanten*)
REFERENCE: Whetstone et al. 1997b.

Nandina domestica Thunb., (domesticated, frequently used as a house plant), SACRED-BAMBOO, HEAVENLY-BAMBOO. Evergreen shrub to ca. 1.5 m tall; leaves 2–3 pinnately compound, 50–100 cm long; leaflets lanceolate to elliptic-lanceolate, glabrous, often with reddish coloration; inflorescence a panicle to 20 cm long; flowers 3-merous, with whitish or cream-colored perianth 6–8 mm long; stamens 6; anthers opening by a longitudinal slit; fruits 6–9 mm in diam., red. Widely cultivated and escapes into sandy woods; Grayson Co., also Dallas Co. (E. McWilliams, pers. comm.) and Tarrant Co. (R. O'Kennon, pers. obs.). May–Jun. Native from India to e Asia.

PODOPHYLLUM MAY-APPLE, MANDRAKE

A genus of ca. 5 species native to e North America and the Himalayas to e Asia; sometimes segregated into the Podophyllaceae. (Greek: *podos*, foot, and *phyllon*, a leaf, probably referring to the stout petioles of the radical leaf)
REFERENCES: Swanson & Sohmer 1976; George 1997.

Podophyllum peltatum L., (peltate, shield-shaped), MAY-APPLE, WILD JALAP, AMERICAN-MANDRAKE. Rhizomatous perennial herb to 50 cm tall; leaves 2 (sometimes 1), appearing opposite at summit of stem, long-petioled; leaf blades large, to 30 cm or more in diam., typically peltate, suborbicular, with 5–9 segments, inconspicuously pubescent beneath; flower solitary, nodding, overtopped by the leaves; perianth white (rarely rose), of 12–15 broad tepals, the 6 outer shorter (= sepals) than the inner, falling early, the inner (= petals) 2–3.5 cm long; stamens twice as many as the petals; anthers opening by a longitudinal slit; fruit an ovoid, many-seeded, fleshy berry 3–5 cm long, yellow to purplish at maturity. Open woods and thickets, sandy soils; e TX w to Hunt Co. in South Sulphur River drainage (Mahler 1988), locally w in Red River drainage to Grayson

Co. (Buckner Preserve), and to Kaufman Co. in Trinity River drainage. Late Mar–mid-Apr. The unripe fruit is toxic but reported to be edible when ripe and is sometimes used to make preserves; all other parts of the plant are toxic, apparently including the seeds of the ripe fruit; there are over 15 biologically active compounds including the poisonous podophyllin, a bitter resin containing lignins and flavonols; podophyllin has strong cathartic and antineoplastic properties. Children have been poisoned by the unripe fruit and poisoning has resulted from the misuse of medicinal preparations. Eating large quanities of the plant or repeated applications of the resin to the skin can be fatal; workers handling the rootstocks sometimes develop dermatitis and conjunctivitis. The plant has long been used medicinally by Native Americans for warts; recently it has been used against venereal warts and testicular cancer (Muenscher 1951; Kingsbury 1964; Morton 1977; Mabberley 1987; Turner & Szczawinski 1991; Leung & Foster 1996). A rust fungus (*Puccinia podophylli* Schwein.) sometimes causes leaf lesions (J. Hennen, pers. comm.). ☠

BETULACEAE BIRCH OR HAZELNUT FAMILY

Ours monoecious shrubs to small or large trees; leaves alternate, simple, deciduous, short-petioled; leaf blades pinnately straight-veined, serrulate or doubly serrate; stipules deciduous; perianth small or absent; staminate flowers in spreading or drooping catkins (= aments); pistillate flowers in spike-like or cone-like inflorescences; pistil 1; ovary inferior; fruit a 1-seeded samara (2-winged) or nutlet (unwinged).

A small (ca. 125 species in 6 genera (Furlow 1997)), mainly n temperate family of wind-pollinated trees and shrubs with some on tropical mountains; species are variously important economically for timber, as a source of edible nuts, as ornamentals, and as aids in soil nitrification and stabilization; the two additional genera not occurring in Texas are the north temperate *Corylus* (source of hazelnuts and filberts) and the Asian *Ostryopsis*. The family was previously often split into two families, Betulaceae (*Alnus* and *Betula*) and Corylaceae (*Carpinus, Ostrya, Corylus, and Ostryopsis*); most modern authorities now recognize these groups at the subfamilial level (Furlow 1997). (subclass Hamamelidae)

FAMILY RECOGNITION IN THE FIELD: shrubs or trees with alternate, simple, sharp-toothed leaves, early deciduous stipules, and male flowers in long *catkins*; fruits 1-seeded, in cone-like or spike-like, obviously *bracteate* infructescences.

REFERENCES: Little 1971; Furlow 1990, 1997; Kubitzki 1993a.

1. Leaves truncate (= as if cut or squared off) or broadly cuneate (= wedge-shaped) at base; fruits winged, subtended by a small (3–8 mm long) bract and grouped with bracts into a cone-like structure.
    2. Trees with shaggy, conspicuously exfoliating bark; lower surface of leaves often grayish white, densely hairy when young; pistillate inflorescences solitary; bracts 3-lobed, 6–8 mm long, papery and eventually deciduous; buds not stalked ______ **Betula**
    2. Shrubs or small trees with ± smooth bark; lower surface of leaves green, glabrous or nearly so even when young; pistillate inflorescences in groups of 2–3; bracts not lobed, 3–4 mm long, becoming woody and persistent; buds stalked ______ **Alnus**
1. Leaves rounded to subcordate at base; fruits wingless (but wing-like bract can be present), enclosed in or subtended by conspicuous sac-like or 3-lobed, wing-like bracts 8 mm or more long (well-developed bracts usually at least 15 mm long), these ± densely clustered together in a spike-like inflorescence.
    3. Fruit subtended (not enclosed) by a 3-lobed bract; bark smooth, gray, the trunk appearing like rippling muscles; lower surface of mature leaves glabrous except for some hairs in the axils of the veins; none of the main side veins near the base of the leaf forked ______ **Carpinus**

3. Fruit enclosed by a pouch- or sac-like unlobed bract; bark rough, shreddy, brownish, the trunk not appearing rippled; lower surface of mature leaves pubescent; some of the main side veins near the base of the leaf forked ______ **Ostrya**

ALNUS ALDER

A genus of ca. 25 species (Furlow 1997) ranging from the n temperate zone s to Assam, se Asia, and the Andes of South America. ALDERS have a symbiotic relationship with species of the actinomycete *Frankia* (filamentous gram positive bacteria) resulting in root nodule formation and fixation of atmospheric nitrogen (Furlow 1997); as a result they can be important ecologically. Some species also produce useful timber. (Latin name for alder)
REFERENCE: Furlow 1979.

Alnus serrulata (Aiton) Willd., (somewhat serrate or saw-toothed), HAZEL ALDER, COMMON ALDER, SMOOTH ALDER. Shrub or small tree 2–5+ m tall; bark grayish brown or blackish gray to reddish brown; leaf blades elliptic to obovate, 1.6–12 cm long, usually widest beyond middle, cuneate to broadly cuneate basally, usually rounded to acute at apex, usually serrulate and occasionally somewhat undulate; staminate flowers 3 per bract; pistillate inflorescences cone-like, ca. 15–20 mm long, the bracts becoming woody and persistent. Along streams and wet areas; mainly deep e TX, disjunct w to Milam Co. near the e margin of nc TX, also in OK just n of the Red River. Mar–Apr.

BETULA BIRCH

A genus of ca. 35 species of n temperate and boreal zones of the n hemisphere (Furlow 1997); related to *Alnus* but the catkins shatter when ripe; some provide timber for furniture and plywood; the bark of *B. papyrifera* Marshall (PAPER BIRCH) is impervious to water and was used by Native Americans for canoes, baskets, cups, and wigwam covers. (Latin name for birch)

Betula nigra L., (black), RIVER BIRCH, RED BIRCH. Tree to 30 m tall; bark soft, shaggy, pinkish to tan, freely shedding; leaf blades roughly triangular to ovate, truncate to broadly cuneate at base, acute at apex, conspicuously doubly serrate at least beyond middle; staminate bracts ovate to suborbicular; staminate flowers 3 per bract, with calyx; pistillate inflorescences (= aments) up to 25–35 mm long; pistillate bracts divided from apex into 3 oblong-linear lobes, papery and eventually deciduous. Along streams and in low woods; Lamar Co. in Red River drainage, also Henderson and Limestone cos. on e edge of nc TX, also Fannin Co. (Little 1971); mainly se and e TX. Mar–Apr.

CARPINUS HORNBEAM, IRONWOOD

A genus of ca. 25 species (Furlow 1997) of the n temperate zone, especially e Asia; some provide good timber for turnery and tools; the very hard wood is used for such things as mallet heads, tool handles, and levers. According to Peattie (1948), the common name is derived from horn, for toughness, and beam, an old word for tree, related to the German, *baum*, tree. (Latin: *carpinus*, hornbeam, possibly derived from Latin: *carpentum*, a Roman horse-drawn vehicle with wheels made from its hard wood—Furlow 1997)
REFERENCES: Furlow 1987a, 1987b.

Carpinus caroliniana Walter, (of Carolina), AMERICAN HORNBEAM, BLUE-BEECH, WATER-BEECH, MUSCLETREE, LECHILLO, LEANTREE. Small tree with very hard wood to ca. 10 m tall; bark smooth, neither shedding nor peeling; leaf blades usually elliptic to ovate, sharply serrate, inconspicuously doubly so; staminate flowers 1 per bract, without calyx; staminate bracts acute, not awned; pistillate bracts with a long terminal lobe and 2 small lateral lobes near base. Rich or low woods and along streams; e TX w in Red River drainage to Lamar Co., also Henderson Co., also Hopkins Co. (Little 1971). Mar–May. This species is easily distinguished by its smooth gray bark and the rippled, muscle-like appearance of the stems. *Carpinus caroliniana* has an interesting geographic distribution—occurring mainly in the e U.S. but with disjunct populations in the mountains of Mexico and Guatemala (Miranda & Harrell 1950). This pattern is the result of a middle to late Miocene (Miocene epoch—24.6–5.1 mya) extension (during a period of climatic cooling) of deciduous forest and associated fauna (particularly amphibians) into Mexico. Subsequently, during Pliocene (Pliocene epoch—5.1–2 mya) and later times as the climate warmed and dried, these deciduous forests became disjunct, surviving in Mexico only in isolated pockets of appropriate microclimate in the highlands (Miranda & Sharp 1950; Graham 1993).

OSTRYA HOP-HORNBEAM, IRONWOOD

A genus of ca. 5 species (Furlow 1997) ranging from the n temperate zone to Central America; some provide timber. The hard wood has been used for such things as sleigh runners, airplane propellers, and mallet heads (Furlow 1997). (Greek: *ostrys*, shell, alluding to the inflated floral bracts)

Ostrya virginiana (Mill.) K. Koch, (of Virginia), IRONWOOD, EASTERN HOP-HORNBEAM, AMERICAN HOP-HORNBEAM, WOOLLY AMERICAN HOP-HORNBEAM, LEVERWOOD. Small tree with very hard wood to ca. 20 m tall; bark roughened, shreddy, brownish; leaf blades usually elliptic to ovate, sharply doubly serrate; staminate bracts abruptly narrowed to a spine tip (= awned); staminate flowers 1 per bract, without calyx; pistillate bracts pouch-like. Lowland or upland woods; e TX with a disjunct collection in the East Cross Timbers at Handley, Tarrant Co., in 1910, but not found there recently (Mahler 1988). Late Feb–Mar.

BIGNONIACEAE CATALPA OR TRUMPETVINE FAMILY

Ours shrubs, trees, or woody vines; leaves opposite or the uppermost alternate, simple or pinnately compound, entire, toothed, or lobed; flowers rather large and showy, in terminal panicles or spike-like racemes, or in axillary clusters; calyces short, 2-lipped or unequally 4- to 5-toothed; corollas campanulate to funnelform, radially symmetrical to bilabiate, 5-lobed; stamens 2 or 4; pistil 2-carpellate; style and stigma 1; fruit a capsule; seeds numerous, winged.

A medium-sized (750 species in 109 genera), mainly tropical family centered in n South America; nearly all species are woody, ranging from numerous lianas to trees and shrubs, or rarely herbs; it includes showy tropical ornamentals such as *Crescentia* (CALABASH-TREE), *Jacaranda*, *Kigelia* (SAUSAGETREE), *Spathodea* (AFRICAN-TULIPTREE, FLAMETREE), and *Tabebuia* (POUI) as well as some valuable for timber including *Tabebuia*. *Tecoma stans* (L.) Juss. ex Kunth, (genus: from tecomaxochitl, a Mexican name; species: erect, upright), TRUMPET-FLOWER, a yellow-flowered small shrub (< 1 m tall), known mainly from the Trans-Pecos, was reported by Hatch et al. (1990) from vegetational area 4 (Fig. 2); this is apparently based on a collection near San Antonio (Correll & Johnston 1970) well to the s of nc TX. (subclass Asteridae)

FAMILY RECOGNITION IN THE FIELD: shrubs, trees, or woody vines with large showy flowers in racemes, panicles, or axillary clusters; corollas sympetalous, tubular, bilaterally symmetrical, with 2 or 4 separate, epipetalous stamens; capsules usually large, woody, with winged seeds. The family shares a number of characters with the Scrophulariaceae (e.g., corolla shape, number of stamens) and the two are apparently related; however, all nc TX scrophs are herbs.

REFERENCE: Shinners 1961.

1. Corollas bright yellow or greenish yellow to orange-red; leaves all opposite, mostly compound (usually 2–13 leaflets); trailing to high-climbing vines; fruits 8–17 cm long.
   2. Tendrils present; leaflets (1–)2, entire, evergreen; flowers in axillary inflorescences ________ **Bignonia**
   2. Tendrils absent; leaflets 7–13, sharply toothed, deciduous; flowers in terminal inflorescences ________ **Campsis**

Podophyllum peltatum [LAM, NIC]

Alnus serrulata [CO1]

Betula nigra [SA3]

Carpinus caroliniana [SA3]

Ostrya virginiana [SA3]

1. Corollas white to rosy lavender with orange or purple markings inside; uppermost leaves usually alternate; leaves simple; shrubs or trees; fruits to 45 cm long.
   3. Leaves subsessile, the blades linear-lanceolate, glabrous, unlobed ________ **Chilopsis**
   3. Leaves petioled, the blades ovate, pubescent beneath, the larger lobed ________ **Catalpa**

BIGNONIA CROSSVINE, QUARTERVINE

A monotypic genus of the se United States. (Named for the Abbé Jean-Paul Bignon, 1662–1743, court-librarian at Paris and a friend of Tournefort)

Bignonia capreolata L., (winding or twining), CROSSVINE, QUARTERVINE. High-climbing, woody, evergreen vine; leaves typically 2-foliate, terminated by a several-branched, disk-bearing tendril; leaflets cordate at base; petioles 1–2 cm long; flowers in axillary clusters; corollas red-orange, lighter inside, 4–5 cm long; capsules linear, flattened, 10–17 cm long. Moist woods in se and e TX, cultivated and spreads or escapes in nc TX; Dallas and Grayson cos. Mar–Jul. [*Anisostichus capreolata* (L.) Bureau]

CAMPSIS TRUMPET-CREEPER

A genus of 2 species of adventitious root-climbers; both are cultivated as ornamentals; one occurs in e North America and the other in e Asia.

This disjunct e Asia-e North America distribution pattern is an interesting one to plant geographers. In the geologic past, dispersal between the Eurasian and North American continents was possible, and the combined area is considered a single "Holarctic" biogeographic region. The fossil record shows that many plants had distributions across the Northern Hemisphere—temperate forests, for example, occurred very broadly and reached their maximum extension in the mid-Tertiary (the Tertiary extended from 65 million years ago to 5 mya). This widespread flora has been referred to as the Arcto-Tertiary flora or the Tertiaro-mesophytic flora. Geohistorical events from the mid-Tertiary to the present have included alterations in the shapes of the northern land masses, fluctuations in sea levels, mountain building, and profound changes in the climate. As a result, there have been great changes in both the composition and the disposition of the flora and the ranges of many plant have been greatly restricted (e.g., eliminated from Europe and w North America). A significant number now survive in only two areas, e North America and e Asia. The genus *Campsis* is such an example (other nc TX examples include *Carya*, *Menispermum*, *Penthorum*, *Phryma*, *Sassafras*, *Saururus*, *Triadenum*, *Triosteum*, and *Veronicastrum*) (Li 1952b; Little 1970; Graham 1972; Boufford & Spongberg 1983; Hamilton 1983; Hsü 1983; Wu 1983; Ying 1983; Cox & Moore 1993; Graham 1993a).

(Greek: *campsis*, curvature, from the curved stamens)

Campsis radicans (L.) Seem. ex Bureau, (rooting), COMMON TRUMPET-CREEPER, COWITCH VINE, TRUMPET-HONEYSUCKLE. Shrubby vine climbing to 10 m or more, with aerial rootlets; leaves odd-pinnate; flowers in terminal panicles; calyx teeth about 1/5 as long as the tube; corollas 2–2.5 cm long, with 5 short recurved lobes, orange to orange-red with orange-red to scarlet lobes; capsules cylindric-oblong, 8–12 cm long. Stream banks, disturbed ground, along fences, also cultivated; e 1/2 of TX. Jun–Aug. This species is sometimes planted as an ornamental but can get out of control and become problematic. The flowers are visited by and presumably pollinated by ruby-throated hummingbirds (*Archilochus colubris*) (James 1948). Dermatitis has been reported in some individuals from handling the leaves or flowers (Muenscher 1951; Burlage 1968). ☠ /81

CATALPA INDIAN CIGARTREE, CIGARTREE

A genus of 11 species of e North America (including the West Indes) and e Asia; this disjunct distribution pattern is discussed under the genera *Campsis* (Bignoniaceae) and *Carya*

(Juglandaceae); cultivated as ornamentals and used for timber. (The Native American name)
REFERENCE: Weniger 1996.

Catalpa speciosa (Warder) Warder ex Engelm., (showy, good-looking), NORTHERN CATALPA, CATAWBA-TREE, CIGARTREE, HARDY CATALPA. Deciduous tree to 30 m tall; leaf blades ovate to ovate-oblong, 15–30 cm long, truncate to cordate at base, acuminate at apex; petioles to 15 cm long; inflorescence a showy, few-flowered panicle; corollas to ca. 50 mm long, white, with yellow lines and brown-purple markings inside; fruit a long (20–45(–60) cm), cylindric, cigar-like capsule ca. 1.5 cm thick; seeds flat, 30–40 mm long, 4–5(–8) mm wide. Cultivated, roadsides, along streams; escapes in e 1/2 of TX. May. Native of Mississippi Valley. When damaged by herbivores, the leaves produce extrafloral nectar which attracts insects that deter the leaf-eating herbivores (Mabberley 1987).

CHILOPSIS DESERT-WILLOW, MIMBRE

A monotypic genus of the sw United States and Mexico; the branches are used to make baskets. (Greek: *chil*, lip, and *opsi*, appearance, from flower shape resembling lips)
REFERENCES: Fosberg 1936; Henrickson 1985.

Chilopsis linearis (Cav.) Sweet, (narrow, with sides nearly paralled), DESERT-WILLOW, FLOWERING-WILLOW, WILLOW-LEAF CATALPA, DESERT-CATALPA, FLOR DE MIMBRE, BOW-WILLOW. Shrub or small tree to 10 m tall; leaves opposite or alternate; leaf blades linear-lanceolate, entire; inflorescences terminal racemes to 30 cm long; flowers sometimes sterile (J. Stanford , pers. comm.); corollas variable in color, usually white with purple lower lip and yellow lines in throat to rarely both lips white or purplish red with purple stripes; fruits to 30 cm long, usually ca. 6 mm thick. Cultivated and long persists, planted along highways, escapes; Tarrant Co., also long persisting (ca. 100 years) in Hamilton Co. (J. Stanford, pers. comm.). May–Sep. Native of Mexico and sw U.S. e to wc TX.

BORAGINACEAE FORGET-ME-NOT OR BORAGE FAMILY

Ours pubescent, bristly-hispid, or glabrous annual or perennial herbs; leaves alternate, simple, entire, usually sessile; flowers terminal or axillary, solitary, in small cymes, or uncurling, 1-sided, spike-like or raceme-like inflorescences (= scorpioid cymes); sepals 5, barely united at base; corollas 5-toothed or -lobed, rotate, tubular, funnelform, or salverform; stamens 5, attached near base of corolla or higher in tube; pistil 2-carpellate; ovary superior, usually 4-lobed; style and stigma 1; fruit usually breaking into 4 one-seeded achene-like mericarps ("nutlets"), sometimes reduced to fewer by abortion.

A large (2,300 species in 130 genera) family of mainly temperate and subtropical herbs to shrubs, trees, and vines including a number of ornamentals such as *Heliotropium* (HELIOTROPE), *Mertensia* (VIRGINIA BLUEBELLS) and *Myosotis* (FORGET-ME-NOT); some, including *Borago officinalis* L. (BORAGE) and *Symphytum officinale* L. (COMFREY), have been used for flavorings or medicinally; most members of the family are characterized by a 1-sided inflorescence that uncoils as it matures; alkaloids are often present. The Boraginaceae are related to and sometimes lumped with the tropical family Ehretiaceae (Judd et al. 1994). There are also affinities with the Hydrophyllaceae and Verbenaceae. *Ehretia anacua* (Terán & Berland.) I.M. Johnst., (genus: for Georg Dionysius Ehret, 1708–1770, illustrator; sp: *anacua* or *anacahuite*, Mexican names for the plants), SUGARBERRY, KNOCKAWAY, ANAGUA, a tree to 15 m tall with orange or dark yellow drupes ca. 5–8 mm in diam., was cited by Hatch et al. (1990) for vegetational area 4 (Fig. 2). While planted further n, it is apparently native only as far n as Travis Co. just s of nc TX (Correll & Johnston 1970). This genus of 50 species of tropical and warm areas is treated in the Boraginaceae by some authorities and in the Ehretiaceae by others. (Asteridae)

FAMILY RECOGNITON IN THE FIELD: herbs with alternate, coarse, rough-hairy leaves and round stems; flowers usually in *scorpioid* cymes (= 1-sided, uncurling inflorescences), with sympetalous, radially symmetrical corollas and 5 stamens; fruits of 4 mericarps.
REFERENCES: Johnston 1954a, 1966; Al-Shehbaz 1991.

1. Flowers in naked, spike-like or raceme-like cymes, none or only the lowest in the axils of leaves or bracts; corollas very small in all species keying here except the rare *Cynoglossum*.
   2. Flowers sessile, crowded, remaining less than 5 mm apart in age; ovary unlobed ____ **Heliotropium**
   2. Flowers sessile OR short- to rather long-pedicelled, becoming widely spaced, usually 5–30 mm or more apart during or after flowering; ovary 4-parted or lobed.
      3. Calyces with ± appressed hairs; corollas usually pale to deep blue (rarely white), 8–12 mm broad; basal leaves large, to 26 cm long ____ **Cynoglossum**
      3. Calyces with widely spreading or reflexed hairs; corollas white or yellow (rarely slightly bluish), very small; leaves small, < 7.5 cm long.
         4. Flowers sessile; calyx hairs all straight and sharp-pointed; axis of spike with both wide-spreading (usually conspicuous with a hand lens) and appressed hairs.
            5. Corollas white with greenish yellow eye, < 3 mm long; calyx lobes remaining close together and thus appearing fused even at maturity, only slightly spreading apart apically; mericarps essentially smooth but with a groove along 1 side ____ **Cryptantha**
            5. Corollas light yellow, 4–5 mm long; calyx lobes often spreading and clearly separated at maturity; mericarps with toothed ridges and a conspicuously warty surface, without a groove ____ **Amsinckia**
         4. Flowers pedicelled; calyx hairs (at least some) hooked; axis of racemes (except near base) with uniformly incurved, subappressed hairs ____ **Myosotis**
1. Flowers in the axils of leaves or bracts (these sometimes small); corollas of various sizes, very small to large.
   6. Corollas strongly bilaterally symmetrical, funnelform, with the upper side longer, bright blue (rarely rose or white), showy, 10–20 mm long; stamens conspicuously exserted from the corolla, separate; rare introduced species ____ **Echium**
   6. Corollas radially symmetrical, of various colors and sizes; stamens not exserted (except exserted in *Borago* where they form a cone around the style as in *Solanum*); widespread native and introduced species.
      7. Corolla limb (= open face of corolla) 10–24 mm wide, pure white with yellow center OR blue or purplish; only in West Cross Timbers and e along rivers OR an escaped cultivar.
         8. Corollas pure white with yellow center, with a distinct narrow tube 8–11 mm long; stamens not exserted; lower leaf blades ca. 18 mm or less wide; stem hairs mostly appressed; native in West Cross Timbers and e along rivers ____ **Heliotropium**
         8. Corollas blue or purplish, rotate to bell-shaped, without a distinct tube; stamens strongly exserted, conspicuous; lower leaf blades 30–80 mm wide; stem hairs spreading at ± right angle to stem; escaped cultivar ____ **Borago**
      7. Corolla limb < 10 mm wide, OR if more then yellow or orange-yellow; native and introduced species widespread in nc TX.
         9. Corollas tubular (lobes not reflexed to form a flattened limb), white to green; leaf blades conspicuously 5–7 veined, the veins visible at a glance ____ **Onosmodium**
         9. Corollas salverform or funnelform (lobes reflexed to form a somewhat flattened limb), variously colored; leaf blades with only the central vein conspicuous.
            10. Corollas yellow to orange-yellow, 11–48 mm long (tube plus lobes); plants perennial ____ **Lithospermum**
            10. Corollas variously colored but not yellow (except sometimes in very center), usually < 7.5 mm long, to 13 mm in one species with blue or purple flowers; plants usually annual.

11. Stems with wide-spreading hairs.
    12. Corollas 4–5.5 mm long, white with a yellow center; fruits (= mericarps) not prickly or spiny, easily visible at maturity ____ **Lithospermum**
    12. Corollas 3 mm or less long, blue to white, sometimes with a yellow center; fruits either prickly or spiny and easily visible at maturity OR not prickly or spiny and completely obscured from view by the calyx lobes.
        13. Calyces rather coarsely pubescent with slender ± appressed hairs only; calyx lobes not significantly obscuring the fruit from view; fruits prickly or spiny, the prickles sometimes partly or almost completely united into a wing or cup-like structure ____ **Lappula**
        13. Calyces prickly-hispid with spiny, conspicuously spreading hairs and pubescent with slender ones as well; calyx lobes completely obscuring the fruit from view; fruits essentially smooth, not prickly ____ **Cryptantha**
11. Stems with appressed hairs only.
    14. Upper stem leaves short-petioled or with slender petiolar bases; fruiting calyx lobes conspicuously unequal; flowering late May and after ____ **Heliotropium**
    14. Upper stem leaves sessile, often slightly clasping; fruiting calyx lobes ± equal in length; flowering early May or before ____ **Buglossoides**

AMSINCKIA FIDDLENECK, TARWEED

A genus of 15 species of w North America and w temperate South America. (Named for Wilhelm Amsinck, a burgomaster of Hamburg, who early in the 19th century gave important support to the botanical garden of that city)

REFERENCES: MacBride 1917; Ray & Chisake 1957a, 1957b, 1957c.

Amsinckia menziesii (Lehm.) A. Nelson & J.F. Macbr., (for Archibald Menzies, 1754–1842, British naval surgeon and botanist who accompanied Vancouver on his voyage of Northwest Pacific exploration, 1790–1795), SMALL-FLOWER FIDDLENECK. Bristly annual 30–60 cm tall; stems simple or freely branched, usually decumbent; inflorescences spike-like, becoming greatly elongated, with few or no bracts; fruiting calyces 6–8 mm long, the lobes narrowly to broadly elongate; corollas light yellow, narrowly funnelform, 4–5 mm long, the lobes small; mericarps 4, triangular, 2.5–3 mm long, the surface conspicuously ridged and roughened. Grasslands and dry areas; included based on citation of vegetational areas 4 and 5 (Fig. 2) by Hatch et al. (1990); mainly c TX. Mar–May. Native of nw U.S. and adjacent Canada. [*A. micrantha* Suksd.]

BORAGO

A genus of 3 species native to the Mediterannean region, Europe, and Asia; cultivated as ornamentals. (Possibly from Latin: *burra*, a hairy garment, alluding to the leaves which are usually densely hairy)

Borago officinalis L., (medicinal), BORAGE. Coarsely hairy annual; stems ascending to erect, 20–70 cm tall; leaves alternate, entire, obovate to oblong; inflorescences open, leafy, cymose; calyces of 5 linear segments; corollas blue or purplish, large, the lobes 8–12 mm long and somewhat reflexed; anthers 5, connivent in an erect cone-like arrangement, 5–8 mm long; ovary 4-lobed; mericarps 4. Cultivated and escapes, weedy area; Tarrant Co. (R. O'Kennon pers. obs.); we are not aware of other TX localities. Spring–fall. Native to Mediterranean region. This species has been used as a flavoring, medicinally, and as an ornamental (Woodland 1997). *Borago* flowers are an example of the "vibrator" or "buzz" pollination syndrome; pollinators (such as bumblebees) shake the anthers by vibrating their thoracic flight muscles at a certain frequency; this sets up a resonance in the anthers or the space they enclose and the otherwise inaccessible pollen is released from the anthers and collected by the insect (Barth 1985; Proctor et al. 1996). The turned back (reflexed) corollas and exposed anther-cone of "vibrator"-type flowers may be an adaptation to minimize dampening of vibration resonance or it may be an adaptation related to microclimate in the flower (e.g., to keep the pollen in a dry powdery condition so that it is easily dispersed) (Corbet et al. 1988; Proctor et al. 1996).

BUGLOSSOIDES

A temperate Eurasian genus of 15 species; it is sometimes lumped into *Lithospermum* (e.g., Mabberley 1997). (From the old genus name, *Buglossum*, and the Greek, *-oides*, resembling)

Buglossoides arvensis (L.) I.M. Johnst., (pertaining to cultivated fields). Erect annual, 20–70 cm tall; stem leaves lanceolate to linear; flowers in leafy bracted racemes which at maturity are loosely flowered; corollas white (rarely yellowish or purplish); mericarps brown, tuberculate, wrinkled, or pitted, ca. 3 mm long. Ditch banks, roadsides, disturbed sites; se and e TX w to West Cross Timbers, also Edwards Plateau. Mar–May. Native of Europe. Sometimes recognized in the genus *Lithospermum* [as *L. arvense* L.] (e.g., Jones et al. 1997).

CRYPTANTHA

A genus of 100 species of w North America. (Greek: *krypto*, to hide, and *anthos*, flower, from cleistogamous flowers of some species)
REFERENCE: Johnston 1925.

Cryptantha texana (A. DC.) Greene, (of Texas), TEXAS CRYPTANTHA. Prickly-hispid annual with erect to decumbent stems to 40 cm long; flowers in uncurling, 1-sided inflorescences to ca. 15 cm long, without bracts; calyces with conspicuous sharp hairs; corollas white with greenish yellow eye, < 3 mm long; mericarp usually 1. Loose sandy soils; s Texas n to Somervell Co.; endemic to TX. Apr–May.

CYNOGLOSSUM HOUND'S-TONGUE, BEGGAR'S-LICE

A temperate and warm area, especially Old World genus of ca. 75 species; some have alkaloids; a few are cultivated as ornamentals and the European *C. officinale* L. (HOUND'S TONGUE) was formerly used medicinally. (Greek: *kyon* or *cyno*, dog, and *glossa*, tongue, alluding to the strap-shaped, lumpy leaves)

Cynoglossum virginianum L., (of Virginia), BLUE COMFREY, BLUE HOUND'S-TONGUE, WILD COMFREY. Erect pubescent perennial, 30–80 cm tall; basal leaves large, to ca. 26 cm long, elliptic-oblong, with long petioles to ca. 7 cm long; stem leaves smaller, sessile and some ± clasping; inflorescences branched, long pedunculate; corollas blue to whitish, ca. 8–12 mm broad; mericarps 5.5–7 mm long, covered with bristles and sticking to hair or clothing. Sandy woods, low ground; Dallas Co.; mainly e TX. Mar–Apr. According to Burlage (1968), reputed to be poisonous.

Cynoglossum zeylanicum (Hornem.) Thunb. ex Lehm., (of Ceylon, Ceylonese), native to Asia, is reported by Hatch et al. (1990) from vegetational area 4 (Fig. 2) and e TX; we have seen no TX specimens of this species. It differs from *C. virginianum* in having smaller mericarps (2.5–4 mm long) and smaller corollas (< 6 mm broad).

ECHIUM VIPER'S-BUGLOSS

An Old World genus of 60 species; a number are cultivated as ornamentals, some were used medicinally, and some contain pyrrolizidine alkaloids (Kingsbury 1964). (A plant name used by Dioscorides, from Greek: *echis*, viper; from a supposed resemblance of the mericarps to a viper's head)

Bignonia capreolata [RAD]

Campsis radicans [REE]

Chilopsis linearis [SA3]

Catalpa speciosa [SA3]

Amsinckia menziesii [PNW]

Borago officinalis [BEN, EN2]

Buglossoides arvensis [REE]

Echium vulgare L., (common), BLUEWEED, BLUEDEVIL, VIPER'S-BUGLOSS. Rough, bristly, biennial herb; stems erect, 30–90 cm tall; leaves alternate; lower leaves oblanceolate, broadly stalked, forming a non-persistent rosette; stem leaves reduced upward, the middle ones linear-lanceolate, 3–9 cm long, sessile; inflorescences panicle-like, the side branches one-sided, bracteate; corollas bright blue (rarely rose or white), showy, 10–20 mm long; stamens 5, 4 conspicuously exserted, the fifth included or barely exserted; mericarps 1–4, ca. 2 mm long, much shorter than the persistent calyces. Roadsides, waste places; Burnet Co. on sw margin of nc TX (portion of a single plant was collected by Chuck Sexton in 1997, pers. comm.); other TX locations not known. Native of Eurasia. This species can be an obnoxious weed in some parts of e North America (Fernald 1950a). Reported to contain poisonous pyrrolizidine alkaloids, consolicine and cynoglossine; toxicity is associated with the use of *Echium* in herbal teas; severe liver damage can result; the bristly hairs can also produce dermatitis (Muenscher 1951; Burlage 1968; Lampe & McCann 1985).

HELIOTROPIUM HELIOTROPE, TURNSOLE

Ours annual to perennial, glabrous or pubescent herbs; flowers in terminal, bractless, 1-sided, uncoiling cymes or in bracted cymes or solitary in the upper axils and at the tips of branches; corollas funnelform to salverform, white to blue or purplish; mericarps 4, free or in pairs, sometimes seemingly 2.

A tropical and temperate genus of ca. 250 species; a number are cultivated as ornamentals; some contain pyrrolizidine alkaloids and can cause severe liver damage; because toxicity is associated with their use as herbal teas (Lampe & McCann 1985), they should not be used in teas or ingested in other ways. (Ancient name from Greek: *helios*, the sun, and *trope*, a turn; ancient writers believed the flowers turned toward the sun; or because some species flower at summer soltice)

1. Flowers numerous, in 1-sided, bractless, conspicuously uncurling spike-like cymes.
   2. Plant glabrous and glaucous; leaves succulent ____ **H. curassavicum**
   2. Plant pubescent, not glaucous; leaves not succulent.
      3. Leaves with a very distinct petiole 4–10 cm long; leaf blades ovate to elliptic, 20–100 mm wide ____ **H. indicum**
      3. Leaves without a distinct petiole, sessile or nearly so; leaf blades oblong to oblanceolate, 18 mm or less wide ____ **H. amplexicaule**
1. Flowers solitary in leaf axils and at branch tips or relatively few in a leafy-bracted cyme that does not uncurl.
   4. Corolla limb (= open face of flower) 4–8 mm broad; leaf blades linear or linear-lanceolate, 1–3(–5) mm or less wide ____ **H. tenellum**
   4. Corolla limb 10–22 mm broad; leaf blades lanceolate to elliptic or ovate, to 18 mm wide ____ **H. convolvulaceum**

Heliotropium amplexicaule Vahl, (stem-clasping), VIOLET HELIOTROPE. Pubescent perennial 20–50 cm tall; inflorescence terminal on an essentially bractless peduncle 1–10 cm long, of 2–5 uncoiling, 1-sided, spike-like cymes; corollas funnelform, blue or purple (rarely white), the limb 4–8 mm across, the tube ca. 5 mm long. Cultivated and escaped to fields and waste places; Bell Co. and Milam–Bell Co. line; mainly e TX and Edwards Plateau. Apr–Aug. Native of Argentina and Uruguay.

Heliotropium convolvulaceum (Nutt.) A. Gray, (resembling *Convolvulus*—bindweed), BINDWEED HELIOTROPE, Annual to ca. 40 cm tall; flowers axillary, not in well-defined inflorescences; corollas conspicuous, white with yellow throat, fragrant, with broad limb that is obscurely if at all lobed and narrow tube 8–11 mm long. Loose sandy soils; Panhandle to Trans-Pecos, e to Palo

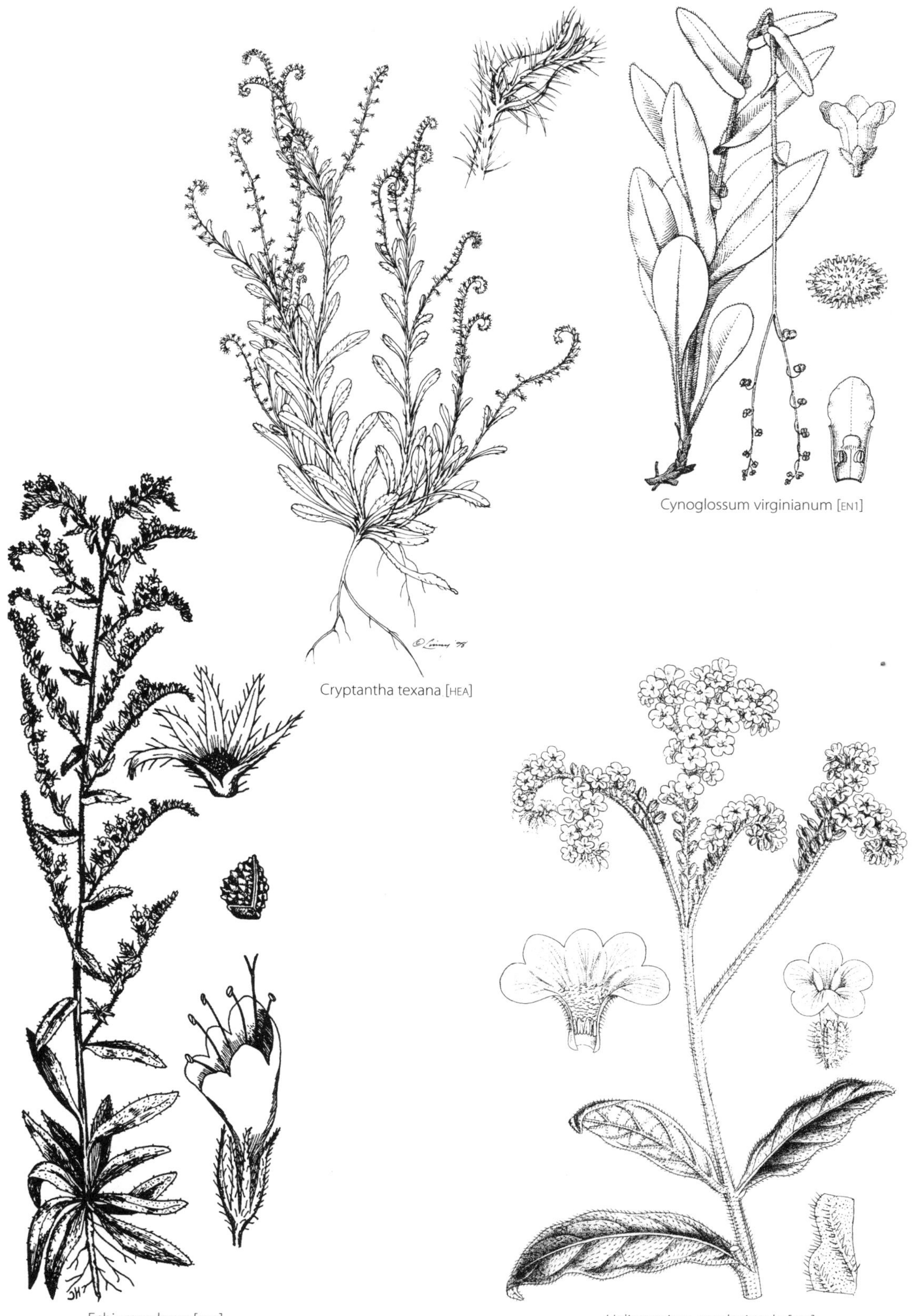

Cynoglossum virginianum [EN1]

Cryptantha texana [HEA]

Echium vulgare [MUE]

Heliotropium amplexicaule [CUR]

Pinto and Wichita cos. and e along Red River to Grayson and Lamar cos., also e along Brazos River to McLennan Co. (Correll & Johnston 1970). Late Jun–Sep. The hairs of this plant can be quite irritating.

Heliotropium curassavicum L., (of Curaçao in s West Indies), SALT HELIOTROPE, CHINESE-PULSEY, QUAILPLANT, SEASIDE HELIOTROPE, COLA DE MICO. Usually rhizomatous perennial with prostrate to half decumbent rubbery stems; corollas usually white or bluish, with yellow eye, small, the limb 4–7 mm across. Low alkaline places, saline situations; Archer, Brown, Dallas, Grayson, and Young counties, also Tarrant Co. (Mahler 1988); mainly Rolling Plains s and w to w TX, also coastal. Jun–Aug. The Spanish common name, COLA DE MICO, meaning monkey's tail, is presumably in reference to the uncoiling inflorescences. The species is considered native from the s U.S. s through the West Indes and Central America s to South America (Correll & Johnston 1970). Toxicity has been reported from the use of this plant in herbal teas (Mulligan & Munro 1990). ☠

Heliotropium indicum L., (of India), INDIA HELIOTROPE, TURNSOLE, ALACRANCILLO. Densely pubescent annual to 1 m tall; corollas blue or violet, rarely white, the limb to 4.5 mm across, the throat glabrous outside; fruits 2-lobed; mericarps smooth, strongly ribbed. Moist disturbed areas of creeks and ponds; se and e TX w to Grayson and Tarrant cos.; also w to Bexar Co. Jun–Oct. Despite the specific epithet, this species is considered to be native to the warmer parts of the New World; it is also present [introduced?] in the Old World tropics (Correll & Johnston 1970); Hatch et al. (1990) treated it as native to TX. Poisonous due to the presence of pyrrolizidine alkaloids (Burlage 1968; Lampe & McCann 1985). ☠

Heliotropium tenellum (Nutt.) Torr., (tender, soft), PASTURE HELIOTROPE. Erect annual to ca. 50 cm tall, appressed-pubescent; flowers solitary at branch tips and in upper axils, together sometimes appearing like a loose, bracted raceme; corollas white with a yellowish eye, the limb 5–6 mm across; mericarps 4, ribbed, at first paired and the fruit conspicuously 2-lobed. Limestone outcrops; most of e 1/2 of TX. May–Oct.

Two additional species, *H. procumbens* Mill., (prostrate), and *H. racemosum* Rose & Standl., (with flowers in racemes), cited by Hatch et al. (1990) for vegetational area 4 (Fig. 2), probably occur only to the s of nc TX. *Heliotropium procumbens* resembles *H. indicum* but has white corollas with the throat sparingly pubescent outside, while *H. racemosum* resembles *H. convolvulaceum* but has the corollas clearly lobed, the lobes triangular and acute, and the flowers in well-defined racemose inflorescences.

LAPPULA STICKSEED

A genus of 40 temperate Eurasian species and 5 in North America. (Latin: *lappa*, a bur, presumably in reference to the fruit)

REFERENCE: Johnston 1924.

Lappula occidentalis (S. Watson) Greene, (western). Annuals 10–50 cm tall, hispid-villous; corollas pale blue to white, sometimes with yellow eye, small, the tube about 1.2 mm long; mericarps 4, smooth with margins of a single row of prickles, the prickles separate or united into wings. Sandy or gravelly open ground. Late Mar–May.

1. Prickles of fruits partly or almost completely joined into spiny-margined wings or a cup-like structure ____ var. **cupulata**
1. Prickles of fruits separate or nearly so ____ var. **occidentalis**

var. **cupulata** (A. Gray) L.C. Higgins, (cup-like), HAIRY STICKSEED, CUPSEED. Mills, Palo Pinto, and Wichita cos. w and s to w TX, local e to Fort Worth (Tarrant Co.) where probably introduced. [*L.*

Heliotropium convolvulaceum [IPL]

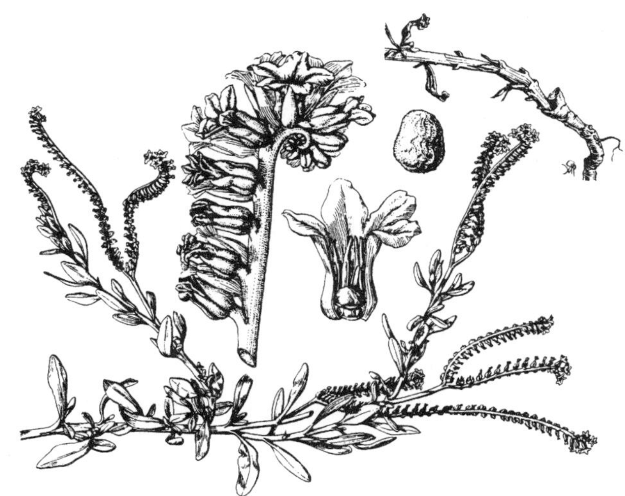
Heliotropium curassavicum [MAS]

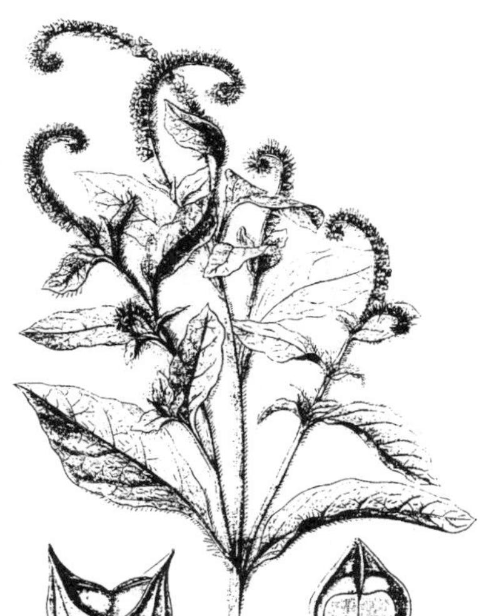
Heliotropium indicum [EN2]

Heliotropium tenellum [GLE]

Lappula occidentalis var. cupulata [BB2]

Lithospermum caroliniense [GLE]

Lithospermum incisum [HO1]

redowskii of authors, not (Hornem.) Greene var. *texana* (Scheele) Brand, *L. redowskii* of authors, not (Hornem.) Greene var. *cupulata* (A. Gray) M.E. Jones, *L. texana* (Scheele) Britton]

var. **occidentalis** (S. Watson) Rydb., FLAT-SPINE STICKSEED. Mainly Palo Pinto and Wichita cos. s and w to w TX. [*L. redowskii* of authors, not (Hornem.) Greene, *L. redowskii* of authors, not var. *occidentalis* (S. Watson) Rydb.]

LITHOSPERMUM PUCCOON, GROMWELL

Ours low, pubescent perennials or annuals with rather showy yellow to orange-yellow flowers (or flowers small and white with yellow centers in one species rare on s margin of nc TX) in leafy-bracted terminal inflorescences; mericarps smooth or pitted, sometimes with a constriction above the base.

A genus of 45 species in temperate regions of the world except Australia; some are cultivated as ornamentals or used medicinally; the roots of some U.S. species were used by Native Americans and early settlers as a source of a red dye; it is still used today in dying weaver's wool (Ajilvsgi 1984). (Greek: *lithos*, stone, and *sperma*, seed, from the hard mericarps)
REFERENCES: Johnston 1952; Govoni 1973.

1. Corollas white with yellow centers; corolla tubes < 2 mm long; plants annual; on s margin of nc TX **L. matamorense**
1. Corollas yellow to orange-yellow; corolla tubes 7–37 mm long; plants perennial; widespread in nc TX.
   2. Corolla lobes toothed or ruffled; corolla tubes 13–35 mm long; stems appressed-pubescent; leaves usually acute **L. incisum**
   2. Corolla lobes entire; corolla tubes 7–14 mm long; stems spreading-pubescent; leaves usually obtuse **L. caroliniense**

Lithospermum caroliniense (Walter ex J.F. Gmel.) MacMill., (of Carolina), PUCCOON, CAROLINA GROMWELL, CAROLINA PUCCOON. Erect perennial 30–100 cm tall; corollas orange-yellow, the limb to ca. 22 mm across; basal leaves often dried up or absent by flowering time; upper stem leaves 3–7 times as long as wide, to ca. 10 mm wide; mericarps 3–3.5 mm long, smooth or pitted, lustrous, white. Sandy woods; mainly se and e TX w to West Cross Timbers, also in Panhandle. Apr–May.

Lithospermum incisum Lehm., (incised, cut), NARROW-LEAF GROMWELL, NARROW-LEAF PUCCOON. Erect perennial to ca. 40 cm tall; basal leaves sometimes dried up or absent by flowering time; stem leaves usually 3–6 mm wide; upper stem leaves 8–20 times as long as wide; spring inflorescences with usually sterile flowers, the corollas showy, lemon- or deep-yellow, the limbs 9–20 mm across; late spring and summer inflorescences with minute, very fertile, cleistogamous flowers with permanently closed corollas 1–2.5 mm long; mericarps 2.5–3 mm long, smooth or pitted, lustrous, whitish. Prairies, open woods, and roadsides; throughout TX. Late Mar–Apr. After the production of conspicuous flowers in the spring, plants of this species typically produce large numbers of small cleistogamous flowers that are extremely fertile (Johnston 1952). The roots were the source of a red dye for Native Americans and settlers (Tveten & Tveten 1993). /96

Lithospermum matamorense DC., (of Matamoros, Mexico), ROUGH GROMWELL. Decumbent to erect annual; stems to 40 cm long; leaves obtuse or retuse apically, gradually reduced up the stem; basal leaves largest, to 9 cm long and 20 mm wide, often present at flowering time; corolla limb 4–7 mm across; mericarps 2.5–3 mm long, whitish or brownish, pitted. Thickets, open woods, floodplains, and weedy areas; Burnet Co. (Balcones Canyonlands Nat. Wildlife Refuge, C. Sexton, pers. comm.) on s margin of nc TX; also San Saba Co. just s of nc TX, Sexton also indicates a Travis Co. location just s of nc TX; mainly Edwards Plateau to s TX.

MYOSOTIS SCORPION-GRASS, FORGET-ME-NOT

Ours small hairy annuals; inflorescences racemose, becoming loosely flowered; calyces hairy; corollas inconspicuous, usually white (or slightly bluish), 2–3.5 mm long; mericarps 4, ovoid to ellipsoid, smooth and shiny.

A genus of ca. 100 species of temperate areas and tropical mountains; a number are cultivated as ornamentals and some have corollas changing from pink to blue in color. (Greek: *myos*, of a mouse, and *oto*, ear, from shape of the leaves or the soft leaves in some species)

1. Pedicels obliquely spreading from base, straight or slightly curved; racemes elongating rapidly, the calyces soon becoming 5–30 mm apart; stem hairs mostly 1–1.5 mm long; fruiting calyces sometimes > 5 mm long; most mature calyx hairs hooked ____ **M. macrosperma**
1. Pedicels closely ascending or erect in their basal part, rather abruptly outcurved near tip; racemes elongating gradually, the calyces (except lowest) becoming 1–8 mm apart; stem hairs 0.5–1.1 mm long; fruiting calyces < 5 mm long; only a few of the mature calyx hairs hooked ____ **M. verna**

Myosotis macrosperma Engelm., (large-seeded), SPRING FORGET-ME-NOT. Plant erect, 20–50 cm tall. Sandy or silty woods, stream banks, fencerows, and roadsides; se and e TX w to West Cross Timbers, also Edwards Plateau. Apr–early May.

Myosotis verna Nutt., (of Spring), EARLY SCORPION-GRASS, SOUTHERN FORGET-ME-NOT. Sandy open woods, roadsides, and old fields; se and e TX w to West Cross Timbers, also Edwards Plateau. Apr–early May.

ONOSMODIUM MARBLESEED, FALSE GROMWELL

Ours conspicuously hairy, leafy perennials from a taproot; leaves alternate, entire, usually conspicuously 5-7 veined; flowers usually numerous in uncurling, 1-sided, terminal or branched racemes or occasionally axillary; corollas tubular, slightly enlarged just below junction of tube and 5 erect lobes, greenish white to whitish or yellowish white, 9–20 mm long; fruit usually of (by abortion) 1(–2–4) mericarp(s) 5 mm or less long, conspicuously white to whitish brown.

A genus of 7 species native to Mexico, the United States, and closely adjacent Canada (Turner 1995a). We are following the recent treatment by Turner (1995a) for nomenclature of *Onosmodium*. The conspicuous mericarps give rise to the common name MARBLESEED. (Named from a likeness to the genus *Onosma*—an Old World genus of Boraginaceae, from Greek: *osnos*, ass, and *osme*, smell, alluding to the roots)
REFERENCES: MacKenzie 1906; Johnston 1954a, 1954b; Turner 1995a.

1. Pedicels of flowers (both at flowering time and later) short, 0.5–5(–7) mm long; leaves with short strigose (= appressed) hairs on lower and upper surfaces under longer pubescence, at least near veins; larger corollas mostly 9–13 mm long; bracts lanceolate, usually not greatly enlarged; widespread in nc TX ____ **O. bejariense**
1. Pedicels of flowers (both at flowering time and later) elongate, usually 5–15 mm long, the lower ones often 10–15 mm long; leaves completely lacking short strigose hairs, having only erect or ascending bristles ca. 1 mm long on the leaves; larger corollas mostly (11–)17–20 mm long; bracts ovate or elliptic, much enlarged; mainly sc TX n to Williamson Co. on extreme s margin of nc TX ____ **O. helleri**

Onosmodium bejariense DC. ex A. DC., (presumably for Bexar County on se edge of the Edwards Plateau, location of San Antonio). Erect or ascending often suffruticose herb usually 0.4–1.1 m tall; mid-stem leaves lanceloate, oblanceolate to elliptic, 6–12 cm long, 2–4 cm wide, sessile or nearly so; calyces 4–9 mm long; larger corollas whitish or greenish white; mature styles extending from the corollas for 5–25 mm; mericarps mostly 3–5 mm long. According to Turner (1995a), who excluded *O. molle* from this species (see note below), the earliest name is *O. bejariense*.

1. Mid-stem hairs widely spreading, the vestiture mostly 2–4 mm high; mericarps mostly 3–4 mm long; widespread in nc TX ____ var. **bejariense**
1. Mid-stem hairs appressed or ascending, the vestiture mostly 1–2 mm high; mericarps mostly 4 5 mm long; in TX known only from Tarrant and Wichita cos. ____ var. **occidentale**

var. **bejariense**, BEJAR MARBLESEED. Grasslands or openings in forests, silty or silty-clay soils, limestone outcrops; se and e TX w to West Cross Timbers and Edwards Plateau; by far the most common nc TX MARBLESEED. Turner's treatment (1995a) differs significantly from previous ones (e.g., Correll & Johnston 1970) in that nearly all nc TX MARBLESEEDS are treated as *O. bejariense* var. *bejariense*. Apr–Jul. [*O. molle* Michx. var. *bejariense* (DC. ex A. DC.) Cronquist, *O. molle* subsp. *bejariense* (DC. ex A. DC.) Cochrane]

var. **occidentale** (Mack.) B.L. Turner, (western), WESTERN MARBLESEED. Shady areas; deep clay or rocky banks; according to Turner (1995a), this variety is known in TX only in Tarrant and Wichita cos. May–Jul. [*O. occidentale* Mack.]

var. *hispidissimum* (Mack.) B.L. Turner, (most hispid or bristly), ROUGH MARBLESEED. According to Turner (1995a), this variety, which can be distinguished from the 2 varieties occurring in nc TX by its shorter corollas (mostly 6–10 mm long versus mostly (11–)17–20 mm) and mericarps mostly flared at base (versus mericarps mostly tapered to the base), is known only to the e of TX (ne U.S. sw to Louisiana); Correll and Johnston (1970) reported an 1880 collection from Dallas–however, given that this is several hundred miles w of the range of this variety, an error is suspected. [*O. hispidissimum* Mack.]

Onosmodium helleri Small, (for its discoverer, Amos Arthur Heller, 1867–1944, Pennsylvania botanist and collector of w plants), HELLER'S MARBLESEED. Erect herb 30–60 cm tall; stems pubescent with both reflexed and spreading coarse hairs, the vestiture 1–2 mm high; mid-stem leaves narrowly to broadly elliptic, mostly 8–16 cm long, (2–)3–8 cm wide, sparsely pilose on upper and lower surfaces, the hairs rough to the touch; calyces 6–10 mm long; larger corollas creamy white; styles extending beyond the corollas for 6–12 mm; mericarps 3–3.6 mm long. Juniper oak woodlands, on crumbly, rather bare limestone soils, typically on slopes; Williamson Co. (Turner 1995a), also Burnet Co. (Balcones Canyonlands Nat. Wildlife Refuge, C. Sexton, pers. comm.); narrowly endemic to ca. 8 counties of the Edwards Plateau and s Lampasas Cut Plain. Mar–May. (TOES 1993: V) ⚠

Onosmodium molle Michx., (soft). According to Turner (1995a), this species is limited to the cedar glades of c Tennessee and closely adjacent Kentucky and Alabama. Texas MARBLESEEDS previously treated as *O. molle* by a number of authors are now considered part of *O. bejariense*.

BRASSICACEAE (CRUCIFERAE) MUSTARD FAMILY

Ours annual or perennial herbs; leaves basal or alternate, simple or pinnately compound, entire, toothed, or lobed; flowers in terminal or axillary racemes; sepals 4; petals 4 (or more numerous in cultivated plants), in a cross-like arrangement, commonly with an elongate claw and an abruptly spreading blade, equal or unequal (2 larger, 2 smaller), or petals absent; stamens 2–6 (when 6, 4 long and 2 short); pistil 1; ovary superior; fruit a silique (= dry, dehiscent, variously shaped, many-seeded, 2-valved capsule with valves splitting from the bottom and leaving a false partition known as a replum).

Lithospermum matamorense [HEA]

Myosotis macrosperma [GLE]

Onosmodium bejariense var. bejariense [HEA]

Myosotis verna [GLE]

Onosmodium helleri [HEA]

A large (3,250 species in 365 genera), economically important, cosmopolitan, but predominently n temperate family of mostly herbs or rarely shrubs; sulfur-containing mustard oil glucosides are often present and frequently cyanogenic compounds as well; these can result in digestive problems or even death in livestock (Kingsbury 1964; Blackwell 1990). Many *Brassica* species are cultivated and yield a variety of foods (e.g., cabbage, turnip, mustard) and oils (e.g., rapeseed or canola oil); other genera (e.g., *Lobularia*-SWEET-ALYSSUM, *Lunaria*-HONESTY or MONEYPLANT) are cultivated as ornamentals, for landscapes, and for dried arrangements; some are aggressive weeds. Petal measurements in the treatments include both blade and claw (if present). Brassicaceae are closely related to Capparaceae and appear to represent an herbaceous clade within that family. From a cladistic standpoint they should be lumped with Capparaceae to form a more inclusive monophyletic family, which based on nomenclatural rules, should be called Brassicaceae (Judd et al. 1994). (subclass Dilleniidae)

FAMILY RECOGNITION IN THE FIELD: herbs with alternate or basal leaves and flowers with *4 petals* in a cruciform or *cross* arrangement (flower color variable, but typically yellow or white), stamens usually *4 long and 2 short* or 2 or 4, and characteristic fruits (siliques) borne in racemes. Similar to Capparaceae but that family often has somewhat bilaterally symmetrical flowers (not cross-like), inflorescence bracts (these absent in Brassicaceae), stamens never 4+2, and fruits lacking a transverse partition (this present in Brassicaceae).

REFERENCES: Rollins 1981, 1993; Al-Shehbaz 1984, 1985a, 1985b, 1986b, 1987, 1988a, 1988b, 1988c; Judd et al. 1994.

1. Plants with auricled-clasping upper stem leaves.
  2. Stems glabrous.
    3. Petals 7–15 mm long including claw; fruits 3–11 cm long, much longer than broad.
      4. Petals deep yellow, with claw shorter to slightly longer than the blade; mature fruits 3–10 cm long; lower leaves often pinnately lobed or divided ____ **Brassica**
      4. Petals yellowish white, with claw 3–4 times as long as the blade; mature fruits 8–11 cm long; leaves neither lobed nor divided ____ **Conringia**
    3. Petals 1–5 mm long including claw OR petals absent; fruits < 2 cm long, often nearly as broad as long.
      5. Petals white; fruits flat, orbicular, 8–15 mm wide, deeply notched at apex ____ **Thlaspi**
      5. Petals greenish yellow or yellow, withering whitish OR petals absent (then sepals yellow); fruits not as above.
        6. Petals 3–4 mm long including claw; upper leaves widest at base; fruits roughly inverted triangular (widest distally, also with a small beak) ____ **Myagrum**
        6. Petals 0–2.5 mm long including claw; upper leaves narrow at base; fruits narrowly cylindric ____ **Rorippa**
  2. Stems pubescent or pilose, at least in lower part.
    7. Petals 0–2.5 mm long including claw; fruits inverted triangular-heart-shaped ____ **Capsella**
    7. Petals 4–10 mm long including claw; fruits ± round.
      8. Pedicels loosely pubescent; petals rich yellow or orange-yellow; leaves pinnatifid to entire ____ **Lesquerella**
      8. Pedicels glabrous; petals light yellow or yellowish white; leaves entire or merely serrate ____ **Camelina**
1. Plants without auricled-clasping upper leaves.
  9. Petals light or greenish yellow to deep yellow or red-orange OR petals absent and sepals yellow.
    10. Petals 0–3 mm long including claw, not or slightly exceeding the sepals.
      11. Leaves simple (but can be lobed or pinnatifid), or only the lower compound and with lobed leaflets.
        12. Larger leaf blades with main lobes obtuse; young fruits 2–3 times as long as thick, ovoid, oblong, or cylindric ____ **Rorippa**

12. Larger leaf blades with lobes acute; young fruits 5–10 times as long as thick, narrowly linear _______ **Sisymbrium**

11. Leaves all pinnately compound, the leaflets usually deeply once or twice pinnatifid and thus appearing highly dissected _______ **Descurainia**

10. Petals 3–15 mm long including claw, 1.5–2.5 times as long as the sepals.

13. Pedicels usually gray or silvery, with minute, appressed or subappressed, simple, stellate, or half-scaly pubescence (use lens).

14. Petals with wide claw shorter than or equaling the blade; mature fruits globose or nearly so, < 10 mm long _______ **Lesquerella**

14. Petals with slender claw longer than the blade; mature fruits much longer than wide, 40–120 mm long _______ **Erysimum**

13. Pedicels glabrous or with few loose hairs.

15. Pedicels elongating rapidly and becoming 2–many times as long as the calyces before the petals wither; plants ill-smelling; leaves crowded at base of plant; petals 5–8 mm long including claw; mature fruits 20–35 mm long, 1–2 mm in diam. _______ **Diplotaxis**

15. Pedicels elongating gradually, slightly shorter to slightly longer than the calyces before the petals wither; plants not with the above combination.

16. Stem leaves (except lowest) unlobed or deeply lobed only in basal half; large terminal portion of blade as wide as rest of leaf or wider, with convex (= outcurved) sides; sepals 3–11 mm long; petals bright yellow.

17. Buds when ready to open 4–10 mm long; mature fruits 10–100 mm long, linear or narrowly cylindric.

18. Beak of fruits terete or narrowly conical, not containing a seed; valves of fruits with 1 most prominent nerve, the other nerves much weaker; leaves not clasping OR clasping; stems usually glabrous (or hirsute in *B. nigra*); fruits variable in length, 10–100 mm long _______ **Brassica**

18. Beak of fruits large, flat or conspicuously angled, usually containing 1 seed; valves of fruits with 3 parallel nearly equal veins; leaves not clasping; stems hispid-pubescent at least in lower part; fruits 35 mm or less long _______ **Sinapis**

17. Buds when ready to open 2–4 mm long; mature fruits (including beak) 3–9 mm long, of 2 distinct segments, the upper segment larger and nearly globose, the lower cylindrical _______ **Rapistrum**

16. Stem leaves deeply lobed; terminal portion of blade narrower than the rest of the leaf OR widely triangular with deeply indented sides; sepals 2–5 mm long; petals sulfur or light yellow.

19. Larger leaf blades with obtuse lobes; lower flowers usually with pinnatifid bracts; fruits 2–4 cm long; pedicels and fruits spreading-ascending; upper stem and axis of raceme usually pubescent _______ **Erucastrum**

19. Larger leaf blades with acute lobes; lower flowers usually without bracts; fruits 1–10 cm long; pedicels and fruits closely appressed to spreading-ascending; upper stem and axis of raceme glabrous _______ **Sisymbrium**

9. Petals yellowish white, white, pink, blue-purple, or purple OR petals absent and sepals green or partly reddish or purple-brown.

20. Fruits indehiscent, separating transversely into 1- or 2-seeded segments.

21. Petals blue-purple, 7–11 mm long including claw; mature fruits 1–2 mm wide; stigmas slender, minute _______ **Chorispora**

21. Petals white to pale purple, 15–20 mm long including claw; mature fruits 6–10 mm wide; stigmas broad, emarginate _______ **Raphanus**

20. Fruits dehiscent longitudinally by valves or manner of dehiscence unclear.

22. Fruits ending in 2 distinct very conspicuous horns (horns ca. 3–6 mm long), densely pubescent, grayish, 70–100 mm long; petals 15–20 mm long including claw, lilac or purple ____ **Matthiola**
22. Fruits not ending in 2 distinct horns, usually glabrous (rarely pubescent), 2–90 mm long; petals 1.5–20 mm long including claw, variously colored.
    23. Plants glabrous; sepals deep purple, not spreading, together urn-shaped; petals 12–20 mm long including claw; fruits 6–10 cm long, to 2 mm wide ____ **Streptanthus**
    23. Plants without the above combination; petals 7 mm or less long including claw (except in *Raphanus* which has much wider fruits than *Strepthanthus* and in *Iodanthus* which has fruits only 2–4 cm long).
        24. Petals 15–20 mm long including claw, white to pale purple; mature fruits 40-50 mm long, 6–10 mm wide, round in cross-section ____ **Raphanus**
        24. Petals 14 mm or less long including claw (usually 7 mm or less), variously colored; mature fruits 2–90 mm long, up to 3 mm wide, round or flat in cross-section.
            25. Petals 7–14 mm long including claw, white to lavender; fruits 2–4 cm long ____ **Iodanthus**
            25. Petals 7 mm or less long including claw, variously colored; fruits variable in length, 2–90 mm long.
                26. Ovaries and young fruits more than twice as long as wide; fruits variable in length, 2–90 mm long.
                    27. Stems pubescent in upper part ____ **Draba**
                    27. Stems glabrous in upper part.
                        28. Petals 1.5–3 mm long including claw; stem leaves compound or deeply lobed; fruits 2–30 mm long.
                            29. Leaves compound, the leaflets distinctly narrowed at very base; fruits < 1 mm wide; lower leaves with 1–5(–6) leaflets on each side ____ **Cardamine**
                            29. Leaves lobed almost to midrib, appearing nearly compound but the rachis wing-margined, the lobes not narrowed at very base; fruits ca. 1.2–2 mm wide; lower leaves with 5–14 divisions on each side ____ **Sibara**
                        28. Petals 3–7 mm long including claw; stem leaves simple (can be fairly deeply lobed) OR compound (if compound the plant aquatic or semi-aquatic); fruits 10–90 mm long.
                            30. Stem leaves few or none, the leaves mostly basal; fruits 20-35 mm long; flowers lavender ____ **Diplotaxis**
                            30. Stem leaves conspicuously present; fruits 4–9 cm long OR if shorter the plant aquatic or semiaquatic; flowers white or lavender-pink.
                                31. Leaves simple; fruits 40–90 mm long; plants erect ____ **Arabis**
                                31. Leaves compound; fruits 10–20 mm long; plants floating, creeping, or ascending ____ **Rorippa**
                26. Ovaries and young fruits less than twice as long as wide; fruits 8 mm long or less.
                    32. Ovaries and fruits inverted triangular-heart-shaped; axis of inflorescence and pedicels greatly elongating, the flowers becoming widely spaced and long-pedicelled (pedicels 8–15 mm long) ____ **Capsella**
                    32. Ovaries and fruits not inverted triangular-heart-shaped; axis of inflorescence and pedicels elongating gradually and moderately, the flow-

ers remaining rather crowded and short-pedicelled (pedicels to ca. 10 mm long at most).

33. Inflorescences terminal; fruits flattened, elliptic or orbicular in outline, the surfaces smooth ________ **Lepidium**

33. Inflorescences lateral; fruits not flattened, swollen, with 2 subglobose lobes, the surfaces wrinkled ________ **Coronopus**

ARABIS ROCKCRESS

Annuals or perennials, glabrous or pubescent with simple hairs; seeds winged.

A genus of 180 species in the n temperate zone and the Mediterranean to tropical African mountains; a number are cultivated as ornamentals. (Name from the country, Arabia, according to Linnaeus)

REFERENCES: Hopkins 1937; Rollins 1941.

1. Lower stem leaves sessile, entire or slightly toothed; petals white to yellowish white, 3–5 mm long, slightly longer than the sepals; fruits drooping; stems glabrous or nearly so ________ **A. canadensis**
1. Lower leaves petioled, deeply lobed; petals lavender-pink, 5–7 mm long, nearly twice as long as the sepals; fruits erect; stems densely spreading-pubescent below ________ **A. petiolaris**

Arabis canadensis L., (of Canada), SICKLEPOD. Perennial to 90 cm tall; fruits 50–90 mm long, 2–3 mm wide, curved. Woods; Grayson and Tarrant cos., also Dallas Co. (Mahler 1988); e TX w to East Cross Timbers, also Edwards Plateau. May–Jun.

Arabis petiolaris (A. Gray) A. Gray, (with a petiole or leaf stalk), BRAZOS ROCKCRESS. Annual to 90 cm tall; fruits 40–80 mm long, 3–4 mm wide, nearly straight. Sandy or rocky open ground; Bell, Burnet, Coleman, and McLennan cos., also Young Co. (Mahler 1988); s and w parts of nc TX s to Edwards Plateau; endemic to TX. Apr–May.

BRASSICA

Ours all introduced annuals; leaves pinnately dissected or lobed, at least on lower portion of the stem; petals yellow, rather large and showy.

An Eurasian and Mediterranean genus of 35 species of herbs, but sometimes woody; includes many crop plants such as TURNIP (*B. rapa* L.), MUSTARD (*B. nigra* (L.) Koch), *B. oleracea* (source of many foods as listed below), and *B. napus* L., the seeds yielding canola oil or rapeseed oil. Some species can be poisonous to livestock due to the presence of mustard oil glycosides (Stephens 1980). (The Latin name of CABBAGE)

REFERENCES: Sun 1946; Lemke & Worthington 1991.

1. Lower stems glabrous or nearly so; mature fruits 3–6 cm or more long; leaves clasping or not so; beak of fruits 4 mm long or longer.
   2. Sepals 4–6 mm long.
      3. Upper leaves auricled-clasping ________ **B. rapa**
      3. Upper leaves not auricled-clasping ________ **B. juncea**
   2. Sepals 7–11 mm long ________ **B. oleracea**
1. Lower stems hirsute; mature fruits 1–2 cm long; leaves not clasping; beak of fruits 1–4 mm long ________ **B. nigra**

Brassica juncea (L.) Czern., (resembling *Juncus*—rush), CHINESE MUSTARD, LEAF MUSTARD, INDIAN MUSTARD. Plant to 2 m tall, essentially glabrous, somewhat glaucous; leaves petiolate or the upper sessile; siliques 3–6 cm long, with beak 4–8 mm long. Roadsides and fields; se and e TX w to Blackland Prairie, also one collection is known from Tarrant Co. in the Fort Worth Prairie. Apr–

May, sporadically to Sep. Native of e Europe and w Asia. This is the most common leaf mustard used for greens.

Brassica nigra (L.) W.D.J. Koch, (black), BLACK MUSTARD. Plant to 1.5(–2) m tall, hirsute below, glabrous above, green or slightly glaucous above; leaves all petioled, the lower pinnatifid; siliques quadrangular because of the stout midveins of the valves, closely appressed, with beak 1–4 mm long. Weedy areas; included based on citation of vegetational area 4 (Fig. 2) by Hatch et al. (1990); scattered in TX. May–Aug. Native of Eurasia. The seeds are a main source of table mustard. When the crushed seeds are mixed with "must" of old wine the result is "mustum ardens" or mustard; the flavor results from the pungent allyl isothiocynate released by enzymatic activity within 10 minutes of adding liquid to the powdered mustard seeds (Mabberley 1997).

Brassica oleracea L., (of the vegetable garden, a potherb used in cooking). Plant 0.3–0.9 m tall, essentially glabrous, glaucous; upper leaves subsessile or wing-petioled, slightly clasping; petals 10–25 mm long; siliques up to 10 cm long, with beak 4–8(–10) mm long. Cultivated, in modified varieties, as BROCCOLI, BRUSSELS-SPROUTS, CABBAGE, CAULIFLOWER, COLLARDS, ORNAMENTAL KALE, and KOHLRABI; reseeds in gardens; included based on citation of vegetational areas 4 and 5 (Fig. 2) by Hatch et al. (1990); widely scattered in TX. Apr–Jun, sporadically later. Derived from plants native to coastal Europe.

Brassica rapa L., (the Latin name for turnip), TURNIP, BIRD'S RAPE, RAPE. Plant 0.4–2 m tall; stems glabrous, usually glaucous; petals 6–10 mm long; siliques 3–7 cm long, with beak 8–15 mm long. Commonly cultivated for its root and greens; frequent as an escape and maintaining itself on roadsides or in waste places, especially in sandy soils; Dallas, Delta, Grayson, Henderson, Hill, and Tarrant cos.; mainly se and e TX w to East Cross Timbers. Mid-Mar–early May. Native of Europe. [*B. campestris* L.] Cultivated plants have a greatly enlarged taproot; wild ones vary from with noticeable thickening to none. This species is now often planted with winter WHEAT and OATS for forage after grain harvesting; it is thus becoming widespread (J. Stanford, pers. comm.); Burlage (1968) reported it can cause abortion and a decrease in milk flow in cattle.

CAMELINA

Annual or biennial [?]; pubescence stellate and simple; flowers white to pale yellow; fruits round, beaked.

A genus of 6–7 species native to Europe and the Mediterranean to c Asia. McGregor (1985c) gave a detailed review of the genus in central North America. (Greek: *chamai*, dwarf, and *linon*, flax, from inhibition of flax plants)
REFERENCES: McGregor 1984c, 1985c; Akeroyd 1993.

1. Basal leaf rosette usually withered at flowering time; petals 2–5(–6) mm long; lower stems usually with long simple hairs and an understory of branched hairs, the simple hairs sometimes absent or rare; stems unbranched or with ascending or erect branches; pedicels 10–15(–20) mm long in fruit ____ **C. microcarpa**
1. Basal leaf rosette green at flowering time; petals (5–)6–9 mm long; lower stems ± densely covered with long, simple, unbranched hairs, branched hairs absent or rare; stems usually much branched, with spreading or ascending branches; pedicels 7–10(–14) mm long in fruit ____ **C. rumelica**

Camelina microcarpa Andrz. ex DC., (small fruited), LITTLEPOD, FALSE FLAX, SMALLSEED FALSE FLAX. Annual or biennial[?] to 75 cm tall; stems with long spreading pubescence, at least below; leaves (except lowest) sessile; leaf blades auricled-clasping, oblong- or triangular-lanceolate, with long ± spreading pubescence (often stellate), ciliate marginally; fruiting raceme elongate, usually rather dense; petals pale yellow; fruits ca. 5–7 mm in diam.; $2n = 40$. Railroads, road-

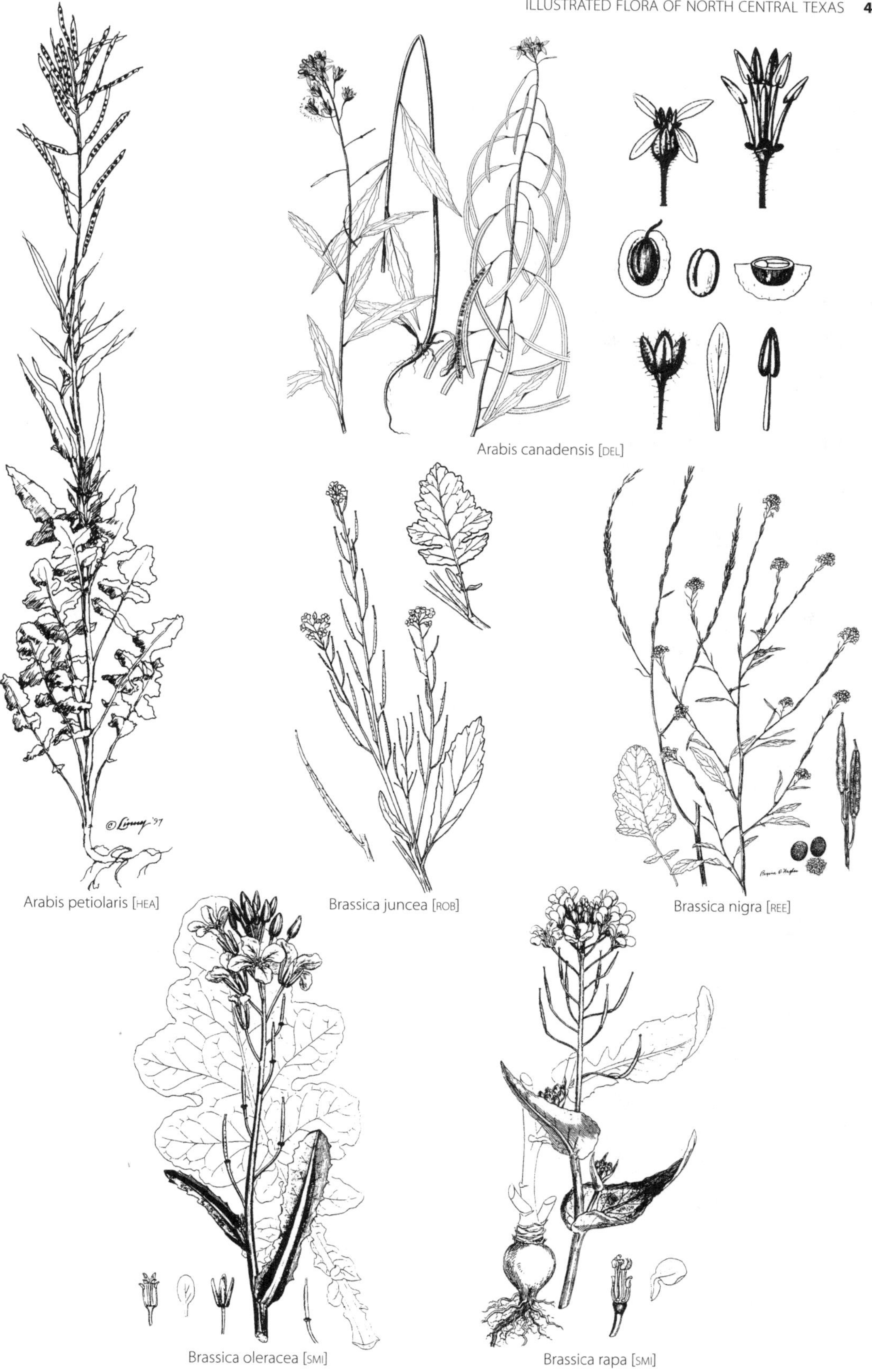

Arabis canadensis [DEL]

Arabis petiolaris [HEA]

Brassica juncea [ROB]

Brassica nigra [REE]

Brassica oleracea [SMI]

Brassica rapa [SMI]

sides, and waste ground; various soils; Collin, Dallas, Grayson, Shackelford, and Tarrant cos.; se and e TX w to nc TX and Edwards Plateau. Late Mar–May. Native of Europe.

Camelina rumelica Velen., (of Roumelia, se Europe). Erect annual or biennial[?] 14–30(–60) cm tall, similar to *C. microcarpa*; fruiting raceme elongate, lax (less dense); petals white to pale yellow; fruits ca. 5–8 mm in diam.; $2n = 12, 26$. Along railroads; Denton and Wise cos. (locally abundant); in TX only known from nc part of state and Hemphill and Ochiltree cos. in the Panhandle (McGregor 1985c). Late Mar–Apr. Native of Europe.

CAPSELLA SHEPHERD'S-PURSE

A temperate and warm area Eurasian genus of 5 species. (Diminutive of Latin: *capsa*, a box)

Capsella bursa-pastoris (L.) Medik., (shepard's purse, the fruits resembling the scrotum of a sheep used as a purse by shepherds), SHEPHERD'S-PURSE, PICKPOCKET. Inconspicuously pubescent annual to 60 cm tall; basal leaves forming a rosette, petioled, the blades pinnatifid; stem leaves mostly sessile, usually auricled-clasping; flowers usually self-pollinated; petals white or yellowish, rarely absent; fruits inverted triangular-heart-shaped (= obcordate-triangular), flattened, 5–8 mm long, 3.5–6 mm wide. Roadsides, lawns, and disturbed sites; common, especially in cities and towns; essentially throughout TX. Reverchon (1880) indicated that it was seen for the first time in Dallas Co. in 1865. Feb–May. Native of Europe, but now a cosmopolitan weed. Used medicinally by Chinese for eye disease and dysentery; apparently used by humans for 1,000s of years—collected seeds have been found at an ancient site nearly 8,000 years old (Mabberley 1987). The seeds of SHEPHERD'S PURSE appear to be carnivorous; a chemical released by the seed appears to attract organisms such as protozoans, nematodes, or mobile bacteria; subsequently, an enzyme secreted by the seed digests the protein of the victim and allows the seed to absorb the material (Pietropaolo & Pietropaolo 1986).

Two forms, treated as species by European authors, occur in nc TX but appear to intergrade. They are about equally common, and do not appear to be distinct: *C. bursa-pastoris*, with petals 1.7–2.5 mm long, markedly exceeding the sepals; and *C. rubella* Reut., with petals 1–1.7 mm long, shorter than to slightly exceeding the sepals.

CARDAMINE BITTERCRESS

Ours annual or biennial with fibrous roots; basal rosette present; leaves pinnatifid or pinnately lobed (simple in an e TX species); fruits linear, compressed; seeds in single row in each locule, wingless.

A genus of 200 species of temperate areas and tropical mountains; it includes cultivated ornamentals and weeds. (A Greek name, *kardamon*, used by Dioscorides for some cress with medicinal uses, perhaps in treating heart problems; presumably from *cardia* or *kardia*, heart)

1. Petioles of lower stem leaves ciliate, at least basally; stamens 4; basal leaves larger and more conspicuous than the relatively few stem leaves ____ **C. hirsuta**
1. Petioles of stem leaves glabrous; stamens 6; basal leaves and lower and middle stem leaves ± similar.
   2. Lateral leaflets narrow; terminal leaflet usually 1–4 mm wide, usually not conspicuously larger than laterals; nearly all leaflets distinct, not decurrent along rachis; stems glabrous throughout; widespread in e part of nc TX ____ **C. parviflora**
   2. Lateral leaflets broad; terminal leaflet usually 4–17 mm wide, usually conspicuously larger than laterals; distal leaflets often slightly decurrent along rachis; stems hispid basally; known in nc TX only from Dallas Co. ____ **C. pensylvanica**

Cardamine hirsuta L., (hairy), HAIRY BITTERCRESS. Annual similar to *C. parviflora*; stems 10–30 cm tall; rosette leaves 3–8 cm long; stem leaves usually smaller; petals white, 2–3 mm long;

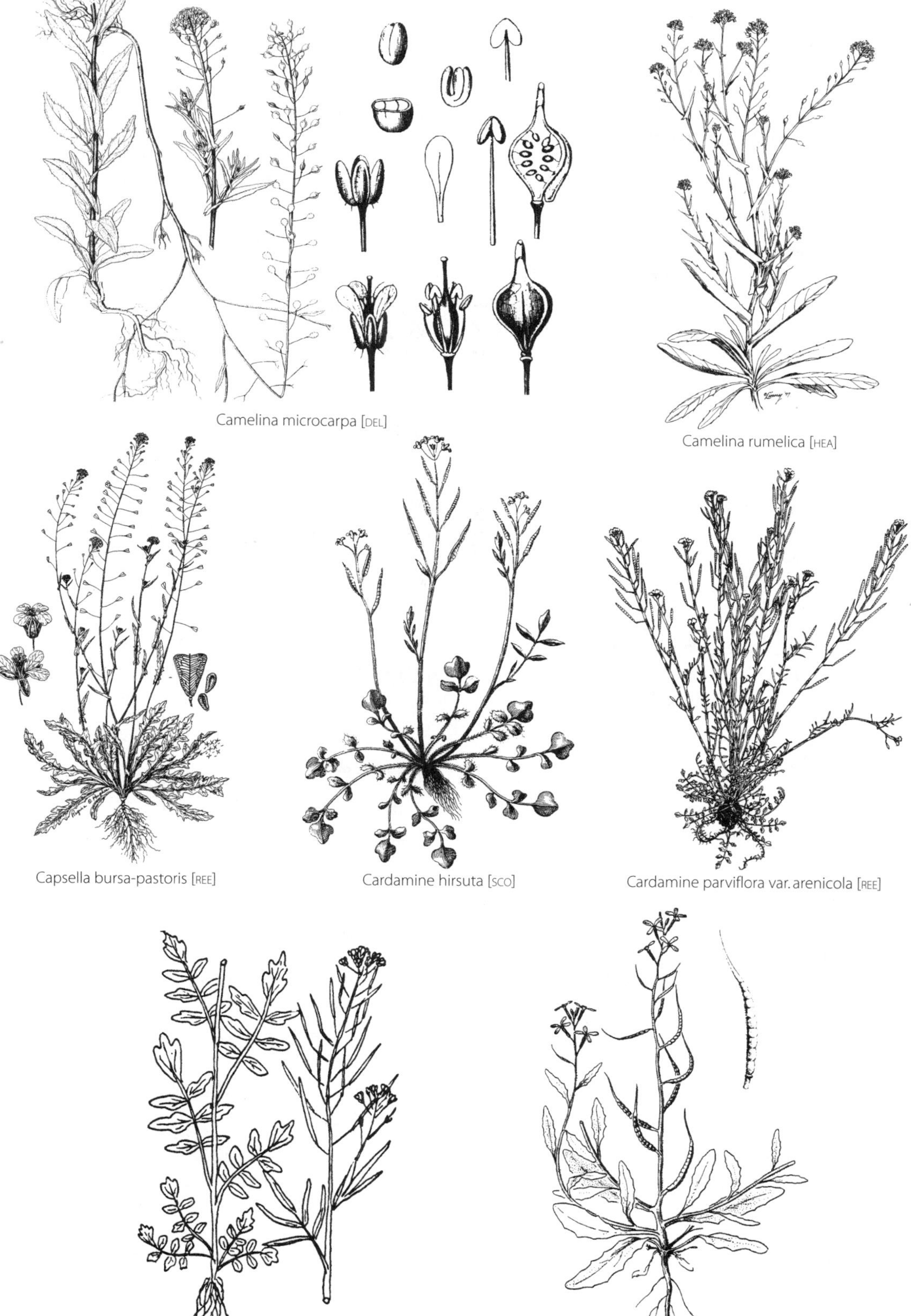

Camelina microcarpa [DEL]

Camelina rumelica [HEA]

Capsella bursa-pastoris [REE]

Cardamine hirsuta [SCO]

Cardamine parviflora var. arenicola [REE]

Cardamine pensylvanica [RYD]

Chorispora tenella [EN2]

styles 0.5 mm or less long; mature fruits 15–25 mm long, ca. 1 mm wide; beak 0.5–1 mm long. Roadsides, weedy areas; Lamar Co. in Red River drainage (Carr 1994); mainly se and e TX. Feb–Apr. Native of Europe.

Cardamine parviflora L. var. **arenicola** (Britton) O.E. Schulz, (sp.: small-flowered; var.: growing in sandy places), SAND BITTERCRESS. Glabrous, somewhat delicate annual to 25(–40) cm tall; leaves mostly 2–4 cm long; petals white, 1.5–3.5 mm long; style 0.2–0.7 mm long; mature fruits beakless, 20–30 mm long. Damp thickets, ditches, and stream banks; se and e TX w to Grayson Co., also Edwards Plateau. Late Feb–Apr.

Cardamine pensylvanica Muhl. ex Willd., (of Pennsylvania), BITTERCRESS. Annual or biennial similar to *C. parviflora*; plant to 75 cm tall, glabrous except for hispid basal region of stem; leaves mostly 4–8 cm long; petals white, 1.5–4 mm long; style 0.5–2 mm long; mature fruits short beaked. Lawns, roadsides, and weed in nursery; Dallas Co; rare; mainly se and e TX. Mar–Apr, Nov. Brown and Marcus (1998) in a recent paper tentatively indicated that some TX specimens (including the Dallas Co. collection) previously identified as *C. pensylvanica* are actually *C. debilis* D. Don, an introduced Old World species. Schulz (1903), in his monograph of the genus, treated both of these similar taxa as members of the same species (a variable *C. flexuosa* With.). Rollins (1993) indicated that *C. debilis* has fruits < 1 mm wide and fibrillose roots, while *C. pensylvanica* has fruits > 1 mm wide and the roots mostly thicker and scarcely fibrillose. Examination of the limited material available does not allow us to make a firm determination and consequently we are tentatively continuing to call this entity *C. pensylvanica*. However, the nearly glabrous basal region of the stem (typical of *C. debilis*), makes us suspicious that Dallas collections may belong to *C. debilis* as suggested by Brown and Marcus. Further work on this complex is needed.

Cardamine bulbosa (Schreb. ex Muhl.) Britton, Sterns, & Poggenb., (bulbous), (BULB BITTERCRESS, SPRING CRESS), a tuberous perennial with entire or shallowly dentate, simple leaves, occurs in se and e TX just to the e of nc TX. Jones et al. (1997) recognized this species as [*C. rhomboidea* (Pers.) DC.].

CHORISPORA BLUE-MUSTARD

An e Mediterranean and c Asian genus of 13 species. (Latin: *choris*, asunder, and *spora*, seed, from fruit breaking apart between seeds)

Chorispora tenella (Pall.) DC., (tender, soft), BLUE-MUSTARD. Annual, sparsely stipitate-glandular, 20–50 cm tall; lower leaves sharply incised; sepals 3.5–6 mm long; petals blue-purple, 7–12 mm long, clawed, with claws longer than blade, the blade 3–5 mm long; fruits indehiscent, breaking at maturity into 2-seeded segments, 3–4.5 cm long including beak; beak ca. 1–1.5 cm long. Disturbed areas; in TX apparently known only from Dallas, Deaf Smith, El Paso, and Tarrant cos. Reported as new for TX by Lipscomb (1984). Mar–May. Native of Asia.

CONRINGIA

A genus of 6 species native from the Mediterranean and Europe to c Asia. (Named for Hermann Conring, 1606–1681, professor at Helmstadt, Germany)

Conringia orientalis (L.) Andrz., (eastern), TREACLE HARE'S-EAR, HARE'S-EAR-MUSTARD, HARE'S-EAR. Glabrous glaucous annual to 110 cm tall; basal rosette absent; leaves oblong to ovate or oblong-lanceolate, entire, cordate clasping; petals yellowish white; fruits linear, 4-angled, 8–11 cm long, 2–3 mm in diam. Along railroads or less often in disturbed areas; gravelly or sandy soils; Dallas, Denton, Grayson, and Navarro cos., also Tarrant Co. (Mahler 1988); e TX w to East Cross Timbers. Apr–early May. Native of se Europe.

CORONOPUS WARTCRESS

A nearly cosmopolitan genus of 10 species including weeds. (Greek: *korone*, crow, and *pous*, foot, from the deeply cleft leaves)

Coronopus didymus (L.) Sm., (in pairs, as of stamens), SWINE WARTCRESS. Rank-scented annual, usually decumbent, slightly pubescent or glabrous; stems 10–40 cm long; leaf blades deeply pinnatifid, almost compound; flowers in mostly lateral spike-like racemes, borne nearly from base to tip of stems; sepals green with white margins; petals white, shorter than the sepals; fruits with 2 lobes, very small, to ca. 1.7 mm long, 2–2.3 mm broad, wrinkled, notched at apex. Lawns or disturbed areas in cities, towns, chiefly in sandy soils; Dallas and Milam cos.; mainly se and e TX. Apr–May. Native of s South America.

DESCURAINIA TANSY-MUSTARD

Annuals or biennials with pinnately compound leaves, the leaflets pinnatifid or again compound, usually pubescent with branched hairs; petals small, yellow to whitish, clawed, slightly shorter to slightly longer than the greenish yellow sepals.

A genus of 40 species of temperate and cool n hemisphere areas and s Africa. (Named for Francois Déscourain, 1658–1740, French apothecary and botanist)
REFERENCES: Detling 1939; Shinners 1949f.

1. Fruits 1.5–2 mm wide, usually 5–12 mm long, ca. as long as to shorter than pedicels, subclavate (= somewhat club-like); seeds usually in 2 rows, rarely 1 ____ **D. pinnata**
1. Fruits 0.5–1.5 mm wide, 8–30 mm long, usually distinctly longer than pedicels, linear; seeds always in 1 row.
   2. Stem leaves 2–3 times pinnate; petals 2–2.5 mm long, shorter than or equaling the relatively long (2–2.5 mm) sepals; fruits 10–30 mm long; upper stems and inflorescences without glandular pubescence (non-glandular pubescence may be present) ____ **D. sophia**
   2. Stem leaves once pinnate; petals 1.5–2 mm long, longer than the short (1–1.5 mm) sepals; fruits 8–15 mm long; upper stems and inflorescences glandular-pubescent ____ **D. incana**

Descurainia incana (Bernh. ex Fisch. & C.A. Mey.) Dorn subsp. **viscosa** (Rydb.) Kartesz & Gandhi, (sp.: hoary, quite gray; subsp.: sticky, clammy). Plant to ca. 120 cm tall. Open areas; included based on citation for vegetational area 5 (Fig. 2) by Hatch et al. (1990); also Trans-Pecos. Jun-Sep. [*D. richardsonii* subsp. *viscosa* (Rydb.) Detling, *Sophia viscosa* Rydb.]

Descurainia pinnata (Walter) Britton, (pinnate, feathery). Plant 5–80 cm tall. Open stream bottoms, roadsides, and waste ground, often sandy soils. Mar–Apr. Oil can be obtained from the seeds. There is apparently introgression among the varieties, which are frequently difficult to distinguish. Subspecies *pinnata* and subsp. *halictorum* are particularly difficult to separate and are here recognized as distinct with some hesitation. Reportedly poisonous to livestock with symptoms similar to those of selenium poisoning; animals can become blind, wander aimlessly, and lose the ability to swallow; the condition is sometimes referred to as blind staggers or paralyzed tongue (Kingsbury 1964; Stephens 1980; James et al. 1992).

1. Leaf blades moderately to densely pubescent, gray-green.
   2. Terminal segment of middle stem leaves elliptic-lanceolate to ovate or rhombic, toothed or lobed like the lateral segments, resembling them in size and shape ____ subsp. **pinnata**
   2. Terminal segment of middle stem leaves oblong-lanceolate to linear, entire or toothed, longer and less divided than the lateral segments ____ subsp. **halictorum**
1. Leaf blades glabrous (petioles and lower rachis sometimes pubescent), green ____ subsp. **brachycarpa**

subsp. **brachycarpa** (Richardson) Detling, (short-fruited). Stems with rather abundant gland-tipped hairs but little or no branched pubescence; lobing of leaf blades as in subsp. *halictorum*. W part of nc TX w to Panhandle and s to Edwards Plateau. [*D. pinnata* var. *brachycarpa* (Richardson) Fernald]

subsp. **halictorum** (Cockerell) Detling, (hoary, quite gray). Stems with branched hairs. W part of nc TX nw to Panhandle and s to Edwards Plateau. [*D. pinnata* var. *osmiarum* (Cockerell) Shinners]

subsp. **pinnata**. PINNATE TANSY-MUSTARD. Stems with branched hairs. Nearly throughout TX.

Descurainia sophia (L.) Webb ex Prantl, (old generic name), FLIXWEED TANSY-MUSTARD. Plant 15–80 cm tall. Roadsides and waste ground; various soils; Denton and Grayson cos.; mostly w 1/2 of TX. Apr–May. Native of Europe. The seeds are sometimes used like mustard.

DIPLOTAXIS

A genus of 27 species native to Europe and the Mediterranean to nw India. (Greek: *diplous*, double, and *taxis*, row or arrangement, alluding to the biserrate seeds or with two shields)

Diplotaxis muralis (L.) DC., (of walls), STINKING WALLROCKET, SANDROCKET, WALLROCKET. Ill-scented, inconspicuously hispid annual to perennial, to 60 cm tall; leaves crowded near base, rather thick and fleshy, the blades toothed or pinnatifid; flowers few in elongated racemes; petals 5–8 mm long, sulfur yellow to lavender; fruits 20–35 mm long, 1–2 mm in diam. Road margins, limestone areas; Collin, Dallas, Ellis, Grayson, and Hamilton cos.; also Edwards Plateau; first collected in TX in 1947 (Cory 1948a). Mar–Sep. Native of Europe.

DRABA WHITLOW-GRASS

Ours small annuals, pubescent, the hairs simple or branched; petals white or absent.

A genus of ca. 300 species of n temperate and boreal areas and mountains of South America; some are cultivated as rock-garden ornamentals; it is the largest genus in the family. (Greek: *drabe*, acrid; applied by Dioscorides to some cress)
REFERENCES: Fernald 1934; Hartman et al. 1975.

1. Mature fruits 2–4(–5) mm long, 0.6–1 mm wide; leaves of main stem more than twice as long as wide, narrowly oblong or oblanceolate, entire; plants pubescent with sessile branched hairs ____ **D. brachycarpa**
1. Mature fruits 5–15 mm long, 1–3 mm wide; leaves of main stem (at first all nearly basal) usually not more than twice as long as wide, entire or toothed; plants with either simple hairs or stalked branched hairs (use hand lens or conspicuous with a dissecting scope).
   2. Leaves entire; fruits erect or strongly ascending on glabrous pedicels; fruits usually glabrous ____ **D. reptans**
   2. Leaves with 1 or more pairs of coarse teeth; fruits widely spreading on pubescent pedicels; fruits often with short hairs.
      3. Fruits lanceolate-elliptic, usually 5–8 mm long, 2–3 mm wide; stem leaves 4–10, scattered up the stem; stems with largely simple pubescence ____ **D. platycarpa**
      3. Fruits elliptic to oblong, 8–16 mm long, 1–3 mm wide; stem leaves 1–7 (or apparently none), crowded near base; stems with spreading, stalked, branched pubescence ____ **D. cuneifolia**

Draba brachycarpa Nutt. ex Torr. & A. Gray, (short-fruited), SHORT-POD DRABA. Stems erect, to ca. 20 cm tall; basal leaves to ca. 15 mm long; petals absent or 1.2–3 mm long; fruits glabrous or hispid. Woods, roadsides, and open ground; sandy soils; e TX w to Rolling Plains and Edwards Plateau. Feb–Mar.

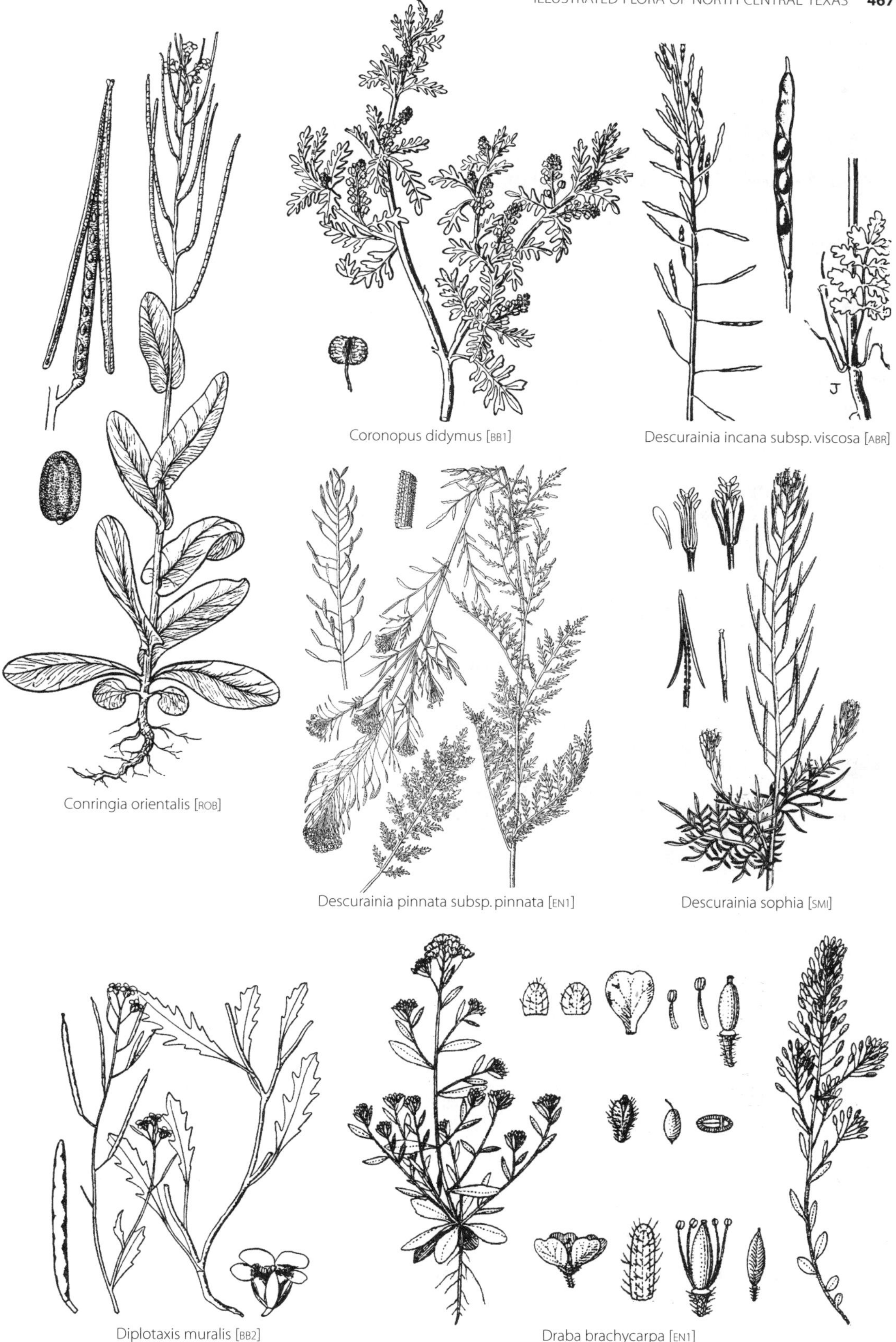

Conringia orientalis [ROB]

Coronopus didymus [BB1]

Descurainia incana subsp. viscosa [ABR]

Descurainia pinnata subsp. pinnata [EN1]

Descurainia sophia [SMI]

Diplotaxis muralis [BB2]

Draba brachycarpa [EN1]

Draba cuneifolia Nutt. ex Torr. & A. Gray, (wedge-leaved), WEDGE-LEAF DRABA, WHITLOW-WORT. Similar to *D. platycarpa*; stems erect, to ca. 30 cm tall. Sandy and gravelly soils; nearly throughout TX. Late Feb–Apr.

Draba platycarpa Torr. & A. Gray, (broad-fruited), BROAD-POD DRABA. Petals 3–3.5 mm long, rarely absent. Rocky slopes, bare spots in prairies, disturbed sites, sand or calcareous clay; e Blackland Prairie w to Rolling Plains and Edwards Plateau. Mar–Apr, main period 1 or 2 weeks later than *D. cuneifolia*.

Draba reptans (Lam.) Fernald, (creeping), CAROLINA DRABA. Petals absent or 3–4 mm long. Open woods, fields, and roadsides, sandy or rocky soils; Blackland Prairie w to High Plains. Feb–Mar. [*D. reptans* (Lam.) Fernald var. *micrantha* (Nutt.) Fernald]

ERUCASTRUM

A genus of 20 species native to the Mediterranean, c and s Europe, and Macaronesia. (Latin: resembling *Eruca*, a Mediterranean genus whose name derives from the Latin name for an edible species)
REFERENCE: Luken et al. 1993.

Erucastrum gallicum (Willd.) O.E. Schulz, (of Gaul or France), ROCKETSALAD, ROCKETWEED, DOG-MUSTARD. Annual to 60(–80) cm tall; leaf blades deeply pinnatifid, the lobes coarsely toothed, obtuse; petals 7–8 mm long, pale yellow; fruits 2–4 cm long, 1–2 mm wide. Roadsides and railroads; Grayson Co., also Denton, Hill, and Tarrant cos. (Mahler 1988); nc TX s to Edwards Plateau; first collected in the U.S. in Wisconsin in 1903 and in Texas in 1926 (Luken et al. 1993). Late Mar–Apr. Native of Europe.

ERYSIMUM WALLFLOWER

Annuals or perennials, gray with minute, appressed, simple or both simple and stellate hairs; leaves linear-lanceolate to oblong, entire, toothed, or pinnatifid; petals long-clawed.

A genus of ca. 200 species of the Mediterranean, Europe, Asia, and North America; it includes weeds and cultivated ornamentals. (Greek: *eryomai*, help or save, from supposed medicinal properties of some species or *eryo*, to draw out, alluding to the blistering properties of some species)
REFERENCE: Rossbach 1958.

1. Petals greenish yellow, 5–8 mm long; pedicels 2–4 mm long ____ **E. repandum**
1. Petals yellow to orange-red, 15–25 mm long; pedicels 5–10 mm long.
   2. Petals yellow (rarely orange-yellow); mature fruits widely spreading, usually 8–12 cm long; plants often 40 cm or less tall ____ **E. asperum**
   2. Petals yellow to orange or orange-red; mature fruits erect (rarely merely ascending), usually 5–8 cm long; plants usually 40–100 cm tall ____ **E. capitatum**

Erysimum asperum (Nutt.) DC., (rough), WESTERN WALLFLOWER. Perennial to 40(–75) cm tall. Limestone outcrops, hillsides, prairies; Lampasas and Wise cos., also Limestone Co. (Mahler 1988); nc TX w to Plains Country. Apr–May. /89

Erysimum capitatum (Douglas ex Hook.) Greene, (headed), PLAINS ERYSIMUM, WESTERN WALLFLOWER, PRAIRIE-ROCKET. Coarse biennial or perennial [?] to 40–100 cm tall; young plants without mature fruits are often very similar to *E. asperum*. Rocky outcrops, roadcuts, dry streambeds, open wooded slopes; Limestone Co.; Post Oak Savannah w through much of TX. Apr–Jun.

Erysimum repandum L., (with wavy margins), SPREADING ERYSIMUM, BUSHY WALLFLOWER. Annual to ca. 40 cm tall; fruits 4–7 cm long. Railroads and roadsides, various soils; Callahan,

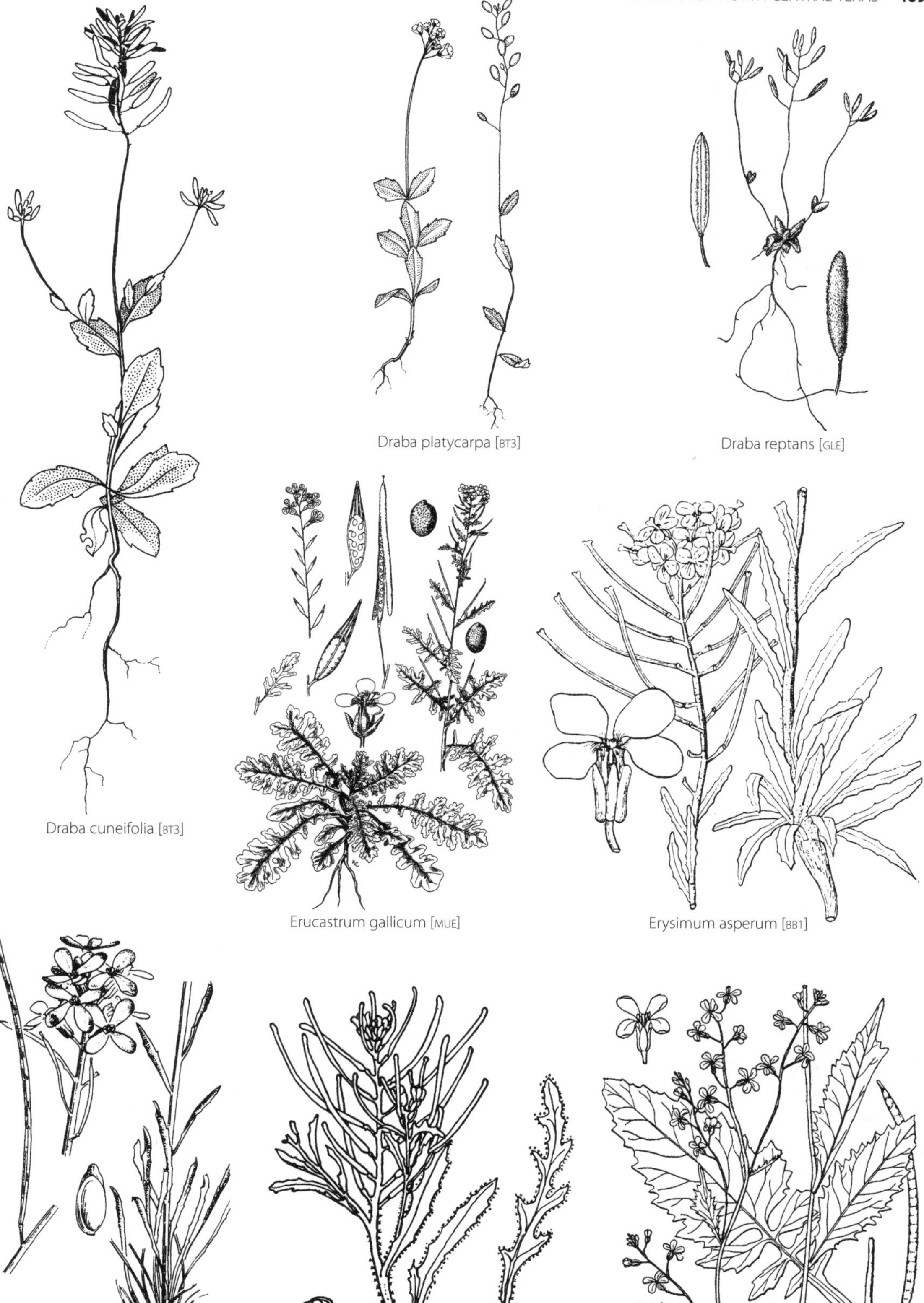

Draba platycarpa [BT3]

Draba reptans [GLE]

Draba cuneifolia [BT3]

Erucastrum gallicum [MUE]

Erysimum asperum [BB1]

Erysimum capitatum [ABR]

Erysimum repandum [BB2]

Iodanthus pinnatifidus [BB2]

Cooke, Dallas, Grayson, and Somervell cos., also Denton Co. (Mahler 1988); nc TX s to Edwards Plateau and se TX, apparently spreading; first collected in TX in Pecos Co. in 1944 (Cory 1947, 1948a, 1950). Apr–May. Native of Europe.

IODANTHUS

A genus of 4 species, 1 in e North America and 3 in Mexico. (Greek: *iodes*, violet-colored, and *anthos*, flower)
REFERENCE: Rollins 1942.

Iodanthus pinnatifidus (Michx.) Steud., (pinnately cut), PURPLE-ROCKET. Perennial, glabrous or nearly so; stems erect, 30–100 cm tall; leaves ovate to lanceolate, usually sharply toothed; lower leaves 5–20 cm long, usually with winged petioles with small lobes and basal auricles clasping the stem; upper leaves sessile and cuneate; sepals erect, often purplish, 3–5 mm long; petals white to lavender; fruits 2–4 cm long, linear. Low and rich woods; included based on citation of vegetational area 4 (Fig. 2) by Hatch et al. (1990); c and e TX. Apr–Jun.

LEPIDIUM PEPPERWEED, PEPPER-GRASS

Our species annuals; leaves mostly toothed or pinnatifid; flowers small; petals white or absent.

A cosmopolitan, but especially temperate genus of ca. 140 species. The fruits of some have a tasty, peppery flavor; a number have been cultivated as salad plants and are usually eaten at cotyledon or seedling stage. (Greek: *lepidion*, little scale, alluding to the fruit)
REFERENCES: Hitchcock 1936, 1945; Al-Shehbaz 1986a.

1. Uppermost leaves deeply toothed or lobed; stem leaves rather uniform in size, deeply pinnatifid; plants prostrate to ascending, usually much branched near base ____ **L. oblongum**
1. Uppermost leaves entire or nearly so; stem leaves gradually much reduced upward, the lower leaves toothed to deeply pinnatifid, the upper shallowly pinnatifid to entire; plants ± erect, usually single stemmed below, branched above.
    2. Stems glabrous or minutely and inconspicuously pubescent.
        3. Lower leaves usually bipinnatifid; fruits 1–2(–2.5) mm broad, ovate to broadly elliptic; plants ill-smelling ____ **L. ruderale**
        3. Lower leaves usually incised to pinnatifid, not bipinnatifid; fruits 2–3.3 mm broad, elliptic to orbicular; plants usually not ill-smelling.
            4. Pedicels about as long as fruits or shorter; petals rudimentary, shorter than sepals OR petals absent; fruits 2–3.5 mm long ____ **L. densiflorum**
            4. Pedicels usually slightly longer than the fruits; petals typically conspicuous, usually as long as or longer than sepals; fruits 2.5–4.2 mm long ____ **L. virginicum**
    2. Stems moderately to densely hispid-pilose, the hairs conspicuous (with a hand lens) and spreading at right angles ____ **L. austrinum**

Lepidium austrinum Small, (southern), SOUTHERN PEPPERWEED. Similar to *L. virginicum*; petals frequently absent. Disturbed areas; mainly Blackland Prairie w to Rolling Plains and s to s TX. Late Mar–May. Apparently hybridizes introgressively with *L. virginicum*.

Lepidium densiflorum Schrad., (densely-flowered), PRAIRIE PEPPERWEED, GREEN-FLOWER PEPPER-GRASS. Plant somewhat erect, to ca. 50 cm tall; basal leaves usually deeply serrate-incised, rarely somewhat pinnatifid; fruits ovate or rarely suborbicular. Sandy soils in disturbed areas; Brown Co. on western margin of nc TX; scattered nearly throughout TX. Feb–Jun.

Lepidium oblongum Small, (oblong), VEINY PEPPERWEED. Stems prostrate to ascending, much branched from base, to ca. 20 cm long, minutely pubescent; basal leaves pinnatifid to

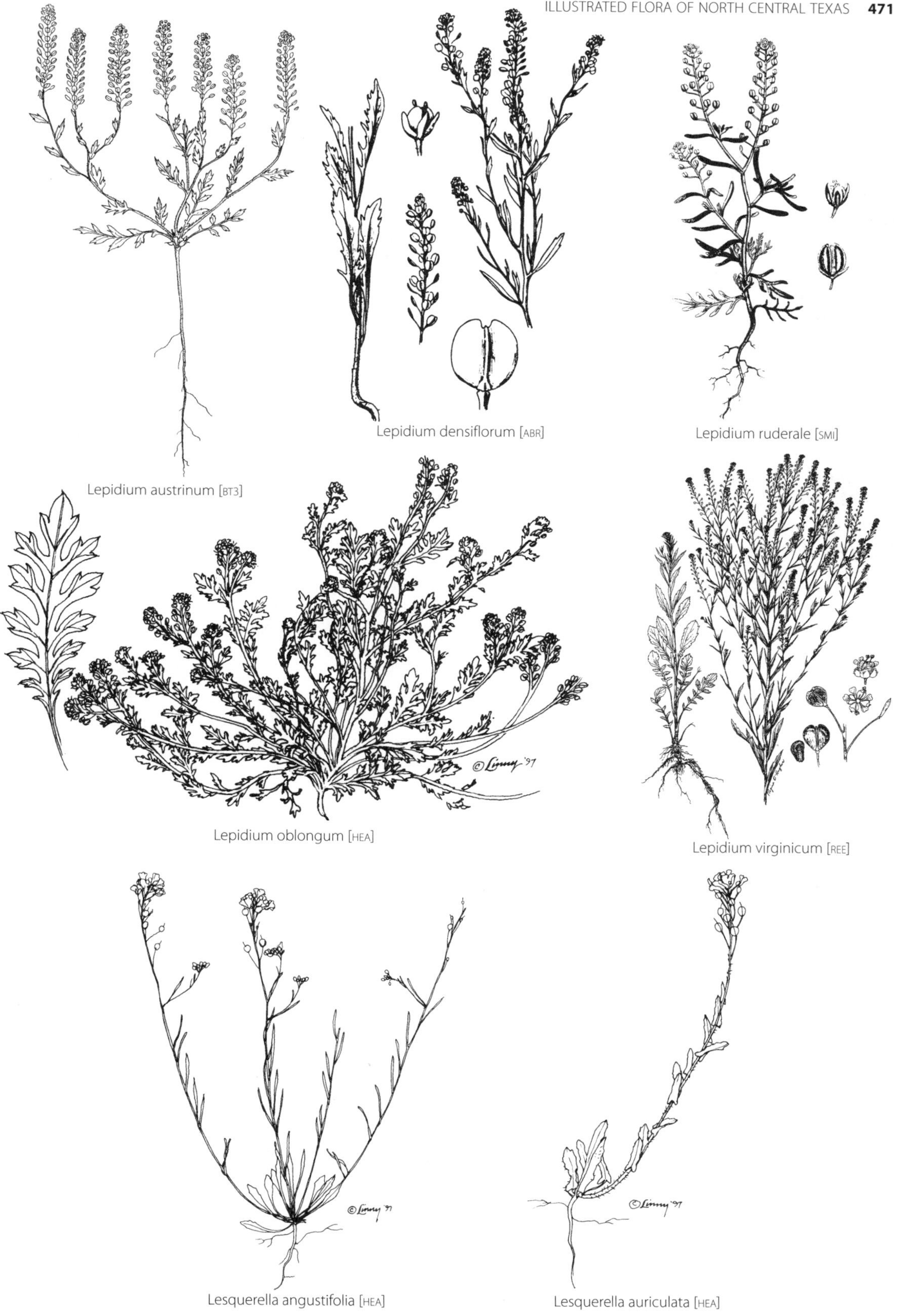

Lepidium austrinum [BT3]

Lepidium densiflorum [ABR]

Lepidium ruderale [SMI]

Lepidium oblongum [HEA]

Lepidium virginicum [REE]

Lesquerella angustifolia [HEA]

Lesquerella auriculata [HEA]

bipinnatifid; petals minute or absent. Sandy or silty ground; Tarrant Co. (Mahler 1988); mainly Rolling Plains and Edwards Plateau w to w TX. Mid-Mar–Apr.

Lepidium ruderale L., (of rubbish). Stems to ca. 30 cm tall; basal leaves usually bipinnatifid, rarely only pinnatifid. Roadsides, waste places, and disturbed areas; included based on citation for vegetational areas 4 and 5 (Fig. 2) by Hatch et al. (1990); scattered in e 1/2 of TX. Mar–Jul. Native of Europe.

Lepidium virginicum L., (of Virginia), VIRGINIA PEPPER-GRASS, POORMAN'S-PEPPER, AMERICAN PEPPER-GRASS, LENTEJILLA. Plant erect, to 60 cm tall; basal leaves usually pinnatifid; inflorescences at least sparingly pubescent; pedicels terete; petals usually longer than the sepals, linear-oblanceolate to obovate, rarely shorter or absent; fruits usually nearly orbicular, less often broadly elliptic to obovate. Usually sandy soils, disturbed areas; throughout TX. Mar–May. Two varieties, var. *medium* (Greene) C.L. Hitchc. and var. *virginicum*, are sometimes recognized (Kartesz 1994). Variety *medium* supposedly differs in having the inflorescences usually glabrous and the pedicels slightly inflated. However, consistent differences in nc TX material could not be found.

LESQUERELLA BLADDERPOD, WILD MUSTARD

Low, pubescent or scurfy annuals or perennials; hairs usually branched or stellate, the rays sometimes fused and the hairs then semi-peltate or peltate and scale-like; leaves (except lowest) sessile or subsessile, entire or toothed or pinnatifid; flowers in racemes; petals yellow; fruits globose or nearly so.

A North American genus of 40 species. (Named for Leo Lesquereux, 1805–1889, distinguished American bryologist and paleobotanist)

REFERENCES: Payson 1921; Rollins 1955, 1956; Rollins & Shaw 1973; Clark 1975; Nixon et al. 1983.

1. Plants annual (rarely biennial), not from a woody base; petals (including claw) usually 4–8 mm long; fruits 3–6 mm long (to 7 mm in *L. gordonii*).
  2. Stem leaves distinctly auriculate.
    3. Lower stems pubescent with appressed branched hairs; stems 20–70 cm long; sepals ca. 4–6 mm long ____ **L. grandiflora**
    3. Lower stems with spreading simple hairs; stems to ca. 22 cm long; sepals ca. 3–4 mm long ____ **L. auriculata**
  2. Stem leaves not auriculate or only obscurely so.
    4. Fruiting pedicels uniformly recurved (= bent downward) and the fruits therefore pendant; buds when ready to open (yellow-green in color, petals not yet visible) 2.5–3.5 mm long ____ **L. recurvata**
    4. Fruiting pedicels various but not uniformly recurved; buds when ready to open 4–6 mm in diam.
      5. Pedicels sigmoid (curved into an S-shape); mainly West Cross Timbers w to w TX ____ **L. gordonii**
      5. Pedicels straight or simply curved, not sigmoid; distribution various.
        6. Inflorescence dense, the flowers crowded at end of stem; inflorescence closely subtended by upper leaves even after fruit has formed; stellate hairs ca. 0.3–0.4 mm across (with a dissecting scope), distinctly hair-like; in nc TX known only from s Grand Prairie ____ **L. densiflora**
        6. Inflorescence quickly elongating, not dense, the flowers not crowded at end of stem; inflorescence separated a short distance from the uppermost leaves by the time a few flowers have opened; stellate hairs ca. 0.1–0.2 mm across, so small as to appear almost scale-like; widespread.
          7. Stem leaves 4–10 cm long, 1–3.5 mm wide; ovules 2 per locule; fruits with a minute stipe (0.5 mm or less long) or none connecting it to the pedicel; in nc TX only known from Red River Co. ____ **L. angustifolia**

7. Stem leaves 1–5 cm long, 2–20 mm wide; ovules 8–10 per locule; fruits with a stipe (ca. 0.75–2 mm long) connecting it to the pedicel; widespread ______ **L. gracilis**

1. Plants perennial from a woody base; petals usually 8–12 mm long; fruits 4.5–8 mm long.
   8. Plants densely pubescent and grayish green but not conspicuously silvery-gray; inflorescence dense and crowded at end of stem; rays (= branches) of stellate hairs free or fused basally only (use dissecting scope); fruits with stipe ca. 0.5–1 mm long; style 4–8 mm long ______ **L. engelmannii**
   8. Plants conspicuously silvery-gray pubescent; inflorescence usually elongated and loose (at least with age); rays of stellate hairs fused together from their base to 1/2 their length; fruits usually without a stipe or occasionally with a very short stipe; style 3–5 mm long ______ **L. fendleri**

Lesquerella angustifolia (Nutt. ex Torr. & A. Gray) S. Watson, (narrow-leaved), NARROW-LEAF BLADDERPOD. Annual with stems to ca. 40 cm long; stems and leaves scurfy with peltate or semi-peltate hairs. Rocky limestone soils; central Red River Co. in extreme ne part of nc TX and nearby OK; rare. Apr–May.

Lesquerella auriculata (Engelm. & A. Gray) S. Watson, (eared, with an ear-shaped appendage), EAR-LEAF BLADDERPOD. Annual densely hirsute with long simple hairs and shorter several-rayed ones beneath; stems erect to decumbent, to 22 cm long; basal leaves dentate to lyrate, sometimes nearly entire; cauline leaves 1–4 cm long, 3–10 mm wide, sessile and auriculate; fruits 4–6 mm long, globose to slightly longer than broad. Prairies and disturbed soils; Kaufman Co. (Reverchon collection in 1903, "sands, common"), also Navarro Co. (Rollins & Shaw 1973); ne Blackland Prairie e to e TX. Mar–May.

Lesquerella densiflora (A. Gray) S. Watson, (densely flowered), DENSE-FLOWER BLADDERPOD. Annual or biennial with erect to decumbent stems to ca. 40 cm long; stems and leaves rather densely pubescent with loosely appressed, stellate hairs; leaves rather numerous and crowded, elliptic- or oblong-lanceolate; fruits 2.5–4.5 mm long. Sandy or gravelly ground; Burnet and Somervell cos., also Brown, Comanche, Eastland, Hood, Palo Pinto (Rollins & Shaw 1973), and McLennan (Mahler 1988) cos.; mainly s Grand Prairie s to Edwards Plateau and w to Rolling Plains; endemic to TX. Late Mar–Apr.

Lesquerella engelmannii (A. Gray) S. Watson, (for George Engelmann, 1809–1884, German-born botanist and physician of St. Louis), ENGELMANN'S BLADDERPOD. Perennial with several stems to ca. 50 cm tall; stems and leaves silvery with dense, appressed, stellate pubescence; stem leaves linear to narrowly oblong-oblanceolate; raceme at first short, umbel-like; fruits 5.5–5.8 mm long; ovules 5–8 per locule. Rocky limestone slopes; Cooke, Dallas, Ellis, Erath, Parker, Tarrant, and Wise cos., also Lampasas Co. (Rollins & Shaw 1973); Blackland Prairie to West Cross Timbers s to s TX. Apr–May.

Lesquerella fendleri (A. Gray) S. Watson, (for August Fendler, 1813–1883, one of the first botanists to collect in New Mexico and Venezuela), FENDLER'S BLADDERPOD, POPWEED. Perennial to 40 cm tall, branched from base, conspicuously silvery-gray pubescent by stellate hairs with rays fused to about middle; fruits 4.6–8 mm long, globose to broadly ellipsoid or ovoid; ovules 10–16 per locule. Roadsides, open sandy or rocky, often calcareous soils; Shackelford Co., also Brown and Coryell cos. (Rollins & Shaw 1973); w part of nc TX w to w TX. Apr.

Lesquerella gordonii (A. Gray) S. Watson, (for James Gordon, d. ca. 1781, correspondent of Linnaeus and London nurseryman), GORDON'S BLADDERPOD, POPWEED, BEADPOP. Stellate pubescent annual; stems prostrate, decumbent or erect, to 40 cm long, usually branched; fruits 4–7 mm long, ca. globose to broadly ellipsoid. Sandy and gravelly soils; Brown, Burnet, Callahan, Shacklelford, Somervell, and Young cos., also Comanche, Eastland, Jack, and Johnson cos. (Rollins & Shaw 1973); mainly West Cross Timbers w to w TX. Mar–Jun. The dry fruits pop when stepped on giving rise to the common name (Kirkpatrick 1992).

Lesquerella gracilis (Hook.) S. Watson, (graceful). Annual with erect to decumbent stems to 50 cm tall, younger parts gray with appressed, stellate pubescence, older parts sparsely pubescent and green; fruits 3–5 mm in diam. Prairies, roadsides, and disturbed ground, calcareous clay. Late Mar–May.

1. Fruits globose to ellipsoid ____ subsp. **gracilis**
1. Fruits obpyriform (= pear-shaped with attachment at narrow end), usually with a distinct basal shoulder ____ subsp. **nuttallii**

subsp. **gracilis** (Hook.) S. Watson, WHITE BLADDERPOD, LAX BLADDERPOD, PEAR-FRUIT BLADDERPOD, CLOTH-OF-GOLD. Blackland Prairie and Grand Prairie; nc to sc TX.

subsp. **nuttallii** (Torr. & A. Gray) Rollins & E.A. Shaw, (for Sir Thomas Nuttall, 1786–1859, English-American botanist). Dallas, Grayson, and Kaufman cos. (Rollins & Shaw 1973); ne TX. [*L. nuttallii* (Torr. & A. Gray) S. Watson, *L. gracilis* (Hook.) S. Watson var. *repanda* (Nutt.) Payson]

Lesquerella grandiflora (Hook.) S. Watson, (large-flowered), BIG-FLOWER BLADDERPOD. Annual (biennial?) with erect to decumbent stems to 70 cm long, densely pubescent with often 5-rayed hairs; basal leaves dentate to bipinnatifid; fruits 4–6 mm long, globose or slightly longer than broad. Loose sandy soils; included based on citation for vegetational area 4 (Fig. 2) by Hatch et al. (1990); n to at least Travis Co. just s of nc TX (Rollins & Shaw 1973); s and c TX; endemic to TX. Mar–Apr.

Lesquerella recurvata (Engelm. ex A. Gray) S. Watson, (curved backward), SLENDER BLADDERPOD. Resembling smaller, delicate forms of *L. gracilis*; stems scurfy with peltate or semi-stellate hairs; leaves stellate-pubescent with subappressed hairs. Limestone outcrops and gravelly calcareous prairies; mainly Blackland and Grand prairies from Ellis, Hood, and Johnson cos. s to c TX; endemic to TX. Mar–Apr.

MATTHIOLA STOCK

An Old World genus of 55 species native to w Europe, the Mediterranean region, and Macaronesia; some are cultivated as ornamentals. (Named for Pietro Andrea Matthiola, 1500–1577, Italian botanist and physician)

Matthiola longipetala (Vent.) DC., (long-petaled), EVENINGSTOCK, PERFUMEPLANT, GILLIFLOWER, STOCK. Annual or biennial with dense, grayish, branched pubescence; leaves linear-lanceolate to lanceolate, entire to dentate; flowers sessile, lilac or purple, fragrant; petals 1.5–2 cm long; fruits 7–10 cm long, forked apically into 2 conspicuous horns. Escapes from cultivation, weedy areas; Bell Co., also Edwards Plateau. Spring–summer. Native of the Old World. [*M. bicornis* (Sibth. & Sm.) DC.]

MYAGRUM

A monotypic genus native of the Mediterranean area and c Europe to India. (Greek: *myo*, mouse, and *agra*, trap)

Myagrum perfoliatum L., (perfoliate, with leaves surrounding the stem). Glabrous annual to 70(–100) cm tall; stem leaves sessile, auricled-clasping, mostly entire, oblong- or triangular-lanceolate; flowers crowded, in spike-like racemes; petals light yellow, 3–4 mm long; fruits roughly triangular (club-shaped, much wider distally), ca. 5–6 mm long. Fields and roadsides in black clay; Dallas, Delta, Fannin, Grayson, and Johnson cos., also Denton (C. Taylor, pers. obs.) and Rockwall (VDB) cos. and Bryan Co., OK (C. Taylor, pers. obs.); also Edwards Plateau; thought to have been introduced into Texas (Delta Co.) with vetch seed from Oregon in 1949 (Cory 1950); abundant and spreading. Apr–May. Native of the Mediterranean area and c Europe to Iran.

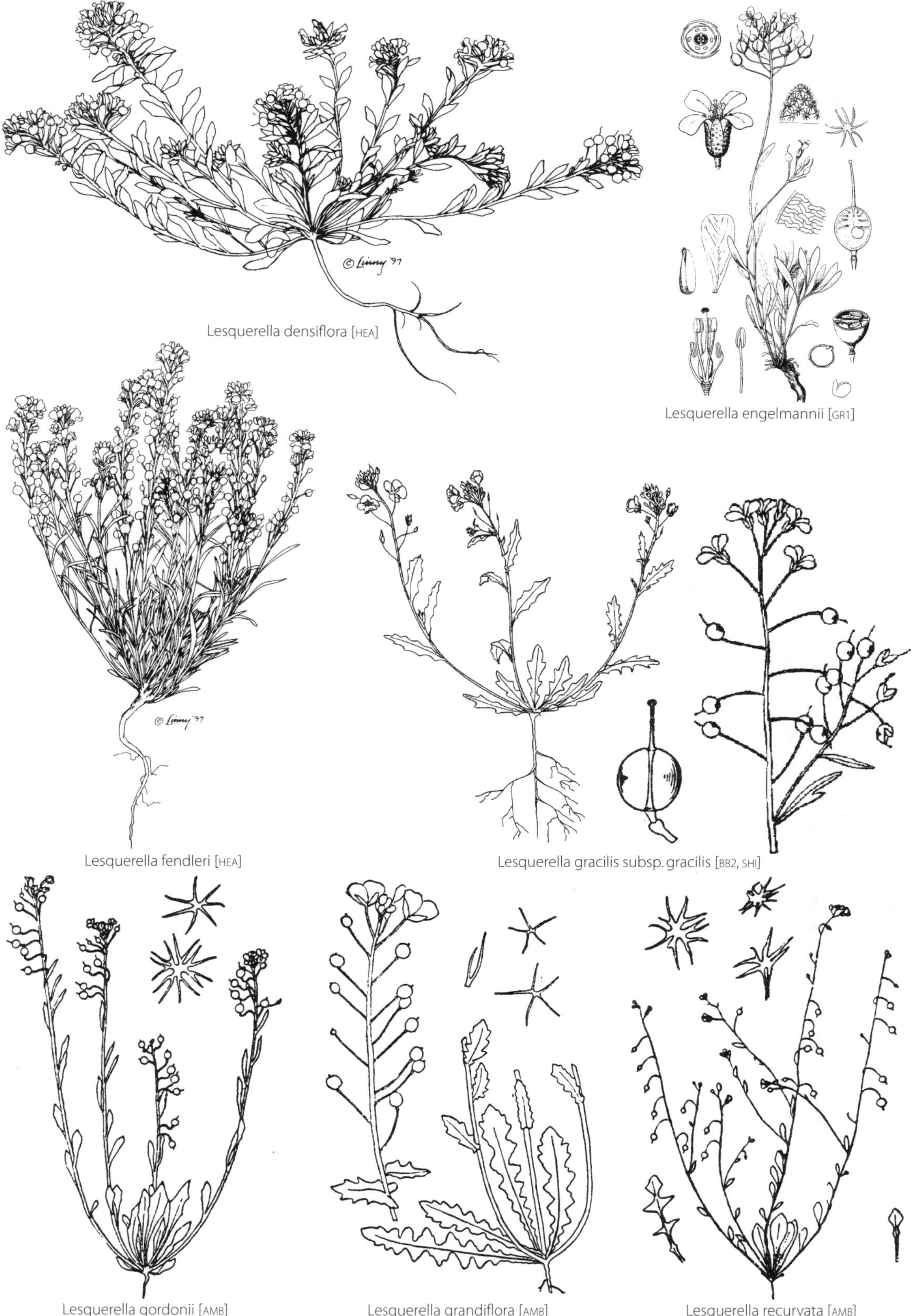

Lesquerella densiflora [HEA]

Lesquerella engelmannii [GR1]

Lesquerella fendleri [HEA]

Lesquerella gracilis subsp. gracilis [BB2, SHI]

Lesquerella gordonii [AMB]

Lesquerella grandiflora [AMB]

Lesquerella recurvata [AMB]

RAPHANUS RADISH

A genus of 3 species native to w and c Europe and the Mediterranean region to c Asia. (Greek: *raphanos*, quickly appearing, alluding to the rapid germination; classical Greek name for radish)

Raphanus sativus L., (cultivated), RADISH, RABANO. Sparsely hispid or nearly glabrous annual 25–125 cm tall, with swollen taproot; larger leaves pinnatifid, with large terminal lobe; petals white to rosy or lavender, 15–22 mm long; fruits 4–5 cm long, 6–10 mm in diam. Commonly cultivated throughout TX and sometimes a transitory escape; included based on citation of vegetational areas 4 and 5 (Fig. 2) by Hatch et al. (1990). Cultivated since time of Assyrians and probably of hybrid origin (Mabberley 1987). Exact place of origin unknown; European or Asiatic.

RAPISTRUM

An Old World genus of 2 species native of c Europe, the Mediterranean region, and w Asia. (Latin name of the wild rape)
REFERENCE: Lemke & Worthington 1991.

Rapistrum rugosum (L.) All., (rugose, wrinkled). Annual, resembling species of *Brassica*, glabrous or sparsely hispid; stems 0.3–0.8 m tall; lower leaves pinnatifid to coarsely dentate; stem leaves few, smaller, dentate; petals 5–10 mm long, yellow; pedicels 2–5 mm long; fruits 3–9 mm long including beak, of 2 segments; upper fruit segment nearly globose, ca. 3 mm in diam., one-seeded, larger than the lower segment, abruptly contracted to the beak; lower fruit segment one-seeded or seedless, cylindrical, appearing like a thickened extension of the pedicel; beak of fruit 1–3 mm long. Fields and roadsides; Grayson and Kaufman cos. s to Bell and Milam cos.; Blackland Prairie; also se and e TX and Edwards Plateau; apparently increasing. Apr–early May. Native of the Mediterranean region.

RORIPPA YELLOWCRESS

Annuals or perennials; our species glabrous or nearly so; leaves simple or compound, toothed or pinnatifid; petals white or yellow, small or absent; seeds not winged.

A nearly cosmopolitan genus of 80 species. (From the old Saxon word, *rorippen*, for plants of this genus)
REFERENCES: Green 1962; Stuckey 1972.

1. Petals white; leaves pinnately compound ____ **R. nasturtium-aquaticum**
1. Petals yellow OR absent; leaves simple (but may be deeply pinnatifid).
    2. Fruits (siliques) sessile or pedicels to 1.5 mm long; petals absent or rarely 1 present ____ **R. sessiliflora**
    2. Fruits pedicellate with pedicels 2 mm long or longer; petals present.
        3. Pedicels about equal to fruits in length; fruits usually 3–9 mm long; seeds 0.5–0.9 mm long; vesicular (= sac-like) hairs absent, the plants glabrous or with pointed hairs ____ **R. palustris**
        3. Pedicels much shorter than fruits; fruits usually 10–15 mm long; seeds 0.4–0.5 mm long; vesicular hairs present on lower part of stem or foliage ____ **R. teres**

Rorippa nasturtium-aquaticum (L.) Hayek, (aquatic nasturtium, from Latin, *nasi tortium*, a twisted nose, from the plant's pungent qualities), WATERCRESS. Aquatic to semiaquatic perennial; stems floating, creeping, or ascending, rooting at the nodes; leaflets oblong- to rhombic-orbicular, entire or shallowly and bluntly toothed; petals 3–4 mm long. Shallow, fairly clear water, in streams or springs, sometimes stranded in wet mud; Bell, Burnet, Dallas, Grayson, and Tarrant cos., also Johnson Co. (R. O'Kennon, pers. obs.); widely scattered in TX. Apr–Jun. Native of Europe. [*Nasturtium officinale* W.T. Aiton, *Sisymbrium nasturtium-aquaticum* L.]

Myagrum perfoliatum [GLE, LAM]

Matthiola longipetala [HEA]

Raphanus sativus [LAM]

Rapistrum rugosum [ALL]

Rorippa nasturtium-aquaticum [REE]

Rorippa palustris subsp. fernaldiana [GLE]

Rorippa palustris (L.) Besser subsp. **fernaldiana** (Butters & Abbe) Jonsell, (sp.: marsh-loving; subsp.: for Merritt Lyndon Fernald, 1893–1950, author of *Gray's Manual of Botany*, 8th ed.), BOG YELLOWCRESS, BOG MARSHCRESS. Annual to 80 cm tall, usually erect, branched above, essentially glabrous, without vesicular hairs; lower leaves lobed; upper leaves toothed; petals usually 1–2 mm long; fruits short-cylindrical to ellipsoid, usually 3–9 mm long, the replum becoming twisted when dry. Margins of lakes and streams; Cooke and Dallas cos.; mainly se and e TX w to Grand Prairie and Edwards Plateau. Apr–Sep. [*R. palustris* (L.) Besser var. *fernaldiana* (Butters & Abbe) Stuckey, *R. islandica* of TX auth., not (Oeder) Borbás]

Rorippa sessiliflora (Nutt.) Hitchc., (stalk-less-flowered), STALK-LESS YELLOWCRESS. Glabrous summer, fall, or winter annual; stems to 50 cm tall, erect, single to much-branched; leaves entire, repand or somewhat toothed; sepals greenish yellow; petals absent or rarely 1 present; fruits thick, can be over 2 mm wide, 6–10 mm long, sessile or on pedicels 0.5–1.5 mm long. Shores of streams and lakes; Bell, Cooke, Dallas, and Hopkins cos., also Brown (HPC) and Tarrant (R. O'Kennon, pers. obs.) cos.; se and e TX w to Grand Prairie and Edwards Plateau. Apr–May, less commonly Aug–Sep.

Rorippa teres (Michx.) Stuckey, (terete, circular in cross-section), TANSY-LEAF YELLOWCRESS. Annual with erect to decumbent, usually branched stems to 40 cm long; vesicular hairs obovate to clavate on lower part of stems and upper leaf surfaces of lower leaves; leaves deeply pinnatifid; petals 1(–2) mm long; pedicels 2–5 mm long; fruits (5–)10–15(–18) mm long. Wet thickets, stream banks, sandy soils; Dallas Co. (Cedar Hill-R. O'Kennon, pers. obs.), also cited for vegetational area 4 (Fig. 2) by Hatch et al. (1990); mainly se and e TX w to Edwards Plateau and Trans-Pecos. Mar–Apr, sporadically later.

SIBARA

A genus of 10 species native to e and s North America. (An anagram of *Arabis*)
REFERENCE: Rollins 1947.

Sibara virginica (L.) Rollins, (of Virginia), VIRGINIA SIBARA. Annual to 35 cm tall, similar in appearance to *Cardamine parviflora*; stems often numerous, sparsely pubescent toward base; basal rosette of leaves present; basal leaves pinnately dissected with 5–14 pairs of lateral divisions; stem leaves similar but reduced; racemes 10–20 cm long; petals white or with tinge of pink, 2.5–3 mm long; fruits 15–30 mm long, ca. 1.2–2 mm wide, flattened. Wet thickets, ditches, disturbed areas; se and e TX w to Edwards Plateau and Rolling Plains. Feb–Apr.

SINAPIS MUSTARD

Our species introduced annuals; leaves not clasping, at least the lower pinnatifid; petals yellow, rather large and showy; beak of silique large, usually containing a seed.

A genus of 7 species of Europe and the Mediterranean region; sometimes included in *Brassica*. As in the related genus *Brassica* (e.g., *Brassica nigra*), the seeds of *Sinapis alba* are a source of edible mustard. ☠ Some species can be poisonous to livestock due to the presence of mustard oil glycosides (Stephens 1980). (Latin: *sinapis*, mustard, from flavor of seeds)
REFERENCES: Sun 1946; Lemke & Worthington 1991.

1. Leaves all pinnatifid; fruits bristly, 3–4 mm in diam., the beak 1–2 times longer than the fruit body; seeds 4–8 per fruit ____ **S. alba**
1. Middle and upper leaves merely toothed; fruits glabrous or sparsely bristly, 1.5–2.5 mm in diam., the beak shorter than or equal to fruit body in length; seeds 7–13 per fruit ____ **S. arvensis**

Sinapis alba L., (white), WHITE MUSTARD. Rough hairy annual to 70 cm tall; siliques 15–35 mm long, the beak 10–20(–rarely more) mm long, flat, often curved. Weedy areas; included based on citation of vegetational areas 4 and 5 (Fig. 2) by Hatch et al. (1990); scattered in TX. Probably native of Mediterranean region. [*Brassica hirta* Moench] Cultivated for its mustard-producing seeds and for greens.

Sinapis arvensis L., (pertaining to cultivated fields), FIELD MUSTARD, CHARLOCK. Annual to 80 cm tall; stems spreading pubescent; siliques 20–45 mm long, the beak quadrangular-flattened, 6–15 mm long. Fields, roadsides, and railroads, various soils; se and e TX w to West Cross Timbers and Edwards Plateau. Feb–May. Native of Europe. [*Brassica kaber* (DC.) L.C. Wheeler] Poisoning and death have been reported in cattle, chickens, horses, and pigs from eating large quantities of the plant or seeds (Mulligan & Munro 1990)

SISYMBRIUM

Our species small to robust annuals, glabrous or sparsely pubescent with simple hairs; leaves unevenly pinnatifid, often with large apical section; petals light yellow.

A genus of 90 species of Eurasia, the Mediterranean region, s Africa, North America, and the Andes. (Latinized from an ancient Greek name for some plant of this family)
REFERENCE: Payson 1922a.

1. Terminal lobe of upper leaves linear-oblong or linear; petals 5–10 mm long; pedicels spreading-ascending; mature fruits 3.5–10 cm long ____ **S. altissimum**
1. Terminal lobe of upper leaves triangular; petals 2.5–4 mm long; pedicels spreading-ascending OR closely appressed; mature fruits 1–5 cm long.
   2. Pedicels spreading-ascending in age, 4–10 mm long; stems glabrous; mature fruits 3–5 cm long ____ **S. irio**
   2. Pedicels closely appressed in age, 1–2 mm long; stems sparingly hispid-pubescent at least below; mature fruits 1–2 cm long ____ **S. officinale**

Sisymbrium altissimum L., (tallest), TUMBLE-MUSTARD. Plant to 1.5 m tall; lower leaves and stems sparsely pilose. Railroads and roadsides; Dallas Co.; scattered in TX. Native of Europe and w Asia.

Sisymbrium irio L., (Latin name for a siliquose plant), ROCKET-MUSTARD, LONDON-ROCKET. Plant to 0.6 m tall. Fields, roadsides, and waste ground, various soils; Dallas and Brown cos.; widespread in TX. Feb–May, Nov on further s. Native of Europe.

Sisymbrium officinale (L.) Scop., (medicinal), HEDGE-MUSTARD, TANSY-MUSTARD. Plant to 1.2 m tall. Fields, thickets, and waste ground, various soils; Cooke, Denton, and Grayson cos., also McLennan and Navarro cos. (Mahler 1988); widespread in TX. Apr–Jun. Native of Europe. [*S. officinale* var. *leiocarpum* DC.]

STREPTANTHUS TWISTFLOWER

A genus of 35 species of w and s North America. Calluses on the leaf margins of some species mimic pierid butterfly eggs, reducing egg deposition and thus herbivory by larvae (Buck et al. 1993). Some species are known to hyperaccumulate nickel (Kruckeberg & Reeves 1995). (Greek: *streptos*, twisted, and *anthos*, flower, from the wavy-margined petals)

Streptanthus hyacinthoides Hook., (resembling *Hyacinthus*—hyacinth), SMOOTH TWISTFLOWER. Erect, glabrous annual to 1 m tall; basal leaves absent; leaves linear-lanceolate; sepals deep purple; petals 12–20 mm long, lavender with dark veins to dark purple; mature fruits 6–10 cm long, to 2 mm wide. Open woods and roadsides, loose sand; w part of Blackland Prairie and West Cross Timbers (Comanche and Young cos.) (Mahler 1988); e TX w to nc TX, also Edwards Plateau. May–Jun, sporadically to Aug.

THLASPI PENNYCRESS, FRENCHWEED

A genus of 60 species of n temperate areas and n hemisphere mountains; some European and Asian species accumulate metals (nickel and zinc). (Greek: *thlaein*, to crush, from the flattened silicle)
REFERENCE: Payson 1926.

Thlaspi arvense L., (pertaining to cultivated fields), FIELD PENNYCRESS, MITHRIDATE-MUSTARD, FRENCHWEED, FANWEED, PENNYCRESS. Glabrous annual to 50 cm tall; leaves (except lowest) sessile and clasping, oblong to triangular-lanceolate, entire or toothed; petals white, 2–5 mm long; fruits conspicuous, flat, orbicular, 8–15 mm wide, notched at apex. Railroads and roadsides, various soils; Denton, Grayson, Henderson, and Tarrant cos.; nc TX s to Edwards Plateau. Mar–early May. Native of Europe. Formerly used medicinally (Mabberley 1997); reported to have caused poisoning and death of cattle when fed in large amounts in hay; photosensitization has also been reported (Mulligan & Munro 1990). ☠

BUDDLEJACEAE BUDDLEJA FAMILY

A small (120 species in 8 genera), tropical and warm area, especially e Asian family of trees, shrubs, or rarely herbs with most species in the ornamentally important genus *Buddleja*. Family name from *Buddleja*, BUTTERFLY-BUSH, a genus of ca. 100 species of warm areas, especially e Asia; some are used medicinally or as fish poisons. *Buddleja davidii* Franch., a Chinese species widely cultivated in Europe and milder climate regions of the United States, is extremely attractive to butterflies. (Named for Reverend Adam Buddle, 1660–1715, amateur botanist) (subclass Asteridae)
FAMILY RECOGNITION IN THE FIELD: the only nc TX species is a small bushy-branched herb with pposite linear leaves and small, 4-merous, white flowers.
REFERENCES: Moore 1947; Rogers 1986; Jensen 1992.

POLYPREMUM POLLYPRIM

A monotypic genus of warm areas of the Americas; previously treated in the Loganiaceae. (Name altered from the Greek: *polypremnos*, many-stemmed)

Polypremum procumbens L., (procumbent, prostrate), JUNIPER-LEAF, POLLYPRIM. Bushy-branched, reclining to erect herb to 30 cm high; perennial but flowering as an annual; leaves opposite, narrowly linear, to 3 cm long, rarely > 2 mm wide; flowers solitary or in a terminal leafy cyme, 4-merous; corollas white, almost rotate, 1.8–2.3 mm long, divided 1/3 their length; capsules ovoid, notched at apex, ca. 1.5–2.5 mm long. Sandy woods, old fields, and disturbed ground; se and e TX w to West Cross Timbers and Edwards Plateau, also Trans-Pecos. Jun–Oct.

CABOMBACEAE WATER-SHIELD FAMILY

Aquatic, rhizomatous, perennial herbs rooting in mud; vegetative tissues with conspicuous air chambers; flowers axillary, solitary, on thick peduncles; sepals and petals each usually 3(–4); stamens 3–36; pistils (1–)2–18, simple; ovaries superior; fruits indehiscent.

A very small (6 species in 2 genera—Wiersema 1997c) pantropical and warm temperate family of aquatics. It has often been lumped with the Nymphaeaceae, but differs in having free carpels. *Cabomba* has flower parts in 3s like those of monocotyledons. Some taxonomists refer to the Cabombaceae as "paleoherbs" (a group including Aristolochiales, Piperales, and Nymphaeales) and believe them to be an early branch off the evolutionary line leading to monocots; this view is supported by characters such as the 3-merous flowers and molecular data which place the paleoherbs as the immediate sister group of the monocots (Chase et al.

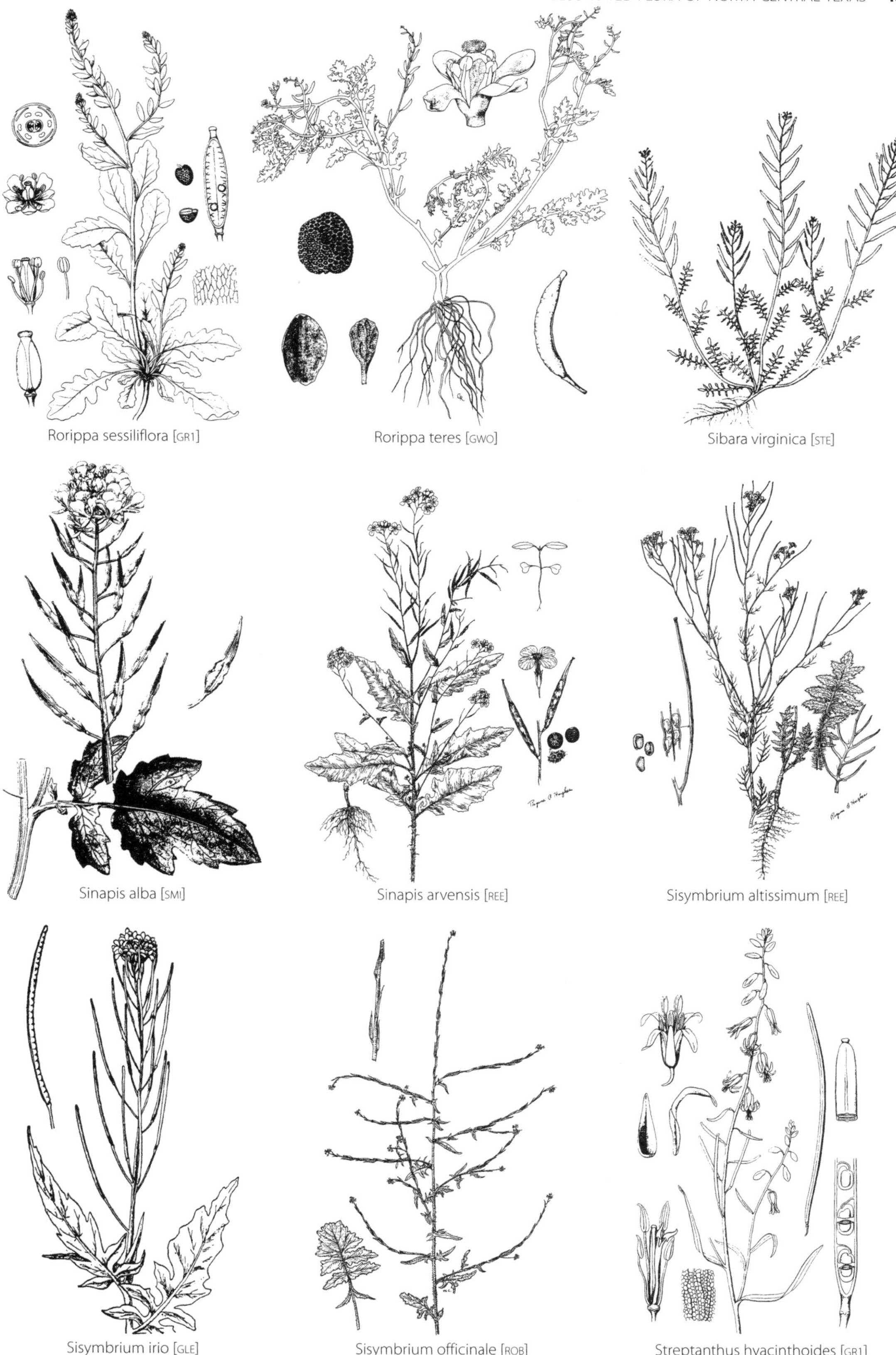

Rorippa sessiliflora [GR1]

Rorippa teres [GWO]

Sibara virginica [STE]

Sinapis alba [SMI]

Sinapis arvensis [REE]

Sisymbrium altissimum [REE]

Sisymbrium irio [GLE]

Sisymbrium officinale [ROB]

Streptanthus hyacinthoides [GR1]

1993) (see Fig. 41 in Appendix 6). In the past *Brasenia* and *Cabomba* were often included in the Nymphaeaceae (e.g., Correll & Johnston 1970). (subclass Magnoliidae)
FAMILY RECOGNITION IN THE FIELD: aquatics with *peltate* floating leaves and either with conspicuous mucilage on underwater surfaces or with much dissected submerged leaves; perianth usually 3-merous; carpels free.
REFERENCES: Wood 1959; Williamson & Schneider 1993a; Wiersema 1997c.

1. Plants with only undivided, alternate, peltate, floating leaves, the blades 13.5–11(–13.5) cm long; lower surfaces of leaf blades and petioles coated with a heavy, very conspicuous layer of jelly-like mucilage; stamens 18–38(–51); perianth purplish, 10–20 mm long ____ **Brasenia**
1. Plants with much dissected, opposite (rarely whorled), submerged leaves and also usually small (blades ca. 1–3 cm long), alternate, peltate, floating leaves; lower surfaces of leaf blades and petioles without conspicuous mucilage layer, the tissues only barely mucilaginous; stamens 3–6; perianth usually whitish (rarely purplish or yellowish), 4–12 mm long ____ **Cabomba**

BRASENIA PURPLE WEN-DOCK, WATER-SHIELD

A monotypic genus of America, Africa, India, temperate e Asia, and Australia. Submerged parts are conspicuously covered with mucilaginous jelly. (Named for Christopher Brasen, 1734–1774, Moravian missionary and plant collector in Greenland and Labrador—Wiersema 1997c)
REFERENCE: Osborn & Schneider 1988.

Brasenia schreberi J.F. Gmel., (for Johann Christian Daniel von Schreber, 1739–1810, German botanist), PURPLE WEN-DOCK, WATER-SHIELD, SCHREBER'S WATERSHED. Rooted aquatic; jelly-like mucilage layer thick and obvious; leaves long-petiolate; leaf blades elliptic to broadly oval, 3.5–11(–13.5) cm long, 2–6.5 cm wide, centrally peltate, entire or slightly crenate; flowers on peduncles to 15 cm long, emergent when open; perianth 12–20 mm long, the petals slightly longer and narrower than the sepals; petals without proximal auricles; pistils 2–18; fruits 5–10 mm long, with 1–2 seeds. Ponds, lakes, slow streams; Lamar Co. in Red River drainage (Carr 1994); mainly e TX. May–Jul. [*B. peltata* Pursh] Cultivated as an ornamental and the young leaves are eaten in Japan (Mabberley 1987). Reported to be wind-pollinated (Wiersema 1997c). Williamson and Schneider (1993b) gave an uncited reference indicating that *Brasenia* has phytotoxic properties which contribute to its dominance in some situations; it is possibly useful as an allelopathic way of controlling aquatic weeds.

CABOMBA FANWORT

A genus of 5 species (Wiersema 1997c) of mostly tropical to warm areas of the Americas. (Probably an aboriginal name, possibly from Guiana)
REFERENCES: Fassett 1953; Ørgaard 1991.

Cabomba caroliniana A. Gray, (of Carolina), CAROLINA FANWORT, FANWORT. Rooted aquatic; stems to 2 m or more long; submersed leaves opposite or rarely whorled, the blades palmately 2–3 times dissected into linear-filiform segments; floating leaves alternate, few, the blades peltate, entire, linear-elliptic, 1–3 cm long; flowers solitary on axillary peduncles; perianth 4–12 mm long; sepals 3; petals 3; sepals and petals similarly colored; petals proximally with yellow, nectar-bearing auricles; pistils (1–)2–4; fruits 1–3-seeded, 4–7 mm long. Ponds, lakes, quiet streams; included based on citation of vegetational area 4 (Fig. 2) by Hatch et al. (1990); mainly se and e TX, also Edwards Plateau. Apr–Jul. *Cabomba* is sometimes used in aquaria (Mabberley 1987).

CACTACEAE CACTUS FAMILY

Fleshy or soft-woody, succulent, green-stemmed, usually armed shrubs to small trees; leaves in ours absent, or present as fleshy points on new growth; specialized, axillary, cushion-like bud

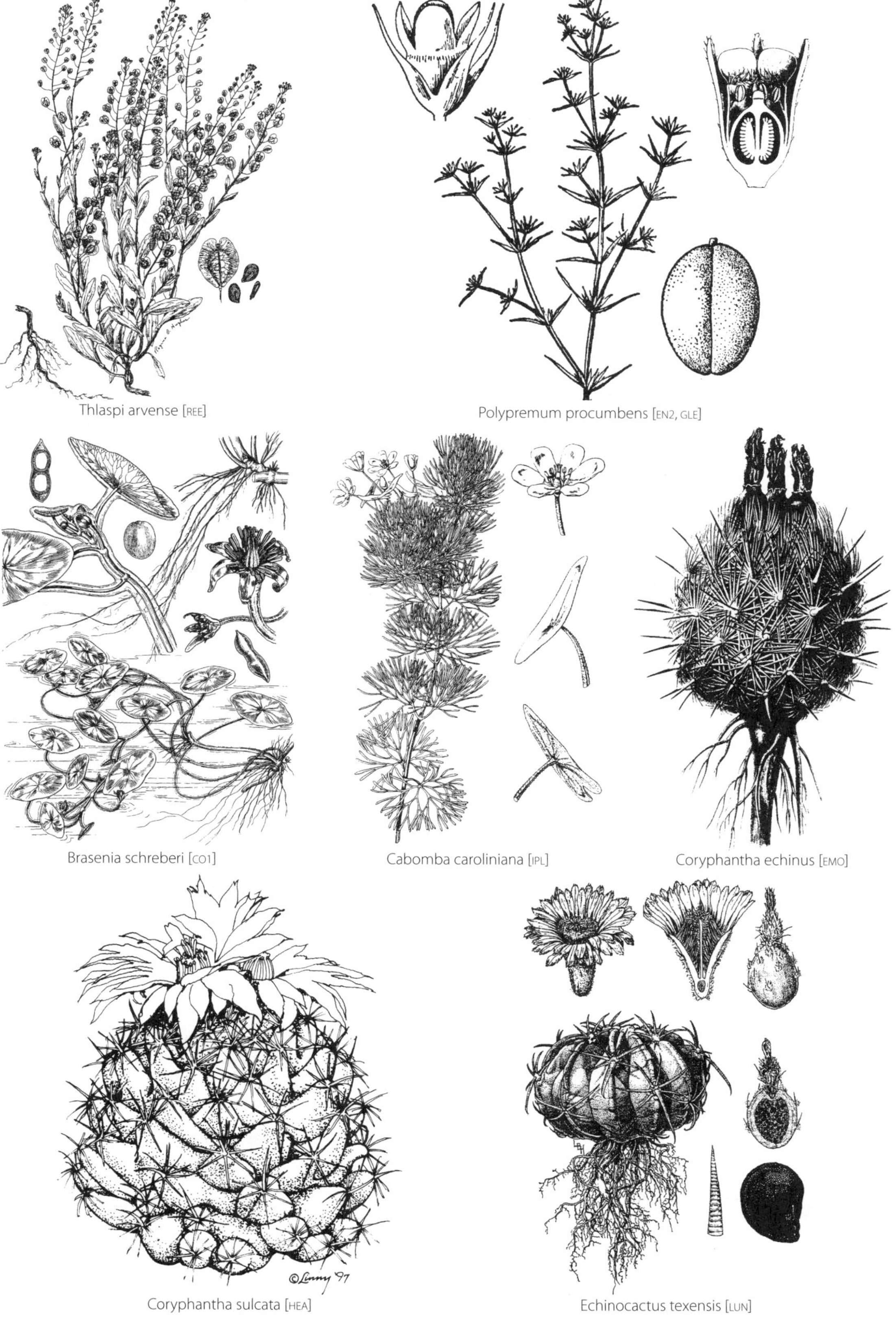

Thlaspi arvense [REE]

Polypremum procumbens [EN2, GLE]

Brasenia schreberi [CO1]

Cabomba caroliniana [IPL]

Coryphantha echinus [EMO]

Coryphantha sulcata [HEA]

Echinocactus texensis [LUN]

areas (= areoles) usually bearing spines and sometimes barbed hairs or bristles (= glochids); flowers solitary or crowded, sessile, closed at night, often showy; sepal-like perianth parts 5 or more, grading into the many, thin-textured petal-like structures, all joined at base to form a cup or tube (= hypanthium) on summit of the thick, pedicel-like, inferior ovary; stamens many; fruit a dry or fleshy berry.

A medium-large (1,400 species in 97 genera) family of stem-succulent xerophytes native almost exclusively to the New World, but now widely naturalized; some have become problematic invaders (e.g., *Opuntia* in Australia); the family includes a number of epiphytes. Xerophytic adaptations include a thick cuticle, large volume to surface ratio, widespread shallow root system (to take in any available rain), crassulacean acid metabolism (CAM photosynthesis—allows night absorption and storage of CO_2 thereby reducing water loss through transpiration during the day—Crosswhite 1984), and spines (condense dew and protect from herbivores). Numerous species are utilized as ornamentals and some are endangered due to overcollecting (e.g., "cactus rustling"); cacti are protected by the Convention of International Trade in Endangered Species (CITES) (Barthlott & Hunt 1993). Many species have alkaloids (e.g., *Lophophora williamsii* (Lem. ex Salm-Dyck) J.M. Coult.—PEYOTE, which contains the hallucinogen, mescaline, and is used in religious ceremonies by Native Americans).

Because of their extreme morphological specializations, the taxonomic placement of the Cactaceae has been problematic; however, biochemical and molecular markers indicate a relationship with the Caryophyllidae (e.g., Cactaceae are characterized by betalain pigments like most other members of this subclass—Cronquist and Thorne 1994). A study by Downie and Palmer (1994) suggested the family may be derived from Portulacaceae. On the basis of molecular data, Hershkovitz and Zimmer (1997) also indicated that the family is derived from within the Portulacaceae and that the molecular divergence "between pereskioid cacti and the genus *Talinum* (Portulacaceae) is less than that between many Portulacaceae genera." The ancestral condition within the family is clearly observable in the small subfamily Pereskioideae composed of leafy shrubs and trees with only slightly succulent stems. Family name conserved from *Cactus*, a generic name rejected in favor of *Mammillaria* (Farr et al. 1979). (Greek: *cactus*, a spiny or prickly plant) (subclass Caryophyllidae)
FAMILY RECOGNITION IN THE FIELD: leafless stem *succulents* with watery sap, areoles (= pad-like axillary buds unique to the family) typically bearing *spines*, and flowers showy, with numerous perianth parts and numerous stamens.
REFERENCES: Britton & Rose 1919–1923, 1937, 1963; Schulz & Runyon 1930; Marshall & Bock 1941; Benson 1969, 1982; Weniger 1970, 1984; Grant & Grant 1979a; Barthlott & Hunt 1993; Behkne & Mabry 1994; Hershkovitz & Zimmer 1997.

1. Stems jointed, cylindrical OR flattened; glochids (= small barbed hairs) present at areoles ____ **Opuntia**
1. Stems not jointed, variously shaped but never flattened; glochids absent.
  2. Stems vertically ribbed.
    3. Central spine (of an areole) strongly hooked at end, 12–38 mm long; flowers yellow with red basally ____ **Thelocactus**
    3. Central spine absent OR present, if present curved downward, not hooked; flowers variously red, red and yellow, pink to light purple, OR reddish basally and then orange and pink to white apically.
      4. Spines, if present, not cross-ribbed; ovaries with slender spines; filaments white, green, yellow, or pink; stems usually multiple, cylindric, 10 cm or less in diam. ____ **Echinocereus**
      4. Spines stout, cross-ribbed (= annulate); ovaries spineless; filaments red; stems usually single, hemispheric, to 30 cm in diam. ____ **Echinocactus**
  2. Stems tuberculate (= with nipple-like protrusions), not ribbed.
    5. Flowers variously colored, located on new growth near the stem apex at the base of the

upper side of a tubercle; above ground portion of plant not flat, the small stems hemispheroidal to globose, ovoid, obovoid, or cylindroid; stems sometimes solitary but often forming clumps; tubercles usually with a longitudinal groove on the upper side.

6. Fruits green at maturity; seeds brown; central spines present and differentiated from radials; flowers yellow to yellow with pink basally to dark purplish pink.
   7. Flowers dark purplish pink; sepal-like structures fringed with long hairs (= fimbriate); seeds reticulate; radial spines 12–40; central spines 3–10 per areole, the lower central spine shorter than the upward-directed other central spines or absent ______ **Escobaria**
   7. Flowers yellow or yellow with pink basally; sepal-like structrures not fimbriate; seeds smooth and shining, sometimes minutely punctate; radial spines 6–26; central spines 1–4 per areole, the lower central spine longer than the upward-directed other central spines ______ **Coryphantha**
6. Fruits red at maturity; seeds black; central spines usually none (sometimes 1(–2), if present not differentiated from radials); flowers greenish to yellowish (rarely bronzish) ______ **Escobaria**

5. Flowers pink, white, cream, or some mixture, located on old growth of preceding seasons below the stem apex, between the tubercles and not obviously connected with them; above ground portion of plant ± flat or slightly convex; stems usually solitary, sometimes clustered; tubercles without a longitudinal groove on the upper side ______ **Mammillaria**

CORYPHANTHA

Stems solitary or branching basally and forming mounds; ribs none; with spirally arranged tubercles; areoles apical on the tubercles; flowers and fruits produced at base of upper side of tubercles, connected to the areole by a groove on the tubercle; flowers ca. 50–62 mm in diam.; fruits fleshy, green at maturity, smooth, ellipsoid; seeds 1–2 mm in greatest diam., brown, smooth and shiny, sometimes minutely punctate.

A genus of 45 species of sw North America; some are cultivated as ornamentals. While some authors (e.g., Benson 1982; Jones et al. 1997) include the morphologically similar genus *Escobaria* in *Coryphantha*, we are following Hunt (1978), Kartesz (1994), and J. Kartesz (pers. comm. 1997) in recognizing *Escobaria* as a distinct genus (see *Escobaria* for additional information). Castetter et al. (1975) gave characters separating the genera. (Greek: *corypha*, top, and *anthus*, flower, from the position of the flowers at the apex of the stem)
REFERENCES: Craig 1945; Castetter et al. 1975.

1. Petal-like structures yellow; radial spines 16–26; seeds minutely punctate ______ **C. echinus**
1. Petal-like structures yellow, pink basally; radial spines 6–8; seeds not punctate ______ **C. sulcata**

Coryphantha echinus (Engelm.) Britton & Rose, (prickly), RHINOCEROS CACTUS. Spines brown, usually down-curved; central spines 3 or 4, ca. 12–17 mm long; radial spines 16–26, to 28 mm long on upper side of areole, 12 mm long on lower; flowers yellow; fruits green, ca. 2.5 cm long. Limestone soils; Cooke Co., (Benson 1982); mainly Trans-Pecos. Jun–Jul. [*C. cornifera* (DC.) Lem. var. *echinus* (Engelm.) L.D. Benson, *Mammillaria echinus* Engelm.] We are following Jones et al. (1997) and J. Kartesz (pers. comm. 1997) for nomenclature of this species.

Coryphantha sulcata (Engelm.) Britton & Rose, (sulcate, furrowed), PINEAPPLE CACTUS, FINGER CACTUS. Spines at first yellow and pink or pink, later overlain by gray or white; central spines 1–3 per areole, spreading, curving, the longer 9–12 mm long; radial spines 6–8 per areole; petal-like structures yellow apically, pink basally; fruits green, to 3 cm long, fleshy at maturity; seeds brown, smooth and shining. Prairies, sandy or gravelly soils; Denton and Tarrant cos., also Brown, Comanche (HPC), Mills, and Lampasas (Benson 1982) cos.; nc TX s to sc TX. May–Jun. [*Mammilaria sulcata* Engelm.] /85

ECHINOCACTUS BARREL CACTUS

A genus of 6 species of ribbed cacti often with long spines native to sw North America; some are cultivated as ornamentals. While now treated as a small genus, *Echinocactus* formerly included most cacti with ribbed stems (Barthlott & Hunt 1993). (Greek: *echinos*, hedgehog, and *cactus*, name for another spiny plant)
REFERENCE: Maddox 1986.

Echinocactus texensis Hopffer, (of Texas), HORSECRIPPLER, DEVIL'S-HEAD CACTUS, DEVIL'S-PINCUSHION, MANCA CABALLO. Stems solitary, green, hemispheroidal, 12.5–20 cm long, up to 30 cm in diam.; ribs 13–17; spines stout, cross-ribbed; central spine 1 per areole, curved downward; radial spines 5–7 per areole; petal-like structures basally reddish, then orange and pink to white apically, with midribs purplish to violet; filaments red; anthers yellow; fruits red, fleshy becoming dry, splitting irregularly with deciduous scales. Sandy or limestone soils; Coryell, Tarrant, Williamson, and Young cos. (Benson 1982); nc TX s and w to s TX and Trans-Pecos. Apr–May [*Homalocephala texensis* (Hopffer) Britton & Rose] Grazing animals can be badly injured by the strong spines and older plants are reported capable of puncturing pickup tires; as a result this species has been eliminated from many areas (Wills & Irwin 1961; Kirkpatrick 1992). ▣/88

ECHINOCEREUS HEDGEHOG CACTUS

Stems branching near ground level, cylindroid, with 5–18 ribs; spines smooth (not cross-ribbed); fruits on old growth of preceding year above areoles; fruits fleshy at maturity, bearing slender spines.

A genus of 47 species of sw North America; all are cultivated. (Greek: *echinos*, hedgehog, and genus *Cereus* from Latin: *cereus*, a wax taper or candle)
REFERENCES: Taylor 1985; Miller 1988; Ferguson 1989.

1. Flowers red or red and yellow (no blue mixture); stems usually 6.2–10 cm in diam.; ribs 5–7(–12); radial spines 4–6 ____ **E. coccineus**
1. Flowers pink to light purple; stems 2.5–5 cm in diam.; ribs 12–18; radial spines 22–32 ____ **E. reichenbachii**

Echinocereus coccineus Engelm. var. **paucispinus** (Engelm.) D.J. Ferguson, (sp.: scarlet; var.: few spined), CLARET-CUP CACTUS, RED-FLOWER HEDGEHOG. Stems few, 15–20 cm long; central spines absent; radial spines nearly straight, gray, 3–3.8 cm long; flowers 25–38 mm long; fruits red. Rocky limestone soils; Lampasas (Benson 1982), Burnet, and Williamson (Ferguson 1989) cos.; sw part of nc TX w and sw to Trans-Pecos. Mid-Mar–Jul. [*E. triglochidiatus* Engelm. var. *paucispinus* (Engelm.) W.T. Marsh.] ▣/88

Echinocereus reichenbachii (Terscheck ex Walp.) F. Haage, (for Heinrich Gottlieb Ludwig Reichenbach, 1793–1979, German naturalist), LACE CACTUS, WHITE LACE CACTUS, BROWN LACE CACTUS, HEDGEHOG CACTUS. Stems solitary or branching, 7.5–15(–22.5) cm long; areoles narrow, 3 mm long; spines obscuring the stem, central spines usually absent (rarely 1–3), if present, much smaller than the radials; radial spines arched, 4.5–6 mm long, straw-colored to pale gray, distally pink to reddish; flowers 20–60 cm long; fruits green with conspicuous, deciduous, woolly hairs from the areoles. Gravelly, rocky, or sandy soils, especially limestone; Hood, Johnson, Palo Pinto, and Williamson cos., also Brown (HPC), Burnet, Jack, McLennan, Parker, Young (Benson 1982), and Tarrant (R. O'Kennon, pers. obs.) cos.; nc TX s to Edwards Plateau and w to Panhandle. May–Jun. ▣/88

ESCOBARIA NIPPLE CACTUS

Stems solitary or branching basally and forming mounds; ribs none; with spirally arranged tubercles; areoles apical on the tubercles; flowers and fruits produced at base of upper side of tu-

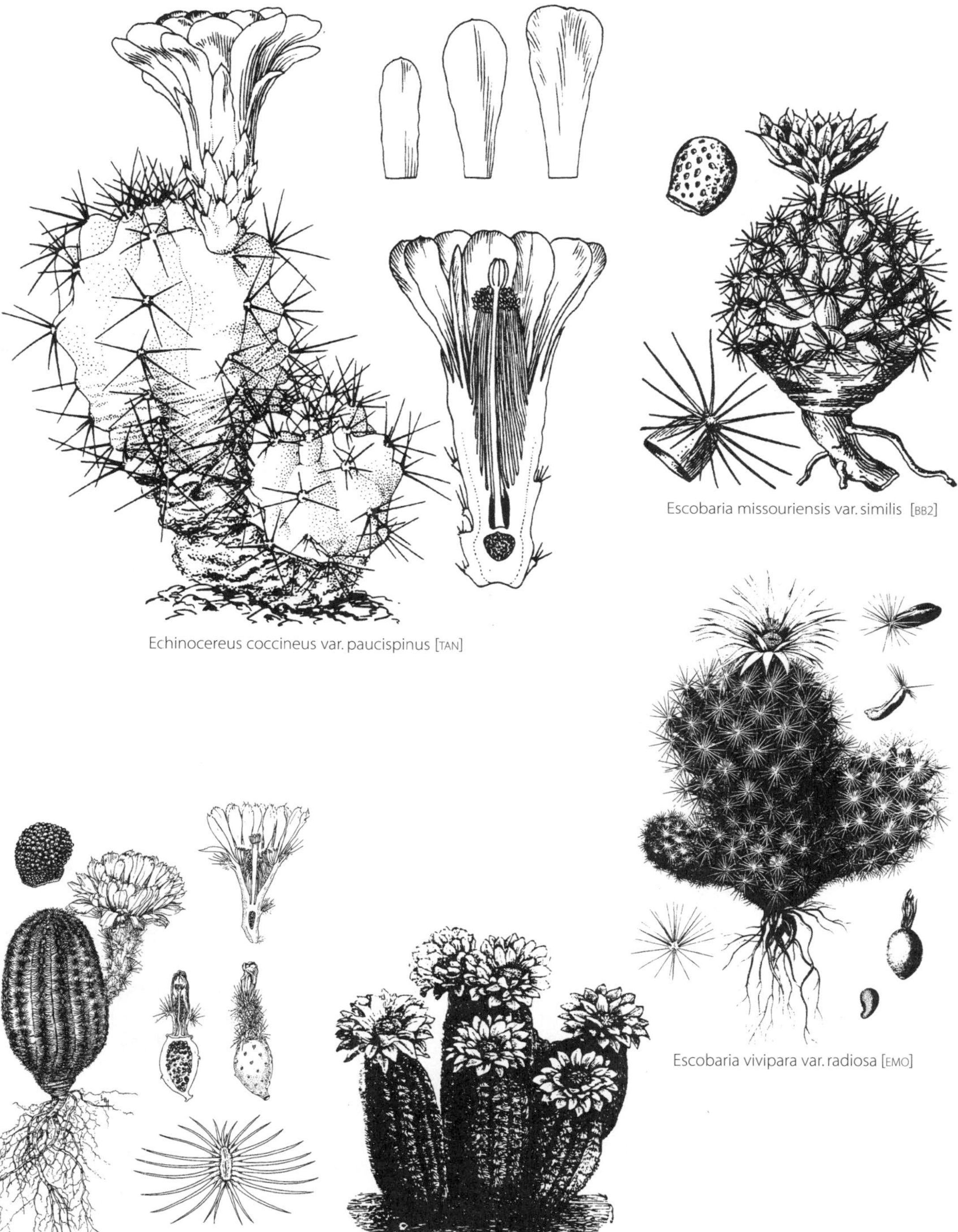

Escobaria missouriensis var. similis [BB2]

Echinocereus coccineus var. paucispinus [TAN]

Escobaria vivipara var. radiosa [EMO]

Echinocereus reichenbachii [BEL, GAR]

bercles, connected to the areole by a groove on the tubercle; flowers 25-56 mm in diam.; fruits fleshy, red or green at maturity, smooth, globose to ellipsoid; seeds 1–2 mm in greatest diam., brown or black, punctate or reticulate.

A genus of 15 species of sw North America to s Canada and Cuba. This genus is morphologically similar to *Coryphantha* and has often been lumped with it (e.g., Benson 1982; Jones et al. 1997). Hunt (1978), however, believed "... that *Escobaria* (with *Neobesseya*) represent an evolutionary lineage independent of *Coryphantha*, that is to say that they are distinct phyletic groups at an approximately analogous stage." We are with some hesitation following Hunt (1978), Kartesz (1994), and J. Kartesz (pers. comm. 1997) in recognizing *Escobaria*. The following dichotomy separating the two genera by Barthlott & Hunt (1993) raises questions about their distinctiveness:

1. Seedcoat cells tabular; outer tepals entire, or areoles with glands ______ *Coryphantha*
1. Seedcoat cells tabular-concave or par-concave (pitted); outer tepals fringed; areolar glands absent ______ *Escobaria*

Because of the difficulty of separating the two using the above dichotomy, other characters are used in our key to genera with *Escobaria* coming out in two places. Castetter et al. (1975) discussed characters separating the genera. (Named for the Escobar brothers, Romulo and Numa, Mexicans of the early 20th century)
REFERENCES: Britton & Rose 1919–1923, 1937, 1963; Craig 1945; Castetter et al. 1975; Hunt 1978; Taylor 1978; Fischer 1980.

1. Flowers greenish to yellowish (rarely bronzish); fruits red at maturity; seeds black; central spines usually none (sometimes 1(–2), if present not differentiated from radials); ______ **E. missouriensis**
1. Flowers dark purplish pink; fruits green at maturity; seeds brown; central spines 3–4 per areole, differentiated from radials ______ **E. vivipara**

Escobaria missouriensis (Sweet) D.R. Hunt, (of Missouri), PLAINS NIPPLE CACTUS. Spines yellowish becoming dark gray, pubescent; central spines usually none, if present, undifferentiated from radial spines; radial spines 12–15 per areole, 10–20 mm long; fruits red, 10–20 mm long; seeds black, punctate. Spring–?

1. Flowers ca. 4.4–5 cm in diam. and length; sepal-like structures not fimbriate; fruits ca. 1 cm in diam., globose; petal-like structures abruptly long acuminate ______ var. **robustior**
1. Flowers ca. 5–6.2 cm in diam. and length; outer sepal-like structures fimbriate; fruits 1.5–2 cm long, globular to ellipsoid; petal-like structures acuminate-attenuate ______ var. **similis**

var. **robustior** (Engelm.) D.R. Hunt, (robust, stout). Fruits globose, ca. 1 cm in diam. Grasslands; Denton, Williamson, and Young cos. (Benson 1982); also Hale Co. in Panhandle and Bexar Co. in c TX (Benson 1982). [*Coryphantha missouriensis* (Sweet) Britt. & Rose var. *robustior* (Engelm.) L.D. Benson] While some authors have put *Escobaria missouriensis* var. *robustior* into synonymy under *Coryphantha sulcata* (e.g., Kartesz 1994; Jones et al. 1997), the individuals in nc TX traditionally placed in var. *robustior* (e.g., Benson 1982; Mahler 1988) seem much closer to *Escobaria missouriensis* var. *similis* than to *C. sulcata*; J. Kartesz (pers. comm. 1997) agrees that var. *robustoir* should be treated as a var. of *Escobaria missouriensis*.

var. **similis** (Engelm.) N.P. Taylor, (similar). Fruits globose to ellipsoid, 1.5–2 cm long. Grasslands; Grayson Co., also Dallas, Kaufman, Lampasas, and Tarrant cos. (Benson 1982); nc TX to c TX (Bexar Co.) and w to Wichita Co. in Rolling Plains. [*Coryphantha missouriensis* (Sweet) Britton & Rose var. *caespitosa* (Engelm.) L.D. Benson, *Escobaria missouriensis* var. *caespitosa* (Engelm.) D.R. Hunt, *Mammilaria similis* Engelm.] /89

Escobaria vivipara (Nutt.) Buxb. var. **radiosa** (Engelm.) D.R. Hunt, (sp.: producing live young, freely producing asexual propagating parts; var.: with many rays), SPINY-STAR, PINCUSHION

CACTUS, BEEHIVE NIPPLE CACTUS. Spines relatively dense, obscuring stems to varying degrees; central spines usually white basally but tipped with pink, red, or black, 3–4 per areole, 1.5–2.2 cm long, straight; radial spines white to pink, 20–40 per areole, 1.2–1.9 cm long; fruits green, 1.9–2.5 cm long; seeds brown, reticulate. Grasslands and woodlands, limestone soils; Hood and Wise cos., also Brown, Hamilton (HPC), Montague (S. Lusk, pers. comm.), and Young (Benson 1982) cos.; mainly w 1/2 of TX. May–Jun. [*Coryphantha vivipara* (Nutt.) Britton & Rose. var. *radiosa* (Engelm.) Backeberg, *Mammilaria vivipara* (Nutt.) Haw. var. *radiosa* Engelm.] ▣/90

MAMMILLARIA FISHHOOK CACTUS, PINCUSHION CACTUS, NIPPLE CACTUS

A genus of ca. 150 species ranging from the sw U.S. to Colombia and Venezuela, but especially in Mexico; they are low, often tuft-forming and a number are cultivated as ornamentals. (Latin: *mamilla*, nipple, in reference to the shape of the tubercles)
REFERENCES: Craig 1945; Hunt 1971.

Mammillaria heyderi Muehlenpf., (for Herr Heyder, 1808–1884, a noted cultivator of cacti in Berlin), FLATTENED MAMMILLARIA, NIPPLE CACTUS, LITTLE-CHILIS, BIZNAGA DE CHILITOS. Plant low-growing spiny pincushion; sap milky; stems typically solitary to rarely clustered, turbinate to subglobose, but the above ground portion ± flat to shallowly convex, 7.5–10(–15) cm in diam.; tubercles prominent, ± conical from a subpyramidal base, 9–12 mm long; areoles apical on the tubercles; spines dense but not obscuring the stem; central spines 0–2(–4) per areole, the longer ones 3–9 mm long, straight; radial spines 6–22 per areole, the longer ones 6–16 mm long; flowers pink, white, cream, or some mixture, not connected by a groove to the areoles; fruits red, 12–40 mm long, 6–9 mm broad. Limestone soils in grasslands or dry areas, typically in partial shade; Brown Co. (Stanford 1971); also San Saba Co. just sw of nc TX; w margin of nc TX s to s TX and w to Trans-Pecos. May–Jun. [*M. gummifera* Engelm. var. *applanata* (Engelm.) L.D. Benson, *M. heyderi* var. *applanata* Engelm.] The fruits are reported to be edible (Kirkpatrick 1992). ▣/97

OPUNTIA PRICKLY-PEAR, CHOLLA

Mat- or clump-forming perennials to shrubs or small trees; stems jointed, either cylindrical or flattened; areoles usually with spines and small, brown, retrorsely barbed hairs (= glochids); young stems with a small, single, fleshy, vestigial, early deciduous leaf below each areole; fruits dry or fleshy, sometimes sweet and edible.

A genus of ca. 200 species native from Massachusetts and British Columbia s to Straits of Magellan and the Galápagos Islands; stems are either flattened or cylindrical (the latter known as CHOLLAS); some are tree-like; a number have edible fruits (PRICKLY-PEARS); the flattened stems of others are eaten as a vegetable; some serve as hosts for the cochineal insect (*Dactylopius coccus*), the source of a red dye; alkaloids, including mescaline, can be present. In some parts of TX during periods of drought, flame-throwers are used to burn the spines off *Opuntia* species so they can be used as an emergency food for livestock. Some species of *Opuntia* have been introduced into the Old World and have become problematic invaders in certain areas (e.g., Australia). A number of other *Opuntia* species including *O. stricta* (Haw.) Haw. var. *stricta* (SPINELESS PRICKLY-PEAR), a spineless ornamental 0.6–2 m tall, and *Opuntia imbricata* (TREE CHOLLA), a large (1–3 m tall) cylindrical species, are cultivated in nc TX and long persist. ☠ In addition to the obvious spines, glochids (= minute barbed hairs or bristles) are also present and can cause dermatitis; they can be extremely irritating; gloves should be worn when handling the plants (Spoerke & Smolinske 1990). The PRICKLEY-PEAR was designated the **state plant of Texas** by the 74th state legislature; all members of subgenus *Opuntia* (with flat stems) are considered the state plant, while those of subgenus *Cylindroopuntia* (with cylindrical stems) are not (Jones et al. 1997). (Greek name for a spiny plant (not Cactaceae) that grew near Greek town of Opus (Opuntis)

REFERENCES: Grant & Grant 1979b, 1979c, 1982a, 1982b; Grant & Hurd 1979; Grant et al. 1979; Leuenberger 1991, 1993.

1. Joints (= stem segments or sections) cylindrical (CHOLLAS) (Subgenus *Cylindroopuntia*).
 2. Plants to only 10 cm tall, mat-like or clump-like; spines with epidermis separating into a sheath only at apex (sheath short), 6–12(–15) per areole; flowers yellow; joints narrowed gradually toward their base, noticeably broader apically ____ **O. schottii**
 2. Plants to several meters tall, usually shrubby; spines with entire epidermis separating into a thin paper-like sheath (sheath as long as spine), 1–10 per areole; flower color various including yellow; joints ± equal in diam. throughout their length except where attached.
  3. Plants golden-tan from sheath color; spines 6–10 per areole, reddish brown; flowers usually green (centers yellowish and the apices and exteriors often tinged with red or brown); terminal joints 9–12 mm in diam. ____ **O. tunicata**
  3. Plants not golden-tan; spines 1–4 per areole, gray or grayish pink; flowers yellow, green, bronze, lavender, purple, or reddish; terminal joints 3–10 mm in diam.
   4. Flowers green, yellow, or bronze; terminal joints 3–4.5 mm in diam.; fruits not tuberculate ____ **O. leptocaulis**
   4. Flowers lavender, purple, or reddish; terminal joints 5–10 mm in diam.; fruits strongly tuberculate (= with conspicuous raised areas) ____ **O. kleiniae**
1. Joints flattened (PRICKLY-PEARS) (Subgenus *Opuntia*).
 5. Plants trees 3–5(–7) m tall with a main trunk 0.6–1.2 m tall (but plants can be very dense and ± as wide as tall); larger terminal joints (22.5–)30–60 cm long; fruits 5–10 cm long; escaped cultivar in sw part of nc TX ____ **O. ficus-indica**
 5. Plants prostrate to sprawling or suberect shrubs 7.5 cm to 3(–3.5) m tall, a main trunk if present very short; larger terminal joints 3.8–34(–40) cm long (except 45–90(–120) cm long in the locally rare cultivar, *O. engelmannii* var. *linguiformis*); fruits 2.1–7 cm long; native nc TX taxa (except var. *linguiformis*) widespread in the area.
  6. Spines usually needle-like, not flattened, elliptic to circular in cross-section; plants prostrate, most or all of the flattened joints on the ground; spines usually present only on the uppermost part of the joints or absent.
   7. Spines 0–1 per areole, 1.9–3 cm long, 0.5–0.7 mm in diam.; spines gray or brownish; all roots fibrous; joints green; seed margins smooth and regular, ca. 0.5 mm broad ____ **O. humifusa**
   7. Spines 1–6 per areole, usually 3.8–5.6 cm long, 0.25–0.5 mm in diam.; spines white or pale gray; main roots tuberous; joints glaucous; seed margins irregular, ca. 1 mm broad ____ **O. macrorhiza**
  6. Spines, at least some larger ones, flattened basally, narrowly elliptic in cross-section; plants suberect to sprawling shrubs to nearly prostrate, many of the joints often separated by 1 or more others from the ground; spines often present on at least the upper half of the joints.
   8. Spines cream to yellow; joints green; petal-like structures yellow ____ **O. engelmannii**
   8. Spines not all cream to yellow, usually dark brown; joints bluish green; petal-like structures yellow or yellow with red bases ____ **O. phaeacantha**

Opuntia engelmannii Salm-Dyck, (for George Engelmann, 1809–1884, German-born botanist and physician of St. Louis). Large, suberect to sprawling shrub 1–3(–3.5) m high; joints green; spines cream to yellow, (1–)3–6 per areole, usually on all but lower areoles, 1.2–4(–5) cm long, erect; flowers yellow; fruits purple, 2.1–7 cm long. May–Jun.

1. Larger terminal joints usually obovate to orbicular, 15–25(–30) cm long ____ var. **lindheimeri**
1. Larger terminal joints elongate, lanceolate to linear-lanceolate, 45–90(–120) cm long ____ var. **linguiformis**

var. **lindheimeri** (Engelm.) B.D. Parfitt & Pinkava, (for Ferdinand Jacob Lindheimer, 1801–1879, German-born TX collector), TEXAS PRICKLY-PEAR, NOPAL PRICKLY-PEAR. Sandy, gravelly, or alluvial soils; grasslands; Mills Co., also Bell, Coryell, Ellis, Lampasas, and Navarro cos. (Benson 1982); c Blackland Prairie s to s TX and w to Trans-Pecos (also Arbuckle Mountains in OK). [*Opuntia lindheimeri* Engelm.] According to Crosswhite (1980), during pioneer days the fruits were eaten fresh, made into preserves or syrup, or boiled and fermented into "colonche"; the young pads were also eaten; however, eating excessive amounts of older pads can cause oxalic acid poisoning. ☠ ▣/101

var. **linguiformis** (Griffiths) B.D. Parfitt & Pinkava, (tongue-shaped), COW'S-TONGUE, COW-TONGUE PRICKLY-PEAR, LENGUA DE VACA. Cultivated as an ornamental because of the tongue-like joints, escapes; Brown (Stanford 1971) and Comanche (HPC) cos.; also e Edwards Plateau and s TX. [*O. lindheimeri* var. *linguiformis* (Griffiths) L.D. Benson] According to Benson (1982), "... this cultivated and escaped variety is known only from plants descended from those collected near San Antonio, Texas, by Griffiths in 1906."

var. *engelmannii*, ENGELMANN'S PRICKLY-PEAR, native to the s and w of nc TX, is distinguished by the often chalky white or pale straw-colored spines. It is used in Mexico as a source of nopalitos—the tender young green pads, which have not yet formed spines, are cut into strips; they are boiled several times and used as a food (Kirkpatrick 1992); nopalitos can sometimes be found fresh or canned in TX supermarkets. The ripe fruits are known as tunas; in order to preserve them, Native Americans squeezed the juice out and dried them in the sun (Kirkpatrick 1992); a candy-like jelly known as queso de tuna is made from the fruits in Mexico. [*O. phaecantha* Engelm. var. *discata* (Griffiths) L.D. Benson & Walk.]

Opuntia ficus-indica (L.) Mill., (Indian fig), INDIAN-FIG. Tree 3–5(–7) m tall with a main trunk 0.6–1.2 m tall; larger terminal joints green, broadly to narrowly obovate or oblong; spines none to abundant, 1–6 per node, flattened basally, the longer ones 1.2–2.5(–4) cm long; flowers yellow or orange-yellow, externally pink-tinged; fruits yellow, orange, red, or purplish, edible at maturity. Cutivated and escapes; Lampasas Co. (Benson 1982), also Brown Co. (Stanford 1971); scattered in s and w TX. Probably native of Mexico. Spineless forms have been valued for their fruits since prehistoric times and were probably spread by trading; this species has become naturalized in many parts of the world and is a pest in some areas (Benson 1982). ☜

Opuntia humifusa (Raf.) Raf., (spreading on the ground), EASTERN PRICKLY-PEAR. Low clump- or mat-forming perennial 7.5–10 cm high; joints green or reddish purple in winter, 3.8–10 cm long, elliptic, orbicular, or obovate, often noticeably wrinkled; spineless or spines 1 per areole in the upper areoles; spines 25–38 mm long; flowers yellow; fruits purple or red, 2.5–3.8 cm long; seeds 4.5 mm in diam., 1.5 mm thick. Open dry areas; Denton, Grayson, Henderson, Kaufman, and Limestone cos., also Brown (HPC), Dallas, and Tarrant (Benson 1982) cos.; se and e TX w to nc TX and e Edwards Plateau. Apr–Jun. [*O. compressa* (Salisb.) J.F. Macbr.]

Opuntia kleiniae DC., (resembling *Kleinia* in the Asteraceae), KLEIN'S CHOLLA, CANDLE CHOLLA, TASAJILLO. Bush or shrub to 2 m tall; larger terminal joints 10–30 cm long, 0.5–1 cm in diam.; tubercules prominent; spines 1–4 per areole; flowers lavender to purple or reddish; fruits red to green and red, to 2 cm long. Rocky soils of hillsides, deserts, and grasslands; Eastland and Lampasas cos. (Benson 1982); sw edge of nc TX s and w to Trans-Pecos. May.

Opuntia leptocaulis DC., (thin-scaled), DESERT CHRISTMAS CACTUS, PENCIL CACTUS, PENCIL CHOLLA, CHRISTMAS CHOLLA, TASAJILLO, TESAJO, RAT-TAIL CACTUS, SLENDER-STEM CACTUS. Bush or erect shrub 50–70 cm tall; joints 30–40 cm long; terminal joints usually 2.5–7.5 cm long, 3–4.5 mm in diam.; tubercules almost absent; spines 1 per areole, 2.5–5 cm long; flowers green, yellow, or bronze; fruits bright red, ca. 12 mm long. Clay or alluvial soils; Archer, Bell, Brown, Coleman, Milam, Palo Pinto, and Parker cos., also Clay, Hood, Johnson, and McLennan cos. (Benson 1982); Blackland Prairie w through w 2/3 of TX. Apr–May.

Opuntia macrorhiza Engelm., (large-rooted), PLAINS PRICKLY-PEAR, GRASSLAND PRICKLY-PEAR, CHAIN PRICKLY-PEAR, TUBEROUS-ROOT PRICKLY-PEAR. Low, clump-forming perennial 7.5–12.5 cm high; joints glaucous, bluish green, 5–10 cm long, orbicular to obovate; spines 1–6 per areole, mostly in upper areoles, the longer spines 3.8–5.6 cm long, slender; flowers yellow, usually with base red-tinged; fruits reddish purple, 25–38 mm long; seeds 4.5 mm in diam., 1.25–2.25 mm thick. Sandy or rocky soils; nearly throughout TX, less so in Edwards Plateau and e Texas. May–Jun.

Opuntia phaeacantha Engelm., (dark-thorned), PRICKLY-PEAR, ENGLEMANN'S PRICKLY-PEAR, PURPLE-FRUIT PRICKLY-PEAR. Large, prostrate or sprawling to suberect perennial 30–90+ cm high; larger terminal joints bluish green, with some purplish pigmentation in cold weather, 10–34(–40) cm long; spines usually dark brown, deflexed, 4–7 cm long, long and arching, some definitely flattened; flowers yellow or the bases reddish; fruits purplish, 31–62 mm long. May–Jun.

1. Spines over most of joint; spines 5–8 per areole above, 1–4 below; plants low-growing, semi-prostrate, spreading, forming clumps or rosettes of joints ____ var. **camanchica**
1. Spines usually on upper one-half, one-third, or less of joint; spines 1–3 per areole; plants ascending and spreading, relatively larger, to 90+ cm tall ____ var. **major**

var. **camanchica** (Engelm. & J.M. Bigelow) L.D. Benson, (of Camanche, CA). Sandy soils; Lampasas Co. (Benson 1982); mainly Panhandle.

var. **major** Engelm., (greater), BROWN-SPINE PRICKLY-PEAR. Rocky, gravelly, or sandy soils of hillsides; c part of nc TX (Denton and Ellis cos.) w through w 2/3 of state.

Opuntia schottii Engelm., (for Richard van der Schott, d. 1819, head gardener at Austrian palace at Schönbrunn), CLAVELLINA, DEVIL CHOLLA, SCHOTT'S CHOLLA, DOG CHOLLA. Clump- or mat-forming cholla to 10 cm tall, the clumps or mats 1–3 m in diam.; terminal joints 4–6 cm long, 1.5–2.5 cm in diam.; areoles (at least the uppermost) spiny; spines 6–12(–15) per areole; flowers yellow; fruits yellow, 38–56 mm long. Rocky soils; Brown Co. (Benson 1982), known from only one large population w of Brownwood (J. Stanford, pers. comm.); mainly extreme w and s TX. Spring.

Opuntia tunicata (Lehm.) Link & Otto var. **davisii** (Engelm. & J.M. Bigelow) L.D. Benson, (sp.: coated; var.: for Jefferson Davis who was Secretary of War when the Whipple expedition explored w TX), GREEN-FLOWER CHOLLA, JEFF DAVIS' CHOLLA, JUMPING CHOLLA, JUMPING CACTUS. Golden tan (from the color of spine sheaths) shrub 30–40(–100) cm tall, bushy or clump-forming; joints strongly woody; terminal joints 5–15 cm long, 0.9–1.2 cm in diam.; spines 6–10 per areole, ca. 3.8 cm long, barbed, the sheaths 1.5–2 mm in diam., loose; fruits red, ca. 30 mm long. Sandy soils; Brown (HPC) and Coryell (Benson 1982) cos.; scattered in w 2/3 of TX. Jun. The sheaths conspicuously glisten in the sun giving the plants a "blond" appearance; the common names JUMPING CHOLLA and JUMPING CACTUS were earned by the ease with which the joints break off; they are easily attached to animals and thus dispersed (Kirkpatrick 1992); this results in vegetative reproduction. ▣/101

Opuntia edwardsii Grant & Grant, a small subshrub 20–45 cm tall, native from the Edwards Plateau to the Panhandle, occurs in Travis Co. just to the s of nc TX. It is in the *O. phaeacantha* group and can be distinguished from *O. phaeacantha* and *O. lindheimeri* using the following characters: joints 11–18 cm long, blue-green; spines white or ashy gray, deflexed, distributed over the entire joint, 1.1–4.0 cm long (Grant & Grant 1979b). It would not be unexpected to find this species in the Lampasas Cut Plain.

THELOCACTUS

A genus of 11 species of Mexico and the sw U.S.; small cacti with large flowers; some are cultivated as ornamentals. (Greek: *thele*, a nipple, and *cactus*, name for another spiny plant; the ribs of some species have a nipple-like appearance)

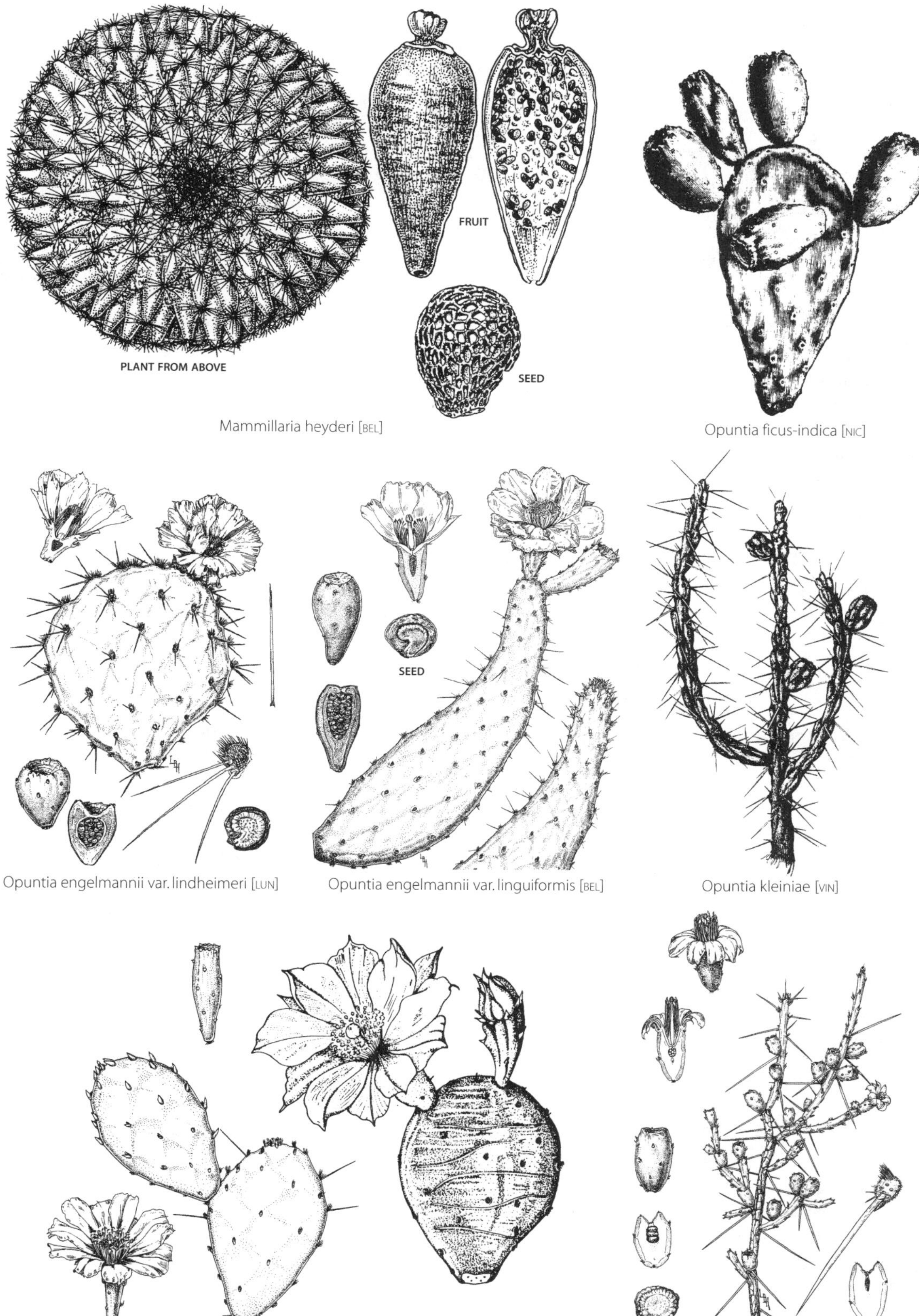

Mammillaria heyderi [BEL]

Opuntia ficus-indica [NIC]

Opuntia engelmannii var. lindheimeri [LUN]

Opuntia engelmannii var. linguiformis [BEL]

Opuntia kleiniae [VIN]

Opuntia humifusa [BEL, LUN]

Opuntia leptocaulis [BEL]

Thelocactus setispinus (Engelm.) E.F. Anderson, (bristly spines), HEDGEHOG CACTUS. Minature barrel cactus with stems solitary to sometimes several to numerous, ovoid or cylindroid, 3.8–10(–20) cm long, 3.8–5 cm in diam., vertically ribbed; ribs ca. 13; spines smooth, not cross-ribbed, dense but not obscuring stem; main central spine strongly hooked, 12–38 mm long; 1–3 smaller, straight, central spines also present; radial spines 12–15 per areole, ± straight; petal-like structures clear yellow, red basally, the largest 20–25 mm long; fruits red, globular, ca. 9–12 mm in diam. Black or clay soils, grasslands, mesquite thickets; Brown Co. (Stanford 1971) near w margin of nc TX; also San Saba and Travis cos. just sw and s of nc TX; mainly e Edwards Plateau s to se and s TX. [*Echinocactus setispinus* Engelm., *Ferocactus setispinus* (Englem.) L.D. Benson] Benson (1982) treated this species in the genus *Ferocactus*. Early spring–late fall.

CALLITRICHACEAE WATER STARWORT FAMILY

A very small (17 species), nearly cosmopolitan family with a single genus. Based on molecular data, Reeves and Olmstead (1993) indicated that the Callitrichaceae should be considered part of the Scrophulariaceae. (subclass Asteridae)

FAMILY RECOGNITION IN THE FIELD: very small wet area annuals with tiny, entire, opposite or apparently whorled leaves often forming rosettes at the stem tips, minute, solitary, axillary flowers without a perianth, and often flattened, slightly heart-shaped fruits.

REFERENCE: Reeves & Olmstead 1993.

CALLITRICHE WATERWORT, WATER STARWORT, WATER-CHICKWEED

Diminutive, glabrous annuals of shallow water or damp ground; leaves opposite or seemingly whorled, linear to oblanceolate, entire; flowers minute, axillary, usually sessile or short-pedicelled, unisexual; perianth absent; flowers consisting of a single stamen or a single pistil; pistil of 2 carpels; fruits somewhat flattened laterally, often slightly heart-shaped and appearing to have 2 lobes, eventually splitting into 4 achene-like mericarps.

Some species are grown as aquarium plants, while others are sensitive to pollution and can be used as ecological indicators—that is, their performance can be used to predict the presence of pollutants. Both aerial and underwater (= hydrophily or more specifically, hypohydrophily) pollination systems are known to occur in *Callitriche*; it is the only genus in which both of these systems have been confirmed (Philbrick 1993; Philbrick & Osborn 1994). *Callitriche* is also one of only two dicot genera for which hydrophily has been documented; the other is *Ceratophyllum* in the Ceratophyllaceae (Philbrick 1991, 1993). Microscopic examination of the small fruits is necessary for definitive identification to species. The only 2 species to occur with frequency in nc TX are *C. heterophylla* and *C. nuttallii*; additional distinguishing characters between these 2 are given in the descriptions. The key to species is adapted from Fassett (1951). (Greek: *callos*, beautiful, and *thrix*, hair, from the slender stems)

REFERENCES: Fassett 1951; Philbrick & Jansen 1991; Philbrick & Anderson 1992; Philbrick 1993; Philbrick & Osborn 1994.

1. Plants either mostly terrestrial or on mud; fruits wider than high; stigmas 0.3–1 mm long; anthers 0.1–0.2 mm wide; flowers without bracts; leaves ± uniform in shape.
   2. Fruit base greatly thickened or gibbose due to the 2 segments pushing against each other at base; fruits 0.3–0.8 mm wide __________ **C. peploides**
   2. Fruit base not thickened or gibbose; fruits 0.5–1.2 mm wide.
      3. Fruits pedicelled; stigma ca. 0.8 mm long; wing of thin margin of carpels turned outward at right angles or revolute and appearing like a thickened margin (under magnification) __________ **C. nuttallii**
      3. Fruits usually nearly sessile; stigma 0.2–0.4 mm long; carpels scarcely winged, under high magnification a very narrow wing seen __________ **C. terrestris**

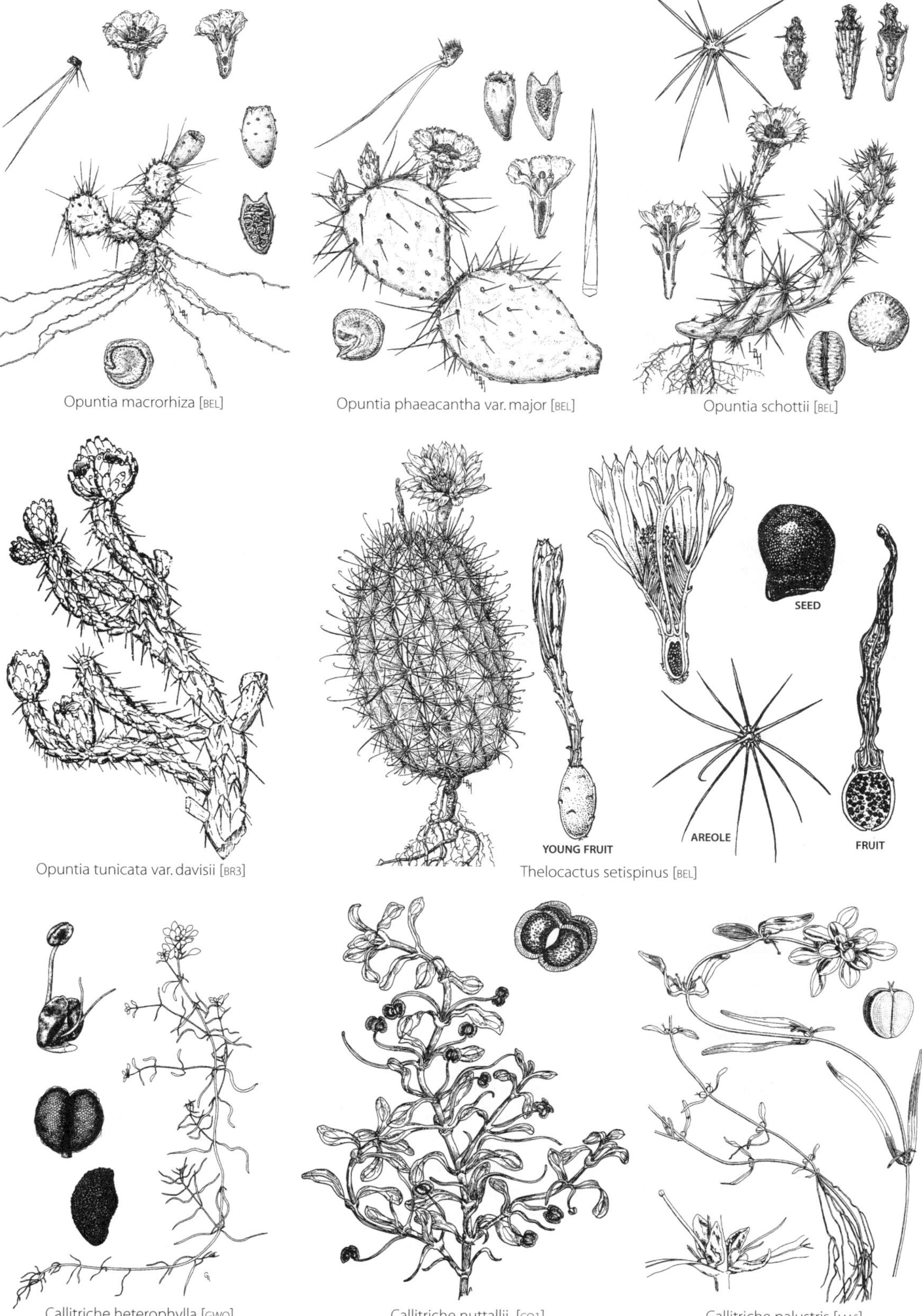

Opuntia macrorhiza [BEL]

Opuntia phaeacantha var. major [BEL]

Opuntia schottii [BEL]

Opuntia tunicata var. davisii [BR3]

Thelocactus setispinus [BEL]

Callitriche heterophylla [GWO]

Callitriche nuttallii [CO1]

Callitriche palustris [MAS]

1. Plants completely submersed OR with terminal rosette of leaves floating OR stranded on mud; fruits as high as wide or slightly higher; stigmas 0.7–6 mm long; anthers 0.3–1.5 mm wide; flowers with 2 whitish inflated bracts at base; leaves often of various shapes.
   4. Fruits thickest at base; margins of carpels with a definite scarious wing; rare in nc TX (reported but no specimens seen) **C. palustris**
   4. Fruits thickest just below middle; margins of carpels usually wingless; commonly found in nc TX **C. heterophylla**

Callitriche heterophylla Pursh, (various-leaved), LARGER WATERWORT. Flowers subtended by 2 small bracts (resembling stipules); fruits sessile or nearly so, very slightly 2-lobed. In quiet, clear water; Burnet, Dallas, Grayson, Henderson, and Lamar cos.; nearly throughout TX. Late Mar–Apr.

Callitriche nuttallii Torr., (for its discoverer, Sir Thomas Nuttall, 1786–1859, English-American botanist), NUTTALL'S WATERWORT. Moss-like; flowers without bracts; fruits in age with well-developed pedicels, deeply lobed. In damp silty or sandy places; se and e TX w to Navarro and Lamar cos. on the e margin of nc TX; also Edwards Plateau. Late Mar–Apr.

Callitriche palustris L., (marsh-loving), COMMON WATERWORT. Fruits sessile, very slightly 2-lobed. In quiet water; included based on citation of vegetational areas 4 and 5 (Fig. 2) by Hatch et al. (1990); mainly c and e TX. Mar–Jun. [*C. verna* L.]

Callitriche peploides Nutt., (resembling *Peplis*, now included in *Lythrum*—loosestrife), MAT WATERWORT. Fruits sessile, moderately 2-lobed. Growing on mud, wet sand, and moist lawns; Tarrant Co.; mainly se, e, and c TX. Feb–May.

Callitriche terrestris Raf., (of or growing in ground), ANNUAL WATERWORT. Fruits sessile or nearly so, moderately 2-lobed. Damp to wet open areas of lawns, fallow fields, paths, or similar areas; included based on citation of vegetational area 4 (Fig. 2) by Hatch et al. (1990); mainly se, e, and c TX. Mar–Jun.

CAMPANULACEAE BLUEBELL, BELLFLOWER OR HAREBELL FAMILY

Ours annual, biennial, or perennial herbs; leaves alternate, simple, entire or toothed, sessile or the lower petioled; stipules absent; flowers axillary or terminal, solitary or in racemes or spike-like inflorescences; calyces with 5 acute or acuminate lobes; corollas sympetalous at least at base, radially symmetrical and 5-lobed OR bilaterally symmetrical and 2-lipped OR (in cleistogamous forms of *Triodanis*) vestigial or absent; stamens 5; pistil 1; style 1; stigma 1; ovary inferior, with placentation axile; fruit a capsule; seeds numerous.

A medium-large (2,000 species in 82 genera) cosmopolitan family of herbs or rarely shrubs; it includes a number of ornamentals in genera such as *Campanula*, *Downingia*, and *Lobelia*; ☠ a number of species contain toxic alkaloids. The family is split by some authorities, with the subfamily Lobelioideae sometimes recognized as a distinct family, the Lobeliaceae (e.g., Lammers 1992); Cronquist (1988), however, included the Lobelioideae and treated the family in the broad sense. Recent molecular analyses (e.g., Michaels et al. 1993; Olmstead et al. 1993) indicate that Campanulaceae are phylogenetically reasonably closely related to Asteraceae. *Sphenoclea zeylanica* Gaertn. (PIEFRUIT, CHICKENSPIKE), here treated in the Sphenocleaceae, was previously placed in the Campanulaceae; according to N. Morin (pers. comm.), it is not closely related to Campanulaceae. Family name from *Campanula*, BELLFLOWER, an ornamentally important genus of 300 species of herbs or rarely shrubs native to the n temperate zone, especially the Mediterranean, and tropical mountains. *Campanula reverchonii* A. Gray, (for Julien

Reverchon, 1837–1905, a French-American immigrant to Dallas and important botanical collector of early TX), BASIN BELLFLOWER, is endemic to granite areas of the Central Mineral Region just s of nc TX. It is a small (to 30 cm tall) annual with radially symmetrical, light blue corollas 9–13 mm long; it can be distinguished from *Triodanis* species by its long-peduncled flowers (at least those of the main axes), narrowly funnelform corollas (tube portion 4–7 mm long), and cleistogamous flowers absent. (Diminutive of Latin: *campana*, bell, in reference to the shape of the flowers) (subclass Asteridae)

FAMILY RECOGNITION IN THE FIELD: herbs with alternate simple leaves and sometimes milky sap, inferior ovaries maturing into capsules with many seeds, showy, bilaterally (2-lipped) or radially symmetrical corollas, and sometimes fused anthers.

REFERENCES: McVaugh 1943, 1961; Diggs 1982; Rosatti 1986; Shetler & Morin 1986; Lammers 1992.

1. Corollas bilaterally symmetrical and 2-lipped (the dorsal lip with 2 lobes separated by a deep cleft, the ventral lip with 3 lobes); stamens united into a tube; flowers in terminal racemes or spike-like inflorescences; capsules opening by apical valves (Subfamily Lobelioideae) ____ **Lobelia**
1. Corollas radially symmetrical or vestigial or absent; stamens separate; flowers axillary, sessile, 1–3 per axil; capsules opening by 1–3 lateral pores (these positioned from near middle of the capsule to near its apex) (Subfamily Campanuloideae) ____ **Triodanis**

LOBELIA

Perennial, biennial, or annual herbs often with milky juice; inflorescences often 1-sided; calyces 5-cleft, radially symmetrical; corollas bilaterally symmetrical, split nearly to base between 2 lobes of the dorsal lip, the tube fenestrate (= with lateral openings) or not so; stamens 5; filaments united above into a tube; anthers united into a tube; ovary 2-carpellate.

A genus of ca. 300 species of tropical and warm areas, especially the Americas; also a few in temperate regions; many have toxic alkaloids including lobelamine and lobeline (Krochmal et al. 1972; Hardin & Arena 1974) and are strongly poisonous; according to some sources (e.g., Hewyood 1993), even smelling *Lobelia tupa* L. of Chile may cause sickness; some have been used medicinally; lobeline has an effect on the central nervous system similar to nicotine (Blackwell 1990); a number are cultivated as ornamentals. Some *Lobelia* species can be hybridized (e.g., *L. cardinalis* × *L. siphilitica* = *L.* ×*speciosa* Sweet) (e.g., Bowden 1964a, 1982); these are rarely found in nature (McVaugh 1943) and are unknown from nc TX. (Named for Matthias de l'Obel, 1538–1616, Flemish herbalist/botanist and physician to James I of England)

REFERENCES: McVaugh 1936, 1940; Bowden 1959a, 1960a, 1960b, 1961, 1964a, 1982; Krochmal et al. 1972; McGregor 1985d; Lammers 1993; Thompson & Lammers 1997.

1. Corollas bright red; flowers (including calyx) 30–50 mm long; filaments including tube portion (but not anther tube) (15–)19–33 mm long ____ **L. cardinalis**
1. Corollas blue or violet; flowers 10–33 mm long; filaments including tube portion 2–15 mm long.
    2. Flowers 15–33 mm long; stems glabrous OR short hirsute or puberulent; filaments including tube portion 6–15 mm long; corolla tube fenestrate (= with 2 lateral slit-like openings as well as the dorsal split).
        3. Stems nearly glabrous to sparsely pubescent; flowers 23–33 mm long; corollas white-striped in throat; corolla tube often 5–10 mm wide; filaments including tube portion 10–15 mm long ____ **L. siphilitica**
        3. Stems densely short hirsute or puberulent throughout; flowers 15–24 mm long; corollas not white-striped in throat; corolla tube 3–4 mm wide; filaments including tube portion 6–7 mm long ____ **L. puberula**

2. Flowers 10–15 mm long; stems glabrous or nearly so; filaments including tube portion 2–4 mm long; corolla tube not fenestrate **L. appendiculata**

Lobelia appendiculata A. DC., (with an appendage), EARFLOWER. Slender annual or biennial 0.15–0.9(–1.1) m tall; stems glabrous or with a few basal hairs; leaves oblong-lanceolate, obtuse, subentire or toothed; middle stem leaves usually clasping or subsessile with broad base, ± glabrous; flowers in loose spikes to 32 cm long; calyx lobes linear-lanceolate, usually densely bristly-ciliate, sometimes only near tips, with conspicuous basal auricles that are flat, ciliate, lanceolate, 1–3 mm long; inflorescences ± secund (= 1-sided); corollas light violet or light lavender-blue to white; anther tube 2–2.5 mm long, bluish gray. Sandy open woods, fields, and roadsides; Fannin, Henderson, and Lamar cos.; se and e TX w to e margin of nc TX. May–Jun.

Lobelia cardinalis L., (cardinal, in reference to the flower color), CARDINAL-FLOWER. Perennial to 1.3(–2) m tall, with basal offshoots; raceme terminal, to 50 cm long; calyx lobes foliaceous, to 14 mm long and 6 mm wide; corollas bright crimson red (rarely white or pink); corolla tube fenestrate; anther tube 4–5.5 mm long, bluish gray; capsules 8–10 mm long. Low moist areas, stream banks; throughout TX. Mainly Sep–Oct. [*L. cardinalis* subsp. *graminea* (Lam.) McVaugh var. *phyllostachya* (Engelm.) McVaugh] A BRIT/SMU sheet annotated by McVaugh as *L. cardinalis* var. *phyllostachys* is known from Dallas. Bowden (1960b) indicated that plants from TX are difficult to classify and apparently represent hybrid swarms between *L. cardinalis* subsp. *cardinalis* and *L. cardinalis* subsp. *graminea* var. *phyllostachya*. As can be seen from these examples, infraspecific taxa are often recognized in this species (e.g., Kartesz 1994; Jones et al. 1997). However, because of apparently unstable character combinations, we are following Brooks (1986b) and Thompson and Lammers (1997) in not recognizing infraspecific taxa. Thompson and Lammers (1997) concluded that all populations of the *L. cardinalis* complex (sometimes divided into four species) occurring from s Canada to n Colombia should be recognized as a single species without infraspecific taxa; they further concluded that *L. cardinalis* is most closely related to *L. siphilitica*, with which it is known to hybridize. The flowers of this species are visited by and presumably pollinated by ruby-throated hummingbirds (*Archilochus colubris*) (James 1948). The entire plant is poisonous to humans and livestock, sometimes fatally so, due to a mixture of pyridine alkaloids including lobeline; the sap can also irritate the skin; it was at one time used to treat conditions including syphilis and worm infestations but because of severe illness or even death resulting from overdoses, its use has long been discontinued (Krochmal et al. 1972; Turner & Szczawinski 1991). ☠ ▣/96

Lobelia puberula Michx., (somewhat pubescent), DOWNY LOBELIA, PURPLE-DEWDROP. Perennial; stems to 1.6(–2.7) m tall; racemes to 50 cm long; calyx lobes with small basal auricles; corollas blue to purple (rarely white); anther tube 3–3.5 mm long, bluish gray; capsules 4–7 mm long. Wooded areas to prairies or fields, usually in wet places; Henderson and Milam cos., also Lamar Co. (Carr 1994); se and e TX w to e margin of nc TX. Aug–Dec. [*L. puberula* var. *mineolana* Wimmer; *L. puberula* var. *simulans* Fernald] While varieties are sometimes recognized (e.g., Kartesz 1994; Jones et al. 1997), we are not distinguishing infraspecific taxa in this variable species. Lobeline and three other alkaloids are present in quantities similar to those in *L. cardinalis* (Krochmal et al. 1972). ☠

Lobelia siphilitica L. var. **ludoviciana** A. DC., (sp.: syphilitic, from supposed medicinal value; var.: of Louisiana), BIG BLUE LOBELIA, BLUE CARDINAL-FLOWER, GREAT LOBELIA, LOUISIANA LOBELIA. Nearly glabrous perennial with basal offshoots; stems to 1.3 m tall; racemes to 50 cm long; calyx lobes foliaceous, to 14 mm long, with broad basal auricles 2–5 mm long; corollas blue, white striped in throat; filament tube 12–15 mm long; anther tube 4–5 mm long, bluish gray; capsules 8–10 mm long. Low moist areas, prairies, woodlands; collected by Reverchon in Dallas Co. (Mahler 1988), also a Grayson Co. population (no longer extant) was observed for a number of years by

Callitriche peploides [CO1]

Callitriche terrestris [CO1]

Lobelia appendiculata [HEA]

Lobelia cardinalis [GLE]

Lobelia puberula [CO1]

Lobelia siphilitica var. ludoviciana [STE]

Sally Crosthwaite (pers. comm.); e TX w to nc TX. Aug–Sep. This species was used by Native Americans and early settlers as a treatment for syphilis (Larsen 1940); according to Steyermark (1963), it was also formerly used medicinally as an emetic and to relieve spasms of the air passages in cases of laryngitis and asthma. However, it is poisonous when taken in overdoses (Steyermark 1963); lobeline and two other alkaloids are present (Krochmal et al. 1972). ☠ /96

Lobelia spicata Lam., (spicate, with spikes), PALE-SPIKE LOBELIA, HIGHBELIA, is known just to the e and n of nc TX in e TX and OK. It can be distinguished from *L. appendiculata* (the most similar nc TX species) using the following characters: stems densely short pubescent at base; middle stem leaves not clasping, narrowed to base, with pubescence on both surfaces; inflorescences usually not secund; calyx lobes glabrous or ciliate near their tips, with basal auricles that are usually deflexed, glabrous, and filiform or shortly triangular. Reported to be poisonous (Burlage 1968). ☠

TRIODANIS VENUS'-LOOKING-GLASS

Annuals with 1 or several, usually unbranched, densely leafy stems to ca. 1 m tall (usually shorter); leaves mostly sessile, often clasping, usually < 3(–7) cm long, the upper gradually smaller; flowers sessile; early flowers (cleistogamous, self-pollinated, in lower leaf axils) often without corolla or corolla vestigial, maturing fruit very early; chasmogamous flowers with corollas radially symmetrical, rotate, rather deeply 5-lobed, blue-purple to purple or rarely white; capsules subcylindrical, opening by lateral pores (like window shades), the pores positioned from near the middle of the capsule to its apex; seeds lenticular (quadrangular in 1 species) in cross-section.

A genus of 7 species, 6 in North America (1 extending to South America) and 1 in the Mediterranean region; previously lumped into the European and Mediterranean genus *Legousia* (= *Specularia*). All species, except the single one from the Old World (*T. falcata* (Ten.) McVaugh), occur in nc TX. (Greek: *tri*, three, and *odontos*, a tooth, possibly in reference to the three calyx lobes of some flowers)

REFERENCES: McVaugh 1945, 1948; Fernald 1946b; Bradley 1975.

1. Leaves (not bracts subtending flowers) with pubescence on the upper surface; calyx lobes 5, all in a given flower ± alike, usually 8–15 mm long; capsules of both cleistogamous and chasmogamous flowers usually with 3 locules and thus 3 pores; pores of capsule opening first at apex with covering of pore curling toward capsule base; in nc TX known only Bell and Williamson cos. near s margin of area ____ **T. coloradoensis**
1. Leaves glabrous on upper surface (or with a few bristles near tips); calyx lobes 3–5, often 3 in cleistogamous flowers or sometimes 1 or 2 much smaller than others, usually < 10 mm long (often much less); capsules of cleistogamous flowers usually with 1 or 2 locules and thus usually 1 or 2 pores (chasmogamous flowers have 3 locules and 3 pores); pores of capsule opening first at base with covering of pore curling toward capsule apex; widespread in nc TX.
   2. Floral leaves (= bracts subtending flowers) narrow, oblong-lanceolate or lanceolate to nearly linear, 3.5–10 times as long as wide, not clasping; capsules of cleistogamous flowers often curved, subulate, usually 8–12(–20) mm long, dehiscent by fractures near apex or by a single pore near apex; capsules of chasmogamous flowers 15–20(–25) mm long ____ **T. leptocarpa**
   2. Floral leaves broader, rhombic-lanceolate to semi-orbicular to reniform, 0.5–3 times as long as wide, not clasping to strongly clasping; capsules of cleistogamous flowers straight, not subulate, usually < 8(–12) mm long, dehiscent by 2(–3) pores near capsule apex or middle; capsules of chasmogamous flowers usually 12 mm or less long.
      3. Pores of capsules near apex.

4. Floral leaves longer than wide, mostly < 1 cm wide; seeds 0.4–0.65 mm long; widespread in nc TX ____ **T. perfoliata** var. **biflora**
4. Floral leaves wider than long to about as long as wide, mostly > 1 cm wide; seeds 0.7–1 mm long; rare, in nc TX known from only from n edge of area ____ **T. lamprosperma**

3. Pores of capsules near middle.
   5. Pores linear or very narrowly oblong, 0.2–0.5 mm wide, the cartilage (which rolls out resulting in pore) before dehiscence with only very narrow scarious margins ____ **T. holzingeri**
   5. Pores oblong-ovate or broader, 0.5–1.5 mm wide, the cartilage with broad scarious margins (relative to cartilage width).
      6. Stems and underside of leaves glabrous, scabrous, or main veins pilose; seeds smooth or roughened by minute points ____ **T. perfoliata** var. **perfoliata**
      6. Stems and underside of leaves densely hirsute or pilose; seeds reticulate (= with network of ridges) ____ **T. texana**

Triodanis coloradoensis (Buckley) McVaugh, (of the Colorado River area), COLORADO VENUS'-LOOKING-GLASS, TEXAS VENUS'-LOOKING-GLASS, LINDHEIMER'S VENUS'-LOOKING-GLASS. Upper and middle leaves oblanceolate to elliptic, to 7 cm long and 15 mm wide, sessile; lower leaves broader, short petiolate; floral bracts lanceolate, attenuate; corollas of chasmogamous flowers 9–18 mm long; capsules from cleistogamous flowers 11–18 mm long, those from chasmogamous flowers to 23 mm long; pores of capsules 1.5–2.5 mm below attachment of calyx lobes, 2–4 mm long, 0.8–1.7 mm wide; seeds 0.8–1 mm long. Dry hillsides, bluffs, rocky ledges, woods, gravel bars, floodplains; Bell and Williamson cos. (McVaugh 1945, 1961) on s margin of nc TX, also Fort Hood (Bell or Coryell cos.—Sanchez 1997); Edwards Plateau and adjacent areas; endemic to TX. Apr–early Jun. [*Campanula coloradoense* Buckley, *Legousia coloradoensis* Briq., *Specularia coloradoensis* Buckley ex Small, *S. lindheimeri* Vatke] This species is most similar to the Mediterranean *T. falcata*. (Ten.) McVaugh.

Triodanis holzingeri McVaugh, (for John Mitchell Holzinger, 1853–1929, German-born bryologist, who first noted its distinctions). Leaves ovate to elliptic or obovate, marginally crenate; floral bracts broader than leaves, ovate; corollas of chasmogamous flowers 7–9(–11) mm long; pores of capsules the narrowest of any nc TX species; seeds 0.3–0.7 mm long. Sandy open woods and open ground; throughout much of TX. May–Jun. [*Specularia holzingeri* (McVaugh) Fernald]

Triodanis lamprosperma McVaugh, (with shining seeds). Leaves broadly elliptic to obovate or ovate, sometimes broadly so, apically acute, marginally crenate to nearly entire; floral bracts often wider than long, to 25 mm wide; corollas of chasmogamous flowers 9–12 mm long; pores of capsules broadly elliptic, 0.7–1.5 mm wide; seeds 0.7–1 mm long. Stream bottom woods; Fannin and Grayson cos.; rare; ne TX w in Red River drainage to nc TX. May–Jun. [*Specularia lamprosperma* (McVaugh) Fernald]

Triodanis leptocarpa (Nutt.) Nieuwl., (slender-fruited), SLIMPOD VENUS'-LOOKING-GLASS, SLENDER VENUS'-LOOKING-GLASS. Leaves narrowly elliptic to lanceolate or oblanceolate, apically acute to obtuse, marginally crenate; floral bracts narrower than leaves, apically acute to acuminate (vs. acute to obtuse in other nc TX species except *T. coloradoensis*); corollas of chasmogamous flowers mostly 7–10 mm long; capsules dimorphic: capsules of chasmogamous flowers ± straight, linear (usually longer and relatively narrower than in other nc TX species), with 1 apical pore (vs. 2 or 3 in other nc TX species); capsules of cleistogamous flowers curved, subulate; seeds 0.7–1 mm long. Rocky limestone outcrops or less often sandy soils; nc TX and Edwards Plateau. May–Jun. [*Specularia leptocarpa* (Nutt.) A. Gray]

Triodanis perfoliata (L.) Nieuwl., (with leaf surrounding the stem), HEN-AND-CHICKENS. Corollas of chasmogamous flowers 5.5–12 mm long; seeds 0.5–0.6 mm long. Eroding or disturbed areas.

Bradley (1975) indicated that the following 2 taxa hybridize, that a gradient exists between them, and that extensive cleistogamy in var. *biflora* results in a partial isolating mechanism that allows individuals of that variety to be identified; he concluded that recognition at the varietal level is most appropriate. See key to species to separate the 2 varieties.

var. **biflora** (Ruiz & Pav.) T.R. Bradley, (two-flowered), SMALL VENUS'-LOOKING-GLASS. Leaves not clasping, longer than wide, apically acute to rarely obtuse, marginally crenate; floral bracts longer than wide, mostly < 1 cm wide; cleistogamous flowers at nearly all nodes, each stem usually with 1 open terminal flower; pores of capsules oval to nearly round, ca. 1 mm in diam. Brown, Denton, and Grayson cos., McVaugh (1961) cited numerous nc TX cos.; nearly throughout TX. Apr–Jun. [*T. biflora* (Ruiz & Pav.) Greene, *Specularia biflora* (Ruiz & Pav.) Fisch. & C.A. Mey.]

var. **perfoliata**, CLASPING VENUS'-LOOKING-GLASS. Leaves and floral bracts usually clasping, about as wide or wider than long, apically broadly acute to rounded, marginally crenate to serrate; each stem typically with more than 1 open flower; pores of capsules elliptic, 0.5–1.5 mm wide. Collin and Grayson cos., McVaugh (1961) cited numerous nc TX cos.; nearly throughout TX. Apr–May. [*Specularia perfoliata* (L.) A. DC.]

Triodanis texana McVaugh, (of Texas). Leaves ovate to elliptic or obovate, apically rounded to obtuse (rarely acute), marginally crenate; corollas of chasmogamous flowers 7–14 mm long; pores of capsules oval; seeds 0.4–0.6 mm long. Sandy open woods and open ground; Denton Co., also Dallas, Erath, and Milam cos. (McVaugh 1945, 1961); e TX w to nc TX, also Edwards Plateau; endemic to TX. Apr–May. According to Shetler and Morin (1986), this species has seed morphology unique within the genus (quadrangular in cross-section, different surface texture). [*Specularia texana* (McVaugh) Fernald]

Individuals intermediate between a number of the taxa above are known. These include those that combine the characters of *T. perfoliata* var. *perfoliata* and *T. perfoliata* var. *biflora*; *T. perfoliata* and *T. texana*; and *T. perfoliata* and *T. holzingeri* (McVaugh 1945). McVaugh (1945) listed two nc TX collections (Bell Co., Erath Co.) for the last pair mentioned.

CANNABACEAE HEMP FAMILY

The Cannabaceae is a very small family (3–5 species in 2 genera—Kubitzki 1993)) of the n temperate zone to se Asia; the species are erect or twining, wind-pollinated herbs with pyridine alkaloids; the family is closely related to the Moraceae and Urticaceae. From a cladistic standpoint these families should be lumped to form a more inclusive monophyletic family, which based on nomenclatural rules should be called Urticaceae (Judd et al. 1994). The beer flavoring HOPS comes from *Humulus*. The family is sometimes referred to as the Cannabidaceae; however, the name Cannabaceae is conserved. (subclass Hamamelidae)

FAMILY RECOGNITION IN THE FIELD: the only nc TX species is an introduced annual herb with palmately compound leaves with 3–9(–11) coarsely toothed leaflets and small flowers.

REFERENCES: Miller 1970; Kubitzki 1993b; Judd et al. 1994; Small 1997.

CANNABIS HEMP

A monotypic genus of c Asia; long in cultivation. (From the Greek name, *kannabis*, thought by some to come from the Persian name, *kanab*, or the Arabic, *kinnab*)

REFERENCES: Emboden 1974; Small & Cronquist 1976; Small et al. 1976.

Cannabis sativa L., (cultivated), HEMP, MARIJUANA. Erect annual herb to 4 m tall; stems angled, minutely pubescent; leaves palmately compound, the lower opposite, the upper alternate; leaf-

Triodanis holzingeri [HEA]

Triodanis coloradoensis [HEA]

Triodanis lamprosperma [HEA]

Triodanis leptocarpa [BB2]

lets 3–9(-11), narrowly lanceolate, coarsely and sharply toothed, scabrous above, pubescent beneath; stipules lance-linear; inflorescence a narrow, terminal panicle with leafy bracts, or the flowers in mostly axillary, small, spike-like clusters; flowers unisexual, the sexes on separate plants; staminate flowers on pedicels 0.5–3 mm long, the perianth greenish, 5-parted; pistillate flowers sessile, enclosed by sepal-like bracts; achenes 2–5 mm long. Found in the city of Dallas in Jun 1876 by Reverchon (Mahler 1988); scattered localities in TX. Jul–Oct. Native of c Asia. [*C. sativa* subsp. *indica* (Lam.) E. Small & Cronquist] While infraspecific taxa are sometimes recognized (e.g., Jones et al. 1997), according to Small (1997), "the chemical and morphologic distinctions by which *Cannabis* has been split into taxa are often not readily discernible, appear to be environmentally modifiable, and vary in a continuous fashion." For these reasons we are not recognizing taxa below the specific level. Cultivated for use as a psychoactive drug and as a source of the fiber known as hemp which has numerous uses including rope, paper, cloth, and nets; due to demand for rope, *Cannabis* was a major economic crop in the American colonies; during World War II, because fiber supplies were low, the U.S. government encouraged and subsidized the cultivation of *Cannabis* for hemp. It is one of the world's oldest cultivated plants having been domesticated for ca. 8500 years; the active ingredients are tetrahydrocannabinol (THC) and related phenolic resins; it is controlled in the U.S. by the Comprehensive Drug Abuse Act of 1970 and is used illegally as a recreational drug; medicinally, *Cannabis* is used in the treatment of glaucoma and in combating the nausea experienced during cancer chemotherapy; there are adverse effects from concentrated doses or heavy usage and the pollen can be a cause of hay fever; a single inflorescence has been estimated capable of producing more than 500 million pollen grains (Tippo & Stern 1977; Stephens 1980; Fuller & McClintock 1986; Mabberley 1987; Blackwell 1990; Turner & Szczawinski 1991; Kubitzki 1993b; Small 1997).

CAPPARACEAE CAPER OR SPIDER-FLOWER FAMILY

Ours annuals with alternate, palmately compound leaves; leaflets entire or nearly so; flowers in terminal, bracted racemes; sepals 4; petals 4, ± unequal, clawed; stamens 6 to many; pistil 1; ovary superior; fruit a capsule.

A medium-sized (650 species in 39 genera) mainly tropical and subtropical family with a few in arid temperate regions; many species are woody and most produce mustard-oil glucosides and in some cases alkaloids; capparids are similar in many respects to members of the Brassicaceae. The Capparaceae are closely related to the Brassicaceae and are probably paraphyletic when treated separately. From a cladistic standpoint the two should be lumped to form a more inclusive monophyletic family, which based on nomenclatural rules, should be called Brassicaceae (Judd et al. 1994). Family name from *Capparis* (CAPER), a genus of 250 species of shrubs, scramblers, or trees of warm areas of the world. The spice capers are the flower buds of *Capparis* species. The family is sometimes referred to as the Capparidaceae; however, the name Capparaceae is conserved. (Greek: *kapparis*, capers) (subclass Dilleniidae)

FAMILY RECOGNITION IN THE FIELD: annuals (many woody in other parts of the world) with palmately compound leaves (3–7 leaflets); similar to Brassicaceae (e.g., flowers in racemes, 4 sepals, 4 petals) but with somewhat bilaterally symmetrical flowers (not cross-like as in Brassicaceae), stamens never 4+2 (the condition in the Brassicaceae), inflorescence bracts often present (these absent in Brassicaceae), and fruits lacking a transverse partition (this present in Brassicaceae).

REFERENCES: Iltis 1957, 1958 [1959]; Ernst 1963a; Judd et al. 1994.

1. Stems glandular-pubescent, at least in upper part (use lens); petals white or yellowish, sometimes pinkish or purple-tinged or pink to purple; mature capsules 40–100 mm long, very much longer than wide; stamens sometimes much exceeding the petals.
   2. Leaflets 3; plants unarmed; stamens 6–20, of unequal lengths, exposed in bud; petals white to

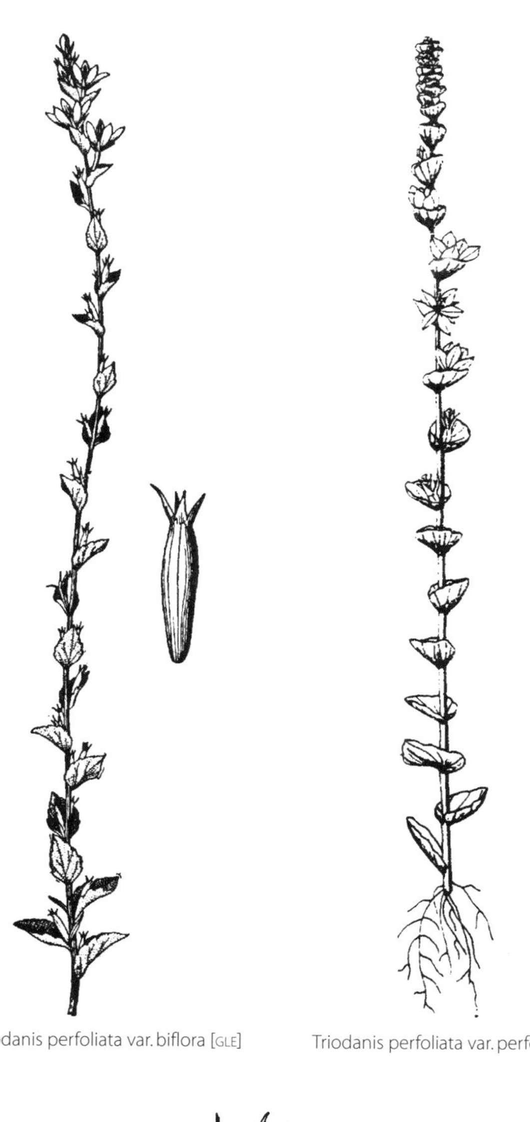

Triodanis perfoliata var. biflora [GLE]

Triodanis perfoliata var. perfoliata [STE]

Cannabis sativa [REE]

Cleome hassleriana [BL2]

Triodanis texana [HEA]

Cleomella angustifolia [IPL]

yellowish, sometimes pinkish-tinged or purplish-tinged, 5–16 mm long; fruits without a stipe or on a short stipe (to ca. 14 mm long) beyond the 10–40 mm long pedicel _____ **Polanisia**

2. Leaflets 5 or 7; plants armed with prickles; stamens 6, of equal length, covered in bud by the overlapping petals; petals pink to purple (rarely white), 20–40 mm long; fruits on an elongate slender stipe 30–80 mm long (beyond the 20–40 mm long pedicel) _____ **Cleome**

1. Stems glabrous; petals deep yellow; mature capsules 5–10 mm long, ca. as wide as long; stamens equaling or barely exceeding the petals _____ **Cleomella**

CLEOME SPIDER-FLOWER, BEEPLANT

A genus of 150 species of tropical and warm areas of the world; a number are cultivated as ornamentals, for medicinal uses, or for edible seeds. (Name of unknown origin; applied early to some mustard-like plant)
REFERENCE: Iltis 1959.

Cleome hassleriana Chodat, (for Emile Hassler, 1861–1937, Swiss botanical collector and physician who settled in Paraguay), SPIDERPLANT, SPIDER-FLOWER, SPINY SPIDER-FLOWER, PINK-QUEEN. Plant robust, 1–2 m tall, armed, strongly scented; leaflets oblanceolate-elliptic to nearly elliptic, to 12 cm long; racemes dense, to 1 m long; petals showy; anthers ca. 1 cm long; capsules linear-cylindric, to 4 mm thick, divergent to deflexed. Cultivated and apparently escapes; included based on citation of vegetational area 4 (Fig. 2) by Hatch et al. (1990), also s TX. Apr–Oct. Native of South America.

CLEOMELLA RHOMBOPOD

A genus of 10 species of sw North America. (Diminutive of *Cleome*)
REFERENCE: Payson 1922b.

Cleomella angustifolia Torr., (narrow-leaved), NARROW-LEAF RHOMBOPOD. Plant glabrous, erect, ca. 0.2–2(–2.6) m tall; leaves 3-foliolate; leaflets linear-elliptic, 2–8 mm wide; petals 4–6 mm long; capsules 5–9 mm wide, rhomboid-obdeltoid (like a distorted triangle); seeds 3–6 per capsule. Draws, ditch bottoms, and roadsides, sandy soils; Clay, Jack, and Montague cos.; West Cross Timbers w to Rolling Plains and s to Edwards Plateau. Jul–Oct.

POLANISIA CLAMMYWEED

Herbaceous glandular-pubescent annuals with strong odor; leaves 3-foliolate, petiolate; petals (in ours) with a long claw prominent nectary gland present in flowers; capsules elongate, erect, dehiscing apically by valves; seeds numerous.

A North American genus of 6 species; thought to be related to xerophytic African *Cleome* species and often lumped into that genus. (Greek: *polys*, many, and *anisos*, unequal, in reference to the characters by which the stamens differ from those of *Cleome*)
REFERENCES: Iltis 1958, 1966.

1. Leaflets lanceolate or elliptic to rhombic-orbicular or ovate, 10–30 mm wide; petals slightly unequal, entire or notched; stamens 9–30 mm long; capsules 5–10 mm wide _____ **P. dodecandra**

1. Leaflets linear, 1–5 mm wide; petals very unequal (upper 2 much larger than lower), with deeply ragged margin; stamens 6–12 mm long; capsules 1.5–3.5(–5) mm wide _____ **P. erosa**

Polanisia dodecandra (L.) DC. subsp. **trachysperma** (Torr. & A. Gray) H.H. Iltis, (sp.: twelve-stamened; subsp.: rough-seeded), CLAMMYWEED. Plant 20–100 cm tall; leaflets oblanceolate, 20–40 mm long; petals 5–16 mm long, white or sometimes with purplish tinge; stamens 6–20, pink to purple; nectary gland solid, truncate to shallowly concave, 1–2 mm long, yellowish, the apex orange to orange-red; capsules 20–70 mm long; seeds many. Sandy stream banks, roadsides, and disturbed sites; widespread in TX, apparently native from West Cross Timbers westward (Mahler 1988), local e to Bell, Dallas, Denton, Falls, Grayson, and McLennan cos. May–Oct. [*Polanisia trachysperma* Torr. & A. Gray] Touching the foliage transfers a strong-smelling substance that makes the hands sticky or clammy. ☒/103

Polanisia erosa (Nutt.) H.H. Iltis, (jagged as if gnawed, in reference to petal margin), LARGE CLAMMYWEED, LARGE CRESTPETAL. Plant 20–60 cm tall; leaflets linear to oblanceolate, 10–40 mm long; upper (larger) petals 6–11 mm long, white to pale-yellow, sometimes with pink tinge, the claw purplish pink; stamens 6–15, pink; nectary gland tubular, truncate, 3.5–5.5 mm long, 0.9 mm in diam., yellow, drying pink-purple, persistent in fruit at base of gynophore; capsules 20–60 mm long, with 6–36 seeds; stipe (= gynophore) (4–)7–14 mm long. Loose sand; Dallas, Limestone, Parker, and Tarrant cos.; se and e TX w to West Cross Timbers. Late May–Oct. Sometimes segregated into the genus *Cristaltella* [as *C. erosa* Nutt.].

CAPRIFOLIACEAE HONEYSUCKLE FAMILY

Herbs, shrubs, trees, or woody vines; leaves opposite, sessile or petioled, simple or pinnately compound, entire, toothed, or lobed; flowers axillary or terminal, solitary, in pairs, or in spikes, whorls, umbels, or cymes; calyx lobes minute to large, 3–5; corollas 2-lipped or radially symmetrical and 4–5-lobed, funnelform or tubular to rotate; stamens 4 or 5; pistil 1; style and stigma 1; ovary half to wholly inferior; fruit in ours a drupe or berry.

A small (420 species in 15 genera), nearly cosmopolitan family of shrubs or small trees or less frequently lianas or herbs; it includes a number of ornamentals such as *Abelia*, *Weigela*, *Lonicera*, and *Symphoricarpos*. Based on a cladistic analysis, Judd et al. (1994) suggested segregating *Sambucus*, *Viburnum*, and relatives into the family Adoxaceae. They further suggested including the Valerianaceae and Dipsacaceae with the remaining Caprifoliacaeae to form a more inclusive monophyletic Caprifoliaceae. Family name from *Caprifolium*, a genus now treated as *Lonicera*. (Latin: *capra*, goat, and *folius*, leaved, from the foliage being used as goat fodder) (subclass Asteridae)

FAMILY RECOGNITION IN THE FIELD: shrubs or vines (1 herb) with similarities to the Rubiaceae (e.g., opposite leaves, ± inferior ovaries) but Caprifoliaceae usually lack stipules and typically have larger showier flowers with the flower parts mostly in 5s (Rubiaceae have stipules and flower parts often in 4s).

REFERENCES: Ferguson 1966a; Brooks 1986c; Donoghue et al. 1992; Judd et al. 1994.

1. Leaves pinnately compound; fruits 4–6 mm in diam., purple-black, numerous — **Sambucus**
1. Leaves simple; fruits not as above.
  2. Plants erect herbs to 80 cm tall; flowers usually solitary in the leaf axils — **Triosteum**
  2. Plants woody, either vines, shrubs, or small trees; flowers either not axillary, or if axillary then not solitary.
    3. Leaves usually with distinct teeth; corollas rotate or shallowly funnelform, radially symmetrical; flowers small and many, in terminal cymes; fruits 10–15 mm long, dark blue — **Viburnum**
    3. Leaves usually without teeth (however, the morphologicaly different juvenile leaves (= those on new growth) can have teeth, lobing, or wavy edges); corollas funnelform to long-tubular, nearly radially symmetrical to 2-lipped; flowers few or many, axillary or terminal, in pairs or spikes or whorls; fruits differing from above in either size or color or both.
      4. Corollas 10–50 mm long, often bilaterally symmetrical; leaves often > 4 cm long; plants shrubs or twining or trailing vines; fruits with several seeds — **Lonicera**
      4. Corollas 2.5–5 mm long, nearly radially symmetrical; leaves usually < 4 cm long; plants small shrubs to only 2 m tall; fruits with 2 seeds — **Symphoricarpos**

LONICERA HONEYSUCKLE

Vines or shrubs; leaves short-petioled or sessile, entire; flowers axillary or terminal, rather large and showy; fruit a several-seeded berry.

A genus of 180 species of the n hemisphere s to Mexico and the Philippines; many are cultivated as ornamentals and a number have become serious invasive weeds; some have very fragrant flowers. (Named for Adam Lonitzer, 1528–86, German naturalist and herbalist)
REFERENCES: Rehder 1903; Luken & Thieret 1995, 1996.

1. Uppermost leaves united around stem; flowers terminal, in heads or in whorls forming spikes; twining, trailing or climbing vines.
   2. Corollas 25–50 mm long, partly or wholly red, purple, or pink ____ **L. sempervirens**
   2. Corollas 12–18 mm long, creamy white ____ **L. albiflora**
1. Uppermost leaves separate, short-petioled; flowers axillary or both axillary and terminal, mostly in 2s; vines or shrubs.
   3. Twining, trailing or climbing vines; corolla tube > 15 mm long ____ **L. japonica**
   3. Shrubs; corolla tube < 15 mm long.
      4. Leaves acuminate apically; branches hollow, with brown pith ____ **L. maackii**
      4. Leaves rounded or acute apically (but short apiculate); branches with solid white pith ____ **L. fragrantissima**

Lonicera albiflora Torr. & A. Gray, (white-flowered), WHITE HONEYSUCKLE, BUSHY HONEYSUCKLE. Evergreen, low, shrubby vine with arched to twining branches; young stems usually glabrous (rarely with sparse pubescence); leaves sessile or subsessile; leaf blades variable in shape, elliptic to ovate, suborbicular, rhombic, or obovate, (1.5–)2–4(–6.5) cm long, ± coriaceous, apically usually rounded (rarely broadly acute); upper surface of leaf blades gray-green or yellow-green, the lower surface variously glaucous; typically only the terminal pair of leaves perfoliate; fruits reddish orange, 5–15 mm in diam. Limestone outcrops; Blackland Prairie (Austin Chalk) s and w to w TX. Apr–May. [*L. albiflora* var. *dumosa* (A. Gray) Rehder] The fruits are reported to contain a substance that induces vomiting (Powell 1988).

Lonicera fragrantissima Lindl. & Paxton, (very fragrant), SWEET-BREATH-OF-SPRING. Semi-evergreen shrub to ca. 3 m tall; leaves short petioled; leaf blades broadly oval, to ca. 8 cm long, strongly apiculate; flowers fragrant, several pairs in leaf axils; corollas ca. 16 mm long, creamy white, sometimes tinged with pink, the tube ca. 5–6 mm long; fruits orange to red, 8–10 mm long. Widely cultivated and escapes, forest margins; Collin Co. (Heard Museum property), also Tarrant Co. (R. O'Kennon, pers. obs.); we do not know of other TX reports. Jan–Apr. Native of China.

Lonicera japonica Thunb., (of Japan), JAPANESE HONEYSUCKLE. Trailing or twining vine, evergreen; young stems usually densely pubescent to almost glabrous; leaves short-petioled; leaf blades oblong-ovate or oblong-lanceolate, those on new spring or summer growth (= juvenile leaves) often pinnatifid; flowers very fragrant; corollas 25–40 mm long, creamy white or white tinged with purple, becoming yellow with age, eventually changing to brownish, the tube pubescent outside, the upper lip with at least 2 of its lobes united more than half way; fruits black, 5–8 mm in diam. Cultivated, escaped, and locally established in disturbed areas, woods, and thickets; se and e TX w to Rolling Plains and Edwards Plateau. Can be an invasive pest overtaking the habitat of native species. Mar–Jul. Native of Asia.

Lonicera maackii (Rupr.) Maxim., (for Richard Maach, 1825–1886, Russian naturalist), AMUR HONEYSUCKLE, BUSH HONEYSUCKLE, TREE HONEYSUCKLE, MAACK'S HONEYSUCKLE. Shrub to ca. 6 m tall, deciduous but leafing out very early and holding leaves late; leaves short-petioled; leaf blades to ca. 8 cm long, acuminate; flowers fragrant, in axillary pairs; corollas white becoming

Polansia dodecandra subsp. trachysperma [GR1]

Polansia erosa [HEA]

Lonicera albiflora [VIN]

Lonicera fragrantissima [CUR]

Lonicera japonica [REE]

Lonicera maackii [SID]

Lonicera sempervirens [BT3]

yellow, ca. 20 mm long; fruits dark red, 2–4 mm in diam. Widely cultivated and escapes, calcareous slopes and forest margins; Dallas Co., also Tarrant Co. (R. O'Kennon, pers. obs.); first referred to as a weed in the U.S. in Chicago in 1924 and now considered problematic in some areas of the e U.S. (Luken et al. 1995). Mar. Native of c and ne China, the Amur River and Ussuri River valleys, Korea, and Japan (Luken & Thieret 1996).

Lonicera sempervirens L., (evergreen), CORAL HONEYSUCKLE, TRUMPET HONEYSUCKLE, EVERGREEN HONEYSUCKLE. Glabrous, evergreen, twining vine; young stems glabrous or nearly so; leaves sessile or subsessile; leaf blades ovate to elliptic or obovate, 3–6(–7) cm long, 2–4 cm wide, subcoriaceous, apically obtuse to acute; upper surface of leaf blades olive-green or dark green, the lower surface lighter than the upper, often glaucous; terminal 1–2(–several) pairs of leaves perfoliate; corollas tubular, usually red outside, orange inside, shallowly lobed, lobes nearly equal; fruits red or reddish orange, 6–10 mm in diam. Stream banks or hillside woods, also cultivated; se and e TX w to West Cross Timbers. Mar–Apr. The flowers of this species are visited by and presumably pollinated by ruby-throated hummingbirds (*Archilochus colubris*) (James 1948). /96

SAMBUCUS ELDERBERRY

A temperate and subtropical genus of 9 species of shrubs and small trees often with alkaloids and extrafloral nectaries; some are cultivated as ornamentals or used for their edible fruit; however, some species have toxic fruit. While we are treating *Sambucus* in the Caprifoliaceae, Bolli (1994) indicated that even though many authors regard *Sambucus* as most closely related to *Viburnum*, the genus is isolated and he recommended that it be treated in the unigeneric family Sambucaceae. (The Latin name, perhaps from the Greek: *sambuce*, an ancient musical instrument; because of the readily removed tubes of bark, these were used for flutes and whistles)
REFERENCE: Bolli 1994.

Sambucus nigra L. var. **canadensis** (L.) Bolli, (sp.: black; var.: of Canada), COMMON ELDERBERRY, AMERICAN ELDERBERRY. Coarse perennial, developing pithy-woody stems, becoming shrubby, 1–4 m tall; leaflets 5–7(–11), broadly lanceolate, abruptly acuminate, finely and sharply toothed, glabrous to densely soft-pubescent beneath; flowers small, in broad, flat-topped corymbs; corollas creamy white, ca. 5 mm wide; fruit a 3-stoned, purple-black, berry-like drupe 4–6 mm in diam. Stream bottoms and ditch banks, in shade or sun; mainly e 1/2 of TX, scattered further w. May–Jun, sporadically to Sep. [*S. canadensis* L., *S. canadensis* L. var. *submollis* Rehder] While this species has long been recognized as *S. canadensis* (e.g., Correll & Johnston 1970; Kartesz 1994; Jones et al. 1997), we are following Bolli's (1994) revision of the genus for nomenclature of this species; he recognized it as being composed of 6 subspecies indicating "... it is important to emphasize the close relationship and morphological similarity of all subspecies of *S. nigra*." Even though the ripe fruits are edible when cooked (Lampe & McCann 1985) and used in making wines and jellies, the roots, stems, leaves, flowers, and unripe fruits contain a poisonous alkaloid and cyanogenic glycoside (Hardin & Arena 1974); during frontier days the leaves were dried and used as an insecticide (Crosswhite 1980).

SYMPHORICARPOS CORAL-BERRY, SNOWBERRY

A genus of 17 species of deciduous shrubs of North America and China; some are cultivated as ornamentals. The fruits of some species are reportedly toxic if ingested in quantity (Lampe & McCann 1985). (Greek: *symphorein*, to bear together, and *carpos*, fruit; from the clustered fruits)
REFERENCE: Jones 1940.

Symphoricarpos orbiculatus Moench, (orbicular, round), INDIAN-CURRANT, CORAL-BERRY,

BUCKBRUSH. Short, rhizomatous, inconspicuously pubescent, erect to ascending shrub to ca. 2 m tall; leaves very short-petioled; leaf blades elliptic-orbicular, entire or with a few blunt teeth, usually < 4 cm long, rarely to 6 cm; flowers in small dense axillary spikes, rather inconspicuous; corollas funnelform, 3–5 mm long, greenish white, sometimes partly brown-red; drupes usually with 2 stones, coral-red to pink or purple-tinged, 5–7 mm long, long persisting. Woods and thickets; se and e TX w to West Cross Timbers, also Edwards Plateau; increasing under disturbance. Jun–Aug. Reported to be poisonous (Burlage 1968). ☠

TRIOSTEUM FEVERWORT, HORSE-GENTIAN

A genus of 5–6 species of perennial herbs of e Asia and e North America; this disjunct distribution pattern is discussed under the genera *Campsis* (Bignoniaceae) and *Carya* (Juglandaceae). (Greek: *tri*, three, and *osteon*, bone, alluding to 3 bony nutlets)
REFERENCE: Lewis & Frantz 1973.

Triosteum angustifolium L., (narrow-leaved), YELLOW-FLOWERED HORSE-GENTIAN. Perennial herb to ca. 80 cm tall; stems long-hirsute, the hairs 1–4 mm long; leaves narrowly obovate to lanceolate, to ca. 15 cm long, apically acuminate; flowers usually solitary in the leaf axils, subtended by a pair of bracts; calyx lobes 7–13 cm long; corollas tubular, greenish or yellowish white or yellowish, 12–15 mm long; fruit a pale orange drupe 5–7 mm in diam., usually with 3 nutlets. Wooded areas, thickets; Grayson Co. in Red River drainage, also Lamar Co. (Carr 1994); mainly e TX. Mar–May.

VIBURNUM ARROW-WOOD

A genus of 150 species of small trees and shrubs of temperate and warm areas, especially Asia and North America; a number of species are important ornamental shrubs with conspicuous flowers and ornamental fruits; some species have the marginal flowers of the inflorescence sterile and enlarged; ☠ fruits of different species are variously edible or poisonous. (The classical Latin name for *V. lantana* L.)
REFERENCES: Egolf 1962; Donoghue 1983.

Viburnum rufidulum Raf., (somewhat rufid, reddish), SOUTHERN BLACKHAW, DOWNY VIBURNUM, NANNY-BERRY, RUSTY BLACKHAW. Shrub or small tree to 10 m tall, deciduous; leaves reddish-pubescent on petiole and along midrib beneath; leaf blades finely toothed, unlobed, elliptic-lanceolate to orbicular; flowers numerous, in terminal, compound, dense umbels; corollas rotate or shallowly funnelform, white; drupes blue-black, glaucous, 10–15 mm long, 1-seeded. Rocky or sandy woods; widespread in TX, particularly the e 1/2. Apr. Correll and Johnston (1970) and Cox and Leslie (1991) indicated that the fruit pulp is sweet and edible with a raisin-like taste.

CARYOPHYLLACEAE PINK OR CARNATION FAMILY

Ours annual, biennial, or perennial herbs; stems often with swollen nodes; leaves opposite, simple, entire; stipules scarious or absent; flowers solitary or inflorescences cymose, panicle-like, or capitate; sepals or tepals (4–)5; petals 0–5 (or more in cultivated forms), often notched ("pinked") at apex, frequently differentiated into claw and blade, sometimes with a crown of appendages (projections or scales) at junction of claw and blade; stamens (1–)5–10; pistil 1; ovary superior; placentation usually free-central or basal; fruit a capsule dehiscing apically by valves or teeth or an utricle.

A large (ca. 2,200 species in ca. 86 genera—Bittrich 1993), cosmopolitan, but especially temperate and warm n hemisphere family of mostly herbs or rarely shrubs or small trees. It includes ornamentals such as *Dianthus* (CARNATION, SWEET-WILLIAM) and *Gypsophila* (BABY'S-

BREATH). The family is unusual in its subclass in having anthocyanin rather than betalain pigments (Cronquist & Thorne 1994); however, molecular analyses link it with other members of the Caryophyllales (Giannasi et al. 1992; Downie & Palmer 1994). The common name PINK is an old name probably referring to the notched or "pinked" petals (as in pinking shears). Family name conserved from *Caryophyllus* Mill., a genus now treated as *Dianthus* (the name *Dianthus* was published earlier and thus has priority in terms of nomenclature). In pre-Linnaean times, some authorities (e.g., Tournefort) referred to all PINKS as belonging to the genus *Caryophyllus*; Linnaeus, however, used the name for *Syzygium aromaticum* (L.) Merr. & L.M. Perry, CLOVES or CLOVE TREE, of the Myrtaceae; *Caryophyllus* L. has been rejected in favor of *Syzygium* (Farr et al. 1979). Linnaeus did use "caryophyllus" as an epithet for one of the familiar carnations also known as the CLOVE PINK (Bailey 1938). (Greek: *caryon*, nut, and *phyllon*, leaf, possibly in reference to the capsular or utricular fruit being subtended by bracts—R. Rabeler, pers. comm.) (subclass Caryophyllidae)

FAMILY RECOGNITION IN THE FIELD: herbs with opposite, simple, entire leaves, *swollen nodes*, and separate, often *notched petals* frequently with claw and blade; fruit often a toothed or valved capsule.

REFERENCES: Larsen 1986; Rabeler & Thieret 1988; Giannasi et al. 1992; Bittrich 1993; Behnke & Mabry 1994; Downie & Palmer 1994.

1. Flowers or several-flowered head-like clusters subtended by 2–3 pairs of conspicuous, broadly ovate-oblong or obovate, scarious (= dry, not green) bracts; rare introduced species ___ **Petrorhagia**
1. Flowers or inflorescences without bracts OR bracts very different from those above (e.g., in *Dianthus* bracts linear or lance-linear, not scarious); widespread native and introduced species.
  2. Stipules present.
    3. Leaves elliptic, ovate, or obovate AND plants prostrate or decumbent; lower leaves usually in whorls of 4 or apparently so; introduced species rare in nc TX, known only from s part of Lampasas Cut Plain ___ **Polycarpon**
    3. Leaves linear to filiform, needle-like, elliptic, or oblanceolate, if elliptic or oblanceolate, then plants neither prostrate nor decumbent; lower leaves not in whorls of 4, instead either opposite OR needle-like and crowded; including species widespread in nc TX.
      4. Petals absent or minute; sepals green to yellow or brown, sometimes white apically; stipules ovate-lanceolate to linear, 2–6 times as long as wide; widespread in nc TX.
        5. Sepals acute to acuminate or with minute awn-point at apex, entire; fruit a single-seeded utricle ___ **Paronychia**
        5. Sepals rather abruptly narrowed to an awn-like tip about half their total length, the outer 3 with a distinct bristle-tooth on each side from about the middle; fruit a several-seeded capsule ___ **Loeflingia**
      4. Petals present, white or pink; stipules triangular, not over twice as long as wide; rare in nc TX ___ **Spergularia**
  2. Stipules absent.
    6. Sepals separate or united only at base; petals not differentiated into claw and blade.
      7. Petals split more than half way to base (corollas apparently of 10 petals) ___ **Stellaria**
      7. Petals entire OR divided less than half way to base OR absent.
        8. Petals absent.
          9. Leaf blades very narrow, < 1 mm wide, glabrous; capsules opening apically by 4–5 valves; calyces without a red band at base; styles equal in number to capsule valves ___ **Sagina**
          9. Leaf blades usually > 1 mm wide, glabrous or pubescent; capsules opening apically by 6 valves or 10 teeth; calyces with a red band at base OR not so; styles half as many as capsule valves or teeth.
            10. Capsules opening apically by 6 valves (opening to ca. middle of capsule); styles

usually 3; plants ± glabrous or with inconspicuous pubescence, the pubescence if present not glandular ___ **Stellaria**

10. Capsules opening apically by 10 teeth (opening only at very apex); styles usually 5; plants with obvious pubescence, the pubescence usually glandular, at least in the inflorescences ___ **Cerastium**

8. Petals present.

11. Styles 3; petals 2–15 mm long; capsules opening apically by 3 or 6 valves.

12. Petals 2–4 mm long, shorter than the sepals, acute, entire at apex; sepals 2–4 mm long; capsules opening apically by 6 valves ___ **Arenaria**

12. Petals 5–15 mm long, longer than the sepals, toothed or shallowly notched at apex; sepals 2.5–8 mm long; capsules opening apically by 3 valves ___ **Minuartia**

11. Styles 5 (occasionally 4); petals ca. 1–7 mm long; capsules opening apically by 10 teeth or 4–5 valves.

13. Petals toothed or notched at apex; leaves ± pubescent, oblong or lanceolate to elliptic or obovate; capsules opening apically by 10 teeth ___ **Cerastium**

13. Petals entire; leaves glabrous, narrowly linear; capsules opening apically by 4–5 valves ___ **Sagina**

6. Sepals united 1/3 or more (may have separate bracts outside), often forming a campanulate, funnelform, cylindric or inflated calyx tube; petals often, but not always, clawed.

14. Flowers in dense head-like cymes; calyces each with 2–several bracts at base; petals white to variously colored, often minutely dotted with white ___ **Dianthus**

14. Flowers in very open to congested cymes; calyces without bracts at base; petals white to variously colored but if colored not minutely dotted with white.

15. Calyx lobes 15–45 mm long, longer than calyx tube; styles (4–)5; petals usually purplish red, 24–36 mm long including claw, without a crown of projections or scales on upper (inner) side at junction of blade and claw ___ **Agrostemma**

15. Calyx lobes 0.5–5 mm long, shorter than calyx tube; styles 2–3(–4); petals white to pinkish or purplish, 5–38 mm long including claw, without OR with a crown.

16. Leaves not clasping; calyces 10–20-nerved, not angled, cylindric to campanulate to funnelform in fruit; petals white to pinkish or purplish, with a crown of projections or scales on upper (inner) side at junction of blade and claw OR without a crown.

17. Styles 3(–4); inflorescences open; calyx tube 5–10 mm long; petals with crown absent or minute; capsules opening apically by 6 teeth ___ **Silene**

17. Styles usually 2; inflorescences congested; calyx tube 12–20 mm long; petals crowned with an appendage at top of the claw; capsules opening apically by 4 teeth ___ **Saponaria**

16. Leaves clasping; calyces strongly 5-angled or winged, inflated in fruit, ovoid; petals lavender-pink, without a crown ___ **Vaccaria**

AGROSTEMMA CORN-COCKLE

A temperate Eurasian genus of 2 species. (Greek: *agros*, field, and *stemma*, a crown or garland, Linnaeus apparently believing it was suitable for such a use)

Agrostemma githago L., (an old generic name for *Agrostemma*), COMMON CORN-COCKLE. Erect taprooted annual or biennial 0.3–1 m tall; stems simple or few-branched, densely pubescent with appressed hairs; leaves linear to lanceolate, 4–12 cm long, 1–10 mm wide, sessile, entire; stipules absent; flowers 1 per node, on elongate peduncles to 20 cm long; calyx tube 10-ribbed, 10–18 mm long; calyx lobes 15–45 mm long, longer than the tube, linear or lance-linear; petals 24–36 mm long including claw, usually purplish red, with small black dots along veins near base; capsules 18–22 mm long, opening by (4–)5 teeth, many-seeded. Disturbed areas; a 1998 Parker Co. collection is the first documented for nc TX. Mar–Jul. Native of Europe. The seeds are possibly poisonous due to saponins (Mabberley 1997). ☠

ARENARIA SANDWORT, CHICKWEED

Our species small annuals; flowers terminal, solitary or cymose; petals white, entire, shorter than the sepals; fruit a capsule opening by 6 valves.

A n temperate genus of ca. 150 species (Bittrich 1993) including some cultivated ornamentals. Some species previously included in *Arenaria* are now treated in *Minuartia.* (Latin: from *arena*, sand, in which many of the species grow)

REFERENCES: Maguire 1947, 1951; Shinners 1962c; McNeill 1980; Wofford 1981.

1. Sepals glabrous; pedicels 2–15 mm long in flower, up to 35 mm in fruit; on limestone ____ **A. benthamii**
1. Sepals minutely scabrous-pubescent; pedicels 1–3 mm long in flower, up to 10 mm in fruit; in sand ____ **A. serpyllifolia**

Arenaria benthamii Fenzl ex Torr. & A. Gray, (for George Bentham, 1800–1884, English taxonomist and president of Linnaean Society), HILLY SANDWORT. Plant to 50 cm tall; leaves oblong-lanceolate or oblanceolate to elliptic-lanceolate, to 2 cm long, usually smaller. Limestone outcrops, usually in shade of shrubs or small trees; Coryell, Hood, Lampasas, and Johnson cos.; widespread in TX. Late Mar–May.

Arenaria serpyllifolia L., (with leaves like *Thymus serpyllum*—thyme), THYME-LEAF SANDWORT. Plant to 20 cm tall; leaves elliptic or ovate, to 7 mm long, sessile. Sandy roadsides, waste areas, or stream bottoms; Grayson, Henderson, and Kaufman cos., also Tarrant Co. (Mahler 1988); local, apparently spreading; naturalized in parts of e 1/2 of TX. Late Mar–May. Native of Europe.

CERASTIUM CHICKWEED

Small annuals or perennials; petals white, shallowly notched, or petals sometimes absent; capsules cylindrical, exceeding the calyces in age, scarious, with 10 prominent teeth when opened, the whole suggesting a corolla.

An almost cosmopolitan genus of ca. 100 species (Bittrich 1993) including some cultivated annuals and many weeds. (Greek: *cerastes*, horned, alluding to the shape of the slender and often curved fruit)

REFERENCES: Good 1984; Turner 1995d [1996]; Rabeler & Reznicek 1997.

1. Capsule 1–2 times as long as the sepals; petals if present about equaling the sepals (sometimes withering very early); sepal pubescence of dense, glandular or non-glandular hairs, these long (some ca. 1/3 or more as long as sepal width).
    2. Sepals with long, spreading hairs up to and projecting beyond the apex; upper bracts of inflorescence entirely herbaceous, without scarious margins; inflorescence or its main divisions remaining compact OR open and diffuse.
        3. Inflorescence or its main divisions remaining compact, the pedicels mostly shorter than the sepals, 0.5–3(–5) mm long ____ **C. glomeratum**
        3. Inflorescence or its main divisions open and diffuse, the pedicels once to twice as long as the sepals, 5–12 mm long ____ **C. brachypetalum**
    2. Sepals with subappressed to spreading hairs becoming shorter and stopping just below apex; upper bracts of inflorescence with scarious margins; inflorescence open and diffuse in age.
        4. Flowers 8–10 mm wide; petals deeply notched (to 1/2 way to base); sepals pilose, seldom glandular-hairy; stamens 10; capsules mostly over 8 mm long; sepals 5–6 mm long in flower;

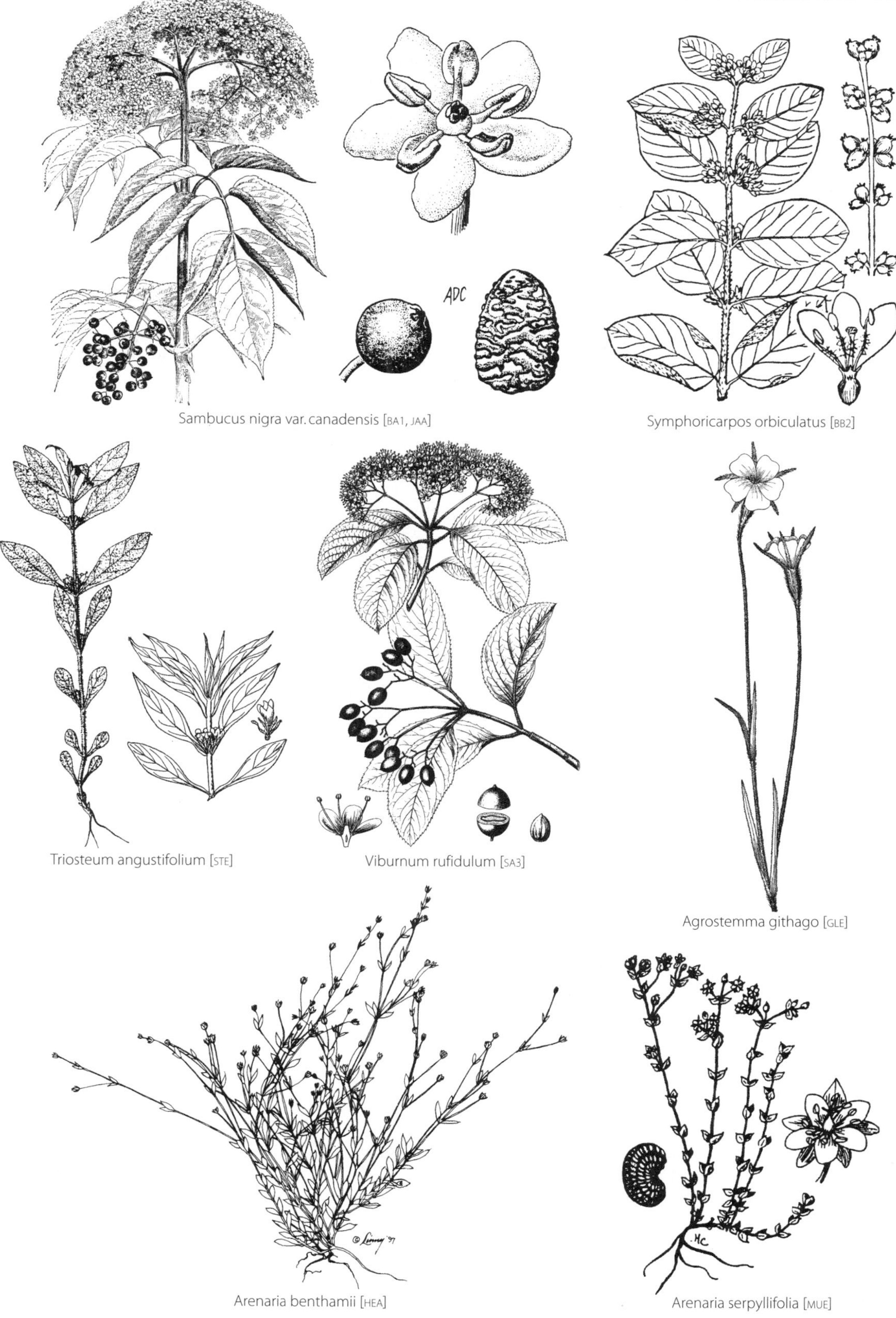

Sambucus nigra var. canadensis [BA1, JAA]

Symphoricarpos orbiculatus [BB2]

Triosteum angustifolium [STE]

Viburnum rufidulum [SA3]

Agrostemma githago [GLE]

Arenaria benthamii [HEA]

Arenaria serpyllifolia [MUE]

plants perennial, often with non-flowering basal shoots ____ **C. fontanum**

4. Flowers 5–6 mm wide; petals shallowly notched (to 1/4 way to base); sepals glandular-hairy; stamens 5 (rarely 10); capsules < 8 mm long; sepals 4–5 mm long in flower; plants annual ____ **C. pumilum**

1. Capsule 2–3 times as long as the sepals; petals often 1.5–2 times as long as the sepals; sepal pubescence of sparse, glandular hairs, these short (much < 1/3 as long as sepal width).
   5. Leaves usually 30 mm or less long; pedicels straight or only slightly curved at maturity, usually equaling or shorter than the capsules in fruit, 0.5–1.25 times the length of the flowering calyces, to 3 times the calyx length at maturity (ca. 2–13 mm long); widespread in nc TX ____ **C. brachypodum**
   5. Leaves often > 30 mm long; pedicels sharply curved or hooked just below the calyx at maturity, usually much longer than the capsules in fruit, 1–3 times the length of the flowering calyces, to 5 times or more the calyx length at maturity (longer pedicels 10–55 mm long); rare if present in nc TX ____ **C. nutans**

Cerastium brachypetalum Pers., (short-petaled), GRAY CHICKWEED. Annual; stems erect, covered with long, shiny, mostly non-glandular hairs; leaves spatulate, elliptic, or ovate; tip of pedicel often bent in fruit. Roadside ditches; a Red River Co. collection (*Rabeler 1333*) made in April of 1998 is the first report for TX; the species should be expected elsewhere in nc TX. Apr. Native of Europe.

Cerastium brachypodum (Engelm. ex A. Gray) B.L. Rob., (short-stalked), SHORT-STALK CHICKWEED. Annual; stems erect, glandular-pubescent with hairs spreading at right angles; leaves glabrous to rather densely viscid-pubescent; pedicels usually 10 mm or less long, reflexed before and after flowering, erect or ascending in flower and again in mature fruit, not sharply curved or hooked just below calyx; sepals 3–4.5 mm long; capsule 6–12 mm long. Prairies, disturbed sites; widespread in e 1/2 of TX. Mar–early Apr.

Cerastium fontanum Baumg. subsp. **vulgare** (Hartm.) Greuter & Burdet, (sp.: pertaining to springs or fountains; subsp.: common), COMMON MOUSE-EAR. Apparently annual in nc TX, though perennial farther n (Larsen 1986; R. Rabeler, pers. comm.); leaves oblong-elliptic to lanceolate or oblanceolate. Sandy roadsides, disturbed sites; Denton, Grayson, and Henderson cos., also Tarrant Co. (R. O'Kennon, pers. obs.); scattered in TX. Apr–May. Native of Europe. [*C. holosteoides* of authors, not Fr., *C. triviale* Link, *C. vulgatum* L.]

Cerastium glomeratum Thuill., (glomerate, clustered). Glandular-pilose annual, erect or with decumbent stems; leaves orbicular ovate to obovate. Roadsides, disturbed sites; widespread in e 1/2 of TX. Mar–early Apr. Native of Europe. [*C. viscosum* of authors, not L.]

Cerastium nutans Raf., (nodding), POWDERHORN CHICKWEED, NODDING CHICKWEED. Erect or decumbent annual; stems finely glandular-pubescent; pedicels 10–55 mm long, sharply curved or hooked just below calyx. Moist or rich woods, open areas; included based on citation of vegetational area 4 (Fig. 2) by Hatch et al. (1990); supposedly scattered mostly in e 1/2 of TX; Turner (1995d [1996]) indicated that he had seen no collections from the state. Apr–Jun.

Cerastium pumilum Curtis, (dwarf), DWARF MOUSE-EAR CHICKWEED, CURTIS' MOUSE-EAR CHICKWEED. Annual resembling a small plant of *C. fontanum*. Roadside ditches and open disturbed areas; a Kaufman Co. collection (*Reznicek 10336*, BRCH, MICH) was the first report for TX (Rabeler & Reznicek 1997); in April of 1998 the species was found in Cooke, Denton, Fannin, Grayson, Lamar, and Red River cos. (e.g., *Rabeler & Diggs 1321*); this is an easily overlooked species; it is to be expected in other parts of nc TX and possibly elsewhere in the state. Apr–May. Native of Europe and sw Asia.

Cerastium brachypetalum [COS]

Cerastium brachypodum [BB2]

Cerastium fontanum subsp. vulgare [REE]

Cerastium glomeratum [BB2, GLE]

Cerastium nutans [GR1]

Cerastium pumilum [COS]

Dianthus armeria [GLE]

DIANTHUS PINK, CARNATION

Annual to perennial herbs; flowers in ours in dense terminal cymes; calyces with cylindrical tube, subtended by 2–several, linear or lance-linear bracts; petals 5, long clawed, without a crown; fruit a capsule.

A Eurasian to s African genus of ca. 300 species (Bittrich 1993) with many cultivated ornamentals including CARNATIONS (*D. caryophyllus* L., *D.* ×*allwoodii* Hort. Allwood); CARNATION has religious significance extending back beyond the Dark Ages (Baumgardt 1982). The common name PINK is an old name probably referring to the notched or "pinked" petals (as in pinking shears). (Greek: *dios*, Jove, Jupiter, or Zeus, chief of the Greek gods, and *anthos*, flower; Jove's flower or divine flower, from beauty or fragrance of flowers)

1. Leaves 2–8 mm wide, linear; stems usually pubescent to glabrate; plants annual or biennial; cymes dense, but not head-like ____ **D. armeria**
1. Leaves 10–18 mm wide, lanceolate to oblanceolate; stems glabrous; plants perennial; cymes usually head-like ____ **D. barbatus**

Dianthus armeria L., (the Latin name for a kind of *Dianthus*), DEPTFORD PINK. Plant 20–80 cm tall, dichotomously branched above; leaves grass-like, olive-green; calyx tube 10–15 mm long, closely subtended by elongate bracts ca. equaling the tube; petals pink to rosy or purplish, dotted with white, drying purplish, long clawed, the blades 4–5 mm long; capsules dehiscent by 4 valves. Disturbed areas; Grayson and Tarrant cos; in TX apparently only in nc and ne parts of state; first reported for TX by Lipscomb (1984). May–Aug. Native of Eurasia.

Dianthus barbatus L., (barbed, bearded), SWEET-WILLIAM. Plant 20–60 cm tall, glabrous; calyx tube 10–12 mm long; petals whitish to pink, dark red, or spotted. Cultivated and possibly escapes; included based on citation of vegetational area 4 (Fig. 2) by Hatch et al. (1990) probably based on a Hunt Co. collection at TAES which is questionably identified as *D. barbatus* (R. Rabeler, pers. comm.); in TX apparently only in nc part of state. Jun–Aug. Native of Eurasia.

LOEFLINGIA

A North American and Mediterranean genus of 7 species (Bittrich 1993). (Named for P. Loefling, 1729–1756, Swedish botanist and explorer)
REFERENCE: Barneby & Twisselmann 1970.

Loeflingia squarrosa Nutt., (with recurved tips), SPREADING LOEFLINGIA. Small, minutely viscid-pubescent, erect annual to ca. 15 cm tall; stems much branched, the plants often globose in shape, about as broad as tall; leaves crowded, almost needle-like, 4–6 mm long; flowers inconspicuous, axillary, sessile, usually solitary or few together in fascicles; sepals 5, resembling the leaves; petals absent or minute; capsules slender, 3-valved. Loose, dry sand; Bell, Dallas, Hood, Parker, and Tarrant (Fort Worth Nature Center) cos.; nearly throughout TX. Late Apr–May. [*L. squarrosa* subsp. *texana* (Hook.) Barneby & Twisselm., *L. texana* Hook.] Inconspicuous and more rarely collected than its widespread occurrence would suggest.

MINUARTIA SANDWORT

Annuals or perennials; inflorescences loosely cymose or racemose; petals white, notched at apex; fruit a capsule opening by 3 valves.

A genus of ca. 120 species (Bittrich 1993) ranging from the Arctic to Mexico, Ethiopia and the Himalaya Mts., also 1 species in Chile. Some are cultivated as ornamentals. Previously included in *Arenaria*. (Named for Juan Minuart, 1693–1768, Spanish botanist and pharmacist)
REFERENCES: Shinners 1949d; Maguire 1951; Wofford 1981; Rabeler 1992.

1. Perennials with stiff, almost needle-like leaves, with axillary tufts of smaller leaves; inflorescences glabrous ____ **M. michauxii**
1. Annuals with flat, soft leaves, without axillary tufts of smaller leaves; inflorescences usually glandular-pubescent (use lens).
   2. Sepals usually elliptic-lanceolate, acute, strongly 3–5 ribbed.
      3. Sepals 5-ribbed; corollas scarcely exceeding the calyces; leaves 1.5 mm or less wide; plants usually < 15 cm tall; seeds dull reddish brown ____ **M. patula**
      3. Sepals 3-ribbed; corollas conspicuously exceeding the calyces, showy; leaves often 1.5–3 mm wide; plants often larger (to 30 cm tall); seeds black, shiny ____ **M. muscorum**
   2. Sepals usually ovate, obtuse, not prominently ribbed ____ **M. drummondii**

Minuartia drummondii (Shinners) McNeill, (for its discoverer, Thomas Drummond, 1780–1835, Scottish botanist and collector in North America), DRUMMOND'S SANDWORT. Simple or sparingly branched annual to ca. 20 cm tall; stems and inflorescences mostly glandular-pubescent; leaves to 35 mm long, 2–7 mm wide; pedicels 10–25 mm long; flowers in dichotomous racemes, showy; petals to 15 mm long. Sandy or sandy clay soils, roadsides, fencerows, or waste areas; Kaufman, Navarro, and Tarrant cos.; mainly se and e TX, also Edwards Plateau. Mar–May. [*Arenaria drummondii* Shinners]

Minuartia michauxii (Fenzl) Farw. var. **texana** (B.L. Rob.) Mattf., (sp.: for André Michaux, 1746–1803, French botanist and explorer of North America), ROCK SANDWORT. Densely tufted perennial, glabrous; stem leaves rather crowded in basal 1/3–2/3 of plant, to 15 mm long, ca. 0.5 mm wide; inflorescences loosely cymose, to ca. 30-flowered; pedicels 5–15(–25) mm long; petals 5–8 mm long. Limestone outcrops, rocky areas; Coryell, Hood, Johnson, Lampasas, Parker, and Wise cos., mainly Grand Prairie, also Grayson Co., also Dallas Co. (*Reverchon* in 1872), but not found there recently; nc TX and Edwards Plateau w to Panhandle. Apr–Jun. [*Arenaria stricta* Michx. var. *texana* B.L. Rob., *Arenaria texana* (B.L. Rob.) Britton]

Minuartia muscorum (Fassett) Rabeler, (mossy). Branched annual similar to but often larger than *M. patula*; seeds black, microscopically foveolate (Wofford 1981). Roadsides, stream bottoms, often in sandy soils; Delta, Fannin, Hopkins, Hunt, and Kaufman cos.; e TX w to Blackland Prairie. Apr–May. We are following Rabeler (1992) in recognizing this species; he indicated there has long been confusion between this species, *A. muriculata*, and *A. patula* var. *robusta*, all three of which represent the same entity. [*Arenaria patula* var. *robusta* (Steyerm.) Maguire, *M. patula* var. *robusta* (Steyerm.) McNeill, *Arenaria muriculata* McNeill, *Minuartia muriculata* (Maguire) McNeill, *Stellaria muscorum* Fassett]

Minuartia patula (Michx.) Mattf., (spreading), PITCHER SANDWORT. Branched annual; leaves to 40 mm long; inflorescences open cymose, sparingly and minutely glandular-pubescent; pedicels to 50 mm long; seeds reddish brown, not microscopically foveolate. Damp sandy ground; Limestone Co.; e TX w to e Blackland Prairie. Mar. [*Arenaria patula* Michx.]

PARONYCHIA NAILWORT, WHITLOW-WORT

Annuals or perennials; scarious stipules usually conspicuous; flowers few or many, in terminal cymes; sepals 5, cupped or folded at apex; petals absent or essentially so; fruit a 1-seeded utricle.

A cosmopolitan genus of ca. 110 species (Bittrich 1993) including some cultivated ornamentals. This is the largest genus of Caryophyllaceae in TX (13 species—fide Turner 1983). The scaly appearance of *Paronychia* species (due to the stipules) caused the plant to be used historically in the treatment of whitlow, a condition which caused the fingernails to look scaly—this is an example of the Doctrine of Signatures, an ancient belief that a plant that resembled a portion of the human body (a sign or signature) gave clues to its use; i.e., was useful in treating an illness of the body structure it resembled (Ajilvsgi 1984). (Greek: *para*, close to, and *onyx*, nail, alluding to the original use of the plant to treat whitlow, an inflammation of the finger, especially beneath the nail)

REFERENCES: Core 1941; Turner 1983, 1995b.

1. Plants annual, arising from a single, usually unbranched taproot; leaves needle-like to oblanceolate or narrowly elliptic, sometimes flat, from < 1 to 5 mm wide, acuminate, acute, or obtuse (if 1 mm or less wide then known only from extreme s part of nc TX).
   2. Leaves elliptic to oblanceolate, 2–5 mm wide; e part of nc TX w to Dallas Co.
      3. Calyces (below point where sepals separate) pubescent with hooked hairs; sepals ca. 1.5 mm long, with prominent, white, scarious margins, dilated into a white hood with a tiny awn/horn from back ___ **P. drummondii**
      3. Calyces glabrous or nearly so; sepals 0.8–1.2 mm long, with very narrow or obscure margins, not hooded ___ **P. fastigiata**
   2. Leaves narrowly linear, 1 mm or less wide; extreme s part of nc TX ___ **P. lindheimeri**
1. Plants perennial, arising from a thickened, often woody base or branched, persistent root-stock; leaves needle-like, 1 mm or less wide, acuminate.
   4. Plants with stems minutely pubescent; sepals 2–3 mm long, usually obscurely nerved; calyces with pubescence, the hairs longer near base (below point where sepals separate); West Cross Timbers w to w TX ___ **P. jamesii**
   4. Plants glabrous or nearly so; sepals ca. 3 mm long, prominently 3-nerved; calyces glabrous to short pubescent but without longer hairs near base; mainly Grand Prairie and Blackland Prairie ___ **P. virginica**

Paronychia drummondii Torr. & A. Gray, (for its discoverer, Thomas Drummond, 1780–1835, Scottish botanist and collector in North America), DRUMMOND'S NAILWORT. Annual to 25 cm tall, branching above into wide-spreading branches, often wider than tall; leaves to 25 mm long and 2–5 mm wide; sepals red-brown with white margins and apex. The white, apical hoods on the sepals distinguish this from all other *Paronychia* species in nc TX . Loose sand; collected at Dallas by Reverchon, not found there recently (Mahler 1988), also Henderson and Milam cos. near e margin of nc TX; mainly se and e TX endemic to TX and LA. May–Sep. [*P. drummondii* subsp. *parviflora* Chaudhri]

Paronychia fastigiata (Raf.) Fernald, (with crowded erect branches), CLUSTER-STEM NAILWORT, FORKED CHICKWEED. Annual to 30 cm tall, branched well above base, low and spreading; leaves flat, oblanceolate to linear-elliptic, to 15 mm long and 3 mm wide; sepals slightly mucronate. Dry woods or sandy openings; Dallas Co. (Turner 1983); mainly scattered localities in e TX. Jun-Aug.

Paronychia jamesii Torr. & A. Gray, (for Dr. Edwin James, 1797–1861, American botanical explorer of the Rocky Mt. area), JAMES' NAILWORT. Perennial branching from woody base, to ca. 30 cm tall; stems minutely pubescent; leaves gray-green; sepals light yellow, 2–3 mm long including the terminal awn-like cusp. Rocky, sandy, open ground; Archer, Clay, and Coleman cos.; according to Turner (1983) this species occurs in the w 1/2 of TX w of a line through Clay, Jack, Palo Pinto, Erath, Comanche, and Brown cos. Jun–Sep. [*P. jamesii* Torr. & A. Gray var. *praelongifolia* Correll]

Paronychia lindheimeri Engelm. ex A. Gray, (for Ferdinand Jacob Lindheimer, 1801–1879, German-born TX collector), LINDHEIMER'S NAILWORT. Annual to 30 cm tall, much branched from near base; leaves to 15 mm long; calyces short pubescent at base, the sepals 1.7–2 mm long, terminating in an awn-like cusp. Rocky, sandy, or gravelly areas; Burnet and Williamson cos. (Turner 1983); mainly e Edwards Plateau with isolated stations in e TX and the Trans-Pecos. Jul–Nov. [*P. chorizanthoides* Small] Stanford (1976) recognized *P. chorizanthoides* (endemic to c TX) based on the following distinctions:

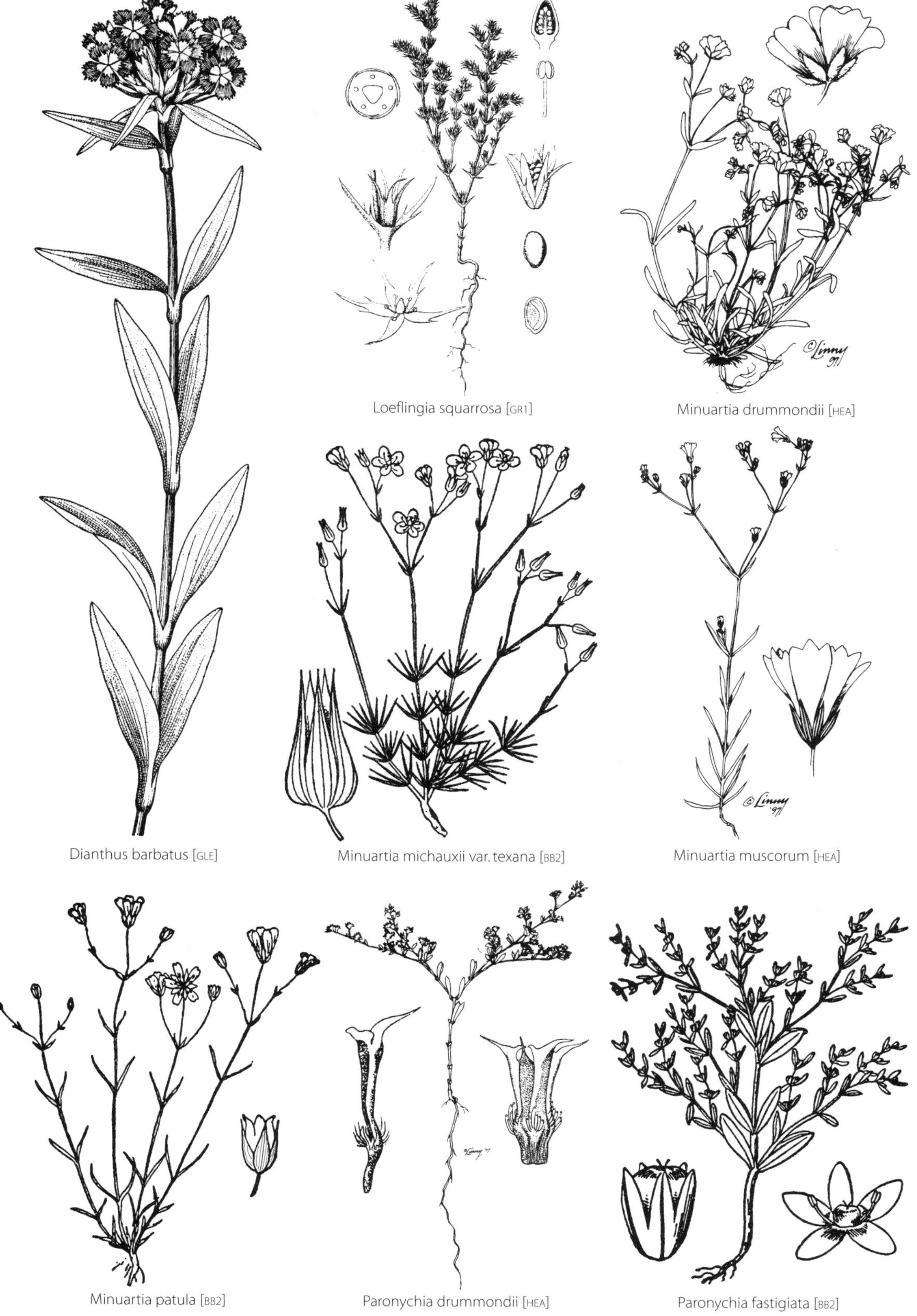

Loeflingia squarrosa [GR1]

Minuartia drummondii [HEA]

Dianthus barbatus [GLE]

Minuartia michauxii var. texana [BB2]

Minuartia muscorum [HEA]

Minuartia patula [BB2]

Paronychia drummondii [HEA]

Paronychia fastigiata [BB2]

1. Flowers mostly clustered; calyces (including awns) over 2 mm long; plants pubescent or puberulent; endemic to central Texas _____ *P. chorizanthoides*
1. Flowers mostly solitary and separate; calyces (including awns) usually 2 mm long or less; [plants glabrous or at most minutely scabrous]; Edwards Plateau with disjunct stations in east Texas and Trans-Pecos _____ *P. lindheimeri*

Paronychia virginica Spreng., (of Virginia), PARKS' NAILWORT, BROOM NAILWORT. Perennial branching from woody base, to 40 cm tall, glabrous; leaves green or yellowish green; sepals yellowish, greenish yellow to brownish, ca. 3 mm long, with a short cusp. Limestone outcrops; mainly Blackland Prairie and Grand Prairie sw to Edwards Plateau; according to Turner (1983, 1995b) this species occurs e of a line through Montague, Jack, Parker, Erath, and Mills cos. Aug–Oct. [*P. parksii* Cory, *P. virginica* Spreng. var. *scoparia* (Small) Cory]

PETRORHAGIA

An Old World genus of 33 species (R. Rabeler, pers. comm.; including *Kohlrauschia* and *Petrorhagia* sensu stricto as recognized by Bittrich 1993) ranging from the Canary Islands and Mediterranean region to Kashmir. Some are cultivated as ornamentals. (Greek: *petros*, rock, and *rhagas*, chink or fissure, from principal habitat of some species)
REFERENCES: Shinners 1969; Rabeler 1985.

Petrorhagia dubia (Raf.) G. López & Romo., (doubtful), CHILDING-PINK. Erect annual to ca. 90 cm tall; stem leaves linear, to ca. 5 cm long and 2 mm wide; inflorescences terminal, with conspicuous bracts subtending flowers or clusters; flowers solitary or in several-flowered head-like clusters; calyces 10–13 mm long; petals pink or purplish, 10–14 mm long. Roadsides and fields; Cooke and Grayson cos., also Limestone Co. (HPC); weedy in e TX, also Hatch et al. (1990) cited vegetational area 7 (Edwards Plateau); first reported for TX by Shinners (1969) who suggested it was possibly introduced with *Lolium perenne* (RYE GRASS) by the Texas Highway Department. Rabeler (1985) indicated that introduction with *Trifolium incarnatum* (CRIMSON CLOVER) was also a possibility. Apr–Jun. Native of the Mediterranean region. [*P. velutina* (Guss.) P.W. Ball & Heywood, *P. prolifera* of TX auth., not (L.) P.W. Ball & Heywood]

POLYCARPON POLYCARP

A genus of 18 species, 16 in Europe and the Mediterranean, and 2 in South America (Bittrich 1993). (Greek: *poly*, many, and *carpus*, fruit)

Polycarpon tetraphyllum (L.) L., (four-leaved), FOUR-LEAF MANYSEED. Plant small, prostrate, glabrous, much-branched, usually annual; stems to ca. 15 cm long; leaves opposite or mostly in whorls of four, 2–8(–15) mm long, 1–4(–8) mm wide, apically obtuse to rounded, entire, subsessile or on petioles to 1.2 mm long; stipules scarious, ovate, 1–3 mm long; flowers numerous, in dense cymes; bracts scarious; pedicels 0.5–3 mm long; sepals 5, 1.5–2.5 mm long; petals 5, white, 0.5–0.8 mm long, oblanceolate; stamens usually 3–5; styles partly united; capsules 3-valved, many-seeded. Sandy or silty soils, openings; Fort Hood (Bell or Coryell cos.—Sanchez 1997); se and s TX w to Edwards Plateau and n to Post Oak Savannah and s part of nc TX. Mar–Jul. Native of Europe and the Mediterranean.

SAGINA PEARLWORT

A n temperate and tropical mountain genus of ca. 25 species (Bittrich 1993) of usually tufted herbs. (Latin: *sagina*, fodder or fattening; previously applied to *Spergula* which was used as early forage)
REFERENCES: Crow 1978, 1979.

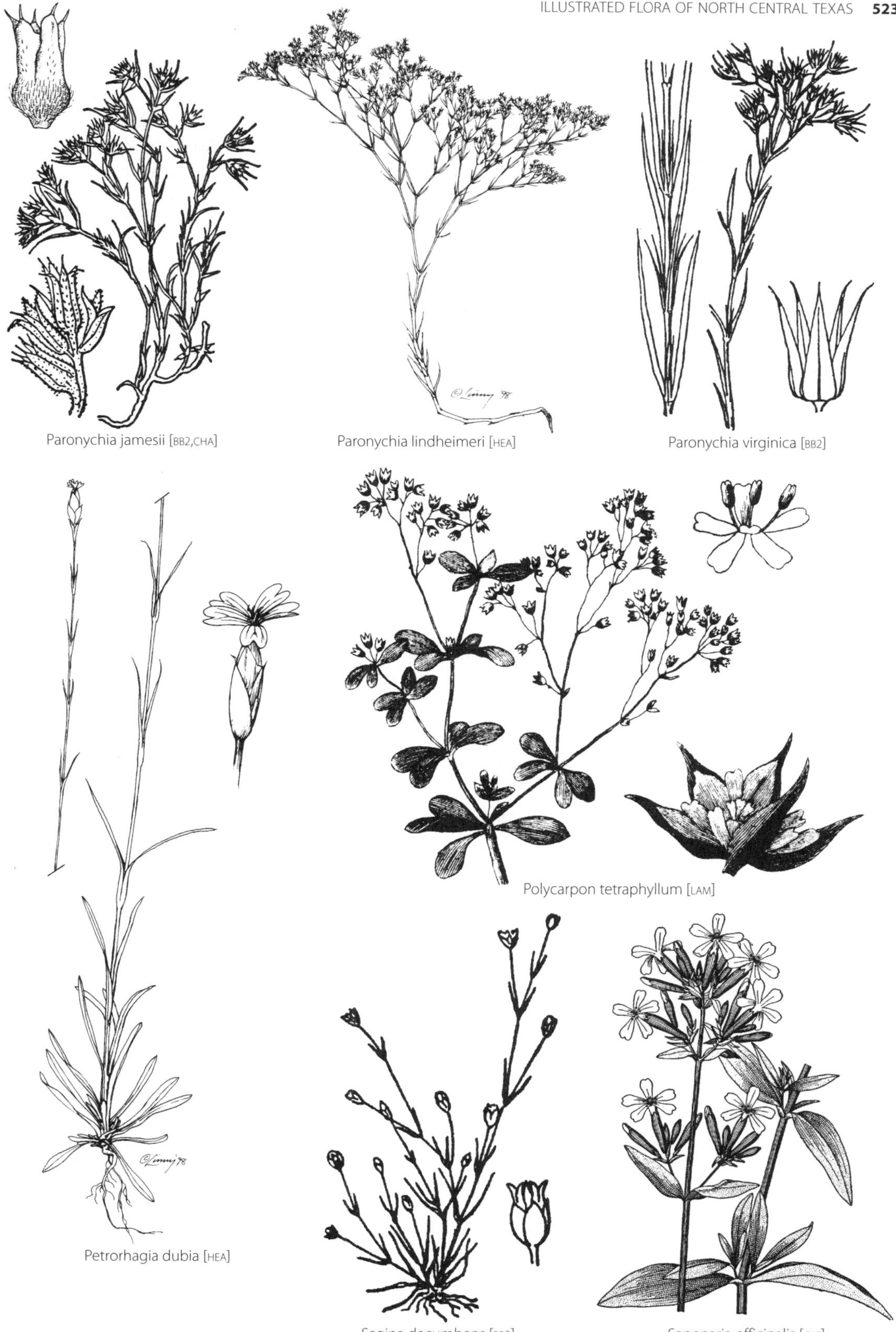

Paronychia jamesii [BB2,CHA]

Paronychia lindheimeri [HEA]

Paronychia virginica [BB2]

Polycarpon tetraphyllum [LAM]

Petrorhagia dubia [HEA]

Sagina decumbens [BB2]

Saponaria officinalis [GLE]

Sagina decumbens (Elliott) Torr. & A. Gray, (trailing with tips upright), TRAILING PEARLWORT. Inconspicuous, small, to ca. 10(–17) cm tall, erect or partly decumbent annual, largely glabrous or with some glandular hairs above; leaves linear to linear-subulate, 3–15 mm long, very narrow, < 1 mm wide; flowers axillary and terminal, long pedicelled (pedicels 3–25 mm long); sepals 4–5, 1.4–2.5 mm long; petals none or 1–5, entire, white, usually slightly shorter than the sepals to sometimes slightly longer; capsules with 4–5 valves. Sandy open woods, roadsides, disturbed sites; se and e TX w to West Cross Timbers, also Edwards Plateau and Trans-Pecos. Late Mar–early May.

SAPONARIA BOUNCING-BET, SOAPWORT

A temperate Eurasian genus of ca. 40 species (Bittrich 1993). (Latin: *sapo*, soap, referring to the mucilaginous juice of *S. officinalis* forming a lather with water)

Saponaria officinalis L., (medicinal), BOUNCING-BET, SOAPWORT, FULLER'S-HERB. Erect glabrous perennial to 30–80(–150) cm tall; leaves 3–10 cm long, 1–5 cm wide, prominently 2–5 nerved; flowers showy; calyx tube 12–20 mm long; calyx lobes ca. 2–5 mm long; petals 5, white or pink; petal claw prominent, 10–20 mm long; petal blade 8–18 mm long, notched apically; stamens 10; capsules 10–12 mm long, 4–6 mm wide. Open areas and waste places; Tarrant Co.; also e TX. Jun–Sep. Native of Europe. Formerly used as a soap substitute, when the rootstocks were beaten in water (Baumgardt 1982); all parts of the plant contain saponins; toxic and potentially lethal to livestock but rarely eaten because it is distasteful (Kingsbury 1964; Burlage 1968; Stephens 1980).

SILENE CATCHFLY, CAMPION

Erect annuals or perennials; flowers in terminal panicles or cymes; calyces tubular to campanulate or funnelform; petals often clawed; fruit in ours a 6-toothed capsule.

A n hemisphere genus of nearly 700 species (Bittrich 1993) of mostly herbs. A number are cultivated as ornamentals. (Name adopted by Linnaeus from earlier authors; probably from mythical Greek *Silenus*, intoxicated foster father of Bacchus being described as covered with foam, from sticky secretions of many species)

REFERENCES: Hitchcock & Maguire 1947; Maguire 1950.

1. Middle stem leaves opposite; stem leaves to ca. 15 mm wide; petals usually 5–10 mm long, 2-lobed ____ **S. antirrhina**
1. Middle stem leaves in whorls of 4; stem leaves to ca. 40 mm wide; petals usually 10–20 mm long, fimbriately 8–12-lobed ____ **S. stellata**

Silene antirrhina L., (possibly for likeness to *Antirrhinum*-snapdragon, whose flowers are said to resemble a snouted dragon; from Greek: *Anti*, against, opposed to, like, and *Rhis*, nose or snout), SLEEPY CATCHFLY. Annual to 1.2 m tall (usually much smaller); leaves linear- or oblong-lanceolate; flowers small, open only in afternoon and evening; calyx tube 6–9 mm long; calyx lobes 0.5–1.6 mm long; petals white to partly (back or crown) or wholly rose-lavender or red-violet, occasionally absent; crown minute or absent. Open woods, prairies, disturbed sites; throughout much of TX. Late Apr–early Jun. Reported to be poisonous (Burlage 1968).

Silene stellata (L.) W.T. Aiton, (starry), WHORLED SILENE, STARRY CAMPION, WIDOW'S-FRILL. Perennial to 1.2 m tall; leaves lanceolate to elliptic or ovate; calyx tube 5–10 mm long; calyx lobes 2–5 mm long; petals large, white, showy, with wide, ragged apex, the claw not well-differentiated from blade; crown absent. Woods and thickets, sandy or clayey soils; in nc TX known only from Dallas, Denton, and Grayson cos.; also e TX. Late May–early Jul.

SPERGULARIA SANDSPURRY

Annuals; leaves linear to filiform; stipules scarious; flowers usually numerous in cyme inflorescences, pink or whitish, 5-merous; styles 3; capsules 3-valved.

A cosmopolitan genus of ca. 25 species (Bittrich 1993) including a number of halophytes. (Derived from the name of a related genus, *Spergula*)
REFERENCE: Rossbach 1940.

1. Plants glabrous; sepals 0.8–1.6 mm long; capsules 1.4–2.6 mm long; inflorescences generally 4–7+ times compound ____ **S. platensis**
1. Plants glandular-pubescent (at least branches of inflorescence); sepals 2.5–5 mm long; capsules 3–6.5 mm long; inflorescences simple or 1–3+ times compound ____ **S. salina**

Spergularia platensis (Cambess.) Fenzl, (from the district of the Río de la Plata in South America). Glabrous low annual often forming mats; leaves thread-like, 1–3 cm long, ca. 1 mm wide; flowers in loose terminal cymes; stipules triangular, 1.5–3.5 mm long, ca. as long as broad or a little longer; sepals broadly lanceolate; petals white, small or rarely none; stamens 5. Collected at Dallas by Reverchon in April, 1880, in wet sandy soil, later reported by him as "common," but not found there since; also s TX and Edwards Plateau. Native of s South America.

Spergularia salina J. & K. Presl, (growing in salty places), SALT-MARSH SANDSPURRY, MARSH SANDSPURRY. Erect or diffuse fleshy annual ca. 5–20 cm tall; leaves linear, 0.8–2.5(–4) cm long, to 1.5 mm wide; stipules conspicuous, membranous, triangular, 2–4 mm long, ca. as long as broad or a little longer; flowers in leafy bracted cymes, the inflorescences often making up much of the plant; sepals ovate, exceeding the pink or white petals; stamens 2–5. In brackish or saline soils; Brown (HPC) and Tarrant (Mahler 1988) cos.; mostly coastal and along the Rio Grande. Apr–Sep. Native of Europe. [*S. marina* (L.) Griseb.]

STELLARIA CHICKWEED, STARWORT, STITCHWORT

Ours low annual herbs; stipules absent; flowers solitary or in terminal cymes of usually 3–7, 5-merous; petals split more than half way to base (corollas thus appearing to be of 10 petals) or petals absent; stamens 1–10; styles usually 3; capsules dehiscent by 6 valves.

A cosmopolitan genus of ca. 150–200 species (Bittrich 1993) including some cultivated ornamentals. The key to species is modified from Rabeler and Reznicek (1997) with additional characters from Morton (1972). We are following Kent (1997) for nomenclature of *S. pallida*. (Latin: *stella*, a star, in allusion to the star-shaped flowers)
REFERENCES: Whitehead & Sinha 1967; Morton 1972; Rabeler 1988; Kent 1997; Rabeler & Reznicek 1997.

1. Flowers usually open, with white petals; plants green, rarely yellowish green; sepals 3.5–7 mm long, uniformly green; seeds dark brown, (0.8–)0.9–1.3 mm in diam., the surface covered with wavy, blunt papillae; capsules 4.5–9 mm long; stamens 3–10; anthers red-violet ____ **S. media**
1. Flowers usually cleistogamous, apetalous; plants yellowish green; sepals seldom over 3 mm long, often with a red basal band; seeds yellowish brown, 0.5–0.8 mm in diam., the surface covered with acute papillae; capsules 3–4 mm long; stamens usually 1–3(–5); anthers gray-violet ____ **S. pallida**

Stellaria media (L.) Vill., (intermediate, the middle), TENPETAL, COMMON CHICKWEED, CHICKWEED STARWORT. Annual, possibly overwintering in protected places; stems prostrate to ascending, usually pubescent in lines; lower leaves petioled, upper sessile; leaf blades oblong-lanceolate to ovate-lanceolate, widely tapered to abruptly narrowed at base; sepals 3.5–7 mm long; petals white, slightly shorter than the sepals, 2-lobed nearly to base and thus appearing as 10 petals; capsules usually pointing downward, equal or exceeding the calyces by 1–2 mm; $2n$ = 40, 42, 44 (Morton 1972). Widespread weed of stream bottoms, lawns, disturbed sites; nearly throughout TX. Feb–Apr. Native of Europe. This now cosmopolitan weed is usually autogamous. Reported to cause digestive disorders in sheep (Burlage 1968).

Stellaria pallida (Dumort.) Crép., (pale), LESSER CHICKWEED. Annual similar to *S. media*; differing as described in the key; $2n$ = 22 (Morton 1972). Disturbed areas, typically in sand; a Hopkins Co. collection (*Reznicek 10361*, BRCH, MICH) near the e edge of nc TX is the first report for this species from TX (Rabeler & Reznicek 1997); in April of 1998 the species was found in Denton and Grayson cos. (e.g., *Rabeler & Diggs 1320*);because of its similarity to *S. media*, this is an easily overlooked species; it is to be expected in other parts of nc TX and possibly elsewhere in the state, especially early in the spring. Spring. Native of Europe. [*S. media* subsp. *pallida* (Dumort.) Asch. & Graebn., *S. apetala* of authors, not Ucria] First documented in North America at Kitty Hawk, NC in 1969 (Morton 1972).

VACCARIA

A genus of 1 or 4 species (Bittrich 1993) of Eurasia and the Mediterranean. (Latin: *vacca*, cow, from use as fodder or prevalence in pastures)

Vaccaria hispanica (Mill.) Rauschert, (Spanish), COWCOCKLE, COWHERB. Erect, glabrous, usually wide-branched annual to 60 cm tall; leaves ovate- to broadly oblong-lanceolate, auricled-clasping; flowers in terminal, open cymes, long-pedicelled; calyces tubular in flower, ovoid, inflated in fruit, sharply 5-angled, the tube 8–14 mm long, the lobes 1.5–3 mm long; petals long-clawed, lavender-pink, showy; capsules dehiscent by 4 valves. Roadsides and disturbed areas, at margins of grain fields; also cultivated as an ornamental; Bosque, Collin, Dallas, Denton, Grayson, Rockwall, and Tarrant cos., also Brown, Erath (Stanford 1976), and Fannin (VDB) cos.; nc TX s to Edwards Plateau. Late Apr–early Jun. Native of Eurasia and the Mediterranean. [*Saponaria vaccaria* L., *V. pyramidata* Medik.]

CELASTRACEAE BITTERSWEET OR STAFFTREE FAMILY

Ours shrubs, small trees, or woody vines; leaves simple, alternate or opposite, deciduous; stipules minute, falling early; inflorescences terminal or axillary; pedicels jointed; flowers small, radially symmetrical, 4- or 5-merous, perfect or unisexual; stamens inserted on margin of a prominent disk; ovary superior; fruit a loculicidal capsule; seeds arillate.

A medium-large (1,300 species in 88 genera), mainly tropical and subtropical to temperate family of mostly trees, shrubs, or woody vines; laticifers are usually present. The African *Catha edulis* (Vahl) Endl. (KHAT, QAT, CAFTA) is cultivated for the leaves which are chewed by Moslems as a daily stimulant in countries such as Ethopia, Somalia, and Yemen; in the U.S. it is considered a controlled substance and was of concern when the U.S. deployed troops to Somalia in 1993 (Baker 1993). (subclass Rosidae)

FAMILY RECOGNITION IN THE FIELD: shrubs, small trees, or woody vines with simple leaves, small flowers with a *disk* beneath or ± surrounding the ovary, and often leathery capsules containing seeds usually with *brightly colored arils*.

REFERENCES: Brizicky 1964a; Lundell 1969b.

1. Leaves alternate; twining, high-climbing, woody vine; flowers in terminal racemes or panicles, 5-merous ____ **Celastrus**
1. Leaves opposite; erect shrub or small tree; flowers in axillary cymes, 4-merous ____ **Evonymus**

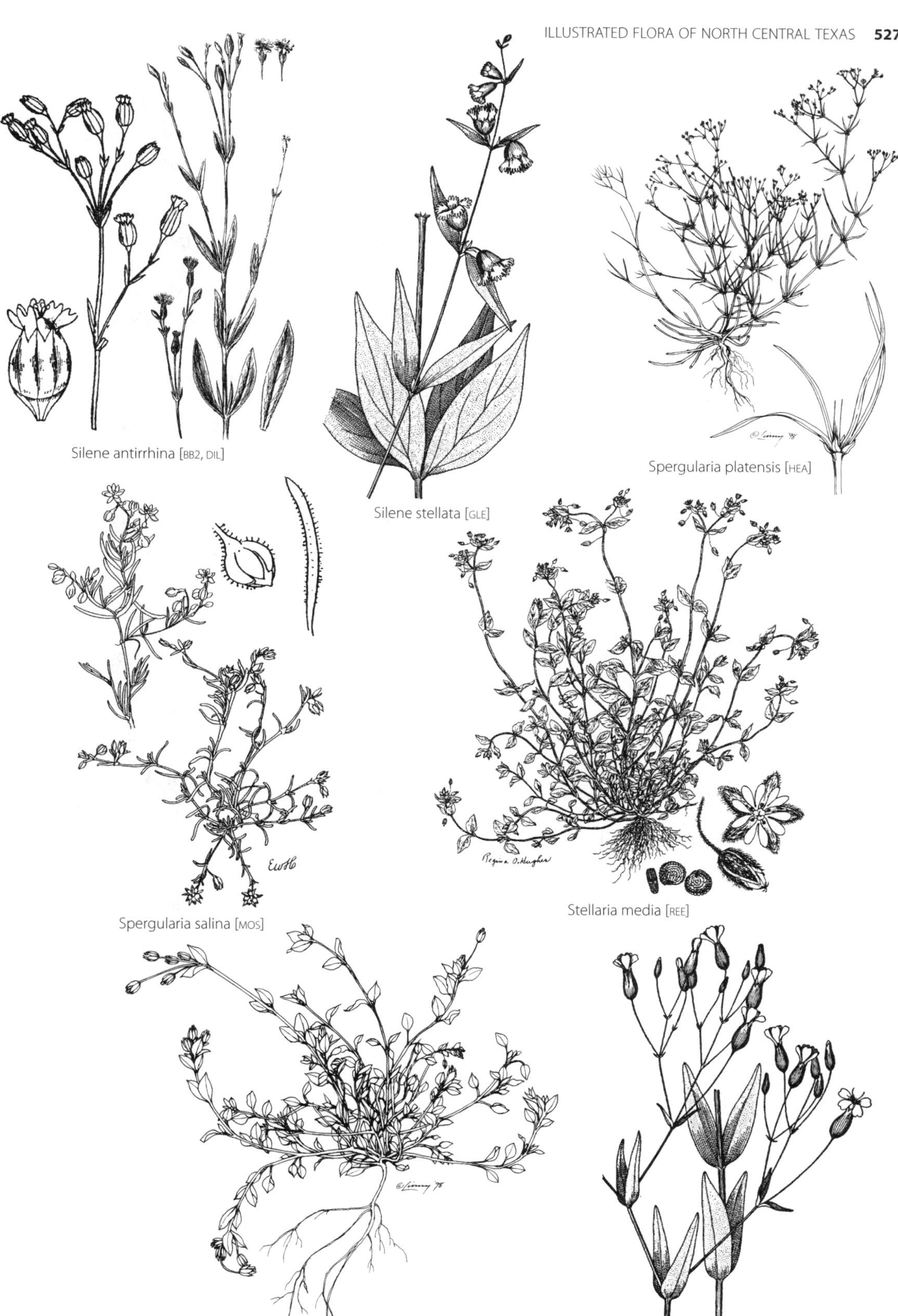

Silene antirrhina [BB2, DIL]

Silene stellata [GLE]

Spergularia platensis [HEA]

Spergularia salina [MOS]

Stellaria media [REE]

Stellaria pallida [HEA]

Vaccaria hispanica [GLE]

CELASTRUS BITTERSWEET

A genus of 32 species native from the tropics to warm temperate areas; some are cultivated as ornamentals or used medicinally. (From *celastros*, an ancient Greek name for some evergreen tree)

Celastrus scandens L., (scandent, climbing), CLIMBING BITTERSWEET, AMERICAN BITTERSWEET. Twining, climbing shrub/liana to 18 m; main stem to 2.5 cm in diam.; leaves elliptic, ovate to obovate, 5–11 cm long, 2.5–6 cm wide, serrulate-crenulate; petioles ca. 1–2(–3) cm long; inflorescences ca. 3–8 cm long; flowers unisexual, greenish, small, ca. 4–5 mm across; fruits globose, orange or orange-yellow, 3-valved, 6–12 mm in diam.; seeds with conspicuous scarlet to crimson aril. Woods; native to n U.S. and in TX to mountains of Trans-Pecos, apparently introduced in Tarrant Co.; cultivated for the colorful fruits and arils. Apr–Jun. Possibly poisonous to humans and livestock (Kingsbury 1964; Blackwell 1990).

EVONYMUS SPINDLETREE, STRAWBERRY-BUSH

A genus of 177 species of evergreen or deciduous trees or shrubs native to n temperate areas, especially Asia; also found in Australia; many are cultivated as ornamentals for their bright autumn foliage, evergreen leaves, or showy capsules with scarlet to orange, bird-dispersed, arillate seeds; the seeds of many species have cardiotoxic glycosides and the leaves, bark, and seeds of some species are poisonous (Hardin & Arena 1974). The genus is sometimes spelled *Euonymus*. (Greek: *eu*, good, and *onoma*, name, but used ironically, the plants having had the bad reputation of poisoning cattle)
REFERENCE: Lundell 1941.

Evonymus atropurpurea Jacq., (dark purple). Erect shrub or small tree to 8 m tall; bark smooth, gray or green; leaves lanceolate, acute, finely and sharply toothed, with distinct petioles 5–20 mm long; inflorescences pedunculate, of usually 5–19 flowers; flowers perfect, 4-merous, greenish purple to dark-red; fruit a pinkish to reddish or purple, usually deeply 4-lobed, smooth capsule ca. 15 mm in diam.; seeds with scarlet aril. Limestone soils, stream bottom woods. Late Apr–early Jun. This species is the source of wahoo root bark, formerly used medicinally, and a medicinal resin, called euonymin; euonymin and the fruits are apparently toxic (Duke 1985); Burlage (1968) reported the plant to be a violent purgative, dangerous to livestock and children, with symptoms ranging from vomiting to mental symptoms and loss of consciousness.

1. Lower surface of leaf blades with persistent pubescence, at least on the veins (use lens); leaf blades elliptic to ovate, acute or abruptly short-acuminate ____ var. **atropurpurea**
1. Lower surface of leaf blades glabrous; leaf blades lanceolate to elliptic, long attenuate at apex ____ var. **cheatumii**

var. **atropurpurea**, WAHOO, EASTERN WAHOO, BURNING-BUSH, INDIAN ARROW-WOOD, SPINDLETREE. Cooke, Dallas, Grayson, Kaufman, Lamar, Tarrant, and Wise cos., also Bell and Delta cos. (Little 1976 [1977]); in TX only in nc part of state.

var. **cheatumii** Lundell, (for E.P. Cheatum, a colleage of C.L. Lundell at Southern Methodist Univ.), WAHOO. Only known from Dallas Co. (Urbandale); endemic to nc TX. According to Lundell (1969b), "A local endemic population, the stand of which was decimated by a scale insect in 1944." We have seen only five sheets of this variety and all of these were collected in the same area in the period 1940–1944. While we are following Lundell (1969b), Correll and Johnston (1970), Kartesz (1994), and Jones et al. (1997) in recognizing this variety, it is unknown if the population still exists and it is not completely clear that it should be given varietal recognition. Lundell (1941) indicated that intermediate forms exist in Dallas Co.

Evonymus americana L., (of America), (STRAWBERRY-BUSH, BURSTING-HEART, BROOK EVONYMUS), with 5-merous flowers, tuberculate capsules, and leaves sessile or nearly so (petioles < 5 mm long), occurs in e TX just e of nc TX.

CERATOPHYLLACEAE HORNWORT FAMILY

A very small cosmopolitan family of 1 genus and 6 species (Les 1997). *Ceratophyllum* is an extremely reduced (e.g., no roots) and highly specialized aquatic. Some molecular analyses have indicated that *Ceratophyllum* is the sister group of all other angiosperms and suggested the genus is a vestige of an ancient angiosperm lineage that diverged early from the line leading to other living angiosperms (Chase et al. 1993; Qiu et al. 1993); a more recent molecular analysis placed *Ceratophyllum* as the sister group to the monocots (Soltis et al. 1997). Its unsealed carpels, among other distinctive characters, makes the genus anomalous among flowering plants (Qui et al. 1993). (subclass Magnoliidae)

FAMILY RECOGNITION IN THE FIELD: rootless aquatics floating below water's surface, with whorled leaves *forked* into nearly thread-like segments.

REFERENCES: Wood 1959; Les 1988a, 1993, 1997; Endress 1994.

CERATOPHYLLUM HORNWORT, COON-TAIL

Perennial, bushy-branched, rootless, aquatic herbs, floating completely submerged; leaves whorled, sessile, 1–4 times dichotomously forked into nearly thread-like segments; flowers solitary in 1 axil of a whorl, sessile, without perianth but with involucre of 8–12 minute bracts, unisexual, both sexes on the same plant; stamens 4–10; pistil 1; style persistent; ovary superior.

In some parts of the world bilharzia-carrying snails and malaria-carrying mosquito larvae are sheltered by these floating aquatics; they can also choke waterways and disturb operations of hydroelectric plants; on the positive side, they oxygenate the water and provide cover for baby fishes. The flowers are water-pollinated (= hydrophilous) with the anthers breaking off and floating to the surface of the water; the pollen grains are then released at the surface and slowly sink to the female flowers—pollination thus occurs underwater; technically this type of pollination is known as hypohydrophily (pollination below the water surface) in contrast to epihydrophily (pollination at the water surface). *Ceratophyllum* is one of only two dicot genera for which hydrophily has been documented; the other is *Callitriche* in the Callitrichaceae (Cox 1988; Philbrick 1991, 1993; Philbrick & Osborn 1994). When sterile, HORNWORTS are often confused with other aquatics, such as *Myriophyllum*, *Najas*, or the algae *Chara*; however, they can be recognized by their forked leaves. (Greek: *ceras*, a horn, and *phyllon*, leaf, from stiff leaf divisions)

REFERENCES: Lowden 1978; Godfrey & Wooten 1981; Les 1986a, 1986b, 1988b, 1988c, 1989; Chase et al. 1993; Qiu et al. 1993; Schneider & Carlquist 1996.

1. Leaves usually forked 1 or 2(–3) times, usually with 2–4 ultimate segments, the segments with 4–5 conspicuous, widely spaced teeth on 1 side (the teeth visible to the naked eye); achenes without spines except at very base _____ **C. demersum**
1. Leaves usually forked 2–4 times, usually with 3–8 ultimate segments, the segments entire or obscurely serrate (the teeth not visible to the naked eye); achenes with spines along lateral margins and at base _____ **C. echinatum**

Ceratophyllum demersum L., (under water). Leaves 5–12 per whorl, 1–3 cm long, ultimate segments to ca. 0.5 mm wide, to ca. 1 mm at the teeth; involucral bracts 1–2 mm long; style to ca. 10 mm long; achenes 4–6 mm long. Streams, lakes, and ponds; se and e TX w to Cooke, Dallas, Fannin, and McLennan cos., also Edwards Plateau. By far the most common of the 2 species in nc TX. May–Sep. This species reproduces primarily asexually; it grows prolifically and in some situations can be a serious weed (Les 1997).

Ceratophyllum echinatum A. Gray, (prickly). Similar to *C. demersum*; style to ca. 6 mm long; achenes 5–7 mm long. Streams, lakes, and ponds; included based on citation of vegetational area 4 (Fig. 2) by Hatch et al. (1990) for *C. muricatum* into which they lumped *C. echinatum*; no other nc TX specimens or citations found; mainly se and e TX, also Edwards Plateau. Summer. Jones et al. (1997) treated this taxon as *C. muricatum* Cham. subsp. *australe* (Griseb.) Les. However, Les (1997) indicated that *C. muricatum* is known in North America only from coastal se United States, while *C. echinatum* occurs in e TX w to near the edge of nc TX.

CHENOPODIACEAE GOOSEFOOT OR PIGWEED FAMILY

Ours annual, biennial, or perennial herbs with stems often ± succulent; leaves alternate (rarely opposite), simple, the blades entire or rather coarsely or irregularly and bluntly toothed or lobed, often ± succulent; stipules absent; flowers 1–many, glomerate in spikes or panicles of spikes or axillary, perfect or unisexual, very small; sepals 0–5, green, sometimes with white to yellowish or pink margins; petals absent; stamens 0–5; pistil 1; ovary usually superior; fruit a 1-seeded indehiscent or irregularly rupturing urticle.

A medium-large (1,300 species in 103 genera) cosmopolitan family, especially of desert or dry areas; most are herbaceous with some shrubs or rarely small trees; many are halophytes (= capable of living in areas of high salt concentrations) or xerophytes. The family includes food plants such as *Beta vulgaris* L. (BEET, SUGAR BEET, and SWISS CHARD), *Chenopodium quinoa* Willd. (the Andean crop QUINOA), and *Spinacia oleracea* L. (SPINACH) as well as a number of agricultural weeds. They exhibit the unusual, reddish, nitrogen-containing pigments known as betalains (characteristic of most Caryophyllidae—Cronquist & Thorne 1994) which derive their name from the genus *Beta*). Many species have Kranz anatomy and the associated C4 photosynthetic pathway, an integrated set of anatomical and physiological adaptations for hot dry conditions. Molecular evidence indicated the family is most closely related to Amaranthaceae (Downie & Palmer 1994). ☠ Kühn (1993) cited references indicating that many weedy Chenopodiaceae produce allergy-causing pollen that can give rise to asthma and rhinitis. (subclass Caryophyllidae)

FAMILY RECOGNITION IN THE FIELD: herbs often with ± succulent stems and leaves; leaves usually alternate, simple; flowers very small, inconspicuous, apetalous, greenish, not subtended by scarious bracts, often in spikes; filaments separate. Amaranthaceae have similar flowers (e.g., very small, lacking petals), but have scarious (= dry, papery), often colored bracts below each flower and united filaments.

REFERENCES: Standley 1916; Reed 1969a; Kühn 1993; Wilken 1993a; Behkne & Mabry 1994; Downie & Palmer 1994.

1. Stem leaves petioled or with narrowed, petiolar base; leaf blades usually > 2 mm wide (often much greater), not bristle-tipped.
   2. Root greatly thickened, often pinkish or reddish in color; fruits strongly adherent in clusters; introduced cultivar (BEETS) that possibly escapes ______ **Beta**
   2. Roots not as above; fruits not adherent (but can be in clusters); native and introduced species widespread in nc TX.
      3. Flowers all unisexual, the staminate and pistillate often in different inflorescences or in the axils of different leaves; pistillate flowers subtended by 2 bracts, these fused 1/2 or more their length and enclosing the fruit.
         4. Leaves and stems glabrous, green; leaf blades 4–10 cm long; lower leaves with conspicuous petioles 10–50 mm long; stigmas 4–5 ______ **Spinacia**
         4. Leaves and stems (at least when young) covered with scales giving the plant a grayish appearance, sometimes glabrate; leaf blades 1–5 cm long; lower leaves often sessile or nearly so OR with petioles to 20 mm long; stigmas 2–3 ______ **Atriplex**

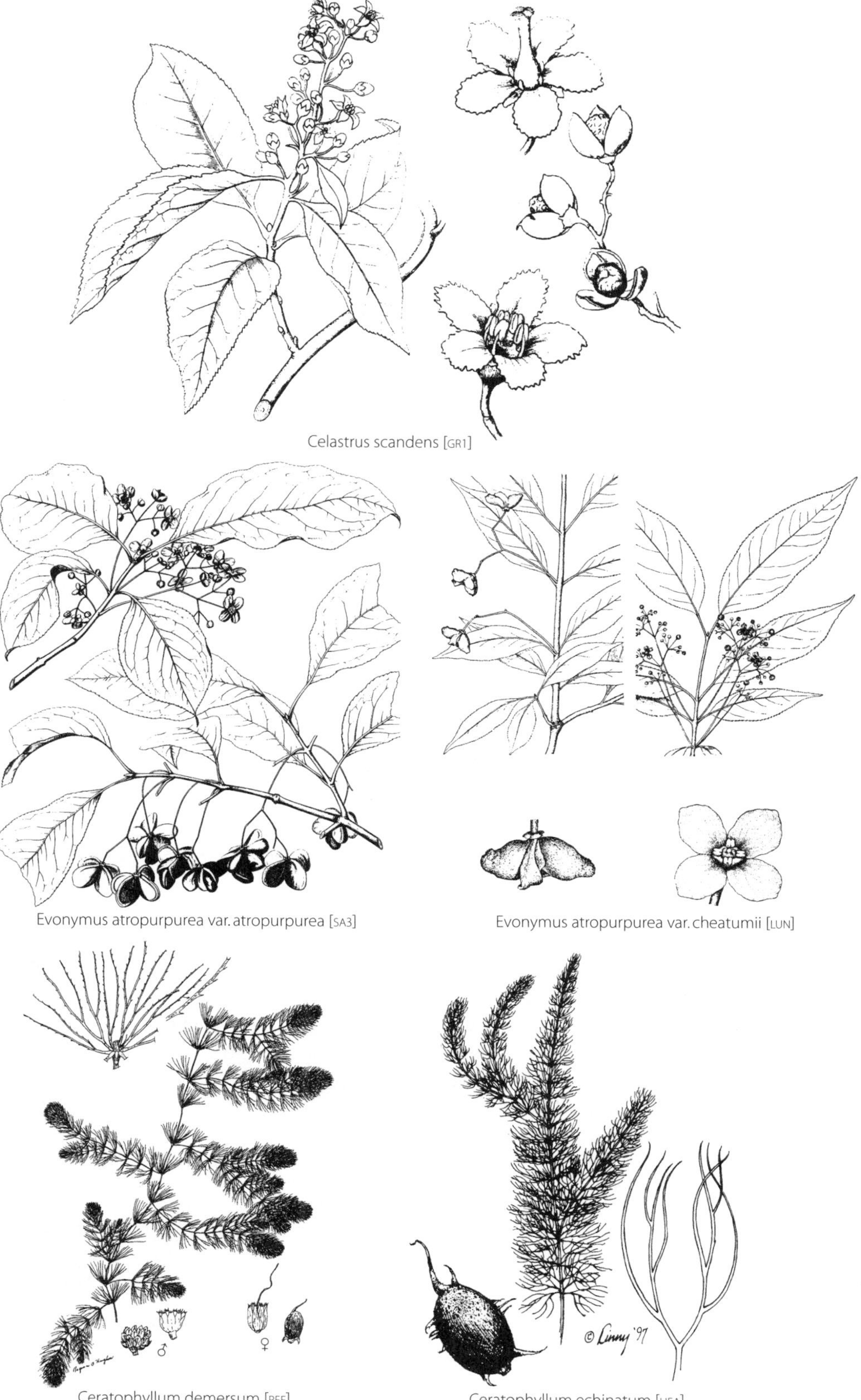

Celastrus scandens [GR1]

Evonymus atropurpurea var. atropurpurea [SA3]

Evonymus atropurpurea var. cheatumii [LUN]

Ceratophyllum demersum [REE]

Ceratophyllum echinatum [HEA]

3. Flowers generally with both sexes (some can be unisexual); flowers not subtended by 2 fused bracts; fruit enclosed only by sepals.
   5. Leaf blades various, but usually not arrowhead-shaped; sepals 4–5, ovate-lanceolate or oblong-ovate; stamens 3–5 per flower.
      6. Plants with silky or woolly pubescence (sometimes glabrate except around flowers); fruiting sepals with a wing running across their backs.
         7. Fruiting sepals each with a separate wing across the back; leaf blades entire; plants not tumbleweed-like ____ **Kochia**
         7. Fruiting sepals not with a separate wing each, rather the calyces with an irregularly lobed but continuous wing going all the way around; leaf blades sinuate-dentate; plants with roundish, tumbleweed-like appearance ____ **Cycloloma**
      6. Plants with glandular or mealy-appearing pubescence (sometimes glabrate); fruiting sepals rounded on back or keeled lengthwise, not winged ____ **Chenopodium**
   5. Larger leaf blades typically arrowhead-shaped with 2 basal lobes; sepal 1, linear-oblanceolate or linear-obovate; stamen 1 per flower ____ **Monolepis**

1. Stem leaves sessile or clasping, widest at base; leaf blades very narrow (2 mm or less wide), short, stiff, bristle-tipped ____ **Salsola**

ATRIPLEX SALTBUSH, ORACHE

Annual or perennial, monoecious or dioecious herbs or sometimes ± woody; surfaces usually gray-scurfy, at least when young; leaves alternate or alternate above and opposite below; leaf blades entire; staminate flowers ebracteate, with 5-parted perianth and 5 stamens; pistillate flowers subtended by 2 bracteoles, without perianth or perianth of minute squamellae; fruit enclosed by bracts.

A genus of ca. 300 species of temperate and warm areas; a number are used as salt-tolerant forage; many are edible and some were used by Native Americans. The Asian *A. hortensis* L. (ORACHE) is widely cultivated for its leaves, which are used like spinach. (Ancient Latin name)
REFERENCES: Hall & Clements 1923; Brown 1956; Freeman et al. 1984; McGregor 1986.

1. Leaf blades widest below middle; fruiting bracts not winged (but sometimes with tubercles or appendages), 4–8 mm long, usually not united to apex; plants annual monoecious herbs 0.15–0.6 m tall ____ **A. argentea**
1. Leaf blades usually widest at or beyond middle; fruiting bracts very conspicuously winged lengthwise, (4–)8–10(–15) mm long, united nearly to apex; plants usually dioecious ± woody perennials 0.4–1.5(–2.5) m tall ____ **A. canescens**

Atriplex argentea Nutt., (silvery). Annual monoecious herb with a ± globose shape; stems erect, 0.15–0.6 m tall, branched from base; leaves alternate or sometimes opposite below; leaf blades 2–5 cm long, triangular-ovate to rounded-ovate; staminate flowers in upper axils or in short dense spikes or the staminate and pistillate mixed in axils near middle of the plant; bracts subtending fruits 2–8 mm wide, united at least 1/2 their length, subentire to irregularly laciniate, not winged but the faces sometimes with appendages or tubercles to 2 mm long. Alkaline areas; w margin of nc TX w to w TX. Jul–Sep.

1. Upper leaves usually subsessile to petioled, the lowest opposite; margins of leaf blades entire ____ subsp. **argentea**
1. Upper leaves sessile, the lowest alternate; margins of leaf blades entire or irregularly dentate ____ subsp. **expansa**

subsp. **argentea**. SILVER-SCALE SALTBUSH, SILVER ORACHE. Apparently only to the w of nc TX; included based on citation of vegetational area 5 (Fig. 2) by Hatch et al. (1990), probably based on a Wichita Co. collection.

subsp. **expansa** (S. Watson) H.M. Hall & Clem., (expanded), SPREADING SALTBUSH, FOGWEED. Clay Co. near extreme w margin of nc TX.

Atriplex canescens (Pursh) Nutt., (gray-pubescent), FOUR-WING SALTBUSH, CHAMIZA. Dioecious (rarely monoecious) perennial herb; stems erect, 0.4–1.5(–2.5) m tall; leaves alternate, sessile or nearly so; leaf blades 1–5 cm long, linear-spatulate to narrowly oblong, entire, thickish, the upper and lower surfaces both gray-scurfy, becoming glabrous; staminate flowers in glomerules in dense spikes arranged in terminal panicles; pistillate flowers in short axillary spikes; bracts subtending fruits with prominent longitudinal wings (total of 4 wings per fruit); wings each ca. 2–9 mm wide. Saline or alkaline soils; Callahan and Shackelford cos., also Brown Co. (HPC); w margin of nc TX s and w to w TX. Apr–Oct. Native Americans ground the seeds into a "baking powder" to use in making bread (Powell 1988). Individuals of this species are sexually labile, i.e., able to change from one sex to the other. Sex change (generally from female to male) is apparently associated with stress such as unusually cold or drought and, "... appears to confer a survival advantage to the individual" (Freeman et al. 1984). Reported to cause bloat in sheep (Burlage 1968).

BETA BEET

A genus of 11–13 species of Europe and the Mediterranean region. (Celtic: *bett*, red, presumably in reference to the betalain pigments)

Beta vulgaris L., (common), BEET, SUGAR BEET, SWISS CHARD. Glabrous annual or biennial; stems to 0.6–1.2 m tall, green or often red; basal leaves in a rosette, long-petioled, reduced up the stem; flowers in glomerules, these axillary or in terminal spike-like panicles, perfect; sepals 5; stamens 5; ovary sunken in a disk. Cultivated; included based on reference that it is occasionally escaped in TX (Correll & Johnston 1970). Spring–fall. Native of s Europe and adjacent areas. Cultivated since the time of the Assyrians for its edible roots, for the roots as a source of sugar (up to 20% by weight), and for its edible leaves; the roots have high concentrations of the red pigments known as betalains (Mabberley 1987); ca. half the world's sugar supplies are derived from BEET; a different subspecies (subsp. *cicla* (L.) Koch), SWISS CHARD, SPINACH BEET, is cultivated for the edible leaves (Mabberley 1987; Hyam & Pankhurst 1995).

CHENOPODIUM LAMB'S-QUARTERS, GOOSEFOOT, PIGWEED

Weedy annual (rarely biennial or perennial) herbs, some strongly aromatic, often with mealy-coated or glandular foliage; leaves alternate; flowers crowded in small bunches (= glomerules), axillary or in terminal panicled spikes, usually perfect; calyces usually white-mealy or glandular-pubescent outside; petals absent; stamens usually 5; fruit a small urticle with thin pericarp that is easily separable or firmly attached; because of the thinness of the pericarp, the seed and fruit are essentially the same size; embryo partly or completely encircling the endosperm.

A genus of ca. 100 species of mostly weedy herbs of temperate areas; some are cultivated as "grains," as ornamentals, or for medicinal uses. *Chenopodium quinoa* Willd. (QUINUA, QUINOA), of Andean South America, was an important pre-Colombian "grain" and is currently making a resurgence; various QUINUA products (e.g., cereals, pastas) are at present widely available in health food stores; see Wilson (1990) for information on the origin of QUINUA. *Chenopodium* species are often difficult to distinguish because of vegetative variability and the small size of the taxonomically important reproductive characters (e.g., fruits, calyces). Mature fruits are extremely helpful in identification. Because of the difficulties involved, several species can be

reached more than one way in the key. (Greek: *chen*, a goose, and *pous*, foot, in allusion to the shape of the leaves of some species)
REFERENCES: Aellen & Just 1943; Wahl 1954; Reed 1969a; Crawford 1975; Crawford & Wilson 1986; Dorn 1988.

1. Leaves and inflorescences glandular (the glands stalked or sessile) and often pubescent with non-glandular hairs; leaves coarsely toothed or pinnately lobed; plants usually with a strong odor.
   2. Sepals densely glandular, the glands stalked or sessile; leaf blades mostly 4(–4.5) cm or less long; rare if present in nc TX.
      3. Leaves conspicuously lobed, the sinuses usually cutting more than 1/2 way to the midrib; glands on calyces short-stalked; seeds mostly horizontal within calyces ____ **C. botrys**
      3. Leaves quite coarsely toothed (2–4 teeth per side) but not deeply lobed; glands on calyces sessile; seeds mostly vertical within calyces ____ **C. pumilo**
   2. Sepals usually glabrous or inconspicuously glandular; leaf blades mostly (4–)6–15 cm or more long; weedy species widespread in nc TX ____ **C. ambrosioides**
1. Leaves and inflorescences glabrous or mealy but neither glandular nor pubescent; leaf blades entire, dentate, or lobed; plants usually without a strong odor (sometimes with an odor).
   4. Leaf blades linear to oblong, usually less than 15(–18) mm wide, 3 times as long as wide or longer.
      5. Leaf blades with 3 or more veins from near base, 4–18 mm wide, typically wider toward the middle or base, not of the same width for the whole length of the blade, the sides usually not parallel.
         6. Lower surface of leaf blades densely mealy, usually conspicuously whitened; sepals completely enclosing the mature fruit; inflorescences usually dense, stout ____ **C. pratericola**
         6. Lower surface of leaf blades glabrous or very sparsely mealy below, green in color; sepals barely curling over the margins of the mature fruit; inflorescences usually loose, open, slender, and often nodding ____ **C. standleyanum**
      5. Leaf blades 1-veined, usually 1–3(–6) mm wide, of about the same width for the whole length of the blade, the sides nearly parallel.
         7. Plants densely mealy; fruits 0.9–1.2 mm in diam.; seeds 0.8–1.1 mm in diam.; calyces open in fruit ____ **C. leptophyllum**
         7. Plants nearly glabrous; fruits 1.3–1.6 mm in diam.; seeds 1.2–1.5 mm in diam.; calyces closed in fruit ____ **C. pallescens**
   4. Leaves ovate, deltoid to rhombic, wider than 15 mm, 1–3 times as long as wide.
      8. Leaf blades ca. as wide as long, 1–3.5 cm long, essentially entire; plants with semi-prostrate habit ____ **C. vulvaria**
      8. Leaf blades longer than wide, 2–10 (rarely more) cm long, at least the lower often toothed at very base; plants with various habits.
         9. Glomerules with mature fruits and flowers simultaneously.
            10. Leaf blades large, 50–150 mm wide, with 1–4, large, conspicuous teeth per side; fruits 1.5–2.5(–2.7) mm in diam. ____ **C. simplex**
            10. Leaf blades small, usually 20 mm or less wide (reports up to 50 mm[?]), entire or with a few low teeth; fruits 1–1.5(–1.6) mm in diam. ____ **C. standleyanum**
         9. Glomerules either flowering or fruiting but usually not both (*C. simplex* and *C. standleyanum* can also be keyed here).
            11. Leaf blades except uppermost deltoid; seeds (pericarp removed) dull; rare in nc TX ____ **C. murale**
            11. Leaf blades ovate to rhombic, gradually reduced upward; seeds (pericarp removed) shiny; including species widespread in nc TX.
                12. Leaf blades large, 50–150 mm wide, with 1–4 large prominent teeth per side; fruits 1.5–2.5(–2.7) mm in diam. ____ **C. simplex**

12. Leaf blades not as above; fruits up to 1.5 mm in diam.
    13. Pericarp with alveolae giving the fruit surface a "honeycombed" appearance (under a lens or scope); fresh plant with unpleasant odor ___ **C. berlandieri**
    13. Pericarp without alveolae, the fruit surface not "honeycombed" or this not discernable (*C. berlandieri* can also be reached this way); fresh plant with OR without unpleasant odor.
        14. Lower surface of leaf blades densely white-mealy.
            15. Plants flowering ca. 2nd week in Sep.; fruits 0.9–1.2 mm in diam.; inflorescences of small delicate glomerules, typically diffuse and often pendulous at maturity ___ **C. missouriense**
            15. Plants flowering Apr–Sep; fruits 1–1.5 mm in diam.; inflorescences often of relatively large glomerules, usually dense but sometimes diffuse, typically erect to ascending or spreading.
                16. Sepals broadly keeled; pericarp "honeycombed" (quite clear under dissecting scope); fresh plants with unpleasant odor ___ **C. berlandieri**
                16. Sepals with keel absent or poorly developed; pericarp not "honeycombed"; fresh plants without unpleasant odor ___ **C. album**
        14. Lower surface of leaf blades not densely white-mealy.
            17. Sepals sharply keeled, completely enclosing the mature fruit; glomerules of flowers usually in dense, erect inflorescences; pericarp "honeycombed," rather adherent to the seed; fresh plants with unpleasant odor ___ **C. berlandieri**
            17. Sepals rounded, not sharply keeled, barely curling over the margins of the mature fruit; glomerules of flowers in open, sometimes nodding inflorescences; pericarp not "honeycombed," easily separable from seed; fresh plants without unpleasant odor ___ **C. standleyanum**

Chenopodium album L., (white), LAMB'S-QUARTERS, PIGWEED. Plant annual, pale-green; stems erect to 0.6–3 m tall, solitary; lateral branches well-developed, ascending and compacted; nodes and infructescences without dark pigmentation; leaf blades moderately to densely mealy; blades of lower leaves 1.5 times as long as wide or longer; flowers in large glomerules in axillary or terminal, dense, paniculate spikes; sepals densely mealy, enclosing the mature fruit; fruits 1.1–1.5 mm in diam., the pericarp not "honeycombed"; seeds shiny, black. Around old homesteads and abandoned farms, disturbed areas; Denton Co., also Dallas and Grayson cos. (Reed 1969a); widespread in TX. Apr–Sep. Apparently native of Eurasia. As is the case in the Great Plains (Crawford & Wilson 1986), this species is probably less abundant in nc TX than generally assumed because individuals of the similar *C. berlandieri* are often identified as *C. album*. However, the "honeycombed" pericarp of *C. berlandieri* clearly distinguishes it from *C. album*. It was used medicinally by Native Americans and by pioneers as a potherb; however, the plants may contain toxic quantities of nitrate and oxalic acid; poisoning and death in cattle, horses, and pigs have been reported (Burlage 1968; Schmutz & Hamilton 1979; Mulligan & Munro 1990). ☠ ☜

Chenopodium ambrosioides L., (resembling *Ambrosia*—ragweed), MEXICAN-TEA, WORMSEED, EPAZOTE, SPANISH-TEA, WORMSEED LAMB'S-QUARTERS. Plant annual or perennial, strongly aromatic; stems erect or ascending, 0.3–1 m tall; leaf blades densely yellow-glandular to nearly glabrous; calyces completely enclosing the fruit; fruits 0.6–1.0 mm in diam.; seeds blackish. Disturbed habitats, sandy soils of woodlands; se and e TX w to West Cross Timbers and Edwards Plateau. Aug–Nov. Weed and medicinal herb native to tropical America; used as a condiment in Mexican cooking and as an antiflatulent. An antihelminthic (= agent that expels worms) oil is obtained from the seeds with the active ingredient being a terpene; the plant has been variously used medicinally including in the treatment of malaria; the oil is toxic and overdoses have been fatal to humans and animals (Burlage 1968; Kingsbury 1964; Leung & Foster 1996).

Chenopodium berlandieri Moq., (for Jean Louis Berlandier, 1805–1851, French botanist who explored TX, NM, and Mexico, and one of the first botanists to make extensive collections in TX), PIT-SEED GOOSEFOOT. Plant annual, typically ill-scented; stems erect, 0.4–1.5 m tall; leaf blades often densely mealy when young, later glabrate, irregularly sinuate-dentate; inflorescences typically dense, erect, although sometimes more diffuse; sepals densely mealy, completely enclosing the fruit; seeds 0.8–1.5 mm in diam., shiny, black. Dry disturbed habitats. Jul–Sep.

1. Leaf blades definitely 3-lobed ____ var. **sinuatum**
1. Leaf blades usually not 3-lobed.
   2. Inflorescences leafy; seeds 1.2–1.5 mm in diam. ____ var. **zschackei**
   2. Inflorescences sparsely leafy; seeds 0.8–1.3 mm in diam.
      3. Sepals sparsely mealy, the lobes green, white-margined; leaf blades not prominently toothed; seeds 1.3 mm in diam. ____ var. **boscianum**
      3. Sepals densely mealy, the lobes uniform in color; leaf blades ± toothed; seeds 0.8–1 mm in diam. ____ var. **berlandieri**

var. **berlandieri**. Leaf blades 1.2–3 cm long, irregularly sinuate-dentate; sepals strongly keeled. Brown, Coleman, and Milam cos., also Bell, Denton, Tarrant, and Young cos. (Reed 1969a); widespread in TX.

var. **boscianum** (Moq.) Wahl, (for its discoverer, Louis Augustin Guillaume Bosc, 1759–1828, French naturalist). Leaf blades 2–6 cm long, triangular-oblong; tepals slightly keeled. Tarrant Co. (*Ruth 321*); also e TX.

var. **sinuatum** (Murray) Wahl, (wavy-margined). Leaf blades ca. 2.5 cm long; inflorescences somewhat leafy; tepals with small keel; seeds 1.0–1.3 mm in diam. Palo Pinto Co.; mainly Plains Country w of nc TX.

var. **zschackei** (Murray) Murray ex Asch., (for Georg Zschacke, 1867–1937, German botanist). Leaf blades variable but typically larger than in other varieties; tepals strongly keeled. Brown, Denton, Grayson, Montague, Shackelford, and Wise cos., also Stephens and Tarrant cos. (Reed 1969a); e TX w to nc TX.

Chenopodium botrys L., (a bunch of grapes), JERUSALEM-OAK, FEATHER-GERANIUM. Annual with strong, not unpleasant scent; stems erect, 0.2–0.6 m tall, densely glandular-viscid throughout; sepals densely glandular-pubescent, imperfectly enclosing the fruit; stamens 5; fruits 0.6–0.8 mm in diam.; seeds dull, dark-brown. Disturbed habitats; collected by Reverchon (*819*) in 1879 (Dallas Co.); apparently otherwise unknown in TX. Jul–Oct. Native to Eurasia.

Chenopodium leptophyllum (Moq.) Nutt. ex S. Watson, (slender-leaved), SLIM-LEAF GOOSEFOOT, NARROW-LEAF GOOSEFOOT, NARROW-LEAF LAMB'S-QUARTERS. Annual to ca. 0.9 m tall; leaf blades 1–6 cm long, densely white-mealy; sepals densely white-mealy, barely covering the fruit; fruits 0.9–1.2 mm in diam.; seeds shiny, black. Dry places, slopes; included based on citation of vegetational area 5 (Fig. 2) by Hatch et al. (1990); mainly w 1/2 of TX. Jun–Sep.

Chenopodium missouriense Aellen, (of Missouri). Annual; stems to ca. 1.5 m tall (often over 1 m), solitary; lateral branches usually well-developed, spreading, and flexous at maturity; nodes and infructescences often with dark pigmentation; leaf blades coarsely toothed, those of lower leaves < 1.5 times as long as wide; inflorescences flexuous and arching, of relatively small, delicate glomerules, usually diffuse, typically pendulous at maturity; sepals slightly mealy, the keel not well-developed; fruits 0.9–1.2 mm in diam. Disturbed habitats; Dallas, Grayson, Limestone, and Tarrant cos., also McLennan Co. (Reed 1969a); in TX apparently only in nc part of the

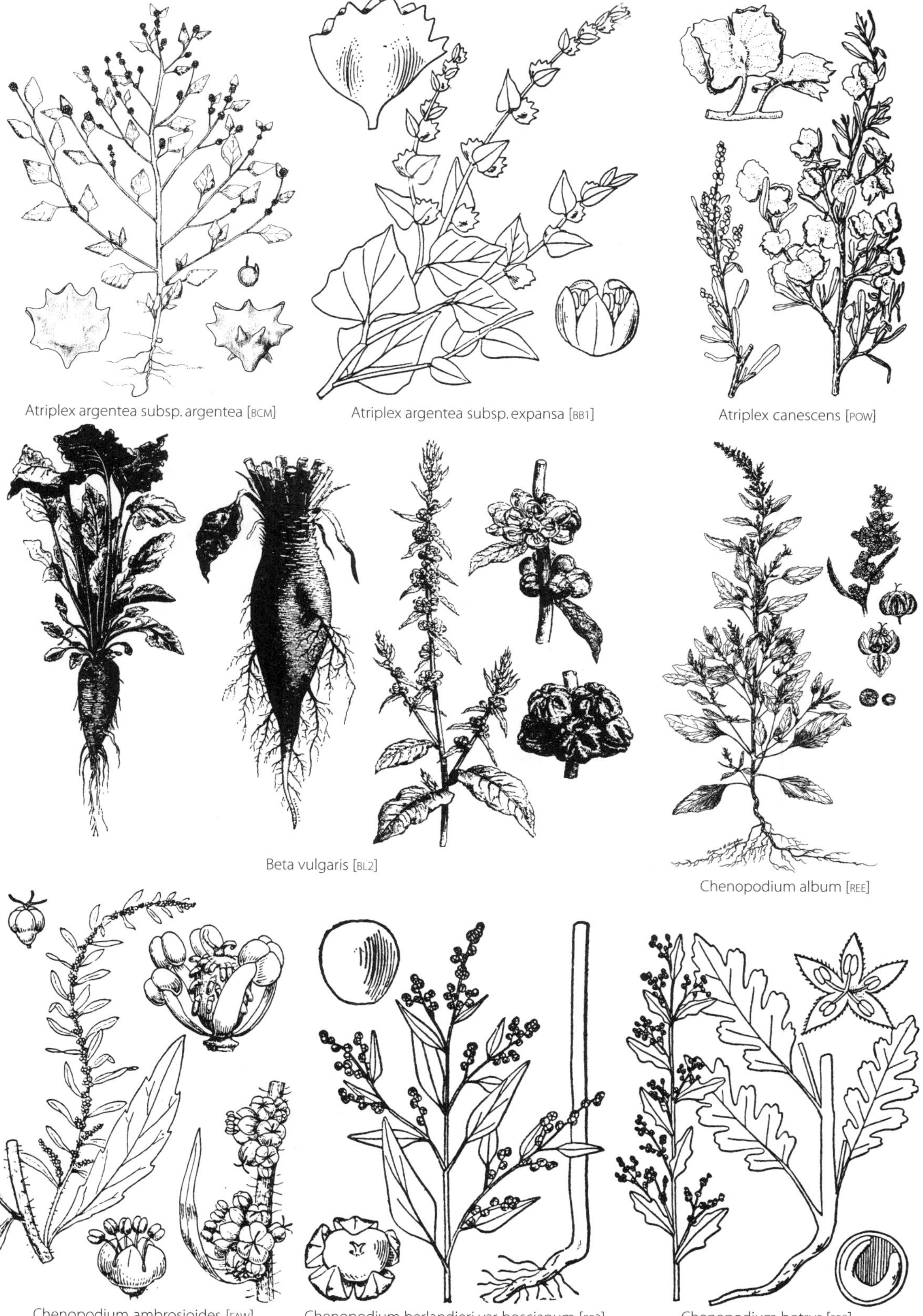

Atriplex argentea subsp. argentea [BCM]

Atriplex argentea subsp. expansa [BB1]

Atriplex canescens [POW]

Beta vulgaris [BL2]

Chenopodium album [REE]

Chenopodium ambrosioides [FAW]

Chenopodium berlandieri var. boscianum [BB2]

Chenopodium botrys [BB2]

state. Flowering restricted to ca. 2nd week in Sep. This taxon is lumped by some authorities (e.g., Kartesz 1994; Jones et al. 1997) with *C. album* [as *C. album* var. *missouriense* (Aellen) Bassett & Crompton]; however, we are following Reed (1969a), Correll and Johnston (1970), and Crawford and Wilson (1986) in recognizing it as a separate species.

Chenopodium murale L., (of walls), NETTLE-LEAF GOOSEFOOT, SOWBANE. Annual; stems erect or decumbent, 0.1–0.6 m long; leaf blades thin, ovate to rhombic-ovate, coarsely sinuate-dentate; sepals partly covering the fruit; seeds 1.2–1.5 mm in diam., dull, black. Disturbed habitats. Collected "in the streets of Dallas" in Jun 1874 and Jun 1875 by Reverchon; not found there since (Mahler 1988); widely scattered in TX. Native of Old World.

Chenopodium pallescens Standl., (rather pale), LIGHT GOOSEFOOT. Annual; stems erect, 0.3–0.6 m tall; leaf blades linear, 1.5–4 cm long, 1.5–3(–6) mm wide, 1-nerved; petioles 2–5 mm long; flowers in large glomerules; sepals slightly mealy, completely enclosing the fruit; fruits 1.3–1.6 mm in diam.; seeds shiny, black. Open floodplains; Parker Co. (Mineral Wells State Park, *Lipscomb 2393*); known from a few scattered localities in TX. Aug–Oct.

Chenopodium pratericola Rydb., (living in meadows or prairies), THICK-LEAF GOOSEFOOT. Annual; stems 0.2–0.9 m tall; leaf blades densely mealy beneath; petioles ca. half as long as the blades; sepal lobes white-margined; sepals completely enclosing the fruit; fruits 1–1.5 mm in diam.; seeds shiny, black. Stream banks, disturbed sites, sandy soils; Dallas, Denton, Jack, Parker, Tarrant, and Young cos., also Erath, Grayson, and Palo Pinto cos. (Reed 1969a); widespread mainly in w 2/3 of TX e to East Cross Timbers. Late May–Oct. [*C. desiccatum* A. Nelson var. *leptophylloides* (Murray) Wahl]

Chenopodium pumilo R. Br., (dwarf), RIDGED GOOSEFOOT. Annual, both glandular and pubescent; stems prostrate to ascending, 0.2–0.4 m long; leaf blades ovate or oblong to lanceolate, with lower surface conspicuously glandular; petioles ca. 1/2 as long as blades. Waste places; included based on citation of vegetational areas 4 and 5 (Fig. 2) by Hatch et al. 1990; Reed (1969a) cited only 1 county in TX; Hatch et al. (1990) cited vegetational areas 1 through 10 [?]; the only TX specimen we have seen is from Upshur Co. in e TX. Nov. Native of Australia.

Chenopodium simplex (Torr.) Raf., (simple, unbranched), MAPLE-LEAF GOOSEFOOT, BIG-SEED GOOSEFOOT. Annual; stems glabrous, erect, 0.6–1.3 m tall; leaf blades with 1–4 large teeth on each side; panicles open, leafless; sepals partly covering the fruit; fruits 1.5–2.5(–2.7) mm in diam.; seeds umbonate-lenticular, tapering from raised center to margin, shiny, black. Disturbed woodlands; Bell, Collin, Dallas, and Tarrant cos.; nc TX s and w to w TX. Jun–Aug. [*C. gigantospermum* Aellen]

Chenopodium standleyanum Aellen, (for Paul Carpenter Standley, 1884–1963, student of floras of sw U.S., Mexico, and Central America), STANDLEY'S GOOSEFOOT. Annual; stems erect or arched to 1(–2.5) m long, usually much smaller; leaf blades to 8 cm long, usually 20 mm or less wide (reports up to 50 mm); inflorescence a slender, often nodding terminal panicle; sepals slightly mealy; fruits 1–1.5(–1.6) mm in diam.; seeds shiny, black. Disturbed wooded habitats, floodplains; Dallas, Grayson, and Tarrant (*Ruth*) cos., also Bell and Brown cos. (Reed 1969a); e TX w to nc TX, also Edwards Plateau.

Chenopodium vulvaria L., (old generic name). Annual with bad odor, strongly white-mealy; stems erect or ascending, branched at base, 0.1–0.5 m tall; sepals densely mealy, rounded on back, completely enclosing the fruit; seeds ca. 1 mm in diam., dull, black. Included based on citation of vegetational areas 4 and 5 (Fig. 2) by Hatch et al. (1990); Reed (1969a) indicated that TX is cited in the literature, but cited no specimens; Hatch et al. (1990) cited vegetational areas 1 through 10 [?]. We have seen no TX specimens. Native of Europe.

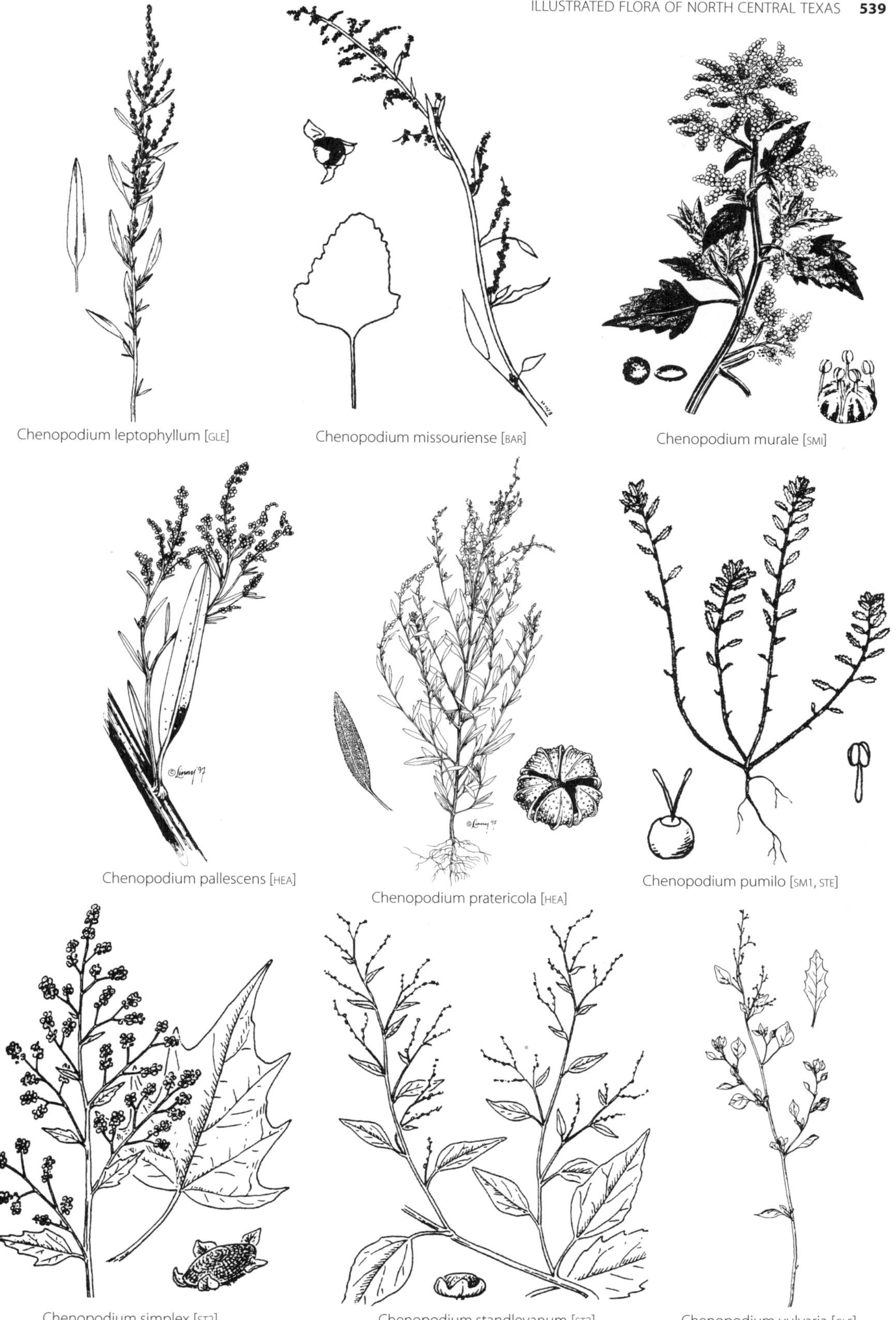

Chenopodium leptophyllum [GLE]

Chenopodium missouriense [BAR]

Chenopodium murale [SMI]

Chenopodium pallescens [HEA]

Chenopodium pratericola [HEA]

Chenopodium pumilo [SM1, STE]

Chenopodium simplex [ST2]

Chenopodium standleyanum [ST2]

Chenopodium vulvaria [GLE]

CYCLOLOMA WINGED-PIGWEED

A monotypic c and w North American genus. (Greek: *cyclos*, a circle, and *loma*, a border or fringe, from the encircling wing of the calyx)

Cycloloma atriplicifolium (Spreng.) J.M. Coult., (with leaves resembling *Atriplex*—saltbush), WINGED-PIGWEED, TUMBLE-RINGWING, PLAINS TUMBLEWEED. Bushy annual 15–100 cm tall and broad, with habit of a tumbleweed, glabrous or with thin pubescence; leaves sessile or short-petioled; leaf blades 2–8 cm long, oblong-lanceolate, irregularly sinuately toothed or pinnately lobed; flowers sessile and rather widely spaced in terminal paniculate spikes, perfect, some pistillate; sepals 5; calyces with a conspicuous wing; stamens 5; styles 2–3. Loose sandy soils, roadsides and disturbed sites; more common in w part of nc TX; widespread in TX. Jun–Oct. The seeds were eaten by Native Americans (Mabberley 1987). Used medicinally by Native Americans (Hopi) for conditions including fever, rheumatism, and headache (Burlage 1968).

KOCHIA SUMMER-CYPRESS

A genus of ca. 20 species of w North America and Eurasia (Wilken 1993a); sometimes lumped into the genus *Bassia* (Kühn 1993; Mabberley 1997). (Named for Wilhelm Daniel Joseph Koch, 1771–1849, German botanist and physician)
REFERENCE: Blackwell et al. 1978.

Kochia scoparia (L.) Schrad., (broom-like), MEXICAN FIREWEED, BELVEDERE SUMMER-CYPRESS, SUMMER-CYPRESS, MOCK CYPRESS, MEXICAN FIREBUSH. Annual, woolly to glabrate; stems erect, 0.3–2(–4) m tall, often with reddish or purplish streaking; leaves 2–7(–10) cm long, 0.5–8(–12) mm wide, linear to lanceolate, ± sessile to petioled; inflorescences dense, leafy, spike-like, some plants with flowers nearly throughout; flowers mostly perfect, some pistillate; calyces with 5 lobes, each with a wing; stamens 5; styles 2(–3). Waste places; Grayson Co.; scattered in TX but mainly in Plains Country to the w of nc TX. Jun–Aug. Native of Eurasia. A cultivated form is sometimes grown as an ornamental for its purplish red autumn foliage. Considered a weed but high in protein and readily grazed by livestock; ingestion may, however, cause photosensitization in cattle, sheep, and horses; reported to contain saponins (Burlage 1968; McGregor 1986).

MONOLEPIS POVERTY-WEED

A genus of 6 species of n and e Asia, North America, and temperate South America; some were formerly eaten. (Greek: *monos*, solitary, and *lepsis*, scale, from sepal number in most species)

Monolepis nuttalliana (Schult.) Greene, (for Sir Thomas Nuttall, 1786–1859, English-American botanist), NUTTALL'S MONOLEPIS, POVERTY-WEED. Glabrous or sparsely mealy annual with prostrate to erect stems 5–35 cm long; foliage pale green; lower leaves long-petioled; leaf blades fleshy, 1–6.5 cm long, rhombic or lanceolate, with 1 or few coarse teeth or often a pair of lobes toward base, giving the leaf an arrowhead shape; flowers axillary, sessile, crowded, in upper part of stem and branches, mostly perfect or a few pistillate; the single bract-like sepal 1.5–2.5 mm long; styles 2. Stream bottoms, roadsides, disturbed sites; widespread in TX but more common in the w 1/2 of nc TX and to the w. Mar–May.

SALSOLA

A cosmopolitan genus of 150 species primarily of coastal or other saline habitats. Nomenclature for *Salsola* follows Mosyakin (1996). (Latin: *salsus*, salty, from habitat of some species)
REFERENCES: Beatley 1973; Mosyakin 1996.

Salsola tragus L., (Greek: *tragos*, a goat), RUSSIAN-THISTLE, TUMBLEWEED. Bushy, glabrous annual 15–100 cm or more tall and broad, with habit of a tumbleweed; stems usually with reddish or

Cycloloma atriplicifolium [BB2]

Kochia scoparia [REE]

Monolepis nuttalliana [BB2]

Salsola tragus [REE, USB]

Spinacia oleracea [HEA]

purple streaks; leaves filiform, 2–8 cm long, only 1–2 mm wide; flowers in interrupted terminal spikes, enclosed or subtended by a pair of stiff, spine-pointed bracts 3–8 mm long, perfect; sepals 5; stamens 3–5; stigmas 2–3; fruit with a prominently winged apex, 3–6 mm wide including wing. Roadsides, disturbed sites; Clay and Jack cos. in West Cross Timbers, also Parker Co. (R. O'Kennon, pers. obs.), also Grayson Co. in Red River drainage; widespread in w 1/2 of TX. Jun–Oct. Native of e Europe and Asia that is now a drought resistent agricultural pest in parts of North America; possibly introduced into the U.S. in South Dakota in 1873 or 1874 in flax seed imported from Russia (Mosyakin 1996; Tellman 1997); this species rapidly became a noxious weed in the w U.S. According to Tellman (1997), "The newly built railroad was an ideal vehicle for spreading tumbleweed throughout the west and the tumbleweed's early distribution pattern shows it moving outward along railways and roadways." [*S. australis* R. Br., *S. kali* L. subsp. *tenuifolia* Tausch, *S. kali* L. subsp. *tragus* (L.) Celak., *S. pestifer* A. Nelson, *S. tragus* subsp. *iberica* Sennen & Pau] The bushy, roughly globe-shaped plants break off at ground level in late fall; the whole plant, when blown by the wind, rolls or tumbles across open areas and disperses the seeds (Kirkpatrick 1992). Reported to contain the alkaloids, salsolidine and salsoline, and to be poisonous to livestock (Burlage 1968). ☠ ☞

SPINACIA SPINACH

A genus of 4 species of sw Asia and n Africa. (Latin: *spina*, spine, referring to spine-tipped bracts around the fruits)

Spinacia oleracea L., (of the vegetable garden, a potherb used in cooking), SPINACH. Glabrous annual; stems erect, 30–45 cm tall; basal leaves in a rosette, long-petioled, reduced up the stem; leaf blades oval to triangular-ovate to hastate, entire or sinuate-dentate; flowers usually unisexual; staminate flowers with 4–5 sepals and 4–5 stamens; pistillate flowers without sepals, subtended by 2 usually spine-tipped bracts that grow together and enclose the fruit; stigmas 4–5; fruiting bracts 2–4 mm long. Cultivated for the edible leaves; included based on citation of vegetational areas 4 and 5 (Fig. 2) by Hatch et al. (1990); occasionally escapes throughout TX. Spring–summer. Native of sw Asia. The tissue contains high levels of oxalic acid which combines with calcium and can cause kidney problems or calcium deficiencies; occasional consumption rarely causes ill effects (Schmutz & Hamilton 1979). ☠ ☞

CISTACEAE SUN-ROSE OR ROCK-ROSE FAMILY

Ours low perennial herbs; leaves alternate or opposite, sessile, simple, linear to elliptic or oblanceolate, entire; flowers small, terminal or in upper leaf axils, solitary, racemose, or in panicled, compact clusters; sepals 3, united at base, bractless or with 1–2 narrow bracts [sepals?] attached on calyx-base (bracts sepal-like in some species, more often resembling reduced upper leaves); petals 3 or 5, opening in sunshine, lasting for one day only (or flowers cleistogamous); stamens 3-many; pistils 2- to 3-carpellate; ovary superior; fruit a capsule.

A small (175 species in 8 genera), mainly n temperate and warm area, especially Mediterranean family of shrubs, subshrubs, or herbs; some are important as ornamentals, for example in rock gardens. Family name from *Cistus*, ROCK-ROSE, a Mediterranean genus of 17 species of shrubs. The leaves of some *Cistus* species exude an aromatic resin, laudanum, used in making perfumes and formerly in medicines (Woodland 1997). (Greek: *kistos*, classical name for these plants) (subclass Dilleniidae)

FAMILY RECOGNITION IN THE FIELD: perennial, sometimes stellate-pubescent herbs with simple entire leaves; petals convolute (= twisted), quickly falling; ovary one-celled, many-seeded, superior; stamens sometimes numerous.

REFERENCES: Brizicky 1964b; Wilbur 1969.

1. Petals 5 (except in late-appearing cleistogamous flowers), yellow, 4–12 mm long; plants pubescent with stellate hairs _____ **Helianthemum**
1. Petals 3, usually reddish to purplish (can also vary to yellowish white or greenish), 1.5–3 mm long; plants pubescent with simple hairs OR largely glabrous _____ **Lechea**

HELIANTHEMUM SUN-ROSE, ROCK-ROSE, FROSTWEED

Ours perennial herbs with stellate pubescence on stems, leaves, and in inflorescence; leaves alternate; flowers dimorphic; cleistogamous flowers often present, with 3–8 stamens, without petals; chasmogamous flowers with 15–36 stamens, with 5 conspicuous yellow petals.

A genus of ca. 110 species native from Europe to the Sahara, ne Africa to c Asia, and North and South America; a number are cultivated as ornamental shrublets. (Greek: *helios*, the sun, and *anthemon*, flower, the flowers of some tending to open only in bright sunshine)
REFERENCES: Wilbur & Daoud 1964; Daoud & Wilbur 1965.

1. Leaves usually 6–8 mm wide; petals 6–12 mm long; calyces (3–)4–6.6 mm long (3–4.2 mm long in cleistogamous flowers); capsules 4–6 mm long, with 12–35 seeds _____ **H. georgianum**
1. Leaves usually < 4(–7) mm wide; petals 4–6.4 mm long; calyces 1–4 mm long (< 2 mm long in cleistogamous flowers); capsules 1.3–4 mm long, with 1–6 seeds _____ **H. rosmarinifolium**

Helianthemum georgianum Chapm., (of Georgia), GEORGIA SUN-ROSE, GEORGIA ROCK-ROSE, HOARY SUN-ROSE. Stems 10–40 cm tall; basal leaves present or absent, similar to stem leaves; stem leaves elliptic to oblanceolate, 10–28(–35) mm long, usually 6–8 mm wide; pedicels of chasmogamous flowers 3–15 mm long; pedicels of cleistogamous up to 3(–6) mm long; chasmogamous sepals 3.6–6.6 mm long; seeds 12–35 per capsule. Sandy open woods, roadsides; se and e TX w to West Cross Timbers and Edwards Plateau. Late May–Jun. ⊠/91

Helianthemum rosmarinifolium Pursh, (with leaves like *Rosemarinus*—rosemary), ROSEMARY SUN-ROSE. Stems 13–50 cm tall; basal leaves absent; stem leaves linear to narrowly oblanceolate, 5–14 times as long as wide, 2–4(–7) mm wide; pedicels of chasmogamous flowers usually 10–22 mm long; pedicels of cleistogamous up to 3 mm long; seeds 1–6 per capsule. Sandy open woods; se and e TX w to e Rolling Plains and Edwards Plateau. Mid-May–Jun (cleistogamous flowers Jun–Jul).

LECHEA PINWEED

Ours perennial herbs lacking stellate pubescence; petals 3, usually reddish to purplish (can be yellowish white or greenish); stamens (3–)6–15(–25); style absent; seeds 2–6 per fruit.

A genus of 17 species native to the Americas. (Named for Johan Leche, 1704–1764, Swedish botanist)
REFERENCES: Hodgdon 1938, 1966; Wilbur & Daoud 1961.

1. Mid-stem leaves linear-lanceolate to almost thread-like, not over 2 mm wide; stems inconspicuously appressed-pubescent.
   2. Pedicels 2–3 mm long (up to 4 or 5 mm in fruit), widely spreading or reflexed; seeds 6 per fruit _____ **L. san-sabeana**
   2. Pedicels 0.5–1.5 mm long (up to 2.5 mm in fruit), ascending or loosely appressed; seeds usually 3(–5) per fruit _____ **L. tenuifolia**
1. Mid-stem leaves lanceolate, 3–9 mm wide; stems spreading-pilose _____ **L. mucronata**

Lechea mucronata Raf., (mucronate, pointed), HAIRY PINWEED, PINE PINWEED. Stems usually 30–90 cm tall; flowers few–several in a compact cluster; seeds (2–)3(–4) per fruit. Sandy woods, old fields, and roadsides; se and e TX w to e Rolling Plains and Edwards Plateau. Jun–Jul. [*L. villosa* Elliott]

Lechea san-sabeana (Buckley) Hodgdon, (of San Saba Co., TX), SAN SABA PINWEED, DRUMMOND'S PINWEED. Similar to *L. tenuifolia*; stems 15–35 cm tall; flowers spaced apart. Sandy open woods; Bosque, Brown, Jack, Montague, Parker, Tarrant, and Wise cos., also Dallas Co. (Mahler 1988); se and e TX w to West Cross Timbers and Edwards Plateau; endemic to TX. Apr–Jun.

Lechea tenuifolia Michx., (slender-leaved), NARROW-LEAF PINWEED. Stems 12–40 cm tall; flowers spaced apart. Sandy woods; se and e TX w to West Cross Timbers and Edwards Plateau, also Trans-Pecos. May–Jul.

CLUSIACEAE (GUTTIFERAE) ST. JOHN'S-WORT OR GARCINIA FAMILY (INCLUDING HYPERICACEAE)

Ours annual or perennial herbs or small shrubs; foliage often pellucid or black punctate/dotted; leaves opposite; leaf blades entire; flowers solitary or in cymose inflorescences, perfect, radially symmetrical; sepals (2–)4 or 5; petals 4 or 5, free, usually yellow or orange-yellow or in one rare nc TX species pinkish or reddish; stamens few–numerous, sometimes in groups; pistil 1; ovary superior; fruit a capsule.

A medium-large (1,370 species in 45 genera), mainly tropical but also n temperate family of trees, shrubs, lianas, and herbs usually with yellow or brightly colored resinous sap; a number of species are important sources of woods, gums, pigments, drugs, oilseeds, and fruits. *Garcinia mangostana* L. (MANGOSTEEN) with delicious arils, and *Mammea americana* L. (MAMMEY-APPLE) are well known tropical fruits. The alternate family name, Guttiferae, means "drop-bearing" and refers to the resinous sap, oil vesicles, glands, and dots seen in many species. Our species are sometimes treated in the Hypericaceae, a family now generally lumped with the Clusiaceae. Family name from *Clusia*, a genus of ca. 145 species of dioecious trees and shrubs of tropical and warm areas of the Americas; they can be epiphytic or in some cases stranglers with roots surrounding and damaging the host. (Named for Charles de l'Ecluse (Carolus Clusius), 1526–1609, of Arras, France, botanist, gardener, and author; he introduced the tulip to the Netherlands) (subclass Dilleniidae)
FAMILY RECOGNITION IN THE FIELD: herbs or small shrubs with mostly opposite, simple, entire, *dotted* (dots visible when leaves are held up to a light) leaves; flowers with 4 or 5 separate colorful (yellow to orange-yellow, pinkish, or reddish) petals which often have a slight contortion or twist; stamens often in bundles, sometimes numerous.
REFERENCES: Adams 1973; Wood & Adams 1976.

1. Petals yellow or orange-yellow; stamens few–numerous, in bundles or not so, without glands at base; widespread in nc TX ____ **Hypericum**
1. Petals pinkish or reddish; stamens 9, in 3 bundles of 3, with 3 orange glands at base alternating with stamen bundles; on e margin of nc TX ____ **Triadenum**

HYPERICUM ST. JOHN'S-WORT

Ours annual or perennial herbs or small shrubs often punctate with resin or oil dots or glands on leaves or other parts (the dots visible on surfaces or using transmitted light); leaves simple, sessile; flowers solitary or in dichotomously cymose inflorescences; sepals (2–)4 or 5; petals 4 or 5, yellow or orange-yellow; stamens few–numerous, often in small bunches; ovary 2- to 5-carpellate.

A genus of ca. 370 species of trees, shrubs, and herbs native to temperate regions and tropical mountains. Many are used as cultivated ornamentals while some have medicinal uses. A number of *Hypericum* species, especially *H. perforatum*, contain hypericin, a reddish, fluorescent, multiple-ringed phenolic substance concentrated in the glandular dots on the leaves.

Helianthemum georgianum [HEA]

Helianthemum rosmarinifolium [HEA]

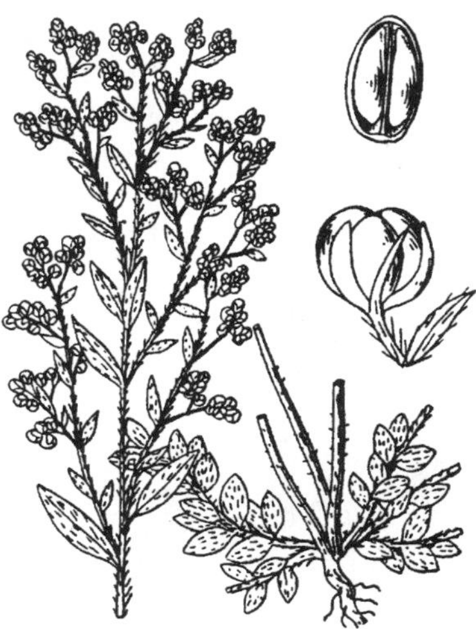
Lechea mucronata [BB2]

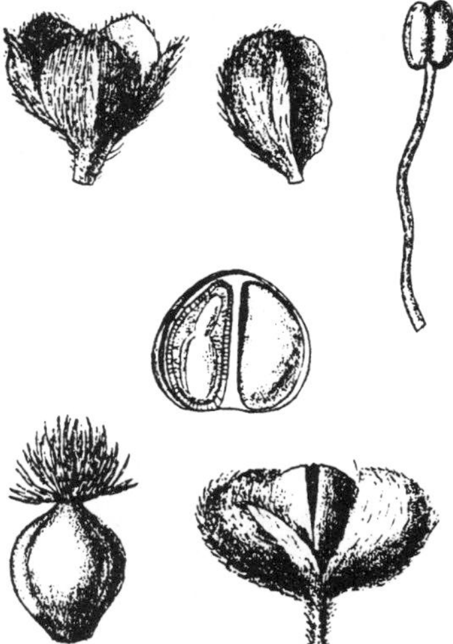
Lechea san-sabeana [EN1]

Lechea tenuifolia [BB2]

Hypericum crux-andreae [GR1]

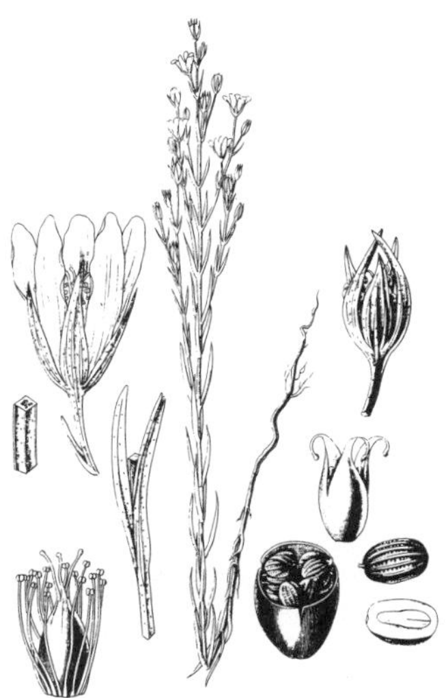
Hypericum drummondii [ANO]

When the plant is eaten, hypericin, whose numerous double bonds absorb UV radiation, increases skin sensitivity and can cause photosensitization and consequent dermatitis in light-skinned livestock; symptoms include swelling, blistering, and lesions as well as more serious reactions including death. Contact may also cause dermatitis in humans (Muenscher 1951; Lewis & Elvin-Lewis 1977; Stephens 1980; Crompton et al. 1988; Turner & Szczawinski 1991). (Greek: *hyper*, above, and *eikon*, picture; some species were hung above pictures to ward off evil spirits)

REFERENCES: Adams & Robson 1961; Adams 1957, 1962; Robson 1980.

1. Petals 4; sepals 2 or 4 and unequal (2 large and leaf-like, 2 smaller or absent); stamens free, not in clusters.
   2. Inner 2 sepals minute or absent; petals ca. 6–10 mm long, to 4(–7) mm wide; styles 2; leaves 9 mm or less wide; widespread in nc TX ____ **H. hypericoides**
   2. Inner 2 sepals smaller than outer but well-developed, 7–14 mm long, to 4 mm wide; petals 10–18 mm long, 7–15 mm wide; styles 3–4; leaves (at least some) 9–17(–20) mm wide; e margin of nc TX ____ **H. crux-andreae**
1. Petals 5; sepals 5 and equal; stamens clustered in 3–5 groups OR free, not in clusters.
   3. Sepals densely and conspicuously black gland-dotted (visible under a hand lens); petals conspicuously black dotted over their surfaces; stamens numerous, usually 20 or more; plants with perennial roots.
      4. Main stem mostly unbranched, with branches only near the summit; leaves obtuse, rounded, or even retuse at apex; sepals 2.5–4 mm long; petals 4–7.5 mm long; styles 2–4 mm long, mostly persistent ____ **H. punctatum**
      4. Main stem mostly with small axillary branches throughout; leaves acute or narrowly obtuse at apex; sepals 4–7 mm long; petals 8–14 mm long; styles 6–10 mm long, mostly soon withering ____ **H. pseudomaculatum**
   3. Sepals not densely and conspicuously black gland-dotted (can have obscure light-colored glands; petals not black dotted or in 1 species with black dots limited to the margins only; stamens few (6–20) or numerous; plants annual or perennial.
      5. Leaves flat, 2 mm or more wide (usually much more), usually 3 or more nerved, spreading.
         6. Petals 5–12 mm long, with or without black dots along margins; stamens > 12 per flower; ovary and capsule with or without prominent, elongate, yellow-amber oil vesicles visible under a hand lens.
            7. Petals usually with black dots along margins, 7–12 mm long; ovary and capsule with prominent, elongate, yellow-amber oil vesicles visible under a hand lens; leaves usually rather small, 10–15(–25) mm long, 2–5(–11) mm wide ____ **H. perforatum**
            7. Petals without black dots along margins, 5–9 mm long; ovary and capsule without prominent oil vesicles; leaves larger, usually 30–70 mm long, 5–15 mm wide ____ **H. sphaerocarpum**
         6. Petals 6 mm or less long, without black dots along margins; stamens 12 or fewer per flower; ovary and capsule without prominent oil vesicles visible.
            8. Main stem with scattered spreading branches; middle and upper leaves rounded at apex; capsule with rounded end (but tipped by persistent style ca. 1 mm long); bracts leaf-like in much of inflorescence ____ **H. mutilum**
            8. Main stem unbranched or nearly so; middle and upper leaves usually tapering to apex; capsule with pointed end (and tipped by persistent style); bracts reduced, narrow and awl-like in much of inflorescence ____ **H. gymnanthum**
      5. Leaves subulate (= awl-shaped) or linear, 1.5 mm or less wide, 1-nerved, appressed or strongly ascending.
         9. Leaves 6–20 mm long, linear; capsules ovoid, slightly longer than calyces; sepals 3–5 mm long ____ **H. drummondii**

9. Leaves < 4 mm long, scale-like; capsules elongate, lance-subulate, much longer than calyces; sepals ca. 2 mm long ______ **H. gentianoides**

Hypericum crux-andreae (L.) Crantz, (St. Andrew's cross), ST. PETER'S-WORT. Glabrous, essentially evergreen, erect or suberect small shrub 30–100 cm tall; upper leaves almost clasping; flowers axillary or in small clusters, pedicelled, subtended by 2 lanceolate bractlets; sepals 4, the 2 larger sepals 10–15 mm long and ca. as wide, broadly ovate to suborbicular; petals light yellow. Low woods, pond margins, and other moist areas; Limestone Co. at extreme e edge of nc TX; mainly se and e TX. Jul–Sep. [*Ascyrum stans* Michx. ex Willd.]

Hypericum drummondii (Grev. & Hook.) Torr. & A. Gray, (for its discoverer, Thomas Drummond, 1780–1835, Scottish botanist and collector in North America), NITS-AND-LICE, DRUMMOND'S ST. JOHN'S-WORT. Erect much branched annual to ca. 80 cm tall; flowers solitary in upper axils; petals orange-yellow, 2.5–4.5 mm long; stamens to ca. 12. Dry sandy or gravelly soils in fields or open woods; in nc TX either in the Cross Timbers or on the sandy soils at the extreme e margin of the area; se and e TX w to Rolling Plains and Edwards Plateau. Jul–Sep.

Hypericum gentianoides (L.) Britton, Sterns, & Poggenb., (resembling *Gentiana*—gentian), PINEWEED ST. JOHN'S-WORT, ORANGE-GRASS, PINEWEED. Erect much branched annual to ca. 60 cm tall; stems and erect branches thread-like and wiry; flowers minute; petals 1.5–4 mm long; stamens few. Dry gravelly or sandy soils; Palo Pinto and Parker cos., also Lamar Co. (Carr 1994); se and e TX w to Rolling Plains and Edwards Plateau. May–Sep.

Hypericum gymnanthum Engelm. & A. Gray, (naked flower), CLASPING ST. JOHN'S-WORT. Erect perennial to ca. 90 cm tall; leaves cordate-clasping; petals 3–6 mm long; stamens 10–12. Sandy soils; included based on citation of vegetational areas 4 and 5 (Fig. 2) by Hatch et al. (1990); mainly se and sc TX. Jun–Jul.

Hypericum hypericoides (L.) Crantz, (resembling *Hypericum*—St. John's wort; this species was originally treated in the genus *Ascyrum*), ST. ANDREW'S-CROSS, ST. PETER'S-WORT. Glabrous, low, essentially evergreen shrub with punctate leaves; stems erect-ascending to decumbent, 30–100+ cm tall; flowers axillary, solitary or in small clusters, pedicelled, subtended by 2 narrow bracteoles; sepals 2 or 4, the 2 larger sepals ovate to elliptic, ca. 5–12(–15) mm long and ca. 3–10 mm broad; petals light yellow. Sandy woods. May–Sep. Often segregated into the genus *Ascyrum*. The following two subspecies seem to intergrade.

1. Erect and freely branched above the ground ______ subsp. **hypericoides**
1. Decumbent with several basal stems ______ subsp. **multicaule**

subsp. **hypericoides**. Se and e TX w to West Cross Timbers and Edwards Plateau. [*Ascyrum hypericoides* L. var. *oblongifolium* (Spach) Fernald]

subsp. **multicaule** (Michx. ex Willd.) Robson, (many-stemmed). Se and e TX w to West Cross Timbers. [*Ascyrum hypericoides* L. var. *multicaule* (Michx. ex Willd.) Fernald]

Hypericum mutilum L., (cut-off, the Linnean type being merely a cut-off fragment of a plant), DWARF ST. JOHN'S-WORT. Erect annual or perennial to 90 cm tall; leaves clasping basally; flowers very small; petals ca. 2–3 mm long, light yellow; stamens 6–12, free. Edge of water or other wet areas; Henderson and Hopkins cos. near e margin of nc TX, also Lamar Co. in Red River drainage (Carr 1994); mainly se and e TX and Edwards Plateau. May–Oct.

Hypericum perforatum L., (perforated, having or appearing to have small holes), COMMON ST. JOHN'S-WORT, KLAMATHWEED, TIPTONWEED, GOATWEED, EOLA-WEED, AMBER, ROSIN-ROSE. Erect, rhizomatous, branched perennial to 70(–150) cm tall; petals orange-yellow, with black oil dots usually present along margins, 7–12 mm long; stamens numerous, united below into 3–5 groups; capsules with numerous yellow-amber oil vesicles. Roadsides; originally seen in TX (Lipscomb 1984) between Bowie and Decatur along Hwy 287 (Wise and Montague cos.); now also in Grayson Co. along Hwy 82, also Tarrant Co.; possibly introduced with CROWN-VETCH by the Highway Department (Mahler 1988). Jun–Sep. Native of Europe. Eating the plant can cause photosensitization and consequent dermatitis in light-skinned livestock; contact may also cause dermatitis in humans—see discussion in generic synopsis. This species is also used medicinally as a possible herbal treatment for depression (Leung & Foster 1996). ☠

Hypericum pseudomaculatum Bush, (false *Hypericum maculatum*; maculatum = spotted), FALSE SPOTTED ST. JOHN'S-WORT. Erect perennial to ca. 80 cm tall; petals conspicuously black-dotted. Fields; included based on citation of vegetational area 5 (Fig. 2) by Hatch et al. (1990); mainly e TX. May–Jun.

Hypericum punctatum Lam., (dotted), SPOTTED ST. JOHN'S-WORT. Erect perennial to ca. 100 cm tall; petals pale yellow, streaked with numerous black oil dots; stamens united below into 3–5 groups; capsules ovoid, 4–6 mm long, with numerous yellow-amber oil vesicles. Edges of woodlands; Denton, Grayson, Hunt, Lamar, Limestone, and Tarrant cos., e TX w to nc TX. Jun–Jul. Reported to be poisonous to horses (Burlage 1968). ☠

Hypericum sphaerocarpum Michx., (globe- or round-fruited), ROUND-FRUIT ST. JOHN'S-WORT. Erect perennial; stems 30–60 cm tall from thin woody rhizome; leaves usually 3–7 cm long, 5–15 mm wide, obtuse to acute; sepals 3–5 mm long; stamens numerous, free. Prairies, embankments; Fannin Co. in Red River drainage and Collin Co. near the Heard Museum; mainly e TX. May–Jul.

TRIADENUM ST. JOHN'S-WORT

A genus of 6–10 species of e Asia and e North America; this disjunct distribution pattern is discussed under the genera *Campsis* (Bignoniaceae) and *Carya* (Juglandaceae). (Greek: *tri*, three, and *aden*, gland, referring to the three large glands alternating with three sets of stamens)

Triadenum walteri (J.G. Gmel.) Gleason, (for Thomas Walter, 1740–1789, British-American botanist and Carolina planter). Perennial herb with rhizomes; stems erect, 40–100 cm tall; leaves 2–15 cm long, 1–5 cm wide, obtuse or rounded apically; petioles 3–15 mm long; flowers usually in axillary cymules or solitary; sepals 5; petals 5, 4–7 mm long, pinkish or reddish; filaments fused 1/2–2/3 of their lengths; ovary 3-carpellate; capsules 7–11 mm long. In or on edge of water or in moist woods; se and e TX w to Milam Co. near extreme e margin of nc TX. Aug–Oct. [*Hypericum walteri* J.G. Gmel.]

CONVOLVULACEAE MORNING-GLORY FAMILY

Ours annual or perennial, herbaceous, some with milky sap; twining and climbing vines or stems trailing to erect; leaves alternate, simple or palmately compound, entire or lobed or with few, coarse teeth, exstipulate; flowers perfect, axillary, solitary or few in small cymes (the cymes head-like in *Jacquemontia*); calyx lobes or sepals 5; corollas sympetalous, funnelform to campanulate, salverform, or rotate, 5-toothed, -angled, or -lobed; stamens 5, distinct, attached at base or on tube of corolla; pistil usually 2-carpellate; ovary superior, styles 1–2; fruit a capsule with usually 1–4 seeds.

A medium-large family of 55 genera (Austin 1998) and 1,600 species. It is a cosmopolitan family of herbaceous climbers, lianas, herbs, shrubs, or rarely trees; a number are showy ornamentals; *Ipomoea batatas* (L.) Lam. (SWEET-POTATO) is widely cultivated; some are problematic weeds (e.g., *Convolvulus*). (subclass Asteridae)

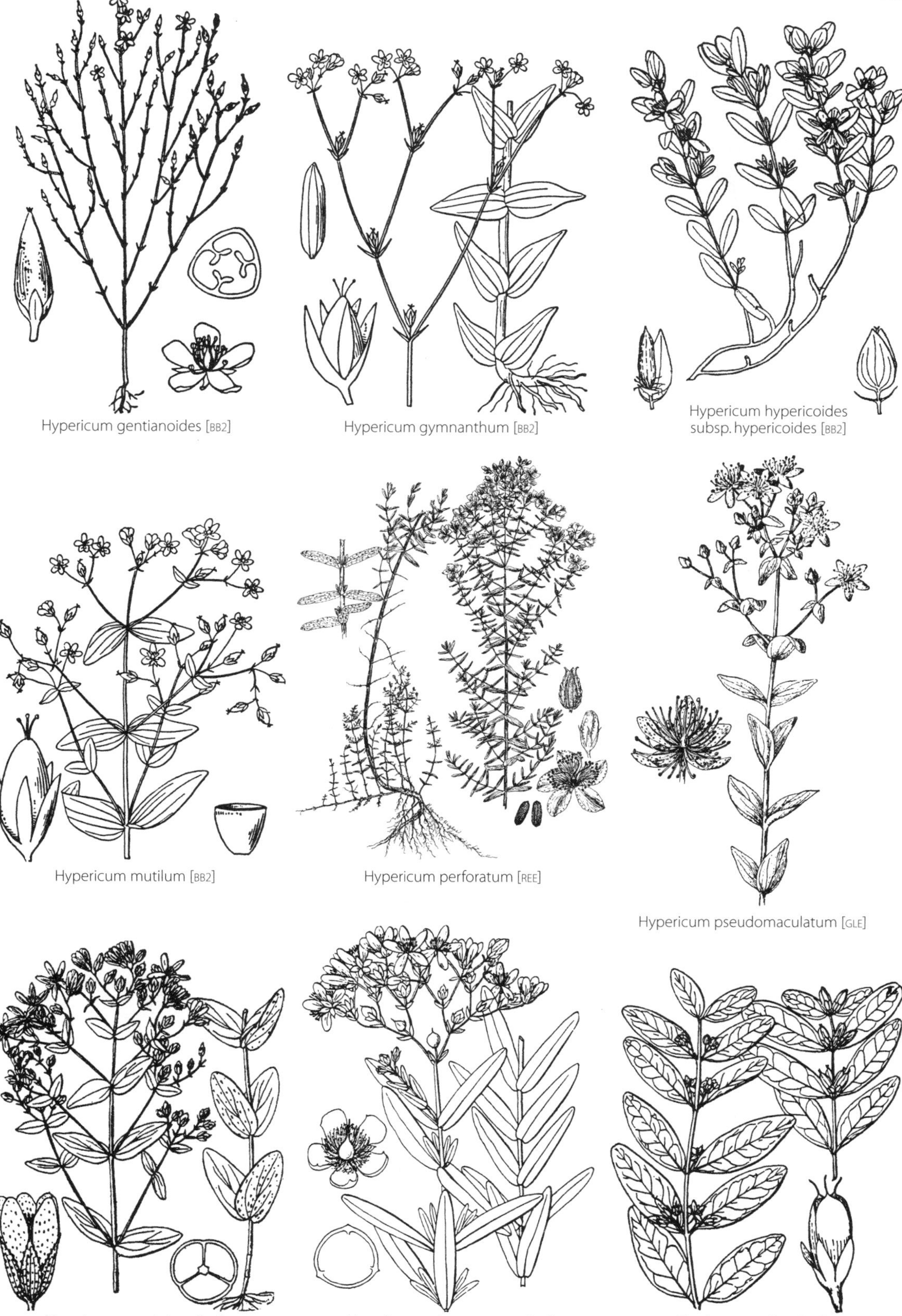

Hypericum gentianoides [BB2]

Hypericum gymnanthum [BB2]

Hypericum hypericoides subsp. hypericoides [BB2]

Hypericum mutilum [BB2]

Hypericum perforatum [REE]

Hypericum pseudomaculatum [GLE]

Hypericum punctatum [BB2]

Hypericum sphaerocarpum [BB1]

Triadenum walteri [BB2]

FAMILY RECOGNITION IN THE FIELD: *twining* herbs (2 species erect to decumbent) with alternate, sometimes cordate leaves; sap sometimes milky; corollas often showy, sympetalous, radially symmetrical, variable in shape but often funnelform to campanulate; ovary superior.
REFERENCES: Wilson 1960c; Austin 1986a, 1990, 1998.

1. Parasitic vines, yellowish brown; leaves absent ____ **Cuscuta** (see Cuscutaceae)
1. Plants green, chlorophyllous; leaves present.
   2. At least some leaf blades usually kidney-shaped, on petioles longer than the blades; open corollas 1.5–7 mm broad; corollas light green or greenish white, deeply lobed, the lobes longer than the tube; capsules with 2 distinct lobes ____ **Dichondra**
   2. Leaf blades usually not kidney-shaped, on petioles shorter to longer than the blades; open corollas 8–70 mm or more broad; corollas variously colored including white, not deeply lobed; capsules without 2 distinct lobes.
      3. Stigmas or stigma lobes 2 or 4 per flower, variously shaped (filiform to linear or oblong- or elliptic-flattened) but neither globose nor subglobose; corollas either 25(–30) mm or less long OR if longer (in *Calystegia*) then completely white.
         4. Flowers numerous in a densely crowded, leafy-bracted, head-like inflorescence; sepals with conspicuous, long (ca. 2–3 mm), tawny hairs easily visible to the naked eye; corollas blue ____ **Jacquemontia**
         4. Flowers solitary or in 2s or 3s; sepals glabrous to pubescent, but without long hairs conspicuous to the naked eye; corolla color various including blue.
            5. Erect to decumbent herbs with short stems to only 35 cm long; flowers sessile or short-pedicelled; pedicels shorter than calyces; corollas rotate to shallowly campanulate, white or pale to deep lavender; stigmas 4 per flower ____ **Evolvulus**
            5. Vines with elongate stems that can reach 1–2 m in length; flowers on long pedicels or peduncles exceeding calyces; corollas funnelform; variously colored including white or lavender; stigmas 2 per flower.
               6. Leaves narrowly linear to linear, 1–3 mm wide, entire; corollas white ____ **Stylisma**
               6. Leaves variously shaped but not narrowly linear to linear, much > 3 mm wide, entire to toothed or lobed; corollas white to pink or pink with red eye to white or pink inside, with the outside wholly pinkish or brownish lavender or with broad bands of those colors.
                  7. Calyces ebracteate; corollas 1.2–3 cm long, usually with some color, sometimes all white ____ **Convolvulus**
                  7. Calyces immediately subtended by 2 large floral bracts; corollas 4–5 cm long, white ____ **Calystegia**
      3. Stigma globose and unlobed OR stigma with 2–3 globose or subglobose lobes; corollas usually > 25 mm long (< 25 mm in 2 red-flowered *Ipomoea* species and in the white-flowered *Ipomoea lacunosa*), usually not completely white.
         8. Leaves with 5–7 palmate lobes that are toothed to pinnatifid; corollas white with purple-red center ____ **Merremia**
         8. Leaves not 5–7 palmately lobed with the lobes toothed to pinnatifid; corollas variously colored, usually not as above ____ **Ipomoea**

CALYSTEGIA HEDGE-BINDWEED

A genus of ca. 25 species widespread in the temperate zones (D. Austin, pers. comm.); it includes cultivated ornamentals and pernicious weeds. (Greek: *kalyx*, cup, cover, calyx, and *stege*, cover, alluding to the two bracts that subtend the flowers and conceal the calyx)
REFERENCES: Brummitt 1965, 1981; Lewis & Oliver 1965; Austin 1986a, 1992; Austin et al. 1997.

Calystegia macounii (Greene) Brummitt, (for John Macoun, who collected the type specimen in

1895). Semi-erect or sparsely twining, pubescent, perennial herb; leaf blades 2–6 cm long, 1–5 cm wide, basally cordate to subsagittate, marginally entire; petioles 0.5–40 mm long; flowers axillary, solitary, on peduncles usually 3–5 cm long; floral bracts 2, enclosing calyces, foliaceous, strongly inflated, 20–25 mm long, 10–15 mm wide; sepals 10–12 mm long, 5–7 mm wide; corollas funnelform, 4–5 cm long, white; style 1. Along a stream bank in limestone area, typically sandy soils near waterways in open grasslands or openings in woodlands; Cooke Co. (*Shinners 13256*, BRIT/SMU); otherwise unknown in TX; the presence of *C. macounii* in TX was only recently recognized (Austin et al. 1997). May–Jun. [*Convolvulus interior* House, *Convolvulus macounii* Greene]

Calystegia sepium (L.) R. Br. subsp. *angulata* Brummitt, (sp.: growing along hedges; subsp.: angular), TRAILING HEDGE-BINDWEED, *Calystegia sepium* (L.) R.Br. subsp. *limnophila* (Greene) Brummitt, (swamp-loving), and *Calystegia silvatica* (Kit.) Griseb. subsp. *fraterniflora* (Mack. & Bush) Brummitt, (sp.: pertaining to the woods; subsp.: with brotherly flowers, these sometimes in pairs), are also known from TX (Hatch et al. 1990; Austin et al. 1997). Because of past confusion involving TX *Calystegia* taxa and because of the possibility of these additional taxa being found in nc TX, the following key (from Austin et al. 1997) is included.

1. Leaf sinus quadrate (= nearly square); blade tissues not beginning for 2–5(–10) mm from petiole attachment (only vascular tissue near attachment) ______ *C. silvatica* subsp. *fraterniflora*
1. Leaf sinus V- or U-shaped; blade tissues beginning at the point of petiole attachment.
   2. Basal lobes of leaf blades with 1–2 small, tooth-like angles; plants normally glabrous or with a few trichomes on petioles ______ *C. sepium* subsp. *angulata*
   2. Basal lobes of leaf blades usually rounded; plants normally pubescent on all vegetative parts.
      3. Plants twining throughout; leaf blades narrow (mostly 3:1 length to width ratio) ______ *C. sepium* subsp. *limnophila*
      3. Plants semi-erect or sparsely twining; leaf blades broad (mostly 2:1 length to width ratio) ______ *C. macounii*

CONVOLVULUS BINDWEED

Trailing to decumbent or twining perennials; leaves petioled; leaf blades entire, toothed, or lobed, basally truncate to cordate, hastate, or sagittate; flowers axillary, solitary or in cymes of 2 or 3; corollas campanulate; style 1; stigmas 2, linear, ± flattened, slightly acute; capsules 1–4-seeded.

A cosmopolitan, but especially temperate genus of ca. 100 species; some have alkaloids. A number are cultivated as ornamentals, while others can be pernicious weeds. (Latin: *convolvere*, to entwine, in reference to the twining habit of some species)
REFERENCE: Lewis & Oliver 1965.

1. Calyces 3–5 mm long, inconspicuously pubescent or glabrous; corollas merely 5-angled; plants sparsely pubescent; leaf blades entire except for basal lobes, ovate to ovate-lanceolate to elliptic, often almost as wide as long ______ **C. arvensis**
1. Calyces 6–12 mm long, densely pubescent; corollas with margin having 5 sharp points or projections; plants densely gray-pubescent; leaf blades from entire to toothed or deeply lobed, variously shaped, usually much longer than wide, sometimes with the main part very long and narrow ______ **C. equitans**

Convolvulus arvensis L., (of fields), POSSESSION-VINE, BINDWEED, FIELD BINDWEED, COMMON BINDWEED, CORNBIND. Perennial from deep creeping root, forming extensive beds; stems to 1 m or more long, decumbent or twining; calyx lobes elliptic-orbicular; corollas white to pink inside, outside with broad bands or wholly pinkish or brownish lavender, 1.2–2.5 cm long. Roadsides, disturbed sites, a problematic weed in gardens and difficult to eradicate because of the deep root; throughout much of TX. May–Jul, less freely later. Native of Eurasia.

Convolvulus equitans Benth., (riding a horse, like legs of a rider around a horse), GRAY BINDWEED, TEXAS BINDWEED. Perennial from taproot, not forming extensive beds; stems to 2 m long, trailing to decumbent; leaves very variable; calyx lobes auricled basally to not auricled; corollas white, pink, or pink with red eye, 1.5–3 cm long. Prairies, disturbed areas; nearly throughout TX except extreme e. May–Oct.

DICHONDRA PONY-FOOT

Small, mat-forming, sparsely pubescent, creeping or trailing perennials rooting at the nodes; leaves long-petioled; leaf blades orbicular-ovate to reniform, entire or notched at apex; flowers small, 1–2 per axil, long-pedicelled but pedicels shorter than petioles; corollas with lobes longer than the tube, light green or greenish white; styles 2, 1 from each of the 2 nearly separate carpels; capsules with 2 or 4 seeds.

A genus of 16 species (D. Austin, pers. comm.) native to tropical and warm areas of the world. (Greek: *di*, two or double, and *chondros*, a grain, from the deeply lobed fruit)
REFERENCES: Tharp & Johnston 1961; Johnston 1963a.

1. Pedicels straight; calyx lobes 2–3 times as long as wide; corollas nearly as long as calyces; widespread in nc TX ____ **D. carolinensis**
1. Pedicels abruptly recurved near summit; calyx lobes 1.5–2 times as long as wide; corollas almost 1/3 longer than calyces; found only in extreme s part of nc TX ____ **D. recurvata**

Dichondra carolinensis Michx., (of Carolina). Plant to 12 cm high; pedicels during flowering 1/3–2/3 as long as petioles; calyx lobes oblong or oblanceolate, separate nearly to base, slightly exceeding the corolla. Damp sandy or silty ground, also a common lawn weed; Bell, Dallas, Denton, and Grayson cos.; se and e TX w to East Cross Timbers. Apr. [*Dichondra repens* J.R. Forst. var. *carolinensis* (Michx.) Choisy]

Dichondra recurvata Tharp & M.C. Johnst., (curved backward). Plant to 17 cm high; pedicels during flowering 1/10 to 1/2 as long as petiole; flowering calyces 2.5–3.2 mm long, shorter than corollas. Roadsides, brushy creekbanks; Mills and Williamson cos., also Bell and Burnet cos. (Johnston 1963a); s part of nc TX s to c TX; endemic to nc and c TX.

EVOLVULUS

Ours small pubescent perennials with several erect to ascending stems from a small woody base; stems decumbent to erect; leaves sessile, entire, narrowly lanceolate to oblong-lanceolate, small, 3 cm or less long, 1–8 mm wide; flowers solitary, axillary; corollas 8–18 mm across, rotate or shallowly campanulate; styles 2, each with 2 branches, the 4 stigmas linear or filiform; capsules 1–4-seeded.

A genus of 98 species of warm and tropical areas of the Americas; 2 species extend to the Old World. (Latin: *evolvere*, to untwist or unroll, alluding to the non-climbing habit compared with other members of the family)
REFERENCES: Van Ooststroom 1934; Perry 1935.

1. Leaves densely pubescent on both surfaces; corollas pale to deep lavender; sepals lanceolate to narrowly lanceolate, 4–5 mm long, spreading pilose ____ **E. nuttallianus**
1. Leaves glabrous above or nearly so; corollas white; sepals oblong lanceolate, 3–5 mm long, appressed pilose ____ **E. sericeus**

Evolvulus nuttallianus Schult., (for Sir Thomas Nuttall, 1786–1859, English-American botanist), HAIRY EVOLVULUS, NUTTALL'S EVOLVULUS, SHAGGY EVOLVULUS, DWARF-MORNING-GLORY. Stems 5–

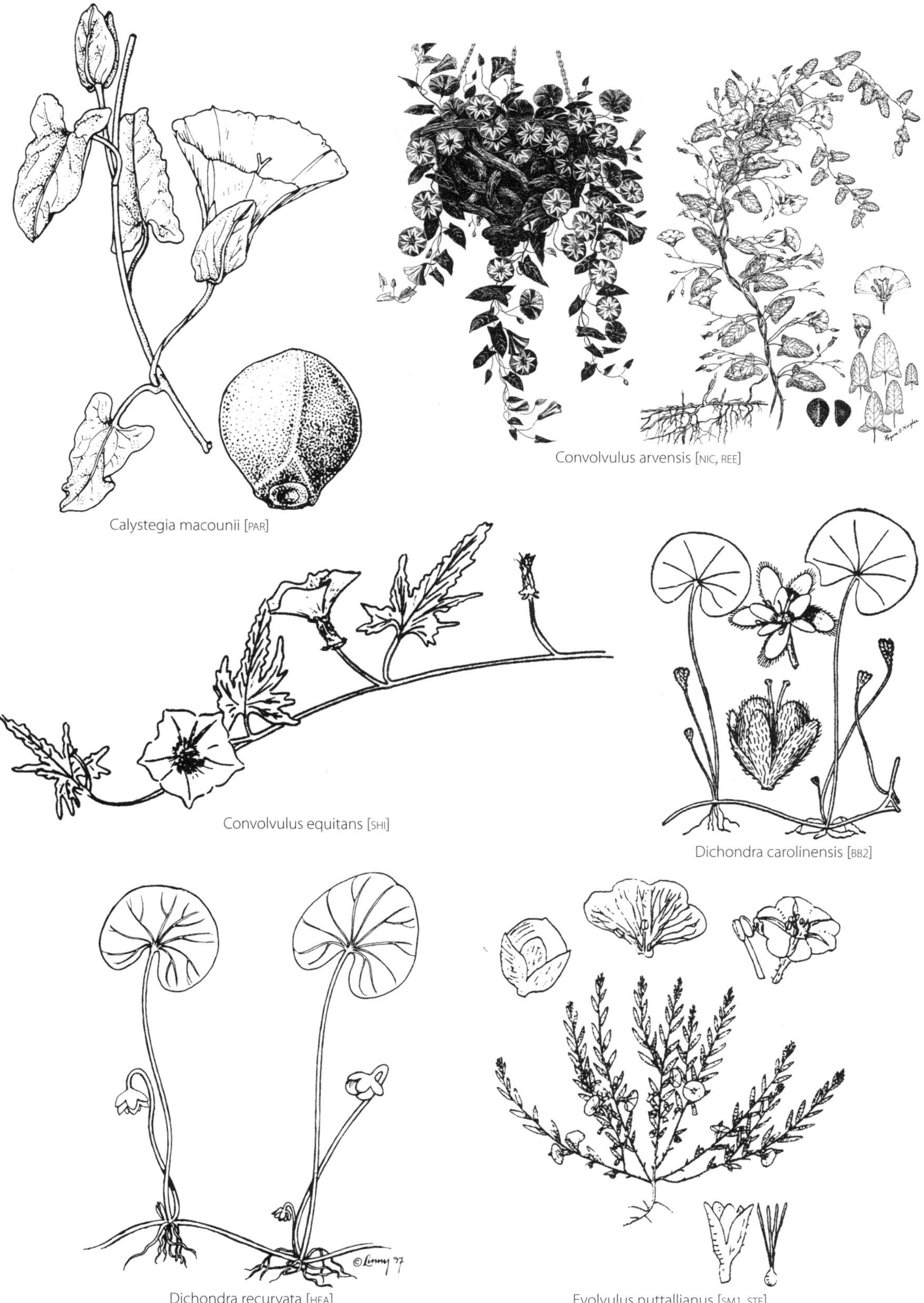

Calystegia macounii [PAR]

Convolvulus arvensis [NIC, REE]

Convolvulus equitans [SHI]

Dichondra carolinensis [BB2]

Dichondra recurvata [HEA]

Evolvulus nuttallianus [SM1, STE]

25 cm long, densely villous. Sandy or rocky limestone soils; nearly throughout TX except extreme e. Apr–Oct. [*Evolvulus pilosus* Nutt.]

Evolvulus sericeus Sw., (silky), SILKY EVOLVULUS. Stems 6–35 cm long, densely villous; leaves densely pubescent on lower surface in contrast to the upper surface. Sandy, mixed sandy and gravelly, or rocky limestone soils, on hillsides, roadsides, and in open woods; nearly throughout TX except extreme e part.

IPOMOEA MORNING-GLORY, MOONFLOWER

Ours trailing or twining annual or perennial vines or rarely a shrubby perennial, mostly flowering in summer and fall; leaves petioled; leaf blades variable in shape, often ± cordate at base, entire to lobed or apparently compound; flowers axillary, solitary or in 2- to ca. 5-flowered cymose inflorescences; corollas funnelform, funnelform-campanulate, or salverform, usually large and showy, often open for less than 8 hours; style 1; stigma solitary, unlobed, globose or with 2–3 globose or subglobose lobes; capsules usually with 1–4 seeds.

A huge genus of ca. 600–700 species (Austin & Huáman 1996) of climbing vines or shrubs native to tropical and warm temperate regions. Many species are cultivated as ornamentals for their flowers; some escape and become troublesome weeds; ☠ others are toxic and hallucinogenic due to ergoline alkaloids; a Mexican species was used in religious ceremonies by the Aztecs (Schmutz & Hamilton 1979). The genus includes the SWEET-POTATO or YAM as discussed below. According to Strausbaugh and Core (1978), the common name, MORNING-GLORY, is derived from the tendency of the flowers of some species to open at night or in diffuse light (frequently in the morning). The rust fungus *Coleosporium ipomoeae* Burrill sometimes forms conspicuous yellow lesions on the leaves of *Ipomoea* species in nc TX; it is a heterecious rust (= uses more than one host to complete its life cycle) that also infects *Pinus taeda* (LOBLOLLY PINE), but it can survive using only MORNING-GLORIES (J. Hennen, pers. comm.) (Greek: *ips*, a worm, and *homois*, resembling, from the twining habit)

REFERENCES: O'Donell 1959; Jones & Deonier 1965; Austin 1976, 1978, 1988; Woolfe 1992; Bohac et al. 1995; Austin & Huáman 1996.

1. Corollas red, orange-red, or red with yellow tube (white in some cultivated forms), 2–3.5 cm long, salverform, the tube abruptly flared near the summit; stamens exserted—in other words, projecting beyond plane of limb (= open face of corolla); corolla limb 2 cm or less wide.
   2. Leaf blades pinnately divided into very narrow, linear, almost thread-like segments ca. as wide as the midrib ____ **I. quamoclit**
   2. Leaf blades entire to angle-toothed, not pinnately divided ____ **I. coccinea**
1. Corollas variously colored, usually not red, of variable length, often longer than 4 cm long, funnelform or funnelform-campanulate, the tube gradually expanded from below middle; stamens included within corolla, not exserted; corolla limb usually > 2 cm wide (except in the small white-flowered *I. lacunosa*).
   3. Leaves palmately compound with 3–7 leaflets (blades divided to base) ____ **I. wrightii**
   3. Leaves entire to variously and sometimes deeply divided, but not divided to base (thus not compound).
      4. Pedicels and peduncles with reflexed hairs; corollas variable in color but in some light blue to blue with white or yellowish center.
         5. Corollas purple or blue-purple to pink or white; sepals 6–17 mm long, lanceolate-elliptic or lanceolate-oblong, with slightly narrowed straight tips shorter to slightly longer than sepal body, glabrous to pubescent; leaf blades variable but often unlobed ____ **I. purpurea**
         5. Corollas usually light blue to blue with white or yellowish center, sometimes drying purplish pink (varying to all lavender or in cultivated forms other colors); sepals 12–24 mm

long, with very narrow, elongate, usually curved, tail-like tips much longer than ovate sepal body, at least the body densely long hirsute; leaf blades variable but usually 3-lobed ______ **I. hederacea**

4. Pedicels and peduncles with spreading to ascending hairs or glabrous; corollas variable in color but neither light blue nor blue with white or yellowish center.
   6. Leaf blades linear-lanceolate to linear, 2–8 mm wide; petioles 1–7 mm long ______ **I. leptophylla**
   6. Leaf blades triangular to cordate-ovate or sagittate, usually wider than 8 mm; petioles usually 10–80 mm long.
      7. Corollas white or white with a tinge of pink, 1.6–2(–2.5) cm long ______ **I. lacunosa**
      7. Corollas white with purple-red center, or pink to purplish, or pink to purplish with dark center, or rarely all white, 2.8–8 cm long.
         8. Corollas 5–8 cm long, white (rarely pink) with purple-red center; sepals rounded or nearly so at apex; leaf blades ovate-lanceolate to broadly ovate or cordate, entire or pandurate (= contracted at sides so as to be fiddle-shaped).
            9. Leaf blades pandurate or not so, 2–9 cm wide, broadly ovate to cordate in outline, cordate to rounded-truncate at base; widespread in nc TX ______ **I. pandurata**
            9. Leaf blades not pandurate, 1–4 cm wide, narrowly ovate to narrowly ovate-lanceolate, rounded to cuneate at base; extremely rare, in nc TX known only from Cooke and Montague cos. ______ **I. shumardiana**
         8. Corollas 3–5 cm long, pink to purplish or pink to purplish with dark center or rarely all white; sepals pointed at apex; leaf blades ovate to cordate-ovate or nearly triangular, entire or 3–5-lobed.
            10. Sepals about equal in length; root not tuberous; corollas pink to purplish with dark center or rarely all white; stems not rooting at the nodes; leaf blades ovate to cordate-ovate, entire or 3–5-lobed; native species widespread in nc TX ______ **I. cordatotriloba**
            10. Sepals unequal in length, the outer 3/4–5/6 as long as the inner; root tuberous; corollas pink to purplish; stems rooting at the nodes; leaf blades nearly triangular in outline, usually entire or 3–5-lobed; cultivated plant introduced to nc TX, possibly persisting but unlikely to be found in the wild ______ **I. batatas**

Ipomoea batatas (L.) Lam., (vernacular name for this species from Chibchan-speakers of Colombia; from it is derived the English word potato—Austin 1988), SWEET-POTATO, YAM. Perennial (but cultivated as an annual) vine, trailing or twining, from soft tuberous root; sepals acuminate; corollas funnelform to funnelform-campanulate, pink to purplish, 4–7 cm long. Widely cultivated and possibly persisting; included based on citation of vegetational areas 4 and 5 (Fig. 2) by Hatch et al. (1990); mainly e TX. May–Sep. Probably native of tropical America, now widely grown in tropical and warm areas for its edible root; it is cultivated in more than 100 countries (Woolfe 1992) and is especially important in Japan (Mabberley 1987) and China; in terms of production, SWEET POTATO is the world's seventh largest food crop (Woolfe 1992; Bohac et al. 1995). Contreras et al. (1995) gave information about the SWEET POTATO in Mexico. True YAMS are in the genus *Dioscorea* (Dioscoreaceae).

Ipomoea coccinea L., (scarlet), SCARLET-CREEPER, SCARLET MORNING-GLORY, STARGLORY, RED MORNING-GLORY. Low-climbing annual vine; leaf blades broadly ovate, entire to dentate with 3–5 teeth; sepals obtuse with a narrow awl-like appendage from just below the apex; corollas salverform, orange-red or red with yellow tube, 2–2.5(–4) cm long, the limb 1.7–1.9 mm wide; stamens exserted, 2.7–3 cm long. Cultivated and escapes; McLennan Co.; otherwise known from coastal TX and Edwards Plateau. Jul–Nov. Native of se U.S. Austin (1975) considered this species adapted for pollination by ruby-throated hummingbirds (*Archilochus colubris*).

Ipomoea cordatotriloba Dennst., (cordate or heart-shaped and three-lobed). Perennial (but flowering the first year), twining and low-climbing vine from branched root; sepal tips acute or acuminate; corollas funnelform to funnelform-campanulate, purple-rose with dark eye, rarely white, 2.8–5.5 cm long. Stream bottoms, fencerows, disturbed sites. Jun–Sep. According to Austin (1976), the following two varieties occupy different habitats in TX; var. *torreyana* is found mainly in prairie and plain areas, while var. *tricocarpa* is found mainly in deciduous forest; some intermediates are known where the two habitat types meet.

1. Sepals pubescent or pilose, at least at apex or on margins; leaves and stems glabrous to pilose ____ var. **cordatotriloba**
1. Sepals glabrous; leaves and stems glabrous ____ var. **torreyana**

var. **cordatotriloba**, SHARP-POD MORNING-GLORY, WILD MORNING-GLORY, PURPLE MORNING-GLORY, TIEVINE. Dallas, Hill, Limestone, Parker, and Tarrant cos., also Bell, Ellis, Falls, Grayson, and Johnson cos. (Austin 1976); se and e TX w to nc TX. [*Ipomoea trichocarpa* Elliott] ▣/94

var. **torreyana** (A. Gray) D.F. Austin, (for John Torrey, 1796–1873, coauthor with Asa Gray of *The Flora of North America*), COTTON MORNING-GLORY. Coryell, Parker, and Tarrant cos., also Bell, Brown, Dallas, Hood, Johnson, Lampasas, McLennan, Mills, Palo Pinto, Somervell, and Williamson cos. (Austin 1976); se and e TX w to nc TX. [*Ipomoea trichocarpa* Elliott var. *torreyana* (A. Gray) Shinners]

Ipomoea hederacea Jacq., (resembling *Hedera*—English ivy), IVY-LEAF MORNING-GLORY. Annual twining vines; stems retrorsely pubescent; leaf blades usually 3-lobed to rarely unlobed or 5-lobed; sepals with tail-like tips much longer than ovate sepal body, at least the body densely long hirsute; corollas funnelform, usually light blue or blue with white or yellowish center (varying to all lavender or in cultivated forms other colors), sometimes drying puplish pink, 2.5–5 cm long. Disturbed areas, roadsides; Bell, Dallas, Grayson, and Rockwall cos.; scattered localities in TX. Jul–Oct.

Ipomoea lacunosa L., (with air spaces, holes or pits, from the loosely reticulate venation of the leaves), PITTED MORNING-GLORY, SMALL WHITE MORNING-GLORY, WHITE MORNING-GLORY. Annual often twining vine from taproot, sparsely to densely hispid-pubescent; leaf blades unlobed or angulate or lobed; sepals acuminate; corollas funnelform, white or white with a tinge of pink, 1.5–2(–2.5) cm long. Disturbed areas; Dallas and Grayson cos., also Tarrant Co. (Mahler 1988); se and e TX w to nc TX, also Edwards Plateau. Sep–Oct.

Ipomoea leptophylla Torr., (thin-leaved), BUSH MORNING-GLORY. Perennial herb; stems erect to decumbent, bushy-branched, to ca. 1.2 m tall; leaf blades lanceolate to linear, unlobed; petioles 1–7 mm long; corollas funnelform to funnelform-campanulate, lavender-pink with dark center or completely purple-red, 5–9 cm long; Clay and Montague cos.; West Cross Timbers s and w to w TX.

Ipomoea pandurata (L.) G. Mey., (fiddle-shaped), BIG-ROOT MORNING-GLORY, WILD POTATO-VINE, MAN-OF-THE-EARTH, WILD POTATO, WILD SWEET-POTATO, INDIAN-POTATO. Perennial twining or trailing vine; juice milky; leaf blades cordate-ovate, often fiddle-shaped (= pandurate); sepal tips obtuse; corollas funnelform, white with purple-red eye, 5–8 cm long. Sandy roadsides and disturbed ground; Cooke, Dallas, Grayson, and Tarrant cos., also Montague Co. (S. Lusk, pers. comm.) e TX w to nc TX, also Edwards Plateau. Jun–Sep. The starchy, enormously thickened roots of old plants can weight up to 25 pounds (Wills & Irwin 1961); they were reportedly used by Native Americans as food; they are also reported to be poisonous, to have a milky resinous juice, and to contain the glucoside ipomoein (Burlage 1968). ☠ ▣/94

Ipomoea purpurea (L.) Roth, (purple), MEXICAN MORNING-GLORY, WOOLLY MORNING-GLORY, COMMON MORNING-GLORY. Annual, low- to high-climbing, twining vine; leaf blades usually unlobed

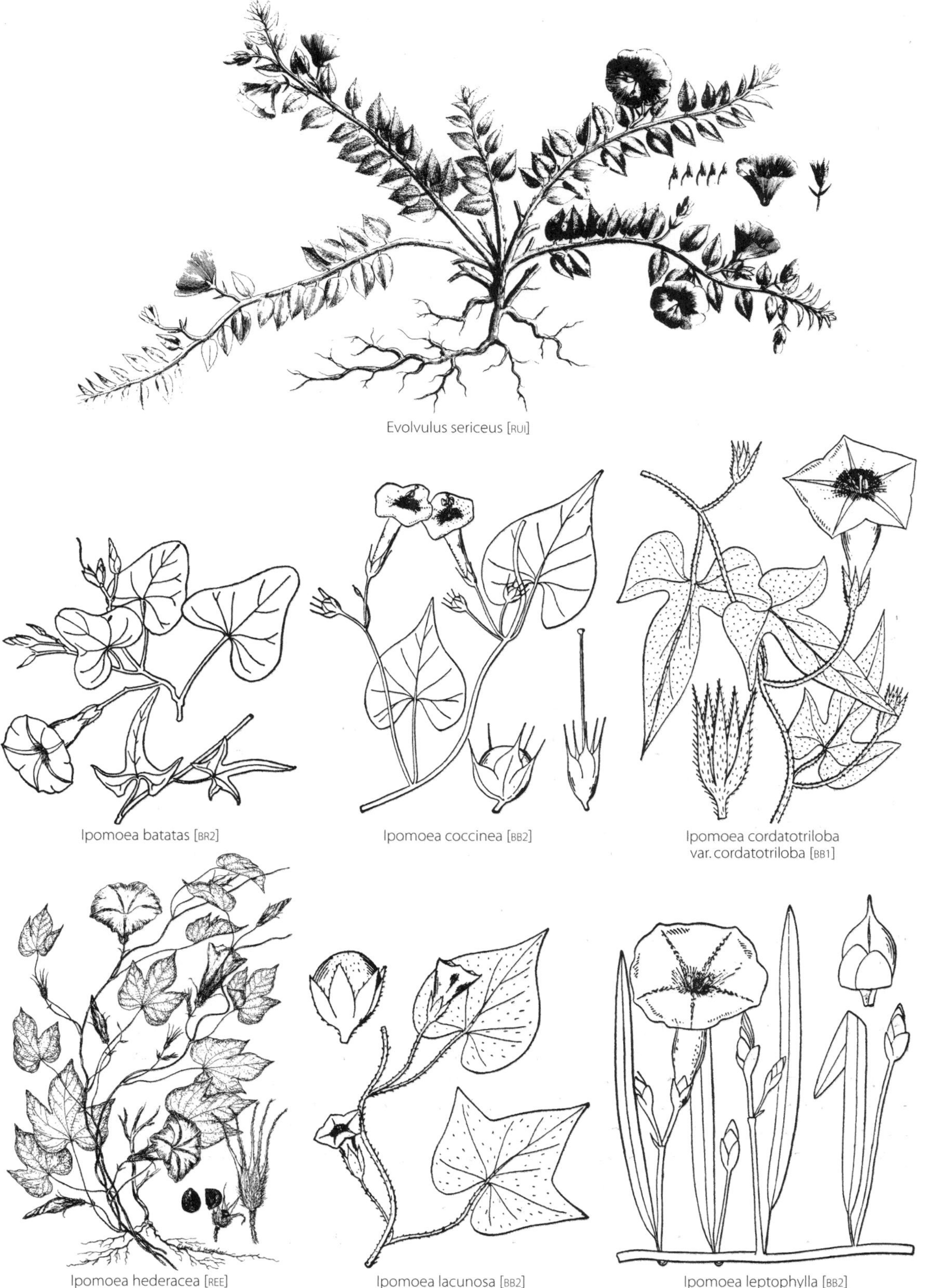

Evolvulus sericeus [RUI]

Ipomoea batatas [BR2]

Ipomoea coccinea [BB2]

Ipomoea cordatotriloba var. cordatotriloba [BB1]

Ipomoea hederacea [REE]

Ipomoea lacunosa [BB2]

Ipomoea leptophylla [BB2]

to rarely 3- or 5-lobed; corollas funnelform, blue-purple to reddish or white (variable in cultivated forms), 2.5–5(–7) cm long. Stream banks, disturbed or waste areas; Dallas Co.; mainly w part of TX. [*I. purpurea* (L.) Roth var. *diversifolia* (Lindl.) O'Donell] Reported to be poisonous (Burlage 1968).

Ipomoea quamoclit L., (native Mexican name), CYPRESSVINE. Low-climbing, twining, annual vine; leaf blades pinnately divided into very narrow, linear, almost thread-like segments about as wide as the midrib (0.2–1.5 mm wide); corollas salverform, deep red (white in some cultivated forms), 2–3 cm long, limb 1.8–2 cm wide; stamens exserted, 2.5–3 cm long. Cultivated and spreads; Grayson Co.; also Blackland Prairie and Post Oak Savannah. Jul–Nov. Native of se U.S.

Ipomoea shumardiana (Torr.) Shinners, (for Benjamin Franklin Shumard, 1820–1869, state geologist of TX in 1860), NARROW-LEAF MORNING-GLORY. Perennial vine similar to *I. pandurata* but leaf blades often narrower, 3–8 cm long; corollas funnelform, white (rarely pink) with purple-red throat, 5–8 cm long. Possibly a hybrid between *I. leptophylla* and *I. pandurata* (Correll & Johnston 1970). Gravelly roadside prairie; rare; Montague Co. (*S. Lusk s.n.*, 1997), also Cooke Co. (Correll & Johnston 1970); also Edwards Plateau; endemic of TX, OK, and KS. Jun–Aug. The recent Montague Co. collections are the only ones known from nc TX for many decades. [*Convolvulus shumardianus* Torrey, *I. carletoni* Holz.]

Ipomoea wrightii A. Gray, (for Charles Wright, 1811–1885, TX collector), WRIGHT'S MORNING-GLORY. Low climbing vine with palmately compound leaves. Alluvial or damp soils; Grayson (Hagerman Nat. Wildlife Refuge) and Rockwall cos.; mainly c and s TX. Jun–Oct. Native probably of India.

JACQUEMONTIA

A genus of 80–100 species of tropical and warm areas, especially the Americas; some are cultivated as ornamentals. (Named for Victor Jacquemont, 1801–1832, botanical explorer)

Jacquemontia tamnifolia (L.) Griseb., (with leaves resembling *Tamnus* of Dioscoreaceae), HAIRY CLUSTERVINE. Annual; stems twining, to 2 m long, but can flower while still small and erect; leaves long-petioled; leaf blades ovate-elliptic to ovate, 3–12 cm long, 2–9 cm wide, the larger ones basally cordate, acute to acuminate at apex; flowers many, densely arranged in a leafy-bracted, pedunculate, head-like inflorescence; peduncles shorter to longer than leaves; bracts and sepals with conspicuous long hairs; bracts lanceolate to elliptic; sepals lance-linear; corollas 12–16 mm long, blue; style 1; stigmas 2, ±flattened; capsules 4-seeded. Cultivated, escaped in Dallas Co.; native to se and e TX. Jul–Oct. [*Ipomoea tamnifolia* L., *Thyella tamnifolia* (L.) Raf.]

MERREMIA

A primarily tropical genus of ca. 70 species including some that are serious weeds in plantations; others are cultivated as ornamentals or used medicinally. (Named for Blasius Merrem, died 1824, German naturalist)

Merremia dissecta (Jacq.) Hallier f., (dissected, deeply cut), ALAMO-VINE, CORREHUELA DE LAS DOCE. Low-climbing to trailing perennial vine; leaf blades palmately deeply 5- or 7-lobed, 4–15 cm long, ca. as wide, the lobes toothed to pinnatifid, obtuse; peduncles 1–2-flowered; corollas white with purple-red center, 3.5–5 cm long. Open and disturbed areas, banks of stream; included based on citation of vegetational areas 4 and 5 (Fig. 2) by Hatch et al. (1990); c TX, native at least as far n as Travis Co. May–Nov. Sometimes lumped into the genus *Ipomoea* [as *I. sinuata* Ortega].

STYLISMA BONAMIA

A genus of 6 species native to s and e U.S. (Derivation of generic name unexplained by origi-

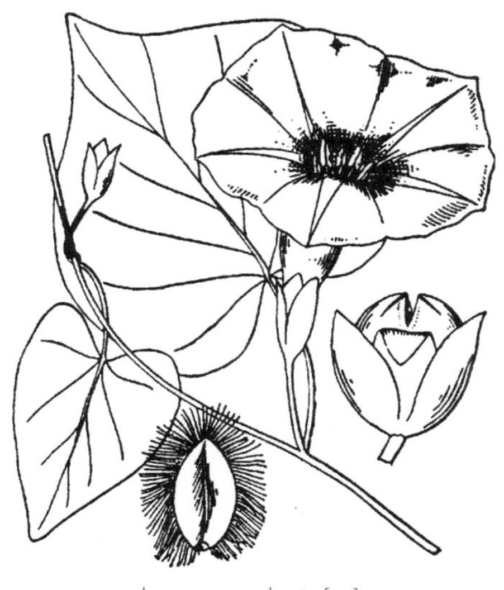

Ipomoea pandurata [BB2]

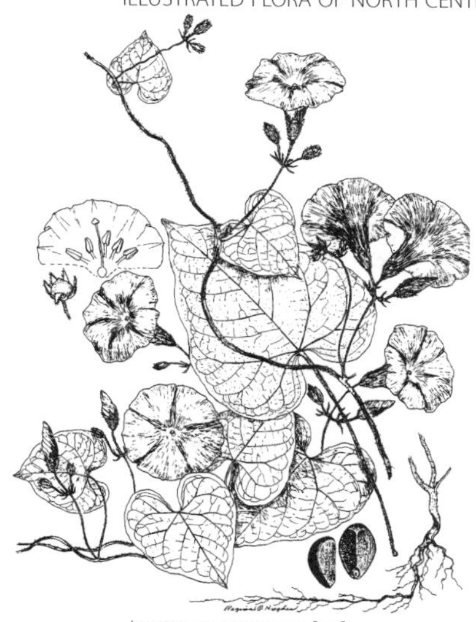
Ipomoea purpurea [REE]

Ipomoea quamoclit [LAM]

Ipomoea shumardiana [USH]

Ipomoea wrightii [WIG]

Jacquemontia tamnifolia [DIL]

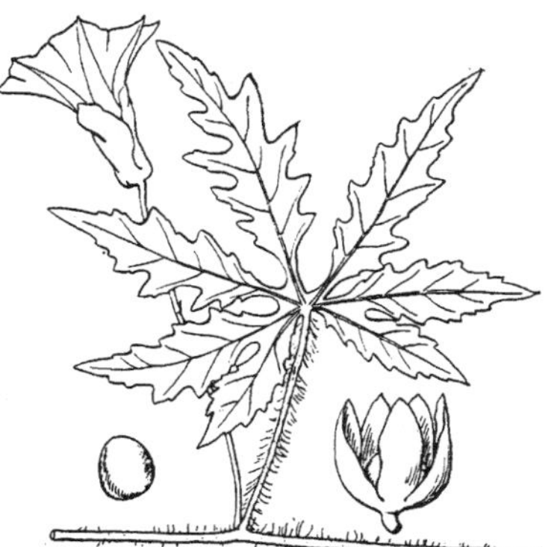
Merremia dissecta [BR2]

Stylisma pickeringii var. pattersonii [BB2]

nal author, but possibly from Greek: *stylus*, a pillar, stake, or column, in reference to the style)
REFERENCES: Fernald & Schubert 1949; Shinners 1962a; Myint 1966; Lewis 1971.

Stylisma pickeringii (Torr. ex M.A. Curtis) A. Gray var. **pattersonii** (Fernald & B.G. Schub.) Myint, (sp.: for Charles Pickering, 1805–1878, botanist and physician on Wilkes expedition in the Columbia River area; var.: for Harry Norton Patterson, 1853–1919, American printer, botanist, and explorer), BIG-POD BONAMIA. Vine-like, pubescent, perennial herb with long, prostrate or trailing to ascending stems to 2 m long; leaves sessile or nearly so; leaf blades entire, narrowly linear to linear, 1–3 mm wide; flowers 1–3(–5) together, on short pedicels at tips of long axillary peduncles; peduncles with leafy bracteoles at tip; bracteoles 1.5–2.5 cm long; sepals hoary pubescent on back; corollas white, 10–18 mm long, funnelform to campanulate; filaments glabrous or nearly so; styles 2, fused nearly to base of the capitate stigmas; capsules with 1–2 seeds. Sandy open woods; Parker and Tarrant cos.; mainly se and e TX. May–Aug. [*Breweria pickeringii* (Torr. ex M.A. Curtis) A. Gray var. *pattersonii* Fernald & B.G. Schub.]

Hatch et al. (1990) cited 2 additional species, *S. aquatica* (Walter) Raf. (growing in or near water) (PURPLE BONAMIA) and *S. villosa* (Nash) House (covered with soft hairs) (HAIRY BONAMIA), for vegetational area 4 (Fig. 2); both of these apparently occur only to the s and e of nc TX. *Stylisma aquatica* can be distinguished by its filaments glabrous or nearly so, lavender corollas, and styles divided halfway or more; *S. villosa* has filaments densely spreading-pilose in lower part, sepals pilose on the back, and white corollas. Another species, *S. humistrata* (Walter) Chapm., (stretched out on the ground), extends up the Sabine River to Wood Co. just e of nc TX. It has peduncles only minutely bracted, the elliptic-oblong leaves short-petioled, filaments densely spreading-pilose in lower part, sepals glabrous on the back, and white corollas.

CORNACEAE DOGWOOD FAMILY

A small (90 species in 12 genera—Mabberley 1987), mainly n temperate family rare in the tropics and subtropics; most are trees and shrubs or rarely rhizomatous herbs; several are cultivated including the ornamentally important genus *Cornus* (DOGWOOD). The Nyssaceae, here recognized as a distinct family, is sometimes lumped into the Cornaceae (e.g., Burckhalter 1992; Mabberley 1997). Recent molecular analyses (Xiang et al. 1993, 1998) suggest that *Cornus*, Nyssaceae, Hydrangaceae, and Loasaceae, as well as several other groups, form a "cornaceous clade"; however, they do not support the lumping of Nyssaceae into Cornaceae. (subclass Rosidae)

FAMILY RECOGNITION IN THE FIELD: shrubs or small trees with opposite simple leaves with *arcuate veins* (= pinnate veins arching toward leaf tip) and small 4-merous flowers in open terminal inflorescences or in some cases in heads subtended by showy bracts; inferior ovary developing into a drupe fruit. The "*Cornus* test" discussed below is also distinctive.
REFERENCES: Coulter & Evans 1890; Rickett 1945a; Ferguson 1966c; Ziang et al. 1998.

CORNUS DOGWOOD, CORNEL

Ours shrubs or small trees; leaves opposite, short-petioled, simple, entire, with prominent pinnate veins arching toward the leaf tip; flowers very small, perfect, in open inflorescences or conspicuously bracteate heads terminal on branches or branchlets; calyx lobes or sepals 4; petals 4, creamy white or yellow-green; stamens 4; filaments longer than the anthers; pistil 1; ovary inferior; fruit a small drupe.

A genus of ca. 65 species of trees, shrubs, and rhizomatous herbs native mainly to n temperate areas with a few taxa in South America and Africa. A number of *Cornus* species are widely cultivated for the showy, bract-surrounded inflorescences that resemble individual flowers. An interesting field character which can be used to identify *Cornus* species is the "*Cornus* test"—if a leaf is very *gently* torn into 2 pieces (apical part separated from basal), the primary veins remain connected by delicate whitish threads which represent the slinky-like, unraveled, spiral thickenings of the vascular tissue (Zomlefer 1994). (Latin: *cornu*, a horn, alluding to the hardness of the wood, the European *C. sanguinea* L. having long been used for skewers by butchers, whence SKEWERWOOD in English provinces and DAGWOOD from the old English: *dagge*, a dagger or sharp pointed object)

REFERENCES: Wilson 1964 [1965]; Ferguson 1966a, 1966b; Eyde 1987; Xiang et al. 1993, 1996 [1997].

1. Flowers in open inflorescences, with no large bracts; petals creamy white; fruits white ________ **C. drummondii**
1. Flowers in small compact inflorescences (heads) subtended by 4 large (to 5 cm long and 3 cm wide), white (rarely pink), petal-like bracts; petals yellow-green; fruits usually red ________ **C. florida**

Cornus drummondii C.A. Mey., (for its discoverer, Thomas Drummond, 1780–1835, Scottish botanist and collector in North America), ROUGH-LEAF DOGWOOD. Shrub or bushy small tree to ca. 5 m tall; twigs sometimes reddish; leaf blades pale and densely spreading pubescent or less often appressed pubescent beneath; calyx lobes shorter than the tube; petals 3.5–5.5 mm long. Stream banks, hillsides, woodlands, and thickets; e 1/2 of TX. May. [*C. priceae* Small] During the winter, the naked reddish twigs are often conspicuous.

Cornus florida L., (free flowering, producing abundant flowers), FLOWERING DOGWOOD, EASTERN DOGWOOD. Small tree with wide-spreading branches, to ca. 12 m tall; young branches usually greenish but sometimes reddish; leaf blades glabrous or silky-pubescent beneath, often scarlet in fall; inflorescence resembling a large 4-petaled flower. Wooded areas; Cooke, Grayson, Fannin, and Lamar cos. in Red River drainage, also Limestone Co. (J. Stanford, pers. comm.) and reported by Mrs. Stillwell from one Dallas locality (Mahler 1988); common in sandy woods in e TX to the e of nc TX; mainly e and c TX. This native understory species is widely cultivated as an ornamental. Late Mar–Apr. Dogwood anthracnose, a fungal disease caused by a member of the genus *Discula*, is currently damaging FLOWERING DOGWOODS in the e U.S. (Mielke & Daughtrey 1990). The wood is extremely shock resistant and was therefore widely used historically for applications including golf club heads, chisel handles, and the shuttles used in the textile industry (Peattie 1948). The fruits are bird-dispersed and have a high lipid content (up to 35% dry weight) which makes them particularly attractive to migratory birds which have high energy demands (Stiles 1984).

CRASSULACEAE STONECROP, ORPINE OR SUCCULENT FAMILY

Ours annual or perennial, often succulent herbs; leaves alternate, opposite, or whorled, simple, entire, slightly to very thick and fleshy; stipules absent; flowers axillary or usually terminal, solitary or in 1-sided racemes or in cymes; sepals 3–5(–7); petals 0–5; stamens 3–10; pistils 3–5(–7); ovaries superior; fruit a follicle or capsule.

A medium-large (1,100 species in 33 genera) nearly cosmopolitan, especially s African family of usually succulent shrubs, herbs, and treelets with crassulacean acid metabolism (CAM photosynthesis—allows night absorption and storage of CO_2 thereby reducing water loss through transpiration during the day). A number are cultivated as ornamentals including species of *Crassula* (e.g., JADE PLANT), *Echeveria*, *Kalanchoe* (MOTHER-OF-THOUSANDS), and *Sedum* (STONECROPS). Crassulaceae show similarities to Saxifragaceae and the two families are thought by some to be related (e.g., Heywood 1993). (subclass Rosidae)

FAMILY RECOGNITION IN THE FIELD: often *succulent* herbs (*Penthorum* not succulent) with radially symmetrical flowers with separate petals and usually *separate* (at least above) carpels of

the same number as the sepals. The somewhat similar Saxifragaceae are not succulent and typically have fused carpels fewer in number than sepals.
REFERENCES: Britton & Rose 1905; Spongberg 1978.

1. Leaves 5–15 cm long, lanceolate to elliptic-lanceolate, serrate, alternate, scattered along the stem; plants upright perennials to 80 cm tall; petals absent or inconspicuous ______ **Penthorum**
1. Leaves 2.5 cm or less long, linear to linear-oblong, entire, opposite OR if alternate, then crowded; plants small annuals to 30 cm tall (usually less); petals present, often conspicuous.
   2. Flowers minute, < 2 mm across, axillary, solitary, white to greenish white (rarely pinkish); leaves opposite; stamens as many as calyx segments ______ **Crassula**
   2. Flowers ca. 7–12 mm across, in terminal raceme-like cymes, yellow, white, pink, or lavender; leaves alternate; stamens twice as many as calyx segments ______ **Sedum**

CRASSULA

A nearly cosmopolitan, but especially tropical and s African genus of ca. 200 species of succulent herbs, shrubs, and treelets including many cultivated ornamentals; many have xeromorphic features. (Diminutive of Latin: *crassus*, thick, alluding to the succulent texture)
REFERENCES: Cody 1954; Tölkin 1977.

Crassula aquatica (L.) Schönl., (growing in or near water), WATER PIGMYWEED. Inconspicuous, glabrous, green or reddish, moisture-loving annual with decumbent to erect stems to 10 cm long; leaves opposite, linear to linear-oblong, to ca. 7 mm long, with united bases; flowers 3–4-merous; some pedicels exceeding the leaves; follicles ca. 1.5–2 mm long. Damp bare spots, pond margins, pathways, mudflats, sandy soils; collected by Reverchon in Dallas Co., also Navarro Co.; probably often overlooked; mainly se and e TX, also Edwards Plateau. Mar–Apr. Sometimes put in the genus *Tillaea* [as *T. aquatica* L.]. Relatively long-pedicelled plants, such as those in nc TX, are sometimes segregated as *C. drummondii* (Torr. & A. Gray) Fedde (e.g., Kartesz 1994), leaving those with sessile or short pedicellate flowers in *C. aquatica*; we are following Correll and Johnston (1970) and Hatch et al. (1990) in lumping [*C. drummondii*] into *Crassula aquatica*.

PENTHORUM DITCH-STONECROP

A genus of 1–3 species of e and se Asia and e North America; this disjunct distribution pattern is discussed under the genera *Campsis* (Bignoniaceae) and *Carya* (Juglandaceae). *Penthorum* is sometimes put into the Saxifragaceae; other authors have separated it into its own family the Penthoraceae. (Greek: *pente*, five, and *horos*, a mark, from the often symmetrically 5-parted flower)

Penthorum sedoides L., (resembling *Sedum*—stonecrop). Rhizomatous and stoloniferous, largely glabrous, succulent, perennial herb to 80 cm tall; leaves alternate, short-petioled or sessile; leaf blades lanceolate to elliptic-lanceolate, 50–100(–150) mm long, sharply toothed; flowers yellowish green, in curved, 1-sided, spike-like, panicled cymes; sepals usually 5(–7); petals usually absent or inconspicuous; stamens 10; pistils 5(–7), united below; fruit a 5-angled capsule. Stream bottom thickets or damp open ground; Dallas, Fannin, Grayson, Henderson, Hunt, Lamar, and Tarrant (Fort Worth Nature Center) cos.; se and e TX w to nc TX. Jul–Sep.

SEDUM STONECROP, ORPINE

Ours glabrous annuals; leaves alternate, crowded, terete or nearly so; sepals and petals 4–5; stamens 8–10; pistils 4–5; fruit a dry, many-seeded follicle with tapered beak. (Classical Latin name for several succulent plants, from *sedo*, to sit, alluding to the manner in which many species affix themselves to rocks or walls)

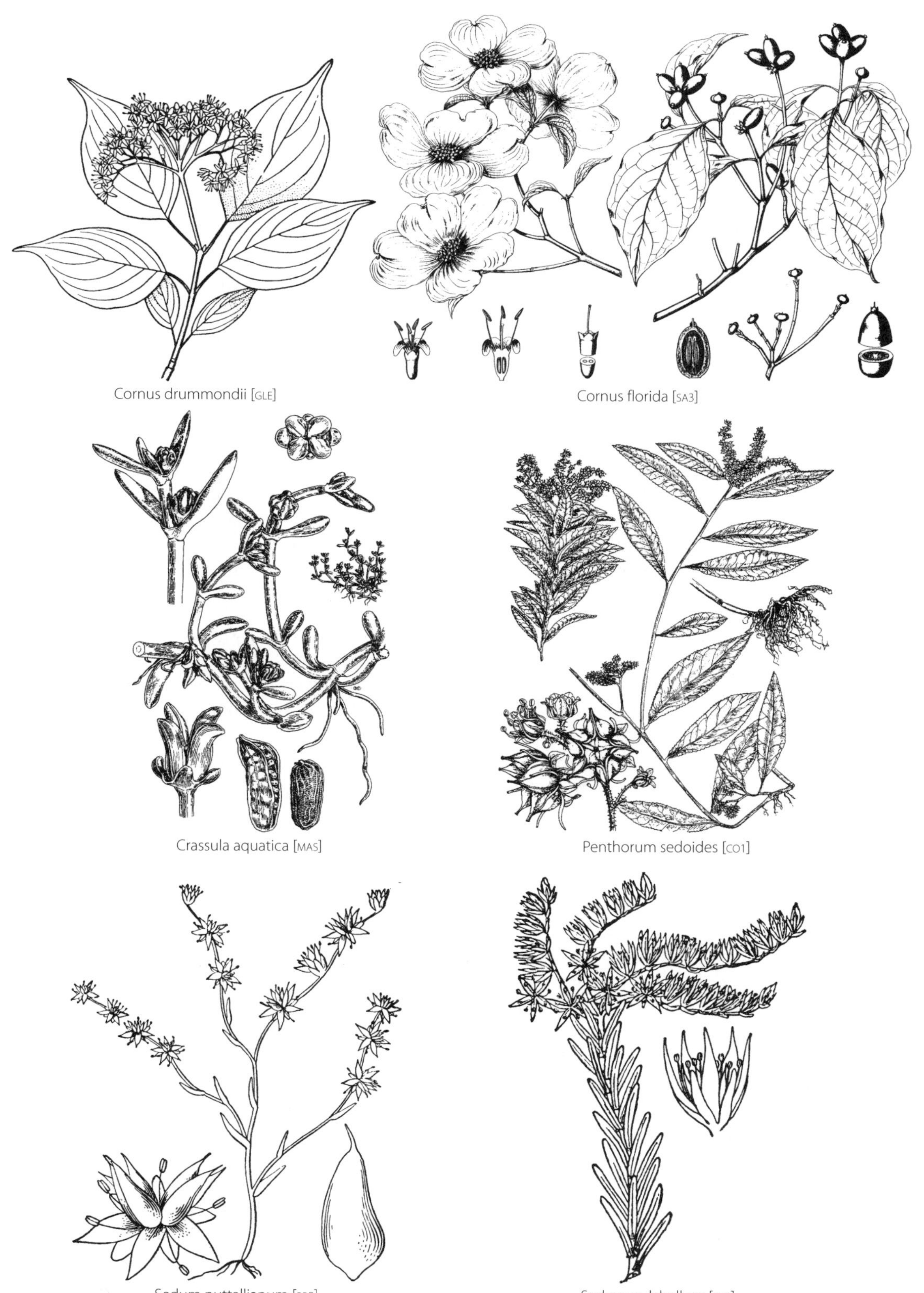

Cornus drummondii [GLE]

Cornus florida [SA3]

Crassula aquatica [MAS]

Penthorum sedoides [CO1]

Sedum nuttallianum [BB2]

Sedum pulchellum [BB2]

A genus of ca. 280 species of n temperate areas, tropical mountains, Madagascar, and Mexico; it incudes many cultivated, ornamental, succulent herbs and shrublets.
REFERENCES: Clausen 1975; Nesom & Turner 1995; White et al. 1998.

1. Petals yellow; flowers ca. 7 mm across; follicles widely divergent in fruit ____ **S. nuttallianum**
1. Petals white to pink or lavender; flowers 8–12 mm across; follicles ascending in fruit ____ **S. pulchellum**

Sedum nuttallianum Raf., (for Sir Thomas Nuttall, 1786–1859, English-American botanist), YELLOW STONECROP. Glabrous pale green annual to 13(–20) cm tall; leaves linear-oblong, nearly as thick as wide; inflorescences with 2–5 branches, the flowers remote. Loose sandy or eroding limestone slopes, shallow soils, or on sandstone outcrops; Post Oak Savannah w to West Cross Timbers and s to Edwards Plateau. May–Jun.

Sedum pulchellum Michx., (beautiful), TEXAS STONECROP, ROCK-MOSS, WIDOW'S-CROSS. Stems ascending, 10–30 cm tall; leaves linear, terete; inflorescences with 2–7 branches, the flowers closely spaced. Moist rocky areas; Bell Co. (Fort Hood—Sanchez 1997), also Coryell, Grayson, Lamar, and Limestone cos. (White et al. 1998); also far e TX and e Edwards Plateau. Mar–May.

CUCURBITACEAE

SQUASH, GOURD OR CUCUMBER FAMILY

Herbaceous annual or perennial vines; monoecious or dioecious; stems with tendrils, trailing or climbing; leaves usually palmately lobed or compound; flowers unisexual, radially symmetrical; corollas 5-lobed, yellow to greenish yellow, yellow-orange, or white; stamens 3 or 5, free or variously united; ovary inferior, usually 3-carpellate; fruit in ours a berry (sometimes with a firm rind and then called a pepo).

A medium-sized family (775 species in 119 genera) of mainly tropical and warm area (few temperate) vines containing numerous economically important food plants including SQUASHES, PUMPKINS, AND MELONS (see following treatments); the fruits and/or flowers of a number of species are edible; other species provide ornamental and useful gourds. (subclass Dilleniidae)
FAMILY RECOGNITION IN THE FIELD: herbaceous *vines* with *tendrils* and alternate, palmately-veined, usually lobed or compound, often conspicuous leaves; flowers usually yellow to white, unisexual; ovary inferior.
REFERENCES: Jeffrey 1975, 1990; Bates et al. 1990; Lira et al. 1997 [1998].

1. Corollas over 5 cm long or across; fruits > 5 cm long or broad, smooth.
  2. Corollas light yellow to yellow or yellow-orange; leaf without disk-shaped gland at base of leaf blade; fruits globose to obovoid or cylindric.
    3. Fruits globose to obovoid, to ca. 9 cm long, fleshy inside, with a hard rind; corollas yellow or yellow-orange; flowers all solitary; anthers united ____ **Cucurbita**
    3. Fruits cylindric, curved or straight, 30–60 cm long, fibrous inside, the rind becoming dry and papery; corollas light yellow to yellow; staminate flowers racemose; anthers free ____ **Luffa**
  2. Corollas white; leaf with disk-shaped gland on each side of the base of the leaf blade at junction with petiole; fruits variously shaped (bottle, dumbell, club, crook-necked) ____ **Lagenaria**
1. Corollas < 5 cm long or across (often much less); fruits 5 cm or less long or broad (except much larger in *Cucumus melo*, MUSKMELON, and *Citrullus lanatus*, WATERMELON), smooth or prickly.
  4. Stems and petioles glabrous.
    5. Flowers greenish yellow; fruits unarmed, orange to red when ripe; leaves nearly unlobed to deeply 3–5-lobed but not divided into separate leaflets ____ **Ibervillea**
    5. Flowers white; fruits conspicuously armed with long slender spines, green; leaves divided into 3–7 distinct leaflets ____ **Cyclanthera**

4. Stems and/or petioles pubescent (use lens).
   6. Fruits smooth, reddish, 12–14 mm long, with 3–6 seeds; corollas greenish white ____ **Cayaponia**
   6. Fruits prickly or smooth, not reddish, small to very large, with seeds 1–very numerous; corollas yellow to greenish yellow to white (if white then fruits with only 1 seed).
      7. Corollas about 4 cm across, yellow; tendrils 2–3-forked; fruits very large (to 60 cm or more in length), smooth; seeds obovoid, flat, often 10–15 mm long; leaves very deeply lobed, the large central lobe itself ± pinnately lobed ____ **Citrullus**
      7. Corollas 2.5 cm or less across, white to yellow or greenish yellow; tendrils usually unbranched (if branched then fruits prickly); fruits 5 cm or less long (except in *Cucumis melo*), smooth or prickly; seeds variously shaped, 10 mm or less long (to 12 mm in *Cucumis melo*); leaves usually angled or shallowly lobed, or if deeply lobed, then the central lobe itself not pinnately lobed.
         8. Tendrils branched; fruits ovoid, pointed, 1–1.5 cm long, with a single seed, usually with prickly bristles; corollas white to cream ____ **Sicyos**
         8. Tendrils unbranched; fruits either not pointed or > 3 cm long, many-seeded, smooth or prickly; corollas yellow or greenish yellow.
            9. Corollas < 8 mm across; leaves mostly 5-lobed (the basal 2 lobes sometimes small); fruits green often with lighter patterning (like a baby watermelon), turning purplish black when ripe, smooth, 1–2 cm long ____ **Melothria**
            9. Corollas > 10 mm across; leaves either unlobed (but angled) or 3-lobed; fruits never purplish blackish, either prickly or much > 2 cm long ____ **Cucumis**

CAYAPONIA

A mainly tropical and subtropical New World genus of 60 species (1 in tropical Africa and Madagascar) (Jeffrey 1990). (A Brazilian name)

Cayaponia quinqueloba (Raf.) Shinners, (five-lobed). Monoecious, rhizomatous, climbing perennial; leaves long-petiolate; leaf blades 3-angled or -lobed, 5–10 cm long; tendrils simple or branched; pistillate flowers and fruits on short stalks to 4 mm long or nearly sessile; corollas greenish white; stamens 3; ovary 3-carpellate with 1 or 2 ovules in each cell; fruits reddish, ovoid to ellipsoid, somewhat fleshy, 12–14 mm long; seeds 3–6. River bottoms and along streams; included based on citation for vegetational areas 4 and 5 (Fig. 2) by Hatch et al. (1990); Post Oak Savannah w to nc TX and s to Edwards Plateau. Jun–Aug.

CITRULLUS

A genus of 4 species native to tropical and s Africa and probably also Asia. (Diminutive of *Citrus*, from the supposed resemblance of the fruits to those of that genus of Rutaceae)
REFERENCES: Bailey 1930b; Hara 1969a.

Citrullus lanatus (Thunb.) Matsum. & Nakai var. **lanatus**, (woolly), WATERMELON, SANDÍA. Monoecious annual with branched tendrils; stems long-trailing, prostrate; leaf blades once or twice deeply lobed or dissected; ovary densely lanate; fruits green or mottled or striped, very large, the flesh sweet, succulent, edible, usually red but can be orange, yellow, or white; seeds numerous, obovoid, variously colored. Commonly cultivated and found as a waif around picnic areas, trash heaps, roadsides, and waste areas; Grayson, Lamar, Palo Pinto, and Somervell cos.; scattered mainly in e 1/2 of TX. May–Nov. Native of Africa. [*C. citrullus* (L.) H. Karst, *C. vulgaris* Schrad.] This species has long been in cultivation; it was known in Egypt in the Bronze Age and probably much earlier (Zohary 1982); seeds were found in the tomb of Tutankhamun (Hepper 1990).

Citrullus lanatus var. *citroides* (L.H. Bailey) Mansf., (resembling *Citrus*), (CITRON or PRESERVING MELON), with smaller fruits having hard white flesh that is not edible raw, is also cultivated for making preserves.

The name, *Citrullus colocynthis* (L.) Schrad., (from the classical name), was misapplied to the WATERMELON by Hatch et al. (1990) and apparently Jones et al. (1997). *Citrullus colocynthis*, an Old World species commonly known as the BITTER-APPLE or BITTER-CUCUMBER, is a perennial with small fruits (< 8 cm in diam.) and a sparsely hispid ovary; it is used as a purgative; we have no evidence that it is cultivated or escaped in nc TX. For discussion of nomenclature see Hara (1969a).

CUCUMIS MELON

Annual, trailing or climbing, pubescent vines; usually monoecious; tendrils unbranched; leaf blades entire or somewhat lobed; flowers usually short-stalked and hidden by foliage; corollas yellow, bell-shaped to rotate, deeply 5-parted; fruits fleshy, prickly or smooth; seeds numerous.

An Old World tropical genus of ca. 30 species, mainly of tropical and s Africa (Jeffrey 1990); cultivated since early times. It includes *C. sativus* L. (CUCUMBER). (Name from classical Latin word for cucumber)
REFERENCE: Kirkbride 1993.

1. Leaf blades with 3 prominent lobes; ovaries and fruits prickly; flowers ca. 13 mm across (staminate sometimes larger); seeds < 5 mm long ____ **C. anguria**
1. Leaf blades angled, but usually not distinctly lobed; ovaries and fruits not prickly; flowers ca. 25 mm across; seeds ca. 12 mm long ____ **C. melo**

Cucumis anguria L., (Greek name for cucumber), BUR GHERKIN, WEST INDIAN GHERKIN. Leaf blades to 9 cm long, deeply 3-lobed, with sinuses rounded and lobes obtuse; fruits oval or oblong, furrowed, prickly, ca. 5 cm long, on crooked peduncles. Occasionally cultivated, escapes [?]; included based on citation for vegetational areas 4 and 5 (Fig. 2) by Hatch et al. (1990); nc to s TX. Aug–Sep. Native of Africa. Kirkbride (1993) named the wild plants var. *longaculeatus* J.H. Kirkbr., (long sharp-pointed, in reference to the fruit prickles), with the cultivars treated as var. *anguria*. Since Hatch et al. (1990) did not distinguish varieties and we have seen no nc TX material, we are uncertain which variety is present in nc TX. The two can be distinguished as follows (modified from Kirkbride 1993):

1. Prickles on fruits 0.8–1.9 mm long, appearing as small bumps or warts; apical hyaline bristle of prickles on hypanthium of female flowers 2–3 times longer than the opaque base ____ var. *anguria*
1. Prickles on fruits 4–10(–15) mm long, appearing as prickles; apical hyaline bristle of prickles on hypanthium of female flowers shorter than the opaque base ____ var. *longaculeatus*

Cucumis melo L., (melon), MUSKMELON, MELON, CANTELOPE. Leaf blades to 13 cm wide, rounded at apex; fruits mostly globular or ellipsoid with musky fragrance, the flesh usually yellowish or orangish to green; corolla tube of male flowers 0.8–2 mm long; corolla tube of female flowers 0.8–2.8 mm long; seeds slender, white. Commonly cultivated and found as a waif around picnic areas, trash heaps, roadsides, and waste areas; Dallas and Lamar cos.; mainly e 1/2 of TX. May–Oct. Native of Asia.

Cucumis sativus L., (cultivated), is the cultivated CUCUMBER; it has ±prickly fruit, corolla tube of male flowers 3.4–4.9 mm long, corolla tube of female flowers 3.5–6.5 mm long, and leaves angled or shallowly 3-lobed, with a pointed middle lobe and with ±acute sinuses. It is native to s Asia.

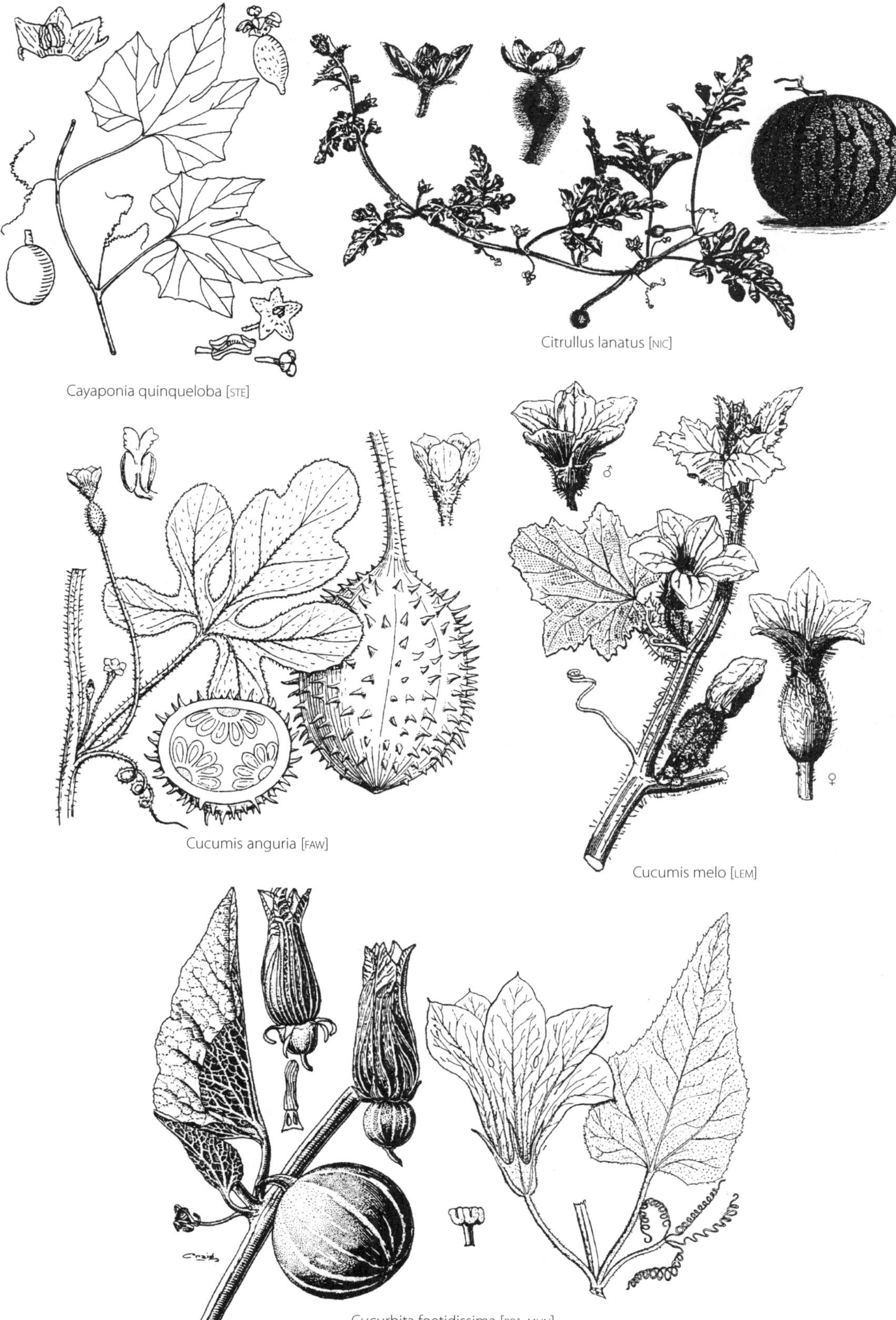

Cayaponia quinqueloba [STE]

Citrullus lanatus [NIC]

Cucumis anguria [FAW]

Cucumis melo [LEM]

Cucurbita foetidissima [BB1, MUN]

CUCURBITA GOURD, SQUASH

Annuals or perennials with long running stems; leaf blades large, scabrous, 15–30 cm long; flowers yellow, solitary in axils; staminate flowers with long pedicels; fruits smooth, fleshy, indehiscent, with a hard rind; seeds usually numerous.

A genus of ca. 20 species (Jeffrey 1990) of tropical and warm parts of the Americas including a number of widely cultivated food plants. The genus was important in pre-Colombian Mesoamerica as part of the maize/beans/squash agricultural system. (Classical Latin name for a type of gourd)

REFERENCES: Bailey 1929, 1930b; Whitaker & Bemis 1964, 1975; Rhodes et al. 1968; Bemis et al. 1970; Heiser 1979; Decker & Wilson 1986, 1987; Andres 1987; Decker 1988; Decker-Walters 1990; Nee 1990.

1. Leaf blades usually longer than wide, triangular-ovate, to 30 cm long, angled to weakly lobed, scabrous, extremely ill-smelling especially when bruised ____ **C. foetidissima**
1. Leaf blades usually nearly as wide as long, broadly ovate to reniform, angled to deeply lobed, to 15 cm long, pubescent, not noticeably ill-smelling ____ **C. texana**

Cucurbita foetidissima Kunth, (very fetid, very bad-smelling), BUFFALO GOURD, STINKING GOURD, FOETID GOURD, CALABAZILLA, CHILICOTE, CALABACILLA LOCA, CHILICOYOTE. Rank-growing, ill-smelling perennial; stems widely running, to 6 m or more long; leaf blades irregularly and finely toothed, thickish, often gray-green, rough to the touch; corollas to 10 cm long; fruits globose, green with lighter stripes, lemon-yellow at maturity, to 7.5 cm broad. Disturbed areas; nearly throughout TX, more common in w 1/2. Apr–Jul. The seeds were eaten by Native Americans; the plant was also used as soap (contains a saponin) and medicinally (Heiser 1979). During frontier days, the leaves and roots were used as a purgative (Crosswhite 1980). This species is being investigated as a source of starch and oil (Heiser 1993). ▣/86

Cucurbita texana A. Gray, (of Texas), TEXAS GOURD. Annual; leaves, at least some, distinctly lobed, relatively thin, green, the margins sharply serrate; corollas to 7 cm long; fruits obovoid to subglobose, green with lighter stripes or nearly white, to 9 cm long and 6 cm broad, bitter. Along streams and tributaries, weedy areas; Bell and Lamar (in a seasonally wet ox-bow off the Red River and in weedy pastures) cos., also Denton, Hamilton, Milam, Navarro (TOES 1993), and Hamilton (HPC) cos.; scattered in e 1/2 of TX. Jul–Oct. This species was thought to be endemic to TX (Correll & Johnston 1970), but naturally occurring populations have apparently rarely been found in several other states (Decker-Walters 1990; Nee 1990). According to M. Nee (pers. comm.) and Decker-Walters (1990), *C. texana* is involved in the origin of *C. pepo*. Because of the ability of *C. texana* and the various *C. pepo* cultivars to hybridize (Whitaker & Bemis 1964), nomenclature of the group is unsettled. Names such as [*C. pepo* subsp. *ovifera* (L.) D.S. Decker var. *texana* (Scheele) D.S. Decker] have been proposed; since there is no current concensus, we are following tradition and recognizing *C. texana* at the specific level. (TOES 1993: V) ⚠ ▣/86

Cucurbita pepo L., (one-celled, many-seeded, pulpy fruit), PUMPKINS and SQUASH (including SUMMER CROOK-NECK, SUMMER STRAIGHT-NECK, ZUCCHINI, SCALLOP, PATTYPAN, and ACORN), was brought into cultivation by Native Americans. This species, with leaves mostly strongly lobed, can be distinguished from *C. texana* by its very variable (in the different cultivated forms), but usually larger and conspicuously differently shaped and colored fruits. While there has been controversy over whether *C. texana* populations are naturally occurring or represent escapes from cultivation of *C. pepo* var. *ovifera* (L.) Alef., (egg-bearing), (ORNAMENTAL GOURDS—with very variable fruits), electrophoretic evidence indicates *C. texana* populations are distinct from *C. pepo* var. *ovifera*; *C. texana* is fairly homogeneous genetically with the low levels of genetic diversity probably related to endemism and small population size (Decker & Wilson 1987); hybridization between the two can occur (Heiser 1979). It now seems likely that *C. texana* is a natural element of the flora (Heiser 1993). Two other *Cucurbita* species are also cultivated. Both have foliage soft, or at least not harsh to the touch, in contrast to the rough or harsh foliage of *C. pepo*; they also both have the leaves mostly not lobed:

Cucurbita moschata (Duchesne ex Lam.) Duchesne ex Poir., (musky), WINTER CROOK-NECK SQUASH, BUTTERNUT SQUASH (native of tropical America). *Cucurbita moschata* has the leaf blades broadly ovate to ±triangular in outline and peduncles long, furrowed, and flared at junction with the fruit.

Cucurbita maxima Duchesne, (largest), AUTUMN SQUASH, WINTER SQUASH, HUBBARD SQUASH, TURBAN SQUASH (probably native of South America). *Cucurbita maxima* has the leaf blades circular to reniform in outline and peduncles short and spongy, nearly cylindric, and not flared at attachment to the fruit.

CYCLANTHERA

A mainly Neotropical genus of ca. 30 species (Jeffrey 1990). (Greek: *cyclos*, circle, and *anthera*, anther, possibly in reference to the united stamens)

Cyclanthera dissecta (Torr. & A. Gray) Arn., (dissected, deeply cut), CUT-LEAF CYCLANTHERA, BUR-CUCUMBER. Annual monoecious climber with slender stems to 3 m or more long; tendrils simple to trifid; leaves 3- to 7-foliolate, to 6 cm long; staminate and pistilate flowers both from the same axils; staminate flowers in racemes or panicles; stamens united into a central column; pistillate flowers solitary; corollas rotate, 5-parted, white, to 7 mm across; fruits peduncled, 2–3 cm long, ovoid, armed with slender smooth spines, rupturing irregularly, with several seeds. Rocky soils; Bell and Palo Pinto cos.; widespread in TX. May–Oct.

IBERVILLEA GLOBEBERRY

A genus of 5 species native to the s U.S. and Mexico (Jeffrey 1990). (Derivation of generic name not explained by original author, however, possibly named for Iberville, a parish in LA)
REFERENCE: Kearns 1994.

Ibervillea lindheimeri (A. Gray) Greene, (for Ferdinand Jacob Lindheimer, 1801–1879, German-born TX collector), BALSAM GOURD, LINDHEIMER'S GLOBEBERRY, BALSAM-APPLE. Perennial dioecious climber with turnip-like taproot; leaves petioled; leaf blades mostly deeply to shallowly 3-lobed (rarely unlobed or 5-lobed), to 12 cm wide, rather fleshy, with swollen-based hairs; corollas greenish yellow; staminate flowers 5–8 per raceme, tubular, glandular-puberulent, 6–8 mm long; stamens 3; pistillate flowers solitary, funnelform, ca. 10 mm long; ovary 3-carpellate; berry globose, ca. 3 cm in diam., reddish orange when ripe, when immature green with lighter patterning (resembling a small watermelon); seeds numerous. Prairies, woodlands, thickets; Bell, Callahan, Cooke, Grayson, McLennan, and Palo Pinto cos., also Brown, Kaufman (HPC), Parker, Somervell, and Tarrant (R. O'Kennon, pers. obs.) cos. ; mainly nc to c TX, also w to Plains Country; endemic to TX and s OK. Apr–Sep. /93

LAGENARIA BOTTLE GOURD, GOURD

A genus of 6 species of tropical Africa and Madagascar with 1 species widespread in rest of tropics. (Greek: *lagenos*, flask, referring to the shape and use of the fruits)
REFERENCE: Heiser 1979.

Lagenaria siceraria (Molina) Standl., (intoxicating, perhaps derived from use of a *Lagenaria* species in making an intoxicating drink), BOTTLE GOURD, WHITE-FLOWER BOTTLE GOURD, CALABASH

GOURD. Viscid-pubescent, monoecious or rarely dioecious, climbing or trailing annual; tendrils branched; leaf blades trianglular-ovate, shallowly palmately 3–5 lobed or unlobed, 15–30 cm wide, cordate basally; flowers 1–2 from leaf axils; corollas white, 5–10 cm across; fruits to 30 cm long or more, variously shaped, without spines or prickles but sometimes with knobs or ridges; seeds numerous. Cultivated for the fruit; the tough pericarp is used in making containers (cups, dippers, bowls, etc.) and decorations; escapes to weedy areas; Dallas Co.; also scattered in se and e TX. ?–Oct. Native of the Old World tropics. [*L. vulgaris* Ser.] This species has long been in cultivation; BOTTLEGOURDS have been found in Egyptian tombs dating to ca. 3500–3000 BC and in New World sites dating to ca. 7000 BC; apparently they dispersed by floating across the Atlantic from Africa (Zohary 1982).

LUFFA VEGETABLE-SPONGE, LOOFAH

A genus of 7 species, 4 of the Old World tropics and 3 of the Neotropics; several are variously cultivated for the fibrous interior of the fruit, for food, or medicinally (Heiser & Schilling 1990). (From the Arabic name)

REFERENCES: Heiser 1979; Heiser & Schilling 1990.

Luffa aegyptiaca Mill., (of Egypt), VEGETABLE-SPONGE, ESTROPAJO. Monoecious strong vine with branched tendrils; leaf blades 3–7-lobed, deltoid to nearly orbicular, 12–30 cm long; corollas deeply 5-lobed, 5–10 cm across, light yellow to yellow; anthers free; fruits large and conspicuous, ± cylindrical, with light furrows and stripes but without ribs or angles, the interior conspicuously fibrous; seeds numerous, narrowly winged. Cultivated for use as a sponge (dried vascular system of fruit); escapes to weedy areas; Montague Co.; also s TX. ?–Oct. Native of the Old World. [*L. cylindrica* (L.) M. Roem.]

MELOTHRIA MELONETTE

A genus of 12 species native to the New World (Jeffrey 1990). (Altered from *melothron*, an ancient Greek name for some fruiting vine)

Melothria pendula L., (pendulous, hanging), DROOPING MELONETTE, CREEPING-CUCUMBER, MELONCITO. Slender, usually monoecious, climbing perennial; stems glabrous or nearly so; tendrils simple; leaf blades orbicular, 5-angled or -lobed, 3–7 cm long, with pubescence at least on the veins beneath; cordate at base; petioles pubescent, the hairs sometimes stiff; flowers axillary, small; corollas yellowish green, 5-angled or -lobed; staminate flowers in racemose or corymbose inflorescences, the corollas campanulate; stamens 3; anthers free to barely united; pistillate flowers usually solitary or few in a cluster, peduncled, the corollas rotate; ovary 3-carpellate; fruits smooth, green often with lighter patterning (like a baby watermelon), turning purplish black, ovoid, 1–2 cm long; seeds numerous, white. Wooded areas, exposed sites; se and e TX w to Rolling Plains and Edwards Plateau. May–Nov. The fruits are reported to act as a drastic purgative (Burlage 1968).

SICYOS BUR-CUCUMBER

A genus of ca. 40 species of the New World, Hawaiian Islands, and Australasia (Jeffrey 1990). (Greek: *sicyus*, cucumber or gourd)

Sicyos angulatus L., (angled), ONE-SEED BUR-CUCUMBER, WALL BUR-CUCUMBER. Glandular-pubescent, monoecious, climbing annual; tendrils divided into 4 segments; leaves petioled, the petioles to 8 cm long; leaf blades to 20 cm long, suborbicular in outline, 5-angled or -lobed; staminate flowers in axillary racemes or corymbs; pistillate flowers usually from the same axils in long-peduncled capitate clusters; corollas white to cream; fruits yellowish, 1–1.5 cm long, ovoid, villous with prickly bristles, with a single seed. Wooded areas; Bell, Cooke, Denton, Tarrant, and

Cucurbita texana [HEA]

Cyclanthera dissecta [BB2]

Ibervillea lindheimeri [EN1]

Lagenaria siceraria [KER, LAM]

Luffa aegyptiaca [EN1, EN2]

Melothria pendula [GLE]

Sicyos angulatus [DIL]

Williamson cos.; e and nc TX, also Edwards Plateau. May–Sep. Sometimes cultivated as a climbing screen in landscapes (Mabberley 1997).

CUSCUTACEAE DODDER FAMILY

A small (ca. 145 species), cosmopolitan family of often brightly colored, ± chlorophyll-less, twining parasites. *Cuscuta* is the only genus. Related to the Convolvulaceae and sometimes treated as a monogeneric tribe in that family; however, Cuscutaceae can be readily distinguished by their parasitic habit, absence of chlorophyll, and lack of contact with the soil after parasitizing a host (Tyrl et al. 1994). (subclass Asteridae)
FAMILY RECOGNITION IN THE FIELD: orange to yellow or whitish, essentially leafless, parasitic herbs with *thread-like* stems *twining* on other plants; flowers small, white to yellowish.
REFERENCES: Austin 1986b; Gandhi et al. 1987.

CUSCUTA DODDER, LOVEVINE

Ours glabrous annual herbs, ± without chlorophyll, with thread-like, orange, yellow, to whitish, irregularly twining stems; growing as parasites on other plants and attached by haustoria invading the host tissue; losing their chlorophyll upon contact with host plant as a seedling; essentially leafless, the leaves reduced to functionless scales; flowers small, usually numerous, in loose or dense cymose clusters, pedicelled or sessile; calyces of 3–5 lobes or sometimes of separate or nearly separate sepals, the lobes or sepals often overlapping laterally; corollas globose, white to yellowish, 3- to 5-lobed, with a ring of small, flat, fringed or fimbriated scales inside at base opposite the stamens; stamens 3–5, attached above the scales; ovaries superior, 2-celled; stigmas linear to capitate; fruit a capsule.

This genus of ca. 145 species includes a number of problematic parasitic weeds which can seriously reduce the yield of some crops. Other common names are ANGEL'S-HAIR, TANGLEGUT, WITCHES'-SHOELACES, DEVIL'S-GUT, STRANGLEVINE, and SCALD. In the Sinhalese language of Sri Lanka, the name of a *Cuscuta* species means "plant without beginning or end" (Austin 1979a). According to Austin (1979a), the name DODDER, "... is an ancient one, having come from its Frissian (Middle English) [Frisian (Low German tongue closely related to Anglo-Saxon)] origin without change. The word dodder is said to signify a 'tangle of threads' in reference to the intertwined stems of the plant." The species are often difficult to identify, especially in spring when just beginning to flower. (Arabic: *kushkut* or *kusat*, a tangled wisp of hair—Austin 1979a)
REFERENCES: Yuncker 1932, 1961; Austin 1979a, 1979b, 1980.

1. Styles partly or completely united; capsules very large, 5–7(–10) mm long; typically parasitizing woody plants; stems coarse ____ **C. exaltata**
1. Styles completely separate; capsules usually smaller, usually much < 5 mm long; typically parasitizing herbaceous or woody plants; stems fine to coarse.
   2. Flowers closely subtended by enlarged bracts which resemble the sepals (usually at least 1 bract per flower); calyces of separate or nearly separate sepals; flowers ca. 4–5 mm long.
      3. Flowers on pedicels (bracts along the pedicels), in loose paniculate clusters ____ **C. cuspidata**
      3. Flowers sessile, in dense clusters.
         4. Bracts with acute, recurved tips; stems disappearing early in the season (other than those portions bearing flowers); flowers in very thick, rope-like clusters ____ **C. glomerata**
         4. Bracts with obtuse, erect tips; stems persisting; flowers clustered, sometimes densely so, but the clusters not rope-like ____ **C. compacta**
   2. Flowers with only minute scale-like bracts not resembling the sepals (an enlarged bract can sometimes subtend a group of flowers); calyces of at least partly fused sepals (at base); flowers 1.5–4 mm long.

5. Calyces and corollas of most flowers 3–4-lobed, rarely 5-lobed; rare in nc TX, reported from only 2 counties.
   6. Sepals reaching the sinuses of corolla; corolla lobes acute with inflexed tips ____ **C. coryli**
   6. Sepals shorter than the corolla tube, not reaching the sinuses; corolla lobes obtuse, not inflexed ____ **C. cephalanthi**
5. Calyces and corollas of most flowers 5-lobed, rarely 4-lobed; including species widespread in nc TX.
   7. Calyx lobes subacute to acute or acuminate, triangular-ovate to triangular-lanceolate in shape ____ **C. indecora**
   7. Calyx lobes obtuse, ovate in shape.
      8. Corolla lobes acute to acuminate, with the tips inflexed ____ **C. pentagona**
      8. Corolla lobes obtuse to acute, the tips not inflexed.
         9. Flowers 2–4 mm long, without granular or glandular coating; calyx lobes overlapping at base; styles 0.8–1.8 mm long when fully extended, slender; capsules often slightly longer than wide; widespread in e half of nc TX ____ **C. gronovii**
         9. Flowers ca. 1.5–2 mm long, with granular or glandular coating; calyx lobes not or scarcely overlapping at base; styles 0.3–0.7 mm long, rather thick; capsules not longer than wide; rare in nc TX, reported only from Dallas Co. ____ **C. obtusiflora**

Cuscuta cephalanthi Engelm., (of *Cephalanthus*—buttonbush), BUTTONBUSH DODDER. Flowers ca. 2 mm long; calyx lobes obtuse. Found at Dallas by Reverchon (Yuncker 1961); not collected recently in nc TX. Jul–Oct.

Cuscuta compacta Juss. ex Choisy, (compact, dense), COMPACT DODDER. Flowers 4–5 mm long; calyces subtended by 3–5 bracts; corolla lobes obtuse. Stream bottoms; Dallas Co. (Reverchon); also e TX. Jul–Oct.

Cuscuta coryli Engelm., (of *Corylus*—hazel or filbert; often found on members of that genus), HAZEL DODDER. Flowers ca. 2 mm long; calyx lobes acute; scales inside corolla (below stamens) rudimentary, unfringed, bifid (with a wing on either side of filament). Related to and similar to *C. indecora*; differs in usually 4(–5)-parted flowers and the rudimentary bifid scale. Williamson Co. (Yuncker 1961); also Robertson Co. in Post Oak Savannah (Yuncker 1961); Hatch et al. (1990) reported only vegetational areas 3 and 4 (Fig. 2) for TX. Aug–Sep.

Cuscuta cuspidata Engelm., (cuspidate, with a sharp stiff point), CUSP DODDER, CUSPIDATE DODDER. Pedicels distinctive, with 1 or more bracts just below the calyces; flowers ca. 4 mm long; calyx lobes obtuse or acute; corolla lobes obtuse to acute. Parasitic on a number of herbaceous species, but according to Yuncker (1961), it seems to prefer members of the Asteraceae including *Ambrosia*, *Baccharis*, *Helianthus*, *Iva* and *Liatris*. Chiefly low open ground; Dallas, Denton, Grayson, and Limestone cos; widespread in TX. Aug–Oct.

Cuscuta exaltata Engelm., (exalted, very tall), TREE DODDER. Flowers 4–5 mm long, sessile or subsessile; calyx and corolla lobes obtuse; capsules circumscissile. Limestone areas; typically parasitizing woody plants, often oaks; Dallas and Johnson (Cleburne State Park) cos., also Bell (Yuncker 1961) and Coryell (Fort Hood—Sanchez 1997) cos.; also se TX and Edwards Plateau. Sep–Oct. This species is the only member of its subgenus in North America (Yuncker 1932).

Cuscuta glomerata Choisy, (with glomerules, clustered into ± rounded heads), CLUSTER DODDER, GLOMERATE DODDER. Stems disappearing leaving thick, yellow, rope-like masses of flowers around the host stems and thus easily recognized by growth form; flowers 4–5 mm long; calyx and corolla lobes acute or obtuse. Stream bottoms; Hunt and Lamar cos., also Dallas Co. (Yuncker 1961); also Edwards Plateau and Panhandle. Jul–Sep.

Cuscuta gronovii Willd. ex Schult., (for Jan Frederick Gronovius, 1690–1762, teacher of Linnaeus and author of "Flora Virginica"), GRONOVIUS' DODDER, COMMON DODDER. Flowers ca. 2–4 mm long; calyx lobes overlapping; corolla lobes obtuse. Stream bottoms, low ground; Hunt, Lamar, and Rockwall cos., also Dallas and Milam cos. (Yuncker 1961); se and e TX w to nc TX. Aug–Nov. [*C. gronovii* var. *latiflora* Engelm.]

Cuscuta indecora Choisy, (not ornamental). Flowers 2–2.5 mm long; calyx lobes shorter or longer than corolla tube; corolla lobes acute with inflexed tips; scales inside corollas (below stamens) fringed, not bifid. Prairies, roadsides, and stream bottoms. Jun–Oct.

1. Calyx lobes triangular-ovate, subacute, shorter than the corolla tube ____ var. **indecora**
1. Calyx lobes narrowly triangular-lanceolate, acute or acuminate, equaling or exceeding the corolla tube ____ var. **longisepala**

var. **indecora**, SHOWY DODDER, PRETTY DODDER, LARGE-SEED DODDER. Widespread in TX.

var. **longisepala** Yunck., (long-sepaled), LONG-SEPAL DODDER. Dallas, Ellis, Grayson, and Hill cos.; scattered in e 1/2 of TX.

Cuscuta obtusiflora Kunth var. **glandulosa** Engelm., (sp.: obtuse- or blunt-flowered; var.: glandular), RED DODDER. Flowers 1.5–2 mm long, often with granular or glandular coating; calyx lobes obtuse, not overlapping; corolla lobes acute to obtusish; capsules depressed globose, not longer than wide. Sandy or rocky ground, often growing on *Polygonum* species (Yuncker 1961); Dallas Co. (Yuncker 1961); Hatch et al. (1990) cited regions 1–10 (throughout TX), but this is possibly an error. May–Sep. [*C. glandulosa* (Engelm.) Small]

Cuscuta pentagona Engelm., (five-angled). Flowers 1.5–3 mm long. May–Oct. According to Austin (1986b), parasitizing a variety of hosts including cultivated Fabaceae (e.g., CLOVER and ALFALFA). Daniel Austin (pers. comm.) questions the recognition of varieties indicating that even specific determination is often difficult.

1. Calyx lobes slightly auricled and overlapping at base, sometimes forming angles giving the calyces an angulate appearance; withered corollas often only at base of capsules ____ var. **pentagona**
1. Calyx lobes not auricled or overlapping at base, the calyces not angulate; withered corollas enveloping the capsules.
   2. Flower parts and pedicels usually not papillate or only slightly so ____ var. **glabrior**
   2. All flower parts (including ovaries and capsules) and pedicels ± densely papillate ____ var. **pubescens**

var. **glabrior** (Engelm.) Gandhi, R.D. Thomas & S.L. Hatch, (smooth, without hairs). Widespread in TX. [*C. glabrior* (Engelm.) Yunck.]

var. **pentagona**, FIELD DODDER, FIVE-ANGLED DODDER. Bell, Coryell, Dallas, Denton, Hunt, and Limestone cos.; se and e TX w to Lampasas Cut Plain and Edwards Plateau. [*C. campestris* Yunck.]

var. **pubescens** (Engelm.) Yunck., (downy, pubescent). Bell, Burnet, and Dallas cos. (Yuncker 1961); also Edwards Plateau and Trans-Pecos. [*C. glabrior* (Engelm.) Yunck. var. *pubescens* (Engelm.) Yunck.]

DIPSACACEAE TEASEL FAMILY

A small (290 species in 11 genera), Old World (Eurasia and Africa, especially Mediterranean) family of herbs or subshrubs including a number of ornamentals. Teasels, formerly used in raising nap on cloth, are obtained from *Dipsacus*. The Dipsacaceae are closely related to the Caprifoliaceae and appear to represent an herbaceous clade within that mainly woody family. From a cladistic standpoint they should be lumped to form a more inclusive monophyletic

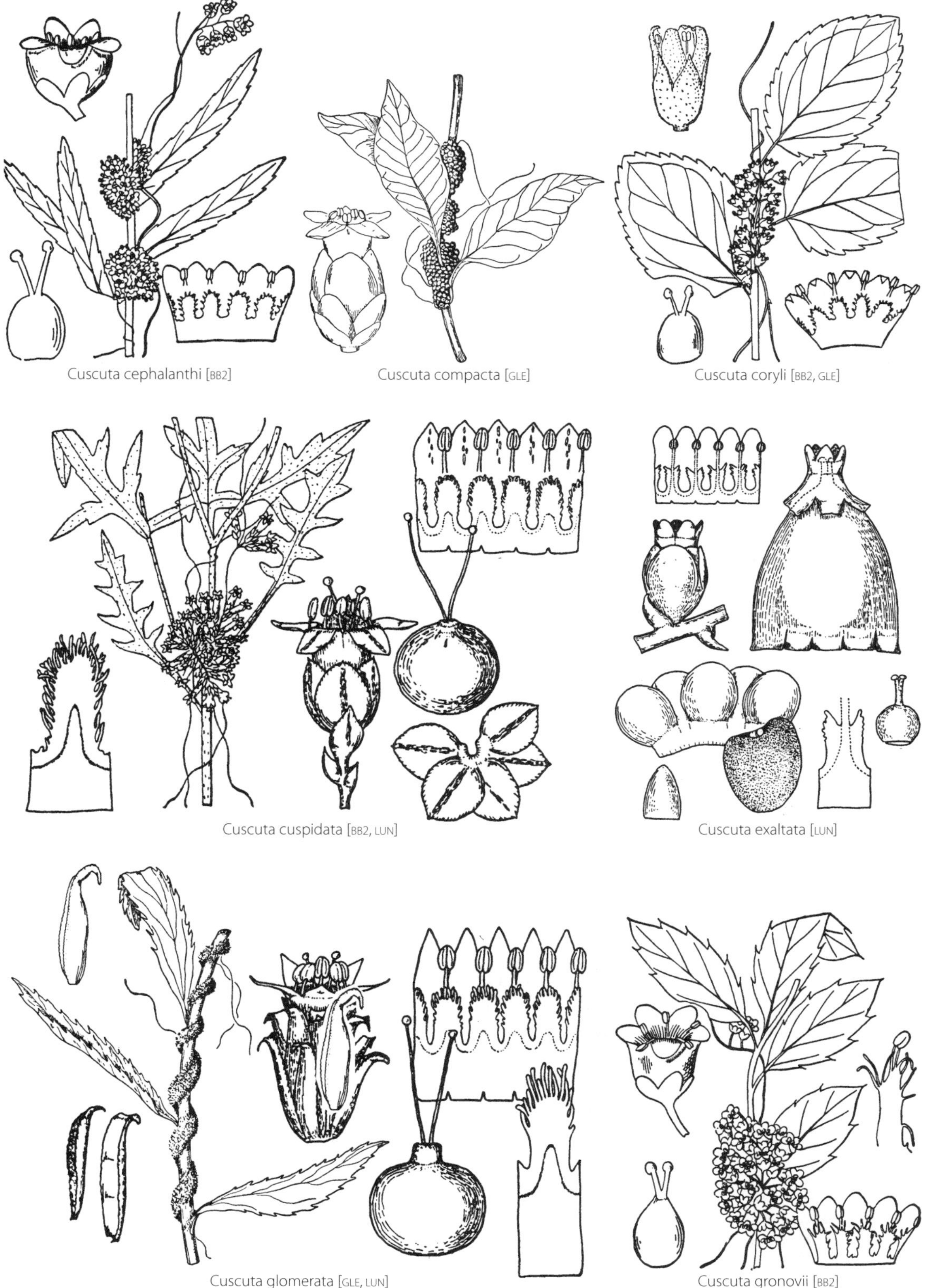

Cuscuta cephalanthi [BB2]

Cuscuta compacta [GLE]

Cuscuta coryli [BB2, GLE]

Cuscuta cuspidata [BB2, LUN]

Cuscuta exaltata [LUN]

Cuscuta glomerata [GLE, LUN]

Cuscuta gronovii [BB2]

Caprifoliaceae (Judd et al. 1994). Family name from *Dipsacus*, TEASAL, a genus of 15 species of Eurasia, the Mediterranean, and tropical African mountains; the rigid-bracted heads of *D. sativus* (L.) Honck. were long used to raise the nap on cloth—hence the name TEASEL. (Greek: *dipsakos*, the classical name for teasel) (subclass Asteridae)

FAMILY RECOGNITION IN THE FIELD: the only nc TX species is an introduced annual herb with opposite leaves and small flowers in *involucrate heads* (resembling Asteraceae); corollas sympetalous, bilaterally symmetrical, with 4 separate epipetalous stamens; ovary inferior. Superficially similar to Asteraceae; however, Asteraceae can be distinguished by their flowers with 5 stamens fused by their anthers.

REFERENCES: Ferguson 1965; Donoghue et al. 1992; Judd et al. 1994.

SCABIOSA

A genus of 80 species native to temperate Eurasia, the Mediterranean region, mountains of e Africa, and s Africa; herbs usually with an umbrella-like epicalyx functioning in wind dispersal of the fruit. (Latin: *scabies*, itch; the rough leaves were used medicinally to treat skin complaints)

Scabiosa atropurpurea L., (dark purple), PINCUSHIONS, SWEET SCABIOUS. Annual herb 20–60 cm tall; leaves opposite; basal leaves simple, dentate; upper leaves pinnately parted, the lobes entire to dentate; stipules absent; flowers in peduncled involucrate heads resembling some in the Asteraceae; involucral bracts narrowly lanceolate, herbaceous, distinct; involucel subtending each flower; corollas white to rose, lilac, or dark purple, bilaterally symmetrical; calyces pappus-like, of 5 bristles; corollas 5-lobed, the lobes unequal; stamens 4, distinct, epipetalous; ovary inferior; achenes with 5 persistent conspicuous calyx bristles. Garden escape currently rapidly spreading and becoming a problematic invasive weed in fields, roadsides, and prairies; Collin, Dallas, Fannin, Grayson, Hopkins, Hunt, and Lamar cos.; apparently at present in TX only in nc part of the state. Jun–Sep. Native of Europe.

DROSERACEAE SUNDEW FAMILY

A small (ca. 113 species in 4 genera) cosmopolitan family of insectivorous herbs including *Dionaea muscipula* J. Ellis, (fly-catching), (VENUS'-FLYTRAP), from the se U.S. *Aldrovanda*, *Dionaea*, and *Drosophyllum* are all monotypic with *Drosera* having most of the species in the family. In *Drosera* the hairs are motile, and entrap and digest insects using secreted proteolytic enzmes and ribonucleases; in *Dionaea* the 2 halves of the leaves swing together trapping the victim; *Aldrovanda* is a rootless aquatic with insect-trapping leaves like *Dionaea*. Most species are found in nitrogen poor habitats such as bogs or wet sandy areas. As with most carnivorous plants, nutrients (especially nitrogen), rather than calories, are gained through carnivory. A molecular study using 18S ribosomal DNA sequences suggested that Droseraceae and Nepenthaceae (an Old World carnivorous family) are the sister group to the Caryophyllidae (Soltis et al. 1997). (subclass Rosidae)

FAMILY RECOGNITION IN THE FIELD: the only nc TX species is an herb with small *rosettes* of leaves; leaves conspicuously broader towards tip, with *tentacle-like hairs* secreting glistening droplets of sticky fluid which trap insects; flowers on a scape.

REFERENCE: Wood 1960.

DROSERA SUNDEW

A cosmopolitan genus (especially s hemisphere) of ca. 110 species of insectivorous herbs typically found in wet areas. (Greek: *droseros*, dewy; the glands of the leaves exude drops of a clear glutinous fluid, glittering like dew drops and giving rise to both the generic and common names)

REFERENCES: Wynne 1944; Shinners 1962d; Wood 1966; Pietropaolo & Pietropaolo 1986.

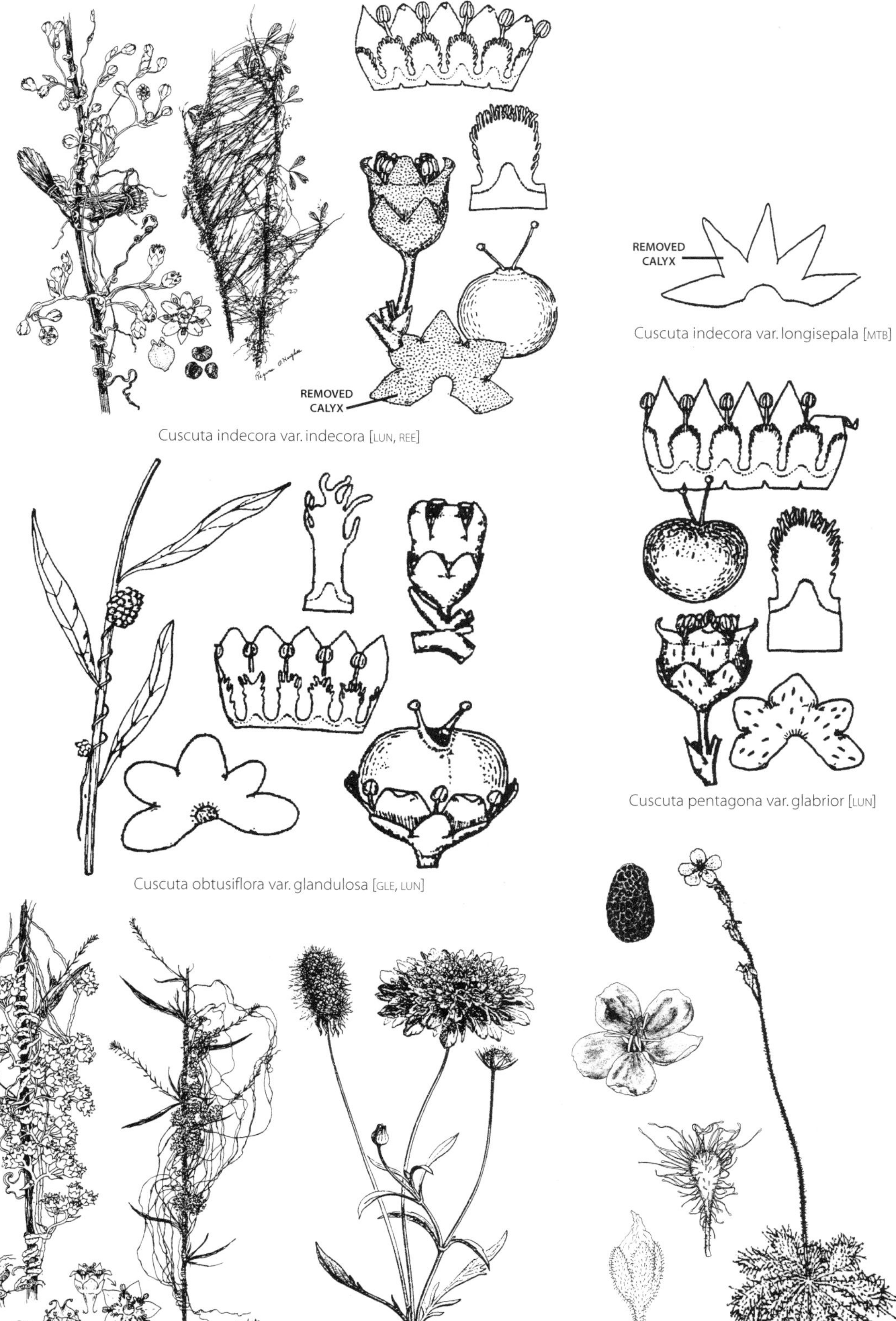

Cuscuta indecora var. indecora [LUN, REE]

Cuscuta indecora var. longisepala [MTB]

Cuscuta obtusiflora var. glandulosa [GLE, LUN]

Cuscuta pentagona var. glabrior [LUN]

Cuscuta pentagona var. pentagona [REE]

Scabiosa atropurpurea [BL1]

Drosera brevifolia [GWO]

Drosera brevifolia Pursh, (short-leaved), ANNUAL SUNDEW. Small insectivorous herb; leaves in a basal rosette, fiddle-head-like in bud, suborbicular (with petiole longer than blade) to nearly spatulate with enlarged terminal portion, to 15 mm long, with conspicuous, red, gland-tipped, motile hairs which secrete a clear sticky fluid used in trapping insects; stipules absent or vestigial; scape to 12 cm tall, with gland-tipped hairs except at base; flowers up to 6 per scape, radially symmetrical, usually 5-merous, hypogynous; sepals ovate, 2.5–4 mm long; petals pink or roseate, 2.5–8 mm long; ovary superior; capsules 3.5–4 mm long; seeds black, with pits in 10–12 rows. Deep sand in woods or in open bogs; included based on citation of vegetational area 4 (Fig. 2) by Hatch et al. (1990); se and e TX w to at least c Henderson Co. (s of Athens) near e margin of nc TX. Feb–Jun. [*D. annua* E.L. Reed, *D. leucantha* Shinners] This *Drosera* species, 2 *Utricularia* species (BLADDERWORTS—Lentibulariaceae), and *Sarracenia alata* A.W. Wood (PITCHERPLANT or YELLOW-TRUMPETS—Sarraceniaceae) are the only carnivorous plants in nc TX.

EBENACEAE PERSIMMON OR EBONY FAMILY

The Ebenaceae is a medium-sized (485 species in 2 genera) family of the tropics and warm areas with a few temperate species; most are trees; some have edible fruits (e.g., *Diospyros*—PERSIMMONS, JAPANESE PERSIMMONS, DATE-PLUMS, VELVET-APPLES), while others are valuble for timber including EBONY, a black, hard, heavy wood obtained from species of *Diospyros*. Family name conserved from *Ebenum*, a genus now treated as *Diospyros* (the name *Diospyros* was published earlier and thus has priority in terms of nomenclature) (Derivation either from Greek: *ebenos*, name used by Hippocrates for a leguminous plant or possibly the ebony tree, or from Latin: *ebenus*, ebony or black; the heartwood of some species is strikingly black in color) (subclass Dilleniidae)
FAMILY RECOGNITION IN THE FIELD: shrubs or small trees with alternate, simple, entire, exstipulate leaves; milky juice absent; flowers small, axillary, usually of a single sex, with connate petals and superior ovaries; fruit a berry with several large seeds.
REFERENCES: Wood & Channell 1960; Spongberg 1977; Morton et al. 1996 [1997].

DIOSPYROS PERSIMMON, EBONY

Shrubs or small trees; leaves alternate, nearly sessile to short-petioled; leaf blades simple, oblong or elliptic, entire; flowers rather small, axillary, solitary or few together, short-pedicelled, drooping, imperfectly unisexual, the pistillate larger, with 8 empty filaments, the staminate smaller, with 16 stamens; calyces subrotate, deeply 4-lobed, enlarging in age; corollas urceolate to campanulate or salverform, creamy or yellowish white, somewhat fleshy, with 4, short, auricled, laterally overlapping lobes; stamens attached to corolla near its base, with wide, pilose filaments; pistil 1; ovary superior; style divided half way or more into 4 branches; fruit a berry with several large seeds.

A mainly tropical genus of ca. 475 species including a number important for fruit and timber. The common name is derived from *pasimenan*, the Native America Lenape word for the tree (Peattie 1948). *Diospyros ebenum* J. König ex Retz., native of India and Sri Lanka, is the EBONY of commerce; its wood is black and extremely hard; it is used in inlays, for piano keys, and in other uses where its color and density are important. EBONY is so dense that it is one of relatively few woods that will not float in water. *Diospyros kaki* L.f., (JAPANESE PERSIMMON, KAKI), of e Asia, produces large fruits that are widely available in U.S. supermarkets. (Greek: *dios*, of Zeus or Jove, and *pyros*, grain, alluding to the edible fruits)
REFERENCE: Hiern 1873.

1. Ovaries and fruits pubescent; fruits black; leaves 2–5 cm long, subsessile; s part of nc TX; bark ± smooth, peeling in thin layers, grayish ____ **D. texana**

1. Ovaries and fruits glabrous; fruits yellowish to orangish; leaves 7–15 cm long, with distinct petioles (ca. 7–10 mm long); widespread in nc TX; bark deeply divided into small blocks or plates, dark brown or blackish _______ **D. virginiana**

Diospyros texana Scheele, (of Texas), TEXAS PERSIMMON, MEXICAN PERSIMMON, BLACK PERSIMMON, CHAPOTE PERSIMMON. Tree to ca. 30 m tall; leaves permanently pubescent; fruits ca. 2 cm in diam., sweet and edible when ripe. Rocky woods, slopes, open areas, and along streams; s part of nc TX in Bell, Brown, and McLennan cos., also Milam Co. (Little 1976); widespread in TX, especially w 2/3. Feb–Jun. The black juice of the fruits was used during frontier days to dye buckskin or leather black (Crosswhite 1980).

Diospyros virginiana L., (of Virginia), COMMON PERSIMMON, EASTERN PERSIMMON. Shrub or small tree to ca. 16 m tall; heartwood hard and dark brown; leaves glabrous or nearly so at maturity; fruits to 5 cm in diam., ripening in fall and some often long persistent on the tree (numerous fruits have been observed in January), sweet and edible when ripe. Woods, old fields, and clearings; se and e TX w to Rolling Plains and Edwards Plateau. Apr–Jun. The tough hard wood has been used for such things as billard balls, golf-club heads, and textile weaving shuttles (Cox & Leslie 1991). The unripe fruits are notoriously astringent; however, despite anecdotes to the contrary, a frost is not necessary for ripening. The fruits are an important wildlife food; they were reportedly used to make a beer by Native Americans in the e U.S. (Hedrick in Heiser 1993).

ELAEAGNACEAE OLEASTER FAMILY

A small (45 species in 3 genera) family mainly native to temperate and warm areas of the n hemisphere to tropical Asia and Australia; most are shrubs to small trees with silvery or golden scales; they are typically armed and usually have nitrogen-fixing root nodules; several genera are used as ornamentals including *Elaeagnus* and *Hippophae*. (subclass Rosidae)
FAMILY RECOGNITION IN THE FIELD: alternate-leaved, usually spiny, introduced shrubs with very distictive, peltate or stellate, *silvery scales* conspicuously covering the leaves and young twigs; flowers 4-merous.
REFERENCE: Graham 1964.

ELAEAGNUS SILVER-BERRY OLEASTER

A genus of ca. 40 species native to Europe, Asia, and North America including cultivated ornamental shrubs; some species are important for their edible fruits. (Greek: *elaia*, olive, and *agnos*, the name of the CHASTE-TREE, *Vitex agnus-castus*, from *hagnos*, pure; the name was originally applied to a willow with massed white fruit)

Elaeagnus pungens Thunb., (piercing, sharp-pointed), THORNY ELAEAGNUS. Spreading shrub to ca. 4 m tall, usually spiny; branchlets brown; leaves evergreen, alternate, simple, elliptic to nearly ovate, 4–10 cm long, 2–3.5 cm wide, the lower surface densely covered with silvery, lobed, peltate scales and dotted with a few reddish brown scales giving the surface a striking dirty silver appearance, the margins entire, undulate; petioles 6–12 mm long; flowers usually 1–3 in the leaf axils, fragrant, with a tubular hypanthium 7–10 mm long; sepal lobes 4, petal-like, shorter than the hypanthium; petals absent; stamens 4, on short filaments attached in throat of hypanthium; carpel 1; ovary superior but hypanthium persistent and making it seem inferior; fruits drupe-like, 10–15 mm long, covered with scales and tipped with the persistent hypanthium, initially brown, eventually red. Widely cultivated and escapes; Dallas (spreading along White Rock Creek) and Tarrant cos.; we are not aware of other TX localities. Late fall. Native of Japan and China.

Elaeagnus angustifolia L., (narrow-leaved), OLEASTER, RUSSIAN-OLIVE, a Eurasian species with deciduous leaves, silvery-white or silvery-gray branchlets, and flowering in spring, is also cultivated in nc TX.

ELATINACEAE WATERWORT FAMILY

Ours small annual herbs of wet areas; leaves simple, opposite, entire, with paired membranous stipules between them; flowers 1–few in the leaf axils, small, radially symmetrical, 2–5-merous; sepals and petals both present; stamens 1 or 2 times as many as the petals; pistil 1; ovary superior, 2–5 celled; fruit a subglobose capsule; seeds pitted.

A small (34 species in 2 genera), temperate and especially tropical family including a number of aquatics. (subclass Dilleniidae)
FAMILY RECOGNITION IN THE FIELD: small wet area herbs with opposite, simple, stipulate leaves and inconspicuous flowers.
REFERENCE: Tucker 1986.

1. Plants erect to ascending, 10–40 cm tall; stems with short glandular pubescence; sepals and petals usually 5 each; sepals with thick green midrib and scarious margins, with a small sharp tooth at tip ______ **Bergia**
1. Plants creeping and mat-forming (some stems can be ascending to erect), 10 cm or less tall; stems glabrous; sepals and petals 2–3 each; sepals without distinct midrib, obtuse ______ **Elatine**

BERGIA

A genus of 24 species native to warm areas of the world. (Named for Peter Bergius, 1723–1817, Swedish botanist and student of Linnaeus)

Bergia texana (Hook.) Walp., (of Texas), TEXAS BERGIA. Annual, usually branched at base, the whole plant ± glandular-puberulent; leaves to 30 mm long, elliptic to oblong, acute, serrulate; flowers 1–3 in the leaf axils, on short pedicels ca. 1 mm long; sepals 5, 2–4 mm long; petals 5, white, not exceeding the sepals; stamens 5 or 10; fruits to ca. 3 cm wide. Muddy pond margins, along creeks; Cooke, Dallas, Denton, Falls, and Parker cos., also locally abundant around stock tanks in Brown Co. (Stanford 1976); widespread in TX. May–Oct.

ELATINE WATERWORT

Low, creeping, mat-forming, almost moss-like annuals, often rooting at the nodes; flowers 1–2 per node.

A genus of 10 species native to tropical and temperate areas; some are used as aquarium plants. (Greek: *elatino*, fir-like, from a European species that suggests such a plant in miniature)
REFERENCES: Duncan 1964; Kaul 1986c.

1. Leaves without a notch at tip, usually obovate to narrowly oblong-ovate (rarely linear-spatulate); largest seeds with 9–15 pits in each row ______ **E. brachysperma**
1. Leaves often with a slight notch at tip, linear to narrowly spatulate; largest seeds with 16–25 pits in each row ______ **E. triandra**

Elatine brachysperma A. Gray, (short-seeded), SHORT-SEED WATERWORT. Forming small mats to ca. 5 cm across, the branches ascending; leaves to 6 mm long and 2 mm wide; sepals 2 or with a third reduced; petals 3, pinkish. On mud or in shallow water; Comanche Co. (Correll & Correll 1972), also cited for vegetational areas 4 and 5 (Fig. 2) by Hatch et al. (1990); also se and c TX to Rolling Plains. Mar–Oct. [*E. triandra* Schkuhr var. *brachysperma* (A. Gray) Fassett]

Elatine triandra Schkuhr, (with three anthers or stamens), AMERICAN WATERWORT. Forming mats, the branches can be ascending to erect; leaves 3–6(–12) mm long, 1–3 mm wide; perianth greenish or pinkish, usually with 3 sepals and 3 petals. On mud or in shallow water; found at Dallas by Reverchon in Mar 1874, apparently not collected there since; se and c TX to nc TX and Rolling Plains. Mar–Oct. We are following Kaul (1986c) and Hatch et al. (1990) in lumping [*E. americana* (Pursh) Arn., *E. triandra* var. *americana* (Pursh) Fassett].

ERICACEAE HEATH OR BLUEBERRY FAMILY

Ours shrubs, trees, or mycoparasitic herbs; leaves simple, alternate; petals united (separate in *Monotropa*); stamens twice as many as corolla lobes; anthers awned or awnless, opening by terminal pores; pollen grains in tetrads (singly in *Monotropa*); ovary superior or inferior; fruit a capsule or berry.

A large (4,500 species in 160 genera (Stevens 1995)), cosmopolitan family of usually shrubs with some herbs and trees; the majority of tropical taxa are montane. Most species are found in nutrient-poor, acidic habitats and have a relationship with mycorrhizal fungi. The family includes many important ornamentals (e.g., AZALEAS and RHODODENDRONS in the genus *Rhododendron*, HEATHS in the genus *Erica*, *Arctostaphylos* (BEARBERRY and MANZANITA), and *Kalmia* (MOUNTAIN-LAUREL), as well as edible *Vaccinium* species (BLUEBERRIES and CRANBERRIES) and *Gaylussacia* species (HUCKLEBERRIES). The leaves of other species (e.g., *Kalmia*—LAMBKILL, SHEEPKILL, *Leucothoe*—FETTERBUSH, and *Lyonia*) are poisonous; *Kalmia* species can be the source of poison honey, when visited by bees (Hardin & Brownie 1993). The family has sometimes been split with groups such as the mycoparasitic *Monotropa* and its relatives raised to family rank (e.g., Kartesz 1994; Jones et al. 1997). However, molecular studies and cladistic analyses (Judd & Kron 1993; Kron & Chase 1993; Kron 1996) support the inclusion of the Monotropaceae, Pyrolaceae, Epacridaceae, and Empetraceae to form a monophyletic Ericaceae. We are therefore treating the Ericaceae in the broad inclusive sense. Family name from *Erica*, HEATH or HEATHER, a genus of ca. 665 species of shrubs and small trees native primarily to s Africa (650 species there with 520 endemic to the Cape region), and also in tropical African mountains, the Mediterranean, Macaronesia, and Europe; *Erica* species cover large areas of moorlands in Europe. (Greek: *ereike*, heather, from *ereiko*, to break; an infusion from the leaves was supposed to break bladder stones) (subclass Dilleniidae)

FAMILY RECOGNITION IN THE FIELD: usually shrubs or trees (1 genus of reduced mycoparasitic herbs) with alternate, simple, rather leathery, often evergreen leaves and usually *urn-shaped*, cylindric, or campanulate corollas (many funnel-shaped in species outside nc TX); anthers opening by terminal pores.

REFERENCES: Small 1914a, 1914b; Wood 1961; Judd & Kron 1993; Kron & Chase 1993; Luteyn 1995; Stevens 1995.

1. Plants mycoparasitic herbs lacking chlorophyll (plant white to tinged with rose), 5–30 cm tall; leaves reduced to scale-like bracts ____ **Monotropa**
1. Plants shrubs or trees with green photosynthetic leaves with well-developed blades.
   2. Flowers in panicles; fruit a berry with a distinctive roughened-tuberculate surface; ovary superior; plant a small tree or rarely arborescent shrub, 4–6(–10 m or more) tall; leaves evergreen ____ **Arbutus**
   2. Flowers in racemes or umbel-like clusters; fruit a dry capsule OR a berry with a smooth surface; ovary superior and developing into a capsule OR inferior and developing into a berry; plants varying from shrubs < 2 m tall to small trees to 8 m tall; leaves essentially evergreen OR deciduous.
      3. Fruit a berry (blueberry-like) crowned by the small persistent calyx teeth; corollas 4–6 mm long, open campanulate; ovary inferior; shrubs or small trees to 8 m tall; leaves essentially evergreen ____ **Vaccinium**

3. Fruit a dry capsule; corollas 8–13 mm long, urceolate-cylindric; ovary superior; shrubs < 1(–2) m tall; leaves deciduous ______ **Lyonia**

ARBUTUS MADRONE, MADROÑO

A genus of 10 species, with 3 in Europe, the Middle East, and North Africa, 1 on the Canary Islands, and 6 in the New World (Sørensen 1995). (Latin name for strawberry tree, probably originally applied to *A. unedo* L., strawberry tree)
REFERENCE: Sørensen 1995.

Arbutus xalapensis Kunth, (for its type locality near Jalapa or Xalapa, in the Mexican state of Veracruz), TEXAS MADRONE, MADROÑO, LADY'S-LEG, NAKED-INDIAN. Small tree or rarely arborescent shrub, 4–6(–10 m or more) tall; bark pinkish to brick red, peeling off in large smooth flakes, the naked branches distinctive; leaves alternate, evergreen, coriaceous, ovate to oval or oblong-elliptic, to 10 cm long and 45 mm wide, entire to serrate, glabrate with age; inflorescence a panicle; flowers perfect; calyx lobes 5, pinkish white; corollas ovoid-urceolate, white, often pink-tinged, ca. 7 mm long; stamens included, 10; anthers ca. 1.5 mm long, with a pair of slender reflexed spurs on the back, opening by terminal pores; ovary superior; fruit a ± spherical berry, red to yellowish red, 7.5–9 mm in diam., fleshy, with ovules 2–several per locule, the fruit surface roughened-tuberculate. Wooded rocky hills, slopes, canyons, plains; three occurrences are known n of the Colorado River in Travis Co. (Balcones Canyonlands Nat. Wildlife Refuge, C. Sexton, pers. comm.) just s of the nc TX border; included because of possibility of occurrence in extreme s part of nc TX; Edwards Plateau to Trans-Pecos. Feb–Apr. [*A. texana* Buckley, *A. xalapensis* var. *texana* (Buckley) A. Gray] According to Sørensen (1995), the TX plants with small glabrous leaves are part of a north-south cline involving a gradual change from plants with large pubescent leaves found in s Mexico and n Central America. The TX plants have often been recognized as a distinct species or variety, but we are following Sørensen (1995) in lumping them into *A. xalapensis* despite their morphological distinctiveness. Fruits edible (Correll & Johnston 1970).

LYONIA

A genus of 36 species native to e and se Asia, e North America including Mexico, and the Greater Antilles (Judd 1995); some are cultivated as ornamental shrubs. (Named for John Lyon, 17??–1818, early American botanist and explorer of the s Alleghenies)
REFERENCES: Judd 1995; Kron & Judd 1997 [1998].

Lyonia mariana (L.) D. Don, (of Maryland), STAGGERBUSH. Shrub usually < 1 m tall (rarely to 2 m); leaves elliptic-oblong to elliptic-lanceolate or narrowly obovate, entire, to 11 cm long and 5 cm wide, deciduous; flowers in umbel-like fascicles (= clusters) along leafless old branches; corollas 8–13 mm long, white or pinkish; ovary superior; capsules ca. 7 mm long. Usually moist, sandy, wooded areas; se and e TX w to nc TX and Edwards Plateau; included based on citation of vegetational areas 4 and 5 (Fig. 2) by Hatch et al. (1990). Mar–Jun. The leaves are thought to be poisonous to young grazing animals; the poisonous principle is apparently andromedotoxin, a resinoid, or arbutin, a glycoside (Hardin & Brownie 1993). ☠

MONOTROPA

Obligately mycoparasitic (= indirectly parasitize plants via fungi) fleshy herbs lacking chlorophyll; plants waxy white or yellow or reddish, becoming tinged with pink or red, and eventually drying dark; leaves reduced to scale-like bracts; inflorescences nodding in flower, becoming erect in fruit; flowers (3–)5(–6)-merous; calyces similar to or different from corollas or occasionally absent; corollas cylindric or slightly flared distally, of separate petals; stamens 8 or 10; pollen grains shed singly (usually in tetrads in the Ericaceae); fruit a capsule.

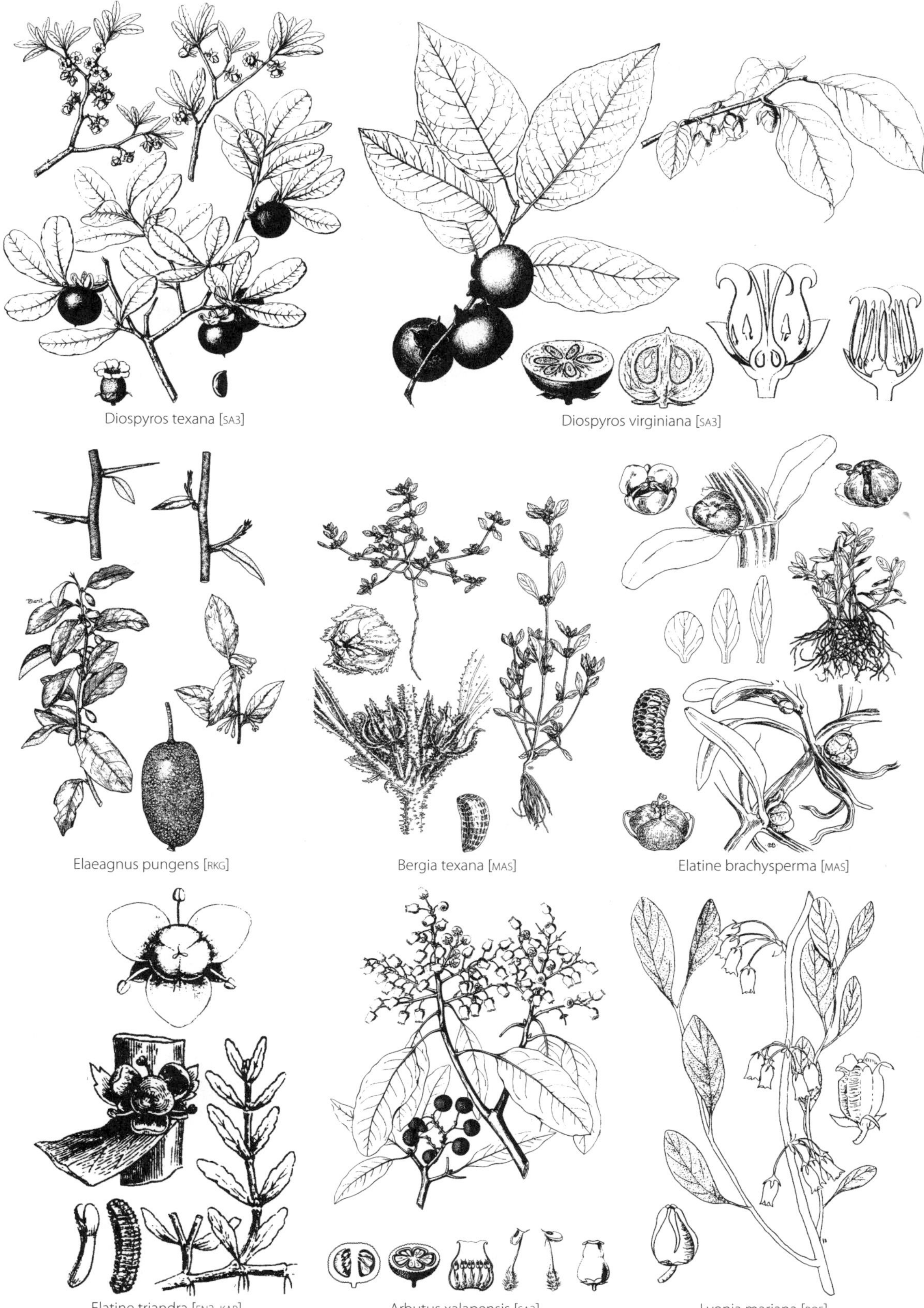

Diospyros texana [SA3]

Diospyros virginiana [SA3]

Elaeagnus pungens [RKG]

Bergia texana [MAS]

Elatine brachysperma [MAS]

Elatine triandra [EN2, KAR]

Arbutus xalapensis [SA3]

Lyonia mariana [ROE]

A widespread n temperate genus (Europe, Himalayas, Japan, North and Central America) of 2 species with disjunct populations in the tropics (Wallace 1995). While Kartesz (1994) and Jones et al. (1997) placed *Monotropa* in the Monotropaceae, we are following Stevens (1995) and Wallace (1995) in treating the Monotropoideae as a subfamily of the Ericaceae. (Greek: *monos*, one, and *tropos*, turn, from the summit of the flowering stem nodding, turned to one side)
REFERENCES: Correll 1965; Stevens 1995; Wallace 1975, 1993, 1995.

1. Flowers several (rarely reduced to only 1) per inflorescence; inflorescences (as well as rest of plant) at flowering time ranging in color from light yellow to reddish, drying brownish; plants often downy or finely pubescent; stigma often subtended by a ring of stiff hairs ____ **M. hypopithys**
1. Flower solitary per inflorescence, never more; inflorescences (as well as rest of plant) at flowering time usually waxy white or rarely tinged pink or orange-red or combinations of these, drying black; plants glabrous; stigma not subtended by a ring of stiff hairs ____ **M. uniflora**

Monotropa hypopithys L., (old generic name from Greek, *hypo*, under, and *pitys*, pine), AMERICAN PINESAP. Plant 5–30 cm tall; leaves scale-like, 5–12 mm long; inflorescence usually a several-flowered raceme; pedicels slender, 2–20 mm long, to 30 mm in fruit; sepals ± unlike petals, sometimes absent; petals 8–17(–20) mm long; styles 1–2 mm wide; capsules 6–10 mm long. Humus in woods; included based on citation (Correll 1965) of Lamar Co. specimen (*Correll & Correll 27488*); mainly e TX, also disjunct to Trans-Pecos. Apr–Jul. [*M. latisquama* Rydb.]

Monotropa uniflora L., (one-flowered), INDIAN-PIPE. Plant 5–30 cm tall; leaves scale-like, 5–12(–15) mm long; sepals ± like petals; petals 10–20 mm long; styles 2–5 mm wide; capsules 7–11 mm long. Forest floor humus, under pines and hardwoods; Fannin Co. (Talbot property) in Red River drainage, also Lamar Co. (Carr 1994); mainly e TX. Apr–Jul. ▣/100

VACCINIUM BLUEBERRY

A large genus (ca. 450 species) native to circumpolar areas, Europe, North America, tropical areas of the Americas, c and se African mountains, Madagascar, Japan, and tropical Asia to Malaysia. It includes deciduous or evergreen shrubs, small trees, and lianas, many of which have edible fruits including BLUEBERRY, BUCKBERRY, HUCKLE-BERRY, BLUETS, BILBERRY, WHORTLE-BERRY, BLAEBERRY, CRANBERRY, COWBERRY, LINGBERRY, LINGENBERRY, and FOXBERRY; ☠ at least several Latin American species have toxic fruits. The CRANBERRY is *V. macrocarpon* Aiton, native to e North America. Edible BLUEBERRIES are cultivated in the sandy soils of n Fannin Co. by the Walker family. (Latin name variously applied to *Vaccinium myrtillus* L. or *Hyacinthus*; origin obscure, possibly from *bacca*, berry, or *vaccinus*, of cows)
REFERENCES: Camp 1945; Vander Kloet 1988.

Vaccinium arboreum Marshall, (tree-like), FARKLE-BERRY, SPARKLE-BERRY. Shrub or small tree to 8 m tall; leaves obovate to oblong-elliptic, entire or with some teeth, to ca. 7 cm long and 3.5 cm wide, essentially evergreen; flowers in loose leafy-bracted racemes, the bracts much smaller than the leaves; corollas 4–6 mm long, white; ovary inferior; berries black or reddish black, often dryish and mealy to moist. Sandy soils, open woods, thickets, fields; Henderson and Limestone cos. on e margin of nc TX, also w in Red River drainage through Lamar to Fannin and Grayson cos.; the Grayson population (Preston Bend Park) is isolated on an outcrop of Trinity sand ca. 30 miles west of any other known TX site; mainly se and e TX. Mar–May. [*V. arboreum* var. *glaucescens* (Greene) Sarg.]

EUPHORBIACEAE SPURGE FAMILY

Ours annual or perennial herbs, shrubs, or small trees with watery or milky juice; leaves alternate or less often opposite, simple (sometimes so deeply lobed as to appear compound, occasionally some leaves divided into palmate leaflets), entire, toothed, or lobed, with or without stipules; flowers axillary or terminal, solitary, in racemes, spikes, or heads, various in structure, unisexual or bisexual; perianth absent, or present and of 1 type of part only, or of both sepals and petals; stamens 1–many; pistil 1, usually 3-celled and forming a 3-angled or 3-lobed ovary (a few species with these parts reduced to 2 or 1); ovary superior; fruit a capsule (sometimes considered a capsular-schizocarp), usually 3-seeded, or in 1 species of *Croton*, an achene.

A huge (8,100 species in 313 genera), cosmopolitan, but especially tropical family; vegetatively they vary from herbs to shrubs, lianas, large trees, or succulents; often there are specialized cells or tubes with milky or colored latex. A number are economically important including *Aleurites* species (TUNG-OIL), *Euphorbia pulcherrima* Willd. ex Klotzsch (POINSETTIA), *Hevea brasiliensis* (Juss.) Müll.Arg. (RUBBER, PARÁ RUBBER), *Manihot esculenta* Crantz (MANIOC, CASSAVA, or TAPIOCA), and *Ricinus communis* L. (CASTOR-BEAN); many other species are used as ornamentals. The latex of *Hevea brasiliensis*, a native of the Amazon Basin, is obtained ("rubber tapping") by making sloping incisions in the bark and collecting the white liquid in cups attached below the incisions; upon drying in the air or when coagulated by acid, the latex takes on its well known elastic properties. ☠A number of species are poisonous due to the presence of alkaloids and cyanogenic glycosides; others have diterpenoids that can cause irritant dermatitis and may act as co-carcinogens (= promote the action of "sub"-carcinogenic doses of known carcinogens) (Kinghorn 1979; Lampe 1986). In dry areas of the Old World, a number of *Euphorbia* species are xerophytically adapted and convergent vegetatively with Cactaceae. The common name SPURGE is apparently derived from the Latin word *purgare*, to purge or cleanse; a number of members of the family are regarded as having cathartic or purgative properties (Tveten & Tveten 1993). (subclass Rosidae)

FAMILY RECOGNITION IN THE FIELD: ours mostly herbs (a few small shrubs and 1 tree) often with *milky sap* and leaves usually alternate or less often opposite (mostly in *Chamaesyce*); *ovary 3-celled*, superior, typically developing into a capsule with 3 lobes or sections; flowers often unisexual, sometimes highly reduced.

REFERENCES: Wheeler 1943; Punt 1962; Webster 1967, 1994a, 1994b; Jensen et al. 1994; Kapil & Bhatnagar 1994; Nowicke 1994; Seigler 1994; Steinmann & Felger 1997.

1. Trees 3–10 m tall; leaf blades rhombic-ovate, 3–7(–10) cm long, with petioles usually longer than blades, with 2 glands at juncture of blade and petiole; rarely escaped introduced species ____ **Sapium**
1. Herbs or small shrubs, usually 1 m or less tall, often much less, rarely to ca. 2 m or more tall (*Croton alabamensis*); leaves not as above; mostly native species.
   2. Leaf blades palmately lobed, 6–60 cm wide.
      3. Plants with extremely painful stinging hairs; leaf blades with petioles attached basally, 6–15 cm wide; perianth white ____ **Cnidoscolus**
      3. Plants without stinging hairs; leaf blades peltate, 10–60 cm wide; perianth greenish or purplish ____ **Ricinus**
   2. Leaf blades entire to toothed, not palmately lobed, usually < 6 cm wide.
      4. Flowers with scaly or green perianth parts, borne in various sorts of inflorescences, but not in cup-like involucres; sap usually not milky (milky only in *Stillingia*).
         5. Plants with stellate hairs or lepidote scales on at least some parts, these often giving the foliage a distinctive silvery to yellowish or brownish color (the hairs and scales, while visible to the naked eye, are striking with a hand lens) ____ **Croton**
         5. Plants with simple (unbranched) or malpighiaceous (attached at middle) hairs or glabrous; foliage green.
            6. Flowers or inflorescences axillary or at least appearing so (uppermost sometimes subtended only by very reduced leaves); sap not milky.

7. Flowers solitary or in small umbel-like clusters; leaves 5–30 mm long (to 45 mm in the rare *Reverchonia* restricted to deep sand); seeds 2 per locule.
   8. Plants of deep sand; calyx lobes conspicuous, purplish or pinkish, with a central greenish strip, 1.5–2.9 mm long; capsules 7–9.8 mm wide; extremely rare if present in w part of nc TX ______ **Reverchonia**
   8. Plants of various soils including sand and limestone; calyx lobes not as above; capsules 1.7–3.2(–7.5) mm wide (3.2 mm or less except in the rare *Andrachne*); widespread in nc TX.
      9. Herbs (*P. polygonoides* becoming lignified at base) to ca. 0.5 m tall; leaf blades elliptic to narrowly oblanceolate, 2–8(–11) mm wide; pedicels absent or up to 7 mm long; male and female flowers usually, but not always, on the same plant; flowers apetalous; capsules 1.7–3.2 mm wide; including species widespread in nc TX ______ **Phyllanthus**
      9. Shrubs up to ca. 1 m tall; leaf blades elliptic-oblong to orbicular, 6–15 mm wide; pedicels 5–12(–22) mm long; male and female flowers on separate plants; flowers with petals; capsules 7–7.5 mm wide; one species rare in nc TX ______ **Leptopus**
7. Flowers few to many in spikes or spike-like racemes; leaves 10–112 mm long; seeds 1 per locule.
   10. Leaf blades often but not always rather triangular (sometimes narrowly so) in shape, truncate to cordate or hastate basally; plants with stinging hairs mixed with soft spreading hairs ______ **Tragia**
   10. Leaf blades variously shaped but not triangular, usually not truncate, cordate, or hastate basally (except cordate in *Acalypha ostyifolia*); plants without stinging hairs.
       11. Leaf blades entire; female flowers not surrounded by conspicuous folded bract; stipules absent; some malpighian (= attached at center) hairs usually present ______ **Ditaxis**
       11. Leaf blades toothed; female flowers surrounded by a conspicuous, toothed or lobed, folded bract; stipules present; hairs not malpighian ______ **Acalypha**

6. Inflorescences terminating the stems (may be over-topped by leaves or leafy bracts); sap milky or not so.
   12. Sap milky; female flowers exposed ______ **Stillingia**
   12. Sap not milky; female flowers ± concealed by a conspicuous, toothed or lobed, folded bract ______ **Acalypha**

4. Flowers lacking a perianth; staminate flowers and a single central pistillate flower together congested within small cup-like involucres (= cyathia) which simulate flowers, the cyathia with prominent glands and often petal-like appendages on the rim; sap milky.
   13. Leaf bases asymmetrical; leaves all opposite; stems often (but not always) prostrate; main stem short, shorter than branches (stems branching basally) ______ **Chamaesyce**
   13. Leaf bases symmetrical; leaves opposite or alternate, but at least some usually alternate; stems erect or ascending, not prostrate; main stem prominent, usually longer than branches (stem often branching apically) ______ **Euphorbia**

ACALYPHA COPPERLEAF, THREE-SEEDED MERCURY

Ours annual or perennial herbs; sap not milky; leaves alternate, petioled; blades entire or toothed, often turning reddish in fall; stipules minute; inflorescences spicate; staminate and pistillate flowers on the same or separate plants; staminate flowers: subtended by a small bract; calyces 4-parted; petals absent; stamens 4–8, united basally; pistillate flowers: each surrounded by a conspicuous folded bract that is variously toothed or lobed; calyces 3-parted; petals absent; ovary 3-carpellate; style branches often highly divided; seeds carunculate.

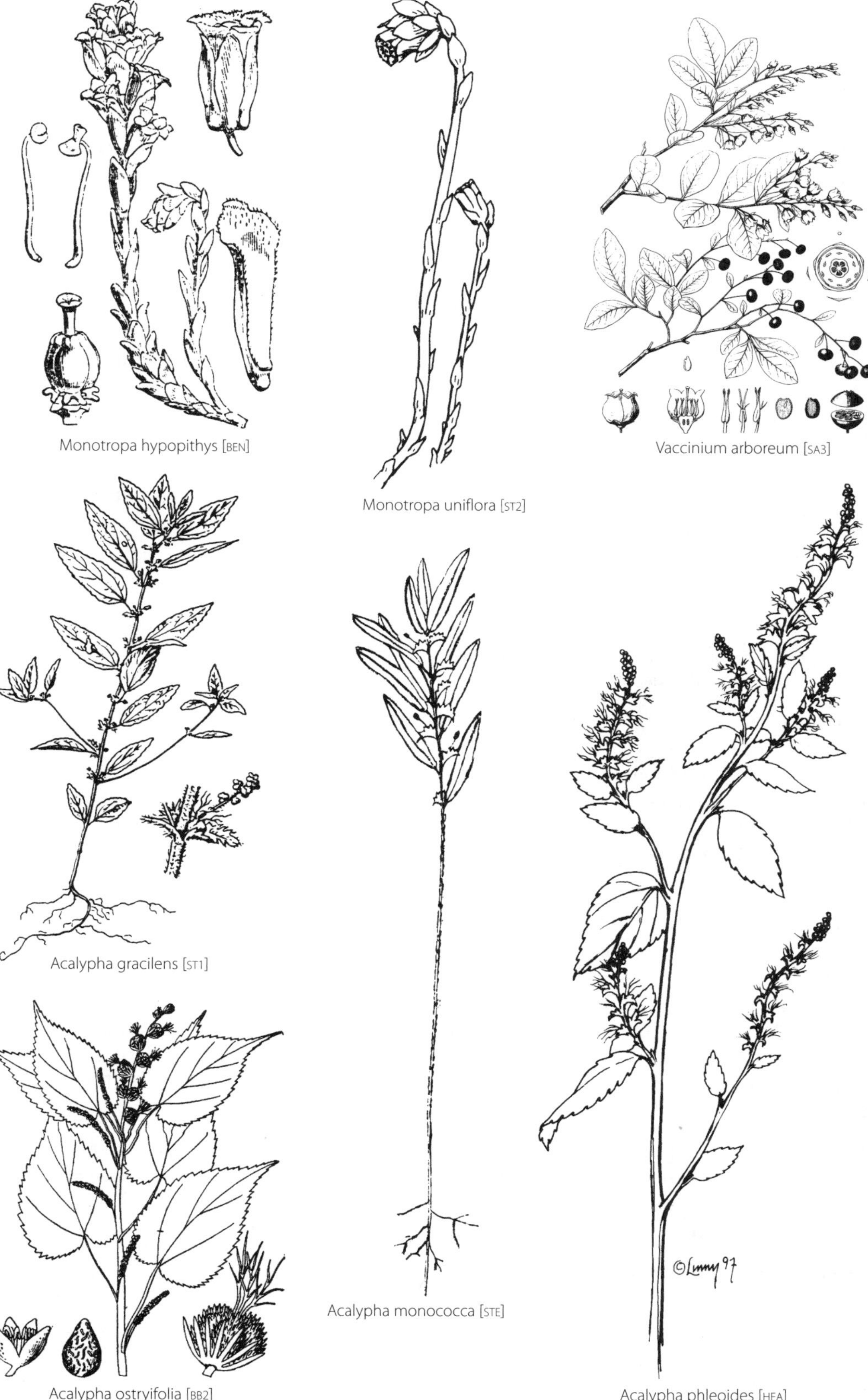

Monotropa hypopithys [BEN]

Monotropa uniflora [ST2]

Vaccinium arboreum [SA3]

Acalypha gracilens [ST1]

Acalypha monococca [STE]

Acalypha ostryifolia [BB2]

Acalypha phleoides [HEA]

A genus of ca. 430 species of tropical and warm areas; most species are shrubs with some trees and herbs. (Greek: *acelephe*, nettle, alluding to the similarity of the leaves to those of stinging nettle, *Urtica dioica* L.)

1. Leaf blades deeply divided, small, 5–12 mm long; staminate and pistillate flowers on different plants; extreme s part of nc TX ____ **A. radians**
1. Leaf blades variously toothed but not divided, > 12 mm long (usually much greater); staminate and pistillate flowers on the same plant, either on the same OR on separate spikes; widespread in nc TX.
   2. Base of leaf blades slightly cordate; leaf blades ovate; staminate and pistillate flowers borne on separate spikes ____ **A. ostryifolia**
   2. Base of leaf blades not cordate; leaf blades linear-lanceolate to ovate or rhombic; staminate and pistillate flowers borne on the same spike.
      3. Spikes terminal; plants perennial ____ **A. phleoides**
      3. Spikes axillary; plants annual.
         4. Fruits 1-seeded ____ **A. monococca**
         4. Fruits 3-seeded.
            5. Leaf blades linear to lanceolate or narrowly ovate, the margins entire to slightly crenate; petiole usually < 1/4 as long as blade ____ **A. gracilens**
            5. Leaf blades lanceolate, elliptic, narrowly ovate, or broadly rhombic, the margins crenate to serrate; petiole 1/4 the length of blade to as long as blade.
               6. Pistillate bracts (5–)7- to 9-(–11) lobed, the lobes oblong-lanceolate, usually sparsely pubescent (long glandular hairs may be present, but without long, non-glandular hairs); leaf blades ovate or elliptic to broadly rhombic; ____ **A. rhomboidea**
               6. Pistillate bracts with (8–)10–14(–16) lanceolate lobes, usually with dense long-spreading hairs; leaf blades narrowly rhombic to broadly lanceolate ____ **A. virginica**

Acalypha gracilens A. Gray, (graceful), SLENDER COPPERLEAF. Erect annual 10–60 cm tall, simple or freely branched, usually at base; leaf blades linear to lanceolate or narrowly ovate, 4–21 mm wide; styles 1–2 mm long, white or pinkish. This species is similar to *A. monococca* but usually occurs on deeper soils under more mesic conditions, typically in areas with more vegetation, usually in sandy soils; Bell and Grayson cos., also Hunt and Lamar cos. (M. Mayfield, pers. comm.); se and e TX w to ne part of nc TX. Jun–Oct. [*A. gracilens* A. Gray var. *delzii* Lill. W. Mill.]

Acalypha monococca (Engelm. ex A. Gray) Lill.W. Mill. & Gandhi, (one-berried), SLENDER ONE-SEED COPPERLEAF. Erect annual similar to *A. gracilens*; leaf blades linear to lanceolate, 3–13 mm wide; fruits with only 1 seed. This species is similar to *A. gracilens* but usually occurs in areas with more sun and shallower soils, typically in sandy soils; Grayson and Jack cos.; se and e TX w to West Cross Timbers and Edwards Plateau. Jun–Oct. [*A. gracilens* subsp. *monococca* (Engelm. ex A. Gray) G.L. Webster, *A. gracilens* var. *monococca* Engelm. ex A. Gray] We are following G. Levin (pers. comm.), who is treating *Acalypha* for the *Flora of North America*, in recognizing this taxon at the specific level; he indicated that there is no evidence of hybridization with *A. gracilens*.

Acalypha ostryifolia Riddell, (with leaves resembling *Ostrya*—hop-hornbeam), HOP-HORNBEAM COPPERLEAF. Erect annual to 50 cm tall, usually freely branched; leaves long-petioled, nearly glabrous, 1.5–5.3 cm wide; pistillate spikes terminal; staminate spikes axillary. Stream banks or bottoms, disturbed areas, roadsides; widespread in TX, but mainly nc TX s and w to w TX. Late May–Oct.

Acalypha phleoides Cav., (resembling *Phleum*—timothy or cat-tail grass), LINDHEIMER'S COPPERLEAF. Perennial; stems several to many from a woody root, branched, spreading to erect, to 60 cm long, pilose with long, wide-spreading hairs or short-pubescent with curled hairs or glabrous; leaf blades lanceolate or ovate-lanceolate, sharply and evenly toothed; styles conspicuously reddish. Rocky slopes, chiefly limestone; Trans-Pecos and Edwards Plateau e and n to Bell and Somervell cos. (Mahler 1988). May–Oct. [*A. lindheimeri* Müll.Arg.] We are following G. Levin (pers. comm.) for nomenclature of this species. He indicated that it will be treated as *A. phleoides* in the forthcoming treatment of *Acalypha* for *Flora of North America.*

Acalypha radians Torr., (radiating outward), ROUND-CROTON, YERBA DE LA RABIA. Perennial; stems ±decumbent, 20–40 cm tall, densely pubescent with short stiff and long-spreading hairs; leaf blades reniform to orbicular, 8–12 mm wide; petioles 4–16 mm long, usually as long as or longer than blades; bracts 3.5–10 mm long, with 7–13 lobes; capsules with 3 seeds. Dry sandy or gravelly areas; Burnet Co. (Correll & Johnston 1970), also cited by Hatch et al. (1990) for vegetational areas 4 and 5 (Fig. 2); extreme s part of nc TX e and s to s TX. Summer–fall.

Acalypha rhomboidea Raf., (rhomboidal, ± diamond-shaped with unequal sides), RHOMBOID COPPERLEAF. Erect annual 10–60 cm tall; stems simple or freely branched, densely pubescent above with recurved hairs, sparsely pubescent below; petioles usually 1/2 to as long as blades; pistillate bracts cut about half way into lobes. Sandy soils; Dallas Co., also Tarrant Co. (Mahler 1988); mainly se TX. Jul–Oct. [*A. virginica* L. var. *rhomboidea* (Raf.) Cooperr.] While this species is cited for vegetational areas 1 through 5 (Fig. 2) by Hatch et al. (1990), according to M. Mayfield (pers. comm.), it occurs mainly in se TX; however, the identification of the Dallas specimen was confirmed by G. Levin. This taxon is sometimes treated as a variety of *A. virginica* (e.g., Kartesz 1994; Jones et al. 1997). However, we are following G. Levin (pers. comm.), who is treating *Acalypha* for *Flora of North America*, in recognizing it at the specific level. According to M. Mayfield (pers. comm.), *A. rhomboidea* and *A. virginica* grow together without intergadation and are "very distinct species."

Acalypha virginica L., VIRGINIA COPPERLEAF. Erect annual 10–50 cm tall; stems simple or sparsely branched; sparsely to densely pubescent with recurved and long, soft, spreading hairs; petioles usually 1/4–1/2 as long as blades; pistillate bracts cut half or more their length into lobes; styles white. Sandy open woods or low areas; se and e TX w to West Cross Timbers, also Edwards Plateau. Jul–Oct.

CHAMAESYCE CREEPING SPURGE

Annual or perennial herbs; sap milky; leaves opposite, 2-ranked, folding together at night or in bad weather; leaf blades entire or toothed, usually asymmetrical at base, nearly sessile or short-petioled; stipules ±scarious or papery, connected or united; cyathia and fruits as in *Euphorbia*—flowers greatly reduced, minute, unisexual (consisting only of a single pistil or a single stamen on a short pedicel), in small cup-like involucres (= cyathia); each cyathium containing one pistillate and several staminate flowers mixed with minute bracts; the cyathia also have fleshy glands on their rims; these glands in some species have small, petal-like gland-appendages—the cyanthium is then termed a pseudanthium because of the resemblance to standard flowers; pistillate pedicel elongating in age, the 3-locular capsule exserted from the involucre. In the past there has not always been clarity in the terms used for the measurement of glands and gland-appendages. We are following terminology suggested by M. Mayfield (pers. comm.)—in the key and descriptions the width of the gland signifies the measurement in the tangential dimension relative to the cyathium—in our species this is typically the longest dimension of the gland; the length of the gland signifies the measurement in the radial dimension relative to the cyathium; the length of the gland-appendage signifies the measurement in the radial dimension relative to the cyathium—this means from attachment at the gland to the apex of the gland-appendage.

A ±cosmopolitan segregate of *Euphorbia*; often included in that genus as a subgenus; some Hawaiian species are arborescent (Wheeler 1941); the genus is estimated to have 300+ species

(M. Mayfield, pers. comm.). ☠As in *Euphorbia*, the latex of some species is apparently toxic and can cause dermatitis or other reactions in sensitive individuals; inflammation and large blisters can develop; the eyes are particularly sensitive. While livestock are supposedly sometimes accidentally poisoned by eating *Chamaesyce* species mixed with other species, animals generally will not eat the plants (Wheeler 1941; Muenscher 1951; Kingsbury 1964; Lampe 1986). Because pubescence, stipule, fruit, and seed characters are critical in determining species of *Chamaesyce*, a hand lens is essential. While Jones et al. (1997) lumped *Chamaesyce* with *Euphorbia*, we are following Kartesz (1994), Webster (1994b), J. Kartesz (pers. comm. 1997), and G. Webster (pers. comm. 1997) in maintaining it as a separate genus. (Ancient Greek name for a kind of prostrate plant; presumably involving the root *chamai*, on the ground or low growing)
REFERENCES: Wheeler 1941; Burch 1966.

1. Capsules, ovaries, underside of leaf blades, and usually stems glabrous.
   2. Larger leaf blades linear or narrowly oblong, 4.5–10 times as long as wide ________ **C. missurica**
   2. Larger leaf blades oblong or elliptic to triangular or orbicular, less than 4.5 times as long as wide.
      3. Larger leaves 13–35(–rarely more) mm long (including petioles); plants ascending, 15–75 cm tall.
         4. Capsules 1.9–2.3 mm long; branches off main stem 1–4 mm thick; cyathia in small clusters or solitary; leaf blades often with a central reddish or purplish splotch; stems pubescent in lines; native species widespread and abundant in nc TX ________ **C. nutans**
         4. Capsules ca. 1.3 mm long; branches off main stem ca. 1 mm thick; cyathia mostly strongly and densely clustered in glomerules; leaf blades usually without reddish or purplish splotch; stems glabrous; in nc TX known only as introduced weed in Tarrant Co. ___ **C. hypericifolia**
      3. Larger leaves 4–13 mm long; plants usually prostrate or nearly so (except sometimes ascending in *C. fendleri* and *C. glyptosperma*), usually 1–10 cm tall.
         5. Stipules united into a single, triangular, entire or ragged-margined, white scale on each side of the stem; stems often rooting at nodes.
            6. Cyathial glands 0.5–1 mm wide (tangential dimension; oblong in shape, the length or radial dimension much less); gland-appendages usually conspicuous, 0.8–1.5 mm long (radial dimension), 1–3 times wider than glands are wide (tangential dimension); plants perennial; staminate flowers 12–30 per cyathia; capsules 1.3–2.3 mm long ________ **C. albomarginata**
            6. Cyathial glands < 0.4 mm wide (oblong in shape, the length or radial dimension much less); gland-appendages usually inconspicuous, 0.2–0.6 mm long, about as wide as the glands; plants annual; staminate flowers 3–8(–10) per cyathia; capsules 1–1.2 mm long ________ **C. serpens**
         5. Stipules cut into linear segments, or reduced to a linear-lanceolate, not noticably white segment; stems not rooting at nodes.
            7. Plants annual from a taproot; gland-appendages white or pink; stems usually prostrate or nearly so.
               8. Leaf blades oblong-ovate to orbicular, less than twice as long as wide; petioles absent or very short, usually less than 1 mm long; gland-appendages 0.75–1.2 mm long (radial dimension), showy ________ **C. cordifolia**
               8. Leaf blades oblong or oblong-elliptic, twice as long as wide or longer; petioles of larger leaves up to 1.3 mm long; gland-appendages 0.5–0.75 mm long, not particularly showy.
                  9. Seeds transversely ridged; leaves usually minutely toothed at least near apex (under a hand lens); gland-appendages usually wider than long (tangential dimension greater than radial dimension) ________ **C. glyptosperma**
                  9. Seeds smooth; leaves usually entire; gland-appendages usually longer than wide ____ **C. geyeri**

Acalypha rhomboidea [ST1]

Acalypha virginica [BB2]

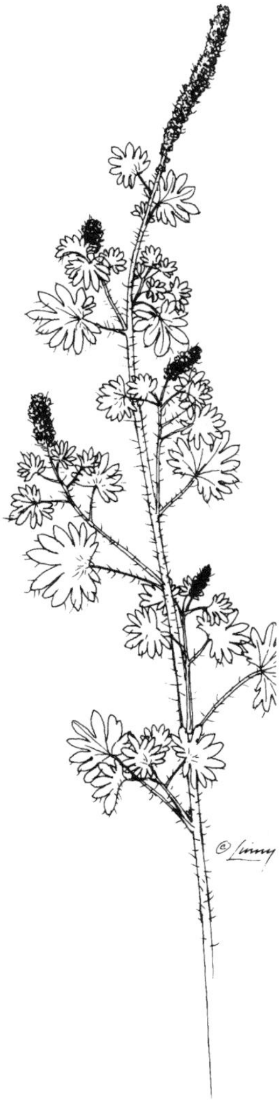

Acalypha radians [HEA]

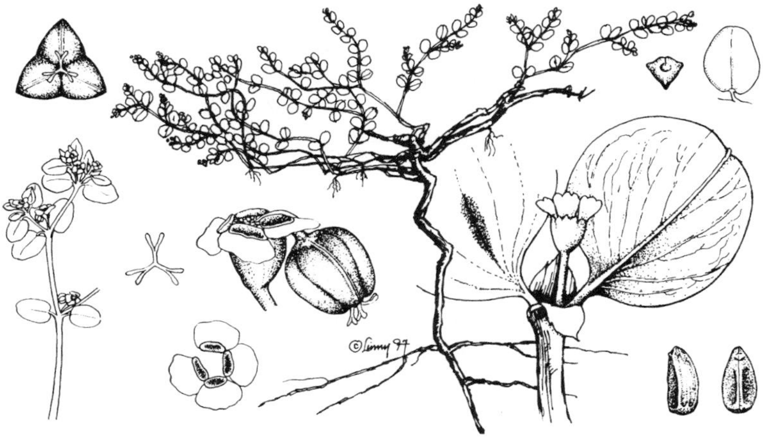

Chamaesyce albomarginata [HEA, RHO]

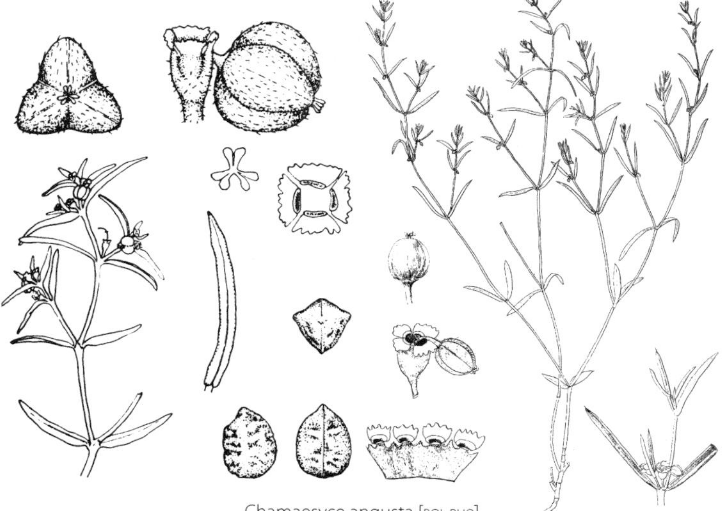

Chamaesyce angusta [BOI, RHO]

7. Plants perennial from branching roots; gland-appendages absent or yellow-green to red-brown (rarely white); stems prostrate to spreading or ascending ______ **C. fendleri**

1. Capsules, ovaries, underside of leaf blades, and stems sparsely to densely pubescent.
   10. Upper leaf blades linear or nearly so, > 9 times as long as wide; plants erect perennials from woody taproots; rare in nc TX, known locally only from the Lampasas Cut Plain ______ **C. angusta**
   10. Upper leaf blades variously shaped but not linear, ca. 1–6 times as long as wide; plants annual or perennial, prostrate to erect; including species common and widespread in nc TX.
       11. Larger leaf blades oblong to oblong-elliptic or oblong-obovate; stems prostrate to erect; including species common and widespread in nc TX.
           12. Leaf blades ca. 3 cm long or longer ______ **C. hirta**
           12. Leaf blades 1 cm long or less.
               13. Styles bifid for from 1/4 to nearly their entire length; leaf blades usually toothed, but neither prominently nor sharply, glabrous on upper surface or sparsely pubescent with hairs up to 0.5 mm long; seeds transversely rugose; stem hairs 0.2–0.7 mm long, sometimes with a few scattered longer ones.
                   14. Capsules loosely appressed-pubescent, the hairs rather uniformly scattered over the surface; leaf blades often with reddish or purplish central splotch; pubescence of ter minal internodes loosely upwardly appressed ______ **C. maculata**
                   14. Capsules spreading-pubescent, the hairs mainly on the angles of the capsules; leaf blades without reddish or purplish central splotch; pubescence of terminal internodes widely spreading or reflexed ______ **C. prostrata**
               13. Styles unbranched; leaf blades prominently and sharply toothed, at least near apex, sparsely to densely pubescent above with hairs 0.5–1.3 mm long; seeds punctately pitted and mottled; stem hairs 0.4–1.5 mm long ______ **C. stictospora**
       11. Larger leaf blades broadly or narrowly triangular; stems spreading to erect; rare, only in s part of nc TX.
           15. Stems and leaf blades with long (ca. 0.8–1.5 mm), conspicuous, soft hairs visible to the naked eye; leaf blades broadly triangular, thin, reddish or brownish beneath, with margins flat; styles bifid, the divisions terete, not reddish; capsules 1.3–1.9(–2.1) mm long; seeds 1–1.3 mm long ______ **C. villifera**
           15. Stems and leaf blades with very short (< 0.2 mm long), stiff hairs invisible to the naked eye; leaf blades narrowly triangular, thick, gray-green or blue-green, with margins inrolled downward; styles bifid, the divisions clavate, deep red; capsules 1.9–2.3 mm long; seeds 1.5–1.8 mm long ______ **C. lata**

Chamaesyce albomarginata (Torr. & A. Gray) Small, (white-margined), WHITE-MARGIN EUPHORBIA. Prostrate, mat-forming perennial; leaf blades orbicular to oblong, entire; styles bifid nearly to base. Stream bottoms, roadsides, disturbed areas; Callahan, Coleman, Cooke, Shackelford, and Young cos.; nc TX s and w to w TX. Late Apr–Oct. [*Euphorbia albomarginata* Torr. & A. Gray] According to M. Mayfield (pers. comm.), this species is found almost exclusively w of the West Cross Timbers; the Cooke Co. specimen, identified by M. Mayfield, is the easternmost record of which we are aware.

Chamaesyce angusta (Engelm.) Small, (narrow), BLACK-FOOT EUPHORBIA. Perennial; stems erect, 6–50 from a woody crown, 10–45 cm tall; leaf blades entire; lower leaf blades elliptic to linear-oblong, 7–25(–38) mm long; higher leaf blades narrower, linear, > 9 times as long as wide; stipules minute, distinct; cyathia at the upper nodes, solitary; gland-appendages white, 2–3 times as wide as gland width; styles bifid only at apex or 1/3 the distance to the base. Comanche Co. (Wheeler 1941), also Burnet Co. on the s margin of nc TX (Balcones Canyonlands Nat. Wildlife Refuge, C. Sexton, pers. comm.); also Travis Co. just s of nc TX

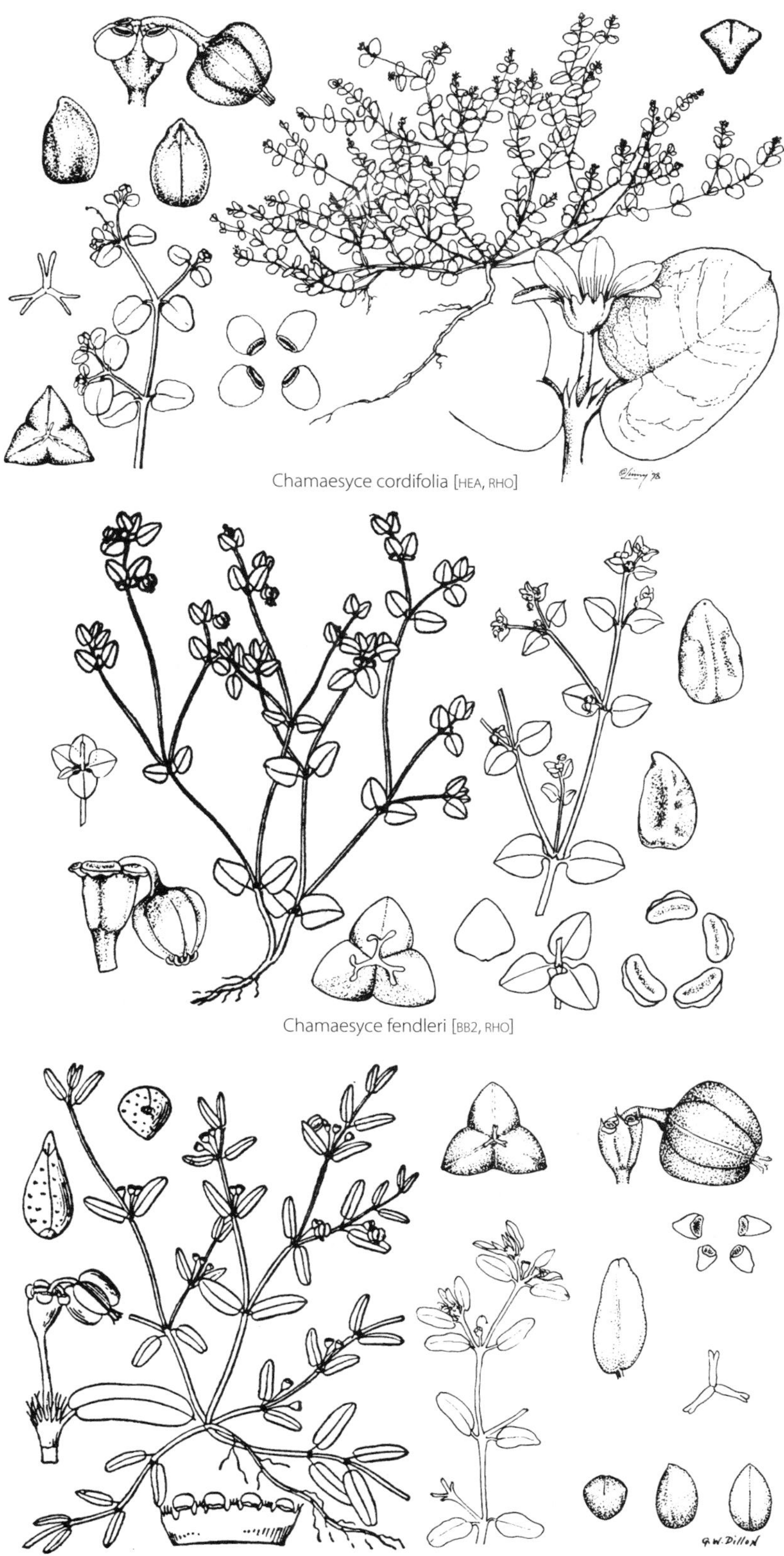

Chamaesyce cordifolia [HEA, RHO]

Chamaesyce fendleri [BB2, RHO]

Chamaesyce geyeri [BB2, RHO]

(Correll & Johnston 1970); mainly Edwards Plateau to Trans-Pecos; endemic to TX. Spring–fall. [*Euphorbia angusta* Engelm.]

Chamaesyce cordifolia (Elliott) Small, (with heart-shaped leaves), HEART-LEAF SPURGE, HEART-LEAF EUPHORBIA. Usually prostrate annual; stems to 65 cm long; leaf blades entire; styles bifid to base. Loose sandy soils, open woods; Denton, Henderson, Limestone, and Tarrant cos.; se and e TX w to East Cross Timbers. Jun–Oct. [*Euphorbia cordifolia* Elliott]

Chamaesyce fendleri (Torr. & A. Gray) Small, (for August Fendler, 1813–1883, one of the first botanists to collect in New Mexico and Venezuela). Perennial; stems rather wiry, prostrate to spreading or ascending, to 20 cm long; roots becoming woody; leaf blades orbicular-ovate to oblong, entire, acute; styles bifid at least 1/2 their length. Rocky or sandy soils; Bell, Bosque, and Tarrant cos., also Brown and Hamilton cos. (HPC); w part of nc TX s and w to Edwards Plateau and Trans-Pecos. Apr–Jun, sporadically later. [*Euphorbia fendleri* Torr. & A. Gray]

Chamaesyce geyeri (Engelm.) Small, (for its discoverer Carl Andreas Geyer, 1809–1853, Austrian botanist), GEYER'S EUPHORBIA. Annual; stems prostrate; gland-appendages usually longer than wide, 0.5–0.75 mm long, acute or obtuse, usually entire; styles bifid 1/3–1/2 their length. Sandy open woods, disturbed areas; Hood and Parker cos. in West Cross Timbers; mainly Plains Country and Trans-Pecos. Jun–Oct. [*Euphorbia geyeri* Engelm.]

Chamaesyce glyptosperma (Engelm.) Small, (carved seed), RIDGE-SEED EUPHORBIA. Annual; stems ascending to decumbent, rarely prostrate, to 50 cm long; leaf blades finely toothed towards apex; gland-appendages usually wider than long, 0.1–0.6 mm long, usually with uneven or shallowly toothed or lobed margin; styles bifid ca. 1/3–1/2 their length. Mostly sandy soils, stream bottoms, open woods, and disturbed areas; widespread in TX, in nc TX mainly East Cross Timbers westward, locally to the e in Dallas Co. (Mahler 1988). Late May–Oct. [*Euphorbia glyptosperma* Engelm.]

Chamaesyce hirta (L.) Millsp., (hairy), PILL-POD EUPHORBIA. Pubescent annual with flowering stems erect; stems with long, yellowish, jointed hairs; leaves ca. 3 cm long, elliptic-oblong; petioles 1–2 mm long; styles bifid 1/2–2/3 their length. Weed in flower bed; Dallas Co., also coastal Texas. Reported as new for TX by Lipscomb (1984). Jun. [*Euphorbia hirta* L.] Native to Africa. Used as an arrow poison ingredient in Zaire and medicinally in many countries in Africa (Neuwinger 1996); reportedly causes cardiac and respiratory depression (Burlage 1968).

Chamaesyce hypericifolia (L.) Millsp., (with leaves like *Hypericum*—St. John's wort), TROPICAL EUPHORBIA. Plant to ca. 50 cm; main stem erect, the plant thus resembling *C. nutans*; branches erect to ascending; leaves usually without a central reddish or purplish splotch; glomerules of cyathia conspicuous, the gland-appendages white to pinkish; styles bifid to about the middle. Weed along sidewalks, and in flower gardens, nurseries, and landscapes; Tarrant Co. (introduced weed in Ft. Worth); first seen in Tarrant Co. in 1994 and now spreading; mainly s TX. Summer. Native to extreme s TX. [*Euphorbia hypericifolia* L.]

Chamaesyce lata (Engelm.) Small, (broad, wide), HOARY EUPHORBIA. Perennial < 20 cm tall, rhizomatous, the rhizomes slender; plant mostly glabrous or stems and leaves with pubescence; stems subdecumbent or spreading, to 35 cm long; leaf blades minutely pubescent, entire, less than 12 mm long; styles bifid nearly to base. Sandy or rocky soils; Coleman, Coryell (Wheeler 1941), Erath, and Hamilton (Stanford 1976) cos.; mainly Plains Country and Edwards Plateau w to Trans-Pecos. [*Euphorbia lata* Englem.] According to M. Mayfield (pers. comm.), this species is found almost exclusively w of the West Cross Timbers.

Chamaesyce maculata (L.) Small, (spotted, from the spot on the leaf), SPOTTED EUPHORBIA, SPOTTED SPURGE. Prostrate or rarely suberect annual, not rooting at the nodes; stems to 50 cm long;

Chamaesyce glyptosperma [BB2, RHO]

Chamaesyce hirta [BB2, RHO]

Chamaesyce hypericifolia [BR2]

Chamaesyce lata [BB2, RHO]

Chamaesyce maculata [REE, RHO]

Chamaesyce missurica [GLE, RHO]

leaf blades finely toothed, oblong, often with a reddish or purplish central splotch; gland-appendages white to red; styles bifid 1/4–1/3 their length. Sandy soils, stream banks, woods, disturbed sites; throughout most of TX. [*C. supina* (Raf.) Moldenke, *Euphorbia maculata* L., *Euphorbia supina* Raf.] Reported to be toxic to livestock; when injested by sheep, photosensitization, swelling of the head, and emaciation can result (Burlage 1968). ☠

Chamaesyce missurica (Raf.) Shinners, (of Missouri), PRAIRIE SPURGE, MISSOURI SPURGE. Annual with erect main stem to 1 m tall and wide-spreading branches to 50 cm or more long; leaves with short petioles 1–3.2 mm long; leaf blades linear to oblong, 1–5 mm wide, 4.7–14 times as long as wide, symmetrical or slightly asymmetrical at base, entire; gland-appendages white (rarely pinkish), prominent, ca. 0.5–2.5 mm long; styles bifid 1/2 their length. Loose sandy soils, limestone outcrops; throughout TX e of Trans-Pecos. Late May–Oct. [*C. missurica* (Raf.) Shinners var. *calcicola* Shinners, *Euphorbia missurica* Raf.]

Chamaesyce nutans (Lag.) Small, (nodding), EYEBANE. Annual with (1)-several low-spreading to erect stems to 1 m long; stems pubescent in lines; gland-appendages white to red; leaf blades asymmetrically oblong, often with a central reddish or purplish splotch; styles bifid ca. 1/2 their length. Stream bottoms, roadsides, and disturbed soils; throughout TX. May–Oct. The name *C. maculata* has often been incorrectly applied to this species. [*C. maculata* of authors, not (L.) Small, *Euphorbia maculata* of authors, not L., *E. nutans* Lag.] Toxic and has caused death in lambs through photosenstization (Kingsbury 1964); the common name also suggests that care should be taken with this species. ☠

Chamaesyce prostrata (Aiton) Small, (prostrate, flat to the ground), PROSTRATE EUPHORBIA. Annual similar to *C. maculata*; leaf blades varying to broadly elliptic, without central reddish or purplish splotch; styles bifid to base or nearly so. Stream banks, prairies, disturbed sites, chiefly clay soils; throughout most of TX. Jun–Oct, sporadically earlier. [*Euphorbia prostrata* Aiton, *Euphorbia chamaesyce* of authors, not. L.] Can be toxic to cattle (Kingsbury 1964). ☠

Chamaesyce serpens (Kunth) Small, (crawling), MAT EUPHORBIA, HIERBA DE GOLONDRINA. Annual resembling *C. albomarginata*; stems slender, almost thread-like, often rooting at the nodes; styles notched. Stream bottoms, prairies, clay flats, disturbed sites; throughout TX. Jul–Oct. [*Euphorbia serpens* Kunth]

Chamaesyce stictospora (Engelm.) Small, (straight seed), SLIM-SEED EUPHORBIA. Annual; stems prostrate, to 30 cm long; leaf blades varying to broadly elliptic, without central red spot; styles unbranched. Limestone slopes, prairies, disturbed sites, clay soils; mainly East Cross Timbers s and w to w TX, locally e to Dallas Co. Jun–Oct. [*Euphorbia stictospora* Engelm.]

Chamaesyce villifera (Scheele) Small, (bearing soft hairs), HAIRY EUPHORBIA. Perennial but flowering the first year, commonly behaving as an annual; stems ascending to erect, to 40 cm tall; leaf blades entire or very minutely toothed, 3–10 mm long; styles bifid 1/2–2/3 their length. Limestone slopes, dry uplands; Bell Co.; s part of nc TX s to Edwards Plateau, also Trans-Pecos. Late May–Oct. [*Euphorbia villifera* Scheele]

CNIDOSCOLUS BULL-NETTLE, MALA MUJER

A mainly tropical American genus of ca. 75 species, often with stinging hairs. The leaves of some species are eaten as greens. (Greek: *cnide*, nettle, and *scolopes*, prickle or sting, alluding to the stinging hairs)

REFERENCE: McVaugh 1944.

Cnidoscolus texanus (Müll.Arg.) Small, (of Texas), TEXAS BULL-NETTLE, MALA MUJER, TREAD-SOFTLY. Perennial with milky sap, to 80(–100) cm tall from a large (to 20 cm thick), deep,

Chamaesyce nutans [REE, RHO]

Chamaesyce prostrata [BR2]

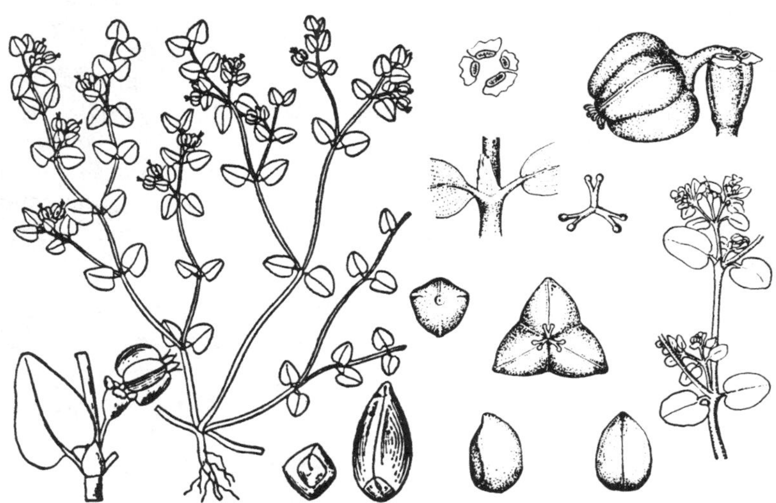

Chamaesyce serpens [BB2, RHO]

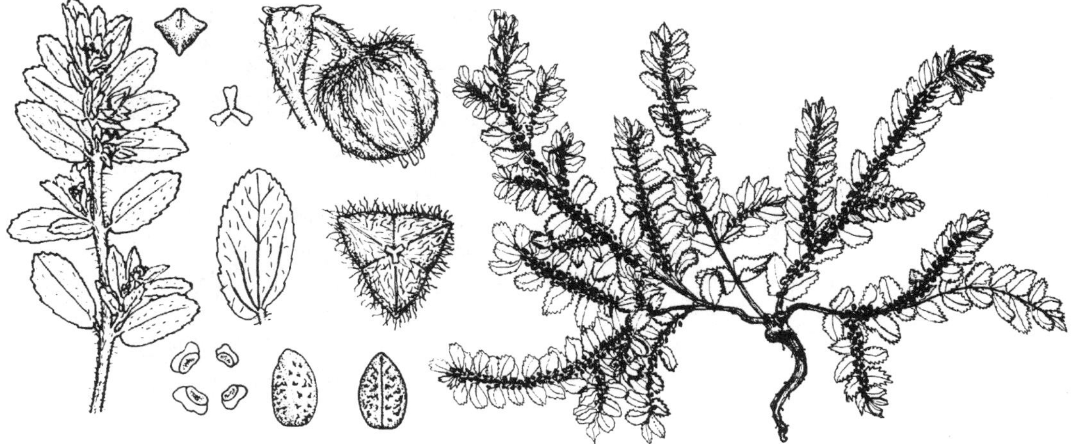

Chamaesyce stictospora [BT3, RHO]

branching root; stems densely and leaves more sparsely hispid with pale, extremely painful, stinging hairs; leaves alternate, long-petioled; leaf blades ca. 6–15 cm wide, cordate, palmately deeply 3–5-lobed, the lobes coarsely toothed or again lobed; stipules inconspicuous, narrow, toothed; flowers in terminal, cymose, pedunculate inflorescences slightly shorter to slightly longer than the leaves; male and female flowers in the same inflorescence; perianth large, white, showy, sweet-scented, of a single whorl of parts; perianth of male flowers with subcylindrical tube 15–20 mm long, slightly longer than the (4–)5 subrotate lobes; perianth of female flowers 10–17 mm long, 5-parted essentially to base; capsules 15–20 mm long, hispid, with 3 large seeds, 14–18 mm long. Sandy open woods, fields, disturbed areas; nearly throughout TX. Apr–Jul, less freely to Sep. [*Jatropha texana* Müll.Arg.] The seeds are reported to be edible; however, there are reports that they may contain some cyanide. If the foliage is touched, glass-like hairs break off in the skin and act like hypodermic needles; they release a toxin which causes an intense burning sensation; this type of effect is known as contact urticaria; the stinging hairs can penetrate even heavy clothing such as jeans; subsequent to the sting, the affected skin can be red, swollen, and irritated for a number of days. According to Lampe (1986), only four families (Euphorbiaceae, Hydrophyllaceae, Loasaceae, and Urticaceae) have stinging hairs—nc TX has stinging representatives of all of these except the Hydrophyllaceae. ☠ ▣/85

CROTON

Ours annual or perennial, often aromatic herbs or shrubs with stellate hairs or peltate scales on the epidermis of at least some parts; sap not milky; leaves alternate or subopposite, or those under the flowers apparently whorled, entire or toothed; stipules small, usually falling early; flowers solitary or in spike-like racemes or head-like clusters, unisexual, the sexes on the same or separate plants; staminate flowers with (4–)5(–6) sepals united at base; petals as many or none; stamens 5–20; pistillate flowers with 5–6(–9) sepals united at or near base; petals minute or none; pistil 1, usually 3-carpellate; fruit a 1- to usually 3-seeded, globose or subglobose capsule, or in *C. michauxii* an achene.

A huge genus (ca. 750 species) of tropical and warm areas ranging from temperate herbs to tropical rain forest trees. Oil of croton, one of the most purgative substances known, is obtained from an Old World species, *C. tiglium* L. Many contain alkaloids and are used medicinally; an Ecuadorian species is currently being investigated for medicinal uses. *Croton* species are distasteful and rarely eaten by livestock; they thus increase under overgrazing; ☠some species are toxic. The widely cultivated horticultural "crotons" with variegated foliage belong to the Euphorbiaceae genus *Codiaeum*, a group of 15 species native from Malesia to the Pacific. (Greek: *croton*, a tick; the Greek name of the CASTOR-OIL-PLANT of this family, from similarity of the seed to a tick)

REFERENCES: Ferguson 1901; Johnston 1959; Ginzbarg 1992; Webster 1992, 1993; Aplet et al. 1994.

1. Leaf blades serrate; small, distinctive, whitish, saucer-shaped gland present on each side of the undersurface of the base of the leaf blade where the petiole attaches (use hand lens) ______ **C. glandulosus**
1. Leaf blades usually entire or rarely minutely serrate; leaf glands not present.
   2. Plants shrubby, the stems woody; in nc TX known only from Coryell, Bell, and Johnson cos.
      3. Upper and lower surfaces of mature leaf blades strikingly different in appearance, the upper green and essentially glabrous (rarely a few scales present), the lower silvery-white with reddish brown tinge due to it being covered with silvery-white peltate scales (some reddish brown), the actual leaf surface not visible; leaf blades obtuse ______ **C. alabamensis**
      3. Upper and lower surfaces of mature leaf blades with stellate pubescence, more dense on lower surface, the upper surface green, the lower grayish green; leaf blades acute ______ **C. fruticulosus**
   2. Plants herbaceous, the stems not woody; widespread in nc TX.
      4. Leaf blades with markedly different types of indumentum above and below, stellate-pu-

bescent above, stellate-scaly beneath (scales silvery or brownish); leaf blades 8 mm wide or less (often considerably less); calyces ca. 1 mm long; fruit an achene to ca. 3 mm long; not collected in nc TX since 1882 ____ **C. michauxii**

4. Leaf blades with similarly coated upper and lower surfaces, both pubescent or both scaly (upper sometimes much less so than lower); leaf blades variable, often > 8 mm wide; calyces usually > 1 mm long; fruit a capsule 4–9 mm long; widespread in nc TX.
   5. Plants dioecious; all flowers without petals; leaf blades 15–35 mm long, to only 10(–12) mm wide, usually 4–5 times as long as wide ____ **C. texensis**
   5. Plants monoecious; at least the staminate flowers with petals; leaf blades usually either longer, wider, or < 4 times as long as wide.
      6. Styles 2, the ultimate branches 4; mature capsules 1-seeded; leaf blades 10–25 mm long ____ **C. monanthogynus**
      6. Styles 3, the ultimate branches 6 or more; mature capsules 3-seeded; leaf blades 10–80(–100) mm long.
         7. Ultimate style branches 6; pistillate sepals subtending mature fruits 4–5 mm long; leaf blades suborbicular (typically) to oblong-lanceolate, often obtuse at apex ____ **C. lindheimerianus**
         7. Ultimate style branches ca. 10 or more; pistillate sepals subtending mature fruits 5–10 mm long; leaf blades ovate to lance-elliptic, often acute at apex ____ **C. capitatus**

Croton alabamensis E.A. Sm. ex Chapm. var. **texensis** Ginzbarg, (sp.: of Alabama; var.: of Texas), TEXABAMA CROTON. Monoecious shrub to ca. 2 m or more tall; younger leaves and young twigs more reddish brown due to greater number of pigmented scales; inflorescence a terminal 6–14-flowered raceme. Limestone canyon slopes, in understory of forests or in full sun; endemic to Coryell and Travis cos. (narrowly endemic to nc TX and immediately adjacent area); first discovered in 1989; recently described by Ginzbarg (1992). Aplet et al. (1994) gave additional information. Feb–Mar. The only other variety, variety *alabamensis*, is known only from 3 cos. in Alabama and Tennessee. (TOES 1993: V)

Croton capitatus Michx., (capitate, headed), WOOLLY CROTON, HOGWORT. Monoecious annual herb to 0.2–1(–1.5) m tall, densely stellate tomentose; leaf blades usually acute to obtuse; flowers in terminal spike-like racemes; styles 3, two or three times 2-parted. Sandy or calcareous prairies, pastures, roadsides. Jun–Oct. While generally avoided, cattle can be poisoned by eating the plant fresh or in hay (Boughton 1931; Muenscher 1951).

1. Petioles of middle and upper leaves ca. equal; leaf blades often blunt, not strongly tapered from base; seeds uniform in color ____ var. **capitatus**
1. Petioles decreasing in length from middle of stem upward; leaf blades mostly acute, strongly tapered from base; seeds mottled ____ var. **lindheimeri**

var. **capitatus**. Dallas, Grayson, and Tarrant cos., also Denton (J. Quayle, pers. comm.), Fannin, and Clay cos. (Johnston 1959); also ne TX and Rolling Plains.

var. **lindheimeri** (Engelm. & A. Gray) Müll.Arg., (for Ferdinand Jacob Lindheimer, 1801–1879, German born TX botanist). Hunt and Denton cos., also Johnson, Kaufman, Hill, Limestone, and McLennan cos. (Johnston 1959); se and e TX w to nc TX and Edwards Plateau.

Croton fruticulosus Torr., (shrubby and dwarf), ENCINILLA, HIERBA LOCA. Monoecious shrub to ca. 1 m tall; flowers in terminal racemes. In brush on limestone uplands; Bell Co., also Johnson (R. O'Kennon, pers. obs.) and Williamson (Johnston 1959) cos. and Fort Hood (Bell or Coryell cos.—Sanchez 1997); w part of nc TX s to Edwards Plateau, also Trans-Pecos. May–fall.

Croton glandulosus L., (glandular), TROPIC CROTON. Monoecious annual, normally erect, to 60 cm tall, widely branched, stellate-pubescent; gland present on each side of the undersurface of the base of the leaf blade where the petiole attaches; capsules 4.5–5.5 mm long. Stream bottoms or disturbed habitats. Jun–Oct.

1. Plants ca. 10–20 cm tall; stellate hairs on stem with central branch usually shorter than radials; larger leaves about 2.5 cm long; glands at base of leaf blade 0.1–0.4 mm thick apically ______ var. **lindheimeri**
1. Plants usually > 25 cm tall; stellate hairs on stem with a long, erect, central branch usually longer than radials; larger leaves over 3 cm long; glands at base of leaf blade 0.5–0.8 mm thick apically ______ var. **septentrionalis**

var. **lindheimeri** Müll.Arg., (for Ferdinand Jacob Lindheimer, 1801–1879, German born TX botanist), LINDHEIMER'S CROTON. Clay and Tarrant cos., also Bell, Erath, and Parker cos. (Johnston 1959); widespread in TX.

var. **septentrionalis** Müll.Arg., (northern), NORTHERN CROTON. Se and e TX w to Rolling Plains and e Edwards Plateau.

Croton lindheimerianus Scheele, (for Ferdinand Jacob Lindheimer, 1801–1879, German born TX botanist), THREE-SEED CROTON. Monoecious annual; leaf blades densely stellate pubescent, thick, usually obtuse to acute; the 3 styles bifid to base. Sandy ground; widespread but local over much of TX, in nc TX mainly in West Cross Timbers (Archer, Clay, Eastland, Jack, Palo Pinto, and Young cos.), also to the e in Grayson Co. (RR yard). Jul–Oct.

Croton michauxii G.L. Webster, (for André Michaux, 1746–1803, French botanist and explorer of North America), NARROW-LEAF RUSHFOIL. Erect, slender, loosely branched, monoecious annual to 60(–90) cm tall; stems coated with stellate hairs, the rays of which are united at the base and therefore scale-like, silvery or brownish; leaves alternate or subopposite, very short-petioled; leaf blades linear-lanceolate to oblong-elliptic; flowers minute, unisexual, both sexes on the same plant; staminate flowers: in slender, short, terminal, minutely bracted spikes; sepals 3–5; petals as many, narrower, white; pistillate flowers: mostly solitary at base of staminate spikes; calyces unequally 3–5-lobed; petals absent; style branches 3, inconspicuously bifrucate near apex; fruit an achene 2.5–3 mm long. Sandy open areas; collected at Dallas by Reverchon in 1882; mainly e TX. Jun–Oct. This taxon, previously treated as [*Crotonopsis linearis* Michx.], was recently lumped into the genus *Croton* (Webster 1992). According to Webster (1992), "... the main generic character of *Crotonopsis*—unicarpellate gynoecium and indehiscent fruit—represents merely the end-point in a reduction series from the 3-carpellate gynoecium of most *Croton* species through the 2-carpellate gynoecium of *Croton monanthogynus* Michx. to the 1-carpellate gynoecium of *Crotonopsis*. Consequetly, there appears to be a much stronger case for treating *Crotonopsis* as a section of *Croton* than as an independent genus." [not *C. linearis* Jacq.]

Croton monanthogynus Michx., (one female flower), DOVEWEED, PRAIRIE-TEA, ONE-SEED CROTON. Closely stellate pubescent monoecious annual to 50 cm tall, widely branched; leaf blades rhombic-ovate to elliptic or lanceolate; capsules ca. 4 mm long. Rocky or eroding ground, roadsides and waste places, particularly calcareous soils; throughout most of TX. Jun–Nov.

Croton texensis (Klotzsch) Müll.Arg., (of Texas), TEXAS CROTON, SKUNKWEED. Dioecious annual herb 20–80 cm tall; leaf blades linear-lanceolate to nearly ovate-oblong, usually 15–35 mm long, (3–)4–5(–6) times as long as wide, to ca. 10(–12) mm wide, densely stellate pubescent, more so below; calyces 2–4 mm across; capsules 4–6 mm long, globose to globose-ovoid, often slightly warty, stellate-tomentose. Sand or sandy loam; widespread in TX. Jun–Oct. While generally avoided, cattle can be poisoned by eating the plant fresh or in hay (Muenscher 1951). ☠

DITAXIS WILD MERCURY

Perennials, herbaceous from a woody crown, with pubescence of hairs attached at the middle (= malpighian); sap not milky; foliage sometimes darkening upon drying; leaves alternate; leaf

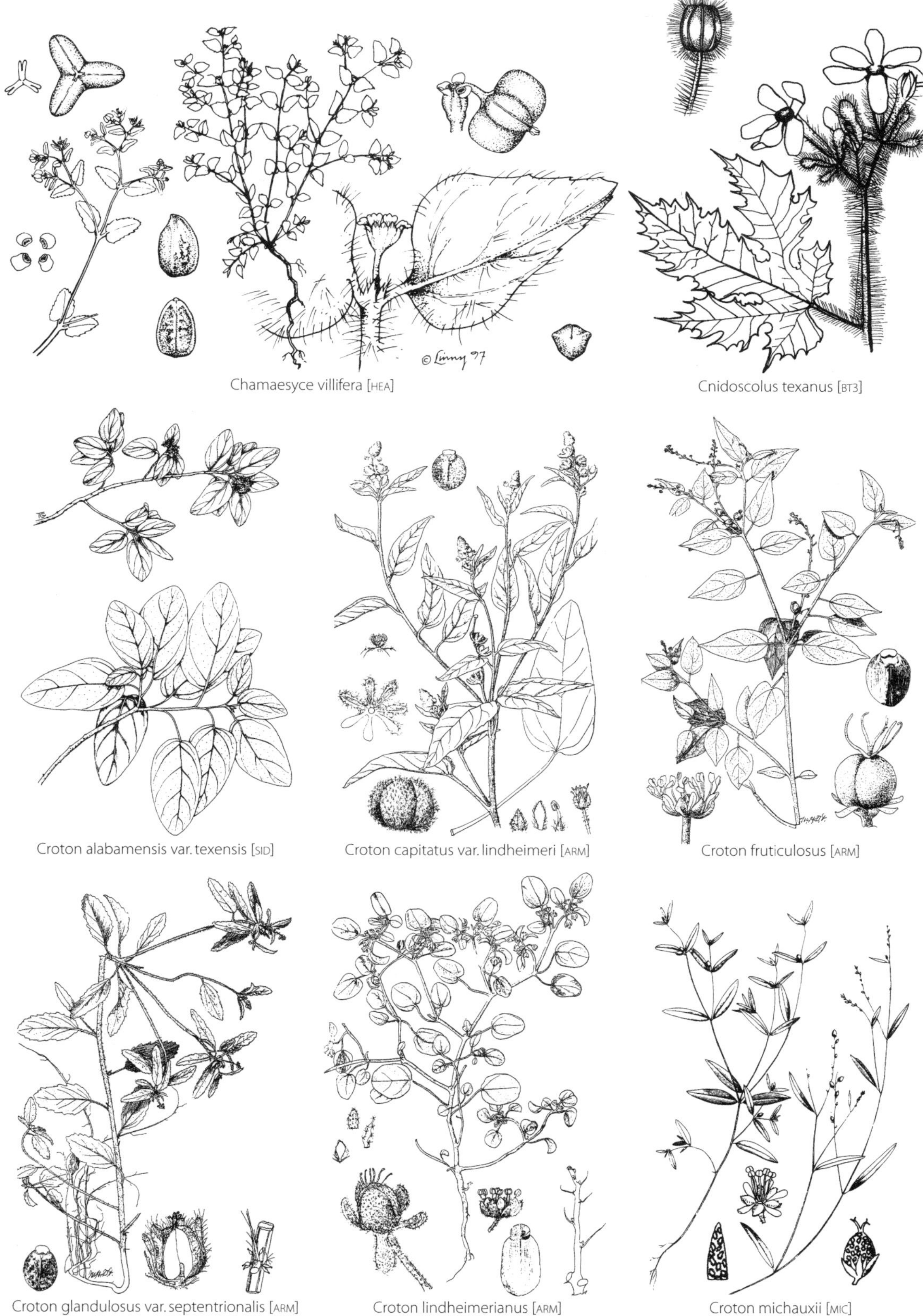

Chamaesyce villifera [HEA]

Cnidoscolus texanus [BT3]

Croton alabamensis var. texensis [SID]

Croton capitatus var. lindheimeri [ARM]

Croton fruticulosus [ARM]

Croton glandulosus var. septentrionalis [ARM]

Croton lindheimerianus [ARM]

Croton michauxii [MIC]

blades entire; inflorescences racemose, axillary; flowers unisexual; staminate flowers: sepals 5; petals 5; glands 5; stamens 7–10, the filaments coherent; pistillate flowers: sepals 5; petals 5 or absent; glands 5, ovary 3-carpellate; ovules 1 per cell; styles 3, bifid; capsules 3-seeded; seeds ecarunculate.

An American genus of 40–50 species ranging from the U.S. to Argentina (Webster 1994a). The species have often been treated in a broadly circumscribed *Argythamnia* (e.g., Correll & Johnston 1970; Ingram 1980; Hatch et al. 1990; Kartesz 1994). However, we are following Webster (1994b) who considered *Ditaxis* a distinct genus, as did Takahashi et al. (1995) and Steinmann and Felger (1997). According to Takahashi et al. (1995), *Ditaxis* "... has pollen that can be easily distinguished from all remaining Euphorbiaceae (and probably from all dicotyledons) by the unusual aperture configuration and the shape of the grain." If treated broadly to include *Chiropetalum* and *Ditaxis*, *Argythamnia* is an American, mainly tropical genus of 78–88 species. (Greek: *di*, two, separate, and *taxis*, arrangement, presumably in reference to the separate male and female flowers)

REFERENCES: Shinners 1956c; Ingram 1967, 1980; Waterfall 1971 [1972]; Webster 1994b; Takahashi et al. 1995.

1. Inflorescences shorter than leaves; plants with much-branched, trailing or spreading stems _______ **D. humilis**
1. Inflorescences longer than leaves; plants with several, unbranched, erect to ascending stems.
   2. Glands of pistillate and staminate flowers truncate, square to rectangular _______ **D. aphoroides**
   2. Glands of pistillate and staminate flowers acute, linear _______ **D. mercurialina**

Ditaxis aphoroides (Müll.Arg.) Pax, (resembling *Aphora*, a generic synonym used by Nuttall), HILL COUNTRY WILD MERCURY, SHRUBBY DITAXIS. Plant densely villous; stems usually erect; leaves elliptic to ovate, sessile above, veins prominent below; usually dioecious; staminate flowers: petals obovate to cuneate, ca. 4–5 mm long; glands wider than long, truncate; stamens 8-10; pistillate flowers: petals absent or rudimentary; glands thick, squarish, truncate; styles erect, bifid half the free length. Sandy or rocky limestone soils; Brown Co., also Mills Co. (TOES 1993); also on Edwards Plateau in Blanco, Gillespie and Kerr cos., also Bandera, Kimble (HPC), Hays, Kendall (Mahler 1988), Comal, and Uvalde (R. Roberts, pers. comm.) cos.; current populations not known from Brown, Hays, or Kendall cos. (Mahler 1988); Reverchon's collection from Williams Ranch (reportedly in Brown Co.—Apr 1882) represents the most n collection; narrowly endemic to Edwards Plateau and the sw part of nc TX. Mar–Apr. [*Argythamnia aphoroides* Müll.Arg.] (TOES 1993: V) ⚠

Ditaxis humilis (Engelm. & A. Gray) Pax var. **humilis**, (low-growing, dwarf), LOW WILD MERCURY, LOW DITAXIS. Stems usually trailing or spreading; leaves sparsely to densely villous, elliptic to narrowly obovate, sessile or nearly so above, with veins prominent on lower surface; monoecious; staminate flowers: petals narrowly lanceolate to oblanceolate; glands linear; stamens 8 or 9; pistillate flowers: petals to 0.5 mm long or absent; glands linear; styles erect and bifid half the free length. Rocky or disturbed calcareous clay soils; throughout much of TX, mainly Blackland Prairie s and w to w TX. Apr–Oct. [*Argythamnia humilis* (Engelm. & A. Gray) Müll.Arg.]

var. *leiosperma* Waterfall, (smooth-seeded), was named in the genus *Argythamnia* by Waterfall (1971 [1972]) apparently based on a single seed character (the name has apparently not been recognized in *Ditaxis*). It seems better lumped with var. *humilis*. Kartesz (1994) recognized the variety but it was not recognized by Jones et al. (1997) nor more recently by Kartesz (pers. comm., 1997). The two can be separated as follows:

1. Seeds with distinctly rugose-roughened areas _______ var. *humilis*
1. Seeds basically smooth, with several whitish slighly roughened encircling bands _______ var. *leiosperma*

var. *laevis* (Torr.) A. Heller, (smooth), SMOOTH WILD MERCURY, SMOOTH DITAXIS, with foliage en-

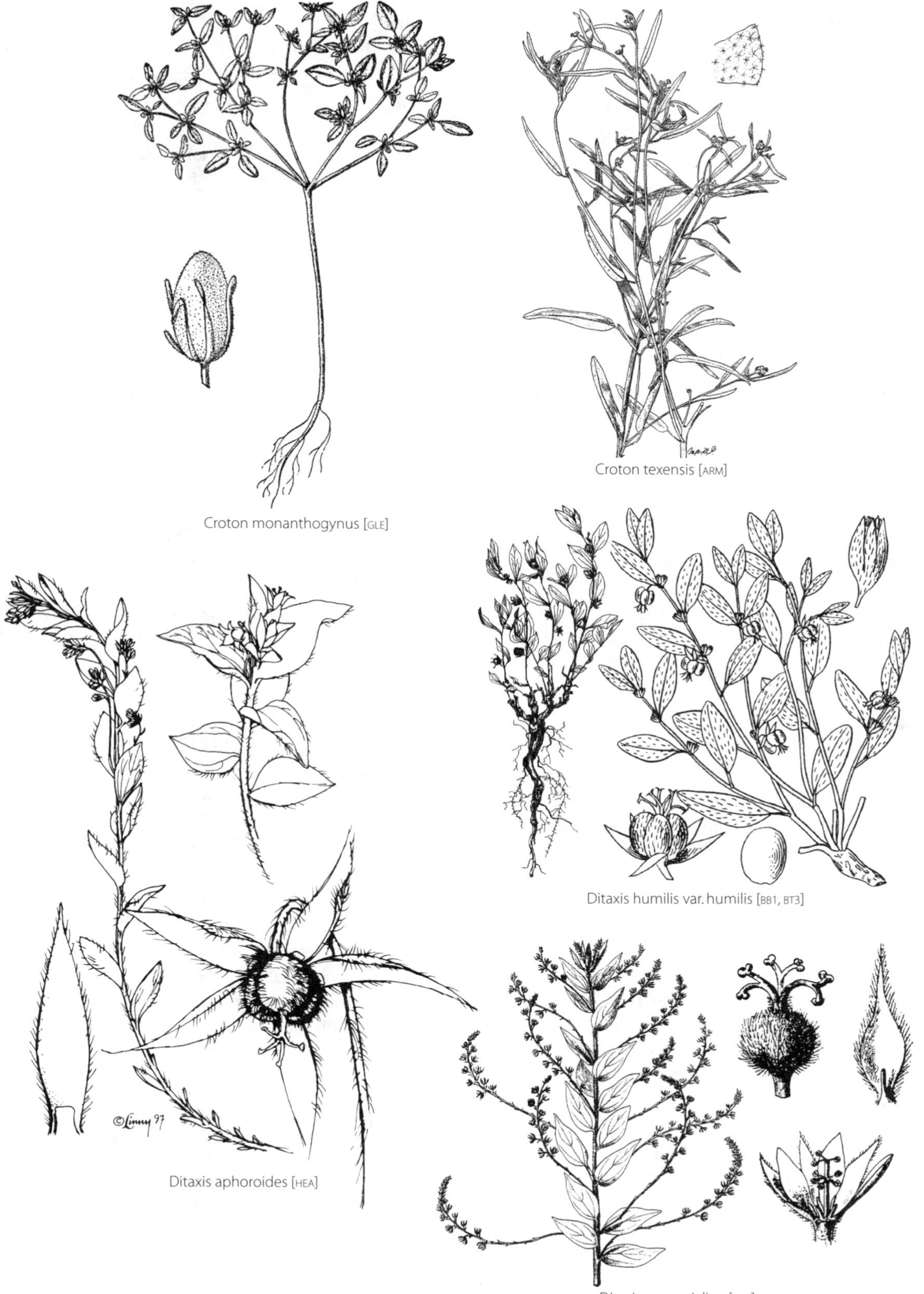

Croton monanthogynus [GLE]

Croton texensis [ARM]

Ditaxis humilis var. humilis [BB1, BT3]

Ditaxis aphoroides [HEA]

Ditaxis mercurialina [EN1]

tirely glabrous, rather succulent, and becoming brittle upon drying, is known from the Edwards Plateau and Trans-Pecos. [*Argythamnia humilis* var. *laevis* (Torr.) Shinners]

Ditaxis mercurialina (Nutt.) J.M.Coult., (Mercury, messenger of the gods), TALL WILD MERCURY, TALL DITAXIS. Plant dark green, villous; monoecious; staminate flowers: petals oblanceolate to broadly so, ca. 3 mm long; glands linear; stamens usually 8; pistillate flowers: petals usually absent; glands linear; styles spreading, bifid near apex only. Sandy or rocky limestone soils, prairies; Bell, Collin, Dallas, Grayson, and Tarrant cos.; nc TX w to Rolling Plains and sw to Edwards Plateau and Trans-Pecos. Apr–Jun. [*Aphora mercurialina* Nutt., *Argythamnia mercurialina* (Nutt.) Müll.Arg.]

EUPHORBIA SPURGE

Annuals or perennials with milky sap; leaves alternate or opposite; stipules absent, or represented by minute glands; inflorescences dichotomously or trichotomously branched, with alternate or whorled leafy bracts; cyathia and fruits as in *Chamaesyce*—flowers greatly reduced, minute, unisexual (consisting only of a single pistil or a single stamen on a short pedicel), in small cup-like involucres (= cyathia); each cyathium containing one pistillate and several staminate flowers mixed with minute bracts; the cyathia also have fleshy glands on their rims; these glands in some species have small, petal-like gland-appendages—the cyanthium is then termed a pseudanthium because of the resemblance to standard flowers; pistillate pedicel elongating in age, the 3-locular capsule exserted from the involucre. In the past there has not always been clarity in the terms used for the measurement of glands and gland-appendages. We are following terminology suggested by M. Mayfield (pers. comm.)—in the key and descriptions the width of the gland signifies the measurement in the tangential dimension relative to the cyathium; the length of the gland signifies the measurement in the radial dimension relative to the cyathium; the length of the gland-appendage signifies the measurement in the radial dimension relative to the cyathium—this means from attachment at the gland to the apex of the gland-appendage.

When considered broadly, *Euphorbia* is a huge, cosmopolitan, especially warm area genus of ca. 2,000 species—the second largest genus of flowering plants; we are treating the segregate, *Chamaesyce*, with 300+ species, as a distinct genus. Euphorbias are basically monoecious or dioecious herbs, succulents, shrubs, or trees with milky latex. Many are important ornamentals including *E. pulcherrima* Willd. ex Klotzsch (POINSETTIA, named for Joel Poinsett, the first U.S. minister to Mexico, who imported the plant into the U.S. in 1829—Spoerke & Smolinske 1990); the inflorescences are sometimes surrounded by colorful bracts (these often large and photosynthetic and possibly better referred to as leaves). Some Old World species superficially resemble Cactaceae and function there as ecological equivalents of that family. ☠ As in *Chamaesyce*, the latex of all species is apparently toxic and can cause dermatitis or other reactions in sensitive individuals; inflammation and large blisters can develop; the eyes are particularly sensitive. While livestock are sometimes accidentally poisoned, animals generally will not eat the plants; the toxins are complex terpenes such as euphorbol; some are suspected of acting as co-carcinogens (= promote the action of "sub"-carcinogenic doses of known carcinogens); ingestion of euphorbias can produce severe gastric distress; e.g., POINSETTIA has caused death in children (Wheeler 1941; Muenscher 1951; Kingsbury 1964, 1965; Kinghorn 1979; Lampe & McCann 1985; Lampe 1986). *Euphorbia antisyphilitica* Zucc., CANDELILLA, of the Trans-Pecos, is the source of Candelilla wax, one of the the highest quality waxes known. The wax is obtained by boiling the stems in water and dilute acid and then scooping off the floating wax. Unfortunately, in some areas populations of this species have been greatly reduced (Powell 1988). (Named for Euphorbus, 1st century AD physician to King Juba of Mauritania, who used latex from some species for medicinal purposes)

There is no general agreement on whether *Euphorbia* should be recognized as one large genus or divided into a number of segregates. Subgenus *Chamaesyce* is now frequently treated as a separate genus—as done here following Kartesz (1994), Webster (1994b), G. Webster (pers. comm. 1997), and M. Mayfield (pers. comm.). Four nc TX species, *E. longicruris, E. roemeriana, E. spathulata, and E. tetrapora*, are sometimes treated in the genus *Tithymalus*. They have flat or convex involucral glands and bracts forming a whorl beneath the umbelliform, 3–5 rayed, symmetrical inflorescence (= pleiochasium). *Euphorbia cyathophora, E. davidii*, and *E. dentata* are segregated by some authors into the genus *Poinsettia*, and are characterized by usually having a solitary involucral gland without appendges. The four nc TX species with usually petal-like gland-appendages, *E. bicolor, E. corollata, E. hexagona*, and *E. marginata*, are in subgenus *Agaloma*, which could likewise be recognized at the generic level.
REFERENCES: Norton 1900; Wheeler 1936, 1941; Krochmal 1952; Richardson 1968; Johnston 1975; Subils 1984; Park 1998.

1. Leaves sessile or with narrowed, winged, petiolar base (leaves sometimes absent from flowering plants).
    2. Involucres with yellow-green to brownish glands but without conspicuous gland-appendages; stem leaves widest near apex.
        3. Leaves entire; fruits not warty; involucral glands with 2 pointed, horn-like appendages.
            4. Uppermost leafy floral bracts partially fused, sometimes to nearly half their length (thus appearing perfoliate) ____ **E. roemeriana**
            4. Uppermost leafy floral bracts essentially free to their base.
                5. Uppermost leafy floral bracts bluntly triangular, nearly as long as wide or longer; inflorescences usually appearing relatively open; seeds with relatively few pits in rows to nearly smooth; plants at maturity usually green in color; growing in sandy soil ____ **E. tetrapora**
                5. Uppermost leafy floral bracts suborbicular or reniform, wider than long; inflorescences usually appearing very crowded; seeds with numerous small pits ± evenly distributed over the seed surface; plants at maturity tannish or coppery in color; growing in limestone areas ____ **E. longicruris**
        3. Leaves finely toothed, at least around apex; fruits warty; involucral glands rounded, entire ____ **E. spathulata**
    2. Involucres with white gland-appendages, these usually conspicuous and petal-like (but very small and whitish to greenish white in *E. hexagona*); stem leaves mostly widest near middle or of uniform width.
        6. Stem leaves opposite, 1–6 mm wide; bracts never white-margined; gland-appendages small, 0.5 mm or less long (in radial dimension—see genus description) ____ **E. hexagona**
        6. Stem leaves alternate (can be opposite or whorled just below inflorescence), usually > 6 mm wide; bracts white-margined OR not so; gland-appendages conspicuous, petal-like, 1.5–4 mm long.
            7. Floral bracts and upper leaves not white-margined (leaves sometimes absent from flowering plants); middle and upper stem leaves usually 12 mm wide or less; capsules glabrous ____ **E. corollata**
            7. Floral bracts and upper leaves prominently white-margined; middle and upper stem leaves usually > 12 mm wide; capsules pubescent.
                8. Uppermost (just below inflorescence), white-margined, leaf-like bracts linear, usually 5–10 times as long as broad; main stem leaves lanceolate to narrowly elliptic, usually 3–5 times as long as wide, pubescent ____ **E. bicolor**
                8. Uppermost, white-margined, leaf-like bracts narrowly lanceolate to ovate, < 4 times as long as broad; main stem leaves elliptic, oblong, or ovate, usually < 2.5 times as long as wide, often glabrous ____ **E. marginata**

1. Leaves with distinct petioles 1–25 mm long.
   9. Involucral glands usually solitary, lacking appendages; larger leaf blades usually toothed or lobed; upper stem leaves with linear to narrowly lanceolate, elliptic, or broadly ovate blades usually (2–)4–45 mm wide; petioles 2–25 mm long.
      10. Main stem leaves all or mostly alternate, essentially glabrous above; upper leaves sometimes fiddle-shaped ____ **E. cyathophora**
      10. Main stem leaves all or mostly opposite, pubescent above; upper leaves not fiddle-shaped.
         11. Leaf blades usually widest at the middle, linear-elliptic to broadly elliptic; petioles often nearly as long as the blades at midstem; trichomes of the lower leaf surfaces stiff, strongly tapered with a broad basal cell; seeds angular in cross-section, unevenly tuberculate, 2.0–2.7 mm long ____ **E. davidii**
         11. Leaf blades usually widest below the middle, lanceolate to trullate (= trowel-shaped); petioles not usually more than a third as long as the blade at midstem; trichomes of the lower leaf surfaces weak, filiform, lacking a broad basal cell; seeds rounded in cross-section, evenly tuberculate, 1.7–2.0(–2.2) mm long ____ **E. dentata**
   9. Involucral glands 4 or 5, with conspicuous or inconspicuous, white or pinkish gland-appendages; larger leaf blades entire; upper stem leaves with linear to lanceolate or narrowly oblong blades 1–6 mm wide; petioles 1–3.5 mm long
      12. Involucres with 4 glands; gland-appendages often conspicuous, white or pinkish, 0.5–2.5 mm long (in radial dimension—see genus description); capsules 2–2.5 mm long; stipules 1–1.5 mm long; stems several from at or near base (this *Chamaesyce* species often has the leaf blades symmetrical basally and is thus included in the *Euphorbia* key to prevent confusion) ____ see **Chamaesyce missurica**
      12. Involucres with 5 glands; gland-appendages very small and inconspicuous, 0.5 mm or less long, whitish or greenish white; capsules 3–5 mm long; stipules absent or represented by minute glands; stem 1 from base ____ **E. hexagona**

Euphorbia bicolor Engelm. & A. Gray, (two-colored), SNOW-ON-THE-PRAIRIE. Erect, softly pilose annual to 130 cm tall; middle stem leaves 3–5 times as long as wide; involucral glands usually 5; gland-appendages petal-like, 2–3 mm long. Prairies, roadsides, and waste ground; mainly Grand Prairie and Blackland Prairie e to e TX, also Montague Co., in nc TX sw to Johnson Co. While generally quite distinct, in the vicinity of Johnson Co. (and probably other areas as well) where the ranges of *E. bicolor* and *E. marginata* overlap, intermediate individuals can be found. Jul–Oct. The fruits are ballistic (= seeds thrown catapult-like from the fruits). The milky sap is caustic and can cause a skin rash (M. Mayfield, pers. comm.). ☠

Euphorbia corollata L., (corolla-like), FLOWERING SPURGE, TRAMP'S SPURGE. Perennial with solitary or few stems to 90 cm tall; leaves alternate, linear to elliptic-oblong; involucral glands 5, cupped; gland-appendages petal-like, 1.5–4 mm long. Clay soils; Collin, Cooke, Dallas, Fannin, Grayson, Lamar, and Montague cos.; e TX w to n part of nc TX, May–Sep. The rootstock reportedly contains a poisonous substance, euphorbon (Burlage 1968). ☠

Euphorbia cyathophora Murray, (cup-bearing), WILD POINSETTIA, FIRE-ON-THE-MOUNTAIN, PAINTED EUPHORBIA, PAINTED SPURGE, PAINTEDLEAF. Nearly glabrous annual to 75 cm tall; leaf blades extremely variable in shape, elliptic to oblong-ovate or ovate, varying to linear-lanceolate or even linear, sometimes lobed and appearing fiddle-shaped, usually finely toothed to entire, the uppermost leaves and the floral bracts usually rosy or rosy and whitish at base; involucral glands usually solitary, cupped, without appendages. Stream banks, open areas; Bell, Dallas, Grayson, Hunt, and Kaufman cos., also Denton (Mahler 1988) and Tarrant (F. Norris, pers. comm.) cos.; widely scattered nearly throughout TX. May–Sep.

Euphorbia davidii Subils, (for David L. Anderson of Villa Mercedes, Argentina, who in 1984 col-

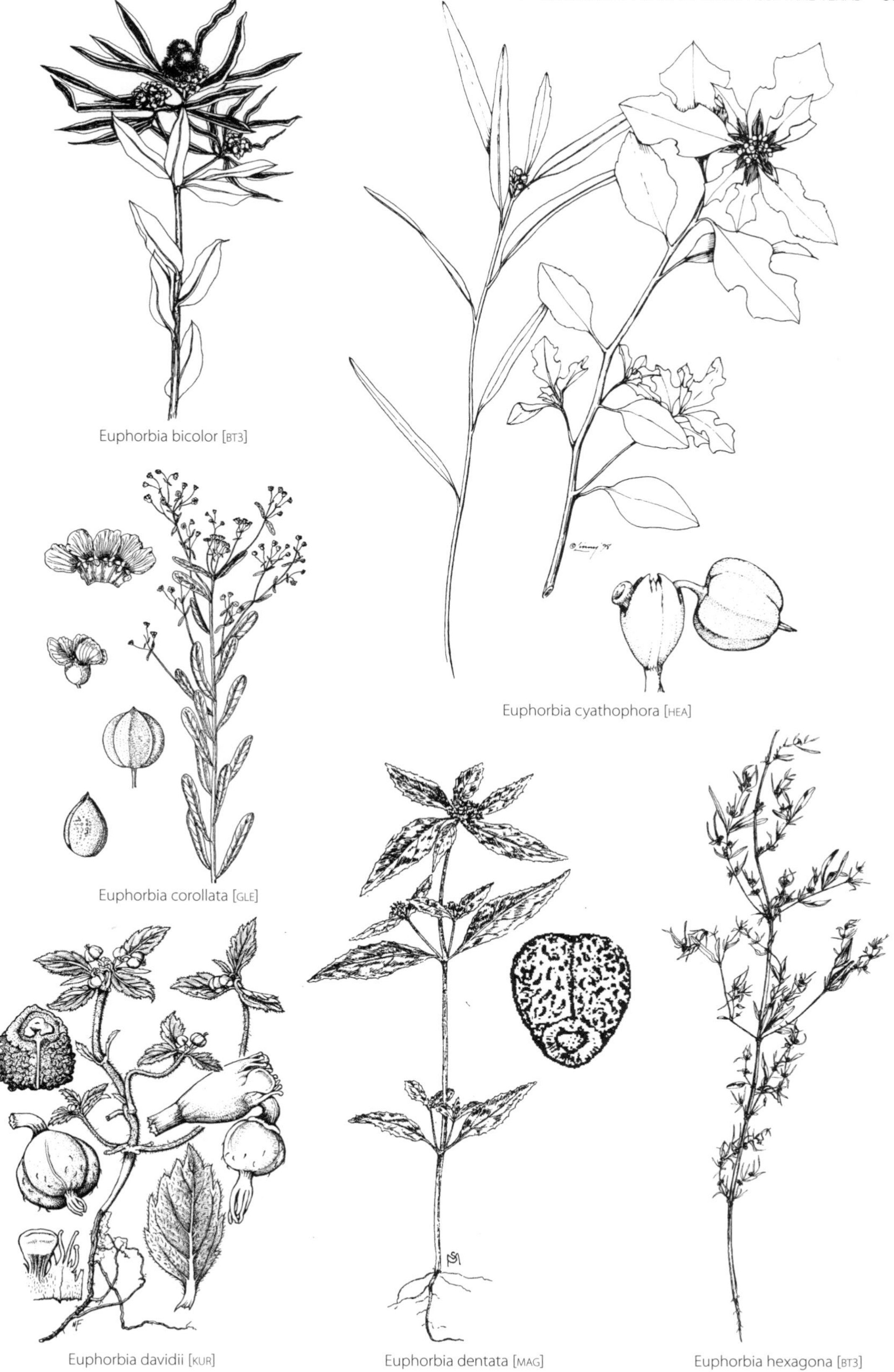

Euphorbia bicolor [BT3]

Euphorbia cyathophora [HEA]

Euphorbia corollata [GLE]

Euphorbia davidii [KUR]

Euphorbia dentata [MAG]

Euphorbia hexagona [BT3]

lected one of the paratypes). Similar to *E. dentata* but more robust, with hairs on the upper stems and young growth coarse, slightly recurved, and stiff (not to the touch so much as to the eye); involucral glands usually solitary, cupped, without appendages; capsules somewhat ovoid or broader at the base, tapering to the apex; seeds angular-ovoid with unevenly and roughly sculptured surface, with a suggestion of 2 transverse ridges in well-formed seeds on the dorsal surface, 2.2–2.7 mm long, with caruncle 1.0–1.4 mm wide; $n = 28$ (tetraploid). Disturbed sites; Eastland Co. (TEX—M. Mayfield, pers. comm.). Specimens of this species have long been misidentified as the rather similar and more common *E. dentata*. Consequently, its range within TX is unclear but it is probably more common in w TX; according to M. Mayfield (pers. comm.), it is to be expected as a weed in other parts of nc TX area. This description and the dichotomy in the key separating *E. davidii* and *E. dentata* are from M. Mayfield (pers. comm.).

Euphorbia dentata Michx., (toothed), TOOTHED SPURGE. Rather sparsely pubescent annual to 75 cm tall, with long weak hairs on the young growth and abaxial leaf surfaces; leaf blades narrowly oblong-lanceolate to rhombic-ovate, rather coarsely and shallowly toothed; involucral glands usually solitary, cupped, without appendages; capsules smoothly rounded, depressed spherical; seeds very rounded to roundly ovoid, the dorsal surface evenly tuberculate with low sculpturing, 1.7–2.0(–2.2) mm long with caruncle ca. 0.8 mm or less wide; $n = 14$ (diploid). Stream bottoms, fields, roadsides and disturbed sites; throughout TX. May–Oct, commonly two generations in one year.

Euphorbia hexagona Nutt. ex Spreng., (hexagonal, six-angled), GREEN SPURGE, SIX-ANGLE EUPHORBIA. Minutely and sparsely pubescent annual, normally erect, to 80 cm tall; leaves opposite, entire; involucral glands 5, with small, inconspicuous gland-appendages. Loose sandy open areas, stream banks or bottoms; Grayson, Somervell, and Tarrant cos.; nc TX and Plains Country. Jun–Oct.

Euphorbia longicruris Scheele, (long-legged), WEDGE-LEAF EUPHORBIA. Plant to 25 cm tall; leaves obovate with long-tapering base; branches or their internodes mostly remaining short, usually not exceeding the very wide floral bracts until late fruiting stage (the floral bracts are upright or ascending and enclose the involucres); involucral glands 4, cresent-shaped, with an erect horn at each end, without appendages. Limestone outcrops; Bell, Brown, Coryell, Tarrant, Wise, and Young cos., also collected in 1880 in Navarro Co.; Edwards Plateau and e Plains Country n and e to nc TX. Late Mar–early Jun.

Euphorbia marginata Pursh, (margined), SNOW-ON-THE-MOUNTAIN. Pilose to glabrous annual to ca. 200 cm tall; involucral glands usually 5, cupped; gland-appendages petal-like, 2–4 mm long. Calcareous uplands, stream bottoms, and low areas; w TX e to a line from Bell to Cooke cos., cultivated further e. Jul–Oct. Reported to produce evil-tasting poisonous honey and to be toxic to livestock; the sap causes skin eruptions and has been used to brand cattle in preference to a hot iron (Kingsbury 1964, 1965; Burlage 1968); care should thus be taken to avoid the sap. ☠ ▣/90

Euphorbia roemeriana Scheele, (for Ferdinand Roemer, 1818–1891, geologist, paleontologist, and explorer of TX), ROEMER'S SPURGE, ROEMER'S EUPHORBIA. Annual to 30 cm tall; leafy bracts of the floral branches paired, opposite, semi-circular, partially fused—sometimes to nearly half their length; involucral glands with an erect horn at each end, without appendages. Calcareous soils; Bosque and Montague cos. (Mahler 1988), also Williamson Co. (M. Mayfield, pers. comm.); rare in nc TX, mainly Edwards Plateau; endemic to TX. Spring. Mark Mayfield (pers. comm.) indicated that this species occurs mainly to the s of nc TX and that the Montague Co. citation is probably erroneous.

Euphorbia spathulata Lam., (spoon-shaped), WARTY EUPHORBIA. Glabrous annual to 55 cm tall; leaves numerous, oblanceolate, obtuse; leafy floral bracts mostly broader than long; involucral

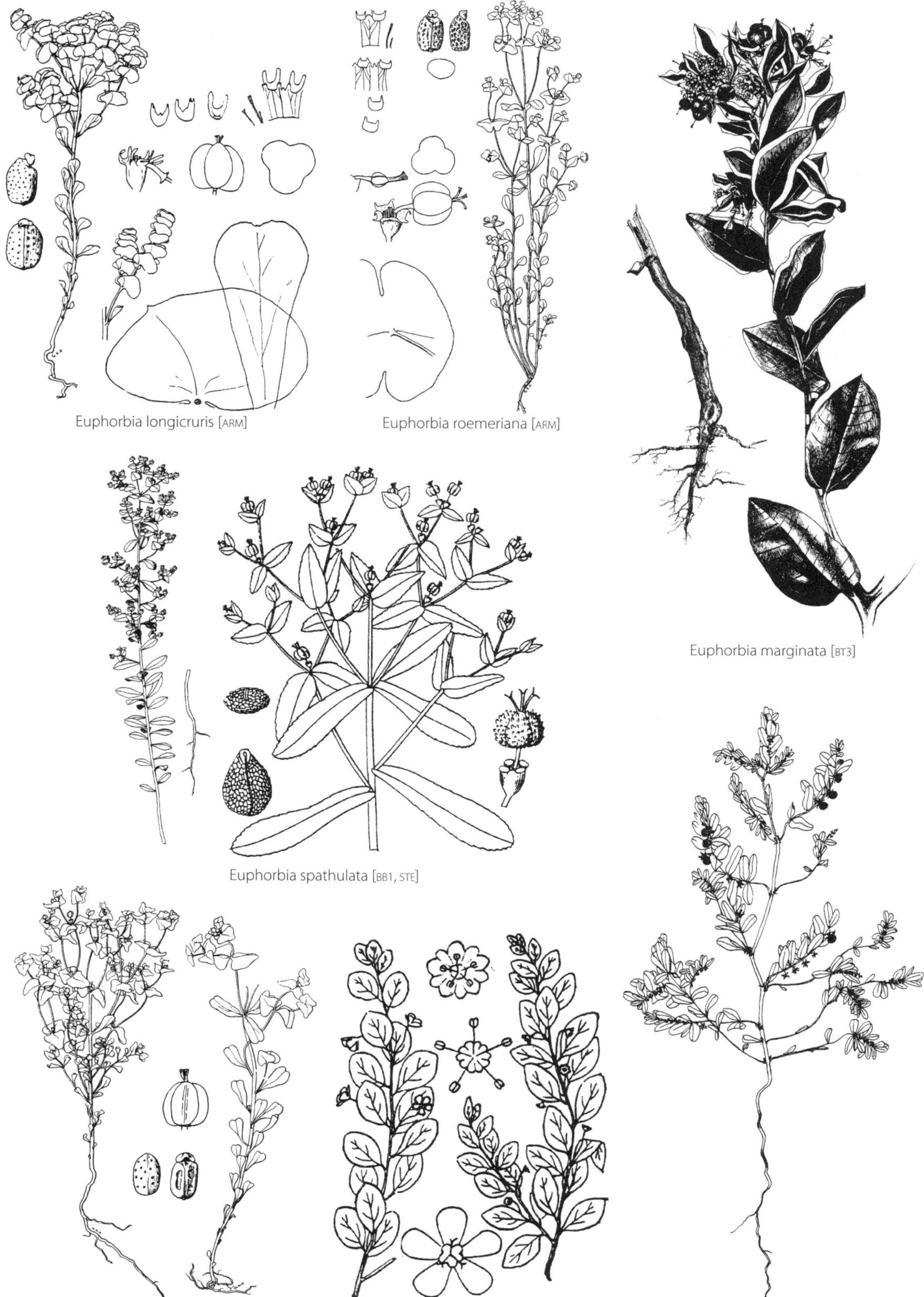

Euphorbia longicruris [ARM]

Euphorbia roemeriana [ARM]

Euphorbia marginata [BT3]

Euphorbia spathulata [BB1, STE]

Euphorbia tetrapora [ARM]

Leptopus phyllanthoides [BB2]

Phyllanthus abnormis [BT3]

glands 4, minute, without horns or appendages; style branches about as long as the unbranched portion. Prairies, disturbed areas; nearly throughout TX. Late Mar–May.

Euphorbia tetrapora Engelm., (four-pored), WEAK EUPHORBIA. Slender, to 20 cm tall; leaves narrowly oblanceolate; branches of the inflorescence or their internodes longer than the floral bracts; involucral glands 4, with an erect horn at each end, without appendages. Sandy woods and fields; e TX w to West Cross Timbers. Late Mar–May.

LEPTOPUS

A genus of ca. 10 species widely scattered in the Old World and North America (Webster 1994b); previously submerged in *Andrachne* (e.g., Correll & Johnston 1970; Kartesz 1994; Jones et al. 1997). Webster (1994b) indicated that *Leptopus* differs from *Andrachne* (sensu stricto) in having anatropous ovules and that it represents a connecting link between *Andrachne* and the small tropical American genus *Astrocasia*. He assigned it to a separate subtribe in the tribe Phyllantheae. (Greek: *lepto*, fine or slender, and *pus*, foot, presumably meaning with slender or thin stalks)
REFERENCE: Webster 1994b.

Leptopus phyllanthoides (Nutt.) G.L. Webster, (resembling *Phyllanthus*—leaf flower), MAIDENBUSH. Woody-based perennial or small shrub to 1 m tall; sap not milky; leaves numerous, alternate, subsessile; leaf blades entire, sparsely pilose or glabrous; stipules oblong-lanceolate; staminate and pistillate flowers usually on separate plants; perianth with both sepals and petals; capsules 4–5 mm long, 7–7.5 mm across, with 2 seeds in each of the 3 cells. Ravines or stream banks on limestone; collected by Reverchon in Johnson Co., also McLennan Co. (Mahler 1988); nc TX s to Edwards Plateau. May–Jun. [*Andrachne phyllanthoides* (Nutt.) J.M. Coult., *Savia phyllanthoides* var. *reverchonii* (J.M. Coult.) Pax & K. Hoffm.] We are following Webster (1994b) in treating this species in *Leptopus*.

PHYLLANTHUS LEAF-FLOWER

Ours low, glabrous annuals or perennials; sap not milky; leaves many, rather crowded, alternate, short-petioled or subsessile, entire; stipules narrowly triangular-lanceolate; flowers unisexual, both sexes usually on the same plant; petals absent; fruit a usually ballistically dehiscent capsule; seeds usually 2 in each of the 3 locules.

A genus of ca. 600 species of herbs, shrubs, and trees of tropical and warm areas; some are used ornamentally, others medicinally. (Greek: *phyllon*, leaf, and *anthos*, flower, because the flowers in a few species are borne upon leaf-like dilated branches)
REFERENCES: Webster 1956, 1970.

1. Leaves spirally arranged; plants perennial with base of plant and root becoming thickened and woody ____ **P. polygonoides**
1. Leaves distichous (= in 2 ranks); plants annual.
   2. Pedicels of female flowers (3–)4–7 mm long; stamens 5; introduced weed known in nc TX only from Tarrant Co. ____ **P. tenellus**
   2. Pedicels of female flowers < 0.5–3(–3.5) mm long; stamens 2–3; native, including species widespread in nc TX.
      3. Main stems with leaves and flowers; stamens 3; filaments free; seeds verruculose (= with extremely minute warts—use dissecting scope) ____ **P. caroliniensis**
      3. Main stems with leaves reduced to scales, the leaves and flowers only on specialized deciduous side branches; stamens 2 or 3; filaments entirely connate; seeds not verruculose.
         4. Leaves not minutely rough-scabrid beneath to the touch; female flowers with pedicels 1–3(–3.5) mm long, distal on the branches, the male flowers proximal; stamens usually 2

(3 in largest flowers); seeds striatulate (= with very fine lines), not transversely ribbed; pistillate calyx lobes ovate to obovate; native species widespread in TX _______________ **P. abnormis**

4. Leaves minutely rough-scabrid beneath to the touch; female flowers and capsules subsessile (pedicels 0.5 mm or less long), proximal on the branches, the male flowers distal; stamens 3; seeds conspicuously transversely ribbed (use lens); pistillate calyx lobes linear-oblong to lanceolate; introduced species known in nc TX only from Tarrant Co., mainly se and e TX _______________ **P. urinaria**

Phyllanthus abnormis Baill., (abnormal), DRUMMOND'S LEAF-FLOWER. Annual but basal stem and root can sometimes be thickened and appear perennial; leaf blades not oblique at base; stems 10–50 cm tall; leaves 3–10 mm long, 1–4 mm wide; pedicels of female flowers 1–3(–3.5) mm long; fruiting calyx lobes scarcely over 1 mm long; stamens 2 in most flowers; filaments entirely connate; style branches not capitate; capsules 2.3–2.7 mm wide. Loose sandy soils; widespread in TX, in nc TX mainly East Cross Timbers w, once found in railroad gravel at Dallas. May–Oct. Reported to be poisonous to livestock; symptoms include listlessness, ceaseless walking, diarrhea, exhaustion, kidney degeneration, liver cirrhosis, and death (Kingsbury 1964; Burlage 1968). ☠

Phyllanthus caroliniensis Walter, (of Carolina), CAROLINA LEAF-FLOWER. Plant 10–40 cm tall; leaves 5–20(–30) mm long, 4–10(–16) mm wide; pedicels of female flowers 0.5–1(–1.5) mm long; pistillate calyx lobes linear-lanceolate or narrowly spatulate; capsules 1.5–2 mm wide. Weedy areas; Henderson Co., also Bosque, Erath, Hamilton (Stanford 1976), Lamar (Carr 1994), and Tarrant (Webster 1970) cos.; se and e TX w to nc TX and Edwards Plateau. Jun–Nov.

Phyllanthus polygonoides Nutt. ex Spreng., (resembling *Polygonum*—knotweed), KNOTWEED LEAF-FLOWER. Plant ca. 10–50 cm tall; leaves on the main stems and branches, usually 5–10 mm long, 1.5–5 mm wide; usually monoecious but sometimes dioecious; stamens 3, the filaments united about half way into a column; pedicels of female flowers 2.5–7 mm long; capsules 2.7–3.2 mm wide. Limestone outcrops, rarely in sandy clay; nearly throughout TX. Apr–Jul, Sep–Oct.

Phyllanthus tenellus Roxb., (tender, delicate). Plant 20–50 cm tall; main stems with leaves reduced to scales, the leaves and flowers on specialized deciduous side branches; leaves 6–25 mm long, 4–11 mm wide; filaments free; capsules 1.7–1.9 mm wide. Yard in Fort Worth (Tarrant Co.); otherwise known in TX only in se part of the state; a 20th century introduction to the se U.S. (Webster 1970). Throughout growing season. Native to the Mascarene Islands (Indian Ocean).

Phyllanthus urinaria L., (presumably in reference to urine), PEEWATER LEAF-FLOWER. Plant 15–50 cm tall; leaves 6–25 mm long, 2–9 mm wide; monoecious; capsules ca. 2–2.2 mm wide. Yard weed; Tarrant Co.; mainly se and e TX. Flowering throughout the growing season. Native of tropical e Asia. Not detected in the continental U.S. until 1944 (Webster 1970). Reported to contain alkaloids and to be used as a fish poison (Burlage 1968). ☠

An additional species, *P. niruri* (Kunth) G.L. Webster, (derivation unknown), is reported by Hatch et al. (1990) for vegetational area 4 (Fig. 2) but apparently occurs only to the s of nc TX (Webster 1970). It differs from *P. abnormis* in having the leaf blades oblique at base, filaments connate halfway or less, style branches capitate, and fruiting calyx lobes larger (3–3.5 mm long).

REVERCHONIA

A monotypic genus of sandy areas of the sw U.S. and adjacent Mexico. It is closely related to *Phyllanthus* (Webster 1994b). (This unusual genus was named in honor of Julien Reverchon, 1837–1905, a French-American immigrant to Dallas and important botanical collector of early TX. Many of his specimens are in the herbaria at BRIT and the Missouri Botanical Garden)

REFERENCE: Webster & Miller 1963.

Reverchonia arenaria A. Gray, (of sand or sandy places), SAND REVERCHONIA. Annual monoecious herb to 0.5 m tall, glabrous; sap not milky; main stem glaucous-white; lateral branches prominent; leaf blades elliptic to narrowly oblong-elliptic or nearly linear (15–)20–40(–45) mm long, 1.8–9 mm wide, apiculate at apex; petioles 1–3 mm long; inflorescences small cymules in the axils of leaves of lateral branches (not occurring on main stem), the cymules usually with a single female flower and 4–6 male flowers; flowers conspicuous; male flowers: calyx lobes 4, purplish to pinkish with a central green strip, 1.5–2.5 mm long; disk with an I-beam-like outline; stamens 2; female flowers: pedicel 1.5–2 mm long in fruit, elongating up to 8.7 mm in fruit; calyx lobes 6(–5), purplish to pinkish with a central green strip, to 2.9 mm long; disk flat and roundish or 6-angled in outline; capsules ± spheroidal and flattened at the ends, conspicuous, 7–9.8 mm in diam.; seeds 4.4–6.6 mm long. Deep sand; Webster and Miller (1963) cited a Tarrant Co. collection (Fort Worth, Mar. 1890, *Bodin* (US)) as "very dubious"; Trans-Pecos and Plains Country e to Wichita Co. just to the w of the nw part of nc TX; included because of the Tarrant Co. specimen and because of the possibility of occurrence in sandy areas along the Red River in the nw part of nc TX. Summer–fall. Reported to be toxic to livestock (Kingsbury 1964) and in experimental animals to cause acute liver and kidney damage (Burlage 1968). ☠

RICINUS CASTOR-BEAN

A monotypic genus native from e and ne Africa to the Middle East, now naturalized throughout the tropics. (Latin: *ricinus*, tick, because of the resemblance of seed to a tick)

Ricinus communis L., (common, general), CASTOR-BEAN, CASTOR-OIL-PLANT, PALMA CHRISTI, HIGUERILLA. Large, coarse, herbaceous, monoecious annual (in nc TX) 1–5 m tall; sap watery; leaves alternate, with long petioles to 30 cm long; leaf blades peltate, with (5–)7–9(–11) palmate lobes, these serrate with the teeth gland-tipped; inflorescences terminal, raceme- or panicle-like, the staminate flowers usually near the base, the pistillate flowers near apex; sepals usually 5; petals 0; stamens numerous; styles 3, plumose, red; capsules round, 12–20 mm in diam., usually covered with soft, dark spines; seeds 8–11 mm long, glabrous, with mottled surface. Cultivated for its foliage and rarely escapes; roadfill, waste areas; Henderson Co. near e edge of nc TX, also Palo Pinto Co. (*Reverchon*) and Tarrant Co. along Trinity River (R. O'Kennon, pers. obs.); mainly s TX. Flowering throughout the growing season. Native to Africa and Middle East. CASTOR-BEAN has a very long history of cultivation; it has been found in 6000 year-old Egyptian tombs (Zohary 1982). All parts of the plant, but especially the seeds, contain many alkaloids and toxic principles including a lectin (type of phytotoxin or toxic protein), ricin, one of the most toxic compounds known. If eaten, 1–8 seeds are fatally poisonous; they are also the source of castor oil, used medicinally in the past; the toxins are not oil soluble. Castor oil is also used as an industrial lubricant. Livestock can be poisoned from eating the leaves or seeds (Muenscher 1951; Kingsbury 1964; Hardin & Arena 1974; Stephens 1980; Mabberley 1987). ☠

SAPIUM

A genus of ca. 90–100 species (Webster 1994b) native to tropical and warm areas to Patagonia; the majority are neotropical. According to Webster (1994b), "*Triadica*, accepted by some Asian workers, is distinctively different from the neotropical species in such features as its non-arillate seeds; further investigation may show that it should be generically separate from *Sapium*." If *Triadica* is recognized, our single introduced species will have to be treated in that segregate genus. (Latin name used by Pliny for a resinous pine or fir, alluding to the greasy latex of these plants)

Sapium sebiferum (L.) Roxb., (tallow-bearing), CHINESE TALLOW TREE, VEGETABLE TALLOW TREE. Rapidly growing monoecious tree usually 3–10 m tall, unarmed, essentially glabrous, with milky sap; branches often slender and drooping; leaves alternate, the blades rhombic-ovate, 3–

Phyllanthus caroliniensis [BB2]

Phyllanthus polygonoides [SHI]

Phyllanthus tenellus [HO5]

Phyllanthus urinaria [LAM]

Reverchonia arenaria [HEA, RHO]

Ricinus communis [BL1, LAM]

7(–10) cm long, entire; petioles usually longer than blades, with 2 glands at juncture of blade and petiole; flowers in terminal spike-like inflorescences (3–)5–15 cm long; petals absent; staminate flowers in clusters at the upper nodes; pistillate flowers solitary at the nodes; fruit a 3-lobed capsule ca. 1–4 cm long and about as broad; seeds 7–8 mm long, long persistent on the placenta after the capsule walls have fallen, chalky-white. Widely cultivated as ornamental shade tree, rarely escapes in nc TX; Dallas-Fort Worth area is near limit of cold hardiness as cold damage is often observed; Grayson Co. (edge of Waterloo lake on sandy soil), also Dallas (H.S. Moss Park, Whiter Rock Lake) and Tarrant cos.; more commonly escapes in se TX where it can be a problematic invader of native prairies. Aug–Nov. Native of China and Japan. [*Triadica sebifera* (L.) Small] This species was introduced into the U.S. in South Carolina in the late 1770s and is now widespread; it displaces native vegetation and is considered one of the most serious invasive exotics in the U.S.; it apparently releases compounds that modify soil chemistry and affect the establishment of native species (Flack & Furlow 1996; Jubinsky & Anderson 1996). The fatty covering of the seed is used for candlewax and soap (Mabberley 1987). Burlage (1968) reported the milky sap as poisonous. ☠ ☞

STILLINGIA QUEEN'S-DELIGHT, QUEEN'S-ROOT

Glabrous unarmed perennial herbs usually with several to many stems from a woody root; sap milky; leaves numerous, crowded, alternate, subsessile, evenly and finely glandular-toothed; flowers in terminal, sessile, spike-like inflorescences (which become over-topped by whorls of leafy branches developed from beneath them), unisexual with both sexes in the same inflorescence, borne in the axils of minute triangular-ovate bracts, the bracts with fleshy, saucer- or cup-shaped, gland-like stipules at each side, the glands larger than the bracts; staminate flowers above, on a fleshy axis which falls after flowering; pistillate flowers few, at base; gynobase (= lower portion of ovary) becoming thick and hard, persistent after the seeds fall, triangular; fruit a shallowly 3-lobed, 3-seeded capsule; seeds with a prominent caruncle (outgrowth).

A genus of ca. 30 species of tropical and warm America, Madagascar, e Malesia, and Fiji. ☠ Feeding experiments with *S. treculeana* (Müll. Arg.) I.M. Johnst., of s TX and the Edwards Plateau, indicated that the plant may be highly toxic due to hydrocyanic acid; small amounts of leaves and stems were lethal to sheep. (Named for Dr. Benjamin Stillingfleet, 1702–1771, English naturalist)

1. Stem leaves lanceolate or elliptic, 9–30 mm wide; capsules ca. 12 mm long and broad; plants of sandy soil ______ **S. sylvatica**
1. Stem leaves linear, 2–5 mm wide; capsules ca. 6 mm long and broad; plants of limestone habitats ______ **S. texana**

Stillingia sylvatica Garden ex L., (forest-loving), QUEEN'S-DELIGHT. Stems ascending to erect, to ca. 1 m tall; gynobase lobes 6 mm long; seeds ca. 8 mm long (not counting caruncle). Sandy prairies, open woods, or open ground; Grayson, Lamar, Montague, Parker, and Tarrant cos.; nearly throughout TX. May–Jun. Used medicinally, but overdoses cause toxic symptons; the latex is reported to have vesicant (= blister-causing) properties (Duke 1985; Lampe 1986). ☠

Stillingia texana I.M. Johnst., (of Texas), TEXAS STILLINGIA. Stems decumbent to erect, to 65 cm long; gynobase lobes 3–3.5 mm long; seeds ca. 5 mm long (not counting caruncle). Gravelly or rocky, calcareous soils; nc TX and the Edwards Plateau. Mid-May–Jun. The latex is reported to have vesicant (= blister-causing) properties (Lampe 1986). ☠

TRAGIA NOSEBURN

Perennial herbs, erect to trailing or twining, usually with stinging hairs; sap not milky; leaves alternate; leaf blades toothed; stipules foliaceous, lanceolate to ovate; flowers in peduncled, axil-

Sapium sebiferum [EN1]

Stillingia sylvatica [BB1,BT3, GLE]

Stillingia texana [BT3]

Tragia betonicifolia [RHO]

Tragia brevispica [RHO]

Tragia ramosa [SHI]

Tragia urticifolia [BT3]

lary (at least appearing so), spike-like, minutely bracted racemes, apetalous, unisexual, most often both sexes in the same inflorescence; staminate flowers above, minute; pistillate below, solitary or few, larger; staminate flowers with 3–4(–6) sepals; stamens usually equaling the number of calyx lobes; pistillate flowers usually with 5–6 narrow sepals; styles 3; fruit an explosively dehiscent capsule.

A genus of 100 species of tropical and warm areas of the world; usually with stiff, nettle-like stinging hairs. ☠All nc TX species have the stinging hairs which can be quite painful if allowed to come into contact with skin. According to Lampe (1986), only four families (Euphorbiaceae, Hydrophyllaceae, Loasaceae, and Urticaceae) have stinging hairs—nc TX has stinging representatives of all of these except the Hydrophyllaceae. (Name from *Tragus*, Latin name for early herbalist, Hieronymous Bock, 1498–1554)
REFERENCE: Miller & Webster 1967.

1. Persistent base of the staminate pedicel usually nearly as long as or longer than its subtending bract; stigmatic surfaces papillate; styles connate 1/3–1/2, with a slight constriction where they join the ovary _____ **T. urticifolia**
1. Persistent base of the staminate pedicel conspicuously shorter than its subtending bract; stigmatic surfaces smooth to papillate; styles free or connate, without a constriction where they join the ovary.
   2. Calyx lobes (pistillate) as long or longer than pistil at anthesis; staminate flowers 14–75 per raceme, compactly arranged on axis _____ **T. betonicifolia**
   2. Calyx lobes (pistillate) shorter than pistil at anthesis; staminate flowers 2–20 per raceme, not compactly arranged.
      3. Larger leaf blades usually broadly triangular, usually 1–2.5 times as long as wide; stigmatic surfaces papillate; styles connate basally to 1/3 their length _____ **T. brevispica**
      3. Larger leaf blades usually linear-lanceolate to ovate, usually 2–4 times as long as broad; stigmatic surfaces not papillate; styles connate 1/3–1/2 or more their length _____ **T. ramosa**

Tragia betonicifolia Nutt., (with leaves like *Betonica*—betony, now = *Stachys* in Lamiaceae), BETONY NOSEBURN. Stems to 50 cm tall, erect, decumbent, or trailing; leaves ovate to triangular-lanceolate, usually cordate to truncate at base; pistillate calyx lobes 6, 1.8–3 mm long at anthesis, 3–5 mm long in fruit; styles connate basally; stigmatic surfaces papillate. Sandy soils; se TX and Post Oak Savannah w to Rolling Plains, also Edwards Plateau. May–Jun, Sep–Oct. [*T. urticifolia* Michx. var. *texana* Shinners] Stinging hairs present. ☠

Tragia brevispica Engelm. & A. Gray, (short-spiked), SHORT-SPIKE NOSEBURN. Stems to 100 cm long or longer, erect to trailing and twining; pistillate calyx lobes 6, 1.3–2 mm long at anthesis, 1.8–3.5 long in fruit. Calcareous soils, open woods and prairies; nc TX s and w to s TX and Edwards Plateau. Apr–Oct. Stinging hairs present. ☠

Tragia ramosa Torr., (branched), CATNIP NOSEBURN. Stems to 50 cm tall, erect, decumbent, trailing, with tendency toward twining; leaf blades usually truncate basally; pistillate calyx lobes 5–7, 0.8–2.5 mm long at anthesis, 1.5–3 mm long in fruit. Disturbed areas; Post Oak Savannah s and w through nc TX to w TX. Apr–Oct. [*T. nepetifolia* Cav. var. *leptophylla* (Torr.) Shinners] Stinging hairs present. ☠

Tragia urticifolia Michx., (with leaves like *Urtica*—nettle), NETTLE-LEAF NOSEBURN. Stems to 65 mm tall, erect to decumbent; pistillate calyx lobes (5–)6, 1.3–2.2 mm long at anthesis, 2–3 mm long in fruit. Sandy soils, fields, and open woods; Henderson Co. near extreme e margin of nc TX; mainly se and e TX. Spring–fall. Jones et al. (1997) treated TX material as *T. urticifolia* var. *texana* Shinners. We are following Kartesz (1994) who recognized *T. urticifolia* but put the var. *texana* into synonymy under *T. betonicifolia*. According to M. Mayfield (pers. comm.), this species occurs only in sandy habitats to the e of nc TX. Stinging hairs present. ☠

Tragia amblyodonta (Müll.Arg.) Pax & K. Hoffm., (blunt-toothed), DOG-TOOTH NOSEBURN, was cited for vegetational area 5 (Fig. 2) by Hatch et al. (1990), but according to M. Mayfield (pers. comm.) it occurs in s and w TX and does not come close to nc TX. In the above key it would come out closest to *T. brevispica*. The two can be distinguished as follows:

1. Mature leaves usually densely pubescent with stinging hairs; plants grayish green in appearance; leaf blades usually truncate to sagittate (rarely cordate) basally; staminate flowers 3–60 per raceme ____ *T. amblyodonta*
1. Mature leaves usually only sparsely pubescent with stinging hairs; plants green in appearance; leaf blades truncate to cordate basally; staminate flowers 3–6(–10) per raceme ____ *T. brevispica*

FABACEAE (LEGUMINOSAE) LEGUME, BEAN, OR PULSE FAMILY

Plants herbaceous or woody; leaves basal or alternate (sometimes crowded and appearing subopposite or whorled), compound or apparently simple (with only 1 leaflet); leaflets entire, or in a few genera toothed or lobed; stipules present (except *Lotus* and *Crotalaria*), falling early in many species; flowers solitary or in racemes, panicles, spikes, heads, or umbel-like clusters; sepals 5, separate or united; petals 1–5, equal or in most genera unequal; stamens 5 to many; pistil 1; fruit a legume, developed from a 1-celled superior ovary with 1–many ovules and parietal placentation, in general opening along both sutures.

A huge (16,400 species in 657 genera), cosmopolitan, vegetatively variable family ranging from herbs to rain forest canopy trees; Ballenger et al. (1993) suggested there are ca. 20,000 species, while Cronquist (1993) indicated 18,000 species. Nearly a third of all species are in 6 large genera: *Acacia*, *Astragalus*, *Cassia*, *Crotalaria*, *Indigofera*, and *Mimosa*. The Fabaceae is the third largest angiosperm family in terms of numbers (after Asteraceae and Orchidaceae), and in importance to humans is second only to the Poaceae. It is also extremely important ecologically because of the symbiotic nitrogen-fixing *Rhizobium* bacteria associated with the roots of many species. The family includes many important timber trees, ornamentals, and particularly protein-rich food plants including *Arachis* (PEANUTS), *Cicer* (CHICK-PEAS), *Glycine* (SOYBEANS), *Lens* (LENTILS), *Phaseolus* (BEANS), and *Pisum* (PEAS); in fact, seeds of legumes are the world's most important source of vegetable protein for man and animals (Isely 1990). ☠ There are often toxic, defensive, non-protein amino acids in the seeds or vegetative tissues as well as alkaloids. For example, the tropical *Abrus precatorius* L. (PRECATORY-BEAN or ROSARY-PEA) has striking red and black seeds sometimes used in necklaces; they contain abrin, a protein so toxic that a single chewed seed is enough to kill a human (Kingsbury 1964). The family, here recognized as having three subfamilies, is sometimes split into three questionably monophyletic families (e.g., Cronquist 1993). Recent evidence (Ballenger et al. 1993) suggested that the Caesalpinioideae is a basal paraphyletic assemblage from which the monophyletic Mimosoideae and Papilionoideae are derived. The tribal arrangement followed here is from Polhill and Raven (1981). The Fabaceae is the third largest family in the nc TX flora (after Asteraceae and Poaceae), with 176 species. Family name conserved from *Faba*, a genus of a single widely cultivated species, *F. vulgaris* Moench, BROAD BEAN, of the Mediterranean region; it is now usually lumped into *Vicia* as *V. faba* L. (Latin: *faba*, bean) (subclass Rosidae)

FAMILY RECOGNITION IN THE FIELD: characteristic most helpful in field recognition is the *legume* fruit—a 1-chambered, pod-like, often bean-like fruit; other helpful clues include leaves with 3 leaflets and pea-like flowers, pinnately compound leaves with flowers in heads with obviously exserted stamens, or stamens in a 9 (fused) + 1 (separate) arrangement.

REFERENCES: Rydberg 1919–1920, 1923–1929; Britton & Rose 1928, 1930; Turner 1959; Isely 1973,

1975, 1981, 1982, 1986a, 1990; Elias 1974; Robertson & Lee 1976; Isely & Polhill 1980; Summerfield & Bunting 1980; Allen & Allen 1981; Duke 1981; Polhill & Raven 1981; Gunn 1983, 1984, 1991; Stirton 1987; Stubbendieck & Conard 1989; Ballenger et al. 1993.

Key To Subfamilies Of Fabaceae

1. Flowers usually small and individually inconspicuous, arranged in dense heads or clusters, regular (= radially symmetrical); corollas usually sympetalous and so small as to be ± inevident; stamens 5–numerous, their filaments exserted beyond the petals and often showy; leaves twice pinnately compound ____ **Mimosoideae**
1. Flowers whether small or large usually individually conspicuous, arranged in various types of inflorescences, usually zygomorphic (= bilaterally symmetrical); corollas usually easily seen, of separate petals except 2 petals (= keel) fused along the margin in many; stamens 1–10, usually not exserted beyond the petals (except in *Caesalpinia*); leaves simple or palmately or once or twice pinnately compound (3–numerous leaflets).
   2. Corollas of 5 distinct petals, these not differentiated into a larger standard (= banner), wings, and keel; sepals separate or nearly so; stamens externally visible and free from one another, not enclosed by a keel; leaves simple or once or twice pinnately compound (usually with > 3 leaflets except in *Cercis* and 2 species of *Senna*); equivalent of standard (= adaxial petal) inside of lateral petals in bud ____ **Caesalpinioideae**
   2. Corollas of 3 separate petals (larger standard, 2 wings) and 2 petals fused to form a keel (corollas rarely of only 1 petal); sepals at least partially fused; stamens usually enclosed by keel and usually not externally visible, their filaments usually fused or free in a few species; leaves simple, once pinnately compound (often with 3 leaflets OR with many leaflets), or palmately compound; standard outside of lateral petals in bud ____ **Papilionoideae**

Subfamily 1. Mimosoideae

Leaves twice even-pinnately compound; stipules thread-like or minute, often falling early; flowers in heads or spikes; sepals united; petals 4 or 5, equal, separate or united, rather inconspicuous; stamens 5 to many, separate or the filaments all united toward base, exceeding the corolla.

Key To Genera Of Mimosoideae

1. Trees or woody shrubs or if herbaceous perennials, then stamens numerous.
   2. Stamens numerous.
      3. Stamens 23–36 mm long, pinkish to reddish; introduced species ____ **Albizia**
      3. Stamens less than 20 mm long, white to yellow; native species ____ **Acacia**
   2. Stamens 10 or less.
      4. Leaflets 15 mm or more long; fruits (5–)7–20 cm long, about as thick as broad, without prickles; flowers yellowish white ____ **Prosopis**
      4. Leaflets less than 15 mm long; fruits 1.5–5 cm long, flattened (sometimes contorted), with or without prickles; flowers pink to purple ____ **Mimosa**
1. Herbaceous perennials; stamens 10 or less.
   5. Plants armed with sharp recurved prickles capable of causing pain ____ **Mimosa**
   5. Plants unarmed or with numerous, weak, hair-like or bristle-like structures not painful to the touch.
      6. Flowers pink or purplish; fruits 15–20 mm long, with conspicuous appressed hairs ____ **Mimosa**
      6. Flowers whitish or yellow; fruits of variable lengths, sometimes much longer than 20 mm, usually glabrous or essentially so.

7. Flowers whitish; fruits sessile; leaves with a petiolar gland between the lowest pair of pinnae (the gland sometimes minute) ____ **Desmanthus**
7. Flowers yellow; fruits stipitate (= stalked); leaves without a petiolar gland ____ **Neptunia**

SUBFAMILY 2. CAESALPINIOIDEAE

Leaves once or twice compound, even- or odd-pinnate (reduced to 1 leaflet and apparently simple in *Cercis*); stipules very small, often falling early; sepals united or separate; petals 5, slightly or markedly unequal; stamens 5 or 10, separate.

KEY TO GENERA OF CAESALPINIOIDEAE

1. Trees or shrubs.
   2. Leaves simple; flowers purple-red (rarely white); flowering Mar–Apr ____ **Cercis**
   2. Leaves once or twice pinnately compound; flowers yellow, orange, greenish yellow, or greenish white; flowering at various times of the growing season.
      3. Leaflets < 10 mm long.
         4. Plants armed with sharp spines; leaves with 2(–4) pinnae (each with leaflets), but each pinna appearing as a once pinnate leaf due to the absence of a common petiole; stamens small, inconspicuous ____ **Parkinsonia**
         4. Plants unarmed; leaves with 11–29 pinnae (each with leaflets); stamens very long, 70–90 mm, bright red ____ **Caesalpinia**
      3. Leaflets 15–50 mm long.
         4. Plants conspicuously armed with straight or branched thorns OR unarmed; leaves once or twice pinnately compound with 18–numerous leaflets; corollas inconspicuous, 4–10 mm long, greenish yellow or greenish white; trees to ca. 15(–35) m tall.
            5. Plants usually armed (cultivated forms sometimes unarmed); leaflets usually 4–15 mm wide, lanceolate to narrowly elliptic, rounded or ± blunt at apex, often crenulate, sometimes entire; native species widespread in nc TX ____ **Gleditsia**
            5. Plants unarmed; leaflets usually 15–30 mm wide, ovate, rather abruptly acuminate at apex, entire; cultivated species rarely escaping in nc TX ____ **Gymnocladus**
         4. Plants unarmed; leaves once pinnately compound with 4–24 leaflets; corollas conspicuous, 10–20 mm long, yellow to orangish; shrubs to 3.5 m tall ____ **Senna**
1. Herbaceous annuals or perennials.
   6. Leaflets glandular-dotted beneath; leaves twice compound with 5–7 pinnae (each with leaflets) ____ **Pomaria**
   6. Leaflets not glandular-dotted beneath; leaves once OR twice compound, if twice compound, then with 5–11(–13) pinnae (each with leaflets).
      7. Leaves twice compound; leaflets 3–8 mm long ____ **Hoffmanseggia**
      7. Leaves once compound; leaflets usually much longer (7–170 mm long).
         8. Leaflets 2 ____ **Senna**
         8. Leaflets 12 to numerous.
            9. Leaflets 2.5 cm or more long; petiolar glands slender or stipitate or absent ____ **Senna**
            9. Leaflets 2 cm or less long; petiolar glands disc-shaped ____ **Chamaecrista**

SUBFAMILY 3. PAPILIONOIDEAE

Leaves once compound with 3-many leaflets (simple or apparently so in species of *Baptisia* and *Crotalaria*), odd- or even-pinnate, or palmate; stipules various; sepals united (barely so in *Indigofera*); petals 5 (except in *Amorpha* and sometimes *Eysenhardtia*) and very unequal: uppermost petal (= standard) usually largest, two lateral (= wings) smaller and separate, two lowest (= keel) smaller and united except at base; stamens 5–10; filaments separate or united.

KEY TO TRIBES AND GENERA OF PAPILIONOIDEAE

1. Stamens distinct.
   2. Leaves pinnately compound; trees or shrubs ______ Sophoreae (only **Sophora**)
   2. Leaves simple or palmately compound; herbs ______ Thermopsideae (only **Baptisia**)
1. Stamens united, at least basally.
   3. Leaves simple; petals yellow ______ Crotalarieae (only **Crotalaria**)
   3. Leaves compound; petals variously colored including yellow.
      4. Leaves palmately compound with 4–7 leaflets; stamens monadelphous; petals deep blue (rarely white) ______ Genisteae (only **Lupinus**)
      4. Leaves pinnately compound with 3–many leaflets OR if palmately compound, then stamens diadelphous; stamens monadelphous or diadelphous; petals variously colored including blue.
         5. Leaflets dentate, the teeth usually easily visible but very small and often visible only near the leaflet tips in *Trifolium arvense* and *T. pratense*; stipules adnate to petioles (Trifolieae).
            6. Fruits coiled or curved, sometimes prickly; stems usually 4-angled ______ **Medicago**
            6. Fruits straight, never prickly; stems terete (= round).
               7. Inflorescence an elongated slender raceme, loose enough that the axis is easily visible to the naked eye; petals deciduous, the mature fruit visible ______ **Melilotus**
               7. Inflorescence a globose to cylindrical, often head-like spike or raceme, so dense that the axis is not easily visible; petals withering and persistent, concealing the mature fruit from view ______ **Trifolium**
         5. Leaflets entire to dentate; stipules distinct OR if united to petioles then leaflets entire.
            8. Tendrils present in place of terminal leaflets; leaflets entire to dentate (Vicieae).
               9. Corollas yellow; leaflets absent, the leaf consisting only of a tendril (however, leaf-like stipules present) ______ **Lathyrus**
               9. Corollas white to blue, lavender, pink-purple, purple, or bicolored but not yellow; leaflets 2–numerous.
                  10. Leaflets 2; stems winged ______ **Lathyrus**
                  10. Leaflets 4–18; stems not winged (but can be angled or edged).
                      11. Leaflets usually 30–60 mm long, 10–30 mm wide; style flattened; rare species possibly on e margin of nc TX ______ **Lathyrus**
                      11. Leaflets usually smaller, 5–35 mm long, 1.5–15 mm wide; style ± round in cross-section; widespread species common throughout nc TX ______ **Vicia**
            8. Tendrils absent; leaflets entire.
               12. Leaves pinnately compound (3–numerous leaflets) OR palmately compound with (3–)5–7 leaflets; leaflets usually glandular-punctate; fruits 1-seeded (Amorpheae and Psoraleeae).
                   13. Shrubs.
                       14. Petals white to pale yellow, 4–5; leaflets (3–)5–12 mm long; in nc TX only in extreme s part in Bell and Williamson cos. ______ **Eysenhardtia**
                       14. Petals bluish to purplish, 1 or 5; leaflets 1.5–80 mm long; widespread in nc TX.
                           15. Petal 1 (standard only); leaflets (3–)10–80 mm long; small to large shrubs 0.3–3.5 m tall ______ **Amorpha**
                           15. Petals 5; leaflets 1.5–3.5(–5) mm long; small shrubs 0.3–1.2 m tall (only *D. fructescens*) ______ **Dalea**
                   13. Herbaceous plants.
                       16. Wing and keel petals attached laterally or apically on staminal tube; leaves pinnately compound with 3–numerous leaflets ______ **Dalea**
                       16. All petals basally attached; leaves palmately compound (3–7 leaflets) or pinnately compound with 3 leaflets.

17. Fruits conspicuously cross-wrinkled; leaves pinnately compound with 3 leaflets; extreme e or ne part of nc TX ______ **Orbexilum**
17. Fruits smooth, not cross-wrinkled; leaves pinnately compound with 3 leaflets (1 species) or palmately compound with 3–7 leaflets (sometimes reduced to 1 or 2 near top of stem); widespread in nc TX.
18. Fruits enclosed in enlarging calyces with beak exserted (in 1 species calyces only ca. 1/2 as long as fruits); pericarp (= fruit wall) thin, usually papery ______ **Pediomelum**
18. Fruits exserted above calyces, the calyces only at very base of the fruits; pericarp thick, coriaceous ______ **Psoralidium**
12. Leaves pinnately compound (3–numerous leaflets, often 3) OR palmately compound with 4 leaflets; leaflets not glandular-punctate; fruits 1- to several-seeded.
19. Fruits segmented, the segments separating at maturity OR fruits 1-seeded (or in 1 species unsegmented and borne underground) (Aeschynomeneae, Coronilleae, and Desmodieae).
20. Leaves palmately or pinnately compound with 4 leaflets; flowers yellow.
21. Leaves palmately compound (all leaflets arising from a single point); leaflets 1–3 cm long; fruits borne above ground, breaking part into 4–6 segments; each flower enclosed by 2 conspicuous bracts ______ **Zornia**
21. Leaves pinnately compound (2 leaflets separated from the other 2 by a rachis); leaflets 2–6 cm long; fruits developing underground, not breaking apart into segments; flowers not enclosed in bracts ______ **Arachis**
20. Leaves odd-pinnately compound with 3–many leaflets; flowers variously colored including yellow.
22. Leaves pinnately compound with 9–25 leaflets; fruit (loment) segments 4-angled to terete ______ **Coronilla**
22. Leaves pinnately compound with 3 leaflets; fruit segments usually flattened.
23. Fruits of 2–6 segments, usually with hooked hairs causing them to stick to hair or clothing; leaflets usually stipellate ______ **Desmodium**
23. Fruits of 1(–2) segment, without hooked hairs; leaflets estipellate.
24. Petals orange or orange-yellow (very rarely white); stipules adnate to petioles; stamens monadelphous; anthers of 2 kinds ______ **Stylosanthes**
24. Petals purplish or pinkish to white; stipules free from petioles; stamens diadelphous; anthers of 1 kind.
25. Stipules nearly thread-like; leaflets not striate; plants perennial ______ **Lespedeza**
25. Stipules ovate or nearly so, striate; leaflets striate with conspicuous parallel lateral veins; plants annual ______ **Kummerowia**
19. Fruits non-segmented, several-seeded.
26. Inflorescences umbellate; leaves pinnately compound with 5 leaflets (the lower 2 stipular in position); petals yellow, marked with red ______ Loteae—in part (only **Lotus**)
26. Inflorescences racemose OR of solitary flowers; leaves pinnately compound with 3–numerous leaflets; petals variously colored.
27. Plants herbaceous twining vines (except in introduced *Glycine* and *Erythrina* and a small annual *Lotus*); leaves with 3 leaflets (except *Apios* and *Galactia* which have 5–7 leaflets) (Phaseoleae).
28. Leaves with 5–7 leaflets.

29. Lower 2–3 pairs of leaflets attached at different places; leaflets 2–10 cm long; corollas at least somewhat brownish red ____________ **Apios**
29. Lower 2 pairs of leaflets attached at the same place (leaves almost palmate but with the terminal leaflet short-stalked); leaflets 1–3 cm long; corollas mostly lavender (standard often with some white) ____________ **Galactia**

28. Leaves with 3 leaflets.
30. Plants not twining vines.
31. Corollas red, very long (30–53 mm long); flowers in a terminal raceme ____________ **Erythrina**
31. Corollas white to pink, violet, or purple (rarely yellowish with pink tinge), 8 mm or less long; flowers axillary, either solitary or in racemes.
32. Leaflets 1–2 cm long; fruits not densely bristly ____________ **Lotus**
32. Leaflets 3–15 cm long; fruits densely bristly ____________ **Glycine**
30. Plants twining vines.
33. Standard (= banner) greatly exceeding keel and wings (ca. 2 times as long), 20–60 mm long.
34. Calyx tube short, ca. 4 mm long; calyx lobes linear, longer than the tube; fruits ca. 7–12(–14) cm long, sessile or nearly so ____________ **Centrosema**
34. Calyx tube cylindrical, 10–20 mm long; calyx lobes ovate, much shorter than the tube; fruits 3–6(–8) cm long, stipitate, the stipe 10–20 mm long ____________ **Clitoria**
33. Standard ± equaling or shorter than keel and wings, 4–25 mm long.
35. Keel of corollas strongly incurved ____________ **Strophostyles**
35. Keel of corollas essentially straight.
36. Corollas yellow; fruits 11–20 mm long ____________ **Rhynchosia**
36. Corollas white, pink, or purplish; fruits 15–55 mm long (except subterranean fruits).
37. Plants high climbing vines with stems often many meters long; corollas 15–25 mm long; fruits densely and conspicuously long hairy at a glance ____________ **Pueraria**
37. Plants relatively low climbing vines, the stems to ca. 2 m long; corollas 8–15 mm long; fruits ± glabrous to pubescent, but not densely and conspicuously long hairy.
38. Bracts subtending pedicels ± as wide as long (2–5 mm long), easily visible to the nake eye; calyx teeth nearly equal, ca. 1/2 as long as tube or less; leaflets gradually narrowed to ± acute apex, the larger leaflets 2–10 cm long, 18–70 mm wide; stipules 3–8 mm long, persistent; fruits 6–12 mm wide; corollas lilac or whitish ____________ **Amphicarpaea**
38. Pedicels without conspicuous bracts ± as wide as long; calyx teeth unequal, ca.

as long as tube; leaflets usually abruptly narrowed to a rounded or obtuse apex, the larger leaflets 2–4(–6) cm long, 10–25(–35) mm wide; stipules 1–3 mm long, soon deciduous; fruits 3–6 mm wide; corollas pink or rose ______ **Galactia**

27. Plants erect, non-twining (except in woody *Wisteria*); leaves pinnately compound with (5–)7–40(–52) leaflets (Galegeae, Indigofereae, Loteae in part, Robinieae, and Tephrosieae).
    39. Trees, shrubs, or woody vines; flowers white or rose, reddish purple, or violet-blue; plants sometimes with stipular spines or conspicuously hispid young stems.
        40. Plants trees or shrubs (not vine-like); racemes axillary; young stems either hispid or with stipular spines ______ **Robinia**
        40. Plants woody vines or vine-like shrubs; racemes terminal; young stems neither hispid nor with stipular spines ______ **Wisteria**
    39. Herbaceous plants; flowers variously colored; plants without either stipular spines or hispid young stems.
        41. Leaves even-pinnately compound.
            42. Flowers 6–9 mm long; peduncle 5–12 cm long; fruits 2.5–8 cm long, oblong to ellipsoid, not winged, with 2 seeds ______ **Glottidium**
            42. Flowers 10–20 mm long; peduncle 1–5 cm long; fruits usually 5–20 cm long, either winged or narrowly linear, usually with 3–40 seeds ______ **Sesbania**
        41. Leaves odd-pinnately compound.
            43. Leaflets mostly alternate along leaf rachis; petals reddish orange with tints of pink or salmon; pubescence of hairs attached at middle ______ **Indigofera**
            43. Leaflets opposite along leaf rachis; petals variously colored, but not reddish orange; pubescence of hairs attached basally or submedianly.
                44. Main racemes terminal or if apparently lateral, then opposite a leaf; standard orbicular; fruits flattened, with neither suture intruding; style with fine stiff hairs ______ **Tephrosia**
                44. Main racemes axillary; standard obovate; fruits subcylindric or somewhat triangular in cross-section to inflated, with 1 suture intruding; style glabrous.
                    45. Leaves all basal or crowded near base, the leafless peduncles more than twice as long as the leaf-bearing lower portion of the stem; flowers 15–26 mm long; keel tipped with a pointed slender beak ______ **Oxytropis**
                    45. Leaves all basal (1 species with flowers 8.5–14 mm long) OR leaves usually extending well up the stem; keel acute or rounded, not beaked ______ **Astragalus**

Acacia

Herbaceous perennials, spiny shrubs, or small trees; leaves twice even-pinnately compound; inflorescences of many-flowered heads or spikes; flowers very small, white to yellow; petals inconspicuous, separate nearly to base; stamens numerous, 20–40 per flower, serving as the main attractant structure for the flower.

Acacia is a huge genus of 1,200 species of tropical and warm areas, especially Australia; most are trees and shrubs of dry regions where they can be important components of the vegetation. In Mexico, Central America, and in Africa, ants have a mutualistic relationship with species of *Acacia*, protecting the plant in exchange for shelter and/or food. Many species are of economic importance as sources of timber, fuel, forage, tanbark, gums, scents, and as cultivated ornamentals. *Acacia senegal* (L.) Willd., of arid tropical Africa, is the source of gum arabic used in lozenges, adhesives (e.g., postage stamps), watercolors, bakery products, and medicines (Morton 1977). (Greek: *akis*, a sharp point, alluding to the thorns of many species) (subfamily Mimosoideae, tribe Acacieae)
REFERENCES: Isely 1969, 1973; Beauchamp 1980; Clarke et al. 1989.

1. Flowers in globose heads; plants unarmed or armed.
   2. Plants unarmed; herbs or subshrubs to ca. 1 m tall, usually less; pinnae 6–16 pairs (these with numerous leaflets) per leaf; fruits flat, 6–10 mm wide; filaments creamy white; petiolar gland absent ___ **A. angustissima**
   2. Plants armed; shrubs or small trees 1–4 m tall; pinnae 1–6 pairs per leaf; fruits either round in cross-section OR flat and 15–30 mm wide; filaments creamy white or bright yellow; petiolar gland usually present.
      3. Plants with straight, paired, stipular spines; pinnae 2–6 pairs per leaf; filaments bright yellow; fruits ± round in cross-section ___ **A. farnesiana**
      3. Plants without paired, stipular spines, instead armed along the internodes with recurved prickles; pinnae usually 1–3 pairs per leaf; filaments creamy white; fruits flat in cross-section ___ **A. roemeriana**
1. Flowers in elongate spikes or racemes; plants armed ___ **A. greggii**

Acacia angustissima (Mill.) Kuntze var. **hirta** (Nutt.) B.L. Rob., (sp.: most narrow; var.: hairy), FERN ACACIA, PRAIRIE ACACIA, WHITE-BALL ACACIA. Perennial with creeping, woody root; leaflets sensitive, folding together on being touched, in rain, or at night; heads solitary or few in axils of upper leaves, also terminal, forming erect, ± leafy racemes of heads; fruits 5–7 cm long. Common in prairies in calcareous clay, occasional to frequent in sandy soils; widespread in TX. Late May–early Jul. [*A. hirta* Nutt.] This is the most common *Acacia* in nc TX.

Acacia farnesiana (L.) Willd., (for Cardinal Odoardo Farnese, 1573—1626, of Rome, who grew the plant in his gardens), HUISACHE, SWEET ACACIA. Shrub or small tree 2–4 m tall; petioles usually with a small circular gland; flowers sweet-scented; fruits (2–)3–8 cm long. Chiefly in sandy or silty ground; Bell Co. (Fort Hood—Sanchez 1997), reportedly wild n to McLennan Co. (Mahler 1988), also Hopkins [?] (Turner 1959) and Tarrant (R. O'Kennon, pers. obs.) cos.; common in s Texas. Mar–Apr. [*Acacia minuta* (M.E. Jones) Beauch. subsp. *densiflora* (Alexander ex Small) Beauch., *A. smallii* Isely] Previously used as a source of fragrant oils for perfumes (Correll & Johnston 1970), valued as a honey plant in s TX (Wills & Irwin 1961), and used medicinally (Crosswhite 1980). The common name HUISACHE is derived from the Nahuatl (language of the Aztecs), *huitz-axin* (Crosswhite 1980). Texas populations of this species have sometimes been treated as *A. minuta* subsp. *densiflora* (Beauchamp 1980; Kartesz 1994), *A. minuata* (orthographic error) (Jones et al. 1997), or *A. smallii* (Powell 1988; Isely 1973, 1990); however, we are following Clarke et al. (1989) and Powell (1998) in treating it as *A. farnesiana*.

Acacia greggii A. Gray, (for Josiah Gregg, 1806–1850, who collected in Mexico and died in the wilderness in n CA). Shrub or small tree (1–)2–3 m tall; stems with recurved prickles usually along the internodes or occasionally at a node; pinnae 1–3 pairs per leaf; flowers sweet-scented; filaments creamy white; fruits 5–8 cm long, 15–20 mm wide. Possibly poisonous to animals due to the high cyanide content of the foliage (Kingsbury 1964). ☠

1. Leaflets 3–6 mm long; inflorescences 2–5 cm long; fruits usually strongly constricted, often twisted ____ var. **greggii**
1. Leaflets 5–9(–12) mm long; inflorescences often exceeding 5 cm; fruits straight-margined or some somewhat constricted, not twisted ____ var. **wrightii**

var. **greggii**, CATCLAW, CATCLAW ACACIA, DEVIL'S-CLAW. Rocky, gravelly, or sandy soils; Brown, Coleman, Shackelford, and Stephens cos., also Young Co. (Mahler 1988); West Cross Timbers s and w to w TX. Apr–May.

var. **wrightii** (Benth.) Isely, (for Charles Wright, 1811–1885, TX collector), CATCLAW, WRIGHT'S ACACIA, JOINT-VETCH, UÑA DE GATO, HUISACHILLO. Very similar to var. *greggii*. Similar habitats as var. *greggii*; Callahan and Shackelford cos.; West Cross Timbers s and w to w TX. Apr–early Jun. [*Acacia wrightii* Benth.]

Acacia roemeriana Scheele, (for Ferdinand Roemer, 1818–1891, geologist, paleontologist, and explorer of TX), ROEMER'S ACACIA, CATCLAW. Shrub 1–2(–3) m tall with weak stems; flowers fragrant; fruits 5–10 cm long, 15–30 mm wide. Rocky or sandy areas; Brown Co.; w margin of nc TX s and w to w TX. Spring or later according to rains. [*A. malacophylla* Benth.] Jack Stanford (pers. comm.) indicated that flowers of this species dissected by him and U.T. Waterfall had five separate carpels.

Calliandra conferta A. Gray, (genus: Greek: *kalos*, beautiful, and *andros*, male; sp.: compact, closely crowded together), a very small shrub native to s TX, occurs n to Travis Co. just s of nc TX. It is similar to *Acacia* species but can be distinguished by the combination of its small size (10–30 cm tall), unarmed stems, and leaves with only 2(–4) leaflets. The pollen of this species is quite distinctive—in large, golden, tear-shaped groups of usually 8 cells (= octads) (Stanford 1966).

ALBIZIA

A genus of 118 species of trees, shrubs, or lianas of warm areas of Asia, Africa, and South America. Species are variously used for timber, gums, shade for coffee or tea, or as ornamentals. (Named for Filippo delgi Albizzi, 18th century Italian naturalist, who introduced the genus into European culture) (subfamily Mimosoideae, tribe Igneae)
REFERENCES: Isely 1970b, 1973.

Albizia julibrissin Durazzo, (modification of Persian name), MIMOSA, SILKTREE. Broad-headed, fast growing, short-lived, small, unarmed tree with flat top; leaves twice even-pinnately compound, the ultimate leaflets folding together at night or in rain, or on being touched; flowers in small, short-peduncled, hemispherical heads; heads clustered, conspicuous and showy due to mass of stamens; corollas narrowly funnelform, greenish to yellowish, inconspicuous, with united petals; stamens usually 20–40; filaments ca. 23–36 (rarely more) mm long, united and white in basal 1/4, separate and pink to rose-lavender or reddish above, serving as the main attractant structure for the flower; fruits flat, ca. 8–18 cm long. Commonly cultivated, occasionally escaped; can aggressively invade native habitats; Grayson Co., also Dallas, Parker, and Tarrant cos. (O'Kennon & Diggs, pers. obs.); widespread in TX. Late May–early Jun. Native of s Asia.

AMORPHA

Unarmed erect shrubs; leaves once odd-pinnately compound; leaflets many, oblong or elliptic, often gland-dotted; stipules slender, falling when the leaves expand; flowers many, in terminal, solitary or panicled, slender, dense, erect, spike-like racemes (young tips often nodding); calyces 5-toothed, often gland-dotted; corollas usually of one petal (standard), sometimes with very reduced additional ones; stamens 10; fruits gland-dotted.

A North American genus of 15 species. (Greek: *amorphos*, shapeless deformed, from the absence of four of the petals) (subfamily Papilionoideae, tribe Amorpheae)
REFERENCE: Wilbur 1975.

1. Inflorescences (15–)20–40 cm long; leaflets 30–80 mm long, (15–)20–30 mm wide; se and e TX to very e margin of nc TX ______ **A. paniculata**
1. Inflorescences (2–)7–15(–25) cm long; leaflets (3–)10–50 mm long, (2–)4–15(–20) mm wide; widespread in nc TX.
   2. Leaves sessile or short-petioled, the petioles up to 5(–8) mm long; shrubs less than 1 m tall; leaflets (3–)10–18(–25) mm long; fruits often canescent, 3–4(–5) mm long (not counting conspicuous beak ca. 2 mm long), 1-sided, but ± straight except for beak; all calyx lobes acute ______ **A. canescens**
   2. Leaves with petioles (10–)20–30(–50) mm long; shrubs 1–3.5 m tall; leaflets (15–)20–50 mm long; fruits usually glabrous, 5–7 mm long (not counting inconspicuous beak < 1 mm long), definitely curved; adaxial calyx lobes often rounded ______ **A. fruticosa**

Amorpha canescens Pursh, (gray-pubescent), LEADPLANT. Rhizomatous, usually conspicuously canescent, gray-green, low shrub; leaflets 11–41(–49); racemes usually numerous near stem apices, often forming a dense compound cluster; standard 4.5–6 mm long, bright violet. Sandy prairies; Montague Co. in West Cross Timbers; also Panhandle, se, and sc TX. May–Jul.

Amorpha fruticosa L., (shrubby, bushy), FALSE INDIGO, BASTARD INDIGO, INDIGO-BUSH AMORPHA. Much-branched shrub to 3 m tall, dark-green; leaflets 9–21(–31), pubescent at least on veins on lower surface; racemes solitary or in clusters of 2–4; calyces ca. 3–5 mm long; standard 5–6 mm long, dark blue to reddish purple. Stream banks, chiefly limestone areas; widespread in TX. Apr–May. [*A. fruticosa* var. *angustifolia* Pursh] Reported as poisonous to livestock and to contain alkaloids (Burlage 1968). ☠

Amorpha paniculata Torr. & A. Gray, (with flowers in panicles), PANICLED AMORPHA. Shrub 1–3 m tall; leaflets 15–19, tomentose, often densely so and with conspicuous raised veins on the lower surface; racemes clustered 5–10 together in a compound inflorescence exserted above the leaves; standard purple; calyx lobes all acute or acuminate; fruits 6–8 mm long. Wooded and wet areas; Henderson Co. on the far e margin of nc TX; mainly se and e TX. May–Jun.

Amorpha roemeriana Scheele, (for Ferdinand Roemer, 1818–1891, geologist, paleontologist, and explorer of TX), occurs n to Travis Co. just s of nc TX. It is similar to *A. paniculata* but differs in having leaflets with inconspicuous veins beneath, 15–40 mm long, and an inflorescence 10–20 cm long. [*A. texana* Buckley]

AMPHICARPAEA HOG-PEANUT

A genus of 3 species of e Asia, North America, and Africa; some have cleistogamous flowers giving rise to subterranean fruits. Sometimes spelled *Amphicarpa.* (Greek: *amphi*, of both kinds, and *carpos*, fruit, in allusion to the two different fruit types) (subfamily Papilionoideae, tribe Phaseoleae)
REFERENCE: Turner & Fearing 1964.

Amphicarpaea bracteata (L.) Fernald, (bracteate, with bracts), SOUTHERN HOG-PEANUT. Annual with twining or sprawling stems 0.3–2 m long; leaves alternate, pinnately compound with 3 leaflets; leaflets 2–10 cm long, 18–70 mm wide, entire; petioles 2–10 cm long; stipules 3–8 mm long; chasmogamous flowers in axillary racemes of 1–17 flowers on peduncles 1–6 cm long; calyces 4–5 mm long, with 4 lobes; corollas lilac or whitish; stamens 10, diadelphous; fruits 1.5–4 cm long, flattened; racemes of cleistogamous, inconspicuous, often apetalous flowers near or underground producing fleshy, often subterranean fruits 6–12 mm in diam. Wooded areas; Bell, Dallas, and Grayson cos.; se and e TX w to nc TX. May–Sep. [*A. bracteata* var. *comosa* (L.)

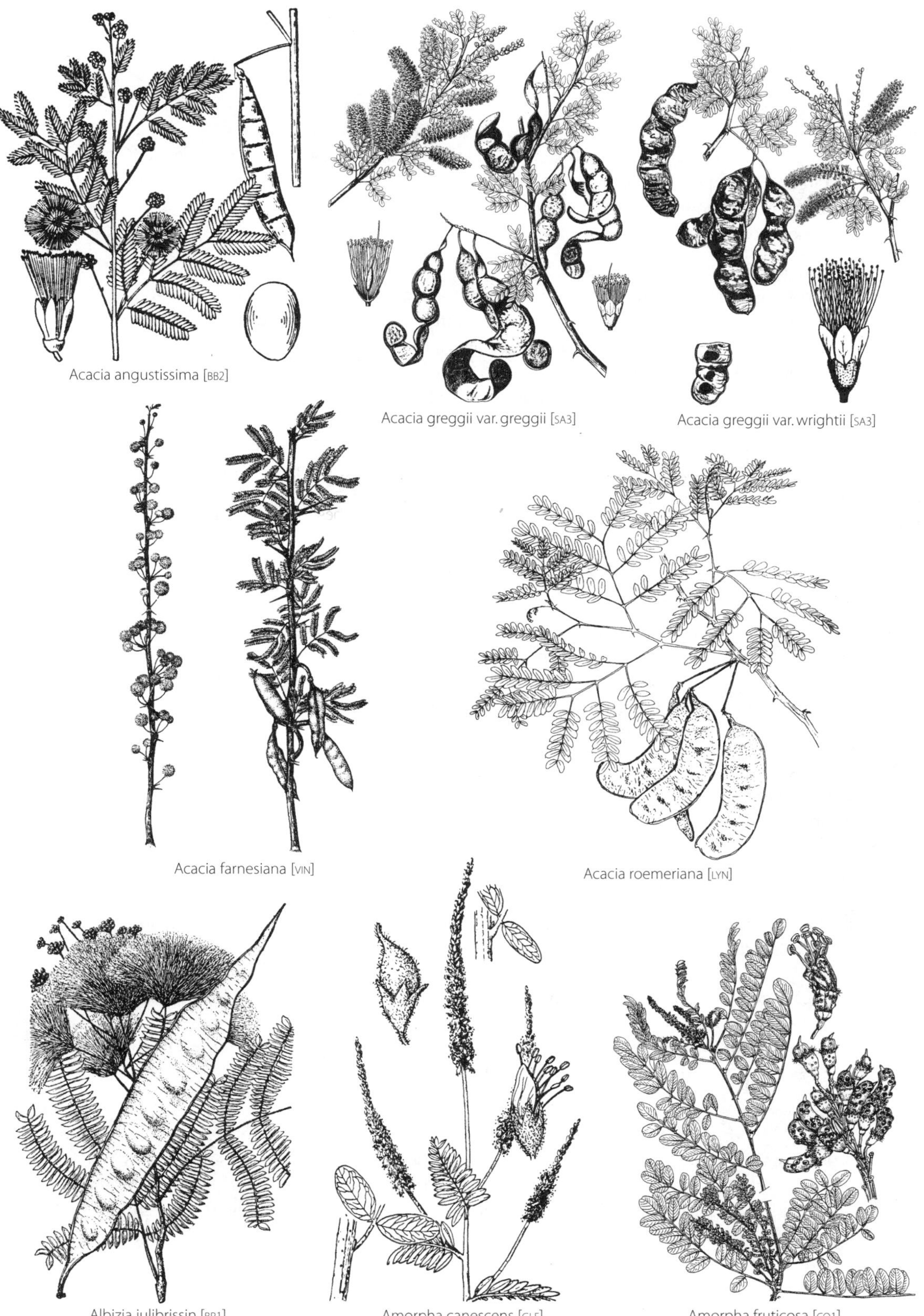

Acacia angustissima [BB2]

Acacia greggii var. greggii [SA3]

Acacia greggii var. wrightii [SA3]

Acacia farnesiana [VIN]

Acacia roemeriana [LYN]

Albizia julibrissin [BR1]

Amorpha canescens [GLE]

Amorpha fruticosa [CO1]

Fernald] The underground fleshy fruits, when cooked, are edible and were used by Native Americans (Mabberley 1987).

APIOS GROUNDNUT, POTATO-BEAN

A genus of 10 species of e Asia and North America. The habit (vine), characteristic leaves, and striking flower color make the single nc TX member of this genus easily recognizable in the field. (Greek: *apios*, pear, from the somewhat pyriform tuberous enlargement of the rhizomes) (subfamily Papilionoideae, tribe Phaseoleae)
REFERENCE: Seabrook & Dionne 1976.

Apios americana Medik., (of America), GROUNDNUT, AMERICAN POTATO-BEAN. Perennial from tuberous-enlarged rhizomes to 6 cm in diam.; stems annual, twining and climbing, to 4 m long, sparsely to densely pubescent; leaves once pinnately compound; leaflets 5–7, ovate- or elliptic-lanceolate, 2–10 cm long, pubescent, especially beneath, rounded at base; petioles 15–70 mm long; stipules nearly thread-like, 4–7 mm long, soon deciduous; flowers slightly sweet-scented, in axillary, short-peduncled, spike-like racemes; pedicels 2–6 mm long; standard cupped or hooded, brown-red with pale back; wings down-curved, brown-red or purple-red; keel sickle-shaped, brownish or dull red; stamens 10, diadelphous; fruits linear, 5–12 cm long, 4–7 mm wide, slightly flattened. Damp woods and thickets; Cooke, Denton, Fannin, Grayson, and Henderson cos.; e TX w to East Cross Timbers, rare in Edwards Plateau. Jun–Aug. The cooked tubers were eaten by Native Americans and early explorers (McGregor 1986; Heiser 1993). /78

ARACHIS PEANUT

A South American genus of 69 species (Krapovickas & Gregory 1994). The rich yellow to nearly orange flowers, conspicuous stipules, and readily identified subterranean fruits make the single species occasionally escaped in nc TX easily recognizable in the field. (Greek: *a*, without, and *rachos*, branch) (subfamily Papilionoideae, tribe Aeschynomeneae)
REFERENCES: Hermann 1954b; Gregory et al. 1980; Krapovickas & Gregory 1994.

Arachis hypogaea L., (underground), COMMON PEANUT, PEANUT, GOOBER, GROUNDNUT, MONKEYNUT, EARTHNUT. Annual with erect or decumbent, villous stems; leaves pinnately compound, usually with 4 leaflets; leaflets 2–6 cm long; stipules prominent, linear-subulate, 10–50 mm long; flowers axillary; pedicel (really sterile lower portion of fruit) eventually elongating and pushing the fruit underground where it matures; corollas yellow (standard can be nearly orange), 1–1.5 cm long; stamens 10, monadelphous; fruits indehiscent, with 1–3 seeds, 1–5 cm long, 1–2 cm wide. Sandy areas along rivers, weedy areas; Comanche, Dallas, Grayson, and Somervell cos., also Hood Co. (R. O'Kennon, pers. obs.); also Post Oak Savannah. Jun–Oct. Native of South America. PEANUT is thought to be of allopolyploid origin (Mabberley 1987). Leaf lesions caused by the rust fungus *Puccinia arachidis* Speg. can sometimes be found (J. Hennen, pers. comm.). This is an important crop locally in areas of sandy soils (e.g., annual Peanut Festival in Whitesboro, in w Grayson Co.); it is cultivated for its edible seeds and for oil. Peanut meal can be toxic as a result of contamination with aflatoxin, produced by the fungus *Aspergillus flavus* Link: Fr. (Lewis & Elvin-Lewis 1977).

ASTRAGALUS LOCOWEED, MILK-VETCH

Ours herbaceous, unarmed, annuals or perennials, prostrate or decumbent to wide-spreading or erect or nearly so; leaves once odd-pinnately compound, stipulate, estipellate; flowers in peduncled, terminal or axillary heads or head-like or spike-like racemes; petals various shades of pink, lavender, or purple to white or yellowish; stamens 10, diadelphous; fruits inflated, bladdery, or linear to narrowly oblong, not inflated.

A huge genus of 1,750 species; *Astragalus* is considered one of the largest genera of vascular plants. It occurs mainly in the n temperate zone, especially w North America and c and w Asia and extends to Chile, n India, and the mountains of tropical Africa; it is notoriously difficult taxonomically. Some species concentrate selenium in their tissues (POISON-VETCHES); such plants can cause a toxic, sometimes fatal response in livestock—when symptoms include emaciation, hoof deformity, and lameness, the condition has been referred to as alkali disease; when blindness, excitement, and other nervous symptoms occur, the condition has been referred to as blind staggers; other *Astragalus* species (LOCOWEEDS) are apparently toxic due to indolizidine alkaloids (e.g., swainsonine) and cause locoism; symptoms in livestock include intoxication, paralysis, respiratory problems, and sudden death; affected horses become excessively excited and wild and are dangerous to ride (Kingsbury 1964, 1965; Blackwell 1990; James & Welsh 1992; Ralphs 1992). Other species, however, are cultivated for hay or forage. *Oxytropis*, sometimes included in *Astragalus*, is here treated as a separate genus. (Ancient Greek name of a leguminous plant; perhaps from *astragalus*, ankle bone or dice, possibly alluding to rattling of seeds within fruit) (subfamily Papilionoideae, tribe Galegeae)

REFERENCES: Barneby 1964; Isely 1983–1986.

1. Leaves all basal or crowded near base of stem, the leafless peduncles more than twice as long as the leaf-bearing lower portion of the stem; leaflets, above and beneath, densely gray-pubescent; pubescence consisting of hairs attached above their bases so that they have a long and a short arm (= malpighian or dolabriform hairs); plants perennial.
   2. Leaflets linear or narrowly lanceolate, 4–15 times as long as wide; keel tipped with a pointed, slender beak; corollas 15–26 mm long; fruits 7–12 mm long (without beak) ______ see **Oxytropis lambertii**
   2. Leaflets lanceolate to elliptic-orbicular, 1.5–4 times as long as wide; keel acute or rounded, not beaked; corollas 8.5–14 mm long; fruits 12–37 mm long (without beak) ______ **A. lotiflorus**
1. Leaves extending well up the stem, the peduncles about equaling or shorter than the leaf-bearing portion of stem; leaflets glabrous to sparsely pubescent above OR else plant an annual; pubescence consisting of hairs attached at their bases (not malpighian or dolabriform—except in *A. canadensis*).
   3. Plants perennial, with rhizomes or woody taproot.
      4. Plants erect and tall (0.4–1.6 m); stipules dry, papery, deciduous; inflorescences densely spike-like, the flowers usually 20 or more; stems and leaves with hairs attached above their bases so that they have a long and a short arm ______ **A. canadensis**
      4. Plants prostrate to ascending-erect, usually < 0.4 m tall; stipules herbaceous, persistent; inflorescences with 1–20 flowers; stems and leaves with hairs attached at their bases.
         5. Leaflets acute or obtuse, pubescent beneath, at least when young; mature fruits 1.8–27 mm wide.
            6. Calyces 7–10 mm long, the tube longer than lobes; corollas 12–20 mm long; mature fruits 10–27 mm wide.
               7. Ovaries and fruits glabrous; plants without slender rhizomes; mature fruits 11–27 mm wide ______ **A. crassicarpus**
               7. Ovaries and fruits densely gray-pubescent; old plants with slender rhizomes; mature fruits 10–13 mm wide ______ **A. plattensis**
            6. Calyces 2.5–5 mm long, the tube about equaling lobes; corollas 5–12 mm long; mature fruits 1.8–3.5 mm wide ______ **A. nuttallianus**
         5. Leaflets notched or truncate, glabrous beneath; mature fruits 1.8–7 mm wide.
            8. Inflorescences with 6–20 flowers, spicate or racemose; calyces with short white or white and black hairs; mature fruits 3.5–7 mm wide ______ **A. distortus**
            8. Inflorescences often with 1–6 flowers (sometimes more), head-like or subumbellate; calyces varying from glabrous to villous, without black hairs; mature fruits 1.8–3.5 mm wide ______ **A. nuttallianus**

3. Plants annual, with slender taproot.
   9. Leaflets of all leaves notched to truncate apically.
      10. Peduncles with wide-spreading hairs; keel longer than the wings; wings whitish; standard and keel violet or white-tipped or margined; corollas 4–5.2 mm long; fruits 5.5–9 mm long ___ **A. reflexus**
      10. Peduncles with appressed or ascending hairs or glabrous; keel shorter than the wings; petals variously colored; corollas (4–)4.8–18 mm long; fruits 12–37 mm long.
         11. Corollas (12–)13–18 mm long; wings wholly white in apical 2/3; ovules 8–12; style pubescent; mature fruits 3.5–6.5 mm wide, stipitate, the stipes 1–3 mm long ___ **A. lindheimeri**
         11. Corollas 5–12 mm long; wings colored to apex along lower edge; ovules 10–26; style glabrous; mature fruits 1.8–3.5 mm wide, sessile or nearly so, the stipes < 1 mm long.
            12. Ovules 20–26; mature fruits 20–37 mm long, straight or gently and evenly curved for their whole length; standard 8.3–12 mm long; occasional in Blackland Prairie and West Cross Timbers ___ **A. leptocarpus**
            12. Ovules 10–18; mature fruits 12–26 mm long, usually curved near base, otherwise straight or nearly so; standard (4–)4.8–10 mm long; widespread throughout nc TX ___ **A. nuttallianus**
   9. Leaflets of middle and upper leaves acute to obtuse at apex.
      13. Calyx teeth about equal in length to calyx tube; mature fruits glabrous or sparsely hairy, 12–26 mm long, curved near base or straight, spreading, ascending, or deflexed, not erect; calyces half as long as corollas or less; ovules 10–18; widespread throughout nc TX ___ **A. nuttallianus**
      13. Calyx teeth 1.5–2.5 times as long as calyx tube; mature fruits densely pilose-hairy, 7–13 mm long, straight or nearly so, erect; calyces over half as long as corollas to nearly as long as corollas; ovules 5–9; limited to the s and sw margins of nc TX ___ **A. wrightii**

Astragalus canadensis L., (of Canada), CANADA MILK-VETCH. Perennial, inconspicuously pubescent or glabrous, usually tall and robust; stems 0.4–1.6 m tall, erect, leafy, branched above; petals greenish white, light yellowish green, yellowish white, or dull straw-colored; fruits 10–15 mm long, 4–5.2 mm wide. Rocky or sandy thickets; Cooke, Dallas, and Grayson cos., also Lamar Co. (Mahler 1988); se and e TX w to the n part of nc TX. Jun–early Jul.

Astragalus crassicarpus Nutt., (thick-fruited), GROUND-PLUM, BUFFALO-PLUM, POMME DE PRAIRIE, GROUND-PLUM MILK-VETCH. Perennial; stems 10–60 cm long, decumbent, ascending, or suberect; flowers often showy; petals lilac, pink-purple, or purple. Native Americans ate the immature, succulent, plum-like fruits; care must be taken because there are several closely related poison MILK-VETCHES or LOCOWEEDS. The two varieties below can be separated by the following key modified from Barneby (1964).

1. Stems arising singly or few together from slender, widely creeping, subterranean caudex-branches, forming loosely matted or colonial growths; ovules 34–50; c TX n to extreme s part of nc TX ___ var. **berlandieri**
1. Stems arising together from the crown of the taproot or shortly forking, determinate caudex at or just below soil-level; ovules 52–68; widespread in nc TX ___ var. **crassicarpus**

var. **berlandieri** Barneby, (for Jean Louis Berlandier, 1805–1851, French botanist who explored TX, NM, and Mexico, and one of the first botanists to make extensive collections in TX). Similar to var. *crassicarpus*; petals lilac or pink-purple. Soils derived from limestone or clay, gravelly areas, prairies; Williamson Co. (Correll & Johnston 1970); mainly Edwards Plateau e to coastal plain; endemic to TX. Late Apr–summer.

var. **crassicarpus**. GROUND-PLUM, MILK-VETCH, INDIAN-PEA. Petals purple or lilac; fruits globose to ovoid or obovoid, sometimes superficially plum-like in appearance, up to 1.5 times as long as

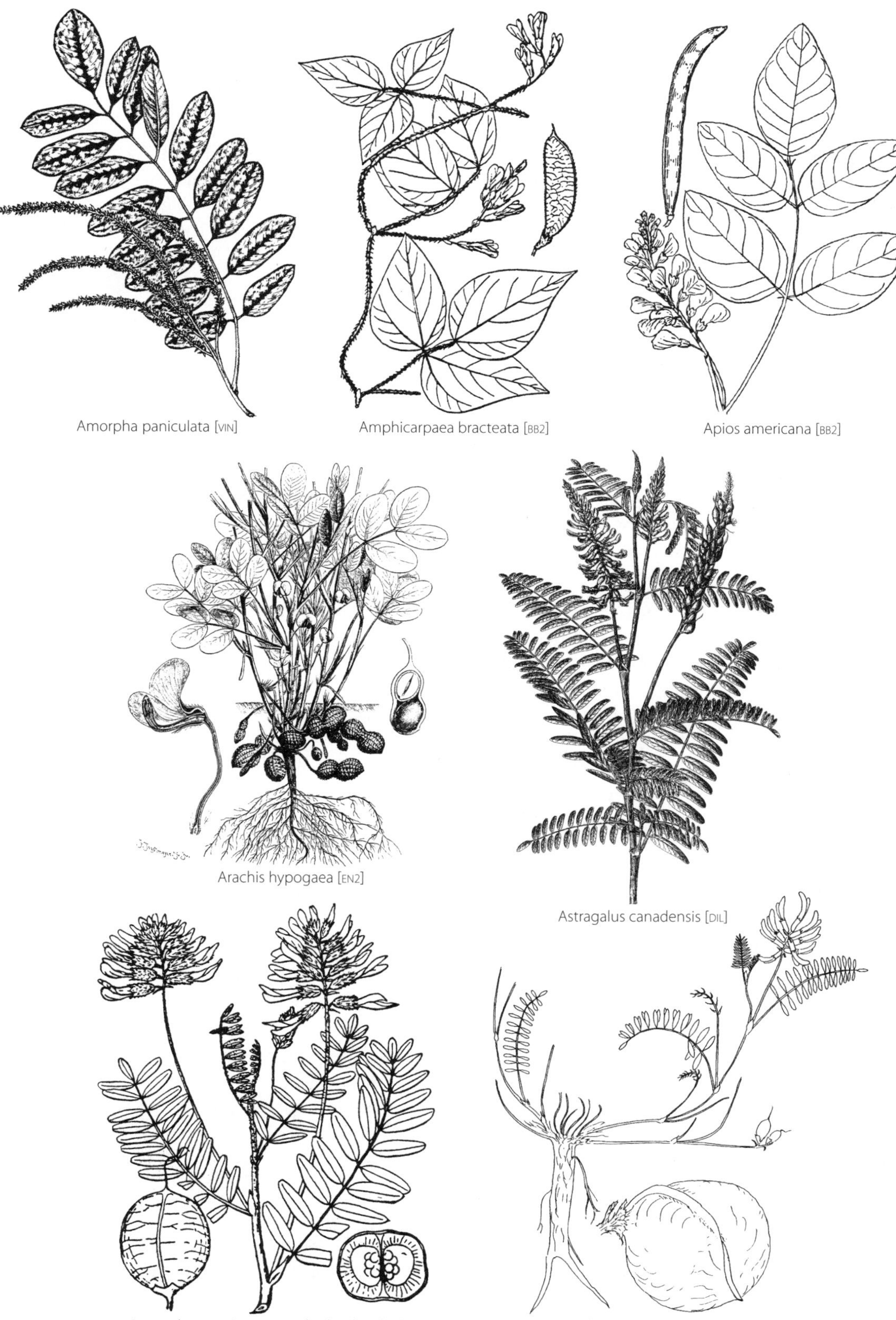

Amorpha paniculata [VIN]

Amphicarpaea bracteata [BB2]

Apios americana [BB2]

Arachis hypogaea [EN2]

Astragalus canadensis [DIL]

Astragalus crassicarpus var. berlandieri [BB2]

Astragalus crassicarpus var. crassicarpus [BT3]

wide, 15–27 mm long, 12–25 mm wide. Clayey or rocky prairies; mainly Blackland and Grand prairies; also Panhandle to Trans-Pecos. Mar–Jul.

var. *trichocalyx* (Nutt.) Barneby, (with a hairy calyx), with greenish white or creamy petals, occurs in open woods just to the e of nc TX.

Astragalus distortus Torr. & A. Gray, (twisted, from the fruit shape). Low perennial; stems decumbent to prostrate, 10–35 cm long; calyces with white or mixed black and white hairs, 3.1–6.3 mm long; corollas 8.2–15.3 mm long, variable in color, pink-purple, lilac to whitish, sometimes with markings; fruits 13–25 mm long. The 2 varieties below can be distinguished by the following key modified from Barneby (1964).

1. Flowers relatively large, the standard 11–15.5 mm long, the keel (7–)7.4–9.3 mm long; fruits deeply sulcate both dorsally and ventrally; ovules 26–37; only known n of Red River to the n of nc TX __ var. **distortus**
1. Flowers smaller, the standard 8.2–12 mm long, the keel 5.5–7.2 mm long; fruits shallowly sulcate dorsally but not or only obscurely so ventrally; ovules 16–28; only known s of Red River ____ var. **engelmannii**

var. **distortus**. OZARK MILK-VETCH, BENT-POD MILK-VETCH. Prairies and post oak woodland; Cross Timbers just n of Red River in OK; not reported in TX; included because of proximity and possiblity of occurrence. Late Mar–Jul.

var. **engelmannii** (E. Sheld.) M.E. Jones, (for George Engelmann, 1809–1884, German-born botanist and physician of St. Louis), ENGELMANN'S MILK-VETCH. Sandy soils; Hunt, Henderson, and Kaufman cos., also Tarrant Co. (Barneby 1964); in much of e 1/2 of TX; not reported n of Red River in OK. Mar–May.

Astragalus leptocarpus Torr. & A. Gray, (thin- or slender-fruited), SLIM-POD MILK-VETCH, BODKIN MILK-VETCH. Inconspicuously appressed-pubescent or glabrous annual; stems erect, ascending, or decumbent, 3–20(–35) cm long; racemes with 2–7(–12) flowers; petals purple-blue drying violet, the center of standard and upper sides of wings and keel often white; fruits straight or slightly curved, linear, 2.2–3.1(–3.5) mm wide. Sandy open woods or calcareous clays; Bell, Dallas, Denton, Henderson, Limestone, and Parker cos.; se and e TX w to nc TX. Apr. *Astragalus leptocarpus* "is closely related to *A. nuttallianus* and might logically be treated as forming part of that polymorphic complex; for apart from the ordinarily longer pod enclosing from one to four extra pairs of ovules and seeds, it possesses no single attribute which cannot be matched somewhere in *A. nuttallianus*. In practice it is easily distinguished from sympatric varieties of *A. nuttallianus*: from var. *trichocarpus* by its glabrous ovary; from var. *pleianthus* by its retuse leaflets; from var. *macilentus* by the acute keel-tip; and from var. *nuttallianus*, which it most nearly resembles in its green, glabrescent foliage and in shape of the leaflets, by the nearly always longer and more amply proportioned flower" (Barneby 1964).

Astragalus lindheimeri Engelm. ex A. Gray, (for Ferdinand Jacob Lindheimer, 1809–1879, German-born TX collector), LINDHEIMER'S MILK-VETCH, BUFFALO-CLOVER, MILK-VETCH. Inconspicuously appressed-pubescent annual; stems decumbent to ascending, to ca. 35 cm long; inflorescences racemose, with 2–8 crowded flowers; flowers sweet-scented, large and very showy; petals bicolored, purple-blue, with center of standard, upper side and tips of wings, and tip of keel white, turning violet upon drying; fruits 17–27 mm long, flattened, curved-oblong. Sandy, silty, or rocky ground; West Cross Timbers, also Dallas and Navarro cos. (Barneby 1964); also Rolling Plains and Edwards Plateau and in sw part of Oklahoma along Red River (Barneby 1964). Late Mar–early May.

Astragalus lotiflorus Hook., (with flowers like *Lotus*—deer-vetch, trefoil), LOTUS MILK-VETCH. Plant low, tufted, ± scapose, spreading to erect, to 18 cm tall, usually densely pubescent with hairs attached above base so that they have a long and a short arm (= malpighian or

Astragalus distortus var. distortus [BB2]

Astragalus distortus var. engelmannii [HEA, JME]

Astragalus leptocarpus [BT3]

Astragalus lindheimeri [BT3]

Astragalus lotiflorus [BT3]

Astragalus nuttallianus var. austrinus [BT3]

Astragalus nuttallianus var. macilentus [BT3]

Astragalus nuttalianus var. trichocarpus [COC]

dolabriform hairs); chasmogamous and cleistogamous inflorescences both produced, commonly on different plants; inflorescences with (3–)5–17 flowers; petals greenish white to yellowish or tinged with red or purple or purple-tipped or -margined. Limestone and sandstone outcrops; n Blackland Prairie and Grand Prairie w to Plains Country and Trans-Pecos. Late Mar–Apr. Barneby (1964) considerd the species to be a polymorphic indivisible entity. [*A. lotiflorus* var. *reverchonii* (A. Gray) M.E. Jones]

Astragalus nuttallianus DC., (for Sir Thomas Nuttall, 1786–1859, English-American botanist), TURKEY-PEA, SMALL-FLOWER MILK-VETCH. Glabrous to pubescent annual (but can also be keyed as a perennial in the key); stems erect to decumbent, to 35(–45) cm long; petals varying in color from various shades of pinkish to purplish, often with variable white markings, to whitish; fruits 12–26 mm long, 1.8–3.5 mm wide. This is an extremely variable species with a number of often difficult to distinguish varieties occurring in or near nc TX. They are distinguished in the key modified from Barneby (1964); distributions also follow Barneby (1964).

1. Leaflets of all leaves notched or truncate-emarginate.
   2. Flowers small, the standard 4.3–7.3(–7.6) mm long, the keel-tip triangular-acute or sharply deltoid; raceme axis not elongating or scarcely so, not over 0.8(–1) cm long in fruit; widespread in nc TX _____ var. **nuttallianus**
   2. Flowers larger, the standard 8.5–13 mm long, the keel-tip obtusely rounded; raceme axis mostly elongating and usually 1–3 cm long in fruit; just to the s of nc TX _____ var. **macilentus**
1. Leaflets not all notched or truncate-emarginate, elliptic or ovate and obtuse to subacute in at least some upper or often all leaves.
   3. Keel obtusely rounded at tip; racemes elongating, the axis 1–3 cm long in fruit; fruits commonly glabrous; just to the s of nc TX _____ var. **macilentus**
   3. Keel triangular or narrowly deltoid at tip, acute or subacute; racemes not or scarcely elongating, the axis 0.8(–1.2) cm or less long in fruit; fruits glabrous or with pubescence; widespread in nc TX.
      4. Flowers relatively large, the standard mostly 7–9 mm long; racemes mostly with 4–10 flowers; fruits consistently glabrous _____ var. **pleianthus**
      4. Flowers small, the standard mostly 5–7 mm long; racemes mostly with 1–5(–6) flowers; fruits glabrous or with pubescence.
         5. Fruits hirsutulous, the hairs 0.5–1 mm long; leaflets commonly 11–13 _____ var. **trichocarpus**
         5. Fruits glabrous or if pubescent the hairs shorter, appressed, 0.5 mm long or less; leaflets commonly 7–11 _____ var. **austrinus**

var. **austrinus** (Small) Barneby, (southern). Dry plains and hillsides; from West Cross Timbers s and w to w TX. Mar–May.

var. **macilentus** (Small) Barneby, (thin, lean). Rocky and disturbed areas, often on calcareous soils; just to the s of nc TX in s Burnet and Travis cos.; c to w TX. Mar–May. Isely (1983–1986) treated this variety as a separate species indicating that although partially sympatric with *A. nuttallianus*, it does not intergrade with that species. [*A. macilentus* (Small) Cory]

var. **nuttallianus**. NUTTALL'S MILK-VETCH, TURKEY-PEA, PEAVINE. Disturbed ground, mostly limestone; Blackland and Grand Prairies and Cross Timbers; also South TX Plains, Edwards Plateau, and Rolling Plains. Mar–Jun.

var. **pleianthus** (Shinners) Barneby, (many-flowered). Calcareous soils; prairies, open woodlands; Blackland Prairie sw to Edwards Plateau; endemic to TX. Mar–May. [*A. pleianthus* (Shinners) Isely] Kartesz (1994) and Jones et al. (1997), following Isely (1983–1986), treated this taxon as a separate species; Isely (1983–1986) indicated that "Though sympatric with several of its congeners, it seemingly does not blend with any of them." However, we are following Barneby (1964) in maintaining it at the rank of variety.

var. **trichocarpus** Torr. & A. Gray, (hairy-fruited), SOUTHWESTERN MILK-VETCH. Calcareous or sandy soils; Archer, Young, Palo Pinto, Erath, Somervell, and McLennan cos. (Barneby 1964); scattered mainly in e 1/2 of TX. Mar–May.

Astragalus plattensis Nutt., (of the Platte River), PLATTE RIVER MILK-VETCH, GROUND-PLUM. Resembling *A. crassicarpus*, usually largely decumbent, more pubescent throughout; flowers sweet-scented; standard yellowish white to light or deep lavender or rosy lavender; wings similar or deeper-colored; keel lavender to purple or red-purple with dark tip; ovary pubescent. Sandy or rocky prairies or open woods; from e edge of East Cross Timbers s to South TX Plains and nw to Panhandle. Late Mar–mid-Jul.

Astragalus reflexus Torr. & A. Gray, (reflexed, bent sharply backward), TEXAS MILK-VETCH, DROOPING MILK-VETCH. Rather sparsely pilose, prostrate to erect annual; stems to 35 cm long; racemes with 3–10 flowers; corollas 4–5.2 mm long; petals bicolored, bluish violet or reddish violet with white; fruits 5.5–9 mm long, 2.5–5.5 mm wide. Prairies, roadsides, calcareous or clay soils; Bell, Dallas, Hill, McLennan, Somervell, and Tarrant cos.; nc and c TX mainly between the Trinity and Colorado rivers from Dallas-Ft. Worth to Austin, also disjunct to Cameron Co. in far s TX; endemic to TX. Feb–May.

Astragalus wrightii A. Gray, (for Charles Wright, 1811–1885, TX collector), WRIGHT'S MILK-VETCH. Gray-pilose, erect annual to 35 cm tall; inflorescences head-like, 3–7-flowered; corollas 5.2–6.2 mm long; petals reddish violet, lilac, or whitish with lilac tinge; fruits 7–13 mm long, 2.5–3.5 mm wide. Sandy or gravelly ground; Bell, Burnet, Lampasas, and Williamson cos., also Brown Co. (Barneby 1964); Edwards Plateau and s and w parts of nc TX; endemic to TX. Apr.

BAPTISIA WILD INDIGO, FALSE INDIGO

Herbaceous perennials with woody crown and tough roots; foliage sometimes darkening upon drying; leaves very short-petioled, palmately compound with 3 leaflets (or uppermost with only 1 or 2); stipules large to small, persistent or falling early; flowers terminal, large, solitary or in erect to horizontal or drooping racemes or solitary in the leaf axils; corollas variously colored; stamens 10, separate; fruits rounded or subcylindric, becoming woody, often with a beak.

A genus of 17 species endemic to the e U.S.; nc TX species can usually be easily recognized to genus by the combination of the palmately 3-foliate leaves, free stamens, and inflated fruits (Isely 1990). *Baptisia tinctoria* (L.) R. Br. ex W.T. Aiton was formerly used as a dye plant; others are used as ornamentals or medicinally; a number contain alkaloids. ☠ The plants are avoided by grazing animals and some species are reportedly toxic to livestock. Hybridization is well known in *Baptisia* (e.g., Turner & Alston 1959; Alston & Turner 1962, 1963). Kosnik et al. (1996) described a *Baptisia* hybrid complex in nc TX including a new hybrid. In nearly every case where 2 or more parental species have been found growing together in nc TX, hybrid individuals have also been found. The key to species below includes the hybrids known to occur within nc TX. (Greek: *baptis*, dye or dip, from the economic use of some species which yield a poor indigo dye) (subfamily Papilionoideae, tribe Thermopsideae)
REFERENCES: Larisey 1940; Turner & Alston 1959; Alston & Turner 1962, 1963; Kosnik et al. 1996.

1. Petals cream-colored; inflorescences held distinctly below the level of the leaves; vegetative structures densely pubescent ____ **B. bracteata**
1. Petals blue-violet to yellow, orange, or brick-red, but not cream; inflorescences at least partially erect and held above the level of the leaves OR flowers solitary in the upper leaf axils; vegetative structures glabrous to moderately pubescent.

2. Flowers solitary in the upper leaf axils (at least some) or in short terminal racemes with 1–4 flowers; petals yellow ______ **B. nuttalliana**
2. Flowers in racemes of 5–20 or more flowers; petals yellow, orange, brick-red, or blue-violet.
   3. Petals blue-violet (keel sometimes whitish); stems at base thick (> 5 mm diam.); plants usually growing individually; inflorescences one to a few per plant.
      4. Petals intensely to moderately blue-violet; inflorescences vertical (strictly erect); foliage glabrous and glaucous ______ **B. australis**
      4. Petals moderately to pale blue-violet; inflorescences angled about 45°; foliage slightly pubescent, not glaucous ______ **B. ×bicolor**
   3. Petals yellow or multicolored; stems at base thin (< 5 mm diam.); plants usually growing in dense clusters; inflorescences usually numerous per plant.
      5. Petals yellow, all ± the same color; foliage glabrous to moderately pubescent.
         6. Petals vividly yellow; racemes nearly vertical (strictly erect); foliage glabrous ______ **B. sphaerocarpa**
         6. Petals pale to medium yellow; racemes angled about 45°; foliage moderately pubescent ______ **B. ×bushii**
      5. Petals brick-red and yellow to orangish to blue-violet and yellow, different petals of the same flower often of different colors; foliage glabrous ______ **B. ×variicolor**

Baptisia australis (L.) R. Br. ex W.T. Aiton var. **minor** (Lehm.) Fernald, (sp.: southern; var.: smaller), BLUE WILD INDIGO, WILD BLUE-INDIGO, BLUE-INDIGO. Plant usually growing individually, tall (usually 46–74 cm), of strictly erect posture, with a single thick (> 5 mm diam.), glaucous stem rising from the ground several cm before branching; branches generally few and rigid; leaves small (center leaflet 18–34 mm long) and glabrous; stipules small (4–11 mm long); flowers borne on a usually single, rigidly erect raceme; floral bracts quickly deciduous; pedicels 5–15 mm long; corollas large (27–36 mm long), blue-violet of variable intensity; fruits usually much longer than wide, 30–60 mm long, 20–30 mm wide, with a distinct persistent beak that is noticeably widened at base, black at maturity, glabrous. Clay soils, prairies, pastures; Fannin, Grayson, and Montague cos. in Red River drainage, also Collin and Hunt cos., also Dallas Co. (Mahler 1988); otherwise only known in TX from the Panhandle and Titus Co. to the e. Apr. [*B. minor* Lehm., *B. minor* var. *aberrans* Larisey, *B. texana* Buckley, *B. vespertina* Small ex Rydb.] According to Ajilvsgi (1984), the sap turns purple when exposed to air; it is still used in making a blue dye for dying wool for use in weaving. ⊠/80

Baptisia australis var. *australis* is cited by Hatch et al. (1990) for vegetational area 4 (Fig. 2) and Jones et al. (1997) cited it for TX. However, J. Kartesz (pers. comm. 1997) indicated there is not a TX record. Isely (1990) separated the 2 varieties as follows;

1. Legume more nearly symmetric, tending to be oblong, moderately inflated, 1.2–1.5(–2) cm in diam.; stems commonly 1–1.5 m tall, with ascending branches ______ var. *australis*
1. Legume usually strongly asymmetric, ovoid to oblong-lanceoloid, much inflated, 2–3 cm in diam.; stems usually 0.5–1 m long or tall, with spreading, divaricate branches ______ var. *minor*

Baptisia ×bicolor Greenm. & Larisey [*B. australis* × *B. bracteata*], (two-colored), TWO-COLOR WILD INDIGO. Plant usually growing individually, intermediate to both parents in numerous characters, of medium height (30–60 cm tall), generally branching at or near the ground; stems and foliage moderately pubescent; leaves and stipules of moderate length (center leaflet 27–51 mm long; stipules 4–10 mm long); flowers borne on 1-several racemes angled about 45° above the ground; floral bracts quickly deciduous; pedicels of 10–18 mm long; corollas large (27–33 mm long), moderately to pale blue-violet. Prairies, pastures, soils often intermediate between clay and sand or in areas where soil types are in close proximity; Fannin and Grayson cos. in Red River drainage; we are unaware of other TX locations. Apr. ⊠/80

Astragalus plattensis [BB1]

Astragalus reflexus [BT3]

Astragalus wrightii [BT3]

Baptisia australis var. minor [STE]

Baptisia bracteata var. leucophaea [STE]

Baptisia nuttalliana [COC]

Baptisia sphaerocarpa [BT3]

Baptisia bracteata Muhl. ex Elliott var. **leucophaea** (Nutt.) Kartesz & Gandhi, (sp.: with bracts; var.: dusky-white), PLAINS WILD INDIGO. Plant usually growing individually, low (21–36 cm tall), often wider than tall, branching at or near the ground, with all vegetative parts densely pubescent; leaves and stipules large (center leaflet 30–60 mm long; stipules 10–40 cm long); flowers on reclining or hanging racemes; floral bracts persistent and large (10–30 mm long); pedicels long (15–35 mm long); corollas large (25–38 mm long); fruits (2–)3–4(–5) cm long, 1.5–2.5 cm wide, tapering to a slender beak (beak rather wide at base), black, pubescent or rarely later glabrate. Sandy open woods, prairies, pastures, and roadsides; e TX w to East Cross Timbers. Late Mar–Apr. [*B. bracteata* var. *glabrescens* (Larisey) Isely, *B. leucophaea* Nutt., *B. leucophaea* var. *glabrescens* Larisey] The plants turn blackish or silver-gray late in the summer, break off, and tumble in the wind (Ajilvsgi 1984). ⊠/80

Baptisia ×bushii Small [*B. bracteata* × *B. sphaerocarpa*], (for Benjamin Franklin Bush, 1858–1937, amateur botanist of Missouri), BUSH'S WILD INDIGO. Plant usually growing individually, intermediate to both parents in numerous characters, low (26–36 cm tall), branching at or near the ground; stems and foliage moderately pubescent; leaves and stipules moderate to large (center leaflet 42–60 mm long; stipules 7–34 mm long); flowers on several racemes angled about 45° above the ground; floral bracts quickly deciduous; pedicels 7–15 mm long; corollas medium-sized (22–29 mm long), pale to medium yellow. Usually on sandy soils; Fannin and Grayson cos. in Red River drainage; also e TX. Apr. [*B. ×intermedia* Larisey, *B. ×stricta* Larisey] ⊠/80

Baptisia nuttalliana Small, (for Sir Thomas Nuttall, 1786–1859, English-American botanist), NUTTALL'S WILD INDIGO. Plant erect, to 120 cm tall, with well-developed individuals spherical and bush-like in form, puberulent, then glabrate; leaflets 25–60(–80) mm long; stipules very small, deciduous; pedicels 2–5 mm long; corollas 17–20 mm long, yellow; fruits ovoid to subspheroid, usually 8–13 mm long, (5–)8–13 mm wide, abruptly beaked, black, glabrate. Sandy open woods, pastures, and roadsides; Tarrant Co., also Lamar (Carr 1994) and McLennan (Turner 1959) cos.; mainly e TX. Apr–May. According to Isely (1990), often forming extensive poorly fruiting stands. This is the only nc TX *Baptisia* species with mostly axillary flowers.

Baptisia sphaerocarpa Nutt., (globe- or spherical-fruited), GREEN WILD INDIGO, YELLOW BUSH-PEA. Plant often growing in a dense cluster with a number of apparently identical individuals in close proximity, suggesting vegetative spread, tall (40–70 cm) and of generally erect posture, with 1-many small, flexible, glabrous stems, branching close to the base and often; leaves large (center leaflet 36–67 mm long); upper leaves often reduced to 1–2 leaflets; stipules small (2–5(–6.1) mm long) to absent; flowers borne on 1-many flexible vertical racemes; floral bracts quickly deciduous; pedicels short (1–7 mm long); corollas small (17–25 mm long), intensely yellow; fruits subspheroid, 1.4–1.8 cm in diam., usually light brown (Isely (1990) also reported black), glabrous; in nc TX the small, nearly round, light brown fruits immediately distinguish *B. sphaerocarpa* from the other 2 common parental species with larger, elongate, black fruits. Sandy or silty clay; e TX w to w Blackland Prairie (Grayson Co.). Apr–May. [*B. viridis* Larisey] ⊠/80

Baptisia ×variicolor Kosnik, Diggs, Redshaw, & Lipscomb, [*B. australis* × *B. sphaerocarpa*], (variably colored), VARICOLORED WILD INDIGO. Plant often growing in a dense cluster with a number of apparently identical individuals in close proximity, suggesting vegetative spread, intermediate to both parents in numerous characters, but vegetatively often more similar to *B. sphaerocarpa* than *B. australis*, tall (40–76 cm) and of generally erect posture, with one to many small, flexible, glabrous stems, branching at or near the base and often; leaves medium-sized (center leaflet 30–50 mm long); stipules small (3–15 mm long) to absent; flowers borne on 1-many flexible vertical racemes; floral bracts quickly deciduous; corollas of intermediate size (22–27 mm long), brick-red and yellow to orangish to blue-violet and yellow. Prairies, pastures, soils often intermediate between clay and sand or in areas where soil types are in close proximity; known in TX only in Fannin and Grayson cos. in the Red River drainage; also probably in s OK. Apr–May. Described in 1996 (Kosnik et al.). ▣/80

Baptisia alba (L.) Vent. var. *macrophylla* (Larisey) Isely, (sp.: white; var.: large-leaved), with white flowers in erect racemes, is cited by Hatch et al. (1990) for vegetational area 4 (Fig. 2). It probably occurs only to the s and e of nc TX.

CAESALPINIA RUSH-PEA

A genus of ca. 138 species (excluding *Pomaria*—Simpson 1998) of trees, shrubs, or hook-climbers with extra-floral nectaries; they are found in the tropics, warm areas of the Americas, and Namibia. A number of species are cultivated as ornamentals or used as a source of tannin; ☠ the leaves of *C. pulcherrima* (L.) Sw. (PRIDE-OF-BARBADOS) have been used as a fish poison in Guatemala and Panama and the seeds have been used to poison criminals (Lewis & Elvin-Lewis 1977). Several Texas species (*C. drummondii*, *C. jamesii*) previously treated in *Caesalpinia* are now placed in *Hoffmannseggia* and *Pomaria* respectively (Simpson 1998; B. Simpson, pers. comm.) (Named for Andrea Caesalpini, 1519–1603, Italian botanist physician to Pope Clement VIII and director of the botanical garden at Bologna; his work, *De Plantis*, influenced Linnaeus; his herbarium, dating from the 1500s, is preseved in Florence—Porter 1967) (subfamily Caesalpinioideae, tribe Caesalpinieae)
REFERENCES: Isely 1975; Simpson & Miao 1997; Simpson 1998.

Caesalpinia gilliesii (Hook.) Wall. ex D. Dietr., (for John Gillies, 1792–1834, Scottish physician and botanist who collected in Argentina, Brazil, and Chile), BIRD-OF-PARADISE, POINCIANA, POP-BEAN BUSH. Unarmed shrub or tree to ca. 5 m tall; leaves twice pinnately compound with 11–29 pinnae (each with many leaflets), the pinnae paired or not so; inflorescences densely glandular-pubescent with yellow glands; sepals separate to base, the lower broader than others; petals 5, yellow, 25–30 mm long; stamens 10, 70–90 mm long, much longer than petals; filaments free, bright red; fruits flat, 6–12 cm long, 1.5–2 cm wide, glandular-pubescent. Cultivated for its showy flowers and escapes, dry habitats, pastures; Callahan, Parker, and Young cos., also Brown (HPC) and Tarrant (R. O'Kennon, pers. obs.) cos.; w part of nc TX s and w to w TX. Feb–Jul. Native to Argentina and Uruguay. [*Poinciana gilliesii* Wall. ex Hook.] The seeds are reported to be useful medicinally due to antitumor activity (Mabberley 1987) and the green pods to be severely irritating to the digestive tract (Kingsbury 1964). ☠

CENTROSEMA BUTTERFLY-PEA

A genus of 35 species of warm areas of the Americas; some are used as green manures under crops such as rubber and coconut. *Centrosema* and *Clitoria* are the only U.S. members of the family with large resupinate flowers (= turned upsided down or inverted, with standard on lower side of the flower) (Isely 1990). (Greek: *centron*, a spur, and *sema*, a standard) (subfamily Papilionoideae, tribe Phaseoleae)

Centrosema virginianum (L.) Benth., (of Virginia), BUTTERFLY-PEA. Perennial vine; stems glabrous or inconspicuously pubescent, trailing or twining, to 1.6 m long; leaves pinnately compound; leaflets 3, oblong-ovate to lanceolate, scabrous-pubescent above, minutely pubescent or nearly glabrous and with prominent raised veins beneath; flowers axillary, solitary (rarely in 2s); calyces campanulate, the tube ca. 4 mm long; corollas 20–40 mm long; petals lavender to violet-blue, the standard with reddish central markings and whitish base, rarely all petals white; stamens 10, diadelphous; fruits linear, beaked, 7–12(–14) cm long, ca. 4 mm wide. Sandy woods, roadsides; se and e TX w to Lampasas Cut Plain and Edwards Plateau, also to West Cross Timbers (Erath Co.—Correll & Johnston 1970). Jun–Sep. ▣/83

CERCIS

A n temperate genus of 6 species extending s to ne Mexico; they are deciduous trees with flowers appearing along the branches and trunk (= cauliflory) before the leaves have expanded; some are cultivated as ornamentals. The tribe Cercideae is considered among the most basal elements in the family (Ballenger et al. 1993). (Ancient Greek name applied perhaps to a popular, but also to *C. siliquastrum* L., JUDASTREE; possibly from *kerkis*, a weaver's shuttle, alluding to the large woody fruits; the common name reflects the tradition that this was the tree from which Judas hanged himself) (subfamily Caesalpinioideae, tribe Cercideae)
REFERENCE: Hopkins 1942.

Cercis canadensis L., (of Canada), REDBUD, JUDASTREE. **State tree of Oklahoma**. Small thornless tree; leaves apparently simple (with 1 leaflet), opening after the flowers, cordate basally; flowers pedicillate in clusters of 2–6; calyces asymmetrically cup-shaped, 5-toothed; petals purple-red (very rarely white), ca. 9–12 mm long, very unequal, the flower simulating a papilionaceous form; stamens 10, all fertile, separate; fruits slightly stipitate, flattish, 4–10 cm long, 8–18 mm wide. Mar–Apr. Hybridization apparently occurs where both varieties are found. Our cultivated redbuds are mostly mongrel types derived from the two native varieties, especially the second (Mahler 1988). Cox and Leslie (1991) reported that the buds, flowers, and young pods are edible—particularily sauteed; however, Burlage (1968) indicated that REDBUD is poisonous due to a saponin (no further details as to which organ, etc., were given). ☠

1. Plants usually single-trunked; blades of mature leaves thin (but firm), cordate, about as long as wide or slightly longer, pointed, dull green on both surfaces ____ var. **canadensis**
1. Plants often multi-trunked; blades of mature leaves thick and leathery, cordate to cordate-reniform, wider than long, rounded or barely pointed, rich deep green, shiny, distinctly glaucous above ____ var. **texensis**

var. **canadensis**, EASTERN REDBUD. Stream bottoms and lower slopes, sandy or silty ground; mainly e and nc TX. ❀/83

var. **texensis** (S. Watson) M. Hopkins, (of Texas), TEXAS REDBUD. Rocky limestone slopes; Blackland Prairie w to West Cross Timbers and s to Edwards Plateau. [*C. occidentalis* Torr. & A. Gray] The National Champion TEXAS REDBUD (largest recorded in the U.S.) is located in Dallas Co. (American Forestry Association 1996).

CHAMAECRISTA SENNA

Erect annuals; leaves 2-ranked, once even-pinnately compound; petioles with a sessile to stalked, ±disc-shaped gland; petals yellow, some of them sometimes with a reddish spot at base; stamens 5 or 10, all fertile, separate; anthers basifixed, apically dehiscent by pores; fruits elastically dehiscent.

A genus of 265 species; pantropical but mainly South America to s U.S.; sometimes lumped into the genus *Cassia* (e.g., Isely 1975), but differing in characters such as anthers apically dehiscent by pores, filaments not sigmoid, and fruits elastically dehiscent. Irwin and Barneby (1982) gave a detailed key separating *Cassia*, *Chamaecrista*, and *Senna*. (Greek: *chamai*, on the ground or low, and *crista*, crest) (subfamily Caesalpinioideae, tribe Cassieae)
REFERENCES: Turner 1955; Isely 1975; Irwin & Barneby 1982.

1. Flowers and fruits distinctly pedicellate (pedicels 5–10 mm long); flowers 10–17 mm long; stamens 10 ____ **C. fasciculata**
1. Flowers and fruits essentially sessile (pedicels rarely to 3 mm long); flowers about 5(–8) mm long; stamens usually 5 ____ **C. nictitans**

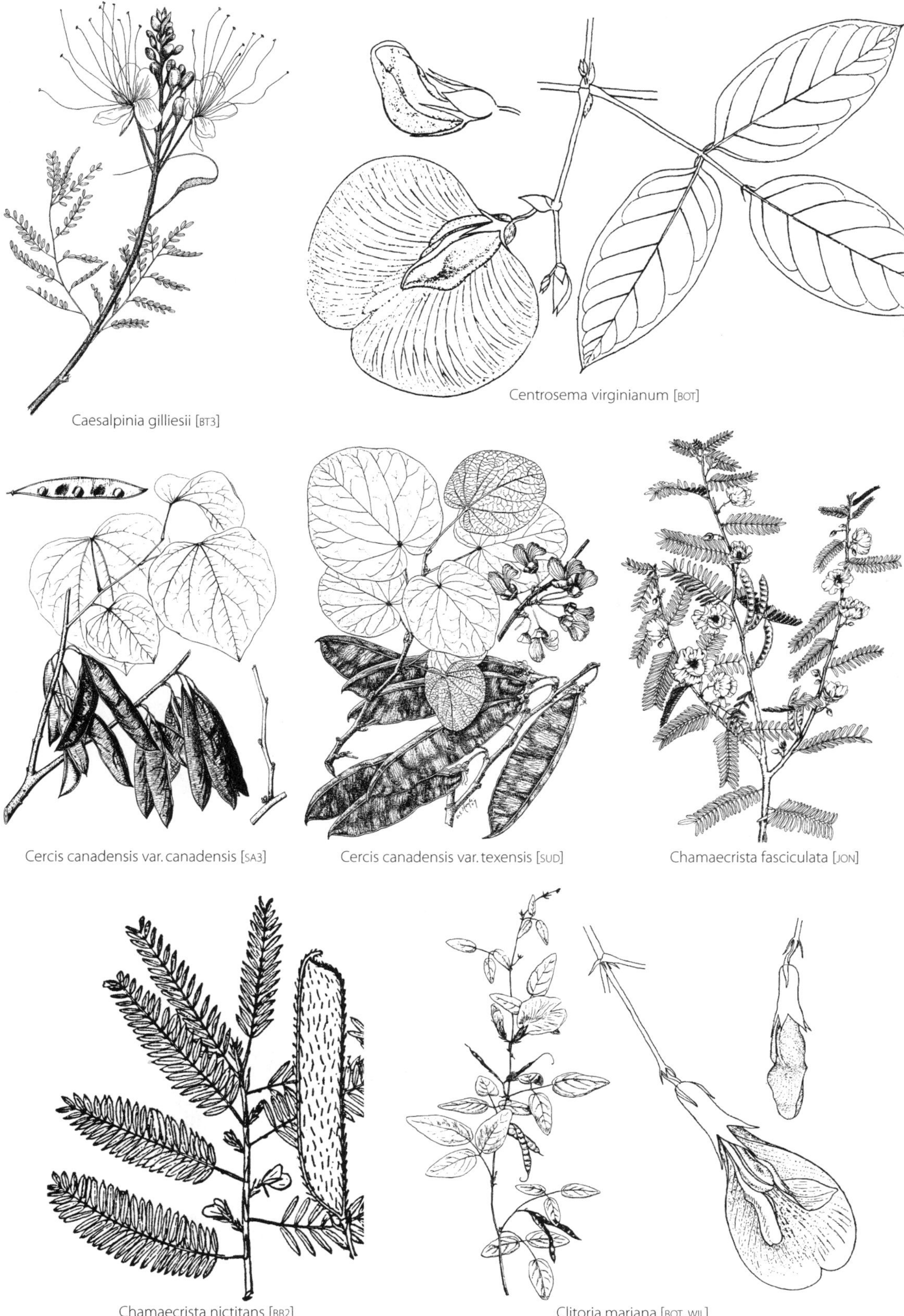

Caesalpinia gilliesii [BT3]

Centrosema virginianum [BOT]

Cercis canadensis var. canadensis [SA3]

Cercis canadensis var. texensis [SUD]

Chamaecrista fasciculata [JON]

Chamaecrista nictitans [BB2]

Clitoria mariana [BOT, WIL]

Chamaecrista fasciculata (Michx.) Greene, (fascicled, clustered), PARTRIDGE-PEA, SHOWY PARTRIDGE-PEA, BEEFLOWER, PRAIRIE SENNA. Annual to ca. 1 m tall; leaves 2-ranked, usually with 8–20 pairs of leaflets, very slightly touch sensitive; petioles with a conspicuous (using hand lens) sessile or short-stalked disk-like gland ca. 1.5 mm across, the gland located from near the middle of the petiole to somewhat closer to the lower pair of leaflets; petals yellow, the upper 4 with a reddish spot at base, the lower larger; anthers yellow and/or purple; fruits usually 5–7 mm long. Open woods or prairies, disturbed areas, often on sandy soils; widespread in TX, but mainly e 1/2. Jun–Oct. [*Cassia chamaecrista* L., *Cassia fasciculata* Michx., *Cassia fasciculata* var. *ferrisiae* (Britton ex Britton & Rose) Turner, *Cassia fasciculata* var. *puberula* (Greene) J.F. Macbr., *Cassia fasciculata* var. *robusta* (Pollard) J.F. Macbr., *Cassia fasciculata* var. *rostrata* (Wooton & Standl.) B.L. Turner] The petiolar glands are small, orangish, and nectar-producing; ants utilize the nectar (Ajilvsgi 1984). The green plant, hay, and seeds have been considered toxic to animals (Kingsbury 1964). ☠ ▣/83

Chamaecrista nictitans (L.) Moench, (blinking, moving), SENSITIVE-PEA, SENSITIVE PARTRIDGE-PEA. Annual usually less than 0.5 m tall; leaves 2-ranked, somewhat touch sensitive; petiolar gland usually stalked or sometimes subsessile; petals yellow; anthers pinkish to rose; fruits 2.5–5 cm long. Sandy open woods, disturbed areas; Denton, Fannin, Grayson, Limestone, and Tarrant cos.; se and e TX w to East Cross Timbers. Sep–Oct. [*Cassia nictitans* L.] The leaves are reported to act as a cathartic (Burlage 1968). ☠

CLITORIA PIGEON-WINGS, BUTTERFLY-PEA

A genus of 60 species primarily of the tropics, especially of the Americas. The flowers are inverted so that the anthers and pistil touch visiting insects on the back. Some are cultivated as climbing ornamentals and others used as aphrodisiacs, possibly based on the ancient idea known as the Doctrine of Signatures. *Centrosema* and *Clitoria* are the only U.S. members of the family with large resupinate flowers (= turned upsided down or inverted, with standard on lower side of the flower) (Isely 1990). (Derivation from the small keel relative to the large standard, suggesting mammalian clitoris) (subfamily Papilionoideae, tribe Phaseoleae)
REFERENCES: Fantz 1977, 1991.

Clitoria mariana L., (of Maryland), SPOONFLOWER, ATLANTIC PIGEON-WINGS. Perennial herb; stems glabrous or minutely pubescent above, erect to low-spreading or trailing, to 1 m long; leaves pinnately compound; leaflets 3, ovate-elliptic to oblong-lanceolate, glabrous, much paler below; stipules linear-lanceolate; flowers solitary or few on short, axillary peduncles; flowers inverted, with standard at base; corollas very large, the standard 3.5–6 cm long, 3–4 cm wide; petals violet-blue or lavender-blue, the standard with darker or reddish markings toward center and pale or whitish base; stamens 10, diadelphous. Sandy open woods; se and e TX w to West Cross Timbers and Edwards Plateau. Late May–Sep.

CORONILLA CROWN-VETCH

A genus of 9 species of the Atlantic Islands, the Mediterranean, and Europe. Some are cultivated as ornamentals with strongly scented flowers, others for erosion control. The single naturalized nc TX species can be easily recognized in the field by the combination of its pinkish flowers in umbels and its segmented fruits. (Diminutive of Latin: *corona*, a crown, alluding to the inflorescence) (subfamily Papilionoideae, tribe Cornilleae)
REFERENCE: Lassen 1989.

Coronilla varia L., (variable), CROWN-VETCH. Perennial herb with erect to ascending or trailing stems 30–50(–100) cm long; leaves once odd-pinnately compound with 9–25 leaflets, essentially sessile; umbels axillary, on peduncules 5–15 cm long; flowers 5–15(–20) per umbel; calyces

2–3 mm long; petals pinkish (can dry lavender), 7–13 mm long; stamens 10, diadelphous; ovary glabrous; fruits glabrous, linear, ±cylindrical with 4 angles, 1.5–5.5 cm long, stipitate, with 3–7(–12) disarticulating segments. Roadsides, planted for erosion control, apparently spreads; originally reported for TX from Montague Co. (Lipscomb 1984), now also Grayson and Tarrant cos.; Hatch et al. (1990) cited only vegetational area 4 (Fig. 2) for TX. May–Jul. Native of Europe and the Mediterranean area. According to McGregor (1986), the seeds are reported to be poisonous. ☠ ☜

CROTALARIA RATTLEPOD

Ours annual or perennial herbs with stems sometimes winged above; leaves sessile or subsessile, apparently simple in our species; flowers usually in axillary or terminal, 2-many-flowered racemes, long-pedicelled; sepals united up to half way; petals yellow; stamens 10, monadelphous; fruits much inflated, with the seeds rattling inside when dry.

A genus of ca. 600 species of the tropics and subtropics, mostly (511 species) in Africa and Madagascar. Some are used as sources of fodder or fiber; ☠ others can cause poisoning in livestock or humans, apparently due to pyrrolizidine alkaloids (e.g., monocrotaline) (Kingsbury 1964; Fuller & McClintock 1986; Turner & Szczawinski 1991; Hardin & Brownie 1993). (Greek: *crotalon*, a rattle, from the loose seeds rattling in the coriaceous inflated pods) (subfamily Papilionoideae, tribe Crotalarieae)

REFERENCES: Senn 1939; Windler 1974.

1. Larger leaves 20 mm or more wide; standard 15 mm long or longer, much longer than the calyx; escaped cultivated species.
   2. Bracts in inflorescence conspicuous, ovate, 5–12 mm long; stipules ovate, 5 mm or more long; calyces glabrous or essentially so; leaves 5–15 cm long ____ **C. spectabilis**
   2. Bracts in inflorescence small (3 mm or less long) or absent, linear to awl-shaped if present; stipules bristle-like, very small, < 2 mm long or absent; calyces with short appressed hairs; leaves 4–8 cm long ____ **C. retusa**
1. Larger leaves 15 mm or less wide; standard < 15 mm long, shorter than or equal to the calyx; native species.
   3. Stems appressed-pubescent or nearly glabrous, the hairs 0.3–1 mm long; leaves usually 4–10 mm wide; rare or possibly absent from nc TX ____ **C. purshii**
   3. Stems ± spreading-pilose, the hairs usually 1–2 mm long; leaves 8–15 mm wide; widespread in nc TX ____ **C. sagittalis**

Crotalaria purshii DC., (for Frederick Traugott Pursh, 1774–1820, German explorer, collector, horticulturist, and author), PURSH'S CROTALARIA. Erect perennial to 40(–50) cm tall; leaves linear- or oblong-lanceolate; racemes with (2–)4–6 flowers; fruits 2.5–4 cm long. Senn (1939) cited a Tarrant Co. collection; Hatch et al. (1990) cited vegetational areas 4 and 5 (Fig. 2); Turner (1959) and Correll and Johnston (1970) questioned the occurrence of *C. purshii* in Texas. Windler (1974), who treated the unifoliolate crotalarias of North America, indicated that the w most location was in e LA. We have seen no TX specimens. Late May–Aug.

Crotalaria retusa L., (retuse, notched slightly at a rounded apex). Erect annual; stems 30–90 cm tall; racemes 10–30 cm long; petals yellow or yellow streaked with red, 20–25 mm long; fruits 2.5–5 cm long. Cultivated and escapes, apparently not persistent; Senn (1939) cited a Tarrant Co. specimen; no recent TX collections are known. Summer–fall. Native to the Asian tropics. Significant losses have occurred in fowl due to the alkaloid monocrotaline in the seeds (Kingsbury 1964); poisoning is similar to that caused by *C. spectabilis* (Burlage 1968). ☠ ☜

Crotalaria sagittalis L., (sagittate, arrow-like), ARROW CROTALARIA. Annual or perennial; stems

10–50 cm tall; stipules conspicuous; racemes with 2–4 flowers; fruits 2–4 cm long. Sandy areas; se and e TX w to West Cross Timbers. Apr–Sep. According to McGregor (1986), there are reports of fatal poisoning in horses from eating fresh or dried plants; Burlage (1968) reported this species as the cause of "bottom disease" or "Missouri Bottom Disease." "The condition is characterized by slow emaciation, weakness, stupor, fatal hemorrhages and degenerative changes in the liver and spleen." ☠

Crotalaria spectabilis Roth, (spectacular, remarkable, showy), SHOWY CROTALARIA. Erect annual 50–100(–200) cm tall; racemes 15–50 cm long, with up to 20–50 flowers; standard yellow, can have veins darker with purple streaks, 15–25 mm long; fruits 3–5 cm long. Cultivated and escapes, sandy or weedy areas and roadsides; Dallas and Hood cos.; se and e TX w to nc TX and Edwards Plateau. Spring–Oct. Native of Asia. This species was introduced into the U.S. in 1921 for use as a cover crop and soil enricher; however, its use was abandoned with the advent of mechanized harvesting of crops and the associated danger of grain contamination by SHOWY CROTALARIA seeds (Morton 1982). The plants, hay, and seeds are toxic to animals and humans due to the presence of pyrrolizidine alkaloids including monocrotaline; liver damage and heart failure can result and animal fatalities are known; human poisoning has resulted from eating the seeds or drinking a "medicinal" tea; severe liver damage can occur (Muenscher 1951; Morton 1982). ☠

DALEA PRAIRIE-CLOVER

Ours unarmed herbaceous perennials or low shrubs; leaves once odd-pinnately compound; leaflets few to many, resin-dotted; stipules slender, falling early; flowers in terminal spikes, these often dense and cone-like; bracts often resin-dotted; petals usually 5, separate; stamens 5–10, monadelphous, attached to staminal tube, either laterally or apically; fruits indehiscent. We are including *Petalostemon* which is separated by some authors; according to Correll and Johnston (1970) who recognized it, *Petalostemom* is "... not at all well-delimited from *Dalea* but technically keyed out on the basis of the insertion of 4 of the petals at the rim of the 'stamen tube' instead of along the length of it."

A New World genus of ca. 160 species native from Canada to Argentina, but especially in Mexico and the Andes. They mainly occur in dry or desert areas. (Named for Samuel Dale, 1659–1739, English botanist and physician who practiced at Essex) (subfamily Papilionoideae, tribe Amorpheae)

REFERENCES: Shinners 1949b, 1949c, 1953f; Wemple & Lersten 1966; Wemple 1970; Barneby 1977; Meeson 1977.

1. Small shrubs with clearly woody branches; petals purple ____ **D. frutescens**
1. Herbaceous perennials; petals purple to pink, white, or yellow.
  2. Calyx teeth triangular or lanceolate, broadly or narrowly acute, much shorter than or equaling the tube; stamens usually 5; 1 petal inserted near rim of floral cup, the other 4 at end of the "stamen tube."
    3. Calyces glabrous outside or nearly so; petals white.
      4. Floral bracts below buds or very young flowers nearly equaling or exceeding the calyces (but bracts deciduous); spikes 14–70(–100) mm long in flower, elongate and cylindric; leaflets 10–30 mm long ____ **D. candida**
      4. Floral bracts below buds or very young flowers 1/3–2/3 as long as the calyces; spikes 7–16 mm long in flower, globose or subglobose; leaflets 6–12 mm long ____ **D. multiflora**
    3. Calyces densely pubescent or pilose outside, at least on the teeth; petals white, pink, or purple.
      5. Leaves long pilose; stems ± pilose; rare on extreme e margin of nc TX ____ **D. villosa**

Coronilla varia [WIL]

Crotalaria purshii [RHO]

Crotalaria retusa [WIL]

Crotalaria sagittalis [RHO]

Crotalaria spectabilis [GRE]

Dalea aurea [MNY]

Dalea candida var. candida [MIC, MNY]

Dalea candida var. oligophylla [MIC, MNY]

5. Leaves glabrous; stems glabrous or short pubescent; widespread in nc TX.
   6. Leaflets 13–41(–49) per leaf, the larger lanceolate to oblong or elliptic, 2–5 times as long as wide; petals white ____ **D. phleoides**
   6. Leaflets 3–9(–11) per leaf, the larger linear or linear-oblong to linear-lanceolate, 5–20 times as long as wide; petals pink to purple.
      7. Spikes crowded but not densely cone-like, the inflorescence axis visible in part after flowering, at least in pressed specimens; calyx teeth as long as the tube or nearly so and distinguishable with the naked eye; leaflets usually 7–9(–11) per leaf; stems decumbent; rare endemic known only from Hood, Parker, and Wise cos. ____ **D. reverchonii**
      7. Spikes permanently dense and cone-like, the flowers even when pressed completely concealing the inflorescence axis; calyx teeth shorter than the tube and usually not easily distinguishable with the naked eye; leaflets 3–7 per leaf; stems ascending to erect; including species widespread in nc TX.
         8. Cone-like spikes 7–10 mm thick in fruit (or in flower excluding corollas); pubesence of calyces of two types, that of tube retrorsely descending, that of teeth antrorsely ascending; s Wise Co. s and w ____ **D. tenuis**
         8. Cone-like spikes usually 10–14 mm thick in fruit (or in flower excluding corollas); pubescence of calyces antrorsely ascending; including species widespread in nc TX.
            9. Calyx tube largely or wholly glabrous in sharp contrast to the pubescent teeth; peduncles 4–12 cm long; widespread in nc TX ____ **D. compacta**
            9. Calyx tube densely pubescent, the pubescence essentially the same as on the teeth; peduncles variable in length but often < 4 cm long; only in n part of nc TX ____ **D. purpurea**

2. Calyx teeth with bristle-like tips slightly shorter to much longer than the tube; stamens usually 7–10; 1 petal attached near rim of floral cup, the other 4 at various places on the "stamen tube," but not at its end.
   10. Petals white; flowers widely spaced in slender, drooping spikes; plants usually with a single main stem from the base (this can be quite branched); leaflets completely glabrous ____ **D. enneandra**
   10. Petals yellow or purple; flowers crowded in rather slender to thick, spreading to erect spikes; plants usually with several main stems from the base; leaflets glabrous or pubescent.
       11. Petals purple; leaflets glabrous or pubescent.
           12. Leaflets with dense pubescence; spikes narrow, 3–8 mm wide in fruit (or in flower excluding corollas); stems prostrate; on sandy soils on n margin of nc TX ____ **D. lanata**
           12. Leaflets completely glabrous; spikes 10–13 mm wide in fruit (or in flower excluding corollas); stems decumbent to ascending; on limestone on w margin of nc TX ____ **D. lasiathera**
       11. Petals yellow; leaflets all pubescent.
           13. Main leaves of the stem with 5–7(–9) leaflets (mostly 5).
               14. Spikes (12–)14–21 mm wide in fruit (or in flower excluding corollas), 20–50 mm long; stems usually erect, 30–75 cm long; standard (blade and claw) 6.3–8.6 mm long; yellow petals staying yellow even when old or dry; w Blackland Prairie w through most of nc TX ____ **D. aurea**
               14. Spikes 7–13(–15) mm wide in fruit (or in flower excluding corollas), 10–30 mm long; stems often spreading, 5–35 cm long; standard 4.4–5.5 mm long; yellow petals often fading to pink or brown; on the sw margin of nc TX ____ **D. nana**
           13. Main leaves of the stem with 3 leaflets (some upper leaves with 1 or 2) ____ **D. hallii**

Dalea aurea Nutt. ex Pursh, (golden, in reference to the flowers), GOLDEN DALEA, SILK-TOP DALEA. Stems spreading to usually erect, 30–75 cm long; leaflets 5(–7), 10–20 mm long, pubescent; petals yellow. Silty or gravelly limestone soils; Blackland Prairie (Grayson Co.) s and w to w TX. Late May–Jul, sporadically to Sep. ▣/86

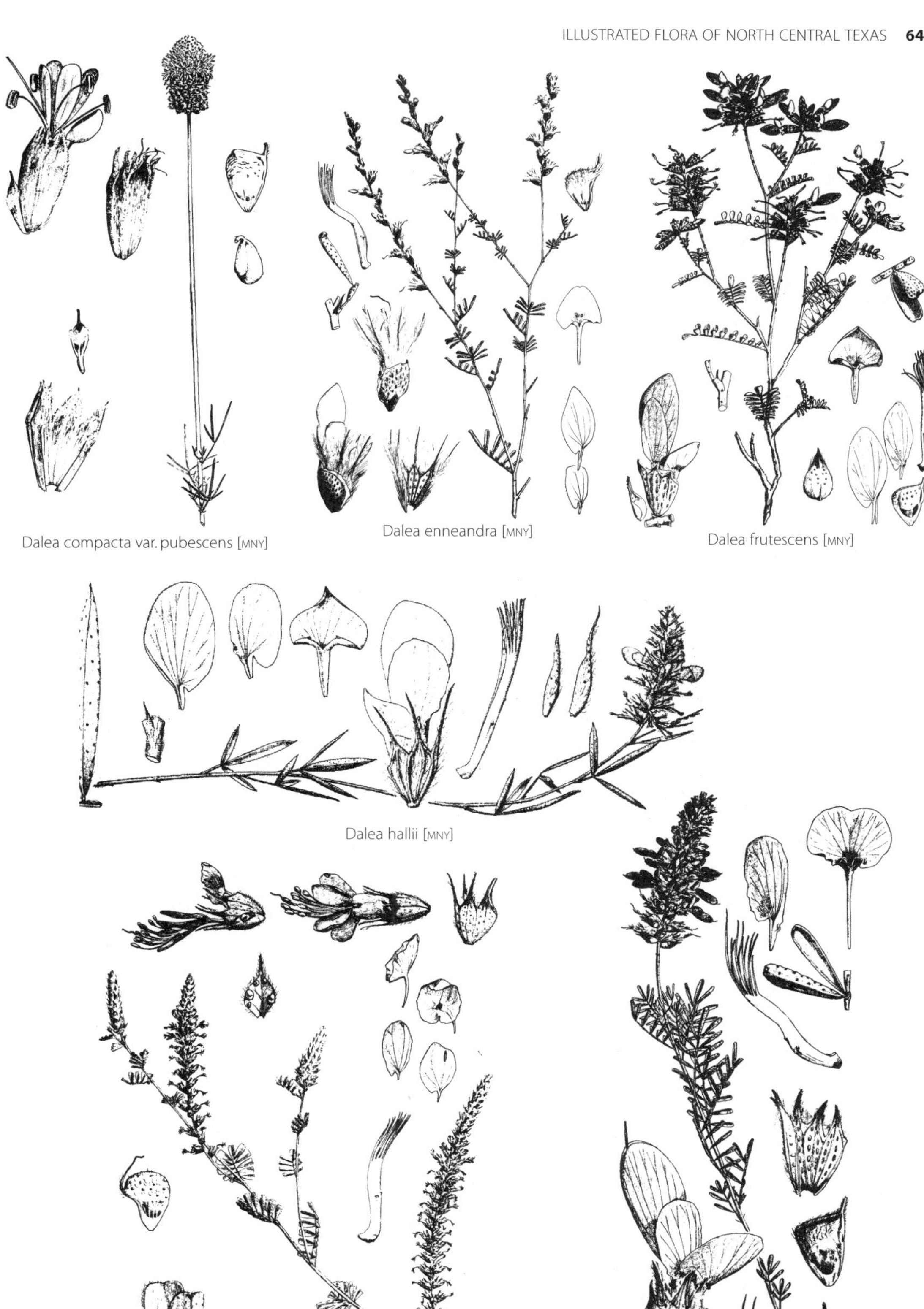

Dalea compacta var. pubescens [MNY]

Dalea enneandra [MNY]

Dalea frutescens [MNY]

Dalea hallii [MNY]

Dalea lanata [MNY]

Dalea lasiathera [MNY]

Dalea candida Willd., (shining or pure white, in reference to the flowers). Stems spreading or ascending, 30–100 cm long; leaflets 5–9, 10–30 mm long, linear-lanceolate or oblanceolate, acute to obtuse, glabrous; petals white; calyces usually with a ring of small glands near top. Prairies and open woods; sandy, rocky, or clayey soils. Late May–early Jul. The 2 varieties intergrade (Turner 1959) and are questionably distinct.

1. Most lateral leaflets > 2 mm wide; apical leaflet of larger leaves 15–32 mm long, 3–8 mm wide (much smaller on second-growth stems of injured plants); spikes remaining dense during and after flowering, the axis concealed ____ var. **candida**
1. Most lateral leaflets < 2 mm wide; apical leaflet of larger leaves 7–20 mm long, 1–6 mm wide; spikes becoming loose, exposing the axis ____ var. **oligophylla**

var. **candida**. Dallas, Fannin, Grayson, Hopkins, and Lamar cos., also Denton Co. (J. Quayle, pers. comm.); e TX w to Rolling Plains (Turner 1959). [*Petalostemon candidus* Michx.]

var. **oligophylla** (Torr.) Shinners, (few-leaved). Panhandle to Trans-Pecos, e to Rolling Plains (Wichita Co.) just w of nc TX; there is a single report (Montague Co. in the West Cross Timbers) from nc TX (Turner 1959). [*Petalostemon candidus* Michx. var. *oligophyllus* (Torr.) F.J. Herm.]

Dalea compacta Spreng. var. **pubescens** (A. Gray) Barneby, (sp.: compact, dense; var.: pubescent, downy), SHOWY PRAIRIE-CLOVER, PRAIRIE-CLOVER. Stems erect, 30–70 cm tall; leaflets 3–7, 10–25 mm long, glabrous or with sparse pubescence; spikes usually 10–14 mm thick (more slender on small, new shoots of mowed plants); petals purplish. Prairies, on clay and limestone; Blackland Prairie w to e Rolling Plains. Jun–early Jul. [*D. helleri* Shinners, *Petalostemon pulcherrimus* (A. Heller) A. Heller]

Dalea enneandra Nutt., (with nine stamens), BIG-TOP DALEA. Stems 20–150 cm tall, bushy-branched above, from a tough, woody, orange root; leaflets usually 5–9, narrowly linear to oblong, 5–12 mm long, glabrous; calyx teeth conspicuously white-pilose; keel pale yellow; standard and wings white. Rocky, sandy, or silty soils; Blackland Prairie w to the Panhandle. Jun–Jul. [*D. laxiflora* Pursh, *D. enneandra* var. *pumila* (Shinners) B.L. Turner] While recently lumped by most authorities, *D. enneandra* var. *pumila*, originally named by Shinners who observed the plants for many years in the field, likely deserves varietal recognition. The plants are consistently short (20–30 cm tall) and compact, with shorter, denser inflorescences, and overall in the field have a different aspect from typical *D. enneandra*. These plants are known from Ellis and Hill cos. in nc TX and from Gillespie and Kerr cos. on the Edwards Plateau.

Dalea frutescens A. Gray, (shrubby, bushy), BLACK DALEA. Small shrubs, 30–80(–120) cm tall; leaflets 13–17, 1.5–3.5(–5) mm long, glabrous; flowers in short, few-flowered spikes; bracts conspicuously gland-dotted; calyx lobes shorter than tube, the tube glabrous; petals purplish. Rocky, disturbed habitats; Montague, Dallas, and Bell cos. s and w to c TX and the Trans-Pecos. Apr–Sep. [*D. frutescens* var. *laxa* B.L. Turner] *Pilostyles thurberi* A. Gray (THURBER'S PILOSTYLES), a member of the Rafflesiaceae, parasitizes *D. frutescens* in nc TX; its vegetative structures are ± entirely within the tissues of the host plant with the visible portions bud-like; only the small flowers and sometimes a few subtending scale-like leaves are externally visible.

Dalea hallii A. Gray, (for its collector, Elihu Hall, 1822–1882, collected in TX and also botanized with Parry in the Rocky Mts.), HALL'S DALEA. Stems trailing to erect, to 35 cm long, appressed-pubescent; leaflets 3, 10–25 mm long, pubescent below; calyces red-brown, the tube and teeth pilose; petals yellow. Eroding limestone slopes; Blackland Prairie and Grand Prairie s to c TX; endemic to TX. Late May–Jun, repeating in Sep. Type from Dallas (Barneby 1977).

Dalea lanata Spreng., (woolly), WOOLLY DALEA. Perennial from a tough orange root, with ascending to trailing stems 30–70 cm long; stems and leaves gray-pubescent; leaflets 9–13, 4–12 mm

Dalea multiflora [MNY]

Dalea nana var. carnescens [MNY]

Dalea nana var. nana [MNY]

Dalea phleoides var. microphylla [MNY]

Dalea purpurea [MIC, MNY]

long; petals purple or red-purple. Gravelly or sandy soils; Grayson Co. in Red River drainage; extreme s TX, Panhandle to Trans-Pecos, e in Red River drainage to nc TX. Jul–Sep.

Dalea lasiathera A. Gray, (with woolly awns, of the calyx teeth), PURPLE DALEA. Stems decumbent to ascending, 10–35 cm tall; leaflets 7–13, 5–15 mm long, glabrous; bracts conspicuously gland-dotted to the naked eye; calyx teeth always shorter than tube, pilose but not plumose; petals purple. Limestone hillsides; Brown Co. on w margin of nc TX; c and w TX n to w edge of nc TX. Mar–Jun. /86

Dalea multiflora (Nutt.) Shinners, (many-flowered), ROUND-HEAD DALEA, WHITE PRAIRIE-CLOVER. Bushy-branched; stems erect or ascending, 30–60 cm long; leaflets 3–9, 6–12 mm long, glabrous, linear or narrowly oblanceolate; petals white. Prairies, on limestone or calcareous clay; Blackland Prairie w to Rolling Plains and s to se TX. Mid-Jun–mid-Jul (mowed plants as late as Sep). [*Petalostemon multiflorus* Nutt.]

Dalea nana Torr. ex A. Gray, (dwarf), DWARF DALEA. Stems decumbent to ascending, 10–30 cm tall; leaflets 5–9, 5–15 mm long, pubescent; petals yellow. Sandy, rocky, or gravelly areas, prairies, roadsides; s and w TX n to sw part of nc TX. Late Mar–Sep.

1. Bracts narrower, narrowly ovate, lanceolate, or elliptic-acuminate, 1.2–2 mm wide; spikes permanently dense and cone-like, the flowers in each vertical rank contiguous or nearly so, the axis concealed; calyx tube vase-shaped, 1.6–2.4 mm in diam.; usually on rocky or gravelly areas, particularly limestone ____ var. **carnescens**
1. Bracts broadly ovate, obovate, or elliptic-acuminate, 2–4 mm wide; spikes relatively loose, at least with age, the flowers in each vertical rank at least 1 mm apart, the axis partly visible in pressed specimens; calyx-tube campanulate, 2.2–3 mm in diam.; usually on sandy soils, not on limestone ____ var. **nana**

var. **carnescens** Kearney & Peebles, (derivation not given in type description, possibly from Latin: *carneus*, flesh-colored, and *escens*, becoming, from the corolla described as fading to reddish). Brown and Lampasas cos.; mainly Edwards Plateau to Trans-Pecos. [*D. nana* var. *elatior* A. Gray ex B.L. Turner]

var. **nana**. Callahan Co.; mainly s and w TX.

Dalea phleoides (Torr. & A. Gray) Shinners var. **microphylla** (Torr. & A. Gray) Barneby, (sp.: resembling *Phleum* of the Poaceae, in reference to the narrow spike; var.: small-leaved), LONG-BRACT PRAIRIE-CLOVER. Stems erect or ascending, 20–60 cm tall; leaflets usually 25–41(–49) per leaf, 4–6 mm long, pilose or glabrate with age; peduncles conspicuously glandular; spikes becoming elongate; axis of spike and calyx tube, at least at base and often throughout, pilosulous with spreading-incurved hairs; petals white. Sandy open woods and open ground; Limestone, Parker, and Tarrant cos., also Brown (HPC) and Montague (Barneby 1977) cos.; se and e TX w to Rolling Plains (Young Co.—Mahler 1988) and Edwards Plateau. Jun–early Jul, repeating in September. [*D. drummondiana* Shinners, *Petalostemon phleoides* Torr. & A. Gray var. *microphyllus* (Torr. & A. Gray) Barneby]

var. *phleoides*, GLANDULAR PRAIRIE-CLOVER, SLIM-SPIKE PRAIRIE-CLOVER, is known just to the e of nc TX. It can be distinguished by the leaflets 13–21(–25) per leaf and the axis of the spike and the exterior of calyces glabrous or almost so (but glands often conspicuous on the peduncle).

Dalea purpurea Vent., (purple), PURPLE PRAIRIE-CLOVER, PRAIRIE-CLOVER. Stems erect to ascending, 20–90 cm tall, simple or sparingly branched above; leaflets usually 5, 8–20 mm long, linear, glabrous or villous; spikes, at least well-developed ones, usually 10–13 thick in fruit; petals purple. Sandy prairies; Montague Co., also Red River and Wichita cos. (Mahler 1988), rarely s to

Tarrant Co. (Mahler 1988); also in the Trans-Pecos (Correll & Johnston 1970). Jun–early Jul. [*Petalostemon purpureus* (Vent.) Rydb.]

Dalea reverchonii (S. Watson) Shinners, (for Julien Reverchon, 1837–1905, a French-American immigrant to Dallas and important botanical collector of early TX), COMANCHE PEAK PRAIRIE-CLOVER. Similar to *D. tenuis*; stems 10–30 cm long, decumbent; leaflets usually 7–9(–11), 5–10 mm long, glabrous; spikes 10–70 mm long in fruit, essentially sessile (peduncles to ca. 1.5 cm long); calyx teeth rather short pubescent; petals rose to magenta-purple. Limestone with sandy surface; type locality—Comanche Peak (Hood Co.); only known current populations are in Parker and Wise cos; narrowly endemic to nc TX. May–Jun, sporadically afterward depending upon rainfall. [*Petalostemon reverchonii* S. Watson] (TOES 1993: V) ⚠

Dalea tenuis (J.M. Coult.) Shinners, (slender, thin). Stems erect or ascending, 15–50 cm tall, glabrous or inconspicuously pubescent, widely branched; leaflets 3–5, 5–12 mm long, glabrous or nearly so; spikes 10–30 mm long in fruit, on slender peduncles 1.5–15 cm long; petals pink-purple. Eroding limestone slopes; sw Fort Worth Prairie (from s Wise Co.) s through Lampasas Cut Plain to Edwards Plateau and w to e Rolling Plains; endemic to TX. Jun. [*D. stanfieldii* (Small) Shinners, *Petalostemon tenuis* (J.M. Coult.) A. Heller]

Dalea villosa (Nutt.) Spreng. var. **grisea** (Torr. & A. Gray) Barneby, (sp.: villous, soft hairy; var.: gray). Stems usually erect, 20–70 cm tall; leaflets 9–17, 6–12 mm long; spikes short-peduncled; petals pink or pink-purple. Milam Co. (Turner 1959) on e margin of nc TX; mainly se and e TX. Spring–summer. [*Petalostemon griseus* Torr. & A. Gray]

Dalea emarginata (Torr. & A. Gray) Shinners, (with a shallow notch at the end), an annual with a slender taproot cited by Hatch et al. (1990) for vegetational area 4 (Fig. 2), probably only occurs s of nc TX n to Llano Co.

Dalea pogonathera A. Gray, (with bearded awns, in reference to the plumose calyx teeth), BEARDED DALEA, HIERBA DEL CORAZÓN, similar to *D. lasianthera*, but with calyx lobes usually longer than the tube and plumose, is cited by Hatch et al. (1990) for vegetational area 5 (Fig. 2) but apparently comes e only as far as Wichita Co., in the Rolling Plains to the w of nc TX.

DESMANTHUS BUNDLE-FLOWER

Unarmed herbaceous or semi-shrubby perennials; root often woody; leaves twice even-pinnately compound, with a petiolar gland between the lowest pair of pinnae (gland sometimes minute); flowers in peduncled, axillary heads; petals separate, linear; stamens 5 or 10; filaments white or yellowish white, separate, serving as the main attractant structure for the flower; fruits flattened, 1–several-seeded, dehiscent. Many species have nyctinastic (= nighttime or "sleep") leaf movements (Luckow 1993).

A genus of 24 species of warm areas of the Americas. We are following Luckow (1993) for nomenclature of *Desmanthus*. (Greek: *desme*, a bundle, and *anthos*, a flower, presumably from the flowers clustered in heads) (subfamily Mimosoideae, tribe Mimoseae)
REFERENCES: Turner 1950a, 1950b; Isely 1970a, 1973; Luckow 1993.

1. Young stems minutely pubescent all around; peduncles ± equaling or exceeding the subtending leaves _____ **D. velutinus**
1. Young stems glabrous except on the angles; peduncles shorter than the subtending leaves.
    2. Leaves with 2–4(–5) pairs of pinnae (each with numerous leaflets); stamens 10.
        3. Stipules conspicuously pubescent; nectaries on petiole large, wider than petiole, somewhat flattened; fruits falcate; taproot red, cylindrical or napiform _____ **D. acuminatus**
        3. Stipules usually glabrous; nectaries on petiole small, orbicular; fruits straight; taproot brown, cylindrical _____ **D. virgatus**

2. Leaves (except smallest) with 5–18 pairs of pinnae; stamens 5.
   4. Heads 4–12-flowered; fruits nearly straight, at maturity 2–3 mm wide, 35–70 mm long, usually ca. 7 times or more longer than wide, readily dehiscent along both sutures _______ **D. leptolobus**
   4. Heads 20–70-flowered; fruits strongly sickle-shaped, at maturity 4.5–7 mm wide, 15–32 mm long, 3–4 times longer than wide, dehiscent along both sutures but tardily so on one side _______ **D. illinoensis**

Desmanthus acuminatus Benth., (tapering to a long narrow point), SHARP-POD BUNDLE-FLOWER. Stems sprawling to decumbent or ascending, usually 20–60 cm long; petioles 3–6 mm long; heads with 6–13 flowers; fruits ca. 30–50 mm long. Gravelly or sandy ground; n to Bell, Burnet, and McLennan cos. in s part of nc TX, also Hamilton (HPC) and Williamson (Luckow 1993) cos.; s part of nc TX s to sc TX; endemic to TX. Apr–May(–fall). [*Desmanthus virgatus* (L.) Willd. var. *acuminatus* (Benth.) Isely] While this taxon is often treated as a variety of *D. virgatus* (e.g., Kartesz 1994; Jones et al. 1997), we are following Luckow (1993) in treating it as a distinct species; she cited numerous differences.

Desmanthus illinoensis (Michx.) MacMill. ex B.L. Rob. & Fernald, (of Illinois), ILLINOIS BUNDLE-FLOWER, ILLINOIS DESMANTHUS, PRAIRIE-MIMOSA, PRICKLEWEED. Stems erect or spreading, 20–150 cm tall; petioles 2–10 mm long; fruits 3–4 times as long as broad (vs. at least 7 times in our other species), tightly clustered. Ditches, stream bottoms, fields, roadsides, and low areas, often clay soils; nearly throughout TX. Late May–Jun, sporadically to Sep.

Desmanthus leptolobus Torr. & A. Gray, (thin-lobed), PRAIRIE BUNDLE-FLOWER, PRAIRIE-MIMOSA. Stems prostrate to suberect, 60–100 cm long; leaflets essentially linear or narrowly elliptic; petioles 2–5 mm long. Prairies and open ground, clayey, rocky, or less often sandy soils, often a weed in lawns; Post Oak Savannah w through nc TX to Rolling Plains and e Edwards Plateau. Late May–Jun.

Desmanthus velutinus Scheele, (velvety). Stems spreading to erect, 20–50 cm long; leaves with 3–7 pairs of pinnae; petioles 4–12 mm long; flowers 15–33 per head; stamens 10; fruits straight, ca. 30–80 mm long. Rocky limestone outcrops; Bell, Bosque, Dallas, Hill, and Williamson cos., also Collin Co. and many cos. in the West Cross Timbers (Luckow 1993); nc TX s and w to w TX. May–Jun and Sep–Oct, sporadically Jul–Aug.

Desmanthus virgatus (L.) Willd., (twiggy). Stems prostrate to erect, to 1.5 m tall; foliage on living plants blue-green, glaucous; petioles usually 1–3(–5) mm long; heads with 3–22 flowers; fruits straight or slightly falcate, 22–88 mm long. Disturbed areas; included based on citation of vegetational areas 4 and 5 (Fig. 2) by Hatch et al. (1990); Rio Grand Plains and sc TX n to at least Travis Co. (Correll & Johnston 1970; Luckow 1993). Apr–Nov. [*D. depressus* Humb. & Bonpl. ex Willd., *D. virgatus* var. *depressus* (Humb. & Bonpl. ex Willd.) B.L. Turner] While TX material is sometimes treated as *D. virgatus* var. *depressus* (Kartesz 1994; Jones et al. 1997), we are following Luckow (1993) in lumping this variety. She indicated that the holotype of *D. virgatus* (at LINN) corresponds to what has traditionally been called *D. virgatus* var. *depressus*.

Desmanthus reticulatus Benth., (netted), NET-LEAF BUNDLE-FLOWER, cited by Hatch et al. (1990) for vegetational area 4 (Fig. 2), is endemic to s and c TX and apparently extends n only as far as Travis Co. (Correll & Johnston 1970; Luckow 1993) just s of nc TX. It is distinguished by having the lower surface of the leaflets with raised, somewhat reticulate veins (not so in nc TX species), leaves with 1–4 pairs of pinnae, stems pubescent all around, and the fruits on peduncles 80–150 mm long (4–60 mm long in nc TX species).

DESMODIUM TICK-CLOVER, BEGGAR'S-LICE, BEGGAR'S-TICKS

Ours perennial herbs; leaves pinnately compound; leaflets 3 (except *D. psilophyllum* with 1

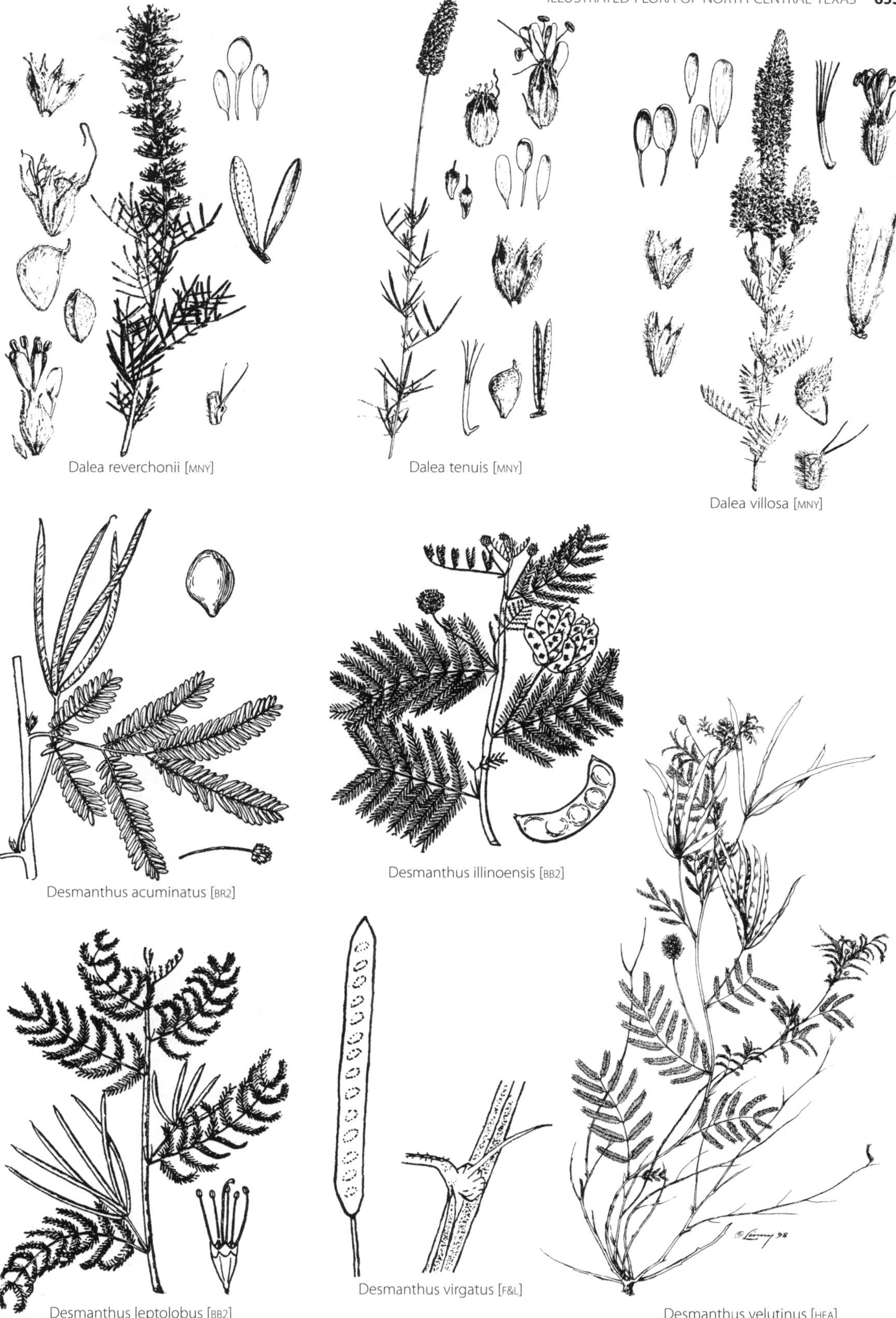

Dalea reverchonii [MNY]

Dalea tenuis [MNY]

Dalea villosa [MNY]

Desmanthus acuminatus [BR2]

Desmanthus illinoensis [BB2]

Desmanthus leptolobus [BB2]

Desmanthus virgatus [F&L]

Desmanthus velutinus [HEA]

leaflet); stipules broad or narrow, persistent or falling early; flowers small, usually many, in erect, narrow racemes or panicles; petals pink, lavender, red-purple, purple, or whitish, often drying bluish or a striking blue-green; fruits (loments) flat, constricted into ca. 2–6 one-seeded segments, usually with small hooked hairs. In nc TX the genus is easily identified in the field because it is the only native group with flat fruits breaking into segments.

A genus of 450 species of warm areas of the world, especially e Asia, Brazil, and Mexico. The fruits fall into 1-seeded segments which stick to hair or clothing, hence the common names. Species are variously used as fodder, green manure, or medicinally. (Greek: *desmos*, a chain, alluding to the connected segments of the fruit giving it a chain-like appearance) (subfamily Papilionoideae, tribe Desmodieae)
REFERENCES: Schubert 1950; Isely 1983.

1. Leaves apparently simple (with 1 leaflet) ____ **D. psilophyllum**
1. Leaves compound, with 3 leaflets.
    2. Calyces hardly lobed, merely with 5 nearly equal teeth, these not more than 1/2 as long as the tube; incisions between fruit segments cutting almost completely through fruit; stipe of fruit exserted from calyx, 3 times or more as long as calyx; leaflets estipellate; stamens monadelphous.
        3. Stems usually branched; leaves dispersed over the stem; petals white; leaflets acute or slightly acuminate; inflorescences usually 20 cm or less long ____ **D. pauciflorum**
        3. Stems unbranched below inflorescence; leaves clustered, almost in a whorl; petals usually pink to purple; leaflets conspicuously and often rather abruptly acuminate; inflorescences 30–80 cm long ____ **D. glutinosum**
    2. Calyces 2-lobed (the upper lobe 2-toothed, the lower lobe with the central tooth longer than the 2 laterals), the teeth very unequal, slightly shorter to much longer than the tube (lower lobe with central tooth longer); incisions between fruit segments various, usually cutting 1/2–7/8 of way through fruit, not completely to band of tissue at upper margin; stipe of fruit usually not exserted from calyx or only slightly so (up to 2.5 times as long as calyx); leaflets stipellate; stamens diadelphous.
        4. Either inflorescence axis pilose (long straight hairs) as well as having short hooked hairs OR leaves with large conspicuous pale or whitish area along midvein; stipules ovate (but with elongate tip), 3–8 mm wide at base and semi-clasping; pedicels long, in flower 5–15 mm long; lower surface of leaf blades with pubescence of hooked hairs (lens or scope necessary).
            5. Leaflets with large conspicuous pale to whitish areas along midvein; inflorescence axis mainly with short hooked hairs; petals white with greenish or yellowish tinge, sometimes with lavender base ____ **D. tweedyi**
            5. Leaflets without pale to whitish areas along midvein; inflorescence axis pilose in addition to having short hooked hairs; petals pinkish purple or white ____ **D. canescens**
        4. Inflorescence axis not pilose (only with short hooked hairs); leaves without pale to whitish areas along midvein; stipules awl-shaped or lanceolate, 0.5–3 mm wide, not clasping; pedicels long or short, in flower 1–15 mm or more long; lower surface of leaf blades without hooked hairs (or with only a few along the veins).
            6. Fruits mostly with 2(–3) segments, these rounded below; flowers small, 3–6 mm long; plants glabrate or pubescent, not conspicuously villous.
                7. Leaves subsessile (petioles 3 mm or less long); leaflets narrow, narrowly oblong to linear, 4–10 times as long as wide; pedicels 5 mm or less long ____ **D. sessilifolium**
                7. Leaves usually petiolate (petioles 1–25 mm long); leaflets usually broadly ovate to elliptic, < 3.5 times as long as wide (very rarely to 5 times as long as wide); pedicels 3–20 mm long.
                    8. Lateral leaflets of middle and lower leaves nearly as long as petioles; stems and leaves essentially glabrous or sparsely puberulent with short hooked hairs ____ **D. marilandicum**

8. Lateral leaflets of middle and lower leaves distinctly longer than petioles; stems and leaves pilose and/or densely puberulent with short hooked hairs.
    9. Stems and leaves sparsely to densely pilose (hairs ± straight); terminal leaflets 9–30 mm long, usually 2.5 times as long as wide or less, similar to lateral leaflets _______ **D. ciliare**
    9. Stems and leaves pubescent with short hooked hairs, but without pilose hairs or these rare; terminal leaflets (25–)30–75 mm long, 2–3.5 times as long as wide, usually longer and narrower than lateral leaflets _______ **D. obtusum**

6. Fruits mostly with 3–6 segments, these usually obtusely angled or if rounded then the plant at least somewhat villous; flowers usually 6–9 mm long.
    10. Terminal leaflets rather broad, narrowly ovate to broadly ovate or rhombic to deltoid, 1–2.2 times as long as wide; plants in general and especially lower surface of leaflets villous (usually ± velvety to the touch); stipules often brick-red in color; bracts villous.
        11. Fruits nearly straight below, with (3–)4–5(–6) segments, these usually 5–8 mm long and obtusely angled; terminal leaflet usually at least 2/3 as wide as long, rhombic to deltoid _______ **D. viridiflorum**
        11. Fruits usually curved, with 2–4 segments, these 4–5 mm long and often rounded below; terminal leaflet ca. 1/2 as wide as long, elliptic-ovate _______ **D. nuttallii**
    10. Terminal leaflets narrow, usually lanceolate to narrowly oblong, usually 2.5–10 times as long as wide; plants in general, including lower surface of leaflets with sparse pubescence, not velvety to the touch; stipules not reddish; bracts glabrate or sparsely hairy _______ **D. paniculatum**

Desmodium canescens (L.) DC., (gray pubescent), HOARY TICK-CLOVER. Stems erect or ascending, 0.5–1.2(–2) m tall; leaflets 5–10 cm long, 1.5–2 times as long as wide, petioles 19–100 mm long; pedicels 8–13 mm long; corollas 9–13 mm long, purplish or pinkish becoming greenish. Sandy woods; Dallas and Henderson cos.; mainly e TX. Sep.

Desmodium ciliare (Muhl. ex Willd.) DC., (ciliate, fringed), LITTLE-LEAF TICK-CLOVER. Stems ascending to erect, 0.4–1.5 m tall; leaflets 1.5–3 cm long, 1.5–2.3(–5) times as long as wide; petioles 1–3(–5) mm long; pedicels 3–8 mm long; corollas 3.5–5 mm long, lavender-purple or pink; fruits with 1–2(–3) segments, each segment 4–5 mm long. Sandy woods and openings; se and e TX w to East Cross Timbers. Aug–Oct.

Desmodium glutinosum (Muhl. ex Willd.) A.W. Wood, (glutinous, sticky). Stems erect, 0.3–0.8(–1.3) m tall; leaflets 5–10 cm long, usually 1–1.5 times as long as wide; petioles 30–80 mm long; leaves crowded just below inflorescence; pedicels 3.5–5.5 mm long; corollas 5–7 mm long, pink to pink-purple, drying bluish (rarely white); fruit segments 2–3, each segment 7.5–9 mm long. Woods; Dallas, Grayson, and Lamar cos.; mainly e TX. Jun–Jul.

Desmodium marilandicum (L.) DC., (of Maryland), MARYLAND TICK-CLOVER. Stems ascending or erect, 0.3–1 m tall; leaflets 1.5–2.5(–4) cm long, 1.2–1.5(–2) times as long as wide; petioles 10–25 mm long; pedicels 8–15 mm long; corollas ca. 5 mm long, lavender to red-violet, often drying bluish green; fruit segments 2(–3), each segment 3.5–4.5 mm long. Open woods; included based on citation of vegetational area 5 (Fig. 2) by Hatch et al. (1990); mainly e TX. Aug–Oct.

Desmodium nuttallii (Schindl.) B.G. Schub., (for Sir Thomas Nuttall, 1786–1859, English-American botanist), NUTTALL'S TICK-CLOVER. Stems ascending to erect, 0.3–1 m tall; terminal leaflet often largest; leaflets 2–8 cm long, 1.5–2(–2.2) times as long as wide; petioles 10–30 mm long; pedicels 4–10 mm long; corollas 6–7 mm long, purple or pink; fruit segments 2–4, each segment 4–5 mm long. Sandy open woods; included based on citation of vegetational area 5 (Fig. 2) by Hatch et al. (1990); mainly e TX. Jun–Sep.

Desmodium obtusum (Muhl. ex Willd.) DC., (blunt), RIGID TICK-CLOVER, STIFF TICK-CLOVER.

Stems ascending-erect, 0.5–1.5 m tall; leaflets (2.5–)4–6(–7.5) cm long, 2–3.5 times as long as wide, petioles 3–12 mm long; pedicels 4–10 mm long; corollas 4.5–6 mm long, pink-purple (rarely white); fruit segments 2–3, each segment 3–5 mm long. Sandy woods; included based on citation of collection near Tarrant-Denton County line (Turner 1959); mainly e TX. Oct. [*D. rigidum* (Elliott) DC.]

Desmodium paniculatum (L.) DC., (with flowers in panicles), PANICLED TICK-CLOVER. Stems 0.3–1(–1.5) m tall, ascending-spreading to erect, usually with few, long, wide-spreading branches; leaflets 2–5(–6) cm long, usually 3–8 times as long as wide, glossy and dark green above; petioles (10–)20–50 mm long; pedicels usually 6–12 mm long; corollas 6–7(–8) mm long, lilac to purple; fruit segments 3–5, each segment 3.5–7 mm long. Low woods, fencerows, and open ground; se and e TX w to Rolling Plains and Edwards Plateau. Jun–Sep. [*D. dichromum* Shinners] is lumped with *D. paniculatum*; it is known only from a single collection n of Mineral Wells, Palo Pinto Co.

Desmodium pauciflorum (Nutt.) DC., (few-flowered), FEW-FLOWER TICK-CLOVER. Stems ascending or spreading, 0.2–0.6 m long; leaflets 3–7 cm long, 1.2–1.5 times as long as wide; petioles 30–70 mm long; pedicels 2–7 mm long; corollas 4.5–6 mm long, white; fruiting segments 1–2(–3), each segment 9–10 mm long. Woods; included based on citation of vegetational area 5 (Fig. 2) by Hatch et al. (1990); mainly e TX. Jul.

Desmodium psilophyllum Schltdl., (slender- or naked-leaved), WRIGHT'S TICK-CLOVER, SIMPLE-LEAF TICK-CLOVER. Stems ascending to erect, to 0.5(–0.9) m tall, minutely and inconspicuously pubescent; leaves apparently simple, the single leaflet 3.3–5.5 cm long, 1–2 cm wide; petioles 6–9 mm long; flowers in a loose raceme; pedicels 4–10 mm long; corollas pink or lavender-pink; fruit segments 4–4.5 mm long. Stream banks and rocky slopes; Bell Co. (Mahler 1988) in s part of nc TX; Trans-Pecos to Edwards Plateau and n to edge of nc TX. May. [*D. wrightii* A. Gray]

Desmodium sessilifolium (Torr.) Torr. & A. Gray, (sessile-leaved), SESSILE-LEAF TICK-CLOVER. Stems ascending-erect to erect, 0.5–1(–2) m tall, branched above, densely spreading-pubescent; leaves sessile or subsessile; leaflets (3–)4–8 cm long, 5–10 mm wide, 4–10 times as long as wide, densely soft-pubescent beneath, with prominent, raised veinlets; pedicels 2–5 mm long; corollas ca. 5 mm long, pale lavender to reddish purple to nearly white, withering greenish; fruit segments 2(–3), each segment 4.5–6 mm long. Sandy or rocky open woods, roadsides, and fields; e 1/2 of TX. Late May–Jul, less freely to Sep.

Desmodium tweedyi Britton, (for Frank Tweedy, 1854–1937, topographic engineer and botanical collector in the nw U.S.), TWEEDY'S TICK-CLOVER. Stems erect, to 1.25 m tall, short-pilose; leaflets 3–12 cm long, ovate to rhombic-elliptic, with raised veinlets beneath, forming a prominent network, densely spreading-pubescent beneath; petioles 40–90 mm long; pedicels 10–22 mm long; corollas white; fruit segments 6–8 mm long. Thickets in limestone areas; Blackland Prairie w to Rolling Plains and Edwards Plateau. Jun–Jul.

Desmodium viridiflorum (L.) DC., (green-flowered), VELVET-LEAF TICK-CLOVER. Stems erect or ascending, 0.3–1 m tall; leaflets 5–12(–15) cm long, 1–1.5(–1.9) times as long as wide; petioles 20–50 mm long; pedicels 3–9 mm long; corollas 7–8 mm long, purple to pink or light lavender; fruit segments (3–)4–5(–6), each segment usually 5–8 mm long. Woods; Dallas, Grayson, and Henderson cos.; se and e TX w to nc TX. Sep–Oct.

ERYTHRINA CORAL-BEAN, COLORÍN

A genus of 112 species of warm areas of the world. The species have red to orange flowers and are apparently all bird-pollinated. Species are variously used as shade trees, ornamentals with showy flowers (CORAL TREES), for their colorful seeds used as beads, or to shade coffee on

Desmodium canescens [BB2]

Desmodium ciliare [WIL]

Desmodium glutinosum [WIL]

Desmodium marilandicum [WIL]

Desmodium nuttallii [WIL]

Desmodium obtusum [WIL]

Desmodium paniculatum [WIL]

Desmodium pauciflorum [WIL]

Desmodium psilophyllum [HEA]

plantations. Some have extrafloral nectaries to attract ants which guard the plant against herbivores. ☠ The seeds of all species are reported to be poisonous; some are used as fish poisons (Burlage 1968). (Greek: *erythros*, red, referring to the color of the flowers) (subfamily Papilionoideae, tribe Phaseoleae)
REFERENCES: McClintock 1953; Krukoff & Barneby 1974.

Erythrina herbacea L., (herbaceous, not woody), CORAL-BEAN, EASTERN CORAL-BEAN, CHEROKEE-BEAN. Subshrub or shrub in nc TX dying back to the ground in winter; stems prickly, 0.5–2+ m tall; leaves alternate, pinnately compound with 3 leaflets; leaflets hastately 3-lobed to widely deltoid, (2–)4–13 cm long, glabrous or nearly so; inflorescence a terminal, long pedunculate raceme; pedicels 3–9 mm long; calyces ± tubular, 5–11.5 mm long; corollas scarlet, very showy, narrow and elongate; standard 30–53 mm long; wings and keel 5.5–13 mm long; stamens 10, usually diadelphous; fruits 7–15(–21) cm long, 1.2–1.6 cm wide, constricted between the seeds, the stipe 1.5–4.5 cm long; seeds scarlet, 5–13 mm long. Cultivated as an ornamental, apparently escaped in Fort Worth (Tarrant Co.); native to sandy woods of se and e TX to the se of nc TX. Apr–Jun. Austin (1975) considered this species adapted for pollination by ruby-throated hummingbirds (*Archilochus colubris*). The seeds have been used as beads; they contain numerous toxic alkaloids (e.g., erysodine, erysopine) and are used as a rat or dog poison in Mexico (Schmutz & Hamilton 1979; Morton 1982); according to Burlage (1968), the poison is similar in action to curare. ☠ ▣/89

EYSENHARDTIA KIDNEYWOOD, BEEBRUSH

A genus of 10 species native from the sw U.S. to Guatemala. (Named for C.W. Eysenhardt, M.D., professor at Univ. of Konigsberg, Prussia) (subfamily Papilionoideae, tribe Amorpheae)

Eysenhardtia texana Scheele, (of Texas), TEXAS KIDNEYWOOD, VARA DULCE. Largely glabrous unarmed shrub to 3 m tall; leaves crowded, once pinnately compound; leaflets many (15–47), (3–) 5–12 mm long, oblong, gland-dotted, aromatic, with the aroma of tangerine rinds; stipules thread-like, persistent; flowers in terminal, erect, solitary or panicled, slender, spike-like racemes, sweet-scented; petals 4–5, white to pale yellow, separate; stamens 10, diadelphous. Limestone hillsides; Bell and Williamson cos. in s part of nc TX, also Fort Hood (Bell or Coryell cos.—Sanchez 1997); s part of nc TX s to s TX and w to Trans-Pecos. Apr–Sep. [*E. angustifolia* Pennell] Dyes have been obtained from the wood and the wood is fluorescent in water (Powell 1988); the wood, according to Crosswhite (1980), "when leached in water turned the water orange, fluorescing blue against a black background. This was used as a diuretic [during frontier days]."

GALACTIA MILK-PEA

Prostrate perennials with trailing or twining stems, pubescent to nearly glabrous; leaves pinnately compound; leaflets 3 (5 in one species rare in nc TX); stipules thread-like, falling early; flowers in axillary, uncrowded, spike-like racemes; petals pink, roseate, pink-purple, or lavender.

A genus of 140 species of warm areas of the world, especially the Americas. Many have latex (rare in Fabaceae), hence the common name. (Greek: *gala*, milk; Patrick Browne originally stating that it has "milky branches") (subfamily Papilionoideae, tribe Phaseoleae)
REFERENCES: Vail 1895; Rogers 1949; Duncan 1979.

1. Leaves (at least most) with 5 leaflets; corollas lavender (standard often with some white); in nc TX known only from Brown Co. on w margin of area ______ **G. heterophylla**
1. Leaves with 3 leaflets; flowers pink to roseate or pink-purple; widespread in nc TX.
   2. Leaflets thick, leathery, with densely appressed pubescence on both surfaces but particularly

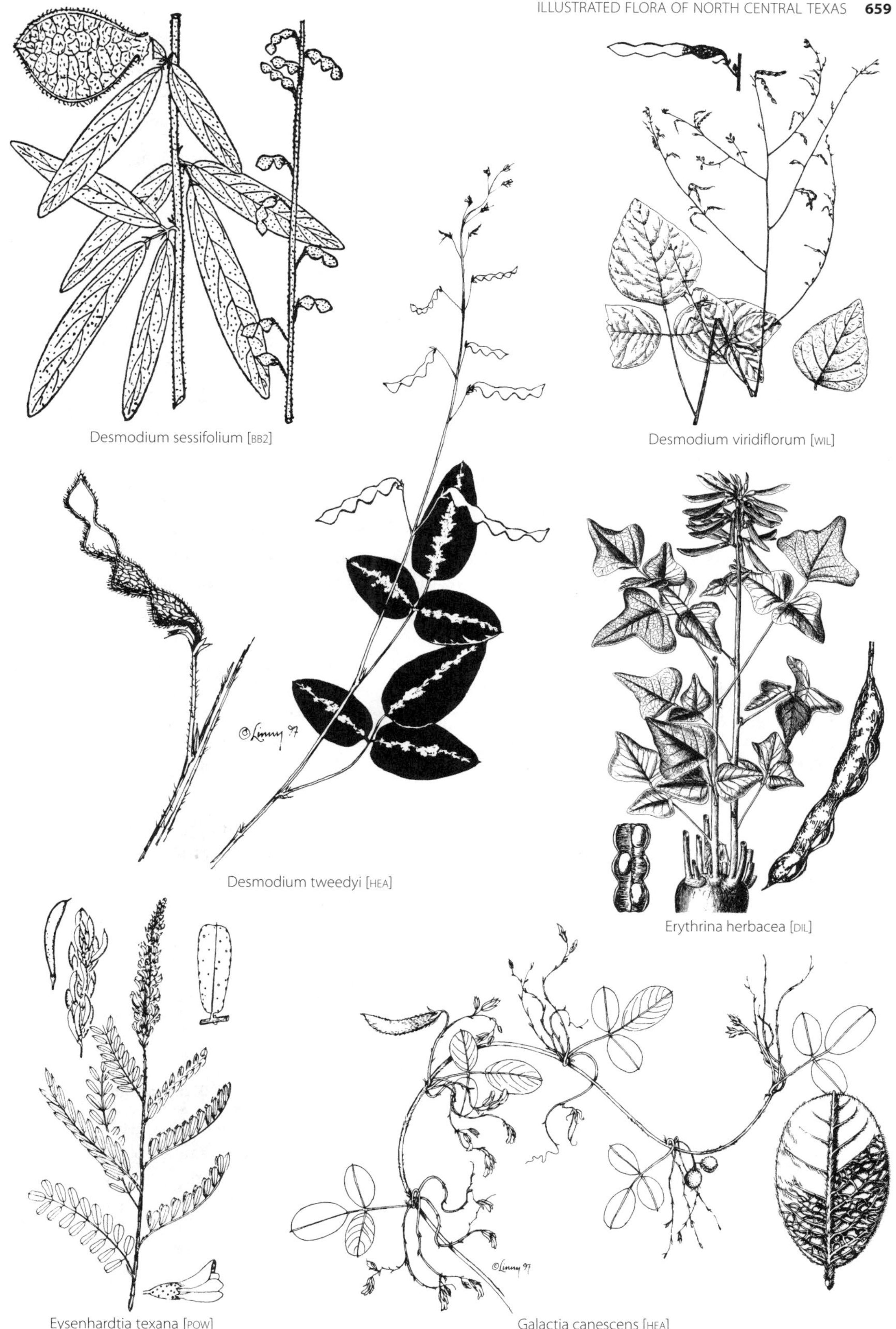

Desmodium sessifolium [BB2]

Desmodium viridiflorum [WIL]

Desmodium tweedyi [HEA]

Erythrina herbacea [DIL]

Eysenhardtia texana [POW]

Galactia canescens [HEA]

beneath, prominently reticulate-veined beneath to the naked eye; racemes 2–7 per node; stems trailing but not twining; rare in nc TX ______ **G. canescens**

2. Leaflets thin, with pubescence ± spreading to nearly glabrate, not prominently reticulate beneath; racemes 1–2 per node; stems usually twining; widespread in nc TX ______ **G. volubilis**

Galactia canescens (Scheele) Benth., (gray-pubescent), HOARY MILK-PEA. Stems appressed-pubescent, to 2 m long; leaflets elliptic to orbicular-ovate; corollas 9–12 mm long, pinkish, developing into fruits above ground; cleistogamous flowers and small, peanut-like, 1-seeded fruits also produced underground. Sandy soils; South TX Plains disjunct n to Brazos River terrace (Hood Co.—R. O'Kennon, pers. obs., and Somervell Co.—Correll & Johnston 1970), also in deep sand near Azle in Parker and Tarrant cos.; endemic to TX. May–Jun.

Galactia heterophylla A. Gray, (various-leaved). Stems to 0.6 m long; leaflets 5, usually oblanceolate, the lower 2 pairs attached at the same spot, the terminal leaflet with a short (1–4 mm long) stalk; cleistogamous flowers and underground fruits not present. Shallow or gravelly soils; Brown Co. (HPC); w edge of nc TX s to Edwards Plateau, s, and se TX; endemic to TX. Apr–Sep. [*G. grayi* Vail]

Galactia volubilis (L.) Britton, (twining around a support), DOWNY MILK-PEA. Stems short villous to subappressed-pubescent, to 1–2 m long; leaflets ovate to oblong-lanceolate, 1.5–4.5(–5) cm long; corollas 8–12(–14) mm long, pink to roseate or pink-purple, fading pale; fruits 2–6 cm long. Dry open woods, thickets, semi-open areas; se and e TX w to nc TX and Edwards Plateau. Late May–Sep. [*G. mississippiensis* (Vail) Rydb.] Duncan (1979) had a different view of the correct nomenclature of this taxon and treated it as *G. regularis* (L.) Britton, Sterns, & Poggenb.

GLEDITSIA HONEY-LOCUST

Polygamous or dioecious trees or shrubs, usually armed with straight or branched thorns; leaves alternate, deciduous, once or twice pinnately compound, sometimes a leaf partly once pinnate, partly two times pinnate; leaflets entire or crenulate; flowers small, in lateral, catkin-like racemes appearing with the leaves; petals 3–5, resembling the calyx lobes, inconspicuous, yellowish or greenish yellow; stamens usually 6–10, separate; fruits stalked, flattened.

A genus of 14 species, with 2–3 e North America, 1 South America, 1 Caspian area, and the rest India and Japan to New Guinea. They are trees usually with stout, axillary, branched thorns. Species are variously used as cultivated ornamentals, for shade, timber, or as hedges. (Simplified and Latinized from name of Johann Gottlieb Gleditsche, 1714–1786, a botanist contemporary with Linnaeus) (subfamily Caesalpinioideae, tribe Caesalpinieae)
REFERENCE: Michener 1986.

1. Fruits with only 1(–3) seeds, 3–5(–8) cm long, ± ovate, neither twisted nor contorted; mature leaf axes, stalks of leaflets, and leaflets glabrous or with a few hairs; seeds not surrounded by a sugary pulp ______ **G. aquatica**

1. Fruits many-seeded, 10–40 cm long, elongate, oblong, often twisted or contorted; mature leaf axes, stalks of leaflets, and often midribs on lower surfaces of leaflets pubescent; seeds surrounded by a sugary pulp ______ **G. triacanthos**

Gleditsia aquatica Marshall, (growing in or near water), WATER HONEY-LOCUST, WATER-LOCUST. To ca. 25 m tall, armed or unarmed; thorns if present simple or few-branched; leaves resembling *G. triacanthos* except for pubescence; fruits 2–3.5 cm wide; seeds 10–15 mm in diam. Swampy areas, along streams, bottomland woods; Lamar Co. in Red River drainage (Carr 1994); mainly se and e TX and Edwards Plateau. May–Jun.

Gleditsia triacanthos L., (three-thorned), COMMON HONEY-LOCUST, HONEYSHUCK. To 30(–45) m

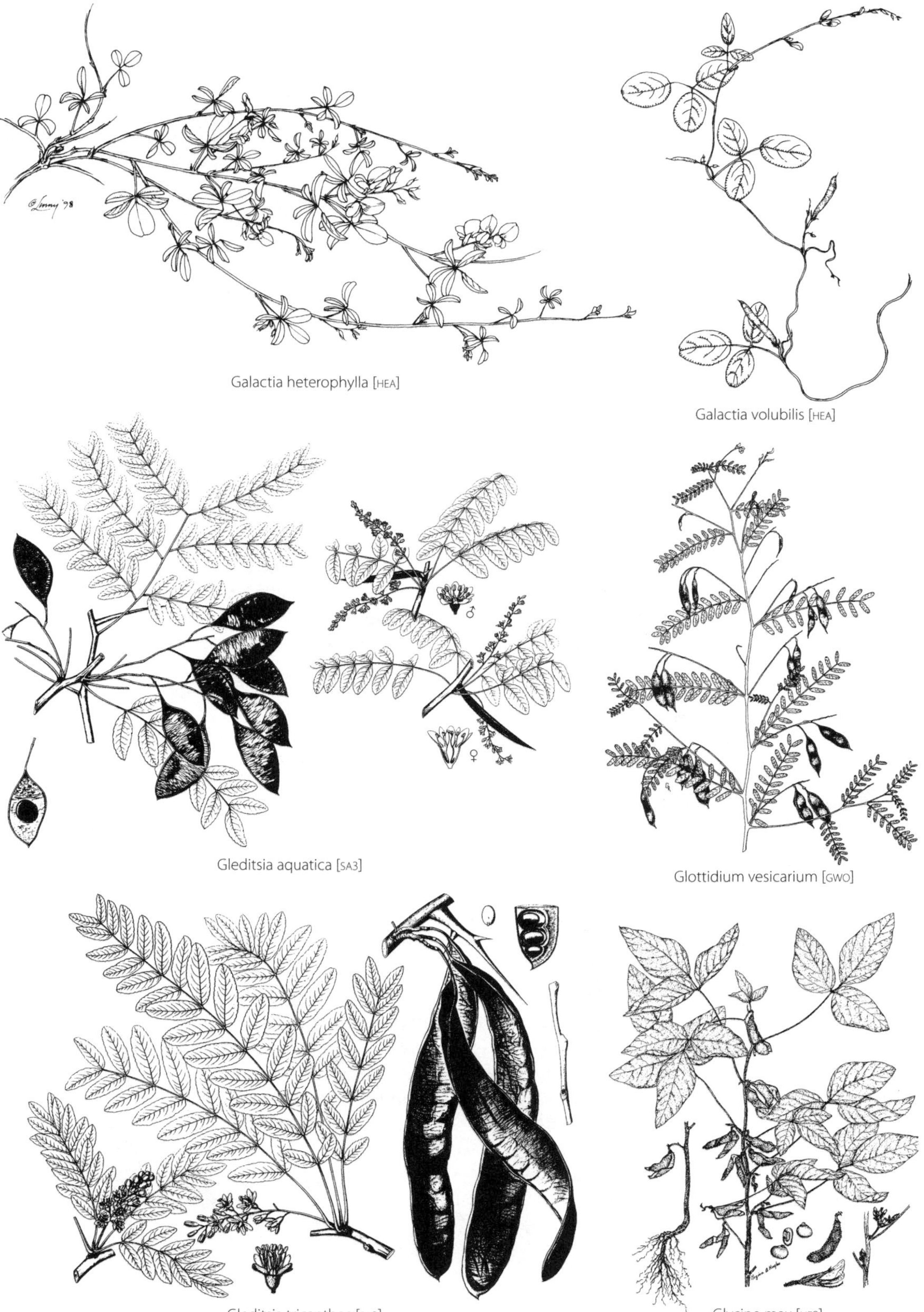

Galactia heterophylla [HEA]

Galactia volubilis [HEA]

Gleditsia aquatica [SA3]

Glottidium vesicarium [GWO]

Gleditsia tricanthos [SA3]

Glycine max [HE2]

tall, with long, stout, branched thorns on trunk and mostly simple ones on branches, rarely thornless; thorns sometimes very conspicuous, 6–15(–40) cm long; once pinnate leaves with 10–14 pairs of leaflets; twice pinnate leaves with 2–8 pairs of pinnae; leaflets 1.5–3(–3.5) mm long; petals 4–6 mm long; fruits 2–4 cm wide; seeds < 10 mm in diam. Stream bottoms, also weedy invader of disturbed sandy slopes or open ground, becoming problematic in some areas; se and e TX w to West Cross Timbers; also scattered in Rolling Plains and Edwards Plateau. Apr. [*G. triacanthos* var. *inermis* (L.) C.K. Schneid.] The very long thorns observed on some individuals are among the most striking examples of physical plant defense seen on any species in nc TX; Native Americans used them for fishing-spear tips (Cox & Leslie 1991). The thornless condition observed in some plants is unstable and appears to vary with age (Michener 1986). The sweet fruits (containing up to 30 percent sugar) are eaten by cattle, and Native Americans ate the honey-like substance in the young pods (Cox & Leslie 1991); according to Crosswhite (1980), "Early settlers using them for food called them 'Honey-Shucks,' eating the young pods and considering the older ones too bitter."

GLOTTIDIUM BLADDERPOD, BAGPOD

A monotypic genus of the se U.S.; previously treated in *Sesbania*. It can be recognized in the field by its large size, slender racemes of small flowers, and somewhat inflated fruits. (Presumably from Greek: *glotti*, tongue) (subfamily Papilionoideae, tribe Robinieae)

Glottidium vesicarium (Jacq.) R.M. Harper, (bladder-like), BLADDERPOD, BAGPOD. Unarmed annual herb to 4 m tall; leaves once pinnately compound; leaflets numerous; flowers in axillary racemes, reddish brown to orange or yellowish or tinged with pink or red; body of fruits oblong, 2.5–8 cm long, 1.5–2 cm wide, acuminate at both ends, the valves of fruits separating at maturity into a thicker outside layer and a papery-membranous, thinner, inside layer; stipe of fruits 1–1.5 cm long, 1–1.5 mm thick. In damp soils; se and e TX w to West Cross Timbers and Edwards Plateau. Aug–Sep. [*Sesbania vesicaria* (Jacq.) Elliott] The seeds and green leaves are reportedly poisonous to livestock due to the presence of saponins (Kingsbury 1964; Mabberley 1987; Hardin & Brownie 1993).

GLYCINE

A genus of 18 species native from Asia to Australia including the important crop plant SOYBEAN. (Greek, *glyco*, sweet, alluding to the sweet taste of the tubers of a species of *Apios* at one time included in *Glycine*) (subfamily Papilionoideae, tribe Phaseoleae)
REFERENCES: Hermann 1962; Hymowitz & Newell 1981.

Glycine max (L.) Merr., (old name), SOY-BEAN, SOYA-BEAN. More or less erect, bushy, annual herb 0.3–2 m tall, densely long-hairy; leaves pinnately compound with 3 entire leaflets; flowers in rather inconspicuous axillary inflorescences of (1–)5–8(–12) flowers; calyces 4–7 mm long, densely pubescent; petals 4.5–7 mm long, white to pink, violet, or purple; stamens diadelphous; fruits (2–)4–8 cm long, with conspicuous, dense, bristly pubescence, pendant, with 2–4 seeds. Waif along railroads, waste places; extremely widely cultivated; Grayson Co.; probably widespread as a waif in TX. Summer. Native of e Asia. An old and very important crop cultivated primarily for its seeds which are extremely rich in protein and also as a source of oil; SOYBEAN is also the source of soy sauce and bean curd or tofu. It was apparently derived in ne China ca. 11th century BC from *Glycine soya* Siebold & Zucc.; first cultivated in the U.S. in 1924; perhaps the world's most important pulse and currently one of the major agricultural crops of the U.S.; in this country grown mainly as an oil seed crop (Mabberley 1987; Isely 1990).

GYMNOCLADUS

A genus of 5 species, 1 in e North America and 4 in e and se Asia; this disjunct distribution pattern is discussed under the genera *Campsis* (Bignoniaceae) and *Carya* (Juglandaceae). The species are trees with the fruits opening along a partial suture like a follicle (an ancestral characteristic). While quite distinct (e.g., spines absent), *Gymnocladus* shows some similarities to *Gleditsia* (e.g., plants polygamous or dioecious, seeds with copious endosperm, roots lacking nodule formation, and calyx similarities—Lee 1976). (Greek: *gymnos*, naked, and *clados*, branch, referring to the branches being naked when the deciduous leaves are shed) (subfamily Cesalpinioideae, tribe Caesalpinieae)
REFERENCE: Lee 1976.

Gymnocladus dioicus (L.) K. Koch, (dioecious, having the sexes on separate plants), KENTUCKY COFFEE TREE. Unarmed, dioecious or polygamous tree to 23(–35) m tall; leaves alternate, irregularly twice pinnately compound, with 3–7 pairs of pinnae, the pinnae with (3–)4–7 pairs of leaflets in addition to a terminal leaflet; leaflets 2–7 cm long, usually 15–30 mm wide, ovate, rather abruptly acuminate at apex, entire; petioles often 20–40 cm long; inflorescences racemose to paniculate; flowers usually functionally imperfect; hypanthium 6–12(–18) mm long; perianth radially symmetrical, of 2 similar 5-merous whorls, greenish white; stamens 10; fruits asymetrically oblong, (5–)10–18(–25) cm long, 3–5 cm wide, woody at maturity; seeds few, embedded in a pulp. Cultivated and long persisting; apparently rarely escaping into native woodland; Tarrant Co. (S. Urshel, pers. comm.). May–Jun. Native to the e and c U.S. s to OK just n of nc TX. There are problems with the generic name *Gymnocladus* Lamarck and N. Harriman (pers. comm.) has submitted a manuscript (Harriman, forthcoming) to *Taxon* to conserve the name with a conserved gender; according to Harriman (forthcoming), there is an earlier available name making *Gymnocladus* illegitimate. Secondly, there is a problem with the gender of *Gymnocladus*; Lamarck treated the name as feminine, but according to the *International Code of Botanical Nomenclature* (Greuter et al. 1994), the name should be masculine. Until this matter is resolved, we are following Harriman's recommendation and treating this species as *Gymnocladus dioicus*. Poisoning in animals (potentially fatal) has been reported from eating the foliage or sprouts and in humans from eating the fruit pulp or seeds; the toxic substance is reported to be a quinolizidine alkaloid or cytisine (Kingsbury 1964; Turner & Szczawinski 1991; Hardin & Brownie 1993). ☠

HOFFMANNSEGGIA RUSH-PEA

A genus of 28 species mostly native to the New World from sw U.S. to Chile, with 3 species in s Africa; Simpson (1998) indicated that future study will probably result in the transfer of several South American species to *Pomaria*. Some sw U.S. species have tubers that are edible when roasted. (Named for J. Centurius, 1766–1849, Count of Hoffmannsegg, Germany) (subfamily Caesalpinioideae, tribe Caesalpinieae)
REFERENCES: Isely 1975; Simpson & Miao 1997; Simpson 1998.

Hoffmannseggia glauca (Ortega) Eifert, (whitened with a coating or bloom), SICKLE-POD RUSH-PEA, MESQUITEWEED, INDIAN RUSH-PEA, HOG-PEANUT, HOG-POTATO, CAMOTE DE RATÓN. Perennial herb to 30 cm tall; stems glabrous or pubescent; leaves twice pinnately compound; pinnae 5–11(–13); leaflets 4–11 pairs per pinna, not glandular-dotted beneath; petiole and rachis glandular; racemes glandular, 10–20 cm long; petals 5, yellow, clawed; stamens 10, separate, glandular; fruits falcate, 2–4 cm long, 5–8 mm broad. Roadsides and disturbed areas; Montague and Palo Pinto cos.; West Cross Timbers s and w to w TX. Late Apr–Sep. [*H. densiflora* Benth.] According to Ajilvsgi (1984), this species forms underground tubers that were roasted by Native Americans and have sometimes been fed to hogs. ▣/93

Hoffmannseggia drummondii Torr. & A. Gray, (for Thomas Drummond, 1780–1835, Scottish botanist and collector in North America), DRUMMOND'S RUSH-PEA, was cited for vegetational area 4 (Fig. 2) by Hatch et al. (1990); however, it occurs only well to the s of nc TX (B. Simpson, pers. comm.). The leaves with 3 pinnae (each pinna with 4–5 pairs of leaflets) distinguish this species from *H. glauca*. Apr–Sep. [*Caesalpinia drummondii* (Torr. & A. Gray) Fisher] While this species has often been treated in *Caesalpinia* (e.g., Isely 1975; Kartesz 1994; Jones et al. 1997), molecular evidence shows it should be included in *Hoffmannseggia* (B. Simpson, pers. comm.).

INDIGOFERA INDIGO

A genus of ca. 700 species of tropical and warm areas of the world. Several tropical species (e.g., *I. tinctoria* L.) are a source of the blue dye indigo, now largely replaced by synthetics. Some species are cultivated as ornamentals. (Latin: *indigus*, indigo, and *fero*, to bear, in reference to the dye) (subfamily Papilionoideae, tribe Indigofereae)

Indigofera miniata Ortega, (cinnabar-red), SCARLET-PEA. Appressed-pubescent perennial from tough taproot; stems ascending to prostrate, to 50 cm long, usually freely branched; leaves once pinnately compound; leaflets 5–9; stipules nearly thread-like; flowers in axillary and terminal, peduncled, spike-like racemes; sepals divided nearly to base, acuminate; petals light brick-red to salmon-rose, the standard abruptly bent back from near base; fruits 10–40 mm long, 2–3 mm in diam. Sandy or rocky prairies and open woods. Mid-May–Jun, less freely to Sep.

1. Petals 8–11(–20) mm long ____ var. **leptosepala**
1. Petals 6–8 mm long ____ var. **miniata**

var. **leptosepala** (Nutt. ex Torr. & A. Gray) B.L. Turner, (thin-sepaled), WESTERN SCARLET PEA, WESTERN INDIGO, SCARLET-PEA. Common, increases under disturbance; e 1/2 of TX. Intergrades with the following variety. ▣/94

var. **miniata**, COAST INDIGO, SCARLET-PEA. Brown and Comanche cos. (Stanford 1971) in w part of nc TX, also cited by Hatch et al. (1990) for vegetational area 4 (Fig. 2).

Indigofera suffruticosa Mill., (somewhat shrubby), INDIGO, a native of tropical America and one of the sources for the precursor of the blue dye indigo, is cited by Hatch et al. (1990) for vegetational area 4 (Fig. 2). It apparently occurs to the s and se of nc TX. This species can be distinguished from *I. miniata* by its size (0.5–2 m tall), erect habit, and greater number of leaflets (9–15).

KUMMEROWIA LESPEDEZA, BUSH-CLOVER

Low or prostrate annuals; leaves palmately compound with 3 leaflets; leaflets striate with parallel veins; stipules ovate, striate, 3–8 mm long; flowers axillary, solitary or few, shorter than the leaves.

An Asian genus of 2 species; in the past lumped into *Lespedeza*, but differing in its annual habit, striate leaflets, conspicuous stipules, and solitary flowers (Isely 1990). (Named in 1912 for Professor Kummerov Posnaniensi) (subfamily Papilionoideae, tribe Desmodieae)
REFERENCE: Isely 1948.

1. Stem hairs antrorse (= pointing up toward apex of stem, the free end of the hair distal to the attached end); leaflet margins not conspicuously ciliate ____ **K. stipulacea**
1. Stem hairs retrorse (= pointing down toward base of stem); leaflet margins usually conspicuously spreading-ciliate ____ **K. striata**

Kummerowia stipulacea (Maxim.) Makino, (having stipules), KOREAN BUSH-CLOVER, KOREAN LESPEDEZA. Stems erect or ascending, 10–30(–60) cm long, freely branched; main stem leaves with petioles 2–10 mm long; calyces glabrous or nearly so; corollas 6–7 mm long; standard violet with dark base; wings white; keel white with red-black tip; fruits 3 mm long, covered by calyx for 1/2 the fruit length. Cultivated and naturalized, sandy soils; Fannin and Grayson cos.; mainly e TX; introduced into the U.S. in 1919 and used for pasture, hay, and as a cover crop (McGregor 1986); there are reports of death in cattle due to uncontrollable hemorrhaging (Lewis & Elvin-Lewis 1977). Jul–Sep. Native of Korea. [*Lespedeza stipulacea* Maxim.]

Kummerowia striata (Thunb.) Schindl., (striated, striped), JAPANESE BUSH-CLOVER, COMMON LESPEDEZA, JAPANESE LESPEDEZA. Stems freely branched, prostrate to erect, pubescent or in age nearly glabrous, 10–40(–50) cm long; main stem leaves with petioles 1–3 mm long; calyx lobes usually ciliate; corollas 4.5–6 mm long; standard rose-violet; wings white; keel white, often with purple tip; fruits 3–4 mm long, covered by calyx for 1/2–3/4 the fruit length. Cultivated and naturalized, sandy roadsides, pastures, and open woods; se and e TX w to East Cross Timbers; introduced into the U.S. in 1846 and used for hay, pasture, and as a cover crop (McGregor 1986). Jun–Sep. Native of China and Japan. [*Lespedeza striata* (Thunb.) Hook. & Arn.]

LATHYRUS PEAVINE, VETCHLING

Annual or perennial, trailing, sprawling, or ascending to climbing herbs with winged stems (except in *L. venosus*); leaves pinnately compound, the rachis terminating in a tendril that is often branched; leaflets in nc TX species 2 (except in *L. aphaca* and *L. venosus*), estipellate; stipules semi-sagittate, usually persistent and conspicuous; racemes few- to many-flowered, axillary, pedunculate; calyces essentially symmetrical basally; corollas white, pink-purple, lavender-blue, blue-purple, or bicolored; stamens 10, diadelphous; style dilated and flattened, pubescent on upper surface (= adaxial side); fruits elastically dehiscent, several-seeded, essentially linear, usually somewhat flattened.

A genus of 160 species of the n temperate zone, mountains of e Africa, and temperate South America; they are usually climbers with branched tendrils. Many are used as ornamentals and a number as fodders. Eating the seeds of some species can lead to lathyrism (including spinal cord degeneration and paralysis of the legs); heavy losses of human life have occurred under conditions where other food was not available (e.g., Africa); poisonous due to the presence of toxic nonprotein amino acids (Tampion 1977; Schmutz & Hamilton 1979; Fuller & McClintock 1986). According to Kupicha (1981), the genus is closely related to *Vicia*; with 2 rare exceptions (*L. aphaca* and *L. venosus*), nc TX *Lathyrus* species can be readily distinguished by the combination of winged stems and leaves with only 2 leaflets (*Vicia* has stems unwinged and leaves with 4–18 leaflets). (Greek: *la*, very, and *thoures*, stimulant or passionate, the name due either to the seeds which were said to have irritant properties or from the reputation of the first described species as an aphrodisiac) (subfamily Papilionoideae, tribe Vicieae)
REFERENCES: Shinners 1948a; Hitchcock 1952; Kupicha 1981, 1983; Jones & Reznicek 1997.

1. Corollas yellow; leaflets absent, the leaf consisting only of a tendril; stipules leaf-like, 5–40 mm wide; stems not winged ____ **L. aphaca**
1. Corollas various shades of blue to purple, pink, or white; leaflets 2–14; stipules various, usually < 10 mm wide; stems winged OR not winged (in 1 species).
    2. Leaflets 2; stems winged.
        3. Corollas 18–25 mm long, usually pink-purple to white; racemes with 5 or more flowers; fruits 6–9(–12) cm long; petioles with conspicuous wings ca. 3–4 mm wide on each side; plants perennial ____ **L. latifolius**
        3. Corollas 6–13 mm long, lavender-blue, pink-purple, blue-purple, or bicolored; racemes with 1–4 flowers; fruits 2.5–4.5 cm long; petioles wingless or with wings to ca. 1 mm wide on each side; plants annual.

4. Corollas 6–9 mm long, pale or light lavender-blue; fruits and enlarging ovaries glabrous; mature fruits 3–5 mm wide ____ **L. pusillus**
4. Corollas 10–13 mm long, with standard pink-purple or blue-purple with white eye, the wings paler, and the keel nearly white; fruits and enlarging ovaries hirsute; mature fruits 6–8 mm wide ____ **L. hirsutus**

2. Leaflets 8–14; stems not winged ____ **L. venosus**

Lathyrus aphaca L., (name used by Pliny for a leguminous plant), Sprawling, suberect, or slightly scandent annual; stems glabrous, 0.3–0.6(–1) m long; leaflets absent (leaves of seedlings with 1 pair of leaflets); tendrils unbranched; stipules 1–5 cm long, ovate to broadly lanceolate, hastate; racemes with 1(–2) flowers; calyces 6–10 mm long; corollas yellow, 10–12(–18) mm long; fruits 2–4 cm long, glabrous. Roadsides or other open areas; a recent collection from Kaufman Co. is the first record of this species from TX (Jones & Reznicek 1997). Late spring–summer. Native of Europe. A report for Tennessee by Beardsley & Brown (1972) was the first report for the se U.S.

Lathyrus hirsutus L., (hairy), ROUGH-PEA, SINGLETARY-PEA, SINGLETARY VETCHLING, CALEY-PEA. Ascending or sprawling, glabrous or sparsely pubescent annual; stems 0.2–1 m long; leaflets 2–7 cm long; stipules 10–18 mm long, 1–2 mm wide, linear; racemes with 1–4 flowers; calyces 5–7 mm long; fruits hirsute with pustulate hairs 1 mm or more long. Occasionally cultivated for pasture or soil improvement, roadside escape; Limestone Co. on e margin of nc TX, also Denton Co. (Shinners 1948a), also Tarrant Co. (R. O'Kennon, pers. obs.); mainly e TX. May. Native of the Mediterranean region. Reported as poisonous by Burlage (1968).

Lathyrus latifolius L., (broad-leaved), PERENNIAL SWEET-PEA, EVERLASTING-PEA. Glabrous perennial; stems climbing, 0.8–2 m long, broadly winged; leaflets to 4–8(–15) cm long; tendrils branched; stipules large, 3–5 cm long, 4–10 mm wide, lanceolate to ovate; racemes with 5–12(–15) flowers; calyces 10–11 mm long; corollas 18–22(–25) mm long, usually pinkish purple but varying to white or even red or striped; fruits 7–10 mm wide, glabrous. Cultivated in TX as an ornamental, persists and escapes?; Grayson Co., also observed on a fencerow in Tarrant Co. (R. O'Kennon, pers. obs.) Spring. Native of Europe. Seeds poisonous.

Lathyrus pusillus Elliott, (very small), LOW PEAVINE. Sprawling or low-climbing, glabrous or slightly pubescent annual; leaflets 2.5–6 cm long; stipules 1–3 cm long, usually 1–5 mm wide, lanceolate to lance-ovate, the upper lobe 2–3 times as long as the lower; racemes with 1–3 flowers; calyces 5–8 mm long. Sandy or rarely clayey soils, open woods, roadsides; se and e TX w to West Cross Timbers. Apr. Reported to cause livestock poisoning (Kingsbury 1964).

Lathyrus venosus Muhl. ex Willd., (veiny). Perennial; stems 0.4–1 m long, erect or scandent; leaflets 3–8 cm long, 1–3 cm wide; stipules linear-lanceolate to lanceolate, (0.5–)1–2.5(–3.5) cm long, usually < 10 mm wide, 1/6–1/2 as long as leaflets; racemes with 5–15(–25) flowers; calyces 6–14 mm long, with unequal lobes; corollas 12–22 mm long, bluish or purplish; fruits 3–6 cm long, 5–8 mm wide. Open woods; included based on citation of vegetational area 4 (Fig. 2) by Hatch et al. (1990); also Post Oak Savannah. Apr–May. [*L. venosus* var. *intonus* Butters & H. St. John]

LESPEDEZA BUSH-CLOVER

Perennial herbs; leaves rather small, numerous, and crowded, short-petioled, pinnately compound with 3 leaflets; leaflets entire; stipules inconspicuous, slender, linear to nearly thread-like; flowers usually small, axillary or terminal, in pairs or in head-like or loose racemes or panicles; petals white to cream to purplish or pinkish (on open flowers; cleistogamous flowers also produced by some species); stamens 10, diadelphous; fruits flattened, 1-seeded, indehiscent,

Gymnocladus dioicus [SA3]

Hoffmannseggia glauca [LAM]

Indigofera miniata var. leptosepala [BB2]

Kummerowia stipulacea [STE, WIL]

Kummerowia striata [RCA, STE, WIL]

Lathyrus aphaca [BEN]

Lathyrus hirsutus [BEN]

Lathyrus latifolius [STE, WIL]

usually ovate or rounded; style elongate on chasmogamous fruits (though easily broken off), recurved tightly on cleistogamous fruits.

A genus of 40 species of temperate North and South America, tropical and e Asia, and Australia; some are cultivated for forage, fodder, or green manure. *Lespedeza* species are well known to hybridize (Turner 1959; Clewell 1966, 1968; McGregor 1986) and because of this some individuals are very difficult to definitively identify. A population including *L. virginica*, *L. procumbens*, and numerous hybrid individuals is known from Grayson Co. (Named for Vincente Manual de Céspedes, Spanish Govenor of East Florida, on account of the hospitality provided in the late 1700s to Michaux during his explorations there; later misspelled, probably by Michaux's editor as de Lespedez) (subfamily Papilionoideae, tribe Desmodieae)
REFERENCES: Clewell 1966, 1968.

1. Plants basically prostrate (procumbent to trailing) for most of their length, sometimes weakly ascending; petaliferous flowers long-pedunculate, exceeding subtending leaves.
   2. Stem hairs spreading ____ **L. procumbens**
   2. Stem hairs appressed.
      3. Main stems usually completely trailing; fruits 3–5 mm long; stipules 1.5–3(–4) mm long; found on sandy soil ____ **L. repens**
      3. Main stems weakly ascending or erect for up to 15 cm and then trailing; fruits 4.5–7 mm long; stipules 2.5–8 mm long; found on rocky limestone soil.
         4. Stems weakly ascending; petioles of main leaves 20–40 mm long; main stem much-branched; leaflets usually green or light green beneath; rare in n part of nc TX ____ **L. violacea**
         4. Stems erect for up to 15 cm and then trailing; petioles of main leaves 5–20 mm long; main stem unbranched or nearly so; leaflets gray-green to whitish beneath; c TX n through nc TX ____ **L. texana**
1. Plants erect or strongly ascending; petaliferous flowers short-pedunculate, not greatly exceeding leaves (except longer in *L. violacea*).
   5. Flowers white or cream-colored, sometimes with a purplish spot; calyces equal to or longer than the mature fruits; flowers either 1–4 in axillary clusters or in spike-like globose or short-cylindric inflorescences of 10–45 flowers.
      6. Leaflets spatulate to cuneate (= wedge-shaped); flowers 1–4 in axillary clusters; wing petals and keel ± equal in length ____ **L. cuneata**
      6. Leaflets not spatulate to cuneate; flowers in spike-like, globose or short-cylindric inflorescences; wing petals longer than keel.
         7. Leaflets narrowly elliptic to oblong; well-developed terminal leaflets > 2 times as long as wide; rachis (stalk of terminal leaflet) longer than petiole on well-developed leaves; calyx teeth 7–13 mm long; standard 7–12 mm long ____ **L. capitata**
         7. Leaflets mostly ovate to elliptic; well-developed terminal leaflets usually 1–2 times as long as wide; rachis equal to or shorter than petiole; calyx teeth 5–8 mm long; standard 6–8 mm long ____ **L. hirta**
   5. Flowers pink to purple or violet; calyces 1/2 or less as long as mature fruits; flowers in axillary, racemose inflorescences with 4–14 flowers (not appearing spike-like).
      8. Leaflets narrow, linear to linear-oblong, 3–8 times as long as wide ____ **L. virginica**
      8. Leaflets broader, elliptic or ovate, 1–3 times as long as wide.
         9. Upper surface of leaflets and stems glabrous or with sparse pubescence; lower surface of leaflets with sparse pubescence; keel longer than wing petals, extending 1–2 mm past wings; racemes usually longer than subtending leaves ____ **L. violacea**
         9. Upper surface of leaflets and stems conspicuously appressed-pubescent or pilose; lower surface of leaflets with dense pubescence; keel ± equal in length to wing petals, included within the wings; racemes usually shorter than subtending leaves ____ **L. stuevei**

Lathyrus pusillus [WIL]

Lathyrus venosus [WIL]

Lespedeza capitata [BB1]

Lespedeza cuneata [WIL]

Lespedeza hirta [MIC]

Lespedeza procumbens [MIC]

Lespedeza repens [WIL]

Lespedeza stuevei [WIL]

Lespedeza capitata Michx., (capitate, headed), ROUND-HEAD BUSH-CLOVER, ROUND-HEAD LESPEDEZA. Stems erect or arched, 0.5–1.2(–2) m long, densely spreading-pubescent; leaflets 2–5 cm long; corollas 7–12 mm long; fruits 4–5 mm long. Sandy open woods or open ground; Grayson Co., also Lamar Co. (Turner 1959); e and nc TX and Rolling Plains. Sep.

Lespedeza cuneata (Dum.Cours.) G. Don., (wedge-shaped), SERICEA, SERICEA LESPEDEZA, CHINESE BUSH-CLOVER. Plant bushy-branched; stems 0.5–2 m tall; leaflets 1–2 cm long, the larger leaflets truncate or slightly notched; corollas 6–9 mm long; fruits 2.5–3 mm long. Cultivated for erosion control, escaped and now a problematic invader, especially on sandy soils; e TX w to East Cross Timbers. Jul–Oct. Native of e and c Asia.

Lespedeza hirta (L.) Hornem., (hairy), HAIRY BUSH-CLOVER, HAIRY LESPEDEZA. Stems low-spreading to erect or over-arched, 0.5–1.8 m long, densely spreading-pubescent; leaflets 1.5–4 cm long; corollas 6–8 mm long; fruits 4–7 mm long. Sandy woods; Denton and Grayson cos.; mainly se and e TX. Jul–Sep.

Lespedeza procumbens Michx., (procumbent, prostrate), TRAILING BUSH-CLOVER, TRAILING LESPEDEZA. Stems prostrate or nearly so, to 1 m long, usually freely branched; leaflets oblong-elliptic to suborbicular, spreading-pubescent beneath; stems and peduncles pilose; flowers 8–12 per raceme; petals rosy lavender to purple. Sandy open woods or open ground; e TX w to West Cross Timbers. Late May–Sep. Similar to *L. repens* which, however, has stems and peduncles sparsely short-appressed pubescent and typically 4–8 flowers per raceme.

Lespedeza repens (L.) Barton, (creeping), CREEPING BUSH-CLOVER, CREEPING LESPEDEZA. Stems partly decumbent to spreading or ascending, usually freely branched, to 1 m long; leaves ± of one size, though reduced upwards; leaflets oblong-lanceolate or oblanceolate to obovate or short-elliptic, appressed-pubescent beneath; stipules 1.5–3(–4) mm long; calyces 1/4 or more the length of fruits developed from cleistogamous flowers; corollas purple. Sandy soils, woods and fencerows; se and e TX w to East Cross Timbers. Late May–Sep. This species is sometimes difficult to distinguish from *L. violacea* which, however, has the calyces < 1/4 the length of the fruits from cleistogamous flowers, stipules up to 2.5–6 mm long, and leaves of at least 2 distinct sizes due to smaller leaves in the axils of larger leaves.

Lespedeza stuevei Nutt., (for its discoverer, W. Stüwe, 1875–?, Prussian pharmacist and botanist), TALL BUSH-CLOVER, STUEVE'S BUSH-CLOVER. Stems spreading to erect or arched, 0.5–1.5 m tall; stems and lower surface of leaflets densely spreading- or occasionally appressed-pubescent; leaflets 0.5–3 cm long; corollas 5–8 mm long, lavender-pink to rose-purple; fruits 4–7 mm long, hairy. Sandy open woods, fencerows, roadsides; Dallas, Grayson, and Tarrant cos., also Turner (1959) gave many other citations from nc TX; e 1/2 of TX. Late May–Sep.

Lespedeza texana Britton, (of Texas), TEXAS BUSH-CLOVER, TEXAS LESPEDEZA. Similar to *L. repens*; stems to 1.5 m long; leaflets to 3.5 cm long, grayish green beneath, coriaceous; stipules (3–)4–8 mm long; corollas purple; fruits 4–7 mm long. Clay soils, on or near limestone outcrops; Bell, Dallas, Fannin, Grayson, Hood, and Somervell cos.; Blackland Prairie and Grand Prairie; also Rolling Plains and Edwards Plateau. Late May–Sep.

Lespedeza violacea (L.) Pers., (violet), VIOLET LESPEDEZA, PRAIRIE-CLOVER. Stems weakly erect to trailing, usually branched, 0.2–0.7 m long, pubescence appressed; leaves petioled, of two sizes, small leaves in axils of larger leaves; leaflets 2–5 cm long; calyces 3–6 mm long; corollas purple; fruits 3.5–7 mm long. Open woodland along edges of openings; Grayson and Dallas cos. (Turner 1959); rare in TX, Hatch et al. (1990) cited only vegetational area 5 (Fig. 2). A Hunt Co. population with single sessile flowers in the leaf axils appears most similar to *L. violacea* (*Sanders 3449*, BRIT). Jul–Aug.

Lespedeza virginica (L.) Britton, (of Virginia), SLENDER BUSH-CLOVER, SLENDER LESPEDEZA. Stems erect, 0.3–1 m tall; stems and leaflets appressed-pubescent to short pilose; corollas 6–8 mm long, purple; fruits 4–5 mm long. Sandy open woods, fencerows, roadsides; e 1/2 of TX. Jul–Sep.

Lespedeza intermedia (S. Watson) Britton, (intermediate), according to Correll and Johnston (1970), was once collected in Tarrant Co. (in 1910). This e U.S. species, otherwise unknown in TX, is similar to *L. stuevei* but differs in having appressed-pubescent stems, leaflets glabrous above, and fruits glabrate.

LOTUS DEER-VETCH, TREFOIL

Annuals or perennials; leaves very short-petioled or sessile, pinnately compound, with 5 leaflets (lower 2 stipule-like in position) or with 3 leaflets or upper leaves often reduced to 1 leaflet (leaves thus appearing palmately compound or simple); stipules apparently absent (or reduced to glands); flowers solitary in the leaf axils or in peduncled umbels; fruits dehiscent.

A genus of 100 species of the n temperate zone; some are cultivated as ornamentals or for forage. (From the classical Greek name *lotus*, used for a number of different plants; restricted by Linnaeus to this genus) (subfamily Papilionoideae, tribe Loteae)
REFERENCES: Ottley 1944; Isely 1981.

1. Petals yellow, marked with red; plants trailing perennials; leaves with 5 leaflets (the lower pair stipule-like in position); flowers in long-peduncled, head-like umbels ____ **L. corniculatus**
1. Petals white to rosy or lavender-pink; plants erect annuals; leaves with 3 leaflets or the upper leaves reduced to 1 leaflet; flowers solitary in the leaf axils ____ **L. unifoliolatus**

Lotus corniculatus L., (horned), BIRD-FOOT TREFOIL, BIRD-FOOT DEER-VETCH, EGGS-AND-BACON. Sparsely pubescent or glabrous perennial; umbels with (2–)4–8 flowers; corollas 10–14 mm long; fruits 1.5–3.5 cm long. Cultivated and escaped into disturbed and weedy areas; Grayson Co.; also Post Oak Savannah. May–Aug. Native of Eurasia. Used for fodder, but can be cyanogenic and poisonous (Mabberley 1987).

Lotus unifoliolatus (Hook.) Benth., (with only one leaflet), PURSH'S DEER-VETCH. Moderately short-pilose annual to 80 cm tall; calyces nearly equaling the corollas; corollas 5–7(–8) mm long. Sandy open ground; widespread in much of e 1/2 of TX. Mid-May–early Jul. We are following Kartesz (1994) and Jones et al. (1997) in treating this species (which has long been known as *L. purshianus*) as *L. unifoliolatus*. [*L. purshianus* (Benth.) Clem. & E.G. Clem.]

LUPINUS BLUEBONNET, LUPINE, LUPIN

A genus of 200 species widespread geographically including the Andes, Rockies, Mediterranean, tropical African highlands, and e South America; often in areas of temperate climate. A number are cultivated as ornamentals, as fodder, or as green manure. Many have alkaloids (e.g., lupinine) and can cause poisoning in humans or livestock (Kingsbury 1964; Schmutz & Hamilton 1979). A South American species is cultivated as a food crop by descendants of the Incas.

In 1901, the Texas legislature was in the process of adopting a state flower. "... in the House, debates were flying fast and furious as one legislator launched his appeal for his favorite, to be followed by more eloquent protestations of the virtues of yet another. Phil Clement of Mills pleaded the case of the open cotton boll, which he likened to 'the white rose of commerce.' John Nance Garner, later the vice-president of the United States, jousted in behalf of the prickly-pear cactus flower. ...Then up to the podium strode John M. Green of Cuero. As Green made his appeal for the beautiful bluebonnet, calls came from the floor asking, 'What the devil is a bluebonnet?'. ...'You must mean '*el conejo*'.' 'The rabbit' was a name used by the Mexicans because the waving white tip reminded them of the bobbing tail of a cottontail rabbit. 'No, no, no,' roared another. 'He's referring to what some have called 'buffalo clover'.'...At this point a group of stalwart Texas women rose to the cause. ...The National Society of the Colonial Dames of America in the State of Texas ... had originated the idea of using the bluebonnet as the state flower and they were not going to let their favored blossom be left by the roadside for a cactus bloom or cotton boll just because a bunch of representatives didn't know what it was . ... A bluebonnet painting was sent for, and one painted by Miss Mode Walker of Austin was carried into the chamber. We are told by Mary Daggett Lake that 'deep silence reigned for an instant. Then deafening applause fairly shook the old walls.' The bluebonnet had won hands down." (Andrews 1986). Unfortunately, due to confusion about common names and the fact that the legislators probably didn't know there were six bluebonnet species in the state, *Lupinus subcarnosus*, which some felt was the least attractive of the bluebonnet species, was officially designated as the state flower, rather than the more beautiful and widespread *L. texensis*. As a result, "For seventy years the argument kicked up dust in the halls of the state Capital until the politically astute representatives ... decided to correct their oversight. In 1971, in order to make certain that they would not be caught in another botanical trap, they covered all their bases ... by offering an additional resolution that would include 'any other variety of Bluebonnet not heretofore recorded'." (Andrews 1986). As a result, because there are six *Lupinus* species in TX, there are six state flowers.

(From the ancient Latin: *lupinus*, wolf, from the mistaken idea that the plants rob soil of nutrients) (subfamily Papilionoideae, tribe Genisteae)
REFERENCES: Shinners 1953c; Erbe 1957; Turner 1957; Andrews 1986; Turner & Andrews 1986.

Lupinus texensis Hook., (of Texas), TEXAS BLUEBONNET, BUFFALO-CLOVER, TEXAS LUPINE. **One of the six *Lupinus* species which are the state flowers of Texas**. Winter annual to 60 cm tall; stems pilose with appressed or ascending hairs; leaves long-petioled, palmately compound; leaflets 4–7, lanceolate or oblancolate, glabrous above or nearly so, obtuse to acute or abruptly short-pointed apically; flowers in erect, terminal racemes, the white, pointed or acute raceme tips conspicuous from a distance (due to incompletely expanded and silvery-white hairy buds); calyces 6–8 mm long; petals deep blue (rarely white); standard with white center (changing to magenta red) bearing yellow-green dots; wing petals not inflated, nearly straight when viewed from front; fruits conspicuously hairy, ca. 2.5–4.2 cm long. Rocky limestone soils, clay; sc to nc TX mainly on the Blackland and Grand prairies and Edwards Plateau; now widely cultivated and planted along highways in much of TX and adjacent OK; endemic to TX (can now be found in s OK). Apr–May. Apparently unpalatable to cattle. /97

Lupinus subcarnosus Hook., (somewhat fleshy), TEXAS BLUEBONNET, FLESHY-LEAF LUPINE, occurs in deep sandy soils just to the se of nc TX. It can be distinguished by its racemes without distinct white tips (the buds at the rounded raceme tips have yellowish gray or brownish hairs), leaflets rounded, very obtuse or even indented apically, and wing petals inflated (cheek-like), light blue. Several Brown Co. and Comanche Co. specimens (HPC) of *L. texensis* have inflated wing petals and are thus similar to *L. subcarnosus*.

MEDICAGO MEDIC, MEDICK, BUR-CLOVER, ALFALFA, BURWEED

Annual or perennial herbs; stems prostrate to erect; leaves pinnately compund; leaflets 3, with small, sharp teeth, at least in apical portion; flowers in axillary or terminal, peduncled, head-like or short cylindrical, spike-like racemes; petals yellow (in all nc TX species except 1), blue, purple, or rarely white; stamens 10, diadelphous; fruits spirally coiled to curved.

An Old World genus of 85 species native to Europe, the Mediterranean, Ethiopia, and s Africa. A number are cultivated as ornamentals, for fodder or green manure. All nc TX species are introduced from the Old World. (Greek: *medice*, the name of alfalfa, because it came to the Greeks from Media) (subfamily Papilionoideae, tribe Trifolieae)
REFERENCES: Wagner 1948; Lesins & Lesins 1979; Small & Jomphe 1989.

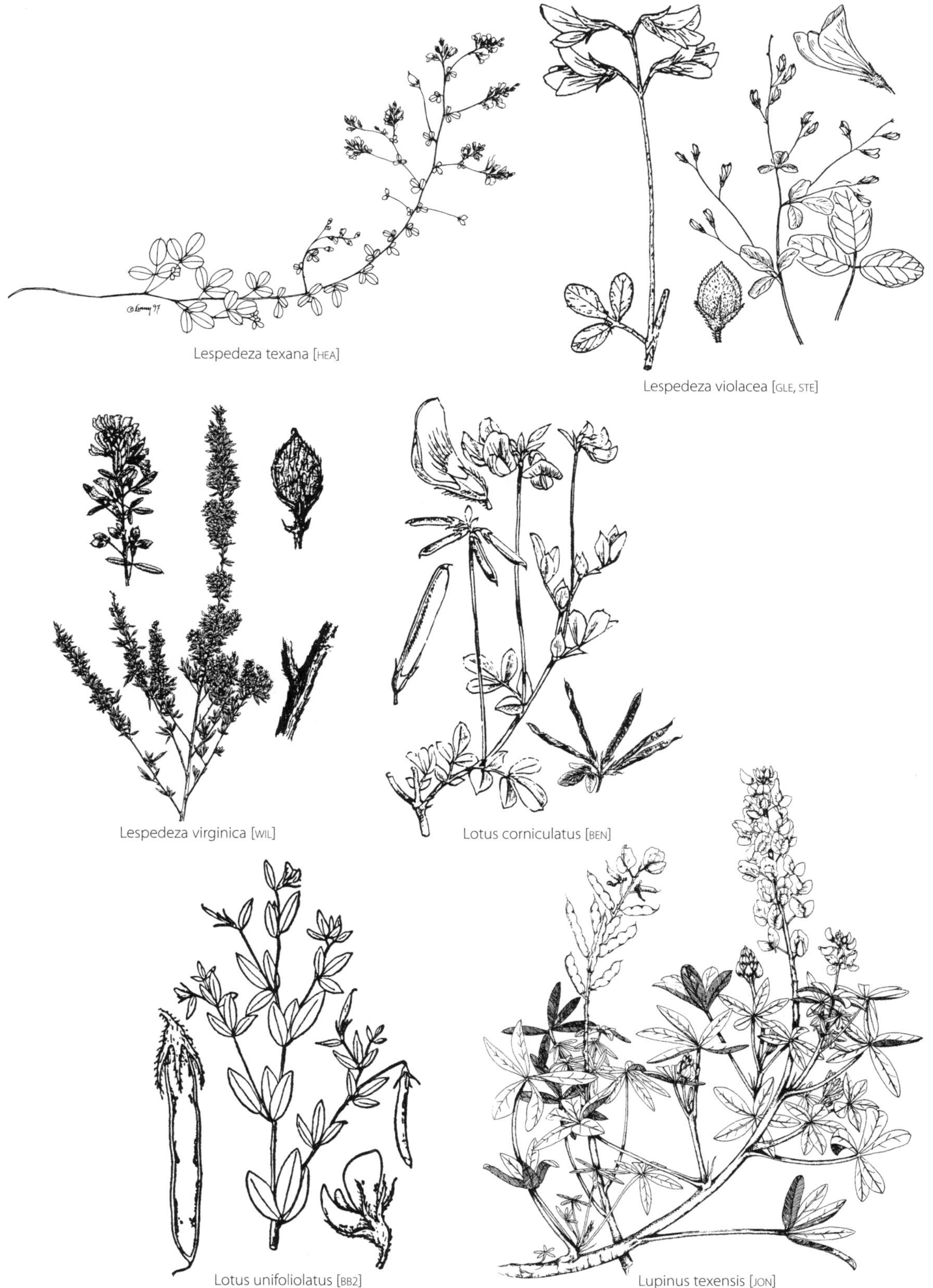

Lespedeza texana [HEA]

Lespedeza violacea [GLE, STE]

Lespedeza virginica [WIL]

Lotus corniculatus [BEN]

Lotus unifoliolatus [BB2]

Lupinus texensis [JON]

1. Petals purple to pale blue, rarely white, 6–12 mm long; ovaries and fruits without prickles; plants perennial ____ **M. sativa**
1. Petals yellow, 2–5 mm long; ovaries and fruits with or without prickles; plants annual.
   2. Stipules (2 at base of each leaf) shallowly toothed; stems and leaflets pubescent.
      3. Ovaries and fruits without prickles; flowers 8–20(–50) per raceme; petals 1.5–2 mm long ____ **M. lupulina**
      3. Ovaries and fruits prickly; flowers 5–10 per raceme; petals 2.5–4 mm long ____ **M. minima**
   2. Stipules deeply toothed to deeply lobed; stems sparsely pubescent or glabrous; leaflets glabrous.
      4. Stipules of upper leaves not divided more than half way to base; leaflets with prominent, central, purple-red spot ____ **M. arabica**
      4. Stipules of upper leaves divided more than half way to base; leaflets without central spot.
         5. Pedicels shorter than the calyx lobes, some flowers subsessile; mature fruits prickly or roughened with short points, loosely curled, barrel-shaped, 4–6 mm in diam. ____ **M. polymorpha**
         5. Pedicels equaling or exceeding the calyx lobes; mature fruits smooth, tightly coiled and flattened, disk-shaped, 10–15(–20) mm in diam. ____ **M. orbicularis**

Medicago arabica (L.) Huds., (of Arabia), SPOTTED BUR-CLOVER, SPOTTED MEDIC. Stems erect to decumbent, 10–60 cm long; leaflets 10–25 mm long; terminal leaflet very short-stalked; flowers 1–5 per head-like raceme; petals ca. 4–5 mm long; fruits prickly, with 4–7 coils, 5–6 mm in diam. (excluding prickles); prickles of fruits 2–3 mm long, recurved. Sandy pastures, lawns, and roadsides, chiefly in low ground; se and e TX w to East Cross Timbers; first TX record 1913 (Wagner 1948). Late Mar–Apr. Native of s Europe and sw Asia. ☜

Medicago lupulina L., (resembling *Humulus lupulus*—hops, apparently based on a resemblance of the clusters of tiny fruits to those of hops), BLACK MEDICK, NONE-SUCH, HOP-CLOVER, YELLOW TREFOIL. Stems ascending or trailing, to 100 cm long; leaflets 10–20 mm long; racemes 7–10 mm long, with 8–20(–50) flowers; fruits with 1 partial coil, curved, kidney-shaped, 2–3 mm in diam. Planted for pasture and soil improvement; established as a weed on roadsides, in lawns, and waste places; se and e TX w to Grand Prairie and Edwards Plateau, also West Cross Timbers (Erath Co.—Turner 1959); first TX record 1937 (Wagner 1948). Apr–May, less commonly to Jul. Native of Eurasia. Said by some to be the Irish SHAMROCK; however, it is more likely that either *Trifolium dubium* or *T. repens* is actually the SHAMROCK (Nelson 1991). ☜

Medicago minima (L.) L., (least, smallest), BUR-CLOVER, SMALL MEDIC, SMALL BUR-CLOVER. Similar to *M. lupulina* except for the fruits and usually short stems 10–30(–50) cm long; fruits with 3–5 coils, ca. 3–5 mm in diam. (excluding prickles); prickles of fruits 1.5–3 mm long, hooked at tip. Lawns, pastures, and roadsides; e 1/2 of TX; first TX record 1914 (Wagner 1948). Apr–May, rarely again in Sep. Native of Europe. ☜

Medicago orbicularis (L.) Bartal., (orbicular, round), BUTTON-CLOVER, BUTTON MEDIC. Similar to *M. polymorpha*; stems 10–50 cm long; leaflets 8–18 mm long; flowers 1–5 per head-like raceme; petals ca. 3 mm long; fruits with 4–6 coils, 10–15(–20) mm in diam. Lawns and roadsides, various soils; Callahan, Dallas, Ellis, Grayson, Hamilton, Lampasas, and Tarrant cos., mainly nc TX w to Rolling Plains and s to c TX; first TX record 1915 (Wagner 1948). Apr–early Jun. Native of the Mediterranean region. ☜

Medicago polymorpha L., (of many forms), CALIFORNIA BUR-CLOVER, BUR-CLOVER. Stems decumbent to ascending, 5–50 cm long; leaflets 6–15 mm long; terminal leaflet rather long-stalked; flowers (1–)3–5(–8) per head-like raceme; petals 2.5–4 mm long; fruits coiled 2–5 times, 4–6 mm in diam. (excluding prickles); prickles of fruits usually well-developed, 2–3 mm long, hooked at tips, rarely absent. Sandy or less commonly clayey pastures, lawns, and roadsides; widespread

Medicago arabica [BB2]

Medicago lupulina [REE]

Medicago minima [BEN]

Medicago orbicularis [MOR]

Medicago polymorpha [GLE]

Medicago sativa [BB2, GLE]

Melilotus albus [BEN]

Melilotus indicus [MOR]

in TX; first TX record 1900 (Wagner 1948). Apr–early May. Native of Eurasia. [*M. polymorpha* var. *vulgaris* (Benth.) Shinners, *M. hispida* Gaertn.]

Medicago sativa L., (cultivated), ALFALFA, LUCERNE. Glabrous perennial, 30–100 cm tall; leaflets 10–30 mm long; racemes 10–40 mm long, crowded, with 10–20(–30) flowers; fruits loosely coiled in 1 or 2 turns, 4–5 mm in diam. Cultivated and persisting, locally established as an escape on roadsides; nearly throughout TX; first TX record 1915 (Wagner 1948). Late Apr–Sep. Native of sw Asia. Long cultivated for fodder and silage (Mabberley 1987); it is considered by some to be the most important forage plant in the world (Isely 1990). ALFALFA hay can sometimes contain blister beetles (*Epicauta* species) which, if ingested, are extremely dangerous and even potentially fatal to horses as well as other livestock (Hardin & Brownie 1993). Such beetles have been found in ALFALFA hay being fed to horses in nc TX.

MELILOTUS SWEET-CLOVER, MELILOT

Annuals or biennials with erect or spreading, branched stems; leaves pinnately compound; leaflets 3, with small, sharp teeth, at least in apical portion; flowers in erect, elongated, spike-like racemes; petals yellow or white; stamens 10, diadelphous.

A genus of 20 species of fragrant herbs native to temperate and subtropical Eurasia and n Africa; cultivated for forage, hay, and green manure; also valued as a bee plant for honey production; similar to *Medicago*. All of the nc TX species are introduced. SWEET-CLOVERS can be recognized in the field by the combination of their slender racemes of small yellow or white flowers, indehiscent 1–2-seeded fruits, and often fragrant herbage. Sweet-clover poisoning, a hemorrhagic disease of cattle in which the animal bleeds to death, is caused by ingestion of improperly cured or molded sweet-clover hay (coumarin glycosides release coumarin which becomes converted to dicoumarin, a toxic substance—dicoumarin prevents blood clotting; it is now used medicinally as an anticoagulant) (Kingsbury 1964; Lewis & Elvin-Lewis 1977). (Greek: *meli*, honey, and *lotus*, some leguminous plant) (subfamily Papilionoideae, tribe Trifolieae)

REFERENCES: Hennen 1951; Isely 1954, 1990; Stevenson 1969.

1. Corollas 1.7–2.4(–3) mm long, yellow; stipules with wide, clasping, short-auricled base; mature fruits 1.5–2.5 mm long ____ **M. indicus**
1. Corollas 3.6–5.2 mm long, yellow or white; stipules narrow-based, or with wing-like extension part way around stem, but not free auricles; mature fruits 2.5–4 mm long.
   2. Corollas yellow; standard about as long as wings ____ **M. officinalis**
   2. Corollas white; standard markedly longer than wings ____ **M. albus**

Melilotus albus Medik., (white), WHITE SWEET-CLOVER, WHITE MELILOT, HUBAM-CLOVER (an annual strain). Annual or biennial; stems 0.5–2 m tall; crushed leaves fragrant; very similar to *M. officinalis* and lumped by some into that species (e.g., Kartesz 1994); we are following Isely (1990) in recognizing it as a separate species. Roadsides, railroads, and disturbed sites; widely scattered in TX. May–Jun. Native of Europe and Asia. This species is known to cause sweet-clover poisoning in cattle.

Melilotus indicus (L.) All., (of India), SOUR-CLOVER, YELLOW SOUR-CLOVER, ALFALFILLA. Annual 0.1–0.6 m tall; racemes with 10–60 flowers; fruits flattened, nearly orbicular to ovoid. Roadsides, pastures, and disturbed sites; widely scattered across TX. Apr–May. Native of the Mediterranean region and s Asia.

Melilotus officinalis (L.) Lam., (medicinal), YELLOW SWEET-CLOVER, YELLOW MELILOT. Annual or biennial 0.4–1(–2) m tall; crushed leaves fragrant; stipules lanceolate, usually 5–8 mm long; racemes with 30–70 flowers; fruits ovoid, 2.5–4 mm long, 2–2.5 mm wide. Roadsides and railroads, disturbed areas; n half of Blackland Prairie w to Grand Prairie, scattered elsewhere in TX. Late Apr–May, sporadically later. Native of Europe and Asia. The leaves have a characteristic, vanilla-like odor when crushed or dried and have been used in making sachets (Ajilvsgi 1984). This species is known to cause sweet-clover poisoning in cattle.

MIMOSA CATCLAW, SENSITIVE-BRIAR

Ours prostrate, low spreading, or trailing, herbaceous perennials to 4 m long or erect shrubs, usually armed; leaves twice even-pinnately compound, the ultimate leaflets folding together at night, in rain, or on being touched, sensitive to touch or not so; flowers sweet-scented, in our species in pink to purple (rarely white), globose heads; petals separate or united; stamens usually 8–12, the filaments separate, showy, serving as the main attractant structure for the flower.

A genus of 480 species of tropical and warm areas, especially the Americas. The group includes herbs, shrubs, lianas, and trees, often with stipular spines. Some *Mimosa* species contain the toxic amino acid, mimosine, and if eaten can cause hair loss, gastritis, and cataracts in animals (Spoerke & Smolinske 1990). As treated here, *Mimosa* includes *Schrankia*, the SENSITIVE-BRIARS; however, Stanford (1966) and J. Stanford (pers. comm.) indicated that pollen morphology warrants further study and possibly justifies generic recognition for *Schrankia*. We are following Turner (1994d) for nomenclature of the *M. quadrivalis* complex; he treated the 4 nc TX taxa as species arguing that they are isolated geographically, do not appreciably intergrade, and when occasionally growing together apparently do not hybridize; Barneby (1991) treated them as varieties of *M. quadrivalis*. The leaves of a number of *Mimosa* species (previously treated in *Schrankia*) are touch sensitive and display rapid movements upon being touched; these movements result from pressure changes in the pulvinules (= tiny swollen joints) at the base of each leaflet and the larger pulvini (singular: pulvinus) at the base of each leaf (Wills & Irwin 1961); this is possibly an adaptation to reduce water loss or a defense against herbivores. (Greek: *mimos*, mimic, referring to the sensitive leaves) (subfamily Mimosoideae, tribe Mimoseae)

REFERENCES: Isely 1971a, 1971b, 1973; Barneby & Isely 1986; Barneby 1991; Turner 1994d.

1. Plants prostrate, sprawling, or trailing; stems herbaceous; fruits with conspicuous hairs or distinctly prickly over the surface.
   2. Plants with numerous weak, hair-like or bristle-like structures not painful to the touch; recurved prickles absent; mature fruits 10–20 mm long, noticeably flattened, with numerous appressed hairs on the surface ____ **M. strigillosa**
   2. Plants conspicuously armed with numerous recurved prickles capable of causing pain; mature fruits 20–120 mm long, not noticeably flattened OR flattened in *M. roemeriana*, usually covered with erect, recurved to straight prickles (*M. quadrivalvis* complex).
      3. Leaflets with raised midrib and side-veins beneath.
         4. Fruits 4–12 cm long with acute or beaked apex; peduncles in late flower or fruit 2–7(–10) cm long; heads in bud stage with bracts not at all protruding or only slightly so; widespread in nc TX ____ **M. nuttallii**
         4. Fruits 1–4 cm long with round apex; peduncles in late flower or fruit (4–)8–12(–20) cm long; heads in bud stage with bracts protruding (can be seen with naked eye); if present in nc TX only on extreme se margin, mainly se and e TX ____ **M. hystricina**
      3. Leaflets smooth or only midrib raised beneath.
         5. Mature fruits flattened at maturity, 3–6 times as wide as thick, 4–6 mm wide; lower part of stems rounded or 5-sided; young stems puberulent or less often glabrous; usually on calcareous soils; widespread in nc TX ____ **M. roemeriana**
         5. Mature fruits usually 4-sided to nearly terete, essentially unflattened, at most 2 times as wide as thick, 2–4 mm wide; lower part of stems mostly distinctly 4-sided; young stems

glabrous; usually on sandy soils; in nc TX known only from Milam Co. on extreme se margin of area ________ **M. latidens**

1. Plants erect shrubs, never prostrate, sprawling, or trailing; stems woody; fruits glabrous or with recurved prickles limited to the margins.
   6. Leaves with 1–4(–5) pairs of pinnae per leaf; mature fruits 6–8 mm broad, often without spine-like structures on the margins, sometimes with an occasional recurved prickle; filaments definitely pink, fading pale ________ **M. borealis**
   6. Leaves with 4–8 pairs of pinnae per leaf; mature fruits usually 3–4 mm broad, often with numerous, conspicuous, recurved prickles on the margins; filaments white, tinged with pink ________ **M. aculeaticarpa**

Mimosa aculeaticarpa Ortega var. **biuncifera** (Benth.) Barneby, (sp.: prickly-fruited; var.: bearing two hooks), CATCLAW, WAIT-A-BIT, WAIT-A-MINUTE. Rounded or spindly shrub ca. 1 m tall; stems straight to often zig-zag in appearance; petals united halfway or more; fruits 35 mm or less long. Calcareous or sandy rocky open areas, also alluvial soils; Bell, Palo Pinto, Stephens, and Williamson cos., also Brown Co. (HPC); nc TX s and w to w TX. May–Aug. [*M. biuncifera* Benth.]

Mimosa borealis A. Gray, (northern), CATCLAW, FRAGRANT MIMOSA, PINK MIMOSA. Rounded shrub usually ca. 1(–2.5) m tall; stems ± straight; flowers very fragrant; petals distinct; fruits 25–50 mm long. Sandy or rocky open woods or open ground; Grand Prairie s and w to w TX. Apr–May.

Mimosa hystricina (Small) B.L. Turner, (porcupine-like, bristly), BRISTLY SENSITIVE-BRIAR. Stems sprawling or trailing, 2–4 m long; pinnae usually 4–5 pairs per leaf; ultimate leaflets folding together at night or on being touched; petals united nearly halfway; filaments conspicuous, pink; fruits subterete, densely prickly. Sandy wooded areas; included based on citation of vegetational area 4 (Fig. 2) by Hatch et al. (1990); mainly se and e TX, probably only to the se of nc TX (Turner 1994d). Feb–May. [*M. quadrivalvis* var. *hystricina* (Small) Barneby, *Schrankia hystricina* (Small) Standl.]

Mimosa latidens (Small) B.L. Turner, (broad-toothed), KARNES SCHRANKIA. Stems sprawling or trailing; pinnae usually (1–)2–3 pairs per leaf; ultimate leaflets folding together at night or on being touched; stipules 1–3(–4) mm long; petals united nearly halfway; filaments conspicuous, pink. Sandy soils or sandy loams; Milam Co. (Turner 1994d) on se margin of nc TX; mainly se and s TX. Apr–Sep. [*M. quadrivalvis* var. *latidens* (Small) Barneby, *Schrankia latidens* (Small) K. Schum., *Schrankia microphylla* of TX auth., not (Dryand) J.F. Macbr.]

Mimosa nuttallii (DC.) B.L. Turner, (for Sir Thomas Nuttall, 1786–1859, English-American botanist), CATCLAW SENSITIVE-BRIAR, CATCLAW SCHRANKIA, SHAME-BOY, NUTTALL'S SENSITIVE-BRIAR. Stems sprawling or trailing, usually 0.6–1.2 m long; pinnae usually 4–8 pairs per leaf; ultimate leaflets folding together at night or on being touched; petals united nearly halfway; filaments conspicuous, pink; fruits 4-sided or nearly terete, densely or sparsely prickly. Open woods, prairies, and roadsides, various but usually sandy or silty alluvial soils; widespread in TX. Apr–Jun. [*M. quadrivalvis* L. var. *nuttallii* (DC.) Barneby, *Schrankia nuttallii* (DC.) Standl., *Schrankia uncinata* of TX auth., not Willd.]

Mimosa roemeriana Scheele, (for Ferdinand Roemer, 1818–1891, geologist, paleontologist, and explorer of TX), ROEMER'S SENSITIVE-BRIAR, ROEMER'S SCHRANKIA. Stems sprawling or trailing, usually 0.3–1 m long; pinnae usually 4–5 pairs per leaf; ultimate leaflets folding together at night or on being touched; stipules 3–6 mm long; petals united nearly halfway; filaments conspicuous, pink. Rocky slopes on limestone, occasional on sandy soils; mainly nc TX and Edwards Plateau. Apr–Jul. [*M. quadrivalvis* var. *platycarpa* (A. Gray) Barneby, *Schrankia roemeriana* (Scheele) Blank.] ▣/99

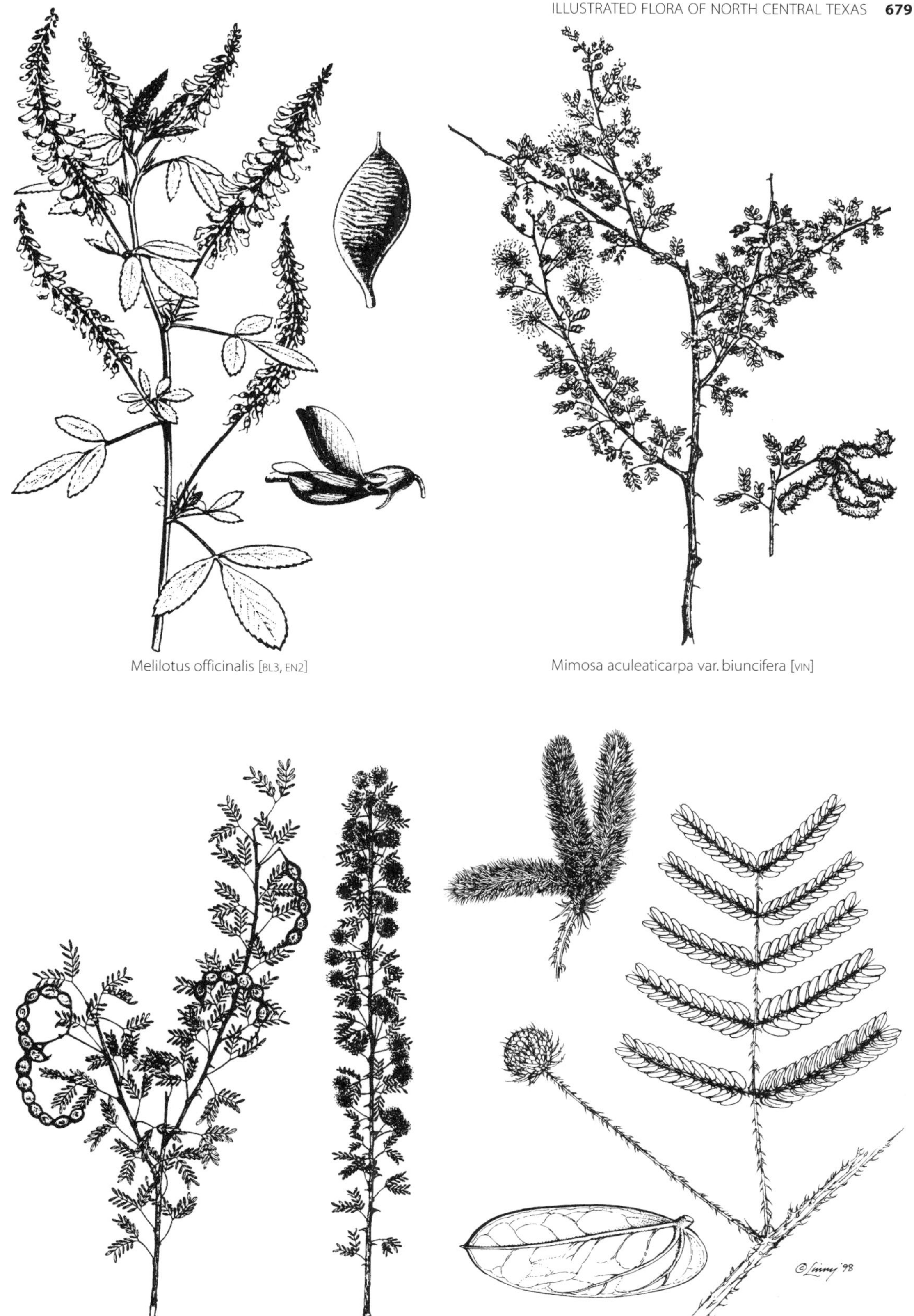

Melilotus officinalis [BL3, EN2]

Mimosa aculeaticarpa var. biuncifera [VIN]

Mimosa borealis [VIN]

Mimosa hystricina [HEA]

Mimosa strigillosa Torr. & A. Gray, (somewhat strigose, with stiff bristles), POWDERPUFF, VERGONZOSA, HERBACEOUS MIMOSA. Stems sprawling, 1–2(–4) m long, with numerous stiff bristle-like structures, not painful to the touch; pinnae usually 4–6 pairs per leaf; filaments pink to purple. Open, often somewhat sandy areas; Dallas and Tarrant cos.; mainly e TX. May–Oct.

NEPTUNIA

A genus of 11 species native to tropical and warm areas of the world, especially Australia and America; some have sensitive leaves (leaflets folding together on being touched) as in *Mimosa*. (Named for Neptune, in Roman religion a god who had to do with perpetuity of springs and streams; sometimes referred to as god of the seas; so named because some species are aquatic) (subfamily Mimosoideae, tribe Mimoseae)
REFERENCES: Turner 1951; Isely 1970a, 1973.

Neptunia lutea (Leavenw.) Benth., (yellow), YELLOW-PUFF, YELLOW NEPTUNIA. Perennial, unarmed (in contrast to the armed sensitive species of *Mimosa* previously treated in *Schrankia*); stems prostrate, vine-like, usually densely and minutely pubescent, to 2 m long; foliage blue-green; leaves twice even-pinnately compound, sensitive—the leaflets folding together on being touched; leaflets 8–18 pairs per pinna; flowers in short heads ca. 2 cm long on erect, axillary peduncles; heads (in bud) with 30–60 flowers, all flowers in the heads with stamens all alike and bearing anthers; calyces 1–2 mm long; petals separate almost or quite to base; stamens 10; filaments separate, deep yellow, serving as the main attractant structure for the flower; fruits 2.5–5 cm long, 10–15 mm wide, the stipe 4–15 mm long. Disturbed ground, various soils (some preference for sandy clay); in much of the e 1/2 of TX. Mid-May–Jun, sporadically later.

Neptunia pubescens Benth. (TROPICAL NEPTUNIA) var. *microcarpa* (Rose) Windler and var. *pubescens* are both reported from vegetational area 4 (Fig. 2) by Hatch et al. (1990). This primarily se coastal plain species apparently only reaches ne to Travis Co. just s of nc TX. The species is distinguished from *N. lutea* as follows: flowers in upper part of head with stamens bearing anthers, the lower flowers smaller and with staminodes; heads (in bud) with 20–30 flowers; stipe of fruits 0–4(–5) mm long; leaflets 14–43 pairs per pinna; calyces 2–2.7 mm long. Variety *pubescens* has stipe of fruits longer than calyx, fruits tapering to stipe, and leaves with 3–6 pairs of pinnae; var. *microcarpa* has stipe usually shorter than calyx, fruits rounded to stipe, and leaves usually with 2–3 pairs of pinnae.

ORBEXILUM SNAKEROOT

Perennial herbs; leaves pinnately compound with 3 leaflets; leaflets gland-dotted; racemes dense, spike-like; fruits cross-wrinkled, nearly beakless.

A genus of 8 species native to the U.S. and Mexico (Grimes 1990); it was previously placed in *Psoralea*. (Derivation of generic name unknown, not indicated by original author) (subfamily Papilionoideae, tribe Psoraleeae)
REFERENCES: Isely 1986a; Grimes 1990.

1. Corollas grayish lavender or lilac (rarely white), 4–7 mm long; calyces 2–3 mm long ______ **O. pedunculatum**
1. Corollas deep purple, 7–10 mm long; calyces 3.2–4 mm long ______ **O. simplex**

Orbexilum pedunculatum (Mill.) Rydb., (with a flower stalk), SAMPSON'S SNAKEROOT, BOBSROOT SNAKEROOT. Plant erect, to 0.6(–0.8) m tall; leaflets 40–70 mm long, 10–20 mm wide, elliptic to lanceolate; racemes 4–10 cm long; floral bracts and calyces eglandular. Damp or dry sandy open ground and open woods; Hopkins and Lamar cos. in ne part of nc TX; mainly se and e TX. Apr–May. [*Psoralea pedunculata* (Mill.) Vail, *Psoralea psoralioides* (Walter) Cory var. *eglandulosa* (Elliott) Freeman, *O. pedunculatum* var. *eglandulosum* (Elliott) Isely]

Mimosa latidens [HEA]

Mimosa nuttallii [RYD]

Mimosa roemeriana [HEA]

Mimosa strigillosa [COC]

Neptunia lutea [HEA, SM1]

Orbexilum simplex (Nutt. ex Torr. & A. Gray) Rydb., (simple, unbranched), SINGLE-STEM SNAKE-ROOT. Plant erect, to 1 m tall, minutely and inconspicuously pubescent; leaflets 20–70 mm long, 5–15 mm wide, linear-lanceolate or lanceolate; racemes 2–5 cm long; calyces purple, gland-dotted. Damp sandy woods or open ground; Limestone Co. on e margin of nc TX; mainly se and e TX. May–Jun. [*Psoralea simplex* Nutt. ex Torr. & A. Gray]

OXYTROPIS LOCOWEED, PURPLE LOCO, CRAZYWEED

A genus of 300 species of the n temperate zone, particularly c Asia; some are harmful to livestock and a number are cultivated as ornamentals; sometimes lumped into *Astragalus*. (Greek, *oxys*, sharp, and *tropis*, keel, alluding to the keel of the flowers) (subfamily Papilionoideae, tribe Galageae)
REFERENCE: Barneby 1952.

Oxytropis lambertii Pursh, (for Alymer Bourke Lambert, 1761–1842, from whose cultivated plants Pursh described it), LOCOWEED, PURPLE LOCO, LAMBERT'S CRAZYWEED, WHITE LOCO, LAMBERT'S LOCO, CRAZYWEED, POINT LOCO. Perennial pubescent herb, often forming colonies; plant tufted, ±scapose, erect, to 35 cm tall; pubescence of hairs attached above base so that they have a long and a short arm (= malpighian or dolabriform hairs); leaves once odd-pinnately compound, often dimorphic; leaflets 7–19, sometimes falcate; stipules 7–24 mm long; flowers 10–25 per raceme, with sweet carnation scent; corollas 15–26 mm long, light to deep purple-red to almost white, withering to blue-violet; fruits 8–15(–25) mm long, the beak 3–7 mm long, strigose. Limestone outcrops; Blackland Prairie and Grand Prairie; also Rolling Plains and Edwards Plateau. Apr–May. [*Astragalus lambertii* (Pursh) Spreng. var. *abbreviatus* (Greene) Shinners, *Oxytropis lambertii* var. *articulata* (Greene) Barneby] Some authorities (e.g., Jones et al. 1997) treat TX material as var. *articulata* (Greene) Barneby. As in some *Astragalus* species, toxins (apparently indolizidine alkaloids—e.g., swainsonine) can cause fatal loco poisoning or locoism in livestock; horses are particularly susceptible; symptoms include trembling, abortion, tendency to become excessively excited and wild, and eventually, inability to eat, paralysis, and death; all parts of the plant are toxic and the poison is cumulative; (Kingsbury 1964; Burlage 1968; James & Welsh 1992; Ralphs 1992). ☠ ▣/101

PARKINSONIA

A genus of 29 species with 25 in drier parts of the Americas, 1 in s Africa, and 3 in ne Africa. (Named for John Parkinson, London apothecary and botanical author, 1567–1650) (subfamily Caesalpinioideae, tribe Caesalpinieae)

Parkinsonia aculeata L., (prickly), RETAMA, PALOVERDE, HORSE-BEAN, JERUSALEM-THORN, MEXICAN PALOVERDE. Shrub or small tree to ca. 12 m tall; branches green, well-armed with sharp spines; leaves twice pinnately compound with 2(–4) pinnae (each with leaflets), but each pinna appearing as a once pinnate leaf due to the absence of a common petiole; pinnae with flattened green rachis; leaflets very small, 2–5(–9) mm long; flowers in racemes; petals 8–13 mm long, yellow; stamens 10, separate, inconspicuous; fruits usually 5–10 cm long, constricted between the seeds. Low sandy or gravelly limestone areas; s part of TX n to Williamson Co. (Correll & Johnston 1970) on s margin of nc TX, also Fort Hood (Bell or Coryell cos.—Sanchez 1997); also spreading from cultivation in Tarrant Co., also Brown Co. (HPC). Spring–Fall. Native of tropical America n to TX; Correll and Johnston (1970) indicated this species is perhaps adventive in TX and possibly native to South America; Elias (1989) said it is native from s TX s through Mexico to South America. This species was used in a famous forensic botany case in 1993. A PALOVERDE tree was instrumental in linking a murder suspect to an Arizona crime site where the suspect allegedly dumped the body of a victim. Plant geneticist Tim Helentjaris of the University of Arizona demonstrated that two seed pods found in the back of suspect's truck came from a specific PALOVERDE tree scraped by the truck at the crime scene. This example is important because it was the first in which plant DNA was used in a criminal case (Mestel 1993; Yoon 1993).

PEDIOMELUM SCURF-PEA

Perennials with tough or swollen (tuberous) root, caulescent or acaulescent; leaves once pinnately or palmately compound with 3–7 leaflets, often gland-dotted; corollas blue to lavender-purple, brick-red, or salmon-pink (rarely pink); stamens 10, monadelphous; fruits gland-dotted or eglandular, not cross-wrinkled, enclosed by enlarged, often gland-dotted calyces, dehiscent by transverse rupture and with the base of the fruit persistent on the receptacle; beak of fruit prominent, not cross-wrinkled.

A genus of 21 species native from c Mexico n through the U.S. to sc Canada (Grimes 1990); it was previously placed in *Psoralea*. With the exception of *P. hypogaeum* and *P. rhombifolium*, nc TX species usually have conspicuously gland-dotted leaves, fruits, and often calyces and bracts. (Greek: *pedion*, plain, and *melo*, apple) (subfamily Papilionoideae, tribe Psoraleeae)
REFERENCES: Shinners 1951a; Ockendon 1965; Isely 1986a; Grimes 1990.

1. Leaves pinnately compound with 3 leaflets, not gland-dotted; stems normally prostrate; petals brick-red to salmon-pink (rarely pink) ____ **P. rhombifolium**
1. Leaves palmately compound with 3–7 leaflets, gland-dotted (except in *P. hypogaeum*); stems erect or ascending or plants acaulescent; petals blue to lavender or purple.
   2. Inflorescence a slender loose raceme or slender, interrupted, spike-like raceme, < 2 cm thick; calyces 2–9 mm long; bracts up to 5 mm long; root thick, elongate.
      3. Plants eglandular except for upper surfaces of leaflets; pedicels 1–3 mm long; petioles of middle and upper stem leaves 6 mm or more long (often much more); bracts 2–10 mm long ____ **P. digitatum**
      3. Plants gland-dotted on both leaf surfaces, stipules, bracts, and calyces; pedicels 3.5–10 mm long; petioles of middle and upper stem leaves 5 mm or less long; bracts 1.5–3.5 mm long ____ **P. linearifolium**
   2. Inflorescence a dense, spike-like raceme 2–4 cm thick; calyces 8–17 mm or more long; bracts 4–15 mm or more long; root thick, elongate or often tuberous-enlarged, sometimes conspicuously so.
      4. Leaves all basal or nearly so at flowering time; fruits long-beaked (beak 7–19 mm long); foliage eglandular or inconspicuously gland-dotted ____ **P. hypogaeum**
      4. Leaves distributed well up the stem; fruits short-beaked (beak 2–6 mm long); foliage usually conspicuously gland-dotted.
         5. Stems densely pilose with widely spreading hairs; leaf blades long-pubescent on margins and midrib beneath ____ **P. latestipulatum**
         5. Stems and leaf blades glabrous or appressed-pubescent.
            6. Floral bracts ovate-orbicular, abruptly contracted to a narrow sharp point, the body about as wide as long (6–13 mm wide); inflorescences with 3–7 flowers, with only 2 nodes ____ **P. reverchonii**
            6. Floral bracts ovate- or oblong-lanceolate, gradually acute or acuminate, longer than wide (1–6 mm wide); inflorescences many-flowered, usually with 3 or more nodes.
               7. Peduncles shorter than or equal to subtending petioles; stipules 2.5–7 mm wide, elliptic-lanceolate to falcately oblanceolate or orbicular, the largest often in the middle of the plant ____ **P. latestipulatum**
               7. Peduncles longer than subtending petioles; stipules (except quickly deciduous ones of very basal leaves) usually 2.5 mm or less wide, linear-lanceolate to lanceolate.
                  8. Leaflets (except those of quickly deciduous basal leaves) linear-lanceolate or nar-

rowly oblong-lanceolate, 7–14 times as long as wide, 35–95 mm long, 3–9(–13) mm wide ____ **P. cyphocalyx**

8. Leaflets broadly lanceolate to rhombic, elliptic, or obovate, 2–3(–5) times as long as wide, 20–50 mm long, (6–)10–24 mm wide ____ **P. cuspidatum**

Pediomelum cuspidatum (Pursh) Rydb., (with a sharp stiff point), TALL-BREAD SCURF-PEA. Stems decumbent or ascending, occasionally erect, to 80 cm long, freely branched; leaves and fruits gland-dotted; flowers often with sweet-clover scent, 12–20 mm long; petals blue or purplish; fruit body 6–8 mm long, completely enclosed in calyx, the beak to 2 mm long. Clayey, rocky, or sandy prairies; Blackland Prairie s and w to w TX. Late Apr–May. [*Psoralea cuspidata* Pursh]

Pediomelum cyphocalyx (A. Gray) Rydb., (bent or curved calyx), TURNIP-ROOT SCURF-PEA. Stems usually solitary, ascending or erect, to 1 m tall, simple or sparingly branched above; leaves and fruits gland-dotted; flowers similar to *P. cuspidatum*; fruit body 5–6 mm long, completely enclosed in calyx, glabrous, the beak pubescent, to ca. 4 mm long. Limestone outcrops; Bell and Hamilton cos.; also Burnet, Lampasas, Parker, and Wise cos. (Ockendon 1965); Grand Prairie s to Edwards Plateau; endemic to c and nc Texas. Late May–Jun. [*Psoralea cyphocalyx* A. Gray]

Pediomelum digitatum (Nutt. ex Torr. & A. Gray) Isely, (finger- or hand-like), PALM-LEAF SCURF-PEA. Plant stiffly and openly branched from lower down, 30–80 cm tall; root not tuberous-enlarged; leaves and fruits gland-dotted; leaflets linear-oblanceolate, (2–)3–8 mm wide, gray and densely appressed-pubescent beneath, nearly glabrous above; spikes or spike-like racemes erect or ascending, interrupted; fruit body 5–6 mm long, the beak to 4 mm long. Sandy or clayey soils; e TX w to Rolling Plains and Edwards Plateau. Mid-May–early Jul. [*Psoralea digitata* Nutt. ex Torr. & A. Gray, *Psoralea digitata* var. *parvifolia* Shinners]. Variety *parvifolia*, supposedly with narrower linear leaflets, those of the middle leaves only 2–4 mm wide, is recognized by some authorities; we are following Grimes (1990), who indicated there is no discrete break in leaflet width and did not recognize var. *parvifolia*. It occurs in e TX w to Henderson and Limestone cos. at the extreme e margin of nc TX.

Pediomelum hypogaeum (Nutt. ex Torr. & A. Gray) Rydb., (underground). Plant to 25 cm tall, eglandular or with obscure glands; leaflets elliptic to rhombic or oblanceolate; flowers in short, dense spikes; petals lavender to purple; fruit body 5–6.5 mm long, the beak 7–19 mm long, projecting well past calyx. Sandy or rocky areas, roadsides, prairies, woods. Mar–May.

1. Peduncles (= stalk of inflorescence) 1.2–3.5 cm long, < 1/2 as long as the petiole of the subtending leaf ____ var. **hypogaeum**
1. Peduncles (1.7–)4–14 cm long, > 1/2 as long as the petiole of the subtending leaf.
   2. Peduncles and petioles with appressed or closely ascending hairs; leaflets narrow, 4–11(–13) mm wide, 3–4 times as long as wide ____ var. **scaposum**
   2. Peduncles and petioles with wide-spreading hairs; leaflets broad, (8–)13–37 mm wide, 2 times as long as wide or less ____ var. **subulatum**

var. **hypogaeum**, EDIBLE SCURF-PEA, PRAIRIE-POTATO, POMME BLANCHE, POMME-DE-PRAIRIE. Parker Co. (Grimes 1990); scattered in w 1/2 of TX. [*Psoralea hypogaea* Nutt. ex Torr. & A. Gray, *Psoralea scaposa* (A. Gray) J.F. Macbr. var. *breviscapa* Shinners] The tuber-like root was an important food source for Native Americans (Ajilvsgi 1984).

var. **scaposum** (A. Gray) Mahler, (with scapes). C TX n to Fort Worth Prairie and West Cross Timbers; type locality near Fredricksburg; endemic to TX. [*Psoralea hypogaea* var. *scaposa* A. Gray, *Psoralea scaposa* (A. Gray) J.F. Macbr.]

var. **subulatum** (Bush) J.W. Grimes, (awl-shaped). Cooke and Denton cos., also Grayson, Navarro (Ockendon 1965), McLennan (Grimes 1990), and Tarrant (R. O'Kennon, pers. obs.) cos.; type locality

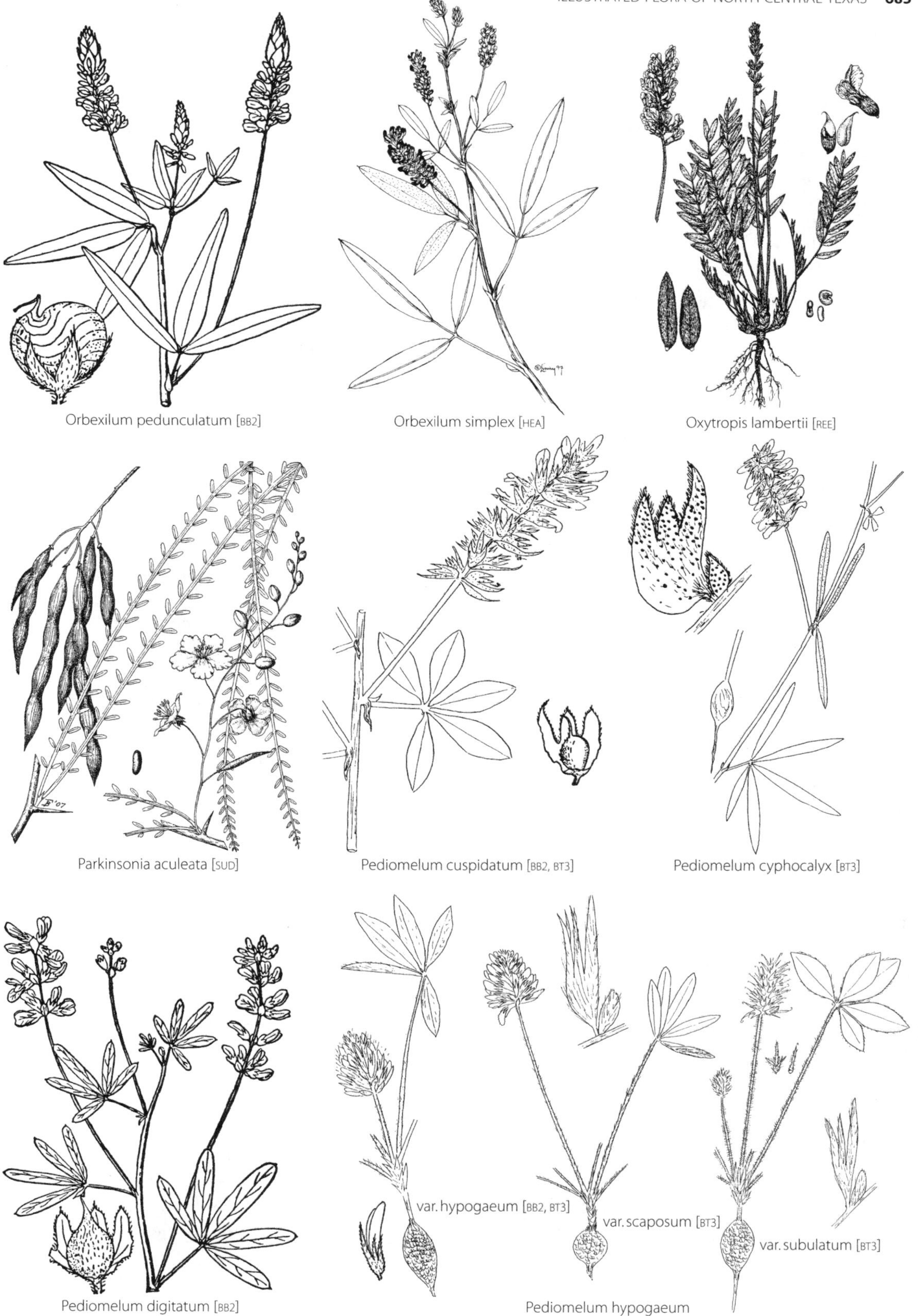

Orbexilum pedunculatum [BB2]

Orbexilum simplex [HEA]

Oxytropis lambertii [REE]

Parkinsonia aculeata [SUD]

Pediomelum cuspidatum [BB2, BT3]

Pediomelum cyphocalyx [BT3]

Pediomelum digitatum [BB2]

Pediomelum hypogaeum

in Dallas Co.; se and e TX w to East Cross Timbers. [*Psoralea subulata* Pursh, *Psoralea subulata* var. *minor* Shinners]

Pediomelum latestipulatum (Shinners) Mahler, (with broad stipules). Stems erect, 7–18(–45) cm tall; rootstock usually turnip-shaped; leaves and fruits gland-dotted; leaflets usually elliptic to oblanceolate, 5–15 mm wide; flowers 18–26 mm long; petals blue to lavender or purple; fruit body 5–6.5 mm long, the beak narrow, 5–6 mm long. Rocky or sandy prairies; Grand Prairie southwestward. Apr–mid-May.

1. Stems with appressed pubescence ______ var. **appressum**
1. Stems with erect pubesence ______ var. **latestipulatum**

var. **appressum** (Ockendon) Ghandi & L.E. Br., (pressed close to, lying flat against). According to Grimes (1990), while var. *appressum* is larger overall than var. *latestipulatum*, there is overlap in the size characters. The difference in pubescence is the only consistent character separating the varieties. Williamson Co. (Grimes 1990); mainly Edwards Plateau; type is from Travis Co.; endemic to TX. [*Psoralea latestipulata* Shinners var. *appressa* Ockendon]

var. **latestipulatum**. Brown, Coleman, Comanche, Mills, Parker, Stephens, and Tarrant cos., also Callahan, Eastland, and Wise cos. (Ockendon 1965); mainly Grand Prairie w to Rolling Plains; endemic to TX. [*Psoralea latestipulata* Shinners]

Pediomelum linearifolium (Torr. & A. Gray) J.W. Grimes, (linear-leaved). Plant to 170 cm tall, branched above, glabrate to inconspicuously appressed-pubescent; leaves and fruits gland-dotted; leaflets linear, 1–4(–6) mm wide; racemes slender, loose, drooping; flowers 8–10 mm long; petals blue to purplish; fruits 8–10.5 mm long including a broad beak to 3.5 mm long. Rocky prairies or open ground, chiefly on limestone; w Blackland Prairie w and n to Panhandle. Late May–Jun, sporadically later. [*Psoralea linearifolia* Torr. & A. Gray, *Psoralidium linearifolium* (Torr. & A. Gray) Rydb.]

Pediomelum reverchonii (S. Watson) Rydb., (for Julien Reverchon, 1837–1905, a French-American immigrant to Dallas and important botanical collector of early TX), ROCK SCURF-PEA. Plant rather bushy-branched, erect, to 120 cm tall; branches with appressed pubescence; leaflets lanceolate to elliptic, 5–9(–14) mm wide; bracts large, 18–20 mm long, 8–12 mm wide, conspicuously gland-dotted; flowers 10–15 mm long; petals blue-lavender; fruit body 7–8 mm long, the beak 3–4 mm long. Limestone outcrops; Cooke, Hood, and Tarrant cos. (type locality—Hood Co., *Reverchon*); also Johnson, Montague, Parker, and Wise cos. (Grimes 1990); w part of nc TX; narrowly endemic to nc TX and sc OK. Jun–Sep. [*Psoralea reverchonii* S. Watson]

Pediomelum rhombifolium (Torr. & A. Gray) Rydb., (with rhomboid leaves), ROUND-LEAF SCURF-PEA. Plant sparsely to densely gray-pubescent; stems normally prostrate, to 100 cm long; root not tuberous-enlarged; leaflets and fruits eglandular; leaflets variable in shape, ca. 10–30 mm wide; peduncles axillary, erect, 2–12 cm long; bracts inconspicuous; flowers 5–8(–10) mm long; fruit body 8–11 mm long, the beak 6–8 mm long. Sandy stream banks, roadsides, disturbed areas; e 1/2 of TX w through Rolling Plains; endemic to TX, s most OK and w most LA. May–Jul, sporadically to Sep. [*Psoralea rhombifolia* Torr. & A. Gray]

POMARIA

A genus of 12 species with 9 occurring in the U.S. from Kansas southward and in Mexico to the state of Hidalgo; also 3 species occur in South Africa (Simpson 1998). We are following Simpson and Miao (1997) and Simpson (1998) in recognizing this segregate of *Caesalpinia*; future study will probably result in the transfer of several South American species of

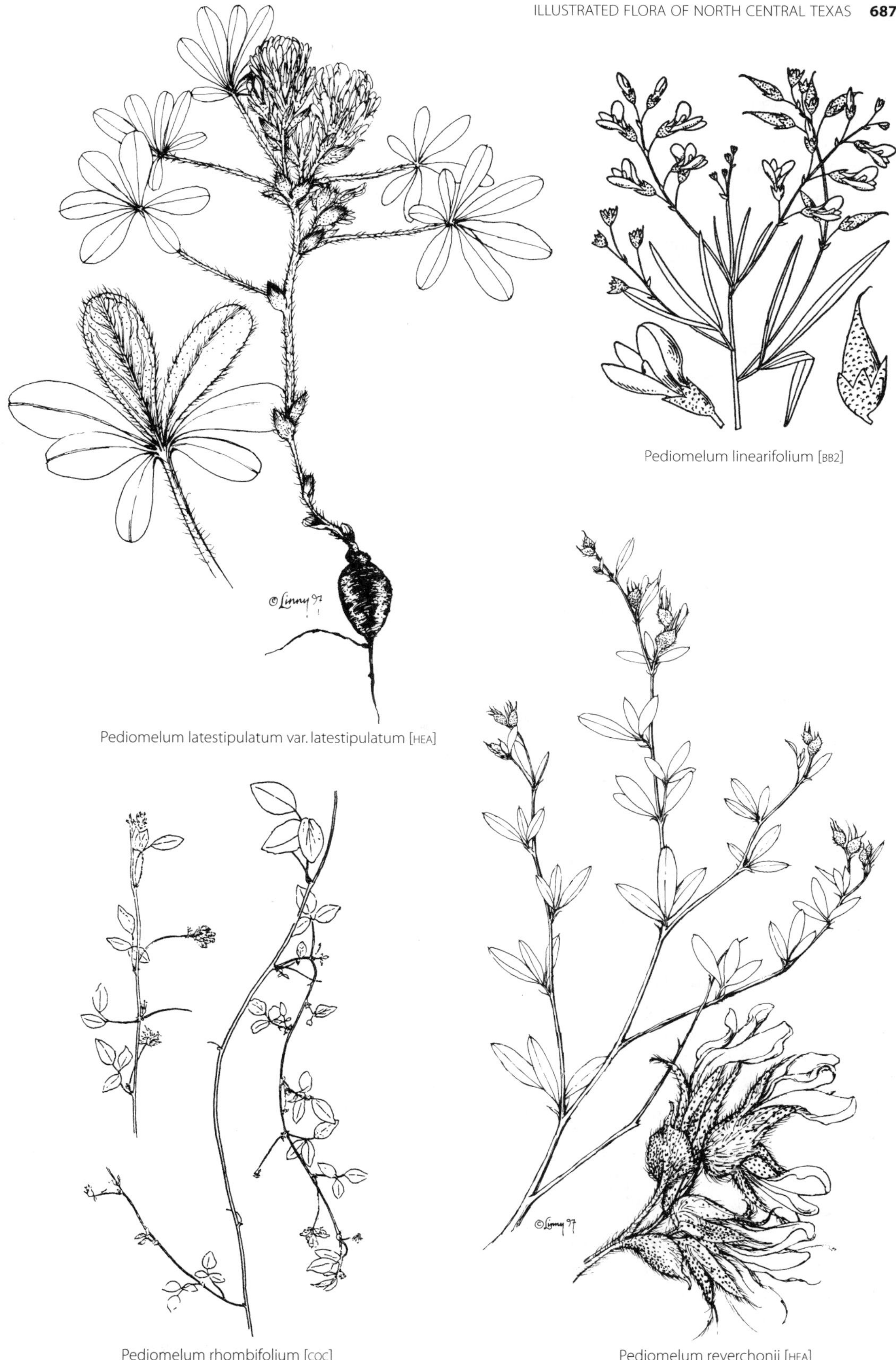

Pediomelum latestipulatum var. latestipulatum [HEA]

Pediomelum linearifolium [BB2]

Pediomelum rhombifolium [COC]

Pediomelum reverchonii [HEA]

Hoffmannseggia to *Pomaria* (Simpson 1998). (Named in 1799 by Cavanilles for Domini Pomár, physician of Phillip III) (subfamily Caesalpinioideae, tribe Caesalpinieae)
REFERENCES: Isely 1975; Simpson & Miao 1997; Simpson 1998.

Pomaria jamesii (Torr. & A. Gray) Walp., (for Edwin James, 1797–1861, American botanical explorer of the Rocky Mt. area). Unarmed perennial herb to 40 cm tall, from a thick woody root; stems appressed-pubescent, glandular; leaves twice pinnately compound with 5–7 pinnae; each pinna with 5–10 pairs of leaflets; leaflets 2.5–7 mm long, with orange sessile glands on under surface (glands black upon drying); stipules linear to lanceolate, ±entire; flowers zygomorphic; calyces and corollas glandular; petals 5, yellow, with red markings or sometimes red at base; stamens 10, to 6 mm long, shorter than the petals; filaments free; fruits 20–25 mm long, 8–10 mm wide; seeds 2. Loose sandy soils, disturbed areas; Hamilton Co., also Brown (HPC) and Comanche (Turner 1959; Simpson 1998) cos.; Trans-Pecos to Panhandle, e to Rolling Plains and sw part of nc TX. May–Sep. [*Caesalpinia jamesii* (Torr. & A. Gray) Fisher, *Hoffmannseggia jamesii* Torr. & A. Gray] While this species has usually been treated in *Caesalpinia* (e.g., Correll & Johnston 1970; Isely 1975; Kartesz 1994; Jones et al. 1997), we are following Simpson (1998) in treating it in *Pomaria*.

PROSOPIS MESQUITE

A mainly American genus of 44 species with some in sw Asia and Africa; mostly trees or shrubs, usually spiny. (From the Greek name for a kind of prickly fruit, such as the head of *Arctium lappa* L.—burdock, for obscure reasons) (subfamily Mimosoideae, tribe Mimoseae)
REFERENCES: Isely 1972, 1973; Solbrig & Bawa 1975; Solbrig & Cantino 1975; Burkart 1976.

Prosopis glandulosa Torr., (glandular), HONEY MESQUITE, MESQUITE, ALGAROBA. Shrub or small tree, spreading by deep roots; branches with small, stout, straight spines at enlarged nodes; leaves twice even-pinnately compound; leaflets glabrous; flowers in drooping, short-peduncled, catkin- or spike-like racemes, yellowish white; petals separate nearly or quite to base; stamens 10, separate, serving as the main attractant structure for the flower; fruits (5–)7–20 cm long, nearly as thick as broad, constricted between seeds. Various soils; throughout most of TX, common and locally abundant from Post Oak Savannah w, rare and local further e; frequently preserved as a semi-cultivated tree about houses. Late Apr–May, less freely to Jul or later. [*P. juliflora* (Sw.) DC. var. *glandulosa* (Torr.) Cockerell] This species was originally uncommon and restricted to stream banks and rocky slopes; it has greatly increased in abundance since settlement and is now a problematic invader of abused grasslands, particularly under abusive grazing regimes. It is well known for its deep roots which can go to depths of 100 feet or more in search of water (Cox & Leslie 1991). Large, fist-sized, swollen galls are sometimes formed on the stems by the rust fungus *Ravenelia holwayi* Dietel; leaf lesions can also be present; the rust has a complex life cycle involving 5 different spore types (J. Hennen, pers. comm.). Smoke from the burning wood is widely used to flavor foods and the seeds of MESQUITE have long been eaten by livestock and humans; Native Americans made bread from the pods and an intoxicating beer by fermenting the meal (Standley 1922b; Powell 1988); however, large amounts can cause poisoning in livestock (Kingsbury 1964). Furniture and flooring are made from the hard reddish brown wood. ☠

PSORALIDIUM SCURF-PEA

A genus of 3 species widespread in North America from s Canada to n Mexico (Grimes 1990); it was previously placed in *Psoralea*. *Psoralidium linearifolium* is here treated in *Pediomelum*. (Diminutive of *Psoralea*, from Greek: *psoraleos*, warty or scurfy, alluding to the glandular, dotted leaves of some species of that genus) (subfamily Papilionoideae, tribe Psoraleeae)
REFERENCES: Shinners 1951a; Isely 1986a; Grimes 1990.

Pomaria jamesii [HEA]

Prosopis glandulosa [SA2, SUD]

Psoralidium tenuiflorum [GLE]

Pueraria montana var. lobata [REE]

Psoralidium tenuiflorum (Pursh) Rydb., (slender-leaved), SLIM-LEAF SCURF-PEA, WILD ALFALFA, SCURVY-PEA, SLENDER SCURFY-PEA. Much-branched perennial herb with erect or ascending stems 40–60(–120) cm tall; stems, lower surfaces of leaf blades and inflorescences strigose with conspicuous whitish hairs; leaves with 3 leaflets, rarely 5 below; leaflets 10–50 mm long, up to 6(–12) mm wide, narrowly elliptic to oblong, conspicuously gland-dotted on both surfaces; stipules gland-dotted; racemes erect to drooping, with flowers numerous and crowded; calyces conspicuously gland-dotted; corollas usually 5–7 mm long; fruits 7–9 mm long, densely gland-dotted, not cross-wrinkled, exerted from the calyces, the calyx remnants basal only. Sandy or rocky prairies, open woods, and roadsides; widely scattered across TX. Apr–Jul, sporadically later. [*Psoralea tenuiflora* Pursh] Reported to be poisonous to livestock (Muenscher 1951). ☠

PUERARIA KUDZU, KUDZUVINE

A genus of 17 species of twiners with extra-floral nectaries native to tropical and e Asia. (Named for Marc Nicolas Puerari, 1766–1845, Swiss botanist) (subfamily Papilionoideae, tribe Phaseoleae)
REFERENCES: Frankel 1989; Ward 1998.

Pueraria montana (Lour.) Merr. var. **lobata** (Willd.) Maesen & Almeida, (sp.: pertaining to mountains; var.: lobed), KUDZU, KUDSU, KUDZUVINE, JAPANESE ARROWROOT. Perennial vine with trailing or high-climbing, villous stems to 20 or even 30 m long; foliage killed back to the ground during the winter in nc TX; leaves pinnately compound with 3 leaflets; leaflets ovate-rhombic to ovate or nearly rotund, entire to often 2–3 lobed, pubescent below, 5–20 cm long; petioles very long, often as long as rest of the leaf; inflorescence an axillary raceme; pedicels 2–8 mm long; flowers fragrant; corollas 15–25 mm long, violet-purple or reddish purple; stamens monadelphous; fruits linear-oblong, 4–5 cm long, conspicuously tawny to reddish brown villous. Roadsides, waste places; Grayson and Tarrant cos., also Lamar Co. (G. Diggs, pers. obs.); se and e TX w to nc TX. Late summer–fall. Native of China and Japan. [*Pueraria lobata* (Willd.) Ohwi] This species has been used for its edible root, as a green manure, for fodder, and in misguided attempts to control erosion or serve as a ground cover. It was introduced to the U.S. in 1876 by the Japanese as a gift at the Philadelphia Centennial Exposition and was endorsed and encouraged by the U.S. Department of Agriculture as a soil binder and fertilizer (Frankel 1989). It grows extremely rapidly (a foot or more a day) and invades and literally covers up native vegetation including large trees. In some areas of the se U.S. it is probably the most problematic noxious alien invader; KUDZU now covers more than one million acres in the southeastern U.S., is sometimes referred to as the "Scourge of the South," and is outlawed in several states (Frankel 1989). Recent studies with animals suggested that isoflavones and isoflavone glycosides from KUDZU may be useful in the treatment of alcohol abuse (Keung & Vallee 1993; Leung & Foster 1996). For an explanation of the nomenclature of this species, see Ward (1998).

RHYNCHOSIA SNOUT-BEAN

Ours trailing or twining perennials with deep, often partly tuberous-thickened or branched roots; leaves pinnately compound in our species (1-foliate in a species to the se of nc TX); leaflets 3, with resin droplets beneath; stipules small; flowers axillary, solitary or spike-like racemes; petals yellow or orangish, sometimes tinged with brown or red.

A mainly tropical genus of ca. 300 species; they can usually be recognized by the combination of ± yellow corollas, small typically 2-seeded fruits, and conspicuously glandular foliage (Isely 1990). (Greek: *rhynchos*, snout or beak, presumably in allusion to the somewhat beak-like form of the keel) (subfamily Papilionoideae, tribe Phaseoleae)
REFERENCE: Grear 1978.

1. Calyces 1.5–6 mm long, divided ca. 1/2 way to base, shorter than the corollas; corollas 3.5–7 mm long.
   2. Flowers 1–3 in the axils of the leaves ____ **R. senna**
   2. Flowers in conspicuous elongate racemes of 5–15 flowers ____ **R. minima**
1. Calyces 8–14 mm long, divided nearly to base, ca. equal to or longer than the corollas; corollas 8.5–13 mm long ____ **R. latifolia**

Rhynchosia latifolia Nutt. ex Torr. & A. Gray, (broad-leaved), BROAD-LEAF SNOUT-BEAN, BROAD-LEAF RHYNCHOSIA. Plant short-pilose; stems to 100 cm long; leaflets 12–80 mm wide, suborbicular to rhombic-ovate or elliptic-lanceolate; flowers in elongate racemes 3–30 cm long; corollas yellow; fruits ca. 13–20 mm long, 7–9 mm wide, shortly ovate-oblong. Sandy woods and roadsides; se and e TX w to West Cross Timbers (Parker Co. at Azle). Late May–Jun, Sep, sporadically Jul–Aug.

Rhynchosia minima (L.) DC., (least, smallest), LEAST SNOUT-BEAN. Plant minutely pubescent; racemes usually 4.5–16.5 cm long; corollas 4–7 mm long, yellow, sometimes with brownish tinge; fruits 12–20 mm long, 3–4.5 mm wide, scimitar-shaped (= like a curved sword). Clay soils; included based on citation of vegetational area 4 (Fig. 2) by Hatch et al. (1990); the n most specimen we have seen is from Travis Co.; mainly coastal plain to c TX. Apr–Dec. [*R. minima* var. *diminifolia* Walraven]

Rhynchosia senna Gillies ex Hook. var. **texana** (Torr. & A. Gray) M.C. Johnst., (sp.: from the Arabic name; var.: of Texas). Plant minutely pubescent throughout; stems to 70 cm long; leaflets 4–18 mm wide, ovate-lanceolate or elliptic to narrowly oblong; calyces half as long as the corollas; corollas yellow, often tinged with red or brown; fruits to 11–19 mm long, scimitar-shaped. Gravelly or rocky limestone soils; Bell, Burnet, and Somervell cos., also Dallas and Palo Pinto cos. (Correll & Johnston 1970); c part of nc TX s to Edwards Plateau and w to Trans-Pecos. Jun–Sep. [*R. senna* var. *angustifolia* (A. Gray) Grear, *R. texana* Torr. & A. Gray]

Rhynchosia americana (Houst. ex Mill.) Metz is cited by Hatch et al. (1990) for vegetational area 4 (Fig. 2), probably based on a Fayette Co. record (Correll & Johnston 1970) to the se of nc TX. This species differs from all nc TX species of *Rhynchosia* in having a single reniform leaflet.

ROBINIA LOCUST

Trees or shrubs; leaves once odd-pinnately compound; leaflets 7–19, entire; flowers in drooping axillary racemes; calyces campanulate; fruits linear-oblong, compressed to flat.

A North American genus of 4 species of deciduous trees and shrubs usually with stipular spines and extrafloral nectaries; some are cultivated as ornamentals. (Named for Jean Robin, 1550–1629, herbalist to Henry IV of France, and his son Vespasian Robin, 1579–1662, who first cultivated the locust in Europe) (subfamily Papilionoideae, tribe Robinieae)
REFERENCES: Isely & Peabody 1984; Peabody 1984.

1. Young stems and leaves densely and conspicuously hispid (= bristly); petals rose to reddish purple; fruits densely hispid (but rarely developed); usually unarmed shrub ____ **R. hispida**
1. Young stems and leaves glabrous; petals whitish; fruits glabrous; tree usually armed with stipular spines ____ **R. pseudoacacia**

Robinia hispida L., (hispid, bristly), BRISTLY LOCUST, ROSE-ACACIA, MOSSY LOCUST. Shrub to 3 m tall (rarely small tree); stipules not spiny; racemes with 3–ca. 10 flowers; flowers inodorous, very showy; petals 2–3 cm long; fruits 5–8 cm long, often not formed. Cultivated, persisting and spreading around old homesites; Grayson Co.; Hatch et al. (1990) cited only vegetational areas 3 and 4 (Fig. 2). Spring. Introduced from the e U.S. This variable species has been used in some areas in soil conservation plantings and in reclamation of strip mine spoils (Isely 1990).

Robinia pseudoacacia L., (false *Acacia*), BLACK LOCUST, FALSE ACACIA, BASTARD ACACIA. Tree becoming 20–30 m tall, forming colonies by root sprouts; bark furrowed; racemes with ca. 10–35 flowers; flowers fragrant, 1.5–2.5 cm long; fruits 5–10 cm long. Sandy roadsides and fencerows; Denton, Grayson, Johnson, and Tarrant cos.; se and e TX w to East Cross Timbers, also Edwards Plateau. Mar—May. Native of e U.S. This species is very difficult to eradicate once established because of the root sprouts. The wood is one of the strongest and most durable in North America (Peattie 1948; Cox & Leslie 1991). The leaves exhibit nyctinastic (nighttime or "sleep") movements, the leaflets drooping on their petiolules (Peattie 1948). The inner bark, young leaves, and seeds are toxic to all classes of livestock and humans; children have been poisoned by sucking on fresh twigs or eating inner bark or seeds; toxins include a lectin (= type of phytotoxin or toxic protein), known as robin, that agglutinates red blood cells, and robitin, a glycoside (Hardin & Arena 1974; Lewis & Elvin-Lewis 1977; Fuller & McClintock 1986). ☠

SENNA

Ours annual or perennial, unarmed herbs or shrubs; leaves once even-pinnately compound; petiolar glands present or absent; corollas yellow or orange; stamens 10, unequal, usually not all fertile, separate; anthers basifixed, dehiscing by apical pores; fruits not elastically dehiscent.

A genus of ca. 250 species (Isely 1990) or ca. 350 species (Mabberley 1997); pantropical in distribution but mostly in the New World; extending northward in xeric regions (Isely 1990). Sometimes lumped into the genus *Cassia*, but *Senna* differs in characters such as anthers dehiscent by apical pores, filaments not sigmoid, and bracteoles absent (Irwin & Barneby 1982). Irwin and Barneby (1982) gave a detailed key separating *Cassia*, *Chamaecrista*, and *Senna*. Reported to have the "vibrator" or "buzz" pollination syndrome; pollinators (such as bumblebees) shake the anthers by vibrating their thoracic flight muscles at a certain frequency; this sets up a resonance in the anthers or the space they enclose and the otherwise inaccessible pollen is released from the terminal pores of the anthers and collected by the insect (Barth 1985; Proctor et al. 1996). ☠ Some are poisonous or important medicinally (source of pharmaceutical senna) due to their anthraquinone glycosides; others are cultivated as ornamentals. (Ancient Arabic name of these plants, *sana*) (subfamily Caesalpinioideae, tribe Cassieae)
REFERENCE: Irwin & Barneby 1982.

1. Leaflets 2 (1 pair) per leaf.
  2. Leaflets linear, 1–3 mm wide; flowers solitary ____ **S. pumilio**
  2. Leaflets lanceolate, 7–12 mm wide; flowers 2–5 per peduncle ____ **S. roemeriana**
1. Leaflets 4–numerous (2 or more pairs) per leaf.
  3. Leaflets usually 4–6 per leaf; mature fruits either terete OR 16–20 cm long.
    4. Plants not malodorous; leaflets oblanceolate to lanceolate or lanceolate-elliptic, acute to acuminate at apex; petiolar gland (between lowest leaflets) erect; mature fruits 6–8 cm long, 6–9 mm wide ____ **S. corymbosa**
    4. Plants malodorous; leaflets obovate, apex rounded and apiculate; petiolar gland appressed; mature fruits 16–20 cm long, 4–5.5 mm wide ____ **S. obtusifolia**
  3. Leaflets 8–numerous per leaf; mature fruits neither terete nor as long as 16 cm.
    5. Petioles without glands; mature fruits 10–20 cm long, winged ____ **S. alata**
    5. Petioles with gland(s) between each pair of leaflets or below lowest pair of leaflets (near base of petiole); mature fruits 4–12 cm long, not winged.
      6. Petiolar glands occurring between each pair of leaflets; mature fruits 4–6 cm long ____ **S. lindheimeriana**
      6. Petiolar glands solitary, below lowest pair of leaflets (usually near base of petiole); mature fruits 7–12 cm long.
        7. Plants annual, malodorous; flowers usually 2–3(–5) per raceme; mature fruits brown

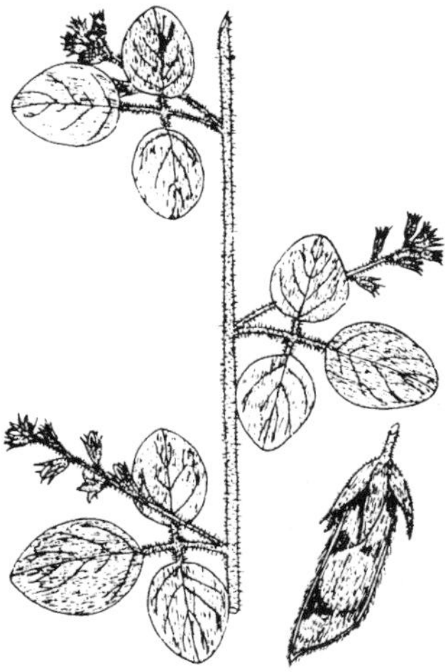

Rhynchosia latifolia [STE]

Rhynchosia minima [FAW]

©Lenny97

Rhynchosia senna var. texana [HEA]

Robinia hispida [LAM]

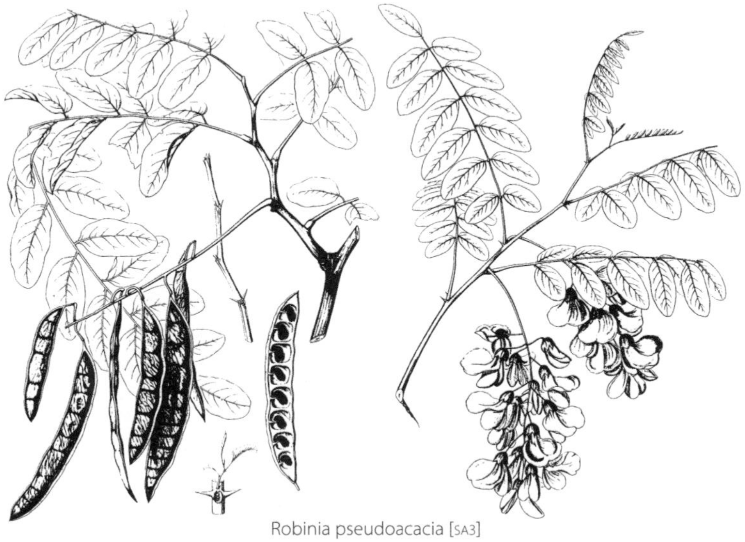

Robinia pseudoacacia [SA3]

Senna alata [WIG]

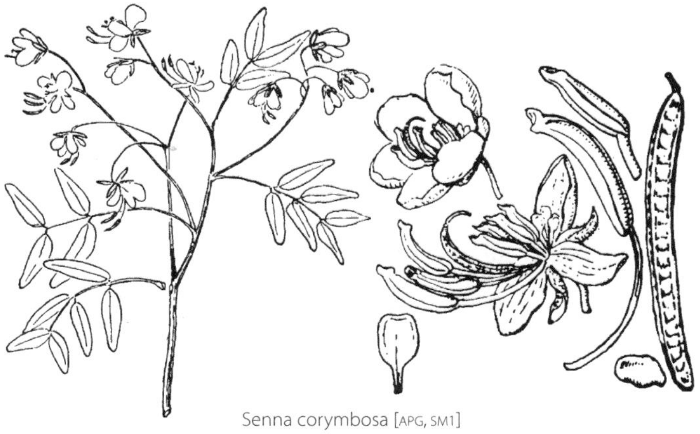

Senna corymbosa [APG, SM1]

with lighter margins, somewhat raised over the seeds but without distinct septations between the seeds ______ **S. occidentalis**

7. Plants perennial, not malodorous; flowers (4–)5–numerous per raceme; mature fruits black, flat, with cross septations (visible externally as very distinct lines across the fruit) ______ **S. marilandica**

Senna alata (L.) Roxb., (winged), EMPEROR'S-CANDLESTICKS. Perennial shrub to 2–3 m tall, usually dying back to the groundin winter in TX; leaflets 6–12 pairs per leaf; upper 3 stamens much reduced; lowest 3 stamens with larger anthers; our only *Senna* with winged fruit. Cultivated and apparently escapes; included based on citation for vegetational areas 4 and 5 (Fig. 2) by Hatch et al. (1990); sc to c TX. Aug–Oct. Native of tropical America. [*Cassia alata* L.]

Senna corymbosa (Lam.) H.S. Irwin & Barneby, (with corymbs), ARGENTINE SENNA. Glabrous perennial shrub or small tree to 3.5 m tall; evergreen where not killed back by cold; leaflets 2–3 pairs per leaf, oblanceolate to lanceolate or lanceolate-elliptic, acute to acuminate apically; upper 3 stamens much reduced; lowest 3 stamens with long arching filaments. Cultivated and apparently escapes; included based on citation for vegetational areas 4 and 5 (Fig. 2) by Hatch et al. (1990); sc to c TX. Aug–Sep. Native of Argentina and Uruguay. [*Cassia corymbosa* Lam.]

Senna lindheimeriana (Scheele) H.S. Irwin & Barneby, (for Ferdinand Jacob Lindheimer, 1801–1879, German born TX botanist), LINDHEIMER'S SENNA. Erect perennial herb 1–2 m tall; leaflets 5–8 pairs per leaf; upper 3 stamens much reduced; fruits compressed, 4–6 cm long, 6–8 mm wide. Limestone soils; Burnet Co.; mainly Edwards Plateau to Trans-Pecos. [*Cassia lindheimeriana* Scheele] The leaves are reported to be toxic but not fatal to grazing animals; they also act as a strong laxative (Kingsbury 1964; Ajilvsgi 1984).

Senna marilandica (L.) Link, (OF MARYLAND), WILD SENNA, MARYLAND SENNA. Erect perennial herb from woody root, to 1(–2) m tall, essentially glabrous; leaflets (5–)6–10 pairs per leaf; petiolar glands usually conical; upper 3 stamens much reduced; lowest 3 stamens with somewhat larger anthers; fruits 7–11 cm long, 8–11 mm wide, flattened, the segments rectangular. Disturbed areas, sandy fields, and open woods; Bell, Dallas, Grayson, and Johnson cos.; se and e TX w to nc TX. Aug–Sep. [*Cassia marilandica* L.] The leaves were used as a cathartic by Native Americans; glycosides are present and probably also saponins (Burlage 1968).

Senna obtusifolia (L.) H.S. Irwin & Barneby, (blunt-leaved), SICKLE-POD, COFFEEWEED. Erect annual to 1(–1.5) m tall, usually smaller, essentially glabrous; leaflets usually 3 pairs per leaf; petiolar glands slender; upper 3 stamens much reduced; lowest 3 stamens with very large anthers; fruits terete or 4-angled, usually sickle-shaped. Sandy soils, disturbed sites; Dallas, Grayson, and Milam cos.; se and e TX w to Blackland Prairie. Jul–Sep. [*Cassia obtusifolia* L.] Glycosides, alkaloids, and other toxins are present which can poison cattle and possibly other livestock if the leaves, stems, or seeds are eaten; fatalities can result (Hardin & Brownie 1993).

Senna occidentalis (L.) Link, (western), COFFEE SENNA, STYPICWEED, BRICHO. Erect annual to 1–2 m tall; leaflets 4–6 pairs per leaf, lanceolate to ovate, acute or acuminate at apex; upper 3 stamens much reduced; lowest 3 stamens with elongate filaments; fruits 8–12 cm long, 7–10 mm wide. Roadsides and disturbed sites; Travis Co. (Turner 1959) at the s margin of nc TX; included based on citation of vegetational area 4 (Fig. 2) by Hatch et al. (1990); se and e TX, also Edwards Plateau. Aug–Nov. Geographic origin unclear (Irwin & Barneby 1982), possibly pantropical, n to s U.S. [*Cassia occidentalis* L.] Use as a coffee substitute in some areas has been reported; slightly toxic to livestock (Correll & Johnston 1970).

Senna pumilio (A. Gray) H.S. Irwin & Barneby, (dwarf), DWARF SENNA, PYGMY SENNA. Small perennial to 0.2 m tall, usually shorter, nearly glabrous; root tuberous; leaflets 1 pair per leaf; upper 3 stamens much reduced; fruits inflated, subglobose, indehiscent, 10–15 mm long, 7–10 mm

Senna lindheimeriana [VIN]

Senna marilandica [WIL]

Senna occidentalis [WIL]

Senna obtusifolia [DIL]

Senna pumilio [HEA]

Senna roemeriana [HEA]

wide. Rocky limestone soils; Shackelford Co., also Brown (HPC), Mills, and Throckmorton (Turner 1959) cos.; w edge of nc TX s and w to w TX; sometimes abundant and showy. Apr–early May. [*Cassia pumilio* A. Gray]

Senna roemeriana (Scheele) H.S. Irwin & Barneby, (for Ferdinand Roemer, 1818–1891, geologist, paleontologist, and explorer of TX), TWO-LEAF SENNA. Perennial 0.3–0.6 m tall, from thick, woody root; stems and leaves pubescent, gray-green; leaflets 1 pair per leaf; upper 3 stamens much reduced; fruits turgid, 20–30 cm long. Limestone outcrops; Blackland Prairie s and w to w TX. Apr–May, sparingly Jun–Aug, and again in Sep. [*Cassia roemeriana* Scheele] Agricultural agents indicate that this species can cause fatal poisoning in grazing livestock. ☠ ▣/105

Senna pendula (Humb. & Bonpl. ex Willd.) H.S. Irwin & Barneby var. *glabrata* (Vogel) H.S. Irwin & Barneby, a less hardy species closely related to *S. corymbosa*, can be distinguished from that species by its obovate leaflets rounded apically. This tropical American taxon is reported by Hatch et al. (1990) from vegetational area 5 (Fig. 2); we have seen no nc TX material and persistence seems unlikely. [*Cassia bicapsularis* sensu Correll & Johnston, not L.]

SESBANIA

Annual or perennial herbs or subshrubs, unarmed; stems long, green, glabrous, unbranched below; leaves once even-pinnately compound, up to 30 cm long; leaflets numerous, 2–3 cm long, glabrous; flowers in axillary racemes; calyces campanulate, the tube broader than long, the lobes shorter than tube; corollas yellow or orange-yellow, often tinged with red; keel petals auriculate.

A genus of 50 species of warm and usually wet areas of the world; it includes herbs, shrubs, and trees, some cultivated as ornamentals; including *Daubentonia*. *Glottidium*, recognized here as a distinct genus, is sometimes treated as part of *Sesbania*. (Either from the Arabic: *sesban*, name for *Sesbania sesban* (L.) Merrill, or possibly from Persian, *sisaban*, rope, fiber, or a kind of tree) (subfamily Papilionoideae, tribe Robinieae)
REFERENCES: Isely 1986a; McVaugh 1987.

1. Flowers 10–30 per raceme; fruits 5–6 cm long, ca. 10 mm wide, 4-winged ____ **S. drummondii**
1. Flowers 2–6 per raceme; fruits 10–20 cm long, 3–4 mm wide, not winged ____ **S. herbacea**

Sesbania drummondii (Rydb.) Cory, (for its discoverer, Thomas Drummond, 1780–1835, Scottish botanist and collector in North America), RATTLEBUSH, POISON-BEAN, COFFEE-BEAN, DRUMMOND'S SESBANIA. Subshrub, lower part becoming woody; plant to 3 m tall; flowers yellow or orange-yellow, often with red lines; fruits 4-sided and conspicuously 4-winged, apically short-beaked, the stipe 1–1.5 cm long; seeds usually 3–7 per fruit. Moist low areas; Denton Co., also Comanche (Stanford 1976) and Williamson (Correll & Johnston 1970) cos.; mainly se and e TX, chiefly in Coastal Plain, also Edwards Plateau. [*Daubentonia drummondii* Rydb.] The seeds become loose and rattle when moved and are poisonous to sheep, goats, and cattle when eaten, apparently due to the presense of saponins; affected animals develop diarrhea, weakness, lethargy, and sometimes die; the seeds are also poisonous to humans (Sperry et al. 1955; Kingsbury 1964; Ellis 1975; Lewis & Elvin-Lewis 1977; Crosswhite 1980). ☠

Sesbania herbacea (Mill.) McVaugh, (herb-like, with succulent stems), COFFEE-BEAN, BEQUILLA, COLORADO RIVER-HEMP, SESBANE. Large annual to 4 m tall, openly branched; flowers yellow; fruits narrowly linear, glabrous, the sutures thickened, with beak 5–10 mm long, the stipe 0–4 mm long; seeds 30–40 per fruit. Disturbed areas; Dallas, Denton, Ellis, Grayson, Kaufman, and Tarrant cos.; se and e TX w to nc TX, also Edwards Plateau. Jul–Oct. [?*Darwinia exaltata* Raf., ?*S. exaltata* (Raf.) Rydb. ex A.W. Hill, ?*S. macrocarpa* Muhl. ex Raf.] This species has long gone under the name *S. exaltata* (e.g., Kartesz 1994; Jones et al. 1997). However, we are following McVaugh

(1987) who made the combination, *S. herbacea*, based on an older name, [*Emerus herbacea* Mill.], published in 1768. The species was important as a source of fiber for Native Americans (Mabberley 1987). The seeds have been reported to be poisonous (Hardin & Brownie 1993). ☠

SOPHORA

Ours unarmed shrubs or small trees; leaves once odd-pinnately compound; stipules very small, falling early; flowers in terminal or lateral racemes of 4–15 flowers, large and showy; stamens 10, separate; fruits torulose or moniliform.

A genus of 45 species, mostly in n temperate areas and the tropics. A number are cultivated as ornamentals including the Chinese and Korean *S. japonica* L. (PAGODA TREE). (Arabic: *sophera*, used for a tree with pea-like flowers) (subfamily Papilionoideae, tribe Sophoreae)
REFERENCES: Rudd 1972; Merrill 1977.

1. Leaves deciduous, thin; leaflets 13–19 (on well-developed leaves); standard white changing to pink, keel rose- to lavender-pink; fruits 7–8 mm thick, black at maturity; widespread in nc TX ________ **S. affinis**
1. Leaves evergreen, coriaceous; leaflets 5–11; petals all deep violet-blue; fruits 10–20 mm thick, brown at maturity; n to s part of nc TX ________ **S. secundiflora**

Sophora affinis Torr. & A. Gray, (related), EVE'S-NECKLACE, TEXAS SOPHORA. Usually a small tree; racemes drooping; flowers showy, but obscured by new leaves; corollas 11–14 mm long; fruits usually with 2–6 fertile segments, irregularily moniliform (= constricted at intervals and resembling a string of beads), hence the common name; seeds brown, ca. 5 mm long. Rocky limestone slopes and ravines; Blackland Prairie and Grand Prairie s to Edwards Plateau. Apr–early May. The seeds are reported to be poisonous (Correll & Johnston 1970). ☠ ▣/106

Sophora secundiflora (Ortega) Lag. ex DC., (with flowers on only one side of a stalk), MOUNTAIN-LAUREL, TEXAS MOUNTAIN-LAUREL, MESCAL-BEAN, FRIJOLITO. Usually a shrub; corollas 15–16 mm long; fruits torulose, knobby with 1–4 fertile segments; seeds 10–15 mm in diam. Rocky limestone slopes; Fort Hood (Bell or Coryell cos.—Sanchez 1997), also McLennan (Mahler 1988) and Williamson (Little 1976) cos; s part of nc TX s and w to w TX; also cultivated. Mar. One of the most beautiful native plants of Texas; unfortunately poisonous to livestock, and the flowers and bright red seeds poisonous to humans; one seed is reported to be enough to cause fatal poisoning. Before the use of peyote, MESCAL-BEAN was used as a hallucinogen in sw North America by Native Americans, but is very toxic (due to cytisine and related quinolizidine alkaloids) and fatal in excess (Standley 1922b; Kingsbury 1964; Lampe & McCann 1985; Mabberley 1987). The name MESCAL-BEAN is supposedly derived from "use (misuse) in adulterating weak whisky or mescal liquor, a small amount making the liquor so potent that over-indulgence brought on dizziness, disorientation and sometimes death" (Crosswhite 1980). Merrill (1977) gave extensive ethnographic and archeological information on this plant. ☠

STROPHOSTYLES WILD BEAN, FUZZY-BEAN

Annual twining or trailing vines, glabrous to pubescent; leaves pinnately compound with 3 leaflets, petiolate, stipulate; leaflets ovate to linear, entire or 3-lobed, stipellate; racemes axillary, long pedunculate, few-flowered; petals short-clawed, usually pinkish to purplish; keel strongly and conspicuously incurved or sickle-shaped; stamens 10, diadelphous; fruits linear, sessile, dehiscent. Nc TX species are easily recognized in the field by the combination of long peduncles and curved keels.

A North American genus of 3 species. (Greek: *strophe*, a turning, and *stylos*, a pillar or style, apparently in reference to the style which is curved to conform with the incurved or sickle-shaped keel surrounding it) (subfamily Papilionoideae, tribe Phaseoleae)

REFERENCE: Stace & Edye 1984.

1. Flowers 9–14 mm long; leaflets usually < 2 times as long as wide, often somewhat 3-lobed; fruits ca. 50–100 mm long; seeds 5–10 mm long, hairy ______ **S. helvula**
1. Flowers 5–8 mm long; leaflets usually 2 times or more as long as wide, usually not lobed; fruits ca. (15–)30–45 mm long; seeds 3–4 mm long, glabrous ______ **S. leiosperma**

Strophostyles helvula (L.) Elliott, (pale yellow), AMBERIQUE-BEAN, TRAILING WILD BEAN. Stems glabrous to pilose; leaflets ovate, sparsely strigose, 3–6.5 cm long, often 3-lobed or with an indentation on one or both sides; peduncles (5–)10–30 cm long; calyces ca. 4–5 mm long including longest lobe; keel incurved, darkened apically; seeds covered with a thick, felty layer. Sandy open ground; se and e TX w to West Cross Timbers and Edwards Plateau. Jul–Oct. [*Phaseolus helvula* L. (original spelling by Linnaeus of the specific epithet)]; sometimes spelled *S. helvola*; see Isely 1986b for explanation.

Strophostyles leiosperma (Torr. & A. Gray) Piper, (smooth-seeded). Stems pilose; leaflets linear to narrowly ovate, pilose, 2.5–5 cm long, usually not lobed, rarely with an indentation on one side or with very shallow lobing; peduncles 3–10(–11) cm long; calyces 2–4 mm long; keel incurved; seeds shiny. Sandy open ground; widespread in TX. Jun–Sep. [*Phaseolus leiospermus* Torr. & A. Gray]

STYLOSANTHES PENCIL-FLOWER

A genus of 25 species of tropical and warm areas of the world. The single nc TX species is easily recognized in the field by the combination of conspicuous stipules and orange to orange-yellow flower color. (subfamily Papilionoideae, tribe Aeschynomeneae)
REFERENCES: Mohlenbrock 1957 [1958], 1958.

Stylosanthes biflora (L.) Britton, Sterns, & Poggenb., (two-flowered), SIDE-BEAK PENCIL-FLOWER. Perennial herb from a tough, deep, vertically or obliquely branched root; stems decumbent to erect, 10–50(–60) cm long; leaves short-petioled, pinnately compound; leaflets 3, narrowly lanceolate or oblanceolate to elliptic, acute, with short, bristle-point; stipules narrow, united ca. 2/3–3/4 the length of the petiole, the apical 3–8 mm free; flowers solitary or few together, terminal, closely subtended by reduced leaves; petals orange to orange-yellow (rarely white); standard 5–9 mm long; wings and keel 3.5–4.5 mm long; stamens 10, monadelphous; fruits 3–5 mm long, with 2 segments, the lowest usually sterile and pedicel-like. Sandy open woods; se and e TX w to West Cross Timbers (Montague and Parker cos.); also Edwards Plateau. May–Oct. [*S. biflora* var. *hispidissima* (Michx.) Pollard & C.R. Bell]

TEPHROSIA CATGUT, HOARY-PEA

Clump-forming perennials with tough roots; stems densely short-pilose; leaves once odd-pinnately compound; leaflets many, conspicuously pilose, often with prominent bristle-tip; stipules very slender; flowers in terminal or lateral, spike-like, short or elongate racemes; petals large and showy; fruits dehiscent.

A genus of 400 species mainly in seasonal areas of the tropics, especially Africa. Some *Tephrosia* species are the source of the alkaloid rotenone, used as a fish poison and insecticide; use of *Tephrosia* species as a fish poison developed independently in the Americas, Africa, Asia, and Australia; typically the pounded leaves, branches, or roots are thrown into a body of water and the fish rise stunned to the surface (Wood 1949). (Greek: *tephros*, ash-colored or hoary, presumably from the dense pubescence of this color in some species) (subfamily Papilionoideae, tribe Tephrosieae)
REFERENCE: Wood 1949.

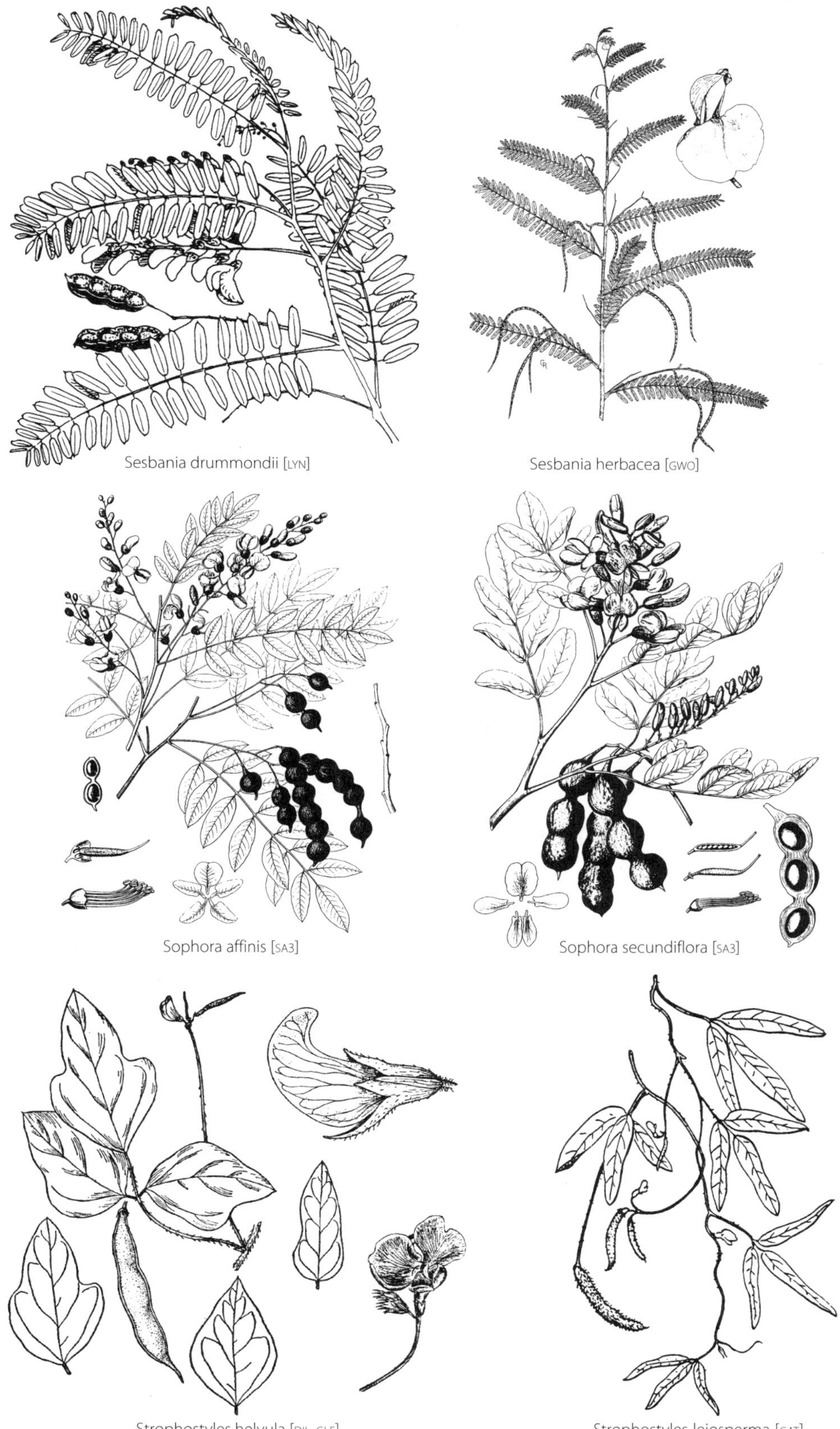

Sesbania drummondii [LYN]

Sesbania herbacea [GWO]

Sophora affinis [SA3]

Sophora secundiflora [SA3]

Strophostyles helvula [DIL, GLE]

Strophostyles leiosperma [GAT]

1. Leaflets narrow, linear (narrowly to broadly), 3–6 times as long as wide; mature fruits 5.5 mm or less wide; widespread in nc TX.
   2. Corollas bicolored, the standard light yellow, the wings and keel partly or wholly pink or red; flowers crowded in thick, short racemes 3–10 cm long; stems ascending or erect, 25–70 cm tall; leaflets usually acute at apex (also bristle-tipped) ____ **T. virginiana**
   2. Corollas of 1 color, all petals white to yellowish white or whitish pink, changing to rosy or purple-red; flowers loosely spaced in long-peduncled, slender, elongate racemes 10–40 cm long; stems spreading to largely decumbent, 30–100 cm long; leaflets usually obtuse to truncate (also bristle-tipped) ____ **T. onobrychoides**
1. Leaflets broad, obovate or elliptic to suborbicular, only 1–2 times as long as wide; mature fruits 7 mm or more wide; limited to the extreme s margin of nc TX ____ **T. lindheimeri**

Tephrosia lindheimeri A. Gray, (for Ferdinand Jacob Lindheimer, 1809–1879, German-born TX collector), LINDHEIMER'S TEPHROSIA, ROUND-LEAF TEPHROSIA. Leaflets (5–)7–15(–19), 11–37 mm long; flowers 7–21 mm long; corollas rose-purple, the standard with a white spot near base; fruits 25–50 mm long, 7–8.5 mm wide. Sandy areas; Burnet Co.; s margin of nc TX s to s TX; endemic to TX. Apr–Sep.

Tephrosia onobrychoides Nutt., (resembling *Onobrychis*—another genus of Fabaceae), MULTI-BLOOM TEPHROSIA. Leaflets (11–)13–25(–29), 17–60 mm long; flowers 15–20 mm long; fruits 35–85 mm long, 3.5–5 mm wide. Sandy open woods or open ground; Dallas, Grayson, Henderson, and Milam cos., also Lamar (Carr 1994) and Tarrant (Wood 1949) cos.; se and e TX w to nc TX. Mostly Jun–Jul, less freely to Sep. The light colored flowers open in the evening and close the next morning; before closing, they change color (Ajilvsgi 1984) becoming rosy or purple-red.

Tephrosia virginiana (L.) Pers., (of Virginia), GOAT'S-RUE, VIRGINIA TEPHROSIA, DEVIL'S-SHOESTRING, CATGUT. Leaflets (9–)15–25(–31), 11–33 mm long; flowers 14–21 mm long; fruits 25–55 mm long, 3.5-5.5 mm wide. Sandy open woods or open ground; mostly n and e TX. Mostly middle and late May(–Jun). Rotenone is present in the roots; the plant was used medicinally by Native Americans and early settlers (Ajilvsgi 1984).

TRIFOLIUM CLOVER

Annual or perennial herbs; leaves alternate, palmately or pinnately compound; leaflets 3, rounded or notched at apex, entire or with small, sharp teeth, usually with well-developed petioles; flowers sessile or pedicellate, in terminal or axillary, sessile to long-peduncled, very dense, head-like or spike-like inflorescences; petals usually withering and persistent long after flowering; stamens 10, diadelphous; fruits 1–4-seeded.

A genus of 238 species of bee-pollinated herbs of temperate and subtropical areas except Australia. A number of species are agriculturally important as forage, fodder, hay, in crop rotation, and as bee plants; some species are cyanogenic; a number are used as ornamentals, especially *T. incarnatum* in nc TX. (Latin: *tres*, three, and *folius*, leaf, from the three leaflets) (subfamily Papilionoideae, tribe Trifolieae)

REFERENCES: McDermontt 1910; Hennen 1950; Isely 1951; Hermann 1953; Brown & Peterson 1984; Zohary & Heller 1984.

1. Petals pale to bright yellow or greenish yellow; terminal leaflet on a stalk longer than those of the lateral leaflets (leaves pinnately compound).
   2. Petals not striate-sulcate; inflorescences 5–20-flowered; flower clusters 5–8 mm in diam.; flowers 2.5–3 mm long, with standard 1–2 mm wide; petioles of middle stem leaves usually shorter than leaflets ____ **T. dubium**
   2. Petals striate-sulcate (with lines in shallow grooves–use lens); inflorescences 20–40-flowered;

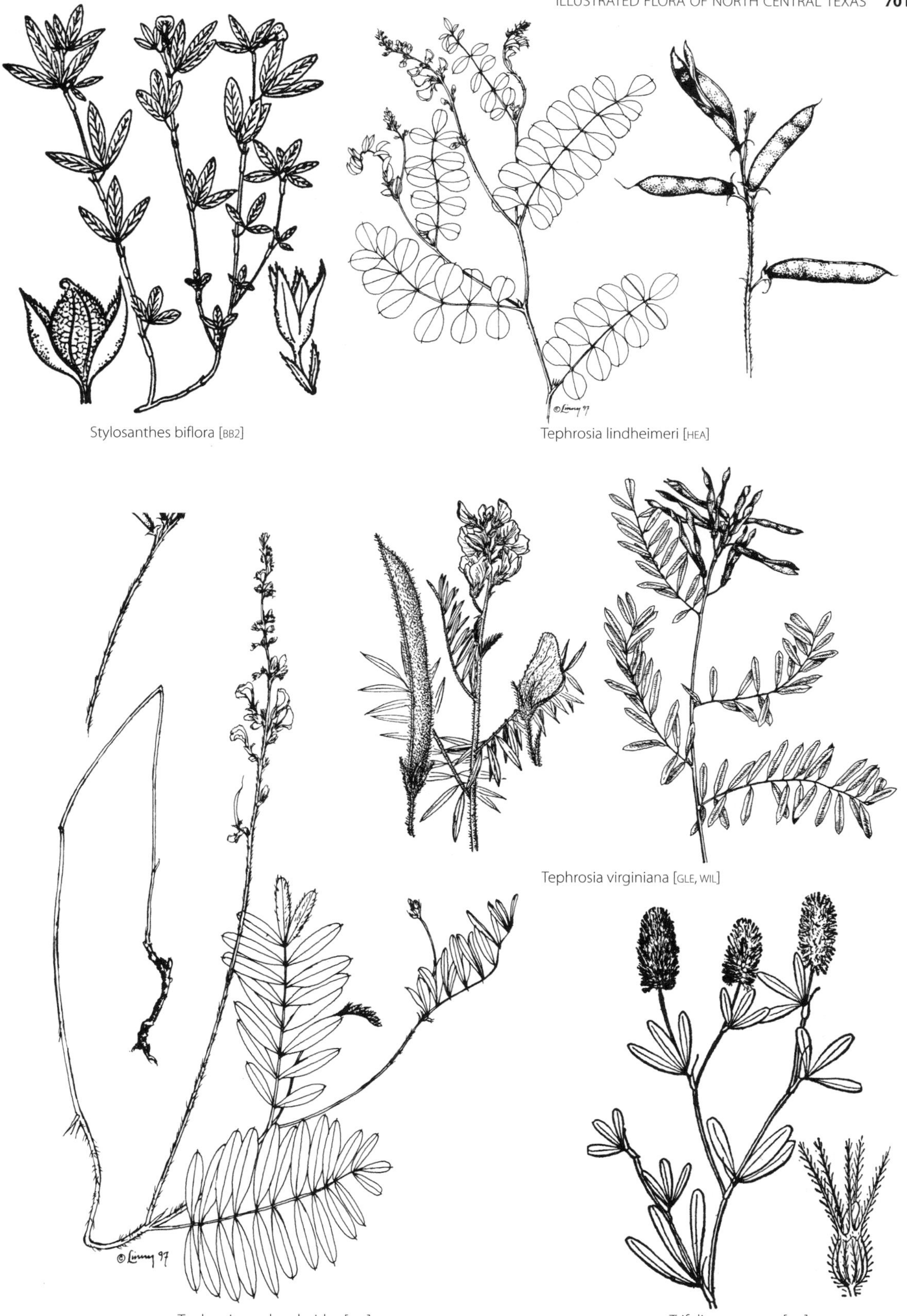

Stylosanthes biflora [BB2]

Tephrosia lindheimeri [HEA]

Tephrosia virginiana [GLE, WIL]

Tephrosia onobrychoides [HEA]

Trifolium arvense [BB2]

flower clusters 8–15 mm in diam.; flowers 2.5–5 mm long, with standard 2–4 mm wide; petioles of middle stem leaves usually longer than leaflets ________ **T. campestre**

1. Petals pink to purplish, red, or white; terminal leaflet sessile or nearly so (leaves palmately compound).
   3. Flower clusters sessile or nearly so at the ends of the main branches (clusters immediately subtended by a leaf; peduncles can be up to 5 mm long); corollas usually pinkish or reddish purple ________ **T. pratense**
   3. Flower clusters on peduncles (5–)10 mm or more long; corollas variously colored, white to pink, purplish, or scarlet-red.
      4. Calyces greatly inflated, contracted at mouth, with extremely long (ca. 4–5 mm) awn-like teeth; corollas white, becoming pinkish ________ **T. vesiculosum**
      4. Calyces not inflated OR if inflated then with much shorter teeth; corollas variously colored.
         5. Leaflets narrow, > 3 times as long as wide; stems, leaves, and calyces conspicuously silky pubescent; calyces usually longer than corollas ________ **T. arvense**
         5. Leaflets broader, < 3 times as long as wide (usually < 2 times as long as wide); stems, leaves, and calyces not all conspicuously silky pubescent; corollas usually as long as or longer than calyces (often much longer).
            6. Corollas a striking scarlet-red; stems ascending, not rooting at the nodes; flower clusters spike-like, at least 2 times as long as wide ________ **T. incarnatum**
            6. Corollas not red OR if red then with stems creeping and rooting at the nodes; flower clusters globose or nearly so, ca. as wide as long.
               7. Individual flowers when fully opened sessile or nearly so and corollas white, turning pink; in nc TX known only from extreme ne part (Lamar Co.) ________ **T. lappaceum**
               7. Individual flowers when fully opened sessile and lavender-pink OR on pedicels 1–4(–8) mm long and variously colored; widespread in nc TX.
                  8. Individual flowers when fully opened sessile or nearly so (pedicels 1 mm or less long); corollas lavender-pink; calyces conspicuously inflated on one side at maturity ________ **T. resupinatum**
                  8. Individual flowers when fully opened on pedicels 1–4(–8) mm long; corollas variously colored, white to pink or red; calyces not inflated.
                     9. Corollas red; plants producing solitary cleistogamous flowers near base which push into the soil and produce underground fruits; stems creeping ________ **T. polymorphum**
                     9. Corollas white to yellowish white or some petals pinkish; plants not producing underground fruits; stems creeping or decumbent to erect.
                        10. Stems creeping and rooting at the nodes; flower clusters short-racemose, the pedicels arising over a 2–5 mm distance at the end of the peduncle; lobes of calyx equal to or shorter than tube ________ **T. repens**
                        10. Stems decumbent, ascending, or erect, not rooting at the nodes; flower clusters umbellate, the pedicels all arising at the very end of the peduncle; lobes of calyx mostly longer than the tube.
                            11. Calyx lobes broadly oblong, 1–1.5(–2.5) mm wide, some of them nearly as wide as long, usually acute to slightly acuminate ________ **T. bejariense**
                            11. Calyx lobes very narrow, linear or nearly so, 0.5 mm or less wide, all of them 2–3 times as long as wide, very acuminate ________ **T. carolinianum**

Trifolium arvense L., (pertaining to cultivated fields), RABBIT-FOOT CLOVER, OLD-FIELD CLOVER. Soft-pilose erect annual 10–40(–45) cm tall; leaflets 8–25 mm long, the margins with teeth small and visible only near the leaflet tips; flowers sessile in short, dense spikes; spikes 5–25 mm long, 10–15 mm in diam.; calyces prominently pilose, giving the whole inflorescence a silky-hairy appearance; corollas rose, pinkish, or white, ca. 4 mm long, usually exceeded by the calyx lobes. Sandy roadsides, pastures, disturbed areas; Grayson and McLennan cos., also

Trifolium bejariense [ZO1]

Trifolium campestre [GLE]

Trifolium carolinianum [BB2]

Trifolium dubium [GLE]

Trifolium incarnatum [USC]

Trifolium lappaceum [ZO1]

Trifolium polymorphum [ZO1]

Lamar Co. (Carr 1994); also e TX. May–early Jun. Native of Europe. Reported to produce dermatitis and photosensitization in domestic animals (Burlage 1968).

Trifolium bejariense Moric., (for San Antonio de Béxar— now San Antonio in Bexar County on se edge of the Edwards Plateau), BEJAR CLOVER. Annual; stems glabrous or nearly so; leaflets usually 5–10 mm long; at least some calyx lobes as broad as long, very unequal; corollas slightly longer than calyces, white to yellowish white. Open wooded areas, prairies, in sandy or sandy clay soils; Hunt, Kaufman, and Navarro cos. (Turner 1959), also Lamar Co. (Hennen 1950); e TX w to nc TX; endemic to TX. Spring.

Trifolium campestre Schreb., (of the fields or plains), LARGE HOP CLOVER, LOW HOP CLOVER, HOP TREFOIL. Erect to decumbent annual 10–40 cm tall; leaflets 6–15 mm long. Waste areas and roadsides; Grayson and Lamar cos. in Red River drainage; mainly se and e TX. Apr–May. Native of Europe.

Trifolium carolinianum Michx., (of Carolina), CAROLINA CLOVER. Annual (perennial ?); stems glabrous or sparsely pubescent; leaflets 4–10(–15) mm long; flower clusters 10–15 mm in diam. at flowering; calyx lobes 2–3 times as long as wide; corollas 1–2 times as long as calyces, 5–7 mm long, yellowish white, changing to brown or according to some, purplish. Fields, roadsides; Henderson Co., also Dallas Co. (Hennen 1950), also Fannin, Hunt, Navarro, and Williamson cos. (Turner 1959); se and e TX w to nc TX. Mar–May.

Trifolium dubium Sibth., (doubtful), LEAST HOP CLOVER, SMALL HOP CLOVER, SHAMROCK. Erect to decumbent annual 5–25(–35) cm tall; leaflets 5–12 mm long. Waste areas and roadsides; Grayson, Hopkins, Lamar, and Tarrant cos.; se and e TX w to East Cross Timbers. Apr–May. Native of Europe. This species and *T. repens* are the two species usually considered to be the Irish SHAMROCK (Nelson 1991).

Trifolium incarnatum L., (flesh-colored), CRIMSON CLOVER, ITALIAN CLOVER. Soft-pilose annual 20–45(–80) cm tall; leaflets 10–40 mm long; spikes elongated, cylindrical, 10–25 mm in diam., tapering apically, showy; flowers sessile; calyx tubes densely long hairy; corollas 8–12 mm long. Sandy roadsides and fields; Grayson Co.; se and e TX w to nc TX. Apr–May. Native of Europe. This species is often planted along roadsides because of its scarlet-red corollas. Overripe crimson clover hay, with numeous stiff hairs on the pedicels and calyces, can be dangerous to horses; indigestible phytobezoars (= hairballs) can form in the stomach and death can result from impaction (Kingsbury 1964; Burlage 1968).

Trifolium lappaceum L., (perhaps from resemblance to bur or burdock from Latin, *lappa*, bur or burdock), LAPPA CLOVER. Annual; stems decumbent to erect, 10–40 cm long, glabrous to thinly pubescent; leaflets 5–25 mm long; inflorescences globose, ca. 1–1.3 cm in diam. at flowering, in fruit bur-like, 1.5–2 cm in diam.; corollas ±equal to calyces in length, 7–8 mm long. According to Isely (1990), this species can be distinguished by its calyx tube which is 20-ribbed and glabrous, with pilose lobes. Weedy areas and roadsides; Lamar Co. in Red River drainage (Carr 1994); mainly se and e TX; first reported for TX by Brown and Peterson (1984). Spring–summer. Native of Mediterranean region.

Trifolium polymorphum Poir., (of many forms), PEANUT CLOVER. Glabrous or sparsely pubescent creeping perennial with cleistogamous flowers and long-peduncled clusters of open flowers about twice as high as the leaves; corollas 2–4 times as long as calyces; called PEANUT CLOVER because of fruits developing underground from cleistogamous flowers. Sandy prairies, open woods, and roadsides; Denton, Dallas, Kaufman, Johnson, McLennan, Milam, and Tarrant cos. (Hennen 1950); no recent collections seen; se and e TX w to nc TX. Apr–early May. [*Trifolium amphianthum* Torr. & A. Gray]

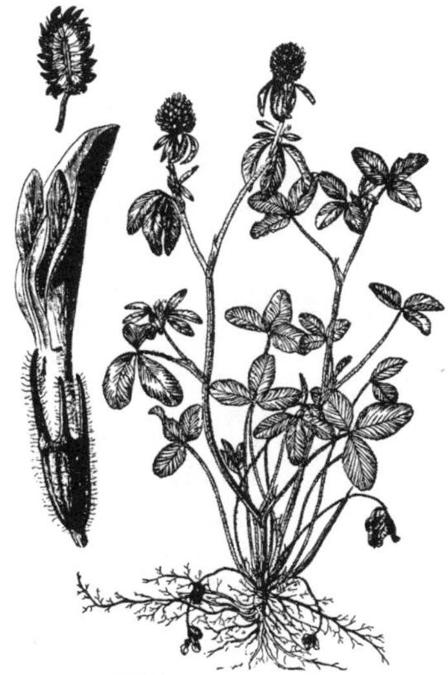

Trifolium pratense [BL3]

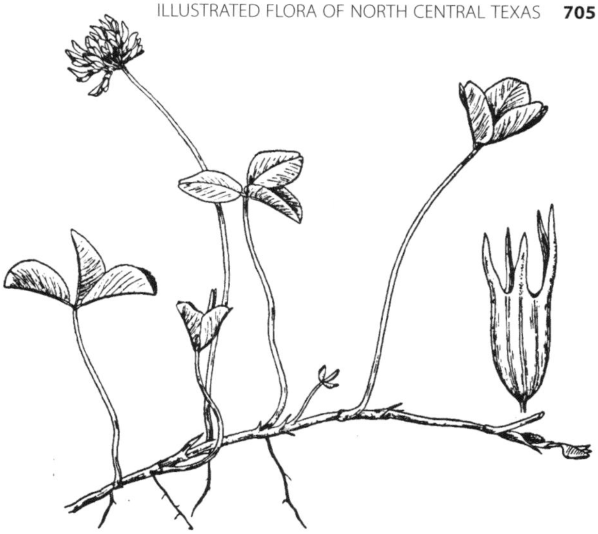

Trifolium repens [GLE]

Trifolium vesiculosum [HEA]

Trifolium resupinatum [BEN, WIL]

Trifolium pratense L., (of meadows), RED CLOVER. Perennial or biennial; stems usually with spreading pubescence; leaflets 10–30(–60) mm long, marginally with very small teeth to nearly entire, often with a reddish or darkened spot; flowers sessile or nearly so, in subglobose to ovoid clusters 1–3 cm long; corollas 12–20 mm long. Roadsides; Grayson and Red River cos. in Red River drainage, scattered escape in e TX w to nc TX, also Edwards Plateau. May–Jul. Native of Europe. Said by some to be the Irish SHAMROCK; however, it is more likely that either *Trifolium dubium* or *T. repens* is actually the SHAMROCK (Nelson 1991). Reported to cause slobbering, bloating, stiffness of gait, diarrhea, emaciation, and abortion in cattle; a saponin is present (Burlage 1968).

Trifolium repens L., (creeping), WHITE CLOVER, DUTCH CLOVER. Glabrous perennial, often mat-forming; leaflets usually 10–20(–30) mm long; petioles long (5–20 cm); flower clusters 10–30 mm in diam.; corollas 6–12 mm long, white or pinkish. Lawns, roadsides, and disturbed areas; Dallas, Grayson, Lamar, and Tarrant cos.; se and e TX w to nc TX and Edwards Plateau. Apr–May, sporadically later. Native of Europe. This species and *T. dubium* are the two species usually considered to be the Irish SHAMROCK (Nelson 1991). It can be toxic to livestock due to the presence of a cyanogenic glycoside (Lewis & Elvin-Lewis 1977).

Trifolium resupinatum L., (upside-down), PERSIAN CLOVER, REVERSED CLOVER. Largely glabrous annual with erect or partly decumbent stems 10–45 cm long; leaflets usually 10–20 mm long; heads hemispherical, 5–10 mm in diam. at flowering, enlarging to 15–20 mm in fruit; flowers resupinate; calyces pilose, papery, prominently reticulate veined, with awn-like teeth < 2 mm long, inflated at one side in age, the heads bur-like; corollas 4–6 mm long, lavender-pink. Roadsides and lawns; Denton, Grayson, and Lamar cos.; se and e TX w to nc TX. Apr–May. Native of the Mediterranean region.

Trifolium vesiculosum Savi, (bladder-like), ARROW-LEAF CLOVER. Annual; stems usually 15–50(–70) cm long; leaflets 15–30(–60) mm long; flower clusters globose, ovoid, or oblong, 20–60 mm long, 20–35 mm wide; corollas white, becoming pink, 1.5–2 times as long as calyces. Sandy open areas; Grayson and Tarrant cos., also Wise Co. (B. Lipscomb, pers. obs.); also e TX; first reported for the U.S. from LA and MS by Thieret (1969b). May–Aug. Native of s Europe.

VICIA VETCH

Ours usually annual, decumbent, trailing to climbing herbs; leaves once pinnately compound, the rachis tip usually terminating in a simple or branched tendril, stipulate; leaflets estipellate, 2–9 pairs, entire, opposite to alternate on the rachis; flowers 1–2 in leaf axils or in few to many-flowered, axillary, spike-like racemes; calyces persistent, symmetrical or gibbous basally; corollas white to blue, lavender, pink-purple, purple, or bicolored; stamens 10, diadelphous; style nearly terete, bearded apically on side adjacent to standard or in a tuft about the apex; fruits elastically dehiscent, linear to oblong, flattened, 2- to several-seeded.

A mostly n temperate genus of 140 species with some in South America, Hawaii, and tropical e Africa; possibly paraphyletic; they are mainly scrambling herbs often with alkaloids. Some are used as forage, fodder, or as green manure; others for their edible seeds including the Old World *V. faba* L. (BROAD-BEAN, FIELD-BEAN, HORSE-BEAN), important before introduction of New World *Phaseolus* species; some species, however, are toxic; the uncooked seeds of *V. faba* can cause hepatitis in some people of Italian or Jewish ancestry due to a genetic predisposition to a biochemical deficiency of red blood cells. In nc TX, VETCHES are recognized in the field by the combination of their leaves with the rachis ending in a tendril, numerous small leaflets, and unwinged stems. (The classical Latin name, from *vincire*, to bind, alluding to the clasping tendrils) (subfamily Papilionoideae, tribe Vicieae)
REFERENCES: Shinners 1948a; Hermann 1960; Kupicha 1976; Lassetter 1978, 1984; Olwell 1982.

1. Flowers sessile or nearly so in the axils of the leaves, solitary or paired ____ **V. sativa**
1. Flowers 1–many in peduncled racemes, the peduncles nearly equal to much longer than the leaves.
   2. Calyces strongly swollen on one side basally; fruits 8–10 mm wide; racemes with 8–40 flowers; corollas 10–18 mm long ____ **V. villosa**
   2. Calyces symmetrically tapered basally; fruits 4–7 mm wide; racemes with 1–19 flowers; corollas 4–8 mm long.
      3. Racemes 3–19-flowered ____ **V. ludoviciana**
      3. Racemes 1- or 2-flowered.
         4. Larger leaves with 2–8 leaflets; ovules or seeds 7–12; fruits 4–4.5 mm wide; calyces 2.2–2.8 mm long ____ **V. minutiflora**
         4. Larger leaves with 8–18 leaflets; ovules or seeds 3–8; fruits 4–8 mm wide; calyces 2.8–3.7 mm long ____ **V. ludoviciana**

Vicia ludoviciana Nutt., (of Louisiana), DEER PEA VETCH. Glabrate or sparsely pubescent annual; stems 10–100 cm long; corollas 4.5–8 mm long. The species is quite variable and has been divided into 2 intergrading subspecies (Lassetter 1984).

1. Flowers opening before elongation of peduncle; peduncle 1/8–1/2 as long as its subtending leaf when all flowers have opened, elongating rapidly as pods develop, ultimately equaling or slightly exceeding the leaf, 0.5–3.5 cm long in flower, 2–8.5 cm in age; racemes with 1–5(–6) flowers; corollas pinkish white to light lavender; leaflets (7–)11–15(–17) ____ subsp. **leavenworthii**
1. Flowers opening after elongation of peduncle; peduncle more than 1/2 as long as to exceeding its subtending leaf when all flowers have opened, elongating slightly in age (but tip of raceme often shriveled then), mostly 3–8.5 cm long in flower, about the same in age; racemes with 1–19 flowers; corollas pinkish white to deep bluish purple; leaflets 7–10(–13) ____ subsp. **ludoviciana**

subsp. **leavenworthii** (Torr. & A. Gray) Lassetter & C.R. Gunn., (for its discoverer, Melines Conklin Leavenworth, 1796–1862, s U.S. botanist), LEAVENWORTH'S VETCH. Rocky, clayey, or occasionally sandy prairies, roadsides; se and e TX w to e Rolling Plains and Edwards Plateau. Apr–May. [*V. leavenworthii* Torr. & A. Gray]

subsp. **ludoviciana**. Rocky or sandy soils, open woods and roadsides; se and e TX w to e Rolling Plains and Edwards Plateau. Late Mar–early May. [*V. exigua* Nutt., *V. leavenworthii* var. *occidentalis* Shinners, *V. ludoviciana* var. *laxiflora* Shinners, *V. ludoviciana* var. *texana* (Torr. & A. Gray) Shinners]

Vicia minutiflora A. Dietr., (small-flowered), SMALL-FLOWER VETCH, PYGMY-FLOWER VETCH. Glabrous annual; stems 20–80 cm long; leaflets of middle and upper leaves linear or narrowly oblong; corollas 5–6(–7) mm long, pale blue to lavender; fruits glabrous or sparsely to densely pubescent. Sandy open woods, fencerows, and roadsides; se and e TX w to East Cross Timbers. Mar–Apr. Including [*V. reverchonii*], with fruits densely pubescent (Olwell 1982). [*V. micrantha* Nutt. ex Torr. & A. Gray, *V. reverchonii* S. Watson]

Vicia sativa L., (cultivated), COMMON VETCH. Annual, glabrous or (especially younger parts) ± pubescent; stems 30–100 cm long; leaflets (6–)8–14; flowers solitary or paired; fruits 3.5–8 mm wide. Apr–May. Native of Europe and the Mediterranean region. This is a complex group, often treated as several species and many varieties; according to Isely (1990), determination of variety can be ambiguous.

1. Corollas 10–18 mm long, pink-purple to white; calyces 7–11(–12) mm long; leaflets on upper leaves essentially linear, 1.5–4 mm wide, subacute to truncate or slightly notched at apex; seeds 2.5–4 mm in diam., mottled or black; widespread in nc TX ____ subsp. **nigra**

1. Corollas 18–25(–30) mm long, usually pink-purple; calyces 10–12(–15) mm long; leaflets on upper leaves narrowly oblong, 3–9 mm wide, slightly or strongly notched at apex; seeds 6–8 mm in diam., black; rare in nc TX _______ subsp. **sativa**

subsp. **nigra** (L.) Ehrhend., (black), NARROW-LEAF VETCH. Leaflets 4–10 times as long as wide. Frequently cultivated and escapes to roadsides and weedy areas; se and e TX w to East Cross Timbers. [*V. angustifolia* L., *V. sativa* var. *nigra* L., *V. sativa* var. *segetalis* (Thuill.) Ser.] Extrafloral nectaries are often present on the stipules; the nectaries are small, ± round, depressed, deep purple areas; under appropriate conditions glistening droplets of nectar can be seen in the nectaries; ants have been observed feeding at the nectaries in nc TX (G. Diggs, pers. obs.).

subsp. **sativa**. Leaflets 2–5(–7) times as long as wide. Rarely cultivated; found once as a weed in field near Dallas in May 1946. Extrafloral nectaries are often present on the stipules (Hermann 1960). Reported in some instances to be poisonous to livestock and humans due to lethal concentrations of a cyanogenic glycoside in the seeds (Kingsbury 1964).

Vicia villosa Roth, (villose, soft-hairy), Annual; stems 50–100 cm long; leaflets (10–)14–18; racemes usually 1-sided; calyces 5–6 mm long; corollas violet or bicolored (rarely white). Both of the following subspecies are cultivated for hay, forage, soil improvement, and erosion prevention; however, poisoning and death in cattle have been reported. Native of Eurasia.

1. Stems glabrous or sparsely and inconspicuously appressed-pubescent; flowers usually 10–20 per raceme _______ subsp. **varia**
1. Stems densely and conspicuously villous, the hairs spreading or ascending; flowers usually more than 20 per raceme _______ subsp. **villosa**

subsp. **varia** (Host) Corb., (varied), WINTER VETCH, WOOLLY-POD VETCH. Flowers in dense or loose racemes, showy. Commonly cultivated and escapes; sandy soils, roadsides; se and e TX w to West Cross Timbers. Late Apr–May. [*V. dasycarpa* Ten., *V. villosa* var. *glabrescens* W.D.J. Koch]

subsp. **villosa**. HAIRY VETCH, RUSSIAN VETCH, WINTER VETCH. Flowers closely crowded in dense racemes. Cultivated and escapes; mostly sandy soils; e TX w to West Cross Timbers. Late Apr–May.

Vicia caroliniana Walter, (of Carolina), CAROLINA VETCH, PALE VETCH, WOOD VETCH, with racemes of 8–20 or more flowers, corollas 8–12 mm long, white to lavender tinged, and fruits 4–5 mm wide, is cited by Hatch et al. (1990) for vegetational area 4 (Fig. 2). It apparently occurs only to the s of nc TX.

WISTERIA

High climbing woody vines or vine-like shrubs with twining branches; plants unarmed; leaves once odd-pinnately compound; leaflets entire; racemes terminal, drooping, with numerous, large, showy, usually violet-blue (rarely white) flowers; calyces campanulate; fruits compressed, stipitate.

A genus of ca. 6 species native to e Asia and e North America; this disjunct distribution pattern is discussed under the genera *Campsis* (Bignoniaceae) and *Carya* (Juglandaceae). The Asiatic species are sometimes split as the segregate *Rehnsonia* (Stritch 1984). All parts of *Wisteria* species should be considered poisonous, especially the fruits and seeds (Blackwell 1990) due to the presence of the glycoside wisterin (Schmutz & Hamilton 1979; Lampe & McCann 1985); as little as 2 seeds are enough to cause serious illness in children (Hardin & Arena 1974). (Named for Professor Caspar Wistar, 1760–1818, anatomist of Philadelphia) (subfamily Papilionoideae, tribe Tephrosieae)
REFERENCES: Stritch 1984; Valder 1995.

Vicia ludoviciana var. leavenworthii [HE1]

Vicia ludoviciana var. ludoviciana [HE1]

Vicia minutiflora [HE1]

Vicia sativa subsp. nigra [HE1]

Vicia sativa subsp. sativa [HE1]

Vicia villosa subsp. varia [HE1]

Vicia villosa subsp. villosa [HE1]

Wisteria sinensis [WIL]

Zornia bracteata [WIL]

Wisteria sinensis (Sims) DC., (Chinese), CHINESE WISTERIA. High climbing vine or vine-like shrub with twining branches, to ca. 20 m long; young stems densely short pubescent; leaves with 7–13 leaflets; racemes 15–30 cm long, the flowers opening nearly simultaneously; pedicels (10–)15–25 mm long; corollas ca. 2–2.7 cm long; ovary and fruit velvety pubescent. Widely cultivated and long persists around old home sites; escapes?; included based on citation of vegetational area 4 (Fig. 2) by Hatch et al. (1990). Mostly Apr–May. Native of China. Poisonous. Two other species are also cultivated in nc Texas (Shinners 1958a):

Wisteria floribunda (Willd.) DC., JAPANESE WISTERIA, native of Japan, has the ovary and fruit velvety pubescent, (13–)15–19 leaflets per leaf, flowers gradually opening from base to tip of raceme, and corollas 1.5–2 cm long. Poisonous (Lampe & McCann 1985).

Wisteria fructescens (L.) Poir., (fruitful, fruit-bearing), [=*W. macrostachya* (Torr. & A. Gray) Nutt. ex B.L. Rob. & Fernald], native to se and e TX, has the ovary and fruit glabrous and pedicels usually ca. 4–10 mm long. Presumably poisonous.

ZORNIA

A genus of 86 species of warm areas of the world. (Named for Johannes Zorn, 1739–1799, a German apothecary) (subfamily Papilionoideae, tribe Aeschynomeneae)
REFERENCE: Mohlenbrock 1961.

Zornia bracteata J.F. Gmel., (bracteate, with bracts), BRACTED ZORNIA, VIPERINA. Glabrous perennial herb from a deep, tough, vertically or obliquely branched root; stems prostrate to decumbent or ascending, 20–80 cm long; leaves palmately compound; leaflets 4, usually 1–3 cm long, inconspicuously punctate; petioles as long as the leaflets; flowers widely spaced in long-peduncled, axillary spikes of 3–10 flowers, each flower with a pair of large, conspicuous, elliptic-ovate bracts 7–15 mm long; petals 9–14 mm long, yellow to orangish; stamens 10; fruits 10–20 mm long, bristly, breaking apart into 2–6 segments. Sandy, low, open woods or open ground; Tarrant Co., also Erath and Parker cos. (Turner 1959); se and e TX w to nc TX and Edwards Plateau. Apr–Jun. The combination of 4 leaflets, large paired bracts, and bristly fruits make this species distinctive (Isely 1990).

FAGACEAE OAK OR BEECH FAMILY

An economically very important family of 600–800 species in 9 genera (Nixon 1997a); cosmopolitan in distribution except in the tropics and s Africa; often vegetational dominants in n hemisphere temperate forests; usually wind-pollinated trees and shrubs, most with large amounts of tannins. Many species are utilized as a source of hardwood lumber, as ornamental trees, or for their edible nuts. North American genera include *Castanea* (CHESTNUT), *Castanopsis* (CHINQUAPIN), *Fagus* (BEECH), *Lithocarpus* (TAN OAK, TAN-BARK OAK), and *Quercus* (OAK). *Castanea dentata* (Marshall) Borkh. (AMERICAN CHESTNUT), of the e U.S., was virtually wiped out in the first half of the 20th century by chestnut blight, caused by an introduced Old World fungus (*Cryphonectria parasitica* (Murrill) Barr [*Endothia parasitica* (Murrill) H.W. & P.J. Anderson]). The botanical journal, *Castanea*, published by the Appalachian Botanical Club, is named after *C. dentata* (Brooks 1937). The distribution of the s hemisphere genus *Nothofagus* (SOUTHERN-BEECH), in New Guinea, New Caledonia, temperate Australia, New Zealand, and temperate South America, in part reflects the different distribution of the continents during Tertiary times caused by plate tectonics; while this genus has traditionally been placed in Fagaceae, molecular evidence indicates that it should be recognized in its own family, Nothofagaceae (Manos et al. 1993). Family name from *Fagus*, BEECH, a mainly n temperate genus of 10 species of deciduous monoecious trees. (Classical Latin name, derived from Greek: *figos*, an oak with edible acorns, probably from Greek *fagein*, to eat—Nixon 1997a) (subclass Hamamelidae)

FAMILY RECOGNITION IN THE FIELD: trees or shrubs with male flowers in *catkins*; fruit an *acorn* (= a nut subtended or enveloped by a cupule—involucre or cup of wholly or partly fused bracts); leaves alternate, simple, often lobed or toothed, with straight pinnate veins; buds clustered at tips of twigs. The cupule is characteristic of the family; in the genus *Castanea* (CHESTNUTS and CHINQAPINS—1 species in e TX) it is spiny and completely surrounds the nuts.
REFERENCES: Forman 1966; Elias 1971b; Nixon 1984, 1997a; Jones 1986; Kaul 1986a; Kubitzki 1993c; Manos et al. 1993.

QUERCUS OAK

Trees or shrubs with alternate (often closely bunched), simple, deciduous or evergreen (in nc TX only *Quercus fusiformis* and *Q. virginiana*) leaves; stipules very slender or minute, falling early; flowers in nc TX species appearing in early spring; unisexual, both sexes borne on the same plant (monoecious), with 5- to 6-parted calyces, apetalous; staminate flowers in clustered spreading to drooping catkins; stamens usually 4–10; pistillate flowers solitary or several together in short spikes; pistil 1; ovary inferior; each pistillate flower subtended by a cupule (= an involucre of numerous bracts/scales that harden at maturity into a cup subtending or partly enveloping the 1-seeded fruit); seed enclosed in a shell forming a nut; shell of nut glabrous or tomentose on inner surface (the nut and its cupule are together known as an acorn); flowering in spring, the flowers (in deciduous species) appearing just before or with the leaves.

A genus of ca. 400 species (Nixon 1997b) of the n temperate zone s to Malesia and Colombia at higher elevations. OAKS are important as a source of timber, as ornamentals, for tannins, and for cork. Cork is derived from the thick bark of *Q. suber* L. (CORK OAK) of s Europe and n Africa. Acorns, after they were treated to remove bitter tannins, were an important food source for a number of Native American groups; species in the white oak group were preferred because of their lower tannin content (Martin et al. 1951). The windborne pollen contributes to allergies. Numerous insect galls (= abnormal swellings or growths produced by plants in response to attack by a parasite) are found on nc TX OAKS and are frequently observed; in fact, no other plant group is known to have more insect galls than OAKS; hundreds of different kinds are known (Felt in Strausbaugh & Core 1978). OAK species are not easily identifiable before the leaves are fully expanded, by which time the flowers are largely past (last year's leaves still present on live oaks; some old dead leaves often still present on others). The following key is for use with full-grown leaves of flowering or fruiting branches, not sucker-shoots, stump-sprouts, or shade forms. Hybridization is well known in *Quercus* (Palmer 1948), and hybrids occur between a number of nc TX species (e.g., *Q. marilandica* × *Q. velutina; Q. marilandica* × *Q. shumardii*); therefore, individuals difficult to identify should not be a surprise. The 17 nc TX species make this the largest genus of woody plants found in the area; it is also the largest woody genus in adjacent Oklahoma (Tyrl et al. 1994). ☠ Toxicity to cattle due to the presence of tannins can cause significant loss in some areas, particularly during drought years; mostly new foliage is involved, but mature leaves and acorns can also cause poisoning if eaten in large quantities; the compounds actually causing the toxic effects are apparently low molecular weight phenolic compounds produced as the result of biodegradation of high molecular weight tannins (Kingsbury 1964; Zhicheng 1992). (Classical Latin name for the ENGLISH OAK, *Quercus robur* L., from some central European language)

Oaks in c TX and in parts of nc TX are being killed by oak wilt, caused by the fungus *Ceratocystis fagacearum* (Bretz) J. Hunt. Live oaks and members of the red oak group are most susceptible. Live oaks show variability in response to infection with most individuals dying within 3–6 months while some survive indefinitely with crown loss. In the most susceptible red oaks, death can occur in a matter of weeks after initial symptons are seen. Transmission seems to occur by root grafts and in some cases by insects (beetles) (Appel 1995; MacDonald

1995; Reisfield 1995). Appel and Billings (1995) gave detailed information on this disease. REFERENCES: Trelease 1924; Palmer 1948; Müller 1951; Little 1971; Hardin 1975; Jensen 1977; Elias 1980; Axelrod 1983; Appel & Billings 1995; Dorr & Nixon 1985; Simpson 1988; Nixon 1997b.

1. Leaf blades with midrib and side veins not exserted bristles (blade or its lobes may have a minute point not formed by an exserted rib or nerve); bark pale, often scaly; inner surface of acorn shell glabrous; cup scales usually thickened at base and loosely appressed at apex; fruits maturing first autumn after spring flowering (**"White" oak group**).
  2. Lower surface of blades mostly glabrous except along main veins or in their axils (or hairs so inconspicuous that surface appears glabrous).
    3. Leaf blades shallowly or deeply lobed ____ **Q. alba**
    3. Leaf blades entire (without teeth) or irregularly toothed.
      4. Leaves evergreen, with a coarse, hard to leathery texture.
        5. Leaf blades in general relatively broad to near base and often rounded to broadly so basally; mature acorns fusiform, much longer than broad, usually extended from cup for more than 1/2 their length, (17–)20–30(–33) mm long, the apex acute; widespread in nc TX (mainly East Cross Timbers westward) ____ **Q. fusiformis**
        5. Leaf blades in general tapering to base and cuneate (= wedge-shaped) basally; mature acorns usually ovoid, not much longer than broad, usually not extended from cup for more than 1/2 their length, 15–20(–25) mm long, the apex rounded or blunt; probably only to the e of nc TX ____ **Q. virginiana**
      4. Leaves deciduous, not coarse (this is included because some Red/Black group oaks with unlobed leaves have only 1 nerve exserted as a bristle at the leaf tip–this is sometimes lost and they are thus apparently without bristles) ____ see Red/Black oak group–dichotomy 13
  2. Lower surface of blades pale with soft and felty or close and firm pubescence.
    6. Petioles 0.7–4 cm long, thinly and loosely pubescent or glabrous.
      7. Leaf blades deeply and unevenly lobed.
        8. Terminal leaf lobe only slightly larger than others; lobes usually pointed; mature acorn nearly completely (>3/4) enclosed in the thin, shell-like cups, the cup opening usually < 1/2 the diam. of the acorn; cups 3 cm or less in diam. ____ **Q. lyrata**
        8. Terminal leaf lobe much larger than others; lobes usually rounded; mature acorn rarely more than half-enclosed in cup, the cup opening ca. same diam. as acorn; cups large, 3–6 cm in diam. ____ **Q. macrocarpa**
      7. Leaf blades rather evenly toothed (shallowly or deeply), not lobed ____ **Q. muehlenbergii**
    6. Petioles 0.1–0.7 cm long, sparsely to densely short-pubescent.
      9. Leaf blades 3–10 cm wide, usually lobed more than half way to midrib; pubescence on lower surface rather open, in widely separated stellate tufts; veinlets of lower surface raised, the network of veinlets prominent.
        10. Twigs glabrous or nearly so; plant a shrub (small tree) to ca. 7 m tall ____ **Q. margarettiae**
        10. Twigs persistently densely tomentulose or velvety; plant a small to large tree ____ **Q. stellata**
      9. Leaf blades 1–4 cm wide, entire to toothed or shallowly and evenly lobed; pubescence on lower surface close and dense, concealing the surface; veinlets faint or invisible.
        11. Leaves deciduous, entire to shallowly lobed; acorn nearly cylindrical, with wide, blunt apex ____ **Q. sinuata**
        11. Leaves evergreen, entire or with 1–3 teeth near base or apex; acorn usually tapering gradually to apex, ovoid or fusiform.
          12. Leaf blades in general relatively broad to near base and often rounded to broadly so basally; mature acorns fusiform, much longer than broad, usually extended from cup for more than 1/2 their length, (17–)20–30(–33) mm long, the apex acute; widespread in nc TX (mainly East Cross Timbers westward) ____ **Q. fusiformis**

12. Leaf blades in general tapering to base and cuneate (= wedge-shaped) basally; mature acorns usually ovoid, not much longer than broad, usually not extended from cup for more than 1/2 their length, 15–20(–25) mm long, the apex rounded or blunt; probably only to the e of nc TX ____ **Q. virginiana**

1. Leaf blades with main veins (at least the midrib) exserted as bristles (these sometimes broken off in age–species with unlobed leaves will have only one bristle at the leaf tip); bark often dark, furrowed but rarely scaly; inner surface of acorn shell with hairs; cup scales usually scarcely thickened at base and tightly appressed at apex; fruits maturing second autumn after spring flowering (**"Red/Black" oak group**).
 13. Leaves entire.
  14. Leaves linear-lanceolate to narrowly elliptic, at least 5 times as long as wide ____ **Q. phellos**
  14. Leaves elliptic to oblanceolate, obtriangular to club-shaped, rarely more than 3 times as long as wide.
   15. Leaves usually broadest close to middle; twigs and lower (abaxial) leaf surfaces uniformly and persistently gray-tomentose ____ **Q. incana**
   15. Leaves usually broadest close to apex; twigs and lower leaf surfaces glabrate or nearly so or with brownish or yellowish pubescence, not gray-tomentose.
    16. Base of leaf blade narrowly cuneate (tapering to a narrow point); petioles nearly absent or sometimes present; twigs and petioles glabrous or nearly so; leaves glabrous beneath except for tufts of stellate hairs in the axils of the veins ____ **Q. nigra**
    16. Base of leaf blade rounded to subcordate; petioles usually 5 mm or more long; twigs and petioles usually with brownish pubescence; leaves sometimes with some brownish pubescence beneath (especially near base of the midrib) ____ **Q. marilandica**
 13. Leaves variously toothed or lobed.
  17. Leaves at most irregularly toothed or shallowly lobed.
   18. Leaves neither 3-lobed apically nor so much broader apically than basally as to be distinctly club-shaped, leaves usually broadest near middle ____ **Q. incana**
   18. Leaves apically 3-lobed (often obscurely so), much broader apically than basally and distinctly club-shaped.
    19. Twigs and petioles glabrous or nearly so; leaves glabrous except for axillary tufts of stellate hairs ____ **Q. nigra**
    19. Twigs and petioles brownish- or yellowish-pubescent; leaves variously pubescent beneath.
     20. Base of leaf blade broadly cuneate (wedge-shaped, narrowed to a point) to rounded; leaves densely and uniformly brownish- or yellowish-tomentulose beneath ____ **Q. falcata**
     20. Base of leaf blade rounded to subcordate; leaves variously pubescent beneath (especially near the attachment of the petiole) but never uniformly tomentulose ____ **Q. marilandica**
  17. Leaves regularly toothed or lobed, the lobes often large.
   21. Main lateral lobes simple (not divided into smaller lobes), usually with only 1(–2) bristle-tipped tooth, broadly to narrowly triangular in shape.
    22. Terminal lobe elongate, often prominent, oblong, acuminate apically, usually with 2–several lateral teeth ____ **Q. falcata**
    22. Terminal lobe scarcely more prominent than the lateral lobes or nearly absent, not oblong, apically acute to rounded, usually with 0–1 lateral teeth ____ **Q. marilandica**
   21. Main lateral lobes divided into several tiny lobes, usually with 2–several bristle-tipped teeth, variously shaped, usually not triangular.
    23. Leaves densely brownish- or yellowish-tomentulose beneath.

24. Terminal lobe often greatly elongated and much more prominent than lateral lobes and/or leaf bases sometimes broadly U-shaped ____ **Q. falcata**
24. Terminal lobe usually not much more prominent than lateral lobes; leaf bases variously obtuse to rounded or cordate but not broadly U-shaped ____ **Q. velutina**

23. Leaves variously pubescent or glabrate but not densely tomentulose.
25. Terminal lobe often greatly elongated and much more prominent than lateral lobes and/or leaf bases sometimes broadly U-shaped ____ **Q. falcata**
25. Terminal lobe usually not much more prominent than lateral lobes; leaf bases variously obtuse to rounded or cordate but not broadly U-shaped.
26. Terminal buds large (6–10 mm long), often quadrangular, densely brownish- or yellowish- or gray-tomentose or strigose; in nc TX found only in counties adjacent to the Red River (w to Grayson Co.) ____ **Q. velutina**
26. Terminal buds not over 5(–6) mm long, spindle-shaped to lanceolate or narrowly ovoid, sparsely pubescent or glabrous; nearly throughout nc TX.
27. Acorn cups 16–25 mm broad, abruptly narrowed at base, enclosing acorn by 1/3 or less (nearly flat with abruptly raised sides to shallowly cup-shaped); acorns 20–25 mm long, usually flat at base; leaf blades mostly 8–18 cm long, the lower surfaces at maturity with conspicuous (to the naked eye) tufts of tomentum in the vein axils; mostly Blackland Prairie and to the e ____ **Q. shumardii**
27. Acorn cups 12–18 mm broad, rounded or tapered at base, enclosing 1/3–1/2 the acorn (deeply cup-shaped); acorns 14–20 mm long, usually round at base; leaf blades mostly 4–11(–15) cm long, the lower surfaces at maturity glabrous or with minute (often only detectible with magnification) tufts of tomentum in the vein axils; mostly East Cross Timbers and to the w ____ **Q. buckleyi**

Quercus alba L., (white), WHITE OAK, STAVE OAK, RIDGE WHITE OAK, FORKED-LEAF WHITE OAK. Large tree. Stream bottom woods; once collected in Dallas Co. (Mahler 1988); mainly far e TX w to Red River Co. (Little 1971). The wood has long been valued for staves used in making barrels that would hold liquids—such as wine and liquor (Peattie 1948).

Quercus buckleyi Nixon & Dorr, (for Samuel Botsford Buckley, 1809–1884, state geologist of TX and plant collector), TEXAS RED OAK, SPANISH OAK, SPOTTED OAK, ROCK OAK. Small tree usually to only 10 m (rarely more) tall; leaf blades appearing more finely and deeply cut than in *Q. shumardii*. Limestone outcrops and slopes or in stream bottoms, much less often on sandy soils. This species has long incorrectly gone under the name *Q. texana*. Dorr and Nixon (1985) pointed this out and named this species, which occurs in nc and c TX mainly from the East Cross Timbers westward, *Q. buckleyi*. The name *Q. texana* is now restricted to a species extending w only as far as extreme e TX (much to the east of nc TX). *Quercus buckleyi* and the similar *Q. shumardii* (occurring mainly from the w edge of the Blackland Prairie e to e TX) hybridize along a narrow zone of overlap from the Cooke and Grayson co. area near the Red River s to the vicinity of San Antonio (Bexar Co.) (Simpson 1988). To the w of this hybrid zone "pure" individuals of *Q. buckleyi* can be found, while to the e "pure" *Q. shumardii* occurs. In the hybrid zone, specific determination is often not possible. This taxon is possibly better treated as a variety of *Q. shumardii* as intended by Shinners (1956b) who made the combination, *Q. shumardii* Buckley var. *microcarpa* (Torr.) Shinners. However, the type of this variety is equal to *Q. gravesii* Sudworth (Dorr & Nixon 1985). [*Q. shumardii* var. *microcarpa* in part, excluding type, *Q. shumardii* var. *texana* in part, excluding type, *Q. texana* of authors, not Buckley]

Quercus falcata Michx., (sickle-shaped), SOUTHERN RED OAK, SPANISH OAK, SWAMP RED OAK, SWAMP SPANISH OAK, CHERRY-BARK OAK, BOTTOM-LAND RED OAK, THREE-LOBE RED OAK. Large tree;

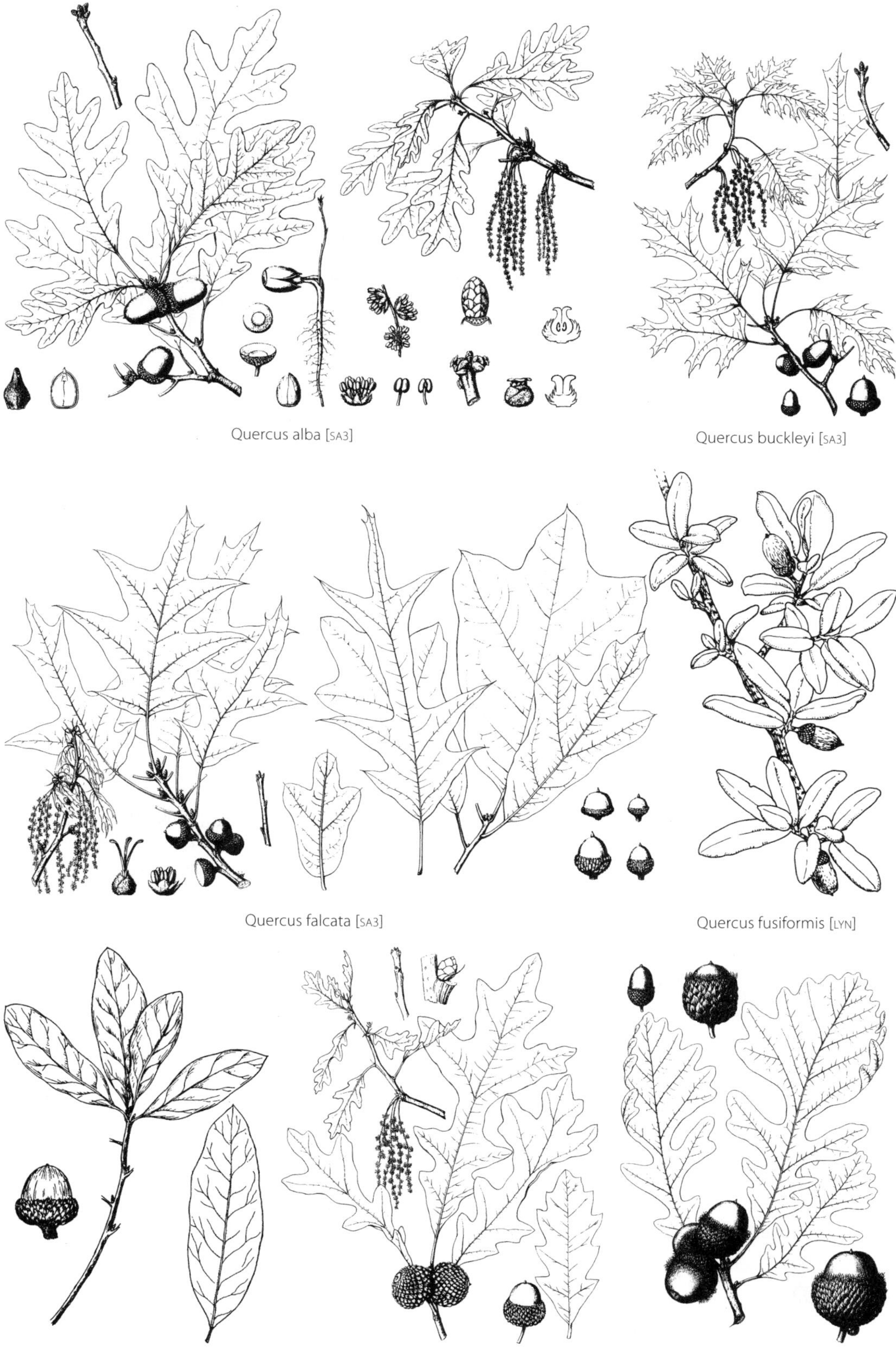

Quercus alba [SA3]

Quercus buckleyi [SA3]

Quercus falcata [SA3]

Quercus fusiformis [LYN]

Quercus incana [GLE]

Quercus lyrata [SA3]

Quercus macrocarpa [SA3]

leaves quite variable, sometimes entire with 3 apical lobes, usually with long, falcate (sickle-shaped) lobes with deep, rounded sinuses, terminal lobe usually elongated. Moist to wet forests; Fannin and Lamar cos. in Red River drainage in ne part of nc TX, also Delta Co. (Little 1971); mainly e TX. ▣/104

Quercus fusiformis Small, (spindle-shaped), PLATEAU LIVE OAK, ESCARPMENT LIVE OAK, SCRUB LIVE OAK, WEST TEXAS LIVE OAK, LIVE OAK. Evergreen; small shrub to rather large tree, spreading by root sprouts. Possibly only a more xeric and cold tolerant subspecies of the more widespread *Q. virginiana*. Limestone outcrops, well-drained soils, mainly xeric habitats; primarily East Cross Timbers w to West Cross Timbers and sw to Edwards Plateau, also n into the Arbuckle and Wichita Mountains of s OK; also cultivated. [*Quercus virginiana* Mill. var. *fusiformis* (Small) Sarg.] Jones et al. (1997b) treat this species as a variety of *Q. virginiana*. While we are following Nixon (1997b) in treating it as a separate species, there appears to be extensive hybridization and introgression with *Q. virginiana*; see discussion under that species.

Quercus incana W. Bartram, (hoary, quite gray), BLUEJACK OAK, SANDJACK OAK, UPLAND WILLOW OAK, CINNAMON OAK, SHIN OAK, TURKEY OAK, HIGHGROUND OAK. Shrub or small tree to ca. 8 m tall; leaves distinctly gray-tomentose on lower (abaxial) surfaces. Sandy uplands; Kaufman Co.; scattered in se, e and c TX.

Quercus lyrata Walter, (lyre-shaped), OVERCUP OAK, SWAMP POST OAK, SWAMP WHITE OAK, WATER WHITE OAK. Medium size tree. Swamps, stream banks or other areas with frequent standing water, often acidic soils; Henderson Co., also Fannin (Little 1971) and Lamar (Simpson 1988) cos. in Red River drainage; mainly se and e TX. OVERCUP OAK is sometimes confused with *Q. macrocarpa* (BUR OAK) which is found on at least moderately drained calcareous soils; the 2 species are thus extremely different in terms of habitat.

Quercus macrocarpa Michx., (large-fruited), BUR OAK, MOSSY-CUP OAK, PRAIRIE OAK, MOSSY-OVERCUP OAK. Large tree; nuts and cups (3–6 cm wide) largest of all nc TX species. Stream bottoms, lower slopes, upland woods; usually in at least moderately drained places; in or near limestone areas; se and e TX w to West Cross Timbers and Edwards Plateau. This species is well known for its large acorns and thick, fire-resistant bark.

Quercus margarettiae Ashe ex Small, (for Margaret Henry Wilcox, later married to William Willard Ashe, 1872–1932, who named the species), SAND POST OAK, DWARF POST OAK, SCRUBBY POST OAK, RUNNER OAK. Shrub (rarely a small tree) to ca. 7 m tall. Restricted to deep sandy soils; Eastland and Palo Pinto cos., also Comanche, Grayson, and Tarrant cos. (Simpson 1988); e TX w to nc TX and Edwards Plateau. [*Q. stellata* Wangenh. var. *margarettiae* (Ashe ex Small) Sarg.] Jones et al. (1997) treated this species as a variety of *Q. stellata*. While we are following Nixon (1997b) in treating it as a separate species, he noted that, "Populations of post oak in east Texas (the Cross Timbers region) on sands and gravels exhibit characteristics intermediate between *Q. stellata* and *Q. margaretta*[e]." He further indicated that these populations have been referred to as *Q. drummondii* Liebm., DRUMMOND'S POST OAK, and "perhaps the Texas material is best treated as a nothospecies, *Q.* ×*drummondii*."

Quercus marilandica Münchh., (of Maryland), BLACKJACK OAK, BLACKJACK, BARREN OAK, JACK OAK, BLACK OAK. Small to large tree, smaller westward, often coarse and densely branched to near base; leaves nearly entire varying to having 3 shallow lobes near apex (rarely with a basal lobe on each side), narrowest basally, wedge-shaped or club-shaped in general outline. Sandy or occasionally gravelly and silty soils; chiefly upland; se and e TX w to West Cross Timbers and Edwards Plateau.

Quercus muehlenbergii Engelm., (for Gotthilf Henry Muhlenberg. 1753–1815, German-educated minister and pioneer botanist of Pennsylvania), CHINQAPIN OAK, CHESTNUT OAK, SWAMP CHEST-

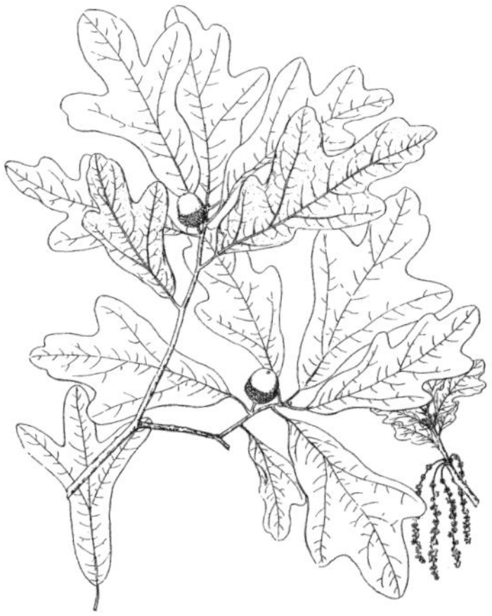

Quercus margarettiae [SA3]

Quercus muehlenbergii [SA3]

Quercus nigra [SA3]

Quercus marilandica [LYN, SA3]

Quercus phellos [SA3]

Quercus shumardii [SA2]

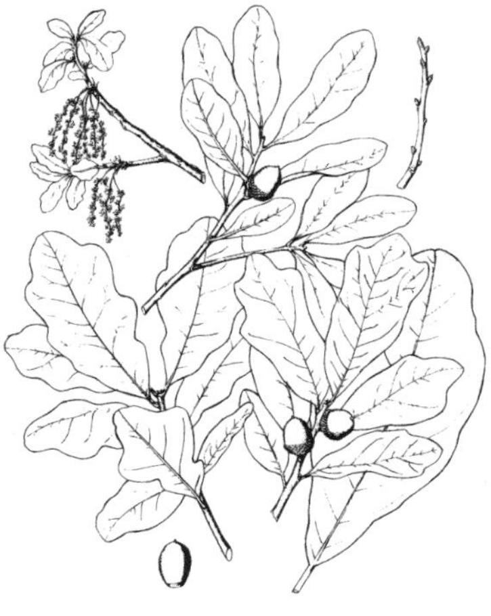

Quercus sinuata var. breviloba [SA3]

NUT OAK, YELLOW OAK, ROCK OAK. Medium to large tree; leaves shallowly to deeply rather evenly toothed, not lobed. Uplands, creek bottoms, ravines, on limestone or calcareous soils; e TX w to nc TX and s and w to Edwards Plateau and Trans-Pecos. The common name derives from the similarity of the leaves to those of CHINQAPIN, *Castanea pumila* Mill., another member of the Fagaceae native to the e U.S. as far w as e TX.

Quercus nigra L., (black), WATER OAK, POSSUM OAK, SPOTTED OAK, DUCK OAK, PUNK OAK. Large tree with dark gray or blackish bark, smooth in upper part. Stream bottom woods; Fannin, Kaufman, Henderson, Lamar, and Limestone cos., also Ellis Co. (Mahler 1988), also Grayson Co. (possibly introduced); se and e TX w to e part of nc TX.

Quercus phellos L., (Greek for cork oak—*Q. suber*), WILLOW OAK, PIN OAK, PEACH OAK, SWAMP WILLOW OAK, WILLOW-LEAF OAK. Medium to large tree; leaves usually linear-lanceolate to lanceolate (oblanceolate to narrowly ovate or narrowly obovate), 5–12(–16) cm long, 1–2.5(–4) cm wide. Moist forests; Fannin and Lamar cos.; e TX w in Red River drainage to ne part of nc TX.

Quercus shumardii Buckley, (for Benjamin Franklin Shumard, 1820–1869, state geologist of TX in 1860), SHUMARD'S OAK, SHUMARD'S RED OAK, RED OAK, SWAMP RED OAK, SPOTTED OAK. Small to large tree to 30 m or more tall. Leaf blades divided about 0.5–0.6 of distance to midrib. Moist forests, chiefly in stream bottoms or drainage ways; mainly w edge of Blackland Prairie e to e TX. Similar to *Q. buckleyi*; see note under that species. Very susceptable to oak wilt. [*Q. schneckii* Britt., *Q. shumardii* var. *schneckii* (Britt.) Sarg.] While var. *schneckii* is sometimes recognized (e.g., Kartesz 1994; Jones et al. 1997), we are following Nixon (1997b) in lumping this variety that differs in having more deeply rounded cups covering ca. 1/3 of the nut (vs. shallow cups covering ca. 1/4 of the cup).

Quercus sinuata Walter, (with wavy margins). Includes the following 2 varieties differing greatly in growth form and habitat.

1. Shrubs to small often multi-trunked trees usually to only 6(–12) m tall, sometimes forming thickets; in limestone upland areas; widespread in nc TX ____ var. **breviloba**
1. Large trees to 20 m or more tall; in bottomland forests; e margin of nc TX ____ var. **sinuata**

var. **breviloba** (Torr.) C.H. Müll., (short-lobed), BIGELOW'S OAK, SCRUB OAK, SHIN OAK, WHITE OAK, SCALY-BARK OAK. Often on limestone outcrops or rocky areas; Blackland Prairie (on the Austin Chalk) w through West Cross Timbers and Edwards Plateau. [*Q. san-sabeana* Buckley, *Q. breviloba* (Torr.) Sarg.]

var. **sinuata**. BASTARD OAK, DURAND'S WHITE OAK, WHITE OAK, BLUFF OAK. Moist woods; mainly e TX, in nc TX only reported from Navarro Co. (Simpson 1988). [*Q. durandii* Buckley]

Quercus stellata Wangenh., (stellate, star-shaped, from the stellate hairs), POST OAK, IRON OAK, CROSS OAK. Usually rather small (medium) tree; leaves with 2–4 lobes on each side, the main lobe on each side usually rather large and perpendicular to midvein giving the leaf a cross-like appearance. Sandy or rarely gravelly-silty ground, chiefly in uplands; eastern 2/3 of TX; this is the commonest oak species in nc TX and the vegetational dominant in many areas. Post oak tree-ring chronologies extending from about 200 to 300 years have been obtained from North Central Texas sites, with individual trees dating back to 1681 (e.g., Stahle et al. 1985). According to J. Stanford (pers. comm.), hybrids of this species and *Q. marilandica* can be found in the Lampasas Cut Plain.

Quercus velutina Lam., (velvety, from the young foliage), BLACK OAK, YELLOW OAK, QUERCITRON OAK, QUERCITRON, SMOOTH-BARK OAK, YELLOW-BARK OAK. Medium to large trees. Sandy ground, upland or lowland; Fannin, Grayson, and Lamar cos.; e TX w to nc TX in Red River drainage.

Quercus virginiana Mill., (of Virginia), LIVE OAK, ENCINO, COAST LIVE OAK, VIRGINIA LIVE OAK. Small to very large evergreen tree. In the se part of nc TX, *Q. virginiana* and *Q. fusiformis* form a complicated hybrid complex (Simpson 1988). True examples of *Q. virginiana* probably only occur naturally well to the se of nc TX. Even though helpful in separating these two species, leaf shape appears quite variable. While used extensively in landscaping, *Q. virginiana* is much more sensitive to low temperature than the closely related *Q. fusiformis*. Most individuals of *Q. virginiana* were severely damaged or killed in the Dallas-Fort Worth Metroplex during the severe winter of 1983. *Quercus fusiformis*, however, survived without problems much further n. Both of these species are being significantly affected by oak wilt.

FUMARIACEAE FUMITORY FAMILY

Ours low annuals with clear sap, glabrous and often glaucous; leaves alternate and/or basal, pinnately compound, with deeply divided leaflets; flowers in spike-like racemes, bilaterally symmetrical; sepals 2, minute, scale-like; petals 4, pale to bright yellow or orange-yellow or lavender to purple, unequal, slightly united at base but mostly falling separately; outer 2 petals dissimilar, the uppermost longest, its base prolonged into a spur, its wide apex bent up or back, ± keeled or hooded; stamens 6, in 2 bundles of 3 each, the filaments of each bundle partly to completely connate; carpels 2; fruit an elongate dehiscent capsule or a subglobose to obovoid indehiscent capsule.

A medium-sized (ca. 450 species in 19 genera), mainly n temperate family of herbs of Eurasia and North America (Stern 1997a); they contain alkaloids, though in smaller amounts than the related Papaveraceae. Some are used as ornamentals including *Corydalis* species and *Dicentra* species (BLEEDINGHEART, DUTCHMAN'S-BREECHES). The family is related to the Papaveraceae and Lidén (1986) and Judd et al. (1994) lumped within the Papaveraceae those genera (*Corydalis* and *Fumaria* in nc TX) often separated into the Fumariaceae (e.g., Kartesz 1994; Jones et al. 1997; Kiger 1997a; Mabberley 1997; Stern 1997a). This lumping of the families was based on morphological and molecular analyses (e.g., Chase et al. 1993; Judd et al. 1994) which indicated the Fumariaceae was derived from within the Papaveraceae sensu stricto. However, a more recent study (Hoot et al. 1997 [1998]) supported the monophyly of both families; we are therefore recognizing both the Fumariaceae and the Papaveraceae. According to Stern (1997a), " ... although a few taxa are morphologically intermediate, the members of the Fumariaceae generally are quite distinct from those of Papaveraceae in several respects, including floral symmetry, sap character, and stamen number and fusion." (subclass Magnoliidae)
FAMILY RECOGNITION IN THE FIELD: herbs with *pinnately compound* leaves with deeply divided leaflets, watery sap, and *bilaterally symmetrical* flowers with 1 *spurred* petal; stipules absent; sepals 2; petals 4; fruit a capsule.
REFERENCES: Ernst 1962; Gunn 1980; Lidén 1986, 1993; Chase et al. 1993; Judd et al. 1994; Stern 1997a; Hoot et al. 1997 [1998].

1. Petals pale to bright yellow or orange-yellow; fruits elongated cylindrical, 10–45 mm long, with many seeds, dehiscent; seeds with elaisomes ____ **Corydalis**
1. Petals light to deep purple; fruits subglobose–obovoid, ca. 2.5 mm long, 1-seeded, indehiscent; seeds without elaisomes ____ **Fumaria**

CORYDALIS SCRAMBLED-EGGS, FITWEED, FUMITORY, FUMEWORT

Leaves with 2–3 orders of leaflets and lobes; racemes initially congested, soon elongating; petals pale to bright yellow or orange-yellow, one conspicuously spurred; fruit an elongate dehiscent capsule; seeds many, with elaisomes.

A genus of ca. 100 species (Stern 1997b) of the n temperate zone and tropical African mountains. Many are cultivated as ornamentals; some have edible tubers; many contain isoquinoline alkaloids (e.g., aporphine, protoberberine) and are toxic to livestock (Kingsbury 1964). A number are difficult to distinguish without mature fruits and seeds. (Greek: *korydallis*, name of the crested lark, possibly from the resemblance of the spur to the hind claw of the bird, or from a resemblance of the shape of the flower to the bird's head)
REFERENCES: Ownbey 1947, 1951; Stern 1997b.

1. Ovaries and capsules coated with thick, short, white hairs; spurred petal of fully developed flower 16–22 mm long, with crest conspicuous and marginal wing very broad ____ **C. crystallina**
1. Ovaries and capsules glabrous; spurred petal 9–18 mm long, with crest inconspicuous OR conspicuous, the marginal wing narrow to medium or narrow.
   2. Spurred petal 9–15 mm long, the spur 3–7 mm long; greatest diam. of seeds 1.2–1.6 mm; plants often with inconspicuous racemes of 1–5 cleistogamous flowers ____ **C. micrantha**
   2. Spurred petal 14–18 mm long, the spur 4–9 mm long; greatest diam. of seeds 1.8–2.2 mm; plants usually without cleistogamous flowers.
      3. Seeds finely roughened (under magnification); mature capsules 20–35 mm long; bracts subtending flowers to 17 mm long; spurred petal crested with a fold or not crested ____ **C. curvisiliqua**
      3. Seeds faintly dotted or smooth except on margins; mature capsules 12–20 mm long; bracts subtending flowers usually 4–10 mm long; spurred petal not crested ____ **C. aurea**

Corydalis aurea Willd. subsp. **occidentalis** (Engelm. ex A. Gray) G.B. Ownbey, (sp.: golden; subsp.: western), GOLDEN CORYDALIS. Stems usually prostrate with age, usually 10–25 cm long; racemes usually longer than leaves, bracts much reduced upward; petals bright yellow, the spurred petal 14–18 mm long, with hood not crested, the spur 5–9 mm long; seeds ca. 2 mm in diam. Sandy soils; Brown, Callahan, and Shackelford cos.; also Comanche Co. (Ownbey 1947); w part of nc TX s and w to w TX. Late Mar–Jun. [*C. aurea* var. *occidentalis* Engelm. ex A. Gray, *C. montana* Engelm.] Reported as occasionally toxic to livestock due to alkaloids (Burlage 1968).

Corydalis crystallina Engelm., (crystalline, transparent), MEALY CORYDALIS. Stems erect or ascending, 20–40 cm tall; racemes longer than the leaves; bracts 5–12 mm long; petals bright yellow, sometimes with inconspicuous brown-red marking, the hood always crested; fruits 14–18 mm long; seeds ca. 2 mm in diam., minutely roughened. Sandy open ground; Post Oak Savannah w to East Cross Timbers, also Edwards Plateau. Late Mar–early May.

Corydalis curvisiliqua Engelm., (with curved pods). Stems ascending, 10–40 cm long; foliage glaucous; bracts to 15 mm long and 6 mm wide; flowers bright yellow, usually ca. 12 per inflorescence; spurred petal 15–18 mm long, the spur 7–9 mm long and often somewhat globose at the tip. Mar–May.

1. Mature fruits 26–34 mm long, usually abruptly acute; bracts (lowermost) 10 mm or less long, much reduced upward; hood (= end of spurred petal) crestless or with a crest (= fold); on w and s margins of nc TX ____ subsp. **curvisiliqua**
1. Mature fruits 20–25 mm long, gradually tapered; bracts (lowermost) 10–15 mm long, conspicuous, somewhat reduced upward; hood with a well-developed crest; widespread in nc TX ____ subsp. **grandibracteata**

subsp. **curvisiliqua**, CURVE-POD CORYDALIS. Disturbed areas, sandy soils; Young Co. on w margin of nc TX and Travis Co. just s of nc TX s to s Edwards Plateau; endemic to TX.

subsp. **grandibracteata** (Fedde) G.B. Ownbey, (large-bracted). Sandy or silty ground; Clay, Collin, Denton, Montague, and Shackelford cos.; also Brown (HPC), Archer, Dallas, and Navarro (Ownbey 1947) cos.; Blackland Prairie w to e Rolling Plains, also Edwards Plateau. [*C. curvisiliqua* var. *grandibracteata* Fedde]

Quercus stellata [LYN, SA3]

Quercus velutina [SA3]

Quercus virginiana [SA3]

Corydalis aurea subsp. occidentalis [BB2]

Corydalis curvisiliqua subsp. curvisiliqua [HEA]

Corydalis micrantha (Engelm. ex A. Gray) A. Gray, (small-flowered). Stems erect to ascending, to 30(-60) cm tall; bracts 5-8 mm long; flowers pale yellow; spurred petal with low crest usually present on hood, the spur 4-7 mm long; cleistogamy (= having closed self-pollinated flowers) is more common in this species than in any other *Corydalis*. Sandy or silty clay ground. Mar-Apr.

1. Racemes of normal flowers elongated, much longer than the leaves, with 3-20 flowers, these becoming widely spaced; spur not globose at tip; fruits 15-25 mm long (including beak) _____ subsp. **australis**
1. Racemes of normal flowers often short, not much longer than the leaves, with 3-12 flowers, these crowded; spur usually somewhat globose at tip; fruits usually 10-15 mm long (including beak) _____ subsp. **micrantha**

subsp. **australis** (Chapm.) G.B.Ownbey, (southern), SOUTHERN CORYDALIS. Stems semi-erect or ascending, 20-40(-60) cm tall, weak, not strongly striate when dry; foliage green to glaucous. Se and e TX w to East Cross Timbers, and in Brazos River valley w to Somervell Co. (Mahler 1988). [*C. micrantha* var. *australis* (Chapm.) Shinners]

subsp. **micrantha**, SMALL-FLOWER CORYDALIS, SLENDER FUMEWORT. Stems erect or ascending, 15-25 cm tall. Bell, Hood, and Tarrant cos.; also Dallas Co. (Ownbey 1947); e TX w to nc TX.

subsp. *texensis* G.B. Ownbey, (of Texas), TEXAS CORYDALIS. This endemic subspecies was cited by Hatch et al. (1990) for vegetational area 4 (Fig. 2). It is similar to subsp. *australis* but differs in having fruits 25-30 mm long, glaucous foliage, and stems usually stout and strongly striate (= lined) when dry. Based on distributional information from Ownbey (1947) and from specimens available, this subspecies apparently occurs n only to Travis Co. just s of nc TX.

FUMARIA FUMITORY, EARTH-SMOKE

A genus of ca. 50 species native to Eurasia, Africa, and Atlantic islands, with greatest diversity in the w Mediterranean area (Boufford 1997a); a number contain alkaloids. (Latin: *fumus*, smoke; according to Pliny, the juice of an Old World species causes the eyes to water as if exposed to smoke (Tveten & Tveten 1993), or perhaps from the nitrous odors of the fresh roots)
REFERENCE: Boufford 1997a.

Fumaria officinalis L., (medicinal), DRUG FUMITORY, COMMON FUMITORY. Glabrous, glaucous annual; stems spreading or ascending; leaves with 3-4 orders of leaflets and lobes; plant similar in foliage and flower structure to *Corydalis*, but with lavender or purple corollas 6-9.5 mm long and short, subglobose to obovoid fruits. Roadsides, railroads, and disturbed sites; Dallas Co., also Tarrant Co. (Mahler 1988); rare in nc TX, more frequent in c and s TX. Spring-summer. Native of Europe.

GARRYACEAE SILKTASSEL FAMILY

A very small family of 1 genus and 13 species of w North America, Central America, and the West Indies. They are dioecious evergreen trees and shrubs with highly toxic alkaloids. The Garryaceae were previously thought to be closely related to Cornaceae and were sometimes treated in that family (e.g., Correll & Johnston 1970; Hatch et al. 1990); however, recent molecular studies (Xiang et al. 1993) did not support a close relationship to Cornaceae. (subclass Rosidae)
FAMILY RECOGNITION IN THE FIELD: the only nc TX species is an evergreen shrub or small tree with opposite, simple, entire leaves and small, inconspicuous, apetalous, unisexual flowers in catkin-like pendulous inflorescences; fruit a blue drupe.
REFERENCE: Xiang et al. 1993.

GARRYA SILKTASSEL

A genus ranging from Washington to Panama and the West Indies; some are cultivated as ornamentals and used medicinally; some species have bark containing at least 5 alkaloids including delphinine otherwise known only from *Aconitum* and *Delphinium* in the Ranunculaceae. (Named for N. Garry, 1782[?]–1856, first secretary of Hudson Bay Co. and friend of David Douglas—Daniel 1993)
REFERENCES: Coulter & Evans 1890; Dahling 1978.

Garrya ovata Benth. subsp. **lindheimeri** (Torr.) Dahling, (sp.: ovate, egg-shaped; subsp.: for Ferdinand Jacob Lindheimer, 1801–1879, German-born TX collector), LINDHEIMER'S SILKTASSEL, MEXICAN SILKTASSEL. Dioecious, low shrub or rarely a small tree, 1–4 m tall; leaves evergreen, opposite, petiolate; leaf blades oblong-elliptic to broadly elliptic, ovate, or obovate, ca. 4.5–8 cm long and 2.5–5 cm wide, entire, the lower surfaces densely pubescent with curled or crinkly hairs, whitish gray or gray-green; inflorescences racemose, pendulous, branched, fasciculate; flowers without petals; staminate flowers pedicelled, with 4 stamens; pistillate flowers ± sessile; ovaries inferior; fruits blue, glaucous, globose drupes 6–10 mm in diam. Rocky slopes and bottoms in juniper-oak woodland; Burnet and Williamson cos. (Balcones Canyonlands Nat. Wildlife Refuge, C. Sexton, pers. comm.); endemic to Edwards Plateau and adjacent Lampasas Cut Plain. Mar–early April. [*G. lindheimeri* Torr.]

GENTIANACEAE GENTIAN FAMILY

Ours low, glabrous annual or short-lived perennial herbs; leaves opposite, sessile or subsessile, simple, entire, without stipules; flowers terminal, solitary or in cymose inflorescences; calyces often with prominently ribbed tube, usually with 4–5 teeth or lobes; corollas salverform, rotate or campanulate, usually 4–5-lobed; stamens as many as corolla lobes; ovary superior, 2-carpellate; fruit a many-seeded, 2-valved capsule.

A medium-large (1,225 species in 78 genera), cosmopolitan family of mainly herbs or shrubs or rarely small trees; they usually accumulate bitter iridoid substances; many are showy and frequently used as ornamentals including species of *Centaurium*, *Eustoma*, *Exacum*, *Gentiana*, and *Sabatia*. *Nymphoides*, previously placed in this family, is here treated in the Menyanthaceae. Family name from *Gentiana*, GENTIAN, a temperate, arctic, and montane genus of 361 species of usually perennial herbs. (Named for Gentius, 2nd century king of Illyria, who is said to have discovered the medicinal properties of *G. lutea* L. roots) (subclass Asteridae)
FAMILY RECOGNITION IN THE FIELD: herbs with opposite, simple, glabrous leaves without stipules; calyces bell-shaped or tubular; corollas showy, salverform to rotate or bell-shaped, radially symmetrical with the same number of epipetalous stamens alternating with the corolla lobes.
REFERENCES: Wood & Weaver 1982; Mésezáros et al. 1996.

1. Corollas pink to rose, rarely white, salverform or rotate, the lobes < 24 mm long; style filiform, usually deciduous from capsule; anthers coiled after dehiscence.
   2. Corollas rotate, the tube much shorter than the lobes; leaves 5–26 mm wide; corolla lobes 11–23 mm long; anthers recurved or revolute (circinately coiled) after dehiscence ____ **Sabatia**
   2. Corollas salverform, the tube equal to or longer than the lobes; leaves 1–13 mm wide (4 mm or less wide except in *C. calycosum* which is rare in nc TX); corolla lobes 3.5–13 mm long; anthers spirally curved after dehiscence ____ **Centaurium**
1. Corollas blue-purple with darker center, rarely largely white or rose, campanulate, the lobes 30 50 mm long; style usually stout and persistent; anthers not coiled ____ **Eustoma**

CENTAURIUM CENTAURY

Erect, usually much-branched annuals; flowers numerous, showy; calyces and corollas 4- or 5-parted; corollas usually pink to rose (rarely white), often with a light area at very base, the tube elongate, slender; anthers spirally coiled or twisted after dehiscence.

A mainly n hemisphere genus of ca. 50 species (J. Pringle, pers. comm.) with several in South America and Australia; some are cultivated as ornamentals. ☠ Some are toxic to livestock though eaten only when other forage is scarce (Kingsbury 1964). Material for the key and descriptions was obtained in part from an unpublished manuscript being prepared for the Flora of North America by J. Pringle. (An old name, variously applied by the herbalists, from Latin: *Centaurus*, Centaur, mythical discoverer of its medicinal qualities)
REFERENCES: Hess 1968; Melderis 1972; Turner 1993a; Holmes & Wivagg 1996.

1. Corolla lobes 1–6(–7) mm long, 1.5(–1.7) mm or less wide, much shorter than the corolla tube; anther sacs (when coiled at maturity) 0.6–1.1 mm long.
   2. Flowers on pedicels (1–)2–12 mm long; inflorescence rather open (but there can be numerous flowers), not flat-topped.
      3. Pedicels usually ca. as long as calyces; corolla lobes 3–6(–7) mm long; stem leaves linear to linear-lanceolate, 1–3(–4) mm wide; native species widespread in nc TX ____ **C. texense**
      3. Pedicels shorter than calyces; corolla lobes 1–4 mm long; stem leaves lance-ovate to lanceolate, 2–7 mm wide; introduced species known in nc TX only from extreme e margin of area ____ **C. pulchellum**
   2. Flowers sessile or subsessile; inflorescence dense, flat-topped ____ **C. floribundum**
1. Corolla lobes 6–13 mm long, usually > 1.5 mm wide, nearly equal in length to the corolla tube; anther sacs (when coiled at maturity) 1.5–2.5 mm long.
   4. Stem leaves typically linear to very narrowly linear-oblanceolate, 1–3 mm wide; corolla tube at first about equaling, finally twice the calyx length (calyx measured to end of calyx lobes); plants much branched from the very base; widespread in nc TX ____ **C. beyrichii**
   4. Stem leaves oblong to oblong-elliptic or narrowly lanceolate, 3–13 mm wide; corolla tube shorter than to slightly longer than calyx (calyx measured to end of calyx lobes); plants usually with one stem from base, branched well above base but not much-branched from the very base (but sometimes somewhat branched from very base); rare in nc TX ____ **C. calycosum**

Centaurium beyrichii (Torr. & A. Gray) B.L. Rob., (for Heinrich Kral Beyrich, 1796–1834, Prussian botanist who collected in GA, SC, and TX), ROCK CENTAURY, MOUNTAIN-PINK. Plant to ca. 30 cm tall; without numerous glands; stem leaves to 3 cm long and 3 mm wide; flowers numerous, densely packed in the much-branched inflorescence; corollas pink, rarely white; seeds dark brown. Limestone gravel; Grand Prairie s and w to w TX. May–Aug. [*Erythraea beyrichii* Torr. & A. Gray] Turner (1993a) indicated that *C. beyrichii* "... sometimes occurs near or with *Centaurium calycosum* and the occasional hybrid between these probably occurs." This species is poisonous to livestock; severe gastroenteritis, organ damage, and death can result; however, it is eaten only when other forage is scarce (Kingsbury 1964; Burlage 1968). ☠ ▣/83

Centaurium calycosum (Buckley) Fernald, (with large calyx), BUCKLEY'S CENTAURY, ROSITA. Plant to ca. 60 cm tall, usually much smaller; stem leaves to 6 cm long and 13 mm wide; inflorescence open; corollas pink to rose-colored, rarely white; seeds light brown. Moist soils, along streams, prairies, and hillsides; Coleman Co., also Parker Co. (Turner 1993a); w part of nc TX s and w to w TX. Apr–Jun(–Aug). [*Erythraea calycosa* Buckley] This species is much less common in nc TX than the other native species. It is toxic to livestock but somewhat less so than *C. beyrichii* (Kingsbury 1964). ☠

Centaurium floribundum (Benth.) B.L. Rob., (flowering abundantly), JUNE CENTAURY. Plant 5–

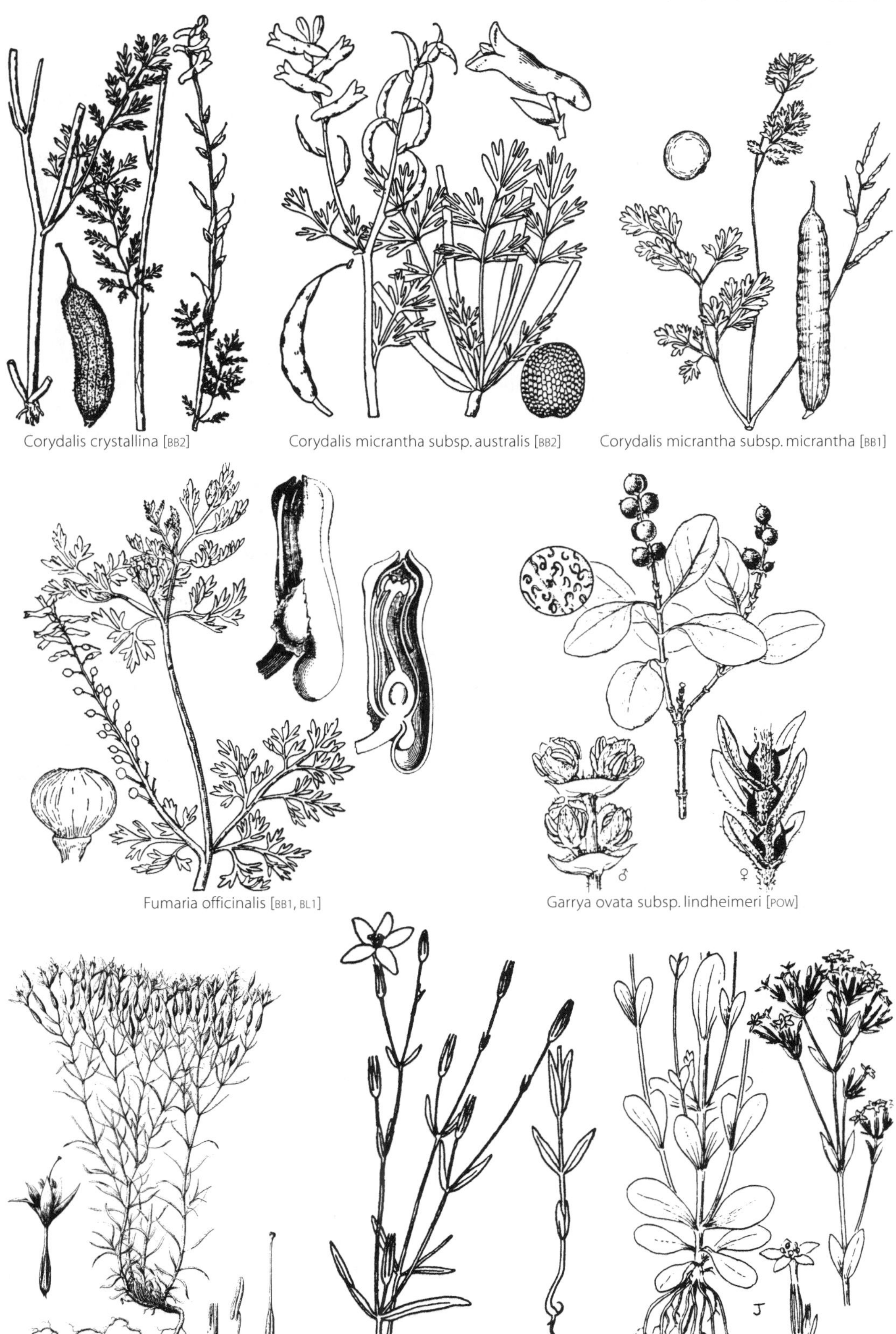

Corydalis crystallina [BB2]

Corydalis micrantha subsp. australis [BB2]

Corydalis micrantha subsp. micrantha [BB1]

Fumaria officinalis [BB1, BL1]

Garrya ovata subsp. lindheimeri [POW]

Centaurium beyrichii [RBM]

Centaurium calycosum [BB2]

Centaurium floribundum [ABR]

50(–90) cm tall; stem leaves 1–2.5 cm long, 2–7(–9) mm wide; corollas pink, the lobes 2–5 mm long; seeds brown. Moist open areas, stream banks, open woods; Ellis, Falls, Hill, Johnson, Limestone, McLennan, Milam, and Navarro cos. (Holmes & Wivagg 1996, cited as *C. muehlenbergii*); nc and e TX. Summer–fall. Native of the w U.S.A., naturalized and spreading in TX. [*Erythraea floribunda* Benth.] While some authorities have lumped *C. floribundum* with *C. muehlenbergii* (e.g., Holmes & Wivagg 1996), J. Pringle (pers. comm.) indicated that it appears to be a distinct species and that specimens from TX cited by Holmes and Wivagg (1996) as *C. muehlenbergii* are actually *C. floribundum*. According to J. Pringle (pers. comm.), "True *C. muehlenbergii*, as indicated by studies of the type, is characterized by proportionately elongate, relatively few-flowered inflorescences with predominantly monochasial branching, with some or all flowers being distinctly pedicellate. *Centaurium floribundum*, in contrast is characterized by flat-topped, densely many-flowered inflorescences with predominantly dichasial branching, with all of the flowers sessile or subsessile."

Centaurium pulchellum (Sw.) Druce, (pretty, beautiful but small). Plant 10–26 cm tall; stem leaves 1–2 cm long, 2–7 mm wide; inflorescence an open compound dichasium, not flat topped; pedicels 2–5 mm long; calyces 5–9 mm long; corolla tube exceeding calyx; corollas pink, the lobes 1–4 mm long; seeds light brown. Roadsides, fields; Limestone Co. at the extreme e margin of nc TX (Holmes & Wivagg 1996); mainly e TX; this species was first reported for TX by Correll and Johnston (1972). Apr–Jun. Native of Europe. [*Erythraea pulchella* (Sw.) Fr.]

We are following Holmes and Wivagg (1996) in treating this taxon as *C. pulchellum*; however, there may be problems with the identification of TX material. James Pringle (pers. comm.) indicated, "... Anton A. Reznicek (in annotations, TRT) has suggested that some of the plants naturalized in North America identified as *C. pulchellum* may actually be *C. tenuiflorum* (Hoffsgg. & Link) Fritsch ex E. Janchen, a European species abundantly naturalized in parts of Australia but thus far not reported in print for North America." Pringle indicated that further study is thus needed. Melderis (1972), in *Flora Europaea*, separated the two as follows:

1. Stem usually dichotomously branched in the lower part, with patent branches; cauline internodes usually 2–4; flowers long-pedicellate, usually in a lax dichasial cyme ______ *C. pulchellum*
1. Stem branched in the upper part, with strict branches; cauline internodes usually 5–9; flowers shortly pedicellate, usually in a dense dichasial cyme ______ *C. tenuiflorum*

Centaurium texense (Griseb.) Fernald, (of Texas), TEXAS CENTAURY, LADY BIRD'S CENTAURY. Stem leaves linear to linear-lanceolate, to 2.5 cm long, 1–3(–4) mm wide; inflorescence ± open; pedicels ca. as long as calyces; corollas light pink; corolla tube at first shorter, finally longer than calyx; corolla lobes 3–6(–7) mm long; seeds light brown. Dry limestone areas; Bell, Bosque, Burnet, Dallas, Denton, Hood, Montague, and Tarrant cos., also Brown (HPC) and Somervell (R. O'Kennon, pers. obs.) cos.; Post Oak Savannah w to nc TX and Edwards Plateau. Jun–Aug. [*Erythraea texensis* Griseb.] According to Correll and Johnston (1970), the common name is for Mrs. Lyndon B. Johnson, who made a special effort to collect seeds of this species from limestone hills south of Johnson City and planted them along her ranch's airfield runway.

Centaurium glanduliferum (Correll) B.L. Turner, (bearing glands), [*C. beyrichii* var. *glanduliferum* Correll], is cited by Hatch et al. (1990) for vegetational area 4 (Fig. 2); we have seen no nc TX material of this mainly w TX taxon. It can be distinguished from the related *C. beyrichii* by the numerous minute glands, the leaves especially being densely gland-covered, seeds blackish or very dark brown, and plant size (to ca. 15 cm tall). We are following Turner (1993a), Jones et al. (1997), and J. Kartesz (pers. comm. 1997) in recognizing this taxon at the specific level.

EUSTOMA
BLUEBELLS, PRAIRIE GENTIAN, CATCHFLY GENTIAN

A genus of a single species (J. Pringle, pers. comm.) of s North America to n South America. (Greek: *eu*, good, and *stoma*, mouth, from corolla tube)
REFERENCE: Shinners 1957.

Eustoma russellianum (Hook.) G. Don, (derivation either for the Dukes of Bedford, whose family name is Russell, or for Mr. Russell of Falkirk), BLUEBELL GENTIAN, BLUEBELLS, TEXAS BLUEBELLS, SHOWY PRAIRIE GENTIAN, LIRA DE SAN PEDRO, PURPLE PRAIRIE GENTIAN. Erect glaucous annual or short-lived perennial to 70 cm tall; leaves ovate to elliptic-oblong or elliptic-lanceolate, 1.5–8 cm long, noticeably 3-veined; flowers solitary or in few-flowered inflorescences, 5(–6)-merous; corollas blue-purple with darker center, rarely largely white, white with yellow center, white with deep red center, or pink, very large, deeply divided, the lobes 30–40 mm long, 15–24 mm wide, at least 3 times as long as tube; anthers ca. 10 mm long, not coiled; stigma 2-lobed, yellow, ca. 8 mm wide. Low spots in prairies, low open ground, or disturbed sites; throughout most of TX. Late Jun–Aug. [*E. grandiflorum* (Raf.) Shinners] This is one of the most striking wildflowers in nc TX. It is closely related to *E. exaltatum* (L.) Salisb. ex G. Don; according to J. Pringle (pers. comm.), there are not two distinct categories of flower size—a character often used in the past to separate *E. exaltatum* and *E. russellianum*. Instead, considerable intergradation in flower size exists; however, plants with flowers in the size range traditionally associated with *E. russellianum* do tend to prevail in some areas. Kancheepuram Gandhi (pers. comm.) is currently studying these taxa to determine if *E. russellianum* is most appropriately recognized at the species level or as a subspecies/variety of *E. exaltatum*. Until that study is completed, we are following the traditional approach and treating this taxon at the specific level. According to Stanford (1976), "Numerous color forms occur in Texas. Dried material is often misleading in that the water-soluble pigments at the corolla base are sometimes dispersed throughout the petals, resulting in a wide array of petal color." Shinners (1957) and Stanford (1976) listed the following color forms: forma *grandiflorum* [forma *russellianum*]—typical, blue-purple with darker center, forma *fisheri* (Standl.) Shinners—pure white; forma *bicolor* (Standl.) Shinners—white with deep red center (dried material usually with purple-tinged lobes); forma *roseum* (Standl.) Shinners—pink; and forma *flaviflorum* (Cockerell) Shinners—white with bright yellow center (dried material with entire corolla yellow). /90

SABATIA ROSE GENTIAN

Erect glabrous annuals; leaves mostly cauline; flowers in terminal cymose inflorescences, showy, in ours usually 5-merous; corollas usually pink to rose (rarely white) with a yellow or greenish yellow triangular spot at the base of each corolla lobe, these spots together forming a yellow or greenish yellow star in the center of the corolla; anthers bright yellow, recurved or revolute (circinately coiled) after dehiscence.

A genus of 19 species (J. Pringle, pers. comm.) of North America and the West Indies; some are cultivated as ornamentals. Material for the key and descriptions was obtained in part from an unpublished manuscript being prepared for the *Flora of North America* by J. Pringle. (Named for Liberato Sabbati, born 1714?, Italian botanist and surgeon in Rome)
REFERENCES: Wilbur 1955; Perry 1971; Bell & Lester 1980.

1. Calyx tube essentially smooth, wingless, 1–3.5 mm long, covering < 1/3 of the corolla tube; stems slightly winged; upper nodes usually with 2 side branches per node ____ **S. angularis**
1. Calyx tube prominently 5-ribbed or winged, 4–8 mm long, covering 2/3 or more of the corolla tube; stems not winged; upper nodes usually with 1 side branch per node.

2. Calyces shorter than corollas in mature bud; yellow eye of corollas with sharply defined red border; known from s and w parts of nc TX ______ **S. formosa**
2. Calyces usually ca. as long as or longer than corollas; yellow eye of corollas with adjacent reddish zone grading into pink but usually without sharply defined red border; known from n and e parts of nc TX ______ **S. campestris**

Sabatia angularis (L.) Pursh, (angular, angled), ROSE-PINK, SQUARE-STEM ROSE GENTIAN, BITTER-BLOOM. Plant (10–)30–80 cm tall; leaves sub-orbicular to cordate-ovate; corolla lobes to ca. 20 mm long, ca. 4–6(–8) mm wide, pink to rose or rarely white, with yellow or greenish yellow spot at base usually bordered by a dark red line. Prairies, woods; e TX w in Red River drainage to Cooke, Grayson, and Lamar cos. Jun–Aug.

Sabatia campestris Nutt., (of the fields or plains), PRAIRIE ROSE GENTIAN, MEADOW-PINK, TEXAS-STAR, ROSE-PINK. Plant 10–37(–50) cm tall; leaves oblong-elliptic to broadly ovate-elliptic; largest leaves usually near mid-stem; basal rosette usually absent at flowering time; pedicels 18–91 mm long; corolla lobes ± obovate, usually widest distally (but narrowed at very apex), 13–25 mm long (mean ca. 19), ca. 7–11(–15) mm wide, rose to pale pink (rarely white), basally with a 3–4 mm long by 1–1.5 mm wide yellow or greenish yellow triangular spot, this often bordered by a white band and sometimes by a reddish area (this not sharply defined); width of white band at base of corolla lobes > 1 mm when present. Chiefly in clay soils but also in sand, open ground, edge of woods, roadsides; e 1/2 of TX. May–early Jul. ▣/104

Sabatia formosa Buckley, (handsome, beautiful), BUCKLEY'S SABATIA. Similar to *S. campestris*; plant 7–28 cm tall; leaves 7–25 mm long; largest leaves generally at or near the stem base, with basal rosette typically persisting at flowering time; pedicels 20–62 mm long; corolla lobes ± broadly lanceolate to elliptic-rhombic, widest near the middle (narrower in shape overall than in *S. campestris*), 9–20 mm long (mean ca. 15), magenta-rose (typically more deeply pigmented than in *S. campestris*), with patterning similar to *S. campestris* but yellow or greenish yellow area with sharply defined red border and width of white band at base of corolla lobes < 1 mm. Loose sandy soils, but sometimes in clay, typically in more xeric situations than *S. campestris*; Comanche and Eastland cos. (Bell & Lester 1980); widespread in e 1/2 of TX, but not known from extreme ne part of the state. Late Mar–Apr. While *S. formosa* has often been considered an early flowering morph of *S. campestris*, we are following J. Pringle (pers. comm. of treatment forthcoming in Flora of North America) and Bell and Lester (1980) in distinguishing *S. formosa* from *S. campestris*. Electrophoretic as well as morphological evidence supports the separation; hybrids between the two have been observed but are limited by differences in flowering time (Bell & Lester 1980).

GERANIACEAE GERANIUM FAMILY

Ours annual or biennial, pubescent herbs; leaves alternate or subopposite, simple or compound, toothed or lobed; stipules ovate to lanceolate, somewhat papery; flowers axillary or terminal, solitary or in peduncled pairs, congested cymose inflorescences, or umbels; sepals 5; petals 5; stamens 5 or 10, the filaments united basally; pistil 5-carpellate, the united styles and central column forming a prolonged beak; ovary superior; carpels 1-seeded, variously separating at maturity.

A medium-sized (700 species in 11 genera) mainly temperate (few tropical) genus of mostly herbs or small shrubs including the ornamental GERANIUMS (most in the South African genus *Pelargonium*); many have glandular hairs containing aromatic essential oils and some have been used medicinally. (subclass Rosidae)

FAMILY RECOGNITION IN THE FIELD: herbs with lobed or compound leaves and a distinctive

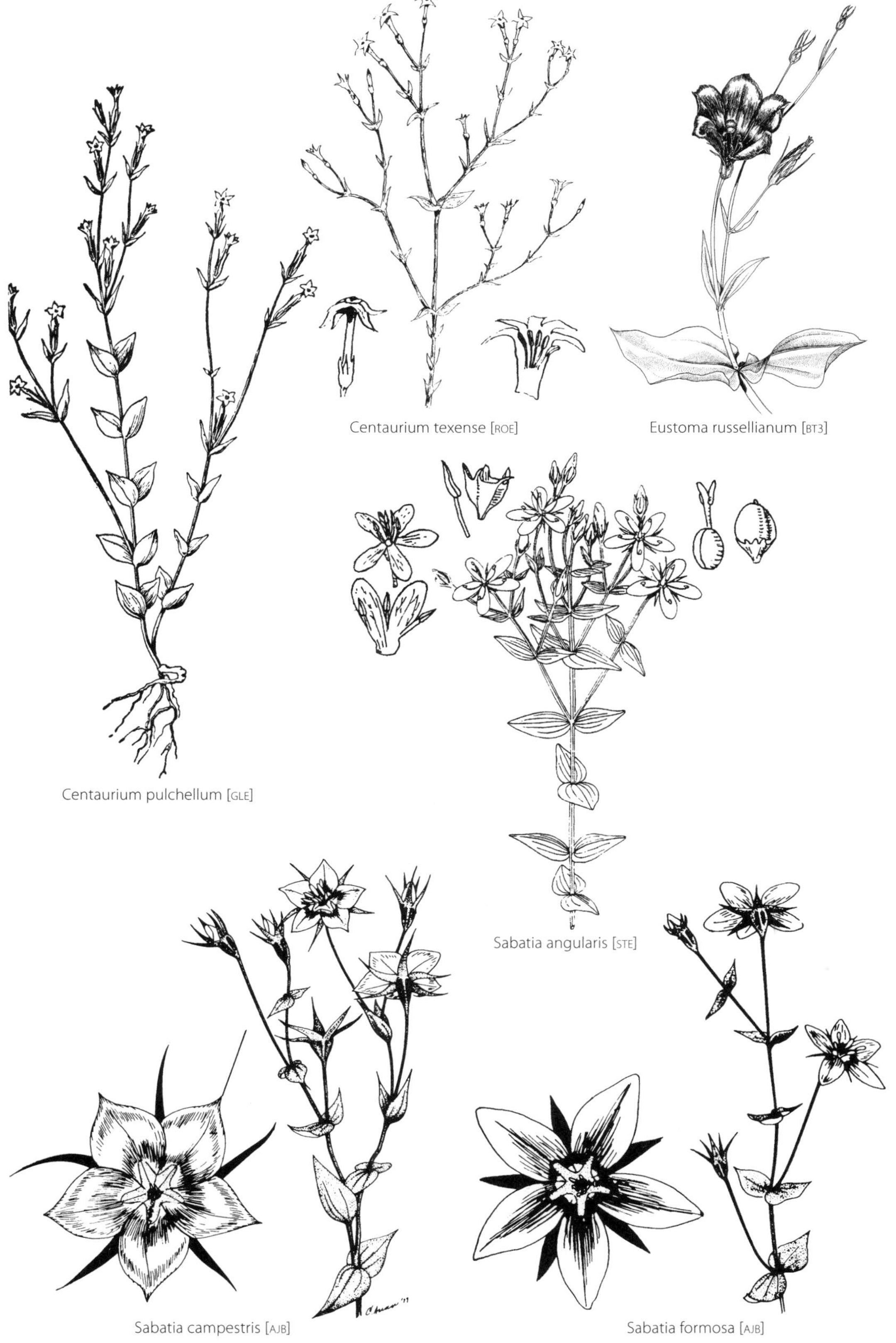

Centaurium texense [ROE]

Eustoma russellianum [BT3]

Centaurium pulchellum [GLE]

Sabatia angularis [STE]

Sabatia campestris [AJB]

Sabatia formosa [AJB]

birdbill-like, *long-beaked pistil* that separates into segments in fruit; flowers radially symmetrical, with 5 separate petals.
REFERENCES: Hanks & Small 1907; Robertson 1972; Price & Palmer 1993.

1. Fertile stamens (= those with anthers) 10; beak short (< 15 mm long); basal rosette usually absent; leaf blades usually ± as wide as long, deeply palmately or ternately divided (the ultimate segments usually ± narrow in appearance); seeds noticeably reticulate; stylar portion of the carpels remaining attached at the beak apex, the beak separating from the base first, nearly glabrous inside, the beak segments merely outwardly coiled ____ **Geranium**
1. Fertile stamens 5; beak long, usually 20–50(–75) mm; basal rosette present; leaf blades usually slightly to much longer than wide, palmately lobed (lobes rounded and crenate, not narrow in appearance) or pinnately compound; seeds smooth; beak separating at the apex first, with long hairs inside, the beak segments tightly twisted when dry ____ **Erodium**

ERODIUM STORK'S-BILL

Ours annuals or biennials; stems partly decumbent to ascending, to 40 cm long; leaves petiolate, the blades usually longer than broad; inflorescences axillary, long-peduncled, umbellate; petals deeply colored, opening in the morning, usually falling by late afternoon.

A genus of ca. 60 species of Europe, the Mediterranean to c Asia, temperate Australia, and tropical s America. The beak segments ("awns") coil and uncoil in response to moisture and drive the seeds into the ground; this can be easily observed in the laboratory using water to cause uncoiling and 95% ethanol to cause drying and recoiling. (Greek: *erodios*, a heron, from the long bill-like beak of the fruit)

1. Leaves pinnately compound with leaflets finely pinnately dissected; petals to 6–8 mm long, pink to pinkish purple ____ **E. cicutarium**
1. Leaves simple, usually palmately lobed; petals 10 mm or more long, purple to red ____ **E. texanum**

Erodium cicutarium (L.) L'Hér. ex Aiton, (resembling *Cicuta*—water hemlock), FILAREE, PIN-CLOVER, CALIFORNIA FILAREE, ALFILARIA, ALFILERILLO. Stems ± pilose with flattened, whitish hairs; umbels with 2–8 flowers. Roadsides and disturbed areas; throughout most of TX. Late Feb–May. Native of Europe.

Erodium texanum A. Gray, (of Texas), STORK'S-BILL, TEXAS FILAREE, HERONBILL. Stems minutely pubescent; sinuses between leaf lobes usually to less than 2/3 the distance to the petiole. Rocky or sandy prairies, disturbed sites; widespread in TX. Mar–Apr. /88

GERANIUM CRANE'S-BILL

Ours annuals or biennials, pubescent; leaves essentially orbicular in outline, deeply palmately or ternately lobed (usually divided 3/4 or more of distance to petiole), long-petioled; flowers usually in pedicellate pairs on short peduncles borne from the leaf axils; inflorescences sometimes congested, cymose; sepals aristate; petals white to pink or reddish pink.

A genus of 300 species of the temperate zone and tropical mountains; many are cultivated ornamentals, particularly as ground-covers. (Old Greek name, from *geranos*, crane, the long beak thought to resemble the bill of that bird)
REFERENCES: Fernald 1935; Jones & Jones 1943.

1. Sepals ca. 6–8 mm long, ovate-lanceolate to elliptic-lanceolate, pubescent with long spreading hairs on nerves and margins; stem pubescence of short and long hairs, at least some spreading.
   2. Petals pale pink or white with pink veins; carpel body with non-glandular hairs 0.8–1.5 mm

Erodium cicutarium [BT3]

Erodium texanum [GR1]

Geranium carolinianum [BB2]

Geranium dissectum [BEN]

Geranium texanum [HEA]

long; beak pubescent, with only a few hairs glandular; style branches yellow, 1 mm or more long; widespread in nc TX ____ **G. carolinianum**

2. Petals dark pink or dark reddish pink; carpel body with gland-tipped hairs 0.2–0.5 mm long; beak densely pubescent with gland-tipped hairs; style branches purplish, ca. 0.5 mm long; rare in nc TX, known only from Dallas Co. ____ **G. dissectum**

1. Sepals 4–5 mm long, orbicular-ovate, glabrous except for short appressed hairs on the nerves; stem pubescence short, appressed, retrorse (= pointing down) ____ **G. texanum**

Geranium carolinianum L., (of Carolina), CRANE'S-BILL, CAROLINA CRANE'S-BILL, CAROLINA GERANIUM. Winter annual or biennial, beginning to flower with stems only 3 cm tall; stems ascending, ultimately to 80 cm long; leaf blades sub-orbicular, deeply lobed and toothed; flowers usually in pairs; petals ca. as long as sepals. Woods, fields, waste places, various soils; throughout most of TX. Mar–early May.

Geranium dissectum L., (dissected, deeply cut). Winter annual or biennial; similar to *G. carolinianum*; stems ascending, to 60 cm long; ultimate leaf segments linear; petals ca. 4–6 mm long. Fields, waste areas; Dallas Co., mainly ne TX to the e of nc TX. Apr–May. Native of Europe.

Geranium texanum (Trel.) A. Heller, (of Texas), TEXAS GERANIUM. Winter annual similar to *G. carolinianum*; petals white or white with pinkish veins. Disturbed soils; widespread in TX, but mainly Blackland Prairie w to Rolling Plains and Edwards Plateau. Mar–Apr, usually flowering before *G. carolinianum*. [*G. carolinianum* L. var. *texanum* Trel.]

GROSSULARIACEAE CURRANT FAMILY

The Grossulariaceae is a small (330 species in 24 genera), cosmopolitan family of trees and shrubs; some are armed, while others accumulate aluminium or are cynogenic. The family is sometimes included in the Saxifragaceae. Family name conserved from *Grossularia*, a genus now treated as *Ribes*, GOOSEBERRY (the name *Ribes* was published earlier and thus has priority in terms of nomenclature). (Latin: *grossula*, gooseberry, apparently from *grossus*, an unripe fig, from a resemblance of the fruits to small unripe figs) (subclass Rosidae)

FAMILY RECOGNITION IN THE FIELD: the only nc TX species is a shrub with alternate, simple, lobed leaves; flowers with 5 yellow to red petals; fruit a berry.

REFERENCES: Coville & Britton 1908; Spongberg 1972.

RIBES CURRANT, GOOSEBERRY

A genus of 150 species of often spiny shrubs of temperate regions of the n hemisphere and the Andes; many species have edible fruits (gooseberries and currants) and some are cultivated as ornamentals. In some areas of the U.S. *Ribes* species have been eliminated and their cultivation prohibited because they serve as the alternate host of white pine blister rust, an introduced fungus that attacks white pines. (According to Fernald (1950a), said by A. DeCandolle to come from the old Danish colloquial, *ribs*, for the red currant; according to other sources from the Arabic: *ribas*, acid tasting, alluding to the fruit)

REFERENCES: Berger 1924; Sinnott 1985.

Ribes aureum Pursh var. **villosum** DC., (sp.: golden; var.: villous, with soft hairs), CLOVE CURRANT, BUFFALO CURRANT. Erect, unarmed shrub to ca. 2 m tall; leaves simple, alternate; leaf blades 3–8 cm wide, nearly as wide as long, 3–5-lobed, the lobes usually dentate; inflorescence a raceme of 3–10 flowers, usually nodding; flowers fragrant, 5-merous; calyces yellow, with tube 10–15 mm long; petals yellow to red; ovary inferior; fruit an ovoid to globose berry 7–10 mm long, yellowish turning black. Rocky slopes, sandy bluffs; included based on citation of vegetational area 5

(Fig. 2) by Hatch et al. (1990); mainly w 1/2 of TX. Feb–May. [*R. odoratum* H. Wendl.] The fruits of this species were formerly used with buffalo meat to make pemmican (Mabberley 1987).

HALORAGACEAE WATER-MILFOIL FAMILY

Glabrous perennial herbs growing in water or on wet ground; leaves alternate or whorled, sessile, simple (but in most so deeply divided as to appear compound), sharply toothed or pinnately lobed with almost thread-like segments; flowers minute, sessile, axillary or in terminal spikes, unisexual or perfect; calyces 3–4-lobed; petals 0 or 4; stamens 3–8; ovary inferior; fruits nut-like, 3–4-celled.

A small (145 species in 9 genera), cosmopolitan, but especially s hemisphere family of usually aquatic or wet area herbs or sometimes shrubs or small trees. The family is sometimes interpreted to include the well-known tropical genus *Gunnera*. It is sometimes referred to as the Haloragidaceae; however, the name Haloragaceae is conserved. Family name from *Haloragis*, a genus of 26 species of Australia, New Caledonia, New Zealand, Rapa, and Juan Fernandez. (Greek: *hals*, the sea, a lump of salt, or salt, and *rhax*, *rhagos*, berry or grape) (subclass Rosidae)
FAMILY RECOGNITION IN THE FIELD: aquatic or wet area herbs; leaves alternate or whorled, at least the submerged ones pinnately divided into numerous segments and *feather-like* in appearance; flowers minute.

1. Leaves mostly or all whorled, the ones out of the water usually reduced and bract-like; stamens 4 or 8; carpels 4; sepals 4 **Myriophyllum**
1. Leaves alternate, the ones out of water well-developed; stamens 3; carpels 3; sepals 3 **Proserpinaca**

MYRIOPHYLLUM WATER-MILFOIL, PARROT'S-FEATHER

Stems and leaves mostly submersed except in *M. aquaticum*; leaves whorled, the submersed leaves pinnately divided into thread-like segments; flowers in terminal spikes held above the surface of the water or in the axils of leaves, small, unisexual or perfect; petals 0 or 4; stigmas 4; fruits small (1–3 mm long), subglobose to ovoid, 3-angled.

A cosmopolitan genus of 60 species, especially in Australia; they resemble *Ceratophyllum* but the leaves are pinnately divided, not forked. Some are cultivated as aquarium plants. (Greek: *myrios*, numberless, and *phyllon*, leaf, alluding to the innumerable leaf divisions)
REFERENCES: Löve 1961; Jarman 1968; Correll & Correll 1972; Reed 1977; Aiken 1981; Godfrey & Wooten 1981.

1. Whorls of leaves on lower and middle parts of stem usually 1 cm or more apart.
   2. Many of the leafy vegetative branches emersed (= out of the water), sometimes trailing on mud or wet ground and erect-ascending; leaf segments usually 6 mm or less long; flowers in clusters in axils of leaves essentially like the sterile leaves **M. aquaticum**
   2. Most of the leafy vegetative branches submersed (or sometimes stranded out of the water when water low); leaf segments usually 6–20 mm long; flowers in whorls in a slender terminal spike, the individual whorls not subtended by leaf-like bracts, the bracts very small and shorter than to slightly exceeding the flowers.
      3. Lowermost bracts usually slightly exceeding the flowers and fruit, pectinate to serrate or nearly entire; middle stem leaves usually < 2.5 cm long, with 12 or more segments on each side, the segments usually 15 mm or less long; stem thickened below the inflorescence to almost double the width of the lower stem; abundant in several nc TX lakes and probably being more widely spread at present **M. spicatum**
      3. Lowermost bracts shorter than flowers and fruit, entire to serrate; middle stem leaves

usually 3 cm or more long, with 11 or fewer segments on each side, the segments often > 15 mm long; stem not thickened below the inflorescence; rare if present in nc TX; probably not occurring in TX _____ **M. sibiricum**

1. Whorls of leaves on lower and middle parts of stem usually much < 1 cm apart.
   4. Emersed bracts or bracteal leaves very deeply pinnately dissected or lobed (and thus comb-like in appearance), as long as flowers or fruits or to only 2–3 times as long—the spikes thus appearing ± naked; anthers (4–)8; fruits 2–3 mm long; submersed foliage leaves with a total of 18–26 thread-like segments _____ **M. verticillatum**
   4. Emersed bracts or bracteal leaves merely serrate OR with long serrations (and thus nearly comb-like in appearance), many times as long as the flowers and fruits—the bracts thus obvious at a glance; anthers 4; fruits 1–1.8 mm long; submersed foliage leaves with a total of 8–20 thread-like segments.
      5. Submersed foliage leaves with a total of 12–20 thread-like segments; emersed bracts or bracteal leaves with only small serrations _____ **M. heterophyllum**
      5. Submersed foliage leaves with a total of 8–10 thread-like segments; emersed bracts or bracteal leaves with long (relative to leaf size) serrations, usually nearly comb-like _____ **M. pinnatum**

Myriophyllum aquaticum (Vell.) Verdc., (growing in or near water), PARROT'S-FEATHER, WATER-FEATHER. Creeping and freely branching, forming masses around ponds or along stream banks; leaves all whorled, with 20 or more linear-filiform divisions; upper leaves often yellow, suggesting flowers from a distance; flowers axillary, unisexual, apparently seldom produced; the species is dioecious (= sexes on different plants); only pistillate flowers are known from North America and no seed set is known to occur (Aiken 1981). Common aquarium plant, cultivated and escaped, ponds, lakes, streams; Grayson Co., also Denton (G. Dick, pers. comm.) and Tarrant (C. Owens, pers. comm.) cos.; se and e TX w to Rolling Plains and Edwards Plateau. Mar–May. Native of Brazil. [*M. brasiliense* Camb., *M. proserpinacoides* Gillies ex Hook. & Arn.]

Myriophyllum heterophyllum Michx., (variously leaved). Submersed leaves 2–5(–6) cm long, usually in whorls of 5–6; surface of fruits papillose. Fort Hood (Bell or Coryell cos.—Sanchez 1997), also Palo Pinto and Wise cos. (Mahler 1988); mainly Rolling Plains, Edwards Plateau and se TX. Apr–Aug.

Myriophyllum pinnatum (Walter) Britton, Sterns, & Poggenb., (feather-like, having leaflets arranged on each side of a common stalk), GREEN PARROT'S-FEATHER. Plant variable, either submersed or predominantly terrestrial; submersed leaves to ca. 3 cm long, usually in pseudowhorls of 3–4(–5) and often some alternate or subopposite; surface of fruits tuberculate. Ponds, lakes, streams; found at Dallas by Reverchon, not seen there recently (Mahler 1988); mainly se and e TX, also Edwards Plateau. Mar–Aug. Note that the illustration we use for *M. pinnatum* is the same as figure 570 from Correll and Correll (1972); however, the figure captions for *M. pinnatum* and *M. heterophyllum* were inadvertantly switched in Correll and Correll (1972).

Myriophyllum sibiricum Kom., (of Siberia). Plant rarely branching near the water surface; developing oblanceolate turions (= swollen scaly structures that store carbohydrates and act as a means of propagation); scales (= modified leaves) of turions shorter, thicker, and darker green than regular leaves; stamens 8. Included based on citation of vegetational areas 4 and 5 (Fig. 2) by Hatch et al. (1990); supposedly widespread in TX; we have seen no material for nc TX and Aiken (1981) indicated that TX records of this species (as *M. exalbescens*) were probably based on misidentifications; J. Kartesz (pers. comm. 1997) also indicated that it is not in TX. [*M. exalbescens* Fernald, *M. spicatum* L. var. *exalbescens* (Fernald) Jeps.] Despite the epithet, this species is native to North America as well as the Old World (McClintock 1993).

Myriophyllum spicatum L., (spicate, with spikes), EURASIAN WATER-MILFOIL. Often branching prolifically near the water surface; not developing turions; stamens 8. Lakes, ponds; Grayson

Ribes aureum var. villosum [GLE]

Myriophyllum aquaticum [MAS]

Myriophyllum heterophyllum [REE]

Myriophyllum pinnatum [GWO]

Myriophyllum spicatum [REE]

Myriophyllum sibiricum [REE]

Myriophyllum verticillatum [MAS]

(Lake Ray Roberts) and Lamar (Pat Mayse Lake) cos.; distribution in other parts of TX not known; probably fragments spread by powerboats. Native of Eurasia. Flowering material from nc TX not seen. This species can be a problematic invader of aquatic habitats in some areas (Jarman 1968); Reed (1977) discussed its spread in North America. In Texas, EURASIAN WATER-MILFOIL is considered a "harmful or potentially harmful exotic plant" and it is illegal to release, import, sell, purchase, propagate, or possess this species in the state (Harvey 1998). Jones et al. (1997) treated TX material of this species as *M. spicatum* var. *exalbescens* which is apparently a synonym of *M. sibiricum* (Kartesz 1994).

Myriophyllum verticillatum L., (whorled). Developing turions; submersed leaves to 30(–45) mm long, the divisions to 28 mm long. Lakes; included based on citation of vegetational area 4 (Fig. 2) by Hatch et al. (1990); we have seen no material for nc TX; ne TX. Apr–Jun.

PROSERPINACA MERMAID-WEED

A genus of 2–3 species of e North America to the West Indies; unusual for a dicot in having 3-merous flowers. (Possibly from Latin: *prospero*, to creep, referring to the creeping stems of most species)

Proserpinaca palustris L. var. **amblyogona** Fernald, (sp.: of marshes; var.: with rounded angles), MARSH MERMAID-WEED. Stems to 1 m or more long, the lower stems usually creeping and rooting at the nodes, the upper stems ascending or suberect; leaves alternate; submerged leaves finely pinnatifid with a total of 8–14 divisions, grading into merely sharply toothed upper ones; upper leaves to 7 cm long and 10 mm wide; flowers usually solitary or in clusters of 2–5 in axils of upper serrate leaves, perfect, the parts in threes; petals absent; fruits small (ca. 2–4 mm long), 3-angled. Shallow water, shorelines, other wet areas; Henderson Co.; se and e TX w to extreme e margin of nc TX. Late May–Jul.

Proserpinaca pectinata Lam., (comb-like), which occurs in e TX to the e of nc TX, has the upper leaves deeply pinnately divided into narrow segments like the lower leaves.

HAMAMELIDACEAE WITCH-HAZEL FAMILY

A small family of ca. 100 species in ca. 31 genera (Meyer 1997); mainly temperate and subtropical, especially e Asian trees and shrubs including *Hamamelis*, the source of medicinal witch-hazel. Many of the genera (12–14) are monotypic and the family may thus represent relics of an ancient group. The subfamily Altingioideae, which includes *Liquidambar*, is sometimes recognized as a discrete family, the Altingiaceae. Species are variously used for timber, as a source of medicinal extracts or scents, or cultivated as ornamentals. Family name from *Hamamelis*, WITCH-HAZEL, a genus of 5–6 species of e North America and e Asia. (Greek name for an unidentified pear-shaped fruit) (subclass Hamamelidae)

FAMILY RECOGNITION IN THE FIELD: the only nc TX species is a tree with alternate, simple, deeply palmately 5–7 lobed (*star-like* in shape), toothed, fragrant leaves with stipules or at least stipule scars; fruits woody 2-valved capsules (in ours grouped together in stalked *ball-like clusters*).

REFERENCES: Ernst 1963b; Goldblatt & Endress 1977; Endress 1989, 1993; Meyer 1997.

LIQUIDAMBAR SWEETGUM

A genus of 3–4 species (Meyer 1997) of monoecious, deciduous trees native to se North America, Central America, the e Mediterranean, and e Asia; the species are valuable for timber, for the aromatic balsam (storax) used medicinally and in scents, and as cultivated ornamentals with spectacular fall colors. The Asian *L. orientalis* Mill. is the source of Levant storax, the balm (of Gilead) of the Bible. (According to Fernald (1950a), a mongrel name, from Latin:

liquidus, fluid, and Arabic: *ambar*, amber, in allusion to the fragrant terebinthine juice or gum which exudes from the tree)
REFERENCES: Duncan 1959; Bogle 1986; Hoey & Parks 1994.

Liquidambar styraciflua L., (flowing with storax or gum), SWEETGUM, REDGUM. Monoecious tree to 40 m or more tall, often exuding a gum (chewed by some); small branches often with corky ridges; leaves alternate, simple, deciduous; leaf blades to ca. 18 cm long and 12 cm wide, deeply palmately 5–7 lobed, star-like in shape, fragrant if crushed, quite showy in fall, yellow to orange, crimson, or purplish, the lobes acuminate, usually serrate; petioles to 12 cm long; flowers unisexual, without petals; staminate flowers in bracteate globose masses on racemose inflorescences to 6 cm long; pistillate flowers in peduncled globose heads that develop into woody globose clusters (to 3 cm in diam.) of 2-beaked capsules. Wet areas, low or rich woods; native w in Red River drainage to Lamar Co., widely cultivated further w, escaped in vacant lot in Fort Worth (Tarrant Co.); mainly e TX. Mar–May. SWEETGUM is particularly noted for its colorful fall foliage; it is also the source of a fragrant gum (copalm balm, American copalm, storax) obtained from the inner bark after wounding or gashing; it has a long history of usage including being smoked with tobacco and burned as an incense by the Aztecs; it has also been used for a chewing gum, as an antiseptic, and medicinally in the treatment of skin diseases and dysentery; it was much used by Confederate doctors (Standley 1922a; Peattie 1948; Mabberley 1987). SWEETGUM is a leading commercial hardwood species in e TX (Cox & Leslie 1991). Populations of this species (sometimes treated as a different variety or species) occur in the mountains of Mexico and Central America; their occurrence is presumably a relict of a once more widespread distribution. It is odd to observe such a familiar tree in the floristically very different montane forests of Central America.

HIPPOCASTANACEAE BUCKEYE FAMILY

A very small (15 species in 2 genera) family of mainly n temperate trees and shrubs with some to n South America and se Asia; the family is best known for the ornamental, *Aesculus hippocastanum* L. (HORSE-CHESTNUT). The evergreen genus *Billia* has 2 species native from s Mexico to tropical South America. The Hippocastanaceae is closely related to the Sapindaceae and appears to represent a clade within that family. From a cladistic standpoint they should be lumped to form a more inclusive monophyletic Sapindaceae (Judd et al. 1994). Family name conserved from *Hippocastanum*, a genus now treated as *Aesculus* (the name *Aesculus* was published earlier and thus has priority in terms of nomenclature). (Greek: *hippos*, horse, and *kastana*, chestnut; also the Latin name for horse-chestnut) (subclass Rosidae)
FAMILY RECOGNITION IN THE FIELD: shrubs or small trees with opposite, *palmately compound* leaves; flowers bilaterally symmetrical, in *showy panicles*; seeds large, in a leathery capsule.
REFERENCES: Hardin 1957a, 1957b; Brizicky 1963b; Judd et al. 1994.

AESCULUS BUCKEYE, HORSE-CHESTNUT

Ours shrubs or small trees; leaves palmately compound, deciduous; leaflets toothed; inflorescence a large, showy, terminal panicle with perfect and unisexual, bilaterally symmetrical flowers; sepals 5; petals 4–5, with a claw and blade; stamens usually 7; ovary superior; fruit a 3-celled, (1–)2–3-seeded leathery capsule; seeds large.

A genus of 13 species of North America, India, e Asia, and se Europe; some are cultivated as ornamentals, while others are used as fish poisons. ☠ The raw seeds of *Aesculus* species are poisonous to humans, domestic animals, and livestock; the early leaves are also apparently poisonous; a strychnine-like effect is produced with the toxic principle apparently a glycoside, aesculin; in severe cases coma and death from respiratory paralysis can result; BUCKEYE honey

may also cause poisoning (Kingsbury 1964; Burlage 1968; Schmutz & Hamilton 1979; Cheatham & Johnston 1995; Turner & Szczawinksi 1991). The toxic seeds and roots were used by Native Americans to stupefy fish (Tyrl et al. 1994). *Aesculus hippocastanum* L. (HORSE-CHESTNUT), native from the Balkans to the Himalayas, is a widely used ornamental tree. Hybridization and introgression are known to occur in *Aesculus* in the e U.S. (e.g., DePamphilis & Wyatt 1989). (Ancient Latin name of some OAK or mast-bearing tree)
REFERENCES: Hardin 1957c, 1957d; Wyatt & Lodwick 1981; McGregor 1984a; DePamphilis & Wyatt 1989.

1. Leaflets 7–11; flowers creamy yellow to greenish yellow; calyces about 6 mm long; petals subequal; stamens exserted to almost twice the corolla length; fruits spiny or rarely smooth ____ **A. arguta**
1. Leaflets 5(–7); flowers red; calyces usually 8–16 mm long; petals unequal, the upper pair longer and narrower with minute spatulate blade and the claw often equaling the lateral petals in length, the lateral petals with a wide and nearly rounded blade; stamens included or only slightly exserted beyond the upermost petals; fruits smooth ____ **A. pavia**

Aesculus arguta Buckley, (sharp-toothed), WHITE BUCKEYE, TEXAS BUCKEYE, WESTERN BUCKEYE. Shrub or shrubby small tree to ca. 7(–12) m tall; leaflets to 12 cm long and 4 cm wide; corollas to 15 mm long. Limestone slopes and sandy open woods; Bell, Comanche, Cooke, Grayson, McLennan, and Tarrant cos., also Hamilton Co. (HPC); e TX w to Rolling Plains and Edwards Plateau. Apr. [*A. glabra* Willd. var. *arguta* (Buckley) B.L. Rob.] Poisonous. ☠

Aesculus pavia L. var. **pavia**, (for Peter Paaw, 1564–1617, Dutch botanist), RED BUCKEYE, SCARLET BUCKEYE, FIRECRACKER-PLANT. Large shrub or small tree to 10 m tall; leaflets to 17 cm long and 7 cm wide; calyces rarely < 8 mm long; corollas to 30 mm long. Forests, thickets, along streams; Callahan, Dallas (apparently cultivated), Hopkins, and Williamson cos. (Wyatt & Lodwick 1981); mainly se and e TX, rarely further w. Mar–May. Pollinated by ruby-throated hummingbirds (*Archilochus colubris*) (James 1948; Wyatt & Lodwick 1981). Poisonous. ☠ /77

Aesculus pavia L. var. *flavescens* (Sarg.) Correll, (yellowish), PALE BUCKEYE, TEXAS YELLOW BUCKEYE, PLATEAU YELLOW BUCKEYE), a yellow-flowered variety, is endemic to the Edwards Plateau. According to Wyatt and Lodwick (1981), there is limited introgression between the varieties along the Balcones Escarpment; however, a number of morphological characters separate the two and Wyatt and Lodwick (1981) concluded that they are distinctive taxa appropriately recognized as varieties. Poisonous. ☠

HYDRANGEACEAE HYDRANGEA FAMILY

The Hydrangeaceae is a small (190 species in 17 genera), mainly n temperate family extending to Malesia. Species range from trees to shrubs, lianas, or herbs including the ornamentally important *Hydrangea* and *Philadelphus*. The family has often been included in the Saxifragaceae. *Philadelphus* and related genera are sometimes segregated as Philadelphaceae (Quibell 1993). Recent molecular analyses (Xiang et al. 1993, 1998) suggest that *Cornus*, Nyssaceae, Hydrangaceae, and Loasaceae, as well as several other groups, form a "cornaceous clade." ☠ Some members of the family (e.g., *Hydrangea* species) contain the cyanogenic glycoside, hydrangin (Spoerke & Smolinske 1990). Family name from *Hydrangea*, a genus of 23 species of shrubs native from the Himalaya to Japan, the Philippines, and America. (Greek: *hydor*, water, and *aggos*, jar, or *aggeion*, vessel, in reference to the cup-shaped fruits) (subclass Rosidae)
FAMILY RECOGNITION IN THE FIELD: the only nc TX species is a shrub with opposite, simple, unlobed, exstipulate leaves, flowers with *4 showy white petals*, and *numerous stamens*; fruit a capsule.
REFERENCES: Small & Rydberg 1905b; Spongberg 1972.

PHILADELPHUS MOCK ORANGE

A n temperate genus of 65 species of shrubs with white, usually strongly scented flowers. A number of *Philadelphus* species and their hybrids are cultivated as ornamentals. (Named either for Ptolomy Philadelphus, 283–247 B.C., King of Egypt, or from Greek name meaning brotherly love from *phileo*, love, and *adelphus*, brother)
REFERENCE: Hu 1954–1956.

Philadelphus pubescens Loisel., (pubescent, downy). Shrub to ca. 3 m tall; leaves simple, opposite, deciduous, the blades ovate to ovate-elliptic, 5–8 cm long, 2.5–5 cm wide, apically acuminate, dentate to nearly entire; inflorescences racemose, of 5–7 flowers; flowers ca. 2.5 cm wide, showy; petals 4, white; stamens numerous; ovary partly inferior; fruit a capsule. Wooded river bluffs; included based on citation of vegetational area 4 (Fig. 2) by Hatch et al. (1990); also Red River Co. (Correll & Johnston 1970); in TX apparently limited to ne part of the state. Apr–Jun.

HYDROPHYLLACEAE WATERLEAF FAMILY

Low annuals, biennials, or perennials; leaves alternate, simple or pinnately compound, entire, toothed, or lobed; flowers radially symmetrical, perfect, axillary or terminal, solitary or in spike-like, uncurling, one-sided cymes; sepals 5, linear or oblong-lanceolate, barely united at base; corollas rotate, funnelform, or almost salverform, 5-lobed or -angled, with or without outgrowths variously called appendages, scales, or glands near the attachment of the filaments; stamens 5, attached to base of corolla tube, the filament bases rarely expanded on both sides into wings or appendages; pistil 2-carpellate; ovary superior; styles or stigmas 2; capsules usually many-seeded.

A small (270 species in 18 genera), subcosmopolitan (especially dry w North America) family of herbs and small shrubs including a number of ornamentals. Many either have glandular hairs and are odorous or are rough hairy. Family name from *Hydrophyllum*, WATERLEAF, a North American genus of 8 species. (Greek: *hydor*, water, and *phyllon*, leaf, in reference to the high water content of the young leaves) (subclass Asteridae)
FAMILY RECOGNITION IN THE FIELD: usually bristly herbs typically with *scorpioid cymes* (= un curling, 1-sided inflorescences); corollas sympetalous; leaves alternate, frequently toothed, lobed, or compound; fruit a many-seeded capsule (the similar Boraginaceae has a fruit of 4 or fewer nutlets).
REFERENCES: Constance 1939; Wilson 1960b.

1. Plant growing at edge of ponds or streams or in other aquatic habitats; stems with conspicuous spines painful to the touch ____ **Hydrolea**
1. Plant of terrestrial habitats; stems not spiny.
    2. Leaves toothed, lobed, or compound; placentation parietal.
        3. Flowers mostly solitary and axillary or sometimes a few in a terminal cyme; pedicels often much longer than the calyces; upper leaves distinctly petioled.
            4. Corollas 8–12 mm long, ca. 1.5–2 times as long as the calyces; calyx lobes alternating with auriculate appendages 1–2 mm long; seeds with an elaiosome (= appendage) ____ **Nemophila**
            4. Corollas 4 mm or less long, shorter than or equal to the calyces; calyx lobes not alternating with auriculate appendages; seeds without an elaiosome ____ **Ellisia**
        3. Flowers many, in terminal, 1-sided, uncoiling (= scorpioid) cymes; pedicels shorter to longer than the calyces; upper leaves sessile (petioled in 1 species) ____ **Phacelia**
    2. Leaves entire, simple; placentation appearing axile ____ **Nama**

ELLISIA AUNT LUCY

A monotypic North American genus. (Named for John Ellis, 1711–1776, Irish-born merchant, naturalist, and correspondent of Linnaeus)
REFERENCE: Constance 1940.

Ellisia nyctelea (L.) L., (nocturnal), AUNT LUCY, WATERPOD. Annual 5–40 cm tall, ± hispid-pilose; leaf blades 3–8 cm long, 1–2(–3) cm wide, deeply pinnatifid, the divisions mostly coarsely toothed; flowers solitary, axillary; pedicels 4–15(–28) mm long; calyces 3–4 mm long at anthesis, becoming accrescent; corollas white to bluish- or lavender-tinged, rather small and inconspicuous, with minute corolla appendages alternating with filament bases; stamens included; capsules globose, 5–6 mm in diam. Stream bottom thickets; in TX known only from Denton Co. Apr. Reportedly introduced to Texas (Correll & Johnston 1970) from n U.S.

HYDROLEA

A mainly tropical genus of 11 species of semi-aquatics (Davenport 1988). (Derivation of generic name uncertain, presumably from Greek: *hydor*, water, in reference to aquatic habitat)
REFERENCE: Davenport 1988.

Hydrolea ovata Nutt. ex Choisy, (ovate, egg-shaped), HAIRY HYDROLEA. Erect, spiny, rhizomatous, pubescent perennial to 1 m tall; spines 1–2 per node, 5–12 mm long; leaves entire, oblong to ovate, 1.5–7 cm long, 10–25 mm wide; calyces shorter than corollas; corollas rotate-campanulate, blue (rarely white), showy, 11–17 mm long, the lobes 5–9 mm wide; corolla appendages absent; capsules globose, 4.5–5.5 mm long. In water or wet areas; Cooke, Fannin, Grayson, and Lamar cos. in Red River drainage, also Henderson, Hunt, Limestone, and Milam cos. on e edge of nc TX, also Dallas (Davenport 1988) and Kaufman (R. O'Kennon, pers. obs.) cos.; se and e TX w to nc TX. Sep–Oct. [*Nama ovata* (Nutt. ex Choisy) Britton] ▣/93

NAMA

Ours small, pubescent or hispid annuals; leaves simple, entire; flowers solitary, axillary or few together in small terminal cymes; corolla appendages absent; stamens included, filament bases without or rarely with minute appendages; placentation appearing axile but actually extruded parietal (J. Bacon, pers. comm.).

A genus of 45 species of sw U.S. and tropical America with 1 species in Hawaii. (Greek: *nama*, a stream or spring, referring to habitat of the first described species)
REFERENCES: Hitchcock 1933a, 1933b; Bacon 1974, 1984.

1. Leaves linear or linear-oblong to oblanceolate, 1–8 mm wide, not decurrent; plant erect or spreading; corollas pink to purple, 8–15 mm long ____ **N. hispidum**
1. Leaves oblanceolate or obovate, 5–35 mm wide, conspicuously decurrent at base forming wings on stem; plant largely prostrate; corollas white, 6–7 mm long ____ **N. jamaicense**

Nama hispidum A. Gray, (hispid, bristly), SANDBELL, ROUGH NAMA. Plant hispid pubescent; stems 10–50 cm tall; corollas funnelform-campanulate. Sandy open woods or open ground; in nc TX from Bell and Dallas cos. to the w; widespread in TX. Late Apr–Jun.

Nama jamaicense L., (of Jamaica), FIDDLELEAF, FIDDLELEAF NAMA. Plant with pubescence softer than in *H. hispidum*; stems 10–50 cm long; corollas nearly tubular. Sandy or silty ground, roadsides or disturbed places; Bell and Williamson cos., also Johnson Co. (Mahler 1988); se and e TX w to s part of nc TX and Edwards Plateau. Apr–May, earlier farther s.

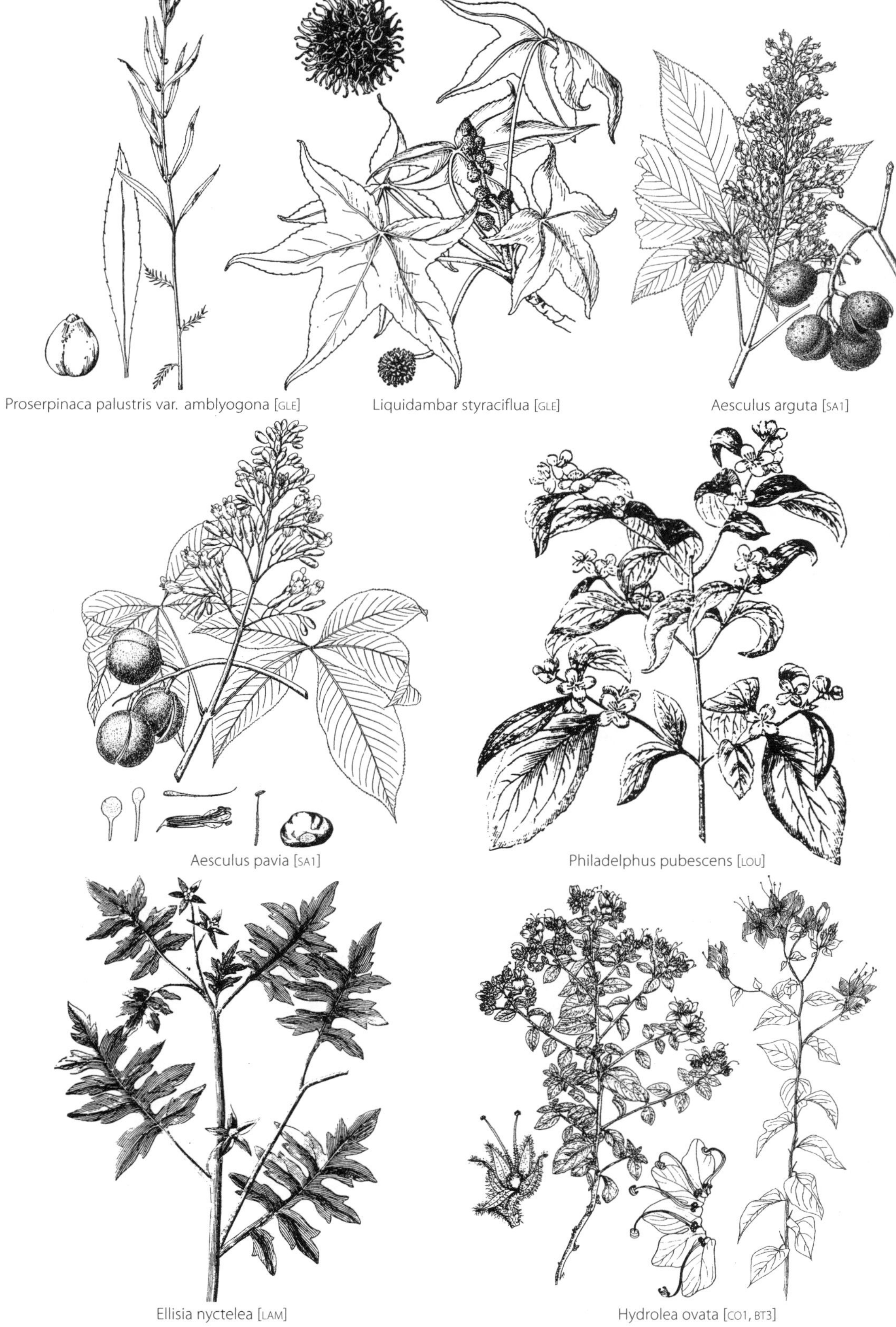

Proserpinaca palustris var. amblyogona [GLE]

Liquidambar styraciflua [GLE]

Aesculus arguta [SA1]

Aesculus pavia [SA1]

Philadelphus pubescens [LOU]

Ellisia nyctelea [LAM]

Hydrolea ovata [CO1, BT3]

NEMOPHILA BABY BLUE-EYES

A genus of 11 species of w and se North America including cultivated ornamental annuals. (Greek: *nemos*, glade, and *phileo*, to love)
REFERENCE: Constance 1941.

Nemophila phacelioides Nutt., (resembling *Phacelia*, another genus of Hydrophyllaceae), BABY BLUE-EYES, LARGE-FLOWER NEMOPHILA, FLANNEL-BREECHES. Annual 20–60 cm tall, hispid-pilose, at least on younger parts; leaf blades 6–8 cm long, 2–5 cm wide, deeply pinnatifid or compound, usually with 9–11 mostly entire divisions; flowers solitary, axillary or few in small terminal cymes; pedicels very long, usually 15–60 mm or more; corollas rotate, blue-lavender with pale center, rather large and showy, 8–12 mm long, with broad fimbriate corolla appendages alternating with filament bases; stamens included; capsules globose, 5–9 mm in diam., equaled or exceeded by the accrescent calyx. Stream bottom woods, sandy or silty ground; Bell, Dallas, Grayson, McLennan, and Williamson cos., also Ellis, Lampasas (Constance 1941), Hamilton (HPC), and Johnson (R. O'Kennon, pers. obs.) cos.; se and e TX w to nc TX, also Edwards Plateau. Apr–May. /100

PHACELIA

Glabrous to hispid-pilose annuals or biennials, ours without stinging hairs; leaves toothed to once or twice pinnately compound; corollas rotate to campanulate, pale to deep lavender or violet-blue, often white centered, rarely white, with scale-like corolla appendages or glands alternating with filament bases; stamens included to exerted.

A genus of 150 species mostly of w North America, but also e U.S. and South America. Typically bee-pollinated with anthers that turn inside out at maturity; many are cultivated as ornamentals. The bristly stinging hairs of some species to the w of TX (e.g., nine Arizona species) can cause dermatitis in sensitive individuals; the substances causing the reaction are low molecular weight compounds (haptens) that upon penetrating the skin, combine with skin proteins and form antigens (Fuller & McClintock 1986; Lampe 1986; Wilken et al. 1993). According to Lampe (1986), only four families (Euphorbiaceae, Hydrophyllaceae, Loasaceae, and Urticaceae) have stinging hairs—nc TX has stinging representatives of all of these except the Hydrophyllaceae. The scale-like corolla appendages alternating with the filament bases apparently act as little flaps which shield some nectar from easy access; in the process of obtaining this hard-to-reach nectar, bees jar the anthers which shower them with pollen by explosively turning inside-out (Wills & Irwin 1961). (Greek: *phacelos*, a fascicle, alluding to the flowers clustered in raceme-like scorpoid cymes)
REFERENCES: Voss 1937; Constance 1949, 1950; Atwood 1975; Wilken 1986b; Wilken et al. 1993; Moyer & Turner 1994.

1. Upper leaves petioled; leaves compound with toothed leaflets ____ **P. congesta**
1. Upper leaves sessile; leaves simple or the lower compound.
   2. Plants viscid-glandular (= sticky due to gland-tipped hairs); main stem stout, at base usually > 3 mm thick; stamens exserted; filaments each with a pair of scaly basal appendages, but glands absent between filament bases; seeds 4 per capsule ____ **P. integrifolia**
   2. Plants not viscid-glandular; main stem delicate, at base usually 3 mm or less thick; stamens included; filament bases alternating with glands bordered by minute flaps; seeds 4–20 per capsule.
      3. Pedicels and stems pubescent.
         4. Lowest stem leaves sessile or short-petioled, the petioles broadly winged; basal leaves usually shallowly toothed or shallowly lobed; fruiting pedicels erect, often shorter than calyces ____ **P. strictiflora**

4. Lowest stem leaves often long-petioled, the petioles narrowly margined; basal leaves deeply pinnately lobed or pinnate, usually appearing compound or nearly so; fruiting pedicels spreading-ascending to reflexed, often longer than calyces.
   5. Stem leaves toothed or shallowly lobed (usually not more than 1/2 way to midvein); ovules 6–12 per placenta; seeds usually 10–15 per capsule ____ **P. patuliflora**
   5. Stem leaves deeply lobed to pinnatifid (often more than 1/2 to midvein); ovules typically 4 per placenta; seeds 6–8 per capsule ____ **P. hirsuta**

3. Pedicels and stems glabrous ____ **P. glabra**

Phacelia congesta Hook., (arranged very closely together), BLUE-CURLS, SPIKE PHACELIA. Variously pubescent annual or biennial, usually somewhat viscid-glandular; corollas 4–7 mm long; stamens exserted, the filaments with a pair of scaly basal appendages; capsules with 4 seeds. Gravelly or sandy soils; Cooke, Montague, Somervell, and Tarrant cos., also Dallas Co. (Mahler 1988); nc TX s and w to w TX. Apr–Jun, sporadically later.

Phacelia glabra Nutt., (smooth, hairless), SMOOTH PHACELIA. Annual 5–40 cm tall; leaves deeply lobed or pinnatifid; stamens included, the filament bases alternating with glands bordered by minute flaps; seeds 4–8 per capsule. Prairies, forest margins, sandy loam; Kaufman and Hopkins cos.; e TX w to ne part of nc TX. Apr.

Phacelia hirsuta Nutt., (hairy), HAIRY PHACELIA. Densely pubescent annual 10–50 cm tall; corollas 5–8 mm long; stamens included, the filament bases alternating with glands bordered by minute flaps; capsules with 6–8 seeds. Sandy areas, forest margins and openings; Collin (railroad fill), Hunt, Lamar, and Limestone cos.; se and e TX w to nc TX and Edwards Plateau. Apr.

Phacelia integrifolia Torr., (entire-leaved), GYP PHACELIA, CRENATE-LEAF PHACELIA. Viscid annual or biennial; leaves 2–7 cm long, crenate to shallowly pinnatifid, the cauline ones usually sessile; corollas 4–7 mm long; stamens exserted, the filaments with a pair of scaly basal appendages; capsules with 4 seeds. Sandy or rocky areas especially on gypsum or limestone; included based on citation of vegetational area 5 (Fig. 2) by Hatch et al. (1990); mainly w and nw TX. Mar–May. Reported to cause skin inflammation (Burlage 1968), but apparently not due to stinging hairs. ☠

Phacelia patuliflora (Engelm. & A. Gray) A. Gray, (spreading flower). Densely to finely pubescent annual 8–60 cm tall; stamens included, the filament bases alternating with glands bordered by minute flaps; capsules usually with 10–15 seeds. Sandy areas, woods, and terraces. Moyer and Turner (1994) recognized the following 2 varieties as well as a variety (var. *austrotexana*) from s TX.

1. Peduncles with small, well-defined glandular hairs ca. 0.1 mm long in addition to non-glandular hairs; branches decumbent; fruiting pedicels spreading to reflexed; calyx lobes obtuse ____ var. **patuliflora**
1. Peduncles without glandular hairs; branches ascending; fruiting pedicels spreading-ascending in age; calyx lobes acute ____ var. **teucriifolia**

var. **patuliflora**, SAND PHACELIA. Burnet Co., also Falls Co. (Constance 1949); s part of nc TX s to s TX. Apr–May.

var. **teucriifolia** (I.M. Johnst.) Constance, (with leaves like *Teucrium*—germander). Rich open woods; found by A. Ruth in Tarrant Co., also Coleman Co. (Constance 1949); otherwise s TX, particularly w and n Edwards Plateau. Apr–May.

Phacelia strictiflora (Engelm. & A. Gray) A. Gray, (erect- or upright-flowered). Densely pubescent annual 5–30 cm tall, with a basal rosette; leaves linear-oblong to orbicular, with 1–6 pair of teeth or lobes; stem leaves sessile; corollas 7–10 mm long; stamens included, the filament bases alternating with glands bordered by minute flaps; capsules with 10–20 seeds. Sandy or gravelly areas. Mar–May. The four varieties discussed below are sympatric and intergrade (Constance

1949; Wilken 1986b). Constance (1949) suggested the variability within this species may be due to introgressive hybridization with *P. hirsuta*.

1. Stems with widely spreading or spreading-ascending hairs ____ var. **strictiflora**
1. Stems with appressed or closely ascending or incurved hairs.
   2. Stem leaves deeply pinnatifid; lower leaves pubescent on both sides; basal rosette usually withering early.
      3. Stems relatively thicker, 1–3 mm thick on well-developed plants; leaves linear-oblong, crowded during flowering (overlapping 1/4–1/2); leaf lobes acute; widespread in nc TX ____ var. **connexa**
      3. Stems slender, mostly 0.2 –1 mm thick; leaves oblong-ovate, not very crowded; leaf lobes usually obtuse; in nc TX known only from Grayson Co. ____ var. **robbinsii**
   2. Stem leaves toothed or shallowly pinnatifid; lower leaves glabrous or nearly so on lower surfaces; basal rosette usually present ____ var. **lundelliana**

var. **connexa** Constance, (connected, closely related). Dallas, Grayson, Limestone, Tarrant, and Wise cos.; se and e TX w to West Cross Timbers.

var. **lundelliana** Constance, (for Cyrus Longworth Lundell, 1907–1994, botanist and founder of the Texas Research Foundation, Renner, the institution that published the *Flora of Texas* and the *Manual of the Vascular Plants of Texas*). Clay, Hood, Montague, Parker, and Somervell cos., also Navarro Co. (Constance 1949); Blackland Prairie w to Rolling Plains, also Edwards Plateau.

var. **robbinsii** Constance, (named for the G.T. Robbins, collector of the type specimen). Grayson Co. (Constance 1949) and OK.

var. **strictiflora**. Collin and Dallas cos., also Milam Co. (Constance 1949); se and e TX w to nc TX; endemic to TX.

JUGLANDACEAE WALNUT FAMILY

Trees with alternate, pinnately compound leaves; leaflets often asymmetrical, often gland-dotted beneath and aromatic; stipules absent; flowers appearing with the leaves, unisexual, the sexes on the same plant (monoecious); staminate flowers in drooping catkins, with 3–many stamens; pistillate flowers solitary or in short spikes, with separate pistils; ovaries inferior; fruits usually hard, nut-like, covered by a fibrous-fleshy to firm, dry husk (bract and bractlets of involucre); seed 1 per fruit, 2–4 lobed. The species exhibit mast fruiting, a type of cyclical fruiting with years of heavy fruit production occurring irregularly. This is possibly an adaptation to reduce seed predation by making the seeds an unpredictable resource. The seeds are nevertheless an important wildlife food.

A small (59 species in 8 genera) mainly temperate and warm n hemisphere family extending to South America and Malesia. Juglandaceae are wind-pollinated, resinous and aromatic, deciduous trees valued for their fine timber, as ornamentals, and for their nuts. Species producing edible nuts include PECANS, BLACK WALNUTS, and ENGLISH or PERSIAN WALNUTS (*Juglans regia* L.). (subclass Hamamelidae)

FAMILY RECOGNITION IN THE FIELD: trees with *alternate, pinnately compound* leaves, often aro matic; branches stout, round, with conspicuous leaf-scars and several buds per node; male flowers in *catkins*; fruits nut-like.

REFERENCES: Elias 1972; Stone 1993, 1997a.

1. Leaflets 2–8 pairs per leaf, with a single terminal leaflet; terminal 3 leaflets equal to or larger than the lateral leaflets; pith of branchlets continuous, not chambered; husk around nut at least partly splitting into sections; surface of nut smooth or reticulate, not furrowed; staminate catkins clus-

Nama hispidum [PHY]

Nama jamaicense [BR2]

Nemophila phacelioides [CUR]

Phacelia glabra [HEA]

Phacelia hirsuta [AAA]

Phacelia congesta [GBN]

Phacelia patuliflora var. patuliflora [HEA]

tered, mostly attached to short, new-growth peduncles; staminate flowers lacking true perianth but with 1–3 perianth like bractlets at one side; stamens 3–8 **Carya**

1. Leaflets 5–11 pairs per leaf, with or without a single terminal leaflet; terminal 2–3 leaflets usually smaller than the middle lateral leaflets; pith of branchlets chambered (= separating into thin plates); husk around nut not splitting into sections; surface of nut with irregular furrows; staminate catkins solitary, attached directly to twigs of the previous year; staminate perianth 4- or 5-parted; stamens 8–40 **Juglans**

CARYA HICKORY, PECAN

Husk of nut splitting vertically into sections, firm, dry, usually 4-valved.

A genus of 14 species of e North America (s into Mexico), with a few in e Asia.

This disjunct e Asia-e North America distribution pattern is an interesting one to plant geographers. In the geologic past, dispersal between the Eurasian and North American continents was possible, and the combined area is considered a single "Holarctic" biogeographic region. The fossil record shows that many plants had distributions across the Northern Hemisphere—temperate forests, for example, occurred very broadly and reached their maximum extension in the mid-Tertiary (the Tertiary extended from 65 million years ago to 5 mya). This widespread flora has been referred to as the Arcto-Tertiary flora or the Tertiaro-mesophytic flora. Geohistorical events from the mid-Tertiary to the present have included alterations in the shapes of the northern land masses, fluctuations in sea levels, mountain building, and profound changes in the climate. As a result, there have been great changes in both the composition and the disposition of the flora and the ranges of many plant have been greatly restricted (e.g., eliminated from Europe and w North America). A significant number now survive in only two areas, e North America and e Asia. The genus *Carya* is such an example with other nc TX examples including *Campsis*, *Menispermum*, *Penthorum*, *Phryma*, *Sassafras*, *Saururus*, *Triadenum*, *Triosteum*, and *Veronicastrum* (Li 1952b; Little 1970; Graham 1972; Boufford & Spongberg 1983; Hamilton 1983; Hsü 1983; Wu 1983; Ying 1983; Cox & Moore 1993; Graham 1993a).

Carya species are important for timber and nuts (pecans, hickory nuts); the nuts were widely used as food by Native Americans (Heiser 1993). The wood of HICKORIES is hard, heavy, and very tough; it is excellent for tool handles because of the combined strength and shock resistance (Stone 1997b). (Greek: *carya* or *káryon*, nut, kernel; ancient name of the WALNUT)

REFERENCES: Grauke et al. 1986; Brummitt 1988; Manaster 1994; Stone 1997b.

1. Bud scales valvate (edges touching but not overlaping) in pairs, usually 4–6; seams of the fruit husk with prominent narrow wings or keels; fruit husk thin, ca. 1 mm or less thick; leaflets usually without tufts of hairs beneath (except these often present in *C. cordiformis*).
   2. Leaflets (7–)9–17; lateral leaflets falcate (= sickle-shaped), the base usually asymmetrical, some times conspicuously so; terminal leaflet lanceolate to oblong-lancelate to ovate-lanceolate, usually broadest at middle or below, with stalk 8–25 mm long to nearly sessile; buds usually brown or reddish brown (rarely yellowish brown).
      3. Teeth of leaflets usually prominent; bark ridged or with appressed scales or exfoliating with small plate-like scales; fruits (nut with husk) oblong-cylindric, usually not flattened laterally, ca. 25–80 mm long; lateral petiolules 0–7 mm long; midribs adaxially mostly glabrous, rarely hirsute near base; nut longer than wide, smooth; fascicles of staminate catkins sessile or nearly so; widespread in nc TX **C. illinoinensis**
      3. Teeth of leaflets usually inconspicuous; bark exfoliating in long strips or plate-like scales; fruits subglobose to obovoid, somewhat compressed or flattened laterally, to 40 mm long; lateral petiolules 0–2 mm long; midribs adaxially villous near base; nut as wide as long, furrowed or wrinkled; fascicles of staminate catkins with definite stalks; only on extreme e edge of nc TX **C. aquatica**
   2. Leaflets (5–)7–9; lateral leaflets straight to slightly falcate, the base essentially symmetrical or

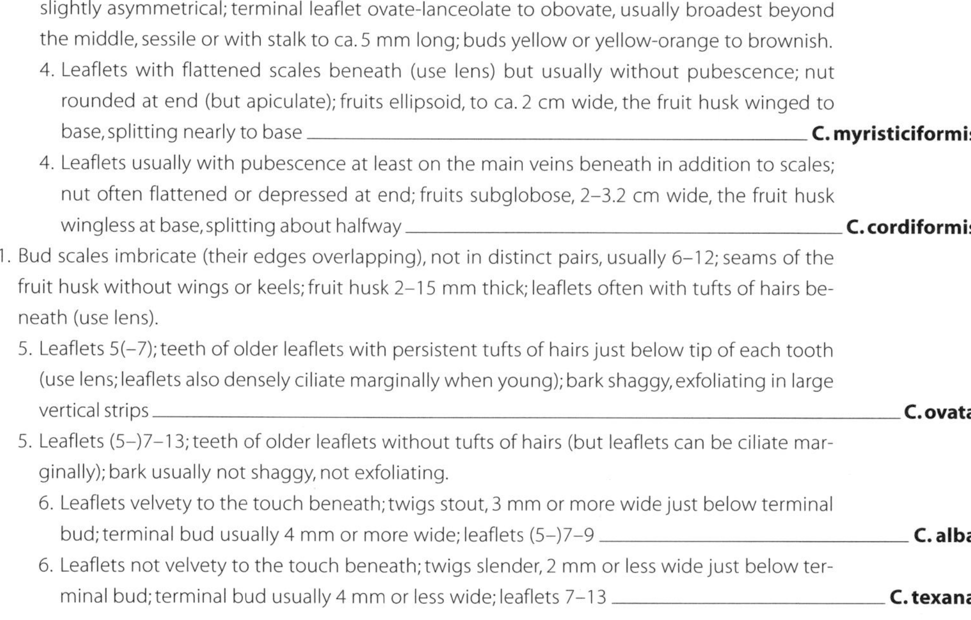

slightly asymmetrical; terminal leaflet ovate-lanceolate to obovate, usually broadest beyond the middle, sessile or with stalk to ca. 5 mm long; buds yellow or yellow-orange to brownish.

4. Leaflets with flattened scales beneath (use lens) but usually without pubescence; nut rounded at end (but apiculate); fruits ellipsoid, to ca. 2 cm wide, the fruit husk winged to base, splitting nearly to base ____ **C. myristiciformis**
4. Leaflets usually with pubescence at least on the main veins beneath in addition to scales; nut often flattened or depressed at end; fruits subglobose, 2–3.2 cm wide, the fruit husk wingless at base, splitting about halfway ____ **C. cordiformis**

1. Bud scales imbricate (their edges overlapping), not in distinct pairs, usually 6–12; seams of the fruit husk without wings or keels; fruit husk 2–15 mm thick; leaflets often with tufts of hairs beneath (use lens).
   5. Leaflets 5(–7); teeth of older leaflets with persistent tufts of hairs just below tip of each tooth (use lens; leaflets also densely ciliate marginally when young); bark shaggy, exfoliating in large vertical strips ____ **C. ovata**
   5. Leaflets (5–)7–13; teeth of older leaflets without tufts of hairs (but leaflets can be ciliate marginally); bark usually not shaggy, not exfoliating.
      6. Leaflets velvety to the touch beneath; twigs stout, 3 mm or more wide just below terminal bud; terminal bud usually 4 mm or more wide; leaflets (5–)7–9 ____ **C. alba**
      6. Leaflets not velvety to the touch beneath; twigs slender, 2 mm or less wide just below terminal bud; terminal bud usually 4 mm or less wide; leaflets 7–13 ____ **C. texana**

Carya alba (L.) Nutt. ex Elliott, (white), MOCKERNUT HICKORY, WHITE-HEART HICKORY, WHITE HICKORY, HARD-BARK HICKORY. Fruits 30–50 mm long, with husk ca. 3–15 mm thick; shell of nut usually very thick; kernel sweet. Dry to moist woods; Lamar Co., also Delta and Hopkins cos. (Little 1971); e TX w to e edge of nc TX. Apr. [*C. tomentosa* (Lam. ex Poir.) Nutt.] We are following Kartesz (1994), Jones et al. (1997), and J. Kartesz (pers. comm. 1997) in recognizing this taxon nomenclaturally as *C. alba*; Stone (1997b) treated it as *C. tomentosa*.

Carya aquatica (F. Michx.) Nutt., (growing in or near water), WATER HICKORY, SWAMP HICKORY, BITTER PECAN, WATER PIGNUT. Leaflets glabrous beneath; buds usually dark red-brown; fruits winged to base, often in clusters of 3–4; nut somewhat flattened laterally; kernel bitter. Stream banks and wet woods; Kaufman and Lamar cos., also Delta, Hopkins, Hunt cos. (Little 1971); se and e TX w to e part of nc TX. Apr.

Carya cordiformis (Wangenh.) K. Koch, (in the form of a heart), BITTER-NUT, BITTER-NUT HICKORY, PIGNUT. Leaflets pubescent beneath; buds yellow or orange-yellow; nut depressed or obcordate at apex, smooth to uneven; kernel bitter. Stream banks and wet woods; Delta, Falls, Hopkins, Hunt, and Lamar cos. (Little 1971) on e margin of nc TX, also disjunct w to Dallas Co. (*Reverchon*); mainly se and e TX. Apr.

Carya illinoinensis (Wangenh.) K. Koch, (of Illinois), PECAN, NOGAL MORADO, NUEZ ENCARCELADA. **State tree of Texas**. Buds usually brown (rarely yellowish brown); fruits 25–80 mm long, 12–25 mm in diam.; nut usually not flattened laterally, ± pointed at both ends; kernel sweet. Stream bottoms or slopes; also cultivated, both as a shade tree and for the nuts; mainly e 1/2 of TX, scattered further w. Apr. [*Carya pecan* (Marshall) Engl. & Graebn., *Hicoria pecan* (Marshall) Britton] PECAN was adopted as the state tree by the Texas State Legislature in 1919 (Jones et al. 1997). The wind borne pollen can be a significant hay fever causing allergen. The first pecan cultivars were selected in 1846; currently there are over 500 named cultivars including thin-shelled pecans; in addition to being edible, the nuts provide an oil used in cosmetics (Mabberley 1987). All PECAN cultivars with Indian names were developed at the USDA Field Station in Brownwood, TX (J. Stanford, pers. comm.). The National Champion PECAN (largest recorded in the U.S.) is located in Weatherford, Parker Co. (American Forestry Association 1996);

it is located 3 miles n of downtown Weatherford on State Highway 51 (H. Garrett, pers. comm.). Manaster (1994) gave extensive information on the PECAN.

Carya myristiciformis (F. Michx.) Nutt., (in the form of *Myristica*—nutmeg), NUTMEG HICKORY, NOGAL, BITTER WATERNUT. Fruits usually solitary, ca. 35 mm long; nut rounded and apiculate at both ends; kernel sweet. Stream banks and swampy areas; scattered localities in e TX, w to nc TX in river drainages in Delta, Fannin, and Hunt cos. (Little 1971), also Lamar Co. Apr.

Carya ovata (Mill.) K. Koch, (ovate, egg-shaped), SHAG-BARK HICKORY, SCALY-BARK HICKORY, UPLAND HICKORY, RED-HEART HICKORY, SHELLBARK. Bark separating into large, conspicuous, light gray plates remaining attached at the middle; fruits 25–50(–60) mm long, with husk 4–15 mm thick; nut often 3-ridged or -angled, usually with thin shell; kernel sweet. Rich woods, often moist areas; se and e TX w to Red River Co. (Little 1971), also a disjunct collection from Parker Co. (West Cross Timbers). The wood is very strong and tough (Steyermark 1963) and was valued for smoking meat such as hams (Peattie 1948).

Carya texana Buckley, (of Texas), BLACK HICKORY, TEXAS HICKORY, BUCKLEY'S HICKORY, OZARK HICKORY. Variable in size and shape of leaflets and nuts; buds rusty; fruits 30–50 mm long, with husk 2–4 mm thick; nut acute at both ends, rough and pitted; kernel bitter. Lowland or upland woods, sandy soils or occasional on limestone; se and e TX w to East Cross Timbers. Apr.

JUGLANS WALNUT, NOGAL

Husk of nut fibrous-fleshy, indehiscent.

A genus of 21 species native from the Mediterranean to e Asia and North America s to the Andes; this is one of the few n temperate tree genera extending s to South America (*J. neotropica* Diels is the source of an important carving wood in Ecuador). Species are variously used as cultivated ornamentals, for the edible nuts (these were widely used by Native Americans—Heiser 1993), and for timber. The wood of most species is durable, dark-colored, and highly prized for woodworking. *Juglans* species produce juglone, an allelopathic compound that inhibits the growth of other plants; it is found in the leaves, bark, and fruits; ☠ this compound can also cause a toxic reaction in horses when walnut shavings or leaves are used in stalls (Hardin & Brownie 1993). (Classical Latin name for *J. regia* L., from *jovis*, of Jupiter, and *glans*, acorn, nut, or gland; thought by some to refer to the male organs of Jupiter)
REFERENCES: Sudworth 1934; Wittemore & Stone 1997.

1. Mature fruits 35 mm or larger in diam.; nuts with conspicuously irregular grooves and ridges; margins of leaflets conspicuously serrate to serrate-crenate; se and e TX w to Grand Prairie and further w in Red River drainage ____ **J. nigra**
1. Mature fruits 35 mm or less in diam.; nuts with regular lengthwise grooves and ridges; margins of leaflets subentire to serrate-crenate; from w Cross Timbers and Lampasas Cut Plain s and w to w TX.
    2. Fruits 25–35 mm in diam.; leaflets usually 9–15; mature leaflets usually 15–35 mm wide ____ **J. major**
    2. Fruits usually 20 mm or less in diam.; leaflets (15–)17–23; mature leaflets usually 15(–17) mm or less wide ____ **J. microcarpa**

Juglans major (Torr.) A. Heller, (greater, larger), ARIZONA WALNUT, ARIZONA BLACK WALNUT, NOGAL SILVESTRE. Tree to ca. 15 m tall; leaflets usually 9–15 (rarely more), kernel edible. Along streams and in floodplains; Hood Co., also Lampasas Co. (Little 1976) and Fort Hood (Bell or Coryell cos.—Sanchez 1997); mainly Edwards Plateau and Trans-Pecos. Apr.

Juglans microcarpa Berland., (small-fruited), TEXAS WALNUT, LITTLE WALNUT, TEXAS BLACK WALNUT, DWARF WALNUT, NOGAL, NOGALILLO, NOGALITO. Shrubby small tree to ca. 6 m tall; kernel ed

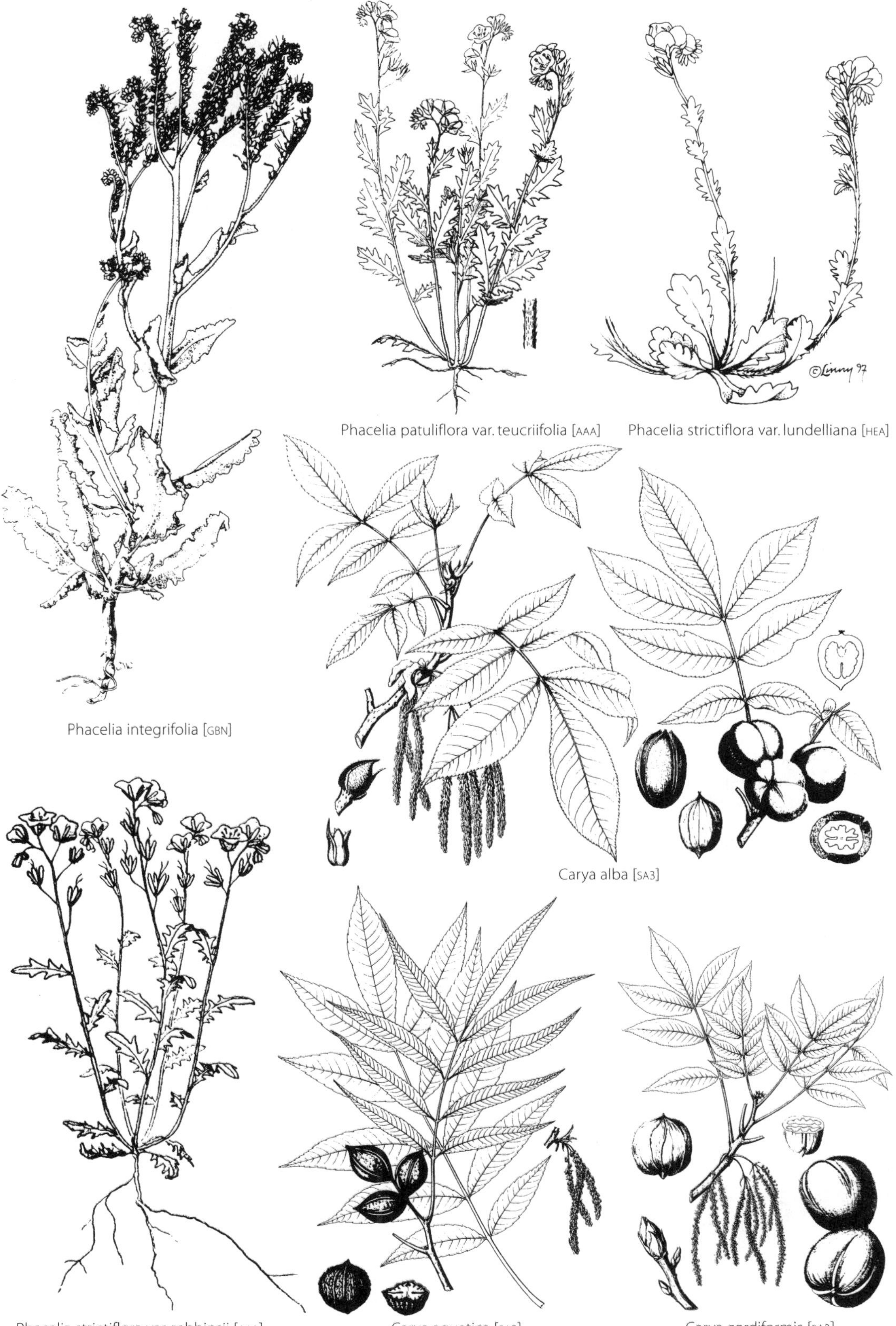

Phacelia patuliflora var. teucriifolia [AAA]

Phacelia strictiflora var. lundelliana [HEA]

Phacelia integrifolia [GBN]

Carya alba [SA3]

Phacelia strictiflora var. robbinsii [AAA]

Carya aquatica [SA3]

Carya cordiformis [SA3]

ible. Limestone outcrops and gravelly stream bottoms; McLennan Co. (Mahler 1988), also West Cross Timbers (Jack Co.—Simpson 1988 and Palo Pinto Co.—Little 1976); mainly Edwards Plateau and Trans-Pecos. Apr. The National Champion LITTLE WALNUT (largest recorded in the U.S.) is located in Denton Co. (American Forestry Association 1996).

Juglans nigra L., (black), BLACK WALNUT. A broad-headed tree; leaflets usually 11–23; kernel edible. Stream bottom woods on calcium rich soils; se and e TX w to Grand Prairie, also w in the Red River drainage to Wilbarger Co. in the Rolling Plains (Little 1971), also Edwards Plateau; sometimes cultivated. Late Apr. The chocolate to purplish brown heartwood is one of the most valuable North American woods; it is used in cabinets, furniture, and gunstocks; because a single tree can be worth thousands of dollars, they are sometimes "rustled." This species exhibits allelopathy, the inhibition of one plant by another via the release of chemicals into the environment. Rain dripping from the leaves has a strong influence on the types of plants capable of growing beneath the trees (Brooks in Daubenmire 1974); according to H. Garrett (pers. comm.), members of the Solanaceae (e.g., tomatoes) in particular are adversely affected by proximity to BLACK WALNUT trees. Because of the possible allelopathic effects, some organic gardeners are careful concerning the use of BLACK WALNUT leaves as mulch. According to J. Stanford (pers. comm.), BLACK WALNUT sawdust is sometimes effective in killing fire ants. A brown dye can be obtained from the husk of the fruits and was used by Native Americans (Burlage 1968). BLACK WALNUT is reported to be capable of causing contact dermatitis (Lampe & McCann 1985). ☠

KRAMERIACEAE RATANY OR KRAMERIA FAMILY

A very small (15 species), New World family containing only a single genus; it has at times been placed in the Fabaceae or Polygalaceae, and is apparently related to the Polygalaceae. Chromosome evidence presented by Turner (1958) argued against a relationship with the Fabaceae. (subclass Rosidae)

FAMILY RECOGNITION IN THE FIELD: the single nc TX species is a prostrate or trailing herb with simple, alternate, entire leaves; flowers bilaterally symmetrical, purplish or reddish, superficially orchid-like; fruits pea-sized, spiny.

REFERENCES: Britton 1930; Robertson 1973; Simpson 1989.

KRAMERIA RATANY

A genus of 15 species of hemiparasitic shrubs, trees, or herbs native from the sw U.S. to Argentina and Chile, especially in dry areas. (Possibly named for J. Kramer, 1700s, Austrian army physician)

REFERENCES: Musselman 1975 [1976], 1977; Simpson & Skvarla 1981.

Krameria lanceolata Torr., (lanceolate, lance-shaped), TRAILING RATANY, CRAMERIA, PRAIRIE BUR, SANDBUR. Perennial herb; roots woody; stems prostrate or trailing, to 0.2–1(–1.8) m long; leaves alternate, entire, linear to elliptical, to 2 cm long, minutely spiny-tipped; peduncles 2–3 cm long, 2-bracted; flowers solitary, axillary, bilaterally symmetrical, superficially orchid-like; sepals 4–5, unequal, colored, 8–10 mm long; petals 5, the upper 3 long-clawed and often united, purplish or reddish, the lower 2 thick, sessile; stamens 4; filaments united below; anthers opening by terminal pores; ovary 1- or 2-celled, superior; fruits indehiscent, subglobose, 6–9 mm in diam., 1-seeded, woolly, spiny, the spines retrorsely scabrous. Rocky prairies, nearly throughout TX. Apr–Nov. [*K. secundiflora* of authors, not DC.] Reported to be hemiparasitic; forming haustoria on the roots of a broad range of host plants (Musselman 1975 [1976], 1977). ▣/95

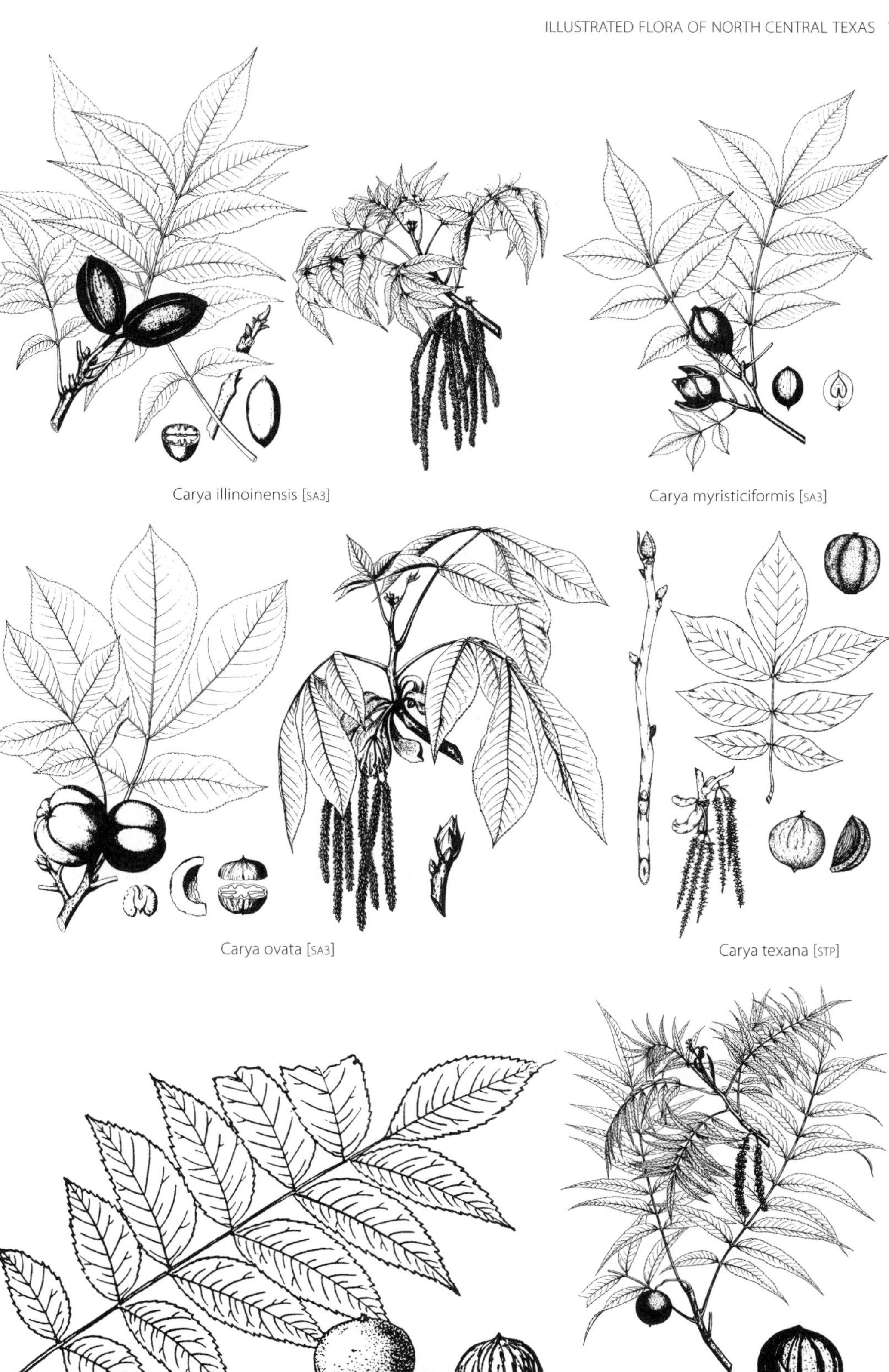

Carya illinoinensis [SA3]

Carya myristiciformis [SA3]

Carya ovata [SA3]

Carya texana [STP]

Juglans major [SA2]

Juglans microcarpa [SA3]

LAMIACEAE (LABIATAE) MINT FAMILY

Annual or perennial, often aromatic herbs or shrubs; stems square; leaves opposite, simple, entire, toothed, or lobed; flowers axillary or terminal, solitary or in whorls, spikes, racemes, panicles, or small heads; calyces 2- to 5-toothed or -lobed; corollas usually bilaterally symmetrical, commonly 2-lipped; stamens 2 or 4, attached to the corolla tube; pistil 1, 2-carpellate, deeply 4-lobed; ovary superior; style 1, simple or 2-branched apically; fruit usually of 4 one-seeded nutlets.

A large (6,700 species in 252 genera), cosmopolitan, but especially Mediterranean to c Asian family of mostly herbs or small shrubs including many ornamentals (e.g., *Salvia*—SAGE, *Solenostemon*—COLEUS) and numerous aromatic herbs including *Lavendula* (LAVENDER), *Mentha* (MINT), *Nepeta* (CATNIP), *Ocimum* (BASIL), *Origanum* (MARJORAM and OREGANO), *Rosmarinus* (ROSEMARY), *Salvia* (SAGE), and *Thymus* (THYME); many species have also been used medicinally. Often there are short-stalked epidermal glands containing volatile essential oils giving rise to characteristic aromas. The family is related to the Verbenaceae. Judd et al. (1994) argued that as traditionally circumscribed, the Verbenaceae are paraphyletic and the Lamiaceae polyphyletic; taxa traditionally recognized as MINTS seem to be derived independently from a number of different Verbenaceae clades. They indicated that in order for the families to be monophyletic, the Verbenaceae should be restricted to the Verbenoideae (those taxa with racemose or spicate inflorescences), while the Lamiaceae should be expanded to include most of the Verbenaceae. However, recent molecular studies (Wagstaff & Olmstead 1997) do not support the monophyly of a clade composed of Lamiaceae sensu lato and Verbenaceae sensu stricto. Until the phylogeny of this group is more clearly resolved, we are treating these families in the traditional manner. (subclass Asteridae)

FAMILY RECOGNITION IN THE FIELD: usually aromatic herbs or shrubs with *square stems* and *opposite leaves*; corollas sympetalous, *2-lipped*; stamens 4 (in 2 sets) or 2; ovary usually 4-lobed, with a basal style; fruits usually of 4 one-seeded nutlets. Similar families can be distinguished as follows: Boraginaceae have alternate leaves and radially symmetrical corollas; Verbenaceae have the style usually terminal on the ovary; Scrophulariaceae have an unlobed ovary and many-seeded fruits.

REFERENCES: Cantino & Sanders 1986; Cantino 1992; Harley & Reynolds 1992; Abu-Asab & Cantino 1993; Judd et al. 1994; Wagstaff & Olmstead 1997.

1. Calyces with a conspicuous cap-like or shield-like projection on 1 side, bilabiate, with both lips entire ____ **Scutellaria**
1. Calyces without projections, radially symmetrical or bilabiate with 1 or both lips toothed.
   2. Corollas bilaterally symmetrical and appearing 1-lipped, the upper lip inconspicuous and hardly discernible or lobes of upper lip on the margins of the lower lip.
      3. Corollas white to cream or lavender with darker markings; middle lobe of lower lip of corollas 2–3 times as long as lateral lobes; leaves of flowering stems sharply or coarsely toothed or deeply lobed; flowering stems 7–150 cm tall ____ **Teucrium**
      3. Corollas blue, rarely white; middle lobe of lower lip of corollas only slightly longer than lateral lobes; leaves of flowering stems entire or indistinctly and bluntly toothed; flowering stems 30 cm or less tall ____ **Ajuga**
   2. Corollas definitely 2-lipped, the upper lip clearly discernible OR corollas nearly radially symmetrical with 4 or 5 ± equal lobes.
      4. Fertile (anther-bearing) stamens 2.
         5. Flower clusters in the axils of well-developed typical leaves usually longer than the flower clusters; corollas < 10(–13) mm long (often much less).

6. Corollas very bilaterally symmetrical, ± conspicuously 2-lipped; calyces bilabiate; plants aromatic **Hedeoma**
6. Corollas nearly radially symmetrical, funnelform or nearly salverform, with small nearly equal lobes or teeth, not distinctly 2-lipped; calyces radially symmetrical or nearly so; plants aromatic OR not so.
7. Flowers essentially sessile in dense axillary whorls; leaves 1.5–15 cm long; plants not aromatic **Lycopus**
7. Flowers in short-stalked axillary cymes; leaves 1–4(–5) cm long; plants with a mint like aroma **Cunila**
5. Flower clusters in spikes or racemes, the clusters not subtended by well-developed leaves, OR in terminal heads or head-like whorls and subtended by large, often colored bracts unlike the leaves; corollas typically > 10 mm long (often much more).
8. Flowers in dense heads or head-like whorls with involucres of large, often colored bracts immediately below the heads; calyces radially symmetrical **Monarda**
8. Flowers in elongate spikes or racemes; calyces bilabiate **Salvia**
4. Fertile (anther-bearing) stamens 4 (2 are sometimes shorter).
9. Flowers opposite, 1 per axil of each floral bract.
10. Calyces deeply 2-lipped, the upper lip either with 3 lobes much larger than the 2 lobes of the bottom lip (in *Warnockia*) OR upper lip with lobes fused for almost their entire length and ± indistinct (in *Brazoria*).
11. Stems pubescent below inflorescences; corollas mostly 15–22(–25) mm long; uppermost lobe of corollas split for a portion of its length and erect; nutlets pubescent; calyces often with a tuft of long hairs at base; plants growing on sand, rare if present in nc TX **Brazoria**
11. Stems glabrous below inflorescences; corollas 8.5–12 mm long; uppermost lobe of corollas entire and sub-galeate; nutlets glabrous; calyces minutely pubescent; plants widespread in nc TX on gravelly or thin clayey soils on limestone **Warnockia**
10. Calyces with 5 slightly unequal, distinct lobes, not deeply 2-lipped **Physostegia**
9. Flowers whorled or compactly panicled, > 1 per axil of each floral bract.
12. Main stem leaves and leaves (leaf-like bracts) subtending flowers (at least some) deeply lobed.
13. Plants tall, to 200 cm high; leaves (leaf-like bracts) subtending flowers long-petioled; stem leaves all deeply lobed; calyx lobes ± spine-tipped **Leonurus**
13. Plants to only 45 cm high; leaves (leaf-like bracts) subtending flowers sessile; stem leaves (at least some) merely crenate; calyx lobes not spine-tipped **Lamium**
12. Main stem leaves and leaves (leaf-like bracts) subtending flowers not lobed.
14. Stems gray with soft, matted or incurled hairs.
15. Flowers in dense axillary whorls; calyces (8–)10-toothed, the teeth often tipped by hooked spines; leaf blades wedge-shaped to subcordate at base; corollas 5–6 mm long, white **Marrubium**
15. Flowers in terminal compact panicles; calyces 5-toothed, the teeth straight or slightly incurved; leaf blades cordate at base; corollas 7–12 mm long, white to pale purple, with dark purple dots **Nepeta**
14. Stems glabrous or with spreading pubescence.
16. Inflorescences generally appearing axillary, the flower clusters in axils of leafy bracts very similar to regular leaves.
17. Flowers and fruits sessile; leaf blades crenate to deeply lobed, never entire **Lamium**
17. Flowers and fruits on pedicels (these can be short); leaf blades entire or crenate or rarely serrulate, never lobed.

18. Calyces 2-lipped, upper and lower teeth unequal.
    19. Leaf blades 5 mm or less wide; calyces ca. 3 mm long; plant with pennyroyal odor; s part of nc TX ____ **Calamintha**
    19. Leaf blades 4–25 mm wide; calyces 2.7–6 mm long in flower, enlarging to 4.6–8.9 mm long in fruit; plant without pennyroyal odor; ne part of nc TX ____ **Trichostema**
18. Calyces nearly radially symmetrical, the teeth all equal or nearly so.
    20. Leaves reniform (= kidney-shaped) to suborbicular; corollas (7–)10–20(–22) mm long; stems (except short erect flowering stems) trail ing, creeping, and rooting at the nodes ____ **Glechoma**
    20. Leaves linear-elliptic to lanceolate or ovate; corollas 1.5–4.5 mm long; stems erect ____ **Trichostema**

16. Inflorescences generally terminal in appearance, the flower clusters in dense glomerules, loose corymbs, or spike-like inflorescences, not in axils of leaf-like bracts.
    21. Flowers numerous, in 1–3 large globose clusters 40–60 mm thick, these encircling the stem; corollas orange-yellow or orange-red, 20–25 mm long, very pubescent ____ **Leonotis**
    21. Flowers not in inflorescences as above; corollas not as above.
        22. Flower clusters in dense head-like glomerules or in ± flat-topped inflorescences; corollas nearly radially symmetrical, with small nearly equal lobes or teeth, not distinctly 2-lipped ____ **Pycnanthemum**
        22. Flower clusters in elongate spike-like inflorescences; corollas distinctly 2-lipped.
            23. Inflorescences so dense that main flowering stalk is not visible among the flowers; calyces 6–11(–15) mm long at flowering time; corollas 10–15(–20) mm long ____ **Prunella**
            23. Inflorescences less dense, loose or interrupted in places so main flowering stalk is usually visible; calyces 1.2–6 mm long at flowering time (can elongate in fruit); corollas 1.7–14 mm long (6 mm or less long in all except 1 species of *Stachys*).
                24. Calyces distinctly 2-lipped, the lower lobes different from the upper, ca. 3 mm long in flower, elongating in fruit to 8–12 mm long; petioles 10–80 mm long; corollas ca. as long as or only slightly longer than calyces; leaves sometimes distinctly crinkly in appearance, sometimes with striking purplish coloration ____ **Perilla**
                24. Calyces nearly radially symmetrical, the lobes all ± the same, only 1.2–6 mm long in flower or fruit, not elongating; petioles (except lowermost) 0–20 mm long; corollas longer than calyces, often much longer; leaves not distinctly crinkly, usually without striking purplish coloration.
                    25. Lower leaves with petioles nearly as long as or longer than blades; plants without aromas as described below ____ **Stachys**
                    25. Lower leaves sessile or with blades many times longer than petioles; plants strikingly aromatic with peppermint, spearmint, or apple odors OR not so.
                        26. Corollas 10–14 mm long, with upper lip hooded over stamens; petioles 10–20 mm long; calyces 5–6 mm long; plants without aromas as described below ____ **Stachys**

26. Corollas 1.7–5 mm long, the upper lip not hooded over the exserted stamens; leaves sessile or with petioles to 15 mm long; calyces 1.2–3.4 mm long; plants strikingly aromatic with peppermint, spearmint, or apple odors ________ **Mentha**

AJUGA BLUE-BUGLE

An Old World genus of 50 species, mainly temperate but extending to lowland Malesia; some are medicinal, others cultivated as ornamentals. The upper lip of the corolla is very short or absent. (Greek: *a*, without, and *zugos* (Latin *jugum*), yoke, from the seeming absence of the upper corolla lip)

Ajuga reptans L., (creeping), CARPET AJUGA, BUGLE, BUGLEWEED. Low mat-forming perennial with basal rosette and leafy stolons; flowering stems erect, to 30 cm tall; flowers in whorls of 2–6 in leafy-bracted, spike-like racemes; calyces 4–6 mm long; corollas blue to purplish, bilabiate but appearing ± 1-lipped, the upper lip much shorter (0.5–1.5 mm long) than the lower lip (5–10 mm long); lower lip with middle lobe bilobed, much broader than and only slightly longer than lateral lobes; stamens 4. Cultivated as a ground cover, becoming weedy; Dallas and Grayson cos., also Tarrant Co. (R. O'Kennon, pers. obs.); Hatch et al. (1990) cited only vegetational area 4 (Fig. 2) for TX. Mar–Apr. Native of Europe. Duke (1985) referenced sources indicating that this species is a narcotic hallucinogen and that it is known to have caused fatalities.

BRAZORIA BRAZOS MINT

A genus of 2 species endemic to Texas (Turner 1996). The related monotypic *Warnockia scutellarioides* was previously treated in this genus. (Named for its habitat on the Brazos River, the longest river in Texas; derived from the Spanish name-*Brazos de Dios* or Arms of God)
REFERENCES: Lundell 1945, 1968, 1969a; Shinners 1953d; Turner 1996.

Brazoria truncata (Benth.) Engelm. & A. Gray var. **truncata**, (cut off square), BLUNT-SEPAL BRAZORIA, RATTLESNAKE WEED. Plant usually 20–35 cm tall; stems simple or branched from base; leaves oblanceolate to spatulate; mature spikes densely flowered, with lower internodes mostly 5–6 mm long; lower lip of calyces inflexing after anthesis, with a tuft of long hairs reaching 2 mm in length; corollas 15–22(–25) mm long, the 2 upper lobes lavender, the 3 lower lobes very pale lavender; $2n = 28$. On loose sandy soils; included based on citation for vegetational area 4 (Fig. 2) by Hatch et al. (1990); this Texas endemic, probably occurs only s of nc TX n to the sw part of Burnet Co. (Turner 1996). Apr–May.

var. *pulcherrima* (Lundell) M.W. Turner, (very handsome), endemic to e TX just e of nc TX can be distinguished as follows: mature spikes loosely interrupted, with lower internodes mostly 8–13 mm long; lower lip of calyces lightly pubescent to canescent at base, with most hairs reaching only 0.2 mm in length, occasional hairs reaching 1.0 mm long (Turner 1996). This variety is possibly worthy of specific recognition as originally treated by Lundell (1968).

CALAMINTHA SAVORY, CALAMINT

A genus of ca. 12 species native from Europe to c Asia and the New World (estimate of number of species from Shinners 1962e and Mabberley 1987). Previously recognized in the genus *Satureja* and sometimes lumped into *Clinopodium* (e.g., Mabberley 1997) (Greek: *kallos*, beautiful, and *minthe*, mint)
REFERENCES: Dewolf 1955; Shinners 1962e; Epling & Játiva 1964, 1966.

Calamintha arkansana (Nutt.) Shinners, (of Arkansas), OZARK SAVORY, ARKANSAS CALAMINT. Stoloniferous perennial, also with erect stems 10–40 cm tall; aromatic with pennyroyal odor; stolon leaves ovate to elliptic; cauline leaves ± linear, essentially entire, to 25 mm long and 5 mm wide, gland-dotted; flowers in cymes; calyces strongly ribbed, ca. 3 mm long; corollas bluish, 10 mm long, 2-lipped; stamens 4; style branches curled. Calcareous outcrops; Bell and Burnet cos. in s part of nc TX; se and e TX w to nc TX and Edwards Plateau. Apr–Aug. [*Satureja arkansana* (Nutt.) Briq.]

CUNILA DITTANY, STONE-MIST

A New World genus of 15 species native from e North America to Uruguay. (An ancient Latin name for some fragrant plant, transferred to this American genus)

Cunila origanoides (L.) Britton, (resembling *Origanum*—another mint genus including oregano), MARYLAND STONE-MIST, AMERICAN DITTANY. Glabrous perennial herb to ca. 40 cm tall, with woody base; leaves essentially sessile, ovate, 1–4(–5) cm long, rounded to cordate basally, gland-dotted, aromatic; flowers in terminal or axillary cymes usually shorter than subtending leaf; calyces radially symmetrical, ca. 3 mm long, strongly 10-nerved, the throat villous; corollas purplish to white, nearly radially symmetrical, 6–8 mm long, pubescent; stamens 2; filaments straight, exceeding corolla. Dry open woods; included based on citation of vegetational areas 4 and 5 (Fig. 2) by Hatch et al. (1990); Post Oak Savannah w to nc TX. Sep–Oct. [*C. mariana* L.] Used as a culinary herb and source of medicinal cunila oil (Mabberley 1987).

GLECHOMA GROUND-IVY

A temperate Eurasian genus of ca. 10 species. (Old Greek name *glechon*, for a kind of mint)

Glechoma hederacea L., (resembling *Hedera*—English ivy), GROUND-IVY, GILL-OVER-THE-GROUND, RUNAWAY-ROBIN. Perennial herb with creeping stems rooting at the nodes; flowering stems erect, 5–20(–40) cm tall; leaf blades reniform to suborbicular, 1–2.5(–4.5) cm long, coarsely crenate, cordate at base, green or with purplish coloration, gland-dotted on lower surface, petiolate; petioles on creeping stems long, to 10 cm; flowers 2–7 in axillary and terminal cymules; calyces 4–7 mm long; corollas bilabiate, (7–)10–20(–22) mm long, blue to violet or with white streaks (rarely white), the lower lip speckled with red-purple; stamens 4. Weedy areas and waste places; Dallas Co., also Tarrant Co. (R. O'Kennon, pers. obs.); not cited for TX by Hatch et al. (1990). Apr–Jun. Native of Europe. Formerly used as a medicinal tea and added to beer on long voyages (Mabberley 1987); toxic and even potentially fatal to horses if ingested in large amounts (Kingsbury 1964).

HEDEOMA MOCK PENNYROYAL, FALSE PENNYROYAL

Small, pubescent, aromatic annuals or perennials; leaves short-petioled; leaf blades linear to elliptic, entire or indistinctly toothed, gland-dotted on 1 or both surfaces; flowers axillary or both axillary and terminal, solitary or in condensed axillary cymes of 2–12, the overall appearance sometimes spike-like; calyces narrow, curved, with bristle-like teeth; corollas bilabiate, small, white to pink, lavender, or blue; fertile stamens 2.

A genus of 38 species of sw North America and South America. (Name altered from *hedyosman*, an ancient name of mint, from Greek: *hedys*, sweet, and *osme*, scent)
REFERENCES: Epling & Stewart 1939; Irving 1976, 1979, 1980.

1. Plants annual; calyx teeth not convergent (instead straight to spreading or reflexed); corolla tube not inflated.
   2. Leaves ovate to elliptical, 3.5–8 mm wide, petiolate, inconspicuously toothed or entire; upper lip of corollas bent at right angle to tube; calyx tube saccate for up to 1/3 its length ____ **H. acinoides**
   2. Leaves linear, 1–3 mm wide, sessile, entire; upper lip of corollas ± straight, not bent at right angle to tube; calyx tube markedly saccate for 3/4 its length ____ **H. hispida**

Juglans nigra [MIC, SA3]

Krameria lanceolata [GR1]

Ajuga reptans [LAM]

Brazoria truncata var. truncata [JON]

Calamintha arkansana [BB2]

Cunila origanoides [STE]

Glechoma hederacea [LAM]

1. Plants annual or perennial (sometimes woody based with shoots from previous season often persistent); EITHER calyx teeth strongly convergent and completely or incompletely closing the calyx opening at maturity OR corolla tube markedly inflated.
   3. Plants with aroma of peppermint; leaves usually more than 3 times as long as wide, bright green; upper and lower calyx teeth usually strongly convergent (± completely closing the calyx opening at maturity); corollas 7–11 mm long, blue; annuals or herbaceous to woody-based perennials averaging < 25 cm tall _____ **H. drummondii**
   3. Plants with aroma of camphor or lemon; leaves usually less than 3 times as long as wide, gray green or dark green; upper and lower calyx teeth incompletely closing opening; corollas 8–15 mm long, white or lavender; woody-based perennials usually averaging > 25 cm tall _____ **H. reverchonii**

Hedeoma acinoides Scheele, (resembling *Acinos*—another genus of Lamiaceae), SLENDER HEDEOMA. Delicate annual, ephemeral, 5–30 cm tall; taproot slender; basal leaves often purplish; petioles 1–4 mm long; calyces 5–6 mm long, the lower 1/3 of the tube saccate; corollas pink, 9–13 mm long, not inflated. Rocky limestone soils; Palo Pinto Co., also Brown, Hamilton (HPC), Parker, Somervell (R. O'Kennon, pers. obs.), and Tarrant cos. (Irving 1980); s and w parts of nc TX, also c and coastal TX. Apr–May. Dried plants are used by some as an aromatic tea (R. O'Kennon, pers. obs.).

Hedeoma drummondii Benth., (for its discoverer, Thomas Drummond, 1780–1835, Scottish botanist and collector in North America), DRUMMOND'S HEDEOMA, DRUMMOND'S FALSE PENNYROYAL. Annual or robust perennial 15–45 cm tall; leaves usually essentially linear, 5–11 mm long, 1–4 mm wide, entire; calyces finely hirsute, 5–6 mm long, the tube saccate for ca. 2/3 of its length; corollas blue, 7–11 mm long. Disturbed habitats; McLennan and Palo Pinto cos., also Parker, Somervell, and Tarrant cos. (R. O'Kennon, pers. obs.); nc TX s and w to w TX. Jun–Sep. According to Irving (1980), *H. drummondii* can hybridize with *H. reverchonii*.

Hedeoma hispida Pursh, (hispid, bristly), ROUGH HEDEOMA, ROUGH FALSE PENNYROYAL. Coarse annual 9–40 cm tall; taproot slender; leaves essentially sessile or subsessile; calyces 5–6 mm long; corollas dimorphic: cleistogamous corollas small, blue or white, ca. 5.3 mm long, scarcely exserted from the calyces; chasmogamous corollas larger, blue, 6–7 mm long, well-exserted from calyces. Prairies, sandy soils; se and e TX w to Rolling Plains. Apr–May.

Hedeoma reverchonii (A. Gray) A. Gray, (for Julien Reverchon, 1837–1905, a French-American immigrant to Dallas and important botanical collector of early TX), REVERCHON'S FALSE PENNYROYAL, ROCK HEDEOMA. Robust, woody-based perennial 15–60 cm tall, with shoots persisting from previous year; leaves elliptic-oblong, 6–14 mm long, 2.2–5 mm wide.

1. Plants lemon-scented; leaves gray green; calyces 6–7 mm long; corollas 10–15 mm long, the tube conspicuously dilated upward _____ var. **reverchonii**
1. Plants camphor-scented; leaves dark green; calyces 5–6 mm long; corollas 8–10 mm long, the tube only slightly dilated _____ var. **serpyllifolium**

var. **reverchonii**, ROCK HEDEOMA. Corollas white or lavender. Open, exposed, calcareous outcrops; e TX w to Blackland and Grand prairies s to c TX. Jun–Sep. [*H. drummondii* Benth. var. *reverchonii* A. Gray]

var. **serpyllifolium** (Small) R.S. Irving, (with leaves like *Thymus serpyllum*—thyme). Corollas mainly white, occasionally lavender. Open, exposed, calcareous outcrops; sw part of nc TX s and w to w TX. Jun–Sep. [*H. drummondii* Benth. var. *serpyllifolium* (Small) R.S. Irving, *H. serpyllifolium* Small]

LAMIUM DEAD-NETTLE

Ours low, ± pubescent annuals often rooting at lower nodes; stems decumbent at base, erect to flexuous above, 10–45 cm tall; lower leaves rather long-petioled; leaf blades triangular-ovate, bluntly and coarsely toothed; flowers in dense whorls subtended by leaf-like bracts; corollas 10–20 mm long, 2-lipped, pinkish purple to reddish purple, rarely white, with darker markings on lower lip, the upper lip hooded, with reddish pubescence; stamens 4; mericarps yellowish to olive-brown mottled with white scaly areas.

A genus of ca. 40 species of n Africa and Eurasia; called DEAD-NETTLES because non-flowering stems resemble *Urtica* species. (Old Latin name of a nettle-like plant mentioned by Pliny, presumably from resemblance of leaves to those of nettles)
REFERENCE: Mennema 1989.

1. Leaf-like bracts (those just below flowers) sessile and clasping; calyces densely villous; leaves, including those below flowers, coarsely crenate to lobed ______ **L. amplexicaule**
1. Leaf-like bracts (those just below flowers) petioled; calyces glabrous or sparsely hairy; leaves, including those subtending flowers, crenate, not lobed ______ **L. purpureum**

Lamium amplexicaule L., (stem-clasping), HENBIT, DEAD-NETTLE. Bracts (leaf-like) usually wider than long, ± horizontal; plant producing cleistogamous flowers (appearing to be in bud, the corollas unexpanded) from Nov–Feb. Extremely abundant in gardens, lawns, roadsides, disturbed areas; throughout TX. Feb–May, sporadically throughout the year. Native of Europe and the Mediterranean to Iran.

Lamium purpureum L., (purple), RED DEAD-NETTLE, PURPLE DEAD-NETTLE. Bracts (leaf-like) usually longer than wide, often reflexed; leaves and bracts deep green or often tinged with purple. Less common than HENBIT; disturbed areas; Dallas and Grayson cos., also Tarrant Co. (R. O'Kennon, pers. obs.); e TX w to nc TX. Late Mar–Apr(–Sep). Native of Europe and the Mediterranean.

A number of varieties are sometimes recognized in this species (e.g., Mennema 1989); Jones et al. (1997) listed two varieties for TX, var. *incisum* (Willd.) Pers. and var. *purpureum*; nc TX material appears to be var. *purpureum*. Mennema (1989) separated the two as follows:

1. Leaves irregularly and usually deeply incised, especially the upper ones ______ var. *incisum*
1. Leaves regularly and mostly faintly crenate ______ var. *purpureum*

LEONOTIS

A genus of 15 species of tropical Africa with 1 extending to Asia and the New World. (Greek: *leon*, lion, and *otis*, ear; the corolla has been likened to a lion's ear)

Leonotis nepetifolia (L.) W.T. Aiton, (with leaves like *Nepeta*—another mint genus including catnip), NEP-LEAF LION'S-EAR, LION'S-HEAD. Robust annual herb 0.5–2 m tall; stems erect, canescent; leaves ovate, 5–12 cm long, 4–10 cm wide, petiolate; inflorescence of 1–3 dense, globose flower clusters 4–6 cm thick, these encircling the stem; calyces curved, to 15–20 mm long; corollas bilabiate, 20–25 mm long, conspicuous, orange-yellow to scarlet; stamens 4. Roadsides and in waste places; included based on citation of vegetational areas 4 and 5 (Fig. 2) by Hatch et al. (1990); naturalized in parts of e 1/2 of TX. Jun–Nov. Native of s Africa.

LEONURUS MOTHERWORT

Erect, aromatic, biennial or perennial herbs; leaves usually palmately 3–5(–7)-parted or -lobed; flowers in numerous close whorls in axils on terminal, leafy-bracted, spike-like inflorescences; calyces 5- or 10-ribbed, with 5 ± equal spiny teeth; corollas bilabiate; fertile stamens 4.

A temperate Eurasian genus of 3 species containing alkaloids. In the past, used as a tea for mothers at child birth, hence the common name (Strausbaugh & Core 1978). (Greek: *leon*, lion, and *oura*, tail, alluding to the hairy flowers)

1. Calyces with 5 ribs and 5 angles; upper lip of corollas densely long-hairy; stem leaves shaped somewhat like a maple leaf, with 3–5(–7) conspicuous large lobes ____ **L. cardiaca**
1. Calyces usually with 10 ribs and scarcely any angles; upper lip of corollas short-hairy; stem leaves deeply 3-parted, the divisions cleft so that the ultimate divisions are narrow ____ **L. sibiricus**

Leonurus cardiaca L., (for the heart), MOTHERWORT, COMMON MOTHERWORT, LION'S-TAIL. Perennial; stems 0.6–2 m tall, freely branched, glabrous or retrorsely pubescent on angles; leaves all cauline, gradually reduced upwards; leaf blades 5–12 cm long, palmately 3–5(–7)-lobed or the smallest unlobed (the lobes toothed or lobed), basally cuneate, rounded, truncate or cordate, apically obtuse to acute or acuminate; petioles 1–5 cm long; bracts similar to the leaves but slightly smaller; calyces 3–8 mm long, 5-ribbed; corollas 6–12 mm long, exceeding calyces, white to pink with purple spots. Disturbed areas; Montague Co. (photographed in 1983 by R. O'Kennon, photo at BRIT); possibly otherwise unknown in TX since not in Hatch et al. (1990) or Jones et al. (1997); however, Steyermark (1963) cited TX without locality; known from e OK (Taylor & Taylor 1994). May–Aug. Native of Eurasia. The leaves can cause contact dermatitis in susceptible individuals and the fragrant lemon-scented oil can cause photosensitization; grazing animals can have their mouths injured by the sharp teeth of the calyces (Steyermark 1963; Duke 1985).

Leonurus sibiricus L., (of Siberia), SIBERIAN MOTHERWORT. Biennial; stems to nearly 2 m tall, retrorsely pubescent; leaf blades deeply palmately 3-parted, the divisions 2–7 cleft and incised, with minute golden glands, reduced upwards, long-petioled; calyces ca. 7 mm long, (5–)10-ribbed; corollas ca. 10–16 mm long, ca. 2 times as long as calyces, rose-pink to purplish. Open and waste areas; included based on citation of vegetational areas 4 and 5 (Fig. 2) by Hatch et al. (1990), also Tarrant Co. (R. O'Kennon, pers. obs.); naturalized in parts of e 1/2 of TX. Apr–Aug. Native of Asia.

LYCOPUS WATER-HOREHOUND, BUGLEWEED

Erect perennial herbs with stolons; not fragrant; leaves linear-elliptic to nearly ovate, subentire to toothed or pinnatifid, gland-dotted, usually petiolate to subsessile; bracts similar to leaves, many times longer than dense whorls of small flowers in their axils; corollas white, 4–5-lobed, 1.8–5 mm long, sometimes with spots; fertile stamens 2.

A n temperate and Australian genus of 4 species. (Greek: *lycos*, wolf, and *pous*, foot, for some fancied likeness in the leaves)
REFERENCES: Hermann 1936; Henderson 1962.

1. Calyces equal to or shorter than the nutlets, the calyx lobes acute to obtuse at tips ____ **L. virginicus**
1. Calyces much longer than the nutlets, the calyx lobes acuminate to subulate (awl-shaped) at tips.
   2. Lower and sometimes upper leaves usually distinctly incised to pinnatifid at least at base; corollas with 4 lobes, 2.5–3.5 mm long; w to at least w part of Blackland Prairie ____ **L. americanus**
   2. Leaves merely toothed; corollas with 5 lobes, 2.5–5 mm long; only in extreme e part of nc TX ____ **L. rubellus**

Lycopus americanus Muhl. ex Barton, (of America), AMERICAN BUGLEWEED, WATER-HOREHOUND. Plant stoloniferous without tubers; stems 0.3–0.9 m tall, glabrous or sparingly appressed-pubescent with dark hairs; leaves lanceolate to nearly ovate, to ca. 12 cm long and 3(–6) cm wide; lower leaves incised or pinnatifid, sometimes only serrate; calyx teeth with long subulate tips; corollas white, barely longer than calyces. Low moist or wet areas; Collin, Dallas, Grayson, Parker, and Tarrant cos., also Fort Hood (Bell or Coryell cos.—Sanchez 1997); also Panhandle. Aug–Nov.

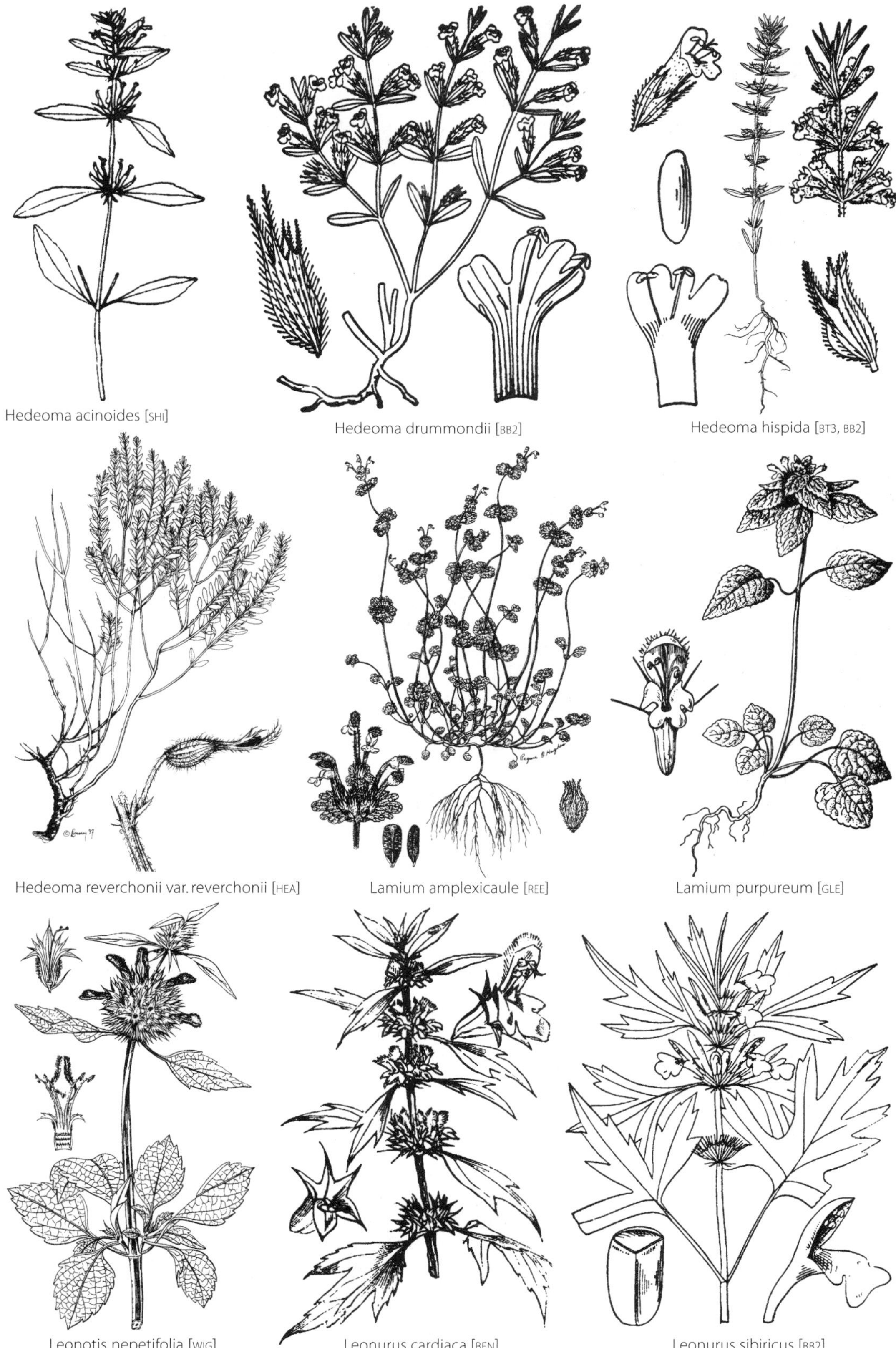

Hedeoma acinoides [SHI]

Hedeoma drummondii [BB2]

Hedeoma hispida [BT3, BB2]

Hedeoma reverchonii var. reverchonii [HEA]

Lamium amplexicaule [REE]

Lamium purpureum [GLE]

Leonotis nepetifolia [WIG]

Leonurus cardiaca [BEN]

Leonurus sibiricus [BB2]

Lycopus rubellus Moench, (reddish), WATER-HOREHOUND, ARKANSAS BUGLEWEED, TAPER-LEAF BUGLEWEED, STALKED BUGLEWEED. Stems 0.4–1.2 m tall, from stolons and slender tubers; leaves to 5–15 cm long and 1.5–5 cm wide; calyx teeth acuminate and sharp-pointed, scarcely subulate-tipped; corollas white, often with purple spots. Wet areas; se and e TX w to Hopkins and Milam cos. on e edge of nc TX. Aug–Dec. [*L. rubellus* var. *arkansanus* (Fresen.) Benner, *L. rubellus* var. *lanceolatus* Benner]

Lycopus virginicus L., (of Virginia), VIRGINIA BUGLEWEED. Stems to 0.3–0.9 m tall, from stolons, mostly without tubers; leaves sometimes purple-tinged, to 15 cm long and 5 cm wide, coarsely toothed; calyx teeth lanceolate to triangular; corollas whitish, 1.8–2.2 mm long, 4-lobed; mature nutlets usually longer than and concealing the calyces. Wet areas; se and e TX w to Henderson Co. near the extreme e margin of nc TX. Aug–Dec.

MARRUBIUM HOREHOUND

A genus of 30 species of Europe, the Mediterranean, and Asia. (Name used by Pliny, from Hebrew: *marrob*, bitter juice)

Marrubium vulgare L., (common), COMMON HOREHOUND, WHITE HOREHOUND, MARRUBIO. Perennial aromatic herb 0.3–0.7(–1) m tall, with very conspicuous white tomentum and often stellate hairs; old plants with woody root; leaves broadly ovate to suborbicular, 1.5–5 cm long, toothed, wrinkled-veiny, petioled; flowers in widely spaced dense axillary clusters; calyces 4–5 mm long, usually with 10 teeth, some of these hooked at the tip; corollas bilabiate, 5–6 mm long, white, often with rose-purple dots; stamens 4. Eroding or disturbed limestone soils, waste places; Bell Co. in s part of nc TX, also Somervell Co. (common at Fossil Rim Wildlife Center—R. O'Kennon, pers. obs.); naturalized in scattered localities throughout TX. Reverchon (1880) indicated that it appeared in Dallas Co. in the vicinity of cattle and sheep lots after a few years of settlement. Apparently, it requires large amounts of nitrogen—it is often abundant near livestock pens and beneath large trees where livestock stand in the shade and their droppings add to soil fertility (J. Stanford, pers. comm.). Apr–Jul, less freely later. Native of Eurasia, Mediterranean, and Macaronesia. Brooks (1986a) indicated that there may be some taxonomic confusion with this species. It is used as a flavoring and medicinally in remedies such as cough drops; it was formerly much used medicinally as a tea and in sweets and liqueurs (Mabberley 1987). Ranchers detest this species because the dried calyces get in wool and mohair and greatly reduce their value.

MENTHA MINT

Conspicuously aromatic perennial herbs from rhizomes or stolons; leaves simple, sessile or petiolate, toothed, usually gland-dotted; flowers small, in whorls in axils of leaves or in terminal spike-like inflorescences; corollas bilabiate, purplish to whitish or pink; stamens 4, equal.

A temperate Old World genus of 25 species of aromatic herbs long cultivated for use as flavorings. (Latin name; possibly from Minthe of Theophrastus, a nymph fabled to have been changed by Proserpine into mint)

1. Upper stem leaves sessile; leaf blades 1–2 times as long as wide, crenate-serrate, rounded in outline at tip (except for a pointed tooth); leaf blades and calyx tubes densely pubescent ____ **M. ×rotundifolia**
1. Upper stem leaves sessile to petiolate; leaf blades 2–3.5 times as long as wide, sharply serrate, acute to acuminate at tip; leaf blades and calyx tubes glabrous or nearly so.
   2. Calyces 1.5–2 mm long; leaves sessile to short-petioled (petioles 0–3 mm long); spikes slender, including corollas < 10 mm wide, ± interrupted ____ **M. spicata**
   2. Calyces 2.5–3.4 mm long; petioles 4–15 mm long; spikes thicker, including corollas 10–15 mm wide, mostly dense, with few interruptions ____ **M. ×piperita**

Lycopus americanus [MAS]

Lycopus rubellus [GLE]

Lycopus virginicus [ST1]

Marrubium vulgare [LAM]

Mentha spicata [CO1]

Mentha xpiperita [GLE]

Mentha xrotundifolia [SMI]

Mentha ×piperita L. [*M. aquatica* × *M. spicata*], (like the pepper vine, pepper-like), PEPPERMINT. Plant with leafy stolons; stems erect to decumbent, 0.3–0.9 m tall, often purplish; leaves lanceolate to ovate-lanceolate, 3–8 cm long, acute, sharply serrate; calyces to 3.4 mm long; corollas 3–5 mm long, rose-purplish to lavender or white. Wet areas; Grayson and Parker cos.; also Tarrant Co. (R. O'Kennon, pers. obs.); mainly c to w TX. Jun–Oct. Native of Europe. Some authorities (e.g., Brooks 1986a) suggest it is possibly a hybrid of *M. arvensis* and *M. aquatica*. PEPPERMINT is the source of menthol, a widely used flavoring in things such as chocolate, crème de menthe, tea, ice cream, and toothpaste (Mabberley 1987); the fresh leaves are also frequently used to flavor beverages and foods.

Mentha ×rotundifolia (L.) Huds. [*M. longifolia* × *M. suaveolens*], (round-leaved), APPLEMINT, ROUND-LEAF MINT. Plant with leafy stolons; stems usually erect, 0.45–1(–1.5) m tall; leaves broadly elliptic, 2.5–5 cm long, crenate-serrate; calyces ca. 2 mm long; corollas ca. 4 mm long, white or pink. Roadsides, waste places; included based on citation of vegetational areas 4 and 5 (Fig. 2) by Hatch et al. (1990); mainly c and w TX. May–Sep. Native of Europe.

Mentha spicata L., (with spikes), SPEARMINT. Plant stoloniferous; stems to 0.3–0.7(–1.2) m tall, often purplish; leaves oblong-lanceolate to ovate-lanceolate, acute to acuminate, 3–7(–9) cm long, sharply serrate; calyces 1.5–2 mm long; corollas 1.7–3 mm long, whitish to lavender or pinkish. Ditches, around lakes, other moist areas; Bell Co., also Brown (HPC), Dallas, and Tarrant (R. O'Kennon, pers. obs.) cos.; mainly c and w TX. Native of Europe. Jun–Oct. The fresh leaves are frequently used to flavor beverages and foods; an oil is also obtained from SPEARMINT and used in flavoring a variety of items including chewing gum.

MONARDA HORSE MINT, BEEBALM

Ours erect, aromatic, annual or perennial herbs; leaf blades usually toothed, gland-dotted; flowers large, sessile or nearly so, in leafy-bracted, head-like floral clusters, the floral clusters solitary or arranged in interrupted spikes; floral clusters subtended by large conspicuous bracts and the individual flowers also usually subtended by small, inconspicuous, linear or subulate bracteoles; calyces tubular, prominently ribbed, with 5, short, erect, stiff, acute or spine-tipped teeth, hairy or not in the throat; corollas bilabiate, elongate, with a long tube and an erect to sickle-shaped upper lip, the lower lip wider, 3-lobed; stamens 2.

A North American of genus of ca. 19 species (Turner 1994e); a number are used as teas, herbs to flavor foods, or medicinally; others are cultivated as ornamentals; some are frequently components of commercial wildflower mixes. We are following a recent treatment by Turner (1994f) for nomenclature of *Monarda*. (Named for Nicolas Monardes, 1493–1588, Spanish physician and botanist who wrote on medicinal and other useful plants of the New World; he authored a book translated into English in 1577 under the title *Joyful Newes out of the Newe Founde Worlde*.)

REFERENCES: McClintock & Epling 1942; Fernald 1944; Shinners 1953b; Scora 1967; Turner 1994f.

1. Flower heads solitary (rarely 2) at ends of branches; middle and lower leaf blades abruptly narrowed to truncate or subcordate base; corolla tube gradually expanded; upper lip of corollas erect or slightly arching; stamens exserted beyond upper lip.
   2. Corollas nearly white or pale lavender or pale pink, with purple or red-purple dots; broad leaf-like bracts subtending floral clusters long-pilose on the margins, 2–8 mm wide; calyx teeth with conspicuously stalked glands (use lens); uppermost stem leaves 3–14 mm wide; mid-stem leaves with petioles mostly 1–3 mm long ____ **M. russeliana**
   2. Corollas white or lavender, not purple- or red-purple-dotted; broad leaf-like bracts minutely pubescent on the margins, 5–17 mm wide; calyx teeth without stalked glands; uppermost stem leaves 7–20 mm wide; mid-stem leaves with petioles 3 mm or more long.

3. Nodes and often lower stems with long, spreading hairs; pedicels 2–5 mm long; corollas cream-white; mid-stem leaves with petioles mostly 3–7 mm long, or if somewhat longer then ± pilose with spreading hairs ____ **M. lindheimeri**
3. Nodes and stems nearly glabrous or with minute appressed or incuved hairs; pedicels 1–2 mm long; corollas lavender (rarely white); mid-stem leaves with petioles mostly 8 mm or more long, never pilose with spreading hairs ____ **M. fistulosa**

1. Flower heads (1–)2–7 in an interrupted spike; middle and lower leaf blades usually gradually narrowed at base; corolla tube narrow for most of its length and then abruptly expanded; upper lip of corollas sickle-shaped; stamens not exserted.
4. Bracts subtending floral clusters and calyx teeth both ending in long, bristle-like tips usually 2 mm or more long; calyx teeth (3–)5–10 times as long as wide (including bristle-like tips); corollas white to pink or lavender, with or without purple spots.
5. Bracts all ± the same in shape, oblong, usually bent outward near the middle revealing the pubescent inner surface, variable in color but often purplish, abruptly narrowed into a bristle tip; corollas white to pink or lavender with purple spots ____ **M. citriodora**
5. Bracts dissimilar, the outer ones broad, progressing to narrow inner ones, usually not bent outward, or if so, the inner surface glabrous or nearly so, variable in color but often whitish or yellowish to green, sometimes tinged with purple, gradually tapering into a terminal bristle; corollas white to pink or lavender, usually without spots.
6. Calyx teeth stout, stiff, 3–6 mm long, often reddish or purplish; bracts elliptic, with veins strongly raised; widespread in nc TX ____ **M. clinopodioides**
6. Calyx teeth slender, not stiff, 1–3 mm long, mostly greenish; bracts ovate, the veins slightly raised; mainly w TX, questionably e as far Somervell Co. in West Cross Timbers ____ **M. pectinata**
4. Bracts subtending floral clusters acute to acuminate but not bristle-tipped; calyx teeth triangular to triangular-lanceolate, not bristle-tipped, 1–3 times as long as wide; corollas yellow to cream, white, or rarely pink, usually with red-brown spots ____ **M. punctata**

Monarda citriodora Cerv. ex Lag., (lemon-scented), LEMON BEEBALM, LEMON MINT, HORSE MINT. Pubescent annual 15–80 cm tall; floral bracts and calyces hispid-margined; bracts with a spinose bristle 2–5 mm long, the inner surface often purple; calyx teeth with bristle-like tips 2–7 mm long; corollas white to pink or lavender with purple dots, with tubes 7–19 mm long. Prairies, savannahs, roadsides; throughout TX. Mainly May–Jul. According to Kirkpatrick (1992), citronellol, used as a perfume and insect repellent, is obtained from this species. ▣/99

Monarda clinopodioides A. Gray, (from resemblance of the bracteal leaves to those of *Clinopodium*, now treated as *Pycnanthemum incanum* (L.) Michx.), BASIL BEEBALM. Pubescent annual 15–55 cm tall; bracts and calyx teeth hispid-margined, the inner bracts 3–7 mm wide; corollas white to pink or lavender, sometimes with purple markings, with tubes 11–15 mm long. Sandy open ground; mainly e 1/2 of TX. May–Jun.

Monarda fistulosa L. var. **mollis** (L.) Benth., (sp.: hollow; var.: soft, with soft hairs), WILD BERGAMOT, LONG-FLOWER HORSE MINT. Finely pubescent perennial 70–150 cm tall; calyx teeth acuminate, 1–2 mm long; corollas lavender (rarely white), with tubes 15–24 mm long. Stream banks, hillsides, open ground, prairies, or woods; se and e TX w to West Cross Timbers. May–Jul. Sometimes treated as [*M. fistulosa* subsp. *fistulosa* var. *mollis*]. [*M. mollis* L.] ▣/99

Monarda lindheimeri Engelm. & A. Gray ex A. Gray, (for Ferdinand Jacob Lindheimer, 1801–1879, German-born TX collector), LINDHEIMER'S BEEBALM. Finely pubescent perennial 30–65 cm tall; calyx teeth acuminate, 1–2 mm long; corollas creamy white, with tubes 11–19 mm long. Open woods, sandy or gravelly limestone soils; Grayson Co. in Red River drainage, also Fannin Co. (Turner 1994f); mainly se and e TX. May. [*M. hirsutissima* Small]

Monarda pectinata Nutt., (comb-like), PLAINS BEEBALM. Pubescent annual 15–50 cm tall; floral bracts and calyces hispid-margined; corollas white to pink or lavender, with or without purple markings, with tubes 8–14 mm long. Panhandle to Trans-Pecos, e to Somervell Co. (according to McClintock and Epling 1942, who cited this record with question mark); Turner (1994f) mapped this species only in the w 1/2 of TX far to the w of nc TX and questioned the McClintock and Epling citation. May–Jun, sporadically later.

Monarda punctata L., (spotted), SPOTTED BEEBALM, PERENNIAL SANDY-LAND-SAGE, DOTTED MONARDA, HORSE MINT, YELLOW HORSE MINT. Annual or short lived perennial, usually freely branched, to 100 cm tall; corollas yellow to cream, white or rarely pink, usually with reddish brown dots. Sandy open woods or open ground; in much of nc TX this species is an excellent indicator of sandy soil. Late May–Jul, sporadically to Sep. This noticeably aromatic species contains thymol, an antiseptic substance; it is not grazed by cattle and sometimes forms extensive stands (Ajilvsgi 1984). The following are recognized by some authorities (e.g., Kartesz 1994) as varieties within subsp. *punctata*.

1. Leaf blade pubescence consisting mainly of long (0.3–1 mm), erect or ascending hairs along the midvein on the lower surface; corollas white to cream, with reddish brown dots; in nc TX only known from Lamar Co. in Red River drainage ____ var. **lasiodonta**
1. Leaf blade pubescence on both surfaces of very short (0.1–0.2 mm), appressed or incurved hairs; corollas yellow, white, cream, or pink, with reddish brown dots; widespread in nc TX.
   2. Calyx teeth usually longer than broad, often 2 times as long as broad, usually acuminate; corollas yellow; bracts green to yellowish green; leaves 5–7 cm long; widespread in nc TX ____ var. **intermedia**
   2. Calyx teeth ca. as broad as long, usually acute; corollas white to cream or pink; bracts pale or whitish green; leaves 3–4 cm long; w part of nc TX ____ var. **occidentalis**

var. **intermedia** (E.M. McClint. & Epling) Waterf., (intermediate). Leaf blades 10–23 mm wide; calyx teeth narrow, elongate, acuminate. Nc TX w to Rolling Plains and e Edwards Plateau; endemic to TX. This is by far the most common variety in nc TX. [*M. punctata* subsp. *intermedia* E.M. McClint. & Epling] /99

var. **lasiodonta** A. Gray, (woolly-toothed), PLUMETOOTH BEEBALM. Leaf blades commonly 4–7 cm wide. Lamar Co. in Red River drainage (Turner 1994f); mainly se and e TX. [*M. lasiodonta* (A. Gray) Small]

var. **occidentalis** (Epling) E.J. Palmer & Steyerm., (western), WESTERN BEEBALM. Leaf blades 8–13 (–27) mm wide; calyx teeth broadly triangular-lanceolate. According to Turner (1994f), this variety only occurs to the w of nc TX; Scora (1967), however, cited a number of nc TX counties: Bell, Dallas, Hamilton, Johnson, Milam, and Tarrant; specimens from Brown, Comanche, and Montague cos. appear to match this variety; nc TX w to w TX. [*M. punctata* subsp. *occidentalis* Epling]

Monarda russeliana Nutt. ex Sims, (for Thomas Russell, 1793–1819, surgeon and collector with Thomas Nuttall), RUSSELL'S BEEBALM. Perennial 30–60(–80) cm tall; stems slender, rather weak; calyx teeth ca. 2 mm long; corollas white or pale lavender, with purple dots. Sand or sandy loam, woods, roadsides, fields; Hopkins, Lamar, and Red River cos., also Fannin Co. (Turner 1994f); ne corner of nc TX and adjacent n part of e TX.

Monarda stanfieldii Small, (named for Stanfield who collected it at San Marcos, TX in 1897), [*Monarda punctata* var. *stanfieldii* (Small) Cory, *M. punctata* subsp. *stanfieldii* (Small) Epling]. According to Turner (1994f) who recognized it at the specific level, this is a well-marked taxon largely confined to the granitic sands along the middle course of the Colorado River; it occurs immediately to the s of nc TX. Its calyx (mouth closed by a dense mass of white hairs) and glabrous corolla tube distinguish it from *M. punctata* (calyx opening merely ciliate; corolla tube pubescent).

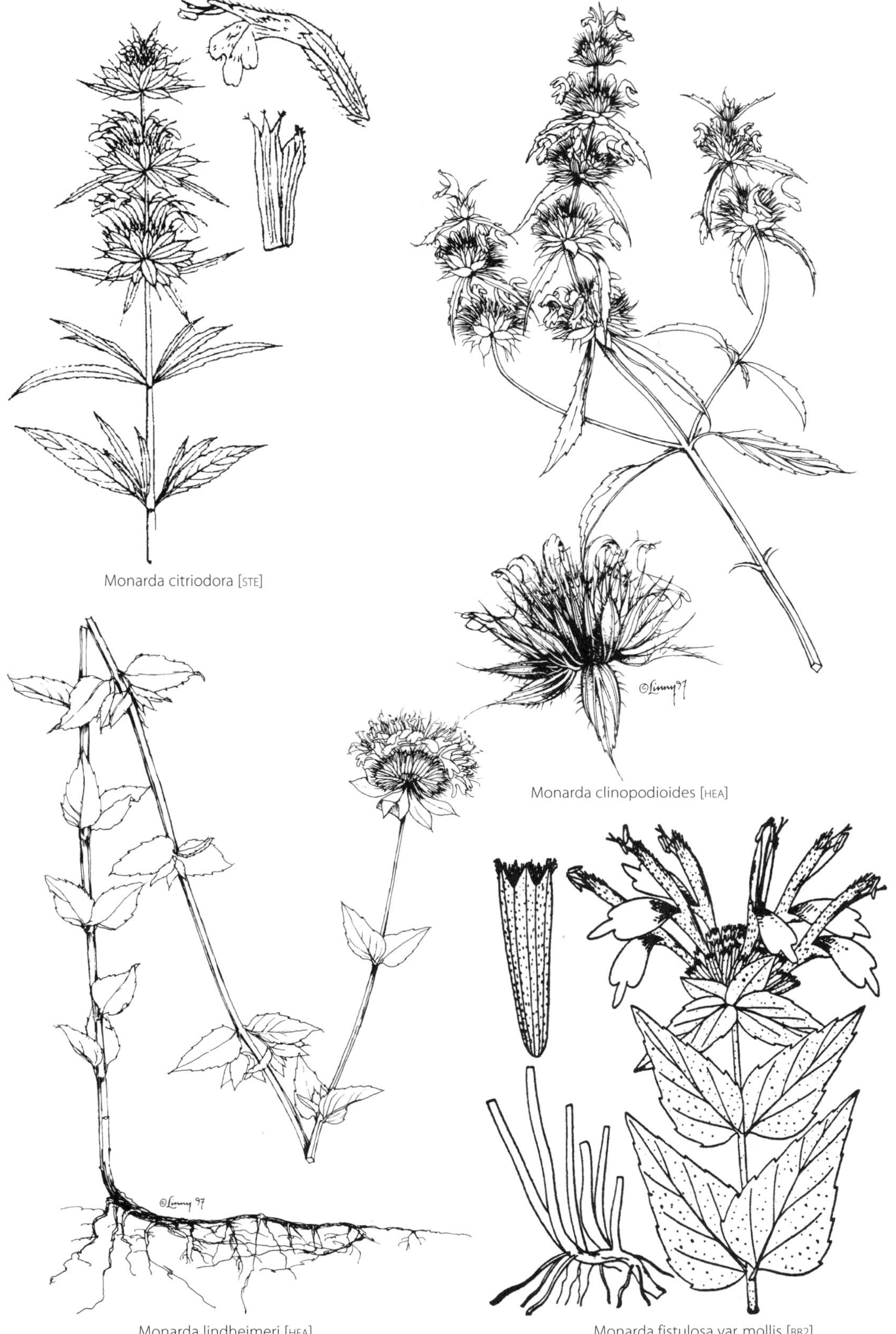

Monarda citriodora [STE]

Monarda clinopodioides [HEA]

Monarda lindheimeri [HEA]

Monarda fistulosa var. mollis [BB2]

Mondara viridissima Correll, (very green), [*M. punctata* subsp. *punctata* var. *viridissima* (Correll) Scora]. This is a mostly fall flowering entity that occurs just to the se of nc TX. According to Turner (1994f), who recognized it at the specific level, this taxon is largely confined to areas of Carrizo sands in sc TX. It has 2 or more flowers heads in an interrupted spike and can be distinguished from nc TX taxa by the combination of its narrow (mostly 4–6 mm wide), linear-lanceolate leaves and stem hairs short and usually spreading at right angles to the stem.

NEPETA CATNIP, CAT MINT

A genus of ca. 250 species of temperate Eurasia, n Africa, and tropical African mountains, usually in dry habitats; some are strongly attractive to cats. (Latin name, possibly from Nepete, an Etruscan city, or Nepi in Italy)

Nepeta cataria L., (old generic name, pertaining to cats), CATNIP, CATNEP. Pubescent, perennial, erect or ascending herb 0.3–1 m tall; leaves pale green, petioled; leaf blades triangular-ovate, coarsely toothed, with minute golden glands on lower surface; flowers in dense or interrupted, spike-like inflorescences; calyces 5–7 mm long; corollas bilabiate, 7–12 mm long, white with red-violet dots; stamens 4. Disturbed areas and waste places; reported by Reverchon as "adventive in Dallas Co." (Mahler 1988), also Henderson and McLennan cos. (Mahler 1988); e TX w to nc TX and Edwards Plateau. May–Sep. Native of Eurasia. While CATNIP is sometimes used as a herbal remedy, toxicity was reported in a 19-month-old child who displayed a "drugged" appearance after ingesting CATNIP (Osterhoudt et al. 1997).

PERILLA

A genus of ca. 6 species native from India to Japan. (Possibly from the Hindu name)

Perilla fructescens (L.) Britton, (shrubby, bushy), COMMON PERILLA, BEEFSTEAK-PLANT. Aromatic, annual, coarse herb, often with purplish coloration; stems 0.2–1 m tall; leaf blades ovate to suborbicular, 4–15 cm long, 3–10 cm wide, coarsely serrate to crenate, sometimes ± crinkly in appearance, acute to acuminate at apex, with minute glands on lower surface; petioles 1–8 cm long; flowers solitary in axils of very small bracts, together forming spike-like, elongate, somewhat 1-sided racemes 5–15 cm long; calyces 2–3 mm long at flowering, elongating to 8–12 mm long in fruit, the base slightly inflated and villous outside; corollas bilabiate, 5-lobed, white to lavender, 2.5–3 mm long, ca. as long as or only slightly longer than calyces; stamens 4, ca. equal. Wet areas, low woods, along streams, low pastures; Grayson Co., also Lamar Co. as an abundant pasture weed (G. Diggs, pers. obs.); mainly se and e TX. Jul–Nov. Native from Himalaya to e Asia. Much cultivated in the Old World as an ornamental, for the leaves and seed which are eaten in Asia, and for an oil (yegoma) like linseed oil used to waterproof papers, in paints, and in printing inks (Duke 1985; Mabberley 1987). Duke (1985) cited references of toxicity including dermatitis; pulmonary edema, respiratory distress, and even death can result from ingestion by cattle and horses (Hardin & Brownie 1993).

PHYSOSTEGIA

OBEDIENT-PLANT, LION'S-HEART, FALSE DRAGON'S-HEAD

Largely glabrous, erect, perennial herbs; leaves, except lowest, sessile, slightly clasping, oblong to lanceolate or elliptic, toothed; flowers in terminal, spike-like racemes with small floral bracts; corollas bilabiate, usually large and showy, pale to deep lavender to reddish violet or white, the throat mottled white with dark dots; stamens 4.

A North American genus of 12 species including cultivated ornamentals; almost all the species are showy and are potential ornamentals. Flowers if pushed to one side remain in place, hence the common name OBEDIENT-PLANT (not one generally used in TX). (Greek: *physa*, bladder,

and *stege*, covering, in allusion to the calyx which can become somewhat inflated)
REFRERENCES: Lundell 1959, 1960, 1969a; Cantino 1982.

1. Several pair of small sterile bracts usually present and conspicuous below lowest flower of inflorescence; plants flowering Aug–Dec ____ **P. virginiana**
1. Sterile bracts absent below lowest flower of inflorescence; plants flowering mid-Mar–Jul, rarely to Aug.
   2. Corollas less than 2 cm long (usually 1.5 cm or less); flowering calyces usually 3–5(–6) mm long.
      3. Corollas 5–7 mm long; calyces 3–4 mm long ____ **P. micrantha**
      3. Corollas 10–15(–19) mm long; calyces 5–6 mm long ____ **P. intermedia**
   2. Corollas, at least most, over 2 cm long; flowering calyces 5–10 mm long.
      4. Leaves conspicuously broad, to 75 mm wide; flowering late Jun–Jul, rarely to Aug ____ **P. digitalis**
      4. Leaves narrow, 25 mm or less wide; flowering Apr–Jun, rarely to Jul.
         5. Corollas usually pale lavender to whitish; lower petiolate leaves usually absent at flowering time; leaves usually 12 mm or less wide; flowering mid-May–Jun, rarely to Jul; limited to extreme s part of nc TX ____ **P. angustifolia**
         5. Corollas deep lavender to reddish violet; lower petiolate leaves usually present at flowering time; leaves to 25 mm wide; flowering Apr–mid-May; widespread in Blackland Prairie ____ **P. pulchella**

Physostegia angustifolia Fernald, (narrow-leaved). Stems to 2 m tall, usually much less; rhizomes short and vertical; lowest 1–4 stem nodes bearing petiolate leaves deciduous by anthesis; leaves in nc TX usually < 12 mm wide, linear to narrowly lanceolate or oblanceolate, sharply serrate, usually clasping stem; corollas usually 22–33 mm long, usually pale lavender to whitish (rarely brighter lavender), spotted and sometimes streaked inside with purple; nutlets usually 2–3 mm long. Low moist areas, roadside ditches; s part of Lampasas Cut Plain (Lampasas Co.), also s part of Blackland Prairie (Burnet Co.) (Cantino 1982); also Edwards Plateau. Mid-May–Jun (Jul). [*P. edwardsiana* Shinners]

Physostegia digitalis Small, (of the finger, possibly from the flowers being somewhat like the finger of a glove, or from resemblance to flowers of the genus *Digitalis*—foxglove), FINGER LION'S-HEART, FALSE DRAGON'S-HEAD. Stems stout, to 2 m tall; leaves large, to 22 cm long, oblong to elliptic-oblong; calyces 8–10 mm long; corollas 20–25 mm long, pale lavender to whitish, often with reddish purple dots. Sandy open areas; included based on citation of vegetational area 4 (Fig. 2) by Hatch et al. (1990); possibly on extreme e edge of nc TX; mainly se and e TX. Jun–Jul(–Aug).

Physostegia intermedia (Nutt.) Engelm. & A. Gray, (intermediate), INTERMEDIATE LION'S-HEART. Stems 0.3–1.5 m tall; rhizomes producing horizontal secondary and tertiary rhizomes to 40 cm long; lower 3–8 pairs of stem leaves petiolate, usually deciduous by anthesis; leaves widely spaced, at least some clasping the stem, linear-lanceolate to linear, the larger usually 3–15 mm wide, entire or teeth few; corollas lavender, spotted and streaked inside with purple; nutlets usually 2–2.5 mm long. Low moist areas, roadside ditches; Dallas, Denton, and Rockwall cos., also Bell (Fort Hood—Sanchez 1997), Cooke (Mahler 1988), and Fannin (Lundell 1969a) cos.; se and e TX w to nc TX. Late Mar–Late Jul.

Physostegia micrantha Lundell, (small-flowered). Similar to *P. intermedia*; stems to ca. 0.9 m tall, usually much smaller; leaves to ca. 12 cm long and 11 mm wide; corollas pinkish to light lavender or white. Wet bottomlands; originally known only from a single population from Titus Co. in e TX; also from Collin Co. in the Blackland Prairie; also a population has been observed in Red River Co. (A. Crosthwaite, pers. comm.). May–Jun. The species has also been reported from McCurtain Co., OK (Correll & Correll 1972), but no specimens were located when a status report was prepared by Tyrl et al. (1978). Cantino (1982) indicated that since this taxon was based on a single population, since it resembled *P. intermedia* except in having smaller flowers with aborted anthers, and since typical *P. intermedia* individuals were also present in the population, it should be lumped with *P. intermedia*. However, more recently (in 1995), a large population (100s to a thousand individuals) of consistently small-flowered individuals closely resembling the type collection of *P. micrantha* was located in a roadside wetland in Collin Co. by R. O'Kennon; in early 1998 this population was destroyed by the TX Dept. of Transportation. While final disposition of this entity is unclear, we are here recognizing it at the specific level. Kartesz (1994) and Jones et al. (1997) lumped it with *P. intermedia*.

Physostegia pulchella Lundell, (pretty, beautiful), BEAUTIFUL FALSE DRAGON'S-HEAD. Stems to 1.4 m tall; rhizomes unbranched, vertical, to 6 cm long; lowest 1–4 pairs of stem leaves petiolate, some usually present at anthesis; stem leaves usually sharply serrate to base; base of sessile leaves auriculate clasping; corollas spotted or streaked inside with purple; nutlets 2.2–3 mm long. Low moist areas, roadsides ditches; se and e TX w to Grayson and Collin cos., also Cooke and Denton cos.; this is the most common *Physostegia* of the Blackland Prairie; endemic to TX. Apr–mid-May. /103

Physostegia virginiana (L.) Benth. subsp. **praemorsa** (Shinners) P.D. Cantino, (sp.: of Virginia; subsp.: as though the end was bitten off). Stems to ca. 1.3 m tall; rhizomes short, usually unbranched, vertical; lower petiolate leaves usually early deciduous, to 13 mm wide; stem leaves narrower (to ca. 6 mm wide), the margins sharply serrate; frequently sterile floral bracts (without flowers) present (up to 40 pairs) below lowest flower; corollas reddish violet, lavender, or white, usually spotted and streaked with purple inside; nutlets usually 2.9–3.8 mm long. Prairies, limestone glades, roadside ditches; the holotype was collected in Fannin Co., also two nc TX localities (without county) were mapped by Cantino (1982); se and e TX w to nc TX, also Trans-Pecos. Aug–Dec. [*P. praemorsa* Shinners, *P. serotina* Shinners]

subsp. *virginiana*, native to the e U.S., is cultivated in nc TX and possibly escapes (1 individual has been found in Collin Co.); apparently also escaping in e TX. It can be distinguished by its much broader leaves (those of the stem 3–55 mm wide).

PRUNELLA HEALALL, SELFHEAL

A genus of 4 species of the n temperate zone and nw Africa. (From the pre-Linnean name *brunella*, which may have been derived from the German: *breaume*, quinsy, a throat infection these plants were supposed to cure)

Prunella vulgaris L. subsp. **lanceolata** (Barton) Hultén, (sp.: common; subsp.: lanceolate, lance-shaped), COMMON SELFHEAL, CARPENTER-WEED. Low perennial; stems erect to half decumbent or prostrate, spreading-pilose; leaves petioled; leaf blades oblong-lanceolate, 3–7 cm wide, indistinctly toothed or entire, usually cuneate to attenuate at base; flowers in dense, cylindrical, terminal spikes 2–5(–7) cm long; bracts green or often with purple; calyces green or purple, 6–11(–15) mm long; corollas bilabiate, 10–15(–20) mm long, lavender to blue-purple (rarely white), with pale or white center on lower lip, the upper lip hooded. Sandy ground, damp woods; se and e TX w to West Cross Timbers, more common eastward. Apr–May. [*Prunella vulgaris* var. *lanceolata* (Barton) Fernald]

subsp. *vulgaris* including [*P. vulgaris* var. *hispida* Benth.], with wider, ovate leaves rounded basally, also occurs in vegetational area 4 (Fig. 2) according to Hatch et al. (1990). We have seen no nc TX material of this subspecies. Brooks (1986a) indicated that within this cosmopolitan species, subsp. *vulgaris* represents European plants and that the presence of intermediate individuals makes taxonomic recognition of infraspecific taxa questionable.

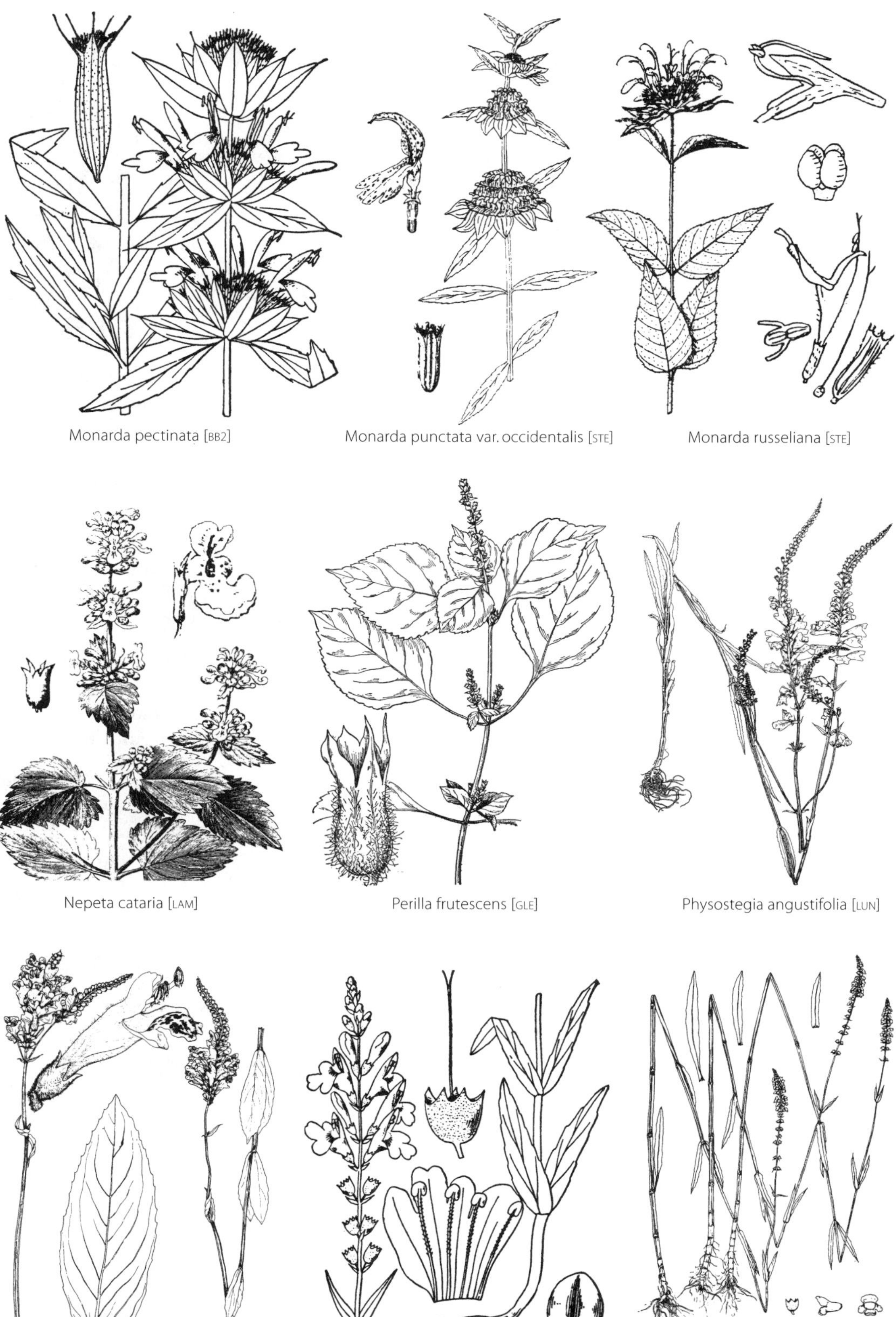

Monarda pectinata [BB2]

Monarda punctata var. occidentalis [STE]

Monarda russeliana [STE]

Nepeta cataria [LAM]

Perilla frutescens [GLE]

Physostegia angustifolia [LUN]

Physostegia digitalis [CO1]

Physostegia intermedia [BB2]

Physostegia micrantha [LUN]

PYCNANTHEMUM MOUNTAIN MINT

Erect perennial herbs with a strong mint aroma; foliage gland-dotted; bracts often whitish; flowers crowed in terminal head-like clusters or looser corymbs with branches visible; corollas much larger than calyces, stamens 4, didynamous.

A North American genus of 17 species including some used medicinally and as flavorings. (Greek: *pycnos*, dense, and *anthemon*, flower, from the densely-flowered inflorescences) REFERENCE: Grant & Epling 1943.

1. Leaf blades linear, usually 4.5 mm or less wide, entire ____ **P. tenuifolium**
1. Leaf blades narrowly lanceolate to ovate, > 5 mm wide, subentire to toothed, the teeth often scattered.
   2. Leaf blades rounded to subcordate at base, sessile or nearly so; flowers in very tight globose heads, so tightly clustered that the branches holding the flowers or small groups of flowers cannot be seen; calyces ± radially symmetrical, all of the teeth ca. the same length, not 2-lipped ____ **P. muticum**
   2. Leaf blades wedge-shaped at base, on petioles 2–6(–12) mm long; flowers in crowded inflorescences but loose enough that the stalks supporting individual flowers or small groups of flowers are clearly visible; calyces 2-lipped with some teeth longer than others ____ **P. albescens**

Pycnanthemum albescens Torr. & A. Gray, (whitish), WHITE-LEAF MOUNTAIN MINT, WHITE MOUNTAIN MINT. Stems 0.4–0.8(–1.5) m tall, white canescent above; leaves ovate, ovate-lanceolate, or elliptic, 2.5–7 cm long, 10–25 mm wide, all but the lower strongly whitened; bracts strongly whitened; calyces 3.5–5 mm long; corollas 5–7.5 mm long, whitish or pale lavender with purple spots on lower lip. Low woods, along streams; Fannin Co. (Talbot property) in Red River drainage, also Lamar Co. (Carr 1994); mainly se and e TX. Jul–Nov.

Pycnanthemum muticum (Michx.) Pers., (blunt, pointless), CLUSTER MOUNTAIN MINT. Stems to 1.1 m tall; leaves ovate-lanceolate to ovate, 2.5–8 cm long, 15–40 mm wide; bracts often whitened; calyces 3–5 mm long; corollas 4–6 mm long, white to purplish with purple spots. Dry open woods; Dallas Co.; rare in TX where it is limited to the nc part of the state; widespread in the e U.S. Jul–Nov.

Pycnanthemum tenuifolium Schrad., (slender-leaved), SLENDER MOUNTAIN MINT, NARROW-LEAF MOUNTAIN MINT, SLENDER-LEAF MOUNTAIN MINT. Plants often forming dense colonies from horizontal roots; stems 0.4–1.1 m tall, glabrous or puberulent on the angles; leaves 1.5–6 cm long, 1–4.5 mm wide, sessile, not whitened; flowers in dense or open corymbs; calyces radially symmetrical, 3.7–5 mm long; corollas 5–7 mm long, whitish or pale lavender, with purple spots on lower lip. Moist open areas, limestone; Cooke, Grayson, Henderson, and Lamar cos.; se and e TX rarely w to nc TX. May–Oct.

SALVIA SAGE

Annual or perennial herbs or shrubs, often aromatic; leaves often gland-dotted; flowers in terminal, spike-like racemes, opposite or whorled, with minute to large leafy bracts; calyces bilabiate; corollas strongly bilabiate, purple, blue, red, or white, often showy; stamens 2.

A genus of 900 species of the tropics to temperate regions, especially of the Americas, the Sino-Himalaya area, and sw Asia. Some are used medicinally or as culinary herbs; *Salvia officinalis* L. (COMMON SAGE, GARDEN SAGE), native to s Europe and the Mediterranean, is widely grown and used as an herb; it is the primary flavoring used in sausage and in many stuffings and dressings. Many other *Salvia* species, particularly ones with red flowers, are used as ornamentals. The seeds of some species of *Salvia* (known to Native Americans as Chia) were used

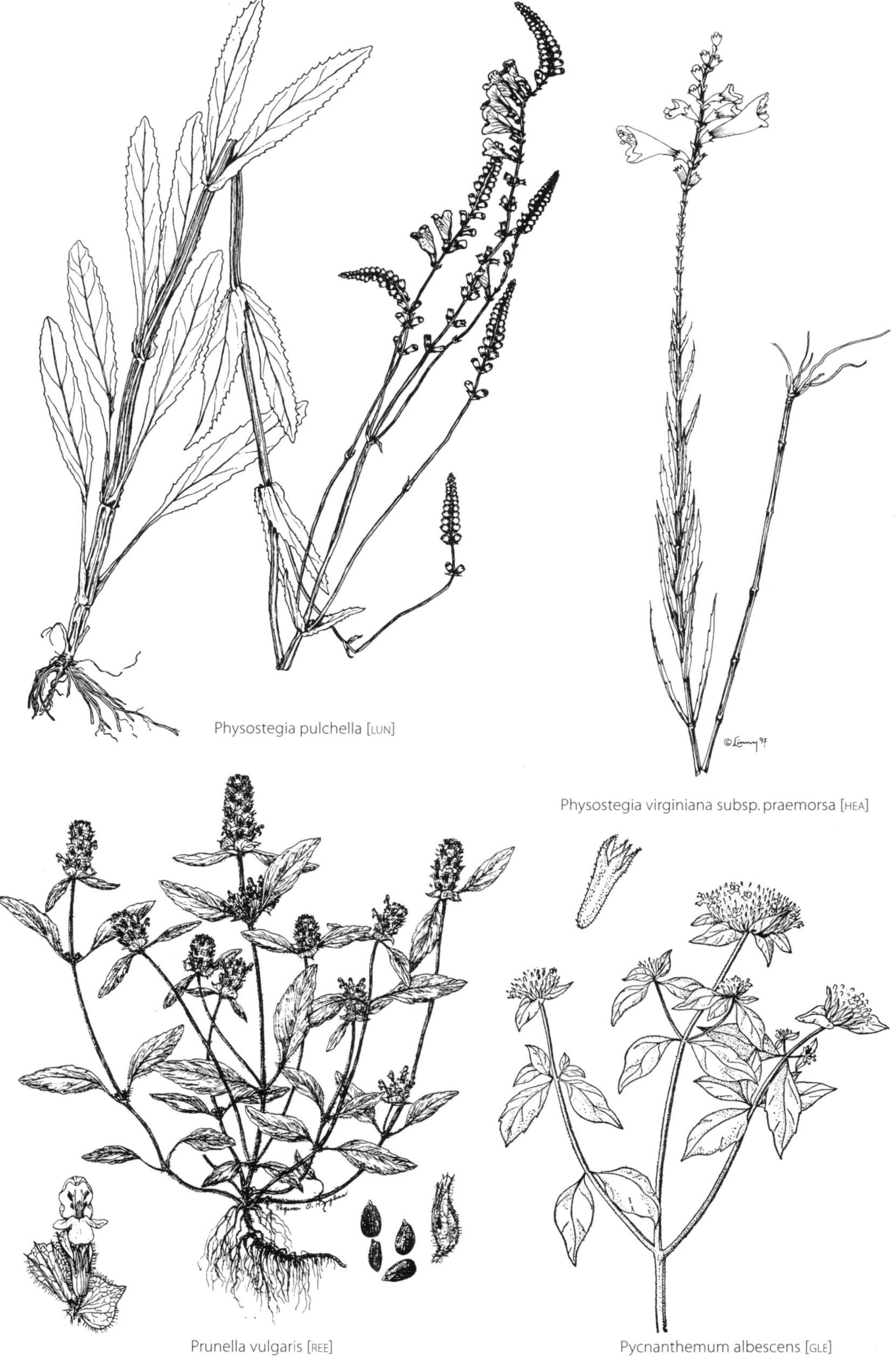

Physostegia pulchella [LUN]

Physostegia virginiana subsp. praemorsa [HEA]

Prunella vulgaris [REE]

Pycnanthemum albescens [GLE]

for food; the seeds, which are rich in protein and easily digested fats, were toasted and ground to make a flour (Powell 1988). The 2 anthers have a rocker action, with pollinating bees, etc. having pollen pressed onto their backs or heads from the fertile part of the anther as a result of pushing against a sterile projection from the other end. (Old Latin name, from *salvare*, to save or heal, alluding to the medicinal properties of many of the species).
REFERENCES: Epling 1938–1939; Whitehouse 1949.

1. Floral bracts large and leafy, exceeding the calyces; calyx throat with dense ring of bristly white hairs inside.
   2. Upper lip of corolla ca. 5–6.5 mm long, shorter than the undivided portions of the corolla (tube and throat) which together are ca. 10–13 mm long; stems usually with 4–6 pairs of leaves below inflorescences; inflorescences with few flowers open at one time ____ **S. texana**
   2. Upper lip of corolla 10–12 mm long, longer than undivided portions of the corolla (tube and throat) which together are ca. 8–10 mm long; stems with 8–11 pairs of leaves below inflorescences; inflorescences with 10–30 flowers open at one time ____ **S. engelmannii**
1. Floral bracts small, not exceeding the calyces; calyx without dense ring of bristly hairs inside.
   3. Leaves (at least some, often the lowest) deeply divided, lobed, or compound.
      4. Corollas blue or violet; leaves mostly in a basal rosette (1 or 2 pairs of small leaves can be on the flower stalk), at least some usually deeply divided or lobed, but not with distinct leaflets; on sandy soils in ne part of nc TX ____ **S. lyrata**
      4. Corollas scarlet red; leaves abundant on the stem as well as basal, the lower ones often with 3–4 distinct leaflets, the terminal leaflet much larger; on limestone in s part of nc TX ____ **S. roemeriana**
   3. Leaves entire or at most toothed, neither divided, lobed, nor compound.
      5. Basal leaves present at flowering time; leaves mostly in a basal rosette (1 or 2 pairs of small leaves can be on the flower stalk) ____ **S. lyrata**
      5. Basal leaves usually withered by flowering time; stem leaves numerous, crowded.
         6. Plants shrubs with definitely woody stems; corollas red ____ **S. greggii**
         6. Plants herbaceous, the stems not woody; corollas purple, violet-blue, blue, or red.
            7. Flowers mostly opposite (one per axil and thus 2 per node); corollas 6–10.5 mm long; plants annual ____ **S. reflexa**
            7. Flowers whorled (usually > 1 per axil and thus usually > 2 per node); corollas 14–22 mm long; plants perennial (but may flower the first year).
               8. Calyces with matted, felty, white or purplish hairs; calyx teeth very short, virtually unnoticeable due to the matted hairs; corollas purple or violet-blue ____ **S. farinacea**
               8. Calyces with appressed or spreading hairs, the hairs not matted; calyx teeth clearly visible to the naked eye, ca. 1/4–1/3 the length of the calyx; corollas red or blue.
                  9. Corollas deep red; leaf blades deltoid-ovate, truncate to cordate basally, abruptly contracted to the long petioles; petioles 10–35 mm long ____ **S. coccinea**
                  9. Corollas blue; leaf blades linear to lanceolate or oblong, long tapering to the short petioles; petioles 0–15 mm long ____ **S. azurea**

Salvia azurea Michx. ex Lam. var. **grandiflora** Benth., (sp.: azure, sky-blue; var.: large-flowered), BLUE SAGE, AZURE SAGE, PITCHER SAGE. Perennial herb, occasionally flowering when as little as 15 cm tall; stems to 1.5 m tall, pubescent with reflexed hairs; leaves 3–10 cm long, 1–4 cm wide; calyces 4.5–10 mm long; corollas 15–25 mm long, deep blue to light blue with whitish center (rarely white). Rocky, clayey, or sandy prairies; throughout most of TX. Jun–Oct. ▣/105

Salvia coccinea Buc'hoz ex Etl., (scarlet), TROPICAL SAGE, TEXAS SAGE, MIRTO, SCARLET SAGE, INDIAN-FIRE, MEJORANA. Perennial; stems to 1 m tall; leaves to 7 cm long and 5 cm wide; calyces 6–9 mm long, usually tinged with red, the lobes of the calyx nearly half the tube length; corollas 16.5–22 mm long, bright or deep red. Sandy woodlands; Tarrant Co.; scattered in e 1/2 of TX. Feb–Nov.

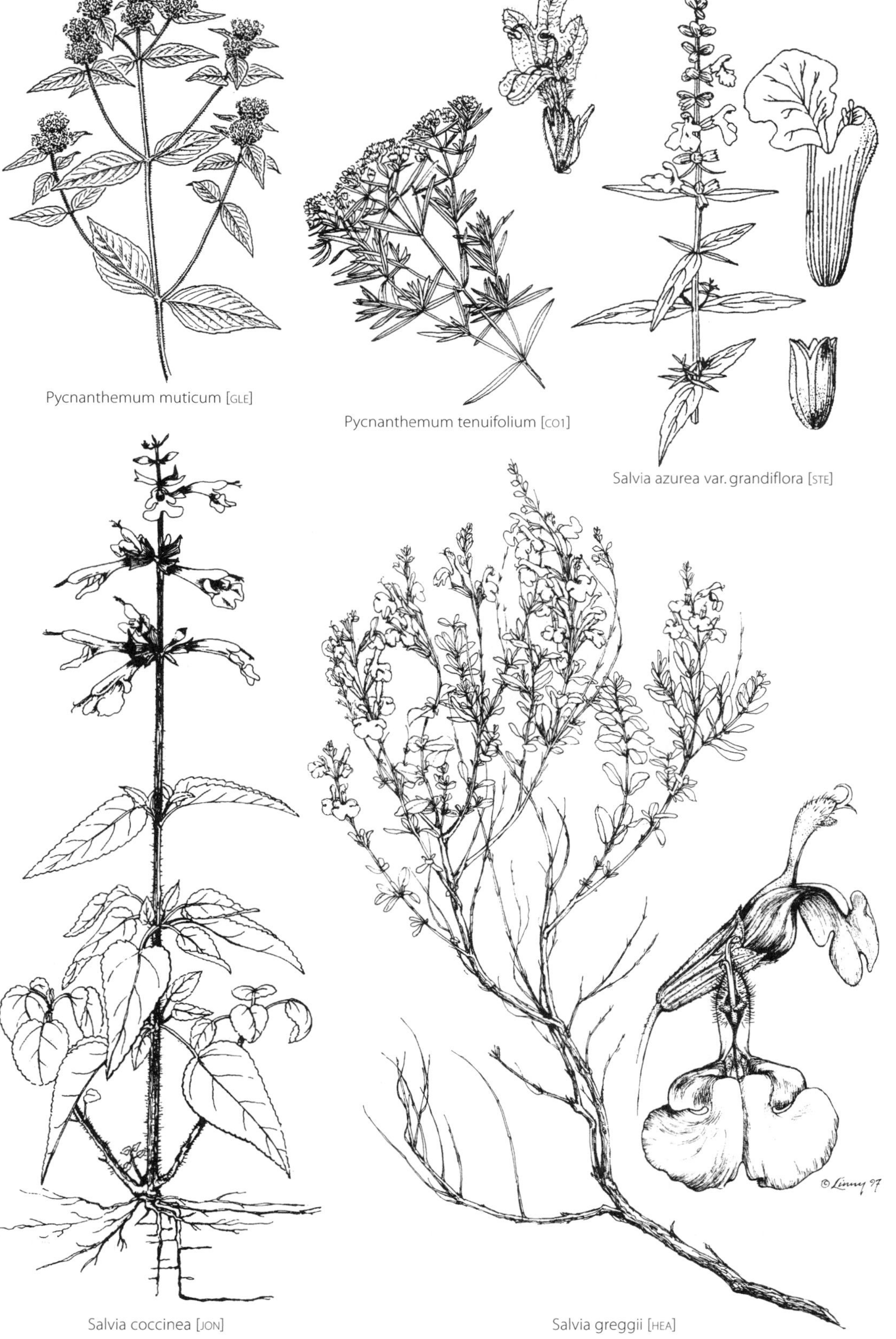

Pycnanthemum muticum [GLE]

Pycnanthemum tenuifolium [CO1]

Salvia azurea var. grandiflora [STE]

Salvia coccinea [JON]

Salvia greggii [HEA]

Salvia engelmannii A. Gray, (for George Engelmann, 1809–1884, German-born botanist and physician of St. Louis), ENGELMANN'S SAGE. Perennial herb from a woody root; stems to ca. 0.4 m tall; leaves linear-lanceolate, to 8 cm long and 1 cm wide, entire; corollas to ca. 22 mm long, gaping, pale lavender-blue, the throat with darker veins, the lower lip with 2 white areas towards base. Lime-stone prairies, often very shallow soils; Bell, Dallas (Whitehouse 1949), and Denton cos. w to West Cross Timbers and Lampasas Cut Plain and s to Edwards Plateau; endemic to TX. Apr–May.

Salvia farinacea Benth., (mealy, powdery), MEALY SAGE, MEALY-CUP SAGE. Minutely pubescent perennial herb 0.25–1 m tall, flowering the first year, developing a woody root; leaves lanceolate or oblong-lanceolate, to 10 cm long and 3 cm wide, shallowly toothed or entire; calyces with matted, felty, white or purplish hairs; corollas to 25 mm long, purple or violet-blue. Limestone prairies and rock outcrops; in nc TX mainly Dallas and Wise cos. s and w; widespread in TX. Apr–Jul. Widely planted as a native wildflower. /105

Salvia greggii A. Gray, (for Josiah Gregg, 1806–1850, who collected in Mexico and died in the wilderness in n CA), AUTUMN SAGE, GREGG'S SAGE. Much-branched, shrubby perennial to 0.9 m tall; leaves coriaceous, obovate to elliptic, small, 10–25 mm long; calyces 10–15 mm long; corollas 25–30 mm long, red. Native in rocky soils of c, w, and s TX probably s of nc TX; included based on citation of vegetational areas 4 and 5 (Fig. 2) by Hatch et al. (1990); a showy ornamental cultivated in nc TX that long persists and possibly escapes. Mar–May.

Salvia lyrata L., (lyre-shaped), CANCERWEED, LYRE-LEAF SAGE. More or less pilose, usually scapose, perennial herb 0.25–0.65(–0.8) m tall, with prominent basal leaves; stems with 1–2 pairs of reduced leaves below the simple or few-branched inflorescences; corollas 20–30 mm long, light blue-lavender with darker blue markings. Sandy woods, low ground; Hunt Co. and w in Red River drainage to Grayson Co., also Dallas Co. (S. Wasowski, pers. comm.); mainly se and e TX. Apr.

Salvia reflexa Hornem., (bent sharply backward), ROCKY MOUNTAIN SAGE, LANCE-LEAF SAGE. Minutely pubescent annual 0.2–0.6(–0.7) m tall; leaves lanceolate to linear-lanceolate, to 5 cm long and 12 mm wide, toothed or entire, rather short-petioled; calyces 4–8 mm long; corollas pale blue to whitish, 6–10.5 mm long. Ditches, disturbed soils; Dallas and Tarrant cos. s and w to w TX. May–Oct. Extremely aromatic with an aroma reminiscent of the genus *Mentha*. Reported to be toxic to cattle, possibly through nitrate poisoning (Kingsbury 1964).

Salvia roemeriana Scheele, (for Ferdinand Roemer, 1818–1891, geologist, paleontologist, and explorer of TX), CEDAR SAGE. Herbaceous perennial to ca. 0.7 m tall; foliage often conspicuously white hairy; leaves (upper) or terminal leaflet (on lower leaves) suborbicular to reniform-cordate, coarsely crenate or with an undulate margin, 2.5–5 cm wide; racemes elongate; calyces to 15 mm long, glabrous inside; corollas 25–35 mm long. Limestone rocks, cedar brakes, wooded areas; Bell Co. in s part of nc TX; mainly Edwards Plateau, also Trans-Pecos. Mar–Aug. Widely planted as a landscape plant, escapes in the landscape, and persists.

Salvia texana (Scheele) Torr., (of Texas), TEXAS SAGE. Perennial 0.1–0.4 m tall, from a woody root; stems spreading-pilose; leaves subsessile, lanceolate, to 6 cm long and 2 cm wide, entire or slightly toothed; calyces hispid-pilose outside; corollas ca. 16.5–21.5 mm long, purplish blue, the lower lip 6.5–8.5 mm long, with 2 prominent white areas toward base. Limestone prairies or rock outcrops; Dallas and Denton cos. s and w to w TX. Apr–May.

SCUTELLARIA SKULLCAP, HELMET-FLOWER

Annual or perennial herbs, sometimes with woody root or base, not aromatic; leaves often gland-dotted; flowers in the axils of ± reduced upper leaves or floral bracts; calyces with a shield-like (or skullcap-like) protrusion (= scutellum), hence the common name; corollas blue

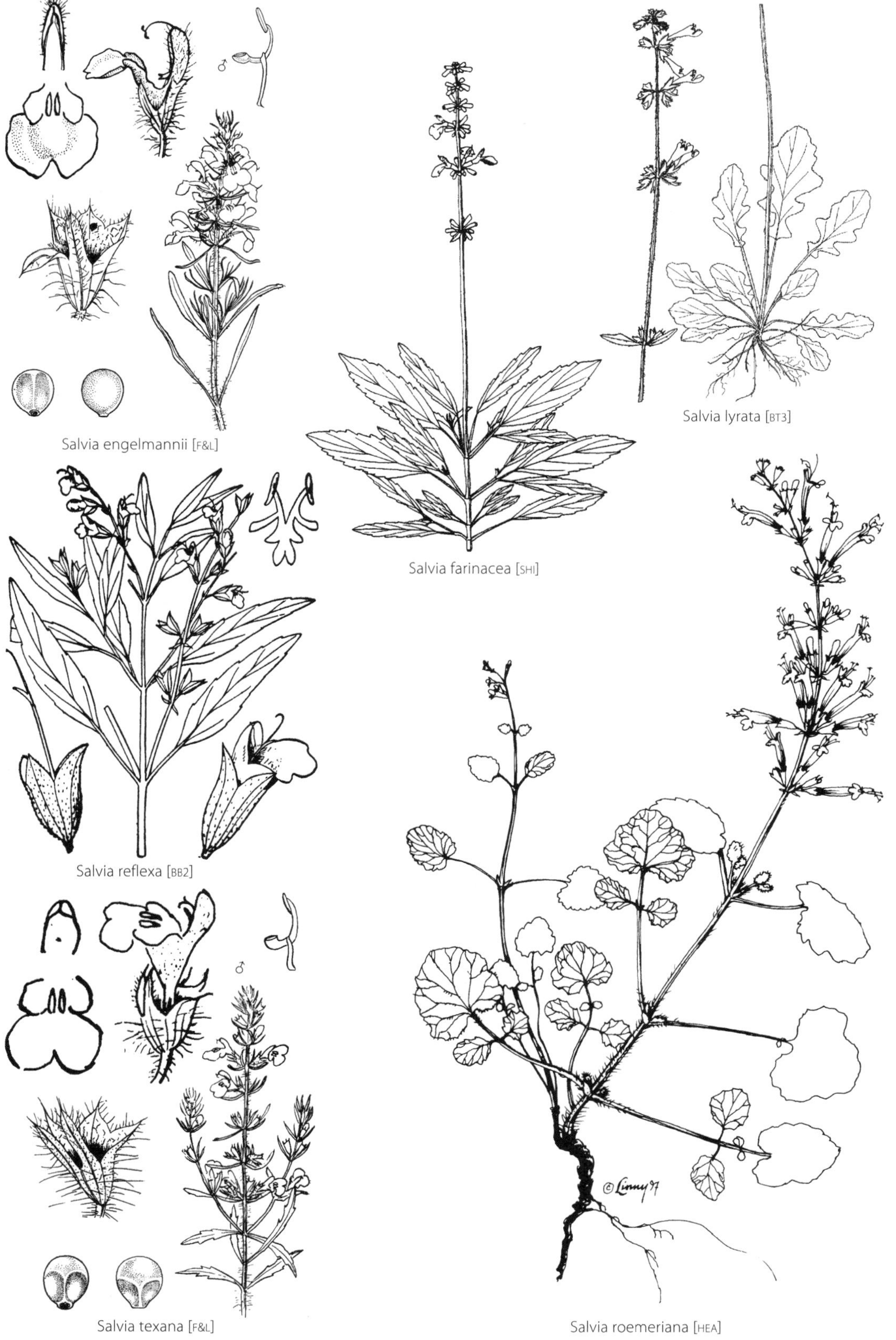

Salvia engelmannii [F&L]

Salvia lyrata [BT3]

Salvia farinacea [SHI]

Salvia reflexa [BB2]

Salvia texana [F&L]

Salvia roemeriana [HEA]

to purple or violet (rarely white), with ± prominent white markings toward base of lower lip, the upper lip hooded; stamens 4, didynamous.

A genus of 350 species, cosmopolitan except s Africa; some are used medicinally or cultivated as ornamentals. (Latin: *scutella*, a small dish or shield, in allusion to the protrusion from the calyx)

REFERENCES: Leonard 1927; Epling 1939, 1942; Lane 1983, 1986; Paton 1990; Goodman & Lawson 1992; Turner 1994b.

1. Stem leaves distinctly petioled, the longest leaves including petiole 3 cm or more long; leaf blades toothed, with cordate to truncate bases; plants usually much > 30 cm tall.
   2. Stems minutely pubescent, the hairs inconspicuous; upper leaves becoming gradually reduced in size as they grade rather smoothly into the leafy floral bracts; leaf blades glabrous on upper surfaces.
      3. Corollas large, 13–22 mm long; leaf margins minutely ciliate; calyces 4–4.5 mm long in flower, 5–6 mm long in fruit ______ **S. cardiophylla**
      3. Corollas small, 5–9 mm long; leaf margins not ciliate; calyces 2–2.7 mm long in flower, 3–4 mm long in fruit ______ **S. lateriflora**
   2. Stems prominently spreading or recurved pubescent, the hairs obvious to the naked eye; upper leaves only slightly reduced, not grading smoothly into the floral bracts—instead with an abrupt transition to the much smaller floral bracts; leaf blades pubescent on upper surfaces ______ **S. ovata**
1. Stem leaves petioled OR sessile, the largest leaves including petiole 1–3(–4) cm long; leaf blades entire or few-toothed, the bases tapering to truncate or subcordate; plants usually < 30 cm tall.
   4. Middle stem leaves with subcordate or truncate bases; plants with rhizomes; nutlets with peg-like, cylindrical, blunt projections ______ **S. parvula**
   4. Middle stem leaves with rounded to tapering bases; plants with a taproot; nutlets not as above, with short tubercles or lamellae.
      5. Calyces spreading-pubescent or pilose, with long hairs conspicuous with a lens; stems (especially upper stems) with relatively long, spreading, often gland-tipped hairs; lower leaves with petioles 4–18 mm long, the petioles mostly narrowly margined; corollas 5–11 (–13) mm long; plants annual, the stems hardened but neither enlarged nor distinctively woody ______ **S. drummondii**
      5. Calyces short-pubescent with inconspicuous hairs; stems with minute spreading or appressed hairs without gland-tips; lower leaves sessile or with petioles to 4 mm long, the blades decurrent as wings on the petioles; corollas (11–)12–23 mm long; plants perennial, the stems (at least on older plants) enlarged and distinctively woody.
         6. Stems with minute pubescence of downwardly curved and appressed hairs (use lens); leaf pubescence usually downcurved; calyces without gland-tipped hairs or sometimes with a few (as well as sessile glands); widespread in nc TX ______ **S. wrightii**
         6. Stems with minute pubescence of spreading hairs, sometimes somewhat retrorse but not downcurved; leaf pubescence spreading or ascendent; calyces usually with minute gland-tipped hairs (as well as sessile glands); on extreme w edge of nc TX ______ **S. resinosa**

Scutellaria cardiophylla Engelm. & A. Gray, (with heart-shaped leaves), HEART-LEAF SKULLCAP. Perennial, flowering the first year, 25–80 cm tall; leaf blades ± triangular, 2–5 cm long; corollas 13–22 mm long. Sandy woods; Denton, Henderson, Kaufman, and Limestone cos.; se and e TX w to East Cross Timbers, more common to the e. Jun–Jul, sporadically later.

Scutellaria drummondii Benth., (for its discoverer, Thomas Drummond, 1780–1835, Scottish botanist and collector in North America), DRUMMOND'S SKULLCAP. Annual or short-lived perennial 20–30 cm tall, old plants forming small, bushy, many-stemmed clumps; upper stems glandular pubescent; leaves 1–3(–4) cm long; corollas violet; nutlets fuscous or black, covered with ±

imbricated lamellae. Apr–May. The foliage is reported to be toxic (Kirkpatrick 1992). While this species has traditionally been recognized as a single variable species (e.g., Epling 1942; Kartesz 1994); Turner (1994b) separated 2 weakly differentiated varieties as occurring in nc TX: ☠

1. Pubescence of middle and lower stems to some considerable extent glandular, the longer-spreading hairs bearing minute capitate glands at their apices; extreme e margin of nc TX, mainly se and e TX ______ var. **drummondii**
1. Pubescence of middle and lower stems mostly of eglandular hairs; widespread in nc TX (plants with ± eglandular stems also occur in the n part of nc TX as a result of apparent hybridization between var. *edwardsiana* and *S. wrightii*) ______ var. **edwardsiana**

var. **drummondii**. On sandy or sandy-loam soils; Milam Co. (Turner 1994b) on extreme e edge of nc TX; mainly se and e TX w to e edge of Edwards Plateau.

var. **edwardsiana** B.L. Turner, (of the Edwards Plateau). On limestone; nc TX s and w to w TX; this is the variety occurring over the vast majority of nc TX. Turner (1994b) noted that intergrades between the two varieties occur along the e edge of the Edwards Plateau; he also noted that occasional plants intermediate between *S. drummondii* var. *edwardsiana* and *S. wrightii* have been observed in the n part of nc TX (e.g., Collin, Dallas, and Grayson cos.) and adjacent OK. They have calyces and nutlets of *S. drummondii*, but intermediate vestiture and habit; the stems have very short downcurved hairs and only a few if any glandular hairs (Turner 1994b).

Scutellaria lateriflora L., (lateral-flowered), MAD-DOG SKULLCAP, VIRGINIAN SKULLCAP, SIDE-FLOWERING SCULLCAP. Delicate perennial with rhizomes and stolons; stems to 60(–100) cm tall; leaves thin, ovate, 3–11 cm long, crenate to serrate; flowers mainly in 1-sided racemes from the axils of leafy bracts that are reduced upwards; corollas blue-violet. Low woods; Lamar Co. in Red River drainage; e TX, also Panhandle. Jul–Sep.

Scutellaria ovata Hill, (ovate, egg-shaped), EGG-LEAF SKULLCAP. Perennial to ca. 85(–100) cm tall; leaves to 13 cm long, cordate-ovate, crenate-serrate, long-petiolate; flowers in ± distinct racemes from upper nodes; bracts 4–9 mm long, 3–6 mm wide; corollas 17–23 mm long, blue to violet, with a whitish lower lip. Sandy or rocky slopes, in woods or thickets. Apr–Jul. While recognizing it is a variable species, Lane (1986) questioned the need for infraspecific taxa.

1. Stems glandular throughout with both spreading glandular and non-glandular hairs ______ subsp. **bracteata**
1. Middle and lower parts of the stems below the inflorescences with only decurved or retrorse or even subappressed non-glandular hairs ______ subsp. **mexicana**

subsp. **bracteata** (Benth.) Epling, (having bracts), TUBER SKULLCAP. Burnet and Dallas cos.; se and e TX w to nc TX and Edwards Plateau.

subsp. **mexicana**, (of Mexico). Bell Co., also Dallas Co. (Epling 1942); se and e TX w to nc TX and Edwards Plateau.

subsp. *ovata* was reported by Hatch et al. (1990) for vegetational area 4 (Fig. 2); we have found no specimens or other citations for this subspecies in nc TX.

Scutellaria parvula Michx., (very small). Small herbaceous perennial 8–30 cm tall, developing horizontal rhizomes with bead-like bulbous thickenings; leaves deltoid-ovate to ovate-oblong; corollas 7–8 mm long, blue. Prairies and open woods. Late Mar–May.

1. Stems glabrous or with a few non-glandular curled or appressed hairs ______ var. **missouriensis**
1. Stems definitely hairy with glandular or spreading non-glandular hairs.
   2. Hairs along stem ca. 1/3 as long as stem is wide; lower leaf surfaces rather evenly covered with sessile golden glands and also with long hairs mainly along the veins; lateral veins of the leaf usually not noticeably connected ______ var. **parvula**

2. Hairs along stem ca. 1/2 as long as stem is wide; lower leaf surfaces without sessile glands, only with long hairs mainly along the veins; lateral veins of the leaf often connected and forming a ± continuous vein just below the leaf margin _______ var. **australis**

var. **australis** Fassett, (southern). Se and e TX w to w Blackland Prairie, also Erath Co. (Epling 1942). [*S. australis* (Fassett) Epling] This is by far the most common variety in nc TX.

var. **missouriensis** (Torr.) Goodman & C.A. Lawson, (of Missouri). Included based on citation of vegetational areas 4 and 5 (Fig. 2) by Hatch et al. (1990); se and e TX w to nc TX. [*S. leonardii* Epling, *S. parvula* var. *leonardii* (Epling) Fernald] Goodman and Lawson (1992) explained the need to replace the varietal epithet *leonardii* with *missouriensis*, an earlier overlooked name.

var. **parvula**, SMALL SKULLCAP. Dallas Co., also Tarrant Co. (Epling 1942); se and e TX w to nc TX.

Scutellaria resinosa Torr., (full of resin), RESIN-DOT SKULLCAP, SHORT-LEAF SKULLCAP, RESINOUS SKULLCAP. Stiff low perennial usually ca. 15–20 cm tall with woody taproot and base, usually with many stems; leaves (7–)9–20 mm long; foliage with small sessile glands that glisten under a lens; calyces with minute spreading hairs; corollas 17–22 mm long, deep violet-blue; nutlets black and minutely, evenly, and closely tuberculate. Prairies, slopes, rocky or sandy areas; Callahan Co. (Turner 1994b) on w margin of nc TX w to Panhandle. Apr–Oct. Closely related to *S. wrightii*; see discussion under that species.

Scutellaria wrightii A. Gray, (for Charles Wright, 1811–1885, TX collector), WRIGHT'S SKULLCAP. Perennial ca. 15–20 cm tall usually with several stems from a woody base; similar to *S. resinosa*; flower size variable, the corollas 12–23 mm long (rarely more), deep violet-blue; nutlets black and minutely, evenly, and closely tuberculate. We are following Turner (1994b) in recognizing this species that has sometimes been lumped with *S. resinosa* (e.g., Lane 1983; Hatch et al. 1990; Kartesz 1994). Turner (1994b), agreeing with Epling (1942), indicated the 2 are largely allopatric and they are readily distinguished by pubescence; intermediates have not been found. The stem pubescence is very minute and requires a strong lens or scope to properly observe; however, it appears to be definitive in separating the 2 species. Calcareous areas or sandy loam, prairies, slopes, and open woods; Blackland Prairie w to Rolling Plains and Edwards Plateau. Mar–Jul. [*S. brevifolia* (A. Gray) A. Gray, *S. integrifolia* L. var. *brevifolia* A. Gray, *S. resinosa* var. *brevifolia* (A. Gray) Penland] A showy plant in full bloom. /105

STACHYS BETONY, HEDGE-NETTLE

Ours annual to perennial herbs; leaves simple; flowers in interrupted inflorescences; corollas strongly 2-lipped; stamens 4, didynamous.

A genus of 300 species of herbs and shrubs with alkaloids; native to temperate and warm areas except Australasia and tropical mountains. Some are cultivated as ornamentals; others provide edible tubers or are used medicinally. (Greek: *stachys*, spike or ear of corn, alluding to the inflorescence)

REFERENCES: Epling 1934; Nelson 1981; Mulligan & Munro 1989 [1990]; Turner 1994c.

1. Leaf blades 4 cm or less long; lower leaves with petioles nearly as long as or longer than blades; uppermost leaves sessile; leaf margins crenate; leaf apices obtuse; corollas 5–6 mm long; plants annual or biennial _______ **S. crenata**
1. Leaf blades to 13+ cm long; lower leaves with petioles many times shorter than blades; uppermost leaves with at least a short petiole; leaf margins dentate or serrate; corollas 10–14 mm long; plants perennial _______ **S. tenuifolia**

Stachys crenata Raf., (crenate, scalloped), SHADE BETONY. Plant slender, pubescent, usually several-stemmed, these erect or decumbent, to 30 cm long; leaves petioled, the petioles reduced

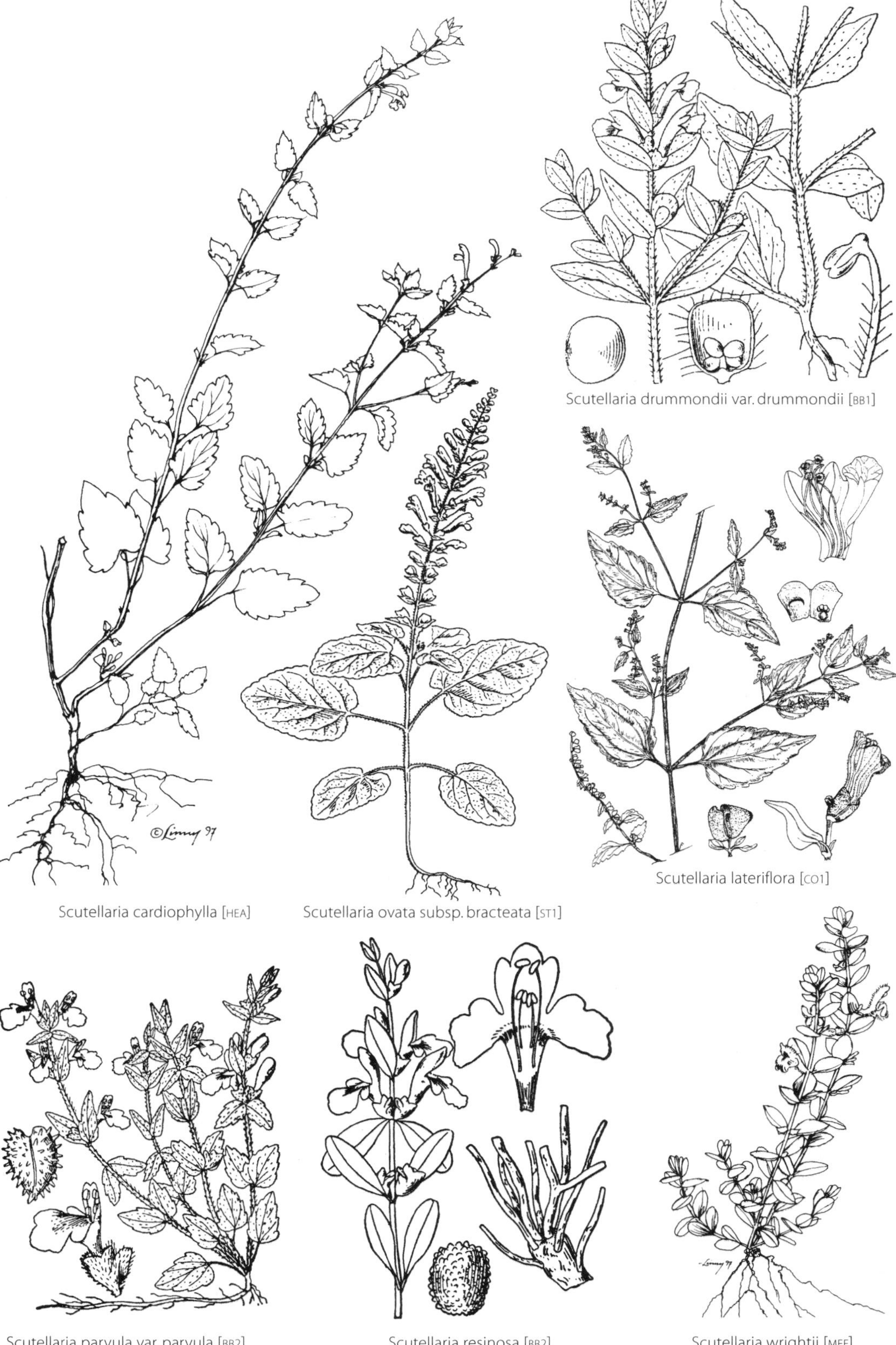

Scutellaria drummondii var. drummondii [BB1]

Scutellaria lateriflora [CO1]

Scutellaria cardiophylla [HEA]

Scutellaria ovata subsp. bracteata [ST1]

Scutellaria parvula var. parvula [BB2]

Scutellaria resinosa [BB2]

Scutellaria wrightii [MEE]

above; leaf blades ovate or oblong-ovate, to 4 cm long, the upper gradually much smaller, passing into floral bracts; flowers few, in whorls, the lower in leaf axils, the upper short-bracted; calyces 3–5 mm long; corollas light lavender-pink to pink or bluish (rarely white). Ditches, damp woods, weedy areas; n to Dallas Co., also Brown (Stanford 1976) and Denton (Mahler 1988) cos.; se and e TX w to nc TX and Edwards Plateau. Apr–May. [*S. agraria* of authors, not Cham. & Schltdl.]

Stachys tenuifolia Willd., (slender-leaved), SLENDER-LEAF BETONY, SMOOTH HEDGE-NETTLE. Stems erect, to 1.3 m tall, glabrous or with a few hairs on the angles; leaf blades linear to linear-lanceolate or narrowly ovate, the main ones 6 cm or more wide; middle and lower leaves with blades 4–11 times longer than petioles; inflorescence a few-flowered, interrupted spike; corollas variously colored including white with hints of pink/purple. Low woods, stream banks; Lamar Co. in Red River drainage; otherwise in TX known only from deep e TX. Aug–Nov.

Stachys coccinea Ortega, (scarlet), TEXAS BETONY, is becoming popular in nc TX as a xeric landscape plant where it persists and spreads. This native of the Trans-Pecos can be distinguished from the species above by the large (18–24 mm long), showy, scarlet corollas. The authority for this species is often given as Jacq.; however, Turner (1994c) pointed out that it should be Ortega.

TEUCRIUM GERMANDER, WOOD-SAGE

Perennial herbs; leaves simple, serrate to pinnatifid, often gland-dotted; flowers rather showy, in terminal, spike-like racemes, with small or large leafy bracts; corollas bilabiate but appearing ± 1-lipped; upper lip of corollas much shorter than the lower lip, divided into 2 lobes each equal to or smaller than the lateral lobes of the lower lip; lower lip of corollas with middle lobe much larger than lateral lobes; stamens 4.

A cosmopolitan, especially Mediterranean genus of 100 species; the corollas appear to have a single 5-lobed lip. Various species are used medicinally or as cultivated ornamentals. (Greek: *teucrion*, name used by Dioscorides for some related plant, possibly from Teucer, the Trojan king who used the plant medicinally)
REFERENCE: McClintock & Epling 1946.

1. Leaf blades toothed, conspicuously pubescent beneath; corollas lavender with dark dots, the tube 4–7 mm long ________ **T. canadense**
1. Leaf blades, at least the lower, deeply lobed to pinnatifid, glabrous; corollas white to cream, usually with pink or purple markings toward base, the tube 1–2 mm long.
   2. Largest lobes of corollas 5–8 mm long; leaf blades with few wide lobes, the lobes mostly 2 mm or more wide; plants 15–70 cm tall; e part of nc TX ________ **T. cubense**
   2. Largest lobes of corollas 9–13 mm long; leaf blades deeply cut, the ultimate segments narrow (1–1.5 mm wide); plants 5–20(–30) cm tall; West Cross Timbers and Lampasas Cut Plain westward ________ **T. laciniatum**

Teucrium canadense L., (of Canada), AMERICAN GERMANDER, WOOD-SAGE. Rhizomatous perennial 30–150 cm tall; leaves ovate to lanceolate, 6–10 cm long, 2–4 cm wide, the lower surfaces silvery pubescent; inflorescences silvery from numerous appressed hairs; bracts about as long as calyces or slightly longer; flowers subsessile or on pedicels to ca. 2 mm long; calyces 5–9 mm long at flowering, usually silvery pubescent; corollas 10–18 mm long, lavender (rarely white). Stream bottom woods and low open ground; Blackland Prairie westward. Jun–Aug. This is by far the most abundant *Teucrium* species in nc TX.

Teucrium cubense Jacq. var. **laevigatum** (Vahl) Shinners, (sp.: of Cuba; var.: smooth), ANNUAL GERMANDER. Perennial (also possibly annual), ± bushy herb; stems usually several from a taproot, to 70 cm tall; median leaves mostly 3–5-lobed nearly to the midrib; floral bracts 3-parted nearly to the base, often much longer than calyces; flowers on pedicels 4–12 mm long; calyces 5–10 mm long at flowering; corollas 6–15 mm long. Low open ground; Kaufman Co. (possibly introduced); mainly s and sw TX; endemic to TX. Jun–Oct. [*T. cubense* subsp. *laevigatum* (Vahl) E.M. McClint. & Epling]

Teucrium cubense var. *cubense* is cited by Hatch et al. (1990) for vegetational area 4 (Fig. 2); this taxon apparently only occurs to the s and e of nc TX. It differs in having the median leaves crenate to lobed only ca. 1/2 way to midrib and in having the floral bracts 3-lobed only to the middle to entire.

Teucrium laciniatum Torr., (laciniate, torn), CUT-LEAF GERMANDER. Perennial; plant forming beds from oblique, creeping roots; stems many, tufted, 5–20(–30) cm tall; leaves cut nearly to midrib into narrow lobes; bracts longer than calyces, quite similar to the leaves; pedicels 3–8 mm long; flowers with strong, spicy, sweet scent; calyces 8–13 mm long at flowering; corollas 14–22 mm long. Sandy or gravelly prairies and roadsides; Grand Prairie (Tarrant Co.) and Lampasas Cut Plain (Bell and Hamilton cos.), also Brown Co. (HPC); w part of nc TX and Edwards Plateau w to w TX. Apr–May. Jones et al. (1997) lumped this species with *T. cubense* var. *laevigatum*; we are following Kartesz (1994) and J. Kartesz (pers. comm. 1997) in recognizing it at the specific level. /107

TRICHOSTEMA BLUE-CURLS

Ours annual herbs; leaves simple, usually entire or sometimes irregularly serrate; flowers in axillary cymes of 1–7 flowers, together appearing racemose or paniculate; corollas not exserted from calyces, 5-lobed, bilabiate, the lower single-lobed lip usually longer than the upper lip made up of 4 equal lobes; stamens 4.

A North American genus of 16 species including cultivated ornamental herbs and shrubs. (Greek: *thrix*, hair, and *stema*, stamen, from the capillary filaments)
REFERENCES: Lewis 1945; Abu-Asab & Cantino 1989.

1. Calyces radially symmetrical, the 5 teeth ± equal; stamens straight or only slightly arched, 2.3–4.2 mm long, not greatly exserted; stems to 0.5 m tall; w as far as Tarrant Co. ____ **T. brachiatum**
1. Calyces strongly 2-lipped, the lower 2 teeth ca. 1/3 as long as the partially fused upper 3 teeth (calyces inverting at maturity so the 3 longer teeth become lowermost); stamens arched, 6–16 mm long, exserted; stems to 1 m tall; in nc TX known only from Lamar Co. in extreme ne part of area ____ **T. dichotomum**

Trichostema brachiatum L., (branched at right angles), FLUX-WEED, FALSE PENNYROYAL. Stems 0.15–0.5 m tall, usually freely branched, pubescent; leaves linear to elliptic, 10–50 mm long, 4–16(–20) mm wide, entire or sometimes irregularly serrate, sessile to with a petiole 1–5 mm long; inflorescence a 1–3-flowered axillary cyme, usually glandular; pedicels 1–15(–20) mm long; flowering calyces 2.5–4 mm long, enlarging in fruit to 3.5–8 mm, with minute stalked glands; corollas 1.5–4.5 mm long, not much longer than calyces, bluish to rose-pink; stamens 4; nutlets pubescent and glandular. Sandy or limestone sites, disturbed areas; Bell, Grayson, and Johnson cos., also Tarrant Co. (Mahler 1988); also Edwards Plateau. Jul–Oct. [*Isanthus brachiatus* (L.) Britton, Sterns, & Poggenb.]

Trichostema dichotomum L., (forked in pairs), FORKED BLUE-CURLS, BASTARD PENNYROYAL. Stems to 1 m tall, typically branched, pubescent; leaves oblong to ovate, 15–60 mm long, 5–25 mm wide, entire, narrowed to petioles 2–15 mm long; cymes 3–7-flowered; flowering calyces 2.7–6 mm long, enlarging in fruit to 4.6–8.9 mm long; corollas bluish; nutlets glabrous. Open woods, stream banks, sandy soils; Lamar Co. in Red River drainage (Carr 1994); mainly se and e TX. Jul–Oct.

WARNOCKIA

A monotypic genus endemic to Texas and one site in the Arbuckle mountains of s Okla-

homa (also possibly a site in Mexico) (Turner 1996). Previously treated in the genus *Brazoria* (Kartesz 1994). (Named for Dr. Barton Holland Warnock, 1911–1998, professor of Biology, Sul Ross State Univ., Alpine, TX, and avid student and collector of the flora of the Trans-Pecos)
REFERENCES: Shinners 1953d; Lundell 1969a; Turner 1996.

Warnockia scutellarioides (Engelm. & A. Gray) M.W. Turner, (resembling *Scutellaria*—skullcap), PRAIRIE BRAZORIA, RATTLESNAKE-FLOWER. Erect annual herb 7–45(–75) cm tall, simple or widely branched; leaves, except lowest, sessile, oblong or oblanceolate, to 70 mm long and 23 mm wide, slightly clasping, toothed toward apex; flowers in terminal spike-like racemes, with small floral bracts 3–6 mm long; calyces 3–5 mm long, to 7.9 mm long at maturity, minutely pubescent, closed at maturity; corollas rosy, pinkish, or lavender, the throat paler or white with purple dots, 8.5–12 mm long; stamens 4; mature nutlets conspicuously 3-angled, sulfur yellow maturing to brown overlain with gold; $2n = 10$. Gravelly or thin clayey soils on limestone; Coryell Co., also Dallas, Erath, Hamilton, Tarrant (Lundell 1969a), Bell, Johnson (Shinners 1953b), Bosque, Hill, McLennan, Parker, Williamson (Turner 1996), and Somervell (R. O'Kennon, pers. obs.) COS.; scattered in e 2/3 of TX. Mid-Apr–early Jun. While we are following Turner (1996) in recognizing this taxon at the generic level, it is possibly only worthy of subgeneric recognition in *Brazoria*. [*Brazoria scutellarioides* Engelm. & A. Gray]

LAURACEAE LAUREL FAMILY

Ours shrubs and small trees, conspicuously aromatic; leaves alternate, simple, deciduous, entire or lobed; flowers small, in clusters, imperfect or perfect, greenish yellow or yellowish; sepals 6; petals 0; staminate flowers with 9 stamens in 3 rows; pistillate flowers with rudimentary stamens and 1 pistil; ovary superior; fruit a 1-seeded, red or blue drupe.

A large (2,000–3,000 species in ca. 50 genera—van der Werff 1997a) family of mostly tropical and subtropical areas, especially se Asia and Brazil; they are mostly aromatic evergreen trees and shrubs (however, nc TX species are deciduous). Economically important members include *Cinnamomum* species (source of cinnamon and camphor), *Laurus nobilis* L. (TRUE or BAY LAUREL—source of bay leaf; also used for the Greek and Roman crown or wreath of laurel—hence resting on your laurels, baccalaureate, poet laureate), and the Central American *Persea americana* Mill. (AVOCADO). AVOCADO, with many cultivars, has been cultivated since 8,000 BC. for the highly nutritious fruit which is rich in oils and vitamins A, B, and E; the oil is also used in cosmetics (Rohwer 1993a). Fruits of Lauraceae are important foods for tropical frugivorous birds such as bell-birds, toucans, and quetzals; the whole fruit is often swallowed and in a short time the seed regurgitated unharmed (Snow 1981; Rohwer 1993a). Family name from *Laurus*, a genus of two species of evergreen shrubs and trees, one of the Mediterranean region, the other of Macaronesia. (Classical Latin name for *L. nobilis*) (subclass Magnoliidae)
FAMILY RECOGNITION IN THE FIELD: *aromatic* shrubs or small trees with alternate simple leaves; flowers small; petals absent; anthers opening by valves or flaps; fruit a small drupe.
REFERENCES: Wood 1958; Rohwer 1993a; van der Werff & Richter 1996; van der Werff 1997a.

1. Leaves completely unlobed, pinnately veined with 5–7 pairs of nerves originating from along the midvein; young twigs brownish; flowers appearing before the leaves ____ **Lindera**
1. Leaves often with 1 or 2 conspicuous side lobes (rarely up to 4), with 3 main veins from near the base of the blade; young twigs yellowish green; flowers appearing with the young leaves ____ **Sassafras**

LINDERA SPICEBUSH, WILD ALLSPICE, FEVERBUSH

A genus of ca. 100 species (Wofford 1997) of aromatic trees primarly of tropical and temperate Asia, with 2 in e North America. (Named for Johann Linder, 1676–1723, Swedish botanist)
REFERENCE: Wofford 1997.

Stachys crenata [BT3, SM2]

Stachys tenuifolia [BB2, GLE]

Teucrium canadense [CO1]

Teucrium cubense [CO1]

Teucrium laciniatum [BB2]

Trichostema dichotomum [DOR]

Trichostema brachiatum [LAM]

Warnockia scutellaroides [PSE]

Lindera benzoin (L.) Blume var. **pubescens** (E.J. Palmer & Steyerm.) Rehder, (sp.: from an Arabic word meaning aromatic gum; var.: pubescent, downy), SPICEBUSH. Dioecious or polygamo-dioecious shrub or small tree to ca. 5 m tall, much branched; young branches pubescent; leaf blades (4–)6–15 cm long, 2–6 cm wide, mostly obovate to elliptic to ovate, entire, cuneate or tapering at base, glaucous and pubescent below, very aromatic with a spicy odor; petioles ca. 10 mm long; flowers and fruits along the branches and at their tips, in numerous nearly sessile clusters at the nodes; flowers fragrant; sepals 6, yellow; pistillate flowers with ca. 12–18 rudimentary stamens; drupes red, 6–10 mm long, 3–7 mm wide, on pedicels 3–5 mm long. Rich woods, along streams; Bell Co. on s margin of nc TX and Lamar Co. in Red River drainage; mainly Edwards Plateau and e TX. Mar–Apr. SPICEBUSH is so aromatic that even brushing against the plant makes you aware of its presence. The leaves were formerly used to make a tea and the fruits as an allspice substitute (Mabberley 1987). The fruits are bird-dispersed (Moore & Willson 1982) and have a high lipid content (ca. 37%) which makes them particularly attractive to migratory birds which have high energy demands (Stiles 1984).

SASSAFRAS

A genus of 3 species with 2 in e Asia and 1 in e North America; this disjunct distribution pattern is discussed under the genera *Campsis* (Bignoniaceae) and *Carya* (Juglandaceae). (Probably derived by French or Spanish settlers from a Native American name for the plant)
REFERENCES: Rehder 1920; van der Werff 1997b.

Sassafras albidum (Nutt.) Nees, (white), SASSAFRAS. Dioecious, usually small tree with spicy aromatic bark; leaves ovate to elliptic in outline, entire, unlobed or 2–3 (rarely more) -lobed, 6–12(–18) cm long, 2–8(–10) cm wide, aromatic when crushed, densely pubescent beneath, at least when young; flowers and fruits at the branch tips; flowers in stalked, branched clusters; sepals 3–4 mm long, greenish yellow or yellow; pistillate flowers with 6 rudimentary stamens; fruit a subglobose to ovoid, blue-black drupe ca. 6–10(–15) mm long, on an elongate (3–4 cm) reddish pedicel, the upper part of which grades into a reddish, cup-like structure enclosing the base of the fruit. Forest margins, old fields, fencerows; Fannin and Lamar cos. in Red River drainage, also Delta and Hopkins cos. (Little 1971); mainly se and e TX. Mar–Apr. [*S. albidum* var. *molle* (Raf.) Fernald] The blue-black fruit contrasting with the reddish pedicel is an example of a bicolor fruit display, thought to be more effective than a single color at attracting birds which act as dispersal agents (Willson & Thompson 1982). Dried SASSAFRAS leaves are the file of Creole file gumbo; the species is also used for light timber and the oil medicinally, including killing lice and for insect bites; the bark of the roots has been used to flavor root beers and to make sassafras tea. However, the plant contains safrole (an allylbenzene) which is considered by the FDA to cause cancer; it also has weak hepatotoxic and mutagenic potential; its use as a flavoring is now prohibited and interstate marketing of SASSAFRAS for sassafras tea has been banned by the FDA (Fuller & McClintock 1986; Mabberley 1987; Duke 1985; McGuffin et al. 1997). ☠

LENTIBULARIACEAE BLADDERWORT FAMILY

A small (245 species in 3 genera), cosmopolitan family of aquatic or moist area herbs. The Lentibulariaceae as a whole, and *Utricularia* in particular, are well known as insectivorous/carnivorous; this family and the unrelated Droseraceae (SUNDEWS) and Sarraceniaceae (PITCHER-PLANTS) are the only such groups within nc TX. As with other carnivorous plants, nutrients (especially nitrogen), rather than calories, are obtained through carnivory. Some authorities believe the Lentibulariaceae is derived from the Scrophulariaceae. Family name conserved from *Lentibularia*, a genus now treated as *Utricularia* (the name *Utricularia* was published earlier and thus has priority in terms of nomenclature). (Latin: *lens*, *lentis*, shaped

like the seed of the lentil, lens-shaped; presumably in reference to the lens-like shape of some part of the plant, possibly the bladders) (subclass Asteridae)
FAMILY RECOGNITION IN THE FIELD: aquatic or wet area insectivorous herbs with small, *bladder like traps*; leaves or their segments *thread-like* [these "leaves" may actually be stem tissue]; corollas 2-lipped, yellow.
REFERENCE: Barnhart 1916.

UTRICULARIA BLADDERWORT

Ours small, glabrous, perennial herbs, floating in clear water or stranded or buried in mud or wet sand, bearing small bladder-like traps for capturing small aquatic organisms; leaves alternate or whorled, linear-filiform or mostly cut into thread-like segments (while leaves are often recognized in this group, all vegetative structures may actually be stem tissue, some of which can be leaf-like—Godfrey & Wooten 1981); flowers solitary or racemose, on naked, erect peduncles; calyces 2-lipped; corollas yellow, 2-lipped, spurred at base; stamens 2; pistil 1; ovary superior; style and stigma 1; fruit a capsule.

A cosmopolitan (especially tropical) genus of 180 species of aquatics, epiphytes, and twiners. They possess bladders with trap doors; when triggered by a microsopic animal, the prey is sucked into the bladder, the door closes, and nutrients are absorbed from the victim. The trap is set by removal of most of the water in the bladder resulting in lower pressure inside the bladders than outside. When a prey touches one or more of the trigger hairs on the door, the door opens and water rushes in carrying the prey. Trapping is extremely rapid, usually within 1/50 of a second (Pietropaolo & Pietropaolo 1986). *Utricularia* species can become noxious aquatic weeds in rice fields in some parts of the world (Woodland 1997). (Latin: *utriculus*, a little bladder, from the bladder-like traps)
REFERENCES: Rossbach 1939; Reinert & Godfrey 1962; Pietropaolo & Pietropaolo 1986; Taylor 1989.

1. Leaves minute and linear (not branched) or none; plant usually terrestrial (can be aquatic); inflorescences to 35 cm tall; corollas 15–25 mm high (from tip of spur to tip of upper lip) and nearly as wide ______ **U. cornuta**
1. Leaves 1–3 times finely dichotomously branched; plant forming mats in water or on mud; inflorescences to 15 cm tall (usually shorter); corollas 6–12 mm high, 6–8 mm wide ______ **U. gibba**

Utricularia cornuta Michx., (horned), HORNED BLADDERWORT. Leaves inconspicuous, usually underground, usually seen only by carefully washing away soil; flowers 1–several (rarely up to 9), the spur 7–12 mm long, conspicuous; seeds reticulate, not winged. Wet soils or bogs or at edge of water; included based on citation of vegetational area 4 (Fig. 2) by Hatch et al. (1990); mainly e TX w to at least Henderson Co. near e margin of nc TX, also Edwards Plateau. May–Sep.

Utricularia gibba L., (swollen on one side), CONE-SPUR BLADDERWORT. Leaves usually conspicuous on the stems, forked 1–3 times; flowers 1–4, the spur inconspicuous; seeds smooth, broadly winged. Shallow water and mud; Bell, Dallas, Fannin, Grayson, Henderson, and Tarrant cos.; se and e TX w to Rolling Plains and Edwards Plateau. May–Aug. Including [*U. biflora* Lam.] previously separated on the basis of the leaves forking 2–3 times. /108

LINACEAE FLAX FAMILY

A small (250 species in 14 genera), cosmopolitan family of herbs, shrubs, lianas, and trees, sometimes cyanogenic. Some yield timber, edible fruit, or are cultivated as ornamentals; *Linum usitatissimum* L. is the source of flax (fibers used in linen, fine writing paper, and cigarette paper) and linseed oil. (subclass Rosidae)

FAMILY RECOGNITION IN THE FIELD: herbs with alternate, simple, entire leaves; flowers with 5 easily lost petals; stamens connate basally around the superior ovary; fruit a capsule.
REFERENCES: Small 1907b; Robertson 1971; Rogers 1984.

LINUM FLAX

Ours annual or perennial herbs; leaves alternate, opposite or whorled, sessile, simple, entire; stipular glands present or not so; flowers in terminal cymes or sometimes in racemose or paniculate inflorescences; sepals 5; petals 5, falling early, often showy, blue, white, yellow, or yellow with red at base (or red in 1 introduced species rare in nc TX); stamens 5; staminodes 5; ovary 5-carpellate, superior; capsules globose, dehiscing in 5 or 10 segments.

A genus of ca. 180 species of temperate and subtropical areas, especially the Mediterranean. Many species are cultivated as ornamentals; some as a source of fibers. Use of a hand lens to examine cilia and glands on the margins of sepals is often necessary for definitive identification. The petals of some species fall off at the slightest disturbance or in very hot weather (Kirkpatrick 1992). (Latin: *linum*, classical name of flax; this word is also the source of the words linen, linseed, and lingerie—Ajilvsgi 1984)
REFERENCES: Rogers 1963, 1964, 1968; Mosquin 1971.

1. Petals blue or rarely white.
  2. Sepals (at least alternate ones) ciliate-margined, 6–8 mm long; pedicels erect after flowering; stigmas slender, not capitate; introduced species ____ **L. usitatissimum**
  2. Sepals entire-margined or nearly so, 3.5–5 mm long; pedicels becoming reflexed after flowering; stigmas capitate; native species ____ **L. pratense**
1. Petals yellow or orange, often partly red-brown, OR red (in one introduced species rare in nc TX).
  3. Flowers red; introduced species known in nc TX only from s and w part of area ____ **L. grandiflorum**
  3. Petals yellow or orange, often partly red-brown; widespread native species.
    4. Styles separate or united only at base; fruits ultimately dehiscing into 10 one-seeded segments.
      5. Outer sepals entire; stipular glands not present at bases of leaves.
        6. Petals 5–8 mm long; margins of inner sepals with conspicuous stalked glands; mature fruits in dried specimens usually adhering to the plant; leaf blades narrowly lanceolate or narrowly oblanceolate, 3.5 mm or less wide ____ **L. medium**
        6. Petals 2.5–4.5 mm long; margins of inner sepals glandless or with very inconspicuous glands; mature fruits in dried specimens usually soon shattering; leaf blades elliptic to oblanceolate or obovate, to 10 mm wide ____ **L. striatum**
      5. All sepals with glandular teeth; stipular glands present at bases of most or all leaves.
        7. Plants annual, typically with 1 main stem and a rather small taproot; styles united at the base ____ **L. sulcatum**
        7. Plants perennial, typically with several stems from a hardened or enlarged base; styles completely separate.
          8. Leaves lanceolate to oblanceolate or broader, some of the lower ones in whorls of 4; petals 3–6 mm long; probably only to the w of nc TX ____ **L. schiedeanum**
          8. Leaves linear, the lower ones alternate or opposite; petals 6–10 mm long; widespread in nc TX ____ **L. rupestre**
    4. Styles united at least beyond the middle and often to near the summit; fruits dehiscing along the false septa into 5 two-seeded segments.
      9. Leaves 5–10 mm long; petals usually 6.5–12 mm long; sepals entire or fringed, not glandular-toothed; flowers few, mostly terminating leafy branches.
        10. Upper leaves and bracts sparsely but conspicuously ciliate-margined (use hand lens); upper stems and pedicels short hirsute; petals 6.5–8 mm long ____ **L. imbricatum**

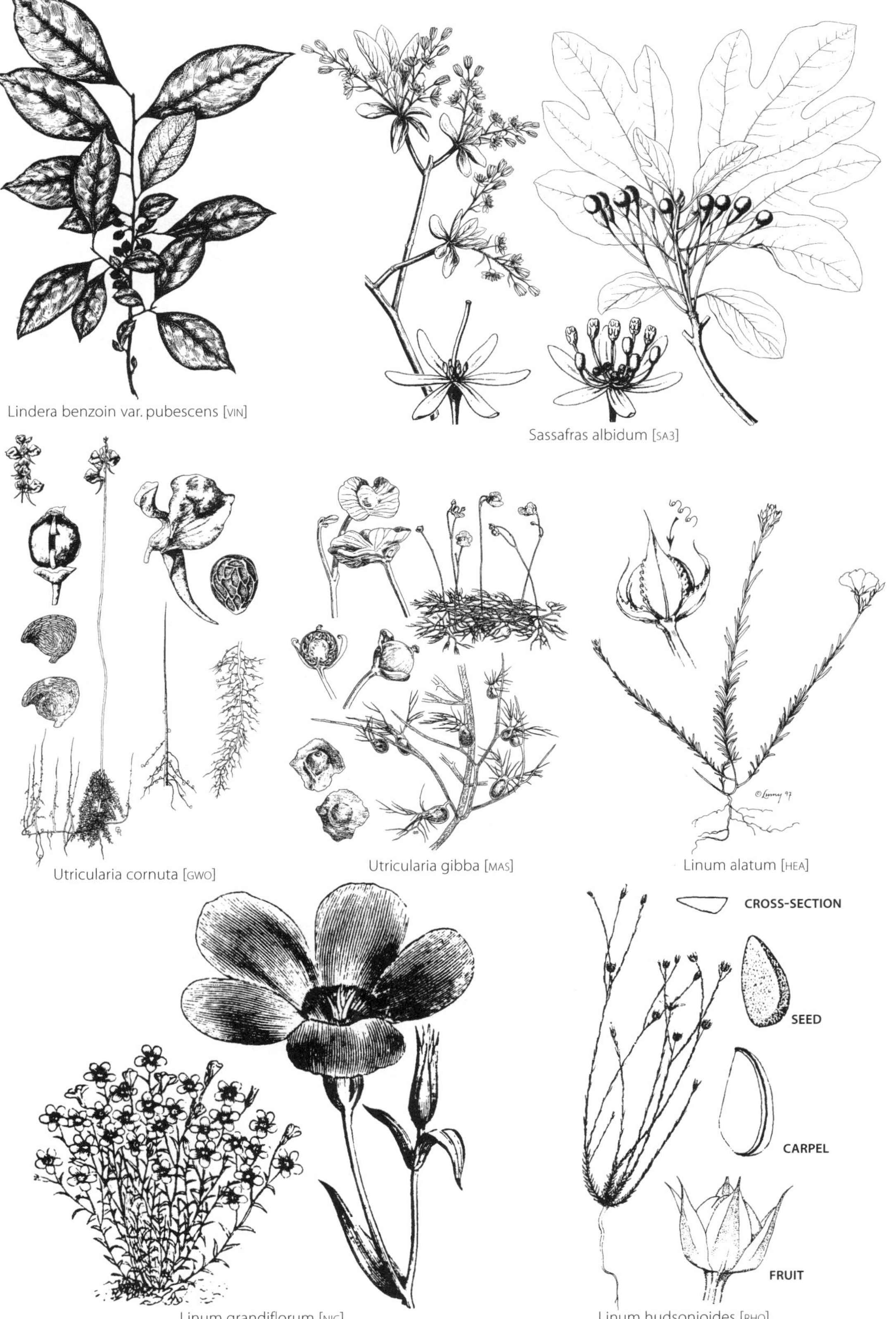

Lindera benzoin var. pubescens [VIN]

Sassafras albidum [SA3]

Utricularia cornuta [GWO]

Utricularia gibba [MAS]

Linum alatum [HEA]

Linum grandiflorum [NIC]

Linum hudsonioides [RHO]

10. Upper leaves and bracts not ciliate-margined; upper stems and pedicels with hairs only on the angles; petals 8–12 mm long ____ **L. hudsonioides**

9. Leaves 10–30 mm long; petals usually 10–18 mm long; sepals glandular-toothed; flowers in racemose or paniculate inflorescences.

11. Outer sepals ovate, the broad scarious margins irregularly crenate, each of the coarse teeth bearing a delicate gland; in nc TX only on the e margin (Milam Co.) ____ **L. alatum**

11. Outer sepals lanceolate or narrower, the margins not scarious or narrowly so, regularly (though sometimes sparsely) minutely serrate with gland-tipped teeth; widespread in nc TX ____ **L. rigidum**

Linum alatum (Small) H.K.A. Winkl., (winged). Glabrous annual 10–40 cm tall; leaves linear to linear-lanceolate, alternate or lowest opposite; petals 10–18 mm long, yellow with reddish base. Open sandy areas; Milam Co. (Rogers 1968); mainly se TX. Mar–Jul.

Linum grandiflorum Desf., (large-flowered), FLOWERING FLAX. Erect annual 30–60 cm tall; leaves oblong to lanceolate, 10–20 mm long; sepals ciliate; petals 15–30 mm long, various shades of red; anthers blue. Roadsides; Fort Hood (Coryell Co.—Sanchez 1997), also Brown and Lampasas (J. Stanford, pers. comm.), and Tarrant (R. O'Kennon, pers. obs.) cos.; according to J. Stanford (pers. comm.), this species has been introduced along roadsides in c TX. Native of n Africa.

Linum hudsonioides Planch., (resembling *Hudsonia*—beach-heath in the Cistaceae). Annual 5–30 cm tall, glabrous except on angles of upper stem; leaves linear to linear-lanceolate, 10 mm or less long; petals yellow, with or without a brick-red base. Sandy or gravelly areas; Bosque, Jack, and Wise cos., also Clay, Dallas (Rogers 1968), Brown, Callahan, Eastland, Erath, Hood, and Mills (Rogers 1963) cos.; Post Oak Savannah s and w to w TX. Mar–Sep.

Linum imbricatum (Raf.) Shinners, (overlapping in regular order like tiles), TUFTED FLAX. Glabrous, erect to spreading annual 3–30 cm tall, usually with several stems, sparingly branched above; leaves conspicuously overlapping; petals yellow, with or without red-brown base. Sandy or rocky open ground; Bell, Burnet, Cooke, Hamilton, Hunt, and Montague cos., also Grayson (Rogers 1968), Dallas, McLennan, and Navarro (Rogers 1963) cos.; nc to s TX. Apr–early May.

Linum medium (Planch.) Britton var. **texanum** (Planch.) Fernald, (sp.: intermediate; var.: of Texas), TEXAS FLAX, SUCKER FLAX. Erect glabrous perennial 20–80 cm tall; leaves narrowly lanceolate, 10–25 mm long; petals yellow. Open areas, often in sandy soils; Fannin, Hunt, Kaufman, Milam, Parker, and Tarrant cos., also Lamar Co. (Carr 1994); se and e TX w to nc TX. Mar–Aug.

Linum pratense (J.B. Norton) Small, (of meadows), MEADOW FLAX, NORTON'S FLAX. Glabrous annual; stems erect to partly prostrate, to 60 cm long, sometimes flowering when quite small; leaves linear to linear-lanceolate, 10–20 mm long; pedicels becoming reflexed after flowering; petals blue or rarely white, 5–14 mm long; styles mostly 3 mm long or less; capsules 4–6 mm long. Rocky limestone to sandy soils, prairies or disturbed ground; widespread in TX, mainly Blackland Prairie w to Panhandle and Edwards Plateau. Late Mar–May.

Linum rigidum Pursh, (stiff). Glabrous, erect or ascending annual to 40 cm tall, usually widely branched above; flowers showy; sepals 6–12 mm long; petals (9–)12–16(–19) mm long; fruits 3.5–4.7 mm long. Rocky or sandy areas, disturbed prairies or open ground. Mid-Apr–early Jun. This species, reported to contain a saponin, has been implicated in livestock poisoning (Lewis & Elvin-Lewis 1977).

1. Fruits thick-walled, opaque, broadly ovoid, tapering abruptly to the flattened base; usually with stipular glands ____ var. **berlandieri**

1. Fruits thin-walled, translucent (dark seeds commonly evident through the wall), elliptic, the base rounded; without stipular glands ____ var. **rigidum**

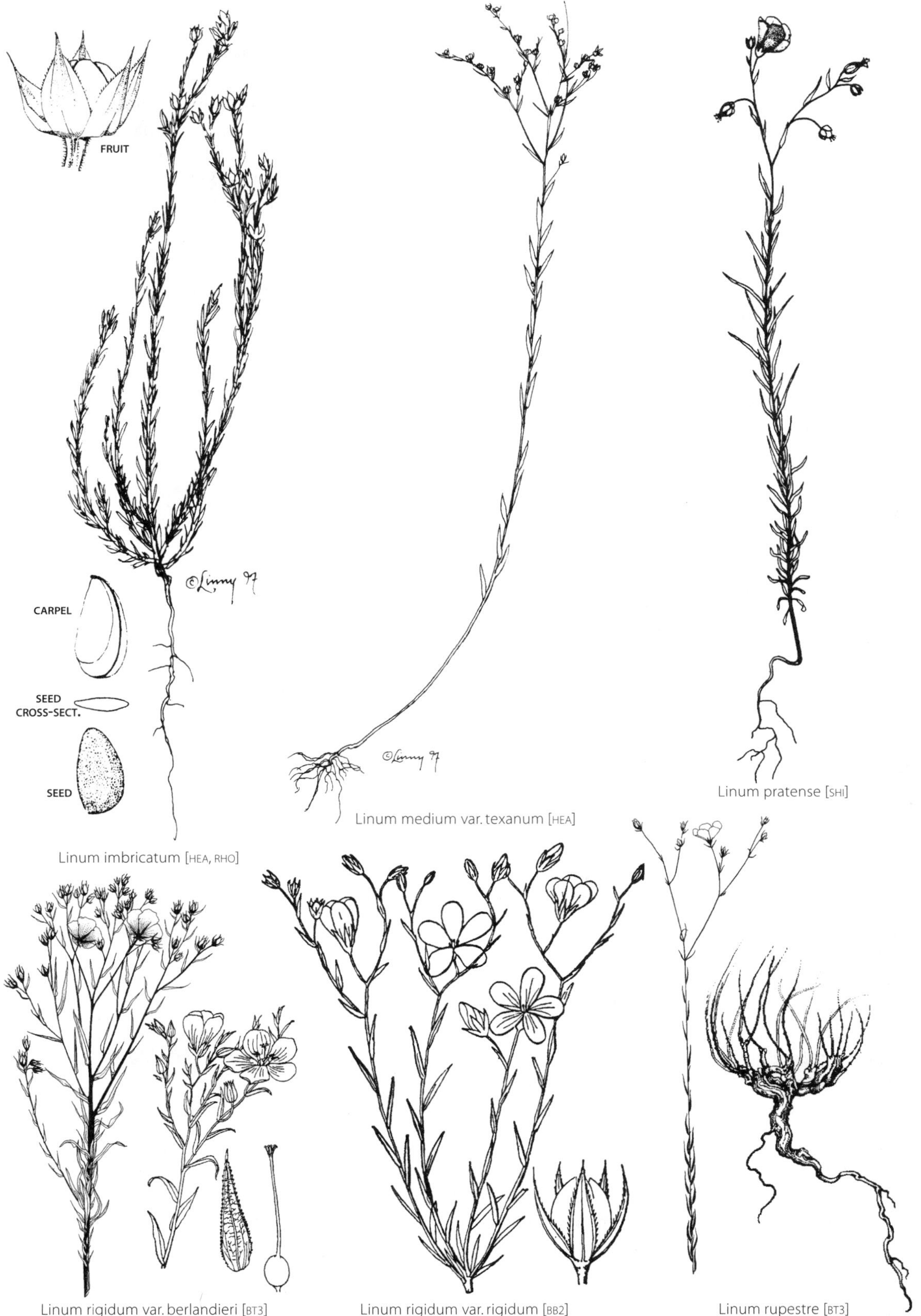

Linum imbricatum [HEA, RHO]

Linum medium var. texanum [HEA]

Linum pratense [SHI]

Linum rigidum var. berlandieri [BT3]

Linum rigidum var. rigidum [BB2]

Linum rupestre [BT3]

var. **berlandieri** (Hook.) Torr. & A. Gray, (for Jean Louis Berlandier, 1805–1851, French botanist who explored TX, NM, and Mexico, and one of the first botanists to make extensive collections in TX), BERLANDIER'S FLAX. Leaves linear-lanceolate or lanceolate, acute, 1–4.1 mm wide, usually with stipular glands; petals usually 13–16 mm long, deep yellow with broad, deep brown-red basal zone; sepals 7–9 mm long in flower. Widespread in TX, but mainly Blackland Prairie s and w throughout most of TX. [*L. berlandieri* Hook.] While often treated as a separate species (e.g., Kartesz 1994; Jones et al. 1997), because of hybridization and introgression, we follow Rogers (1968) and McGregor (1986) in keeping this taxon as a variety of *L. rigidum*. ▣/95

var. **rigidum**, STIFF-STEM FLAX. Leaves linear, acute or acuminate, 1–2 mm wide, without stipular glands; petals mostly 12–14 mm long, coppery yellow or orange, red-lined or with short, pale to deep brown-red zone at base; sepals 6–7.5 mm long in flower. Montague, Tarrant, and Wise cos.; mainly nc TX nw to Panhandle.

Linum rupestre (A. Gray) Engelm. ex A. Gray, (rock-loving), ROCK FLAX. Glabrous perennial 20–70 cm tall; leaves linear or nearly so; petals clear yellow. Limestone outcrops; Bell, Hood, Johnson, Parker, Somervell, and Tarrant cos.; widespread in TX, but mainly Trans-Pecos to Edwards Plateau, ne to nc TX. May–Jun.

Linum schiedeanum Schltdl. & Cham., (for Christian J.W. Schiede, d. 1836, German-born physician who collected plants in Mexico). Perennial to 39(–68) cm tall; petals lemon-yellow. Open or semi-shaded areas, calcareous soils; reported as occurring in vegetational area 5 (Fig. 2) by Hatch et al. (1990), but according to Rogers (1968) limited to the Trans-Pecos. Jun–Aug.

Linum striatum Walter, (striated, striped), RIGID FLAX. Glabrous perennial 30–100 cm tall, usually with several stems from base; petals yellow. Open or semi-shaded wet areas; included based on citation of vegetational area 4 (Fig. 2) by Hatch et al. (1990); mainly se and e TX. May–Aug.

Linum sulcatum Riddell, (furrowed), GROOVED FLAX. Glabrous erect annual 20–80 cm tall, loosely branched above; leaves linear to linear-lanceolate; petals 5–10 mm long, pale yellow. Sandy open woods and prairies; Hopkins and Kaufman cos., also Montague and Tarrant cos. (Mahler 1988); se TX and Post Oak Savannah w to West Cross Timbers and Edwards Plateau. Jun–Oct.

Linum usitatissimum L., (most useful), COMMON FLAX, CULTIVATED FLAX, LINAZA. Erect, glabrous annual 35–90 mm tall; leaves linear to linear-lanceolate, 12–30 mm long; pedicels erect after flowering; petals 11–15 mm long, blue or rarely white; sepals (at least alternate ones) ciliate-margined; stigmas slender, not capitate; capsules 6–8 mm long. Cultivated and escapes; included based on citation of vegetational areas 4 and 5 (Fig. 2) by Hatch et al. (1990); widely scattered in TX. Apr–May. Native of Mediterranean region. The flax of commerce is a fiber derived from this species; the fibers have great tensile strength and are used for such things as linen, thread, carpets, and paper; the seeds yield linseed oil, used in food-processing, paints, varnish, printing inks, and water-proofing (Mabberley 1987). FLAX is thought by some to be one of the oldest known textile plants; it was used in Egypt as long ago as the fourth millennium BC (Hepper 1992) and linen objects were found in the tomb of Tutankhamun (Hepper 1990). The leaves and seed chaff contain a poisonous cyanogenic glycoside, linamarin; livestock fatalities from eating the plant are known to have occurred (Kingsbury 1964). ☠ ☞

Linum lewisii Pursh, (for Meriwether Lewis, 1774–1809, explorer, naturalist, and leader of expedition with William Clark), LEWIS' FLAX, BLUE FLAX, PRAIRIE FLAX, native to the Trans-Pecos, is used in some wildflower mixes and has been planted in Parker Co.; it is apparently persisting. It can be distinguished by the combination of its blue or rarely white petals 10–15 mm long, inner sepals entire, styles usually 4 mm or more long, stigmas capitate, fruiting pedicels spreading or recurved, and its cespitose perennial habitat.

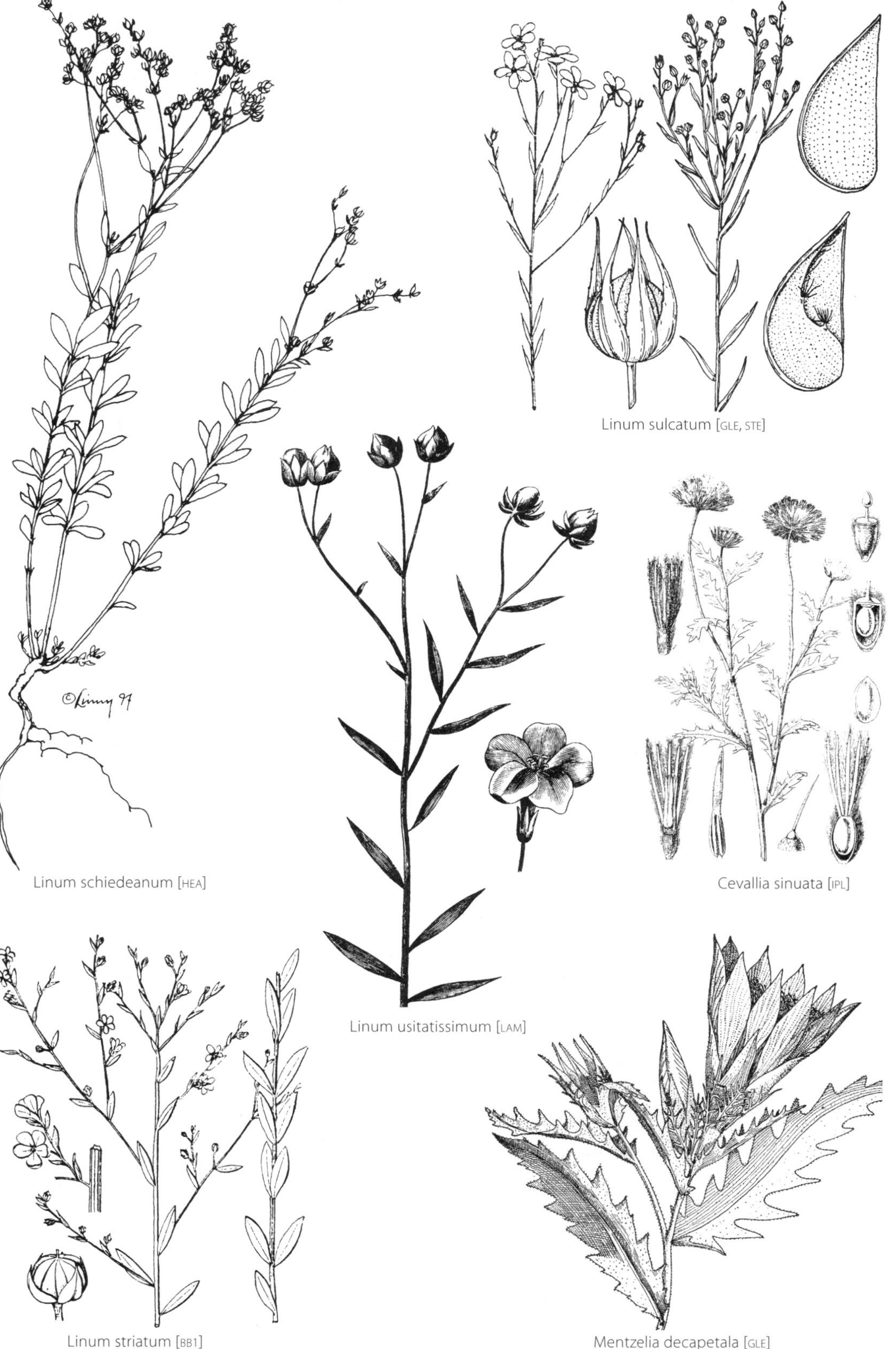

Linum sulcatum [GLE, STE]

Linum schiedeanum [HEA]

Linum usitatissimum [LAM]

Cevallia sinuata [IPL]

Linum striatum [BB1]

Mentzelia decapetala [GLE]

LOASACEAE
STICKLEAF OR BLAZINGSTAR FAMILY

Brittle-stemmed annuals or perennials often with barbed or stinging hairs, with papery or flaky bark on older parts; leaves usually alternate (sometimes subopposite), sessile or short-petioled, simple, toothed or lobed, in our species scabrous-pubescent or hispid; flowers terminal, solitary, corymbose, or in head-like clusters; calyces with cylindrical tube and 5 slender lobes; petals 5 or ca. 10, attached to summit of calyx tube; stamens 5 to many, with or without prominent appendages, attached next to the petals, or inside narrow to petal-like staminodia (= modified sterile stamens); ovary inferior, unilocular, with 1–5 placentae, with 1-many ovules; fruit a capsule or in *Cevallia* achene-like.

A small (260 species in 14 genera), largely herbaceous (some shrubs or small trees) family mainly of tropical and subtropical America with some in Africa and Arabia. They often have coarse, multicellular, silicified or calcified hairs, these sometimes stinging or often glandular; the hairs can have harpoon-like barbs. It is a family whose phylogenetic affinities have been unclear; recent molecular analyses (Xiang et al. 1993, 1998) suggest that *Cornus*, Nyssaceae, Hydrangaceae, and Loasaceae, as well as several other groups, form a "cornaceous clade." Family name from *Loasa*, a genus of 105 species of herbs and subshrubs usually with stinging hairs; native from Mexico to South America, especially in the mountains. (Name derived from a South American native name) (subclass Dilleniidae)
FAMILY RECOGNITION IN THE FIELD: usually herbs with alternate leaves; rough, *barbed or stinging hairs* present; stamens variable in number but often numerous; petals 5 or often more than 5.
REFERENCES: Ernst & Thompson 1963; Thompson & Powell 1981; Kaul 1986d.

1. Flowers in tight, head-like clusters on long peduncles; stamens 5, with prominent, spatulate, inflated appendages extending beyond the anthers; plants with stinging hairs; fruits indehiscent, 1-seeded ______ **Cevallia**
1. Flowers solitary or cymose; stamens 10 to numerous, without appendages; stinging hairs absent; fruits dehiscent, 2–many-seeded ______ **Mentzelia**

CEVALLIA STINGING CEVALLIA, SHIRLEY'S-NETTLE

A monotypic genus of sw North America. (Derivation of generic name unknown)

Cevallia sinuata Lag., (with wavy margins), STINGING CEVALLIA, SHIRLEY'S-NETTLE. Low, clump-forming, suffrutescent perennial to 60 cm tall; stems and leaves tomentose and armed with long stinging hairs; leaves oblong, pinnately lobed or coarsely toothed, to 5 cm long and 2.5 cm wide, sessile; flowers opening in the morning; calyx lobes and petals narrowly linear, pilose, ± yellow to orange or reddish, very similar to each other. Open areas, rocky or gravelly ground; Clay Co. (Mahler 1988); West Cross Timbers s and w to w TX. Jun–Jul. The stinging hairs look like tiny glass trees when examined with a hand lens; they contain formic acid and can produce a severe skin rash (Ajilvsgi 1984). According to Lampe (1986), only four families (Euphorbiaceae, Hydrophyllaceae, Loasaceae, and Urticaceae) have stinging hairs—nc TX has stinging representatives of all of these except the Hydrophyllaceae. /83

MENTZELIA STICKLEAF

Herbaceous perennials or biennials; hairs barbed but not stinging; stems whitish; leaves lobed or entire, petiolate or sessile; flowers 5-merous; stamens 10 to numerous; petal-like modified stamens sometimes present.

A genus of 60 species of warm areas of the Americas including some cultivated ornamen-

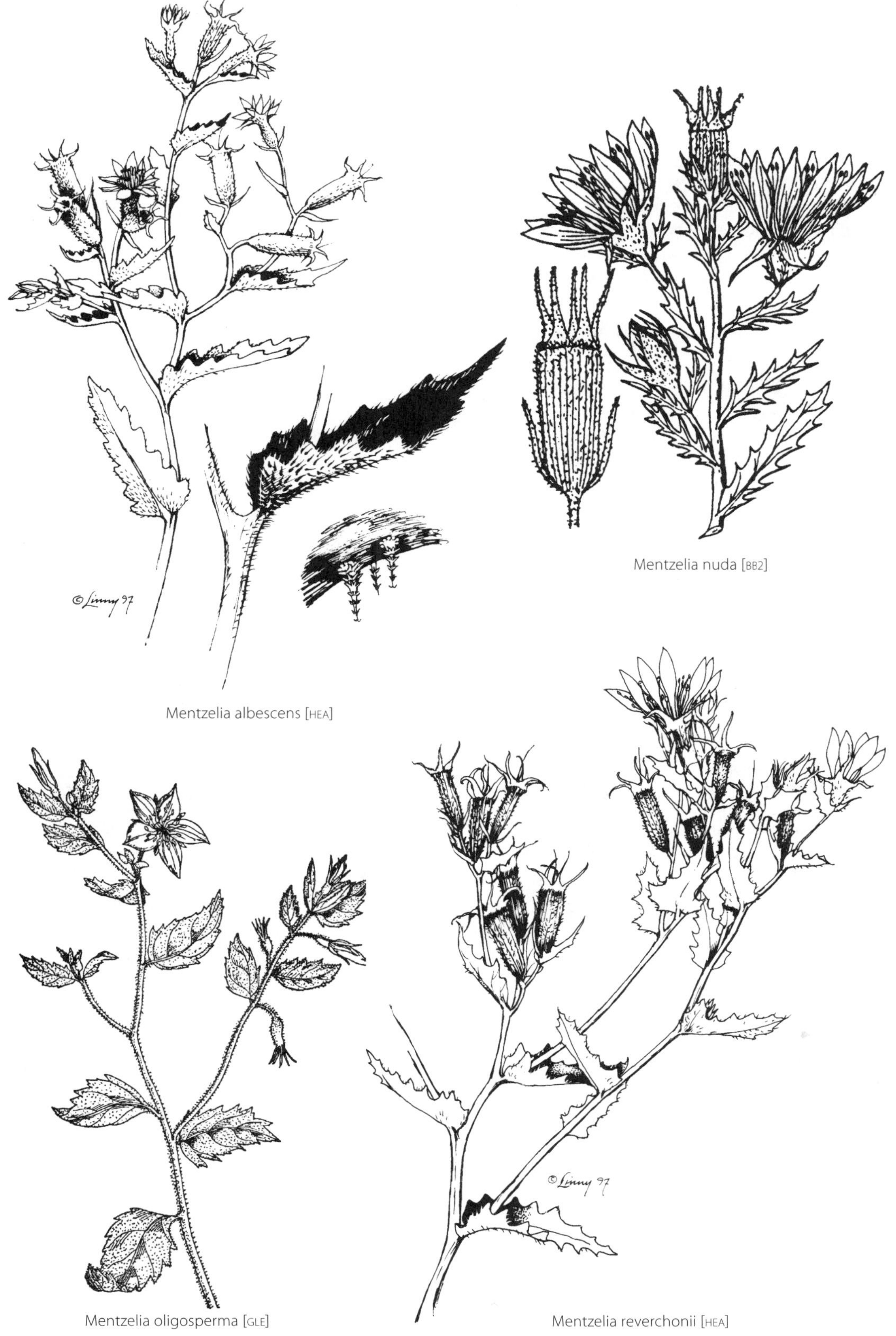

Mentzelia nuda [BB2]

Mentzelia albescens [HEA]

Mentzelia oligosperma [GLE]

Mentzelia reverchonii [HEA]

tals. The leaves of some species have a pubescence that gives them a sandpaper-like feel and causes them to attach to clothes and animal fur—hence the common name; this is problematic for sheep ranchers and can lower the market value of the fleece (Wills & Irwin 1961). (Named for Christian Mentzel, 1622–1701, German botanist)

REFERENCES: Osterhout 1902; Darlington 1934.

1. Petals 5, usually orange (rarely yellow), 8–15 mm long; seeds not winged; filaments narrow; staminodia (= modified sterile stamens) absent; flowers opening in early morning ____ **M. oligosperma**
1. Petals (including petal-like modified stamens) more than 5, yellow to white, 6 mm or more long (except in the small-flowered *M. albescens*, the petals are 15 mm or more long); seeds winged; filaments of outer fertile stamens sometimes broadened; staminodia sometimes present; flowers opening in late afternoon or early evening.
   2. Petals light yellow to yellow, 6–30 mm long; calyx lobes 3–12 mm long; styles 4–17 mm long.
      3. Petals 6–11 mm long, light yellow, lanceolate to ± ovate; calyx lobes 3–6 mm long, styles 4–6 mm long ____ **M. albescens**
      3. Petals 15–30 mm long, yellow, spatulate; calyx lobes 10–12 mm long; styles 11–17 mm long ____ **M. reverchonii**
   2. Petals white to cream, 25–80 mm long; calyx lobes 10–40 mm long; styles 18–60 mm long.
      4. Petals 25–40 mm long; calyx lobes 10–19 mm long; styles 18–30 mm long ____ **M. nuda**
      4. Petals 40–80 mm long; calyx lobes 25–40 mm long; styles 50–60 mm long ____ **M. decapetala**

Mentzelia albescens (Gillies & Arn.) Griseb., (whitish), WAVY-LEAF MENTZELIA. Stems to 0.6 m tall; flowers opening in late afternoon; stamens 40 or less, with 5 of the outer stamens petaloid (these together with 5 true petals make the flowers appear to have 10 petals); capsules cylindric, 20–30 mm long; seeds flattened, winged. Disturbed sites; Archer and Parker cos. in w part of nc TX, Hatch et al. (1990) also cited vegetational areas 4 and 5 (Fig. 2); mainly Rolling Plains and Edwards Plateau w to w TX. May–Aug.

Mentzelia decapetala (Pursh ex Sims) Urb. & Gilg ex Gilg, (ten-petaled), TEN-PETAL MENTZELIA. Stems to 1 m tall; flowers large, opening about 1 hour after sunset; petals ca. 10 (the inner 5 smaller and representing petaloid stamens), 10–20 mm wide; stamens numerous; capsules 3–4.5 cm long; seeds flattened, winged. Disturbed habitats; Dallas Co. (on Austin Chalk on Cedar Hill); nc TX w to Panhandle. Usually Jun–Aug.

Mentzelia nuda (Pursh) Torr. & A. Gray, (naked), BRACTLESS MENTZELIA, SAND-LILY, POOR-MAN'S-PATCHES, STARFLOWER. Stems to 1 m tall, usually unbranched below, branched above; flowers opening in late afternoon; bracts subtending flowers serrate to laciniate; petals ca. 10, 3–10 mm wide; stamens numerous; narrow staminodia (= modified sterile stamens) present, some nearly as long as the petals; capsules cylindrical, 18–30 mm long; seeds flattened, winged. Usually sandy soils, prairies; Palo Pinto and Young cos., also Dallas Co. (Mahler 1988); nc TX s and w to w TX. Jun–Oct. The leaves adhere tenaciously to clothing or hair and can be problematic to sheep ranchers because of their tendency to stick in sheep's wool (Kirkpatrick 1992). While [var. *stricta* (Osterh.) H.D. Harr.] is sometimes recognized (e.g., Kartesz 1994; Jones et al. 1997), we are following Correll and Johnston (1970), Kaul (1986d), and Hatch et al. (1990) in lumping this variety. Jones et al. (1997) listed only var. *stricta* for TX. If varieties are recognized, nc TX plants best fit var. *stricta*. [*M. stricta* (Osterh.) G.W. Stevens, *Nuttallia nuda* (Pursh) Greene, *Nuttallia stricta* (Osterh.) Greene] Rydberg (1932) separated the 2 (as species of *Nuttallia*) as follows:

1. Plant branched below; flowers subtended by solitary entire bracts ____ var. *nuda*
1. Plant simple below; flowers subtended by several toothed bracts ____ var. *stricta*

Mentzelia oligosperma Nutt. ex Sims., (few-seeded), STICKLEAF, CHICKENTHIEF, REGAHOSA.

Rounded semi-woody perennial; stems 0.2–0.6(–1) m tall; roots enlarged; flowers opening in early morning; petals 5, 3–4 mm wide, usually orange (reported as varying to yellow by Kaul 1986d); stamens 15–40, all fertile; styles 7–10 mm long; capsules cylindrical, 7–13 mm long; seeds oblong, 3-angled. Limestone soils; Blackland Prairie s and w to w TX, also se TX. Late May–Jun. This is the most common *Mentzelia* in nc TX.

Mentzelia reverchonii (Urb. & Gilg) H.J. Thomps. & Zavort., (for Julien Reverchon, 1837–1905, a French-American immigrant to Dallas and important botanical collector of early TX), PRAIRIE STICKLEAF, BUENA MUJER. Stems to 1 m tall; flowers opening in late afternoon; petals apparently more than 10, yellow; stamens numerous; narrow staminodia present; capsules cylindrical, 15–30 mm long; seeds flattened, winged. Limestone or gravel soils; Archer, Shackelford, and Young cos.; apparently widespread in TX, but mainly extreme w part of nc TX s and w to w TX. May–Sep. The leaves cling tightly to clothing or animal hair due to the presence of barbed hairs; this has given rise to the Spanish common name BUENA MUJER meaning good woman in English (Ajilvsgi 1984).

LOGANIACEAE STRYCHNINE OR LOGANIA FAMILY

Ours annual or perennial herbs or high-climbing, twining, woody vines; leaves opposite, sessile or short-petioled, simple, entire or nearly so; leaf bases connected around the stem by united short stipules or a stipular ridge; flowers sessile or short-pedicelled, terminal or axillary, solitary or in cymes or cymose spike-like inflorescences; corollas funnelform to salverform or nearly tubular, 5-lobed; sepals 5, united at least basally; stamens 5; pistil 2-carpellate; ovary superior; fruit a capsule or separating into 2 carpels at maturity.

A medium-sized (ca. 570 species in 29 genera) tropical to temperate family of herbs, shrubs, trees, and vines. ☠ The family contains numerous extremely poisonous and medicinal plants (due to alkaloids, iridoids, and saponins) such as *Strychnos nux-vomica* L. (source of strychnine); some were used as arrow poisons in South America and as ordeal poisons in Africa; the family also includes ornamentals such as *Gelsemium* (CAROLINA or YELLOW JESSAMINE). *Polypremum*, here treated in the Buddlejaceae, is sometimes placed in the Loganiaceae. Family name from *Logania*, a genus of 15 species native from Australia to New Caledonia and New Zealand. (Named for James Logan, 1674–1751, Irish botanist and writer, William Penn's agent in the U.S., and governor of Pennsylvania) (subclass Asteridae)
FAMILY RECOGNITION IN THE FIELD: herbs or vines with opposite simple leaves; stipules present; corollas sympetalous, radially symmetrical; fruit a capsule developing from a superior ovary.
REFERENCES: Moore 1947; Rogers 1986; Jensen 1992.

1. Plant an erect or spreading herb < 1 m tall; corollas white, white suffused or lined with pink outside, or red outside and yellowish inside.
   2. Corollas 10–50 mm long, white, white suffused or lined with pink outside, or red outside and yellowish inside; fruits separating into 2 carpels at maturity, falling from persistent base; calyces ca. 5–10 mm long or longer; stems pubescent or glabrous ____ **Spigelia**
   2. Corollas 1.5–2.5 mm long, white; fruits 2-horned but not separating into 2 carpels at maturity, persistent; calyces ca. 1 mm long; stems glabrous ____ **Mitreola**
1. Plant a twining, high-climbing woody vine; corollas yellow ____ **Gelsemium**

GELSEMIUM YELLOW JESSAMINE

A genus of 3 species, 2 s U.S. to Guatemala and 1 se Asia to w Malesia. They contain alkaloids and have been used medicinally and ☠ in murder and suicide. (Italian: *gelsemino*, jessamine)
REFERENCES: Duncan & Dejong 1964; Ornduff 1970.

Gelsemium sempervirens A. St.-Hil., (evergreen), CAROLINA JESSAMINE, POOR-MAN'S-ROPE, EVENING TRUMPET-FLOWER, YELLOW JESSAMINE. Perennial twining vine; leaves evergreen, opposite; leaf blades ovate to lanceolate or elliptic, to 75 mm long and 30 mm wide, petioles ca. 5 mm long; flowers axillary, solitary or 2–6 in cymes, very fragrant; corollas showy, yellow, funnelform, 5-lobed, 25–35 mm long; capsules 1.4–2 cm long; seeds winged. Wooded areas and forest margins; Dallas Co., apparently spreading or escaped from cultivation, also Lamar Co. (Carr 1994); mainly se and e TX. Feb–Apr. Numerous alkaloids including sempervirine, gelsemine, and gelsemoidine are found throughout the plant; a tea made from as few as three leaves has been reported to cause death; children have been poisoned by chewing on leaves or sucking nectar; honeybees can also be poisoned and honey derived from the flowers can be lethal; animals may also be poisoned (Muenscher 1951; Hardin & Arena 1974; Westbrooks & Preacher 1986; Leung & Foster 1996). ☠

MITREOLA MITERWORT, HORNPOD

A mainly tropical genus of 6 species. (Greek: *mitra*, mitre or cap, in reference to similarity of the 2-horned fruit to a certain kind of cap)
REFERENCE: Nelson 1980.

Mitreola petiolata (J.F. Gmel.) Torr. & A. Gray, (with a petiole or leaf-stalk), LAX HORNPOD. Glabrous annual herb to ca. 75 cm tall; leaves ovate-elliptic to elliptic-lanceolate, 2–8 cm long, tapering to a petiolate base; flowers in long peduncled cymes; corollas white; capsules 3–4 mm long, the surface smooth or with a few scattered papillae. Wet, sandy areas, ponds, streams; Bell, Burnet, Dallas, and Tarrant cos.; e 1/2 of TX. Jun–Nov. Previously treated in the genus *Cynoctonum* [as *C. mitreola* (L.) Britton].

Mitreola sessilifolia (J.F. Gmel.) G. Don, with leaves broadly oval, sessile, and rarely more than 2 cm long, occurs in se and e TX just to the e of nc TX.

SPIGELIA PINKROOT, WORM-GRASS

Perennial herbs; flowers solitary or in short 1-sided inflorescences; corollas tubular-funnelform or funnelform, with small lobes; calyces with slender lobes; anthers linear; fruits separating into 2 carpels at maturity.

A genus of 50 species of tropical and warm areas of the Americas; species have variously been used as a vermifuge, medicinally, ☠ or as a criminal poison. (Named for Adrian van der Spiegel, Latinized Spigelius, 1578–1625, Dutch professor of anatomy at Padua, who was perhaps the first to give directions for preparing an herbarium)
REFERENCE: Henrickson 1996.

1. Corollas white or white suffused or lined with pink externally, 10–13.5 mm long; plant 5–15(–19) cm tall, much branched from base; leaf blades 1.2–3.5 cm long, 3–13 mm wide, tapering at base ____ **S. hedyotidea**
1. Corollas red outside, yellowish inside, 35–50 mm long; plant 15–80 cm tall, usually with only 1 stem from base; leaf blades 5–12 cm long, 10–60 mm wide, broadly rounded at base ____ **S. marilandica**

Spigelia hedyotidea A. DC., (sweet-ear), PRAIRIE PINKROOT. Plant bushy-branched, low, 5-15(-19) cm tall; stems purplish tinged near base; flowers terminal and in leaf axils, 2 per node; corollas funnelform; anthers and style included; fruit of 2 nearly spheroid cocci, the pair 2.5-3.7 mm long, 5-6 mm wide, glabrous. Limestone outcrops and gravelly soils; Brown (HPC), Bell, McLennan, and Palo Pinto (Henrickson 1996) cos.; s and w parts of nc TX s and w to s TX and Edwards Plateau. May. [*Coelostylis lindheimeri* (A. Gray) Small, *S. lindheimeri* A. Gray] We are following Henrickson (1996) for nomenclature on *Spigelia hedyotidea*.

Spigelia marilandica L., (of Maryland), INDIAN-PINK. Stems erect, unbranched or sparsely so; leaves sessile, entire; flowers showy, in short, 1-sided, terminal inflorescences; corollas tubular-funnelform; anthers and style exserted; fruits 4–6 mm long, 6–10 mm wide. Rich woods; included based on citation of vegetational areas 4 and 5 (Fig. 2) by Hatch et al. (1990); mainly e TX. May–Oct. Austin (1975) considered this species adapted for pollination by ruby-throated hummingbirds (*Archilochus colubris*). An alkaloid, spigiline, is present and has been used medicinally as an antihelmintic or vermifuge; convulsions and poisoning have been reported from overdoses (Muenscher 1951; Blackwell 1990). ☠

LYTHRACEAE
CRAPE-MYRTLE OR LOOSESTRIFE FAMILY

Annual or perennial herbs, shrubs, or small trees; leaves opposite or alternate, sessile or subsessile, simple, entire; flowers terminal or axillary, solitary or in cymes, spike-like racemes, or panicles; flowers perigynous; hypanthium cup-shaped or tubular, calyx lobes 4–6; petals 4–6, attached near summit of hypanthium, lavender or pink to red or white, clawed; stamens 6 to many; pistil 2- to 6-celled; ovary superior, free in the hypanthium; fruit a capsule.

A medium-sized (600 species in 27 genera) family of mostly herbs with some shrubs and trees found mainly in the tropics, but with some temperate species; alkaloids are often present. A number of taxa provide dyes (e.g., *Lawsonia*–HENNA), while others are ornamentals including *Cuphea* (CIGAR-FLOWER), *Lythrum* (LOOSESTRIFE), and *Lagerstroemia* (CRAPE-MYRTLE). (subclass Rosidae)

FAMILY RECOGNITION IN THE FIELD: herbs (or 1 introduced species a shrub or small tree) with simple entire leaves and often 4-angled stems; flowers with a *hypanthium*; petals usually 4 or 6, often *crumpled* like crepe paper, particularly in bud; stamens free, inserted inside the hypanthium below the petals, often unequal in length; fruit a capsule.

REFERENCES: Shinners 1953a; Graham 1975, 1986.

1. Plant a shrub or small tree to ca. 7 m tall; leaves alternate; petals ca. 10–20 mm long (including narrow claw); introduced showy ornamental ____ **Lagerstroemia**
1. Plant herbaceous (stem can be woody at base) < 1.5 m tall (usually much less); leaves opposite, alternate, or whorled; petals 7 mm or less long; native species (also 1 introduced).
   2. Flowers borne in nearly all axils of well-developed stem leaves; petals 0–4, 2.5 mm or less long; fruits subglobose, ca. as wide as long.
      3. Flowers 1 per leaf axil (therefore 2 per node); leaves above mid-stem short-petioled and/or with narrowed base; fruits opening along distinct sutures ____ **Rotala**
      3. Flowers (1–)3–many per leaf axil; leaves above mid-stem sessile, auriculate to cordate at base; fruits opening irregularly ____ **Ammannia**
   2. Flowers borne in axils of reduced leaves along the terminal part of the stem in spike-like inflorescences with reduced leaves or bracts; petals usually 6, 3–7 mm long; fruits cylindric, ca. 2 times as long as wide ____ **Lythrum**

AMMANNIA TOOTHCUP

Glabrous, usually widely branched annual herbs to 60(–100) cm tall; leaves opposite; inflorescences of sessile or stalked axillary cymes of (1–)3–15 flowers; flowers 4-merous; petals small (to 2.5 mm long), quickly deciduous; stamens 4–8; capsules irregularly dehiscent.

A cosmopolitan genus of 25 species, mainly of wet habitats. (Named for Paul Ammann, 1634–1691, German botanist)

REFERENCES: Graham 1979, 1985.

1. Plant delicate in appearance; peduncles filiform, 3–9 mm long; flowers (1–)3–12(–15) per axil, commonly 7; petals deep rose-purple; leaves membranous; capsules (1–)1.5–3(–3.5) mm in diam.; lowest branches much shorter than main stem ____ **A. auriculata**
1. Plant robust in appearance; peduncles stout, 0–4(–9) mm long; flowers usually 1–5 (rarely more) per leaf axil; petals deep rose-purple or pale lavender; leaves fleshy or membranous; capsules 3.5–6 mm in diam.; lowest branches sometimes nearly equal in length to the main stem.
   2. Flowers mostly 1–3 per leaf axil; peduncles essentially absent (inflorescences thus sessile); petals pale lavender; anthers pale yellow; fruits 4–6 mm in diam. ____ **A. robusta**
   2. Flowers mostly 3–5 per leaf axil; peduncles stout, 0–4(–9) mm long; petals deep rose-purple; anthers deep yellow; fruits 3.5–5 mm in diam. ____ **A. coccinea**

Ammannia auriculata Willd., (eared), EAR-LEAF AMMANNIA. Plant to ca. 0.8 m tall; pedicels elongated; anthers deep yellow. In muddy ground about lakes, ponds, and ditches; Denton and Tarrant cos., also Comanche Co. (HPC); nc TX w to Rolling Plains and Edwards Plateau. Jul–Oct.

Ammannia coccinea Rottb., (scarlet), PURPLE AMMANNIA, TOOTHCUP. Plant to 1 m tall, usually widely branched; leaves membranous to fleshy, oblong- or linear-lanceolate, with widened base; petals deep rose-purple, sometimes with a deeper purple spot at base. In muddy ground about lakes, ponds, and ditches; e 1/2 of TX. Jul–Oct, occasionally as early as May. Graham (1979) has shown that *A. coccinea* is a morphologically intermediate polyploid hybrid ($n = 33$) derived from *A. auriculata* ($n = 16$) and *A. robusta* ($n = 17$). She further indicated that all occur in apparently identical, intermittently open, wet habitats (e.g., margins of drying ponds), and occur in mixed populations. Based on her range maps, all 3 taxa are widespread in nc TX. As might be expected with hybridization, there is overlap in a number of characters making correct identification of some specimens problematic.

Ammannia robusta Heer & Regel, (stout, strong). Plant to ca. 1 m tall; lowest branches often nearly equal in length to main stem; leaves fleshy; pedicels none; easily distinguished when fresh by the pale lavender petals (these sometimes with a deep rose spot at base of midvein or with rose-purple midvein). In muddy ground about lakes, ponds, and ditches; Falls and Williamson cos. (Graham 1979); nc to se TX. Jul–Oct. [*A. coccinea* Rottb. subsp. *robusta* (Heer & Regel) Koehne]

LAGERSTROEMIA CRAPE-MYRTLE

A genus of 53 species of the tropics from Asia to Australia. Some species are cultivated as ornamentals, others used for timber. (Named for Magnus von Lagerstrom, 1691–1759, Swedish merchant and friend of Linnaeus)

Lagerstroemia indica L., (of India), COMMON CRAPE-MYRTLE, CRESPÓN. Shrub or small tree to ca. 7 m tall, with smooth grayish bark; leaves alternate, deciduous, suborbicular to obovate, to ca. 7 cm long, entire, sessile or nearly so; inflorescence a terminal panicle to 20 cm long; flowers very showy, perfect, radially symmetrical, pedicillate; calyces campanulate, 6–9-lobed, 7–10 mm long; petals usually 6, to 2 cm long, purple or pink to white, conspicuously slender-clawed with orbicular-cordate blades, the margins crisped; stamens numerous (36–42); fruit a capsule ca. 1 cm long. Widely cultivated, persisting around old homesites, rarely escaping; Lamar and Tarrant cos. Summer–fall(–late fall). Native of Asia.

LYTHRUM PURPLE LOOSESTRIFE

Rhizomatous perennial herbs with woody bases; stems 4-angled; leaves opposite, alternate, or whorled; inflorescence a terminal spike or spike-like; flowers 1–2 per axil or clustered in cymules; calyces glabrous, the lobes alternating with appendages; petals (5–)6, light lavender or bluish purple to reddish purple.

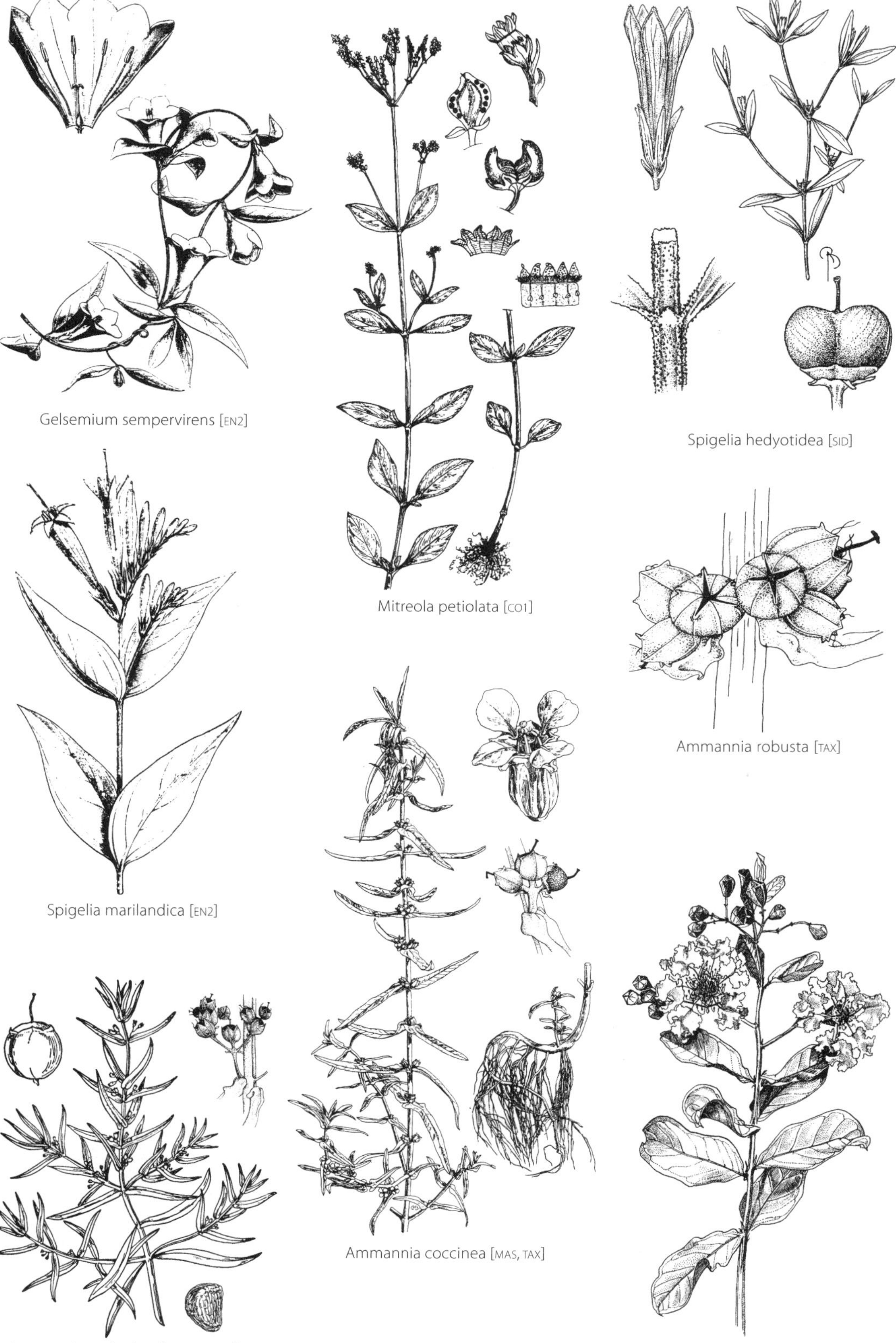

Gelsemium sempervirens [EN2]

Mitreola petiolata [CO1]

Spigelia hedyotidea [SID]

Ammannia robusta [TAX]

Spigelia marilandica [EN2]

Ammannia coccinea [MAS, TAX]

Ammannia auriculata [GLE, MAS, TAX]

Lagerstroemia indica [BT3]

A cosmopolitan genus of 36 species. (Name used by Dioscorides for *L. salicaria* L., from Greek: *lythron*, blood, alluding to the color of the flowers of some)
REFERENCES: Shinners 1953a; Stuckey 1980.

1. Stamens 6(–8); upper stem leaves usually alternate (rarely subopposite); flowers solitary or paired in the axils (many together forming a spike-like inflorescence).
   2. Middle stem leaves lanceolate to oblanceolate to narrowly elliptical (the upper leaves/bracts smaller, but usually ± similar in shape), cuneate (= wedge-shaped) basally, green, membranous ____ **L. alatum**
   2. Middle stem leaves narrowly linear to linear-oblong, oblong, or lanceolate-oblong (the upper leaves/bracts smaller, usually linear or nearly so), rounded to nearly auriculate basally, gray-green, glaucous, sometimes somewhat fleshy ____ **L. californicum**
1. Stamens usually 12; upper stem leaves opposite or whorled; flowers clustered in cymules in the axils (many together forming a showy spike-like inflorescence) ____ **L. salicaria**

Lythrum alatum Pursh var. **lanceolatum** (Elliott) Rottb. & A. Gray ex Rothr., (sp.: winged; var.: lanceolate, lance-shaped), LANCE-LEAF LOOSESTRIFE, WINGED LOOSESTRIFE. Plant to 1.2 m tall; lower stem leaves opposite to subopposite; upper stem leaves supopposite to mostly alternate; petals 3–6 mm long, purple. Low moist areas; se and e TX w to East Cross Timbers, also Edwards Plateau. Late May–Jul. [*Lythrum lanceolatum* Elliott]

Lythrum californicum Torr. & A. Gray, (of California), CALIFORNIA LOOSESTRIFE, HIERBA DEL CÁNCER, WINGED LOOSESTRIFE. Plant 0.2–1 m tall; lower stem leaves opposite to subopposite; upper stem leaves subopposite to mostly alternate; petals 4–6 mm long, purple. Low moist areas; Parker and Tarrant cos.; also Montague Co. (Mahler 1988); nc TX s and w to w TX. Late May–Jul. The differences between *L. californicum*, apparently rare in nc TX, and the common *L. alatum* var. *lanceolatum*, seem rather tenuous. Graham (1986) suggested they may ultimately be viewed as parts of a single, widespread, variable species consisting of several geographic races. Shinners (1953a) suggested that there is apparently hybridization and introgression between *L. californicum* and *L. alatum* var. *lanceolatum* [as *L. lanceolatum*] in a wide belt through c and n TX and OK.

Lythrum salicaria L., (resembling *Salix*—willow), PURPLE LOOSESTRIFE. Plant to 1.2 m tall; leaves opposite or whorled, cordate to rounded basally; inflorescences showy. Cultivated and observed in a drainage area adjacent to a flower bed in Dallas Co. (R. O'Kennon, pers. obs.); it is included because of the possibility of escape and spread in nc TX; in some parts of the ne U.S. *L. salicaria* can aggressively invade native marshlands eliminating native species; dense stands covering thousands of acres are sometimes formed with even tenacious natives such as *Typha* species being excluded. PURPLE LOOSESTRIFE is often cited as one of the most detrimental cases of habitat alteration by an exotic species in the U.S.; it was introduced in New England in the early 1800s and by 1995 was known in every state but Florida; because of its potential as a pest, it has been declared a noxious weed in several states with laws banning its distribution and cultivation; this species should not be planted (Stuckey 1980; Graham 1986; Yatskievych & Spellenberg 1993; Flack & Furlow 1996). ?–Jun–Jul. Native of Eurasia.

ROTALA

A genus of 44 species of temperate to tropical areas of the world, typically in wet habitats. (Name an incorrect diminutive of Latin: *rota*, a wheel, from the whorled leaves of the first described species)

Rotala ramosior (L.) Koehne, (very branched), TOOTHCUP. Small glabrous annual to ca. 40 cm tall; leaves opposite, linear to narrowly oblong-lanceolate; petals usually 4, pink to purple-red,

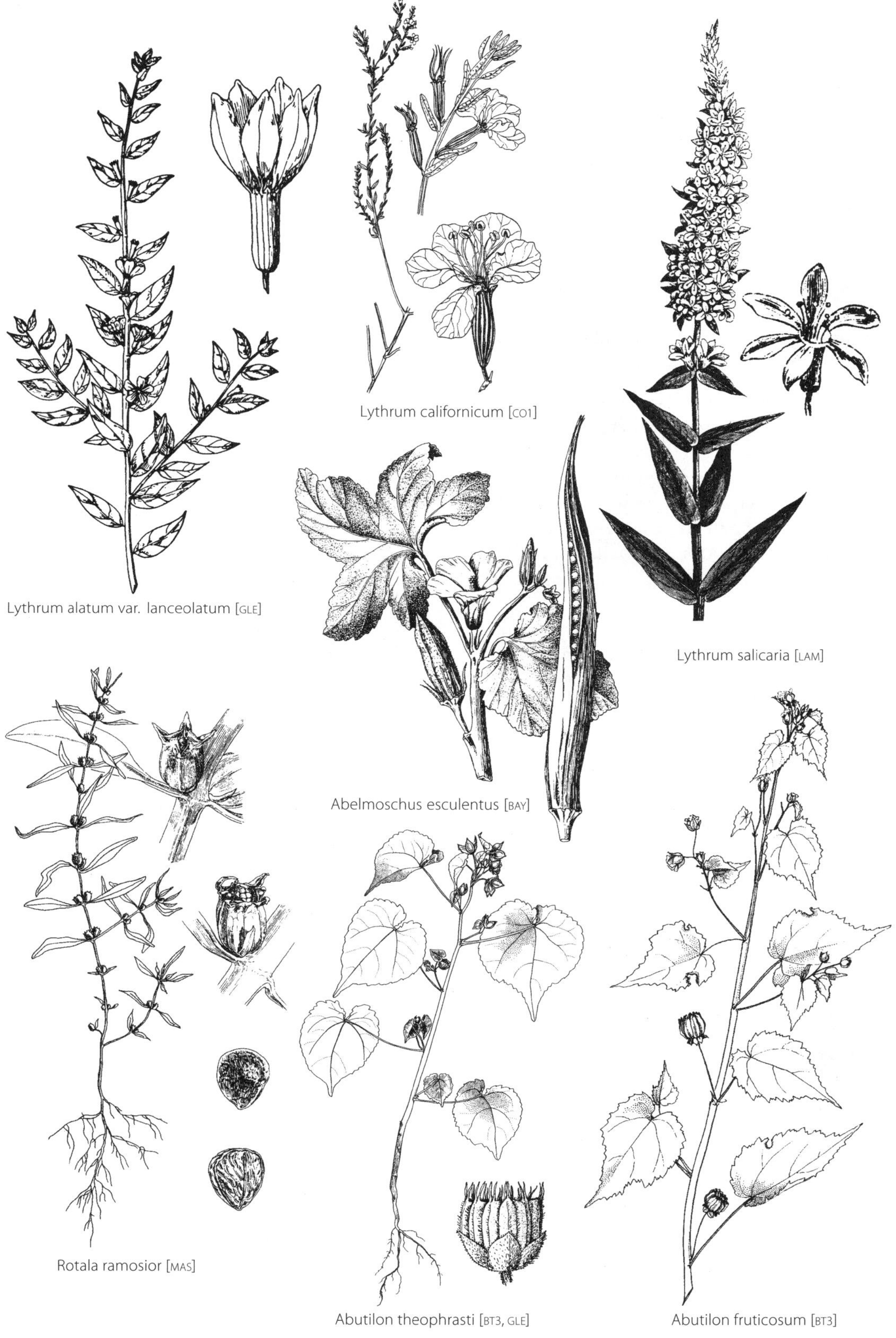

Lythrum californicum [CO1]

Lythrum alatum var. lanceolatum [GLE]

Lythrum salicaria [LAM]

Abelmoschus esculentus [BAY]

Rotala ramosior [MAS]

Abutilon theophrasti [BT3, GLE]

Abutilon fruticosum [BT3]

scarcely exceeding calyx lobes; sepals 4; fruit a many-seeded capsule. Damp, sandy or silty ground, sometimes shallow water; se and e TX w to West Cross Timbers, also Edwards Plateau. Late Jun–Oct.

MALVACEAE MALLOW FAMILY

Ours annual or perennial herbs or shrubs often with mucilaginous sap, usually with simple, forked, stellate or lepidote pubescence; leaves alternate, usually palmately nerved, simple or rarely palmately compound, entire, toothed or lobed; stipules present, sometimes falling early; flowers solitary or clustered in the leaf axils or terminal in spike-like racemes or corymbs, radially symmetrical; sepals 5, united at least near base, often subtended by an involucel of separate or united bracts (called epicalyx); petals 5, free from each other but often fused to base of the stamen tube; stamens numerous; anthers free; filaments united into a tube at least near base and thus surrounding the ovary and styles, separate apically; pistil with 3–more carpels; ovary superior; fruit a capsule or schizocarp, rarely a berry.

A medium-large (1,800 species in 111 genera), cosmopolitan, but especially tropical family of herbs, shrubs, and some trees; there are usually stellate hairs and mucilage. It includes several economically important plants, *Gossypium* species (COTTON) and *Abelmoschus esculentus* (L.) Moench (OKRA), as well as a number of showy ornamentals (e.g., *Alcea*—HOLLYHOCK, *Hibiscus*, and *Malva*—MALLOW). The family is related to Sterculiaceae, Tiliaceae, and the tropical woody Bombacaceae which includes BAOBAB (*Adansonia*) and KAPOK (*Bombax* and *Ceiba*) (Judd et al. 1994). The European *Althaea officinalis* L. (MARSH-MALLOW) yielded a mucilaginous sap that was converted into a foamy sweet confection—the original marshmallows. (subclass Dilleniidae)

FAMILY RECOGNITION IN THE FIELD: usually herbs (2 species of shrubs) with alternate, usually *palmately veined* leaves and often with stellate pubescence; flowers with a conspicuous and unique *stamen tube* with numerous free anthers; petals separate, usually attached basally to the stamen tube; involucel of bracts often subtending the calyx.

REFERENCES: Hanson 1920; Kearney 1951; Fryxell 1988, 1997; La Duke & Doebley 1995.

1. Plants a woody shrubs to 3 m tall; petals bright red, 20–35 mm long; fruits berry-like until fully mature, with a fleshy outer layer, red in color ____ **Malvaviscus**
1. Plants herbaceous (except shrubby in 1 species of *Hibiscus* and 1 species of *Pavonia*); petals variously colored and sized; fruits not berry-like (either a capsule or schizocarp).
   2. Plants small shrubs to ca. 1(–1.5) m tall; petals deep rose-pink; number of style branches and stigmas (10) twice the number of carpels (5) ____ **Pavonia**
   2. Plants usually herbaceous (1 species of *Hibiscus* shrubby); petals variously colored and sized; number of style branches and stigmas equal the number of carpels.
      3. Fruit a capsule, the 3–5 carpels united into a compound ovary, opening apically at maturity to dehise the seeds, the carpels remaining attached to one another and the axis after the seeds have been shed (seeds within the compound ovary as a whole).
         4. Sepals subtended by an involucel of 3 bracts; bracts ovate in outline, 15–25 mm or more wide, divided beyond the middle into numerous (7–13) slender teeth; seeds densely hairy; foliage usually gland-dotted and punctate (use lens). ____ **Gossypium**
         4. Sepals subtended by an involucel of usually 8 or more bracts; bracts linear, < 3 mm wide, not divided beyond the middle into numerous teeth; seeds not hairy; foliage neither gland-dotted nor punctate.
            5. Mature capsules < 4 cm long; leaves usually < 9 cm wide; calyces symmetrical, 5-merous, persistent in fruit (involucel of bracts also present just below calyx) ____ **Hibiscus**
            5. Mature capsules 7.5–20 cm long; leaves usually > 9 cm wide; calyces asymmetrical,

spathe-like, splitting laterally at flowering time, deciduous in fruit (involucel of bracts also present just below calyx) ______ **Abelmoschus**

3. Fruit a schizocarp, the 5–many individual carpels (called mericarps) loosely united into a ring around a central axis and at maturity separating and dispersing individually (seeds inside the individual carpels can be either dehiscent or indehiscent).
   6. Style branches not abruptly larger at very tip, thread-like or club-like, with the stigmas decurrent along their inner side.
      7. Involucel of (1–)3 separate bracts or bracts absent.
         8. Pedicels of lower flowers shorter than the petioles of their subtending leaves or bracts; carpels beakless; petals often deeply notched at apex; taproot not thickened ______ **Malva**
         8. Pedicels of lower flowers equaling or exceeding the petioles of their subtending leaves or bracts; carpels usually with a beak; petals not deeply notched at apex; taproot often much thickened ______ **Callirhoe**
      7. Involucel of 6–9 bracts fused at base ______ **Alcea**
   6. Stigmas usually distinctly and abruptly larger than style branches, limited to very tip of style branch or nearly so.
      9. Carpels greatly inflated, the walls thin and papery; petals whitish to pale yellow to yellow-orange, ca. 8–12 mm long ______ **Herissantia**
      9. Carpels not inflated, the walls not thin and papery; petals variously colored and of various sizes.
         10. Petals yellow or orange.
            11. Carpels sharply differentiated into 2 parts, the upper part without seeds, smooth, the lower part seed-containing, reticulate; w part of nc TX (Tarrant Co. and w) ______ **Sphaeralcea**
            11. Carpels not differentiated into upper and lower parts; widespread in nc TX.
               12. Involucel bracts absent; silvery lepidote tomentum absent; leaves symmetrical; widespread in nc TX.
                  13. Ovules usually 3 in each carpel; leaves either large (to 15 cm or more in diam.) and suborbicular OR broadly ovate, cordate basally, acute to acuminate apically, and up to 10 cm long (often much smaller) ______ **Abutilon**
                  13. Ovules 1 in each carpel; leaves various but not as above.
                     14. Calyces accrescent (=enlarging with age), becoming 10–12 mm long, forming a wing-angled, loose, globose, membranous covering over the fruit; mature carpels indehiscent ______ **Rhynchosida**
                     14. Calyces not as above; mature carpels apically dehiscent ______ **Sida**
               12. Involucel of 1–3 bracts OR foliage with silvery lepidote tomentum OR leaves asymmetrical (sometimes all 3); uncommon in nc TX.
                  15. Leaves symmetrical (= the 2 sides shaped the same); calyces inflated at maturity; silvery lepidote tomentum absent; involucel bracts 3, ovate, ca. 3–4 mm wide or lanceolate, 0.6–1 mm wide; carpels (9–)10–14(–16) ______ **Malvastrum**
                  15. Leaves asymmetrical (= the 2 sides shaped differently); calyces not inflated at maturity; silvery lepidote tomentum usually present; involucel bracts 0–3, if present linear, < 1 mm wide; carpels usually 6–10 ______ **Malvella**
         10. Petals various shades of pink, lavender, purple, or red.
            16. Leaves asymmetrical; silvery lepidote tomentum usually present (stellate pubescence can also be present) ______ **Malvella**
            16. Leaves symmetrical; silvery lepidote tomentum absent (pubescence of simple, paried, or stellate hairs).

17. Petals 4–6 mm long; calyces and leaves glabrous or with pubescence of simple or paired hairs ______ **Modiola**
17. Petals 7–22 mm long; calyces and leaves velvety-or felty-pubescent with dense stellate hairs (use hand lens) ______ **Sphaeralcea**

ABELMOSCHUS

A genus of 15 species of the Old World tropics; previously treated as part of *Hibiscus*. (Arabic: *abu-l-mosk*, father of musk, referring to the musk-scented seeds or Arabic: *kabb-el-misk*, musk seed)
REFERENCES: Bates 1968; Charrier 1984.

Abelmoschus esculentus (L.) Moench, (edible), OKRA, GUMBO, GOBO, GOMBO, LADY'S-FINGER, BANDAKAI, GOBBO. Stout annual to ca. 2 m or more tall, glabrate to bristly; leaves very large, often > 30 cm across, palmately compound to divided to scarcely lobed; flowers solitary in the upper axils; involucel bracts 8–12, linear; petals white to yellow, purple to red at base; capsules tender when young, becoming tough with age, ± long cylindrical. Widely cultivated and rarely escapes; Milam Co. Summer–fall. Native of tropical Asia. [*Hibiscus esculentus* L.] The immature capsule is the vegetable okra; the soup, gumbo, derives its character from the mucilaginous quality of the capsule.

ABUTILON INDIAN-MALLOW

Ours annual or perennial herbs with stellate pubescence; leaf blades toothed, cordate basally; flowers solitary, axillary, pedunculate, or inflorescence slightly paniculate; petals yellow; carpels separating at maturity.

A genus of ca. 160 species (Fryxell 1997) of tropical and warm areas of the world; many are cultivated as ornamentals; some are used medicinally or as sources of fiber. (Arabic name for a species of *Malva*; or probably compounded from the Arabic: *abu*, father of, and Persian: *tula* or *tulha*, mallow—Fryxell 1997)
REFERENCES: Kearney 1955b; Fryxell 1983.

1. Carpels 5–9 per fruit, 6–9 mm long, essentially awnless; leaves to 5(–9.5) cm wide; calyces 3–5 mm long, parted ca. 1/2 way, spreading to reflexed in fruit ______ **A. fruticosum**
1. Carpels 10–15 per fruit, 10–18 mm long, with awns (beaks) 2–6 mm long; leaves to 16(–21) cm wide; calyces 8–14 mm long, 5-parted nearly to base, accrescent, not reflexed ______ **A. theophrasti**

Abutilon fruticosum Guill. & Perr., (shrubby, bushy), INDIAN-MALLOW, PELOTAZO. Perennial to 85 cm tall from a woody root, finely and densely gray-pubescent with stellate hairs; leaf blades ovate to triangular, (2–)4–10(–16) cm long; peduncles 1–3 cm long; petals yellow, 4–6(–10) mm long. Slopes, prairies, limestone outcrops; in nc TX more common to the w; Blackland Prairie s and w to w TX. Jun–Oct. This species (with 6–9 carpels) has in the past been incorrectly treated by TX authors as *A. incanum* (Link) Sweet, a species with 5 carpels which occurs further west (P. Fryxell, pers. comm.). [*A. texense* Torr. & A. Gray]

Abutilon theophrasti Medik., (for the Greek naturalist, Theophrastus, 372–287 B.C., considered to be Aristotle's finest student and the founder of botany based on his detailed descriptions of plants growing in the botanical gardens of Athens; his *History of Plants* is the oldest botanical work in existence—Porter 1967; Simpson & Ogorzaly 1986), INDIAN-MALLOW, VELVET-LEAF BUTTERPRINT, CHINGMA. Annual to 2 m tall, velvety with soft stellate pubescence throughout; leaf blades ovate to nearly orbicular, usually larger than in *A. fruticosum*, (3–)5–15(–17.5) cm long; peduncles 1.5–5 cm long; petals yellow, 6–15 mm long. Disturbed or weedy areas; Collin, Dallas, Grayson, Johnson, and Tarrant cos.; scattered in TX. Jun–Oct. Native of Eurasia.

ALCEA HOLLYHOCK

An Old World genus of 60 species (Fryxell 1997) native from the Mediterranean to c Asia; a number are cultivated as ornamentals. (Greek: *alkaia*, used for a kind of mallow; from *akce*, remedy, in reference to its medical uses or *alke*, strength, in reference to its vigorous growth—Fryxell 1997)

Alcea rosea L., (rose-colored), HOLLYHOCK, AMAPOLA GRANDE. Biennial or perennial with stellate pubescence; stems mostly unbranched, to 3 m tall; leaf blades suborbicular, 8–30 cm wide, cordate basally, long-petioled; inflorescence a terminal spike-like raceme; bracts 6–9 below the calyx; flowers showy; corollas white to pink, red, or purple, to ca. 10 cm wide; fruit a schizocarp, the carpels beakless, 25 or more. Cultivated and occasionally escapes; Grayson Co. May–Sep (–frost). Native of Asia. [*Althaea rosea* (L.) Cav.]

CALLIRHOE WINECUP, POPPY-MALLOW

Annuals or perennials from a slender or in perennial species an enlarged root; petals large (to 23 mm long) and extremely showy, purple-red to pink or white; carpels separating at maturity.

A North American genus of 9 species; some are cultivated as ornamentals or valued as wildflowers. Gynodioecy is known in a number of species (Dorr 1990). (Named after Callirhoe, in Greek mythology, daughter of the river god Achelous)
REFERENCES: Waterfall 1951; Bates et al. 1989; Dorr 1990.

1. The 5-lobed calyces with involucel of 3 bracts at or close to base ____ **C. involucrata**
1. The 5-lobed calyces without bracts.
   2. Calyces hispid-pubescent outside; pedicels 0.5–4 cm long; carpels pubescent at least on the beaks; petals white to rarely pink ____ **C. alcaeoides**
   2. Calyces glabrous or sparsely pubescent on the ribs outside; pedicels 3–14 cm long; carpels glabrous; petals deep red-purple to light pink or white.
      3. Plants perennial with a swollen root; mature fruits 6–10 mm wide; carpel beak small, making up < 1/3 of the total length of the carpel; back of mature carpel body not or only slightly prolonged over the base of the beak; widespread in nc TX ____ **C. pedata**
      3. Plants annual with a slender taproot; mature fruits 4–6 mm wide; carpel beak large and hollow, making up 1/3 or more of the total length of the carpel; back of carpel body prolonged ca. 1 mm into a conspicuous whitish "collar" covering the base of the beak; s and w parts of nc TX ____ **C. leiocarpa**

Callirhoe alcaeoides (Michx.) A. Gray, (resembling *Alcea*—hollyhocks), PLAINS WINECUP, PLAINS POPPY-MALLOW, LIGHT POPPY-MALLOW, CLUSTERED POPPY-MALLOW. Erect to spreading perennial to 55 cm tall, pubescent with 4-rayed hairs or glabrous; flowers crowded in terminal corymbs; petals white, rarely pink. Prairies; calcareous clay; nc TX and Edwards Plateau. Apr–May.

Callirhoe involucrata (Nutt.) A. Gray, (with an involucre), PURPLE POPPY-MALLOW. Stems curved-ascending or trailing, to 70(–80) cm long; flowers axillary, solitary; petals reddish purple; pedicels 3-10(–21) cm long. Sandy, eroding, disturbed ground, roadsides. Late Apr–Jun. Some authorities lump the following 2 varieties (e.g., Thompson & Barker 1986).

1. Sinuses between lobes of the leaf extending to within 5–10 mm of the petiole; stipules usually large, 5–15 mm long, 5–14 mm wide; carpels strigose ____ var. **involucrata**
1. Sinuses between lobes of the leaf extending to within 2–4 mm of the petiole; stipules 2–7(–10) mm long, 1–5 mm wide; carpels glabrous or with varying amounts of strigose pubescence ____ var. **lineariloba**

var. **involucrata**, LOW WINECUP, LOW POPPY-MALLOW, PURPLE POPPY-MALLOW, BUFFALO-ROSE,

WINECUP, PURPLE-MALLOW. Dallas, Denton, and Grayson cos.; throughout TX except Trans-Pecos. ▣/81

var. **lineariloba** (Torr. & A. Gray) Torr., (with linear lobes), SLIM-LOBE POPPY-MALLOW, GERANIUM POPPY-MALLOW, WINECUP, COWBOY-ROSE. Se and e TX w to West Cross Timbers, also Edwards Plateau. Dorr (1990) indicated this is an exceedingly variable taxon with a number of morphological types described as species; however, he further indicated there is a gradual intergradation of populations and thus did not recognize formal taxa.

Callirhoe leiocarpa R.F. Martin, (smooth-fruited), TALL WINECUP, TALL POPPY-MALLOW. Erect annual (rarely biennial) to ca. 85 cm tall; glabrous or slightly pubescent with 4-rayed hairs; flowers solitary, petals to ca. 23 mm long, red-purple to light pink. Prairies, roadsides, wooded areas; Bell and Shackelford cos. on s and w margins of nc TX, also Brown Co. (HPC); mainly c and s TX. Apr–Aug.

Callirhoe pedata (Nutt. ex Hook.) A. Gray, (footed, with leaf lobes at the foot of the leaf), FINGER POPPY-MALLOW. Perennial; stems erect to reclining, to ca. 100 cm tall, glabrous to strigose pubescent; flowers solitary or in loose corymbose inflorescences; petals usually reddish purple, sometimes white or pink; a population at Tandy Hills Park in Fort Worth had white flowered individuals out numbering those with reddish purple flowers 100 to 1—B. Benz, pers. comm. Limestone outcrops or rocky prairies; mainly Blackland Prairie w to West Cross Timbers and s to Edwards Plateau. Apr–May. [*C. digitata* Nutt. var. *stipulata* Waterf.]

GOSSYPIUM COTTON

A genus of ca. 50 species (Wendel & Albert 1992) of warm temperate and tropical areas of the world. Four COTTON species, cultivated for at least 5,000 years, seem to have been domesticated independently; this is an example of multiple domestication otherwise unknown among crop plants (Wendel & Albert 1992). The hairs covering the seeds and the oil extracted from the seeds are important products. (Latin, *gossypion*, cotton plant)
REFERENCES: Watt 1907; Saunders 1961; Fryxell 1968, 1979a, 1992; Kohel & Lewis 1984; Wendel & Albert 1992; Seelanan et al. 1997.

Gossypium hirsutum L., (hairy), UPLAND COTTON, COTTON BELT COTTON, WEST INDIAN COTTON, ALGODÓN. Annual or perennial to ca. 1.5 m tall; foliage commonly glandular-dotted and punctate; leaf blades usually broadly palmately 3–5 lobed; involucel bracts 3, ovate, ca. 3–5.5 cm long, lacerate with 7–13 slender teeth to 2.5 cm or more long; corollas whitish to yellow, fading to pinkish purple, to 7.5 cm long; capsules (boll) dehiscent, 3–5 celled, to nearly 3.5 cm long, beaked, each cell with 5–11 seeds covered with a close tomentum (fuzz) and a loose, woolly tomentum (lint) yielding the cotton of commercial importance. Widely cultivated in TX and occasional as a waif; Milam Co. Summer. This tetraploid cultivar was presumably brought into cultivation in Mexico or Central America; it has been found in Mexican archaeological sites dating to 3,400 B.C. (Simpson & Ogorzaly 1986). COTTON cultivation was historically very important on the Blackland Prairie and was an important component of the economic development of nc TX; one result was the nearly complete conversion of the native prairie ecosystem into cultivated fields. COTTON is still widely cultivated in TX, particularly in the w part of the state. A toxic dihydroxyphenol, gossypol, found in the seedlings and seeds, must be removed before the seeds can be used for animal food (Duke 1985); Duke (1985) referenced a source indicating hogs have died from eating the raw seeds; in women it is an abortifacient, and chronic herbal use may cause sterility in men (McGuffin et al. 1997) ☠ ☞

HERISSANTIA

A genus of 4 species occurring from the s U.S. to tropical America and also (probably introduced) in tropical Asia and Australia (Fryxell 1973); sometimes lumped with *Abutilon* (e.g.,

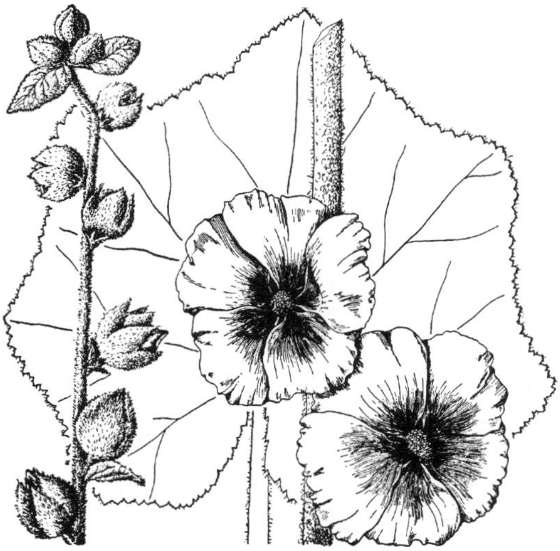

Alcea rosea [GLE]

Callirhoe alcaeoides [BB2]

Callirhoe involucrata var. involucrata [GR1]

Callirhoe pedata [BT3]

Callirhoe leiocarpa [HEA]

Mabberley 1997). (Named for Louis Antoine Prospère Herissant, 18th century French physician, naturalist, and poet)
REFERENCES: Brizicky 1968; Fryxell 1973.

Herissantia crispa (L.) Brizicky, (finely waved, closely curled), NET-VEIN HERISSANTIA. Trailing perennial with stems to 1 m or more long, velvety-tomentulose; leaves broadly ovate, 2–7 cm long, deeply cordate basally, usually abruptly acuminate apically, often prominently net-veined, crenulate; involucel bracts absent; calyx lobes 4–6 mm long; petals ca. 2 times as long as calyces, whitish to pale yellow or yellowish orange; carpels ca. 12, thin and papery, greatly inflated at maturity, usually long hirsute; fruits globose, 1–2 cm in diam. Brushy or rocky areas; included based on citation of vegetational areas 4 and 5 (Fig. 2) by Hatch et al. (1990); c, s, and w TX. Flowering throughout the growing season. [*Abutilon crispum* (L.) Medik.]

HIBISCUS ROSE-MALLOW, MARSH-MALLOW

Annuals or perennials; leaves alternate, stipulate; flowers often large and showy, solitary, axillary, peduncled; involucel bracts usually 8 or more; ovary of 5 permanantly united carpels; styles 5, stigmas capitate or peltate; fruit a loculicidal capsule.

A genus of ca. 300 species of warm temperate to tropical areas of the world. Species are variously used for fiber and medicine and a number are cultivated as ornamentals including some with very large showy flowers. *Abelmoschus esculentus* (OKRA) is sometimes treated in the genus *Hibiscus*. (Greek: *ibiscos*, old name used for some large mallow)
REFERENCE: Fryxell 1980.

1. Plants woody shrubs or small trees; seeds long-ciliate ____ **H. syriacus**
1. Plants herbaceous; seeds not ciliate.
   2. Plants low annuals to 0.6(–1) m tall with some branches becoming prostrate; plants hispid; leaves deeply incised or divided; calyces bladdery-inflated, loose around the capsule ____ **H. trionum**
   2. Plants erect perennials to 2.5 m tall; plants glabrous or pubescent; leaves toothed or lobed but not deeply incised or divided; calyces not bladdery-inflated, filled by the capsules.
      3. Stems, lower leaf surfaces, and calyces glabrous or nearly so; middle and upper leaf blades often hastate (= arrowhead-shaped) basally ____ **H. laevis**
      3. Stems, lower leaf surfaces, and calyces covered with fine stellate pubescence; leaf blades rounded or cordate to broadly cuneate basally ____ **H. moscheutos**

Hibiscus laevis All., (smooth), HALBERD-LEAF ROSE-MALLOW, SCARLET ROSE-MALLOW, HALBERD-LEAF HIBISCUS. Calyces closely enclosing the capsules; petals pink or whitish with crimson to purple blotch at base, 5–8 cm long; capsules glabrous. In low wet areas and shallow water; Dallas, Grayson, and Tarrant cos; se and e TX w to Rolling Plains and Edwards Plateau. May–Nov. [*H. militaris* Cav.] ▣/92

Hibiscus moscheutos L., (musky), Stems to 2.5 m tall; leaves lanceolate to ovate, some angled or obscurely lobed, to ca. 22 cm long; petals 5–10 cm long. Wet areas; mainly e and se TX. Jun–Oct.

1. Upper surface of leaf blades permanently pubescent; capsules densely pubescent ____ subsp. **lasiocarpos**
1. Upper surface of leaf blades glabrous or becoming so; capsules glabrous ____ subsp. **moscheutos**

subsp. **lasiocarpos** (Cav.) Blanch., (woolly-fruited), WOOLLY ROSE-MALLOW, SWAMP ROSE-MALLOW, MALLOW-ROSE, ROSE-MALLOW. Leaves long petiolate (petioles to 10 cm long); petals white to rose with crimson or deep purplish red blotch at base. Grayson Co., also Dallas Co. (hybrids with *H. laevis*); e TX w to Rolling Plains and Edwards Plateau. [*H. lasiocarpos* Cav.]

subsp. **moscheutos**. COMMOM ROSE-MALLOW, SWAMP ROSE-MALLOW, MALLOW-ROSE, WILD COTTON. Petioles to ca. 5 cm long; petals white or creamy-yellow with a crimson-purple base. Included

Gossypium hirsutum [WAT]

Herissantia crispa [SBM]

Hibiscus laevis [CO1]

Hibiscus moscheutos subsp. lasiocarpos [G&F]

Hibiscus moscheutos subsp. moscheutos [GR1]

Hibiscus syriacus [GLE]

Hibiscus trionum [GLE]

Malva neglecta [BT3, REE]

based on citation of vegetational area 4 (Fig. 2) by Hatch et al. (1990); mainly se and e TX. Jones et al. (1997) did not list this subspecies for TX; J. Kartesz (pers. comm. 1997) does; we have seen no specimens from TX and its status in the state is unclear.

Hibiscus syriacus L., (of Syria), ROSE-OF-SHARON, ALTHAEA. To ca. 7 m tall; foliage glabrous; leaves ovate to rhombic-ovate in outline, 2–8(–12) cm long, usually 3-lobed, toothed; flowers axillary, subtended by 6–10 linear to lanceolate involucel bracts; petals 5–7 cm long, white to pink, reddish or lavender with a dark red base (flowers sometimes doubled—with more than the normal 5 petals). Commonly cultivated, long persists, and spreads by seeds; Tarrant Co. (R. O'Kennon, pers. obs.); widely cultivated in TX. Jun–Sep. Native of e Asia.

Hibiscus trionum L., (old generic name), FLOWER-OF-AN-HOUR, VENICE-MALLOW. Calyces bladdery, loosely enclosing the capsules; petals pale yellow or whitish, with a purple or brown eye, 2.5–4 cm long; capsule pubescent. In disturbed habitats; Tarrant Co, also Grayson Co. (K. Haller, pers. comm.); nc TX w to Rolling Plains and s to Edwards Plateau. Aug–Sep. Native of c Africa. /92

MALVA MALLOW, CHEESE WEED

Annuals, biennials or perennials; leaves mostly long-petioled; blades orbicular to suborbicular or reniform, shallowly palmately lobed and toothed, cordate basally; petals purple-red to pale blue or white; carpels indehiscent.

A genus of ca. 40 species of Europe, the Mediterranean, temperate Asia, and tropical African mountains; some are cultivated as ornamentals, while others are used as a leaf vegetable. (Classical Latin name for mallow, from Greek: *malachos*, soothing or soft, alluding to its medicinal properties or its soft mucilaginous quality, or *malos*, tender, soft, woolly)
REFERENCES: Morton 1937; Ray 1995.

1. Petals purple-red, 15–30 mm long, 3–5 times as long as the sepals; involucel bracts oblong to ovate or obovate ____ **M. sylvestris**
1. Petals pale blue to white, 4–12 mm long, < 2.4 times as long as sepals; involucel bracts linear to linear-lanceolate or oblong.
   2. Petals ca. 2 times as long as sepals; mature carpels smooth or faintly reticulate ____ **M. neglecta**
   2. Petals 2 times as long as sepals; mature carpels rugose-reticulate (with a hand lens)
      3. Calyces not enlarged or only barely so with age, mostly closed over the fruits; pedicels 10–. 25 mm long; claw of petals pilose ____ **M. rotundifolia**
      3. Calyces much-enlarged with age and widely spreading under the fruits; pedicels 3–10(–15) mm long; claw of petals glabrous ____ **M. parviflora**

Malva neglecta Wallr., (neglected, overlooked), COMMON MALLOW, CHEESES. Plant at first erect, becoming partly decumbent; stems to ca. 1 m long; leaves 3–6 cm wide, shallowly 5–9-lobed; flowers fascicled in leaf axils; petals 6–12 mm long; carpels 12–15. Waste ground and roadsides; Dallas, Grayson, and Jack cos., also Brown, Colemen, and Hamilton cos. (HPC); established in various localities mainly in w 2/3 of TX. Late Apr–Jun, sometimes repeating Sep–Oct. Native of Eurasia and n Africa.

Malva parviflora L., (small-flowered), LITTLE MALLOW. Glabrous or sparsely pubescent annual similar to *M. neglecta* and *M. rotundifolia*; stems erect or ascending; leaf blades suborbicular-cordate to reniform or angulate-lobed, to 7 cm long and 12 cm wide, usually wider than long; flowers fascicled in leaf axils; petals 4–5 mm long; carpels ca. 10. Weedy areas; Brown Co. on w margin of nc TX (*B. Ellebracht s.n.*, 19 Mar 1998, HPC); according to label data, possibly introduced with wheat seeds from Kansas; also s, se, and e TX, Edwards Plateau, and Trans-Pecos. Mar–Jul. Native from Spain and North Africa to India.

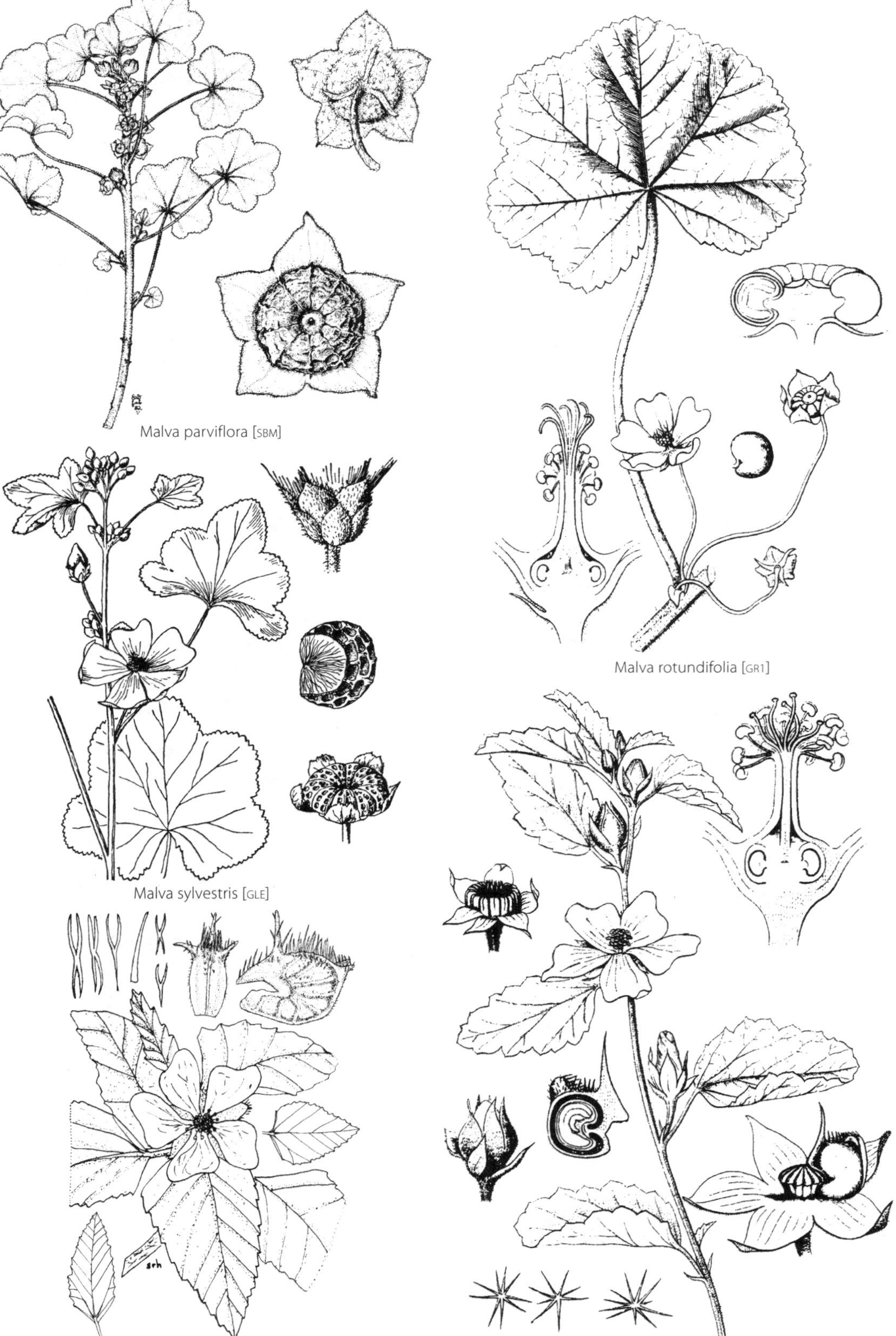

Malva parviflora [SBM]

Malva rotundifolia [GR1]

Malva sylvestris [GLE]

Malvastrum coromandelianum [RHO]

Malvastrum aurantiacum [GR1]

Malva rotundifolia L., (round-leaved), RUNNING MALLOW, ROUND-LEAF MALLOW, COMMON MALLOW, DWARF MALLOW, NORTHERN MALLOW. Similar to *M. neglecta*; stems decumbent to ascending; carpels 8–15. Found as a farm-yard weed n of Denton in May 1950, also Tarrant Co. (Dec 1998); nc TX and w TX. May–Dec. Native of Europe.

Malva sylvestris L., (of woodland), HIGH MALLOW. Erect biennial or perennial 0.2–1 m tall; leaves orbicular to reniform, shallowly 3–7-lobed; flowers fascicled in upper leaf axils or in a spike-like inflorescence; carpels ca. 10. Cultivated and escapes, waste places; Brown, Dallas, and Tarrant cos.; e TX w to nc TX, also s TX. May–Oct. Native of Eurasia.

MALVASTRUM FALSE MALLOW

Stellate-pubescent annuals or perennials; leaves petioled; leaf blades symmetrical; flowers solitary, axillary or rarely on reduced 2–3-flowered, apical, axillary branches; involucels of 3 bracts present; petals of varying shades of golden-yellow or orange-yellow; carpels (9–)10–14(–16), indehiscent.

A genus of 14 species of tropical and warm areas of the world. (Possibly from genus name *Malva* and Greek: *aster*, star, presumably in reference to the fruit or possibly from *Malva* and the diminutive suffix *astrum*, indicating a resemblance, hence the common name FALSE MALLOW—Fryxell 1997)

REFERENCES: Kearney 1955a; Brizicky 1966b; Fryxell & Hill 1977; Hill 1980b, 1982.

1. Stems and leaves with sub-lepidote 6–10-rayed stellate hairs usually with basally united rays; petals (12–)15–16 mm long, 10–13(–15) mm wide; mericarps (4–)5(–6) mm long, with a prominent medial-apical cusp 1.5–2.3 mm long and 2 contiguous flattened obtuse cusps at the distal margin ____ **M. aurantiacum**
1. Stems and leaves with simple or 2–4-rayed stellate hairs, the rays usually not united basally; petals (6–)8–10(–13) mm long, (4.5–)6–7(–9) mm wide; mericarps (3–)3.5–4 mm long, with a prominent or small medial-apical cusp 0.2–1.2 mm long and 2 divergent pointed cusps at the distal margin ____ **M. coromandelianum**

Malvastrum aurantiacum (Scheele) Walp., (orange-red), WRIGHT'S FALSE MALLOW. Perennial; stems rigid, 0.4–1 m tall, from a woody base; leaf blades subcordate to broadly ovate to oblong, (18–)30–40(–55) mm long, usually unlobed or very infrequently with 2 obscure lateral lobes, crenate-dentate, obtuse in general outline apically (but often with a small, pointed, apical tooth); bracts of involucel ovate to subcordate, adnate to base of calyx; petals golden-yellow to pale orange-yellow, drying to pinkish purple; carpels (12–)14(–16). Stream banks and pastures; Bell, Dallas, and Tarrant cos.; mainly c and s TX; endemic to TX. Apr–Jul (Oct). [*M. wrightii* A. Gray]

Malvastrum coromandelianum (L.) Garcke, (of the Coromandel Coast of se India), THREE-LOBE FALSE MALLOW. Herbaceous annual or perennial, 0.2–0.6(–1) m tall; leaf blades ovate to lanceolate, (17–)30–40(–65) mm long, unlobed or infrequently with 3 lobes, dentate to serrate, usually acute apically; bracts of involucel linear to narrowly spatulate; petals pale golden-yellow to orange-yellow; carpels (9–)10–14(–15). Alkaline soils; disturbed areas; Fort Hood (Bell or Coryell cos.—Sanchez 1997); also P. Fryxell (pers. comm.) indicated that this species is a common weed in Austin, just to the s of nc TX and therefore likely in the s part of nc TX; mainly s TX and Edwards Plateau n to edge of nc TX, also Callahan Co. (Hill 1982) just w of nc TX. Apr–Nov. A pantropical weed extending into temperate regions (Correll & Johnston 1970; Fryxell 1988).

MALVAVISCUS TURK'S-CAP

A mainly tropical American genus of 3 species. (From genus name *Malva*, and Latin: *viscus*, glue, but the meaning is unclear since *Malvaviscus* is not viscid—Fryxell 1997; an alternative

possibility is that the name is a combination of *Malva* and *Hibiscus*)
REFERENCES: Schery 1942; Turner & Mendenhall 1993.

Malvaviscus arboreus Dill. ex Cav. var. **drummondii** (Torr. & A. Gray) Schery, (sp.: tending to be woody, tree-like; var.: for its discoverer, Thomas Drummond, 1780–1835, Scottish botanist and collector in North America), DRUMMOND'S WAX-MALLOW, TEXAS-MALLOW. Shrub to 3 m tall; leaf blades 4–9 cm long, cordate-orbicular, as wide as long, shallowly and evenly 3-lobed or 3-angled, rather closely toothed; flowers in upper leaf axils; bracts of involucel linear-spatulate; petals bright red, 2–3.5 cm long, the corolla somewhat contorted and tube-like, the petals spreading only above; number of style branches and stigmas (10) twice the number of carpels (5); fruits 5-celled, the carpels berry-like, indehiscent, red, reported to be edible (Standley 1923b; Crosswhite 1980). Bell and Limestone cos., also Brown and McLennan cos. (Turner & Mendenhall 1993), also escaped in Dallas Co. (E. McWilliams, pers. comm.); native n to s part of nc TX, mainly s and e TX n to nc TX; also cultivated as a perennial further n and freezing back in most winters. While this taxon is often recognized as a distinct species (e.g., Kartesz 1994), we are following Turner and Mendenhall (1993) and J. Kartesz (pers. comm. 1997) in treating it as a variety of *M. arboreus*. Jun–Jul. [*M. drummondii* Torr. & A. Gray] This is the only temperate zone member of the genus. Austin (1975) considered this species adapted for pollination by ruby-throated hummingbirds (*Archilochus colubris*).

MALVELLA

Low perennial herbs with decumbent, prostrate to ascending stems to 40 cm long; foliage with silvery scales (lepidote) to free-rayed stellate hairs; leaves variable in shape, often asymmetrical, mostly palmately 5-nerved from base; flowers axillary; sepals subtended by an involucel of 0–3 bracts; petals 10–17 mm long, white to sulfur-yellow to rose; carpels usually 6–10.

A genus of 4 species, 3 of the Americas and 1 from the Mediterranean; previously included in *Sida*. The key to *Malvella* species is adapted from Fryxell (1974). (Diminutive of genus name *Malva*)
REFERENCES: Clement 1957; Fryxell 1974.

1. Leaf blades rounded to subacute at apex, reniform, wider than long, with stellate pubescence predominating; involucel of 1–3 bracts present; calyx lobes ovate ____ **M. leprosa**
1. Leaf blades acute at apex, ovate, triangular to narrowly triangular, longer than wide, predominantly or exclusively silvery-lepidote (= covered with scales); involucel usually absent but sometimes present; calyx lobes ovate-cordate to cordate.
   2. Leaf blades ovate-triangular, 1–2(–3) times as long as wide, often with at least some stellate pubescence mixed with silvery scales, the margins toothed to apex; involucel sometimes present; calyx lobes ovate-cordate ____ **M. lepidota**
   2. Leaf blades narrowly elongate-triangular, (2–)2.5–5(–6) times as long as broad, indument solely silvery-lepidote, the margins entire except for few hastate teeth at base; involucel absent; calyx lobes cordate ____ **M. sagittifolia**

Malvella lepidota (A. Gray) Fryxell, (with small scrufy scales), SCURFY SIDA. Leaf blades 10–30(–45) mm long, to 50 mm wide; petals white to cream to pale brownish yellow, rose upon drying. Rocky or silty soils along irrigation canals and in depressions; Hamilton Co. (HPC—in that county known from a single collection); also Hatch et al. (1990) cited vegetational areas 4 and 5 (Fig. 2); mainly c to w TX. Mar–Oct. [*Sida lepidota* A. Gray, *Sida leprosa* (Ortega) K. Schum. var. *depauperata* (A. Gray) Clement]

Malvella leprosa (Ortega) Krapov., (scrufy), ALKALI SIDA, DOLLAR WEED, ALKALI-MALLOW. Leaf blades to 40 mm long and 50 mm wide; petals white to cream or rose, brownish on drying.

Flats and rocky areas; included based on citation of vegetational areas 4 and 5 (Fig. 2) by Hatch et al. (1990); mainly c to w TX. Mar–Oct. [*Sida hederacea* (Douglas ex Hook.) Torr. ex A. Gray, *Sida leprosa* var. *hederacea* (Douglas ex Hook.) K. Schum.]

Malvella sagittifolia (A. Gray) Fryxell, (arrow-leaved). Leaves to 54 mm long, 2–5(–10) mm wide; petals yellow or white, often suffused with red. Flats and rocky areas; included based on citation of vegetational areas 4 and 5 (Fig. 2) by Hatch et al. (1990); mainly c to w TX. Throughout growing season. [*Sida leprosa* var. *sagittifolia* (A. Gray) Clement, *Sida sagittifolia* (A. Gray) Rydb.]

MODIOLA

A monotypic New World genus native from Virginia to Argentina. (Latin: *modiolus*, hub or body of a wheel, in allusion to the fruit)

Modiola caroliniana (L.) G. Don, (of Carolina), CAROLINA MODIOLA. Creeping perennial herb, sparsely hispid-pilose; leaf blades orbicular-ovate, shallowly to deeply palmately 3–5-lobed and toothed, to 6 cm long and 4 cm wide; flowers axillary, solitary, rather small; petals 4–6 mm long, salmon to purplish red; fruits of 15–30 carpels. Disturbed and waste areas; found as a garden weed at Dallas in 1949, also Denton and Limestone cos., also Brown (Stanford 1976), Parker, and Tarrant (R. O'Kennon, pers. obs.) cos.; common in s and se Texas, scattered elsewhere in the state. Apr–Jun. Reported as poisonous to livestock (Burlage 1968).

PAVONIA

A genus of ca. 150 species of herbs, subshrubs, or shrubs (rarely arborescent) (Fryxell 1997) of tropical and warm areas; some are cultivated as ornamentals and others for their fibers. (Named for José Antonio Pavón, 1754–1844, Spanish botanist and physician, traveler in Peru, and one of the authors of *Flora Peruviana*—Fryxell 1997)
REFERENCE: Fryxell 1979b.

Pavonia lasiopetala Scheele, (woolly-petaled), WRIGHT'S PAVONIA. Small shrub to ca. 1(–1.5) m tall; stems densely to sparsely stellate-pubescent; leaves alternate, petioled; leaf blades ovate-cordate, usually acute apically, 2–5(–7.5) cm long, stellate-pubescent on both surfaces; flowers solitary, axillary, on pedicels 2–5 cm long, opening in early morning and lasting only 1 day; involucel of 5(–8) linear bracts; calyx lobes ovate, acuminate, conspicuously nerved, the nerves green and the intervening tissue whitsh; petals deep rose-pink, 12–25 mm long; styles 10; carpels 5; fruits 8–9 mm in diam. Cultivated and spreading in landscapes; Tarrant Co.; native in rocky woods of Edwards Plateau and s TX. Flowering throughout the growing season. /102

RHYNCHOSIDA BEAKED SIDA

A genus of 2 species of the Americas; previously treated in the genus *Sida*; Fryxell (1978) discussed generic relationships of *Rhynchosida*, *Sida*, and related genera. (Presumably from Greek: *rhyncho*, snout, and the genus name *Sida*)
REFERENCE: Fryxell 1978.

Rhynchosida physocalyx (A. Gray) Fryxell, (bladder calyx, alluding to the somewhat inflated calyx), SPEAR-LEAF SIDA. Perennial from a fleshy-woody rootstock; stems spreading or decumbent, to ca. 40 cm long; foliage with stellate pubescence; leaf blades suborbicular to oblong, to 6 cm long and 5 cm wide, obtuse to broadly rounded apically, cordate basally; flowers solitary in the leaf axils; calyx lobes cordate, with an apical mucro, ca. 8 mm long in flower, later enlarging and covering the fruit, petals ca. as long as calyx, yellow or buff in color; fruits of 10–14 carpels, disk-like, blackish at maturity. Rocky or sandy soils, prairies, waste places; Brown, Coleman,

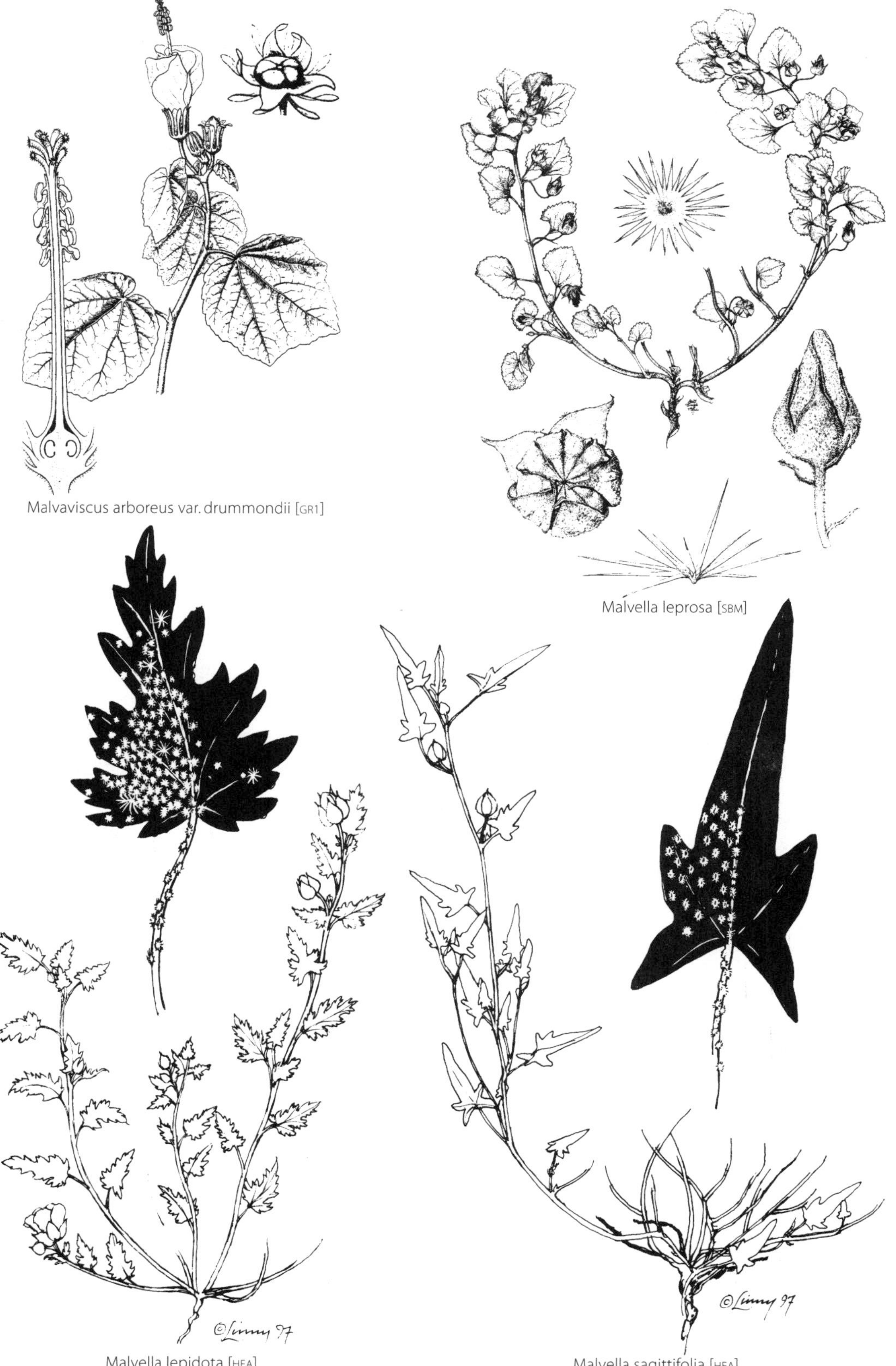

Malvaviscus arboreus var. drummondii [GR1]

Malvella leprosa [SBM]

Malvella lepidota [HEA]

Malvella sagittifolia [HEA]

Stephens, McLennan, Mills, Palo Pinto, Tarrant, and Young cos., also Hamilton Co. (HPC); mainly w part of nc TX s and w to w TX. Apr–Oct. [*Sida hastata* A. St.-Hil., *Sida physocalyx* A. Gray] ▣/104

SIDA

Ours prostrate to erect annuals or perennials with stellate pubescence; flowers axillary, petals light yellow to orange-yellow; carpels 5, dehiscent at apex, usually apiculate with 2 beaks 0–1 mm long.

A genus of ca. 100 species (Fryxell 1997) native to tropical and warm areas, especially the Americas; some are sources of fiber. The botanical journal *Sida, Contributions to Botany*, published by the Botanical Research Institute of Texas, is named after this genus; Lloyd Shinners chose the abbreviated one word title for simplicity in reference citations and because it would be familar to botanists all over the world (Mahler 1973b). (Greek name used by Theophrastus for some similar plant)
REFERENCES: Kearney 1954; Clement 1957; Fryxell 1978, 1985.

1. Stems prostrate; small spine-like structure absent; pedicels much longer than the petioles; petals yellow to orange-yellow ____ **S. abutifolia**
1. Stems erect; small (ca. 1 mm long) spine-like structure present at the base of the petiole of well-developed leaves; pedicels shorter than the petioles; petals pale yellow ____ **S. spinosa**

Sida abutifolia Mill., (with leaves like *Abutilon*—Indian mallow), SPREADING SIDA. Perennial from a woody rootstock; stems to 50 cm long; foliage stellate-pubescent; leaf blades linear to ovate or oblong, to 35 mm long, usually much smaller, crenate-dentate; calyces ca. 5 mm long; petals much longer than calyces; carpels apiculate varying to having 2 prominent points. Limestone outcrops, rocky prairies, and roadsides; in nc TX mainly Blackland Prairie and Grand Prairie; widespread in TX. Apr–Oct. [*Sida filicaulis* Torr. & A. Gray]

Sida spinosa L., (spiny), PRICKLY SIDA, PRICKLY-MALLOW. Minutely stellate-pubescent, branched annual to ca. 100 cm tall; leaves ovate to narrowly lanceolate, to 55 mm long and 30 mm wide, crenate-dentate; calyces 5–7 mm long; petals slightly longer than calyces; carpels opening into 2 prominent beaks. Disturbed and waste areas; se and e TX w to East Cross Timbers. May–Nov.

SPHAERALCEA GLOBE-MALLOW, FALSE MALLOW

Perennials with stellate pubescence; flowers axillary and in terminal, spike-like racemes; carpels sharply differentiated into 2 parts, the lower part seed-containing, indehiscent, reticulate, the upper part seedless, smooth.

A genus of ca. 40 species (Fryxell 1997) of arid areas of the Americas; some are cultivated as ornamentals. (Greek: *sphaera*, sphere or globe, and the genus name *Alcea*—hollyhock, from the commonly spherical fruit)
REFERENCES: Kearney 1935; La Duke & Northington 1978.

1. Larger leaf blades toothed or with only 2 prominent basal lobes; calyces subtended by linear-lanceolate to thread-like involucel bracts, these soon falling.
    2. Leaf blades oblong-lanceolate to linear-lanceolate, toothed, but unlobed; most pedicels shorter than calyces ____ **S. angustifolia**
    2. Leaf blades triangular-lanceolate, with 2 basal lobes; longer pedicels as long as or longer than calyces ____ **S. hastulata**
1. Larger leaf blades compound or very deeply several-lobed (typically 5 narrow lobes); calyces usually not subtended by involucel bracts ____ **S. coccinea**

Modiola caroliniana [GR1]

Pavonia lasiopetala [GR1]

Rhynchosida physocalyx [MAR, SBM]

Sida spinosa [REE]

Sida abutifolia [BT3]

Sphaeralcea angustifolia (Cav.) G. Don subsp. **cuspidata** (A. Gray) Kearney, (sp.: narrow-leaved; subsp.: with a cusp or sharp stiff point), NARROW-LEAF GLOBE-MALLOW, POINT-SEED GLOBE-MALLOW, COPPER-MALLOW. Stems spreading to erect, to 100 cm long; petals 7–20 mm long, variable in color, salmon (Tarrant Co. specimen), reddish to orangish to lavender; carpels 10–15. Sandy or rocky soils; Tarrant Co. (near Fort Worth stockyards), also Brown Co. (Kearney 1935 and HPC); mainly w 1/2 of TX. Late May–Oct. [*S. angustifolia* var. *cuspidata* A. Gray] ▣/106

Sphaeralcea coccinea (Nutt.) Rydb., SCARLET GLOBE-MALLOW, RED FALSE MALLOW. Stems decumbent to ascending, to 35(–50) cm long; plant spreading by branching roots; petals scarlet, 10–20 mm long; carpels 10–14. Sandy or gravelly open ground, roadsides; Jack and Montague cos., also Brown Co. (HPC) and Fort Hood (Bell or Coryell cos.—Sanchez 1997); mainly West Cross Timbers s and w to w TX. Apr–Jun, sporadically to Oct. The stellate trichomes of this species were used as evidence in a forensic botany case involving a 1989 plane crash near Ruidoso, New Mexico (Fish vs. Beech Aircraft Corp.); trichomes obtained from the stored wreckage of the airplane engine were argued to be evidence for faulty engine design; i.e., that the design had allowed plant material to be sucked into the engine, thus causing the crash. However, botanists were able to show that the trichomes were introduced into the engine post-crash and could not have caused the accident; their evidence included information showing the trichomes could not have survived the heat of the crash (melted aluminium was present), that the trichomes were from *S. coccinea* which was common at the storage site, and that associated with the trichomes there were dead bees, pollen, and chewed *Sphaeralcea* leaves suggesting that the presence of the plant material was the result of nest building activities by bees in the wreckage during post-crash storage (Blaney 1995; Bates et al. 1997; Brunk 1997; Linddell 1997; Rozen & Eickwort 1997).

Sphaeralcea hastulata A. Gray, (somewhat spear-shaped). Stems usually decumbent, 15–40 cm long, forming bushy clumps; petals orange to rose-orange or scarlet, 11–22 mm long. Sandy, silty, gravelly, and rocky ground; Trans-Pecos to s Rolling Plains, spreading e along roadsides in Erath Co. in w nc TX, and along railroad in Tarrant Co. (Mahler 1988), also Brown Co. (HPC). Apr–May, sporadically to Oct.

MELASTOMATACEAE
MELASTOME OR MEADOW-BEAUTY FAMILY

A large (4,950 species in 188 genera), mainly tropical and warm area family with many species in South America; most are shrubs or herbs, less often trees or lianas. Some yield timber or dyes or are used as cultivated ornamentals. Family name from *Melastoma*, a genus of ca. 70 species native from Indomalesia to the Pacific. (Greek: *melas*, black, and *stoma*, mouth, in reference to the mouth being stained black by the fruits of some species) (subclass Rosidae)

FAMILY RECOGNITION IN THE FIELD: herbs with opposite, simple, palmately veined leaves with the *main veins parallel*; stems often 4-angled; *petals* 4, whitish rose to rose to purple; filaments geniculate (= elbow-shaped); anthers opening by apical pores, with sterile appendages.

REFERENCE: Wurdack & Kral 1982.

RHEXIA MEADOW-BEAUTY, DEER-GRASS

Perennial herbs often with long glandular hairs; stems above the middle with 4 well-defined sides (faces); leaves opposite, distinctly palmately 3-veined, serrate-ciliate; stipules absent; flowers showy, solitary or in cymose inflorescences; calyx lobes 4; petals 4, lavender-rose to lavender or white, obovate; stamens 8; anthers yellow, conspicuous, basifixed, typically with a basal appendage, often curved or straight, dehiscing by a pore; hypanthium cylindrical at flowering, ±

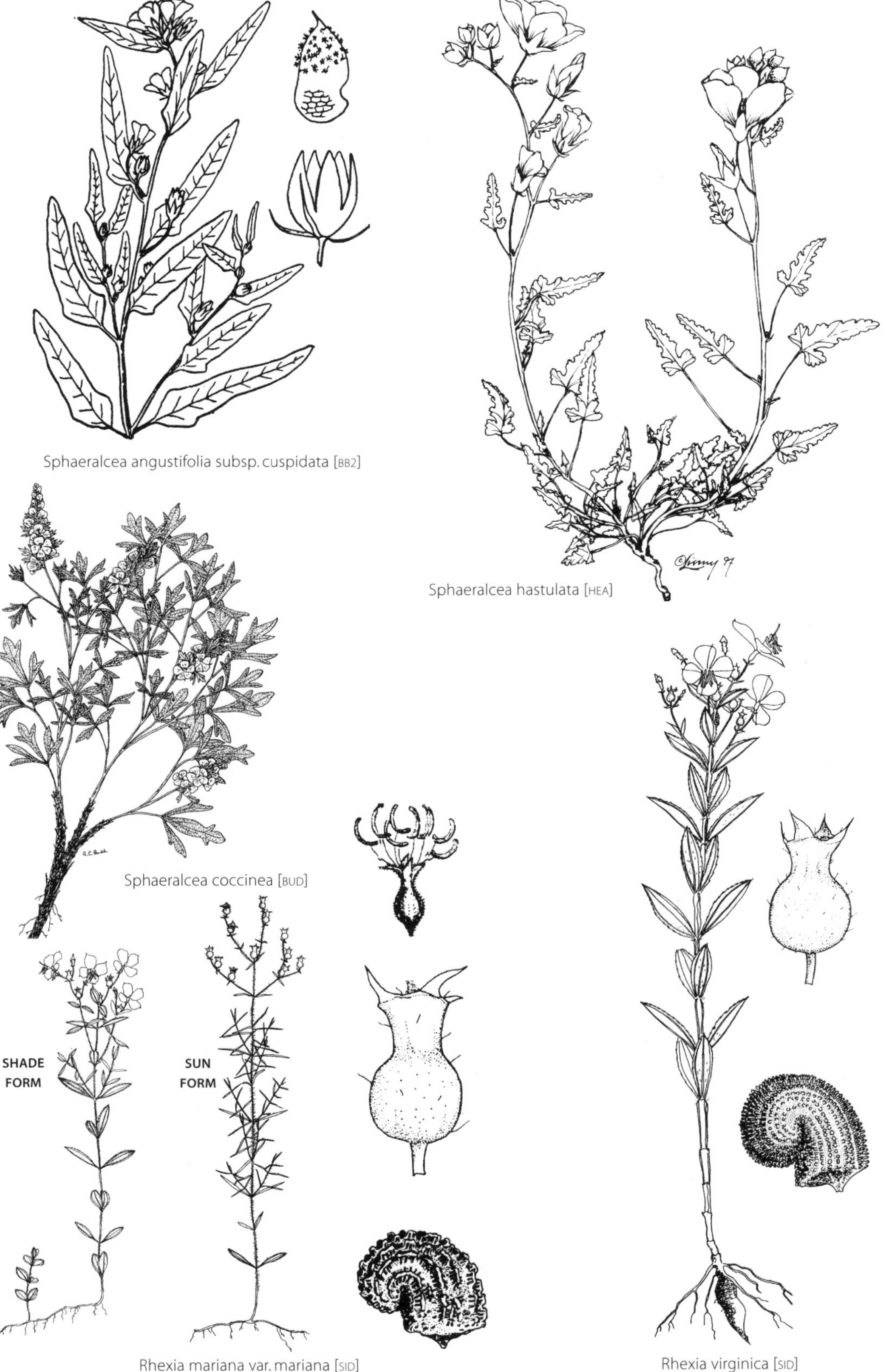

Sphaeralcea angustifolia subsp. cuspidata [BB2]

Sphaeralcea hastulata [HEA]

Sphaeralcea coccinea [BUD]

Rhexia mariana var. mariana [SID]

Rhexia virginica [SID]

urceolate at maturity and enclosing all or most of the capsule, narrowed distally to form a neck-like portion, usually with glandular-stipitate hairs; fruit a globose or subglobose capsule.

A North American genus of 13 species including cultivated ornamental herbs. (Name used by Pliny for a member of the Boraginaceae, which was reputed to be useful in curing ruptures; from Greek; *rhexis*, a breaking or rupture, for breaking or bursting forth of the entrails of victims; why adopted for its modern use unknown)
REFERENCES: James 1956; Eyde & Terri 1967; Kral & Bostick 1969.

1. Leaves short-petiolate; stems not conspicuously winged, with one pair of the 4 stem faces flat to concave and much narrower than the convex or rounded other pair; petals lavender to white or rarely lavender-rose; plants from elongate stoloniferous rhizomes ____ **R. mariana**
1. Leaves sessile or nearly so; stems usually with conspicuous wings to ca. 2 mm wide, with the 4 stem faces flat to convex and essentially equal; petals lavender-rose; plants from spongy-thickened or tuberiferous rootstocks ____ **R. virginica**

Rhexia mariana L. var. **mariana**, (of Maryland), MARYLAND MEADOW-BEAUTY. Stems to ca. 0.8 m tall, branched or unbranched; leaf blades linear to elliptic to lanceolate or ovate (variable depending on whether in sun or shade—see illustration), to 6.5 cm long and 2 cm wide; calyx lobes to 1–3 mm long; petals 10–25 mm long, lavender to white, rarely lavender-rose; anthers 6–10 mm long; hypanthium 6–10 mm long. Moist or wet, usually open areas; Lamar Co. in Red River drainage, also Limestone and Milam cos. near e margin of nc TX, also Grayson Co. (S. Crosthwaite, pers. obs.); se and e TX w to e part of nc TX. May–Sep.

Rhexia mariana var. *interior* (Pennell) Kral & Bostick, (inland), [*R. interior* Pennell], has the 4 stem faces subequal (and thus using that character in the key above it would key to *R. virginica*), stem angles sharp or narrowly winged, rhizomes stout, leaves short-petiolate, flowers bright lavender-rose, and hypanthium 10–13 mm long. This variety is known from se and e TX and from s OK just n of nc TX (Kral & Bostick 1969); its occurrence would not be surprising in nc TX in the Red River drainage or on the e margin of nc TX.

Rhexia virginica L., (of Virginia), COMMON MEADOW-BEAUTY. Stems to 1 m tall, usually branched; leaves ovate to ovate-lanceolate or elliptic, to 10 cm long and 3.5 cm wide (usually smaller); calyx lobes to 1.5–3 mm long; petals 10–25 mm long, lavender-rose; anthers 4–7 mm long; hypanthium usually 8–10 mm long. Seeps or wet areas; included based on citation of vegetational area 4 (Fig. 2) by Hatch et al. (1990); mainly se and e TX; much less common in TX than *R. mariana*. Jun–Oct.

MELIACEAE MAHOGANY FAMILY

A medium-sized (565 species in 51 genera), mainly tropical family with a few in the subtropics; most are trees or rarely shrubs; the bark is typically bitter and astringent. The family includes the genus *Swietenia* (MAHOGANY), the source of a valuable tropical wood used for furniture including cabinets and formerly in shipbuilding. (subclass Rosidae)
FAMILY RECOGNITION IN THE FIELD: the single species introduced in nc TX is a tree with large, alternate, *bipinnately compound* leaves; flowers radially symmetrical, usually *lavender*, in a panicle; fruit a small drupe.
REFERENCES: Wilson 1924; Miller 1990.

MELIA CHINA-BERRY, UMBRELLA-TREE

An Old World, mainly tropical genus of 3 species. (Classical Greek name for the ash, which has similar leaves)

Melia azedarach L., (derivation unknown), CHINA-BERRY, PRIDE-OF-INDIA, CAVELÓN, PARÁISO, CHINA-TREE, WHITE-CEDAR, CEYLON MAHOGANY. Bushy-topped small tree to ca. 15 m tall; leaves alternate, bipinnately compound, 30 cm or more long; ultimate leaflets rhombic- or ovate- or elliptic-lanceolate, toothed, to 6(–8) cm long and 3 cm wide; flowers many, in terminal panicles over-topped by leaves; sepals 5, pubescent with simple or stellate hairs; petals 5, narrowly oblanceolate, 7–13 mm long, lavender (rarely whitish); filaments united into a narrow, purplish, cylindrical tube 6–10 mm long with a many-toothed summit; ovary superior; fruit a yellowish drupe ca. 15 mm in diam. Commonly cultivated, established as an escape in floodplain forests, thickets, and forest margins and almost appearing native; se and e TX w to nc TX and Edwards Plateau. Mid-Apr–mid-May. Native of Himalaya Mountains and e Asia. Cultivated as an ornamental and used as a timber tree in its native habitat; the bark and leaves have been used medicinally and as an insecticide (Mabberley 1987); poisoning is known from ingesting the fruits, and fatalities in livestock and humans have been reported; 6–8 fruits are reported to have caused death in a young child; leaves, bark, and flowers are also toxic; toxins include tetranotriterpene neurotoxins and unidentified resins which cause digestive tract irritation and degeneration of the liver and kidneys (Kingsbury 1964; Hardin & Arena 1974; Lampe & McCann 1985; Turner & Szczawinksi 1991). ☠

MENISPERMACEAE MOONSEED FAMILY

Ours perennial vines; leaves alternate, simple; leaf blades entire or usually few-toothed or -lobed, palmately veined at base; stipules absent; flowers very small, in axillary and terminal, racemose panicles, unisexual, the sexes on different plants (dioecious); sepals 4–9; petals 4–8 or absent; stamens 6–many, the filaments enlarged at base; pistils 2–4; ovaries superior; fruit a drupe with the bony endocarp typically curved, laterally compressed, and often sculptured.

A medium-sized (ca. 525 species in ca. 78 genera—Rhodes 1997), mainly tropical and warm area family of primarily lianas, vines, and scandent shrubs, or rarely trees or herbs; ☠ most have bitter, poisonous sesquiterpenoids and alkaloids; Kessler (1993b) considers the accumulation of some alkaloids unparalleled by any other angiosperm family. A number of species are used medicinally or as fish-poisons, sweeteners, or contraceptives; the alkaloid tubocurarine, traditionally used as a component of curare arrow or dart poisons by Amazonian natives, and now to relax muscles during surgery, is obtained from the South American *Chondrodendron tomentosum* Ruíz & Pav. (subclass Magnoliidae)

FAMILY RECOGNITION IN THE FIELD: *vines* with alternate, simple, palmately-veined leaves; flowers small, inconspicuous, unisexual, the female with 2–4 pistils; fruit a drupe, the endocarp (= hard seed-like inner part of fruit) often *curved and sculptured*.

REFERENCES: Ernst 1964a; Barneby & Krukoff 1971; Loconte & Estes 1989; Kessler 1993b; Rhodes 1997.

1. Leaves usually densely soft-pubescent beneath, at least some usually longer than wide, often without lobes or with wavy margins or 3–5-lobed; fruits red at maturity; stamens (in staminate flowers) 6 **Cocculus**
1. Leaves nearly glabrous to sparsely pubescent beneath, usually ca. as wide as long, usually 3–7-lobed or-angled; fruits blue to bluish black or black; stamens (in staminate flowers) 12–24.
    2. Leaves peltate (the petiole attached to lower leaf surface slightly back from the margin); fruits globose, 6–15 mm in diam.; flowers with 4–8 petals **Menispermum**
    2. Leaves not peltate (petiole attached to margin of leaf); fruits ellipsoid, ca. (15–)20–25 mm long; flowers with petals absent **Calycocarpum**

CALYCOCARPUM CUPSEED, WILD SARSAPARILLA

A monotypic genus of e North America. (Greek: *calyx*, cup, and *carpos*, fruit, presumably from the shape of the stone)

Calycocarpum lyonii (Pursh) A. Gray, (for John Lyon, a Scottish gardener and botanist who died on an expedition in Tennessee between 1814 and 1818), CUPSEED, WILD SARSAPARILLA. High climbing vine; leaves to 20 cm long, ca. the same width to wider, with 3–5 acuminate lobes, cordate basally, thin; inflorescences to ca. 20 cm long; flowers ca. 5 mm across; sepals 6–9, petaloid, greenish white; petals 0; stamens (in staminate flowers) 12; fruits blackish at maturity, the stone deeply scooped out on one side, cup-like in shape, and with irregularly toothed margins. Low woods near stream; Lamar Co. in the Red River drainage; this record is the westernmost locality known in TX; previously known only from deep e TX. May–Jun.

COCCULUS SNAILSEED, CORALVINE, CORALBEAD

A genus of 8 species of tropical and warm areas excluding South America and Australia; some contain alkaloids and are used medicinally; others have edible fruits or are cultivated as ornamentals. (An old name, diminutive of Greek: *coccus*, berry)

Cocculus carolinus (L.) DC., (of Carolina), CAROLINA SNAILSEED, CORALBERRY, CORALBEAD, CAROLINA MOONSEED, RED-BERRY MOONSEED. Becoming a high-climbing vine, but often beginning to flower when only 15–30 cm long, sometimes prostrate or scandent; leaf blades extremely variable in shape, toothing, lobing, and size, to ca. 12 cm long (usually much smaller), widely cordate, truncate or cuneate basally, leathery; inflorescences to ca. 15 cm long; sepals 6–9; petals (5–)6, yellow; fruits 5–8.5 mm in diam., globose, red, the stone ridged with a central depression and roughly snail-shaped. Thickets, roadsides, disturbed sites; nearly throughout TX. Jun–Jul, sporadically to Oct. Often abundant and by far the most common member of the family in nc TX. Some individuals with completely unlobed leaves superficially resemble *Smilax* (Smilicaceae); *Cocculus* can easily be distinguished by the lack of both armature and tendrils, its leaf pubescence, and its snail-shaped stones. Reported to contain alkaloids (Burlage 1968).

MENISPERMUM MOONSEED

A genus of 3 species of e North America (including Mexico) and e Asia (Rhodes 1997); this disjunct distribution pattern is discussed under the genera *Campsis* (Bignoniaceae) and *Carya* (Juglandaceae). (Greek: *men*, moon, and *sperma*, seed, from the shape of the seed)

Menispermum canadense L., (of Canada), MOONSEED, YELLOW PARILLA. Vine to ca. 3–4 m; leaf blades 3–15 cm long, 5–7 palmately lobed or angled, orbicular-ovate or orbicular-reniform, ± cordate at base, thin or firm; inflorescences to ca. 15 cm long; sepals 4–8; petals 4–8; stamens (in staminate flowers) 12–24; fruits globose, blue to blue-black, the stone flattened and somewhat crescent (moon)-shaped. Damp woods; collected by Reverchon in Oak Cliff (Dallas), not found there since; otherwise only reported in TX from the Edwards Plateau. May–Jun. This species is cultivated as an ornamental and the rhizomes are used medicinally (Mabberley 1987); the fruits are dangerous due to the presence of isoquinoline alkaloids, including dauricine, a compound with curare-like action; they can be mistaken for grapes and deaths have been reported (Hardin & Arena 1974; Tampion 1977; Lampe & McCann 1985; Turner & Szczawinski 1991). ☠

MENYANTHACEAE BUCKBEAN OR BOGBEAN FAMILY

A small (40 species in 5 genera), nearly cosmopolitan family of aquatic or wet area herbs. *Nymphoides* is sometimes lumped into the Gentianaceae; however, the Menyanthaceae is now

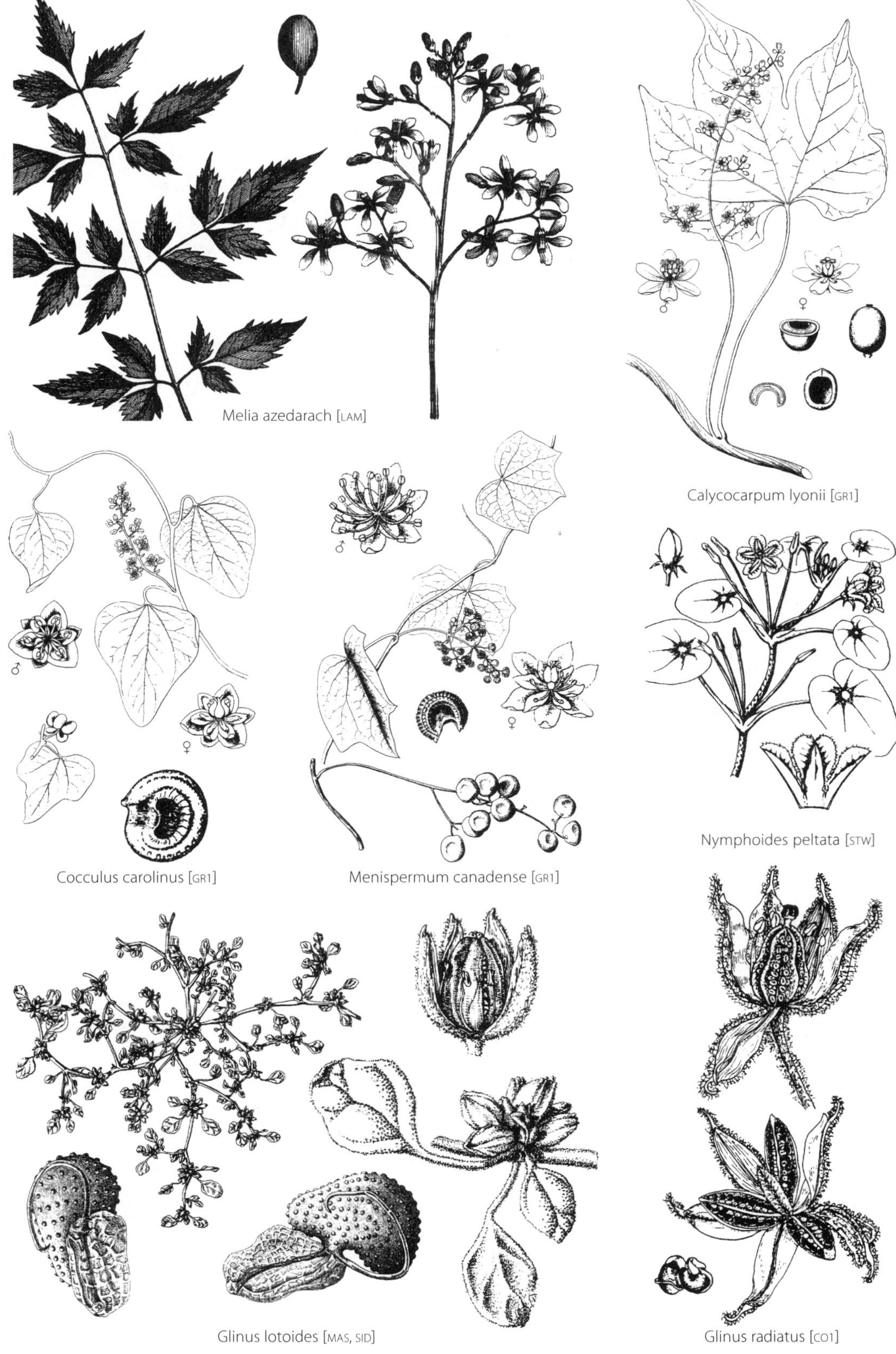

Melia azedarach [LAM]

Calycocarpum lyonii [GR1]

Cocculus carolinus [GR1]

Menispermum canadense [GR1]

Nymphoides peltata [STW]

Glinus lotoides [MAS, SID]

Glinus radiatus [CO1]

generally considered to be allied with other Asteridae and not closely related to the Gentianaceae. Family name from *Menyanthes*, BUCKBEAN or BOGBEAN, a circumboreal genus represented by a single, rhizomatous, herbaceous, perennial species. (Either from Greek: *men*, moon or month, and *anthos*, flower, in reference to the length of the flowering period, or from *menanthos*, moonflower, the name used by Theophrastus for a plant growing on Lake Orchomenos) (subclass Asteridae)

FAMILY RECOGNITION IN THE FIELD: the single nc TX species is an *aquatic* with suborbicular floating leaves and *yellow flowers in umbels*; corollas short tubular below the 5 lobes, with stamens attached at base. Other superficially somewhat similar plants (Cabombacae, Nymphaeaceae, Nelumbonaceae) have separate petals.

REFERENCE: Wood 1983b.

NYMPHOIDES FLOATING-HEART

A cosmopolitan genus of 20 species; some have edible tubers, medicinal seeds, or are cultivated as ornamentals. (From the genus name *Nymphaea*, from Greek: *nymphe*, water nymph, and *eidos*, form, like, or appearance, alluding to the resemblance)

REFERENCE: Ornduff 1966.

Nymphoides peltata (J.G. Gmel.) Kuntze, (shield-shaped), YELLOW FLOATING-HEART, WATER-FRINGE. Perennial, submersed, aquatic herb with floating leaves; leaves opposite, petiolate, 2 or more per long, petiole-like stem; leaf blades suborbicular, coarsely undulate-dentate, to 15 cm long and wide, cordate basally; flowers in umbels; pedicels often 6 cm or more long; calyces 5-parted; corollas bright yellow, 2–3 cm broad, the 5 lobes somewhat fringed, also with basal crests of hairs; anthers 4–5 mm long; ovary superior; capsules beaked, to 2.5 cm long. Cultivated and escaped, quiet water of streams and ponds; Dallas Co. (J. Stanford, pers. comm.), also Hatch et al. (1990) cited vegetational areas 4 and 5 (Fig. 2); ne and nc TX. Jul–Sep. Native of Eurasia.

MOLLUGINACEAE CARPETWEED FAMILY

Ours annual, nonsucculent or scarcely succulent herbs; leaves simple, alternate to apparently whorled, sometimes unequal at the nodes, entire; flowers solitary or clustered in the leaf axils, small, radially symmetrical, perfect, hypogynous, inconspicuous; sepals 5; petals absent; fruit a many-seeded capsule dehiscent by valves.

A small (130 species in 13 genera), usually herbaceous family mainly of tropical and warm areas, especially s Africa; it was formerly included in the Aizoaceae and some of the genera have been placed in the Ficoidaceae. The family is unusual in its subclass in having anthocyanin rather than betalain pigments (Cronquist & Thorne 1994); however, molecular evidence places it within the Caryophyllales (Downie & Palmer 1994). (subclass Caryophyllidae)

FAMILY RECOGNITION IN THE FIELD: prostrate to ascending herbs with simple, alternate to whorled, entire leaves; flowers small, inconspicuous, radially symmetrical, lacking petals; fruit a many-seeded capsule.

REFERENCES: Thieret 1966; Bogle 1970; Endress & Bittrich 1993; Behkne & Mabry 1994; Downie & Palmer 1994.

1. Plants with branched or stellate pubescence; leaf blades 8–15 mm wide, narrowly to broadly obovate; flowers essentially sessile; seeds with conspicuous bladder-like appendage _____ **Glinus**
1. Plants glabrous; leaf blades narrow, usually 3–5(–10) mm wide, spatulate to oblanceolate to narrowly obovate; flowers on filiform pedicels 5–15 mm long; seeds without an appendage _____ **Mollugo**

GLINUS

Pubescent annuals with habit resembling *Mollugo*; flowers in small, dense, few (ca. 5–10)-flowered clusters in the leaf axils; seeds with a distinct bladder-like appendage (= strophiole) and a persistent, long slender funiculus.

A genus of ca. 12 species of tropical and warm areas of the world. The following 2 species are very similar; according to Godfrey and Wooten (1981), "There appears to be no concensus as to whether plants naturalized in the U.S. are representatives of 2 species." (Greek: *glinos*, sweet juice; the usage not evident)

1. Seeds minutely tuberculate (use scope if possible), 0.4–0.6 mm long, 0.3–0.4 mm wide; stamens (3–)5–10 ______ **G. lotoides**
1. Seeds usually smooth, 0.4–0.5 mm long, 0.25–0.3 mm wide; stamens 3–5 ______ **G. radiatus**

Glinus lotoides L., (resembling *Lotus*—deer-vetch, trefoil). Plant ± gray-green from branched or stellate pubescence; stems prostrate or ascending, branched at base, ca. 10–30 cm long; leaves alternate to apparently whorled, 0.5–3 cm long, ovate to orbicular, narrowed to a slender petiole; sepals 4–7 mm long; capsules ellipsoid to ovoid, loculicidal, ca. 4 mm long. Waste places, exposed mud of dried-up lake bottoms; Denton and Grayson cos.; also Post Oak Savannah. Jul–Sep. Introduced from the Old World. An excellent illustration of the distinctive seed, including its strophile and funiculus, was provided by Thieret (1966).

Glinus radiatus (Ruiz & Pav.) Rohrb., (with rays). Very similar to *G. lotoides*; leaves tending to be narrower. Muddy or sandy soils; included on the basis of citation for vegetational area 4 (Fig. 2) by Hatch et al. (1990); mainly se and e TX. Apr–Oct. Apparently native to tropical America, possibly n to s U.S. Bogle (1970) suggested that these 2 similar species need further study.

MOLLUGO CARPETWEED

A genus of ca. 35 species of tropical and warm areas of the world. (Old name for *Galium mollugo* L. of Rubiaceae, transferred to this genus, probably due to the similarly whorled leaves)

Mollugo verticillata L., (whorled), GREEN CARPETWEED, INDIAN-CHICKWEED. Glabrous nonsucculent annual with prostrate to ascending stems to 20(–50) cm long; leaves apparently whorled, 3–6(–8) together, to ca. 3 cm long, unequal, the broadest leaves usually basal; flowers small, pedicellate, 2–5 from each node; sepals to 2.5 mm long, green to white with green back; stamens usually 3(–5); seeds without a distinct appendage (= strophiole). Waste ground, disturbed low areas, cultivated sites; throughout TX. Late May–Oct. Apparently native of tropical America. In the past, placed by some workers in the Ficoidaceae.

MORACEAE MULBERRY OR FIG FAMILY

Shrubs, trees, or herbaceous plants usually with ± milky juice; leaves usually rather short-petioled, usually alternate, simple, entire or toothed or lobed; stipules slender, falling when the leaves open or soon after; flowers unisexual, the sexes on the same or separate trees (plants thus monoecious or dioecious), in catkin-like clusters, heads, or hollow receptacles; perianth 2–6-parted or perianth absent from pistillate flowers; stamens 1–6; pistil 1; fruits achenes surrounded by the fleshy perianths, all the fruits of a pistillate inflorescence united into a multiple fruit (= syncarp), or fruits inside a fleshy receptacle (= synconium).

A medium-large family of nearly 1,100 species in 38 genera (Wunderlin 1997); they are primarily tropical and warm area (a few temperate) trees, shrubs, lianas, stranglers, and herbs, usually having laticifers with milky sap. It includes *Artocarpus* (JACKFRUIT and BREADFRUIT) and *Ficus* (FIGS, including tropical STRANGLER FIGS). *Artocarpus altilis* (Parkinson) Fosberg,

BREADFRUIT, is famous because plants of it were being transported on the sailing ship Bounty, when a mutiny occurred against Captain Bligh. The Moraceae are closely related to the Urticaceae and Cannabaceae and probably paraphyletic when treated separately. From a cladistic standpoint these families should be lumped to form a more inclusive monophyletic Urticaceae (Judd et al. 1994). (subclass Hamamelidae)

FAMILY RECOGNITION IN THE FIELD: woody plants (1 herb) with alternate simple leaves; sap usually *milky*; flowers reduced, unisexual, with a 4-parted perianth; fruits usually *multiple* or inside a *receptacle*.

REFERENCES: Rohwer & Berg 1993; Judd et al. 1994; Wunderlin 1997.

1. Plant an erect annual herb < 1 m tall ____ **Fatoua**
1. Plant a perennial tree, shrub, or creeping vine.
   2. Flowers and fruits hidden from view inside a hollow receptacle; plant a shrub (rarely a small tree) or creeping vine; terminal vegetative bud surrounded by a pair of stipules; rare introduced species persisting or spreading around homesites ____ **Ficus**
   2. Flowers not inside a hollow receptacle; plant a tree (rarely a shrub); terminal vegetative bud scaly, not surrounded by a pair of stipules; widespread native and introduced species.
      3. Leaf blades entire; branches usually with short thorns (the thorns sometimes absent); multiple fruits large, to 15 cm in diam., green to greenish yellow ____ **Maclura**
      3. Leaf blades toothed and/or lobed; branches thornless; multiple fruits small, to 2–3 cm in diam. or less, variously colored.
         4. Twigs glabrous or nearly so; bark scaly or furrowed; multiple fruits cylindric; leaves alternate; buds with 3–6 scales ____ **Morus**
         4. Twigs pubescent; bark smooth; multiple fruits globe-shaped; leaves alternate, sometimes opposite; buds with 2–3 scales ____ **Broussonetia**

BROUSSONETIA PAPER-MULBERRY

A genus of 8 species native to tropical and warm Asia and Madagascar; anthers are explosive (as in *Urtica*, but rare in Moraceae in the strict sense). (Named for Auguste Broussonet, 1761–1807, of Montpellier, French physician and naturalist)

Broussonetia papyrifera (L.) L'Hér. ex Vent., (paper-bearing), PAPER-MULBERRY. Dioecious small tree to 16 m tall; sap milky; leaves broadly ovate, unlobed or sometimes lobed, coarsely dentate marginally, rounded to cordate at base; petioles to 10 cm long; staminate catkins 4–8 cm long, slender, pendulous; multiple fruits orange-red, 2–3 cm in diam., the red achenes protruding. Dallas, Grayson, and Tarrant cos.; ornamental naturalizing in se and e TX w to nc TX and Edwards Plateau. Mar–Apr. Native of China and Japan. Tapa or kapa cloth and fine paper are obtained from the inner bark (Mabberley 1987; Rohwer & Berg 1993). This species can spread by rhizomes and be a problematic invader.

FATOUA CRABWEED

An herbaceous genus of 2 species (Wunderlin 1997) native from Madagascar to e Asia, n Australia, and New Caledonia. (Derivation of generic name unknown)

REFERENCES: Thieret 1964; Massey 1975.

Fatoua villosa (Thunb.) Nakai, (soft hairy), MULBERRY-WEED, HAIRY CRABWEED. Monoecious annual to 80 cm tall; sap not milky; stems finely pubescent, erect, seldom branched, resembling a member of the Urticaceae; leaves alternate, petioled, broadly ovate, 3-nerved at base, cordate to truncate basally, crenate to dentate; staminate and pistillate flowers mixed in axillary, pedunculate glomerules; staminate perianths 4-merous; stamens 4, exserted; pistillate perianths 6-lobed; achenes compressed-trigonous. Weed in lawns, flower beds, and nurseries; first observed

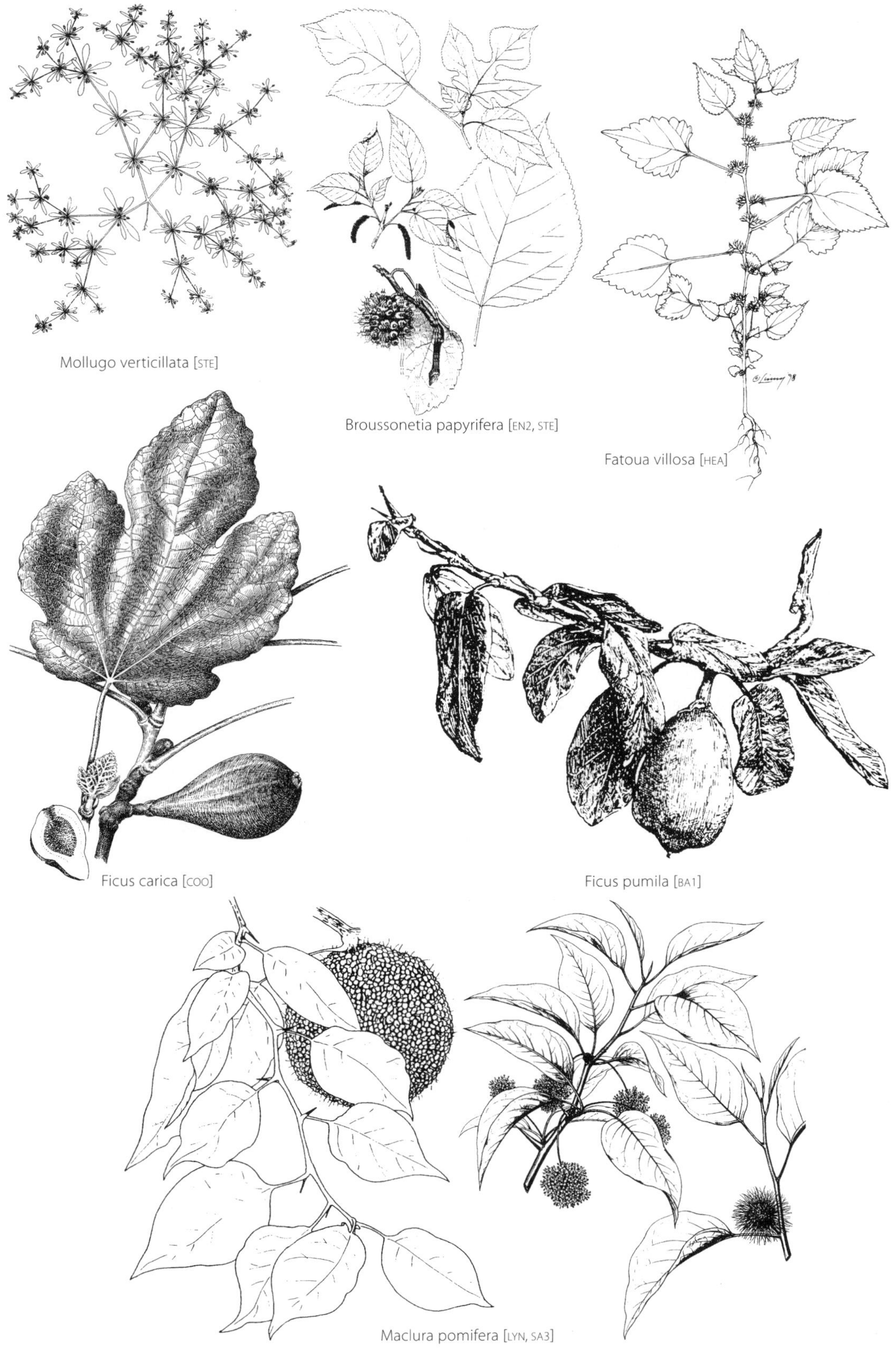

Mollugo verticillata [STE]

Broussonetia papyrifera [EN2, STE]

Fatoua villosa [HEA]

Ficus carica [COO]

Ficus pumila [BA1]

Maclura pomifera [LYN, SA3]

in TX in Dallas in 1978 (R. O'Kennon, pers. obs.), first vouchered TX collection 1980 (Lipscomb 1984); now common in portions of nc TX; first reported for North America by Thieret (1964) in se U.S.; apparently spreading westward. Jul–Sep. Native of e Asia. [*Urtica villosa* Thunb.]

FICUS FIG

Ours shrubs (rarely small trees) or root-clinging vines; sap milky; flowers unisexual inside a receptacle (the receptacle fleshy at maturity with the whole structure called a fig) with a small apical opening.

A large genus of ca. 750 species (Wunderlin 1997) of tropical and warm areas, especially Indomalesia to Australia; the genus is diverse vegetatively, ranging from trees, shrubs, lianas, and epiphytes to stranglers; some spread by aerial roots descending and becoming additional trunks—they can thus occupy several hectares. FIGS often have intricate, species-specific pollination mutualisms with wasps. Many species are cultivated as ornamentals, as sources of rubber, or for fiber, paper, timber, fruits, or medicines. *Ficus religiosa* L. (PEEPUL, BO-TREE), native from India to se Asia, is sacred to Hindus and Buddhists because Buddha is said to have had the true insight beneath one. The Indomalesian *F. elastica* Roxb. ex Hornem. (RUBBER-PLANT, INDIAN RUBBER-TREE) yields a latex-containing sap that was formerly used as a source of rubber. Currently, most natural rubber is obtained from *Hevea brasiliensis* (Juss.) Müll.Arg. (RUBBER, PARÁ RUBBER), a South American member of the Euphorbiaceae. The latex of *Ficus* species contains protein-digesting enzymes as well as other toxins; some species including COMMON FIG can cause severe dermatitis; some also contain photosensitizers which can result in phytophotodermatitis (Lampe & McCann 1985; Fuller & McClintock 1986; Turner & Szczawinksi 1991). (Classical Latin name for *Ficus carica*, the edible FIG)

1. Plants shrubs (rarely small trees); leaves 3–5 lobed, very large (10–20 cm long and nearly as wide) **F. carica**
1. Plants creeping vines; leaves not lobed, often small (< 2.5 cm long), with larger leaves (to 5–10 cm long) on fruiting branches **F. pumila**

Ficus carica L., (Latin for Caria, a district in Asia Minor where neither fig nor papaya originated), COMMON FIG, HIGUERA, FIG TREE. Deciduous shrub, sometimes killed back during severe winters; leaves large, usually 3–5 lobed (rarely unlobed), thick, scabrous, usually cordate basally; figs axillary, solitary, spherical or pear-shaped, fleshy, 5–8 cm long, greenish or brownish violet. Cultivated and long persists around old homesites in nc TX but probably does not naturalize; Grayson Co., also Fort Hood (Bell or Coryell cos.—Sanchez 1997); also se TX. May–Aug. Native of w Asia. This is the cultivated, edible FIG; Egyptians were growing it by 4,000 BC (Mabberley 1987) and dried figs were uncovered at a Neolithic Age (5,000 BC) site on the western slopes of the Judean mountains (Zohary 1982). The latex can cause dermatitis or phytophotodermatitis in some people (Burlage 1968; Lampe & McCann 1985).

Ficus pumila L., (dwarf), CREEPING FIG, CLIMBING FIG, CREEPING RUBBER-PLANT. Climbing on walls, etc. by roots; leaves dimorphic, those on creeping vegetative stems sessile or short petioled, cordate-ovate, small (< 2.5 cm long), oblique at bases; leaves on erect fruiting branches elliptic to oblong, larger, (to ca. 5–10 cm long); figs yellowish, pear-shaped, to ca. 5 cm long. Cultivated in Fort Worth (Tarrant Co.) and spreading from cultivation. Native of e Asia. [*F. repens* Hort.] The latex can cause phytophotodermatitis (Lampe & McCann 1985).

MACLURA BOIS D'ARC, HORSE-APPLE, OSAGE-ORANGE

A monotypic genus endemic to TX, OK, and AR. This is one of only 11 genera of trees restricted to the e U.S. (only three of these, *Asimina*, *Maclura*, and *Taxodium*, occur in nc TX)

(Little 1983). It is sometimes (e.g., Mabberley 1997) treated as including *Cudrania* and *Plecospermum*; if so the genus has 12 species ranging from Indomalesia to Australia, Africa, and America. (Named for William Maclure, 1763–1840, early American geologist)
REFERENCES: Lipscomb 1992; Laushman et al. 1996; Weniger 1996.

Maclura pomifera (Raf.) C.K. Schneid., (pome-bearing), BOIS D'ARC, HORSE-APPLE, OSAGE-ORANGE, NARANJO CHINO, BOW-WOOD, HEDGE-APPLE. Broad-headed small tree often with arching branches; wood yellow to orange; roots with conspicuous orange coloration; sap milky; leaf blades elliptic to ovate-lanceolate, acuminate, entire, glabrous above, rather sparsely pubescent beneath; staminate flowers in short-peduncled, dense, short, oblong-ovoid clusters; pistillate flowers on separate trees (plants thus dioecious), in globose heads with long, conspicuous styles; fruits multiple, superficially resembling a greenish or greenish yellow large orange with a wrinkled surface, to 12 cm or more in diam., the individual achenes completely enclosed by fleshy calyces. Stream bottoms, lower slopes, waste places; a weedy invader of disturbed uplands, chiefly in limestone areas; mainly ne to nc TX and s to c TX; native to a relatively small area in TX and adjacent OK and AR (Little 1971). According to Weniger (1996), based on field notes of pre-1860 surveys, the native range of BOIS D'ARC in TX was apparently limited to 12 counties in nc and extreme ne TX from Dallas and Grayson cos. on the w, e to Bowie Co. and s to Kaufman Co. Apr–early May. This species was used by Native Americans (e.g., Osage Indians) to make bows (French: *bois d'arc*, wood of the bow) and war-clubs; before the invention of barbed wire, it was widely cultivated for hedges and fencerows—such a hedge was said to be "horse high, bull-strong and pig tight." According to some sources, the inventor of barbed wire, who had a BOIS D'ARC fence, saw the thorns on the plant and had the idea of "putting steel thorns" on twisted wire (Conrad 1992). The wood is one of the most decay resistent in North America and was widely used in nc TX as piers in pier and beam houses; the roots are a source of yellow dye and the leaves have been used to feed silkworms (Mabberley 1987; Conrad 1992). The fruits were used historically in TX to repel insects such as cockroaches (Conrad 1992) and such usage continues to the present. While horses relish them, the large fruits can cause death in horses and ruminants by lodging in the esophagus (Burlage 1968). Some humans develop dermatitis from contact with the milky sap of stems, leaves, and fruits (Muenscher 1951). ☠

MORUS MULBERRY

Deciduous monoecious or dioecious trees (rarely shrubs); leaf blades ovate, lobed or unlobed, serrate or dentate, 3–5 nerved from base, truncate to cordate basally, subobtuse to acute or acuminate apically; flowers unisexual; both sexes in separate catkin-like inflorescences; calyces 4-parted; stamens 4; edible multiple fruits resembling a blackberry or raspberry, composed of numerous achenes covered by sweet, juicy, white to dark purple calyces.

A genus of 10 species (Wunderlin 1997) of deciduous trees of the subtropics; it includes cultivated ornamental and fruit trees. (Latin: *morum*, classical name for MULBERRY)

1. Mature leaf blades at most slightly scabrous on one side, usually > 8 cm long; petioles 1 cm or more long.
   2. Leaf blades glabrous except for small tufts of hairs in axils of main veins on lower surface; fruits usually white to pinkish; leaves usually variously lobed ____ **M. alba**
   2. Leaf blades soft-pubescent beneath, sparsely pubescent or glabrous above; fruits red to dark purple; leaves lobed or often unlobed ____ **M. rubra**
1. Mature leaf blades scabrous with very short, stiff hairs on both surfaces, 7 cm or less long; petioles usually < 1 cm long ____ **M. microphylla**

Morus alba L., (white), WHITE MULBERRY, RUSSIAN MULBERRY, SILKWORM MULBERRY, MORAL BLANCO. Small tree to ca. 15 m tall; leaves to 20 cm long. Established in thickets in limestone areas, local; Dallas, Denton, Grayson, and Tarrant cos., also McLennan Co. (Mahler 1988); also cultivated; self-sowing and apt to become a yard weed; naturalized at scattered localities across TX. Early and mid-Apr. Native of e Asia. The leaves are the main food of silkworms (larvae of the silkmoth—*Bombyx mori*) (Correll & Johnston 1970; Rohwer & Berg 1993). Fruits edible when ripe (Correll & Johnston 1970). The cultivated FRUITLESS MULBERRY is a cultivar of *M. alba.*

Morus microphylla Buckley, (small-leaved), MEXICAN MULBERRY, TEXAS MULBERRY, MOUNTAIN MULBERRY. Shrub or small tree to ca. 7 m tall. Rocky slopes; Brown, Dallas, and Palo Pinto cos., also Coleman, Comanche, Erath (Little 1976) and Somervell (Fossil Rim Wildlife Center—R. O'Kennon, pers. obs.) cos.; nc TX s and w to w TX. Apr. Fruits edible when ripe (Correll & Johnston 1970) and used by Native Americans (Powell 1988).

Morus rubra L., (red), RED MULBERRY, MORAL. Becoming a medium to large tree; leaves unlobed or lobed, to 20 cm long. Stream bottoms, less often on slopes; se and e TX w to e Rolling Plains and Edwards Plateau. Late Mar–Apr. While it is sometimes recognized as a variety (e.g., Kartesz 1994; Jones et al. 1997), we are following Wunderlin (1997) in lumping [*M. rubra* var. *tomentosa* (Raf.) Bureau]. Fruits edible when ripe (Correll & Johnston 1970). The milky sap from leaves and unripe fruits can cause dermatitis, hallucinations, and central nervous system disturbances (Schmutz & Hamilton 1979).

MYRICACEAE WAX-MYRTLE OR BAYBERRY FAMILY

The Myricaceae is a small (55 species in 3 genera) nearly cosmopolitan family of aromatic shrubs and trees; 2 of the genera are monotypic. The family appears to have a relationship with the Juglandaceae (Kubitzki 1993d). (subclass Hamamelidae)

FAMILY RECOGNITION IN THE FIELD: the only nc TX species is a shrub with alternate, simple, *gland-dotted, fragrant* leaves; flowers small, inconspicuous, in spike-like catkins; fruits small, drupaceous, usually with a *white waxy coating.*

REFERENCES: Baird 1968; Elias 1971b; Kubitzki 1993d; Wilbur 1994; Bornstein 1997.

MYRICA WAX-MYRTLE

A subcosmopolitan genus of ca. 50 species (Bornstein 1997) of shrubs, usually with nitrogen-fixing bacteria; it includes species used as sources of wax, for their edible fruit, or as cultivated ornamentals. (Greek, *myrike*, name of the tamarisk or some other fragrant shrub, perhaps from *myrizein*, to perfume; the name later transferred to this aromatic genus)

Myrica cerifera L., (wax-bearing), SOUTHERN WAX-MYRTLE, CANDLE-BERRY, TALLOW-SHRUB, WAXBERRY, SPICEBUSH, BAYBERRY, SWEET-OAK. Evergreen shrub 0.3–7 m tall; leaves alternate, short-petioled or sessile with narrowed base; leaf blades oblanceolate, entire or irregularly few-toothed in apical part, glabrous, sprinkled with fine yellow glandular dots, especially beneath, fragrant when crushed; flowers unisexual, in short, chiefly lateral, spike-like catkins, without perianth but with small bracts; staminate and pistillate flowers on separate plants (= dioecious); ovary superior; mature fruits drupaceous, 2–3 mm in diam., usually with a white waxy coating. Boggy ground, or a weedy invader in drier areas; se and e TX w locally to Henderson Co. on e margin of nc TX and along Brazos River to McLennan Co. (Little 1976 [1977]), also Dallas Co. (Mahler 1988) [introduced?]. Late Mar–Early Apr. Wilbur (1994) argued that the genus *Myrica* should be split into *Morella* and *Myrica* sensu stricto; Jones et al. (1997) accepted this viewpoint and recognized the SOUTHERN WAX-MYRTLE as [*Morella cerifera* (L.) Small]. We are following the traditional approach advocated by Bornstein (1997) and retaining this species in a broadly defined *Myrica*. The wax-like coating on the fruits is rich in palmitic acid and can be used to make candles and scented bayberry soap (Mabberley 1987). This is an attractive shrub widely used in landscapes.

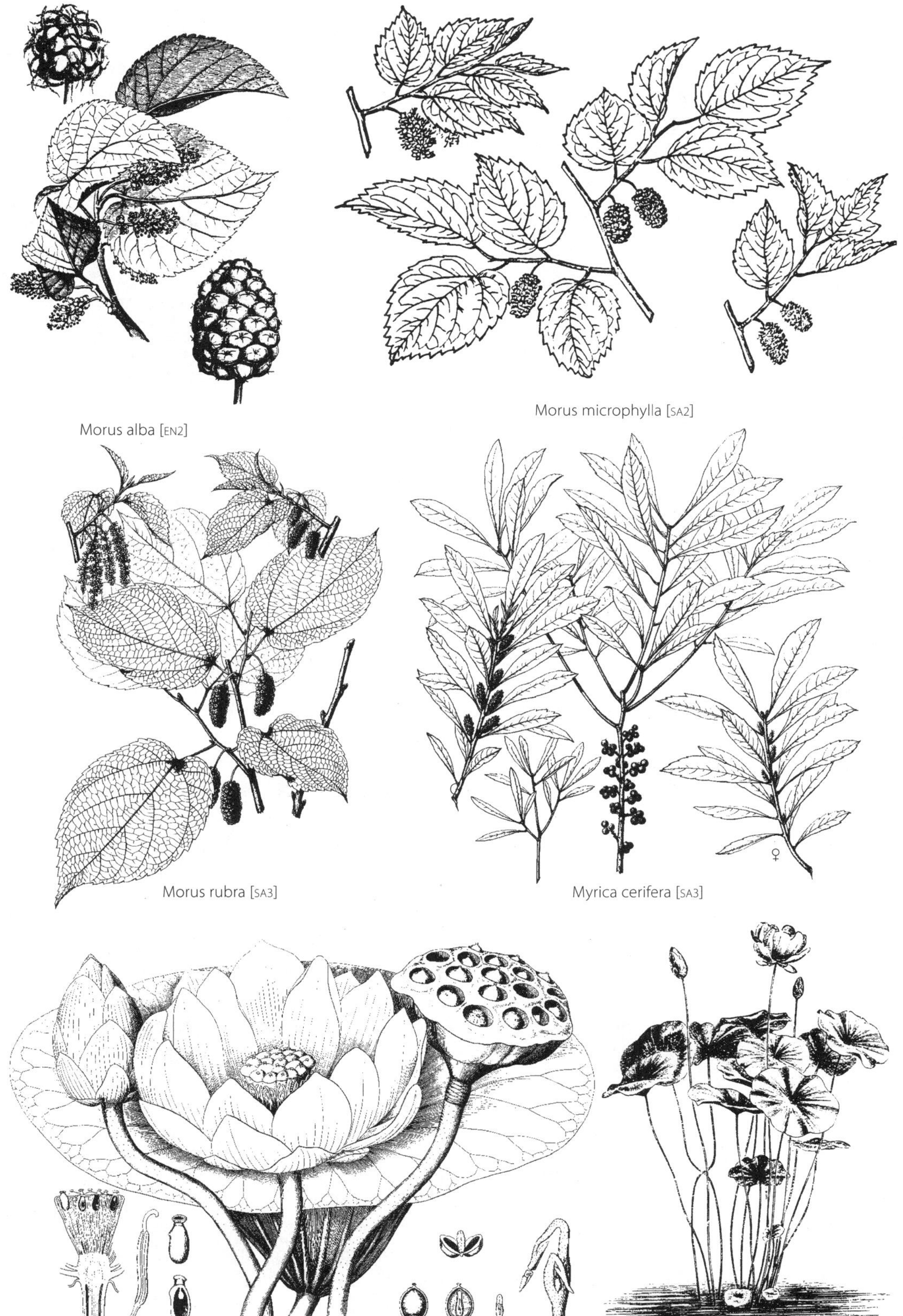

Morus alba [EN2]

Morus microphylla [SA2]

Morus rubra [SA3]

Myrica cerifera [SA3]

Nelumbo lutea [BA1, GR1]

NELUMBONACEAE LOTUS OR LOTUS-LILY FAMILY

The Nelumbonaceae is a family of a single genus with only 2 species; it has sometimes been placed in the Nymphaeaceae, but is profoundly different in terms of its chemistry (Williamson & Schneider 1993b) and 3-aperturate pollen (1-aperturate in Nymphaeaceae) (Cronquist 1993). According to Meacham (1994), this is a classic case of an ancient taxon with an unusual combination of characters; he indicated it is more closely related to Nymphaeales than the Ranunculidae (with which it has sometimes been linked). However, in the analyses of Hambry and Zimmer (1992) and Chase et al. (1993), *Nelumbo* did not appear related to Nymphaeales. The modern concensus is that it is a distinct taxon deserving familial rank (Williamson & Schneider 1993b). (subclass Magnoliidae)

FAMILY RECOGNITION IN THE FIELD: aquatic, water-lily-like herbs with large *peltate* leaves (not peltate in the somewhat similar Nymphaeaceae), large conspicuous flowers, and acorn-like fruits in pits in a distinctive, enlarged, *inverted-cone-like receptacle.*

REFERENCES: Wood 1959; Williamson & Schneider 1993b; Meacham 1994; Wiersema 1997a.

NELUMBO LOTUS, SACRED-BEAN

A genus of 2 species, 1 in s and e North America and the West Indies to Colombia, 1 in lower Volga and s and se Asia to tropical Australia. The Old World *Nelumbo nucifera* Gaertn., SACRED LOTUS, EGYPTIAN-BEAN, with fragrant, pink/red flowers, is sacred in India, Tibet, and China; its seeds are viable for several hundred years in the mud of rivers; this species, easily distinguished by flower color, is cultivated in nc TX as an ornamental; it is also widely cultivated in Asia as a food crop for its edible tubers and embryos (Williamson & Schneider 1993b). (Name used in Sri Lanka for *N. nucifera*)

REFERENCES: Hall & Pendfound 1944; Ward 1977.

Nelumbo lutea (Willd.) Pers., (yellow), YELLOW LOTUS, LOTUS, YELLOW NELUMBO, WATER-CHINQUAPIN, POND-NUT, YOUQUEPEN. Robust, aquatic, perennial herb rooted in mud; rhizomes of two types: slender, elongated, 6–8 mm in diam. or thick, banana-like in appearance, 8–20 mm in diam.; leaf blades nearly orbicular, to ca. 70 cm in diam., floating or usually held above the water, glabrous; petioles to 1 m or more long, attached to center of lower surface of blade (= peltate), the blade not notched; flowers solitary, held well above water on long peduncles, large, to 25 cm wide, not noticeably fragrant or only mildly so, opening in morning and closing at night; petals many, large, yellowish white or yellowish cream; stamens numerous (ca. 200), spirally arranged; pistils numerous, sunken in pits in the nearly flat upper surface of the greatly enlarged (to ca. 10 cm in diam.), inverted-conical, erect receptacle that becomes dry, hard, and brown by the time the fruits ripen; fruits acorn-like, hard, indehiscent, 1-seeded, ca. 1 cm in diam. Lakes, ponds; se and e TX w to West Cross Timbers and Edwards Plateau; said to have been spread by Indians in prehistoric times. Jun–Sep. The storage tubers along the rhizomes, as well as the seeds, are edible and were used by Native Americans (Wiersema 1997a); the dried receptacles are often used in floral arrangements. Once established, this species can be aggressive and hard to eradicate. ▣/100

NYCTAGINACEAE FOUR-O'CLOCK FAMILY

Ours annual or perennial herbs or shrubby low perennials (*Selinocarpus*); stems often with at least the lower nodes swollen; leaves opposite, simple, entire or indistinctly toothed, somewhat thick or fleshy, sessile or petioled, usually the 2 at each node unequal; stipules absent; inflorescences axillary or terminal; involucres enclosing 1–several flowers, in some cases an involucre resembling a calyx present below the corolla-like calyx (the whole structure thus superficially resembling a typical flower with both calyx and corolla); flowers perfect; calyces 3- to 5-lobed, corolla-like, sometimes very showy; corollas absent; stamens 1–9; pistil 1; ovary superior but apparently inferior (tightly enclosed by base of perianth); fruit an anthocarp, an accessory fruit composed of the persistent base of the perianth tube enclosing the indehiscent achene—the whole structure is referred to in the following treatment as the fruit.

A small (390 species in 30 genera), mainly tropical and warm area family, especially in the Americas with a few in the temperate zone; they are betalain-containing herbs, shrubs, and trees. Some are edible, used medicinally, or cultivated as ornamentals including *Mirabilis* and the widely grown tropical ornamental genus *Bougainvillea* with brightly colored bracts subtending the 3-flowered inflorescences. Molecular evidence indicates the family is related to Phytolaccaceae (Downie & Palmer 1994). (subclass Caryophyllidae)
FAMILY RECOGNITION IN THE FIELD: usually herbs (1 shrubby perennial) with opposite leaves and often swollen nodes; single perianth whorl often subtended by an involucre; ovary superior, developing into a nut-like, 1-seeded achene enclosed by the perianth tube.
REFERENCES: Standley 1918; Reed 1969c; Bogle 1974; Bittrich & Kühn 1993; Behkne & Mabry 1994; Downie & Palmer 1994.

1. Flowers with an involucre of bracts (apparently with calyx, the corolla is absent in this family and the colorful perianth parts are actually the corolla-like calyx) or in bracted heads; fruits winged or not winged.
   2. Flowers with calyx-like involucre of united bracts ____ **Mirabilis**
   2. Flowers or heads with separate bracts.
      3. Flowers in pedunculate heads of 8–many flowers; leaf blades usually 3–9 cm long; fruits with or without wings; petioles 10–85 mm long.
         4. Stigmas linear; fruits with at least narrow wings; flowers usually > 15 per head; stamens (3–)5 ____ **Abronia**
         4. Stigmas capitate; fruits not winged, merely 1-ribbed; flowers 8–15 per head; stamens 5–8 ____ **Nyctaginia**
      3. Flowers solitary or in pairs in the leaf axils; leaf blades 1.2–2.5 cm long; fruits with conspicuous thin wings; petioles 3–25 mm long ____ **Selinocarpus**
1. Flowers neither involucrate nor in bracted heads; fruits not winged.
   5. Flowers solitary, axillary; perianth very long, 80–170 mm (with very slender elongate tube), white tinged with purple or pink; plants prostrate or sprawling ____ **Acleisanthes**
   5. Flowers in numerous umbel-like or head-like clusters of ca. 2–6 at the ends of slender peduncles OR in racemes arranged in panicles, terminal; perianth tiny, 1–1.5 mm long, white to pink, purple, or reddish; plants procumbent to ascending or erect ____ **Boerhavia**

ABRONIA SAND-VERBENA

Perennial herbs; stems viscid-pubescent; leaves opposite, the pair usually unequal; flowers perfect, in conspicuous heads, the heads long pedunculate, each subtended by 5(–7) distinct bracts; perianth viscid, with long slender tube and short funnelform limb; stamens 5; stigma linear; fruits turbinate (= top-shaped) or biturbinate, deeply lobed or winged.

A North American genus of 33 species ranging from sw U.S. to n Mexico; the roots, when ground, of some species were formerly eaten by Native Americans; some are cultivated as ornamentals. (Greek: *abros*, delicate or graceful, referring to the bracts)
REFERENCE: Galloway 1975.

1. Leaf blades elliptic to orbicular; fruits glabrous or with a few hairs apically, 6–9 mm long, 3–4.5 mm wide, never rugose; involucral bracts rounded to acutish at apex ____ **A. ameliae**
1. Leaf blades ovate-oblong to narrowly triangular-ovate; fruits pubescent, 5–12 mm long, 2.5–7 mm wide, somewhat rugose; involucral bracts acute or acuminate at apex ____ **A. fragrans**

Abronia ameliae Lundell, (for Amelia Anderson Lundell, 1908–1998, wife of Cyrus Longworth Lundell who described the species), AMELIA'S SAND-VERBENA. Stems spreading, to 60 cm long; leaf blades 3–8 cm long, 2–6 cm wide; perianth 18–25 mm long, orchid-color (from label data) to bright red-magenta, the limb ca. 10 mm wide; fruits top-shaped. Sandy woodlands or deep sand; included based on citation of vegetational area 4 (Fig. 2) by Hatch et al. (1990); mainly s 1/2 of TX; endemic to TX. Mar–Jun. /77

Abronia fragrans Nutt. ex Hook., (fragrant), SAND-VERBENA, SWEET SAND-VERBENA, SNOWBALL, LASATER'S-PRIDE. Stems erect or procumbent, freely and widely branched, 25–100 cm long; leaf blades usually 2–9 cm long; flowers sweet-scented; perianth 18–32 mm long, the limb deep lavender (varying to pink or white westward), 5–10 mm wide; fruits usually biturbinate (= narrowing both ways from the middle). Loose sand; Panhandle, South TX Plains, Rolling Plains, and West Cross Timbers (Mahler 1988); Hatch et al. (1990) also cited vegetational area 5 (Fig. 2). Late Apr–Jun, sporadically to Oct. [*A. speciosa* Buckley—type from Ft. Belknap, Young Co.] The flowers are open from late afternoon through the night (Kirkpatrick 1992).

ACLEISANTHES TRUMPETS

A genus of 7 species of sw North America. (Greek: *a*, without, *cleis*, close, and *anthus*, flower, alluding to absence of involucre)

REFERENCE: Smith 1976.

Acleisanthes longiflora A. Gray, (long-flowered), ANGEL-TRUMPETS, YERBA-DE-LA-RABIA. Glabrous perennial; stems brittle, prostrate, freely branched, mat-forming or sprawling; leaves opposite; leaf blades 15–50 mm long, 3–35 mm wide, deltoid to linear-lanceolate; flowers axillary, sessile or subsessile, opening in afternoon and throughout the night, erect, fragrant, white tinged with purple or pink, the funnelform limb 15–20 mm across, the tube long (80–170 mm), slender, 1.5–2 mm in diam.; fruits 5–10 mm long, 5-angled. Sandy or rocky ground, often a roadside weed; Coleman Co. (*Reverchon*) on the extreme w margin of nc TX, also Burnet Co. (Smith 1976) on s margin of nc TX; mainly s and w TX. Apr–Oct, after rains. The flowers are some of the longest found among TX plants; the species is nocturnal—the flowers open at dusk and are apparently pollinated by moths (Wills & Irwin 1961; Ajilvsgi 1984).

BOERHAVIA SPIDERLING

Annual or perennial herbs, our species with minutely glandular-pubescent stems and opposite, unequal, glabrous, entire or sinuate leaves often lighter in color beneath; perianth funnelform, corolla-like, small (1–1.5 mm long); stamens 1–5; fruits 5-angled.

A genus of ca. 50 species of warm areas of the world. (Named for H. Boerhave, 1668–1738, Dutch botanist)

REFERENCES: Woodson et al. 1961; Procher 1978.

1. Fruits minutely spreading-pubescent; perianth deep red or purple-red; plants perennial with prostrate to ascending stems ____ **B. diffusa**
1. Fruits glabrous (can be sticky); perianth white to pink or pink-lavender; plants annual usually with ascending to erect stems (can be decumbent).
   2. Flowers in numerous small umbel-like or head-like clusters of 2–6 at the ends of long slender peduncles, not in racemose inflorescences; stamens equal to or slightly longer than perianth; mature fruits truncate or nearly so at tip, 3–4 mm long; widespread in nc TX ____ **B. erecta**
   2. Flowers in elongate racemose inflorescences, never in small umbel-like or head-like clusters; stamens included within perianth; mature fruits rounded at tip, ca. 2.5 mm long; possibly in the extreme w part of nc TX ____ **B. spicata**

Abronia fragrans [USH]

Abronia ameliae [HEA]

Acleisanthes longiflora [EMO]

Boerhavia diffusa [IVE]

Boerhavia diffusa L., (diffuse, spreading), SCARLET SPIDERLING. Stems to 3 m long; leaf blades 1.5–5.5 cm long, 0.8–5 cm wide; flowers in numerous small umbel-like or head-like clusters at the ends of slender peduncles; perianth deep red or purple-red; fruits 2.5–4 mm long, rounded at tip. Rocky, gravelly, or sandy ground; widespread in TX. May–Oct. While *B. coccinea* is often recognized as a separate species (e.g., Kartesz 1994; Jones et al. 1997), we are following Woodson et al. (1961) and Procher (1978) in putting [*B. coccinea* Mill.] into synonymy with *B. diffusa*. Woodson et al. (1961), for example, said that, "A rather extensive examination of herbarium specimens has revealed no tangible differences . . . ." Further, R. Spellenberg (pers. comm.) recently indicated that he could not see the differences between the two once an array of variation was considered. He suggested that the two seem to differ primarily in flower color, ". . . in the New World [*B. coccinea*] the deep wine color predominates, in the Old World [*B. diffusa*] pink to white predominates."

Boerhavia erecta L., (erect, upright), ERECT SPIDERLING. Stems erect to decumbent, 20–120 cm long; leaf blades 2–8 cm long, 1–5 cm wide; perianth white or tinged with pink or purple. Along roadsides or disturbed sites; mainly e 1/2 of TX. Jun–Sep.

Boerhavia spicata Choisy, (with spikes), SPICATE SPIDERLING. Stems erect to decumbent, 20–60 cm long; leaf blades 1–4.5 cm long, 4–25 mm wide; flowers in short, dense to remotely-flowered racemes; perianth pink to pink-lavender or whitish. Sandy or rocky soils; included based on citation of vegetational area 5 (Fig. 2) by Hatch et al. (1990); mainly c and w TX. Jul–Aug.

Boerhavia intermedia M.E. Jones, (intermediate), similar to *B. erecta*, was reported from Dallas Co. by Reed (1969c). According to his key, this species has small fruits (2.2–2.7 mm long). A sheet at BRIT/SMU from Dallas, annotated by Reed as *B. intermedia*, has fruits 3 mm long and is possibly *B. erecta*. *Boerhavia intermedia* is otherwise known in TX only from the w part of the state. The following key to separate the two is modified from R. Spellenberg (pers. comm.):

1. Pedicels (of a given few-flowered group) not all attached at the same point; half mature fruits turbinate at apex ____ *B. erecta*
1. Pedicels all attached at the same point (the few-flowered groups thus ± umbels); half mature fruits ± flat at apex ____ *B. intermedia*

MIRABILIS FOUR-O'CLOCK

Perennial herbs with stout, pithy or woody taproot; foliage sometimes glaucous or whitened; leaves opposite; flowers solitary or few in 5-lobed, calyx-like, saucer- or cup-shaped to cylindrical, sometimes colored involucres, these axillary and solitary, or terminal and panicled (both types produced by the same plant, often on different stems at different times, the two phases utterly dissimilar); perianth of 5 fused sepals, extremely corolla-like, salverform to funnelform-campanulate, lasting part of one day; true corollas absent; fruits 5-ribbed or -angled, ± smooth to tuberculate.

A genus of 54 species, mainly of warm areas of the Americas, especially sw North America, and 1 species in the Himalayas; some are cultivated as ornamentals. ☠ *Mirabilis* species contain the alkaloid trigonelline; poisoning has occurred in children (Schmutz & Hamilton 1979). We are following the recent treatment by Turner (1993b) for nomenclature of *Mirabilis*. The common name results from some species opening their flowers in late afternoon (Woodland 1997). (Latin: *mirabilus*, wonderful)
REFERENCES: Shinners 1951g; Turner 1993b; Le Duc 1995.

1. Perianth 3–6 cm long, salverform; involucre tubular-campanulate, divided 1/3–2/3, the lobes erect; fruits smooth or slightly 5-angled, but not ribbed ____ **M. jalapa**

1. Perianth 1.5 cm or less long, variously shaped; involucre rotate to funnelform-campanulate, divided 1/3 or less, the lobes spreading or ascending; fruits with 5 prominent ribs.
   2. Fruits glabrous or with a few short appressed hairs, the ribs on the fruit surface roughened but not conspicuously tuberculate; plants essentially glabrous throughout (involucres can have a few short appressed hairs); growing in deep loose sands in far w part of nc TX ______ **M. glabra**
   2. Fruits conspicuously pubescent, the ribs on the fruit surface conspicuously tuberculate OR not so; plants pubescent, at least involucres or inflorescences, the pubescence sometimes viscid; growing in various, often calcareous soils; widespread in nc TX.
      3. Blades of stem leaves narrowly linear to linear-lanceolate, 1–5(–10) mm wide, ca. 15–30 times as long as wide; ribs on the fruit surface roughened but without tubercles (at least in nc TX) ______ **M. linearis**
      3. Blades of stem leaves narrowly lanceolate to oblong-elliptic, 3–100 mm wide, 1–12 times as long as wide; ribs on the fruit surface usually with conspicuous elongate or cylindrical tubercles (visible to the naked eye).
         4. Upper and lower internodes densely appressed-puberulent; robust herbs 0.6–2 m tall; growing deep sandy soils ______ **M. gigantea**
         4. Upper internodes glabrous or slightly pubescent, the lower glabrous (rarely short pubescent) OR internodes spreading-pilose; herbs 0.3–0.8 m tall; growing in various, often calcareous soils.
            5. Well-developed leaf blades 3–30 mm wide, gradually tapering to the petioles OR truncate to cordate; flowers variously arranged but often in rather open inflorescences.
               6. Blades of stem leaves linear or lanceolate to oblong-elliptic, tapered at base, 3–30 mm wide, ca. 3–12 times as long as wide ______ **M. albida**
               6. Blades of stem leaves elliptic to ovate or triangular, widely V-shaped to truncate or cordate at base, 10–30 mm wide, the larger ones 1–4 times as long as wide (mostly less than 3 times) ______ **M. latifolia**
            5. Well-developed leaf blades mostly 40–100 mm wide, broadly obtuse, truncate or cordate at base; flowers mostly in rather congested terminal clusters ______ **M. nyctaginea**

Mirabilis albida (Walter) Heimerl, (white), WHITE FOUR-O'CLOCK. Plant usually ± glabrous except for viscid-pubescent inflorescence and upper stem; stems erect, 0.20–1.5 m tall, whitish; leaf blades bright green above, glaucous or whitened below; perianth 8–10 mm long, pink to rose or whitish; fruits ca. 5 mm long. Dry soils, rocky or sandy prairies, open woods, open areas; se and e TX w to Rolling Plains and across Edwards Plateau to Trans-Pecos; this is the commonest *Mirabilis* in nc TX. May–Jul and Sep–Oct. Turner (1993b) indicated this species is exceedingly variable vegetatively, but uniform in terms of fruit characters; the anthocarps are markedly tuberculate, including on the 4–5 ribs, irregularly pubescent with tufted hairs ca. 0.5 mm long, and in addition with a minute layer of much shorter glandular hairs. Turner (1993b) considered *M. eutricha* to be a form with longer stem-hairs; he considered *M. dumetorum*, here treated as *M. latifolia*, to represent broad-leafed, pubescent-stemmed individuals. [*M. albida* var. *lata* Shinners, *M. eutricha* Shinners]

Mirabilis gigantea (Standl.) Shinners, (gigantic), GIANT FOUR-O'CLOCK. Stems erect, 0.6–2 m tall; leaves almost sessile or petioles to 10 mm long; larger leaf blades to 80 mm wide; perianth ca. 10 mm long, rose-pink to light purple; fruits ca. 5 mm long, conspicuously pubescent with tufted hairs, with ribs tuberculate. Sandy open ground or open woods; East and West cross timbers and on sandy river terraces within the Blackland Prairie (McLennan Co., also Dallas Co. (Shinners 1951g; Turner 1993b)); Hatch et al. (1990) also reported it from vegetational areas 3 and 7 (Fig. 2); endemic to TX. Apr–Oct.

Mirabilis glabra (S. Watson) Standl., (smooth, hairless), TALL FOUR-O'CLOCK. Stems erect, 0.6–1.5

m tall, essentially glabrous throughout, glaucous below; leaves often whitish below; fruits 5-angled. Deep loose sands; Callahan Co., also Comanche Co. (Shinners 1951g; Turner 1993b); w part of nc TX w to w TX. May–Nov. [*M. exaltata* (Standl.) Standl.]

Mirabilis jalapa L., (Latin for Xalapa, a town in Veracruz, Mexico; the drug jalap was at one time mistakenly thought to be derived from this plant), FOUR-O'CLOCK, MARVEL-OF-PERU, COMMON FOUR-O'CLOCK, FALSE JALAP. Erect, essentially glabrous perennial with large tuberous root; stems 0.5–1 m tall; upper leaves nearly sessile; leaf blades ovate to ovate-deltoid, 4–14 cm long, 2–8.5 cm wide; petioles 5–50 mm long; involucres herbaceous, 7–15 mm long, the lobes ovate to lanceolate; flowers open ca. 4 pm and close the following morning; perianth very large and extremely showy, purplish red, white, yellow, orange, or varigated, 3–6 cm long, the tube 2–5 mm in diam., the limb 20–35 mm wide; fruits 7–10 mm long. Commonly cultivated; questionably native populations on limestone outcrops; Bell and Dallas cos., also Brown and Hamilton (HPC) cos.; also Edwards Plateau and Trans-Pecos; probably naturalized from tropical America. May–Nov. We are following Le Duc (1995) in lumping [*M. jalapa* L. subsp. *lindheimeri* Standl., *M. lindheimeri* (Standl.) Shinners] with *M. jalapa*. Cultivated by the Aztecs prior to the Spanish conquest as a medicine and for its flowers (Emmant in Le Duc 1995). Seeds and roots are reported to cause digestive disturbances (Hardin & Arena 1974). ☠ ☜

Mirabilis latifolia (A. Gray) Diggs, Lipscomb, and O'Kennon, comb. nov. BASIONYM: *Oxybaphus nyctagineus* (Michx.) Sweet var. *latifolius* A. Gray, Bot. Mex. Bound. Surv. 174. 1859; *Allionia latifolia* (A. Gray) Standl., Contr. U.S. Natl. Herb. 12:350. 1909. TYPE: TEXAS. Travis Co.: near Austin, 1849, *C.Wright 603* (LECTOTYPE: GH—designated by Turner 1993b), (broad-leaved). Resembles a small-leaved *M. nyctaginea*; erect or decumbent and growing over other plants; stems 0.18–1.15 m long; upper stems with short incurved hairs; leaf blades ± ovate, the larger 2–5 cm long; petioles 5–12 mm long; inflorescences widely branched; fruits 4-angled, some fruits 5-angled on type specimen. Various soils, disturbed sites; Dallas, Denton, and Grayson cos., also Burnet and Williamson cos. (Turner 1993b); nc TX and Edwards Plateau, also e TX. Jun–Oct. Turner (1993b) lumped this taxon, which has long gone under the name of *M. dumetorum*, with *M. albida* but indicated that he is ". . . not especially sure of my relegation of *M. dumetorum* to synonymy." Because of this uncertainty and the rather distinctive leaf shape, we are following Shinners (1951g) in recognizing this taxon at the specific level. Turner (1993b) pointed out that the combination *M. latifolia* must be used if the entity is given specific recognition. [*M. dumetorum* Shinners]

Mirabilis linearis (Pursh) Heimerl, (narrow, with sides nearly parallel), LINEAR-LEAF FOUR-O'CLOCK. Stems erect to procumbent, 0.2–2 m long, usually glaucous, often very whitish; leaves usually glaucous, at least below; inflorescence and involucres densely viscid-pubescent; perianth ca. 10 mm long, purple-red to lavender-pink; fruits 4.5–5 mm long, pubescent, the ribs on the fruit surface roughened but without tubercles. Sandy or rocky soils; widespread in TX but mainly w 2/3. May–Jun and Aug–Oct. Richard Spellenberg (pers. comm.) indicated that *M. linearis* often has tuberculate fruits and that the lack of tubercles (a character used above in the key to species) will not serve as a distinction for material outside nc Texas; however, material from nc TX consistently has non-tuberculate ribs.

Mirabilis nyctaginea (Michx.) MacMill., (night-blooming), WILD FOUR-O'CLOCK. Stems several, erect to ascending, 0.3–1.2 m tall, glabrous or nearly so; leaf blades ovate-lanceolate to cordate; involucres becoming enlarged and colored in fruit; flowers 3–5 per involucre; perianth campanulate, pink to purple or reddish (rarely white), the limb 8–18 mm wide; fruits 4–6 mm long, with small tubercles, pubescent. Prairies and thickets, calcareous clay; from Blackland Prairie w to Plains Country. Late Apr–May. ▣/99

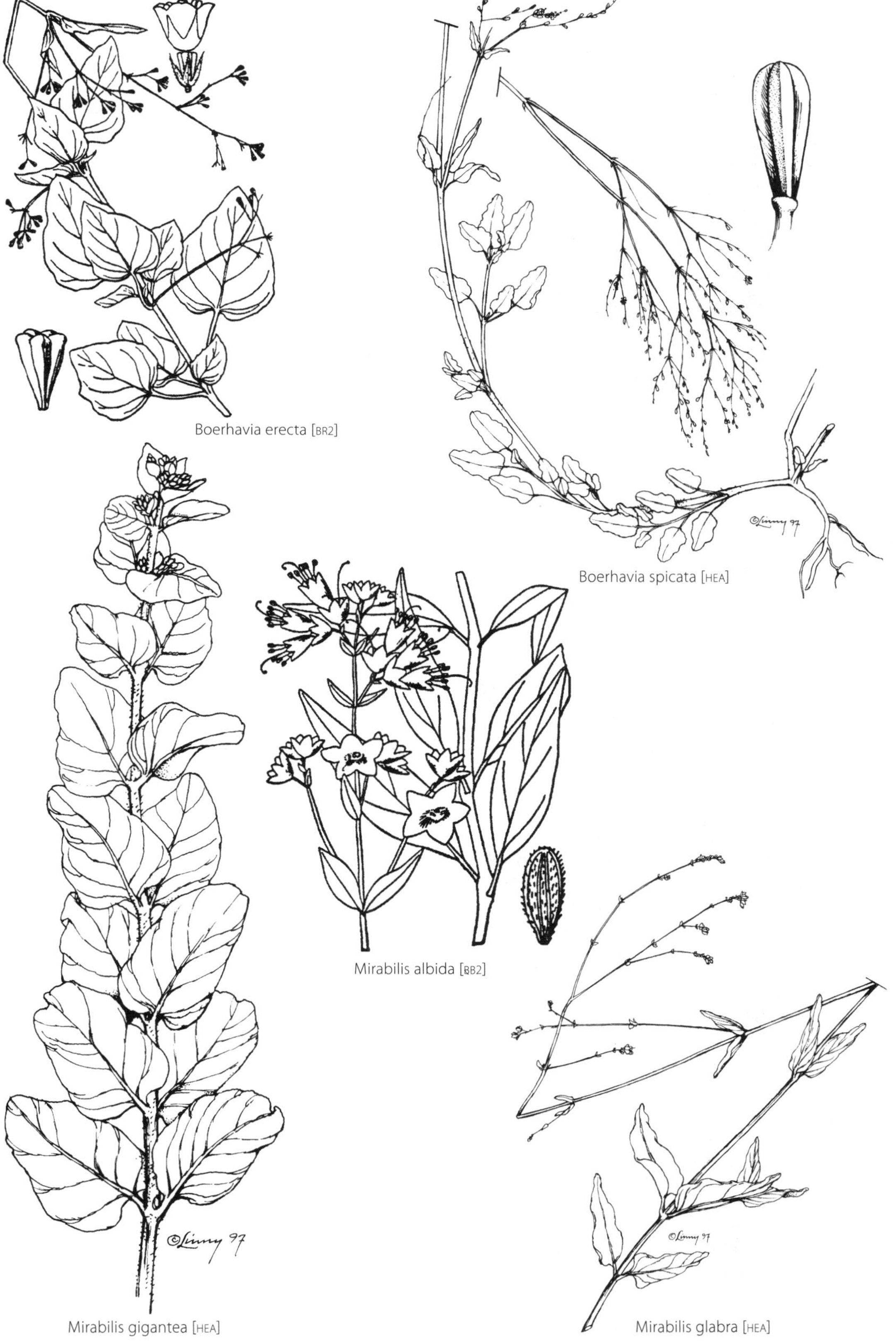

Boerhavia erecta [BR2]

Boerhavia spicata [HEA]

Mirabilis albida [BB2]

Mirabilis gigantea [HEA]

Mirabilis glabra [HEA]

NYCTAGINIA SCARLET MUSKFLOWER, DEVIL'S-BOUQUET

A monotypic genus of sw North America. (Greek: *nyktos*, night, and *gignomai*, becoming)

Nyctaginia capitata Choisy, (capitate, headed), SCARLET MUSKFLOWER, DEVIL'S-BOUQUET. Perennial herb from woody taproot to 4.5 cm in diam., flowering the first year; stems viscid-puberulent or glabrate with age; leaves opposite; leaf blades 4–9 cm long, 6–55 mm wide, entire or sinuate; flowers in clusters 5 cm or more in diam.; perianth funnelform, the limb pink to salmon-rose to deep red, 10–14 mm wide; stamens longer than the perianth; fruits 5–6 mm long, 4 mm in diam., glabrous, many-ribbed. Sandy soils; included based on citation of vegetational area 5 (Fig. 2) by Hatch et al. (1990); mainly s 1/2 of TX. May–Oct, especially after wet periods. Cultivated as an ornamental. Wills and Irwin (1961) indicated that as with many members of this family, the flowers of SCARLET MUSKFLOWER open in late afternoon and close early the next day or sometimes later in cloudy weather. The flowers have a strong, ± offensive odor (Ajilvsgi 1984). ⊠/100

SELINOCARPUS MOONPOD

A tropical American genus of 10 species. (Greek: *selinon*, parsley, and *carpus*, fruit)
REFERENCE: Fowler & Turner 1977.

Selinocarpus diffusus A. Gray, (diffuse, spreading), SPREADING MOONPOD. Shrubby low perennial; stems erect to decumbent, 10–30 cm tall; leaves opposite; leaf blades 12–25 mm long, 6–15 mm wide, fleshy; upper leaves not reduced; perianth tubular-funnelform, 35–45 mm long, densely glandular-hirtellous externally, pale greenish yellow; perianth tube very slender; perianth limb ca. 15 mm wide; stamens 5; fruits 6–7 mm long with 5 wings 2–3 mm long. Sandy or gypseous soils; included based on citation of vegetational area 5 (Fig. 2) by Hatch et al. (1990); nc TX s and w to w TX. Spring and summer.

NYMPHAEACEAE WATER-LILY FAMILY

Rhizomatous and stoloniferous aquatic perennial herbs; leaves long-petioled, simple, entire, not peltate; flowers solitary; receptacle ± obconical or club-shaped, the flower parts (except pistils) closely spiral; in ours sepals 4–6(–14), sometimes grading into the numerous petals, which in turn may grade into staminodes; stamens many; pistils many; ovaries superior; fruits in ours berry-like, many-seeded.

A small family of ca. 50 species in 6 genera (Wiersema & Hellquist 1997); it is a cosmopolitan group of aquatic herbs often with alkaloids; a number of species are cultivated including the huge (can support a human's weight) *Victoria amazonica* (Poepp.) J.C. Sowerby (QUEEN VICTORIA'S WATER-LILY), native to the Amazon basin. The family has a number of characters linking it to monocots (e.g., lack of a cambium, scattered vascular bundles) and Cronquist (1993) suggested that the early dicots from which monocots arose were probably something like modern Nymphaeales; he considered the Nymphaeales as the probable sister group of the monocots. Some taxonomists refer to the Nymphaeaceae as "paleoherbs" (a group including Aristolochiales, Piperales, and Nymphaeales) and believe them to be an early branch off the evolutionary line leading to monocots (Zomlefer 1994). This view is supported by molecular data which place the "paleoherbs" as the immediate sister group of the monocots (Chase et al. 1993) (see Fig. 41 in Appendix 6). According to Mabberley (1987), "... there is good evidence that the group is a successful, specialized relic of the stock which existed before the monocotyledons were recognizably distinct from dicotyledons." *Nelumbo* (LOTUS), as well as *Brasenia* (PURPLE WEN-DOCK) and *Cabomba* (FANWORT), previously placed in this family, are here treated in the Nelumbonaceae and Cabombaceae respectively. (subclass Magnoliidae)
FAMILY RECOGNITION IN THE FIELD: water-lilies with long-petioled, floating, simple leaves that

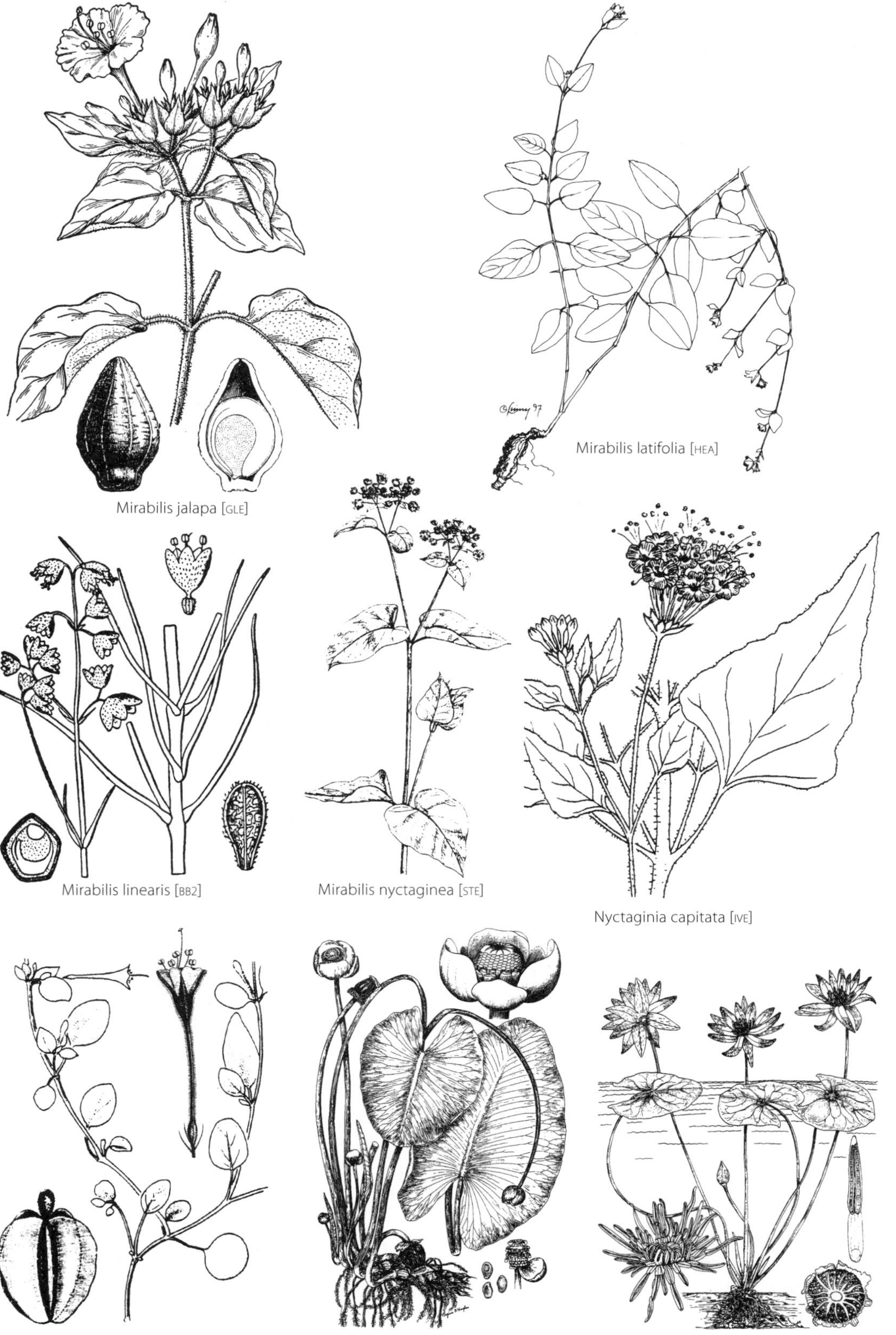

Mirabilis jalapa [GLE]

Mirabilis latifolia [HEA]

Mirabilis linearis [BB2]

Mirabilis nyctaginea [STE]

Nyctaginia capitata [IVE]

Selinocarpus diffusus [ABR, EN2]

Nuphar advena [REE]

Nymphaea elegans [CO1]

are *not peltate* (peltate in Nelumbonaceae); carpels united (free in Cabombaceae); flower parts numerous, spirally arranged.
REFERENCES: Wood 1959; Schneider & Williamson 1993; Wiersema & Hellquist 1997.

1. Sepals usually 6, suborbicular, about as long as wide; petals inconspicuous, 0.6–0.9 cm long, slender, resembling broad staminal filaments; perianth subglobose, the flower appearing somewhat closed and ball-like; leaf venation essentially pinnate ____ **Nuphar**
1. Sepals 4, oblong-lanceolate, 2.5–4 times as long as wide; petals showy, 2–10 cm long, not filament-like; perianth widespreading, the flower appearing conspicuously open; leaf venation essentially palmate ____ **Nymphaea**

NUPHAR SPATTER-DOCK, COW-LILY, YELLOW POND-LILY

A genus of 10–12 species (Wiersema & Hellquist 1997) of n temperate and cold areas; the plants are water-lilies typically with alkaloids and flowers usually held above the water; the seeds often have a slimy pericarp with air-bubbles which are thought to aid in dispersal. (Derivation either from Arabic (Persian): *ninufar*, pond-lily, or Arabic (Egyptian): *nilufar*, water-lily—Paclt 1998)
REFERENCES: Beal 1956; Wiersema & Hellquist 1994; Paclt 1998.

Nuphar advena (Aiton) W.T. Aiton in Aiton & W.T. Aiton, (newly arrived, adventive, not native), YELLOW COW-LILY, SPATTER-DOCK. Leaf blades floating or emergent, elliptic- or orbicular-ovate, to 30 cm long, glabrous above, glabrous or pubescent below; petiole attached at base of deep notch or sinus in blade; petioles and peduncles with numerous minute air cavities; flowers to 45 mm across; sepals (4–)6(–14), in 2 rows, petaloid, green on backs, usually yellow (rarely green or red-tinged) inside, 12–24 mm long; "petals" (apparently outer stamens) numerous, small and inconspicuous, stamen-like or scale-like, yellow or reddish, inserted with the stamens; stamens numerous; anthers 3–7 mm long; pistils numerous, united; fruits broadly ovoid, slightly con-stricted below the stigmatic disk, ca. 3–5 cm broad, maturing underwater; seeds not arillate. Lakes, ponds; Bell, Dallas, and McLennan cos., also Parker and Tarrant cos. (R. O'Kennon, pers. obs.); mainly e TX and Edwards Plateau. Jun–Oct. [*Nuphar lutea* (L.) Sm. subsp. *advena* (Aiton) Kartesz & Gandhi, *N. lutea* subsp. *macrophylla* (Small) E.O. Beal, *Nymphaea advena* Aiton] Cultivated as an ornamental. While sometimes treated as *N. lutea* subsp. *advena* (e.g., Kartesz 1994), we are following Wiersema and Hellquist (1994, 1997) in recognizing this taxon as *N. advena.*

NYMPHAEA WATER-LILY, WATER-NYMPH

Leaf blades floating or slightly elevated, elliptic- or orbicular-ovate to suborbicular, usually glabrous, often reddish or purplish beneath; petiole attached at base of deep notch in blade; flowers closed at night; sepals 4; petals many, showy, the inner petals grading into stamens; anthers ca. 4–6 mm long; fruits ovoid to depressed globose, in ours ca. 1.5–3 cm long, maturing underwater; seeds arillate. A number of cultivated types have become established in artificial ponds and lakes, where deliberately introduced; most are of hybrid origin; their identity can only be approximated with the key.

A cosmopolitan genus of 35–40 species of water-lilies (Wiersema 1997b). Many species and hybrids are cultivated as ornamentals; the rhizomes and seeds of some species are edible. The seeds have a spongy aril which traps air-bubbles, causing the seed to float and aiding in dispersal. (Greek: *nymphe*, water nymph, from the habitat)
REFERENCES: Conrad 1905; Wiersema 1987, 1988, 1997b.

1. Petals pale to deep yellow; stolons present ____ **N. mexicana**
1. Petals white to blue, lavender, or rosy; stolons absent.

2. Leaf blades about as long as wide (length usually not over 1 cm greater than width); petals white to rosy, usually > 25 per flower; flowers floating; sepals abaxially uniform in color, not flecked with short dark lines **N. odorata**
2. Leaf blades distinctly longer than wide (length usually 2–3 cm greater than width); petals blue to lavender-blue or pale violet (rarely white), usually 6–10 per flower; flowers usually raised on peduncle above surface of water; sepals abaxially flecked with short dark lines **N. elegans**

Nymphaea elegans Hook., (elegant), BLUE WATER-LILY, SENORITA WATER-LILY, LAMPAZOS. Stolons elongate, spongy, at the terminal nodes developing clusters of curved, fleshy, overwintering roots that resemble tiny bananas; leaves usually 6–10(–20) cm wide; flowers small, the petals ca. 2–4 cm long. In ponds or other bodies of water; McLennan Co.; mainly s TX. May–Oct.

Nymphaea mexicana Zucc., (of Mexico), YELLOW WATER-LILY, BANANA WATER-LILY, LAMPAZO AMARILLO. Leaves to ca. 20 cm wide; flowers small in the original (non-hybrid) form, with petals 2–4 cm long; hybrids with petals up to 10 cm long, but size may vary in the same colony according to depth of water and time of year; petals usually ca. 25 per flower; sepals abaxially uniform in color, not flecked with short dark lines. Ponds, lakes or other bodies of water; Hood Co. in Lake Granbury (probably a hybrid with *N. odorata*). Jul–Oct, occasionally earlier. Native of s TX, Mexico, and Florida.

Nymphaea odorata Aiton, (fragrant), WHITE WATER-LILY, AMERICAN WATER-LILY, FRAGRANT WATER-LILY, ALLIGATOR-BONNET, POND-LILY, NINFA ACUÁTICA. Leaves usually 9–25 cm wide; flowers sweet-scented, large, with white petals, the outer ones 4.5–9 cm long; hybrids with smaller or larger flowers, with white to rose-pink petals. Dallas, Grayson, and Hood cos., also Comanche (HPC), Erath, and Hamilton (Stanford 1976) cos.; native and frequent in se and e Texas, probably only introduced farther w. May–Sep. [*Nymphaea odorata* var. *villosa* Casp.] ▣/101

NYSSACEAE SOUR-GUM FAMILY

The Nyssaceae is a very small (10 species in 3 genera—treated as a separate family by Mabberley (1987) with numbers updated following Mabberley 1997) family of trees and shrubs of e North America (s to Central America) and e Asia; it is related to Cornaceae, and Burckhalter (1992) and Mabberley (1997) placed *Nyssa* in that family. Recent molecular analyses (Xiang et al. 1993, 1998) suggest that *Cornus*, Nyssaceae, Hydrangaceae, and Loasaceae, as well as several other groups, form a "cornaceous clade"; however, they do not support the lumping of Nyssaceae into Cornaceae. Some species accumulate cobalt and can be used as cobalt indicators. (subclass Rosidae)

FAMILY RECOGNITION IN THE FIELD: the only nc TX species is a tree with alternate, simple, exstipulate leaves; flowers small, greenish, in axillary clusters; fruit a blue-black drupe.

REFERENCES: Rickett 1945b; Eyde 1959, 1963, 1966; Eyde & Barghoorn 1963; Little 1971; Burckhalter 1992; Wen & Stuessy 1993; Xiang et al. 1993.

NYSSA TUPELO, SOUR-GUM

A genus of 8 species with 3 in s North America, 1 Central America, 1 in China, and 1 in Indomalesia; some are used for timber or cultivated as ornamentals for their showy fall colors. (From Nyssa, a water nymph in Greek mythology, because the first described species grows in water)

Nyssa sylvatica Marshall, (of woodland, forest-loving), BLACK-GUM, BLACK TUPELO, SOUR-GUM, PEPPERIDGE, COTTON-GUM. Tree to 30 m tall; leaves alternate, deciduous, the blades obovate to oblanceolate or elliptic (rarely suborbicular), to ca. 10(–15) cm long and 6(–10) cm broad, entire or toothed above middle, bright crimson or burgundy in fall, with petioles to 2(–2.5) cm long; inflorescences usually axillary, pedunculate; staminate inflorescences umbellate or umbellate-racemose; pistillate flowers small, greenish, solitary or in clusters of 2–8; fruit a 1-seeded, blue-black, ellipsoid to subglobose drupe 10–15 mm long, on peduncles to 7 cm long. Swamps, low woods and open woods; Fannin and Lamar cos. in Red River drainage, also Delta, Hopkins, Henderson, Navarro, and Limestone cos. (Little 1971) on e margin of nc TX; mainly e TX w to e part of nc TX. Apr–May. The common name TUPELO is apparently derived from the Native American Creek language: *eto*, tree, and *opelwv*, swamp (Peattie 1948).

OLEACEAE OLIVE FAMILY

Perennial shrubs or trees (nearly herbaceous in *Menodora*); leaves opposite, simple or compound, entire, toothed or lobed; stipules none; flowers axillary, lateral, or terminal, perfect or unisexual, with or without perianth; calyces and corollas, when present, synsepalous and sympetalous, each 4-lobed (except 10–14 calyx lobes and 5–6 corolla lobes in *Menodora*); stamens usually 2 or 4; pistil 1, 2-locular; ovary superior; fruit a drupe, capsule, or samara.

A medium-sized (615 species in 24 genera), nearly cosmopolitan, but especially Asian family of mostly trees and shrubs including timber trees and the economically important *Olea europaea* L. (OLIVE); OLIVE has been in cultivation north of the Dead Sea for its oil and drupes since at least 3700–3600 BC; the foliage was an ancient sign of good will—the olive branch. The family also includes a number of ornamentals including some commonly cultivated in nc TX such as *Forsythia* (GOLDENBELLS) species (shrubs with yellow flowers in early spring) and *Syringa* (LILAC) species. Family name from the genus *Olea*, a genus of 20 species of evergreen trees and shrubs of the Old Word tropics and warm temperate regions. (Greek: *elaia*, olive) (subclass Asteridae)

FAMILY RECOGNITION IN THE FIELD: shrubs or trees (1 species nearly herbaceous) with opposite, simple or compound leaves; calyces and corollas (if present) usually 4-merous; stamens usually 2 or 4, epipetalous; ovary superior.

REFERENCES: Wilson & Wood 1959, Hardin 1974.

1. Leaves compound; fruit a samara with a prominent terminal wing OR a small berry (in the introduced *Jasminum*); plants trees (*Fraxinus*) or shrubs (*Jasminum*).
   2. Leaves pinnately compound with 5–9 leaflets; plants native trees; flowers in drooping panicles; corollas absent ____ **Fraxinus**
   2. Leaves with 3 leaflets; plants introduced shrubs; flowers borne singly; corollas yellow, showy ____ **Jasminum**
1. Leaves simple (unlobed or pinnately lobed); fruit a drupe or unwinged capsule; plants small to large shrubs (rarely small trees) or nearly herbaceous.
   3. Fruit a thin-walled capsule; plants nearly herbaceous, small, to 25 cm or less tall; leaves entire to pinnately lobed; corollas with 5–6 lobes ____ **Menodora**
   3. Fruit a small drupe; plants woody, usually much more than 25 cm tall; leaves entire or minutely or indistinctly toothed, not lobed; corollas with 4 lobes or absent.
      4. Corollas absent; flowers in clusters or small panicles in the axils of leaves; leaf margins usually minutely or indistinctly toothed ____ **Forestiera**
      4. Corollas present, bright or creamy white; flowers in panicles terminating the branches; leaf margins entire ____ **Ligustrum**

FORESTIERA

Ours deciduous shrubs, sometimes half vine-like, or small trees; leaves short-petioled; leaf blades minutely or indistinctly toothed; flowers appearing before the leaves, lateral, in sessile clusters with sepal-like bud-scales, or in small panicles, unisexual, the sexes usually on separate plants (= dioecious); sepals minute; fruit a dark blue to purple to blackish, glaucous drupe.

A genus of 15 species of the Americas, especially sw North America; some are cultivated as

Nymphaea mexicana [GWO]

Nymphaea odorata [REE]

Nyssa sylvatica [SA3]

Forestiera pubescens var. pubescens [LYN, POW]

Forestiera acuminata [GLE]

Fraxinus americana [SA3]

ornamentals. (Named for Charles Le Forestier, deceased ca. 1820, French physician and naturalist and Poiret's first teacher in botany)
REFERENCE: Johnston 1957.

1. Leaf blades usually acuminate, 3–9 cm long, on petioles 4–20 mm long; pistillate inflorescence a short panicle 10–20 mm long with 18–32 flowers; mature drupes 9–18 mm long; leaves and twigs usually glabrous; filaments 4–7 mm long ______ **F. acuminata**
1. Leaf blades acute or obtuse, 2.5–4 cm long, on petioles 1–7 mm long; pistillate inflorescence umbel-like or a corymbose cluster, of 3–12 flowers; mature drupes 5–7 mm long; leaves and twigs glabrous or pubescent; filaments 1–4 mm long ______ **F. pubescens**

Forestiera acuminata (Michx.) Poir., (tapering to a long narrow point), SWAMP-PRIVET, TEXAS ADELIA. Shrub or small tree; lenticels conspicuous on twigs. Low woods; Lamar and Navarro cos., also Bell, Collin, and Dallas cos. (Little 1976 [1977]); mainly se and e TX. Mar.

Forestiera pubescens Nutt., (pubescent, downy), SPRING-HERALD, HERALD-OF-SPRING DEVIL'S-ELBOW, ELBOW-BUSH, STRETCH-BERRY, CHAPARRAL. Shrub, often with looping, half vine-like stems. Limestone outcrops, pastures, brushy prairies. Jan–Mar.

1. Leaves glabrous ______ var. **glabrifolia**
1. Leaves pubescent ______ var. **pubescens**

var. **glabrifolia** Shinners, (smooth- or hairless-leaved), SMOOTH-LEAF FORESTIERA, EVERGREEN FORESTIERA. Blackland Prairie w to West Cross Timbers and Lampasas Cut Plain, also Rolling Plains.

var. **pubescens**. Blackland Prairie (Dallas, Ellis, and Grayson cos.) s and w to w TX.

FRAXINUS ASH

Deciduous, often dioecious trees; branchlets, petioles, and leaflets usually glabrous or nearly so in ours; leaves odd-pinnately compound, the leaflets entire or with rather widely spaced teeth; flowers in drooping panicles (at first compact) in late March and April, appearing before or with the leaves; corollas absent; species hardly identifiable until fully expanded leaves are present.

A primarily n temperate genus of 65 species with a few extending to the tropics. Some cultivated as ornamentals for bright fall foliage or valuable for timber; the elastic wood of some was formerly used for making wheels; it is also used for making tool handles and sports equipment, such as baseball bats or hockey sticks. The ash flowergall mite, *Aceria fraxiniflora* Felt, can infect the male inflorescences of ashes in nc TX turning them into masses of lumpy distorted galls (Solomon et al. 1993). (The classical Latin name for ASH)
REFERENCE: Miller 1955.

1. Wing of fruit not decurrent on fruit body (wing ± ending where body of fruit begins); petiolules of lateral leaflets 3–14 mm long; blades of leaflets usually pale glaucous below, abruptly narrowed, usually asymmetrical basally, short-decurrent on petiolules.
   2. Leaflets 5–9, typically 7; blades usually 2.5–3.5 times as long as wide, 6–15 cm long, normally widely spreading; fruits 2.7–5 cm long ______ **F. americana**
   2. Leaflets 5–7, typically 5; blades usually 1.75–2.5 times as long as wide, usually < 6(–8) cm long, often drooping; fruits 1.5–2.7(–3) cm long ______ **F. texensis**
1. Wing of fruit decurrent over half way on fruit body (wing extending along body of fruit); petiolules of lateral leaflets 0–7 mm long; blades of leaflets lighter green below but not noticeably pale, usually gradually narrowed basally and long-decurrent (sometimes with blade tissue to petiolule base and leaflet thus appearing sessile) ______ **F. pennsylvanica**

Fraxinus americana L., (of America), WHITE ASH, FRESNO. Large tree to 40 m tall; usually glabrous throughout. Stream bottom woods or on slopes; e TX w to Collin, Grayson, Johnson, and McLennan cos., also Hamilton (HPC) and Tarrant (Little 1971) cos. Feb–Mar. The wood of this species is famous as the source of baseball bats, tennis racquets, and hockey sticks (Peattie 1948).

Fraxinus pennsylvanica Marshall, (of Pennsylvania), GREEN ASH, RED ASH. Tree to 20 m tall; in nc TX usually glabrous throughout; leaflets 5–7; fruits 3–7.5 cm long, lanceolate to oblong-ovate. Stream bottom woods or on slopes; e TX w to Rolling Plains. Feb–Mar. [*F. pennsylvanica* var. *integerrima* (Vahl) Fernald]

Fraxinus texensis (A. Gray) Sarg., (of Texas), TEXAS WHITE ASH, TEXAS ASH. Small tree to ca. 16 m; glabrous throughout; leaflets suborbicular-ovate to obovate or sometimes narrowly elliptic (usually rather wide). Limestone slopes, bluffs; Grayson Co. s and w through the Edwards Plateau; endemic to TX and s OK. Closely related to *F. americana*, sometimes difficult to distinguish, and possibly only a subspecies or variety of that species. Feb–Mar. [*F. americana* L. subsp. *texensis* (A. Gray) G.N. Mill., *F. americana* L. var. *texensis* A. Gray]

JASMINUM JASMINE, JESSAMINE

A mainly Old World genus of ca. 200 species (1 American) of tropical areas with a few in temperate regions; the species are deciduous and evergreen shrubs and lianas. The flowers of some are used in scent-making and perfuming tea. (Latinized form of the Persian name: *yasmin*)

Jasminum nudiflorum Lindley, (naked flower), WINTER JASMINE. Deciduous shrub to 3(–5) m tall, rambling; branches long, slender, glabrous, 4-angled; leaves opposite, with 3 leaflets; leaflets ovate to oblong-ovate, 1–3 cm long, glabrous, ciliate; flowers borne singly in axils of previous year's leaves (2 per node), appearing before the leaves; pedicels ca. 6 mm long, covered with bracts; corollas yellow, (5–)6-lobed, 2–3 cm across; corolla lobes ca. 1/2 as long as the tube; stamens 2; fruit a blackish berry. Cultivated and escapes; chalky roadside; Collin Co.; we are not aware of other TX localities. Feb. Native of China.

LIGUSTRUM PRIVET, HEDGEPLANT

Evergreen or deciduous shrubs or small trees; leaves entire; flowers in dense, narrow, terminal panicles, perfect; corollas bright or creamy white; stamens usually exserted from corolla tube; fruit a small black or blackish blue drupe.

A genus of ca. 40 species native to Europe, n Africa, and e and se Asia to Australia. PRIVETS are widely used as ornamentals, often as hedges; all are tolerant of city pollution. However, they are problematic weeds in that they escape from cultivation, naturalize, and displace native plants. The flowers typically have a heavy scent, offensive to some because of ammonia undertone due to presence of timethylamine—the result is a somewhat fishy smell. The fruits and leaves of all species growing in nc TX should be considered poisonous due to syringin (an irritant glycoside) and alkaloids; children have been fatally poisoned by eating the fruits and the leaves are poisonous to livestock; some species can cause dermatitis in individuals trimming hedges or bushes (Schmutz & Hamilton 1979; Lampe & McCann 1985; Turner & Szczawinksi 1991). (The classical Latin name of *L. vulgare* L.)

1. Twigs glabrous; larger leaf blades 6–15 cm long, evergreen.
   2. Larger leaf blades usually 8–15 cm long, with 6–8 ± distinct veins on each side of midrib, acuminate; petioles of larger leaves 10–20 mm long; tube of corolla equaling lobes in length ____ **L. lucidum**
   2. Larger leaf blades usually 6–10 cm long, with 4–5 indistinct veins on each side of midrib, short

acuminate to nearly obtuse; petioles of larger leaves 6–12 mm long; tube of corolla slightly longer than the lobes **L. japonicum**

1. Twigs pubescent; larger leaf blades 2–6(–7) cm long, evergreen in mild winters or somewhat deciduous.
   3. Flowers sessile or subsessile; corolla tube as long as lobes; leaves often tapering at base **L. quihoui**
   3. Flowers distinctly pedicelled; corolla tube shorter than lobes; leaves cuneate at base, usually not tapering **L. sinense**

Ligustrum japonicum Thunb., (of Japan), WAX-LEAF LIGUSTRUM, JAPANESE PRIVET, WAX-LEAF PRIVET. Similar to *L. lucidum* but usually smaller, to ca. 3 m or more tall; leaves glabrous, leathery. Cultivated and escapes; Dallas and Tarrant cos. (R. O'Kennon, pers. obs.). Jun–Jul. Native of Korea and Japan. Poisonous.

Ligustrum lucidum W.T. Aiton, (bright, shining, clear), GLOSSY PRIVET, CHINESE PRIVET, NEPAL PRIVET, WAX-LEAF PRIVET, WHITE WAX TREE, TREE PRIVET. Shrub or small tree to ca. 10 m; leaves glossy, glabrous, (6–)10–15 cm long. Cultivated and escapes; Dallas and Grayson cos., also Fort Hood (Bell or Coryell cos.—Sanchez 1997) and Johnson, Parker, and Tarrant cos. (R. O'Kennon, pers. obs.). Late summer. Native of China and Korea. Poisonous.

Ligustrum quihoui Carriére, (for Antoine Quihou, 1820–1889, French horticulturalist), QUIHOU'S PRIVET. Shrub to ca. 2 m tall; branches nearly horizontal; leaves very short-petioled, with dark green, usually lanceolate or oblanceolate, glabrous blades; flowers in whorl-like, separated clusters, at tips of branches and on paired side branchlets, forming an open panicle; corolla tube about equaling the lobes; stamens exserted from corolla tube. Cultivated and naturalized; Dallas and Grayson cos., also Tarrant Co. (R. O'Kennon, pers. obs.); also e TX. Late May–Jul. Native of China. Poisonous.

Ligustrum sinense Lour., (Chinese), CHINESE PRIVET, TRUENO DE SETO. Shrub to ca. 4 m; leaves short-petioled; leaf blades rhombic-elliptic to ovate, usually pubescent on midrib beneath; flowers in compact panicles; corolla tube shorter than lobes; stamens exserted from corolla tube. Cultivated, persisting, and naturalized; Collin, Dallas, Grayson, and Red River cos., also Comanche, Erath (Stanford 1976), and Tarrant (R. O'Kennon, pers. obs.) cos.; in TX mainly from nc TX e and s. Late April–May. Native of China. A problematic invader of native habitats; in some areas it has become the dominant understory shrub. Poisonous.

Ligustrum vulgare L., (common), COMMON PRIVET, a native of the Mediterranean area similar to *L. sinense*, is cultivated in nc TX and possibly escapes. It differs in having the stamens included within the corolla tube and the leaves glabrous on the midrib beneath. Poisonous.

MENODORA

A genus of ca. 25 species mainly of warm areas of the Americas with 3 species in s Africa. (Perhaps from Greek: *mene*, moon, and *dory*, spear or shaft, from appearance of fruit on stiff pedicel) REFERENCES: Steyermark 1932; Turner 1991b.

Menodora heterophylla Moric. ex DC., (various-leaved), REDBUD, TWINPOD, LOW MENODORA. Small, largely glabrous perennial to 25 cm tall forming patches from branching and creeping roots; leaves sessile or subsessile, varying from entire to deeply pinnatifid with sharp lobes or teeth, linear-lanceolate to elliptic-obovate in outline; pedicels 5–8 mm long, recurved in fruit; flowers solitary, terminal or axillary, perfect; unopened flower buds bright red; calyces divided to base into narrowly linear lobes; corollas to 15 mm long, with short, narrowly cylindrical tube and broadly funnelform limb, yellow within, ± red without, showy; fruit a circumscissile, distinctly 2-celled capsule 6–10 mm long. Rocky, silty, or sandy ground; Brown and Stephens cos.,

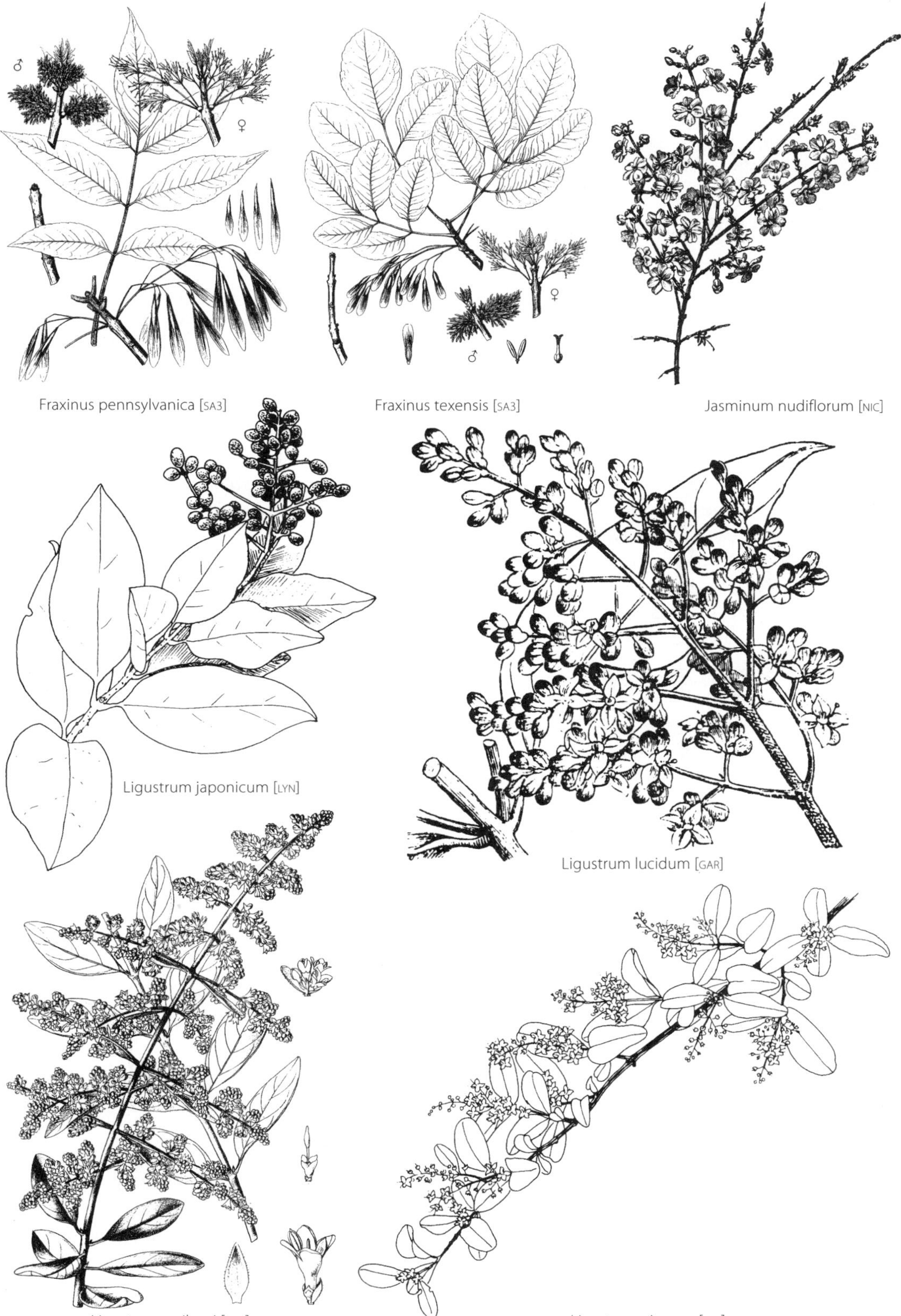

Fraxinus pennsylvanica [SA3]

Fraxinus texensis [SA3]

Jasminum nudiflorum [NIC]

Ligustrum japonicum [LYN]

Ligustrum lucidum [GAR]

Ligustrum quihoui [CUR]

Ligustrum sinense [LYN]

also Burnet and Mills cos. (Turner 1991b); w margin of nc TX s through Edwards Plateau to s TX. Apr–Jun, sporadically to Sep (as early as Jan in s Texas).

Menodora longiflora A. Gray, (long-flowered), SHOWY MENODORA, native to c and w TX, extends as far north as McCullough, Llano, and Travis cos. just s and w of nc TX (Turner 1991b). This species can be distinguished by the leaves mostly entire (occasionally the lowest are 2–3-lobed), flowers numerous in terminal inflorescences, pedicels erect in fruit, and bright yellow corollas 30–60 mm long.

ONAGRACEAE EVENING-PRIMROSE FAMILY

Annual or perennial herbs or half-shrubs; leaves basal, alternate, or opposite, simple, entire, toothed or pinnately lobed; flowers basal, axillary, terminal, solitary, in spikes, or panicles; hypanthium of a short or elongate, cylindrical tube; calyx lobes or sepals (3–)4(–5); petals (0–)4 (–5), attached at summit of hypanthium (= floral tube); stamens (4–)8(–10); pistil usually 4-carpellate; ovary inferior; fruit a dehiscent capsule or an indehiscent, 1-seeded, nut-like capsule.

A medium-sized (650 species in 18 genera) family, cosmopolitan in distribution but especially of temperate and warm areas of the Americas. The family consists of herbs and shrubs or more rarely trees and includes a number of ornamentals such as *Clarkia*, *Epilobium* (FIREWEED), *Fuchsia*, and *Oenothera*. Family name conserved from *Onagra*, a genus now treated as *Oenothera* (the name *Oenothera* was published earlier and thus has priority in terms of nomenclature) (Greek: *onagra*, evening-primrose) (subclass Rosidae)
FAMILY RECOGNITION IN THE FIELD: herbs (sometimes woody-based) often with showy flowers with parts usually in *4s* and a *conspicuously inferior* ovary; *tube-shaped* hypanthium often present; stamens typically 8.
REFERENCES: Munz 1961, 1965; Raven 1964; Conti et al. 1993; Hoch et al. 1993.

1. Petals equal, not clawed, yellow to white or pink, or petals absent; flowers radially symmetrical, basal, axillary, or in terminal inflorescences.
   2. Hypanthium (= floral tube) not prolonged beyond the ovary, the flower parts (petals, stamens) at the summit of the ovary; sepals persistent in fruit ____ **Ludwigia**
   2. Hypanthium much prolonged beyond the ovary, pedicel-like, the flower parts well-separated from the ovary; sepals deciduous in fruit.
      3. Stigmas deeply lobed or branched, not peltate ____ **Oenothera**
      3. Stigmas entire to only slightly lobed, peltate ____ **Calylophus**
1. Petals slightly unequal, clawed, white to pink or reddish; flowers usually bilaterally symmetrical (*Gaura*) OR radially symmetrical (*Stenosiphon*), in slender, solitary or panicled racemes.
   4. Petals white to pink or reddish, usually all projecting upward, 1.5–15 mm long; anthers red, purple, or brown, ca. 1–6 mm long; hypanthium nearly cylindrical; leaves abruptly reduced in the inflorescence; filaments often with a scale at base (except in *G. parviflora*) ____ **Gaura**
   4. Petals pure white, evenly spaced, 4–6 mm long; anthers creamy white (withering brownish), less than 2 mm long; hypanthium thread-like; leaves gradually reduced in the inflorescence; filaments without a scale at base ____ **Stenosiphon**

CALYLOPHUS EVENING-PRIMROSE, SUNDROPS

Herbaceous or woody based perennials or rarely annuals; flowers radially symmetrical, 4-merous, axillary, opening near sunrise or near sunset; petals yellow, but often with UV markings visible only to certain insects; stamens 8; stigma peltate, the lobes 4, shallow; fruit a cylindric capsule.

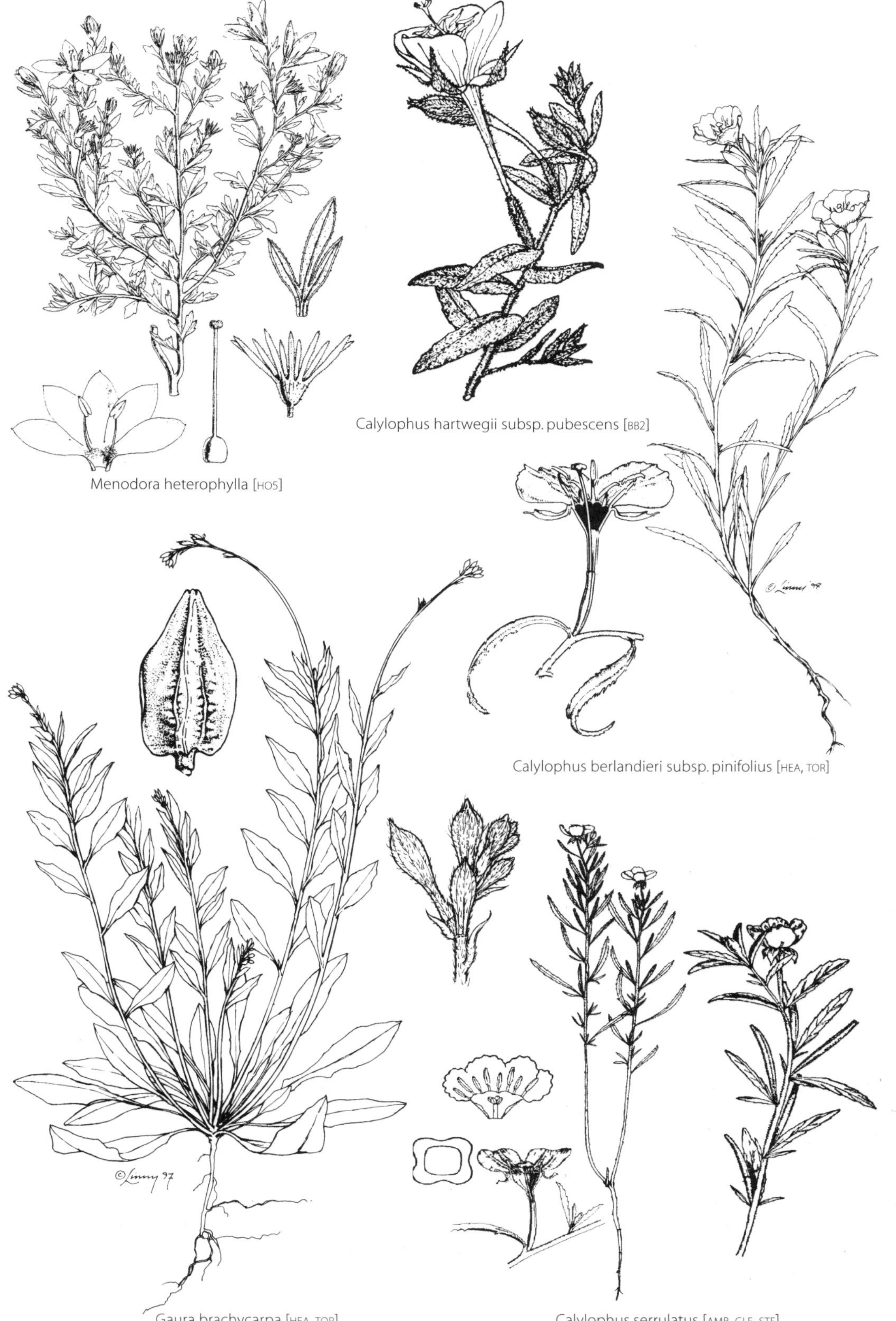

Menodora heterophylla [HO5]

Calylophus hartwegii subsp. pubescens [BB2]

Calylophus berlandieri subsp. pinifolius [HEA, TOR]

Gaura brachycarpa [HEA, TOR]

Calylophus serrulatus [AMB, GLE, STE]

A North American genus of 6 species previously included in *Oenothera*; segregated by Raven (1964). (Greek: *calyx*, cup, and *lophus*, crest or tuft)
REFERENCES: Raven 1964; Shinners 1964a; Towner 1977.

1. Hypanthium 5–16(–20) mm long; flower buds somewhat 4-angled (due to keeled sepals); filaments of 2 lengths, some ca. 2 times as long as others; widespread in nc TX.
   2. Stigma exserted to tip of outer anthers or beyond; flowers relatively larger, the petals usually 9–21 mm long; pollen 85–100% fertile ____ **C. berlandieri**
   2. Stigma extends only to tip of inner anthers near apex of hypanthium; flowers relatively smaller, the petals usually 6–12 mm long; pollen 30–80% aborted ____ **C. serrulatus**
1. Hypanthium 20–50 mm long; flower buds not 4-angled (sepals not keeled); filaments nearly equal in length; limited to sw part of nc TX ____ **C. hartwegii**

Calylophus berlandieri Spach, (for Jean Louis Berlandier, 1805–1851, French botanist who explored TX, NM, and Mexico, and one of the first botanists to make extensive collections in TX), BERLANDIER'S EVENING-PRIMROSE, SQUARE-BUD DAY-PRIMROSE. Annual or perennial to 50 cm tall, similar to *C. serrulatus* except for floral characters; leaves linear to oblanceolate, subentire to denticulate; flowers self-incompatible, opening near sunrise. May–Jun. Hybridization sometimes occurs between the 2 subspecies (Towner 1977).

1. Plants ± bushy, the stems several–many, nearly decumbent to ascending, 10–40 cm tall; largest leaf blades on stem 1–4 cm long; hypanthium and stigma yellowish, never black; rare in nc TX ____ subsp. **berlandieri**
1. Plants usually not bushy, the stems 1–several, suberect to erect, 30–80 cm tall; largest leaf blades on stem 2.5–9 cm long; hypanthium (inside) and stigma sometimes black; widespread in nc TX ____ subsp. **pinifolius**

subsp. **berlandieri**, HALF-SHRUB SUNDROPS, DRUMMOND'S SUNDROPS, BEACH EVENING-PRIMROSE. Prairies, sandy, gravelly, and limestone soils in relatively dry areas; in nc TX known only from Palo Pinto and Wise cos. (Towner 1977) in West Cross Timbers; widespread in TX, but mainly s TX and Panhandle. Mar–Nov. [*C. drummondianus* Spach subsp. *berlandieri* (Spach) Towner & P.H. Raven, *Oenothera serrulata* Nutt. subsp. *drummondii* (Torr. & A. Gray) Munz]

subsp. **pinifolius** (Engelm. ex A. Gray) Towner, (with leaves like *Pinus*—pine). Prairies, sandy and rocky areas; Post Oak Savannah and se TX w to Edwards Plateau and Red Plains. Mar–Jul. [*C. drummondianus* Spach, *C. serrulatus* (Nutt.) P.H. Raven var. *spinulosus* (Nutt. ex Torr. & A. Gray) Shinners, *Oenothera serrulata* subsp. *pinifolia* (Engelm. ex A. Gray) Munz]

Calylophus hartwegii (Benth.) P.H. Raven subsp. **pubescens** (A. Gray) Towner & P.H. Raven, (sp.: for Theodore Hartweg, 1812–1871, Royal Horticultural Society collector in CA and Mexico; subsp.: pubescent, downy), GRAND PRAIRIE EVENING-PRIMROSE. Bushy perennial to ca. 40 cm tall; leaves linear to oblong-lanceolate, entire to denticulate, abruptly narrowed to truncate or slightly clasping basally; flowers self-incompatible, opening in afternoon or near sunset; petals 10–35 mm long; ovary with long spreading hairs. Open sandy areas or on limestone; Brown, Callahan, Comanche, Erath, Hamilton, Lampasas, Mills, and Tarrant cos.; w part of nc TX s and w to w TX. Apr–Jun, sporadically later. This species is made up of 5 subspecies, only 1 of which occurs in nc TX. [*C. hartwegii* var. *pubescens* (A. Gray) Shinners, *Oenothera greggii* A. Gray, *Oenothera greggii* var. *lampasana* (Buckley) Munz]

Calylophus serrulatus (Nutt.) P.H. Raven, (with minute teeth), YELLOW EVENING-PRIMROSE, DAY-PRIMROSE. Similar to *C. berlandieri* but self-compatible; difficult to distinguish without floral characters; flowers opening in the morning; anthers shedding pollen directly on stigma. Open, sandy or rocky soils; w Blackland Prairie w to Panhandle and s to near Gulf coast. May–Jun. [*Oenothera serrulata* Nutt.]

GAURA BUTTERFLY-WEED, WILD HONEYSUCKLE

Annual or perennial herbs (can be woody at base); leaves reduced upward; basal leaves often lyrate; inflorescence a spicate raceme or group of these; flowers opening near sunrise or sunset, (3–)4-merous, usually strongly bilaterally symmetrical; hypanthium well-developed; petals clawed, usually white, becoming pink or red with age; stamens 8; stigma (3–)4-lobed; fruit an indehiscent, woody, nut-like capsule, sometimes stipitate.

A North American genus of 21 species; some are cultivated for cut flowers. The key is adapted from Raven and Gregory (1972a) with modifications from G. Hoggard (1998). (Greek: *gauros*, proud, showy, or majestic, from the sometimes showy flowers)

REFERENCES: Munz 1938; Raven & Gregory 1972a, 1972b; Carr et al. 1986, 1988, 1990; G. Hoggard 1999.

1. Flowers 3-merous; fruits 3-angled.
   2. Sepals 4.5–6 mm long; petals 3.5–5 mm long ____ **G. triangulata**
   2. Sepals 9–15 mm long; petals 8–12.5 mm long ____ **G. brachycarpa**
1. Flowers 4-merous; fruits 4-angled.
   3. Flowers small: anthers ca. 1 mm long, oval; sepals 2–5.8 mm long; petals 1.5–4 mm long; fruits weakly angled ____ **G. parviflora**
   3. Flowers relatively larger: anthers 1.5–6 mm long, linear; sepals 4.5–17 mm long; petals 3.5–15 mm long; fruits usually distinctly angled or winged.
      4. Fruits narrowed to a slender stipe 2–8 mm long (stipe ± pedicel-like in appearance); stems densely villous OR glabrous to with sparse pubescence; plants perennial from a woody base or rhizome.
         5. Body of fruits tapering to a slender stipe, the stipe noticably broader near attachment to body; stems usually glabrous to with sparse pubescence; widespread in nc TX ____ **G. sinuata**
         5. Body of fruits abruptly narrowing to a slender stipe, the stipe filiform and ± equal in diam. its entire length; stems usually conspicuously and densely villous, the hairs 2–3 mm long; in nw part of nc TX ____ **G. villosa**
      4. Fruits subsessile OR with a thick cylindrical stipe; stems pubescent or pilose, sometimes densely so; plants perennial from a woody base or rhizome OR annual from a taproot.
         6. Plants perennial from a branching woody base or aggressively rhizomatous; fruits rather abruptly constricted to a thick cylindrical stipe ca. 1/2 the length of the fruit body and 1/4–1/2 the width of the fruit body (the stipe can appear like an abruptly narrowed bottom part of a sessile fruit).
            7. Fruits relatively small, only 4–9 mm long, 1–3 mm wide, covered with dense, short, appressed pubescence (use hand lens); bracts 2–5 mm long; plants from a branching woody base ____ **G. coccinea**
            7. Fruits relatively larger, 7–13 mm long, 3–5 mm wide, usually glabrous or with a few scattered hairs; bracts 2–8 mm long; plants aggressively rhizomatous ____ **G. drummondii**
         6. Plants annual or biennial from a taproot (and 1 perennial species); fruits subsessile, not abruptly constricted to a thick stipe.
            8. Sepals with long, erect hairs; plants perennial; flowers opening near sunrise and withering in the afternoon; mainly to the s of nc TX ____ **G. lindheimeri**
            8. Sepals glabrous or with appressed hairs; plants annual; flowers opening near sunset and withering the next morning; widespread in nc TX.
               9. Fruits with wings on the angles and conspicuous furrows between the angles; plants usually < 1.2 m tall, typically branched from base; flowering in spring (Feb–Jun).
                  10. Buds and backs of sepals glabrous; bracts 4.5–7 mm long; fruits ellipsoidal, broader near middle ____ **G. suffulta**

10. Buds and backs of sepals pubescent; bracts 2–4 mm long; fruits pyramidal, broader near base ____ **G. brachycarpa**

9. Fruits angled but without wings; plants typically 1–4 m tall, usually not branched at base; flowering in summer and fall (Jun–Dec) ____ **G. longiflora**

Gaura brachycarpa Small, (short-fruited), PLAINS GAURA. Low annual to 0.65(–0.85) m tall; hypanthium 6.5–12 mm long; petals 8–12.5 mm long, pink to rose-red or white; fruits (3–)4-angled, 5.5–10 mm long, sessile or with stipe 1 mm long. Sandy open areas; Post Oak Savannah w to West Cross Timbers and s to s TX. Mar–May(–Jun).

Gaura coccinea Nutt. ex Pursh, (scarlet), SCARLET GAURA, WILLOW-HERB GAURA, SMOOTH GAURA, WILD HONEYSUCKLE, BEE-BLOSSOM. Low perennial 0.2–0.5(–1) m tall; petals 3–8 mm long, white, withering to rosy or maroon; fruits abruptly constricted to a thick cylindrical stipe. Sandy or rocky prairies; Burnet, Callahan, Erath, Hamilton, Palo Pinto, and Parker cos., also Brown (HPC), Bell, and Williamson (Munz 1961) cos.; s and w parts of nc TX s and w to w TX. Apr–Jun. [*G. coccinea* var. *glabra* (Lehm.) Munz, *G. coccinea* var. *parviflora* (Torr.) Rickett, *G. odorata* Sessé ex Lag.]

Gaura drummondii (Spach) Torr. & A. Gray, (for its discoverer, Thomas Drummond, 1780–1835, Scottish botanist and collector in North America), SWEET GAURA, SCENTED GAURA. Perennial spreading by rhizomes to form large patches; stems usually several, 0.2–0.6(–1.2) m tall; petals 6–10 mm long, pink to red, rarely white; fruits 7–13 mm long, abruptly constricted to cylindrical stipe. Rocky or sandy open ground; Dallas, Erath, and Tarrant cos.; e 1/2 of TX. Apr–Oct. This species has long incorrectly gone under the name of *G. odorata* (Raven & Gregory 1972a). [*Gaura odorata* sensu Mahler and various authors, not Sessé ex Lag.]

Gaura lindheimeri Engelm. & A. Gray, (for Ferdinand Jacob Lindheimer, 1801–1879, German born TX botanist), WHITE GAURA, MUNZ GAURA. Villous, clumped perennial, 0.5–1.5 m tall; flowers opening near sunrise; petals 10.5–15 mm long, white fading to light or deep pink; fruits 6–9 mm long, subsessile. Prairies, sometimes cultivated; included based on citation of vegetational area 4 (Fig. 2) by Hatch et al. (1990); mainly se and sc TX. Apr–Jul.

Gaura longiflora Spach, (long-flowered), TALL GAURA, KEARNEY'S GAURA. Large annual or biennial typically 1–4 m tall; inflorescence pubescent or glandular-pubescent; hypanthium 4–13.5 mm long; petals 6.5–15 mm long; fruits 4.5–7 mm long, subsessile. Open disturbed areas; Dallas, Grayson, Hunt, and Lamar cos., also Wise Co. (Munz 1961); se and e TX w to n part of nc TX. Jun–Dec. [*G. biennis* var. *pitcheri* Torr. & A. Gray, *G. filiformis* Small]

Gaura parviflora Douglas ex Lehm., (small-flowered), LIZARD-TAIL GAURA, VELVET-LEAF GAURA, SMALL-FLOWER GAURA, DOWNY GAURA. Tall annual from a large taproot, 0.15–2(–3) m tall, densely glandular-pubescent and villous with long, wide-spreading hairs; leaves entire or toothed; petals brick red to white; fruits 5–11 mm long, subsessile. Disturbed soils, stream bottoms; widespread in TX. Apr–Oct.

Gaura sinuata Nutt. ex Ser., (with wavy margins), WAVY-LEAF GAURA. Brittle-stemmed perennial 0.2–1.5 m tall, spreading by rhizomes and often forming mats; inflorescence simple or open-panicled, on a long, naked peduncle (to 20 cm tall) with a basal whorl of leafy branches; buds appressed-pubescent; petals 7–14.5 mm long, pink to red or white; fruit body 8–15 mm long; stipe 2–8 mm long. Sandy or rocky prairies or disturbed soils; widespread in TX. Apr–Oct.

Gaura suffulta Engelm. ex A. Gray, (supported or propped), ROADSIDE GAURA, WILD HONEYSUCKLE, BEE-BLOSSOM, KISSES. Winter annual; stems erect or with decumbent bases, 0.25–1.2 m tall, spreading-pilose; leaves entire, toothed, or lobed; inflorescence glabrous or with a few long hairs near base; petals 10–15 mm long, white, withering to rosy; fruits 4.5–8 mm long, sessile or

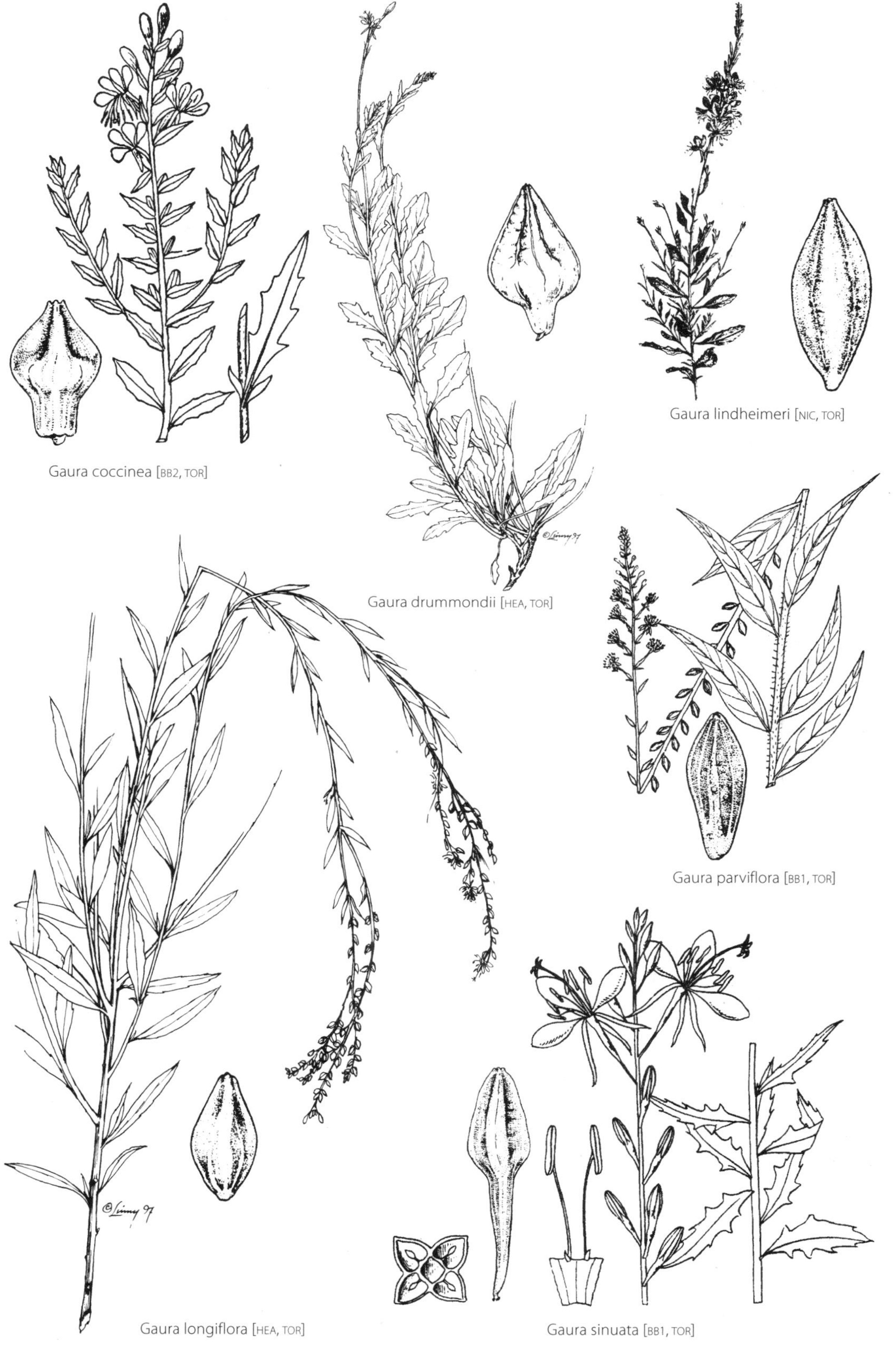

Gaura coccinea [BB2, TOR]

Gaura drummondii [HEA, TOR]

Gaura lindheimeri [NIC, TOR]

Gaura parviflora [BB1, TOR]

Gaura longiflora [HEA, TOR]

Gaura sinuata [BB1, TOR]

on short stipe to 2.2 mm long. Prairies, roadsides, disturbed sites; e TX w to Rolling Plains, also Edwards Plateau. Feb–May(–Jun).

Gaura triangulata Buckley, (triangular, 3-angled). Slender low annual to 0.6 m tall; hypanthium 4–5.5 mm long; fruits 7–9 mm long, sessile. Sandy, open areas; West Cross Timbers and Rolling Plains; type locality in Young Co. (Buckley 1861 [1862]). Mar–Jun. [*G. tripetala* Cav. var. *triangulata* (Buckley) Munz]

Gaura villosa Torr., (soft hairy), WOOLLY GAURA. Perennial 0.6–1.8 m tall, with many erect branches from woody base, below inflorescence usually densely villous; leaves narrowly lanceolate to very narrowly elliptic or linear, sinuate-dentate to subentire; inflorescence strigulose, glandular-pubescent, or hirtellous; flowers opening near sunset and withering the next morning; petals 8.5–13 mm long, white; anthers 2.5–4.5 mm long; fruit body 9–18 mm long; stipe 2–8 mm long. Dunes and sandy flats; Cooke, Montague, and Somervell cos., also Hood Co. (R. O'Kennon, pers. obs.); nw part of nc TX w through Plains Country. Mar–Jun.

LUDWIGIA SEEDBOX, WATER-PRIMROSE, FALSE LOOSESTRIFE

Usually perennial herbs of damp ground or shallow water; stems creeping or floating to erect; leaves entire or minutely toothed; flowers 4- or 5-merous, axillary; sepals persistent in fruit; petals yellow, sometimes conspicuous, or absent; stamens as many as or twice the number of sepals; stigma capitate or globose; fruit a capsule.

A cosmopolitan, especially American genus of 82 species of wet habitats. Some are cultivated as ornamentals, including in aquaria. Some species were formerly segregated into the genus *Jussiaea*; that genus was named by Linnaeus for the brothers Antoine Laurent (1686–1758) and Bernard (1699–1777) de Jussieu, French botanists and important early proponents of the "natural system" of classification that grouped related plants; this was a significant step forward from Linnaeus' "artificial system" of classification. (*Ludwigia* is named for Christian Gottlieb Ludwig, 1709–1772, German botanist and physician)
REFERENCES: Munz 1944; Raven 1963; Ramamoorthy & Zardini 1987; Peng 1988, 1989; Zardini & Raven 1992.

1. Leaves opposite.
   2. Petals absent; capsules with broad green bands at the corners, sessile — **L. palustris**
   2. Petals present; capsules without green bands, on pedicels 0.3–1.5 mm long — **L. repens**
1. Leaves alternate.
   3. Stems conspicuously 4-winged (with 4 narrow bands or wings of leaf-like tissue along the stems) — **L. decurrens**
   3. Stems not winged.
      4. Sepals 5; petals 5; plant prostrate, floating, or erect.
         5. Pedicels 10–60(–80) mm long; branches prostrate or floating; seeds not falling from capsule, fused in large mass of endocarp — **L. peploides**
         5. Pedicel 2–20 mm long; branches erect; seeds readily falling from capsule, not fused in a large mass of endocarp (each seed with a small horseshoe-shaped piece of endocarp) — **L. leptocarpa**
      4. Sepals 4; petals 4 or absent; plants erect or nearly so.
         6. Stamens 8; capsules 17–45 mm long; known locally only from Bell Co. in s part of nc TX — **L. octovalvis**
         6. Stamens 4; capsules 2–10(–12) mm long; widespread in nc TX.
            7. Capsules nearly globose, on short pedicels 3–5 mm long; petals present — **L. alternifolia**
            7. Capsules subcylindric or elongate obpyramidal (not globose), sessile or subsessile; petals present OR absent.

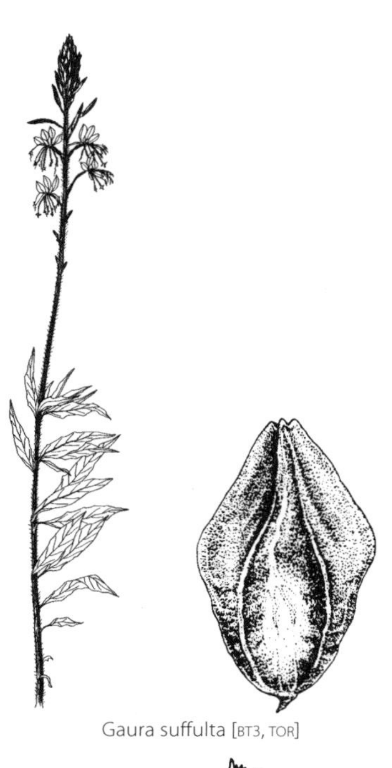

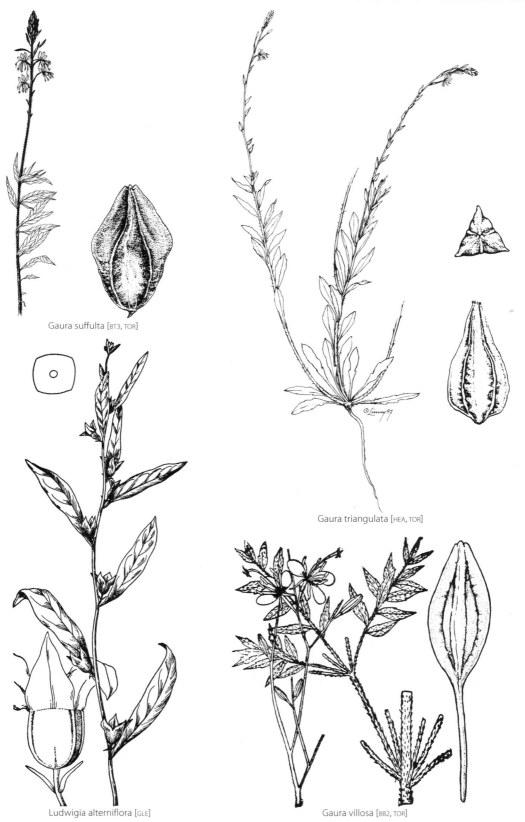

Gaura suffulta [BT3, TOR]

Gaura triangulata [HEA, TOR]

Ludwigia alterniflora [GLE]

Gaura villosa [BB2, TOR]

8. Petals present; leaf blades linear, 1.5–5(–8.5) mm wide, subsessile or with petioles to 5 mm long ______ **L. linearis**
8. Petals absent; leaf blades narrowly elliptic, 4–20 mm wide, with petioles 2–10 mm long ______ **L. glandulosa**

Ludwigia alternifolia L., (with alternate leaves), SEEDBOX, RATTLE-BOX, BUSHY SEEDBOX. Plant erect, to 1 m tall, subglabrous or rather minutely pubescent, at least in upper part; leaf blades lanceolate; petioles usually 3-7 mm long; flowers (at least upper ones) in the axils of reduced leaves or bracts; calyx lobes 4; petals 8-10 mm long, showy, but falling easily, lasting less than one day; capsules 5-6 mm long, on pedicels 3-5 mm long. Wet places; Denton, Fannin, Grayson, and Tarrant cos., also Lamar Co. (Carr 1994); se and e TX w to East Cross Timbers. Jun–Sep, rarely as early as Apr.

Ludwigia decurrens Walter, (running down the stem, from the leaf-like tissue along the stems), PRIMROSE-WILLOW, UPRIGHT PRIMROSE-WILLOW. Plant subglabrous, erect, to 2 m tall; leaves subsessile; leaf blades lanceolate to elliptic; sepals and petals 4; petals 8–12 mm long; stamens 8; capsules 10–20 mm long, on a pedicels 0–10 mm long. Wet places; Denton, Fannin, and Hopkins cos., also Dallas (Munz 1961) and Lamar (Carr 1994) cos.; se and e TX w to nc TX. Jun–Oct.

Ludwigia glandulosa Walter, (glandular), TORREY'S SEEDBOX, CREEPING SEEDBOX, CYLINDRIC-FRUIT LUDWIGIA. Plant glabrous to slightly hairy, to 1 m tall; leaf blades lanceolate to elliptic; petioles 2–10 mm long; petals absent; capsules 2–8 mm long, sessile. Wet places; Denton, Grayson, and Montague cos.; se and e TX w to nc TX. Jun–Oct.

Ludwigia leptocarpa (Nutt.) H. Hara, (thin-fruited), ANGLE-STEM WATER-PRIMROSE. Similar to *L. octovalvis*; plant hairy, to 1.5 m tall; leaf blades lanceolate to broadly so; petioles 2–35 mm long; flowers 5-merous, rarely with only 4 sepals and petals; petals 5–11 mm long; stamens usually 10; capsules 15–50 mm long, on pedicels 2–20 mm long; seeds 1–1.2 mm long, in a horseshoe-shaped piece of endocarp. Wet places; Dallas, Johnson, and Tarrant cos.; se and e TX w to East Cross Timbers. Aug–Oct.

Ludwigia linearis Walter, (narrow, with sides nearly parallel), NARROW-LEAF SEEDBOX, LINEAR-LEAF LUDWIGIA. Plant glabrous to densely minutely strigillose or puberulent; stems erect, often well-branched, to 1 m or more tall; petals 3–6 mm long; capsules 5–10(–12) mm long, sessile; seeds 0.45–0.65 mm long. Wet places; Johnson and Lamar cos. (Peng 1989); se and e TX w to nc TX.

Ludwigia octovalvis (Jacq.) P.H. Raven, (eight-valved), SHRUBBY WATER-PRIMROSE, NARROW-LEAF WATER-PRIMROSE. Plant nearly glabrous to hairy, to 1 m or more tall; leaf blades narrowly lanceolate to narrowly ovate; petioles 0–10 mm long; petals 5–16 mm long; capsules 17–45 mm long, on pedicels 0–10 mm long; seeds 0.6–0.75 mm long, not in a horseshoe-shaped piece of tissue. Wet places; Bell Co. in s part of nc TX; mainly s 1/2 of TX. Jul–Oct.

Ludwigia palustris (L.) Elliott, (of marshes), MARSH-PURSLANE, AMERICAN SEEDBOX, MARSH SEEDBOX. Plant ascending or creeping and rooting at nodes; leaf blades elliptic-lanceolate, acute, entire or subentire; petioles 2–10 mm long; flowers sessile; petals absent; calyx lobes 4; capsules 2–8 mm long. Wet places; Fannin and Hopkins cos., also Lamar (Carr 1994) and Somervell (R. O'Kennon, pers. comm.) cos. and Fort Hood (Bell or Coryell cos.—Sanchez 1997); se and e TX w to nc TX, also Edwards Plateau and Trans-Pecos. Jun–Oct. Jones et al. (1997) did not list this species for TX.

Ludwigia peploides (Kunth) P.H. Raven, (resembling *Peplis* which is now treated in *Lythrum* in the Lythraceae), WATER-PRIMROSE, SMOOTH WATER-PRIMROSE, FLOATING EVENING-PRIMROSE, VERDOLAGA DE AGUA, PRIMROSE-WILLOW. Plant trailing in shallow water or creeping in mud, with flowering branches somewhat ascending, glabrous; leaf blades oblong-elliptic or -lan-

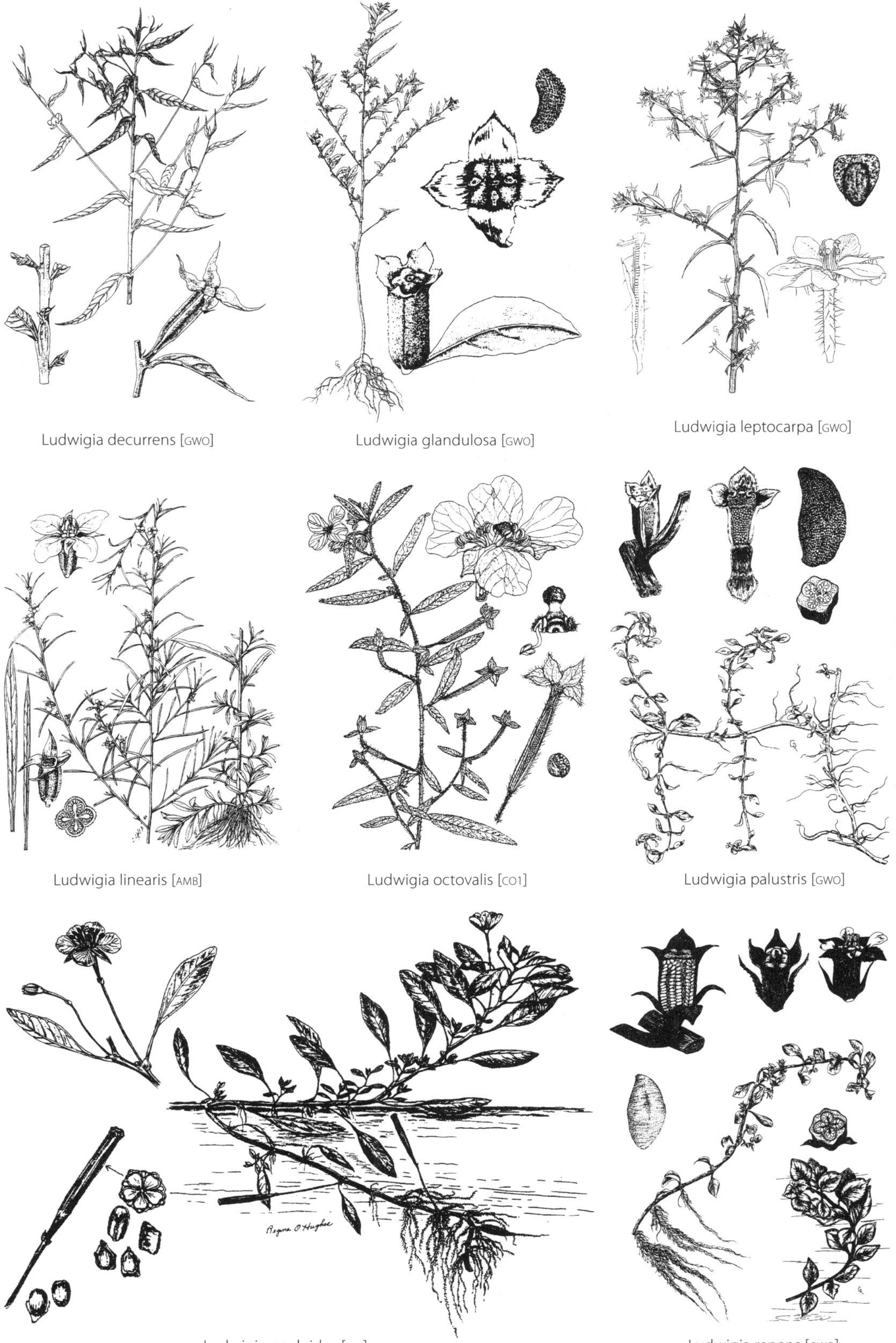

Ludwigia decurrens [GWO]

Ludwigia glandulosa [GWO]

Ludwigia leptocarpa [GWO]

Ludwigia linearis [AMB]

Ludwigia octovalis [CO1]

Ludwigia palustris [GWO]

Ludwigia peploides [REE]

Ludwigia repens [GWO]

ceolate, entire; petioles 2–40 mm long; flowers axillary, solitary; calyx lobes 5; petals 7–14(–24) mm long, showy; capsules 10–40 mm long. Mid-May–Oct. Ponds, tanks, streams, other wet areas; se and e TX w to West Cross Timbers and Edwards Plateau. We are following McGregor (1986) in lumping [*L. peploides* subsp. *glabrescens* (Kuntze) P.H. Raven and *Jussiaea repens* L. var. *glabrescens* Kuntze]; also [*L. peploides* var. *glabrescens* (Kuntze) Shinners]. Jones et al. (1997) listed only subsp. *glabrescens* as occurring in TX. ▣/97

Ludwigia repens J.R. Forst., (creeping), ROUND-LEAF SEEDBOX, FLOATING PRIMROSE-WILLOW, CREEPING PRIMROSE-WILLOW, FLOATING WATER-PRIMROSE, CREEPING WATER-PRIMROSE. Plant trailing in shallow water or creeping in mud, rooting at the nodes, glabrous or puberulent; leaf blades narrowly elliptic to subrotund; petioles 3–25 mm long; flowers 4-merous; petals 4–5 mm long; capsules 3.3–7.5 mm long, on pedicels 0.3–1.5 mm long. Damp ground or shallow water; Bell and Williamson cos., also Dallas and Tarrant cos. (Munz 1961); widespread in TX. Jul–Sep. [*L. natans* Elliott, *L. natans* var. *rotundata* (Griseb.) Fernald & Griscom]

OENOTHERA EVENING-PRIMROSE, SUNDROPS

Annual or perennial herbs; flowers in leaf axils or in ± distinct inflorescences, radially symmetrical, 4-merous, opening near sunset or near sunrise; hypanthium well-developed, extending beyond the ovary; calyx lobes reflexed, not persistent in fruit; petals yellow (these sometimes drying or fading to pinkish or reddish), white to pink or rose-purple; stamens 8; stigma deeply 4-lobed; fruit a dehiscent or nut-like and indehiscent capsule.

A genus of 124 species of the Americas, especially in temperate areas. The genus is well known for its complicated genetics including reciprocal translocation and structural heterozygotes; many species have unusual meiotic chromosome configurations (Dietrich & Wagner 1988). Oil of *Oenothera* is an important ingredient used in many cosmetics including lipstick. Many are cultivated as ornamentals; the flowers often open and are scented in the evening for moth pollination. (Greek: *oinotheras*, name used by Theophrastus for some species of *Epilobium*, possibly from wine-scenting (*oeno*, wine), in allusion to an ancient use of the roots)
REFERENCES: Gates 1958; Straley 1977; Ellstrand & Levin 1980; Raven et al. 1979; Wagner 1983, 1986; Dietrich & Wagner 1987, 1988; Harte 1994; Dietrich et al. 1997.

1. Hypanthium 0.2–5 cm long; ovaries and fruits neither deeply sharp-angled nor winged.
  2. Petals light to deep yellow (withering white or reddish).
    3. Flowers many, in dense, terminal, head-like or finally elongate spikes.
      4. Leaf blades 10–60 mm wide; petals obovate to obtriangular, usually broadest near apex; capsules 3–6 mm thick at base; ovules and seeds horizontal in locule, sharply angled ____ **O. biennis**
      4. Leaf blades 3–15 mm wide; petals broadly elliptic to nearly rhombic, usually broadest near middle; capsules 2–4 mm thick at base; ovules and seeds ascending in locule, not sharply angled.
        5. Free sepal tips 2–6 mm long; flower buds with wide-spreading hairs to nearly glabrous, the hairs pustulate-based (= swollen at base)—use dissecting scope; mature lower buds usually extending past youngest buds at end of spike ____ **O. heterophylla**
        5. Free sepal tips usually 0.5–1.5 mm long; flower buds with appressed hairs, the hairs not pustulate; mature lower buds usually not extending past youngest buds at end of spike ____ **O. rhombipetala**
    3. Flowers solitary, or few in loose terminal spikes.
      6. Hypanthium as long as the ovary or longer; fruits cylindrical, usually > 10 mm long.
        7. Stem leaves nearly entire, with only a few small teeth; plants biennial, 0.5–2 m tall; ovules and seeds horizontal in locule, sharply angled; mature capsules 3–6 mm in diam. at base ____ **O. biennis**

7. Stem leaves sinuate-dentate to sinuate-pinnatifid, rarely nearly entire; plants annual, 0.1–0.6(–1) m tall; ovules and seeds ascending in locule, not sharply angled; mature capsules 2–4 mm in diam. at base.
    8. Petals 5–22 mm long; sepals 5–12(–15) mm long; stigma surrounded by anthers at flowering time; anthers 2–6 mm long ____ **O. laciniata**
    8. Petals 25–40 mm long; sepals 15–30 mm long; stigma elevated above anthers at flowering time; anthers 4–11 mm long ____ **O. grandis**

6. Hypanthium shorter than the ovary; fruits club-shaped, the lower part narrower, the upper part swollen, 10 mm or less long.
    9. Flowers terminal, borne above the leaves; leaf blades (except lowest) linear, 0.2–1.2(–3) mm wide; hypanthium 1–2 mm long; sepals 1.5–2 mm long, without free tips; petals 3–5(–7) mm long; capsules usually 4–6 mm long ____ **O. linifolia**
    9. Flowers lateral and finally terminal, in the axils of leaves or leafy bracts; leaf blades oblong-lanceolate to oblong-linear, mostly 2–10 mm wide; hypanthium 4–10 mm long; sepals 4–10 mm long, with free tips 1–1.5 mm long; petals 5–15 mm long; capsules 5–15 mm long ____ **O. spachiana**

2. Petals white to rosy lavender with yellowish base ____ **O. speciosa**

1. Hypanthium 2–12.5 cm long (5 cm or more long except in *O. triloba*); ovaries and fruits deeply sharp-angled and winged (except neither deeply sharp-angled nor winged in *O. jamesii*).
    10. Plants essentially stemless or with suberect or trailing stems to 0.5 m or less long; ovaries and capsules deeply sharp-angled and winged, often conspicuously so.
        11. Plants winter annuals; petals 1–2 cm long; capsules winged mostly above middle, borne at base of plant ____ **O. triloba**
        11. Plants perennials; petals 2–5 cm long; capsules winged their entire length, borne at base of plant or along the stems.
            12. Wings of capsules 4–6 mm wide; capsules 2–3 cm long; plant essentially stemless or caespitose (clumped) with a short stem; leaf blades nearly entire to pinnatifid; limited to extreme w part of nc TX (Callahan Co.) ____ **O. coryi**
            12. Wings of capsules 7–20 mm wide; capsules 2–6.5 cm long; plant usually with distinct stems to 50 cm long; leaf blades nearly entire to with scattered small teeth; widespread in nc TX ____ **O. macrocarpa**
    10. Plants with erect stems usually 1–3 m tall; ovaries and capsules neither deeply sharp-angled nor winged ____ **O. jamesii**

Oenothera biennis L., (biennial, living two years and flowering in the second), COMMON EVENING-PRIMROSE. Erect biennial to 2 m tall; inflorescence a terminal spike, often with short branches; flowers opening near sunset; hypanthium 2–5 cm long; buds with free sepal tips 1–4 mm long; petals yellow; capsules cylindric, tapering towards apex, 1.4–2.5 cm long. Disturbed wooded and weedy areas; Cooke, Grayson and Hopkins cos., also Tarrant Co. (Mahler 1988); e TX w to nc TX. Aug–Oct.

Oenothera coryi W.L. Wagner, (for Victor Louis Cory, 1880–1964, TX botanist at TX A&M and Southern Methodist universities). Perennial from a woody taproot; plant essentially stemless or caespitose with a short stem; pubescence hirsute and strigulose; leaf blades nearly entire to remotely pinnately lobed; flowers opening near sunset; hypanthium (5.5–)7.5–10(–12.5) cm long; buds with free sepal tips 0.7–1.2 mm long; petals broadly obovate, 3.5–4.3 cm long, yellow, fading orange. Open grasslands or disturbed areas; Callahan Co. in extreme w part of nc TX (Wanger 1986); nc TX w to Panhandle; endemic to TX. Apr–May. [*O. brachycarpa* A. Gray var. *typica* sensu Munz, not Munz (as to the type)]

Oenothera grandis (Britton) Smyth, (large, big). Annual to 0.6(–1) m tall; flowers opening near sunset; hypanthium 2.5–5 cm long; petals deep yellow; capsules (1–)2.5–3.5(–5) cm long, cylindric. Open sandy areas; Dallas and Limestone cos. w to Panhandle and s to s TX. Apr–Jun.

Oenothera heterophylla Spach, (various-leaved), VARIABLE EVENING-PRIMROSE. Annual or short-lived perennial to ca. 0.7 m tall; flowers opening near sunset; hypanthium 2.5–3 cm long; petals yellow; sepals 1.5–3 cm long; petals 2.5–3.5 cm long; capsules 1.3–2.5 cm long, cylindric. Sandy open woods; Williamson Co., also Dallas, Henderson, Hopkins, and Limestone cos. (Dietrich & Wagner 1988); se and e TX w to Blackland Prairie, also Edwards Plateau. May–Sep.

Oenothera jamesii Torr. & A. Gray, (presumably for Edwin James, 1797–1861, surgeon-naturalist, first botanical collector in CO and first known botanical collector in TX, with Major Long's expedition to the Rocky Mts. in 1819–1820), TRUMPET EVENING-PRIMROSE. Robust biennial, appressed pubescent; stems erect, to 3 m tall; inflorescence unbranched or few-branched; flowers opening near sunset, very large; hypanthium 6–11 cm long; sepals 4–6 cm long, in bud with free tips 3–6 mm long; petals yellow, fading reddish, 3.5–5 cm long; capsules 2–5 cm long, cylindric. Stream banks or other moist situations; Bell, Coryell (Fort Hood—Sanchez 1997), Somervell, Tarrant (R. O'Kennon, pers. obs.), and Williamson (Balcones Canyonlands Nat. Wildlife Refuge, C. Sexton, pers. comm.) cos.; mainly Edwards Plateau w to Trans-Pecos. Jul–Oct.

Oenothera laciniata Hill, (laciniate, torn), CUT-LEAF EVENING-PRIMROSE, DOWNY EVENING-PRIMROSE, SINUATE-LEAF EVENING-PRIMROSE. Annual or short-lived perennial to ca. 0.5 m tall; flowers opening late afternoon to late morning; hypanthium 1.5–3.5 cm long; petals light yellow; capsules 2–5 cm long. Disturbed soils; nearly throughout TX. Apr–Jun.

Oenothera linifolia Nutt., (with leaves like *Linum*—flax), THREAD-LEAF SUNDROPS. Slender, low (to 0.3 m tall), glabrous, small-flowered annual; leaf blades entire, filiform, to 3 mm wide (usually less); flowers opening near sunrise; hypanthium 1.5–2 mm long; sepals 1.5–2 mm long; petals 3–5 mm long, yellow. Sandy open woods; Kaufman and Lamar cos., also Dallas Co. (Mahler 1988); se and e TX w to nc TX. Apr–May.

Oenothera macrocarpa Nutt., (large-fruited), MISSOURI-PRIMROSE. Perennial with minutely gray-pubescent to glabrous, suberect to trailing stems to 0.5 m long or nearly acaulescent; leaves green to silvery or gray with dense, appressed pubescence; flowers opening near sunset; hypanthium 5–12 cm long; sepals 2–4 cm long, in bud with free tips 1.5–8 mm long; petals 2–5 cm long, yellow, sometimes fading reddish; capsules 2–6.5 cm long, conspicuously 4-winged. Limestone outcrops or prairies. Apr–Jun.

1. Plants completely glabrous ____ subsp. **oklahomensis**
1. Plants with appressed pubescence on either stems and buds or on whole plant.
   2. Leaf blades linear-lanceolate to broadly lanceolate, greenish, only the youngest tissue gray hairy; widespread in nc TX ____ subsp. **macrocarpa**
   2. Leaf blades ovate to broadly lanceolate, conspicuously gray-hairy, the entire plant with thick pubescence; possibly e to w margin of nc TX ____ subsp. **incana**

subsp. **incana** (A. Gray) W.L. Wagner, (hoary, quite gray), MISSOURI-PRIMROSE. While subsp. *incana* is generally found to the w of nc TX, a BRIT/SMU sheet from Callahan Co. on our w margin was annotated by W.L. Wagner as subsp. *macrocarpa* intermediate to subsp. *incana*. Edwards Plateau to nw TX.

subsp. **macrocarpa**, FLUTTER-MILL. Nc TX to Edwards Plateau. [*O. missouriensis* Sims] ❀/101

subsp. **oklahomensis** (Norton) W.L. Wagner, (of Oklahoma), OZARK SUNDROPS, GLADE-LILY. Cooke Co.; mainly Rolling Plains to Panhandle. [*Megapterium oklahomense* Norton, *O. macrocarpa* var. *oklahomensis* (Norton) Reveal, *O. missouriensis* Sims var. *oklahomensis* (Norton) Munz]

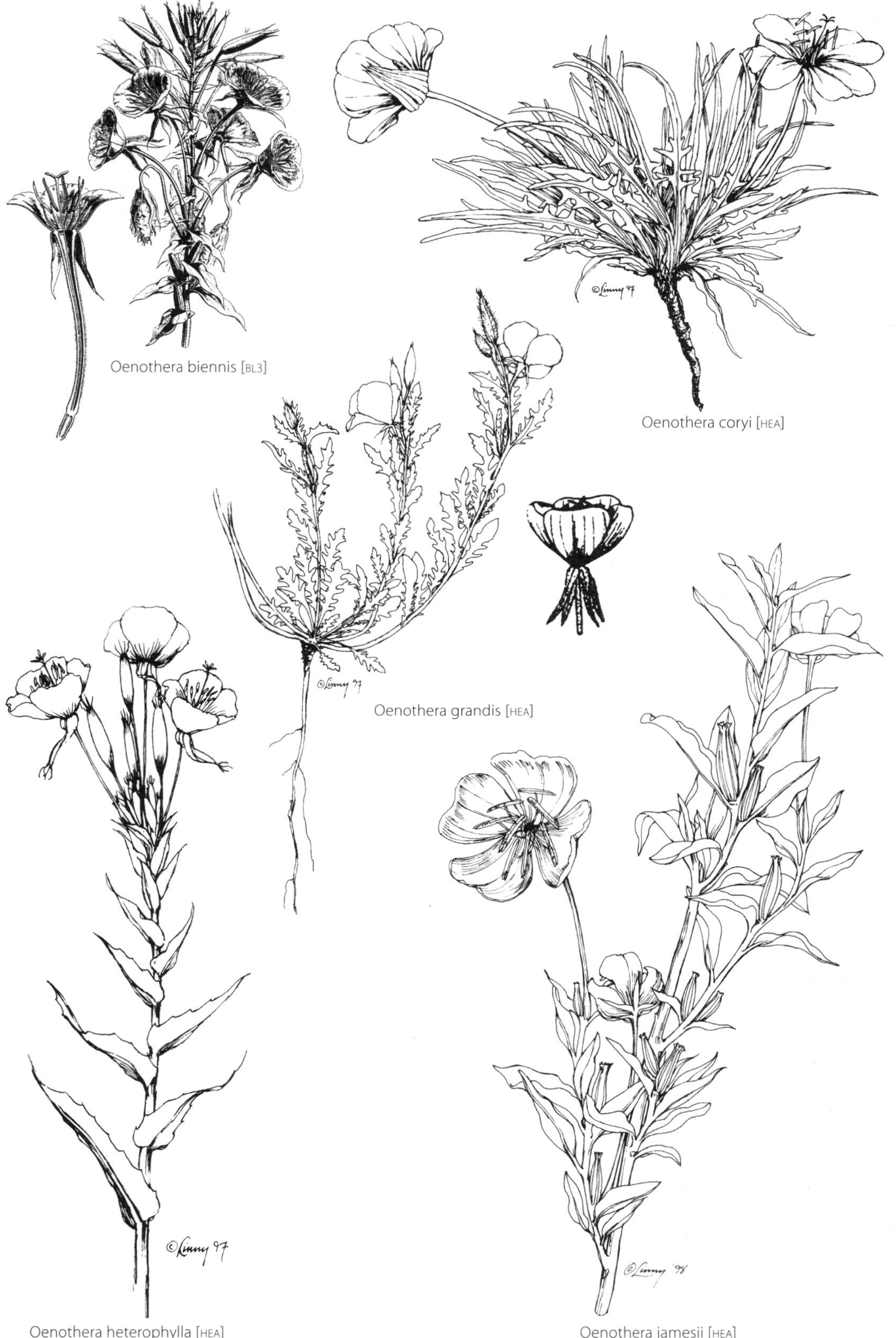

Oenothera biennis [BL3]

Oenothera coryi [HEA]

Oenothera grandis [HEA]

Oenothera heterophylla [HEA]

Oenothera jamesii [HEA]

Oenothera rhombipetala Nutt. ex Torr. & A. Gray, (with rhomboid or diamond-shaped petals), FOUR-POINT EVENING-PRIMROSE. Biennial 0.3–1(–1.5) m tall; flowers opening near sunset; hypanthium 2.5–3 cm long; sepals 1.5–3 cm long; petals yellow, 1.5–3.5 cm long; capsules cylindric, 1.3–2.5 cm long. Sandy disturbed areas; n 1/2 of Texas. May–Sep.

Oenothera spachiana Torr. & A. Gray, (for Édouard Spach, 1801–1879, French (Alsatian) botanist), SPACH'S EVENING-PRIMROSE. Low (to 0.5 m tall), appressed-pubescent annual; leaf blades entire or nearly so; flowers opening near sunrise; hypanthium 4–10 mm long; petals yellow. Sandy areas, often open woods; Denton, Grayson, Hunt, Navarro, and Tarrant cos., also Kaufman and Lamar cos. (Munz 1961); se and e TX w to East Cross Timbers. Apr–May.

Oenothera speciosa Nutt., (showy, good-looking), SHOWY-PRIMROSE, BUTTERCUP, TEXAS-BUTTERCUP, SHOWY EVENING-PRIMROSE, MEXICAN EVENING-PRIMROSE, MEXICAN-PRIMROSE, WHITE EVENING-PRIMROSE, AMAPOLA DEL CAMPO. Low perennial usually with several pubescent stems; leaf blades toothed to deeply pinnatifid; flowers opening in evening and morning, white to pink or rose-purple, also turning reddish or rose-purple with age; hypanthium 1–2 cm long; sepals 1.5–3 cm long; petals 2.5–4 cm long; capsules clavate, 1–1.5 cm long, the terminal portion ribbed. Low disturbed areas, especially roadsides; nearly throughout TX. Groups of thousands of individuals make extremely vivid late spring displays; this species is one of our showiest and most abundant wildflowers. According to McGregor (1986), in the Great Plains the white, evening-opening individuals are diploid, while the rose-purple, morning-opening individuals are tetraploid. Apr–Jul. ▣/101

Oenothera triloba Nutt., (three-lobed), STEMLESS EVENING-PRIMROSE, THREE-LOBED-PRIMROSE. Nearly stemless winter annual with numerous basal rosette leaves varying from entire to deeply lobed; flowers opening near sunset; hypanthium 2–10 cm long; sepals 1–1.8 cm long; petals pale yellow; capsules obpyramidal, 1–2 cm long, 4-winged apically, borne at base of plant. Grassy areas, disturbed soils, lawn weed; Blackland Prairie w to Rolling Plains and s to Edwards Plateau. Mar–Apr.

Oenothera brachycarpa A. Gray, (short-fruited), which occurs just to the w of nc TX, is similar to *O. coryi*. It differs as follows: leaves with a terminal lobe; hypanthium (9–)12–21(–22) mm long; buds with free sepal tips 1–7 mm long; petals broadly rhombic-obovate.

Oenothera mexicana Spach, (of Mexico), and *O. fulfurriae* W. Dietr. & W.L. Wagner, (for Falfurrias, where the type was collected, in Brooks Co. in se TX), cited by Hatch et al. (1990) for vegetational areas 4 and 5, and 4, respectively, apparently occur only to the se of nc TX (Dietrich & Wagner 1987, 1988). *Oenothera brachycarpa* A. Gray (WRIGHT'S EVENING-PRIMROSE, SHORTPOD EVENING-PRIMROSE) and *O. pubescens* Willd. ex Spreng., cited by Hatch et al. (1990) for vegetational area 5, and 4 and 5, respectively, apparently occur only to the w and s of nc TX (Wagner 1986; Dietrich & Wagner 1988).

STENOSIPHON FALSE GAURA

A monotypic genus of the c and s U.S. (Greek: *stenos*, slender, and *siphon*, a tube)

Stenosiphon linifolius (Nutt. ex E. James) Heynh., (with leaves like *Linum*—flax), FALSE GAURA, FLAX-LEAF STENOSIPHON, TALL GAURA. Glabrous, brittle-stemmed, slender biennial or perennial 0.6–3 m tall, only rarely branched below inflorescence; leaves narrowly lanceolate, 3–8(–10) cm long, 4–28 mm wide, sessile, acute, entire; flowers in panicled terminal spikes; petals 4, white, 4–6 mm long; stamens 8; fruit a 1-seeded capsule 2–4 mm long, ca. 2 mm thick. Limestone outcrops, clay soils; in nc TX Dallas to Grayson cos. w through West Cross Timbers; nc TX s to Edwards Plateau and nw to Panhandle. May–Jul, less freely to Oct.

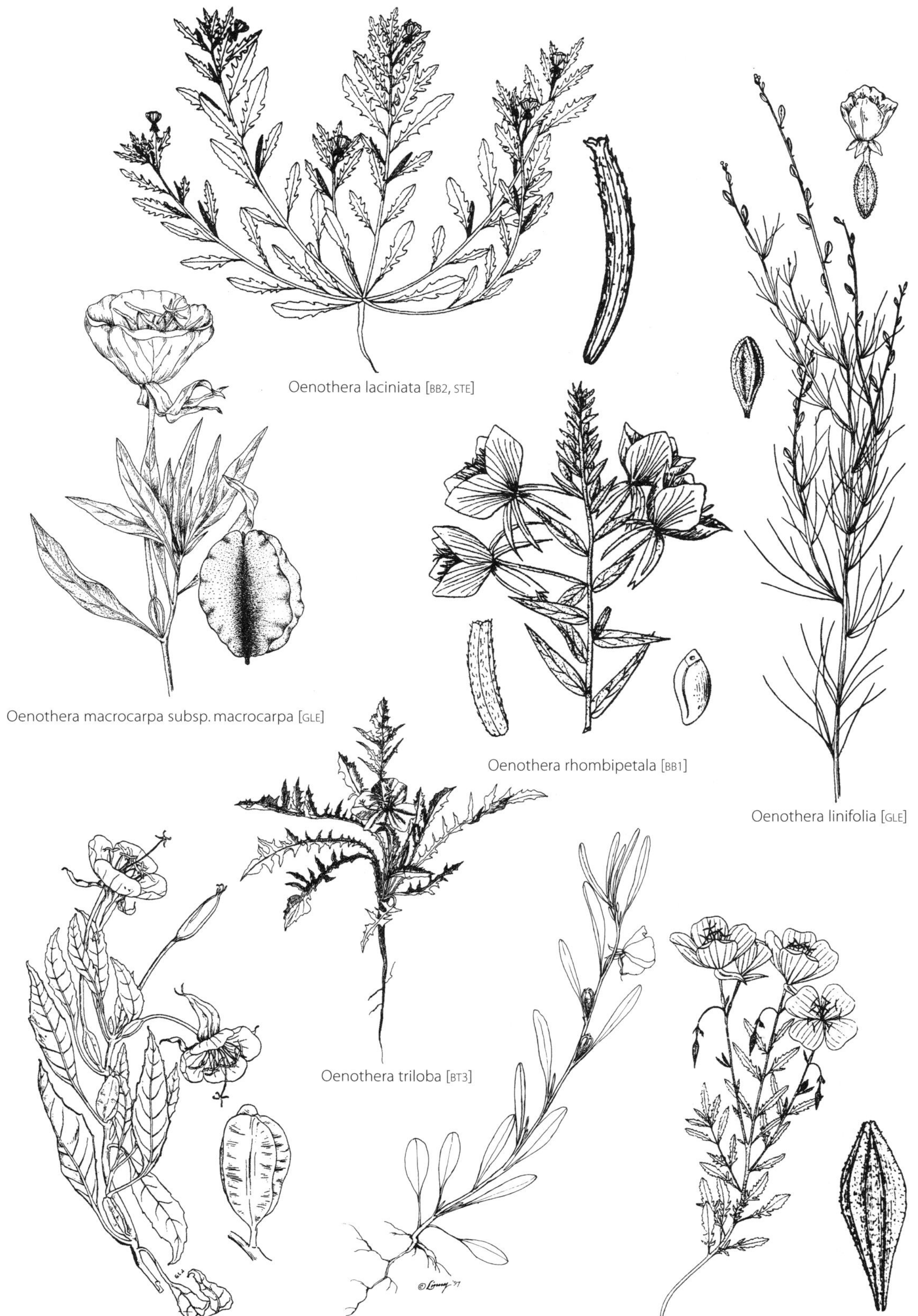

Oenothera laciniata [BB2, STE]

Oenothera macrocarpa subsp. macrocarpa [GLE]

Oenothera rhombipetala [BB1]

Oenothera linifolia [GLE]

Oenothera triloba [BT3]

Oenothera macrocarpa subsp. oklahomensis [ARM]

Oenothera spachiana [HEA]

Oenothera speciosa [BB2, STE]

OROBANCHACEAE BROOMRAPE FAMILY

A small (210 species in 15 genera) family of root parasites lacking chlorophyll. The species occur in the n hemisphere, especially in temperate regions and the Old World subtropics; some can be parasitic pests on cultivated crops. This family is similar in many respects to, and is sometimes included in, the Scrophulariaceae (which also includes some parasitic or hemiparasitic taxa). Two other genera, *Epifagus* (BEECHDROPS) and *Conopholis* (SQUAWROOT), occur in other parts of TX. (subclass Asteridae)

FAMILY RECOGNITION IN THE FIELD: glandular-pubescent *root parasites* with short, fleshy, erect stems and scale-like leaves; plants *lacking chlorophyll* (and therefore not green), variously colored; corollas bilabiate, with a slightly curved tube.

REFERENCES: Thieret 1969a, 1971.

OROBANCHE BROOMRAPE

Perennial, fleshy, herbaceous plants without chlorophyll, parasitic on vascular plant roots, yellowish to brownish or purplish, glandular-pubescent; stems fleshy; leaves alternate, sessile, scale-like; flowers solitary on axillary pedicels or in spikes; corollas with a slightly curved tube, bilabiate; stamens 4, didynamous; fruit a 2-valved capsule; seeds numerous, minute.

A genus of 150 species of n temperate and warm areas of the world; a number are restricted to certain families as hosts, rarely to just 1 species; HEMP, TOBACCO, TOMATO, EGGPLANT, COTTON, and many legumes are parasitized by *Orobanche* species (Thieret 1969a). The inflorescences of some were eaten by Native Americans. (Greek: *orobas*, vetch, and *anchein*, to strangle, from the parasitic habit)

REFERENCES: Munz 1931; Achey 1933; Collins 1973.

1. Flowers solitary on 2–12 naked, erect, long (3–12 cm) pedicels from leaf axils; bracts absent at base of calyces; stems to 10 cm tall ____ **O. fasciculata**
1. Flowers sessile in dense spikes or on pedicels shorter than the corollas; bracts 1(–2) at the base of calyces; stems 5–30(–50) cm tall ____ **O. ludoviciana**

Orobanche fasciculata Nutt., (fascicled, clustered), CHESTER BROOMRAPE. Parasitic on various species including Asteraceae; lower portion of plant becoming subligneous to woody; calyces 5–8.5 mm long, the lobes triangular; corollas purple (rarely yellowish), to ca. 20(–30) mm long, with small semiorbicular lobes. Bare limestone prairies; Montague Co. (Correll & Johnston, 1970); mainly w TX. Mar–May. [*O. fasciculata* Nutt. var. *subulata* Goodman]

Orobanche ludoviciana Nutt. subsp. **multiflora** (Nutt.) Collins, (sp.: of Lousiana; subsp.: many-flowered), LARGE-FLOWER BROOMRAPE. Parasitic on members of Asteraceae; calyces 8–17(–19) mm long, exceeding the fruits, the lobes lance-linear to attenuate; corollas 15–35 mm long, pale purple to rose or yellow, the lobes obtusely rounded. Sandy areas, sandy prairies, and gypsum soils; Bell and Tarrant (Ft. Worth Nature Center) cos; also Dallas (S. Lusk, pers. comm.) and Hamilton (HPC) cos.; widespread in TX. (Mar–)Apr–Jul(–Sep). Two varieties of *O. multiflora* separated as follows are here put in synonymy: calyces 12–17 mm long; corollas 25–35 mm long, the upper lip purple, 9–12 mm long (var. *multiflora*) vs. calyces 8–10 mm long; corollas 20–25 mm long, the lips rose purple or lighter, 5–7 mm long (var. *pringlei*). [*O. ludoviciana* Nutt. var. *multiflora* (Nutt.) Beck., *O. multiflora* Nutt. var. *multiflora*, *O. multiflora* Nutt. var. *pringlei* Munz] Thieret (1969a) indicated that Texas material of this species is perplexingly variable.

Orobanche ludoviciana subsp. *ludoviciana*, currently regarded as a Great Plains taxon, is known from w TX. It differs from subsp. *multiflora* in having the calyces shorter than to equaling the fruits and the corolla lobes triangularly acute.

OXALIDACEAE WOODSORREL FAMILY

A medium-sized (775 species in 6 genera) family of some small trees and shrubs, but mostly herbs with tubers or bulbs; they occur from the tropics to a few in temperate areas; oxalates are usually accumulated and the leaves are often folded together at night. The family contains *Averrhoa carambola* L. (CARAMBOLA, STARFRUIT), a tropical Asian tree cultivated for its fruits; the fruits, which resemble giant fleshy *Oxalis* fruits, are now commonly available in supermarkets. (subclass Rosidae)

FAMILY RECOGNITION IN THE FIELD: herbs with palmately compound, *clover-like* leaves and sour sap; flowers radially symmetrical, 5-merous, usually either yellow or ± pink; stamens united basally; fruit a capsule.

REFERENCES: Small 1907a; Robertson 1975.

OXALIS WOODSORREL

Ours annual or perennial herbs with sour sap due to the presence of oxalic acid; leaves basal or alternate, palmately compound, superficially resembling those of clover, leaflets 3, crescent-shaped, obovate, or obcordate, downwardly folded together at night or in bad weather; stipules small or absent; flowers in peduncled, axillary or basal, umbel-like cymes, closing at night and in bad weather, nodding before and after blooming; sepals 5; petals 5; stamens 10; pistil 5-carpellate; pedicels often reflexed in fruit; capsules narrowly cylindrical.

A genus of 700 species, cosmopolitan in distribution, but especially South America and s Africa. Many species have tubers and are thus difficult to eradicate, weedy pests. A number of *Oxalis* species are cultivated as ornamentals and the tubers of one, the Peruvian *O. tuberosa* Molina, OCA, are eaten as a vegetable. The Irish SHAMROCK is said by some to be *O. acetosella* L.; however, it is more likely that either *Trifolium dubium* or *T. repens* is actually the SHAMROCK (Nelson 1991). The leaves of some species have an acidic sour taste and are eaten by some people; ☠ however, they possess soluble oxalates which are toxic to humans and can cause colic, coma, and even death in animals if consumed in large quantities (Lewis & Elvin-Lewis 1977). The leaves of *Oxalis* species exhibit sleep movements with the leaflets folding downward at dusk or in cloudy weather (Wills & Irwin 1961). (Greek name for sorrel, from *oxys*, acid, alluding to the taste of the leaves)

REFERENCES: Wiegand 1925; Eiten 1955, 1963; Shinners 1956d; Lourteig 1979; Turner 1994e.

There is considerable confusion regarding the North American yellow-flowered species of *Oxalis* section *Corniculatae* DC. both in terms of the number of taxa to recognize and the appropriate names to use. A variety of authors cited above have worked on the problem, but more work on this complex is clearly needed. Within nc TX there are at least two entities, 1) *O. corniculata* var. *corniculata*, a creeping plant with dark green-purplish leaves and visible stipules (fused with base of petioles), which is apparently found mainly in disturbed or cultivated areas around houses and landscapes, and 2) another entity with erect or sometimes creeping stems, lighter green leaves, and stipules inconspicuous to absent, which is widespread throughout nc TX. Turner (1994e) decided that this second entity (which has long gone under the name of *O. dillenii*) should be called *O. corniculata* var. *wrightii*. John Kartesz (pers. comm. 1997) has concluded that *O. dillenii* should be lumped with *O. stricta* and that *O. stricta* has nomenclatural priority. While we are unsure regarding the rank at which these two entities are best recognized, we do not want to cause further confusion by making additional combinations. As a result we are following J. Kartesz.

1. Plants stemless; leaves all basal; flowers pink to rose, violet, or pinkish purple (rarely white).
   2. Under surface of leaflets and pedicels pubescent; stems arising from a woody crown, with rhizomes and root-tubers; introduced ornamental ____ **O. articulata**

2. Under surface of leaflets and pedicels glabrous; stems arising from a scaly bulb; native species.
   3. Leaflets shallowly notched (apical lobes not longer than main part of blade); tubercles mostly 2 at tip of each sepal, usually distinct or sometimes confluent; throughout nc TX. ____ **O. violacea**
   3. Leaflets deeply lobed, crescent-shaped (apical lobes longer than main part of blade); tubercles up to 6 in a mass at tip of each sepal; mainly in s 1/2 of nc TX ____ **O. drummondii**

1. Plants leafy-stemmed; flowers yellow.
   4. Stems creeping and rooting at the nodes (but peduncles erect); leaves usually deep green with purplish pigmentation; stipules (which are fused to the petiole base) noticeable, together with base of petiole forming a short broad flange, often brownish or purplish ____ **O. corniculata**
   4. Stems usually ascending or ± erect OR creeping; leaves light green; stipules obscure or absent ____ **O. stricta**

Oxalis articulata Savigny subsp. **rubra** (A. St.-Hil.) Lourteig, (sp.: articulated, jointed; subsp.: red), WINDOWBOX WOODSORREL. Perennial; leaflets broadly obcordate (rarely shallowly notched), red-spotted beneath toward the margins or sometimes over the entire surface; sepals with tubercles at tips; petals 10–15 mm long, pink to rose (rarely white), with veins sometimes darker; capsules 6–8 mm long. Cultivated and persists or spreads; Tarrant Co. (R. O'Kennon, pers. obs.); also se and e TX. Apr–May. Native of Brazil. [*O. rubra* A. St.-Hil.]

Oxalis corniculata L., (horned), CREEPING LADIES'-SORREL, AGRITO, JOCOYOTE. Stems creeping and rooting at the nodes; peduncles usually from axils of leaves at rooted nodes; sepals without tubercles at tip; petals 4–8 mm long; capsules 8–20(–25) mm long. Disturbed or cultivated areas around houses and landscapes; Tarrant Co., also Fort Hood (Bell or Coryell cos.—Sanchez 1997); also se and s TX and Edwards Plateau. Radford et al. (1968) suggested many plants are found that are intermediate in growth habit between *O. corniculata* and *O. dillenii* (which we are treating as *O. stricta*); however, locally we have been able to distinguish the two. While *O. corniculata* is often considered native (e.g., Hatch et al. 1990), its place of origin is uncertain; it is possibly native to the Australasian region (Eiten 1963).

Oxalis drummondii A. Gray, (for its discoverer, Thomas Drummond, 1780–1835, Scottish botanist and collector in North America), DRUMMOND'S OXALIS. Glabrous perennial; bulb scales 3-ribbed; scapes to 30 cm tall, glabrous, with 4–8 flowers; sepals 5–7 mm long, the apical tubercles ± confluent; petals 15–23 mm long, pinkish purple; capsules 7–9 mm long. Sand or limestone soils, woodlands or prairies; Coryell, Dallas (*Reverchon*), and Somervell cos., also Brown Co. (HPC) and Fort Hood (Bell or Coryell cos.—Sanchez 1997); South TX Plains and Edwards Plateau, e to Gonzales Co., n to nc TX; endemic to TX. Flowering only in fall, with or rarely without leaves.

Oxalis stricta L., (upright, erect), SHEEP-SHOWERS, GRAY-GREEN WOODSORREL, DILLEN'S OXALIS. Perennial without rhizomes to ca. 25 cm tall, flowering the first year; taproot slender or stout; stems erect to creeping and mat-forming; leaflets glabrous to pilose above; sepals 3–7 mm long, without tubercles at tip; petals 5–12 mm long; capsules 8–25 mm long, glabrous to canescent. Sandy open woods, clayey prairies, disturbed areas; widespread in TX. Apr–Jun, less freely to Oct. [*O. corniculata* L. var. *wrightii* (A. Gray) B.L. Turner, *O. dillenii* Jacq., *O. dillenii* var. *filipes* (Small) Eiten, *O. dillenii* var. *radicans* Shinners] While we are lumping the two species, *O. stricta* and *O. dillenii* have often been separated (e.g., Radford et al. 1968; Correll & Johnston 1970; McGregor 1986) using characters such as the following:

1. Inflorescence umbellate, rarely with an occasional cymose branch; septate hairs not present on any part of plant; capsules appressed-pubescent ____ *O. dillenii*
1. Inflorescence cymose, the branches often only bracteate by abortion of the lateral flowers; septate hairs present on stems, petioles, or pedicels; capsules glabrous or with a few nonappressed septate hairs ____ *O. stricta*

Stenosiphon linifolius [BB1]

Orobanche fasciculata [HO1]

Orobanche ludoviciana subsp. multiflora [ABR]

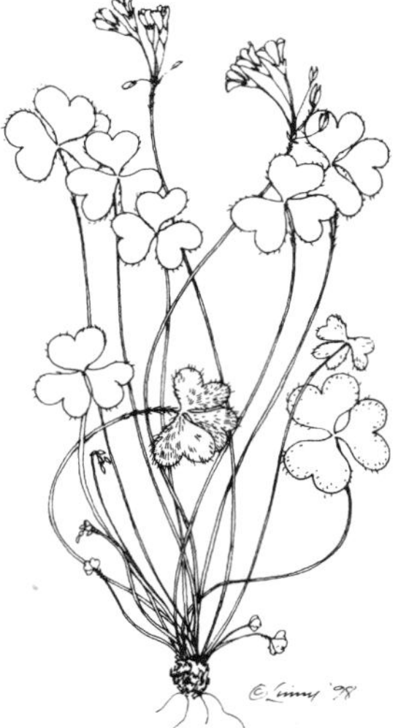

Oxalis articulata subsp. rubra [HEA]

Oxalis corniculata [PHY]

Oxalis drummondii [HEA]

Oxalis stricta [DIL]

Oxalis violacea [GR1]

Oxalis violacea L., (violet), VIOLET WOODSORREL. Glabrous perennial; bulb scales 3-ribbed; scapes to 30(–40) cm tall, with 4–19 flowers; sepals 4–6 mm long; petals 14–20 mm long, violet to pinkish purple (rarely white); capsules 4–6 mm long. Sandy open woods, rarely in prairie clay; se and e TX w to West Cross Timbers. Late Mar–early May, repeating sparingly Sep–Oct. and then without leaves.

PAPAVERACEAE POPPY FAMILY

Ours herbaceous annuals or biennials (*Eschscholzia* can be perennial from a taproot), usually with milky or colored sap or latex (sap watery in *Eschscholzia*); leaves alternate and/or basal, from wavy margined to toothed, lobed, deeply pinnatifid, or dissected, prickly or not so; stipules absent; flowers terminal or axillary, usually solitary (can be cymosely arranged in *Eschscholzia*), conspicuous, radially symmetrical, perfect; sepals 2–3, enclosing the bud until anthesis and then falling; petals usually 4–6, often showy; stamens numerous, distinct; pistil 1, of 2-many carpels; ovary superior; fruit a capsule deshicent by valves or pores.

A small (230 species in 23 genera) mainly n temperate family; most are herbs or more rarely shrubs; usually with latex, often containing isoquinoline alkaloids. *Papaver somniferum* (OPIUM POPPY) produces a milky latex, that when dried is known as opium (see details under that species). A number of species are used as ornamentals including the widely cultivated, orange- or yellow-flowered *Eschscholtzia californica* Cham. (CALIFORNIA POPPY), various *Papaver* species (POPPY), and *Sanguinaria canadensis* L. (BLOODROOT). The family is related to the Fumariaceae and Lidén (1986) and Judd et al. (1994) included within the Papaveraceae those genera (*Corydalis* and *Fumaria* in nc TX) often separated into the Fumariaceae (e.g., Kartesz 1994; Jones et al. 1997; Kiger 1997a; Mabberley 1997; Stern 1997a). This lumping of the families was based on morphological and molecular analyses (e.g., Chase et al. 1993; Judd et al. 1994) which indicated the Fumariaceae was derived from within the Papaveraceae sensu stricto. However, a more recent study (Hoot et al. 1997 [1998]) supported the monophyly of both families; we are therefore recognizing both the Fumariaceae and the Papaveraceae. (subclass Magnoliidae)
FAMILY RECOGNITION IN THE FIELD: herbs with alternate and/or basal leaves and usually *milky or colored* juice; stipules absent; flowers with 2–3 sepals, radially symmetrical showy corollas with 4–6 petals, and *numerous* stamens; fruit a capsule either armed with prickles, dehiscing by pores located below a stigmatic disk, or linear and opening by valves.
REFERENCES: Ernst 1962; Gunn & Seldin 1976; Gunn 1980; Chase et al. 1993; Grey-Wilson 1993; Kadereit 1993; Judd et al. 1994; Kiger 1997a; Hoot et al. 1997 [1998].

1. Plants prickly; petals white, lavender, or yellow; capsules usually armed with prickles ____ **Argemone**
1. Plants not prickly; petals red, orange (sometimes with dark spot at base) or yellow (orange at base) to white or purple; capsules unarmed.
   2. Capsules obovate to subglobose, narrowed at base, < 3 times as long as wide, dehiscing through pores just below the flattened apical stigmatic disk; latex whitish ____ **Papaver**
   2. Capsules very long and narrow, linear, the sides ± parallel, > 10 times as long as wide, opening from above by 2 valves, eventually opening nearly to base OR opening from base; stigmatic disk absent; latex yellow or clear.
      3. Petals red (drying blackish) with a blackish spot at base OR infrequently pale orange with dark spot; leaves deeply pinnatifid with 4–8 pairs of lobes; foliage with sparse to dense covering of coarse whitish hairs; sepals distinct; receptacle neither hollow nor with a rim; latex yellow ____ **Glaucium**
      3. Petals yellow to orangish, usually with orange area at base; leaves pinnately dissected into numerous narrowly oblong or ± linear segments; foliage glabrous; sepals fused; receptacle hollow, cup-like, and closely surrounding the ovary base, with a prominent spreading rim; latex clear ____ **Eschscholzia**

ARGEMONE PRICKLY-POPPY

Usually glaucous prickly annuals or biennials, our species with pale yellow to reddish orange latex becoming brownish black upon drying; leaves sessile (except lowest), often clasping, toothed or pinnatifid, sometimes with a bluish cast, often mottled or lighter along the veins; flowers large and showy, in ours 4–12 cm in diam., cymose and crowded, or solitary; sepals 2–3, often prickly and with a conspicuous horn; petals (4–)6, in ours white, yellow, or lavender, falling early; fruit a valvate capsule, usually armed.

A genus of 32 species (Ownbey 1997) of North and South America, the West Indies, and Hawaii. The leaves and seeds of *Argemone* species contain isoquinoline alkaloids (e.g., berberine, protopine, sanguinarine); poisoning has occurred when food grains became contaminated with PRICKLY-POPPY seeds (Kingsbury 1964; Hardin & Arena 1974); some have been used medicinally; others are cultivated as ornamentals. (Latin: *argema*, cataract of the eye, alluding to former medicinal use of these or other plants with this name)
REFERENCES: Ownbey 1958, 1997; Stermitz et al. 1969.

1. Petals yellow; fruits usually not densely prickly, the prickles usually all large and rather even-sized (be careful not to confuse the prickly sepals, which form the covering of the flower buds, with fruits) ____ **A. mexicana**
1. Petals white or lavender; fruits usually densely prickly, usually with various sized prickles.
   2. Petals lavender ____ **A. polyanthemos**
   2. Petals white (or flower color unknown).
      3. Lower leaf surfaces minutely hispid or prickly between as well as on the primary and secondary veins; stems with closely spaced prickles; latex reddish orange when fresh ____ **A. aurantiaca**
      3. Lower leaf surfaces usually prickly only on the primary and secondary veins, sometimes nearly smooth; stems usually with more widely spaced prickles or nearly smooth; latex yellow or reddish orange when fresh.
         4. Largest prickles on capsules compound (with few to many smaller prickles arising from basal portion of prickles), 15–35 mm long; latex reddish orange when fresh ____ **A. aurantiaca**
         4. Largest prickles on capsules simple, 4–10(–12) mm long; latex yellow when fresh.
            5. Lower stem leaves divided to ca. 2/3 the distance to the midrib, the middle and upper less so, the latter not greatly reduced nor widely separated, the upper strongly auricled-clasping; all leaves very glaucous, thick and leathery; inflorescences leafy; sepal horns 6–10(–15) mm long ____ **A. polyanthemos**
            5. Lower, middle, and sometimes upper stem leaves divided ca. 4/5 the distance to the midrib, the upper much reduced, widely spaced, slightly clasping; all leaves green or rather glaucous, thin; inflorescences with reduced leafy bracts; sepal horns 4–6(–10) mm long ____ **A. albiflora**

Argemone albiflora Hornem. subsp. **texana** G.B. Ownbey, (sp.: white-flowered; subsp.: of Texas), WHITE PRICKLY-POPPY. Annual or biennial to ca. 1.5 m tall; leaves usually entirely smooth (or with a very few prickles) on the upper surface; capsules to ca. 4 cm long, the surface visible or sometimes partly obscured by prickles. Rocky or sandy soils, weedy areas; Dallas, Limestone, and Somervell cos., also Brown (HPC), Grayson, Lamar, and Tarrant (Ownbey 1958) cos.; e TX w to Grand Prairie, probably introduced w to Rolling Plains (Baylor Co.). Mar–Jul, sporadically later. [*A. albiflora* Hornem. var. *texana* (G.B. Ownbey) Shinners]

Argemone aurantiaca G.B. Ownbey, (orange-red). Annual or biennial to ca. 0.8 m tall; capsules 4–5 cm long. Fields, pastures or hilly areas, often rocky or sandy substrates; McLennan Co., also Bell Co. (Ownbey 1958) and Fort Hood (Bell or Coryell cos.—Sanchez 1997); endemic to sc TX n to s part of nc TX; also to Taylor Co. to the w of nc TX. Mar–Aug.

Argemone mexicana L., (of Mexico), DEVIL'S-FIG, YELLOW PRICKLY-POPPY, CARDO SANTO, CHICALOTE, MEXICAN-POPPY, THORN-APPLE. Annual to ca. 0.8 m tall; latex bright yellow; capsules 2.5–4.5 cm long. Fields, roadsides, and waste areas; sc TX n to at least Travis Co. (Ownbey 1958) just to the s of nc TX; included based on citation of vegetational area 4 (Fig. 2) by Hatch et al. (1990). Mar–Apr. Native to West Indies and probably Central America and Florida. The yellow latex has been used to treat skin diseases (Schmutz & Hamilton 1979); the leaves and seeds are poisonous if eaten; poultry can also be poisoned by the vegetative parts and seeds (Hardin & Arena 1974; Hardin & Brownie 1993). ☠

Argemone polyanthemos (Fedde) G.B. Ownbey, (many-flowered). Annual or biennial to ca. 1.2 m tall; leaves without prickles on upper surface; petals white or lavender; capsules to ca. 3 cm long, usually less prickly than in *A. albiflora*, the surface clearly visible. *Argemone polyanthemos* appears to hybridize with *A. albiflora* subsp. *texana* where their ranges overlap in nc TX and many specimens can only be identified with reservations (Ownbey 1958). Ownbey (1958), however, concluded based on morphological and distributional grounds that the two warrant recognition as separate species. Rocky or sandy soils, prairies, fields, weedy areas; Brown, Clay, Hill, Palo Pinto, Parker, and Tarrant cos.; mainly w part of nc TX w and n through n TX; rare and apparently introduced eastward; according to Ownbey (1958), e to Bell, Dallas, and Fannin cos. Apr–Jul, sporadically later. [*A. intermedia* of authors, not Sweet, *A. intermedia* Sweet var. *polyanthemos* Fedde]

ESCHSCHOLZIA CALIFORNIA-POPPY, GOLD-POPPY

A genus of 12 species of the w United States and nw Mexico (Clark 1997); some are cultivated as ornamentals. (Named for Johann F.G. von Eschscholtz, 1793–1831, Estonian physician and biologist who traveled with Chamisso [who named the genus] on the Romanzoff (or Kotzebue) Expedition to the Pacific Coast—Clark 1997)
REFERENCES: Greene 1905; Lewis & Snow 1951; Ernst 1964b; Clark & Jernstedt 1978; Clark 1978, 1993, 1997.

Eschscholzia californica Cham. subsp. **californica,** (of California), CALIFORNIA-POPPY. Erect to spreading annual or perennial (from taproot) 5–60 cm tall, glabrous and sometimes glaucous; sap watery; leaves alternate, basal and cauline, 1–4 times deeply pinnately dissected into numerous narrowly oblong or ± linear segments; cotyledons usually 2-lobed; inflorescences usually 1-flowered (can be cymose); peduncles 5–15(–20) cm long; receptacles obconic, expanded, hollow, cup-like, and closely surrounding the ovary base, the spreading rim of the receptacular cup prominent; sepals 2, fused into a hood-like structure that is pushed off as a unit by the opening petals; petals 4, yellow to orangish, usually with orange area at base, 2–6 cm long; stamens numerous; capsules 3–9 cm long, dehiscent from the base. Widely cultivated as an ornamental; recently (May 1998) found escaped in an open, grassy, weedy area along railroad tracks; Tarrant Co. (Fort Worth); this is the first Texas collection of which we are aware (*Lipscomb 3495, Diggs & McCullough*). May–Jun. Native from n Mexico through CA to s WA. This striking and morphologically variable species is the state flower of California (Clark 1993) and has become naturalized and weedy in various countries (Jepson 1925; Fuller & McClintock 1986) including New Zealand where it can form ± pure stands (Mabberley 1997). The latex is reported to be mildly narcotic and to have been used by Native Americans for toothache (Mabberley 1997); the entire plant is toxic to humans due to isoquinoline alkaloids which can depress respiration; thirteen different alkaloids have been found in this species (Fuller & McClintock 1986; Clark 1993, 1997). ☠

Eschscholzia californica subsp. *mexicana* (Greene) C. Clark, (of Mexico), MEXICAN GOLD-POPPY, AMAPOLA DE CAMPO, is native to the Trans-Pecos of Texas. It has the spreading rim of the receptacular cup often inconspicuous and the cotyledons unlobed.

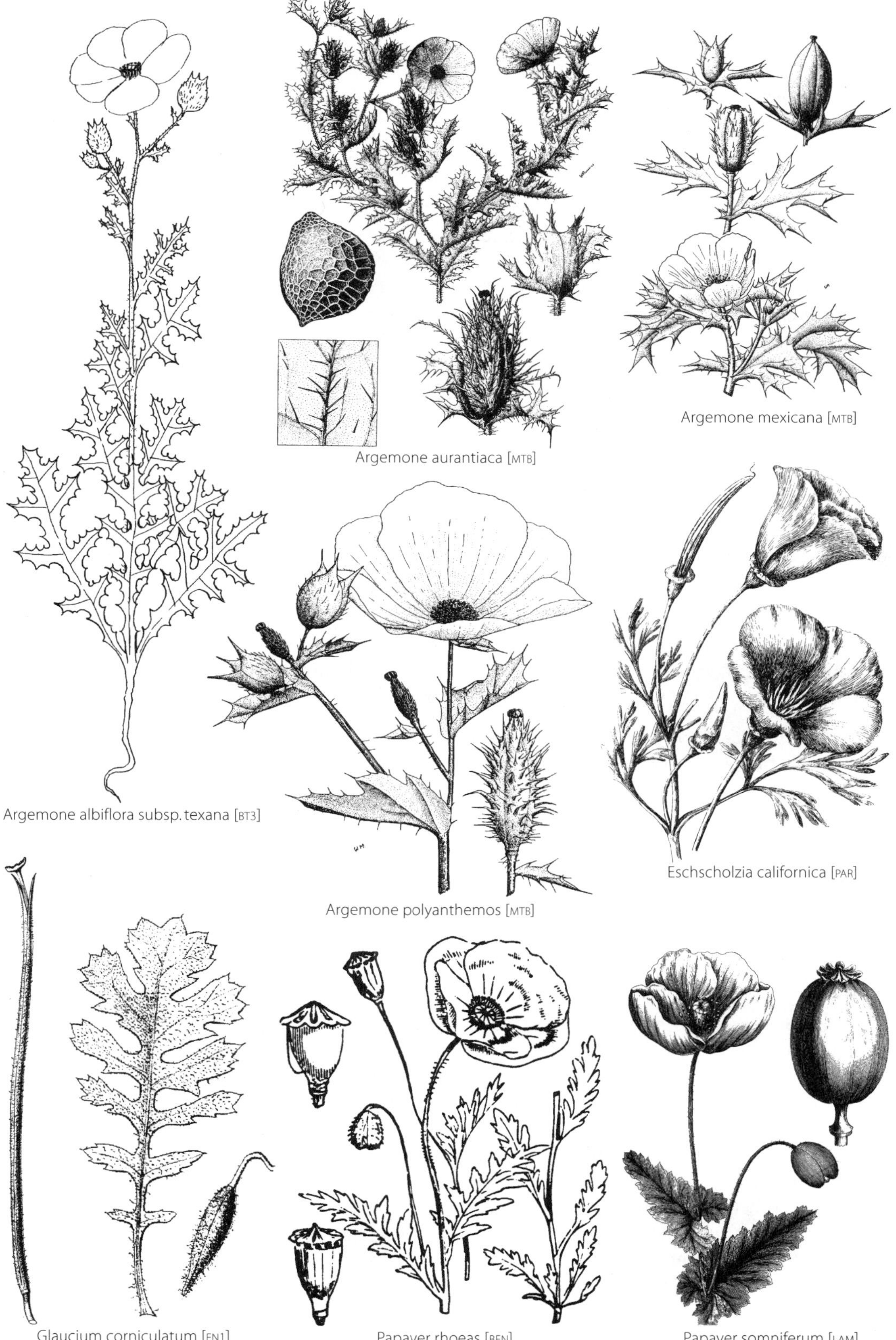

Argemone aurantiaca [MTB]

Argemone mexicana [MTB]

Argemone albiflora subsp. texana [BT3]

Argemone polyanthemos [MTB]

Eschscholzia californica [PAR]

Glaucium corniculatum [EN1]

Papaver rhoeas [BEN]

Papaver somniferum [LAM]

GLAUCIUM HORNED-POPPY

A genus of ca. 23 species of Europe and sw and c Asia; a number contain alkaloids. (Greek, *glaucos*, gray-green or bluish gray, alluding to the color of the foliage)
REFERENCES: Kiger 1997b; Kirkpatrick & Williams 1998.

Glaucium corniculatum (L.) Rudolph, (horned), RED HORNED-POPPY. Annual or biennial herbs with yellow latex; foliage with sparse to dense covering of coarse whitish hairs; stems 30–90 cm tall; leaves 4–20 cm long, 1.5–8 cm wide, deeply pinnatifid with 4–8 pairs of lobes; rosette leaves petiolate, often withered by flowering time; cauline leaves sessile; flowers 1 per peduncle, terminal or in leaf axils; peduncles 1–3(–7) cm long; sepals 2; petals 4, red (drying blackish) with a blackish spot at base or infrequently pale orange with dark spot, broadly obovate, 2–3.5(–4) cm long; capsules linear, straight to slightly curved, 15–20 cm long, to 4–6 mm in diam., pubescent with appressed hairs. Weedy areas; Tarrant Co.; otherwise only known in TX from Garza, Gillespie, Kerr, San Saba, and Travis cos. Apr–May. Native of Europe. The first documented TX collection was by Toney Keeney in 1986 in San Saba Co.; it was listed for TX by Jones et al. (1997) with details reported by Kirkpatrick and Williams (1998).

PAPAVER POPPY

Ours herbaceous annuals to ca. 1 m tall with milky latex; leaves with wavy margins to deeply pinnatifid; flowers solitary, axillary or terminal, often showy, perfect; peduncles usually > 10 cm long; sepals 2; petals 4(–6), showy; ours with subglobose to broadly ovoid or obovoid capsules dehiscing through pores just below the flattened apical stigmatic disk.

A genus of 70–100 species (Kiger & Murray 1997) of Europe, Asia, s Africa, Cape Verde Islands, and w North America. Many species are cultivated as ornamentals or for their alkaloids (e.g., see OPIUM POPPY below); ☠ all should be considered potentially poisonous to livestock (Kingsbury 1964). (The ancient name, from Latin: *pappa*, milk, alluding to the milky latex)
REFERENCES: Duke 1973; Kiger 1973, 1975, 1985; Novák & Preininger 1987; Kiger & Murray 1997.

1. Stem leaves not cordate clasping; plants densely hirsute, not glaucous ____ **P. rhoeas**
1. Stem leaves cordate clasping; plants glabrous, glaucous (= with a white or gray-silvery coating) ____ **P. somniferum**

Papaver rhoeas L., (an old Greek name), CORN POPPY, FIELD POPPY, SHIRLEY POPPY, AMAPOLA. Leaves deeply pinnatifid with the divisions lobed or incised; corollas usually red with black towards center to purple or white, ca. 8 cm in diam.; capsules obovoid to subglobose to 2 cm long. Cultivated and escaped; Collin, Dallas, Denton, and Grayson cos., also Burnet, Comanche, Erath, and Mills cos. (HPC); Post Oak Savannah w to nc TX and s to Edwards Plateau. Apr–Jun. Native of Eurasia and n Africa. Poisonous alkaloids are present and have been responsible for poisoning in livestock (Burlage 1968; Kingsbury 1964). ☠

Papaver somniferum L., (sleep-bearing or producing), OPIUM POPPY, COMMON POPPY. Leaves with wavy margin or coarsely toothed or shallowly lobed; corollas white to red or purple, to 8 cm or more in diam.; capsules subglobose, 1.5–5(–9) cm long. Cultivated and escaped; Clay, Dallas, and Grayson cos., also Mills and Lampasas cos. (HPC); nc TX and Edwards Plateau. May–Jun. Native of Eurasia. The latex obtained by sliting the unripe capsules of this species is known as opium; it is the source of ca. 25 different alkaloids, including morphine and codeine, widely used as medicines and narcotics; overdoses can result in death due to respiratory failure. Morphine, unmatched by synthetics, is one of the best pain relievers known; unfortunately it is addictive. Heroin is synthetically derived from morphine by acetylation; it is even more addictive than morphine (Duke 1973; Schmutz & Hamilton 1979; Fuller & McClintock 1986; Mabberley 1987). This species has been used for thousands of years; Sumerian tablets ca. 6,000 years old that were found in Mesopotamia mention the opium poppy (Talalaj & Talalaj 1991). Poppy seeds, used in baking and as birdseed, are also obtained from this species; while not narcotic, poppy seeds are reported to have caused positive drug tests.

PASSIFLORACEAE PASSION-FLOWER FAMILY

A medium-sized (575 species in 17 genera) family mainly of tropical to warm temperate areas, especially in the Americas; the family consists of vines, trees, shrubs, and herbs, often with alkaloids. It is thought to be related to the mainly tropical family Flacourtiaceae. The family also shares similarities with the Cucurbitaceae and Loasaceae. (subclass Dilleniidae)
FAMILY RECOGNITION IN THE FIELD: herbaceous tendriled *vines*; flowers with an elaborate *fringed corona* overlaying a rather flat ring of petals and sepals; ovary superior, raised on a stalk surrounded by the united filaments.
REFERENCES: Killip 1938; Brizicky 1961a.

PASSIFLORA PASSION-FLOWER

Ours perennial climbing or trailing vines with axillary, simple tendrils; leaves alternate, simple, palmately 3-lobed, sometimes with petiolar glands; stipules slender; flowers axillary, solitary, on long, 1-flowered peduncles (these jointed and often bracted), open during the morning, (3-)5-merous, radially symmetrical, perigynous with floral cup, with a crown (= corona) of many, filament-like segments with united bases attached to floral cup inside the petals, the outer row equaling the petals in length, the inner rows shorter; stamens 5; filaments united into a sheath around a stalk (= gynophore) supporting the single pistil; fruits berry-like; seeds arillate.

A genus of 430 species of tropical and warm areas of the Americas, Indomalesia, and the Pacific region. A number of species have edible fruits and several are cultivated commercially—the edible part is the pulpy aril; others are cultivated for their unusual, sometimes very showy flowers. Jørgensen et al. (1984) provided excellent illustrations of the complex variation in flowers of *Passiflora*. PASSION-FLOWER leaves are fed on by larvae of the butterfly family Heliconiidae; the heliconiids are distasteful to birds and other predators apparently due to substances obtained from the leaves (Howe 1975); *Passiflora* species have evolved a number of interesting anti-herbivore mechanisms including leaf shape variation (to escape detection by egg-laying female butterflies), extrafloral nectaries (to attract ants and thus discourage herbivores), egg mimics that apparently discourage egg-laying, hooked hairs, and toxins (Gilbert 1980). All four nc TX *Passiflora* species are fed on by *Dione [Agraulis] vanillae*, the Gulf Fritillary, a member of the Heliconiidae (Scott 1986). The common name, related to the scientific, was derived from Catholic missonaries in South America who used the flowers as a lesson on the Crucifixion of Jesus (e.g., 3 stigmas = 3 nails, 5 anthers = 5 wounds; corona = crown of thorns). (An adaptation of *flos passionis*, a translation of *fior delle passione*, the Italian name early applied to the flower from a fancied resemblance of its parts to the implements of the Crucifixion)
REFERENCE: Jørgensen et al. 1984.

1. Flowers large and showy; petals bluish lavender, 25–35 mm long; fruits to 50 mm long; leaf lobes finely toothed ____ **P. incarnata**
1. Flowers relatively small and inconspicuous; petals greenish yellow, 3–8 mm long, OR absent and sepals greenish; fruits 8–15 mm long; leaf lobes entire or with a few large, lobe-like teeth.
   2. Petiolar glands absent; leaf blades shallowly to deeply lobed, the leaf lobes broadly triangular to ovate, oblong, obovate, or oblanceolate, usually blunt or rounded at apex (also often mucronate), the lower and upper surfaces both glabrous; petals present; peduncles 10–40 mm long; seeds with wrinkled transverse ridges; including species widespread in nc TX

3. Leaf blades lobed from ca. 1/4–1/2 their length; peduncles without bracts; sepals 5–10 mm long; outer corona filaments not knobbed at tips; widespread in nc TX — **P. lutea**
3. Leaf blades lobed from (1/3–)1/2–2/3 their length; peduncles with small narrow bracts 1–3 mm long, these sometimes early deciduous; sepals 10–12 mm long; outer corona filaments knobbed at tips; known in nc TX only from the Lampasas Cut Plain — **P. affinis**

2. Petiolar glands present and obvious at juncture of petiole and blade; leaf blades deeply lobed, the leaf lobes narrow, usually linear to narrowly oblong or narrowly lanceolate, acute to acuminate at apex, the lower surfaces glabrous but the upper surfaces with pubescence; petals absent; peduncles 3–8 mm long; seeds reticulate; rare in nc TX — **P. tenuiloba**

Passiflora affinis Engelm., (similar, related), BRACTED PASSION-FLOWER. Vine; leaf blades 2–10 cm long, 3–14 cm wide, sometimes similar to those of *P. lutea*, but usually more deeply lobed, with a few ocelli (small yellowish dots that apparently act as egg mimics and thus discourage egg-laying—Gilbert 1980); petioles 1–3.5 cm long, glandless; peduncles 1–3 cm long, with remote bracts; petals linear, narrower than sepals; corona filaments purplish near base; fruits subglobose, 8–10 mm long, black-purple when ripe; seeds ca. 3 mm long. Stream bottoms in limestone areas; Bell Co. (Fort Hood—Sanchez 1997); Edwards Plateau n to s Lampasas Cut Plain. Jun–Aug. ⊞/102

Passiflora incarnata L., (flesh-colored), MAYPOP PASSION-FLOWER, PASIONARIA, APRICOT VINE. Vine to several m long; leaf blades 6–15 cm long, 7–15 cm wide, pubescent beneath, the lobes finely toothed, pointed; petioles to 8 cm long, with 2 glands near juncture with leaf blades; petals bluish lavender, rarely white; fruits ovoid to subglobose, orange-yellow when ripe; arils edible; seeds 4–5 mm long. Stream bottoms, fencerows, and disturbed ground; Cooke, Dallas, Grayson, Henderson, Kaufman, and Rockwall cos.; se and e TX w to East Cross Timbers. Jun–Aug. This species was long cultivated by Native Americans for the edible fruits (arils) (Mabberley 1987). ⊞/102

Passiflora lutea L., (yellow), YELLOW PASSION-FLOWER. Vine to ca. 5 m long; leaf blades 3–7 cm long, 2–10 cm wide, usually wider than long; petioles to 5 cm long, glandless; peduncles 1.5–4 cm long, without bracts; peduncles to 100 mm long; petals ± linear, narrower than sepals; fruits globose-ovoid, 8–15 mm long, black-purple when ripe; seeds 4.5–5 mm long. Stream bottom or hillside woods, climbing into trees; Bell, Dallas, Grayson, McLennan, and Montague cos., also Lamar (Carr 1994), Hood, Somervell, and Tarrant (R. O'Kennon, pers. obs.) cos.; se and e TX w to West Cross Timbers, also Edwards Plateau. Jun–Jul. [*P. lutea* var. *glabriflora* Fernald]

Passiflora tenuiloba Engelm., (with narrow, slender lobes), SPREAD-LOBE PASSION-FLOWER, BIRD-WING PASSION-FLOWER. Vine; glands at juncture of petiole and leaf blades saucer-shaped, sessile, 1–1.2 mm in diam.; leaf blades of 2 different forms: 1) with 2 long narrow lateral lobes and a much shorter midlobe; 2) a western form to the w of nc TX with the lobes themselves lobed, the leaf blades thus appearing to have up to 7 lobes; sepals 6–10 mm long, ca. 2 mm wide, greenish; fruits globose, 8–15 mm in diam., black when ripe; seeds 3–4 mm long. Typically in open limestone areas, generally prostrate, sometimes climbing into low shrubs; Killip (1938) cited a Reverchon collection from Dallas Co., also known from Travis Co. just s of nc TX; mainly Edwards Plateau to s TX and Trans-Pecos. Apr–Oct. Killip (1938) indicated that the leaf blades have a few ocelli (small yellowish dots that apparently act as egg mimics and thus discourage egg-laying—Gilbert 1980) ⊞/102

PEDALIACEAE SESAME FAMILY

A small (85 species in 17 genera) family of herbs and shrubs of tropical and warm areas, especially coastal and arid regions. The family includes the Old World tropical *Sesamum*

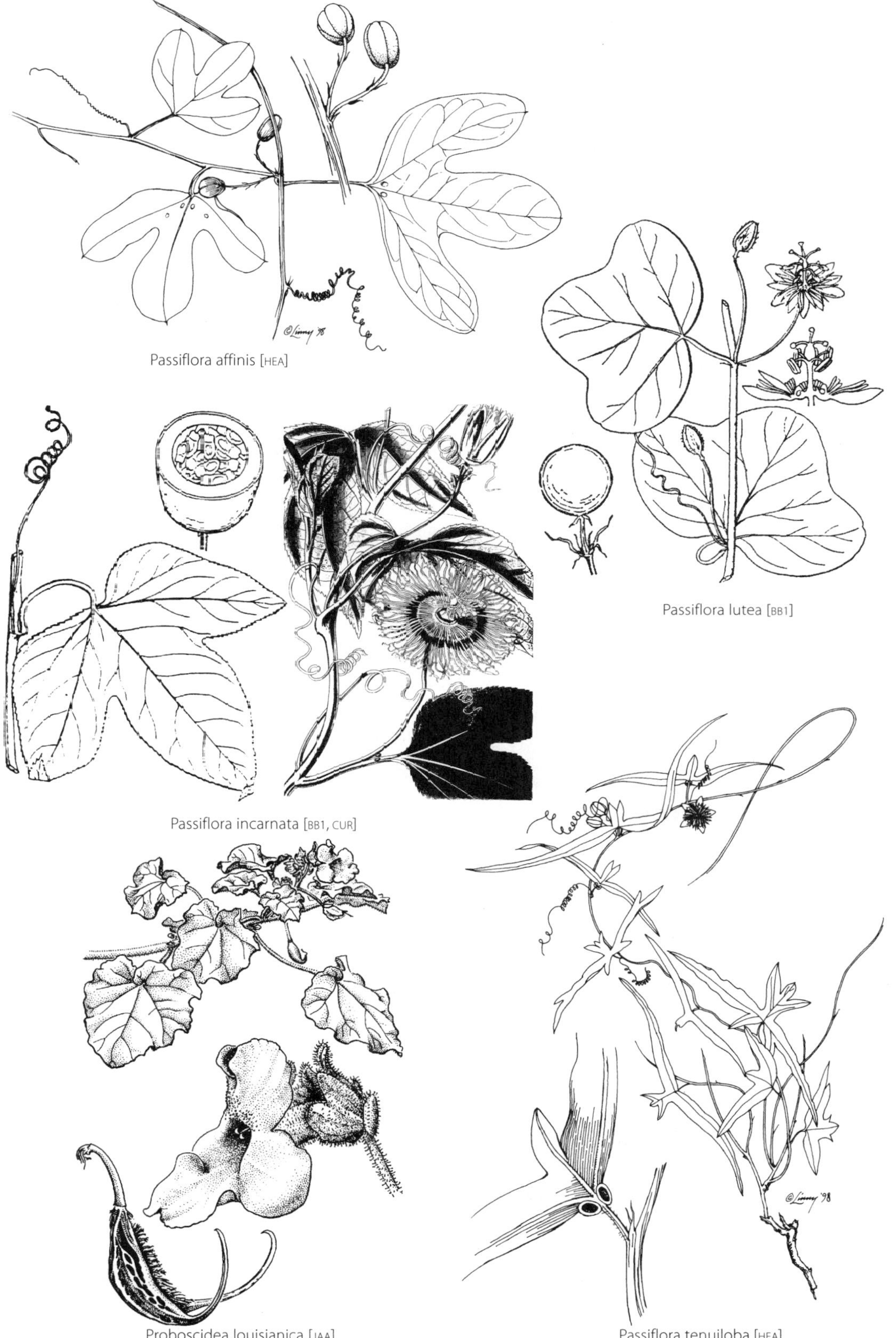

Passiflora affinis [HEA]

Passiflora lutea [BB1]

Passiflora incarnata [BB1, CUR]

Proboscidea louisianica [JAA]

Passiflora tenuiloba [HEA]

indicum L., SESAME, SESAMUM, GINGELLY, the source of sesame seeds and oil. *Proboscidea* and close relatives are sometmes segregated as the Martyniaceae, a family of 3 genera and 13 species of New World herbs (e.g., Hutchinson 1979; Bretting & Nilsson 1988). Family name from *Pedalium*, a monotypic genus of the Old World tropics based on *P. murex* L. whose leaves are used as a vegetable and seeds medicinally. (Name possibly from Greek: *pedalion*, *pedalium*, a rudder or kind of plant) (subclass Rosidae)

FAMILY RECOGNITION IN THE FIELD: the only nc TX species is a *glandular-pubescent* herb with large, tubular-cylindric-campanulate corollas and unique capsules which split and have *2 conspicuous curved horns.*

REFRERENCES: Thieret 1977a; Bretting & Nilsson 1988; Manning 1991.

PROBOSCIDEA DEVIL'S-CLAW, UNICORN-PLANT, CINCO LLAGAS

A genus of 9 species of warm areas of the Americas; some are cultivated as ornamentals or for their edible fruits. (Greek: *proboscis*, snout, alluding to the long curved beak of the fruit)

REFERENCES: Thieret 1976; Bretting 1983.

Proboscidea louisianica (Mill.) Thell., (of Louisiana), COMMON DEVIL'S-CLAW, UNICORN-PLANT, COW-CATCHER, RAM'S-HORN, MULE-GRAB. Low, viscid-pubescent, ill-scented annual densely covered with glandular hairs; leaves alternate or opposite; leaf blades ovate to suborbicular on densely short pubescent petioles to 20 cm long, cordate basally; flowers long-pedicelled, in terminal racemes with small, inconspicuous bracts; sepals 5, united, unequal; corollas to 55 mm long, 5-lobed and somewhat bilatiate, tubular-cylindric for 2–5 mm, then becoming broadly campanulate, dull white to lavender, mottled purple and yellow inside; fertile stamens 4, didynamous; pistil 2-carpellate; fruit a capsule to 10 cm long, 2–3 cm thick, at maturity splitting to form 2 curved horns sharply curved at ends to form hooks, often dispersed by hooking around the ankles of small livestock. Loose sandy soils; widespread in TX. May–Sep. [*Martynia louisianica* Mill.] The fruits get tangled in the wool of sheep and can interfere with shearing; they have even been known to clamp shut the jaws of sheep resulting in starvation; however, the young pickled fruits are edible and the plants are commercially cultivated for this purpose (Thieret 1977a; Mabberley 1987). ▣/104

PHRYMACEAE LOPSEED FAMILY

A monotypic family native to Asia and North America. *Phryma* has been placed in the Verbenaceae by a number of authorities (e.g., Cronquist 1981; Kartesz 1994). However, Chadwell et al. (1992) concluded that when a variety of characters including pollen morphology are considered, a position near, but not within subfamily Verbenoideae is best supported. Chadwell et al. (1992) further indicated that definitive familial placement of *Phryma* cannot be determined until the broader problem of family delimitation in the Lamiales as a whole is addressed. In a recent molecular study of Labiatae and Verbenaceae (Wagstaff & Olmstead 1997), *Phryma* "... emerges as a distinct lineage, not clearly associated with any other group within the Lamiales s.l." In order to emphasize its distinct phylogenetic position, we are recognizing it at the family level. (subclass Asteridae)

FAMILY RECOGNITION IN THE FIELD: the only species is a Verbenaceae-like herb with opposite leaves and small flowers in spike-like inflorescences; in fruit the calyces (enclosing the fruits) are *reflexed* and *parallel* to the inflorescence axis.

REFERENCES: Thieret 1972; Chadwell et al. 1992; Wagstaff & Olmstead 1997.

PHRYMA LOPSEED

A genus of a single species native to Asia (India to Japan) and e North America; this disjunct

distribution pattern is discussed under the genera *Campsis* (Bignoniaceae) and *Carya* (Juglandaceae). *Phryma* is sometimes divided into an Asian species and a North American species but the two are distinguished by only minor differences such as length of the upper calyx lobes, shape of the upper lip of the corollas, and pubescence (Hara 1969); Hara indicated that this species "... must be of very ancient origin." (Derivation of generic name unexplained by Linnaeus)
REFERENCE: Hara 1969.

Phryma leptostachya L., (slender-spiked), LOPSEED. Pubescent or glabrous, perennial herb 30–100 cm tall; leaves opposite, the median ones rather long-petioled, the upper ones shorter-petioled; leaf blades thin, ovate to lanceolate, (3–)5–16 cm long, 3–10 cm wide, irregularly coarsely toothed; flowers minutely bracted, opposite, in slender, long-peduncled, spike-like inflorescences terminating stem and upper, axillary, leafless branches; pedicels < 1 mm long; calyces with 3 linear, acuminate upper teeth as long as the tube (2–3 mm), and 2 shorter (0.3–0.5 mm), wide, lower teeth; in fruit the calyces are closed, become reflexed, and lie parallel to the inflorescence axis; corollas 2-lipped, ca. 8 mm long, purplish to light lavender, whitish inside, the lower lip much larger and lighter in color; stamens 4; pistil 1; style 1; stigma 2-lobed; fruits achenes or achene-like, enclosed in calyces. Stream bottom woods; Bell, Dallas, Grayson, Lamar, and Tarrant cos., also Fort Hood (Bell or Coryell cos.—Sanchez 1997); e TX w to nc TX. Jun.

PHYTOLACCACEAE POKEWEED FAMILY

Ours annual or perennial herbs, sometimes with woody base; leaves alternate, simple, entire; flowers in axillary or terminal, peduncled, leafless racemes; calyces rotate, small, white to greenish, pink, rose, or reddish purple; petals absent; fruit in ours a berry.

A small (65 species in 18 genera) mainly tropical and warm area family, especially of the Americas; usually herbs but also shrubs, trees, or woody climbers. Like most Caryophyllidae, they produce betalain pigments rather than anthocyanins (Cronquist & Thorne 1994). Some are used medicinally, as sources of dye, or as potherbs. Molecular evidence (Downie & Palmer 1994) indicated the family is polyphyletic and that *Phytolacca* and *Rivina* are related to Nyctaginaceae. (subclass Caryophyllidae)
FAMILY RECOGNITION IN THE FIELD: herbs with alternate, simple, entire leaves and usually drooping, terminal or axillary racemes of small flowers that are white to greenish or variously colored with some reddish pigmentation; fruit a *brightly colored* berry; foliage usually rather succulent, often strongly reddish- or purplish-tinted and/or with a bloom.
REFERENCES: Thieret 1966; Nowicke 1969; Brown & Varadarajan 1985; Rogers 1985; Rohwer 1993b; Behkne & Mabry 1994; Downie & Palmer 1994.

1. Fruits dark purple, 6–10 mm in diam.; sepals 5; stamens 10; plants usually much more than 1 m tall, coarse annuals ____ **Phytolacca**
1. Fruits red or orange, 2–3.5 mm in diam.; sepals 4; stamens 4; plants usually < 1 m tall, perennials from a woody root, sometimes with woody stem-base ____ **Rivina**

PHYTOLACCA POKEWEED

A genus of 25 species native to tropical and warm regions of the world; species vary from trees to shrubs or herbs, and typically possess alkaloids; ☠ some are poisonous, while a number are used as vegetables or as cultivated ornamentals. (Greek: *phyton*, plant, and either Latin: *lacca*, derived from the Hindi *lakh*, for the dye obtained from the lac insect (Tveten & Tveten 1993), or French: *lac*, lake, from the crimson (lake) color of the fruits)

Phytolacca americana L., (of America), POKEWEED, POKEBERRY, POKE, INKBERRY. Plant to 3 m tall;

stems reddish or purplish; leaf blades lanceolate or elliptic-lanceolate, to 25 cm long and 10 cm wide; flowers perfect; sepals petaloid, white or pink, 2–3 mm long, persistent in fruit; fruits usually of 10 carpels; seeds 1 per carpel, shiny, glabrous, somewhat flattened. Stream bottom woods and thickets or occasionally in disturbed ground; throughout much of TX. Late Jun–Sep. The word "poke" is possibly derived either from an American Indian word *pak*, meaning blood, in reference to the fruits, from which they obtained a brillant purplish red dye (Core 1967) or from *pocan* or *puccoon*, probably Algonquin for a plant that contains dye (Rogers 1985). The fruits are dispersed by birds which are presumably attracted by the bicolor displays—the reddish stems and preripe fruits constasting with the black ripe fruits (Willson & Thompson 1982; McDonnell et al. 1984). All parts of the plant should be considered poisonous and potentially lethal, especially the roots, purple rind of the stem, and seeds; however, the very young leaves and shoots are used by some as a cooked vegetable ("poke salat" or "sallet"—Peacock 1982; Rogers 1985) after proper preparation including several boilings; the boiled water is toxic. Toxins include phytolaccatoxin and related triterpenoid saponins; an additional mitogenic (= causing cell division) compound, apparently a type of protein lectin, is present resulting in some sources recommending that the plant should not be eaten. Further, if the plant is handled, the mitogens can be absorbed through skin abrasions, potentially causing serious blood aberrations; the plant **should not be handled** except with gloves. A dye, used to color ink, wine, and sweets, has been obtained from the fruits (Hardin & Arena 1974; Lewis & Elvin-Lewis 1977; Lampe & McCann 1985; Rogers 1985; Mabberley 1987; Turner & Szczawinski 1991). ☠

RIVINA PIGEON-BERRY

A monotypic genus native from the s U.S. to the American tropics. Brown and Varadarajan (1985) argued that *Rivina* and related taxa should be segregated as the family Petiveriaceae. (Named for A.Q. Rivinus, 1652–1723, German botanist)

Rivina humilis L., (low-growing, dwarf), PIGEON-BERRY, ROUGEPLANT, BLOODBERRY, CORALITO. Plant to ca. 85 cm tall; leaf blades ovate to elliptic-ovate or rhombic, to 15 cm long and 9 cm wide; sepals whitish, greenish, or rose to reddish purple, 2–2.5 mm long; ovary unilocular with a solitary basal ovule; seed 1 per fruit, hairy (Thieret 1966). Stream bottom woods and thickets, chiefly in limestone areas; throughout much of TX, rare e. Jun–Oct. Cultivated as an ornamental; the fruits have been used as a source of a red dye (Mabberley 1987); the leaves and roots are poisonous and the fruits have been implicated in non-fatal poisonings (Morton 1982; Lampe & McCann 1985). ☠

PLANTAGINACEAE PLANTAIN FAMILY

A small (275 species in 3 genera), cosmopolitan (more temperate and tropical montane) family of wind-pollinated, usually scapose herbs or a few shrubs; the family is of little economic importance except several are problematic weeds. Based on molecular data, Reeves and Olmstead (1993) indicated that the Plantaginaceae should be considered part of the Scrophulariaceae. (subclass Asteridae)

FAMILY RECOGNITION IN THE FIELD: herbs with *rosettes* of basal, ± *parallel-veined* leaves and *leafless* peduncles terminated by slender spikes of small, densely packed, membraneous, 4-merous flowers.

REFERENCES: Rosatti 1984; Reeves & Olmstead 1993.

PLANTAGO PLANTAIN, RIBWORT

Ours small annual or perennial scapose herbs; leaves basal, subsessile or petioled; blades entire, toothed or pinnately lobed; flowers bracted, in heads or spikes terminating leafless, simple peduncles; sepals 4 or 3 (2 abaxial sepals fused to form 1); corollas thin, dry, and membraneous, white or yellowish, 4-lobed, the lobes varying from erect to reflexed; stamens 4 or 2, often exserted; pistil 1; fruit a circumscissile capsule.

A cosmopolitan genus of ca. 270 species of herbs and a few shrubs including many weeds. The seeds of some are mucilaginous when wet and are efficient laxatives—i.e. *P. afra* L. (PSYLLIUM), native from the Mediterranean to India. Because of the small size of the flowers and the importance of pubescence characters, a hand lens is often necessary for positive identification of *Plantago* species. (The Latin name, from *planta*, footprint or sole of the foot, alluding to the leaves of *P. major* that are broad and flat and pressed to the ground)

REFERENCES: Shinners 1950d, 1967; Bassett 1966, 1973; Bassett & Crompton 1968; Rahn 1974; R. Hoggard 1999.

1. Leaves deeply pinnately lobed — **P. coronopus**
1. Leaves entire or with few teeth or shallow lobes.
   2. Floral bracts scarious-margined or wholly herbaceous, with herbaceous, acute or acuminate tips formed by the midvein, without long-acuminate, scarious tips; abaxial 2 sepals not fused, each with 1 midvein (thus 4 sepals present); plants usually annual (except *P. rugelii* and *P. major*).
      3. Sepals and bracts glabrous.
         4. Leaf blades less than 4 mm wide, linear to filiform, inconspicuous; stamens 2; plants annual.
            5. Corolla lobes erect in age (look at tip of capsule); seeds 10–30 per capsule, somewhat asymmetrical, 0.6–0.8 mm long — **P. heterophylla**
            5. Corolla lobes often becoming conspicuously reflexed in age; seeds (2–)4(–8) per capsule, symmetrical, (0.5–)0.8–1.7 mm long — **P. elongata**
         4. Leaf blades 3–13 cm wide, broadly elliptic to ovate or ovate-cordate, large and conspicuous; stamens 4; plants perennial.
            6. Capsules circumscissile below middle, the upper part ca. 2 times as long as lower, (2.1–)2.5–5 mm long; petioles purple at least at base; floral bracts usually narrowly lance-triangular; sepals acute, (1.5–)2–3.5 mm long; seeds 4–10 per capsule — **P. rugelii**
            6. Capsules circumscissile near middle, 2.2–2.6 mm long; petioles green, rarely with slight purplish tinge; floral bracts usually broadly ovate; sepals rounded, 1.6–1.8 mm long; seeds 10–25 per capsule — **P. major**
      3. Sepals and bracts pubescent or pilose.
         7. Hairs on middle part of flowering stem (= peduncle) spreading at right angles.
            8. Bracts at base of spike keeled and clasping; corolla lobes usually erect and folded together before and after flowering, 0.8–3 mm long; plants usually not turning dark on drying.
               9. Bracts 1–2.8 mm long; abaxial sepals (2 on side away from axis) rounded at apex, without green midvein extending beyond scarious margins; corolla lobes 0.8–2.3 mm long; mature seeds yellow-brown to black, (0.8–)1–1.7 mm long, without a transparent margin — **P. virginica**
               9. Bracts (2.2–)2.8–5.2(–5.4) mm long; abaxial sepals acuminate at apex, with green midvein extending beyond scarious margins; corolla lobes 2–3 mm long; mature seeds red to reddish black; (1.7–)2–2.8 mm long, with pronounced transparent margin — **P. rhodosperma**
            8. Bracts at base of spike neither keeled nor clasping; corolla lobes at first erect, spreading to reflexed during and after flowering, (2.7–)3–4 mm long; plants usually turning noticeably dark on drying — **P. helleri**
         7. Hairs on middle part of flowering stem closely ascending or appressed.
            10. Floral bracts triangular-ovate, broadly scarious-margined except at apex (each scarious margin 1/4–1/3 entire width of bract), never conspicuous at a glance.

11. Leaf blades glabrous on upper surface, with thin pubescence on lower surface; corolla lobes cordate-ovate, ca. 0.5 mm longer than wide, 2.5–3 mm long ______ **P. wrightiana**
11. Leaf blades woolly-pubescent on upper and lower surfaces; corolla lobes suborbicular, ca. as wide as long, to ca. 2–2.5 mm long ______ **P. hookeriana**

10. Floral bracts narrowly triangular-lanceolate to linear-lanceolate, narrowly scarious-margined at base, otherwise herbaceous, inconspicuous OR sometimes conspicuously elongate at a glance.

12. Leaf blades glabrous on upper surface; bracts in middle of spike 3–6 times as long as the calyces; plants usually turning noticeably dark on drying ______ **P. aristata**
12. Leaf blades densely pubescent on upper surface; bracts in middle of spikes shorter than or up to 4 times as long as the calyces; plants usually not turning dark on drying ______ **P. patagonica**

2. Floral bracts scarious except at base and in center, ovate and abruptly narrowed to long-acuminate, scarious tip extending beyond the midvein; abaxial 2 sepals fused to form 1 with 2 veins (thus apparently only 3 sepals present); plants perennial ______ **P. lanceolata**

Plantago aristata Michx., (bearded, with an awn), BUCKTHORN, BOTTLEBRUSH PLANTAIN, BRACTED PLANTAIN. Annual to ca. 25 cm tall; spikes very dense, to 15 cm long but usually much shorter, 6–10 mm thick (excluding bracts); floral bracts narrow, elongate, conspicuous. Sandy open woods; se and e TX w to West Cross Timbers. May–Jun.

Plantago coronopus L., (presumably named for the resemblance of the leaves to those of *Coronopus*, swine cress, in the Brassicaceae), BUCK-HORN PLANTAIN. Pubescent annual or biennial; leaves closely spreading on the ground or ascending, linear to lanceolate in outline, 4–25 cm long, with spreading-ascending, acute, ± linear lobes; inflorescences sometimes numerous (up to 65 observed), 5–50 cm long including the peduncle terminated by a narrowly cylindric, dense spike to ca. 12.5 cm long; bracts usually not surpassing the flowers, broadly scarious-margined at base, the keel prolonged into an acuminate tip; petals acute. Landscape weed or in weedy areas; Tarrant Co., this collection (*O'Kennon 14221*, 1998) is the first report for Texas (O'Kennon et al. 1998). Late Apr–Jul–? In addition to the conspicuously different leaves, this species differs from all other PLANTAINS occurring in Texas in having short hairs on the corolla tubes and 3- or 4-locular capsules; other members of the genus occurring in the state have the corolla tubes glabrous and 2-locular capsules.

Plantago elongata Pursh, (elongate), SLENDER PLANTAIN. Glabrous annual, 2–18 cm tall; leaves narrowly linear, entire; spikes loosely flowered with part of axis often exposed, 2–10 cm long. Damp or disturbed bare ground; Denton, Grayson, and Tarrant cos., also Comanche Co. (HPC); mainly se and e TX w to Rolling Plains and Edwards Plateau. Apr–May. [*Plantago pusilla* Nutt.]

Plantago helleri Small, (for Amos Arthur Heller, 1867–1944, Pennsylvania botanist and plant collector of w America), CEDAR PLANTAIN. Annual; flowering stalks with conspicuous, long, spreading hairs; spikes short (to ca. 4 cm long), very thick (8–12 mm); corolla lobes conspicuous. Limestone outcrops; Johnson and Tarrant cos. w and sw to w TX. Apr–Jun.

Plantago heterophylla Nutt., (various-leaved), SLIM-SPIKE PLANTAGO. Annual similar to *P. elongata*. Open ground, roadsides, often sandy soils; Burnet, Dallas, Denton, Hopkins, and Hunt cos., se and e TX w to nc TX, also Edwards Plateau. Mar–May.

Plantago hookeriana Fisch. & C.A. Mey., (for William Jackson Hooker, 1785–1865, director of Kew Gardens), TALLOW-WEED, HOOKER'S PLANTAIN. Annual similar to *P. wrightiana*; leaves linear to narrowly oblanceolate; spikes dense, to 12 cm long and to ca. 8 mm thick. Sandy, gravelly or rocky open areas; Bell and Henderson cos.; se and e TX w to the Trans-Pecos, mainly in the s part of TX. Apr–Jul.

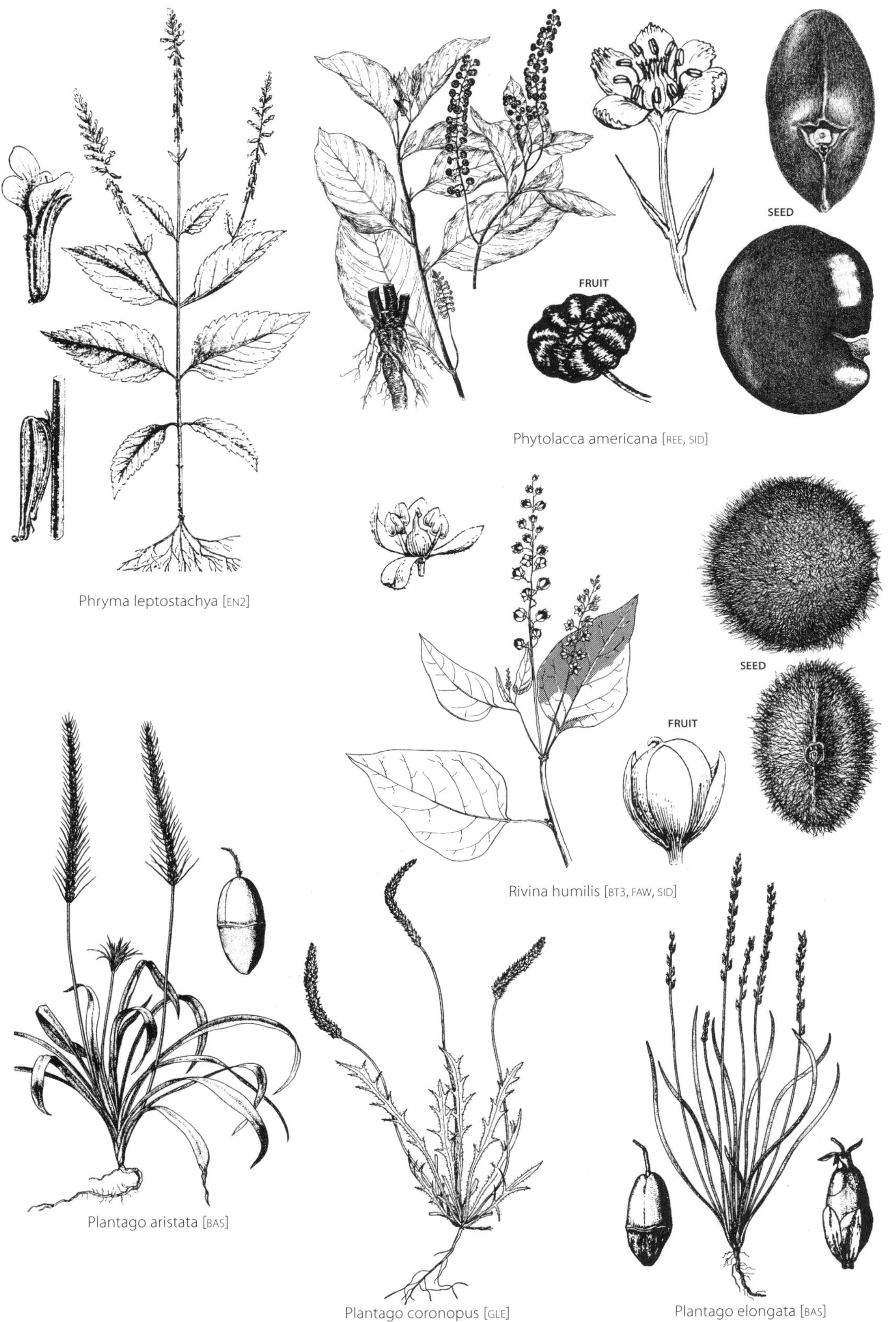

Phytolacca americana [REE, SID]

Phryma leptostachya [EN2]

Rivina humilis [BT3, FAW, SID]

Plantago aristata [BAS]

Plantago coronopus [GLE]

Plantago elongata [BAS]

Plantago lanceolata L., (lanceolate, lance-shaped), BUCKHORN, BUCKHORN PLANTAIN, RIBBON-GRASS, ENGLISH PLANTAIN, RIBWORT, RIB-GRASS, RIPPLE-GRASS. Perennial or biennial; leaves narrowly to broadly lanceolate, prominently ribbed lengthwise; spikes very dense, short and ovoid when young, elongating and cylindric, 1.5–5(–8) cm long, to ca. 10 mm thick. Lawn weed; Dallas and Grayson cos., also Brown Co. (HPC); introduced in scattered localities in TX. Apr–May. Native of Europe.

Plantago major L., (greater, bigger), DOORYARD PLANTAIN, COMMON PLANTAIN, WHITEMAN'S-FOOT, BROAD-LEAF PLANTAIN, LANTEN, GREAT PLANTAIN. Perennial; leaves petioled, large, to 30 cm long including petiole, ovate to elliptic, usually strongly ribbed; spikes loosely flowered to dense, typically dense above, the axis often exposed below, to 22 cm long and to ca. 8 mm thick. Low areas; Dallas Co.; introduced in scattered localities in TX. May–Oct. Native of Europe.

Plantago patagonica Jacq., (of Patagonia), BRISTLE-BRACT PLANTAIN. Annual, gray- or yellowish pubescent with woolly or cottony hairs; corolla lobes 1–2 mm long; spike (excluding bracts) 4–8 mm thick. Sandy or gravelly stream bottoms, roadsides, and fields; nearly throughout TX. Apr–Jun. [*Plantago patagonica* Jacq. var. *breviscapa* (Shinners) Shinners, *Plantago patagonica* Jacq. var. *gnaphalioides* (Nutt.) A. Gray, *Plantago patagonica* Jacq. var. *spinulosa* (Decne.) A. Gray, *P. purshii* Roem. & Schult., *P. purshii* Roem. & Schult. var. *breviscapa* Shinners, *P. purshii* Roem. & Schult. var. *spinulosa* (Decne.) Shinners] A variable species previously divided into a number of varieties.

Plantago rhodosperma Decne., (red-seeded), TALLOW-WEED, RED-SEED PLANTAIN. Annual; leaves frequently toothed or lobed; spikes dense, to 20 cm long and to ca. 10 mm thick. Prairies, pastures, roadsides, and disturbed areas; throughout TX but mainly Blackland Prairie westward. Apr–May.

Plantago rugelii Decne., (for its discoverer, Ferdinand Rugel, 1806–1879, German-born planter and explorer in se U.S.), COMMON PLANTAIN, RUGEL'S PLANTAIN. Perennial similar to *P. major*; leaves petioled, large, broadly ovate to elliptic, to 13 cm wide, prominently and usually 5-ribbed; spikes slender, loosely flowered to dense, the axis sometimes visible below, to 22(–50) cm long and to ca. 8(–12) mm thick. Low, moist disturbed areas; Grayson and Lamar cos., also Dallas Co. (Lipscomb 1978b); also far e TX. Jun–Sep.

Plantago virginica L., (of Virginia), PALE-SEED PLANTAIN, DWARF PLANTAIN, HOARY PLANTAIN. Pubescent annual similar to *P. rhodosperma*, 5–30 cm tall; leaves lanceolate to elliptic, entire or rarely toothed; spikes dense to interrupted, to 20(–25) cm long and to ca. 8 mm thick. Sandy open woods, roadsides, and fields; se and e TX w to East Cross Timbers, also Edwards Plateau. Mar–May.

Plantago wrightiana Decne., (for Charles Wright, 1811–1885, TX collector), WRIGHT'S PLANTAIN. Annual, 12–40 cm tall, pubescent; leaves linear-lanceolate; spikes dense, to 10 cm long and to ca. 9 mm thick; corollas conspicuous. Rocky or sandy open ground; Blackland Prairie s and w to w TX. Late Apr–Jun. [*P. hookeriana* Fisch. & C.A. Mey. var. *nuda* (A. Gray) Poe]

PLATANACEAE SYCAMORE OR PLANETREE FAMILY

A very small monogeneric family of ca. 8 species (Kaul 1997) of wind-pollinated trees often with visually striking bark; several are widely cultivated; with the exception of 3 Old World species, all are North American (Kaul 1997). (subclass Hamamelidae)

FAMILY RECOGNITION IN THE FIELD: the only TX species is a tree with large, shallowly palmately lobed or coarsely toothed, alternate leaves and usually *conspicuously exfoliating* bark; buds completely sheathed by the petiole base; flowers in spherical heads; fruits in *dangling ball-like heads.*

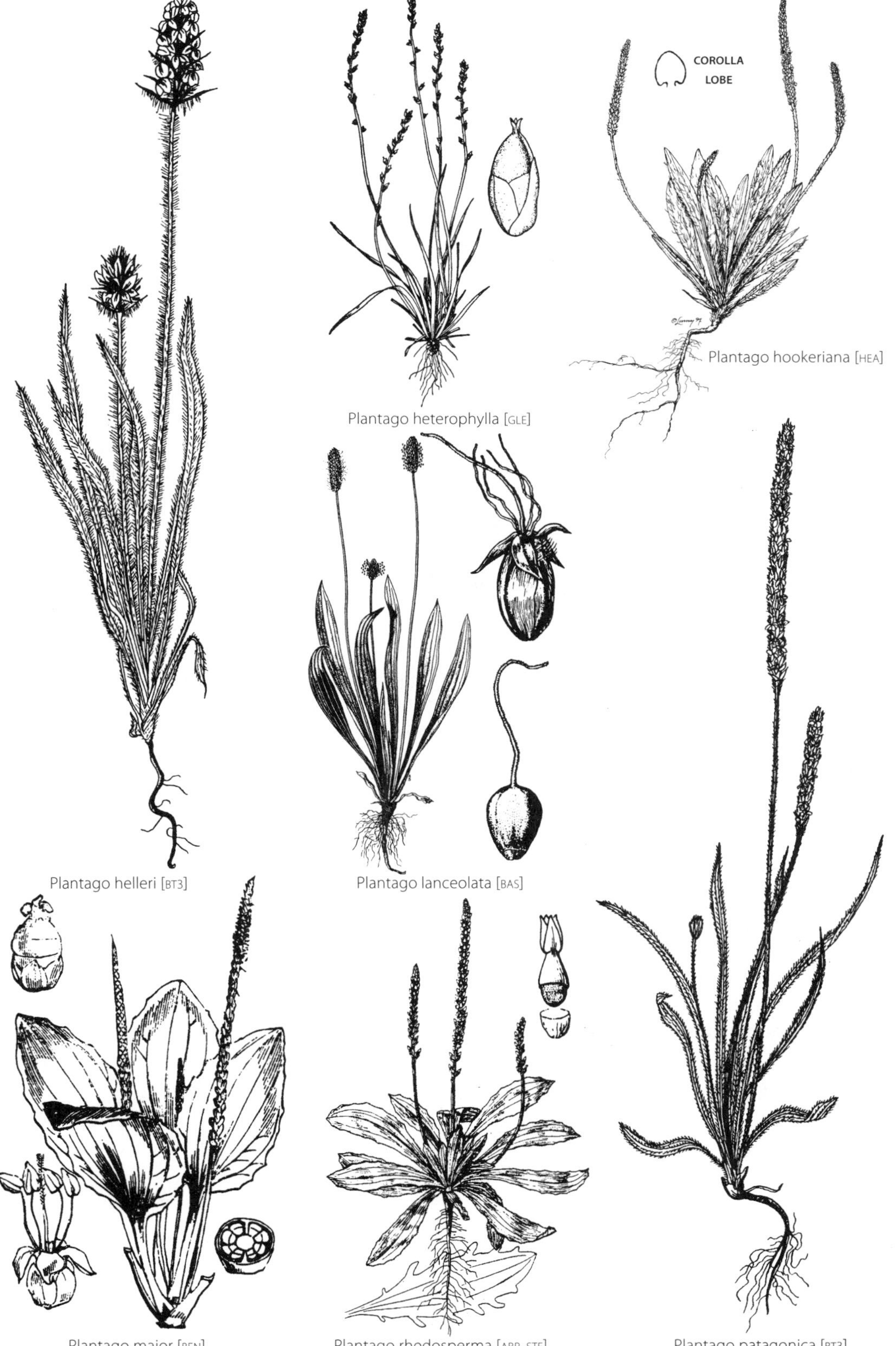

Plantago heterophylla [GLE]

Plantago hookeriana [HEA]

Plantago helleri [BT3]

Plantago lanceolata [BAS]

Plantago major [BEN]

Plantago rhodosperma [ABR, STE]

Plantago patagonica [BT3]

REFERENCES: Boothroyd 1930; Ernst 1963b; Schwarzwalder & Dilcher 1991; Kubitzki 1993e; Manos et al. 1993; Kaul 1997.

PLATANUS SYCAMORE, PLANETREE, BUTTONWOOD

A n hemisphere genus; some are important as ornamentals or as a source of timber for furniture, pulp, or paneling. The wood is resistant to splitting, making it useful for butcher blocks and buttons, thus the old vernacular name BUTTONWOOD (Kaul 1997). (The ancient name from Greek: *platys*, broad, apparently referring to the large leaves)

Platanus occidentalis L., (western), AMERICAN SYCAMORE. Broad-headed tree with smooth or flaky bark mottled white, tan, and green, the trunk thus often quite striking in appearance; leaves deciduous, alternate, long-petioled, the petiole hollow at base and completely surrounding the axillary bud; leaf blades simple, large, to ca. 35 cm long and wide, truncate to cordate at base, shallowly palmately lobed or coarsely toothed, the lobes or teeth acute or acuminate, the blade surfaces at first coated with soft, whitish, stellate hairs, becoming sparsely pubescent in age; stipules large, conspicuous, united around the twigs, falling when leaves expand; flowers unisexual, in separate, long-peduncled, drooping, globular heads on the same tree; sepals and petals minute; stamens and styles several per flower, long and conspicuous; pistillate heads 3–4 cm in diam. in fruit; fruits indehiscent, single-seeded, surrounded at base by a tuft of tawny hairs. Stream bottoms; more common in some areas as a planted tree than a wild one; se and e TX w to Grand Prairie and Edwards Plateau. Late Mar–early Apr.

POLEMONIACEAE PHLOX OR POLEMONIUM FAMILY

Low annual, biennial, or perennial herbs or suffrutescent; leaves alternate or opposite, sessile or short-petioled, simple, entire to very deeply pinnately lobed; stipules absent; flowers radially symmetrical or slightly bilaterally symmetrical; calyces 5-toothed or -lobed, the tube scarious between the green ribs; corollas 5-lobed, rotate to tubular-funnelform to salverform; stamens 5, epipetalous; pistil 3-carpellate; ovary superior; style 3-lobed; fruit a capsule.

A small (290 species in 20 genera), American (especially w North America) and Eurasian family; it is a vegetatively variable group of usually herbs or shrubs, lianas, or small trees; it contains a number of showy ornamentals including *Gilia*, *Phlox*, and *Polemonium* (JACOB'S-LADDER). The family is apparently closely related to the Hydrophyllaceae. Family name from *Polemonium*, a genus of 25 species of annual and perennial herbs with alternate pinnate leaves native from the n temperate zone s to Mexico and Chile. (Greek: *polemonium*, name used by Dioscorides, perhaps for Polemon, an Athenian philosopher, or possibly from *polemos*, war) (subclass Asteridae)

FAMILY RECOGNITION IN THE FIELD: usually herbs; flowers usually radially symmetrical with 5 lobed calyx, 5-lobed corolla often with a *conspicuous tube*, 5 stamens attached high inside the corolla tube and alternating with the corolla lobes, and usually a 3-celled superior ovary with a 3-lobed style.

REFERENCES: Grant 1959; Wilson 1960b; Grant & Grant 1965; Wherry 1966.

1. Leaves incised to pinnately divided, sometimes appearing compound, all alternate; stamens exserted from corolla tube.
   2. Plants 20–200 cm tall; corollas deep red, rarely yellow or white, with speckled throat; corolla tube 20–25 mm long, ca. 2 times or more as long as corolla lobes; widespread in nc TX ______ **Ipomopsis**
   2. Plants 50 cm or less tall; corollas whitish to lavender, or blue-violet with a yellow eye; corolla tube < 6 mm long, shorter than to ca. as long as corolla lobes; rare in nc TX ______ **Gilia**
1. Leaves entire or nearly so, all opposite OR opposite below, becoming alternate above; stamens not or barely exserted from the corolla tube ______ **Phlox**

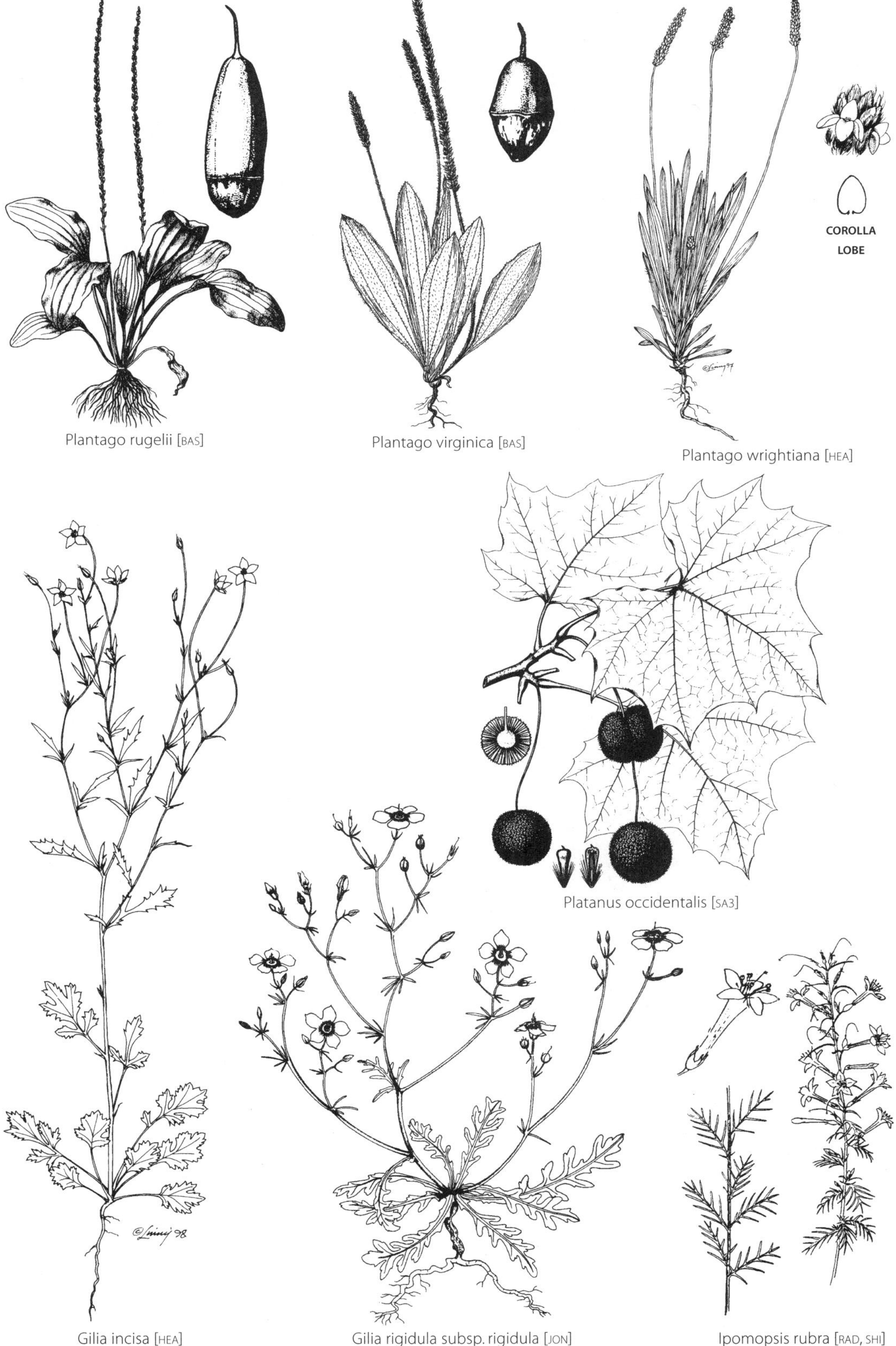

Plantago rugelii [BAS]

Plantago virginica [BAS]

Plantago wrightiana [HEA]

Platanus occidentalis [SA3]

Gilia incisa [HEA]

Gilia rigidula subsp. rigidula [JON]

Ipomopsis rubra [RAD, SHI]

GILIA

Glandular-pubescent annuals to perennials; leaves alternate; flowers solitary, in pairs, or loosely glomerate; corollas in ours subrotate to rotate.

A New World genus of ca. 25 species, especially in w North America; some are cultivated as ornamentals. (for Felipe Gil, 18th century Spanish botanist)
REFERENCES: Shinners 1963; Turner 1994a.

1. Lower stem leaves with long slender petioles and with wide, flat, incised-toothed blades or blade segments; basal leaves often simple, serrate to deeply so; corollas lavender to whitish; sepals 4–5 mm long; corollas 4–7(–8) mm long ____ **G. incisa**
1. Lower stem leaves sessile or subsessile or with petioles short and winged, the blades pinnatifid to pinnate, the segments needle-like to narrowly oblong; basal leaves pinnatifid to pinnate; corollas violet-blue to purple, with a yellow eye; sepals usually 5–7(–9) mm long (some poorly developed flowers can have shorter sepals); corollas typically (7–)8–15 mm long ____ **G. rigidula**

Gilia incisa Bentham, (incised, cut deeply into irregular lobes), SPLIT-LEAF GILIA. Annual, biennial, or perennial to 50 cm tall; leaves conspicuously reduced up the stem; leaf blades to 15 mm wide; flowers solitary or paired, on peduncles to 35 mm long; corollas lavender to whitish; corolla tube 3–4 mm long. Rocky, gravelly, silty, or sandy soils; Bell Co. in s part of nc TX, also Burnet Co. (Turner 1994a); mainly s and w TX. Late Mar–early Jun.

Gilia rigidula Bentham, (somewhat stiff or rigid). Suffrutescent perennial, to 25 cm tall, diffusely branched at base; leaves with 2–7 segments, the segments 2–12 mm long; flowers solitary to loosely glomerate; corollas distinctive—violet-blue to purple, with a yellow eye; capsules 3–5 mm long. Dry, sandy to rocky prairies.

1. Leaf segments mostly stiffly linear, needle-like (= acerose), sharp-tipped; w and n parts of nc TX ____ subsp. **acerosa**
1. Leaf segments typically broader, usually not needle-like; s part of nc TX ____ subsp. **rigidula**

subsp. **acerosa** (A. Gray) Wherry, (needle-like). Plant usually < 10 cm tall; leaves to 20 mm long. Callahan Co., also Collin and Shackelford cos. (Turner 1994a); the Collin Co. locality is well to the east of all other known sites for this species; mainly w 1/2 of TX. (Late Mar–)Apr–Sep. [*G. acerosa* (A. Gray) Britton] Wilken (1986a) lumped this subspecies.

subsp. **rigidula**, PRICK-LEAF GILIA. Plant to 25 cm tall; leaves to 35 mm long; peduncles to 25 mm long. Burnet Co., also Milam Co. (Turner 1994a); s part of nc TX s to s TX and w to Edwards Plateau. Late Mar–May(–Oct).

IPOMOPSIS

A genus of 26 species of w North America and Florida, with 1 species in South America. Some are showy and cultivated as ornamentals. (Greek: *ipo*, to impress, and *opsis*, appearance, alluding to the showy flowers)
REFERENCES: Grant 1956; Shinners 1963.

Ipomopsis rubra (L.) Wherry, (red), STANDING-CYPRESS, TEXAS PLUME, RED GILIA, INDIAN-PLUME. Glabrous winter annual or biennial with a prominent basal rosette; stems normally unbranched, erect; leaves alternate, numerous, crowded, deeply pinnately divided into linear to filiform segments, appearing compound, 4–8 cm long, with ca. 10–15 segments 5–30 mm long and ca. 1 mm wide; flowers in showy terminal spikes; corollas tubular-funnelform, deep red, rarely yellow or white, with speckled throat, the tube 20–25 mm long, the lobes 9–11 mm long; stamens bluish; capsules 8–10 mm long. Sandy, gravelly, or rocky ground; e 1/2 of TX. May–Jul. [*Gilia rubra* (L.) A. Heller] This is one of the most striking wildflowers in nc TX; it is the only member of the genus adapted to relatively moist climates (Wherry 1966). Ruby-throated hummingbirds (*Archilochus colubris*) pollinate the flowers (Grant & Grant 1965). ▣/94

PHLOX

Annuals or perennials; leaves opposite or opposite below and alternate above; flowers radially symmetrical, solitary or in cymes; corollas salverform with conspicuously distinct tube and lobes.

A genus of 67 species of North America and ne Asia; many are cultivated as ornamentals. *Phlox cuspidata* and *P. drummondii* are known to hybridize and form hybrid swarms (Correll & Johnston 1970). Lepidopterans (butterflies & moths) pollinate the slender-tubed flowers (Grant & Grant 1965). (Greek: *phlox*, flame, an ancient name of *Lychnis* of Caryophyllaceae, transferred to this genus)

REFERENCES: Nelson 1899; Kelly 1915; Bogusch 1929; Whitehouse 1939, 1945; Wherry 1955, 1966; Erbe & Turner 1962.

1. Plants annual from a slender taproot, easily pulled from ground with roots attached.
   2. Corolla lobes shorter than the tube; corolla tube not constricted at junction with corolla lobes; corollas without conspicuous yellow eye.
      3. Middle and upper stem leaves 3–6 times as long as wide, oblong-ovate or oblong-lanceolate; corolla lobes usually > 7 mm wide; corolla color variable; foliage pubescence typically conspicuous to the naked eye ____ **P. drummondii**
      3. Middle and upper stem leaves 7–10 times as long as wide, oblong-lanceolate to linear; corolla lobes usually < 7(–8) mm wide; corollas purple to pink, with faintly striate pale eye; foliage pubescence typically inconspicuous ____ **P. cuspidata**
   2. Corolla lobes longer than the tube; corolla tube constricted at junction with corolla lobes; corollas with conspicuous yellow eye ____ **P. roemeriana**
1. Plants perennial from a crown or branching root; stems typically breaking off near ground level when pulled.
   4. Corolla lobes obtuse to apiculate, not notched; plants (20–)30–60(–75) cm tall; widespread in nc TX ____ **P. pilosa**
   4. Corolla lobes notched; plants 20 cm or less tall; known in nc TX only from Dallas Co. ____ **P. oklahomensis**

Phlox cuspidata Scheele, (with a cusp or sharp stiff point), POINTED PHLOX, LARGE-FLOWER PHLOX. Stems 5–55 cm tall; usually smaller and more finely pubescent than *P. drummondii*; leaves to 35 mm long and 5 mm wide; corolla tube 8–15 mm long, pilose, some hairs glandular. Sandy or silty clay prairies and open woods; se and e TX w to East Cross Timbers, also Somervell Co. (sandy terrace of Brazos River), also Hood Co. (R. O'Kennon, pers. obs.). Apr–Jun. [*P. cuspidata* var. *grandiflora* Whitehouse, *P. cuspidata* var. *humilis* Whitehouse]

Phlox drummondii Hook., (for its discoverer, Thomas Drummond, 1780–1835, Scottish botanist and collector in North America), PRIDE-OF-TEXAS. Erect or half-decumbent annual, 10–50 cm tall, densely pubescent with flattened, jointed, coarse hairs; corolla color variable, the tube 12–16 mm long. Sandy open ground. Apr–May.

1. Corolla color varying conspicuously within colonies, ranging from red to pink, purple, violet, lavender, white, or even pale yellow ____ subsp. **drummondii**
1. Corolla color not noticably variable within colonies, either red, lavender, or purplish.
   2. Corollas bright red, with a dark red eye-ring or star ____ subsp. **wilcoxiana**
   2. Corollas lavender or purplish, eye various.
      3. Eye pale or white with at most a narrow-rayed purple star; many leaves only 4–5 times as long as wide; native to nc TX ____ subsp. **mcallisterii**

3. Eye well-filled by a dark red star or ring; main (lower) leaves 5–10 times as long as wide (averaging 7); native to the s of nc TX, but widely cultivated ____ subsp. **drummondii**

subsp. **drummondi**, DRUMMOND'S PHLOX. Including [var. *peregrina* Shinners] which is very variable in color and which is cultivated and escaped or planted by the highway department. Grayson Co. (Hagerman National Wildlife Refuge) [probably planted]; native to sc TX s of nc TX. /103

subsp. **mcallisterii** (Whitehouse) Wherry, (for Dr. Frederick McAllister, born 1876, professor of botany at Univ. of TX). Stems erect; leaves thin; corollas lavender with white eye or whitish lines around center. Open oak woods and roadsides; e TX w to Rolling Plains and Edwards Plateau. [*P. drummondii* var. *mcallisterii* (Whitehouse) Shinners]

subsp. **wilcoxiana** (Bogusch) Wherry, (". . . from the geologic formation upon which the [sub]species is found most abundantly"—Bogusch 1929). Main leaves 5–10 times as long as wide (averaging 7). Milam Co. (Wherry 1966; Correll & Johnston 1970); se, e, and s TX and Edwards Plateau, mainly s of nc TX; endemic to TX. [*P. drummondii* var. *wilcoxiana* (Bogusch) Whitehouse] /103

Phlox oklahomensis Wherry, (of Oklahoma). Rhizomatous to stoloniferous, subshrubby perennial, 8–20 cm tall and broad, pubescent; leaves linear to lanceolate, to 6 cm long and 5 mm wide; flowers fine glandular to glandless; sepals 7–10 mm long; corolla tube 8–12 mm long; corolla lobes white or tinged with red, with apical notch 1–4 mm deep; styles united to middle, 2–3 mm long. The only known collection for Texas was from sandy woods in ne Dallas Co. (Wherry 1966; Correll & Johnston 1970); also KS and OK.

Phlox pilosa L., (with long soft hairs), PRAIRIE PHLOX. Pubescent perennial from branching roots; stems (20–)30–60(–75) cm tall; leaves 3–9 mm wide; corollas purple, pink, lavender, or rarely white, the tube 10–20 mm long. Apr–May. Wherry (1966) noted that the subspecies of this markedly variable species intergrade and that further study is needed to determine their real relationships; intermediates between the subspecies should be expected.

1. Sepals subulate with awn 1.5–3 mm long; main leaves 40–90 mm long; widespread in nc TX ____ subsp. **pilosa**
1. Sepals linear-subulate with awn 1–2 mm long; main leaves 30–60 mm long; s and w parts of nc TX.
   2. Stems typically simple, moderately glandular; main leaves 45–60 mm long; s and w parts of nc TX ____ subsp. **latisepala**
   2. Stems typically branched, copiously glandular; main leaves 30–45 mm long; in nc TX known only from Bell and Williamson cos. ____ subsp. **riparia**

subsp. **latisepala** Wherry, (broad-sepaled), ROUGH PHLOX. Open woods on dry slopes, sometimes in grasslands, often over limestone; Bell, Bosque, Burnet, Comanche, Hood, Lampasas, McLennan, Tarrant, and Williamson cos. (Wherry 1966); endemic to Edwards Plateau and adjacent areas. [*P. asper* E. Nelson, *P. pilosa* var. *asper* (E. Nelson) Wherry ex Gould] According to Wherry (1966), this subspecies intergrades with both subspecies *pilosa* and subspecies *riparia*.

subsp. **pilosa**, DOWNY PHLOX. Herbage glandular, at least above. Prairies and open woods; se and e TX w to West Cross Timbers and Edwards Plateau. This is the commonly seen subspecies in nc TX.

subsp. **riparia** Wherry, (of river banks), TEXAS PHLOX. Gravelly areas and talus slopes; Bell and Williamson cos. (Wherry 1966); endemic on and near Edwards Plateau.

Phlox roemeriana Scheele, (for Ferdinand Roemer, 1818–1891, geologist, paleontologist, and explorer of TX), GOLD-EYE PHLOX, ROEMER'S PHLOX. Plants to 35 cm tall, densely pubescent with

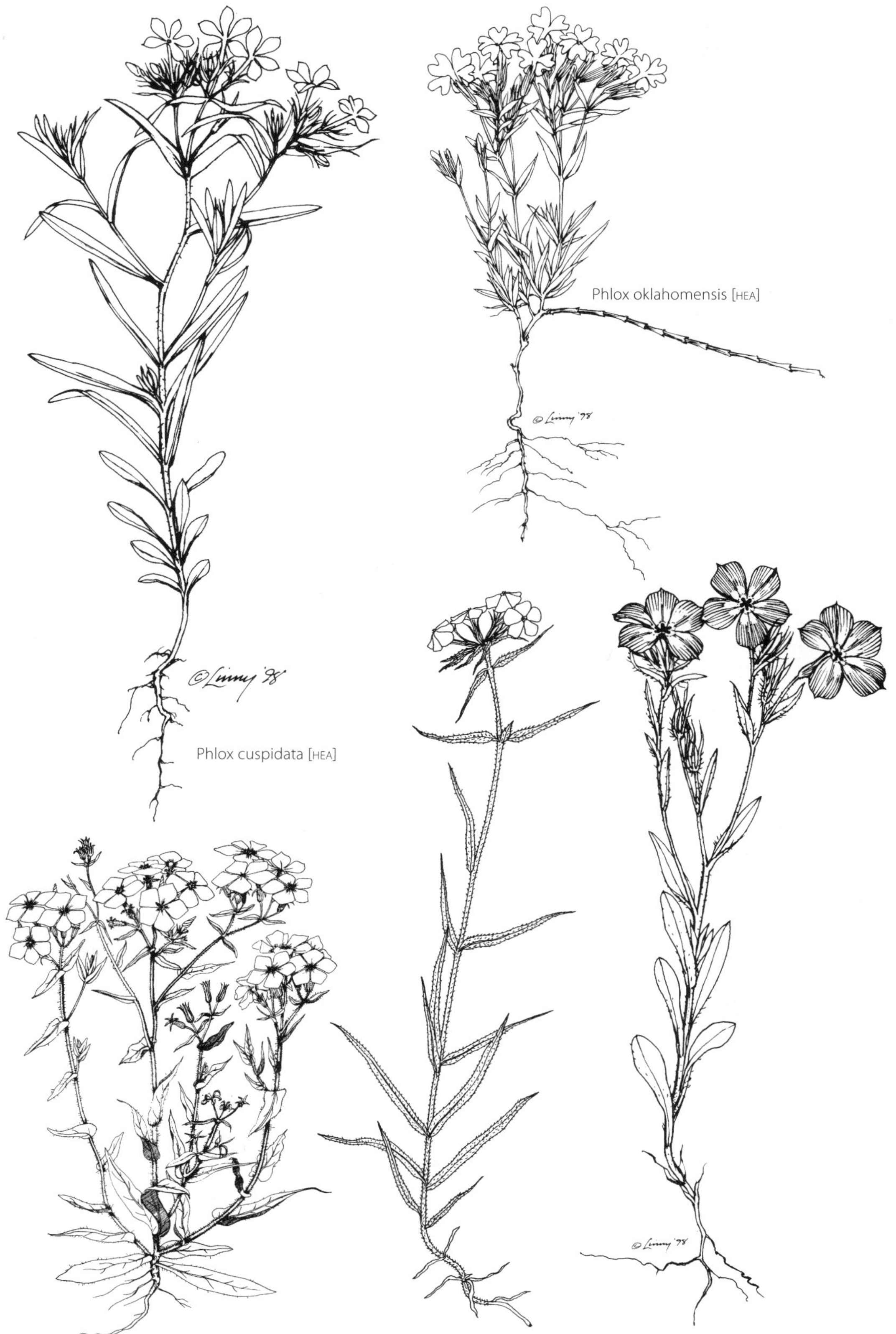

Phlox oklahomensis [HEA]

Phlox cuspidata [HEA]

Phlox drummondii subsp. drummondii [JON]

Phlox pilosa subsp. pilosa [BT3]

Phlox roemeriana [HEA]

flattened, jointed, coarse hairs; corollas purple, pink, or rarely white, with conspicuous yellow eye bordered by white, the tube 9–13 mm long, the lobes averaging 14 mm long; seeds 3–5 in each cell. Rocky slopes, limestone areas; Bell, Bosque, Brown, Burnet, and McLennan cos., also Comanche, Hood, Johnson (Whitehouse 1945), Mills, Williamson (Wherry 1966), and Hamilton (HPC) cos.; s and sw parts of nc TX s to Edwards Plateau and w to Plains Country; endemic to TX. Late Mar–early May. /103

POLYGALACEAE MILKWORT FAMILY

A medium-sized (950 species in 17 genera), vegetatively variable, nearly cosmopolitan family whose flowers superficially resemble those of the unrelated Fabaceae. Species range from trees to shrubs, lianas, or herbs; some are parasites and extra-floral nectaries are often present. A relationship with the Malpighiaceae-Vochysiaceae complex has been suggested. (subclass Rosidae)

FAMILY RECOGNITION IN THE FIELD: herbs (1 species woody-based) with a distinctive, bilaterally symmetrical perianth—appearing *pea-like*—with 5 sepals (2 larger and petal-like) and 3 petals (1 often crested); anthers dehiscing through terminal pores or slits.

REFERENCES: Blake 1924; Graham & Wood 1965; Miller 1971b.

POLYGALA MILKWORT

Ours small annual or perennial herbs or 1 species a woody-based perennial (*P. lindheimeri*); sap not milky; leaves alternate, opposite, or whorled, sessile or subsessile, simple, entire; flowers in terminal or axillary, spike-like or head-like racemes; sepals 5, unequal, 2 longer and petal-like; petals 3, united at base, unequal, the lower one keeled, with variously shaped or crested apex; filaments united into a tubular sheath split down upper side; anthers 8; pistil 2-carpellate; ovary superior; capsules dehiscent; seeds usually arillate.

A subcosmopolitan genus of ca. 500 species of trees, shrubs, and herbs. Once thought to increase the yield of cow's milk—hence the common name. A number of species are used ornamentally or medicinally (e.g., the North American *P. senega* L. (SENEGA-ROOT or SENECA-SNAKE-ROOT) was used for treating snakebite.) (Greek: *polys*, much, and *gala*, milk; some European species were thought to enhance milk production in cows)

REFERENCES: Blake 1916; Pennell 1931, 1933; Smith & Ward 1976; Wendt 1979; McGregor 1986.

1. Stems densely pubescent; plants with woody base; keel (= lower of the 3 petals) with a conic or cylindric beak, not crested ____ **P. lindheimeri**
1. Stems glabrous; plants without woody base (or in *P. alba* a woody taproot); keel with a fimbriate crest.
   2. Leaves whorled or apparently so.
      3. Inflorescences slender, 2.2–4.5 mm wide, tapered to a point; flowers usually whitish and greenish (rarely purple-tinged) ____ **P. verticillata**
      3. Inflorescences thick, 10–17 mm wide, blunt or rounded at end; flowers rosy purple to greenish (rarely white) ____ **P. cruciata**
   2. Leaves alternate.
      4. Plants perennial; stems usually several from base.
         5. Flowers white with greenish center (but can have purple crest); inflorescences in flower 4–8 mm thick; plants without cleistogamous flowers underground ____ **P. alba**
         5. Flowers pink to pink-purple (rarely white); inflorescences in flower 9–14 mm thick; plants with whitish cleistogamous flowers underground ____ **P. polygama**
      4. Plants annual; stems usually single from base.

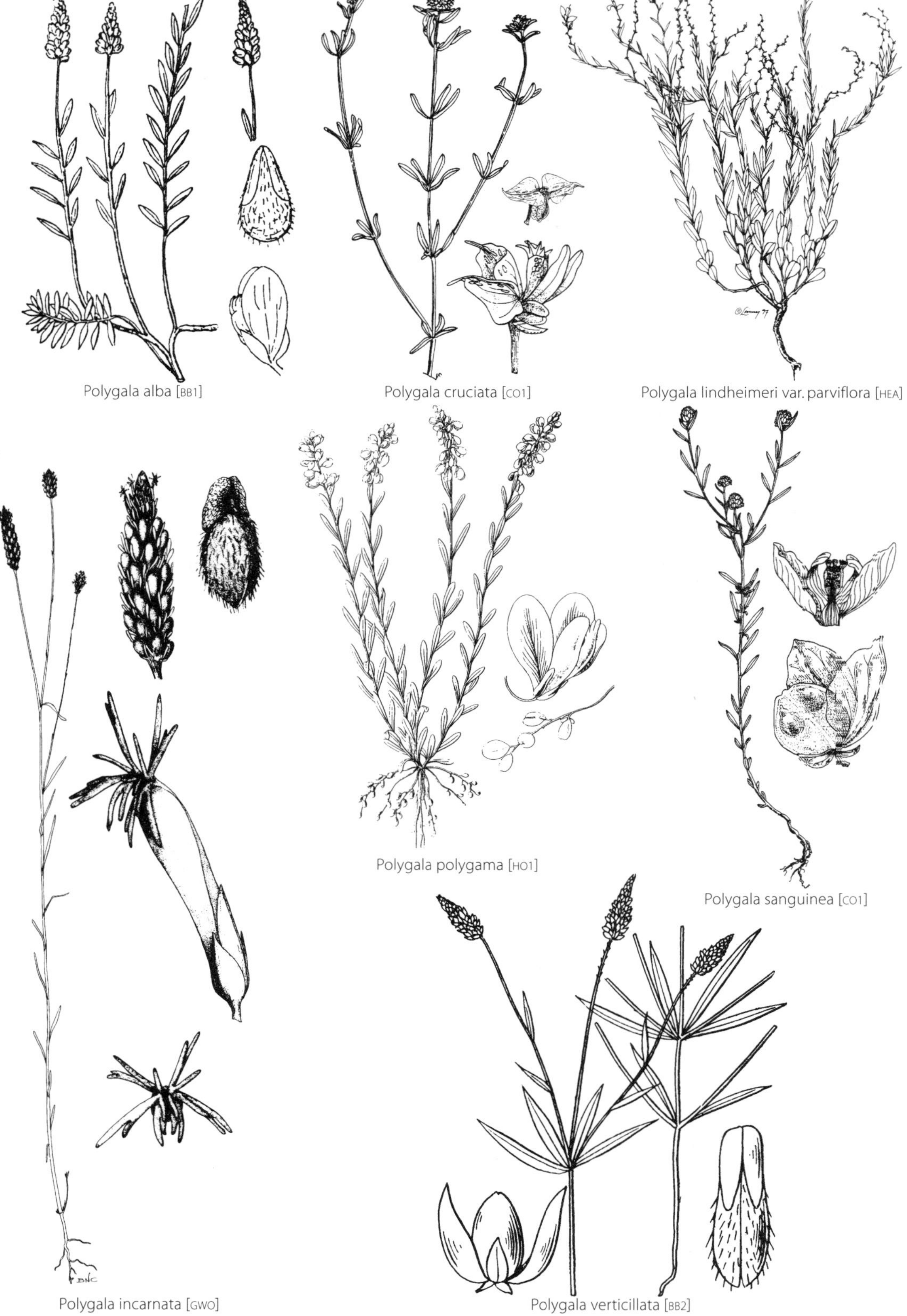

Polygala alba [BB1]

Polygala cruciata [CO1]

Polygala lindheimeri var. parviflora [HEA]

Polygala polygama [HO1]

Polygala sanguinea [CO1]

Polygala incarnata [GWO]

Polygala verticillata [BB2]

6. Leaves 0.3–1 mm wide, 4–12 mm long; petals and staminal tube united into a conspicuous narrow tube-like trough 5 mm long; stem and leaves glaucous ______ **P. incarnata**
6. Leaves 1–4.5 mm wide, 7–39 mm long; petals and staminal tube not united into a conspicuous narrow tube-like trough; stem and leaves not glaucous ______ **P. sanguinea**

Polygala alba Nutt., (white), WHITE MILKWORT. Ascending or erect perennial from a stout vertical rootstock, to 45 cm tall; leaves narrow, crowded toward base of stems; flowers white with greenish center (but crest can be purple). Rocky and sandy areas; nearly throughout TX. Mid-Apr–Jun.

Polygala cruciata L., (cross-like), MARSH MILKWORT. Annual to 40 cm tall; leaves in whorls of 3 or 4 or scattered; flowers rosy purple to greenish (rarely white). Bogs, seepage slopes, and savannahs; included based on citation of vegetational area 4 (Fig. 2) by Hatch et al. (1990); mainly e TX. May–Sep.

Polygala incarnata L., (flesh-colored), PINK MILKWORT, SLENDER MILKWORT. Glaucous, erect, simple or sparingly branched annual to 50 cm tall, with few leaves; petals rose-purplish. Low sandy or silty open ground or open woods; Cooke, Grayson, Kaufman, Milam, Montague, and Wise cos., also Lamar Co. (Carr 1994); se and e TX w to West Cross Timbers. May–Jul.

Polygala lindheimeri A. Gray, (for Ferdinand Jacob Lindheimer, 1809–1879, German-born TX collector). Stems several–many from a woody base, decumbent to erect, to 28 cm long; stems and leaves with spreading or incurved hairs; flowers pink to purple or lavender (rarely whitish).

1. Leaves essentially similar throughout, narrowly elliptic to oval or rarely suborbicular, the pubescence densely spreading or sometimes spreading-incurved (like on the stem) ______ var. **lindheimeri**
1. Leaves mostly linear to lanceolate but sometimes with the lower ones broader, the puberulence merely incurved (like on the stem) ______ var. **parviflora**

var. **lindheimeri**, SHRUBBY MILKWORT. Brushy limestone hills and bluffs; sw TX ne to Burnet Co. near s edge of nc TX. Mar–Oct.

var. **parviflora** Wheelock, (small-flowered), ROCK MILKWORT. Limestone outcrops or gravelly areas; Burnet, Coleman, and Lampasas cos., also Brown (HPC) and Hood (Mahler 1988) cos.; mainly Trans-Pecos to Edwards Plateau ne to sw part of nc TX. Jun–Jul. [*P. tweedyi* Britton ex Wheelock]

Polygala polygama Walter var. **obtusata** Chodat, (sp.: with sexes mixed, having flowers with both male and female parts, presumably for the cleistogamous flowers; var.: obtuse, blunt), BITTER MILKWORT, RACEMED MILKWORT. Stems ascending to erect, 15–30 cm tall; flowers pink to rose-purplish (rarely lighter to white), also with racemes of whitish cleistogamous (= non-opening, self-pollinating) flowers from base of plant growing underground or at soil surface. Bogs, low sandy soils, and along streams; se and e TX disjunct w to Parker and Tarrant cos. Apr–Jun. Jones et al. (1997) and J. Kartesz (pers. comm. 1997) treated all TX representatives of this species as var. *obtusata* (but spelled *obtusa* in Jones et al.).

Polygala sanguinea L., (blood-red), BLOOD MILKWORT. Erect annual 10–40 cm tall; flowers rose to purple or green (rarely white). Moist woods, prairies; Lamar Co. in Red River drainage; mainly e TX. May–Jul.

Polygala verticillata L., (whorled), WHORLED MILKWORT. Erect annual to 30 cm tall, widely branched; leaves in whorls of 4–5, rarely the upper alternate; flowers white or partly green, rarely purplish-tinged. Sandy open woods; se and e TX w to e edge of West Cross Timbers and Edwards Plateau. Mid-Apr–Jul. [*P. verticillata* L. var. *ambigua* (Nutt.) A.W. Wood, *P. verticillata* L. var. *isocycla* Fernald, *P. verticillata* L. var. *sphenostachya* Pennell] We are following Hatch et al.

(1990) in not recognizing infraspecific taxa in *P. verticillata*. Kartesz (1994) and Jones et al. (1997) recognized three varieties in TX. The three, which can intergrade, are sometimes separated using characters such as the following (Pennell 1931; McGregor 1986):

1. Leaves all or mostly whorled; plants with spreading branches; flowers and fruits crowded, the racemes ± conical, 0.5–2 cm long.
   2. Plants 5–20 cm tall; leaves 1–2.5 cm long; capsules to 1.8 mm long ____ var. *isocycla*
   2. Plants 20–30 cm tall; leaves 2–3 cm long; capsules 1.8–2.4 mm long ____ var. *sphenostachya*
1. Upper leaves alternate; plants less branched above; flowers and fruits less crowded (the lower ones remote), the racemes long-cylindric, slender, 1–5 cm long ____ var. *ambigua*

POLYGONACEAE KNOTWEED FAMILY

Ours annuals or perennials, mostly herbs or a few vines, these rarely woody; leaves alternate, simple, entire or with very small, blunt teeth (sometimes with basal lobes), with scarious stipules united to form a tubular sheath known as an ocrea (plural: ocreae) (ocreae absent from *Eriogonum*, apt to shrivel or break apart in others); perianth usually of (4–)5–6 tepals, in 1 or 2 rows, equal or the outer shorter, colored alike; stamens 1–9; pistil 1; ovary superior, unilocular, with a single basal ovule; fruit a trigonous or lenticular achene.

A medium-large (1,100 species in 46 genera), nearly cosmopolitan, but chiefly n temperate family of shrubs, trees, lianas, and many herbs. The nodes are often swollen and the family is unusual in its subclass in having anthocyanin rather than betalain pigments. The ocrea is usually an excellent field character. Species are variously used as edible plants, for timber, as a source of tanning material, or as cultivated ornamentals; some are problematic weeds. The achenes of *Fagopyrum esculentum* Moench (BUCKWHEAT) are used for food as are the young petioles of *Rheum rhabarbarum* L. (RHUBARB, PIEPLANT); however, RHUBARB leaf blades contain oxalates and anthraquinone glycosides and are toxic (Spoerke & Smolinske 1990). A number of characters including the often 3-merous flowers of Polygonaceae suggests to some that, "The ancestry of Caryophyllidae [including Polygonaceae] may lie in or near Ranunculaceae." (Cronquist 1993) (3-merous flowers are known in a number of Ranunculaceae). (subclass Caryophyllidae)

FAMILY RECOGNITION IN THE FIELD: herbs (and a few vines) often with *enlarged* nodes; leaves alternate, simple, with a *tubular, membranous or papery sheath* (= ocrea) enclosing the base of the petiole; flowers small; fruit an often 3-angled or lens-shaped achene.

REFERENCES: Graham & Wood 1965; Horton 1972; McNeill 1981; Brandbyge 1993.

1. Plant not a vine, OR if an herbaceous vine, then without tendrils.
   2. Ocreae (stipules) absent; flowers short-pedicelled in small, funnelform involucres of whorled, partially united bracts; leaves with white to gray pubescence beneath, sometimes strikingly so ____ **Eriogonum**
   2. Ocreae present (often falling from older parts); flowers not involucrate, the inflorescences various; leaves without white to gray pubescence beneath (however, some hairs can be present).
      3. Branches appearing to emerge from internodes (due to short coalescence of base of branch and stem); plants heath-like with many, small, usually linear leaves 5–20 mm long; inflorescences branched, panicle-like ____ **Polygonella**
      3. Branches emerging directly from nodes; plants not heath-like; leaves not linear OR if linear then the inflorescences not panicle-like.
         4. Outer tepals markedly shorter than the inner tepals in flower, the outer greatly enlarged in fruit; leaf blades lobed basally in several species, unlobed in other species ____ **Rumex**

4. Outer tepals equaling the inner tepals in flower, not greatly enlarged in fruit; leaf blades never lobed basally ______ **Polygonum**

1. Plant an herbaceous or woody vine with conspicuous tendrils.
   5. Flowers pink or white; leaf blades usually deeply cordate at base; fruiting perianth usually 5–10 mm long, without a flattened wing-like base; introduced cultivar ______ **Antigonon**
   5. Flowers greenish or yellow-green; leaf blades truncate or subcordate at base; fruiting perianth enlarging to 20–30 mm long, with a flattened wing-like base; native in low woods ______ **Brunnichia**

ANTIGONON CORALVINE, MOUNTAIN-ROSE

A genus of 2–3 species native to Mexico and Central America. (Greek: probably from *anti*, against, opposite, and *gonia*, an angle, or *gony*, knee, in reference to the geniculate nodes)

Antigonon leptopus Hook. & Arn., (slim- or slender-stalked), QUEEN'S-WREATH, CORALVINE, CORALLITA, CONFEDERATE-VINE, COAMECATL. Perennial climbing vine to 10 m high; roots enlarged; leaf blades 2–13 cm long, entire, cordate-ovate, acuminate at apex; flowers pendulous in branched showy inflorescences. Cultivated and volunteers mainly in s and e TX; included based on citation of vegetational area 4 (Fig. 2) by Hatch et al. (1990). Summer–fall. Native of Mexico. The tuberous roots are reported to be edible with a nut-like flavor (Mabberley 1987).

BRUNNICHIA EARDROP-VINE

A genus of 3 species, 1 in America and 2 in Africa (Graham & Wood 1965). The wing-like perianth base apparently aids in dispersal. (Named for M.T. Brünnich, a Norwegian naturalist of the 18th century)

Brunnichia ovata (Walter) Shinners, (ovate, egg-shaped), EARDROP-VINE, AMERICAN BUCKWHEAT-VINE, LADIES'-EARDROPS. Glabrous or minutely pubescent, high-climbing to trailing perennial vine; leaf blades ovate, 5–15 cm long, truncate to slightly cordate basally, acute to acuminate apically; ocreae absent; flowers greenish or yellow-green in terminal, panicled, spike-like racemes, these spreading to drooping; fruiting perianths pinkish and showy, with a conspicuous, flattened, wing-like base; achenes 6–7 mm long. Stream bottom woods; Ellis, Henderson, Hunt, Lamar, and Kaufman cos.; se and e TX w to e part of nc TX. Late Jun–Aug. [*B. cirrhosa* Gaertn.]

ERIOGONUM WILD BUCKWHEAT

Annuals or perennials; stems and at least undersurfaces of leaf blades conspicuously gray or white (sometimes with brownish tinge) with cottony, matted pubescence; inflorescence a cymose panicle; flowers short-pedicelled in small, funnelform involucres of whorled, partially united bracts; stamens 9.

A genus of 240 species of mostly w and s North America; taxa range from Alaska to central Mexico and e to the Appalachian Mountains and Florida (Reveal 1989). The species vary from shrubs to herbs and cushion plants and differ from most of the family in lacking ocreae. Several species are cultivated as ornamentals or for their edible leaves and roots; one species, *E. ovalifolium* Nutt., is used as a silver indicator in Montana. (Greek: *erion*, wool, and *gonu*, knee, from the woolly stems and leaves and the swollen nodes)
REFERENCES: Stokes 1936; Reveal 1968, 1989.

1. Perianth (tepals) pubescent externally; involucres 4–6 mm long; basal and lower stem leaves 10–20+ cm long; plants perennial ______ **E. longifolium**
1. Perianth glabrous externally; involucres 2–4 mm long; basal and lower stem leaves 7 cm or less long; plants annual or biennial.

2. Involucres densely white to gray pubescent, 2.5–4 mm long; outer perianth segments obovate ________ **E. annuum**
2. Involucres with only scattered hairs, 2–2.5 mm long; outer perianth segments oblong-cordate ________ **E. multiflorum**

Eriogonum annuum Nutt., (annual), ANNUAL WILD BUCKWHEAT. Annual or biennial 0.5–2 m tall; leaves to 7 cm long; inflorescences elongated with few long branches, usually open; perianth glabrous, dull white or rarely pinkish, maturing to a dark red-brown, the outer tepals obovate; achenes glabrous. Grasslands, disturbed areas; throughout most of TX. May–Nov.

Eriogonum longifolium Nutt., (long-leaved), LONG-LEAF WILD BUCKWHEAT. Perennial herb 1–2 m tall, with thick taproot; basal and lower stem leaves 10–20 (rarely more) cm long; involucres, perianth, and achenes densely white- to silvery-pubescent externally, the perianth glabrous and yellow internally. Sandy soils and calcareous clay; throughout most of TX. Jun–Aug. [*E. longifolium* Nutt. var. *lindheimeri* Gand., *E. longifolium* var. *plantagineum* Engelm. & A. Gray]

Eriogonum multiflorum Benth., (many-flowered), HEART-SEPAL WILD BUCKWHEAT. Annual (biennial?) herb 0.5–2 m tall; leaves to 4 cm long; inflorescences compacted with several short branches; perianth glabrous, white, maturing with red-brown midribs, the outer tepals cordate; achenes glabrous. Sandy soils; e 1/2 of TX. Aug–Nov.

POLYGONELLA JOINTWEED

A genus of 9 species of e North America. (Diminutive of *Polygonum*, a related genus)
REFERENCE: Horton 1963.

Polygonella americana (Fisch. & C.A. Mey.) Small, (of America), SOUTHERN JOINTWEED, SMALL JOINTWEED. Perennial with numerous short leafy branches, prostrate to erect, to nearly 1 m tall, from long, woody taproots; leaves linear to linear-spatulate, 5–20 mm long, to ca. 1 mm wide, about as thick as wide; ocreae not ciliate; inflorescences panicle-like; flowers 1 per ocrea; tepals 5, white, to 3 mm long, the outer 2 becoming reflexed, the inner enlarging to 4 mm long in fruit and becoming pinkish brown. In sand; mainly se and e TX, disjunct w to Parker and Tarrant cos. in nc TX; also reported from Edwards Plateau and Plains Country. Jun–Sep(–Oct). The common name is apparently derived from the conspicuous swollen nodes or "joints" (Ajilvsgi 1984).

POLYGONUM KNOTWEED, SMARTWEED (INCLUDING *PERSICARIA* AND *TOVARA*)

Plants erect or twining vines, often of moist or wet areas; leaves alternate; joint present between petiole and base of leaf blade or not so; flowers in inconspicuous axillary fascicles or in axillary and/or terminal, spike-like or raceme-like inflorescences; tepals usually white or pink; stamens 4–8; achenes lenticular or trigonous.

A nearly cosmopolitan, especially n temperate genus of ca. 172 species; it is often split into 3 genera: *Persicaria* (ca. 150 species), *Polygonum* sensu stricto (20 species), and *Tovara* (2 species) (Mabberley 1997); representatives of all three are found in nc TX and they are separated in the key to species. Some are cultivated as ornamentals, used medicinally, eaten as vegetables, or are problematic weeds. Several have been reported to cause dermatitis in susceptible individuals (Muenscher 1951; Stephens 1980); the common name SMARTWEED is possibly derived from the tendency of the sap to causing "smarting" when gotten on the skin (Kirkpatrick 1992). (Greek: *poly*, many, and *gonu*, knee or joint, from the swollen nodes of the stem)
REFERENCES: Small 1894, 1895; Stanford 1925; Fassett 1949; Li 1952a; Mertens & Raven 1965; Mitchell 1968; McDonald 1980; Wolf & McNeill 1986.

1. Flowers in dense terminal spike-like or raceme-like inflorescences OR axillary and inconspicuous, in either case, neither widely separated nor reflexed in age; plants erect to ascending or twining or prostrate and rooting at nodes; styles 2–3, falling early or not hooked at tip; tepals (4–)5–6.
   2. Plants twining or trailing vines; leaf blades sagittate to cordate to truncate basally (section *Polygonum*).
      3. Outer 3 tepals minutely keeled but not winged, becoming up to 3.5 mm long; achene dull black, closely enclosed by but scarcely exceeded by perianth _______ **P. convolvulus**
      3. Outer 3 tepals conspicuously winged (the wings 0.25–2 mm wide), becoming up to 8 mm long; achene glossy black, closely enclosed by and much exceeded by the perianth _______ **P. scandens**
   2. Plants not twining or trailing vines, rather the stems erect to ascending or prostrate; leaf blades never sagittate, cordate, or truncate basally.
      4. Flowers single or in 2s or 3s at the nodes; joint present between petiole and base of leaf blade; leaf blades 30 mm or less long (part of section *Polygonum*).
         5. Achenes striated (surface roughened with fine lines), unequally trigonous, dull to mildly shiny; tepals 5; stems usually prostrate, sometimes ascending _______ **P. aviculare**
         5. Achenes smooth or striated only on the margins, equally trigonous, shiny; tepals 5 or 6; stems erect or ascending.
            6. Achenes black, marginally striated; stem 4-angled or winged below ocreae; tepals 5; leaves usually with 2 pleats parallel to the midrib _______ **P. tenue**
            6. Achenes reddish brown, generally entirely smooth; stem angles inconspicuous; tepals 6; leaves without pleats _______ **P. ramosissimum**
      4. Flowers in dense terminal spike-like or raceme-like inflorescences; joint not present between base of leaf blade and petiole; leaf blade usually much more than 30 mm long (section *Persicaria*).
         7. Ocreae (= stipules) without cilia at their summits or cilia less than 1 mm long (but surface of ocreae can be strigose).
            8. Tepals reddish pink; raceme 1 per stem or a shorter 2nd one sometimes present, terminal on the stems; ocreae and leaf blades strigose OR glabrous (note: ocreae are without cilia at summit) _______ **P. amphibium**
            8. Tepals green to white or pink; racemes several–numerous, both terminal and lateral; ocreae and leaf blades glabrous or nearly so.
               9. Peduncles with numerous stalked glands (visible under a hand lens); racemes erect or nearly so; tepals without forked veins (anchor-shaped) near apex (visible under hand lens). _______ **P. pensylvanicum**
               9. Peduncles eglandular or glandular but the glands not stalked; racemes nodding OR erect; tepals with OR without forked veins (anchor-shaped) near apex (if present, visible under hand lens).
                  10. Racemes usually nodding; tepals not glandular, with forked veins (anchor-shaped) near apex (visible under hand lens); achenes flat or concave on one or both faces; plants annual; widespread in nc TX _______ **P. lapathifolium**
                  10. Racemes erect; tepals glandular (visible under hand lens), without forked veins; achenes convex on both faces; plants perennial; in nc TX known only from w and s parts of nc TX _______ **P. densiflorum**
         7. Ocreae with cilia 1.5 mm long or more at their summits.
            11. Tepals greenish white, glandular, the glands yellow (visible under hand lens) _______ **P. punctatum**
            11. Tepals variously colored, eglandular.
               12. Ocrea cilia 3 mm or less long; achenes lenticular or trigonous, longer than wide; racemes usually less than 3(–4) cm long _______ **P. persicaria**

Antigonon leptopus [HO3]

Brunnichia ovata [REE]

Eriogonum annuum [BB1]

Eriogonum longifolium [BB2]

Eriogonum multiflorum [IPL]

Polygonella americana [BB1]

Polygonum amphibium var. emersum [MAS]

Polygonum aviculare [REE]

Polygonum convolvulus [REE]

12. Ocrea cilia over 3 mm long; achenes trigonous, as wide as long; racemes over 3 cm long.
    13. Ocrea cilia 6–22 mm long; leaf blades 10–60 mm wide ____ **P. setaceum**
    13. Ocrea cilia usually 5 mm or less long; leaf blades 8–15 mm wide ____ **P. hydropiperoides**

1. Flowers in slender terminal spikes, becoming widely separated and reflexed in age; plants erect perennials forming clumps from short rhizomes; styles 2, quickly elongating, persisting as 2 slender beaks with hooked tips; tepals 4 ____ **P. virginianum** (*Tovara*)

Polygonum amphibium L. var. **emersum** Michx., (sp.: amphibious, growing on land or in water; var.: raised above water level), WATER SMARTWEED. Perennial, rhizomatous, to ca. 1 m tall; ocreae usually eciliate; leaf blades 1–6 cm wide; tepals without forked veins near apex; achenes lenticular, biconvex, turgid, 2.8–4 mm long. Low moist areas; Grayson and Parker cos., also Comanche Co. (HPC); in much of TX, particularly e TX. Jun–Oct. Two intergrading varieties of *P. amphibium* are sometimes recognized in North America: var. *stipulaceum* Coleman, with stems and leaves glabrous, plants floating or sprawling on shore, and flowering while floating or recently stranded; and var. *emersum* Michx., with stems and leaves glabrous to hirsute, plants erect, and flowering from aerial shoots or shoots strongly emergent from water. The species, including the type variety, also occurs in Eurasia (Mitchell 1968; Kartesz 1994; J. Kartesz pers. comm. 1997). All nc TX plants seem to be var. *emersum*. [*Persicaria coccinea* (Muhl. ex Willd.) Greene, *Polygonum coccinum* Muhl. ex Willd.]

Polygonum aviculare L., (pertaining to small birds, which eat the young leaves and achenes), PROSTRATE KNOTWEED, KNOTWEED. Annual or weak perennial, prostrate or ascending (rarely erect); leaf blades narrowly oblong to narrowly elliptic to linear; flowers solitary or in 2s or 3s at some nodes; pedicels shorter than ocreae; tepals 5, greenish with white or pink margins. Disturbed areas; throughout most of TX. May–Nov. Probably native to Europe. Formerly used as a tea in treatment of asthma; now a bad weed, in places resistant to herbicides (Mabberley 1987).

Polygonum convolvulus L., (old generic name for twiners, to twine around), DULL-SEED CORNBIND, CORN BINDWEED, BLACK BINDWEED, CLIMBING-BUCKWHEAT, WILD BUCKWHEAT. Annual, herbaceous, twining vine to ca. 1.2 m long; tepals greenish, sometimes purple-spotted; achenes trigonous. Disturbed sites; Collin, Dallas, Grayson, Rockwall, and Tarrant cos.; widespread in TX. Apr–Sep. A European weed.

Polygonum densiflorum Meisn., (densely flowered), SNOUT SMARTWEED. Perennial 0.6–2 m tall; stems relatively coarse, usually 7–15 mm in diam. at base; ocreae eciliate; racemes 4–8 cm long; tepals greenish white to white or pink, gland-dotted; achenes lenticular, biconvex. Low moist areas; Brown Co. (HPC) on w margin of nc TX, also Fort Hood (Bell or Coryell cos.—Sanchez 1997); also cited by Hatch et al. (1990) for vegetational area 4 (Fig. 2); widespread in s 1/2 of TX. Jun–Nov. [*Persicaria densiflora* (Meisn.) Moldenke, *Persicaria portoricensis* (Bertero ex Small) Small]

Polygonum hydropiperoides Michx., (resembling *P. hydropiper*, that name meaning water-pepper), SWAMP SMARTWEED, WATER-PEPPER. Annual or perennial to ca. 1 m tall; inflorescences usually 4–8 cm long; tepals white to pink; stamens included; achenes trigonous. Low moist areas; widespread in TX. Jun–Nov. [*Persicaria hydropiperoides* (Michx.) Small]

Polygonum lapathifolium L., (with leaves like *Rumex lapathum*—dock), WILLOW SMARTWEED, CURLTOP SMARTWEED, PALE SMARTWEED. Annual to 1 m or more tall; ocreae eciliate, strongly ribbed; tepals white to pinkish white, with distinctive and obvious (using lens or scope) forked, anchor-like veins near apex; stamens included; achenes lenticular, concave on one side. Low moist areas; throughout TX. Apr–Dec. Probably introduced from Europe. [*Persicaria lapathifolia* (L.) Gray] The leaves of this species have a pronounced tendency to stick to the newspapers in which they were pressed.

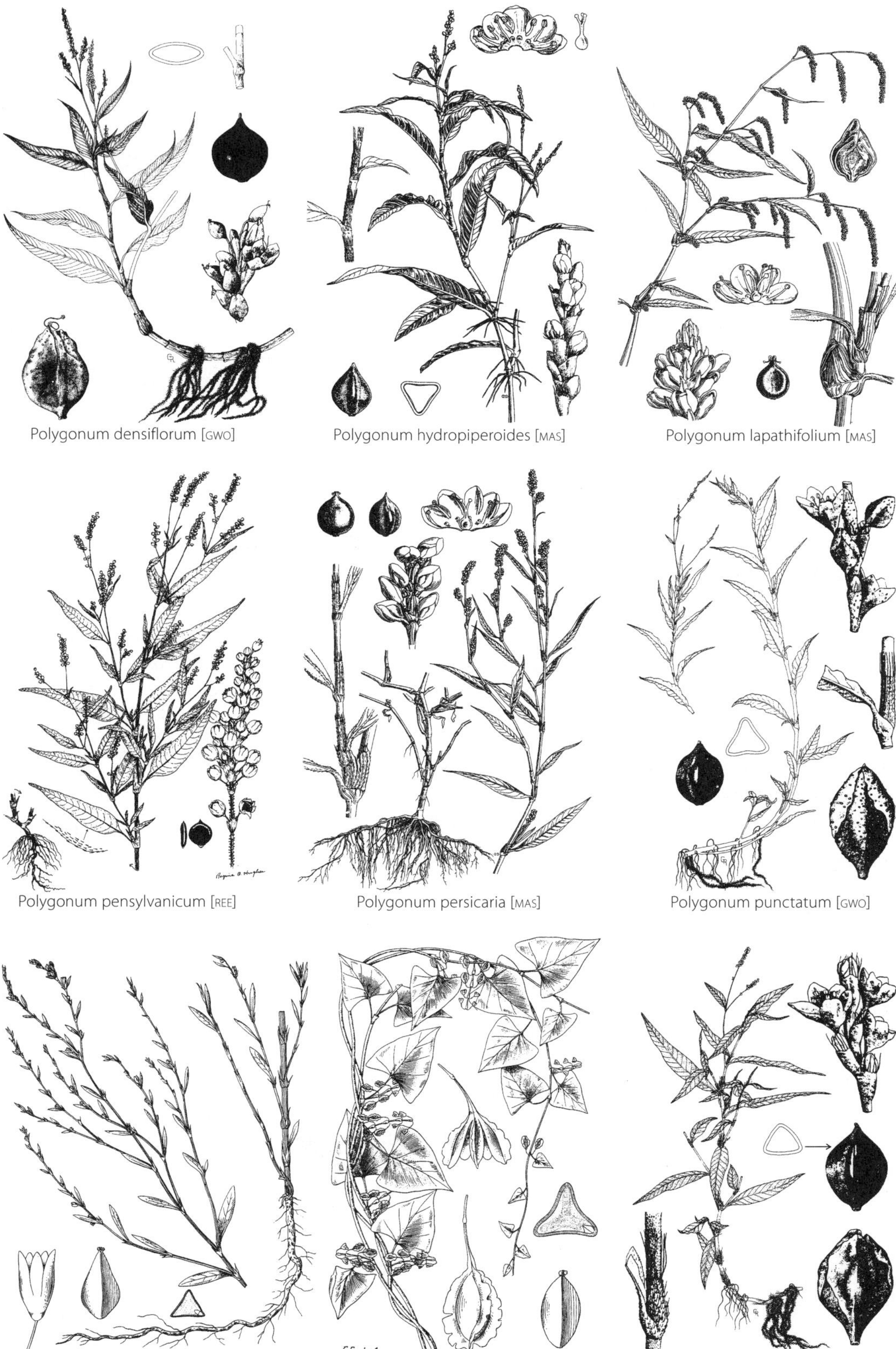

Polygonum densiflorum [GWO]

Polygonum hydropiperoides [MAS]

Polygonum lapathifolium [MAS]

Polygonum pensylvanicum [REE]

Polygonum persicaria [MAS]

Polygonum punctatum [GWO]

Polygonum ramosissimum [SM2]

Polygonum scandens var. cristatum [SM2]

Polygonum setaceum [GWO]

Polygonum pensylvanicum L., (of Pennsylvania), SMARTWEED, PINK SMARTWEED, PENNSYLVANICA SMARTWEED. Annual to 1.5 m tall; stem nodes often cherry-red; ocreae usually eciliate; inflorescences erect; tepals white-pink to pinkish; stamens included; achenes lenticular, flat or concave on both faces or ridge on one surface. Low moist areas; nearly throughout TX. May–Jan. [*Persicaria bicornis* (Raf.) Nieuwl., *Persicaria pensylvanica* (L.) M. Gomez, *Polygonum bicorne* Raf.]

Polygonum persicaria L., (old generic name, said to come from the leaves resembling those of peach or *persica* in Latin), LADY'S-THUMB, MOCO DE QUAJOLOTE, HEARTS-EASE. Erect annual to 1 m tall; ocrea cilia usually less than 3 mm long; peduncles usually glabrous; racemes numerous; tepals white to pink or dusky pink; stamens included; achenes lenticular or trigonous. Low moist areas, included based on citation of vegetational areas 4 and 5 (Fig. 2) by Hatch et al. (1990); supposedly throughout TX; we have seen no nc TX material. Jun–Dec. Introduced from Europe. [*Persicaria vulgaris* Webb & Moq.]

Polygonum punctatum Elliott, (dotted), WATER SMARTWEED, DOTTED SMARTWEED. Erect or ascending annual or perennial 0.1–1 m tall; ocrea bristles usually less than 6 mm long; inflorescences erect to arching, slender, to 10 cm long, the flowers not crowded, irregularly spaced, especially below, the peduncles glandular; tepals with yellowish glands, greenish white; stamens included; achenes usually trigonous. Wet areas; widespread but more common in e 1/2 of TX. Feb–Dec. [*Persicaria punctata* (Elliott) Small] Infraspecific taxa are sometimes recognized in this species (Fassett 1949; Kartesz 1994; Jones et al. 1997); Jones et al. (1997) apparently treated all TX material as var. *confertiflorum* (Meisn.) Fassett). We are following Correll and Correll (1972), Godfrey and Wooten (1981), Kaul (1986b), and Hatch et al. (1990) in not recognizing infraspecific taxa.

Polygonum ramosissimum Michx., (much-branched), BUSHY KNOTWEED. Erect or ascending branched annual to 0.8(–1.2) m tall; leaf blades lanceolate to linear; flowers solitary or in 2s or 3s at some nodes; tepals yellowish green, 6. Low, moist, disturbed sites; Clay, Grayson, and Montague cos.; widespread in TX. Jun–Nov. Infraspecific taxa are sometimes recognized in this variable species (e.g., Small 1894; Kartesz 1994; Jones et al. 1997); however, we are following Correll and Correll (1972) and Kaul (1986b) in not recognizing taxa below the rank of species. Kaul (1986b) noted that the species is highly variable and that, "The complex needs critical study to determine the role of environment, season, and genetics in such phenotypic plasticity."

Polygonum scandens L. var. **cristatum** (Engelm. & A. Gray) Gleason, (sp.: climbing; var.: crested), THICKET KNOTWEED, FALSE BUCKWHEAT. Perennial twining or trailing vine to 5 m long, herbaceous; tepals greenish to whitish; achenes trigonous. Disturbed open areas of woodlands, woods margins; Cooke, Dallas, Grayson, Hunt, and Tarrant cos.; e TX w to nc TX, also Edwards Plateau. Aug–Oct. [*Polygonum cristatum* Engelm. & A. Gray]

Polygonum setaceum Baldwin, (bristle-like), Perennial to 2 m tall; ocrea cilia 6–22 mm long; peduncles strigose to glabrous; tepals white to pink; stamens included; achenes trigonous, the faces flat. Low moist areas; Dallas, Henderson, Hopkins, and Grayson cos.; se and e TX w to nc TX, also Edwards Plateau. Jun–Oct. Sometimes treated as a variety of *P. hydropiperoides* (e.g., Correll & Correll 1972); in some cases the two are distinguished with difficulty; McDonald (1980) gave evidence supporting the recognition of the 2 as distinct species. Jones et al. (1997) recognized TX members of this species as *P. setaceum* var. *interjectum* Fernald. [*Persicaria setacea* (Baldwin) Small, *Polygonum hydropiperoides* Michx. var. *setaceum* (Baldwin) Gleason]

Polygonum tenue Michx., (slender), PLEAT-LEAF KNOTWEED. Erect annual 20–30(–40) cm tall; leaves linear; tepals green with white (rarely pinkish) margins. Sandy woodlands; Denton, Grayson, and Hunt cos.; rare in e 1/2 of TX. Sep–Nov.

Polygonum virginianum L., (of Virginia), JUMPSEED. Sparsely appressed-pubescent to almost glabrous perennial 0.3–1.5 m tall; leaf blades elliptic-lanceolate to ovate-lanceolate; perianth greenish white to white; achenes lenticular, strongly biconvex. In woods, stream bottoms or lower slopes; se and e TX w to Grayson and Tarrant cos., also Edwards Plateau. Jul–Oct. Sometimes recognized in the segregate genus *Tovara* [as *T. virginiana* (L.) Raf.]. The hooked beaks of the achenes are unique in the genus and possibly aid in dispersal by animals (Li 1952a).

RUMEX DOCK, SORREL

Annuals, biennials, or perennials; inflorescence a terminal usually elongate panicle; flowers perfect, or unisexual and the plants dioecious or monoecious; perianth of 6 tepals, the inner 3 usually becoming greatly enlarged in age (and called valves), often with a grain-like tubercle on back, the outer 3 remaining small; stamens 6; fruit a 3-angled (= trigonous) achene enclosed by the valves.

A temperate, especially n temperate genus of 200 species. Some are variously weeds, used for their edible leaves, or a source of tannin; others contain soluble oxalates and are potentially fatally toxic if eaten in large quantities by livestock, especially sheep (Lewis & Elvin-Lewis 1977). (Classical Latin name for *R. acetosella*, sorrel)
REFERENCES: Rechinger 1937; Sarkar 1958.

1. Leaf blades (except sometimes the upper ones) with lobes basally; tepals (valves) without tubercles; flowers usually unisexual, the sexes usually on separate plants (dioecious).
   2. Plants annual, with taproot; pistillate tepals greatly enlarged in age (2–3 mm long) and enclosing the achene; membraneous clasping bracts in upper part of inflorescence ovate, equaling or mostly shorter than the pedicels when flowers are fully open; widespread in nc TX ____ **R. hastatulus**
   2. Plants perennial, with rhizomes or stolons; pistillate tepals not enlarging (1–1.5 mm long), much shorter than the achene; membraneous clasping bracts in upper part of inflorescence ovate-lanceolate, equaling or exceeding the pedicels when flowers are fully open; rare if present in nc TX ____ **R. acetosella**
1. Leaf blades tapered, rounded or cordate basally, without lobes; tepals with or without tubercles; flowers perfect or if unisexual the sexes not on separate plants (monoecious).
   3. Flowers in widely spaced whorls, the whorls mostly separated by internodes 3–10 times as long as the pedicels; inflorescences not dense in fruit, usually broad and open, with simple primary branches long, spreading-ascending; tepal margins often with teeth longer than wide; lower leaf blades usually cordate at base ____ **R. pulcher**
   3. Flowers in crowded whorls, the middle whorls of main branches separated by internodes 1–3 times as long as the pedicels; inflorescences becoming dense in fruit, much longer than broad, branches erect or closely ascending; tepal margins entire or denticulate, the teeth if present wider than long; lower leaf blades cuneate to rounded or rarely subcordate at base.
      4. At least 1 tepal usually with tubercle at base; inner fruiting tepals not becoming very large, 6 mm or less long or wide; outer fruiting tepals not reflexed in fruit; achenes to ca. 3 mm long; widespread in nc TX, typically in moist soils.
         5. Leaf blades (at least lower ones) with rather minutely toothed and/or obviously crisped margins, usually wavy as well, dark green, not glaucescent; axillary shoots usually absent below inflorescence; tepal tubercles usually broadly elliptic to ovate ____ **R. crispus**
         5. Leaf blades with entire, flat or often wavy margins, pale green or glaucescent; axillary shoots often present below inflorescence; tepal tubercles usually narrowly elliptic ____ **R. altissimus**
      4. Tepals without tubercles; inner fruiting tepals becoming very large, including basal lobes (8–)13–17 mm long, 7–12 mm wide; outer fruiting tepals reflexed in fruit; achenes 4–5 mm long; rare in nc TX growing in dry sandy areas ____ **R. hymenosepalus**

Rumex acetosella L., (old generic name meaning little sorrel), SHEEP SORREL, SORREL, SOUR-GRASS. Glabrous perennial 10–50 cm tall; pedicel jointed just below flower; calyces scarcely enlarged in fruit. Disturbed sites; included based on citation for vegetational area 4 (Fig. 2) by Hatch et al. (1990), probably based on a Travis Co. record; very rare if present in nc TX; also Edwards Plateau. Apr–May. Native of Europe.

Rumex altissimus Wood, (very tall, tallest), PALE DOCK, SMOOTH DOCK, PEACH-LEAF DOCK. Glabrous perennial or biennial to 160 cm tall; pedicels not much longer than tepals (up to 2 times as long); tepal margins entire. Disturbed moist areas; widespread in much of TX. Apr–Jun.

Rumex crispus L., (crisped, curled), CURLY DOCK, CURLY-LEAF DOCK. Perennial 20–100(–160) cm tall; tepal margins entire or denticulate. Disturbed moist areas; widespread in TX. Apr–May. Native of Europe.

Rumex hastatulus Baldwin, (somewhat spear-shaped), HEART-WING SORREL, ENGELMANN'S DOCK. Glabrous perennial 15–100 cm tall; pedicels jointed below middle. Sandy open woods, disturbed sites; e 1/2 of TX. Apr–Jun.

Rumex hymenosepalus Torr., (with membranous sepals), CANAIGRE, WILD RHUBARB, TANNER'S DOCK. Glabrous perennial from cluster of tuber-like roots; stems 30–120 cm tall; leaf blades tapered at base; inflorescences dense; tepals becoming pinkish, entire. Dry sandy areas; Brown Co. (J. Stanford, pers. comm.), also Hatch et al. (1990) cited vegetational area 5 (Fig. 2); mainly Edwards Plateau and Plains Country w to w TX. Spring. The tubers are rich in tannin and were formerly used to treat leather (Mabberley 1987).

Rumex pulcher L., (handsome, beautiful), FIDDLE DOCK. Perennial 30–120 cm tall; fruiting tepals varying from spiny-toothed to subentire. Damp or disturbed ground; se and e TX w to Grand Prairie and Edwards Plateau. May. Native of Europe.

Rumex verticillatus L., (whorled), SWAMP DOCK, cited by Hatch et al. (1990) for vegetational area 4 (Fig. 2), apparently occurs only to the s of nc TX. Its pedicels, much longer than the tepals ((2–)2.5–5 times as long), distinguish it from the similar *R. altissimus.*

PORTULACACEAE PURSLANE FAMILY

Ours annual or perennial herbs; leaves ± fleshy, flattened to terete, alternate or opposite, simple, entire; flowers solitary or in corymbose panicles or racemes; sepals 2, separate, sometimes falling early; petals usually 4–5(–6), separate; stamens 5–many; pistil 1; ovary superior or half inferior; fruit a capsule, 3-valved or circumscissile.

A small (380 species in 32 genera), cosmopolitan, especially w North American family of usually succulent herbs and shrubs with mucilaginous tissues and betalain pigments; it includes various ornamentals. A number of species have specialized Kranz ("wreath") anatomy and the associated C_4 photosynthetic pathway, adaptations for hot dry conditions (the C_4 pathway allows for a more efficient capture of CO_2 thereby reducing the degree to which the stomata have to be open and thus the amount of water lost through transpiration). The family is possibly polyphyletic or paraphyletic and may be ancestral to Cactaceae (Rodman 1990; Downie & Palmer 1994); a recent study by Hershkovitz and Zimmer (1997) indicated on the basis of molecular data that the Cactaceae is derived from within the Portulacaceae and that the molecular divergence "... between pereskioid cacti and the genus *Talinum* (Portulacaceae) is less than that between many Portulacaceae genera." The Portulacaceae also seems related to the Basellaceae (Rodman et al. 1984; Carolin 1993). Plant anatomists/morphologists consider the perianth to be of a single series, with the petal-like structures actually sepals and the sepal-like

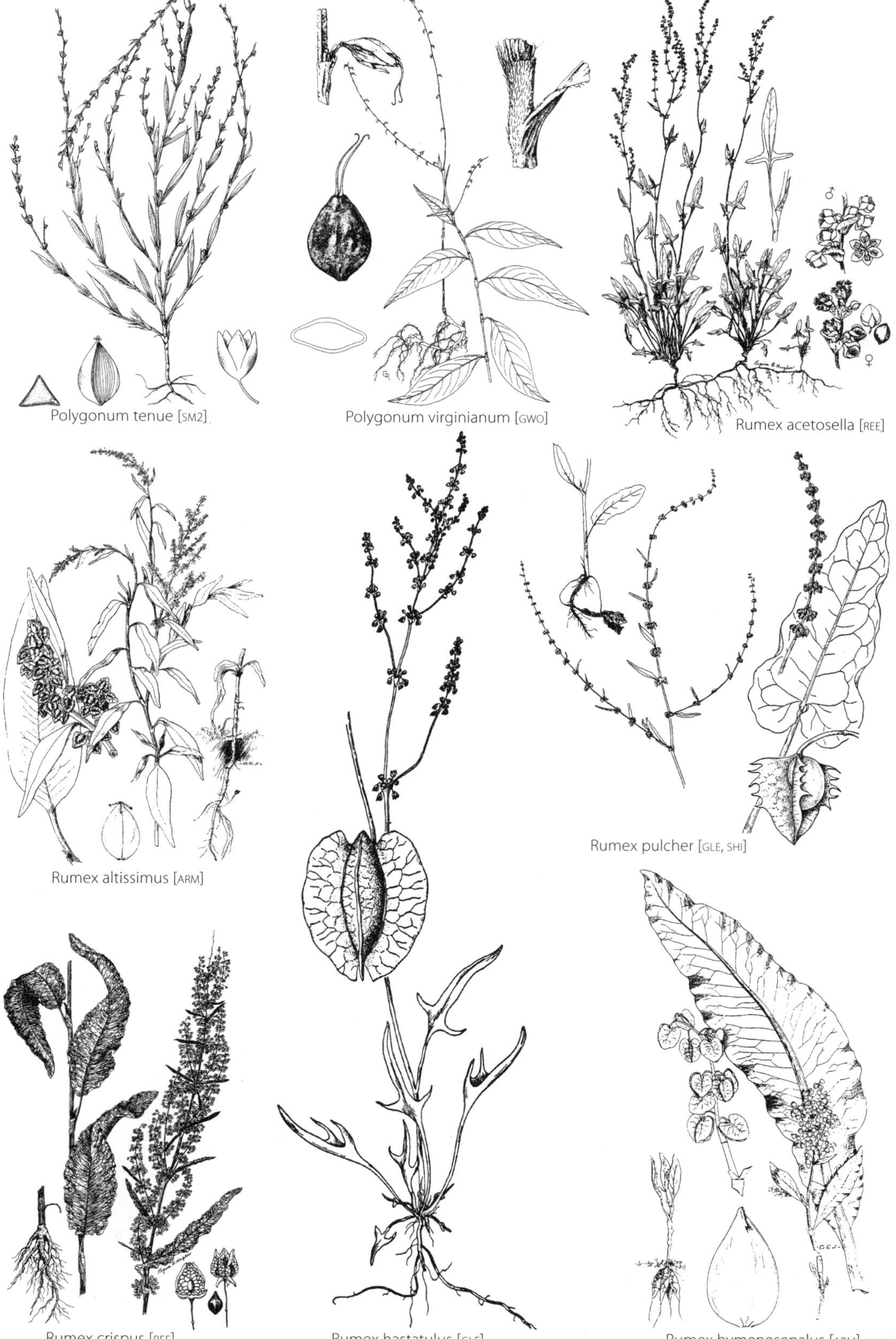

Polygonum tenue [SM2]

Polygonum virginianum [GWO]

Rumex acetosella [REE]

Rumex altissimus [ARM]

Rumex pulcher [GLE, SHI]

Rumex crispus [REE]

Rumex hastatulus [GLE]

Rumex hymenosepalus [ARM]

structures actually bracts (Tyrl et al. 1994); for convenience we are following most taxonomists and using the terms sepals and petals. (subclass Caryophyllidae)
FAMILY RECOGNITION IN THE FIELD: herbs with stems or leaves *fleshy* and often mucilaginous; leaves simple, entire; flowers with *2 sepals* and 4–5(–6) petals; fruit a capsule.
REFERENCES: Rydberg 1932; Bogle 1969; Barker 1986a; Carolin 1993; Behkne & Mabry 1994; Downie & Palmer 1994; Hershkovitz & Zimmer 1997.

1. Flowers on leafless pedicels or peduncles; ovary superior; fruit a 3-valved capsule splitting from the top (= apex) down.
   2. Stem leaves several–many, alternate (but stem often short and leaves thus crowded), 7 cm or less long, 3.5 mm or less wide; flowers axillary or in a terminal branched inflorescence; sepals deciduous; petals pink to purple, orange, or reddish ______ **Talinum**
   2. Stem leaves 2, opposite or nearly so, ca. 9–15 cm long, 5–15 mm wide; flowers in a terminal unbranched raceme; sepals persistent in fruit; petals white to pink with pink to purplish veins ______ **Claytonia**
1. Flowers sessile or subsessile, subtended by leaves or leafy bracts; ovary half-inferior; fruit a circumscissile capsule (= top coming off like a lid) ______ **Portulaca**

CLAYTONIA SPRING-BEAUTY

A mainly North American genus of 24 species with some in e Asia; a number are cultivated as ornamentals or used for their edible corms; some have ant-dispersed seeds. (Named for John Clayton, 1686–1773, one of the earliest American botanists; he sent Johan F. Gronovius the specimens for his *Flora Virginica* (published in 1739), one of the first books on American botany—Strausbaugh & Core 1978)
REFERENCES: Rothwell 1959; Rothwell & Kump 1965; Lewis et al. 1967; Lewis 1976; Lewis & Semple 1977; Doyle 1983, 1984.

Claytonia virginica L., (of Virginia), VIRGINIA SPRING-BEAUTY. Perennial from a soft-woody, globose, small corm; stems 1 to many, thread-like and oblique underground, half-decumbent to ascending above ground; exposed part of plant 5–30 cm long; basal leaves present (somewhat fleshy) as well as 2 stem leaves; inflorescence a terminal raceme with 6–15 flowers; flowers closed at night and in cloudy weather; petals 5, ca. 9–14 mm long; stamens 5. Sandy open woods, prairies, disturbed sites; se and e TX w to East Cross Timbers, locally in Red River drainage w to Archer Co., in Brazos River drainage w to Bosque Co. (Mahler 1988); also Edwards Plateau. Late Feb–early Apr. Cultivated as an ornamental; chromosome number varies from $2n = 12$ to almost 200, with different numbers even within the same plant (Rothwell 1959; Rothwell & Kump 1965; Lewis et al. 1967; Lewis & Semple 1977; Mabberley 1987); the bulb-like corms were apparently eaten by Native Americans (Ajilvsgi 1984). The seeds have elaiosomes and are dispersed by ants (Handel 1978). /84

PORTULACA PURSLANE

Ours low, often prostrate, succulent annuals; leaves alternate or subopposite or the uppermost apparently whorled, sessile or subsessile, flat or terete; flowers opening in late morning or afternoon, closing at night; petals usually 5; stamens 8–many; fruits circumscissile with a bottom valve (like a pot) and an upper valve (like a lid), with or without a wing at the line where the valves separate.

A tropical and warm area genus of 40 species of succulent, trailing, usually annual herbs. Some are used as potherbs, medicinally, or as cultivated ornamentals (e.g., *P. grandiflora* Hook., ROSE-MOSS, MOSS-ROSE, of n South America with variously colored, often double flowers). (Presumably diminutive of Latin: *porta*, a gate or door, from the fruit lid)

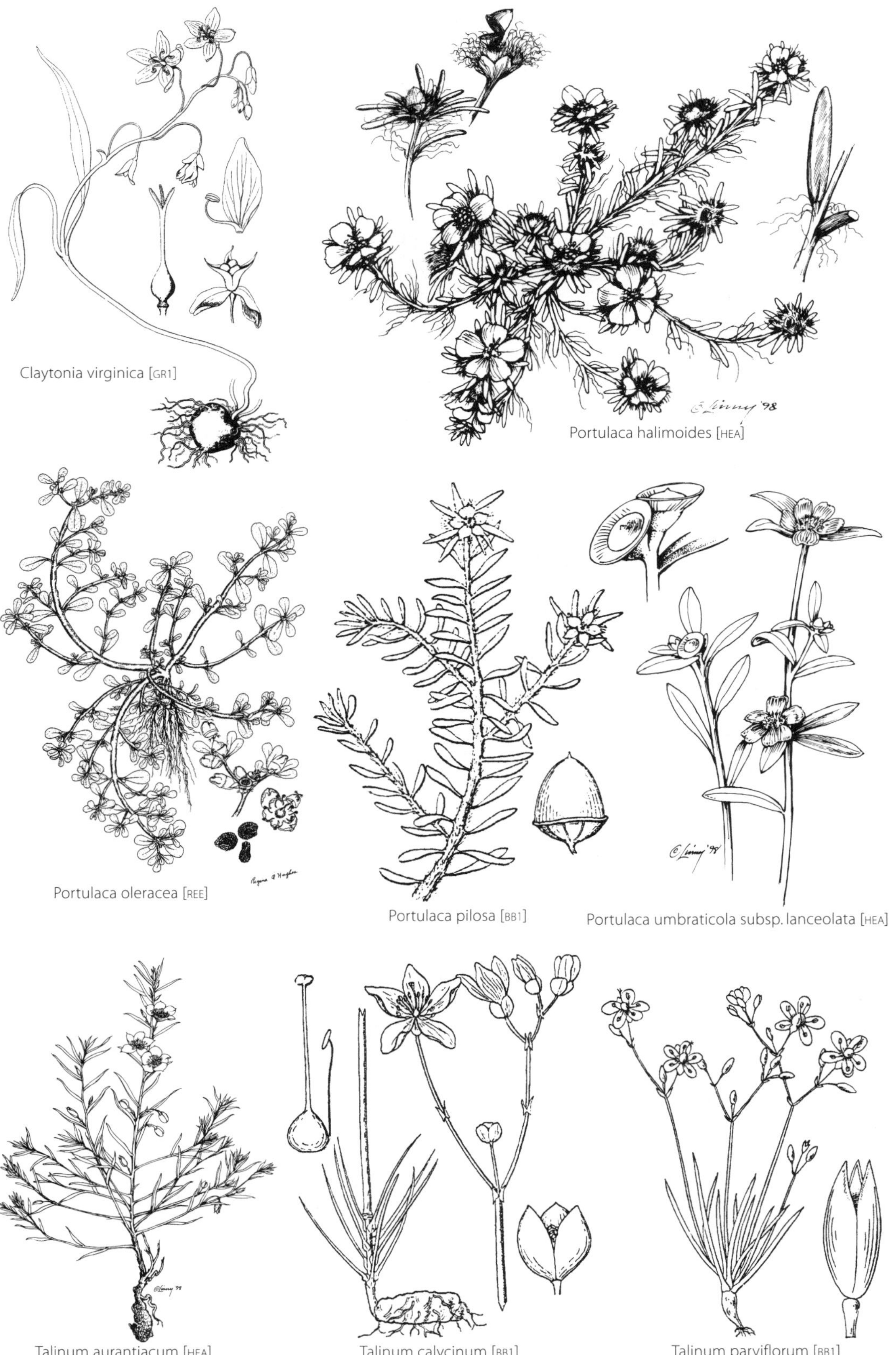

Claytonia virginica [GR1]

Portulaca halimoides [HEA]

Portulaca oleracea [REE]

Portulaca pilosa [BB1]

Portulaca umbraticola subsp. lanceolata [HEA]

Talinum aurantiacum [HEA]

Talinum calycinum [BB1]

Talinum parviflorum [BB1]

REFERENCES: Legrand 1962; Matthews & Levins 1985a, 1985b; Matthews 1986; Matthews & Ketron 1991; Matthews et al. 1992a, 1992b, 1993, 1994.

1. Stem leaves terete to subterete (definitely not flat), linear or linear-oblanceolate; leaf axils and inflorescence conspicuously villous with long, white, kinky hairs; flowers subtended by an involucral ring of 6–10 small leaves.
   2. Petals red-purple, 3–6(–7.5) mm long; capsules 2.5–3.5 mm in diam.; common in parts of nc TX ____ **P. pilosa**
   2. Petals yellow to bronze, < 3 mm long; capsules 1.5–2 mm in diam.; included here on basis of only one citation, apparently rare in nc TX ____ **P. halimoides**
1. Stem leaves thick but flat, oblong-lanceolate to oblong-elliptic or obovate; leaf axils and inflorescence glabrous or with inconspicuous hairs; flowers subtended by 2–4 small leaves.
   3. Capsules with only a line around the wide rim at the point of dehiscence; leaves oblanceolate or obovate, the upper 2–4 times as long as wide; petals yellow ____ **P. oleracea**
   3. Capsules with projecting, scarious wing 0.4–0.8 mm wide around the wide rim under the point of dehiscence; leaves oblong-lanceolate or oblanceolate, the upper 3–5 times as long as wide; petals yellow tipped with red or copper ____ **P. umbraticola**

Portulaca halimoides L., (resembling *Halimium* of the Cistaceae), SINKER-LEAF PORTULACA. Sandy or gravelly areas; included on basis of citation of vegetational area 4 (Fig. 2) by Hatch et al. (1990); mainly c and w TX. Mar–Nov. [*P. parvula* A. Gray]

Portulaca oleracea L., (of the vegetable garden, a potherb used in cooking), COMMON PURSLANE, PURSLANE, PUSLEY, VERDOLAGA. Petals yellow, 3–10 mm long. Dried-up lake beds, disturbed areas; Grayson Co.; nearly throughout TX. Jul–Sep. Cosmopolitan weed of uncertain origin but probably native to the Old World. While this species is often divided into many subspecies (e.g., Kartesz 1994), we are following Matthews et al. (1993) in not recognizing infraspecific taxa; they concluded *P. oleracea* is a polymorphic species not divisible based on seed surface characters (which were used to divide it in the past); chromosome number is variable, ranging from $2n$=18 to 36, 45, 52, or 54 (Matthews et al. 1993). This species can be an aggressive weed. It has been long cultivated, eaten as a potherb and salad, and used as fodder; however, the tart taste comes from oxalic acid, which in large quantities can be toxic (Matthews & Levin 1985a; Tveten & Tveten 1993). ☠ ☜

Portulaca pilosa L., (with long soft hairs), CHISME, SHAGGY PORTULACA. Loose sandy or gravelly soils; Burnet and Grayson cos., also Brown Co. (HPC); nearly throughout TX. Late Jun–Oct. Matthews et al. (1992b) concluded that [*P. mundula* I.M. Johnst.] is a synonym.

Portulaca umbraticola Kunth subsp. **lanceolata** (Engelm.) J.F. Matthews & Ketron, (sp.: growing in shady places; subsp.: lance-shaped), WING-POD PORTULACA, CHINESE-HAT. Similar to *P. oleracea* but distinguished by the winged capsules and flower color. Typically in sandy soils, disturbed areas; Grayson Co., also Brown and Comanche cos. (HPC); nearly throughout TX. Jul–Sep. [*P. lanceolata* Engelm.] Matthews and Ketron (1991) recognized three subspecies within this variable species. Subspecies *coronata* (Small) J.F. Matthews & Ketron has solid yellow petals, a chromosome number of $2n$=36, and is apparently endemic to South Carolina and Georgia. Subspecies *lanceolata*, "... is characterized by yellow petals tipped with red (copper), a chromosome number of $2n$=54, and a distribution generally west of the Mississippi River." Subspecies *umbraticola* " ... is variable in flower color: pink, purple, red or yellow. It has a chromosome number of $2n$=18" (Matthews & Ketron 1991); subspecies *umbraticola* is known from South America and the Antilles.

TALINUM FLAMEFLOWER

Succulent perennials with pithy root; leaves alternate or opposite, sometimes crowded on a short stem; flowers axillary or in terminal cymose panicles, open in late afternoon and evening or in cloudy weather; petals 5 (rarely more); stamens 4–30 or more.

A genus of 39–40 species of the Americas and s Africa; some are cultivated as ornamentals and as potherbs. (Derivation obscure, possibly from an African vernacular name)

1. Flowers in terminal inflorescences; petals pink to purple (rarely reddish); leaves nearly as thick as wide, linear-lanceolate, usually rather crowded on a short stem (stems 10 cm or less long); widespread in nc TX.
   2. Petals 4–9 mm long; stamens 4–10; capsules 3–5 mm long; usually on sand ____ **T. parviflorum**
   2. Petals 9–15 mm long; stamens 25 or more; capsules 5–8 mm long; usually on limestone ____ **T. calycinum**
1. Flowers solitary, axillary; petals orange or reddish; leaves flat (but fleshy), linear-lanceolate to obovate, alternate or opposite, well separated up the entire length of the elongate stem; extreme w margin of nc TX ____ **T. aurantiacum**

Talinum aurantiacum Engelm., (orange-red), ORANGE FLAMEFLOWER. Plant 12–35 cm or more tall, stiffly branched, becoming woody toward base; sepals 6–9 mm long; petals 9–13 mm long; capsules 5–7 mm long. Sandy or rocky ground; Callahan, Stephens, and Throckmorton cos., also Brown Co. (HPC); extreme w part of nc TX s and w to w TX. Late Apr–Sep. /107

Talinum calycinum Engelm., (calyx-like), ROCK-PINK. Stems ca. 3–10 cm; plant including inflorescence to ca. 33 cm tall; sepals 4.5–8 mm long. Gravelly or rocky limestone soils; n Fort Worth Prairie—Cooke, Parker, and Wise cos., also Montague Co. (R. O'Kennon, pers. obs.), also nw Grayson Co. on Goodland limestone outcrop; nc and nw TX. May–Jul. /107

Talinum parviflorum Nutt., (small-flowered), PRAIRIE FLAMEFLOWER, DWARF FLAMEFLOWER. Plant with short stem or nearly acaulescent, including inflorescence to ca. 19 cm tall; sepals 2.7–4 mm long. Bare sandy or sandy clay soils; throughout most of TX. May–Sep.

PRIMULACEAE PRIMROSE FAMILY

Ours low, glabrous, annual or perennial herbs with fibrous roots; leaves basal, alternate, opposite, or whorled, sessile or indistinctly petioled, simple, entire; flowers axillary or terminal, solitary or in racemes or umbels; calyces 5-lobed; corollas moderately to very deeply 5-lobed, funnelform to rotate or with reflexed lobes; stamens 5; pistil 1; ovary superior; placentation free-central; fruit a dehiscent capsule.

A medium-sized (825 species in 22 genera), nearly cosmopolitan, but mainly n temperate family of herbs or rarely subshrubs; it includes a number of ornamentals such as *Cyclamen* (CYCLAMENS) and *Primula* (PRIMROSES). *Anagallis* and *Cyclamen* species are used medicinally, but also contain a poisonous substance. Primulaceae are related to the woody tropical family Myrsinaceae (Judd et al. 1994). Family name from *Primula*, a genus of 400 species of perennial herbs native to the n hemisphere and s to Java, New Guinea, and s South America. (Latin: *primus*, first, in reference to the early flowers of some species) (subclass Dilleniidae)
FAMILY RECOGNITION IN THE FIELD: herbs with simple entire leaves, sometimes in a basal rosette; corollas 5-lobed, radially symmetrical, short to long tubular; stamens epipetalous, opposite the corolla lobes; ovary superior; style and stigma single.
REFERENCE: Channell & Wood 1959.

1. Flowers in an umbel terminating a scape; leaves basal only (but the umbel can have leafy bracts at base).

2. Plants dwarf annuals 1–7 cm tall; corollas white, funnelform to subrotate, 0.8–1.5 mm long or broad, shorter than the calyces; stamens distinct, neither forming a cone nor exserted ________ **Androsace**
2. Plants perennials 30–60 cm tall; corollas lavender or purple-rose with yellow eye, the lobes reflexed, 12–28 mm long, much longer than the calyces; stamens with filaments joined below, the anthers forming a conspicuous exserted cone ________ **Dodecatheon**

1. Flowers in leaf axils (1 per leaf) OR in racemes; well-developed leaves extending up stem.
3. Corollas yellow; leaves opposite or whorled ________ **Lysimachia**
3. Corollas white to pink, scarlet, or blue; leaves alternate or opposite.
4. Flowers in terminal racemes with numerous flowers; corollas white; capsules valvate ________ **Samolus**
4. Flowers axillary, solitary per leaf axil; corollas white to pink, scarlet, or blue; capsules circumscissile ________ **Anagallis**

ANAGALLIS PIMPERNEL

Annuals; leaves sessile or subsessile; corollas rotate; capsules globose or subglobose, circumscissile.

A genus of ca. 28 species of Europe, African mountains, and South America, with 1 pantropical. (Greek: *anagelao*, to laugh or delight; from belief they dispelled sadness)

1. Leaves opposite; corollas usually scarlet, rarely blue, with purple eye ring; flowers on pedicels exceeding the leaves ________ **A. arvensis**
1. Leaves alternate; corollas white or pinkish; flowers sessile or nearly so ________ **A. minima**

Anagallis arvensis L., (pertaining to cultivated fields), SCARLET PIMPERNEL, POORMAN'S-WEATHERGLASS, COMMON PIMPERNEL, HIERBA DEL PÁJARO, SHEPHERD'S -WEATHERGLASS. Erect to sprawling, usually freely branched annual; leaves sessile, elliptic or ovate; flowers 5-merous; corollas ca. 4–6 mm wide, closed at night or in cloudy weather, usually lasting one day; capsules ca. 4 mm in diam. Sandy disturbed ground; Dallas, Grayson, and Lampasas cos.; e 1/2 of TX. Native of Europe. The flowers quickly close at the approach of bad weather; thus the common name WEATHERGLASS. Formerly used medicinally (Mabberley 1987); apparently poisonous due to the presence of triterpene saponins (Cheatham & Johnston 1995).

Anagallis minima (L.) E.H.L. Krause, (lesser, smaller), CHAFFWEED, FALSE PIMPERNEL. Inconspicuous small annual; leaves obovate; flowers sessile or nearly so, 4 or occasionally 5-merous; corollas minute, ca. 1 mm wide; capsules to 2 mm in diam. Damp sandy ground; Dallas and Denton cos., also Lamar Co. (Carr 1994); se and e TX w to nc TX, also Edwards Plateau. Apr–May. Native of Europe. Sometimes segregated into the genus *Centunculus* [as *C. minimus* L.]

ANDROSACE ROCK-JASMINE

A n temperate genus of 100 species, often xerophytic; some are cultivated as rock garden ornamentals. (Greek: *andros*, male, and *sakos*, buckle, alluding to the shape of the anthers)
REFERENCE: Robbins 1944.

Androsace occidentalis Pursh, (western), WESTERN ROCK-JASMINE. Dwarf annual 1–7 cm tall; leaves in a basal rosette, usually elliptic-oblanceolate, to 2 cm long and 6 mm wide, white-pubescent above; scapes with stellate hairs; corollas ca. 2.5 mm wide, white; stamens distinct; capsules 5-valved. Sandy or silty ground; Blackland Prairie w to w TX. (Jan–)Feb–Apr. Very small, usually overlooked, and probably more widespread than reported.

DODECATHEON SHOOTING-STAR, AMERICAN COWSLIP

A genus of 13 species, mainly North American with 1 in e Siberia; related to *Primula* but with reflexed corollas and self-pollination possible. Many are cultivated as ornamentals because of the showy, unusually shaped flowers. Members of this genus exhibit the "vibrator" or

"buzz" pollination syndrome; pollinators (such as bumblebees) shake the anthers by vibrating their thoracic flight muscles at a certain frequency; this sets up a resonance in the anthers or the space they enclose and the otherwise inaccessible pollen is released from the terminal pores of the anthers and collected by the insect (Macior 1964; Barth 1985; Proctor et al. 1996). The turned back (reflexed) corollas and exposed anther-cone of "vibrator"-type flowers may be an adaptation to minimize dampening of vibration resonance or it may be an adaptation related to microclimate in the flower (e.g., to keep the pollen in a dry powdery condition so that it is easily dispersed) (Corbet et al. 1988; Proctor et al. 1996). (Greek: *dodeca*, twelve, and *theos*, god, name given by Pliny to the related PRIMROSE, which was believed to be under the care of the twelve superior gods)
REFERENCES: Fassett 1944; Gleason 1952; Macior 1964.

Dodecatheon meadia L., (for Richard Mead, 1673–1754, English physician), COMMON SHOOTING-STAR, AMERICAN COWSLIP. Perennial scapose herb to ca. 60 cm tall; leaves to 20 cm long and 4 cm wide, usually tinged with red at base (all nc TX specimens we have observed have the leaves gradually tapering to the base—this corresponds with var. or subsp. *media* if infraspecific taxa are recognized); scapes erect; inflorescence an umbel of few–many flowers; flowers extremely showy, nodding at anthesis; corollas lavender or purple-rose with yellow eye, the lobes reflexed, 12–28 mm long, much longer than the calyces; filaments connate below; anthers forming a conspicuous slender cone; capsules 7.5–18 mm long, 5-valved. Prairies, calcareous clay, on slopes or in low ground; Grand Prairie (Cooke, Montague, Tarrant, and Wise cos.), also Grayson Co.; e TX w to Grand Prairie and Edwards Plateau. Apr–May. Also cultivated as an ornamental. We are following Correll and Johnston (1970) and Hatch et al. (1990) in not recognizing infraspecific taxa in nc TX; Kartesz (1994) and Jones et al. (1997) recognized 2 subspecies in TX (subsp. *brachycarpum* (Small) R. Knuth and subsp. *meadia*). Gleason (1952) distinguished the 2 (as varieties) as follows: ▣/87/frontispiece

1. Leaves cordate at base or abruptly narrowed into the petiole; anthers (4–)5.5–6.6(–7) mm long; capsules 7.5–10 mm long ________ var. *brachycarpum*
1. Leaves gradually tapering to the base; anthers (6.5–)7–8(–10) mm long, capsules 10.5–18 mm long ________ var. *meadia*

LYSIMACHIA LOOSESTRIFE

A temperate and warm area genus (especially the Himalayas, China, and Europe) of ca. 150 species of herbs or rarely shrubs; some are used locally as medicines or cultivated as ornamentals. (Greek *lysis*, releasing, and *mache*, strife; according to tradition, King Lysimachos of Thrace (ca. 360–281 BC), upon being chased by a maddened bull, was said to have pacified it by waving a LOOSESTRIFE plant)
REFERENCE: Coffee & Jones 1980.

Lysimachia lanceolata Walter, (lanceolate, lance-shaped), LANCE-LEAF LOOSESTRIFE. Perennial reproducing vegetatively by rhizomes; stems (10–)20–90 cm tall, erect; petioles of middle stem leaves pubescent from node to base of blade; stem leaves opposite or whorled, 3–14(–16) cm long, 4–16 mm wide, linear to lanceolate, entire, short-petiolate; basal leaves often persistent as a rosette; flowers solitary per leaf axil, on long pedicels 10–40 mm long; corollas deeply 5-lobed, the lobes 5–9 mm long, yellow; stamens 5; 5 small staminodia present; capsules 2–4.5 mm in diam., valvate. Moist to dry woods, slopes; Lamar Co. in Red River drainage; in TX known only from three other counties (Bowie, Cass, Red River) in extreme ne part of the state. May–Jun.

Lysimachia radicans Hook., (rooting), TRAILING LOOSESTRIFE, occurs in e TX just to the e of nc TX. It can be distinguished by its habit (reclining or trailing and rooting at the nodes), corolla lobes only 2–5 mm long, and by the petioles of the middle stem leaves pubescent only along the basal portion.

SAMOLUS BROOKWEED, WATER-PIMPERNEL

Perennial or annual [?], somewhat succulent herbs; leaves in basal rosettes; alternate stem leaves also present, entire; flowers pedicellate, in racemes, these often paniculately arranged; corollas 5-lobed, white in our species; ovary partially inferior; capsules 5-valved.

A cosmopolitan genus of 15 species, especially in wet areas. (Ancient name probably of Celtic origin, said to refer to curative properties of the genus in diseases of cattle and swine)
REFERENCE: Henrickson 1983.

1. Racemes on long stout peduncles longer than the stems, no bracts close to the lowest flowers; corollas 4–6 mm wide, the lobes usually shorter than the tube; capsules 3–4 mm in diam. _____ **S. ebracteatus**
1. Racemes sessile or nearly so, leaf-like bracts near the lowest flowers; corollas 2–3 mm wide, the lobes longer than the tube; capsules 2–3 mm in diam. _____ **S. valerandi**

Samolus ebracteatus Kunth subsp. **cuneatus** (Small) R. Knuth, (sp.: without bracts; subsp.: wedge-shaped), LIMEROCK BROOKWEED. Plant to 60 cm tall; leaves obovate to oblanceolate to broadly spatulate; staminodia not present. Wet limestone, seepage areas, or along streams; Bell and Palo Pinto cos.; nc TX s and w to w TX. Mar–Oct. [*S. cuneatus* Small]

Samolus valerandi L. subsp. **parviflorus** (Raf.) Hultén, (sp: derivation unexplained by Linnaeus; subsp.: small-flowered), THIN-LEAF BROOKWEED. Plant to 60 cm tall, typically much smaller; leaves obovate, oblanceolate, elliptic to spatulate; 5 staminodia present in sinuses of corolla lobes. Along creeks, ditches, or seepage areas; throughout TX. Apr–Oct. [*S. parviflorus* Raf.]

RAFFLESIACEAE RAFFLESIA FAMILY

A small family (50 species in 9 genera) of chlorophyll-less stem or root parasites of tropical areas with a few in temperate regions. The vegetative portion of the plant is ± thread-like and occurs within the host tissues like fungal mycelia. Family name from *Rafflesia*, a genus of 16 rare species in a few localities in Borneo, Sumatra, Java, Peninsula Malaysia, Thailand, and the Philippines; *Rafflesia arnoldii* R. Br., an endangered Sumatran parasite, has the largest flowers of any known plant; a single flower can be 97 cm (over 38 inches) in diam. and weigh 7 kg (over 15 pounds). The flowers smell of carrion and are visited by and probably pollinated by flies (Mat Salleh 1991). While placed in the Rosidae by Cronquist (1981), a relationship with Aristolochiaceae or some other Magnoliidae group seems more likely (Meijer 1993). (Named for Sir Thomas Stamford Raffles, the founder Governor of Singapore during the British East India Company's rule in southeast Asia; he was present when the first species was collected in Sumatra in 1818) (subclass Rosidae)
FAMILY RECOGNITION IN THE FIELD: the only nc TX species is a *stem parasite* with the *only structures visible* outside the host being the small, *bud-like* flowers and a few subtending, tiny, scale-like leaves.
REFERENCES: Mat Salleh 1991; Meijer 1993.

PILOSTYLES

A genus of ca. 11 species mainly of the New World (CA and TX to the tropics) but with 2 in sw Australia, 1 in Iran, and 1 in tropical Africa. They parasitize all three subfamilies of Fabaceae. (Greek: *pilos*, cap, and *stylus*, column, stake, or style; presumably from the shape of the style/stigma)
REFERENCE: Rose 1909.

Pilostyles thurberi A. Gray, (for George Thurber, 1821–1890, botanist and quartermaster with the U.S.-Mexican Boundary Survey), THURBER'S PILOSTYLES. Parasite with vegetative structures ± en-

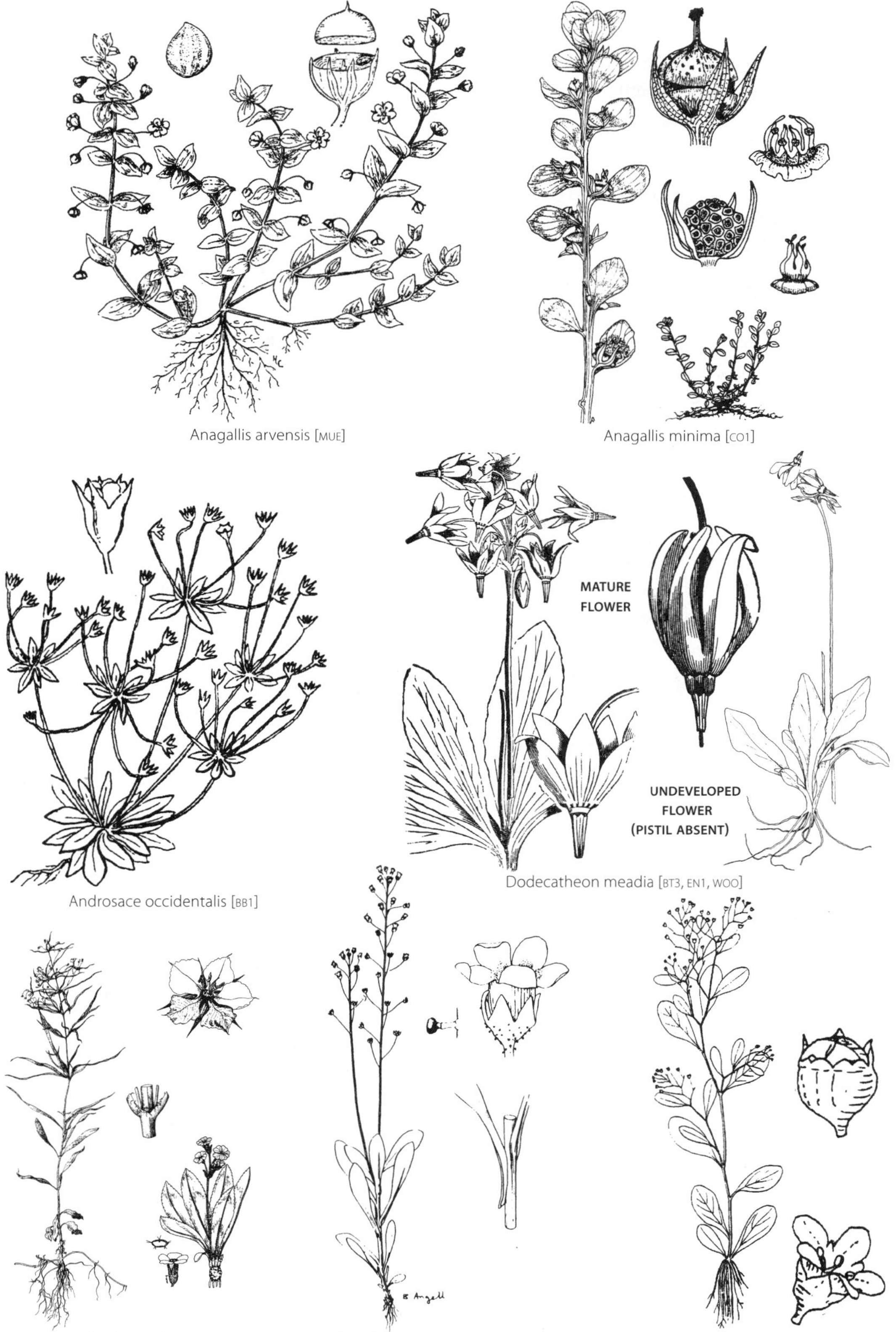

Anagallis arvensis [MUE]

Anagallis minima [CO1]

Androsace occidentalis [BB1]

Dodecatheon meadia [BT3, EN1, WOO]

Lysimachia lanceolata [CO1]

Samolus ebracteatus subsp. cuneatus [SWN]

Samolus valerandi subsp. parviflorus [STE]

tirely within the tissues of the host plant with only the small flowers and sometimes a few subtending scale-like leaves visible externally (complete visible portion of plant bud-like); floral bracts 2–7, broadly ovate, 1–1.5 mm long; flowers unisexual, solitary per short (to 1.5 mm long) peduncle, brownish; calyx segments 4–5, similar to bracts, 1–1.5 mm long; petals absent; staminate flowers with many sessile anthers in 1–3 series around a column that is expanded at tip into a disk ca. 1 mm in diam.; pistillate flowers with stigmas in ring along disk margin; ovary inferior; fruit a globose capsule 1–1.5 mm in diam; seeds numerous. Parasitic on *Dalea frutescens* (Fabaceae) in nc TX and also on *D. formosa* to the s and w; Bell, Brown, Comanche, Hamilton, and Mills cos. (HPC); sw part of nc TX s and w to w TX. May–Jul. [*P. covillei* Rose]

RANUNCULACEAE BUTTERCUP FAMILY

Ours annual or perennial herbs or seldom woody climbers (*Clematis*); leaves basal, alternate, opposite, or apparently whorled, simple or compound, entire, toothed, or lobed; flowers solitary or in spikes, racemes, or panicles, perfect or unisexual, often with a prominent, elongating receptacle, the flower parts closely spiral; perianth parts few to many, alike or differentiated into sepals and petals, the petals often producing nectar (sometimes called honey leaves); stamens usually many; pistils few to many; ovary superior; fruit in ours an achene or follicle.

A large family of 2,500 species in ca. 60 genera (Whittemore & Parfitt 1997); it is a mainly temperate and boreal group of usually herbs with some lianas and small shrubs. The family contains numerous ornamentals including species of *Aconitum* (MONKSHOOD), *Anemone*, *Aquilegia*, *Caltha* (MARSH-MARIGOLD), *Clematis*, *Delphinium*, and *Ranunculus*. A number are highly poisonous (e.g., *Aconitum*—WOLFSBANE, MONKSHOOD) due to the presence of alkaloids or other toxins such as glycosides or saponins. (subclass Magnoliidae)

FAMILY RECOGNITION IN THE FIELD: herbs (or 1 genus of woody climbers) with *cone- or dome-shaped* receptacles with *numerous spirally arranged* stamens and few to numerous *separate* pistils; leaves often variously divided or compound, the bases often sheathing.

REFERENCES: Keener 1977; Loconte & Estes 1989; Ziman & Keener 1989; Duncan & Keener 1991; Tamura 1993; Whittemore & Parfitt 1997.

1. Flowers with 1 or 5 large spurs; fruit a follicle.
   2. Perianth radially symmetrical, reddish to rose and yellow; conspicuous spurs 5, 1 from each of the petal-like perianth parts **Aquilegia**
   2. Perianth bilaterally symmetrical, white to blue, pink, or purple; conspicuous spur 1, from the upper perianth part of the outer whorl.
      3. Plants native perennials; upper leaves petioled (petioles sometimes short); carpels and later follicles 3(–4); lowest floral bracts not conspicuously dissected, with 1–few segments; inner series of perianth (= honey leaves) of 4 separate structures (2 upper included within spur of outer series, 2 lower short-clawed) **Delphinium**
      3. Plants introduced annuals; upper leaves sessile or nearly so; carpels and later follicles 1; lowest floral bracts conspicuously dissected into many filiform segments; inner whorl of perianth of 2 fused structures **Consolida**
1. Flowers with minute spurs or none; fruit an achene.
   4. Flowering stems with opposite or whorled leaves or leafy bracts; perianth of sepals only (these petal-like).
      5. Stems with 1 whorl of leafy bracts (at first close to the flower, somewhat calyx-like); plants erect herbs, small, to ca. 30 cm tall; styles short, < 3 mm long; sepals numerous, (7–)10–20 (–30) **Anemone**
      5. Stems with opposite leaves; plants climbing or scrambling vines, the stems > 30 cm long except when young; styles often very long, 10–100 mm long; sepals 4(–6) **Clematis**

4. Flowering stems with alternate leaves or none; perianth of sepals and petals OR of sepals only (these petal-like).
   6. Leaves all basal, entire ____ **Myosurus**
   6. Leaves alternate up the stems, varying from entire to toothed, lobed, or compound.
      7. Perianth of 4 or 5 sepals, falling after the flowers open; sepals 5 mm or less long, whitish to purplish, rather inconspicuous; flowers all unisexual or some perfect; leaves 2–3 times distinctly ternately compound ____ **Thalictrum**
      7. Perianth of 10 or more sepals and petals, in 2 or more rows (some sepals may be very small and inconspicuous); perianth parts 1–21 mm long, usually yellow, red, or red-purple (rarely green), often showy and conspicuous; flowers perfect; leaves various, often lobed or compound but not 2–3 times ternately compound.
         8. Petals yellow or rarely green; leaves various; widespread native and naturalized species ____ **Ranunculus**
         8. Petals red to red-purple; leaves much-dissected into numerous linear segments; rare escaped cultivar ____ **Adonis**

ADONIS PHEASANT'S-EYE

A genus of ca. 35 species (Parfitt 1997) of temperate Eurasia. ☠ *Adonis* species contain glycosides with marked cardiac activity; some are used medicinally with effects similar to digitalin; lethal poisoning in livestock has been reported (Kingsbury 1964); some are cultivated as ornamentals. (Named for Adonis of Greek mythology, lover of Aphrodite and the god of plants, from whose blood the plant allegedly grew; based on the blood red flowers)
REFERENCES: Heyn & Pazy 1989; Parfitt 1997.

Adonis annua L., (annual), PHEASANT'S-EYE. Erect, glabrous annual with much-branched stems 20–70 cm tall; leaves alternate, much-dissected, the numerous segments linear; petioles with clasping base; flowers solitary, terminal, short-pedicelled; sepals 5–8, thin; petals 6–10, red to red-purple with dark base, 8–15 mm long; stamens 15–20; pistils many; achenes in cylindric heads 1–2 cm long, the individual achenes 3–5 mm long, glabrous. Cultivated as an ornamental, naturalized along roadsides, in waste places, and disturbed soils; Dallas, Denton, Grayson, McLennan, Williamson, and Wise cos., also Bosque Co. (Mahler 1988); mainly n part of e 1/3 of TX. Reverchon (1880) stated that it was imported by French colonists in 1855. Late Mar–May(–later). Native of Europe. Reportedly poisonous due to a glycoside (Burlage 1968). ☠

ANEMONE WINDFLOWER

Ours small perennial herbs to ca. 30 cm tall from a tuberous rootstock; basal leaves long-petioled, palmately compound, with toothed, lobed, or compound leaflets; flowers usually solitary in ours, 1.5–4.3 cm across, terminating an elongate scape; scapes naked except for a single set of apparently opposite or whorled, sessile, somewhat leaf-like, deeply divided bracts; perianth parts many, of 1 type (sepals), the sepals petal-like; stamens and pistils numerous; receptacle greatly elongating in age forming an elongate head-like structure with numerous achenes; achenes nearly covered with long, white, woolly hairs.

A nearly cosmopolitan, especially n temperate and arctic genus of ca. 150 species (Dutton et al. 1997); a number are cultivated as ornamentals, while others are used medicinally. ☠ Some species contain the toxic irritating glycoside protoanemonin (Cheatham & Johnston 1995). Several fungal rusts (*Tranzschelia cohaesa* (Long) Arthur, *Tranzschelia ornata* López-Franco & J.F. Hennen—type specimen from Sherman, TX) are known to infect *Anemone* species. *Tranzschelia ornata* is heterecious—its life cycyle requiring 2 hosts, *Anemone berlandieri* and *Prunus mexicana* (MEXICAN PLUM); it causes a systemic infection of the *Anemone* and is manifested by reddish brown spots on the leaves of *P. mexicana* (J. Hennen, pers. comm.). (Derivation

probably from Greek: *anemos*, wind; possibly from *Naaman*, Semitic name for Adonis, whose blood, according to myth, produced *Anemone coronaria* L.—Dutton et al. 1997)
REFERENCES: Britton 1891; Joseph & Heimberger 1966; Keener 1975a; Hoot et al. 1994; Keener & Dutton 1994; Dutton et al. 1997.

1. Involucre (of a single set of sessile, leaf-like, deeply divided bracts) above middle of otherwise naked scape at flowering time; scape densely pubescent below involucre; body of achenes 2.7–3.5 mm long, the style ca. 1/2 the length of the achene, slightly S-shaped, often largely concealed by the achene pubescence; all basal leaf blades lobed or dissected differently from involucral bracts; plants without rhizomes ____ **A. berlandieri**
1. Involucre below middle of otherwise scape at flowering time; scape nearly glabrous below involucre; body of achenes 1.5–2.5(–3) mm long, the style ca. the length of the achene, straight, often clearly projecting beyond the achene pubescence; one or more basal leaves dissected similarly to involucral bracts; plants with slender rhizomes ____ **A. caroliniana**

Anemone berlandieri Pritz., (for Jean Louis Berlandier, 1805–1851, French botanist who explored TX, NM, and Mexico, and one of the first botanists to make extensive collections in TX), TEN-PETAL ANEMONE. Nonrhizomatous from a tuberous root; sepals usually white within, blue-lavender to violet without, 7–15(–20) mm long; fruiting receptacle 11–28 mm long. Prairies, in calcareous clay, often a lawn weed; throughout much of TX. Late Feb–mid-Apr. [*Anemone decapetala* of authors, not Ard. var. *heterophylla* (Torr. & A. Gray) Britton, *A. heterophylla* Nutt. ex Torr. & A. Gray] /78

Anemone caroliniana Walter, (of Carolina), CAROLINA ANEMONE. Rhizomatous from a small globose tuber; flowering stem glabrous or nearly so below the leafy bracts; sepals white to pink or lavender-blue within, pink to violet without, 10–22 mm long; fruiting receptacle 8–19(–22) mm long. Sandy open woods, roadsides, and prairies; se and e TX w to East Cross Timbers; in Red River drainage w to Rolling Plains; also e Edwards Plateau. Late Feb–mid-Apr.

Anemone edwardsiana Tharp, (of the Edwards Plateau), TWO-FLOWER ANEMONE, typically with 2, but sometimes up to 10 flowers per scape (in contrast to nc TX species which usually have a single flower), is endemic to the Edwards Plateau n to Travis Co. just s of nc TX. It has involucral bracts 2-tiered (vs. 1-tiered in nc TX species), basal leaves usually 1-ternate, and (8–) 10–20 sepals 2–3 mm wide.

Anemone okennonii Keener & B.E. Dutton, (for Robert J. O'Kennon, 1942–, TX collector and co-author of this book), O'KENNON'S ANEMONE, is found to the sw of nc TX; endemic to TX. This species, which was recently named (Keener & Dutton 1994), can be distinguished from *A. edwardsiana* by the basal leaves 2–3 ternate and the 7–11 sepals 3–4.5 mm wide. [*A. tuberosa* Rydberg var. *texana* Enquist & Crozier]

AQUILEGIA COLUMBINE

A n temperate genus of ca. 70 species (Whittemore 1997c) with long perianth spurs and alkaloids. A number are cultivated as ornamentals for their showy flowers. (Perhaps from Latin: *aquila*, eagle, from supposed resemblance of the curved spurs to claws, or possibly from *aqua*, water, and *legere*, to collect, from the evident fluid at the base of the hollow spurs)
REFERENCES: Payson 1918; Munz 1946; Whittemore 1997c.

Aquilegia canadensis L., (of Canada), COMMON COLUMBINE, AMERICAN COLUMBINE, WILD COLUMBINE, CANADIAN COLUMBINE. Perennials 0.2–0.8(–1) m tall; leaves basal and alternate along the stems, palmately 2(–3)-compound; ultimate leaflets fan-shaped, toothed and lobed; flowers rather long-pedicelled, solitary and terminal on ascending branches, nodding, showy, 2–5 cm long (from end of stamens to base of spur); perianth parts in 2 rows, the sepals 5, red or rose, the

Pilostyles thurberi [EMO]

Adonis annua [BB1]

Anemone berlandieri [HO3]

Anemone caroliniana [BB2]

Aquilegia canadensis [GEN]

Clematis crispa [GR1, GWO]

Clematis drummondii [VIN]

Clematis pitcheri [BB2]

petals 5, red or rose at base passing to yellow at tips, with long, conspicuous basal spurs 18–25 mm long; stamens exserted; follicles 5 per flower, erect, 1.5–2(–3) cm long, usually slender-beaked. Limestone cliffs above streams; Bell and Williamson cos. in s part of nc TX s to Edwards Plateau. Late Mar–early May. [*A. canadensis* L. var. *latiuscula* (Greene) Munz] The flowers of this species are visited by and presumably pollinated by ruby -throated hummingbirds (*Archilochus colubris*) (James 1948). /78

CLEMATIS VIRGIN'S-BOWER

Ours mostly low- to high-climbing or sprawling, herbaceous or slightly woody, perennial vines (but can flower when small); leaves opposite, pinnately compound, the rachis sometimes ending in a tendril; leaflets entire, toothed, or lobed; dioecious or with perfect flowers; flowers frequently nodding, often solitary or 2–7 in cymose clusters or (1–)3–12 in cymose or paniculate inflorescences; pedicels often conspicuously elongate; perianth parts of 1 type (sepals); sepals 4(–6), all petal-like, often with densely white-pubescent margins; stamens numerous; pistils numerous; achenes with long, sometimes plumose, persistent style (often referred to as a beak).

A genus of ca. 300 species (Pringle 1997) of lianas or herbs of the n temperate zone, s temperate zone, and tropical African mountains. The genus contains the only woody members of any size in the family; a number of the liana species and their hybrids are important ornamentals, often with very showy flowers. Several are used medicinally; ☠ the plants should be considered poisonous due to the acrid juice (containing anemonin, a dilactone derived from protoanemonin) that can cause skin, mouth, or internal inflammations; all are apparently distasteful to animals though some are suspected of having caused livestock mortality; fortunately, mouth irritation usually prevents the ingestion of fatal doses (Kingsbury 1964; Lampe & McCann 1985; Spoerke & Smolinske 1990; Turner & Szczawinksi 1991). (From *clematis*, a name used by Dioscorides for a climbing plant with long and lithe branches, from Greek: *clema*, twig or tendril)

REFERENCES: Erickson 1943; Hara 1975; Keener 1975b; Dennis 1976; Keener & Dennis 1982; Essig 1992; Moreno & Essig 1997; Pringle 1997.

1. Sepals white or whitish, mostly < 15 mm long, thin, neither fleshy nor leathery; perianth rotate, not at all tube-like in appearance, the sepals wide-spreading from their bases; flowers (1–)3–12 in cymose or paniculate inflorescences.
   2. Leaflets entire or rarely with an entire lobe; mature styles relatively short, 1–3(–6) cm long; flowers bisexual; pistils 5–10 per flower; plants glabrous ____ **C. terniflora**
   2. Leaflets conspicuously coarsely toothed; mature styles very long, 4–10 cm long; flowers unisexual; pistils 35–90 per flower; plants usually pubescent to almost glabrous ____ **C. drummondii**
1. Sepals partly to wholly lavender to blue-purple or red, mostly > 15 mm long, thick and fleshy or leathery; perianth bell-shaped to ovoid or urn-shaped, at least bottom part of sepals appressed together and tube-like in appearance; flowers often solitary or 2–7 in cymose clusters.
   3. Sepals partly or wholly lavender to blue-purple; stems and leaves glabrous but not glaucous; apices of leaf blades rounded to acute; widespread in nc TX.
      4. Tips of sepals with broad, thin, ruffled margins 2–6 mm wide distally; leaflets thin, not conspicuously reticulate-veined; rare in nc TX ____ **C. crispa**
      4. Tips of sepals without broad, thin, ruffled margins (margins can be up to 1 mm wide); leaflets thick and firm, ± conspicuously reticulate-veined (= with a network of noticeable veins), the secondary and sometimes tertiary veins prominently raised on adaxial surface; including a species widespread in nc TX.
         5. Achene tails plumose (= feathery, with long, slender, spreading, lateral hairs; however, tails can appear appressed-pubescent if achene is not mature), 4–6 cm long at maturity; pedicels usually with a pair of bracts at base (similar to the leaves except they are simple

rather than compound); leaflets finely reticulate-veined, even the tertiary and quaternary veins prominently raised on the abaxial blade surface, the ultimate closed areoles mostly < 2 mm in longer dimension; possibly on e or s margins of nc TX. **C. reticulata**

5. Achene tails partially glabrous to long, silky, appressed-pubescent, but not plumose, 1–3.5 cm long; pedicels usually with a pair of compound leaves at base; leaflets less finely reticulate-veined, the tertiary and quarternary veins scarcely or not raised on the abaxial blade surface, the ultimate closed areoles mostly > 2 mm in longer dimenson; widespread in nc TX **C. pitcheri**

3. Sepals definitely red; stems and leaves glabrous and ± glaucous (especially on lower surfaces); apices of leaf blades rounded to obtuse or emarginate (and usually with a small mucro); Edwards Plateau n to Lampasas Co. in Lampasas Cut Plain **C. texensis**

Clematis crispa L., (crisped, curled), CURLY CLEMATIS, BLUE-JASMINE. Climbing or ascending vine, but can flower when quite small; leaves with 5–9(–11) leaflets; pedicels usually with a pair of compound leaves at base; flowers solitary, terminal, bell-shaped; sepals 25–50 mm long, blue-purple to lavender at base, rosy purple to almost white at tips, distally strongly spreading to recurved; sepal margins broad, thin, 2–6 mm wide distally; mature styles 2–3.5 cm long. Stream banks, low open woods, and roadsides; rare nw to Dallas and Tarrant cos. (Mahler 1988), also Williamson Co. (Correll & Johnston 1970); mainly se and e to c TX. Apr–Sep.

Clematis drummondii Torr. & A. Gray, (for its discoverer, Thomas Drummond, 1780–1835, Scottish botanist and collector in North America), TEXAS VIRGIN'S-BOWER, OLD-MAN'S-BEARD, BARBAS DE CHIVATO, LOVE-IN-THE-MIST. Climbing vine, finely pubescent or almost glabrous; leaves usually with 5 or 7 leaflets; inflorescences usually axillary, of 3–12 flowers, cymose or paniculate, or flowers solitary or paired; flowers rotate; sepals (7–)9–13(–15) mm long, white or whitish, wide-spreading, not recurved distally; mature styles 4–9 cm long. Sandy or rocky ground, along fencerows; Grayson Co., also Dallas Co. (Mahler 1988); nearly throughout TX except extreme e part. Jul–Sep. A number of the common names are derived from the plumose styles persistent on the achenes; a cluster of these has the appearance of a feathery ball (Ajilvsgi 1984).

Clematis pitcheri Torr. & A. Gray, (for its discoverer, Zina Pitcher, 1797–1872, physician, botanist, and mayor of Detroit), LEATHER-FLOWER, BLUEBELL, PITCHER'S CLEMATIS, PITCHER-FLOWER. Climbing ± herbaceous or somewhat woody vine; leaves usually with 3 or 5 leaflets, some simple; inflorescences axillary, 1–7-flowered; flowers ovoid to urn-shaped; sepals 10–35 mm long, dull-purple to brick-red-purple on outer surface, deeply colored or greenish on inner surface, recurved at tips; mature styles 1–3 cm long. Stream bottom thickets, open woods, and fencerows; widespread in TX. Late Apr–early Jul. This is the most common *Clematis* species in nc TX.

Clematis reticulata Walter, (netted), NET-LEAF CLEMATIS. Climbing herbaceous or slightly woody vine; leaves with 3–9 leaflets; inflorescences axillary, 1–3-flowered; flowers urn-shaped; sepals 15–25 mm long, purplish red to mauve or pink-lavender, recurved at tips; mature styles 4–6 cm long. Sandy soils of fields and thickets; e TX w to Red River and Travis cos. on e and s margins of nc TX; included because of likelihood of occurrence in nc TX. Apr–Jul.

Clematis terniflora DC., (flowers in threes), SWEET-AUTUMN CLEMATIS. High climbing or spreading vine to 3 m or more long, glabrous or nearly so; leaflets usually 5; inflorescences axillary, of 3–12 flowers, cymose or paniculate; flowers rotate; sepals 9–22 mm long, white, wide-spreading, not recurved distally; mature styles 1–3(–6) cm long. Frequently cultivated as an ornamental and escapes; Dallas and Tarrant cos., also Brown Co. (J. Stanford, pers. comm.); se, e, and c TX. Jul–Sep(–Dec). Native of Japan. [*C. dioscoreifolia* H. Lév. & Vaniot, *C. maximowicziana* Franch. & Sav., *C. paniculata* Thunb., not J.F. Gmel.] While showy, this species can spread extensively and has the potential of becoming problematic.

Clematis texensis Buckley, (of Texas), SCARLET CLEMATIS, RED LEATHER-FLOWER. Herbaceous or somewhat woody climbing vine to 3 m or more long; leaves usually with 9–11 leaflets; inflorescences axillary, 1–7-flowered; flowers ovoid to urn-shaped; sepals 20–30 mm long, red, recurved at tips; mature styles 5–7 cm long. Rocky stream banks; Bosque Co., also Bell, Coryell (HPC), and Lampasas (Erickson 1943) cos; also Travis Co. just s of nc TX; Edwards Plateau n to Lampasas Cut Plain; endemic to TX. Late Mar–May. According to Pringle (1997), this is the only truly red-flowered species of *Clematis*. /84

CONSOLIDA LARKSPUR

A genus of ca. 40 species (Warnock 1997b) native from the Mediterranean to c Asia; sometimes included in *Delphinium*, but with a number of well-defined differences (Keener 1976a). A number are cultivated as ornamentals. Like *Delphinium*, *Consolida* species contain poisonous diterpenoid alkaloids (e.g., delphinine) which can cause death in livestock (Kingsbury 1964; Schmutz & Hamilton 1979). (Latin: *consolatio*, consolidate, alleviation, or to become solid or firm, from reported ability to heal wounds)
REFERENCES: Chater 1964; Keener 1976a; Warnock 1997b.

Consolida ajacis (L.) Schur, (for Ajax, the Greek hero, because of the shield-like marks on the petals), ROCKET LARKSPUR, ANNUAL LARKSPUR, ESPUELA DE CABALLERO. Annual similar to *Delphinium* (but flower structure different); stems to ca. 1 m tall; leaves basal and cauline, divided into numerous linear to filiform (= thread-like) segments except on lower leaves; bracteoles short, usually not reaching base of flower; flowers usually blue, sometimes violet, purple, pink, or white; spur 13–18 mm long; perianth segments 10–14(–20) mm long; stamens in 5 spirally arranged series; follicles 15–20 mm long, ca. 5 mm wide; seeds black. Commonly cultivated and escapes to roadsides and disturbed sites; Bell, Brown, Dallas, Grayson, Somervell, and Tarrant cos.; nc TX to e and c TX. Apr–Jun. Native of the Mediterranean area. [*C. ambigua* (L.) P.W. Ball & Heywood] Often recognized in the genus *Delphinium* [as *D. ajacis* L. or *D. ambiguum* L.].

Consolida orientalis (J. Gay) Schrödinger, (oriental, eastern), native to s Europe and n Africa, is also cultivated and known from the Edwards Plateau; it could possibly be found escaped in nc TX. It can be distinguished from the similar *C. ajacis* as follows: bracteoles (on pedicels) long, reaching to or beyond base of flower; spur (8–)10–12 mm long; flowers purple to purplish violet; and seeds reddish brown. [*Delphinium orientale* J. Gay]

DELPHINIUM LARKSPUR

A genus of ca. 300 species; mainly n temperate but also the arctic, subtropical areas, and African mountains (Warnock 1997a); many are cultivated as ornamental herbaceous perennials. In the w U.S. LARKSPURS are second only to the LOCOWEEDS as a cause of fatal cattle poisoning; they contain alkaloids such as delphinine and ajacine; while of variable toxicity, all species should be considered poisonous (Kingsbury 1964; Stephens 1980). The common name LARKSPUR is apparently an old English name referring to the resemblance of the spur to the elongate hind claw of the European crested lark (Tveten & Tveten 1993). (Latin: *delphinus*, dolphin, in allusion to the shape of flowers, which are sometimes not unlike the classical figures of dolphins)
REFERENCES: Marsh & Clawson 1916; Ewan 1945, 1951; Keener 1976a; Kral 1976; Warnock 1981, 1987a, 1995, 1997a; Brooks 1982; Olsen et al. 1990.

Delphinium carolinianum Walter, (of Carolina). Herbaceous perennial 0.3–1.5 m tall; stems simple or rarely branched; leaves mainly basal to mainly cauline, palmately deeply lobed or compound; flowers in terminal, spike-like racemes (rarely branched), bilaterally symmetrical; perianth parts in 2 series, all petal-like, the sepals (outer series) 5, showy, with dorsal green spot below apex, the upper with an elongate spur, the petals (inner series) 4, smaller, the upper 2

Clematis reticulata [LOU, TOR]

Clematis terniflora [G&F]

Clematis texensis [HEA]

Consolida ajacis [STE]

with spurs enclosed by large spur from the upper sepal, the lower pair short-clawed; stamens many, in 8 spirally arranged series; follicles 3(–4), erect. Extensive morphological variation and overlap make clear separation of taxa within the *D. carolinianum* complex difficult. While subsp. *carolinianum* can usually be fairly easily distinguished, the other 2 subspecies are more difficult. We are following Warnock's (1981, 1997a) treatment of these taxa. He (1981) noted that "In regions where they occur sympatrically [like nc TX], characters separating the subspecies become obscure and some plants are difficult to place subspecifically."

1. Basal leaves usually absent or rare at flowering; leaf segments usually < 2 mm wide; uppermost petiole usually < 1 cm long; flowers usually deep blue, blue-violet to purple, rarely white; extreme ne part of nc TX ____ subsp. **carolinianum**
1. Basal leaves usually present at flowering; leaf segments usually > 2 mm wide; uppermost petiole usually > 1 cm long; flowers white to blue; widespread in nc TX.
   2. Leaves usually with (3–)5 or more major divisions, these often much further subdivided; flowers white to light lavender; roots ± horizontal with several major branches; widespread in nc TX ____ subsp. **virescens**
   2. Leaves usually with 3 major divisions, these with relatively few subdivisions; flowers blue to white; roots usually ± vertical, often without major branches; s and e parts of nc TX ____ subsp. **vimineum**

subsp. **carolinianum**, WILD BLUE LARKSPUR, CAROLINA LARKSPUR, BLUE LARKSPUR. Sandy open woods and roadsides, dry to rather damp ground; Hopkins Co., also Fannin Co. (Warnock 1981); extreme ne TX. Apr–Jun.

subsp. **vimineum** (D. Don) M.J. Warnock, (with long slender shoots, like *Salix*—willows or osiers), PINEWOODS LARKSPUR, BLUE LARKSPUR, GULF COAST LARKSPUR. Leaf segments often 4 mm or more wide; petioles 3–15 cm long. Clay or sandy soils, open to semi-open grassland; Lampasas, McLennan, Navarro, and Williamson cos. (Warnock 1981); s and se TX nw to s and e parts of nc TX. Mar–Jun. [*D. vimineum* D. Don]

subsp. **virescens** (Nutt.) R.E. Brooks, (light green), PRAIRIE LARKSPUR, PLAINS LARKSPUR, WHITE LARKSPUR, PENARD'S LARKSPUR, RABBIT-FACE. Glandular pubescence often present; leaf segments mostly 2–4 mm wide. Grasslands; ne TX w and s to Panhandle and Edwards Plateau; this is by far the most common subspecies in nc TX. Apr–Jun. [*D. carolinianum* Walter subsp. *penardii* (Huth) M.J. Warnock, *D. virescens* Nutt., *D. virescens* var. *macroceratilis* (Rydb.) Cory, *D. virescens* var. *penardii* (Huth) L.M. Perry] Brooks (1982) split this subspecies into 2, subsp. *penardii* and subsp. *virescens*, separated as follows: leaves mostly equally distributed and cauline, few basal; upper stem and rachis covered with yellow pustulate trichomes—subsp. *virescens*; versus distinct basal rosette of leaves present with cauline leaves few; upper stem canescent and sparsely pustulate hairy; rachis canescent—subsp. *penardii*. We are following Warnock (1997a) in lumping the two. ▣/87

MYOSURUS MOUSETAIL

A genus of 15 species (Whittemore 1997b) of temperate areas; characterized by an elongate receptacle. (Greek: *myos*, of a mouse, and *oura*, a tail, from the long slender fruiting receptacle)
REFERENCES: Campbell 1952; Stone 1959; Whittemore 1997b.

Myosurus minimus L., (least, smallest), TINY MOUSETAIL, MOUSETAIL. Very small glabrous annual 4–16.5 cm tall; leaves basal, linear-oblanceolate, entire; leafless flowering stem at first shorter, eventually longer, than the leaves; flower solitary, terminal, radially symmetrical; sepals 5, reflexed, greenish; petals often 5, linear, whitish (rarely pinkish), 2–3.5 mm long, often soon deciduous, sometimes absent; carpels numerous, usually more than 100 on a greatly elongating, narrowly cylindrical receptacle (2–5 cm long in fruit) that resembles a mouse's tail; achenes 0.9–

Delphinium carolinianum subsp. carolinianum [DEL]

Delphinium carolinianum subsp. vimineum [CUR]

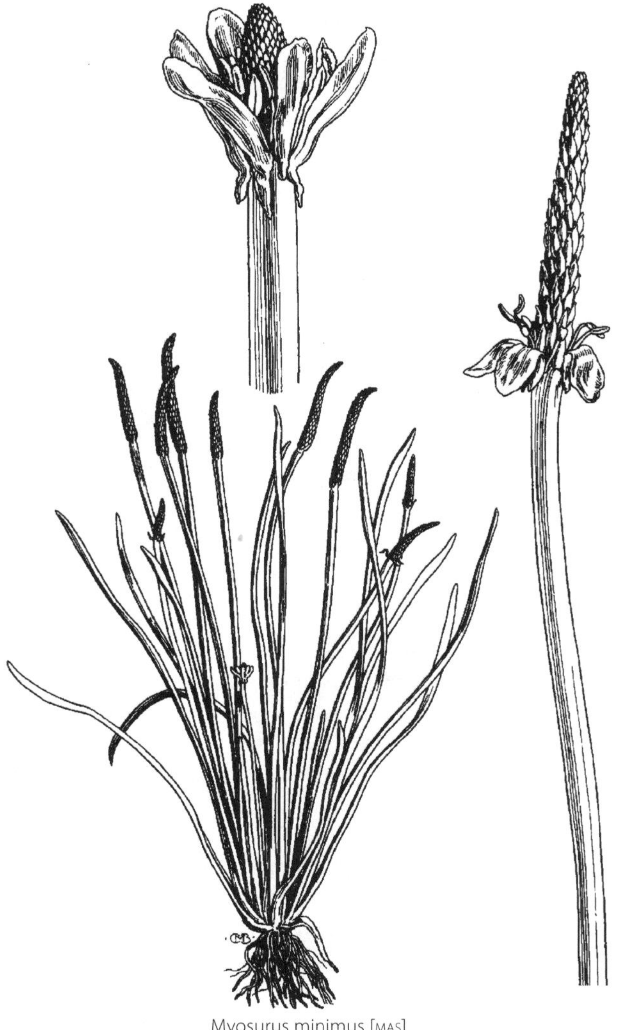
Myosurus minimus [MAS]

Delphinium carolinianum subsp. virescens [BB1]

Ranunculus abortivus [BB2]

2 mm long, usually with short beak. Damp sand or clay soils; Dallas, Denton, Grayson, and Hunt cos.; throughout much of TX, particularly the e 1/2. Mar–Apr. Inconspicuous and easily overlooked; it can occur with and superficially resemble *Plantago elongata.* ▣/100

RANUNCULUS BUTTERCUP, CROW'S-FOOT, PAIGLE

Small annual or perennial herbs; leaves simple or compound, entire, toothed, or lobed; petioles with widened margins toward base, not distinctly stipulate; flowers solitary and terminal or in upper leaf axils; sepals 3–5, green; petals (honey leaves) (1–)usually 5 or more, in ours usually yellow to greenish, often showy; stamens and pistils usually numerous; fruit an achene, these aggregated in clusters or head-like groups.

A genus of ca. 300 species of herbs with yellow, greenish, white, or red flowers native nearly worldwide except in lowland tropics (Whittemore 1997a). ☠ All species are apparently poisonous and generally avoided by livestock; they contain the toxic glycoside protoanemonin which is irritating to the skin and mucous lining of the digestive tract; cows poisoned by BUTTERCUPS produce bitter or reddish milk (Schmutz & Hamilton 1979; Blackwell 1990). A number are cultivated as ornamentals and some are weeds. (Diminutive of Latin: *rana*, frog; name applied by Pliny to these plants, many of which grow in wet habitats)

REFERENCES: Benson 1940, 1948, 1954; Keener 1976b; Duncan 1980; Keener & Hoot 1987; Whittemore 1997a.

1. Stem leaves simple, entire or slightly toothed ____ **R. pusillus**
1. Stem leaves (except uppermost) compound or deeply lobed.
  2. Petals greatly exceeding the sepals, 7–21 mm long; plants perennial.
    3. Uppermost leaves or bracts similar in appearance to lower leaves, compound; leaflets rhombic-ovate to ovate-lanceolate, toothed and often lobed; stems often rooting at the nodes; roots fleshy-fibrous, slender ____ **R. hispidus**
    3. Uppermost leaves or bracts often quite different in appearance from lower leaves, simple, the blades oblong-lanceolate to linear-lanceolate, entire or with few, irregular teeth or lobes (rarely uppermost leaves compound with narrow, mostly entire leaflets); stems not rooting at the nodes; roots fleshy-fibrous and some tuberous-thickened.
      4. Stems with appressed or ascending hairs; petals 5(–9), 7–12 mm long; sepals 5–7; achenes 1.8–2.2 mm wide; ne part of nc TX ____ **R. fascicularis**
      4. Stems with long hairs spreading at right angles; petals (8–)10–22, 10–21 mm long; sepals 7–10; achenes 2.8–3.4 mm wide; s and w parts of nc TX ____ **R. macranthus**
  2. Petals shorter than or equaling the sepals, 1–8 mm long; plants annual.
    5. Achenes 5–5.5 mm long at maturity including a conspicuous beak ca. 2–2.5 mm long; faces of achenes with stout prickles ____ **R. muricatus**
    5. Achenes ca. 1–2 mm long at maturity, the beak inconspicuous (absent or < 1 mm long); faces of achenes glabrous or with minute hooked bristles/prickles.
      6. Leaf blades sparsely or densely short pilose, with acute teeth or lobes ____ **R. parviflorus**
      6. Leaf blades glabrous, with obtuse teeth or lobes or none.
        7. Blades of lower leaves mostly or all divided half way to base or less; achenes ca. (1–)1.5 mm across ____ **R. abortivus**
        7. Blades of lower leaves all divided nearly or quite to base; achenes 0.8–1 mm across ____ **R. sceleratus**

Ranunculus abortivus L., (abortive, i.e., with reduced styles, petals, etc.), LITTLE-LEAF BUTTERCUP, KIDNEY-LEAF BUTTERCUP, EARLY WOODS BUTTERCUP. Glabrous, erect or nearly so, to 35 cm tall; petals 2.5–3.5 mm long; achenes smooth, 1.4–1.6 mm long, 10–35 in an ovoid head. Open areas, woods, damp shady ground; Dallas, Grayson, Hopkins, and Hunt cos., also Lamar Co. (Carr

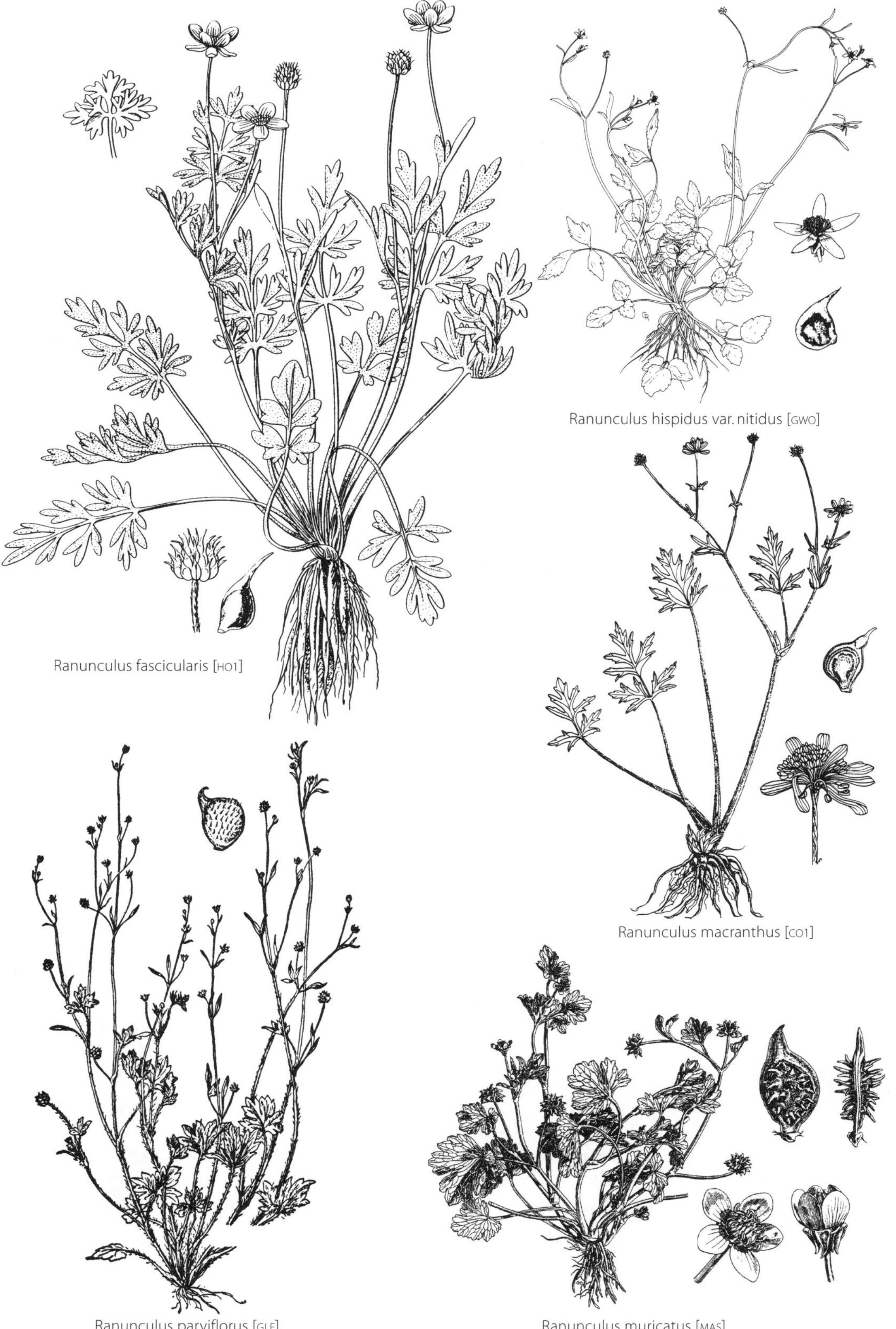

Ranunculus fascicularis [HO1]

Ranunculus hispidus var. nitidus [GWO]

Ranunculus macranthus [CO1]

Ranunculus parviflorus [GLE]

Ranunculus muricatus [MAS]

1994); e TX w to nc TX. Mar–Jun. Reportedly poisonous to livestock, with symptoms including gastrointestinal irritation, salivation, diarrhea, blindness, convulsions, and death (Burlage 1968). ☠

Ranunculus fascicularis Muhl. ex Bigelow, (fascicled, in clusters), TUFTED BUTTERCUP, PRAIRIE BUTTERCUP, EARLY BUTTERCUP. Stems erect or spreading, to 30 cm long; achenes 2–2.8 mm long. Sandy open woods and open ground; Grayson, Henderson, Hopkins, Kaufman, and Lamar cos.; se and e TX w to e Blackland Prairie. Late Feb–early Apr. [*R. fascicularis* Muhl. ex Bigelow var. *apricus* (Greene) Fernald]

Ranunculus hispidus Michx. var. **nitidus** (Chapm.) T. Duncan, (sp.: bristly; var.: shining), BRISTLY BUTTERCUP, MARSH BUTTERCUP. Stems low-spreading or trailing, glabrous or sparsely pilose, to 60 cm long; petals 8–12 mm long. Stream bottoms and ditch banks; Dallas, Grayson, Kaufman, Lamar, and Rockwall cos.; se and e TX w to nc TX. Late Feb–Mar. [*Ranunculus carolinianus* DC.]

Ranunculus macranthus Scheele, (large-flowered), SHOWY BUTTERCUP, LARGE BUTTERCUP. Stems erect to reclining, to 50(–100) cm tall; achenes 2.2–4.2 mm long. Damp, silty or sandy clay ground; Bell, Burnet, McLennan, Parker, Somervell, and Williamson cos.; s and w parts of nc TX s to s TX and w to Trans-Pecos. Late Mar–early May.

Ranunculus muricatus L., (roughened by means of hard points), ROUGH-SEED BUTTERCUP. Stems reclining to erect, to 50 cm long; petals 4–7 mm long; achenes 5–5.5 mm long, with stout prickles, 10–20 in a cluster. Grassy or wet areas; Dallas, Grayson, and Tarrant cos.; mainly se and e TX. Mar–May. Native of the Old World.

Ranunculus parviflorus L., (small-flowered), STICKTIGHT BUTTERCUP. Beginning to flower when almost stemless; stems becoming as much as 30 cm long, trailing to erect; flowers inconspicuous, obscured by the leaves; petals 1–2 mm long; achenes 1.5–2 mm long, papillate and usually with minute hooked prickles, 10–20 in a globose head. Damp sandy ground, in woods, roadsides, and sometimes a lawn weed; Dallas and Hunt cos.; se and e TX w to nc TX. Mar–Apr. Native of Europe. Reported to have toxic properties similar to *R. abortivus* (Burlage 1968). ☠

Ranunculus pusillus Poir., (very small), WEAK BUTTERCUP. Ascending, glabrous and glaucous annual to 50 cm tall; stems usually rooting at lowest nodes; petals 1–3(–5), 1.5–2.5 mm long; achenes 1–1.2 mm long. Damp sand or silt, wet areas; Bell, Dallas, Grayson, Hopkins, and Red River cos., also Burnet (Correll & Johnston 1970) and Lamar (Carr 1994) cos.; se and e TX w to nc TX, also Edwards Plateau. Mar–Apr. While often recognized (Kartesz 1994; Jones et al. 1997), we are following Whittemore (1997a) in lumping [*R. pusillus* var. *angustifolius* (Engelm. ex Engelm. & A. Gray) L.D. Benson]. This species is often inconspicuous and easily overlooked.

Ranunculus sceleratus L., (cursed, growing in vile places), BLISTER BUTTERCUP, ROUGE BUTTERCUP, CURSED CROW'S-FOOT, CELERY-LEAF BUTTERCUP. Resembling *R. abortivus*, but distinguished as in the key, more freely branched, and usually in different habitats; stems to 1 m long; petals 2–5 mm long; achenes not spiny, ± smooth, ca. 1–1.2 mm long, 40–300 in a cylindrical head. Ditches, sandy shores of lakes, ponds, and streams; Bell, Grayson, and Tarrant cos., also Hamilton (HPC) and Parker (R. O'Kennon, pers. obs.) cos.; nc and s TX, Edwards Plateau, and Rolling Plains. May–Jun. The acrid sap is reported to cause blisters on human skin and the plant is reportedly poisonous to livestock (Burlage 1968; Correll & Correll 1972). Moerman in Whittemore (1997a) indicated that *R. sceleratus* was used by Native Americans as a poison for arrow points. ☠

THALICTRUM MEADOW-RUE

Ours herbaceous, usually dioecious perennials; leaves 2 or 3 times ternately compound, the numerous ultimate leaflets usually 3-lobed or several-toothed at apex; petioles with clasping

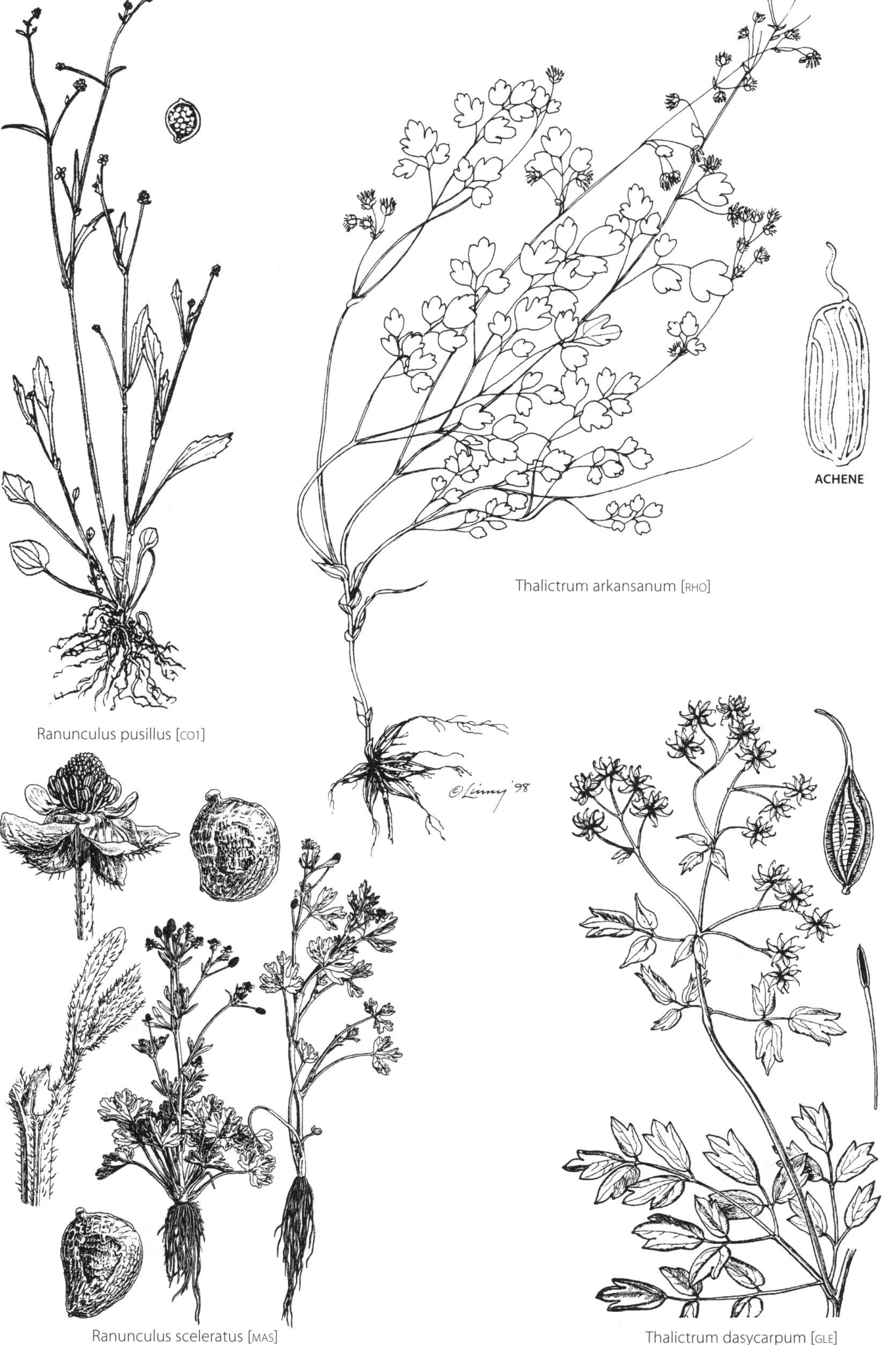

Thalictrum arkansanum [RHO]

Ranunculus pusillus [CO1]

Ranunculus sceleratus [MAS]

Thalictrum dasycarpum [GLE]

bases; flowers in terminal panicles, drooping; perianth parts in 1 series (sepals); sepals 4–5, whitish to purplish; stamens many; pistils few–15; fruit an achene.

A genus of 120–200 species (Park & Festerling 1997) of herbs with alkaloids native to the n temperate zone, tropical areas of the Americas, and s Africa. Some are used medicinally and as ornamentals. (Greek: *thaliktron*, classical name used by Dioscorides)
REFERENCES: Boivin 1944; Park & Festerling 1997.

1. Plants decumbent, 15–30(–40) cm tall; stems 0.5–1.1 mm thick below middle; roots few, tuberous; ultimate leaflets 1.5 cm or less long; inflorescences few-flowered; leaflets glabrous ______ **T. arkansanum**
1. Plants erect, 60–150(–200) cm tall; stems 3–6 mm thick below middle; roots many, coarsely fibrous; ultimate leaflets 2–5.5 cm long; inflorescences many-flowered; leaflets typically pubescent on lower (= abaxial) surfaces ______ **T. dasycarpum**

Thalictrum arkansanum B. Boivin, (of Arkansas), ARKANSAS MEADOW-RUE. Roots ribbed, brown; lobes of leaflets often obtuse to rounded and often notched or toothed; perianth parts whitish, those of male flowers 2–3 mm long, of female flowers 1–1.5 mm long; filaments pink, 2–3 mm long; stigma 1.3–3 mm long; mature achenes sessile, ellipsoid, 3.5–4.5 mm long, 10–12-nerved. Damp sandy woods; rare in e Red River drainage, Grayson (Mahler 1988) and Red River cos., also Lamar Co. (TOES 1993); limited in TX to ne part of state; narrowly endemic to ne TX and adjacent AR and OK. Late Mar–early Apr. Park and Festerling (1997) indicated that *T. arkansanum* is poorly known, that it is closely related to *T. texanum* and *T. debile* Buckley (of se U.S.), and that it should possibly be considered a variety of *T. debile*. Robert Kral (pers. comm.) indicated that *T. arkansanum* is nearly indistinguishable from *T. debile*. Further study is need to clarify the relationships of these species. (TOES 1993: V) ⚠

Thalictrum dasycarpum Fisch. & Avé-Lall., (thick-fruited), PURPLE MEADOW-RUE. Stems often purple; leaflets sometimes glaucous on lower surface, entire or usually 3-lobed, the lobes usually acute and entire; perianth parts 3–5 mm long, purplish to whitish; filaments white, 4–7 mm long; stigma 2–5 mm long; mature achenes sessile or nearly so, (2.5–)3.8–5.5 mm long. Damp thickets and open woods, sandy or silty ground; Hunt Co., also Dallas Co. (*Hall* 2, 1872 cited in Boivin 1944); se and e TX rare w to the Panhandle. Apr–May. [*Thalictrum dasycarpum* Fisch. & Avé-Lall. var. *hypoglaucum* (Rydb.) B. Boivin]

Thalictrum texanum (A. Gray) Small, (of Texas), HOUSTON MEADOW-RUE, native to se TX, is cited by Hatch et al. (1990) for vegetational area 4 (Fig. 2). This species, which differs from *T. arkansanum* in being erect, having shorter stigmas (0.5–1 mm long), ovoid achenes, and black roots (when dry), apparently only occurs to the s and e of nc TX.

RHAMNACEAE BUCKTHORN FAMILY

Half-woody to woody shrubs, vines, or small trees; leaves usually alternate (opposite in 1 species rare in nc TX), short-petioled, simple, entire or toothed; stipules small, falling early; flowers small, radially symmetrical, usually perfect, axillary or terminal, solitary or few together, or many in compact, umbel-like racemes or small panicles; calyces (4–)5-lobed; petals (4–)5 or absent, green, yellowish, or white (rarely pinkish), usually clawed or spatulate and often hood-shaped, delicate, inconspicuous (except in *Ceanothus*); stamens 5, alternate with the sepals, often hooded by the petals, attached around an often prominent, fleshy disk; pistil 1; ovary superior; fruit a drupe or capsule-like.

A medium-sized (900 species in 49 genera), cosmopolitan, but especially tropical and warm area family of trees, shrubs, climbers, and rarely herbs including some ornamentals (e.g., *Ceanothus*). A number of species contain anthraquinone glycosides and alkaloids and have been used medicinally; others have been used as dyes, for edible fruit, or as ornamentals. (subclass Rosidae)

FAMILY RECOGNITION IN THE FIELD: woody plants usually with alternate (opposite in 1 species) simple leaves and sometimes with *3 main veins* or with *secondary veins strikingly parallel*; stipules small; flowers usually small, 5-merous, with a hypanthium; stamens 5, opposite the petals.

REFERENCES: Brizicky 1964c; Johnston & Johnston 1969.

1. Plants twining, climbing woody vines; fruits ca. 2 times as long as wide ______ **Berchemia**
1. Plants shrubs or small trees; fruits usually nearly as wide as long.
    2. Leaves opposite; known only on extreme w margin of nc TX ______ **Karwinskia**
    2. Leaves alternate; including species widespread in nc TX.
        3. Petals absent; plants armed (branches ending in stout thorns); limited to extreme w and s parts of nc TX ______ **Condalia**
        3. Petals present (may be small); plants armed (in *Ziziphus*) OR unarmed in all other nc TX species; widespread in nc TX.
            4. Plants usually armed, the branches with terminal thorns or axillary spines; leaves 3-veined from the base; petals green to yellowish ______ **Ziziphus**
            4. Plants unarmed; leaves pinnately veined or with one main vein, not 3-veined from the base (except in *Ceanothus* with usually white or rarely pinkish petals); petals green to yellowish to white (rarely pinkish).
                5. Petals white (rarely pinkish), becoming much longer than the calyx lobes; flowers in umbel-like inflorescences at the end of elongate axillary peduncles or at the end of leafy branches, conspicuously surpassing the leaves; leaves usually 3-veined from base; enlarged cup and disk below fruit persistent on the pedicel as a roughly triangular structure ______ **Ceanothus**
                5. Petals green to yellowish, smaller than or equaling the calyx lobes; flowers axillary, single or in small clusters or glomerules, inconspicuously borne among the leaves; leaves usually pinnately veined or with 1 vein; without persistent cup and disk.
                    6. Leaves 3–14 cm long; fruit juicy at maturity, a drupe; widespread in nc TX.
                        7. Flowers 1–3 in the axils of new growth, at least some unisexual; leaf blades 3–7 (–9) cm long; calyx lobes, petals, and stamens 4; fruits with 2 stones ______ **Rhamnus**
                        7. Flowers (1–)3–8 in umbel-like axillary clusters, perfect; leaf blades 5–15 cm long; calyx lobes, petals, and stamens 5; fruits with 3 stones ______ **Frangula**
                    6. Leaves < 3 cm long; fruit dry at maturity, a slow-maturing capsule; found only in extreme s and possibly w parts of nc TX ______ **Colubrina**

BERCHEMIA SUPPLEJACK, RATTANVINE

A genus of ca. 12 species of twiners, some used as ornamentals; native from e Africa to e Asia, and w North America. (Presumably named for Berthout van Berchem, 17th or 18th century Dutch or French botanist)

Berchemia scandens (Hill) K. Koch, (climbing), SUPPLEJACK, RATTANVINE, ALABAMA SUPPLEJACK. High climbing, twining, woody vine (= liana), sometimes forming a vining shrub if unsupported; stems conspicuously smooth, reddish brown or yellowish brown, with age to greenish or grayish; leaf blades oblong-elliptic, 3–6(–8) cm long, glabrous, entire or nearly so, pale or glaucous beneath, conspicuously pinnately veined, the veins parallel; inflorescence a small raceme or narrow panicle at the ends of lateral branches; flowers 5-merous, small (ca. 2 mm wide), yellowish or greenish; fruit a small, ellipsoidal, blue-black drupe ca. 6–8 mm long, juicy at maturity. Thickets or woods, stream bottoms or slopes; much of the e 1/2 of TX. Late Apr–early May. The smooth, greenish or grayish older stems make this vine easily recognizable in nc TX forests. In winter, the reddish brown or yellowish brown naked twigs are also quite conspicuous.

CEANOTHUS BUCKBUSH

Ours unarmed, weak shrubs with leaves alternate, deciduous, usually 3-ribbed (= nerved or veined) basally; inflorescences dense, umbel-like panicles; petals in ours with claw and hooded blade each nearly 1 mm long; ovary 3(–4) celled; fruits capsule-like, the upper portions abscising and leaving the cup and disk (roughly triangular and shallowly bowl-like in shape) persistent on the pedicel.

A North American genus of 55 species of shrubs and small trees, especially common in California, particularly in the chaparral. A number of species are cultivated for their showy inflorescences; nitrogen-fixing Ascomycete fungi occur in the roots. The burl-like "rootstocks" or "grubs" of nc TX species become greatly enlarged and pioneers had to dig these up prior to cultivation (Correll & Johnston 1970). (From a Greek name used by Theophrastus for a spiny plant not of this genus)
REFERENCES: Parry 1889; Brandegee 1894.

1. Inflorescences on elongate, axillary, essentially naked peduncles; leaves usually ovate to broadly elliptic, usually pilose above, usually acute _____ **C. americanus**
1. Inflorescences terminating leafy branches; leaves narrowly elliptic or elliptic-lanceolate, glabrous or nearly so above, obtuse or barely acute _____ **C. herbaceus**

Ceanothus americanus L., (of America), REDROOT, NEW JERSEY-TEA, JERSEY-TEA. Erect shrub to ca. 1 m tall; fruits 3(–4) -lobed, ca. 4–5 mm across. Sandy soils; woodlands, prairies; Grayson, Henderson, and Lamar cos.; much of the e 1/2 of TX. Apr. The dried leaves have been used as a tea and also medicinally by Native Americans; they were also used by early settlers during the Revolutionary War when Asian tea could not be obtained (Correll & Correll 1969). [*C. americanus* L. var. *pitcheri* Torr. & A. Gray] We are following Kartesz (1994) and J. Kartesz (pers. comm. 1997) in lumping var. *pitcheri*; Jones et al. (1997) treated all TX material of this species as var. *pitcheri*.

Ceanothus herbaceus Raf., (herbaceous, not woody), REDROOT, INLAND CEANOTHUS, FUZZY CEANOTHUS. Similar to *C. americanus*; usually nearly glabrous. Limestone outcrops, gravelly prairies; from Post Oak Savannah s and w across most of TX. Apr. [*C. herbaceus* var. *pubescens* (Torr. & A. Gray ex S. Watson) Shinners]

COLUBRINA SNAKEWOOD

A genus of 31 species of tropical and warm parts of the world, especially the New World. ☠ Some species have been used as a fish poison and others medicinally. (Latin derived from French for serpent tree, for the snake-like shape of some species)
REFERENCES: Johnston & Johnston 1969; Johnston 1962, 1971.

Colubrina texensis (Torr. & A. Gray) A. Gray, (of Texas), TEXAS SNAKEWOOD, TEXAS COLUBRINA, HOG-PLUM. Unarmed rounded shrub 1–2 m tall; leaves small, ovate, thinly tomentose beneath, 1-nerved; flowers small, ca. 5–6 mm across, yellow-green, borne on short, twig-like condensed spur shoots, 1 per axil but appearing glomerate (some leaves also crowded together with flowers on the spur shoots); ovary 3-celled; fruits drupe-like, with stony endocarp becoming tardily dehiscent and capsule-like. Slopes or rocky hillsides; Brown and Burnet cos. (Johnston & Johnston 1969) on sw margin of nc TX s and w to w TX. May–Jul.

CONDALIA

A genus of 18 species of warm areas of the New World. (Named for Antonio Condal, native of Barcelona, Spanish explorer of South America)

Condalia hookeri M.C. Johnst., (for Sir W.J. Hooker, 1785–1875, director of Kew Gardens or his son, Sir Joseph D. Hooker, 1817–1911, traveler in the Himalaya, who succeeded his father as director of Kew), BLUEWOOD, BRASIL, BRAZIL, CAPUL NEGRO. Armed shrubs to ca. 2 m tall; leaves obovate (10–)15–20 mm long, usually rounded apically to rarely slightly acute, basally acuminate, entire or rarely with 2–4 inconspicuous teeth; flowers inconspicuous; fruit a globose, dark blue or black, fleshy drupe 5–6 mm in diam. Thickets; Brown Co. (Stanford 1971), also Williamson Co. (Correll & Correll 1969; Little 1976); w and s margins of nc TX s to s TX. Summer–Sep. [*C. obovata* Hook.] The dense heartwood is a brilliant red color and has been used to make ornamental objects (Standley 1923a; Correll & Correll 1969); a blue dye was obtained in frontier days from the wood (Standley 1923a; Crosswhite 1980). The fruits were used during frontier days to make jelly (Crosswhite 1980).

FRANGULA BUCKTHORN

A genus of ca. 40 species primarily in the New World but with some in e Asia, Europe, North Africa, and the Azores (Brizicky 1964c); it is often treated as a subgenus of *Rhamnus* but differs in having naked winter buds and flowers 5-merous and bisexual. ☠ Like *Rhamnus*, some species contain purgatives (anthraquinones). *Frangula purshiana* (DC.) Cooper [*Rhamnus purshiana* DC.], native to the nw U.S. and adjacent Canada, is commonly known as CASCARA SAGRADA; it is the source of commercial purgatives, with the annual sale of the bark valued at many tens of millions of dollars. According to Heiser (1993), it "... is generally thought to be the world's most widely used carthartic." (Latin: *frangere*, to break, because the wood of some species breaks easily)

Frangula caroliniana (Walter) A. Gray, (of Carolina), CAROLINA BUCKTHORN, INDIAN-CHERRY, YELLOWWOOD, POLECAT-TREE. Large shrub or small tree to ca. 6 m tall, unarmed; buds without scales, densely brown-hairy; leaf blades oblong-lanceolate, finely or indistinctly toothed, subacute, pinnately veined, at maturity glabrous beneath except on main veins to soft-pubescent over the lower surface; flowers small, yellow-green, in compact, axillary, umbel-like clusters shorter to slightly longer than the adjacent petiole; calyces 3–4 mm long; petals 1–1.3 mm long, appearing with emerging leaves or after the leaves have emerged; fruit a small drupe ca. 5–8 mm in diam., black at maturity, going through a red phase making the bicolored infructescence rather showy. Ravines and stream bottoms; in much of the e 1/2 of TX. May–early Jun. [*R. caroliniana* Walter] Given the close relationship to *Rhamnus* and other toxic species of *Frangula*, the bark, leaves, and fruits should be assumed to have toxins; reported to have laxative effects (Pammel 1911). ☠

KARWINSKIA

A genus of 16 species ranging from the sw U.S. to Bolivia and the West Indies. Some are used for timber, others medicinally. (Named for Wilhelm Friedrich von Karwinski, 1780–1855, Bavarian botanist who collected in Mexico and Brazil)

Karwinskia humboldtiana (Schult.) Zucc., (named for Alexander von Humboldt, 1769–1859, German naturalist, traveller, and geographer; explored South America and Mexico from 1799–1804), COYOTILLO, HUMBOLDT'S COYOTILLO. Unarmed shrub usually 1–2 m tall, usually glabrous; leaves opposite, 3–7(–8) cm long, oblong or elliptic-oblong, entire or rarely slightly crenate, with numerous, conspicuously parallel secondary veins (leaves thus resembling those of *Berchemia*); secondary veins when viewed on abaxial (= lower) leaf surface often with a conspicuous alternating pattern of light and dark areas; inflorescence a few-flowered axillary cyme; flowers small, inconspicuous; petals present; fruit a globose drupe, black at maturity. Dry plains and prairies; Brown Co. (HPC); w margin of nc TX s to s TX and w to Trans-Pecos. Summer–fall. The fruits (flesh) are reported to be edible but experimental evidence showed some toxicity; the stones in particular are dangerous—they contain a toxic substance that paralyzes the motor nerves of vertebrates; paralysis of the limbs results in humans and domestic animals; there may be a lengthy (weeks) lag period between poisoning and onset of symptoms; in Mexico, COYOTILLO is used in treating tetanus and the leaves and roots in treating fevers; however, the foliage can be fatally toxic to livestock and all parts of the plant should be considered poisonous (Standley 1923a; Vines 1960; Kingsbury 1964; Correll & Johnston 1970; Mabberley 1987). ☠

RHAMNUS BUCKTHORN

A genus of 125 species of usually deciduous trees and shrubs ranging from the n hemisphere s to Brazil and s Africa; often treated as including *Frangula*, which is here recognized as a distinct genus. ☠ A European species, *R. cathartica* L. (PURGING BUCKTHORN), is the source of a powerful purgative (hydroxymethylanthraquinones); the toxins are present in the fruits and bark (Kingsbury 1964; Lampe & McCann 1985). A number of species, including *R. lanceolata*, are known to be alternate hosts for the fungus, *Puccinia coronata* Corda, crown rust of OATS (Brizicky 1964c). (Greek: *rhamnos*, the classical name for these plants)
REFERENCE: Gleason 1947

Rhamnus lanceolata Pursh subsp. **glabrata** (Gleason) Kartesz & Gandhi, (sp.: lanceolate, lance-shaped; subsp.: smooth, hairless), LANCE-LEAF BUCKTHORN. Large polygamodioecous shrub to small tree, unarmed; leaves and young branches glabrous; buds covered with scales; leaves lanceolate-elliptic, finely serrulate; flowers small, 4-merous, yellowish green, with petals 0.5–1.3 mm long, appearing before or with the emerging leaves, the leaves if present at flowering time only ca. 1 cm long; male flowers (1–)2–3 per axil; female flowers solitary; fruit a black drupe ca. 6 mm in diam. Low woods and blackland prairie; Hunt Co. (Clymer Meadow, *O'Kennon 13247*, BRIT); mainly e TX. Spring. [*R. lanceolata* var. *glabrata* Gleason] Subspecies *glabrata* has not been previously reported for TX; the subspecies supposedly occurring in the state (Jones et al. 1997; J. Kartesz, pers. comm. 1997) is subsp. *lanceolata* with the leaves and young branches pubescent at anthesis and the leaves permanently pubescent beneath, at least on the veins. However, all TX material we have observed is essentially glabrous and matches the description of subsp. *glabrata*. Given that the only difference between the subspecies is apparently pubescence (Gleason 1947 as varieties), the recognition of taxa below the level of species is questionable. The fruits are reported to have the same poisonous properties as *R. cathartica* (Pammel 1911). ☠

ZIZIPHUS

Shrubs or small trees, nearly glabrous, usually armed; leaves alternate, 3-veined, petioled; flowers usually axillary, solitary or in cymes; petals present; fruit a drupe with woody 2-celled stone.

A genus of 86 species of trees and shrubs native to tropical and warm areas of the world. The related *Paliurus spina-christi* Mill. [*Ziziphus spina-christi* (L.) Willd.] (CHRIST'S-THORN, CROWN-OF-THORNS), a Mediterranean species, is reputed to have been used as Christ's crown of thorns; it is a naturalized pernicious weed in Gillespie Co. on the Edwards Plateau where eradication efforts are being undertaken (O'Kennon 1991). (Arabic: *zizouf*, the name for the lotus fruit of antiquity, *Z. lotus* Desf.)
REFERENCES: Thomas 1936; Johnston 1963b.

1. Leaf blades usually 3 cm or less long, gray-green; branches terminating in straight thorny tip; fruits pea-like, 1 cm or less long; branches glaucous ____ **Z. obtusifolia**
1. Leaf blades usually (2–)3–5 cm long, glossy green; branch tips not spiny, spines axillary when present, hooked or curved; fruits date-like, > 1.5 cm long (often much longer); branches not glaucous ____ **Z. zizyphus**

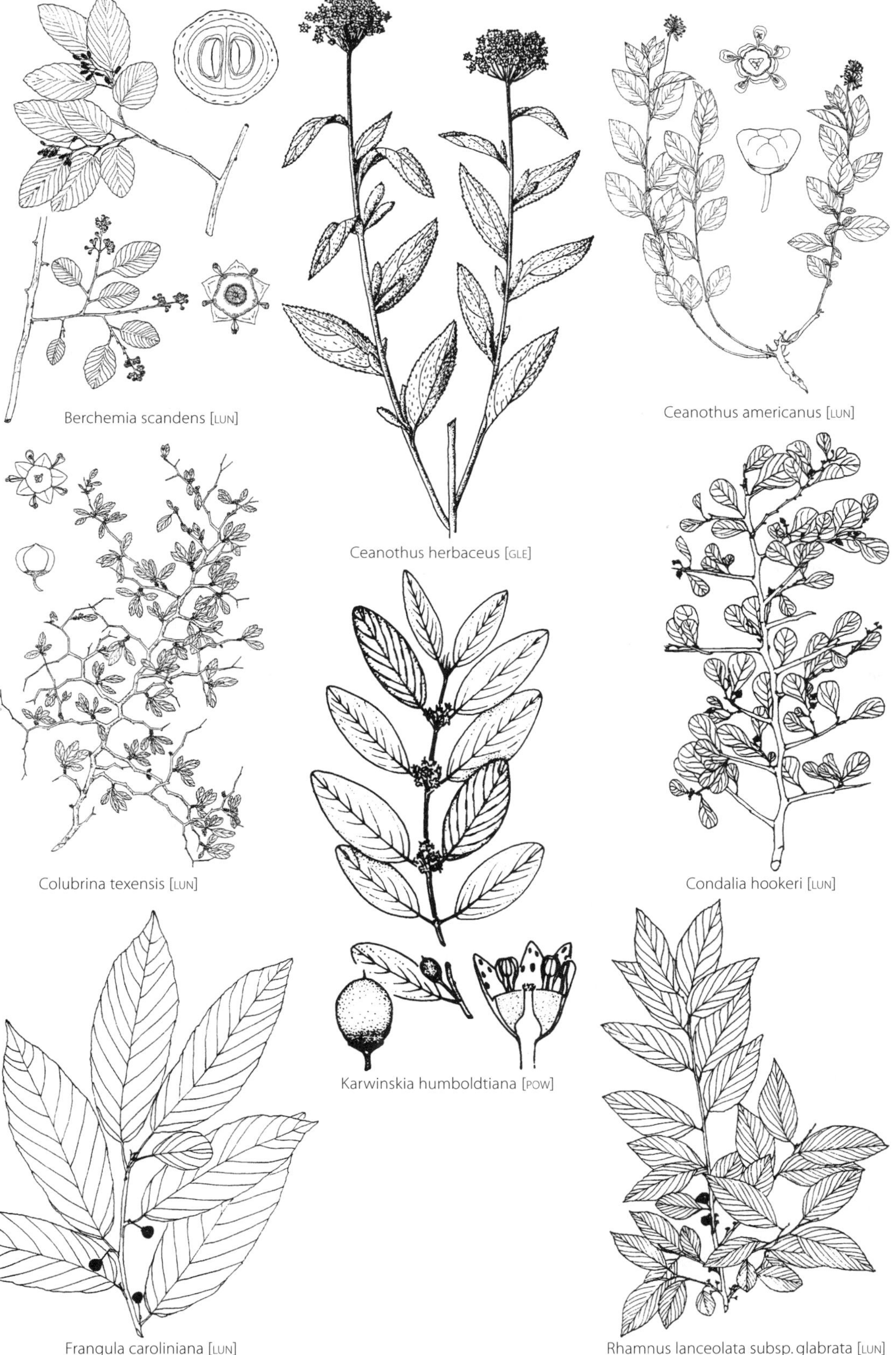

Berchemia scandens [LUN]

Ceanothus herbaceus [GLE]

Ceanothus americanus [LUN]

Colubrina texensis [LUN]

Karwinskia humboldtiana [POW]

Condalia hookeri [LUN]

Frangula caroliniana [LUN]

Rhamnus lanceolata subsp. glabrata [LUN]

Ziziphus obtusifolia (Hook. ex. Torr. & A. Gray.) A. Gray, (blunt-leaved), LOTEBUSH, GUMDROP-TREE, GUMDROPBUSH, BLUETHORN, CLEPE. Densely and stiffly branched shrub to ca. 2 m tall; branches glaucous with a grayish or whitish, waxy bloom; leaves very variable, small, usually < 4 cm long, typically much smaller, oblong to elliptic to nearly linear, entire or indistinctly and bluntly toothed (leaves on sucker or root shoots ovate, bluntly toothed), very obtuse to slightly indented at apex, glabrous, often grayish green; flowers in small clusters or short, raceme-like, usually axillary inflorescences; petals ca. 1 mm long, early deciduous. Rocky or sandy ground; Brown, Clay, Coleman, Jack, and Palo Pinto cos., also Callahan and Shackelford cos. (Correll & Correll 1969); mainly West Cross Timbers s and w to w TX, also further e in Bell, Burnet, and Williamson cos. (Correll & Correll 1969). April and later according to rains. [*Condalia obtusifolia* (Hook. ex Torr. & A. Gray) Weberb.] *Condalia hookeri* is superficially similar to *Z. obtusifolia* because of its branch tips ending in thorns and its small leaves. Its bright green leaves broader towards the apex, non-glaucous branches, and lack of petals distinguish it from *Z. obtusifolia*.

Ziziphus zizyphus (L.) H. Karst., (from an Arabic name), JUJUBE, JAPANESE-APPLE, COMMON JUJUBE, CHINESE JUJUBE, CHINESE-DATE. Large shrub or tree to 12 m tall; at least some branches usually with axillary spines; leaf blades ovate-lanceolate to ovate, shallowly and rather bluntly toothed, glabrous or nearly so; flowers on new growth, in very small axillary clusters or solitary. Occasionally cultivated, tending to spread from root-sprouts, and to self-sow; this species can form thorny thickets; Dallas, Grayson, and Tarrant cos., also Brown and Hamilton cos. (HPC) and Fort Hood (Bell or Coryell cos.—Sanchez 1997); se and e TX w to nc TX and Edwards Plateau. Early May. Native of se Europe and s Asia. [*Z. jujuba* Mill.] The flesh of the ripe fruit is edible raw or cooked and has a date-like taste; for this reason the species is widely cultivated in some parts of the world; it was cultivated in China many centuries B.C. (Thomas 1936). The National Champion JUJUBE (largest recorded in the U.S.) is located in the Fort Worth Botanic Gardens (American Forestry Association 1996). This species can spread and become problematic.

ROSACEAE ROSE FAMILY

Plants herbaceous or woody, armed or unarmed; leaves basal or alternate (sometimes closely bunched), simple or compound, entire or usually toothed or lobed; stipules small to large and leafy, or minute and falling when leaves open, or absent; flowers terminal or axillary, solitary, in racemes, panicles, corymbs, or umbel-like clusters, radially symmetrical or nearly so, with a floral cup (= hypanthium) formed by fusion of bottom portion of flower parts (in taxa with inferior ovaries this appearing like outer ovary wall); sepals 5, sometimes subtended by bracteoles; petals 5; stamens 10–many, inserted with the petals on the edge of the floral cup; pistils separate or united, 1–many; ovary superior or inferior; fruits diverse, in ours achenes, drupes, pomes or an aggregates of achenes or drupelets.

A large (2,825 species in 95 genera), economically important, subcosmopolitan, but especially temperate and warm n hemisphere family of herbs, shrubs, and trees including many ornamentals (e.g., *Cotoneaster, Photinia, Pyracantha, Rosa,* and *Spiraea*) and temperate fruits such as APPLES (*Malus*), ALMONDS, CHERRIES, PLUMS, PEACHES (*Prunus*), BLACKBERRIES, RASPBERRIES (*Rubus*), and STRAWBERRIES (*Fragaria*). (subclass Rosidae)
FAMILY RECOGNITION IN THE FIELD: usually alternate, simple or compound, usually toothed leaves, typically with *stipules*; flowers radially symmetrical, bisexual, 5-merous, with *hypanthium* and *10 to many, free* stamens.
REFERENCES: Rydberg 1908, 1913, 1918; Robertson 1974; Phipps et al. 1990, 1991; Dickinson & Campbell 1991a, 1991b; Campbell et al. 1991; Robertson et al. 1991.

1. Annual or perennial unarmed herbs (base can be somewhat woody); leaves compound.
  2. Leaves pinnately compound with more than 3 leaflets.
    3. Petals absent; sepals 4; flowers numerous in dense subglobose to cylindric heads ____ **Sanguisorba**
    3. Petals 5, white or yellow; sepals 5; flowers solitary or few on long pedicels or in elongate spike-like racemes.
      4. Petals white; styles persistent on achenes and elongating into conspicuous beaks 4–7 mm long; hypanthium without hooked bristles; flowers solitary or few on long pedicels ____ **Geum**
      4. Petals yellow; achenes without beaks; hypanthium with hooked bristles; flowers in spike-like racemes ____ **Agrimonia**
  2. Leaves 3-foliate or palmately compound.
    5. Flowers with only 5 pistils; fruits follicles (dehiscent) with 2–4 seeds ____ **Porteranthus**
    5. Flowers with numerous pistils; fruits achenes (indehiscent) with a single seed inside.
      6. Styles persistent, elongating and forming a conspicuous beak 4–7 mm long on the fruits; petals white ____ **Geum**
      6. Styles usually deciduous, neither elongating nor forming a beak; petals white or yellow.
        7. Leaves with 3 leaflets; receptacles becoming enlarged, red, and conspicuously spongy or fleshy; petals white or yellow.
          8. Petals white; bracts below calyces similar in size to sepals, not toothed apically ____ **Fragaria**
          8. Petals yellow; bracts below calyces much larger than sepals, with 3(–5) apical teeth ____ **Duchesnia**
        7. Leaves palmately compound with 5 or more leaflets; receptacles neither becoming enlarged nor red nor spongy nor fleshy; petals yellow ____ **Potentilla**
1. Perennial shrubs or trees or if somewhat herbaceous, then armed with thorns or prickles; leaves simple OR compound.
  9. Leaves simple, plant armed or unarmed.
    10. Ovary superior; style 1; fruit a drupe with a single stone ____ **Prunus**
    10. Ovary inferior; styles 2–5; fruit a pome with several separate seeds or seed-containing carpels.
      11. Plants usually armed with conspicuous woody spines; fruits 2 cm or less in diam.; mature carpels hard and bony; leaves deciduous with toothed margins or evergreen with entire margins.
        12. Leaves deciduous, the margins toothed; petals 6 mm or more long (often much longer) ____ **Crataegus**
        12. Leaves evergreen, the margins entire; petals ca. 4 mm long ____ **Pyracantha**
      11. Plants unarmed (if rarely armed, then fruits much > 2 cm in diam.); fruits small (6–12 mm in diam.) to very large; mature carpels not hard and bony; leaves deciduous with toothed margins.
        13. Flowers small, 8 mm or less across; inflorescences broad panicles 10–16 cm across; fruits 5–6 mm in diam.; leaves evergreen, somewhat leathery ____ **Photinia**
        13. Flowers larger, much > 8 mm across (petals ca. 10 mm or more long); inflorescences short racemes or umbel-like clusters; fruits 6 mm or more in diam., often much more; leaves deciduous, not leathery.
          14. Inflorescences umbel-like clusters; pedicels 15–30 mm long; leaf blades not cordate basally; fruits either small (ca. 6 mm in diam.) or large and much > 12 mm in diam. ____ **Pyrus**
          14. Inflorescences short racemes; pedicels 17 mm or less long; leaf blades usually slightly cordate basally; fruits small, 12 mm or less in diam. ____ **Amelanchier**
  9. Leaves compound, plant armed.
    15. Hypanthium globose to urn-shaped, with a constricted opening, the achenes concealed inside (the hypanthium is termed a hip, is smooth in outline, and typically red or reddish orange); ROSES ____ **Rosa**

15. Hypanthium flat or hemispheric, the ovules and drupelets conspicuously exposed (the cluster of druplets is commonly termed a blackberry or dewberry and is lumpy in outline and red to dark purple or black); BLACKBERRIES and DEWBERRIES ____ **Rubus**

AGRIMONIA AGRIMONY, GROOVEBUR, COCKLEBUR, HARVEST-LICE

Pubescent, rhizomatous, erect, perennial herbs to 2 m tall; leaves pinnately compound, the leaflets dimorphic, the large primary leaflets sharply toothed, alternating with very small ones; stipules leafy, toothed or lobed, adnate to the sides of the petiole; inflorescences terminal, spike-like racemes; flowers 5-merous, small; petals (2–)3–4 mm long, yellow; stamens 5–15; hypanthium with small but conspicuous hooked bristles above, hardened at maturity, 1.5–5 mm long, containing an achene; calyx lobes incurved after flowering and forming a beak on the hypanthium.

A genus of 15 species mainly in n temperate areas s to c and s Africa. All 3 nc TX species are limited to the ne part of nc TX. (Possibly a corruption of the genus name *Agremonia* used by the Greeks and derived from *argema*, a fleck in the eye; formerly used as a cure for eye disease)

1. Leaves with 11–19(–25) leaflets, these usually lanceolate or nearly so, sharply toothed and sharply pointed at apex (additional very small leaflets also present); lower stems conspicuously long hairy ____ **A. parviflora**
1. Leaves with 3–9 leaflets, these usually elliptic to obovate, either the teeth or the apex ± blunt or rounded (additional very small leaflets also present); lower stems either long hairy or glabrous or with scattered hairs.
   2. Axis of inflorescence and lower surfaces of leaflets without resin dots (but with pubescence); lower stems long hairy; leaves with 3–7 leaflets ____ **A. microcarpa**
   2. Axis of inflorescence and lower surfaces of leaflets with resin dots (use hand lens); lower stems glabrous or with scattered hairs; leaves with 5–9 leaflets ____ **A. rostellata**

Agrimonia microcarpa Wallr., (small-fruited), SLENDER GROOVEBUR, SMALL-FRUIT AGRIMONY. Stems 0.3–0.6(–1.2) m tall; leaflets coarsely toothed, rounded at summit; lower surface of leaflets velvety hairy; inflorescences loosely flowered, minutely pilose. Sandy woods, chiefly low ground; Lamar Co. (Mahler 1988); e TX w in Red River drainage to nc TX. Jul–Oct.

Agrimonia parviflora Aiton, (small-flowered), MANY-FLOWER GROOVEBUR, MANY-FLOWER AGRIMONY. Stems usually stout, typically in clumps, 0.3–2 m tall; undersurface of leaflets with resin dots and sparingly hairy, especially on the veins; axis of inflorescence with resin dots, finely pubescent with short hairs and often with long spreading hairs; flowers crowded; hypanthium ribbed. Edge of low woods, thickets; Grayson Co.; mainly extreme e TX and n Panhandle. Jul–Oct. This species usually has a much stouter appearance than the other 2 nc TX species.

Agrimonia rostellata Wallr., (with a small beak), WOODLAND GROOVEBUR, WOODLAND AGRIMONY. Stems usually slender, typically solitary or 2–3 together, 0.3–0.6(–1) m tall; leaflets coarsely and ± bluntly serrate, the lower surfaces glabrous or some with hairs on the veins (in addition to the resin dots); axis of inflorescence glabrous or nearly so except for resin dots. Moist woods, sandy or clayey soils; Grayson Co.; e TX w in Red River drainage to nc TX. Jul–Sep.

AMELANCHIER

JUNE-BERRY, SUGAR-PLUM, SHAD-BUSH, SERVICE-BERRY, SARVICE-BERRY

A n temperate genus of 33 species; some are cultivated as ornamentals or used for their edible fruit. (French: *amélanchier*, name for *A. ovalis* Medik.)

REFERENCES: Wiegand 1912; Jones 1946; Hess 1968.

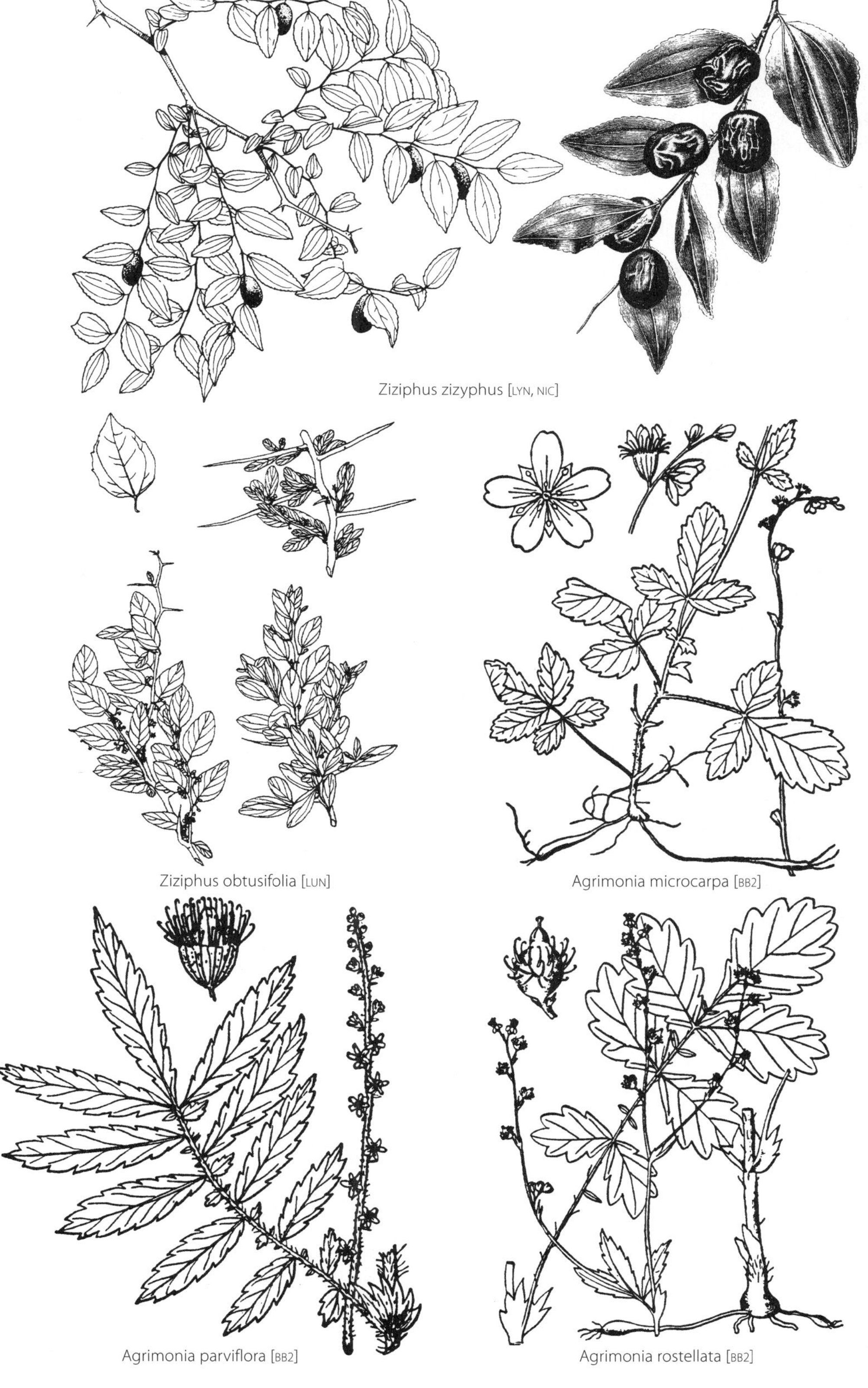

Ziziphus zizyphus [LYN, NIC]

Ziziphus obtusifolia [LUN]

Agrimonia microcarpa [BB2]

Agrimonia parviflora [BB2]

Agrimonia rostellata [BB2]

Amelanchier arborea (F. Michx.) Fernald, (tree-like), JUNE-BERRY, SERVICE-BERRY, SHAD-BERRY, COMMON SERVICE-BERRY, DOWNY SERVICE-BERRY. Unarmed shrub or tree to 8(–20) m tall, usually with 1 trunk; leaves deciduous, simple, ovate to elliptic, oblong or obovate, 4–10 cm long, acute or acuminate at apex, slightly cordate or rounded basally, sharply serrate, often doubly so; petioles 1–3.5 cm long; racemes terminal, 3–5 cm long, pendant, 3–15-flowered; flowers 5-merous; hypanthium ± adnate to ovary; petals white, sometimes roseate, 10–20 mm long, 7 mm or less wide; stamens numerous; ovary wholly or partially inferior; fruit a dryish berry-like pome 6–12 mm in diam., reddish to reddish purple, edible. Wooded areas; included based on citation of vegetational area 4 (Fig. 2) by Hatch et al. (1990); mainly e TX. Mar–May. The fruits were eaten by Native Americans and the wood is one of the heaviest in North America (Steyermark 1963).

CRATAEGUS HAWTHORN, REDHAW, THORN, THORN-APPLE

Deciduous shrubs or small trees, usually armed with thorns; branches often crooked; leaves alternate, simple, subentire, toothed, lobed, or dissected; stipules falcate, falling when the leaves expand, or large, leafy, and persistent on summer shoots; inflorescence a simple or compound, flattened, cyme-like panicle of few–many flowers or flowers solitary; flowers 5-merous, with rank or rank-sweet scent; petals white, occasionally pink in bud; stamens 5–20; ovary inferior; fruit a small pome with 1–5 bony nutlets (= pyrenes), the mature fruits of all nc TX species are some shade of red with the exception of the black- or bluish-fruited *C. brachyacantha.* The references to leaves in the descriptions and keys refer only to those leaves on short-shoots, that is those on previous years' woody growth (flowers are only produced on short-shoots). Leaves on long-shoots (= the current year's growth) are often variable and can lead to misinterpretations.

A n temperate genus of ca. 180 species (J. Phipps, pers. comm.) of usually thorny deciduous shrubs or small trees. *Crataegus* is one of the most taxonomically complex genera in North America, apparently because of hybridization, polyploidy, and apomixis; over 1,100 species have been named with the vast majority of these probably not warranting specific recognition. Because of the lack of consistent distinguishing characters, we have lumped a number of taxa. The genus is being actively researched at the present time; e.g., a new species, *C. nananixonii* J.B. Phipps & O'Kennon, was named in 1997 from Nacogdoches Co. in e TX (Phipps & O'Kennon 1997); that species was named for Elray S. Nixon, a prominent TX botanist retired from Stephen F. Austin State Univ.; the first part of the name, nana, refers to the dwarf stature of the plant. The fruits of many *Crataegus* species are used for preserves and jelly; the large-fruited e TX MAYHAW (*C. opaca* Hook. & Arn.) is particularly prized for making jelly. All species are alternate hosts of cedar apple rusts (*Gymnosporangium* spp.) and because of the abundance of cedars (*Juniperus* spp.) in nc TX, infected HAWTHORNS are commonly seen. (Greek: *kratos*, strength, alluding to the tough wood or thorns)

REFERENCES: Palmer 1925, 1946, 1960; Kruschke 1965; Phipps & Muniyamma 1980; Phipps 1983, 1988, 1997; Phipps et al. 1990.

1. Leaf blades deeply incised to dissected, the incisions cutting 1/2 the distance to the midrib or more, the overall appearance of the leaf parsley-like; main veins of well-developed leaf blades running to the sinuses as well as to the tips of the lobes; limited to the extreme ne portion of nc TX ____ **C. marshallii**
1. Leaf blades toothed to lobed but not deeply incised, definitely not parsley-like in appearance; main veins of well-developed leaf blades running only to the lobe tips (except also to the sinuses in *C. phaenopyrum* and *C. spathulata*); widespread throughout nc TX.
    2. Leaf blades with a broad base, either rounded, truncate, or cordate.
        3. Lower surface of leaf blades often velvety-pubescent to the touch, the pubescence sometimes lost with age and lower surface thus only slightly pubescent at maturity; calyces

densely woolly-pubescent outside; flowers 20–23 mm across; fruits 13–18 mm wide; widespread native species **C. mollis**

3. Lower surface of leaf blades essentially glabrous, not velvety-pubescent to the touch; calyces glabrous outside; flowers ca. 13 mm across; fruits 4–6 mm wide; rare escape from cultivation **C. phaenopyrum**

2. Leaf blades with a narrowed base, from wedge-shaped to tapering.

4. Thorns short, usually < 2 cm long; ripe fruits black, bluish when immature; found only in extreme ne corner of nc TX (Red River Co.) **C. brachyacantha**

4. Thorns large, usually > 4 cm long, rarely absent; ripe fruits red, never bluish or black at any stage of maturation; widespread throughout nc TX.

5. Main veins of well-developed leaves running to the sinuses as well as to the tips of the lobes; bark usually ± smooth and mottled, flaking **C. spathulata**

5. Main veins of well-developed leaves running only to the lobe tips; bark uniformly rough, not flaking OR ± smooth and mottled, flaking (only in *C. viridis*).

6. Calyces densely woolly-pubescent outside; leaf blades often ovate or deltoid in outline, often widest below the middle (*C. engelmannii* can have the calyces densely woolly-pubescent, but the leaf blades are typically spatulate to obovate and widest beyond the middle; it can therefore be keyed either here or under 6 below).

7. Leaf blades coarsely toothed to often shallowly lobed, the lower surface often velvety-pubescent to the touch; pubescence toward base of midrib on lower surface of blades matted or curled, partly directed forward or loosely appressed; leaf blades mostly ovate or deltoid, usually widest at or below the middle, < 2 times as long as wide, often nearly as wide as long, short-decurrent on petiole, the petiole barely or not at all winged for most of its length; stamens ca. 20 **C. mollis**

7. Leaf blades finely toothed, not lobed, the lower surface not velvety-pubescent to the touch; pubescence toward base of midrib on lower surface of blades short and straight, spreading nearly at right angles; leaf blades spatulate to obovate, widest beyond the middle, normally 2 times or more as long as wide, long-decurrent on the petiole, the petiole noticeably winged for most of its length; stamens ca. 10 **C. engelmannii**

6. Calyces glabrous or pubescent outside; leaf blades mostly spatulate to narrowly obovate, wedge-shaped or oblong-ovate, usually widest beyond the middle (leaf blades at branch tips sometimes broader).

8. Leaf blades coarsely toothed to shallowly lobed, often > 3 cm wide, with tufts of hairs beneath in the axils of main veins (often inconspicuous, rarely present only on new growth), otherwise glabrous, light to dark green, not very glossy above, at maturity firm but not thick; bark usually smooth and mottled, flaking; thorns short (typically ca. 3(–5) cm long), straight, usually few in number, sometimes absent (rarely numerous or large) **C. viridis**

8. Leaf blades usually finely toothed, not lobed, usually < 3 cm wide, glabrous beneath or evenly pubescent along the veins, dark green, very glossy above, at maturity thick and leathery; bark uniformly rough, not flaking; thorns long (at least some typically 4–7 cm or more long), slightly curved, usually abundant.

9. Foliage and inflorescences glabrous or nearly so; calyces glabrous or nearly so.

10. Leaf blades on new growth (at branch tips) elliptic or lanceolate; branches stiffly and uniformly horizontal **C. crus-galli**

10. Leaf blades on new growth suborbicular; branches irregularly horizontal **C. reverchonii**

9. Foliage and inflorescences pubescent while young and usually throughout the season; calyces pubescent outside **C. engelmannii**

Crataegus brachyacantha Sarg. & Engelm., (short-thorned), BLUEBERRY HAWTHORN. Tree to 15 m tall, armed with numerous, short, usually somewhat curved thorns; leaves to 5 cm long and 25 mm wide; flowers ca. 8 mm across; petals initially white, becoming orangish with age; fruits 8–13 mm wide, black when ripe, bluish when immature, glaucous. Stream margins, roadsides; a sterile specimen (*Whitehouse 21448a*, BRIT/SMU) from s Red River Co., identified by Shinners as *C. brachyacantha*, seems to match this species. It is otherwise found only further e in e TX. Apr. Fruiting Aug–Sep.

Crataegus crus-galli L., (cock-spur), COCKSPUR HAWTHORN, BUSH'S HAWTHORN. Glabrous shrub or small tree 2–6(–8) m tall; thorns conspicuous, 3–8 cm long; flowers ca. 10–15 mm across; fruits short oblong to slightly obovoid (rarely subglobose), 8–10 mm wide, greenish or dull red. Limestone bluffs and hilltops; se and e TX w to West Cross Timbers and Edwards Plateau. Late Apr. Fruiting Oct–Nov. [*C. bushii* Sarg., *C. cherokeensis* Sarg., *C. pyracanthoides* (Aiton) Beadle]

Crataegus engelmannii Sarg., (for George Engelmann, 1809–1884, German-born botanist and physician of St. Louis). Similar to and possibly not specifically distinct from *C. crus-galli*; basically a pubescent type of *C. crus-galli*; fruits subglobose or short oblong, 6–8 mm wide, dull-crimson. Sandy upland or lowland woods; e TX w to e edge of Blackland Prairie (Mahler 1988). Apr. Fruiting Oct–Nov. [*C. berberifolia* Torr. & A. Gray is possibly a synonym (Phipps et al. 1988)]

Crataegus marshallii Eggl., (for botanist Humphrey Marshall, 1722–1801), PARSLEY HAWTHORN. Shrub or small tree to 8 m tall; leaves conspicuously incised and sharply serrate; flowers 10–15 mm across; fruits oblong to obovoid, 4–8 mm thick, bright red. Acidic sandy soils, woods, roadsides; Lamar Co. in Red River drainage; mainly se and e TX. Mar–Apr. Fruiting Sep–Oct.

Crataegus mollis Scheele, (soft, with soft hairs), REDHAW, SUMMERHAW, DOWNYHAW. Tree 3–12 m tall with scattered stout thorns to nearly thornless; leaf blades large, (3–)5–7(–10) cm long, variable in shape, usually wedge-shaped basally but sometimes rounded to even truncate (*C. mollis* can therefore be reached 2 ways in the key), tomentose to slightly pubescent beneath; flowers 20–23 mm across; fruits subglobose (rarely oblong or obovoid), 13–18 mm wide, scarlet or bright crimson. Stream bottoms or hillside woods and thickets; se and e TX w to West Cross Timbers and Edwards Plateau. Early and mid-Apr. Fruiting Aug–Oct. [*C. brachyphylla* Sarg.]. *Crataegus dallasiana* Sarg. [probably = *C. brazoria* Sarg.], known from Dallas and Ellis cos., will key to *C. mollis* in our key. This entity is possibly a hybrid between *C. mollis* and *C. viridis*, and is intermediate in a number of characters between these 2 species.

Crataegus phaenopyrum (L.f.) Medik., (with the appearance of a pear—*pyrus* in Latin means pear), WASHINGTON-THORN. Small tree 7–8(–12) m tall; leaves often 3-lobed or with 2–3 pairs of lateral lobes, the sinuses between the lobes usually < 1/2 the distance to the midrib (rarely but never consistently deeper); flowers 10–13 mm across; fruits shining scarlet. The 4–6 mm diam. fruits are the smallest of any *Crataegus* in nc TX. It is also the only nc TX species in which the calyces are deciduous from the mature fruits, resulting in the tops of the nutlets being exposed. Cultivated and escapes; Tarrant Co.; native further e in the e U.S. May–Jun. Fruiting fall.

Crataegus reverchonii Sarg., (for Julien Reverchon, 1837–1905, a French-American immigrant to Dallas and important botanical collector of early TX), REVERCHON'S HAWTHORN. Shrub or bushy small tree to 6 m tall; leaves normally glabrous, sometimes pubescent on stump sprouts; flowers ca. 10–15 mm across; fruits subglobose, ca. 12 mm wide, light scarlet. Similar to *C. crus-galli* except for leaf shape differences and possibly not specifically distinct. Rocky or sandy ground, on slopes or in drainage ways; mainly Blackland Prairie w to West Cross Timbers, also Edwards Plateau. Middle and late Apr. Fruiting Sep–Oct.

Crataegus spathulata Michx., (spoon-shaped, in reference to the leaves), LITTLE-HIP HAWTHORN, PASTURE HAWTHORN. Shrub or tree 5–7 m tall; branchlets usually thorny and horizontal; leaves

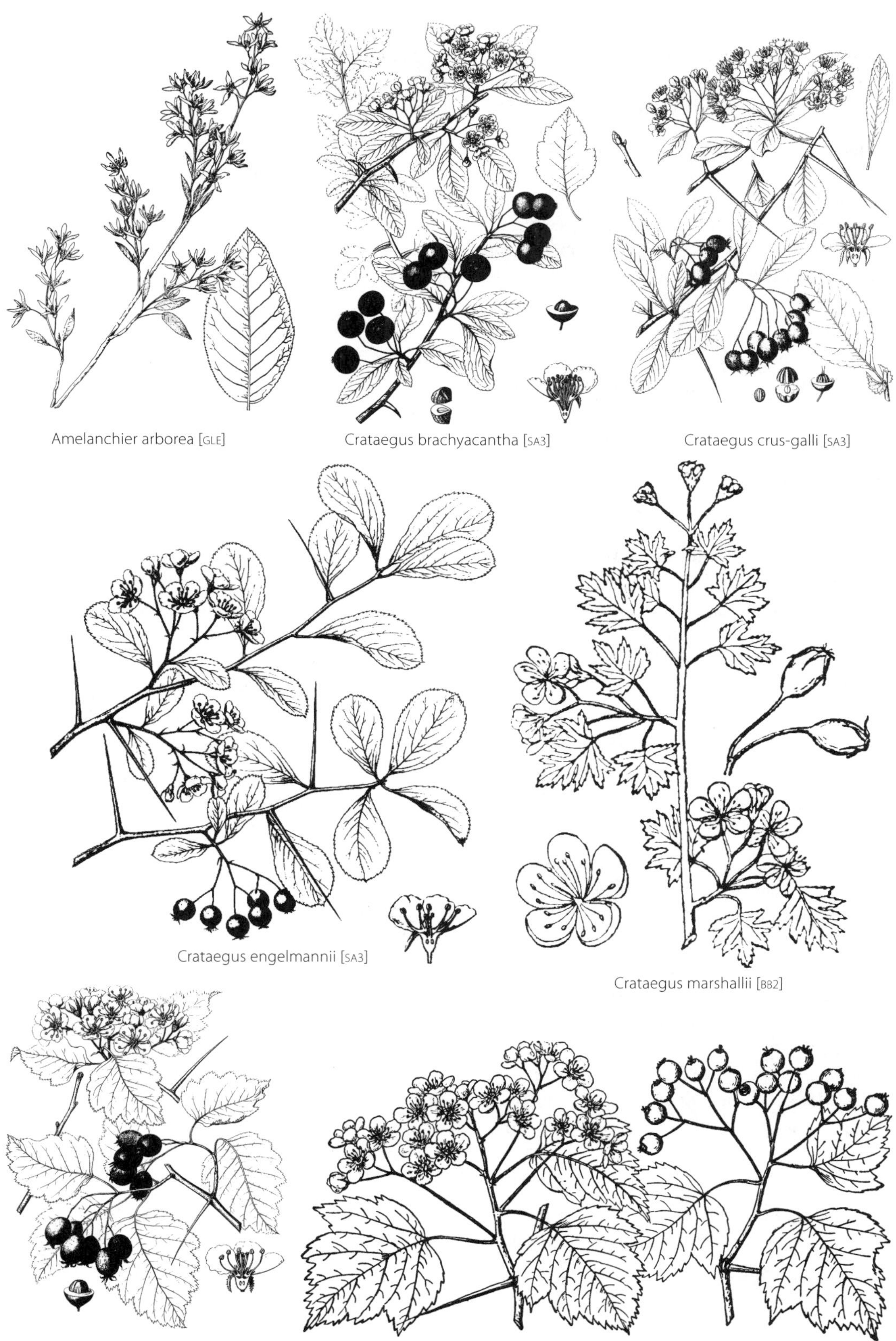

Amelanchier arborea [GLE]

Crataegus brachyacantha [SA3]

Crataegus crus-galli [SA3]

Crataegus engelmannii [SA3]

Crataegus marshallii [BB2]

Crataegus mollis [SA3]

Crataegus phaenopyrum [SA2]

of flowering branchlets usually 10–20 mm long, 10 mm or less wide, narrowly obovate with several coarse rounded teeth or small lobes beyond middle or near apex, gradually narrowed to entire base; flowers 6–8 mm across; fruits subglobose, 4–7 mm wide, red. Sandy woods, roadsides; se and e TX w to Bell, Grayson, Hunt, and Navarro cos. Apr. Fruiting Oct–Nov.

Crataegus viridis L., (green), GREENHAW. Small tree 3–6(–12) m tall with thorny to nearly thornless branches; leaf blades variable in shape and often asymmetrical, with tufts of hair beneath in axils of main veins; flowers ca. 12–15 mm across; fruits subglobose, 5–8 mm wide, red or orange-red. Stream bottoms, fields, slopes; se and e TX w to Rolling Plains and Edwards Plateau. Late Mar–Apr. Fruiting Sep–Nov. [*C. anamesa* Sarg., *C. antiplasta* Sarg., *C. glabriuscula* Sarg.] The state champion *C. viridis* is in Harry S. Moss Park in Dallas (R. May, pers. comm.).

DUCHESNEA INDIAN-STRAWBERRY, MOCK STRAWBERRY, SNAKEBERRY

A genus of 2 species of s Asia (Mabberley 1987); sometimes lumped with *Potentilla* (e.g., Mabberley 1997); similar to *Fragaria* (STRAWBERRY) but with yellow flowers, dry receptacles, and toothed bracts associated with calyces. (Named for Antoine Nicolas Duchesne, 1747–1827, French botanist and early monographer of *Fragaria*)

Duchesnea indica (Andr.) Focke, (of India), INDIAN MOCK STRAWBERRY, YELLOW-STRAWBERRY. Perennial stoloniferous herb; leaves compound with 3 serrate-crenate leaflets, resembling those of STRAWBERRY; petioles long, to 30 cm; flowers solitary on axillary peduncles 3–10 cm long, 5-merous; calyx lobes alternating with 5 larger leafy 3(–5)-toothed bracts; petals yellow; stamens 20; pistils numerous; "fruit" an aggregate of achenes, the receptacle enlarged, spongy and red at maturity, not juicy, resembling a small STRAWBERRY, not poisonous but tasteless and not worth eating. Weedy, often wet areas, alleys in cities; naturalized mainly se and e TX w to at least Red River Co., also Dallas (E. McWilliams, pers. comm.) and Tarrant (R. O'Kennon, pers. obs) cos. Mar–Aug. Native of India. Sometimes cultivated in hanging baskets (Mabberley 1987).

FRAGARIA STRAWBERRY

A genus of ca. 12 species of herbs with rooting runners and usually red, fleshy receptacles; native to the n temperate zone and Chile. (Latin: *fraga*, the classical name for strawberry, from *fragrans*, fragrant, alluding to the fruit)

Fragaria virginiana Duchesne subsp. **grayana** (Vilm. ex J. Gay) Staudt, (sp.: of Virginia; subsp.: for Asa Gray, 1810–1888, botanist at Harvard and preeminent American plant taxonomist), VIRGINIA STRAWBERRY, WILD STRAWBERRY. Low perennial scapose herb with rhizomes and long stolons; leaves compound; leaflets 3, elliptic to obovate, toothed; petioles to 30 cm long; inflorescence an umbel-like cyme; flowers 5-merous, perfect or unisexual, the pistillate smaller than staminate; petals white, 6–15 mm long; stamens usually 20; pistils numerous; "fruit" an aggregate of minute dry achenes, these embedded in pits on the surface of a swollen, fleshy, red, edible, conical receptacle 5–20 mm in diam. (thus an accessory "fruit"—with fleshy parts derived from organs other than the pistil). Fields, prairies, forest margins, wooded areas; Grayson Co.; mainly e TX w to nc TX. Mar–Apr. The cultivated STRAWBERRY, *F. ×ananassa*, is a hybrid between *F. virginiana* and *F. chiloensis* (L.) Mill., native to w North America and South America, respectively.

GEUM AVENS

A temperate and cold area genus of ca. 40 species; some are cultivated as ornamentals, others have glucosides and are used medicinally. (Classical Latin name for these plants)

Geum canadense Jacq., (of Canada), WHITE AVENS. Perennial rhizomatous herb, 30–100(–120) cm tall, usually with rather densely pilose stem (rarely glabrous) and sparsely appressed-pilose

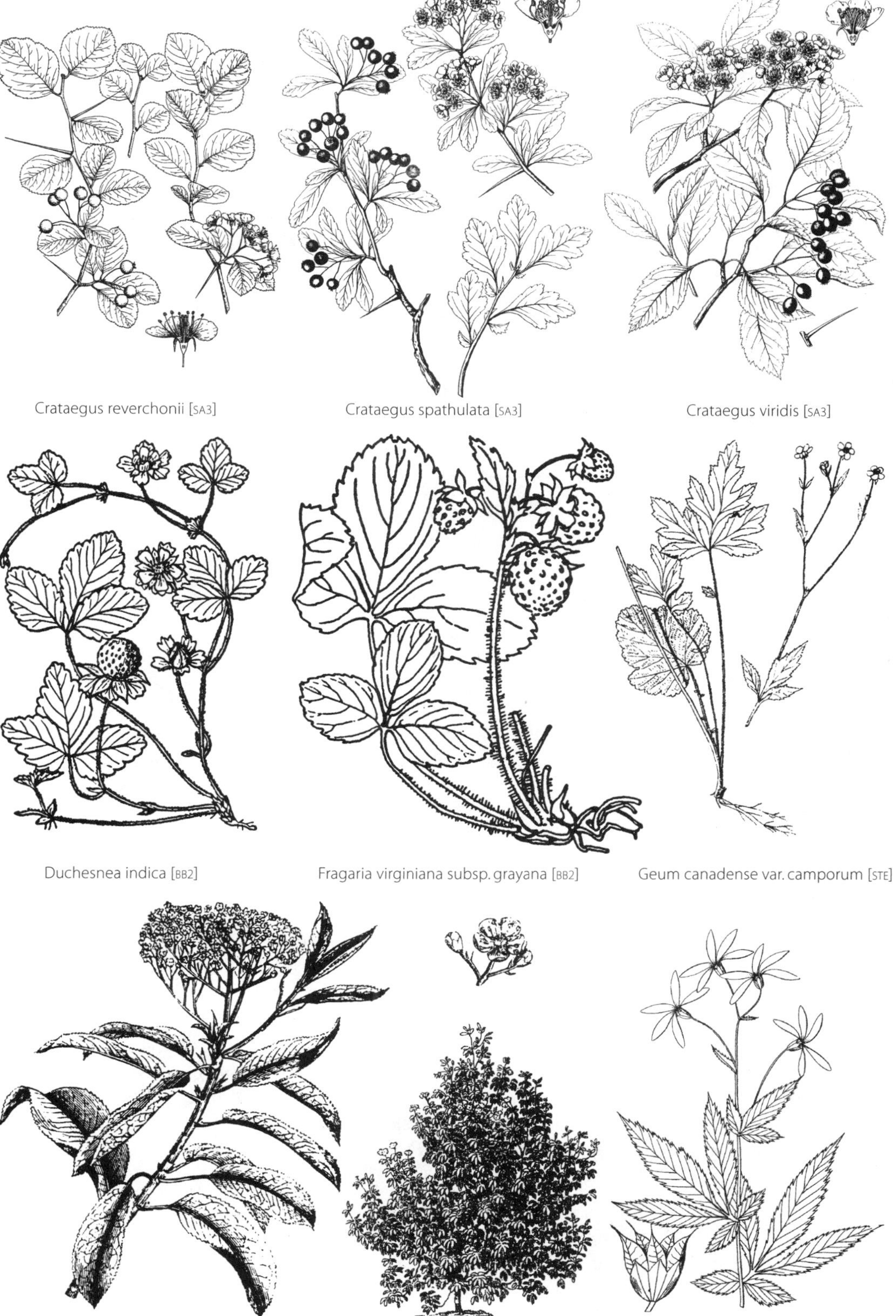

Crataegus reverchonii [SA3]

Crataegus spathulata [SA3]

Crataegus viridis [SA3]

Duchesnea indica [BB2]

Fragaria virginiana subsp. grayana [BB2]

Geum canadense var. camporum [STE]

Photinia serratifolia [LOU]

Porteranthus stipulatus [MTB]

leaves; basal leaves long-petioled, usually pinnately compound with 3 large terminal leaflets and 2–4 much smaller ones, or apparently palmate (only the terminal leaflets developed); stem leaves with shorter petioles, with 3 leaflets or the uppermost leaves simple; leaflets sharply toothed; stipules of stem leaves leafy, toothed or lobed; flowers solitary, long pedicellate, terminal on arched-spreading branches, 5-merous; petals white, fading yellowish, 5–9 mm long; stamens numerous; achenes numerous, in a spherical head-like aggregate 1–2 cm in diam., with persistent conspicuous beaks (= 4–7 mm long styles). Woods and thickets, various soils; e 1/2 of TX. Apr–May(–Jun).

1. Terminal leaflet of middle stem leaves usually acute; carpels broadly ovate, 3–4 mm long _____ var. **camporum**
1. Terminal leaflet of middle stem leaves usually obtuse; carpels narrowly obovate to wedge-shaped, 2–3 mm long _____ var. **texanum**

var. **camporum** (Rydb.) Fernald & Weath., (of the plains). E 1/2 of TX. This is the common variety in nc TX.

var. **texanum** Fernald & Weath., (of Texas). Included based on citation of vegetational areas 4 and 5 (Fig. 2) by Hatch et al. (1990); we have observed no specimens clearly distinguishable as this variety from nc TX; supposedly e 1/2 of TX.

PHOTINIA

A genus of 65 species native from the Himalayas to Japan and Sumatra and e North America and Central America; a number are cultivated as evergreen or deciduous shrubs including use as hedge plants. (Greek: *photos*, light, referring to the shiny leaves of some species)
REFERENCE: Robertson et al. 1991.

Photinia serratifolia (Desf.) Kalkman, (serrate-leaved). Evergreen shrub or tree to ca. 12 m tall; leaves simple, oblong, large, to ca. 20 cm long, serrate, somewhat leathery, dark green above, yellowish green beneath, reddish when young; petioles 2–3 cm long; panicles 10–16 cm across; flowers 5-merous, small (6–8 mm across); petals white; fruits globose, 5–6 mm in diam., red. Widely cultivated and escaping particularly in sandy soils; Grayson Co.; distribution in other parts of TX not known. Late spring–summer. Native of China, Taiwan, Japan, and Sumatra. [*P. serrulata* Lindl.]

PORTERANTHUS INDIAN-PHYSIC

A North American genus of 2 species, formerly recognized as *Gillenia*. (Named for Thomas Conrad Porter, 1822–1901, Professor at Lafeyette College)

Porteranthus stipulatus (Muhl. ex Willd.) Britton, (having stipules), INDIAN-PHYSIC, AMERICAN IPECAC. Erect perennial herb; stems 0.4–1(–1.2) m tall; leaves alternate, compound; leaflets 3, lanceolate, acuminate, 4–8 cm long; leaflets of lower leaves deeply incised or laciniate; leaflets of upper usually sharply double-serrate; petioles 0–10 mm long; stipules 1–2(–3) cm long, 1–1.5 cm wide, leaf-like, ovate to orbicular; inflorescences leafy, panicle-like, each branch terminating in 1–7 long-pedicelled flowers; flowers 5-merous; petals white or pinkish, 10–15 mm long; stamens 20; pistils 5; follicles 6–8 mm long. Wooded slopes; Lamar Co. in Red River drainage; mainly e TX. Apr–Jun. Sometimes recognized in the genus *Gillenia* [as *G. stipulata* (Muhl. ex Willd.) Nutt.]. The glycosides in the roots and bark were formerly used medicinally (Mabberley 1987); they are potentially poisonous (Burlage 1968).

POTENTILLA CINQUEFOIL, FIVE-FINGER

Pubescent annual, biennial, or perennial herbs; leaves in our species palmately compound with 5–7 sharply toothed leaflets; stipules conspicuous, 0.5–3 cm long; flowers 5-merous; calyces subtended by 5 bractlets (epicalyx); petals yellow; stamens 20–30; achenes numerous, in head-like arrangement on a prolonged receptacle.

A genus of ca. 500 species of herbs and shrubs with an epicalyx of bractlets alternating with the sepals; it is primarily a n temperate and boreal group, with a few in s temperate zone. Some are cultivated as ornamentals or used medicinally. (Diminutive of Latin: *potens*, powerful, originally applied to *P. anserina* L., from its reputed medicinal value)
REFERENCE: Shinners 1955.

1. Pedicels shorter than or equaling the calyces in flower (up to 3 times as long in age); flowers many in a terminal, cymose inflorescence; petals light sulfur yellow ____ **P. recta**
1. Pedicels 5–12 times as long as the calyces in flower; flowers axillary, solitary; petals deep yellow ____ **P. simplex**

Potentilla recta L., (upright), SULFUR CINQUEFOIL. Erect annual, biennial, or perennial to 70(–80) cm tall; leaflets 3–8 cm long, coarsely serrate; petals 7–11 mm long. Weedy areas; found at Polytechnic (now in Fort Worth), Tarrant Co. in May 1929, more recently (1995) at Fort Worth Nature Center; also known from Cass Co. (Correll & Johnston 1970) in e TX. Native of Europe. Reported to be poisonous (Burlage 1968).

Potentilla simplex Michx., (unbranched, simple), OLDFIELD CINQUEFOIL. Perennial from short thick rhizomes; stems initially erect to ascending, later widely spreading, arching and rooting at tips to produce small tubers which become the rhizomes for the following year; leaflets 2–7 cm long, toothed; petals 4–7 mm long. Sandy open woods, upland or lowland; Lamar Co. in Red River drainage; mainly e TX. Apr.

PRUNUS CHERRY, PEACH, PLUM

Deciduous or evergreen (1 species) shrubs or small trees, sometimes rhizomatous and thicket-forming; leaves simple; leaf blades entire to sharply or bluntly toothed; petioles often with glands; stipules paired, small, soon deciduous; flowers in fascicles, umbels, racemes, or solitary, appearing before, with, or after the leaves, often numerous and showy; petals white to pink, rosy lavender, or red; stamens ca. 15–20; fruit a drupe, frequently edible, the exocarp fleshy or juicy, the endocarp hard, indehiscent.

A temperate (especially n) and tropical mountain genus of 200+ species including important fruit and nut trees such as ALMONDS, APRICOTS, CHERRIES, NECTARINES, PEACHES, and PLUMS. *Prunus dulcis* (Mill.) D.A. Webb, ALMOND, native to w Asia, is the most widely grown of all "nuts." The genus also has widely used ornamentals (e.g., the Eurasian *P. laurocerasus* L.—CHERRY-LAUREL) and is valued for its beautiful wood used in furniture, turnery, and cabinet-making. All parts of some species (e.g., *P. persica*, *P. serotina*), particularly the seeds, twigs, and leaves may contain the cyanogenic glycoside amygdalin which breaks down to release dangerous levels of hydrogen cyanide (HCN); when the leaves are damaged, an enzymatic reaction releases hydrogen cyanide; death can result from animals eating the leaves; caution should therefore be taken where *Prunus* species occur in pastures. Children have been fatally poisoned by swallowing the cyanide-producing stones while eating wild cherries. HCN disrupts the metabolic pathway by which oxygen is utilized in the mitochondria of cells; death from HCN poisoning occurs within 12–20 minutes of exposure (Sperry et al. 1955; Hardin & Arena 1974; Schmutz & Hamilton 1979; Stephens 1980; Blackwell 1990; Turner & Szczawinski 1991; McGuffin et al. 1997). A single crushed leaf is sufficient to kill an insect in a jar (J. Phipps, pers. comm.) (Classical Latin name for a plum tree)
REFERENCES: Wight 1915; McVaugh 1951.

1. Plants with mature leaves (with or without flowers or fruits).
   2. Non-rhizomatous trees, usually a solitary individual, not forming thickets; fruits in racemes, solitary, or in *P. mexicana* in fascicles (= clusters) or umbels.
      3. Leaves evergreen, ± leathery, entire or with remotely spaced teeth — **P. caroliniana**
      3. Leaves deciduous, not leathery, with small teeth regularly and closely spaced.
         4. Lower surface of leaf blades with a mustache of hairs at base of midrib, otherwise glabrous (sometimes entirely glabrous in var. *eximia* on s margin of nc TX); flowers and fruits in an elongate raceme; fruits red when immature, becoming purple-black at maturity — **P. serotina**
         4. Lower surface of leaf blades entirely glabrous or with scattered pubescence, lacking distinct mustache at base of midrib; flowers and fruits solitary or in fascicles or umbels; fruits yellow, red, or purplish.
            5. Leaf blades glabrous beneath, often folded lengthwise, 4–8 times as long as wide; fruits solitary, large (ca. 5–8 cm in diam.), and tomentose to velvety; flowers sessile or subsessile — **P. persica**
            5. Leaf blades with scattered pubescence beneath, not folded lengthwise, ca. 1–3 times as long as wide; fruits in fascicles or umbels, smaller (1–3 cm long), glabrous, with a waxy bloom; flowers on pedicels 10–17 mm long
               6. Leaf blades 6–12 cm long, 3–6 cm wide, apically abruptly acuminate, basally rounded or subcordate, marginally sharply and often doubly serrate; fruits ellipsoid to ± globose, 2–3 cm long; widespread in nc TX — **P. mexicana**
               6. Leaf blades usually 4–7(–7.5) cm long, < 3(–3.5) cm wide, apically acute or gradually acuminate, basally broadly cuneate to rounded, marginally finely serrate; fruits subglobose, 1–2 cm long; rare in nc TX — **P. umbellata**
   2. Rhizomatous shrubs or trees usually forming thickets of numerous individuals; fruits in fascicles (clusters) or umbels, not in racemes.
      7. Leaf blades mostly (6–)7–10 cm long; shrub or tree to 10 m tall — **P. munsoniana**
      7. Leaf blades mostly < 7 cm long; shrubs to 4 m tall.
         8. Leaf blades not folded lengthwise, spreading or partly drooping, the lower surface densely pubescent, the apex obtuse to acute; young branchlets (current year's growth) usually pubescent; petioles pubescent — **P. gracilis**
         8. Leaf blades folded lengthwise, drooping, the lower surface pubescent only on veins, the apex acute to acuminate; young branchlets glabrous or rarely pubescent; petioles glabrous to pubescent.
            9. Leaf blades acute or acuminate, 1–2 cm wide, the leaf teeth each tipped with a large permanent gland; fruits 2–2.5 cm long; on sandy soils — **P. angustifolia**
            9. Leaf blades acuminate, usually (1.5–)2–4 cm wide, the leaf teeth with or without a small deciduous gland; fruits 1.3–2 cm long; usually on calcareous clay soils or limestone outcrops — **P. rivularis**
1. Plants with flowers; mature leaves usually absent (except in evergreen *P. caroliniana*).
   10. Flowers in racemes.
      11. Racemes ca. 3 cm long, axillary; petals 2–3 mm long; leaves evergreen — **P. caroliniana**
      11. Racemes (4–)6–15 cm long, terminating short branches; petals 2.5–6 mm long; leaves deciduous — **P. serotina**
   10. Flowers solitary, fascicled, or in umbels.
      12. Petals 15–25 mm long, lavender-pink or red; flowers sessile or subsessile; ovary velvety hairy — **P. persica**
      12. Petals 3–11 mm long, white or light pink; flowers pedicelled; ovary glabrous.
         13. Petals 3–7 mm long; flowers to ca. 15 mm across; leaf teeth usually ± obtuse, tipped with a gland (use hand lens on young leaves if present); calyx lobes usually subglabrous to pubescent below the middle on inner surfaces; rhizomatous shrubs or trees, usually forming thickets.

14. Calyx lobes entire; pedicels usually 2–6 mm long ______ **P. angustifolia**
14. Calyx lobes glandular-toothed; pedicels 5–15 mm long.
    15. Hypanthium glabrous or pubescent only at summit; pedicels glabrous.
        16. Shrubs 0.5–2(–3) m tall; calyx lobes oblong-lanceolate or ovate-lanceolate, shorter than hypanthium, 1.5–2 mm long; widespread in nc TX ______ **P. rivularis**
        16. Trees or shrubs 3–10 m tall; calyx lobes ovate or oblong-ovate, as long as hypanthium, 2–4 mm long; rare in nc TX ______ **P. munsoniana**
    15. Hypanthium and pedicels densely pubescent ______ **P. gracilis**

13. Petals 7–11 mm long; flowers 15–23 mm across; leaf teeth sharp, without a gland; calyx lobes usually densely pubescent on inner surfaces; non-rhizomatous usually solitary tree, not forming thickets.
    17. Hypanthium sparsely to densely pubescent; pedicels of open flowers 6–11(–13) mm long; calyx lobes usually densely pubescent on inner surfaces and sparsely to densely pubescent on outer surfaces; tree to 12 m tall; widespread in nc TX ______ **P. mexicana**
    17. Hypanthium glabrous to sparsely pubescent; pedicels of open flowers 9–17 mm long; calyx lobes usually densely pubescent on inner surfaces, but sparsely pubescent on outer surfaces; small tree to 6 m tall; mainly e TX, rare in nc TX ______ **P. umbellata**

Prunus angustifolia Marshall, (narrow-leaved), CHICKASAW PLUM, SANDHILL PLUM. Shrub (rarely a small tree) to ca. 4 m tall; branchlets usually zigzag, sometimes spine-tipped; leaf blades 2–6(–8) cm long, folded lengthwise; inflorescence a compact umbel of 2–4 flowers; flowers expanding with the leaves; calyces glandless; petals white to creamy-white, 3.5–6 mm long; fruits red or yellow, 2–2.5 cm long. Sandy open woods, roadsides, and fencerows; e 2/3 of TX. Late Feb–early Apr. The fruits of this species can be delicious and may be eaten raw or made into jelly or preserves; Native Americans are reported to have dried the fruits on hot rocks so that they could be stored for future use (Kirkpatrick 1992).

Prunus caroliniana (Mill.) Aiton, (of Carolina), LAUREL CHERRY, CAROLINA LAUREL CHERRY. Tree to ca. 12 m tall with evergreen leaves; leaf blades thick, glabrous, 5–12 cm long, to 4 cm wide, dark green and shiny above, paler below; axillary racemes ca. 3 cm long, densely flowered; pedicels 3–4 mm long; sepals ca. 1 mm long; petals cream, 2–3 mm long; fruits black, shiny, ripening in fall and long persistent. Low areas; cultivated more in the past than at present, escaping; Grayson and Tarrant cos., also Dallas Co. (E. McWilliams, pers. comm.); native to se and e TX disjunct w to McLennan Co. (Little 1976 [1977]); according to Mahler (1988) penetrating nc TX from e TX along rivers. Feb–Apr.

Prunus gracilis Engelm. & A. Gray, (graceful), SAND PLUM, OKLAHOMA PLUM. Small shrub to ca. 1.5 m tall; mature leaf blades oblong-elliptic or lanceolate, 2–5 cm long, 10–25 mm wide, stiff, prominently veiny and densely soft-pubescent beneath; flowers abundant and showy, expanding with the leaves, in clusters of 2–4(–8); petals white, 5–6.5 mm long; fruits 1.5–1.8 cm long, yellow-red to red. Sandy open woods or open ground; e 2/3 of TX. Mar–early Apr. The fruits are edible and were much dried by Native Americans for winter use (Mabberley 1987).

Prunus mexicana S. Watson, (of Mexico), WILD PLUM, MEXICAN PLUM, BIG-TREE PLUM. Tree to 12 m tall; leaf blades 6–12 cm long, 3–6 cm wide, obovate to oblong-ovate to ovate; flowers appearing before or with young leaves, in clusters of 2–4(–6); petals white; fruits 2–3 cm long, purplish red, with a bloom, seem to never ripen (or ripen quite late). Woods and thickets, various soils; se and e TX w to West Cross Timbers; also Edwards Plateau. Late Feb–early Apr. The National Champion MEXICAN PLUM (largest recorded in the U.S.) is located in Hood Co. (American Forestry Association 1996).

Prunus munsoniana W. Wight & Hedrick, (for Thomas Volney Munson, 1843–1913, who developed numerous grape varieties and is credited with saving the French wine industry in the

1870s from the root disease, grape phylloxera), WILDGOOSE PLUM, MUNSON'S PLUM. Shrub or tree to 10 m tall; similar to *P. rivularis* and doubtfully separable from it. Shinners identified Grayson, Lamar, and Rockwall co. specimens as *P. munsoniana*; Little (1976 [1977]) also mapped Burnet, Clay, Collin, Fannin, Hunt, and Lampasas cos.; said by Sargent (1922) to occur w to Clay and Lampasas cos.; mainly nc TX and Edwards Plateau. Possibly only a larger phase of *P. rivularis*; Simpson (1988) questioned this species. Mar.

Prunus persica (L.) Batsch, (Latin for peach), PEACH, DURAZNO. Small tree 3–10 m tall; leaf blades oblong-lanceolate, 7–15 cm long, finely toothed, glabrous; flowers appearing before the leaves, usually solitary, 2.5–3.5 cm across; fruits 5–8 cm in diam. Commonly cultivated, occasional as an escape on roadsides and in waste places; Bosque, Brown, Grayson, and Hill cos.; e TX w to nc TX and Edwards Plateau. Mar–early Apr. Native of e Asia. Second only to apple as the most widely grown tree fruit worldwide (Mabberley 1987). An introduced Old World rust fungus, *Tranzschelia discolor* (Fuckel) Tranzschel & Litv., forms lesions on the leaves of PEACH (J. Hennen, pers. comm.). The frost-damaged leaves and seeds have high concentrations of hydrogen cyanide and are potentially fatal to livestock (Burlage 1968; Schmutz & Hamilton 1979).

Prunus rivularis Scheele, (growing by streams), THICKET PLUM, HOG PLUM, CREEK PLUM. Shrub with leaf blades 5–6(–7) cm long, usually (1.5–)2–4 cm wide, folded lengthwise; flowers with unfolding leaves, in clusters of 2–4(–8); petals white, 5–6.3 mm long; fruits 1.3–2 cm long, yellow with reddish areas or rarely red. Usually calcareous clay soils or limestone outcrops; Post Oak Savannah w to Rolling Plains and Edwards Plateau. Late Feb–early Apr. The fruits are generally quite tart and not as sweet as those of *P. angustifolia*.

Prunus serotina Ehrend., (late-ripening), BLACKCHERRY, WILD BLACKCHERRY, RUM CHERRY, CAPULIN. Tree 10–15(–30) m tall; leaf blades 3.5–15 cm long, 2–5 cm wide; racemes (4–)6–15 cm long; petals white, 2.5–6 mm long; fruits 0.7–1(–1.2) cm in diam., dark red becoming purple-black, sweet or bitter. Mar–Apr. Highly prized for its beautiful wood, used for furniture and cabinet-making; also used as a flavoring for rum and brandy and the bark was formerly used medicinally (Mabberley 1987). After storms, some ranchers watch for fallen branches since cattle can be poisoned from eating wilted leaves which contain hydrocyanic or prussic acid (cyanide); poisoning in humans has been reported from ingesting seeds, chewing twigs, and from making tea from the leaves (Mulligan & Munro 1990).

1. Plants from Lampasas Cut Plain and Edwards Plateau; lower surface of leaf blades sometimes entirely glabrous; leaf blades relatively more coarsely toothed (2nd floral leaf with about 5 teeth per cm of margin); petioles relatively longer, those of the 2nd floral leaf ca. 15 mm long but varying from 12–20 mm ____ var. **eximia**
1. Plants from n part of Blackland Prairie and Red River drainage; lower surface of leaf blades with a mustache of hairs at base of midrib; leaf blades relatively more finely toothed (floral leaves often with ca. 7 teeth per cm of margin); petioles relatively shorter, those of the 2nd floral leaf ca. 10–11 mm long but varying from 4–17 mm ____ var. **serotina**

var. **eximia** (Small) Little, (out of the ordinary, distinguished), ESCARPMENT BLACKCHERRY. Burnet and Williamson cos. on s margin of nc TX (Balcones Canyonlands Nat. Wildlife Refuge, C. Sexton, pers. comm.); otherwise known only from the Edwards Plateau; endemic to TX. [*P. serotina* subsp. *eximia* (Small) McVaugh]

var. **serotina**, BLACKCHERRY, WILD BLACKCHERRY. Woods, thickets; Fannin, Grayson, Hunt, Kaufman, and Lamar cos., also Dallas Co. (Little 1971); se and e TX w to ne part of nc TX.

Prunus umbellata Elliott, (with umbels), FLATWOOD PLUM. Small tree to 6 m tall; leaf blades usually 4–7(–7.5) cm long, < 3(–3.5) cm wide, elliptic-oblong to elliptic or obovate; flowers before or

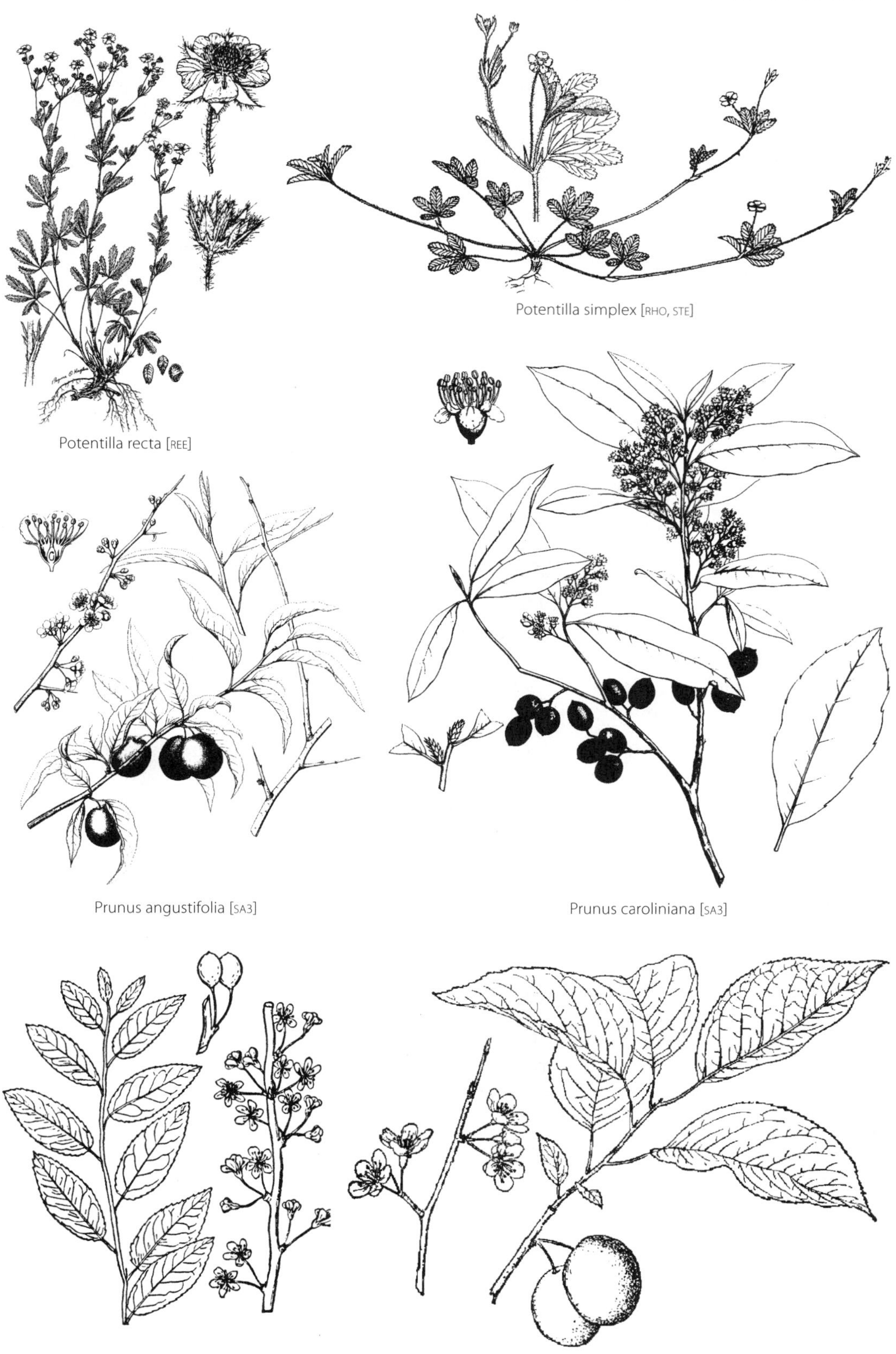

Potentilla recta [REE]

Potentilla simplex [RHO, STE]

Prunus angustifolia [SA3]

Prunus caroliniana [SA3]

Prunus gracilis [BB2]

Prunus mexicana [SA2]

with young leaves, in clusters of 2–4(–5); petals white; fruits 1–2 cm long, red to yellow or dark purple, with a bloom. Forests, forest margins; Brown (HPC), Coleman (Correll & Johnston 1970), and Henderson (Little 1976 [1977]) cos.; mainly e TX. Feb–Apr. While this species is usually easy to distinguish from *P. mexicana* when leaves are present, separating the two with only flowers is more difficult. It may be that this species is only present on the extreme e margin of nc TX. [*P. mitis* Beadle, *P. umbellata* var. *tarda* (Sarg.) W. Wight]

PYRACANTHA FIRE-THORN

A genus of 9 species native to Asia and se Europe; it includes cultivated ornamentals. *Pyracantha* is similar to the Old World genus *Cotoneaster* (widely cultivated), but is distinguished by the presence of thorns and toothed leaves (Robertson et al. 1991) (Greek: *pyr*, fire, and *acantha*, a spine or thorn)
REFERENCE: Robertson et al. 1991

Pyracantha koidzumii (Hayata) Rehder, (for Gen'ichi Koidzumi, 1883–1953, Japanese botanist). Much branched, thorny, evergreen shrub to ca. 4 m tall; leaves simple, entire, ca. 2–3 cm long, rounded apically, clustered at tips of short branchlets; inflorescence a small corymb scarcely longer than the leaves; flowers 5-merous, small; petals ca. 4 mm long, white; stamens 20; ovary half-inferior; fertile ovules 2; fruit an orange-scarlet or red pome 6–8 mm in diam. Widely cultivated and escapes; Tarrant Co.; we are not aware of other TX localities. Spring. Native of Taiwan.

PYRUS PEAR

Ours trees with simple, alternate, deciduous, serrate to crenate leaves; inflorescences corymb- or umbel-like, appearing with or before the leaves; flowers 5-merous, showy; petals white or with a pink tinge; stamens numerous; styles 2–5; fruit a pome with numerous, small, hard stone (grit) cells; seeds black or nearly so.

An Eurasian and Mediterranean genus of ca. 25 species; a number of species formerly treated in *Pyrus* are now placed in the related *Malus* (APPLE) or *Sorbus* (MOUNTAIN-ASH); *Pyrus* is currently restricted to trees with stone (sclerencyma) cells in the fruits (giving the flesh a gritty texture). The genus includes important fruit trees and showy ornamentals. (Classical Latin name for a pear)

1. Fruits globose, ca. 1 cm across; styles 2(–3); inflorescences glabrous; flowers 2–2.5 cm across ____ **P. calleryana**
1. Fruits usually pear-shaped, much > 1 cm across; styles usually 5; inflorescences villous to nearly glabrous; flowers 2.5–3 cm across ____ **P. communis**

Pyrus calleryana Decne., (for J.M.M. Callery, 1810–1862, missionary and botanist in Korea and China), CALLERY PEAR, BRADFORD PEAR. Small tree; leaves broadly ovate to ovate, crenate, very colorful (yellows to reds, often mixtures) in autumn. Widely cultivated in nc TX in form of cultivar Bradford or cultivar Aristocrat; escaping to weedy areas; Grayson Co. (weedy creekbank), also naturalized in Dallas Co. (E. McWilliams, pers. comm.). Spring. Native of China.

Pyrus communis L., (common, general), COMMON PEAR, PERA. Long-lived, usually small, sometimes spiny tree; leaves ovate to elliptic-ovate, to ca. 8 cm long, acute to acuminate apically, rounded to cuneate basally, crenate-serrulate; petioles 2–5 cm long; inflorescences umbel-like, appearing with first leaves; flowers 5-merous, with unpleasant odor; petals white, rarely with pinkish tint, broadly oblong, ca. 10 mm long, > 7 mm wide; stamens numerous; fruit a large edible pome with numerous small, hard stone (grit) cells. Cultivated and according to Correll and Johnston (1970) escaping in e TX; in nc TX persisting for many decades around old home sites; a Henderson Co. collection is probably an escape, also Brown and Callahan (HPC) cos. The seeds

Prunus munsoniana [SA2]

Prunus serotina var. serotina [SA2]

Prunus persica [BL1, VIN]

Prunus serotina var. eximia [HEA]

Prunus rivularis [LYN]

Prunus umbellata [SA3]

contain amygdalin, a cyanogenic glycoside (Lewis & Elvin-Lewis 1977). Apr–May. Native of Europe and w Asia.

ROSA ROSE

Woody perennials, shrubby or vine-like, with prickles and often bristles as well; leaves 3-foliate or odd pinnately compound; leaflets sharply toothed; stipules paired, narrow, usually fused with petiole for part of their length; flowers mostly large and showy, solitary or in corymbose or paniculate inflorescences; petals usually 5 (numerous in cultivated forms), white, pink, or rose (red or yellow in cultivated forms); stamens numerous; ovaries numerous; hypanthium (called a hip) with a constricted opening, becoming fleshy and berry-like at maturity and often colored (e.g., red) (technically an accessory "fruit"—with fleshy parts derived from organs other than the pistil); fruits (inside enlarged hypanthium) achenes.

A genus of 100–150 species native to the n temperate zone and tropical mountains; generally prickly shrubs, sometimes climbing or trailing, often with showy flowers. Species have been variously used as a source of essential oils, essences, or medicinally. The genus is particularly important ornamentally having been cultivated since ancient times (e.g., Romans); there are currently thousands of cultivars. Tyler, in Smith Co., just e of nc TX, is a major center of rose cultivation. Unfortunately, except in the song, there is no native yellow rose of TX; the famous rose in the song refers to a young lady, not a flower. The armature of roses is technically referred to as prickles; they are epidermal outgrowths lacking vascular tissue and can be rather easily popped off the stems by lateral pressure. True thorns (e.g., in *Crataegus*) are modified branches (with vascular tissue) and true spines are modified leaf tissue (also with vascular tissue). (Classical Latin name for these plants, but the name is ancient and without a clear origin; possibly from Celtic: *rhod*, red—Tveten & Tveten 1993)

REFERENCES: Rydberg 1920, 1923.

1. Hypanthium densely velvety-pubescent; calyx closely subtended by an involucre of large, dissected, pubescent bracts (falling soon after flowering) ____ **R. bracteata**
1. Hypanthium glandular- or hispid-pubescent, not velvety; calyx without an involucre of bracts.
    2. Styles united into a column, conspicuously exerted from the opening of the hypanthium and nearly equaling the stamens (look among the stamens).
        3. Stipules with widely spaced, long, narrow teeth; prickles usually only found just below the stipules; leaves of flowering branchlets with 7–9 leaflets ____ **R. multiflora**
        3. Stipules entire or with minute teeth; prickles scattered along the stem; leaves of flowering branchlets with 3–5 leaflets.
            4. Petals pink, 2–3.5 cm long; stipules fused to petiole for > 1/2 stipule length; hypanthium glabrous or with short gland-tipped hairs ____ **R. setigera**
            4. Petals white, 3–4 cm long; stipules free for > 1/2 stipule length; hypanthium conspicuously long-bristly ____ **R. laevigata**
    2. Styles separate, not exerted or only slightly so, the stigmas often forming a dense brush-like structure closing the opening of the hypanthium, much shorter than the stamens.
        5. Leaves with 3, rarely 5 leaflets 3–6 cm long; hypanthium with conspicuous long bristles up to 4–5 mm long; flowers 6–8 cm across ____ **R. laevigata**
        5. Leaves with 5–11 leaflets 1–4 cm long; hypanthium glabrous, glandular-pubescent, or with short bristles to 2 mm long; flowers 3–5.5 cm across.
            6. Leaflets with resin dots or glandular hairs on the lower surfaces; stems with conspicuously hooked prickles; outer sepals deeply pinnately divided
                7. Leaflets usually glandular on both sides, suborbicular to broadly elliptic, usually obtuse in outline apically (but with a small apical tooth); styles pubescent; sepals spreading or erect, persistent until full ripening of fruit ____ **R. eglanteria**

Pyracantha koidzumii [LIH]

Pyrus calleryana [YUN]

Pyrus communis [VIN]

Rosa arkansana var. suffulta [BB2]

Rosa bracteata [VIN]

Rosa carolina [DIL]

Rosa eglanteria [BB2]

Rosa foliolosa [BA1]

7. Leaflets eglandular above, elliptic to ovate, acute or short acuminate apically; styles glabrous or nearly so; sepals reflexed, soon deciduous ______ **R. micrantha**

6. Leaflets without either resin dots or glandular hairs on the lower surfaces; stems with prickles and bristles straight or nearly so; outer sepals entire.

8. Pedicels and hypanthium with stalked glands; sepals with stalked glands, spreading after flowering, deciduous.

9. Flowers rose-colored; largest leaflets 8–19 mm wide; sepals 16–22 mm long; petals 20–25(–32) mm long ______ **R. carolina**

9. Flowers white or rose-colored; largest leaflets 3–8 mm wide; sepals 12–16 mm long; petals 15–20 mm long ______ **R. foliolosa**

8. Pedicels and hypanthium without stalked glands; sepals pubescent but without stalked glands, upright after flowering, usually persistent ______ **R. arkansana**

Rosa arkansana Porter var. **suffulta** (Greene) Cockerell, (sp.: of Arkansas; var.: supported or propped), SUNSHINE ROSE, PRAIRIE WILD ROSE. Shrub to 2 m tall; stems with numerous slender prickles and bristles; leaflets 7–11, thick in texture; flowers few–many in corymbs, ca. 4 cm across; petals pink. Twice collected along railroad right-of-way in West Cross Timbers (Jack and Young cos.—Mahler 1988), also Comanche Co. (Stanford 1976); cited by Hatch et al. (1990) only from Area 5. May. [*Rosa suffulta* Greene]

Rosa bracteata J.C. Wendl., (with bracts), MACARTNEY ROSE. Stems trailing or arching, forming dense tangles to 3 m high, with curved, flattened, broad-based, paired prickles below the stipules; leaves evergreen, dark green; leaflets 5–9, leathery, glossy above, glabrous or pubescent on midrib beneath; flowers solitary or few together, subsessile at tips of leafy branchlets, 5–7 cm across; petals white; hips 2–2.5 cm long. Cultivated and escapes; Jack Co. in West Cross Timbers and Tarrant Co. in East Cross Timbers; se and e TX w to nc TX and Edwards Plateau; apt to freeze back in severe winters. May–Jun, sporadically to Nov. Native of China.

Rosa carolina L., (of Carolina), CAROLINA ROSE, PASTURE ROSE. Shrub 0.2–0.7(–1) m tall; stems with ± straight, usually slender prickles and bristles; leaflets 5–9; flowers mostly solitary, 35–55 mm across; hips red. Roadsides, woods; Lamar Co. in Red River drainage; mainly e TX. May–Jul.

Rosa eglanteria L., (Latinization of the old English and French name), SWEET BRIAR ROSE, SWEET-BRIER, EGLANTINE. Shrub to ca. 2(–3) m tall, similar to *R. micrantha*; stems with curved prickles, sometimes mixed with bristles; leaves with 5–7 leaflets; leaflets 1–3 cm long, resinous aromatic; pedicels and hypanthium glandular-hispid; flowers 3–5 cm across; petals bright pink; hips orange to scarlet. Cultivated and escapes, disturbed grassland; Bell Co. (Fort Hood—Sanchez 1997); we are not aware of other TX localities. Summer–fall. [*R. rubiginosa* L.] The leaves when crushed have an odor like fermenting apples (Radford et al. 1968). Native of Europe.

Rosa foliolosa Nutt. ex Torr. & A. Gray, (full of leaves, profusely-leaved), WHITE PRAIRIE ROSE, LEAFY ROSE. Dwarf, rhizomatous shrub to 0.5 m tall; prickles few, very small, slender, straight or nearly so; leaflets glabrous or pubescent on veins beneath, 7–11; stipules glandular-ciliate; flowers usually solitary, short-pedicelled, ca. 4 cm across; petals white or rarely light pink. Prairies and open thickets or roadsides, calcareous clay or less often sandy soils; Blackland Prairie w to Grand Prairie; mainly nc TX s to Edwards Plateau. Mid-May–early Jul. [*R. ignota* Shinners]

Rosa laevigata Michx., (smooth), CHEROKEE ROSE. Plant high-climbing, to 5 m; stems with curved, flattened, broad-based prickles, without bristles; leaves evergreen, usually with 3(–5) acute to acuminate leaflets; flowers large, fragrant; petals 3–4 cm long, usually white (rarely rose-colored); styles united, only exerted ca. 1 mm (therefore this species can be reached 2 ways in key); hips red, 3–4 mm long. Cultivated and escapes; included based on citation of vegeta-

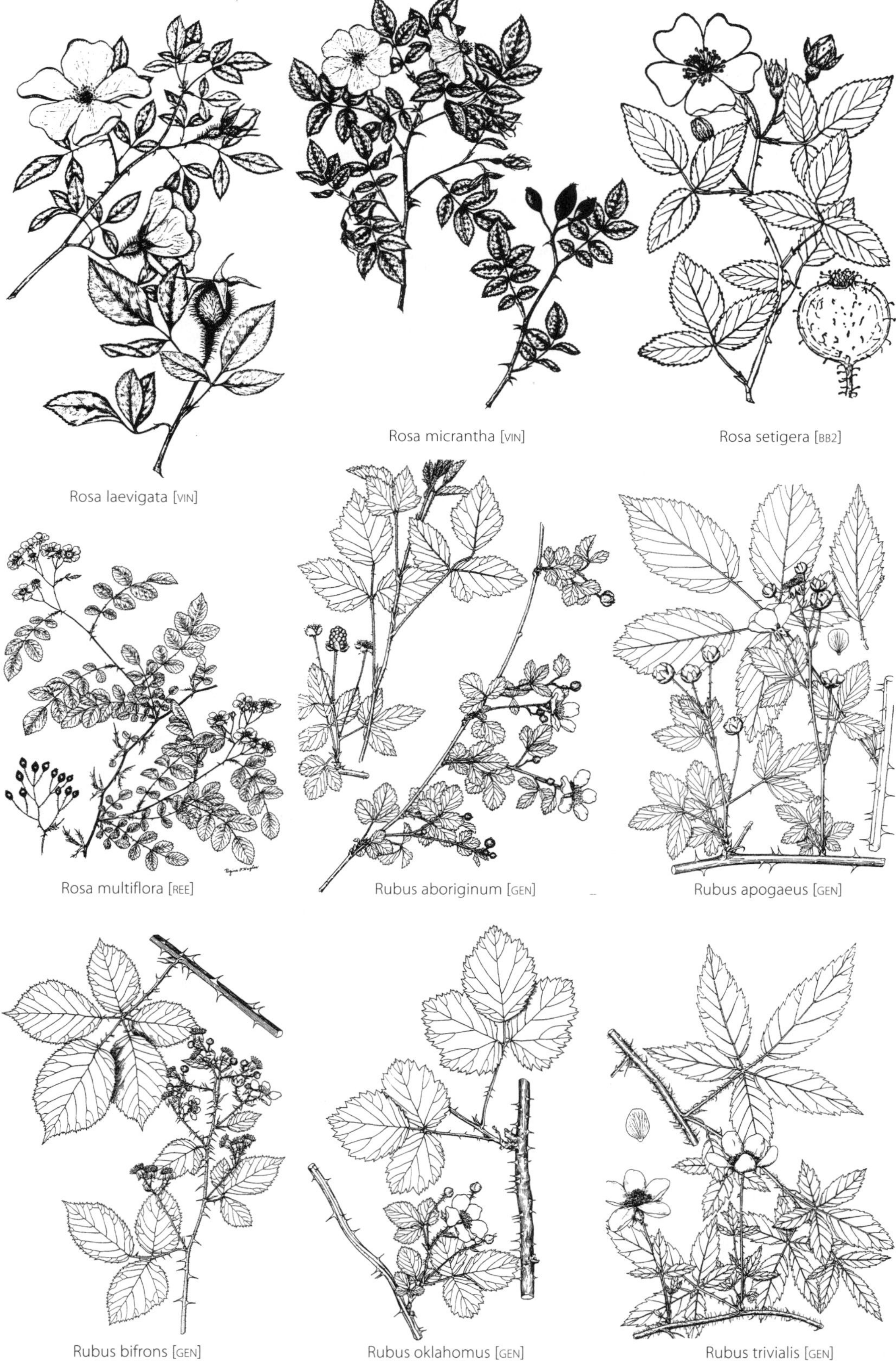

Rosa laevigata [VIN]

Rosa micrantha [VIN]

Rosa setigera [BB2]

Rosa multiflora [REE]

Rubus aboriginum [GEN]

Rubus apogaeus [GEN]

Rubus bifrons [GEN]

Rubus oklahomus [GEN]

Rubus trivialis [GEN]

tional areas 4 and 5 (Fig. 2) by Hatch et al. (1990); mainly se and e TX, also Edwards Plateau. Spring. Native of China.

Rosa micrantha Borrer ex Sm., (small-flowered). Shrub to 2 m tall; stems with curved prickles, without bristles; leaves with 5–7 leaflets; leaflets 1.5–3 long, not strongly aromatic; pedicels glandular-hispid; hypanthium glabrous or sparsely glandular-hispid; flowers ca. 3 cm across; petals pink to white. Cultivated and escapes; included based on citation of vegetational areas 4 and 5 (Fig. 2) by Hatch et al. (1990); mainly se and e TX, also Edwards Plateau. Spring. Native of Europe, nw Africa, and sw Asia. Summer–fall.

Rosa multiflora Thunb. ex Murr., (many-flowered), JAPANESE ROSE, MULTIFLORA ROSE. Shrub; stems erect and arching to trailing, with stout, recurved, flattened, broad-based prickles; stipules pectinate; flowers usually many, in rounded or pyramidal inflorescences; petals white (rarely pink), 7–15 mm long; hips red, 6–9 mm long. Cultivated and escapes, weedy areas; Dallas Co., also Fort Hood (Bell or Coryell cos.—Sanchez 1997); mainly e TX. May–Jun. Native of Asia.

Rosa setigera Michx. var. **tomentosa** Torr. & A. Gray, (sp.: bristle-bearing; var.: densely woolly, with matted hairs), PRAIRIE ROSE, CLIMBING ROSE, CLIMBING PRAIRIE ROSE, FUZZY ROSE. Climbing, sprawling, or trailing shrub; stems to 5 m or more long, with remote, curved, flattened, broad-based prickles, occasionally also with bristles; leaves with 3–5 leaflets; leaflets 3.5–5(–10) cm long, tomentose below; stipules entire or remotely toothed; flowers 5–15 in a corymb, 4–7 cm across, roseate, fading to white; hips red, 8–12 mm long. Thickets, open woods; Collin, Dallas, Fannin, Grayson, Hopkins, Hunt, Kaufman, and Lamar cos.; e TX w to n part of nc TX, also Edwards Plateau. Apr–Jul.

RUBUS DEWBERRY, BLACKBERRY, BRAMBLE

Usually woody, ± prickly perennials, often reproducing vegetatively by suckers or rooting stem tips; the mostly biennial stems flowering the second year, then dying; mature plants have both first year, usually unbranched stems (= primocanes), and second year, flowering, branched stems (= floricanes); leaves petioled, sometimes partly simple on floricanes to usually pinnately or approximately palmately compound (central leaflet longer-stalked than the others), with 3–5 leaflets; stipules slender; flowers showy, solitary, racemose, corymbose, or panicled; petals usually white or pink (frequently pink in bud); stamens numerous; carpels many, separate, on an elongating receptacle; "fruit" an aggregate of red to black, 1-seeded juicy druplets, the whole structure commonly called a "berry."

A cosmopolitan, especially n hemisphere genus of 250 species, plus innumerable apomictic lines. Many are cultivated for the fruits or as ornamentals; the fruits are an important wildlife food (Martin et al. 1951); those of some species go from green to red to finally black. Both ripe (black) and pre-ripe (red) fruits are often present at the same time—this is an example of a bi-color fruit display, thought to be more effective than a single color at attracting birds which act as dispersal agents (Willson & Thompson 1982). The popular distinction (originating in the North) between blackberries (with erect or arched stems) and dewberries (with trailing stems) does not fit nc TX plants. *Rubus* taxonomy is notoriously difficult because of hybridization, polyploidy, and apoximis (= production of seeds without fertilization). Careful note of growth form (e.g., stem upright vs. trailing) is often important in making a definitive identification. (Classical Latin name for a blackberry or bramble)
REFERENCES: Bailey 1925, 1932, 1941–45; Mahler 1979.

1. Lower surface of leaflets whitish or grayish with dense, closely matted pubescence; petals usually pink, rarely white; inflorescence an open panicle; pedicels with small stout prickles ____ **R. bifrons**
1. Lower surface of leaflets green to gray, glabrous to with dense spreading pubescence; petals

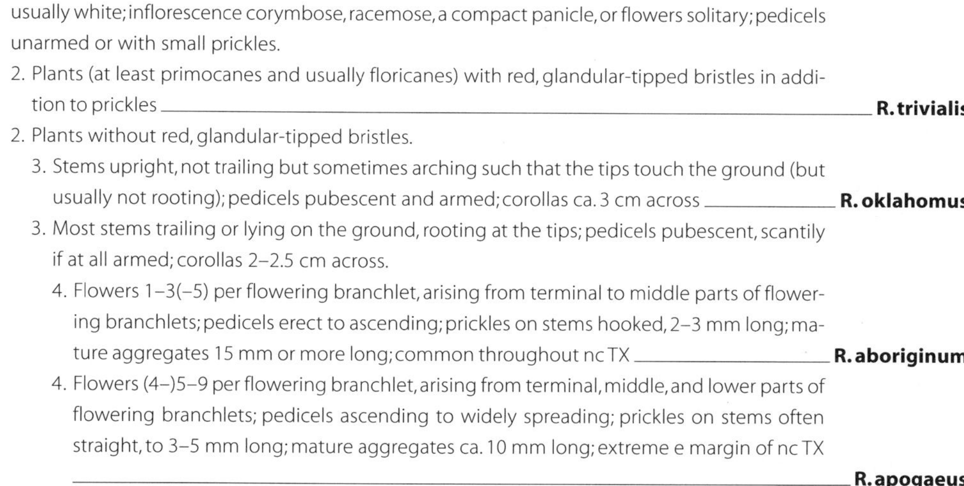

usually white; inflorescence corymbose, racemose, a compact panicle, or flowers solitary; pedicels unarmed or with small prickles.

2. Plants (at least primocanes and usually floricanes) with red, glandular-tipped bristles in addition to prickles ____ **R. trivialis**
2. Plants without red, glandular-tipped bristles.
   3. Stems upright, not trailing but sometimes arching such that the tips touch the ground (but usually not rooting); pedicels pubescent and armed; corollas ca. 3 cm across ____ **R. oklahomus**
   3. Most stems trailing or lying on the ground, rooting at the tips; pedicels pubescent, scantily if at all armed; corollas 2–2.5 cm across.
      4. Flowers 1–3(–5) per flowering branchlet, arising from terminal to middle parts of flowering branchlets; pedicels erect to ascending; prickles on stems hooked, 2–3 mm long; mature aggregates 15 mm or more long; common throughout nc TX ____ **R. aboriginum**
      4. Flowers (4–)5–9 per flowering branchlet, arising from terminal, middle, and lower parts of flowering branchlets; pedicels ascending to widely spreading; prickles on stems often straight, to 3–5 mm long; mature aggregates ca. 10 mm long; extreme e margin of nc TX ____ **R. apogaeus**

Rubus aboriginum Rydb., (derivation not given in type description but presumably meaning native or indigenous). Most stems trailing or lying on the ground, rooting at the tips; lower surface of leaflets usually densely soft-pubescent or velvety to the touch to sparsely pubescent; flowers 20–25 mm across; aggregates 15 mm or more long. Sandy woods, fencerows, and roadsides; e 1/2 of TX. Apr.

Rubus apogaeus L.H. Bailey, (derivation not given in type description but possibly from the floricanes rooting at the apex or apogee). Plant low, forming mounds, a few canes can be erect but growth in general bends toward the ground with the floricanes long-running and rooting at the tips; prickles nearly straight, 3–5 mm long; flowers 25 mm or less across; aggregates ca. 10 mm long. Roadsides, grassy areas, fencerows; Navarro Co.; mainly e TX; Hatch et al. (1990) also cited vegetational areas 5 and 7 (Fig. 2); endemic to TX. Apr–May. [*R. uncus* L.H. Bailey]

Rubus bifrons Vest ex Tratt., (two-fronded). Stems low arching with tips reaching the ground, making clumps to 1.5 m high; flowers 20–25 mm across; aggregates to 20 mm long. Thickets, roadsides, pastures; Henderson, Lamar, and Milam cos. on e edge of nc TX, also Fort Hood (Bell or Coryell cos.—Sanchez 1997); naturalized mainly in e TX. May–Jun. Native of Europe.

Rubus oklahomus L.H. Bailey, (of Oklahoma). Plant usually reaching 1–2 m in height, very prickly, the prickles straight or curved, 3–6 mm long; leaflets densely to sparsely soft-pubescent beneath; flowers large and showy, ca. 3 cm across, 3–5 per lateral branchlet. Thickets and fencerows, in stream bottoms or on slopes; Post Oak Savannah w to Rolling Plains and Edwards Plateau. Apr.

Rubus trivialis Michx., (common, ordinary), SOUTHERN DEWBERRY, ZARZAMORA. Stems trailing or low-arching, rooting at the tips, with recurved prickles; leaves half-evergreen; leaflets glabrous beneath or with pubescence only along the main veins, marginally without glands; flowers usually 1–3 per lateral branchlet; petals 15–25 mm long; aggregates 10–30 mm long. Roadsides, fencerows, and thickets, various soils; e TX w to West Cross Timbers, occasionally farther w. Late Mar–Apr. Jones et al. (1997) did not list *R. trivialis* for TX. This species is sometimes infected by "orange rust," the rust fungus *Arthuriomyces peckianus* (Howe) Cummins & Y. Hirats., which causes a systemic infection resulting in a witches' broom-like, dense, abnormal branching pattern and also orange masses of spores covering the leaves (J. Hennen, pers. comm.).

Rubus flagellaris Willd., (whip-like, having long, thin, supple shoots like whips), NORTHERN DEWBERRY, is an eastern species cited for vegetational areas 1 and 5 (Fig. 2) by Hatch et al. (1990);

we have seen no unambiguous specimens of this taxon from nc TX. The species differs from *R. aboriginum* in having the lower surface of the primocane leaves glabrous or with pubescence only on the veins, neither soft nor velvety to the touch.

Rubus riograndis L.H. Bailey, (of the Rio Grand River), [*R. trivialis* Michx. var. *duplaris* (Shinners) Mahler, *R. duplaris* Shinners] occurs in the Post Oak Savannah vegetational area just to the e of nc TX (Mahler 1979). It is similar to *R. trivialis* but differs in having scattered stalked glands along the margins of the leaflets and with the lower surface of primocane leaflets soft pubescent.

SANGUISORBA BURNET

A n temperate genus of ca. 10 species including *S. officinalis* L. (BURNET), an Eurasian species cultivated as an ornamental and for its edible leaves. (Latin: *sanguis*, blood, and *sorbeo*, to drink up or absorb; it was said to stop bleeding or have styptic properties)

Sanguisorba annua (Nutt. ex Hook.) Torr. & A. Gray, (annual), PRAIRIE BURNET. Glabrous erect annual to 40 cm tall; leaves pinnately compound, the 7–15 leaflets divided nearly to midrib; stipules similar to leaflets; inflorescence a dense, short, thick, cylindrical spike or spike-like head to 2(–3) cm long, on naked peduncles terminating the stems; flowers with calyx-like involucre of 4 broad, scarious bracts; calyces with 4 broad lobes, the lobes 2–3 mm long, green, petaloid, with scarious margins; petals absent; fruit usually a solitary (rarely 2) achene enclosed in the dry, 4-angled hypanthium. Sandy or gravelly prairies, pastures, and roadsides, often in moist areas; Post Oak Savannah w to Rolling Plains, also Edwards Plateau; apparently spreading eastward. Late Mar–May. Probably native only from West Cross Timbers westward (Mahler 1988).

RUBIACEAE COFFEE OR MADDER FAMILY

Ours annual or perennial herbs or in 1 species (*Cephalanthus*) shrubs or small trees; stems square or multiple-angled (rounded in woody plants); leaves opposite or whorled, sessile or petioled, simple, entire; stipules scarious or herbaceous, persistent or deciduous, or stipules modified into additional leaves in whorls; flowers axillary or terminal, solitary or in heads, cymes, or panicles; calyces 2–8-toothed or -lobed, or entire-margined; corollas funnelform, salverform, or rotate, with (3–)4(–6) lobes; stamens 3–6; pistil 1; styles and stigmas 2–4; ovary usually inferior; fruit a capsule or schizocarp.

A huge (10,200 species in 630 genera) cosmopolitan family concentrated in the tropics and subtropics; tropical species are typically shrubs or trees while most temperate taxa are herbaceous; some are rarely aquatic, epiphytic, or inhabited by ants; the group includes a broad range of chemical defenses including alkaloids. The family is the source of coffee (seeds of African species of *Coffea*—with the alkaloid caffeine) and the malaria treatment quinine, an alkaloid derived from the bark of South American species of *Cinchona*. *Cinchona* bark (also known as Peru bark or Jesuits' bark) was long valued by South American natives and is still the source of quinine used medicinally and in flavoring tonic water. The family also includes ornamentals such as *Gardenia*, *Ixora*, and *Pentas*. Family name from *Rubia*, MADDER, a genus of 60 species native to the Mediterranean area, Africa, temperate Asia, and the Americas. (Latin: *ruber*, red, in reference to the red dye obtained from the roots of *R. tinctoria* L. of s Europe and w Asia) (subclass Asteridae)

FAMILY RECOGNITION IN THE FIELD: usually herbs (1 woody) with opposite or whorled, simple, entire leaves; *stipules* present (sometimes modified into additional "leaves"); small often *4-merous* flowers; 2-celled *inferior* ovary.

REFERENCES: Standley 1918–1934; Darwin 1976; Rogers 1987.

1. Plants woody shrubs or small trees; flowers in peduncled globose heads ca. 15–35 mm in diam. ____ **Cephalanthus**
1. Plants herbaceous, annual or perennial; flowers not borne as above.
   2. Leaves opposite or in 3s; stipules present, at least on young growth.
      3. Stipules not bristly-margined; flowers terminal, solitary or corymbose; fruit usually a capsule with numerous seeds.
         4. Corollas rotate (tube essentially absent), white, ca. 1 mm long; flowers axillary and subsessile; flowering May–Aug ____ **Oldenlandia**
         4. Corollas salverform or funnelform (tube obvious), variously colored or sometimes white, 4 mm or more long OR if smaller (1–3 mm long) and white then flowering in early spring; flowers not axillary and subsessile, either terminal or if axillary then on stalks 3–15 mm long.
            5. Plants perennial from a conspicuous stout taproot; stems 5–50 cm tall; flowers in leafy cymes ____ **Hedyotis**
            5. Plants annual with fine taproot or fine fibrous roots; stems to ca. 30 cm tall, often much smaller; flowers in cymes OR solitary in the leaf axils OR 1 or 2 terminating the stem ____ **Houstonia**
      3. Stipules bristly-margined (bristles usually conspicuous at the nodes); flowers in axillary pairs or whorls or terminal leafy-bracted clusters or heads; fruit usually separating into 1–4 one-seeded carpels.
         6. Flowers in mostly terminal, leafy-bracted clusters or heads ____ **Richardia**
         6. Flowers axillary.
            7. Flowers 2–6 per node (1–3 per leaf axil); hairs present on some part of the stems, leaves, or fruit; corollas 4–10 mm long, the tube conspicuous and much longer than calyx lobes ____ **Diodia**
            7. Flowers 10–30 per node; all parts of plant glabrous (but note bristly stipules); corollas 2–3 mm long, scarcely longer than calyx lobes ____ **Spermacoce**
   2. Leaves in whorls of 4 or more; stipules absent (actually modified into leaves, the "leaves" thus in whorls).
      8. Flowers solitary or in cymes or panicles, the inflorescences not involucrate; corollas rotate (tube obscure), white, yellow-green, or reddish brown ____ **Galium**
      8. Flowers in sessile or subsessile, few-flowered, involucrate heads, the involucral bracts resembling leaves; corollas funnelform with conspicuous tube ca. as long as or longer than lobes, pink to lavender to blue.
         9. Corollas pink to lavender; fruits 2–7 mm long, with short appressed bristles; sepals lanceolate, persistent in fruit; leaves 4–6 per whorl, obovate to lanceolate or elliptic, acute to acuminate at apex; widespread in nc TX ____ **Sherardia**
         9. Corollas bright blue to blue-violet; fruits 2–3 mm long, smooth; sepals minute, not persistent; leaves 6–9 per whorl, linear, obtuse at apex; known in nc TX only from Dallas Co. ____ **Asperula**

ASPERULA

A genus of ca. 90 species of Eurasia (especially the Mediterranean) and Australia. Some are cultivated as ornamentals or used as source of dye. (Latin: *asper*, rough, referring to the roughly hairy stems)

Asperula arvensis L., (pertaining to cultivated fields), WOODRUFF. Annual to ca. 30 cm tall, nearly glabrous; stems 4-angled; leaves 6–9 per whorl, linear, obtuse apically; flowers in sessile terminal heads; calyces inconspicuous; corollas bright blue to blue-violet, ca. 4(–6.5) mm long, funnelform, the tube equal or exceeding the (3–)4 lobes; fruits 2–3 mm long, smooth, of 2 one-seeded nutlets. Disturbed areas; Dallas Co.; in TX apparently only known from the Blackland Prairie. Mar–May. Introduced from Old World.

CEPHALANTHUS BUTTONBUSH

A genus of 6 species (3 New World, 2 Asia, 1 Africa) of the tropics n to North America. (Greek: *cephale*, head, and *anthos*, flower, alluding to the globose flower clusters)
REFERENCE: Ridsdale 1976.

Cephalanthus occidentalis L., (western), COMMON BUTTONBUSH, HONEY-BALLS, GLOBEFLOWER. Shrub or small tree 1–5(–15) m tall; trunk rarely to 30 cm in diam.; leaves opposite or in 3s(–4s), short-petioled; leaf blades elliptic-lanceolate, entire; stipules triangular, falling early; flowers 4–5-merous, in terminal and axillary, naked-peduncled, conspicuous, globose heads; corollas white; stamens 4, exserted; fruits 4–8 mm long, splitting into indehiscent segments. Stream banks, ditches, and damp woods; throughout TX. Jun–Jul, sporadically later. [*C. occidentalis* var. *californicus* Benth., *C. occidentalis* var. *pubescens* Raf.] Cultivated as an ornamental but should be used with caution since it can become invasive; also used medicinally (Mabberley 1987); the leaves are reported to contain glucosides and are possibly poisonous to animals and humans; symptoms include vomiting, convulsions, chronic spasms, and muscular paralysis (Muenscher 1951; Burlage 1968; Crosswhite 1980). /82

DIODIA BUTTONWEED

Erect to procumbent, annual or perennial herbs; leaves sessile; stipules bristly-margined; flowers sessile, axillary, usually 4-merous; corollas tubular-funnelform, rather small; fruits splitting into 2–3 indehiscent carpels or sometimes not splitting.

A genus of ca. 30 species of tropical and warm areas of the Americas and Africa. (Greek: *diodos*, a thoroughfare, alluding to the species often growing by the wayside)

1. Stems usually pubescent; corollas pinkish purple (rarely white), 4–6 mm long; fruits not ribbed, ca. 2.5–4 mm long; calyx teeth 4 ____ **D. teres**
1. Stems glabrous; corollas white, 7–10 mm long; fruits strongly 6-ribbed, 5–8 mm long; calyx teeth usually 2 ____ **D. virginiana**

Diodia teres Walter, (terete, circular in cross-section), POOR-JOE, BUTTONWEED. Erect annual with prostrate or ascending branches to ca. 80 cm long; leaves linear to lanceolate. Sandy open woods, fields, and roadsides; se and e TX w to West Cross Timbers and Edwards Plateau. Jun–Oct.

Diodia virginiana L., (of Virginia), VIRGINIA BUTTONWEED, LARGE BUTTONWEED. Perennial; stems spreading or procumbent, rooting, to ca. 60 cm long; leaves linear-lanceolate to broadly lanceolate. Along streams or in marshy ground; Dallas, Grayson, Hopkins, Red River, and Tarrant cos., also Lamar Co. (Carr 1994); se and e TX w to nc TX. Jun–Sep.

GALIUM CLEAVERS, BEDSTRAW

Annuals or perennials with whorled, sessile, linear-lanceolate to oblanceolate or elliptic, entire leaves; flowers very small, terminal or axillary, solitary or in cymes or small panicles; calyx lobes not evident; corollas rotate, 3- to 4-lobed, white to green or brown-red; stamens usually 4; fruit a usually bristly or smooth schizocarp that when ripe splits into 2 seed-like, indeshiscent, 1-seeded segments.

A cosmopolitan genus of ca. 300 species of usually slender herbs typically with square stems and whorled leaves and stipules. Some with fragrant foliage were formerly used to stuff

Sanguisorba annua [RBM]

Asperula arvensis [GLE]

Cephalanthus occidentalis [SA3]

Diodia teres [BT3, CO1, GLE]

Diodia virginiana [GLE, LAM]

Galium aparine [REE]

Galium circaezans [BB1]

Galium obtusum [STE]

mattresses—hence the common name BEDSTRAW. The common name CLEAVERS comes from the hooked hairs of some that cause the plants to stick or cleave to clothing (Tveten & Tveten 1993). According to early Christian tradition, *G. vernum* L., OUR LADY'S BEDSTRAW, was abundant about Bethlehem and was part of the "straw" used for cattle-bedding in stables; it supposedly formed the bed for the infant Jesus in the manger (Hausman 1950). (Greek: *gala*, milk, from use of some species in curdling milk)
REFERENCES: Weatherby & Blake 1916; Puff 1976, 1977.

1. Leaves in whorls of 6–9, the blades widest beyond middle ____ **G. aparine**
1. Leaves in whorls of 4–6, the blades widest near middle or below.
   2. Leaves scabrous or pubescent, at least on margins and midrib beneath.
      3. Flowers in leaf axils, subsessile, the very short pedicels quickly becoming reflexed; leaves 4–10 mm long ____ **G. virgatum**
      3. Flowers in inflorescences at the ends of stems or branches, pedicellate, the pedicels not reflexed OR sessile or nearly so along branches of inflorescence; leaves variable in length, 5–50 mm long.
         4. Flowers on distinct pedicels; leaves usually 25 mm or less long.
            5. Middle stem leaves 5–13 mm long; pedicels becoming 5–30 mm long; leaves not minutely punctate on lower surfaces ____ **G. texense**
            5. Middle stem leaves 10–25 mm long; pedicels becoming 2–5 mm long; leaves usually minutely punctate on lower surfaces (use hand lens) ____ **G. pilosum**
         4. Flowers sessile or nearly so along branches of inflorescence (occasionally solitary at ends of branchlets); larger leaves usually > 25 mm long ____ **G. circaezans**
   2. Leaves glabrous or with scabrous margins only.
      6. Leaves abruptly sharp-pointed; fruits densely bristly ____ **G. triflorum**
      6. Leaves obtuse; fruits smooth, glabrous.
         7. Corollas 2–3 mm across, 4-lobed; leaves mostly 4 per node, rarely 5 or 6 ____ **G. obtusum**
         7. Corollas < 2 mm across, 3-lobed; leaves mostly 5 or 6 per node, sometimes 4 ____ **G. tinctorium**

Galium aparine L., (old generic name, long interpreted to mean catch, cling, or hold onto), CLEAVERS, CATCHWEED BEDSTRAW, GOOSEGRASS CLEAVERS, SWEETHEARTS. Weak-stemmed annual, scabrous with small, hooked hairs, catching on fingers and clothes; stems to ca. 100 cm long; flowers on small, axillary branchlets; corollas white; fruits bristly. Shady or damp ground, thickets; widespread in TX. Mar–Apr. Formerly used as a coffee substitute in Ireland (Mabberley 1987); the dried and roasted fruits are reported to be the best coffee substitute in North America (McGregor & Brooks 1986).

Galium circaezans Michx., (resembling *Circaea*—enchanter's nightshade, in the Onagraceae), WOODS BEDSTRAW, CROSS-CLOVER, WILD LICORICE. Clump-forming perennial with stems to ca. 45 cm tall; leaves 7–18 mm wide; flowers often 1 cm or more apart in inflorescence; corollas greenish or whitish; fruits bristly. Rocky or sandy woods; Bell, Cooke, Dallas, Denton, Grayson, and Tarrant cos., also Lamar Co. (Carr 1994); e TX w to nc TX, also Rolling Plains and Edwards Plateau. Apr–Oct.

Galium obtusum Bigelow, (blunt), BLUNT-LEAF BEDSTRAW. Rather delicate annual or perennial, usually freely branched; stems to 80 cm tall; leaves to 6 mm wide; corollas white, usually 4-lobed; fruits smooth. Damp sandy woods or open ground; Hunt Co. (Mahler 1988); se and e TX w to e part of nc TX. Apr–Jun.

Galium pilosum Aiton, (with long soft hairs), HAIRY BEDSTRAW. Clump-forming low perennial; stems erect to ascending, to 90 cm tall, coarse, up to 1 mm thick near base, 1.2 mm thick near middle; corollas brown-red, rarely greenish yellow or whitish; fruits bristly. Sandy woods; se and e TX w to West Cross Timbers, also Trans-Pecos. May–Jun. Varieties based on differences in

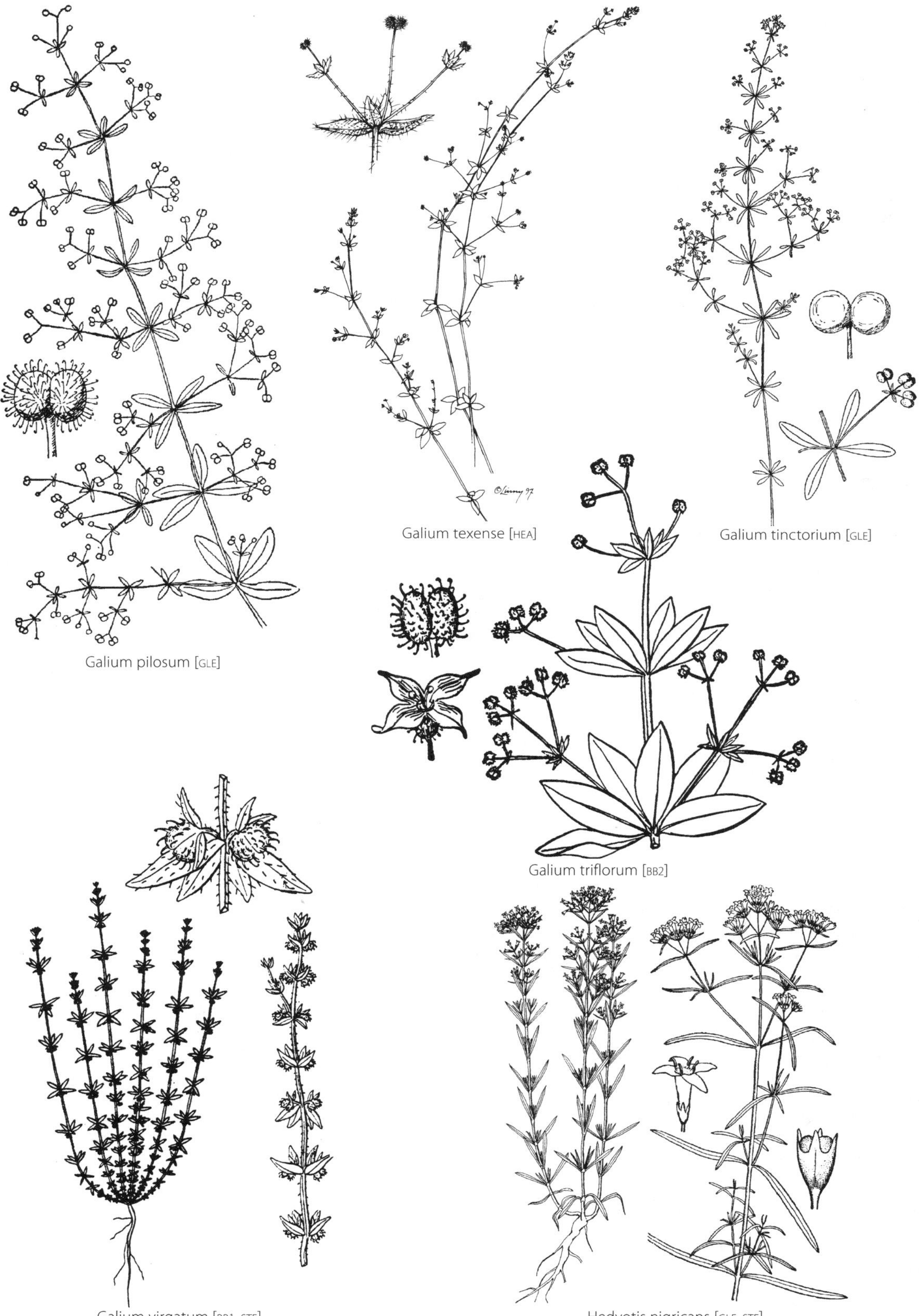

Galium pilosum [GLE]

Galium texense [HEA]

Galium tinctorium [GLE]

Galium triflorum [BB2]

Galium virgatum [BB1, STE]

Hedyotis nigricans [GLE, STE]

stem pubescence are sometimes recognized (e.g., Weatherby & Blake 1916; Kartesz 1994; Jones et al. 1997); however, we are not recognizing infraspecific taxa. [*G. pilosum* var. *laevicaule* Weath. & S.F. Blake; *G. pilosum* var. *puncticulosum* (Michx.) Torr. & A. Gray]

Galium texense A. Gray, (of Texas), TEXAS BEDSTRAW. Annual [or perennial?]; stems to 30 cm tall, weak, slender, up to 0.5 mm thick near base, 1 mm thick near middle; leaves usually < 8 mm long; flowers on axillary branchlets and terminal; corollas white to yellow; fruits bristly. Gravelly limestone slopes, under trees and shrubs; Bell, Johnson, Lampasas, Palo Pinto, and Somervell cos., also Comanche and Hamilton cos. (HPC); s and c TX n to nc TX. Apr–May.

Galium tinctorium (L.) Scop., (used in dying, of the dyers), DYE BEDSTRAW, STIFF MARSH BEDSTRAW. Perennial; leaves to 2 cm long, usually 1–2(–4) mm wide; flowers on distinct pedicels; corollas white; fruits smooth. Swamps or moist areas; Collin, Grayson, Hopkins, Hunt, and Limestone cos., also Lamar Co. (Carr 1994); se and e TX w to nc TX, also Rolling Plains. Mar–Aug.

Galium triflorum Michx., (three-flowered), FRAGRANT BEDSTRAW, SWEET-SCENTED BEDSTRAW. Perennial with slender creeping rootstocks; flowering the first year; stems to 1 m or more long; corollas greenish or whitish; fruits bristly. Damp sandy woods; collected at Dallas by Reverchon (Mahler 1988), not found there recently; scattered in e 1/2 of TX. Jun–Jul.

Galium virgatum Nutt., (twiggy), SOUTHWEST BEDSTRAW. Erect annual, usually with several short, stiff stems to ca. 40 cm tall, usually smaller; corollas light green to yellowish or whitish; fruits bristly or rarely smooth. Eroding or disturbed ground, especially in calcareous clay; widespread in TX. Apr–May. This is the only nc TX species with axillary subsessile flowers.

HEDYOTIS BLUET, STAR-VIOLET

A genus of 110 species of tropical and warm areas, especially the Old World (130 species if *Houstonia* is lumped into *Hedyotis* as is sometimes done); some are cultivated as ornamentals; Terrell (1991) recognized 21 species in North America. The genus is sometimes treated as including *Houstonia* and *Oldenlandia* which are here recognized as separate genera. (Greek: *hedys*, sweet, and *ous*, ear, significance not explained)
REFERENCES: Shinners 1949g; Terrell 1975, 1986b, 1991; Turner 1995c [1996].

Hedyotis nigricans (Lam.) Fosberg, (black, some plants blackening upon drying), PRAIRIE BLUETS, STAR-VIOLET, FINE-LEAF BLUETS. Glabrous perennial; stems 5–50 cm tall; leaves linear to oblanceolate, 1–4 cm long, 0.5–4 mm wide; flowers in leafy cymes; corollas funnelform, 4-lobed, white to lavender-pink, 5–8 mm long; capsules 2.5–3 mm long. Limestone outcrops, prairies; throughout TX. May–Oct. [*Gentiana nigricans* Lam., *H. nigricans* (Lam.) Fosberg var. *filifolia* (Chapm.) Shinners, *Houstonia nigricans* (Lam.) Fernald] Turner (1995c [1996]) recognized 5 varieties in this species, of which 3 occur in TX. Only the type variety occurs in nc TX.

HOUSTONIA BLUET

Ours annual herbs; flowers axillary, 1 or 2 terminating the stem, or in cymes; flowers 4-merous; corollas salverform or funnelform, colorful or white; fruit a capsule.

A genus of 20 species native to the U.S., Canada, and Mexico (Terrell 1996); sometimes included in the genus *Hedyotis*. (Named for Dr. William Houstoun, 1695–1733, an English surgeon and botanist who collected in tropical America)
REFERENCES: Shinners 1949g; Lewis 1958 [1959]; Terrell 1975, 1986a, 1991, 1996; Terrell et al. 1986.

1. Calyx lobes long and quite narrow, apically subulate (= awl-shaped); pedicels recurved in fruit; flowers heterostylous (pin and thrum flowers with different length styles—see description); rare in nc TX, known only from exteme s margin (Burnet Co.) ______ **H. humifusa**

1. Calyx lobes subtriangular, ovate, lanceolate, or narrowly lanceolate, apically obtuse, acute, or sharply acute; pedicels usually straight or in 1 species recurved in fruit; flowers homostylous (= all styles equal in length); widespread in nc TX.
   2. Flowers either in few-flowered terminal cymes OR solitary per axil on pedicels recurved in fruit; corollas 1–3 mm long, usually white, tips of lobes sometimes pink or whole corollas rarely light pink; very rare in nc TX.
      3. Capsules usually hirtellous (use hand lens), on pedicels reflexed or bent at their base; corollas longer than calyx lobes; flowers solitary per axil ____ **H. subviscosa**
      3. Capsules glabrous, on erect peduncles; corollas ca. as long as calyx lobes; flowers in few-flowered terminal cymes ____ **H. parviflora**
   2. Flowers solitary per leaf axil or 1 or 2 terminating the often diminutive stem; pedicels straight in fruit; corollas 2–12.5 mm long, purple, blue-violet, lilac, light violet, pink, or white; widespread in nc TX.
      4. Corollas white with yellow center, 2–5.5 mm long; corolla tube 0.8–2.5 mm long, ca. as long as calyx lobes and ± concealed by them; corolla lobes 0.8–3 mm long. ____ **H. micrantha**
      4. Corollas white to pink, light violet, purple, blue-violet, or lilac, with yellow to brownish or reddish center, 3.5–12.5 mm long; corolla tube 2–8 mm long, usually longer than calyx lobes (sometimes much longer) and not concealed by them; corolla lobes 1–7 mm long.
         5. Corollas white to light pink or light violet, with yellow center, the tube 3.5–8 mm long, densely pubescent within; plants small, the stems to only 3(–7) cm long; leaves 0.3–3 mm wide, mostly oblanceolate ____ **H. rosea**
         5. Corollas usually purple to blue-violet or lilac, with reddish or brownish center or more rarely lavender to pink or white with yellowish or brownish yellow center, the tube 2–5.5 mm long, glabrous or sparsely pubescent within; plants larger, the stems 1.5–15(–25) cm long; leaves usually (0.5–)4–5(–9) mm wide, mostly elliptic, narrowly elliptic, ovate, or spatulate ____ **H. pusilla**

Houstonia humifusa (A. Gray) A. Gray, (sprawling on the ground), MAT BLUETS, Stems 1–15 cm tall; internodes 3–6; leaves 5–30(–45) mm long, 0.5–4(–6) mm wide, sessile; flowers heterostylous (= having styles of different lengths; in this case, distylous—2 different style lengths): pin flowers with anthers sessile and included, the styles long and stigmas thus exserted; thrum flowers with anthers exserted, the styles short and stigmas thus included; calyx lobes 1–4.7 mm long; corollas 3.5–10 mm long, usually purplish or pink, sometimes white, the tube 1–2 times longer than lobes. Sandy areas, sometimes over gypsum; Burnet Co. on s margin of nc TX (Terrell 1996); mainly c and w TX. Mar–Jun(–Oct). [*Hedyotis humifusa* A. Gray]

Houstonia micrantha (Shinners) Terrell, (small-flowered), SOUTHERN BLUETS. Vegetatively similar to *H. pusilla*; flowers are needed to distinguish the 2; stems 1–11 cm tall, usually with 2–4 internodes; basal leaves absent or few; $n = 16$. Grassy areas, roadsides, bottomlands, woodlands; included based on range map of Terrell (1996); apparently mainly e TX w to e part of nc TX. Feb–Apr. *Houstonia micrantha*, *H. pusilla*, and *H. rosea* can all be found growing together, but apparently do not hybridize, possibly due to differences in chromosome number (Terrell 1996). [*Hedyotis crassifolia* Raf. var. *micrantha* Shinners, *Hedyotis australis* W.H. Lewis & D.M. Moore]

Houstonia parviflora Holz ex Greenm., (small-flowered). Stems 3–15(–20) cm tall; internodes usually 3–8; leaves 5–22(–30) mm long, 0.5–3.5(–4.2) mm wide; corollas white with tips of lobes sometimes pink, 0.8–2.5 mm long. Grassy areas and bare spots; included based on citation of vegetational areas 4 and 5 (Fig. 2) by Hatch et al. (1990); mainly c and s TX; according to Terrell (1996), this species extends north only to Travis Co. just s of nc TX; endemic to TX. Mid-Feb–mid-Apr. [*Hedyotis greenmanii* Fosberg]

Houstonia pusilla Schoepf, (very small), TINY BLUET, LOW BLUET, STAR-VIOLET, SMALL BLUET,

SOUTHERN BLUET, INNOCENCE, ANGEL-EYES, QUAKER-LADIES. Stems with 2–5 internodes; basal leaves absent or few; *n* = 8. Grassy areas, roadsides, bottomlands, open woods; se and e TX w to West Cross Timbers, also Edwards Plateau. Feb–Apr. [*Hedyotis crassifolia* Raf.]

Houstonia rosea (Raf.) Terrell, (rose-colored), ROSE BLUET. Stems with 2–4 internodes; basal leaves sometimes present, similar but slightly larger than the stem leaves; *n* = 7. Grassy areas, roadsides, open woods; se and e TX w to West Cross Timbers, also Edwards Plateau and Trans-Pecos. Feb–early Apr. [*Hedyotis rosea* Raf.]

Houstonia subviscosa (C. Wright ex A. Gray) A. Gray, (somewhat sticky), NODDING BLUET. Stems erect to spreading, 3–20(–30) cm tall; internodes 4–9 or more; leaves 5–25 mm long, 0.2–1.5(–3) mm wide; corollas white or rarely light pink, 1.5–3 mm long. Sandy soils, prairies, roadsides; Milam Co. on the e margin of nc TX, also Erath Co. (Terrell 1996); in parts of e 1/2 of TX; endemic to TX. Mar–May(–Jun). [*Hedyotis subviscosa* (C. Wright ex A. Gray) Shinners]

Houstonia longifolia Gaertn., (long-leaved), LONG-LEAF HOUSTONIA or SLENDER-LEAF BLUET, is cited by Hatch et al. (1990) (as *H. nuttalliana* Fosberg) as occurring in vegetational area 5 (Fig. 2); however, according to Terrell (1996), this species only occurs to the e and n of TX. It is a perennial—either cespitose or with slender rhizomes—in contrast to all other nc TX *Houstonia* species which are annuals; it also has flowers several to many in cymes.

OLDENLANDIA

A genus of ca. 300 species of annual or perennial herbs of warm to tropical areas of the world, especially Africa; some are important as sources of dyes. (Named for H.B. Oldenland, ?-1699, a Danish botanist and superintendent of the Dutch East India Company's garden in Cape Town)
REFERENCES: Terrell 1975, 1990, 1991; Verdcourt 1976.

Oldenlandia boscii (DC.) Chapm., (for its discoverer, Louis Augustin Guillaume Bosc, 1759–1828, French naturalist). Perennial with prostrate or spreading stems to 30 cm long; leaves linear, 10–25 mm long, 1–3 mm wide, glabrous; flowers 4-merous, solitary or in small clusters; corollas white, the lobes can be tipped with pink, shorter than calyx lobes; capsule to 2.5 mm wide; seeds numerous. Edges of streams and ponds, ditches; Henderson and Milam cos. on the e margin of nc TX; mainly se and e TX, also Edwards Plateau. May–Aug. [*Hedyotis boscii* DC.]

RICHARDIA MEXICAN-CLOVER

Ours annual or perennial herbs; flowers in mostly terminal, leafy-bracted clusters or heads; corollas funnelform or salverform, white; fruits usually separating into 3 indehiscent, 1-seeded segments.

A mainly tropical American genus of 15 species. (Named for Richard Richardson, 1663–1741, British botanist and physician at Yorkshire)
REFERENCE: Lewis & Oliver 1974.

1. Corollas 5–6 mm long, usually (4–)6-lobed; leaf blades ovate to elliptic-lanceolate, 20–60 mm long, up to ca. 19 mm wide; internodes often much > 2 cm long ____ **R. scabra**
1. Corollas 2–4 mm long, 3–4-lobed; leaf blades linear to lanceolate, 25 mm or less long, ca. 5 mm or less wide; internodes usually < 2(–3) cm long ____ **R. tricocca**

Richardia scabra L., (rough), ROUGH MEXICAN-CLOVER. Short-pilose annual to 85 cm tall; older plants becoming bushy-branched, partly decumbent; corollas usually (4–)6-lobed. Weed of roadsides and gardens; Dallas Co.; mainly se and e TX, also Edwards Plateau. Jun–Oct. Native of Brazil.

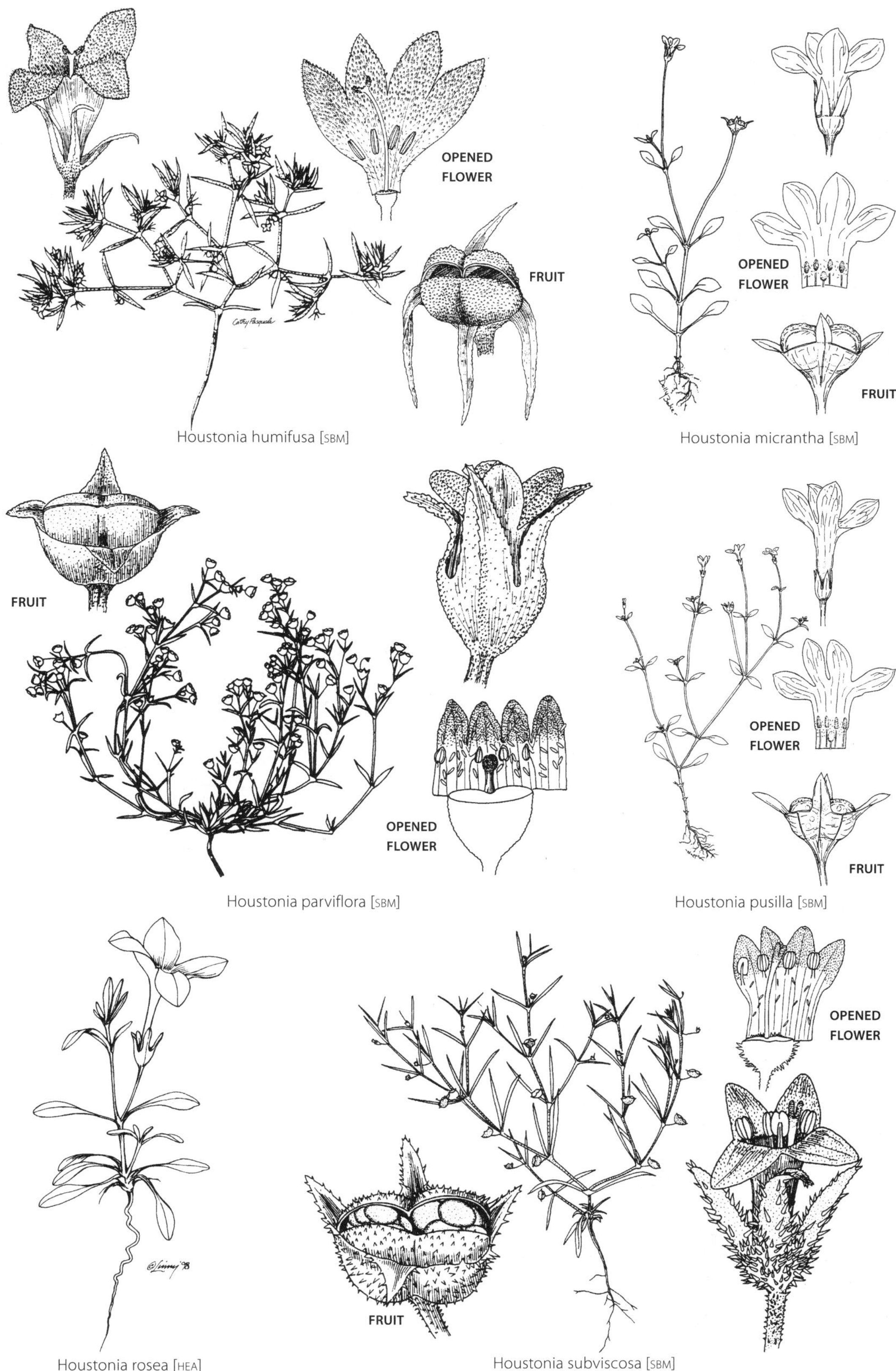

Houstonia humifusa [SBM]

Houstonia micrantha [SBM]

Houstonia parviflora [SBM]

Houstonia pusilla [SBM]

Houstonia rosea [HEA]

Houstonia subviscosa [SBM]

Richardia tricocca (Torr. & A. Gray) Standl., (with three berries, from the fruit separating into three segments), PRAIRIE BUTTONWEED, PRAIRIE MEXICAN-CLOVER. Low perennial often forming mats 30 cm or more across; corollas 3–4-lobed. Brushy areas or open woods, in sand or sandy-clay; Bell and McLennan cos.; mainly s TX n to s part of nc TX. Mar–Nov. Sometimes treated in the genus *Diodia* [as *D. tricocca* Torr. & A. Gray].

SHERARDIA SHERARD

A monotypic genus native to Europe, the Mediterranean, and w Asia. (Named for Dr. William Sherard, 1659–1728, patron of Johann Jacob Dillenius and friend of John Ray)

Sherardia arvensis L., (pertaining to cultivated fields), SPURWORT, HERB SHERARD, FIELD-MADDER. Annual with slender reddish roots; stems prostrate or decumbent, to ca. 40 cm long; leaves acute to acuminate apically; flowers 4–8 in terminal heads subtended by involucres of 8–10 lanceolate leaf-like bracts; sepals 4–6; corollas pink to lavender, 4–5 mm across, the tube twice the length of the 4 or 5 lobes; stamens 4 or 5; fruits ca. 4 mm long, of 2 one-seeded segments, with short appressed bristles. Disturbed areas, lawn weed, rapidly spreading and becoming a lawn pest; Dallas, Grayson, Red River, and Williamson cos., also Hunt, Somervell, and Tarrant cos. (R. O'Kennon, pers. obs.); se and e TX w to nc TX. Mar–May. Introduced from the Old World.

SPERMACOCE BUTTONWEED, BUTTONPLANT

A genus of 150 species of warm areas of the Americas. (Greek: *sperma*, seed, and *acoce*, a point, probably from the pointed calyx teeth on the fruit)

Spermacoce glabra Michx., (smooth, hairless), SMOOTH BUTTONWEED. Rhizomatous, glabrous, low perennial; stems freely branching, erect or partly decumbent, to 60 cm or more long; leaves short-petioled; leaf blades lanceolate to elliptic, to 8 cm long, to 1–2 cm wide; corollas white, 4-lobed, 2–3 mm long; fruits 2–4 mm long, splitting into 2 carpels, 1 open on inner surface, the other closed. Stream banks, ditches, and damp ground; collected at Dallas by Reverchon, but not found so far w recently; mainly se and e TX. May–Oct.

RUTACEAE CITRUS OR RUE FAMILY

Perennial herbs, shrubs, or small trees; leaves alternate, simple or compound; leaf blades entire or toothed, usually firm, thickish, aromatic, with oil dots (glands) usually visible when backlit; flowers terminal or lateral, perfect or unisexual, solitary or in racemes, cymes, or short broad panicles; sepals 3–5; petals 3–5; stamens 3–20; pistils 1–3, ovary superior, raised on a disk-like projection of the receptacle; fruits various.

A medium-large (1,800 species in 156 genera), cosmopolitan, but especially tropical and warm temperate family of mainly trees and shrubs or rarely herbs including the economically important citrus fruits (genus *Citrus* native to se Asia and Malay Peninsula); they are sometimes thorny and often have bitter terpenoids or alkaloids; the tissues are usually aromatic due to the presence of essential oils. Some species are cultivated for their essential oils (e.g., *Ruta* for oil of RUE) or as ornamentals. Old World *Dictamus albus* L. (GASPLANT, BURNINGBUSH) has so many glands releasing volatile inflammable oil vapors that in hot still weather the vapors can be ignited with a flash. Citrus fruits, including GRAPEFRUIT, LEMON, LIME, and ORANGE, are a type of specialized berry known as a hesperidium; the individual segments that we eat represent single carpels. The TEXAS RED GRAPEFRUIT, *Citrus* ×*paradisi* (L.) Macfad. [*C. maxima* × *C. sinensis*] (cultivar "Ruby" (redblush)), was designated the **state fruit of Texas** in 1993; while this cultivar was developed in TX, the parents are introduced (Jones et al. 1997). ☠ *Citrus* species contain furocoumarins that can photosensitize the skin making it much more sensitive to sun-

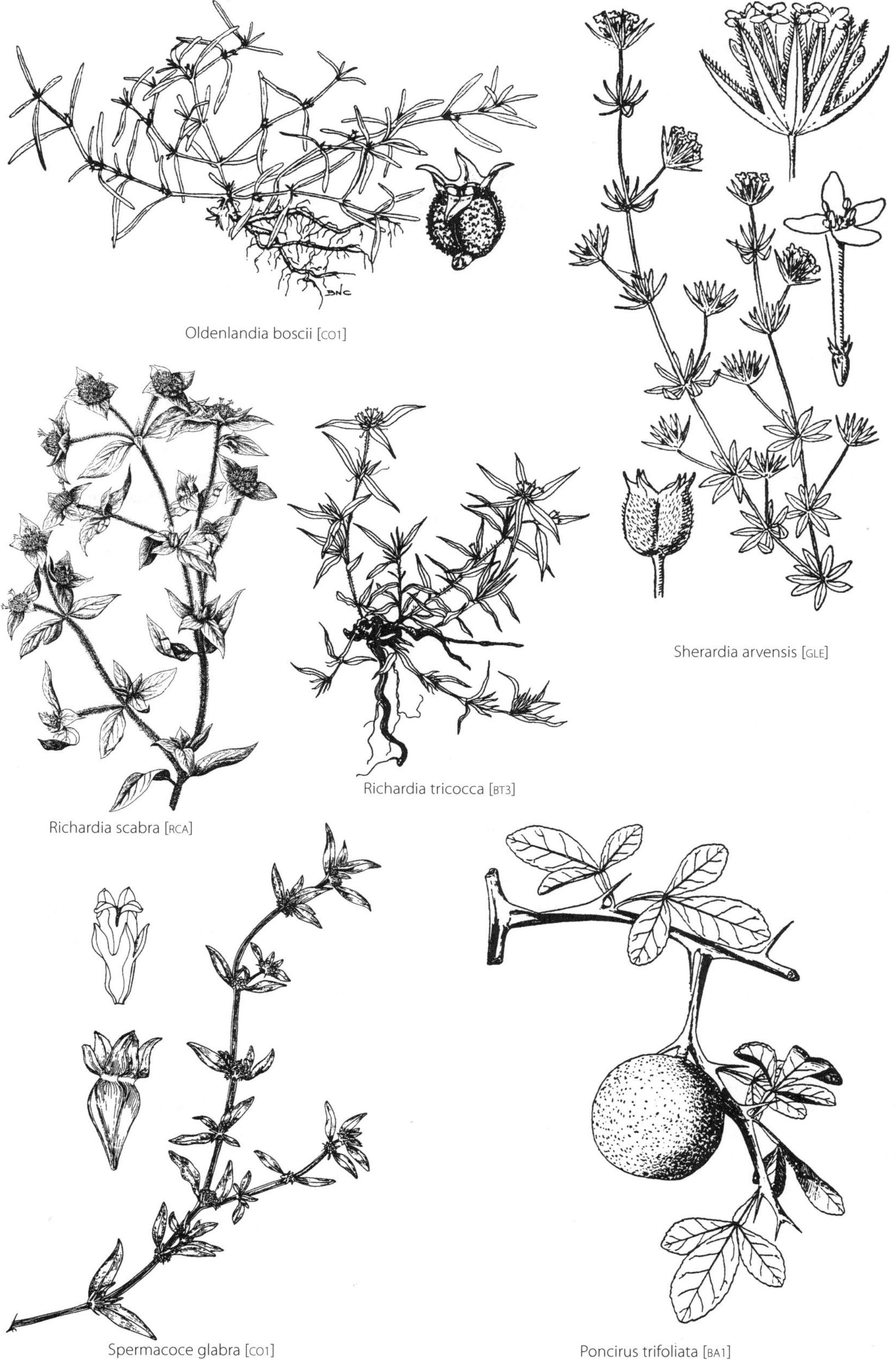

Oldenlandia boscii [CO1]

Sherardia arvensis [GLE]

Richardia tricocca [BT3]

Richardia scabra [RCA]

Spermacoce glabra [CO1]

Poncirus trifoliata [BA1]

light; the result is that sun exposure can sometimes result in reddening, swelling, and blistering (Fuller & McClintock 1986; D.M. Eggers-Ware, pers. comm.); after handling citrus fruits, the hands should be washed before extensive sun exposure; other Rutaceae (e.g., *Ptelea, Thamnosma*) are also capable of causing such phytophotodermatitis (Lampe 1986; L. Woodruff, pers. comm.). According to Lampe (1986), "... the affected skin becomes pigmented and may remain so for many months. Sometimes precise leaf patterns can be seen as dark tattoos on the skin." Family name from *Ruta*, RUE, a genus of 7 species native from Macaronesia and the Mediterranean to sw Asia; it has been cultivated since ancient times for its strong flavor and for medicinal uses (due to the presence of ethereal oils); ☠ however, it can cause dermatitis in some individuals. (Latin: *ruta*, rue, a bitter herb, bitterness, unpleasantness, in reference to the bitter taste) (subclass Rosidae)

FAMILY RECOGNITION IN THE FIELD: usually shrubs or trees (1 can be herbaceous) with alternate, scented, *gland-dotted* leaves (glands easily seen when the leaves are *held up to a strong light*).

REFERENCES: Wilson 1911; Brizicky 1962a.

1. Herbs or subshrubs usually less than 30 cm tall, unarmed; leaves simple, linear ____ **Thamnosma**
1. Shrubs or small trees, armed with thorns or spines OR unarmed; leaves compound, the leaflets broader than linear.
   2. Leaves pinnately compound with 5–17 leaflets; fruit a small, leathery, 1-seeded follicle 5–9 mm long ____ **Zanthoxylum**
   2. Leaves with 3 leaflets; fruit either flat and winged OR like a small orange.
      3. Plants unarmed; fruit a flat, circular, wind-dispersed samara with a thin wing all around, 22–28 mm across ____ **Ptelea**
      3. Plants usually armed with stout straight thorns; fruit orange-like, round or subglobose, ca. 40 mm in diam. ____ **Poncirus**

PONCIRUS BITTER ORANGE, TRIFOLIATE ORANGE

A monotypic genus of c and n China; related to, hybridizes with, and sometimes placed in the Asiatic genus *Citrus* which includes the cultivated citrus fruits such as GRAPEFRUIT, LEMON, LIME, and ORANGE. (French: *poncire*, for a kind of citron)

Poncirus trifoliata (L.) Raf., (three-leaved), BITTER ORANGE, TRIFOLIATE ORANGE. Usually armed shrub or small tree, the armature often conspicuous; thorns often flattened basally; leaves with 3 leaflets; leaflets elliptic or obovate, gland-dotted and aromatic; flowers solitary, perfect; petals 4–5, whitish; stamens 15 or more; fruits like a small orange. Cultivated, escaped and naturalized in low woods at base of bluffs along the Red River, Grayson Co., also along White Rock Creek in Dallas Co. (R. O'Kennon, pers. obs.) and in Tarrant Co. (L. Woodruff, pers. comm.); mainly e and se TX. Mar–Apr. Native of c and n China. [*Citrus trifoliata* L.] Fruits fragrant, acidic; reportedly used for marmalade; also reported to be toxic possibly due to saponins (Mabberley 1987; Lampe & McCann 1985). ☠

PTELEA WAHOO, HOPTREE

A North American genus of ca. 11 species of trees and shrubs containing alkaloids. (Greek: *ptelea*, name for an elm, used by Linnaeus for this genus with superficially similar winged fruits)

REFERENCE: Bailey 1962.

Ptelea trifoliata L., (three-leaved), HOPTREE, WAFER-ASH, SKUNKBUSH, COLA DE ZORRILLO. Shrub or small tree 1–4 m tall; leaflets 3, broadly to narrowly elliptic-lanceolate, entire, gland-dotted, aromatic; flowers in terminal panicles, mostly unisexual by abortion of one sex, usually with vestigial parts of other sex; petals 4–5, greenish yellow, small. Hillsides, rocky ground, sandy woods, and woods along streams; nearly throughout TX. Apr. The fruits can be used as a hops substitute in brewing beer. Reported to contain an alkaloid and a poisonous saponin (Burlage 1968); the species is also known to cause phytophotodermatitis (Lampe 1986—see discussion in family synopsis). There are a number of infraspecific taxa which intergrade and are sometimes distinguished with difficulty; the following 2 occur within nc TX: ☠

1. Leaflets glabrate, with glands 0.15–0.25 mm across; fruits often 3-carpellate; seed part of fruit often 2(–3) mm thick, sometimes located below middle of wings of fruit ____ subsp. **angustifolia** var. **persicifolia**
1. Leaflets pubescent, with glands < 0.1 mm across; fruits mostly 2-carpellate; seed part of fruit usually < 1 mm thick, located near center or above middle of wings of fruit ____ subsp. **trifoliata** var. **mollis**

subsp. **angustifolia** (Benth.) V.L. Bailey var. **persicifolia** (Greene) V.L. Bailey, (subsp.: narrow-leaved; var.: peach-leaved). Burnet, Coryell, Dallas, Grayson, McLennan, and Tarrant cos., also Erath, Hood, and Palo Pinto cos. (Bailey 1962); nc TX s to Edwards Plateau. [*P. persicifolia* Greene]

subsp. **trifoliata** var. **mollis** Torr. & A. Gray, (var.: soft, soft-hairy), WOOLLY HOPTREE. Bell, Tarrant, and Williamson cos., also Comanche (HPC) and McLennan (Bailey 1962) cos.; se and e TX and Edwards Plateau n to nc TX. [*P. mollis* M.A. Curtis, *P. tomentosa* Raf., *P. trifoliata* var. *mollis* Torr. & A. Gray]

THAMNOSMA DUTCHMAN'S-BREECHES

A genus of 11 species of herbs or shrubs of sw North America, s and c Africa, s Arabia, and the island of Socotra (L. Woodruff, pers. comm.). (Greek: *thamnos*, shrub or bush, and *osma*, smell or odor)

Thamnosma texanum (A. Gray) Torr., (of Texas), DUTCHMAN'S-BREECHES, TEXAS DESERT-RUE, RUDA DEL MONTE. Perennial; leaves strongly gland-dotted, aromatic; inflorescence a raceme or racemose cyme; flowers 4-merous, perfect; petals yellow or purple and creamy white; fruit a leathery 2-lobed capsule 3–7 mm long, the 2 lobes resembling the inflated legs of a dutchman's breeches. Rocky slopes, grasslands, on limestone; Bell, Shackelford, and Somervell cos., also Brown (Stanford 1971), Burnet, Coleman, and Dallas (L. Woodruff, pers. comm.) cos.; s and w parts of nc TX, s and w to w TX. Feb–Apr (also in fall). While the epithet is often spelled *texana*, according to L. Woodruff (pers. comm.) who is studying the genus, the correct spelling is *texanum*. This species contains coumarins and can have a phototoxic effect on sheep (Oertli et al. 1983; Dominguez et al. 1984); it can also cause phytophotodermatitis in susceptible humans (L. Woodruff, pers. comm.). ☠

ZANTHOXYLUM PRICKLY-ASH, PEPPERBARK

Ours shrubs or small trees, dioecious or polygamous; older bark with short, sharp spines from conspicuous, cushion-like bases, the younger parts with slender-based, curved stipular spines; leaves odd-pinnately compound; leaflets 5–17, toothed to almost entire, gland-dotted; rachis often with prickles; flowers small, in short, wide, usually terminal panicles, 5-merous; petals yellow-green; carpels 2–5, each maturing into a dry follicle splitting open to reveal a single glossy black seed.

A genus of ca. 250 species of aromatic trees and shrubs native to the Americas, Africa, Asia, and Australia. Chewing a small amount of bark from the twigs or a leaflet gives the tongue a tingling, numb sensation; some have been used medicinally; during frontier days, TX species were used for toothache or applied externally for rheumatism (Crosswhite 1980). The name has sometimes been incorrectly spelled as "Xanthoxylum." (Greek: *xanthos*, yellow, and *xylon*, wood)
REFERENCES: Fosberg 1958, 1959; Brizicky 1962b; Porter 1976.

1. Leaflets of flowering branches 9–17, 3.5–9 cm long, acute or acuminate; leaves of flowering branches 10–30 cm long ____ **Z. clava-herculis**

1. Leaflets of flowering branches 5–11, 1–3.5 cm long, obtuse or subacute; leaves of flowering branches 2.5–12 cm long ______ **Z. hirsutum**

Zanthoxylum clava-herculis L., (club of Hercules), PRICKLY-ASH, SOUTHERN PRICKLY-ASH, TICKLETONGUE, TOOTHACHETREE, PEPPERBARK, HERCULES'-CLUB. Shrub, becoming a round-headed tree, 1.5–7 m tall; twigs and leaf rachises pubescent to glabrous; leaflets with indistinct, forward-pointing teeth; inflorescences 6–15 cm long, naked (= without flowers) in basal 1/4 to 1/2; ovaries 1, rarely 2 or 3. Sandy or rocky ground; se and e TX w to Grand Prairie. Apr–early May.

Zanthoxylum hirsutum Buckley, (hairy), PRICKLY-ASH, TICKLETONGUE, TOOTHACHETREE. Shrub or bushy tree to 5 m tall; twigs and leaf rachises densely to sparsely spreading-pubescent (latter extreme is the common form in nc TX); leaflets with shallow, rounded teeth; inflorescences 1–7 cm long, naked in basal 1/4 or less, commonly flower-bearing or branching nearly from base; ovaries 1 or often 2. Sandy or rocky ground; Bosque, Brown, Burnet, Montague, Parker, and Tarrant cos., also Grayson (Little 1976 [1977]) and Hamilton (HPC) cos.; mainly w part of nc TX w to Rolling Plains and sw to w TX. Apr–early May. According to Porter (1976), this species appears to hybridize with *Z. clava-herculis* where their ranges overlap in c and n TX.

SALICACEAE WILLOW FAMILY

Shrubs or trees; bark bitter; leaves alternate, simple, entire or toothed, deciduous, often with basilaminar glands (= glands at base of leaf blade); stipules minute and falling early (or on summer shoots or suckers leafy and persistent); flowers each in axil of a small bract, grouped in erect to pendulous catkins (= aments), unisexual, the sexes on separate plants (= dioecious); perianth absent; basal disk or 1–2 glands present; stamens (1–)2–many; pistil 1, ovary superior; fruit a many-seeded capsule; seeds tufted with long, silky, white hairs.

A medium-sized (ca. 435 species in 2 genera) nearly cosmopolitan, but mainly n hemisphere family of trees and shrubs, sometimes creeping, with wind-dispersed seeds. The number of species is questionable due to hybridization in *Salix*. Species are variously used for timber, pulp, medicinally, or as ornamentals. While *Salix* and *Populus* are somewhat different morphologically, their biochemical similarity and evolutionary relationship is reflected in the fact that the cottonwood borer (*Plectrodera scalator*) utilizes species of both genera for food (Linsley & Chemsak 1984); this is large (25–40 mm body length) long-horned beetle with a striking black and white checkered pattern; the conspicuous antennae are often nearly as long as or longer than the body. The Salicaceae are related to the tropical and subtropical family Flacourtiaceae (Judd et al. 1994). (subclass Dilleniidae)
FAMILY RECOGNITION IN THE FIELD: shrubs or trees with alternate simple leaves with petioles often flattened or with glands at junction of petiole and blade; flowers unisexual, apetalous, in catkins; fruit a small capsule with *hairy, wind-dispersed* seeds.

1. Leaf blades nearly as wide as long or wider, rhomboid-ovate to deltoid-ovate or deltoid-reniform, abruptly narrowed to rounded-truncate, subcordate, or widely cuneate base; leaf buds covered by several imbricate bud scales; floral bracts with deeply and unevenly cut margins; flowers with saucer-like basal disk; stamens usually many; stigma with several wide lobes or fleshy branches ______ **Populus**
1. Leaf blades many times longer than wide, linear to lanceolate, usually gradually narrowed at base; leaf buds covered by a single bud scale; floral bracts entire or toothed; flowers without fleshy basal disk (but with 1 or 2 enlarged glands); stamens usually 2–8; stigma 2-lobed ______ **Salix**

POPULUS COTTONWOOD, ÁLAMO, POPLAR, ASPEN

Fast growing, often short-lived trees with soft wood; petioles sometimes conspicuously laterally compressed (= flattened), especially just below leaf blades, or petioles round; aments usually pendulous (= drooping), scentless; stamens up to ca. 60.

A n temperate genus of 35 species of wind-pollinated trees. A number are used as ornamental shade and street trees, for shelter-belts, or for pulp, plywood, and excelsior; they are among the fastest growing trees in the n temperate zone. The common name POPLAR is said to have been derived from their growing around public squares and meeting-places (Mabberley 1997). The Spanish common name for POPLAR is ÁLAMO. It is thought by some that the historically famous Alamo of San Antonio got its name from a nearby grove of *Populus. Populus tremuloides* Michx. (QUAKING ASPEN), native to far w TX, may be the most widespread tree in North America, ranging from n Mexico throughout much of the U.S. and Canada to Alaska (Correll & Johnston 1970); its leaves quake or tremble in the slightest breeze due to the flexible flattened petioles; COTTONWOOD leaves do the same to a certain extent. (Classical Latin name for plants of this genus)
REFERENCES: Correll 1961b; Sudworth 1934; Eckenwalder 1977, 1986; Sokal et al. 1986.

1. Mature leaves conspicuously persistently white tomentose (= with densely matted hairs) on lower surface; young twigs and buds white tomentose; petioles rounded or nearly so ____ **P. alba**
1. Mature leaves glabrous on lower surface; young twigs and buds not white tomentose (at most pubescent); petioles distinctly flattened.
   2. Leaf blades usually truncate (= squared-off) or cordate at base, with the same light green color on both surfaces, often with 2 or more prominent glands on upper surface where blade joins petiole; trees with an open crown, not columnar-shaped; native species ____ **P. deltoides**
   2. Leaf blades usually widely cuneate (= wedge-shaped) at base, bright green above and paler on lower surface, without glands on upper surface where blade joins petiole; trees with a very slender, erect, columnar-shaped crown; persistent or escaped cultivar ____ **P. nigra**

Populus alba L., (white), WHITE POPLAR, SILVER-LEAF POPLAR, ÁLAMO BLANCO, ABELE. Tree to ca. 30 m tall with open crown, spreading by root sprouts; leaf blades deltoid-ovate, with small lobes or coarsely toothed, conspicuously white tomentose beneath; petioles 2–4 cm long; our plants apparently all pistillate, but not maturing seeds. Commonly cultivated, persists, and spreads vegetatively; in some instances its aggressive spread can be problematic; Grayson Co., also Erath, Lamar (Correll 1961b), Dallas, and Tarrant (R. O'Kennon, pers. obs.) cos.; e TX w to nc TX, also Trans-Pecos. Mar–May. Native of Europe and Asia. The bark of this species can be rather striking—that of the lower trunk is light gray and furrowed, that of the upper trunk and branches white and smooth with gray areas where branches were formerly attached.

Populus deltoides Bartram ex Marshall, (triangular), COTTONWOOD. Tree to 30 m or more tall with open crown; leaf blades deltoid (= triangular) to deltoid-ovate or deltoid-reniform, finely to coarsely crenate-serrate at least in middle portion of margins, glabrous; petioles 2.5–9 cm long; fruits 3–15 mm long. Along streams, in ditches or draws, around stock tanks, invading bare ground that is damp at seed-time; also cultivated. Mostly mid-Mar–early Apr. Under good conditions COTTONWOODS can grow extremely rapidly—up to 5 feet of height per year; however, like many fast-growing species, they are relatively short-lived (Cox & Leslie 1991). The leaves flutter in the breeze (due to the flexible flattened petioles) like those of QUAKING ASPEN.

1. Leaf blades usually with 10–20 coarse or fine teeth on each side; bud scales glabrous; glands present on upper blade surface where petiole attaches ____ subsp. **deltoides**
1. Leaf blades usually with 5–10 coarse teeth on each side; bud scales pubescent to rarely glabrous; glands absent or present on upper blade surface where petiole attaches ____ subsp. **monilifera**

subsp. **deltoides**, EASTERN COTTONWOOD, ÁLAMO, ALAMO COTTONWOOD. Roughly e 1/2 of TX w to Wichita Co. in Rolling Plains (Correll 1961b; Simpson 1988).

subsp. **monilifera** (Aiton) Eckenw., (bearing a necklace), TEXAS COTTONWOOD, PLAINS COTTONWOOD. Panhandle s and e to Clay, Eastland, Stevens, and Young cos., also in Cooke and Montague cos. It forms hybrid swarms with subsp. *deltoides* (Simpson 1988). Hatch et al. (1990) indicated this taxon comes e to the Blackland Prairie. [*P. deltoides* var. *occidentalis* Rydb., *P. sargentii* Dode, *P. texana* Sarg.]

Populus nigra L., (black), BLACK POPLAR, LOMBARDY POPLAR, SAUCE. Tree to 30 m tall with columnar crown (sometimes very narrow), spreading by root sprouts; leaf blades rhomboid-ovate to broadly deltoid-ovate, finely crenate-serrate along most of margin, glabrous; petioles 3–5 cm long. Commonly cultivated, persists, and spreads vegetatively (plants in nc TX apparently male only); included based on citation of vegetational area 4 (Fig. 2) by Hatch et al. (1990); scattered in TX. Mar–May. Native of Europe and Asia. This species is commonly attacked by pests (including cotton root rot and long-horned beetles) and will die out in a few years (J. Stanford, pers. comm.).

SALIX WILLOW, OSIER, SALLOW

Shrubs and trees typically of moist to wet areas; aments erect, ascending or spreading, usually not pendulous (= drooping).

A genus of ca. 400 species (Mabberley 1997; however, Argus 1997 estimated ca. 450 species), primarily in cold and temperate areas of the n hemisphere, with a few in the s hemisphere. WILLOWS vary from 30 m trees to dwarf shrubs only 1–2 cm tall in tundra regions. Glands in the flowers secrete nectar and attract insect pollinators. Native peoples in both the Old and New worlds chewed *Salix* twigs as a pain reliever. Salicylic acid, similar to the now synthesized acetylsalicylic acid (aspirin), is found in the twigs and was used medicinally for pain relief by native inhabitants. Species are variously cultivated as ornamentals for their form (weeping) or branch color (sometimes yellow or orange), used in basketry because of their pliable branches, or cut for timber. (Classical Latin name for plants of this genus)
REFERENCES: Sudworth 1934; Ball 1961; Dorn 1976; Argus 1986, 1997; Brunsfeld et al. 1992.

1. Plants with flowers and immature leaves.
   2. Aments only 1 per peduncle; stamens 2–12; mature capsules 1.5–6 mm long; pistils glabrous; trees to 20 m or more tall or non-colonial shrubs; widespread in nc TX.
      3. Twigs not conspicuously elongate or pendulous; stamens 4–12 but usually 6; fruiting aments 20–100 mm long, 10–20 mm wide; mature capsules 4.5–6 mm long; bud apex sharp-pointed; bud scale margins free and overlapping; floral bracts deciduous after flowering; native species.
         4. Young unfolding leaves green on both sides; widespread species probably in every county in nc TX ____ **S. nigra**
         4. Young unfolding leaves pale to white-glaucous beneath; known in nc TX only from Hood, Johnson, and Somervell cos. ____ **S. caroliniana**
      3. Twigs conspicuously elongated and pendulous ("weeping willow"); stamens 2; fruiting aments 10–20 mm long, 5–10 mm wide; mature capsules 1–2.5 mm long; bud apex blunt; bud scale margins fused; floral bracts persistent after flowering; introduced species ____ **S. babylonica**
   2. Aments (especially staminate) often 2–4 per peduncle, some developing later at base of the first; stamens 2; mature capsules 5–9 mm long; pistils glabrous or pubescent; colonial shrubs often forming thickets; in nc TX known only from sandy bottoms of Red and Sulfur rivers ____ **S. exigua**
1. Plants past flowering, with mature leaves.
   5. Leaves linear-lanceolate to lanceolate, up to 10–12 times longer than wide, the margins usually with ± closely and evenly spaced teeth; petioles sometimes with glands or dots near base of blade; trees to 20 m or more tall or non-colonial shrubs; widespread in nc TX.

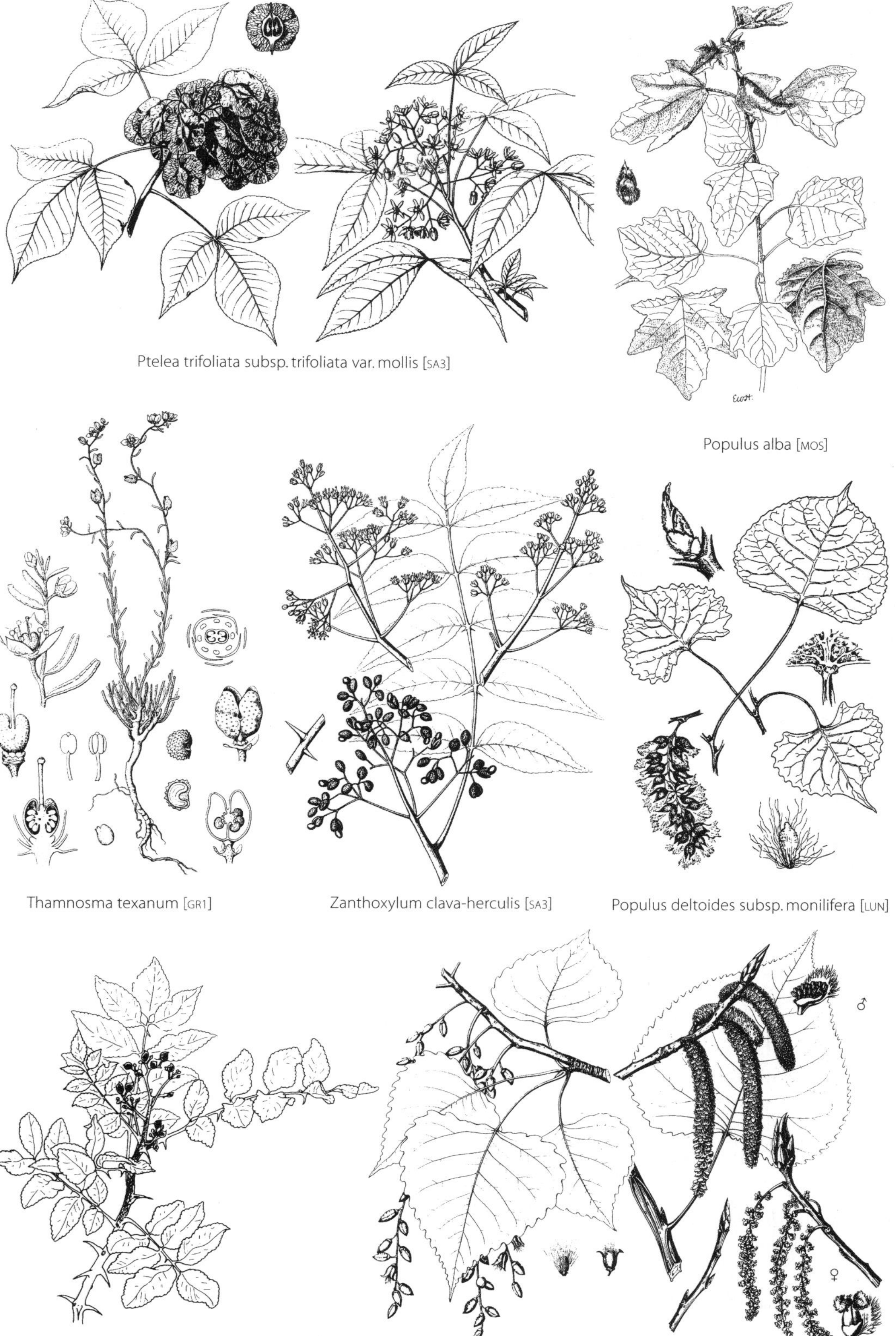

Ptelea trifoliata subsp. trifoliata var. mollis [SA3]

Populus alba [MOS]

Thamnosma texanum [GR1]

Zanthoxylum clava-herculis [SA3]

Populus deltoides subsp. monilifera [LUN]

Zanthoxylum hirsutum [LYN]

Populus deltoides subsp. deltoides [SA3]

6. Leaves green beneath (upper and lower leaf surfaces ± similar in appearance); widespread native species probably in every county in nc TX ________ **S. nigra**
6. Leaves pale to white-glaucous beneath (upper and lower surfaces conspicuously dissimilar); native species known in nc TX only from Hood, Johnson, and Somervell cos. OR escaped cultivated "weeping willow."
    7. Twigs not conspicuously elongated or pendulous (not weeping), reddish brown to yellowish brown, usually pubescent; leaf blades dark green above; bud apex sharp-pointed; bud scale margins free and overlapping ________ **S. caroliniana**
    7. Twigs conspicuously elongated and pendulous (="weeping"), yellowish to yellowish brown, glabrous; leaf blades yellowish green above; bud apex blunt; bud scale margins fused ________ **S. babylonica**

5. Leaves linear to linear-lanceolate, usually 8–20 times longer than wide, the margins subentire to with widely and unevenly spaced teeth; petioles without glands near base of blade; colonial shrubs often forming thickets; in nc TX known only in sandy bottoms of Red and Sulphur rivers ________ **S. exigua**

Salix babylonica L., (of Babylon), WEEPING WILLOW, BABYLON WEEPING WILLOW. Tree to ca. 10(–20) m tall; twigs/branchlets very long, slender and pendulous (weeping); buds flat-beaked, blunt; petioles eglandular or with 2 small dots near base of blade. Widely cultivated, persisting and possibly escaping; included based on citation of vegetational area 4 (Fig. 2) by Hatch et al. (1990), also Hunt Co. (R. O'Kennon, pers. obs.). Mar–Apr. Native to Asia with center of distribution in China. Linnaeus named this species based on the mistaken assumption that it was the willow of Babylon (actually a poplar) referred to in the Bible (Moldenke & Moldenke 1952; Vines 1960).

Salix caroliniana Michx., (of Carolina), COASTAL PLAIN WILLOW, CAROLINA WILLOW, LONG-PEDICEL WILLOW, SOUTHERN WILLOW, WARD'S WILLOW. Non-colonial shrub or tree to ca. 12 m tall; trunks 1–few; twigs reddish brown to yellowish brown; bud apex sharply pointed; bud scale margins free and overlapping; petioles with glandular dots or processes near base of blade; stipules usually prominent and persistent; stamens 4–12, usually 6. Along rocky streams; Hood, Johnson, and Somervell cos. (Little 1976 [1977]), also Edwards Plateau. Mar–Apr. This species is closely related to *S. nigra* and hybridization occurs in some areas where their ranges overlap; Argus (1986) concluded that because they are clearly separable over most of their ranges and because of the limited nature of hybridization, the 2 should be recognized as separate species.

Salix exigua Nutt., (little, small, poor), SANDBAR WILLOW, COYOTE WILLOW, NARROW-LEAF WILLOW, LONG-LEAF WILLOW, TEXAS SANDBAR WILLOW, TARAY, RIVERBANK WILLOW, OSIER WILLOW, SHRUB WILLOW, BASKET WILLOW. Multiple-stemmed rhizomatous shrub to 4(–5.5) m tall, colonial and often forming dense thickets; twigs yellowish brown to reddish brown; bud apex blunt; leaf blades at first silvery with appressed pubescence, later glabrous, linear or linear-lanceolate, 3–10 mm wide (to 15 mm wide on new shoots), the margins subentire or usually with teeth unevenly and widely spaced (teeth near middle of leaf blade 1.2–7 mm apart); petioles absent or up to 3 mm long (to 6 mm long on new shoots), eglandular; floral bracts deciduous after flowering. Damp sandy river bottoms; in nc TX in counties along the Red River—Grayson Co., also Clay, Cooke, and Montague cos. (Little 1976 [1977]), also Delta Co. along the Sulphur River; widely scattered in TX. Apr., sporadically later. Numerous infraspecific taxa of *S. exigua* and *S. interior* have been lumped into *S. exigua*. [*S. interior* Rowlee]

Salix nigra Marshall, (black), BLACK WILLOW, SAÚZ, LINDHEIMER'S BLACK WILLOW, SWAMP WILLOW, WESTERN BLACK WILLOW, SCYTHE-LEAF WILLOW, PUSSY WILLOW. Tree to 20 m or more tall; twigs reddish brown to yellowish brown; bud apex sharply pointed; bud scale margins free and overlapping; leaf blades lanceolate, 5–20 mm wide, non-glaucous beneath (rarely thinly glaucous), the margins with teeth evenly close-set (teeth near middle of leaf blade 0.3–2.1 mm apart); petioles 2–6 mm long (to 15 mm long on new shoots), usually with glands near base of blade; stipules usually small and caducous; stamens 3–5. Stream banks, ditches, tanks, and low ground; widespread in TX, probably in every county in nc TX. Late Mar–Apr. [*S. nigra* Marsh var. *lindheimeri* C.K. Schneid.]

SANTALACEAE SANDALWOOD FAMILY

A medium-sized (540 species in 34 genera), nearly cosmopolitan, especially tropical and warm dry area family of hemiparasitic trees, shrubs, and herbs; they usually parasitize the roots of other species. Family name from *Santalum*, a genus of 8–9 species native from Indomalesia to Australia, Hawaii, and Juan Fernandez. It includes the Asian *Santalum album* L., SANDALWOOD-TREE, known for its fragrant timber and sandal oil used medicinally, in making perfumes, cosmetics, and soaps, and as incense for Buddhist, Hindu, and Muslim religions. (Persian: *shandal*, name for the sandalwood-tree) (subclass Rosidae)

FAMILY RECOGNITION IN THE FIELD: the single TX species is a herbaceous perennial root hemiparasite (green and photosynthetic); leaves alternate, simple, entire; flowers small, with a single white perianth whorl; fruits small, drupe-like.

COMANDRA BASTARD TOADFLAX

A genus of a single species composed of 4 subspecies, 3 in North America, 1 Mediterranean. (Greek: *come*, hair, and *aner*, man, in allusion to the hairs of the sepal lobes which adhere to the anthers)

REFERENCE: Piehl 1965.

Comandra umbellata (L.) Nutt. subsp. **pallida** (A. DC.) Piehl, (sp.: with umbels; subsp.: pale), BASTARD TOADFLAX, WESTERN COMANDRA. Glabrous and glaucous, low (to 30 cm tall), herbaceous, rhizomatous, perennial hemiparasite; parasitizing the underground parts of other plants via modified roots (= haustoria); leaves alternate, sessile, lanceolate or narrowly oblong-lanceolate, green and photosynthetic; flowers in small, compact, terminal, rounded, corymbose inflorescences; sepals 5, petaloid, white, united below into a green tube, the lobes with epidermal hairs on inner surface that adhere to the anthers; petals 0; stamens 5, with a 5-lobed fleshy disk around their bases; pistil 1; ovary half inferior; fruits ovoid to nearly round, indehiscent, 1-seeded, drupe-like, ca. 4–8 mm long. Sandy or rocky open ground; Cooke and Montague cos. in nw part of nc TX s and w to w TX. Late Mar–early May. [*C. pallida* A. DC.] This species is known to parasitize over 200 species of hosts (Piehl 1965); its fruits are reported to be sweet and edible (Mabberley 1987).

SAPINDACEAE SOAPBERRY FAMILY

Shrubs, trees, or climbing vines; leaves alternate, compound; leaflets entire, toothed, or lobed; ours with flowers in lateral fascicles or in terminal panicles, radially symmetrical or bilaterally symmetrical, perfect or unisexual; sepals 4–5; petals 4–5, white to pink or purplish pink; stamens 7–10; ovary superior; fruit a capsule or berry.

A medium-large (1,450 species in 131 genera) family of tropical and warm areas with a few in temperate regions; it includes trees, shrubs, lianas, and herbaceous climbers; ☠ toxic saponins are usually present. Several are important for their fruits (e.g., *Nephelium* (RAMBUTAN) and *Litchi* (LYCHEE)) or for timber; the family also includes ornamentals such as the frequently cultivated, *Koelreuteria paniculata* Lam., GOLDENRAIN-TREE or CHINATREE, with odd-pinnate leaves, coarsely toothed leaflets, bright yellow flowers, and an inflated, membranous capsule.

While the nc TX species of this family (including the cultivated *Koelreuteria*) are morphologically very dissimilar (e.g., annual vines to large trees), their biochemical similarity and evolutionary relationship is reflected in the fact that the soapberry bug (*Jadera haematoloma*) utilizes seeds of all the species (except possibly *Ungnadia*—no information available) as a food source (K. Walker, pers. comm.). The Sapindaceae are closely related to the Hippocastanaceae and Aceraceae and are probably paraphyletic when treated separately. The genus *Aesculus*, here treated in the Hippocastanaceae, is often included in the Sapindaceae. From a cladistic standpoint these families should be lumped to form a more inclusive monophyletic Sapindaceae (Judd et al. 1994). (subclass Rosidae)

FAMILY RECOGNITION IN THE FIELD: trees, shrubs, or vines with alternate compound leaves; stamen number variable, but often 8; fruit a berry or capsule, sometimes inflated.

REFERENCES: Brizicky 1963b; Judd et al. 1994.

1. Vines; fruit a membranous, inflated, balloon-like, 3-lobed capsule; flowers white, in small axillary inflorescences with tendrils ____ **Cardiospermum**
1. Trees or shrubs; fruit a fleshy berry or a somewhat woody 3-lobed capsule; flowers pink to purple or white (if white, in large terminal inflorescences); no tendrils present.
   2. Flowers white, in large, dense, terminal panicles; leaflets entire; leaves usually even-pinnate (the terminal leaflet usually absent); fruit an amber or yellowish, 1-seeded, fleshy berry persisting over the winter ____ **Sapindus**
   2. Flowers pink to purplish pink, in lateral fascicles (= clusters); leaflets sharply toothed; leaves odd-pinnate (with a terminal leaflet); fruit a somewhat woody capsule ____ **Ungnadia**

CARDIOSPERMUM BALLOONVINE, HEARTSEED

A genus of 14 species of climbers with inflated, balloon-like fruits, mostly of the tropics, especially in the Americas. (Greek: *cardia*, heart, and *sperma*, seed, referring to the white, heart-shaped spot on the seed)

Cardiospermum halicacabum L., (old generic name, as *Halicacabus*), COMMON BALLOONVINE, FAROLITOS, HEARTSEEDVINE. Annual vine sprawling and trailing, with axillary tendrils; leaves alternate, usually twice ternately compound; leaflets toothed and lobed; flowers about 4 mm long; sepals 4, 2 large, 2 small; petals 4, obovate, often unequal; stamens 8; fruits membranaceous, inflated (balloon-like), 3-lobed, 3-celled, 3–4.5 cm broad; seeds 5 mm in diam., black at maturity, with conspicuously lighter tissue attaching them inside the fruit. Disturbed areas, sometimes extremely abundant; in e 1/2 of TX. Jun–Nov. Cultivated as an ornamental vine.

SAPINDUS SOAPBERRY

A genus of ca. 13 species of tropical and warm areas of the world; all are cultivated as ornamentals. (Latin: *sapo*, soap, and *indicus*, Indian, alluding to the use of the fruits of *S. saponaria* as a soap by Native Americans)

Sapindus saponaria L. var. **drummondii** (Hook. & Arn.) L.D. Benson, (sp.: soapy; var.: for its discoverer, Thomas Drummond, 1780–1835, Scottish botanist and collector in North America), WESTERN SOAPBERRY, WILD CHINABERRY, JABONCILLO. Broad-topped tree to 10(–15) m tall; leaves usually even- or occasionally odd-pinnately compound; leaflets 10–18(–19), curved-lanceolate, glabrous to soft-pubescent beneath; flowers radially symmetrical, white; stamens 8 or 10; fruits globose, ca. 13 mm in diam., with amber or yellowish, translucent, foul-tasting flesh around the single seed; seed 8–9 mm long. Stream bottoms, forest margins, disturbed areas; scattered throughout TX. Late May–Jun. [*S. drummondii* Hook. & Arn.] The plant contains saponins (with soap-like properties) and the fruits have been used as a soap substitute; the fruits are considered poisonous and there are reports of contact dermatitis in some individuals from han-

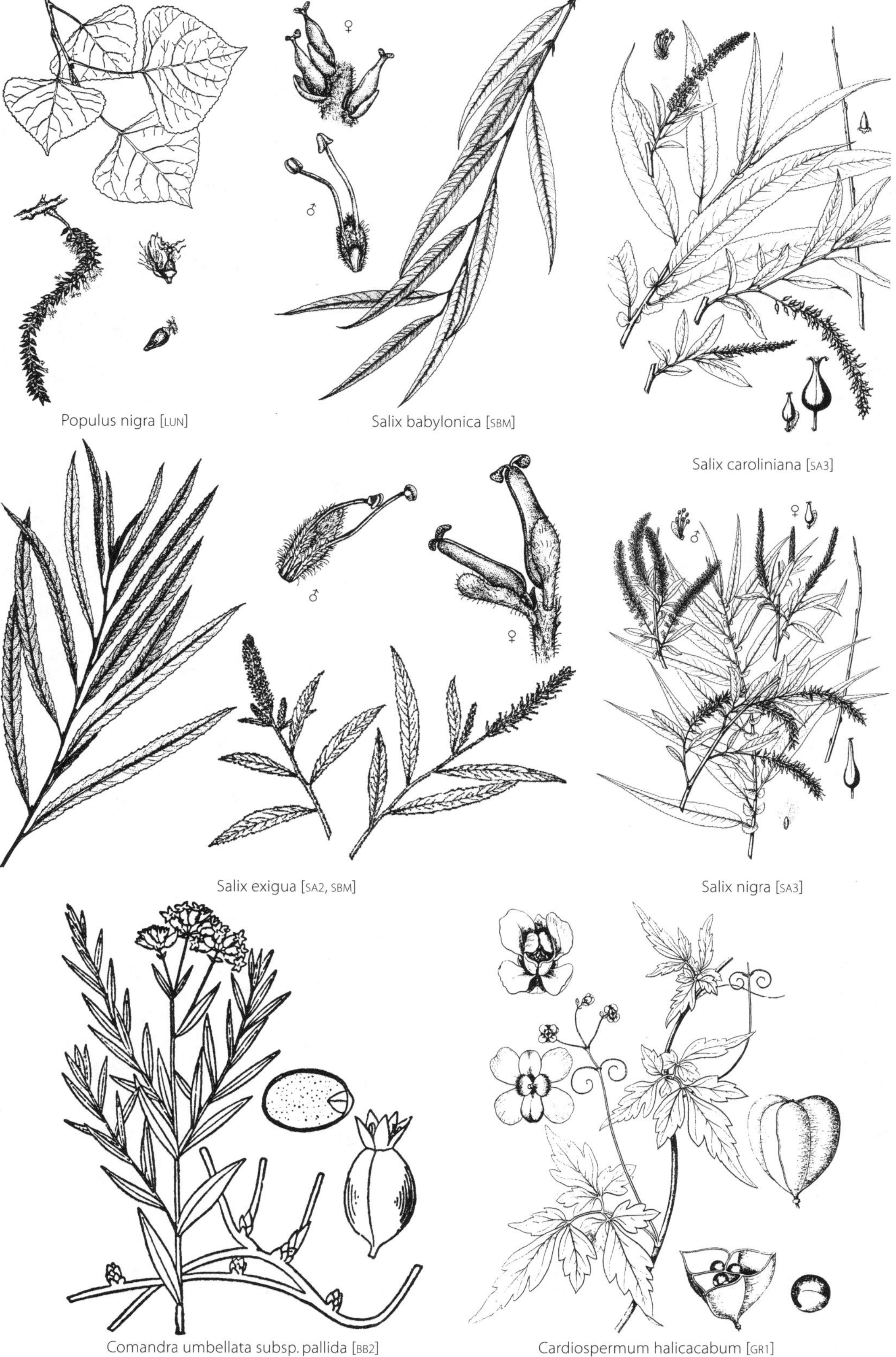

Populus nigra [LUN]

Salix babylonica [SBM]

Salix caroliniana [SA3]

Salix exigua [SA2, SBM]

Salix nigra [SA3]

Comandra umbellata subsp. pallida [BB2]

Cardiospermum halicacabum [GR1]

dling the fruits; the crushed fruits have been used to poison fish (Lampe & McCann 1985; Fuller & McClintock 1986; Lampe 1986; McGregor 1986). ☠

UNGNADIA MEXICAN-BUCKEYE, TEXAS-BUCKEYE

A monotypic genus of s North America. (Named for Baron Ungnad, who in 1576 introduced the COMMON HORSE-CHESTNUT to w Europe by sending seeds to Clusius at Vienna)

Ungnadia speciosa Endl., (showy, good-looking), MEXICAN-BUCKEYE, TEXAS-BUCKEYE, MONILLA. Shrub or small tree rarely to 10 m tall; leaves with (3–)5–9, ovate-lanceolate, prominently veined, glabrous leaflets (the apical 2–3 leaflets sometimes partly united); flowers appearing before or with the leaves, fragrant; petals 4 or 5, pink to purplish pink, drying darker, obovate, with pilose claw; stamens 7–10, anthers red; fruits 3-lobed, 3.5–5 cm thick, pale green, often with some reddish coloring; seeds 10–15 mm in diam., dark-brown to blackish. Limestone outcrops, rocky areas; Bell, Cooke, and Dallas cos., also Brown, Burnet, Coleman, Hamilton, Mills (HPC), Comanche, Coryell, McLennan, Tarrant, and Williamson (Little 1976) cos.; nc TX s and w across much of the w 1/2 of TX. Apr. Seeds poisonous (Burlage 1968; Correll & Johnston 1970). ☠ ▣/108

SAPOTACEAE
CHICLE, SAPOTE, OR SAPODILLA FAMILY

A medium-sized (975 species in 53 genera) mainly tropical family of trees and shrubs with a few in temperate regions; a well-developed latex system bearing white latex is usually present. A number of the tropical species are cauliflorous (= flowers on the trunk) and bat-pollinated. Old World *Palaquium* species were famous as the source of gutta-percha, a non-elastic rubber-like substance (polymer of isoprene) once widely used in industry and still in dentistry for temporary fillings. Family name conserved from *Sapota*, a genus now treated as *Manilkara*, a group of 70 tropical species with milky latex; the genus includes *M. zapota* (L.) P. Royen, SAPODILLA, the source of chicle (the elastic component of early chewing gums) and an edible fruit. (Name derived from the South American native name for sapodilla) (subclass Dilleniidae)

FAMILY RECOGNITION IN THE FIELD: the single nc TX species can be recognized as shrubs or small trees with leaves simple, entire, oblanceolate to elliptic, alternate or bunched on spur shoots; branchlets usually spine-pointed; flowers small, in axillary clusters; fruits small, purplish black, drupe-like.

REFERENCES: Cronquist 1946; Wood & Channell 1960; Pennington 1990, 1991.

SIDEROXYLON CHITTAMWOOD, IRONWOOD

A tropical genus of 75 species; the New World taxa are sometimes segregated as the genus *Bumelia*. *Sideroxylon sessiliflorum* (Poir.) Aubrév. of Mauritius was supposedly dispersed by the now extinct dodo [flightless bird]; turkeys have been force-fed the seeds and germination has been improved. (Greek: *sidero*, iron, and *xylon*, wood, in reference to the hard wood)

REFERENCES: Clark 1942; Cronquist 1945.

Sideroxylon lanuginosum Michx. subsp. **oblongifolium** (Nutt.) T.D. Penn., (sp.: woolly, downy; subsp.: oblong-leaved), CHITTAMWOOD, COMA, GUM BUMELIA, WOOLLY-BUCKTHORN, GUM-ELASTIC. Shrub or small tree to 15 m or more tall, usually with spine-pointed branchlets; leaves alternate or bunched on spur branchlets, very short-petioled, simple; leaf blades oblanceolate to elliptic, entire, obtuse, stiff and leathery, thin to densely white to gray or tawny cobwebby pubescent beneath; flowers axillary, small, in umbel-like clusters; calyces cup-like, with 5 orbicular-ovate, widely overlapping sepals; corollas slightly exceeding the calyces, yellowish white, with cylindrical tube and 5 orbicular-ovate main lobes overlapping and alternating with short spreading ones; calyces and corolla tubes densely pubescent; stamens 5, not exserted; staminodes 5, alternating with corolla lobes and nearly equaling the corolla lobes in length; pistil 1; ovary superior; fruits obovoid to broadly ellipsoid or subglobose, usually purplish black, 7–12 mm long. Woods, stream banks, hillsides, and rocky areas; e 1/2 of TX. May–Jul. [*Bumelia lanuginosa* (Michx.) Pers. var. *oblongifolia* (Nutt.) R.B. Clark, *S. lanuginosum* subsp. *albicans* (Sarg.) Kartesz & Gandhi] According to Tyrl et al. (1994), this species is a good bee-tree (in terms of honey production). Burlage (1968) reported that the fruits are edible but cause digestive disturbances and dizziness if eaten in quantity. Because of the apparant lack of consistent differences, we follow Jones et al. (1997) in including subsp. *albicans* within subsp. *oblongifolium*; Barker (1986b) also lumped var. *albicans* with var. *oblongifolia*. Clark (1942) separated the 2 (as varieties) as follows:

1. Pubescence dense, whitish tomentose; pedicels slender; seeds ellipsoid ____ var. *albicans*
1. Pubescence sparser, whitish becoming tawny; pedicels stoutish; seeds obovoid ____ var. *oblongifolia*

SARRACENIACEAE PITCHER PLANT FAMILY

A very small (14 species in 3 genera) family of pitcher plant-type carnivorous plants of mainly North America and the n part of South America. Insects are lured into the pitchers by strong odors, nectar secretion, and usually reddish coloration on the pitchers. Inside the pitcher is a slick "slide zone," down-pointing hairs to prevent escape, and a pool of liquid containing enzymes which digest the insect after drowning. As with other carnivorous plants, nutrients (especially nitrogen), rather than calories, are obtained through carnivory. (subclass Magnoliidae)
FAMILY RECOGNITION IN THE FIELD: obvious *pitcher-like* leaves containing liquid that drowns insects; flowers solitary, down-facing, at the end of a naked scape.
REFERENCES: Wood 1960; Thanikaimoni & Vasanthy 1972; Bayer et al. 1996.

SARRACENIA PITCHER PLANT, TRUMPET

A genus of 8 species of e North America. (Named for Dr. Michel Sarrasin de l'Étang, 1659–1734, Quebec collector and physician)
REFERENCES: Bell 1949; McDaniel 1971; Schnell & Krider 1976; Pietropaolo & Pietropaolo 1986.

Sarracenia alata A.W. Wood, (winged), YELLOW TRUMPET, PITCHER PLANT. Insectivorous, rhizomatous perennial herb; leaves yellowish green, hollow, tubular-trumpet-shaped, to 70 cm long, rigidly erect, partially filled with liquid (which digests insects), with a ridge on 1 side and a terminal hood; flowers solitary, nodding at end of a long naked scape; sepals 5, 4–5 cm long; petals 5, greenish yellow, drooping, 5–6 cm long, to 4 cm wide near the rounded tips; stamens numerous; ovary superior; style disk expanded; fruit a capsule. Boggy areas; included based on label information for *Xyris baldwiniana* (*Correll & Wasshausen 27469*, BRIT/SMU) indicating occurrence of PITCHER PLANT in c Henderson Co. near e margin of nc TX. Mar–Apr. [*S. sledgei* Macfarl.] PITCHER PLANT, two *Utricularia* species (BLADDERWORTS, Lentibulariaceae), and *Drosera brevifolia* Pursh (ANNUAL SUNDEW—Droseraceae) are the only carnivorous plants in nc TX.

SAURURACEAE LIZARD'S-TAIL FAMILY

Rhizomatous perennial herbs; leaves alternate, simple, entire, with a distinct petiole; inflorescences dense, spike-like, subtended by involucral bracts or not so, with numerous small bisexual flowers, each flower subtended by a very small bract; perianth absent; stamens 3–8; ovary superior or inferior; fruit a capsule, sometimes somewhat succulent.

The Saururaceae is a very small (6 species in 4 genera) family of perennial, often aromatic herbs of temperate and subtropical North America and e Asia; some are cultivated as ornamentals. It is closely related to the mostly tropical family Piperaceae (PEPPER family, source of black pepper) (Judd et al. 1994), but the two families are apparently both monophyletic (Tucker et al. 1993). Saururaceae and Piperaceae have a number of characters linking them to monocots (e.g., scattered vascular bundles and monosulcate pollen). Some taxonomists refer to the family as "paleoherbs" (a group including Aristolochiales, Piperales, and Nymphaeales) and believe them to be an early branch off the evolutionary line leading to monocots (Zomlefer 1994). This view is supported by molecular data which place the paleoherbs as the immediate sister group of the monocots (Chase et al. 1993) (see Fig. 6 in Appendix 41). (subclass Magnoliidae)

FAMILY RECOGNITION IN THE FIELD: often wet area herbs with alternate, simple, entire leaves; flowers small, individually inconspicuous, but numerous in *dense, spike-like* inflorescences; colored bracts sometimes subtending inflorescences.

REFERENCES: Wood 1971; Cheng-Yih & Kubitzki 1993; Tucker et al. 1993; Buddell & Thieret 1997.

1. Leaves mostly basal (stem with a few small leaves subtended by sheathing bract-like leaves); leaf blades truncate or cordate basally, ± pinnately veined; inflorescence subtended by 6–8 white or reddish involucral bracts ____ **Anemopsis**
1. Leaves mostly cauline (scattered up the stem); leaf blades cordate basally, ± palmately veined; inflorescence either not subtended by bracts OR subtended by 4 white bracts.
   2. Spike-like inflorescence 2.5–5 cm long, subtended by 4 white involucral bracts ca. 13 mm long; leaf blades 5–ca. 8 cm long; stems < 40 cm tall; stamens 3 ____ **Houttuynia**
   2. Spike-like inflorescence 10–20(–30) cm long, not subtended by involucral bracts; leaf blades 5–15 cm long; stems 50–100(–150) cm tall; stamens 3–8 ____ **Saururus**

ANEMOPSIS YERBA MANSA

A monotypic genus of the sw U.S. and Mexico; because of the conspicuous bracts, the inflorescence resembles a single flower. (From the genus name *Anemone* and Greek: *opsis*, likeness or similarity, alluding to the resemblance of the inflorescence to a flower of *Anemone* in the Ranunculaceae)

Anemopsis californica (Nutt.) Hook. & Arn., (of California), YERBA MANSA. Perennial herb with creeping, aromatic rootstocks; stems ca. 30 cm tall; leaves elliptic-oblong, to 15 cm long, with petioles ± as long as blades; inflorescence a dense, terminal spike 1.5–4 cm long, on a scape-like stalk; involucral bracts to 30 mm long, white or reddish; stamens 6(–8); carpels 3; fruits dry, dehiscent. Cultivated and escapes in wet areas; Dallas Co. (R. May, pers. comm.); Edwards Plateau, Trans-Pecos, and Panhandle. May–Jul. According to R. May (pers. comm.), seeds germinate exactly 5 weeks from the day they are put in water.

HOUTTUYNIA

A monotypic genus native from Japan s to mountains of Nepal and Java. (Named for Maarten Houttuyn, 1720–1798, Dutch naturalist and physician at Amsterdam)

Houttuynia cordata Thunb., (heart-shaped), Leaves ovate, basally cordate, 5-nerved from base, gland-dotted; inflorescence to ca. 4 cm long in flower, to 5 cm long in fruit; involucral bracts white, ca. 13 mm long; flowers small; stamens 3; stigmas 3; fruits dry, opening apically. Used as a ground cover, escaping cultivation in Tarrant Co. (Fort Worth Botanic Garden). May–?. Native of e Asia. The shoots are eaten as a vegetable in China (Mabberley 1987).

SAURURUS LIZARD'S-TAIL

A genus of 2 species, 1 e North America, 1 e Asia; this disjunct distribution pattern is discussed under the genera *Campsis* (Bignoniaceae) and *Carya* (Juglandaceae). (Greek: *sauros*, a lizard, and *oura*, a tail, in allusion to the slender tail-like inflorescence)

REFERENCES: Hall 1940; Baldwin & Speese 1949.

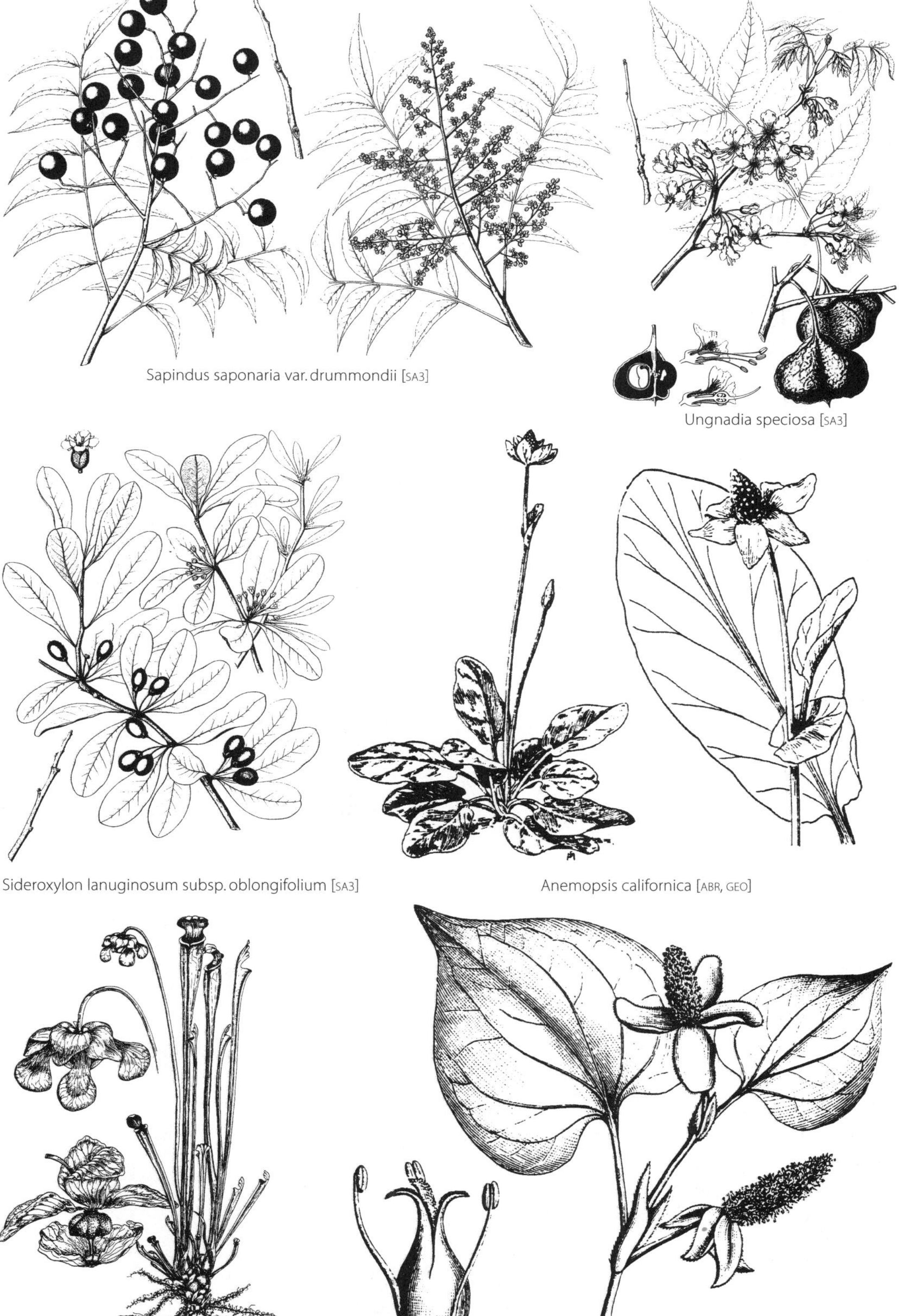

Sapindus saponaria var. drummondii [SA3]

Ungnadia speciosa [SA3]

Sideroxylon lanuginosum subsp. oblongifolium [SA3]

Anemopsis californica [ABR, GEO]

Sarracenia alata [CO1]

Houttuynia cordata [EN2]

Saururus cernuus L., (nodding), LIZARD'S-TAIL, WATER-DRAGON, SWAMP-LILY. Stems erect, 50–100(–150) cm tall, with well-developed, adventitious roots on lower half of stem; rhizomes long-creeping, aromatic; leaf blades ovate, 5–9 cm wide, cordate basally, usually longer than petioles; inflorescences terminal and/or opposite the leaves, 10–15 mm in diam., erect, the tip often drooping; peduncle 3–8 cm long; flowers small, numerous (to 300 per inflorescence), crowded, whitish; carpels 3–5; fruits somewhat fleshy, ca. 3 mm in diam., indehiscent. In water or muddy soils, swampy areas, oxbow or other lakes, streams; Henderson Co. on e margin of nc TX, also Lamar Co. in Red River drainage, also escaping cultivation in Tarrant Co. (Fort Worth Botanic Garden); se and e TX and Edwards Plateau. May–Aug.

SAXIFRAGACEAE SAXIFRAGE FAMILY

Ours annual or perennial herbs, not succulent; leaves basal or alternate, simple, entire, toothed or lobed; flowers solitary, in panicles, or cymes; sepals 5; petals 5; stamens 5–10; pistils united or partly so; ovary superior or partly inferior; fruit a capsule or follicle-like.

A medium-sized (660 species in 35 genera), almost cosmopolitan, but especially n temperate and cold area family; most are perennial herbs and many are cultivated as ornamentals. *Ribes*, here treated in the Grossulariaceae, and *Philadelphus*, here treated in the Hydrangaceae, are sometimes placed in the Saxifragaceae. The genus *Penthorum*, here treated in the Crassulaceae, is sometimes put into the Saxifragaceae (e.g., Correll & Johnston 1970; Mahler 1988); other authors have separated it into its own family the Penthoraceae. Saxifragaceae show similarities to Crassulaceae and the two families are thought by some to be related (e.g., Heywood 1993). Saxifragaceae are also apparently related to the Rosaceae. (subclass Rosidae)
FAMILY RECOGNITION IN THE FIELD: herbs often with a basal rosette and leafless flowering stem; similar to the Rosaceae (e.g., radially symmetrical, 5-merous flowers with separate petals) but *stipules usually absent*; stamens only as many as or twice as many as the petals; carpels fewer than sepals. The somewhat similar Crassulaceae can be distinguished by their typically succulent habit and the usually separate (at least above) carpels of the same number as the sepals.
REFERENCES: Small & Rydberg 1905a; Spongberg 1972; Morgan & Soltis 1993; Soltis et al. 1993; Johnson & Soltis 1995.

1. Flowers in an elongate inflorescence; plants perennial, > 2 cm tall, with basal rosette and erect, leafless, flowering stem; leaves (including petioles) usually much > 8 mm long, not entire, the margins undulate, toothed or lobed.
    2. Leaf blades at maturity usually > 3 cm long (to 15 cm), basally cordate; inflorescences to ca. 100 cm tall; fertile stamens 5; ovary 1-celled ____ **Heuchera**
    2. Leaf blades at maturity usually < 3 cm long (to 4 cm, but usually much smaller), abruptly narrowed at base, but not at all cordate; inflorescences to only ca. 20 cm tall; fertile stamens 10; ovary 3–4-celled ____ **Saxifraga**
1. Flowers solitary; plants annual, very small and inconspicuous, to ca. 2 cm tall; leaves to 8 mm long, entire ____ **Lepuropetalon**

HEUCHERA ALUMROOT

A North American genus of 55 species; some are cultivated as tufted ground-covers including *H. sanguinea* Engelm. (CORALBELLS) of sw North America. (Named for Johann Heinrich Heucher, 1677–1747, German botanist and professor of medicine)
REFERENCES: Rosendahl et al. 1936; Wells 1984.

Heuchera americana L., (of America), ALUMROOT. Scapose perennial; leaves basal, long-petiolate; leaf blades suborbicular to ovate, with 6–9 rounded lobules, also toothed, basally cordate; flowering stem leafless, glabrous to with pubescence; panicle glandular-puberulent; flowers ca. 3–5

Saururus cernuus [CO1]

Heuchera americana [LAM]

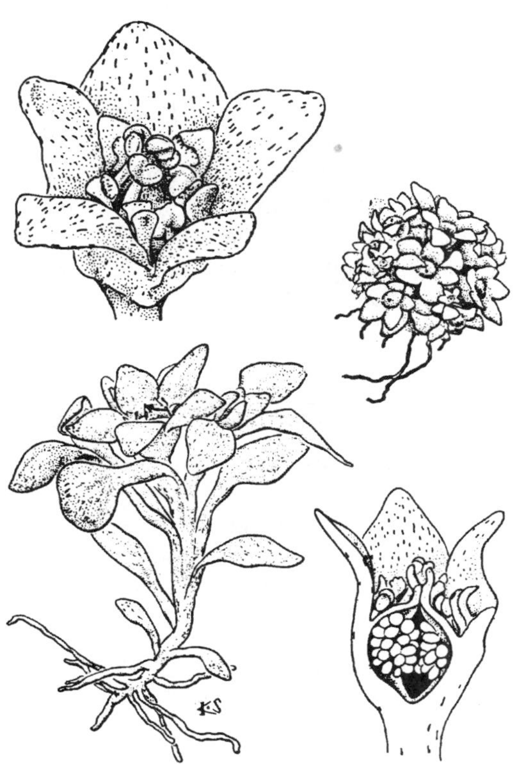

Lepuropetalon spathulatum [JAA]

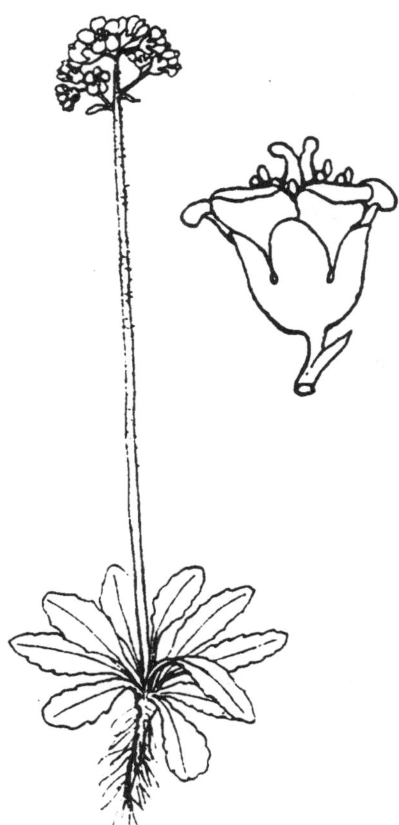

Saxifraga texana [BB2, STE]

mm long; petals greenish to pinkish to purplish; fruit a capsule ca. 3–7 mm long. Woods; included based on citation of vegetational area 4 (Fig. 2) by Hatch et al. (1990); ne TX. Apr–Jun. [*H. americana* L. var. *brevipetala* Rosend., Butters & Lakela]

LEPUROPETALON

A monotypic genus of the se U.S., Mexico, Chile, and Uruguay; some authorities recognize it as a monotypic family. (Greek: *lepro*, scaly, and *petalo*, petal, from the small, scale-like petals)
REFERENCES: Ward 1987; Ward & Gholson 1987.

Lepuropetalon spathulatum Elliott, (spoon-shaped). Very small, inconspicuous, glabrous, yellow-green annual to 2 cm tall; leaves alternate, 3–8 mm long, oblanceolate or obovate to spatulate, often with reddish glands; flowers solitary, terminal or nearly so; calyx lobes ovate, 1–2 mm long, appearing almost like the leaves; petals minute, white, or apparently absent; stamens 5; fruit a capsule. Damp sand or silt; Fannin, Grayson, Kaufman, and Lamar cos., also Hopkins and Hunt cos. (Ward & Gholson 1987); se and e TX w to ne part of nc TX, also Edwards Plateau. Mar. This species is very inconspicuous and seldom collected; it is one of the smallest terrestrial flowering plants; it is abundant on granite derived soils of the Central Mineral Region just s of nc TX (J. Stanford, pers. comm.).

SAXIFRAGA SAXIFRAGE, ROCKFOIL

A genus of ca. 440 species of the n temperate zone to the arctic, with a few to Thailand, South America, and n African mountains; usually perennial or some annual herbs many of which are cultivated as ornamentals. According to Soltis et al. (1993), molecular data indicate that *Saxifraga* may be polyphyletic, being made up of two phylogenetically well-separated clades. (Latin: *saxum*, a stone, and *frangere*, to break, the name early applied through the Doctrine of Signatures to European species bearing granular bulblets, which were supposed to dissolve kidney stones)
REFERENCE: Gornall 1987.

Saxifraga texana Buckley, (of Texas), TEXAS SAXIFRAGE. Scapose perennial; leaves basal, long-petiolate; leaf blades rhombic- or elliptic-lanceolate to ovate, abruptly narrowed basally, the margins undulate to slightly toothed; flowering stems sparsely to densely pubescent; inflorescence of cymules aggregated into a head-like cluster, sometimes becoming looser at maturity; petals white or pink, 2–3 mm long; carpels partly united, dehiscing along the inner face of each carpel. Prairies or pastures, sandy or blackland soils; Fannin, Hopkins, and Kaufman cos., also Lamar Co. (Carr 1994); se and e TX w to ne part of nc TX. Mar. [*Saxifraga reevesii* Cory]

SCROPHULARIACEAE
FOXGLOVE, FIGWORT, OR SNAPDRAGON FAMILY

Ours annual or perennial herbs; leaves basal, alternate, or opposite, simple or compound, entire, toothed, or lobed; stipules absent; flowers axillary or terminal, solitary or in whorls, spikes, racemes, or panicles; sepals separate or united; corollas nearly radially symmetrical and 4- to 5-lobed, to very bilaterally symmetrical and 2-lipped (= bilabiate); stamens 2, 4 (in some genera there is also a staminode), or rarely 5 and slightly unequal; pistil 2-carpellate; ovary superior; style 1; stigmas 1 or 2; locules mostly 2; capsules many-seeded.

A large (5,100 species in 269 genera), cosmopolitan, but especially temperate and tropical mountain family of mostly herbs with some trees and shrubs. A number of genera are grown as ornamentals including *Antirrhinum* (SNAPDRAGONS), *Calceolaria* (SLIPPER-FLOWERS), *Digitalis* (FOXGLOVE), *Leucophyllum* (e.g., *L. frutescens* (Berland.) I.M. Johnst., TEXAS PURPLE-SAGE, CENIZA), and *Penstemon* (BEARDTONGUES). *Digitalis purpurea* L., COMMON FOXGLOVE, is also the source of the heart drugs digitalin and digoxin used to treat conditions including congestive heart failure by increasing the force of systolic contractions; its therapeutic use has been traced back to the 10th century; ☠ all parts of the plant are potentially fatally poisonous to humans and animals due to the presence of cardiac glycosides (Morton 1977). Some species of Scrophulariaceae are hemiparasites and can be serious weeds in certain parts of the world (e.g., *Striga*-WITCHWEED). Many scrophs have flowers similar to those of mints; however, the capsular fruits clearly distinguish the Scrophulariaceae. The family also has affinities to the Gesneriaceae (AFRICAN-VIOLET FAMILY) and Bignoniaceae (CATALPA FAMILY). While we are treating it traditionally, according to Reeves and Olmstead (1993) and Olmstead and Reeves (1995), the Scrophulariaceae is polyphyletic; some members of the family as traditionally viewed are in a clade more closely related to the Acanthaceae, Lamiaceae, and Verbenaceae. Reeves and Olmstead (1993) and Olmstead and Reeves (1995) further indicated that the Plantaginaceae and Callitrichaceae are in a clade with some Scrophulariaceae; this would suggest that these families be included in the Scrophulariaceae. The uncertainty seen here in regard to the boundaries of families is indicative of the current dynamic nature of the study of plant evolution driven in part by recent work in molecular systematics. (subclass Asteridae)

FAMILY RECOGNITION IN THE FIELD: herbs similar to the Lamiaceae (MINTS) (e.g., the usually bilaterally symmetrical, often 2-lipped corollas and stamens usually 2 or 4) but differing in usually having *many-seeded capsule* fruits (versus fruits usually of 4 one-seeded nutlets in the Lamiaceae); the corollas and stamen number distinguish the family from the somewhat similar Solanaceae which has radially symmetrical corollas and 5 stamens.

REFERENCES: Pennell 1920, 1935; Thieret 1967; De-Yuan 1983; Holmgren 1986; Reeves & Olmstead 1993; Olmstead & Reeves 1995.

1. Upper floral bracts colored, conspicuous, red to purple, pink, orange, yellow, or white, nearly concealing the flowers ___ **Castilleja**
1. Upper floral bracts neither conspicuously colored nor concealing the flowers.
    2. Plant a twining vine ___ **Maurandya**
    2. Plant not a twining vine.
        3. Stem leaves alternate or whorled (in circle of 3–7).
            4. Stem leaves whorled ___ **Veronicastrum**
            4. Stem leaves alternate.
                5. Corollas rotate (= more or less flat, without an elongate tube), only slightly bilaterally symmetrical; fertile (= anther-bearing) stamens 5 ___ **Verbascum**
                5. Corollas not rotate (not flat, the tube obvious), strongly bilaterally symmetrical (divided into upper and lower lips); fertile stamens 4.
                    6. Corollas with a basal spur; leaves entire in outline (neither deeply parted nor pinnatifid).
                        7. Plants glabrous to puberulent OR glandular-pubescent in inflorescences only; leaves ± sessile; inflorescences quite distinct; corollas 8–33 mm long (including 2–17 mm long spur); corolla spur straight or curved.
                            8. Corollas pale bluish violet or bluish purple with whitish palate, 8–22 mm long including spur; spur very slender; lower lip of corollas longer than upper; plants annual; capsules 2–3.5 mm long ___ **Nuttallanthus**
                            8. Corollas yellow with orangish yellow palate, (19–)25–33 mm long including spur; spur stout; lower lip of corollas shorter than upper; plants perennial; capsules 5–12 mm long ___ **Linaria**
                        7. Plants glandular-pubescent nearly throughout; leaves short petiolate, at least below; inflorescences indistinct, the bracts not greatly different from leaves; corollas 4.5–9 mm long (including 1–3 mm long spur); corolla spur straight ___ **Chaenorrhinum**
                    6. Corollas without a spur; leaves deeply pinnately parted to pinnatifid ___ **Pedicularis**

3. Stem leaves opposite (bracts subtending flowers may be alternate).
  9. Leaf blades pinnatifid, at least on the lower leaves.
    10. Corollas either pale lavender OR pink to lavender with purple spots inside throat.
      11. Corollas 3–5 mm long, shorter than to scarcely exceeding calyces; calyces 2.8–5.5 mm long; plants small, to only ca. 20 cm tall ____ **Leucospora**
      11. Corollas 20–35 mm long, greatly exceeding calyces; calyces 10–16 mm long; plants to 100 cm tall ____ **Agalinis**
    10. Corollas yellow.
      12. Plants to 2 m tall; upper lip of corollas not hooded; leaves conspicuous on the stem; flowers and fruits loosely arranged with spaces separating them in the inflorescences; widespread in nc TX.
        13. Corollas 14–16 mm long; pedicels 1–4 mm long; capsules 6–11 mm long ____ **Dasistoma**
        13. Corollas ca. 40 mm long; pedicels 5–14 mm long; capsules 15–20 mm long ____ **Aureolaria**
      12. Plants 0.4 m or less tall; upper lip of corollas hooded; leaves mostly basal; flowers and fruits crowded in dense spike-like racemes; in nc TX limited to Red River drainage ____ **Pedicularis**
  9. Leaf blades entire or toothed, not pinnatifid.
    14. Plants usually mat-forming, rooting at nodes.
      15. Calyx lobes or sepals 4; corollas with 4 lobes; fruits somewhat flattened, often with a notch at the tip ____ **Veronica**
      15. Calyx lobes or sepals 5; corollas with 5 obvious or obscure lobes, the lobes either part of a 2-lipped corolla or all nearly equal; fruits not flattened, tapering to rounded at the tip, without a notch.
        16. Calyx tube shorter than lobes; corollas almost radially symmetrical, white (sometimes with yellow throat), pale blue, or lilac ____ **Bacopa**
        16. Calyx tube equal to or longer than lobes; corollas 2-lipped, yellow ____ **Mimulus**
    14. Plants erect or nearly so, not rooting at the nodes.
      17. Corollas with a small spur (but spur conspicuous when examined closely); corollas 4.5–9 mm long; rare introduced species ____ **Chaenorrhinum**
      17. Corollas without a spur; corollas variously sized but often much larger than 9 mm long; native and introduced, often widespread species.
        18. Glandular-pubescent annuals; inflorescence a dense spike-like raceme with showy flowers; corollas 20–25 mm long, white or white-lavender; calyx lobes 4, unequal ____ **Bellardia**
        18. Plants without the above combination.
          19. Flowers sessile (= without any stalk); corollas subsalverform with a very narrow, curved to straight tube 7–15 mm long; inflorescence a terminal spike well-separated from the leaves by a long, nearly naked peduncle ____ **Buchnera**
          19. Flowers on pedicels (these may be either conspicuous or short); corollas and inflorescence not as above.
            20. Calyx lobes or sepals 4; fruits often but not always with a notch at the tip ____ **Veronica**
            20. Calyx lobes or sepals 5 (apparently 7 in *Gratiola* including 2 bractlets); fruits tapering to rounded at the tip, without a notch.
              21. Fertile (= anther-bearing) stamens 2; calyces divided to base into separate sepals or nearly so; calyx tube absent or not evident.

22. Corollas blue to lavender; sterile (= without anthers) stamens 2, conspicuous; pedicels without bractlets just below calyces (calyces of 5 sepals) ______ **Lindernia**
22. Corollas white to pale lavender to yellow; sterile stamens absent or minute; pedicels with 2 bractlets just below calyces (calyces thus appearing to be of 7 sepals) ______ **Gratiola**

21. Fertile (= anther-bearing) stamens 4; calyces divided to base OR not divided to base; calyx tube absent OR present and well-developed.
    23. Plants to only ca. 20 cm tall; most of the leaves in a basal rosette or near the stem base; corollas blue with yellow to white area on lip bordered by reddish brown ______ **Mazus**
    23. Plants not as above.
        24. Flowers 2–5 in axillary whorls; lower lip of corollas with a keel-like or pouch-like central fold, each half notched at tip, purple with white base ______ **Collinsia**
        24. Flowers 1 per leaf or bract axil (bracts can be extremely reduced); lower lip of corollas not as above.
            25. Plants 0.6–3 m tall; leaves with well-developed petioles 15–60(–80) mm long; inflorescence a large open panicle of small flowers; corollas 5–10 mm long, brown or reddish brown with pale green ______ **Scrophularia**
            25. Plants usually 1 m or less tall (rarely to 2 m); leaves sessile or on petioles to only 25 mm long; inflorescence not as above; corollas 6–60 mm long, variously colored but never brown or reddish brown with pale green.
                26. Corollas 10–35 mm long, pink to lavender, purple, blue, red, or rarely white; leaves not glandular-punctate.
                    27. Calyx segments (= sepals) divided nearly to base; staminode (= sterile 5th stamen) usually conspicuous ______ **Penstemon**
                    27. Calyx segments united for part (often most) of their length into a tube; staminode absent.
                        28. Leaf blades or leaf segments usually filiform to linear, sometimes wider; lower lip of corolla without a raised palate, the corolla throat round and open; calyx tube rounded, not 5-angled ______ **Agalinis**
                        28. Leaf blades ovate-lanceolate to ovate; lower lip of corolla with 2 ridges forming a raised palate that nearly closes the throat of the corolla; calyx tube strongly 5-angled ______ **Mimulus**
                26. Corollas 6–12 mm long, yellow or white, often lined with purple; leaves minutely glandular-punctate (use lens) ______ **Mecardonia**

AGALINIS GERARDIA, AGALINIS

Erect annuals frequently blackening upon drying, often hemiparasitic (with green photosynthetic tissues but also parasitizing roots of other plants); leaves opposite or subopposite, sessile, linear to ovate in outline, entire to auricled to pinnately divided into linear segments; inflorescences racemose or spicate; bracts leafy below, reduced upwards; calyces nearly radially symmetrical; corollas bilaterally symmetrical, weakly bilabiate, pink to purple (rarely white), often with yellow lines and darker purple or red-purple spots in throat; stamens 4, didynamous; filaments pubescent; capsules usually globose or nearly so.

A genus of 40 species of tropical and warm areas of the Americas; ± parasitic on the roots of other plants. Two nc TX species were previously segregated into the genus *Tomanthera*; all were at one time placed in the no longer recognized genus *Gerardia*. (Greek: *aga*, wonder, and Latin: *linum*, flax; superficially some species thought to resemble flax)
REFERENCES: Pennell 1928, 1929; Canne 1979 [1980], 1980, 1981.

1. Leaf blades ± linear, usually entire (or sometimes lobed at base in *A. heterophylla*); calyces 3–10 mm long, the lobes 0.4–6.5 mm long, linear to subulate, slightly longer to much shorter than the tube; corollas usually with 2 yellow lines within throat; widespread in nc TX.
   2. Flowering pedicels 4 mm or less long.
      3. Calyx lobes as long as or longer than calyx tube, the lobes 3.5–6.5 mm long ____ **A. heterophylla**
      3. Calyx lobes much shorter than calyx tube, the lobes to only 2 mm long ____ **A. fasciculata**
   2. Flowering pedicels 5 mm or more long.
      4. Upper lobes of corollas very small, < 1/2 the length of the lower lobes; corollas 23–27 mm long; widespread in East and West cross timbers and Red River drainage; only in sandy soils ____ **A. homalantha**
      4. Upper lobes of corollas > 1/2 the length of the lower lobes; corollas 10–23 mm long; in nc TX known only from extreme ne part (Lamar Co.) or extreme s part (Burnet Co.); in various substrates.
         5. Upper corolla lobes reflexed spreading, not arching over the stamens; calyx tube reticulate-venose (use lens or scope); plants not darkening upon drying; known in nc TX only from Lamar Co. ____ **A. gattingeri**
         5. Upper corolla lobes arching over the stamens; calyx tube not reticulate-venose; plants often darkening or blackening upon drying; in nc TX known only from Lamar Co. or Burnet Co.
            6. Racemes relatively short, with 3–4 pairs of pedicels; inside of corollas pubescent in a narrow line below the posterior sinus; in nc TX known only from Burnet Co. on extreme s margin of area ____ **A. edwardsiana**
            6. Racemes relatively elongated, with more than 4 pairs of pedicels; inside of corollas glabrous below the posterior sinus; in nc TX known only from Lamar Co. in ne corner of area ____ **A. tenuifolia**
1. Leaf blades lanceolate to ovate in outline (can be highly divided), either usually with a pair of auricles at base or pinnately divided into linear segments; calyces 10–16 mm long, the lobes 6–10(–13) mm long, ovate to widely lanceolate, longer than the tube; corollas without yellow lines within throat; rare in nc TX.
   7. Leaf blades usually with a pair of lobes (= auricles) at base, otherwise entire; stems between the leaves (= internodes) with numerous long, spreading hairs; corollas 20–25(–27) mm long ____ **A. auriculata**
   7. Leaf blades pinnately divided into linear segments; stems retrorsely hispid, but without numerous long, spreading hairs; corollas (23–)25–33 mm long ____ **A. densiflora**

Agalinis auriculata (Michx.) S.F. Blake, (eared), EAR-LEAF GERARDIA, EARED FALSE FOXGLOVE. Plant 20–80 cm tall; stems 4-angled; leaf blades lanceolate to ovate-lanceolate, the uppermost basally auricled; calyx lobes 6–9(–12) mm long; corollas 20–25(–27) mm long, pink to purple with dark purple spots, the upper lobes longer; capsules ovoid, 1.0–1.4 cm long. Prairies, open woods; a Reverchon collection is reported for Tarrant Co. (Mahler 1988); supposedly limited in

TX to nc part. Aug–Sep. [*Gerardia auriculata* Michx., *Tomanthera auriculata* (Michx.) Raf.] (TOES 1993: V) ⚠

Agalinis densiflora (Benth.) S.F. Blake, (densely-flowered), FINE-LEAF GERARDIA. Plant to ca. 50(–80) cm tall; leaves with 3–7 narrowly acute, linear divisions; calyx lobes 7–10 mm long; corollas (23–)25–33 mm long, pink to lavender with purple spots; capsules 8–10 mm long. Prairies, dry limestone soils; Cooke, Coryell, Montague, Palo Pinto, Parker, and Tarrant cos., also Brown and Hamilton cos. (Stanford 1971); nc TX w to Rolling Plains and s to Edwards Plateau. Aug–Oct. [*Gerardia densiflora* Benth., *Tomanthera densiflora* (Benth.) Pennell]

Agalinis edwardsiana Pennell, (of the Edwards Plateau), PLATEAU GERARDIA. Plant 40–80 cm tall; stems often with reddish purple coloration; pedicels to 30 mm long; calyx lobes to 0.5 mm long; corollas 20–23 mm long, pinkish to purplish with 2 yellow lines and many small red-purple dots in throat; capsules 6–7 mm long. Thin soils on limestone; Burnet Co. (Balcones Canyonlands National Wildlife Refuge, C. Sexton, pers. comm.); also Travis Co. just s of nc TX; mainly Edwards Plateau; endemic to TX. Aug–Oct. [*A. edwardsiana* var. *glabra* Pennell, *Gerardia edwardsiana* (Pennell) Pennell]

Agalinis fasciculata (Elliott) Raf., (fascicled, in clustered), BEACH GERARDIA. Plant 40–70(–100) cm tall; pedicels 2–4 mm long at flowering time; calyx tube 3–4 mm long, the lobes acuminate; corollas rose-pink, with 2 yellow lines and many darker purple spots in throat, (16–)20–35 mm long; capsules 4.5–6(–7) mm long. Disturbed areas, woodlands; included based on citation of vegetational areas 4 and 5 (Fig. 2) by Hatch et al. (1990); se and e TX w to nc TX. Sep–Oct.

Agalinis gattingeri (Small) Small ex Britton, (for its discoverer, Augustin Gattinger, 1825–1903, TN botanist), GATTINGER'S GERARDIA. Plant to ca. 70 cm tall; calyx lobes to ca. 1.8 mm long; corollas 12–20 mm long, rose-pink with 2 yellow lines and relatively large darker purple spots in throat; capsules 4–5 mm long. Open woods; Lamar Co. in Red River drainage; mainly se and e TX. Sep–Oct. [*Gerardia gattingeri* Small]

Agalinis heterophylla (Nutt.) Small ex Britton, (various-leaved), PRAIRIE AGALINIS. Plant 30–100 cm tall; pedicels 1–3 mm long; calyx tube 3.5–5 mm long, the lobes acute; corollas 20–32 mm long, deep pink to white and lavender-tinged with 2 yellow lines and many small darker purple spots in throat; capsules 5–8 mm long. Prairies or open woodlands; se and e TX w to Bell, Grayson, and Tarrant cos. Jun–Oct. [*Gerardia heterophylla* Nutt.]

Agalinis homalantha Pennell, (smooth- or flat-flowered), FLAT-FLOWER GERARDIA. Plant 40–70 cm tall; pedicels 10–20 mm long in flower, to 30 mm in fruit; calyx tube 3–3.5 mm long, the lobes 1–1.2 mm long; corollas 23–27 mm long, lavender to pink with 2 yellow lines and small darker purple spots in throat; capsules subglobose, 6 mm long. Sandy soils; East and West cross timbers and Red River drainage; also e TX. Aug–Oct. [*Gerardia homalantha* (Pennell) Pennell] /77

Agalinis tenuifolia (Vahl) Raf. var. **leucanthera** (Raf.) Pennell, (sp.: slender-leaved; var.: white-anthered), SLENDER GERARDIA. Plant to ca. 50 cm tall; calyx lobes usually < 1(–2) mm long; corollas 10–20(–23) mm long, pink to purplish with 2 yellow lines and darker purple spots in throat; capsules 3–7 mm long. Moist open areas and low woods; Lamar Co. in Red River drainage; mainly e TX. Sep–Nov. [*Gerardia leucanthera* Raf., *Gerardia tenuifolia* Vahl var. *leucanthera* (Raf.) Shinners] Reputed to be poisonous to cattle and sheep (Burlage 1968). ☠

Agalinis aspera (Douglas ex Benth.) Britton, (rough), ROUGH GERARDIA, similar to *A. fasciculata* but with the calyx lobes acute to obtuse or nearly rounded, is reported by Hatch et al. (1990) for vegetational areas 4 and 5 (Fig. 2). *Agalinis caddoensis* Pennell, (presumably for the Caddo Lake area on the LA–TX border), STIFF-LEAF GERARDIA, reported from vegetational area 4 (Fig. 2) by Hatch et al. (1990), is similar to *A. tenuifolia* but differs in the upper corolla lobes spreading and not arching over the stamens. We have seen no TX material of these 2 species; nor did Correll and Johnston (1970). Jones et al. (1997) did not list *A. caddoensis* as occurring in TX.

AUREOLARIA FALSE FOXGLOVE, OAKLEECH

A genus of 10 species of the e U.S. with 1 in Mexico; mostly hemiparasitic on Fagaceae, but sometimes on Ericaceae. (Greek: *aureo*, golden, in reference to the striking yellow flowers)
REFERENCES: Pennell 1928; Canne 1980.

Aureolaria grandiflora (Benth.) Pennell var. **serrata** (Torr. ex Benth.) Pennell, (sp.: large-flowered; var.: serrate, saw-toothed), DOWNY OAKLEECH. Erect perennial to 1.5 m tall, hemiparasitic (with green photosynthetic tissue but also parasitizing the roots of other plants); leaves opposite; lower leaf blades ovate in outline, pinnatifid; upper leaf blades abruptly smaller than lower; bracts serrate to entire, oblong, acute; inflorescence a raceme; pedicels slender, 5–14 mm long in fruit; calyx lobes linear to lanceolate, essentially entire; corollas weakly bilabiate, yellow, ca. 4 cm long and quite showy; stamens 4, didynamous; capsules glabrous, ovoid, 1.5–2 cm long. Sandy open woodlands; Cooke, Fannin, Grayson, and Montague cos. in Red River drainage, also Tarrant (Fort Worth Nature Center), and Parker cos., also collected at Dallas by Reverchon (Mahler 1988); se and e TX w to nc TX. May–Aug. Reported to be poisonous (Burlage 1968). ☠ ▣/79

Variety *cinerea* Pennell, (ashy-gray), with upper leaf blades gradually smaller than the lower, bracts entire to shallowly serrate, lanceolate, acuminate-attenuate, and pedicels stout, 5–9 mm long in fruit—is also cited by Hatch et al. (1990) for vegetational area 4 (Fig. 2); we have seen no nc TX material.

BACOPA WATER-HYSSOP

Perennial low herbs of wet areas, succulent; leaves opposite, entire, sessile; flowers axillary, 5-merous; corollas campanulate, slightly bilaterally symmetrical; stamens 4. (Presumed to be a South American aboriginal name)

A genus of 56 species of warm areas, especially the Americas. Some are cultivated in aquaria or used medicinally.
REFERENCE: Pennell 1946.

1. Leaves 1-nerved, narrowed to wedge-shaped or narrow non-clasping base; pedicels with 2 linear bracts just below calyces, much exceeding the subtending leaves; flowers solitary per node ____ **B. monnieri**
1. Leaves palmately many-nerved, not so narrowed basally and ± clasping; pedicels without bracts, usually shorter than the subtending leaves; flowers usually 2–4 per upper node ____ **B. rotundifolia**

Bacopa monnieri (L.) Pennell, (for Loius Guilliame le Monnier, 1717–1799, French botanist, physician, and student of Jussieu), COASTAL WATER-HYSSOP. Plant mat-forming, fleshy, glabrous; leaves spatulate, to ca. 20 mm long, essentially entire; pedicels to 25 mm long; corollas white to pale blue or lilac, 8–10 mm long; capsules 5–7 mm long, shorter than sepals. In low wet areas, ditches, streams; Bell Co., apparently rare in nc TX; scattered in e 1/2 of TX. Apr–Sep.

Bacopa rotundifolia (Michx.) Wettst., (round-leaved), DISC WATER-HYSSOP. Plant mat-forming, on mud or floating; leaves suborbicular to ovate, to 35 mm long; pedicels to 20 mm long; corollas white with yellow throat, 6–10 mm long; capsules ca. as long as sepals. Mud and water of lakes and streams; Denton Co., apparently rare in nc TX; widespread in TX. May–Nov.

BELLARDIA

A monotypic genus native to the Mediterranean. (Named for C.A.L. Bellardi, 1740–1826, Italian botany professor, Turin)
REFERENCES: Lipscomb & Ajilvsgi 1982; Do et al. 1996.

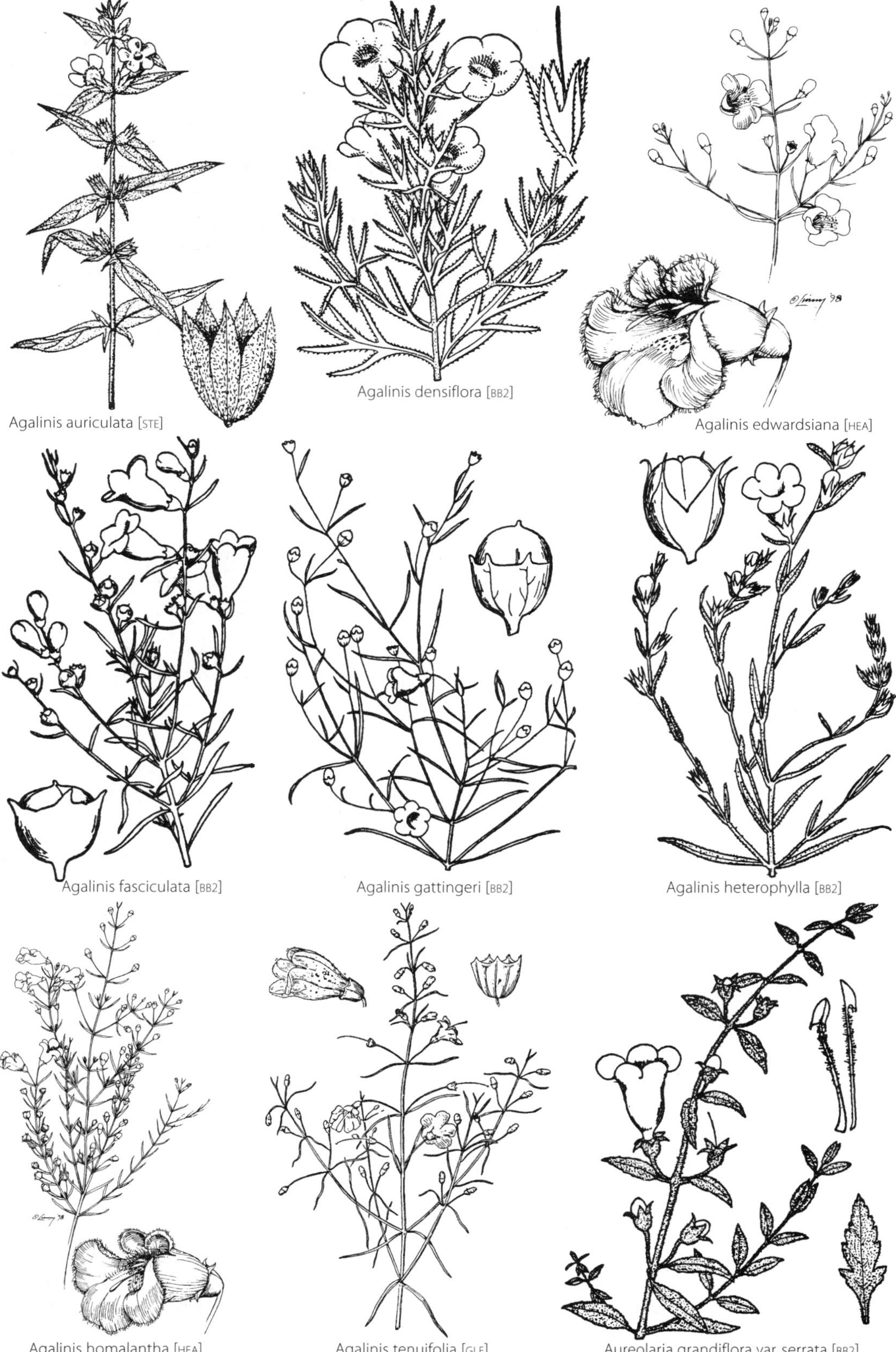

Agalinis auriculata [STE]

Agalinis densiflora [BB2]

Agalinis edwardsiana [HEA]

Agalinis fasciculata [BB2]

Agalinis gattingeri [BB2]

Agalinis heterophylla [BB2]

Agalinis homalantha [HEA]

Agalinis tenuifolia [GLE]

Aureolaria grandiflora var. serrata [BB2]

Bellardia trixago (L.) All., (Latin for germander—*Teucrium* in the Lamiaceae). Erect, usually glandular-pubescent annual 15–70 cm tall; stems unbranched below; leaves opposite; leaf blades linear to linear-lanceolate, 15–90 mm long, 1–15 mm wide, broadly and coarsely toothed; flowers in a terminal, bracteate, dense, spike-like raceme; calyces 8–10 mm long, the lobes 4, unequal, obtuse; corollas 20–25 mm long, white-lavender or white, conspicuous, the lower lip longer; capsules subglobose. Roadsides, open fields, and pastures, often in sandy soils; first collected in TX in Navarro Co. in 1980 (Lipscomb & Ajilvsgi 1982); now also known in nc TX from Henderson, Hill, Kaufman, Limestone, and Milam cos. (Do et al. 1996), also Ellis, Johnson, and Tarrant cos. (R. O'Kennon pers. obs.); other scattered localities are now known further s and e of nc TX; all other known U.S. populations are in California. Apr–May. A recent Old World invader of nc TX.

BUCHNERA BLUEHEARTS

A genus of ca. 100 species of warm areas; 16 are native to the New World; hemiparasitic. (Named for Johann Gottfried Büchner, 1695–1749, German botanist)

Buchnera americana L., (of America), AMERICAN BLUEHEARTS, FLORIDA BLUEHEARTS. Low, slender, scabrous-pilose perennial, hemiparasitic (with green photosynthetic tissues but also parasitizing the roots of other plants); stems erect, usually simple, to ca. 80 cm tall; leaves opposite or subopposite, sessile; leaf blades elliptic to linear-lanceolate, to 10 cm long, entire or with a few coarse teeth, dark green, blackening in drying, the larger prominently 3-ribbed; flowers in terminal, nearly naked spikes; corollas deep purple to lavender-pink or violet (rarely white), subsalverform, unequally 4-lobed, the lobes 3–8 mm long, the tube 7–15 mm long, straight or curved; capsules 5–7 mm long. Sandy soils, prairies, open woods, wet areas; Cooke, Erath, Grayson, Henderson, Lamar, Milam, and Montague cos.; se and e TX w to nc TX, also Edwards Plateau and Trans-Pecos. [*B. floridana* Gand.] Reported as apparently [hemi]parasitic on the roots of other plants (Ajilvsgi 1979).

CASTILLEJA INDIAN PAINTBRUSH, PAINTBRUSH, PAINTEDCUP

Low, pubescent annuals or perennials, hemiparasitic (with green photosynthetic tissues but also parasitizing the roots of other plants); leaves alternate, sessile or subsessile, narrow and entire or with narrow lobes; flowers in terminal, leafy-bracted spikes; upper floral bracts with the terminal portion colored, more conspicuous than the corollas; calyces narrowly tubular, colored at least at tip; corollas inconspicuous, narrow, 2-lipped, the upper lip long, slender, and hooded, the lower lip short and 3-lobed; stamens 4, didynamous; seeds numerous.

A genus of 200 species mainly of w North America but some in e North America, Eurasia, Central America, and the Andes. Interspecific hybridization is well known in the genus and has contributed to its taxonomic complexity (Egger 1994 [1995]). According to Crosswhite (1983), *Castilleja* species are obligately hemiparasitic. Some species concentrate selenium and can have levels that cause poisoning in animals (Crosswhite 1983). The large populations and bright colors make the PAINTBRUSHES some of the most showy wildflowers in nc TX. (Named in 1781 for Domingo Castillejo, 18th century Spanish botanist at Cadiz)
REFERENCES: Nesom 1992a; Egger 1994 [1995].

1. Bracts with conspicuous lateral lobes; plants perennial, with woody root; tips of bracts and calyces purplish pink, purplish red, purple, red, orange, brownish orange, yellowish orange, bright yellow, or greenish yellow (rarely white); leaves with long, narrow lobes ____ **C. purpurea**
1. Bracts entire or with very short lobes near the apex; plants annual, with slender taproot; tips of bracts and calyces orange-red to bright red (very rarely light yellow); leaves (except lowest) entire ____ **C. indivisa**

Castilleja indivisa Engelm., (undivided), TEXAS PAINTBRUSH, ENTIRE-LEAF PAINTBRUSH. Very

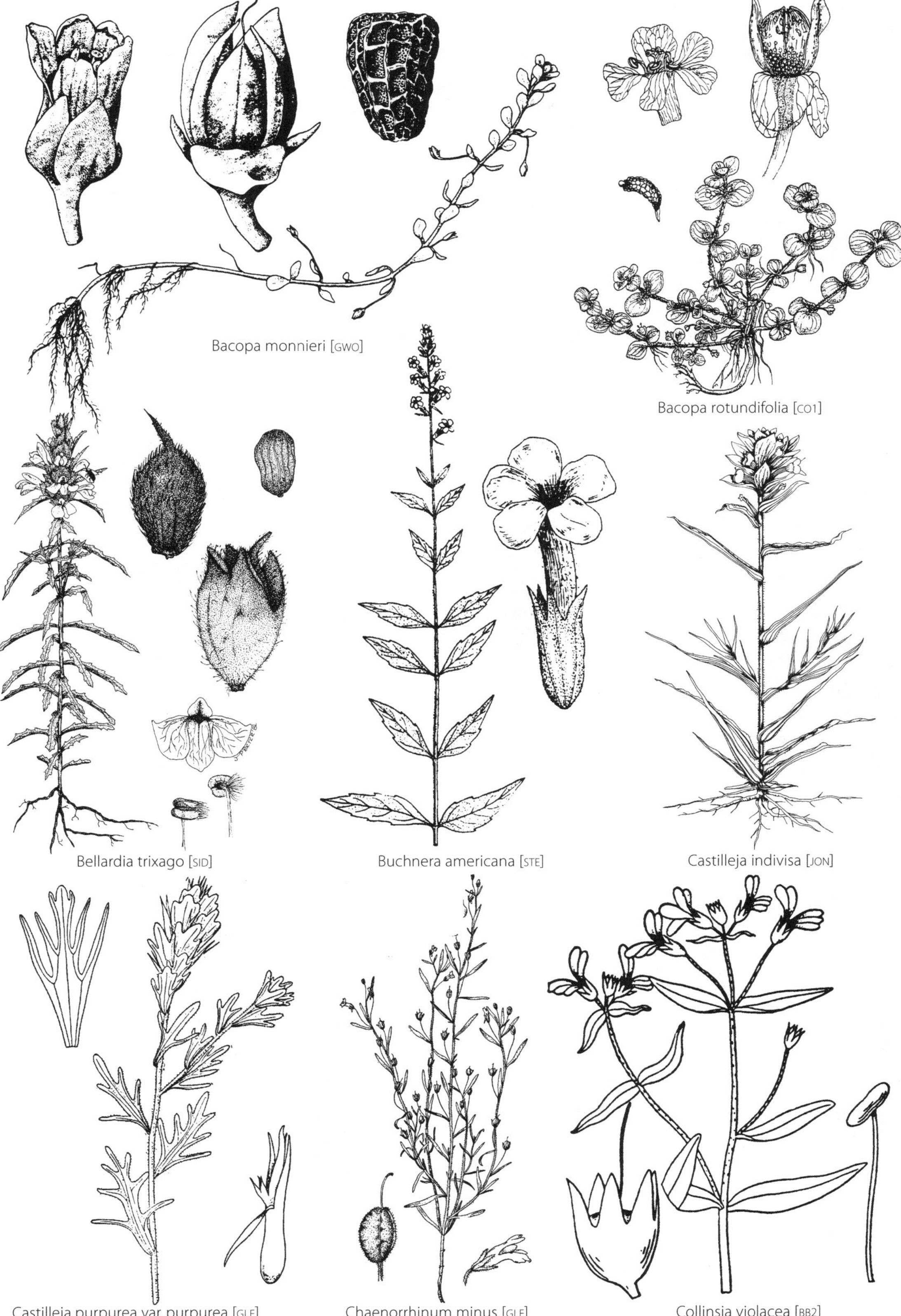

Bacopa monnieri [GWO]

Bacopa rotundifolia [CO1]

Bellardia trixago [SID]

Buchnera americana [STE]

Castilleja indivisa [JON]

Castilleja purpurea var. purpurea [GLE]

Chaenorrhinum minus [GLE]

Collinsia violacea [BB2]

rarely a light yellow individual is seen in a large population of orange-red or red individuals; corollas 20–28 mm long, the hood 6–9 mm long. Sandy or occasionally silty open woods, prairies, disturbed areas; se and e TX w to East Cross Timbers; now widely seeded by the Texas Highway Department. Apr–May. Austin (1975) considered this species adapted for pollination by ruby-throated hummingbirds (*Archilochus colubris*). It is reported to concentrate selenium on certain soils (Crosswhite 1980). ▣/82

Castilleja purpurea (Nutt.) G. Don, (purple), PRAIRIE INDIAN PAINTBRUSH, PRAIRIE PAINTBRUSH, PURPLE PAINTBRUSH. Bracts and calyces very variable in terms of color; corollas 25–40 mm long, the hood 9–13 mm long. A variable species with the following 3 varieties in nc TX.

1. Tips of bracts and calyces bright lemon yellow to greenish yellow; lower lip of corollas ca. 3–7 mm long; w and s parts of nc TX ____ var. **citrina**
1. Tips of bracts and calyces usually ranging from purplish pink to purplish red, purple, red, orangish red, brownish orange, or yellowish orange (rarely lighter); lower lip of corollas ca. 1.5–3(–4) mm long; widespread in nc TX.
   2. Tips of bracts and calyces yellowish orange to brownish orange, orangish red, or red; extreme s part of nc TX ____ var. **lindheimeri**
   2. Tips of bracts and calyces usually purplish pink to purplish red or purple (rarely lighter); widespread in nc TX ____ var. **purpurea**

var. **citrina** (Pennell) Shinners, (*Citrus*—citron- or lemon-colored), YELLOW PAINTBRUSH, CITRON INDIAN PAINTBRUSH, LIPPED INDIAN PAINTBRUSH, CITRON PAINTBRUSH, LEMON PAINTEDCUP, LEMON PAINTBRUSH. Rocky or sandy soils; West Cross Timbers w to Rolling Plains and s to Edwards Plateau. Apr–May. This variety exhibits less color variation than the other two (Nesom 1992a). ▣/82

var. **lindheimeri** (A. Gray) Shinners, (for Ferdinand Jacob Lindheimer, 1801–1879, German-born TX collector), LINDHEIMER'S INDIAN PAINTBRUSH, LINDHEIMER'S PAINTBRUSH. Calcareous gravelly, sandy or clay soils; Lampasas Cut Plain and s Blackland Prairie s to s TX; according to Nesom (1992a), in nc TX this variety occurs only in the s part of the area in Bell, Burnet, Lampasas, and Williamson cos.; endemic to TX. Mar–May. Nesom (1992a) suggested that the color variation within this taxon is a subset of that in var. *purpurea*; however, he further indicated there is an apparent abrupt geographic transition between the two and they can thus be justifiably maintained as weakly delimited varieties. Based on field observations, J. Stanford (pers. comm.) indicated that this is a clearly distinct PAINTBRUSH.

var. **purpurea**, PURPLE PAINTBRUSH. This variety is extremely variable in color; while the tips of bracts and calyces are predominantly purplish pink to purplish red or purple, they vary within a single population to red, reddish orange, burnt orange, peach, light yellow, creamy, and rarely white (Nesom 1992a). Limestone outcrops or prairies; Blackland Prairie w to West Cross Timbers and s to Central Mineral Region. Mar–May. In McCulloch Co., just s of nc TX, 17 color variants have been collected in less than 1/2 mile (J. Stanford, pers. comm.). ▣/82

In some areas, such as the se corner of Montague Co., varieties *citrina* and *purpurea* occur together at the edge of their ranges and numerous colors are present from white to many different undescribed color phases as a result of apparent interbreeding in this vicinity (Mahler 1988).

While the above two *Castilleja* species are quite distinct morphologically and even in different sections of the genus, in nc TX (e.g., Coryell and Hill cos.) hybrids are known between them with introgression and character convergence (Nesom 1992a; Egger 1994 [1995]); hybrids are also known from Grayson Co.

CHAENORRHINUM DWARF-SNAPDRAGON

A genus of 21 species of the Mediterranean area including cultivated ornamentals. Similar to

Linaria but differing in the capsule opening by narrow apical pores. (Greek: *chainen*, to gape, and *rhis*, snout, alluding to the open-mouthed flowers)
REFERENCES: Fernandes 1972; Arnold 1981a, 1981b; Widrlechner 1983; Diggs et al. 1997.

Chaenorrhinum minus (L.) Lange, (lesser, smaller), SMALL-SNAPDRAGON, DWARF-SNAPDRAGON. Erect, glandular-pubescent, often branched annual herb with stems to 40 cm tall; leaves opposite below to opposite or alternate above; leaf blades linear to oblong-lanceolate, 5–20(–35) mm long, 1–3(–8) mm wide, entire; flowers in terminal, indistinct, bracteate racemes; pedicels conspicuous, 3–20 mm long in fruit; calyx lobes 2–5 mm long, linear, obtuse; corollas 4.5–9 mm long, pale lavender or lilac with yellow palate, with tube and 2-lipped limb, with a straight, ± cylindrical spur 1–3 mm long; stamens 4; capsules 3–6 mm long, dehiscing by irregular terminal pores. Roadsides, along railroads; a European weed moving w from the e U.S.; known in Texas only from Fannin (*Taylor & Taylor 10570*, 1972) and Grayson (*Diggs 5748*, 1994) cos. (Diggs et al. 1997). May–Jul. [*Linaria minor* (L.) Desf.] This species was first reported (as *Linaria minor*) in the U.S. from New Jersey in 1874 (Martindale 1876). Widrlechner (1983) indicated that seeds were probably introduced in ship's ballast. Railroads are important in the dispersal of a number of introduced plants (Mühlenbach 1979) and *C. minus* seems an excellent example; Widrlechner (1983) noted that this species, "... has not been found in counties or census districts lacking railroads at the time of introduction to the corresponding state or province." Exclusion experiments showed that the species is capable of self-pollination (Arnold 1981b).

COLLINSIA BLUE-EYED-MARY

A genus of ca. 20 species of North America, especially the w U.S.; some are cultivated as ornamentals. (Named for Zaccheus Collins, 1764–1831, Philadelphia botanist)
REFERENCE: Newsom 1929.

Collinsia violacea Nutt., (violet), VIOLET COLLINSIA. Minutely pubescent annual 5–35(–60) cm tall; stem leaves opposite, sessile, oblong- or ovate-lanceolate, entire or shallowly toothed; floral bracts often whorled; flowers 2–5 in axillary whorls, rather long-pedicelled (pedicels 6–14 mm long, to 25 mm in fruit); corollas bilabiate, showy, 9–13 mm long; lower lip of corollas large, with keel-like or pouch-like central fold, purple with white base, the halves each notched at tip; upper lip of corollas short, white with purple edge and yellow basal protuberance; stamens 4, didynamous, staminode gland-like; capsules globose, 4–5 mm long. Sandy open woods or open ground; Post Oak Savannah w to East Cross Timbers. Mar–May.

DASISTOMA MULLEIN SEYMERIA, MULLEIN FOXGLOVE

A monotypic genus of the se U.S.; hemiparasitic (with green photosynthetic tissues but also parasitizing the roots of other plants). (Greek: *dasys*, hairy, and *stoma*, mouth, referring to the pubescent corolla)
REFERENCES: Pennell 1928; Piehl 1962.

Dasistoma macrophylla (Nutt.) Raf., (large-leaved), MULLEIN SEYMERIA, MULLEIN FOXGLOVE. Perennial to 2 m tall, hemiparasitic on roots of *Aesculus* and other taxa (e.g., *Ulmus americana*) through usually disc-like haustoria (Piehl 1962); stems pubescent above; leaves opposite; leaf blades ovate, 15–35 cm long, 8–22 cm wide, the lower pinnately divided, the upper pinnatifid to toothed or entire; flowers in elongated leafy spike-like racemes, weakly bilabiate; pedicels 1–4 mm long; corollas yellow, 14–16 mm long, pubescent inside; stamens 4, strongly didynamous; capsules globose, ca. 6–11 mm long. Woods along streams and rivers; Dallas and Grayson cos.; also e TX. Jun–Sep. [*Seymeria macrophylla* Nutt.] /86

GRATIOLA HEDGE-HYSSOP

Small, erect, annual or perennial herbs mainly of wet areas; leaves opposite, sessile, narrowly ovate to oblong-lanceolate, entire or slightly toothed; flowers axillary; corollas bilabiate, 5-lobed, white to pale lavender or yellow; fertile stamens 2; capsules many-seeded.

A genus of ca. 20 species of temperate areas and tropical mountains. (Latin: *gratia*, grace, favor, or thanks, from supposed medicinal properties)

1. Corollas golden to orange-yellow; capsules ca. 2 times as long as calyx lobes; stems < 10 cm tall, not fleshy ____ **G. flava**
1. Corollas white, pale lavender to honey-colored or pale yellow, sometimes with yellowish tube; capsules ca. same length as calyx lobes; stems usually > 10 cm tall, thick and fleshy.
   2. Stems glabrous or minutely glandular-puberulent; plants annual (rarely biennial); flowers and fruits pedicellate (pedicels short to long); calyx lobes equal or essentially so (do not confuse with 2 bractlets just below calyx); corollas much longer than calyces.
      3. Pedicels of upper flowers usually 1–5(–8) mm long, shorter than the sepals and adjacent leaves; plants glabrous ____ **G. virginiana**
      3. Pedicels of upper flowers 8–25 mm long, longer than the sepals and soon exceeding adjacent leaves; plants minutely glandular-puberulent ____ **G. neglecta**
   2. Stems with long, spreading pubescence; plants perennial; flowers and fruits subsessile or with very short pedicels; calyx lobes distinctly unequal; corollas only slightly longer than calyces ____ **G. pilosa**

Gratiola flava Leavenw., (yellow), GOLDEN HEDGE-HYSSOP. Annual; leaves linear-oblanceolate, to 15 mm long and 5 mm wide; corollas ca. 12 mm long, the limb golden yellow, the tube orange-yellow. Prairies and fields, sandy soils; included based on citation of vegetational areas 4 and 5 (Fig. 2) by Hatch et al. (1990); se and e TX w to nc TX. Feb–Apr.

Gratiola neglecta Torr., (neglected, overlooked), YELLOW-SEED HEDGE-HYSSOP. Annual to 40 cm tall; leaves rhombic-lanceolate to lanceolate; corollas to 8–12 mm long, the limb white, creamy white, pale yellow, or pale lavender, the tube yellowish. Muddy shores, damp ground, or shallow water; Bell and Lamar cos., also Dallas Co. (Mahler 1988); mainly e TX. Apr–early Jun.

Gratiola pilosa Michx., (with long soft hairs), SHAGGY HEDGE-HYSSOP, HAIRY HEDGE-HYSSOP. Perennial to 75 cm tall; stems square; leaves ovate-lanceolate to elliptic-ovate; flowers subsessile; corollas white or lavender-tinged, 5–9 mm long. Wet areas; Henderson Co. near e margin of nc TX, also collected by Reverchon at Dallas (La Réunion); se and e TX w to nc TX. Apr–Aug.

Gratiola virginiana L., (of Virginia), VIRGINIA HEDGE-HYSSOP. Annual (rarely biennial); base of stems fleshy; leaves lanceolate or elliptic, undulate to sharply serrate; corollas to 15 mm long, white with purplish lines. Muddy shores, damp ground, or shallow water; Grayson Co., also Tarrant Co. (Mahler 1988); se and e TX w to nc TX. Apr–May. Reported to be poisonous to cattle (Burlage 1968).

LEUCOSPORA NARROW-LEAF CONOBEA

A New World genus sometimes considered monotypic (e.g., Correll & Johnston 1970; Mabberley 1987) or as having ca. 4 species (Henrickson 1989) including the 2 in the genus *Schistophragma*; it is sometimes lumped with *Stemodia*, a genus of ca. 55 species (Mabberley 1997). Further work on generic circumscription in this complex appears needed. (Greek: *leukos*, white, and *spora*, seed)
REFERENCE: Henrickson 1989.

Leucospora multifida (Michx.) Nutt., (many times parted), NARROW-LEAF CONOBEA. Low, annual, simple to much-branched herb, 10–15(–20) cm tall; leaves opposite, petioled, triangular-ovate in

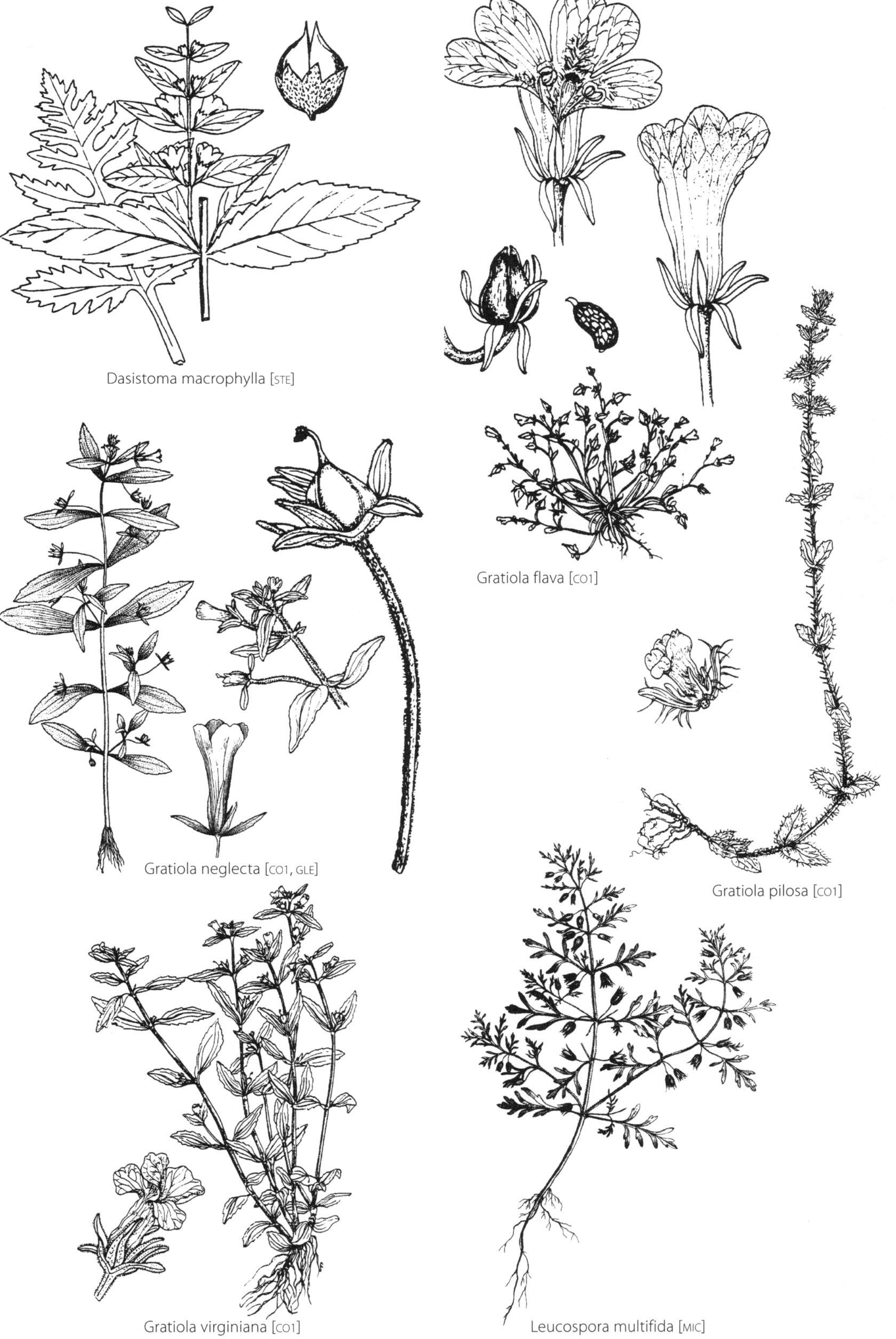

Dasistoma macrophylla [STE]

Gratiola flava [CO1]

Gratiola neglecta [CO1, GLE]

Gratiola pilosa [CO1]

Gratiola virginiana [CO1]

Leucospora multifida [MIC]

outline, pinnately to bipinnately parted, 1–3 cm long; flowers 5-merous, single or in pairs in the leaf axils, on pedicels 5–10 mm long; corollas pink or pale lavender, 3–5 mm long, tubular, shorter than to barely exceeding calyces; stamens 4, didynamous; capsules ovoid, 3.5–4.5 mm long. Low wet areas, city weed; e 1/2 of TX. Jun–Oct. [*Capraria multifida* Michx., *Conobea multifida* (Michx.) Benth., *Stemodia multifida* (Michx.) Spreng.]

LINARIA TOAD-FLAX, SPURRED-SNAPDRAGON

An Old World, primarily temperate, especially Mediterranean genus of 150 species (Sutton 1988); some are used medicinally, others as ornamentals. (From genus *Linum*, FLAX, from the flax-like leaves of some species)

REFERENCES: Munz 1926; Sutton 1988.

Linaria vulgaris Mill., (common), BUTTER-AND-EGGS, COMMON TOAD-FLAX. Strong-scented rhizomatous perennial; stems erect or ascending, 0.3–0.6(–1.3) m tall; leaves alternate or subopposite below, numerous, linear or narrowly linear-lanceolate, 2.5–6 cm long, 2–6(–15) mm wide, subsessile, not clasping; racemes crowded, spike-like; pedicels 1–4 mm long; corollas 2-lipped, yellow with orangish yellow palate, (19–)25–33 mm long including 8–17 mm long spur; spur 2–3 mm wide at base. Cultivated and escaping; Hood (reported as an escape near Center Mills) and Tarrant cos., also Johnson Co. (R. O'Kennon, pers. obs.); in TX reported only from nc part of the state. May. Native of Europe and Asia. Suspected of being poisonous; the glucoside, linariin, is present (Burlage 1968).

LINDERNIA FALSE PIMPERNEL

A genus of 80 species of warm areas, especially the Old World. (Named for Franz Balthasa von Lindern, 1682–1755, German botanist and physician)

Lindernia dubia (L.) Pennell var. **anagallidea** (Michx.) Cooperr., (sp.: doubtful; var.: resembling *Anagallis*—pimpernel in the Primulaceae), CLASPING FALSE PIMPERNEL. Small glabrous annual 9–20(–30) cm tall; leaves opposite, to 20 mm long, 3–5-nerved; lowest leaves narrow-based, subsessile; middle and upper sessile, slightly clasping; flowers solitary in axils of leaves or the upper merely bracted, long-pedicelled; pedicels to 25 mm long, 1–3 times as long as the subtending leaf; corollas ca. 7–9 mm long, pale lavender, pale blue, or white with lavender tube; capsules ca. 4–5 mm long. Mud flats, wet ground, or shallow water; Grayson and Tarrant cos., also Dallas (R. O'Kennon, pers. obs.) and Lamar (Carr 1994) cos.; se and e TX w to Edwards Plateau and Plains Country. Late May–Oct. [*Lindernia anagallidea* (Michx.) Pennell] Because of the lack of consistent characters, Holmgren (1986) suggested var. *anagallidea* should possibly be lumped with var. *dubia*.

MAURANDYA

A genus of 4 species of the sw U.S. and Mexico (Elisens 1985); sprawlers or vines with coiling petioles; sometimes cultivated. They are sometimes (e.g., Mabberley 1987) treated in *Asarina* (16 species, mostly in North America, 1 in Europe) (Named for Catherina Pancratia Maurandy, 18th century botany professor, Cartagena, Spain)

REFERENCES: Munz 1926; Elisens 1985; Sutton 1988.

Maurandya antirrhiniflora Humb. & Bonpl. ex Willd., (with flowers like *Antirrhinium*—snapdragon), SNAPDRAGON-VINE, SNAPDRAGON MAURANDELLA, VINE BLUE-SNAPDRAGON, VIOLET TWINING-SNAPDRAGON. Extensively twining, much-branched, glabrous, herbaceous, perennial vine; leaves subopposite to generally alternate; leaf blades hastate to sagittate, 15–25 mm long, the margins entire, on petioles 10–30 mm long; flowers solitary in axils, on pedicels 10–30(–47) mm long; corollas bilabiate, 20–25 mm long, the tube pale, the lobes violet to bluish to purple (rarely whitish), the lower lip swollen, often with yellowish or whitish area; stamens 4; capsules ca. 6 mm long. Disturbed, sandy, and rocky areas; Burnet Co. (Elisens 1985) near s edge of nc TX, also Fort Hood (Bell or Coryell cos.—Sanchez 1997); also Travis Co. just s of nc TX; mainly s part of TX and Trans-Pecos. Feb–Oct. [*Asarina antirrhiniflora* (Humb. & Bonpl. ex Willd.) Pennell] While Kartesz (1994) recognized this taxon in the monotypic genus *Maurandella* [as *M. antirrhiniflora* (Humb. & Bonpl. ex Willd.) Rothm.], we are following Elisens (1985) and Jones et al. (1997) in treating it in *Maurandya*. Elisens (1985) gave detailed justification. ▣/99

MAZUS

A genus of 10–15 species of mat-forming groundcovers native from Asia to Australia. (Greek: *mazos*, breast, from the swelling in throat of corolla)

Mazus pumilus (Burm. f.) Steenis, (dwarf). Low herb 5–20 cm tall from a basal rosette; leaves opposite, obovate to spatulate, with a few coarse teeth; inflorescence a loose few-flowered raceme; pedicels 4–7 mm long; corollas bilabiate, 7–10 mm long, blue with lower lip yellowish or whitish with reddish brown margins; stamens 4; capsules 3–4 mm long, shorter than calyces. Lawns, weedy areas, roadsides; Dallas Co., also Tarrant Co. (R. O'Kennon, pers. obs.); mainly se and e TX. Mar–Oct. Native of Asia. [*M. japonicus* (Thunb.) Kuntze]

MECARDONIA WATER-HYSSOP

Perennial herbs, glabrous, often drying black; stems 4-angled; leaves opposite, toothed, glandular-punctate; flowers axillary; bractlets 2, at base of pedicel, shorter than floral bracts; flowers 5-merous; corollas bilabiate; capsules many-seeded.

A genus of 10 species of warm areas of the Americas. (Named in 1794 for De Antonio de Meca et Cardona who gave the land for the botanical garden to the Royal College of Surgery of Barcelona)

1. Corollas white, often tinged or lined with purple; 3 outer sepals lanceolate; largest leaves usually > 25 mm long ____ **M. acuminata**
1. Corollas yellow; 3 outer sepals ovate; largest leaves usually < 25 mm long ____ **M. procumbens**

Mecardonia acuminata (Walter) Small, (tapering to a long narrow point), PURPLE MECARDONIA, SAW-TOOTH WATER-HYSSOP. Stems erect to ascending, to ca. 70 cm long; leaves oblanceolate, serrate above middle, 20–40 mm long; pedicels equal to or exceeding the subtending leaves; outer sepals usually < 2 times as wide as inner; corollas 7–10 mm long. Low wet areas; prairies, swamps, and pinelands; Bell and Grayson cos.; se and e TX w to nc TX, also Edwards Plateau. Aug–Oct.

Mecardonia procumbens (Mill.) Small, (prostrate), PROSTRATE MECARDONIA, PROSTRATE WATER-HYSSOP, STALKED WATER-HYSSOP. Stems procumbent to erect-ascending, often branched from base, to ca. 40 cm long; leaves ovate to obovate, 10–25 mm long, serrate above middle; pedicels usually much exceeding the subtending leaf; outer sepals usually > 3 times as wide as inner; corollas yellow with dark veins, 6–12 mm long. Low wet areas or in water; Bell Co.; widespread in TX. Mar–Nov. [*Mecardonia vandellioides* of authors, not (Kunth) Pennell]

MIMULUS MONKEY-FLOWER

Ours perennial, rhizomatous or stoloniferous herbs, glabrous or nearly so; leaves opposite; flowers axillary, 5-merous; corollas bilabiate; stamens 4, didynamous; anthers with sacs divergent; capsules cylindric, many-seeded.

A genus of 150 species of the Americas, s Africa, and Asia; many are cultivated as ornamentals. (Diminutive of Latin: *mimus*, mime, buffoon, or comic, from the face-like corollas of some species) REFERENCE: Grant 1924.

1. Corollas blue, lavender, or white, 20–28 mm long; plants erect; leaf blades to 5–15 cm long, cuneate to tapering at base; calyx teeth equal or nearly so ____ **M. alatus**
1. Corollas yellow, 8–16 mm long; plants creeping and rooting at lower nodes to decumbent; leaf blades to 7 cm long, usually much less, rounded to cordate at base; calyx teeth unequal, one tooth larger than the others ____ **M. glabratus**

Mimulus alatus Aiton, (winged), SHARP-WING MONKEY-FLOWER. Stems erect, 30–70(–100) cm tall, 4-angled, the angles ± winged; leaves broadly ovate to ovate-lanceolate, to 6 cm wide, serrate, petiolate; pedicels 2–8 mm long, to 14 mm in fruit; calyces 11–18 mm long; corollas often with lighter spots on lower lip, the palate nearly closing throat; stamens included; capsules 8–11 mm long. Low wet areas; Collin, Dallas, Fannin, Grayson, and Hunt cos., also Lamar Co. (Carr 1994); se and e TX w to Blackland Prairie. Jun–Nov.

Mimulus glabratus Kunth, (rather smooth, without hairs), ROUND-LEAF MONKEY-FLOWER. Stems weak, to 75 cm long, terete, hollow; leaves broadly ovate to suborbicular, to 6 cm wide, dentate, long-petiolate below to subsessile above; pedicels 10–40 mm long; calyces to 10 mm long; corollas sometimes with red brown spots, the tube slender. Low wet areas or in water; mainly Edwards Plateau and Trans-Pecos. Apr–May.

1. Corollas 12–16 mm long; calyces becoming 10–13 mm long ____ var. **glabratus**
1. Corollas 8–12 mm long; calyces becoming 5–11 mm long ____ var. **jamesii**

var. **glabratus**. Dallas and Williamson cos., also Somervell Co. (Mahler 1988). Jones et al. (1997) did not list this variety for TX.

var. **jamesii** (Torr. & A. Gray ex Benth.) A. Gray, (for Edwin James, 1797–1861, surgeon-naturalist, first botanical collector in CO and first known botanical collector in TX, with Major Long's expedition to the Rocky Mts. in 1819–1820), FREMONT'S MONKEY-FLOWER. Burnet Co.; also Bell Co. (Mahler 1988). [*M. glabratus* var. *fremontii* (Benth.) A.L. Grant, *M. jamesii* Torr. & A. Gray ex Benth.]

NUTTALLANTHUS TOAD-FLAX

Slender annuals or biennials to 70 cm tall, ± glabrous; leaves alternate or subopposite, subsessile, linear, entire; flowers in terminal, spike-like racemes; corollas 2-lipped, with slender, basal, usually curved spur, light lavender-blue with whitish palate.

A genus of 4 species native to North and w South America; it has often been lumped into *Linaria* (Holmgren 1986; Wetherwax 1993). We, however, are following the recent treatment by Sutton (1988) who segregated *Nuttallanthus* (also Kartesz 1994). Seeds viewed under a strong lens or dissecting scope are helpful for definitive identification. (Named for Thomas Nuttall, 1789–1859, British-born botanist, ornithologist, and collector in w North America)
REFERENCES: Munz 1926; Kral 1955; Sutton 1988.

1. Corollas 8–11(–13) mm long, including 2–3.5 mm spur, the lower lip 2–6 mm long; seeds with low, entire longitudinal ridges, the intervening faces ± smooth or with sparse, low tubercles ____ **N. canadensis**
1. Corollas 14–22 mm long, including 6–11 mm spur, the lower lip 6–11 mm long; seeds densely tuberculate, without longitudinal ridges (use strong lens) ____ **N. texanus**

Nuttallanthus canadensis (L.) D.A. Sutton, (of Canada), OLDFIELD TOAD-FLAX, TOAD-FLAX. Capsules 2–3 mm long. Sandy open woods; Dallas Co., also Comanche-Eastland Co. line, McLennan, and Navarro cos. (Kral 1955); apparently rare in nc TX; se and e TX w to nc TX and Edwards Plateau. Mar–Apr. [*Linaria canadensis* (L.) Chaz.]

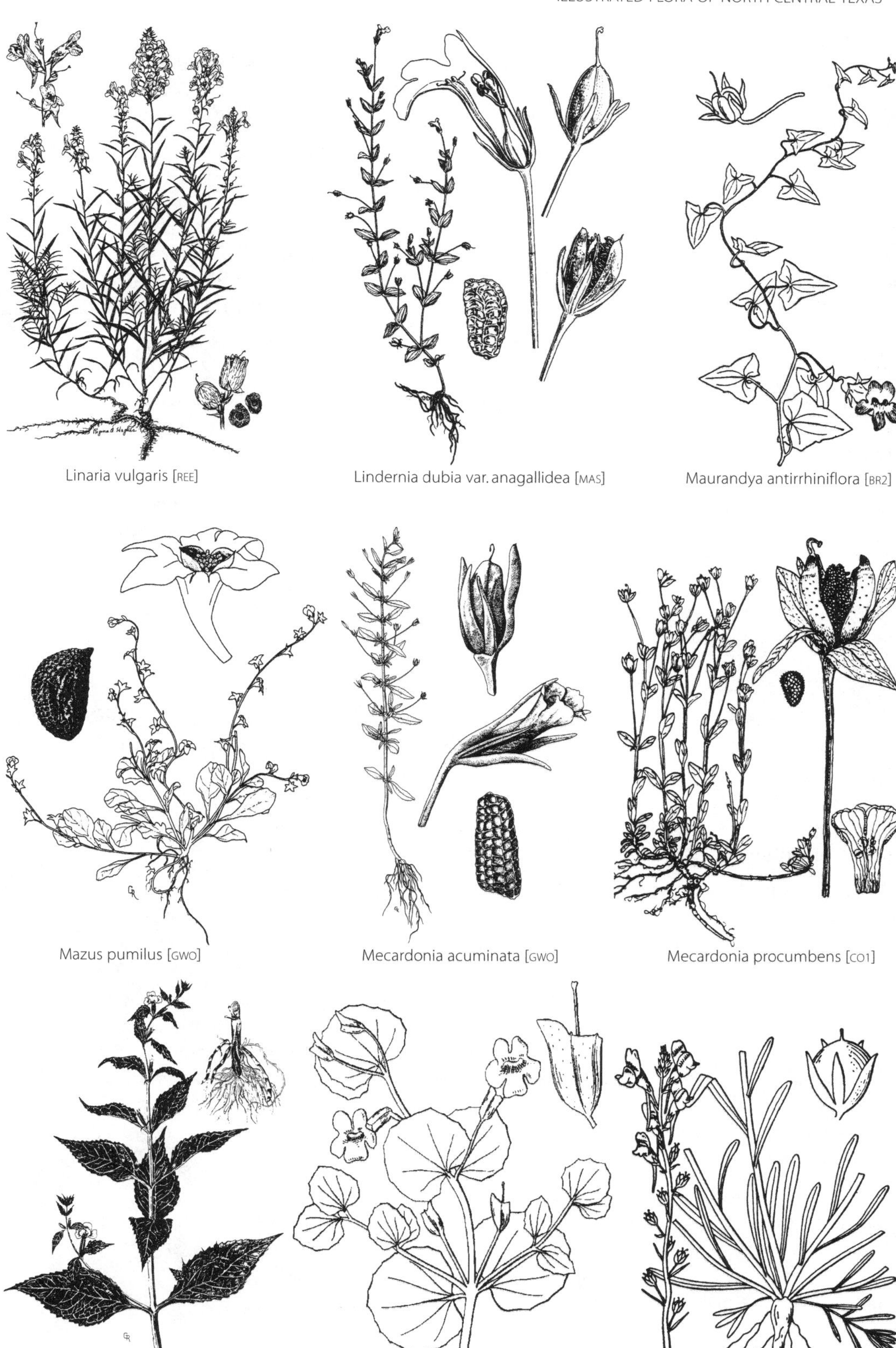

Linaria vulgaris [REE]

Lindernia dubia var. anagallidea [MAS]

Maurandya antirrhiniflora [BR2]

Mazus pumilus [GWO]

Mecardonia acuminata [GWO]

Mecardonia procumbens [CO1]

Mimulus alatus [GWO]

Mimulus glabratus var. jamesii [BB1]

Nuttallanthus canadensis [BB2]

Nuttallanthus texanus (Scheele) D.A. Sutton, (of Texas), TEXAS TOAD-FLAX, TOAD-FLAX. Capsules 2.5–3.5 mm long. Sandy open woods, old fields, and roadsides; widespread in TX, but mainly in e 2/3. Mar–May. This is the common TOAD-FLAX in nc TX. *Nuttallanthus texanus* has sometimes been treated as a variety of *N. canadensis* (e.g., Holmgren 1986; Wetherwax 1993). [*Linaria canadensis* var. *texana* (Scheele) Pennell, *L. texana* Scheele]

PEDICULARIS LOUSEWORT, WOOD-BETONY

A n hemisphere genus of 350+ species with 1 in the Andes; alkaloid-containing hemiparasites (with green photosynthetic tissues but also parasitizing the roots of other plants). At one time LOUSEWORTS were presumed to become lice on sheep coming into contact with them—hence the common name. (Greek: *pediculus*, louse; the presence of these plants in pastures was thought to cause sheep or cattle to become infested with lice)
REFERENCES: Sprague 1962; Piehl 1963.

Pedicularis canadensis L., (of Canada), COMMON LOUSEWORT, EARLY LOUSEWORT, EARLY FERN-LEAF LOUSEWORT, WOOD-BETONY. Perennial herb 10–30(–40) cm tall; leaves mostly basal, pinnately parted to pinnatifid, fern-like in appearance, 4–15 cm long, 8–25(–50) mm wide; inflorescence a short (3–5 cm long), dense, spike-like raceme, later elongating; calyces 7–9 mm long; corollas yellowish, sometimes with lavender, 18–25 mm long, bilabiate, the upper lip hooded; capsules 12–16 mm long. Woods; Fannin (Talbot property) and Lamar cos. in Red River drainage; ne TX. Mar–May. Hemiparasitic on the roots of at least 80 species in 35 families; connections to the host are made by specialized enlargements of the parasite root known as haustoria (Piehl 1963). Jones et al. (1997) treated all TX material of this species as *P. canadensis* subsp. *canadensis* var. *dobbsii* Fernald. Reported to be poisonous to sheep (Burlage 1968).

PENSTEMON BEARDTONGUE

Ours low to rather tall perennials; leaves opposite, toothed or entire, the lower petioled, the upper gradually reduced and sessile, sometimes clasping; flowers in terminal, narrow or spike-like panicles; corollas asymmetrical, distinctly 2-lipped to not so, with cylindrical basal tube, widened limb, and 5 rounded lobes, often quite showy, white to pink, lavender, or red, often with darker lines (= nectar-guides); fertile stamens 4, of 2 lengths; staminode (= sterile stamen) prominent and often bearded; seeds numerous in capsule.

A genus of 250 species of perennial herbs and shrubs of North America with 1 species in ne Asia. Many are showy and cultivated as ornamentals. The common name is derived from the bearded staminode. Crosswhite and Crosswhite (1981) considered *Penstemon* the largest endemic genus of flowering plants in North America. They indicated that 40 of the species are red-flowered and emphasized the geographic correlation between the distribution of red-flowered Penstemons and hummingbirds in w North America. In hummingbird-pollinated species, sucrose-rich nectars are found, while in insect-pollinated species, the nectar is hexose-rich (Mabberley 1997). (Greek *pente*, five, and *stemon*, stamen, the fifth stamen being present and conspicuous, although infertile and thus a staminode—5 stamens are unusual in the Scrophulariaceae whose flowers typically have either 2 or 4 stamens).
REFERENCES: Straw 1966; Crosswhite & Crosswhite 1981.

1. Flowers (except uppermost) borne on axillary or lateral branchlets (these sometimes short).
    2. Corollas 35–60 mm long, widely inflated; sepals densely glandular-pubescent, 7–16 mm long; flowering branchlets mostly shorter than flowers ____ **P. cobaea**
    2. Corollas 34 mm or less long, narrow to moderately inflated; sepals glabrous or sparsely glandular-pubescent, 2–8 mm long; flowering branchlets mostly becoming nearly as long as the flowers or longer.

3. Staminode (= sterile 5th stamen) densely covered with yellow-orange hairs for more than half its length (for 8–10 mm), usually prominently exserted; stems usually puberulent near base; floor of corollas prominently pleated (folded with 2 ridges); corollas gradually enlarging from base ______ **P. laxiflorus**
3. Staminode usually sparsely covered with whitish or yellowish hairs in terminal 1/3, included or reaching the opening of the corolla but not exserted; stems glabrous or nearly so near base; floor of corollas not pleated or only slightly so; corollas gradually enlarging from base OR narrow in basal 1/3 and abruptly enlarged above.
    4. Corollas usually 20–33 mm long, straight and cylindrical in basal 1/3, abruptly enlarged above; sepals 5–7 mm long; mid-stem leaves lanceolate, 15–25 mm wide ______ **P. digitalis**
    4. Corollas 15–20(–23) mm long, gradually enlarging from base; sepals 2–4 mm long; mid-stem leaves broadly elliptic to elliptic lanceolate, 20–40 mm wide ______ **P. tubaeflorus**

1. Flowers whorled or opposite, on the primary axis of the inflorescence.
    5. Corollas white, lavender, pink, rose OR red with white throat with red lines; upper leaves broad, sessile, clasping, but not united around stem.
        6. Inflorescences and stems completely glabrous; leaf blades thick and fleshy, glaucous.
            7. Corollas 35–50 mm long; plants usually 0.5–1.2 m tall; sepals 7–11 mm long at flowering time ______ **P. grandiflorus**
            7. Corollas 14–23(–28) mm long; plants 0.6 m or less tall; sepals 4.5–7 mm long at flowering time ______ **P. fendleri**
        6. Inflorescences and usually stems with pubescence; leaf blades neither thick nor fleshy, not glaucous.
            8. Stem leaves entire or sparsely and irregularly toothed; corollas 1.3–2 cm long, white or nearly so, sometimes with a few colored lines; basal and lower leaves linear, < 6 mm wide; sepals 5–7 mm long ______ **P. guadalupensis**
            8. Stem leaves conspicuously toothed; corollas 3–3.5 cm long, lavender to pink or rose OR red with white throat with red lines; basal and lower leaves lanceolate or oblanceolate, > 6 mm wide; sepals 8–11 mm long ______ **P. triflorus**
    5. Corollas red; upper leaves united around stem ______ **P. murrayanus**

Penstemon cobaea Nutt., (named for the showy-flowered tropical American genus *Cobaea* (Polemoniaceae) which was named for a Jesuit Father, Bernardo Cobo, 1572–1659, Spanish missionary and naturalist in Mexico and Peru), WILD FOXGLOVE, FOXGLOVE, FALSE FOXGLOVE, COBAEA PENSTEMON, BALMONY. Plant 20–50(–65) cm tall, finely and densely pubescent; leaf blades ovate-elliptic to ovate or obovate, sharply toothed or rarely subentire, those at mid-stem usually 25–50 mm wide; calyx lobes lanceolate to lance-ovate; corollas broad, white to light or deep lavender with red-purple lines inside, extremely showy. Limestone prairies and rock outcrops, sandy open woods; mainly Blackland Prairie w to Rolling Plains and Edwards Plateau, also s to coastal plain. Apr–May. One of the most striking wildflowers in nc TX. ⊛/102

Penstemon digitalis Nutt. ex Sims, (of the finger, possibly from the flowers being somewhat like the finger of a glove, or from resemblance to flowers of the genus *Digitalis*—foxglove), SMOOTH PENSTEMON, BEARDTONGUE, SMOOTH BEARDTONGUE. Plant (25–)50–90 cm tall; leaf blades entire to rather finely and sharply toothed; calyx lobes lance-ovate to ovate; corollas pure white except sometimes with faint reddish purple lines, with glandular hairs outside; terminal 6–8 mm of staminode with whitish to yellowish, usually sparse hairs. Low open places in woods, fields, and on roadsides; Dallas and Grayson cos.; e TX w to nc TX. May–Jun. ⊛/102

Penstemon fendleri Torr. & A. Gray, (for August Fendler, 1813–1883, one of the first botanists to collect in New Mexico and Venezuela), FENDLER'S PENSTEMON, PURPLE FOXGLOVE. Plant 15–60 cm tall, glabrous, glaucous; leaf blades entire, those below the inflorescence 10–30 mm wide; sepals

4–7 mm long; corollas 14–20 mm long, lavender, usually lined inside. Rocky or sandy prairies and roadsides; Navarro Co., also Brown Co. (HPC); mainly w 1/2 of TX w of nc TX. Apr–May.

Penstemon grandiflorus Nutt., (large-flowered), LARGE BEARDTONGUE. Plant (35–)60–100(–120) cm tall, glabrous, glaucous; leaf blades entire, those at mid-stem 25–50 mm wide; corollas lavender (rarely white), inflated. Edge of sandy woods; Wise and Callahan cos. in West Cross Timbers; scattered in w 1/2 of TX; more common in n Great Plains. Apr–May. [*Penstemon bradburii* Pursh]

Penstemon guadalupensis A. Heller, (of the Guadalupe), GUADALUPE PENSTEMON. Plant 25–35 cm tall, puberulent; leaf blades linear to lanceolate, 2–10(–18) mm wide; corollas white or nearly so, sometimes with a few colored lines; staminode with a few white hairs. Calcareous soils; Comanche Co. in sw part of nc TX, also Brown Co. (HPC herbarium); endemic to e Edwards Plateau and sw part of nc TX. Mar–May.

Penstemon laxiflorus Pennell, (loose-leaved), BEARDTONGUE. Plant 30–75 cm tall; leaf blades usually toothed to entire, those at mid-stem 4–17 mm wide; sepals lance-ovate to ovate; corollas 17–30 mm long, narrow, lavender or lavender pink to almost white, lined internally with reddish purple nectar guides, with glandular hairs outside; terminal 8–10 mm of staminode densely bearded. Sandy open woods and prairies; se and e TX w to West Cross Timbers and Edwards Plateau. Apr–May. [*P. australis* Small subsp. *laxiflorus* (Pennell) D.E. Bennett] While this taxon has often been treated as *P. australis* subsp. *laxiflorus* (e.g., Kartesz 1994; Jones et al. 1997), we are following J. Kartesz (pers. comm.) in recognizing it at the species level.

Penstemon murrayanus Hook., (for Johann Andreas Murray, 1740–1791, Swedish pupil of Linnaeus and Professor of Medicine and Botany Göttingen), CUP-LEAF PENSTEMON. Plant 50–200 cm tall, glabrous; leaves perfoliate; leaf blades entire, thickened, glaucous, 25–50 mm wide; corollas ca. 30 mm long, bright red (may dry yellowish). Sandy open woods; Henderson Co., also Ellis Co. (Mahler 1988); mainly se and e TX w to e part of nc TX, also Edwards Plateau. Apr–May. Exceedingly showy. Pollinated by hummingbirds (Pennell in Crosswhite & Crosswhite 1981).

Penstemon triflorus A. Heller subsp. **integrifolius** Pennell, (sp.: three-flowered; subsp.: entire-leaved). Plant 30–65 cm tall, puberulent; middle stem leaves with blades toothed, (10–)15–30 mm wide; corollas lavender to pink or rose or red with white throat with red lines; staminode lightly bearded in terminal one-half with yellow hairs. Calcareous soils; Callahan, Coleman, and Burnet cos. in sw part of nc TX; endemic to Edwards Plateau and sw part of nc TX. Apr–May. [*Penstemon helleri* Small]

While we are following Kartesz (1994) and Jones et al. (1997) in lumping *P. helleri*, J. Stanford (1976, pers. comm. 1998) indicated that *P. helleri* may deserve specific recognition; he suggested it and *P. triflorus* occur together and are sometimes confused. These two can be separated as follows (modified from Stanford 1976):

1. Corollas lavender to pink or rose; staminode lightly bearded with yellow hairs for half its length ______ *P. helleri*
1. Corollas red (throat white with red lines); staminode glabrous ______ *P. triflorus*

Penstemon tubaeflorus Nutt., (with trumpet-shaped flowers), TUBE PENSTEMON. Plant 25–100 (150) cm tall, glabrous; leaf blades usually entire to obscurely serrate; calyx lobes lance-ovate to ovate; corollas white, unlined internally, with glandular hairs; terminal 3–4 mm of staminode with sparse yellowish hairs. Open to wooded areas; Lamar Co. in Red River drainage; mainly e TX. May–Jun.

SCROPHULARIA FIGWORT

A genus of 200 species native from the n temperate zone to the American tropics; some are used medicinally or cultivated as ornamentals. (Latin: *scrofula*, disease characterized by swell-

Nuttallanthus texanus [ABR]

Pedicularis canadensis [CO1]

Penstemon cobaea [STE]

Penstemon digitalis [STE]

Penstemon fendleri [IVE]

Penstemon laxiflorus [BT3]

Penstemon grandiflorus [GLE]

Penstemon guadalupensis [FMC]

ing of the neck glands; so called because by the Doctrine of Signatures, the fleshy bulbs on the rhizomes of some species were supposed to cure the disease)

Scrophularia marilandica L., (of Maryland), CARPENTER'S-SQUARE, MARYLAND FIGWORT. Robust herbaceous perennial with square stems to 2(–3) m tall; leaves opposite, simple; leaf blades lanceolate to ovate, 8–15(–20) cm long, 3–7(–10) cm wide, sharply toothed; petioles 1.5–5(–8) cm long; inflorescence a large panicle of small flowers; flowers bilabiate; corollas 5–10 mm long, brown or reddish brown with pale green; fertile stamens 4; staminode brown or brownish purple; capsules 4.5–7 mm long with numerous seeds. Rich woods, thickets; included based on citation of vegetational area 4 (Fig. 2) by Hatch et al. (1990); ne TX. Jul–Sep.

VERBASCUM MULLEIN

Erect biennials; leaves sessile or the lowest with winged-petiolar bases; flowers in terminal, bracted spikes or narrow racemes, open at night and in the morning; corollas with short tube and rotate limb, slightly bilaterally symmetrical, the upper 2 lobes slightly shorter than lower 3; stamens 5; capsules with numerous seeds.

A genus of ca. 360 species mainly of Eurasia with a few to the Ethiopian and e African highlands; some are cultivated as ornamentals; ☠ others are used as fish poisons. The seeds of some have been known to germinate after 100 years. (Classical Latin name for these plants)

1. Flowers long-pedicelled, the pedicels 10–20(–25) mm long; corollas yellow to white, often with purplish center; stem leaves oblong- to ovate-lanceolate, toothed or shallowly lobed, glabrous or with few glandular, non-branched hairs ____ **V. blattaria**
1. Flowers subsessile; corollas yellow; stem leaves oblanceolate to elliptic or obovate, entire or subentire, densely felty-woolly, the hairs stellate or dendritic-branched ____ **V. thapsus**

Verbascum blattaria L., (name used by Pliny, pertaining to a moth), MOTH MULLEIN. Plant 0.5–1.5 m tall, glandular-pubescent above, glabrate below; leaves to 20 cm long, reduced upward, upper partly clasping; corollas 20–30 mm across; capsules glabrous or glandular-pubescent. Roadsides and waste areas; Dallas and Grayson cos., also Lamar (Mahler 1988) and Somervell (R. O'Kennon, pers. obs.) cos.; mainly e TX. May–Jun, rarely later. Native of Eurasia. Reported to be poisonous (Burlage 1968). ☠

Verbascum thapsus L., (classical name, from ancient Thapsus), COMMON MULLEIN, FLANNEL MULLEIN, FLANNEL-PLANT, FLANNEL-LEAF, INDIAN-TOBACCO, JUPITER'S-STAFF, VELVET-DOCK, COWBOY'S-TOILET-PAPER. Plant stout, 0.7–2 m tall, conspicuously and densely grayish felty-woolly due to stellate or dendritic-branched hairs; leaves to 50 cm long; rosette leaves largest, reduced upwards; stem leaves sessile, decurrent; corollas 12–30(–35) mm across; capsules densely tomentose with stellate or dendritic-branched hairs. Roadsides, pastures, and waste areas; nearly throughout TX. May–Jul, rarely later. Native of Eurasia. The dense coating of hairs is particularly interesting under a lens or dissecting scope. The leaves are used medicinally, including in the form of cigarettes for treating asthma (Mabberley 1987).

VERONICA SPEEDWELL

Annuals or perennials; leaves mostly opposite or the upper leaves/bracts alternate; flowers axillary or in terminal racemes, bracted; calyces deeply 4-parted; corollas light blue, lavender, or white, unequally 4-lobed; stamens 2; capsules often with an apical notch and rather heart-shaped.

A genus of ca. 180 species (J. Thieret, pers. comm.) mainly of the n temperate zone with a few on tropical mountains and in s temperate zone. Six of the 7 species occurring in nc TX are

Penstemon murrayanus [CUR]

Penstemon triflorus [FMC]

Penstemon tubaeflorus [STE]

Scrophularia marilandica [BB2]

Verbascum blattaria [REE]

Verbascum thapsus [REE]

Veronica agrestis [BB2]

weedy Old World natives. (Named for St. Veronica, popularly thought to be from Latin: *vera*, true, and Greek: *eikon*, image or picture; an early Christian legend picturing St. Veronica, pitying Christ on the way to Calvary, wiping his face with her handkerchief which received a miraculous true image of his features)
REFERENCES: Pennell 1921; Thieret 1955; Walters & Webb 1972.

1. Flowers in axillary, peduncled racemes; leaves always opposite.
   2. Leaves all ± short-petioled; racemes 6–30-flowered ____ **V. americana**
   2. Leaves, at least the upper ones on flowering stems, sessile and clasping; racemes often with more than 30 flowers ____ **V. anagallis-aquatica**
1. Flowers in terminal racemes OR solitary in the axils of leaves; upper leaves/bracts alternate.
   3. Flowers sessile or very short-pedicelled (pedicels 0.5–2 mm long, shorter than the calyces), in racemes with the subtending leaves/bracts much reduced or modified and at least the uppermost very different from regular leaves; fruiting pedicels much shorter than subtending leaf/bract.
      4. Flowers white; stem leaves (except lowest) sessile, their blades narrowly oblong or oblanceolate, usually > 2 times as long as wide ____ **V. peregrina**
      4. Flowers light blue or lavender-blue; stem leaves (except uppermost) short-petioled, their blades ovate or oblong-ovate, usually < 2 times as long as wide ____ **V. arvensis**
   3. Flowers long-pedicelled (pedicels usually 5–30 mm long, longer than the calyces), solitary in the axils of leaves similar to regular stem leaves (except somewhat smaller); fruiting pedicels somewhat shorter to much longer than subtending leaf.
      5. Corollas relatively large, 8–12 mm across, much exceeding calyces, blue; pedicels much exceeding leaves, 15–30 mm long; 2 lobes of capsule spreading outward ____ **V. persica**
      5. Corollas inconspicuous, 6 mm or less across, only slightly exceeding calyces, blue or white; pedicels shorter than or slightly exceeding leaves, 5–15 mm long; 2 lobes of capsule erect, not spreading.
         6. Corollas all blue; capsules with long glandular hairs and short eglandular hairs; style exceeding notch in capsule ____ **V. polita**
         6. Corollas white with a blue or pink upper lobe; capsules sparsely glandular-hirsute only; style not exceeding notch in capsule ____ **V. agrestis**

Veronica agrestis L., (of or pertaining to fields), WAYSIDE SPEEDWELL. Very similar to *V. polita* and distinguished primarily by characters in key (following Walters & Webb 1972). Included based on citation of vegetational areas 4 and 5 (Fig. 2) by Hatch et al. (1990); they, however, lump *V. polita* in *V. agrestis*. Holmgren (1986) also lumped the 2 species. *Veronica agrestis* is thus questionably present in nc TX. Native of Europe.

Veronica americana Schwein. ex Benth., (of America), AMERICAN BROOKLIME. Perennial similar to *V. anagallis-aquatica*; stems 20–60 cm long; leaves petioled; pedicels to 10 mm long or more in fruit. Shallow water or stream banks; Coryell Co. (Fort Hood—Sanchez 1997); otherwise in TX known only from the Edwards Plateau. Jun–Sep.

Veronica anagallis-aquatica L., (aquatic *Anagallis*—pimpernel), WATER SPEEDWELL, BROOK-PIMPERNEL. Plant typically perennial; stems, sprawling to decumbent or erect, to 100 cm long; leaves sessile and clasping, oblong-lanceolate, shallowly toothed; flowers numerous; pedicels 4–6(–8) mm long; corollas 5–6 mm across, pale lavender to blue; capsules nearly orbicular. Wet ground or running water; Cooke and Tarrant cos.; also Fannin (Correll & Johnston 1970) and Hamilton (HPC) cos.; Post Oak Savannah w to nc TX, also Edwards Plateau. Apr–Jun. Native of Europe and n Asia.

Veronica arvensis L., (pertaining to cultivated fields), COMMON SPEEDWELL, CORN SPEEDWELL, CORNSPERRY, NECKLACEWEED, NECKWEED. Pubescent, erect or semi-decumbent annual 3–20(–

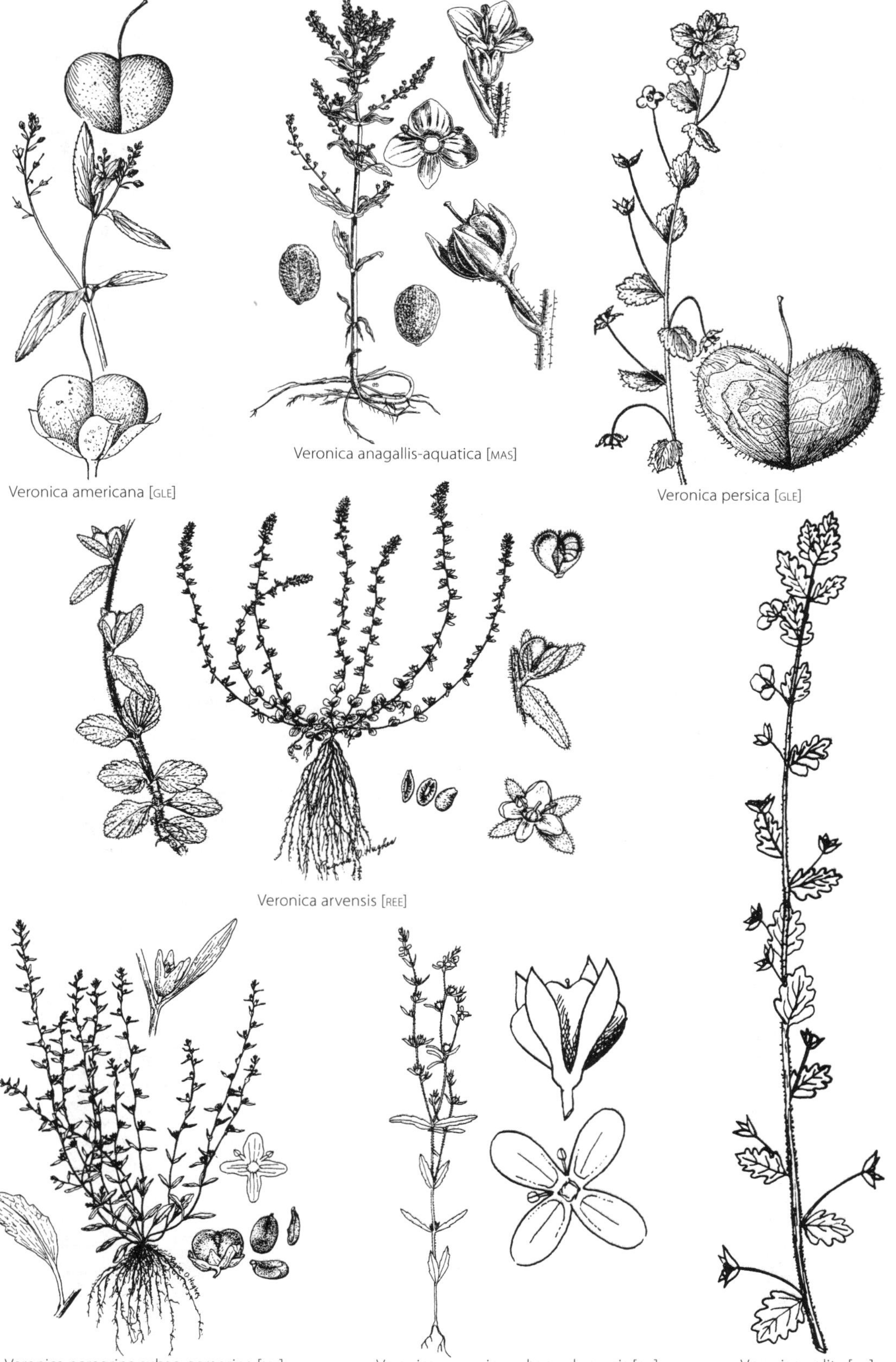

Veronica americana [GLE]

Veronica anagallis-aquatica [MAS]

Veronica persica [GLE]

Veronica arvensis [REE]

Veronica peregrina subsp. peregrina [REE]

Veronica peregrina subsp. xalapensis [STE]

Veronica polita [STE]

30) cm tall; corollas 2–3 mm across. Roadsides, fields, and lawns, chiefly in sandy soils; se and e TX w to Dallas, Denton, Grayson, and Tarrant cos., also Brown Co. (HPC); also Edwards Plateau. Mar–May. Native of Europe. Jack Stanford (pers. comm.) indicated that this species is being spread with ST. AUGUSTINE GRASS.

Veronica peregrina L., (wandering), NECKLACEWEED, NECKWEED, PURSLANE SPEEDWELL. Erect annual 3–30 cm tall; corollas 2–3 mm across. Stream banks, damp woods, roadsides, and disturbed areas. Mar–May.

1. Stems glabrous ______ subsp. **peregrina**
1. Stems glandular-pubescent with short, spreading hairs ______ subsp. **xalapensis**

subsp. **peregrina** Se and e TX w to Dallas and Montague cos.

subsp. **xalapensis** (Kunth) Pennell, (of Jalapa or Xalapa, Mexico), JALAPA SPEEDWELL. Throughout most of TX. [*V. peregrina* var. *xalapensis* (Kunth) Pennell, *V. xalapensis* Kunth]

Veronica persica Poir., (of Persia), PERSIAN SPEEDWELL, BIRD'S-EYE SPEEDWELL. Annual decumbent at base, ascending, to 40 cm tall; corollas violet-blue. Lawns and waste places; Dallas Co., also Tarrant Co. (R. O'Kennon, pers. obs.); mainly c and w TX. Feb–May. Native of Eurasia.

Veronica polita Fr., (elegant, polished). Annual with prostrate or reclining stems to 30 cm long; pedicels in age 5–12 mm long; corollas 3–6 mm across. Lawns and waste places; Dallas and Grayson cos., also Tarrant Co. (R. O'Kennon, pers. obs.); se and e TX w to nc TX. Feb–Mar. Native of Europe. [*V. didyma* Ten. in part] Jones et al. (1997) lumped this species with *V. agrestis.*

VERONICASTRUM CULVER'S-ROOT

A genus of 2 species; 1 e Asia, 1 e North America; this disjunct distribution pattern is discussed under the genera *Campsis* (Bignoniaceae) and *Carya* (Juglandaceae). (From genus *Veronica* and Latin: *astrum*, indicating inferiority or incomplete resemblance, thus false *Veronica*)
REFERENCES: Pennell 1921; Thieret 1955.

Veronicastrum virginicum (L.) Farw., (of Virginia), CULVER'S-PHYSIC, CULVER'S-ROOT. Stout erect perennial herb 0.8–2 m tall, glabrous to villous; leaves in whorls of 3–7; leaf blades lanceolate to ovate-lanceolate, serrate, usually 7–14 cm long and 1–3 cm wide; inflorescence of panicles of spike-like racemes, the terminal one to 15 cm or more long; pedicels < 1 mm long; flowers numerous, crowded; corollas salverform, 4–5.5(–6.5) mm long, white or pinkish, the tube ca. 1 mm in diam.; stamens 2, well-exserted; capsules 3–5 mm long with numerous seeds. Rich woods; Lamar Co. in Red River drainage; ne TX. Jun–Sep. [*Veronica virginica* L.]

SIMAROUBACEAE QUASSIA OR SIMAROUBA FAMILY

A small (110 species in 13 genera) family of mainly tropical trees and shrubs with a few extending to temperate Asia; simaroubilides—triterpenoid lactones—are usually present making the bark, wood, and seeds bitter. Some have been used medicinally, as insecticides, for timber, or as ornamentals. Simaroubaceae are similar to Rutaceae and Meliaceae and possibly similar to the ancestral group from which these families were derived. Family name conserved from *Simarouba*, a genus now treated as *Quassia*, a tropical genus of 35 species. (From the Carribbean name for one of the species) (subclass Rosidae)
FAMILY RECOGNITION IN THE FIELD: the single introduced species in nc TX is easily recognized as a tree with large, alternate, pinnately compound leaves with leaflets having a few, *basal, gland-bearing teeth* (the leaflets are without the pellucid dots found in the similar Rutaceae); flowers small and numerous in panicles.
REFERENCES: Small 1911; Brizicky 1962c.

AILANTHUS TREE-OF-HEAVEN

A genus of 5 trees native from Asia to Australia; fossils of *Ailanthus* are known from North America (Graham 1993a). (From *ailanto*, the Indonesian name for *A. moluccana* DC., meaning reaching for the sky, sky tree, or tree of heaven, in allusion to the height in its native habitat)

Ailanthus altissima (Mill.) Swingle, (very tall, tallest), TREE-OF-HEAVEN, COPALTREE, SMOKETREE. Rapidly growing tree to 20 m tall, propagated by seeds and root sprouts, forming colonies; branches brittle; leaves alternate, short-petioled, glabrous, odd- or occasionally even-pinnately compound with up to 27 leaflets, ill-smelling when crushed; leaflets lanceolate, acuminate, entire except for a few basal teeth, these teeth usually with a prominent gland on the lower surface near apex; flowers many, small, in terminal panicles over-topped by leaves, mostly unisexual, the sexes on separate trees or some perfect; the male flowers have an offensive odor; sepals 5, partly united; petals 5, yellow-green, pilose toward base; stamens 10; ovary 2- to 5-parted into flat 1-celled sections, superior; fruit a schizocarp with 2–5, one-seeded, samara-like, twisted mericarps 3–5 cm long, greenish or yellowish becoming pinkish or reddish brown. Cultivated as an ornamental tree, escaped, becoming weedy; Grayson and Tarrant cos., also Brown, Hamilton (HPC), and Parker (R. O'Kennon, pers. obs.) cos.; mainly e TX w to nc TX, also Edwards Plateau. Apr. Native of China, apparently introduced into the U.S. in Pennsylvania in 1784 (Tellman 1997). Pistillate trees with clusters of colorful fruits can be quite showy. This resilient species can withstand drought, is disease and insect resistant, can survive harsh urban environments, and tolerates smoke—hence one of the common names (McGregor 1986; Cox & Leslie 1991). The species also produces a highly phytotoxic herbicidal material (ailanthone) that has allelopathic effects on other plants (Heisey 1990, 1996). The leaves are toxic if ingested; cases of dermatitis have been reported from contact with the leaves (Fuller & McClintock 1986; Blackwell 1990) ☠

SOLANACEAE NIGHTSHADE OR POTATO FAMILY

Ours annual or perennial herbs or shrubs; leaves usually alternate or subopposite, simple or pinnately compound, entire, bluntly toothed or lobed; stipules absent; flowers axillary or terminal, solitary or in spikes, racemes, panicles, or cymes; calyces usually 5-toothed or -lobed; corollas radially symmetrical, sympetalous, rotate to campanulate, tubular, funnelform, salverform, or urceolate, 5-angled or -lobed; stamens 5; pistil 2-carpellate; ovary superior with axile placentation; style and stigma 1; fruit a many-seeded berry or capsule.

A large ((ca. 2,200 species—M. Nee, pers. comm.) in 94 genera), nearly cosmopolitan, but especially South American family of mainly herbs with a few shrubs, trees, and lianas; it is sometimes treated as having up to 2,950 species (Mabberley 1997); branched hairs and prickles are often present. The family contains many important food crops (e.g., EGGPLANT, GREEN and CHILI PEPPERS, POTATOES, TOMATOES, TOMATILLOS). ☠ It also is rich in alkaloids (e.g., atropine, hyoscyamine, nicotine, scopolamine, and solanine) and many species are variously toxic, hallucinogenic, or medicinal. Examples include *Atropa* (BELLADONA, source of atropine), *Datura* (JIMSON WEED), *Hyoscyamus* (BLACK HENBANE, source of hyoscyamine), *Mandragora* (MANDRAKE) and *Nicotiana* (TOBACCO). Some of the tropane alkaloids (e.g., hyoscyamine) are hallucinogenic and were used by sorcerers in the Middle Ages; other solanaceous alkaloids have been used medicinally (Keeler 1979). (subclass Asteridae)

FAMILY RECOGNITION IN THE FIELD: mostly herbs or some shrubs usually with alternate leaves; corollas sympetalous, *radially symmetrical* (in contrast to the bilaterally symmetrical flowers of the somewhat similar Scrophulariaceae), folded in bud (often folded lengthwise—accordion-like—below the ovary); stamens 5, attached to the corolla; fruit a many-seeded berry or capsule.

REFERENCES: Heiser 1969, 1987; D'Arcy 1986; Nee 1986; Hawkes et al. 1979, 1991; Olmstead & Palmer 1992.

1. Corollas short-campanulate to rotate, or with reflexed lobes, broader than high or long; fruit a berry.
   2. Corollas 5-angled, the margin barely indented; calyces divided 1/10–2/3 to base, enlarged at maturity, about as long as fruits or longer, covering the fruits or nearly so; flowers solitary, axillary.
      3. Calyces divided 1/3–2/3, with ovate, subacute lobes, not greatly inflated at maturity, close fitting to fruits ____ **Chamaesaracha**
      3. Calyces divided less than 1/3, greatly inflated at maturity, very loose-fitting.
         4. Corollas blue-purple (rarely white); flowers erect (but fruits can be nodding) ____ **Quincula**
         4. Corollas greenish or yellowish brown to white; flowers nodding ____ **Physalis**
   2. Corollas 5-lobed, divided 1/4 to base or more; calyces divided from ca. 1/2 to near base, much shorter than mature fruits, not covering the fruits; flowers solitary, racemose, cymose, or in axillary clusters.
      5. Often vine-like shrubs to 3 m tall with drooping branches and greenish purple to pink flowers; leaves spineless; filaments longer than anthers ____ **Lycium**
      5. Herbs (or if a shrub—*Capsicum* only, then flowers white); flowers bluish purple to yellow or white; leaves with spines or spineless; filaments shorter than anthers.
         6. Flowers in racemes or cymes; plants spiny or spineless; corollas bluish purple to yellow or white, from ca. 2–50 mm wide; anthers incurved to erect, usually yellowish; plants branching mainly alternately ____ **Solanum**
         6. Flowers solitary; plant spineless; corollas white, ca. 7 mm wide; anthers erect becoming wide-spreading, usually bluish; plants branching dichotomously ____ **Capsicum**
1. Corollas funnelform, salverform, tubular, or urceolate, longer than broad; fruit a capsule (or a berry in *Margaranthus*).
   7. Corollas cylindric-urceolate (urn-shaped), to 4 mm long; fruit a berry; calyces becoming inflated and enclosing fruits ____ **Margaranthus**
   7. Corollas funnelform, salverform, or tubular, 5–150 mm long; fruit a capsule; calyces neither inflated nor enclosing the fruits.
      8. Corolla lobes tipped by acuminate, almost thread-like points; capsules armed with spines ____ **Datura**
      8. Corolla lobes obtuse to subacute; capsules unarmed.
         9. Leaves < 2 mm wide; corollas blue with yellow eye, salverform; corolla tube < 1 mm in diam., conspicuously filiform ____ **Nierembergia**
         9. Leaves usually > 2 mm wide; corollas variously colored (including white) but not blue with yellow eye, funnelform to salverform; corolla tube > 1 mm in diam., not filiform.
            10. Flowers solitary, axillary; plants glandular-pubescent; capsules 2-valved; corollas variously brightly colored (including white), often with stripes or markings.
               11. Corollas (25–)50–90 mm long; leaves to 80 mm long; plants erect or decumbent ____ **Petunia**
               11. Corollas ca. 5–8 mm long; leaves 5–14 mm long; plants prostrate and rooting at the nodes ____ **Calibrachoa**
            10. Flowers solitary or in racemose to paniculate inflorescences; plants not glandular-pubescent; capsules 4-valved; corollas white (or tinged with rose or lavender or dorsally brown-striped).
               12. Corollas 12–18 mm long; leaves to ca. 1 cm wide ____ **Bouchetia**
               12. Corollas 45–75 mm long; leaves much more than 1 cm wide ____ **Nicotiana**

BOUCHETIA

A genus of 3 species ranging from the s U.S. to Brazil. (Named for Dominici Bouchet Avenionensis, 1770–1845, French botanist at Montpellier)

Bouchetia erecta DC., (erect, upright), ERECT BOUCHETIA. Ascending much-branched perennial to ca. 23 cm tall; leaves oblong-spatulate to lanceolate or oval, to 5 cm long and 1 cm wide, usually much smaller; corollas funnelform, white, sometimes tinged with lavender, 12–18 mm long;

capsules unarmed, ca. 8 mm long. Prairies and rocky slopes; Bell (Fort Hood—Sanchez 1997) and Burnet (Balcones Canyonlands Nat. Wildlife Refuge, C. Sexton, pers. comm.) cos.; mainly c and s TX; endemic to TX. May–Oct.

CALIBRACHOA

A New World genus of 7–15 species; often incuded in the genus *Petunia*. (Named for Antonio Cal y Bracho, 19th century Mexican botanist and professor, born in Spain)
REFERENCES: Sink 1984a, 1984b; Wijsman & de Jong 1985; Wijsman 1990.

Calibrachoa parviflora (Juss.) D'Arcy, (small-flowered), WILD PETUNIA, SEASIDE PETUNIA. Prostrate glandular-pubescent annual rooting at the nodes, diffusely branched and mat forming; leaves linear-oblong to spatulate, fleshy, 5–14 mm long, entire; flowers solitary, axillary; calyces 5-parted to below middle, the lobes 3–6 mm long at anthesis, to ca. 11 long in fruit; corollas funnelform, ca. 5–8 mm long, purple or reddish violet with yellow or whitish tube, the lobes short; stamens 5, often with 3 shorter and 2 longer filaments; fruit a 2-valved unarmed capsule 2–4 mm long. Moist or wet soils, stream beds, muddy flats; according to Hatch et al. (1990) in regions 4 and 5; the nearest specimens we have seen are from s Burnet (Marble Falls) and Travis cos. just s of nc TX; throughout much of TX. Apr–Sep. This species is known from South America and from Cuba, Mexico, and the s U.S. including TX; Sink (1984b) indicated that it is not clear whether the plants outside South America are indigenous or not. [*Petunia integrifolia* (Hook.) Schinz & Thell., *Petunia parviflora* Juss.]

CAPSICUM CAYENNE PEPPER, CHILI, CHILI PEPPER

A mainly tropical American genus of 10 species. The common name CHILI is derived from Nahuatl (language of the Aztecs), *chilli* (Foster & Cordell 1992). (Greek: *kapto*, to bite, referring to the hot taste of the fruits)
REFERENCES: Irish 1898; Heiser & Smith 1953; Heiser & Pickersgill 1975; Andrews 1984, 1995; Bosland 1993.

Capsicum annuum L. var. **glabriusculum** (Dunal) Heiser & Pickersgill, (sp.: annual; var.: rather or somewhat smooth), BIRD PEPPER, CHILITEPÍN, CHILIPIQUÍN, CHILE PIQUÍN, BUSH REDPEPPER. Small shrub, rather inconspicuously pubescent on young parts; leaf blades ovate to lanceolate, entire, longer than the petioles; calyces subentire; corollas white, ca. 7 mm wide; fruits ovoid to nearly globose, to ca. 15 mm long, reddish or yellowish, pungently aromatic. Cultivated; reported wild as far n as McLennan Co. (Mahler 1988); mainly Edwards Plateau and s TX southward. Jun–Nov. [*C. annuum* L. var. *minus* (Fingerh.) Shinners, *C. annuum* L. var. *aviculare* (Dierb.) D'Arcy & Eshbaugh] We are following Heiser and Pickersgill (1975) and Kartesz (1994) for nomenclature of this taxon; Jones et al. (1997) followed D'Arcy and Eshbaugh (1973) in treating it as *C. annuum* var. *aviculare* (Dierb.) D'Arcy & Eshbaugh. This variety is extremely "hot" or picante. According to Andrews (1998), "Francisco Hernández, the first European to collect American plants systematically, described the tiny Chiltepín in 1615. ... President Thomas Jefferson grew Chiltepines from Texas seed acquired in 1813." Andrews (1998) also indicated that Texas House Concurrent Resolution 82 made the Chiltepín the official **native pepper of Texas** in 1997. The seeds are dispersed by birds which eat the fruits (Andrews 1998).

Other cultivated varieties of *C. annuum* provide the familiar GREEN or BELL PEPPERS, RED PEPPERS, and various HOT or CHILI PEPPERS. CHILI PEPPERS have been cultivated for thousands of years in Mexico. Their pungency is due to a volatile phenolic compound known as capsaicin. In addition to making food "hot," the material has been used both as a torture and as a pain reliever for muscular aches (Heiser 1987). Black pepper comes from a completely unrelated Asian member of the Piperaceae. The JALAPEÑO (*C. annuum* var. *annuum*) was designated as the **state pepper of Texas** by the 74th state legislature (Jones et al. 1997).

CHAMAESARACHA FALSE NIGHTSHADE

Rhizomatous perennial erect to spreading herbs; rhizomes partly woody; flowers solitary or in axillary pairs; calyces 5-lobed, somewhat enlarged but not greatly inflated with age, close-fitting to the fruit, not angled or ribbed; corollas rotate, white to yellowish, often purple-tinged; anthers longitudinally dehiscent; fruit a globose berry. Because hair characters are so important in this genus, a hand lens or dissecting scope is nearly essential for accurate identification.

A genus of 7 species of the sw U.S. and n Mexico. (Greek: *chamai*, on the ground, and *Saracha*, name of a tropical American genus of Solanaceae)
REFERENCES: Averett 1972, 1973, 1979.

1. Stems glabrous to sparsely pubescent with short stellate (branched with several rays) or branched (2 rays) hairs.
   2. Leaves glabrous; pedicels with glandular hairs ____ **C. edwardsiana**
   2. Leaves sparsely pubescent with stellate hairs; pedicels without glandular hairs ____ **C. coronopus**
1. Stem pubescence glandular or hairs predominantly simple or both.
   3. Leaves essentially glabrous; stems purplish or grayish at base ____ **C. edwardsiana**
   3. Leaves with glandular and/or simple hairs; stems green at base.
      4. Leaves rhombic to broadly lanceolate or ovate, entire to undulate or somewhat lobed, with pubescence mainly of short glandular hairs (other hairs also often present) ____ **C. sordida**
      4. Leaves usually with numerous pinnate or irregular lobes, with pubescence of longer usually simple hairs, occasionally mixed with glandular hairs ____ **C. coniodes**

Chamaesaracha coniodes (Moric. ex Dunal) Britton, (cone-like), GROUND SARACHA, FALSE NIGHTSHADE, PROSTRATE GROUND-CHERRY. Plant usually pubescent with simple and occasionally also glandular hairs. Disturbed areas; Lampasas Cut Plain and West Cross Timbers s and w to w TX. Apr–Oct.

Chamaesaracha coronopus (Dunal) A. Gray, (Greek: *korone*, crow, and *pous*, foot, possibly from resemblance to *Coronopus*—wartcress in Brassicaceae), GREEN FALSE NIGHTSHADE. Plant glabrous to sparsely pubescent, hairs short, stellate, scurfy; leaves linear, usually pinnately lobed. Disturbed areas; Bell and Lampasas cos., also Brown Co. (HPC); Lampasas Cut Plain s and w to w TX. Apr–Oct.

Chamaesaracha edwardsiana Averett, (of the Edwards Plateau), PLATEAU FALSE NIGHTSHADE. Stems glabrous to pubescent, hairs stellate, simple, or glandular; leaves essentially glabrous. Disturbed areas; Williamson Co.; s most part of nc TX s to Edwards Plateau. Apr–Oct.

Chamaesaracha sordida (Dunal) A. Gray, (dirty), HAIRY FALSE NIGHTSHADE. Plant densely pubescent with mainly glandular hairs, sometimes mixed with simple hairs. Dry plains, often on limestone; Fort Hood (Bell or Coryell cos.—Sanchez 1997); Hatch et al. (1990) cited both vegetational areas 4 and 5 (Fig. 2); mainly w 2/3 of TX. Mar–Nov.

DATURA JIMSONWEED, THORN-APPLE, STRAMONIUM

Rank-smelling, low but coarse annuals or perennials; leaf blades rhombic-ovate to oblong-elliptic, subentire to coarsely toothed; flowers very large and showy, solitary, terminal but ± overtopped by leafy branches, opening in late afternoon, closing about mid-morning; corollas white to lavender; capsules globular, prickly.

A genus of 9 species of herbs native to s North America, but widely naturalized. ☠ *Datura*

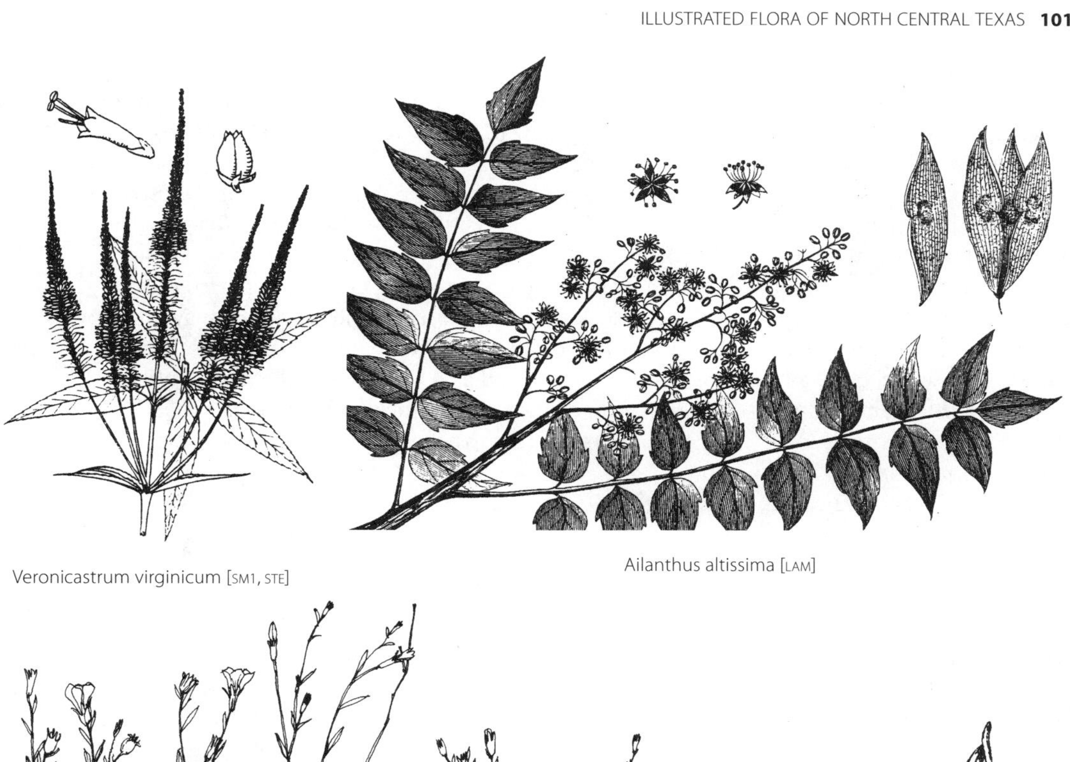

Veronicastrum virginicum [SM1, STE]

Ailanthus altissima [LAM]

Calibrachoa parviflora [AJB]

Bouchetia erecta [HEA]

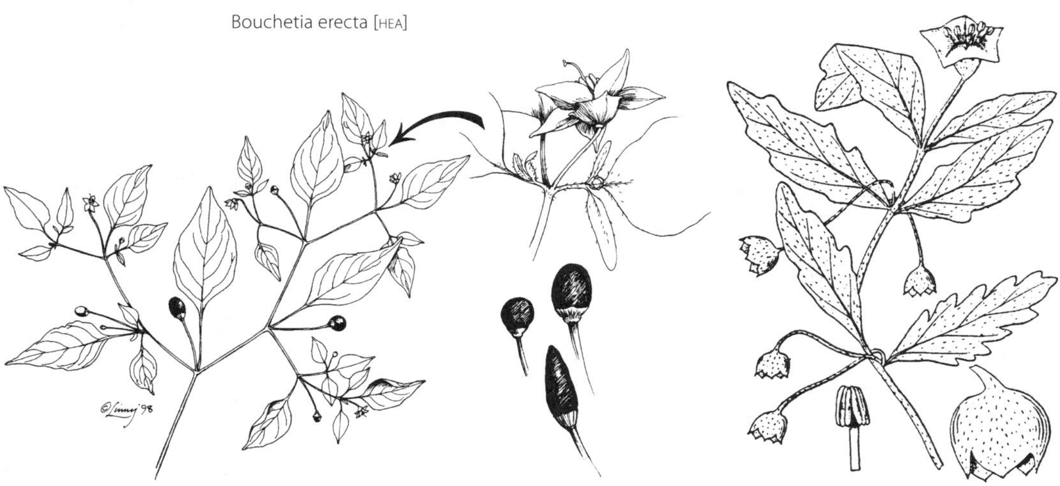

Capsicum annuum var. glabriusculum [HEA]

Chamaesaracha coniodes [BB1]

species contain hallucinogenic and extremely toxic tropane alkaloids (e.g., hyoscamine and hyoscine); all parts of the plant, including the nectar, are poisonous. Fatalities have been reported in Texas in recent years caused by the ingestion of *Datura* (Rivas 1994). The species have been used ritually by some Native Americans including the Aztecs and Algonquins. Some species are also used as ornamentals because of their large, moth-pollinated flowers. *Datura* is sometimes interpreted broadly to include the related South American genus *Brugmansia* (ANGEL'S-TRUMPETS), also with hallucinogenic alkaloids. ☠ *Brugmansia candida* Pers. (ANGEL'S-TRUMPET), is a well-known cause of poisoning in the se U.S.; Greene et al. (1996) reported numerous cases in Florida; the tropane alkaloids block the action of acetycholine at nerve synapses resulting in anticholinergic poisoning; according to Greene et al. (1996), the mnemonic "hot as a hare, dry as a bone, blind as a bat, red as a beet, and mad as a hatter" describes such patients; coma and death can result; *Datura stramonium* causes an identical syndrome of anticholinergic intoxication (Greene et al. 1996). (Altered from the Arabic name, *tatorah*, or the Hindustani, *dhatura*)

REFERENCES: Safford 1921, 1922; Avery et al. 1959.

1. Plants largely glabrous; calyces 3.5–5 cm long; corollas 6–8(–10) cm long and 3–5 cm wide ____ **D. stramonium**
1. Plants finely and densely gray-pubescent; calyces 7–12 cm long; corollas to 15 cm long and to 15 cm wide ____ **D. wrightii**

Datura stramonium L., (old generic name, said to be from Latin: *struma* or *strama*, a swelling), JIMSONWEED, JAMESTOWN WEED, STRAMONIUM, TOLOACHE, COMMON THORN-APPLE. Annual. Farmyards and fields; Coryell Co., also Grayson Co. (S. Crosthwaite, pers. obs.); widely scattered throughout TX. Jun–Oct. Native of tropical America. Used medicinally and as a narcotic by Native Americans including the Zuni Indians (Burlage 1968). However, the species is poisonous; all parts of the plant contain **potentially fatal** tropane alkaloids including hyoscine, hyoscyamine, scopolamine, and atropine; death following ingestion has occurred in all classes of livestock as well as humans; symptoms include dilation of pupils, impared vision, dryness of the skin and mucous membranes, extreme thirst, hallucinations, and convulsions followed by coma and sometimes death (Kingsbury 1964; Burlage 1968; Keeler 1979; Urich et al. 1982). The common name JIMSONWEED results from a corruption of JAMESTOWN WEED; British soldiers sent to Jamestown to quell Bacon's rebellion in 1676 ate *D. stramonium* as greens and were intoxicated for several days (Morton 1982). ☠

Datura wrightii Regel, (for Charles Wright, 1811–1885, TX collector), INDIAN-APPLE, ANGEL-TRUMPET, SACRED DATURA. Pithy-rooted perennial flowering the first year. Sandy or rocky stream bottoms; Bell, Coleman, Denton, Grayson, Parker, and Tarrant cos., also Brown and Hamilton cos. (HPC) and Fort Hood (Bell or Coryell cos.—Sanchez 1997); according to Mahler (1988), probably native only from Hood Co. s and w; widespread in TX. Jun–Oct. [*D. meteloides* of authors, not Dunal] The flowers open at dusk and are pollinated by hawk moths (Wills & Irwin 1961). The plant was used medicinally and as a narcotic by Native Americans including the Aztecs (Burlage 1968); during frontier days small quantities of the leaves were smoked by asthmatics; however, it is poisonous and dangerous or fatal if not used properly (Crosswhite 1980). ☠ /87

LYCIUM MATRIMONY-VINE, WOLFBERRY, DESERT-THORN, SQUAW-BERRY, TOMATILLO, CILINDRILLO

A warm temperate, especially American genus of 100 species of often thorny shrubs. A number of species are native to c and w Texas. (Greek: *lykion*, after Lycia, ancient country of Asia Minor; the name was originally applied to a species of *Rhamnus*)

REFERENCES: Hitchcock 1932; Chiang 1981.

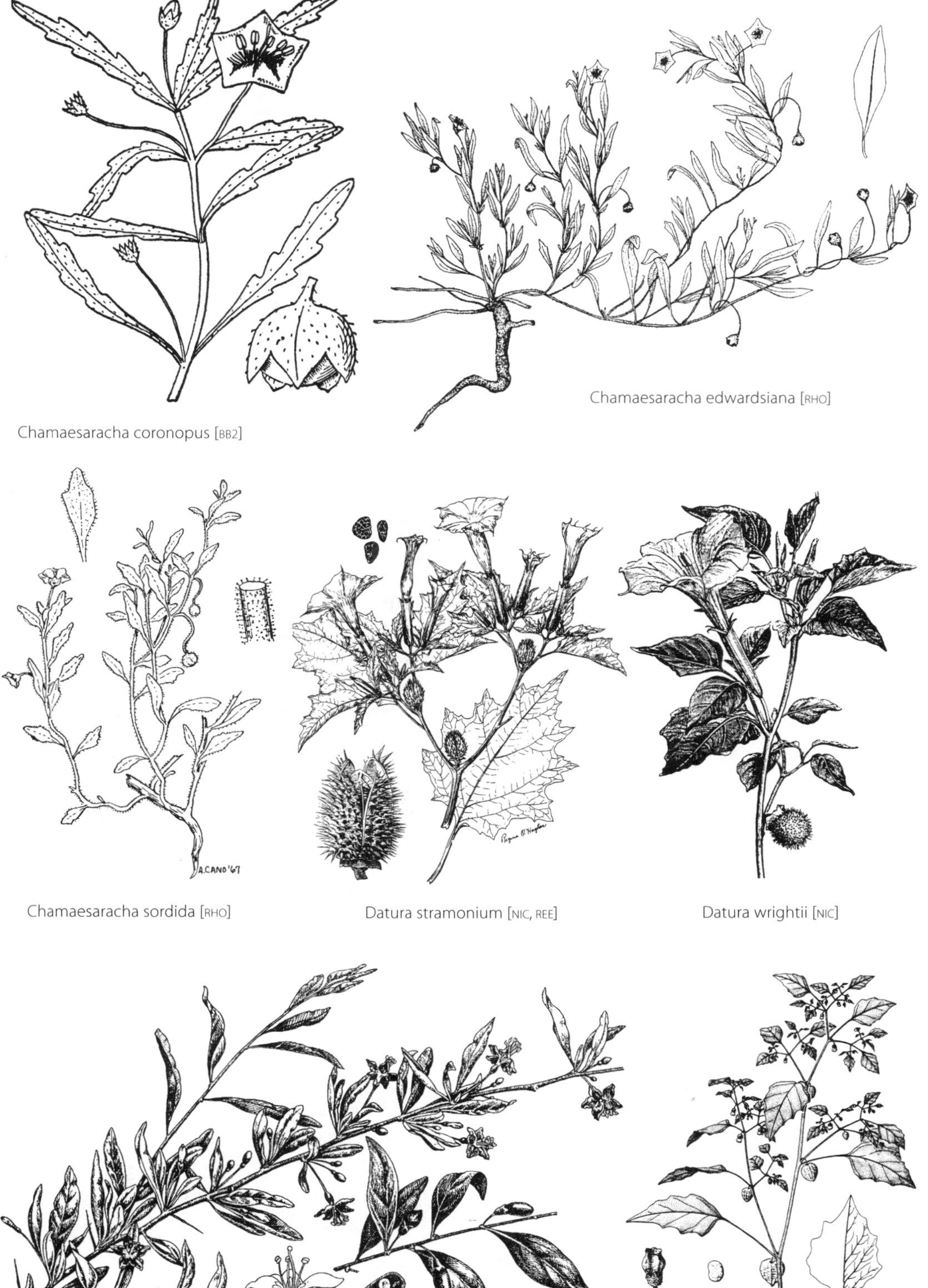

Chamaesaracha coronopus [BB2]

Chamaesaracha edwardsiana [RHO]

Chamaesaracha sordida [RHO]

Datura stramonium [NIC, REE]

Datura wrightii [NIC]

Lycium barbarum [COO]

Margaranthus solanaceus [NEE]

Lycium barbarum L., (foreign), MATRIMONY-VINE, BOX-THORN, FALSE JESSAMINE. Shrub to ca. 3 m tall, often vine-like; branches long, arching, drooping, old growth can be armed at nodes; leaves elliptic to ovate or oblanceolate, 3–5(–7) cm long, glabrous, tapering to the petiole; flowers in axillary clusters of 2–7; corollas greenish purplish, purple, or pinkish to violet or pale lavender, 8–12 mm long; fruit a red or orangish (drying purplish or blackish) ovoid berry 1–2 cm long. Old-fashioned cultivar that occasionally long persists or spreads; Grayson (Hagerman N.W.R., long abandoned homesite) and Cook (hedgerow) cos.; other TX localities not known. May–Aug. Introduced from the Old World. [*L. halimifolium* Mill.] Jones et al. (1997) did not list this species for TX. The leaves are poisonous, possibly due to alkaloids (Lampe & McCann 1985). ☠

MARGARANTHUS NETTED GLOBE-CHERRY

A monotypic genus native from the sw U.S. to Central America. (Possibly from Greek: *margarites*, pearl, and *anthos*, flower)
REFERENCE: Averett 1979.

Margaranthus solanaceus Schltdl., (resembling *Solanum*—nightshade), NETTED GLOBE-CHERRY. Erect nearly glabrous annual to ca. 60 cm tall, resembling *Physalis*; corollas greenish, yellow, or purple, cylindric-urceolate (urn-shaped), to 4 mm long; berry entirely enclosed in the 8–12 mm long, membranous, inflated fruiting calyx. Fields and pastures; included based on citation for vegetational area 4 (Fig. 2) by Hatch et al. (1990); mainly s and w TX. Jul–Nov.

NICOTIANA TOBACCO

A genus of 67 species native to the Americas, s Pacific, Australia, and sw Africa. While *Nicotiana tabacum* L., native of tropical America, is the primary source of TOBACCO products, containing the addictive alkaloid nicotine, probably all species in the genus contain nicotine; ☠ it is often used as a potent insecticide and is fatal to humans in rather small doses (Kingsbury 1964). Exposure to TOBACCO, through smoking, chewing, or other forms of ingestion, is widely considered to be a cause of various cancers including lung cancer. Numerous reports exist in the literature of the tetragenetic effects (= causing birth defects) of TOBACCO (Keeler 1979). *Nicotiana tabacum* and *N. rustica* L. were cultivated by Native Americans in pre-Colombian times. (Named for Jean Nicot, 1530–1600, who sent the seeds of tobacco, *N. tabacum*, to France in 1560)
REFERENCE: Goodspeed 1954.

Nicotiana repanda Willd., (with wavy margins), WILD TOBACCO, FIDDLE-LEAF TOBACCO, TOBACCO CIMARRÓN. Pubescent annual to 0.9 m tall with large basal leaves; stem leaves wing-petioled or sessile, the upper auricled-clasping; calyces divided ca. 1/2 its length; corollas usually white, tubular-funnelform, 4.5–7.5 cm long; to 4 cm in diam.; capsules ca. 1 cm long. Stream bottoms, ravines, thickets, and roadsides; collected at Dallas by Letterman, not found there recently (Mahler 1988); s Texas n to Travis Co. Apr–Jun. The foliage is poisonous and was formerly used for smoking; it contains alkaloids including nicotine (Burlage 1968; Ajilvsgi 1984). ☠

NIEREMBERGIA CUPFLOWER

A genus of 23 species native from Mexico to Chile, especially Argentina; characterized by a long slender corolla tube; a number are cultivated as ornamentals. (Named for John Eusebius Nieremberg, ca. 1595–1658, a Spanish Jesuit and first professor of natural history at Madrid)
REFERENCE: Millán 1941.

Nierembergia hippomanica Miers var. **coerulea** (Miers) Millán, (sp.: derivation not known, possibly Greek, *hippos*, horse and Latin: *manica*, sleeve of a tunic or glove; var.: blue), TEXAS CUPFLOWER. Perennial to ca. 20 cm tall, with non-glandular pubescence; leaves linear to linear-spatulate, < 2 mm wide, entire; flowers solitary, axillary, or in cymose inflorescences; pedicels 3–5 mm long at flowering; calyx tubular, 6–10 mm long, with 10 conspicuous green ribs, the lobes acuminate, > half as long as the calyx tube; corollas with filiform tube 7–10 mm long; corolla limb saucer-shaped, pleated, ca. 2 cm in diam., bright blue with yellow eye; stamens inserted on the upper part of the corolla tube; fertile stamens 4, these in 2 pairs; staminode 1, smaller; anthers yellow; fruit a 2-valved capsule. Several individuals apparently escaped from cultivation were found in an open weedy area of loose sand near the Texas welcome facility along Hwy. 75 just s of the Red River (*Rabeler & Diggs 1316*, 1998) in Grayson Co., also Dallas Co. on a trash pile, also cited by Hatch et al. (1990) for vegetational area 4; mainly se and s TX and Edwards Plateau. Apr. Native of South America. The species is reported to be toxic to cattle (Cabrera 1965).

PETUNIA

A tropical and warm American genus of 35 species; at least 1 South American species is the source of a hallucinogen inducing a sense of flying or levitation; species now treated as *Calibrachoa* have often been lumped with *Petunia. Petunia axillaris* is an important cultivated ornamental; there are over 200 cultivars including some that are striped and doubled. The generic name *Petunia* is conserved (Wijnands et al. 1986; Brummitt 1989). (From *petun*, Brazilian vernacular name for tobacco in the related genus *Nicotiana*)

REFERENCES: Sink 1984a, 1984b; Wijsman & de Jong 1985; Brummitt 1989; Wijsman 1990.

Petunia axillaris (Lam.) Britton, Sterns, & Poggenb., (axillary), COMMON GARDEN PETUNIA. Erect to decumbent glandular-pubescent annual herb; lower (larger) leaves elliptic to ovate, to ca. 80 mm long, entire; flowers axillary, solitary; calyces 5-parted nearly to base; corollas ca. (25–)50–90 mm long, funnelform to salverform, the limb very broad, variously colored, white to deep reddish purple, sometimes with stripes or markings; stamens 5, 1 smaller or rudimentary; fruit a 2-valved unarmed capsule. Widely cultivated and rarely escapes, weedy areas, cracks in sidewalks; Grayson and Tarrant cos. Late spring onward. Native of South America. [*Nicotiana axillaris* Lam, *P. hybrida* Vilm., *P. violacea* of authors, not Juss., *P. ×atkinsiana* D. Don. ex J.W. Loudon, *Stimoryne axillaris* Wijsman]

PHYSALIS GROUND-CHERRY

Glabrous or pubescent, erect or partly decumbent, low annuals or perennials; leaves petioled, alternate, the upper often subopposite; flowers axillary, solitary, drooping; corollas yellow, usually with dark (often brownish purple) eye; calyces inflated in age, enclosing the berries.

A cosmopolitan, but especially American, genus of 80 species. All tissues of some species, but particularly the unripe fruits, contain toxic glycoalkaloids (e.g., solanine); children have been poisoned and cattle have been sickened (Stephens 1980; Lampe & McCann 1985). The TOMATILLO or JAMBERRY (*P. philadelphica* Lam.) is cultivated for its edible fruit, especially in Mexico. *Margaranthus*, with cylindric-urceolate corollas to only 4 mm long, resembles *Physalis*. (Greek: *physa*, bladder, referring to the conspicuously inflated calyx)

REFERENCES: Rydberg 1896; Menzel 1951; Waterfall 1958; Sullivan 1985.

1. Plants with stellate (= star-like with a number of branch hairs from a central point) hairs or jointed hairs some of which are variously branched (use lens).
   2. Flowering calyces 3–10 mm long; fruiting calyces 15–35 mm long; hairs stellate.
      3. Flowering calyces (3–)5–7(–9) mm long; hairs usually relatively sparse, not mat-like, not obscuring leaf surface; pubescence of stellate hairs < 1 mm long only; leaf margins toothed, undulate, or entire ____ **P. cinerascens**
      3. Flowering calyces (6–)7–10 mm long; hairs forming a dense mat at least on the lower surface of the leaf, obscuring the surface; pubescence of stellate hairs < 1 mm long and some-

times branched or unbranched hairs 2–4 mm long; leaf margins coarsely and irregularly few-toothed ____ **P. mollis**

2. Flowering calyces 10–15 mm long; fruiting calyces 30–40 mm long; hairs jointed with some branched ____ **P. pumila**

1. Plants with simple, sometimes glandular hairs (stellate or otherwise branched pubescence small and inconspicuous if any present) to nearly glabrous.
   4. Anthers blue or violet, 1–2.5(–3) mm long; plants annual.
      5. Plants essentially glabrous; flowering calyces ca. 3(–4) mm long; fruiting calyces 10-angled or ribbed ____ **P. angulata**
      5. Plants long pubescent, often viscid; flowering calyces (3–)4–10 mm long; fruiting calyces 5-angled.
         6. Principal leaf blades usually 2–4 cm wide; fruiting calyces 15–30(–32) mm long, the lobes ovate-triangular to lanceolate ____ **P. pubescens**
         6. Principal leaf blades usually 4–8 cm wide; fruiting calyces 30–40 mm long, the lobes usually narrowly lanceolate-attenuate ____ **P. turbinata**
   4. Anthers yellow (sometimes with a bluish tinge or bluish edges, especially in *P. longifolia* var. *subglabrata*), (2–)2.5–5 mm long; plants perennial from deep horizontal root.
      7. Stems usually with short viscid-glandular hairs and long-jointed hairs; leaf blades ovate ____ **P. heterophylla**
      7. Stems eglandular; hairs long or short, incurved; leaf blades ovate or lanceolate.
         8. Stems and leaves villous to short pubescent, at least some of the hairs reflexed or retrorse; calyces with spreading hairs; leaf blades ovate to lanceolate ____ **P. virginiana**
         8. Stems and leaves sparsely pubescent with short antrorse hairs, sometimes nearly glabrous; calyces often with short antrorse appressed hairs in 10 lines or hairs nearly absent; leaf blades linear-lanceolate to lanceolate (rarely ovate) ____ **P. longifolia**

Physalis angulata L., (angular, angled), CUT-LEAF GROUND-CHERRY, SOUTHWEST GROUND-CHERRY, LANCE-LEAF GROUND-CHERRY, PURPLE-VEIN GROUND-CHERRY. Essentially glabrous annual; leaves ovate to ovate-lanceolate; anthers ca. 1–2.5 mm long; fruiting calyces ca. 3(–4) cm long; berry 10–12 mm in diam. Prairies, disturbed areas; nearly throughout TX. May–Oct. [*P. pendula* Rydb., *P. angulata* var. *pendula* (Rydb.) Waterf.]

Physalis cinerascens (Dunal) Hitchc., (becoming ashy-gray), BEACH GROUND-CHERRY. Perennial with stellate pubescence of varying size and density; leaf blades ovate to reniform; anthers yellow, ca. 3 mm long; flowering calyces (3–)5–7(–9) mm long. Disturbed areas; throughout TX. Apr–Oct. [*P. viscosa* L. var. *cinerascens* (Dunal) Waterf.]

Physalis heterophylla Nees, (various-leaved), CLAMMY GROUND-CHERRY. Perennial with deep rootstock; stems with varying proportions of short, viscid-glandular hairs and long-jointed hairs; leaf blades ± ovate; flowering calyces 7–12 mm long; fruiting pedicels 15–40 mm long; anthers yellow, often bluish-tinged, usually 3–4.5 mm long; fruiting calyces 2.5–3 cm long. Sandy, disturbed areas; Bell and Palo Pinto cos.; se and e TX w to nc TX and Edwards Plateau. Apr–Oct. Suspected of being poisonous (Burlage 1968). ☠

Physalis longifolia Nutt., (long-leaved), COMMON GROUND-CHERRY. Perennial, stems often purplish. Suspected of being poisonous (Burlage 1968). ☠

1. Leaf blades usually linear-lanceolate to lanceolate (rarely ovate); anthers yellow ____ var. **longifolia**
1. Leaf blades usually ovate to ovate-lanceolate; anthers light blue or tinged with light blue ____ var. **subglabrata**

var. **longifolia**. Open woods, prairies; Dallas, Grayson, and Rockwall cos.; throughout most of TX. Apr–Oct. [*P. virginiana* var. *sonorae* (Torr.) Waterf.]

Nicotiana repanda [GOO]

Nierembergia hippomanica var. coerulea [HEA]

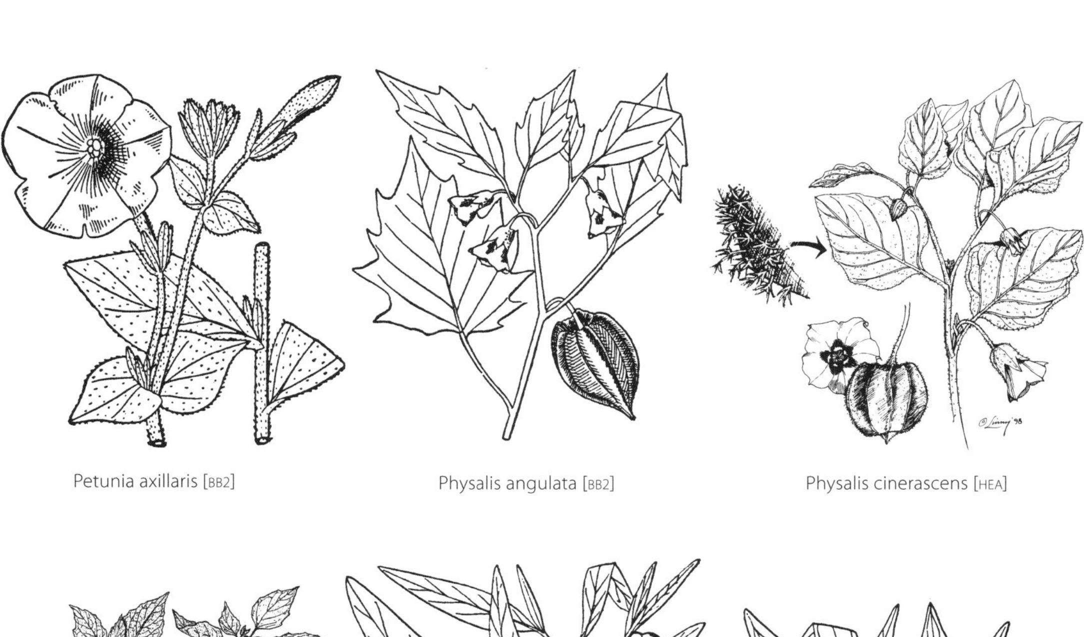

Petunia axillaris [BB2]

Physalis angulata [BB2]

Physalis cinerascens [HEA]

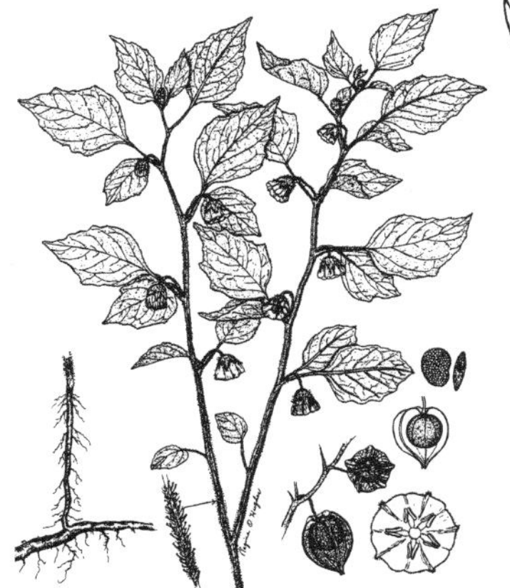

Physalis heterophylla [REE]

Physalis longifolia var. longifolia [BB1]

Physalis longifolia var. subglabrata [BB2]

var. **subglabrata** (Mack. & Bush) Cronquist, (somewhat smooth or hair-less), BLADDER GROUND-CHERRY. Some authors (e.g., McGregor et al. 1986) questioned the recognition of this variety. Included based on citation of vegetational area 5 (Fig. 2) by Hatch et al. (1990); Waterfall (1958) cited Tarrant Co.; se and e TX, also Edwards Plateau. Jun–Oct. [*P. macrophysa* Rydb., *P. virginiana* Mill. forma *macrophysa* (Rydb.) Waterf., *P. virginiana* Mill. var. *subglabrata* (Mack. & Bush) Waterf.]

Physalis mollis Nutt., (soft, soft-hairy), FIELD GROUND-CHERRY. Perennial densely stellate-tomentose; leaves ovate; flowering calyces (6–)7–10 mm long. According to Waterfall (1972), *P. mollis* apparently intergrades with *P. cinerascens*. While we are following Kartesz (1994) in recognizing *P. mollis*, this species may not be specifically distinct from *P. cinerascens*. Sandy soils, roadsides, field, and woods; se and e TX w to West Cross Timbers and Edwards Plateau. May–Jul. [*P. viscosa* L. subsp. *mollis* (Nutt.) Waterf.]

Physalis pubescens L., (pubescent, downy), DOWNY GROUND-CHERRY, LOW HAIR GROUND-CHERRY, TOMATE FRESADILLA. Villous to viscid-glandular or glabrate annuals; leaves ovate; fruiting calyces 1.8–3 cm long, usually hairy, 5-angled; fruiting pedicels 5–13 mm long. Disturbed areas. Apr–Nov. We are following Kartesz (1994) in recognizing 2 varieties. They, however, apparently intergrade and are questionably distinct.

1. Leaf blades usually toothed nearly to the base with 5–8 teeth on each side, seldom translucent ____ var. **integrifolia**
1. Leaf blades with few teeth, 3–4 on each side, or entire, mostly flaccid and translucent ____ var. **pubescens**

var. **integrifolia** (Dunal) Waterf., (entire-leaved). Waterfall (1958) cited Dallas Co.; mainly se and e TX, also Edwards Plateau.

var. **pubescens**. Clay, Cooke, Fannin, and Tarrant cos., also Hamilton (HPC) and McLennan (Waterfall 1958) cos.; se and e TX w to nc TX, also Edwards Plateau.

Physalis pumila Nutt., (dwarf), PRAIRIE GROUND-CHERRY, LOW GROUND-CHERRY. Perennial usually with jointed hairs mostly 1–2 mm long, some 1- to rarely 3-branched, wide-spreading; leaf blades ovate to lanceolate; anthers yellow, 2.5–3 mm long. Prairies, open woods; Dallas, Fannin, Grayson, and Kaufman cos. e TX w to Blackland Prairie. Apr–Jul.

Physalis turbinata Medik., (top-shaped), THICKET GROUND-CHERRY. Annual, similar to *P. pubescens* var. *pubescens*. Open woods, brushy areas; Grayson Co., mainly se, e, and s TX and Edwards Plateau. Jul–Nov. Sometimes lumped into *P. pubescens*.

Physalis virginiana Mill., (of Virginia), VIRGINIA GROUND-CHERRY. Perennial; stems from deep rhizome; pubescence short and retrorse to villous; leaf blades ovate to lanceolate; anthers yellow, 2–4 mm long. Oak woods; Denton and Grayson cos.; e TX w to nc TX. Apr–Jun. Suspected of being poisonous (Burlage 1968). ☠

QUINCULA PURPLE GROUND-CHERRY, CHINESE-LANTERN

A monotypic North American genus related to *Physalis*. (Possibly from Latin: *quin*, five, and *cula*, little, from the five spots on the corolla mentioned in the type description; derivation unexplained by the author)
REFERENCE: Averett 1979.

Quincula lobata (Torr.) Raf., (lobed), PURPLE GROUND-CHERRY, CHINESE-LANTERN-OF-THE-PLAINS. Rhizomatous perennial herb, stems decumbent or low-spreading; plant nearly glabrous in age, young parts granulose-whitened (due to short-stalked white crystalline vesicles which collapse upon drying); leaves petioled; leaf blades oblong-lanceolate or -elliptic, subentire to coarsely toothed or pinnatifid, 4–10 cm long; flowers typically facing upward, conspicuous; corollas rotate, 1.5–2 cm broad, blue-purple typically with deeper purple star-like pattern (rarely white—Stanford (1976) indicated that forma *albiflora* Waterf. is known in TX only from Brown and Val Verde cos.); calyces inflated in age, 1.5–2 cm long at maturity; anthers yellow; berry 4–8(–10) mm in diam. Sandy or gravelly open ground; Montague, Tarrant, and Wise cos. s and w to w TX. Apr–Jun, sporadically during summer, Sep–Oct. Sometimes lumped into the genus *Physalis* [as *P. lobata* Torr.] ⊞/104

SOLANUM NIGHTSHADE

Ours annual or perennial herbs or half-shrubs, glabrous or variously pubescent, some with stellate hairs, some species prickly; flowers in lateral or terminal racemes, cymes, or umbels; anthers usually with terminal pores.

A huge, nearly cosmopolitan genus of 1,200 species (M. Nee, pers. comm.) considered by some to have up to 1,700 species (Mabberley 1997). It includes *S. tuberosum* L. (POTATO), a South American species with edible underground stems (= tubers), and *S. melongena* L. (EGGPLANT), an Old World native with edible fruit. Hundreds of POTATO cultivars are grown in Andean South America; it is an ancient crop brought into cultivation high in the Andes ca. 8,000 years ago and is today the world's fourth most important crop after WHEAT, RICE, and CORN. POTATOES were first grown in Europe in the 16th century (Ireland in 1566), but soon became an important staple; collapse of POTATO cultivation in the middle 1800s as a result of the potato blight (*Phytophthora infestans*) caused tremendous social upheaval in Ireland (see Grun 1990 for a review of POTATO evolution). Recent molecular studies (Bohs & Olmstead 1997; Olmstead & Palmer 1997) indicated that two other economically important genera, *Lycopersicon* (TOMATOES) and *Cyphomandra* (TREE TOMATOES) are nested within the *Solanum* clade and are best considered part of that genus (as is done here for *Lycopersicon*). ☠ Many *Solanum* species contain toxic glycoalkaloids (e.g., solanine); leaves and/or fruits of all species growing in nc TX should be considered potentially fatally poisonous; even the foliage, green tubers (when exposed to light), and sprouts of the edible POTATO are poisonous; animals have died from ingesting POTATO foliage and humans have died from the green tubers (Sperry et al. 1955; Kingsbury 1965; Schmutz & Hamilton 1979; Lampe & McCann 1985). Gaffield et al. (1992) documented that POTATO sprout alkaloids are teratogenic (= cause birth defects) in test animals; high POTATO (and thus solanine) consumption in Britain by pregnant women is alleged to cause spina bifida in babies (Mabberley 1997). *Solanum* flowers are an example of the "vibrator" or "buzz" pollination syndrome; the flowers lack nectar and the abundant pollen is used as a reward; pollinators (such as bumblebees) shake the anthers by vibrating their thoracic flight muscles at a certain frequency; this sets up a resonance in the anthers or the space they enclose and the otherwise inaccessible pollen is released from the terminal pores of the anthers and collected by the insect (Barth 1985; Proctor et al. 1996). The turned back (reflexed) corollas and exposed anther-cone of "vibrator"-type flowers may be an adaptation to minimize dampening of vibration resonance or it may be an adaptation related to microclimate in the flower (e.g., to keep the pollen in a dry powdery condition so that it is easily dispersed) (Corbet et al. 1988; Proctor et al. 1996). (Classical Latin name for these plants, possibly from Latin: *solatium*, soothing, comforting, or quieting, from narcotic properties)

REFERENCES: Muller 1940; Luckwill 1943; Whalen 1979; Schilling 1981; Boyd et al. 1984; Lemke 1991; Bohs & Olmstead 1997; Olmstead & Palmer 1997.

1. Anthers fused into a pointed cone terminated by a sterile tip, opening longitudinally down the inner face; corollas yellow; plants unarmed, with viscid-glandular hairs ______ **S. lycopersicum**
1. Anthers converging around the style but not fused, opening at the tips by pores or slits; corollas usually bluish purple to white (yellow in 1 species); plants unarmed or armed with prickles, usually not viscid-glandular.

2. Leaves pinnately compound or deeply twice-pinnatifid; corollas yellow, violet, or blue.
   3. Corollas yellow; stem hairs stellate; inflorescences stellate-hairy only ____ **S. rostratum**
   3. Corollas violet or blue; stem hairs simple and glandular; inflorescences glandular-villous as well as somewhat stellate-hairy ____ **S. citrullifolium**
2. Leaves entire, toothed, or unevenly lobed; corollas white to purplish or bluish.
   4. Plants usually armed (larger leaves usually prickly on midrib beneath; stems also usually prickly); anthers 5–10 mm long; leaves conspicuously stellate-pubescent (with a lens).
      5. Leaf blades 3–5 times as long as wide, oblong-lanceolate, unlobed to sometimes with a strongly wavy margin, usually silvery or gray-green and often velvety in appearance due to the dense silvery pubescence ____ **S. elaeagnifolium**
      5. Leaf blades 1.5–2.5 times as long as wide, rhombic to oblong-elliptic, often conspicuously lobed, dark green or grayish green, the usually tawny pubescense not so dense as to give a velvety appearance.
         6. Corollas 1.5–2.7 cm across fully open, sparsely pubescent outside; calyces 4–7 mm long; stellate hairs on lower leaf surface sessile, usually with 4–8 rays; flowers usually white, rarely blue-lavender; fruits 1–2 cm in diam. ____ **S. carolinense**
         6. Corollas 2.7–5 cm across fully open, felty-pubescent outside; calyces 7–17 mm long; stellate hairs on lower leaf surface stalked (at least some), usually with 8 or more rays; flowers blue-purple, rarely white; fruits 2.5–3 cm in diam. ____ **S. dimidiatum**
   4. Plants unarmed (without prickles); anthers 1–4 mm long; leaves glabrous or sparsely pubescent.
      7. Larger leaves often deeply 3- to 5-lobed, often roughly triangular in shape, sometimes cordate at base; corolla lobes 4.5–9 mm long; anthers 3–4 mm long; plants half-shrubby or vine-like perennials; mature fruits red ____ **S. triquetrum**
      7. Larger leaves entire or shallowly lobed, neither triangular nor cordate at base; corolla lobes 1–4(–6) mm long; anthers 1–2 mm long; plants annuals; mature fruits black ____ **S. ptychanthum**

Solanum carolinense L., (of Carolina), HORSE-NETTLE, BALL-NETTLE. Perennial with creeping root; leaf blades angled, coarsely toothed, or lobed; corollas white, rarely blue-lavender. Sandy woods, roadsides; se and e TX w to Grayson and Tarrant cos., also Edwards Plateau. May–Oct. Animal and human poisonings have been reported including fatalities in children from eating the berries; chronic poisoning can occur in livestock from eating small quantities over a prolonged period (Kingsbury 1964; Burlage 1968; McGregor et al. 1986; Tveten & Tveten 1993). ☠

Solanum citrullifolium A. Braun, (with leaves of *Citrullus*—watermelon), MELON-LEAF NIGHTSHADE. Annual, very prickly, much branched, to ca. 70 cm tall; corollas very striking violet or blue; one anther enlarged; fruits enclosed by very prickly calyces. Burnet Co. (Lemke 1991); mainly w 1/2 of TX. Jun–Oct. Poisonous. ☠ ▣/106

Solanum dimidiatum Raf., (divided into two dissimilar or unequal parts), WESTERN HORSE-NETTLE, POTATO-WEED. Similar to *S. carolinense* but coarser; corollas blue-purple, rarely white. Prairies, open woods, roadsides; widespread in TX. May–Jul, sporadically to Oct. [*S. perplexum* Small, *S. torreyi* A. Gray] Poisonous, can be lethal if eaten by livestock and suspected of causing birth defects and abortions in livestock. The species is also the cause of the neurological disease in cattle known as "Crazy Cow Syndrome"; symptoms include staggering and incoordination; if frightened, the cow will fall to the ground and struggle to regain its feet; once affected, while often living a normal lifespan, such cattle retain a tendency toward incoordination when disturbed or excited; they are thus accident prone and many have died from drowning; the disease occurs yearly in c and w TX in Concho, Mason, Nolan, Real, Runnels, and Taylor cos. (Menzies et al. 1979; Casteel & Bailey 1992). ☠ ▣/106

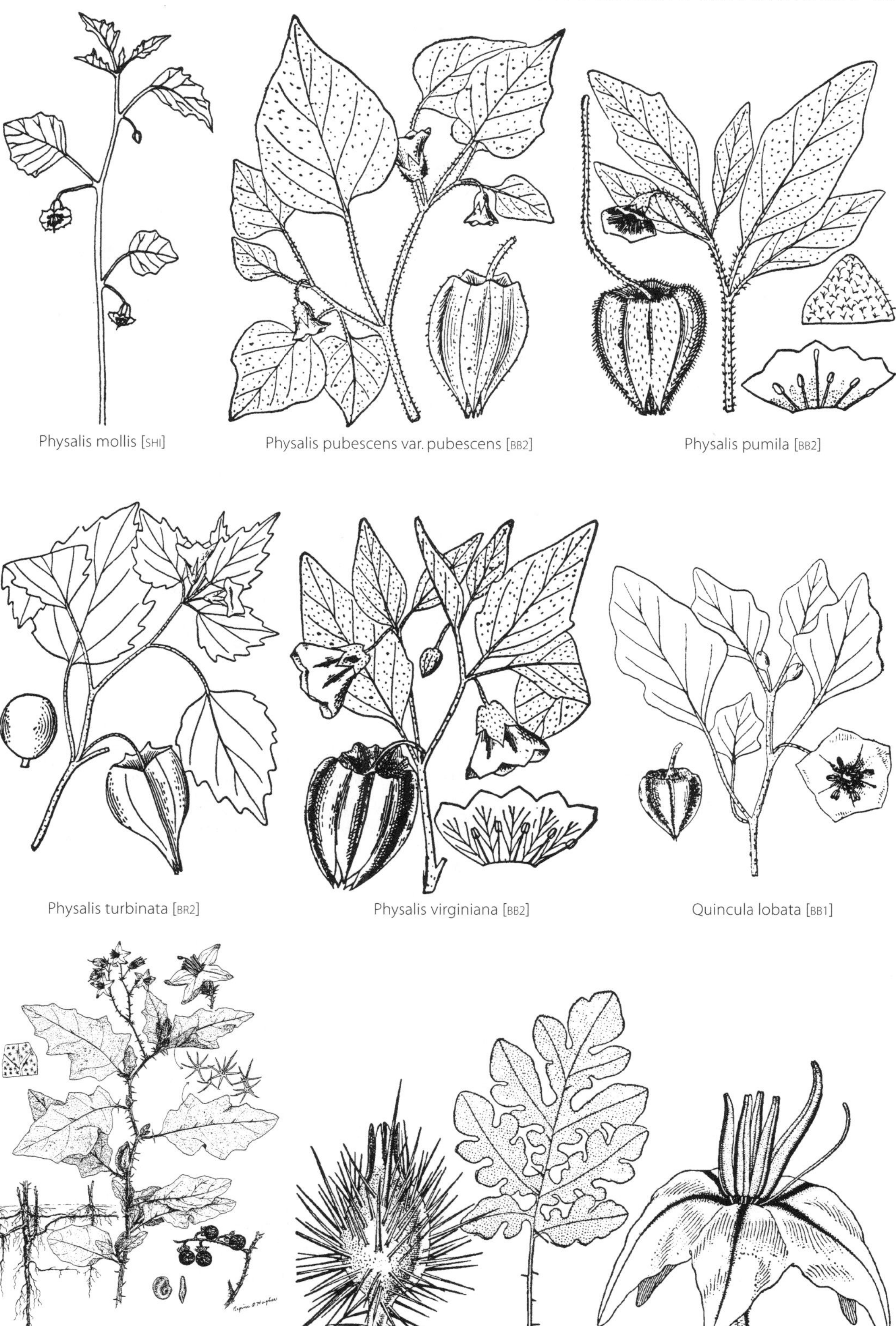

Physalis mollis [SHI]

Physalis pubescens var. pubescens [BB2]

Physalis pumila [BB2]

Physalis turbinata [BR2]

Physalis virginiana [BB2]

Quincula lobata [BB1]

Solanum carolinense [REE]

Solanum citrullifolium [GLE]

Solanum elaeagnifolium Cav., (with leaves like *Elaeagnus* in the Elaeagnaceae or Oleaster family), TROMPILLO, SILVER-LEAF NIGHTSHADE, BULL-NETTLE, WHITE HORSE-NETTLE, WHITEWEED, IRONWEED. Perennial with creeping root; silvery-canescent throughout by many-rayed stellate hairs; stem varying from unarmed to densely prickly; corollas blue-purple, rarely white. Roadsides and waste places; throughout TX. May–Jun, less freely to Oct. Can be lethally poisonous if eaten by livestock due to the presence of the toxic glycoalkaloids solanine and solasonine; the fruits are a source of solasodine, used in the commercial manufacture of steroidal hormones (Kingsbury 1964; Boyd et al. 1984). ☠

Solanum lycopersicum L., (Greek: *lykos*, wolf, and *persikon*, peach, suggesting that the fruit is inferior to a peach), TOMATO, LOVE-APPLE, GOLD-APPLE. Annual herb with unarmed decumbent stems to 2 m or more long; plant glandular-pubescent and strong smelling, also with spreading pubescence; leaves irregularly pinnate or pinnately lobed, pinnatifid, or bipinnatifid, the margins often inrolled; corollas yellow, nearly rotate to with recurved lobes, 1–2 cm broad; fruit a fleshy red or yellow berry; seeds numerous. Commonly cultivated for the edible fruit, sometimes found as a transient escape near dumps or picnic spots; Grayson Co. (open weedy area below abandoned dump), Kaufman Co. (shady riverbank); also Brown Co. (HPC) and Tarrant Co. near refuse dump (R. O'Kennon, pers. obs.). Late spring–fall. Probably native to Andean South America and possibly brought into cultivation there (Rick & Holle 1990) or in Mexico (Heiser 1987). [*Lycopersicon esculentum* Mill.] This species has traditionally been treated in *Lycopersicon* (TOMATO), a genus of 7 species native to w South America and the Galápagos Islands. However, TOMATOES are closely related to *Solanum* and can be hybridized with certain *Solanum* species. Recent molecular studies (Bohs & Olmstead 1997; Olmstead & Palmer 1997) indicated that *Lycopersicon* is nested within the *Solanum* clade and therefore best treated as part of the genus *Solanum*. We are thus following Bohs and Olmstead (1997), Olmstead and Palmer (1997), and J. Kartesz (pers. comm. 1997) in lumping *Lycopersicon* with *Solanum*. The common name TOMATO is derived from Nahuatl (language of the Aztecs), *tomatl* (Rupp 1987). The foliage is poisonous if eaten due to presence of toxic steroidal alkaloids; animal deaths have been reported and human poisonings have occurred from drinking tea made from leaves (Kingsbury 1964, 1965; Keeler 1979; Schmutz & Hamilton 1979). Because it is in the same family as many toxic plants, when originally carried to Europe from the New World, even the fruits of the TOMATO were thought by many to be poisonous; this gave rise to interesting common names such as DEVIL'S WOLF-APPLE. ☠

Solanum ptychanthum Dunal, (folded flower), AMERICAN NIGHTSHADE, BLUE-FLOWER BUFFALO-BUR, HIERBA MORA NEGRA. Annual, larger plants widely bushy-branched; leaf blades thin, lanceolate or elliptic, entire or bluntly toothed (rarely shallowly lobed); corollas usually white; mature fruits 5–9 mm in diam. Stream bottom woods; throughout TX. Jun–Sep. The plant referred to here as *S. ptychanthum* has long commonly gone under the name *S. americanum*; however, because of nomenclatural rules, the correct name is apparently *S. ptychanthum* (Schilling 1981). [*S. americanum* of authors, not Mill., *S. nigrum* of authors, not L.] While the mature fruits of some strains are supposed to be edible when cooked, animal and human poisonings have been reported, probably due to solanine and saponins in immature fruits or foliage; both nervous system and digestive system symptons have been reported (Kingsbury 1964; Morton 1982; Turner & Szczawinski 1991). ☠

Solanum rostratum Dunal, (rostrate, beaked), BUFFALO-BUR, KANSAS-THISTLE, MALA MUJER. Annual, very prickly, usually much-branched, to 70 cm tall, often low and broad; corollas yellow; 4 uppermost anthers yellow, 6–8 mm long, lowermost anther suffused with purple, enlarged, 10–14 mm long; fruits enclosed by prickly calyx. Overgrazed pastures, disturbed ground, various soils; throughout TX. Jun–Oct. The common name is presumably derived from the tendency of the prickly calyces enclosing the fruits to become tangled in the hair of buffaloes or cattle; seed dispersal is thus accomplished (Kirkpatrick 1992). Poisonous and the burs may cause mechanical injury (Kingsbury 1964). ☠

Solanum triquetrum Cav., (three-edged or -cornered), TEXAS NIGHTSHADE, HIERBA MORA. Clump-forming, woody-based perennial from branching roots; longer stems twining; corollas usually white, sometimes pale blue; mature fruits 10–15 mm in diam. Stream banks, fencerows, and disturbed areas; Brown and Grayson cos.; also Dallas, Erath (Mahler 1988) and Somervell (R. O'Kennon, pers. obs.) cos. and Fort Hood (Bell or Coryell cos.—Sanchez 1997); nearly throughout TX. Apr–May. While information on toxicity was not found, this species should be considered poisonous. ☠

SPHENOCLEACEAE
SPHENOCLEA OR CHICKENSPIKE FAMILY

A very small (2 species in a single genus) tropical family of annuals of wet places; treated by some in the Campanulaceae (e.g., Godfrey & Wooten 1981; Hatch et al. 1990; Taylor & Taylor 1994) and by others in its own family (e.g., Cronquist 1988; Lammers 1992; Kartesz 1994; Jones et al. 1997). According to N. Morin (pers. comm.), it is not closely related to Campanulaceae. (subclass Asteridae)

FAMILY RECOGNITION IN THE FIELD: the single species occurring in TX can be recognized as a wet area herb with alternate simple leaves and small, white to yellowish flowers in erect cylindrical spikes.

REFERENCES: Rosatti 1986; Lammers 1992.

SPHENOCLEA

A tropical genus of 2 species, 1 native to w Africa, the other pantropical (originally Africa?). (Derivation of generic name not explained by original author, possibly from Greek: *spheno*, wedge)

Sphenoclea zeylanica Gaertn., (of Ceylon), PIEFRUIT, CHICKENSPIKE, GOOSEWEED. Glabrous, usually branched, annual herb; stems erect, 0.5–1+ m tall, hollow, spongy- or corky-thickened when growing in water and producing numerous fibrous roots from the nodes; leaves alternate, simple; leaf blades elliptic, to ca. 12 cm long and 5 cm wide, marginally entire, apically acute, basally cuneate; petioles 5–20 mm long; inflorescences erect, terminal, cylindrical spikes 1.5–10 cm long, 0.5–1 cm in diam., so dense that the rachis is usually not visible; peduncles to 10 cm long; floral bracts 2–3 mm long and ca. 1 mm wide, not evident without close inspection because of the denseness of the spikes; flowers small, numerous, 5-merous, sessile; calyx lobes green, ca. 1.5 mm long, enlarged to 5 mm long in fruit; corollas sympetalous, radially symmetrical, white to yellowish, ca. 2.5 mm long, the lobes obtuse; stamens inserted near middle of corolla tube, distinct, deciduous with corolla soon after anthesis; anthers roundish; ovary inferior; capsules circumscissile, 2-locular; seeds 0.4–0.5 mm long. Wet places; can be a weed in areas of rice cultivation; included based on citation of vegetational area 4 (Fig. 2) by Hatch et al. (1990); possibly only to the se of nc TX; mainly se and e TX. Aug–Nov. Probably native of the Old World tropics, possibly Africa.

STERCULIACEAE CHOCOLATE OR CACAO FAMILY

Ours trees or small shrubs; leaves alternate, simple; flowers radially symmetrical; sepals united at least at base; petals absent or present; stamens connate at base or higher; ovary superior; fruit a capsule or apparently follicle-like.

A medium-large (1,500 species in 67 genera) family chiefly in tropical and warm areas with

a few in temperate regions; most are trees and shrubs or rarely herbs or lianas; stellate pubescence is usually present. The seeds of a tropical American member of the family, *Theobroma cacao* L., are the source of chocolate and cocoa, containing stimulating alkaloids including theobromine; African species of *Cola* provide flavoring and caffeine for various beverages. The family is related to the Bombacaceae (a mostly tropical family), Malvaceae, and Tiliaceae. Family name from *Sterculia*, a tropical genus of 200 species of monoecious or polygamous trees. (Named for Sterculius, the Roman god of privies, in reference to the dung-like odor of the flowers and leaves of some species; Latin: *stercus*, dung) (subclass Dilleniidae)
FAMILY RECOGNITION IN THE FIELD: trees or small ± herbaceous shrubs with alternate, simple, stipulate leaves; the 2 species occurring in nc TX are very different and best recognized individually.
REFERENCES: Brizicky 1966a; Whetstone 1983.

1. Trees to ca. 20 m tall; leaf blades 10–30 cm long, 3–5-lobed, the lobes entire; petals absent; sepals petaloid, yellow later becomming reddish, 8–12 mm long; flowers in elongate panicles often 50 cm or more long _____ **Firmiana**
1. Shrubs 2 m tall or less; leaf blades to 6 cm long (typically smaller), unlobed, serrate; petals 5, pink or violet with yellowish base, ca. 7 mm long; flowers solitary or in small axillary cymes _____ **Melochia**

FIRMIANA

A genus of 12 species native to the Old World tropics; some are used as ornamentals or for timber. (Named for Karl Josef von Firmian, 1716–1782, Austrian statesman who was Governor of Lombardy and patron of the Padua botanical garden)
REFERENCES: Kostermans 1957, 1989.

Firmiana simplex W. Wight, (of one piece or series as opposed to compound, or simple, unbranched), CHINESE PARASOL-TREE, CHINESE BOTTLETREE, JAPANESE VARNISHTREE, PHOENIXTREE. Deciduous tree to nearly 20 m tall; leaves alternate, simple, conspicuously petiolate; leaf blades palmately 3–5 lobed, 10–30 cm long, often a little wider, cordate to sagittate basally, palmately veined; inflorescence a panicle; sepals 5; petals none; stamens 25–30; fruit of 5 stalked, follicle-like structures to ca. 13 cm long; seeds globose, 7–10 mm in diam., maturing on the open margins of the carpels. Cultivated and spreading from seed; downtown Fort Worth and Fort Worth Botanic Garden, Tarrant Co.; also escaping along Town Creek, near downtown Fredricksburg, Gillespie Co. on the Edwards Plateau. May–Jul. Native of e Asia. [*F. platanifolia* (L. f.) Schott & Endl.]

MELOCHIA BROOMWOOD

A mainly tropical, especially American genus of 54 species. (Derivation of generic name unknown)
REFERENCE: Goldberg 1967.

Melochia pyramidata L., (pyramid-shaped), ANGLE-POD MELOCHIA. Shrub, often ± herbaceous, to ca. (1.5–)2 m tall, glabrous or sparsely pubescent, superficially resembling a *Sida* (in the related Malvaceae); leaf blades ovate to lanceolate, with 5 veins from base but ± pinnately-veined in appearance; petioles to ca. 12 mm long; involucre absent; calyces 5-lobed, the lobes long acuminate; stamens 5; styles filiform; capsules 5-celled, 5 winged, somewhat star-shaped. Sandy or rocky soils, often in disturbed places or waste areas; lawn weed in Dallas, apparently brought in with sod (*Shinners 15288*), also weed in landscape in Tarrant Co.; native to the s 1/2 of TX. Apr–Nov.

STYRACACEAE STORAX FAMILY

A small (160 species in 11 genera) family of trees and shrubs with resinous bark; they range from tropical and warm temperate areas of the New World to the Mediterranean, se Asia, and

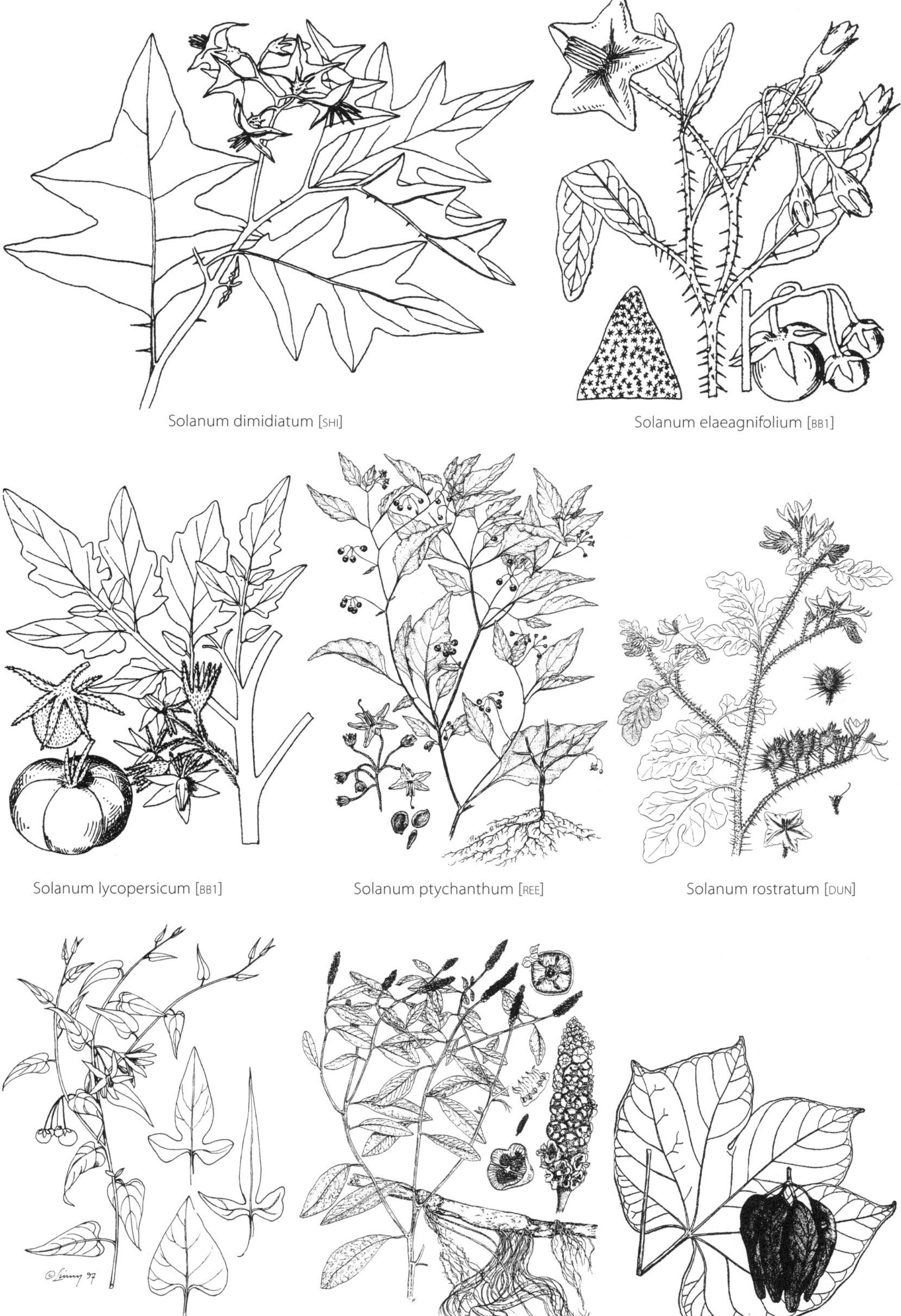

Solanum dimidiatum [SHI]

Solanum elaeagnifolium [BB1]

Solanum lycopersicum [BB1]

Solanum ptychanthum [REE]

Solanum rostratum [DUN]

Solanum triquetrum [HEA]

Sphenoclea zeylanica [CO1]

Firmiana simplex [BR1]

w Malesia. The small genus *Halesia* (SILVERBELL, SNOWDROPTREE, BELLTREE), with two species native to e TX, includes showy ornamental shrubs; its 4-lobed corollas and winged fruits easily distinguish *Halesia* from *Styrax* (with 5-lobed corollas and unwinged fruits). The family is possibly related to Ebenaceae and Sapotaceae. (subclass Dilleniidae)

FAMILY RECOGNITION IN THE FIELD: the only nc TX species is a shrub or small tree with alternate, simple, entire, sometimes slightly lobed, exstipulate leaves; flowers white, showy, in ± drooping, few-flowered racemes.

REFERENCES: Wood & Channell 1960; Spongberg 1976.

STYRAX STORAX, SNOWBELL, SILVERBELL

A genus of 120 species of trees and shrubs of the Mediterranean, se Asia, w Malesia, and tropical America n to the se U.S. The aromatic resins (benzoin, gum benjamin, storax) of some species, obtained by wounding the bark, are used medicinally (in friar's balsam, as an antiseptic, inhalant, and expectorant), in perfumes, and as incense in churches; other species are used ornamentally. (Greek: *styrax*, ancient name for *S. officinalis* L., storax gum tree, native to the Mediterranean and source of storax, a fragrant resin used in incense)

REFERENCE: Cory 1943.

Styrax platanifolius Engelm. ex Torr., (with leaves like *Platanus*—sycamore), SYCAMORE-LEAF STYRAX, SYCAMORE-LEAF SNOWBELL. Many-branched shrub or small tree to 4 m tall; leaves alternate, simple, deciduous, petioled; leaf blades strongly reticulate-veined, broadly ovate to suborbicular, to ca. 10 cm long and wide, glabrous or nearly so, marginally undulate to angulate or with short lobes, basally truncate to subcordate, apically obtuse to abruptly acute; petioles 5–10(–13) mm long; flowers perfect, in short, 3–5-flowered, ± drooping, axillary racemes, these often appearing terminal; pedicels 8–12(–18) mm long; corollas with 5 petals united at base; petals white, showy, 12–14(–15) mm long, 4–6 mm wide; stamens 10; fruits globose or subglobose, 7–8 mm in diam., apically 3-valved; seed usually 1 per fruit. Along streams in limestone areas; Bell (Fort Hood—Sanchez 1997) and Burnet cos.; mainly on the Edwards Plateau; endemic to TX. Apr–May.

TAMARICACEAE SALT-CEDAR OR TAMARIX FAMILY

A small (78 species in 4 genera), mainly temperate and subtropical Old World family of trees and shrubs; many are halophytes or xerophytes; the leaves are small, often scale-like, and frequently have salt-excreting glands. (subclass Dilleniidae)

FAMILY RECOGNITION IN THE FIELD: the only TX genus is an introduced group of shrubs or small trees with leaves alternate, very small, sessile, usually *scale-like* (giving the plant a *gymnosperm-like appearance*); flowers small, in spicate racemes; petals persistent after withering.

REFERENCE: Crins 1989.

TAMARIX SALT-CEDAR, TAMARISK

Ours introduced shrubs or small trees, much branched with slender branches, often evergreen; leaves alternate, entire, very small, scale-like, sessile (plants thus superficially gymnosperm-like in appearance); flowers small, 5-merous (our species) in spicate racemes usually clustered into panicles; petals white to pink to rosy-lavender (rarely reddish); anthers attached on or below a fleshy nectary-disk; ovary superior; fruit a capsule; seeds terminating in a tuft of hairs.

An Eurasian and African genus of 54 species including cultivated ornamentals. Some species produce manna (a sweet hardened exudate) resulting from puncture of stems by scale-insects; it can be observed in the Dead Sea area and is considered to be the Biblical manna (Baum 1978); others produce galls from which tannin is obtained for treating leather. *Tamarix* species

are deep-rooted and weedy and are problematic invaders in many parts of the w U.S.; they crowd out native species along watercourses and are said to lower the water table along streams and irrigation canals through transpiration. They can form nearly monotypic stands which are considered by some to be "biological deserts" with low animal and plant diversity. Nuserymen in the early 1800s apparently made the first introductions of TAMARISKS into the U.S.; the U.S. Dept. of Agriculture also introduced a number of these Old World species (from the late 1860s to 1915) in a misguided attempt for use in erosion control and as windbreaks (Cox & Leslie 1991; Brock 1994; Luken & Thieret 1997). The common name SALT CEDAR is derived from the plants' ability to tolerate salt and their CEDAR (*Juniperus*) -like appearance. The species are distinguished with difficulty. (Classical Latin name, possibly from Tamaris River in Spain)
REFERENCES: Baum 1966, 1967, 1978; Brock 1994.

1. Staminal filaments arising from the tips of the disk lobes; disk lobes longer than wide; petals caducous (= falling early) ____ **T. gallica**
1. Staminal filaments appearing to arise from between the disk lobes, inserted under the disk near the margin or inserted between the lobes; disk lobes wider than long; petals persistent.
   2. Petals ovate to elliptic; sepals ± entire; leaves oblong to narrowly lanceolate, acute; flowers of racemes on green branches with 1–2 of the filaments inserted between lobes of disk ____ **T. chinensis**
   2. Petals obovate; sepals eroded to irregularly denticulate; leaves ovate, acute to acuminate; all filaments inserted below disk near margin ____ **T. ramosissima**

Tamarix chinensis Lour., (Chinese), CHINESE TAMARISK. Bark brown to black-purple. *Tamarix chinensis* and *T. ramosissima* are often separated with difficulty; Baum (1967) indicated that some specimens cannot be distinguished and that hybridization may be a possibility. Cultivated and escapes, sandy soils; Cooke and Palo Pinto cos. along Red and Brazos rivers; also se TX. May–Jul. Native of Mongolia, China, and Japan.

Tamarix gallica L., (of Gaul or France), FRENCH TAMARISK, ROMPEVIENTOS, SALT-CEDAR. Bark blackish brown to deep purple. Cultivated and escapes in some areas; sandy or silty stream bottoms. May–Jul, sporadically later. Native to s Europe. Mahler (1988) considered this the only species in nc TX. However, based on the work of Baum (1967), nc TX material seems to be *T. chinensis* and *T. ramosissima*. According to Baum (1967), while rare in the U.S., *T. gallica* is known from Texas. While no definitive collections of *T. gallica* have been seen from nc TX, it is included here for clarity and because it is cited as occurring in vegetational areas 4 and 5 (Fig. 2) by Hatch et al. (1990).

Tamarix ramosissima Ledeb., (much-branched), SALT-CEDAR. Bark reddish brown. Cultivated and escapes, in sand and silt, along rivers, reservoirs, disturbed areas; Brown, Grayson, Lamar, and Wise cos.; also se TX and Edwards Plateau. Jun–Jul, Oct. Native of Eurasia.

TILIACEAE LINDEN FAMILY

The Tiliaceae is a medium-sized (680 species in 46 genera), subcosmopolitan, but mainly tropical family of trees, shrubs, or rarely herbs with a few temperate taxa; stellate pubescence or peltate scales are often present and the leaves are usually palmately veined. Some are important as sources of timber, as ornamentals, or for fiber. Asian species of *Corchorus* are the source of jute (gunny), the fiber used in sacking, twine, carpeting, and paper. The family is related to the Bombacaceae (a mostly tropical family), Malvaceae, and Sterculiaceae. (subclass Dilleniidae)
FAMILY RECOGNITION IN THE FIELD: the only nc TX species is a tree with few-flowered inflorescences attached by a peduncle to the middle of a distinctive, conspicuous, leafy, *strap-shaped bract*; leaves alternate, simple, toothed, asymmetric at base; fruits round, nut-like, pea-sized.
REFERENCE: Brizicky 1965b.

TILIA BASSWOOD, LINDEN, LIME

A n temperate genus of 45 species with fragrant flowers on an inflorescence emerging from a prominent bract; a number are important as ornamentals or for timber; bees make a superb honey from the flowers of some. According to Strausbaugh and Core (1978), the inner bark is very tough and useful in making fabrics or baskets—thus the name bastwood (= basswood). (Classical Latin name)
REFERENCES: Sargent 1918; Jones 1968; Hardin 1990.

Tilia americana L. var. **caroliniana** (Mill.) Castigl., (sp.: American; var.: of Carolina), CAROLINA BASSWOOD, FLORIDA BASSWOOD, KENDALL BASSWOOD, FLORIDA LINDEN. Large tree to ca. 30 m tall; young twigs with tomentum; leaves deciduous, simple, alternate; leaf blades broadly ovate, to 13 cm long, stellate-tomentose or sparsely stellate-pubescent on abaxial (= lower) surface, basally asymmetric, truncate to cordate, marginally coarsely dentate; petioles to 35 mm long; flowers fragrant, white or yellowish, 5-merous, 10 mm or less long; peduncle adnate to middle of a distinctive, narrow, foliaceous bract, the bracts to 12 cm long and 3 cm wide; ovary superior; fruits globose, nutlike, ca. 6 mm in diam. Low woods; Fannin (Simpson, 1988) and Lamar (Little 1976 [1977]) cos. in ne part of nc TX; mainly se and e TX, also Edwards Plateau. Apr–Jun. The leaves are superficially similar to those of *Morus rubra*, RED MULBERRY, (which has leaves often lobed, usually symmetrical or nearly so basally). CAROLINA BASSWOOD can be distinguished by its never lobed, usually basally asymmetrical leaves.

ULMACEAE ELM FAMILY

Small to large, deciduous (in nc TX) trees; leaves alternate, simple, entire or serrate, often asymmetrical basally, short-petioled; stipules very small, falling when the leaves open; flowers in short, lateral racemes, fascicles (= clusters or bundles), or solitary, perfect or unisexual, small, greenish, not conspicuous; perianth 4- to 9-parted; stamens 4–9; pistil 1; ovary superior; fruit in ours a samara or drupe.

A small (ca. 150 species in ca. 18 genera—Sherman-Broyles et al. 1997a) family of tropical and temperate, especially n temperate, trees and shrubs. According to Todzia (1993), the family is related to other members of the Urticales (e.g., Moraceae, Urticaceae); it appears to be basal in the order and paraphyletic as usually treated (Judd et al. 1994). (subclass Hamamelidae)
FAMILY RECOGNITION IN THE FIELD: trees with alternate, simple, pinnately veined, toothed leaves with *asymmetric bases*; flowers small, greenish, inconspicuous; fruits small, either flattened and winged, or rounded drupes.
REFERENCES: Elias 1970; Giannasi 1978; Todzia 1993; Judd et al. 1994; Sherman-Broyles et al. 1997a.

1. Flowers opening with the leaves in spring; leaves with 3 veins from base (2 basal lateral veins more prominent than other lateral veins), the veins curved; leaf margins entire or once serrate; fruit a small rounded drupe ____ **Celtis**
1. Flowers open before the leaves (except fall flowering species); leaves with 1 main vein from base (basal lateral veins no more prominent than other lateral veins), the veins essentially straight, parallel; leaf margins doubly or once serrate; fruit a flattened and winged samara ____ **Ulmus**

CELTIS HACKBERRY, SUGARBERRY, NETTLE-TREE

A genus of ca. 60 species (Sherman-Broyles 1997b), mostly of the tropics, but also in temperate areas; the fruits are often edible. HACKBERRIES nearly always exhibit galls or other disease manifestations on the leaves or twigs. The fruits are popular as a food for winter birds such as cedar waxwings (Martin et al. 1951). Sherman-Broyles et al. (1997b) indicated that the group is

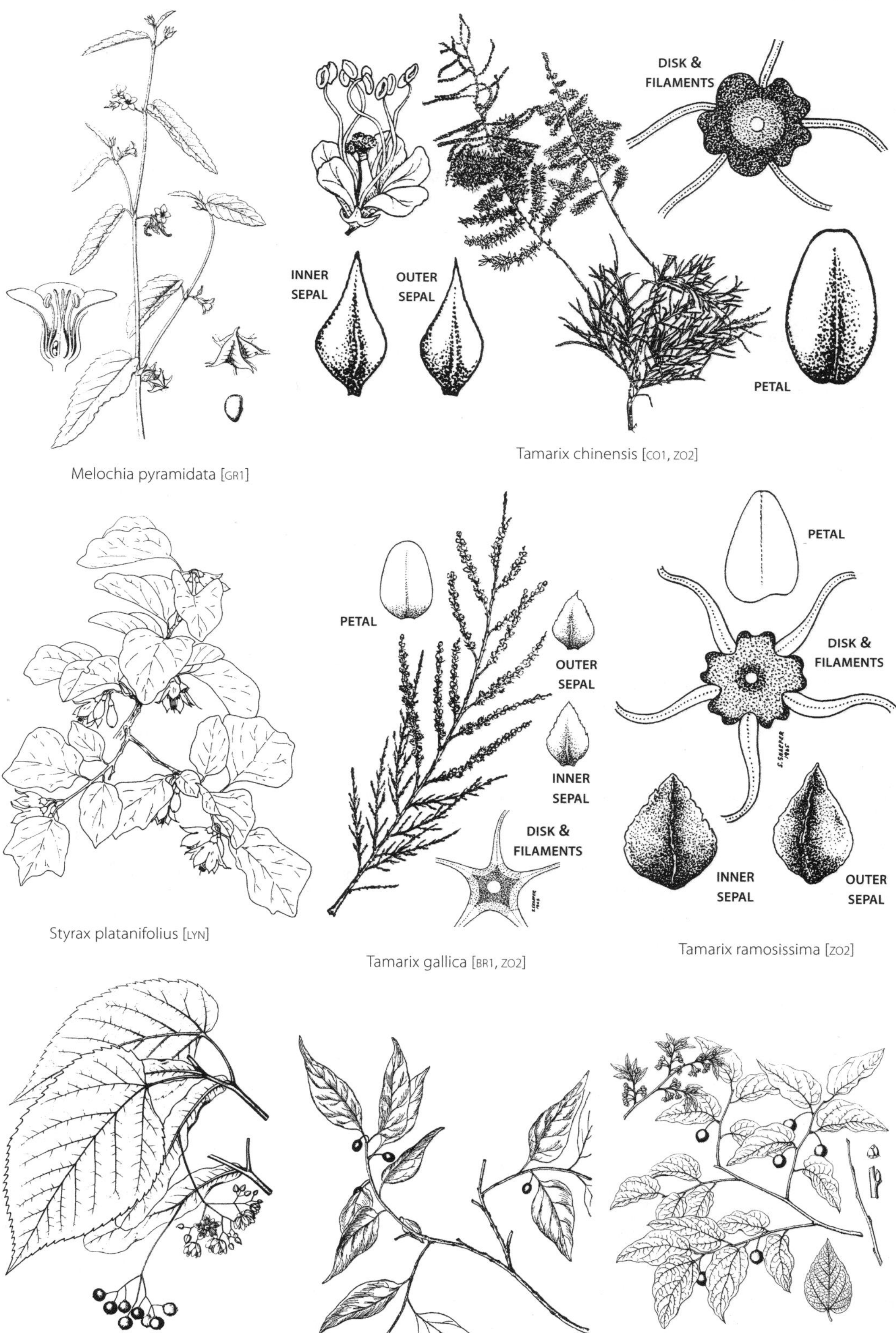

Melochia pyramidata [GR1]

Tamarix chinensis [CO1, ZO2]

Styrax platanifolius [LYN]

Tamarix gallica [BR1, ZO2]

Tamarix ramosissima [ZO2]

Tilia americana var. caroliniana [SA3]

Celtis laevigata var. laevigata [GLE]

Celtis laevigata var. reticulata [SA3]

taxonomically complex and in need of revision. A conspicuous structure known as witches' broom (a dense mass of deformed twigs) is sometimes formed on HACKBERRIES by an infection involving both an insect and a non-rust fungus (J. Hennen, pers. comm.). (Latin: *celtis*, a kind of lotus; name used by Pliny for *Celtis australis* L., the lotus of the Ancients, with sweet fruits) REFERENCE: Sherman-Broyles et al. 1997b.

Celtis laevigata Willd., (smooth), SUGARBERRY, SUGAR HACKBERRY, PALO BLANCO. Small to medium-large tree with gray, smooth or conspicuously warty-roughened bark; leaf blades ovate or lanceolate, 3–9 cm long, entire or few-toothed, cuneate to cordate basally, obtuse to acuminate apically; flowers small, both staminate and perfect present; calyces deeply lobed; fruits small, round, smooth drupes with a thin layer of flesh around a hard stone, orange, brown, dull or dark red to reddish black on pedicels equaling or exceeding the subtending petioles; the fruits persist after the leaves fall. Stream bottoms, slopes, rock hillsides. Late Mar–mid-Apr. The 3 varieties given below are not always clearly distinguishable. Varieties *laevigata* and *reticulata* can usually be told apart without difficulty. However, individuals that appear to be var. *texana* often seem intermediate between the other 2 varieties and thus difficult to place definitively. Variety *texana* may well not warrant recognition. Also, in e TX var. *laevigata* is commonly thinner-leaved than in most of nc TX, intergrading westward with the stiffer-leaved var. *texana*. Sherman-Broyles et al. (1997b) recognized var. *reticulata* as a separate species and lumped var. *texana* with var. *laevigata*. We are following Kartesz (1994) and J. Kartesz (pers. comm. 1997) in recognizing all three varieties. Additional taxonomic work is needed to clarify the situation.

1. Upper surface of mature leaf smooth or nearly so; petioles essentially glabrous ______ var. **laevigata**
1. Upper surface of mature leaf slightly to harshly scabrous (rarely nearly smooth); petioles pubescent.
   2. Upper surface of leaf usually harshly scabrous; leaf base cordate to occasionally oblique; apex of mature leaves obtuse to acute (rarely subacuminate) ______ var. **reticulata**
   2. Upper leaf surface usually slightly scabrous (rarely nearly smooth); leaf base usually cuneate to rounded; apex of mature leaves acute to acuminate ______ var. **texana**

var. **laevigata**, Mature leaf blades (4–)6–8(–15) cm long, entirely glabrous or pubescent on main veins beneath; fruits usually 5–8 mm in diam. Mainly e 2/3 of TX.

var. **reticulata** Torr., (netted), NET-LEAF HACKBERRY, PALO BLANCO. Leaf blades 2–4.5(–7) cm long, rather thick and stiff, dark green above, pale and with prominent raised veins (= strongly reticulate veined) and pubescent beneath; fruits usually (5–)8–10 mm in diam., dark red or reddish black, on pedicels exceeding the subtending petioles. Denton, Hill, McLennan, Parker, and Tarrant cos., also Bell, Coryell (Fort Hood—Sanchez 1997), Brown (HPC), Clay, Cooke, Dallas, Henderson, and Montague (Little 1976) cos.; mainly w 1/2 of TX. Late Mar–mid-Apr. [*Celtis reticulata* Torr.]

var. **texana** (Scheele) Sarg., (of Texas). Mature leaves usually pubescent on veins beneath; fruits usually 5–8 mm in diam. Clay, Cooke, McLennan, and Young cos., also Bell Co. (Fort Hood—Sanchez 1997 as *C. tenuifolia*); w part of nc TX s to the Edwards Plateau and w to w TX. Jones et al. (1997) did not list this variety for TX.

ULMUS ELM

Bark usually with deep furrows (except *U. parvifolia*); leaves short-petioled; leaf blades pinnately veined, sharply toothed, with slightly to strongly asymmetrical base; flowers small, perfect; calyces funnelform; fruit a flat, rounded samara with central seed surrounded by a thin wing.

A genus of 20–40 species Sherman-Broyles 1997) occuring from the n temperate zone s to Central America and se Asia (Wiegrefe et al. 1994); some are important ornamental shade or street

trees while others are used for timber; many are susceptible to the beetle-borne fungus which causes Dutch Elm Disease. (Classical Latin name for elm)
REFERENCES: Sherman-Broyles et al. 1992; Wiegrefe et al. 1994; Sherman-Broyles 1997.

1. Plants with flowers and fruits only (in spring).
   2. Fruits glabrous marginally; flowers and fruits sessile or subsessile, in dense fascicles, neither drooping nor in racemes.
      3. Fruits pubescent over seed (in center of fruit); calyces pubescent ____ **U. rubra**
      3. Fruits glabrous over entire surface; calyces glabrous ____ **U. pumila**
   2. Fruits ciliate marginally; flowers and fruits either drooping on elongate pedicels OR in short racemes.
      4. Surface of fruits glabrous (but marginally ciliate); twigs without corky wings; fruits ovate; flowers and fruits in loose fascicles, drooping on elongate pedicels; calyces shallowly lobed, slightly asymmetric ____ **U. americana**
      4. Surface of fruits pubescent (as well as marginally ciliate); twigs usually with corky wings; fruits lanceolate to oblong-elliptic; flowers and fruits in short racemes, not drooping; calyces deeply lobed, symmetric ____ **U. alata**
1. Plants with mature leaves (including fall flowering species).
   5. Leaves mostly once serrate (teeth without a break or cut); leaf bases symmetrical to slightly asymmetrical; upper surface of leaf blades smooth or nearly so; twigs without corky wings; escaped introduced species.
      6. Plants flowering and fruiting in late summer–fall; bark mottled with orangish to salmon areas, thin and flaky ____ **U. parvifolia**
      6. Plants flowering and fruiting in spring; bark not mottled with orangish to salmon areas, not thin and flaky ____ **U. pumila**
   5. Leaves usually doubly serrate (teeth with a break or cut on one side); leaf bases usually strongly asymmetrical; upper surface of leaf blades smooth to scabrous (= like sandpaper); twigs with OR without corky wings; native species.
      7. Leaf blades mostly 1.5–2 times as long as wide, 2.5–12 cm long; twigs without corky wings.
         8. Leaf blades glabrous to densely soft-pubescent beneath, glabrous and smooth or slightly scabrous above; buds glabrous or slightly pale pubescent ____ **U. americana**
         8. Leaf blades densely soft-pubescent beneath, densely and harshly scabrous above; buds with long yellowish or rusty hairs ____ **U. rubra**
      7. Leaf blades mostly 2–2.5 times as long as wide, 2–7 cm long; twigs with OR without corky wings.
         9. Leaf blades acute or short-acuminate, smooth or slightly scabrous above; twigs usually with corky wings; plants flowering in spring ____ **U. alata**
         9. Leaf blades obtuse or subacute, coarsely scabrous above; twigs with or without corky wings; plants flowering in fall ____ **U. crassifolia**

Ulmus alata Michx., (winged), WINGED ELM, CORK ELM, WAHOO ELM. Small tree; leaf blades smooth or slightly scabrous above, acute or short acuminate; twigs usually conspicuously corky-winged; samaras ovate-elliptic to oblong, ca. 8 mm long. Sandy ground, lowlands or uplands; se and e TX w to East Cross Timbers, also Erath Co. (Little 1971); also Edwards Plateau. Mar.

Ulmus americana L., (of America), AMERICAN ELM, WHITE ELM. Large tree; leaf blades glabrous or in nc TX frequently ± pubescent beneath and slightly scabrous above; samaras elliptic, ca. 1 cm long. Stream bottoms; mainly se and e TX w to West Cross Timbers and Edwards Plateau. Feb–early Mar. This species is particularly susceptable to Dutch elm disease, caused by an introduced European ascomycete fungus, *Ophiostoma ulmi* (Buisman) Nannf. [*Ceratocystis ulmi* (Buisman) C. Moreau]. This fungus, the spores of which are carried by bark beetles, blocks the vascular tissue and thus kills infected trees; it was first discovered in North America in Colorado in the 1930s (Sherman-Broyles 1997).

Ulmus crassifolia Nutt., (thick-leaved), CEDAR ELM, OLMO. Similar to *U. alata*; leaf blades thicker, stiffer, harshly scabrous above, obtuse or subacute; samaras oblong, ca. 1 cm long. Uplands or lowlands; e 1/2 of TX. This is our only native fall-flowering species. Under conditions of fire supression, it can be an aggressive invader of native prairie remnants.

Ulmus parvifolia Jacq., (small-leaved), CHINESE ELM, LACE-BARK ELM. Small to medium tree; leaf blades 20–ca. 60 mm long, similar to those of *U. pumila*; samaras suborbicular to round-ovate, ca. 1 cm long, glabrous. Cultivated and rarely escapes; Dallas Co. (R. O'Kennon, pers. obs.). Late summer–fall. Native of China and Japan.

Ulmus pumila L., (dwarf), SIBERIAN ELM, DWARF ELM, ASIATIC ELM, CHINESE ELM. Small to medium tree or shrub; leaf blades 20–75 mm long, smooth or nearly so above, short-pointed; samaras suborbicular to round-obovate, 1–1.5 cm in diam., glabrous. Widely cultivated and possibly escaping; included based on citation of vegetational area 5 (Fig. 2) by Hatch et al. (1990). Early spring. Introduced from n China and e Siberia.

Ulmus rubra Muhl., (red), SLIPPERY ELM, RED ELM. Medium sized tree; leaf blades harshly scabrous above; samaras suborbicular to obovate or broadly elliptic, 1–2 cm long. Stream bottoms; e TX w to West Cross Timbers (Comanche Co.—Little 1976); also Edwards Plateau. Mar. [*U. fulva* Michx.] The mucilaginous, slimy or slippery inner bark was used medicinally in the past and gave rise to the common name.

URTICACEAE NETTLE FAMILY

Ours annual or perennial herbs sometimes with stinging hairs; leaves alternate or opposite, simple, entire or toothed, petioled; stipules small, narrow; inflorescences terminal or axillary; flowers very small, often unisexual, sometimes perfect; perianth greenish, 2- to 4-parted; stamens as many as perianth parts; pistil 1; ovary superior; fruit an achene.

A medium-large family (1,050 species in 48 genera) of mainly wind-pollinated herbs, shrubs, and lianas (a few trees) occuring from the tropics to a few in temperate areas. The Urticaceae are closely related to the Moraceae and Cannabaceae. From a cladistic standpoint these families should be lumped to form a more inclusive monophyletic Urticaceae (Judd et al. 1994). *Fatoua*, an herbaceous alternate-leaved member of the Moraceae, is often confused with various Urticaceae, particularly *Boehmeria*. *Fatoua* differs from *Boehmeria* in having its flowers in axillary, pedunculate glomerules. (subclass Hamamelidae)
FAMILY RECOGNITION IN THE FIELD: herbs with simple usually toothed leaves with stipules and with or without stinging hairs; flowers *very small, greenish*, usually unisexual, in *string-like clusters or tufts at the nodes*; sap clear.
REFERENCES: Miller 1971a; Bassett et al. 1974; Friis 1993; Judd et al. 1994; Boufford 1997b.

1. Leaves alternate.
   2. Margins of leaves entire; flower clusters sessile in the axils of leaves _____ **Parietaria**
   2. Margins of leaves toothed; inflorescences of numerous small compact clusters along leafless lateral branches (but these often terminated by leaves) _____ **Boehmeria**
1. Leaves opposite.
   3. Plants with stinging hairs; flowers in axillary globular heads usually on short slender peduncles _____ **Urtica**
   3. Plants without stinging hairs; flowers not arranged as above.
      4. Inflorescences short paniculate (branched), from leaf axils of main stem; leaf blades glabrous _____ **Pilea**

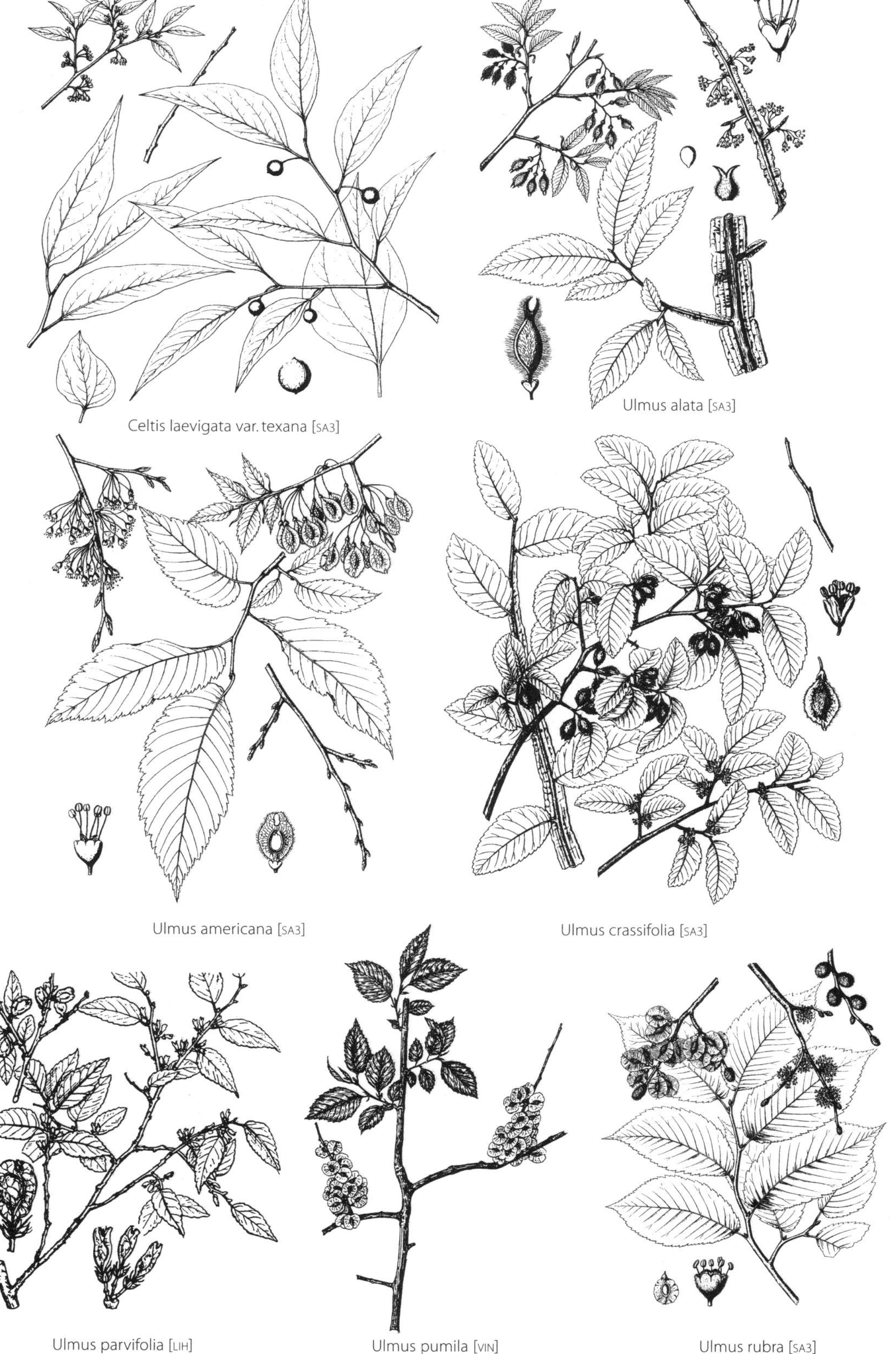

Celtis laevigata var. texana [SA3]

Ulmus alata [SA3]

Ulmus americana [SA3]

Ulmus crassifolia [SA3]

Ulmus parvifolia [LIH]

Ulmus pumila [VIN]

Ulmus rubra [SA3]

4. Inflorescences spike-like (unbranched), of numerous small compact clusters along leafless lateral branches (but these often terminated by leaves); leaf blades glabrous to pubescent or scabrous ____ **Boehmeria**

BOEHMERIA FALSE NETTLE

A genus of ca. 80 species of tropical and n subtropical areas; monoecious or dioecious trees to herbs without stinging hairs; *B. nivea* (L.) Gaudin (RAMIE, CHINA-GRASS) of tropical Asia is cultivated for its long fibers used in rope and Chinese linen. (Named for George Rudolph Boehmer, 1723–1803, professor of botany and anatomy at Wittenberg, Germany)

Boehmeria cylindrica (L.) Sw., (cylindrical), BOG-HEMP, SMALL-SPIKE FALSE NETTLE, BUTTON-HEMP. Monoecious or dioecious, erect perennial to ca. 1.2 m tall; leaves long-petioled; leaf blades ovate to ovate-lanceolate, to ca. 15 cm long and 8 cm wide, serrate, 3-nerved from the base; inflorescences spike-like (unbranched), of numerous, small, compact clusters along leafless lateral branches (but these often terminated by leaves); achenes ovate to round-ovate, 1–1.5 mm in diam., minutely winged, slightly beaked. Wet areas, low woods along streams; Cooke, Dallas, Grayson, Lamar, and Tarrant cos., also Fort Hood (Bell or Coryell cos.—Sanchez 1997); mainly se and e TX, scattered to the w. Jun–Oct.

PARIETARIA PELLITORY

A genus of 20–30 species primarily in temperate and tropical regions (Boufford 1997b). (Ancient Latin name, from *paries*, wall, from the habitat of the first described species)

Parietaria pensylvanica Muhl. ex Willd., (of Pennsylvania), PENNSYLVANIA PELLITORY. Low monoecious or polygamous annual; leaf blades rhombic-lanceolate, minutely scabrous-pubescent, sticking to fingers and clothes; flowers in small axillary clusters, with prominent calyx-like green bracts. Sandy, silty, or rocky ground, in shade; se and e TX w to West Cross Timbers and Edwards Plateau, also Trans-Pecos. Apr–Jun, also occasionally again in Oct. The following 2 varieties are sometimes recognized; they intergrade and in many cases the characters separating them do not seem consistent.

1. Stems commonly much branched at base, densely pubescent with short, spreading, hooked hairs and few to many long, straight ones ____ var. **obtusa**
1. Stems simple or sparsely branched, sparsely to rather densely pubescent with curled or hooked hairs, occasionally with a few long straight ones ____ var. **pensylvanica**

var. **obtusa** (Rydb. ex Small) Shinners, (blunt). Apparently mainly Edwards Plateau and Trans-Pecos. [*P. obtusa* Rydb. ex Small]

var. **pensylvanica**, HAMMERWORT. Most, if not all, nc TX material seems to be of this variety.

PILEA CLEARWEED, RICHWEED

A genus of ca. 400 species of tropical and subtropical areas worldwide except Australia and New Zealand (Boufford 1997b); monoecious and dioecious herbs without stinging hairs; some used ornamentally including *P. microphylla* (L.) Liebm. (ARTILLERY-PLANT, GUNPOWDER-PLANT) of tropical America whose anthers eject pollen explosively. (Latin: *pileus*, cap, alluding to the first described species' enlarged sepal, which partly covers the achene like the felt cap of the Romans) REFERENCE: Fernald 1936.

Pilea pumila (L.) A. Gray, var. **deamii** (Lunell) Fernald, (sp.: dwarf; var.: for Charles C. Deam, 1865–1953, botanist of Indiana), CLEARWEED, CANADA CLEARWEED. Low monoecious annual to ca.

0.4(–0.7) m tall; stems watery, translucent; leaves long-petioled; leaf blades ovate, serrate; achenes ovate, 1.3–2 mm long, pale green, unspotted or with purple markings, unwinged and unbeaked. Moist woods, seepage areas and along streams; Fannin Co.; mostly e TX. Jun–Nov.

URTICA NETTLE

A genus of 45 species (Boufford 1997b); nearly cosmopolitan, but especially n temperate in distribution; usually herbaceous, monoecious or dioecious, and usually with alkaloids and stinging hairs; ☠ the hairs pierce the skin and inject histamine (causing itching), a neurotoxin (probably a sodium channel toxin), and acetylcholine (causing a burning sensation). (Classical Latin name, from *uro*, to burn, alluding to the stinging hairs)
REFERENCES: Woodland et al. 1976; Woodland 1982.

Urtica chamaedryoides Pursh, (presumably from resemblance to *Teucrium chamaedrys* L.—germander, in the Lamiaceae), STINGING NETTLE, HEART-LEAF NETTLE, ORTIGUILLA. Low monoecious annual with square stems; stems and leaves hispid with stinging hairs; leaf blades subcordate to lanceolate; flowers in small axillary clusters, the upper clusters forming a spike-like panicle with reduced leaves or bracts. Shady or damp ground; se and e TX w to West Cross Timbers and Edwards Plateau. Mar–May. Boufford (1997b) lumped [*U. chamaedryoides* var. *runyonii* Correll] from s TX. Jones et al. (1997) recognized var. *runyonii*. If touched, glass-like hairs on the foliage of this species break off in the skin and act like hypodermic needles; they release toxins which cause an intense burning sensation; this type of effect is known as contact urticaria. According to Lampe (1986), only four families (Euphorbiaceae, Hydrophyllaceae, Loasaceae, and Urticaceae) have such stinging hairs—nc TX has stinging representatives of all of these except the Hydrophyllaceae. ☠

VALERIANACEAE VALERIAN FAMILY

A small (300 species in 10 genera), primarily herbaceous (rarely shrubby) family nearly cosmopolitan in distribution, but especially in n temperate areas and the Andes. Many have a characteristic "wet dog" odor due to the presence of valerianic acid and its derivatives. The Valerianaceae are closely related to the Caprifoliaceae and appear to represent an herbaceous clade within that mainly woody family. From a cladistic standpoint they should be lumped to form a more inclusive monophyletic Caprifoliaceae (Judd et al. 1994). Family name from *Valeriana*, a genus of ca. 250 species of herbs and shrubs with alkaloids, native to the n temperate zone, s Africa, and the Andes; some species of *Valeriana* have been used medicinally (e.g., *V. officinalis* L. to treat nervous disorders—Woodland 1997) (see *Valerianella* for derivation) (subclass Asteridae)
FAMILY RECOGNITION IN THE FIELD: herbs with simple, opposite, often clasping leaves; flowers small, often crowded, in terminal, *dichotomously-branched*, leafy-bracted inflorescences; corollas sympetalous, usually white; ovary obviously inferior, developing into a small but *distinctive fruit*—3-celled but only one cell maturing a seed (hand lens needed).
REFERENCES: Ferguson 1965; Donoghue et al. 1992; Judd et al. 1994.

VALERIANELLA LAMB'S-LETTUCE, CORNSALAD

Ours small annual herbs, dichotomously branched; often with a distinctive "wet dog" odor when dry; leaves, except lowest, sessile, often clasping, simple, opposite, entire or with few coarse, basal teeth or lobes; flowers small, terminal, in compact, head-like, dichotomous, leafy-bracted cymose inflorescences; calyx lobes absent; corollas usually white (rarely pale pinkish), funnelform to salverform, subequally 5-lobed, the tube saccate with nectary; stamens 3; pistil 3-carpellate, inferior, of 1 fertile locule (often somewhat flattened) and 2 sterile (= empty) locules (the sterile locules are sometimes inflated); fruits dry, 1-seeded, indehiscent, usually < 3 mm long.

A genus of 50 species of the n temperate zone s to n Africa; some are used as potherbs; many North American taxa exhibit fruit polymorphisms in which a single population of a species may have up to three different genetically controlled fruit morphs. For *Valerianella* nomenclature we are following Eggers-Ware (1983). Mature fruits examined with a hand lens are necessary for positive identification of most of our taxa. Eggers-Ware has annotated a number of BRIT specimens without mature fruits as indistinguishable between *V. radiata* and *V. woodsiana*. (Diminutive of *Valeriana*, a medieval name, perhaps from Latin: *valere*, to be strong or healthy, possibly alluding to medicinal properties of some species, or said by Linnaeus to honor Publius Aurelius Licinius Valerianus, Roman emperor from 253–260 B.C. and patron and friend of botanists)

REFERENCES: Dyal 1938; Eggers 1969; Eggers-Ware 1983.

1. Flowers showy, the corollas 2.5–4.5 mm across; fruits with conspicuous hairs having a minute hook at their ends OR fruits glabrous or with straight hairs.
   2. Mature fruits broadly ovate to nearly round in outline, with conspicuous hairs having a minute hook at their ends; stems, leaves, and bracts essentially glabrous except for tufts of hairs on each side of leaf base; widespread on limestone in nc TX ____ **V. amarella**
   2. Mature fruits narrowly elliptic or lanceolate in outline, glabrous or rarely with straight hairs; stems, leaves, and bracts with pubescence (on their margins or angles); rare if present in nc TX ____ **V. stenocarpa**
1. Flowers not showy, the corollas 0.6–2 mm across; fruits glabrous or with straight hairs.
   3. Combined width of the 2 sterile locules distinctly wider than the single fertile locule (fertile locule conspicuously exceeded laterally by the widely divergent sterile locules); mature fruits wide, broadly elliptic to nearly round in outline ____ **V. woodsiana**
   3. Combined width of the 2 sterile locules narrower to barely wider than fertile locule (fertile locule narrower to barely exceeded laterally by the sterile locules which are not divergent or only slightly so); mature fruits narrower, narrowly lanceolate to elliptic or ovate in outline ____ **V. radiata**

Valerianella amarella (Lindh. ex Engelm.) Krok, (bitter, in reference to the taste of the leaves), HAIRY CORNSALAD. Plant 15–30 cm tall; stem glabrous; rather showy, usually many flowers open at once, over-topping the small bracts; corollas 3–4.5 mm across; fruits with the 2 sterile locules much smaller than the large fertile locule, the groove between sterile locules narrow, very shallow or inconspicuous. Limestone outcrops; Blackland Prairie–on Austin Chalk (Bell, Ellis, and Grayson cos.) s and w to w TX. Late Mar–May. Hundreds or even thousands of plants of *V. amarella* can make an extremely showy display when in flower on nearly bare limestone outcrops.

Valerianella radiata (L.) Dufr., (having rays), BEAKED CORNSALAD, NARROW-CELL CORNSALAD. Plant 15–60 cm tall; stems often pubescent along angles; upper leaves often lobed-toothed toward base; bracts mostly lanceolate to oblong-obovate, broadly pointed. Damp ground, disturbed areas; se and e TX w to West Cross Timbers. Mar–May. Eggers-Ware (1983) indicates that the 3 taxa of *V. radiata* found in nc TX are distinguishable only by fruit shape. Because they sometimes occur in mixed colonies and because of known genetically controlled fruit polymorphisms in other members of the genus, we are following her in recognizing these taxa at the forma level.

1. Single fertile locule somewhat humped across its width on outer surface; mature fruits narrowly lanceolate to narrowly elliptic in outline, usually more than twice as long as wide ____ **V. radiata** forma **parviflora**
1. Single fertile locule nearly flat across its width on outer surface; mature fruits elliptic to ovate in outline, ca. twice as long as wide or less.

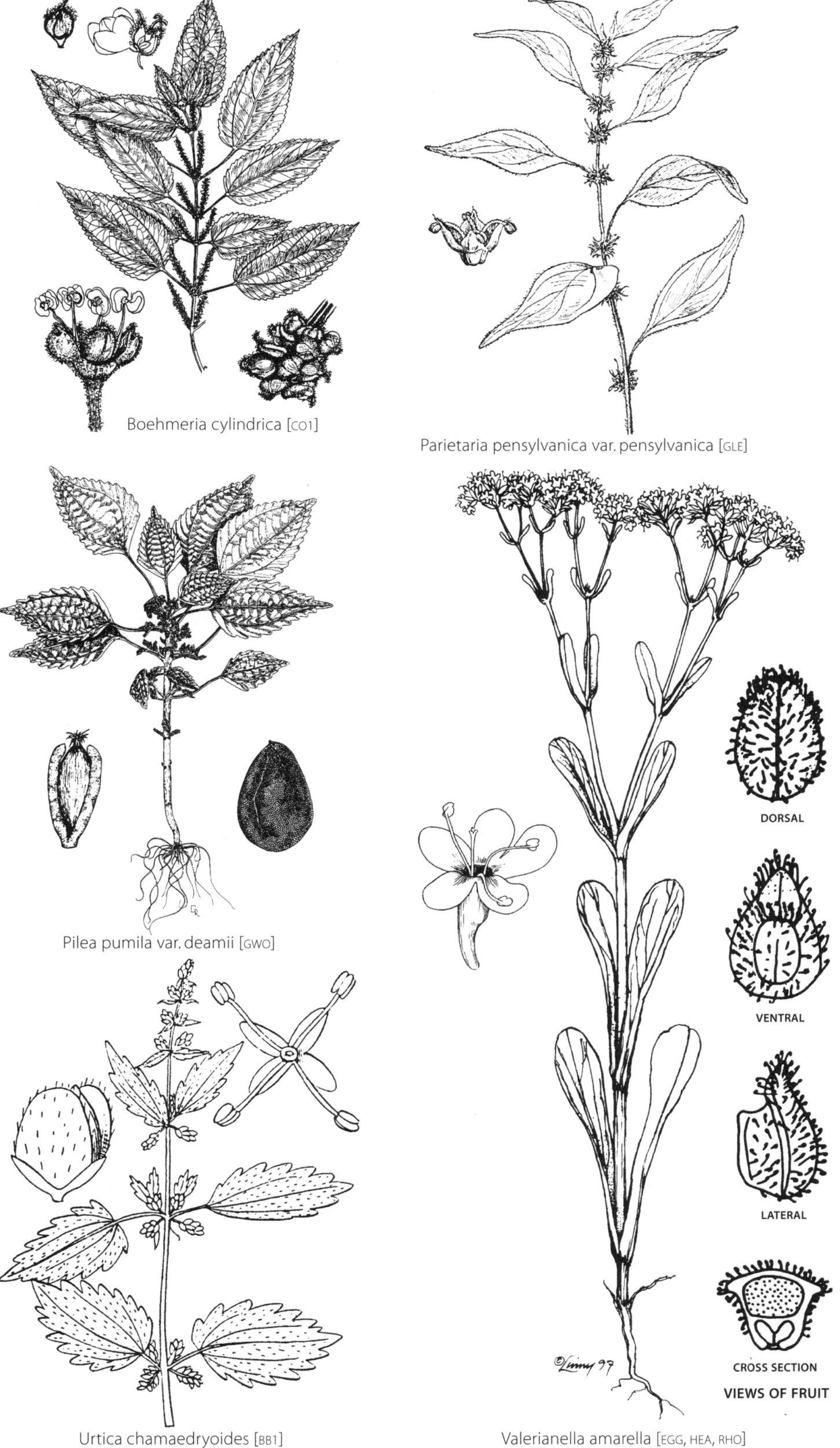

Boehmeria cylindrica [CO1]

Parietaria pensylvanica var. pensylvanica [GLE]

Pilea pumila var. deamii [GWO]

Urtica chamaedryoides [BB1]

Valerianella amarella [EGG, HEA, RHO]

2. The 2 sterile locules slightly spreading apart from each other, with a deep valley-like groove between them; combined width of sterile locules ca. as wide as fertile locule (fertile locule not visible or only barely so when fruit viewed ventrally—from sterile locule side); common in nc TX ______ **V. radiata** forma **radiata**
2. The 2 sterile locules not spreading apart from each other, with only a shallow line-like groove between them; ; combined width of sterile locules distinctly narrower than fertile locule (fertile locule clearly visible when fruit viewed ventrally—from sterile locule side); rare in nc TX, mainly e TX ______ **V. radiata** forma **fernaldii**

forma **fernaldii** (Dyal) Egg.Ware, (for Merritt Lyndon Fernald, 1893–1950, author of *Gray's Manual of Botany*, 8th ed.). Kaufman Co.; otherwise in TX known only from Gonzales and Travis cos. [*V. radiata* (L.) Dufr. var. *fernaldii* Dyal]

forma **parviflora** (Dyal) Egg.Ware, (small-flowered). Se and e TX w to West Cross Timbers. [*V. stenocarpa* (Engelm. ex A. Gray) Krok var. *parviflora* Dyal] This taxon has a hump-backed fruit similar to that of *V. stenocarpa*; however, based on a variety of other floral and vegetative characters (e.g., flower size), it is more appropriate to recognize it as a form of *V. radiata* (Eggers 1969).

forma **radiata**. Denton, Fannin, Grayson, Hopkins, Hunt, and Limestone cos.; se and e TX w to West Cross Timbers. [*V. radiata* (L.) Dufr. var. *radiata*]

Valerianella stenocarpa (Engelm. ex A. Gray) Krok, (narrow-fruited). Plant 10–50 cm tall; stems often pubescent on angles; bracts glabrous or the middle and outer ± hispid-pubescent on margins, mostly linear- to oblong-lanceolate and rather narrowly pointed; corollas 2.5–4 mm across. Margins of thickets or woods, roadsides, fields, and fencerows; included based on Mahler (1988); according to D.M. Eggers-Ware (pers. comm.), this species, which is endemic to sc TX in a six county area along the Balcones Escarpment and on the edge of the Edwards Plateau, probably does not occur in nc TX. Apr–May.

Valerianella woodsiana (Torr. & A. Gray) Walp., (for Joseph Woods, 1776–1864, British botanist and student of this genus), WOOD'S CORNSALAD. Plant 15–50 cm tall; stems often pubescent along the angles. Damp ground, especially sandy soils; Denton, Navarro, and Wise cos.; se and e TX w to nc TX. Apr–May. Eggers-Ware (1983) indicated that *V. woodsiana* can be reliably distinguished from *V. radiata* only by fruit characters and suggested the need for further study to determine whether it is a separate species or another fruit morph of *V. radiata*.

Valerianella florifera Shinners, (flower-bearing), was cited by Hatch et al. (1990) for vegetational area 4 (Fig. 2), probably based on a Fayette Co. record to the se of nc TX. This species, which occurs on sandy soils, has relatively large corollas as in *V. amarella* and *V. stenocarpa*. *Valerianella florifera* differs from all other nc TX species in having margins of floral bracts conspicuously glandular-denticulate (vs. non-glandular or essentially so).

Valerianella texana Dyal, (of Texas), endemic to the Burnet, Gillespie, and Llano Co. area of the Edwards Plateau, is the only other *Valerianella* species in TX. This taxon has the relatively large corollas of *V. amarella* and *V. stenocarpa*, but can be distinguished by its fruits < 1.5 mm long, with 4–6 irregular lines of clavate glands, and with the fertile locule grooved along the midline (vs. fruits > 1.5 mm long, glabrous, with cilia, or with hooked hairs, and the fertile locule not grooved along the midline).

VERBENACEAE VERVAIN OR VERBENA FAMILY

Annual or perennial herbs, shrubs, or small trees; stems square (not distinctly so in woody species); leaves opposite or the uppermost alternate, simple or compound, entire, toothed, or lobed; flowers axillary or terminal, in heads, spikes, or panicles; sepals 5, united basally; corollas sal-

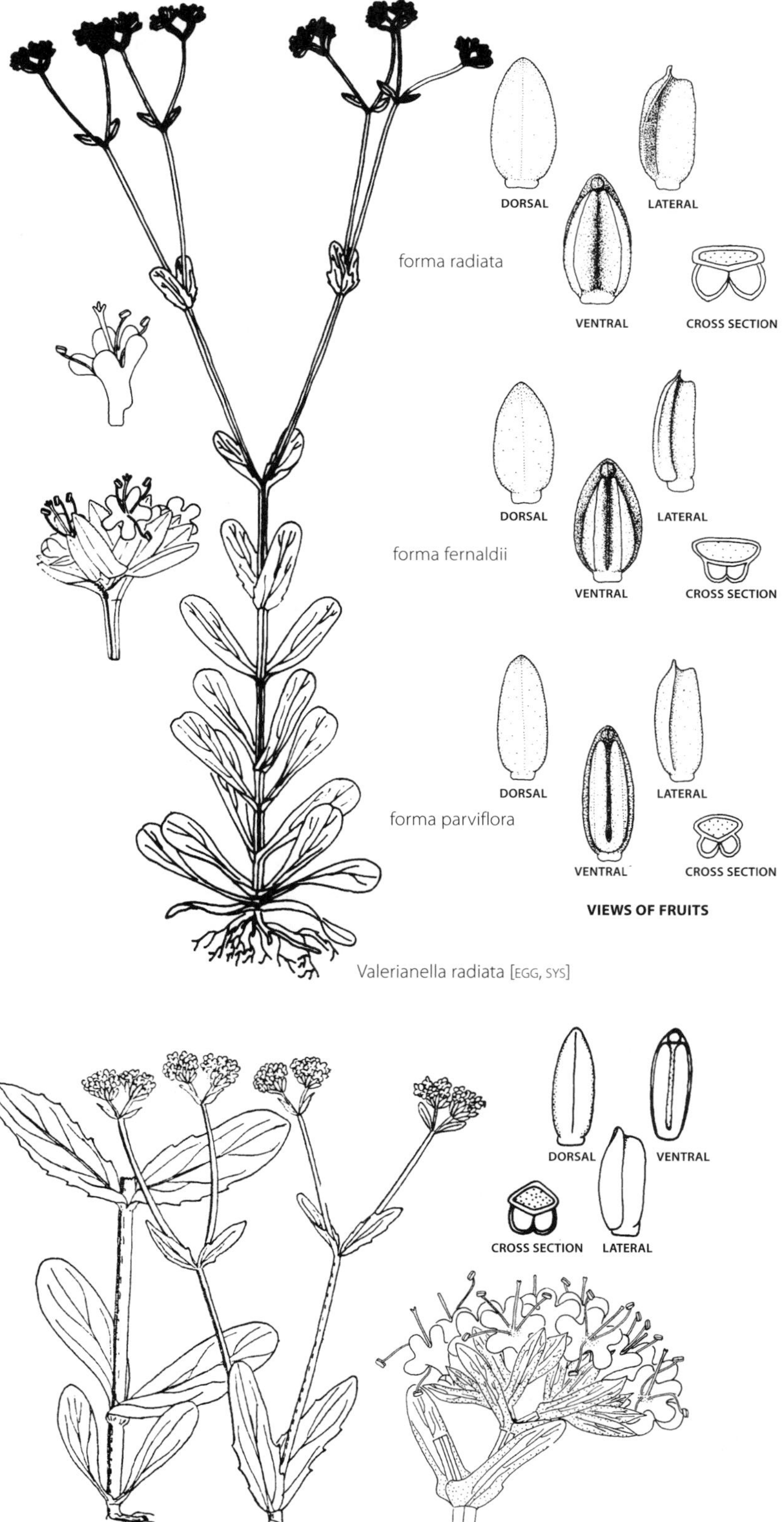

Valerianella radiata [EGG, SYS]

Valerianella stenocarpa [BB1, EGG]

verform or funnelform, 4- to 5-lobed, slightly or markedly bilaterally symmetrical; stamens 4, attached to corolla tube near or below middle; pistil 2-carpellate, usually 2–4-lobed; style 1 and stigmas 1 or 2; ovary superior; fruits dry and separating into 4 one-seeded nutlets or fleshy and drupe-like.

A medium-large (950 species in 41 genera), mainy tropical family with a few species in temperate regions; it has often been treated more broadly to include groups now segregated into other families. It consists of herbs, shrubs, trees, and lianas and includes a number of ornamentals as well as the Asiatic *Tectona grandis* L. f. (TEAK), the source of a valuable, water-resistent wood. The family is related to the Lamiaceae and according to Judd et al. (1997), it appears to be paraphyletic. They suggested that from the cladistic standpoint, the Verbenaceae be limited to those taxa traditionally placed in the Verbenoideae (in nc TX this includes *Aloysia*, *Glandularia*, *Lantana*, *Lippia*, and *Verbena*), with the rest of the family put in a more inclusive, monophyletic Lamiaceae. However, recent molecular studies (Wagstaff & Olmstead 1997) do not support the monophyly of a clade composed of Lamiaceae sensu lato and Verbenaceae sensu stricto. Until the phylogeny of this group is more clearly resolved, we are treating these families in the traditional manner. *Phryma* has been placed in the Verbenaceae by a number of authorities (e.g., Cronquist 1981), but is here treated in its own family (see explanation under Phrymaceae). (subclass Asteridae)

FAMILY RECOGNITION IN THE FIELD: similar to the mint family (usually opposite leaves; stems often square; plants sometimes aromatic; sympetalous corollas) but differs in that the ovary has a single *terminal style* and the flowers are usually individually small.

REFERENCES: Moldenke 1948, 1961b; Judd et al. 1994; Wagstaff & Olmstead 1997.

1. Shrubs or small trees with large, conspicuous, palmately compound leaves with 3–9 distinct leaflets ____ **Vitex**
1. Herbs or shrubs with variously toothed to divided to bipinnatifid leaves, but not as above.
   2. Flowers and fruits in slender spikes; calyces with 3 slender, elongate (2–3 mm long) upper teeth ca. equaling the tube, and 2 short (0.3–0.5 mm long) lower teeth, in fruit the calyces dramatically reflexed and lying parallel to the inflorescence axis; herbs 30–100 cm tall ____ **Phryma** (see Phrymaceae)
   2. Flowers and fruits in heads or in solitary or panicled spikes or slender racemes or in axillary cymes; calyces without teeth as above, in fruit the calyces neither reflexed nor lying parallel to the inflorescence axis; small herbs to shrubs.
      3. Flowers in axillary cymes; stamens prominently exserted; corollas funnelform; shrubs to 3 m tall; leaf blades elliptic to ovate, 8–23 cm long; fruits rose-pink to violet or red-purple, globose, 3–6 mm in diam., in showy clusters ____ **Callicarpa**
      3. Flowers in heads or in solitary or panicled spikes or slender racemes; stamens barely or not exserted; corollas salverform, funnelform, or 2-lipped; herbs or shrubs; leaf blades various; fruits not as above.
         4. Leaf blades toothed, lobed, or deeply pinnatifid; herbs or shrubs, if shrubby, then the fruits drupe-like, with a ± fleshy outer layer; widespread in nc TX.
            5. Flowers in heads or spikes terminating leafy stems or branches; ovary 4-celled, the fruit separating into 4 nutlets.
               6. Nutlets completely enclosed by the twisted-closed calyces, not visible without dissecting calyces, grayish black to black at maturity, with a cavity at base; styles 6–20 mm long, 6 times as long as ovary or longer; inflorescence at flowering time of compact, flat-topped spikes (can elongate later); corollas salverform ____ **Glandularia**
               6. Nutlets barely included in the open calyces, their apices often visible, reddish brown, without a cavity at base; styles 2–3 mm long, 3 times as long as ovary or shorter;

inflorescence at flowering time generally of elongated spikes (can sometimes be compact); corollas usually funnelform (some can be salverform) _____ **Verbena**

5. Flowers in heads or short spikes terminating leafless peduncles arising from the axils of leaves; ovary 2-celled, the fruit with 2 nutlets.
   7. Shrubs to 2 m tall; leaf blades rotund to triangular-lanceolate, abruptly narrowed, rounded, or truncate basally, with distinct petioles; corolla tube 7–10 mm long, the limb 5–9 mm across; calyces nearly truncate with 2 short lateral teeth _____ **Lantana**
   7. Herbs (base can be woody) with trailing or ascending stems rooting at the nodes; leaf blades oblong-lanceolate, oblanceolate, or obovate, sessile or gradually narrowed to short petiolar bases; corolla tube 2–5 mm long, the limb 1.5–4.5 mm across; calyces deeply 2-lobed or divided nearly to the base as 2 bracteole-like segments _____ **Lippia**

4. Leaf blades usually entire; shrubs with small dry fruits; rare in nc TX _____ **Aloysia**

ALOYSIA BEEBUSH

A genus of 37 species native to the Americas. ☠ At least one species is toxic to livestock (Kingsbury 1964). (Named for Maria Louisa Teresa, died 1819, Princess of Asturias, Spain) REFERENCE: Armada & Barra 1992.

Aloysia gratissima (Gillies & Hook.) Tronc., (very pleasing or agreeable), COMMON BEEBUSH, WHITEBRUSH, BEEBLOSSOM, PALO AMARILLO, CEDRÓN, POLEO, CEDRÓN DEL MONTE, NIÑARUPÁ, RESEDA DEL CAMPO, ANGEL FAVORITA, AZMIRILLO, ORGANILLO, ROMERILLO, VARA DULCE, HIERBA DE LA PRINCESA. Much branched shrub to 3 m tall, often with axillary fascicles of smaller leaves; young twigs minutely pubescent; leaves opposite, green, lanceolate to elliptic to narrowly oblong, 3–27 mm long, 2–8 mm wide, entire or with 1–4 teeth per side; flowers very sweet-scented with strong vanilla odor, in peduncled, erect, dense, spike-like racemes 2–7 cm long in the axils of reduced upper leaves, forming narrow panicles; corollas white or tinged with violet, with greenish yellow eye, with tube 3.5 mm long and limb 3.5 mm wide; fruits small, dry, 2-celled. Rocky or gravelly ground; Brown and Coleman cos., also Burnet (Moldenke 1948), Dallas, and Tarrant (Moldenke 1961b) cos.; across much of the s 1/2 of TX rarely n to nc TX. Mar–Oct, according to rains. [*A. gratissima* var. *schulziae* (Standl.) L.D. Benson, *A. lycioides* Cham. var. *schulziae* (Standl.) Moldenke, *A. ligustrina* sensu Moldenke] Reported to produce weakness and death in horses when heavily grazed (Burlage 1968). While two varieties are often recognized (e.g., Kartesz 1994; Jones et al. 1997), we are not distinguishing infraspecific taxa within this species. However, all nc TX material seems to fall within var. *gratissima*. Moldenke (1961b) separated the two varieties (as *A. ligustrina*) as follows: ☠

1. Leaf blades narrowly oblong or elliptic, entire, appressed-puberulent beneath _____ var. *gratissima*
1. Leaf blades more broadly elliptic, obovate, or oblong-elliptic, the larger ones usually 2–8-toothed, densely strigillose beneath and shortly hispidulous along the venation _____ var. *schulziae*

CALLICARPA FRENCH-MULBERRY

A genus of 140 species of the tropics and subtropics; some are used as ornamentals, ☠ others medicinally or as fish poisons. (Greek: *callos*, beauty, and *carpos*, fruit)

Callicarpa americana L., (of America), AMERICAN BEAUTY-BERRY, FRENCH-MULBERRY, BERMUDA-MULBERRY, SOURBUSH, BUNCHBERRY, FILIGRANA DE MAJORCA, FILAGRANA DE PIÑAR, FOXBERRY, PURPLE BEAUTY-BERRY, SPANISH-MULBERRY, TURKEY-BERRY. Soft-woody, stellate-pubescent shrub 0.7–3 m tall, with widely spreading branches; leaves with petioles to 38 mm long; leaf blades elliptic to ovate, 8–23 cm long, sharply toothed, aromatic; flowers in axillary cymes; corollas funnelform, obtusely 4-lobed, lavender-pink, to bluish, reddish, or white, small, the tube 2.6–2.9 mm long; fruits drupaceous, extremely showy, rose-pink to violet or red-purple, globose, 3–6 mm in diam., densely clustered, with a distinctive spicy odor. Woods, especially low ground; se and e TX w to East Cross Timbers, also Fort Hood (Bell or Coryell cos.—Sanchez 1997) in Lampasas Cut Plain; also Edwards Plateau. Jun–Jul. ▣/81

GLANDULARIA VERVAIN, VERBENA

Annual or perennial herbs; leaves opposite; calyces tubular, 5-lobed; corollas 5-lobed, salverform; stamens 4; ovary 2-carpellate, 4-lobed; fruits separating at maturity into 4 one-seeded, linear nutlets; base chromosome number of 5.

While sometimes recognized as a section of *Verbena*, we are following Schnack and Covas (1944), Umber (1979), and Turner (1998) in recognizing *Glandularia* as a distinct genus based on consistent differences including seed morphology, chromosome number, and ratio of style versus ovary length. It is a New World genus of ca. 70 species. (Diminutive of Latin: *glandula*, glandular or glandulose, in reference to the glandular stigmatoid mass on the style branch)
REFERENCES: Perry 1933; Schnack & Covas 1944; Umber 1979; Pruski & Nesom 1992; Turner 1998.

1. Flowers relatively small, the corolla tube 8–10 mm long, only slightly longer than calyx; corolla limb 3–5 mm wide; calyces 6 mm long; nutlets 2–2.8 mm long ______ **G. pumila**
1. Flowers relatively larger, the corolla tube 10–26 mm long, 1.5–2.5 times as long as calyx; corolla limb 7–15 mm wide; calyces 7–13 mm long; nutlets (2–)3–3.5 mm long.
   2. Corolla tube 15–30 mm long, the limb 10–24 mm wide; calyces 8–15 mm long; leaves merely toothed to incised, 3-parted, incised-pinnatifid, or 1-pinnatifid; inflorescences often conspicuously pedunculate.
      3. Leaves relatively little divided, irregularly deeply toothed from apex to base, densely soft pubescent on both surfaces; corolla limb 10–24 mm wide; escaped cultivar ______ **G. ×hybrida**
      3. Leaves much more divided, 3-parted to incised, incised-pinnatifid, or 1-pinnatifid, glabrate or with appressed hairs on both surfaces; corolla limb 10–15 mm wide; native species ______ **G. canadensis**
   2. Corolla tube usually 10–15 mm long, the limb 7–10(–15) mm wide; calyces 7–10 mm long; leaves 1–2-pinnatifid; inflorescences essentially sessile or on a short peduncle ______ **G. bipinnatifida**

Glandularia bipinnatifida (Nutt.) Nutt., (twice pinnately cut), DAKOTA VERVAIN, PRAIRIE VERBENA, PRAIRIE VERVAIN, SWEET-WILLIAM, SMALL-FLOWER VERVAIN, COMMON VERVAIN, WILD VERVAIN, RAGWEED VERVAIN, WESTERN PINK VERVAIN, WESTERN PINK VERBENA, MORADILLA. Perennial; stems ascending, densely hispid-hirsute; leaves variable; bracts usually exceeding calyces; calyces typically with non-glandular pubescence but without glandular pubescence or with only a few such hairs; corollas pink to lavender or purple. Disturbed areas; widespread in TX but mainly w 2/3 of the state. Late Mar–Jun, less freely to Oct. [*Verbena bipinnatifida* Nutt., *V. ciliata* Benth., *V. ciliata* var. *longidentata* L.M. Perry] ▣/91

Glandularia canadensis (L.) Nutt., (of Canada), ROSE VERVAIN. Perennial; stems decumbent to ascending; calyces typically with glandular pubescence in addition to non-glandular pubescence (in contrast to *G. bipinnatifida*); corollas showy, mostly pink to rose, varying to blue, lavender, purple, or white. Sandy open woods, roadsides; Henderson, Limestone, and Milam cos. on e margin of nc TX, also Grayson Co. in Red River drainage, also Denton, Lamar (Moldenke 1948), and Falls (Carr 1994) cos.; mainy se and e TX. Mar–May, sporadically later. [*Verbena canadensis* (L.) Britton, *V.* ×*oklahomensis* Moldenke]

Glandularia ×hybrida (Groen. & Rümpler) G.L. Nesom & Pruski, (hybrid), HYBRID VERBENA, GARDEN VERBENA. Densely hirsute or villous perennial; stems procumbent or ascending, rooting at the nodes; spikes large and showy; flowers fragrant; calyces 8–15 mm long; corollas pink, red, white, yellow, blue, purple, or varigated, often with a white or yellow eye, the tube 15–30 mm long. Cultivated throughout the state as an ornamental and persisting or escaping; Dallas

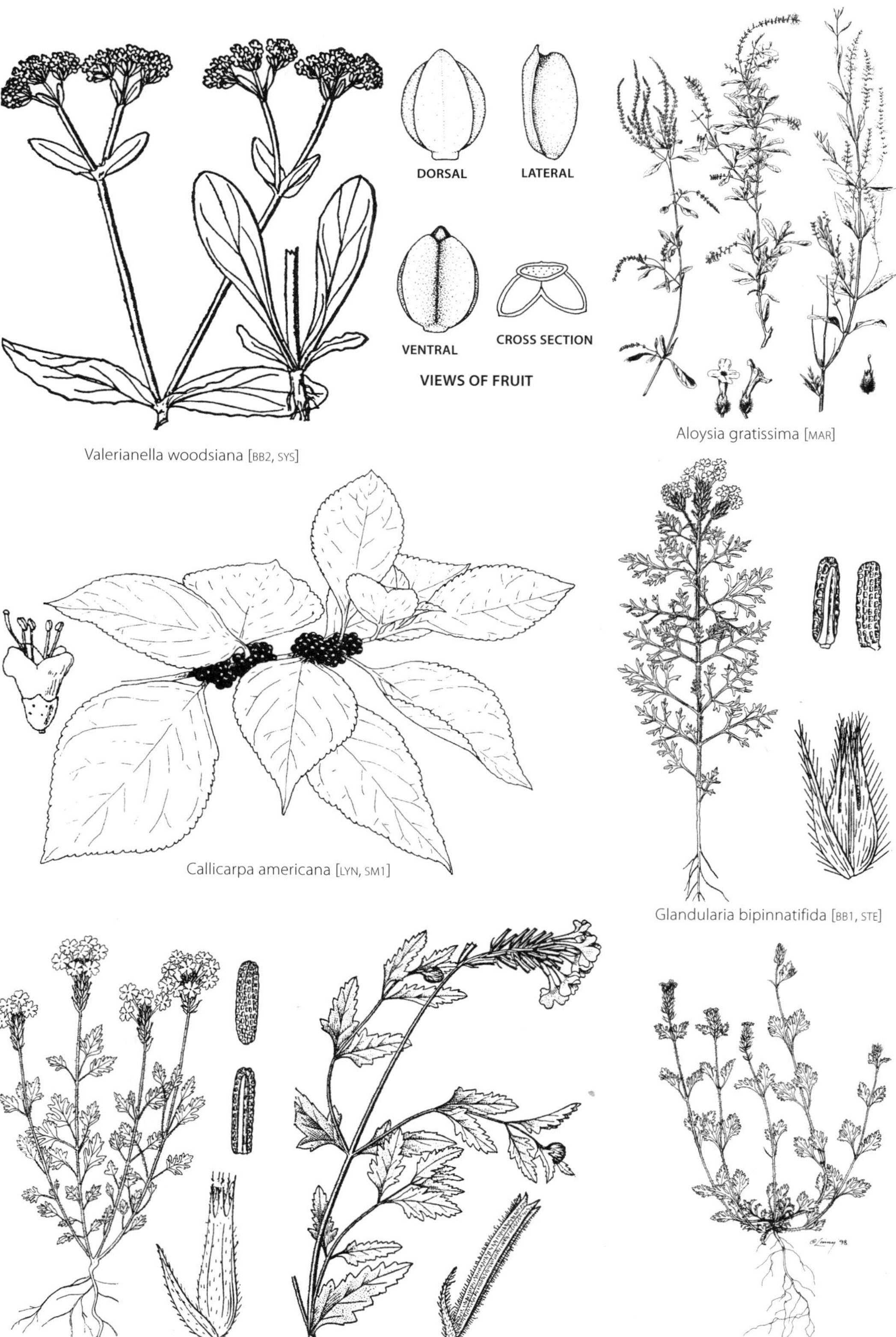

Valerianella woodsiana [BB2, SYS]

Aloysia gratissima [MAR]

Callicarpa americana [LYN, SM1]

Glandularia bipinnatifida [BB1, STE]

Glandularia canadensis [BB1]

Glandularia xhybrida [GLE]

Glandularia pumila [HEA]

Co. (Moldenke 1948). [*Verbena* ×*hybrida* Groen. & Rümpler] Probably a hybrid of *G. peruviana* (L.) Druce with other South American species (Pruski & Nesom 1992). We are following Pruski and Nesom (1992) for nomenclature of this species. Kartesz (1994) and Jones et al. (1997) placed this species in *Verbena* and did not treat it as a hybrid.

Glandularia pumila (Rydb.) Umber, (dwarf), PINK VERVAIN, PINK VERBENA, HAIRY VERBENA, WILD VERBENA. Low annuals; stems usually decumbent-spreading to erect, hirsute, and often finely glandular; leaves three-parted, the divisions variously incised; spikes sessile or on short peduncles; corollas pink to lavender or pale blue, rarely white. Sandy or gravelly soils; widespread in TX, but mainly w 2/3. Late Mar–May. [*Verbena pumila* Rydb.]

Glandularia quadrangulata (A. Heller) Umber, (four-angled), native to the s 1/2 of TX, is cited for vegetational area 4 (Fig. 2) by Hatch et al. (1990). This species, which can be distinguished by its beaked nutlets (unbeaked in all other nc TX species) and usually white corollas, apparently occurs only to the s of nc TX.

LANTANA

Perennial shrubs to ca. 2 m tall, pubescent; leaves opposite or in threes, toothed, often rugose; flowers showy, in dense axillary heads on elongate peduncles; corollas with a slender cylindric tube expanding into a flat limb, radially symmetrical or obscurely 2-lipped, 4–5 parted; stamens 4; fruits drupaceous, usually with a fleshy outer layer, the 2 seed chambers of the hard inner layer remaining fused at maturity.

A genus of 150 species of tropical America and tropical and s Africa; it includes ornamentals, some with edible fruit, and problematic weeds; all parts of some species, particularly the green fruits, are poisonous (Schmutz & Hamilton 1979). (Ancient Latin name for a member of the Caprifoliaceae, *Viburnum lantana* L., which has a similar inflorescence)

1. Corollas orange, deep yellow, or red, the tube to 10 mm long, the limb to 9 mm wide; leaf blades broad, mostly 20–70 mm wide; bracts subtending flowers narrowly triangular-lanceolate to oblanceolate or spatulate; heads large, mostly 2–3 cm wide.
   2. Leaf blades broadly ovate to orbicular, with few coarse teeth mostly 2–5 mm high; bracts subtending flowers oblanceolate to spatulate and persistent into fruit; stem angles with long, hispid hairs ____ **L. urticoides**
   2. Leaf blades cordate-ovate or triangular-ovate, with numerous fine crenate teeth mostly 0.5–1.5 mm high; bracts subtending flowers narrowly triangular-lanceolate (rarely a basal one spatulate or subfoliaceous), usually not persistent; stem angles with short, ± recurved, non-hispid hairs ____ **L. camara**
1. Corollas pale yellow or mostly white to pink, violet, purple, or blue, the tube 4–5 mm long, the limb ca. 3 mm wide; leaf blades narrow, mostly 6–15 mm wide; bracts subtending flowers broadly ovate; heads small, mostly 1–1.5 cm wide ____ **L. achyranthifolia**

Lantana achyranthifolia Desf., (with leaves like *Achyranthes* in the Amaranthaceae), VEIN-LEAF LANTANA, BRUSHLAND LANTANA, HIERBA NEGRA, MEJORANA, YERBA DEL CHRISTO, CARAQUITO BLANCO, CARIACO DE SAN JUAN, FRUTILLA BLANCA, ORGANILLO CIMARRON. Aromatic shrub to 1.5 m tall, pubescent; leaf blades ovate to lanceolate, 5–35 mm long; corollas pale yellow or white often with a yellow or orange eye, fading to pink, violet, purple, or blue; fruits with thin flesh, translucent, nearly dry. Rocky areas, pastures; included based on citation of vegetational area 4 (Fig. 2) by Hatch et al. (1990); probably only to the s of nc TX; s 1/2 of TX. Feb–Nov. [*L. macropoda* Torr.]

Lantana camara L., (South American vernacular name or possibly Latin: *camera*, arch, in reference to the stems), WEST INDIAN LANTANA, ALFOMBRILLA HEDIONA, LARGE-LEAF LANTANA. Unarmed or sparingly armed shrub to 2 m tall; similar to *L. urticoides* but with larger leaves; corolla tube ca. 10 mm long; fruits black. Widely cultivated and escaped; Bell Co.; mainly s and c TX. Flowering nearly throughout the growing season. Native to West Indies, also Central and South America. [*L. camara* var. *mista* (L.) L.H. Bailey] This species can be a problematic weed in subtropical parts of the U.S.; in some areas of the world it invades disturbed habitats, forms impenetrable thickets, and can be a serious pest (Cronk & Fuller 1995). The species is also known to have allelopathic effects (= detrimental to other plants) due to the production of phytotoxic compounds (Casado 1995). It is toxic to livestock, probably due to the presence of a polycyclic triterpenoid; photosensitization may result; the fruits are suspected of being toxic and even lethal to children; contact with the plant can also cause dermatitis (Kingsbury 1964; Morton 1982; Spoerke & Smolinske 1990). ☠

Lantana urticoides Hayek, (resembling *Urtica*—nettle), COMMON LANTANA, TEXAS LANTANA, HIERBA DEL CRISTO, CALICO-BUSH, BUNCHBERRY. Armed to unarmed shrub to 2 m tall; leaf blades ovate to broadly so, very scabrous above; bractlets 4–9 mm long, 1–2 mm wide; corollas yellow or orange, changing to red, the tube 7–10 mm long, the limb 5–9 mm wide; fruits dark blue to black. Cultivated and escaped; open areas, thickets, woods, roadsides; Bell, Coryell, Somervell, and Tarrant cos.; native of c and s TX. Late Apr–Oct. [*L. horrida* of authors, not Kunth, *L. horrida* sensu Moldenke, *L. horrida* Kunth var. *latibracteata* Moldenke] Roger Sanders (pers. comm.) indicated that hybrids between *L. urticoides* and *L. camara* are to be expected. A decotion of the leaves has been used to treat snakebite in Mexico; the plant is reportedly toxic to cattle and sheep; an alkaloid, lantamine, is present (Powell 1988). ☠ /95

LIPPIA FROGFRUIT, FOGFRUIT

Creeping or trailing to ascending, glabrous or inconspicuously pubescent perennials; floral bracts purplish; corollas very small, white to light lavender, pale blue, purple, or rose-purple, with yellow eye changing to orange-red; fruits included in the calyces, dividing into 2 nutlets at maturity, with a thin, dry outer layer.

A tropical African and American genus of ca. 200 species, some are widely naturalized. The nc TX species are all in the portion of the genus that has often been segregated into the genus *Phyla* (e.g., Kartesz 1994; Jones et al. 1997). We, however, are following R. Sanders (unpublished manuscript for the *Generic Flora of the Southeastern U.S.*) in treating these species in *Lippia*. (Named for Agostino Lippi, 1678–1704, Italian naturalist)
REFERENCE: Kennedy 1992.

1. Leaf blades widest at or near middle, toothed from below middle to apex.
   2. Leaf blades not folded like a fan, usually lanceolate or elliptic-lanceolate, widest at or near the middle, the teeth appressed and extending below the widest part of the blade; widespread in nc TX ____ **L. lanceolata**
   2. Leaf blades often slightly folded like a fan, usually ovate to triangular-ovate to rhomboid, widest below the middle, the teeth divergent, not extending below the widest part of the leaf; reported for nc TX but no specimens seen ____ **L. strigulosa**
1. Leaf blades oblanceolate, widest near apex, toothed only near apex.
   3. Leaf blades 4–25 mm wide, acute to rounded or blunt in general outline at apex (not counting apical tooth); small bracts just below heads 2–3 mm long; peduncles 1.5–4 times as long as adjacent leaves in flower; heads or spikes 6–10 mm thick, becoming cylindrical or oblong ovoid, in fruit to 3 cm long; widespread in nc TX. ____ **L. nodiflora**
   3. Leaf blades 2–8 mm wide, acute or subacute in general outline at apex; small bracts just below heads 4–5 mm long; peduncles 0.7–1.5(–2) times as long as the adjacent leaves in flower; heads or spikes 7–12 mm thick, remaining short, globose to subcylindrical, to only 2 cm long; in nc TX reported only from Dallas and Tarrant cos. ____ **L. cuneifolia**

Lippia cuneifolia (Torr.) Steud., (wedge-leaved), WEDGE-LEAF FROGFRUIT. Leaf blades 15–20 mm long, with 1–4 sharp, forward-pointing teeth per side (all beyond the middle) or rarely entire; corollas whitish or purplish, the tube 4–5 mm long, the limb 2–4.5 mm wide. Low grasslands; reported by Moldenke (1961b) from Dallas and Tarrant cos.; mainly w 1/2 of TX. May–Oct. [*Phyla cuneifolia* (Torr.) Greene]

Lippia lanceolata Michx., (lanceolate, lance-shaped), FROGFRUIT, LANCE-LEAF FROGFRUIT, NORTHERN FROGFRUIT. Leaf blades 18–75 mm long, 5–30 mm wide; corollas pale blue, purplish, or white, sometimes with yellow in center. Banks of streams and ponds, moist areas; nearly throughout TX except Trans-Pecos. May–Oct. [*Phyla lanceolata* (Michx.) Greene]

Lippia nodiflora (L.) Michx., (with flowers at nodes), FROGFRUIT, COMMON FROGFRUIT, TURKEY-TANGLE, CAPE-WEED, MAT-GRASS, HIERBA DE LA VIRGIN MARÍA, TEXAS FROGFRUIT, SAW-TOOTH FROGFRUIT, FOGFRUIT, SPATULATE-LEAF FOGFRUIT, WEIGHTY FOGFRUIT, WEDGE-LEAF FROGFRUIT, HOARY FROGFRUIT. Leaf blades 10–72 mm long; corollas white to rose-purple, often with yellow center. Low moist disturbed areas; throughout TX. May–Oct. [*Phyla incisa* Small, *P. nodiflora* (L.) Greene var. *longifolia* Moldenke, *P. nodiflora* var. *reptans* (Kunth) Moldenke, *P. nodiflora* var. *rosea* (D. Don) Moldenke]

Lippia strigulosa M. Martens & Galeotti, (with small or weak appressed hairs), DIAMOND-LEAF FROGFRUIT, TURRE HEMBRA, HIERBA BUENA MONTES. Leaf blades to 75 mm long and 20 mm wide; corollas white, sometimes lavender- or purple-tinged with age, ca. 3 mm long, the limb 1.5 mm wide. Open, usually moist areas; Burnet Co. (Moldenke 1948) near s margin of nc TX and Wood Co. just e of nc TX; mainly s TX. Feb–May. [*Phyla strigulosa* var. *parviflora* (Moldenke) Moldenke, *Phyla strigulosa* (M. Martens & Galeotti) Moldenke var. *sericea* (Kuntze) Moldenke] We are following R. Sanders (pers. comm.) in not recognizing varieties in this species.

Lippia canescens Kunth [*L. nodiflora* var. *canescens* (Kunth) Kuntze, *Phyla canescens* (Kunth) Greene] was recently discovered in a flower bed in Fort Worth (Tarrant Co.—*O'Kennon 14085*, BRIT), probably brought in with nursery stock. While sometimes lumped with *L. nodiflora* (e.g., Wilken 1993b—as *Phyla nodiflora*), this taxon was recognized as a distinct species (in the genus *Phyla*) by Kennedy (1992). It is native to South America and has become established in c Mexico and California; this is the first collection known from TX (identity confirmed by R. Sanders).

Kennedy (1992) separated it from *L. nodiflora* as follows:

1. Floral bracts widely rhombic with straight-tapered membranous upper edges; flowers open in more than a single whorl; calyces glandular with uncinate hairs ____ *L. canescens*
1. Floral bracts widely obtrullate, the upper margins wide-membranous and distinctly undulate and ciliated; flowers presented in a single whorl; calyces eglandular with straight hairs ____ *L. nodiflora*

Lippia graveolens Kunth, (heavily scented), SCENTED LIPPIA, HIERBA DULCE, RED BRUSH LIPPIA, OREGANO CIMARRÓN, ROMERILLO DE MONTE, TÉ DE PAÍS, TARBAY, an aromatic shrub or small tree to 3(–9) m tall, is cited by Hatch et al. (1990) for vegetational area 4 (Fig. 2). This species, found mainly in s TX, apparently occurs only to the s of nc TX. It can be easily distinguished from all other nc TX *Lantana* and *Lippia* species which usually have 1(–2) peduncles per leaf axile (2(–4) per node) and the inflorescence usually projecting well beyond the subtending leaf; it has 2–3 peduncles per leaf axil (4–6 per node) and the inflorescence usually shorter than the subtending leaf.

VERBENA VERVAIN

Usually perennial herbs; stems prostrate to strictly erect; leaves mostly opposite; inflorescences spicate, terminal, often greatly elongate; calyces tubular, 5-lobed; corollas 5-lobed, funnel-shaped (rarely salverform); stamens 4; ovary 2-carpellate, 4-lobed; fruits briefly enclosed by the open calyces, separating at maturity into 4 one-seeded, linear nutlets; base chromosome number of 7.

A tropical and temperate American genus of 200 species; also 2–3 Old World species. *Verbena officinalis* L. (VERVAIN, JUNO'S-TEARS) has bright-eyed flowers and based on the Doctrine of Signatures was used historically in treating eye diseases; it has also been used in witchcraft and ritualistic ceremonies dating back to the Romans and Druids (Baumgardt 1982). *Glandularia*, here treated separately, has often been lumped into *Verbena*. (Latin: *verbena*, ancient name for the common Eurasian vervain, *V. officinalis* L.; possibly derived from Latin: *verbenae*, the foliage of sacred, ceremonial, and medicinal plants such as olive, laurel, and myrtle)
REFERENCES: Perry 1933; Barber 1982.

1. Flowering spikes short and compact, crowded, usually not greatly elongating and never open in fruit, 6 cm or less long (often much less); leaves serrate to coarsely serrate, never divided into segments.
   2. Leaves subcordate and semiclasping at base; corolla tube 3 times as long as calyx or more; plant to ca. 0.6 m tall; bracts noticeably longer than calyx ____ **V. rigida**
   2. Leaves tapering to a wedge-shaped or narrowed subsessile or petioled base; corolla tube slightly longer than calyx; plant to 2.5 m or more tall; bracts equal to or shorter than calyx ____ **V. brasiliensis**
1. Flowering spikes slender and open or compact, greatly elongating and often becoming open in fruit (fruits often widely spaced), often > 6 cm long; leaves varying from crenate to serrate to 3-parted to pinnatifid.
   3. Spikes panicled at the ends of stems and branches, subtended mainly by inconspicuous bracts (at base of spikes); floral bracts (below each flower) inconspicuous.
      4. Middle stem leaves 1- or 2-pinnatifid or 3- or 5-cleft or deeply incised ____ **V. halei**
      4. Leaves (all) merely serrate or crenate.
         5. Leaf blades very scabrous above; corollas blue to pink or lavender; fruiting calyces spreading, with calyx lobes converging or coming together so as to hide the apex of the enclosed fruit ____ **V. scabra**
         5. Leaf blades not markedly scabrous above; corollas white; fruiting calyces ascending, the calyx lobes not converging in a manner that would hide the apex of the enclosed fruit, the fruit apex visible ____ **V. urticifolia**
   3. Spikes solitary or in 3s at the ends of stems and branches OR panicled and subtended by leafy bracts; floral bracts conspicuous or not so.
      6. Leaf blades serrate-dentate or shallowly incised.
         7. Plants coarse; leaf blades ovate or ovate-orbicular, broad, to 44 mm wide; spikes stout, 7–10 mm wide in flower ____ **V. stricta**
         7. Plants slender; leaf blades linear to narrowly elliptic, narrow, 20 mm or less wide; spikes relatively slender, 5–6 mm wide in flower ____ **V. neomexicana**
      6. Leaf blades deeply incised-dentate to pinnatifid or 3-cleft.
         8. Spikes not conspicuously bracteose, the floral bracts not prominent (shorter to only slightly longer than calyces).
            9. Leaf blades (at least the lower) not narrowly elongate, instead, the blades oblong-ovate or obtusely elliptic-ovate, usually 3-parted with the segments incised-dentate.
               10. Leaves distinctly petiolate, the margined petiole often ± as long as the blade; leaf blades ± plicate (= folded fan-like), with venation noticeably whitish near margins; plants not coarsely hairy ____ **V. plicata**
               10. Leaves sessile or at most very short petioled; leaf blades not plicate, with venation not noticeably whitish near margins; plants coarsely hairy (= hirsute) ____ **V. xutha**

9. Leaf blades (at least the lower) narrowly elongate, oblong-lanceolate to spatulate, usually incised-pinnatifid or incised-dentate.
   11. Plants coarse, with a low, ± compact habit; leaves with a broadly margined or semi-clasping subpetiolar base; bracts usually slightly longer than calyces _____ **V. canescens**
   11. Plants more slender, with a taller and open habit; leaves with a narrowly margined petiolar base; bracts mostly equal or shorter than calyces _____ **V. neomexicana**

8. Spikes conspicuously bracteose, the floral bracts prominent (sometimes much longer than the calyces, particularly at base of inflorescence).
   12. Leaf blades ± plicate (folded fan-like); venation noticeably whitish near leaf margins _____ **V. plicata**
   12. Leaf blades neither plicate nor with conspicuous whitish venation near the leaf margins.
      13. Leaves with a broadly margined subpetiolar or semi-clasping base; floral bracts ovate, abruptly acuminate, slightly longer than the flowers, ascending _____ **V. canescens**
      13. Leaves narrowed to a margined petiole; floral bracts linear-lanceolate, much longer than the flowers, often reflexed with age _____ **V. bracteata**

Verbena bracteata Lag. & Rodr., (with bracts), BIG-BRACT VERVAIN, LARGE-BRACT VERVAIN, PROSTRATE VERVAIN. Usually prostrate to decumbent perennial, can flower as an annual; stems spreading-pilose; leaves coarsely toothed and lobed; spikes conspicuously bracteose; calyces 3–4 mm long; corollas inconspicuous, bluish to lavender or purple, the corolla tube slightly longer than calyx, the corolla limb 2.5–3 mm wide. Disturbed areas; widespread in TX. Late Apr–Oct.

Verbena brasiliensis Velloso, (of Brazil), BRAZILIAN VERVAIN. Large, stout perennial, stems erect, 1–2.5 m tall, conspicuously square; leaves elliptic or lanceolate; spikes 0.5–4 cm long; calyces 2.5–3.5 mm or more long; corollas purple or lilac, the corolla tube slightly longer than calyx, the corolla limb 2.5 mm wide. Waste places and along creeks; Grayson (RR yard), Lamar, and Tarrant cos., also Fort Hood (Bell or Coryell cos.—Sanchez 1997); apparently recently introduced into nc TX and now spreading; mainly se and e TX and Edwards Plateau. May–Oct. Native of South America. ☞

Verbena canescens Kunth, (gray-pubescent), GRAY VERVAIN. Perennial from a woody root; stems spreading to erect, to 46 cm tall; foliage canescent; calyces 2–3 mm long; corollas blue to lavender or purple, the corolla tube slightly longer than calyx, the corolla limb 4–6 mm wide. Limestone outcrops; Bell, Brown, Lampasas, Mills, Shackleford, and Williamson cos., also Comanche, Hamilton (HPC), and Somervell (R. O'Kennon, pers. obs.) cos.; widespread in TX. Apr–Oct. [*V. canescens* var. *roemeriana* (Scheele) L.M. Perry]

Verbena halei Small, (for J.P. Hale, ca. 1889, landowner in lower California who collected cacti with Mrs. M.K. Brandegee), SLENDER VERVAIN, BLUE VERVAIN, CANDELABRA VERVAIN, SANDING VERVAIN, TEXAS VERVAIN. Erect perennial, flowering the first year, to 1 m tall; leaves few, petioled; leaf blades oblanceolate or obovate, toothed or lobed; flowers becoming widely spaced; calyces 3–3.5 mm long; corollas lavender-blue (rarely white), the corolla tube only slightly longer than calyx, the corolla limb 6–7 mm wide. Prairies, disturbed areas; throughout TX. Apr–Oct. [*V. officinalis* L. subsp. *halei* (Small) S.C. Barber] ⊛/108

Verbena neomexicana (A. Gray) Small, (of New Mexico), HILLSIDE VERVAIN. Stems erect, to 1 m tall; leaves variable; calyces 3 mm long; corollas blue to lavender or purple, the corolla tube slightly longer than calyx. Rocky areas. Apr–Nov.

1. Leaf blades only rather shallowly incised; floral bracts usually broadly ovate-acuminate; corolla limb ca. 8 mm wide; plants densely canescent-hirtellous _____ var. **hirtella**
1. Leaf blades pinnately parted or almost divided; floral bracts lanceolate-acuminate; corolla limb 4 mm wide; plants hirsute _____ var. **neomexicana**

var. **hirtella** L.M. Perry, (rather hairy), Bell Co.; mainly w 2/3 of TX.

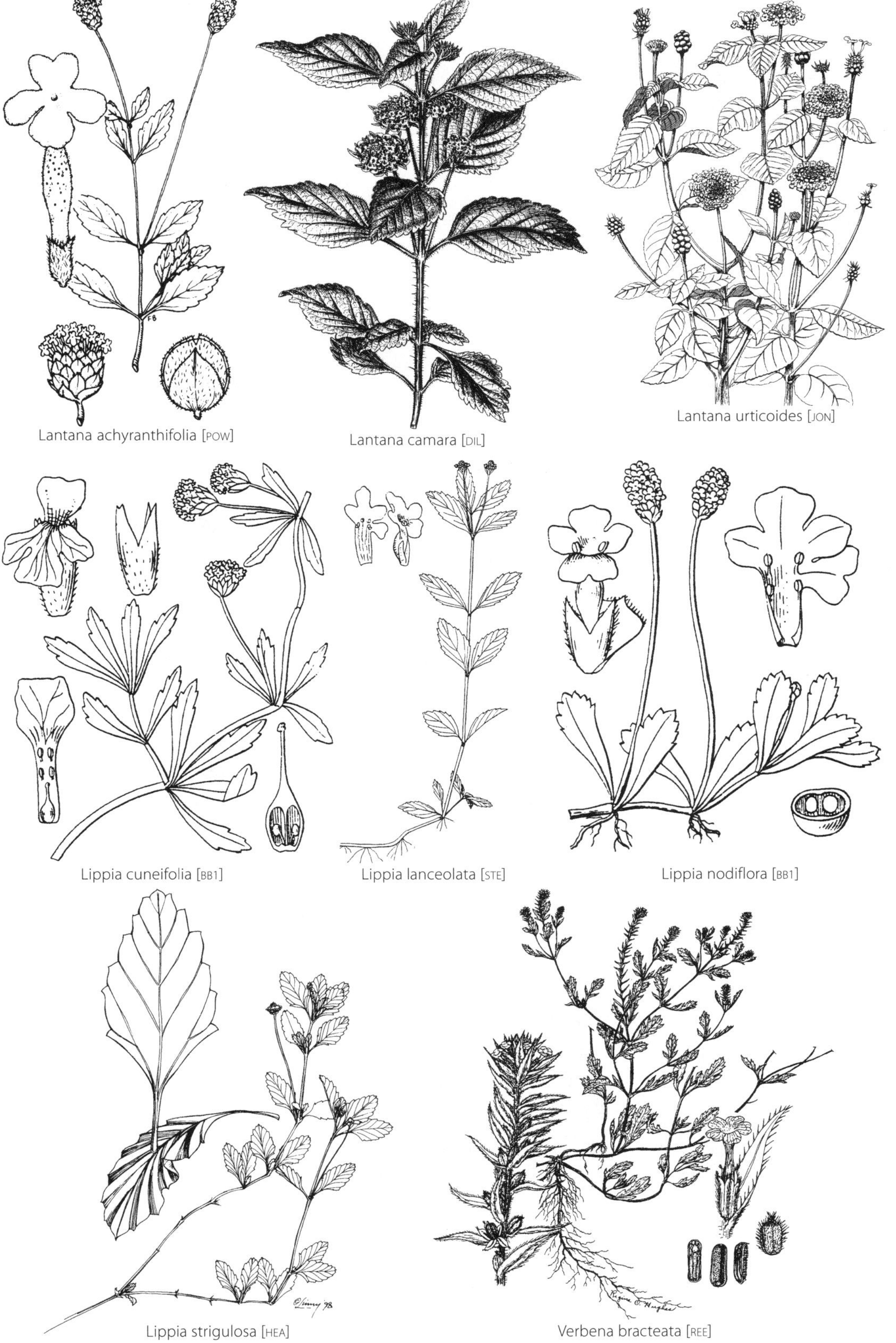

Lantana achyranthifolia [POW]

Lantana camara [DIL]

Lantana urticoides [JON]

Lippia cuneifolia [BB1]

Lippia lanceolata [STE]

Lippia nodiflora [BB1]

Lippia strigulosa [HEA]

Verbena bracteata [REE]

var. **neomexicana**. Included based on citation of vegetational areas 4 and 5 (Fig. 2) by Hatch et al. (1990); mainly w 2/3 of TX.

Verbena plicata Greene, (folded like a fan), FAN-LEAF VERVAIN. Perennial; stems decumbent to ascending; floral bracts equaling or surpassing the calyces; calyces 3.5–4 mm long; corollas blue to lavender or purple, the corolla tube ca. equal in length to calyx, the corolla limb 4–6 mm wide. Open or disturbed areas, sandy or rocky soils; Callahan, Coleman, and Young cos. on w margin of nc TX, also Brown Co. (HPC); mainly Rolling Plains s and w to w TX. Feb–Oct.

Verbena rigida Spreng., (rigid, stiff), VEINY VERVAIN, TUBER VERVAIN. Erect perennial; stems 0.2–0.6 m tall; leaf blades oblong to oblong-lanceolate or narrowly obovate; calyces 4 mm long; corollas purple to pink-purple, the corolla tube 3 times as long as calyx or more, the corolla limb 5–7 mm wide. Pastures, roadsides; Dallas Co.; mainly se and e TX, also Edwards Plateau. Apr–Oct. Native of South America.

Verbena scabra Vahl, (rough), HARSH VERVAIN, WHITE VERVAIN, SANDPAPER VERVAIN. Perennial; stems erect, 0.3–1(–1.5) m tall; leaf blades ovate, extremely scabrous above; spikes slender, elongate, graceful; calyces 2.5–3 mm long; corollas blue to pink or lavender, the corolla tube ca. equaling calyx in length, the corolla limb ca. 2 mm wide. Low areas; included based on citation of vegetational areas 4 and 5 (Fig. 2) by Hatch et al. (1990); supposedly widespread in the state; however, we have seen few TX specimens. Mar–Dec.

Verbena stricta Vent., (upright, erect), WOOLLY VERVAIN, HOARY VERVAIN, MULLEN-LEAF VERVAIN. Stout perennial; stems erect 0.2–2 m tall; leaves sessile or nearly so; leaf blades elliptic-ovate, to 8–20 cm long, irregularly toothed, hirsute-villous and often canescent beneath; flowers densely crowded; calyces (3–)4–5 mm long; corollas blue to purple, the corolla tube slightly longer than calyx, the corolla limb 8–9 mm wide. Open areas, waste places; Collin and Grayson cos., also Cook, Dallas, and Tarrant cos. (Moldenke 1948); se and e TX w to nc TX, also Rolling Plains. Jun–Sep.

Verbena urticifolia L., (with leaves like *Urtica*—nettle), WHITE VERVAIN, NETTLE-LEAF VERVAIN. Perennial; stems coarse, erect 0.5–2.5 m tall; leaf blades ovate, coarsely crenate-serrate, glabrous or sparsely hairy; inflorescences slender, elongate, and graceful; flowers very small; calyces 2 mm long; corollas white, the corolla tube ca. 2 mm long, the corolla limb ca. 2 mm wide. Bottomland woods; Dallas, Denton, Grayson, Tarrant, and Wise cos.; mainly se and e TX w to nc TX. Jun–Oct.

Verbena xutha Lehm., (yellowish, tawny), COARSE VERVAIN, GULF VERVAIN. Coarse and pubescent perennial; stems to 2 m tall; calyces 3–4 mm long; corollas blue to purple, the corolla tube ca. as long as calyx, the corolla limb 5–8 mm wide. Stream bottoms, roadsides; Bell, Dallas, Ellis, Grayson, McLennan, and Milam cos.; widespread in TX. Jun–Oct.

VITEX CHASTETREE

Shrubs or trees with opposite, palmately compound leaves with 3–9 leaflets, these darker above; inflorescences showy, of numerous lavender to lilac, blue, or white flowers; calyces 5-parted; corollas 5-lobed and 2-lipped; stamens 4; fruits drupaceous, ± fleshy.

A genus of 250 species ranging from the tropics to a few in temperate regions; some are used as timber trees and ornamentals. (Classical Latin name for *Vitex agnus-castus*)

1. Leaflets usually entire; inflorescences appearing dense, the adjacent flower clusters (cymules) nearly touching and sessile or nearly so ____ **V. agnus-castus**
1. Leaflets conspicuously toothed to deeply pinnately divided; inflorescences appearing open, the

Verbena brasiliensis [GLE]

Verbena canescens [HUM]

Verbena halei [SHI]

Verbena plicata [HEA]

Verbena rigida [BR2]

Verbena neomexicana var. hirtella [GLE]

Verbena scabra [BR2]

adjacent flower clusters separated from one another and at least the lower ones distinctly stalked ____ **V. negundo**

Vitex agnus-castus L., (ancient name meaning chaste), COMMON CHASTETREE, INDIAN-SPICE, WILD-LAVENDER, HEMPTREE, MONK'S PEPPER-TREE, WILD PEPPER, TRUE CHASTETREE, ABRAHAM'S-BALM, CHASTE-LAMB-TREE, SAGETREE, TREE-OF-CHASTITY. Shrub or small tree to 5 m tall, aromatic; leaves with (3–)5–9 leaflets; leaflets to ca. 12 cm long, dark green above, densely white puberulent below; petioles to 75 mm long; inflorescences to 31 cm long; calyces 2–2.5 mm long; corolla tube 6–7 mm long; corolla limb to 13 mm wide. Widely cultivated as an ornamental in TX, escapes, and naturalizes in moist habitats and waste places; Bell, Dallas, Erath, Hill, Grayson, and Tarrant cos., also Brown Co. (Moldenke 1961b). May–Oct. Native of s Europe. The fruits have been used as a pepper substitute; white-flowered forms have long been considered a symbol of chastity (Mabberley 1987). ☜

The following 2 varieties can apparently be told apart only by flower color.

1. Corollas lavender, lilac or white ____ var. **agnus-castus**
1. Corollas blue ____ var. **caerulea**

var. **agnus-castus**. According to Hatch et al. (1990), widespread in the e 1/2 of TX.

var. **caerulea** Rehder, (cerulean, dark blue). Cited by Hatch et al. (1990) only for vegetational area 4 (Fig. 2).

Vitex negundo L. var. **heterophylla** (Franch.) Rehder, (a Malay word for chastetree; var.: various-leaved), NEGUNO, CUT-LEAF CHASTETREE. Similar to *V. agnus-castus*; leaves with 3–5 leaflets. Cultivated and escapes; Tarrant Co.; also e TX. Summer. Native of Asia. [*V. negundo* L. var. *incisa* (Lam.) C.B. Clarke] ☜

VIOLACEAE VIOLET FAMILY

Small annual or perennial herbs (ours), acaulescent or caulescent; leaves basal, alternate, or opposite, simple (sometimes with very leafy stipules and thus appearing compound), entire, toothed, or lobed; stipules absent, or slender and inconspicuous, or falling early, or large and leafy; flowers solitary, axillary or basal; sepals 5, slightly unequal; petals 5, unequal, the lowest one with widened or swollen base or projecting spur; stamens 5, separate, with large, separate or weakly united anthers and short filaments; pistil 3-carpellate; ovary superior; fruit a 3-valved capsule. Cleistogamous flowers (= closed with rudimentary petals, self-pollinating) are also usually produced, either under or above ground during summer and fall.

A medium-sized (900 species in 20 genera—H. Ballard, pers. comm.) cosmopolitan family of herbs, shrubs or even lianas or trees; alkaloids are often present. The family includes the ornamental VIOLETS and PANSIES (*Viola*). It is possibly related to the mostly tropical Flacourtiaceae. (subclass Dilleniidae)

FAMILY RECOGNITION IN THE FIELD: small, sometimes stemless herbs with simple leaves; flowers bilaterally symmetrical, often nodding, with 5 separate petals, one of which is often *spurred*; stamens often with appendages or spurs; fruit a 3-valved capsule.

REFERENCE: Brizicky 1961b.

1. Leaves sessile to subsessile, linear to linear-lanceolate or broadly lanceolate, entire; sepals not auricled at base; lowest petal merely inflated basally, not spurred; stamens united (eventually separating), the lower 2 not spurred ____ **Hybanthus**
1. Leaves long-petioled (but leafy stipules sessile), the blades lanceolate to triangular to broadly ovate or broadly cordate, entire, toothed, or lobed; sepals with short-auricled base; lowest petal spurred; stamens separate, the lower 2 spurred ____ **Viola**

HYBANTHUS GREEN-VIOLET, NOD-VIOLET

A genus of 150 species of tropical and warm areas of the world. (Greek: *hybos*, hump-backed, and *anthos*, flower)
REFERENCE: Turner & Escobar 1991.

Hybanthus verticillatus (Ortega) Baill., (whorled), NODDING GREEN-VIOLET, WHORLED NOD-VIOLET. Leafy stemmed perennial with deep, branching root; stems usually clustered, to 25(–35) cm tall; leaves opposite or verticillate to alternate above; flowers axillary, nodding on short peduncles; lowest petal yellowish to greenish white, often with rosy or purple spot near the large tip; remaining petals smaller, rosy lavender to brown-red. Prairies, eroding slopes, rocky areas, or disturbed ground; widespread in TX. Apr–Jun, sometimes again in Oct. [*H. linearis* (Torr.) Shinners]

VIOLA VIOLET

Annuals with taproots or usually perennials from a caudex, some species rhizomatous; flowers basal or axillary, white-colored forms (= albinos) present in a number of species; seeds numerous, arillate, typically ant dispersed.

A genus of ca. 550 species (H. Ballard, pers. comm.) of temperate regions, especially n temperate areas and the Andes; it includes the ornamentally important *V.* ×*wittrockiana* Gams (GARDEN PANSY); according to Woodland (1997), this is the largest genus of n temperate [dicot] herbs. Cleistogamous flowers are often present. Like many spring-flowering herbs of temperate deciduous forests, many *Viola* species have seeds with elaiosomes (= oily appendages) and are dispersed by ants which eat the elaiosomes but leave the seeds intact (Beattie et al. 1979; Culver & Beattie 1980; Beattie 1985). The taxonomy of *Viola* is complicated by the extensive morphological variation seen in some species due to growing conditions or season as well as by hybridization in certain groups. According to Ballard (1994), "Taxonomists who have attempted identification of native violets have usually come to see the cloak of romantic folklore and 'innocence' surrounding violets as sheeps' clothing covering a pack of taxonomically incorrigible wolves." (Classical Latin name for a scented flower)
REFERENCES: Brainerd 1911, 1921; Baird 1942; Valentine 1962; Russell 1961, 1965; Lacey 1969; McKinney 1992; Ballard 1994; Gil-ad 1997.

1. Plants with leafy stems (= caulescent); petals yellow OR violet-blue to lavender-blue to cream-white with yellow center; plants annual or perennial.
   2. Stipules large, leafy, deeply lobed; petals violet-blue or lavender-blue to cream-white with yellow center; plants annual with erect taproots _____ **V. bicolor**
   2. Stipules small, entire; petals yellow; plants perennial with short prostrate rhizomes _____ **V. pubescens**
1. Plants stemless (= acaulescent), the leaves all basal; petals light to dark blue-violet, rarely whitish; plants perennial.
   3. Leaf blades divided into 3 or more distinct segments or lobes.
      4. Lateral lobes large, comprising basal one-half to two-thirds of the leaf blade; leaf blades ca. 1–1.5 times as long as wide (in outline around lobes); capsules purple-spotted _____ **V. palmata**
      4. Lateral lobes ± small, restricted to basal one-quarter or one-third of the leaf blade, giving it a sagittate/hastate base; leaf blades ca. 1.5–3 times as long as wide; capsules green _____ **V. sagittata**
   3. Leaf blades completely unlobed (± heart-shaped and marginally crenate to serrate).
      5. Plants usually heavily short-pubescent on both leaf surfaces and along petioles; peduncles up to twice the petiole length; leaves held prostrate (flat on the ground) in life; leaf blades commonly elliptic, rounded at apex; capsules green; limited to sandy soils; apparently rare in nc TX _____ **V. villosa**
      5. Plants glabrous to slightly long-pubescent (rarely heavily pubescent), the hairs most

evident on lower leaf surfaces and summit of petioles; peduncle length variable but if heavily pubescent the peduncles equaling or shorter than the petioles; leaves half erect (off the ground) in life; leaf blades broadly ovate to deltoid-ovate to narrowly triangular, acute to acuminate at apex; capsules purple-spotted; in a variety of soils; widespread in nc TX.

6. Foliage glabrous throughout; leaf blades broadly to narrowly triangular, usually longer than broad (at flowering time), gradually and uniformly tapering to an acuminate apex, truncate to subcordate or cordate at base ____ **V. missouriensis**
6. Foliage usually long-pubescent on lower leaf surfaces and summit of petioles; leaf blades broadly ovate to deltoid-ovate, ca. as broad as long, concavely or abruptly tapering to an acute or short acuminate apex, shallowly to deeply cordate at base ____ **V. sororia**

Viola bicolor Pursh, (two-colored), FIELD PANSY, WILD PANSY, JOHNNY-JUMP-UP, HEARTS-EASE, CUPID'S-DELIGHT. Minutely and inconspicuously pubescent taprooted annual or winter annual; stems 2–15(–25) cm tall; stipules 10–20 mm long, pectinate-palmately lobed; lower leaf blades orbicular to ovate, unlobed; upper leaf blades spatulate to obovate, broadly elliptic, or lanceolate, unlobed; petals violet-blue to light lavender-blue or whitish, paler and at least the lower ones darkly veined toward base, the lowest one ± yellow toward center and base (or rarely the lower petals wholly dark-colored). Sandy soils; e TX w to Rolling Plains and Edwards Plateau. Mar–mid-Apr. [*V. rafinesquii* Greene]

Viola missouriensis Greene, (of Missouri), MISSOURI VIOLET. Petals deep to pale violet-blue with a white center, often bordered with a dark violet area around the center, lines on the lateral and spurred petals dark violet; petal trichomes borne on the lower lateral petals, absent on the spurred petal, cylindrical, gradually widened apically; capsules yellow-green sparsely spotted and dotted with red-purple; seeds dark orange-yellow to strong yellowish brown (Gil-ad 1997). Se and e TX w to West Cross Timbers, also Trans-Pecos. Mid-Mar–mid-Apr. [*V. soria* var. *soria* (Greene) L.E. McKinney] This taxon has long been treated as a distinct species (e.g., Russell 1965); however, McKinney (1992) recognized it as a variety of *V. soria*. We are following Gil-ad (1997) in recognizing it at the specific level; he gave numerous characters (see description above and description of *V. soria* below) distinguishing it and cogently argued that it is a distinct species. Gil-ad (1997) did, however, note that hybrids and introgressants between the two are common and may complicate identification.

Viola palmata L., (palmate), TRI-LOBE VIOLET, LOVELL VIOLET, WOOD VIOLET. Leaf blades variable in shape, with 3–7 lobes, the lobes broad to nearly linear, the central lobe sometimes much larger to little different from the laterals; petals light to dark blue-violet. Dry to bottomland woods; Denton and Grayson cos.; mainly se and e TX, also Edwards Plateau. Mar–Apr. [*V. lovelliana* Brainerd, *V. palmata* var. *dilatata* Elliott, *V. palmata* var. *triloba* (Schwein.) Ging. ex DC., *V. triloba* Schwein., *V. triloba* Schwein var. *dilatata* (Elliott) Brainerd] While varieties are sometimes recognized in this species (e.g., Kartesz 1994), we are following McKinney (1992) in not recognizing infraspecific taxa. Gil-ad (1997) treated this taxon as *V. triloba* Schwein.; however, H. Ballard (pers. comm.) indicated that *V. palmata* is apparently the correct name; until nomenclatural issues are resolved, we are taking the conservative approach and using the name traditionally associated with this species.

Viola pubescens Aiton, (pubescent, downy), YELLOW VIOLET, SMOOTH YELLOW VIOLET. Plant glabrous in nc TX; stems erect or decumbent, 10–45 cm tall; leaf blades reniform-cordate to broadly ovate, unlobed; petals yellow, the lower 3 with purplish brown veins. In woods, low ground; Dallas (on the Austin Chalk), Grayson, Hunt, and Lamar cos.; se and e TX w to nc TX. Apr. [*V. eriocarpon* Schwein., *V. pensylvanica* Michx., *V. pubescens* var. *eriocarpon* Nutt.] If infraspecific taxa are recognized, according to H. Ballard (pers. comm.), nc TX material would be *V. pubescens* var. *scabriuscula* Schwein. ex Torr. & A. Gray.

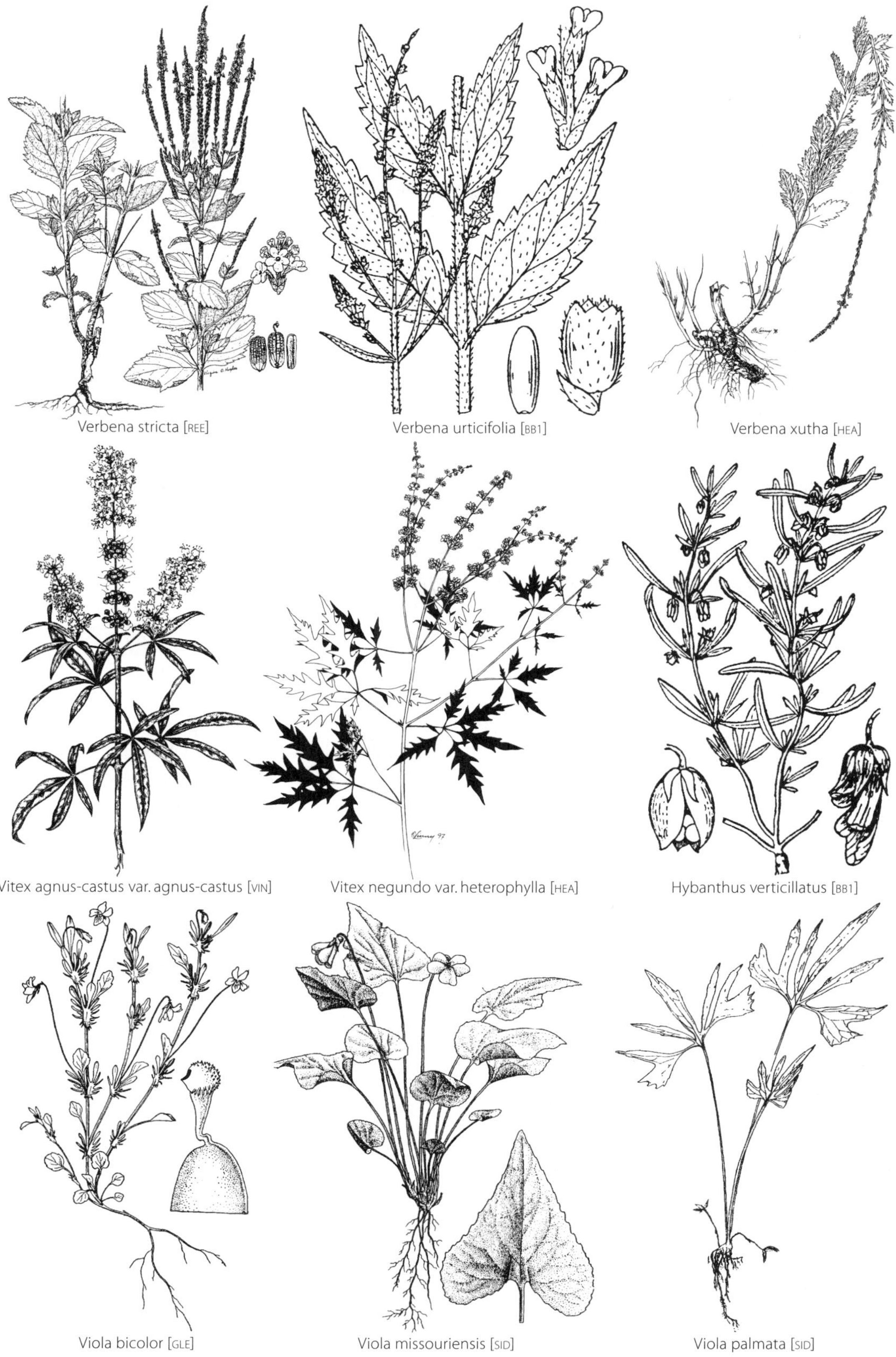

Verbena stricta [REE]

Verbena urticifolia [BB1]

Verbena xutha [HEA]

Vitex agnus-castus var. agnus-castus [VIN]

Vitex negundo var. heterophylla [HEA]

Hybanthus verticillatus [BB1]

Viola bicolor [GLE]

Viola missouriensis [SID]

Viola palmata [SID]

Viola sagittata Aiton, (arrow-like), ARROW-LEAF VIOLET. Leaf blades with 4–6 usually small lobes or large teeth at base, pubescent; petals blue to violet-purple. Dry sandy woods and forest margins; included based on nc TX location (without county) mapped in McKinney (1992), also Lamar Co. (Carr 1994); mainly e TX. Apr–Jun.

Viola sororia Willd., (sisterly, possibly based on similarity to another violet species), SISTER VIOLET, BAYOU VIOLET, DOWNY BLUE VIOLET, BLUE PRAIRIE VIOLET. Petals violet with a white center and violet lines on the lateral and spurred petals; petal trichomes borne on the lateral petals, absent on the spurred petal, cylindrical; capsules blotched with dark overlapping red-purple patches on a yellow-green background; seeds dark grayish brown (Gil-ad 1997). Open woods, waste places, moist areas, often on sandy substrates; Dallas Co., also Montague and Tarrant cos. (as *V. pratincola* sensu Russell, not Greene—Russell 1965) and Hunt and Kaufman cos. (Mahler 1988); scattered mainly in e and c TX. Mar–May. [*V. papilionacea* Pursh, at least in part as to application of name]

Viola villosa Walter, (soft-hairy), CAROLINA VIOLET. Plant very low and compact; leaf blades commonly lying on the ground, dark green with red veins, usually with dense short pubescence on both surfaces but sometimes becoming glabrous late in the year; petals usually dark blue-violet. Sandy woods; Denton and Tarrant cos. (Russell 1965), also Lamar (Carr 1994) and Milam (McKinney 1992) cos.; scattered in e 1/2 of TX. Mar.

VISCACEAE
MISTLETOE OR CHRISTMAS MISTLETOE FAMILY

A medium-sized (385 species in 7 genera), cosmopolitan, but especially tropical and warm area family of photosynthesizing parasites of trees; haustoria actually penetrate and branch in host tissue. The family was formerly recognized as a subfamily in the Loranthaceae. Some are problematic parasites while others are sources of Christmas mistletoe. Family name from *Viscum*, MISTLETOE, a genus of 100 species native from temperate regions to the Old World tropics. (Classical Latin name for mistletoe; possibly related to Latin: *viscos*, sticky) (subclass Rosidae)
FAMILY RECOGNITION IN THE FIELD: the single nc TX species is an *evergreen parasite* on tree branches often deforming the branch at point of attachment; fruits mucilaginous, translucent-whitish; this species becomes very conspicuous on deciduous trees when they lose their leaves for the winter.
REFERENCE: Kuijt 1982.

PHORADENDRON MISTLETOE

An American, especially tropical genus of 190 species. ☠ Various *Phoradendron* species are known to potentially be fatally poisonous to humans and animals; the toxins, present in all parts of the plant, but especially the whitish fruits, apparently include amines and toxic proteins; if eaten, acute gastroenteritis and heart failure can result; because MISTLETOES are widely used as Christmas decorations, care should be taken to prevent access by children or pets; however, some birds (e.g., cedar waxwings, bluebirds) relish the fruits (Martin et al. 1951; Kingsbury 1964; Schmutz & Hamilton 1979; Morton 1982; Hardin & Brownie 1993). (Greek: *phor*, a thief, and *dendron*, tree, from the parasitic habit)
REFERENCE: Wiens 1964.

Phoradendron tomentosum (DC.) Engelm. ex A. Gray, (densely woolly, with matted hairs), MISTLETOE, CHRISTMAS MISTLETOE, INJERTO, HAIRY MISTLETOE. **Official floral emblem of Oklahoma** as designated by the Assembly of the Territory of Oklahoma on 11 February 1893 (Tyrl et al. 1994). Perennial, hemiparasitic (= parasitic but at least partly autotrophic) on tree branches;

whole plant yellow-green; leaves opposite, simple; leaf blades leathery, evergreen, entire, faintly 3-ribbed, obtuse; flowers unisexual, in axillary, simple or branched, somewhat interrupted spike-like inflorescences; the sexes on separate plants; perianth 2–4-parted; ovary inferior; fruit a small 1-seeded drupe with sticky, mucilaginous, translucent-whitish mesocarp. Parasitizing a variety of tree species including *Celtis*, *Maclura*, *Prosopis*, and *Ulmus*, deformaties or death to branches of the host tree often result from the parasitism; MISTLETOE is sometimes abundant on a tree and in winter, when naked branches are visible, it is often very conspicuous; widespread in TX. Oct–Mar. [*P. serotinum* (Raf.) M.C. Johnst. var. *pubescens* (Engelm. ex A. Gray) M.C. Johnst.] This species is often collected locally for use as Christmas mistletoe. Dispersed by birds wiping their beaks and feet on branches to clean off the sticky seeds (Wills & Irwin 1961) or perhaps by the seeds passing through the birds' digestive tracts. Poisonous. ☠

VITACEAE GRAPE FAMILY

Perennial, climbing or shrubby, woody vines (= lianas) with tendrils; leaves alternate, simple or palmately compound, toothed or palmately lobed; flowers small, in axillary, peduncled, racemose, paniculate, or compound umbel-like inflorescences; perianth in *Vitis* falling when the flowers open, in the other genera persistent; sepals (4–)5; petals (4–)5, green to whitish or yellowish; stamens (4–)5, attached around a prominent, fleshy disk, opposite the petals; pistil 1; ovary superior; fruit a 1–4-seeded berry.

A medium-sized (850 species in 14 genera) largely tropical and warm area family of mostly woody climbers usually with tendrils opposite the leaves (rarely succulent treelets or herbs). The most important species economically is the Asian *Vitis vinifera* L. (GRAPE), the source of wine. Other species are used as ornamentals (e.g., *Parthenocissus*). (subclass Rosidae)
FAMILY RECOGNITION IN THE FIELD: woody vines with *tendrils* and inflorescences of small in conspicuous flowers both borne *opposite* the usually large leaves; fruit a *berry* (e.g., grape) with 1–4 seeds.
REFERENCES: Bailey 1934; Brizicky 1965a.

1. All leaves simple.
  2. Leaf blades not fleshy; inflorescences paniculate, racemose, or cymose; flowers 5-merous.
    3. Tendrils with slender, pointed, curling tips; native species.
      4. Petals united at apex, falling together; inflorescence elongate, longer than wide, not dichotomously forking; plants densely pubescent to nearly glabrous; leaf blades cordate at base; pith white; bark usually shreddy ____ **Vitis**
      4. Petals separate, falling individually; inflorescence short, wider than long, dichotomously forking; plants essentially glabrous; leaf blades truncate to cordate at base; pith brown; bark tight, not shreddy ____ **Ampelopsis**
    3. Tendrils with small, disk-like tips; introduced ornamentals ____ **Parthenocissus**
  2. Leaf blades conspicuously fleshy; inflorescences resembling compound umbels; flowers 4-merous ____ **Cissus**
1. Some or all of the leaves divided into separate leaflets.
  5. Leaves with (3–)5–many leaflets, usually not fleshy; inflorescences paniculate, racemose, or cymose; flowers 5-merous; leaflets and petioles usually not falling apart when pressed and dried.
    6. Leaves once palmately compound; leaflets 3–7 per leaf ____ **Parthenocissus**
    6. Leaves 2–3 times pinnately or ternately compound; leaflets usually (9–)11–34 or more per leaf ____ **Ampelopsis**
  5. Leaves with 3 leaflets, conspicuously fleshy; inflorescences resembling compound umbels; flowers 4-merous; leaflets and petioles falling apart when pressed and dried ____ **Cissus**

AMPELOPSIS

Plants bushy to high climbing; leaves simple or compound; tendrils opposite leaves when present, without adhesive disks at tips; flowers small, greenish, perfect; berry dry or pulpy; seeds 1–several.

A genus of 25 species of climbers of temperate and subtropical America and Asia. (Greek: *ampelos*, vine, and *opsis*, appearance, from the habit)

1. Leaves 2–3 times compound; fruits black, 10–15 mm in diam. ____ **A. arborea**
1. Leaves simple; fruits green becoming orange-pink or purplish, eventually turquoise-blue, < 10 mm in diam. ____ **A. cordata**

Ampelopsis arborea (L.) Koehne, (tending to be woody, tree-like), PEPPERVINE. Varying from low, half-woody, and bushy to moderately high-climbing; leaflets usually (9–)11–34 or more per leaf, dark green, oblong-elliptic, to 3–7 cm long, coarsely and sharply toothed, often also lobed, glabrous. Stream bottoms, fencerows, and disturbed areas; Dallas and Grayson cos., also Bell, Brown (HPC), Lamar (Carr 1994) and Tarrant (R. O'Kennon, pers. obs.) cos.; se and e TX w to nc TX and Edwards Plateau. Late Jun–Jul.

Ampelopsis cordata Michx., (heart-shaped), HEART-LEAF AMPELOPSIS, RACOON-GRAPE. Low- to high-climbing vine, closely resembling species of *Vitis*; leaf blades cordate- or triangular-ovate, to 15 cm long and wide, coarsely and sharply toothed, sometimes shallowly lobed, glabrous. Stream bottom woods; se and e TX w to Edwards Plateau and Panhandle. Late May–Jun. [*Cissus ampelopsis* Pers.] The fruits are sometimes confused with grapes, but are not edible (McGregor 1986).

CISSUS COWITCH, POSSUM-GRAPE

A genus of ca. 200 species of mainly vines with tendrils found in tropical and warm parts of the world. (Greek: *kiccos*, classical name of the ivy, alluding to the climbing habit of many species)

Cissus incisa Des Moul., (incised, cut), COWITCH, IVY-TREEBINE, MARINEVINE, MARINE-IVY, HIERBA DEL BUEY. Vines, deciduous or semi-evergreen; pith white; tendrils opposite leaves; leaves alternate, variable, simple to usually palmately 3-folilate, to 8 cm long, reported to have a burnt rubber or sharp nitrogenous odor; leaflets conspicuously fleshy, coarsely toothed, obtuse apically; flowers 4-merous, perfect or unisexual, greenish; stamens 4; disk a 4-lobed cup; style 1, capitate; berries dry, black, 1–4-seeded. Stream banks or in disturbed areas; Grayson, Lamar, and Tarrant cos., also Brown (HPC), Dallas (G. Diggs, pers. obs.), Hood, Parker, Somervell, and Wise (R. O'Kennon, pers. obs.) cos.; throughout most of TX. Late Jun–Jul. According to McGregor (1986), the roots of the related *C. trifoliata* (L.) L. of Arizona and Mexico are poisonous and cause dermatitis (McGregor 1986); this is apparently the result of calcium oxalate crystals in the tissues (Lampe 1986); caution should thus be taken with *C. incisa*. ☠

PARTHENOCISSUS VIRGINIA-CREEPER, WOODBINE

Woody climbing or trailing vines; branched tendrils with adhesive disks or twining tips present; leaves simple to usually palmately compound with 3–7 serrate leaflets; inflorescence a group of cymes; flowers small; calyces of united sepals; petals separate; fruit a black to dark blue, often glaucous berry with thin flesh and 1–4 seeds.

A genus of 10 species of deciduous climbers native to temperate Asia and North America. Several species are used ornamentally. *Parthenocissus* species often display very early fall foliage color (often strikingly red); this is considered to serve as a "foliar fruit flag" which attracts birds that act as dispersal agents for the fall-ripening fruits (Stiles 1984). (Greek: *parthenos*, virgin, and genus name *Cissus*, from *kissos*, ivy; based on the French name vigne-vierge or the English virginia creeper)

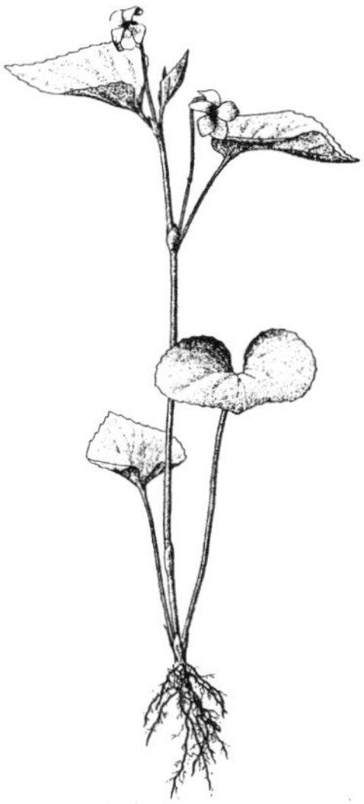

Viola pubescens [SID]

Viola sagittata [SID]

Viola sororia [SID]

Viola villosa [SID]

Phoradendron tomentosum [LYN]

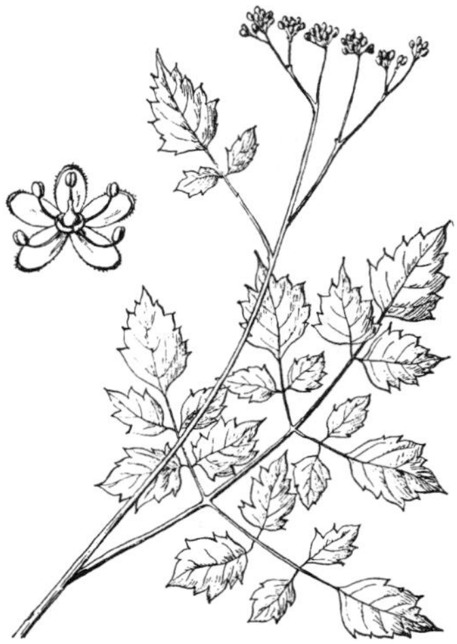

Ampelopsis arborea [GLE]

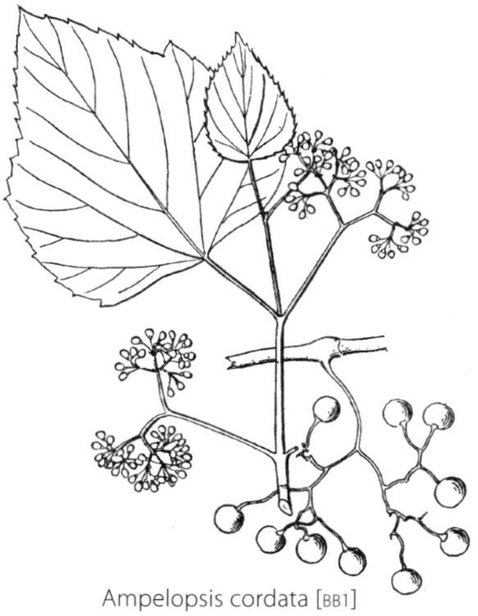

Ampelopsis cordata [BB1]

Cissus incisa [GLE]

1. Leaves simple (but 3-lobed) or with 3 leaflets ____ **P. tricuspidata**
1. Leaves with 5–7 leaflets.
   2. Leaves usually with 7 leaflets (sometimes 5–6), the leaflets usually 3–5(–6) cm long and to ca. 2(–3) cm wide, fleshy-thickened; in nc TX limited to Lampasas Cut Plain ____ **P. heptaphylla**
   2. Leaves usually with 5–6 leaflets, the leaflets usually much more than 5 cm long and 2 cm wide, not fleshy-thickened; widespread in nc TX ____ **P. quinquefolia**

Parthenocissus heptaphylla (Buckley) Britton ex Small, (seven-leaved), SEVENLEAF-CREEPER. Vine to 10 m with long forking tendrils; leaves glossy above; fruits ca. 1 cm in diam.; seeds 1–4. Climbing on vegetation, rocky or sandy soils; Bell, Lampasas, and Williamson cos., also Brown, Hamilton (HPC), and Burnet (C. Sexton, pers. comm.) cos.; endemic to Edwards Plateau and Lampasas Cut Plain of TX. Apr–May (–later?). The texture of the leaves resembles that in *Cissus*.

Parthenocissus quinquefolia (L.) Planch., (five-leaved), VIRGINIA-CREEPER, WOODBINE, AMERICAN-IVY, HIEDRA, PARRA, REDTWIG-CREEPER. Deciduous, high-climbing vine; tendrils branched, tipped by adhesive disks; leaflets to 15 cm long and 5 cm wide, thin herbaceous; flowers 25–200 per inflorescence; berries 5–9 mm in diam.; seeds 1–3(–4), 3.5–4 mm long. Along creeks, wooded areas, also cultivated; e 1/2 of TX. May–Jul. The leaves turn a striking red in the fall. This species does not cause contact dermatitis; sometimes falsely accused because of its association in the same habitat (sometimes on the same tree) as POISON IVY; however, the berries are suspected of being lethally poisonous to children and the tissues of the plant are known to contain microscopic, irritating, needle-like crystals known as raphides (Kingsbury 1964; Burlage 1968; Turner & Szczawinski 1991).

Parthenocissus tricuspidata (Siebold & Zucc.) Planch., (having three cusps or points), JAPANESE-IVY, BOSTON-IVY. High-climbing vine; adhesive disks present; leaves variable, simple on young material, often 3-foliate on older, glossy above; fruits ca. 7 mm in diam. Widely cultivated in TX on buildings and walls; long persists and spreads vegetatively from cultivation; Grayson Co.; wider TX distribution not known. May?–Jun–later? Native of China and Japan. Potentially toxic as in the case of *P. quinquefolia* (Turner & Szczawinski 1991).

VITIS GRAPE

Vine-like shrubs or high climbing polygamo-dioecious vines, often with large woody stems; leaf blades cordate or ovate, variously toothed or lobed; inflorescence a compact compound panicle opposite a leaf; flowers small, fragrant; calyces minute or absent; petals 5, united at apex, falling together; fruit a pulpy berry with 2–4 seeds.

A n hemisphere genus of ca. 65 species. The tendrils are negatively phototropic and force themselves into cracks or crevices in supporting structures; the ends can become enlarged and sticky (Heywood 1993). *Vitis vinifera* L. (GRAPE VINE) is the source of most wine, grape juice, table grapes, and raisins (= dried grapes); some nc TX wild species are edible and can be used in making wine and jelly. North Central Texas is famous in the history of grape cultivation because of the work of T.V. Munson (e.g., Munson 1909) in Grayson Co. Munson experimented extensively with native grapes and developed 300 new varieties including a number still used today. Through some of his disease-resistant varieties, Munson is credited with saving the French wine industry in the 1870s from the root disease known as grape phylloxera (Sperry 1994). The T.V. Munson Memorial Vineyard and Viticulture-Enology Center at Grayson County College continues this legacy and is presently propagating ca. 65 varieties; many of these are also apparently resistent to Pierce's Disease, a bacterial infection problematic in some parts of the U.S. including Texas (W. Martin, pers. comm.). The stems of a number of species have been used in making wreaths. (Classical Latin name for the grape)

REFERENCES: Munson 1909; Bailey 1934; Duncan 1967; Moore 1987, 1991; Gandhi 1989; Mullins et al. 1992.

1. Underside (= abaxial surface) of fully expanded leaf blades ± covered with thin to dense cobwebby, woolly, or erect-spreading pubescence until well after flowering; younger shoots, petioles, and peduncles thinly to densely woolly or pubescent.
  2. Underside of fully expanded leaf blades white to cream with dense, matted hairs completely concealing the surface ____ **V. mustangensis**
  2. Underside of fully expanded leaf blades green, gray, whitish, yellowish, or rusty with rather dense to sparse pubescence, the surface visible.
    3. Underside of leaf blades with long, web-like or matted hairs (these often becoming loosened and curling into small tufts or tangles), sometimes with short straight hairs in addition; grapes without lenticels; infructescences usually with > 25 grapes (sometimes less); widespread in nc TX.
      4. Leaf blades with shallow, acute or rounded, and abruptly small-pointed teeth as wide as high or wider; pubescence gray, yellowish, or rusty; widespread in nc TX.
        5. Leaf blades of flowering shoots shallowly to deeply lobed, the deeply lobed ones entire in the sinuses of the lobes; underside of mature leaf blades glaucous; nodes often glaucous ____ **V. aestivalis**
        5. Leaf blades of flowering shoots unlobed or shallowly lobed, the lobes toothed to base; underside of mature leaf blades not glaucous; nodes not glaucous ____ **V. cinerea** var. **cinerea**
      4. Leaf blades with coarse, uneven, acute or acuminate teeth, the larger teeth longer than wide; pubescence gray or whitish; Grand Prairie and w ____ **V. acerifolia**
    3. Underside of leaf blades with short, straight, erect-spreading hairs, sometimes the main veins with longer matted hairs as well; grapes usually with lenticels; infructescences usually with < 25 grapes; West Cross Timbers and Lampasas Cut Plain s to Edwards Plateau ____ **V. monticola**
1. Underside of fully expanded leaf blades glabrous OR (only along main veins or in their axils) thinly cobwebby or thinly woolly or with short, erect-spreading pubescence; younger shoots, petioles, and peduncles glabrous OR thinly woolly or with short, erect-spreading pubescence, soon becoming largely glabrous.
  6. Tendrils not branched; leaf blades unlobed, 5–10(–12) cm long; bark tight, not shredding; lenticels evident on older stems; e TX w to e part of nc TX ____ **V. rotundifolia**
  6. Tendrils branched; leaf blades lobed or unlobed, 5–18 cm long; bark loosening and shredding with age; lenticels inconspicuous or absent on older stems; widespread in nc TX.
    7. Leaf blades (at least some) deeply lobed.
      8. Margins of leaf blades often ciliolate (= with fringe of short hairs); new branches green to brown; grapes gray-bluish, glaucous; inflorescence axis glabrous to sparsely and loosely long-pubescent ____ **V. riparia**
      8. Margins of leaf blades not fringed; new branches reddish; grapes black, not glaucous; inflorescence axis densely short-pubescent ____ **V. palmata**
    7. Leaf blades entire or at most shallowly lobed.
      9. Leaf teeth usually longer than wide, coarse, uneven, and acute or acuminate; Grand Prairie and West Cross Timbers s to Edwards Plateau ____ **V. acerifolia**
      9. Leaf teeth usually ca. as wide as long or wider, shallow, acute or rounded and often abruptly pointed; including species widespread in nc TX.
        10. Leaf blades usually 8(–10) cm or less long; inflorescences usually 6 cm or less long; tendrils absent or only opposite uppermost leaves or at tips of fertile branches; infructescences typically with < 25 grapes; grapes usually with lenticels; Lampasas Cut Plain s to Edwards Plateau ____ **V. monticola**
        10. Leaf blades usually 8 cm or more long; inflorescences usually much more than 6 cm

long; tendrils present; infructescences typically with > 25 grapes; grapes without lenticels; widespread in nc TX.

11. Petioles, vein axils, and veins on underside of leaf blades glabrous or with short, erect-spreading hairs ______ **V. vulpina**
11. Petioles, vein axils, and often veins on underside of leaf blades with cobwebby or woolly hairs.
    12. Leaf blades often wider than long; leaves developing fully as branches elongate; grapes purplish or blackish; mainly Lampasas Cut Plain s to Edwards Plateau ______ **V. cinerea** var. **helleri**
    12. Leaf blades longer than wide; new growth rapidly produced, the new branch slender and elongate with very immature leaves; grapes black; widespread in nc TX ______ **V. vulpina**

Vitis acerifolia Raf., (with leaves like *Acer*—maple), PANHANDLE GRAPE, BUSH GRAPE, LONG'S GRAPE. Rarely climbing, but covering rocks and shrubs; leaf blades often shallowly lobed as well as irregularly and very sharply toothed; grapes 8–12 mm in diam., black, with heavy bloom, becoming sweet. Stream bottoms and rocky slopes; Cooke Co., also Erath Co. (Mahler 1988); Grand Prairie w to Panhandle. Late Apr–early May. Fruiting Jul–Aug.

Vitis aestivalis Michx., (summer), PIGEON GRAPE. Clump forming to low or high climbing vine; leaf blades suborbicular-ovate, almost as wide as long or wider; grapes 5–20 mm in diam., dark purple or black with thin bloom, taste variable, often sweet. Stream bottom woods, usually on sand. May. Fruiting Sep–Oct.

1. Mature 3- or 4-seeded grapes usually 9–14 mm in diam.; stipules usually > 1.5 mm long; young branches and petioles with whitish or light brownish pubescence; lower leaves of fertile branches obtuse to acute; high climbing vines; widespread in nc TX ______ var. **aestivalis**
1. Mature 3- or 4-seeded grapes usually > 14 mm in diam; stipules usually < 1.5 mm long; young branches and petioles with rusty or reddish pubescence; lower leaves of fertile branches acute to acuminate; clump forming shrubby vines or low climbing vines; e TX w to e margin of nc TX ______ var. **lincecumii**

var. **aestivalis**, SUMMER GRAPE, PIGEON GRAPE. Dallas, Henderson, and Limestone cos., also Hood Co. (Mahler 1988); se and e TX w to nc TX. [*V. lincecumii* Buckley var. *glauca* Munson, *V. lincecumii* Buckley var. *lactea* Small]

var. **lincecumii** (Buckley) Munson, (for Gideon Lincecum, 1793–1874, early TX naturalist and physician; see Lincecum et al. 1997—*Science on the Texas Frontier*), PINEWOODS GRAPE, POST OAK GRAPE, BLUE-LEAF GRAPE. E TX w to Limestone and Milam cos. at e edge of nc TX. [*V. lincecumii* Buckley] The spelling of the epithet was originally given by Buckley (1861 [1862]) as *linsecomii* but has been corrected to *lincecumii* to conform with the family's spelling of the name.

Vitis cinerea (Engelm.) Millardet, (ashy-gray), SUMMER GRAPE, GRAY-BARK GRAPE, SWEET GRAPE. Moderate or high climbing vine; leaf blade (or its lobes) acute, the basal sinus variable; grapes 4–9 mm in diam., blackish or purplish, with a slight bloom. Stream bottom woods. May–early Jun. Fruiting Sep–Nov.

1. Underside of leaf blades cobwebby, woolly, soft-pubescent, or densely canescent; leaf blades usually > 10 cm long; grapes only slightly to not glaucous ______ var. **cinerea**
1. Underside of leaf blades with pubescence when young but eventually nearly glabrous and glossy, although retaining some cobwebby or woolly pubescence on the veins; leaf blades usually < 10 cm long; grapes moderately to heavily glaucous ______ var. **helleri**

var. **cinerea**, SWEET GRAPE, GRAY-BARK GRAPE, PARRA SILVESTRE. Se and e TX w to West Cross Timbers, also Edwards Plateau. [*V. aestivalis* var. *cinerea* Engelm., *V. cinerea* var. *canescens* Engelm.]

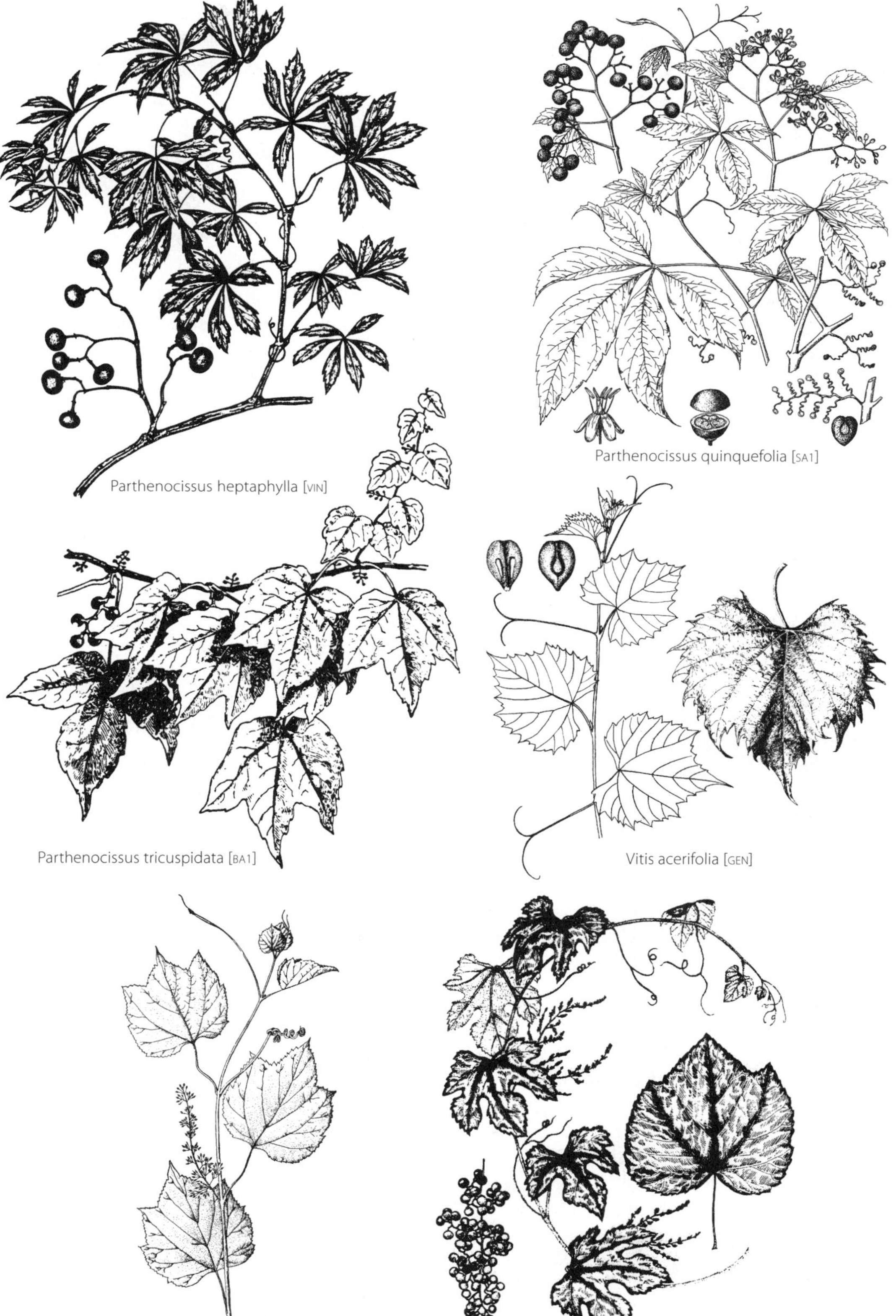

Parthenocissus heptaphylla [VIN]

Parthenocissus quinquefolia [SA1]

Parthenocissus tricuspidata [BA1]

Vitis acerifolia [GEN]

Vitis aestivalis var. aestivalis [GLE]

Vitis aestivalis var. lincecumii [VIN]

var. **helleri** (L.H. Bailey) M.O. Moore, (for Amos Arthur Heller, 1867-1944, Pennsylvania botanist and collector of w American plants), WINTER GRAPE, ROUND-LEAF GRAPE, SPANISH GRAPE, UVA CIMARRONA. Southern part of West Cross Timbers and Lampasas Cut Plain s to Edwards Plateau. [*V. berlandieri* Planch.]

Vitis monticola Buckley, (inhabiting mountains), SWEET MOUNTAIN GRAPE, CHAMPIN GRAPE. Climber; leaf blades relatively small, ca. 5–8(–10) cm long above attachment of petiole, cordate-ovate to suborbicular, acute or obtuse, indistinctly to sharply toothed, often slightly lobed; grapes 6–12 mm in diam., black or rarely red or pinkish, thinly glaucous, sweet. Stream bottoms, limestone areas; Coryell Co., also Bosque and Palo Pinto cos. (Mahler 1988); w part of nc TX, also Edwards Plateau; endemic to TX. Late May. Fruiting Sep–Oct.

Vitis mustangensis Buckley, (according to the type description, "This is called the Mustang grape in Texas, where it is very common." (Buckley 1861 [1862])—thus the epithet is apparently derived from the common name), MUSTANG GRAPE. High climbing vine, often rampant; leaf blades cordate-suborbicular to broadly triangular-ovate, subacute or obtuse, subentire to deeply lobed, becoming glabrous and dark green on upper surface, lower surface with strikingly thick tomentum; grapes 15–20 mm in diam., purple-black to light-colored, without bloom, pungent or with a fiery taste, especially if skin is chewed. Stream bottoms, thickets, fencerows, and disturbed areas, often on sandy soils; se and e TX w to West Cross Timbers and Edwards Plateau. Apr. Fruiting Aug–Sep. [*V. candicans* Engelm. ex A. Gray] This species sometimes literally covers other vegetation. Buckley (1861 [1862]) said that, "It makes an excellent wine; but is little esteemed for eating on account of an acrid juice beneath the skin, which, if swallowed, gives a burning pain in the throat."

Vitis palmata Vahl, (palmate), CATBIRD GRAPE, MISSOURI GRAPE, RED GRAPE. High climber; shoots, flowering branchlets, and petioles red; leaf blades 7–12 cm long above attachment of petiole, ovate, long acuminate; grapes 5–10 mm in diam., black or bluish black, without bloom, sweet at maturity. Low woods; Lamar Co. in Red River drainage; se and e TX. Jun. Fruiting Sep–Oct.

Vitis riparia Michx., (of river banks), RIVER GRAPE, RIVERBANK GRAPE, FROST GRAPE. High climber; petioles glabrous; leaf blades cordate-ovate, prolonged acuminate apically; grapes 8–12 mm in diam., purple-black, with heavy bloom, acidic. Stream bottoms; Trans-Pecos e to Grayson and Van Zandt cos. (Mahler 1988). Early May. Fruiting Aug–Oct.

Vitis rotundifolia Michx., (round-leaved), MUSCADINE GRAPE, SCUPPERNONG, BULLACE GRAPE. Vigorous very high climber; leaf blades relatively small, 5–10(–12) cm long above attachment of petiole, about as wide as long, with broad teeth, usually unlobed; grapes 12–25 mm in diam., purple-black to bronze, without bloom, falling rapidly, flesh musky-tasting. Woods; Henderson, Hopkins, and Lamar cos. in e part of nc TX; mainly se and e TX. May(–Jun). Fruiting Sep–Oct.

Vitis vulpina L., (of the fox), FOX GRAPE. High climber; leaf blades longer than wide, the basal sinus open and broadly U-shaped; grapes 5–10 mm in diam., black, often glaucous. Stream bottoms and hillsides; se and e TX w to West Cross Timbers and Edwards Plateau. Late Apr–mid-May. Fruiting Oct–Nov. [*V. cordifolia* Michx.]

Several hybrids are also known:

Vitis ×*champinii* Planch. [*V. mustangensis* × *V. rupestris*—Moore 1991], (for Aimé Champin, ?-1894, viticulturalist at the agricultural university of Montpellier, France). Leaves only slightly arachnoid pubescent beneath and lacking hirtellous trichomes; grapes not glaucous. Moore (1991) cited Bell, Burnet, and Coryell cos.

Vitis ×*doaniana* Munson ex Viala [*V. acerifolia* × *V. mustangensis*], (for Judge Jonathan Doan of Wilbarger, TX, who started a trading post in 1878—now Doans, TX, and who for years manufac-

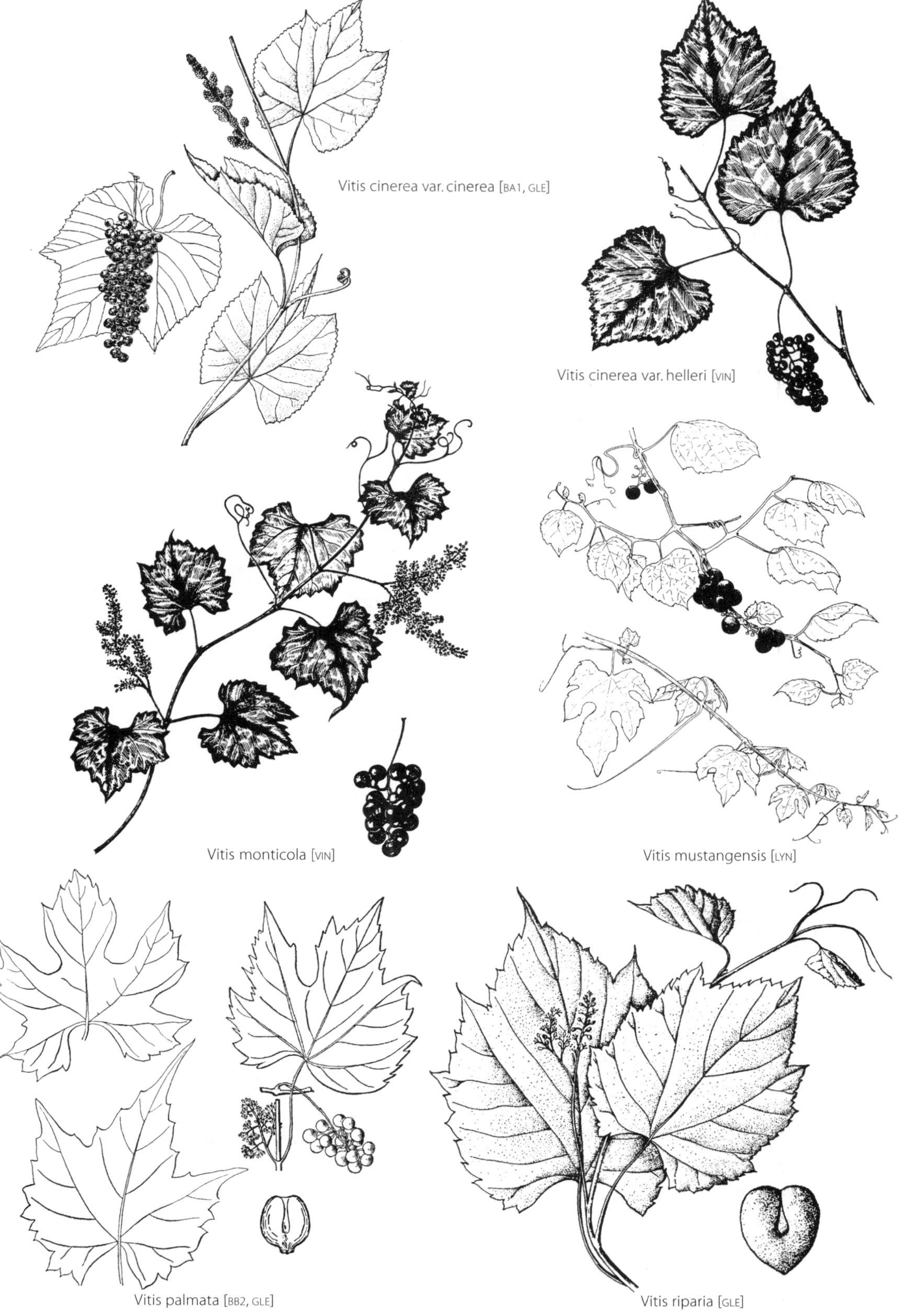

Vitis cinerea var. cinerea [BA1, GLE]

Vitis cinerea var. helleri [VIN]

Vitis monticola [VIN]

Vitis mustangensis [LYN]

Vitis palmata [BB2, GLE]

Vitis riparia [GLE]

tured fine wine from this grape which had been gathered in Greer Co., OK). Leaves moderately to heavily arachnoid pubescent beneath, also with hirtellous trichomes; grapes glaucous. Moore annotated a Montague Co. Whitehouse collection at BRIT/SMU (*15027*) as *V.* ×*doaniana*.

ZYGOPHYLLACEAE CALTROP FAMILY

Ours pubescent to pilose, prostrate to decumbent annuals or evergreen shrubs or small trees; leaves opposite or crowded in fascicles at the nodes, even-pinnately compound; leaflets 3–8 pairs, with asymmetrical bases, entire, folding together at night or in bad weather or in *Guajacum* sometimes in heat of day; stipules linear-lanceolate or subulate; flowers solitary or in small clusters, pedunculate; sepals 5; petals 5, yellow or orange or in *Guajacum* blue to purple, pink, or white; stamens 10 (bristles on ovary sometimes resemble additional filaments); ovary superior, 2–5-carpellate; fruit a schizocarp, 2(–3–4) or 5- or 10-lobed or -loculed, separating at maturity into 5 or 10 sections (mericarps) or a septicidal capsule.

A small (285 species in 27 genera) mostly tropical and warm, especially dry area family of mainly shrubs, herbs, and a few trees; alkaloids or mustard-oils are sometimes present. The family includes *Gaujacum officinale* L. and *Guajacum sanctum* L. (both referred to as LIGNUM VITAE), producing some of the world's hardest wood and a medicinal resin. The family also includes the abundant aromatic *Larrea tridentata* (Sessé & Moç. ex DC.) Coville (CREOSOTE-BUSH) of w TX. Family name from *Zygophyllum*, a genus of 80 species native from the Mediterranean region to c Asia, s Africa, and Australia, often in desert or arid regions. (Greek: *zygo*, yoke, and *phyllon*, leaf, in reference to the paired leaflets) (subclass Rosidae)
FAMILY RECOGNITION IN THE FIELD: opposite, *even-pinnately compound* leaves, the entire, *asym metrically-based* leaflets in 3–8 pairs; fruit a small schizocarp or a capsule.
REFERENCES: Vail & Rydberg 1910; Porter 1972.

1. Prostrate to decumbent annual herbs; petals yellow to orange; fruits ovoid to globose, with stout spines or unarmed, 5- or 10-lobed or -loculed.
   2. Petals lemon-yellow; fruits with two conspicuous stout, spreading prickles on each of the 5 sections; beak of fruit falling with sections ____ **Tribulus**
   2. Petals orange to yellow; fruits roughened or warty but not prickly, separating into 10 sections; beak of fruit persistent after sections fall ____ **Kallstroemia**
1. Erect distinctly woody shrubs or small trees; petals blue to purple, pink, or white; fruits flat, unarmed (but with apiculate tip), 2(–3–4)-lobed or -loculed ____ **Guajacum**

GUAJACUM

A genus of 6 species of trees and shrubs of dry areas in warm parts of the Americas. LIGNUM VITAE, obtained from *G. officinale* L. and *G. sanctum* L., is the hardest commerical hardwood; it was used in the lock gate hinges on the Erie Canal where they lasted for a century. It was also used medicinally, hence the common name meaning wood of life. Previously spelled *Guaiacum*. (From the South American vernacular word *guaiac*, the name for lignum vitae, *G. officinale*)
REFERENCE: Porter 1974.

Guajacum angustifolium Engelm., (narrow-leaved), GUAYACÁN, SOAPBUSH. Evergreen shrub or small tree, 1–7 m tall; leaves opposite or crowded in fascicles at the nodes, folded at night and in heat of day; leaflets 4–8 pairs, 5–15 mm long, 2–3 mm wide; flowers solitary or in small clusters, 12–20 mm in diam., fragrant; petals ca. 10 mm long; capsules 2(–3–4)-lobed, obcordate, 1–2 cm in diam., the margin ± winged, abruptly contracted to an elongate apiculate tip; seeds with scarlet aril. Brushy areas; included based on citation of vegetational area 5 (Fig. 2) by Hatch et al. (1990); mainly s 1/2 of TX. Mar–Sep. [*Porlieria angustifolia* (Engelm.) A. Gray] The bark from the roots has been used to make soap (Powell 1988), hence the common name.

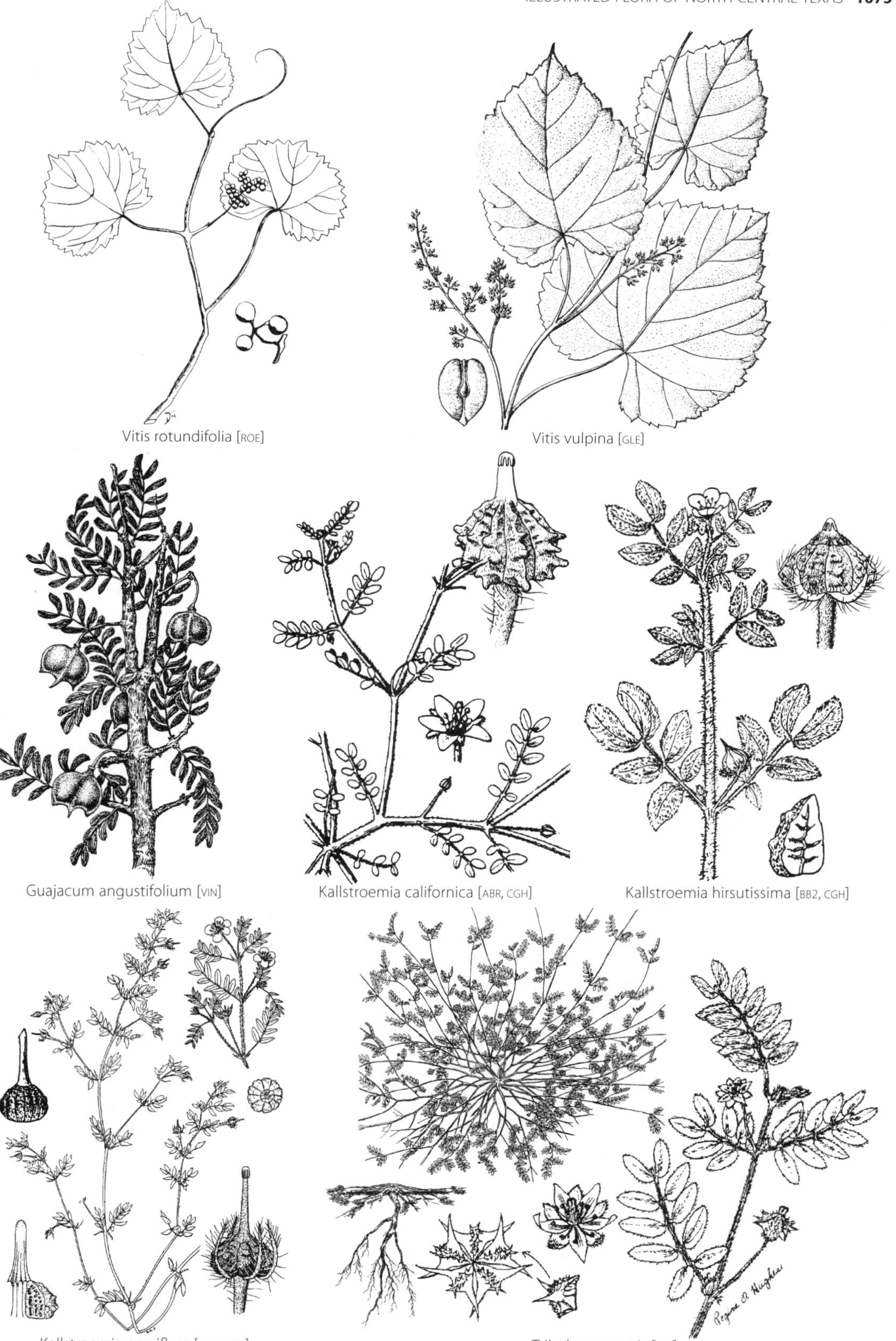

Vitis rotundifolia [ROE]

Vitis vulpina [GLE]

Guajacum angustifolium [VIN]

Kallstroemia californica [ABR, CGH]

Kallstroemia hirsutissima [BB2, CGH]

Kallstroemia parviflora [ARM, BB2]

Tribulus terrestris [REE]

KALLSTROEMIA CALTROP

Prostrate to decumbent (= branch tips ascending) annuals; stems hirsute; leaves opposite, 1 of each pair smaller; flowers solitary, in the axils of the smaller leaf of the pair; intrastaminal glands absent; fruits ovoid, 10-lobed, separating into 10 one-seeded prickleless sections (mericarps); mericarps rugose to tubercled.

A genus of 17 species native to tropical and warm parts of the Americas. The wet mericarps secrete a mucilaginous sheath that may adhere to animals and thus aid in dispersal (Porter 1969). (Named for Kallstroem, obscure scholar and contemporary of Austrian botanist J.A. Scopoli, 1723–1788, author of the genus)
REFERENCE: Porter 1969.

1. Beak of fruit 4–9 mm long, longer than fruit body; petals orange, 5–11 mm long; peduncles equalling or commonly longer than subtending leaves; widespread in nc TX ____ **K. parviflora**
1. Beak of fruit 1–4 mm long, shorter than fruit body; petals yellow, 2–6 mm long; peduncles usually shorter than subtending leaves; rare in nc TX
   2. Beak broadly conical, its base surrounded by a conspicuous ring of short white trichomes; sepals persistent; petals 2–4 mm long, ca. 1.5 mm wide; fruit body 6–8 mm wide ____ **K. hirsutissima**
   2. Beak cylindrical, its base glabrous to pubescent but without a ring of trichomes; sepals deciduous; petals 4–6 mm long, 2.5–3 mm wide; fruit body 3–5 mm wide ____ **K. californica**

Kallstroemia californica (S. Watson) Vail, (of California). Stems 10–65 cm long; leaves 1.5–6 cm long; leaflets up to 12 (3–6 pairs); fruits with 4–5 blunt oblong tubercles up to 1.5 mm long. Disturbed habitats; included based on citation of vegetational area 4 (Fig. 2) by Hatch et al. (1990); mainly s 1/2 of TX. Mar–Nov.

Kallstroemia hirsutissima Vail ex Small, (very hairy), CARPETWEED. Stems 15–70 cm long; leaves 1–4 cm long; leaflets 6–12(–14), in 3–6(–7) pairs); fruits tubercled. Disturbed habitats; Bell and Coryell cos. (Fort Hood—Sanchez 1997); se and s TX w to the Lampasas Cut Plain, Edwards Plateau, and Trans-Pecos. Jun–Nov. Poisonous to sheep, goats, and cattle; it can cause weakness of the hind legs, paralysis, and even death (Kingsbury 1964; Burlage 1968).

Kallstroemia parviflora Norton, (small-flowered), WARTY CALTROP. Stems hirsute, becoming glabrate, to ca. 100 cm long; leaves to ca. 6 cm long; leaflets up to 10(–12), in 3–5(–6) pairs; sepals persistent; petals 3.5–6 mm wide; fruits rugose to tubercled; beak of fruit strongly conic at base. Disturbed habitats; widespread in TX. Apr–Nov.

TRIBULUS CALTROP

A genus of 25 species of tropical and warm areas, especially dry areas of Africa. The common name comes from the Greek, *caltrop*, a pointed weapon placed on the ground to impede cavalry, in reference the armed fruit. (Greek: *tribolus*, three-pointed, apparently referring to the prickly fruit).

Tribulus terrestris L., (of the earth or ground), GOATHEAD, PUNCTUREWEED, PUNCTUREVINE, BULLHEAD, CADILLO, ABROJO DE FLOR AMARILLO. Annual with prostrate stems to > 1 m long; leaves opposite, 1 of each pair smaller, 1–4.5 cm long; leaflets up to 12 (3–6 pairs); flowers solitary, in the axils of the smaller leaf of the pair; peduncles usually shorter than the subtending leaves; petals 3–5 mm long, 2–3 mm wide; intrastaminal glands present; fruits globose, beaked, 5-lobed, separating into 5, 3- to 5-seeded sections with stout prickles. Disturbed sites; throughout most of TX. May–Nov. Native of the Mediterranean region; said by Reverchon to have appeared at Dallas around 1860 (Mahler 1988). The prickly fruits are very painful to both animal and human feet, damage even tires, and are injurious and occasionally fatal to livestock if eaten; they are reported to contain a saponin and to produce photosensitization and swelling of the head and ears (Burlage 1968; Correll & Johnston 1970).

CLASS MONOCOTYLEDONAE

Plants usually herbaceous—in other words, lacking regular secondary thickening (except Palmaceae, Smilacaceae, most Agavaceae, and a few Poaceae); seedlings usually with 1 seed leaf or cotyledon; stems or branches elongating by apical growth and also by growth of basal portion of internodes; leaves when present alternate, whorled, basal, or rarely opposite, elongating by basal growth (readily seen on spring-flowering bulbs whose leaf-tips have been frozen back); leaf blades usually with parallel or concentrically curved veins, these unbranched or with inconspicuous, short, transverse connectives (leaves net-veined or with prominent midrib and spreading side-veins parallel with each other in Alismataceae, Araceae, Smilacaceae, Marantaceae, and some Orchidaceae); perianth with dissimilar inner and outer whorls (petals and sepals), or all parts about alike (tepals), the parianth parts separate or united, commonly in 3s, less often in 2s, rarely in 5s, or perianth of scales or bristles, or entirely absent.

Worldwide, the Monocotyledonae is a group composed of ca. 55,800 species in 2,652 genera arranged in 84 families (Mabberley 1997); 25 of these families occur in nc TX. The monocots appear to be a well-supported monophyletic group derived from within the monosulcate Magnoliidae group of dicots (Chase et al. 1993; Duvall et al. 1993; Qiu et al. 1993). From the cladistic standpoint, the dicots are therefore paraphyletic and thus inappropriate for formal recognition (see explantion and Fig. 41 in Apendix 6). Within the monocots, *Acorus* appears to be the sister group to all other monocots, with the Alismataceae (and *Potamogeton*) being the next most basal group (Duvall et al. 1993).
REFERENCES: Cronquist 1981, 1988, 1993; Dahlgren et al. 1985; Thorne 1992; Chase et al. 1993; Clark et al. 1993; Duvall et al. 1993; Qiu et al. 1993; Reveal 1993a, 1993b; Takhtajan 1997.

ACORACEAE SWEETFLAG OR CALAMUS FAMILY

Acorus has traditionally been placed in the Araceae (e.g., Dahlgren et al. 1985) despite many characters unusual for an aroid. Grayum (1987) gave extensive reasons why the genus should be placed in its own family. The Acoraceae, thus circumscribed, is a very small, Old World and North American family of 2 species (3 if *Gymnostachys* is included). Using cpDNA restriction site analysis, a clade containing *Acorus* and the somewhat similar *Gymnostachys* (also traditionally placed in the Araceae) was resolved as a sister group of all other monocots (Davis 1995). An anaylsis by Duvall et al. (1993) also pointed to *Acorus* as the most basal living lineage of monocotyledons and a more recent molecular study by Soltis et al. (1997) again suggested that *Acorus* is anomalous among monocots. These results all support the recognition of the Acoraceae as separate from the Araceae and suggest further study is needed to determine its phylogenetic position. (subclass Arecidae)
FAMILY RECOGNITION IN THE FIELD: the only member of this family in nc TX is an aromatic herb with sword-like leaves roughly 1 m long and a cylindrical, finger-like spadix diverging laterally from an elongate, spathe-like scape.
REFERENCES: Wilson 1960a; Grayum 1987; Duvall et al. 1993; Davis 1995; Soltis et al. 1997.

ACORUS SWEETFLAG, CALAMUS

A genus of 2 species with iris-like or grass-like leaves; sometimes cultivated for fragrant oils in rhizomes. (Latin name for an aromatic plant or possibly Latin: *acorus*, without pupil, the name used by Dioscorides for an iris used in treating cataracts)
REFERENCES: Buell 1935; Harper 1936.

Acorus calamus L., (ancient name for a reed). DRUG SWEETFLAG, SWEETFLAG, CALAMUS. Aromatic, rhizomatous (thick) perennial herb with erect, linear, sword-shaped, parallel-veined leaves 0.9–1.2 m long, 5–25 mm wide; inflorescence an exposed cylindrical spadix, 4–9 cm long, diverging laterally from an elongate, leaf-like, spathe-like scape; flowers perfect, covering the spadix; perianth of 6 short segments; stamens 6; carpels 2–3; fruit a few-seeded berry. Wet ground or shallow water; Dallas Co., also Denton and Tarrant cos. (Mahler 1988). May–Jun. The geographic origin of *Acorus* has been somewhat confused. The genus is apparently introduced in TX, but was described as native in 1833 (Mahler 1988); Harper (1936) questioned whether *Acorus* is native to the U.S.; Buell (1935), however, concluded that the genus is native to the interior of North America. According to J. Kartesz (pers. comm.), TX plants are introduced from the Old World, with *A. americanus* (Raf.) Raf. extending no further s in the Great Plains than Nebraska and Iowa. Jones et al. (1997) treated TX material as *A. americanus*. This species has been used medicinally since the time of Hippocrates; it was also known from Tutankhamun's tomb; it is used religiously as "oil of holy ointment" for anointing sacred items and referred to in Exodus as SWEET CALAMUS; in Sumatra it is hung up at night to keep evil spirits from children; it is also apparently effective as an insecticide (Mabberley 1997). Duke (1985) referenced sources indicating that oil of calamus is carcinogenic, probably due to the presence of asarone (an allylbenzene) or safrole; McGuffin et al. (1997) indicated asarone is potentially hepatocarcinogenic and can cause chromosome damage in human lymphocytes. ☠ ☜

AGAVACEAE
YUCCA, CENTURY-PLANT OR AGAVE FAMILY

Herbaceous or woody, usually xerophytic perennials from a pithy corm or soft-woody root; leaves usually basal or bunched, narrow, flat to concave or thickened, ± fleshy or leathery, with widened, clasping base; flowering stems with alternate leafy bracts; flowers in racemes or panicles; tepals 6, in 1 or 2 rows; stamens 6; pistil 1; ovary superior or inferior; fruit a capsule.

A medium-sized family (550 species in 18 genera—B. Hess, pers. comm.) mainly of arid or semi-arid tropics and subtropics, especially in the Americas; its taxa have sometimes been treated as Amaryllidaceae, Asteliaceae, Dracaenaceae, Liliaceae, or Nolinaceae; some authorities (e.g., Heywood 1993) have suggested they may be more closely related to taxa in the Liliaceae than to each other. Molecular studies (Bogler & Simpson 1995, 1996) indicated the family as treated here is probably not monophyletic and supported the recognition of Nolinaceae and a more narrowly circumscribed Agavaceae. Ornamentals include species of *Agave*, *Dracaena*, *Sansevieria* (MOTHER-IN-LAW'S-TONGUE), and *Yucca*. Family name from *Agave*, AGAVE, MAGUEY, or CENTURY-PLANT, a genus of 100+ species native from the s United States to tropical South America. *Agave* species are the source of sisal hemp and pulque, a Mexican "beer" distilled to produce mescal and tequila. (Greek: *agave*, noble or admirable, in reference to the handsome appearance when in flower) (subclass Liliidae)

FAMILY RECOGNITION IN THE FIELD: usually *xerophytic*, typically robust perennials with often *elongate narrow* leaves usually basal or crowded near base of stem (or at stem apex in large tree-like YUCCAS) and sometimes *sharp-pointed*; inflorescence a raceme or panicle; fruit a capsule.

REFERENCES: Dahlgren et al. 1985; Bogler & Simpson 1995, 1996.

1. Flowers rosy red or salmon-colored; leaves conspicuously revolute (inrolled) upon drying (the margins nearly touching) ____ **Hesperaloe**
1. Flowers white to greenish or yellowish (can be reddish brown towards tips); leaves not revolute (can be v-shaped in *Manfreda*).
    2. Ovary superior; leaves usually 30 cm long or more OR if shorter with a hard spiny tip, not succulent (but can be thick), often < 4(–8) cm wide (but 3–8 cm wide in the large tree-like YUCCAS).

3. Flowers 13–78+ mm long or broad; leaves 8–80 mm wide; capsules large (much > 1 cm long), the seeds numerous in each cell ____ **Yucca**
3. Flowers 2.5–6 mm long or broad; leaves 2–12 mm wide; capsules small (< 1 cm long), the seeds solitary in each cell ____ **Nolina**
2. Ovary inferior; leaves 10–30 cm long, without a hard spiny tip, succulent, 1–7(–10) cm wide ____ **Manfreda**

HESPERALOE

A genus of 5 species native to sw North America (Starr 1997). (Greek: *hesperos*, western or evening, and the genus name *Aloe*)
REFERENCES: Trelease 1902; Starr 1995, 1997; Pellmyr & Augenstein 1997.

Hesperaloe parviflora (Torr.) J.M. Coult., (small-flowered), RED-FLOWERED-YUCCA, RED HESPERALOE. Leaves numerous, crowded at base of plant, linear, to 1.2 m long; flowering stem to 2.5 m tall, usually few-branched; pedicels to 35 mm long; flowers tubular to oblong-campanulate, 25–35 mm long; stamens shorter than the corolla; style slightly to much exserted; capsules to 3 cm or more long. Rocky slopes, open areas; Mills Co. in sw part of Lampasas Cut Plain, also across the Colorado River in San Saba Co., also spreading from cultivation in Brown Co. (Stanford 1976) and reported by Starr (1997) from Collin Co. [escaped?]; otherwise mostly much further w in sw TX. Mar–Sep. [*Yucca parviflora* Torr.] The striking flower color immediately distinguishes this species from all other nc TX Agavaceae; it is widely used as an ornamental. Pollination is reported to be by hummingbirds as well as bees (Starr 1995); experiments by Pellmyr and Augenstein (1997) showed the species to be self-incompatible and pollinated by black-chinned hummingbirds (*Archilochus alexandri*). /92

MANFREDA FALSE ALOE, AMERICAN-ALOE

A genus of 25 species ranging from the se United States to Honduras and El Salvador (B. Hess, pers. comm.). The species have been variously recognized in *Agave*, *Manfreda*, and *Polianthes*. (Named for Manfred, an ancient Italian writer)
REFERENCES: Shinners 1951f, 1966a; Verhoek-Williams 1975.

Manfreda virginica (L.) Rose, (of Virginia). Glabrous perennial; leaves mostly in a basal rosette, soft, thick-herbaceous, somewhat fleshy; flowers in a spike-like raceme; perianth tubular-funnelform; anthers linear, versatile (= attached near middle); capsules 3-celled, oblong to globose, 14–20 mm long, with numerous flattened seeds. The following key is modified from Shinners (1951f).

1. Leaves 12–18 cm long, (2–)3–8(–10) cm wide, 3–6 times as long as wide, 4–10 per plant; scape 6–10 mm thick near base, 3–5 mm at base of inflorescence; perianth (including ovary) 2.6–3.5 cm long, the lobes 2.5–3 mm wide at base; anthers 13–17(–20) mm long; flowering mid-Jun–mid- in Jul; nc TX mainly in the Blackland Prairie ____ subsp. **lata**
1. Leaves (12–)15–30 cm long, 1–4.5 cm wide, 7–15 times as long as wide, ca. 10 per plant; scape 4–7 mm thick near base, 1.5–3.5 mm at base of inflorescence; perianth (including ovary) 2–2.3 cm long, the lobes 1.5 mm wide at base; anthers 8–10 mm long; flowering mid-Jul–mid-Aug; wooded areas ____ subsp. **virginica**

subsp. **lata** (Shinners) O'Kennon, Diggs, and Lipscomb, comb. nov. BASIONYM: *Agave lata* Shinners, Field & Lab. 19:171–173. 1951. TYPE: TEXAS. Grayson Co.: 4.7 miles south of Sherman, *H.V Daly 61*, 15 Jun 1951, (HOLOTYPE: BRIT/SMU), (broad), WIDE-LEAF FALSE ALOE. Plant 0.6–1.7 m tall (to tip of inflorescence); corm pithy; leaves 4–10, noticeably fleshy, green to bluish gray-green, occasionally with reddish splotches near base, elliptic or broadly lanceolate, deeply concave, glabrous, margins scabrous; pedicels shorter than subtending bracts; flowers spicy-scented; perianth greenish or yellowish with dots or tinge of red-brown toward tips, the lobes

5–8 mm long; filaments green with reddish pigmentation; anthers exserted, cream-colored. Mainly Blackland Prairie; s Grayson (apparently now extinct locally), Hunt, and Kaufman cos., also Parker Co. (R. O'Kennon pers. obs.); otherwise apparently known only from s Oklahoma. Jun–Jul. [*Agave lata* Shinners, *Polianthes lata* (Shinners) Shinners] This taxon was named as a species in the genus *Agave* by Shinners (1951f) and subsequently transferred to *Polianthes* (Shinners 1966a). Verhoek-Williams (1975) placed it in the genus *Manfreda* but lumped it with the more widespread *M. virginica* (L.) Rose. By the time of Verhoek-Williams' study, the Grayson Co. site was apparently no longer in existence and no other TX sites were known. Since that time, several new Blackland Prairie populations with hundreds or even thousands of individuals have been discovered. While there is undoubtedly overlap in most of the characters distinguishing this subspecies from subsp. *virginica*, we agree with Shinners (1951f) that it is a geographically distinct entity; subspecific status appears most appropriate. Detailed taxonomic work on the large Blackland Prairie populations is needed. ▣/98

subsp. **virginica**, (of Virginia), FALSE ALOE, RATTLESNAKE-MASTER, VIRGINIA AGAVE. Leaves lanceolate to somewhat oblong-spatulate, nearly flat. Tarrant Co. (Fort Worth Nature Center; it is not completely certain that this popualtion is native) and Lamar Co. (Carr 1994); mainly se and e TX. Used by Native Americans as an antidote for snakebite, giving rise to the common name.

Manfreda maculosa (Hook.) Rose, (spotted), cited by Hatch et al. (1990) for vegetational area 4 (Fig. 2), is an endemic to sc TX and apparently occurs only to the s of nc TX. It can be distinguished by its longer perianth (including ovary nearly 5 cm long) with longer lobes (10–19 mm long).

Agave americana L., (of America), the CENTURY-PLANT, is cultivated in nc TX and long persists; it has large glaucous-gray leaves with a long (2.5–5 cm) terminal spine and a paniculate inflorescence 5–7 m tall.

NOLINA BEAR-GRASS

Polygamo-dioecious perennials with woody crown; leaves numerous, basal, clustered, linear, with margins smooth or serrulate; panicle pedunculate; perianth small, white, of 6 segments; stamens 6; pistil 1.

A mainly sw North American genus of 30 species; some are used as ornamentals. John Kartesz (pers. comm. 1997) is currently treating *Nolina* in the Liliaceae; other authorities place it in its own family, the Nolinaceae (e.g., Dahlgren et al. 1985); B. Hess (pers. comm.) has indicated that in the forthcoming treatment for Flora of North America, it will be included in the Agavaceae. Molecular analyses (Bogler & Simpson 1995, 1996) indicate that *Nolina*, *Dasylirion*, *Beaucarnea*, and *Calibanus* are a monophyletic group and support the recognition of the Nolinaceae. (Named for P.C. Nolin, an 18th century French agriculturalist)
REFERENCES: Trelease 1911; Bogler & Simpson 1995, 1996.

1. Leaves 4–12 mm wide, flattened in cross-section, the margins strongly serrulate; inflorescence held well above the leaves ____ **N. lindheimeriana**
1. Leaves 2–4(–7) mm wide, roundish with one flattened side in cross-section, the margins smooth to remotely toothed; inflorescence among the leaves ____ **N. texana**

Nolina lindheimeriana (Scheele) S. Watson, (for Ferdinand Jacob Lindheimer, 1809–1879, German-born TX collector), RIBBON-GRASS, DEVIL'S-SHOESTRING, LINDHEIMER'S NOLINA. Perennial 60–180 cm tall, with woody crown; leaves narrow and elongate, flat, soft, with smooth surfaces and serrulate margins with the teeth directed forward. Limestone outcrops, in sun or shade; Bell (J. Stanford, pers. comm.) and Somervell (Mahler 1988) cos., also Fort Hood (Bell or Coryell cos.—Sanchez 1997); mainly Edwards Plateau; endemic to TX. Apr–May.

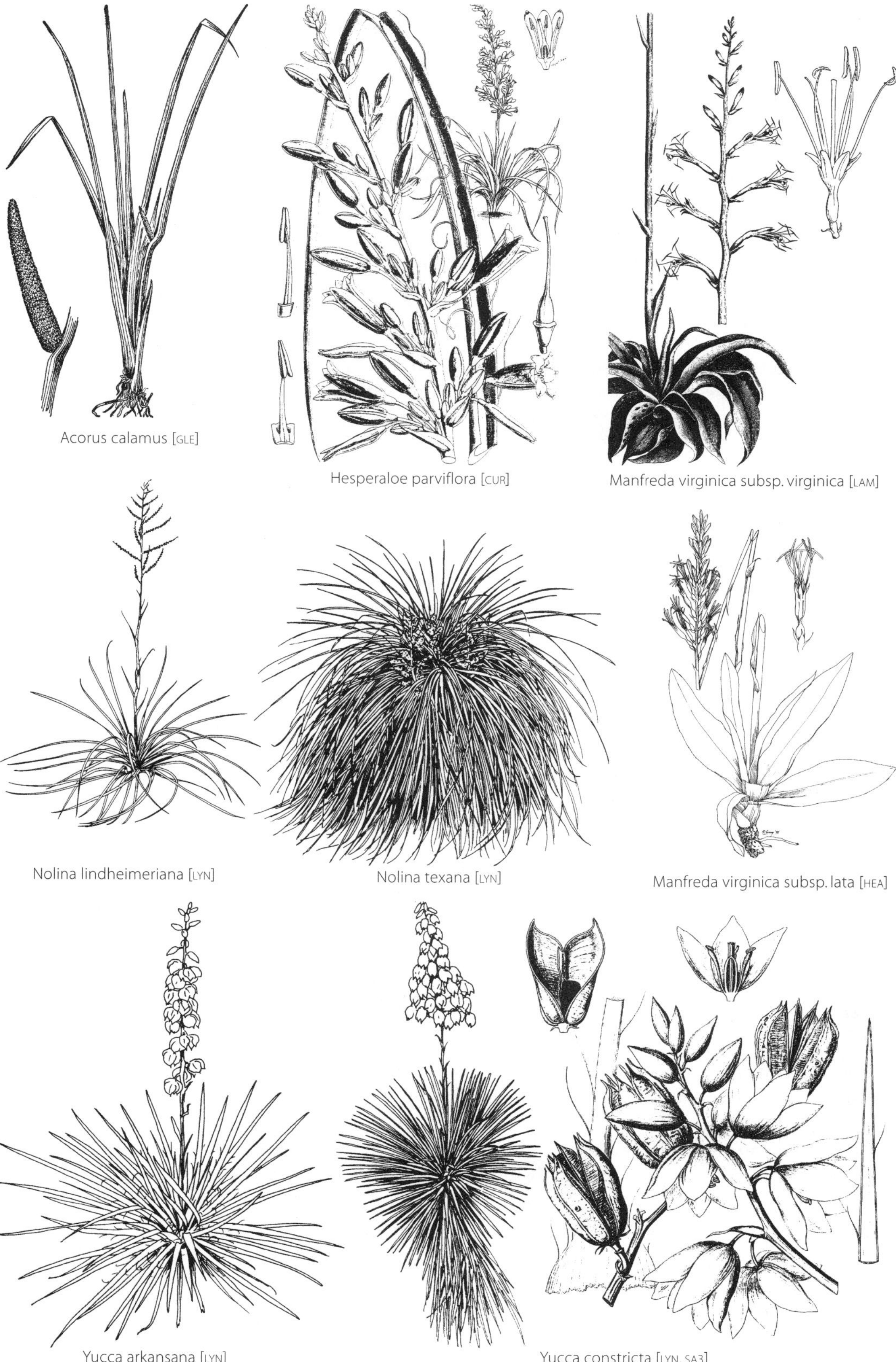

Acorus calamus [GLE]

Hesperaloe parviflora [CUR]

Manfreda virginica subsp. virginica [LAM]

Nolina lindheimeriana [LYN]

Nolina texana [LYN]

Manfreda virginica subsp. lata [HEA]

Yucca arkansana [LYN]

Yucca constricta [LYN, SA3]

Nolina texana S. Watson, (of Texas), SACAHUISTA, BUNCH-GRASS. Perennial 30–60 cm tall; leaves narrow, elongate, almost rounded-triangular in cross-section, the margins smooth or with distant teeth. Rocky soils; Brown (HPC) and Hamilton (Mahler 1988) cos., also Fort Hood (Bell or Coryell cos.—Sanchez 1997); w part of nc TX s and w to w TX. Apr. The flowers are poisonous and potentially fatal to livestock; liver-kidney toxicity and photosensitization are involved (Sperry et al. 1955; Kingsbury 1964). This species is normally avoided by livestock but will be eaten if no other food is available (J. Stanford, pers. comm.). ☠

YUCCA BEAR-GRASS, SPANISH-BAYONET, SOAPWEED

Plants coarse, with one to many crowns of narrow, elongate leaves, in nc TX species these usually in a basal cluster or at ends of very short trunk-like stems or in 2 species at ends of elongate trunk-like stems; flowers in terminal racemes or panicles; flowering stems with wide-based, acute or acuminate, somewhat papery bracts; flowers rather large; perianth drooping, of 6 thick, white to cream-colored or greenish segments (can be tinged with purple in 2 species); fruits dehiscent capsules or in 2 species indehiscent, erect when dry or in 2 species pendant.

A genus of 35 species (B. Hess, pers. comm.) of warm areas of North America. YUCCAS were used by Native Americans as a source of food, fiber, soap, and medicine; the spiny tip was apparently used as a needle, often with the still attached fibers serving as thread (Churchill 1986c); 2,000 year old fiber and twine from YUCCA have been found in Native American ruins in AZ; according to Bell and Castetter (in Webber 1953), "... yucca ranked foremost among the wild plants utilized by the inhabitants of the Southwest. It holds this place because of the great variety of uses to which it could be put and to the wide accessibility of this genus within the Southwest." During World Wars I and II, large amounts of YUCCA were harvested in TN and NM for fiber (Webber 1953). All species are dependant on yucca moths for pollination; if the moths are not present the plants reproduce vegetatively; as a result, large clonal populations are often encountered in the field (K. Clary, pers. comm.). According to Powell (1988), "The Yucca Moth (*Tegeticula = Pronuba*) flies at dusk to a flower where she climbs stamens to collect pollen and pack the pollen in a large ball-like mass under her neck. She then visits another flower where she inserts her ovipositer directly through the ovary wall and deposits 20–30 eggs, one at a time, each directly into an ovule. She then climbs to the stigma of the same flower and spreads the pollen, thus ensuring pollination, subsequent fertilization, and developing seeds that provide nourishment for the moth larvae. Each larva ultimately destroys the seed in which it grows, but there are many undamaged seeds left in the yucca capsule." Baker (1986) described some of the complexities of pollination in *Yucca*. (A native Haitian name)

In addition to the species discussed below, several other YUCCAS are cultivated in nc TX (particularly the sw part) including 3 with trunk-like stems; these are *Y. aloifolia* L. (leaves dark green, stiff, spear-like, without marginal threads; in contrast to all other species, the leaves are arranged along the full length of the trunk instead of in well-defined leaf rosette); *Y. rostrata* Engelm. ex Trel. (leaves glaucous, flexible, the margins denticulate); and *Y. thompsoniana* Trel. (leaves glaucous, flexible, much shorter than in *Y. rostrata*, the margins denticulate) [now sometimes treated as a synonym of *Y. rostrata*]. *Yucca filamentosa* L. and *Y. flaccida* Haw., both with drooping leaves and usually without trunk-like stems or trunk-like stems very short, are also widely used in landscaping in nc TX. They are sometimes lumped as *Y. filamentosa*, but can be distinguished by the leaf margins shredding into threads in *Y. filamentosa*, while not shredding in *Y. flaccida*.

REFERENCES: Trelease 1902; McKelvey 1938, 1947; Webber 1953.

1. Plants with leaves in a crown at ends of 1–4(–8) trunk-like stems 1.5–4.3 m tall; leaves broad, 3–8 cm wide, stiff and spear-like, thickish; fruits indehiscent, eventually drooping; sw margin of nc TX, mainly Edwards Plateau s and w to s TX and Trans-Pecos.

2. Leaves with marginal threads, the margins not denticulate, the apical portion of leaves usually rolled inward so that margins nearly touch; ovary slender for its length, not over 7 mm in diam. at flowering time **Y. torreyi**
2. Leaves without marginal threads, the margins not denticulate, the apical portion of leaves not inrolled; ovary stout for its length, 7–12 mm in diam. at flowering time **Y. treculeana**

1. Plants with leaves in a basal cluster (without visible stems) or at ends of very short trunk-like stems; leaves usually narrower, 0.8–4 cm wide, not as above; fruits dehiscent at maturity, not drooping; widespread in nc TX.
   3. Leaf margins yellowish to dark orangish red or reddish brown, smooth or minutely toothed, not shredding into threads; pistil 3.2–4.5 mm long.
      4. Leaves straight or nearly so, not with strongly inrolled margins, usually pale bluish to sage green, conspicuously glaucous, ± smooth on both surfaces; leaf margins yellowish, flat **Y. pallida**
      4. Leaves twisted, with margins inrolled most of their length, dark green, not glaucous, ± scabrous on both surfaces; leaf margins usually dark orangish red or reddish brown or occasionally yellowish, wavy **Y. rupicola**
   3. Leaf margins whitish, shredding into prominent white threads (these often disappearing late in year); pistil 2–3.2 cm long.
      5. Inflorescence a much branched panicle, beginning well above tips of leaves (separated from them by nearly its own length or more of naked scape); fruits constricted near middle or not so; in nc TX mainly in West Cross Timbers.
         6. Leaves very slender, 8–15 mm wide, 100–200 per strikingly globose rosette; fruits usually conspicuously constricted near middle; on limestone substrates **Y. constricta**
         6. Leaves usually 15–40 mm wide, ca. 50–85 per rosette (rosette not globose in appearance); fruits usually not conspicuously constricted; on sandy substrates **Y. necopina**
      5. Inflorescence usually unbranched and raceme-like or with 1 or 2 short, spreading branches near base (these often soon deciduous), beginning below to just above leaf tips (separated from them by less than its own length of naked scape); fruits not contricted near middle; widespread in nc TX **Y. arkansana**

Yucca arkansana Trel., (of Arkansas), ARKANSAS YUCCA. Leaves 20–60 cm long, the blades 1–2.5 cm wide, the margins at first white, papery with curly fibers; perianth 32–65 mm long, greenish white, globose; capsules ca. 4–7 cm long. Rocky limestone or sandy soils; se and e TX w to eastern Rolling Plains and Edwards Plateau. Late Apr–mid-May.

Hybrids of *Y. arkansana* and *Y. pallida* have been found on limestone in Dallas (McKelvey 1947), at Glen Rose in Somervell Co. (Shinners 1958), and recently in Tarrant County at Tandy Hills Park. The Tandy Hills plants vary from having leaves with curly fibers on the margins to not so, from having leaf margins white to yellowish, and from having inflorescences branched to sparsely branched or unbranched. In general, the plants were from 1–1.5 m tall. This population of hundreds of individuals over a number of acres was quite variable with individuals ranging from much like typical *Y. arkansana* to those much like *Y. pallida* and a full spectrum of intermediates.

Yucca constricta Buckley, (constricted), BUCKLEY'S YUCCA. Usually stemless, rarely with trunk-like stems to 40 cm tall; overall aspect of basal leafy portion almost ball-like in outline; leaves 30–65 cm long, very slender, 100–200 per rosette, very straight but flexible, the margins white or green with fibers that soon erode away; perianth pale greenish white; panicle branches glabrous. Limestone outcrops or rocky prairies; Callahan, Coleman, and Erath cos., also Brown Co. (HPC); West Cross Timbers s and w to w TX; endemic to TX. Apr–Jun.

Yucca necopina Shinners, (unexpected), GLEN ROSE YUCCA. Similar to *Y. arkansana* vegetatively, but taller (1–3 m tall) and with large, much-branched inflorescences held well above the leaves; leaves 50–80 cm long, typically 1.5–4 cm wide, the margins white, with curly fibers; inflorescences completely glabrous; flowers greenish white. Previously known only from a sandy fencerow on Brazos River terrace near Glen Rose, Somervell Co. (Shinners 1958a); recently rediscovered by R. O'Kennon along Brazos River terraces in Hood and Somervell cos. and in deep sand in Parker and Tarrant cos.; these populations number in the hundreds of individuals; apparently endemic to nc TX but should be looked for in s OK. This is the common YUCCA of sandy soils in the West Cross Timbers in nc TX. May–Jun. Shinners thought this to be possibly a hybrid between *Y. pallida* and *Y. arkansana* but unlike an evident hybrid of these 2 species observed nearby (Shinners 1958a). Recent field observations of large numbers of relatively uniform individuals in widely separated populations—in sandy areas where neither *Y. arkansana* or *Y. pallida* typically occur—support the recognition of this entity at the specific level. Molecular evidence (K. Clary, pers. comm.) also supports specific recognition. Bill Hess (pers. comm.)) is treating this species as a synonym of *Y. arkansana* in his treatment of *Yucca* for Flora North America. The closest relative of *Y. necopina* seems to be *Y. louisianensis* Trel. (to be treated as *Y. flaccida* Haw. by B. Hess (pers. comm.), which is distinguished by its usually narrower leaves and pubescent inflorescences. Vines (1960) cited *Y. louisianensis* for Dallas and Fort Worth; these records are likely to be of *Y. necopina*. The genus is in need of more detailed field study in nc Texas. (TOES 1993: V) ⚠ /107

Yucca pallida McKelvey, (pale), PALE YUCCA, PALE-LEAF YUCCA. Plant 1.3–2.5 m tall to tip of inflorescence; leaves 18–35 cm long, 2–4 cm wide, conspicuously glaucous, margins corneous (= horn-like texture); panicle narrowly to widely branched; perianth segments with pale greenish center and white edges. Limestone outcrops or rocky prairies; Grand Prairie and Blackland Prairie (Dallas, on the Austin Chalk); endemic to nc TX or possibly slightly onto the Edwards Plateau. May–Jun.

Yucca rupicola Scheele, (growing on cliffs or ledges), TEXAS YUCCA, TWIST-LEAF YUCCA. Similar to *Y. pallida*; leaves 20–60 cm long; perianth segments whitish or greenish white. Limestone ledges, plains; Bell Co. in s part of nc TX; mainly Edwards Plateau; endemic to TX. Apr–Jun.

Yucca torreyi Shafer, (for John Torrey, 1796–1873, American botanist, physician, and collector of many w North American plants), TORREY'S YUCCA, SPANISH-DAGGER. Stems unbranched or rarely with 2–3 branches; dead leaves reflexed on trunk below leaf crown; leaves 30–110 cm long, light green, stiff, spear-like, with marginal threads; panicle with 10–50% of its total length extending beyond the leaves or rarely entirely within the leaves; perianth subglobose or campanulate, sometimes fully expanding, cream (can be tinged with purple); fruits 7–14 cm long, indehiscent, slightly pulpy. Gravelly soils, grassy and chaparral mesas and slopes; Brown and Burnet cos. (HPC) on sw margin of nc TX; mainly Edwards Plateau and Trans-Pecos. Late Mar–May. This species is sometimes treated as a synonym of the related *Y. treculeana* (e.g., Powell 1988); according to Webber (1953) and Correll & Johnston (1970), the two sometimes hybridize; they can usually be readily distinguished in the field; we are following McKelvey (1938), Kartesz (1994), Jones et al. (1997), and J. Kartesz (pers. comm., 1997) in recognizing them at the specific level. [*Y. baccata* Torr. var. *macrocarpa* Torr., *Y. crassifolia* Engelm., *Y. macrocarpa* (Torr.) Coville]

Yucca treculeana Carr., (for A.A.L. Trécul, 1818–1896, who took plants of this species to France in 1850—Vines 1960), TRECUL'S YUCCA, SPANISH-DAGGER, SPANISH-BAYONET, DON QUIXOTE'S-LANCE, PITA, PALMA PITA, PALMA DE DÁTILES, PALMA LOCA, TEXAS-BAYONET. Stems few-branched, with leaf crown at apex; dead leaves reflexed on trunk below leaf crown (trunks bare of dead leaves on old plants); leaves 50–100 cm long, without marginal threads; panicle with ca. 50–75% of its total length extending beyond the leaves; perianths broadly globose or hemispherical, greenish cream to cream (can be lightly tinged with purple); fruits 5–11.5 cm long, indehiscent, the flesh

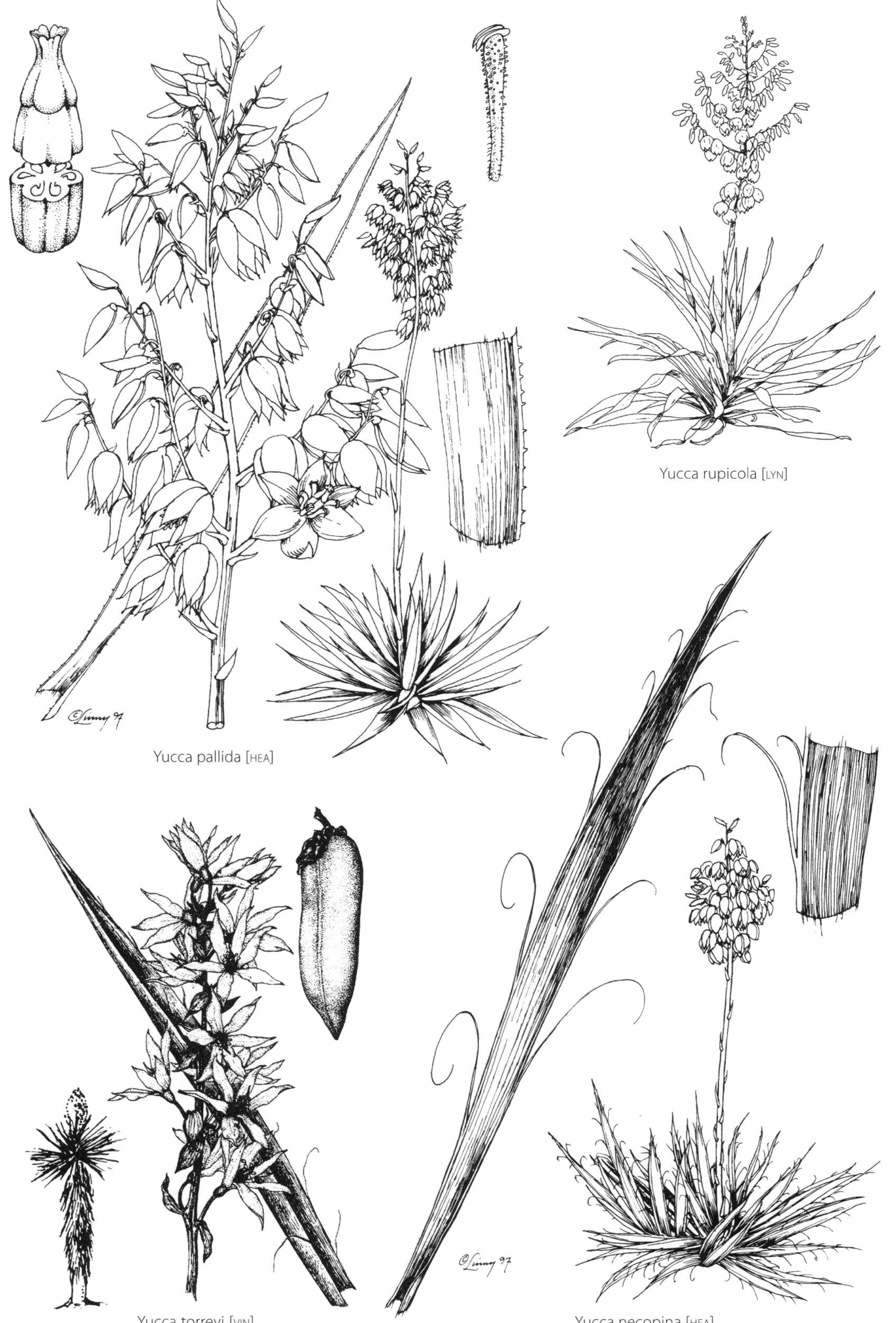

Yucca pallida [HEA]

Yucca rupicola [LYN]

Yucca torreyi [VIN]

Yucca necopina [HEA]

sweetish and succulent. Brushland; Burnet Co. (Buckley in McKelvey 1938; also R. O'Kennon, pers. obs.), also known just s of nc TX in San Saba Co. (Buckley in McKelvey 1938) and Travis Co. (Tharp letter quoted in McKelvey 1938); Trans-Pecos and se and s TX n to Edwards Plateau near s margin of nc TX. Feb–Apr. The spines were used to jab a snake bite and induce bleeding in order to flush away the poison (Vines 1960). Pioneers cooked and prepared the flowers like cabbage and also pickled them (Schulz 1922; McKelvey 1938; Crosswhite 1980). According to Havard (1896), the flesy, banana-like fruits are delicious, contain considerable sugar, and were converted by Chihuahua Indians into a fermented beverage. Long fibers obtained by macerating the leaves were used in the past to make ropes (Crosswhite 1980).

ALISMATACEAE
ARROWHEAD OR WATER-PLANTAIN FAMILY

Wet ground or aquatic, annual or perennial herbs, largely glabrous, with milky sap; leaves basal, sometimes dimorphic with different submerged (linear and bladeless) and emergent (generally with distinct blades) forms; leaf blades entire, linear to ovate-elliptic, or triangular and with basal lobes (= sagittate), longitudinally ribbed (midrib more prominent than others) and with cross-veins; flowers whorled, in scapose racemes or panicles, perfect or imperfect; sepals 3, green; petals 3, white or rarely pink, equal; stamens 6 to many; pistils many, on a swollen or elongating receptacle; ovary superior with basal placentation; fruits achenes in our species.

A small (ca. 75 species in 11 genera) nearly cosmopolitan (Haynes & Holm-Nielsen 1994), but especially n temperate family of aquatic or wet area plants with most species found in the New World. Molecular analyses (Duvall et al. 1993) indicate that the Alismataceae is phylogenetically near the base of the monocotyledons. (subclass Alismatidae)
FAMILY RECOGNITION IN THE FIELD: wet area or aquatic herbs with milky sap and basal, often broad, usually distinctly petiolate leaves; flowers whorled, in scapose racemes or panicles, with 3 green sepals, 3 white or rarely pink petals, and numerous separate carpels (and later achenes).
REFERENCES: Small 1909; Beal 1960; Rogers 1983; Dahlgren et al. 1985; Haynes & Holm-Nielsen 1994.

1. Carpels (and later achenes) in a single ring on the receptacle; stamens 6; leaf blades never sagittate (cuneate to cordate at base); extreme ne part of nc TX ____ **Alisma**
1. Carpels (and later achenes) densely crowded over surface of receptacle forming a head-like mass; stamens usually > 6 (often numerous); leaf blades variable in shape, sometimes sagittate; widespread in nc TX.
   2. Fruiting heads rough in appearance, resembling a bur (due to the conspicuous persistent styles on the achenes); achenes turgid, ribbed or ridged, not membranous-winged; flowers perfect; leaf blades never sagittate ____ **Echinodorus**
   2. Fruiting heads not bur-like (except somewhat bur-like in *S. brevirostra*); achenes flattened, membranous-winged; flowers perfect or imperfect, at least the lower imperfect; leaf blades sagittate OR not so ____ **Sagittaria**

ALISMA WATER-PLANTAIN

A n temperate and Australian genus of 9 species. (Greek: *alisma*, water-plantain)
REFERENCES: Fernald 1946a; Hendrick 1958; Voss 1958; Pogan 1963.

Alisma subcordatum Raf., (slightly cordate), WATER-PLANTAIN, SMALL-FLOWER WATER-PLANTAIN, MUD-PLANTAIN. Emergent perennial; stems erect; leaves basal; leaf blades ovate to elliptic, broadly cuneate to subcordate at base, to 12(–15) cm long and 8(–10) cm wide, long-petioled; inflorescences to 60(–100) cm tall, panicled, with whorled branches; flowers perfect; petals white or pinkish, 1–3 mm long, suborbicular; receptacle flattened, including achenes the whole structure 4 mm wide or less; achenes 1–2 mm long, wingless, smooth, with a single dorsal groove. Shallow water; Lamar Co. (Carr 1994); mainly e TX. Jun–Sep. [*A. plantago-aquatica* L. var. *parviflorum* (Pursh) Torr.]

ECHINODORUS BURHEAD

Ours emergent annuals or perennials; leaves long-petioled; leaf blades with arcuate veins prominent below; inflorescences usually much exceeding the leaves, with flowers in whorls; flowers perfect; petals white; stamens ca. 12–20; fruiting heads rough in appearance, resembling a bur (due to the conspicuous persistent styles on the achenes); achenes turgid, ribbed or ridged, beaked.

A genus of 26 species extending from the n U.S. to Argentina and Chile (Haynes & Holm-Nielsen 1994); some are cultivated as ornamental aquarium plants. (Greek: *echinus*, rough husk, and *doros*, a leather bottle, applied to the ovary, which is in most species armed with the persistent style, forming a sort of prickly head of fruits)
REFERENCES: Fassett 1955; Haynes & Holm-Nielsen 1986.

1. Inflorescences rigidly erect at maturity; veins of sepals smooth; stamens usually 12–15; flowers 6–11 mm across ____ **E. berteroi**
1. Inflorescences erect when young but later becoming prostrate and rooting at nodes; veins of sepals usually with papillose or roughened ridges; stamens usually 20–22; flowers 10–25 mm across ____ **E. cordifolius**

Echinodorus berteroi (Spreng.) Fassett, (for C.G.L. Bertero, 1789–1831, Italian physician who botanized in West Indies), BURHEAD, ERECT BURHEAD. Coarse annual or short-lived perennial with scapes to 80 cm tall; leaves variable, those of mature plants with blades broadly ovate, 3–18 cm long, subcordate to truncate or broadly cuneate at base; secretory tissue visible as pellucid (= clearish, somewhat transparent) line-like markings conspicuous in dried leaves when backlit with strong light; inflorescences with flowers in whorls of 3–9; achenes 2–3 mm long. Mud and shallow water; Dallas, Denton, Grayson, Montague, and Parker cos., also Bell, Coryell (Fort Hood—Sanchez 1997), Brown, Comanche, and Hamilton cos. (HPC); widespread in TX. May–Oct. [*E. berteroi* var. *lanceolatus* (Engelm. ex S. Watson & J.M. Coult.) Fassett, *E. rostratus* (Nutt.) Engelm. ex A. Gray]

Echinodorus cordifolius (L.) Griseb. subsp. **fluitans** (Fassett) R.R. Haynes & Holm-Niels., (sp.: with heart-shaped leaves; subsp.: floating), CREEPING BURHEAD. Annual or short-lived perennial; leaf blades broadly ovate, 2–14(–20+) cm long; pellucid markings usually absent or not conspicuous; inflorescences with flowers in whorls of 5–15, to 1.2 m long, often producing plantlets at tips; achenes ca. 2–3.5 mm long. Mud and shallow water; Tarrant Co., also Brown (Stanford 1971) and Hamilton (HPC) cos. in w part of nc TX and Lamar Co. (Carr 1994) in Red River drainage; also se and e TX and Edwards Plateau. Apr–Nov. We are following Haynes & Holm-Nielsen (1994) and Jones et al. (1997) in recognizing the TX material of this species as subsp. *fluitans*. [*E. fluitans* Fassett]

SAGITTARIA ARROWHEAD

Mostly perennial, aquatic or semi-aquatic, rhizomatous herbs, usually emergent when flowering; leaves varying with environmental conditions (particularly depth of water) and season; leaf blades unlobed or sagittate; petioles long and spongy; submerged leaves typically bladeless; inflorescences usually erect or sometimes procumbent, branched or unbranched; flowers in whorls of 3, pedicelled, bracteate, mostly imperfect or sometimes perfect; petals white; stamens numerous; achenes flattened, membranous-winged, beaked.

A predominantly New World genus of ca. 25 species, ranging from Canada s to Argentina and Chile; 3–4 species also occur in Eurasia (Haynes & Holm-Nielsen 1994); often tuberiferous herbs of aquatic habitats; the tubers are edible in a number of species; some exhibit leaf polymorphism—the submerged leaves ribbon-shaped, the floating ones with ovate blades, the emergent ones with sagittate blades. (Latin: *sagitta*, an arrow, from the leaf shape of some species) REFERENCES: Smith 1895; Bogin 1955; Wooten 1973; Beal et al. 1982.

1. Leaf blades not sagittate, without lobes; filaments usually pubescent OR glabrous in 1 species that is rare in nc TX.
   2. Stalks of fruiting heads recurved; bracts of inflorescence smooth, thinly membranous; filaments pubescent; common and widespread in nc TX ____ **S. platyphylla**
   2. Stalks of fruiting heads ascending or spreading, not recurved; bracts of inflorescence thickened and papillose or coarsely ridged; filaments pubescent OR glabrous; rare, if present in nc TX probably limited to extreme e margin.
      3. Filaments glabrous; leaves ± phyllodial (= petiole and blade indistinct), enlarged-spongy at base, tapering to linear or narrowly lanceolate blade-like portion; achenes 1–1.5 mm long ____ **S. papillosa**
      3. Filaments pubescent; leaves with long petioles and definite blades (elliptic to lanceolate or rarely ovate); achenes 1.8–2.2 mm long ____ **S. lancifolia**
1. Leaf blades usually sagittate, with conspicuous projecting lobes basally; filaments glabrous.
   4. Sepals of pistillate flowers (when in fruit) 5–14 mm long, appressed or spreading; pedicels recurved and noticeably thickened; petioles terete; most flowers perfect; rare if present in nc TX ____ **S. montevidensis**
   4. Sepals of pistillate flowers (when in fruit) 4–7 mm long, reflexed; pedicels ascending or if recurved then not noticeably thickened; petioles angular; few or no flowers perfect; widespread in nc TX.
      5. Basal lobes of leaf commonly twice as long as blade body, typically narrow, usually < 2(–2.5) cm wide ____ **S. longiloba**
      5. Basal lobes of leaf seldom much longer than blade body, narrow to typically much wider, usually much > 2.5 cm wide.
         6. Lower floral bracts triangular-ovate, obtuse or acute, 5–12 mm long; achene beak projecting ± horizontally or slightly downcurved, the wing of the achene extending ± smoothly to upper surface of beak ____ **S. latifolia**
         6. Lower floral bracts lanceolate to narrowly triangular, acuminate, 12–30 mm long; achene beak projecting upward at an angle, the wing of the achene not extending smoothly to upper of the beak, with a definite interruption ("saddle"-like) ____ **S. brevirostra**

Sagittaria brevirostra Mack. & Bush, (short-beaked). Leaves to 0.6 m long; leaf blades to 30 cm long and 20 cm wide (usually smaller); sepals reflexed from fruiting aggregate; achene beak broad-based, up to 1.5 mm long. Rivers, ditches, other wet areas; Dallas, Ellis, and Grayson cos.; also se TX. Jun–Aug.

Sagittaria lancifolia L., (lance-leaved), SCYTHE-FRUIT ARROWHEAD. Leaf blades to 40 cm long and 10 cm wide; bracts of inflorescence striate to strongly papillose, to 35 mm long; sepals reflexed from fruiting aggregate; beak of achene inserted obliquely near apex of achene, to 0.8 mm long, ascending. Swamps, marshes, or other wet areas; included based on citation of vegetational area 4 (Fig. 2) by Hatch et al. (1990); mainly se and e TX; according to R. Haynes (pers. comm.), this is a coastal species that probably does not occur in nc TX. May–Nov. [*S. falcata* Pursh, *S. lancifolia* var. *media* P. Micheli]

Sagittaria latifolia Willd., (broad-leaved), COMMON ARROWHEAD, DUCK-POTATO, WAPATO. Leaves to 1.5 m long; leaf blades to 50 cm long, mostly sagittate, the lobes varying from narrow to del-

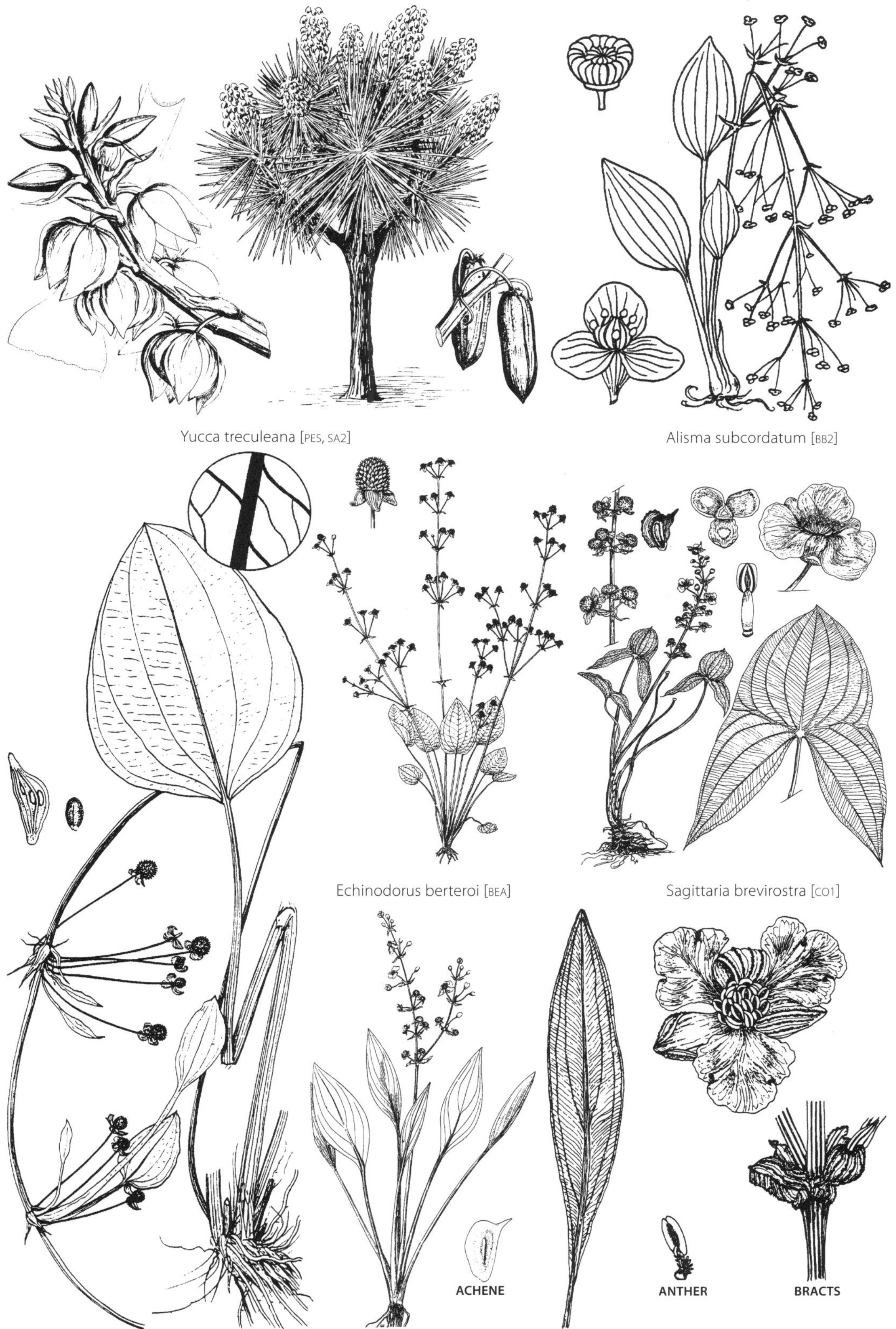

Yucca treculeana [PES, SA2]

Alisma subcordatum [BB2]

Echinodorus berteroi [BEA]

Sagittaria brevirostra [CO1]

Echinodorus cordifolius subsp. fluitans [FLN]

Sagittaria lancifolia [ARM, CO1]

toid; bracts to 15 mm long; sepals reflexed from fruiting aggregate, to 10 mm long; petals to ca. 20 mm long; achene beak broad-based, up to 2 mm long. Lakes, ponds, or other wet areas; Denton, Grayson, and Parker cos.; widespread in TX. May–Aug. The tuberous roots were used as food by Native Americans (Erichsen-Brown 1979).

Sagittaria longiloba Engelm. ex J.G. Sm., (long-lobed), LONG-LOBE ARROWHEAD, LONG-BARB ARROWHEAD, FLECHA DE AGUA. Leaves to 0.8 m long; leaf blades to 22 cm long, 0.5–2.5 cm wide; bracts usually < 15 mm long; sepals 4–7 mm long, reflexed from fruiting aggregate; petals to ca. 14 mm long; achene beak projecting horizontally, triangular, tiny, to 0.15 mm long or obsolete. Ponds, swamps, ditches, or other wet areas; Brown (Mahler 1988) and Hamilton (Stanford 1971) cos.; mainly s and w TX. Apr–Nov. Native Americans and early settlers used the tuberous roots as food; they were called duck potatoes or swan potatoes (Kirkpatrick 1992).

Sagittaria montevidensis Cham. & Schltdl. subsp. **calycina** (Engelm.) Bogin, (sp:. presumably of Montevideo, Uruguay; subsp.: calyx-like), GIANT ARROWHEAD. Leaves to 1 m long; leaf blades 3–40 cm long, 2–25 cm wide; bracts ca. 10 mm long; sepals appressed around fruiting aggregate or spreading; achenes to 2.5 mm long and 1.3 mm wide, the beak horizontal or oblique, narrowly winged, ca. as long as the achene is wide. Lakes, ponds, other wet areas; included based on distribution map (without counties) in Beal et al. (1982) and on citation of vegetational area 4 (Fig. 2) by Hatch et al. (1990), also Brown Co. (HPC); mainly e TX and Edwards Plateau. Jun–Oct. While this taxon is sometimes (e.g., Kartesz 1994) recognized as a separate species [*S. calycina* Engelm.], we are following Haynes and Holm-Nielsen (1994) and Jones et al. (1997) in treating it as a subspecies of *S. montevidensis*.

Sagittaria papillosa Buchenau, (with papillae or nipple-like structures), NIPPLE-BRACT ARROWHEAD. Leaf blades to 25 cm long and 5 cm wide (usually narrower); bracts of inflorescence densely papillose, 3–10 mm long; sepals reflexed from fruiting aggregate; petals to ca. 12 mm long; beak of achene inserted laterally above middle of achene, ca. 0.2 mm long, projecting horizontally or recurving. Swamps, marshes, or other wet areas; included based on citation of vegetational area 4 (Fig. 2) by Hatch et al. (1990); mainly se and e TX. Mar–Nov.

Sagittaria platyphylla (Engelm.) J.G. Sm., (broad-leaved), DELTA ARROWHEAD. Leaf blades 8–20 cm long, 2–8 cm wide, rarely with lateral projection(s) from the base; bracts 3–8 mm long, pistillate pedicels recurved; sepals 4–6 mm long, appressed around fruiting aggregate; beak subulate, 0.3 mm or more long, projecting upward at an angle. Swamps, marshes, ponds, or other wet areas; se and e TX w to Rolling Plains and Edwards Plateau. Apr–Oct. [*S. graminea* Michx. var. *platyphylla* Engelm.] This is the most common *Sagittaria* species in nc TX.

ARACEAE CALLA, ARUM, OR AROID FAMILY

Ours glabrous, herbaceous, rooted perennials with basal, simple or compound, entire leaves or 1 species a free-floating aquatic; inflorescence a fleshy spike (= spadix) with a sometimes highly modified leafy bract (= spathe) subtending or enclosing it; flowers very small, without perianth or with a few minute scales, imperfect; staminate flowers in upper part of spike, with 1–10 stamens, their filaments very short or absent; pistillate flowers in basal part of spike, each with 1 pistil; ovary superior.

A large (2,550 species in 104–105 genera), mainly tropical and subtropical (a few in temperate zones) family of mainly perennial herbs or vines. Many have tissues containing raphides (bundles of microscopic, needle-like calcium oxalate crystals) that can cause injury to the mouth, throat, or hands by puncturing cell membranes; the plants are also often cyanogenic or contain alkaloids, free oxalic acid, or other toxins; some are potentially fatally poisonous (McIntire et al. 1992; Woodland 1997). *Philodendron*, one of the most popular house plants in

Sagittaria latifolia [MAS]

Sagittaria longiloba [CO1]

Sagittaria montevidensis subsp. calycina [CO1]

Sagittaria papillosa [CO1]

Sagittaria platyphylla [AMB]

the U.S., has raphides and questionable unidentified proteins and can cause painful burning to the mouth and throat and contact dermatitis; ingestion of the leaves is highly toxic to cats; the plants should not be left within reach of children or pets (Lampe & McCann 1985; Spoerke & Smolinske 1990). *Dieffenbachia* (DUMBCANE), another common house plant, contains raphides and the alkaloid protoanemonine; ingestion causes swelling of the pharynx and larynx and can result in death through suffocation (Morton 1982); its common name is derived from its ability to paralyze the vocal cords and render people speechless. Food crops include the edible starchy corms of *Colocasia* (TARO) and *Xanthosma* (TANIER); ornamentals include *Anthurium*, *Caladium*, *Dieffenbachia*, *Monstera*, and *Philodendron*. Some species can produce heat in the inflorescences which volatilizes odors to attract pollinators. The genus *Acorus*, traditionally placed in the Araceae, is here recognized in the Acoraceae. Family name from *Arum*, a genus of 26 species of Europe and the Mediterranean area. (Greek: *aron*, the classical name of these plants) (subclass Arecidae)

FAMILY RECOGNITION IN THE FIELD: herbaceous perennials (also 1 floating aquatic) with numerous very small flowers in an often finger-like *spadix* subtended by a sheath-like *spathe*; leaves usually basal, with expanded blades, often with net venation.

REFERENCES: Birdsey 1951; Wilson 1960a; Jacobsen 1985; Bogner & Nicolson 1991; Grayum 1990, 1992.

1. Plants free-floating aquatics; leaves 3–15 cm long; spathes ca. 15 mm long; pistillate flower solitary at base of inflorescence ____ **Pistia**
1. Plants rooted in soil (even when in water); leaves often much longer; spathes much > 15 mm long; pistillate or perfect flowers several to many.
   2. Leaves compound with 3–15 leaflets ____ **Arisaema**
   2. Leaves simple.
      3. Leaf blades narrow (5–25 mm wide), grass-like, parallel-veined ____ **Acorus** (see Acoraceae)
      3. Leaf blades broad (much greater than 25 mm wide), not at all grass-like, net-veined.
         4. Leaves peltate (= the petiole attached on the lower leaf surface away from the margin); leaf blades cordate to sagittate or hastate at base, with a purplish spot on the upper epidermis above where the petiole attaches to the blade; spathes yellow ____ **Colocasia**
         4. Leaves not peltate; leaf blades sagittate at base, without a purplish spot; spathes creamy white ____ **Xanthosoma**

ARISAEMA JACK-IN-THE-PULPIT

A genus of 150 species of e Africa, Arabia, tropical and e Asia, and North America. ☠ All parts of *Arisaema* contain microscopic crystals of calcium oxalate, which if eaten disrupt cells and cause extreme burning and swelling of the mouth and throat (Stephens 1980; Cheatham & Johnston 1995). (Greek: *aris*, a kind of arum, and *haima*, blood, from the red-spotted leaves of some species)

Arisaema dracontium (L.) Schott, (Greek name for a kind of arum, presumably from *draco*, dragon), GREEN-DRAGON, DRAGONROOT. Perennial from a corm; leaves net-veined, divided into (5–)7–15 unequal leaflets (sometimes some of the divisions are not completely separated into leaflets); summit of spathe with margins inrolled, tapering to a slender point; spadix long exserted; fruits reddish orange. Low woods, moist slopes; Dallas, Cooke, Denton, Fannin, Grayson, and Tarrant cos.; se and e TX w to East Cross Timbers and Edwards Plateau. Late Apr–May. The rust fungus *Uromyces ari-triphylli* (Schwein.) Seeler sometimes causes conspicuous lesions on GREEN-DRAGON in nc TX (J. Hennen, pers. comm.). The tissues contain injurious calcium oxalate raphides (Lampe & McCann 1985). ☠ ▣/78

Arisaema triphyllum (L.) Schott, (three-leaved), (JACK-IN-THE-PULPIT, INDIAN-TURNIP), with 3 leaflets, the summit of the spathe arching over the spadix, not inrolled, and the spadix not exserted, occurs in se and e TX to the e of nc TX. The tissues contain injurious calcium oxalate raphides (Lampe & McCann 1985). ☠

COLOCASIA TARO, ELEPHANT'S-EAR

A tropical Asian genus of 8 species of tuberous herbs with peltate leaves; used as ornamentals and for food. ☠ All parts of *Colocasia* species, except the corm (when properly prepared) of TARO, contain calcium oxalate crystals which cause burning and swelling of the mouth and throat (Schmutz & Hamilton 1979). (Arabic: *kolkas*, originally used for the root of a species of *Nelumbo*)

REFERENCES: Arridge & Fonteyn 1981; Wang 1983.

Colocasia esculenta (L.) Schott, (edible), TARO, WILD TARO, KALO, DASHEEN, EDDO, COCOYAM. Large perennial with large tuber-like corm and leaves all basal; easily identified by the large peltate leaves; leaf blades ± ovate, notched at base but not as deep as attachment of petiole, the upper surface often with a velvety sheen; spathe convolute, constricted between the inflated tube and expanded blade, yellow; spadix terminated by a short or long appendage. Cultivated as an ornamental and apparently spreading; Bell Co. (specimens collected in 1997 from a large population along Salado Creek including a flowering individual), also Turtle Creek and White Rock Lake (Dallas Co., R. O'Kennon, pers. obs.); also known from Travis Co. just s of nc TX; its naturalization along the San Marcos River and neighboring areas of sc TX was discussed by Arridge and Fonteyn (1981). Sep. Native of tropical Asia. Widely grown in the tropics for the edible (when appropriately cooked) tuberous corm and young leaves; in Hawaii it is eaten in the form of "poi." However, all parts of the plant except the corms (when properly prepared) contain calcium oxalate crystals and other toxins causing severe burning and swelling of the mouth and throat and even death (Schmutz & Hamilton 1979). ☠

PISTIA WATER-LETTUCE, SHELLFLOWER, WATER-BONNET

A monotypic genus; the leaf hairs produce a water-repellent surface; the leaves are nearly horizontal during the day, but move to a more vertical position at night; the genus seems to evolutionarily link the Araceae to the Lemnaceae. (Greek: *pistos*, water or liquid, referring to its aquatic habitat)

Pistia stratiotes L., (soldier), WATER-LETTUCE, SHELLFLOWER, WATER-BONNET. Monoecious free-floating herb; roots long, feathery, hanging; leaves clustered, entire, gray-green, velvety-hairy, strongly ribbed lengthwise, cuneate to obovate-cuneate, to ca. 3–15 cm long, truncate to emarginate at apex; spathes axillary, inconspicuous, ca. 15 mm long; staminate flowers above, pistillate below, solitary; perianth absent. Streams, lakes, and ponds; apparently spreading from cultivation in creeks in Tarrant Co. (Fort Worth Botanic Garden); also cited for vegetational area 4 (Fig. 2) by Hatch et al. (1990); mainly s TX and Edwards Plateau. Spring. This species can become a serious pest in some regions by occupying areas of open water and in Texas is considered a "harmful or potentially harmful exotic plant"; it is illegal to release, import, sell, purchase, propagate, or possess this species in Texas (Harvey 1998). Widespread in the tropics and subtropics. Despite being called WATER-LETTUCE, this species contains oxalates and possibly other toxins; intense irritation of the mouth, throat, and upper digestive tracts has been reported from eating even small amounts (Morton 1982). ☠

XANTHOSOMA YAUTIA

A tropical American genus of 57 species of herbs with milky sap and sagittate or hastate leaf blades; a number are cultivated as ornamentals and for food. ☠ Nearly all parts of a num-

ber of species are injurious due to the presence of calcium oxalate crystals and possibly a toxic alkaloid; ingestion can cause burning, swelling, and blistering of the mouth, throat, and digestive tract; fatalities have been reported; even tasting small amounts can cause serious reactions (Morton 1982; Lampe & McCann 1985). (Greek: *xanthos*, yellow, and *soma*, body, alluding to the yellow inner tissues of some species)
REFERENCE: Lemke & Schneider 1988.

Xanthosoma sagittifolium (L.) Schott, (arrow-leaved), ELEPHANT'S EAR, TANNIA, TANIA. Large perennial herb from a tuberous corm; aerial stems developing with age (in tropical areas); similar to *Colocasia* but leaves not peltate; leaf blades to 90 cm long from attachment to petiole to tip and ± as wide, with 8–9 pairs of main lateral veins; petioles longer than blades, to ca. 2 m long; spathe convolute, constricted between the inflated tube and expanded blade. Apparently spreading from cultivation; those in water surviving the winter; Bell Co. (Fort Hood-Sanchez 1997), also cited for vegetational read 4 (Fig. 2) by Hatch et al. (1990); also Edwards Plateau. Oct. Native to West Indies. Grown there as an ornamental and in the tropics for its edible (after cooking) corms and specially prepared young leaves; however, the sap is reported to be an irritant and as with other *Xanthosoma* species, the plant should be considered toxic (Kingsbury 1964; Morton 1982). ☠

ARECACEAE (PALMAE) PALM FAMILY

A large (2,650 species in 203 genera), mainly tropical, often conspicuous family of usually unbranched, evergreen trees, shrubs, or lianas containing a number of economically important plants including *Calamus* species (RATTAN), *Cocos nucifera* L. (COCONUT PALM), *Copernicia prunifera* (Mill.) H.E. Moore (CARNAUBA WAX PALM), *Elaeis guineensis* Jacq. (OIL PALM), *Phoenix dactylifera* L. (DATE PALM), *Phytelephas macrocarpa* Ruíz & Pav. (VEGETABLE-IVORY OR TAGUA), *Roystonea regia* (Kunth) O.R. Cook (ROYAL PALM), and *Washingtonia filifera* (L. Linden) H. Wendl. (CALIFORNIA FAN PALM). The world's largest seeds (to 50 cm long) are produced by *Lodoicea maldivica* (J.F. Gmel.) Pers. (SEYCHELLES PALM, DOUBLE-COCONUT, COCO-DE-MER). Because of the variety of species useful for food, fiber, and shelter, the PALM family is considered to be the third most important family to humans, following only the GRASS and BEAN families. The family is unusual in having an unbranched trunk with a single apical bud and a terminal rosette of leaves. The woody stems are completely different in structure and manner of growth (no permanent cambium or enlargement in diam.) from those of dicots. King's-cabbage or heart-of-palm is a food obtained from the apical bud (= apical meristem). Family name from *Areca*, a genus of 60 species native from Indomalesia to tropical Australia and the Solomon Islands. (Name derived from a vernacular name used in Malabar) (subclass Arecidae)
FAMILY RECOGNITION IN THE FIELD: the only nc TX species is a palm usually ca. 1 m tall and has distinctive, large, evergreen, *fan-like* leaf blades divided into numerous segments; flowers very small, numerous, in a panicle.
REFERENCES: Bailey 1961; Moore & Uhl 1982; Dahlgren et al. 1985; Tomlinson 1990; Lockett 1991; Henderson et al. 1995; Zona 1997.

SABAL

A genus of 16 species of dwarf to stout, unarmed palms ranging from the se United States to South America; some are used as a source of thatch. (Possibly derived from an American vernacular name)
REFERENCES: Small 1922; Zona 1990.

Sabal minor (Jacq.) Pers., (smaller), DWARF PALMETTO, BUSH PALMETTO, DWARF PALM, BLUE PALM, PALM, BLUE STEM, SWAMP PALM. Plant about 1 m tall (rarely taller), usually acaulescent with a pithy crown or with a short trunk; leaves long-petioled, with stiff, evergreen, glabrous, fan-like blades longitudinally pleated toward base, divided over half way (sometimes becoming split to base) into many narrowly lanceolate segments; inflorescence a long panicle, its stalk with leafy bracts consisting of closed, tubular basal sheath and grass-like short blade; flowers many, subsessile, very small, with 3-lobed calyx and 3 greenish or brownish petals as long as the calyx; stamens 6; pistil 1; ovary superior; fruits black, 6–13 mm in diam. Stream bottoms; s and se TX n along larger rivers to Dallas and Kaufman cos. in nc TX and Van Zandt and Wood cos. in e TX, also spreading from cultivation in Tarrant Co. (Fort Worth Botanic Garden, R. O'Kennon, pers. obs.). Jun.

BROMELIACEAE PINEAPPLE OR BROMELIAD FAMILY

A relatively large family (2,400 species in 59 genera) of epiphytic or terrestrial xerophytic herbs; mainly limited to tropical to warm temperate areas of the New World. The most important species economically is *Ananas comosus* (L.) Merr. (PINEAPPLE); many other species are cultivated as ornamentals. A number of the epiphytic species are tank epiphytes; these have tightly clasping leaf bases and are thus able to act as well-like "tanks" to store water which is absorbed by specialized roots or hairs. These epiphytes form a "hanging ecosystem" in New World tropical forests and provide habitats for an assortment of animals including frogs. In the field, many species can be recognized by their tough, thickened, usually spiny leaves, often conspicuous inflorescences with colored bracts, and the typically epiphytic or xerophytic habit. Family name from *Bromelia*, a tropical American genus of 48 species. (Named for Olaf Bromel, 1639–1705, Swedish botanist) (subclass Zingiberidae)

FAMILY RECOGNITION IN THE FIELD: gray or gray-green, xerophytic *epiphytes*; leaves with distinctive peltate scales.

REFERENCES: Smith 1938; 1961; Smith & Wood 1975; Dahlgren et al. 1985.

TILLANDSIA BALL-MOSS

Xerophytic epiphytes largely covered with gray scales or trichomes, the plants gray when dry, gray-green when wet; leaves entire, with peltate scales that collect water and nutrients; flowers perfect; stamens 6; ovary superior; fruit a septicidal capsule; seeds with a basal plumose appendage.

A tropical American genus of 380 species of epiphytes typically with leaves in a rosette; some are cultivated as ornamentals. Because of their grayish color, seedlings are sometimes confused with lichens (E. McWilliams, pers. comm.) (Named for Elias Tillands, 1640–1693, professor at Abo, who, as a student crossing directly from Stockholm, was so seasick that he returned to Stockholm by walking more than 1,000 miles around the head of the Gulf of Bothnia and hence assumed his surname (by land); the genus was erroneously supposed by Linnaeus to dislike water)

REFERENCES: Birge 1911; McWilliams 1992, 1995.

1. Plant typically a dense ball-like clump; stems short, 10 cm or less long, completely concealed by the overlapping leaf sheaths; flowers at the end of a scape (= flowering stalk) conspicuously exerted above the leaves ____ **T. recurvata**
1. Plant of slender, wiry, usually curled, elongate, hanging strands, not ball-like; stems elongate, often several meters long, visible between the leaves; flowers sessile, among the leaves ____ **T. usneoides**

Tillandsia recurvata (L.) L., (recurved), BALL-MOSS, BUNCH-MOSS, GALLITOS. Plant rarely > 15 cm tall; roots present; leaves arising close together, curving out, elongate (3–17 cm long), very narrow (to ca. 2 mm wide), covered with scales; scape slender, with 1–2(–5) flowers; flowers bluish. In nc TX, usually epiphytic on LIVE OAKS or further s and w, on rocks, tombstones, and utility wires; Bell, Burnet, Milam, and Williamson cos. (McWilliams 1992); s and sc TX n to s part of nc TX, apparently spreading e and ne (several populations are now known from LA—E. McWilliams, pers. comm.); an introduced population on a single tree was observed in Dallas (McWilliams 1992). Flowering throughout the year. This species has expanded its geographic range in TX over the past 80 years, apparently in response to changing climate; such climate-sensitive species may be able to serve as early indicatiors of projected regional climatic change (McWilliams 1995).

Tillandsia usneoides (L.) L., (like *Usnea*, a lichen that hangs from trees) SPANISH-MOSS, OLD MAN'S-BEARD, LONG-MOSS, BLACK-MOSS, PASTLE, FLORIDA-MOSS. Hanging strands to 3–4(–8) m long; roots absent; leaves thread-like, 2–6 cm long; inflorescence of a single flower; flowers greenish or greenish yellow. Epiphytic or on wires or other supports; native to se, e and c TX; known just to the s of nc TX in Travis Co. (Smith 1961); included because of the possibility of occurrence on extreme s or e margins of nc TX. Feb–Jun. In moist areas in se and e Texas this species hangs in large extremely conspicuous festoons from trees. It occurs from the s U.S. s to Argentina, an incredible distribution stretching across 5,000 miles of latitude. Rarely flowering; distributed by the wind and by birds using it as nest material; the dried plants are used as packing material and in upholstery (Mabberley 1987).

BURMANNIACEAE BURMANNIA FAMILY

A small (160 species in 16 genera) mainly tropical family that occurs n to Japan and the e U.S. and s to New Zealand; small forest herbs including many colorless saprophytic species without chlorophyll. The family is closely related to the Orchidaceae. (subclass Liliidae)
FAMILY RECOGNITION IN THE FIELD: the only nc TX species is a very small herb with a thread-like unbranched stem, scale-like leaves, and a few small terminal flowers.
REFERENCES: Wood 1983a; Dahlgren et al. 1985.

BURMANNIA

A genus of 60 species of tropical and subtropical areas of the world. (Named for Johannes Burmann, 1706–1779, a Dutch botanist)

Burmannia capitata (J.F. Gmel.) Mart., (headed), CAP BURMANNIA. Very small herb with a thread-like, usually unbranched stem 5–20 cm tall; leaves alternate, tiny and scale-like, to 5 mm long, few, widely spaced along the stem; flowers 1–several in a cluster at tip of stem, small, ca. 5 mm long, 0.5–1.5 mm wide, greenish white or cream, sometimes tinged with blue, the six perianth segments fused into a tube, tipped by minute to obsolescent lobes 0.5–1 mm long; stamens 3, almost sessile, attached to the perianth tube; ovary inferior; capsules 2–5 mm long, 3-angled, but not winged (winged in *B. biflora* L., known in TX only from deep e part). Moist woods, bogs; Milam Co. near e margin of nc TX; mainly se and e TX. Aug–Nov.

COMMELINACEAE SPIDERWORT FAMILY

Annual or perennial herbs; leaves alternate or basal, with closed, tubular basal sheaths and broad or narrow, grass-like, entire blades; flowers in small cymose clusters subtended by reduced upper leaves or leafy bracts (if the bracts enclose the flowers they are referred to as spathes); sepals 3; petals 3, thin and delicate, usually lasting half a day or less (flowers open during morning, later in cloudy weather); stamens usually (5–)6, all fertile or some staminodial or lacking; filaments often long-hairy; pistil 1; ovary superior; fruit a capsule.

A medium-sized (650 species in 41 genera—R. Faden, pers. comm.) family of tropical, subtropical, and warm temperate herbs; a number are widely used ornamentals including species

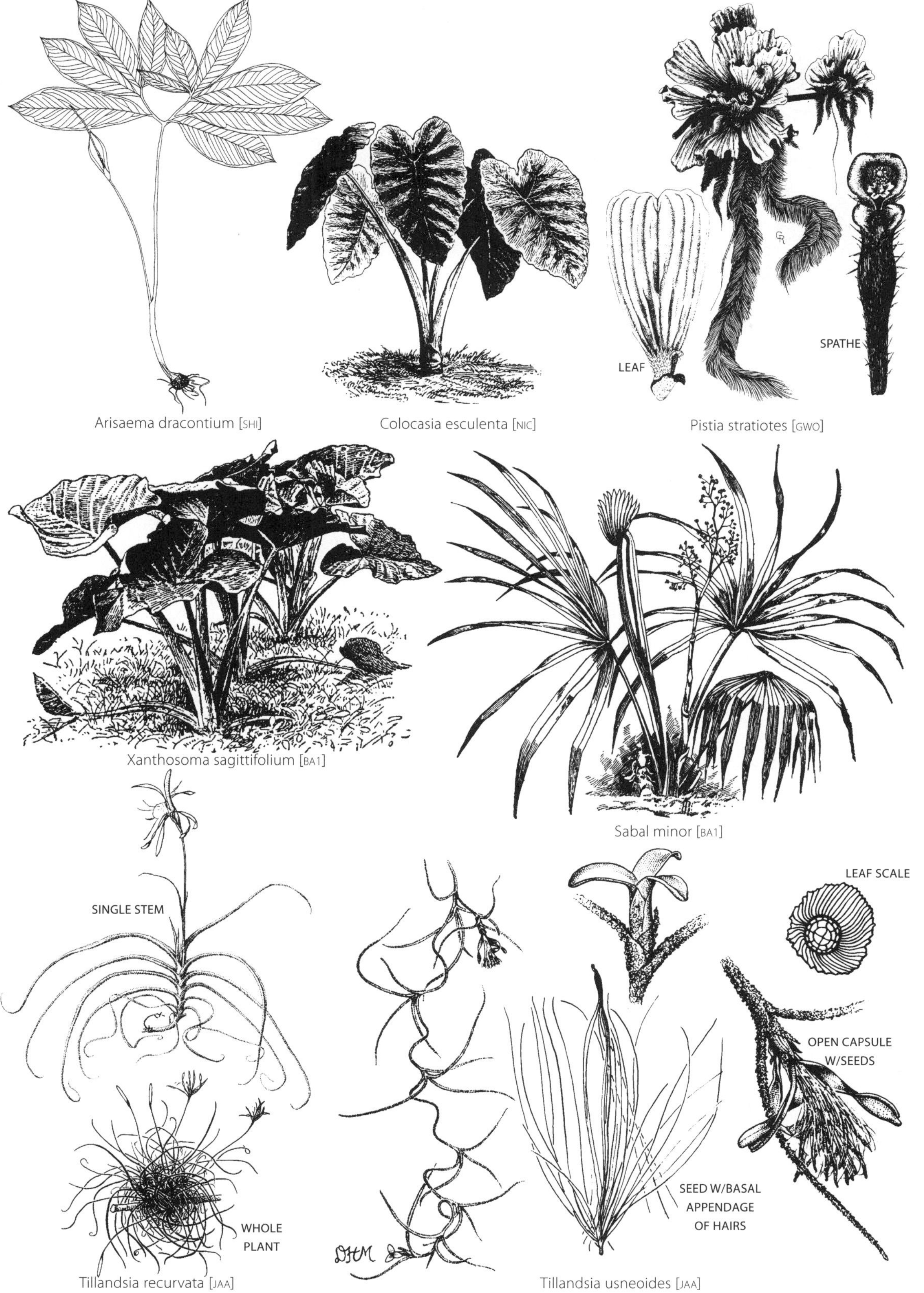

Arisaema dracontium [SHI]

Colocasia esculenta [NIC]

Pistia stratiotes [GWO]

Xanthosoma sagittifolium [BA1]

Sabal minor [BA1]

Tillandsia recurvata [JAA]

Tillandsia usneoides [JAA]

of *Tradescantia* (SPIDERWORT), *Rhoeo* (BOATFLOWER, OYSTERPLANT), and *Zebrina* (WANDERING-JEW); the latter 2 genera are now typically lumped into *Tradescantia* (Hunt 1986). (subclass Commelinidae)

FAMILY RECOGNITION IN THE FIELD: herbs with succulent mucilaginous stems with knotted nodes and usually alternate, ± basally sheathing leaves with strongly parallel veins; petals 3 (1 sometimes smaller), delicate, often blue.

REFERENCES: Poole & Hunt 1980; Faden 1985, 1992; Hunt 1986; Tucker 1989; Faden & Hunt 1991.

1. Inflorescences each subtended or enclosed by a single leafy bract (the bract sometimes conspicuously folded); the 3 petals not all alike, EITHER 2 larger and bluish and 1 one much smaller and white (at least paler than the others) OR one petal slightly smaller than the others; fertile stamens 3 (2–3 staminoidia also present) OR 6 and unequal.
   2. Bract (spathe) folded, the two sides ± parallel, 1.0–3.5 cm long, conspicuously different in shape from stem leaves; fertile stamens 3 (2–3 staminoidia also present); filaments glabrous; petals (upper 2) clawed; foliage not glaucous; widespread in nc TX ____ **Commelina**
   2. Bract flat, not folded, 4–8 cm long, similar to upper stem leaves but shorter and wider; fertile stamens 6; filaments (5 of the 6) bearded; petals not clawed; foliage glaucous; in nc TX restricted to the Lampasas Cut Plain ____ **Tinantia**
1. Inflorescences each subtended by 2(–3) leaves with reduced sheaths but conspicuous long blades; the 3 petals all ± similar; fertile stamens 6, equal. ____ **Tradescantia**

COMMELINA WIDOW'S-TEARS, DAYFLOWER

Annuals or perennials; flowers bilaterally symmetrical; inflorescences enclosed in spathes; petals greatly or slightly unequal, blue or blue and white, the upper 2 larger and clawed; stamens unequal, 3 long (1 of these curved in), 3 slightly shorter, sterile (actually staminoidea).

A genus of ca. 170 species of tropical and warm areas of the world; bees pierce the juicy lobes of the upper 3 sterile anthers to obtain nectar. The common name DAYFLOWER comes from the extremely delicate flowers which open in the morning but are gone by noon on sunny days (Kirkpatrick 1992). (Named for the early Dutch botanists, Commelin, on account of the 2 showy petals and 1 less conspicuous petal. Linnaeus was referring to the three botanists of that name, two of whom, Johan, 1629–1692, and Kaspar, 1667–1731, were conspicuous botanists, while the third "... died before accomplishing anything in Botany")

REFERENCES: Pennell 1916, 1937, 1938; Brashier 1966; Faden 1989, 1993.

1. Leaf-like spathe enclosing flowers open not only on the top margin, but also open down the back margin to where it attaches to its stalk; annuals with fibrous roots.
   2. Two petals blue, 3rd (anterior) petal much smaller and white (or at least paler than the others); leaf blades 15–40 mm wide; leaf sheaths 10–20 mm long; bottom edge of spathes ± straight; stalk of spathes 1–7 cm long; spathes usually pale with contrasting dark green veins; capsules bilocular ____ **C. communis**
   2. All 3 petals blue, 3rd slightly smaller; leaf blades 9–15(–22) mm wide; leaf sheaths 5–10 mm long; bottom edge of spathes usually curved down at tip; stalk of spathes 1–2 cm long; spathes without contrasting veins; capsules trilocular ____ **C. diffusa**
1. Leaf-like spathe enclosing flowers open on the top margin, but with its edges fused together along the back margin; perennials with thickened roots.
   3. Two petals blue, the 3rd much smaller and white; stems usually not erect; spathes scattered along the stem (opposite leaves) and near the stem apex, usually 1 per node; leaf sheath margins inconspicuously ciliate with whitish hairs; leaves with auricles at summits of sheaths; leaf blades 14–35 mm wide ____ **C. erecta**
   3. All 3 petals blue, the lower one slightly smaller; stems usually strictly erect; spathes usually

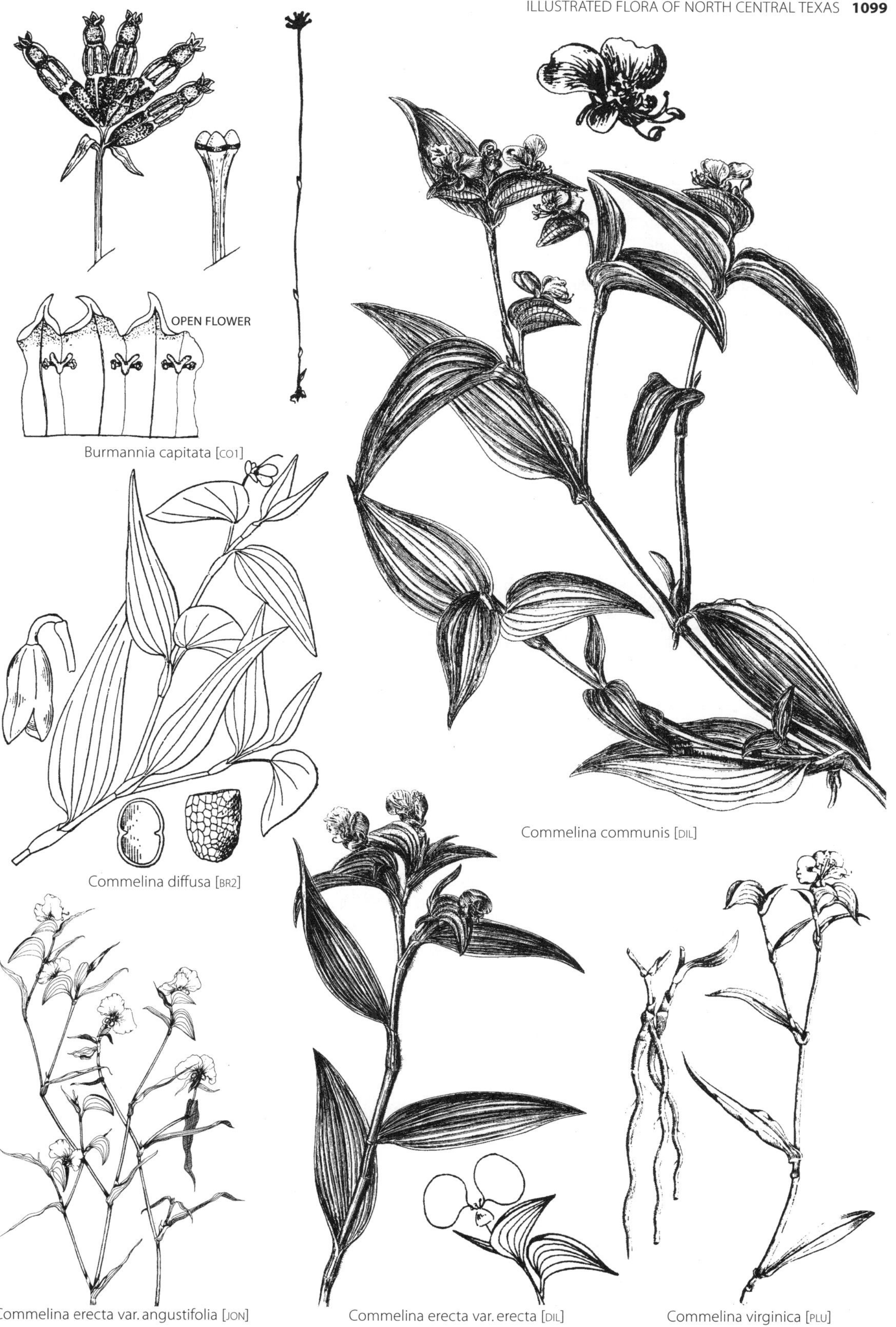

Burmannia capitata [CO1]

Commelina communis [DIL]

Commelina diffusa [BR2]

Commelina erecta var. angustifolia [JON]

Commelina erecta var. erecta [DIL]

Commelina virginica [PLU]

in clusters near the stem apex; leaf sheath margins conspicuously ciliate with reddish hairs; leaves without auricles; leaf blades 20–65 mm wide ____ **C. virginica**

Commelina communis L., (common), COMMON DAYFLOWER. Stems erect, later decumbent, to 4 mm in diam., to 50 cm tall. Stream banks and low thickets, can be a garden weed; Dallas Co.; e TX, also Edwards Plateau. May–Oct. Native of e Asia.

Commelina diffusa Burm. f., (diffuse, spreading), SPREADING DAYFLOWER, CREEPING DAYFLOWER. Stems erect initially, later decumbent, usually not more than 1.5 mm in diam. Low woods; Fannin (Talbot property) and Rockwall cos.; mainly se and e TX. Apr–Nov.

Commelina erecta L., (erect), ERECT DAYFLOWER, HIERBA DE POLLO. Stems erect to decumbent, 10–70(–100+) cm long. Native in various soils, often a weed. May–Jun and Sep–Oct, occasionally Jul–Aug. While often distinguishable, there is considerable variation (Brashier 1966) and overlap between the following varieties; Faden (1992) indicated the three freely intergrade and are of questionable significance.

1. Blades of middle and upper leaves broadly oblong-lanceolate, less than 5 times as long as wide, 1.4–3.2 cm wide ____ var. **erecta**
1. Blades of middle and upper leaves narrowly oblong-lanceolate or linear-lanceolate, more than 5 times as long as wide, 0.5–1.2(–2.0) cm wide.
   2. Spathes (1.3–)1.5–2.0(–2.3) cm long ____ var. **angustifolia**
   2. Spathes (2.2)2.5–2.8(–3.3) cm long ____ var. **deamiana**

var. **angustifolia** (Michx.) Fernald, (narrow-leaved), NARROW-LEAF DAYFLOWER. In habitats as diverse as sandy woods and rocky limestone slopes; throughout TX.

var. **deamiana** Fernald, (for Charles C. Deam, 1865–1953, American botanist). Usually in sandy soils; Hill Co.; e TX w to nc TX.

var. **erecta**. Thickets, stream banks or a weedy invader elsewhere; se and e TX w to Rolling Plains and Edwards Plateau. /85

Commelina virginica L., (of Virginia), VIRGINIA DAYFLOWER. Plant spreading by elongate rhizomes; stems coarse, 3–6 mm in diam. at base, erect (rarely decumbent), to 90 cm tall; leaf blades scabrous when rubbed toward base. Low woods; Hopkins and Lamar cos.; se and e TX w to e part of nc TX. May–Oct.

Commelina caroliniana Walter, (of Carolina), apparently native to India but scattered in the se U.S., is known from Travis Co. just to the s of nc TX (Faden 1993); its presence in nc TX would not be surprising. Faden (1993) separated it from the similar *C. diffusa* as follows:

1. Spathes not at all to slightly falcate; upper cyme usually vestigial (rarely well-developed and 1-flowered); capsules (5–)6–8 mm long; ventral locule seeds 2.4–4.3(–4.6) mm long, smooth to faintly alveolate ____ *C. caroliniana*
1. Spathes usually distinctly falcate; upper cyme in larger spathes usually well-developed and 1–several-flowered; capsules 4–6.3 mm long; ventral locule seeds 2–2.8(–3.2) mm long, deeply reticulate ____ *C. diffusa*

TINANTIA FALSE DAYFLOWER, WIDOW'S-TEARS

A genus of ca. 13 species (R. Faden, pers. comm.) from TX to the American tropics; some cultivated as ornamentals; 2 of the species were previously treated as *Commelinantia*. (Named for Francois A. Tinant, 1908–1858, a forester in Luxembourg—R. Faden, pers. comm.)
REFERENCES: Tharp 1922, 1956; Woodson 1942; Simpson et al. 1986.

Tinantia anomala (Torr.) C.B. Clarke, (anomalous), FALSE DAYFLOWER, WIDOW'S-TEARS. Tufted glabrous annual with erect or spreading stems 20–80 cm long, becoming freely branched, the branches emerging through the back of the leaf sheaths just above the nodes; basal leaves tapered to a long petiole; upper stem leaves sessile or short petioled, the blades ± lanceolate, often somewhat cordate and clasping basally; flowers in elongate cymes; petals not clawed, the 2 upper larger ones 15–18 mm long, lavender-blue, the much smaller lower 1 white; stamens 6, all fertile, very polymorphic; upper 3 stamens, with filaments conspicuously bearded with yellow-tipped hairs, with small anthers (2 lateral upper stamens upright, middle upper stamen less upright); lower 3 stamens curved downward, with larger anthers; filaments of lateral 2 lower stamens with purple hairs; filament of middle lower stamen glabrous; capsules 6–8 mm long. Limestone gravel or rocky crevices, often in some shade; Edwards Plateau ne to McLennan and s Bosque cos. (Mahler 1988), also Fort Hood (Bell or Coryell cos.—Sanchez 1997); endemic to TX or nearly so (1 record from Mexico—Faden 1992). Apr–Jun, rarely later. [*Commelinantia anomala* (Torr.) Tharp] Simpson et al. (1986) gave detailed information on the reproductive biology of this species; it is apparently largely autogamous. /107

TRADESCANTIA SPIDERWORT

Perennial subsucculent herbs; leaf blades linear to oblong-elliptic; petals equal, not clawed, blue to rose, magenta, purple, or white; stamens 6, all fertile and equal; filaments long-pilose with colored hairs.

An American genus of 70 species; a number are cultivated as ornamentals including species previously treated in *Rhoeo* (BOATFLOWER, OYSTERPLANT), and *Zebrina* (WANDERING-JEW). The filaments of most species have long hairs and were used by Robert Brown in 1828 to observe and describe protoplasmic streaming. Hybridization and introgression are well known in *Tradescantia* and complicate the taxonomy of the genus. The common name is possibly derived from the long slender leaves which clasp the stem and dangle like spider legs or from the mucilaginous sap stringing out to resemble a spider's web, and wort, from Anglo-Saxon: *wyrt*, a plant or herb (Tveten & Tveten 1993). (Named for John Tradescant, 1608–1662, gardener to King Charles I of England)

REFERENCES: Bush 1904; Tharp 1932; Anderson & Woodson 1935; Anderson & Sax 1936; Anderson 1954; MacRoberts 1977, 1980; Hunt 1980.

1. Leaf blades broader than their opened flattened sheaths, at least the upper ones ____ **T. edwardsiana**
1. Leaf blades narrower than sheaths or ca. as broad as their opened flattened sheaths.
   2. Upper internodes glabrous.
      3. Sepals glabrous or with only a small tuft of hairs at tips (this is the only nc TX species like this; all others have pubescent sepals) ____ **T. ohiensis**
      3. Sepals sparsely to densely pubescent on back (hairs either eglandular or glandular).
         4. Sepals either completely eglandular-pubescent or with eglandular and glandular pubescence intermixed; bracts pilose to glabrous ____ **T. hirsutiflora**
         4. Sepals usually with glandular pubescence only (sometimes with a tuft of eglandular hairs at tip); bracts glabrous ____ **T. occidentalis**
   2. Upper internodes variously pubescent (puberulent, pilose, or with matted hairs).
      5. Hairs on internodes minute or long and ± straight, wide-spreading.
         6. Sepals with only eglandular pubescence (use lens).
            7. Leaf sheaths glabrous or minutely pubescent; bracts conspicuously saccate, the blade reduced, densely and minutely velutinous ____ **T. gigantea**
            7. Leaf sheaths long pilose or with long matted or tangled hairs; bracts not conspicuously saccate, the blades well-developed, glabrous to ± pilose.

8. Stems 12–49 cm tall, with 2–5 nodes; leaf blades usually edged with purple; sepals relatively firm, dull-green to suffused or edged with rose _______ **T. hirsutiflora**
8. Stems 2–7 cm tall in flower, up to 30 cm in fruit, usually with 1–2 nodes; leaf blades usually edged with pink or purple; sepals somewhat petal-like, usually strikingly purple or rose-colored, occasionally pale green _______ **T. tharpii**

6. Sepals with glandular pubescence and often also eglandular pubescence (use lens).
   9. Leaf sheaths long pilose, at least toward their summits _______ **T. hirsutiflora**
   9. Leaf sheaths glabrous or short pubescent.
      10. Petals ovate, bright blue, occasionally pink; plants 10–30(–45) cm tall; blades of leaves and bracts up to 20 cm long; bracts not conspicuously saccate _______ **T. humilis**
      10. Petals obovate, magenta-pink to blue; plants 20–75 cm tall; blades of leaves and bracts 10–35 cm long; bracts conspicuously saccate _______ **T. gigantea**

5. Hairs on internodes long and matted or tangled, spider-web-like in appearance, ± appressed.
   11. Stems erect or ascending, simple or infrequently branched, 30–105 cm tall; roots conspicuously felty with red-brown hairs easily visible to the naked eye _______ **T. reverchonii**
   11. Stems spreading and ± diffuse, much branched, 10–35 cm long; roots not conspicuously felty to the naked eye _______ **T. subacaulis**

Tradescantia edwardsiana Tharp, (of Edwards Plateau), PLATEAU SPIDERWORT. Stems erect or ascending, puberulent to glabrate, 25–70 cm tall; leaf blades elliptic-lanceolate, 7–30 cm long, 15–65 mm wide, gradually constricted to the sheath, acuminate, minutely puberulent to essentially glabrate; leaf sheaths 7–20 mm wide, nearly glabrous except for the ciliate margins; pedicels minutely and densely puberulent, the hairs sometimes glandular when flowers are in bud stage; sepals glandular-puberulent, sometimes also with eglandular pubescence; petals white to pale blue or lavender, rarely bright pink, broadly ovate. Rich woods, moist alluvial terraces, and ravines; Bell, Collin, Coryell, Dallas, Fannin, and Palo Pinto cos., also Brown and Hamilton cos. (Stanford 1971); Anderson and Woodson (1935) also mapped a collection from just s of the Red River in what is apparently Cooke Co.; also Travis Co. just s of nc TX (Anderson & Woodson 1935); endemic to Edwards Plateau and nc TX. Feb–May.

Tradescantia gigantea Rose, (gigantic), GIANT SPIDERWORT. Stems erect or ascending, branching infrequently, glabrous below, minutely pubescent above; leaf blades glabrous or the upper minutely pubescent; sepals with only eglandular pubescence; petals magenta-pink to blue. Limestone soils; e Edwards Plateau n to Bell Co. on s margin of nc TX, also Burnet Co. (Anderson & Woodson 1935); endemic to TX. Mar–May.

Tradescantia hirsutiflora Bush, (hairy-flowered), HAIRY-FLOWER SPIDERWORT. Stems ± spreading pilose to hirsute or glabrate; sepals with only eglandular pubescence or with both glandular and eglandular; petals bright blue to purplish, rarely pink. Sandy soils; Grayson and Lamar cos. in Red River drainage, also Milam and Henderson cos. on e margin of nc TX; mainly se and e TX, also Edwards Plateau. Mar–Jun.

Tradescantia humilis Rose, (low-growing, dwarf), TEXAS SPIDERWORT. Plant 10–30(–45) cm tall, minutely pubescent or slightly pilose, or largely glabrous; petals bright blue, occasionally pink. Sandy or rocky ground; Bell, Brown, and Williamson cos., also Dallas and Lamar cos. (Anderson & Woodson 1935); mainly sc to nc TX; endemic to TX. Mar–Jun.

Tradescantia occidentalis (Britton) Smyth, (western), PRAIRIE SPIDERWORT. Similar to *T. ohiensis*; 10–90 cm tall; leaf blades averaging narrower (to 20 mm wide); petals bright blue to rose or magenta. Sandy, gravelly, or less often clayey soils, prairies; widespread in w half of nc TX, also Navarro Co. on e margin of nc TX, Anderson and Woodson (1935) also reported Dallas, Kaufman, and McLennan cos. in e part of nc TX; widespread in TX, mainly Blackland Prairie s and w to w TX. Apr–Jun.

Tinantia anomala [HEA, TOR]

Tradescantia edwardsiana [RHO]

Tradescantia hirsutiflora [HEA]

Tradescantia gigantea [HEA]

Tradescantia humilis [HEA]

Tradescantia ohiensis Raf., (of Ohio), OHIO SPIDERWORT. Plant 20–75 cm tall, glabrous and glaucous; leaf blades 3–32(–45) mm wide; petals blue to rose, magenta, or rarely white. Sandy or clayey soils, prairies, meadows, thickets and roadsides; e 1/2 of TX. [*T. canaliculata* Raf.] This is the most common spiderwort in nc TX. According to Faden (1992), it is the most common and widespread species in the U.S.; it hybridizes with a number of other species. Mar–Jun. The roots are reported to contain a poisonous saponin (Ajilvsgi 1984). ☠ /108

Tradescantia reverchonii Bush, (for Julien Reverchon, 1837–1905, a French-American immigrant to Dallas and important botanical collector of early TX), REVERCHON'S SPIDERWORT, GRASS-VIOLET. Stems erect or ascending, simple or infrequently branched, 30–105 cm tall, usually rather densely pilose with somewhat matted or tangled hairs; sepals with eglandular or both eglandular and glandular hairs; petals bright blue (rarely rose or white). Sandy soils, open woods; Falls and Williamson cos., also collected at Dallas by Reverchon (Mahler 1988), but not found there since, also Henderson Co. (Anderson & Woodson 1935); mainly e TX. Mar–Jul.

Tradescantia subacaulis Bush, (almost without a stem), STEMLESS SPIDERWORT. Stems spreading, 10–35 cm long, densely matted-pilose throughout; roots tuberous-thickened; sepals with glandular and eglandular hairs; petals usually bright blue, occasionally pink. Loose sandy soils, open woods, or open ground; Limestone Co., also Navarro Co. (Anderson & Woodson 1935); s TX n to se part of nc TX; endemic to TX. Late Mar–Jun.

Tradescantia tharpii E.S. Anderson & Woodson, (for Benjamin Carroll Tharp, 1885–1964, botanist at Univ. of TX), THARP'S SPIDERWORT. Plant long-pilose throughout, usually densely so; stems rarely branching, often initially acaulescent; sepals with only eglandular pubscence; petals deep rose or purple, sometimes blue. Sandy clay or rarely silty clay soils, rocky prairies, open woods, or open ground; Collin, Denton, Dallas, Tarrant, and Wise cos., also Erath Co. (Anderson & Woodson 1935); Blackland Prairie w to e Rolling Plains and Edwards Plateau. Late Mar–Apr.

CYPERACEAE SEDGE FAMILY

Annual or perennial herbs; culms (= stems) triangular (most commonly), flat, round, square, or multi-angular, with smooth nodes and usually pithy or spongy internodes; leaves with tubular basal sheath (often reduced or absent from upper leaves) closed except at summit (but apt to become split by growth of culm), generally without a scaly ring (ligule) at junction of sheath and blade on upper (inner) side, and with terminal, usually elongate blade (leaves all reduced to inconspicuous sheaths in *Eleocharis* and some *Cyperus* and *Scirpus*); inflorescences various (umbellate in *Cyperus* and *Fimbristylis* and less distinctly so in some other genera); flowers (often referred to as florets) perfect or unisexual, each subtended by a single (rarely 2) scale-like bract (often referred to as floral scales or in this treatment scales of spikelets or simply scales), without perianth or perianth reduced to bristles or small perianth scales, solitary or in spikelets (these often with added empty scale-like bracts at base); stamens 1–3, with anther attached by one end; pistil 1; fruit an achene.

A large, cosmopolitan (especially temperate), taxonomically difficult family of herbs with 4,500–5,000 species in 100–105 genera (Goetghebeur 1987). Because of the often similar vegetative parts and reduced reproductive structures, technical characters requiring at least a hand lens frequently have to be used to distinguish species. Cyperaceae species superficially resemble grasses or rushes; the family is of limited economic importance as wildlife food, for woodland grazing, or for erosion control; in n temperate parts of the world they sometimes replace grasses as forage; in TX in the Hill Country and w part of the state, *Carex emoryi* L. becomes important for livestock during summer months; also some are problematic weeds (S.D.

Tradescantia occidentalis [BB2]

Tradescantia ohiensis [BB2]

© Linny '97

Tradescantia reverchonii [HEA]

© Linny '97

Tradescantia subacaulis [HEA]

Tradescantia tharpii [AND]

Jones, pers. comm.). The monotypic North American genus *Dulichium* (*D. arundinaceum* (L.) Britton (genus: derivation not given by original author; sp.: reed-like)—THREE-WAY SEDGE) occurs in e TX just to the e of nc TX. It resembles *Cyperus* in having 2-ranked spikelets, but differs in having 6–9 perianth bristles subtending the achene (none in *Cyperus*) and the inflorescences being axillary (terminal in *Cyperus*). The Cyperaceae, with 140 species, is the fourth largest family in the nc TX flora (after Asteraceae, Poaceae, and Fabaceae). (subclass Commelinidae)

FAMILY RECOGNITION IN THE FIELD: grass-like or rush-like herbs with solid internodes, round or often 3-*angled* culms ("sedges have edges"), and often 3-ranked leaves without ligules; many (but not all) species grow in wet habitats; flowers small, inconspicuous, without perianth or perianth reduced to bristles or perianth small scales, enclosed by a sac-like perigynium OR subtended by 1 scale-like bract each, and arranged in very reduced spikes/spikelets. The ± similar Poaceae (GRASSES) have hollow internodes, round culms, 2-ranked leaves usually with a ligule, and flowers subtended by 2 scale-like bracts each (lemma and palea); the ± similar Juncaceae (RUSHES) have flowers with a small 6-parted perianth.

REFERENCES: Dahlgren et al. 1985; Goetghebeur 1987; Tucker 1987; Bruhl et al. 1992; Bruhl 1995; Rolfsmeier 1995.

1. Pistil (and later achene) enclosed in a pouch or sac (= perigynium) from which the style is exserted during flowering; flowers imperfect ____ **Carex**
1. Pistil (and later achene) not enclosed in a perigynium (but subtended by flat or concave scales); flowers perfect.
   2. Scales of spikelets (= scale-like bracts also referred to as floral scales or simply scales) 2-ranked (= in two distinct rows), the spikelets ± flattened or square.
      3. Culms (= stems) with leaf blades or leafy bracts at least below the inflorescence; spikelets 1–many-flowered; inflorescences often conspicuously branched, of few–many spikelets; achenes without a tubercle; extremely abundant in nc TX.
         4. Plants small (culms 21(–38) cm or less tall); inflorescence a single unlobed or 3-lobed (but unbranched) terminal structure; spikelets with only 2 scales (also 2 minute, brownish, basal scales) and only 1 bisexual fertile flower; each spikelet with only 1 achene; achenes lenticular (= lens-shaped); styles 2-branched ____ **Kyllinga**
         4. Plants variable in size, but often much larger; inflorescence often obviously branched; spikelets with 4–many scales and 3–many bisexual flowers; each spikelet with > 1 achene; achenes usually trigonous (= 3-sided) OR in a few species lenticular; styles usually 3 branched OR in a few species 2-branched ____ **Cyperus**
      3. Culms without any leaf blades or leafy bracts (leaves reduced to basal sheaths); spikelets 1–3-flowered; inflorescences without any branching whatsoever, of a solitary terminal spikelet; achenes with a tubercle (= persistent style base differing in appearance from body of achene); in nc TX only 1 rare species (*E. baldwinii*) has the scales of spikelets 2-ranked ____ **Eleocharis**
   2. Scales of spikelets spirally arranged, not 2-ranked, the spikelets round.
      5. Scales of spikelets white; spikelets in a terminal, leafy-bracted head, the bracts with white base; style branches 2 ____ **Rhynchospora**
      5. Scales of spikelets variously colored (may rarely be partly white); inflorescences various, but not with white-based bracts; style branches 2 or 3.
         6. Inflorescences with a single spikelet, this terminal at end of culm, the inflorescences thus without any branching whatsoever; culms without leaf blades or leafy bracts; achenes with a tubercle (= persistent style base at apex of achene differing in appearance from body of achene) ____ **Eleocharis**
         6. Inflorescences with 1–many spikelets, these terminal or lateral, the inflorescences thus unbranched OR often conspicuously branched; culms with leaf blades (at least basal ones) or leafy bracts; achenes with or without a tubercle.

7. Achenes with whitish or light grayish, bony or crustaceous, outer layer (pericarp), usually sitting on a distinct hardened ring-like or disk-like pad ____ **Scleria**
7. Achenes with neither whitish, light grayish, bony, nor crustaceous, outer layer (pericarp), nor sitting on a distinct, hardened, ring-like or disk-like pad.
   8. Achenes with prominent tubercle; spikelets 1–few-flowered, the florets with 0–1(–2) pistils (uppermost 1–2 florets without pistil); lower (1–)2–3 scales of spikelets sterile ____ **Rhyncospora**
   8. Achenes without tubercle (except in *Bulbostylis*, which has thread-like leaves); spikelets 1–many-flowered, all florets with pistil; only lowest scale of spiklet sterile (except in *Cladium* which usually has more than 1 sterile lower scale).
      9. Scales of spikelets with prominent bristle-tip longer than width of scale base; perianth of 3 stalked perianth scales, these sometimes alternating with 3 bristles ____ **Fuirena**
      9. Scales of spikelets with short bristle-tip or bristle-tip absent; perianth of 1 thin perianth scale, or of bristles, or periarth absent.
         10. "Spikelets" (actually small spikelet-like spikes or heads of spirally arranged single-flowered spikelets) 1–5(–8) mm long, sessile; inside each scale a single, thin, inconspicuous perianth-like bracteole; plants very small, to only 15 cm tall; scales with 2 or 3 prominent ribs (use strong lens) ____ **Lipocarpha**
         10. Spikelets 2.5–20 mm long, sessile or on distinct pedicels; inside each scale a perianth of bristles or perianth absent; plants small to very large; scales with only 1 prominent rib (the midrib).
            11. Spikelets with a single fertile (= achene-producing) floret subtended by 2–3 empty scales (these lacking achenes but can have stamens); leaf margins scaberulous (= only slightly roughened, almost smooth to the touch) to dangerously spinulose-serrulate (= saw-toothed); inflorescences with numerous spikelets, these in groups of 2–10 at the ends of short branches (= peduncles); perianth absent; rare, known in nc TX only from Dallas and Henderson cos. ____ **Cladium**
            11. Spikelets not as above, with 1–many fertile florets, these subtended by only 1 empty scale; leaf margins various; inflorescences not as above; perianth absent or of bristles; common and widespread in nc TX.
               12. Inflorescences usually either 1-sided or widely spreading or drooping; perianth of bristles (except in 1 small annual species); plants small to very large (to 5 m tall) (segregates of *Scirpus* in the broad sense).
                  13. Inflorescences with a single, erect modified leaf (involucral bract) appearing like a continuation of the culm, the inflorescences thus appearing lateral; culms with 1–3 leaves near base OR without blade-bearing leaves.
                     14. Spikelets 2–10 mm long; plants tufted annuals 2–22(–30) cm tall; culms wiry, < 1.5 mm thick near base; achenes minutely papillose (this can sometimes be obscured by a whitish wax-like layer); perianth bristles absent; inflorescences with 1–3 spikelets ____ **Isolepis**
                     14. Spikelets 5–20 mm long; plants rhizomatous perennials 30–500 cm tall OR tufted annuals or perennials 9–65 cm tall (in the rare *S. saximontanus* known in nc TX only from the Lampasas Cut Plain); culms coarse, 2–20 mm thick near base OR 0.5–1.5 mm thick (in *S. saximontanus*); achenes either smooth OR (in *S. saximontanus*) with prominent, transverse,

wavy ridges; perianth bristles present OR absent (in *S. saximontanus*); inflorescences with 1–150 spikelets ______ **Schoenoplectus**

13. Inflorescences with 2 or more well-developed leaf-like involucral bracts, the inflorescences thus appearing terminal; culms with well-developed leaves.
    15. Spikelets small, 3–10 mm long, 2–4 mm wide, very numerous; achenes < 1.5 mm long; culms mostly obscurely triangular ______ **Scirpus**
    15. Spikelets larger, 10–30(–40) mm long, 6–12 mm wide, few in number; achenes 3–4 mm long; culms sharply triangular ______ **Bolboschoenus**

12. Inflorescences neither distinctly 1-sided, widely spreading, nor drooping; perianth absent; plants 1 m or less tall, often much less.
    16. Leaf blades flat, some or all over 0.8 mm wide; inflorescences with 2 or more leafy bracts; achenes without tubercle.
        17. Spikelets 1-flowered (with only 2 scales per spikelet), sessile in head-like clusters closely subtended by bracts much longer than the heads; style base neither dilated nor fimbriate ______ **Kyllinga**
        17. Spikelets several–many-flowered (scales numerous), on distinct pedicels or if clustered then spikelets conspicuously many-scaled; style base dilated, fimbriate ______ **Fimbristylis**
    16. Leaf blades thread-like, 0.1–0.6 mm wide; inflorescences with 1 prominent bract; achenes with tubercle ______ **Bulbostylis**

BOLBOSCHOENUS BULRUSH

A genus of 6–13 species (Smith 1997a) known from Mesoamerica n to the United States, Eurasia, and Australia. Four species are native to North America n of Mexico, with 1 naturalized. Two species are known in TX with a third suspected (S.D. Jones, pers. comm.). Previously included in *Scirpus* (e.g., Kartesz 1994) and according to some, better treated as a section or subgenus in *Scirpus* in the broad sense; Strong (1993, 1994) argued that *Scirpus* is heterogenous but lumped *Bolboschoenus* with *Schoenoplectus*. We are following Smith (1995, 1997a) and Jones et al. (1997) in recognizing this segregate of *Scirpus* at the generic level. Smith (1995) indicated that *Bolboschoenus* and a number of other segregates of *Scirpus* will be recognized in the forthcoming Cyperaceae treatment for *Flora of North America* (Vol. 11). This approach is supported by the phylogenetic studies of Bruhl (1995) which suggested that *Scirpus* sensu lato is polyphyletic. (Greek: *bulbus*, bulb, and *schoeno*, reed or rush-like)
REFERENCES: Beetle 1942, 1947; Koyama 1962; Strong 1993, 1994; Browning et al. 1995; Smith 1995, 1997a; Smith & Yatskievych 1996.

Bolboschoenus maritimus (L.) Palla, (of the sea), ALKALI BULRUSH, BAYONET-GRASS, PRAIRIE BULRUSH, KOYAMA, SALTMARSH BULRUSH. Perennial 40–150 cm tall, with rhizomes 1–4 mm thick and corms to 20 mm thick; culms sharply triangular, 5–20 mm thick at base; leaves several, well-developed; inflorescences appearing terminal, subtended by several bracts; spikelets 10–30(–40) mm long, 6–12 mm wide, usually few in number (1–)2–15(–40), either all sessile or some sessile and some peduncled; scales of spikelets with midnerve prolonged into a 1–3 mm long point; styles 2(–3) branched; achenes 3–4 mm long, biconvex or nearly so. Wet ground; Grayson Co. (Hagerman National Wildlife Refuge); nc TX w to w TX. Jul–Aug. [*B. paludosus* (A. Nelson) Soó, *B. maritimus* subsp. *paludosus* (A. Nelson) T. Koyama, *Scirpus maritimus* L., *Scirpus maritimus* L. var. *paludosus* (A. Nelson) Kük.] If infraspecific taxa are recognized, all TX material would be subsp. *paludosus* (Smith 1997a). The corms were used as food by Native Americans in w North America (Beetle 1950).

BULBOSTYLIS

Ours tufted annuals or short-lived perennials with fibrous roots; leaves basal, filiform or setaceous; inflorescence a small, simple or compound, umbel-like cyme at end of flowering culm; spikelets several-flowered; scales of spikelets spirally imbricate, keeled; bristles absent; achenes trigonous-obovoid, ca. 1 mm long; style base persistent as a minute tubercle.

A genus of 100 species of tropical and warm areas; previously included in *Fimbristylis*. (Greek: *bolbos*, swelling or bulb, and *stylos*, pillar, column, or style)
REFERENCES: Kral 1971; Jones & Wipff 1992.

1. Umbel-like cymes simple; scales of spikelets truncate at apex, usually notched, the tip of the keel barely reaching base of notch; achenes finely transversely ridged ____ **B. capillaris**
1. Umbel-like cymes sometimes compound (the cyme branches themselves bearing small cymes); scales of spikelets broadly obtuse at apex, the tip of the keel reaching the apex of the scale or slightly exceeding it; achenes with an easily lost, waxy, fine pebbling ____ **B. ciliatifolia**

Bulbostylis capillaris (L.) Kunth ex C.B. Clarke, (hair-like), HAIR-SEDGE. Slender annual 5–35 cm tall; spikelets narrowly ovoid; 3–5 mm long, rarely longer. In loose sand; included based on citation of vegetational areas 4 and 5 (Fig. 2) by Hatch et al. (1990); e 1/2 of TX and Trans-Pecos. Jun–Sep.

Bulbostylis ciliatifolia (Elliott) Fernald, (ciliate-leaved). Annual or short-lived perennial to 40 cm tall; spikelets narrowly ovoid to lance-ovoid or oblong, 2–6 mm long. Sandy areas. Summer–Fall. Kral (1971) separated 2 varieties as follows:

1. Low annuals; inflorescence a simple to rarely compound umbel of few, lance-ovoid spikelets; longest bract of inflorescence seldom exceeding inflorescence; edges of leaves usually hispidulous ____ var. **ciliatifolia**
1. Tall perennials; inflorescence usually of many, oblong or lance-linear spikelets and commonly compound; longest bract of inflorescence commonly longer than inflorescence; edges of leaves usually distinctly tuberculate-scabrid ____ var. **coarctata**

var. **ciliatifolia**. Parker Co.; e TX w to nc TX.

var. **coarctata** (Elliott) Kral, (crowded together). Dallas, Milam, and Parker cos.; se and e TX w to nc TX.

CAREX CARIC SEDGE

Prepared by Stanley D. Jones (BRCH)

Cespitose or rhizomatous grass-like perennial herbs; plants mostly monoecious (all TX taxa), rarely dioecious; culms (= stems) triangular or terete, mostly solid; leaves 3-ranked; sheaths closed; inflorescences of spikes, either unisexual or bisexual, when bisexual they are androgynous (= having the staminate flowers distal to the pistillate), or gynecandrous (= having the pistillate flowers distal to the staminate), when unisexual the staminate spike is terminal with lateral spikes being pistillate or some androgynous; flowers enclosed by a sac-like scale (= perigynium) with an apical orifice from which the style or stigmas protrude; each perigynium subtended by a single pistillate scale; perianth absent; stamens (2–)3; carpels 2 or 3; stigmas 2 or 3; ovary and ovule 1; fruit an achene, lenticular and distigmatic or trigonous and tristigmatic.

A huge genus of ca. 2,000 species, cosmopolitan in distribution, mostly in n temperate and arctic regions of moist to wet habitats. It is the largest genus in the Texas flora with ca. 95 taxa plus 3 introduced species used as cultivars that may persist; similarily it is the largest genus in the Oklahoma flora (Taylor & Taylor 1994). *Carex* is also the largest genus in the flora of nc TX, being represented by 55 species. Forage value for livestock is low but the plants can be of use for wildlife, especially rabbits, rodents, deer, and birds. The genus is also important in preventing soil erosion. (The classical Latin name, of obscure origin; derived by some from the Greek: *keirein*, to cut, on account of the sharp leaves—as indicated in the English name SHEAR-GRASS)

This key is based on mature perigynia and fruit; fruiting material is essential for proper identification. Incomplete veins refer to veins that do not extend the entire length of the perigynium body. County citations for *Carex* are from a number of herbaria including BRCH.
REFERENCES: Mackenzie 1931–35, 1940; Hermann 1954a, 1970; Bryson 1980; Reznicek & Ball 1980; Menapace et al. 1986; Ball 1990; Bernard 1990; Crins 1990; Manhart 1990; Reznicek 1990; Standley 1990; Jones & Hatch 1990; Naczi & Bryson 1990; Jones & Reznicek 1991; Rothrock 1991; Naczi 1992; Reznicek & Naczi 1993; Jones 1994a, 1994b.

1. Achenes 2-sided, plano-convex or unequally biconvex in cross-section, flattened; stigmas 2.
   2. Inflorescence 2–5 cm wide, usually an open panicle of spicate branches; perigynia 6–7 mm long with broad subtruncate, spongy-thickened bases; perigynial wall frequently adhering to the achene; achenes ovate-lanceolate; ventral leaf sheath margins with orange-red dots ____ **C. crus-corvi**
   2. Inflorescence < 1.5 cm wide, a contracted panicle of spicate branches, spicate raceme, or composed of individual lateral or terminal spikes, but not an open panicle; perigynia without broad subtruncate, spongy-thickened bases OR if so then not 6–7 mm long; perigynial wall little to not at all adhering to the achene; achenes variously shaped but not ovate-lanceolate; ventral leaf sheath margins without orange-red dots, OR if dots present, then not in conjunction with the other characters.
      3. Terminal spike solely staminate, or sometimes partly pistillate in *C. crinita*.
         4. Lateral spikes, at least the lower, sessile or nearly so, ascending ____ **C. emoryi**
         4. Lateral spikes on flexuous peduncles, drooping ____ **C. crinita** var. **brevicrinus**
      3. Terminal spike either androgynous or gynecandrous.
         5. Terminal or all spikes androgynous, the staminate flowers often fugacious (= falling or disappearing early) making spikes appear solely pistillate, the spikes variously shaped, but not appearing clavate.
            6. Primary spicate branches usually less than 10, most frequently a primary branch will be comprised of a single spike, sometimes it will rebranch giving rise to secondary spikes.
               7. Leaf sheaths baggy (loose) around the culm ____ **C. gravida**
               7. Leaf sheaths tight (not loose) around the culm.
                  8. Lowest inflorescence bract 5.5–25 cm long, greatly exceeding the inflorescence, two to many times as long as the inflorescence.
                     9. Culms smooth below inflorescence; plants of open bottomlands or floodplain habitats ____ **C. arkansana**
                     9. Culms antrorsely scaberulous below inflorescence; plants of open mesic to submesic woodlands ____ **C. perdentata**
                  8. Lowest inflorescence bract less than 5.5 cm long, not exceeding the inflorescence, or less than two times as long as the inflorescence.
                     10. Beaks of perigynia smooth, not serrated.
                         11. Perigynia ovate-deltoid, veinless ventrally, spongy at base but without a swollen spongy area at base on ventral surface ____ **C. leavenworthii**
                         11. Perigynia ovate-lanceoid, with veins present on ventral surface, at least proximally over an enlarged spongy area at base.

Bolboschoenus maritimus [CO1]

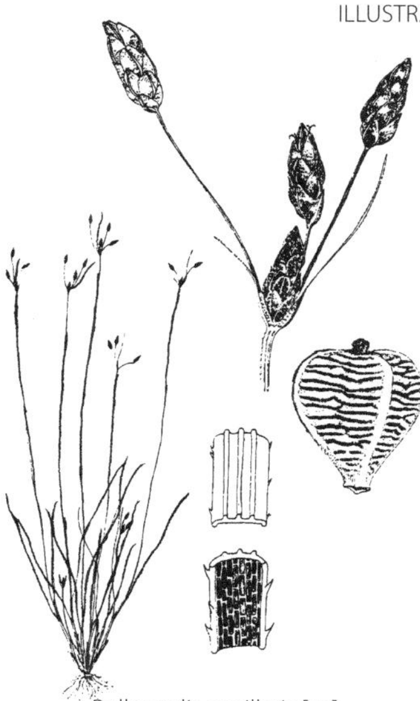

Bulbostylis capillaris [SID]

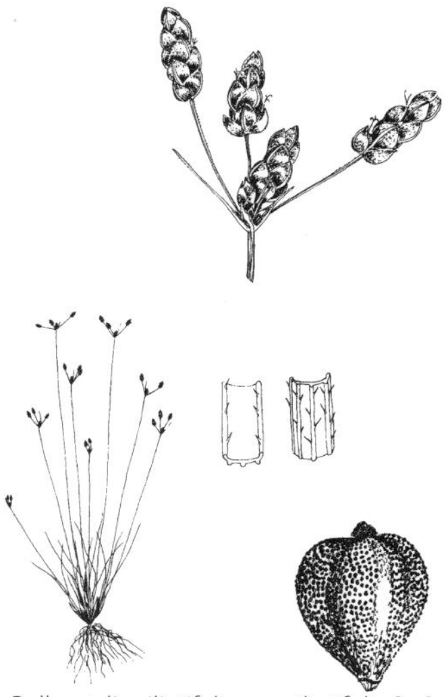

Bulbostylis ciliatifolia var. ciliatifolia [SID]

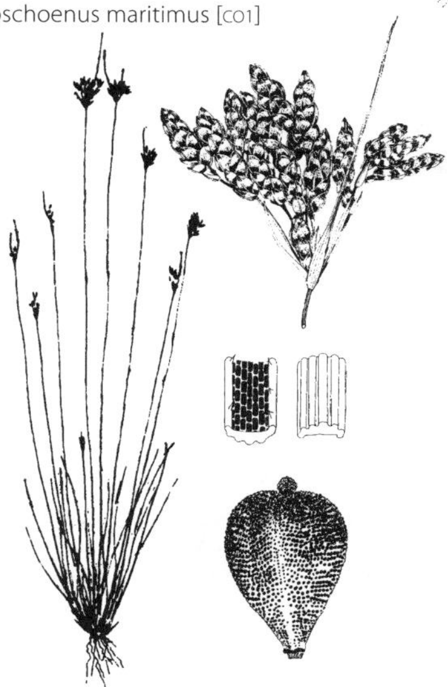

Bulbostylis ciliatifolia var. coarctata [SID]

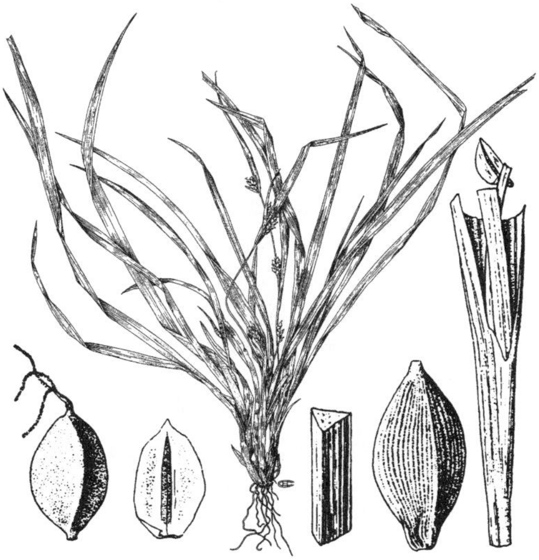

Carex abscondita [MAC]

Carex albicans var. australis [MAC]

Carex albolutescens [BB2]

12. Perigynia 1.3–1.8 mm wide; widest leaf blade 1.5–3 mm wide ____ **C. retroflexa**
12. Perigynia 1–1.3 mm wide; widest leaf blade 1–1.5 mm wide ____ **C. texensis**

10. Beaks of perigynia serrated, not smooth.
13. Perigynia spongy at base, with or without a swollen area basely.
14. Perigynia (1.4–)1.5–2.7(–2.8) mm wide, ovate-deltoid or conspicuously ovate, without a swollen spongy area at base on ventral surface.
15. Perigynia 2.2–3.2(–3.3) mm long, veinless ventrally, with 0(–3) veins dorsally ____ **C. leavenworthii**
15. Perigynia (3.3–)3.4–5.2(–5.6) mm long, with 0–5(–8) narrow veins (ca. 0.1–0.2 mm wide) ventrally, with 0–10(–11) narrow veins dorsally ____ **C. perdentata**
14. Perigynia 0.9–1.8 mm wide, ovate-lanceoid or slightly ovate-oblong, with a swollen spongy area at base of ventral surface ____ **C. socialis**
13. Perigynia not spongy at base.
16. Adaxial and abaxial leaf surfaces smooth, not minutely papillose (not sandpaper-like), except sometimes sparingly so along major veins.
17. Perigynia (3.4–)3.5–4.7 mm long, (2–)2.1–2.7(–3.1) mm wide.
18. Apex of the ventral leaf sheath straight or slightly concave, not callused or only slightly thickened, friable, frequently with scattered reddish dots; dorsal leaf sheath white or pale green with darker green veins with darker green septate-nodules, but not green mottled with white; widest leaves (3–)4–8 mm wide; most culms forming greater than 70° angle with the ground ____ **C. gravida**
18. Apex of ventral leaf sheath concave, callused, not friable, without scattered reddish dots; all dorsal leaf sheaths green OR green with darker green septate-nodule, OR some sheaths green mottled with white; widest leaves 2.5–4.5 mm wide; most culms forming less than a 50° angle with the ground, usually much less ____ **C. austrina**
17. Perigynia 2–3.5 mm long, 1.3–2.3(–2.4) mm wide.
19. Perigynia bodies ovate-deltoid; perigynia beaks 0.3–0.7(–0.8) mm long with a single row of serrations, abruptly arising from the apex of the perigynium; widest leaf blade 1.1–3(–4) mm wide; leaves per fertile culm 2–6(–7); culm width, ca. 2 cm above rootstock, 1–2.4(–3.5) mm wide; pistillate scale (1.3–)1.5–2.2 (–2.5) mm long; pistilate scale awn 0–0.8(–1) mm long; dorsal leaf sheath frequently green mottled with white dots ____ **C. leavenworthii**
19. Perigynia bodies ovate or suborbicular; perigynia beaks 0.8–1.1 mm long with a double row of serrations, gradually tapering from the shoulder of the perigynium; widest leaf blade (1.9)2.5–4.4 mm wide; leaves per fertile culm (4–)5–8; culm width, ca. 2 cm above rootstock, 1.7–3.2(–3.3) mm wide; pistillate scale 1.1–1.7(–1.9) mm long; pistillate scale awn 0–3.2 mm long; dorsal leaf sheath mostly green, infrequently green mottled with white dots ____ **C. cephalophora**
16. Adaxial or both adaxial and abaxial leaf surfaces minutely papillose (sandpaper-like), at least near distal end (this can be detected by placing a leaf between the thumb and index finger and sliding the fingers toward the distal end of the leaf; also easily seen with a hand lens).

20. Inflorescence capitate (short triangular in outline), 12–19 mm long, 9–14 mm wide; leaves conspicuously shorter than culm, (6.5–)8.4–21(–23) mm long; ventral surface of perigynia veinless, the dorsal surface veinless or rarely with 1–4 incomplete narrow veins (ca. 0.1–0.2 mm wide) ____ **C. mesochorea**
20. Inflorescence short-oblong, oblong, or linear, not short triangular in outline, (12–)13.5–47 mm long, 6–18(–28) mm wide, the central axis visible, at least between some spikes, usually the lowest two; leaves shorter or longer than culm, 11.3–46.4(–55.5) mm long; ventral surface of perigynia 0–15 veined, the dorsal surface 0–12-veined.
    21. Ventral surface of perigynia with (5–)6–15 conspicuous broad veins (ca. 0.05 mm wide), the dorsal surface with (0–)1–12 broad veins ____ **C. muehlenbergii** var. **muehlenbergii**
    21. Ventral surface of perigynia with 0–6(–8) narrow veins (ca. 0.01–0.02 mm wide), the dorsal surface with 0–11(–14) narrow veins.
        22. Bodies of pistillate scales 3–4.2(–4.3) mm long, (1–)1.6–2.6(–3) mm wide, the mid-stripe 3-veined, rarely 1-veined; culms usually forming an angle of 50° or less with the ground ____ **C. austrina**
        22. Bodies of pistillate scales (1.5–)1.8–3.1 mm long, (1–)1.2–1.8(–2.2) mm wide, the mid-stripe 1-veined, occasionally 3-veined; culms usually forming an angle of 70° or more with the ground.
            23. Beaks of perigynia 0.2–0.6(–1) mm long, abruptly arising from apex of perigynium; perigynia broadly ovate, (1.5–)2.5–3.8 mm long; dorsal leaf sheaths frequently green mottled white dots ____ **C. muehlenbergii** var. **enervis**
            23. Beaks of perigynia (1–)1.4–1.7(–1.8) mm long, tapering from shoulders or occasionally abruptly arising from apex of perigynia; perigynia ovate or ovate-deltoid, 3.2–5.2(–5.6) mm long; most dorsal leaf sheaths infrequently mottled with white dots ____ **C. perdentata**

6. Primary spicate branches more than 10, often these primary branches will re-branch into secondary branches.
    24. Most leaves equal to or exceeding the culms; perigynia 1–1.8 mm wide, narrowly ovate; beak of the perigynium tapering from the body, the beak 1/2 as long to as long as the body ____ **C. vulpinoidea**
    24. Most leaves shorter than the culms; perigynia 1.6–3 mm wide, narrowly ovate to broadly ovate, orbicular, OR reniform; beak of the perigynium usually arising abruptly from the body but can taper from the body, the beak to ca. 1/2 the length of the body.
        25. Perigynia with red glandular dots, orbicular to reniform, often broader than long; pistillate scales not conspicuous ____ **C. triangularis**
        25. Perigynia without red glandular dots, narrowly ovate to ovate orbicular, rarely broader than long; pistillate scales usually conspicuous ____ **C. fissa**

5. Terminal or all spikes gynecandrous, the spikes frequently appearing clavate.
    26. Perigynia bodies oblanceolate, mostly less than 1.5 mm wide; perigynium wing restricted to upper half of body ____ **C. tribuloides**

26. Perigynia bodies ovate, obovate, orbicular, or reniform, 1.5–6 mm wide; perigynium wing not restricted to upper half of body.
  27. Perigynia with several obvious veins over the achene on the inner (ventral) surface.
    28. Perigynia beaks 1.5–2.5 mm long; perigynia 5–20(–25) per spike; spikes 2–4 per culm ____ **C. hyalina**
    28. Perigynia beaks less than 1.5(–1.8) mm long; perigynia 25–80 per spike; spikes 3–8 per culm.
      29. Perigynium body widest above the middle, the body more or less obovate.
        30. Styles abruptly contorted just above the achene; perigynium beak abruptly tapered to a long tip, spreading at maturity ____ **C. albolutescens**
        30. Styles straight to somewhat sinuous; perigynium beak usually gradually tapering into a peak, appressed, not spreading at maturity.
          31. Achenes 0.8–1.2 mm wide; perigynia 1.6–2.8 mm wide, 4–6(–7) veined over achene abaxially.
            32. Pistillate scales reddish; inflorescences of robust culms arched or nodding, 2.3–8.4 cm long, the spikes strongly separated; spikes clavate, the staminate portion of well-developed spikes 2–11 mm long ____ **C. ozarkana**
            32. Pistillate scales hyaline-white; inflorescences of robust culms erect, 1–4.5 cm long, the spikes slightly separated to congested; spikes rounded to acute at base, the staminate portion of spikes < 2 mm long ____ **C. longii**
          31. Achenes 1.4–1.8 mm wide; larger perigynia 2.5–3.3(–3.4) mm wide, veinless or few-veined abaxially ____ **C. brevior**
      29. Perigynium body widest at or below the middle, the body more or less orbicular.
        33. Larger perigynia 2.5–3.8(–4.2) mm long ____ **C. festucacea**
        33. Larger perigynia 4.2–5.2 mm long ____ **C.** aff. **bicknellii**
  27. Perigynia veinless or rarely with 1–3 faint veins over the achene on the inner (ventral) surface.
    34. Perigynia finely granular-papillose (at 30 x magnification), the bodies reniform, wider than long, (2.6–)3.2–5 mm wide ____ **C. reniformis**
    34. Perigynia smooth, the bodies more or less orbicular or rarely obovate, 1.5–6 mm wide.
      35. Perigynia obovate ____ **C. brevior**
      35. Perigynia more or less orbicular.
        36. Larger perigynia 1.5–3.4 mm wide.
          37. Achenes 1.4–1.8 mm wide, 1.7–2 mm long; larger perigynia 3.4–4.5 mm long, 2.5–3.3(–3.4) mm wide ____ **C. brevior**
          37. Achenes (1–)1.1–1.35 mm wide, 1.3–1.7 mm long; larger perigynia 2.5–3.9(–4.2) mm long, 1.5–2.5 mm wide ____ **C. festucacea**
        36. Larger perigynia 3.7–6 mm wide ____ **C. tetrastachya**

1. Achenes 3-sided, trigonous or obscurely terete in cross-section; stigmas 3.
  38. Styles continuous with the achene, of the same color and texture as the achene, persistent.
    39. Larger perigynia 1 cm long or longer, including beaks.
      40. Pistillate spike outline tending to be globose; perigynia loosely arranged, spreading, drying dark olive-drab green ____ **C. intumescens**
      40. Pistillate spike outline oblong to cylindric; perigynia either loosely arranged or not, drying stramineous, green, or light olive-drab green.

41. Staminate peduncles greatly exceeding the uppermost pistillate spike; perigynia loosely arranged; elongate rhizomes present ____ **C. louisianica**
41. Staminate peduncles shorter than to only slightly exceeding the uppermost pistillate spike; perigynia tightly arranged; elongate rhizomes absent ____ **C. lupulina**

39. Larger perigynia 4–9(–9.5) mm long, including beaks.
42. Larger perigynia 7–9(–9.5) mm long, including beaks; beaks 3–4 mm long ____ **C. lurida**
42. Larger perigynia 4–6.5 mm long, including beaks; beaks 2 mm long or less.
43. Perigynia mostly squarrose to the rachis (at ca. 90° angle); achenes conspicuously obovate, conspicuously minutely granular-papillose; beaks of perigynia abruptly arising from perigynium body ____ **C. frankii**
43. Perigynia ascending along the rachis; achenes broadly elliptic, not conspicuously minutely granular-papillose; beaks of perigynia gradually tapering from perigynium body ____ **C. hyalinolepis**

38. Styles not continuous with the achene, usually not of the same color and/or texture as the achene, not persistent, withering.
44. Perigynia pubescent, at least apically, often minutely so (use 25 x magnification).
45. Perigynia mostly concealed by pistillate scales; pistillate scales 2–3.2 mm long.
46. Spikes borne close together above the middle of the culm, usually exceeding the leaves ____ **C. albicans** var. **australis**
46. Spikes, at least some, borne near the base of the culm, the leaves exceeding the spikes ____ **C. microrhyncha**
45. Perigynia conspicuous, not concealed by pistillate scales; pistillate scales (3–)3.5–4 mm long ____ **C. planostachys**
44. Perigynia glabrous.
47. Spikes, at least some, borne at or near base of plant.
48. Beaks of perigynia conspicuous, 3–3.5 mm long; spikes on long capillary peduncles to 15 cm long; shoot bases reddish ____ **C. basiantha**
48. Beaks of perigynia not conspicuous, less than 0.5 mm long; spikes either not on capillary peduncles or peduncles less than 5 cm long; shoot bases either reddish or white.
49. Staminate spikes sessile; shoot bases white followed by pale brown ____ **C. abscondita**
49. Staminate spikes pedunculate; shoot bases reddish ____ **C. edwardsiana**
47. Spikes borne well above base of plant.
50. Spikes, predominantly gynecandrous.
51. Spikes arising lateral on flexuous peduncles, the spikes drooping.
52. Longest pistillate scales with awns 1.2–5 mm long; perigynia 4.5–6 mm long, ovate-lanceolate ____ **C. davisii**
52. Longest pistillate scales acuminate or short-awned, the awns 0–0.5(–2) mm long; perigynia 3.5–5 mm long, rhomboid or narrowly elliptic, broadest near the middle and tapering to both ends ____ **C. oxylepis**
51. Spikes terminal on culm, the spikes stiff, ascending.
53. Perigynia ascending along the rachis, appearing flattened on the side next to the rachis, the perigynia not appearing inflated ____ **C. complanata**
53. Perigynia spreading (to ca. 90°), or only slightly ascending along the rachis, not flattened on the side next to the rachis, the perigynia inflated.
54. Pistillate scales 3–6 mm long, including awn; achenes 2.1–2.6 mm long; plants of dry upland prairies and open grassy areas ____ **C. bushii**
54. Pistillate scales (2–)2.5–3 mm long, including awn; achenes 1.5–2 mm long; plants most frequently in low wet woods, wooded swamps, river flood plain forests, less frequent in open grassy areas ____ **C. caroliniana**

50. Spikes, at least some, usually the terminal, solely staminate.
  55. Spikes arising lateral on flexuous peduncles, the spikes drooping.
    56. Pistillate spikes (some are androgynous) 7–9 mm wide; rhizomatous perennials; rhizome 3–12 mm thick ____ **C. cherokeensis**
    56. Pistillate spikes (solely pistillate) 2–3 mm wide; tufted perennials ____ **C. debilis**
  55. Spikes terminal or lateral but not on flexuous peduncles, the spikes not drooping.
    57. Most of the beaks of perigynia conspicuously bent.
      58. Pistillate scales brown, reddish brown, or reddish purple on both sides of the mid-vein; plants with creeping rhizomes ____ **C. meadii**
      58. Pistillate scales green or hyaline on both sides of the mid-vein; plants without creeping rhizomes.
        59. Veins on faces of the perigynia raised.
          60. Perigynia 1.6–2.5 mm wide ____ **C. granularis**
          60. Perigynia 1–1.3 mm wide.
            61. Perigynia obovoid, 2.4–2.6 mm long, with a short, abruptly bent beak ____ **C. blanda**
            61. Perigynia fusiform, 3.4–5 mm long, tapering into a curved, more or less elongate beak ____ **C. striatula**
        59. Veins on faces of the perigynia impressed.
          62. Larger perigynia (4–)4.2–6 mm long; style base straight or slightly bent ____ **C. flaccosperma**
          62. Larger perigynia 3.2–4.1 mm long; style base usually conspicuously bent or reflexed ____ **C. glaucodea**
    57. Most of the beaks of perigynia straight, not bent.
      63. Plants with elongate creeping rhizomes ____ **C. microdonta**
      63. Plants tufted, without elongate rhizomes.
        64. Perigynia distichously arranged along the rachis ____ **C. bulbostylis**
        64. Perigynia spirally arranged along the rachis.
          65. Shoot bases brown, often white above the brown.
            66. Perigynia (4–)4.2–6 mm long; style base straight or slightly bent ____ **C. flaccosperma**
            66. Perigynia 3.2–4.1 mm long; style base usually conspicuously bent or reflexed ____ **C. glaucodea**
          65. Shoot bases purplish red.
            67. Perigynia 2–2.6 mm wide, orbicular to suborbicular in cross-section; achene bodies (excluding stipe) 2.2–3 mm long ____ **C. grisea**
            67. Perigynia 1.5–2.3 mm wide, obtusely triangular in cross-section; achene bodies (excluding stipe) 1.8–2.3 mm ____ **C. corrugata**

Carex abscondita Mack., (concealed), HIDDEN-FRUIT CARIC SEDGE. Wet to mesic, shaded deciduous hardwood forests, pine forests, or swamps; in ne part of nc TX in Red River Co., probably in other ne cos.; also Post Oak Savannah, Pineywoods, and upper portion of Gulf Prairies and Marshes. Fruiting Apr–Jun. Section *Careyanae*

Carex albicans Willd. ex Spreng. var. **australis** (L.H. Bailey) Rettig., (sp.: whitish; var.: southern), SOUTHERN BELLOWS-BEAK CARIC SEDGE. Sandy or rocky woods, frequently on slopes in mixed pine-hardwood forests; Dallas, Delta, Hopkins, Lamar, and Red River cos.; also Post Oak Savannah and Pineywoods. Fruiting Apr–Jun. [*C. physorhyncha* Liebm. ex Steud.] Section *Acrocystis*

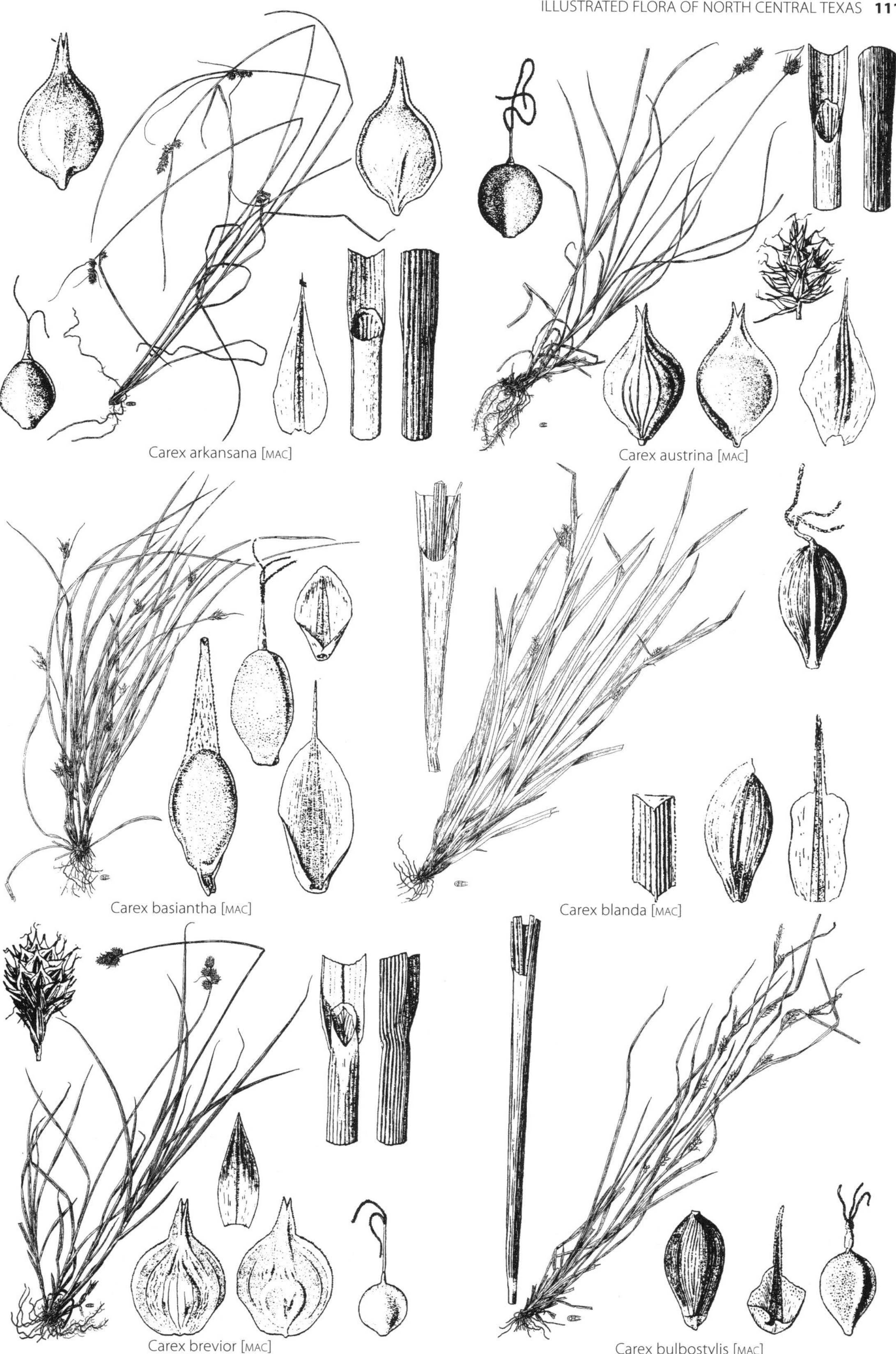

Carex arkansana [MAC]

Carex austrina [MAC]

Carex basiantha [MAC]

Carex blanda [MAC]

Carex brevior [MAC]

Carex bulbostylis [MAC]

Carex albolutescens Schwein., (whitish yellow), WHITISH YELLOW CARIC SEDGE. Wet woods, thickets, and peats; Delta, Hopkins, Lamar, and Milam cos.; also Post Oak Savannah and Pineywoods. Fruiting Apr–Jun(–Aug). Section *Ovales*

Carex arkansana L.H. Bailey, (of Arkansas), ARKANSAS CARIC SEDGE. Openings in low, seasonally wet woods, open bottomlands, damp prairies associated with creeks; Dallas, Hopkins, Kaufman and Red River cos., probably in other ne cos.; also Post Oak Savannah and Pineywoods. Fruiting May–Jun. Section *Phaestoglochin*

Carex austrina (Small) Mack., (southern), SOUTHERN CARIC SEDGE. Obligate heliophyte (= a plant adapted to grow in or tolerate full sun) in open prairies in alfisols, but occasional in mollisols and vertisols, and encroaching on histosols; throughout most of nc TX; also Rolling Plains, Post Oak Savannah, Pineywoods, s to the Coastal Bend Area of the Gulf Prairies and Marshes. Fruiting Mar–early Jul. [*C. muehlenbergii* Schkuhr ex Willd. var. *australis* Olney, *C. muehlenbergii* var. *austrina* Small] Section *Phaestoglochin*

Carex basiantha Steud., (basal-flowered), BASAL-FRUIT CARIC SEDGE. Ravine slopes of mixed pine-hardwood forests and dedicuous woods; usually in rocky or sandy soils; in the ne part of nc TX in Red River Co., and probably in other ne cos.; also Post Oak Savannah and Pineywoods. Fruiting late-Mar–Jul. [*C. willdenowii* of American authors, not Schkuhr ex Willd.] Section *Phyllostachyae*

Carex blanda Dewey, (mild), CHARMING CARIC SEDGE. In dry to mesic woods, bottomlands, slopes, forest edges, and meadows; Bell, Bosque, Collin, Cooke, Dallas, Grayson, Hunt, Johnson, Lamar, Parker, Red River, Tarrant, and Williamson cos. ; also Post Oak Savannah, Edwards Plateau, Pineywoods, and Gulf Prairies and Marshes. Fruiting Apr–May(–Jun). Section *Laxiflorae*

Carex brevior (Dewey) Mack. ex Lunell, (short), SHORT CARIC SEDGE. A heliophyte found in submesic open meadows, prairies, and roadsides; Navarro, Delta, Hopkins, Lamar, Milam, and Red River cos.; also Post Oak Savannah, Pineywoods, and upper portions of Gulf Prairies and Marshes. Fruiting late Apr–Jun. Section *Ovales*

Carex bulbostylis Mack., (bulb-styled), GLOBOSE CARIC SEDGE. Mesic deciduous forests, flood plains and adjacent slopes; usually in neutral soils or slightly acidic or slightly alkaline, loams, sandy loams, sandy clay loams, or clay loams; widespread in nc TX; also Post Oak Savannah, Edwards Plateau, Pineywoods, and upper portions of Gulf Prairies and Marshes. Fruiting Apr–May. [*C. amphibola* Steud. var. *globosa* (L.H. Bailey) L.H. Bailey] Section *Griseae*

Carex bushii Mack., (for its discoverer, Benjamin Franklin Bush, 1858–1937, postmaster in MO and amateur botanist), BENJAMIN BUSH'S CARIC SEDGE. Obligate to facultative heliophyte in open mesic to submesic prairies, open roadsides, and forest edges with sandy soils; Dallas, Denton, Kaufman, Lamar, and Red River cos.; also Post Oak Savannah, Pineywoods, and upper portions of Gulf Prairies and Marshes. Fruiting Apr–May. [*C. caroliniana* Schwein. var. *cuspidata* (Dewey) Shinners] Section *Porocystis*

Carex caroliniana Schwein., (of Carolina), CAROLINA CARIC SEDGE. Facultative sciophyte (= a plant adapted to grow in or tolerate shade) in deciduous woods, usually lower slopes and in bottoms, near wooded streams, sandy soils; in ne part of nc TX in Delta, Hopkins, and Lamar cos.; also Post Oak Savannah, Pineywoods, and upper portions of Gulf Prairies and Marshes. Fruiting Apr–May. Section *Porocystis*

Carex cephalophora Muhl. ex Willd., (bearing heads), HEAD-BEARING CARIC SEDGE. A sciophyte, primarily in alfisols with sandy or sandy loam soils, slopes in mesic to submesic hardwood forests, or mixed hardwood pine forest, or occasionally at wetter sites in entisols along stream

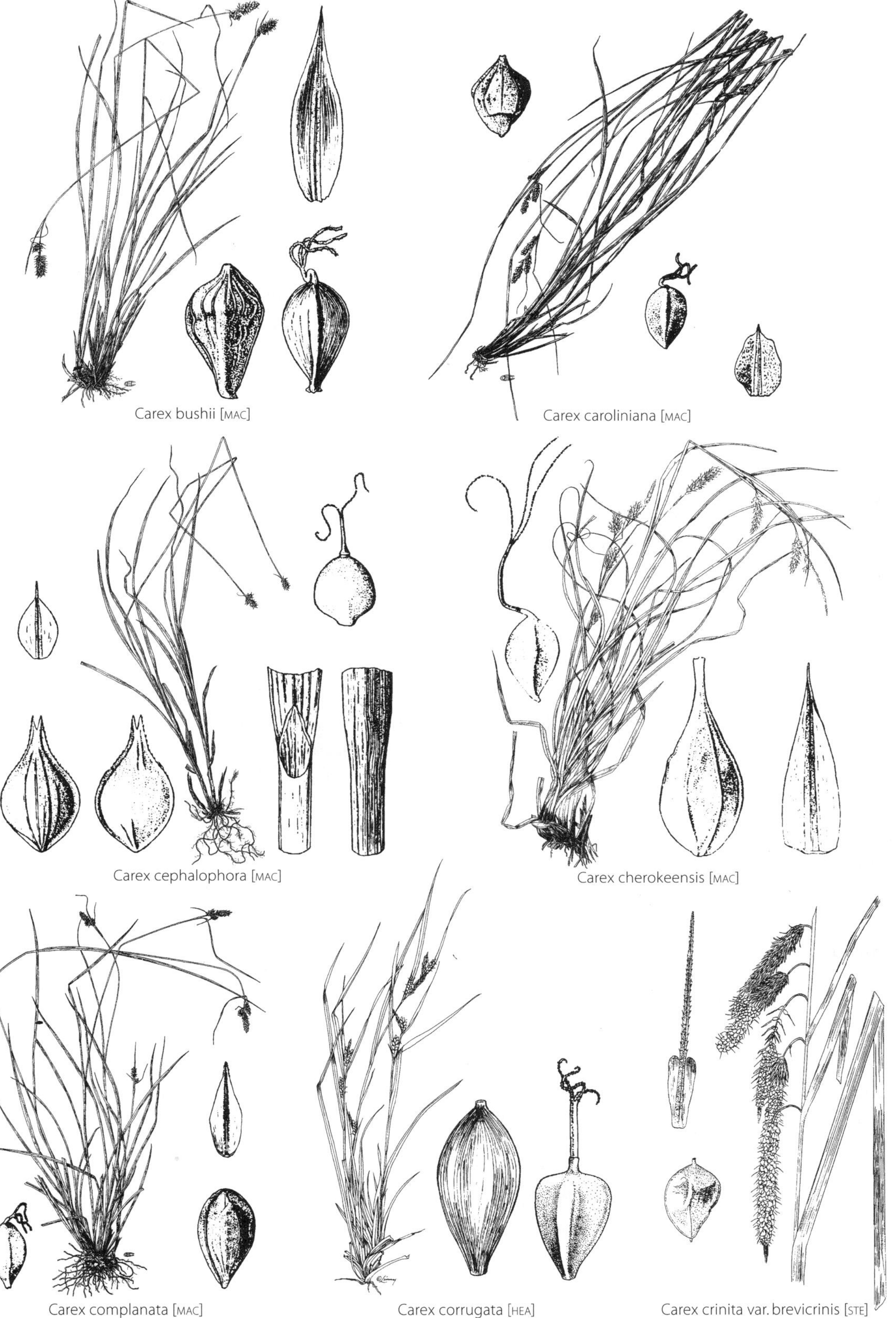

Carex bushii [MAC]

Carex caroliniana [MAC]

Carex cephalophora [MAC]

Carex cherokeensis [MAC]

Carex complanata [MAC]

Carex corrugata [HEA]

Carex crinita var. brevicrinis [STE]

courses, occasionally remaining as remnants in openings, pastures, or roadsides; Ellis and Red River cos.; also Post Oak Savannah, but more frequent in the Pineywoods. Fruiting (late Mar–) late Apr–late Jul(–early Oct). Section *Phaestoglochin*

Carex cherokeensis Schwein., (from its occurrence in "Cherokee country"), CHEROKEE CARIC SEDGE. Low open, damp, deciduous woods with sandy or sandy loam soils, frequently calcareous; Dallas, Hopkins, Hunt, Kaufman, Lamar, Milam, Navarro, Red River, Tarrant, and Wise cos.; also Post Oak Savannah, Pineywoods, and upper portions of Gulf Prairies and Marshes. Fruiting Apr–May(–early Jun). Section *Hymenochlaenae*

Carex complanata Torr. & Hook., (flattened), FLAT-FRUIT CARIC SEDGE. In shade or full sun, usually in open mesic deciduous forests with sandy soils, forest edges, or clear-cuts; in ne part of nc TX in Delta, Hopkins, and Lamar cos.; also Pineywoods, Post Oak Savannah, and upper portions of Gulf Prairies and Marshes. Fruiting Apr–May(–Jun). Section *Porocystis*

Carex corrugata Fernald, (corrugated or wrinkled), WRINKLE-FRUIT CARIC SEDGE. Floodplains of mesic deciduous forests, alluvia, in acidic to alkaline clays, to silt loams; Dallas, Delta, Fannin, Hopkins, Jack, and Lamar cos.; probably in every e county of nc TX; also Post Oak Savannah, Pineywoods, and upper portions of Gulf Prairies and Marshes. Fruiting Apr–May. Section *Griseae*

Carex crinita Lam. var. **brevicrinis** Fernald, (sp.: provided with long hair; var.: short-haired), SHORT-HAIR CARIC SEDGE. Wooded swamps and bottoms; ne part of nc TX in Red River Co., likely in other ne cos.; also Pineywoods and upper portions of Post Oak Savannah. Fruiting May–Jun. Section *Phacocystis*

Carex crus-corvi Shuttlew. ex Kunze, (crow-spur), CROW-FOOT CARIC SEDGE. A heliophyte in wet prairies, depressions, roadside ditches, marshes, and open swamps; widespread in nc TX; also Post Oak Savannah, Edwards Plateau, Pineywoods, Gulf Prairies and Marshes, and s Texas Plains. Fruiting (late Feb–)Apr–Jun(–early Jul). Section *Vulpinae*

Carex davisii Schwein. & Torr., (for Emerson Davis, 1798–1866, amateur student of *Carex*), EMERSON DAVIS' CARIC SEDGE. Rich, deciduous, calcareous woods, forest edges, meadows, and shores; Dallas, Delta, Denton, Hopkins, Hunt, Johnson, Kaufman, and Tarrant cos.; also upper portion of Post Oak Savannah. Fruiting Apr–May(–mid-Jun). Section *Hymenochlaenae*

Carex debilis Michx., (frail), WEAK CARIC SEDGE. Low woods, swamps, forest edges, and especially along creek margins; ne part of nc TX in Delta, Lamar, and Red River cos.; also Post Oak Savannah, Pineywoods, and upper portion of Gulf Prairies and Marshes. Fruiting Apr–May(–Jul). Section *Hymenochlaenae*

Carex edwardsiana E.L. Bridges & Orzell, (of the Edwards Plateau), EDWARDS PLATEAU CARIC SEDGE. Mesic to submesic mixed juniper-hardwood forests and ravine slopes, alkaline clay loams and sandy clay loams; s part of nc TX in Bell Co.; also Edwards Plateau, mostly in the Balcones Canyonlands; endemic to TX. Fruiting Apr–May. [*C. oligocarpa* of Texas authors in part, not Schkuhr ex Willd.] Section *Griseae*

Carex emoryi Dewey, (for Major William Helmsley Emory, 1811–1887, American soldier who worked on U.S./Mexican boundary survey), WILLIAM EMORY'S CARIC SEDGE. A heliophyte along margins of streams, rivers, lakes, ponds, marshes, and open swamps, usually on calcareous soils; Bell, Cooke, Dallas, Milam, and Williamson cos.; also Post Oak Savannah, Pineywoods, Edwards Plateau, and Rolling Plains. Fruiting Mar–May. Section *Phacocystis*

Carex festucacea Schkuhr ex Willd., (fescue-like), FESCUE-LIKE CARIC SEDGE. Damp or wet low areas in woods; Dallas, Delta, Hopkins, Lamar, and Milam cos. in e part of nc TX; also Post Oak

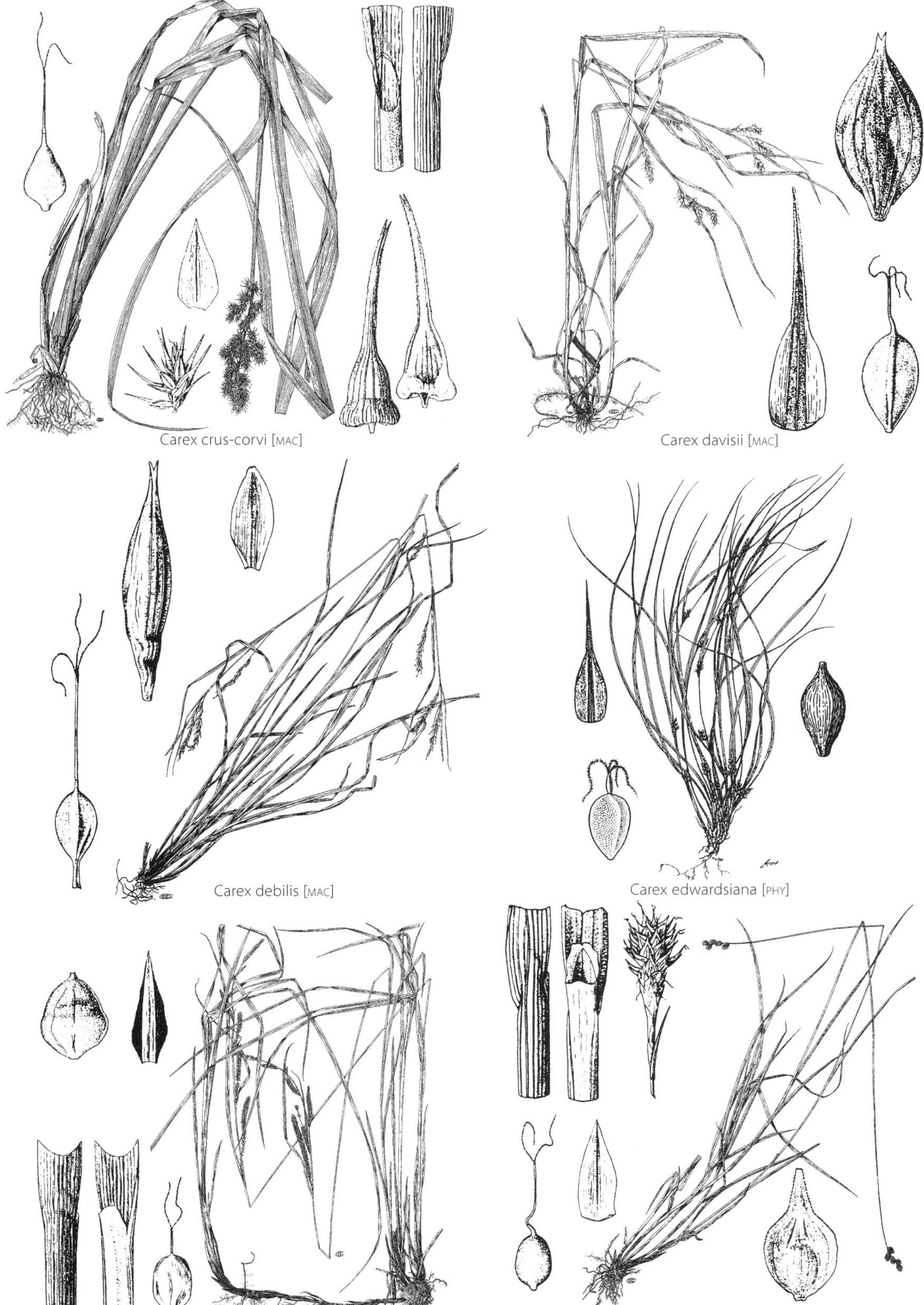

Carex crus-corvi [MAC]

Carex davisii [MAC]

Carex debilis [MAC]

Carex edwardsiana [PHY]

Carax emoryi [MAC]

Carex festucacea [MAC]

Savannah, Pineywoods, and upper portion of Gulf Prairies and Marshes. Fruiting Apr–May(–early Jun). Section *Ovales*

Carex fissa Mack., (split), SHARP-MARGIN CARIC SEDGE. A heliophyte in open, wet, roadside ditches and in open wet areas in floodplains, usually in alluvial clay soils; Dallas, Hopkins, Hunt, Kaufman, and Milam cos.; also Post Oak Savannah and upper portion of Gulf Prairies and Marshes. Fruiting May–Jun. Section *Multiflorae*

Carex flaccosperma Dewey, (flaccid, weak, or soft-seeded or -fruited), FLACCID-FRUIT CARIC SEDGE. Usually found in floodplains in mesic deciduous forests, in acidic silt loams, sandy loams, sandy clay loams, clays, and loams; e and ne parts of nc TX in Delta, Fannin, Henderson, Hopkins, Hunt, Kaufman, and Lamar cos.; also Post Oak Savannah, Pineywoods, and upper portion of Gulf Prairies and Marshes. Fruiting Apr–May. Section *Griseae*

Carex frankii Kunth, (for its discoverer, Joseph Frank, 1782–1835, German botanist, physician, and traveler in U.S.) JOSEPH FRANK'S CARIC SEDGE. Low deciduous woods, bottomlands, and wet meadows, usually in calcareous or neutral soils; Bell, Bosque, Delta, Fannin, Grayson, Henderson, Hunt, Hopkins, Lamar, Milam, and Red River cos.; also Post Oak Savannah, Pineywoods, upper portion of Gulf Prairies and Marshes, Edwards Plateau, and Trans-Pecos. Fruiting May–Sep(–early Nov). Section *Squarrosae*

Carex glaucodea Tuck. ex Olney, (gray-green), GRAY-GREEN-FRUIT CARIC SEDGE. Along edges and in openings of mesic deciduous forests or in ephemeral wet prairies, in acidic to alkaline loams, or clays; ne part of nc TX in Hopkins and Lamar cos., probably in other ne cos.; also upper Post Oak Savannah and upper Pineywoods. Fruiting May–Jun(–early Jul). [*C. flaccosperma* Dewey var. *glaucodea* (Tuck. ex Olney) Kük.] Section *Griseae*

Carex granularis Muhl. ex Willd., (granular, covered with minute grains), GRANULAR CARIC SEDGE. Calcareous rich woods, shores, meadows, and bottomlands, usually in calcareous soils; ne part of nc TX in Hopkins, Hunt, Lamar, and Red River cos.; also Pineywoods and upper Post Oak Savannah. Fruiting May–Jun. [*C. granularis* var. *haleana* (Olney) Porter] Section *Granulares*

Carex gravida L.H. Bailey, (heavy with fruit), HEAVY-FRUIT CARIC SEDGE. A heliophyte in open swales, seepy areas, damp or mesic prairies with calcareous soils; Bell, Burnet, Dallas, Denton, Hamilton, Hunt, Navarro, Palo Pinto, Tarrant, and Wise cos.; also Post Oak Savannah, Rolling Plains, and High Plains. Fruiting May-Jun. [*C. gravida* var. *lunelliana* (Mack.) F.J. Herm., *C. lunelliana* Mack.] Section *Phaestoglochin*

Carex grisea Wahlenb., (gray), INFLATED CARIC SEDGE. Floodplains of mesic deciduous forests, acidic to alkaline sandy loams, loams, sandy clay loams, and clay loams; Dallas, Delta, Ellis, Fannon, Grayson, Hunt, Kaufman, and Palo Pinto cos.; also Post Oak Savannah and Rolling Plains. Fruiting Apr–Jun(–early Jul). [*C. amphibola* Steud. var. *turgida* Fernald] Section *Griseae*

Carex hyalina Boott, (transparent, translucent), FEW-FLOWER CARIC SEDGE, TISSUE SEDGE. A sciophyte of bottomland hardwood forests, usually on secondary flood terraces in wet neutral clay soils but can be slightly acid or slightly alkaline; frequent along the Trinity and Sulphur rivers and their tributaries; less frequent along the Brazos River; Dallas, Delta, Denton, Ellis, Henderson, Hopkins, Hunt, Kaufman, Lamar, Navarro, and Red River cos.; also Pineywoods, upper portion of Gulf Prairies and Marshes, and Post Oak Savannah. Fruiting mid-Mar–mid-May(–mid-Jun). Section *Ovales* (TOES 1993: V) ⚠

Carex hyalinolepis Steud., (with transparent or translucent scales), HYALINE-SCALE CARIC SEDGE. A heliophyte forming massive colonies; found in open roadside ditches, swales, shores, marshes,

Carex fissa [MAC]

Carex flaccosperma [MAC]

Carex frankii [MAC]

Carex glaucodea [MAC]

Carex granularis [MAC]

Carex gravida [MAC]

and open swamps, frequently in black calcareous or neutral clay; Dallas, Delta, Ellis, Hopkins, Navarro, and Red River cos.; also Post Oak Savannah and upper portion of Gulf Prairies and Marshes. Fruiting Apr–May(–early Jul). Section *Paludosae*

Carex intumescens Rudge, (swollen, puffed up), BLADDERY CARIC SEDGE. Swampy woods, bottomland hardwood forests, acidic soils; Red River Co. in ne part of nc TX; also Post Oak Savannah, Pineywoods, and upper portion of Gulf Prairies and Marshes. Fruiting Mar–Sep. Section *Lupulinae*

Carex leavenworthii Dewey, (for its discoverer, Melines Conklin Leavenworth, 1796–1862, s U.S. botanist), MELINES LEAVENWORTH'S CARIC SEDGE. Primarily a heliophyte but grows more robust in shade, primarily in alfisols with sandy or sandy loam soils, occasionally in entisols, histosols, or mollisols, open mesic to submesic sites, occasionally in wetter sites, forest edges, forest openings, pastures, roadsides, lawn weed, appears to do better in recently disturbed sites as a successional species, but persists; widespread in nc TX; also Post Oak Savannah, Pineywoods, upper Gulf Prairies and Marshes, and upper South Texas Plains. Fruiting (mid-Feb–)mid-Mar–early Jul(–late Oct). [*C. cephalophora* Muhl. ex Willd. var. *angustifolia* Boott; *C. cephalophora* var. *leavenworthii* (Dewey) Kük.]. Section *Phaestoglochin*

Carex longii Mack., (for Bayard Henry Long, 1885–1969, of Philadelphia), BAYARD LONG'S CARIC SEDGE. A facultative heliophyte in open damp or wet sites, usually in sandy, agrillaceous or peaty soils; Delta, Hopkins, Lamar, and Milam, cos. in e part of nc TX, undoubtedly in other cos. within nc TX; also Post Oak Savannah, Pineywoods, upper portion of Gulf Prairies and Marshes. Fruiting May–Jul(–Nov). Section *Ovales*

Carex louisianica L.H. Bailey, (of Louisiana), LOUISIANA CARIC SEDGE. Swampy woods, bottomland hardwood forests, acidic soils; in Red River drainage in Fannin Co.; also Post Oak Savannah, Pineywoods, and upper portion of Gulf Prairies and Marshes. Fruiting Apr–Aug. Section *Lupulinae*

Carex lupulina Muhl. ex Willd., (resembling *Humulus lupulus*—hops), HOP CARIC SEDGE. Open swamps, wet ditches, somewhat acidic-neutral to calcareous soils; Hopkins, Lamar, Milam, and Red River cos.; also Post Oak Savannah, Pineywoods, e porton of Edwards Plateau, upper portion of Gulf Prairies and Marshes. Fruiting Apr–Oct. Section *Lupulinae*

Carex lurida Wahlenb., (sallow, pale yellow), SALLOW CARIC SEDGE. Open swales and open swamps; Denton, Lamar, Milam, and Red River cos., expected in other e cos. within nc TX; also Post Oak Savannah, Pineywoods, and upper Gulf Prairies and Marshes. Fruiting late-Apr–early Jul(–Aug). Section *Vesicariae*

Carex meadii Dewey, (for its discoverer, Samuel Barnum Mead, 1798–1880, botanist and physician of CT and IL), SAMUEL MEAD'S CARIC SEDGE, MEAD'S CARIC SEDGE. A heliophyte in open mesic to wet calcareous clay prairies and depressions; in ne part of nc TX in Dallas, Grayson, Kaufman, and Lamar cos.; also upper Gulf Prairies and Marshes. Fruiting late-Mar–mid-May(–early Jun). Section *Paniceae*

Carex mesochorea Mack., (midland), MIDLAND CARIC SEDGE. A facultative to obligate heliophyte growing in entisols near hardwood forest edges or in openings, in pastures, or roadsides, usually on sandy or sandy loam soils, appears to need disturbance as a successional species; the only known station in TX is from Tarrant Co. Fruiting late Mar–late Jul. [*C. cephalophora* Muhl. ex Willd. var. *mesochorea* (Mack.) Gleason] Section *Phaestoglochin*

Carex microdonta Torr. & Hook., (small-toothed), SMALL-TOOTH CARIC SEDGE. A heliophyte in calcareous shores, gravels, meadows, prairies, and glades; Bell, Collin, Coryell, Dallas, Lampasas,

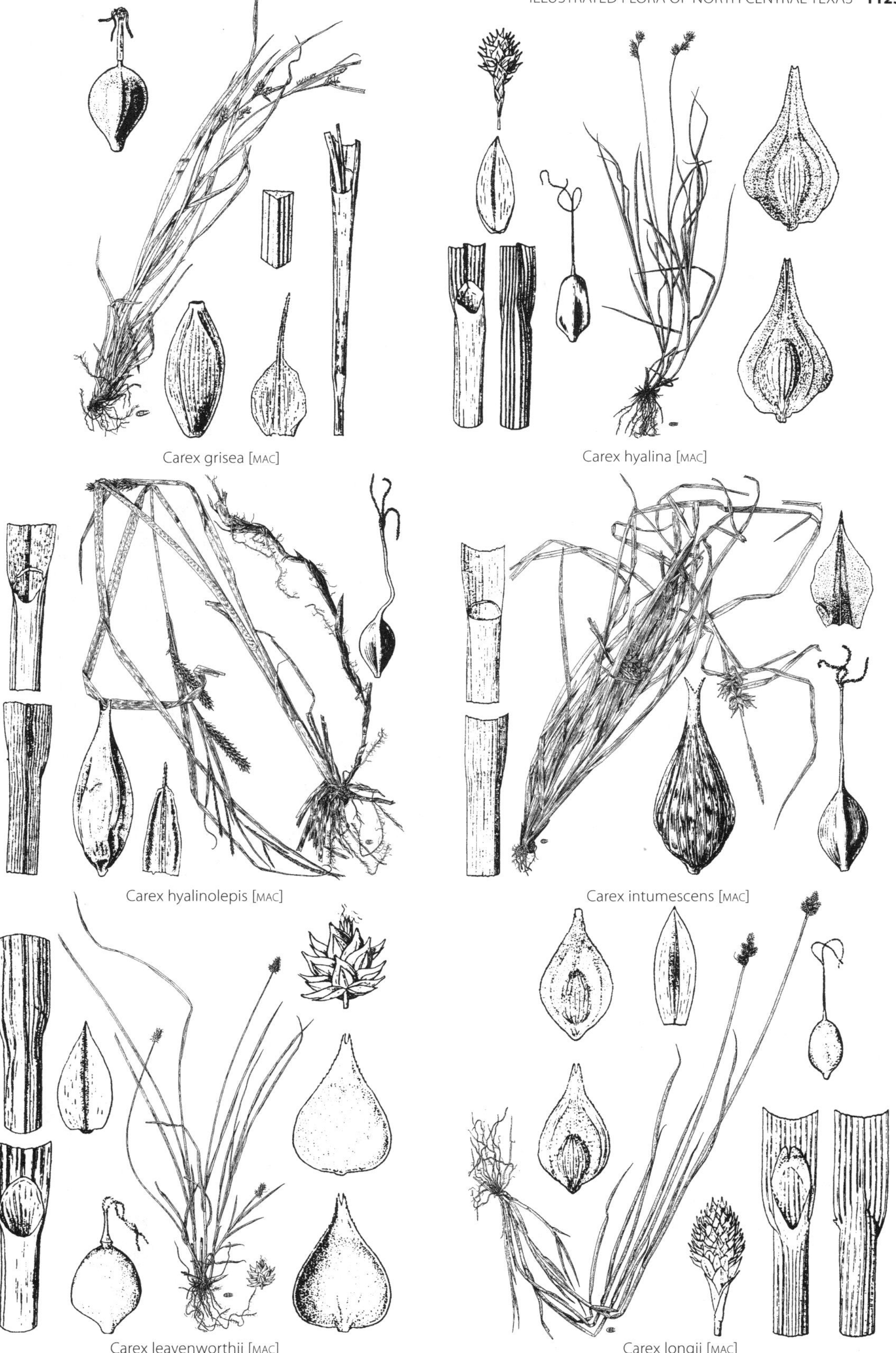

Carex grisea [MAC]

Carex hyalina [MAC]

Carex hyalinolepis [MAC]

Carex intumescens [MAC]

Carex leavenworthii [MAC]

Carex longii [MAC]

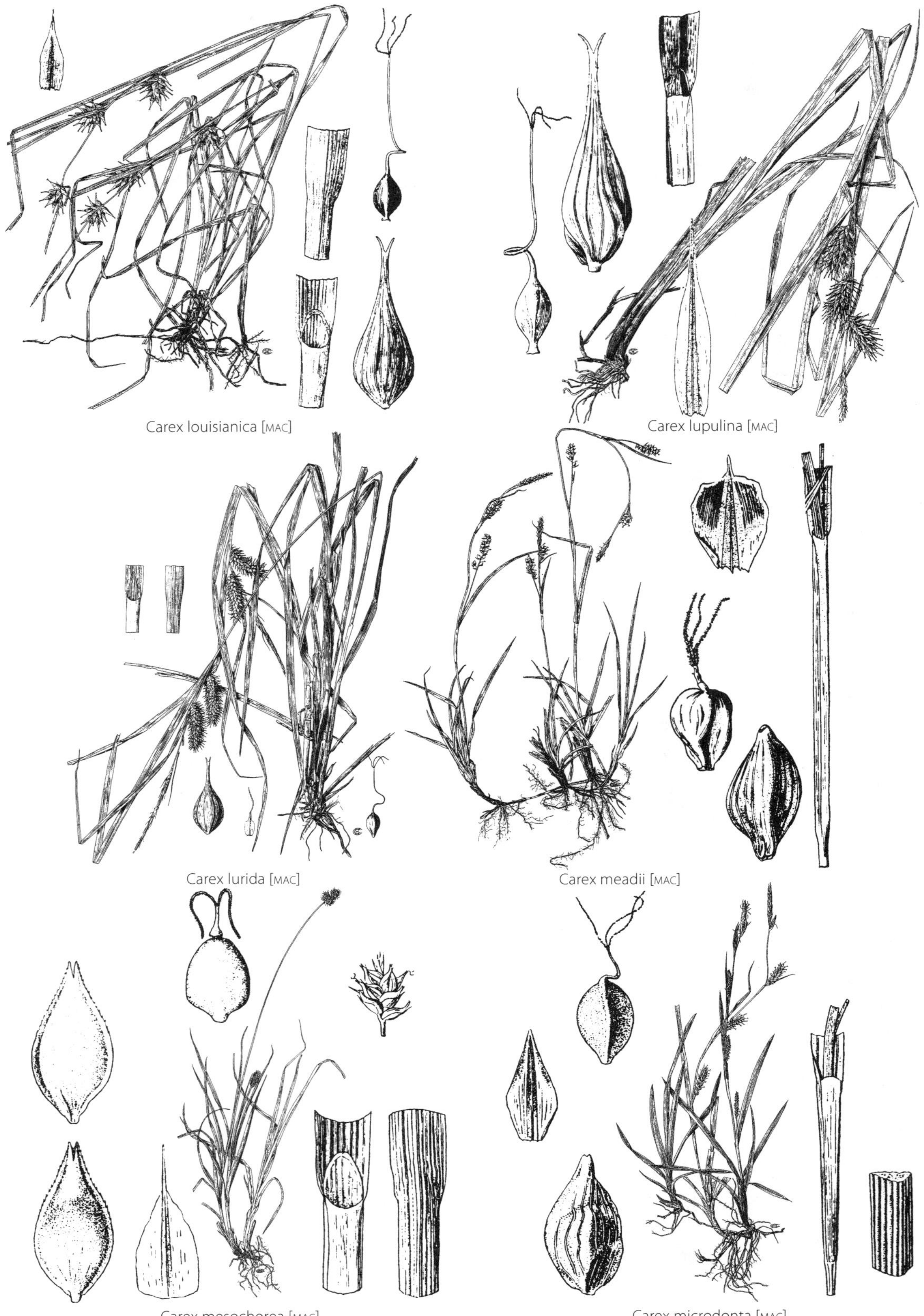

Carex louisianica [MAC]

Carex lupulina [MAC]

Carex lurida [MAC]

Carex meadii [MAC]

Carex mesochorea [MAC]

Carex microdonta [MAC]

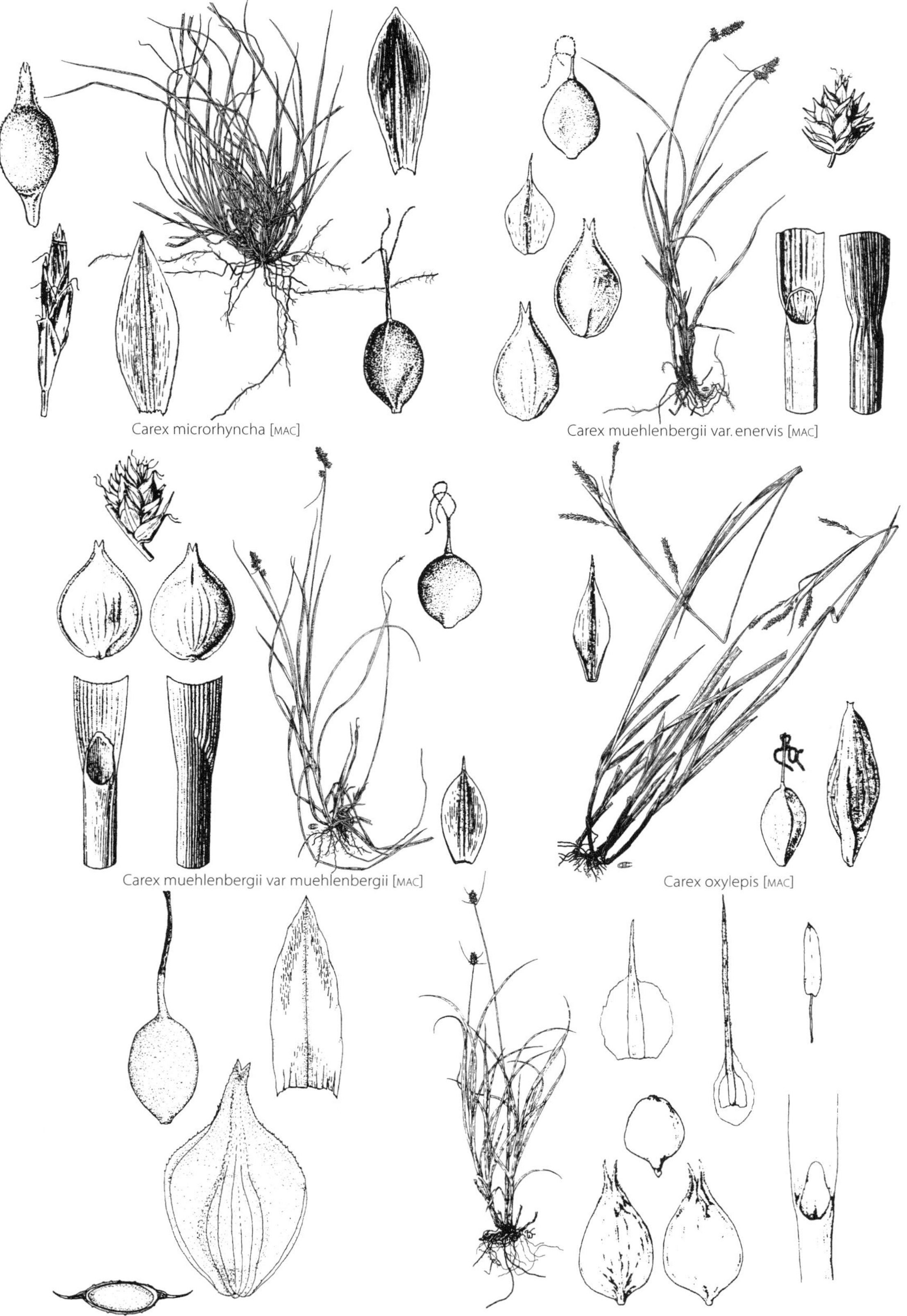

Carex microrhyncha [MAC]

Carex muehlenbergii var. enervis [MAC]

Carex muehlenbergii var muehlenbergii [MAC]

Carex oxylepis [MAC]

Carex ozarkana [BTT]

Carex perdentata [SID]

Milam, Mills, and Tarrant cos.; also Post Oak Savannah, South Texas Plains, Edwards Plateau, and Trans-Pecos. Fruiting late Apr–Jun. Section *Granulares*

Carex microrhyncha Mack., (small-beaked), SMALL-BEAK CARIC SEDGE. Submesic oak-hickory forests or oak-juniper woodlands, at base of trees or in semi-open areas with sandy or gravely sandy soils; Delta, Milam, Parker, and Red River Cos.; also Pineywoods and Post Oak Savannah. Fruiting early Mar–Apr(–May). Section *Acrocystis*

Carex muehlenbergii Schkuhr ex Willd. (for Gotthilf Henry Ernest Muhlenberg, 1753–1815, German-educated Pennsylvania pioneer botanist). See key to species to separate varieties. Section *Phaestoglochin*

var. **enervis** Boott, (nerveless), GOTTHILF MUHLENBERG'S VEINLESS CARIC SEDGE. Obligate to facultative sciophyte, but in some habitats growing as a remnant in full sun, in mesic or submesic hardwood forests (frequently oak-hickory woods), alfisols, less frequent in entisols, vertisols, or histosols, regardless, most frequently found in sandy soils with a humus layer or thin soils over limestone with a humus layer; widespread in nc TX; also Pineywoods, upper Gulf Prairies and Marshes, and e Edwards Plateau. Fruiting Apr–Jul(–Oct). [*C. onusta* Mack., *C. plana* Mack.]

var. **muehlenbergii**, GOTTHILF MUHLENBERG'S CARIC SEDGE. Obligate to facultative heliophyte in entisols of open sand dunes, openings in sandy oak-hickory woods, sandy forest edges, open to semi-open sandstone outcrops, open pine barrens, or on thin soils over limestone, occasionally in alfisols or even histosols; Erath, Henderson, Lamar, Milam, Parker, and Tarrant cos.; also upper portions of Post Oak Savannah and Pineywoods. Fruiting (late Mar–)late Apr–mid-Aug(–early Sep).

Carex oxylepis Torr. & Hook., (sharp-scaled), SHARP-SCALE CARIC SEDGE. Rich moist hardwood forests, frequently along floodplains of forest creeks; ne and se portion of nc TX in Hunt, Lamar, Milam, and Red River cos.; also Post Oak Savannah, Pineywoods, and upper portion of Gulf Prairies and Marshes. Fruiting Mar–Apr(–early Jun). Section *Hymenochlaenae*

Carex ozarkana P. Rothr. & Reznicek, (of the Ozarks), OZARK CARIC SEDGE. A heliophyte in early successional wetlands on mineral soils, often in association with seepage, seepy banks of streams, permanently wet ditches, pond shores, and wet depressions in meadows and pastures; these sites are usually dominated by *Juncus* species; the soils are loamy, ranging from clay loams to silt loams, usually acidic; Hopkins and Lamar cos. in ne part of nc TX; also Upper Post Oak Savannah. Fruiting May. Section *Ovales*

Carex perdentata S.D. Jones, (having teeth), CONSPICUOUSLY-TOOTHED CARIC SEDGE. A facultative sciophyte, primarily in sandy loams, sandstone outcrops, granitic outcrops, or thin soils over limestone, open mesic to submesic hardwood forests, or open hardwood-juniper forest, or woodlands in savannas on granite outcrops; w 2/3 of nc TX; also Edwards Plateau. Fruiting mid-Mar–early Jun. Section *Phaestoglochin* This species, endemic to TX and OK, was described in 1994 (Jones 1994b).

Carex planostachys Kunze, (flat-spiked), CEDAR CARIC SEDGE. Dry oak-juniper or scrub on calcareous soils; Bell, Bosque, Coryell, Dallas, Hamilton, Hill, Hood, Johnson, McLennan, Parker, Somervell, and Tarrant cos.; also Post Oak Savannah, Edwards Plateau, Gulf Prairies and Marshes, and Trans-Pecos. Fruiting Mar–May. Section *Halleranae*

Carex reniformis (L.H. Bailey) Small, (kidney-shaped), KIDNEY-SHAPED CARIC SEDGE. A facultative sciophyte in wet woods and sloughs; Hopkins, Red River, and Tarrant Cos.; also Pineywoods, Post Oak Savannah, and upper portion of Gulf Prairies and Marshes. Fruiting May–Jun. Section *Ovales*

Carex retroflexa Muhl. ex Willd., (bent backward), REFLEXED-FRUIT CARIC SEDGE. Dry rocky or

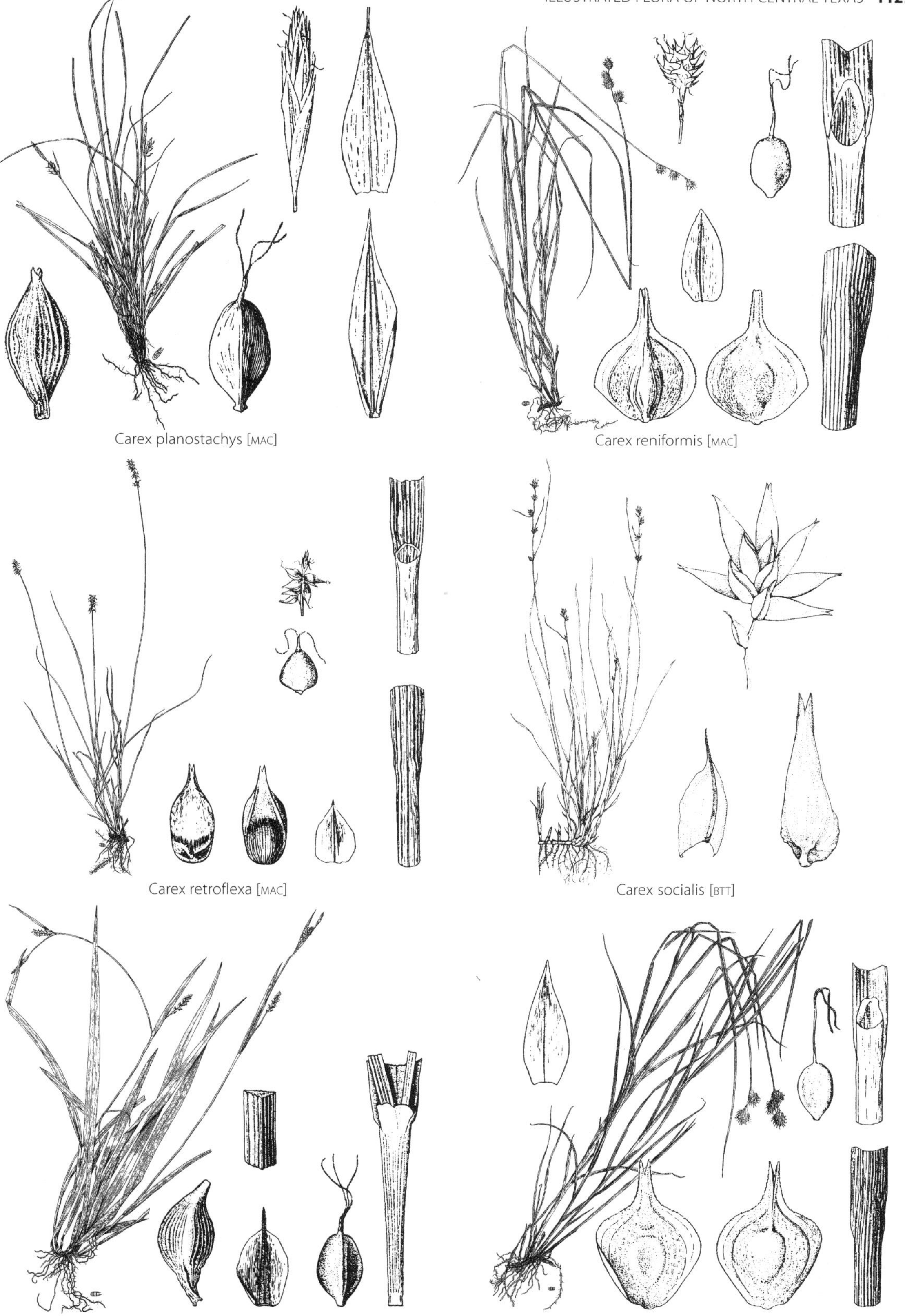

Carex planostachys [MAC]

Carex reniformis [MAC]

Carex retroflexa [MAC]

Carex socialis [BTT]

Carex striatula [MAC]

Carex tetrastachya [MAC]

sandy woods, thickets, and forest edges; widespread in nc TX; also Post Oak Savannah, Pineywoods, upper Gulf Prairies and Marshes, and e Edwards Plateau. Fruiting late Mar–May(–Jun). Section *Phaestoglochin*

Carex socialis Mohlenbr. & Schwegman, (sociable), COMPANION CARIC SEDGE. A facultative sciophyte in clay or sandy clay soils of secondary terraces of river floodplains; in the ne corner of nc TX in Red River Co., probably also in other ne cos., also Henderson Co. on the e margin of nc TX; also Post Oak Savannah, Pineywoods, and upper Gulf Prairies and Marshes. Fruiting late Mar–May. Section *Phaestoglochin*

Carex striatula Michx., (with fine longitudinal lines), FINE-LINE CARIC SEDGE. A facultative sciophyte, frequently on upper slopes of ravines in partial openings of deciduous forests; reported in nc TX from Dallas Co., but probably also in Lamar, Red River, and Delta cos.; also Post Oak Savannah and Pineywoods. Fruiting Mar–May. Section *Laxiflorae*

Carex tetrastachya Scheele, (four-spiked), FOUR-ANGLE CARIC SEDGE. Open, moist to wet sites, wet prairies, roadside ditches, open swamp and marsh edges, most frequent in calcareous soils; widespread in nc TX; also Post Oak Savannah, Gulf Prairies and Marshes, South Texas Plains, Edwards Plateau, and rarely in Pineywoods. Fruiting Mar–May(–early Jun). [*C. brittoniana* L.H. Bailey] Section *Ovales*

Carex texensis (Torr.) L.H. Bailey, (of Texas), TEXAS CARIC SEDGE. In submesic to mesic rocky, or sandy woods and fields; Delta, Fannin, Hopkins, and Lamar cos. in the ne part of nc TX; also Post Oak Savannah, Pineywoods, and upper Gulf Prairies and Marshes. Fruiting mid-Mar–mid-May(–early Jun). [*C. retroflexa* Muhl. ex Willd. var. *texensis* (Torr.) Fernald] Section *Phaestoglochin*

Carex triangularis Boeck., (triangular), TRIANGULAR CARIC SEDGE. A heliophyte in open wet roadside ditches and in open wet areas in floodplains, usually in alluvial clay soils; Delta, Hopkins, Lamar, and Red River cos. in the ne part of nc TX; also Post Oak Savannah, Pineywoods, and upper Gulf Prairies and Marshes. Fruiting May–Jun. Section *Multiflorae*

Carex tribuloides Wahlenb., (resembling *Tribulus*—caltrop), CALTROP CARIC SEDGE. Frequently in the open in bottomlands, swales, swamp margins, and marshes; ne part of nc TX in Delta, Lamar, and Red River cos.; also Post Oak Savannah, Pineywoods, and upper portion of Gulf Prairies and Marshes. Fruiting May–Aug. Section *Ovales*

Carex vulpinoidea Michx., (resembling *Carex vulpina*, with inflorescence like a fox tail), FOX-TAIL CARIC SEDGE. A heliophyte of wet roadside ditches, lakesides, pondsides, and open wet floodplains, usually in clayey soils; ne part of nc TX in Delta, Denton, Fannin, Hopkins, Lamar, and Red River cos.; also Post Oak Savannah and Pineywoods. Fruiting Jun–Aug. Section *Multiflorae.*

Carex affinity *bicknellii* Reznicek [*Ined.*], (the species will not be named *C. bicknellii*, a name already used in *Carex*; however, this taxon has an affinity towards that species). No illustration is available for this as yet unnamed species. A heliophyte in open wet swales and bottoms, depressions, or wet roadside ditches or ones with emphemeral water, usually in sandy soils; in nc TX in Delta, Kaufman, Lamar, Red River, and Tarrant cos.; also Fayette and Harris cos. to the e of nc TX. May–Jun. Section *Ovales*

CLADIUM SAW-GRASS, TWIG-RUSH

Rhizomatous perennials; culms obtusely trigonous; leaves basal and cauline, with well-developed blades; leaf sheaths loose; inflorescences cymosely branched, with numerous spikelets; spikelets with a single fertile floret subtended by 2–3 empty scales (these lacking achenes but

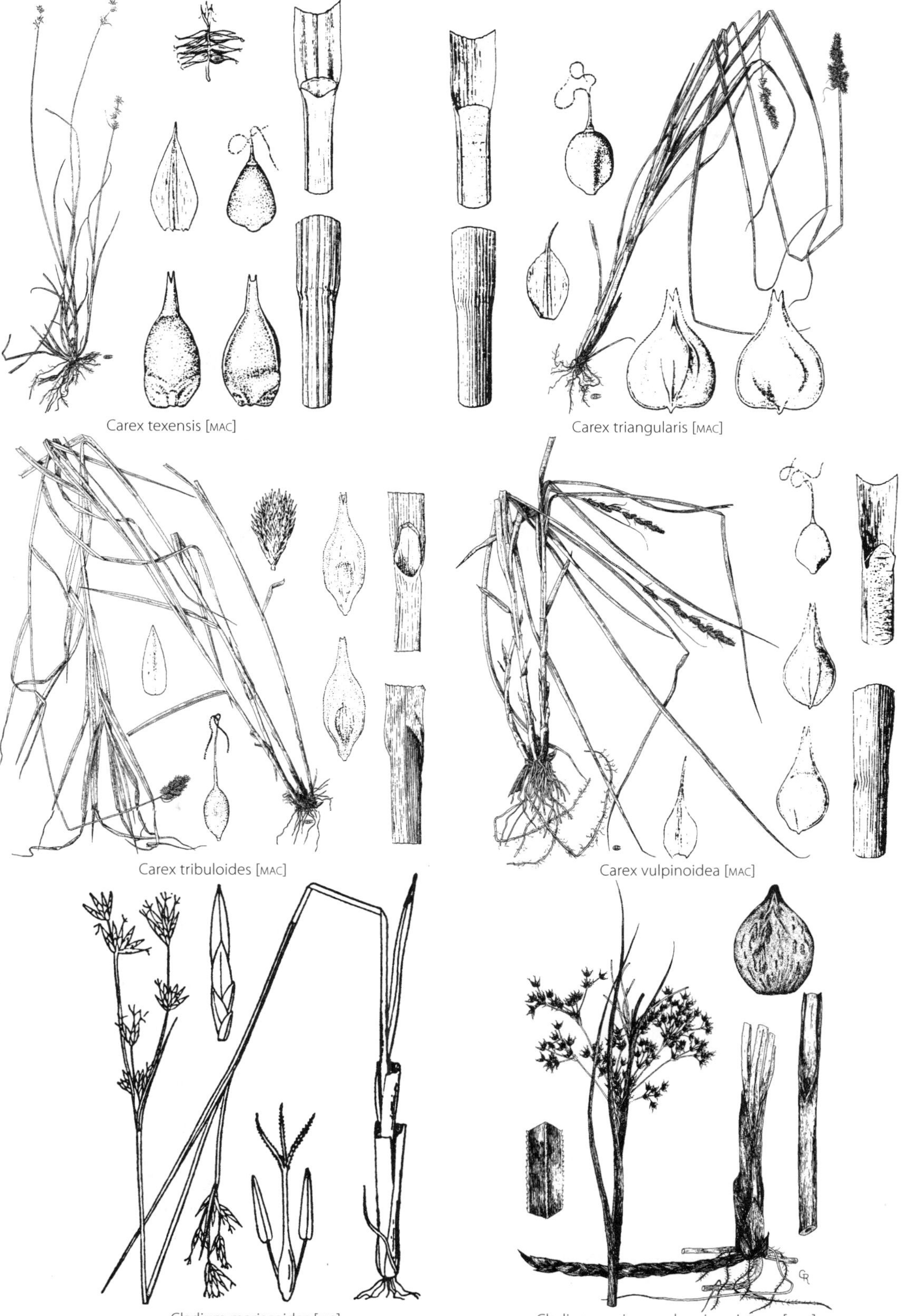

Carex texensis [MAC]

Carex triangularis [MAC]

Carex tribuloides [MAC]

Carex vulpinoidea [MAC]

Cladium mariscoides [BB2]

Cladium mariscus subsp. jamaicense [GWO]

can have stamens); scales of spikelets spirally imbricate; stamens 2; stigmas 2–3; perianth bristles absent; achenes without tubercles.

A genus of 2 species, 1 in North America, the other cosmopolitan. (Greek *cladion*, a branchlet, from the repeatedly branched inflorescence of the first named species) *Cladium* species superficially resemble some *Rhynchospora* taxa but can be easily distinguished by the lack of tubercles on the achenes.

1. Leaf blades 1–3 mm wide, to ca. 0.3 m long, with margins scaberulous (= only slightly roughened, almost smooth to the touch); plants to 1 m tall ____ **C. mariscoides**
1. Leaf blades 5–15 mm wide, to 1 m long, with margins dangerously saw-toothed; plants to 3 m tall ____ **C. mariscus**

Cladium mariscoides (Muhl.) Torr., (resembling *Mariscus*, a segregate now included in *Cyperus*), TWIG-RUSH. Plant 0.4–1 m tall; leaf blades involute, with scaberulous margins; inflorescences 5–30 cm long, slender, ca. 2–5 cm wide, usually of relatively few cymes, the inflorescences much smaller than in *C. mariscus*, with spikelets in groups of 3–10 at the ends of short, erect branches (peduncles); spikelets 3–6 mm long; achenes smooth, short cylindric, apiculate-pointed, truncate basally, 2.5–3.5 mm long. Wet areas; Henderson Co. (Bridges & Orzell 1989) near e margin of nc TX; mainly e TX; very rare in the state. Jul–Sep. [*Mariscus mariscoides* (Muhl.) Kuntze]

Cladium mariscus (L.) J. Pohl subsp. **jamaicense** (Crantz) Kük., (sp.: for resemblance to *Mariscus*; subsp.: of Jamaica), JAMAICAN SAW-GRASS, SAW-GRASS. Plant 1–3 m tall; leaf blades ca. 0.3–1 m long, with dangerously spinulose-serrulate (= saw-toothed) margins and midrib (on lower surface); inflorescences 20–80 cm long, 10–30 cm wide, much-branched, sometimes droopy, with spikelets in groups of 2–6 at the ends of short branches; spikelets 3–5 mm long; achenes with surfaces roughened, obovoid to subglobose, apiculate-pointed or obtuse, contracted basally, 2–3 mm long. Stream or lake margins, wet areas, often in calcareous soils; Dallas Co. (R. O'Kennon, pers. obs.); also se part of the state, Edwards Plateau, and Trans-Pecos (Hatch et al. 1990). Jul–Oct. [*Mariscus jamaicense* (Crantz) Britton]

CYPERUS FLAT SEDGE

Annuals or usually perennials; plants largely glabrous except for scabrous-margined leaves; culms (= stems) triangular; leaves basal or nearly so; inflorescences terminal, head-like or umbel-like, leafy-bracted at base; scales of spikelets 2-ranked (= in two distinct rows), the spikelets ± flattened or square; perianth bristles absent; achenes lenticular or trigonous.

A genus of ca. 300 species of annual or perennial herbs of tropical and warm areas; some are problematic weeds while others are cultivated as ornamentals. The commonly cultivated Old World *Cyperus papyrus* L. (PAPYRUS, PAPER-REED) was used by the Egyptians to make paper at least 5500 years ago; the Greek word for the plant was *papyros* from which our word paper is derived (Hepper 1992); this species was also used to make sandals, ropes, and boats (e.g., Moses in the bulrushes); the Greek word *byblos* was the name for the white pith of PAPYRUS used in making paper (the pith was cut into strips, glued together, and then pressed and dried—Zohary 1982); the word *byblos* became modified into *biblion* and was applied to all scrolls or books and eventually to the Bible (Hepper 1992). *Cyperus* is a taxonomically difficult genus with a number of taxa apparently hybridizing and intergrading morphologically. Intermediates (genetically contaminated individuals) between *C. croceus*, *C. echinatus*, *C. retroflexus*, and *C. retrorsus* are frequently seen. Similar problems occur within other complexes. (*Cypeiros*, the ancient Greek name)

REFERENCES: McGivney 1938, 1941a, 1941b; Corcoran 1941; Marcks 1972, 1974; Baijanth 1975;

Denton 1978; Tucker 1983, 1994; Carter 1984; Carr 1988; Carter & Kral 1990; Schippers et al. 1995; Jones et al. 1996; Carter & Jones 1997.

1. Achenes lenticular (= lens-shaped); styles 2-branched.
  2. Spikelets with only 2 scales (plus 2 minute, brownish, basal scales much smaller than regular scales); achene 1 per spikelet; inflorescences 3–8(–12) mm long ____ see **Kyllinga**
  2. Spikelets with 6 or more scales; achenes several per spikelet; inflorescences variable, often much larger.
    3. Spikelets mostly 1.0–1.9 mm wide, sharp-pointed; achenes 0.4–0.5 mm wide, narrowly oblong to oblong ____ **C. polystachyos**
    3. Spikelets mostly more than 2.0 mm wide, subacute to obtuse, not sharp-pointed; achenes 0.6–0.7 mm wide, usually obovoid, often nearly as broad as long.
      4. Achenes brown to grayish at maturity, with isodiametric cells, not transversely lined, distinctly apiculate apically; rare if present in nc TX ____ **C. lanceolatus**
      4. Achenes black at maturity, with longitudinally elongate cells, usually transversely lined, slightly apiculate apically; known from several counties in nc TX ____ **C. flavescens**
1. Achenes trigonous (= 3-angled); styles 3-branched.
  5. Spikelet axis separating at maturity at the floret nodes ____ **C. odoratus**
  5. Spikelet axis remaining intact or apparently so, the florets either falling separately from the persistent axis OR the entire axis falling as a unit (OR whether unclear in immature *C. odoratus*).
    6. Culms with conspicuous septa (= internal partitions, but visible externally) at intervals of 5–50 mm; leaves usually reduced to just a sheath (bladeless or blades to 2 cm long), the culms thus appearing nearly leafless ____ **C. articulatus**
    6. Culms nonseptate; leaf blades usually present and conspicuous (reduced in *C. haspan* and *C. involucratus*), the culms thus usually appearing leafy.
      7. Scales with strongly recurved (= curved backwards) long acuminate tips; plants with a persistent spice-like odor; scales 7–9 nerved; plants annual, usually 20 cm or less tall ____ **C. squarrosus**
      7. Scales usually incurved to essentially straight or curved back but without long tips; plants usually without a spice-like odor; scales 3–5 nerved; plants perennial OR annual in case of *C. compressus*, *C. difformis*, *C. erythrorhizos*, and sometimes *C. odoratus;* plants of various heights, often much more than 20 cm tall.
        8. Culms rough to the touch, sparsely to densely covered with microscopic retrorse (= down pointing) or antrorse (= up pointing) teeth.
          9. Spikelets usually stalked, in loose heads, the heads usually clustered together; lower spiklets neither markedly reflexed nor parallel to the peduncle; culms with retrorse teeth; largest scales usually 1.1–1.5 mm long; stamen 1; achenes 0.7–0.8 mm long ____ **C. surinamensis**
          9. Spikelets sessile, in dense ovoid heads occurring singly at the ends of elongate peduncles, the peduncles 2–16 cm long; lower spikelets markedly reflexed and ± parallel to the peduncle (appearing to droop around the peduncle); culms with antrorse teeth; scales usually 4–6 mm long; stamens 3; achenes 2.5–3.0 mm long ____ **C. plukenetii**
        8. Culms smooth to the touch, rarely with a few horizontal knobs.
          10. Scales usually slightly to strongly curved outward at the tips (except in *C. reflexus*); stamen 1; spikelet axis essentially wingless, persistent after scales and achenes have fallen.
            11. Leaves usually nodulose (= with knot-like septa visible under a hand lens); scales essentially linear, conspicuously falcate (= sickle-shaped); achenes linear, 1–1.3 mm long ____ **C. pseudovegetus**
            11. Leaves not nodulose; scales ovate, weakly S-shaped OR curved at base but straight at tip; achenes oblong to elliptic, 0.7–1.1(–1.2) mm long.

12. Scales weakly S-shaped, the tips curving outward; plants tufted, without rhizomes; widespread in nc TX ____ **C. acuminatus**
12. Scales curved at base but straight at tip; plants often with short, scaly, creeping rhizomes 1–3 mm thick; in nc TX known only from Denton Co. ____ **C. reflexus**

10. Scale tips not curving outward (but scales may be spreading); stamens usually 3; spikelet axis winged or wingless, persistent or whole spikelet falling as a unit.
13. Most leaves reduced to mere bladeless sheaths or occasionally the uppermost sheaths with short blades very rarely to 10 cm long.
14. Bracts usually 2, 1 of them 0.3–1(–2) times as long as the inflorescence, usually < 13 cm long; culms 0.1–0.7 m tall; native species ____ **C. haspan**
14. Bracts 10–25, often much surpassing the inflorescences, 15–40 cm long; culms 0.3–1.5 m tall; escaped cultivar ____ **C. involucratus**
13. Even the lower leaves with well-developed blades.
15. Scales minute, < 1 mm long; rare in nc TX, known only from Williamson Co. ____ **C. difformis**
15. Scales > 1 mm long; widespread in nc TX.
16. Scales 1.3–2.5(–3.2) mm long; achenes 0.7–1.5(–1.9) mm long; roots reddish OR not so; spikelets in rather loose elongate spikes, the spike axis visible.
17. Roots usually reddish; scales small, 1.3–1.5 mm long; achenes 0.7–0.8 mm long ____ **C. erythrorhizos**
17. Roots not reddish; scales 1.5–2.5(–3.2) mm long; achenes (1–)1.2–1.5(–1.9)mm long ____ **C. odoratus**
16. Scales (2.3–)2.5–5.5 mm long; achenes 1–3 mm long; roots usually not reddish; spikelets in rather loose elongate spikes as above OR in crowded or extremely densely packed short compact heads with the inflorescence axis usually not visible.
18. Achenes 1–1.3 mm long, nearly as thick as long; spikelets 10–24 mm long, 2–3.5(–4) mm wide, conspicuously flattened, digitately arranged (= all arising from about the same point on a very short axis) ____ **C. compressus**
18. Achenes 1.3–3 mm long, much longer than thick; spikelets various, flattened to angled or nearly rounded, pinnate or digitate in dense heads or open spikes.
19. Spikelet axis essentially wingless; spikelets 3–16 mm long with scales 2.5–4.2 mm long, in crowded (but not extremely dense) heads or short spikes.
20. Plants viscid (= sticky); leaves spongy at base, the dried leaves nodulose (= with knot-like septa visible with a hand lens) basally; longer peduncles branched into head-bearing secondary peduncles ____ **C. elegans**
20. Plants nonviscid; leaves neither spongy nor nonseptate basally; secondary peduncles usually absent.
21. Inflorescence usually a single nearly spherical head 1–3 cm long (sometimes with a few peduncles bearing small heads); main head with 15–55 spikelets; widely scattered in nc TX ____ **C. lupulinus**
21. Inflorescence 3–10 cm long, of 4–8 peduncles, each bearing short spikes 1–2 cm long; spikes with 5–18 spikelets each; rare if present in nc TX ____ **C. schweinitzii**

19. Spikelet axis winged; spikelets, scales, and inflorescences various, similar to above OR quite different.
    22. Lower spikelets markedly reflexed and appearing ± drooping around the peduncle; heads cylindric or obovoid; anthers 1 mm or more long; rare in nc TX ________ **C. hystericinus**
    22. Lower spikelets not markedly reflexed OR if somewhat reflexed, then anthers < 1 mm long; heads various; anthers various, < 1 mm long in many species OR 1 mm or more long; widespread in nc TX.
        23. Plants rhizomatous perennials; anthers 1 mm or more in length; spikelet axis persistent.
            24. Bracts 3 or 4, about equaling inflorescence in length; spikelets 3–9 per spike ________ **C. rotundus**
            24. Bracts 5–13, greatly exceeding inflorescence; spikelets 10–50 per spike.
                25. Spikelets reddish brown; achenes 0.4–0.5 mm thick; culms (60–) 75–110 cm tall ________ **C. setigerus**
                25. Spikelets brown to golden-brown; achenes 0.6-0.8 mm thick; culms 15–50(–65) cm tall ________ **C. esculentus**
        23. Plants nonrhizomatous perennials; anthers less than 1 mm long; spikelet axis deciduous at base.
            26. Spikelets usually 12–25 mm long, usually pinnately arranged in often rather loose elongate spikes, the spike axis visible ________ **C. strigosus**
            26. Spikelets 3.5–10 mm long, usually crowded or extemely densely packed into short compact heads (inflorescence axis usually not visible), digitately arranged or if slightly pinnate, then the heads extremely dense.
                27. Inflorescences extremely densely cylindric or subcylindric, usually less than 8(–10) mm broad; spikelets so dense that outline of head is smooth ________ **C. retrorsus**
                27. Inflorescences neither densely cylindric nor subcylindric, mostly > 8 mm in diam., extremely densely globose or subglobose or spikelets crowded in globose or subglobose heads; outline of head appearing somewhat rough or smooth.
                    28. Inflorescences extremely densely globose or subglobose with 100–250 spikelets per head; spikelets so dense that outline of head is smooth ________ **C. echinatus**
                    28. Inflorescences spherical, with spikelets crowded (but not extremely densely), globose or subglobose with 10–70 spikelets per head; outline of head appearing somewhat rough.
                        29. Achenes concave in cross-section; largest scales usually 3 mm or longer;

achenes usually 1.7–2.5 mm long; plants 3–35(–60) cm tall; spikelet axis with thickened and slightly discolored wings clasping the achene __________ **C. retroflexus**

29. Achenes convex in cross-section; largest scales usually 3 mm or less long; achenes usually 1.7 mm or less long; plants 10–80 cm tall; spikelet axis wings neither discolored nor clasping the achene __________ **C. croceus**

Cyperus acuminatus Torr. & Hook. ex Torr., (tapering at tip), TAPER-LEAF FLAT SEDGE. Tufted perennial 10–80 cm tall; culms slender; inflorescences usually compact; sheaths not nodulose; blades 1–4 mm wide; scales weakly S-shaped, the tip with a slight to marked outward curve. Moist areas; nearly throughout TX. Mostly May–Oct.

Cyperus articulatus L., (jointed), CHINTÚL, JOINTED FLAT SEDGE. Perennial 0.5–1.4 m tall with creeping rhizomes, forming colonies; leaves few, basal, reduced to small essentially bladeless sheaths; inflorescences essentially bractless or with very small bracts; spikelets 6–33(–45) mm long. Moist grassland; se TX n to Comal and McLennan cos.; disjunct n to Grayson Co. (extensive colony on edge of small tank on bluff near Red River). May–Oct. The rhizome has been used medicinally (Burkhill 1985).

Cyperus compressus L., (flattened), POORLAND FLAT SEDGE. Tufted annual or occasionally a short-lived perennial. Weed in shrubbery, black clay; Dallas Co.; mainly se and e TX. Oct.

Cyperus croceus Vahl, (saffron-colored, yellow), BALDWIN FLAT SEDGE. Tufted perennial 15–70 cm tall; heads or spikes 8–20 mm broad; spikelets 3–8 mm long; scales with green keel and reddish or yellow-brown sides. Sandy open areas; Grayson, Kaufman, and Tarrant cos., also Milam Co. (S.D. Jones, pers. comm.); mainly e and se TX and Edwards Plateau. May–Oct. [*C. globulosus* of authors, not Aubl.]

Cyperus difformis L., (of unusual or differing forms). Annual 10–50 cm tall; roots red; leaves 2–4 per culm, 1–4 mm wide; heads globose or lobulate; spikelets 4–8 mm long; scales roundish, obtuse, very small, 0.5–0.8 mm long, green with brownish or purplish sides. Unshaded creek beds in perennially wet mud in shallow water over limestone or dolomite, creek banks, lake shores, other wet areas; Williamson Co. (Carr 1988); in TX otherwise known in Travis Co. (Carr 1988); first collected in TX in 1981 and first reported by Carr (1988). Native of Eurasia. Lipscomb (1980) discussed the distribution of *C. difformis* in North America.

Cyperus echinatus (L.) A.W. Wood, (prickly), GLOBE FLAT SEDGE, CYLINDER FLAT SEDGE. Tufted perennial 15–70 cm tall; heads or spikes globose or subglobose, 8–21 mm long, 8.5–18 broad, less than 1/4 longer than broad. Sandy open areas; se and e TX w to West Cross Timbers, also Edwards Plateau. May–Oct. [*C. ovularis* (Michx.) Torr.]

Cyperus elegans L., (elegant), STICKY FLAT SEDGE. Tufted viscid (= sticky) perennial 25–80 cm tall; dried leaves nodulose (= with knot-like septa visible with a hand lens) basally; spikelets mostly in head-like clusters; longer peduncles branched into head-bearing secondary peduncles at ends of branches. Damp soils; known in nc TX only from Erath Co. (West Cross Timbers); mainly s TX. Jun.

Cyperus erythrorhizos Muhl., (red-rooted), RED-ROOT FLAT SEDGE. Tufted annual 0.5–1.4 m tall; fresh roots usually reddish; inflorescences umbel-like; peduncles unequal; spikes several per peduncle; internodes of spikes 0–0.5 mm long; scales relatively small, 1.3–1.5 mm long; achenes

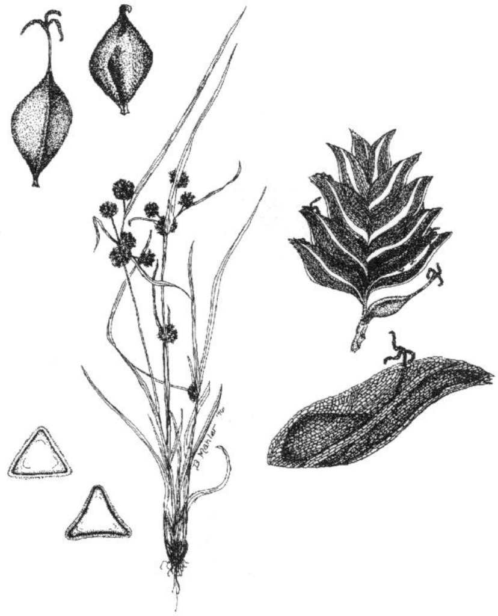

Cyperus acuminatus [BT2]

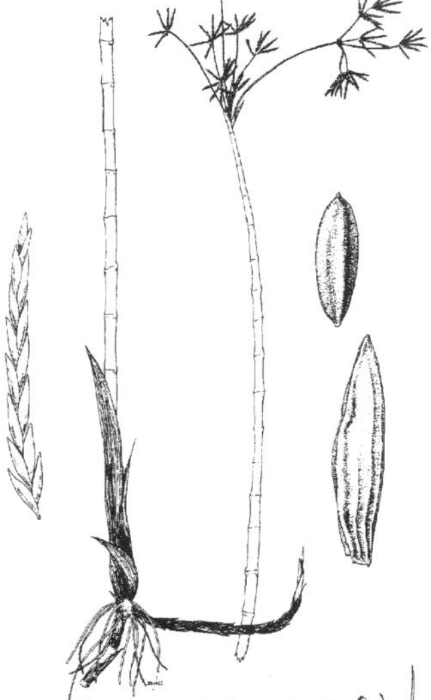

Cyperus articulatus [CO1]

Cyperus compressus [BT2]

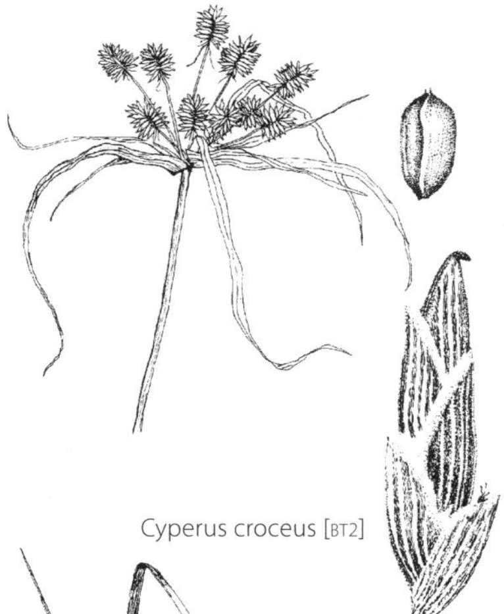

Cyperus croceus [BT2]

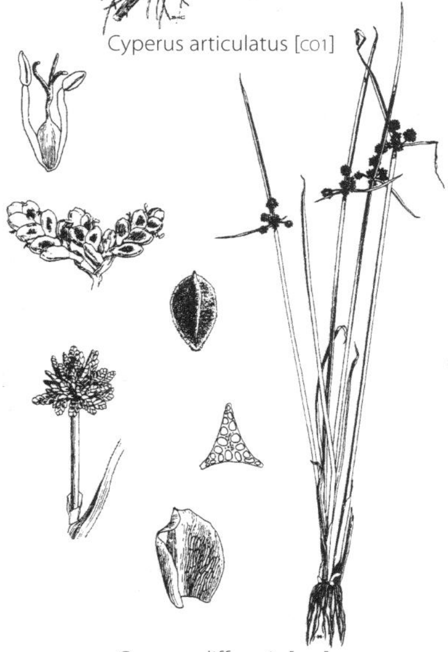

Cyperus difformis [MAS]

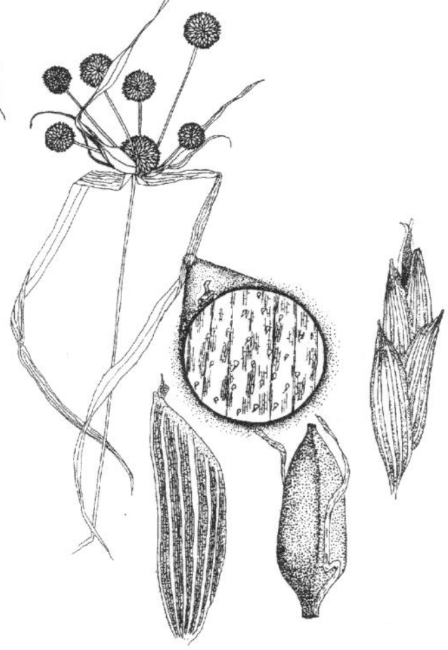

Cyperus echinatus [BT2]

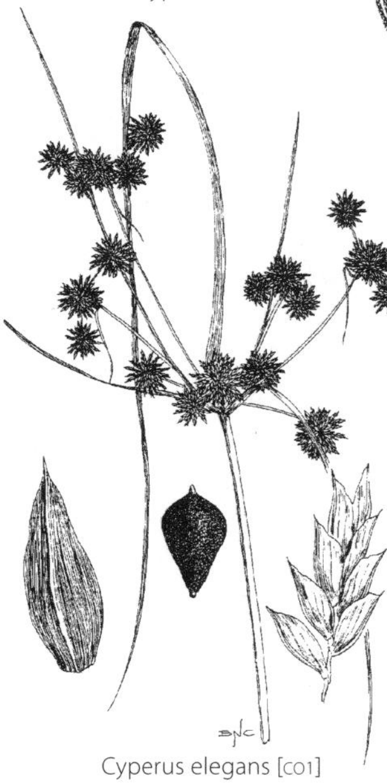

Cyperus elegans [CO1]

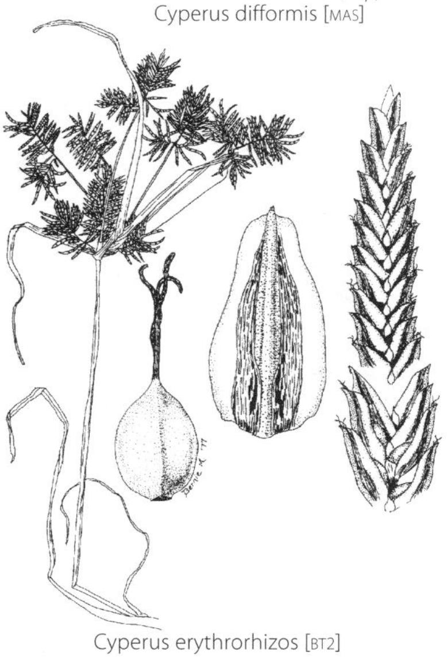

Cyperus erythrorhizos [BT2]

Cyperus esculentus [REE]

0.7–1 mm long. Marshy areas; se and e TX w to Grayson and Parker cos., also Edwards Plateau. Jul–Dec.

Cyperus esculentus L., (edible), YELLOW NUT-GRASS, CHUFA, NORTHERN NUT-GRASS. Perennial, colonial, 15–50 cm tall; rhizomes sometimes with tuber-like thickenings; anther connective prolonged into a red dot 0.05–0.1 mm long. Sandy disturbed soils; scattered nearly throughout TX. Summer–fall. According to Mabberley (1987), native to w Asia and Africa and widely naturalized in New World; however, Tucker (1994) considered it to be cosmopolitan. [*C. esculentus* var. *leptostachyus* Boeck., *C. esculentus* var. *macrostachyus* Boeck.] Schippers et al. (1995) discussed infraspecific variation in this widespread species; it can be a troublesome weed infesting a variety of crops (Holm et al. 1977). Varieties are sometimes recognized (e.g., Jones et al. 1997). According to Crosswhite (1980), the nut-like, edible, tuber-like thickenings were used during pioneer days.

Cyperus flavescens L., (yellowish), YELLOW FLAT SEDGE. Tufted annual 10–25 cm tall; spikelets (1.8–)2–3 mm wide; scales yellow-green to yellowish brown; achenes black, shiny, with rectangular to linear (vertical) cells, the rows of cells marked by horizontal, wavy, usually discolored sutures. Moist sand; Denton, Hamilton, and Tarrant cos.; mainly se and e TX. Jul–Nov. [*C. flavescens* var. *poaeformis* (Pursh) Fernald]

Cyperus haspan L., (the native name in Ceylon), SHEATHED FLAT SEDGE. Tufted perennial 10–70 cm tall; most leaves reduced to bladeless sheaths or rarely with short blades. Moist places; near s edge of nc TX in Burnet Co. (S.D. Jones, pers. comm.) and expected in other parts of nc TX; e and se TX and Central Mineral Region. Jun–Oct. This pantropical to warm temperate species can be a pernicious weed.

Cyperus hystricinus Fernald, (porcupine-like, bristly). Perennial to 1 m tall; rhizomes thick, to 1.5 cm long; leaf blades 4–6 mm wide; spikelets golden brown; lower spikelets markedly reflexed and appearing ± drooping around the peduncle. Xeric sandy soils; Henderson and Tarrant cos. (Tucker 1984); mainly se TX. Summer–fall. [*C. retrofractus* (L.) Torr. var. *hystricinus* (Fernald) Kük.]

Cyperus involucratus Rottb., (with an involucre), UMBRELLA-PLANT, UMBRELLA FLAT SEDGE. Perennial to 1.5 m tall; bracts very numerous (10–25) and very long (15–40 cm). Widely cultivated, persists, and spreads; Tarrant Co. Summer–fall. Native of Old World, probably Africa or Madagascar. [*C. alternifolius* of authors, not L., *C. alternifolius* L. subsp. *flabelliformis* (Rottb.) Kük., *C. flabelliformis* Rottb.] Baijnath (1975) discussed nomenclature for this species.

Cyperus lanceolatus Poir., (lanceolate, lance-shaped). Tufted or mat-forming perennial 5–50 cm tall; spikelets straw-colored or sometimes with an olivaceous tinge, but without darker splotch. Included based on citation of vegetational area 5 (Fig. 2) by Hatch et al. (1990) but possibly not present in nc TX (S.D. Jones, pers. comm.); also Central Mineral Region just s of nc TX (Correll & Johnston 1970). Sep. [*C. lanceolatus* var. *compositus* J. Presl & C. Presl]

Cyperus lupulinus (Spreng.) Marcks, (resembling *Humulus lupulus*—hops), SLENDER FLAT SEDGE. Tufted perennial 15–50 cm tall, with hard swollen culm bases, often developing short knotty rhizomes; inflorescences typically of a single nearly spherical head or sometimes with a few peduncles bearing small heads; spikelets gray-green. Sandy open woods; Grayson, Kaufman, and Tarrant cos.; widely scattered in TX. May–Jun, occasionally to Sep. [*C. filiculmis* of authors, not Vahl—Marks 1974]

Cyperus odoratus L., (fragrant), FRAGRANT FLAT SEDGE. Tufted annual or perennial 5–60 cm tall; spikelets at maturity dull reddish to brownish; spikelet axis separating at maturity at the floret nodes (because this is not always evident in young material, *C. odoratus* can also be reached in

Cyperus flavescens [BT2]

Cyperus haspan [GWO]

Cyperus hystricinus [BB2]

Cyperus involucratus [BA1, BR2]

Cyperus lanceolatus [HEA]

Cyperus lupulinus [BT2]

Cyperus odoratus [BT2]

Cyperus plukenetii [BT2]

the key without recognizing this character); scales relatively small, 1.5–2.5 mm long; achenes (1–) 1.2–1.5(–1.9) mm long. Stream banks or moist areas; abundant in all parts of TX. Jun–Oct. [*C. engelmannii* Steud., *C. ferruginescens* Boeck., *C. odoratus* var. *engelmannii* (Steud.) R. Carter, S.D. Jones, & Wipff, *C. odoratus* var. *squarrosus* (Britt.) S.D. Jones, J. Wipff, and R. Carter] This is one of the most common *Cyperus* species in nc TX. It is often treated as a single variable species (Correll & Johnston 1970; Kartesz 1994; Tucker 1994); Tucker (1983, 1987) for example, indicated that there was not a single consistent character separating *C. engelmannii* from *C. odoratus* and that the two intergrade extensively. Correll and Johnston (1970) indicated concerning *C. odoratus* that "... it is impossible to distinguish segregate taxa." Given that the varieties overlap morphologically and are mostly sympatric, treatment as a variable species without infraspecific taxa is possibly best. Jones et al. (1996), however, recognized the following three varieties indicating that "We find these three taxa closely related and mostly sympatric, but discrete. Although some intermediates exist, they are relatively few. Considering their distinct morphologies, we believe that varietal rank under *C. odoratus* is warranted. ..." The following key is from Jones et al. (1996) modified from O'Neill (1940).

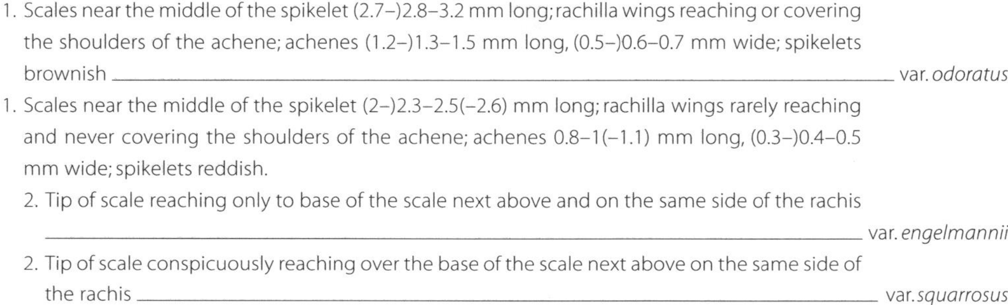

1. Scales near the middle of the spikelet (2.7–)2.8–3.2 mm long; rachilla wings reaching or covering the shoulders of the achene; achenes (1.2–)1.3–1.5 mm long, (0.5–)0.6–0.7 mm wide; spikelets brownish ______ var. *odoratus*
1. Scales near the middle of the spikelet (2–)2.3–2.5(–2.6) mm long; rachilla wings rarely reaching and never covering the shoulders of the achene; achenes 0.8–1(–1.1) mm long, (0.3–)0.4–0.5 mm wide; spikelets reddish.
   2. Tip of scale reaching only to base of the scale next above and on the same side of the rachis ______ var. *engelmannii*
   2. Tip of scale conspicuously reaching over the base of the scale next above on the same side of the rachis ______ var. *squarrosus*

Cyperus plukenetii Fernald, (for Leonard Plukenet, 1642–1704, one of the original describers and illustrators of American plants). Tufted perennial 30–100 cm tall; culm bases hard, swollen, sometimes developing short knotty rhizomes; culms rough to the touch, with minute antrorse teeth; lower spikelets appearing to droop around the peduncle. Sandy woods; Henderson and Limestone cos., also Dallas Co. (S.D. Jones, pers. comm.); se and e TX w to e Blackland Prairie. Jun–Sep.

Cyperus polystachyos Rottb., (many-spiked). Tufted perennial; culms ca. 3–35 cm long; inflorescences variable, with 1–8(–12) main peduncles terminated by clusters of spikelets or clusters sometimes sessile or sometimes inflorescences secondarily branched; spikelets sessile to short-stalked, sometimes diverging ± at right angles from the axis, sometimes ± strongly ascending. Stream banks, moist sand; Hopkins and Limestone cos. on e margin of nc TX; mainly se and e TX. Spring–fall. [*C. polystachyos* Rottb. var. *texensis* (Torr.) Fernald] We are following Jones et al. (1997) in lumping var. *texensis*.

Cyperus pseudovegetus Steud., (false vigorous). Tufted perennial; inflorescenes usually with 3–10 main peduncles terminated by head-like clusters of spikelets, sometimes secondarily branched; spikelets small, 2.5–4 mm long; scales essentially linear, conspicuously sickle-shaped, when spread out 0.6–0.7 mm wide; leaves usually with knot-like septa visible under a hand lens; achenes linear, 1.0–1.3 mm long, maturing brownish. Sandy soils; se and e TX w to Grayson and Parker cos. Jun–Oct. [*C. virens* Michx. var. *arenicola* (Boeck.) Shinners]

Cyperus reflexus Vahl, (bent back), BENT-AWN FLAT SEDGE. Perennial with creeping rhizomes, similar to *C. acuminatus* (which however, has S-shaped scales) and *C. pseudovegetus* (which however, has much narrower scales); scales ca. 1.1 mm wide when spread out, often reddish

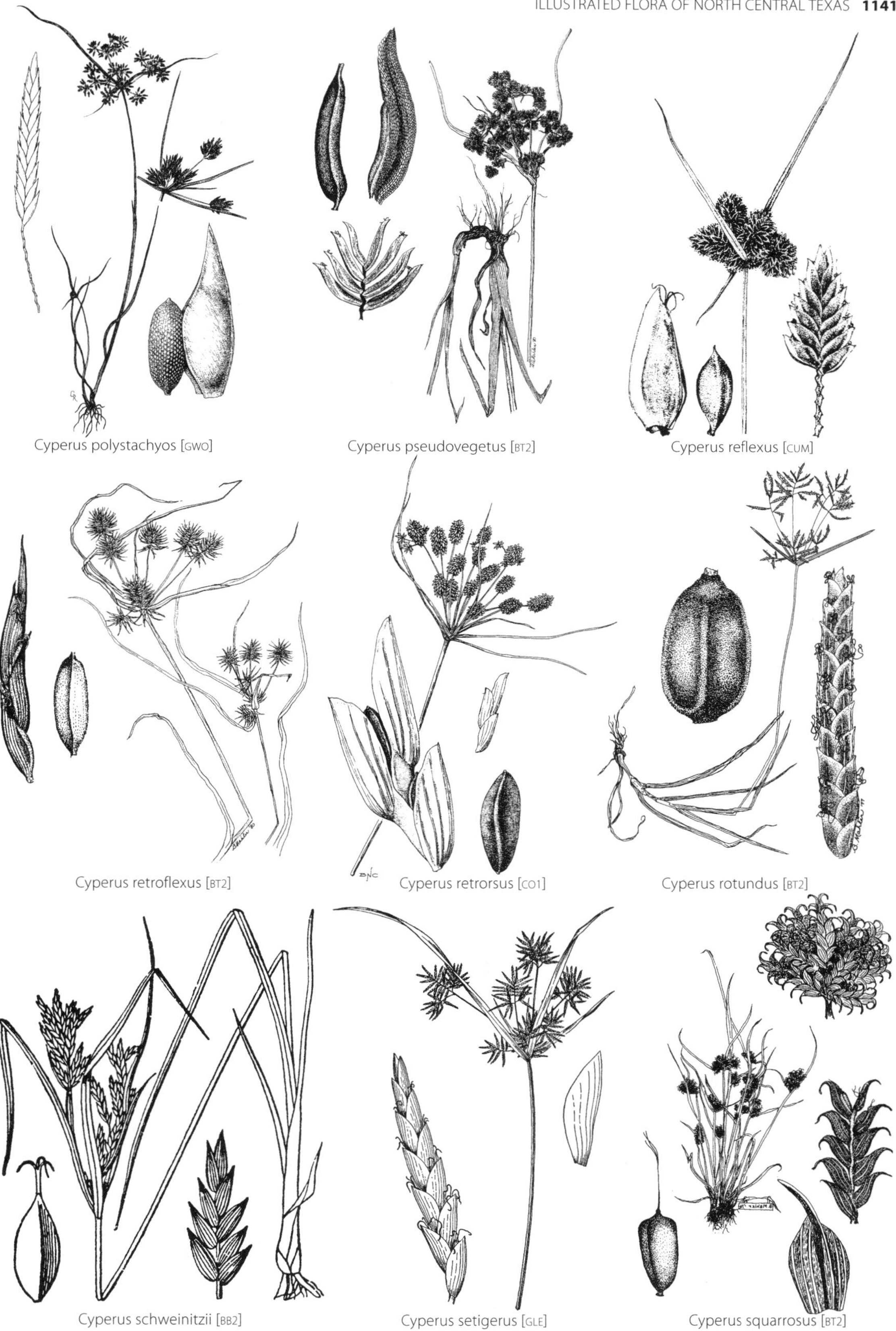

Cyperus polystachyos [GWO]

Cyperus pseudovegetus [BT2]

Cyperus reflexus [CUM]

Cyperus retroflexus [BT2]

Cyperus retrorsus [CO1]

Cyperus rotundus [BT2]

Cyperus schweinitzii [BB2]

Cyperus setigerus [GLE]

Cyperus squarrosus [BT2]

with greenish keels, straight at base but curved at tip; achenes pale brown, 0.9–1.2 mm long. Moist sand; Denton Co., also Bell and Henderson cos. (S.D. Jones, pers. comm.); mostly se and e TX, also Edwards Plateau. Spring–summer. [*C. reflexus* var. *fraternus* (Kunth) Kuntze]

Cyperus retroflexus Buckley, (reflexed), ONE-FLOWER FLAT SEDGE. Tufted perennial usually 3–35(–60) cm tall; scales at maturity often deep red-brown with prominent green keel. Sandy open woods or prairies; throughout TX. May–Oct. [*C. retroflexus* var. *pumilus* (Britt.) R. Carter & S.D. Jones, *C. uniflorus* Torr. & Hook., not Thunb., *C. uniflorus* var. *pumilus* Britt., *C. uniflorus* var. *retroflexus* (Buckley) Kük.] Infraspecific taxa are often not recognized in this species (e.g., O'Neill 1942; Kartesz 1994; Tucker 1994) and given the extensive overlap in morphological characters between the varieties and the lack of geographical isolation, treatment as a variable species without infraspecific taxa is possibly best. However, the following varieties were recognized by Carter and Jones (1997). They indicated that despite "overlap in virtually every characteristic we examined," when combinations of characteristics were used, most specimens could be identified to variety. According to S.D. Jones (pers. comm.), both varieties occur throughout nc TX, but var. *retroflexus* is by far the most common. The following key is from Carter and Jones (1997).

1. Fertile floral scales 1.9–3.0(–3.3) mm long; rachilla wing usually membranaceous throughout; rachilla usually lacking lateral nerves; longest spikelets 2.2–5.8(–8) mm long; terminal sterile floral scale of spikelet often much reduced, less than 2/3 the length of fertile floral scales; longest peduncle less than 2.7(–3.9) cm long; plants diminutive, 3–35(–45) cm tall ____ var. *pumilus*
1. Fertile floral scales (2.8–)3.0–3.9 mm long; rachilla wing usually chartaceous beyond clasped achene angle, border membranaceous; rachilla usually with two lateral nerves, one along each side of median; longest spikelets 4.9–9(–11.3) mm long; terminal sterile floral scale usually not greatly reduced, 2/3 or more the length of fertile floral scales; longest peduncle (0.5–)2.4–6.8 cm long; except for depauperate specimens, plants usually greater than 25(–57) cm tall ____ var. *retroflexus*

Cyperus retrorsus Chapm., (twisted or turned backward). Tufted perennial; spikes extremely densely cylindric or subcylindric, 6–8(–10) mm broad by 7–18 mm long, more than 1/4 longer than broad. Ditches, roadsides, open woods, usually in sandy soils; se and e TX w to West Cross Timbers. Jun–Oct. Because of indistinct boundaries even between species within this complex, we are not recognizing infraspecific taxa within *C. retrorsus*. [*C. ovularis* (Michx.) Torr. var. *cylindricus* (Elliott) Torr., *C. globulosus* Aubl. var. *robustus* (Boeck.) Shinners]

Cyperus rotundus L., (round), NUT-GRASS, NUT SEDGE, COCO-GRASS, PURPLE NUT-GRASS. Deeply rhizomatous perennial 7–50 cm tall, forming colonies; rhizomes at intervals with tube-like thickenings; each cluster of inflorescence with 2–12 spikelets; spikelets 4–40 mm long, dark red-brown to purplish. Disturbed waste or lawn areas; widespread in TX. May–Oct. Native of the Old World. Often a pernicious weed and sometimes referred to as "the world's worst weed" (Mabberley 1987).

Cyperus schweinitzii Torr., (for its discoverer, Lewis David von Schweinitz, 1780–1834, German-born Pennsylvania clergyman and student of fungi), SCHWEINITZ' FLAT SEDGE. Tufted perennial 15–45(–80) cm tall; achenes dark-brown. Deep sands; included based on citation of vegetational area 5 (Fig. 2) by Hatch et al. (1990); also Plains Country and Trans-Pecos (Correll & Johnston 1970). Spring and summer. According to S.D. Jones (pers. comm.), this species does not occur in nc TX.

Cyperus setigerus Torr. & Hook., (bearing bristles). Perennial 40–110 cm tall with shallow creeping rhizomes, forming small colonies; spikelets reddish brown. Stream or pond banks, low areas; widespread in TX. May–Sep.

Cyperus squarrosus L., (with recurved tips), BEARDED FLAT SEDGE. Tufted annuals 1–20 cm tall, with persistent odor, like coffee-and-chicory or curry powder; inflorescences of 1–3 heads, es-

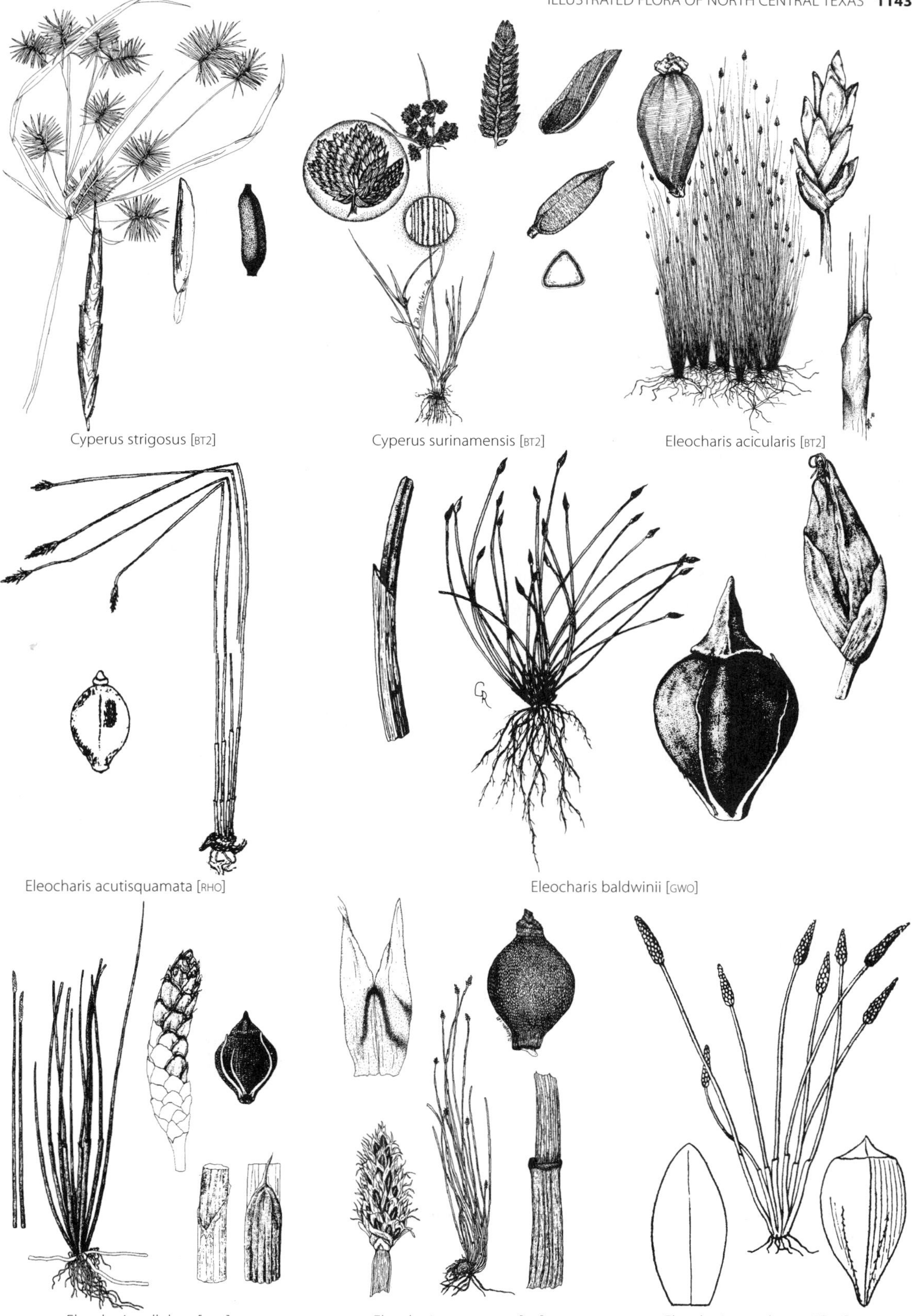

Cyperus strigosus [BT2]

Cyperus surinamensis [BT2]

Eleocharis acicularis [BT2]

Eleocharis acutisquamata [RHO]

Eleocharis baldwinii [GWO]

Eleocharis cellulosa [GWO]

Eleocharis compressa [BT2]

Eleocharis engelmannii [ABR]

sentially sessile; scale tips prominently recurved. Disturbed soils, sand; Bell, Bosque, Dallas, Grayson, Hood, Parker, and Tarrant cos.; nearly throughout TX. Jun–Jul. [*C. aristatus* Rottb.]

Cyperus strigosus L., (strigose, with straight appressed hairs bent at base), FALSE NUT-GRASS. Tufted perennial 10–110 cm tall, without rhizomes; spikelets usually pinnately arranged in often rather loose elongate spikes, the spike axis visible; scales 3.75–4.5 mm long. Low sandy soils; e 1/2 of TX. Jun–Oct.

Cyperus surinamensis Rottb., (of Surinam), TROPICAL FLAT SEDGE. Tufted perennial, short-lived, flowering first year, 10–40(–80) cm tall; culms rough to the touch, with minute retrorse teeth. Low moist areas; Grayson and Tarrant cos., also Dallas Co. (Lipscomb 1978a), also Williamson Co. (S.D. Jones, pers. comm.); s and se TX n to nc and e TX. Jul–Nov.

Cyperus bipartitus Torr., (two-parted), is cited by Hatch et al. (1990) for vegetational area 4 (Fig. 2) but apparently occurs only to the s and e of nc TX. This species can be distinguished from the similar *C. flavescens* as follows: scales yellow with large blotches of chestnut or purplish brown, rarely entirely purplish. [*C. rivularis* Kunth, *C. niger* Ruiz & Pavon var. *rivularis* (Kunth) V.E. Grant]

Cyperus digitatus Roxb., (finger- or hand-like) FINGER FLAT SEDGE, is cited by Hatch et al. (1990) for vegetational area 4 (Fig. 2) but apparently occurs only to the s and e of nc TX. It differs from *C. erythrorhizos* as follows: internodes of spikes 0.6–2 mm long; scales 1.5–2 mm long.

Cyperus ochraceus Vahl, (ochre-colored), is cited by Hatch et al. (1990) for vegetational area 4 (Fig. 2) but apparently occurs only to the s and e of nc TX. This species can be distinguished from *C. pseudovegetus* and *C. reflexus* as follows: scales broader when spread out, 1.5–1.9 mm wide; achenes maturing nearly black, 1.3–1.5 mm long.

ELEOCHARIS SPIKE-RUSH

Culms (= stems) green and glabrous, flat, round, or angled, with leaf sheaths present on basal part; leaf blades absent; inflorescence a single terminal spikelet at the end of a usually elongate culm (hence name SPIKE-RUSH); scales of spikelets spirally arranged (± 2-ranked in 1 species—*E. baldwinii*); perianth of bristles or absent; achenes with a tubercule (= hardened persistent style).

A cosmopolitan genus of ca. 120 species; *E. dulcis* (Burm. f.) Hensch. (WATER-CHESTNUT), native to the Old World tropics, is often used in Chinese cuisine. (Greek: *elos*, marsh, and *charis*, grace; many species being marsh plants)

REFERENCES: Fernald & Brackett 1929; Svenson 1929, 1939, 1953, 1957; Drapalik & Mohlenbrock 1960; Harms 1968, 1972; González-Elizondo & Peterson 1997.

1. Scales of spikelets with long-pointed scarious tip (often split in two).
   2. Culms usually several-angled or -ribbed, not strongly flattened, 0.3–0.8 mm wide; widespread in nc TX ____ **E. acutisquamata**
   2. Culms strongly flattened, to 1.5 mm wide; apparently erroneously reported for nc TX, questionably present in deep e TX ____ **E. compressa**
1. Scales of spikelets with rounded or broadly short-pointed tip (sometimes split).
   3. Spikelets tiny, 1–2 mm thick in flower excluding styles or stamens, 2–7 mm long, with 2–15 scales; culms thread-like.
      4. Scales ± 2-ranked, the spikelets flattish; tubercle ca. 1/3 or more the length of the achene body; rare, known locally only from e margin of nc TX ____ **E. baldwinii**
      4. Scales spirally arranged, the spikelets not flattish; tubercle smaller (1/4 or less the length of the achene body) or not distinct; widespread in nc TX.

5. Achene body with 6–9 ribs or ridges running lengthwise with fine perpendicular lines between ribs **E. acicularis**
5. Achene body smooth, without ribs, ridges, or lines.
    6. Tubercle not differentiated from achene body, blending into achene body so that the junction is not visible (appearing as though tubercle not present); achene body gray to black at maturity, 0.8–1.5 mm long; widespread in nc TX **E. parvula**
    6. Tubercle constricted basally, well-differentiated from achene body, the junction with achene body clearly visible; achene body pearly white to pale greenish gray, 0.5–0.6 mm long; rare, known in nc TX only from Lamar Co. in Red River drainage **E. microcarpa**

3. Spikelets 1.8–5 mm thick in flower, 3–25 mm or more long, with (3–)15–90 or more scales; culms coarsely thread-like, wiry, or thick.
    7. Culms 2–5 mm wide; spikelets of roughly the same diam. at the supporting culm.
        8. Culms conspicuously quadrangular (= square in cross-section) **E. quadrangulata**
        8. Culms not quadrangular.
            9. Culms strongly flattened to ± rounded, 0.9–3 mm wide; spikelets 8–25 mm long; scales usually acute; styles 2-branched; body of achenes 1.2–1.8 mm long, the tubercle basally constricted, distinct from the achene body, the junction between the 2 clear **E. palustris**
            9. Culms ± rounded, 2–5 mm wide; spikelets 19–36(–50) mm long; scales obtuse; styles 3-branched; body of achenes ca. 2 mm long, the tubercle ± appearing as a continuation of the achene body, not at all constricted basally **E. cellulosa**
    7. Culms 2 mm or less wide; spikelets distinctly thicker than the supporting culm.
        10. Plants annuals with only fleshy-fibrous roots, pulled up easily.
            11. Scales definitely acute (= pointed); spikelets lanceolate, acuminate **E. lanceolata**
            11. Scales obtuse; spikelets lanceolate to broadly ovoid to ovoid-cylindric, acute to obtuse.
                12. Tubercle (= hardened persistent style) much narrower than summit of achene body, 0.1–0.2 mm wide, somewhat constricted basally and thus slightly separated from achene body, light-colored; styles 2-branched **E. geniculata**
                12. Tubercle nearly as broad as summit of achene body, 0.5–1 mm wide, in outline merging with achene body, dark-colored; styles 2- or 3-branched.
                    13. Spikelets globose-ovoid or oblong-ovoid and obtuse, mostly 3–12 mm long; bristles mostly surpassing the tubercule; tubercule 1/3–nearly 1/2 as high as achene body **E. obtusa**
                    13. Spikelets mostly cylindrical and acute, usually 6.0–18.0 mm long; bristles shorter than to barely reaching tip of tubercule; tubercle short, not more than 1/4 as high as achene body **E. engelmannii**
        10. Plants perennials with rhizomes, often very difficult to pull up (and underground parts often lost).
            14. Plants often very small, 2–7(–12) cm tall; tubercle not differentiated from achene body, blending into achene body so that the junction is not visible (appearing as though tubercle not present); scales (and thus flowers) 3–8(–20) per spikelet **E. parvula**
            14. Plants usually much larger; tubercle well-differentiated from achene body, the junction with achene body clearly visible; scales (and thus flowers) 15–90(–110) per spikelet.
                15. Spikelets longer, at least (8–)10 mm long in flower, to 25 mm long in age; scales ± acute, 40–100 per spikelet; plants to 120 cm tall; achenes biconvex; style branches 2 **E. palustris**
                15. Spikelets shorter, 3–9 mm long in flower, to 14 mm long in age; scales obtuse, 15–70(–110) per spikeket; plants 50 cm or less tall; achenes trigonous (in species widespread in nc TX) OR biconvex (in species rare in nc TX); style branches 2 or 3.

16. Achenes biconvex, reddish just before maturity, maturing to black or dark purplish black; style branches 2; scales (and thus flowers) 15–25 per spikeket; plant rare, in nc TX known only from Lampasas Co. ______ **E. flavescens**
16. Achenes trigonous, ripening through yellow to brown or dark brown; style branches 3; scales (and thus flowers) 24–70(–110) per spikelet; plant widespread in nc TX ______ **E. montevidensis**

Eleocharis acicularis (L.) Roem. & Schult., (needle-shaped), REVERCHON'S SPIKE-RUSH, NEEDLE SPIKE-RUSH. Small perennial 2–23 cm tall, often forming mats; spikelets with 5–15 scales (and thus flowers), 2–5 mm long; achene body trigonous, 0.5–0.7 mm long, pearly white; tubercle small. Damp soils; Dallas Co.; scattered in TX. May–Jun, sporadically to Oct. [*E. reverchonii* Svenson]

Eleocharis acutisquamata Buckley, (with sharp-pointed, small, scale-like bracts or leaves), SHARP-SCALE SPIKE-RUSH. Rhizomatous perennial, forming mats; rhizomes 2–4(–6) mm thick; culms slender, 8–28 cm tall; spikelets 3–11 mm long, narrowly oblong or cylindric to narrowly elliptic, with 24–44 scales; achene body trigonous, 0.9–1.2 mm long, yellow to golden-brown; tubercle small. Pond margins or low areas of prairies, calcareous soils; mainly nc TX and Edwards Plateau. Late Mar–May.

Eleocharis baldwinii (Torr.) Chapm., (for its discoverer, William Baldwin, 1779–1819, Pennsylvania botanist and physician). Delicate tufted annual or perennial [?], stoloniferous, mat-forming; culms thread-like, 3–20 cm tall; spikelets flattish, 2–7 mm long, with few (2–4(–10)) scales, sometimes proliferating; cleistogamous spikelets often present at base of plant; scales of spikelet ± 2-ranked, folded from the midrib and thus boat-like; achenes trigonous, with body ca. 0.5–0–0.8 mm long, whitish buffy to olive, grayish olive, or brownish olive; tubercle conic-subulate, 0.2–0.3 mm long. Wet areas; Henderson Co., also Milam Co. (S.D. Jones, pers. comm.). Summer–fall. [*E. capillacea* of authors, not Kunth]

Eleocharis cellulosa Torr., (from the cellular surface of the achene), GULFCOAST SPIKE-RUSH. Tufted perennial; culms erect to 80 cm tall; spikelets cylindric, to 1.9–3.6(–5) cm long and 3.5–5 mm thick, with many scales; achene body biconvex, ca. 2 mm long, light brown; tubercle not at all basally constricted, appearing as a continuation of the achene body. Mud; Grayson Co., also Burnet Co. (S.D. Jones, pers. comm.); also se TX and Edwards Plateau. Spring–fall.

Eleocharis compressa Sull., (flattened), COMPRESSED SPIKE-RUSH. Similar to *E. acutisquamata*; rhizomatous perennial 9–20 cm tall, forming mats; spikelets 5–12 mm long, with 20–40 scales, ovoid to narrowly ovoid; achene body trigonous, ca. 1 mm long, yellow to golden-brown; tubercle small. Loamy usually moist soils; reported from Dallas by Svenson (Mahler 1988); e TX[?]. According to S.D. Jones (pers. comm.), *E. compressa* is a species to the e and n that apparently does not reach TX; he indicated that TX material identified as *E. compressa* is actually *E. aqutisquamata*. Brown and Marcus (1998) also did not find any TX material of *E. compressa* and indicated that all specimens examined were referable to *E. aqutisquamata*.

Eleocharis engelmannii Steud., (for George Engelmann, 1809–1884, German-born botanist and physician of St. Louis), ENGELMAN'S SPIKE-RUSH. Included based on citation of vegetational area 4 (Fig. 2) by Hatch et al. (1990); not reported for other parts of TX. [*E. obtusa* var. *detonsa* (A. Gray) Drapalik & Molenbrock; *E. ovata* (Roth) Roem. & Schult. var. *engelmannii* (Steud.) Britton] This species has sometimes been lumped with *E. obtusa* (e.g., Mahler 1988).

Eleocharis flavescens (Poir.) Urb., (yellowish), PALE SPIKE-RUSH. Perennial by elongate rhizomes 0.5–1 mm thick, delicately mat forming; culms 4–35 mm tall; spikelets 3–6 mm long; achene body biconvex, 0.6–1 mm long, reddish just before maturity, maturing to black or dark purplish black; tubercle small. Wet areas; Lampasas Co. in Lampasas Cut Plain (S.D. Jones, pers. comm.); mainly se and e TX and Edwards Plateau. Spring–fall. [*E. flaccida* (Rchb.) Urb.] Similar to and can be confused with the annual *E. geniculata* (which has the summit of the leaf sheaths firm and opaque versus thin-membranous and hyaline in *E. flavescens*).

Eleocharis geniculata (L.) Roem. & Schult., (jointed, bent like the knee), Tufted annual 4–40 cm tall; spikelets 3–6 mm long, ovoid to broadly ovoid, usually of 28–50 scales; achene body biconvex, 0.7–1 mm long, black or purplish black; tubercle small. Moist calcareous soils; in much of TX. Summer and fall. [*E. caribaea* of authors, not (Rottb.) S.F. Blake]

Eleocharis lanceolata Fernald, (lanceolate, lance-shaped), LANCE-SPIKE SPIKE-RUSH. Tufted annual 10–20 cm tall; spikelets with 30–80 scales; achene body biconvex, 0.9–1.1 mm long, brownish; tubercle not basally constricted, merging in outline with achene body. Sandy soils; Grayson (Correll & Johnston 1970) and Lamar (Carr 1994) cos.; also Bowie (Correll & Johnston 1970) and Red River cos.; apparently very rare in TX. Summer. [*E. obtusa* (Willd.) Schult. var. *lanceolata* (Fernald) Gilly]

Eleocharis microcarpa Torr., (small-fruited), SMALL-SEED SPIKE-RUSH. Tufted annual; culms thread-like, 10–30 cm tall; spikelets 1.5–5 mm long, sometimes proliferous; achene body 0.5–0.6 mm long, pearly white to pale greenish gray; tubercle small. Wet areas; Lamar Co. (Carr 1994) in Red River drainage; se and e TX, also Edwards Plateau and Rolling Plains. Spring–fall.

Eleocharis montevidensis Kunth, (presumably of Montevideo, Uruguay), Rhizomatous perennial 10–50 cm tall; spikelets 3–14 mm long, variable in shape, with 24–70(–110) scales; scales often dark brownish purple, obtuse; achene body trigonous, 0.8–1.2 mm long, yellow to golden-brown; tubercle ± conic. Low ground, sandy or clay soils; nearly throughout TX. Late Mar–early Jun. [*E. arenicola* Torr.]

Eleocharis obtusa (Willd.) Schult., (obtuse, blunt), BLUNT SPIKE-RUSH. Tufted annual 3–50 cm tall; spikelets ca. 3–12 mm long, usually broadly ovoid to nearly cylindric, usually with 50–100 scales; achene body biconvex, 0.8–1.5 mm long, pale to dark brown; tubercle nearly as wide as summit of achene. Moist sandy soils, pond margins; e TX w to Grayson and Tarrant cos.; also Edwards Plateau. Late Apr–Oct.

Eleocharis palustris (L.) Roem. & Schult., (of marsh), LARGE-SPIKE SPIKE-RUSH, CREEPING SPIKE-RUSH. Rhizomatous perennial 15–60 cm tall; culms round to strongly compressed; spikelets 8–25 mm long, with 40–100 scales; achene body biconvex, usually more turgid on 1 side, 1.2–1.8 mm long, yellow to golden-brown; tubercle 0.3–0.7 mm long, basally constricted. Wet soils or shallow water, sometimes extremely abundant; nearly throughout TX. May–Jun, sporadically to Oct. [*E. macrostachya* Britton; *E. xyridiformis* Fernald & Brackett] The complex that has been treated as *E. palustris* or *E. macrostachya* is quite confusing and is in need of taxonomic work. We are following Kartesz (1994), Jones et al. (1997), and S.G. Smith (pers. comm.) in using the name *E. palustris*. While one taxon, *E. xyridiformis*, is here lumped, another, *E. erythropoda* Steud. [= *E. calva* Torr.], is possibly a good species. According to Taylor and Taylor (1994), *E. erythropoda* is widely distributed in OK and while we have seen no specimens, it is therefore likely present in nc TX. According to S.D. Jones (pers. comm.), *E. erythropoda* is present in the Rolling Plains and along the TX-LA border. Galen Smith (pers. comm.) indicated that *E. erythropoda* and *E. palustris* intergrade; the two can be distinguished as follows:

1. Culms round in cross-section or nearly so; sterile scale at base of spikelet only 1, this completely encircling the base of the spikelet; spikelets < 15 mm long; anthers 1.2–1.5mm long ____ *E. erythropoda*
1. Culms strongly flattened to round in cross-section; sterile scales at base of spikelets 1–3, the lowest often not encircling the base of the spikelet; spikelets (8–)10–25 mm long; anthers 1.6–2.1 mm long ____ *E. palustris*

Eleocharis parvula (Roem. & Schult.) Link ex Bluff, Nees & Schauer, (very small), DWARF SPIKE-RUSH. Plant often very small, 2–7(–12) cm tall; rhizomes or stolons short, plants forming mats in mud; spikelets often very small, 2–5(–9) mm long; achene body trigonous, 0.8–1.5 mm long including confluent tubercle, gray to black; tubercle scarcely recognizable as separate except under high magnification. In mud; nearly throughout TX. Spring, fall. [*E. parvula* var. *anachaeta* (Torr.) Svenson]

Eleocharis quadrangulata (Michx.) Roem. & Schult., (four-angled), SQUARE-STEM SPIKE-RUSH. Coarse perennial; culms erect, to 80(–150) cm tall, distinctly 4-sided; spikelets cylindric, 20–42 mm long, with 40–90 scales; achene body biconvex, ca. 2 mm long, yellowish brown, reticulate; tubercle longer than broad, constricted basally. Mud, lake margins; Grayson, Lamar, and Tarrant cos., also Denton Co. (G. Dick, pers. comm.) and Fort Hood (Bell or Coryell cos.—Sanchez 1997); mainly e TX. Late spring–fall. [*E. quadrangulata* var. *crassior* Fernald]

FIMBRISTYLIS

Annuals or perennials; leaves filiform to linear; inflorescences of peduncled or sessile spikelets, often subtended by a leafy involucre; scales of spikelets spirally imbricate; flowers perfect; perianth none; stamens 1–3.

A genus of ca. 150–250 species depending on circumscription; some species are used as copper indicators or for their fiber; *Bulbostylis* is sometimes included. (Latin: *fimbria*, a fringe, and Greek: *stylos*, pillar, column, or style, from the style being fringed with hairs in some species)
REFERENCES: Kral 1971; Kolstad 1986a.

1. Styles 3-branched; achenes trigonous or not; spikelets usually 2–7 mm long.
   2. Spikelets lanceolate to linear-oblong, usually 3–7 mm long, apically acute; scales of spikelets acute to acuminate; ligules of short hairs present; achenes trigonous, the surfaces not reticulate or only faintly so, smooth or warty especially at base ____ **F. autumnalis**
   2. Spikelets usually ovoid to nearly round, 2–4 mm long, apically rounded; scales of spikelets obtuse; ligules absent; achenes not trigonous or only obscurely so, obovoid, the surfaces reticulate and usually warty ____ **F. miliacea**
1. Styles 2-branched; achenes lenticular to biconvex or obovoid; spikelets 3–10 mm long.
   3. Spikelets (at least 1 or more) peduncled; achenes 1 mm or more long; plants annual or perennial, to 1 m tall; leaves narrowly linear.
      4. Plants cespitose (= tufted or clumped) annuals; spikelet apex acute; achenes longitudinally and horizontally (less distinct) ribbed, often warty; ligules of short hairs present ____ **F. annua**
      4. Plants perennials with culms solitary or a few together; spikelet apex obtuse to acute; achenes minutely reticulate, usually without either conspicuous ribs or warts; ligules inconspicuous to absent OR of short hairs ____ **F. puberula**
   3. Spikelets sessile, all close together in a capitate cluster; achenes ca. 0.5–0.7 mm long; plants low growing annuals to 15 cm tall; leaves filiform ____ **F. vahlii**

Fimbristylis annua (All.) Roem. & Schult., (annual). Cespitose annual; culms decumbent, ascending, or erect, to 50 cm tall; leaves narrowly linear, 1–2(–4) mm wide; spikelets lance-ovoid to ovoid to oblong, 3–8 mm long, apically acute; achenes 1–1.3 mm long. Weedy areas; Dallas Co.; mainly e TX and Edwards Plateau. Oct. [*F. baldwiniana* (Schult.) Torr., *Scirpus annuus* All.]

Fimbristylis autumnalis (L.) Roem. & Schult., (autumnal, of the fall), SLENDER FIMBRISTYLIS. Cespitose annual, 5–20 cm tall; leaves linear, to 4 mm wide; achenes ca. 1 mm long. Moist or wet, often sandy areas; e TX w to West Cross Timbers; also Edwards Plateau. (Jun–)Jul–Nov. [*Scirpus autumnalis* L.]

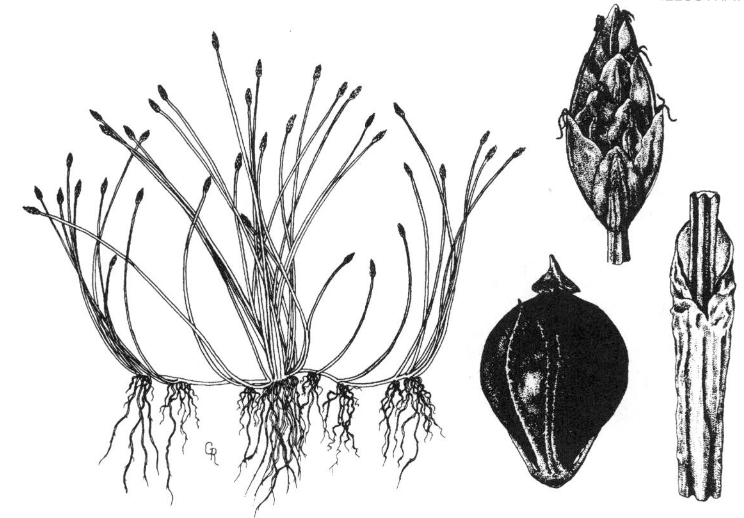

Eleocharis flavescens [GWO]

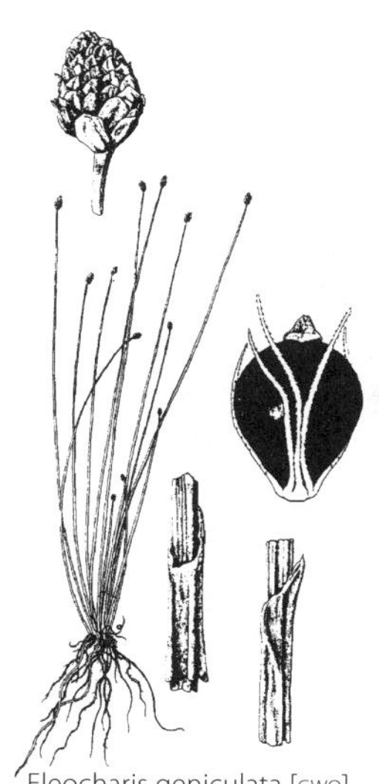

Eleocharis geniculata [GWO]

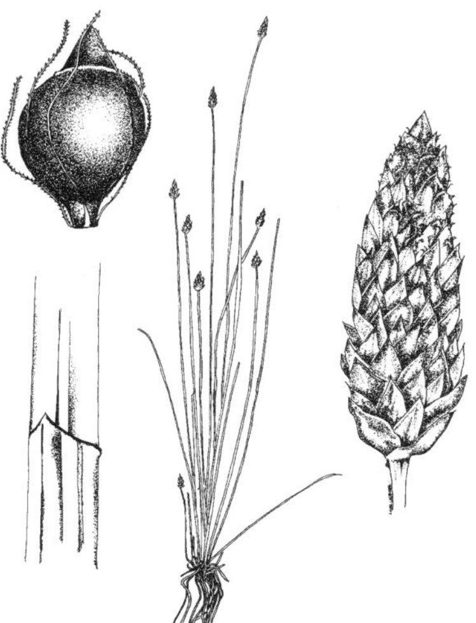

Eleocharis lanceolata [BT2]

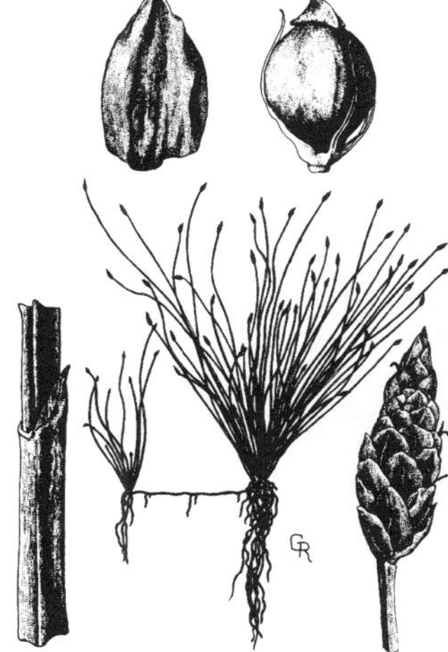

Eleocharis microcarpa [GWO]

Eleocharis montevidensis [GWO]

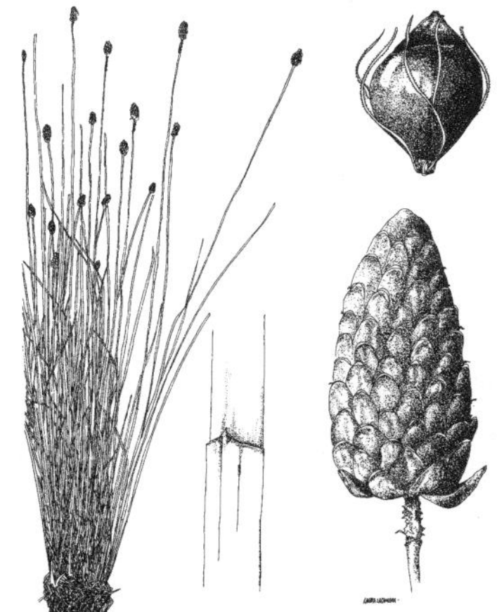

Eleocharis obtusa [BT2]

Eleocharis palustris [BT2]

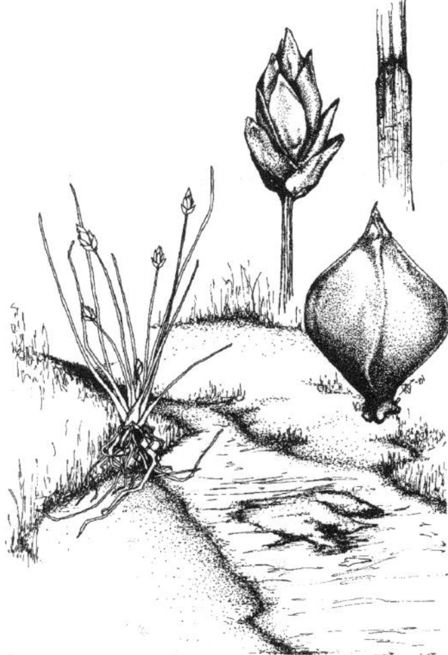

Eleocharis parvula [BT2]

Fimbristylis miliacea (L.) Vahl, (resembling *Milium*—millet), GLOBE FIMBRISTYLIS. Cespitose annual to 50(–100) cm tall; achenes ca. 1 mm long. Sandy areas; Delta, Lamar, Hopkins, and Red River cos. (S.D. Jones, pers. comm.); mainly e TX. Aug.–Oct. Native of Asia. [*Scirpus miliaceus* L.] Kral (1971) indicated, "The history of introduction of this weed into the U.S.A. probably parallels that of rice, in that it is a common species of the rice growing countries of the Orient."

Fimbristylis puberula (Michx.) Vahl, (somewhat pubescent). Glabrous perennial 15–60(–100) cm tall; culms solitary or in small tufts; leaves basal; inflorescences umbel-like; spikelets ± ovoid; achenes to 1.8 mm long. Sandy prairies, open woods, often in wet areas. The following key to varieties is modified from Kral (1971) and Kolstad (1986a).

1. Base of culms rarely bulbous, producing fascicles of slender orangish rhizomes; old leaf bases not persisting as shreddy remnants; outer surface of fertile scales seldom with any puberulence; longest bract of inflorescence usually longer than the inflorescence; ligules inconspicuous or often of short ascending hairs ____ var. **interior**
1. Base of culms bulbous, often jointed together into a stout, knotty rhizome; old leaf bases often persisting as shreddy remnants; outer surface of fertile scales usually with some puberulence; longest bract of inflorescence usually much shorter than inflorescence; ligules inconspicuous to absent ____ var. **puberula**

var. **interior** (Britton) Kral. According to S.D. Jones (pers. comm.), there is a specimen (Hamilton Co.) at TAES; Kral (1971) mapped the species to the s and w of nc TX; mainly Edwards Plateau and Trans-Pecos. Early summer–summer. [*F. interior* Britton]

var. **puberula**. Se and e TX w to Grand Prairie, also Edwards Plateau. This is the common variety in nc TX. Apr–Jul (usually finished flowering by early summer). [*F. drummondii* (Torr. & Hook. ex Torr.) Boeck., *Scirpus puberulus* Michx.]

Fimbristylis vahlii (Lam.) Link, (for Martin Hendriksen Vahl, 1749–1804, Danish botanist and student of Linnaeus), VAHL'S FIMBRISTYLIS. Leaves filiform, < 1 mm wide; spikelets lance-ovoid to linear-ellipsoid or oblong-cylindric, apically obtuse or acute. Usually along lake margins, streams, or in disturbed bottomlands; e TX w to Bell, McLennan, and Wise cos., also Edwards Plateau. Jul–Oct. [*Scirpus vahlii* (Lam.) Link]

FUIRENA UMBRELLA-GRASS

Perennials or annuals to 1 m tall; culms (= stems) obtusely triangular to nearly round; inflorescences condensed, often head-like, of 1–10 ovoid spikelets; scales of spikelets numerous, spirally imbricate, usually with a very conspicuous, often recurved, bristle-like awn; perianth of 3 stalked, scale-like or paddle-like structures often thickened, expanded, or swollen at maturity, often also 3 perianth bristles alternating with the stalked structures; achenes strongly 3-angled.

A genus of 30 species of warm areas. In the past, some authors lumped *Fuirena* into *Scirpus*. (Named for Georg Fuiren, 1581–1628, a Danish botanist)

REFERENCES: Coville 1890; Bush 1905; Svenson 1957; Kral 1978.

1. Upper leaf sheaths with obvious pubescence; scale-like perianth parts (look under scales of spikelets) without a subapical bristle, the apex either conic and swollen or acuminate; rhizomes producing corm-like shoot buds; only near e margin of nc TX.
   2. Scale-like perianth parts with apex thinnish or thickened, acuminate, incurved; bristles of perianth retrorsely barbellate; anthers 1.3 mm or less long; swollen cormous buds of rhizome not separated by narrower internodes ____ **F. squarrosa**
   2. Scale-like perianth parts swollen at maturity, narrowing to the conic and erect, sometimes

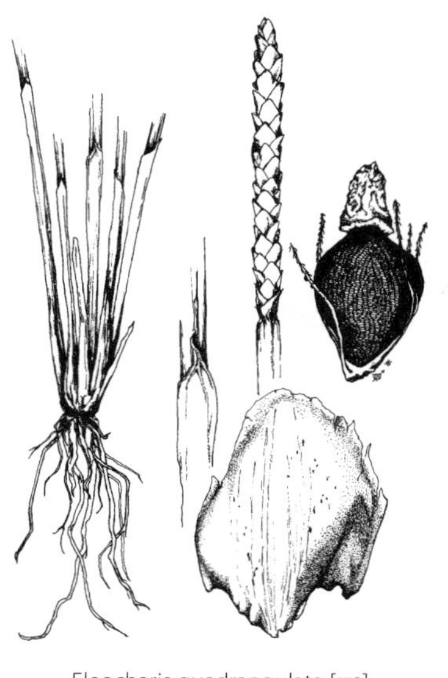
Eleocharis quadrangulata [BT2]

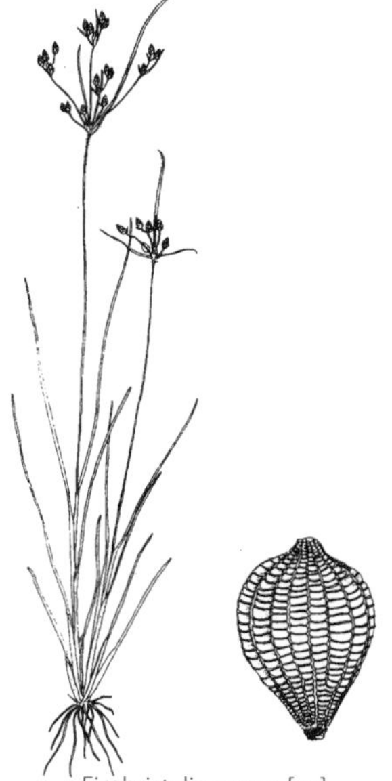
Fimbristylis annua [SID]

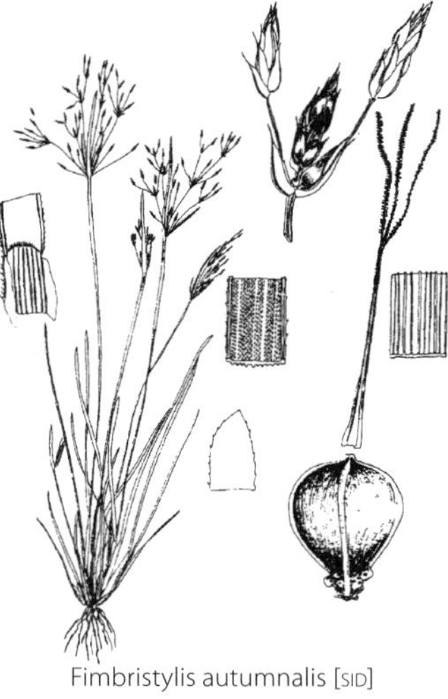
Fimbristylis autumnalis [SID]

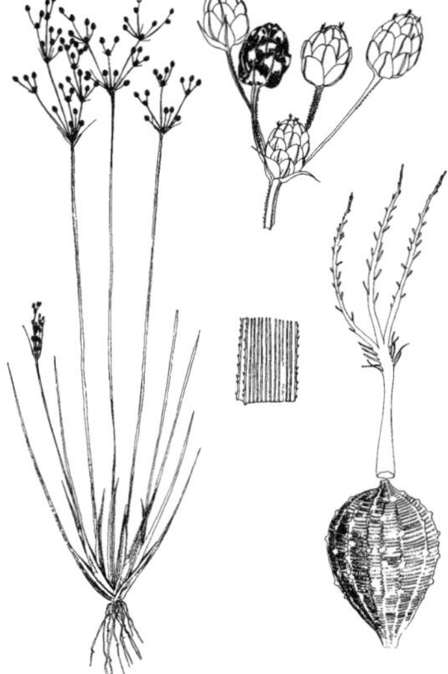
Fimbristylis miliacea [SID]

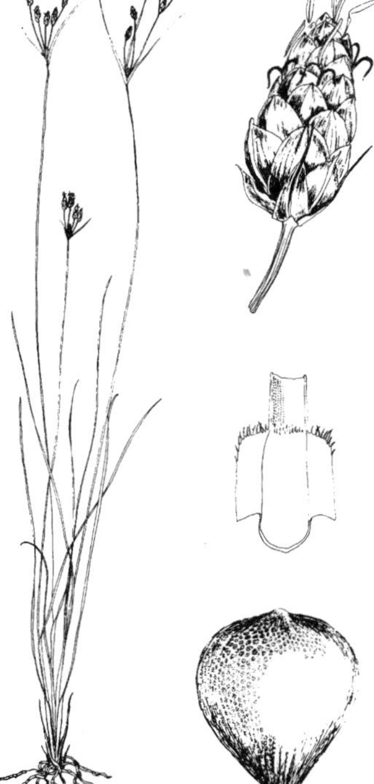
Fimbristylis puberula var. interior [SID]

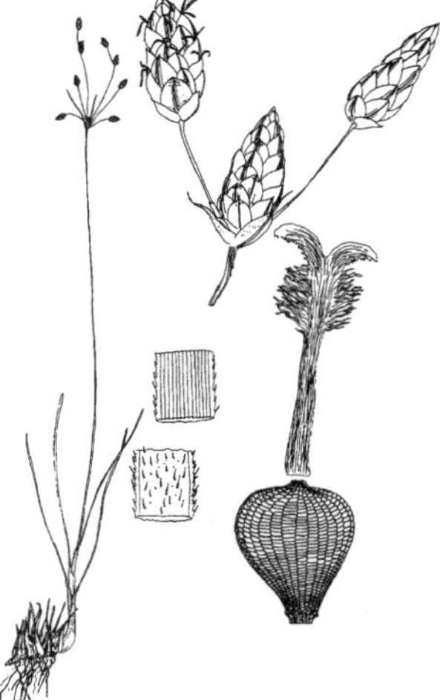
Fimbristylis puberula var. puberula [SID]

Fimbristylis vahlii [SID]

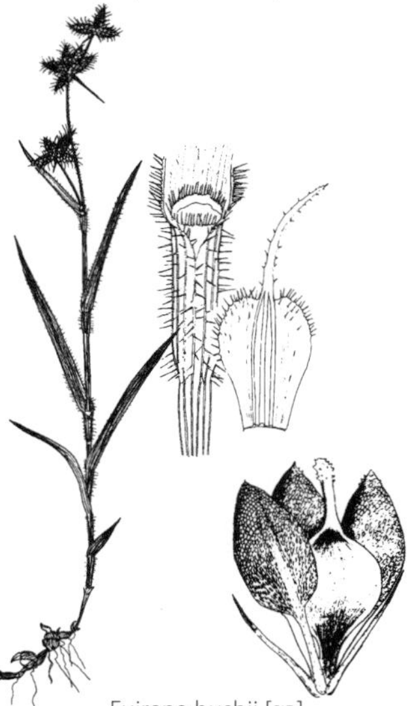
Fuirena bushii [SID]

apiculate apex; bristles of perianth smooth; anthers ca. 2 mm long; swollen cormous buds of rhizome often separated by narrow internodes longer than corm width _______ **F. bushii**

1. Upper leaf sheaths glabrous; scale-like perianth parts usually with a subapical bristle, the apex flattish or notched or turgid and conic; rhizomes simple, lacking corm-like shoot buds; widespread in nc TX _______ **F. simplex**

Fuirena bushii Kral, (for its discoverer, Benjamin Franklin Bush, 1858–1937, postmaster in MO and amateur botanist). Plant to 1 m tall; leaf sheaths strongly hirsute; leaf blades with pubescence; bristles of perianth reaching base of blades of stalked, scale-like perianth parts. Wet, acidic, usually sandy areas; Lamar Co. (BRCH) in Red River drainage; mainly se TX. Jun–Oct. [*F. ciliata* Bush]

Fuirena simplex Vahl, (unbranched), UMBRELLA SEDGE, WESTERN UMBRELLA-GRASS. Plant 0.1–1 m tall; leaf sheaths glabrous or only the lowest hirsute; leaf blades minutely scabrous or glabrous; bristles of perianth reaching at least base of blade of stalked, scale-like perianth part, retrorsely (downwardly) barbed. Aquatic or wet places. Jun–Oct. This is a widespread variable species and is by far the most common species in nc TX. Kral (1978) distinguished two varieties as follows:

1. Plants nonrhizomatous, mostly annual; anthers 0.5–0.6 mm long _______ var. **aristulata**
1. Plants rhizomatous, perennial; anthers 0.9–1.2 mm long _______ var. **simplex**

var. **aristulata** (Torr.) Kral, (bearded or awned). Widespread in TX. [*F. squarrosa* Michx. var. *aristulata* Torr.]

var. **simplex**. Widespread in TX.

Fuirena squarrosa Michx., (with recurved tips). Plant to 1 m tall, usually less; leaf sheaths strongly hispid-hirsute; leaf blades with pubescence; bristles of perianth often reaching middle of blade of stalked scale-like perianth parts. Wet areas, often on sandy substrates; Limestone Co., also Milam Co. (BRCH); mainly se and e TX, also Edwards Plateau and Panhandle. Jun–Oct. [*F. hispida* Elliott]

Fuirena breviseta (Coville) Coville, (short-bristled), is cited by Hatch et al. (1990) for vegetational area 4 (Fig. 2). Kral's (1978) range map, however, showed all localities of *F. breviseta* to be well e and s of nc TX. It can be distinguished by its short perianth bristles, these not reaching the base of the blades of the stalked, scale-like perianth parts; the bristles are also without barbs or weakly, usually upwardly barbed (vs. markedly downwardly barbed in the 3 other nc TX species).

ISOLEPIS BULRUSH

Small tufted annuals 2–25(–30) cm tall, glabrous or nearly so; leaves setaceous, near base of culms; inflorescences of 1–3 sessile spikelets, appearing lateral, with a single, erect, modified leaf (= involucral bract) appearing like a continuation of the culm; flowers (7–)10–30 per spikelet; scales of spikelets keeled, awnless or very short-awned; perianth absent; stigmas 3; achenes trigonous, minutely papillose (this can sometimes be obscured by a whitish, wax-like, surface layer).

A cosmopolitan genus of ca. 70 species (Smith 1997b), mostly of temperate and subtropical climates; when tropical, restricted to mountains at higher elevations. Three species are reported for TX (S.D. Jones, pers. comm.). Previously included in *Scirpus* (e.g., Kartesz 1994) and according to some, better treated as a section or subgenus in *Scirpus* in the broad sense. We are following Smith (1995, 1997b) and Jones et al. (1997) in recognizing this segregate of *Scirpus* at the generic level. Smith (1995) indicated that *Isolepis* and a number of other segregates of *Scirpus* will be recognized in the forthcoming Cyperaceae treatment for *Flora of North America* (Vol. 11).

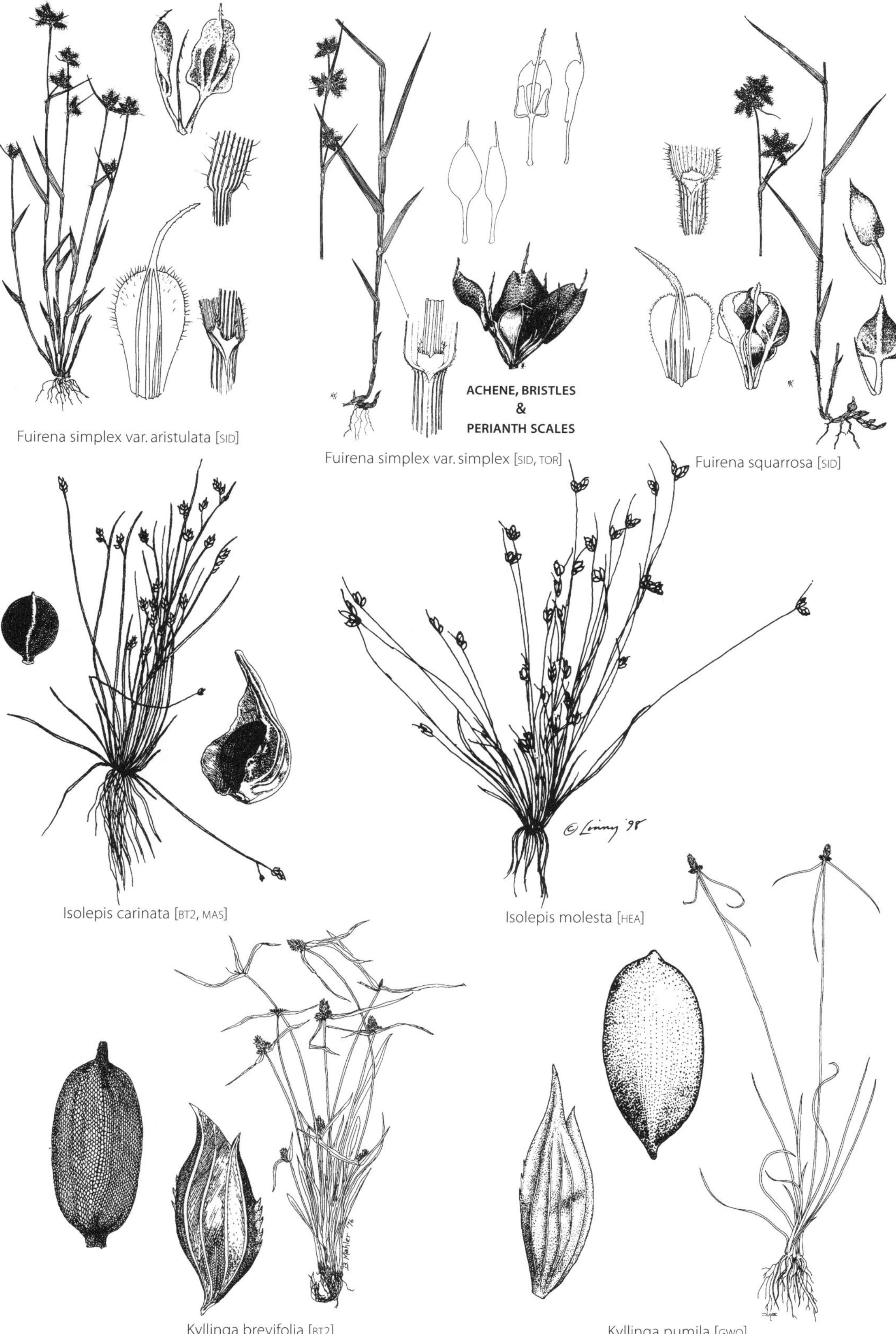

Fuirena simplex var. aristulata [SID]

Fuirena simplex var. simplex [SID, TOR]

Fuirena squarrosa [SID]

Isolepis carinata [BT2, MAS]

Isolepis molesta [HEA]

Kyllinga brevifolia [BT2]

Kyllinga pumila [GWO]

This approach is supported by the phylogenetic studies of Bruhl (1995) which suggested that *Scirpus* sensu lato is polyphyletic. The key to species is modified from Smith (1997b). (Greek: *isos*, equal, and *lepis*, scale)
REFERENCES: Beetle 1947; Johnston 1964; Smith 1995, 1997b; Smith & Yatskievych 1996.

1. Scales from middle of spikelet 1.8–2.0 mm long, short-awned (awns so short that the scales appear merely acuminate); achenes 1–1.5 mm long ____ **I. carinata**
1. Scales from middle of spikelet 1–1.2 mm long, mucronate; achenes 0.7–0.9 mm long ____ **I. molesta**

Isolepis carinata Hook. & Arn. ex Torr., (with a keel), ANNUAL BULRUSH. Tufted annual 4–25(–30) cm tall; involucral bract 5–30 mm long; spikelets usually solitary, sometimes 2(–3), 2–10 mm long, 1.5–2 mm wide. Moist sandy soils; Fannin (Talbot property), Hood, Lamar, Navarro, and Tarrant cos.; mainly se and e TX; also Edwards Plateau. Spring. [*I. koilolepis* Steud., *Scirpus koilolepis* (Steud.) Gleason]

Isolepis molesta (M.C. Johnston) S.G. Smith, (troublesome). Tufted annual 2–20 cm tall; involucral bract 3–10(–25) mm long; spikelets 1–3, 2–8 mm long, 1–1.5 mm wide. Moist sandy soils; based on range map in Smith (1997b), this species apparently occurs in the s part of nc TX; se and e TX w to e part of c TX; it should be looked for in mixed populations with *I. carinata*. Spring. [*Scirpus molestus* M.C. Johnston] Johnston (1964) indicated that while this species seems like a dwarf form of *I. carinata*, clear-cut differences exist and no intermediate specimens are known.

KYLLINGA

Small rhizomatous or tufted annuals or perennials to 21(–38) cm tall; inflorescence a single roundish to 3-lobed congested head or head-like mass, 3–8(–12) mm long; spikelets flat, 1.8–4 mm long, with only a single, fertile, bisexual flower, deciduous as a unit, of only 4 scales, 2 of these normal in size, the basal 2 minute, brownish; achene 1 per spikelet.

A genus of ca. 40–45 species, mostly tropical with the greatest diversity in tropical Africa and Madagascar; 8 occur in the New World (Tucker 1984). *Kyllinga* is sometimes included in the genus *Cyperus*. (Named for P. Kylling, died 1696, Danish Botanist)
REFERENCES: Johnston 1966; Delahoussaye & Thieret 1967; Tucker 1984.

1. Rhizomatous perennials, the culms arising at intervals of 3–10 mm; stamen solitary; roundish inflorescences usually without any visible lobing ____ **K. brevifolia**
1. Densely tufted annuals with 4–20 culms per square cm; stamens paired; roundish inflorescences often slightly 3-lobed ____ **K. pumila**

Kyllinga brevifolia Rottb., (short-leaved), SHORT-LEAF FLAT SEDGE. Perennial with reddish rhizomes to 20 cm long and 1–2 mm thick; culms to ca. 20(–38) cm tall; leaf blades usually 1–3(–10) cm long, 1–4 mm wide; spikes 3–12 mm long; spikelets flat, 2.5–4 mm long. Moist loam; Dallas and Tarrant cos.; mainly se and e TX, also Edwards Plateau. Apr–Nov. [*Cyperus brevifolius* (Rottb.) Endl. ex Hassk.]

Kyllinga pumila Michx., (dwarf, very small), SLENDER-LEAF FLAT SEDGE. Tufted fragrant annual to 21 cm tall; leaf blades 2–11 cm long, 1–1.8(–3) mm wide. Moist sites; Denton Co., also Grayson Co. (Johnston 1966); mainly se and e TX. Sep–Nov. [*Cyperus tenuifolius* (Steud.) Dandy]

LIPOCARPHA

Tufted, delicate, glabrous annuals 1–15 cm tall; leaves basal, 2, one without blade; inflorescence of 1–3 ovoid, sessile, spikelet-like spikes 2–5(–8) mm long, the spikes of numerous single-flowered spikelets whose scales (here called floral scales in contrast to the included hyaline scales) are spirally and imbricately arranged to make the whole spike appear like a single spikelet; in ours inside each floral scale there is 1 inconspicuous hyaline scale or bracteole often split and torn by or adhering to the achene; floral scales with 0–3 prominent veins; inflorescence subtended by 1–3 bracts, 1 of these much larger and appearing like a continuation of the culm, the inflorescence thus appearing lateral; achenes narrowly oblong-obovate to obovate or ovate, 0.5–0.8 mm long, granular, very minutely apiculate.

A small, mainly tropical genus of 8 species with inflorescences and flowers much reduced; possibly derived from *Cyperus*. The species treated here as *Lipocarpha* have historically been segregated into the genus *Hemicarpha*. According to Friedland's (1941) range map, at least two and possibly three difficult to distinguish taxa are to be expected in nc TX. He recognized these as varieties of *Hemicarpa micrantha* and said that in order to key them, "... the spikelet must be boiled and then dissected under a binocular microscope capable of at least a magnification of forty diameters." Tucker (1987) and Kartesz (1994) recognized these taxa as species (*Lipocarpha aristulata* (Coville) G.C. Tucker, *L. drummondii* (Nees) G.C. Tucker, and *Lipocarpha micrantha* (Vahl) G.C. Tucker). While Friedland's range map clearly showed *L. aristulata* and *L. drummondii* in nc TX, the only specimens we have seen from nc TX are annotated *L. micrantha* by Tucker. Using the key from Friedland (1941), we have been completely unable to consistently distinguish the taxa; the following key was developed by S.D. Jones using TX material. (Greek: *lipos*, fat, and *carphos*, chaff, from the thickness of the inner scales of some species) REFERENCES: Friedland 1941; Svenson 1957; Tucker 1987.

1. Mid-upper floral scales of the spikelet mostly tapering into a conspicuous awn 1/4 to longer than the scale body; achenes obovate, maturing dark reddish brown to blackish; inner hyaline scale as long as or longer than the achene and cupped around it, veinless ____ **L. aristulata**
1. Mid-upper floral scales of the spikelet acute-triangular apically, awnless, or, at most with a short mucro; achenes obovate or narrowly obovate, maturing brown to reddish brown, (but not as dark reddish brown to blackish as in *L. aristulata*); inner hyaline scale as long as the achene or longer and cupped around it OR shorter than the achene OR absent, if present, then 3–5-veined.
   2. Apices of floral scales incurved over top of the achenes; achenes not normally readily visible at maturity; achenes narrowly obovate; inner hyaline scale as long as or longer than the achene and cupped around it; when an achene is shed, the scales usually are shed with it; inner hyaline scale not bifurcated ____ **L. drummondii**
   2. Apices of floral scales spreading, barely exceeding the achenes; achenes normally readily visible at maturity; achenes obovate; inner hyaline scale much shorter than achene or absent; when an achene is shed it usually leaves the inner scale behind attached to the rachis; inner hyaline scale usually bifurcated ____ **L. micrantha**

Lipocarpha aristulata (Coville) G.C. Tucker, (bearded or awned). Range map given by Friedland (1941) showed this taxon within nc TX; Hatch et al. (1990) cited vegetational area 4 (Fig. 2); we have seen no nc TX specimens but S.D. Jones (pers. comm.) indicated there is a Cooke Co. specimen at TEX; he also indicated there is a TEX specimen from Wichita Co. just to the w of nc TX; also Edwards Plateau and Trans-Pecos. Friedland (1941) mapped a number of locations from the e 1/2 of TX and one from the Trans-Pecos. Jun–Nov. [*Hemicarpha aristulata* (Coville) Smyth]

Lipocarpha drummondii (Nees) G.C. Tucker, (for its discoverer, Thomas Drummond, 1780–1835, Scottish botanist and collector in North America), COMMON HEMICARPHA. Range map given by Friedland (1941) showed this taxon within nc TX; we have seen no nc TX specimens but S.D. Jones (pers. comm.) indicated there are Johnson Co. and Tarrant Co. specimens at TEX; Friedland (1941) mapped several localities in nc and e TX; also se and e TX and Edwards Plateau. May–Nov. [*Hemicarpha drummondii* Nees, *H. micrantha* (Vahl) Pax var. *drummondii* (Nees) Friedl.]

Lipocarpha micrantha (Vahl) G.C. Tucker, (small-flowered). Moist sandy soils, seasonally wet areas; Bosque, Henderson, Hood, and Parker cos.; e 1/2 of TX; also Trans-Pecos. Mar–Oct. [*Hemicarpha micrantha* (Vahl) Pax, *Hemicarpha micrantha* var. *minor* (Schrad.) Friedl., *Scirpus micranthus* Vahl] Inconspicuous and rarely collected.

RHYNCHOSPORA BEAK-RUSH, HORNED-RUSH

Tufted or clump-forming perennials (rarely annuals), glabrous or with scabrous-margined leaves; culms (= stems) triangular; spikelets narrowly ovoid to fusiform or roundish, brownish, 1-few-flowered, in loose or compact clusters, in spike-like or open panicles; lower (1–)2–3 scales of spikelets sterile; perianth of bristles or absent; achenes with a conspicuous tubercle or "beak" (= hardened and persistent style base) from the apex (hence the name BEAK-RUSH).

A genus of ca. 250 species nearly cosmopolitan in distribution, especially in tropical and warm areas of South America. (Greek: *rhyncos*, a snout, and *spora*, a seed, from the beaked achene)

REFERENCES: Gale 1944; Nixon & Ward 1982; Thomas 1984, 1992.

1. Scales of spikelets white; bracts below inflorescence with white base (much more conspicuous in *R. colorata*).
   2. Bracts exceeding spikelets 1–2(–3) in number, very narrow, most filiform; white spot on bract only at the very base, not longer than spikelets; rhizomes absent; leaf blades 3–15 mm long, ca. 1(–2) mm broad basally, narrower (arcuate-filiform) distally ____ **R. nivea**
   2. Bracts exceeding spikelets 3–6(–7) in number, (1.5–)2.5–5 mm broad at base; white spot on bract conspicuous, usually much longer than spikelets; rhizomes extensively creeping, orangish or whitish; leaf blades 6–25 mm long, 1.2–4 mm broad basally and at least 1 mm broad even ± distally ____ **R. colorata**
1. Scales of spikelets variously colored (not white); bracts below inflorescence not with white base.
   3. Mature spikelets conspicuously long, 15–23 mm long; achene body 3.5–5 mm long; tubercle 13–18 mm long.
      4. Bristles subtending achene shorter than achene body (usually ca. 1/3–2/3 as long) ____ **R. corniculata**
      4. Bristles (at least most) much longer than the achene body (ca. 1.5–2.5 times as long) ____ **R. macrostachya**
   3. Mature spikelets much shorter (2.5–9 mm long); achene body (to ca. 2 mm); tubercle much shorter, 2 mm or less long.
      5. Achene body 0.7–1 mm long, with conspicuous, irregular bone-colored, transverse ridges; tubercle ± as wide as body of achene; perianth absent; all scales of spikelets fertile (developing an achene); plants annual; rare, known locally only from Milam Co. near e margin of nc TX ____ **R. nitens**
      5. Achene body 1.5–2 mm long, surface various, but without conspicuous, bone-colored, transverse ridges; tubercle definitely narrower than body of achene; perianth of 5–6 bristles present; lowest 1(–2) scales of spikelets empty (without an achene); plants perennial; widespread in nc TX.
         6. Spikelets 4.5–6 mm long; bristles conspicuously retrorsely (backward or downward) barbed, much longer than the achene body; achene body usually smooth ____ **R. glomerata**
         6. Spikelets 2.5–4 mm long; bristles antrorsely (forward or upward) barbed or plumose or absent or smooth, shorter than the achene body; achene body usually either honeycombed or finely cross-wrinkled.
            7. Scales of spikelets awned; achene body honeycombed; summit of achene with a smooth collar-like ring fitted against the tubercle ____ **R. harveyi**
            7. Scales of spikelets obtuse or with a very short point; achene body usually finely transversely-wrinkled; summit of achene narrowed up to style (= tubercle) base, without a collar–like ring ____ **R. globularis**

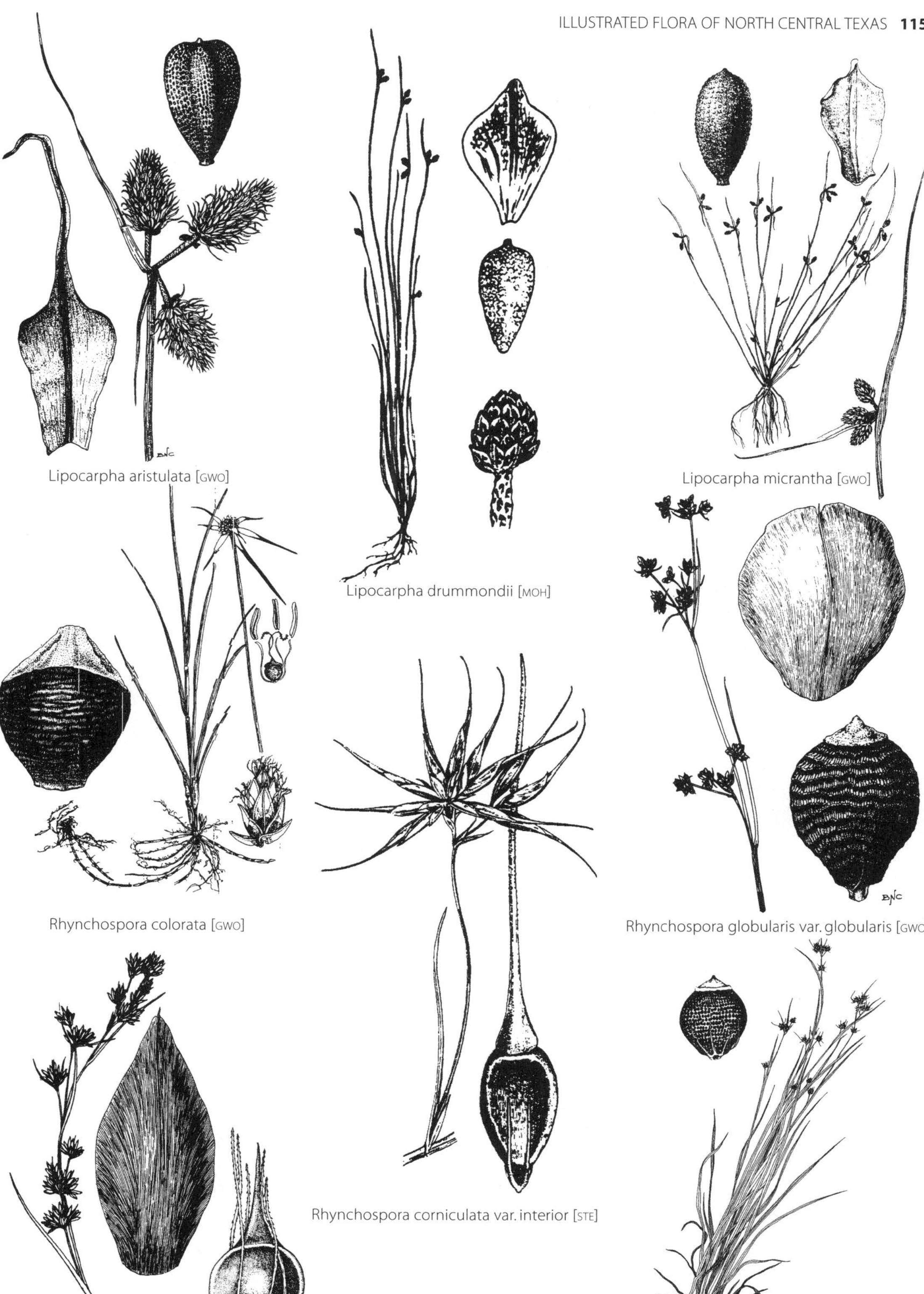

Lipocarpha aristulata [GWO]

Lipocarpha drummondii [MOH]

Lipocarpha micrantha [GWO]

Rhynchospora colorata [GWO]

Rhynchospora globularis var. globularis [GWO]

Rhynchospora corniculata var. interior [STE]

Rhynchospora glomerata [GWO]

Rhynchospora harveyi [BT3, STE]

Rhynchospora colorata (L.) H. Pfeiff., (colored), WHITE-TOP UMBRELLA-GRASS, STAR-RUSH WHITE-TOP SEDGE, UMBRELLA-GRASS. Perennial to ca. 56 cm tall; leaves cauline although sometimes crowded near base; bracts unequal in length, the longer ones mostly 5–15 cm long; white spot on bracts conspicuous, (2.5–)5–20(–25) mm long. Wet places; included on basis of citation of vegetational area 4 (Fig. 2) by Hatch et al. (1990); se TX and s part of e TX w across much of the state. (Spring–)Summer. Previously separated into the genus *Dichromena* [as *D. colorata* (L.) Hitchc.].

Rhynchospora corniculata (Lam.) A. Gray var. **interior** Fernald, (sp.: horned; var.: inland), HORNED BEAK-RUSH, HORNED-RUSH. Rhizomatous or tufted perennial 0.5–1.5 m tall; culms sharply triangular; inflorescence large, with spreading branches; spikelets strikingly elongate (15–23 mm long), usually in clusters of 3–7(–14); achene body flattish, 3.5–5 mm long, 2–3.5 mm wide, with extremely conspicuous, long-subulate tubercle; bristles 3–6, but normally 5, unequal, ca. 2–5 mm long. Mud, edge of ponds, or on decaying logs in water; Grayson, Fannin, Henderson, and Kaufman cos., also Tarrant Co. (R. O'Kennon, pers. obs.); mainly se and e TX, also Edwards Plateau and Trans-Pecos. Spring–Summer (fruit present into fall). Some authorities (eg., Jones et al. 1997) lump this variety.

Rhynchospora globularis (Chapm.) Small var. **globularis,** (globular, of a little ball or sphere), GLOBE BEAK-RUSH. Perennial 15–100 cm tall; leaf blades ca. 1.5 mm wide; spikelets 2.5–4 mm long, 1–3-fruited; bristles to 2/3 as long as achene body; achene body 1.6–2.0 mm long. Damp sandy soils, disturbed sites; Dallas Co.; mainly e TX. May–Sep.

Rhynchospora globularis var. *recognita* Gale is abundant in e TX just to the e of nc TX. Robert Kral, who is treating the genus for the *Flora of North America*, has indicated that he will be recognizing this taxon at the specific level (R. Kral, pers. comm.). Kral indicated that *R. globularis* var. *recognita* is expected in the ne part of nc TX. It is similar to *R. globularis*; the following key separating the two is from Gale (1944) with modifications from R. Kral (pers. comm.).

1. Habit frequently depressed (culms often short, but ranging from 14–68 cm tall); leaves usually 1.5–2 mm wide; branchlets of the cymes terminating in small knobby glomerules; bracts inconspicuous; spikelets 2.5–3 mm long, roundish, dark; achenes 1–1.2 mm wide, 1.3–2.3 mm long (including tubercule), finely cancellate, transversely ridged to rugulose ____ *R. globularis* var. *globularis*
1. Habit robust (culms up to ca. 100 cm tall); leaves usually 2–4 mm wide; branchlets of the cymes usually terminating in dense glomerules; setaceous bracts conspicuous (giving the inflorescence a bristly appearance); spikelets 3–4 mm long, elongate, reddish; achenes 1.2–1.5 mm wide, 1.3–1.6 mm long, coarsely cancellate to striate, tranversely ridged ____ *R. globularis* var. *recognita*

Rhynchospora glomerata (L.) Vahl, (in compact clusters), CLUSTER BEAK-RUSH. Tufted perennial to 1.1 m tall; spikelets usually 2(–3)-fruited; bristles slightly surpassing the tubercle in length; achene body 1.5–2 mm long. Moist sand; Lamar Co. in Red River drainage (Carr 1994); mainly se and e TX. Summer.

Rhynchospora harveyi W. Boott, (for its discoverer, Francis Leroy Harvey, 1850–1900, of NY), HARVEY'S BEAK-RUSH, PLANK BEAK-RUSH. Perennial 20–60 cm tall; leaf blades 1–4 mm wide; inflorescence narrow, of few small clusters; spikelets 2.5–3 mm long, mostly 1-fruited; bristles < 1/2 length of achene body; achene body 1.5–1.8 mm long. Damp sandy soils; disturbed sites; se and e TX w to Grayson and Parker cos., also Edwards Plateau. May–Jul.

Rhynchospora macrostachya Torr. ex A. Gray, (large-spiked), TALL BEAK-RUSH, HORNED BEAK-RUSH. Perennial similar to *R. corniculata* except for the longer bristles subtending the achenes and the plant usually not as tall or coarse; also inflorescence branches usually more stiffly erect rather than spreading and spikelets usually in clusters of 10 or more; bristles usually 6, mostly

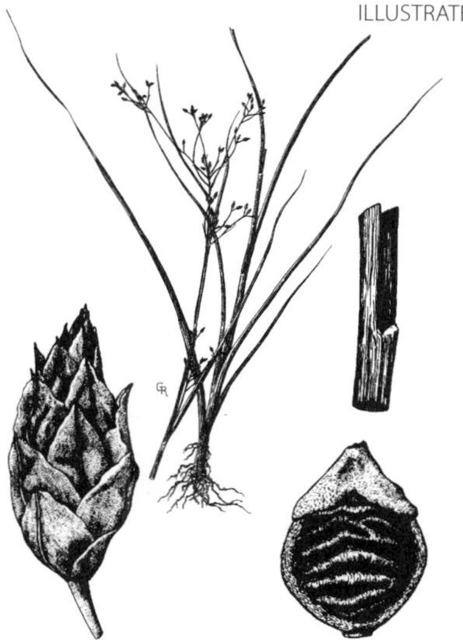

Rhynchospora nitens [GWO]

Rhynchospora nivea [CO1]

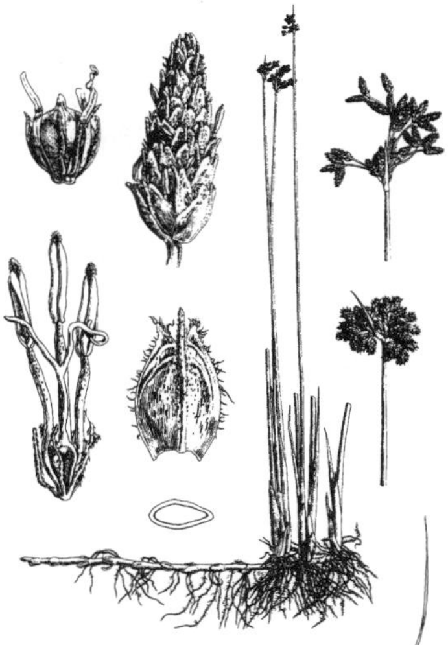

Rhynchospora macrostaycha [CO1]

Schoenoplectus acutus [CO1]

Schoenoplectus californicus [MAS]

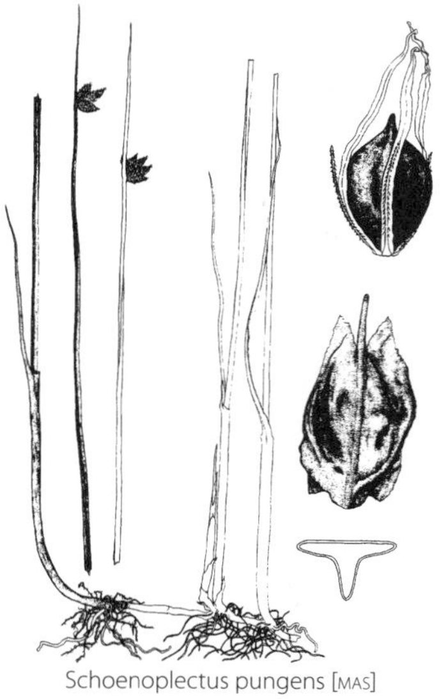

Schoenoplectus pungens [MAS]

Schoenoplectus saximontanus [HEA]

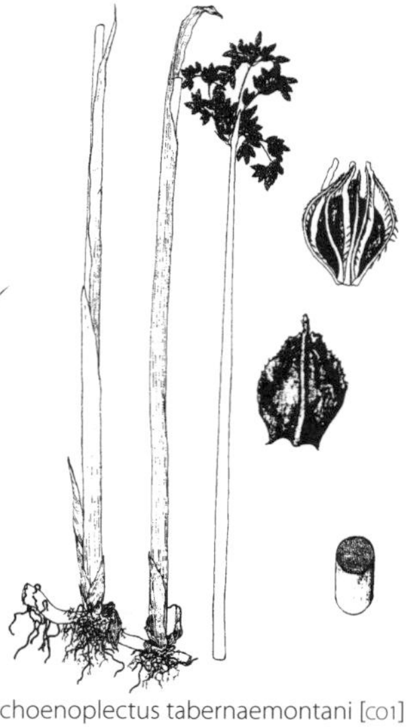

Schoenoplectus tabernaemontani [CO1]

10–12 mm long, antrorsely barbed. Lamar Co. in Red River drainage; mainly se and e TX. Summer. [*R. corniculata* var. *macrostachya* (Torr.) Britton]

Rhynchospora nitens (Vahl) A. Gray, (shining), SHORT-BEAK BALD-RUSH. Glabrous annual 15–80 cm tall; spikelets 4–9 mm long, with numerous scales; perianth absent; achene body ca. as broad as long; tubercle much wider than long. Wet areas; Milam Co. (BRCH); mainly se and e TX. Jul–Aug. Previously segregated into the genus *Psilocarya* [as *P. nitens* (Vahl) A.W. Wood]. [*Psilocarya portoricensis* Britton]

Rhynchospora nivea Boeck., (snowy, white), SNOWY WHITE-TOP SEDGE. Tufted, glabrous perennial 10–30(–40) cm tall; spikelets few-flowered, white; longer bracts 1.7–3.7 cm long; white spot only at the very base of bract. Creek beds on limestone, wet areas; Bell, Brown, Burnet, and Parker cos., also collected along Turtle Creek, Dallas (Austin Chalk), in 1881 or 1882 by Reverchon, noted as "very rare"; not found there since (Thomas 1984; Mahler 1988); also e TX and Edwards Plateau. Previously separated into the genus *Dichromena* [as *D. nivea* (Boeck.) Boeck. ex. Britton, *D. reverchonii* S.H. Wright].

SCHOENOPLECTUS BULRUSH, TULE

Rhizomatous perennials (*S. saximontanus*, which is rare in nc TX, can be annual) of wet areas or in water; plants 0.3–5 m tall (0.09–0.65 m tall in *S. saximontanus*), glabrous or nearly so; culms (= stems) without blade-bearing leaves or with 1–3 blade-bearing leaves near base, sharply triangular or bluntly trigonous to nearly round; inflorescences of one or few sessile spikelets or many in an open panicle, appearing lateral, usually with a single, erect, modified leaf (= involucral bract) appearing like a continuation of the culm; scales of spikelets awnless or nearly so; perianth of bristles present or absent (in *S. saximontanus*); achenes plano-convex to strongly trigonous.

A cosmopolitan genus of ca. 50 species (Smith 1996); 15 are native to North America plus 2 have been introduced. Approximately 10 species occur in TX (S.D. Jones, pers. comm.). Previously included in *Scirpus* (e.g., Kartesz 1994) and according to some, better treated as a section or subgenus in *Scirpus* in the broad sense. We are following Smith (1995, 1996) and Jones et al. (1997) in recognizing this segregate of *Scirpus* at the generic level. Smith (1995) indicated that *Schoenoplectus* and a number of other segregates of *Scirpus* will be recognized in the forthcoming Cyperaceae treatment for *Flora of North America* (Vol. 11). This approach is supported by the phylogenetic studies of Bruhl (1995) which suggested that *Scirpus* sensu lato is polyphyletic. Some species can be ecological dominants in wetlands and provide valuable food and habitat for wildlife (Smith 1996). *Schoenoplectus* species were also extensively used by Native Americans in making baskets, mats, and roof thatching; *S. californicus* (known in South America as TOTORA) is still used for making the famous boats (balsas) seen on Lake Titicaca (Beetle 1950; Heiser 1978). (Greek: *schoeno*, a reed or rush-like, and *plectus*, twine, braid, or plait; alluding to the mat-forming rhizomes of some species)

REFERENCES: Beetle 1941, 1943, 1947; Koyama 1962, 1963; Schuyler 1974; Raynal 1976; Heiser 1978; Strong 1993, 1994; Smith 1995, 1996; Smith & Yatskievych 1996.

1. Plants 0.09–0.65 m tall; culms 0.5–1.5 mm thick near base; perianth bristles absent; achenes with ca. 10–20 prominent (use lens), transverse, wavy, mostly sharp ridges; known in nc TX only from Coryell Co. ______ **S. saximontanus**
1. Plants 0.3–5 m tall; culms 2–23 mm thick near base; perianth bristles present; achenes smooth; widespread in nc TX.
   2. Inflorescences unbranched, usually with 1–5 sessile spikelets; plants 0.3–1.2 m tall; culms sharply triangular, 2–6 mm thick near base ______ **S. pungens**

2. Inflorescences branched, with many, mostly pedicelled spikelets; plants 1–5 m tall; culms usually nearly round to bluntly trigonous, 8–23 mm thick near base.
   3. Achene bristles 2–4 (2–3 different appearing stamens also present), with closely spaced lateral projections; leaf sheaths retrorsely fimbriate-filiferous (= fringed) **S. californicus**
   3. Achene bristles usually 4–6 (2–3 different appearing stamens also present), with well-spaced barbs; leaf sheaths smooth or merely lacerate (= cut as if torn).
      4. Spikelets in glomerules of 3–8, sessile or on short stiff pedicels; scales of spikelets often ca. 5 mm long, with conspicuous elongate reddish glutinous spots (under a hand lens) **S. acutus**
      4. Spikelets usually in glomerules of 2 (sometimes 3 or single) on long lax pedicels; scales of spikelets often 3–4 mm long, nearly smooth (occasionally with a few reddish spots near midrib) **S. tabernaemontani**

Schoenoplectus acutus (Muhl. ex Bigelow) Á. Löve & D. Löve, (acute, sharp-pointed), HARD-STEM BULRUSH, ALKALI TULE. Rhizomatous perennial 1–3 m tall forming extensive colonies; spikelets 8–15 mm long; achenes plano-convex or unequally biconvex. Calcareous mud, usually in water; Grayson and Rockwall cos.; w TX, rare e and nc TX. May. [*Scirpus acutus* Muhl. ex Bigelow] According to Beetle (1950), this species, which was used for mats and roofing, "... was very important in the Indian cultures of western North America."

Schoenoplectus californicus (C.A. Mey.) Soják, (of California), CALIFORNIA BULRUSH, GIANT BULRUSH, TULE, CALIFORNIA TULE. Perennial 1–3+ m tall, from tight subrhizomatous knots; culms bluntly trigonous; spikelets 6–11 mm long; achenes plano-convex or biconvex. Mud or shallow water; nearly throughout TX. Spring and summer. [*Scirpus californicus* (C.A. Mey.) Steud.]

Schoenoplectus pungens (Vahl) Palla, (piercing, sharp-pointed), AMERICAN BULRUSH, SWORD-GRASS, THREE-SQUARE BULRUSH. Rhizomatous perennial 0.3–1.2 m tall; spikelets 5–20 mm long; achenes lenticular to trigonous. Wet soils; Ellis, Montague, and Wise cos.; widespread in TX. Apr–Jul. [*Scirpus pungens* var. *longispicatus* (Britton) S.G. Sm., *Scirpus americanus* Pers. var. *longispicatus* Britton] Varieties *longispicatus* and *pungens* intergrade extensively (Smith 1996); we are therefore following Kolstad (1986b) and Kartesz (1994) in not recognizing infraspecific taxa in this species; Smith (1996) recognized var. *longispicatus* and var. *pungens* in TX and separated the two as follows:

1. Styles trifid; achenes thickly lenticular to bluntly trigonous; floral scales [scales of spikelets] brown to stramineous var. *longispicatus*
1. Styles bifid; achenes lenticular; floral scales brown var. *pungens*

Schoenoplectus saximontanus (Fernald) J. Raynal, (rocky mountain), ROCKY MOUNTAIN BULRUSH. Small annual or perennial 9–65 cm tall; rhizomes inconspicuous, ca. 1 mm thick; culms cylindric, ridged when dry; 0–2 smaller bracts sometimes present in addition to the erect bract which appears like a continuation of the culm; inflorescences with 1–10(–20) spikelets, usually sessile or nearly so, sometimes with 1–2 short branches; spikelets 6–20 mm long; scales with a slightly recurved awn ca. 0.2–0.5 mm long; achenes usually strongly trigonous. Damp soils or emergent from water; Coryell Co. (Fort Hood—Sanchez 1997); in a band from the Panhandle to s TX (Smith 1996). Spring–fall. [*Scirpus bergsonii* Schuyler, *Scirpus saximontanus* Fernald, *Scirpus supinus* L. var. *saximontanus* (Fernald) T. Koyama]

Schoenoplectus tabernaemontani (C.C. Gmel.) Palla, (for Jacob Theodore von Bergzabern, died 1590, Heidelberg botany professor who Latinized his name as Tabernaemontanus), GREAT BULRUSH, SOFT-STEM BULRUSH, GIANT TULE. Rhizomatous perennial 1–5 m tall; spikelets 6–10 mm long; achenes plano-convex. The usually terete fresh culms (just below inflorescences) of this species help distinguish it from the similar *S. californicus* with bluntly trigonous culms. Wet ground, shallow water around lakes; included on basis of citation of vegetational areas 4 and 5 (Fig. 2) by Hatch et al. (1990); widely scattered in TX. May–Jul, occasionally to Oct. [*Scirpus tabernaemontani* C.C. Gmel., *Scirpus validus* Vahl]

SCIRPUS BULRUSH

Perennials of wet areas, with or without rhizomes; plants glabrous or nearly so; culms (= stems) obtusely triangular, with well-developed leaves; inflorescences of many spikelets in open panicles, appearing terminal, with 2 or more well-developed leaf-like involucral bracts; scales of spikelets awnless or essentially so; perianth of bristles.

In the strict sense, a cosmopolitan genus of ca. 30 species; 6 are reported for TX (S.D. Jones, pers. comm.). *Bolboshoenus*, *Isolepis*, and *Schoenoplectus* have traditionally been treated as part of *Scirpus* sensu lato (e.g., Kartesz 1994); according to some authorities these segregates are better treated as sections or subgenera; according to Mabberley (1987), *Scirpus* is a heterogeneous and indivisible genus. If treated in the broad traditional sense, the genus contains ca. 200–300 species (Tucker 1987; Mabberley 1997). We are following Smith (1995, 1996, 1997a, 1997b) and Jones et al. (1997) in recognizing the segregates of *Scirpus* at the generic level. Smith (1995) indicated that the segregates will be recognized in the forthcoming Cyperaceae treatment for *Flora of North America* (Vol. 11). This approach is supported by the phylogenetic studies of Bruhl (1995) which suggested that *Scirpus* sensu lato is polyphyletic. (The Latin name of the bulrush)
REFERENCES: Beetle 1947; Schuyler 1966, 1967; Strong 1993, 1994; Smith 1995; Smith & Yatskievych 1996.

1. Perianth bristles very long, obviously and greatly exceeding the scales of spikelets in length; mature inflorescences appearing almost woolly to the naked eye ____ **S. cyperinus**
1. Perianth bristles shorter than to slightly exceeding the scales of spikelets in length; mature inflorescences not appearing woolly.
   2. Scales of spikelets with prominent green-keeled midribs; spikelets 6–10 mm long; spikelets usually sessile, clustered in glomerules ____ **S. pendulus**
   2. Scales of spikelets without prominent green-keeled midrib; spikelets 2–5 mm long; at least some spikelets often on separate peduncles.
      3. Bristles 0–3, shorter than achenes; bracts usually shorter than inflorescences; each spikelet with 70–150(–200) florets ____ **S. georgianus**
      3. Bristles usually 5 or 6, shorter than or little longer than the achenes; bracts usually as long as or exceeding the inflorescences; each spikelet with 20–40 florets ____ **S. atrovirens**

Scirpus atrovirens Willd., (dark green, from the fuscous spikelets), PALE BULRUSH. Rhizomatous perennial to 1.5 m tall, resembling *S. georgianus*; sheaths and blades cross-septate; spikelets 2–4 mm long. Moist areas; Grayson Co.; mainly e TX and Panhandle. Summer.

Scirpus cyperinus (L.) Kunth, (resembling *Cyperus*), WOOLLY-GRASS BULRUSH. Perennial 0.8–2 m tall, perennating by basal offshoots; spikelets 3–6 mm long. Wet or boggy places; Lamar Co., also Delta and Red River cos. (S.D. Jones, pers. comm.); mainly e TX. Summer. [*C. cyperinus* var. *rubricosus* (Fernald) Gilly]

Scirpus georgianus R.M. Harper, (of Georgia). Tufted perennial, not distinctly rhizomatous, 0.5–1.5 m tall; sheaths and blades usually not cross-septate; spikelets to 4 mm long. Moist or wet soils; included based citation of vegetational area 4 (Fig. 2) by Hatch et al. (1990); mainly e TX. Spring.

Scirpus pendulus Muhl., (pendulous, hanging). Cespitose perennial to 1.5 m tall; spikelets 5–10 mm long. Ditches, streambeds, pond margins; e TX w to Cooke and Palo Pinto cos. (West Cross Timbers). Apr–Jun. [*S. lineatus* of TX authors, not Michx.]

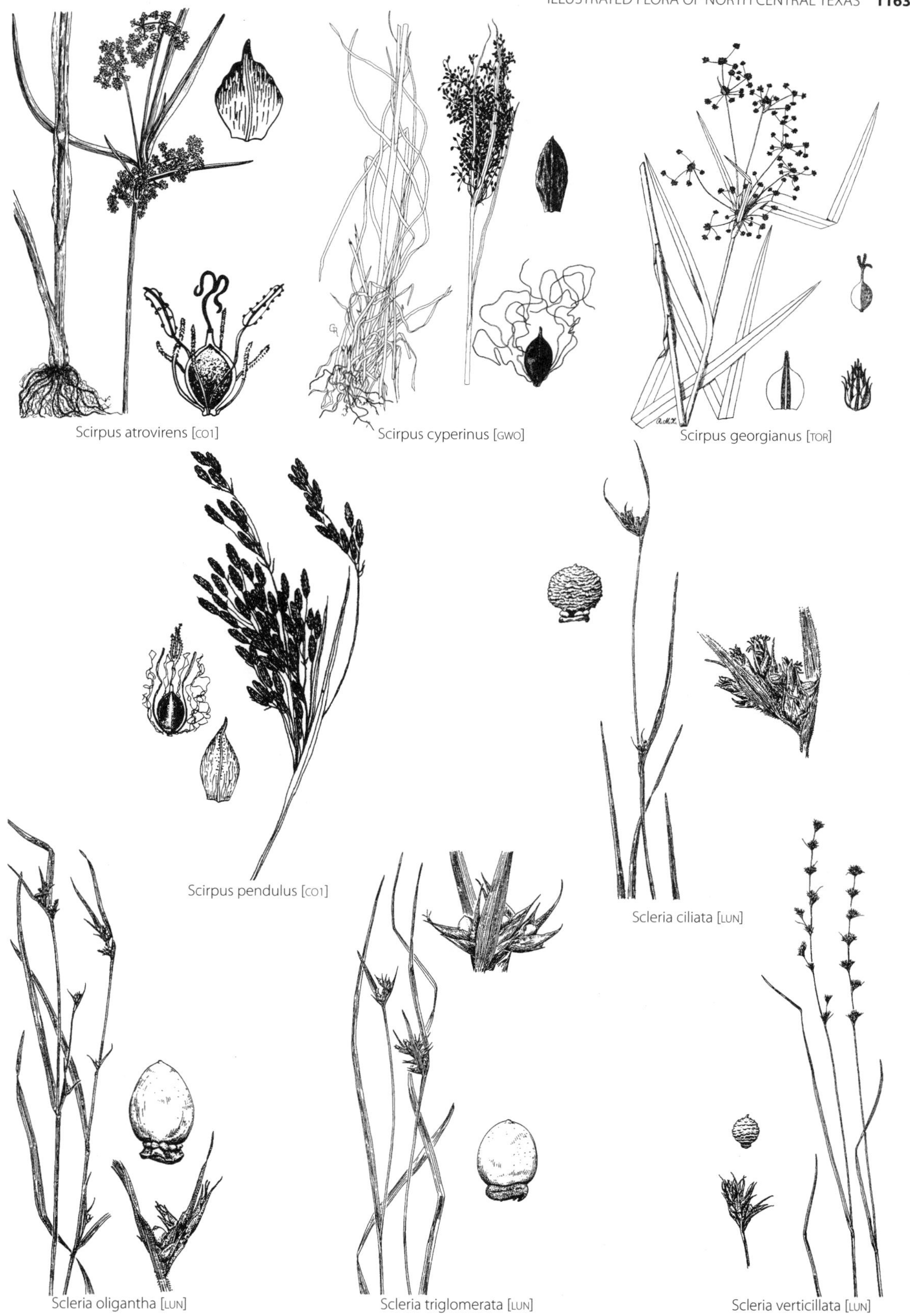

Scirpus atrovirens [CO1]

Scirpus cyperinus [GWO]

Scirpus georgianus [TOR]

Scirpus pendulus [CO1]

Scleria ciliata [LUN]

Scleria oligantha [LUN]

Scleria triglomerata [LUN]

Scleria verticillata [LUN]

SCLERIA NUT-RUSH

Monoecious annuals or perennials usually less than 1 m tall; culms (= stems) sharply triangular, leafy; inflorescences of axillary and terminal, small, compact clusters of few spikelets, leafy-bracted at base; staminate and pistillate spikelets often mixed within a cluster; achenes globose or ovoid to obovoid, usually whitish, bony, or crustaceous, usually on a hardened pad (= hypogynum).

A genus of 200 species of tropical and warm areas. (Greek: *scleria*, hardness, from the hardened fruit)
REFERENCES: Core 1936, 1966; Fairey 1967; Kessler 1987.

1. Plants annual, without rhizomes; achenes wrinkled or roughened, ca. 1 mm long, without pad at base; plants rare or possibly not currently present in nc TX ____ **S. verticillata**
1. Plants perennial with short rhizomes; achenes smooth or warty, 2–4 mm long, with ring-like pad at base; plants widespread in nc TX.
   2. Achene body rough (reticulate, papillose or warty); leafy bracts of inflorescence pubescent and long ciliate ____ **S. ciliata**
   2. Achene body smooth; leafy bracts of inflorescence glabrous OR pubescent and long ciliate.
      3. Pad completely covered with rough white crust, without distinct tubercles; bracts glabrous ____ **S. triglomerata**
      3. Pad not covered with rough white crust, but with 8 or 9 papillose tubercles (use lens); bracts pubescent ____ **S. oligantha**

Scleria ciliata Michx., (ciliate, fringed), FRINGED NUT-RUSH, CILIATE NUT-RUSH. Plant 20–90 cm tall; lower leaf sheaths densely pilose with slightly reflexed hairs. Low sandy woods or open areas; se and e TX w to Dallas and Montague cos. Late Apr–Jul. [*S. ciliata* Michx. var. *elliottii* (Chapm.) Fernald]

Scleria oligantha Michx., (few-flowered), LITTLE-HEAD NUT-RUSH, FEW-FLOWER NUT-RUSH. Perennial 30–60 cm tall; leaf sheaths pubescent or glabrous. Sandy woods, drainage areas; se and e TX w to Dallas and Limestone cos. Late Apr–May.

Scleria triglomerata Michx., (three-clustered), WHIP-GRASS, WHIP NUT-RUSH, TALL NUT-RUSH. Plant 40–100 cm tall, glabrous or nearly so except for scabrous culm angles and leaf margins. Sandy open woods, low areas; se and e TX w to East Cross Timbers. Jun.

Scleria verticillata Muhl. ex Willd., (whorled), LOW NUT-RUSH, WHORLED NUT-RUSH. Plant tufted, slender, 10–60 cm tall; leaf sheaths pilose; leaf blades glabrous. Collected in Dallas Co. by Reverchon (Core 1966), not found in nc TX since; also reported from Post Oak Savannah and Edwards Plateau. Jul–Sep.

DIOSCOREACEAE YAM FAMILY

A medium-sized family (880 species in 8 genera) of mostly herbaceous twining vines from tubers found throughout tropical and warm areas with a few in the n temperate zone; most species are in the genus *Dioscorea*. ☠Raphides are usually present as are steroidal saponins and often lactone alkaloids. (subclass Liliidae)
FAMILY RECOGNITION IN THE FIELD: herbaceous twining vines with alternate, petiolate, cordate ovate leaves with *7–11 main veins*; flowers very small, with 6 stamens and an inferior ovary; fruit a 3-winged capsule.
REFERENCES: Dahlgren et al. 1985; Al-Shehbaz & Schubert 1989.

DIOSCOREA YAM

Dioecious perennials from rhizomes (in nc TX species) or tubers (in some other species); stems to 3 m or more long, leaning on other plants or twining-climbing; leaves alternate, opposite, whorled, or a mixture of these, long-petiolate; leaf blades cordate-ovate, cordate at base, acute to acuminate at apex, with 7–11 main, convergent, parallel veins and numerous cross veins; flowers very small, whitish to greenish yellow, in axillary panicles or spikes; perianth radially symmetrical; petals and sepals 3 each, 0.2–2 mm long; stamens 6; ovary inferior; stigmas 3, bilobed; capsule 3-winged, 1.2–3 cm long, loculicidal, splitting through wings; seeds flat, with a broad wing equal to or wider than body.

A genus of ca. 850 species of tropical and warm areas. *Dioscorea* species (YAMS) are cultivated in some areas as a subsistence crop; ☠proper preparation to remove toxic saponins is necessary. The storage tubers of Mexican species of *Dioscorea* have served as a source of steroidal precursor molecules (e.g., diosgenin) for cortisone and the active ingredients in early birth control pills (Lewis & Elvin-Lewis 1977). *Dioscorea bulbifera* L. (AIR-POTATO), native of tropical Asia and possibly Africa, ☠contains diosgenin, alkaloids, oxalates, and other toxins (Morton 1982); it has both inedible and edible forms (Bailey & Bailey 1976). This species, which is sometimes cultivated in nc TX, can be distinguished from native taxa by the presence of conspicuous, axillary, aerial bulbils. (Named for Dioscorides, Greek naturalist of the first century A.D., who wrote *Materia Medica*, a description of ca. 600 plants used medicinally)

1. Petioles usually pubescent at junction with leaf blade, usually 7 cm or more long on well-developed leaves; leaves in whorls of 4–7 at 1 or more lower nodes ____ **D. quaternata**
1. Petioles glabrous at junction with leaf blade, usually 6 cm or less long on well-developed leaves; leaves of lower nodes at most opposite or with 3 in close proximity ____ **D. villosa**

Dioscorea quaternata J.F. Gmel., (in fours). Rhizomes relatively stout, 10–15 mm thick, often contorted irregularly or with many short branches; leaves of lower nodes whorled, opposite to alternate above; leaf blades 7–15 cm long; capsules usually 2.5–3 cm long; seeds ca. 18 mm wide. Moist rich woods; included based on citation of vegetational area 4 (Fig. 2) by Hatch et al. (1990); ne TX. Apr–Jun. [*D. villosa* var. *glabrifolia* (Bartlett) Fernald]

Dioscorea villosa L., (soft-hairy), WILD YAM, ATLANTIC YAM, COLICROOT. Rhizomes relatively slender, 5–10 mm thick, ± straight, not much branched; leaves except for the lowest all alternate; leaf blades 5–11 cm long; capsules 1.2–2.5 cm long; seeds 7–12 mm wide. Low woods; Lamar Co.; mainly deep e TX. Apr–Jun.

ERIOCAULACEAE PIPEWORT FAMILY

The Eriocaulaceae is a medium-large (1,100 species in 9 genera) family of usually perennial, small, scapose herbs distributed mainly in the tropics and warm areas, especially in the Americas; a few are in temperate regions. (subclass Commelinidae)

FAMILY RECOGNITION IN THE FIELD: wet area herbs with a basal rosette of linear, grass-like leaves; flowers small and inconspicuous, in a small (5–20 mm in diam.) head at the end of an elongate naked peduncle; the head often has a whitish appearance.

REFERENCES: Moldenke 1961a; Kral 1966b, 1989; Dahlgren et al. 1985.

ERIOCAULON PIPEWORT

Clump-forming, usually monoecious perennial or biennial herbs reproducing vegetatively by lateral offshoots or rhizomes; leaves ± all in basal rosette, linear and grass-like in appearance, sheath-like at base; inflorescence a compact hemispherical to globose head terminating an elongate, naked, angled peduncle, the peduncle with a basal sheath; heads with involucral and receptacular bracts; bracts and perianth parts often with conspicuous whitish hairs; flowers small (ca. 3 mm or less long) and crowded together; sepals 2; petals 2; male flowers with 3–6 stamens; carpels 2; ovary superior; fruit a 1-seeded loculicidal capsule.

A genus of ca. 400 species of tropical and warm areas including Japan and North America. In TX the genus occurs mainly in the se and e parts of the state extending w to near the extreme e margin of nc TX. (Greek: *erion*, wool, and *caulos*, a stalk, from the wool at the base of the scape in the first named species)

1. Mature heads 10–20 mm wide, hard and little compressed upon pressing and drying; involucral bracts acute to acuminate at apex; mature scape with 8–12 ridges; leaves 10–40 cm long, exceeding the sheaths of the scape in length **E. decangulare**
1. Mature heads ca. 5 mm wide, soft and compressed upon pressing and drying; involucral bracts rounded at apex; mature scape with 4–7 ridges; leaves 1–6.5 cm long, the sheaths of the scape exceeding all or most of the leaves in length **E. texense**

Eriocaulon decangulare L., (ten-angled), TEN-ANGLE PIPEWORT, PIPEWORT. Mature scapes 30–110 cm long; mature heads subglobose, dull white; lowermost flowers and bractlets reflexed and obscuring the subtending involucral bracts; outer involucral bracts narrowly ovate to lanceolate, 2–4 mm long, straw-colored, with clavate white hairs apically; surface of receptacle villous; receptacular bracts acute to acuminate, sometimes visibly exserted; sepals 2–3 mm long, yellowish white, with clavate white hairs on keel and apex. Meadows, swamps, pond margins, bogs; e Milam Co., also Henderson Co. (Moldenke 1948); mainly se and e TX. Apr–Nov.

Eriocaulon texense Körn., (of Texas), TEXAS PIPEWORT, PIPEWORT. Mature scapes 5–30 cm long, mature head hemispherical, gray except for the white trichomes of the bractlets and perianth parts and the straw-colored outer involucral bracts; outer involucral bracts suborbicular to broadly ovate, ca. 1.5 mm long, at maturity usually hidden by the florets; surface of receptacle with numerous hairs; receptacular bracts acute; sepals ca. 1.5 mm long, dark gray apically, with clavate white hairs on keel and sometimes margins apically. Wet acidic areas, bogs; c Henderson Co., also Milam Co. (Moldenke 1948); mainly se and e TX w to near e margin of nc TX. Moldenke (1961a) indicated Cory recorded the species from the Blackland Prairie. Apr–Jun.

Eriocaulon aquaticum (Hill) Druce, (aquatic), SEVEN-ANGLE PIPEWORT, WHITE-BUTTONS, DUCK-GRASS, [*E. septangulare* With.], was reported for TX in vegetational areas 1 and 4 (Fig. 2) by Hatch et al. (1990). However, Godfrey and Wooten (1979) indicated it occurs in Canada and the n U.S. s only to North Carolina; R. Kral (pers. comm.) confirmed that this species does not occur in TX. According to Kral, "Early reports of *E. aquaticum* (*E. septangulare*) were based on lack of understanding of the similar-looking *E. texense*." This species can be distinguished by the heads which are small (4–5 mm wide) and ± dark in appearance due to usually gray to almost black bracts and perianth parts (only tips of perianth parts and receptacular bracts have whitish hairs); also the receptacles are glabrous.

Ericaulon kornickianum Van Heurck & Müll.Arg., (for Friedrich August Körnicke, 1828–1908, German botanist), SMALL-HEAD PIPEWORT, is known from e TX and could possibly be expected near the e margin of nc TX (R. Kral, pers. comm.). This diminutive species has short scapes (5–8 cm long), mature heads 3–4 mm broad, outer involucral bracts not hidden by bractlets and perianth parts, sepals ca. 1 mm long, and receptacles glabrous.

HYDROCHARITACEAE

FROG'S-BIT, TAPE-GRASS OR WATERWEED FAMILY

Ours submerged, monoecious or dioecious herbs; leaves basal or subopposite, opposite, or in whorls along the stem; inflorescences with a spathe; perianth 3-merous or minute or absent; flowers at the water surface or borne underwater; fruits indehiscent, ripening underwater.

A small (112 species in 16 genera), cosmopolitan, but mainly tropical family of aquatics; it

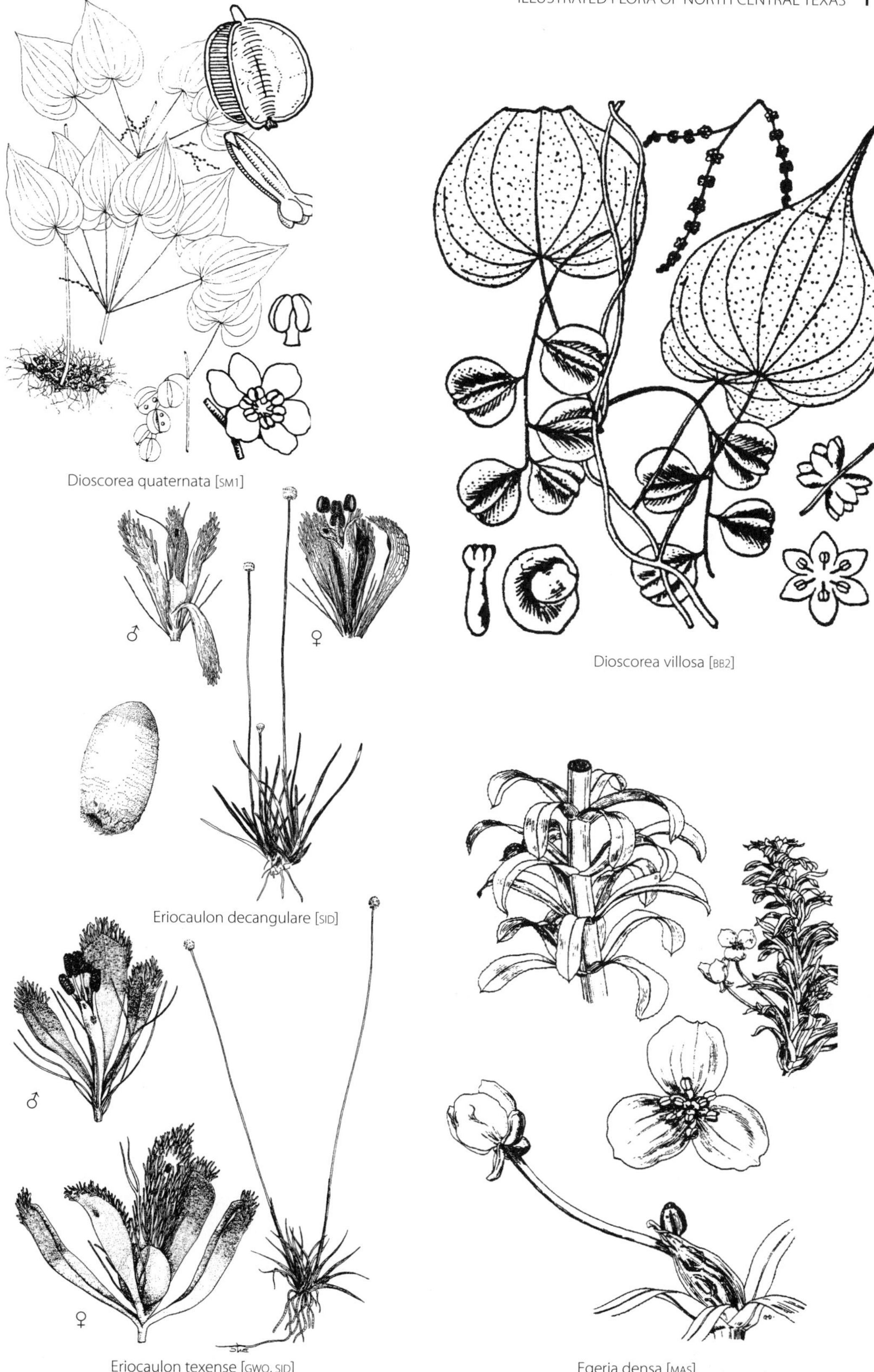

Dioscorea quaternata [SM1]

Dioscorea villosa [BB2]

Eriocaulon decangulare [SID]

Eriocaulon texense [GWO, SID]

Egeria densa [MAS]

includes *Elodea* (WATERWEED, DITCH-MOSS) and *Hydrilla*. The family is notable for its wide variety of pollination mechanisms including male flowers that detatch and float or sail to the female flower, and also anthers that explode and scatter pollen over the water surface; it is one of relatively few families that exhibit hydrophily—water-mediated pollination; both hypohydrophily (pollination under water) and epihydrophily (pollination at the water surface) are known in the family (Philbrick 1991, 1993). A number of species are used as aquarium plants; some have become noxious aquatic weeds. We are following Thorne (1993b) and R. Haynes (pers. comm.) in including *Najas*, a genus previously recognized in its own family, in the Hydrocharitaceae. Family name from *Hydrocharis*, FROGBIT, an Old World genus of 3–6 species of floating aquatics. (Greek: *hydor*, water, and *charis*, graceful, in reference to the habit) (subclass Alismatidae)

FAMILY RECOGNITION IN THE FIELD: submerged *aquatic* herbs with leaves EITHER simple, linear to linear-lanceolate or narrowly oblong, subopposite to opposite or whorled OR long and ribbon-like and clustered at base of plant.

REFERENCES: Dahlgren et al. 1985; Thorne 1993b.

1. Leaves subopposite, opposite, or whorled, distributed along the elongate stem, 4.5 cm or less long, 3(–5) mm or less wide.
   2. Leaves subopposite or opposite (some can occasionally appear whorled where branches arise—if so they are dilated at base); flowers sessile or subsessile, borne underwater; perianth absent or minute, clearish or greenish, virtually indistinguishable without a lens; including extremely abundant native species ____ **Najas**
   2. Leaves in distinct whorls of 3–8, not dilated at base; flowers (male and/or female) borne to the water surface on a thread-like stalk 3–6 cm long; perianth 3–10 mm long, white or translucent, visible with the naked eye; introduced species found in a few lakes in nc TX (currently spreading)
      3. Leaves usually 2–3(–4) cm long, serrulate (teeth scarcely visible to the naked eye) marginally but lacking teeth on the midvein beneath (fresh leaves thus not rough to the touch); male flowers borne to the water surface on a thread-like stalk, with conspicuous white perianth parts 8–10 mm long ____ **Egeria**
      3. Leaves usually 1.5(–2) cm or less long, serrate (teeth visible to the naked eye) marginally and with teeth on the midvein beneath (fresh leaves thus rough to the touch); male flowers detaching and floating to the surface, without a conspicuous white perianth ____ **Hydrilla**
1. Leaves clustered at base of plant on a very short stem, long and ribbon-like, 8–60(–nearly 100) cm long, to 8(–20) mm wide ____ **Vallisneria**

EGERIA

A genus of 2 species of subtropical South America; it has sometimes been lumped with *Elodea* (an American genus of 12 species); *Elodea canadensis* Michx., known from OK but apparently not TX, can be distinguished from the somewhat similar *Egeria densa* and *Hydrilla verticillata* using the following characters: leaves opposite or in whorls of three, mostly 8–15 mm long, leaf margins lacking teeth perceptible to the naked eye, and leaf midveins lacking teeth. *Egeria* has nectariferous flowers and a relatively large perianth (Catling & Wojtas 1986); it is presumably animal-pollinated in contrast to the the many hydrophilous taxa observed in this family. (Named after a Roman goddess of water or mythical water nymph)

REFERENCES: St. John 1965; Cook & Urmi-König 1984; Catling & Wojtas 1986.

Egeria densa Planch., (dense), WATERWEED. Perennial, submerged aquatic, rooting at bottom or if broken, drifting; lowermost leaves opposite or in whorls of 3; rest of leaves in whorls of 4–6(–8); leaves crowded, usually linear-lanceolate, serrulate, transparent, to ca. 2.5(–4) cm long and 3(–5) mm wide; flowers 3-merous; male spathes 2–4-flowered, in upper axils; male flowers reaching the surface on a long thread-like hypanthium 3–6 cm long; petals white, ca. 8–10 mm long, much longer than the sepals; anthers 9; female flowers not observed. Lakes, ponds; Dallas, Grayson (Lake Ray Roberts), and Lamar (Pat Mayse Lake) cos., also Brown (City Reservoir) and Comanche cos. (HPC); mainly e TX and Edwards Plateau; probably spread from lake to lake by boats or boat trailers; commonly called ELODEA even though this is the scientific name of a related genus; widely used in biological laboratories. [*Anacharis densa* (Planch.) Vict., *Elodea densa* (Planch.) Casp.] Native of South America.

HYDRILLA

A monotypic Old World genus introduced into the U.S. and Central America. The male flowers are released underwater, rise to the surface, and float to the female flowers (an example of epihydrophily—see discussion in *Vallisneria* generic synopsis). Cook and Lüönd (1982) indicated that during flower opening the stamens spring upward, the anthers burst, and pollen is scattered in the air, some of this landing on the female flowers and some on the surface of the water; apparently the pollen grains landing on the water are lost in terms of reproduction and only those actually landing on the female flower are involved in pollination. Even though the effective pollen grains never contact the water, the male flowers float and the system can thus be described as epihydrophily. (Diminutive of Greek: *hydra*, water, sea serpent)

REFERENCES: Godfrey & Wooten 1979; Cook & Lüönd 1982; Steward et al. 1984; Langeland 1996.

Hydrilla verticillata (L.f.) Royle, (whorled). Perennial, submersed, dioecious or monoecious aquatic with horizontal stems in the substrate sometimes forming tubers; erect stems branching, capable of ascending as much as 8.5 m (Godfrey & Wooten 1979), growing horizontally near the water surface; leaves in whorls of (2–)3–8, sessile, mostly 1.5(–2) cm or less long, 1.5–2 mm wide, narrowly oblong, serrate marginally (the teeth visible to the naked eye), with a single vein which on the abaxial surface bears conical protrusions tipped with teeth (fresh leaves thus noticeably rough to the touch); staminate flowers detach and float to surface; pistillate flowers reach the surface at the end of an elongate, thread-like floral tube 4–5 cm long; perianth segments 6, colorless, very inconspicuous, 3–5 mm long; fruits 5–6 mm long, ± fusiform. Lakes and other aquatic habitats; rapidly spreading at present in nc TX; probably spread vegetatively from lake to lake by boats or boat trailers and also intentionally by fishermen (L. Hartman, pers. comm.) to "improve" the habitat (this is both illegal and ill-advised since it ultimately degrades the fishery); Cooke, Dallas, Denton, Hopkins, Hunt, and Tarrant cos. (M. Smart, pers. comm.); also se and e TX and Edwards Plateau. Native of the Old World, probably the warmer areas of Asia (Cook & Lüönd 1982). Female plants were first reported in the U.S. in 1960 from South Florida (misidentified as *Elodea canadensis*) (Blackburn et al. in Steward et al. 1984); monoecious plants were first observed in the U.S. in Washington, D.C. in 1982; the female (dioecious) individuals are triploid, while monoecious plants are diploid (Steward et al. 1984). Since its introduction in Florida, this problematic species has spread across the southern U.S. to Texas, California, Washington state, and up the east coast to Delaware and Maryland. The earliest TX collection we are aware of is from 1974 (*Amerson 2097*, BRIT/SMU); the species was first collected in Louisiana in 1973 (Solymosy 1974). *Hydrilla* is a serious invasive pest which can completely dominate aquatic habitats eliminating native species, clogging waterways, and severely curtailing recreational use (Steward et al. 1984; Flack & Furlow 1996; Langeland 1996). In Texas, *Hydrilla* is considered a "harmful or potentially harmful exotic plant" and it is illegal to release, import, sell, purchase, propagate, or possess this species in the state (Harvey 1998). *Hydrilla* can easily be confused with *Egeria densa*.

NAJAS NAIAD, WATER-NYMPH, BUSHY-PONDWEED

Submerged, monoecious or dioecious, aquatic herbs; leaves subopposite, opposite, or crowded and appearing whorled but inserted at barely different levels, sheathing basally, 3 mm wide or less; flowers axillary, few; staminate flowers with 1 stamen; pistillate flowers sessile, 1-carpellate; stigmas 2–4; fruit a 1-seeded nutlet enclosed in a loose coat.

A cosmopolitan genus of ca. 32 species. The stems have reduced xylem lacking vessels; some are serious weeds in rice fields, but make good green fertilizer; *Najas* species are an important source of food for waterfowl and fish; however, they are capable of vigorous growth and may become so dense as to impede water flow and the movement of boats (Tyrl et al. 1994). *Najas* is one of relatively few genera that exhibit hypohydrophily (pollination under water) in contrast to some cases of hydrophily in which pollination occurs at the water surface (epihydrophily; e.g., *Hydrilla, Vallisneria*) (Cox 1988; Philbrick 1991, 1993). While traditionally recognized in its own family, the Najadaceae (e.g., Kartesz 1994; Jones et al. 1997), Thorne (1993b) treated *Najas* in the Hydrocharitaceae, and there is molecular and morphological evidence that it should be lumped with the Hydrocharitaceae (R. Haynes, pers. comm.). Shaffer-Fehre (1991b), for example, put *Najas* in the Hydrocharitaceae based on seed coat anatomy. (Greek: *Naias*, a water-nymph)
REFERENCES: Morong 1893; Taylor 1909b; Clausen 1936; Haynes & Wentz 1974; Haynes 1977, 1979; Stuckey 1985; Lowden 1986; Shaffer-Fehre 1991a, 1991b; Haynes & Hellquist 1996).

1. Leaves usually minutely denticulate (appearing entire to the naked eye) or nearly entire; internodes and back of leaves unarmed ______ **N. guadalupensis**
1. Leaves coarsely and obviously (to the naked eye) toothed; internodes and often back of leaves armed with small spines ______ **N. marina**

Najas guadalupensis (Spreng.) Magnus, (for the Caribbean Island of Guadalupe, from where the type was collected), COMMON WATER-NYMPH, SOUTHERN NAIAD. Monoecious; leaves to 25 mm long and 2 mm wide (usually ca. 10 mm long and 1 mm wide); seeds fusiform, 2–3 mm long, 0.4–0.8 mm wide, areolate, the areolae in rows and easily seen under low magnification. Often very abundant; possibly in nearly every tank, pond, and lake in nc TX, also streams and ditches; throughout TX. Jun–Sep.

Najas marina L., (of the sea), HOLLY-LEAF WATER-NYMPH, SPINY NAIAD. Dioecious; leaves to 45 mm long and 3 mm wide; seeds ovoid, 4–5 mm long, 1.2–2.3 mm wide, smooth or minutely areolate, the areolae irregularly arranged. Lakes and ponds; Somervell Co.; also in TX in Cameron, Tom Green, Travis, and Zapata cos. (Stuckey 1985). May–Jul.

VALLISNERIA TAPE-GRASS, EEL-GRASS

A genus of 2–10 species of tropical and warm temperate areas of the world. Pollination occurs when male flowers break off, float to surface, open, and float to the female flowers which are borne at the surface on long peduncles; the female flowers create a slight depression in the surface of the water and the male flowers fall into this depression (Cox 1988); this type of hydrophily (= water-mediated pollination) occurring at the water surface is known as epihydrophily (in contrast to hypohydrophily or underwater pollination as seen in *Najas*) (Philbrick 1993). Some workers (e.g., Cox 1988) further divide epihydrophily into a category in which pollen is transported just above the surface of the water (e.g., *Hydrilla, Vallisneria*) and a category in which pollen is transported directly on the surface of the water (e.g., *Elodea*) (Named for Antonio Vallisneri, 1661–1730, an Italian botanist)
REFERENCE: Rydberg 1909.

Vallisneria americana Michx., (of America), AMERICAN WILD-CELERY, EEL-GRASS, WATER-CELERY. Submerged stoloniferous aquatic; leaves clustered at base of plant on a very short stem, long and ribbon-like, 8–60(–nearly 100) cm long, to 8(–20) mm wide; male flowers numerous, tiny, breaking from a spathe and free-floating at maturity, with 3 very thin, transparent sepals and 1

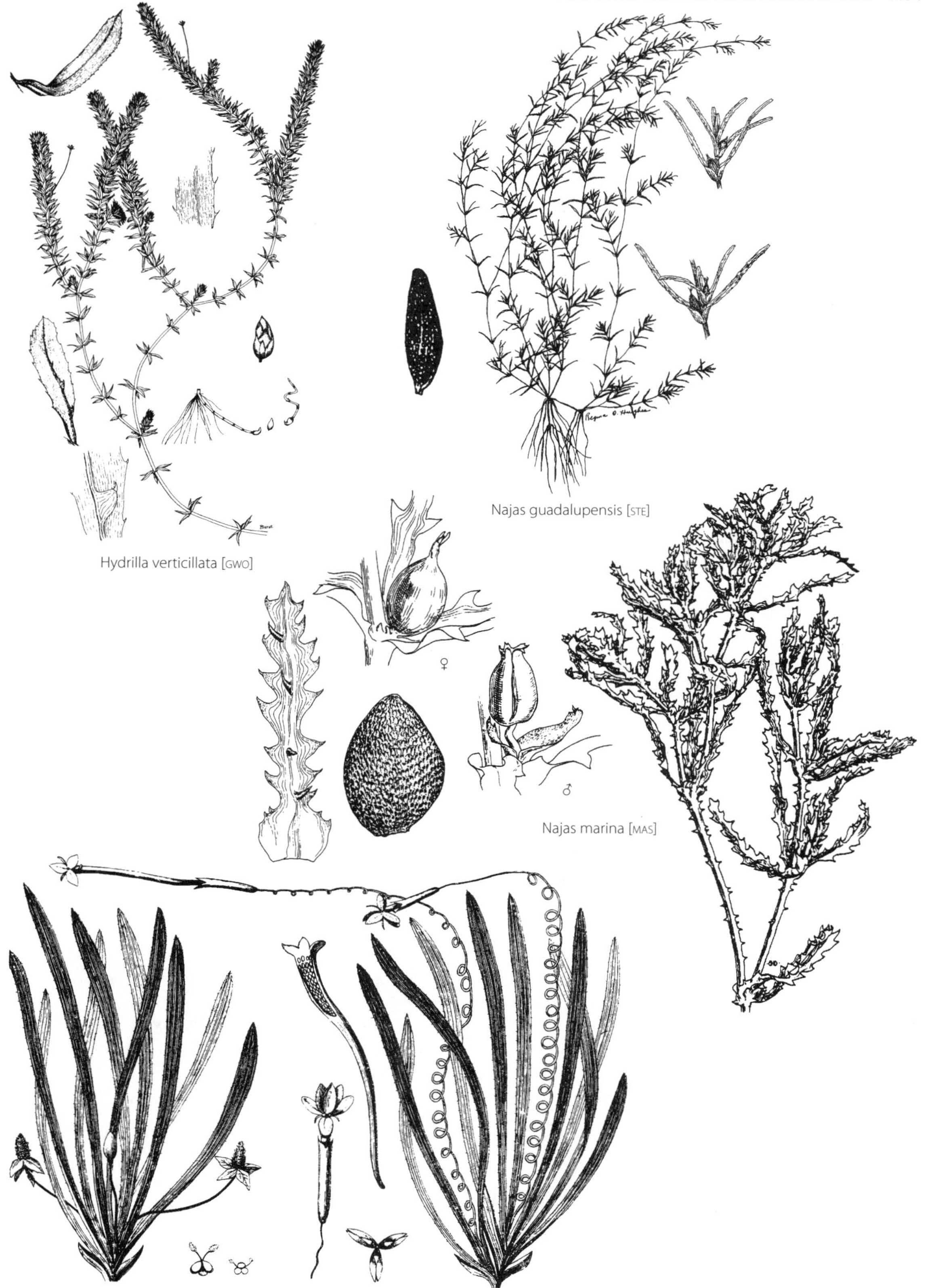

Hydrilla verticillata [GWO]

Najas guadalupensis [STE]

Najas marina [MAS]

Vallisneria americana [LAM]

transparent petal; female flowers solitary in pedunculate spathes at the water surface at flowering time, with 3 thickish persistent sepals 3–4 mm long and 3 rudimentary, soon disintegrating petals to ca. 2 mm long; peduncles to 1 m long, coiling after pollination and pulling the flowers below the surface; fruits cylindric, 8–18 cm long; to ca. 5 mm in diam. Lakes, flowing streams; introduced into Lake Ray Roberts (Grayson Co.) (M. Hackett, pers. comm.) but no nc TX specimens seen; rare in se and e TX w to Edward Plateau. Apr–Jul. [*V. spiralis* of authors, not L.] Where abundant in the n U.S., it is a valuable wildlife food.

IRIDACEAE IRIS FAMILY

Plants annual or perennial herbs from fibrous roots or a bulb, corm, or rhizome; leaves basal or alternate; leaf blades equitant (2-ranked, folded around a stem in the manner of the legs of a rider around a horse; the stem and leaves together having a flattened appearance), or pleated, or concave above, or with prominent midrib; inflorescences various; flowers arising from a spathe of bracts; tepals 6, in 1 or 2 rows, similar or of 2 different sizes or shapes; stamens 3; pistil 1; stigmas petaloid in *Iris* and some other genera; ovary inferior; fruit a capsule.

A medium-large family (1,700 species in 82 genera) of nearly worldwide distribution in tropical and temperate regions, but especially s Africa, e Mediterranean, and Central and South America; it includes many important ornamentals in addition to *Iris*, including *Crocus*, *Gladiolus*, and *Tigridia* (TIGER-FLOWER). The world's most expensive spice, saffron, important as a food coloring, is obtained from the stigmas of *Crocus sativus* L. (subclass Liliidae)
FAMILY RECOGNITION IN THE FIELD: herbs with *equitant* leaves and flowers subtended by bracts, with a 6-parted petaloid perianth, 3 *stamens*, and a 3-celled inferior ovary; the somewhat similar Liliaceae have 6 stamens and a superior or inferior ovary.
REFERENCES: Dahlgren et al. 1985; Goldblatt 1990; Goldblatt & Takei 1997.

1. Leaves flat; rootstock a rhizome or obsolete and roots fibrous (except *Iris xiphium* with a bulb).
   2. Tepals similar in size and positioning (but some can be narrower); style branches filiform, entire; flowers small (ca. 12 mm or less long); plants grass-like vegetatively ____ **Sisyrinchium**
   2. Tepals not similar in size or positioning, the inner whorl erect, the outer whorl reflexed; style branches petaloid, cleft at apex; flowers large (much more than 12 mm long); plants not grass-like, the leaves large ____ **Iris**
1. Leaves plicate (= folded like a fan), occasionally so narrow that pleats not developed; rootstock a bulb.
   3. Tepals (3 inner vs. 3 outer) conspicuously unequal in size, variously patterned, either with very small inner tepals dark (blackish violet) towards base or inner tepals spotted reddish brown over yellow towards base; anthers 6–10 mm long.
      4. Tepals blue or purple-blue, yellow color never present; outer larger tepals lanceolate to broadly so, with a violet halo (= ring) outlining whitish, purple-dotted base; inner tepals much smaller, ca. 8 mm long, blackish violet towards base, acuminate apically; cauline leaves if present reduced to bract-like structures ____ **Herbertia**
      4. Tepals velvety purple to rose-purple, spotted reddish brown over yellow towards base; outer large tepals ovate; inner tepals ca. 15 mm long, obtuse apically; cauline leaves leaf-like towards base of stem, becoming bract-like above ____ **Alophia**
   3. Tepals nearly equal in size and appearance, uniformly blue except lighter to white at very base giving the perianth the appearance of a white "eye"; anthers 11–15 mm long ____ **Nemastylis**

ALOPHIA

A genus of 5 species of perennial herbs native from the s United States to South America; some are used as cultivated ornamentals. (Greek: *a*, without, and *lophos*, crest)

Alophia drummondii (Graham) R.C. Foster, (for its discoverer, Thomas Drummond, 1780–1835, Scottish botanist and collector in North America), PURPLE PLEAT-LEAF. Perennial from a shallow bulb; basal leaves sheathing, narrowly linear to linear-lanceolate, to ca. 30 cm long and 2 cm wide; cauline leaves leaf-like near base of stem, bract-like above; flowering stem to 75 cm tall; flowers few, emerging from the spathes; tepals velvety-purple to rose-purple (rarely white), spotted reddish brown over yellow toward base, to ca. 25 mm long, fugacous; 3 outer tepals somewhat larger than the 3 longitudinally cupped and apically crimped inner; anthers 6–8 mm long; capsules ca. 1.5–5 cm long. Sandy soils, grassy areas, and open woods; Dallas Co. (Reverchon collection), also Milam Co. near e edge of nc TX, also Lamar Co. in Red River drainage (Carr 1994); mainly se and e TX. Mostly May–Jul. [*Nemastylis purpurea* Herb., *Eustylis purpurea* (Herb.) Engelm. and A. Gray, not *A. drummondii* in the sense of Correll & Johnston (1970)]. ▣/77

HERBERTIA

A mainly South American genus of 4 species of perennial herbs with 1 species in TX and LA; some cultivated as ornamentals. (Named for William Herbert, 1778–1847, Dean of Manchester and an authority on bulbiferous plants)
REFERENCE: Goldblatt 1977 [1978].

Herbertia lahue (Molina) Goldblatt subsp. **caerulea** (Herb.) Goldblatt, (sp.: derivation unknown; subsp.: dark blue). Scapose herb from a bulb; leaves with sheathing bases, narrowly linear, to ca. 20 cm long, usually 6 mm or less wide, cauline leaves if present reduced and bract-like; scape to ca. 30 cm tall; spathes with 1–2 flowers; perianth ca. 5 cm across; 3 outer tepals usually pale or dark lavender with patterned base, to 25 mm long; 3 inner tepals much smaller, with upper part violet and lower blackish violet, sometimes with white spots; anthers 7–10 mm long; capsules to 25 mm long. Grasslands and prairies; Denton Co.; also Dallas Co. (probably introduced with sod–R. O'Kennon, pers. obs.); mainly se and e TX; endemic to TX and LA. Mar–May. [*Alophia drummondii* of Correll & Johnston (1970) and Mahler (1988), not (Graham) R.C. Foster, *H. caerulea* (Herb.) Herb., *Trifurcia lahue* (Molina) Goldblatt subsp. *caerulea* (Herb.) Goldblatt] The name *Alophia drummondii* was misapplied by Correll & Johnston (1970) and Mahler (1988) to this species which is correctly placed in *Herbertia* (Goldblatt 1977 [1978]). ▣/92

IRIS FLAG, FLEUR-DE-LIS

Ours perennials from rhizomes (bulb in *I. xiphium*); leaves flat (nearly cylindrical in *I. xiphium*), equitant; flowers usually large and showy, usually mostly bluish, purplish, white, or variously colored in cultivated forms; perianth segments (tepals) of 2 types, the outer (= falls) deflexed or spreading, the inner (= standards) usually erect; style divided into 3 petaloid branches, each overlying a stamen and the claw of a fall; hypanthial tube present.

A mostly n temperate genus of ca. 210 species; *Iris* species are very important ornamentally; e.g., they are the "fleur-de-lis" of French royalty (Tveten & Tveten 1993). Several species are also grown for orris, a powder with the odor of violets; it is obtained from the rhizomes and used in perfumes. ☠Leaves and especially the rhizomes of some species (e.g., *I. germanica*, *I. pseudacorus*) contain an irritating resinous substance (irisin) which can cause severe digestive upset if eaten and dermatitis if handled (Schmutz & Hamilton 1979; Lampe & McCann 1985). There are no *Iris* species native to nc TX; all of the following are Old World taxa that are introduced, persist, and escape. In addition, LOUISIANA IRIS, hybrids of *I. fulva* Ker. Gwal., *I. giganticaerulea* Small, and to a lesser degree *I. brevicaulis* Raf. (all native to LA or adjacent areas), are also cultivated in nc TX, especially in wet areas such as pond margins (e.g., Edith Wolford of Fannin Co.). These are rhizomatous, have flat leaves and beardless falls, and are very variable in color. (Named after Iris, Greek goddess of the rainbow, from flower colors)
REFERENCES: Dykes 1913; Anderson 1936; Foster 1937.

1. Leaves nearly cylindrical, channeled on upper surface; plants with a bulb ____ **I. xiphium**
1. Leaves flat, the two surfaces identical; plants with a thick rhizome.
   2. Petals yellow; standards 4–8 mm wide; falls glabrous ____ **I. pseudacorus**
   2. Petals bluish, purplish, or white (or variously colored in cultivated forms); standards 30–60 mm wide; falls with a beard of multicellular hairs.
      3. Spathes entirely scarious (= dry, papery, and translucent or transparent) at flowering time, 20–35 mm long; hypanthial tube 8–11 mm long ____ **I. pallida**
      3. Spathes herbaceous (= greenish) in the lower half at flowering time, the upper half scarious, 35–55 mm long; hypanthial tube 17–25 mm long ____ **I. germanica**

Iris germanica L., (German), GARDEN IRIS. Perianth bluish violet to white or variously colored in a diversity of cultivated forms; falls 40–60 mm wide; standards 45–60 mm wide. This is the commonly cultivated *Iris* widely planted in TX. It persists indefinitely and escapes widely. Grayson Co., also Dallas (Ed McWilliams, pers. comm.) and Tarrant (R. O'Kennon, pers. obs.) cos. Spring. Probably of hybrid origin from several European species. Toxic (Lampe & McCann 1985).

Iris pallida Lam., (pale), BLUE FLAG, ORRIS. Perianth lilac to violet; falls 30–50 mm wide; standards 30–45 mm wide. Cultivated and escapes; Dallas Co., also Tarrant Co. (R. O'Kennon, pers. obs.). Apr–May. Native to the Old World.

Iris pseudacorus L., (false sweet-flag), YELLOW FLAG. Perianth yellow; falls 20–30 mm wide; standards 4–8 mm wide; hypanthial tube 10–15 mm long. Commonly cultivated and escapes, in ponds and lake margins; Parker and Tarrant (Fort Worth Botanic Garden) cos. (R. O'Kennon, pers. obs). Apr–May. Native to Europe and n Africa. Toxic (Lampe & McCann 1985).

Iris xiphium L., (sword-leaved), DUTCH IRIS, SPANISH IRIS. Flowers usually solitary; perianth blue (to other colors in cultivated forms); falls 18–25 mm wide; standards 15–20 mm wide. Cultivated and escapes; Dallas Co. Apr–May. Native to Europe and n Africa.

NEMASTYLIS CELESTIAL, PLEAT-LEAF, SHELL-FLOWER

A genus of 5 species ranging from the s U.S. to Central America. (Greek: *nema*, a thread, and *stylos*, pillar, column, or style, for the slender style branches)
REFERENCES: Foster 1945; Goldblatt 1975.

Nemastylis geminiflora Nutt., (twin-flowered), PRAIRIE CELESTIAL, PRAIRIE PLEAT-LEAF, CELESTIAL-LILY, PRAIRIE-IRIS. Perennial from a bulb 2–3 cm in diam.; basal leaves (2–)3, 20–40 cm long, 3–6 mm wide; cauline leaves 2 or 3, to 35 cm long, 5–11 mm wide, at least one exceeding the inflorescence; flowering stems 12–30(–46) cm tall; spathes 1–2-flowered; tepals subequal, to 3 cm long, blue, lighter at base giving the appearance of an "eye," all flat or saucer-shaped; filaments free or connate at very base; anthers 11–15 mm long; capsules 15–25 mm long. Prairies or open oak woods; se and e TX w to Rolling Plains and Edwards Plateau. Mar–May. According to Wills and Irwin (1961), the flowers are only open for a few hours; they indicated flower opening occurs in late morning with the perianth parts curling up usually before 3 p.m. /100

Nemastylis nuttallii Pickering ex R.C. Foster, (for Sir Thomas Nuttall, 1786–1859, English-American botanist), is cited by Hatch et al. (1990) for vegetational areas 3 and 4 (Fig. 2); we have seen no TX material and Goldblatt (1975) gave a range map showing only AR, OK, and MO. It can be distinguished by the following characters: filaments united; anthers ca. 4 mm long; cauline leaves < 4 mm wide or bract-like.

SISYRINCHIUM BLUE-EYED-GRASS, GRASS-VIOLET

Annuals or perennials from fleshy-fibrous roots; leaves few, equitant; flowers solitary or in

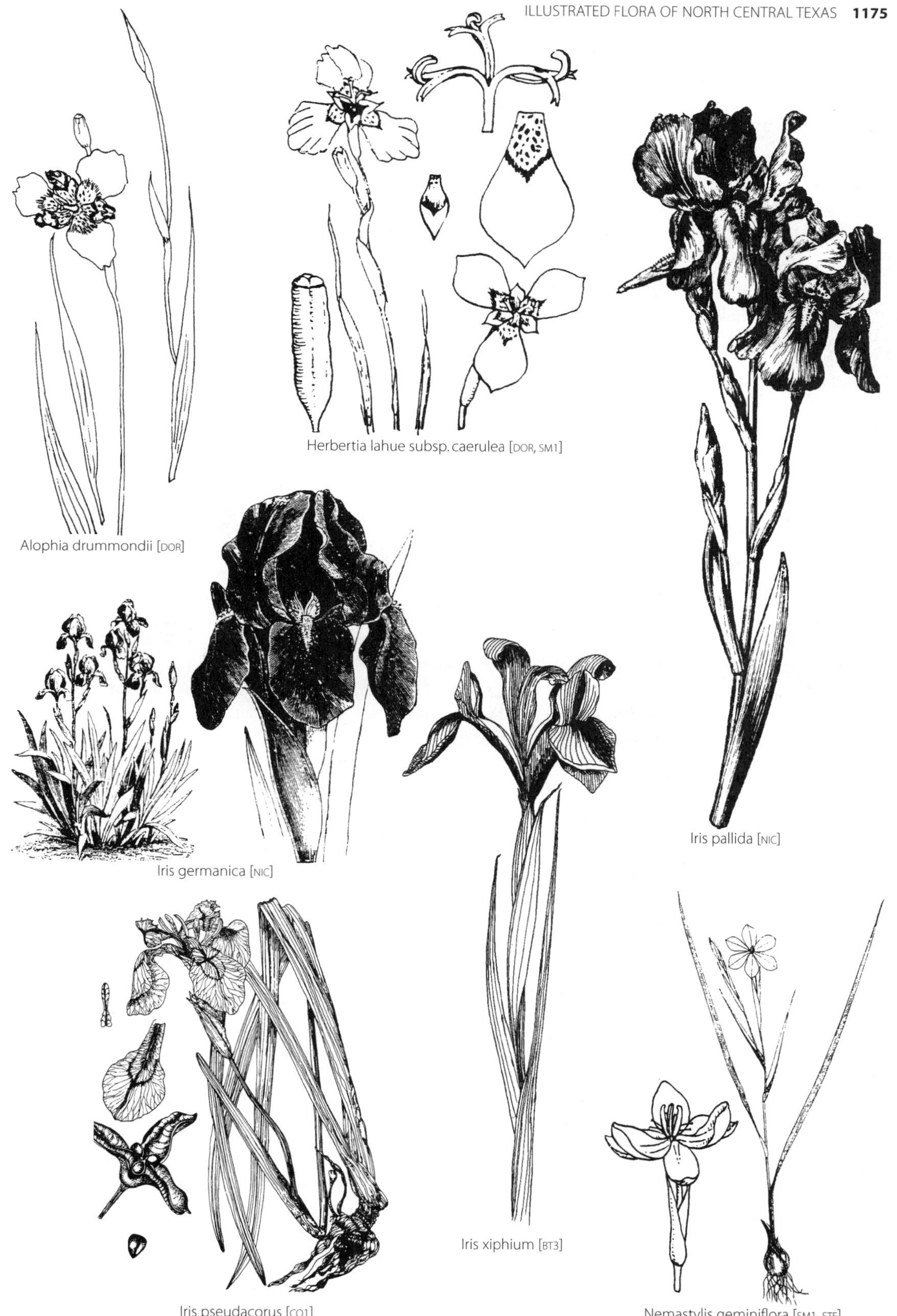

Herbertia lahue subsp. caerulea [DOR, SM1]

Alophia drummondii [DOR]

Iris pallida [NIC]

Iris germanica [NIC]

Iris xiphium [BT3]

Iris pseudacorus [CO1]

Nemastylis geminiflora [SM1, STE]

small, umbel-like clusters, from a pair of leaf-like bracts (referred to as the spathe), closed at night or in cloudy weather; perianth radially symmetrical, the tepals similar or alternating narrow and wide, bluish to purplish, rose or white, often with a yellow "eye"; stamens united into a narrow column.

A mainly American genus (especially Central and South America) of 80 species with 1 in Ireland (possibly naturalized); some are cultivated as ornamentals. It is a taxonomically difficult group that needs detailed study; hybridization sometimes occurs, further complicating the picture. Mosquin (1970) proposed lumping a number of the species, including *S. albidum*, *S. angustifolium*, *S. campestre*, and *S. langloisii*, into a single, widespread, variable taxon. While the flowers are relatively small, the plants are often abundant, making showy displays along roadsides and open areas. The most widely used common name, BLUE-EYED-GRASS, is doubly incorrect, the plants neither having a blue center or "eye" nor being related to grasses. (Name used by Theophrastus for some plant, later transferred to this genus; possibly derived from Greek: *sys*: with, and *rhynchus*, snout or beak; significance not known)

REFERENCES: Bicknell 1901; Shinners 1962b; Mosquin 1970; Goldblatt et al. 1989; Hornberger 1991.

1. Tepals 3–7 mm long, lavender-pink to purple-rose, occasionally white; plants annual ____ **S. minus**
1. Tepals 6.5–12 mm long, white to pale or deep blue or blue-purple; plants perennial.
   2. Bracts immediately subtending flowers (spathe or spathe plus additional leafy bract located so close as to be touching spathe) clearly different in length; flowers white to light blue; stems usually without leaves or leafy bracts except those immediately subtending the flowers; margins of outer spathe bract not fused at base or only slightly so (for 1 mm or less).
      3. Spathes (each composed of 2 bracts) usually paired at stem apex; the pair of spathes immediately subtended by an elongate bract much longer than the spathes ____ **S. albidum**
      3. Spathe (composed of 2 bracts) single at stem apex; spathe not subtended by an elongate bract ____ **S. campestre**
   2. Bracts immediately subtending flowers equal in length or nearly so; flowers intense blue to blue-violet to purple-blue, rarely white; stems with leaves or leafy bracts in addition to those immediately subtending the flowers; margins of outer spathe bract fused at base (for 1.1 mm or more).
      4. Stems winged with each wing equaling stem width and at least 1 mm wide ____ **S. angustifolium**
      4. Stems not broadly winged.
         5. Ovaries usually glabrous; plants drying medium- to olive-green; stems usually numerous ____ **S. pruinosum**
         5. Ovaries minutely pubescent (under a hand lens); plants sometimes drying light-green to glaucous, particularly in *S. chilense*; stems few to numerous.
            6. Leaves 2–4 mm wide; narrower tepals 1.3–3.3 times longer than broad; capsules 4–7 mm long; stems solitary to few ____ **S. chilense**
            6. Leaves 1–2 mm wide; narrower tepals (2.5–)3–5 times longer than broad; capsule 3.5–4.5 mm long; stems usually numerous ____ **S. langloisii**

Sisyrinchium albidum Raf., (white), WHITE BLUE-EYED-GRASS. Plant 20–40 cm tall; stems narrow, broadly winged, simple; leafy bract so close to the spathes as to appear to be the much elongated outer bract of the spathe; paired spathes sessile; flowers white to light blue. In open woods; Milam and Wise cos.; e and nc TX. Mar–Apr.

Sisyrinchium angustifolium Mill., (narrow-leaved), BERMUDA BLUE-EYED-GRASS. Plant 15–30 cm tall; peduncles 1–5 mm longer than the subtending leafy bract; flowers blue to blue-violet, rarely white; capsules 4–5.5 mm long. Sandy woods; Dallas and Lamar cos.; mainly se and e TX, also Edwards Plateau. Mar–May. [*S. bermudiana* of authors, not L.]

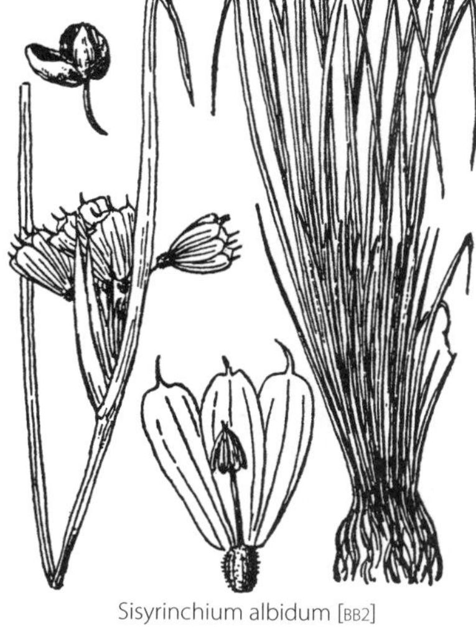

Sisyrinchium albidum [BB2]

Sisyrinchium angustifolium [TOR]

Sisyrinchium campestre [BB2]

Sisyrinchium chilense [CUR]

Sisyrinchium langloisii [HEA]

Sisyrinchium minus [ADD]

Sisyrinchium pruinosum [SHI]

Sisyrinchium campestre E.P. Bicknell, (of the plains or fields), PRAIRIE BLUE-EYED-GRASS. Plant 15–28 cm tall; stems narrow, broadly winged, simple, without leafy bracts or one rarely present; outer bract of spathe ca. 1.5–2(–5) times as long as the inner; flowers white to light blue; capsules 3–4 mm long. Prairies; Delta, Grayson, and Fannin cos.; se and e TX w to n part of nc TX, also Edwards Plateau. Apr–May.

Sisyrinchium chilense Hook., (of Chile), SWORD-LEAF BLUE-EYED-GRASS. Plant 15–50 cm tall; stems solitary or few, slightly to strongly glaucous or gray-green; flowers blue-purple. Prairies; w edge of Blackland Prairie (Bell and Grayson cos.) s and w to w TX. Apr–May. [*S. ensigerum* E.P. Bicknell]

Sisyrinchium langloisii Greene, (for Rev. A.B. Langlois, 1832–1900, French-born clergyman and botanist in LA), PALE BLUE-EYED-GRASS. Plant 12–35 cm tall; stems usually numerous; perianth light violet-blue to purple-blue with yellow "eye," pale outside. Sandy woods; Fannin, Hopkins, Kaufman, McLennan, and Tarrant cos.; se and e TX w to nc TX, also Edwards Plateau. (Mar–)Apr–May.

Sisyrinchium minus Engelm. & A. Gray, (smaller), LEAST BLUE-EYED-GRASS. Plant prostrate to erect, 4–22 cm tall; stems with 1 or 2 leafy nodes and a leafy bract; flowers lavender-pink to purple-rose, occasionally white, rarely yellow; capsules 4–5 mm long. Sandy soils; Bell and Navarro cos., also Dallas (Shinners 1948b) and Tarrant (R. O'Kennon, pers. obs.) cos.; mainly se and c TX. (Mar–)Apr.

Sisyrinchium pruinosum E.P. Bicknell, (with a white glistening coating or bloom), DOTTED BLUE-EYED-GRASS. Plant 9–30 cm tall; stems numerous; leaves 1–3 mm wide; flowers light violet-blue to purple-blue, rarely white or very rarely mottled, with yellow "eye," pale outside; capsules 4–6 mm long. Prairies, disturbed sites, often a lawn weed; se and e TX w to West Cross Timbers, also Edwards Plateau. (Mar–)Apr(–May). Hornberger (1991), Kartesz (1994), and Jones et al. (1997) lumped this taxon into *S. langloisii*; pending a treatment of the group as a whole, we are maintaining it as a separate species. /106

JUNCACEAE RUSH FAMILY

Grass-like or sedge-like annuals or perennials; stems pithy or hollow, round or somewhat flattened on one side, with smooth nodes; leaves basal and/or cauline, with open or closed, tubular, basal sheath smoothly continuous with the blade or with margins or inner lining prolonged at summit into auricles or a scale (resembling the ligule of grasses); leaf blades flat or terete; inflorescences of spike-like racemes, panicles, or heads; perianth of 2 whorls of 3, narrow, scaly, green to red-brown or yellowish perianth parts; stamens 3 or 6; pistil 1; ovary superior; fruit a dry dehiscent capsule.

A small (430 species in 7 genera) family of herbs of temperate and cold areas and tropical mountains primarily in wet or damp habitats. The family is thought by some authorities (e.g., Dahlgren et al. 1985) to be related to the Cyperaceae. (subclass Commelinidae)
FAMILY RECOGNITION IN THE FIELD: grass-like or sedge-like herbs with basal, tufted, linear leaves; flowers inconspicuous with 6 *scaly perianth segments*, clustered but not in spikelets; fruit a *capsule*; the somewhat similar Poaceae and Cyperaceae have 1-seeded fruits, either lack a perianth or have the perianth very reduced, and have flowers in spikelets.
REFERENCE: Dahlgren et al. 1985.

1. Plants glabrous; seeds numerous per capsule; capsules usually 3-celled (sometimes imperfectly so); leaf blades terete (= rounded in cross-section) or flat; leaf sheaths open; widespread in nc TX ____ **Juncus**

1. Plants with pubescence; seeds 3 per capsule; capsules 1-celled; leaf blades flat; leaf sheaths closed; mainly e TX, rare in nc TX ______ **Luzula**

JUNCUS RUSH

Glabrous perennial or annual (only 2 nc TX species) herbs with fibrous roots, some species rhizomatous; leaf blades flattened or terete, sometimes septate; inflorescences variable, sometimes with numerous rebranched branches; flowers solitary on pedicels or often in head-like clusters; seeds numerous and small.

A genus of ca. 300 species, cosmopolitan in distribution but rare in the tropics. (Latin: *iuncus*, name for rush; derived from *iungere*, to join or bind, from use of stems for tying)
REFERENCES: Engelmann 1868; Wiegand 1900.

1. Inflorescence appearing lateral, the "stem" apparently continuing beyond it [actually the "stem" beyond the inflorescence is a stem-like involucral bract]; leafy bracts absent or much shorter than inflorescence.
   2. Capsules obovoid, clearly longer than broad, obtuse, truncate, or even depressed apically; flowers numerous, 30–100 per panicle; upper leaf sheaths without blades; widespread in nc TX ______ **J. effusus**
   2. Capsules spherical (rarely slightly ovoid), ca. as broad as long, apically turgid; flowers few, 2–25 per panicle; upper leaf sheaths with blades; in nc TX known only from Henderson and Lamar cos. ______ **J. coriaceus**
1. Inflorescence terminal, or both terminal and appearing lateral, with either short or long leafy bracts.
   3. Small tufted annuals 5–30 cm tall; leaves ca. 0.5 mm or less broad, inrolled or flat, not nodulose (= without knot-like septa).
      4. Flowering stems usually branched; inflorescence not head-like; leaves inrolled, filiform; bracteoles usually 2–3 beneath each flower; stamens often 6, rarely 3 ______ **J. bufonius**
      4. Flowering stems unbranched; inflorescence of 1–2 head-like clusters which appear sessile; leaves flat; bracteoles solitary beneath each flower; stamens 3 ______ **J. capitatus**
   3. Perennials 8–125 cm tall; leaves various.
      5. Flowers solitary at ends of pedicels or in few-flowered clusters, not forming a dense head-like cluster or glomerule (sometimes clustered in *J. dudleyi*); leaf blades usually thin and flat, or slender and terete, neither equitant (= 2-ranked) nor nodulose (= with knot-like septa); each flower closely subtended by a pair of opposite (or nearly so) bracteoles clearly smaller than the perianth.
         6. Flowers solitary at ends of pedicels (but may be crowded).
            7. Capsules oblong, nearly equaling or exceeding the perianth; sepals 4.5–5 mm long; leaf blades flat or inrolled; capsules distinctly 3-locular ______ **J. brachyphyllus**
            7. Capsules oblong-ovoid to globose-ovoid, shorter than or equaling the perianth; sepals 3–5 mm long; leaf blades flat, inrolled, or terete; capsules apparently 1-locular or incompletely 3-locular.
               8. Leaf auricles (at summit of leaf sheath) 1–2.3(–5) mm long, prolonged, distinctly longer than broad, scarious (= thin and dry) or scarious margined; plants 8–45 cm tall ______ **J. tenuis**
               8. Leaf auricles 0.3–1 mm long, rounded, not longer than broad; firm-membranous or subcoriaceous; plants 20–125 cm tall.
                  9. Leaf blades strongly inrolled, thread-like, grooved on one side or terete; bracteoles underneath flowers acute or acuminate; flowers few and widely scattered ______ **J. dichotomus**
                  9. Leaf blades flat or inrolled; bracteoles underneath flowers obtuse or acute; flowers often rather crowded.

10. Perianth 4–5 mm long; flowers partly solitary, partly in tight clusters of 2–5 ____ **J. dudleyi**
10. Perianth 3.3–4.2 mm long; flowers solitary though crowded ____ **J. interior**

6. Flowers (at least terminal ones) sessile in 2s or 3s, or in small clusters.
11. Perianth 2.5–3.5 mm long; capsule about equaling the perianth; inflorescences often rather open ____ **J. marginatus**
11. Perianth 4–5 mm long; capsule distinctly shorter than perianth; inflorescences of dense, almost head-like clusters ____ **J. dudleyi**

5. Flowers often forming dense heads or head-like clusters or glomerules; leaf blades rather thick and fleshy or spongy, equitant, usually nodulose (except neither equitant nor nodulose in *J. marginatus* and *J. filipendulus* with flat leaves); each flower subtended by a single bracteole.
12. Leaf blades thin and flat, neither equitant nor nodulose.
13. Perianth 2.5–3.5 mm long; capsule about equal to perianth in length; stems 1.5–3 mm broad in broadest dimension; panicle much branched, with 10–80 head-like clusters ____ **J. marginatus**
13. Perianth 4–5 mm long; capsule shorter than perianth; stems 0.5–1 mm broad in broadest dimension; panicle few branched, of 2–5(–10) head-like clusters ____ **J. filipendulus**
12. Leaf blades rather thick and fleshy or spongy, equitant, usually nodulose.
14. Flowers in globose heads 6–18 mm high, 15–60 per head; plants with or without rhizomes.
15. Heads 8–15 mm broad in flower, 10–18 mm in fruit; perianth 3.5–5.5 mm long.
16. Heads closely crowded, few or none widely spaced; branches of inflorescence if present to only 4 cm long; basal leafy bract slightly or greatly exceeding the inflorescence ____ **J. torreyi**
16. Heads mostly well-separated, long-peduncled or separated on the branches by naked internodes longer than diameter of heads; branches of inflorescence 1.5–30 cm long; basal leafy bract shorter to slightly longer than inflorescence.
17. Larger leaf blades 2.5–7.0 mm wide; branches of inflorescence 2–30 cm long.
18. Leaf blades 3–4(–6) mm thick in the larger dimension, with several tough complete septa, tough and resistant to crushing; stems 3–5 mm thick basally; leaves and stems grayish or olivaceous; heads (10–)12–15 mm broad; valves of capsules spreading and free after dehiscence ____ **J. validus**
18. Leaf blades 4–7 mm thick in the larger dimension, with many weak incomplete septa, mostly crushed flat in herbarium specimens; stems 6–10 mm thick basally; leaves and stems greenish; heads 10–12 mm broad; valves of capsules remaining united at their tips after dehiscence ____ **J. polycephalus**
17. Larger leaf blades 0.7–2.0 mm wide; branches of inflorescence 1.5–10(–12) cm long ____ **J. texanus**
15. Heads 6–8 mm broad in flower, 7–10 mm in fruit; perianth 2.7–3.5 mm long.
19. Capsule 1/4–2/3 as long as the perianth, ovoid to obovate, abruptly apiculate ____ **J. brachycarpus**
19. Capsule equaling or exceeding the perianth, slender, tapering uniformly to an elongate apex ____ **J. scirpoides**
14. Flowers in small heads, usually 2–10(–20) (rarely solitary) per head, the heads narrow to hemispherical or nearly globose, 3–6 mm high, 1–8 mm broad; plants without rhizomes.

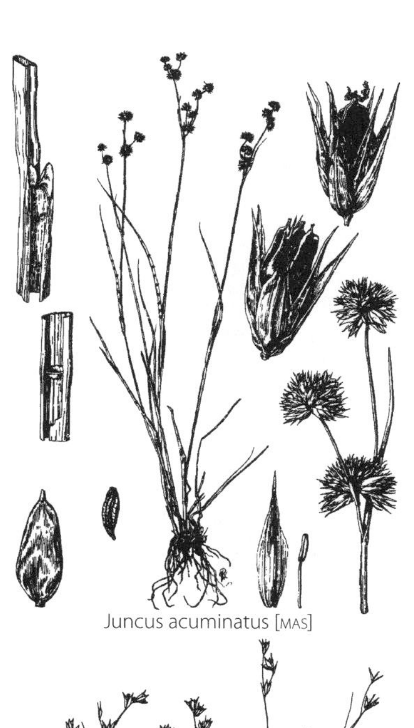

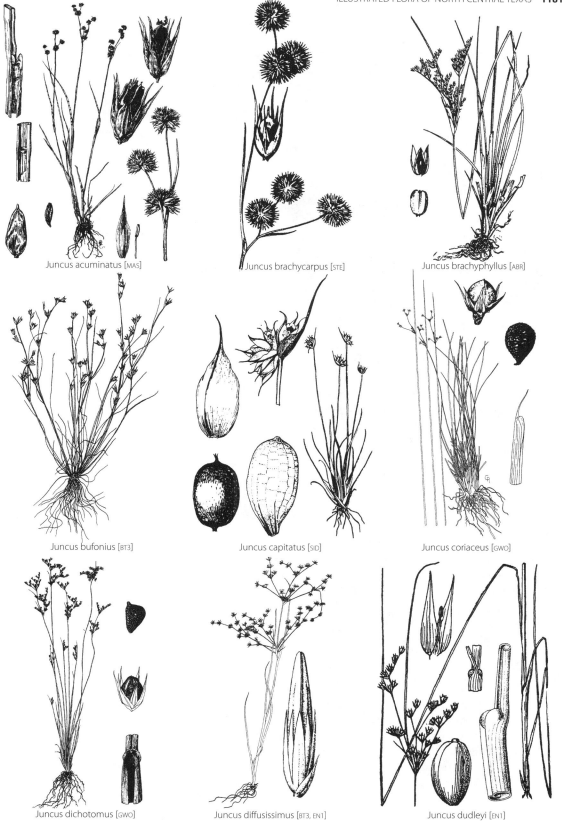

Juncus acuminatus [MAS]

Juncus brachycarpus [STE]

Juncus brachyphyllus [ABR]

Juncus bufonius [BT3]

Juncus capitatus [SID]

Juncus coriaceus [GWO]

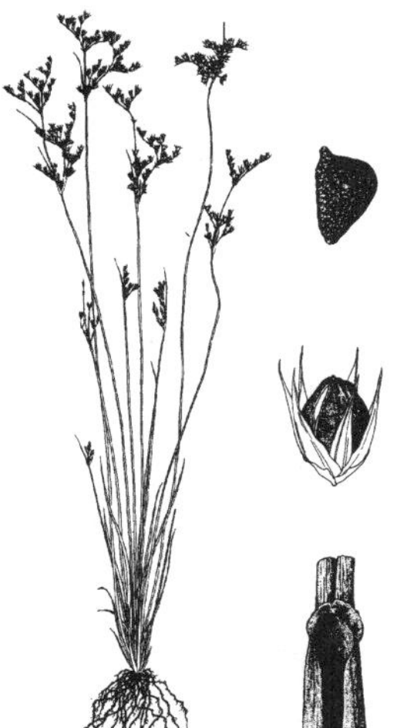

Juncus dichotomus [GWO]

Juncus diffusissimus [BT3, EN1]

Juncus dudleyi [EN1]

20. Heads (or single flowers) 16–200 or more per inflorescence, narrow to hemispherical, 1–6 mm broad (excluding exserted capsules of old flowers), mostly with 1–6 flowers; inflorescences 6–20 cm long, 4–25 cm broad or more.
    21. Capsules distinctly elongate, becoming 3.8–6 mm long, about twice as long as the perianth; larger leaf blades 0.7–2.5 mm wide, not strongly nodulose ____ **J. diffusissimus**
    21. Capsules becoming 2.0–3.2 mm long, equaling or slightly exceeding the perianth; larger leaf blades 2.5–5 mm wide, strongly nodulose ____ **J. nodatus**
20. Heads 6–55 per inflorescence, becoming hemispherical or almost globose, 6–9 mm broad, mostly with 5–10(–20) flowers; inflorescences 2.5–15 cm long, 1.5–10 cm broad ____ **J. acuminatus**

Juncus acuminatus Michx., (long-pointed, tapering to tip), KNOT-LEAF RUSH. Resembling *J. nodatus*, but inflorescences more open. Wet places; Bell and Fannin cos., also Dallas Co. (Mahler 1988) and Parker and Tarrant cos. (R. O'Kennon, pers. obs.); se and e TX w to nc TX, also Edwards Plateau and Trans-Pecos. Apr–Jul.

Juncus brachycarpus Engelm., (short-fruited), WHITE-ROOT RUSH. Plant 15–80 cm tall, rhizomatous or occasionally not so. Damp sandy ground; se and e TX w to West Cross Timbers, also Edwards Plateau. May–Jun, sporadically to Jul.

Juncus brachyphyllus Wiegand, (short-leaved), SMALL-HEAD RUSH. Plant 40–80 cm tall; auricles of leaf sheaths either shorter or longer than wide, chartaceous or subcoriaceous; floral bractlets oblong, obtuse or acute. Low, sandy or rocky, open ground; Bosque, Grayson, Navarro, and Wise cos.; in TX apparently only in nc part of the state. Apr–Jun.

Juncus bufonius L., (pertaining to a toad), TOAD RUSH. Plant 4–18(–30) cm tall; stems often reddish tinged; inflorescence 1/4–3/4 the total height of the plant. Damp sandy soils; Bell, Burnet, Henderson, and Lamar cos.; widespread in TX. Late Apr–Jun.

Juncus capitatus Weigel, (headed), CAPPED RUSH. Plant usually less than 10 cm tall; leaves all basal. Sandy soils; Grayson Co. (Mahler 1988), also Lamar Co. (Carr 1994); mainly e TX, also Edwards Plateau. May–Jun. First reported for TX by Gould in 1962 [1963] from Walker Co; Keeney and Lipscomb (1985) gave additional locations.

Juncus coriaceus Mack., (leathery), LEATHERY RUSH. Plant 30–100 cm tall. Moist sand; Henderson Co., also Lamar Co. (Carr 1994); e TX w to e margin of nc TX. Summer.

Juncus dichotomus Elliott, (2-parted or forked), FORKED RUSH. Plant 35–80 cm tall. Sandy woods or open ground, on slopes or in low places; Hopkins, Lamar, and Navarro cos.; mainly e TX w to e edge of nc TX; scattered elsewhere in the state. May.

Juncus diffusissimus Buckley, (most diffuse), SLIM-POD RUSH. Stems few to many in dense, rounded clumps 15–65 cm tall; inflorescence very open. Damp sandy ground; se and e TX w to East Cross Timbers and Edwards Plateau. May–Jun, less freely to Oct.

Juncus dudleyi Wiegand, (for its discoverer, William Russell Dudley, 1849–1911), DUDLEY'S RUSH. Plant 25–125 cm tall. Low ground; Denton and Tarrant cos. (Mahler 1988); scattered in TX. Jun.

Juncus effusus L. var. **solutus** Fernald & Wiegand, (sp.: loosely spreading; var.: loosened), COMMON RUSH, SOFT RUSH. Perennial forming dense, rounded clumps 70–130 cm tall; stems with rhizomatous base; leaves basal or nearly so, with very small blades or none; inflorescence appearing lateral, a rather compact panicle, almost head-like while young. Wet open ground; Denton, Grayson, Hopkins, and Lamar cos., also Bell (Fort Hood—Sanchez 1997) and Tarrant (R. O'Kennon, pers. obs.) cos.; se and e TX w to East Cross Timbers. Late Apr–Jun.

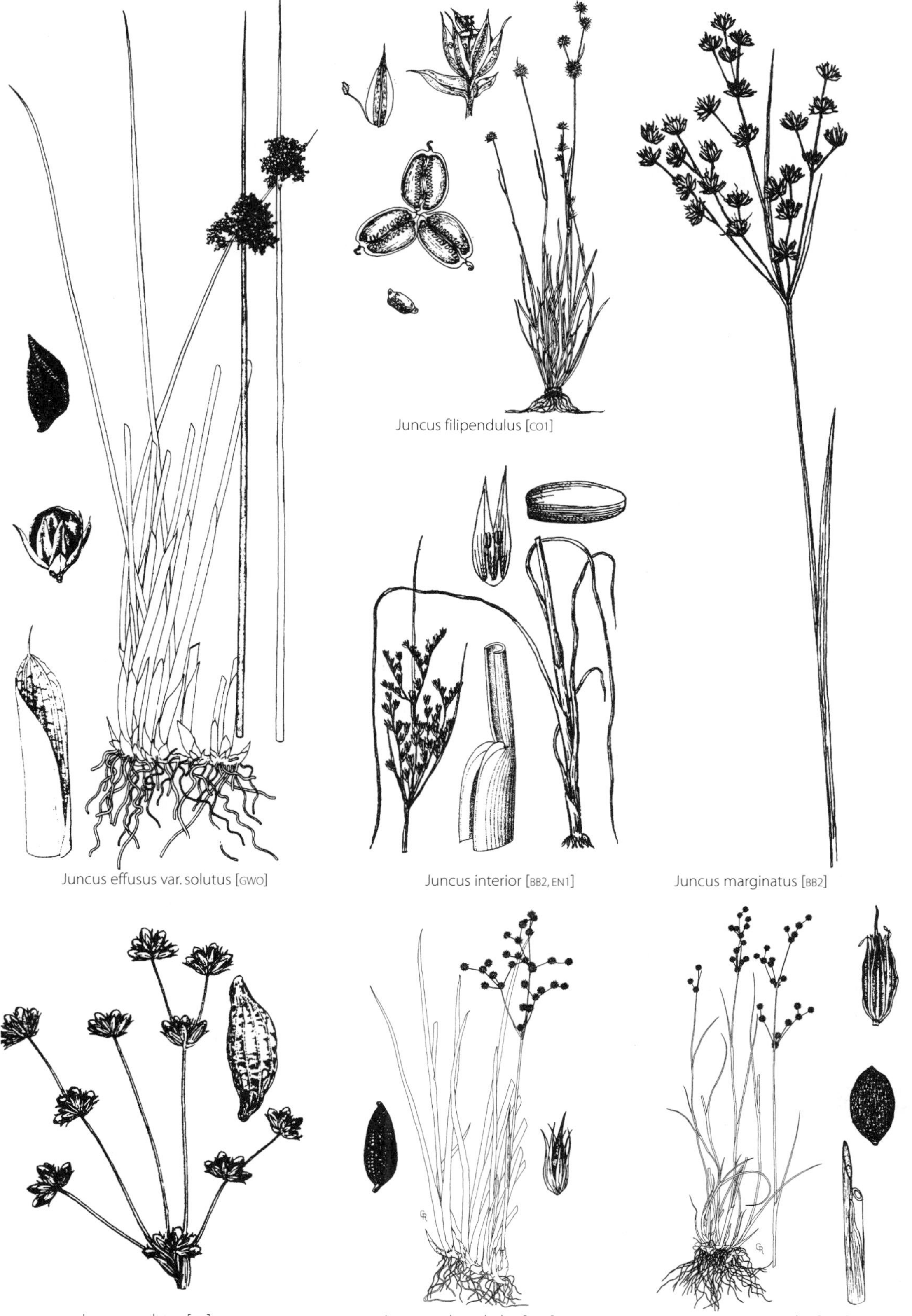

Juncus filipendulus [CO1]

Juncus effusus var. solutus [GWO]

Juncus interior [BB2, EN1]

Juncus marginatus [BB2]

Juncus nodatus [STE]

Juncus polycephalus [GWO]

Juncus scirpoides [GWO]

Juncus filipendulus Buckley, (hanging thread), RING-SEED RUSH. Perennial, tufted or from slightly swollen bases; stems 15–30 cm long. Moist soils or along streams; Bell and Denton cos.; also Edwards Plateau and Rolling Plains. Spring–summer.

Juncus interior Wiegand, (inland), INLAND RUSH. Plant 20–80 cm tall. Chiefly low ground, in open woods or prairies; nearly throughout TX. Late Apr–Jul.

Juncus marginatus Rostk., (margined), GRASS-LEAF RUSH, TWO-FLOWER RUSH, NEEDLE-POINT RUSH. Plant 15–90 cm tall, with swollen stem base, usually short-rhizomatous; inflorescence compact, of many small clusters of 2–12 flowers each; to avoid keying errors, this species which has flowers subtended by only 1 bracteole can be reached either way in the key beginning at dichotomy 5. Sandy old fields, roadsides, moist areas; se and e TX w to West Cross Timbers and Edwards Plateau, also Trans-Pecos. May–Jul, occasionally to Aug. [*J. setosus* (Coville) Small] While var. *setosus* is sometimes recognized (e.g., Jones et al. 1997), we are following Correll and Johnston (1970), Churchill (1986a) and J. Kartesz (pers. comm. 1997) in not recognizing infraspecific taxa in this species.

Juncus nodatus Coville, (with nodes, knots, or swellings), JOINTED RUSH. Stems 40–110 cm tall, forming loose clumps from rhizomatous base; inflorescence rather dense; similar to *J. marginatus* but with nodose leaves and more elongate inflorescences. Damp sandy ground or shallow water; Grayson and Kaufman cos.; se and e TX w to Wichita Co. in Red River drainage (Mahler 1988); scattered elsewhere in TX. Jun–Jul, rarely to Nov.

Juncus polycephalus Michx., (many-headed), FLAT-LEAF RUSH. Tufted perennial from subrhizomatous base; similar to *J. validus*, but with smaller heads. Included based on citation of vegetational area 5 (Fig. 2) by Hatch et al. (1990); mainly se TX, scattered elsewhere. Summer.

Juncus scirpoides Lam., (*Scirpus*-like), NEEDLE-POD RUSH. Very similar to *J. brachycarpus* except for capsules. Damp, sandy ground; in nc TX collected once at w edge of East Cross Timbers in Johnson Co. (Mahler 1988), also Milam Co. near e margin of nc TX; mainly se and e TX, also Edwards Plateau. Late May–Aug.

Juncus tenuis Willd., (slender), SLENDER RUSH, POVERTY RUSH, PATH RUSH. Tufted perennial 8–30(–45) cm tall; panicle 3–6(–9) cm long, about 1/4–1/5 the total height of plant (in contrast to *J. bufonis*). Moist sand in woods; Grayson Co.; mainly se and e TX, also Edwards Plateau. Spring, rarely summer.

Juncus texanus (Engelm.) Coville, (of Texas), TEXAS RUSH. Plant slender, 20–80 cm tall, freely rhizomatous and tuber-bearing; capsules with slender, prolonged, exserted tip. Low ground; Bell, Bosque, Collin, Coryell, Ellis, and Grayson cos.; Blackland Prairie and Grand Prairie; nc TX, s TX, Edwards Plateau, and Trans-Pecos; endemic to TX. Jun–Jul.

Juncus torreyi Coville, (for John Torrey, 1796–1873, coauthor with Asa Gray of "The Flora of North America"), TORREY'S RUSH. Plant rhizomatous, 30–110 cm tall; rhizomes often tuber-bearing; heads 8–10(–11) mm thick. Damp or wet ground; mainly Blackland Prairie s and w to w TX; rare farther e in Red River drainage (Mahler 1988). Jun–Jul, sporadically to Oct.

Juncus validus Coville, (vigorous). Resembling *J. torreyi* but heads well-separated; rhizomes often very short or absent. Damp or wet ground. Jun–Oct.

1. Inflorescences mostly 2–5 cm long, of 6–15 heads; capsules tardily dehiscent ____ var. **fascinatus**
1. Inflorescences 5–25 cm long, of (12–)15–76 heads; capsules promptly completely dehiscent ____ var. **validus**

var. **fascinatus** M.C. Johnst., (fascinating). Included based on citation of vegetational areas 4 and 5 (Fig. 2) by Hatch et al. (1990); in much of e 1/2 of TX; endemic to TX.

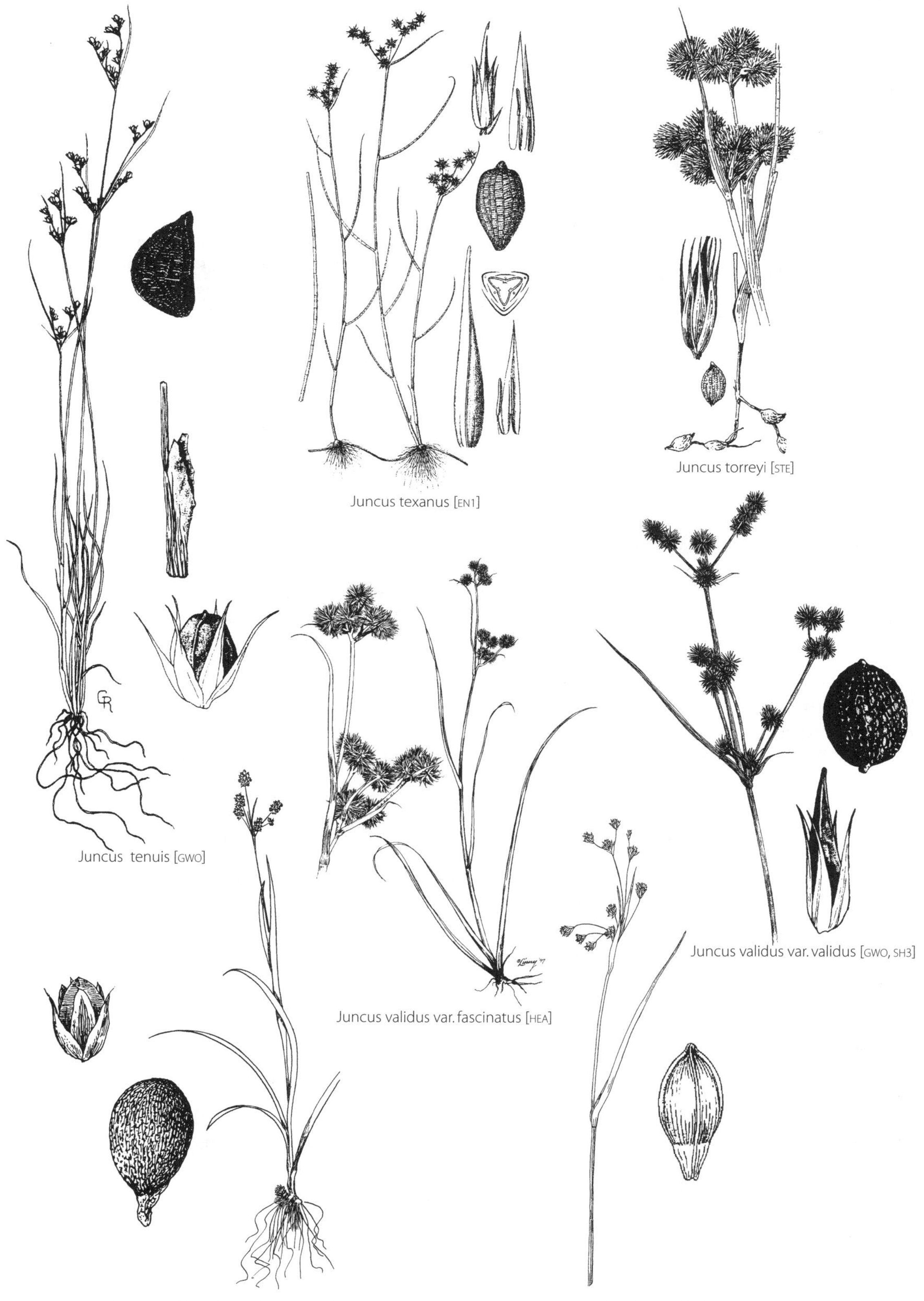

Juncus texanus [EN1]

Juncus torreyi [STE]

Juncus tenuis [GWO]

Juncus validus var. validus [GWO, SH3]

Juncus validus var. fascinatus [HEA]

Luzula bulbosa [SH3, STE]

Luzula echinata [GLE]

var. **validus**. ROUND-HEAD RUSH. Se and e TX w to Grayson and Tarrant cos.; also Edwards Plateau and Panhandle.

LUZULA WOODRUSH

Tufted perennials; leaves flat, grass-like, with pubescence; inflorescences terminal, of 5–10(–13) usually simple branches, with 1–3 leaf-like bracts; flowers (in our species) in few-flowered glomerules; stigmas much longer than styles; seeds 3, with whitish elaiosomes or caruncles (= appendages used in dispersal by ants or other insects) ca. 1/3–2/3 length of seed.

A genus of 115 species, cosmopolitan in distribution but especially in temperate Eurasia. Handel (1978) presented experimental evidence that the seeds of some species are dispersed by ants. (From *Gramen Luzulae*, or *Luxulae*, diminutive of Latin *lux*, light; a name given to one of the species from its shining with dew)

1. Small bulb-like whitish structures (= swollen, reduced leaves) usually present on the rhizomes; glomerules (= flower clusters) cylindric; capsule usually exceeding perianth; elaiosome 0.9–1.3 mm long, 1/2–2/3 length of seed _______ **L. bulbosa**
1. Small bulb-like structures not present on the rhizomes (but base of plant can be swollen); glomerules subglobose or ovoid; perianth usually conspicuously exceeding capsule; elaiosome 0.5–0.6 mm long, ca. 1/3 length of seed _______ **L. echinata**

Luzula bulbosa (A.W. Wood) Smyth, (bulbous), BULB WOODRUSH. Tufted perennial to 45 cm tall; rhizomes slender with whitish, swollen, bulb-like structures (= reduced leaves) ca. 2–4 mm thick; leaves few, 2–7 mm wide, long-hairy marginally; inflorescence branches usually erect or ascending. Sandy soils in forests; Fannin Co. in Red River drainage (Talbot property); Hatch et al. (1990) also cited vegetational area 5 (Fig. 2); mainly e TX. Spring. [*L. campestris* (L.) DC. var. *bulbosa* A.W. Wood]

Luzula echinata (Small) F.J. Herm. var. **mesochorea** F.J. Herm., (sp.: prickley; var.: midland). Similar to *L. bulbosa* but more densely tufted; rhizomes knotty and base of plant sometimes swollen, but rhizomes without bulb-like structures; some of the inflorescence branches usually divergent at right angles. Sandy soils; Lamar Co. (Carr 1994); mainly e TX. Spring. [*L. campestris* (L.) DC. var. *echinata* (Small) Fernald & Wiegand]

LEMNACEAE DUCKWEED FAMILY

Very small or minute annual aquatics, floating free on or in the water, consisting of a flat or solid body (called thallus, frond, or joint) a few mm or less across, not differentiated into stems and leaves, solitary or in small clusters; new plants (= daughter fronds) chiefly produced asexually by budding, often remaining attached to the parent frond by a short stipe; flowering infrequent or at least infrequently observed; inflorescence with 1 or 2 staminate flowers of a single stamen and 1 pistillate flower with a single pistil, produced in 2 lateral pouches at frond base (in *Lemna* and *Spirodela*) or in a cavity on frond surface (in *Wolffia* and *Wolffiella*); flowers minute, imperfect, without perianth (a very minute bract may be present); fruit an urticle.

A small (25 species in 4 genera) family of aquatics thought to be derived from the Araceae; molecular studies link *Lemna* with *Pistia* (a free-floating member of the Araceae) (Duvall et al. 1993). Because of their huge numbers and rapid reproduction, duckweeds can have pronounced influences on aquatic habitats (Hicks 1937); they are important as food for waterfowl and fish, but can also become pests. Lemnaceae have also been used in wastewater treatment systems as biological filters and the plants are sometimes harvested to make a high-protein feed for livestock (Tyrl et al. 1994). Due to their prolific asexual reproduction, Lemnaceae are being studied by NASA as a possible food source for prolonged space voyages. The family is not well-collected or well-studied in nc TX. All nc TX species are found on lakes, ponds, swamps, marshes, or other standing water or stranded on mud; they are prominent chiefly in late summer and fall. This easily recognized family contains the smallest known flowering plants, members of the genus *Wolffia*. While easy to identify to genus, specific determination is often difficult. (subclass Arecidae)

FAMILY RECOGNITION IN THE FIELD: free-floating, very small or minute aquatics (larger plants only a few mm in size).

REFERENCES: Thompson 1897; Hicks 1937; Wilson 1960a; Harrison & Beal 1964; McClure & Alston 1966; Clark & Thieret 1968; den Hartog & van der Plas 1970; Landolt 1980, 1986; Dahlgren et al. 1985.

1. Fronds with 1 or more roots, 2 mm long or longer, oblong to elliptic, ovate, obovate, or orbicular, not long-tapered to apex.
   2. Each frond with 1 root; fronds 2–5 mm long ____ **Lemna**
   2. Each frond with several roots; fronds 2.5–10 mm long (*Lemna* size or much larger) ____ **Spirodela**
1. Fronds rootless, EITHER < 1.4 mm long (pinhead size) OR fronds larger and narrowly lanceolate and distinctly long-tapered from a relatively broad base to a narrow apex.
   3. Fronds < 1.4 mm long, pinhead size, ± distinctly 3-dimensional, never flat except on the upper surface, solitary or usually at most only two fronds attached together, usually floating on the water surface ____ **Wolffia**
   3. Fronds 4–8.4 mm long, flat, membranous, (rarely solitary–)usually 2–many attached together, usually floating below the water surface ____ **Wolffiella**

LEMNA DUCKWEED

Fronds solitary or in clusters of a few, pale green, yellow-green, or green, sometimes with reddish purple pigmentation; base of the single root surrounded by a tubular sheath, not covered by a membranous scale; 2 vegetative reproductive pouches per frond, these lateral near frond base.

A cosmopolitan genus of 7 species (Landolt (1980) indicated 9–13 species). (Classical Greek name for a water plant)

REFERENCES: Landolt 1975; Reveal 1990a.

1. Fronds with 3–5(–7) nerves, without OR with anthocyanin (= red or purple pigmentation) present; upper (= dorsal) surface of fronds with or without prominent protuberances (= papillae); including species widespread in nc TX.
   2. Fronds without any reddish purple pigmentation on either lower or upper surfaces; lower frond surface flat to slightly convex; root sheath with definite wings or appendages; root tips usually sharp-pointed; including species widespread in nc TX.
      3. Upper frond surface with only 1 papilla above the node (= point where root attaches on lower frond surface) and this one typically smaller than the one near the frond tip; widespread in nc TX ____ **L. aequinoctialis**
      3. Upper frond surface very often with 2–3 prominent papillae above the node larger than the one near the frond tip; rare in nc TX, reported only from Dallas Co. ____ **L. perpusilla**
   2. Fronds typically with reddish purple pigmentation, at least on lower surface; lower frond surface slightly to strongly convex or gibbous; root sheath without wings or appendages; root tips mostly rounded; apparently rare in nc TX ____ **L. obscura**
1. Fronds with 1 nerve, without red or purple pigmentation; upper surface of fronds flat, smooth or with obscure median papillae, with no prominent papillae; rare in nc TX.
   4. Fronds very asymmetrical at the base 1$^{1}/_{3}$–3 times as long as wide; nerve reaching at least

$^{3}/_{4}$ of the distance from the node to the apex, usually distinct; plants sometimes submerged ________ **L. valdiviana**

4. Fronds nearly symmetrical at the base, 1–1 $^{2}/_{3}$ times as long as wide; nerve reaching at most $^{2}/_{3}$ of the distance from the node to the apex, sometimes indistinct; plants always floating on surface ________ **L. minuta**

Lemna aequinoctialis Welw., (of equinoctial zone, from equatorial regions). Fronds ovate to lanceolate, 1–6.5 mm long, 0.8–4.5 mm wide, nearly symmetrical, light to medium green; lower surface flat to slightly convex; wing of root sheath 1–2 $^{1}/_{2}$ times as long as wide. Bell, Brown, Dallas, Denton, Grayson, and Tarrant cos.; throughout TX. [*L. trinervis* (Austin) Small] This is the common DUCKWEED in nc TX.

Lemna minuta Kunth, (minute, very small), LEAST DUCKWEED. Fronds ovate to lanceolate, 0.8–4 mm long, 0.5–2.5 mm wide, never pointed, light green; lower surface flat to slightly convex; root sheath not winged. Dallas Co. (Landolt 1986); rare in TX; se and e TX, also Edwards Plateau and Trans-Pecos. [*L. minima* Phil. ex Hegelm., *L. minuscula* Herter, *L. valdiviana* Phil. var. *minima* Hegelm.] According to Landolt (1986), this species is sometimes difficult to distinguish from *L. valdiviana*.

Lemna obscura (Austin) Daubs, (hidden). Fronds broadly ovate to suborbicular, 1 $^{1}/_{5}$–1 $^{2}/_{3}$ times as long as wide, 1–3.5 mm long, 0.8–3 mm wide, green to yellowish green, typically with reddish purple pigmentation, especially on lower surface; upper frond surface with a distinct papilla near tip and several smaller indistinct papillae along midline; lower frond surface slightly to strongly convex or gibbous. The *Lemma gibba-minor-obscura-turionifera* complex is quite confusing. Landolt (1975) indicated that *L. gibba* occurs only to the w of nc TX. Based on incomplete comparative material, but using Landolt's (1980) worldwide key to *Lemna* species and his map (1975), Lamar Co. and Tarrant Co. pigmented populations with the lower frond surfaces convex are apparently *L. obscura*. The name *L. minor* has been misapplied to this taxon in TX in the past (e.g., Jones et al. 1997). Ox-bow lake in Lamar Co. in Red River drainage and Tarrant Co. (Fort Worth Nature Center); Landolt (1975) indicated this species occurs along the Gulf coast and northwards to n TX. [*L. minor* L. var. *obscura* Austin, *L. minor* of TX authors, not L.]

Landolt's (1975) range map indicated that *L. turionifera* could also possibly occur in nc TX; however, he (Landolt 1986) indicated the species is rare in TX and cited no nc TX localities. This species would be identified as *L. obscura* by the key above. The following dichotomy to separate the two is modified from Landolt (1980):

1. Fronds slightly to strongly convex or gibbous beneath; upper frond surface with papilla at apex bigger than others; not forming turions ________ *L. obscura*
1. Fronds nearly flat beneath; upper frond surface with several papillae of ± equal size along the midline; forming small obovate to circular, rootless, dark green to brown turions under unfavorable conditions which sink to the bottom of the water ________ *L. turionifera*

Lemma perpusilla Torr., (very weak and slender). Fronds ovate to lanceolate, 1–4 mm long, 0.8–3 mm wide, 1–1 $^{2}/_{3}$ times as long as wide, light green; apical and central papillae prominent; wing of root sheath 2–3 times as long as wide. Dallas Co. (Landolt 1986); rare in TX; Landolt (1986) gave only the Dallas locality; Hatch et al. (1990) indicated se and s TX.

Lemna valdiviana Phil., (of Valdivia, Chile), VALDIVIANA DUCKWEED, PALE DUCKWEED. Fronds ovate to lanceolate, somewhat falcate to symmetrical except for the oblique base, 1–5 mm long, 0.6–3 mm wide, occasionally somewhat pointed, light green, flat to slightly biconvex; root sheath not winged. Included based on citation of vegetational area 4 (Fig. 2) by Hatch et al. (1990); according to Landolt (1986), rather frequent in TX except in the nw and s parts of the state; however, we have not observed specimens from nc TX.

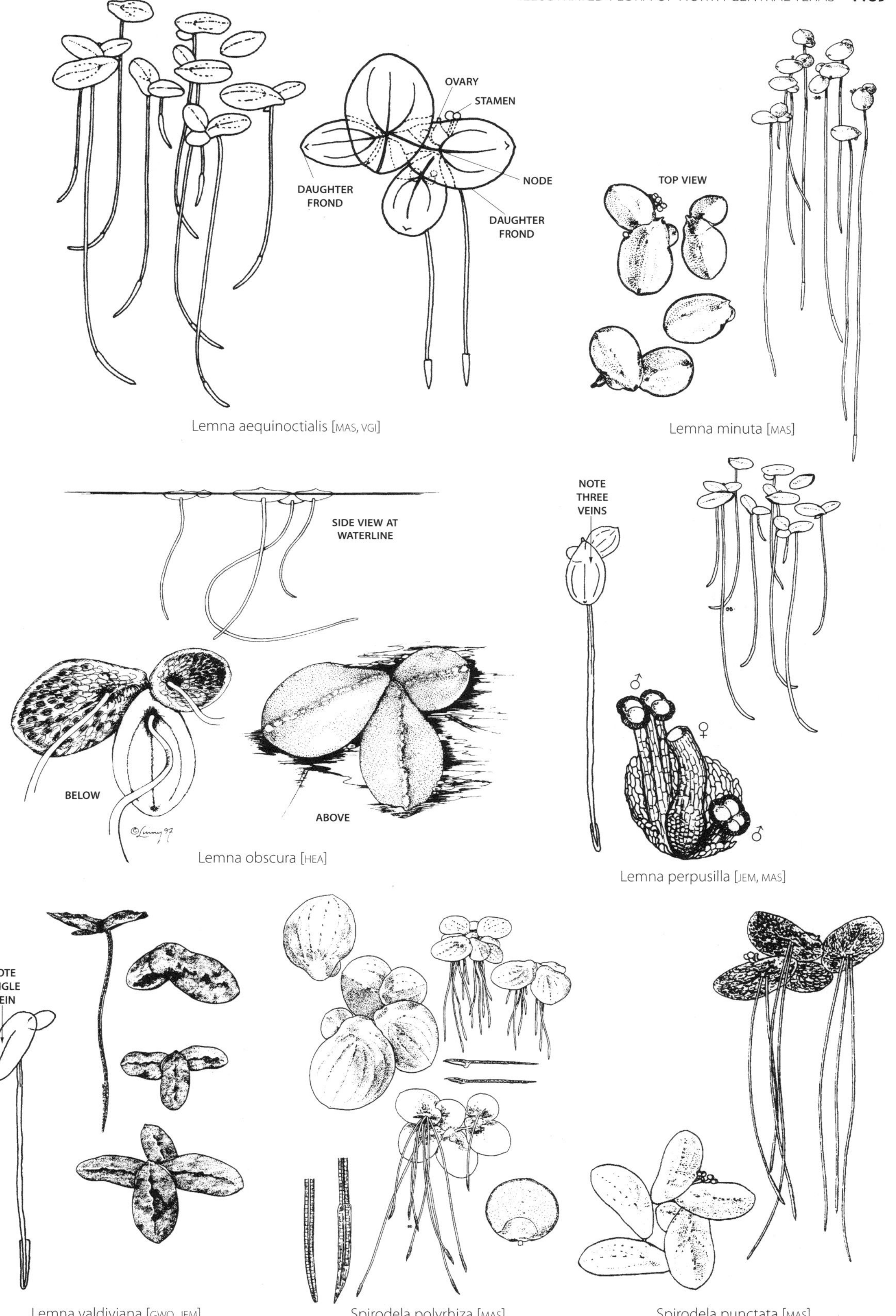

Lemna aequinoctialis [MAS, VGI]

Lemna minuta [MAS]

Lemna obscura [HEA]

Lemna perpusilla [JEM, MAS]

Lemna valdiviana [GWO, JEM]

Spirodela polyrhiza [MAS]

Spirodela punctata [MAS]

SPIRODELA DUCKMEAT

Fronds solitary or in clusters of a few, dark lustrous green on upper surface, usually reddish purple on lower; 1 or more of the roots penetrating a membranous scale (= prophyllum); vegetative reproductive pouches 2 per frond, these lateral near frond base.

A cosmopolitan genus of 4 species. *Spirodela* is the least reduced genus in the family; sometimes grown, using dairy waste water, as a substitute for alfalfa in animal food. (Greek: *speira*, a cord or thread, and *delos*, evident or visible, from the roots)

1. Roots 4–20 per frond; fronds 3–10 mm long, orbicular-obovate to nearly ovate, almost as wide as long, the upper surface often with a conspicuous red dot near center, conspicuously several nerved; widespread in nc TX ______ **S. polyrhiza**
1. Roots 2–5 per frond; fronds 2.5–5 mm long, oblong-ovate to somewhat elliptic-reniform, distinctly longer than wide, the upper surface usually without a conspicuous red dot, not conspicuously nerved; rare in nc TX ______ **S. punctata**

Spirodela polyrhiza (L.) Schleid., (many-rooted), COMMON DUCKMEAT, GREATER DUCKWEED. Fronds usually ca. 3–6 mm wide; only 1(–3) of the roots penetrating the prophyllum. Dallas, Denton, Grayson, Henderson, and Tarrant cos.; nearly throughout TX. This species is the largest of the surface-floating duckweeds.

Spirodela punctata (G. Mey.) C.H. Thomps., (punctate, dotted). Closely resembling some *Lemna* species in size and shape; fronds usually ca. 1–3 mm wide; all of the roots penetrating the prophyllum. Denton Co. (mixed collection with *Lemna aequinoctialis*); rare in se and e TX. [*S. oligorhiza* (Kurtz) Hegelm.] Native to South America, se Asia, Australia, and probably also s Africa. Introduced as an aquarium plant (Godfrey & Wooten 1979).

WOLFFIA WATER-MEAL

Fronds extremely tiny, about the size of a pinhead, solitary or paired, with a single terminal vegetative reproductive pouch; rarely flowering; when in large numbers the plants appear ± like green scum on the water surface.

A tropical to temperate genus of ca. 7 species. *Wolffia* species are the smallest known flowering plants. Concentrations of 1–2 million plants per square yard of water surface can occur (Hicks 1937). (Named for Johann Friedrich Wolff, 1788–1806, German botanist and physician who wrote on *Lemma* in 1801)

1. Fronds usually ellipsoid or broadly ovoid, the upper surface flattened and often (but not always) with a raised, conical papilla in the center; frond epidermis often (but not always) with brownish pigmented cells (seen in dried material); cells of fronds progressively smaller and less inflated from lower to upper surface and fronds thus appearing darker near upper surface when viewed laterally in transmitted light ______ **W. brasiliensis**
1. Fronds usually ± globular, rarely ellipsoid, the upper surface strongly convex, not flattened and without a central papilla; frond epidermis without brownish pigmented cells; cells of fronds uniformly inflated and fronds thus ± uniformly green when viewed laterally in transmitted light ______ **W. columbiana**

Wolffia brasiliensis Wedd., (of Brazil). DOTTED WOLFFIA, POINTED WOLFFIA. Fronds 0.5–1.5 mm long. Fannin, Co.; mainly se and e TX and Edwards Plateau. Even though w of the known range of this species, many individuals from an isolated swamp/beaverpond complex near Telephone in Fannin Co. (L. Talbot property), have definite epidermal pigmentation and papilla. [*W. papulifera* C.H. Thomps., *W. punctata* Griseb.]

Wolffia columbiana H. Karst., (of Colombia), COLOMBIA WOLFFIA, COMMON WOLFFIA. Fronds 0.1–1.4 mm long; upper surface smooth or slightly roughened by a few minute papillae much smaller than the usually central papilla often seen in *W. braziliensis*. Dallas and Lamar cos.; se and e TX w to nc TX and Edwards Plateau.

WOLFFIELLA MUD-MIDGET, BOG-MAT

A genus of ca. 7 species, mostly of tropical and warm areas of the Americas and Africa. (Name a diminutive of *Wolffia*)

Wolffiella gladiata (Hegelm.) Hegelm., (sword-like). Fronds narrowly lanceolate, often falcate, 4–8.4 mm long, 0.5–1.4 mm wide at base, flat, membranous, distinctly long-tapered from a relatively broad base to a narrow apex, usually 5–15 times as long as width at middle of frond, (rarely solitary-)usually 2-many attached, sometimes forming star-like groups, usually floating below the water surface; vegetative reproductive pouch solitary per frond; seldom flowering. Falls Co. on e margin of nc TX (Landolt 1986); mainly se and e TX. [*W. floridana* (Donn. Sm.) C.H. Thomps.]

Wolffiella lingulata (Hegelm.) Hegelm., (strap-shaped), occurs in se TX. This species can be distinguished by its fronds wide tongue-shaped or ovate, up to 1.5–4 times as long as wide, rounded at the tips, solitary to 2(–4) attached.

LILIACEAE LILY FAMILY (including AMARYLLIDACEAE)

Perennial herbs with bulb or rhizome, or with fleshy-fibrous or tuberous roots; leaves basal, alternate, or whorled, sometimes sheathing or with narrow and petiole-like base, or with distinct, tubular basal sheath and grass-like blade; flowers radially symmetrical or nearly so, solitary or in racemes, panicles, corymbs, umbels, or head-like inflorescences, bractless, individually bracted, or inflorescence with an involucre of sheathing bracts; perianth segments (= tepals) 6, in one or two rows, separate or united; stamens 6; pistil 1; ovary superior or inferior; fruit a dry capsule or fleshy berry.

We are following Cronquist (1981, 1988, 1993) and Kartesz (1994) in treating the Liliaceae broadly to include the Alliaceae, Amaryllidaceae, Asparagaceae, Convallariaceae, Hemerocallidaceae, Hyacinthaceae, and Hypoxidaceae, but not the Agavaceae or Smilacaceae. The key to genera separates the species sometimes placed in the Amaryllidaceae from those in a more restricted Liliaceae. Some authors recognize an Amaryllidaceae limited to those taxa with inferior ovaries. The Liliaceae is quite diverse and has been divided into as many as 27 smaller families; however, Cronquist (1993) indicated that while there is significant variation within the group, he has not been able to find a reasonable way of dividing that variation at the familiy level. As treated here, the Liliaceae is a large (4,950 species in 288 genera), cosmopolitan family important for its numerous ornamentals including *Amaryllis*, *Hosta*, *Lilium* (LILY) and *Tulipa* (TULIP); the medically important *Aloe* is sometimes placed in the Liliaceae, while some authorities segregate it into its own family, the Aloeaceae. *Colchicum autumnale* L. (AUTUMN CROCUS), native from Europe to n Africa, is also medicinally important as the source of the alkaloid colchicine, used in treating gout. Some Liliaceae are edible (e.g., *Allium*, *Asparagus*), but many are very toxic due to the presence of alkaloids. *Liriope muscari* (Decne.) L.H. Bailey, (musk), (LILYTURF), with a dense raceme of small violet flowers usually exceeding the foliage, and *Ophiopogon japonicus* (Thunb.) Ker Gawl., (of Japan), (MONDO-GRASS OR MONKEY-GRASS), with racemes of off-white flowers not exceeding the foliage, are both widely cultivated, persist, and spread vegetatively in flower beds in nc TX. Family name from *Lilium*, LILY, a genus of 100 species native from the n temperate zone to the Philippines. (Latin form of Greek: *leirion*, name for Madonna lily—*Lilium candidum* L.) (subclass Liliidae)

FAMILY RECOGNITION IN THE FIELD: bulbous or rhizomatous, perennial herbs; flowers often showy, with 6-parted petaloid perianth, 6 stamens (3 in the somewhat similar Iridaceae), and 3-celled, superior or inferior ovary; fruit a capsule or berry.

REFERENCE: Dahlgren et al. 1985.

1. Perianth pilose, yellow; leaves narrowly linear, pilose (sometimes segregated as the Hypoxidaceae) ____ **Hypoxis**
1. Perianth and leaves not as above.
   2. Ovary inferior or superior; inflorescence with a basal involucre of sheathing bract or bracts; flowers solitary, umbellate or in head-like inflorescences (sometimes segregated as the Amaryllidaceae).
      3. Perianth bright red, to 40 mm long; leaves not present at flowering time; rarely escaped, introduced, ornamental species ____ **Lycoris**
      3. Perianth either not red OR if red then 10 mm long or less; leaves usually (but not always) present at flowering time; widespread native and introduced species.
         4. Ovary superior; perianth (including tube if present) up to 28 mm long (usually much less), white to yellow or variously colored, without a crown; perianth segments without green tips.
            5. Filaments separate; perianth white to yellowish, pink or red, 10 mm or less long.
               6. Umbels with 5–30 or more flowers, or with bulblets; anthers oblong, ca 1 mm long; plants with onion odor; flower stalks (= pedicels) within umbels of ± same length ____ **Allium**
               6. Umbels with 4–12 flowers, without bulblets; anthers linear, ca. 2 mm long; plants without onion odor; flower stalks within umbels of different lengths ____ **Nothoscordum**
            5. Filaments united; perianth lavender-blue or white with pale blue tinge, 16–28 mm long.
               7. Flowers single, rarely 2, with 2 partly or wholly united bracts; perianth white with pale blue tinge, the lobes with a darker central line and brownish tinge on back ____ **Ipheion**
               7. Flowers 1–6, with 2 large and 2 small bracts; perianth, including lobes, lavender-blue ____ **Androstephium**
         4. Ovary inferior; perianth at least 25 mm long OR with a crown OR perianth segments with green tips; perianth white, yelllow, or orange-yellow.
            8. Perianth without a crown (= circle of tissue inside the corolla).
               9. Flowers nodding, usually 2–10; perianth segments tipped with green; introduced species ____ **Leucojum**
               9. Flowers not nodding, usually solitary; perianth segments not tipped with green; native species.
                  10. Perianth white, with tube much more than 30 mm long ____ **Cooperia**
                  10. Perianth orange-yellow, with tube 25–30 mm long ____ **Habranthus**
            8. Perianth with a crown.
               11. Perianth tube 6–8 cm long; crown united with the filaments; perianth white ____ **Hymenocallis**
               11. Perianth tube < 3 cm long; crown not united with the filaments; all or part of perianth usually yellow ____ **Narcissus**
   2. Ovary superior; inflorescence bractless or flowers individually bracted; flowers solitary, racemose, paniculate, or corymbose (sometimes recognized as the Liliaceae in a stricter sense).
      12. Flowers urceolate (perianth segments united almost their whole length and slightly con-

stricted at mouth), 3–6 mm long, dark blue with whitish teeth or rarely all white; flowering stalk to ca. 20 cm tall _______ **Muscari**

12. Flowers not urceolate, 6–120 mm long (except 3–4 mm long in the greenish white flowered *Schoenocaulon*), white to lavender-blue to variously colored; flowering stalk usually much more than 20 cm tall (except in *Ornithogalum*).
   13. Perianth segments united 1/2–2/3 their lengths; rare escape or persistent cultivated species _______ **Hyacinthus**
   13. Perianth segments separate or united up to 1/4 their length; common native and naturalized species.
      14. Inflorescences scapose (= flowers on leafless stalks from the base of the plant); plants without a leafy stem or leafy portion of stem very short, the leaves nearly all basal; fruit a capsule.
         15. Flowers solitary, leaves 2 _______ **Erythronium**
         15. Flowers in racemes, corymbs, or panicles; leaves more than 2.
            16. Perianth segments very large, > 80 mm long, orangish _______ **Hemerocallis**
            16. Perianth segments smaller, < 30 mm long, greenish white, white, lavender-blue, or yellow or orange-yellow.
               17. Flowers greenish white; perianth segments ca. 3–4 mm long; plants both with leaves mostly 4 mm or less wide AND with flowers sessile or nearly so _______ **Schoenocaulon**
               17. Flowers white, lavender-blue, yellow, or yellow-orange; perianth segments 6–20 mm long; plants either with leaves wider than 4 mm wide OR flowers distinctly stalked OR both.
                  18. Perianth segments ca. 7 mm long, yellow or orange-yellow; flowers sessile or nearly so; plants without a bulb _______ **Aletris**
                  18. Perianth segments 6–20 mm long, white or lavender-blue; flowers distinctly pedicellate; plants bulbous.
                     19. Perianth segments white, bright green on back, usually with white margins; flower stalk usually 25 cm or less tall; lower pedicels often very long, to 8.5 cm long _______ **Ornithogalum**
                     19. Perianth segments lavender-blue or white, not green on back; flowering stalk often much more than 25 cm tall; lower pedicels usually < 3 cm long.
                        20. Perianth segments lavender-blue or rarely white; anthers roughly orbicular, ca. 1 mm long _______ **Camassia**
                        20. Perianth segments white; anthers elongate, oblong, ca. 2 mm long _______ **Zigadenus**
      14. Inflorescences axillary (= from axils of the leaves); plants with an erect-arching leafy stem or stem with numerous finely dissected branches; fruit a berry.
         21. Leaves reduced to scales; main stem with numerous finely dissected branches; ripe berry red; perianth 4–6 mm long _______ **Asparagus**
         21. Leaves well-developed, lanceolate-elliptic to broadly elliptic, to 15 cm long and 7 cm wide; main stem unbranched; ripe berry blue-black; perianth 13–20 mm long _______ **Polygonatum**

ALETRIS STAR-GRASS, COLIC-ROOT

An Asian and North American genus of 10 species including some cultivated as ornamentals. (Greek: *aleitris*, a female slave who ground corn, referring to the apparent mealiness of the perianth)

Aletris aurea Walter, (golden), YELLOW STAR-GRASS. Perennial, rhizomatous, scapose herb; leaves flat, lanceolate, to 12 cm long; flowers in a spike-like raceme terminating a ± naked (a few small remote bracts can be present) scape to 80 cm or more tall; pedicels 3 mm or less long; perianth tubular, yellow or orange-yellow. Savannahs, boggy areas; included based on citation of vegetational area 4 (Fig. 2) by Hatch et al. (1990); mainly se and e TX. May–Jul. ▣/77

ALLIUM ONION, GARLIC, LEEK

Bulbous perennials typically with odor of onion or garlic, our species glabrous; leaves with closed, tubular, basal sheath and slender blade; inflorescence subtended by a spathe that divides into 1–3 bracts; flowers white to pink, reddish pink, or purplish, or replaced by bulbils; perianth of 6 similar, ± free segments (= tepals); fruit a capsule.

A n hemisphere genus of ca. 690 species ranging from Eurasia to Africa, Sri Lanka, and Mexico. It includes many species cultivated as ornamentals and for food including ONION, GARLIC, SHALLOT, LEEK, and CHIVES; placed by some authorities in the Amaryllidaceae or the Alliaceae (e.g., Dahlgren et al. 1985). *Allium* species contain sulfur-containing compounds similar to those found in the unrelated mustard family (Blackwell 1990); cows grazing on wild onions are known to give onion-flavored milk and butter (Cheatham & Johnston 1995); while generally edible, gastroenteritis in children or other problems can result from excessive consumption of some species (Lampe & McCann 1985; Cheatham & Johnston 1995). According to Block (1985), when an ONION or GARLIC bulb is cut, low-molecular-weight organic molecules that incorporate sulfur atoms are released. These compounds have a number of biological effects: they are tear-inducing (= lacrimatory; the tear-inducing substance can undergo hydrolysis to form sulfuric acid); certain of the compounds are antibacterial or antifungal; others inhibit blood from clotting. (The ancient Latin name of the garlic, from Celtic: *all*, hot or pungent)
REFERENCES: Fraser 1939; Ownbey 1950b; Ownbey & Aase 1955; Davies 1992; Howard 1994; Mathew 1996; Mes et al. 1997 [1998].

1. Leaf blades cylindrical, hollow; flowering stalk up to more than 1 m tall ____ **A. cepa**
1. Leaf blades flat, folded, or concave; flowering stalk usually 0.5(–0.9) m or less tall (except taller in *A. sativum*).
  2. Umbels with bulbils, often with few or no flowers; the few flowers, if present, rarely producing capsules or seeds.
    3. Flowering stems naked above the subbasal leaves, usually 0.5(–0.9) m or less tall; bulb coats (base of leaves) with fibers that persist as a conspicuous net-like structure enclosing the bulbs; leaf blades 1–5 mm wide; widespread native species ____ **A. canadense** var. **canadense**
    3. Flowering stems with leaves nearly to middle, often 0.5–1.8 m or more tall; bulb coats lacking fibers that persist as a conspicuous net-like structure enclosing the bulbs; leaf blades 5–24(–40) mm wide; persisting or escaped introduced species.
      4. Underground bulb usually with ca. 2–few main cloves and producing numerous bulb lets from the base, the bulblets small (ca. 1 cm long), yellowish or brownish, helmet-shaped or subhemispherical, and typically stalked; stamens equaling or usually slightly exerted beyond the perianth segments; umbels usually large, (3–)5–10 cm in diam. ____ **A. ampelopresum**
      4. Underground bulb without small bulblets, but usually with 5–18+ similar-sized large bulblets (= cloves); stamens included or just equaling perianth segments; umbels usually 5 cm or less in diam. ____ **A. sativum**
  2. Umbels without bulbils; flowers producing capsules and seeds.
    5. Plants 25–180 cm tall, but typically > 50 cm tall; leaves 5–20(–40) mm wide; persisting or escaped introduced species ____ **A. ampelopresum**
    5. Plants 7–90 cm tall, but typically < 50 cm; leaves 1–5(–7) mm wide; widespread native species.

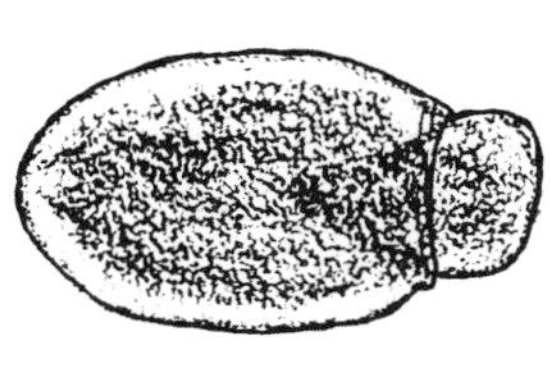

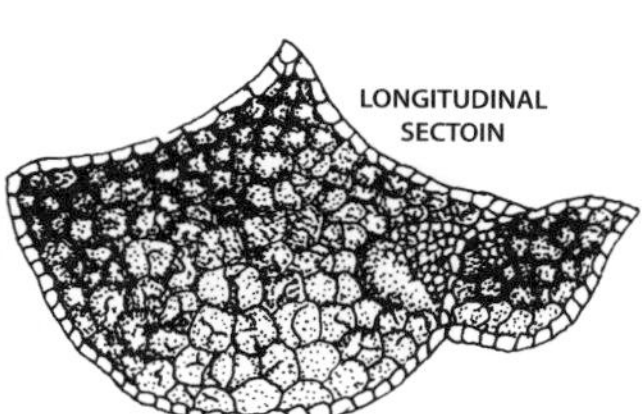

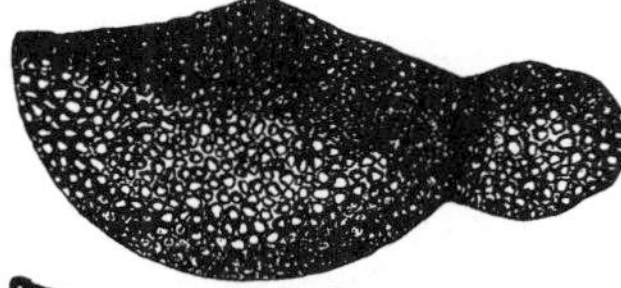

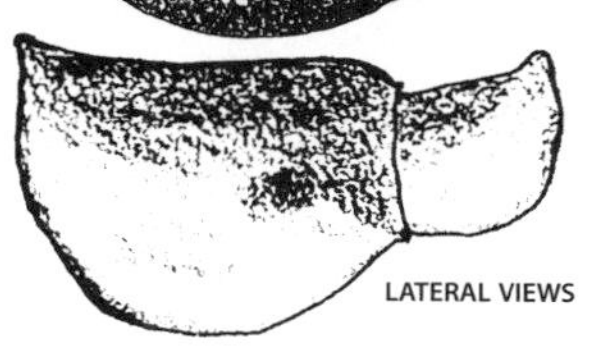

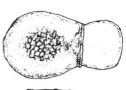

Wolffia brasiliensis [GWO, MIB]

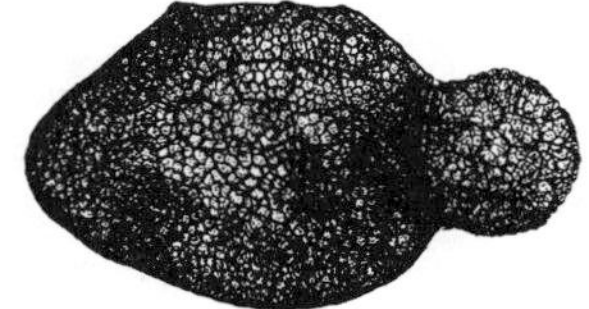

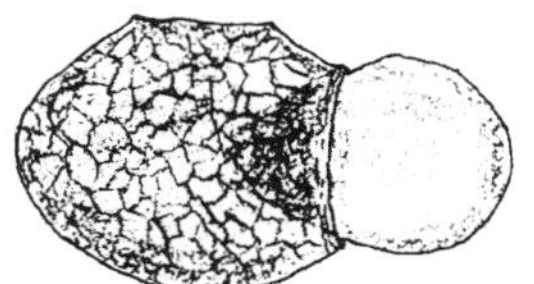

Wolffia columbiana [GWO, MIB]

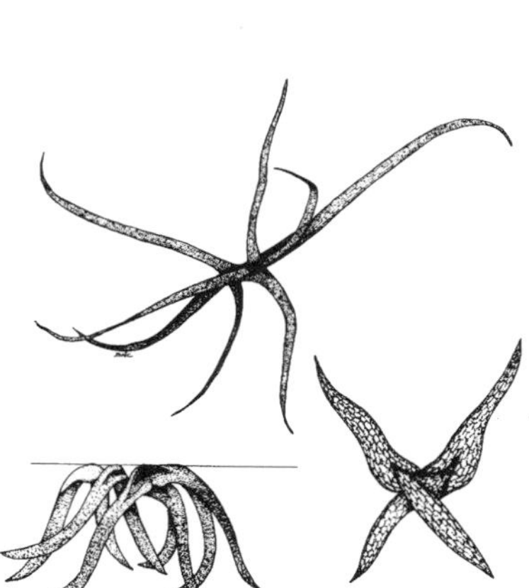

Wolffiella gladiata [GWO]

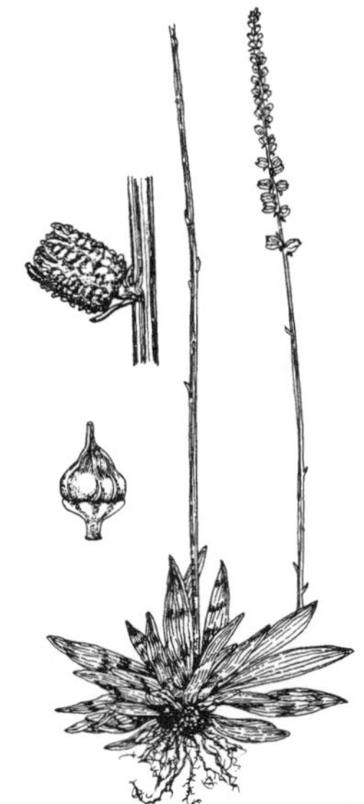

Aletris aurea [CO1]

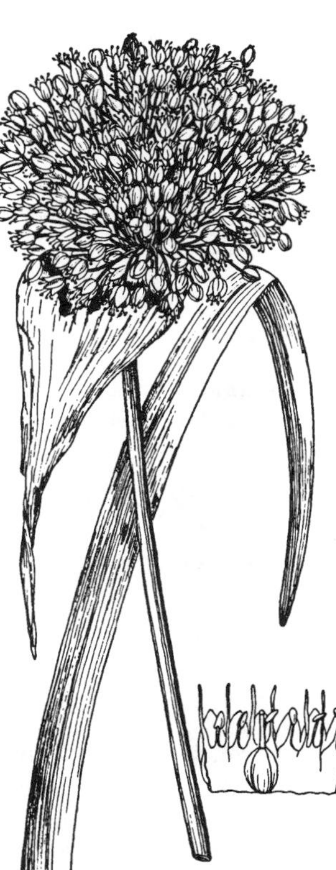

Allium ampeloprasum [GLE]

Allium canadense var. canadense [BB2]

Allium canadense var. mobilense [BB2]

6. Bulb coats lacking fibers that persist as a conspicuous net-like structure enclosing the bulbs; perianth segments deep pink; flowering late summer and fall ________ **A. stellatum**
6. Bulb coats with fibers that persist as a conspicuous net-like structure enclosing the bulbs; perianth segments pink, lavender, purplish red, or white; flowering in spring.
    7. Perianth segments not remaining spreading after flowering, either shriveling or if persistent, then urceolate; involucral bracts with 3–7 nerves (spathe usually divided into 2–3 separate or partly united involucral bracts).
        8. Perianth campanulate or urceolate-campanulate, ultimately withering somewhat and exposing the capsule, the segments 4–7 mm long; net-like structure enclosing bulbs fine or only moderately coarse-meshed.
            9. Perianth usually pinkish or lilac.
                10. Pedicels filiform (= thread-like); plants slender; leaf blades narrower than flowering stem, 0.3–2 mm wide; flowers ± scentless; se and e TX w to East Cross-Timbers ________ **A. canadense** var. **mobilense**
                10. Pedicels stouter; plants more robust; leaf blades narrower to broader than flowering stem, 0.5–7 mm wide; flowers with sweet hyacinth scent; Blackland Prairie w to Rolling Plains ________ **A. canadense** var. **hyacinthoides**
            9. Perianth usually white (rarely pink) ________ **A. canadense** var. **fraseri**
        8. Perianth urceolate, permanently enclosing the capsule, the segments 5–11 mm long; net-like structure enclosing bulbs usually very coarse.
            11. Bulbs at flowering time with a cluster of short-stalked basal bulblets; perianth segments 5–7 mm long, white with pinkish midribs, fading pink; pedicels ca. 2 times- the length of perianth in full flower ________ **A. runyonii**
            11. Bulbs at flowering time without basal bulblets; perianth segments deep rose color, fading purple, 7–11 mm long; pedicels ca. equal in length to perianth in full flower ________ **A. perdulce**
    7. Perianth segments remaining spreading after flowering, becoming dry, papery and rigid; involucral bracts usually 1-nerved (spathe usually divided into 2–3 separate or partly united involucral bracts) ________ **A. drummondii**

Allium ampeloprasum L., (onion of the vineyard), WILD LEEK. Flowering stem 45–200 cm tall, sheathed below, with leaves withering by flowering time; rounded parent bulb with many yellowish bulblets; leaves flat, 30–60 cm long, 15–24(–40) mm wide; umbels 5–10 cm wide; flowers very many, often more than 100 to up to 500 per umbel, white to pink or reddish pink, sometimes with bulbils. Weedy areas; Grayson Co. (extensive roadside colony), also, the map in Cheatham and Johnston (1995) showed 2 nc TX localities; scattered in e 1/2 of TX. Summer. Native of Europe, n Africa, and Asia. *Allium porrum* L., LEEK, probably derived from *A. ampeloprasum,* is also cultivated and probably persists locally. It differs in having the bulb but poorly developed and bulblets few. It is cultivated for the edible fleshy leaf sheaths. Stanford (1971) reported *A. porrum* from Hamilton Co.

Allium canadense L., (of Canada), WILD ONION. Flowering stalk 10–50 cm tall; leaves 1–5(–7) mm wide; flowers usually few or none, intermixed with bulblets and green sprouts, or many and without bulblets or sprouts; perianth white to pink or lavender. The key to varieties of *A. canadense* is included in the key to *Allium* species. In large amounts, this plant has caused death in cattle (Kingsbury 1964); gastroenteritis has been reported in children (Lampe & McCann 1985).

var. **canadense**. CANADA GARLIC, WILD GARLIC, WILD ONION. Flowering stalk 15–50 cm tall; leaf blades 1–5 mm wide; perianth segments when present 4–7 mm long, white to pink. Sandy open woods, fields, roadsides, other open areas; Falls, Henderson, Hopkins, Hunt, and Lamar cos; se and e TX w to East Cross Timbers. Mar–May. [*Allium acetabulum* (Raf.) Shinners] var. **fraseri** Ownbey, (for John Fraser, 1750–1811, Scottish collector in North America). Flowering stalk 20–50 cm tall; leaf blades 1–7 mm wide; perianth segments 4–7 mm long, white (rarely pink). Rocky soils, woods or open areas; Post Oak Savannah w to Rolling Plains and Edwards Plateau. Apr–May. [*A. fraseri* (Ownbey) Shinners]

var. **hyacinthoides** (Bush) Ownbey & Aase, (hyacinth-like). Flowering stalk 15–30(–40) cm tall; leaf blades 0.5–7 mm wide; umbels many-flowered; flowers fragrant; perianth segments 5–7 mm long, pink, thin. Calcareous prairies or infrequently in sandy soils, in sun or shade; Blackland Prairie w to edge of Rolling Plains and Edwards Plateau, mostly to the w of var. *mobilense*. Late Mar–Apr. [*A. hyacinthoides* Bush]

var. **mobilense** (Regel) Ownbey, (of Mobile, Alabama). Flowering stalk 10–30(–50) cm tall; bulb at base of plant sometimes with 1 or 2 bulblets (not in other vars.); leaf blades 0.3–2 mm wide; flowers many, scentless; perianth segments 4–7 mm long, pink (rarely white). Sandy or rocky soils, rarely on limestone or clay, woods and prairies; se and e TX w to East Cross Timbers, also Edwards Plateau. Apr–mid-May. [*A. mobilense* Regel]

var. *ecristatum* (M.E. Jones) Ownbey, (not crested), is cited by Hatch et al. (1990) for vegetational area 4 (Fig. 2) but apparently occurs only to the s of nc TX (Ownbey & Aase 1955). It can be distinguished from the similar var. *hyacinthoides* by its few (5–25)-flowered umbels and thicker perianth segments.

Allium cepa L., (Latin for onion), ONION. Flowering stalk to > 1 m tall; leaves cylindrical with groove on inner surface; pedicels many times longer than flowers; perianth whitish green. Cultivated, persisting, and escaped?; included based on citation of vegetational areas 4 and 5 (Fig. 2) by Hatch et al. (1990), also s TX. May–Jun. Native of w Asia.

Allium drummondii Regel, (for its discoverer, Thomas Drummond, 1780–1835, Scottish botanist and collector in North America), DRUMMOND'S ONION, PRAIRIE ONION. Flowering stalk 7–30 cm tall; leaves 1–3(–5) mm wide; umbels with 10–25 flowers; perianth segments 6–9 mm long, white to pink, lavender, or purple-red (rarely greenish yellow); pedicels ca. 6–18 mm long. Sandy or gravelly, often limestone soils; nearly throughout TX. Mar–May.

Allium perdulce S.V. Fraser, (very sweet). Flowering stalk7–25 cm tall; leaves 1–2 mm wide; flowers sweet-scented; umbel 5–25-flowered; perianth rose-purple. Sandy or gravelly prairies; Archer, Bosque, and Callahan cos., also Clay and Palo Pinto cos. (Ownbey & Aase 1955), w part of nc TX w to Plains Country and s to Edwards Plateau. Mar–early Apr.

Allium runyonii Ownbey, (for H. Everett Runyon, 1881–1968, TX botanist and photographer), RUNYON'S ONION. Flowering stalk 10–35 cm tall; leaves 1–4 mm wide; perianth segments 5–7 mm long, white with pinkish midribs, fading pink. Sandy soils; included based on citation of vegetational areas 4 and 5 (Fig. 2) by Hatch et al. (1990); Ownbey and Aase (1955) gave the distribution as Rio Grande Plains in s TX; endemic to TX. Mar. The cluster of stalked bulblets typically present at the base of the bulb is distinctive among the nc TX species of *Allium*.

Allium sativum L., (cultivated or sown), GARLIC. Plant stout; bulb coats membranous, silky white or pink; bulb of 5–15(+) similar-sized bulblets (= cloves); leaves 6–12, flat, 5–15 mm wide, sheathing the stem; flowers usually aborting before anthesis; perianth segments usually whitish green or pinkish. Cultivated in nc TX and probably persisting and escaping?; included because of likelihood of encounter in abandoned garden spots; May–Jul. Cultigen derived from an Old World species; cultivated since ancient times, probably 3,000 B.C. or earlier; bulbs were found in the tomb of Tutankhamen; used medicinally, in cooking, and worn around the neck to supposedly ward off evil spirits, trolls, and vampires (Rose & Strandtmann 1986; Mabberley 1987; Mathew 1996).

Allium stellatum Nutt. ex Ker Gawl., (star-like), PRAIRIE ONION, WILD ONION, PINK WILD ONION. Flowering stalk 20–70 cm tall; leaves 1–5 mm wide; perianth segments 5–8 mm long, deep pink. Prairies, usually in calcareous soils; Cooke and Tarrant cos., also Lamar Co. (Carr 1994); in TX only in nc part of the state. Jul–Oct.

Allium texanum T.M. Howard, (of Texas). This species was named by Howard (1990) with the type from Bosque Co. and another nc TX collection from Comanche Co. He indicated that "it is distinguished from the closely related *A. fraseri* [*A. canadense* var. *fraseri*] in flowering later, having broadly spiraled, gla[u]cous foliage, taller, more robust habits, larger umbels with flowers having green ovaries, and having membranous coated bulbs or with non-persisting, poorly developed reticulated bulb coats. *Allium texanum* differs from *A. canadense* in its mostly membranous-coated bulbs, floriferous, rather than bulbil-bearing, umbels, and in the individual floral form." We have seen no specimens but based on the description assume it is part of the *A. canadense* complex, possibly a variety of that species. Because of our unfamiliarity with this taxon and our hesitancy to add to nomenclatural complexity by treating it as a var. of *A. canadense*, we are simply including it as a note.

ANDROSTEPHIUM FUNNEL-LILY

An American genus of 2 species; cultivated as ornamentals. (Greek: *andros*, male, and *stephanos*, a crown, in reference to the fused filaments)

Androstephium coeruleum (Scheele) Greene, (dark blue), BLUE FUNNEL-LILY. Glabrous perennial from a fibrous-coated corm, gray-green, 6–25 cm tall; leaves basal, slender, longer than the scape (= flowering stem), exserted from a broad, loose, thinly papery, sheathing bract; scape to 25 cm tall, usually < 15 cm; flowers 1–6 in an umbel-like cluster subtended by membranous bracts, with strong spicy-sweet scent; perianth lavender-blue, 16–24 mm long; filaments ± united to form a tube; capsules ca. 15 mm long. Prairies; Blackland Prairie (on the Austin Chalk) w to Edwards Plateau and Rolling Plains. Late Mar–mid-Apr.

ASPARAGUS

An Old World genus of 130–140 species. (The ancient Greek name)

Asparagus officinalis L., (used in medicine), GARDEN ASPARAGUS. Rhizomatous, dioecious, glabrous perennial usually 1–3 m tall; stems with many very finely dissected branches; leaves reduced to scales; photosynthesis carried out by the stem tissue; flowers axillary, 1–2(–3) per axil, on jointed pedicels, greenish yellow, ca. 4–6 mm long; fruit a red berry ca. 1 cm in diam. Widely cultivated and escapes to sandy areas; Cooke, Grayson, and McLennan cos.; mainly e 1/2 of TX. May–Jul. Native of the coasts of Europe, n Africa, and Asia. Grown since time of the ancient Greeks for the edible young spring shoots (Mabberley 1997). Mature asparagus has caused poisoning in cattle (Kingsbury 1964); the young plants can cause dermatitis and the fruits are suspected of poisoning humans (Schmutz & Hamilton 1979); there are steroid saponins which are molluscicidal (Mabberley 1997).

CAMASSIA WILD-HYACINTH

A genus of 6 species, 5 in North America (1 e U.S., 4 disjunct to w North America—Wood 1970), 1 South American. Some species (e.g., *C. quamash* (Pursh) Greene–CAMASH or QUAMASH of Native Americans) were extensively used as food by Native Americans in the nw U.S. (Kindscher 1987); some also are cultivated as ornamentals. *Camassia* is sometimes segregated with similar genera (e.g., *Hyacinthus, Muscari, Ornithogalum*) into the Hyacinthaceae (Dahlgren et al. 1985). (From the Native American name, *quamash* or *camass*)
REFERENCES: Gould 1942; Ranker & Schnabel 1986.

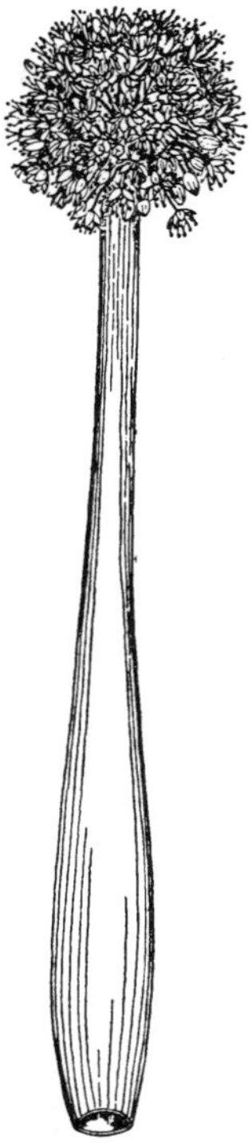

Allium drummondii [BB2]

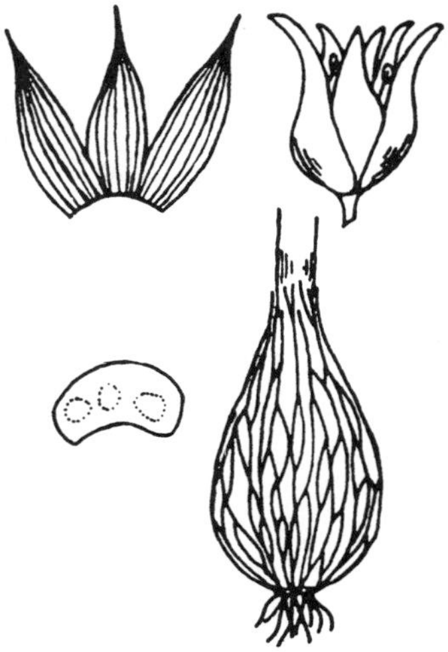

Allium perdulce [TKA]

Allium cepa [GLE]

Allium sativum [ANO]

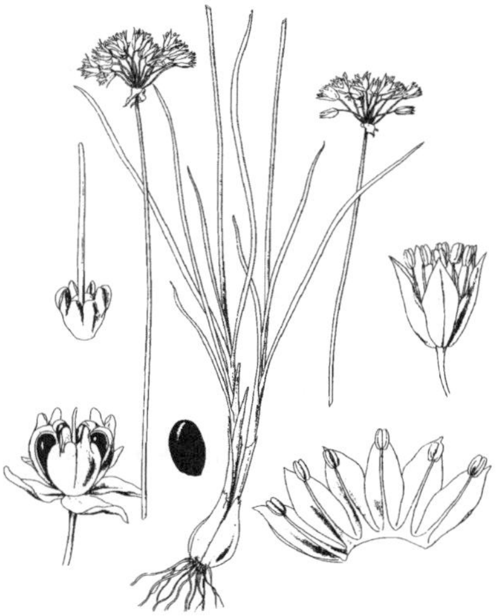

Allium stellatum [HO1]

Allium runyonii [HEA]

Androstephium coeruleum [RYD]

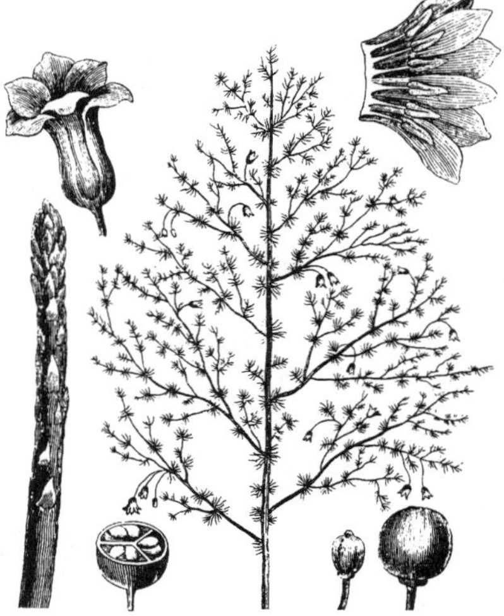

Asparagus officinalis [LAM]

Camassia scilloides (Raf.) Cory, (resembling the genus *Scilla*), WILD-HYACINTH, EASTERN CAMASS, ATLANTIC CAMASSIA, CAMASS, CAMASS-LILY, MEADOW-HYACINTH, MEADOW QUILL, SIKO. Bulbous perennial 15–85 cm tall; leaves crowded at base of stem, their bases clasping and surrounded by a sheathing, somewhat papery bract; inflorescence an elongating erect raceme; flowers sweet-scented; perianth segments 10–14 mm long, lavender-blue, rarely white; fruit a capsule. Open woods and prairies; se and e TX w to West Cross Timbers and Edwards Plateau. Apr–May. The bulb is edible and was an important food source for Native Americans and early settlers; however, it resembles that of the potentially fatal *Zigadenus nuttallii* (DEATH CAMAS) (Ajilvsgi 1984; Kindscher 1987). ▣/81

COOPERIA RAIN-LILY

Plants glabrous, often flowering without leaves, 8–35 cm tall; flowers erect, solitary, subtended by a conspicuous bract; flowering stems appearing quickly after rains; perianth white, sometimes pink-tinted outside, salverform, with an elongate tube and an open limb of similar sepal and petal lobes; flower sessile or short-pedicelled; ovary borne inside the involucre; capsule trilocular.

A small American genus of ca. 6–7 species (Correll & Johnston 1970) sometimes lumped into *Zephyranthes* (e.g., Mabberley 1997) and previously placed in the Amaryllidaceae. We are following Kartesz (1994), Jones et al. (1997), and J. Kartesz (pers. comm. 1997) in maintaining *Cooperia* as a distinct genus. (Named for Daniel Cooper, ?1817–1842, an English botanist)

1. Flower sessile, ovary borne at base of subtending bract; perianth tube (from base to where perianth begins to widen) greatly elongate (3.4–18 cm long); style > 40 mm long ____ **C. drummondii**
1. Flower short-pedicelled, the ovary borne 1/3–1/2 way above base of subtending bract; perianth tube shorter (2.2–4 cm long); style < 35 mm long ____ **C. pedunculata**

Cooperia drummondii Herb., (for its discoverer, Thomas Drummond, 1780–1835, Scottish botanist and collector in North America), CEBOLLETA, RAIN-LILY. Spring leaves 1–3(–5) mm wide; perianth with nearly flat limb about 1/4 as long as the long, slender, pedicel-like basal tube, the lobes 12–20 mm long, obtuse. Prairies, roadsides, often on thin soils over limestone; se and e TX w to West Cross Timbers and Edwards Plateau. Jun–Oct, typically after a rain. [*Zephyranthes brazosensis* (Herb.) Traub, *Zephyranthes herbertiana* D. Dietr.] The flowers open in the late afternoon or evening (Kirkpatrick 1992).

Cooperia pedunculata Herb., (peduncled, stalked, or footed), GIANT RAIN-LILY, PRAIRIE RAIN-LILY, WHITE RAIN-LILY, WIDE-LEAF RAIN-LILY. Spring leaves 4–10 mm wide; perianth with broadly funnelform limb about 1/2 or more as long as the narrowly cylindric basal tube, the lobes 25–30 mm long, with an abrupt small point. Rocky or sandy soils, prairies, roadsides, open woods; Coryell Co., also McLennan Co. (Mahler 1988); s part of nc TX sw to Edwards Plateau, also se and e TX. Apr–Jul, typically a few days after heavy rains. [*Zephyranthes drummondii* D. Don.] According to Wills and Irvin (1961), the flowers "... open slowly around dusk or earlier on cloudy days, the lobes gradually spreading during the night, and appearing fully expanded the next morning. Ordinarily each flower lasts only one day, turning pale pink before withering, but in dull weather withering may not occur until the second day." ▣/85

ERYTHRONIUM DOG-TOOTH-VIOLET, FAWN-LILY

Perennial from deep bulb; runners or stolons present or absent; leaves of flowering plants 2, of sterile plants 1; inflorescence usually a 1-flowered scape; perianth segments 6, in ours white to tinged with varying shades of blue and red, 2–4 cm long; 3 inner perianth segments with a yellow spot at base, stamens 6; pistil superior, 3-carpellate; capsule obovate.

A genus of ca. 20 species of temperate North America and Eurasia; some have edible corms; a number are cultivated as ornamentals. The following key is from Robertson (1966); all nc TX specimens we have observed had mottled leaves; some of these had the mature fruits resting on the ground and some were from prairies; Robertson (1966) indicated that specimens from near Fort Worth were definitely *E. mesochoreum*. Taylor and Taylor (1994) recognized the two taxa as varieties of *E. albidum*. Further work on this complex is needed before specimens can be classified with confidence; we are not fully convinced there are actually two different species in nc TX. (Greek name for the purplish-flowered European species, from *erythros*, red)
REFERENCES: Harper 1945; Ireland 1957; Robertson 1966.

1. Perianth segments reflexed in full bloom; leaves mottled; mature fruits erect to nodding at maturity, held off ground; moist woods ____ **E. albidum**
1. Perianth segments spreading to at most half-reflexed in full bloom; leaves usually not mottled; mature fruits resting on ground; prairies, pastures, dry open woods ____ **E. mesochoreum**

Erythronium albidum Nutt., (white), WHITE DOG-TOOTH-VIOLET, WHITE FAWN-LILY, TROUT-LILY. Leaves abruptly attenuated, mottled, flat to half-folded; perianth reflexed; sterile forms with long runners with a new bulb forming at runner tip; chromosome number $2n = 44$. Usually moist woods; Collin, Cooke, Dallas, Grayson, and Tarrant cos., also Comanche (HPC), Hunt, Kaufman, Red River, and Rockwall (Mahler 1988) cos.; e TX w to nc TX. Feb–Mar. The common name TROUT LILY refers to the leaf mottling which resembles the speckling on a trout (Ajilvsgi 1984).

Erythronium mesochoreum Knerr, (midland). Leaves gradually attenuated, not mottled, conduplicate or occasionally only half-folded; perianth spreading; sterile forms usually without runners, a new bulb forming at base of old one; chromosome number $2n = 22$. Prairies and dry woods; Bell, Bosque, Coryell, Dallas, Grayson, Rockwall, and Tarrant cos. (Mahler 1988); nc TX and Edwards Plateau. (Jan–)Feb–Mar. [*E. albidum* var. *coloratum* Sterns, *E. albidum* var. *mesochoreum* (Knerr) Rickett] Churchill and Bloom in Churchill (1986b) indicated that ants eat the white oil body attached to the seeds and apparently act as dispersal agents. /89

HABRANTHUS COPPER-LILY

A temperate South American genus of 10 species including some cultivated as ornamentals; previously placed in the Amaryllidaceae. (Greek: *habros*, graceful, and *anthos*, a flower)
REFERENCE: Holms & Wells 1980.

Habranthus tubispathus (L'Hér.) Traub, (tube-spathed), ATAMOSCO-LILY, STAGGER-GRASS. Leaves basal; scape to ca. 30 cm tall, with 1 flower; perianth long-funnelform with gradually tapering base, 25–30 mm long, orange-yellow, sometimes with reddish tinge on outer surface; flower on pedicel about twice as long as the subtending involucral bract; ovary borne above the bract; capsule subglobose, ca. 15 mm wide. Moist open areas; Milam Co., also yard weed in Tarrant Co., also McLennan and Williamson cos. (Holmes & Wells 1980); s part of nc TX to se and s TX and Edwards Plateau. Jul–Oct, after rains. [*H. texanus* (Herb.) Herb. ex Steud., *Zephyranthes texana* Hook.] Holmes and Wells (1980) proposed that this species is native to South America and was introduced in the U.S. in the late 1600s or early 1700s possibly by Spanish missionaries.

HEMEROCALLIS DAY-LILY

An ornamentally important genus of ca. 15 species occurring from c Europe to China and Japan; many cultivars are of hybrid origin. Some authorities (e.g., Dahlgren et al. 1985) segregate the genus as the Hemerocallidaceae. (Greek: *hemera*, a day, and *callus*, beauty, referring to the flowers lasting one day)
REFERENCES: Bailey 1930a; Erhardt 1992.

Hemerocallis fulva L., (tawny, dull yellow-brown), TAWNY DAY-LILY, ORANGE DAY-LILY. Perennial from fleshy roots with tuberous swellings; leaves basal, to 1 m long; scape 0.7–1.5 m tall; flowers several, not fragrant; perianth large and showy, usually tawny orange, 8.5–12 cm long, broadly campanulate to funnelform with cylindrical tube; perianth lobes spreading to slightly recurved; margins of inner perianth lobes undulate; fruit a capsule; seeds usually not maturing. Widely cultivated, long persists, and sometimes escapes; Lamar Co., also Grayson Co. (G. Diggs, pers. obs.). Native of e Asia. The dried flowers are used to flavor food in China and Japan (Mabberley 1987).

Hemerocallis lilioasphodelus L., (lily and *Asphodelus,* another genus of Liliaceae), is another cultivated species; this Eurasian native can be distinguished by its bright yellow fragrant flowers and the margins of the perianth lobes plane (not undulate).

HYACINTHUS HYACINTH

A genus of 3 species native from w and c Asia to the Mediterranean; sometimes segregated with similar genera (e.g., *Camassia, Muscari, Ornithogalum*) into the Hyacinthaceae (Dahlgren et al. 1985). (The Greek name)

Hyacinthus orientalis L., (oriental), HYACINTH, GARDEN HYACINTH. Bulbous perennial 10–30 cm tall; leaves crowded at base, ± clasping and sheathing; flowers sweet-scented, short-pedicelled, in an erect, spike-like raceme; perianth narrowly campanulate with recurved lobes, violet-blue to white, rosy, or rarely yellowish. In 1893 Reverchon wrote that a wild strain of this had self-sown and escaped from his garden in Dallas, maintaining itself for several years (Mahler 1988). Modern cultivated varieties sometimes persist about old farms and gardens (Hunt Co.), but are not known to produce seed. Late Feb–Mar. Native of the e Mediterranean region. All parts of the plant, but especially the bulb, can cause poisoning in humans and livestock; toxicity is apparently due to alkaloids such as lycorine (Stephens 1980; Spoerke & Smolinske 1990).

HYMENOCALLIS WHITE SPIDER-LILY, SPIDER-LILY

Glabrous perennials from a large bulb; leaves basal, strap-like; inflorescence an umbel, usually with 6–9 flowers, terminating a naked scape, subtended by 2 or more usually scarious bracts; flowers white, extremely showy, sweet scented; perianth with a long slender tube and linear to narrowly lanceolate, spreading segments; large, conspicuous, cup-like crown (= corona) present and connecting the bases of the filaments; fruit a few-seeded capsule.

A genus of 30–40 species of warm areas of the Americas; a number are cultivated as ornamentals; previously placed in the Amaryllidaceae. The bulbs of some species are poisonous due to the presence of alkaloids such as lycorine and tazettine (Lampe & McCann 1985; Spoerke & Smolinske 1990). (Greek: *hymen*, a membrane, and *callos*, beauty, presumably referring to the crown)

REFERENCES: Shinners 1951d; Sealy 1954.

1. Free portion of filaments 23–35 mm long; crown 33–40 mm long; larger perianth segments usually > 5 mm wide; leaves 18–42 mm wide ____ **H. caroliniana**
1. Free portion of filaments (above attachment to crown) 20 mm or less long; crown 25–35 mm long; larger perianth segments usually 5 mm or less wide; leaves usually < 20(–40) mm wide ____ **H. liriosme**

Hymenocallis caroliniana (L.) Herb., (of Carolina). Plant 35–53 cm tall; perianth segments white, greenish white below, to 10 cm long; crown 3.3–4 cm long. Wet sandy areas; included based on citation of vegetational area 4 (Fig. 2) by Hatch et al. (1990); mainly se and e TX. Mar–May(–Jul).

Camassia scilloides [BB2]

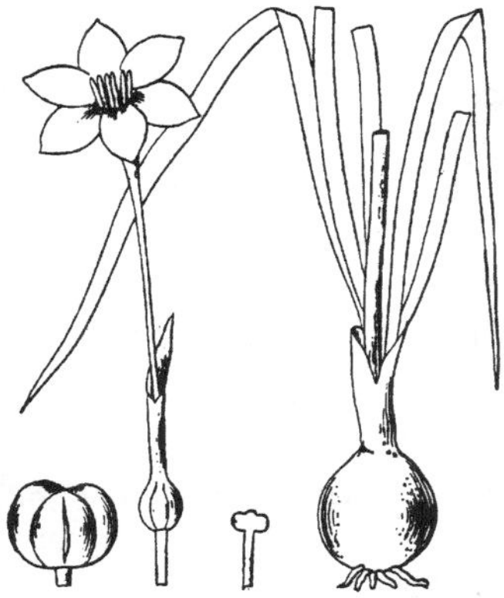

Cooperia drummondii [BB2]

Cooperia pedunculata [CUR]

Erythronium albidum [RYD]

Erythronuim mesochoreum [STE]

Habranthus tubispathus [CUR]

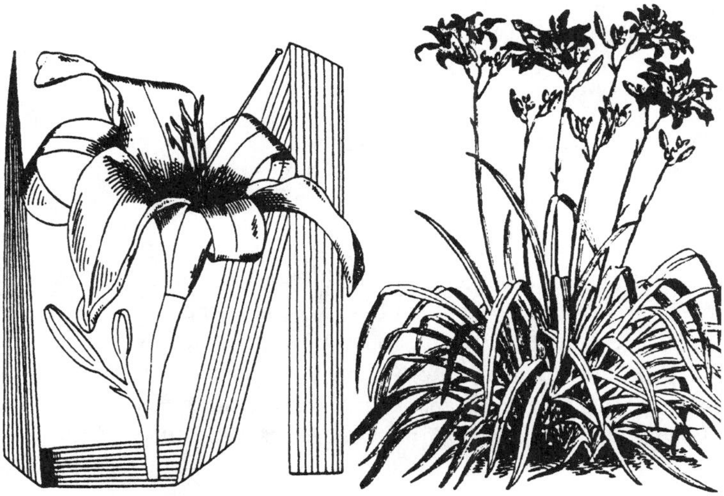

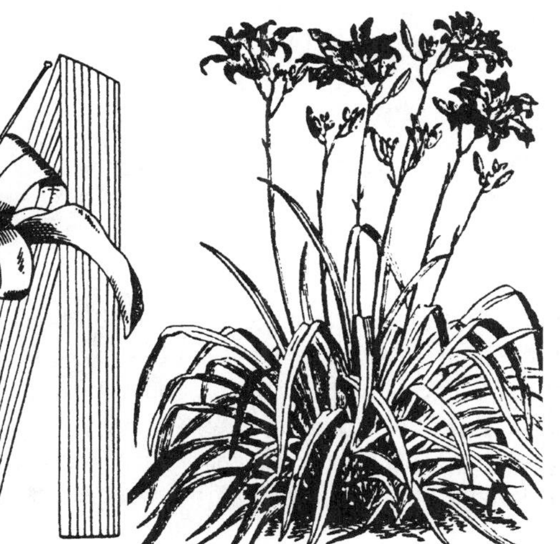

Hemerocallis fulva [BT2, GAR]

Hyacinthus orientalis [NIC]

Hymenocallis liriosme (Raf.) Shinners, (lily-smell or fragrant lily), WESTERN SPIDER-LILY, FRAGRANT SPIDER-LILY. Plant 35–100 cm tall; perianth extremely showy, to ca. 20 cm in diam., snowy white, tinged with yellow in the center, yellowish or greenish on tube; tube 6–8 cm long. Stream bottoms and ditches, sometimes in shallow water; Kaufman and Red River cos. (Correll & Johnston 1970); se and e TX w to Blackland Prairie. Late Mar–May, occasionally to Jul. [*H. eulae* Shinners] ⊠/93

HYPOXIS YELLOW STAR-GRASS, GOLDSTAR

A genus of 150 species of tropical and warm areas of the world, especially the s hemisphere. It is sometimes segregated as the Hypoxidaceae (e.g., Dahlgren et al. 1985; Jones et al. 1997) or placed in the Amaryllidaceae. (Old name taken over by Linnaeus; either from Greek: *hypoxys*, somewhat acidic, or Greek: *hypo*, beneath, and *oxys*, sharp, alluding to the base of the capsule)
REFERENCES: Brackett 1923; Herndon 1992.

Hypoxis hirsuta (L.) Coville, (hairy), YELLOW STAR-GRASS, COMMON GOLDSTAR. Perennial from a small corm; leaves basal, slender, grass-like, with closed, tubular basal sheath; several additional papery sheaths present outside the leaves, the old ones disintegrating and disappearing; flowering stems 6–20 cm tall, from half as long as the leaves to almost as long; flowers usually 2–4, in an umbel-like inflorescence with a pair of opposite, thread-like bracts at its base; perianth yellow, greenish and pubescent outside, nearly rotate when fully open; perianth segments 6, 6–12 mm long, up to 18 mm with age, lanceolate to elliptic; stamens 6; pistil 1; ovary inferior; capsule 2–6 mm long. Open woods, prairies, and roadsides; se and e TX w to Dallas, Grayson, and Wise cos., also Montague Co. (R. O'Kennon, pers. obs.), also Edwards Plateau. Late Mar–early May. [*H. rigida* Chapm., *H. leptocarpa* Engelm.]

IPHEION SPRING-STAR

A South American genus of 10 species of onion-scented herbs (Mabberley 1987) sometimes treated in the genus *Tristagma* (e.g., Mabberley 1997); previously placed by some authorities in the Amaryllidaceae or Alliaceae. (Greek: origin obscure)

Ipheion uniflorum (Lindl.) Raf., (one-flowered), SPRING-STAR. Glabrous perennial, 7–20 cm tall, with onion odor; leaves basal, with long, closed, tubular sheath of thin, scarious texture, and slender green blade; scape with 2 partly or wholly united bracts above the middle, terminated by 1(–2) flowers; perianth 20–28 mm long, white with pale blue tinge, the lobes with darker central line and brownish tinge on back; filaments ± united to form a tube. Cultivated and escaping; Blackland Prairie (Kaufman Co.) and Edwards Plateau. Feb–Mar. Native of Argentina and Uruguay. [*Brodiaea uniflora* (Graham) Engl.]

LEUCOJUM SNOWFLAKE

A genus of 10 species ranging from Europe to Morocco and Iran; related to *Galanthus*; previously placed in the Amaryllidaceae. (Greek: *leucos*, white, and *ion*, a violet, in reference to the delicate fragrance)
REFERENCE: Webb 1980.

Leucojum aestivum L., (summer), GIANT SNOWFLAKE, SUMMER SNOWFLAKE. Bulbous perennial; leaves broadly linear, to 10 cm long and 5–20 mm wide, surrounded at base by tubular sheaths; scapes 20–40(–80) cm tall; umbel with 2–10 nodding flowers, subtended by a membranous bract 3–5 cm long; pedicels 1–4(–7) cm long; perianth campanulate, without elongate perianth tube; corona absent; perianth segments free nearly to base, all about equal, white with green tips, 1.5–3 cm long; anthers opening by longitudinal slits; seeds 5–7 mm long, black. Cultivated

Hymenocallis caroliniana [CO1]

Hymenocallis liriosme [HEA]

Hypoxis hirsuta [BB2]

Ipheion uniflorum [CHT]

Leucojum aestivum [LAM]

Lycoris radiata [SIN]

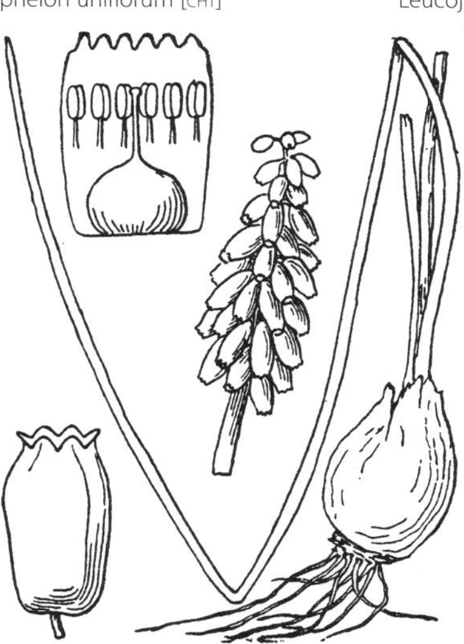

Muscari neglectum [BB2]

and escapes; Dallas Co. Mar. Native of Europe. Two other similar European species are cultivated and persist in nc TX:

Galanthus nivalis L., (genus: Greek: *gala*, milk, and *anthos*, flower; sp.: of snow), SNOWDROP, has the inner tepals much shorter than outer, only the inner green-tipped, and anthers opening at tip.

Leucojum vernum L., (spring), SPRING SNOWFLAKE, has flowers solitary.

LYCORIS

A genus of 11 species native from China and Japan to Burma; cultivated as ornamentals; previously placed in the Amaryllidaceae. The bulbs of *Lycoris* species are poisonous due to the presence of the alkaloid, lycorine (Lampe & McCann 1985). (Named after Lycoris, a beautiful Roman actress and mistress of Marc Antony)
REFERENCE: Ping-Sheng et al. 1994.

Lycoris radiata (L'Hér.) Herb., (with rays), SPIDER-LILY, RED SPIDER-LILY. Bulbous perennial; leaves to ca. 8 mm wide, glaucous, not appearing with the flowers; inflorescence a few-flowered umbel at end of a solid scape; flowers bright red, lacking fragrance; perianth ca. 4 cm long, the lobes reflexed with wavy margins; stamens much-exserted, 2 times as long as perianth; ovary inferior; fruit a few-seeded capsule. Cultivated and long persists or possibly escapes; roadside ditch in Grayson Co. (G. Diggs, pers. obs.). Fall. Native of China and Japan. Poisonous (Lampe & McCann 1985).

MUSCARI GRAPE-HYACINTH

A genus of 30 species ranging from Europe and the Mediterranean to w Asia; some contain alkaloids including colchicine; a number are cultivated as ornamentals and the flowers of some are used in scent making. *Muscari* is sometimes segregated with similar genera (e.g., *Camassia*, *Hyacinthus*, *Ornithogalum*) into the Hyacinthaceae (Dahlgren et al. 1985). (Named in reference to the musky flower odor of some species)

Muscari neglectum Guss. ex Ten., (overlooked), STARCH GRAPE-HYACINTH. Low-growing bulbous perennial; leaves basal, several or many, dark green, about 3–4 mm wide, arched or loosely coiled and spreading, longer than flowering stem; inflorescences scapose, racemose, to ca. 20 cm tall; perianth urceolate, dark blue with whitish teeth (rarely all white), about 3 mm broad, 3.5–6 mm long, with musky odor; pedicels 2–4 mm long; fruit a capsule. Cultivated widely, escaping and naturalizing, yards, fields, and roadsides; Bosque, Dallas, Grayson, McLennan, and Rockwall cos., also Tarrant Co. (R. O'Kennon, pers. obs.); nc, c and e TX. Mar–early Apr. Native of the Mediterranean region. [*M. racemosus* (L.) Lam. & DC.]

NARCISSUS

Bulbous, glabrous, scapose perennials with linear, flat or terete leaves exserted from closed tubular sheaths; inflorescence terminal, subtended by a membranous sheathing bract splitting along 1 side; flowers sweet-scented; perianth united into a tube, with perianth parts spreading to reflexed from the base of the central tubular or cup-shaped crown (= corona), yellow or white; stamens included or barely exserted from the tube; fruit a many-seeded capsule.

A genus of 27 species; a number of species (and their hybrids) of this Old World (European and Mediterranean) genus are cultivated for their early spring flowers; previously placed in the Amaryllidaceae. Several of the most frequently cultivated species and their hybrids, that long persist around old homesites or escape, are treated here. The bulbs have been known to cause poisoning in humans and also in livestock when used as emergency food; even small amounts cause vomiting; bulbs and leaves can cause dermatitis in susceptible individuals; toxins include alkaloids (e.g., narcissine, lycorine) and a glycoside; raphides (= bundles of microscopic, needle-like calcium oxalate crystals) are also present in the bulbs and can cause mechanical injury to the mouth, throat, or hands by puncturing cell membranes (Kingsbury 1964; Schmutz & Hamilton 1979; Spoerke & Smolinske 1990; Turner & Szczawinski 1991). (Greek plant name derived from *narke*, numbness or torpor, from its narcotic properties; in Greek mythology the youth Narcissus fell in love with his own reflection in a pool and was turned into this plant by the gods.)

1. Crown nearly as long as or slightly longer than lobes of perianth; perianth tube (below attachment of perianth lobes) broadly conical, about as long as lobes ____ **N. pseudonarcissus**
1. Crown ca. 2/3 as long as lobes of perianth or less (often much less); perianth tube nearly cylindrical, slightly to much shorter than lobes.
    2. Crown ca. 1/2–2/3 as long as perianth lobes; perianth usually yellow; flower 1 per flowering stalk ____ **N. ×incomparabilis**
    2. Crown 1/3 as long as perianth lobes or less; perianth yellow or white (but crown can be colored); flowers usually 2–10(–15) per flowering stalk or if 1, then perianth white.
        3. Leaves thick, terete (= cylindrical) or nearly so (and grooved on upper surface), green ____ **N. jonquilla**
        3. Leaves flat or nearly so, mostly ± glaucous.
            4. Perianth white; crown red-rimmed; flowers usually 1 per flowering stalk ____ **N. poeticus**
            4. Perianth yellow or white; crown yellow or orange; flowers 2–8(–15) per flowering stalk ____ **N. tazetta**

Narcissus ×incomparabilis Mill. [*N. poeticus* × *N. pseudonarcissus*], (incomparable). Hybrid similar to *N. pseudonarcissus*; leaves 8–12 mm wide; perianth lobes 25–35 mm long; perianth tube 20–25 mm long; corona 13–22 mm long; there are numerous forms including many of the modern "fancy" types (Shinners 1958a). Cultivated; no escaped specimens from nc TX seen; included because it is expected to be found persisting. Mar.

Narcissus jonquilla L., (from Spanish: *junquillo*, rush, for the slender leaves), JONQUIL. Leaves 2–4 mm wide; inflorescences 2–5-flowered; pedicels unequal; flowers fragrant; perianth yellow with deeper yellow crown; perianth lobes 10–15 mm long; perianth tube slender, (17–)20–30 mm long; crown 3–5 mm long. Cultivated and long persists; Grayson and Hunt cos. Feb–Mar. Native of s Spain and Algeria.

Narcissus poeticus L., (pertaining to the poets), PHEASANT'S-EYE, POET'S NARCISSUS. Leaves 5–13 mm wide; flowers very fragrant; perianth lobes 15–30 mm long; perianth tube 20–30 mm long, slender, ± equal in diam. to apex; corona 1–2.5 mm long. Cultivated; no escaped specimens from nc TX seen; included because it is expected to be found persisting. Mar–May. Native from France to Greece. Flowers used in making perfume.

Narcissus pseudonarcissus L., (false *Narcissus*), DAFFODIL, TRUMPET NARCISSUS. Leaves 5–15 mm wide; perianth yellow (rarely white) with crown usually deeper yellow (rarely white); perianth lobes 18–40(–55) mm long; perianth tube broad, gradually flared, about 12–18 mm long; crown 15–50 mm long. Cultivated and long persists; Grayson and Hunt cos. Feb–Mar. Native of Europe.

Narcissus tazetta L., (small cup, from Italian word), POLYANTHUS NARCISSUS. Leaves 5–25 mm wide; flowers fragrant, perianth lobes 8–22 mm long; perianth tube slender, 12–18 mm long tube; crown 3–6 mm long. Cultivated and long persists; according to Shinners (1958a), this species and its hybrids are the most frequent escapes from cultivation in nc TX; included on that basis. Late Feb–Mar. Native of Mediterranean area. PAPER-WHITE NARCISSUS is a large-flowered, all white form of *N. tazetta*.

Narcissus ×*odorus* L. [*N. jonquilla* × *N. pseudonarcissus*], CAMPERNELLE JONQUIL, is occasionally cultivated; this hybrid, with usually 2–4 yellow flowers per flowering stalk, resembles *N. jonquilla* in having thick, grooved leaves, but has slightly larger flowers and the crown ca. 1/2 as long as the perianth lobes.

NOTHOSCORDUM FALSE GARLIC

An American genus of 20 species similar to *Allium* but odorless; previously considered by some authorities to be in the Alliaceae (e.g., Dahlgren et al. 1985) or the Amaryllidaceae. (Greek: *nothos*, false, and *scordon*, garlic)

Nothoscordum bivalve (L.) Britton, (two-valved), CROW-POISON, YELLOW FALSE GARLIC. Bulbous glabrous perennial, 7–40 cm tall; leaves basal, with closed, tubular, basal sheath and slender blade; inflorescences scapose, umbellate, of (3–)6–12 flowers, subtended by 2 scarious bracts; flowers scentless; perianth funnelform, 8–10 mm long, whitish with yellowish base inside and lavender to purple-red midribs on backs of perianth segments. Open woods, prairies, disturbed sites; throughout most of TX; one of our most abundant and widespread native plants. Mostly Mar–early May and late Sep–Oct. Resembling an *Allium* and placed by some in that genus [as *A. bivalve* (L.) Kuntze]; easily distinguished by its longer anthers (2 mm long) and lack of odor.

ORNITHOGALUM

An Eurasian and African genus of ca. 200 species; sometimes segregated with similar genera (e.g., *Camassia*, *Muscari*, *Ornithogalum*) into the Hyacinthaceae (Dahlgren et al. 1985). Some species have alkaloids including colchicine; a number are cultivated as ornamentals; some have edible bulbs; others are toxic. (Greek: *ornis*, a bird, and *gala*, milk; this was an expression used by the ancient Greeks to describe something amazing—Love 1994).

Ornithogalum umbellatum L., (with umbels), STAR-OF-BETHLEHEM. Bulbous perennial 10–35 cm tall; leaves basal, longer than flowering stem; inflorescence a simple corymb; flowers subtended by conspicuous, clasping, papery, acuminate bracts; perianth broadly funnelform or nearly rotate; perianth segments white inside, green with white margins outside; fruit a capsule. Low often shady ground, cultivated, escaped, and naturalized; Dallas and Grayson cos., also Collin Co. (R. O'Kennon, pers. obs.); Blackland Prairie and Edwards Plateau. Apr. Native of Europe. All parts of the plant, particularly the bulbs, are considered poisonous to humans and animals due to the presence of a digitalis-like alkaloid or cardiac glycosides; death in livestock has been reported (Muenscher 1951; Kingsbury 1964; Fuller & McClintock 1986; Blackwell 1990; Spoerke & Smolinske 1990).

POLYGONATUM SOLOMON'S-SEAL

A n temperate (especially sw China) genus of 55 species of herbs with robust horizontal rhizomes; species variously used as ornamentals, for food, or medicinally. *Polygonatum* is sometimes segregated with similar genera into the Convallariaceae (Dahlgren et al. 1985). The common name refers to the scar formed on the rhizome when the stem breaks off at the end of the growing season; it supposedly resembles the official seal of King Solomon (Ajilvsgi 1984). (Greek, *polys*, many, and *gonu*, knee, in reference to the many joints of the rhizome)
REFERENCES: Gates 1917; Ownbey 1944.

Polygonatum biflorum (Walter) Elliott, (two-flowered), GREAT SOLOMON'S-SEAL. Glabrous perennial from knotty rhizomes; stem erect-arching, to ca. 1 m tall; leaves borne along the stem, elliptic-lanceolate to broadly elliptic, to ca. 15 cm long and 7 cm wide; inflorescences axillary, 1–9-flowered; peduncles to ca. 5 cm long; pedicels 0–2 cm long; flowers perfect, pendulous; perianth 13–20 mm long, greenish white; fruit a blue-black berry. Rich, moist, wooded slopes; Dallas

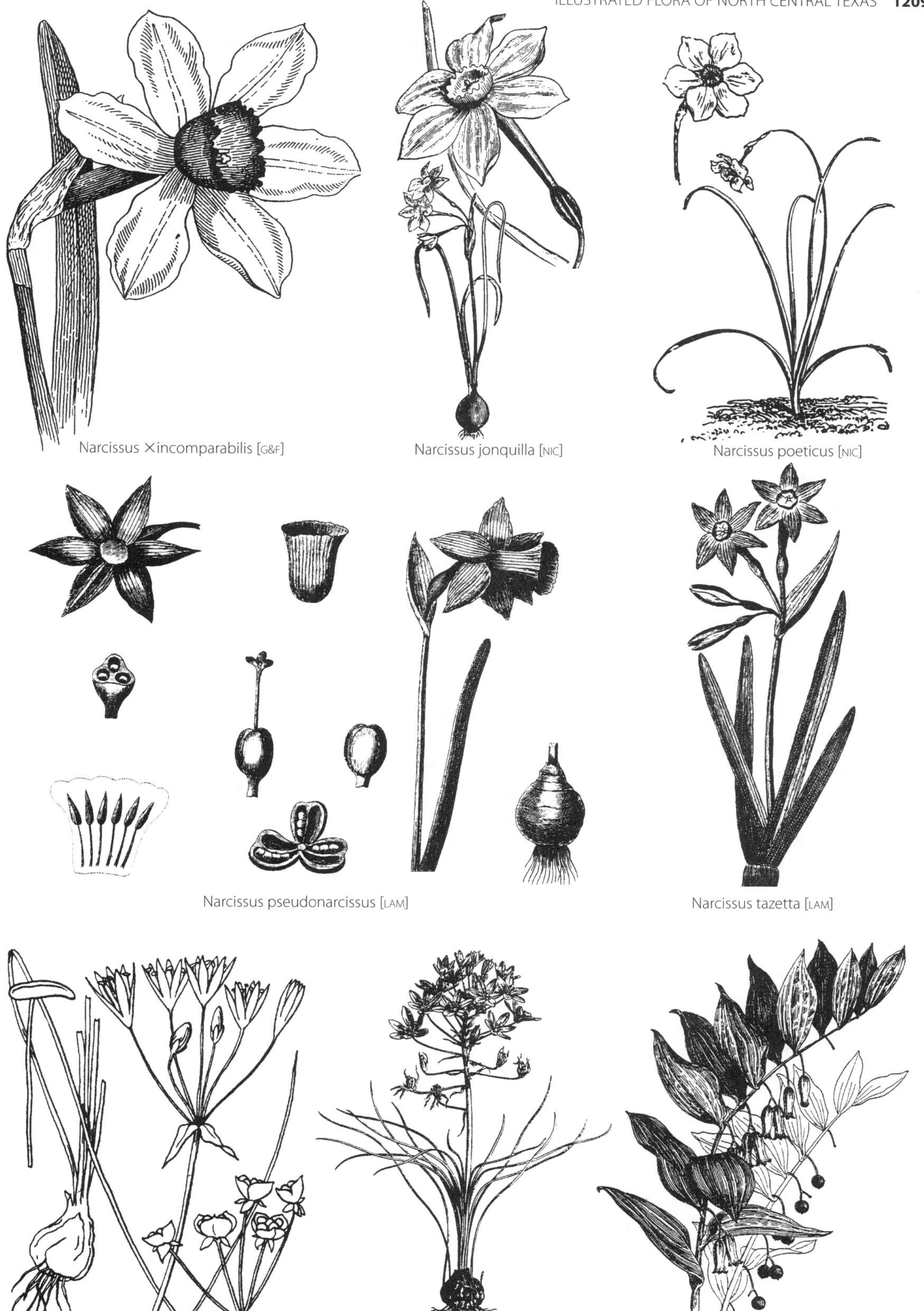

Narcissus ×incomparabilis [G&F]

Narcissus jonquilla [NIC]

Narcissus poeticus [NIC]

Narcissus pseudonarcissus [LAM]

Narcissus tazetta [LAM]

Nothoscordum bivalve [BB2]

Ornithogalum umbellatum [NIC]

Polygonatum biflorum [BAI]

(Spring Creek Preserve in Garland) and Grayson cos., also Lamar Co. (TOES 1993); mainly e TX. Mar–May. (TOES 1993: V) ⚠

SCHOENOCAULON SABADILLA

A genus of 10 species native from the s U.S. to Peru; ☠ some have alkaloids including veratrin; *S. officinale* A.Gray (SABADILLA, CEVADILLA) has seeds that are insecticidal and used in veterinary medicine. (Greek: *schoeno*, reed or rush-like, and *caulos*, a stalk)

Schoenocaulon texanum Scheele, (of Texas), TEXAS SABADILLA. Herbaceous perennials from bulbs; scapes naked, to 55 cm tall; leaves all basal, grass-like, to 60 cm long, mostly 4 mm or less wide; spikes 10–15 mm in diam.; flowers sessile or nearly so, each subtended by a small bract; perianth segments 6, essentially separate, fleshy-thickened or leathery, linear-oblong, with thickish entire margins, 3–4 mm long, ca. 1 mm or less wide, greenish white; stamens 6; filaments 3.5–5 mm long; capsules 10–15 mm long. Limestone soils, rocky grasslands and openings in juniper-oak woodlands; Bell Co., also Burnet and Williamson cos. (Balcones Canyonlands Nat. Wildlife Refuge, C. Sexton, pers. comm.); s margin of nc TX s through Edwards Plateau to Trans-Pecos. Mar–Jul.

Schoenocaulon drummondii A. Gray, (for Thomas Drummond, 1780–1835, Scottish botanist and collector in North America), DRUMMOND'S SABADILLA, GREEN-LILY, the only other member of the genus occuring in TX, is known mainly from the s 1/2 of the state. It can be distinguished by its submembranous, elliptic to ovate-elliptic perianth segments with thin erose margins, a larger spike (15–20 mm in diam.), and its phenology (usually flowering in autumn).

ZIGADENUS DEATH-CAMASS, POISON-SEGO

A genus of 18 species native to North America, the Urals, and e Asia; ☠ many are poisonous due to the presence of alkaloids; a few are cultivated as ornamentals. Previously sometimes spelled *Zygadenus*. (Greek: *zygos*, a yoke, and *aden*, a gland, referring to the glands sometimes being in pairs)

Zigadenus nuttallii (A. Gray) S. Watson, (for Sir Thomas Nuttall, 1786–1859, English-American botanist), NUTTALL'S DEATH-CAMASS, POISON-CAMASS, DEATH-CAMASS. Perennial from bulb, 30–100 cm tall; leaves crowded near base with short, tubular basal sheath (closed but very thin down side opposite blade, easily splitting); inflorescence a raceme or a panicle with a few short branches near base, these flower-bearing nearly to base; pedicels becoming 10–25 mm long; perianth segments 6–9 mm long, white, ovate, abruptly narrow and claw-like at base; fruit a 3-lobed capsule ca. 8–12 mm long. Prairies, open woods; e TX w to West Cross Timbers, also Edwards Plateau. Late Mar–early May. All parts of the plant may be fatally poisonous to cattle, sheep, and horses due to complex steroidal alkaloids (Sperry et al. 1955; Ajilvsgi 1984; Blackwell 1990); *Zigadenus* species have caused poisoning in children from eating the bulbs or chewing the flowers and deaths have been reported from eating the bulbs; toxins present include zygadenine, zygacine, and related alkaloids (Marsh et al. 1915; Schmutz & Hamilton 1979; Stephens 1980; Ajilvsgi 1984; Fuller & McClintock 1986). ☠

MARANTACEAE ARROWROOT FAMILY

A medium-sized (535 species in 29 genera) primarily tropical family of rhizomatous perennial herbs. The stamens are distinctive; all are sterile and modified into staminodes with the exception of one petal-like functional stamen. There are a number of ornamentals including *Calathea* and *Maranta*. Family name from *Maranta*, ARROWROOT or PRAYER-PLANT, a tropical American genus of ca. 20 species. *Maranta arundinacea* L., WEST INDIAN ARROWROOT, has a

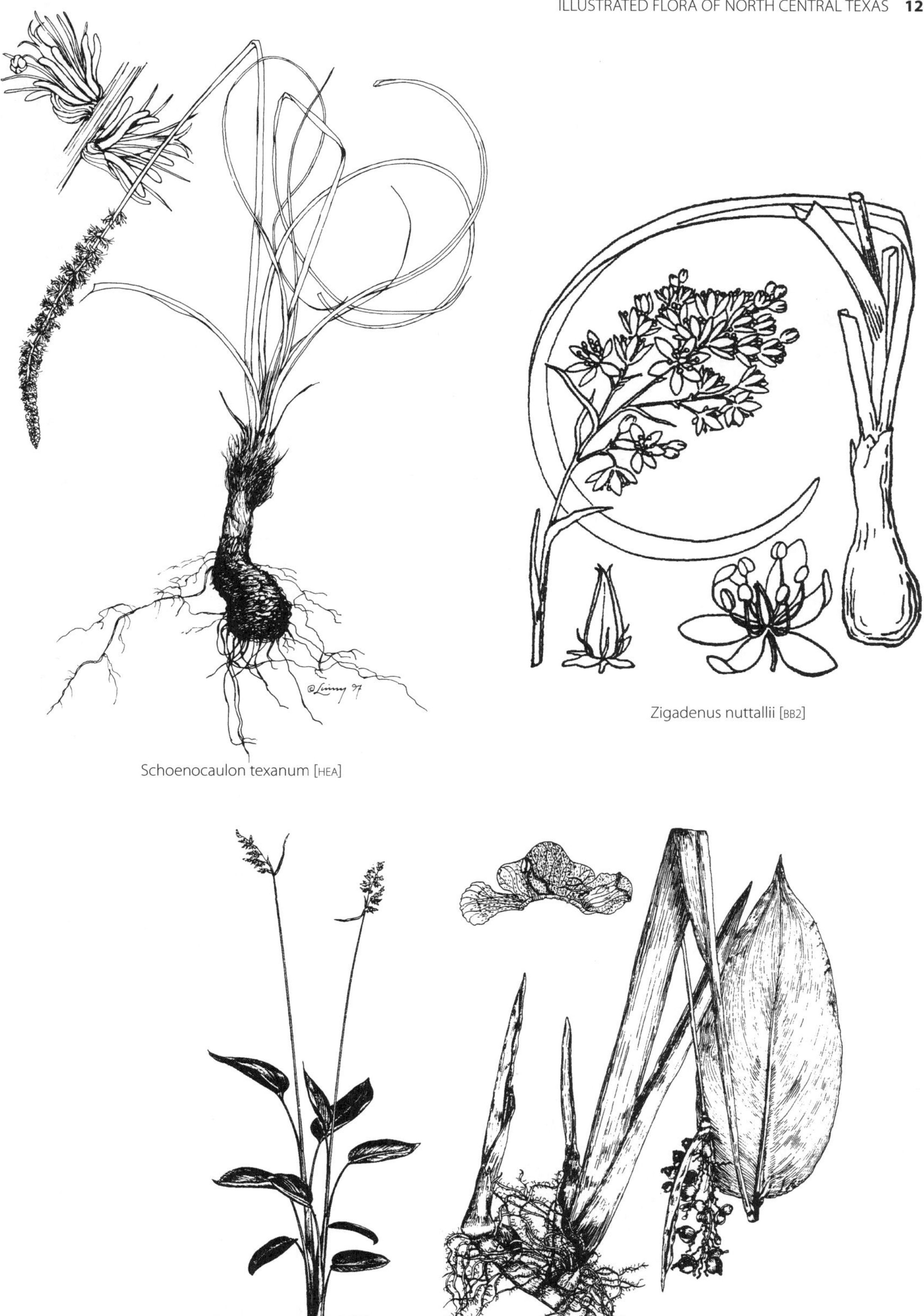

Schoenocaulon texanum [HEA]

Zigadenus nuttallii [BB2]

Thalia dealbata [CO1, NIC]

starchy edible rhizome, the starch being easily digested and used for infants and invalids. (Named for Bartolommeo Maranti, 16th century Venetian botanist) (subclass Zingiberidae)
FAMILY RECOGNITION IN THE FIELD: the only nc TX species is a large rhizomatous herb with broad, *banana-like* leaves with a joint where the long petiole and blade join; inflorescences usually *white-powdery*; flowers purple; androecium of a single fertile stamen and often staminoidea.
REFERENCES: Rogers 1984; Dahlgren et al. 1985.

THALIA

A mostly tropical American (1 in Africa) genus of 7 species. (Named for Johann Thal, a German physician and naturalist who died in 1583)

Thalia dealbata Fraser ex Roscoe, (white-washed), POWDERY THALIA, POWDERED THALIA. Perennial herb 1–2 m tall from strong rhizomes, glabrous and with whitened surfaces; leaves alternate, long-petioled, mostly in basal half of plant, large, banana-like; leaf blades 30–50 cm long, to 20 cm wide, ovate-lanceolate, with prominent midrib and numerous side veins parallel or concentric with each other; petioles to 80 cm long, the base winged and clasping; inflorescence with whitish appearance, the axes zigzag in appearance, a rather small, loose panicle with a prominent basal bract paired with a small one, the bracts typically with conspicuously whitened surfaces; flowers small, usually paired and enclosed by two bracts, sessile; sepals 3, short; petals 3, unequal, purple, 5–10 mm long; staminodes present, petal-like, purple, united, one with cupped, projecting apex exceeding the petals; pistil 1; ovary inferior; fruit a bluish purple, subglobose urticle 10–15 mm in diam. Shallow water, ditches, pond margins, or other wet areas; Dallas Co., also escaping cultivation in Tarrant Co. (Fort Worth Botanic Garden); mainly se and e TX. May.

ORCHIDACEAE ORCHID FAMILY

Ours perennial terrestrial herbs with fleshy-fibrous roots, some species rhizomatous or with swollen, bulbous base, autotrophic and with green leaves or saprophytic and lacking chlorophyll; leaves when present basal or alternate, with sheath or clasping petiolar base; leaf blades with parallel or parallel convergent veins, or ribbed and net-veined, or rather fleshy and not evidently veined; flowers solitary or in spikes or racemes, bilaterally symmetrical; sepals 3, green or colored and petal-like; petals 3, one (the lip or labellum, normally the lowest due to twisting of the flower) slightly or very different from the others in shape or size and often elaborated into lobes, spurs, sacs, fringes, or grooves; nectaries usually present; stamens and pistil united into a central knob-like or column-like structure (the column) with 1 (or 2 in some species in other regions) inconspicuous sessile anther; pollen usually in waxy masses (= pollinia); ovary inferior, in most species twisted and the flower thus resupinate (= inverted 180°, making the lip the lowermost of the petals); placentation parietal; fruit (in ours) a dry dehiscent capsule; seeds very numerous, minute, without endosperm.

This is a huge cosmopolitan group with more species than any other monocot family; estimates range from ca. 17,500–30,000 species (Luer 1975) or even more in 796 genera; Dressler (1981) estimated 19,192 species and Atwood (1986), in an analysis of the size of the family, determined there were 19,128 species and suggested that eventually the total may reach 20,000-23,000 or even an improbable maximum of 25,000. The Orchidaceae is possibly larger even than the Asteraceae and thus possibly the largest family of flowering plants. Orchids are particularly abundant in the tropics where many species are epiphytic; Atwood (1986) indicated epiphytes account for 73% of the family; many have special swollen water and nutrient storage structures called pseudobulbs—a particularly important adaptation for the epiphytic habit.

The family is extremely important horticulturally for its beautiful and intricate flowers; orchid cultivation is very old, dating back nearly 1,000 years in China; well known cultivated genera include *Cattleya* (CORSAGE ORCHID), *Dendrobium*, and *Epidendrum*. In addition to hybridization, tissue culture and other sophisticated techniques are now used in orchid propagation and cultivation. Pollination mechanisms are often incredibly specialized (e.g., pseudocopulation in which the flower mimics a female insect and thus attracts males who carry out pollination); some species even imprison and intoxicate their pollinators. Special relationships with mycorrhizal fungi are often necessary for seed germination. The fruit of *Vanilla planifolia* Jackson, a tropical American spice used by the Aztecs, is the source of the flavoring vanilla. Many orchids have a low tolerance for environmental disturbance and many species are now greatly reduced in number or extinct due to habitat destruction and modification. Family name from *Orchis*, a genus of 33 species native from the n temperate zone to sw China and India. (Greek: *orchis*, testicle, in reference to the shape of the tuberous roots of some species) (subclass Liliidae)

FAMILY RECOGNITION IN THE FIELD: perennial herbs with alternate or basal leaves; reproductive parts united into a central *column*; flowers bilaterally symmetrical, with 3 sepals and 3 petals, 1 of the petals (the *lip*) usually being different from the other 2; fruit a capsule with very numerous nearly microscopic seeds.

REFERENCES: Correll 1937, 1947, 1950, 1961a; Schultes & Pease 1963; Magrath 1973; Luer 1975; Dressler 1981; Rasmussen 1985; Atwood 1986; Burns-Balogh & Funk 1986.

1. Stems with 1 or more leaf blades; plants green and photosynthetic.
   2. Flowers greenish or uniformly orange, neither white nor with any pinkish, reddish, or purplish pigmentation, with a slender spur ca. as long as or longer than the ovary (spur 9–33 mm long).
      3. Flowers greenish; spur 9–14 mm long; lip divided into 3 linear divisions, not fringed ________ **Habenaria**
      3. Flowers orange; spur 20–33 mm long; lip conspicuously fringed ________ **Platanthera**
   2. Flowers either white to white with green or yellow markings OR if not white then variously colored with at least some pinkish to reddish or purplish pigmentation (not orange), without a spur.
      4. Stems with 6–12 leaves below inflorescence, the leaves passing into leafy bracts under the flowers; leaves large, 6–20 cm long, 2–7 cm wide, not at all grass-like; flowers greenish to brownish, yellowish, or pinkish, with purplish or reddish markings ________ **Epipactis**
      4. Stems with 1–6 leaves below inflorescence, the floral bracts if present abruptly differentiated; leaves large OR small and often grass-like; flowers white to white with green or yellow markings OR pink to rose-purple.
         5. Stems with a single leaf inserted ca. halfway up the stem (floral bract also present); lip heavily bearded; flowers 1(–2), not in a conspicuous spiral; lip the lowermost of the petals ________ **Pogonia**
         5. Stems with 1–6 leaves, if 1 then the leaf ± basal; lip bearded or glabrous; flowers 2–numerous; lip either the uppermost of the petals and the flowers not in a conspicuous spiral OR lip the lowermost of the petals and flowers usually in a conspicuous spiral.
            6. Flowers not in a conspicuous easily visible spiral; perianth 7–40 mm broad, pink to rose-pink or rose-purple or rarely white; lip bearded, representing the uppermost of the petals ________ **Calopogon**
            6. Flowers usually arranged in a conspicuous easily visible spiral; perianth 1.5–4 mm broad, white or white with green or yellow markings; lip not bearded, representing the lower most of the petals ________ **Spiranthes**
1. Stems without leaf blades, with sheaths or clasping bracts only; plants green and photosynthetic OR not green (yellowish to reddish, brownish, or purplish) and saprophytic.
   7. Plants green, photosynthetic; flowers sessile; perianth white or white with green or yellow marking; floral bracts slightly longer than the ovary ________ **Spiranthes**

7. Plants not green, saprophytic, the stems usually yellowish to reddish, brownish, or purplish; flowers short-pedicelled; perianth mottled purple, brown, yellow, and white; floral bracts not longer than the ovary.
   8. Stem bracts elliptic-ovate, shorter to slightly longer than broad; floral bracts easily observed, ± conspicuous; perianth either < 10 mm long OR much longer (15–20 mm long); lip with 5–7 longitudinal ridges or crests; rhizomes with annular markings, not coral-like ______ **Hexalectris**
   8. Stem with long, tubular sheaths much longer than broad; floral bracts minute, inconspicuous; perianth < 10 mm long; lip without ridges or crests; rhizomes coral-like, lacking annular markings ______ **Corallorrhiza**

CALOPOGON GRASS-PINK

Erect, scapose herbs from corms; leaf 1 (rarely more), grass-like, sheathing the stem near base; racemes with 2–20 flowers; flower buds waxy; flowers showy, pink to rose-pink or rose-purple, rarely white, not resupinate (the lip therefore the uppermost of the petals); lip bearded with numerous clavellate hairs; pollinia 4; capsules erect.

A North American genus of 5 species (1 described in 1995); sometimes cultivated as ornamentals. (Greek: *calos*, beautiful, and *pogon*, beard, from the colorful tuft of bristles on the lip which is characteristic of the genus)
REFERENCES: Magrath 1989; Goldman 1995.

1. Corm forked; leaf nearly as long as or longer than inflorescence; buds grooved longitudinally; flowers fragrant ______ **C. oklahomensis**
1. Corm spherical, not forked; leaf usually shorter than inflorescence; buds usually smooth; flowers not fragrant.
   2. Flowers usually more than 8 (varying from 4–20), opening in slow succession up the raceme, the blooming period extending over a prolonged period; leaf usually > 5 mm wide (varying from 4–37 mm); column usually 10–20 mm long ______ **C. tuberosus**
   2. Flowers usually 2–5, all opening nearly simultaneously; leaf usually ca. 2 mm wide (varying from 1–4 mm); column 7–8 mm long ______ **C. barbatus**

Calopogon barbatus (Walter) Ames, (barbed), BEARDED GRASS-PINK. Plant 15–30(–45) cm tall; leaf 5–18 cm long; pedicels 6–10 mm long; flowers very closely spaced; lateral sepals grooved longitudinally, recurved; sepals and petals 12–17 mm long; lip 10–13 mm long, 7–10 mm wide; disk of lip pink, the same color as most of flower; stigma flat against column surface. Bogs, marshes, other wet areas; Henderson Co. (Correll & Johnston 1970) near e margin of nc TX; mainly e TX. Apr–May. Goldman (1995) indicated that such TX specimens previously identified as *C. barbatus* are actually small individuals of *C. oklahomensis*; it is therefore probable that *C. barbatus* is not a component of the nc TX flora or even the TX flora.

Calopogon oklahomensis D.H. Goldman, (of Oklahoma). Plant to 35 cm tall; leaf to 35 cm long, 5–15 mm wide; flowers 2–7(–11), opening in rapid succession, with citronella odor; lateral sepals grooved longitudinally, recurved; petals 11–20 mm long; lip 9–15 mm long; disk of lip pinkish, the same color as most of flower, with a basal area of yellow hairs above which is an area of pinkish hairs; stigma flat against column surface. Mesic to damp, acidic, sandy or loamy soils in oak woods, savannas, or bog margins; Lamar Co. (Goldman 1995—specimen at TEX); mainly e TX; one of the most recently described species in nc TX (Goldman 1995). Mar–May. (TOES 1993: IV) ⚠

Calopogon tuberosus (L.) Britton, Sterns, & Poggenb., (tuberous), GRASS-PINK. Plant 20–80(–135) cm tall; leaf to 50+ cm long, 5–50 mm wide; lateral sepals smooth, straight; sepals and petals 12–27 mm long; lip 10–20 mm long, 5.5–13 mm wide; disk of lip white with basal area of white to yellow hairs above which is a region of white, yellow, or orange hairs; stigma usually perpendicular to column surface. Bogs, marshes, other wet areas; Lamar Co. (Correll 1961a) in Red River drainage; mainly se and e TX. May–Jun. [*C. pulchellus* R. Br. ex W.T. Aiton]

CORALLORRHIZA CORALROOT

A n temperate genus of 15 species of saprophytes lacking chlorophyll; some are cultivated as ornamentals. (Greek: *corallion*, coral, and *rhiza*, root, referring to the coral-like rhizome)
REFERENCES: Engel 1997; Freudenstein 1997.

Corallorrhiza wisteriana Conrad, (for its discoverer, Charles Jones Wister, 1782–1865), SPRING CORALROOT, WISTER'S CORALROOT. Plant 10–40 cm tall, from irregular rhizomes; stems yellowish to reddish brown or purple, not bulbous-based; flowers in a small, spike-like raceme; perianth brownish, greenish yellow, or reddish brown with purple dots; lip white with purple to pink markings, 5–6 mm long, with notched apex and entire or denticulate margin; pollinia 4. In leaf mold on limestone hills or on sandy soils; Dallas and Grayson (Hagerman Nat. Wildlife Refuge) cos., also Coryell (Fort Hood—Sanchez 1997) and Fannin (Freudenstein 1997) cos.; mainly se and e TX; also Edwards Plateau. Apr.

Corallorrhiza odontorhiza (Willd.) Nutt., (tooth-rooted), AUTUMN CORALROOT, LATE CORALROOT, has been collected in n Red River Co. (*Magrath 12302*, OCLA) just to the e of nc TX; it differs in having the stems bulbous-based, the lip only 4 mm long with rounded apex and eroded margin, and flowering in the fall.

EPIPACTIS HELLEBORINE

A genus of 22 species ranging from the n temperate zone to the tropics. (Greek: *epipactis*, an ancient name applied by Theophrastus, ca. 350 B.C., to a plant used to curdle milk, possibly hellebore—Luer 1975)

Epipactis gigantea Douglas ex Hook., (gigantic), GIANT HELLEBORINE, STREAM-ORCHIS, STREAM EPIPACTIS, CHATTERBOX. Plant 20–80 cm tall, rhizomatous; leaves clasping stem, broadly elliptic to linear-lanceolate, 6–22 cm long, 2–7 cm wide; inflorescences 2–8-flowered, prominently bracteate; flowers greenish to brownish, yellowish, or pinkish, with purplish or reddish markings; sepals 15–25 mm long; capsules pendent, 2–2.5 cm long. Stream banks in woods; Dallas and Wise cos. (Correll 1961a); also Edwards Plateau and Trans-Pecos. Jun.

HABENARIA FINGER ORCHID

A pantropical and subtropical genus of ca. 600 species depending on circumscription; a number of segregates are sometimes split out (e.g., many species previously placed in *Habenaria* are now treated in *Platanthera*). (Latin: *habena*, reins, thong, or strap, in allusion to the shape of the long, strap-like divisions of the lip of some species)
REFERENCE: Ames 1910.

Habenaria repens Nutt., (creeping), WATER SPIDER ORCHID, CREEPING ORCHID, NUTTALL'S HABENARIA. Plant glabrous, 10–90 cm tall; stems leafy; leaves 3-ribbed, 5–24 cm long, 3.5–20 mm wide; racemes usually densely many-flowered; flowers greenish, small; sepals 3–7 mm long; lip divided into three narrow parts to near lobes, the middle lobe 4–7 mm long, linear, not fringed, the spur 9–14 mm long. Pond margins, ditches, other wet areas, often in water; Milam Co. (Correll 1961a) near extreme e margin of nc TX; Hatch et al. (1990) also cited vegetational area 4 (Fig. 2); mainly se and e TX. May–Nov.

HEXALECTRIS CRESTED-CORALROOT

Plants saprophytic, leafless; rhizomes irregular; stems reddish or purplish; flowers in a spike-like raceme; corolla lip with 5–7 longitudinal crests down the middle of the disc; pollinia 8.

A genus of 7 species of the s U.S. and Mexico. The plants live in symbiosis with a mycorrhizal fungus (Luer 1975). Luer (1975) has excellent color photographs of all three nc TX species. (Greek: *hex*, six, and *alectryon*, cock, referring to the 5–7 crests resembling a cock's comb on the lip) REFERENCES: Luer 1975; Catling & Engel 1993; Engel 1997.

1. Lip with 5–7 longitudinal, purple, non-wavy crests, shallowly 3-lobed, the fissures < 2 mm deep; petals (non-lip) either 11 mm or less long OR larger and yellowish or purplish brown, with brown or purple veins; racemes with up to 25 flowers.
   2. Sepals, petals, and lip 12–20 mm long; petals (non-lip) yellowish or purplish brown, with brown or purple veins; middle lobe of lip usually yellow to white with purple striations ____ **H. spicata**
   2. Sepals, petals, and lip 11 mm or less long; petals (non-lip) pinkish brown; middle lobe of lip bright purple ____ **H. nitida**
1. Lip with 5 conspicuous, longitudinal, yellow, wavy crests, deeply 3-lobed, the fissures > 3 mm deep; petals (non-lip) 15–20 mm long, maroon or deep purple; racemes with 10 flowers or less ____ **H. warnockii**

Hexalectris nitida L.O. Williams, (shining), SHINING HEXALECTRIS. Plant 15–30 cm tall; racemes with up to 20 flowers; flower parts with shiny or polished appearance; middle lobe of lip bright purple, the lateral lobes white. Under juniper trees; Dallas Co.; also known s and w of nc TX from Brewster, Bandera, Kendall, and Taylor cos. (Mahler 1988). Jul. /92

Hexalectris spicata (Walter) Barnhart, (with spikes), CRESTED-CORALROOT, BRUNETTA, BUFF-CREST, COCK'S-COMB, LEAFLESS ORCHID. Plant 15–80 cm tall; racemes with 5–25 flowers; middle lobe of lip usually yellow to white with purple striations; occasionally plants lack purple pigment—the sepals and petals (non-lip) are then mahogany brown with darker brown striations, and the lip is all white except for faint yellowish markings (Luer 1975). While we are following Catling and Engel (1993) in recognizing two varieties, these taxa might better be recognized as forms. The key to varieties is modified from Catling and Engel (1993).

1. Flowers usually not opening (typically auto-pollinating, but sometimes opening); the 5 central veins of the lip with their highest keels raised 0.4–0.7 mm above the lip surface; column without a rostellar flap separating the pollen masses from the stigmatic surface; flowering in nc TX in Jun–Jul ____ var. **arizonica**
1. Flowers opening, the sepals and petals often with recurving tips; the 5 central veins of the lip with their highest keels raised (0.4–)0.7–1 mm above the lip surface; column with a rostellar flap separating the pollen masses from the stigmatic surface; flowering in TX in May–early Jun ____ var. **spicata**

var. **arizonica** (S. Watson) Catling & V.S. Engel, (of Arizona). In rotting wood or leaf litter in oak, pine, or juniper woods over limestone; Palo Pinto Co., also Dallas and Tarrant cos. (Engel 1997); in TX also known from Anderson Co. in e TX, Travis Co. just s of nc TX, and Brewster and Culberson cos. in far w TX. Jun–Jul. Catling and Engel (1993) indicated that most plants seen in the Dallas area and all in Travis Co. have closed flowers that do not open and are apparently auto-pollinating; they suggested this taxon is an auto-pollinating race derived from var. *spicata*; alternatively, they suggested that var. *arizonica* might have a hybrid origin, possibly resulting from a cross between *H. nitida* and *H. spicata* var. *spicata*. [*Corallorhiza arizonica* S. Watson]

var. **spicata**. Oak, hickory, or conifer woods, calcareous sandy or organic soils; Dallas Co., also Bell, Brown, and Coryell cos. (HPC); according to Catling and Engel (1993), this species occurs at the same site (Dallas Nature Center) as *H. spicata* var. *arizonica* and *H. nitida*; TX distribution unclear. May–early Jun. According to J. Stanford (pers. comm.), these delicate-appearing plants have massive underground rhizomes (ca. 10 cm in diam.).

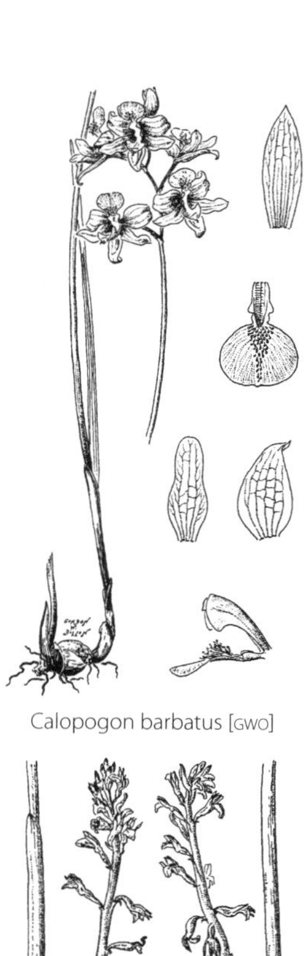
Calopogon barbatus [GWO]

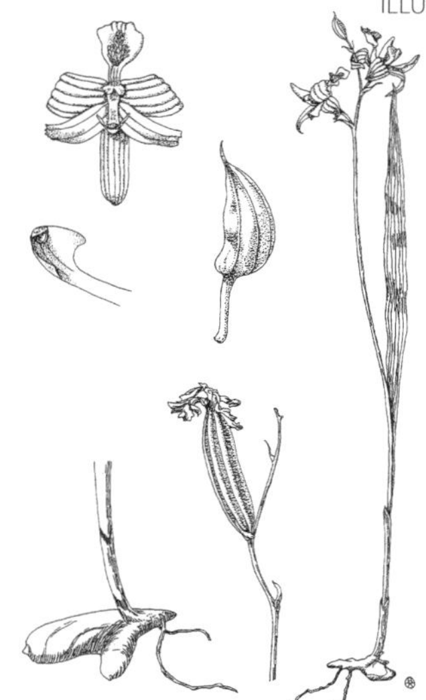
Calopogon oklahomensis [LIN]

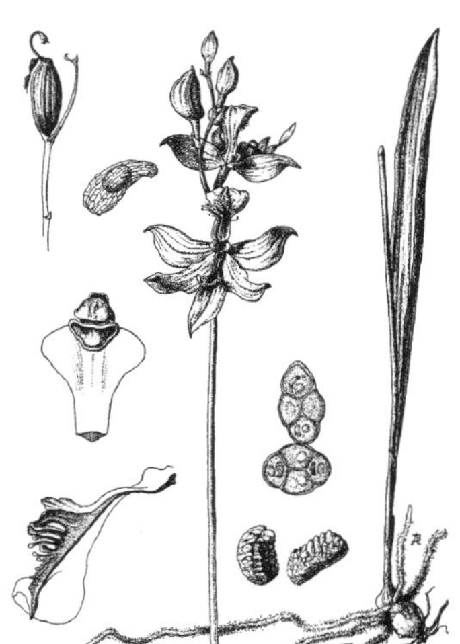
Calapogon tuberosus [LUN]

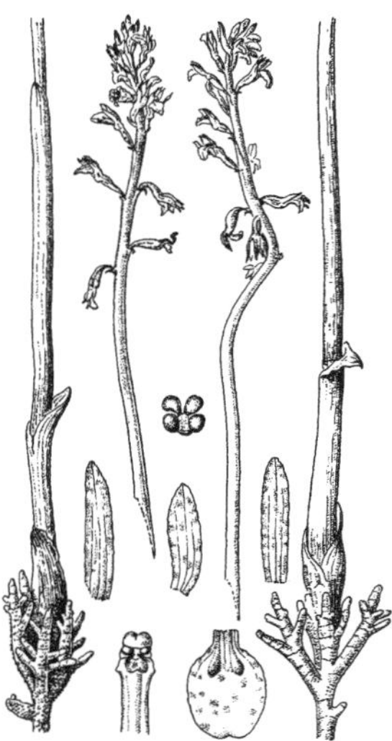
Corallorrhiza wisteriana [LUN]

Epipactis gigantea [LUN]

Habenaria repens [LUN]

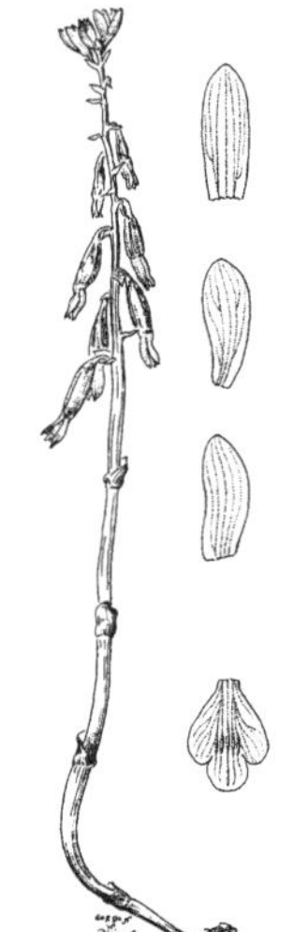
Hexalectris nitida [CO2]

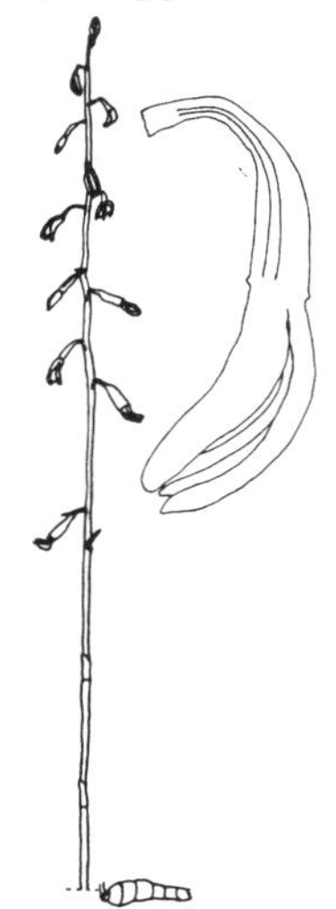
Hexalectris spicata var. arizonica [LIN]

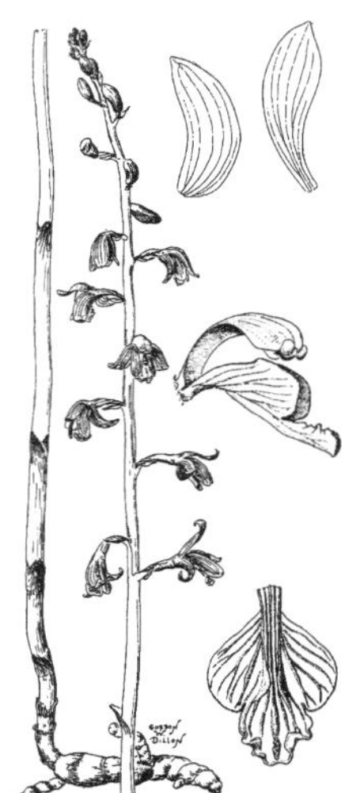
Hexalectris spicata var. spicata [LUN]

Hexalectris warnockii Ames & Correll, (for Barton Holland Warnock, 1911–1998, Trans-Pecos Texas botanist), TEXAS PURPLE-SPIKE. Plant to ca. 30 cm tall; inflorescences with up to 10 flowers; sepals, petals, and lip ca. 15–20 mm long; middle lobe of lip edged in purple or white, with 5 conspicuous yellow crests, 3 of which reach the apex; lateral lobes of lip pale pink with purple veins. Under juniper trees; Dallas Co. (Mahler 1988; Engel 1997); also Edwards Plateau and Trans-Pecos. Jul. ▣/92

PLATANTHERA

FINGER ORCHID, BUTTERFLY ORCHID, FRINGED ORCHID

A northern hemisphere genus of 40 species. In the past this genus has often been lumped with *Habenaria*. Many other *Platanthera* species occur in e TX. (Greek: *platys*, broad or wide, and *anthera*, anther, in reference to the unusually wide anther)
REFERENCE: Ames 1910.

Platanthera ciliaris (L.) Lindl., (ciliate, fringed), YELLOW FRINGED ORCHID, YELLOW FINGER ORCHID. Plant 24–100 cm tall; leaves oblong lanceolate to lanceolate, 7–30 cm long, 0.6–6 cm wide; racemes densely or laxly flowered; flowers very showy, bright to deep orange; spur slender, longer than the ovary; lip 8–12 mm long, 2–3 mm wide, ciliate-fringed, the fringes usually 10 mm or more long. Low woods, near edge of water, moist areas; Henderson and Milam cos. (Correll 1961a) near extreme e margin of nc TX; mainly e TX. Jun–Oct. [*Habenaria ciliaris* (L.) R. Br. ex W.T. Aiton] ▣/103

POGONIA SNAKE-MOUTH

A genus of 2 species of North America and temperate Asia. (Greek: *pogonias*, bearded, referring to the bearded lip)
REFERENCE: Teuscher 1978

Pogonia ophioglossoides (L.) Ker Gawl., (resembling *Ophioglossum*–adder's-tongue fern, in reference to the single leaf), ROSE POGONIA. Plant 20–40 cm tall, from fibrous roots; stems rigidly erect; leaf usually 1, ovate to ovate-lanceolate, 2–12 cm long, 1–3 cm wide; flowers 1(–2) at tip of stem, rose-pink to white, occasionally fragrant, 15–25 mm long, resupinate (the lip therefore the lowermost of the petals); lip heavily yellow-white bearded; capsule 2–3 cm long. Marshy areas, seepage slopes, other wet areas; included based on citation of vegetational area 4 (Fig. 2) by Hatch et al. (1990); se and e TX w to at least c Henderson Co. near e margin of nc TX. Apr–Jul.

SPIRANTHES LADIES'-TRESSES

Roots fleshy, tuberous-thickened; stems with few leaves or none; flowers in a slender twisted or spiraled spike; perianth narrow, almost tubular, white or partly green or yellowish.

A mainly n temperate (few tropical) genus of 30 species; some are cultivated as ornamentals. *Spiranthes* are by far the most common orchids in nc TX. (Greek: *speira*, a coil or spiral, and *anthos*, flower, from the spiraled inflorescence)
REFERENCES: Garay 1980; Catling 1981, 1982, 1983a, 1983b; Sheviak 1982.

1. Axis of spike pubescent (use lens); leaves cauline and/or basal.
   2. Flowers in 1 secund (= 1-sided) or loose spiral; pubescence blunt to capitate (= end enlarged) or pointed.
      3. Lip usually widest near base, often fleshy-thickened; widespread in nc TX.
         4. Lip white with a yellow-green center; pubescence blunt to capitate ____ **S. cernua**
         4. Lip white; pubescence pointed ____ **S. vernalis**

Hexalectris warnockii [LUN]

Pogonia ophioglossoides [LUN]

Platanthera ciliaris [LUN]

Spiranthes cernua [LUN]

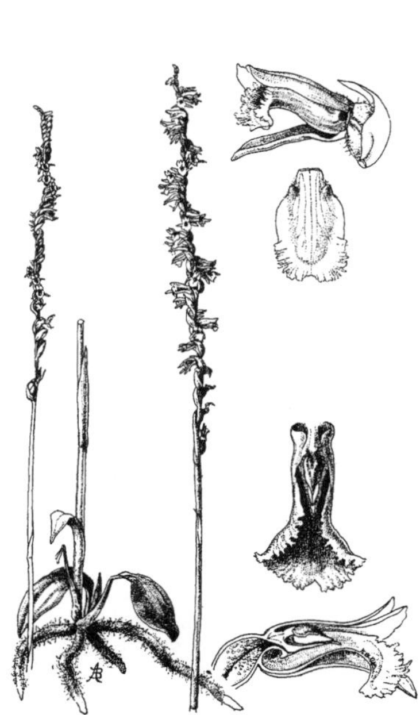

Spiranthes lacera var. gracilis [LUN]

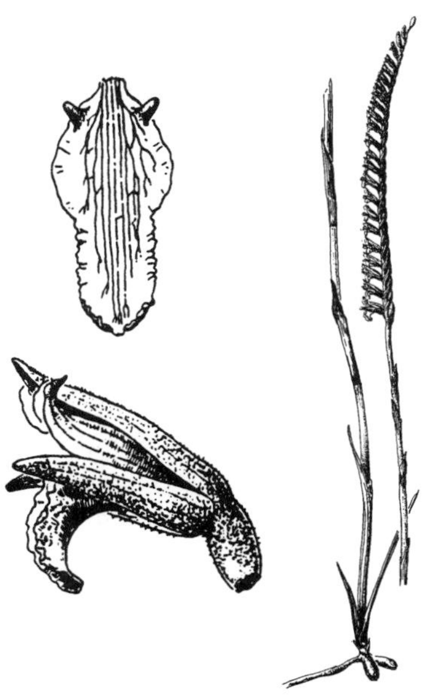

Spiranthes longilabris [CO1]

3. Lip usually widest near apex or lateral margins parallel, thin, membranous; rare near e margin of nc TX ____ **S. praecox**

2. Flowers in 2–4 dense spirals; pubescence blunt to capitate.

5. Perianth 5.5 mm or less long; plants delicate, slender; spike 15 mm or less in diam.; rare in nc TX ____ **S. ovalis**

5. Perianth > 6 mm long; plants usually large, stout (for a *Spiranthes*); spike usually much > 15 mm in diam.; widespread in nc TX.

6. Lip not constricted near the middle, with small basal tuberosities (± short conical knobs); leaves usually absent at flowering time; flowers slender, not inflated; lateral sepals spreading; typically in calcareous soils ____ **S. magnicamporum**

6. Lip slightly to distinctly constricted near the middle, with prominent basal tuberosities (usually somewhat inward-curving); leaves present at flowering time; flowers appearing inflated; lateral sepals appressed; typically in sandy soils ____ **S. cernua**

1. Axis of spike glabrous; leaves all basal.

7. Perianth 5.5–11 mm long; lip 2–6 mm wide; leaves (if present) mostly 5 mm or less wide, erect or ascending, without a petiole, the lower portion sheathing the stem, linear to narrowly lanceolate or oblong-elliptic; flowering Mar–Jun or Oct–Dec.; rare in nc TX.

8. Inflorescences 1-sided (rarely slightly spiraled); lip tapering from broad base to narrow obtuse apex, not marked or veined with green; flowering Oct–Dec ____ **S. longilabris**

8. Inflorescences strongly spiraled; lip usually widest near apex, marked or veined with green; flowering Mar–Jun ____ **S. praecox**

7. Perianth 2–5.5(–7) mm long; lip 1.5–2.5 mm wide; leaves (if present) 6–25 mm wide, spreading, the ± short broad blades with a distinct petiole, ovate to ovate-lanceolate; flowering Jun–Nov; widespread in nc TX.

9. Perianth 4–5.5(–6) mm long; lip with green or yellowish green center; roots several, in a fascicle ____ **S. lacera**

9. Perianth up to 3(–4) mm long; lip white; root 1(–2) ____ **S. tuberosa**

Spiranthes cernua (L.) Rich., (drooping, nodding), NODDING LADIES'-TRESSES, COMMON LADIES'-TRESSES, LADIES'-TRESSES. Plant glabrous below, pubescent above, to 60 cm tall; leaves mostly basal, linear to lanceolate, to 25 cm long and 2.5 cm wide; spike 1.5–2 cm thick; flowers sometimes fragrant; perianth 10–15 mm long; sepals and petals white or nearly so; lip thick, white with yellow-green center, apically recurved and undulate or crenulate; column 3–7 mm long. Prairies, open woodlands, sandy soils; se and e TX w to Rolling Plains and Edwards Plateau. $2n = 60$ (Sheviak 1982). Jul–Nov.

Spiranthes lacera (Raf.) Raf. var. **gracilis** (Bigelow) Luer, (sp.: torn; var.: graceful), SLENDER LADIES'-TRESSES, GREEN-LIP LADIES'-TRESSES. Plant essentially glabrous, 20–60 cm tall; leaves all basal, not always persisting until flowering, short petioled; leaf blades ovate, 1.5–6.5 cm long, 10–25 mm wide; flowers in 1 spiral row; lip white marked with broad green or yellowish green stripe in center, 4–5(–6) mm long, 2–2.5 mm wide, the apex finely lacerate. Sandy woods, prairies; Bell, Denton, and Grayson cos., also Dallas and Kaufman cos. (Correll 1961a); se and e TX w to East Cross Timbers. Jun–Nov. [*S. gracilis* (Bigelow) L.C. Beck] ▣/107

Spiranthes longilabris Lindl., (long-lipped), GIANT SPIRAL ORCHID. Plant ± glabrous throughout, 12–60 cm tall; leaves when present basal, 3–10 cm long, mostly < 5 mm wide; inflorescences with flowers projecting horizontally; flowers white or white tinged with cream, conspicuously open; perianth 6–10 mm long; lip 3–5.5 mm wide near base (which is widest portion), recurved apically, crenate. Low woods, wet open areas; included based on citation of vegetational area 4 (Fig. 2) by Hatch et al. (1990), no other vegetational areas cited for TX; rare if present in nc TX. Oct–Dec.

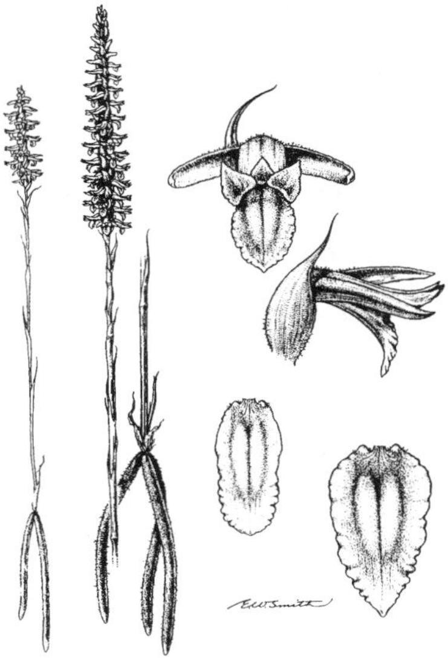

Spiranthes magnicamporum [BML]

Spiranthes ovalis [LUN]

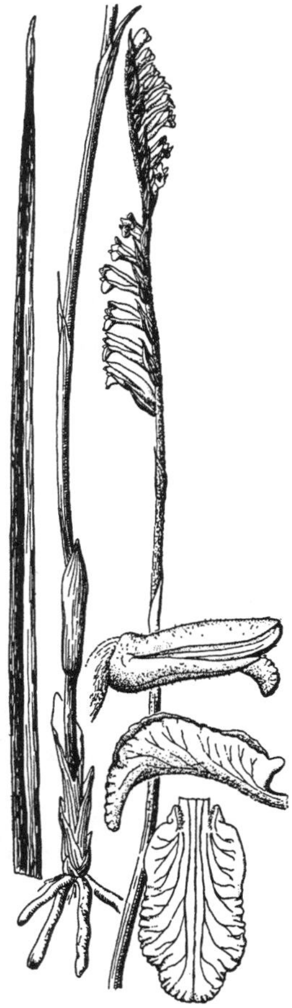

Spiranthes praecox [LUN]

Spiranthes tuberosa [LUN]

Spiranthes vernalis [LUN]

Spiranthes magnicamporum Sheviak, (great field), GREAT PLAINS LADIES'-TRESSES. Plant pubescent, to 60 cm tall; leaves linear-lanceolate, to 14 cm long and ca. 1 cm wide; spike pubescent; flowers fragrant often with strong odor similar to vanilla; perianth 7–11 mm long; lip white with center yellowish and fleshy, slightly crisped apically; column 3 mm long. Upland, calcareous soils, prairies; Grayson, Montague, Parker, and Tarrant cos., also Cooke Co. (H. Garnett, pers. comm.); the only TX citations given by Hatch et al. (1990) were for vegetational areas 4 and 5 (Fig 2). Sep–Nov. $2n = 30$ (Sheviak 1982). Flowers appear slender with spreading sepals as compared to *S. cernua* with inflated tubular flowers and appressed sepals. According to L. Magrath (pers. comm.), *S. magnicamporum* and *S. cernua* may not be specifically distinct; rather they may occupy different ends of a continuum. Sheviak (1982) indicated that while introgression occurs between the two, they differ in chromosome number, are partially genetically isolated, and have different ecological requirements; he recognizes them at the specific level.

Spiranthes ovalis Lindl., (oval), OCTOBER LADIES'-TRESSES, OVAL LADIES'-TRESSES. Plant glabrous below, pubescent above, ± delicate, to only 35(–45) cm tall; leaves 2–4, basal or low on stem, 5–27 cm long, 6–15 mm wide; spike with 2 or 3 tight spirals; flowers small; perianth 3–5.5 mm long, white; lip 4–5.3 mm long, 2.4–4 mm wide, recurved apically, wavy-crenate. Wooded areas; collected by Reverchon in Dallas Co. (Correll 1961a); apparently not collected in TX since that time. Aug–Oct. This is the only small-flowered species of *Spiranthes* in nc TX with the flowers in 2 or more spirals.

Spiranthes praecox (Walter) S. Watson, (precocious), GRASS-LEAF LADIES'-TRESSES. Plant often glabrous throughout (sometimes axis of inflorescence pubescent), 20–75 cm tall; leaves when present up to 7, mostly basal, narrowly linear to filiform, 10–25 cm long, 1–5 mm wide; perianth 5.5–11 mm long, white, usually marked with green; lip thin, 5.5–11 mm long, 2–6 mm wide near apex (which is widest portion), apically mostly wavy and slightly crenate. Low woods, wet open areas; included based on citation of vegetational area 4 (Fig. 2) by Hatch et al. (1990); mainly se and e TX. Mar–Jun.

Spiranthes tuberosa Raf., (tuberous), LITTLE LADIES'-TRESSES, GRAY'S LADIES'-TRESSES. Plant 20–40(–60) cm tall; leaves all basal, ovate, 2.5–6.5 cm long, 6–15 mm wide; flowers small, in 1 spiral row; lip white, 2.3–4 mm long, 1.5–2 mm wide, the apex barely erose. Dry sandy woods; Lamar Co. in Red River drainage; mainly se and e TX. Jun–Oct. [*S. grayi* Ames.]

Spiranthes vernalis Engelm. & A. Gray, (of spring), SPRING LADIES'-TRESSES, TWISTED LADIES'-TRESSES, UPLAND LADIES'-TRESSES. Plant 10–90(–117) cm tall; spike about 1.6 cm thick, the rachis and ovaries with reddish or whitish pointed hairs; perianth 5–10 mm long, usually white; lip 4.5–8 mm long, fleshy, with stout basal tuberosities, recurved, apically crenulate. Sandy woods and prairies, moist areas; se and e TX w to Cooke and Dallas cos.; also Rolling Plains. Late Apr–Jun. [Incl. *S. reverchonii* (Small) Cory]

POACEAE (GRAMINEAE) GRASS FAMILY

Ours herbaceous annuals or perennials or woody in bamboos; roots fibrous; culms (= stems) usually rounded, with prominent, swollen or constricted, solid nodes, the internodes hollow or solid, and with intercalary meristem at base allowing continued elongation; tillers (= basal branches), stolons (= above ground horizontal stems or "runners"), or rhizomes (= underground horizontal stems) often present; leaves alternate, with tubular basal sheath enclosing the culm and usually split to base down one side opposite the blade, and a short or elongate, usually linear, flattened or sometimes involute (= inrolled) blade, with a row of hairs or scaly membrane (= ligule) at junction of blade and sheath on inner (upper) side, often with intercalary meristem near the ligule; inflorescences made up of very reduced branches (called spikelets), these spikelets, which are the basic units of the grass inflorescence, are arranged in various ways (e.g., spikes, racemes, panicles); spikelets composed of 1–numerous flowers, associated scale-like bracts (each flower and its subtending (1–)2 scale-like bracts are collectively called a floret), and a short axis; florets borne 2-ranked, one above another, alternating along the concealed axis (= rachilla), the spikelets usually with an additional pair of empty scale-like bracts (= glumes) at base (in some only 1 glume, or none, or reduced to awns instead); flowers perfect or imperfect, without true perianth, this apparently represented only by 2(–3) minute lodicules which function by swelling and opening the floret; florets usually with 2 alternate, overlapping scale-like bracts (lemma—the outer or lower; palea—the inner or upper, which is usually smaller and more delicate, sometimes absent or not visible); stamens (1–)3(–6), the anthers often dangling outside the floret at anthesis to allow the wind-borne pollen to be carried away, but sometimes retained within the floret; pistil 1, the typically 2(–3) stigmas usually feathery with increased surface area to catch wind-blown pollen; flowers typically open only for a short time (usually during morning) or not at all, wind- or self-pollinated; fruit a caryopsis (= achene-like, 1-seeded fruit with ovary wall adnate to the seed coat), rarely an achene.

Ordinarily dissection is necessary to see the parts of the grass spikelet; because of the small size of the structures involved (e.g., lemmas, paleas), a dissecting scope or at minimum a hand lens is often necessary for definitive identification. In the keys, measurements of glumes, lemmas, or spikelets do not include awns (= beards or bristle-like appendages) if these are present. A number of species of ornamental grasses are cultivated and long persist in nc TX including *Cortaderia selloana* (Schult. & Schult. f.) Asch. & Graebn., *Miscanthus sinensis* Andersson, *Pennisetum* species, and a variety of bamboos.

This huge (ca. 9,500 species in 668 genera) cosmopolitan family, on a worldwide basis, is ecologically the world's most dominant vascular plant family, occurring over vast areas of prairie, plain, steepe, and pampas; it is estimated that 20% of the world's vegetational cover is made up of grasses. The family is also economically the most important, containing all the cereal crops including *Oryza sativa* (RICE), *Triticum aestivum* (WHEAT), and *Zea mays* (CORN), as well as *Saccharum officinarum* (SUGAR CANE). More than 70% of farmland worldwide is devoted to cereals, which provide humans with more than 50% of all calories (Heiser 1990a). Just three plants, RICE, WHEAT, AND CORN provide ca. 45% of total human caloric intake (Chrispeels & Sadava 1977). Most grasses are extremely well-adapted to fire, grazing, trampling, and lawn mowers.

General Characteristics of the Poaceae (Gramineae) Family [jep]

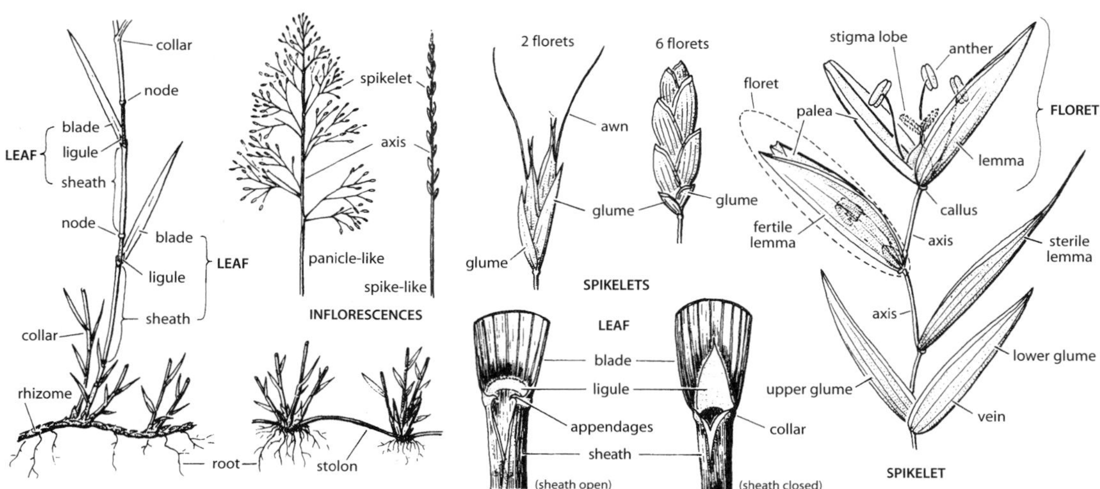

This is due in part to the presence of intercalary meristems (located in the culms just above the nodes and in the leaves near the ligules, thus allowing growth from the base even if terminal parts are damaged) and the tendency to branch ("tiller") or produce stolons or rhizomes near or below ground level. Species worldwide range from tiny annuals (2–3 cm tall) to huge bamboos 40 m (ca. 130 feet) tall. Grasses are wind-pollinated and shed large amounts of pollen; this is a major source of allergic reactions (e.g., hay fever) in humans. The Poaceae is a particularly important part of the nc TX flora; the 249 species present represent about 11% of the total species known for the region and make the Poaceae the second largest family in nc TX (following only the Asteraceae). (subclass Commelinidae)

FAMILY RECOGNITION IN THE FIELD: mostly herbs with 2-ranked leaves having sheathing bases, free blades, and ligules; culms round, with hollow or solid internodes; flowers small, inconspicuous, reduced to stamens and pistils, subtended by 2 scale-like bracts each, and arranged in very reduced spikes called *spikelets*; the ± similar Cyperaceae (SEDGES) have 3-ranked leaves usually without ligules, often 3-sided culms with solid internodes, and flowers subtended by 1 scale-like bract each; the ± similar Juncaceae (RUSHES) have flowers with a small 6-parted perianth.

REFERENCES: Nash 1909, 1912; Nash & Hitchcock 1915; Hitchcock 1931–1939, 1935, 1951; Silveus 1933; Tharp 1952; Shinners 1954; Gould 1957b, 1968a, 1968b, 1973, 1975b, 1978; Box & Gould 1959; deWitt 1981; Stebbins 1981, 1982, 1985; Estes et al. 1982; Gould & Shaw 1983; Campbell 1985; Coughenour 1985; Dahlgren et al. 1985; Redmann 1985; Clayton & Renvoize 1986; Soderstrom et al. 1987; Hignight et al. 1988; Tucker 1988, 1990, 1996; Crins 1991; Allen 1992; Powell 1994; Clark et al. 1995; Kellogg et al. 1996; Peterson et al. 1997.

1. Culms woody, persisting for more than 1 season (bamboos).
    2. Culms terete (= rounded); upper nodes often puberulent; base of midvein on lower leaf surface without a mustache of hairs; branches 3–6 per node; plants native in ne part of nc TX ______ **Arundinaria**
    2. Culms flattened above node on 1 side; upper nodes glabrous; base of midvein on lower leaf surface often with a mustache of hairs along one side; branches usually 2(–3) per node; plants introduced in nc TX ______ **Phyllostachys**
1. Culms not woody, lasting 1 season only.
    3. Plants typically 2–6 m tall, usually reed-like; inflorescence a large panicle up to ca. 60 cm long and 20 cm wide.
        4. Leaves distributed along much of the length of the culm; inflorescences not silvery-white in appearance; leaf blades 0.8–6 cm wide; stout creeping rhizomes present or absent; plants in extensive dense colonies or not so.
            5. Spikelets in pairs of 1 pedicelled and 1 sessile (or nearly so); plants not in dense extensive colonies (but culms can be clumped).
                6. Spikelets with a tuft of long hairs at base ca. as long as spikelets; awn of lemmas 5–25 mm long.
                    7. Awn of lemmas 12–25 mm long; 1 spikelet of each pair sessile, 1 pedicelled; inflorescence branches disarticulating (breaking into sections) at maturity, the pedicels falling with the sessile spikelets ______ **Saccharum**
                    7. Awn of lemmas 5–10 mm long; both spikelets of pairs pedicelled, pedicels 1–5 mm long, unequal, one much shorter; inflorescence branches not disarticulating, the spikelets falling from pedicels at maturity ______ **Miscanthus**
                6. Spikelets without a tuft of long hairs at base; awn of lemmas usually much shorter (usually 1.5 mm or less long) or absent ______ **Sorghum**
            5. Spikelets not in pairs; plants often in extensive dense colonies.
                8. Spikelets 3–5 mm long, glabrous; florets 2 per spikelet (upper perfect, lower staminate); leaf blades 0.3–1.5 cm wide ______ **Panicum** (*P. virgatum*)
                8. Spikelets 10–15 mm long, either with rachilla or lemmas hairy; florets 2–8 per spikelet; leaf blades mostly 1.5–7 cm wide.

9. Lemmas glabrous; rachilla hairy; glumes 8 mm or less long, shorter than the lowest lemma **Phragmites**
9. Lemmas hairy; rachilla glabrous; glumes 10–15 mm long, nearly as long as entire spikelet **Arundo**

4. Leaves mostly at the base of the plant; inflorescences silvery-white in appearance; leaf blades 1.5 cm or less wide; creeping rhizomes absent; culms clumped but plants not forming extensive colonies **Cortaderia**

3. Plants usually < 2(–3+) m tall, not reed-like; inflorescences various, usually much smaller.
10. Spikelets fused with or closely fitted into the axis of the inflorescence, forming a solid cylindrical or flattened spike.
11. Inflorescences as long as or longer than the leafy portion of the culm, with (3–)4–12 branches **Schedonnardus**
11. Inflorescences shorter than the leafy portion of the culm, unbranched or with up to 3 branches.
12. Spikelets (at least upper) awned **Aegilops**
12. Spikelets awnless.
13. Spike-like inflorescences flattened, the axis exposed between the spikelets; plants creeping, with flowering branches only 10–30 cm tall **Stenotaphrum**
13. Spike-like inflorescences (or if branched, the spike-like branches) not flattened or flattened only on one side, the axis not evident; plants not creeping, 30–300 cm or more tall.
14. Inflorescences 2–3 mm thick, unbranched; leaf blades 1.5–4 mm wide **Coelorachis**
14. Inflorescences (if unbranched, or if branched, the inflorescence branches) 5–8 mm thick, unbranched or with 2–3 branches; leaf blades usually 10–25 mm wide **Tripsacum**
10. Spikelets neither fused with nor closely fitted into the axis of the inflorescence (but may be sessile and appressed).
15. Spikelets unisexual, the staminate and pistillate conspicuously different with naked eye observation.
16. Spikelets with 3–6(–10) florets, conspicuously (pistillate) or inconspicuously (staminate) long hairy; plants 0.2–0.6 m tall **Poa** (*P. arachnifera*)
16. Spikelets with 1–2 florets, glabrous or nearly so; plants either 0.3 m or less tall OR 0.5–3+ m tall.
17. Plants low, 30 cm or less tall, mat-forming, usually dioecious (= staminate and pistillate spikelets on different plants) **Buchloe**
17. Plants much larger, not mat-forming, monoecious (= staminate and pistillate spikelets on the same plant).
18. Staminate and pistillate spikelets in separate inflorescences; pistillate spikelets in large "cobs"; whole pistillate inflorescence covered by large, leaf-like, modified leaves or bracts (= shucks) **Zea**
18. Staminate and pistillate spikelets in the same inflorescence; whole inflorescence not covered by leaf-like shucks.
19. Inflorescence unbranched and spike-like OR with 2–3 spike-like **Tripsacum**
19. Inflorescence a much-branched panicle; pistillate spikelets not becoming hard; plants usually growing in shallow water **Zizaniopsis**
15. Spikelets perfect, or if unisexual, not conpsicuously different to the naked eye.
20. Spikelets with 2–many perfect florets.
21. Spikelets pedicelled (pedicels may be short) **KEY A**

21. Spikelets sessile, or nearly sessile and so densely crowded as to conceal the short pedicels ____ **KEY B**

20. Spikelets with 1 perfect floret (staminate or reduced rudimentary florets sometimes present) ____ **Key C**

KEY A

1. Lemmas 1-awned from the back (not from their apex).
   2. Spikelets (excluding awns) 2.5 mm long or less; inflorescence an open panicle with numerous spikelets and main branches bare of spikelets below the middle ____ **Aira**
   2. Spikelets (excluding awns) 4 mm long or more (often much more); inflorescence not as above.
      3. Glumes 1.5–5 mm long, not longer than the lemmas (excluding awns) ____ **Trisetum**
      3. Glumes ca. 20–30 mm long, longer than the lemmas (excluding awns) ____ **Avena**
1. Lemmas awnless or awned from their apex (awn can arise between apical teeth).
   4. Glumes conspicuously longer than individual lemmas (excluding awns) and usually exceeding all of the lemmas.
      5. Spikelets 15 mm or less long (excluding awns).
         6. Lemmas awnless (but with minute mucro < 1 mm long); inflorescences many-flowered, 10–36 cm long ____ **Tridens**
         6. Lemmas awned, the twisted and geniculate awns usually 5–7 mm long; inflorescences few-flowered, to ca. 5 cm long ____ **Danthonia**
      5. Spikelets 20–30 mm long (excluding awns) ____ **Avena**
   4. Glumes not conspicuously longer than individual lemmas.
      7. Spikelets in several very dense 1-sided clusters at the end of branches; the branches bare of spikelets in their lower portions; rare introduced species ____ **Dactylis**
      7. Spikelets not borne as above; common native and introduced species.
         8. Ligule a membrane (may have ciliate margin, or become split or torn).
            9. Lemmas 3-nerved, the nerves often prominent.
               10. Nerves of lemmas pubescent or puberulent OR base of lemmas with long hairs; lemmas with midnerve usually exserted as a short awn or mucro.
                  11. Plants annual; spikelets usually with 2–4 florets; inflorescence a panicle 3–11 cm long with a few rebranched primary branches, the lower branches spreading, bare of spikelets on the lower 1/3–1/2; paleas with long cilia on upper margins ____ **Triplasis**
                  11. Plants perennial; spikelets with 4–12 florets; inflorescences various, but often quite different from above; paleas not ciliate on upper margins.
                     12. Inflorescences 5–35 cm or more long; culms with several nodes above base; leaf margins green ____ **Tridens**
                     12. Inflorescences 2–4 cm long; culms with one node above base; leaf margins white ____ **Erioneuron**
               10. Nerves of lemmas not hairy; base of lemmas without long hairs; lemmas awnless.
                  13. Lemmas < 6 mm long; second glume with 1 nerve; caryopsis (= fruit) not beaked; widespread and abundant in nc TX ____ **Eragrostis**
                  13. Lemmas (3.8–)4.6–10.8 mm long; second glume with 3–5 nerves; caryopsis beaked; rare if present in nc TX ____ **Diarrhena**
            9. Lemmas 5–many-nerved, the nerves prominent or sometimes so obscure as to be almost unnoticeable.
               14. Lemmas awned.
                  15. Spikelets (12–)15–35 mm long, if spikelets at the lower end of this size range then lemmas 2-toothed or minutely notched at apex.
                     16. Lemmas not 2-toothed or notched at apex, rather tapering to a 10–20 mm

awn from an entire apex; inflorescence a spike or spicate raceme of 1–5 large stiffly erect spikelets; spikelets 2–3.5 cm long, usually with 9–18 florets; rare in nc TX _______ **Brachypodium**

16. Lemmas often 2-toothed or minutely notched at apex, with 1–18 mm long awn arising from between teeth or notch; inflorescences not as above; spikelets various; very common in nc TX _______ **Bromus**

15. Spikelets < 15(–18) mm long (often much less), if spikelets at the upper end of this size range then lemmas not 2-toothed or notched at apex.

17. Lemmas awned from (or just below) a conspicuously 2-toothed apex, the awn geniculate _______ **Trisetum**

17. Lemmas awned from an entire or minutely and indistinctly notched apex, the awn straight.

18. Leaf blades 0.1–2.5 mm wide; plants annual; stamens 1, infrequently 3; plants 5–70 cm tall _______ **Vulpia**

18. Leaf blades 2.5–12 mm wide; plants perennial; stamens 3; plants 50–120(–200) cm tall _______ **Festuca**

14. Lemmas awnless.

19. Spikelets 6–20 mm wide, (10–)20–50 mm long, conspicuously flattened, with 1–4 sterile (empty) lemmas below fertile _______ **Chasmanthium**

19. Spikelets much smaller, not conspicuously flattened; spikelets without sterile lemmas below fertile.

20. Lemmas as wide as long, spreading at right angles, inflated with the margins outspread _______ **Briza**

20. Lemmas longer than wide, not spreading at right angles, not inflated, the margins clasping the paleas.

21. Lemmas rounded on back, without prominent keel or raised midnerve except near apex.

22. Upper florets sterile, much reduced to a small "rudiment" usually broadest near apex; glumes thin, papery at least on margins and apices; leaf sheaths with margins united except at summit _______ **Melica**

22. Upper florets ± like the lower; glumes not thin and papery; leaf sheaths open down one side (edges may overlap) except in *Glyceria* which has sheath margins united.

23. Leaf blades mostly 10–20 mm wide, with prominent midrib; caryopsis (= fruit) beaked, turgid, conspicuously exserted from the lemma and palea at maturity _______ **Diarrhena**

23. Leaf blades usually 1–12 mm wide, the midrib not prominent; caryopsis neither beaked, turgid, nor conspicuously exserted.

24. Lemmas with nerves (usually 7) strongly and uniformly developed and equally spaced; leaf sheaths with margins united except at summit _______ **Glyceria**

24. Lemmas with several indistinct nerves or none on each side; leaf sheaths open down one side (edges may overlap).

25. Plants perennial, usually much > 30 cm tall; either spikelets 10–15(–18) mm long OR if smaller then lower panicle branches bare of spikelets in their lower 1/3–1/2 _______ **Festuca**

25. Plants annual, to only 15(–30) cm tall; spikelets 5–7 mm long; panicle branches with spikelets nearly to base _______ **Desmazeria**

21. Lemmas keeled or with prominent midnerve from tip to base.
    26. Plants strongly rhizomatous, of alkaline or alkaline-saline habitats; leaves 2-ranked with sheaths conspicuously overlapping; spikelets unisexual, the similar staminate and pistillate inflorescences on separate plants; spikelets usually 5–20-flowered, 6–18(–28) mm; rare in nc TX, known only from Brown and Comanche cos. ____ **Distichlis**
    26. Plants not rhizomatous or not conspicuously so, of various habitats; leaves not as above; spikelets various; widespread in nc TX.
        27. Spikelets (12–)15–35 mm long ____ **Bromus**
        27. Spikelets 1.5–11(–12) mm long.
            28. Panicles 1–3(–4.5) cm long, overtopped by the leaves; lemmas obtuse, with 5(–7) prominent, raised, almost parallel nerves; culms to only 10(–18) cm long ____ **Sclerochloa**
            28. Panicles 3–25 cm long, exserted above the foliage; lemmas acute to obtuse, without 5 prominent, raised, almost parallel nerves; culms of variable length, often much more than 18 cm long.
                29. Lemmas often puberulent, the base of lemma often with long, kinky hairs; upper glume only slightly wider than lower glume (much less than 2 times as wide) ____ **Poa**
                29. Lemmas glabrous or merely scabrous, the base of lemma lacking hairs; upper glume much wider than lower (2 times as wide or wider).
                    30. Upper glume widest near middle, slightly pointed at apex; main axis of inflorescence puberulent; disarticulation above glumes ____ **Koeleria**
                    30. Upper glume widest near apex, with cupped, blunt apex; main axis of inflorescence glabrous or minutely scabrous; disarticulation below glumes ____ **Sphenopholis**

8. Ligule of hairs entirely, or with membranous base less than half the total length.
    31. Spikelets 6–20 mm wide, (10–)20–50 mm long, conspicuously flattened ____ **Chasmanthium**
    31. Spikelets much smaller, not at all flattened or only slightly so.
        32. Inflorescence an elongate, narrow, stiffly erect, spike-like panicle with a few widely spaced, usually short side branches; lower 1–2 florets sterile; only in extreme ne part of nc TX (Lamar Co.) ____ **Chasmanthium**
        32. Inflorescences various, but usually quite different from above; lower florets fertile; widespread in nc TX.
            33. Lemmas with entire pointed apex; nerves of lemmas not hairy; base of lemmas without long hairs ____ **Eragrostis**
            33. Lemmas with notched or rounded-truncate apex (can be short awned from the notch); nerves of lemmas pubescent or puberulent OR base of lemmas with long hairs.
                34. Plants annual; spikelets usually with 2–4 florets; inflorescence a panicle 3–11 cm long with a few rebranched primary branches, the lower branches spreading, bare of spikelets on the lower 1/3–1/2; paleas with long cilia on upper margins ____ **Triplasis**
                34. Plants perennial; spikelets with 4–12 florets; inflorescences various, but often quite different from above; paleas not ciliate on upper margins.

35. Inflorescences 5–35 cm or more long; culms with several nodes above base; leaf margins green ____ **Tridens**
35. Inflorescences 2–4 cm long; culms with one node above base; leaf margins white ____ **Erioneuron**

KEY B

1. Spikelets borne on opposite sides of the zigzag, usually flattened, main axis of the inflorescence; inflorescence unbranched, consisting of a solitary, 2-sided spike or spike-like raceme.
  2. Glumes 3.6 mm or more wide ____ **Triticum**
  2. Glumes 1.0–3.5 mm wide.
    3. Lemmas minutely spiny-ciliate on keel and margins (the hairs visible with the naked eye and obvious with a lens) ____ **Secale**
    3. Lemmas not minutely spiny-ciliate on keel and margins.
      4. Spikelets solitary at each node of the inflorescence.
        5. Spikelets with 1 glume, oriented with edge (back of lemmas) facing inflorescence axis ____ **Lolium**
        5. Spikelets with 2 glumes, oriented with side facing inflorescence axis ____ **Pascopyrum**
      4. Spikelets 2–3 at each node of the inflorescence.
        6. Lemmas awnless (glumes can have awns up to 5 mm long); plants 30 cm or less tall with leaf blades 3 mm or less wide; inflorescences to 3.5 cm long ____ **Hilaria**
        6. Lemmas conspicuously awned or sometimes awnless; plants usually much > 30 cm tall; leaf blades usually much > 3 mm wide; inflorescences usually > 3.5 cm long.
          7. Spikelets 1-flowered, in groups of 3 per node (lateral ones may be reduced to group of awns) ____ **Hordeum**
          7. Spikelets 2–12-flowered, 2–3 per node.
            8. Lemmas awnless or with awns to only 3 mm long; spikelets 5–12-flowered, solitary or sometimes 2 per middle and lower nodes ____ **Pascopyrum**
            8. Lemmas with prominent awns 5–50 mm long; spikelets 2–6-flowered, usually 2–3 per node ____ **Elymus**

1. Spikelets not on opposite sides of a main axis, rather borne all on 1 side or on all sides of the main axis or its branches; inflorescence usually branched, consisting of 2–many spikes, racemes, or panicles (condensed to open) OR if unbranched then not distinctly 2-sided.
  9. Spikelets in several very dense, 1-sided clusters at the end of branches bare of spikelets in their lower portions; rare introduced species ____ **Dactylis**
  9. Spikelets not borne as above; common native and introduced species.
    10. Plants 5–15(–30) cm tall; inflorescences usually 1–9 cm long and 19 mm or less wide.
      11. Spikelets with 3–4 florets, falling as a unit (disarticulation below glumes); inflorescences usually 1–3(–4.5) cm long, overtopped by the upper leaves; lemmas obtuse, prominently 5-nerved, the lowest lemma ca. 5 mm long ____ **Sclerochloa**
      11. Spikelets with (4–)5–9(–10) florets, not falling as a unit (disarticulation above glumes); inflorescences usually 3–9 cm long, exserted beyond upper leaves; lemmas broadly acute, obscurely nerved, the lowest lemma ca. 2.5 mm long ____ **Desmazeria**
    10. Plants usually much larger; inflorescences usually much larger.
      12. Spikelets not borne on 1 side of the inflorescence axis or its branches, instead spikelets borne crowded on all sides of the axis or on short, crowded branchlets, the whole inflorescence narrow, usually a dense spike or spike-like or head-like.
        13. Inflorescences not dense, the main axis clearly visible at a glance (spikelets crowded on well-separated short side branches); lowermost lemma sterile, not very different from glumes in size or appearance ____ **Chasmanthium**
        13. Inflorescences dense, the main axis often obscured by the numerous spikelets;

lowermost lemma perfect, sometimes quite different from glumes in size or appearance.

14. Lemma nerves pubescent, at least in lower portion, or base of lemma long-hairy; spikelets with 4–11 florets; lemmas rounded at apex or usually notched and thus 2-toothed, sometimes with mid-vein exerted as a mucro ______ **Tridens**
14. Lemma nerves and base of lemma glabrous or nearly so; spikelets with 2–4 florets; lemmas pointed or rounded, not notched.
15. Upper glume widest near middle, slightly pointed at apex; main inflorescence axis puberulent; disarticulation above glumes ______ **Koeleria**
15. Upper glume widest near apex, with cupped, blunt apex; main inflorescence axis glabrous or minutely scabrous; disarticulation below glumes ______ **Sphenopholis**

12. Spikelets borne on 1 side of the inflorescence axis or its branches, often in 2 distinct rows, the inflorescence with 2–many, distinct, short or long, digitately arranged or scattered, spicate branches.
16. Branches of inflorescence distributed (scattered) along main axis, usually only 1 arising per node, not crowded together.
17. Spikelets not crowded on branches, usually not overlapping adjacent spikelet ______ **Eragrostis**
17. Spikelets often rather crowded, usually overlapping adjacent spikelet ______ **Leptochloa**
16. Branches of inflorescence digitate (all arising together at very tip of flowering culm) or nearly so or verticillate (in whorls).
18. Lemmas and glumes awnless ______ **Eleusine**
18. Lemmas and/or glumes with awns.
19. Axis of each spicate inflorescence branch projecting as a stiff point beyond terminal spikelet; second glume mucronate or with a short awn; spikelets with 3–5 perfect florets ______ **Dactyloctenium**
19. Axis of each spicate inflorescence branch not projecting as a stiff point beyond terminal spikelet; second glume neither mucronate nor with a short awn; spikelets with 1–2 perfect florets ______ **Chloris**

KEY C

1. Spikelets in pairs (1 sessile or nearly so and 1 pedicelled, both pedicelled at branch tips), the pedicelled spikelet similar to sessile spikelet or usually reduced or absent (often represented by an empty pedicel); pedicels flattened, with long-ciliate margins.
2. Individual inflorescences usually less than 6 cm long.
3. Inflorescence of a single spike-like raceme per peduncle (each terminal branchlet with a single raceme above the uppermost leaf) ______ **Schizachyrium**
3. Inflorescence a panicle or of 2 or more racemes or spikes per peduncle (each terminal branchlet with 2–many inflorescence branches, racemes, or spikes above the uppermost leaf).
4. Pedicelled spikelets smaller (shorter or narrower, usually rudimentary, vestigial, or absent) than the sessile spikelets, tapered to a narrow apex; extremely widespread and abundant native and introduced species.
5. Pedicels and internodes of inflorescence branches neither strongly flattened nor grooved on both sides (can be slightly flattened or grooved on one side), the central portion thus neither thin nor membraneous ______ **Andropogon**
5. Pedicels and usually upper internodes of inflorescence branches strongly flattened and grooved on both sides, the central portion thus thin to membranaceous, often easily ruptured with a probe ______ **Bothriochloa**

4. Pedicelled spikelets ca. the same size as the sessile spikelets, rounded apically; introduced species known in nc TX only from Coryell Co. in Lampasas Cut Plain ____ **Dichanthium**

2. Individual inflorescences usually over 7 cm long.

6. Pedicelled spikelet absent, represented only by the hairy pedicel; awn of lemmas usually 12–35 mm long ____ **Sorghastrum**

6. Pedicelled spikelet present; awn of lemmas shorter or absent (*Miscanthus* and *Sorghum* or) or 12–25 mm long (*Saccharum*).

7. Spikelets with a tuft of long hairs at base ca. as long as spikelets; awn of lemmas 5–25 mm long.

8. Awn of lemmas 12–25 mm long; 1 spikelet of pairs sessile, 1 pedicelled; spikelets ca. 6–8 mm long; inflorescence branches disarticulating (breaking into sections) at maturity, the pedicels falling with the sessile spikelets ____ **Saccharum**

8. Awn of lemmas 5–10 mm long; both spikelets of pairs pedicelled, pedicels 1–5 mm long, unequal, one much shorter; spikelets 4–5.5 mm long; inflorescence branches not disarticulating, the spikelets falling from pedicels at maturity ____ **Miscanthus**

7. Spikelets without a tuft of long hairs at base; awn of lemmas usually much shorter (usually 1.5 mm or less long) or absent ____ **Sorghum**

1. Spikelets in pairs or not, but usually not in sessile-pedicelled pairs; pedicels glabrous or evenly pubescent on all sides, or pedicels absent.

9. Spikelets surrounded by an involucre of prickly or smooth scales, spines, or bristles.

10. Involucre of scales or spines that are flattened at least at base, these sharply prickly or not so.

11. Involucres smooth, not prickly, not painful to the touch ____ **Buchloe**

11. Involucres sharply prickly, very painful to the touch (SANDBURS) ____ **Cenchrus**

10. Involucre of bristles (not flattened), these not sharply prickly.

12. Bristles remaining when the spikelets fall, usually < 20 mm long ____ **Setaria**

12. Bristles falling with the spikelets, the longer bristles > 20 mm long, often 40–50 mm or more ____ **Pennisetum**

9. Spikelets without a well-developed involucre (may have ring or collar at base).

13. Reduced floret or florets present below the perfect floret (the reduced ones staminate or sterile, sometimes represented only by a sterile lemma); spikelets awnless.

14. Lemmas strongly compressed, sharply keeled; inflorescences spike-like; reduced florets 2 per spikelet, rudimentary or scale-like, their lemmas not similar to the second glume; disarticulation above glumes ____ **Phalaris**

14. Lemmas usually not as above; inflorescences various, spike-like to very different; reduced floret 1 per spikelet, the lemma similar to the second glume; disarticulation below glumes ____ **Key D**

13. Reduced floret below the perfect floret absent (reduced florets found above the perfect or absent or seemingly so); spikelets awnless or awned.

15. Spikelets usually borne along 1 side of the often flattened inflorescence branches, usually in 2 rows, and/or inflorescences of spike-like branches bearing sessile spikelets to very base; spikelets sessile or nearly so.

16. Branches of inflorescence digitate (all arising together at very tip of flowering culm) or nearly so or verticillate (in whorls).

17. Spikelets awnless; inflorescence branches usually 2–5(–7); rudimentary floret absent or present only as a minute scale ____ **Cynodon**

17. Spikelets usually awned (can be awnless in *Eustachys*); inflorescence branches 3–20 or more, variously arranged; rudimentary floret or florets present above perfect floret.

18. Lemma of perfect floret awned, the awns usually conspicuous; widespread and common in nc TX ____ **Chloris**

18. Lemma of perfect floret awnless, sometimes mucronate or minutely aristate; rare in nc TX, mainly to the e __________ **Eustachys**

16. Branches of inflorescence distributed along flowering culm, usually only one arising per node.

19. Plants dwarf, mat-formers to only ca. 30 cm tall; inflorescence branches 1–4, usually 6–14 mm long; plants dioecious, male and female flowers on different plants; spikelets awnless or nearly so __________ **Buchloe**

19. Plants often larger; inflorescence branches 1–80, 8–150 mm long; plants not dioecious, each spikelet with 1 perfect floret; spikelets awned or awnless.

20. Lemmas awnless; leaf blades 3 mm or less wide; spikelets neither densely arranged and comb-like on the side branches nor loosely arranged on the main axis.

21. Inflorescence branches spreading, 2–10(–20) cm long; glumes stiff __________ **Schedonnardus**

21. Inflorescence branches appressed, usually 2–3 cm long; glumes soft __________ **Willkommia**

20. Lemmas usually awned; leaf blades less than to much more than 3 mm wide; if less than 5 mm wide then the spikelets either densely arranged and comb-like on the side branches or loosely arranged on the main axis.

22. Inflorescences mostly wider than long with widely spreading panicle branches mostly 10–20 cm long; leaf blades lanceolate, rather broad in appearance for a grass, abruptly narrowed at base __________ **Gymnopogon**

22. Inflorescences mostly longer than wide with spreading to appressed branches to 15 cm long but usually much shorter; leaf blades linear, narrow in appearance, not abruptly narrowed at base.

23. Spikelets without rudimentary florets; inflorescence branches usually appressed, usually 4–15 cm long __________ **Spartina**

23. Spikelets with 1–several rudimentary or staminate florets above the perfect floret; inflorescence branches spreading, 5 cm or less long __________ **Bouteloua**

15. Spikelets not borne along 1 side of the inflorescence branches; inflorescences without spike-like branches having spikelets to very base; spikelets sessile or pedicelled.

24. Spikelets in clusters of 3 per node (laterals often reduced, staminate or sterile); spikelets sessile.

25. Lemmas awned; inflorescence axis usually disarticulating at maturity, the sections falling with the spikelets; spikes usually 4–10 cm long __________ **Hordeum**

25. Lemmas awnless; inflorescence axis not disarticulating, a cluster of 3 spikelets falling as a unit from the inflorescence axis; spikes 2–3.5 cm long __________ **Hilaria**

24. Spikelets not 3 per node; spikelets sessile or on pedicels.

26. Glumes absent, the caryopsis (= fruit) thus subtended by only 2 scale-like bracts (lemma and palea); inflorescence a panicle with spreading branches.

27. Lemmas hispid or hispidulous, the hairs sometimes conspicuous, awnless, 5-nerved, the lateral nerves marginal; spikelets 2.5–5.5 mm long; aerial culms 3 mm or less thick, 0.7–1.5 m tall __________ **Leersia**

27. Lemmas glabrous, awned, 7-nerved, the lateral nerves not marginal; spikelets 6–8 mm long; aerial culms stout, 5–15 mm thick, 1.5–2.5(–3) m tall __________ **Zizaniopsis**

26. Glumes present, the caryopsis thus subtended by more than 2 scale-like bracts (lemma, palea, glume(s)); inflorescences various.

28. Lemmas hard, closed except in anthesis, permanently enclosing caryopsis,

narrowly fusiform-cylindrical (shaped like a fat needle broadest near middle), with a sharp-pointed callus at base, with 1 or 3 stiff or wiry awns.

29. Lemmas with 1 awn; ring of hairs present at base of awn ____ **Nassella**
29. Lemmas with 3 awns; ring of hairs at base of awn absent ____ **Aristida**

28. Lemmas neither hard nor permanently enclosing caryopsis, open down one side, variously shaped, but not as above, usually without a sharp-pointed callus, with a single awn or awnless.
30. Glumes and lemmas both awnless.
31. Glumes both as long as or longer than lemma.
32. Panicles dense, spike-like, 30 mm or less wide.
33. Fertile lemmas 0.9–1.5 mm long (the single lemma always fertile), glabrous; inflorescences 10–30 mm wide ____ **Agrostis**
33. Fertile lemmas 3–6.8 mm long, pubescent (2 reduced sterile lemmas also present); inflorescences 8–18(–20) mm wide ____ **Phalaris**
32. Panicles open, not spike-like, 20–250 mm wide ____ **Agrostis**
31. Glumes, at least the first, shorter than lemma.
34. Lemmas 1-nerved; fruit dropping from lemma and palea at maturity; seed loose, not fused to ovary wall; ligule a minute ring of hairs with membranous base ____ **Sporobolus**
34. Lemmas 3-nerved, the lateral nerves sometimes faint; fruit enclosed in lemma and palea at maturity; seed fused to ovary wall; ligule a short membrane, usually without hairs ____ **Muhlenbergia**
30. Glumes OR lemmas awned.
35. Inflorescence very dense, spike-like.
36. Glumes awned; lemma awnless or with an awn to ca. 1 mm long arising from the apex of the lemma.
37. Awns of glumes 0.8–1.5 mm long, abrupt from unnotched apex; inflorescences 5–9 mm wide; lemma awnless ____ **Phleum**
37. Awns of glumes 5–9 mm long, from apical notch; inflorescences 10–30 mm wide; lemma with a delicate awn ca. 1 mm long ____ **Polypogon**
36. Glumes awnless; lemma with awn 3–5 mm long arising at or near the base of the lemma ____ **Alopecurus**
35. Inflorescence usually an open or sometimes contracted panicle, not spike-like.
38. Lemmas usually awned from an entire or very minutely cleft apex or awnless; glumes, at least the first, usually shorter than lemma, awnless or awned ____ **Muhlenbergia**
38. Lemmas awned from back, just below tip to near middle; glumes equaling or exceeding lemma, awnless.
39. Panicles contracted, dense, narrow, ca. 10 mm wide; spikelets 3–4 mm long (excluding awns) ____ **Limnodea**
39. Panicles open, very diffuse, > 15 mm wide; spikelets 1.2–2.2 mm long.
40. Awn arising near middle of back of lemma, ca. 2 mm long; florets 2 per spikelet (though only 1 with awned lemma and thus appearing as if only 1 floret) ____ **Aira**
40. Awn arising just below tip of lemma, 4–8 mm long; floret 1 per spikelet ____ **Agrostis**

KEY D

1. Spikelets with 2 well-developed, outer, scale-like bracts (excluding the tough or hard and grain-like fertile lemma), the 3rd if present, much reduced, less than 1/4 as long or 1/2 as wide as the others.
   2. Spikelets long-pedicelled, widely separated in open panicle ____ **Digitaria** (*Leptoloma*)
   2. Spikelets sessile or short-pedicelled, the inflorescence or its branches with closely spaced spikelets, the spikelets often overlapping one another or nearly so.
      3. Pedicels with a prominent collar-like or cup-like structure (modified glume) just under the spikelet (obvious with a lens) ____ **Eriochloa**
      3. Pedicels without collar-like or cup-like structure.
         4. Spikelets with well-developed pedicels, or both sessile and pedicelled together, in panicles or narrow racemes.
            5. Lemmas with silky hairs longer than the spikelet ____ **Digitaria**
            5. Lemmas short-pubescent or glabrous ____ **Sorghum**
         4. Spikelets sessile or subsessile, on a flattened inflorescence branch, the inflorescence consisting of 1 or more spikes or spike-like racemes.
            6. Spikelets lanceolate to oblong or suborbicular, obtuse or acute, 0.8–2.5 times as long as wide; inflorescence branches either only 2 apically or scattered along main inflorescence axis, not digitately arranged; fertile lemma with relatively thick margins, these inrolled over edges of the palea.
               7. Inflorescence with 2–many branches; rounded back of fertile lemma facing the axis of the inflorescence branch; spikelets closely packed and often paired, in 2 or 4 rows, 1.6–4.8 mm long; widespread in nc TX ____ **Paspalum**
               7. Inflorescence with 2–4 branches; rounded back of fertile lemma facing away from the axis of the inflorescence branch; spikelets more widely spaced (only slightly overlapping), not in pairs, in 2 rows, 1.8–2.2(–2.6) mm long; in nc TX only known from extreme e margin ____ **Axonopus**
            6. Spikelets narrowly lanceolate, gradually acute, 3–4 times as long as wide; inflorescence branches digitately arranged or scattered along main inflorescence axis, never only 2 apically; fertile lemma with thin and flat margins, these not inrolled over the palea ____ **Digitaria**
1. Spikelets with 3 or 4 well-developed, outer, scale-like bracts (3rd or 4th may be much shorter than rest, but nearly as wide).
   8. Inflorescences with an easily discernable central axis bearing relatively few (1–ca. 30) discrete unbranched spike-like racemes, these racemes usually discernible at a glance.
      9. Inflorescences usually with only 2 paired spike-like racemes at end of culm; plants of wet habitats ____ **Paspalum** (*P. distichum*)
      9. Inflorescences not as above; plants of various habitats.
         10. Second glume and lower lemma usually with brownish glandular blotches; spike-like racemes usually 2–5; in nc TX known only from Tarrant Co. ____ **Paspalum** (*P. langei*)
         10. Second glume and lower lemma without brownish glandular blotches; spike-like racemes 1–ca. 30; widespread in nc TX.
            11. First glume > 1/2 as long as the spikelet (excluding awns).
               12. First glume awnless ____ **Urochloa**
               12. First glume with an awn 5–10 mm long ____ **Oplismenus**
            11. First glume 1/2 or less as long as the spikelet.
               13. Glumes and lemmas awnless; first glume rounded apically; inflorescence branches winged or not winged.
                  14. Inflorescence branches winged; inflorescences with 2–6 branches ____ **Urochloa**

14. Inflorescence branches not winged; inflorescences with 7–17 branches ____ **Paspalidium**
13. Glumes or lemmas awned or awnless; first glume acute apically; inflorescence branches not winged ____ **Echinochloa**
8. Inflorescences without a conspicuous central axis, the branches usually rebranched, discrete spike-like racemes not easily discernable.
15. Spikelets asymmetric at base, the second glume with a pouch- or sac-like swelling at base; first glume greatly reduced, 1/4 or less as long as second glume; in nc TX known only from Lamar Co. in Red River drainage ____ **Sacciolepis**
15. Spikelets various, but not as above; widespread and extremely common in nc TX ____ **Panicum**

AEGILOPS

A genus of 21 species native from the Mediterranean to c Asia and Pakistan. Related to *Triticum* and one species is thought to have possibly contributed a genome to polyploid WHEAT. (Classical Latin name for wheat or a kind of wild oat) (subfamily Pooideae, tribe Triticeae)
REFERENCES: Gupta & Baum 1986, 1989.

Aegilops cylindrica Host, (cylindrical), JOINTED GOAT GRASS. Annual 35–75 cm tall; leaf sheaths glabrous except on margins; leaf blades with flaring base projecting to form small, sometimes indistinct auricles; ligule a membrane < 1 mm long; inflorescence a slender, cylindrical, pencil-like spike, the spikelets closely fitted into niches in the unbranched, flattened, bilateral axis of the inflorescence; disarticulation usually first in lowest node of inflorescence axis, the spike falling whole; spikelets sessile, 2–5-flowered, the upper florets reduced; glumes rounded, asymmetrical; lemmas and glumes of upper spikelets long-awned, the awns ca. 3–8 cm long. Roadsides and disturbed sites, limestone areas; Dallas, Denton, Grayson, and McLennan cos.; widespread in TX. Apr–Jun. Native of se Europe and adjacent Asia. [*Triticum cylindricum* (Host) Ces., Pass. & Gibson]

AGROSTIS BENT GRASS

Glabrous annuals or perennials; ligule a membrane; inflorescence an open panicle; spikelets 1-flowered; florets separating above glumes; glumes equaling or exceeding the lemma; paleas short or absent; lemmas awned or awnless.

A cosmopolitan, but especially n temperate genus of 220 species; a number are important for use in lawns and pastures. (Old Greek name for grass from *agros*, a field) (subfamily Pooideae, tribe Aveneae)
REFERENCE: Hitchcock 1905.

1. Lemmas awned (awns 4–8 mm long); plants annual ____ **A. elliottiana**
1. Lemmas awnless; plants perennial.
2. Spikelets 1.5–2.1 mm long; lemmas 1–1.3(–1.5) mm long; native and widespread in nc TX ____ **A. hyemalis**
2. Spikelets (at least some) (2–)2.2–3(–3.2) mm long; lemmas 1.3–2 mm long; introduced into nc TX or native only to the extreme ne portion.
3. Panicles with main branches rebranched only towards the tips, the spikelets clustered near the tips; primary panicle branches (at least some) usually 5–15 cm or more long; mainly w TX, rarely introduced elsewhere; spring flowering except in w TX (form of *A. hyemnalis* previously recognized as *A. scabra*) ____ **A. hyemalis**
3. Panicles with main branches rebranched near or below the middle, the spikelets not clustered near the tips; primary panicle branches usually 5 cm or less long, infrequently more; e TX w to Lamar Co.; usually flowering in fall ____ **A. perennans**

Agrostis elliottiana Schult., (for Stephen Elliott, 1771–1830, Carolinian botanist), ELLIOTT'S BENT GRASS, ANNUAL TICKLE GRASS. Annual 10–40 cm tall; leaf blades 1.6 mm or less wide; panicles delicate, becoming very open; spikelets 1.2–2.2 mm long; lemmas 1.1–2 mm long, with awn 4–8 mm long from just below apex, rarely awnless. Sandy soils, disturbed sites; Burnet, Dallas (*Reverchon*, Apr 1876), Denton, Grayson, Hunt, Navarro, and Tarrant cos.; se and e TX w to East Cross Timbers, also Edwards Plateau. Apr–May.

Agrostis hyemalis (Walter) Britton, Sterns, & Poggenb., (of winter), TICKLE GRASS, SPRING BENT GRASS, WINTER BENT GRASS, FLY-AWAY GRASS. Tufted perennial 15–75(–80) cm tall; leaf blades 0.5–3(–4) mm wide; panicles narrow at start of flowering, later becoming very open; primary panicle branches in whorls, at least some 5–15 cm or more long. Open areas, usually sandy soils; Dallas, Denton, Grayson, Hunt, and Lamar cos.; throughout much of TX. Late Apr–early Jun. [*A. scabra* Willd.] Gould (1975b) and Kartesz (1994) recognized *A. scabra* as a distinct species; however, we are following Jones et al. (1997) and S. Hatch (pers. comm.) in lumping this taxon with *A. hyemalis*.

Agrostis perennans (Walter) Tuck., (perennial), AUTUMN BENT GRASS. Clumped perennial 30–80(–100) cm tall; leaf blades 1–6 mm wide. Moist sand along wooded streams, open oak woods; Lamar Co. (Carr 1994); e TX w to ne corner of nc TX. Aug–Oct, occasionally in spring.

AIRA HAIR GRASS

A genus of 10 species of Europe and the Mediterranean region to Iran; also widely distributed as weeds. (An ancient Greek name for some grass, perhaps darnel—*Lolium temulentum* L.) (subfamily Pooideae, tribe Poeae)
REFERENCE: Brown & Peterson 1984.

Aira caryophyllea L. var. **capillaris** Mutel, (sp.: clove-like scent; var.: resembling hair, very slender), ANNUAL HAIR GRASS. Delicate tufted annual to ca. 35 cm tall; leaf blades usually 0.5 mm or less wide; ligules 1.5–4 mm long, white, membranous; panicles delicate, very open and diffuse, 4–12 cm long; spikelets 2-flowered, ca. 2 mm long excluding awn; glumes longer than the lemmas; lower lemma awnless; lemma of upper floret awned from near middle of back, the awn ca. 2 mm long. Sandy open areas; Grayson, Lamar, and Tarrant cos.; mainly e TX. Apr–Jun. Native of Europe. [*A. capillaris* Host, *A. elegans* Willd. ex Roem. & Schult., *A. elegantissima* Schur] Gould (1975b) and Kartesz (1994) treated this taxon as *A. elegans*; however, we are following Jones et al. (1997) and S. Hatch (pers. comm.) in recognizing it as *A. caryophyllea* var. *capillaris*.

Aira caryophyllea L. var. *caryophyllea*, was reported by Carr (1994) as occurring in Lamar Co. This European native is extremely similar to var. *capillaris* except the lower floret as well as the upper has an awn.

ALOPECURUS FOXTAIL

A genus of 36 species of the n temperate zone and South America. (Greek: *alopex*, fox, and *oura*, tail) (subfamily Pooideae, tribe Poeae)

Alopecurus carolinianus Walter, (of Carolina), CAROLINA FOXTAIL. Glabrous, green or usually blue-green annual 7–35 cm tall; ligule a membrane 3.5–7 mm long; inflorescence a dense, cylindrical, spike-like panicle 2–6 cm long and 4–6 mm in diam.; spikelets strongly compressed, 1-flowered; glumes silky-pubescent; lemmas with 3–5 mm awn from back near base, the awn geniculate medianly. Sandy soils; Bell, Grayson, and Wise cos., also Lamar Co. (Carr 1994); se and e TX w to nc TX, also Edwards Plateau. Mar–Jun.

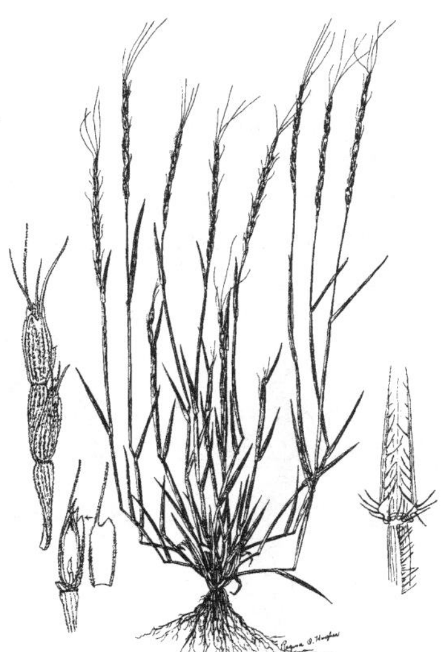

Aegilops cylindrica [REE]

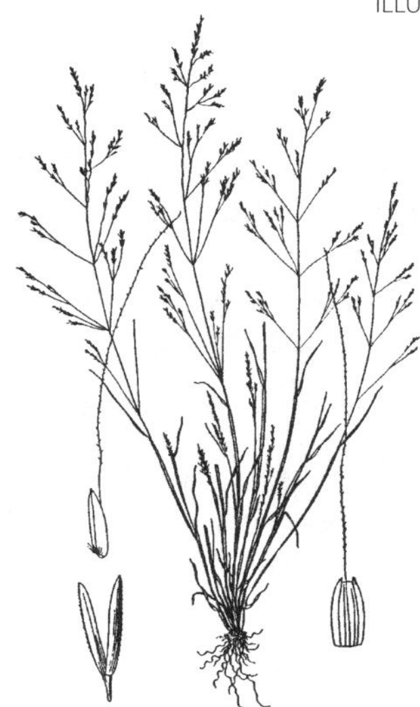

Agrostis elliottiana [USB]

Agrostis hyemalis [HI1]

Agrostis perennans [USB]

Aira caryophyllea var. capillaris [SIL]

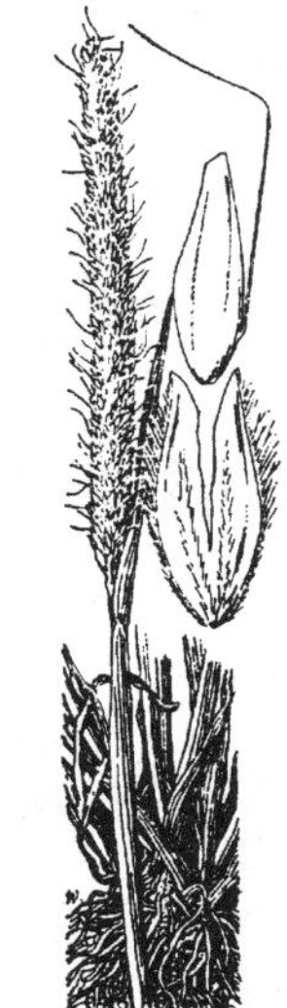

Alopecurus carolinianus [HI1]

Andropogon gerardii [GO1]

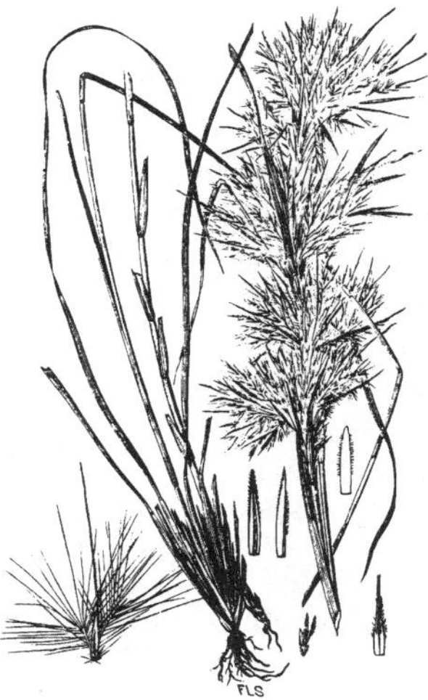

Andropogon glomeratus [USB]

Andropogon gyrans [USB]

ANDROPOGON BLUESTEM, BEARD GRASS

Perennials; culms stiffly erect; ligules membranous; inflorescences paniculate, of 2–several spicate branches; spikelets in pairs: one sessile, perfect, the other pedicellate, staminate, neuter, rudimentary, or absent (represented by pedicel only); disarticulation so that sessile spikelet falls with associated pedicel and section of the inflorescence branch; first glume large, firm, tightly clasping or enclosing second glume; perfect spikelet with 2 florets, the lower neuter, often vestigial; lemma of perfect (upper) floret awned or awnless.

A genus of ca. 100 species of tropical and warm areas of the world. It has sometimes been circumscribed more broadly to include such genera as *Bothriochloa*, *Dichanthium*, and *Schizachyrium*. (Greek: *aner (andr)*, man, and *pogon*, beard, referring to the hairy staminate spikelets) (subfamily Panicoideae, tribe Andropogoneae)
REFERENCES: Gould 1957a, 1967; Campbell 1983, 1986; Barnes 1986; Wipff 1996.

1. Sessile spikelets 7–11 mm long; terminal inflorescence conspicuously exserted beyond bracts, not woolly in appearance; pedicelled spikelets large, well-developed, similar to sessile except awnless ____ **A. gerardii**
1. Sessile spikelets 7 mm or less long; inflorescences either not conspicuously exserted beyond bracts OR if exserted then appearing woolly; pedicelled spikelets rudimentary, vestigial, or absent (= pedicel only).
   2. Each inflorescence with 2 relatively stiff and usually straight branches, the spikelets and hairs usually so dense that the branch axes are not easily visible; inflorescences often conspicuously exserted beyond bracts; sessile spikelets 5–7 mm long ____ **A. ternarius**
   2. Each inflorescence with 2–5 slender, delicate and flexuous branches, the spikelets and hairs often not as dense, the branch axes thus often visible, sometimes easily so; sessile spikelets 5 mm long or less; inflorescences not conspicuously exserted OR exerted in *A. gyrans*.
      3. Upper sheathing bracts of inflorescence inflated-spathe-like; inflorescences usually exerted beyond the bracts; sessile spikelets 4–5 mm long; culms with tufts of long hairs just below sheathing bracts of inflorescence; rare in nc TX, known only from Tarrant Co. ____ **A. gyrans**
      3. Upper sheathing bracts of inflorescence not inflated; inflorescences not conspicuously exerted; sessile spikelets 4 mm or less long; culms glabrous OR with tufts of long hairs just below sheathing bracts of inflorescence; extremely widespread and abundant in nc TX.
         4. Inflorescences crowded apically, broad, broom-like; culms with tuft of long hairs below sheathing bracts of inflorescence ____ **A. glomeratus**
         4. Inflorescences numerous but not clustered apically; culms glabrous below sheathing bracts of inflorescence ____ **A. virginicus**

Andropogon gerardii Vitman, (for Loius Gérard, 1733–1819, French botanist). Plant to 2 m tall, essentially glabrous, often glaucous; inflorescences of 2–7 branches, digitate; racemes 4–11 cm long; sessile spikelets usually scabrous, often glaucous; upper lemma of sessile spikelet awned or awnless (sometimes in subsp. *hallii*); pedicellate spikelets about as large as sessile one, awnless. Aug–Nov. The following key to subspecies is modified from Sutherland (1986) and Wipff (1996).

1. Upper lemma of sessile spikelet awned with an awn (7.5–)10.0–25.0 mm long; rhizomes short or absent, the internodes usually 2 mm or less in length; anthers usually < 3.8 mm long; ligules 0.4–2.5 mm long ____ subsp. **gerardii**
1. Upper lemma of sessile spikelet awnless or with an awn to 8(–11) mm long; rhizomes well-developed, creeping, the internodes often exceeding 20 mm in length; anthers usually > 3.8 mm long; ligules (0.9)3–4.5 mm long ____ subsp. **hallii**

subsp. **gerardii**, BIG BLUESTEM, TURKEYFOOT. In relatively undisturbed areas; throughout TX. One of "big four" tall grasses of the original native prairie along with *Panicum virgatum*, *Schizachyrium scoparium*, and *Sorghastrum nutans*. ▣/78

subsp. **hallii** (Hack.) Wipff, (presumably named for E.A. Hall who collected the type in Nebraska in 1862), SAND BLUESTEM. Sandy soils; a Montague Co. collection is apparently this subspecies, which is generally found to the w of nc TX; Rolling Plains, High Plains, and Trans-Pecos. [*A. gerardii* var. *paucipilus* (Nash) Fernald, *A. hallii* Hack., *A. paucipilus* Nash] The 2 subspecies sometimes hybridize (Barnes 1986) with the hybrids named *A. gerardii* subsp. ×*chrysocomus* (Nash) Wipff (Wipff 1996).

Andropogon glomeratus (Walter) Britton, Sterns, & Poggenb., (clustered), BUSHY BLUESTEM, BUSHY BEARD GRASS. Plant 0.75–1.5 m tall; inflorescences much-branched, crowded, broom-like; inflorescence branches 1.5–3 cm long; sheaths of terminal branchlets of inflorescences narrow, reddish brown; sessile spikelets 3–4.5 mm long; glumes glabrous; pedicellate spikelets rudimentary or absent. Roadsides, low moist areas; throughout TX. Sep–Nov.

Andropogon gyrans Ashe, (going around in circles, concentrically twisted and plaited backward and forward; the significance of the name unclear, not given in the type description), ELLIOTT'S BLUESTEM, ELLIOTT'S BLUE GRASS. Plant 0.3–0.80 m tall; inflorescences not broom-like; inflorescence branches 3–4(–5) cm long; upper sheathing bracts of inflorescence inflated-spathe-like, to 6–10 mm broad, 7–15+ cm long; awns of sessile spikelets 10–15 mm long; pedicellate spikelets rudimentary or nearly so. Usually in partial shade; Tarrant Co. (Campbell 1983); mainly se and e TX. [*A. elliottii* of authors, not Chapman] This species has long incorrectly gone under the name *A. elliottii* (Campbell 1983).

Andropogon ternarius Michx., (in threes), SPLIT-BEARD BLUESTEM, SPLIT-BEARD BEARD GRASS, FEATHER BLUESTEM, SILVERY BEARD GRASS. Plant 0.7–1.2 m tall; culms glabrous or with a tuft of long hairs below bract bearing node; inflorescences on lateral shoots at all upper nodes; inflorescence branches 3–6 cm long; sessile spikelets glabrous, awned; pedicellate spikelets slender, awnless, rudimentary, 2 mm or less long, not wider than pedicel. Sandy soils, woodland or woodland pastures; se and e TX w to West Cross Timbers. Sep–Nov.

Andropogon virginicus L., (of Virginia), BROOMSEDGE BLUESTEM, BROOMSEDGE, VIRGINIA BLUESTEM, YELLOWSEDGE BLUESTEM. Plant 0.50–1 m tall; culm nodes glabrous; inflorescences branched, not densely clustered; inflorescence branches 2–3 cm long; sheathing bracts of inflorescence usually straw-colored; sessile spikelets with awned lemma; pedicellate spikelets rudimentary or absent. Sandy soils; moist areas or slopes; Grayson, Hopkins, Kaufman, Lamar, and Limestone cos.; widespread in e 1/2 of TX. Sep–Nov. This species was inadvertently introduced to the Hawaiian Islands in 1932 and is there considered one of the most threatening alien species; it is a serious invader of native communities and alters the fire and hydrology regimes (Cronk & Fuller 1995).

ARISTIDA THREEAWN

Annuals or perennials; panicles open or contracted; spikelets 1-flowered, usually relatively large (long); glumes 1(–3)-nerved; lemmas hardened, terete, linear, with a sharp-pointed callus at base, with 3 awns; awn column present or absent; caryopsis permanently enclosed within the lemma.

A genus of ca. 230 species of warm areas of the world. The sharp calluses can be problematic for livestock and irritating when in shoes and socks. (Latin: *arista*, a beard or awn) (subfamily Aristidoideae, tribe Aristideae)

REFERENCES: Hitchcock 1924; Holmgren & Holmgren 1977; Allred 1984a, 1984b, 1985a, 1985b, 1986; Sutherland 1986.

1. Lateral awns of lemma much reduced, 1–2 mm long, erect; central awn of lemma 3–8 mm long, deflexed, with a spiral coil at base like a corkscrew ____ **A. dichotoma**
1. Lateral awns of lemma usually well-developed, (2–)4 mm or more long (often much longer), erect to spreading, horizontal, or even deflexed; central awn of lemma (5–)10–36 mm long, deflexed to erect, without a spiral coil at base OR in 1 rare species (*A. basiramea*) with a spiral coil at base.
  2. Awns of lemma, at least central one, spirally coiled at base like a corkscrew OR with a distinct semicircular bend at base.
    3. Lemmas 7–10 mm long (to base of awn); awn column with a well-defined joint at base, separating at the joint at maturity (check mature spikelets); central awn of lemma with a semicircular bend at base ____ **A. desmantha**
    3. Lemmas 4–7 mm long; awn column not jointed basally and not separating at maturity; central awn of lemma spirally coiled at base like a corkscrew ____ **A. basiramea**
  2. Awns of lemma nearly straight or curved, but neither spirally coiled basally nor with a distinct semicircular bend.
    4. Lemmas 16–28 mm long to base of awn ____ **A. oligantha**
    4. Lemmas 15 mm or less long to base of awn.
      5. Leaf sheaths (at least lower ones) lanate pubescent, the hairs cobwebby, kinked, and intertwined ____ **A. lanosa**
      5. Leaf sheaths not lanate pubescent, varying from glabrous to pilose, the hairs if peresent ± straight, not cobbwebby, and usually appressed.
        6. Panicles open, at least lower branches spreading.
          7. Awns of lemma 4–10 cm long; second glume usually 14–25 mm long; lemmas 12–16 mm long (to base of awns) ____ **A. purpurea** var. **longiseta**
          7. Awns of lemma 3–4.5 cm long; second glume usually 15 mm or less long; lemmas usually 10–12 mm long ____ **A. purpurea** var. **purpurea**
        6. Panicles contracted, the branches usually all stiffly appressed along the main axis.
          8. Lemmas narrowing into a slender twisted awn column (sometimes called a beak) 1–4 mm long ____ **A. purpurea** var. **nealleyi**
          8. Lemmas thick to base of awns; awn column absent.
            9. Awns 4–10 cm long ____ **A. purpurea** var. **longiseta**
            9. Awns 3.5 cm or less long.
              10. Plants annual; glumes about equal; central awns 5–30 mm or more long; lemmas (3–)4–9 mm long; lateral awns sometimes reduced and 1/2 or less the length of the central awn OR well-developed ____ **A. longespica**
              10. Plants perennial; glumes about equal OR first glume half to three-fourths as long as second; central awns 15–30 mm long; lemmas 4–15 mm long; lateral awns well-developed, at least 1/2 the length of the central awn.
                11. First (= lower) glume ± as long as to slightly longer than the mostly 5–10 mm long second glume; lemma 4–8 mm long ____ **A. purpurascens**
                11. First glume 1/2–3/4 as long as the 11–15 mm long second glume; lemma ca. 11–15 mm long ____ **A. purpurea** var. **wrightii**

Aristida basiramea Engelm. ex Vasey, (branching from base), FORK-TIP THREEAWN. Annual; panicles contracted; glumes about equal, up to 10 mm long including awn; lemmas 4–7 mm long, central awn 9–15 mm long, conspicuously spirally coiled. Sandy soils; Dallas Co.; known only from Bastrop, Dallas, and Red River cos. in TX (Gould 1975b). Aug–Oct. According to Allred (1986) this species is similar to *A. dichotoma* and could be treated as a variety of that species. The 2 are distinguished by *A. basiramea* having spreading lateral awns usually 4–10 mm long (vs. erect, 1–2 mm long in *A. dichotoma*).

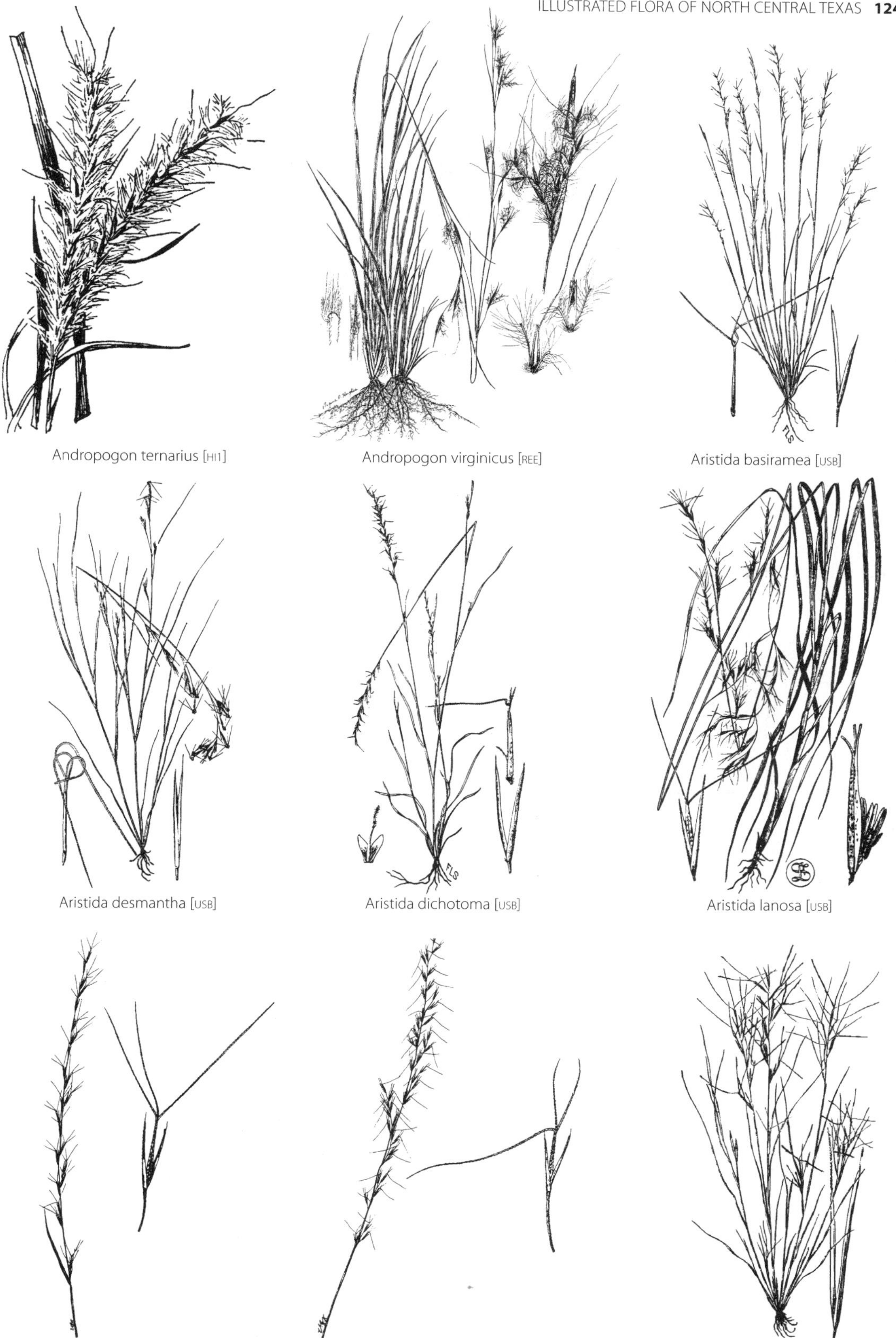

Andropogon ternarius [HI1]

Andropogon virginicus [REE]

Aristida basiramea [USB]

Aristida desmantha [USB]

Aristida dichotoma [USB]

Aristida lanosa [USB]

Aristida longespica var. geniculata [GO1]

Aristida longespica var. longespica [HI1]

Aristida oligantha [USB]

Aristida desmantha Trin. & Rupr., (with clustered flowers), CURLY THREEAWN, WESTERN TRIPLEAWN, WESTERN THREEAWN. Annual; leaf sheaths glabrous to lanate pubescent; panicles loosely contracted, 8–20 cm long; spikelets light yellowish or golden-brown; glumes 1-nerved, awn-tipped, about equal; lemmas 7–10 mm long, the awns about equal, spreading; awn column jointed basally, well-defined. Sandy soils; Dallas (*Reverchon*, 1876; *Stillwell*, 1935), Limestone, Palo Pinto, and Parker cos.; se and e TX w to nc TX. Sep–Nov.

Aristida dichotoma Michx., (2-parted or forked), CHURCH-MOUSE THREEAWN, POVERTY GRASS. Annual; panicles contracted or racemes spike-like; glumes about equal; lemmas 4–6 mm long; central awn 3–8 mm long, spirally coiled basally; lateral awns 1–2 mm long, erect. Sandy soils; Dallas, Grayson, and Lamar cos.; e TX w to Blackland Prairie. Aug–Nov.

Aristida lanosa Muhl. ex Elliott, (woolly), WOOLLY-SHEATH THREEAWN, WOOLLY TRIPLEAWN GRASS, WOOLLY THREEAWN. Perennial; panicles with appressed or somewhat spreading branches; glumes subequal, 9–15 mm long including the awn when present; lemmas 8–9 mm long, without a well-defined awn column; lemma awns only curved below. Similar in appearance to *A. desmantha* which also has leaf sheaths with lanate pubescence. However, *A. desmantha* has lemma awns with a semicircular bend and a distinct awn column well-defined from the lemma. Woods openings, often in sandy soils; Lamar and Montague cos.; mainly e and se TX. Late Aug–Nov.

Aristida longespica Poir., (long-spiked), SLIM-SPIKE THREEAWN. Annual; leaves not in a conspicuous basal tuft; panicles contracted, narrow and spike-like; glumes about equal, (3–)4–9 mm long; central awn erect to reflexed. Two varieties, not always easily distinguished, occur in nc TX; Allred (1986) noted that intermediates between the 2 are not uncommon; however, he further indicated that since the extremes are so strikingly different, distinction at the varietal level seems appropriate. Sandy open areas. Late Aug–Dec.

1. Lemmas usually (3.5–)7–10 mm long; central awn usually (8–)12–36 mm long; lateral awns usually 2/3–3/4 as long as central, usually 6–18 mm long ____ var. **geniculata**
1. Lemmas usually 2.5–7 mm long; central awn usually 5–10(–15) mm long; lateral awns much shorter, usually 1/3–slightly more than 1/2 as long as central, usually 2–5(–8) mm long ____ var. **longespica**

var. **geniculata** (Raf.) Fernald, (jointed), KEARNEY'S THREEAWN, PLAINS THREEAWN. Montague, Parker, and Tarrant cos.; widespread in e 1/2 of TX. Aug–Dec. [*A. intermedia* Scribn. & C.R. Ball]

var. **longespica**. SLIM-SPIKE THREEAWN, SLENDER THREEAWN. Bell, Bosque, Dallas, Grayson, and Parker cos.; se and e TX w to West Cross Timbers.

Aristida oligantha Michx., (few-flowered), OLDFIELD THREEAWN, PRAIRIE THREEAWN, FEW-FLOWER ARISTIDA. Annual; panicles open, much branched; glumes subequal, 18–25 mm long, the second awned, the awn up to 1 cm long; lemmas 16–28 mm long, the awns about equal, 3–7 cm long. Calcareous soils; throughout TX. Jun–Nov. This species apparently has allelopathic effects on other plants; it is also often seen growing on seed harvester ant mounds (J. Stanford, pers. comm.).

Aristida purpurascens Poir., (purplish), ARROW-FEATHER THREEAWN, BROOMSEDGE, ARROW GRASS. Tufted perennial; basal leaves persistent after drying, curly; panicles narrow, contracted; lemmas 4–8 mm long, purple or mottled with purple at maturity; central awns 15–30 mm long. Similar to *A. longespica* except *A. purpurascens* is perennial. Woods openings, borders, and prairies, usually on sandy soils; Dallas, Fannin, Grayson, and Hunt cos.; se and e TX w to Blackland Prairie, also Edwards Plateau. Aug–Nov.

Aristida purpurea Nutt., (purple). Perennial; panicles open or contracted. A highly variable species with 4 varieties in nc TX that have in the past been treated as separate species. Because of extensive intergradation, we are following most recent authors (e.g., Holmgren & Holmgren 1977, Allred 1984b, Sutherland 1986, Kartesz 1994) in treating them as varieties of *A. purpurea*. Most plants may be separated by the characters given in the key to species and varieties.

var. **longiseta** (Steud.) Vasey, (long-bristled), RED THREEAWN, DOGTOWN GRASS, LONG-AWNED ARISTIDA, LONG-AWNED THREEAWN. Panicles contracted or open; glumes unequal, the first 0.5–0.6 as long as second, the second 14–25 mm long; lemmas 12–16 mm long, the awns about equal, 4–10 cm long. Disturbed sites; mainly Blackland Prairie s and w to w TX. Mar–Dec. [*A. longiseta* Steud., *A. reverchonii* Vasey]

var. **nealleyi** (Vasey) Allred, (for Greenleaf Alley Nealley, 1864–1896, botanist), BLUE THREEAWN. Panicles contracted; glumes unequal, the first glume about half as long as second; lemmas slightly longer than second glume; apex of lemma twisted into awn column 1–4 mm long; awns slightly unequal, 15–20(–30) mm long. Rocky, usually limestone soils; Coleman and Somervell cos., also Hood Co. (Mahler 1988); w part of nc TX s and w to w TX. May–Oct. [*A. glauca* (Nees) Walp.]

var. **purpurea**, (purple), PURPLE THREEAWN, PURPLE NEEDLE GRASS. Panicles open, curving, the branches flexuous; glumes unequal, the second 11–15 mm long, up to twice the length of the first; lemmas 10–12 mm long, the awns about equal, 3.5–4.5 cm long. Sandy or rocky soils; Post Oak Savannah s and w to w TX. Apr–Oct.

var. **wrightii** (Nash) Allred, (for Charles Wright, 1811–1885, TX collector), WRIGHT'S THREEAWN, WRIGHT'S TRIPLEAWN GRASS. Panicles contracted; glumes unequal, the second 11–15 mm long, the first 0.5–0.75 as long as second; lemmas equaling or exceeding second glume, the awns about equal, 15–30 mm long, the central occasionally longer than laterals. Calcareous soils; Blackland Prairie s and w to w TX. May–Oct. [*A. wrightii* Nash]

ARUNDINARIA

A genus of ca. 50 species of bamboos native to China, Japan, and the Americas. (Latin: *arundo*, a reed) (subfamily Bambusoideae, tribe Bambuseae)
REFERENCE: Platt & Brantley 1997.

Arundinaria gigantea (Walter) Muhl., (gigantic), GIANT CANE, SOUTHERN CANE, SWITCH CANE. Much branched woody perennial 2–5(–8) m tall from rhizome; leaves variable, the lower often reduced; upper leaf blades usually 15–25 cm long, 2–4(–5.5) cm wide; ligule a firm membrane 1.5 mm or less long; inflorescences racemose or narrowly paniculate; spikelets 1–few, large, 4–7(–8) cm long, ca. 8 mm broad, with 6–12(–13) flowers; lemmas (10–)15–25 mm long. Moist woods or low areas; can form dense stands known as canebrakes; much reduced with the introduction of domestic livestock; Lamar Co. in Red River drainage, also Henderson Co. on e margin of nc TX; mainly se and e TX, also Edwards Plateau and Trans-Pecos. Mostly Apr–May. This is the only native bamboo species in TX.

ARUNDO

A genus of 3 species native to the Mediterranean and Taiwan. (Latin: *arundo*, a reed) (subfamily Panicoideae, tribe Arundineae)

Arundo donax L., (classical name), GIANT REED. Perennial, rhizomatous; extremely large, the culms 2–6 m tall, usually unbranched; leaf blades usually 4–7 cm wide on main culms; panicles 30–60 cm long; branchlets and rachilla joints essentially glabrous; spikelets usually with 2–4 florets; glumes glabrous; lemmas densely hairy. Wet areas of ponds and roadsides; usually tight, clay soils; cultivated along highways; apparently not setting fertile seed, spreads vegetatively; nearly throughout TX. Mostly Sep–Nov. Native to Mediterranean region. This is the "reed" of the Bible and has been used for 5,000 years for pipe instruments (Mabberley 1987).

AVENA OATS

Erect annuals 30–120 cm tall; ligule a whitish membrane 2–4 mm long; inflorescence an open panicle of ca. 8–30 large pendulous spikelets; spikelets 2–4-flowered; glumes ca. 17–30 mm long, longer than the lemmas, acute to acuminate; lemmas awned from near middle of the back or awnless.

A mainly temperate Old World genus of ca. 25 species ranging from Europe to the Mediterranean region and Ethiopia. (Classical Latin name for oats) (subfamily Pooideae, tribe Aveneae)
REFERENCES: Baum 1968, 1977, 1991; Zohary & Hopf 1994.

1. Awn of lemmas geniculate, mostly 2.5–4 cm long; lemmas with stiff, usually reddish brown hairs on dorsal surface; spikelets usually 3–4-flowered ____ **A. fatua**
1. Awn of lemmas not geniculate, irregularly developed, to < 3 cm long or absent; lemmas glabrous; spikelets usually 2-flowered ____ **A. sativa**

Avena fatua L., (simple), WILD OATS. Roadsides and other disturbed areas; Tarrant Co.; scattered in TX. Mostly Apr–May. Native to the Mediterranean area. The pointed callus of the fruit is reported to cause mechanical injuries (Burlage 1968).

Avena sativa L., (cultivated), COMMON OATS, CULTIVATED OATS. Commonly cultivated for grain, often planted to prevent washing on newly graded roadsides, also naturalized and common as a weed on roadsides and other disturbed areas; throughout TX. Mostly Mar–Jun. Native of the Mediterranean area where it has long been a cultivated cereal crop. [*A. fatua* L. var. *sativa* (L.) Hausskn.] While OATS is thought to be a later domesticate than WHEAT, it is one of the most nutritious of the cereals, having a high protein content; this hexaploid species probably evolved from weeds invading the fields of early WHEAT and BARLEY farmers (Heiser 1990a; Zohary & Hopf 1994). It is possibly best treated as a variety (e.g., Jones et al. 1997) or subspecies of *A. fatua*, which is apparently involved in its ancestry (Zohary & Hopf 1994). High nitrate concentrations and fungal contaminants can result in the loss of cattle and horses upon ingesting OAT hay (Lewis & Elvin-Lewis 1977).

AXONOPUS CARPET GRASS

A genus of 35 species of tropical and warm areas of the Americas and Africa. (Greek: *axon*, axis, and *pous*, foot) (subfamily Panicoideae, tribe Paniceae)
REFERENCES: Chase 1938; Black 1963; Crins 1991.

Axonopus fissifolius (Raddi) Kuhlm., (split-leaved), COMMON CARPET GRASS. Perennial, cespitose, stoloniferous, forming carpets but the flower-bearing culms erect; culms 20–75 cm long; leaves essentially glabrous; leaf blades 6–17(–28) cm long, 1.5–7(–9) cm wide, flat, blunt; ligule a membrane ca. 0.3 mm long; inflorescences of 2–4 slender branches, the upper 2 usually paired and spreading at culm tip, the branches ca. 2.5–8(–10) cm long, ca. 1 mm wide, slightly winged; spikelets awnless, 2-flowered, the lower floret staminate or neuter, the upper floret perfect, 1.8–2.2(–2.6) mm long; first glume absent; perfect lemma hardened, glabrous, with inrolled margins, with rounded back facing away from axis. Moist sandy woods, margins of wet areas, roadsides; Henderson, Limestone, and Milam cos., near extreme e margin of nc TX; mostly se and e TX. (Feb) May–Nov(–Dec). [*A. affinis* Chase]

BOTHRIOCHLOA BLUESTEM, BEARD GRASS

Annuals or perennials; ligules membranous; inflorescences paniculate with racemose

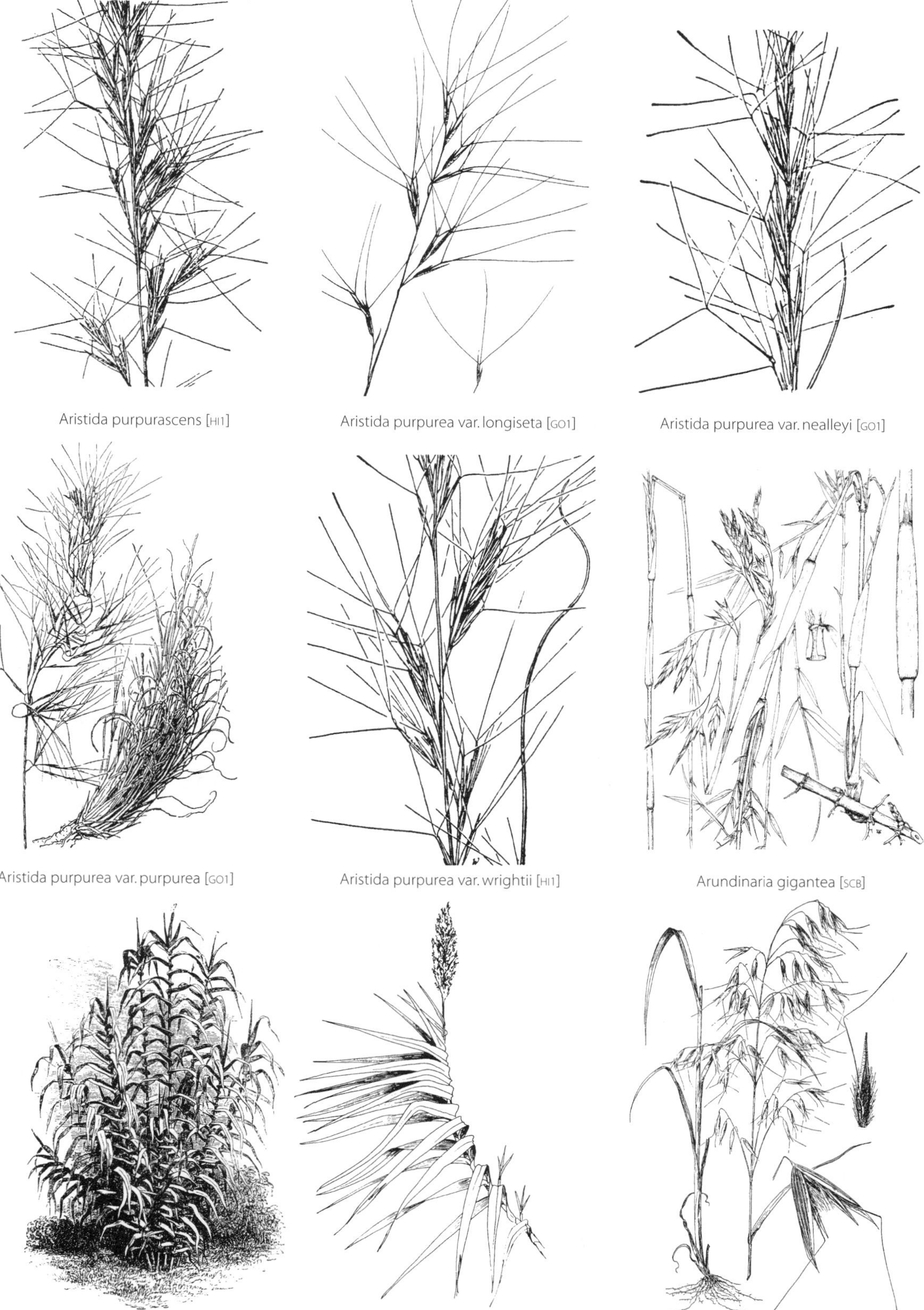

Aristida purpurascens [HI1]

Aristida purpurea var. longiseta [GO1]

Aristida purpurea var. nealleyi [GO1]

Aristida purpurea var. purpurea [GO1]

Aristida purpurea var. wrightii [HI1]

Arundinaria gigantea [SCB]

Arundo donax [LYN, NIC]

Avena fatua [HI1]

branches; spikelets in pairs, one sessile, one pedicelled; pedicels and upper internodes of inflorescence branches with a central groove or broad membranous area; disarticulation at base of sessile spikelet so that associated pedicel and section of inflorescence branch fall with the sessile spikelet; sessile spikelets perfect, 2-flowered, the lower floret sterile, the upper floret fertile; first glume dorsally flattened, not enclosing the second glume; second glume with a rounded median keel; lemma of upper floret usually awned; pedicellate spikelets well-developed, but often much smaller and narrower than sessile spikelets, neuter or staminate, awnless.

A genus of ca. 35 species of warm areas of the world; a number are cultivated for fodder; formerly treated in a more inclusive *Andropogon*. (Greek: *bothrion*, a shallow pit, and *chloa*, grass, from pitted glumes of some species) (subfamily Panicoideae, tribe Andropogoneae)
REFERENCES: Gould 1953, 1957a, 1959b, 1967; Shinners 1956e; Celarier & Harlan 1958; Allred 1981; Allred & Gould 1983.

1. Inflorescences purplish, open, with 2–8 branches; inflorescence axis shorter than branches, the inflorescence appearing nearly digitate; pedicelled spikelets about as large and broad as sessile spikelets ____ **B. ischaemum**
1. Inflorescences whitish silvery, contracted, with panicle branches often numerous; inflorescence axis usually longer than branches, the inflorescence not appearing digitate; pedicelled spikelets conspicuously smaller and narrower than sessile spikelets.
   2. Sessile spikelets 4.5–7.3 mm long; awn of lemma 20–30 mm or more long; first glume of sessile spikelet with or without a glandular pit; culm nodes with a conspicuous ring of short, spreading, white hairs ____ **B. barbinodis**
   2. Sessile spikelets < 4.5 mm long; awn of lemma 18 mm or less long; first glume of sessile spikelet without a glandular pit; culm nodes without a conspicuous ring of short, spreading, white hairs ____ **B. laguroides**

Bothriochloa barbinodis (Lag.) Herter, (bearded at nodes). Panicles included to long exserted; lemma awn of sessile spikelet geniculate. Sandy or rocky limestone soils. Mostly May–Oct.

1. First glume of most sessile spikelets without a glandular pit or depression ____ var. **barbinoidis**
1. First glume of most sessile spikelets with a glandular pit or depression (like a conspicuous "pinhole" in the surface of the glume) ____ var. **perforata**

var. **barbinodis**, CANE BLUESTEM, CANE BEARD GRASS, BRISTLE-JOINT BLUESTEM. Bosque, Brown, Hamilton, Palo Pinto, and Young cos.; Grand Prairie s and w to w TX.

var. **perforata** (Trin. ex E. Fourn.) Gould, (perforated), PINHOLE BLUESTEM, PINHOLE BEARD GRASS, PERFORATED BLUESTEM. Bosque, Palo Pinto, and Tarrant cos.; w part of nc TX s and w to w TX.

Bothriochloa ischaemum (L.) Keng var. **songarica** (Rupr. ex Fisch. & C.A. Mey.) Celerier & Harlan, (sp.: from Greek: *ischaemos*, blood-restraining, from supposed styptic properties; var.: of Dzungaria, central Asia), KING RANCH BLUESTEM, KR BLUESTEM. Plant becoming rhizomatous or stoloniferous with mowing or grazing; lemma awn of sessile spikelet geniculate, 1–1.5 mm long. Calcareous soils, roadsides, fields; throughout TX. Mostly May–Nov. Native of c and e Asia. A pernicious weed crowding out native species.

Bothriochloa laguroides (DC.) Herter. subsp. **torreyana** (Steud.) Allred & Gould., (sp.: like *Lagurus*—hare's-tail grass; subsp.: for John Torrey, 1796–1873, botanist and physician, co-author with Asa Gray, described many w American plants), SILVER BLUESTEM, SILVER BEARD GRASS. Plant 50–130 cm tall; panicles usually exerted. Dry, often sandy soils, increases under disturbance, extremely common along roadsides; throughout TX, more commonly Blackland Prairie and w. Mostly May–Nov. [*Bothriochloa longipaniculata* (Gould) Allred & Gould, *B. saccharoides* (Sw.) Rydb. var. *longipaniculata* (Gould) Gould, *B. saccharoides* (Sw.) Rydb. var. *torreyana*

(Steud.) Gould] While this taxon has been variously treated in the past (e.g., Gould 1975b; Kartesz 1994), we are following Jones et al. (1997) and S. Hatch (pers. comm.) in recognizing it as *B. laguroides* subsp. *torreyana*.

BOUTELOUA GRAMA GRASS

Perennials or annual (*B. barbata*); culms erect, tufted; rhizomes present or absent; leaves mostly basal; leaf blades usually flat, involute apically; inflorescences of 1-many short, spike-like branches; spikelets sessile, with 1 perfect floret and 1 or more staminate or neuter florets above; glumes 1-nerved; lemmas 3-nerved.

An American genus of 24 species ranging from Canada to Argentina, especially in Mexico. It includes a number of valuable native forage grasses; some are also used as ornamentals. All species are C_4 plants with typical Kranz leaf anatomy (Gould 1979); these adaptations allow more effective capture of carbon dioxide and thus less water loss through transpiraton (since stomata do not have to be as open for gas exchange), an advantage in arid environments. (Named for Claudio Boutelou, 1774–1842, a Spanish writer on floriculture and agriculture) (subfamily Chloridoideae, tribe Cynodonteae)

REFERENCES: Griffiths 1912; Featherly 1931; Gould & Kapadia 1962, 1964; Kapadia & Gould 1964a, 1964b; Roy & Gould 1971; Gould 1979; Wipff & Jones 1996.

1. Spikelets usually 3–7 per inflorescence branch, not pectinately arranged on inflorescence branches; inflorescence branches ca. 1.5 cm or less long, falling entire (section *Bouteloua*).
   2. Inflorescence branches 25 to numerous per inflorescence ____ **B. curtipendula**
   2. Inflorescence branches 10 or less per inflorescence ____ **B. rigidiseta**
1. Spikelets usually 8–90 per inflorescence branch, with a striking pectinate (= comb-like) arrangement on inflorescence branches; inflorescence branches 1–5 cm long, not falling entire, the florets separating from persistent glumes (section *Chondrosioides*).
   3. Plants tufted annuals; inflorescence branches 1–3 cm long (usually averaging ca. 2 cm), usually < 3 mm wide including awns ____ **B. barbata**
   3. Plants perennials; inflorescence branches 1.5–5 cm long (typically > 3 cm except in *B. trifida*), usually 3 mm or more wide including awns.
      4. Inflorescence branches ending with a naked extension of branch axis or axis-like rudimentary spikelet extended well beyond the terminal normal spikelet, the inflorescence branches thus with conspicuous, elongated, sharp points (these definitely different from awns).
         5. Tuft of hairs absent at base of the lowermost rudimentary floret of each spikelet; culms 15–40 cm tall, decumbent at base, usually branched, with 4–6 nodes; anthers 2–2.5 mm long; inflorescences 10–30 cm long (above uppermost leaf); inflorescence branches ending with a naked extension of branch axis extended well beyond the terminal normal spikelet ____ **B. hirsuta**
         5. Tuft of hairs present at base of the lowermost rudimentary floret of each spikelet; culms 35–75 cm tall, strictly erect, unbranched, usually with three nodes; anthers ca. 3 mm long; inflorescences 25–40 cm long (above uppermost leaf); inflorescence branches ending with axis-like rudimentary spikelet extending beyond normal spikelets ____ **B. pectinata**
      4. Inflorescence branches ending in a spikelet (this spikelet can be reduced and pointed), without conspicuous, long, sharp points (awns often the terminal most part of the inflorescence branches).
         6. Inflorescence branches usually 1–3 per main inflorescence axis, spreading, 1.5–6 cm long (at least some usually > 3 cm long) ____ **B. gracilis**
         6. Inflorescence branches 3–8 per main inflorescence axis, usually appressed to axis, < 2(–2.5) cm long ____ **B. trifida**

Bouteloua barbata Lag., (barbed), SIXWEEKS GRAMA. Culms to 30 cm; inflorescence branches usually 4–6 per main axis; axis of inflorescence branches not extended beyond spikelets. Dry grasslands, roadsides, and wasteplaces, typically sandy soils; city weed in Tarrant Co.; mainly Rolling Plains s and w to w TX. Apr–Nov.

Bouteloua curtipendula (Michx.) Torr., (short-hanging), SIDE-OATS GRAMA. **State grass of Texas.** Rhizomatous; inflorescence branches usually 25–80, about 1.5 cm long, along 1 elongate main axis terminating a leafy culm; fertile lemma awnless. Jun–Nov. This species is extremely variable morphologically, ecologically, and in terms of chromosome number; it often reproduces apomictically (Gould 1959a). It was adopted as the state grass by the 62nd Texas Legislature in 1971 (Jones et al. 1997).

1. Plants without creeping rhizomes (base can sometimes be knotty); culms in large or small clumps, stiffly erect ____ var. **caespitosa**
1. Plants with creeping rhizomes; culms not in large clumps, decumbent or stiffly erect ____ var. **curtipendula**

var. **caespitosa** Gould & Kapadia, (tufted). Mostly on loose, limey soils; Erath, Palo Pinto, and Tarrant cos.; w part of nc TX s and w to w TX.

var. **curtipendula**. On better soils and little disturbed areas including native prairies; throughout TX. This is the predominant variety throughout nc TX.

Bouteloua gracilis (Willd. ex Kunth) Lag. ex Griffiths, (graceful), BLUE GRAMA. Inflorescence branches 1–3(–4) per main axis, up to 5 cm long, curved; axis of inflorescence branches not extending beyond spikelets; apical rudimentary spikelets absent. Grasslands; Archer and Jack cos.; West Cross Timbers s and w to w TX. Jul–Oct.

Bouteloua hirsuta Lag., (hairy), HAIRY GRAMA. Inflorescence branches 1–4 per main axis, usually 2.3–4 cm long (including branch axis tip), straight to curved; axis of inflorescence branches continuous and extending beyond spikelets, usually scabrous; chromosome number quite variable, $2n = 20–60$ (Roy 1968). Grasslands and a variety of other habitats; throughout TX. Jun–Nov.

Bouteloua pectinata Feath., (comb-like), TALL GRAMA. Inflorescence branches usually 3–5 per main axis, 2.5–4.5 cm long, straight to slightly curved; axis of inflorescence branches not extended but with axis-like rudimentary spikelet extending beyond spikelets; axis-like rudimentary spikelet usually hairy at base, its apex usually visibly bifid under maginification; $2n = 20$. Limestone outcrops, hilltops, well-drained calcareous soils; Bell, Brown, Burnet, Coryell, Grayson, Hood, Johnson, Parker, Somervell, and Williamson cos., also Bosque, Erath, Hamilton, Lampasas, Wise (Roy 1968), and Tarrant (Featherly 1931) cos.; nc TX and Edwards Plateau; endemic to TX and OK. Mostly Jul–Aug, with a much shorter flowering period than *B. hirsuta*. [*B. hirsuta* var. *pectinata* (Feath.) Cory, *B. hirsuta* subsp. *pectinata* (Feath.) Wipff & S.D. Jones] While Wipff and Jones (1996) and Jones et al. (1997) recognized this taxon at the subspecific level, we are following Roy and Gould (1971), whose biosystematic investigation supported its recognition as a separate species. Even though hybridization can occur where the two grow together, they differ in numerous morphological characters and are usually easily distinguished in the field. According to Gould (1979), "... the morphological uniformity of this species contrasts strikingly with the variability observed in populations of plants of *B. hirsuta*." ▣/81

Bouteloua rigidiseta (Steud.) Hitchc., (stiff-awned), TEXAS GRAMA, MESQUITE GRASS. Inflorescence branches 6–8(–10), about 1 cm long, along 1 main axis terminating a leafy culm; fertile lemma 3-awned; single rudimentary floret usually reduced to an awn column, 3-awned. Grasslands; nearly throughout TX. Mar–Oct.

Bouteloua trifida Thurb., (three-parted), RED GRAMA, THREEAWN GRAMA. Inflorescence branches 3–7 per main axis, up to 2(–2.5) cm long, slender and appressed to main axis; axis of inflores-

Avena sativa [NVE]

Axonopus fissifolius [GO1]

Bothriochloa barbinodis var. barbinodis [HI1]

Bothriochloa barbinodis var. perforata [HI1]

Bothriochloa ischaemum var. songarica [GO1]

Bothriochloa laguroides subsp. torreyana [GO1, HEA]

Bouteloua barbata [GO3]

Bouteloua curtipendula var. curtipendula [USB]

Bouteloua gracilis [GO3]

cence branches not extending beyond spikelets; apical rudimentary spikelet absent. Grasslands; Bell Co., also Brown, Comanche, and Hamilton cos. (HPC); throughout much of TX except Blackland Prairie and far e TX. Apr–Nov.

BRACHYPODIUM FALSE BROOM

A genus of 17 species of temperate Eurasia, tropical America, and tropical mountains. (Greek: *brachy*, short, and *podos*, foot, from the short thick spikelets in some species) (subfamily Pooideae, tribe Brachypodieae)

Brachypodium distachyon (L.) P. Beauv., (two-spiked), PURPLE FALSE BROOM. Annual usually 20–50 cm tall; ligules 1–2 mm long, erose apically; inflorescence a spike or spicate raceme with 1–5 large spikelets; spikelets 2–3.5 cm long, usually with 9–18 imbricated florets; lower glume 3- to 7-nerved; upper glume 7- to 9-nerved; lemmas 7-nerved, with 1–2 cm long awn from entire apex; paleas pectinate-ciliate on nerves. Roadsides and disturbed sites; included based on citation of vegetational area 5 by Gould (1975b) and Hignight et al. (1988); also Edwards Plateau. Spring. Native of Europe.

BRIZA QUAKING GRASS, SHAKING GRASS

Annuals (our species) with open panicles of awnless spikelets on long pedicels; spikelets usually with 3–14(–20) florets, the florets crowded and widely spreading; lemmas usually broader than long, rounded apically.

A genus of 20 species of temperate Eurasia and South America including cultivated ornamentals. (Named from Greek word for a kind of grain) (subfamily Pooideae, tribe Aveneae)

1. Spikelets large, mostly 12–25 mm long; inflorescences usually with only 1–6(–12 or more) spikelets ____ **B. maxima**
1. Spikelets small, 2–6 mm long; inflorescences with many spikelets ____ **B. minor**

Briza maxima L., (largest), BIG QUAKING GRASS. Plant to ca. 60 cm tall, glabrous or leaf blades minutely scabrous; ligule of uppermost leaf usually 10 mm or more long; spikelets 12–25 mm long, 8–12 mm wide, with 7–14(–20) florets, not markedly tapered, longer than wide, on long, slender, drooping pedicels. Introduced as a garden ornamental; included based on citation for vegetational area 5 by Hignight et al. (1988); mainly se and e TX. Apr–May. Native of Europe.

Briza minor L., (smaller), LITTLE QUAKING GRASS. Plant to ca. 50 cm tall, glabrous or nearly so; ligules 5–10 mm long; spikelets ca. 2–6 mm wide, with 3–8 florets, markedly tapered toward apex, often almost triangular in shape, about as wide as long, pendulous on long usually kinked pedicels. Open areas in woods, fields, disturbed places, sandy areas; Grayson and Tarrant cos., also Lamar Co. (Carr 1994); mainly se and e TX. Mostly Apr–May. Native to Europe. Reported to contain cyanide (Burlage 1968).

BROMUS BROME GRASS, CHESS

Plants annual or perennial; leaf sheaths closed except at summit; ligule a membrane, often prominent; inflorescence a usually ± 1-sided (barely so in *B. hordeaceus*) panicle or infrequently a raceme; spikelets pedicelled, with 4–numerous florets; lemmas 2-toothed at apex (inconspicuously in *B. catharticus*), usually awned, with awn arising between the teeth, or awnless.

A genus of ca. 100 species of temperate regions and tropical mountains; some are used as ornamentals and for forage. (An ancient Greek name for oats; from *broma*, food) (subfamily Pooideae, tribe Bromeae)
REFERENCES: Wagnon 1952; Soderstrom & Beaman 1968.

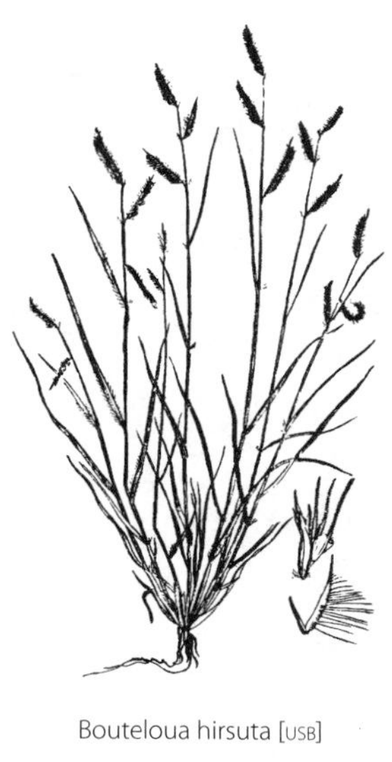

Bouteloua hirsuta [USB]

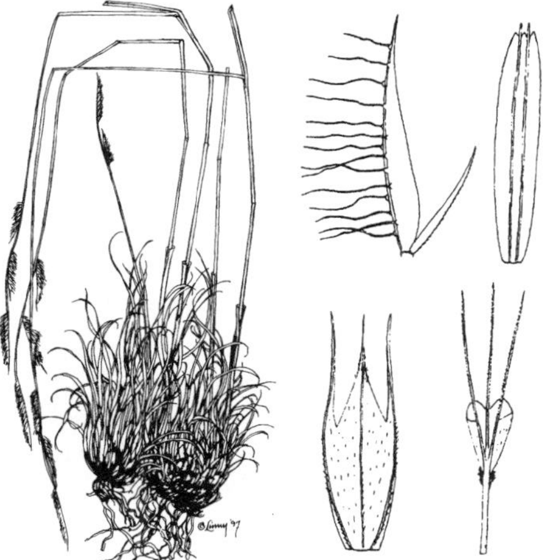

Bouteloua pectinata [BOT, HEA]

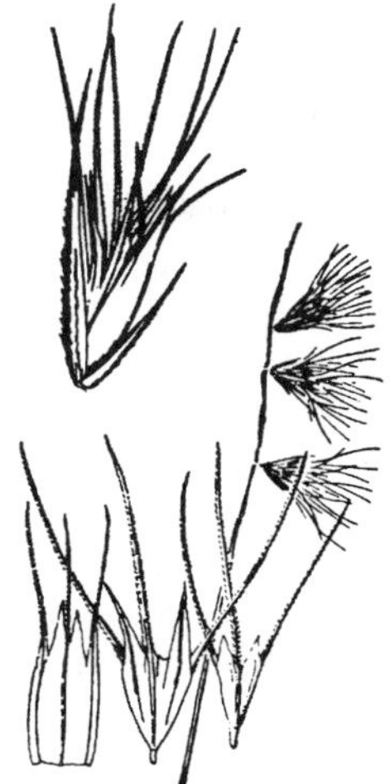

Bouteloua rigidiseta [HI1]

Bouteloua trifida [USB]

Brachypodium distachyon [GLE]

Briza maxima [LAM]

Briza minor [BB2]

Bromus catharticus [USB]

Bromus hordeaceus [USB]

1. Glumes and lemmas sharply keeled, the spikelets strongly flattened; lower lemmas 2.5–3.2 mm wide from keel to margin, awnless or with a short awn 1–3 mm long, with apical teeth 0.1–0.3 mm long **B. catharticus**
1. Glumes and lemmas not sharply keeled, the spikelets not strongly flattened; lower lemmas 1.0–2.7 mm wide from keel to margin, with an awn 4–18 mm long, with apical teeth 0.2–5 mm long (sometimes appressed, inconspicuous).
   2. Apical teeth of lemmas 2–5 mm long, thin, whitish or transparent, lance-linear, acuminate; awn of lemmas 12–18 mm long **B. tectorum**
   2. Apical teeth of lemmas 0.2–2.5 mm long, papery, opaque, triangular or ovate, acute or obtuse; awn of lemmas 4–13 mm long.
      3. Lower glume usually 1-nerved; upper glume narrowly oblong-lanceolate, 0.8–1.3 mm wide from keel to margin, 6–12 mm long; plants perennial, typically found in woods **B. pubescens**
      3. Lower glume 3- or 5-nerved; upper glume broadly lanceolate or oblong-elliptic, 1.2–2 mm wide from keel to margin, 4–8 mm long; plants annual, typically found in disturbed areas.
         4. Glumes pubescent **B. hordeaceus**
         4. Glumes glabrous or slightly pubescent on nerves or near apex.
            5. Palea from 0.8 mm shorter than lemma to equaling or barely exceeding it; mature lemmas mostly 6–7.5 mm long, with margins inrolled; awns usually 4–9 mm long; basal leaf sheath glabrous or inconspicuously hirsute **B. secalinus**
            5. Palea 1–2 mm shorter than lemma; mature lemmas mostly 7–9 mm long, with margins not inrolled; awns usually 8–13 mm long; basal sheaths often conspicuously (with a lens) shaggy-pilose **B. japonicus**

Bromus catharticus Vahl, (cathartic, purgative), RESCUE GRASS, RESCUE BROME, SCHRADER'S GRASS. Winter annual, in green growth from late fall to summer, 10–70 cm tall when in flower; leaf sheaths spreading-pilose; spikelets glabrous. Roadsides, disturbed sites, lawns; throughout TX. Mar–May. Native of s South America. [*B. willdenowii* Kunth, *B. unioloides* Kunth, *Festuca unioloides* Willd.] ☜

Bromus hordeaceus L., (like barley—*Hordeum*), SOFT CHESS. Annual 10–60 cm tall; lower leaf sheaths densely pilose, the upper glabrous; lemmas densely pubescent. Found as a weed at Denton Agricultural Experiment Station in May, 1947; Denton, Limestone, and McLennan cos.; also e TX. Native of Europe. [*B. mollis* of authors, not L., *B. molliformis* J. Lloyd] ☜

Bromus japonicus Thunb. ex Murray, (Japanese), JAPANESE BROME, JAPANESE CHESS, SPREADING BROME. Annual 20–80 cm tall; leaf sheaths densely soft-pubescent or occasionally the uppermost glabrous; leaf blades densely pilose; panicles rather dense, with drooping branches; lemmas glabrous; awns either straight or bent out in age. Roadsides, yards, and disturbed sites; nearly throughout TX. May–Jun. Native of Europe and Asia. ☜

Bromus pubescens Muhl. ex Willd., (downy). Perennial 70–120 cm tall, forming small clumps; leaf sheaths spreading-pilose; spikelets drooping; lemmas usually densely pubescent. Woods and thickets; Bosque, Dallas, Denton, Fannin, Grayson, McLennan, and Tarrant cos., also Johnson Co. (R. O'Kennon, pers. obs.); e TX w to nc TX and Edwards Plateau. May. We are following Gould (1975b) in lumping [*B. nottowayanus* Fernald]; the name has been applied to individuals with 5-nerved first glumes.

Bromus secalinus L., (like rye—*Secale*), RYE BROME. Annual, 20–100 cm tall; leaf sheaths glabrous or (especially lower ones) densely pubescent; leaf blades usually ± pilose; panicles narrow, slightly drooping; lemmas glabrous or inconspicuously pubescent. Roadsides, fields, and disturbed sites; widespread in TX. Apr–Jun. Native of Europe. Similar to and possibly hybridizes with *B. japonicus*. Texas plants of *B. secalinus* have been treated as *B. racemosus* by a number of authors. However, we are following Gould (1975b), Hatch et al. (1990), and Jones et al. (1997) in placing them in *B. secalinus*. [*B. racemosus* of authors, not L.]

Bromus tectorum L., (of houses), CHEAT GRASS BROME. Annual 10–80 cm tall; lower leaf sheaths densely pubescent, the upper glabrous; panicle branches drooping. Roadsides, railroads, and disturbed sites. Apr–May. Native of Europe; in TX since 1945 (Mahler 1988). The awns can cause mechanical injury to grazing livestock (Burlage 1968).

1. Lemmas glabrous or scabrous ____ var. **glabratus**
1. Lemmas soft pubescent ____ var. **tectorum**

var. **glabratus** Spenn., (rather smooth, without hairs), CHEAT GRASS. Included based on citation of vegetational area 4 (Fig. 2) by Gould (1975b) and Hatch et al. (1990); in TX cited only for vegetational area 4. Spring.

var. **tectorum**. CHEAT GRASS, DOWNY BROME, DOWNY CHESS, BRONCO GRASS. Collin, Denton, Grayson, Hill, McLennan, and Wise cos., nc TX s and w to w TX. Apr–May(–Jun).

BUCHLOE BUFFALO GRASS

A monotypic North American genus. (Greek: *bous*, cow or ox, and *chloë*, grass) (subfamily Chloridoideae, tribe Cynodonteae)

Buchloe dactyloides (Nutt.) Engelm., (finger-like), BUFFALO GRASS. Perennial; plants dioecious or occasionally monoecious, stoloniferous forming sod; leaves usually with short, curly blades; ligule a ciliate membrane ca. 0.5 mm long, not auricled; staminate inflorescences elevated above the leaves, with 1–4 spike-like branches 6–14 mm long; staminate spikelets usually 6–12 per branch, sessile, pectinately arranged, 2-flowered; glumes unequal, 1(–2)-nerved; lemmas 3-nerved; pistillate inflorescences bur-like, usually hidden in leafy portion of plant, closely subtended by inflated leaf sheaths, with usually 3–5(–7) one-flowered spikelets, falling entire; inflorescence axis indurate; second glumes indurate, yellowish with apex toothed, green; lemmas 3-nerved, 3-lobed. Grasslands; throughout TX. Apr–Sep. This species is dominant over large areas of the short grass prairie of the Great Plains; it was used by settlers in making sod houses; it is currently increasing in use as a low maintenance, drought resistant yard grass.

CENCHRUS SANDBUR, GRASSBUR

Annuals or perennials, largely glabrous; leaf sheaths compressed and keeled; ligule of hairs; inflorescence spike-like, with zigzag, triangular-flattened axis; spikelets of 2 florets, the lower floret sterile with glume-like lemma, the upper floret fertile with hardened grain-like lemma, enclosed in bur-like involucres with bristles and spines that penetrate the flesh and are quite painful.

A genus of 30 species of warm and dry regions of America, Africa, and India. The "burs" are particularly problematic in areas where sheep are raised. (Modification of the old Greek name, *cenchros*, of *Setaria italica*) (subfamily Panicoideae, tribe Paniceae)
REFERENCES: DeLisle 1963; Crins 1991.

1. Bur with 1 whorl of united flattened spines confined to lower part of bur, subtended by 1–several whorls of shorter, finer bristles ____ **C. echinatus**
1. Bur with more than one whorl of flattened spines, the spines present irregularly throughout the body of the bur, usually not subtended by whorls of bristles.
    2. Bur usually with 8–40 spines, the base of larger spines often to 1.5 mm wide, the base of bur usually without numerous, thin, down-pointing spines ____ **C. spinifex**

2. Bur usually with 45–75 spines, the base of larger spines 1 mm wide or less, the base of bur with numerous, thin, down-pointing spines ____ **C. longispinus**

Cenchrus echinatus L., (finger-like), SOUTHERN SANDBUR, HEDGE-HOG GRASS, CADILLO. Annual with geniculate or trailing culms to 85 cm long; bur short pubescent; spines and bristles retrorsely barbed. Disturbed areas; included based on citations for vegetational area 4 (Fig. 2) by Gould (1975b) and Hatch et al. (1990); mainly se and e TX. Spring–fall.

Cenchrus longispinus (Hack.) Fernald, (long-spined), LONG-SPINE SANDBUR. Annual; culms partly decumbent, 15–65 cm long; bur long-pubescent; spines retrorsely barbed. Sandy or gravelly sites, disturbed areas; Grayson, Montague, and Parker cos., also Tarrant Co. (R. O'Kennon, pers. obs.); Oak-pine s and w to w TX. Jun–Oct.

Cenchrus spinifex Cav., (spiny), COMMON SANDBUR, GRASSBUR. Perennial but flowering the first year; culms partly decumbent, up to 100 cm long; bur glabrous to short-pubescent; spines retrorsely barbed. Sandy or gravelly sites, disturbed areas; throughout TX. May–Oct. [*C. carolinianus* of authors, not Walter, *C. incertus* M.A. Curtis] We are following Jones et al. (1997) and J. Kartesz (pers. comm. 1997) for nomenclature of this species.

CHASMANTHIUM WOOD-OATS

Perennials, ours rhizomatous; leaf blades broad, flat; ligule in ours a minute ciliate membrane; inflorescence an open or contracted panicle (rarely a raceme); spikelets (2–)3–many-flowered, laterally flattened, sometimes conspicuously so; glumes shorter than lemmas; lower 1–2 florets often not seed-bearing; rachilla disarticulating above the glumes and between the florets.

A genus of 6 species of e North America; in the past it was recognized as part of the genus *Uniola*, which includes *U. paniculata* L., a coastal dune species, known as SEA OATS. However, it is now recognized that *Chasmanthium* and *Uniola*, while superficially similar, belong in different subfamilies. (Greek: *chasme*, gaping, and *anthus*, flower, presumably from the form of the spikelets) (subfamily Panicoideae, tribe Centotheceae)

REFERENCES: Yates 1966a, 1966b; Clark 1990; Wipff & Jones 1994 [1995].

1. Inflorescence branches drooping, the spikelets long-pedicelled; spikelets (10–)20–50 mm long, 6–17(–26)-flowered ____ **C. latifolium**
1. Inflorescence branches erect or ascending, the spikelets subsessile or short-pedicelled; spikelets 5–18 mm long, 3–7-flowered ____ **C. laxum**

Chasmanthium latifolium (Michx.) H.O. Yates, (broad-leaved), WILD OATS, BROAD-LEAF WOOD-OATS, CREEK-OATS. Glabrous perennial, 0.4–1.5 m tall, with short rhizomes; culms leafy to 4/5 of their height; leaf blades 8–20(–30) mm wide; spikelets conspicuously flat, very wide (6–20 mm). Along streams and in moist woods; one of our most common woodland grasses; e 1/2 of TX. Jun–Sep. [*Uniola latifolia* Michx.] The large spikelet size makes this an excellent example to use in demonstrating spikelet structure to students; the dried inflorescences are also sometimes used ornamentally in dried flower arrangements.

Chasmanthium laxum (L.) H.O. Yates, (loose), Clumped perennial, glabrous or pubescent; culms usually 0.7–1.5 m tall, leafy < 1/2 their height; leaf blades usually 4–12(–15) mm wide; spikelets flat, 3–5 mm wide.

1. Leaf sheaths essentially glabrous or nearly so; collar of leaf sheaths glabrous; panicle branches ± appressed ____ var. **laxum**
1. Leaf sheaths (at least lower) usually long-pubescent or hirsute (rarely glabrous); collar of leaf sheaths pubescent; panicle branches ± divergent ____ var. **sessiliflorum**

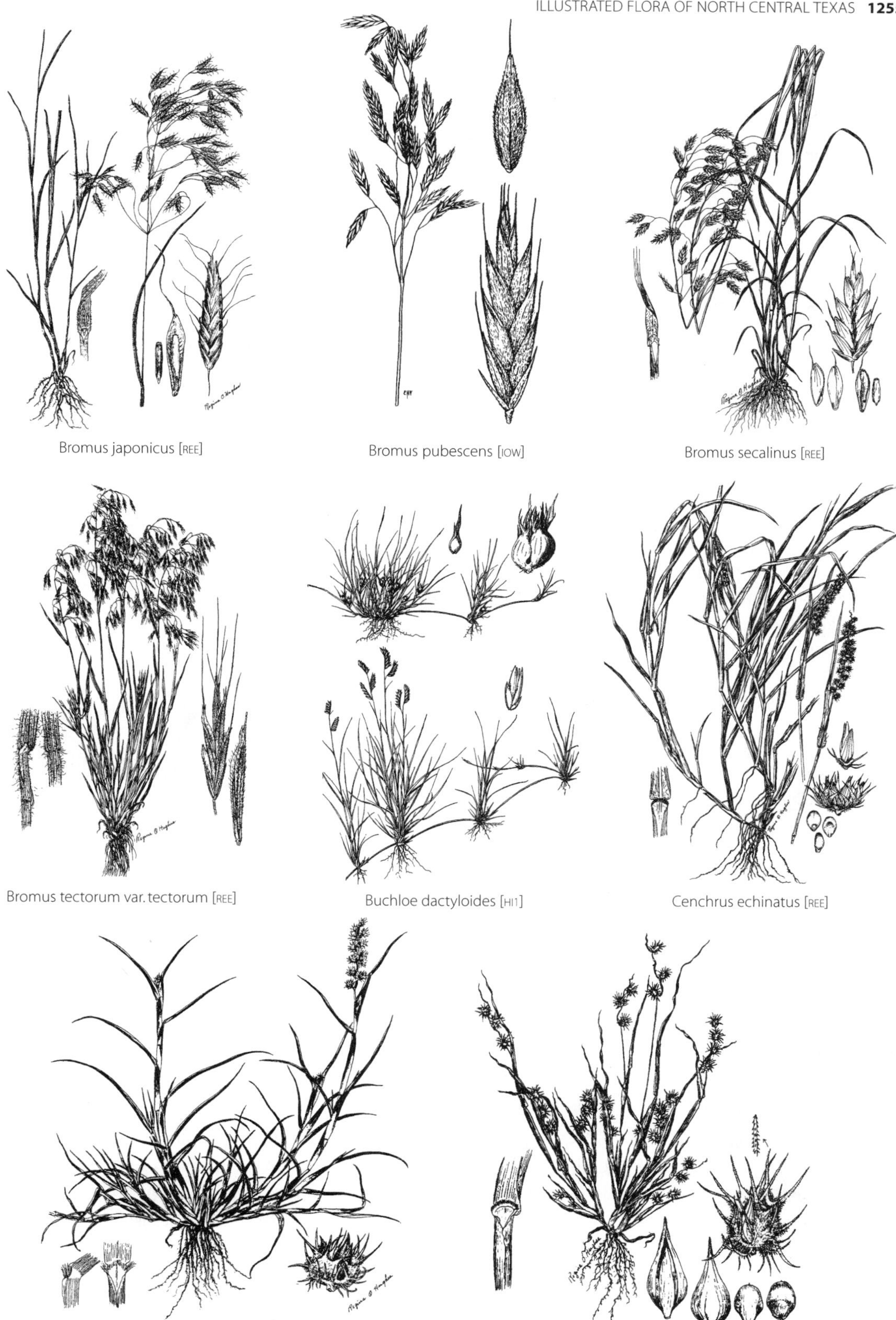

Bromus japonicus [REE]

Bromus pubescens [IOW]

Bromus secalinus [REE]

Bromus tectorum var. tectorum [REE]

Buchloe dactyloides [HI1]

Cenchrus echinatus [REE]

Cenchrus longispinus [REE]

Cenchrus spinifex [REE]

var. **laxum**. Moist, usually sandy areas; Lamar Co. in Red River drainage; mainly se and e TX. Jun–Nov. [*Uniola laxa* (L.) Britton, Sterns, & Poggenb.]

var. **sessiliflorum** (Poir.) Wipff & S.D. Jones, (sessile-flowered), NARROW-LEAF WOOD-OATS. Moist forests and prairie openings, sandy soils; Fannin and Lamar cos. in Red River drainage; mainly se and e TX. Jun–Nov. We are following Wipff and Jones (1994 [1995]) for the nomenclature of this taxon. [*C. laxum* subsp. *sessiliflorum* (Poir.) L.G. Clark, *C. sessiliflorum* (Poir.) H.O. Yates, *Uniola sessiliflora* Poir.]

CHLORIS WINDMILL GRASS

Annuals or perennials; leaf sheaths strongly compressed and sharply keeled; ligule a ciliate membrane; branches of inflorescence digitately or subdigitately arranged at tip of culm or in several whorls; spikelets with 1 perfect floret at base and 1 or more reduced ones above; margins of perfect lemmas often variously pubescent.

A genus of 40 species of tropical and warm areas of the world; it includes some pasture grasses. (Named for Chloris, Greek mother of Nestor, goddess of flowers) (subfamily Chloridoideae, tribe Cynodonteae)
REFERENCES: Nash 1898; Anderson 1974.

1. Branches of inflorescence in more than 1 whorl on main axis ____ **C. verticillata**
1. Branches of inflorescence in 1 whorl only.
  2. Awn of lower lemmas 1.5 mm or less long; lower lemmas 1.5–2 mm long ____ **C. cucullata**
  2. Awn of lower lemmas 2 mm or more long; lower lemmas 2.2–4.2 mm long.
    3. Inflorescences appearing bristly-woolly at arms length; awns 5–15 mm long; upper margins of lower lemmas with a prominent tuft of hairs to ca. 2–3 mm long; plants annual ____ **C. virgata**
    3. Inflorescences not appearing bristly-woolly at arms length; awns 1.5–6 mm long; upper margins of lower lemmas with or without a tuft of hairs; plants perennial.
      4. Sterile or staminate florets (1–)2–4, similar to perfect floret but smaller, often tapering to apex, much longer than wide; upper margins of lower lemmas with a prominent tuft of hairs ____ **C. gayana**
      4. Sterile or staminate floret 1, usually distinctly different from perfect floret, often with a squared-off apex, often nearly as wide as long, sometimes triangular; upper margins of lower lemmas without a prominent tuft of hairs ____ **C. subdolichostachya**

Chloris cucullata Bisch., (hooded), HOODED WINDMILL GRASS, CROWFOOT GRASS, HOODED FINGER GRASS. Perennial; panicle branches 10–20, 2–5 cm long; lower lemma broadly elliptic; sterile floret 1, conspicuously inflated. Sandy soils; mainly East and West cross timbers in nc TX; in much of TX except far e part. Apr–Oct.

Chloris gayana Kunth, (for Jacques Etienne Gay, 1786–1864, French botanist), RHODES GRASS. Perennial; panicle branches 9–30, 8–15 cm long; lemma of lower (perfect) floret 2.5–3.2 mm long, somewhat gibbous (= swollen on 1 side), the awn 1.5–6.5 mm long; sterile florets somewhat gibbous. Cultivated forage grass, escapes along roadsides; included based on citation for vegetational area 5 by Hignight et al. (1988); mainly s TX. May–Dec. Probably native to e Africa.

Chloris subdolichostachya Müll.Hal., (somewhat small-spiked), SHORT-SPIKE WINDMILL GRASS. Perennial, stoloniferous; panicle branches 5-numerous, 3–17 cm long; lemma of lower (perfect) floret 2.2–2.9 mm long, not gibbous, the awn 2–5 mm long; sterile florets variable. Disturbed sandy sites; in much of e 1/2 of TX. May–Oct. [*Chloris latisquamea* Nash] Jones et al. (1997) treated this taxon as *C. ×subdolichostachya* [*C. cucullata* × *C. verticillata*].

Chloris verticillata Nutt., (whorled), TUMBLE WINDMILL GRASS, WINDMILL FINGER GRASS. Peren-

Chasmanthium latifolium [USB]

Chasmanthium laxum var. laxum [USB]

Chasmanthium laxum var. sessiliflorum [USB]

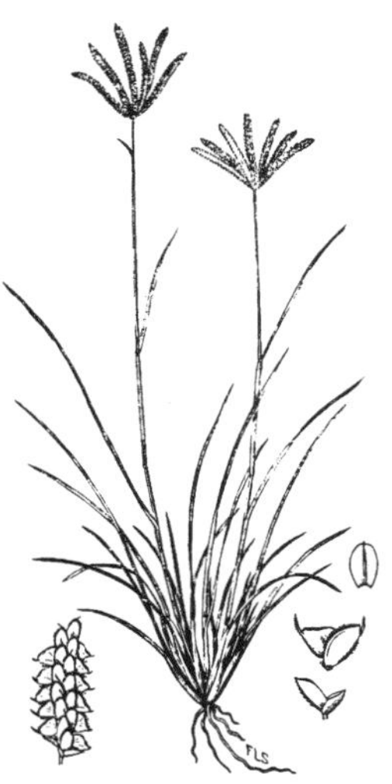

Chloris cucullata [USB]

Chloris gayana [GO1]

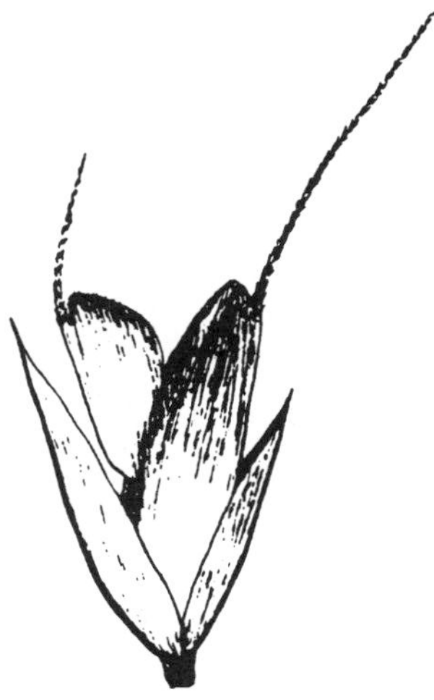

Chloris subdolichostachya [GO1]

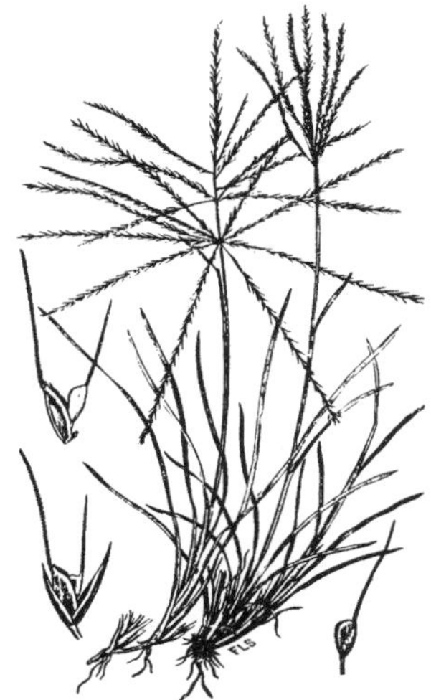

Chloris verticillata [USB]

Chloris virgata [HI1]

Coelorachis cylindrica [HI1]

nial; panicle branches 5–15 cm long, 10–16 per verticil, in 2–5 verticils; lemma of lower (perfect) floret 2–3.5 mm long, not gibbous; sterile floret 1, slightly inflated. Disturbed clay or sandy sites; mainly Blackland Prairie w to w TX. A minor member of original prairie, increasing under disturbance. May–Oct.

Chloris virgata Sw., (twiggy, wand-like), FEATHER FINGER GRASS, SHOWY CHLORIS. Annual; panicle branches 5–10 cm long, 4–20; lemma of lower (perfect) floret 2.5–4.2 mm long, gibbous, producing beaked apical appearance; sterile floret 1, similar to fertile except smaller. Disturbed prairies and roadsides; throughout TX. May–Nov.

Enteropogon chlorideus (J. Presl) Clayton, (genus: ? Greek, *entero*, intestine, and *pogon*, beard; sp.: presumably for resemblance to *Chloris*), (BURY-SEED CHLORIS), previously treated in *Chloris* [as *C. chloridea* (J. Presl) Hitchc.], is cited by Hatch et al. (1990) for vegetational area 4 (Fig. 2), apparently based on a Brazos Co. record (Gould 1975b) to the se of nc TX. This species can be distinguished by the lemma of lower (perfect) floret 4.5–7.5 mm long (versus < 4.5 mm long in all nc TX *Chloris* species) and the plants with underground cleistogamous spikelets at the tips of rhizomes (not present in any nc TX *Chloris* species).

COELORACHIS JOINT-TAIL

A mainly tropical genus of 21 species; related to *Rottboellia*. (Greek: *coelo*, hollow, and *rachis*, spine or backbone, possibly from the niches in the rachis (= inflorescence axis) into which the spikelets fit) (subfamily Panicoideae, tribe Andropogoneae)
REFERENCES: Clayton 1970; Veldkamp et al. 1986.

Coelorachis cylindrica (Michx.) Nash, (cylindrical), CAROLINA JOINT-TAIL. Glabrous perennial with hard-based, solitary or loosely clumped, erect to over-arched culms 25–100 cm long; old plants with short, knotty rhizomes; ligule a short membrane 0.5-1 mm long; inflorescence a very slender (ca. 3 mm thick), elongate cylindrical spike-like raceme (pencil-like but much smaller in diam.) breaking apart at the nodes of the inflorescence axis, the base of each internode with a niche on one side into which the spikelets fit closely; spikelets awnless, in pairs of 1 sessile, fertile (with a pitted glume) and 1 pedicelled, sterile. Prairies and open woods, sandy or clayey soils; widespread in e 1/2 of TX. May–Jun. An important component of certain native sandy prairies, including "mima mound" prairies. [*Manisuris cylindrica* (Michx.) Kuntze, *Mnesithea cylindrica* (Michx.) de Koning & Sosef] Jones et al. (1997) treated this species as *Mnesithea* cylindrica; however, J. Wipff (pers. comm.) has since indicated that the species is better treated in the genus *Coelorachis*.

CORTADERIA PAMPAS GRASS

A genus of 24 species of coarse, clump-forming grasses native to South America, New Zealand, and New Guinea. (From the Argentinian name; a term for cutting) (subfamily Chloridoideae, tribe Danthonieae)

Cortaderia selloana (Schult. & Schult.f.) Asch. & Graebn., (for Friedrich Sellow, 1789–1831, German botanist who collected in South America), PAMPAS GRASS. Dioecious perennial to ca. 3 m tall, forming large clumps to 1 m or more in diam.; leaves mostly basal, the blades usually 0.6–1 m or more long, with scabrous margins; ligule a dense tuft of hairs 3–5 mm long; inflorescence a showy, densely flowered, silvery, feathery panicle 25–100 cm long; spikelets 2–3-flowered, disarticulation above glumes and between florets; pistillate lemmas with long silky hairs; staminate lemmas glabrous. No escaped nc TX specimens have been seen, but this species is widely cultivated throughout TX as a lawn ornamental and long persists. Sep–Nov. Native from Brazil to Argentina and Chile.

CYNODON BERMUDA GRASS

A tropical and warm area genus of 8 species; it includes pasture and lawn grasses. (Greek: *cyon*, dog, and *odous*, tooth, from the close rows of tooth-like spikelets or the hard scales on rhizomes) (subfamily Chloridoideae, tribe Cynodonteae)
REFERENCE: deWet & Harlan 1970.

Cynodon dactylon (L.) Pers., (fingered), BERMUDA GRASS, BAHAMA GRASS. Rhizomatous and stoloniferous perennial 10–40 cm tall; leaf sheaths compressed, keeled, with tufts of hairs at summit; lower leaf sheaths also pilose on back; ligule a short ciliate membrane 0.2–0.5 mm long; branches of inflorescence usually digitately arranged at tip of culm; spikelets with 1 perfect awnless floret, closely overlapping, in 2 rows on a flattened or triangular branch. Cultivated in nc TX since about 1882 for pasture and lawns (Mahler 1988), also a common weed throughout TX. May–Oct. Native probably of Africa, but India has been suggested also (not from Bermuda, despite the common name). Reported to be an important cause of hay fever and potentially poisonous to livestock due to the production of hydrocyanic acid under certain environmental conditions (Lewis & Elvin-Lewis 1977; Fuller & McClintock 1986).

DACTYLIS

A genus of 1–5 Eurasian species (often considered monotypic). (Greek *dactylus*, finger; a name used by Pliny for a grass with digitate spikes or from the crowded spikelets at inflorescence tips) (subfamily Pooideae, tribe Poeae)

Dactylis glomerata L., (clustered), ORCHARD GRASS, COCK'S-FOOT. Erect, densely clumped perennial to 1 m tall; leaf blades 2–10 mm wide; ligules membranous, 2–5 mm long; panicles long exerted, with or without elongated lower branches, of rather dense terminal aggregations of nearly sessile groups of spikelets; spikelets usually with 2–5 flowers, ca. 5–9 mm long. Roadsides, field margins, yard weed; Grayson (first observed 1998) and Tarrant (first observed 1996) cos.; introduced as a forage grass, now a weed in scattered localities in TX. Spring–Summer. Native to Eurasia. This species is now widely naturalized in North America and s Africa; it is an important cause of hay-fever and is complex genetically, there being diploid and tetraploid forms (Mabberley 1997).

DACTYLOCTENIUM CROWFOOT

A genus of 10 species of warm areas of the world. (Greek *dactylos*, finger, and *ctenion*, a little comb, referring to the finger-like arrangement of the comb-like inflorescence branches) (subfamily Chloridoideae, tribe Eragrostideae)
REFERENCE: Peterson et al. 1997.

Dactyloctenium aegyptium (L.) P. Beauv., (of Egypt), CROWFOOT, DURBAN CROWFOOT GRASS, EGYPTIAN CROWFOOT GRASS. Annual with culms mostly 10–60 cm tall, often rooting at lower nodes; ligule a membrane usually 0.1–1 mm long, fringed with short hairs; leaf blades 2–9 mm wide; inflorescences with 2–7(–more) digitately arranged branches usually 1.5–6 cm long; branch tip projecting 1–7 mm beyond terminal spikelets as a sharp point; spikelets sessile, very crowded, in 2 rows along one side of the narrow flattened branch, strongly laterally compressed, usually 3–4 mm long, with 3–5 flowers; second glumes and lemmas usually with short awns. Usually sandy soils, moist areas, disturbed sites; Bell Co.; se and e TX w to s part of nc TX, also Edwards Plateau. (Jul–)Sep–Dec. Native of the Old World tropics.

DANTHONIA POVERTY-OATS

A genus of 100 species native to North America, South America, Europe, Australia, and New

Zealand. (Named for Etienne Danthoine, an early 19th century French botanist of Marseilles) (subfamily Chloridoideae, tribe Danthonieae)
REFERENCE: Darbyshire & Cayouette 1989.

Danthonia spicata (L.) P. Beauv. ex Roem. & Schult., (with spikes), POVERTY-OATS, POVERTY DANTHONIA, POVERTY OAT GRASS. Tufted perennial 25–50 cm tall, with crowded, mostly basal leaves; leaf sheaths pilose near summit; leaf blades persistent, curling in age; ligule a short membrane with long-ciliate margin longer than base; inflorescences narrow, short, spike-like; spikelets erect, 10–13(–15) mm long, with 3-7(–9) florets, on pedicels ca. 1–8 mm long; glumes much longer than lemmas; lemmas awned from between apical acute or acuminate teeth, the awns ca. 5–7 mm long, spirally twisted near base, geniculate. Sandy open woods; Dallas, Grayson, Hunt, and Lamar cos.; mainly e TX. May–Jul.

DESMAZERIA

A genus of 3 species native from Europe to n Africa and Iran; treated here as including *Catapodium*. (Named for J.B. Desmazieres, French botanist) (subfamily Pooideae, tribe Poeae)

Desmazeria rigida (L.) Tutin, (rigid), Glabrous tufted annual 10–30 cm tall; ligule a membrane 1.5–4 mm long; panicles usually 3–10 cm long, narrow, dense, erect; spikelets (4–)5–9(–10)-flowered; glumes and lemmas awnless. Railroads, stock pens, roadsides, and disturbed areas; Hill Co., also Ellis and Wise cos. (Mahler 1988); scattered in e 1/2 of TX. Apr–May. Native of Europe. Sometimes recognized in the genus *Catapodium* [as *C. rigidum* (L.) C.E. Hubb. ex Dony]. [*Scleropoa rigida* (L.) Griseb.]

DIARRHENA BEAKGRAIN

Rhizomatous perennials; culms arching; leaves basal and low cauline; leaf blades flat; ligule a membrane; inflorescence a narrow panicle, long exserted and arching; spikelets with 3–5 florets, the terminal floret reduced and sterile; florets often ± spreading at maturity; glumes 2, awnless; lemmas ± rounded on back, awnless, with a sharp cusp 1–2 mm long at apex; caryopsis (= fruit or grain) with a beak.

A genus of 4 species of e Asia and North America. Subfamilial classification of this genus is problematic with some authorities putting it in the Pooideae and others in the Bambusoideae (Brandenburg et al. 1991a). The following key is from Brandenburg et al. (1991a). (Greek: *dis*, twice, and *arrhen*, male, from the two stamens) (subfamily Pooideae, tribe Diarrheneae)
REFERENCE: Brandenburg et al. 1991a.

1. Callus pubescent on all mature lemmas except first; lemmas widest below the middle and gradually tapering into a cusp at apex, those of first floret 7.1–10.8 mm long; mature fruits 1.3–1.8 mm broad, gradually tapering into a broad, blunt beak ____ **D. americana**
1. Callus glabrous on all mature lemmas; lemmas widest near or above the middle and ± abruptly contracted into cusp at apex, those of the first floret 4.6–7.5 mm long; mature fruits 1.8–2.5 mm broad, abruptly contracted into a bottlenose-shaped beak ____ **D. obovata**

Diarrhena americana P. Beauv., (of America), AMERICAN BEAKGRAIN. Culms ca. 0.6–1.3 m long; leaf blades 7–20 mm wide; ligule a membranous collar 0.5–1.8 mm long; inflorescences 9–30 cm long, with 4–23 spikelets; spikelets 10–20 mm long; lemmas (3.8–)5.3–10.8 mm long, glabrous to scaberulous; anthers (1.7–)2–2.9(–3.5) mm long. This species has long been reported from nc TX based on a Reverchon collection from Dallas (Mahler 1988). However, Dallas material is of the recently named, related species, *D. obovata* (Brandenburg et al. (1991a). *Diarrhena americana*, native to the e U.S., is known from e OK but is apparently unknown from TX (Brandenburg et al. (1991a). Hatch et al. (1990) reported *D. americana* from vegetational area 1

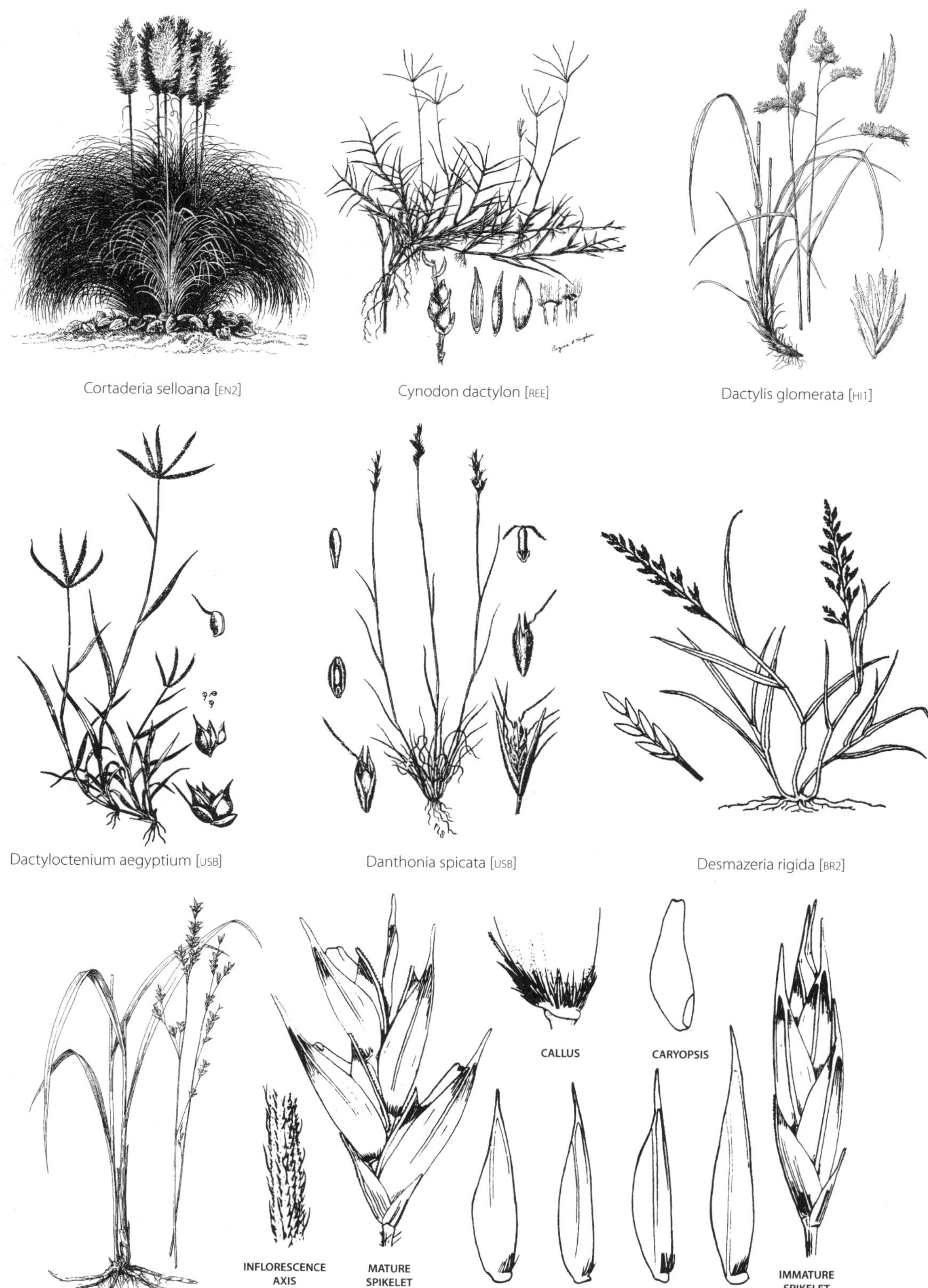

Cortaderia selloana [EN2]

Cynodon dactylon [REE]

Dactylis glomerata [HI1]

Dactyloctenium aegyptium [USB]

Danthonia spicata [USB]

Desmazeria rigida [BR2]

Diarrhena americana [TOR, USB]

(deep e TX), but it is not known which species this citation represents. *Diarrhena americana* is included here to help clarify recent changes in the taxonomy of the genus. Summer–fall.

Diarrhena obovata (Gleason) Brandenburg, (obovate, inversely ovate). Culms ca. 0.5–1.3 m long; leaf blades 6–18 mm wide; ligules 0.2–1 mm long; inflorescences 5–30 cm long, with 4–33 spikelets; spikelets 7–17 mm long; lemmas 4.6–7.5 mm long, glabrous; anthers 1.4–2 mm long. Woodlands; collected by Reverchon in Jun, 1874, in "rich woods, Buzzards Spring" (now in e Dallas and a residential section), not found in nc TX since (cited in Mahler 1988 as *D. americana*); Brandenburg et al. (1991a) indicated that the only TX collection is from Dallas Co. Summer–fall. [*D. americana* var. *obovata* Gleason]

DICHANTHIUM BLUESTEM

An Old World tropical genus of 10 species; formerly treated in a more inclusive *Andropogon*. (presumably from Greek: *dicho*, in two, and *anthus*, flower, in reference to the paired spikelets) (subfamily Panicoideae, tribe Andropogoneae)
REFERENCE: Gould 1967.

Dichanthium annulatum (Forssk.) Stapf, (ringed, ring-like), KLEBERG BLUESTEM. Perennial superficially resembling *Bothriochloa ischaemum* var. *songarica* (KING RANCH BLUESTEM) but with much longer and more consicuous lemma awns; culms both erect and stoniferous; nodes bearded with a conspicuous ring of long white hairs; ligule a short membrane; inflorescences with (2–)3–5(–8) spicate branches, the lowest pair of spikelets on the branches usually without awns and not producing seeds; main inflorescence axis and branches just below spikelets glabrous; pedicels and inflorescence branches without a groove or membranous area; spikelets in pairs, one sessile, awned, the other pedicelled, awnless; pedicelled spikelets ca. the same size as the sessile spikelets, rounded apically; awn of lemma of sessile spikelets ca. 2–2.5 cm long; disarticulation at base of sessile spikelet so that the associated pedicel and section of the inflorescence branch fall with the sessile spikelet. Introduced as a forage grass, disturbed areas; Coryell Co. (Fort Hood—Sanchez 1997); mainly introduced in se and s TX. Flowering throughout the growing season. [*Andropogon annulatus* Forsk.] Native of Africa to India and China.

DIGITARIA CRAB GRASS, FINGER GRASS

Annuals or perennials; ligules membranous; inflorescences often digitate or nearly so (sometimes with branches below the terminal cluster), with few to numerous unbranched spike-like branches or an open, much-branched panicle (*D. cognata*); inflorescence branches winged or unwinged; spikelets 2-flowered, the lower floret staminate or neuter, the upper floret perfect, disarticulating below glumes; first glume minute or absent; second glume well-developed but usually shorter than lemmas; lemma of lower spikelet resembling a glume; lemma margins flat, not inrolled.

A genus of ca. 220 species of tropical and warm areas of the world; many are weedy and some are cultivated for food in Africa. As treated here, *Digitaria* includes *Leptoloma*. (Latin: *digitus*, finger, from the finger-like arrangement of the inflorescence branches) (subfamily Panicoideae, tribe Paniceae)
REFERENCES: Henrard 1950; Ebinger 1962; Gould 1963; Webster 1987; Crins 1991; Wipff & Hatch 1994.

1. Spikelets long-pedicelled (pedicels 2–several times as long as spikelet); inflorescence a much branched and rebranched, open panicle with spikelets far apart ____ **D. cognata**
1. Spikelets subsessile or on short, appressed pedicels; inflorescence of a few main unbranched branches digitately arranged or nearly so (or an unbranched panicle in *D. filiformis*), with spikelets close together.

2. Inflorescence branches winged (wings often as wide as central part of branch); spikelets glabrous to short-pubescent.
   3. Spikelets 1.9–2.2 mm long; fertile lemma dark brown at maturity; mainly e TX, rare in nc TX ____ **D. ischaemum**
   3. Spikelets 2.2–3.6 mm long; fertile lemma light brown or grayish; abundant in nc TX.
      4. Spikelets 2.2–3.3 mm long; second glume 1–1.8 mm long; leaf blades usually densely pubescent; lower lemmas minutely scabrous on lateral nerves (under a dissecting scope) ____ **D. sanguinalis**
      4. Spikelets 2.8–3.6 mm long; second glume 1.5–2.7 mm long; leaf blades glabrous or sparsely pubescent; lower lemmas smooth on lateral nerves ____ **D. ciliaris**
2. Inflorescence branches not winged; spikelets silky-pubescent with long white-silky or purplish hairs OR not so.
   5. Spikelets silky-pubescent (inflorescences appearing conspicuously white-silky or purplish); spikelets 3–4 mm long; plants cespitose perennials; w part of nc TX ____ **D. californica**
   5. Spikelets not silky-pubescent (inflorescences not appearing white-silky or purplish); spikelets 1.5–2.6 mm long; plants tufted annuals; widespread in nc TX.
      6. Spikelets 1.5–1.9(–2) mm long; inflorescence branches usually 8–13 cm long ____ **D. filiformis**
      6. Spikelets 2–2.6 mm long; inflorescence branches usually 13–25 cm long ____ **D. villosa**

Digitaria californica (Benth.) Henrard, (of California), CALIFORNIA COTTONTOP, ARIZONA COTTONTOP. Perennial with erect culms ca. 50–100 cm tall; leaf blades glabrous; inflorescences digitate or nearly so, the branches usually appressed, densely flowered; second glume and margins of lower floret densely long hairy, the hairs 2–4 mm long. Disturbed grasslands; Erath, Lampasas, and Palo Pinto cos.; w part of nc TX s and w to w TX. May–Sep. [*Trichachne californica* (Benth.) Chase]

Digitaria ciliaris (Retz.) Koeler, (ciliate, fringed), SOUTHERN CRAB GRASS. Annual; similar to *D. sanguinalis*, differing in characteristics as enumerated in key. Common weed; throughout TX. Jun–Nov. According to Gould (1975b), presumably introduced from the Old World. [*D. adscendens* (Kunth) Henrard, *D. sanguinalis* var. *ciliaris* (Retz.) Parl.]

Digitaria cognata (Schult.) Pilg., (related to), Clump-forming perennial 20–70 cm tall; culms usually decumbent near base; lower leaf sheaths densely pubescent, upper glabrous; ligule a membrane; panicles open, breaking off and acting as tumbleweeds; spikelets long-pedicelled, narrowly lanceolate, acute; lower glume minute or absent; upper glume and sterile lemma similar. In sandy soils, occasional in rocky or gravelly soils, open woods, fields, and disturbed sites. May–Oct, but chiefly Aug–Oct. Sometimes recognized in the genus *Leptoloma* [as *L. cognatum* (Schult.) Chase]. According to Wipff and Hatch (1994), the only character separating *Leptoloma* and *Digitaria* is inflorescence type and that is unreliable. When examined worldwide, there is no justification for separating the 2 genera. The key and nomenclature for this species follows Wipff and Hatch (1994).

1. Lowermost (sterile) lemma 7-veined (rarely 6-); spikelets glabrous to pubescent; rhizomes absent ____ subsp. **cognata**
1. Lowermost (sterile) lemma 5-veined; spikelets densely pubescent; rhizomes present or absent ____ subsp. **pubiflora**

subsp. **cognata**. FALL WITCH GRASS. Limestone, Montague, Palo Pinto, Parker, and Tarrant cos.; widespread in e 1/2 of TX.

subsp. **pubiflora** (Vasey) Wipff, (with hairy flowers), WESTERN WITCH GRASS. Throughout TX.

Digitaria filiformis (L.) Koeler, (thread-like), SLENDER CRAB GRASS, SLENDER FINGER GRASS. Annual with erect culms 50–80 cm tall; inflorescences paniculate, the primary branches unbranched; second glume usually 3/4 as long as fertile lemma. Sandy fields or woods; Grayson, Parker, and Tarrant cos.; mainly se and e TX. Sep–Oct.

Digitaria ischaemum (Schreb.) Muhl., (Greek: *ischaemos*, blood-restraining, from supposed styptic properties), SMOOTH CRAB GRASS. Tufted annual to ca. 50 cm tall; leaf sheaths glabrous; inflorescences digitate or nearly so, the branches ca. 3–9(–11) cm long; spikelets 1.9–2.2 mm long. Open woods; included on basis of citation for vegetational area 5 (Fig. 2) by Hatch et al. (1990); mainly far e TX. Aug–Nov. Native of Europe.

Digitaria sanguinalis (L.) Scop., (of blood-red color), HAIRY CRAB GRASS, LARGE CRAB GRASS. Annual; culms weak, rooting at nodes; leaves with papilla-based hairs; inflorescences digitate or nearly so, usually with 4–9 branches 6–14 cm long; lemma of lower floret 5-veined, the lateral nerves minutely scabrous above. Weed, much less common in nc TX than *D. ciliaris*; Hood and Wise cos.; also Travis and Wichita cos. just s and w of nc TX; mainly w TX. Jul–Nov. Native of Europe.

Digitaria villosa (Walter) Pers., (softly hairy), SHAGGY CRAB GRASS. Similar to *D. filiformis* and sometimes treated as a variety of that species; annual with erect culms to 125 cm or more tall, much branched at base; second glume usually > 3/4 as long as fertile lemma, the second glume and fertile lemma usually more pubescent than in *D. filiformis*. Disturbed sandy soils; Dallas Co., also Milam and Limestone cos. at the e edge of nc TX, also Lamar Co. (Carr 1994); mainly se and e TX. Aug–Nov. [*D. filiformis* (L.) Koeler var. *villosa* (Walter) Fernald] Jones et al. (1997) treated this species as a variety of *D. filiformis*.

DISTICHLIS SALT GRASS, ALKALI GRASS

A genus of 5 species, 4 in the New World, 1 in Australia. (Greek: *distichos*, two-ranked) (subfamily Chloridoideae, tribe Eragrostideae)
REFERENCE: Peterson et al. 1997.

Distichlis spicata (L.) Greene, (with spikes), SALT GRASS, SPICATE SALT GRASS, INLAND SALT GRASS, DESERT SALT GRASS, COASTAL SALT GRASS, SPIKE GRASS, ALKALI GRASS. Low glabrous perennial 10–35(–70) cm tall from extensive scaly rhizomes; usually dioecious; flowering culms erect, the internodes short with leaf sheaths conspicuously overlapping at least on lower portion of culm; leaves noticeably 2-ranked; leaf blades 2–20 cm long, 1–3 mm wide, often ± involute; ligule a minute membrane < 0.5 mm long; inflorescences contracted spike-like panicles or spike-like racemes 3–8 cm long; spikelets pedicellate, ± similar on male and female plants, usually 5–20-flowered, 6–18(–28) mm long, awnless; disarticulation above glumes and between florets; glumes slightly unequal; lemmas similar to glumes but longer and broader, 3–6 mm long, keeled; pistillate lemmas coriaceous, enclosing the caryopsis; paleas as long as lemmas or nearly so. In w 1/2 of TX in alkaline or alkaline-saline areas, along the coast in saline marshes and flats; Brown Co., also Comanche Co. (Stanford 1971); mainly se TX, also Edwards Plateau and Rolling Plains w to w TX. [*Distichlis spicata* var. *stricta* (Torr.) Scribn.] Nc TX plants fall into var. *stricta*, which is here lumped; according to Gould (1975b), this variety "appears to represent little more than a variable series of inland populations of *Distichlis spicata* growing under a wide range of soil and climatic conditions."

ECHINOCHLOA BARNYARD GRASS

Annuals (our species); ligules absent; panicles dense; spikelets of 2 florets (1 perfect), subsessile, crowded along one side of branches, glabrous, pubescent, or hispid; first glume present, much shorter than second; second glume and lemma of lower (sterile) floret similar; lemma of perfect floret hardened, grain-like, smooth and shiny with inrolled margins.

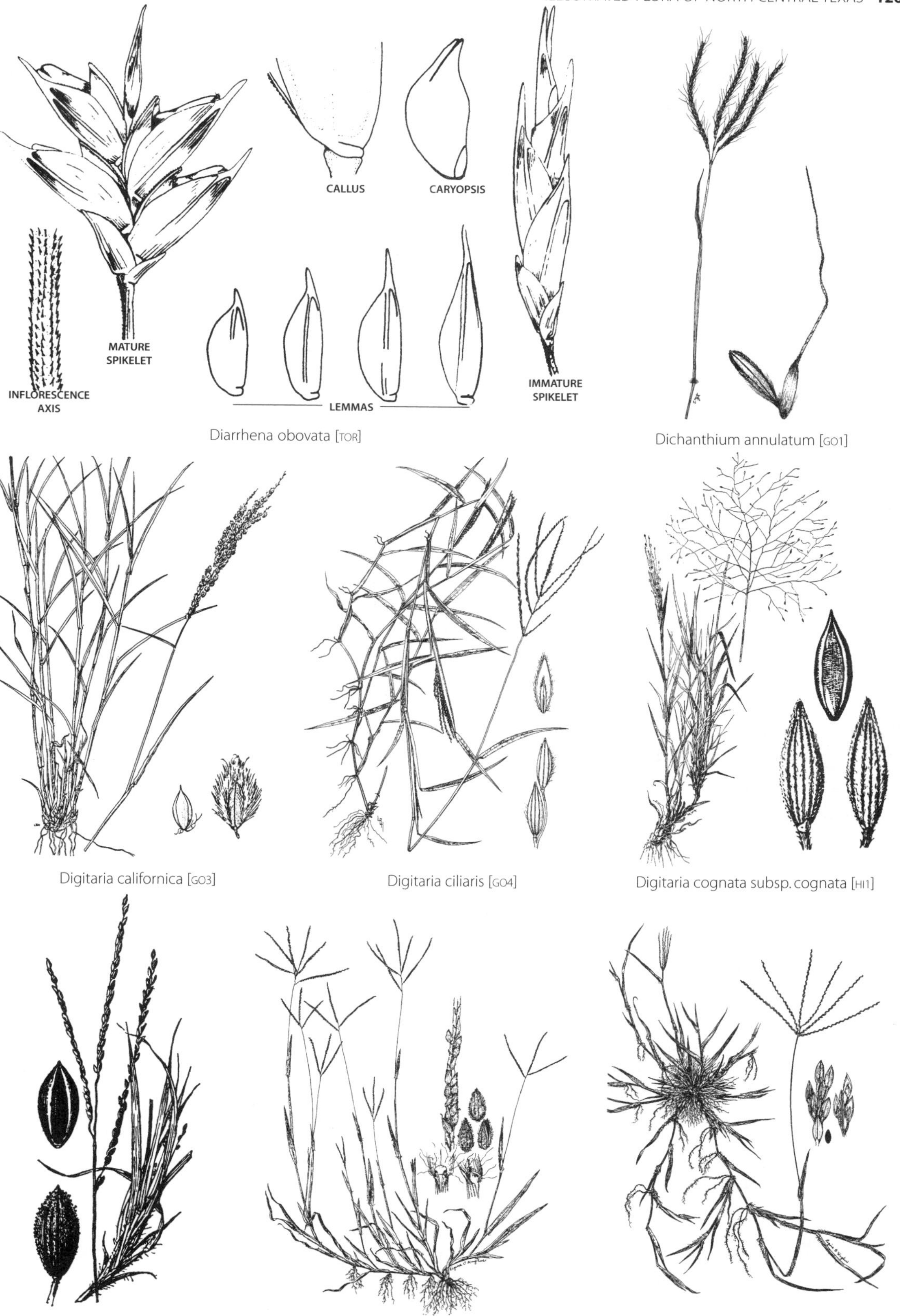

Diarrhena obovata [TOR]

Dichanthium annulatum [GO1]

Digitaria californica [GO3]

Digitaria ciliaris [GO4]

Digitaria cognata subsp. cognata [HI1]

Digitaria filiformis [HI1]

Digitaria ischaemum [REE]

Digitaria sanguinalis [REE]

A genus of ca. 35 species of warm areas of the world. *Echinochloa* species are often extremely variable, making absolute distinctions between taxa difficult. (Greek: *echinus*, sea-urchin, and *chloa*, grass, referring to the bristling awns) (subfamily Panicoideae, tribe Paniceae)
REFERENCES: Hitchcock 1920b; Gould et al. 1972; Crins 1991.

1. Leaf sheaths hirsute or hispid; lemma of lower floret usually with awn 1.5–6 cm long (occasionally short-awned or awnless) ____ **E. walteri**
1. Leaf sheaths glabrous; lemma of lower floret awnless or variously awned.
   2. Palea of lower floret absent or vestigial; spikelets awnless or infrequently with awns to 0.9 cm long ____ **E. crus-pavonis**
   2. Palea of lower floret well-developed; spikelets awnless or awned, the awns if present variable in length, to 6 cm long.
      3. Inflorescence branches 3–7, usually 2(–3) cm or less long; spikelets 2.5–3 mm long, awnless, arranged in 4 regular rows, with hairs without papillate bases; leaf blades 3–6(–9) mm wide ____ **E. colona**
      3. Inflorescence and spikelets without the above combination; leaf blades ca. 3–30 mm wide.
         4. Lemmas with awns usually 1.5–6 cm long; inflorescences usually very large, (10–)20–40 cm or more long and up to 10 cm thick; rare glabrous-sheathed form of this species ____ **E. walteri**
         4. Lemmas awnless or variously awned; inflorescences variable, often smaller; weedy species widespread in nc TX.
            5. Fertile lemma obtuse to broadly acute, with a sharply differentiated, withering, membranous tip; spikelets without papilla-based hairs ____ **E. crus-galli**
            5. Fertile lemma acute to acuminate, tapering to a firm, stiff tip; spikelets usually with some stout, papilla-based hairs ____ **E. muricata**

Echinochloa colona (L.) Link, (farmer), JUNGLE-RICE, SHAMA-MILLET. Spikelets 2.5–3 mm long. Low ground, disturbed areas; throughout TX. Jul–Nov. Native of Old World tropics.

Echinochloa crus-galli (L.) P. Beauv., (cockspur), BARNYARD GRASS. Spikelets 2.8–4 mm long; lower lemma awnless or with an awn to over 5 cm long. Low ground, disturbed areas; throughout TX. Jun–Oct. Apparently introduced from the tropics.

Echinochloa crus-pavonis (Kunth) Schult. var. **macera** (Wiegand) Gould, (sp.: peacock spur; var.: soften). Spikelets 2.8–3.1 mm long. Disturbed areas; nearly throughout TX. Jul–Nov.

Echinochloa muricata (P. Beauv.) Fernald, (with numerous minute short points, roughened). Spikelets green or purple, usually conspicuously echinate with stout, papilla-based hairs. Low moist areas. Jul–Nov.

1. Spikelets 3.5 mm or less long (excluding awn); awn of lemma of lower floret absent to 6 mm long ____ var. **microstachya**
1. Spikelets 3.5 mm or more long (excluding awn); awn of lemma of lower floret usually 6–25 mm long (rarely absent) ____ var. **muricata**

var. **microstachya** Wiegand, (small-spiked). Throughout TX.

var. **muricata**. Widespread in TX.

Echinochloa walteri (Pursh) A. Heller, (for an early Carolinian botanist, Thomas Walter, ?1740–1789). Plant 1–2 m tall; leaf sheaths usually hirsute or hispid with papilla-based hairs, especially toward apex, rarely glabrous; spikelets 3–5 mm long (excluding awn), without or with papilla-based hairs; awn of lemma of lower floret usually 1.5–6 cm long but occasionally short-awned or awnless. Low moist areas; Bell, Grayson, and Kaufman cos.; se and e TX w to Blackland Prairie, also Edwards Plateau. Jul–Nov.

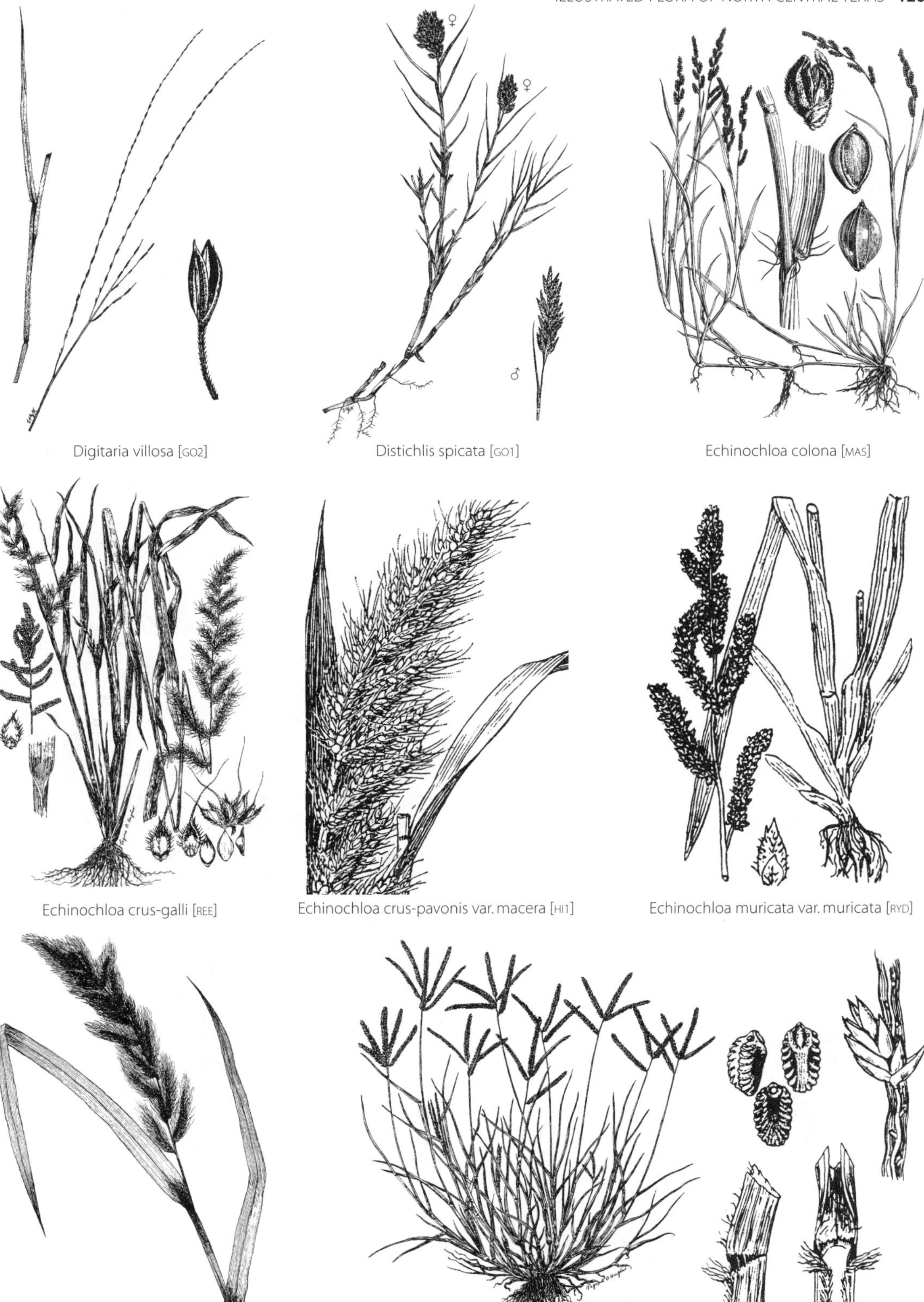

Digitaria villosa [GO2]

Distichlis spicata [GO1]

Echinochloa colona [MAS]

Echinochloa crus-galli [REE]

Echinochloa crus-pavonis var. macera [HI1]

Echinochloa muricata var. muricata [RYD]

Echinochloa walteri [GO1]

Eleusine indica [REE]

ELEUSINE GOOSE GRASS

A genus of ca. 9 species of Africa and South America including *E. coracana* (L.) Gaertn. (FINGER-MILLET), an important grain crop in Africa and India cultivated since the 3rd Millenium BC. (Named from Eleusis, Greek town where Ceres, the goddess of harvests, was worshiped) (subfamily Chloridoideae, tribe Eragrostideae)
REFERENCE: Peterson et al. 1997.

Eleusine indica (L.) Gaertn., (of India), GOOSE GRASS, YARD GRASS, ZACATE GUAIMA. Annual with decumbent to erect culms 15–65 cm long; leaf sheaths compressed, keeled, pilose on margins; ligule a short membrane; branches of inflorescence crowded at tip of culm, usually digitate or with 1–2 branches attached lower; spikelets awnless, crowded, overlapping, in two rows on flattened winged branches. Common weed of gardens, lawns, and disturbed sites; throughout TX. Late Jun–Oct. Native of Old World tropics.

ELYMUS WILD RYE

Ours perennials with solitary or few culms in small clumps; leaf blades auricled at base; ligule a membrane; inflorescence a dense unbranched 2-sided spike with 2–3 sessile spikelets per node, the inflorescence axis in our species remaining intact, disarticulation occurring above glumes and between florets; spikelets with 2–6 florets; glumes and lemmas awned.

A n temperate, especially Asian, genus of ca. 150 species. (Ancient Greek name for millet; from *elyo*, rolled up, from the caryopsis (= fruit) being tightly embraced by the lemma and palea) (subfamily Pooideae, tribe Triticeae)
REFERENCES: Pohl 1959; Bowden 1964b; Church 1967; Runemark & Heneen 1968; Davies 1980; Dewey 1982; Estes & Tyrl 1982.

1. Glumes strongly 1–4-ribbed to base or nearly so, 0.2–1.5 mm wide near middle, not or only slightly bowed out at base, flat, neither yellowish, hardened nor rounded basally, with awn equaling or exceeding the body; lemma awns usually curving outward at maturity, 15–50 mm long ____ **E. canadensis**
1. Glumes strongly 3–8-ribbed above base, 1.0–2.5 mm wide near middle, often strongly bowed out at base, often yellowish, hardened or rounded basally, with awn shorter than or equaling the body; lemma awns straight or slightly curved, 5–25 mm long ____ **E. virginicus**

Elymus canadensis L., (of Canada), CANADA WILD RYE, NODDING WILD RYE. Plant 40–160 cm tall; leaf blades mostly 4–12 mm wide; leaf sheaths glabrous; ligule a membrane 0.5–1 mm long; spikes thick, usually oblique or nodding, (4–)8–21 cm long and 6–18 mm thick (excluding awns). Thickets and open woods, in limestone or sandy clay; in nc TX mainly East and West cross timbers, throughout most of TX. (Mar–)late May–Jun. [*E. canadensis* L. var. *brachystachys* (Scribn. & C.R. Ball) Farw., *E. canadensis* L. var. *villosus* (Muhl.) Shinners]

Elymus virginicus L., (of Virginia), VIRGINIA WILD RYE. Plant 30–140 cm tall; leaf blades 5–15 mm wide; leaf sheaths glabrous or scabrous-pubescent, only the uppermost to all of the leaf sheaths inflated; ligule a minute membrane; spikes rather dense and stiff, usually erect, partly included to well-exserted, usually 3–15 cm long. Thickets or open ground; in much of e 1/2 of TX. (Apr–)May–Aug. [*E. virginicus* L. var. *glabriflorus* (Vasey) Bush] Hybrids with *E. canadensis* are known (Pohl 1959).

ERAGROSTIS LOVE GRASS

Annuals or perennials; ligule a ring of hairs (except in *E. secundiflora*); spikelets in very open or contracted panicles, 2–many-flowered, sometimes with a reddish purple color, often somewhat laterally compressed; lemmas obtuse or acute, awnless, rounded or keeled on back, 3-nerved, the lateral nerves sometimes obscure, glabrous.

A genus of ca. 300 species of temperate and tropical areas of the world; some are cultivated as ornamentals, for fodder, and for edible seeds; some have inflorescences which detach and disperse the seeds by acting as tumbleweeds. (Greek: *eros*, love or god of love, and *agrostis*, a grass) (subfamily Chloridoideae, tribe Eragrostideae)
REFERENCES: Koch 1974, 1978; Witherspoon 1977; Perry & McNeill 1986; Van den Borre & Watson 1994; Peterson et al. 1997.

1. Culms creeping, rooting at nodes; plants mat-forming annuals; flowering culms up to 25 cm tall (usually less).
  2. Plants dioecious; anthers 1.2–2 mm long; lemmas (1.8–)2.6–3.5 mm long; styles long exserted from lemma and palea, often persistent and conspicuous (with a lens) as thread-like structures ____ **E. reptans**
  2. Plants monoecious; anthers 0.5 mm or less long; lemmas 1.5–2(–2.3) mm long; styles not exserted ____ **E. hypnoides**
1. Culms erect; plants clump-forming annuals or perennials; flowering culms usually much more than 25 cm tall.
  3. Panicle branches bearing spikelets to base or nearly so.
    4. Spikelets sessile, widely spaced on long branches ____ **E. sessilispica**
    4. Spikelets sessile or short-pedicelled, closely crowded on short branches.
      5. Spikelets 3–10 mm wide, usually < 2 times as long as wide, strongly compressed; introduced species rare in nc TX ____ **E. superba**
      5. Spikelets up to 5 mm wide (often much narrower), > 2 times as long as wide (often many times), usually slightly to strongly compressed; including abundant and widespread native and introduced species.
        6. Plants perennial; spikelets 3–5 mm wide, strongly flattened; lemmas very acute, 2–5 mm long ____ **E. secundiflora**
        6. Plants annual; spikelets 1–4 mm wide, flattened but not strongly so OR not flattened; lemmas subacute or obtuse, 1.6–2.8 mm long.
          7. Spikelets linear, almost as thick as wide, 1–1.5 mm wide; lemmas not glandular-dotted ____ **E. barrelieri**
          7. Spikelets ovate to lanceolate, much wider than thick, 2–4 mm wide; lemmas usually glandular-dotted on the keels ____ **E. cilianensis**
  3. Panicle branches naked for a distance from base.
    8. Spikelets widely spreading from branches; pedicels longer than spikelets.
      9. Spikelets 2–4(–6)-flowered, 2–4 mm long ____ **E. hirsuta**
      9. Spikelets (4–)5–18-flowered, 4–10 mm long.
        10. Spikelets very narrow, usually 1(–1.3) mm or less wide; lemmas 1.2–1.6 mm long ____ **E. pilosa**
        10. Spikelets 1.5–3 mm wide; lemmas 1.8–3.4 mm long.
          11. Lemmas keeled, the lateral nerves conspicuous; spikelets 4–10(–20) mm long.
            12. Lemmas ca 1.4–2.2 mm long; spikelets oblong to linear; plants with knotty rhizomes ____ **E. spectabilis**
            12. Lemmas 2.4–3.8 mm long; spikelets lanceolate to ovate; plants without rhizomes ____ **E. trichodes**
          11. Lemmas rounded on back, the lateral nerves inconspicuous; spikelets 3–5(–6) mm long ____ **E. intermedia**
    8. Spikelets slightly spreading or appressed along the branches, sessile or on pedicels mostly shorter than spikelets (some pedicels may be longer).

13. Culm bases neither hardened nor enlarged; plants annual.
    14. Closed spikelets (before lemmas spread open) over 1 mm wide; lemmas 2–2.4 mm long; leaf blades 3–7 mm wide; lower inflorescence branches 1–3, if somewhat verticillate then not hair-like ____ **E. pectinacea**
    14. Closed spikelets 1 mm wide or less; lemmas 1.2–1.6 mm long; leaf blades 0.4–2.2 (–3) mm wide; lower inflorescence branches hair-like, usually verticillate (in whorls of 3 or more branches) ____ **E. pilosa**
13. Culm bases hardened or enlarged; plants perennial.
    15. Longer glume 1–1.9 mm long; spikelets 4–5 mm long ____ **E. curtipedicellata**
    15. Longer glume 2.2–5 mm long; spikelets 6–20(–25) mm long.
        16. Spikelets 3–5 mm wide; culms 70 cm or less tall; spikelets 10–40-flowered, often brown-red, densely arranged, in conspicuously overlapping groups ____ **E. secundiflora**
        16. Spikelets 1.4–1.6 mm wide; culms to 150 cm tall; spikelets 7–11-flowered, gray-green, less dense, not much overlapping ____ **E. curvula**

Eragrostis barrelieri Daveau, (for French botanist, Jacques Barrelier, d. 1673), MEDITERRANEAN LOVE GRASS. Low-spreading to erect annual 7–50 cm tall; leaf sheaths glabrous except at top; panicles dense; spikelets gray- or yellow-green to bronze or dark purple-red-brown; lemmas 1.8–2.2 mm long. Roadsides and disturbed sites; nearly throughout TX. Apr–Nov. Native of s Europe.

Eragrostis cilianensis (All.) Vignalo ex Janch., (fringed with hairs), STINK GRASS. Annual often with a strong odor; lemmas 2.2–2.8 mm long, with minute, raised, glandular dots on keel. Disturbed sites; throughout TX. Mostly Aug–Oct. Native of Europe.

Eragrostis curtipedicellata Buckley, (short-stalked), GUMMY LOVE GRASS, SHORT-STALKED LOVE GRASS. Tufted, hard-based perennial 15–75 cm tall; leaf sheaths glabrous except for margins and summit; panicles open, with viscid axis, breaking off in age. Sandy oak woods, fields, roadsides, and open areas; nearly throughout TX. May–Nov.

Eragrostis curvula (Schrad.) Nees, (curved), WEEPING LOVE GRASS. Clump-forming perennial 60–150 cm tall; upper leaf sheaths usually glabrous, the lower pubescent with closely ascending hairs; panicles narrow, ± drooping. Planted on sandy soils for erosion control and range improvement; fields, roadsides; now widespread in TX. May–Jul. Native of Africa. Cattle wait until more palatable species are gone before eating the rather coarse foliage; subsequently they will eat this species nearly to the ground; the hard stubble that remains will hurt their feet making them "tender-footed" and hesitant to walk and properly feed; they have to be removed from such a pasture (J. Stanford, pers. comm.).

Eragrostis hirsuta (Michx.) Nees, (hairy), BIG-TOP LOVE GRASS, STOUT LOVE GRASS. Tufted perennial; spikelets small (to 4 mm long), few-flowered (2–4(–6). Sandy woods and river bottoms; e TX w to West Cross Timbers. Sep–Nov.

Eragrostis hypnoides (Lam.) Britton, Sterns, & Poggenb., (like *Hypnum*—a genus of mosses, moss-like), TEAL LOVE GRASS, SMOOTH CREEPING LOVE GRASS. Annual; culms creeping, rooting at nodes, mat-forming; in aspect similar to *E. reptans*; flowering culms 10–25 cm tall. Mud and sand bars of streams, lakes; Lamar Co. (Carr 1994); mainly se and e TX; also Edwards Plateau. Spring to fall, but mostly fall.

Eragrostis intermedia Hitchc., (intermediate), PLAINS LOVE GRASS. Tufted erect perennial 50–100 cm tall; leaf sheaths glabrous except at top; panicles open; spikelets 5–11-flowered. Sandy woods or less often open ground; mainly e 1/2 of TX, also Trans-Pecos. May–Oct. Many of the specimens we put in *E. intermedia* have been previously placed in *E. lugens* (Correll & Johnston

Elymus canadensis [HI1]

Elymus virginicus [USB]

Eragrostis barrelieri [GO2]

Eragrostis cilianensis [HI1]

Eragrostis curtipedicellata [USB]

Eragrostis curvula [USD]

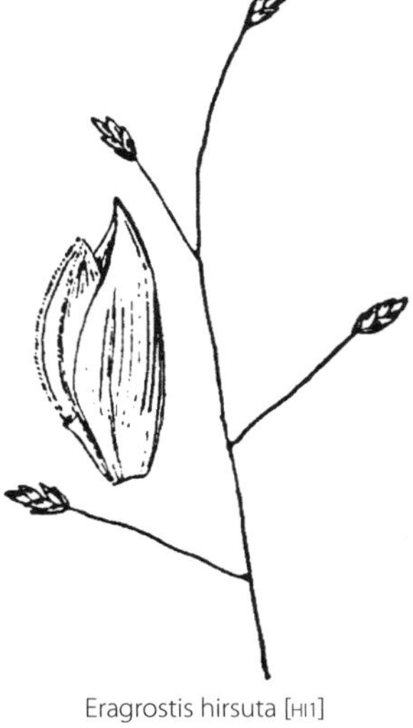

Eragrostis hirsuta [HI1]

Eragrostis hypnoides [GO2]

1970). We are following Gould (1975b) and Hatch et al. (1990) in recognizing *E. lugens* as restricted to vegetational areas 1, 2, 6, and 7 (Fig. 2) in TX.

Eragrostis pectinacea (Michx.) Nees ex Steud., (comb-like). Low-spreading to erect annual 10–60 cm tall; leaf sheaths glabrous except at top; lemmas grayish green. Sandy or clayey roadsides, disturbed areas. Spring–Nov.

1. Pedicels not appressed, 3–10 mm long, averaging > 4 mm ______ var. **miserrima**
1. Pedicels appressed, flexuous, 1–5 mm long, usually averaging < 4 mm ______ var. **pectinacea**

var. **miserrima** (E. Fourn.) Reeder, (possibly from Latin: *miser*, miserable, sickly, or wretched). The characters used to distinguish the varieties given here follow Correll and Johnston (1970), who recognized the taxa at the specific level; however, the distinctions seem weak. Specimens from Grayson, Jack, and Limestone cos. may be examples of var. *miserrima*; scattered in TX. [*E. arida* Hitchc., *E. tephrosanthos* Schult.]

var. **pectinacea**. SPREADING LOVE GRASS. Bell, Dallas, Denton, Erath, Hamilton, Mills, and Montague cos.; widespread in TX. This is by far the more common of the 2 varieties in nc TX. [*E. diffusa* Buckley]

Eragrostis pilosa (L.) P. Beauv., (pilose, with long soft hairs), INDIA LOVE GRASS. Tufted annual 15–45 cm tall; closed spikelets (before lemmas spread open) very narrow, 1 mm wide or less, when open to 1(–1.3) mm wide; lemmas grayish green with reddish or purplish tips. Sandy or sandy clay roadsides, disturbed sites; e TX w to West Cross Timbers, also s TX and Panhandle. Jun–Aug. Native of Europe and Asia.

Eragrostis reptans (Michx.) Nees, (creeping), CREEPING LOVE GRASS. Dioecious annual; culms creeping, rooting at nodes, mat-forming; flowering culms 5–10(–20) cm tall; spikelets usually 0.8–2(–2.5) mm long, with 16–40(–60) florets, crowded, sessile or short-pedicelled, ovate to lanceolate or linear, sometimes curved. Dry lake beds, stream bottoms, weedy areas; Bell, Cooke, Dallas, Denton, Grayson, and Kaufman cos.; e 1/2 of TX. Apr–Nov. Sometimes segregated into the genus *Neeragrostis* [as *N. reptans* (Michx.) Nicora].

Eragrostis secundiflora J. Presl. subsp. **oxylepis** (Torr.) S.D. Koch, (sp.: with flowers on one side of the stalk; subsp.: sharp-scaled), RED LOVE GRASS. Tufted perennial 15–90 cm tall; leaf sheaths glabrous except at top; ligules very short, with membranous base half the total length, obscured by long hairs at base of leaf blade; panicles usually small and compact, occasionally with 1 or 2 long branches; spikelets glaucous or blue-gray to green, or brown-red. Sandy woods or open ground, rarely in limestone gravel; Grayson Co. w through West Cross Timbers; nearly throughout TX, but more common in the e 1/2. May–Nov. Koch (1978) indicated that subsp. *secundiflora* occurs from South America n to Mexico. We are following Koch (1978) for synonymy of this taxon. [*E. beyrichii* J.G. Sm., *E. oxylepis* (Torr.) Torr., *E. oxylepis* (Torr.) Torr. var. *beyrichii* (J.G. Sm.) Shinners]

Eragrostis sessilispica Buckley, (with sessile spikelets), TUMBLE LOVE GRASS. Tufted perennial 30–90 cm tall; leaf sheaths glabrous except at top; panicles becoming loosely coiled and half prostrate, ultimately breaking off and acting as tumbleweeds. Sandy prairies and oak openings; East Cross Timbers w to Panhandle, also s TX. Apr–Sep.

Eragrostis spectabilis (Pursh) Steud., (spectacular), PURPLE LOVE GRASS. Perennial with a knotty, rhizomatous base; leaf sheaths usually pilose on collar and upper margins; panicles open, the branches capillary, wiry. Sandy or disturbed sites; widespread in TX, but mainly e 1/2. Aug–Oct.

Eragrostis superba Peyr., (superb), WILMANN'S LOVE GRASS. Perennial; spikelets 3–10 mm wide, conspicuously compressed, showy; lemmas keeled. Drought resistant species planted as a range

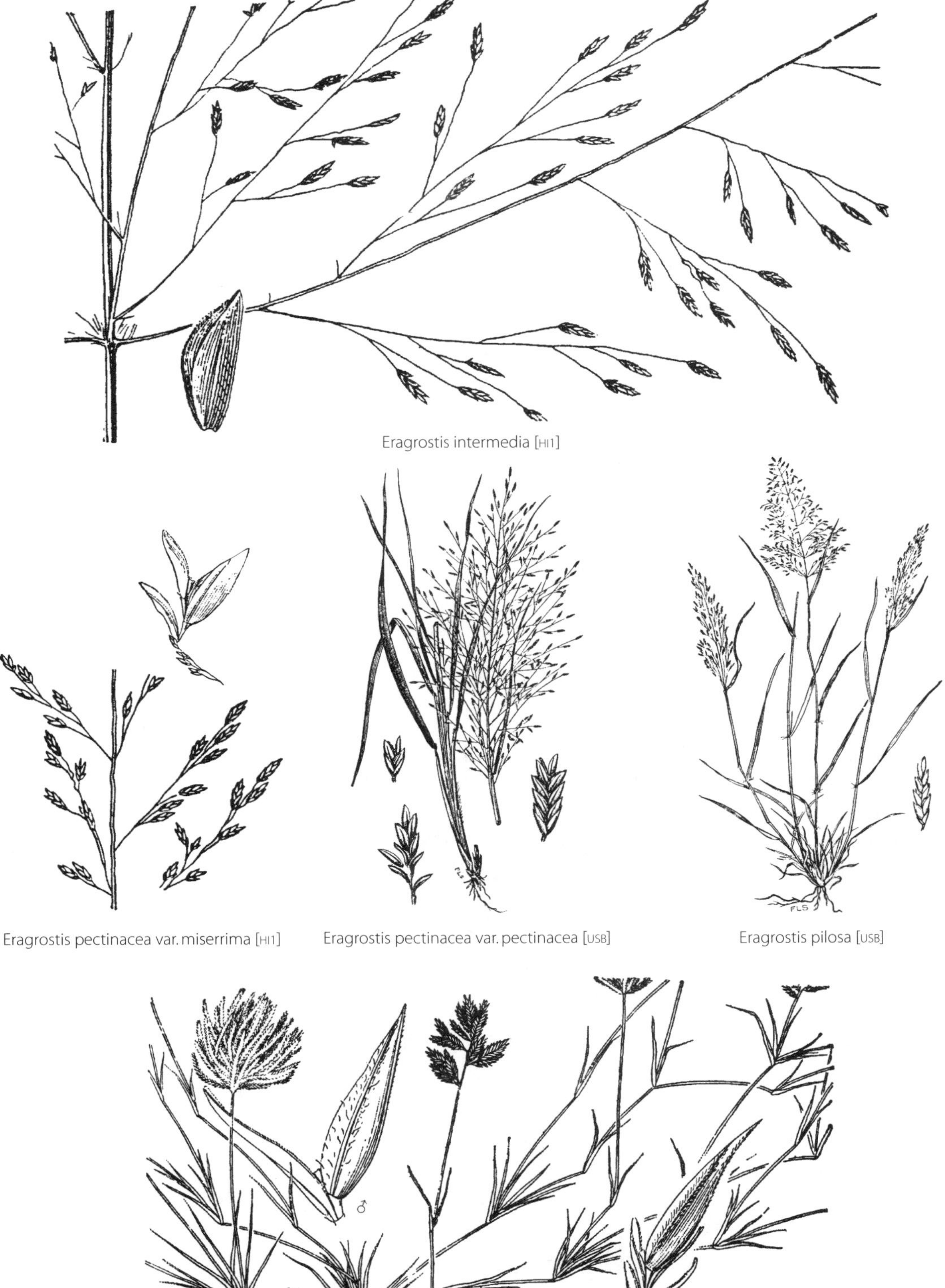

Eragrostis intermedia [HI1]

Eragrostis pectinacea var. miserrima [HI1]

Eragrostis pectinacea var. pectinacea [USB]

Eragrostis pilosa [USB]

Eragrostis reptans [HI1]

grass, persists, and escapes; Parker Co. (R. O'Kennon, pers. obs.), cited for vegetational area 5 by Hignight et al. (1988) and vegetational areas 4, 5, 6, and 7 (Fig. 2) by Hatch et al. (1990). Introduced from Africa.

Eragrostis trichodes (Nutt.) A.W. Wood, (hair-like), SAND LOVE GRASS. Tufted perennial; leaf sheaths pilose at throat, sometimes villous on back or papillose-villous on margins; lemmas with reddish-tinged splotches. Sandy prairies; widespread in TX. Jul–Dec., sporadically in the spring. [*E. pilifera* Scheele, *E. trichodes* var. *pilifera* (Scheele) Fernald] We have not been able to find consistent differences and are thus following Correll and Johnston (1970), Gould (1975b), and S. Hatch (pers. comm.) in lumping var. *pilifera*.

ERIOCHLOA CUP GRASS

Annuals or perennials; ligule of hairs to 1–2 mm long; panicles slender, of erect, spike-like racemes; spikelets with a cup-like collar or ring just under the base (representing the much-reduced lower glume), 2-flowered, the lower staminate or neuter, the upper perfect; upper glume and sterile lemma similar.

A genus of ca. 30 species of tropical and warm areas of the world. (Greek: *erion*, wool, and *chloë*, grass) (subfamily Panicoideae, tribe Paniceae)
REFERENCES: Shaw & Webster 1987; Crins 1991.

1. Pedicels densely short-pubescent; racemes overlapping each other; lemma of perfect floret with an awn 0.5–0.8 mm long; plants annual ____ **E. contracta**
1. Pedicels with silky hairs more than half as long as the spikelets; racemes not or only scarcely overlapping each other; lemma of perfect floret awnless or with a minute awn tip less than 0.5 mm long; plants perennial ____ **E. sericea**

Eriochloa contracta Hitchc., (contracted), PRAIRIE CUP GRASS. Plant 15–75 cm tall, often with many culms, forming bushy clumps, rather densely soft-pubescent. Dry roadsides, disturbed areas; nearly throughout TX. May–Oct.

Eriochloa sericea (Scheele) Munro ex Vasey, (silky), TEXAS CUP GRASS, SILKY CUP GRASS. Plant tufted, 30–100 cm tall, rather densely soft-pubescent. Prairies or open thickets, calcareous clays; nc TX w to Rolling Plains and s to Edwards Plateau. May–Oct. Secondary member of original prairie; it does not withstand grazing.

ERIONEURON WOOLLY GRASS, FLUFF GRASS

A New World genus of 3 species (Valdés-Reyna & Hatch 1997); similar to and previously lumped with *Tridens*. (Greek: *erion*, wool, and *neuro*, nerve, from hairs on nerves of lemmas and paleas) (subfamily Chloridoideae, tribe Eragrostideae)
REFERENCES: Valdés-Reyna 1985; Valdés-Reyna & Hatch 1997; Peterson et al. 1997.

Erioneuron pilosum (Buckley) Nash, (pilose, with long soft hairs), HAIRY TRIDENS, HAIRY ERIONEURON. Tufted perennial; culms 10–30 cm tall, usually with one node above the basal cluster of leaves; ligule of hairs ca. 0.5 mm long; leaf blades 1–2 mm broad, 2–8(–11) cm long, with a thick white margin (visible with a lens); panicles or racemes contracted, 2–4 cm long, 1.5–2 cm broad, of ca. 4–12 large spikelets; spikelets usually 10–16 mm long, with 7–18 closely imbricated florets; glumes 1-nerved, glabrous; lemmas rounded on back, 3-nerved, the nerves and margins conspicuously long-hairy, with awn 1–2 mm long from entire or notched apex. Disturbed sites, prairies; Bell, Dallas, and Grayson cos.; also Parker (R. O'Kennon, pers. obs.) and Tarrant (Mahler 1988) cos.; nc TX s and w to w TX. Apr–Oct. [*Tridens pilosus* (Buckley) Hitchc.]

Eragrostis secundiflora subsp. oxylepis [USB]

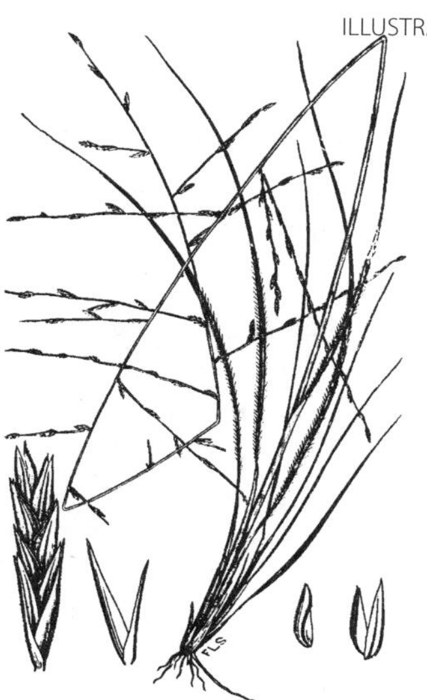

Eragrostis sessilispica [USB]

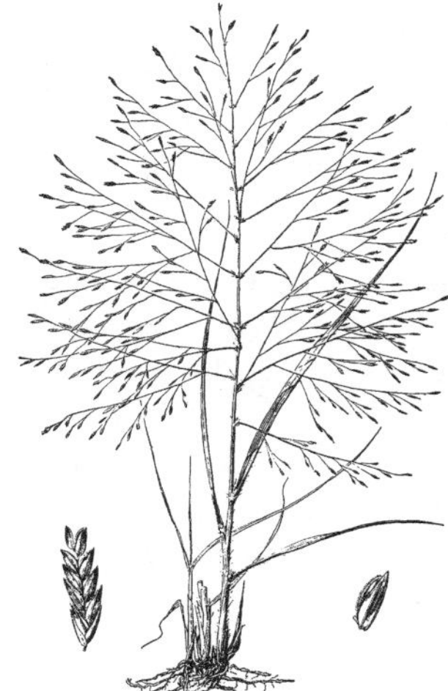

Eragrostis spectabilis [HI1]

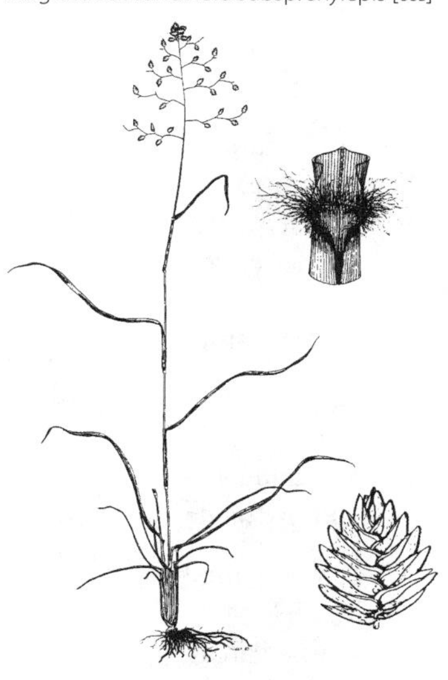

Eragrostis superba [HIG]

Eragrostis trichodes [USB]

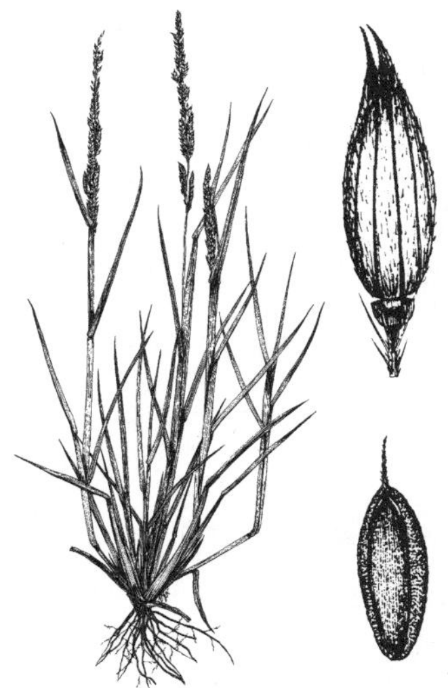

Eriochloa contracta [GO1]

Eriochloa sericea [USB]

Erioneuron pilosum [BB2]

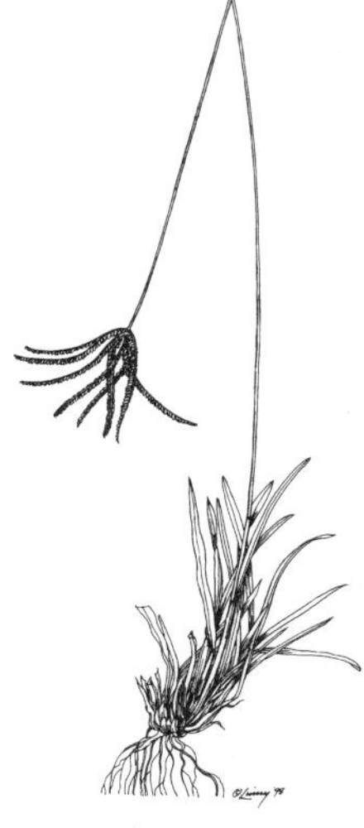

Eustachys retusa [HEA]

EUSTACHYS

A genus of 10 species native to tropical America and Africa. *Eustachys* has sometimes been treated as a section of *Chloris*. (Greek: *eu*, good, and *stachys*, spike, possibly in reference to the spike-like inflorescence branches) (subfamily Chloridoideae, tribe Cynodonteae)
REFERENCE: McKenzie et al. 1987.

Eustachys retusa (Lag.) Kunth, (retuse, notched slightly at apex). Tufted perennial with erect culms 25–90 cm tall; ligule a membrane ca. 0.5 mm long; leaf blades 4–7(–10) mm wide; inflorescences of 3–14 digitately arranged, spike-like branches 4–9 cm long; spikelets crowded, in 2 rows along 1 side of the branches, with 1 perfect lower floret and usually 1 sterile (sometimes staminate) floret above; fertile lemma with margins strongly ciliate; sterile lemma cylindrical to narrowly obtriangular and truncate. Disturbed roadsides; Falls and Limestone cos.; Hatch et al. (1990) only cited vegetational areas 3 and 4 (Fig. 2). May–Sep. Native to South America. [*Chloris argentina* (Hack.) Lillo & Parodi]

FESTUCA FESCUE

Tufted perennials; leaf blades 2.5–12 mm wide; leaf sheaths open; ligule a membrane; panicles open or contracted; spikelets with 2–8 florets; lemmas not toothed at apex, awned or awnless, the backs rounded, not plainly nerved except at apex.

A genus of 450 species of temperate areas and tropical mountains; some are important as pasture and lawn grasses. *Vulpia* (SIXWEEKS GRASS, ANNUAL FESCUE), treated here as a separate genus, has in the past sometimes been included in *Festuca*. Darbyshire (1993) recently suggested that *Festuca* subgenus *Schedonorus*, including *F. arundinacea*, be shifted to the related genus *Lolium*. (Ancient Latin name of some grass) (subfamily Pooideae, tribe Poeae)
REFERENCES: Piper 1906; Aiken & Lefkovitch 1993; Darbyshire & Warwick 1992; Darbyshire 1993.

1. Lower lemmas 5–9 mm long; spikelets 8–20 mm long; panicles contacted, at least above ____ **F. arundinacea**
1. Lower lemmas 3–4.7 mm long; spikelets 4–8 mm long; panicles open ____ **F. subverticillata**

Festuca arundinacea Schreb., (reed-like), TALL FESCUE, ALTA FESCUE. Nearly glabrous perennial 60–120(–200) cm tall, forming clumps; panicles rather narrow and dense; spikelets 5–8-flowered; lemmas usually awnless (or with short awn). In calcareous clay, roadsides, disturbed areas; also cultivated; Grayson, Kaufman, and Rockwall cos., also Lamar Co. (Carr 1994), also observed (R. O'Kennon) in Hunt, Johnson, Parker, and Tarrant cos.; spreading rapidly and probably in numerous nc TX cos.; widespread in TX. Mostly Apr–Jun. Native of Europe. [*F. elatior* L. var. *arundinacea* (Schreb.) Wimm., *Lolium arundinaceum* (Schreb.) Darbysh.] This species can be a problematic invasive weed in native prairie remnants. There are reports of livestock being poisoned (fescue foot or gangrenous fescue poisoning) from eating this species; the symptons are similar to those produced by ingestion of ergot alkaloids and fungi infecting the plant may be at fault (Kingsbury 1964; Hardin & Brownie 1993). Darbyshire (1993) presented evidence that *Festuca* is polyphyletic and that this species is best treated in the genus *Lolium*; however, we are treating it conservatively and for now continuing to recognize it in *Festuca*. Jones et al. (1997) placed it in *Lolium*.

Festuca subverticillata (Pers.) E.B. Alexeev, (slightly whorled), NODDING FESCUE. Glabrous perennial 50–120 cm tall, forming clumps; panicles open, with drooping branches; spikelets 2–5(–6)-flowered; lemmas glabrous, awnless, falling early. Low woods and thickets; Grayson Co.; also Dallas Co. (Mahler, 1988); mainly e TX. Apr–Jun. [*F. obtusa* Biehler]

GLYCERIA MANNA GRASS

Perennials, often with rhizomes or rooting at lower nodes; ligule a membrane; inflorescence an open or contracted panicle (rarely a raceme); spikelets in ours with 3–14(–20) florets, awnless, disarticulating above glumes and between florets; lemmas usually with 7 strong, parallel nerves, rounded on back.

A temperate, but especially North American genus of 40 species; some are good pasture grasses; Native Americans formerly used the fruits of some. (Greek: *glyceros*, sweet, referring to the taste of the seeds of one species) (subfamily Pooideae, tribe Meliceae)

1. Inflorescence branches usually short and stiffly erect or ascending; spikelets 8–30 mm long, with 7–14(–20) florets; first glume 2–3 mm long ____ **G. arkansana**
1. Inflorescence branches (at least lower) long and flexuous; spikelets 5 mm or less long, usually with 3–7 florets; first glume < 1 mm long ____ **G. striata**

Glyceria arkansana Fernald, (of Arkansas), ARKANSAS MANNA GRASS. Plant glabrous; culms ca. 1–1.8 m tall, often rooting near base; ligules ca. 3–6 mm long; inflorescences usually 18–40(–50) cm long, the panicle branches short and rigid; lemmas 2.5–3.5 mm long, hirsute or hirtellous. Wet areas; Fannin Co. in Red River drainage (Talbot property), also Lamar Co. (Carr 1994); mainly e TX. Mar–May. According to Gould (1975b), *G. arkansana* "is close to and weakly differentiated from *G. septentrionalis* Hitchc. (EASTERN MANNA GRASS) and probably should be treated as a variety of that species." Jones et al (1997) treated it as a variety of *G. septentrionalis*, while J. Kartesz (pers. comm. 1997) recognizes it at the species level. *Glyceria septentrionalis* Hitchc., also known from e TX, differs in having longer (3.5–5 mm), scabrous lemmas. [*G. septentrionalis* Hitchc. var. *arkansana* (Fernald) Steyerm. & Kucera]

Glyceria striata (Lam.) Hitchc., (striated, striped), FOWL MANNA GRASS, NERVED MANNA GRASS. Plant glabrous, often with short rhizomes; culms 0.4–0.9 m tall; ligules 1.5–4 mm long; inflorescence a panicle 10–22 cm long with numerous slender, flexuous branches bare of spikelets on the lower 1/3–1/2; spikelets ca. 3–4(–5) mm long; lemmas usually 1.5–2 mm long. Along streams, moist forest margins; Lamar Co. (Carr 1994) in Red River drainage, also Bell Co. (Fort Hood—Sanchez 1997); rare in Trans-Pecos, Edwards Plateau, and e TX. Apr–Aug. Reported as potentially cyanogenic (Burlage 1968). ☠

GYMNOPOGON

A genus of ca. 15 species of warm areas of the Americas and from India to Thailand. (Greek: *gymnus*, naked, and *pogon*, beard, referring to the reduction of the abortive flower to a bare awn) (subfamily Chloridoideae, tribe Cynodonteae)
REFERENCE: Smith 1971.

Gymnopogon ambiguus (Michx.) Britton, Sterns, & Poggenb., (ambiguous), BEARDED SKELETON GRASS. Perennial from short rhizome; culms 25–60 cm long with numerous short overlapping leaves with spreading to deflexed blades to 13 mm wide; ligules minute, membranous or a rim of callus tissue; panicles usually broader than long, 10–25 cm long, with 12–35 widely spreading, spike-like branches; branches usually 10–20 cm long with spikelets to base or bare on basal 2–5 cm; spikelets usually not overlapping, appressed, in 2 rows on 1 side of the slender, slightly flattened branch, with 1 perfect lower floret and 1 rudimentary upper floret; glumes 3–7 mm long; lemmas shorter than the glumes, with a 4–8(–11) mm awn from the minutely notched apex. Usually in sandy soils, typically in shade; Grayson, Hunt, and Montague cos.; also Lamar Co. (Carr 1994); mainly e TX. Late Aug–Nov.

HILARIA

A genus of 10 species ranging from the s U.S. to Guatemala. (Named for A. Saint-Hilaire, 1779–1853, French botanist) (subfamily Chloridoideae, tribe Cynodonteae)
REFERENCES: Cory 1948b; Brown & Coe 1951.

Hilaria belangeri (Steud.) Nash, (named for Charles Paulus Bélanger, 1805–1881), COMMON CURLY-MESQUITE. Low tufted perennial, 10–30 cm tall, sending out slender stolons that produce new tufts; nodes densely bearded with spreading hairs; ligule a membrane 0.5–1 mm long; leaf blades 1–2(–3) mm wide, short and often forming a curly tuft, but sometimes longer and erect; inflorescence a slender spike 2–3.5 cm long, exerted on peduncles, with 4–8 clusters of 3 spikelets each; spikelet clusters usually 4.5–6 mm long; central spikelet fertile, 1-flowered, lateral spikelets staminate, 2-flowered; glumes united below, usually asymmetrical; glumes of central spikelet with awns usually 2.5–5 mm long. Open brushy and rocky areas, usually on calcareous soils; Bell, Brown, and Stephens cos.; nc TX s and w to w TX. (Mar–)Aug–Oct(–Nov).

HORDEUM BARLEY

Annuals (in nc TX); leaves with or without auricles; ligule a membrane; inflorescence a dense, 2-sided spike or spike-like raceme, breaking apart in age except in *H. vulgare*; spikelets 3 per node, each 1-flowered (or lateral spikelets may be reduced to a group of awns), the central spikelet sessile, usually perfect, the 2 lateral spikelets usually short-pedicelled, staminate or sterile (perfect in *H. vulgare*); lemmas usually awned apically.

A n temperate genus of ca. 20 species including *H. vulgare* (cultivated BARLEY). BARLEY is a very old cereal crop originally grown by people of the ancient Near East and is considered along with WHEAT to be one of the oldest cultivated plants; it is currently used primarily in the malting process of beer- and whiskey-making and as animal feed (Heiser 1990a). The awns of some species are minutely barbed and cause physical damage and infection in livestock (Stephens 1980). (Ancient Latin name for BARLEY) (subfamily Pooideae, tribe Triticeae)
REFERENCES: Covas 1949; Bowden 1959b; Dewey 1982; Estes & Tyrl 1982; Baum & Bailey 1986, 1989, 1990; von Bothmer et al. 1991; Zohary & Hopf 1994.

1. Lemmas of lateral spikelets absent or much smaller than lemma of central spikelet; lemma awns 2–10(–15) mm long; leaf sheaths without auricles at summit ____ **H. pusillum**
1. Lemmas of lateral spikelets as large as lemma of central spikelet; lemma awns 10 mm or more long (usually much longer); leaf sheaths with auricles at summit.
   2. Leaf blades 1–6 mm wide; lemmas 1.3–1.9 mm wide across back; lateral spikelets short-pedicelled; glumes of central spikelet ciliate marginally ____ **H. murinum**
   2. Leaf blades mostly 7–18 mm wide; lemmas 2–3 mm wide across back; lateral spikelets sessile; glumes of central spikelet glabrous or scabrous ____ **H. vulgare**

Hordeum murinum L. subsp. **leporinum** (Link) Arcang., (sp.: mouse-gray; subsp.: hare-like), HARE BARLEY. Annual 30–75 cm tall; culms erect or decumbent basally; leaf sheaths glabrous; glumes ciliate marginally; lemma awns 10–35 mm long. Disturbed sites; Dallas, Coleman, and Williamson cos.; throughout TX except extreme e and s. Apr. Native of Europe. [*H. leporinum* Link] The awns can cause mechanical injury to grazing livestock (Burlage 1968).

Hordeum pusillum Nutt., (very small), LITTLE BARLEY, MOUSE BARLEY. Annual 12–60 cm tall; culms erect or with bent or decumbent base; leaf sheaths glabrous; spikes flattened; glumes scabrous, not ciliate; lemma awns 2–10(–15) mm long. Disturbed sites; throughout TX. Apr–May.

Hordeum vulgare L., (common), BARLEY. Glabrous annual; vegetatively resembling WHEAT and RYE; axis of spike remaining intact (vs. abscising at each node in our other species); lemma awns

Festuca arundinacea [USB]

Festuca subverticillata [GO2]

Glyceria arkansana [HI1]

Glyceria striata [HI1]

Gymnopogon ambiguus [USB]

Hilaria belangeri [HI1]

Hordeum murinum subsp. leporinum [MAR]

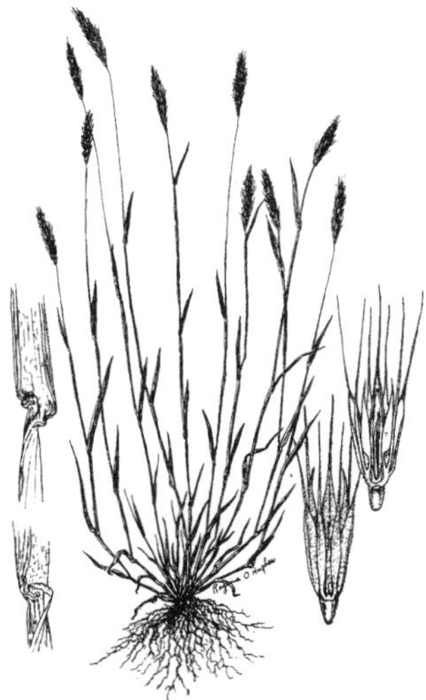
Hordeum pusillum [REE]

Hordeum vulgare [HI1]

long, up to 150 mm. Cultivated throughout much of TX and occasionally occuring as a transitory waif along roadsides, railroads, and field margins; Denton Co., also Tarrant Co. (R. O'Kennon, pers. obs.). Apr–May. Like its wild ancestor, *H. vulgare* subsp. *spontaneum* (C. Koch) Thell., BARLEY is native to Eurasia (Zohary & Hopf 1994). BARLEY is considered one of the two oldest cultivated plants (WHEAT—*Triticum aestivum*, is the other); as such it was probably important in the development of early civilization in the Near East (Heiser 1990a); BARLEY is able to stand drier, less fertile, and more saline conditions than WHEAT. It is the main cereal used in Old World beer production (Zohary & Hopf 1994).

KOELERIA

A temperate genus of ca. 35 species. (Named for Georg Ludwig Koeler, 1765–1807, professor at Mainz and student of grasses) (subfamily Pooideae, tribe Aveneae)
REFERENCE: Arnow 1994.

Koeleria macrantha (Ledeb.) Schult., (large-flowered), JUNE GRASS, PRAIRIE JUNE GRASS. Perennial 35–75 cm tall, forming small clumps; leaves mostly crowded toward base, the lowest with finely pubescent sheaths, the upper glabrous; ligule a short membrane 0.5–1(–2) mm long; inflorescence a narrow panicle, dense, spike-like, with puberulent axis; spikelets shiny, mostly 2–4-flowered, 4–5 mm long, disarticulating above glumes; lemmas usually acute or minutely apiculate. Grasslands; Delta, Denton, Grayson, and Kaufman cos., also Dallas Co. (Mahler 1988); mainly w 1/2 of TX. May–Jun. [*K. cristata* of authors in part, not Pers., *K. pyramidata* of authors in part, not (Lam.) P. Beauv.]

LEERSIA CUT GRASS

Rhizomatous perennials (our taxa); ligule a membrane; spikelets 1-flowered, sessile or subsessile, secund along the distal portions of the branchlets of the inflorescence, compressed laterally, disarticulating from pedicel; glumes absent (the spikelets thus consisting solely of naked florets); lemmas oblong to elliptic, with nerves and keel hispid-ciliate.

A genus of 17 species native to tropical and warm temperate areas; typically marsh grasses sometimes used as fodder; related to *Oryza sativa* L. (RICE), an annual with spikelets 7–10 mm long (vs < 6 mm long in *Leersia*). In se and e TX, RICE is cultivated and is occasionally a weed in moist soils; this native of se Asia is one of the world's most important food crops. While less widely distributed than WHEAT, RICE feeds more people, being the basic food for more than half the world's human population (Heiser 1990a). (Named for Johann Leers, 1727–1774, a German botanist) (subfamily Oryzoideae, tribe Oryzeae)
REFERENCE: Pyrah 1969.

1. Spikelets 3–4 mm wide, nearly as wide as long; leaf blades 5–15(–20) mm wide; in nc TX known only from Lamar Co. in Red River drainage ____ **L. lenticularis**
1. Spikelets 2 mm or less wide, 2–3 times as long as wide; leaf blades 3–10 mm wide; widespread in nc TX.
    2. Lower panicle branches fascicled; spikelets 4–5.5 mm long, 1.5–2 mm wide ____ **L. oryzoides**
    2. Lower panicle branches solitary; spikelets 2.2–3(–3.5) mm long, 1 mm wide ____ **L. virginica**

Leersia lenticularis Michx., (lens-shaped), CATCHFLY GRASS. Culms 0.9–1.5 m long, erect or sprawling and ascending; panicles open, with spikelets clustered towards ends of flexuous, often drooping branches, the branches naked nearly 1/2 their length; spikelets broadly oblong, slightly asymmetrical, mostly 4–5 mm long; keel and nerves of lemmas with conspicuous, stiff hairs giving the appearance of widely spaced comb teeth (use lens). Wet or moist areas along streams and lakes; Lamar Co. at edge of oxbow lake in Red River drainage; mainly se and e TX. Mostly Jul–Nov.

Leersia oryzoides (L.) Sw., (like rice—*Oryza*), RICE CUT GRASS. Plant 1–1.5 m tall; rhizomes slender, creeping; leaf blades 7–10 mm wide; panicle branches usually numerous, naked of spikelets for < 1/2 their length. Marshes, wet places; Bell, Dallas, Grayson, and Limestone cos., also Lamar Co. (Carr 1994); throughout most of TX. Sep–Nov.

Leersia virginica Willd., (of Virginia), WHITE GRASS, VIRGINIA CUT GRASS. Plant 0.5–1.2 m tall; rhizomes scaly, clustered, stouter than culm base; leaf blades 3–8 mm wide; panicle branches few, widely spaced, long and slender, up to 8–12 cm long, naked for > 1/2 their length. Ditches, moist areas; Collin, Dallas, Grayson, and Rockwall cos., also Bell (Fort Hood—Sanchez 1997), Fannin (Mahler 1988), and Lamar (Carr 1994) cos.; se and e TX w to nc TX. May–Nov.

LEPTOCHLOA SPRANGLETOP

Annuals or perennials; inflorescence a panicle of spicate branches; spikelets with 2–12 flowers, the lower 1–2 perfect, the rest staminate or neutral, sessile or nearly so, overlapping in 2 rows on 1 side of a nearly terete branch, disarticulating above the glumes and between florets; lemmas in ours awnless or with a short awn to 1.5 mm long.

A genus of 27 species native to tropical and warm areas of the Americas and Australia; some are used as fodder. (Greek: *leptos*, slender, and *chloa*, grass, from the slender inflorescence branches) (subfamily Chloridoideae, tribe Eragrostideae)
REFERENCES: Snow & Davidse 1993; Snow 1996; Peterson et al. 1997.

1. Lemmas 1–1.5 mm long; leaf sheaths usually papillose-pilose; spikelets 1.4–3 mm long, with 2–4 florets ____ **L. mucronata**
1. Lemmas at least 1.8 mm long; leaf sheaths glabrous to pilose, not papillose; spikelets usually 5–10 mm long, with 3–12 florets.
   2. Tip of lemma appearing chopped off, usually notched, awnless; plants perennial ____ **L. dubia**
   2. Tip of lemma not appearing chopped off, blunt to acute or acuminate (but can be slightly notched), awnless or with an awn 0.5–1.5 mm long; plants annual.
      3. Second glume 3–4.2 mm long; inflorescence branches 4–12(–19) cm long; lemmas lance-elliptic, acute to acuminate, 2.5–4 mm long, usually with awn 0.5–1.5 mm long ____ **L. fascicularis**
      3. Second glume < 3 mm long; inflorescence branches 3–6 cm long; lemmas obovate, blunt, 1.8–3 mm long, awnless or abruptly mucronate ____ **L. uninerva**

Leptochloa dubia (Kunth) Nees, (doubtful), GREEN SPRANGLETOP, TEXAS CROWFOOT. Tufted perennial 25–115 cm tall, erect, unbranched above base (but inflorescences with 2–15 unbranched flexuous, loosely erect or spreading main branches); ligule a dense row of cilia 0.3–1.2 mm long; spikelets usually with 3–8 florets; glumes and lemmas awnless. Rocky slopes, loams; Bell, Johnson, Palo Pinto, Shackelford, and Tarrant cos.; widespread in TX. Spring–fall.

Leptochloa fascicularis (Lam.) A. Gray, (fascicled, clustered), BEARDED SPRANGLETOP, SALT MEADOW GRASS, SALT SPRANGLETOP. Tufted annual usually 50–100 cm tall, moderately branched; ligule a well-developed membrane 2.5–6 mm long; inflorescences with ca. 8–35 stiffly erect or erect-spreading branches; spikelets with 6–12 florets; lemmas with central nerve usually projecting as an awn 0.5–1.5 mm long. Mud, sometimes alkaline or subsaline; Dallas and Grayson cos.; throughout TX. Summer–fall.

Leptochloa mucronata (Michx.) Kunth, (mucronate, with a short and small abrupt tip), RED SPRANGLETOP, SLENDER GRASS. Tufted annual 10–80 cm tall; inflorescence branches few to numerous (to ca. 70), scattered, slender, flexuous; spikelets widely spaced and only slightly overlapping, with 2–4 florets; glumes and lemmas usually awnless. Moist soils and mud; nearly throughout TX. Late spring–fall. [*L. filiformis* (Lam.) P. Beauv.]

Leptochloa uninerva (J. Presl) Hitchc. & Chase, (one-nerved), MEXICAN SPRANGLETOP. Similar to *L. fascicularis*; spikelets more darkly colored at maturity. Mud, sometimes alkaline or subsaline; included based on citation for vegetational area 5 by Hignight et al. (1988); mainly s TX n to Edwards Plateau. Spring–summer(–fall).

LIMNODEA OZARK GRASS

A monotypic genus of the s United States. (Name altered from *Limnas*, an Old World genus of grasses) (subfamily Pooideae, tribe Aveneae)

Limnodea arkansana (Nutt.) L.H. Dewey, (of Arkansas), OZARK GRASS. Annual 15–60(–100) cm tall; leaf blades glabrous or more often hispidulous or hispid on upper and lower surfaces; ligule a lacerate-ciliate membrane 1–2 mm long; inflorescence a narrow but not cylindrical panicle, erect or nodding at tip; spikelets 1-flowered, 3–4 mm long excluding awns; lemmas usually slightly 2-toothed at apex, awned from back at or near base of the 2 teeth, the awn (4–) 8–12 mm long, geniculate. Prairies and disturbed areas, calcareous soils, often in sand; throughout most of TX except extreme w and nw. Mar–Jun.

LOLIUM RYE GRASS

Annuals or perennials; leaves auricled; ligule a membrane; inflorescence an unbranched spike with sessile spikelets in 2 ranks on opposite sides of the inflorescence axis; inflorescence axis remaining intact, disarticulation occuring above glume and between florets; spikelets 1 per node, placed with edge (keels of lemmas) against axis of inflorescence, without a glume on that side, 5–15(–20)-flowered.

An Eurasian genus of 8 species including valuable fodder and lawn grasses. Darbyshire (1993) suggested that *Festuca* subgenus *Schedonorus*, including *F. arundinacea*, be shifted to *Lolium*. (Ancient Latin name for rye grass) (subfamily Pooideae, tribe Poeae)
REFERENCE: Darbyshire 1993.

1. Glumes markedly shorter than spikelet, 5–10 mm long — **L. perenne**
1. Glumes as long as the entire spikelet (excluding awns), 10–20 mm long — **L. temulentum**

Lolium perenne L., (perennial), Annual or short-lived perennial. Cultivated for winter-green lawns and erosion control; escaped to roadsides, disturbed areas; throughout TX. Mar–Jun. Native of Europe. While we are following Kartesz (1994) in recognizing the following 2 subspecies, there appears to be significant intergradation and variation. According to Gould (1975b), "... recognition of two varieties [subspecies] is not satisfactory." Awned and unawned individuals can be found together.

1. Leaf auricles long, slender, pointed; lemmas usually all awned, rarely all awnless, occasionally only a few in upper spikelets awned — subsp. **multiflorum**
1. Leaf auricles mostly short, rounded; lemmas awnless — subsp. **perenne**

subsp. **multiflorum** (Lam.) Husn., (many-flowered), ITALIAN RYE GRASS. More common in cultivation than subspecies *perenne*. Coryell, Ellis, Grayson, Hill, McLennan, and Navarro cos. [*L. multiflorum* Lam., *L. perenne* var. *italicum* sensu Mahler, possibly (A. Braun) D.R. Parnell, *L. perenne* var. *multiflorum* (Lam.) D.R. Parnell] Jones et al. (1997) treated this taxon as *L. perenne* var. *aristatum* Willd.

subsp. **perenne**. PERENNIAL RYE GRASS. Perennial but grown as winter annual. Hill and Grayson cos.

Lolium temulentum L., (drunken), DARNEL, DARNEL RYE GRASS, BEARDLESS DARNEL RYE GRASS, POISON DARNEL. Annual. Roadsides, sandy weedy areas; Grayson, Lamar, and McLennan cos., also

Koeleria macrantha [HI1]

Leersia lenticularis [USB]

Leersia oryzoides [MAS]

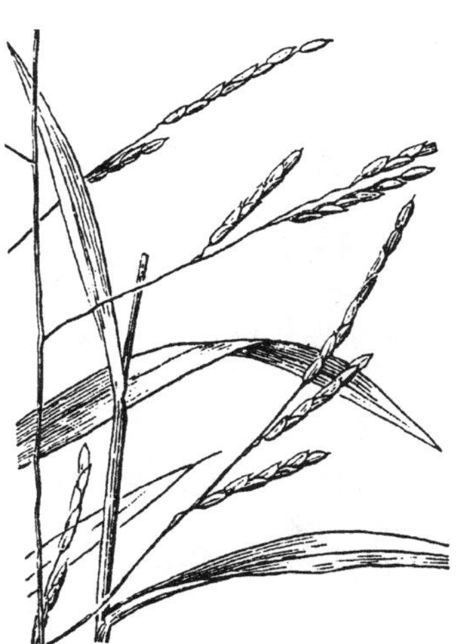

Leersia virginica [HI1]

Leptochloa dubia [USB]

Leptochloa fascicularis [USB]

Leptochloa mucronata [HI1]

Leptochloa uninerva [USB]

Limnodea arkansana [USB]

Fort Hood (Bell or Coryell cos.—Sanchez 1997); throughout much of TX. May–Jun. Native of Eurasia. [*L. temulentum* var. *leptochaeton* A. Braun] Poisonous due to alkaloids produced by fungi infecting grains; livestock can be affected with symptoms including "rye-grass staggers" in sheep; infected grains were formerly used to make an intoxicating beverage (Muenscher 1951; Mabberley 1987; Hardin & Brownie 1993). ☠

MELICA MELIC, MELIC GRASS

Tall perennials; leaf sheaths closed; ligule a membrane; inflorescence a panicle; spikelets usually with (1–)2–3 perfect florets, also usually 2–3 neuter reduced florets ("rudiments") above; glumes broad; lemmas (in our species) awnless; disarticulation below spikelets (in our species).

A genus of 80 species of temperate regions excluding Australia. (Greek: *meli*, honey, or classical name for some plant, possibly with sweet sap, taken up by Linnaeus for this genus) (subfamily Pooideae, tribe Meliceae)
REFERENCE: Boyle 1945.

1. Panicles usually simple (rarely compound); glumes ca. equal in length or second slightly longer; spikelets ± flat-topped and triangular in shape; ligules 1 mm or less long; rudiments broadly obovate to obconic, truncate (appearing abruptly cut off) **M. mutica**
1. Panicles usually compound (branches themselves branched); glumes not equal in length; spikelets neither flat-topped nor triangular in shape; ligules 3–6 mm long; rudiments narrowly obovate or oblong, not truncate **M. nitens**

Melica mutica Walter, (pointless), TWO-FLOWER MELIC, NARROW MELIC. Culms usually 40–80(–100) cm tall from creeping rhizomes; inflorescences 4–16 cm long; spikelets 7–11 mm long; fertile florets usually 2; rudiments broadly obovate to obconic (= like an inverted cone), spreading at an angle from the rachilla. Forest openings on sandy soils; Lamar Co. (Carr 1994) in Red River drainage; mainly se and e TX. Apr–Jun.

Melica nitens (Scribn.) Nutt. ex Piper, (shining), THREE-FLOWER MELIC, TALL MELIC. Rhizomatous, 50–120 cm tall; inflorescences usually 10–26 mm long; spikelets 8–15 mm long, much longer than broad; fertile florets usually 2–3; rudiments not spreading at an angle from the rachilla. Woodland, rocky grasslands; Grayson Co.; also Coleman Co. (Mahler 1988); mainly w 1/2 of TX. Apr–Jun.

MISCANTHUS

A genus of ca. 20 species of the Old World tropics, s Africa, and e Asia; some are cultivated as ornamentals. (Greek: *mischos*, pedicel, and *anthos*, flower, from stalked spikelets) (subfamily Panicoideae, tribe Andropogoneae)

Miscanthus sinensis Andersson, (Chinese), SILVER GRASS, EULALIA. Perennial ca. 1–3(–more) m tall, forming dense, bushy clumps; leaves basal and distributed up the culm, the blades to ca. 1 m long and 2 cm wide, green or variegated (with green and white or yellow bands or stripes) in some cultivated forms, the margins sharply scaberulous; ligule a ciliate membrane 1.5–3 mm long; inflorescence a dense panicle 10–35 cm long, 5–20 cm wide, the branches ascending; inflorescence branches not disarticulating, the spikelets falling from pedicels at maturity; spikelets in pairs, both spikelets of each pair alike, except 1 short-pedicelled, 1 long-pedicelled; pedicels 1–5 mm long; spikelets 4–5.5 mm long, with a tuft of long hairs at base of glumes slightly longer than spikelet; lemmas shorter than glumes, with a twisted, geniculate awn 5–10 mm long. No escaped nc TX specimens have been seen, but this species is cultivated, persists, and possibly escapes. Fall. Native of e Asia.

Lolium perenne subsp. multiflorum [USD]

Lolium perenne subsp. perenne [USB]

Melica mutica [USB]

Lolium temulentum [USB]

Miscanthus sinensis [HI1]

Melica nitens [HI1]

MUHLENBERGIA MUHLY

Ours perennials; ligule a membrane; inflorescences open to contracted; spikelets usually 1-flowered, the floret separating above the glumes; glumes 1-nerved or nerveless, rarely 3–5-nerved; lemmas 3(–5)-nerved, usually awned; paleas 2-nerved.

A genus of ca. 160 species primarily of tropical and warm areas of the Americas, with a few in s Asia. (Named for Gotthilf Henry Ernest Muhlenberg, 1753–1815, distinguished American botanist of Pennsylvania) (subfamily Chloridoideae, tribe Eragrostideae)

REFERENCES: Scribner 1907; Soderstrom 1967; Pohl 1969; Morden & Hatch 1987, 1989, 1996; Crosswhite & Crosswhite 1997; Peterson et al. 1997.

1. Panicles open, usually 2–17 cm or more wide; branches bare of spikelets for a distance from base.
    2. Leaf sheaths compressed-keeled; spikelets on pedicels usually 2–5(–8) mm long; lemma awns 0.5–2 mm long; panicles with branches narrowly spreading, 2–5(–7) cm wide ____ **M. ×involuta**
    2. Leaf sheaths rounded, not keeled; spikelets on pedicels 3–25 mm long, at least some usually 10 mm or more long; lemma awns 0.5–13 mm long; panicles with branches usually widely spreading, 4–17(–20) cm wide.
        3. Awns of lemma usually 5–15 mm long (rarely as little as 2 mm); second (longer) glume < 1/2 as long as lemmas ____ **M. capillaris**
        3. Awns of lemma usually 0.5–4 mm long (rarely to 7 mm); second glume 1/2 as long as lemmas or longer ____ **M. reverchonii**
1. Panicles contracted, usually < 2(–3) cm wide; branches with spikelets nearly to base.
    4. Glumes minute, 0.3 mm or less long, the first often absent; culms decumbent, often rooting at lower nodes ____ **M. schreberi**
    4. Glumes usually much longer; culms usually erect.
        5. Plants without scaly creeping rhizome; leaf sheaths sharply keeled; inflorescences (10–)20–54 cm long, usually 10–15(–30) mm wide; plants with unbranched erect culms, usually 80–100(–150) cm tall ____ **M. lindheimeri**
        5. Plants with scaly creeping rhizome; leaf sheaths not sharply keeled; inflorescences variable in size, to 21 cm long and 10 mm wide, usually smaller; plants either with culms much branched above or not erect, to ca. 90 cm tall, usually much smaller.
            6. Mat-like plants of wet areas; leaf blades mostly 1–2(–3.5) cm long, 0.5–1(–2) mm wide; both glumes and lemmas 2 mm or less long ____ **M. utilis**
            6. Plants not mat-like, of various habitats; leaf blades mostly 4 cm or more long, 1–3 (or more) mm wide; either glumes or lemmas or both > 2 mm long.
                7. Panicles terminal only; glumes narrowly lanceolate; lemmas pilose below, with conspicuous awns 3–10(–18) mm long ____ **M. sylvatica**
                7. Panicles terminal and axillary; glumes lanceolate to ovate; lemmas pilose to glabrous, awnless or with awns to 3 mm long (rarely to 7 mm in *M. bushii*).
                    8. Culms glabrous.
                        9. Glumes overlapping nearly to middle, ovate, abruptly tapering to an awn tip; panicles long-exserted, the peduncles up to 11 cm long ____ **M. sobolifera**
                        9. Glumes not overlapping or overlapping only at base, lanceolate, gradually tapering to an awn tip; panicles not long-exserted, the peduncles only 1–2 cm long.
                            10. Glumes 2 mm or less long; ligules 0.6 mm or less long ____ **M. bushii**
                            10. Glumes 2–4 mm long; ligules 0.8–1.4 mm long ____ **M. frondosa**

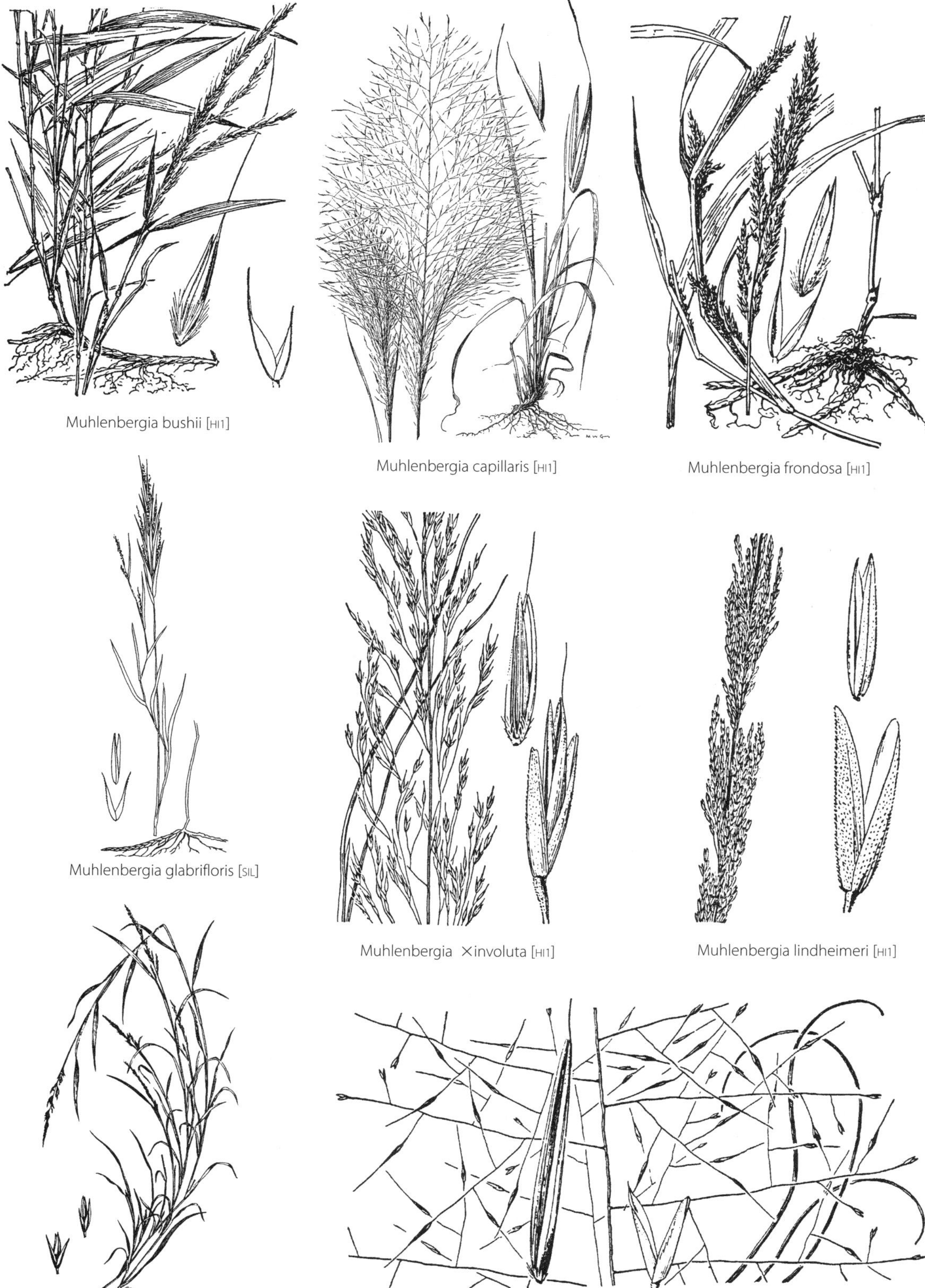

Muhlenbergia bushii [HI1]

Muhlenbergia capillaris [HI1]

Muhlenbergia frondosa [HI1]

Muhlenbergia glabrifloris [SIL]

Muhlenbergia ×involuta [HI1]

Muhlenbergia lindheimeri [HI1]

Muhlenbergia mexicana [USB]

Muhlenbergia reverchonii [HI1]

Muhlenbergia bushii R.W. Pohl, (for Benjamin Franklin Bush, 1858–1937, amateur botanist of Missouri), NODDING MUHLY. Rhizomatous perennial; culms 30–90 cm tall, erect, becoming much-branched above; panicles numerous, slender; lemma awnless or with awn rarely to 7 mm, pilose basally. Rich or low woods; Clay, Denton, Grayson, Hunt, McLennan, and Montague cos., also Tarrant Co. (Silveus 1933); otherwise in TX only known from Van Zandt Co. in e TX. Jul–Oct. [*M. brachyphylla* Bush]

Muhlenbergia capillaris (Lam.) Trin., (hair-like), HAIRY-AWN MUHLY, LONG-AWNED HAIR GRASS, SLENDER MUHLY, GULF MUHLY. Densely tufted perennial; culms erect, 60–100 cm tall; panicles open, diffuse, to 35(–40) cm long, 8–17(–20) cm wide, the branches and pedicels capillary and widely spreading at maturity; lemmas 3.5–4.5(–5) mm long, minutely scabrous, with a few hairs basally, with variable awn (2–)5–15 mm long. Sandy forest openings; Grayson, Henderson, Lamar, and Tarrant cos.; mainly se and e TX. Sep–Oct.

Muhlenbergia frondosa (Poir.) Fernald, (leafy), WIRE-STEM MUHLY. Rhizomatous perennial; culms often becoming decumbent or sprawling, to 100 cm long; panicles numerous, slender; lemmas awnless or with awn to 2 mm long, pubescent basally. Clay soils; in TX known only from Dallas and Grayson cos. (Correll & Johnston 1970). Oct.

Muhlenbergia glabrifloris Scribn., (smooth-flowered), INLAND MUHLY. Rhizomatous perennial similar to *M. frondosa*; panicles numerous, slender; lemmas awnless, glabrous. Rich woods, in TX known only from the Blackland Prairie; type of this species was collected by Reverchon at Dallas (Mahler 1988). Sep.

Muhlenbergia ×involuta Swallen [*M. lindheimeri* × *M. reverchonii*], (rolled inward), CANYON MUHLY. Densely tufted perennial; culms stiffly erect, 60–140 cm tall; panicle branches slender, erect-spreading; spikelets 3–4 mm long; lemmas 3–4 mm long, minutely bifid apically, with awn 0.5–2 mm long from between the minute apical teeth. Prairie draws and openings; Burnet Co.; type collected 20 mi ne of San Antonio; also Bandera, Blanco, Comal, Kendall and Travis cos. to the s of nc TX (Mahler 1988); Edwards Plateau n to very s edge of nc TX; endemic to TX. Fall.

Muhlenbergia lindheimeri Hitchc., (for Ferdinand Jacob Lindheimer, 1801–1879, German born TX botanist), LINDHEIMER'S MUHLY. Large tufted perennial with stiffly erect culms usually 80–100(–150) cm tall; panicles (10–)20–54 cm long, tightly or loosely contracted, densely flowered; panicle branches with flowers nearly to base; lemmas ca. 2–4 mm long, glabrous, scabrous, or puberulent, awnless or infrequently with awn to 3(–4) mm long. Mesic limestone areas and creek banks; Bell Co.; mainly Edwards Plateau n to s part of nc TX. Sep–Dec.

Muhlenbergia mexicana (L.) Trin., (Name given under mistaken idea that the species is Mexican). Rhizomatous perennial; culms 30–90 cm tall, much-branched above; ligules to 1 mm long; panicles slender, the branches densely flowered; spikelets 2–4 mm long; glumes ca. equaling floret, with awn to 1.5 mm long; lemmas 1.3–3.4 mm long, awnless or with awn to 0.5 mm long, pilose basally. Thickets and field borders; a Reverchon collection (*4110*) reported from Granbury, Hood Co. is the only known TX collection (Gould 1975b). Summer–fall.

Muhlenbergia reverchonii Vasey & Scribn., (for Julien Reverchon, 1837–1905, a French-American immigrant to Dallas and important botanical collector of early TX), SEEP MUHLY, REVERCHON'S MUHLY. Densely tufted perennial; culms stiffly erect, 40–80 cm tall; ligules 2–4 mm long; panicles open; glumes to 3 mm long; lemmas glabrous or scabrous, 3.5–5 mm long, with awn 0.5–4(–7) mm long. Calcareous soils, typically in moist or wet areas; mainly Blackland Prairie w to West Cross Timbers and s to Edwards Plateau. Aug–Nov. The inflorescences can have a striking reddish appearance (J. Stanford, pers. comm.).

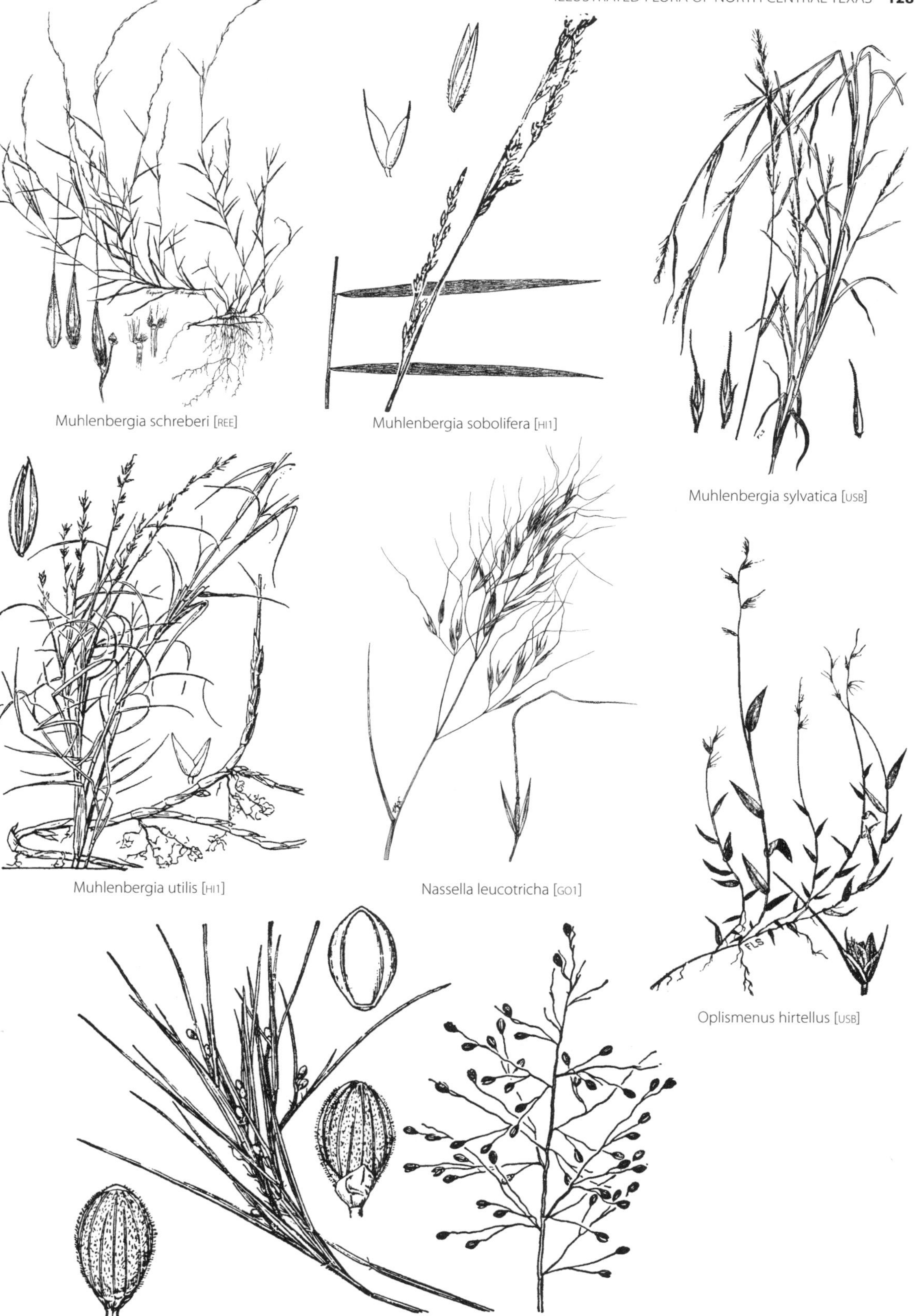

Muhlenbergia schreberi [REE]

Muhlenbergia sobolifera [HI1]

Muhlenbergia sylvatica [USB]

Muhlenbergia utilis [HI1]

Nassella leucotricha [GO1]

Oplismenus hirtellus [USB]

Panicum aciculare var. aciculare [HI1]

Muhlenbergia schreberi J.F. Gmel., (for Johann Daniel Christian von Schreber, 1739–1810, German botanist), NIMBLE-WILL, SATIN GRASS, SCHREBER'S MUHLY. Perennial; culms decumbent below, rooting at lower nodes, usually 10–40(–60) cm tall, much-branched; ligules to 0.5 mm long; panicles contracted; lemmas 2–2.5 mm long, with awn 1.5–5 mm long, pilose basally. Woods and thickets; Bell, Collin, Dallas, Denton, and Fannin cos.; se and e TX w to East Cross Timbers, also Edwards Plateau. Jun–Oct.

Muhlenbergia sobolifera (Muhl. ex Willd.) Trin., (bearing sprouts), ROCK MUHLY, ROCK-DROPSEED. Rhizomatous perennial; culms 40–85(–100) cm tall, much-branched above; ligules to 1 mm long; glumes about equal, overlapping basally; lemmas 2–3 mm long, awnless or with awn to 3 mm long, pubescent basally. Rocky slopes, open woods; Grayson Co., also Brown, Hamilton (HPC), and Dallas (Mahler 1988) cos.; also infrequent in e TX and Edwards Plateau. Sep–Oct. Including [*M. sobolifera* var. *setigera* Scribn.], the type of which was collected by Reverchon (*70*) at Dallas (Mahler 1988).

Muhlenbergia sylvatica Torr. ex A. Gray, (forest-loving), FOREST MUHLY. Rhizomatous perennial; culms decumbent or sprawling below, usually 40–100 cm long, freely branching at middle nodes; internodes puberulent; ligules to 2.5 mm long; panicles terminal, slender, contracted; lemmas 2.2–3.2 mm long, pilose basally, with awns 3–10(–18) mm long. Woods and shaded stream banks; included based on citations of vegetational areas 4 and 5 (Fig. 2) by Gould (1975b) and Hatch et al. (1990); mainly Edwards Plateau. Aug–Sep.

Muhlenbergia utilis (Torr.) Hitchc., (useful), APAREJO GRASS, APAREJO MUHLY. Low rhizomatous perennial; culms 20–40 cm long; panicles narrow, interrupted, loosely flowered, 1–4 cm long; lemmas 1.6–2 mm long, minutely scabrous, scarcely mucronate. Along streams, marshy meadows, usually very wet areas of calcareous soils; Tarrant Co.; mainly Edwards Plateau. Late summer–fall.

NASSELLA

A genus of 79 species of warm and tropical areas of the Americas, especially the Andes; related to and previously treated in *Stipa*. (Latin: *nassa*, a basket with a narrow neck) (subfamily Pooideae, tribe Stipeae)
REFERENCES: Hitchcock 1925; Brown 1952; Barkworth 1990, 1993.

Nassella leucotricha (Trin. & Rupr.) Barkworth, (white-haired), WINTER GRASS, TEXAS WINTER GRASS, SPEAR GRASS, TEXAS NEEDLE GRASS. Perennial, tufted, 25–100 cm tall, green from late fall to summer; ligules variable, from absent to a 1 mm membrane; axillary, cleistogamous florets produced basally in addition to terminal panicles; spikelets 1-flowered; glumes acuminate, 14–18 mm long; lemma 9–12 mm long, with very long (4.5–10 cm), geniculate awn; lemma apex with smooth white neck and ring of hairs around awn base; caryopsis permanently enclosed within the lemma. Prairies, disturbed sites; throughout TX. Apr–May. A minor component of original prairie, increasing under disturbance. The lemma base and rachilla form a callus that is so sharp-pointed that it will easily stick into clothing or skin; wounds can result including those to the mouth of animals (Lipscomb & Diggs 1998). [*Stipa leucotricha* Trin. & Rupr.]

OPLISMENUS

A genus of 7 species of the tropics and warm areas including *O. hirtellus* cultivar *variegatus*, a cultivated greenhouse hanging-basket plant. (Greek: *hoplismos*, a weapon, referring to the awns) (subfamily Panicoideae, tribe Paniceae)
REFERENCES: Hitchcock 1920a; Davey & Clayton 1978; Scholz 1981; Crins 1991.

Oplismenus hirtellus (L.) P. Beauv., (rather hairy), BASKET GRASS. Perennial with culms creeping and rooting at the nodes; ligule a ciliate membrane; leaf blades 1.5–7 cm long, 5–15 mm wide, widely spreading to reflexed; inflorescences of 3–7 widely spaced, very short, spike-like branches ca. 1–6 mm long; spikelets sessile or nearly so, in 2 rows on 1 side of the inflorescence branches, 2-flowered, the lower floret sterile, the upper floret perfect; glumes awned; awn of first glume 5–10 mm long, much longer than awn of second glume; fertile lemma indurate, the margins virtually enclosing the palea. Tarrant Co., also Limestone Co. near the e edge of nc TX; mainly se and e TX. Summer–Oct. Native to New World tropics and subtropics. The nc TX specimens of this species were treated as *O. hirtellus* subsp. *setarius* by Correll and Johnston (1970) and Gould (1975b), while recognized as a separate species, *O. setarius*, by Kartesz (1994). We are following Davey and Clayton (1978) whose worldwide study of the group indicated that the two taxa should be lumped; Crins (1991) and Jones et al. (1997) also treated the two as conspecific. [*O. hirtellus* subsp. *setarius* (Lam.) Mez ex Ekman, *O. setarius* (Lam.) Roem. & Schult.]

PANICUM PANIC GRASS

Annuals or perennials with or without rhizomes; basal rosette leaves not developed (basal leaves few, usually withering by flowering time) or in subgenus *Dichanthelium* basal tuft or rosette of leaves shorter and wider than those of the culms present, produced from fall to spring, persistent; ligule usually a membrane, often ciliate or with a fringe of hairs; panicles much-branched, normally produced in one continuous period of bloom, terminal or both terminal and lateral together or in subgenus *Dichanthelium* terminal panicles on mostly simple culms produced in spring and early summer, and a second crop of more numerous lateral or basal panicles produced in late summer or fall (often of cleistogamous spikelets) (descriptions are of spring phase only, autumnal phase often quite different); spikelets awnless, 2-flowered, the lower floret sterile or staminate, the upper floret perfect; glumes usually both present, the first typically shorter; lower lemma resembling second glume; upper (perfect) lemma firm to hardened, shiny and glabrous with inrolled margins.

A huge genus of > 600 species in 6 subgenera making it the largest genus of grasses (Zuloaga 1987); the species are native from tropical to temperate areas throughout the world and a number are used as fodders, grains, or cultivated ornamentals. The segregate *Dichanthelium* has been recognized by Gould (1974), Kartesz (1994), and others (e.g., Hatch et al. 1990) on the basis of such characters as having an overwintering rosette of short broad leaves, uniformly having the C_3 photosynthetic pathway, and possessing spring chasmogamous inflorescences and later in the season small, axillary, cleistogamous inflorescences. However, the overlap and blurring of these and other characters in Central and South American taxa brings the generic recognition of *Dichanthelium* into question (Zuloaga 1987). We are thus following Lelong (1984), Zuloaga (1987), Webster (1988), Zuloaga et al. (1993), Zuloaga and Morrone 1996, and Jones et al. (1997) in treating *Dichanthelium* (15 nc TX species) and *Steinchisma* (1 species in nc TX) as subgenera within *Panicum*. The other nc TX species of *Panicum* are scattered in 3 of the other 4 subgenera of the genus. Species here treated as *Urochloa* have previously been lumped with *Panicum*; some of these are included in the *Panicum* key as an aid in identification. (Latin: *panus*, an ear of millet) (subfamily Panicoideae, tribe Paniceae)

REFERENCES: Hitchcock & Chase 1910; Silveus 1942; Gould 1974; Allred & Gould 1978; Gould & Clark 1978; Lelong 1984, 1986; McGregor 1985a; Zuloaga 1987; Hansen & Wunderlin 1988; Webster 1988, 1992; Webster et al. 1989; Crins 1991; Zuloaga et al. 1993; Wipff & Jones 1994 [1995]; Zuloaga & Morrone 1996.

1. Basal leaves usually similar to those of the culm, only smaller, basal rosette absent; small axillary panicles absent in fall; plants annual or perennial; sometimes segregated as genus *Panicum* sensu stricto.
   2. Plants annual, without rhizomes, enlarged hard or knotty bases, or densely clumped culms.

3. Inflorescence consisting of several ± secund spike-like primary branches; spikelets subsessile or short-pedicelled; fertile lemma transversely rugose.
    4. Spikelets 5–6 mm long ______ see **Urochloa texana**
    4. Spikelets 2.4–3 mm long ______ see **Urochloa fasciculata**
3. Inflorescence a ± diffuse panicle; spikelets short- or long-pedicelled; fertile lemma smooth.
    5. Spikelets conspicuously warty under a hand lens (second glume and lower lemma verrucose or tuberculate); only at far e margin of nc TX ______ **P. brachyanthum**
    5. Spikelets not warty; widespread in nc TX.
        6. First glume about 1/4 as long as spikelet, obtuse or rounded ______ **P. dichotomiflorum**
        6. First glume > 1/4 as long as spikelet, acute or acuminate.
            7. Spikelets 4.5–5 mm long ______ **P. miliaceum**
            7. Spikelets 1.5–3.5 mm long.
                8. Spikelets 1.8–3.5 mm long, long acuminate at tip; panicles nearly as broad as long; rare in nc TX ______ **P. philadelphicum**
                8. Spikelets 1.5–2(–2.2) mm long, acute to slightly acuminate at tip; panicles usually distinctly longer than broad; includes species widespread in nc TX.
                    9. Palea of lower floret usually absent (do not be confused by lodicules); leaves yellow-green; inflorescences often with reddish or purplish coloration; fertile lemma lacking a crescent-shaped marking at its base; widespread in nc TX ______ **P. capillare**
                    9. Palea of lower floret present; leaves blue-green; inflorescences usually without reddish or purplish coloration; fertile lemma with a crescent-shaped marking at its base; on extreme w margin of nc TX ______ **P. hillmanii**

2. Plants perennial, often with rhizomes, enlarged hard or knotty bases, or culms densely clumped.
    10. Spikelets short-pedicelled along one side of the branch axes, forming appressed spike-like inflorescence branches; inflorescence branches usually not themselves branched; first glume about as long as second ______ **P. obtusum**
    10. Spikelets short- or long-pedicelled, in open or sometimes contracted or congested panicles; inflorescence branches usually themselves branched; first glume usually shorter than second.
        11. Sterile palea enlarged and indurate at maturity, giving the spikelets an expanded appearance (spikelets at maturity gaping open at apex); spikelets 1.8–2.6 mm long, borne toward the ends of the few slender branches (subgenus *Steinchisma*) ______ **P. hians**
        11. Sterile palea usually absent or minute if present; spikelets and branching various.
            12. Plants with conspicuous, creeping, scaly rhizomes.
                13. First glume < 1/2 as long as spikelet; spikelets on short pedicels so appressed as to make the spikelets appear sessile; spikelets 2.8–3.5 mm long; panicles open or contracted, usually sparsely branched ______ **P. anceps**
                13. First glume > 1/2 as long as spikelet; spikelets on relatively long, only slightly appressed pedicels; spikelets 2.8–5 mm long; panicles usually open, much branched ______ **P. virgatum**
            12. Plants without creeping scaly rhizomes.
                14. Spikelets short-pedicelled (appearing nearly sessile) along the usually unbranched inflorescence branches or on short spur branches; spikelets 1.6–2.5 mm long; plants typically of low moist areas ______ **P. rigidulum**
                14. Spikelets either long-pedicelled in open panicles OR if very short-pedicelled then the inflorescence open and its branches rebranched; spikelets 2.1–3.9 mm long; plants of various habitats.
                    15. Culms (at least the lower part) knotty, the conspicuously swollen nodes as much as twice as thick as the middle of the internodes ______ **P. antidotale**

15. Culms not as above.
    16. Lower inflorescence branches in verticels of 3–7, pilose in the axils; inflorescences often nearly half the height of the entire plant; e margin of nc TX ____ **P. bergii**
    16. Lower inflorescence branches usually solitary, glabrescent in the axils; inflorescences usually 6–25 cm long; including species widespread in nc TX.
        17. First glume ca. 1/4 as long as spikelet, abruptly narrowed to an acute apex ____ **P. coloratum**
        17. First glume 1/2 as long as spikelets or longer, gradually narrowed to an acute to acuminate apex.
            18. Leaf sheath margins often ciliate with a line of ascending hairs; nodes spreading-pilose; plants green ____ **P. diffusum**
            18. Leaf sheath margins glabrous or with a tuft of hairs at the summit, not ciliate; nodes appressed-pubescent or glabrous; plants often glaucous ____ **P. hallii**

1. Basal leaves usually different from those of the culm, forming a basal rosette; small axillary panicles on reduced lateral shoots present in fall; plants perennial; (subgenus *Dichanthelium*) sometimes segregated as genus *Dichanthelium*.
    19. Blades of basal and culm leaves all 15–40 times as long as wide, 1–4 mm wide.
        20. Spikelets 2.1–4.2 mm long; leaf sheaths glabrous or pilose with spreading to ascending hairs.
            21. Spikelets 2.1–3.0 mm long ____ **P. linearifolium**
            21. Spikelets 3.0–4.2 mm long ____ **P. depauperatum**
        20. Spikelets 1.7–2.1 mm long; leaf sheaths pilose with widely spreading to slightly reflexed hairs ____ **P. laxiflorum**
    19. Blades of basal and lowest culm leaves 3–15 times as long as wide (of upper culm leaves up to 20 times), 2–30(–35) mm wide.
        22. Spikelets 2.4–4.3 mm long.
            23. Nodes with a dense ring of widely spreading to reflexed hairs.
                24. Spikelets 2.6–3.2 mm long.
                    25. Node lacking a broad glabrous region; leaf blades 5–12 mm wide; culms 0.3–0.7 m tall, slender (< 2 mm thick) ____ **P. malacophyllum**
                    25. Node with a broad, conspicuous, glabrous region (in addition to a ring of hairs); leaf blades 3–30 mm wide; culms usually 0.8–1.5 m tall, stout (usually 2–4 mm thick) ____ **P. scoparium**
                24. Spikelets 3.6–4.2 mm long.
                    26. Leaf blades velvety-pubescent beneath; ligule of hairs 3–4 mm long ____ **P. ravenelii**
                    26. Leaf blades not velvety-pubescent beneath, glabrous or with only scattered pubescence; ligule essentially absent ____ **P. boscii**
            23. Nodes with spreading to ascending hairs, or glabrous.
                27. Leaf blades usually 12 mm or less wide.
                    28. Spikelets narrowly obovate, gradually tapering to a narrow base, 3–3.9 mm long ____ **P. pedicellatum**
                    28. Spikelets ovate, oblong, or slightly obovate, not tapering gradually to a narrow base, 2.4–4 mm long.
                        29. Hairs of ligules 1.5–6 mm long.
                            30. Spikelets 2.4–2.7 mm long ____ **P. acuminatum** var. **villosum**
                            30. Spikelets usually 2.7–4 mm long ____ **P. oligosanthes**
                        29. Hairs of ligules absent or < 1.5 mm long.

31. Spikelets usually 2.7–4 mm long; second glume and lower lemma usually 5-nerved; widespread in nc TX ____ **P. oligosanthes**
31. Spikelets 2.4–2.8 mm long; second glume and lower lemma usually 7-nerved; rare in nc TX, reported only from Dallas and Lamar cos. ____ **P. aciculare**

27. Leaf blades (at least larger ones) usually 13–35 mm wide.
32. Leaf blades velvety-tomentose or puberulent beneath, glabrous or puberulent above; spikelets 2.4–2.8 mm long ____ **P. scoparium**
32. Leaf blades not velvety-tomentose or puberulent beneath (or if so, spikelets 2.8–4 mm long), often glabrous on both surfaces; spikelets 2.4–4 mm long.
33. Spikelets broadly elliptic to obovate, turgid, with heavy broad nerves; leaf blades 5–15 mm wide ____ **P. oligosanthes**
33. Spikelets narrowly elliptic to obovate, neither turgid nor strongly nerved; leaf blades (8–)13–35 mm wide.
34. Leaf blades usually < 10 cm long; leaf sheaths glabrous or pubescent on margins ____ **P. divergens**
34. Leaf blades mostly 10–20(–28) cm long; leaf sheaths, at least lower ones, papillose hispid with spreading hairs ____ **P. clandestinum**

22. Spikelets 1.3–2.4 mm long.
35. Leaf blades glabrous except near base or on margins.
36. Ligules 1.5–6.0 mm long ____ **P. acuminatum** var. **lindheimeri**
36. Ligules 1.3 mm or less long, or apparently absent.
37. Leaf sheaths with widely spreading to slightly reflexed hairs 2.5–3.5 mm long ____ **P. laxiflorum**
37. Leaf sheaths either with ascending hairs up to 2 mm long OR glabrous.
38. Blades of larger culm leaves 3–6 mm wide, narrowed or rounded at base, not auricled.
39. Spikelets 2.0–2.3 mm long, 1.0–1.3 mm wide ____ **P. aciculare**
39. Spikelets 1.3–2.0 mm long, 0.6–1.0 mm wide ____ **P. dichotomum**
38. Blades of larger culm leaves 5–30 mm wide, abruptly narrowed to subcordate base, slightly auricled-clasping ____ **P. sphaerocarpon**
35. Leaf blades sparsely to densely pubescent or pilose over one or both surfaces.
40. Spikelets 2.2–2.4 mm long; species rare in nc TX.
41. Blades of largest culm leaves usually 8–30 mm wide; hairs of ligules 1–1.3 mm long ____ **P. scoparium**
41. Blades of largest culm leaves usually 5–12 mm wide; hairs of ligules 1.5–6 mm long ____ **P. acuminatum** var. **villosum**
40. Spikelets 1–2.1 mm long; species common in nc TX ____ **P. acuminatum**

Panicum aciculare Desv. ex Poir., (bristle-like). Plant 30–75 cm tall; leaves with lower sheaths sparsely pilose, the upper sheaths and blades glabrous or with some pubescence. Sandy woods; Dallas Co. (Mahler 1988), also Lamar Co. (Carr 1994); mainly se and e TX. Apr–Jun, also late summer–fall. The variation within this taxon has sometimes been given no formal recognition (e.g., Gould & Clark 1978) or recognized at the species level (e.g., Lelong 1986); we are following Wipff and Jones (1994 [1995]) in recognizing it at the varietal level. While we have not seen material to determine varietal status of the 2 citations given above, it is possible that both of the following occur in nc TX. Lelong (1986) separated them (as species) as follows:

1. Spikelets 1.7–2.2 mm long; leaf blades 3.5–8 cm long and up to 4 mm wide ____ var. **aciculare**
1. Spikelets 2.4–2.8 mm long; leaf blades 5–15 cm long and up to 7 mm wide ____ var. **angustifolium**

var. **aciculare**. [*Dichanthelium aciculare* (Desv. ex Poir.) Gould & C.A. Clark, *Panicum ovinum* Scribn. & J.S. Sm.]

var. **angustifolium** (Elliott) Wipff & S.D. Jones, (narrow-leaved). [*Dichanthelium angustifolium* (Elliott) Gould, *P. angustifolium* Elliott]

Panicum acuminatum Sw., (long-pointed, tapering to tip). Plant 10–65 cm tall; leaf blades usually 3–12 mm wide.

1. Spikelets 1–2.1 mm long; widespread in nc TX.
   2. Leaf sheaths usually densely long pilose, the hairs on upper part spreading at right angles, 2.5–4 mm long; culms usually pilose with nodes bearded ____ var. **acuminatum**
   2. Leaf sheaths glabrous to with some pubescence (particularly on margins), if present then hairs on upper part ascending, 1–2 mm long; culms glabrous or internodes sparsely pilose ____ var. **lindheimeri**
1. Spikelets 2.2–2.7 mm long; mainly e TX ____ var. **villosum**

var. **acuminatum**. WOOLLY ROSETTE GRASS, WOOLLY PANIC. Plant usually pilose. Sandy open woods; Bell, Denton, and Grayson cos.; widespread in TX. Apr–Jun, again late summer–fall. [*Dichanthelium acuminatum* (Sw.) Gould & C.A. Clark var. *fasciculatum* (Torr.) Freckmann, *Dichanthelium acuminatum* var. *implicatum* (Scribn.) Gould & C.A. Clark, *Dichanthelium lanuginosum* (Elliott) Gould, *P. acuminatum* var. *fasciculatum* (Torr.) Lelong, *P. lanuginosum* Elliott]

var. **lindheimeri** (Nash) Lelong, (for Ferdinand Jacob Lindheimer, 1801–1879, German born TX botanist), LINDHEIMER'S ROSETTE GRASS, LINDHEIMER'S PANIC. Plant nearly glabrous or with sparse pubescence. Sandy or rocky ground, in sun or shade; se and e TX w to West Cross Timbers, also Edwards Plateau. Apr–Jun, again late summer–fall. [*Dichanthelium acuminatum* var. *lindheimeri* (Nash) Gould & C.A. Clark, *Dichanthelium lanuginosum* (Elliott) Gould var. *lindheimeri* (Nash) Fernald, *Dichanthelium lindheimeri* (Nash) Gould, *P. lanuginosum* (Elliott) Gould var. *lindheimeri* (Nash) Fernald, *P. lindheimeri* Nash]

var. **villosum** (A. Gray) Beetle, (soft-hairy), WHITE-HAIRED ROSETTE GRASS, WHITE-HAIRED PANIC. Plant usually 20–60 cm tall; culms and leaf blades pilose; leaf sheaths with erect-spreading hairs to 3 mm long; spikelets 2.2–2.7 mm long. Sandy woods; Parker Co. (Gould 1975b), also Lamar Co. (Carr 1994); mainly e TX. Apr–Jun, also late summer–fall. [*Dichanthelium acuminatum* (Sw.) Gould & C.A. Clark var. *villosum* (A. Gray) Gould & C.A. Clark, *Dichanthelium lanuginosum* (Elliott) Gould var. *villosissimum* (Nash) Gould, *Dichanthelium villosissimum* (Nash) Freckmann, *P. acuminatum* var. *villosum* (A. Gray) Beetle, *P. ovale* Elliott var. *villosum* (A. Gray) Lelong, *P. villosissimum* Nash] While we are following Jones et al. (1997) in recognizing this taxon as a variety, according to J. Wipff (pers. comm.), var. *acuminatum* and var. *villosum* are separated by only one weak character (amount and length of pubescence on the leaf sheaths and blades) which apparently intergrades completely; as a result, he suggests treating var. *villosum* as a synonym of var. *acuminatum*. However, Kartesz (1994) and J. Kartesz (pers. comm. 1997) treated var. *villosum* as a distinct species, *Dichanthelium villosissimum*.

Panicum anceps Michx., (two-edged), BEAKED PANICUM. Perennial 30–100 cm tall with stout rhizomes; ligule a minute membranous collar 0.4 mm or less long. Low moist areas; se and e TX w to Grand Prairie and Edwards Plateau. Jul–Nov. [*P. anceps* var. *rhizomatum* (Hitchc. & Chase) Fernald, *P. rhizomatum* Hitchc. & Chase] Jones et al. (1997) recognized var. *rhizomatum*.

Panicum antidotale Retz., (acting as or of the nature of an antidote), BLUE PANIC. Perennial 0.5–2(–3) m tall. Resembling *P. hallii*, with ligules like that of *P. bergii*. Recommended by some for planting as a forage grass; widely introduced into TX; included based on citation of vegetational area 4 (Fig. 2) by Hatch et al. (1990). Spring–fall. Native of India.

Panicum bergii Arechav., (derivation not known). Tufted perennial 50–100 cm tall; lower nodes with a collar of hairs; leaves with sheaths sparsely pilose, the blades densely so at base; ligule a short membranous base (ca. 0.5 mm long) with fringe of hairs to 2.5 mm long; panicles open, the lower branches in verticils of 3–7, the whole inflorescence disarticulating as a tumbleweed at maturity; spikelets glabrous. Low areas; Navarro Co.; se TX nw to e edge of nc TX. Late Apr–May. Native of s South America. [*P. pilcomayense* Hack.] We are following Jones et al. (1997) and J. Kartesz (pers. comm. 1997) for nomenclature of this species. ☜

Panicum boscii Poir., (for its discoverer, Louis Augustin Guillaume Bosc, 1759–1828, French naturalist). Plant ca. 40–70 cm tall; leaves with sheaths and blades glabrous or with only scattered pubescence; leaf blades to 2.6(–3) cm wide. Wooded or low areas; Lamar Co. (Carr 1994) in Red River drainage; mainly e TX. Mainly Apr–Jul., also fall. [*Dichanthelium boscii* (Poir.) Gould & C.A. Clark]

Panicum brachyanthum Steud., (short-flowered), PIMPLE PANICUM. Glabrous annual with erect to decumbent culms to ca. 100 cm long; panicles few-flowered; spikelets 3–3.6 mm long, conspicuously warty, covered with short stiff hairs. Open, often sandy woods, fencerows, along highways; Lamar Co. (Carr 1994), also Henderson and Milam cos. near the e edge of nc TX; mainly se and e TX. *Panicum verrucosum* Muhl., also with warty spikelets, is known just e of nc TX in e Henderson Co. It has glabrous, much smaller spikelets 1.8–2.6 mm long.

Panicum capillare L., (hair-like), WITCH GRASS. Annual with erect to partly decumbent culms 8–100 cm long; leaf blades usually hirsute or pilose on both surfaces or blades glabrous on upper surface (occasionally merely ciliate marginally below); leaf sheaths with papilla-based hairs; ligules of hairs; panicles very open, large for size of plant, often with reddish or purplish coloration. Disturbed areas and banks of ponds and streams; nearly throughout TX, most common in nc TX. May–Oct. Superficially similar to *Digitaria cognata* which differs in having essentially glabrous leaves and usually lacking a first glume (or glume vestigial). Some authorities (e.g., Jones et al. 1997) recognize a number of varieties for this species.

Panicum clandestinum L., (hidden), DEER-TONGUE, DEER-TONGUE ROSETTE GRASS. Plant 50–150 cm tall. Sandy woods; collected at Dallas by Reverchon in 1875, not found there since (Mahler 1988); mainly e TX. Apr–Jun, again late summer–fall. [*Dichanthelium clandestinum* (L.) Gould]

Panicum coloratum L., (colored), KLEIN GRASS. Tufted perennial usually 60–135 cm tall from often knotty bases; ligule a fringed membrane 0.5–2 mm long including hairs; leaf sheaths glabrous or with papilla-based hairs; leaf blades glabrous or with pubescence. Introduced as a forage grass; Somervell Co. (R. O'Kennon, pers. obs.), also cited for vegetational areas 4 and 5 (Fig. 2) by Hatch et al. (1990); also Post Oak Savannah and Edwards Plateau. May–Sep. Native to Africa. ☜

Panicum depauperatum Muhl., (impoverished), STARVED ROSETTE GRASS. Similar to *P. linearifolium* and possibly intergrading with it; plant to 35 cm tall; leaves pilose to hispid or nearly glabrous, the sheaths common thinly pilose, the blades often glabrous on upper surface. Woods, roadsides; Lamar Co., also Grayson Co. (Mahler 1988); mainly e TX w in Red River drainage. Apr–May. [*Dichanthelium depauperatum* (Muhl.) Gould]

Panicum dichotomiflorum Michx., (with forking inflorescence), FALL PANICUM, SPREADING WITCH GRASS. Coarse annual with culms 1–2 m long, erect or trailing; ligule a 0.5–1 mm long membrane ciliate with hairs ca. 2 mm long; leaf blades usually glabrous (rarely puberulent). Moist, disturbed soils; in nc TX w to Cooke and Tarrant cos.; widepread in TX. Aug–Nov.

Panicum dichotomum L., (2-forked or parted). Plant rather slender, 25–60 cm tall, glabrous except for nodes, summit of leaf sheaths, and margins of leaf blades near base; spikelets very small. Low sandy woods; included based on citation of vegetational areas 4 and 5 (Fig. 2) by

Panicum acuminatum var. acuminatum [HI1]

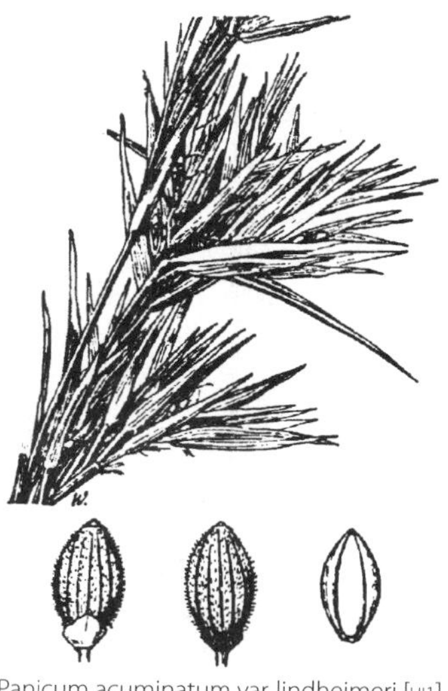
Panicum acuminatum var. lindheimeri [HI1]

Panicum acuminatum var. villosum [USB]

Panicum anceps [USB]

Panicum antidotale [GO1]

Panicum bergii [HEA]

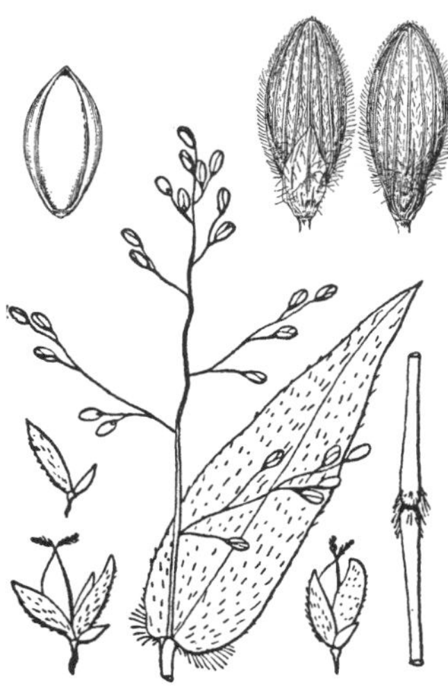
Panicum boscii [BB2]

Panicum brachyanthum [GO1]

Gould (1975b) and Hatch et al. (1990); mainly se and e TX. Apr–Jun, also late summer–fall. [*Dichanthelium dichotomum* (L.) Gould, *P. barbulatum* Michx., *P. lucidum* Ashe, *P. microcarpon* Muhl. ex Elliott, *P. nitidum* Lam., *P. roanokense* Ashe, *P. yadkinense* Nash] Jones et al. (1997) recognized five varieties of this species in TX.

Panicum diffusum Sw., (diffuse, spreading), SPREADING PANICUM. Perennial in small dense tufts; culms slender, spreading (rarely ascending), often branching, usually to ca. 30 cm long but occasionally much longer; nodes pubescent; ligule a membrane ca. 0.5–1 mm long ciliate with hairs 1–2 mm long; leaf blades usually glabrous or spreading pilose; leaf sheaths glabrous or with papilla-based hairs. Disturbed loamy or clayey soils; throughout most of TX. Apr–Nov. Zuloaga and Morrone (1996) indicated that this species is found on Carribean Islands and does not occur in the U.S.; Jones et al. (1997) lumped it with *P. hallii* var. *filipes*. Stephan Hatch (pers. comm.) considers this species to occur in TX and to be distinct from *P. hallii*.

Panicum divergens Kunth, (wide-spreading), VARIABLE ROSETTE GRASS. Culms 40–75 cm tall; larger leaf blades cordate at base, usually < 10 cm long, occasionally up to 15 cm. Sandy woods; Lamar Co. (Carr 1994) in Red River drainage; mainly se and e TX. Apr–Jun, again late summer–fall. [*Dichanthelium commutatum* (Schult.) Gould, *P. commutatum* Schult.] We are following Jones et al. (1997) for nomenclature of this species.

Panicum hallii Vasey, (for its discoverer, Elihu Hall, 1822–1882, American botanist and explorer of Rocky Mts.). Tufted perennial 12–80 cm tall; leaves sparsely pilose to glabrous; ligule a short, fibrous, readily splitting membrane, with fringe of hairs to 1.3 mm long; panicles open, rather small. Prairies, disturbed sites, commonest on limestone or calcareous clay. Apr–Nov.

1. Panicle branches usually > 15; spikelets 2–3 mm long; leaf sheaths without papilla-based hairs ____ var. **filipes**
1. Panicle branches usually < 15; spikelets 3–3.7 mm long; leaf sheaths mostly with papilla-based hairs ____ var. **hallii**

var. **filipes** (Scribn.) F.R. Waller, (slender), FILLY PANICUM. In nc TX known only from Palo Pinto and Tarrant cos.; widely distributed in TX. [*P. filipes* Scribn.]

var. **hallii**. HALL'S PANIC. Nearly throughout TX.

Panicum hians Elliott, (gaping), GAPING PANICUM. Perennial with culms erect or decumbent at base; nodes glabrous or scabrous; leaf sheaths glabrous or with hairs on upper margins; ligule a short glabrous membrane ca. 0.5 mm long; panicles usually 6–20 cm long, the lower 1–3 cm of branches bare of spikelets; spikelets glabrous, 1.8–2.6 mm long, at maturity gaping open at apex; lemma and palea of perfect floret firm, but not tough and hard or grain-like; glumes both present, the first 1/3–1/2 as long as spikelet; lower lemma resembling second glume; upper (perfect) lemma with inrolled margins; palea of the lower (neuter) floret inflated, obovate, often apiculate, distinctive, larger than the lemma and giving the spikelet an expanded or gaping appearance. Low areas, moist soils, often in shade; Dallas, Grayson, and Hunt cos.; se and e TX w to Blackland Prairie and Edwards Plateau. Apr–Oct, typically early in growing season. Treated here in subgenus *Steinchisma* (Zuloaga 1987); sometimes segregated into the monotypic genus *Steinchisma* [as *S. hians* (Elliott) Nash].

Panicum hillmanii Chase, (for Fred Hillman, 1863–1954, botanist at U.S. Dept. of Agriculture), HILLMAN'S PANICUM. Annual 20–65 cm tall; similar to *P. capillare* and possibly conspecific with it; inflorescences usually without reddish or purplish coloration (such coloration often present in *P. capillare*); fertile lemma with a crescent-shaped marking at its base. Disturbed areas; Archer and Brown cos. (McGregor 1985a); mostly n and w TX. Mostly Jul–Oct. The validity of *P. hillmanii* as a distinct species was questioned by Correll and Johnston (1970) and Gould

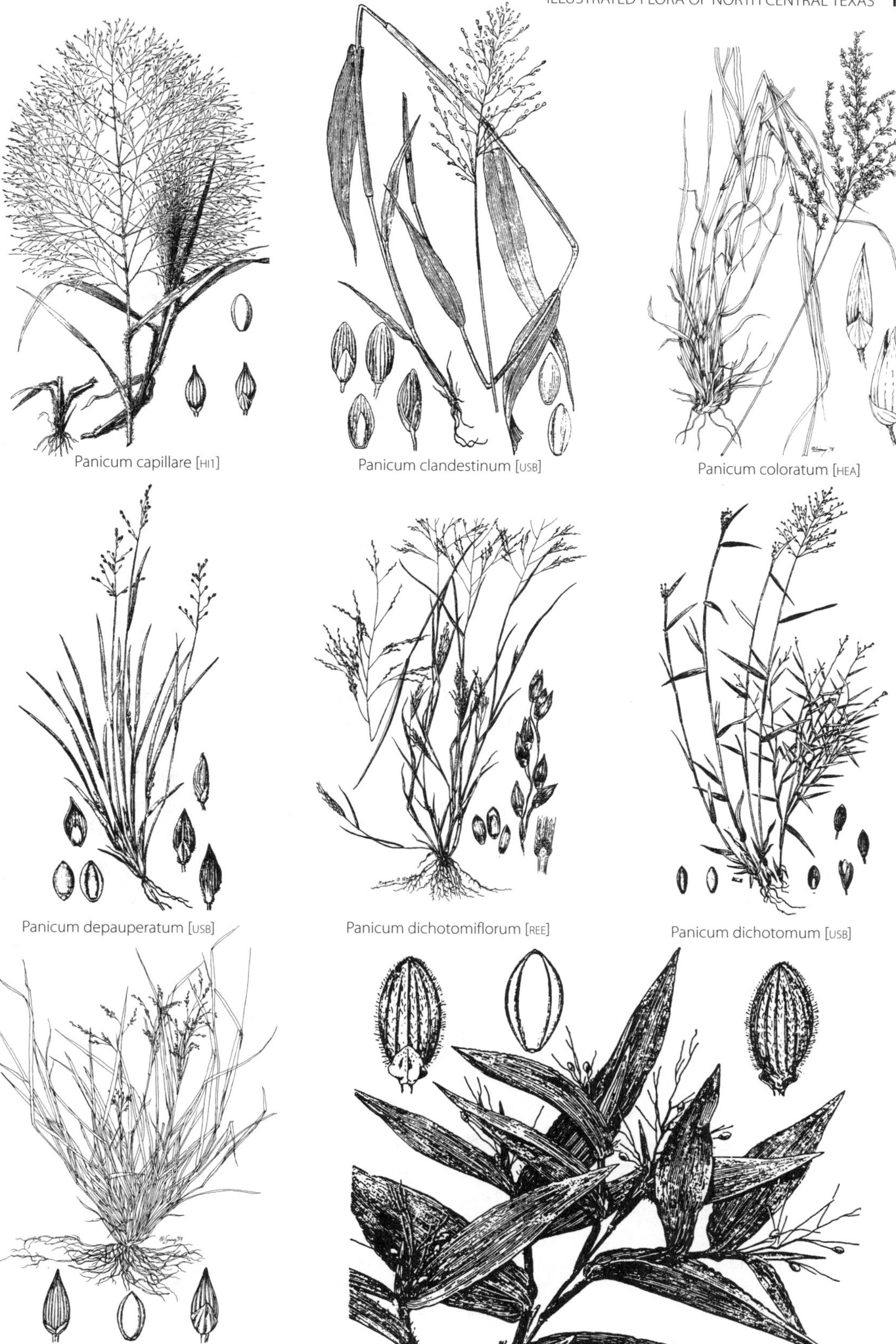

Panicum capillare [HI1]

Panicum clandestinum [USB]

Panicum coloratum [HEA]

Panicum depauperatum [USB]

Panicum dichotomiflorum [REE]

Panicum dichotomum [USB]

Panicum diffusum [HEA, USH]

Panicum divergens [HI1]

(1975b); however, McGregor (1985a) concluded that it is distinct and Zuloaga and Morrone (1996) in a treatment of section *Panicum*, treated it as a distinct species. Jones et al. (1997) gave [*P. capillare* L. subsp. *hillmanii* (Chase) Freckmann & Lelong [ined.]] as a synonym.

Panicum laxiflorum Lam., (loosely-flowered), OPEN-FLOWER ROSETTE GRASS, OPEN-FLOWER PANIC. Plant 12–50 cm tall; leaf sheaths and blades pilose; margins of leaf blades with conspicuously long hairs, at least near base. Sandy woods, especially in low ground; Fannin and Lamar cos.; mainly e TX w to nc TX in Red River drainage. Apr–Jun, again late summer–fall. [*Dichanthelium laxiflorum* (Lam.) Gould]

Panicum linearifolium Scribn. ex Nash, (linear-leaved), SLIM-LEAF ROSETTE GRASS, SLIM-LEAF PANIC. Plant 15–50 cm tall; leaf sheaths pilose or rarely glabrous. Dry sandy woods or open ground; se and e TX w to West Cross Timbers, also Edwards Plateau. Mar–Jun, again late summer–fall. [*Dichanthelium linearifolium* (Scribn. ex Nash) Gould]

Panicum malacophyllum Nash, (soft-leaved), SOFT-LEAF ROSETTE GRASS, SOFT-LEAF PANIC. Plant 25–60 cm tall; leaf sheaths pilose; leaf blades densely soft-pubescent. Sandy or rocky woods; Bosque, Cook, Dallas, Hood, and Lamar cos.; in TX only in nc part. Apr–Jun, again late summer–fall. [*Dichanthelium malacophyllum* (Nash) Gould]

Panicum miliaceum L., (like millet grass—*Milium*), BROOMCORN MILLET, PROSO, HOG MILLET, RUSSIAN MILLET, COMMON MILLET. Coarse annual 20–100 cm tall; leaf sheaths with long papilla-based hairs; leaf blades variously pubescent to glabrous; ligule a fringed membrane ca. 1–3 mm long including the hairs; spikelets plump, broadly ovate to elliptic. Escapes cultivation in disturbed areas; Montague Co.; also reported from nw TX. Jul–Nov. Cultivated in Europe and Asia as a grain crop for humans, here more commonly as animal food. Native to the Old World, probably c Asia (Zohary & Hopf 1994).

Panicum obtusum Kunth, (blunt), VINE-MESQUITE. Perennial 12–50 cm tall, from knotty, rhizomatous base, producing stolons up to 2 m long; culms erect or partly decumbent; nodes densely hairy (on stolons) or nearly glabrous (on erect culms); leaf blades usually ± glabrous; ligule a short membrane 1–2 mm long; inflorescences narrow, spike-like; spikelets swollen, ellipsoidal. Low prairies, roadsides; nearly throughout TX. Jun–Oct. A native weed, sometimes planted for pasture or erosion control.

Panicum oligosanthes Schult., (few-flowered). Culms spreading to erect, 15–65 cm long; ligules 0.1–4.2 mm long. Open woods or open areas. Apr–Jun, again late summer–fall.

1. Lower leaf sheaths pilose with appressed or ascending hairs without swollen bases; spikelets 3–4 mm long; ligules usually 1.6 mm or more long; leaf blades with lower surfaces tomentose or occasionally puberulent ____ var. **oligosanthes**
1. Lower leaf sheaths glabrous or pilose with ascending to spreading hairs from swollen bases; spikelets usually 2.7–3.6 mm long; ligules usually < 1.6 mm long; leaf blades with lower surfaces glabrous to puberulent, never tomentose ____ var. **scribnerianum**

var. **oligosanthes**. Sandy soils; e 1/2 of TX. [*Dichanthelium oligosanthes* (Schult.) Gould]

var. **scribnerianum** (Nash) Gould, (for Frank Lamson Scribner, 1851–1938, agrostologist, U.S. Dept. of Agriculture), SCRIBNER'S ROSETTE GRASS. Sandy soils, occasionally in limestone gravel; throughout TX. [*Dichanthelium oligosanthes* (Schult.) Gould var. *scribnerianum* (Nash) Gould, *P. helleri* Nash]

Panicum pedicellatum Vasey, (stalked), CEDAR ROSETTE GRASS, CEDAR PANIC. Plant 25–60 cm tall; lower leaf sheaths rather sparsely ascending-pilose, the upper pubescent or glabrous except on margins; leaf blades pubescent on upper surface, glabrous or nearly so on lower, with a few long

Panicum hallii var. filipes [USB]

Panicum hallii var. hallii [HI1]

Panicum hians [GO1]

Panicum hillmanii [HEA]

Panicum laxiflorum [USB]

Panicum linearifolium [USB]

Paniucum malacophyllum [USB]

Panicum miliaceum [USB]

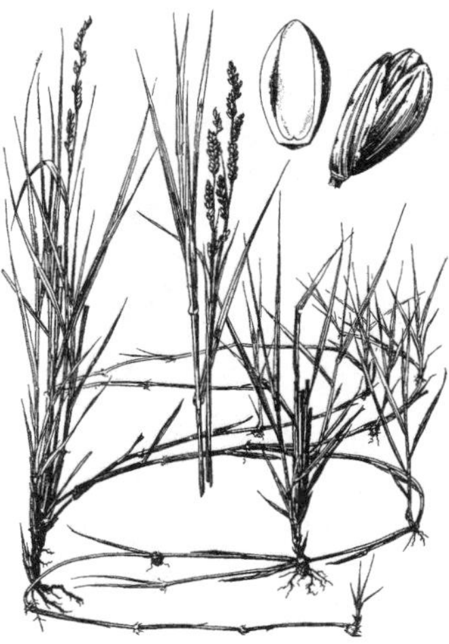
Panicum obtusum [HI1]

marginal hairs near base; ligule of hairs usually 0.3–1 mm long. Rocky limestone slopes; Bell, Bosque, Burnet, and McLennan cos.; s part of nc TX sw through Edwards Plateau. Mar–Jun, occasionally again late summer–fall. [*Dichanthelium pedicellatum* (Vasey) Gould]

Panicum philadelphicum Benth. ex Trin., (of Philadelphia), PHILADELPHIA WITCH GRASS, WOOD WITCH GRASS. Tufted annual 20–50(–60) cm tall; leaf sheaths with papilla-based hairs; leaf blades variously pubescent to nearly glabrous; ligule a fringe of hairs 0.5–1.5 mm long; panicles usually 1/3–1/2 as broad as long. Sandy or gravelly soils; Dallas Co. (*Reverchon 1842*, MO—Zuloaga & Morrone 1996); mainly e TX. Summer–early fall. Jones et al. (1997) treated this taxon as [*P. capillare* var. *sylvaticum* Torr.].

Panicum ravenelii Scribn. & Merr., (for Henry William Ravenel, 1814–1887, botanist and planter of South Carolina). Similar to *P. oligosanthes*; culms 30–80 cm long; leaf sheaths short-pilose; leaf blades densely pubescent beneath, pilose toward base on upper surface and on margins. Sandy woods; Dallas and Lamar cos.; mainly se and e TX. Apr–Jun, again late summer–fall. [*Dichanthelium ravenelii* (Scribn. & Merr.) Gould]

Panicum rigidulum Bosc ex Nees, (somewhat rigid), RED-TOP PANIC. Clump-forming perennial 30–125 cm tall; leaf sheaths glabrous or sometimes hispid; leaf blades glabrous or sparsely hispid; ligule a ragged, ciliate, short membrane, 0.5–1 mm long; spikelets rather crowded along panicle branches. Damp sandy woods and thickets, disturbed sites; Dallas, Lamar, Limestone, and Milam cos.; se and e TX w locally to nc TX and Edwards Plateau. Jun–Oct. [*P. agrostoides* Spreng.] A number of varieties are often recognized in this species (e.g., Lelong 1984; Kartesz 1994; Jones et al. 1997); however, all nc TX material seems to be of the type variety. Lelong (1986) gave a key separating four varieties.

Panicum scoparium Lam., (broom-like), VELVET ROSETTE GRASS. Culms course, tall, to ca. 150 cm; leaf sheaths usually velvety-pubescent; spikelets 2.2–2.8 mm long. Sandy woods and low areas; Lamar Co.; e TX w to ne part of nc TX. May–Jun, also fall. [*Dichanthelium scoparium* (Lam.) Gould]

Panicum sphaerocarpon Elliott, (spherical-fruited), ROUND-SEED ROSETTE GRASS, ROUND-SEED PANIC. Culms 20–80 cm tall; similar to glabrous forms of *P. acuminatum* var. *lindheimeri*; ligule minute or apparently absent; spikelets 1.4–2 mm long. Sandy soils, shaded or open areas; throughout most of TX except nw part. Late Mar–Jun, again late summer–fall. [*Dichanthelium sphaerocarpon* (Elliott) Gould]

Panicum virgatum L., (twiggy, wand-like), SWITCH GRASS. Large rhizomatous perennial 0.6–2(–3) m tall; nodes glabrous; leaf blades 3–15 mm wide; leaf sheaths usually glabrous; ligule a fringed membrane 1.5–3 mm long; panicles large, usually 15–55 cm long; spikelets acuminate-pointed, 2.8–5 mm long. Low moist areas and prairies; throughout TX. Aug–Nov. A member of the original tall grass prairie; considered one of the "big four" tall grasses along with *Andropogon gerardii*, *Schizachyrium scoparium*, and *Sorghastrum nutans*; seen in two growth forms: 1) LOWLAND SWITCH GRASS—very large isolated clumps often nearly 2(–3) m tall, usually in low moist areas, and 2) UPLAND SWITCH GRASS—in dryer sites, shorter not apparently clumped individuals with culms more scattered along the creeping rhizomes. Davis et al. (1995) pointed out additional differences and suggested the variation is possibly worthy of taxonomic recognition.

PASCOPYRUM WESTERN WHEAT GRASS

A monotypic genus of w North America (J. Wipff, pers. comm.). (subfamily Pooideae, tribe Triticeae)

REFERENCES: Dewey 1975, 1983; Löve 1980; Gupta & Baum 1989.

Pascopyrum smithii (Rydb.) Á. Löve, (for its discoverer, Charles Eastwick Smith, 1820–1900),

Panicum oligosanthes var. scribnerianum [USB]

Panicum pedicellatum [USB]

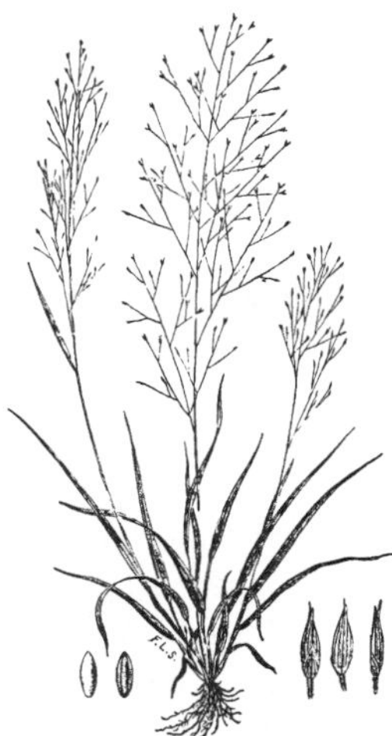
Panicum philadelphicum [USB]

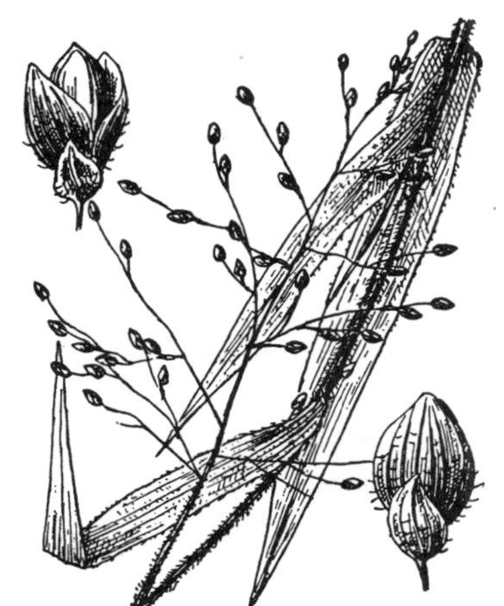
Panicum ravenelii [BB2]

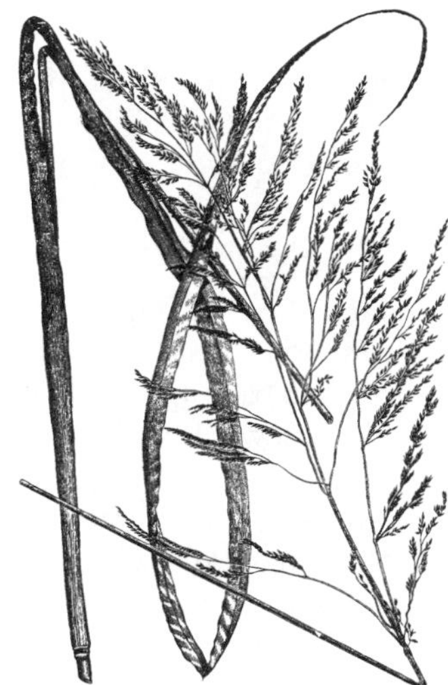
Panicum rigidulum [RCA]

Panicum scoparium [HI1]

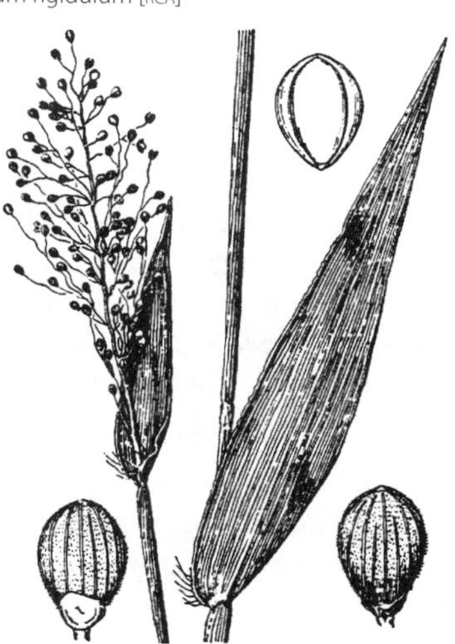
Panicum sphaerocarpon [HI1]

WESTERN WHEAT GRASS, BLUESTEM WHEAT GRASS. Rhizomatous erect perennial 35–100 cm tall; foliage glaucous, blue-green; leaf sheaths glabrous, auricled at summit; leaf blades noticeably scabrous on upper surface; ligule a short membrane; spikes slender, often dense, 6–20 cm long, erect, the internodes of the spike axis flattened but thick; spikelets in a zigzag arrangement on opposite sides of the spike axis, sessile, oriented so that their broadest dimension (not the keels of glumes) is toward the spike axis, awnless or short-awned, 1 or 2 per node, 5–12-flowered, disarticulating above glumes; *n* = 56. Prairies and roadsides, low areas; Archer, Clay, Comanche, Denton, and Jack cos.; nc TX s and w to w TX. [*Agropyron smithii* Rydb., *Elymus smithii* (Rydb.) Gould, *Elytrigia smithii* (Rydb.) Nevski] This species has been variously treated in *Agropyron*, *Elymus*, *Elytrigia*, or *Pascopyrum*. According to Dewey (1975), it is an octoploid that probably originated through hybridization between *Elytrigia dasytachya* and *Elymus triticoides* with subsequent chromosome doubling. While recognized in the genus *Elytrigia* by Dewey (1983), it is quite distinct morphologically from other members of the Triticeae, is unique cytogenetically, and doesn't fit well into any of the traditional genera. Further, there are nomenclatural problems with *Elytrigia* (J. Wipff, pers. comm.). For these reasons, we are following Löve (1980) and Kartesz (1994) in recognizing it in the monotypic genus *Pascopyrum*.

PASPALIDIUM

A mainly tropical genus of ca. 30–40 species with a number of Australian endemics (Crins 1991; Mabberley 1987); related to and sometimes lumped with *Setaria* (e.g., Mabberley 1997). (Greek diminutive of *Paspalum*, in reference to a similarity to that genus) (subfamily Panicoideae, tribe Paniceae)

REFERENCES: Crins 1991; Webster 1995.

Paspalidium geminatum (Forssk.) Stapf, (twin), EGYPTIAN PASPALIDIUM. Erect glabrous perennial 35–80 cm tall, rhizomatous or stoloniferous; panicles with central axis; lateral branches of panicle 7–17, floriferous to base, appressed, with spikelets in 2 rows on the flattened branches; spikelets 2-flowered, the lower floret sterile, with glume-like lemma, the upper floret fertile, with hardened grain-like lemma; first glume broad, rounded to truncate, 1/4–1/3 as long as spikelet; second glume resembling sterile lemma. Shallow water or wet ground; Dallas and Grayson cos., also McLennan Co. (Mahler 1988); mainly se and e TX w to nc TX, also Wichita Co. (Mahler 1988) in e Rolling Plains. May–Aug. Sometimes placed in the genus *Panicum* [as *P. geminatum* Forssk.] or in *Setaria* [as *S. geminata* (Forssk.) Veldkamp] (Webster 1995). Gould (1975b) and Godfrey and Wooten (1979) treated this species as native indicating it occurred in both the Old and New Worlds; however, Crins (1991) considered it introduced and naturalized in the s U.S.; S. Hatch (pers. comm.) agrees with Crins.

PASPALUM

Perennials (our species); ligule a membrane; inflorescences with a central axis and 1-many, spike-like, unilateral branches; spikelets subsessile or short-pedicelled on one side of a flattened branch, lanceolate to nearly circular, flattened on one face, 2-flowered, the lower floret sterile or staminate, the upper floret perfect; lower glume usually reduced or absent; upper glume and sterile lemma similar to each other; fertile lemma firm or hardened, usually smooth and shiny with firm inrolled margins.

A genus of ca. 330 species of tropical and warm areas of the world, especially the Americas. Some are characteristic of the pampas; *P. pyramidale* Nees grows to 15 m tall in the Amazon. Ergot fungi, *Claviceps purpurea* (Fr.: Fr.) Tul., *C. paspali* F. Stevens & J.G. Hall, and related species, are known to grow on a number of Texas grasses including *P. dilatatum* and other *Paspalum* species. These fungi, whose overwintering structures (= sclerotia) replace some grains

in the grass inflorescence, often produce toxic alkaloids (e.g., ergocryptine and ergotamine) chemically similar to LSD; cattle can develop gangrene, have convulsions, or die after ingestion; significant livestock losses have occurred; humans can also be affected (see discussion under the genus *Secale*) (Sperry et al. 1955; Kingsbury 1964, 1965). (Probably from the Greek *paspalos*, millet or meal) (subfamily Panicoideae, tribe Paniceae)
REFERENCES: Chase 1929; Silveus 1942; Banks 1966; Allred 1982; Crins 1991.

1. Spikelets with long hairs around the margin, glabrous or short-pubescent on the faces; glume and sterile lemma abruptly pointed beyond the blunt fruit.
   2. Spikelets 1.8–2.2 mm wide; inflorescence branches (racemes) 2–8 — **P. dilatatum**
   2. Spikelets 1.1–1.4 mm wide; inflorescence branches 7–36 — **P. urvillei**
1. Spikelets glabrous or uniformly short- or long-pubescent; glume and sterile lemma not abruptly pointed beyond the fruit.
   3. Spikelets 3.6–4.8 mm long — **P. floridanum**
   3. Spikelets 1.6–3.5 mm long.
      4. Inflorescence branches 2, paired or less than 1 cm apart (1–2 additional branches occasionally present below); spikelets 2.7–3.5 mm long.
         5. Plants usually of upland habitats; culms erect, not rooting at the nodes; spikelets usually broadly ovate to broadly obovate (rarely elliptic), usually blunt or broadly acute at apex, usually 2.8–3.5 mm long; first glume always absent — **P. notatum**
         5. Plants of wet habitats; culms decumbent, rooting at the nodes; spikelets elliptic, usually tapering to a short acute apex, usually 2.7–3 mm long; minute first glume usually present — **P. distichum**
      4. Inflorescence branches 1–numerous, when 2, the branches 1–2 cm or more apart; spikelets 1.6–3 mm long.
         6. First glume present on some or all spikelets (usually 1/4–1/3 as long as spikelet); second glume and sterile lemma usually with brownish glandular blotches; plants without rhizomes; in nc TX known only from Tarrant Co. — **P. langei**
         6. First glume absent on all spikelets (sterile lemma of lower floret resembles second glume) or minute first glume present (*P. distichum*); second glume and sterile lemma without brownish glandular blotches; plants with or without rhizomes; widespread in nc TX.
            7. Lemma and palea of fertile florets dark brown and shiny at maturity; sterile lemma usually with transverse wrinkles along the margin — **P. plicatulum**
            7. Lemma and palea of fertile florets green, light brown, or straw-colored at maturity; sterile lemma usually without wrinkles.
               8. Spikelets 2.4–3.4 mm long, markedly longer than wide to ca. as wide as long; culms partly decumbent, the longer rooting at the nodes OR erect, not rooting at the nodes.
                  9. Spikelets 1.1–1.4 mm wide, one of each pair with a minute lower glume — **P. distichum**
                  9. Spikelets 1.4–3.2 mm wide, without lower glume.
                     10. Spikelets 1.4–1.8 mm wide, elliptic to obovate, pubescent or glabrous, in pairs in 4 rows or in 2 rows by abortion of upper spikelet of spikelet pairs; culms usually decumbent below and rooting at lower nodes — **P. pubiflorum**
                     10. Spikelets 2–3.2 mm wide, broadly ovate, broadly obovate, or nearly orbicular, glabrous, borne singly in 2 rows; culms erect, not rooting in lower portion — **P. laeve**
               8. Spikelets 1.6–2.2 mm long, ca. as wide as long; culms erect, not rooting at lower nodes — **P. setaceum**

Paspalum dilatatum Poir., (dialated, widened, expanded), DALLIS GRASS, PASPALUM GRASS. Clump-forming perennial; culms low-spreading to erect, 25–120 cm long; lower leaf sheaths pilose, the upper glabrous except at summit; inflorescences of 2–8 branches, the branches 3–8 cm long.

Abundant weed in disturbed sites, lawns, and roadsides; throughout TX. May–Nov. Native of South America. Ergot alkaloids can be present—see generic synopsis.

Paspalum distichum L., (two-spiked), KNOT GRASS, ETERNITY GRASS, JOINT GRASS, FORT THOMPSON GRASS. Perennial 20–60 cm tall, with trailing and rooting culms sometimes several meters long; leaf sheaths usually glabrous or pilose; inflorescences of 2, occasionally 3–4 branches, the branches ca. 2–6 cm long; second glume minutely pubescent. Moist or wet areas along ponds, lakes, and streams; Bell, Dallas, Ellis, and Grayson cos., also Lamar Co. (Carr 1994); throughout TX. Jun–Oct. [*P. distichum* var. *indutum* Shinners] While some authorities recognize var. *indutum* (e.g., Jones et al. 1997), we are following Allred (1982) and J. Kartesz (pers. comm. 1997) in lumping this variety.

Paspalum floridanum Michx., (of Florida), FLORIDA PASPALUM, BIG FLORIDA PASPALUM, BIG PASPALUM. Rhizomatous perennial; culms erect, 1–2 m tall; leaf sheaths and blades nearly or completely glabrous or ± densely hirsute; inflorescences usually of 2–5 branches, the branches 4–13 cm long; spikelets glabrous. Grasslands and open woodlands; se and e TX w to East Cross Timbers, also Edwards Plateau. Aug–Nov. [*Paspalum floridanum* var. *glabratum* Engelm. ex Vasey] Jones et al. (1997) recognized var. *glabratum*.

Paspalum laeve Michx., (smooth). Tufted perennial to ca. 100 cm tall; leaf sheaths glabrous to pilose, inflorescences of (2–)3–6 branches, the branches usually 4–10 cm long. Prairies, open woods, disturbed or moist areas, often on sand. Jul–Oct.

1. Spikelets 2.7–3.2 mm wide, ca. as wide as long ____ var. **circulare**
1. Spikelets 2–2.5 mm wide, conspicuously longer than wide ____ var. **laeve**

var. **circulare** (Nash) Fernald, (circular), ROUND-SEED PASPALUM. Lamar Co. (Carr 1994); mainly se and e TX.

var. **laeve**, FIELD PASPALUM, SMOOTH PASPALUM. Lamar Co. (Carr 1994) in Red River drainage; mainly se and e TX.

Paspalum langei (E. Fourn.) Nash, (for Johann Martin Christian Lange, 1818–1898, Danish botanist), RUSTY-SEED PASPALUM, LANGE'S PASPALUM. Cespitose perennial 30–100 cm tall; leaf blades 7–18 mm broad, often marginally crisped; inflorescences with 2–5 branches, the branches usually 4–8 cm long; spikelets 2.2–2.6 mm long; first glume present on some or all spikelets; second glume and sterile lemma pubescent. Shaded ditchbanks; Tarrant Co.; mainly se and e TX. Apr–Nov.

Paspalum notatum Flüggé, (marked), BAHIA GRASS. Rhizomatous perennial; culms erect, usually 20–75 cm tall; inflorescences usually of 2(–3) spicate branches, the branches 4–12(–15) cm long, paired or one slightly below other; spikelets broadly ovate or broadly obovate (rarely elliptic), glabrous. Introduced as a pasture grass and for erosion control; sandy loam, forest openings, and roadsides. Jun–Nov. Native to Latin America.

1. Spikelets 3.3–4 mm long, 2.3–3 mm wide; inflorescences of 2(–3–4) primary branches ____ var. **latiflorum**
1. Spikelets 2.5–3.2 mm long, ca. 2 mm wide; inflorescences of 4–5 primary branches ____ var. **saurae**

var. **latiflorum** Döll, (broad-flowered). Grayson, Limestone, and Tarrant cos.; mainly e and s TX, also Edwards Plateau.

var. **saurae** Parodi, (derivation unclear, possibly from Greek: *sauros*, lizard). Introduced into the U.S. as PENSACOLA BAHIA GRASS. Possibly present in nc TX but no specimens have been seen.

Paspalum plicatulum Michx., (plicate, folded like a fan), BROWN-SEED PASPALUM, PLAITED PASPALUM. Tufted or clumped perennial 50–100 cm tall; inflorescences usually of 3–10 branches

Panicum virgatum [RCA]

Pascopyrum smithii [USB]

Paspalidium geminatum [HI1]

Paspalum dilatatum [HI1]

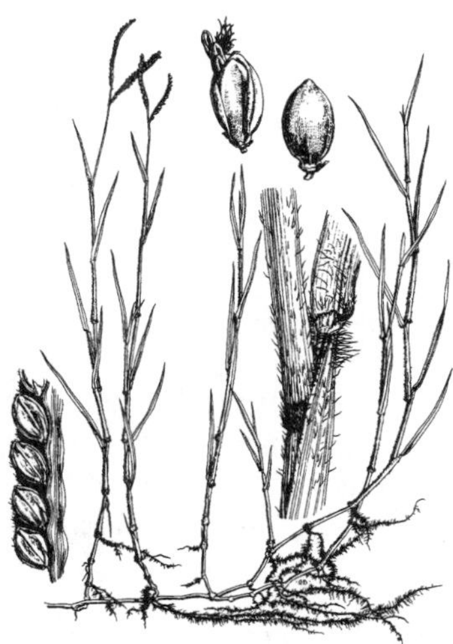
Paspalum distichum [MAS]

Paspalum floridanum [USB]

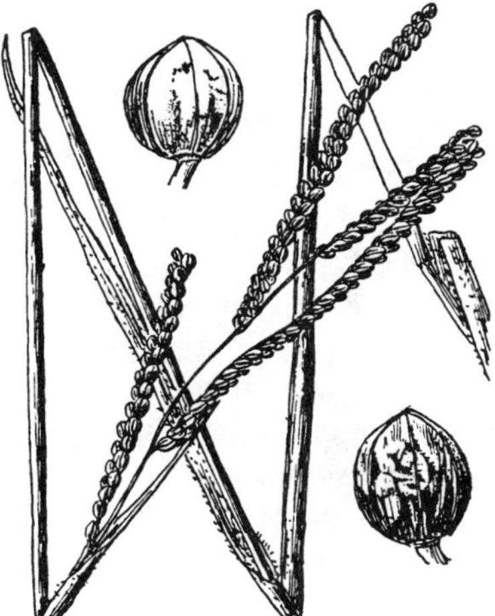
Paspalum laeve var. circulare [BB2]

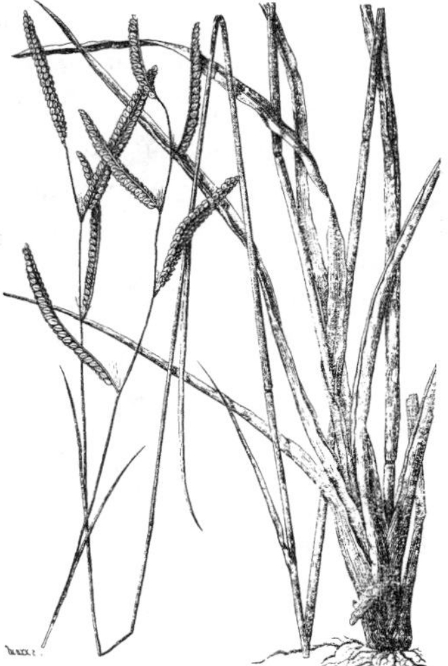
Paspalum laeve var. laeve [RCA]

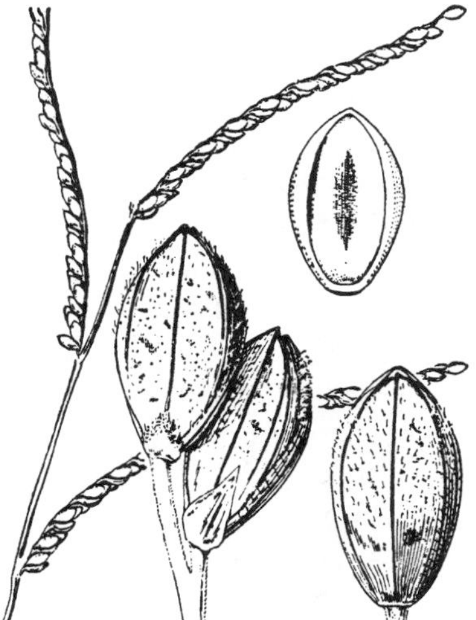
Paspalum langei [HI1]

each commonly 3–10 cm long; spikelets 2.4–2.8 mm long; glume glabrous or minutely pubescent. Sand or sandy loam, often in open woods; Milam Co. at the e edge of nc TX; mainly se and e TX, also Edwards Plateau. Mar–Nov.

Paspalum pubiflorum Rupr., (hairy-flowered). Perennial with culms decumbent below, often rooting at the lower nodes, usually 40–80 cm tall; leaf sheaths glabrous or the lower pilose; inflorescences of 2–5(–7) branches, the branches usually 3–10 cm long; spikelets glabrous or pubescent. Ditches and other moist areas. Mainly May–Nov.

1. Spikelets glabrous ____ var. **glabrum**
1. Spikelets pubescent ____ var. **pubiflorum**

var. **glabrum** Vasey ex Scribn., (smooth, without hairs), SMOOTH-SEED PASPALUM. Collin and Cooke cos., also Parker, Rockwall, and Tarrant cos. (Gould 1975b); reported from vegetational areas 1, 3, 4, 7, and 8 (Fig. 2) by Gould (1975b) and Hatch et al. (1990) but supposedly much less frequent in TX than var. *pubiflorum* (Gould 1975b).

var. **pubiflorum**. HAIRY-SEED PASPALUM, HAIRY-FLOWER PASPALUM. Included based on citation of vegetational areas 4 and 5 (Fig. 2) by Hatch et al. (1990); throughout most of TX.

Paspalum setaceum Michx., (bristle-like), THIN PASPALUM. Clump-forming perennial; culms 20–100 cm long; leaf sheaths and blades pilose to glabrous except on margins, often with fine pubescence with or instead of long hairs; inflorescences of 1–5 branches, the branches 3–17 cm long; spikelets glabrous or pubescent. Disturbed sites, roadsides; throughout TX. May–Oct. [*P. ciliatifolium* Michx., *P. stramineum* Nash, *P. setaceum* Michx. var *caliatifolium* (Michx.) Vasey, *P. setaceum* Michx. var. *stramineum* (Nash) D.J. Banks, *P. setaceum* Michx. var. *muhlenbergii* (Nash) D.J. Banks] Banks (1966) divided this species into a number of varieties.

Paspalum urvillei Steud., (for its discoverer, Jules Sébastien Cesar Dumont d'Urville, 1790–1842, French hydrographer and explorer), VASEY GRASS, URVILLE'S PASPALUM. Clump-forming perennial; culms erect or bent at base, 50–200 cm tall; lower leaf sheaths densely short-bristly with fine, stiff hairs that catch and stick to the skin when touched; upper leaf sheaths glabrous; inflorescences of 7–36 erect branches, the branches 4–10(–14) cm long. Low disturbed areas; Denton, Grayson, Kaufman, and Tarrant cos., also Lamar Co. (Carr 1994); se and e TX w to nc TX, also Edwards Plateau. Late May–Oct. Native of South America.

PENNISETUM

A genus of 130 species of tropical and warm areas of the world; some are used as fodders, lawn grasses, and for their grains. A number of *Pennisetum* species including *P. alopecuroides* (L.) Spreng., *P. americanum* (L.) Leeke, *P. ciliare* (L.) Link, *P. macrostachyum* Trin., and *P. setaceum* (Forssk.) Chiov. (FOUNTAIN GRASS) are cultivated as ornamentals for their large and showy inflorescences and may long persist vegetatively. (Latin: *penna*, a feather, and *seta*, a bristle, referring to the feathery bristles around the spikelets) (subfamily Panicoideae, tribe Paniceae)
REFERENCES: Chase 1921; Crins 1991.

Pennisetum villosum R. Br. ex Fresen., (softly hairy), FEATHERTOP. Perennial, tufted, 20–70 cm tall; leaf sheaths glabrous except on margins and summit; ligule a ciliate membrane with hairs ca. 1 mm long; inflorescence a spike-like contracted panicle, tan or yellowish, feathery, 4–10 cm long, with conspicuous fascicles of bristles 2–5 cm long, the inner bristles plumose with silky hairs; bristles separate to base, disarticulating with spikelets; spikelet one per fascicle of bristles, 8–9 mm long, of 2 florets, the lower floret sterile with glume-like lemma, the upper floret fertile. Cultivated as an ornamental, tending to persist or a transitory escape; Dallas and Parker cos. (R. O'Kennon, pers. obs.); Hatch et al. (1990) also cited vegetational areas 3 and 10 (Fig. 2). Jun–Oct. Native of Africa (Ethiopia).

Paspalum plicatulum [USB]

Paspalum pubiflorum var. glabrum [BB2]

Paspalum notatum var. latifolium [HEA]

Paspalum urvillei [HI1]

Paspalum setaceum [RCA]

Paspalum pubiflorum var. pubiflorum [GO1]

Pennisetum villosum [BA1]

Pennisetum ciliare (L.) Link, (ciliate, fringed with hairs), (BUFFEL GRASS) [*Cenchrus ciliaris* L.], an Old World species introduced into s TX as a forage grass, is cited by Hatch et al. (1990) for vegetational area 4 (Fig. 2) but apparently occurs on to the s of nc TX. It resembles a *Setaria* in having short bristles (4–10 mm long), but these are united at the base and the bur (bristles and spikelet) falls as a unit, whereas in *Setaria* the spikelets disarticulate above the bristles. This species, which was introduced in the 1940s by the Soil Conservation Service, is now a problematic invader of native habitats in some parts of sw North America (Tellman 1997).

PHALARIS CANARY GRASS

Glabrous-erect-tufted annuals; inflorescence a very dense, tightly contracted, ovoid or spike-like panicle; spikelets sessile, strongly laterally compressed, with 1 terminal perfect floret and 2 sterile reduced or scale-like lemmas below; glumes keeled and winged in upper half, ± equal, awnless; fertile lemma appressed-pubescent, awnless; rachilla disarticulating above glumes, the sterile lemmas falling with the fertile one; caryopsis plump.

A genus of ca. 20 species native to Europe, the Mediterranean, n Asia, and the Americas; some are valuable as fodder. (Ancient Greek name for a grass) (subfamily Pooideae, tribe Aveneae)

REFERENCE: Anderson 1961.

1. Caryopses (= fruits) 1.7–2.3 mm long; glumes 4.2–6(–6.7) mm long; reduced lemmas 1.5–2.5 mm long, 1/3–1/2 as long as fertile lemma; fertile lemmas 3–4.7 mm long; native species widespread in nc TX ____ **P. caroliniana**
1. Caryopses 3.9–4.2 mm long; glumes 7–9 mm long; reduced lemmas 2.5–4.5 mm long, more than 1/2 as long as single fertile lemma; fertile lemmas 4.5–6.8 mm long; rare introduced species ____ **P. canariensis**

Phalaris canariensis L., (of the Canary Islands), CANARY GRASS. Culms 25–70(–100) cm tall; ligule a membrane 2–6 mm long; leaf blades 5–15(–20+) cm long, 3–10 mm wide; panicles short, thick, 1.5–3(–4) cm long, 10–18 mm wide; glumes glabrous or sparsely hispid, ± pale with dark green along the lateral nerves. Waste areas, probably coming up from discarded bird seed; Bell, Coryell (HPC), and Brown (Stanford 1971) cos.; also se and e TX and Edwards Plateau. Mar–Jun. Native of the w Mediterranean. Important as a source of commercial bird feed; this is the canary seed of commerce (Gould 1975b; Mabberley 1987).

Phalaris caroliniana Walter, (of Carolina), WILD CANARY GRASS, CAROLINA CANARY GRASS, SOUTHERN CANARY GRASS. Culms usually 25–70(–100) cm tall; ligule a membrane 1–5 mm long; leaf blades 6–15(–20) cm long, 3–10(–13) mm wide; panicles cylindrical to narrowly elliptical in outline, spike-like, usually up to 6(–12) cm long, 8–13(–20) mm wide; glumes with lateral nerves glabrous or scabrous with 5 or less spicules. Moist areas, ravines, disturbed sites; throughout TX. Apr–Jun.

Some individuals of *P. caroliniana* have elongate inflorescences up to 12 cm long (*Whitehouse 15711*, Dallas Co., BRIT/SMU) reminiscent of *P. angusta* Nees ex Trin. The latter species has inflorescences 6–15 cm long, 8–10 mm thick, culms to 1.5 m tall, glumes mostly 3.5–4 mm long, and lateral nerves of glumes scabrous with 9 or more spicules. While known just to the e of nc TX in vegetational areas 1, 2, and 3 (Hatch et al. 1990), we have no definite reports of *P. angusta* from nc TX.

PHLEUM TIMOTHY, CAT-TAIL GRASS

A genus of 15 species of temperate Eurasia, North America, and temperate South America including cultivated forage grasses. (Greek: *phleos*, a kind of reed) (subfamily Pooideae, tribe Poeae)

REFERENCE: Piper & Bort 1915.

Phleum pratense L., (of meadows), TIMOTHY. Clump-forming perennial 30–100(–120) cm tall, glabrous except for scabrous leaf-blades; ligule a membrane 3–6 mm long; inflorescence a slender, dense, cylindrical, pencil-like spike usually 5–9 cm long and 5–9 mm thick; spikelets strongly compressed, 1-flowered; glumes with coarsely ciliate keel, abruptly short awned; lemmas awnless. Collected at Dallas in 1874 by Reverchon (whether wild or cultivated not known), and along railroad at Mineola, Wood Co. in e TX, in Jun, 1900 (Mahler 1988). Native of Eurasia. A cool-climate pasture grass, not now cultivated in nc TX except experimentally; a common cause of hay fever where cultivated.

PHRAGMITES

A cosmopolitan genus of 3 species. (Greek: *phragmites*, growing in hedges, apparently from its hedge-like growth along ditches) (subfamily Panicoideae, tribe Arundineae)
REFERENCE: Clayton 1968.

Phragmites australis (Cav.) Trin. ex Steud., (southern), COMMON REED, REED, NAL, DANUBE GRASS. Rhizomatous perennial; culms 2–4 m tall, usually unbranched; leaf blades mostly 1.5–6 cm wide; panicles densely flowered, 15–40 cm long; spikelets 10–15 mm long, usually with 4–8 florets; rachilla joints long hairy, the hairs to 10 mm or more long; glumes and lemmas glabrous. Wet areas of ponds, low areas, and roadsides, usually tight clay soils; Grayson and Kaufman cos.; throughout TX. Native of North America, South America, Eurasia, Africa, and Australia. Mostly Jul–Nov. [*P. communis* Trin.] The grains were eaten by Native Americans (Mabberley 1987).

PHYLLOSTACHYS BAMBOO

A genus of 55 species of BAMBOOS native from the Himalayas to Japan, especially China; often forming thickets by spreading rhizomes; includes most frequently cultivated hardy BAMBOOS; species variously used for paper, timber, fishing rods, edible shoots, and cultivated ornamentals. (Greek: *phyllum*, a leaf, and *stachys*, a spike, referring to the leafy inflorescences) (subfamily Bambusoideae, tribe Bambuseae)
REFERENCE: Borowski et al. 1996.

Phyllostachys aurea Carriére ex Riviére & C. Riviére, (golden), GOLDEN BAMBOO, FISHPOLE BAMBOO, YELLOW BAMBOO. Woody rhizomatous perennial forming extensive colonies; culms 2–8 m tall, glabrous, green to yellowish, flattened on 1 side above point on node where branches originate; branches 2(–3) per node, 1 of these larger than the other; leaf blades narrowly lanceolate, 5–12(–15) cm long, 1–2 cm wide, basally cuneate to rounded, apically acuminate, glabrous on both surfaces except with short pubescence along midvein near attachment to petiole, the margins entire to scaberulous, with a short but distinct petiole ca. 3–5 mm long; leaf sheaths glabrous except for 2 tufts of bristles at apex; flowers or fruits usually not seen. Widely cultivated in TX; persists and spreads vegetatively; can cover large areas; several extensive colonies spreading on sandy soils in n Grayson Co., also Tarrant Co.; also Bell, Bosque, Callahan, Collin, Dallas, Erath, Kaufman, Limestone, McLennan, Palo Pinto, Parker, and Williamson cos. (Borowski et al. 1996). Borowski et al. (1996) documented the occurrence of this species in TX. We have observed only one flowering collection from nc TX; *Phyllostachys* species generally flower only after intervals of many years. Native of China. [*Bambusa aurea* Hort]

A variety of other exotic bamboos, including *Bambusa* species and additional species of *Phyllostachys* are cultivated in nc TX (e.g., *P. nigra* (Lodd.) Munro, BLACK BAMBOO, with black culms). Our only native bamboo, *Arundinaria gigantea*, is limited to the extreme ne portion of nc TX.

POA BLUE GRASS

Annuals or perennials; tips of leaf blades sometimes cupped; ligule a membrane; panicles open or contracted; spikelets relatively small, usually with 2–6(–10) florets, awnless, disarticulating above glumes and between florets; lemmas acute or obtuse, rounded or keeled on back, often with long, kinky hairs at base.

A genus of 200+ species of temperate and cold areas of the world including tropical mountains; some are important pasture and lawn grasses. (Ancient Greek name for grass or fodder) (subfamily Pooideae, tribe Poeae)
REFERENCE: Marsh 1952.

1. Florets developing into bulbils (often dark purple at base) with lemmas prolonged to 5–15 mm long as if sprouting; culms swollen at base; rare in TX, known only from Denton Co. ____ **P. bulbosa**
1. Florets not developing into bulbils; culms not swollen at base; including species widespread in nc TX.
   2. Plants annual, 3–45 cm tall; keel of glumes not scabrous-ciliate (*P. annua*) or minutely scabrous-ciliate (*P. chapmaniana*).
      3. Lemmas with long, kinky, cobwebby hairs at base, with 3 strong and 2 faint nerves; anthers 0.1–0.2 mm long; rare in nc TX ____ **P. chapmaniana**
      3. Lemmas usually without long hairs at base (but long-pubescent on margins and keel, with 5 strong nerves; anthers 0.5–1.3 mm long; widespread in nc TX ____ **P. annua**
   2. Plants perennial, 15–100 cm tall; keel of glumes minutely to strongly scabrous-ciliate.
      4. Plants with slender creeping rhizomes; lemmas acute or with awn-like tip, pubescent only on keel or marginal nerves; inflorescences tightly contracted to somewhat spreading, ± densely flowered; in various habitats including grasslands and other open areas.
         5. Culms and basal leaf sheaths strongly flattened, sharply keeled ____ **P. compressa**
         5. Culms and basal leaf sheaths terete or only slightly flattened, not sharply keeled.
            6. Lower lemmas 3.9–6.4 mm long; plants dioecious; lowest node of inflorescences usually with (4–)5 branches; native species widespread in nc TX ____ **P. arachnifera**
            6. Lower lemmas usually 2.5–3.2 mm long; plants not dioecious, the florets perfect; lowest node of inflorescences usually with (2–)3–4 branches; introduced species known in nc TX only from Cooke and Lamar cos. ____ **P. pratensis**
      4. Plants without rhizomes; lemmas subacute or obtuse, often pubescent over the surface, at least on lower part; inflorescence branches widely spreading, loosely flowered; in wooded habitats ____ **P. sylvestris**

Poa annua L., (annual, yearly), ANNUAL BLUEGRASS, LOW SPEAR GRASS, DWARF MEADOW GRASS. Glabrous annual; panicles small, compact, erect; lemmas long-pubescent on margins and keel. Disturbed sites; throughout TX. Mar–May, occasionally as early as late Fall. Native of Europe.

Poa arachnifera Torr., (spider-bearing, its long white hairs giving the appearance of a spider web), TEXAS BLUE GRASS. Sod-forming, glabrous perennial 20–60 cm tall; panicles dense, rather narrow, erect; florets unisexual, the sexes on different plants; lemmas of pistillate spikelets with long-ciliate keel and dense basal tuft of long, cobwebby hairs; lemmas of staminate spikelets with keel inconspicuously ciliate toward base and rather sparse basal tuft of cobwebby hairs. Calcareous or sandy clay prairies and oak openings; in much of e 1/2 of TX. Late Mar–Apr. Frequent in original prairie, though not abundant; increases under light disturbance, disappears early under heavy grazing.

Poa bulbosa L., (bulbous), BULBOUS BLUE GRASS. Densely tufted, glabrous perennial 20–50 cm tall; culms swollen at base; florets usually developing into asexual bulbils often dark purple at base; lemmas glabrous, strongly nerved, the tips prolonged as if sprouting. Spreading from cul-

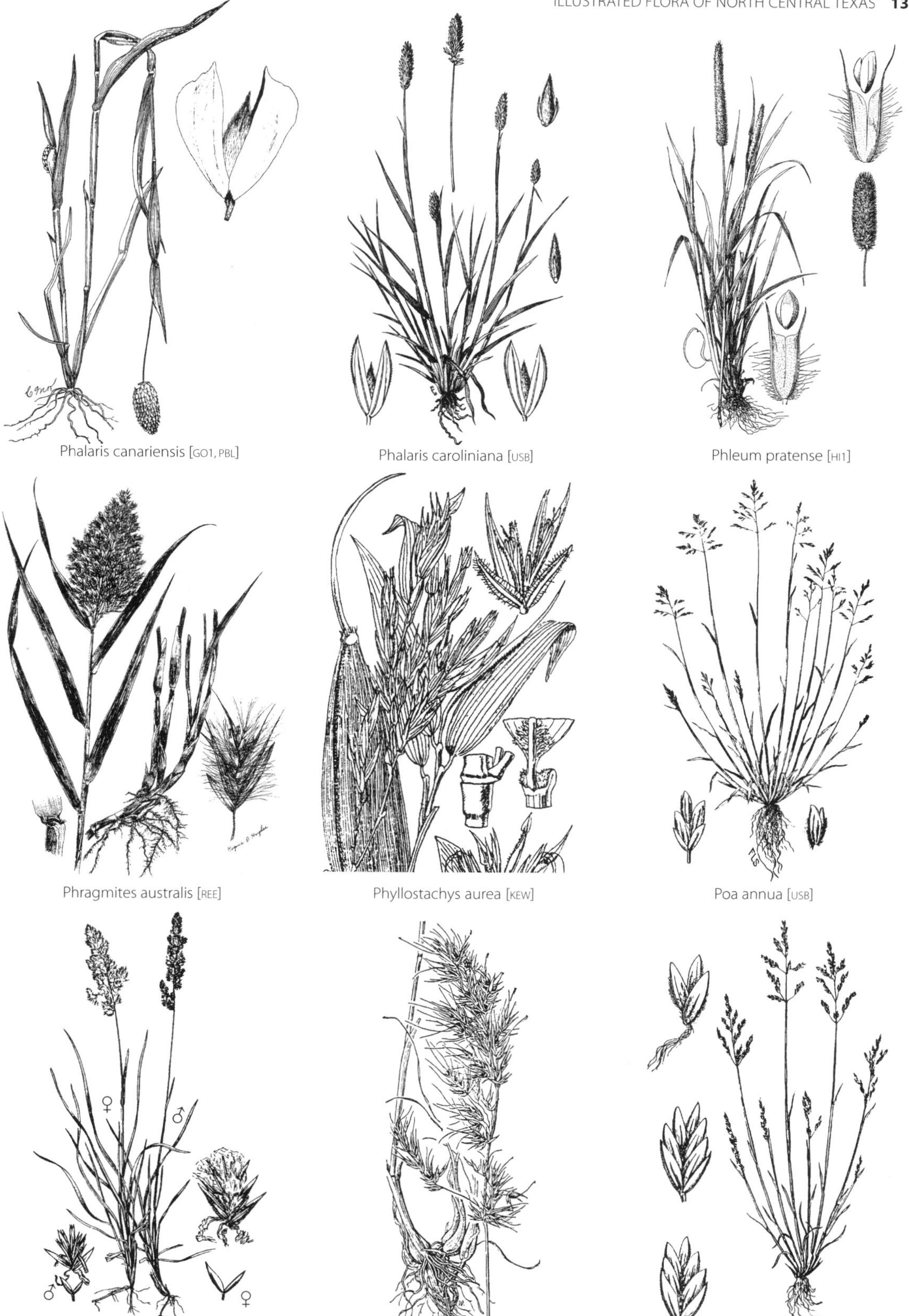

Phalaris canariensis [GO1, PBL]

Phalaris caroliniana [USB]

Phleum pratense [HI1]

Phragmites australis [REE]

Phyllostachys aurea [KEW]

Poa annua [USB]

Poa arachnifera [USB]

Poa bulbosa [HI1]

Poa chapmaniana [USB]

tivation at Denton Agricultural Experiment Station (Shinners 1958a); otherwise unknown in TX. Apr–May. Native of Europe.

Poa chapmaniana Scribn., (for Alvin Wenthworth Chapman, 1809-1899, Florida botanist), CHAPMAN'S POA, CHAPMAN'S BLUE GRASS. Annual to ca. 30 cm tall; similar to *P. annua*, sometimes difficult to distinguish and possibly not specifically distinct from that species; spikelets with cleistogamous florets. Lawns, field, and roadsides; included based on citation of vegetational areas 4 and 5 (Fig. 2) by Gould (1975b); otherwise reported in TX only from e TX (Gould 1975b). Late fall–spring.

Poa compressa L., (flattened), CANADA BLUE GRASS. Rhizomatous perennial 10–60(–80) cm tall; culms and basal leaf sheaths strongly flattened, sharply keeled; panicle branches appressed or only slightly spreading; lemmas with or without long hairs at base. Cultivated as a forage grass, escaping?; included based on citation for vegetational area 5 (Fig. 2) by Gould (1975b); otherwise reported in TX (Gould 1975b; Hatch et al. 1990) only from vegetational areas 6, 9, and 10 (Fig. 2). Apr–Jun. Native of Europe.

Poa pratensis L., (of meadows), KENTUCKY BLUE GRASS, JUNE BLUE GRASS. Sod-forming, glabrous perennial 15–100 cm tall; panicles typically rather loose, with at least the lower branches spreading; lemmas pubescent on keel and marginal nerves, cobwebby at base. Disturbed areas, also appearing occasionally in lawns; Lamar Co., also Cooke Co. (Mahler 1988); scattered in n and w TX. May. Native of the Old World (despite the common name) and possibly also to the northern mountainous areas of North America (Gould 1975b; Sutherland 1986). It is one of the most popular lawn grasses in cooler parts of the country; a number of lawn and pasture strains have been introduced into the U.S. (Gould 1975b; Sutherland 1986). According to Cory (1950), this species was first collected in n TX in 1949.

Poa sylvestris A. Gray, (of woodland), WOODLAND BLUE GRASS, SYLVAN BLUE GRASS. Tufted perennial 40–100 cm tall; panicles open, erect, the lower branches reflexed in age. Woods in limestone ravines, low woods; Fannin and Grayson cos. in Red River drainage, also Dallas Co. (Mahler 1988), not collected there recently; otherwise only far e TX. Apr–May.

POLYPOGON

Annuals or perennials; ligule a membrane; inflorescence a dense, contracted panicle; spikelets small, 1-flowered; glumes falling with rest of spikelet; lemma 1/2–3/4 times as long as the glumes; palea equaling the lemma.

A genus of 10 species of warm temperate areas and tropical mountains. (Greek: *polys*, much, and *pogon*, beard, presumably because of the conspicuous awns of some species) (subfamily Pooideae, tribe Aveneae)

1. Glumes long awned (awns 5–9 mm long); plants annual ____ **P. monspeliensis**
1. Glumes awnless; plants perennial ____ **P. viridis**

Polypogon monspeliensis (L.) Desf., (of Montpellier in s France), RABBIT'S-FOOT, ANNUAL BEARD GRASS, BEARD GRASS, RABBIT'S-FOOT GRASS. Plant 15–75 cm tall, largely glabrous; ligules 4–10 mm long; inflorescences appearing bristly-woolly because of long yellowish awns, ca. 2–15 cm long, 1–2.5 cm broad; spikelets 1.5–2 mm long (not counting conspicuous awns of glumes). Low areas, sometimes in shallow water; throughout TX. Late May–Jul. Native of Europe.

Polypogon viridis (Gouan) Breistr., (green), WATER BENT GRASS. Plant glabrous; culms bent at base or trailing, 15–45(–70) cm long; ligules 2–7 mm long; inflorescences 4–12 cm long, 1–3 cm broad; spikelets 1.3–2(–2.5) mm long, awnless. About springs and in creek beds; Bell and Burnet cos.; s part of nc TX s and w to w TX. Apr–Jul. Native to Mediterranean region. [*Agrostis semiverticillata* (Forssk.) C. Chr., *P. semiverticillatus* (Forssk.) Hyl.]

SACCHARUM PLUME GRASS, SUGARCANE

Large, stout, reed-like perennials; leaf blades elongate, flat; inflorescence a large, dense, often conspicuously hairy, terminal panicle; spikelets all alike, perfect, in pairs, 1 sessile and one pedicellate, 2-flowered, the upper floret fertile; inflorescence branches disarticulating below the spikelets, the pedicels falling with the sessile spikelets; glumes large, usually with a tuft of long hairs at base, the hairs (in our species) about as long as or longer than the spikelets; lemma of upper floret of each spikelet with a long awn.

The genus, as treated here including *Erianthus*, has 35 species of tropical and warm areas of the world. It includes *S. officinarum* L. (SUGARCANE), native to tropical se Asia, which is the source of ca. 1/2 of the world's sugar. (Greek: *sacchar*, sugar, referring to the sweet sap) (subfamily Panicoideae, tribe Andropogoneae)
REFERENCES: Mukherjee 1958; Gandhi & Dutton 1993; Webster & Shaw 1995.

1. Awn of lemma slightly flattened at base, loosely twisted and geniculate (= bent) below middle; culms usually glabrous below the inflorescence; hairs at base of spikelets ca. as long as spikelets; lemma of upper floret 3-nerved ____ **S. brevibarbe**
1. Awn of lemma terete (= round), straight; culms villous (= long hairy) below the inflorescence; hairs at base of spikelets slightly shorter to often much longer than spikelets; lemma of upper floret 1-nerved ____ **S. giganteum**

Saccharum brevibarbe (Michx.) Pers. var. **contortum** (Elliott) R.D. Webster, (sp.: short-bearded; var.: contorted), BENT-AWN PLUME GRASS, BEARD GRASS. Culms usually 1.5–2.5 m tall, the nodes appressed-hispid when young, glabrous with age; ligule a membrane with hairs 1–3 mm long; leaf blades to 80 cm or more long, usually 8–18 mm wide, scabrous; panicles 20–50 (rarely more) cm long; spikelets 6–8 mm long (excluding awns), with basal hairs whitish or twany; lemma awn 15–22 mm long. Usually moist, sandy soils; Hopkins and Lamar cos. in ne part of nc TX; mainly se and e TX. Sep–Nov. We are following Webster and Shaw (1995) for nomenclature of this taxon; it is sometimes recognized in the genus *Erianthus* [as *Erianthus contortus* Baldwin]. [*S. contortum* (Elliott) Nutt.]

Saccharum giganteum (Walter) Pers., (gigantic), SUGARCANE PLUME GRASS. Very large; culms 1–3 m tall in dense clumps, the nodes with a dense ring of hairs 1–6 mm long; ligule and nearby area densely villous; leaf blades to 90 cm or more long, 4–20 mm wide; panicles 10–50 cm long; spikelets 5–8 mm long (excluding awns), with basal hairs abundant, brownish; lemma awn 12–25 mm long. Usually moist sandy soils; Milam and Henderson cos. near e margin of nc TX; mainly se and e TX. Sep–Nov. Sometimes recognized in the genus *Erianthus* [as *E. giganteus* (Walter) P. Beauv.].

SACCIOLEPIS

A genus of 30 species of tropical and warm areas, especially Africa. (Greek: *saccion*, small bag, and *lepis*, scale, referring to the saccate second glume) (subfamily Panicoideae, tribe Paniceae)
REFERENCES: Judziewicz 1990; Crins 1991.

Sacciolepis striata (L.) Nash, (striated, striped), AMERICAN CUPSCALE. Perennial rooting at creeping basal nodes; culms to 1.5 m or more tall; ligule a fringed membrane < 0.5 mm long; inflorescence a narrowly cylindric panicle, 8–25 cm long, ca. 10–15 mm broad; spikelets on short pedicels, glabrous, awnless, basally asymmetrical, 3.5–5 mm long, 2-flowered, the lower staminate, the upper perfect; first glume ca. 1/4 or less as the second; lower lemma about as long as second glume; upper (perfect) lemma hardened, smooth, shiny, ca. 1.5–2 mm long. Moist sand, pond margins; Lamar Co. in Red River drainage; mainly se and e TX. Late Aug–Nov.

SCHEDONNARDUS TUMBLE GRASS, TEXAS CRAB GRASS

A monotypic genus of the s U.S. Mabberley (1997) indicated that this species is a conspicuous feature of deserted towns in Western films. (Greek: *schedon*, near, and *Nardus*, from its resemblance to that genus) (subfamily Chloridoideae, tribe Cynodonteae)

Schedonnardus paniculatus (Nutt.) Trel., (with flowers in panicles), TUMBLE GRASS, TEXAS CRAB GRASS. Tufted perennial 8–50(–70) cm tall or long; leaves crowded toward base of plant; leaf sheaths compressed, keeled, glabrous; leaf blades spirally twisted on drying; ligule a membrane 1–3 mm long; inflorescence as long as or longer than the leafy portion of the culm, spreading to partly decumbent, finally breaking away and acting as a tumbleweed, the main axis becoming loosely coiled, with a few widely spaced spicate branches. Various soils; throughout TX. A minor member of original prairie, increasing under disturbance; rather common on disturbed sites. Apr–Oct.

SCHIZACHYRIUM BLUESTEM

A mainly tropical genus of 60 species, particularly in savannas; formerly treated in a more inclusive *Andropogon*. (Greek: *schizo*, to divide or split, and *achna*, chaff, referring to toothed lemma) (subfamily Panicoideae, tribe Andropogoneae)
REFERENCES: Gould 1967; Carman & Briske 1985; Butler & Briske 1988; Briske & Anderson 1992.

Schizachyrium scoparium (Michx.) Nash, (broom-like), LITTLE BLUESTEM. Perennial 50–200 cm tall, non-rhizomatous, green or glaucous; ligule a firm membrane 1–3 mm long; flowering culms with each leafy branch terminating in a spicate raceme; racemes 2.5–5 cm long; disarticulation at base of sessile spikelets so that associated pedicel and section of raceme fall with sessile spikelet; spikelets in pairs, the sessile spikelets perfect, 6–8 mm long, 2-flowered, the upper fertile; pedicellate spikelets staminate or neuter, narrow, shorter than sessile spikelets. Prairies and woodland openings; nearly throughout TX. Aug–Dec. One of our most important native grasses, often a vegetational dominant, and one of the "big four" tall grass prairie species along with *Andropogon gerardii*, *Panicum virgatum*, and *Sorghastrum nutans*. [*Andropogon scoparius* Michx. var. *frequens* F.T. Hubb., *S. scoparium* (Michx.) Nash var. *frequens* (F.T. Hubb.) Gould]

SCLEROCHLOA HARD GRASS

A monotypic genus of s Europe and w Asia. (Greek: *scleros*, hard, and *chloa*, grass, referring to thick glumes) (subfamily Pooideae, tribe Poeae)
REFERENCE: Brandenburg et al. 1991b.

Sclerochloa dura (L.) P. Beauv., (durable, hard), HARD GRASS. Tufted, glabrous, prostrate to erect annual 3–10(–18) cm tall or long; ligule a membrane 1–2 mm long; inflorescences spike-like, dense, often 1-sided, 1–3(–4.5) cm long; spikelets ca. 3–4-flowered, usually 6–11 mm long; glumes and lemmas blunt, prominently nerved. Roadsides, waste places, or disturbed areas; Dallas and Ellis cos., also sometimes extensive, but local populations were found in Collin, Denton, Fannin, Grayson, and Red River cos. in April 1998 (e.g., *Rabeler & Diggs 1318*); nc TX and Edwards Plateau; first collected in the U.S. in 1895 and in TX in 1944 (Brandenburg et al. 1991b) Apr. Native of Europe.

Poa compressa [USB]

Poa pratensis [USB]

Poa sylvestris [USB]

Polypogon monspeliensis [HI1]

Polypogon viridis [MAS]

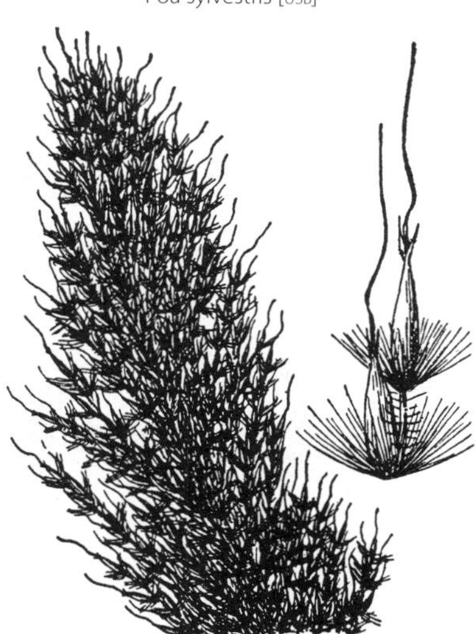

Saccharum brevibarbe var. contortum [BB2]

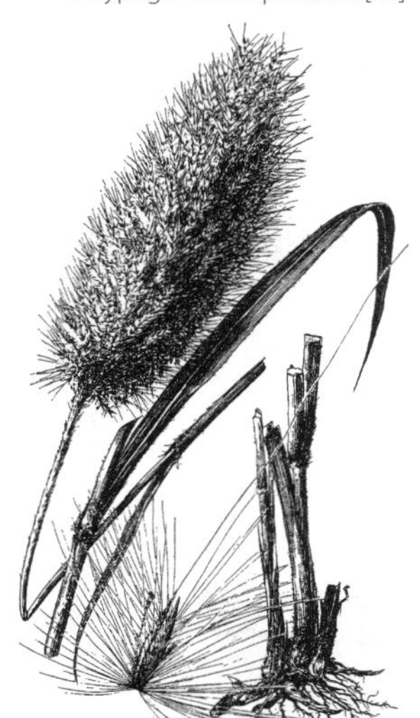

Saccharum giganteum [HI1]

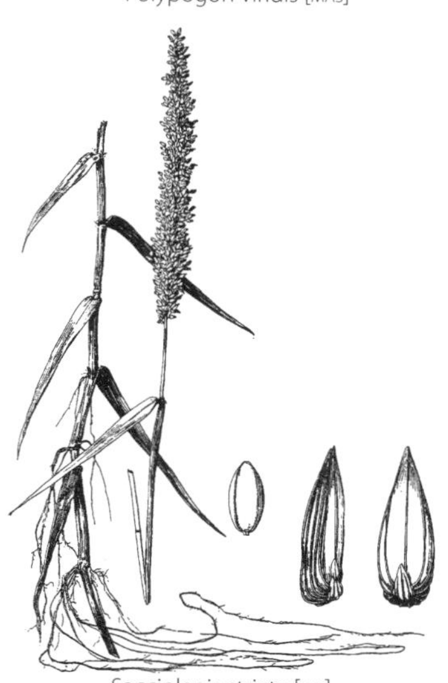

Sacciolepis striata [HI1]

Schedonnardus paniculatus [USB]

SECALE RYE

An Eurasian and s African genus of 3 species including RYE. (Ancient Latin name for rye) (subfamily Pooideae, tribe Triticeae)
REFERENCE: Bowden 1959b.

Secale cereale L., (pertaining to agriculture, from Ceres, goddess of farming), RYE. Annual vegetatively similar to WHEAT (*Triticum*); spikelets usually 2-flowered, both flowers perfect; glumes narrowly lanceolate-subulate, ca. 1–3.5 mm wide, acute or acuminate, apparently 1-nerved; lemmas with awns 5–60 mm long. Occasionally cultivated in much of TX, chiefly in sandy soils; found rarely as a transitory waif on roadsides or in disturbed areas; Grayson and McLennan cos. Apr–May. RYE, an Old World species brought into domestication later than WHEAT or BARLEY, is believed to have evolved from weeds (possibly *S. montanum* Guss.) invading the fields of early WHEAT and BARLEY farmers; large fruited forms were brought into cultivation in e Turkey (Mabberley 1987; Heiser 1990a). RYE and a number of other grasses are susceptible to infection by ergot fungi, e.g., *Claviceps purpurea* (Fr.: Fr.) Tul., which through the production of LSD-like alkaloids can cause hallucinations, psychosis, gangrene of the extremeties (due to vasoconstriction), convulsions, and death in humans and livestock. The condition was referred to in ages past as Saint Anthony's Fire. The reference to fire resulted from the assumption that the burning sensations and blackened (gangrenous) limbs were retribution for sins. Saint Anthony, supposedly with special powers to protect against fire, infection, and epilepsy, was often prayed to for help by those with the condition. Large scale epidemics of ergotism in Europe prior to 1800, from eating bread made with contaminated grain, resulted in 1,000s of deaths; isolated instances still occur where grain purity is not controlled; ergotism is blamed by some for the hysteria that resulted in the Salem witch trials in 17th century Massachusetts (Kingsbury 1964; Caporael 1976; Mabberley 1987; Matossian 1989; Blackwell 1990; Mann 1992).

SETARIA BRISTLE GRASS, FOXTAIL, FOXTAIL-MILLET

Annuals or perennials; ligule a ciliate membrane; panicles usually dense, contracted, spike-like, sometimes less dense with lower branches somewhat spreading (*S. scheelei*); all or at least some spikelets subtended by an involucre of 1–several persistent bristles; spikelets disarticulating above the bristles, awnless, of 2 florets, the lower floret sterile with glume-like lemma, the upper floret fertile with hardened grain-like lemma.

A genus of ca. 150 species of tropical and warm areas; some are used for hay, pasture, silage, or cereal grain. (Latin: *seta*, a bristle, referring to the bristles subtending the spikelets) (subfamily Panicoideae, tribe Paniceae)
REFERENCES: Pohl 1951, 1962; Emery 1957a, 1957b; Rominger 1962; Crins 1991; Webster 1993.

1. Plants very large, usually 1.5–4 m tall; inflorescences very long and thick, 14–45(–50) cm long, 1.5–3 cm in diam.; leaf blades 10–35 mm wide ____ **S. magna**
1. Plants < 1.5 m tall; inflorescences often much smaller in length, diam. or both (can be large in *S. italica*, but plant is much smaller and spikelets larger); leaf blades up to 16(–21) mm broad, usually narrower.
   2. Bristles (short) present at base of terminal spikelets only (sometimes below a few other spikelets); plants resembling a *Panicum*.
      3. Leaf blades 13–20 cm long, usually tapering to an extremely narrow base, often involute; spikelets usually (2.5–)3–4 mm long; bristles shorter to longer than spike lets (rarely > 6 mm long) ____ **S. reverchonii**
      3. Leaf blades 5–13 cm long, not or only slightly narrowing toward base, flat; spikelets 2.5–3 mm long; bristles usually shorter than the spikelets ____ **S. ramiseta**
   2. Bristles present on nearly all spikelets; plants not resembling a *Panicum*.

4. Bristles 1–3 below each spikelet; panicles variable in shape and length, the axis obscured or visible.
  5. Bristles antrorsely scabrous (with up-pointing barbs visible with lens, the free end of the barbs above or distal to the attached end).
    6. Plants annual; bristles on average > 1 per spiklet (1–3 below each spikelet); inflorescences chunky and extremely dense in appearance (axis not visible or visible in only a very few places).
      7. Panicles not lobed or interrupted, usually green, usually 10(–15) cm or less long; spikelets ca. 1.8–2.6 mm long, falling entire; fertile lemma finely rugose ____ **S. viridis**
      7. Panicles often lobed or interrupted, purple or yellow, often large and heavy, up to 30 cm long; spikelets ca. (2.5–)3 mm long, the caryopsis (= fruit) deciduous from glumes and sterile lemma; fertile lemma smooth or nearly so ____ **S. italica**
    6. Plants perennial; bristles mostly 1(–2) below each spikelet; inflorescences usually elongate, not chunky, not extremely dense in appearance (thus much of axis visible).
      8. Leaf blades narrow, 2–8 mm wide; panicles usually cylindric, not tapered, usually 6–15 cm long; bristles mostly 4–15 mm long ____ **S. leucopila**
      8. Leaf blades broad, (5–)9–20 mm wide; panicles usually strongly tapered from a wide base to a narrow apex, (7–)11–35 cm long; bristles (10–)15–35 mm long ____ **S. scheelei**
  5. Bristles retrorsely scabrous (with down-pointing barbs visible with lens, the free end of the barbs below or proximal to the attached end) ____ **S. verticillata**
4. Bristles 4–12 below each spikelet; panicles cylindric, 3–15 cm long, so dense that axis is obscured.
  9. Plants perennial with knotty rhizomes ____ **S. parviflora**
  9. Plants annual with fibrous roots only ____ **S. pumila**

Setaria italica (L.) P. Beauv., (of Italy), FOXTAIL-MILLET, ITALIAN-MILLET, GERMAN-MILLET, HUNGARIAN-MILLET. Annual similar to but larger and coarser than *S. viridis*; leaf blades to 16(–21) mm wide; panicles up to 30 cm long and to 3 cm thick; spikelets ca. (2.5–)3 mm long; bristles to ca. 12 mm long. Denton (damp limestone soil), Grayson (planted for erosion control along Hwy 82), and Tarrant cos., also Parker Co. (R. O'Kennon, pers. obs.); apparently now being widely used to stabilize soil following highway construction; Hatch et al. (1990) cited vegetational areas 4, 7, and 8 (Fig. 2) for TX; Gould (1975b) said that *S. italica* does not persist in the state. Jul–Aug. Related to and interfertile with its wild ancestor *S. viridis*; native of Eurasia; FOXTAIL-MILLET appears to be an old domesticate first brought into cultivation in e Asia; it is used as a cereal grain and sometimes seen as birdseed (Mabberley 1987; Zohary & Hopf 1994).

Setaria leucopila (Scribn. & Merr.) K. Schum., (white-haired). Tufted perennial 20–100 cm tall; bristles mostly 4–15 mm long; fertile lemma finely rugose and with transverse wrinkles. Well-drained soils, sometimes with abundant moisture; Bell Co.; mainly w 1/2 of TX. May–Nov.

Setaria magna Griseb., (large), GIANT BRISTLE GRASS, GIANT FOXTAIL GRASS. Coarse annual; culms to 1–2 cm thick at base; panicles 14–45(–50) cm long and 1.5–3 cm in diam.; bristles 10–20 mm long; spikelets 2–2.5 mm long; fertile lemma smooth and shiny; caryopsis deciduous from glumes and sterile lemma. Wet sand; mainly coastal TX, disjunct w to Tarrant Co. (Trinity River bottom, Fort Worth Nature Center). Oct. This is the largest *Setaria* species in North America.

Setaria parviflora (Poir.) Kerguelén, (small-flowered), KNOT-ROOT BRISTLE GRASS. Perennial 15–150 cm tall, with knotty rhizomes; leaf sheaths slightly compressed, keeled; leaf blades pilose near base; inflorescences dense and stiff; bristles yellowish, tawny, greenish, or purplish, 5–10 mm long; spikelets 2.1–3.0 mm long; fertile lemma rugose. Stream banks, disturbed sites; throughout TX. May–Nov. [*S. geniculata* (Lam.) P. Beauv.]

Setaria pumila (Poir.) Roem. & Schult., (dwarf), YELLOW BRISTLE GRASS. Annual with fibrous roots, similar to *S. parviflora*, differing in its slightly larger spikelets and panicles with relatively fewer spikelets per verticil; bristles yellow at maturity; spikelets 2.7–3.3 mm long. Disturbed soils; nearly throughout TX. Jun–Sep. Native to Europe. [*S. glauca* of authors, not (L.) P. Beauv., *S. lutescens* (Weigel) F.T. Hubbard] This species has long gone under the name *S. glauca*, but because of nomenclatural considerations, *S. pumila* is the appropriate binomial (Clayton & Renvoize 1982). Known to cause mechanical injury to the mouths of livestock; the bristles easily penetrate flesh and remain there because of tiny upwardly directed barbs (Kingsbury 1964).

Setaria ramiseta (Scribn.) Pilg., (branching bristles). Perennial similar to *S. reverchonii*, differing by characters in the key, its occurrence in non-limy soils, and a more s distribution. Sandy loam, dry uplands; s TX n to s parts of region 5 and 8 (Gould 1975b), also Callahan Co. (Correll & Johnston 1970). Spring–Jun, occasionally later. [*Panicum ramisetum* Scribn.]

Setaria reverchonii (Vasey) Pilg., (for Julien Reverchon,1837–1905, a French-American immigrant to Dallas and important botanical collector of early TX), REVERCHON'S BRISTLE GRASS. Tufted erect perennial 20–80 cm tall; culm bases hard, swollen; leaf sheaths pilose on margins and at summit; leaf blades scabrous; ligule a short, membranous-based ring of hairs; fertile lemma conspicuously rugose. Rock outcrops or gravelly soils on limestone; mainly Blackland Prairie (Austin Chalk) s and w to w TX. Apr–Jun, Sep. [*Panicum reverchonii* Vasey]

Setaria scheelei (Steud.) Hitchc., (for Karl Wilhelm Scheele, 1742–1786, German chemist), SOUTHWESTERN BRISTLE GRASS, SCHEELE'S BRISTLE GRASS. Coarse perennial 70–130 cm tall; bristles (10–)15–35 mm long; spikelets 2.1–2.6 mm long; fertile lemma rugose. Fencerows, ravines, open woods, often in shade; Bell and McLennan cos.; s part of nc TX s and w to w TX. May–Nov.

Setaria verticillata (L.) P. Beauv., (whorled), HOOKED BRISTLE GRASS, BUR BRISTLE GRASS, FOXTAIL GRASS. Annual; leaf sheath margins pilose; spikelets 2–2.3 mm long; fertile lemma finely rugulose. Gould (1975b) reported *McCart 9281* from Brown Co.; other TX specimens are from Brewster, Dimmit and El Paso cos. Native to Europe.

Setaria viridis (L.) P. Beauv., (green), GREEN BRISTLE GRASS, GREEN FOXTAIL GRASS. Annual up to 100 cm tall; leaf sheaths slightly compressed; leaf blades mostly 3–10 mm wide; inflorescences rather soft and flexible; bristles usually green; spikelets 1.8–2.6 mm long; fertile lemma finely rugose. Disturbed sites; Blackland Prairie s and w to w TX. May–Jul. Native of Eurasia. While this species is often treated as having a number of varieties (e.g., Kartesz 1994; Jones et al. 1997), we are following Gould (1975b) in not recognizing infraspecific taxa in nc TX.

SORGHASTRUM INDIAN GRASS

Coarse perennials; leaf blades flat, often conspicuously bluish green; inflorescence a long exserted panicle; spikelets in pairs, one sessile and fertile, one vestigial, in ours reduced to a hairy pedicel; disarticulation so that associated pedicel and section of the inflorescence branch fall with sessile spikelet; fertile spikelet 2-flowered, the upper, fertile lemma with a geniculate and twisted awn.

A genus of 17 species of Africa and tropical and warm areas of the Americas. (Named from its resemblance to the genus *Sorghum*) (subfamily Panicoideae, tribe Andropogoneae)
REFERENCE: Davila 1988.

1. Awns of lemmas mostly 23–35 mm long, usually twice-geniculate; spikelets at maturity usually dark brown ____ **S. elliottii**
1. Awns of lemmas usually 12–17 mm long, once-geniculate; spikelets at maturity light brown or straw-colored ____ **S. nutans**

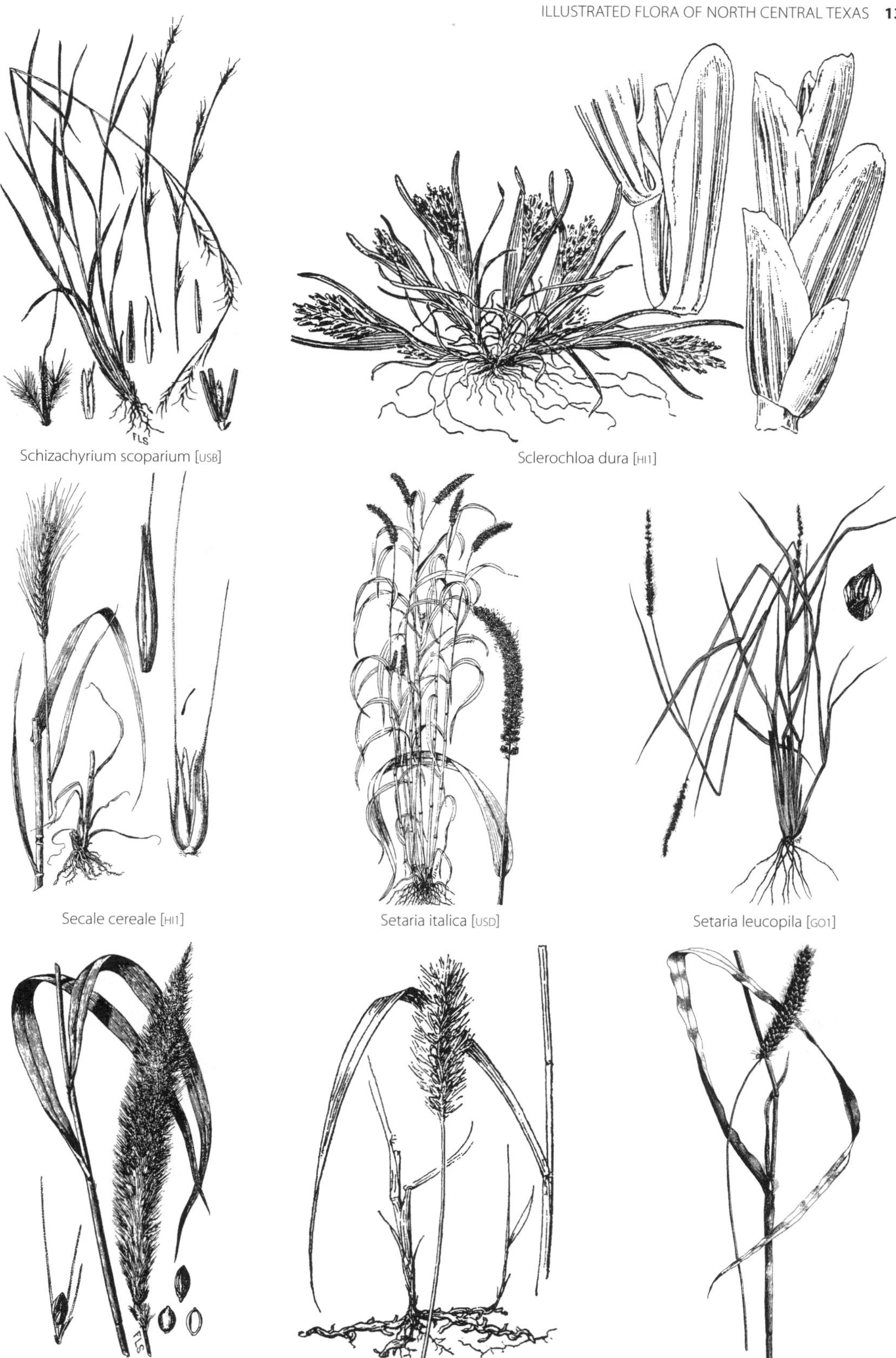

Schizachyrium scoparium [USB]

Sclerochloa dura [HI1]

Secale cereale [HI1]

Setaria italica [USD]

Setaria leucopila [GO1]

Setaria magna [USB]

Setaria parviflora [HI1]

Setaria pumila [RCA]

Sorghastrum elliottii (C. Mohr) Nash, (for Stephen Elliott, 1771–1830, Carolinian botanist), SLENDER INDIAN GRASS, LONG-BRISTLED INDIAN GRASS. Not rhizomatous; culms usually 0.8–1.8 m tall; ligules 1–4 mm long; panicles narrow, looser than in *S. nutans*, sparsely flowered, 15–30 cm long; spikelets usually 5.5–7 mm long. Sandy wooded areas; Fannin Co. (Talbot property), also collected by Reverchon in Dallas Co. in 1876, also Lamar Co. (Carr 1994); mainly se and e TX. Sep–Nov.

Sorghastrum nutans (L.) Nash, (nodding), YELLOW INDIAN GRASS, INDIAN REED. **State grass of Oklahoma** (S. Barber, pers. comm.). Rhizomes short, scaly; culms erect, up to 2.5 m tall; ligule a stiff membrane 2–5 mm long; panicles loosely contracted, up to 30 cm long, yellowish; spikelets usually 6–8 mm long. Moist areas, open woodlands, grasslands; throughout TX. Sep–Nov. One of the dominants in the original tall grass prairie and considered one of the "big four" tall grasses along with *Andropogon gerardii*, *Panicum virgatum*, and *Schizachryium scoparium*; important forage grass and indicator of good range conditions. [*S. avenaceum* (Michx.) Nash]

SORGHUM

Robust annuals or perennials; ligule a ciliate membrane; inflorescence an open or contracted panicle; spikelets in pairs, one sessile and perfect, the other pedicelled and staminate; disarticulation below sessile spikelets so that associated pedicel and section of inflorescence branch fall with spikelet; perfect spikelet 2-flowered, the upper fertile lemma usually awned or awnless.

A mainly Old World genus of 24 species (Mexico, 1 species) including *S. bicolor*, the world's fourth most important cereal after WHEAT, CORN, and RICE; this species, which is more tolerant of drought than most cereals, is of particular importance in Africa and Asia where it feeds millions of people (Heiser 1990a). Some *Sorghum* species, under certain conditions, can be fatally poisonous to livestock due to the presence of hydrocyanic acid (Sperry et al. 1955). (derivation unclear, possibly from *sorgho*, the Italian name of the plant, or perhaps from Latin *syricus*, Syria, and *granum*, grain, the presumed place of origin of *S. halepense*) (subfamily Panicoideae, tribe Andropogoneae)
REFERENCES: Snowden 1936; Celarier 1958; Duvall & Doebley 1990.

1. Plants perennial, with rhizomes.
   2. Plants usually 3–4 m tall; rhizomes short; spikelets 5–6.5 mm long; pedicellate spikelets falling with a part of the pedicel; caryopses (= fruits) 3–3.8 mm long; possibly present in nc TX (no specimens seen) **S. almum**
   2. Plants usually < 2 m tall; rhizome system extensive; spikelets 4.5–5.5 mm long; pedicellate spikelets disarticulating cleanly at the nodes; caryopses 2–3 mm long; extremely abundant in nc TX **S. halepense**
1. Plants annual, without rhizomes **S. bicolor**

Sorghum almum Parodi, (nourishing). Rhizomatous perennial similar to *S. halapense*. Seeded in pastures; included based on citation for vegetational area 5 by Hignight et al. (1988); also s TX and Edwards Plateau. Summer–fall. Native to Argentina. Duvall and Doebley (1990) considered this taxon to be a hybrid between *S. halepense* and a diploid from *S. bicolor* subsp. *arundinaceum* (Desv.) de Wet & J.R. Harlan (subsp. *arundinaceum* is the wild relative of cultivated SORGHUM). Jones et al. (1997) treated this taxon as *S.* ×*almum* Parodi [*S. bicolor* × *S. halepense*].

Sorghum bicolor (L.) Moench, (two-colored). Large succulent annual ca. 0.8–2.5 m tall; leaf blades 1–5 cm or more wide; inflorescences variable. Summer–fall. Probably native of Africa. Poisonous in a manner similar to *S. halepense* (Hardin & Brownie 1993).

1. Panicles compact; lemma of perfect spikelets awnless subsp. **bicolor**
1. Panicles open; lemma of perfect spikelets awned subsp. **drummondii**

Setaria ramiseta [HI1]

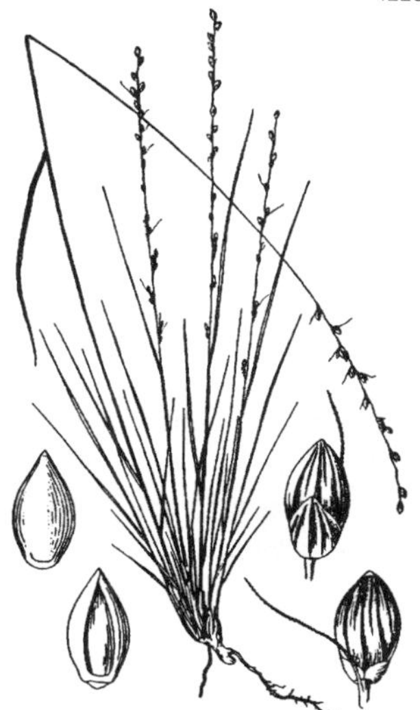
Setaria reverchonii [USB]

Setaria scheelei [GO1]

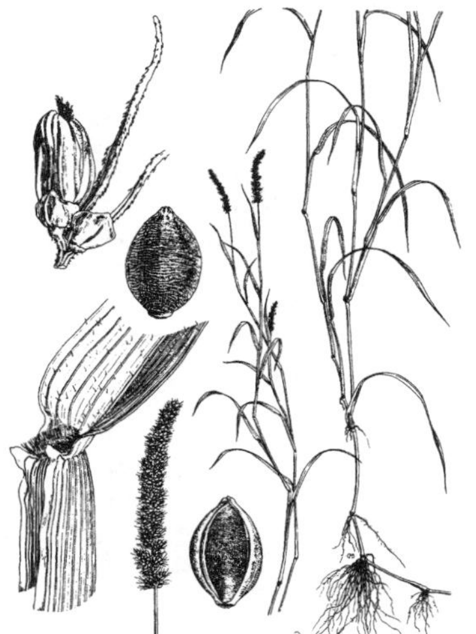
Setaria verticillata [MAS]

Setaria viridis [RCA]

Sorghastrum elliottii [HI1]

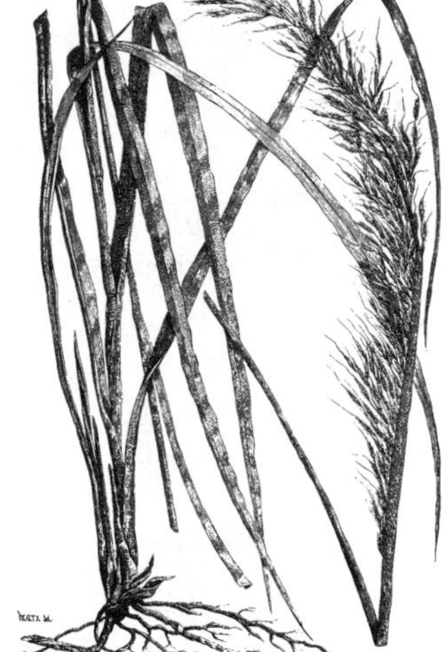

Sorghastrum nutans [HI1, RCA]

Sorghum almum [HEA]

subsp. **bicolor**. SORGHUM, MILO, SORGO, KAFIR, HEGARI, BROOM-CORN, GUINEA-CORN, KAOLANG. Widely cultivated in TX; found as a transitory escape along roadsides, railroads, and field margins; Brown, Grayson, Montague, and Tarrant cos. [*S. vulgare* Pers.] Thought to have been brought into cultivation in Sudan ca. 1000 BC.; cultivated for its grain, sweet juice, broom material, silage, and as a forage plant; a staple in Africa, India, and China; it thrives under drier conditions than appropriate for corn (Mabberley 1987).

subsp. **drummondii** (Nees ex Steud.) de Wet & J.R. Harlan, (for its discoverer, Thomas Drummond, 1780–1835, Scottish botanist and collector in North America). SUDAN GRASS. Awn of lemma ca. 10 mm long. Cultivated and possibly escapes; included based on citation for vegetational area 5 by Hignight et al. (1988). [*S. sudanese* (Piper) Stapf] Jones et al. (1997) treated this taxon as *S.* ×*drummondii* (Steud.) Millsp. & Chase [*S. arundinaceum* × *S. bicolor*]. While this plant is valuable as forage, because of cyanide production at certain stages, caution is advised.

Sorghum halepense (L.) Pers., (of Aleppo or Haleb, a city in n Syria), JOHNSON GRASS. Coarse rhizomatous perennial 25–200 cm tall; leaf sheaths glabrous; leaf blades pilose above at base; ligule a membrane with fringe of hairs; panicles varying from narrow and dense to loose and open, yellow-brown or red-brown to purple-brown; lemma awnless or with awn 1–1.5 mm long. Fields, roadsides, disturbed areas; throughout TX. May–Nov. Native of the Mediterranean region; introduced into TX in the 1880s; now one the most abundant grasses in nc TX and a pernicious invader of native habitats. While normally edible, this species can be poisonous to livestock when young, during dry weather, or after a frost, drought, or period of high temperatures; the poisonous principle is dhurrin, a cyanogenic glycoside (Pammel 1911; Burlage 1968; Hardin & Brownie 1993).

SPARTINA CORD GRASS, MARSH GRASS

A genus of 17 species primarily of coastal America, Europe, and n Africa; typically halophytic (= tolerant of salty or alkaline conditions. (Greek: *spartine*, a cord, possibly from the appearance of the inflorescence branches) (subfamily Chloridoideae, tribe Cynodonteae)
REFERENCE: Mobberley 1956.

Spartina pectinata Link, (comb-like), PRAIRIE CORD GRASS, TALL MARSH GRASS, SLOUGH GRASS. Rhizomatous perennial 1.5–2.5 m tall; ligule a ring of hairs 1–3 mm long; inflorescence branches 8–40, appressed, 4–15 cm long; spikelets 1-flowered, closely placed, sessile, 10–25 mm long including scabrous awns of the glumes; anthers 4–6 mm long. Low moist areas, swales, fresh or brackish water; Comanche, Grayson (Hagerman Nat. Wildlife Refuge), and Fannin cos.; also Rockwall Co. (Wallace Prairie) (Mahler 1988); Blackland Prairie and Edwards Plateau w and n to nw TX. Aug–Oct. Other *Spartina* species are important salt marsh grasses.

SPHENOPHOLIS WEDGE GRASS, WEDGESCALE

A genus of 5 species ranging from Canada to Mexico. (Greek: *sphen*, a wedge, and *pholis*, scale, referring to the shape of the broadly obovate or cuneate second glume) (subfamily Pooideae, tribe Aveneae)
REFERENCE: Erdman 1965.

Sphenopholis obtusata (Michx.) Scribn., (obtuse, blunt), PRAIRIE WEDGESCALE. Annual 15–100 (–120) cm tall; similar in aspect to *Koeleria* but with second glume much more obovate and more blunt at apex; culms single or tufted; leaf sheaths glabrous or pubescent; ligule a membrane 1.5–3 mm long, glabrous or ciliate; inflorescence a narrow panicle, dense, spike-like; inflorescence axis glabrous or minutely scabrous; spikelets shiny, 2–3-flowered, 1.5–5 mm long,

Sorghum bicolor subsp. drummondii [USD]

Sorghum halepense [RCA]

Spartina pectinata [HI1]

Sphenopholis obtusata [HI1]

Sporobolus compositus var. clandestinus [HI1]

Sporobolus compositus var. compositus [USB]

Sporobolus compositus var. drummondii [BB2]

Sporobolus compositus var. macer [HI1]

Sporobolus cryptandrus [USB]

awnless, disarticulating below glumes; apex of second glume conspicuously obovate. Low, open or partly shaded ground, clayey or sandy soils; throughout most of TX. Apr–Jun.

SPOROBOLUS DROPSEED

Annuals or perennials; ligules minute, largely a ciliate fringe on vestigial membranous base; inflorescences spike-like or open-panicled, often partly included within a sheath; spikelets 1-flowered, awnless; glumes shorter than or equaling the lemma, 1-nerved; lemmas usually 1-nerved; fruit falling free from lemma and palea, the seed coat not fused to the pericarp (therefore not a true caryopsis).

A genus of 160 species in the Americas, Asia, Africa, and Europe; some have edible grains. (Greek: *sporos*, seed, and *ballein*, to cast forth, referring to the deciduous grains) (subfamily Chloridoideae, tribe Eragrostideae)

REFERENCES: Clayton 1965; Riggins 1977; Brown 1993; Wipff & Jones 1995; Peterson et al. 1997.

1. Panicles contracted, the branches appressed; plants annual or perennial.
  2. Plants with evident short, creeping, scaly rhizomes ____ **S. compositus** var. **macer**
  2. Plants without creeping, scaly rhizomes.
    3. Culms 1–2 m tall, (2–)3–7 mm in diam. at base; panicles mostly 25–70 cm long; spikelets 2.6–3.2(–4) mm long; mainly far w TX and Panhandle (only Bosque Co. in nc TX) ____ **S. giganteus**
    3. Culms 1.2 m or less tall, 1–3 mm in diam. at base (to 5 mm in *S. asper* which has spikelets 4–7 mm long); panicles 1–30(–40) cm long; spikelet length various; widespread in nc TX.
      4. Plants annual; panicles usually 1–5 cm long.
        5. Florets glabrous.
          6. Glumes shorter than floret; lemmas 1-nerved; lower leaf sheaths not papillose-pilose ____ **S. neglectus**
          6. Glumes longer than floret; lemmas 3-nerved (midnerve more conspicuous); lower leaf sheaths papillose-pilose ____ **S. ozarkanus**
        5. Florets pubescent.
          7. Glumes as long as or longer than floret; lower leaf sheaths and blades papillose-pilose; lemmas 3-nerved (midnerve more conspicuous) ____ **S. ozarkanus**
          7. Glumes shorter than floret; lower leaf sheaths and blades usually not papillose-pilose; lemmas 1- or 3-nerved ____ **S. vaginiflorus**
      4. Plants perennial; panicles usually 5–30 cm long.
        8. Spikelets 1.4–2 mm long; inflorescences mostly exerted from sheaths ____ **S. indicus**
        8. Spikelets 4–7 mm long; inflorescences usually at least partly included within sheaths.
          9. Lemmas glabrous; panicles 5–30 cm long; pericarp gelatinous when wet.
            10. Terminal sheaths 0.8–2(–2.5) mm wide when folded; culms 1–2(–2.5) mm wide near base; primary panicle branches 8–18, not crowded ____ **S. compositus** var. **drummondii**
            10. Terminal sheaths (1.3–)1.5–6 mm wide when folded; culms (1.4–)2–5 mm wide near base; panicle branches 12–35, crowded ____ **S. compositus** var. **compositus**
          9. Lemmas appressed-pubescent; panicles 5–10 cm long; pericarp loose when wet ____ **S. compositus** var. **clandestinus**
1. Panicles open, the branches ascending OR widely spreading; plants perennial.
  11. Pedicels 3–8(–12) mm long; spikelets 4–6(–7.2) mm long; known locally only from Lamar Co. in extreme ne portion of nc TX ____ **S. silveanus**
  11. Pedicels 0.2–ca. 2 mm long; spikelets 1.5–2.8 mm long; widespread in nc TX.
    12. Panicles 3–15(–18) cm long, exserted beyond sheath, the lower branches visibly whorled; sheath summit without collar of hairs; culms 10–30(–50) cm tall ____ **S. pyramidatus**
    12. Panicles 15–30(–40) cm long, partially included within sheath, the branches not visibly

whorled (often hidden within sheath); sheath summit with collar of hairs; culms 35–120 cm tall ______ **S. cryptandrus**

Sporobolus compositus (Poir.) Merr., (compound). Perennial; spikelets in contracted terminal panicles, often with cleistogamous spikelets in axillary panicles; spikelets 4–7 mm long. [*S. asper* (Michx.) Kunth] This species long went under the name *S. asper*, but because of nomenclatural considerations, *S. compositus* is the appropriate binomial (Kartesz & Gandhi 1995). Key to varieties included in key to species.

var. **clandestinus** (Biehler) Wipff & S.D. Jones, (concealed), PURPLE-FLOWER DROPSEED. Spikelets 5–7 mm long. Grasslands, disturbed sites; widespread in e 1/2 of TX. Aug–Oct. [*S. asper* (Michx.) Kunth var. *clandestinus* (Biehler) Shinners, *S. clandestinus* (Biehler) Hitchc.] We are following Wipff and Jones (1995) for nomenclature of this taxon. While it has often been recognized at the specific level (e.g., Kartesz 1994), because var. *clandestinus* differs morphologically from var. *compositus* in only minor ways, Wipff and Jones (1995) argued that it is most appropriately recognized at the varietal level.

var. **compositus**. TALL DROPSEED, LONG-LEAF RUSH GRASS, ROUGH RUSH GRASS. Cespitose, without rhizomes. Grasslands, disturbed sites; widespread in e 1/2 of TX. Sep–Nov.

var. **drummondii** (Trin.) Kartesz & Gandhi, (for its discoverer, Thomas Drummond, 1780–1835, Scottish botanist and collector in North America), MEADOW DROPSEED. Cespitose, without rhizomes. Grasslands, disturbed sites; widespread in e 1/2 of TX. Aug–Nov. Jones et al. (1997) lumped this variety with var. *compositus*.

var. **macer** (Trin.) Shinners, (thin, meager), MISSISSIPPI DROPSEED. Similar to var. *compositus*. Open woods, margins of woods; included based on citation of vegetational area 4 (Fig. 2) by Hatch et al. (1990); mainly e TX. Aug–Nov.

Sporobolus cryptandrus (Torr.) A. Gray, (with hidden flowers), SAND DROPSEED, COVERED-SPIKE DROPSEED. Perennial; summit of sheaths with tufts of long white hairs 2–4 mm long; panicles 15–30(–40) cm long, 2–12 cm wide, usually partially enclosed by the subtending sheath; lemmas as long as spikelet. Grasslands, disturbed sites; throughout TX except far e part. May–Nov. The grains were consumed by Native Americans (Mabberley 1987).

Sporobolus giganteus Nash, (gigantic), GIANT DROPSEED. Large perennial. Loose sand; Bosque Co. (Carr 1989) in Lampasas Cut Plain; mainly far w TX, Edwards Plateau, and Panhandle. Summer–fall.

Sporobolus indicus (L.) R. Br., (of India), SMUT GRASS. Perennial; spikelets 1.4–2 mm long; pericarp mucilaginous, the seed often sticking instead of falling readily. Moist soils, often in disturbed areas; included based on citation of vegetational area 4 (Fig. 2) by Hatch et al. (1990); mainly se and e TX, also Edwards Plateau. Mar–Dec. Native to Asia.

Sporobolus neglectus Nash, (overlooked), PUFF-SHEATH DROPSEED. Annual; erect or decumbent; terminal panicles contracted, 2–5 cm long, often only apical portion exserted from subtending sheath; axillary panicles shorter, almost entirely enclosed by sheaths; sheaths somewhat inflated; spikelets (1.3–)1.6–2.8 mm long; lemmas white or purple-tinged, glabrous. Disturbed sites; Bosque, Grayson, and Mills cos.; nc TX and Edwards Plateau. Aug–Nov. Jones et al. (1997) treated this species as *S. vaginiflorus* var. *neglectus* (Nash) Scribn.

Sporobolus ozarkanus Fernald, (of the Ozarks), OZARK DROPSEED. Annual; culms 4–50 cm tall; spikelets 2.3–3.8(–4.2) mm long. Limestone areas, roadsides; Grayson, Johnson, and Wise cos., also Fort Hood (Bell or Coryell cos.—Sanchez 1997); nc TX and Edwards Plateau. Aug–Oct. Jones et al. (1997) lumped this species with *S. neglectus* (treated by them as *S. vaginiflorus* var. *neglectus*).

Sporobolus pyramidatus (Lam.) Hitchc., (pyramidal), WHORLED DROPSEED. Perennial; panicles 3–15(–18) cm long, becoming pyramidal and exserted at maturity, the lower branches whorled; spikelets 1.5–2 mm long; lemmas 1.2–2 mm long. Open, disturbed sites; mainly Post Oak Savannah s and w through most of TX. Mar–Nov.

Sporobolus silveanus Swallen, (for William Arents Silveus, 1875–1953, TX botanist and attorney), SILVEUS' DROPSEED. Perennial; culms 0.9–1.2 m tall; panicles somewhat open, 20–50 cm long, 10–12(–15) cm wide; spikelets purple, 4–6(–7.2) mm long. Sandy soils, prairies and forest openings; Lamar Co. (Tridens Prairie) in ne corner of nc TX; mainly se and e TX. Sep–Nov.

Sporobolus vaginiflorus (Torr. ex A. Gray) A.W. Wood, (with flowers in the sheaths), POVERTY DROPSEED, SOUTHERN POVERTY GRASS. Annual; panicles terminal and axillary, contracted, 1–4 cm long, 2–5 mm wide, usually partially enclosed within subtending sheaths; lemmas often mottled with dark purple, short appressed-pubescent. Disturbed sites, sandy or clay soils; widespread in e 1/2 of TX. Sep–Nov.

STENOTAPHRUM

A genus of 7 species of tropical and warm areas of the world. (Greek: *steno*, narrow and *taphros*, trench, from grooved inflorescence axis) (subfamily Panicoideae, tribe Andropogoneae)
REFERENCES: Sauer 1972; Crins 1991.

Stenotaphrum secundatum (Walter) Kuntze, (with parts arranged along one side), ST. AUGUSTINE GRASS. Decumbent or ascending, stoloniferous, sod-forming perennial to ca. 30 cm tall; leaf sheaths compressed, keeled, with a few hairs at summit; leaf blades cupped or creased lengthwise; ligule a very short ring of hairs; inflorescences spike-like, with wide, flattened, corky axis; spikelets 4–5 mm long, awnless, 2-flowered, the lower floret staminate or neuter, the upper floret perfect, appressed and sunken into one side of the axis; first glume smaller than second. Commonly planted as lawn-grass, escaping locally; tends to freeze back in very severe winters; Grayson and Tarrant cos.; se and e TX w to nc TX and Edwards Plateau. Jun–Aug. Native to the tropics (South America?), probably not native to the United States.

TRIDENS

Tufted perennials with erect culms; ligule a ciliate membrane; panicles contracted or open; spikelets 4–12-flowered; first glume 1-nerved; lemmas 3-nerved, glabrous or pubescent, rounded on the back, obtuse or acute, usually ± 2-toothed or rounded-truncate at apex, the nerves often slightly mucronate.

A genus of 18 species of warm areas of the Americas (e. U.S. to Argentina) and one species in Angola. Recognized as the genus *Triodia* by Hitchcock (1935). (Latin: *tri*, three, and *dens*, tooth, from the 2-toothed lemma tip often with a mucro from between the teeth) (subfamily Chloridoideae, tribe Eragrostideae)
REFERENCES: Tateoka 1961; Peterson et al. 1997.

1. Panicles open, ± loosely flowered, neither densely flowered nor spike-like, the panicle branches with conspicuous naked (= without spikelets) portions at base.
   2. Leaf blades mostly 3–10 mm wide; panicles mostly 15–35 cm long; plants 60–180 cm tall; spikelets 5–9 mm long ____ **T. flavus**
   2. Leaf blades mostly 1–3 mm wide; panicles mostly 5–16 cm long; plants 20–75 cm tall; spike lets usually 6–13 mm long ____ **T. texanus**
1. Panicles contracted, densely flowered or elongate and spike-like, the panicle branches without

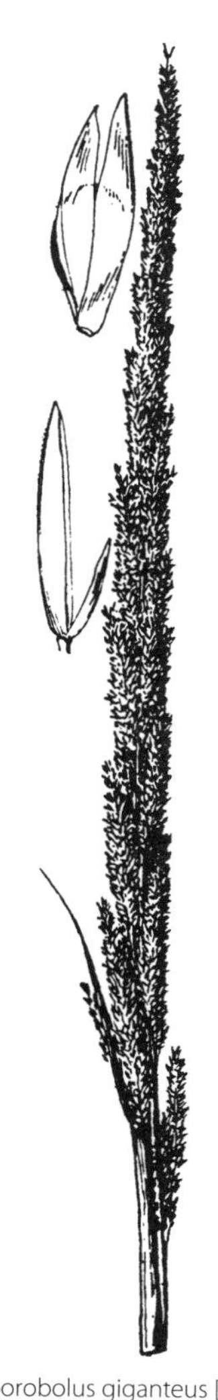

Sporobolus giganteus [HI1]

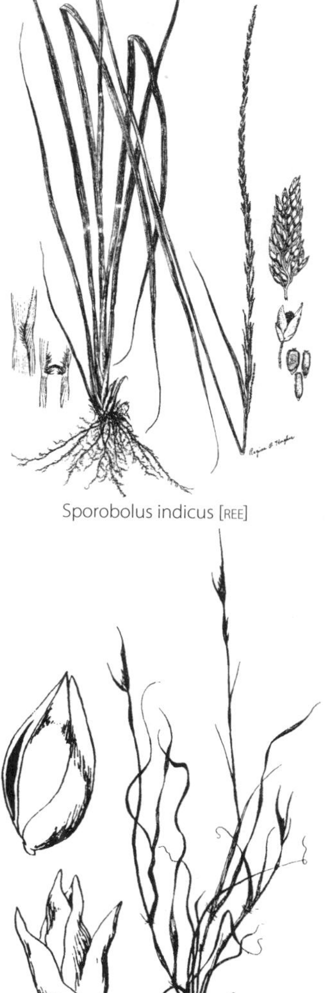

Sporobolus indicus [REE]

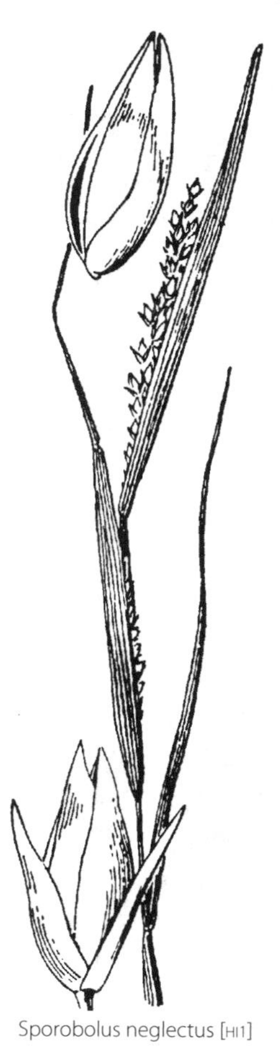

Sporobolus neglectus [HI1]

Sporobolus ozarkanus [ROE]

Sporobolus pyramidatus [GO1]

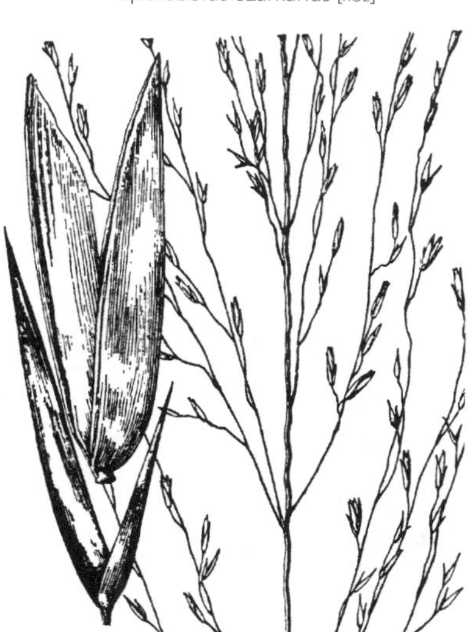

Sporobolus silveanus [HI2]

Sporobolus vaginiflorus [USB]

conspicuous naked basal portions, usually with spikelets nearly to base or apparently so (branches usually tightly appressed at base).

3. Glumes much longer than lemmas, usually as long as the entire spikelet or longer; plants 50–170 cm tall ____ **T. strictus**
3. Glumes slightly longer to shorter than lemmas, much shorter than entire spikelet; plants 20–80(–90) cm tall.
    4. Lemmas glabrous or hairy only at extreme base; lemmas awnless ____ **T. albescens**
    4. Lemmas ciliate or puberulent to well above the base on nerves (at least lower third of lemma with pubescence on nerves); lemmas awnless or awned.
        5. Lemmas awnless; nerves usually with pubescence to well above middle ____ **T. muticus**
        5. Lemmas short awned, the midnerve excurrent (= mucro); nerves with pubescence on lower 1/3–1/2 of lemma ____ **T. congestus**

Tridens albescens (Vasey) Wooton & Standl., (whitish), WHITE TRIDENS, WHITETOP. Plant 30–60 (–90) cm tall, glabrous; panicles 6–30 cm long; spikelets purplish-tinged, thus appearing banded.

Tridens congestus (L.H. Dewey) Nash, (congested, crowded together), PINK TRIDENS. Plant 30–75 cm tall, glabrous; panicles 5–10 cm long; glumes and lemmas thin, papery, usually pink-tinged; lemma apex deeply cleft. Clay, disturbed sites, low moist areas; Grayson Co., also Dallas and McLennan cos. (Mahler 1988); also se TX; endemic to TX. Apr–Nov.

Tridens flavus (L.) Hitchc., (pale yellow), PURPLETOP, REDTOP. Plant 60–180 cm tall, glabrous; lower leaf sheaths laterally compressed and keeled, often giving base of plant a flattish aspect; panicles 15–35 cm or more long, drooping, the branches viscid, flexuous; spikelets green or purplish. Old fields and woods; in much of TX. Aug–Nov.

Tridens muticus (Torr.) Nash, (pointless, cut off). Plant 20–80 cm tall; culm nodes often bearded; panicles narrow, elongate, 7–20(–25) cm long; spikelets short-pedicelled, appearing sessile, not densely crowded, purplish. Dry disturbed sites. Apr–Nov.

1. Second glume 3–7-nerved, usually 6–8 mm long; leaf blades usually 3–4 mm wide ____ var. **elongatus**
1. Second glume 1-nerved, usually ca. 5 mm long or less; leaf blades usually 1–2 mm wide ____ var. **muticus**

var. **elongatus** (Buckley) Shinners, (elongated), ROUGH TRIDENS. Dry disturbed sites; Bell, Denton, Stephens, and Williamson cos.; in much of c and n TX.

var. **muticus**. SLIM TRIDENS. Dry disturbed sites; Bell Co.; in much of TX.

Tridens strictus (Nutt.) Nash, (erect), LONG-SPIKE TRIDENS. Plant 50–170 cm tall, glabrous; panicles 10–36 cm long, the branches erect-appressed; glumes conspicuously longer than rest of spikelet. Sandy or clayey soils, disturbed sites; Cooke, Dallas, Denton, Fannin, Grayson, Lamar, Limestone, and McLennan cos.; se and e TX w to East Cross Timbers and Edwards Plateau. Jul–Nov.

Tridens texanus (Wats.) Nash., (of Texas), TEXAS TRIDENS. Plant 20–75 cm tall; panicles mostly 5–16 cm long, the branches flexuous, bare of spikelets basally; spikelets with 6–12 florets, usually purple or rose-purple at maturity, conspicuous (superficially similar to *Eragrostis secundiflora* subsp. *oxylepis*); lemmas pubescent on nerves below middle; glumes ca. 1/2 as long as lemmas. Plains and dry slopes, often in protection of shrubs; Brown Co. (HPC) on sw margin of nc TX; s and se TX n to Edwards Plateau and edge of nc TX. May–early Jun, late Aug–Nov.

TRIPLASIS

A genus of 3 species occurring from the se United States to Costa Rica. (Greek: *triplasios*, trifarious, threefold, from the tip of the lemma) (subfamily Chloridoideae, tribe Eragrostideae) REFERENCE: Peterson et al. 1997.

Stenotaphrum secundatum [USB]

Tridens albescens [USB]

Tridens congestus [HI1]

Tridens flavus [HI1]

Tridens muticus var. elongatus [USB]

Tridens muticus var. muticus [USB]

Tridens strictus [USB]

Triplasis purpurea (Walter) Chapm., (purple), PURPLE SAND GRASS. Tufted annual; culms 40–100 cm long, spreading-erect or decumbent at base, with 1-flowered cleistogamous spikelets in the axils of enlarged sheaths; ligule a short, dense ring of hairs; panicles open, 3–11 cm long, with a few sparingly rebranched primary branches; spikelets usually with 2–4 florets, 6–10 mm long, usually purple; lemmas notched, mucronate or short awned due to the midnerve extending from between the two lobes of the notch, silky pubescent on the 3 nerves. Sandy soils, forest margins, stream banks, and open areas; Dallas, Grayson, Parker, Somervell, and Tarrant cos.; throughout TX. (Jul–)Sep–Oct(–Nov).

TRIPSACUM

A New World genus of 13 species ranging from the s U.S. to Paraguay, especially Central America. (Supposedly from Greek: *tribein*, to rub, perhaps in allusion to the polished spike-like inflorescence) (subfamily Panicoideae, tribe Andropogoneae)
REFERENCES: Cutler & Anderson 1941; Larson & Doebley 1994.

Tripsacum dactyloides (L.) L., (finger-like), EASTERN GAMMA GRASS. Large, clump-forming, hard-based, rhizomatous perennial 0.5–2(–3) m tall; leaf sheaths glabrous; leaf blades pilose above; ligule a short, membranous-based ring of hairs; lower pistillate portion of inflorescence hard, rounded, cylindrical, breaking up at the nodes into bead-like units, the spikelets 2-flowered, 6–8 mm long, awnless, the upper floret perfect, the lower sterile, the glumes hardened and fused with inflorescence axis and other spikelet parts; terminal portion of inflorescence staminate, unbranched or with 2–3 branches, falling in age, with crowded, paired, unawned, 2-flowered spikelets 6–10 mm long, the glumes papery. Prairies, depressions, or low areas; in nc TX mainly Blackland Prairie w to Denton and Tarrant cos.; throughout TX, but more common in the e part. Late Apr–Jun, less commonly to Oct.

TRISETUM

A temperate genus of ca. 70 species. (Latin: *tres*, three, and *seta*, a bristle, from the awned and 2-toothed lemma) (subfamily Pooideae, tribe Aveneae)

Trisetum interruptum Buckley, (interrupted, not continuous), PRAIRIE TRISETUM. Annual 7–50(–60) cm tall; leaf sheaths pubescent; ligule an asymmetrical, ragged-margined membrane 1.5–2 mm long; panicles narrow, spike-like; spikelets 2–3-flowered, 4–6 mm long (excluding awns); lemmas with 2 slender apical teeth, awned from back at or just below base of the teeth, the awns 5–8 mm long, twisted and geniculate. Disturbed sites; throughout most of TX. Apr–May. Jones et al. (1997) treated this taxon in the genus *Sphenopholis* as *S. interrupta* (Buckley) Scribn.

TRITICUM WHEAT

A genus of 4 species ranging from the Mediterranean to Iran; it includes BREAD WHEAT, the most important temperate cereal (representing ca. 90% of total world WHEAT production). Worldwide, wheats (several species) rank first in grain production and account for more than 20% of total food calories consumed by humans. Wheats are also superior to many other grain crops because of their high (8–14%) protein content (Zohary & Hopf 1994); they are particularly important for bread-making because gluten, the characteristic protein, makes bread dough stick together and gives it the ability to retain gas, thereby making it ideal for making leavened (or risen) bread (Heiser 1990a). WHEAT is the most widely cultivated plant in the world and is considered one of the first two cultivated plants (BARLEY—*Hordeum vulgare*, is the other); as such it was probably important in the development of early civilization in the Near East (Heiser 1990a). (The classical name for wheat) (subfamily Pooideae, tribe Triticeae)
REFERENCES: Bowden 1959b; Gupta & Baum 1986, 1989; Zohary & Hopf 1994.

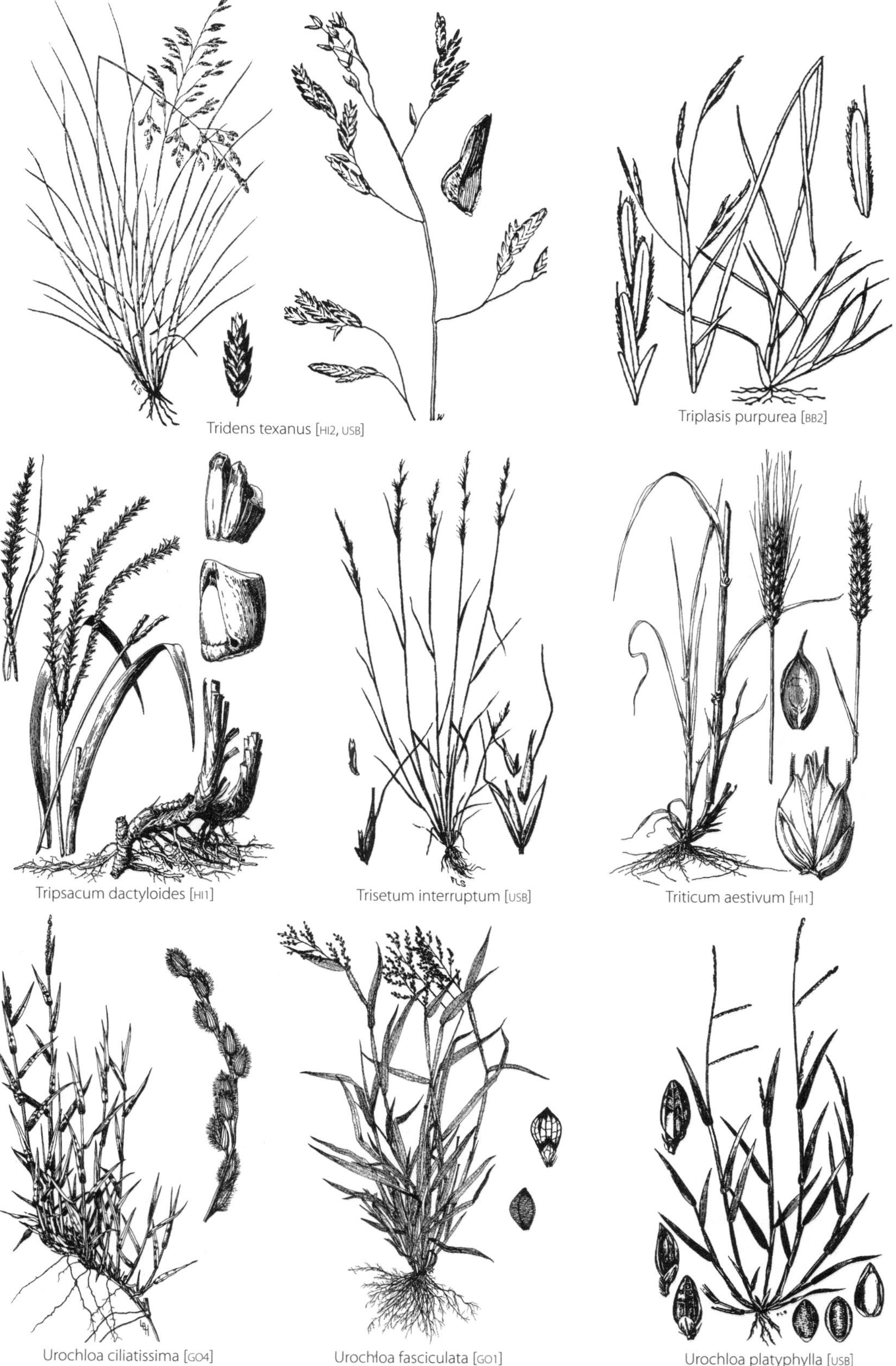

Tridens texanus [HI2, USB]

Triplasis purpurea [BB2]

Tripsacum dactyloides [HI1]

Trisetum interruptum [USB]

Triticum aestivum [HI1]

Urochloa ciliatissima [GO4]

Urochloa fasciculata [GO1]

Urochloa platyphylla [USB]

Triticum aestivum L., (summer), BREAD WHEAT, WHEAT. Glabrous annual 40–100 cm tall; leaf blades prolonged at base into 2 narrow, thin, early-withering, pointed auricles on summit of leaf sheath; ligule a membrane 2–3 mm long; spikelets sessile, solitary at each node, borne on opposite sides of the zigzag spike axis; spikes unbranched, rather stiff, bilateral, terminal, 5–12 cm long and ca. 1 cm thick; spikelets 2–5-flowered with only lower 2 or 3 perfect; glumes broadly ovate, 3.6 mm or more wide, asymmetrical, awnless or awned, 3(–more)-nerved; lemmas broad, slightly keeled, awnless or awned, the awn to 150 mm or more long. Commonly cultivated, in nc TX chiefly in Blackland Prairie and Grand Prairie; frequently seen as a transitory escape along highways, railroads, and waste places; throughout TX. Apr–May. Along with RICE and CORN, WHEAT is one of the three most important food plants for humans worldwide. It is a hexaploid (= 6 sets of chromosomes) species believed to have originated in sw Asia through hybridization between the wild diploid *Aegilops squarrosa* L. and a tetraploid cultivated species, *Triticum turgidum* L. (RIVET WHEAT) (Zohary & Hopf 1994). WHEAT is sometimes infected by the rust fungus *Puccinia graminis* Pers. (black stem rust of wheat) which can cause significant economic losses; this heterecious (= using more than 1 host to complete its life cycle) rust also infects some species in the genus *Berberis* (e.g., the introduced *B. vulgaris* L.—EUROPEAN BARBERRY); as a result, the sale or transport of certain BARBERRY species is illegal in the U.S. and Canada.

UROCHLOA SIGNAL GRASS

Annual or perennial; often rooting at lower nodes; spikelets 2-flowered, the lower sterile or staminate, the upper perfect; glumes usually both present, the first typically shorter; lower lemma resembling second glume; upper (perfect) lemma hardened, glabrous, with inrolled margins.

A genus of ca. 110 species (including most of *Brachiaria*) of tropical, subtropical, and warm areas of the world. All nc TX species have previously been treated in the genus *Brachiaria* (e.g., Kartesz 1994); however, we are following Morrone and Zuloaga (1992) and Jones et al. (1997) in treating them in *Urochloa*; *Brachiaria* (only ca. 1–3 species) is presently considered a small originally Old World genus now more widely introduced (Morrone & Zuloaga 1992). Some *Urochloa* species resemble either *Panicum* or *Paspalum* and at times have been placed in those genera. (Presumably from Greek: *uro*, tail, and *chloa*, grass) (subfamily Panicoideae, tribe Paniceae)

REFERENCES: Webster 1988; Morrone & Zuloaga 1992; Wipff et al. 1993.

1. Spikelets conspicuously arranged on one side of a flattened winged inflorescence branch ca. 2 mm wide; inflorescences of (2–)3–5(–6) widely spaced spike-like branches; plant superficially resembling a *Paspalum* ____ **U. platyphylla**
1. Spikelets not conspicuously arranged on one side of an inflorescence branch, the inflorescence branch neither flattened nor winged; inflorescence branches not widely spaced, sometimes main branches rebranched; plant superficially resembling a *Panicum*.
   2. First glume ca. 2/3–3/4 the spikelet length; spikelets 3–6 mm long; second glume densely long hairy or not so.
      3. Spikelets 5–6 mm long; second glume with scattered hairs, not densely long hairy ____ **U. texana**
      3. Spikelets 3–4.5 mm long; second glume densely long hairy ____ **U. ciliatissima**
   2. First glume 1/4–1/3 the spikelet length; spikelets 2.4–3 mm long; second glume glabrous ____ **U. fasciculata**

Urochloa ciliatissima (Buckley) R.D. Webster, (very fringed with hairs), FRINGED SIGNAL GRASS. Perennial 10–50 cm tall; culms erect or usually decumbent below and rooting at nodes; ligules short, of hairs, ca. 0.5 mm long; panicles few-flowered, with few, short, spreading or ascending branches; first glume glabrous, ca. 3/4 as long as spikelet; upper glume densely long hairy; sterile lemma glabrous, except for pilose margins. Rocky or sandy open sites; Burnet, Erath, McLennan, and Somervell cos.; throughout TX. May–Sep. [*Brachiaria ciliatissima* (Buckley) Chase]

Urochloa fasciculata (Sw.) R.D. Webster, (clustered, growing in bundles), HURRAH GRASS, BROWNTOP, BROWN-TOP SIGNAL GRASS, FIELD GRASS. Annual with spreading to erect, branching culms 10–85 cm long, often rooting at lower nodes; ligules of hairs to ca. 1 mm long; panicles with appressed or erect-spreading branches; spikelets globose-ovoid, with a small point at tip; first glume 1/4–1/3 as long as spikelet; glumes and lower lemma glabrous. Dried-up pond or stream margins, roadsides, disturbed sites; throughout most of TX. May–Oct. [*Brachiaria fasciculata* (Sw.) Parodi, *Panicum fasciculatum* Sw. var. *reticulatum* (Torr.) Beal]

Urochloa platyphylla (Munro ex C. Wright) R.D. Webster, (broad-leaved), BROAD-LEAF SIGNAL GRASS. Coarse annual with decumbent and spreading culm bases, often rooting at lower nodes; inflorescence branches 2–6 cm long, widely spaced, winged; first glume 1/4–1/3 as long as spikelet; glumes and lower lemma glabrous. Disturbed areas, roadsides, ditches; in much of the e 1/2 of TX. Apr–Nov. [*Brachiaria platyphylla* (Munro ex C. Wright) Nash, *Paspalum platyphyllum* Griseb, not Schult.]

Urochloa texana (Buckley) R.D. Webster, (of Texas), TEXAS PANICUM, TEXAS SIGNAL GRASS, TEXAS-MILLET, COLORADO GRASS. Coarse annual, 40–120 cm tall, often creeping and rooting at lower nodes; ligules of hairs to 1 mm long; panicles compact with erect-appressed branches; first glume ca. 2/3 as long as spikelet, strongly 5–7 nerved; second glume and lower lemma with scattered hairs. Moist, disturbed soils; throughout TX. May–Nov. [*Brachiaria texana* (Buckley) S.T. Blake, *Panicum texanum* Buckley]

VULPIA

SIXWEEKS GRASS, SIXWEEKS FESCUE GRASS, ANNUAL FESCUE

Ours annuals; leaf blades 0.1–2.5 mm wide; leaf sheaths open, glabrous; ligule a short membrane; inflorescence a usually contracted, ± 1-sided panicle or spicate raceme; spikelets with 3-many florets; lemmas not toothed at apex, awned or awnless, the backs rounded, not plainly nerved except at apex.

A genus of 22 species found in temperate areas, especially the Mediterranean and w America. *Vulpia* has been treated as part of the genus *Festuca* by some workers. (two possible derivations: named for J.S. Vulpius, pharmacist-botanist of Baden, Germany; or *vulpes*, fox, from the many long awns of the panicle) (subfamily Pooideae, tribe Poeae)
REFERENCES: Lonard & Gould 1974; Darbyshire & Warwick 1992.

1. Lower glume often very small, inconspicuous, 0.5–1.5 mm long, not more than 1/3 as long as upper; inflorescences often not completely exserted from sheath; lemma awns 7.5–22 mm long ____ **V. myuros**
1. Lower glume 1.6–5 mm long, half as long as the upper or more; inflorescences usually well-exserted from sheath; lemma awns 0–12 mm long.
    2. Lemmas awnless or with awn shorter than lemma body; lemma awns 0–7 mm long; spikelets with 5–17 florets ____ **V. octoflora**
    2. Lemmas (except lowest) usually with awn longer than lemma body; lemma awns 3–12 mm long; spikelets with 3–7 florets.
        3. Lemma of lowermost floret 2.5–3.5 mm long; spikelets 3.5–5 mm long (excluding awns); first glume 1.3–2.5 mm long; caryopsis (=fruit) 1.5–2 mm long ____ **V. sciurea**
        3. Lemma of lowermost floret 3.5–7.5 mm long; spikelets 5–10 mm long (excluding awns); first glume 3.5–5 mm long; caryopsis 3.5–5.5 mm long ____ **V. bromoides**

Vulpia bromoides (L.) Gray, (like *Bromus*), BROME SIXWEEKS GRASS. Plant 5–50 cm tall; panicles 5–15 cm long, well-exserted; second glume 4.5–7 mm long; lemmas glabrous or puberulent. Prairies, disturbed habitats; Hunt Co. (Clymer Meadow), also Hamilton Co. (Stanford 1971); mainly e TX. Spring. Native of Europe. [*Festuca bromoides* L., *V. dertonensis* (All.) Gola]

Vulpia myuros (L.) C.C. Gmel., (mouse-tail), RAT-TAIL SIXWEEKS GRASS. Plant 15–70 cm tall; panicles narrow, 3–25 cm long, usually only partially exserted from the sheath; spikelets with 3–7 florets; lemmas scabrous or ciliate, with awns 7.5–22 mm long. Disturbed areas, roadsides; Bell and Hunt cos.; e TX w to e part of nc TX. Apr–May. Native of Europe. [*Festuca myuros* L.]

Vulpia octoflora (Walter) Rydb., (eight-flowered), COMMON SIXWEEKS GRASS. Plant 10–60 cm tall; panicles rather dense, usually narrow, erect or with slightly drooping tip; lemmas glabrous or pubescent. Disturbed sandy or sandy clay soils, limestone gravel, eroding clay; throughout TX. Apr–May. This species is frequently seen growing on seed harvester ant mounds. Three varieties separated by Gould (1975b) as follows occur in nc TX; however, they intergrade and are separated with extreme difficulty.

1. Spikelets excluding awns usually 4.5–5 mm long; awn of lowermost floret 0.3–3 mm long ____ var. **glauca**
1. Spikelets excluding awns 5.5–10 mm long; awn of lowermost floret 2.5–6.5 mm long.
   2. Lemmas sparsely to densely pubescent ____ var. **hirtella**
   2. Lemmas glabrous or slightly scabrous ____ var. **octoflora**

var. **glauca** (Nutt.) Fernald, (whitened with a coating or bloom). Denton, Grayson, Lamar, Navarro, and Somervell cos.; e TX w to Rolling Plains and s to Edwards Plateau. [*Vulpia octoflora* var. *tenella* (Willd.) Fernald]

var. **hirtella** (Piper) Henrard, (somewhat hairy), HAIRY SIXWEEKS GRASS. McLennan and Palo Pinto cos., also Lamar Co. (Carr 1994); nearly throughout TX.

var. **octoflora**, COMMON SIXWEEKS GRASS. Se and e TX w to West Cross Timbers and Edwards Plateau. Apparently the most common variety in nc TX. [*Festuca octoflora* Walter]

Vulpia sciurea (Nutt.) Henrard, (squirrel), SQUIRREL SIXWEEKS GRASS, SIXWEEKS FESCUE. Plant 15–60 cm tall; panicles narrow, elongate, drooping toward tip; second glume 2.5–4 mm long; lemmas pubescent. Loose sandy soils; Bell, Limestone, Parker, and Wise cos.; e TX w to West Cross Timbers and Edwards Plateau. Apr–early May. [*Festuca sciurea* Nutt.]

WILLKOMMIA

A genus of 3 species of the tropics, s Africa, and s United States. (Named for Heinrich Moritz Willkomm, 1821–1895, successively professor of botany at Tharandt, Dorpat, and Prague, and student of Spanish flora) (subfamily Chloridoideae, tribe Cynodonteae)
REFERENCE: Hitchcock 1903.

Willkommia texana Hitchc., (of Texas), TEXAS WILLKOMMIA. Tufted perennial with culms 20–40 cm tall; leaves mostly basal; ligules minute, 0.6 mm or less long; inflorescence a very narrow spike-like panicle 7–18 cm long, 3–10 mm broad, with spikelets in 2 rows along the short (ca. 2–3 cm long) appressed branches; spikelets 1-flowered, sessile, awnless, 3.1–5 mm long, second glume slightly longer than lemma. Bare clay soils; restricted to vegetational areas 1 and 2 according to Gould (1975b) and 2 and 3 (Fig. 2) by Hatch et al. (1990); endemic to TX. However, the description (Hitchcock 1903) indicated that the type is from Ennis (Ellis Co.) in the ec part of nc TX. While we have not seen the type or other specimens from nc TX, the species is included based on the type locality.

Urochloa texana [GO1, USB]

Vulpia bromoides [USH]

Vulpia myuros [USB]

Vulpia octoflora var. octoflora [USB]

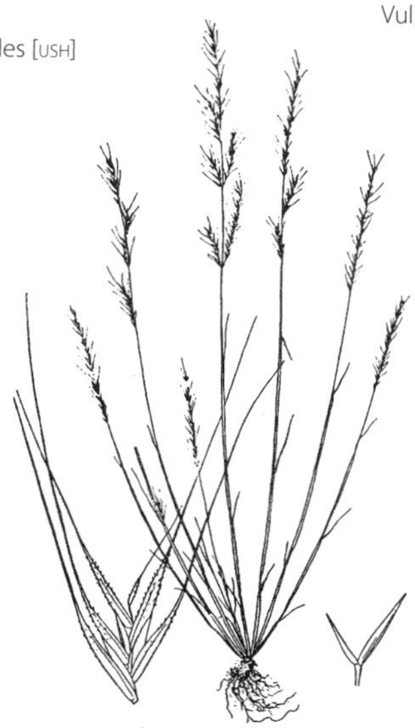

Vulpia sciurea [USB]

Willkommia texana [HI1]

ZEA MAIZE, CORN

A Central American and Mexican genus of 4 species including MAIZE, one of the most important New World crops. (Old Greek name for some grass) (subfamily Panicoideae, tribe Andropogoneae)

REFERENCES: Finan 1948; Iltis 1972, 1983, 1987; Doebley & Iltis 1980; Iltis & Doebley 1980; Crosswhite 1982; Doebley 1990, 1996; Gaut & Doebley 1997.

Zea mays L., (from an aboriginal name), MAIZE, CORN. Large, coarse, monoecious annual to several meters tall with succulent culms; leaves 2-ranked; leaf blades very broad, flat, with auricled base; pistillate spikelets numerous in several rows on a much thickened axis (= cob), the whole inflorescence covered by modified leaves or bracts (= shucks); elongated, unbranched styles (= silk) extending from the inflorescence; staminate spikelets resembling those of *Tripsacum*, in pairs on a terminal panicle (= tassle) with spike-like branches. Widely cultivated and occassional as a transitory escape along roads, disturbed sites, or waste areas; Grayson Co. (G. Diggs, pers. obs.); throughout TX. May–Jul, sporadically to Oct. MAIZE is derived from and interfertile with a wild Mexican grass, TEOSINTE—*Zea mexicana* (Schrad.) Kuntze (TEOSINTE is possibly more appropriately treated taxonomically as *Z. mays* L. subsp. *mexicana* (Schrad.) H.H. Iltis). MAIZE has long been cultivated (since at least ca. 5,600 years ago) by Native Americans from North to South America; but is thought to have originated in Mexico (Mabberley 1987; Heiser 1990a; Doebley 1990); it was very important in pre-Colombian Mesoamerica as part of the maize/beans/squash agricultural system. Along with WHEAT and RICE, MAIZE is one of the three most important food plants for humans worldwide; while world MAIZE production is nearly as great as that for either WHEAT or RICE, a much higher percentage of MAIZE is used for animal food (Chrispeels & Sadava 1977). Cultivars with hard endosperm that explodes when heated are known as popcorns; other cultivars with colored grains were used ritually by Native Americans. The common name MAIZE is derived from a Native American word, *mahiz* (Rupp 1987). The leaves of CORN sometimes have lesions caused by the rust fungus *Puccinia sorghi* Schwein. (J. Hennen, pers. comm.).

ZIZANIOPSIS CUT GRASS

A genus of 5 species of tropical and warm areas of the Americas. (Named from *Zizania*, and Greek: *opsi*, sight or appearance, from resemblance to the genus *Zizania*) (subfamily Oryzoideae, tribe Oryzeae)

Zizaniopsis miliacea (Michx.) Döll & Asch., (pertaining to millet, *Milum*), SOUTHERN WILD RICE, MARSH-MILLET, GIANT CUT GRASS, WATER-MILLET. Coarse, largely glabrous, rhizomatous perennial 1.5–2.5(–3) m tall, forming beds in wet ground or shallow water; leaf blades to 1.2 m long and 35 mm wide, with scabrous, cutting margins; ligule a prominent membrane 6–20 mm long; panicles large, 30–60 cm long, loose, with staminate and pistillate spikelets intermixed; spikelets 1-flowered, without glumes; lemma and palea similar to each other (resembling 2 glumes), strongly ribbed; staminate lemma acuminate; pistillate lemma short-awned. Marshes, creek bottoms, and lakeshores; se and e TX w to Dallas, Collin, Grayson, Hunt, Tarrant, and Williamson cos., also Edwards Plateau. May–Sep.

PONTEDERIACEAE

WATER-HYACINTH OR PICKEREL-WEED FAMILY

Glabrous perennial herbs; leaves basal or alternate, simple, entire, often with or sometimes without a distinct petiole; flowers solitary or in spikes, the inflorescence subtended by a spathe-like bract; sepals 3, colored and petal-like; petals 3, one (the uppermost) differing slightly or greatly in size, shape, or coloration; perianth parts united in lower part to form a very slender basal tube; stamens 3 or 6; pistil 1; ovary superior; fruit a many-seeded capsule or a 1-seeded urticle.

A small (32 species in 6 genera) family of freshwater aquatic herbs of tropical and warm areas, especially in the Americas, with a few in the n temperate zone. Some are problematic weeds; a number are cultivated as ornamentals. (subclass Liliidae)
FAMILY RECOGNITION IN THE FIELD: Rooted or free-floating, wet area or aquatic herbs; flowers often conspicuous, the perianth petaloid; inflorescence a solitary flower or a spike, subtended by a spathe-like bract.
REFERENCES: Dahlgren et al. 1985; Eckenwalder & Barrett 1986; Rosatti 1987.

1. Flowers numerous, in spikes; perianth funnelform; stamens 6.
   2. Perianth 4–6 cm long; petioles inflated; plants typically free-floating ____ **Eichhornia**
   2. Perianth 1–2 cm long; petioles not inflated; plants typically rooted in mud ____ **Pontederia**
1. Flowers solitary; perianth salverform; stamens 3 ____ **Heteranthera**

EICHHORNIA WATER-HYACINTH

A genus of 8 species of rhizomatous or floating aquatics; all native to the New World tropics except 1 African species—Barrett 1988). (Named for Johann Albrecht Friedrich Eichhorn, 1779–1856, of Berlin)
REFERENCE: Barrett 1988.

Eichhornia crassipes (Mart.) Solms, (with a thick stalk), COMMON WATER-HYACINTH. Normally floating on surface of water; with abundant fibrous roots; leaf blades flat, ovate to rhombic or reniform, shorter than the swollen, spongy-inflated petioles, the base broadly cuneate; inflorescence spicate or branched-spicate, well-exserted on a peduncle from a spathe; perianth slightly 2-lipped, with basal tube, the 6 segments free above, showy, bluish lavender, the upper petal with yellow spot surrounded by blue at base; fruit a many-seeded dehiscent capsule. Sometimes cultivated; established as a wild plant in lakes, ponds, and slow streams; Dallas (Mahler 1988) and Denton (G. Dick, pers. comm.) cos., also escaping cultivation in Tarrant Co. (Fort Worth Botanic Garden) (R. O'Kennon, pers. obs.); mainly s and e TX. Late Jun–Sep. Native of Brazil. This species was apparently introduced into the U.S. at the 1884 Cotton States Exposition in New Orleans (Tveten & Tveten 1993); it is an aggressive and problematic weed in some areas, such as Florida, where it can choke waterways; it has an extremely rapid growth rate and is considered by some to be the world's most serious aquatic weed. This species is considered a "harmful or potentially harmful exotic plant" and it is illegal to release, import, sell, purchase, propagate, or possess it in Texas (Harvey 1998). However, this species has the ability to remove large amounts of inorganic pollutants dissolved in water (Woodland 1997) and in some parts of the world is thus useful in wastewater cleanup efforts.

HETERANTHERA MUD-PLANTAIN

Herbs submersed or rooted in mud; leaves linear, ribbon- or grass-like, without a distinct blade or with ovate to elliptic or elliptic-lanceolate blade; flowers solitary (our species), from a spathe; perianth ± radially symmetrical; fruit a many-seeded capsule.

A genus of 12 species of tropical and warm areas of Africa and in the Americas n to North America. (Greek: *hetera*, different, and *anthera*, anther, from the dissimilar anthers of the first described species)

1. Leaves sessile, linear, grass-like, pellucid; plants completely submersed except for flowers; flowers pale yellow; spathe sessile in axils of leaves ____ **H. dubia**
1. Leaves petiolate, with an expanded, ovate to elliptic or elliptic-lanceolate, thickish blade; plants

rooted in mud, forming rosettes and emersed or the leaves floating at the water surface; flowers light blue to purplish blue to white; spathe peduncled ______ **H. limosa**

Heteranthera dubia (Jacq.) MacMill., (dubious), GRASS-LEAF MUD-PLANTAIN, WATER STAR-GRASS. Flowers usually exposed at or above the water surface; stamens all alike; anthers coiled with age. Small streams and quiet waters; included based on citation of vegetational area 4 (Fig. 2) by Hatch et al. (1990); mainly s and c TX. Apr–Jun.

Heteranthera limosa (Sw.) Willd., (of muddy places), BLUE MUD-PLANTAIN. Plant 6–25 cm tall, tufted, becoming rhizomatous; leaves long-petioled; flowering stem with a terminal sheathing spathe; upper petal with a pair of light yellow dots at base; stamens dimorphic, 2 with short yellow anthers, the third with an elongate, light blue or yellow anther; anthers not coiled. Shallow water or wet places; Grayson Co., throughout much of TX but often inconspicuous and rarely collected. Jun–Oct.

PONTEDERIA PICKEREL-WEED

A New World genus of 5 species. (Named for Guilio Pontedera, 1688–1756, professor at Padua)
REFERENCE: Lowden 1973.

Pontederia cordata L., (cordate, heart-shaped), PICKEREL-WEED, WAMPEE. Plant 40–80 cm tall, from thick, short-rhizomatous base; leaves mostly basal, petioled, the petioles with long-clasping basal portion; leaf blades shorter than the petioles, narrowly ovate to triangular-lanceolate, with rounded-truncate to deeply cordate base; inflorescence a slender terminal spike from a spathe; perianth 2-lipped, violet-blue, the upper petal with a central yellow-green spot or pair of spots; fruit a 1-seeded urticle. Margins of lakes, ponds, and streams; Dallas Co. (Mahler 1988), mainly se and e TX, also Edwards Plateau. Jun–Oct. [*P. cordata* var. *lancifolia* (Muhl. ex Elliott) Torr.] Sometimes cultivated as an ornamental; the fruit is said to be edible (Mabberley 1987).

POTAMOGETONACEAE PONDWEED FAMILY

Aquatic rhizomatous herbs; leaves in ours alternate or closely crowded, all submerged or with blades both submerged and floating, sessile or petioled; flowers perfect, in pedunculate axillary spikes; perianth 4-merous; stamens 4; carpels 4, free, sessile; ovaries superior; fruits drupe-like, 1-seeded.

A small (ca. 90 species in 3 genera—R. Haynes, pers. comm.) cosmopolitan family of perennial, rooted aquatics. The third genus in the family is the monotypic, Old World genus *Groenlandia* (with opposite leaves). Molecular analyses (Duvall et al. 1993) indicated a relationship of the Potamogetonaceae with the Alismataceae. (subclass Alismatidae)
FAMILY RECOGNITION IN THE FIELD: aquatics with leaves all submerged or often both submerged and floating—the floating ones ± elliptic and rather leathery with a *waxy, water-repellent* upper surface; flowers small, inconspicuous, 4-merous, in dense, pedunculate, axillary spikes not subtended by bracts.
REFERENCES: Morong 1893; Taylor 1909a; Haynes 1978; Dahlgren et al. 1985; Larsen & Barker 1986.

1. Floating leaves present or absent; stipules free of the leaf base or adnate for < 4(–10) mm, often early deteriorating; submerged leaves translucent, flat, without grooves or channels, of variable width, sometimes less but often much wider than 1 mm (to 45 mm wide); including species widespread in nc TX ______ **Potamogeton**
1. Floating leaves absent; stipules adnate (= fused) to the leaf base for a distance of 10–30 mm, the free portion projecting as a ligule < half as long as the adnate portion; submerged leaves opaque, turgid, channeled, 0.2–1 mm wide; rare in nc TX ______ **Stuckenia**

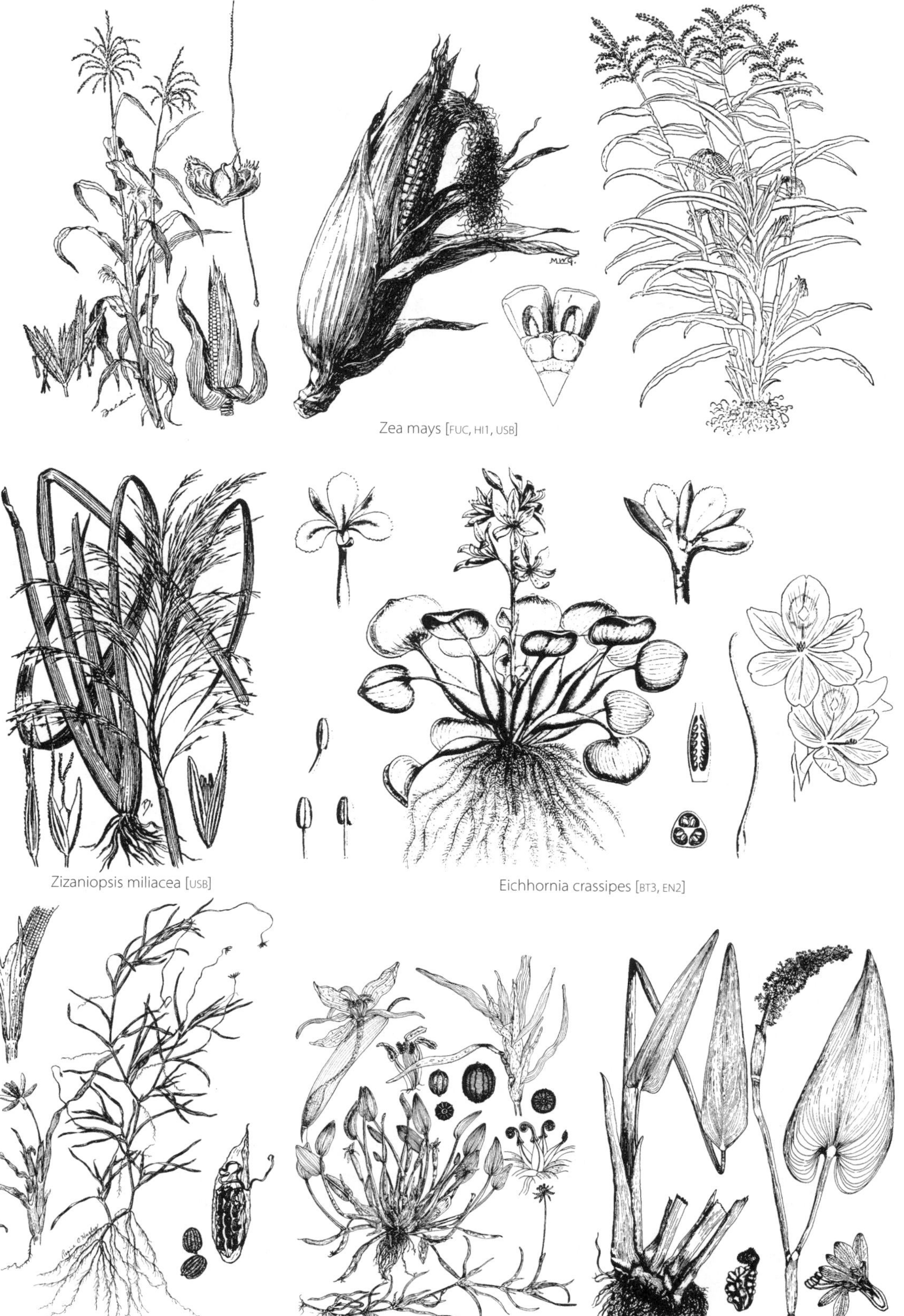

Zea mays [FUC, HI1, USB]

Zizaniopsis miliacea [USB]

Eichhornia crassipes [BT3, EN2]

Heteranthera dubia [REE]

Heteranthera limosa [CO1]

Pontederia cordata [CO1]

POTAMOGETON PONDWEED

Leaves with blades both submerged (filiform to lanceolate, thin and flexuous) and floating (± elliptic and rather leathery with a waxy upper surface) or all leaves submerged, sessile or petioled; flowers in pedunculate axillary spikes, these usually held above the water.

A cosmopolitan genus of ca. 90 species. PONDWEEDS are an important source of food for waterfowl. Because *Potamogeton* species are aquatic and difficult to collect, and because identification can be problematic, the group is neither well-collected nor well known. They are found in lakes, ponds, streams, or other aquatic habitats. Some species of *Potamogeton* exhibit hydrophily or water-mediated pollination (Cox 1988). (Ancient name from Greek: *potamos*, river, and *geiton*, a neighbor, from the aquatic habitat)
REFERENCES: St. John 1916; Fernald 1932; Ogden 1943, 1966; Haynes 1968, 1974, 1986; Reznicek & Bobbette 1976; Stuckey 1979; Catling & Dobson 1985; Haynes & Hellquist 1996.

1. Submerged leaves 2.5 mm or less wide, mostly 20 times or more longer than wide, linear; floating leaves if present with blades < 40 mm long.
   2. Small floating leaves with blades 5–40 mm long usually present; stipules adnate to the base of the submerged sessile leaf blades for mostly 1–4 mm; embryo coil plainly visible through the papery thin walls of the fruit ____ **P. diversifolius**
   2. Leaves all alike, submerged; stipules free of the submerged sessile leaf blades; fruit walls firm, the embryo coil obscured by the wall of the fruit.
      3. Fruits with an undulate to dentate, dorsal ridge or keel; stipules with evident veins appearing as ridges extending the length of the stipules ____ **P. foliosus**
      3. Fruits dorsally smooth and rounded; stipules with veins usually not very evident ____ **P. pusillus**
1. Submerged leaves either > 2.5 mm wide OR < 20 times longer than wide, linear-lanceolate to lanceolate, oblanceolate, oblong, or elliptic; floating leaves if present with blades usually > 40 mm long.
   4. Leaves all submerged, sessile, weakly to strongly clasping at base; leaf margins finely toothed, often undulate-crisped ____ **P. crispus**
   4. Leaves usually not all submerged, floating leaves commonly present by flowering time, occasionally absent; submerged leaves sessile or petiolate, not clasping the stem; leaf margins entire or nearly so, not undulate-crisped.
      5. Floating leaves with petioles 4–20 cm long, the petioles usually longer than blades; submerged leaves sessile or tapering to petioles up to 3.5 cm long (*P. pulcher*) or often much longer, (2–)3–13 cm (*P. nodosus*), acute to blunt-tipped, sometimes disintegrating by fruiting time (*P. nodosus*); widespread in nc TX.
         6. Floating leaf blades cuneate or rounded at base, with 9–21 veins; submerged leaves tapering gradually to a petiole (2–)3–13 cm long; mature fruits usually reddish; widespread in nc TX ____ **P. nodosus**
         6. Floating leaf blades usually cordate, rarely rounded at base, with 21–29 (sometimes more) veins; submerged leaves tapering rather abruptly to a sessile base or short petiole (to 3.5 cm long); mature fruits light-brown to olive-green; rare in nc TX ____ **P. pulcher**
      5. Floating leaves with petioles 2–9 cm long, the petioles usually shorter than blades; submerged leaves sessile or tapering to petioles up to 4 cm long, acute to abruptly acuminate or mucronate, usually persistent; rare in nc TX ____ **P. illinoensis**

Potamogeton crispus L., (crimped), CURLY MUCKWEED. Leaves all submerged, sessile, with finely toothed and ruffled margins, linear-oblong to linear-oblanceolate, oblong, or oblanceolate, 3–10 cm long, 3–15 mm wide. Scattered localities in TX including Dallas, Burnet, Travis, Hays, and Wichita cos., also Grayson and Randall cos. (Ogden 1966). Jun. Native of Europe.

Potamogeton diversifolius Raf., (diverse-leaved), WATER-THREAD PONDWEED. Submerged leaves sessile, linear, 1–8 cm long, 0.3–1.5 mm wide; floating leaves sometimes absent, if present, the blades 5–40 mm long, 5–20 mm wide, the petioles mostly 5–40 mm long. Denton Co., also Burnet, Henderson, Kaufman, Mills (Ogden 1966), and Lamar (Carr 1994) cos.; mainly se TX and Trans-Pecos, scattered elsewhere. May–Sep.

Potamogeton foliosus Raf., (leafy), LEAFY PONDWEED. Leaves all submerged, sessile, linear, 1.3–8.2 cm long, 0.3–2.3 mm wide. Williamson Co. (Ogden 1966); s part of nc TX s and w to Edwards Plateau and Trans-Pecos. May–Oct. [*P. foliosus* var. *macellus* Fernald]

Potamogeton illinoensis Morong, (of Illinois), SHINING PONDWEED, CORNSTALK PONDWEED, ILLINOIS PONDWEED. Submerged leaves sessile or tapering to a petiole to 4 cm long, with blades 5–20 cm long, 10–45 mm wide, elliptic to oblong-elliptic, linear-lanceolate, or lanceolate; floating leaves sometimes absent, if present, the blades 4–8(–19) cm long, 1–4(–7) cm wide. Bell, Burnet, Lampasas, and Williamson cos. in extreme s and sw parts of nc TX (Ogden 1966); mainly c TX, Edwards Plateau, and Trans-Pecos. Apr–Jun.

Potamogeton nodosus Poir., (knotty), LONG-LEAF PONDWEED. Submerged leaves often disintegrated by flowering time or sometimes persistent, tapering to a petiole, with blades 10–20(–30) cm long, 10–20(–35) mm wide, linear-lanceolate to lanceolate-elliptic; floating leaves with blades 4–10(–13) cm long, (1.5–)2–3(–4.5) cm wide. Brown, Cooke, Dallas, Denton, Fannin, Grayson, Hill, McLennan, Parker, and Tarrant cos.; nearly throughout TX. Apr–Jun. This is the most common PONDWEED in nc TX.

Potamogeton pulcher Tuck., (handsome), HEART-LEAF PONDWEED. Submerged leaves sessile or tapering to a short petiole, the blades oblong to lanceolate or linear-lanceolate, to 18 cm long and 35 mm wide (usually smaller), usually persistent, only rarely disintegrating by flowering time; floating leaves with blades 4.5–9(–11) cm long, 2–5.5(–8.5) cm wide. Included based on citation of vegetational areas 4 and 5 (Fig. 2) by Hatch et al. (1990); mainly e TX. Apr–May.

Potamogeton pusillus L., (very small), BABY PONDWEED. Leaves all submerged, sessile, linear, 0.9–6.5 cm long, 0.2–2.5 mm wide. Denton Co., also Dallas Co. (Ogden 1966); scattered mainly in e 1/2 of TX. Late May–Jun.

STUCKENIA

A ± cosmopolitan genus of 4 species previously treated as *Potamogeton* subgenus *Coleogeton* (R. Haynes, pers. comm.); Haynes has further indicated that this segregate will be recognized in the treatment of Potamogetonaceae for *Flora of North America*. (Derivation unknown, not given by original author)

REFERENCES: St. John 1916; Ogden 1966; Haynes 1968, 1986; Larson & Barker 1986; Haynes & Les 1996.

Stuckenia pectinatus (L.) Börner, (comb-like), FENNELL-LEAF PONDWEED, SAGO PONDWEED, SAGO. Rhizomatous, aquatic, perennial herb, often growing in large masses; rhizomes sometimes with tuberous bulblets; stems ca. 1 mm in diam., much branched above, 0.3–1 m long; leaves all submerged, alternate, sessile, filiform to narrowly linear, 3–12(–15) cm long, 0.2–1 mm wide, marginally entire; stipules fused with the leaf base for 10–30 mm and forming a sheath enfolding the stem (leaf thus seemingly arising from apex of sheath), the free portion of the stipules less than half as long as fused portion (i.e., adnate for 2/3 or more of their length); peduncles axillary, 3–25 cm long, flexuous, the inflorescences thus submerged; inflorescence a capitate or cylindrical, often interrupted spike with 2–5(–7) whorls of flowers, in fruit to 5 cm long; fruits 2.5–4 mm long, apiculate. Included based on citation of vegetational area 4 (Fig. 2) by Hatch et al. (1990); widespread in TX. May–Oct. [*Coleogeton pectinatus* (L.) Les & R.R. Haynes, *Potamogeton pectinatus* L.] The fruits and vegetative parts are important wildlife food (Correll & Correll 1972).

SMILACACEAE

GREENBRIER OR CATBRIER FAMILY

A small (320 species in 3 genera) family of the tropics to temperate zones; it has often been lumped into the Liliaceae or even combined with the Dioscoreaceae (Tyrl et al. 1994). (subclass Liliidae)

FAMILY RECOGNITION IN THE FIELD: woody, *prickly* (painfully so) vines with tendrils and alter nate, *net-veined*, ± leathery leaves; flowers (small, inconspicuous) and fruits (small berries) inaxillary *umbels*.

REFERENCES: Dahlgren et al. 1985.

SMILAX GREENBRIER, CATBRIER

Dioecious woody trailers or climbers from tough rhizomes or woody tubers; ± prickly; leaves alternate, short-petioled, bearing tendrils from the petioles; leaf blades several-ribbed (main veins) and net-veined, glabrous or nearly so; flowers in peduncled axillary umbels; perianth green or yellow-green to bronze, small; sepals and petals each 3; stamens 6; pistil 1; ovary superior; fruit in nc TX species a 1–3-seeded blackish or blue-black berry ca. 4–9 mm long.

A genus of ca. 300 species of tropical and temperate areas of the world. Most *Smilax* species are easily recognized in the field as woody vines armed with prickles; they frequently make moving through nc TX forests difficult or painful—hence the common names such as BULLBRIER, HELLFETTER, and DEVIL GREENBRIER. The sarsaparilla of commerce is obtained from a South American species; it was used medicinally as a tonic, for digestive disturbances, or in treating rheumatism; the active substances are steroidal saponins (Dahlgren et al. 1985). *Smilax* species are an important secondary food for white-tailed deer (Martin et al. 1951) (Ancient Greek name of an evergreen oak)

REFERENCE: Coker 1944.

1. Lower surface of leaf blades glaucous (= whitened, silvery, or bluish gray); peduncles longer (usually much longer) than petioles of subtending leaves; stems neither with numerous weak bristle-like dark prickles (only stiff prickles are present) nor leaf blades with indented sides; fruits usually covered with a bloom (= coating of white wax or powder) and bluish, sometimes blackish; only on extreme e margin of nc TX ______ **S. glauca**
1. Lower surface of leaf blades ± the same green color as upper surface; peduncles longer or shorter than petioles of subtending leaves; often either stems with numerous, weak, bristle-like prickles OR leaf blades with indented sides; fruits black, rarely covered with a bloom; widespread in nc TX.
   2. Leaf bases cordate to truncate or rounded; peduncles (= stalk of inflorescence) 1.5 or more times as long as petioles of the subtending leaves, to 70 mm long; fruits usually 1-seeded; widespread in nc TX.
      3. Stems with relatively weak, somewhat bristle-like, usually dark prickles; leaf margins not thickened; leaves drying and fading to an ashy-green color; leaf blades of flowering branches ovate or rounded in outline, the sides ± curved outward, almost never indented, the base rounded to cordate ______ **S. tamnoides**
      3. Stems with rigid, broad-based, pale or only dark-tipped prickles; leaf margins often thickened (as if with a rib (= vein) forming the edge); leaves drying and fading to a tan color; leaf blades of flowering branches triangular to reniform (= kidney-shaped), the sides often in-

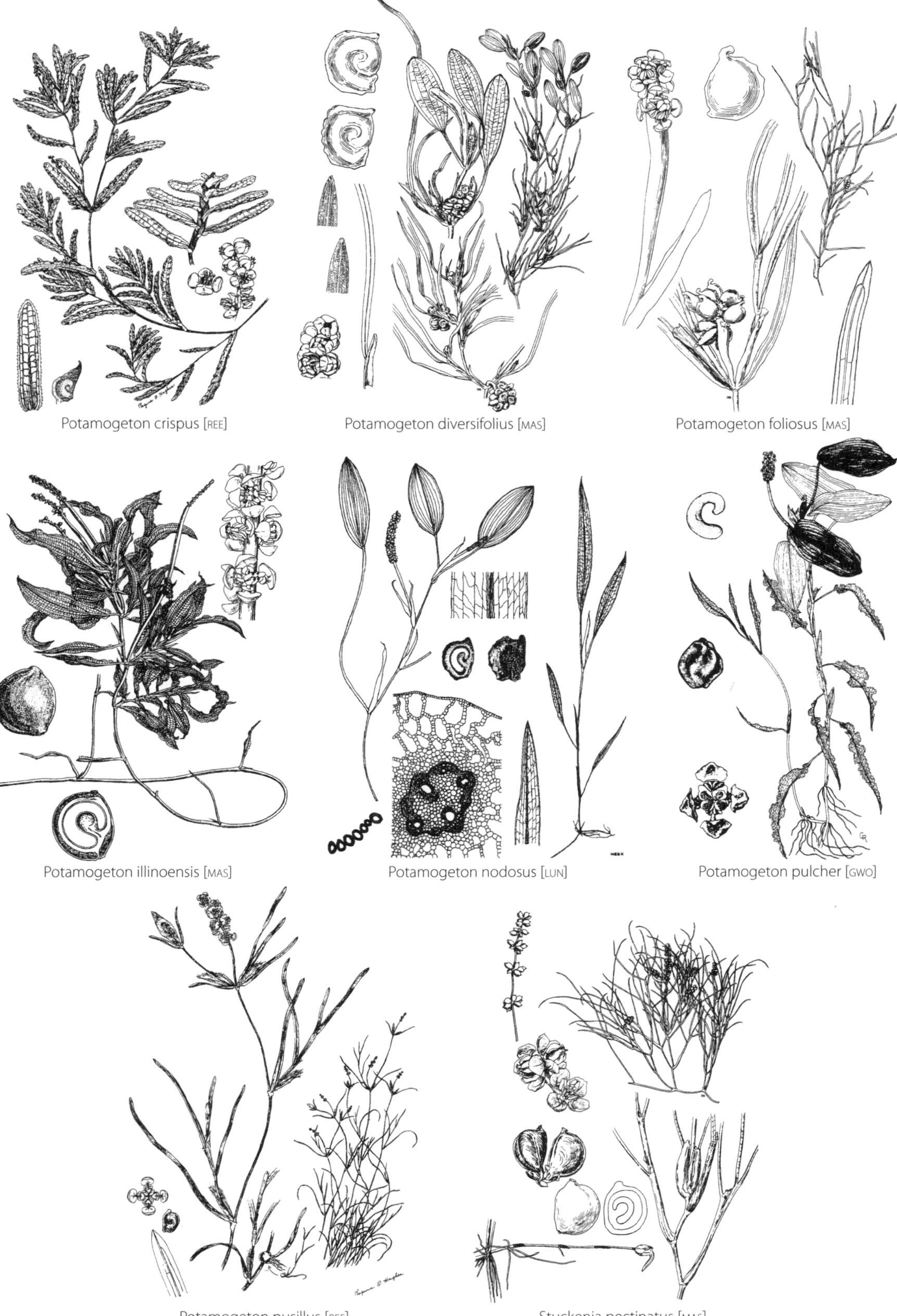

Potamogeton crispus [REE]

Potamogeton diversifolius [MAS]

Potamogeton foliosus [MAS]

Potamogeton illinoensis [MAS]

Potamogeton nodosus [LUN]

Potamogeton pulcher [GWO]

Potamogeton pusillus [REE]

Stuckenia pectinatus [MAS]

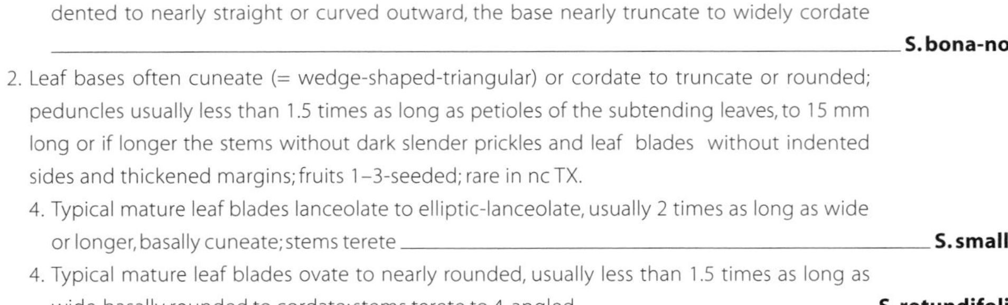

dented to nearly straight or curved outward, the base nearly truncate to widely cordate ______ **S. bona-nox**

2. Leaf bases often cuneate (= wedge-shaped-triangular) or cordate to truncate or rounded; peduncles usually less than 1.5 times as long as petioles of the subtending leaves, to 15 mm long or if longer the stems without dark slender prickles and leaf blades without indented sides and thickened margins; fruits 1–3-seeded; rare in nc TX.
   4. Typical mature leaf blades lanceolate to elliptic-lanceolate, usually 2 times as long as wide or longer, basally cuneate; stems terete ______ **S. smallii**
   4. Typical mature leaf blades ovate to nearly rounded, usually less than 1.5 times as long as wide, basally rounded to cordate; stems terete to 4-angled ______ **S. rotundifolia**

Smilax bona-nox L., (goodnight, from the Spanish: *buenos noches*, for the West Indian species recorded by Clusius), SAW GREENBRIER, FIDDLE-LEAF GREENBRIER, STRETCHBERRY, CHINA-BRIER, BULLBRIER, CATBRIER, ZARZAPARRILLA, TRAMP'S-TROUBLE, FRINGED GREENBRIER. Forming low tangles or climbing on shrubs or trees; leaf blades varying greatly in size and shape; peduncles to 30 mm long; fruits ellipsoid to subglobose, to 6 mm long and 3.5–5 mm wide. Open woods, old fields, pastures, sandy or rocky soils; widespread in the e 1/2 of TX. Apr–May. This is one of the two abundant nc TX GREEN-BRIERS.

Smilax glauca Walter, (whitened with a coating or bloom), SAWBRIER, CATBRIER, CAT GREENBRIER, WILD SARSAPARILLA, GLAUCOUS-LEAF GREENBRIER, SOWBRIER, SARSAPARILLA-VINE. Freely climbing; stems often glaucous, with scattered stiff slender prickles; leaf blades elliptic to ovate or reniform, to 13 cm long and 10 cm wide, the bases rounded to subcordate; peduncles to 30(–38) mm long, usually 1.5–3 times as long as the subtending petiole. Sandy thickets, woods, fields, and along streams; Henderson, Lamar, and Milam cos.; se and e TX w to e margin of nc TX. May–Jun. If the leaf glaucousness is lost, the terete (not 4-angled) stems and peduncles, often much longer than the subtending petioles, can help distinguish this species from *S. rotundifolia*, while the rounded to subcordate leaf bases can separate it from *S. smallii*.

Smilax rotundifolia L., (round-leaved), COMMON GREENBRIER, BULLBRIER, HORSEBRIER. High-climbing and forming thickets; stems terete or 4-angled; peduncles usually to ca. 15 mm long, usually a little shorter to a little longer than the petiole of the subtending leaf; fruits globose, 5–8 mm in diam., usually 12 or less per cluster at maturity. Thickets and woods, moist to dry areas; Bosque Co. (Carr 1989), also Fort Hood (Bell or Coryell cos.—Sanchez 1997); mainly se and e TX, also Edwards Plateau.

Smilax smallii Morong, (for its discoverer, John Kunkel Small, 1869–1938, botanist at NY Bot. Garden), SMALL'S GREENBRIER. Often high climbing; stems armed only below; leaf bases cuneate; peduncles ca. 4–10 mm long; fruits ca. 6 mm in diam. Along creeks, woodlands; Milam and Henderson cos. near e margin of nc TX, also a Reverchon collection from Dallas; mainly se and e TX. May–Jun.

Smilax tamnoides L., (resembling *Tamnus*, a genus of Dioscoreaceae), CHINAROOT, HELLFETTER, BRISTLE GREENBRIER, DEVIL GREENBRIER, HAGBRIER, WILD SARSPARILLA. Usually high-climbing; peduncles 15–65 mm long; fruits to ca. 9 mm long. Stream bottom woods, sandy or less often silty clay soils; se and e TX w to West Cross Timbers and Edwards Plateau. Apr. [*S. hispida* Muhl. ex Torr.] This is one of the two abundant nc TX GREENBRIERS.

Smilax renifolia Small, (kidney-leaved), endemic to the Edwards Plateau, is reported from vegetational area 4 (Fig. 2) by Hatch et al. (1990). However, no definitive specimens have been seen from nc TX. Field observations on the Edwards Plateau (R. O'Kennon) raise doubts about the distinctiveness of *S. renifolia* from *S. bona-nox*. Also, Coker (1944), in a treatment of the woody species of *Smilax* in the U.S., indicated that the type of *S. renifolia* is actually *S. bona-nox*. While

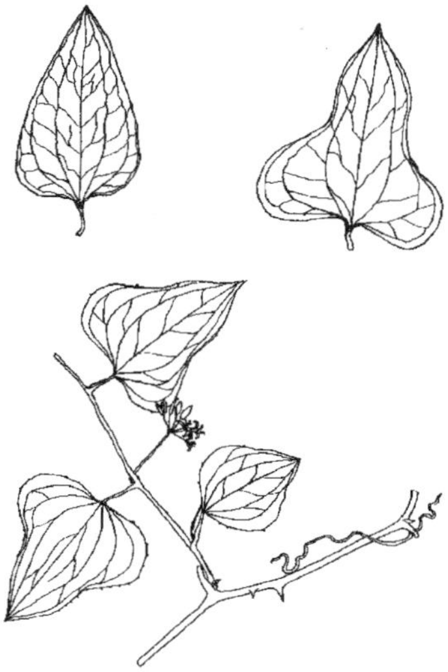

Smilax bona-nox [BT3]

Smilax glauca [G&F]

Smilax smallii [JEM]

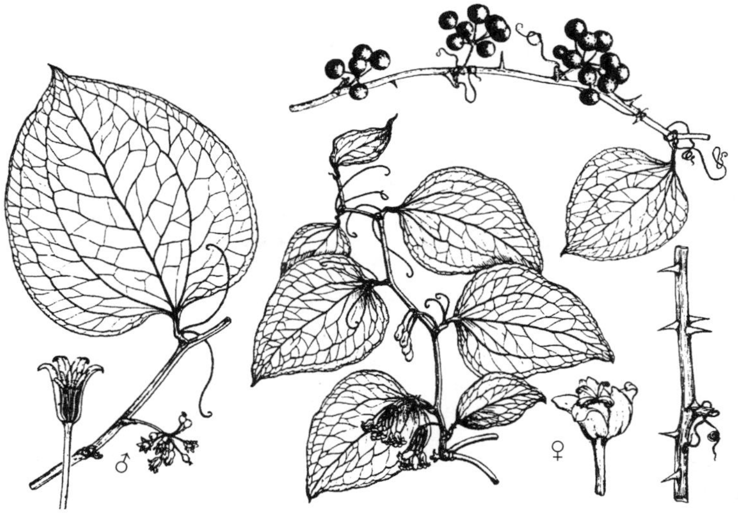

Smilax rotundifolia [JEM]

Smilax tamnoides [JEM]

distinguishing *S. renifolia* by its reniform or deltiod-reniform, mostly broader than long leaf blades (vs. typically panduriform to broadly ovate, usually longer than broad in *S. bona-nox*), Correll and Johnston (1970) likewise indicated *S. renifolia* "... should probably be treated as a geographic variant of *S. bona-nox*, its closest ally."

TYPHACEAE CAT-TAIL FAMILY

A very small (10–12 species), cosmopolitan family represented by a single genus. (subclass Commelinidae)

FAMILY RECOGNITION IN THE FIELD: large, wet area, perennial herbs with elongate, linear, spongy, 2-ranked leaves and dense, felty, brownish, cylindrical inflorescences divided into male and female portions.

REFERENCES: Wilson 1909; Dahlgren et al. 1985; Thieret & Luken 1996.

TYPHA CAT-TAIL

Coarse, wet area perennial herbs to ca. 3 m tall, forming clumps of beds from rhizomes; stems erect, simple, terete; leaves alternate, with closed, tubular sheath continuous with the grass-like blade; flowers in a dense terminal spike, without perianth, imperfect; staminate flowers at summit of spike, consisting of a single stamen (falling early); pistillate flowers below, consisting of a single persistent pistil on a pedicel bearing long hairs; ovary superior; the many crowded flowers making a felty, brownish, cylindrical mass; fruit a minute, wind-dispersed nutlet.

CAT-TAILS provide food and habitat for a variety of animals, but are often considered pests because they spread rapidly and displace other species. Some are eaten by humans and the leaves can be made into mats or other articles. (Named from *typhe*, the old Greek name)

REFERENCES: Hotchkiss & Dozier 1949; Smith 1967; Lee & Fairbrothers 1969; Lee 1975.

1. Staminate and pistillate portions of spike with a gap of 1–4 cm between them; leaves nearly flat to strongly convex (= outwardly curved) on back; stigmas thread-like, nonfleshy, deciduous ______ **T. domingensis**
1. Staminate and pistillate portions of spike touching; leaves flat on back; stigmas lance ovoid, fleshy, persistent ______ **T. latifolia**

Typha domingensis Pers., (of Santo Domingo), NARROW-LEAF CAT-TAIL, SOUTHERN CAT-TAIL, TULE. Plant 2–3 m tall; leaf blades 0.6–1.8 cm wide (fresh), 0.5–1.5 cm wide (dry), light yellowish green; leaf sheath of uppermost leaves tapered to blade, not auricled, with brownish mucilage glands on sheath only, not on inner surface of blade; bracts present on each pistillate flower; inflorescence as tall as or slightly overtopped by the leaves. Often very abundant in ditches, bogs, stock tanks, and lake margins, in shallow water or wet ground; throughout most of TX. Apr–Jul. During pioneer days, the creeping rhizomes "... were eaten, the abundant pollen was mixed with flour for the making of pancakes, and the young female inflorescences were boiled and eaten like miniature roasting ears" (Crosswhite 1980).

Typha latifolia L., (broad-leaved), COMMON CAT-TAIL, BROAD-LEAF CAT-TAIL, TULE ESPADILLA. Plant to ca. 3 m tall; leaf blades 1.0–2.3 cm wide (fresh), 0.6–1.6 cm wide (dry); bracts of pistillate flowers absent or present on a few flowers. Ditches, bogs, stock tanks, and lake margins, in shallow water or wet ground; throughout most of TX. Apr–Jun. These rhizomatous plants often spread to form large stands.

Typha angustifolia L., (narrow-leaved), is cited for e TX to the e of nc TX (Hatch et al. 1990). It supposedly differs from *T. domingensis* in being smaller (1–1.5 m tall), having dark green, narrower leaves, 0.4–1.2 cm wide (fresh), 0.3–0.8 cm (dry), having leaf sheaths auricled, with brownish mucilage glands extending above sheath onto the inner surface of the blade, and with inflorescences much overtopped by the leaves. Jones et al. (1997) lumped *T. angustifolia* with *T. domingensis*; we have been unable to consistently separate the two species with confidence. John Kartesz (pers. comm. 1997) indicated that *T. angustifolia* does not occur in TX.

XYRIDACEAE YELLOW-EYED-GRASS FAMILY

A small (260 species in 5 genera) family of mainly tropical and warm area herbs with a few in temperate regions; they usually occur in wet habitats. (subclass Commelinidae)
FAMILY RECOGNITION IN THE FIELD: grass-like or rush-like, moist or wet area herbs with basal leaves and long naked flowering stalks terminated by small, head-like or cone-like spikes with conspicuous brownish *bracts* subtending the usually yellow flowers.
REFERENCES: Kral 1983; Dahlgren et al. 1985.

XYRIS YELLOW-EYED-GRASS

Perennial, tufted or solitary, grass- or rush-like scapose herbs; leaves basal, linear to filiform; flowering stalks (= scapes) terminated by a head-like or cone-like spike of spirally imbricated brownish, woody bracts; flowers opening in morning, 1 per upper bract; lower bracts usually sterile; sepals 3, dimorphic, the 2 lateral ones keeled and persistent, the outer one covering the flower in bud and deciduous; petals 3, yellow (rarely whitish), composed of a broad blade and a long, narrow claw hidden by the subtending bract, unfolding in the morning; stamens 3; staminodia 3; ovary superior; fruit a dehiscent capsule.

A genus of ca. 240 or more species of tropical and warm areas of the world. A number of other species occur just to the e of nc TX and can be distinguished using the treatment by Kral (1966a). (Greek: *xyris*, name of some plant with 2-edged leaves, from *xyron*, a razor)
REFERENCES: Blomquist 1955; Kral 1960b, 1966a; Bridges & Orzell 1987.

1. Leaves 3–20 mm wide, linear to broadly linear, flat in cross-section; spikes 5–30 mm long with bracts 5–8 mm long; petal blades 3–10 mm long; staminodia bearded with long hairs.
   2. Spikes usually 5–15 mm long, of rather loosely imbricated bracts, the tips of the bracts not appressed; petal blades ca. 3–4 mm long; keel of lateral sepals lacerate (look under bracts) ________ **X. jupicai**
   2. Spikes 10–30 mm long, of tightly imbricated bracts, the tips of the bracts closely appressed; petal blades ca. 8–10 mm long; keel of lateral sepals ciliate-scabrid ________ **X. ambigua**
1. Leaves < 3 mm wide, filiform to linear-filiform, terete, oval, or blocky in cross-section; spikes 4–7 mm long with bracts 4.5 mm or less long; petal blades 3–4(–5) mm long; staminodia beardless ________ **X. baldwiniana**

Xyris ambigua Beyr. ex Kunth, (ambiguous). Solitary or in small tufts; leaves 10–40 cm long, 3–20 mm wide; base of inner leaves with very prominent dark longitudinal veins in sharp contrast to the white or pale intervening tissue; scapes (15–)70–100 cm long. Moist sandy areas, bogs, ditches, lake shores, savannahs; c Henderson Co. near e margin of nc TX; mainly se and e TX. May–Jul.

Xyris baldwiniana Schult., (for its discoverer, William Baldwin, 1779–1819, botanist and physician of Pennsylvania). Growing in large tufts; leaves 10–30 cm long; scapes 20–40(–50) cm long. Moist sandy areas, bogs, ditches, savannahs; c Henderson Co. near e margin of nc TX; mainly e TX. May–Jul. This is the only U.S. species with beardless staminodea.

Xyris jupicai Rich., (derivation unknown). Solitary or in small tufts; leaves 10–60 cm long, 3–10 mm wide; scapes 20–70(–90) cm long. Moist sandy areas, ditches, lakeshores; c Henderson and Limestone cos., also Lamar Co. (Carr 1994); se and e TX w to e edge of nc TX. Jun–Aug. Kral (1966a) considered this species to probably be adventive from Latin America.

ZANNICHELLIACEAE
HORNED-PONDWEED FAMILY

A very small (ca. 12 species in 4 genera) but cosmopolitan family of submerged aquatic herbs; it is one of relatively few families that exhibit hydrophily—water-mediated pollination; in *Zannichellia* pollination actually occurs underwater (hypodydrophily) in contrast to some cases of hydrophily in which pollination occurs at the water surface (epihydrophily; e.g., *Vallisneria* in the Hydrocharitaceae) (Philbrick 1991, 1993). The family was previously treated by some authorities as part of the Potamogetonaceae or in the Zosteraceae (Tyrl et al. 1994). (subclass Alismatidae)

FAMILY RECOGNITION IN THE FIELD: the single nc TX species is a submerged aquatic herb with opposite or apparently whorled, entire, almost *thread-like* leaves and small, curved, stalked fruits.

REFERENCES: Morong 1893; Taylor 1909a; Tomlinson & Posluszny 1976; Dahlgren et al. 1985; Haynes & Holm-Nielsen 1987.

ZANNICHELLIA HORNED-PONDWEED

A cosmopolitan genus of 1–5 species; pollination occurs underwater. (Named for Gian Girolamo Zannichelli, a Venetian botanist, 1662–1729)

REFERENCE: Reese 1967.

Zannichellia palustris L., (marsh-loving), COMMON POOLMAT, HORNED-PONDWEED. Monoecious, submerged, aquatic herb; stems much-branched; leaves opposite but sometimes appearing whorled, very narrow, ca. 0.5 mm wide, to 10 cm long, entire, not sheathing basally; stipules sheathing, membraneous, to 4 mm long; inflorescences axillary, usually 2-flowered (1 flower staminate and 1 pistillate, the 2 together appearing as a single flower); perianth absent; staminate flower of only a single stamen; pistillate flower of (2–)4(–8) carpels; ovaries superior; fruits pedicellate nutlets 2–4 mm long including the beak (= persistent style), oblong, curved, ridged or dentate on back. In water of lakes or streams; McLennan Co., also Denton Co. (G. Dick, pers. comm.); widely scattered in TX. Apr–Jul. The foliage and fruits are important foods for waterfowl (Kaul 1986e).

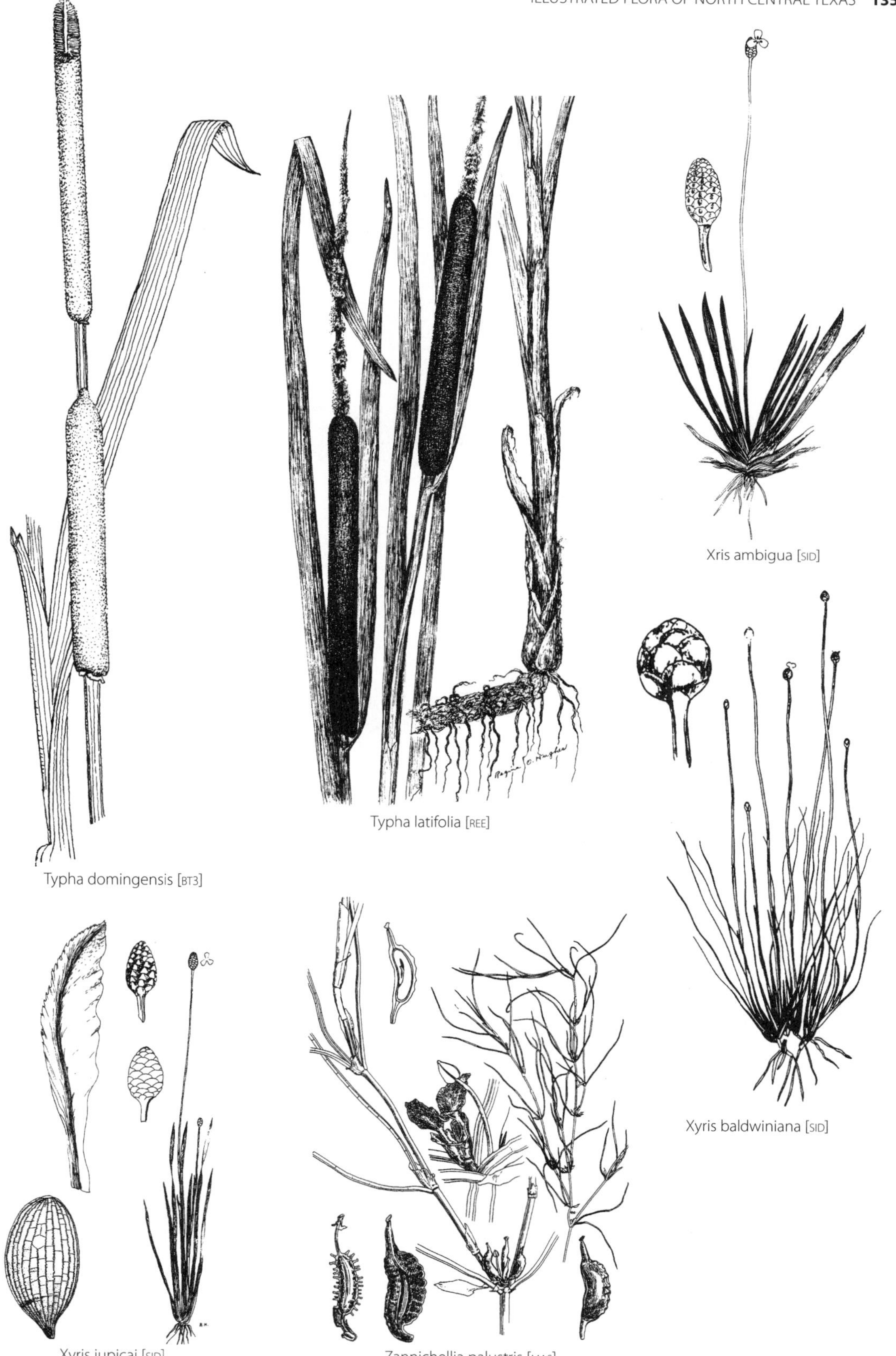

Typha domingensis [BT3]

Typha latifolia [REE]

Xris ambigua [SID]

Xyris baldwiniana [SID]

Xyris jupicai [SID]

Zannichellia palustris [MAS]

Phylogeny/Classification of the Families of Vascular Plants of North Central Texas

This phylogeny/classification is modified from those of Cronquist (1981, 1988), Lellinger (1985), and Hickman (1993). The synopses of subclasses are from Hickman (1993) and Woodland (1997). A classification system including all families of vascular plants can be found in Mabberley (1987, 1997). Figure 36 (from Cronquist 1988) is a diagram of relationships of subclasses of flowering plants.

Ferns and Similar Plants

SPORE-BEARING

Divison LYCOPODIOPHYTA

Order Lycopodiales
- Lycopodiaceae

Order Selaginellales
- Selaginellaceae

Order Isoetales
- Isoetaceae

Divison EQUISETOPHYTA

Order Equisetales
- Equisetaceae

Divison POLYPODIOPHYTA

SUBCLASS OPHIOGLOSSIDAE
- Ophioglossaceae

SUBCLASS OSMUNDIDAE
- Osmundaceae

SUBCLASS SCHIZAEIDAE
- Anemiaceae
- Pteridaceae

SUBCLASS GLEICHENIIDAE
- Polypodiaceae

SUBCLASS HYMENOPHYLLIDAE
- Dennstaedtiaceae
- Thelypteridaceae
- Aspleniaceae
- Dryopteridaceae
- Blechnaceae

SUBCLASS MARSILEIDAE
- Marsiliaceae

SUBCLASS SALVINIIDAE
- Azollaceae

Gymnosperms

"NAKED SEEDS,"
SEEDS NOT ENCLOSED IN AN OVARY;
USUALLY WITH CONES

Divison GNETOPHYTA

Order Ephedrales
- Ephedraceae

Divison PINOPHYTA
(CONIFEROPHYTA)

Order Coniferales
- Cupressaceae
- Pinaceae

Angiosperms

"VESSEL SEEDS,"
SEEDS ENCLOSED IN AN OVARY;
FLOWERS PRESENT

Divison MAGNOLIOPHYTA

Class DICOTYLEDONAE
(MAGNOLIOPSIDA)

SUBCLASS MAGNOLIIDAE
Pistils generally simple; perianth parts and stamens free, generally many, spiralled

Order Magnoliales
- Annonaceae

Order Laurales
- Lauraceae

Order Piperales
- Saururaceae

Order Aristolochiales
- Aristolochiaceae

Order Nymphaeales
- Nelumbonaceae
- Nymphaeaceae
- Cabombaceae
- Ceratophyllaceae

Order Ranunculales
- Ranunculaceae
- Berberidaceae
- Menispermaceae

Order Papaverales
- Papaveraceae

CONTINUED ➤

ANGIOSPERMS *(DICOTS CONTINUED)*

SUBCLASS HAMAMELIDAE

Mostly woody; flowers ± in unisexual catkins, without perianth, typically wind-pollinated

- Order Hamamelidales
  - Platanaceae
  - Hamamelidaceae
- Order Urticales
  - Ulmaceae
  - Cannabaceae
  - Moraceae
  - Urticaceae
- Order Juglandales
  - Juglandaceae
- Order Myricales
  - Myricaceae
- Order Fagales
  - Fagaceae
  - Betulaceae

SUBCLASS CARYOPHYLLIDAE

Mostly herbaceous; petals free (or absent and sepals petal-like, sometimes fused); placentas basal or free-central; stamens developing from inner to outer—centrifugal

- Order Caryophyllales
  - Phytolaccaceae
  - Nyctaginaceae
  - Aizoaceae
  - Cactaceae
  - Chenopodiaceae
  - Amaranthaceae
  - Portulacaceae
  - Basellaceae
  - Molluginaceae
  - Caryophyllaceae
- Order Polygonales
  - Polygonaceae

SUBCLASS DILLENIIDAE

Petals ± free, sometimes fused (if 0, sepals not petal-like); pistil usually compound; placentas generally parietal (or axile); stamens developing from inner to outer; leaves mostly simple

- Order Theales
  - Elatinaceae
  - Clusiaceae (Guttiferae)
- Order Malvales
  - Tiliaceae
  - Sterculiaceae
  - Malvaceae
- Order Nepenthales
  - Sarraceniaceae
  - Droseraceae
- Order Violales
  - Cistaceae
  - Violaceae
  - Tamaricaceae
  - Passifloraceae
  - Cucurbitaceae
  - Loasaceae
- Order Salicales
  - Salicaceae
- Order Capparales
  - Capparaceae
  - Brassicaceae (Cruciferae)
- Order Ericales
  - Ericaceae
- Order Ebenales
  - Sapotaceae
  - Ebenaceae
  - Styracaceae
- Order Primulales
  - Primulaceae

SUBCLASS ROSIDAE

Petals usually free; stamens more than petals or opposite them, developing from outer to inner—centripetal; pistil compound or sometimes simple; placentas most often axile; leaves compound or simple

- Order Rosales
  - Hydrangeaceae
  - Grossulariaceae
  - Crassulaceae
  - Saxifragaceae
  - Rosaceae
- Order Fabales
  - Fabaceae
- Order Proteales
  - Elaeagnaceae
- Order Haloragales
  - Haloragaceae
- Order Myrtales
  - Lythraceae
  - Onagraceae
  - Melastomataceae
- Order Cornales
  - Cornaceae
  - Nyssaceae
  - Garryaceae
- Order Santalales
  - Santalaceae
  - Viscaceae
- Order Rafflesiales
  - Rafflesiaceae
- Order Celastrales
  - Celastraceae
  - Aquifoliaceae
- Order Euphorbiales
  - Euphorbiaceae
- Order Rhamnales
  - Rhamnaceae
  - Vitaceae
- Order Linales
  - Linaceae
- Order Polygalales
  - Polygalaceae
  - Krameriaceae
- Order Sapindales
  - Sapindaceae
  - Hippocastanaceae
  - Aceraceae
  - Anacardiaceae
  - Simaroubaceae
  - Meliaceae
  - Rutaceae
  - Zygophyllaceae
- Order Geraniales
  - Oxalidaceae
  - Geraniaceae
  - Balsaminaceae
- Order Apiales
  - Araliaceae
  - Apiaceae (Umbelliferae)

SUBCLASS ASTERIDAE

Predominantly herbaceous; petals ± fused; stamens equal in number to or fewer than petals and alternate them; pistil compound, generally of 2 carpels

- Order Gentianales
  - Loganiaceae
  - Gentianaceae
  - Apocynaceae
  - Asclepiadaceae
- Order Solanales
  - Solanaceae
  - Convolvulaceae
  - Cuscutaceae

ANGIOSPERMS *(MONOCOTS)*

Menyanthaceae
Polemoniaceae
Hydrophyllaceae
Order Lamiales
Boraginaceae
Lamiaceae (Labiatae)
Verbenaceae
Phrymaceae
Order Callitrichales
Callitrichaceae
Order Plantaginales
Plantaginaceae
Order Scrophulariales
Buddlejaceae
Oleaceae
Scrophulariaceae
Orobanchaceae
Acanthaceae
Pedaliaceae
Bignoniaceae
Lentibulariaceae
Order Campanulales
Sphenocleaceae
Campanulaceae
Order Rubiales
Rubiaceae
Order Dipsacales
Caprifoliaceae
Valerianaceae
Dipsacaceae
Order Asterales
Asteraceae (Compositae)

Class MONOCOTYLEDONAE (LILIOPSIDA)

SUBCLASS ALISMATIDAE

Pistils simple; herbs, ± aquatic

Order Alismatales
Alismataceae
Order Hydrocharitales
Hydrocharitaceae
Order Najadales
Potamogetonaceae
Zannichelliaceae

SUBCLASS ARECIDAE

Inflorescence often of many small flowers, enfolded or subtended by prominent bract(s); pistil usually compound; palm-like to minute aquatics

Order Arecales
Arecaceae (Palmae)
Order Arales
Acoraceae
Araceae
Lemnaceae

SUBCLASS COMMELINIDAE

Flowers small and subtended by chaffy bracts, or sepals and petals unlike; generally wind-pollinated; pistil compound)

Order Commelinales
Xyridaceae
Commelinaceae
Order Eriocaulales
Eriocaulaceae
Order Juncales
Juncaceae
Order Cyperales
Cyperaceae
Poaceae (Gramineae)
Order Typhales
Typhaceae

SUBCLASS ZINGIBERIDAE

Inflorescences often with showy colored bracts; sepals and petals unlike; flowers usually animal-pollinated; pistil compound

Order Bromeliales
Bromeliaceae
Order Zingiberales
Marantaceae

SUBCLASS LILIIDAE

Flowers ± showy, insect-pollinated; sepals and petals generally similar; pistil compound

Order Liliales
Pontederiaceae
Lilaceae
Iridaceae
Agavaceae
Smilacaceae
Dioscoreaceae
Order Orchidales
Burmanniaceae
Orchidaceae

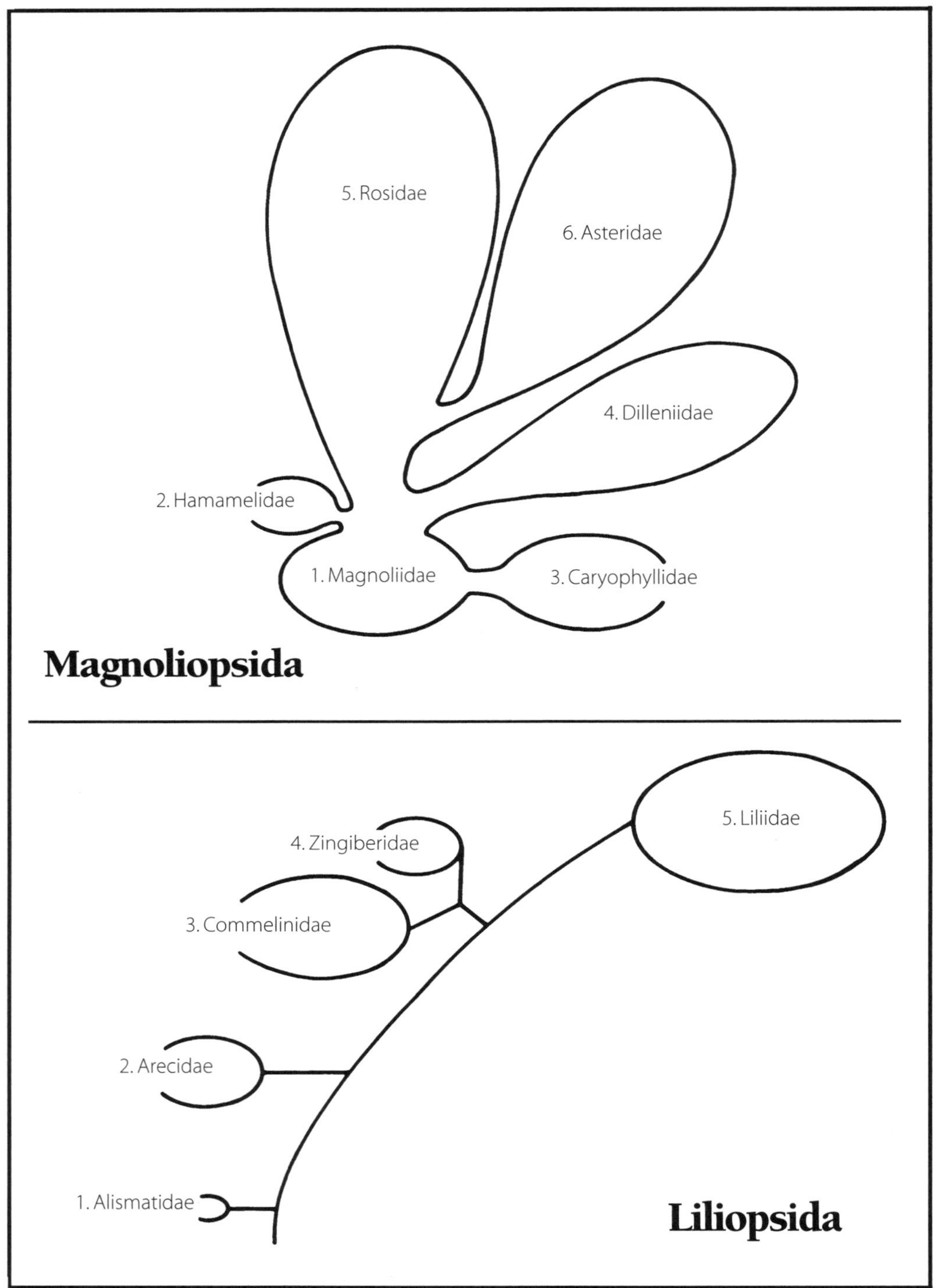

FIG. 36/DIAGRAM OF RELATIONSHIPS OF SUBCLASSES OF FLOWERING PLANTS (FROM CRONQUIST 1988).

Subfamilial and Tribal Phylogeny/Classification of North Central Texas Poaceae (Grass Family)

as recognized by Peterson and Webster (forthcoming)

Subfamily **ARISTIDOIDEAE**

Aristideae

Aristida

Subfamily **BAMBUSOIDEAE**

Bambuseae

Arundinaria
Phyllostachys

Subfamily **CHLORIDOIDEAE**

Cynodonteae

Bouteloua
Buchloe
Chloris
Cynodon
Eustachys
Gymnopogon
Hilaria
Schedonnardus
Spartina
Willkommia

Danthonieae

Cortaderia
Danthonia

Eragrostideae

Dactyloctenium
Distichlis
Eleusine
Eragrostis
Erioneuron
Leptochloa
Muhlenbergia
Sporobolus
Tridens
Triplasis

Subfamily **ORYZOIDEAE**

Oryzeae

Leersia
Zizaniopsis

Subfamily **PANICOIDEAE**

Andropogoneae

Andropogon
Bothriochloa
Coelorachis
Dichanthium
Miscanthus
Saccharum
Schizachyrium
Sorghastrum
Sorghum
Stenotaphrum
Tripsacum
Zea

Arundineae

Arundo
Phragmites

Centotheceae

Chasmanthium

Paniceae

Axonopus
Cenchrus
Digitaria
Echinochloa
Eriochloa
Oplismenus
Panicum
Paspalidium
Paspalum
Pennisetum
Sacciolepis
Setaria
Urochloa

Subfamily **POOIDEAE**

Aveneae

Agrostis
Avena
Briza
Koeleria
Limnodea
Phalaris
Polypogon
Sphenopholis
Trisetum

Brachypodieae

Brachypodium

Bromeae

Bromus

Diarrheneae

Diarrhena

Meliceae

Glyceria
Melica

Poeae

Aira
Alopecurus
Dactylis
Desmazeria
Festuca
Lolium
Phleum
Poa
Sclerochloa
Vulpia

Stipeae

Nassella

Triticeae

Aegilops
Elymus
Hordeum
Pascopyrum
Secale
Triticum

List of Texas Endemics Occurring in North Central Texas

Contributed by Bonnie Amos, Paula Hall, and Kelly McCoy
(Amos et al. 1998) Angelo State University

Dicots

Acanthaceae

Ruellia drummondiana (Nees) A. Gray

Apiaceae

Daucosma laciniatum Engelm. & A. Gray

Asclepiadaceae

Asclepias linearis Scheele
Matelea edwardsensis Correll

Asteraceae

Chaptalia texana Greene
Chrysopsis texana G.L. Nesom
Cirsium engelmannii Rydb.
Palafoxia hookeriana Torr. & A. Gray
Palafoxia reverchonii (Bush) Cory
Pectis angustifolia Torr. var. *fastigiata* (A. Gray) D.J. Keil
Senecio ampullaceus Hook.
Silphium albiflorum A. Gray
Verbesina lindheimeri B.L. Rob. & Greenm.

Boraginaceae

Cryptantha texana (A. DC.) Greene
Onosmodium helleri Small

Brassicaceae

Arabis petiolaris (A. Gray) A. Gray
Lesquerella densiflora (A. Gray) S. Watson
Lesquerella grandiflora (Hook.) S. Watson
Lesquerella recurvata (Engelm. ex A. Gray) S. Watson

Campanulaceae

Triodanis coloradoensis (Buckley) McVaugh
Triodanis texana McVaugh

Celastraceae

Evonymus atropurpurea Jacq. var. *cheatumii* Lundell

Cistaceae

Lechea san-sabeana (Buckley) Hodgdon

Convolvulaceae

Dichondra recurvata Tharp & M.C. Johnst.

Euphorbiaceae

Chamaesyce angusta (Engelm.) Small
Croton alabamensis E.A. Sm. ex Chapm. var *texensis* Ginzbarg
Ditaxis aphoroides (Müll.Arg.) Pax
Euphorbia roemeriana Scheele

Fabaceae

Astragalus crassicarpus Nutt. var. *berlandieri* Barneby
Astragalus nuttallianus DC. var. *pleianthus* (Shinners) Barneby
Astragalus reflexus Torr. & A. Gray
Astragalus wrightii A. Gray
Dalea hallii A. Gray
Dalea reverchonii (S. Watson) Shinners
Dalea tenuis (J.M. Coult.) Shinners
Desmanthus acuminatus Benth.
Galactia canescens (Scheele) Benth.
Galactia heterophylla A. Gray
Lupinus texensis Hook.
Pediomelum cyphocalyx (A. Gray) Rydb.
Pediomelum hypogaeum (Nutt. ex Torr. & A. Gray) Rydb. var. *scaposum* (A. Gray) Mahler
Pediomelum latestipulatum (Shinners) Mahler var. *appressum* (Ockendon) Ghandi & L.E. Br.
Pediomelum latestipulatum (Shinners) Mahler var. *latestipulatum*
Tephrosia lindheimeri A. Gray
Trifolium bejariense Moric.

Fumariaceae

Corydalis curvisiliqua Engelm. subsp. *curvisiliqua*

Garryaceae

Garrya ovata Benth. subsp. *lindheimeri* (Torr.) Dahling

Hydrophyllaceae

Phacelia strictiflora (Engelm. & A. Gray) A. Gray var. *strictiflora*

Lamiaceae

Brazoria truncata (Benth.) Engelm. & A. Gray var. *truncata*
Monarda punctata L. var. *intermedia* (E.M. McClint. & Epling) Waterf.

Physostegia pulchella Lundell
Salvia engelmannii A. Gray
Teucrium cubense Jacq. var. *laevigatum* (Vahl) Shinners

Malvaceae
Malvastrum aurantiacum (Scheele) Walp.

Nyctaginaceae
Abronia ameliae Lundell
Mirabilis gigantea (Standl.) Shinners

Onagraceae
Oenothera coryi W.L. Wagner

Oxalidaceae
Oxalis drummondii A. Gray

Papaveraceae
Argemone aurantiaca G.B. Ownbey

Polemoniaceae
Phlox drummondii Hook. subsp. *wilcoxiana* (Bogusch) Wherry
Phlox pilosa L. subsp. *latisepala* Wherry
Phlox pilosa L. subsp. *riparia* Wherry
Phlox roemeriana Scheele

Ranunculaceae
Clematis texensis Buckley

Rosaceae
Prunus serotina Ehrend. var. *eximia* (Small) Little
Rubus apogaeus L.H. Bailey

Rubiaceae
Houstonia parviflora Holz ex Greenm.
Houstonia subviscosa (C. Wright ex A. Gray) A. Gray

Scrophulariaceae
Agalinis edwardsiana Pennell
Castilleja purpurea (Nutt.) G. Don var. *lindheimeri* (A. Gray) Shinners
Penstemon guadalupensis A. Heller
Penstemon triflorus A. Heller subsp. *integrifolius* Pennell

Solanaceae
Bouchetia erecta DC.

Styracaceae
Styrax platanifolius Engelm. ex Torr.

Valerianaceae
Valerianella stenocarpa (Engelm. ex A. Gray) Krok

Vitaceae
Parthenocissus heptaphylla (Buckley) Britton ex Small
Vitis monticola Buckley

MONOCOTS

Agavaceae
Nolina lindheimeriana (Scheele) S. Watson
Yucca constricta Buckley
✤ *Yucca necopina* Shinners
✤ *Yucca pallida* McKelvey
Yucca rupicola Scheele

Commelinaceae
Tinantia anomala (Torr.) C.B. Clarke
Tradescantia edwardsiana Tharp
Tradescantia gigantea Rose
Tradescantia humilis Rose
Tradescantia subacaulis Bush

Cyperaceae
Carex edwardsiana E.L. Bridges & Orzell

Juncaceae
Juncus texanus (Engelm.) Coville
Juncus validus Coville var. *fascinatus* M.C. Johnst.

Liliaceae
Allium runyonii Ownbey

Poaceae
Muhlenbergia ×*involuta* Swallen [*M. lindheimeri* × *M. reverchonii*]
Tridens congestus (L.H. Dewey) Nash
Willkommia texana Hitchc.

✤ ***Indicates taxa endemic to North Central Texas.***

Illustration Sources

Ref. Code

AAA Proc. Amer. Acad. Arts. 1846–1958. American Academy of Arts and Sciences. Boston, MA.
Reprinted with permission of the American Academy of Arts and Science.

ABR Abrams, L. 1923–1960. Illustrated flora of the Pacific states. Stanford University Press. Stanford, CA. Reprinted from *Illustrated Flora of the Pacific States*, four volumes by Leroy Abrams and Roxana Stinchfield Ferris with permission of the publishers, Stanford University Press. © 1951, 1960 by the Board of Trustees of the Leland Stanford Junior University.

ADD Addisonia. 1916–1964. New York Botanical Garden. Bronx, New York.
Reprinted with permission *Addisonia*, vol. 18, plate 579, Copyright © 1933 The New York Botanical Garden.

AJB Amer. J. Bot. 1914+. Botanical Society of America. Lancaster, PA.
Reprinted with permission of the Botanical Society of America.

ALL Allioni, C. 1785. Flora pedemontana. Joannes Michael Briolus. Torino.

AMB Ann. Missouri Bot. Gard. 1914+. Missouri Botanical Garden. St. Louis.
Reprinted with permission of the Missouri Botanical Garden.

AND Anderson, E. and Woodson, R.E. 1935. The species of *Tradescantia* indigenous to the United States. Arnold Arboretum of Harvard University. Cambridge, MA.
Reprinted with permission of the Arnold Arboretum of Harvard University.

ANO Anonymous. 1821. Medical botany. E. Cox and Sons. London.

APG Apgar, A.C. 1910. Ornamental shrubs of the United States. American Book Company. New York.

ARM Annual Rep. Missouri Bot. Gard. 1889–1912. Missouri Botanical Garden. St. Louis.
Reprinted with permission of the Missouri Botanical Garden.

BA1 Bailey, L.H. 1914–1917. The standard cyclopedia of horticulture. The Macmillan Company. New York.

BAR Bartonia. 1908+. Academy of Natural Sciences. Philadelphia, PA.
Reprinted with permission of the Academy of Natural Sciences.

BAS Bassett, I.J. 1973. The plantains of Canada (monograph no.7). Agriculture Canada. Ottawa.
Reproduced from Agriculture & Agri-Food Canada publications.
Reproduced with the permission of the Minister of Public Works and Government Services Canada 1997.

BAY Baileya. 1953+. L.H. Bailey Hortorium. Ithaca, NY.
Reprinted with permission of the L.H. Bailey Hortorium.

BB1 Britton, N.L. and Brown, A. 1896–1898. An illustrated flora of the northern United States, Canada and the British possessions. Charles Scribner's Sons. New York.

BB2 Britton, N.L. and Brown, A. 1913. An illustrated flora of the northern United States, Canada and the British possessions. Charles Scribner's Sons. New York.

BBS Biltmore Bot. Stud. 1901–1902. Biltmore Herbarium. Biltmore, NC.

BCM Basset, I.J., Crompton, C.W., McNeill, J., Taschereau, P.M. 1983. The genus *Atriplex* (Chenopodiaceae) in Canada (monograph no. 31). Agriculture Canada. Ottawa.
Reproduced from Agriculture & Agri-Food Canada publications.
Reproduced with the permission of the Minister of Public Works and Government Services Canada 1997.

BEA Beal, E.O. and Thieret, J.W. 1986. Aquatic and wetland plants of Kentucky. Kentucky Nature Preserves Commission. Frankfort. Reprinted with permission of Kentucky State Nature Preserves Commission.

BEL Benson, L. 1982. The cacti of the United States and Canada. Stanford University Press. Stanford, CA. Reprinted from *The Cacti of the United States and Canada* by Lyman Benson with permission of the publishers, Stanford University Press © 1982 by the Board of Trustees of the Leland Stanford Junior University.

BEN Bentham, G. 1865 (2nd ed.). Handbook of the British flora. Lovell Reeve and Company. London.

BFA Brotero, F.A. 1816. Phytographia lusitaniae selectior. Typographia Regia. Lisbon.

BL1 Baillon, H. 1871–1888. The natural history of plants. Lovell Reeve and Company. London.

BL2 Baillon, M.H. 1876–1892. Dictionnaire de botanique. Librairie Hachette. Paris.

BL3 Baillon, H. 1866–1895. Histoire des plantes. Librairie Hachette. Paris.

BML Bot. Mus. Leafl. 1932–1986. Botanical Museum, Harvard University. Cambridge, MA.
Reprinted with permission of Botanical Museum, Harvard University.

BOI Boissier, P.E. 1866. Icones Euphorbiarum. Victor Masson et fils. Paris.

BOT Bot. Gaz. (Crawfordsville). 1875+. University of Chicago Press. Chicago, IL.
Reprinted from *Botanical Gazette*, vol. 91, page 104, figs. 1–4, ed. H.C. Cowles, Copyright © 1931 , vol. 18, plate 40, A.F. Foerste, Copyright © 1893, with permission from University of Chicago Press.

BR1 Britton, N.L. 1908. North American trees. Henry Holt and Company. New York.

BR2 Britton, N.L. 1918. Flora of Bermuda. Charles Scribner's Sons. New York.

BR3 Britton, N.L. and Rose, J.N. 1919–1923. The Cactaceae. Carnegie Institute. Washington, DC.

BT2 Lipscomb, B.L. 1999. Previously unpublished original illustrations from the SMU herbarium. Published herein by The Botanical Research Institute of Texas. Fort Worth.

BT3 Shinners, L.H., Whitehouse, E., and P. Mueller. 1999.
Previously unpublished original illustrations from the SMU herbarium.
Published herein by The Botanical Research Institute of Texas. Fort Worth.

BTT Brittonia. 1931/35+. New York Botanical Garden. New York.
Reprinted with permission from *Brittonia*, vol. 21, no. 1, page 78, fig. 1, Copyright ©1969, *Brittonia*, vol. 44, no. 2, page 180, figs. 7, 10, 11, 13, 14, Copyright © 1992, *Brittonia*, vol. 48, no. 1, page 108, fig. 2, Copyright ©1996, The New York Botanical Garden.

BUD Budd, A.C. 1957. Wild plants of the Canadian prairie (publication 983). Agriculture Canada. Ottawa.
Reproduced from Agriculture & Agri-Food Canada publications.
Reproduced with the permission of the Minister of Public Works and Government Services Canada 1997.

CGH Contr. Gray Herb. 1891–1984. Harvard University. Cambridge, MA.
Reproduced with permission of the Library of the Gray Herbarium.

CHA Chaudhri, M.N. 1968. Mededelingen. Botanisch Museum en Herbarium van de Rijksuniversiteit te Utrecht No. 285. Utrecht.
Reprinted with permission of Botanisch Museum en Herbarium.

CHT Chittenden, F.J. 1956. The Royal Horticultural Society dictionary of gardening. Royal Horticultural Society. Oxford.
Reprinted with permission of the Royal Horticultural Society.

CO1 Correll, D.S. and Correll, H.B. 1972. Aquatic and wetland plants of southwestern United States. Environmental Protection Agency. Washington, DC.

CO2 Correll, D.S. 1950. Native orchids of North America north of Mexico. Chronica Botanica Company. Waltham, MA.
Unable to locate copyright owner.

COC Cocks, R.S. 1910. Leguminosae of Louisiana (Bull. no. 1). Louisiana State Museum. New Orleans.

COO Cooper, C.S. and Westell, W.P. 1909. Trees and shrubs of the British Isles native and acclimatised. J.M. Dent and Company. London.

COR Cornut, J.P. 1635. Canadensium plantarum, aliarumque nondum editarum historia. Simon le Moyne. Paris.

CUM Contr. Univ. Michigan Herb. 1939+. University Herbarium. Ann Arbor, MI.
Reprinted with permission of the University of Michigan Herbarium.

CUR Curtis's Botanical Magazine. 1787+. Royal Botanic Gardens, Kew, England.
Reprinted with permission of the Royal Botanic Gardens, Kew.

DAR Darlington, H.T. and Bessey, E.A. 1940. Some important Michigan weeds. Michigan State College Agricultural Experiment Station. East Lansing.

DEL Delessert, B. 1820–1846. Icones selectae plantarum. Fortin, Masson and Company. Paris.

DIL Dillenio, J.J. 1732. Hortus elthamensis seu plantarum rariorum. Sumptibus Auctoris. London.

DOR Dormon, C. 1942. Wild flowers of Louisiana. Department of Conservation. New Orleans.

DUN Dunal, M.F. 1813. Histoire naturelle, medicale et economique des solanum. Strasbourg. London.

EGG Eggers Ware, D.M. A revision of *Valerianella* in North America. Ph.D dissertation.
Reprinted with permission from Donna Eggers Ware.

EMO Emory, W.H. 1857–1859. United States and Mexican boundary survey. United States Government Printing Office. Washington, DC.

EN1 Engler, A. 1900–1953. Das pflanzenreich. Duncker and Humblot. Berlin.
Reproduced with permission of Duncker and Humblot GmbH. Verlagsbuchhandlung. Berlin.

EN2 Engler, A. and Prantl, K. 1887–1915. Die naturlichen pflanzenfamilien. Gebrueder Borntraeger. Stuttgart.
Originally published in "Die Naturlichen Pflanzenfamilien", reprint 1958/60 by J. Cramer in der Gebruder Borntraeger Verlagsbuchhandlung, D–14129 Berlin D–70176 Stuttgart.

F&L	Field & Lab. 1932–1970. Southern Methodist University Press. Dallas, TX.
FAW	Fawcett, W. 1910–1936. Flora of Jamaica. British Museum (Natural History). London.
FLN	Fl. Neotrop. Monogr. 1967+. New York Botanical Garden. New York. Reprinted with permission from *The Alismataceae. Flora Neotropica*, vol. 64, page, 43, fig. 22D, Copyright ©1994, Robert Haynes, and The New York Botanical Garden.
FMC	Heller, A.A. 1895. Botanical explorations in southern Texas during the season of 1894. The New Era Printing House. Lancaster, PA.
FUC	Fuchs, L. 1542. De historia stirpium commentarii insignes. Isingrin. Basel.
G&F	Gard. & Forest. 1888–1897. The Garden and Forest Publishing Co. New York.
GAN	Gandhi, K.N. and Thomas, R.D. 1989. Asteraceae of Louisiana. Sida, Bot. Misc. 4.
GAR	Gartenflora. 1852–1938. Verlag von Ferdinand Enke. Stuttgart, Germany.
GAT	Gates, F.C. 1941. Weeds in Kansas. Kansas State Board of Agriculture. Topeka. Reprinted with permission of the Kansas State Board of Agriculture.
GBN	Great Basin Naturalist. 1939/40+. Brigham Young University. Provo, UT. Reprinted with permission of Brigham Young University.
GEN	Gentes Herbarium. 1920+. L.H. Bailey Hortorium. Ithaca, NY. Reprinted with permission of the L.H. Bailey Hortorium.
GEO	Georgia, A.E. 1916. A manual of Weeds. The Macmillan Company. New York.
GHS	Gentry, H.S. 1972. The Agave family in Sonora (Agricultural Handbook no. 399). United States Department of Agriculture. Washington, DC.
GLE	Gleason, H.A. 1952. The new Britton and Brown illustrated flora of the northeastern United States and adjacent Canada. New York Botanical Garden. New York. Reprinted with permission from the *New Britton and Brown Illustrated Flora of the northeastern United States and Adjacent Canada* by Henry A. Gleason, vol. 1, pages 107, 256, 369, 414, 416; vol. 2, pages 44, 51, 57, 67, 69, 90, 91, 104, 105, 109, 111, 113, 136, 140, 143, 144, 161, 179, 217, 222, 239, 245, 279, 280, 358, 376, 393, 403, 405, 409, 413, 432, 438, 453, 456, 463, 475, 476, 480, 515, 518, 519, 521, 522, 525, 526, 533, 535, 542, 566, 569, 571, 577, 579, 584, 593, 595, 602, 633, 645; vol. 3, pages 51, 55, 61, 93, 95, 115, 119, 128, 129, 132, 134, 158, 163, 181, 184, 188, 190, 200, 212, 222, 224, 231, 235, 236, 245, 253, 272, 273, 277, 279, 280, 281, 284, 289, 313, 317, 319, 338, 341, 343, 349, 351, 366, 369, 392, 399, 409, 411, 431, 433, 453, 466, 471, 473, 476, 497, 498, 501, 502, 510, 536, 537, 541, Copyright © 1952, The New York Botanical Garden.
GO1	Gould, F.W. and Box, T.W. 1965. Grasses of the Texas coastal bend. Texas A&M University Press. College Station. Reprinted with permission from Lucille Gould Bridges and Texas A&M University Press.
GO2	Gould, F.W. 1975. The grasses of Texas. Texas A&M University Press. College Station. Reprinted with permission from Lucille Gould Bridges and Texas A&M University Press.
GO3	Gould, F.W. 1951. Grasses of the southwestern United States. University of Arizona Press. Tucson. Artist: Lucretia B. Hamilton. Copyright © 1951. Reprinted with permission of the University of Arizona Press.
GO4	Gould, F.W. 1978. Common Texas grasses: An illustrated guide. Texas A&M University Press. College Station. Reprinted with permission from Lucille Gould Bridges and Texas A&M University Press.
GOO	Goodspeed, T.H. 1954. The genus *Nicotiana*. Chronica Botanica Company. Waltham, MA. Unable to locate copyright owner.
GR1	Gray, A. 1848–1849. Genera florae americae boreali-orientalis illustrata. James Munroe and Company. New York.
GR2	Gray, A. 1852–1853. Plantae wrightianae. Smithsonian Institution Press. Washington, DC.
GRE	Graham, E.H. 1941. Legumes for erosion control and wildlife (miscellaneous publication no. 412). United States Department of Agriculture. Washington, DC.
GWO	Godfrey, R.K. and Wooten, J.W. 1979–1981. Aquatic and wetland plants of southeastern states. University of Georgia Press. Athens. Reprinted with permission of the University of Georgia Press, Copyright © 1979, 1981.
HAL	Hall, H.M. 1928. The genus *Haplopappus*. Carnegie Institute. Washington, DC.
HE1	Hermann, F.J. 1960. Vetches of the United States – native, naturalized, and cultivated. (Agricultural Handbook no. 168). United States Department of Agriculture. Washington, DC.
HE2	Hermann, F.J. 1962. A revision of the genus *Glycine* and its immediate allies (Technical Bulletin no. 1268). United States Department of Agriculture. Washington, DC.
HEA	Heagy, L. 1999. Original Illustrations. Published herein by The Botanical Research Institute of Texas. Fort Worth. Artist: Linda 'Linny' Heagy. Copyright ©1999.
HI1	Hitchcock, A.S. 1935. Manual of the grasses of the United States. United States Department of Agriculture. Washington, DC.
HI2	Hitchcock, A.S. 1950. Manual of the grasses of the United States (2nd ed.). United States Department of Agriculture. Washington, DC.

HIG Hignight, K.W., Wipff, J.K., and Hatch, S.L. 1988. Grasses (Poaceae) of the Texas cross timbers and prairies. Texas Agricultural Experiment Station. College Station.

HO1 Hooker, W.J. 1829–1840. Flora boreali-americana. Treuttel and Wurtz. Paris.

HO2 Hooker, W.J. 1829–1833. Botanical miscellany. John Murray. London.

HO3 Hooker, W.J. and Arnott, G.A.W. 1830–1841. The botany of Captain Beechey's voyage. Henry G. Bohn. London.

HO5 Hooker's Icones Plantarum. 1867/71+. Williams and Norgate. London.
Reprinted with permission of the Royal Botanic Gardens, Kew.

HUM Humboldt, F.W.H.A. von. 1815–1825. Nova genera et species plantarum. Librairie Greque-Latine-Allemande. Paris.

IOW Iowa State J. Sci. 1959–1972. The Iowa State University Press. Ames, Iowa.
Reprinted with permission of the Iowa State University Press.

IPL Icon.Pl. 1837–1864. Longman, Orme, Brown, Green, and Longmans. London.
Reprinted with permission of the Royal Botanic Gardens, Kew.

IVE Ivey, R.D. 1986. Flowering plants of New Mexico. Robert Dewitt Ivey. Albuquerque, NM.
Reprinted with permission from Robert Dewitt Ivey.

JAA J. Arnold Arbor. 1920–1990. Arnold Arboretum of Harvard University. Cambridge, MA.
Reprinted with permission of the Arnold Arboreum of Harvard University.

JAC Jacquin, N. J. 1763. Selectarum stirpium americanarum historia. Krausiana. Vindobone.

JEM J. Elisha Mitchell Sci. Soc. 1884+. North Carolina Academy of Science. Durham, NC.
Reprinted with permission of North Carolina Academy of Science.

JEP Hickman, J.C. ed. The Jepson manual: Higher plants of California. University of California Press. Berkeley.
Reprinted with permission of the University of California Press.

JME Jones, M. E. 1923. Revision of North American species of *Astragalus*. Marcus E. Jones. Salt Lake City, UT.

JON Jones, F.B. 1982. Flora of the Texas coastal bend (3rd ed.). Welder Wildlife Foundation. Sinton, TX.
Reprinted with permission of the Welder Wildlife Foundation.

KAR Karsten, H. 1891. Abbildungen zur Deutschen flora. R. Friedlander and Sohn. Berlin.

KEM Kew Mag. 1984+. Royal Botanic Gardens, Kew, England.
Reprinted with permission of the Royal Botanic Gardens, Kew.

KER Keraudren, M. 1967. Flore du Cameroun. Museum National d'histoire Naturelle. Paris.
Reprinted with permission of the Museum National d'histoire Naturelle.

KEW Bull. Misc. Inform. 1887–1942. Royal Botanic Gardens, Kew, England.
Reprinted with permission of the Royal Botanic Gardens, Kew.

KIN Kinch, R.C. 1939. Nebraska weeds (Bulletin no. 101). Nebraska Department of Agriculture. Lincoln.

KUR Kurtziana. 1961+. Museo Botanico. Cordoba, Argentina. Reprinted with permission of Museo Botanico.

KVM Kerner von Marilaun, A. 1894–1895. The natural history of plants. Blackie and Son. London.

LAM Lamarck, J.B.A.P.M. 1791–1823. Tableau encyclopedique et methodique des trois regnes de la nature. Chez Pancouke. Paris.

LEM Le Maout, E. and Decaisne J. 1876. A general system of botany (2nd ed.). Longmans, Green, and Company. London.

LIH Li, Hui-Lin. 1963. Woody flora of Taiwan. The Morris Arboretum. Philadelphia, PA.
Reprinted with permission of The Morris Arboretum.

LIN Lindleyana. 1986+. American Orchid Society, Inc. West Palm Beach, FL.
Reprinted with permission from *Lindleyana* vol. 8, no. 3, page 123, figs. 3A–C, Copyright ©1993, V. Engel.
Lindleyana vol. 10, no. 1, page 38, fig. 1, Copyright ©1995, D. Goldman.

LOU Loudon, J.C. 1838. Arbortum et fruticetum Britannicum. Longman, Orme, Brown, Green, and Longmans. London.

LUN Lundell, C.L. 1961–1969. Flora of Texas. Texas Research Foundation. Renner.

LYN Lynch, D. 1981. Native & naturalized woody plants of Austin & the Hill Country. St. Edwards University. Austin, TX.
Reprinted with permission of St. Edward's University and artist: Nancy McGowan.

M&F Moore, R.J. and Frankton, C. 1974. The thistles of Canada. Agriculture Canada. Ottawa.
Reproduced from Agriculture & Agri-Food Canada publications.
Reproduced with the permission of the Minister of Public Works and Government Services Canada 1997.

MAC Mackenzie, K.K. 1940. North American Cariceae. New York Botanical Garden. New York.
Reprinted with permission from *North American Cariceae* by K.K. Mackenzie, vol. 1, plates 25, 26, 33, 34, 35, 36, 37, 41, 44, 46, 49, 61, 64, 78, 165, 167, 172, 175, 176, 182, 186, 199, 215, 233, 253, Copyright ©1940, vol. 2, plates 276, 288, 297, 301, 305, 307, 313, 314, 315, 316, 328, 329, 340, 349, 376, 378, 379, 464, 500, 507, 529, 530, 531, 536, Copyright ©1940, The New York Botanical Garden.

MAG Michigan Agric. Exp. Sta. Bull. 1885+. Michigan State University Press. East Lansing.

MAR Martius, C.F.P. 1840–1906. Flora Brasiliensis. Monachii.

MAS Mason, H.L. 1957. A flora of the marshes of California. University of California Press. Berkeley. Reprinted with permission of the University of California Press.

MAT Mathias, M.E. and Constance, L. 1965. A revision of the genus *Bowlesia* Ruiz & Pav. (Umbelliferae-Hydrocotyloideae) and its relatives. University of California Press. Berkeley.

MEE Meehan, T. 1878–1880. The native flowers and ferns of the United States. L. Prang and Company. Boston, MA.

MEP Parsons, M.E. 1909. The wild flowers of California. Cunningham, Curtiss and Welch. San Francisco.

MGH Mem. Gray Herb. 1917+. Harvard University. Cambridge, MA. Reproduced with permission of the Library of the Gray Herbarium.

MIB Michigan Bot. 1962+. Michigan Botanical Club. Ann Arbor. Reprinted with permission from John Thieret.

MIC Michaux, A. 1820. Flora boreali-americana. Bibliopola Jouanaux Junior. Paris.

MIT Mitt. Bot. Staatssamml. Munchen. 1950+. Botanische Staatssammlung. Munchen. Reprinted with permission of Botanische Staatssammlung.

MNY Mem. New York Bot. Gard. 1900+. New York Botanical Garden. New York. Reprinted with permission from *Memoirs of The New York Botanical Garden* by Rupert C. Barneby, vol. 27, pages 685, 701, 703, 705, 709, 711, 717, 721, 859, 861, 865, 867, 870, plates 46, 54, 55, 56, 57, 58, 62, 64, 133, 134, 136, 137, 139, Copyright ©1977, The New York Botanical Garden.

MOH Mohlenbrock, R.H. 1976. The illustrated flora of Illinois, sedges, *Cyperus* to *Scleria*. Southern Illinois University Press. Carbondale and Edwardsville. Reprinted with permission: Copyright ©1976 Southern Illinois University Press

MOR Moris, J.H. 1837–1859. Flora Sardoa. Ex Regio Typographeo. Torino.

MOS Moss, C.E. 1920. The Cambridge British flora. Cambridge University Press. Cambridge. Reprinted with permission of Cambridge University Press.

MTB Mem. Torrey Bot. Club. 1889/90+. Torrey Botanical Society. New York. Reprinted with permission of the Torrey Botanical Society.

MUE Muenscher, W.C. 1935. Weeds. The Macmillan Company. New York.

MUN Munz, P.A. 1935. A manual of southern California botany. Claremont Colleges. Claremont, CA. Reprinted with permission of the University of California Press.

NEE Nee, Michael. 1986. Flora de Veracruz (fasiculo 49). Instituto de Ecologia. Xalapa. Reprinted with permission of the Instituto de Ecologia.

NIC Nicholson, G. 1885–1888. The illustrated dictionary of gardening. L. Upcott Gill. London.

NVE Nees von Esenbeck, T.F.L. 1843. Genera plantarum florae germanicae. Henry and Cohen. Bonn.

ORA Oregon Agric. Exp. Sta. Bull. 1914+. Oregon State Univ. Agricultural Experiment Station. Corvallis.

PAR Parker, K.F. 1972. An illustrated guide to Arizona weeds. University of Arizona Press. Tucson. Artist: Lucretia B. Hamilton. Copyright ©1972. Reprinted with permission of the University of Arizona Press.

PAX Paxton's Mag. Bot. 1834–1849. W.S. Orr and Co. London.

PBL Pammel, L.H., Ball, C.R. and Lamson-Scribner, F. 1904. The grasses of Iowa part II. Iowa Department of Agriculture. Des Moines.

PES Pesman, M.W. 1962. Meet flora Mexicana. Dale Stewart King. Globe, AZ. Unable to locate copyright owner, author deceased.

PHY Phytologia. 1933+. Michael J. Warnock. Huntsville, TX. Reprinted with permission from Michael J. Warnock.

PLU Plukenetii, L. 1691–1694. Phytographia, sive stirpium illuftriorum & minus cognitarum. Sumptibus Auctoris. London.

PNW Hitchcock, C.L., Cronquist, A., Ownbey, M., and Thompson, J.W. 1955–1969. Vascular plants of the Pacific northwest. University of Washington Press. Seattle. Reprinted with permission of the University of Washington Press.

POW Powell, A.M. 1988. Trees & shrubs of Trans-Pecos Texas. Big Bend Natural History Association, Inc. Alpine, TX. Reprinted with permission from Jim Henrickson and Michael Powell.

PSE Pl. Syst. Evol. 1974+. Springer-Verlag. Vienna. Reprinted with permission of Springer-Verlag.

RAD Radford, A.E., Ahles, H.E., and Bell, C.R. 1968. *Manual of the vascular flora of the Carolinas*. University of North Carolina Press. Chapel Hill. From Manual of the Vascular Flora of the Carolinas by A.E. Radford, H.E. Ahles and C.R. Bell, Copyright ©1968 by the University of North Carolina Press. Used by permission of the publisher.

RBM Marcy. R.B. 1853. Exploration of the Red River of Louisiana. United States Government Printing Office. Washington, DC.

RCA Rep. Commiss. Agric. 1862–1893. United States Government Printing Office. Washington, DC.

REE Reed, C.F. 1970. Selected weeds of the United States (Agricultural Handbook no. 366). United States Department of Agriculture. Washington, DC.

RHO Rhodora. 1899+. New England Botanical Club. Cambridge, MA. Reprinted with permission of the New England Botanical Club.

RKG	Godfrey, R.K. 1988. Trees, shrubs, and woody vines of northern Florida and adjacent Georgia and Alabama. University of Georgia Press. Athens.
ROB	Robbins, W.W., Bellue, M.K., and Ball, W.S. 1951. Weeds of California. California State Department of Agriculture. Sacramento.
ROD	Rodriguesia. 1935+. Jardim Botanico. Rio de Janeiro. Reprinted with permission of Jardim Botanico.
ROE	Roedner, B.J., Hamilton, D.A., and Evans, K.E. 1978. Rare plants of the Ozark Plateau. North Central Forest Experiment Station Forest Service - United States Department of Agriculture. St. Paul, MN.
RUI	Ruiz, H. and Pavon, J. 1794. Florae Peruvianae et chilensis prodromus. en la imprenta de Sancha. Madrid.
RYD	Rydberg, P.A. 1932. Flora of the prairies and plains of central North America. Dover Publications Inc. New York.
SA1	Sargent, C.S. 1902–1913. Trees and shrubs. Houghton, Mifflin and Company. Boston.
SA2	Sargent, C.S. 1905. Manual of the trees of North America (exclusive of Mexico). Houghton, Mifflin and Company. Boston.
SA3	Sargent, C.S. 1890–1902. The silva of North America. Houghton, Mifflin and Company. Boston.
SBM	Syst. Bot. Monogr. 1980+. American Society of Plant Taxonomists. Ann Arbor, MI. Reprinted with permission of the American Society of Plant Taxonomists, G.W. Argus, E.E. Terrell, and A.S. Tomb.
SCB	Smithsonian Contr. Bot. 1969+. Smithsonian Institution Press. Washington, DC.
SCO	Scopoli, J.A. 1771–1772. Flora carniolica. Impensis Joannis Pauli Krauss. Wien.
SHI	Shinners, L.H. 1958. Shinners' spring flora of the Dallas-Fort Worth Area Texas. Lloyd Shinners. Dallas, TX.
SID	Sida, Contributions to Botany. 1962+. Botanical Research Institute of Texas. Fort Worth.
SIL	Silveus, W.A. 1933. Texas grasses classification and descriptions of grasses. W.A. Silveus. San Antonio.
SIN	Chein, P. and Chih-tsun, T. (ed.). 1985. Flora reipublicae popularis Sinicae. Academia Sinica. Beijing. From Flora Reipublicae Popularis Sinicae, Tomus 16(1) with permission of the publisher.
SM1	Small, J.K. 1933. Manual of the southeastern flora. University of North Carolina Press. Chapel Hill. From *Manual of the Southeastern Flora* by J.K. Small. Copyright ©1933 by the University of North Carolina Press, renewed 1961 by Kathryn Small Gerber. Used by permission of the Publisher.
SM2	Small, J.K. 1895. A monograph of the North American species of the genus *Polygonum* (Mem. Dept. Bot. Columbia Coll. Vol. 1). Columbia University Press. New York.
SMI	Smith, J.E. and Sowerby, J. 1790–1814. English botany. Smith, J.E. London.
ST1/2	Strausbaugh, P.D. and Core, E.L. 1978. Flora of West Virginia (2nd ed.). Seneca Books Inc. Grantsville, WV. Reprinted with permission from Seneca Books Inc.
STE	Steyermark, J.A. 1963. Flora of Missouri. Iowa State University Press. Ames. Reprinted with permission of the Missouri Department of Conservation.
STP	Stephens, H.A. 1973. Woody plants of the north central plains. University of Kansas Press. Lawrence. Reprinted with permission of the University Press of Kansas.
STW	Steward, A.N. 1958. Manual of vascular plants of the lower Yangtze Valley River. Oregon State University Press. Corvallis. Reprinted with permission of the Oregon State University Press.
SUD	Sudworth G.B. 1908. Forest trees of the Pacific slope. United States Government Printing Office. Washington, DC.
SWN	SouthW. Naturalist. 1956+. Southwestern Association of Naturalists. San Marcos, TX. Reprinted with permission of the Southwestern Association of Naturalists.
SYS	Syst. Bot. 1976+. American Society of Plant Taxonomists. Laramie, WY. Reprinted with permission of the American Society of Plant Taxonomist, D.M. Eggers Ware, J. J. Furlow, J.C. Semple.
TAN	Taylor, N.P. 1985. The genus *Echinocereus*. Royal Botanic Gardens. Kew, England. Reprinted with permission of the Royal Botanic Gardens, Kew.
TAX	Taxon. 1951+. International Bureau for Plant Taxonomy and Nomenclature. Berlin. Reprinted with permission of the International Bureau of Plant Taxonomy and Nomenclature.
TAY	Taylor, W.C. 1984. Arkansas ferns and fern allies. Milwaukee Public Museum. Milwaukee, WI. Reprinted with permission the Milwaukee Public Museum and W.C. Taylor.
TKA	Trans. Kansas Acad. Sci. 1868/72+. Kansas Academy of Science. Topeka. Reprinted with permission of Kansas Academy.
TOR	Bull. Torrey Bot. Club. 1870+. Torrey Botanical Society. Lancaster, PA. Reprinted with permission of the Torrey Botanical Society.
UCP	Univ. Calif. Publ. Bot. 1902/03+. University of California Press. Berkeley. Reprinted with permission of the University of California Press.
UKS	Univ. Kansas Sci. Bull. 1902–1996. University of Kansas. Lawrence. Reprinted with permission from C.D. Michener and R.C. Jackson.
USB	U.S.D.A. Bull. (1985–1901). 1895–1901. United States Department of Agriculture. Washington, DC.

USC U.S.D.A. Circ. 1895–1901. United States Government Printing Office. Washington, DC.

USD Stefferud, A. (ed.). 1948. Grass: The yearbook of agriculture 1948.
United States Department of Agriculture. Washington, DC.

USG Wheeler, G.M. 1878. Report upon United States geographical surveys west of the one hundredth meridian (vol. 6-Botany).
United States Government Printing Office. Washington, DC.

USH Contr. U.S. Natl. Herb. 1890–1974. United States Government Printing Office. Washington, DC.

UWA Univ. Waterloo Biol. Ser. 1971+. University of Waterloo. Waterloo.
Reprinted with permission of the University of Waterloo and J.C. Semple.

VGI Veroff. Geobot. Inst. ETH Stiftung Rubel Zurich. 1961+. Geobotanischen Institut ETH. Zurich.
Reprinted with permission of Geobotanischen Institut ETH.

VIN Vines, R.A. 1960. Trees, shrubs and woody vines of the southwest. University of Texas Press. Austin.
Reprinted with permission of University of Texas Press, Austin.

WAT Watt, G. 1907. The wild and cultivated cotton plants of the world. Longmans, Green, and Company. London.

WIG Wight, R. 1838–1853. Icones plantarum indiae orientalis. J.B. Pharoah. Madras.

WIL Wilbur, R.L. 1963. The leguminous plants of North Carolina (Technical Bulletin No. 151).
North Carolina Agricultural Experiment Station. Raleigh.

WOO Wood, A. 1895. How to study plants. American Book Company. New York.

YUN South-western Forestry College, Forestry Dept. of Yunnan Provinc. 1991. Iconographia arbororum
Yunnanicorum. Southwest Forestry College. Yunnan, CHINA.
Reprinted with permission of Southwest Forestry College.

ZO1 Zohary, M. and Heller D. 1984. The genus *Trifolium*. Israel Academy of Sciences and Humanities. Jerusalem.
Reprinted with permission of the Israel Academy of Sciences and Humanities.

ZO2 Zohary, M. 1965. Monographic revision of the genus *Tamarix*.
United States Department of Agriculture. Washington, DC.

List of Selected Botanically Related Internet Addresses

The following list of internet addresses is intended to provide an entry point into what is a very large and constantly changing pool of information of interest to botanists; it is by no means an attempt at a comprehensive listing. Some of the addresses will soon be outdated, while others are likely to be constant for significant periods of time. One of the major advantages to botanists of this information explosion is that extensive information is accessible from one's desk even without access to major botanical libraries or institutions. This is particularly important to botanists at small colleges and universities. The addresses below have been obtained from the internet and also from unpublished lists by N.G. Miller, Jorge E. Arriagada, and Rahmona Thompson. Updates to this list can be found at:

http://artemis.austinc.edu/acad/bio/gdiggs/inter.addresses.html

---

Agricultural Research Service Image Gallery (for plant images)
http://www.ars.usda.gov/is/graphics/photos/plants.htm

AltaVista Translation Service (to translate material to or from various languages)
http://babelfish.altavista.digital.com/

American Society of Plant Taxonomists (Home page and to find addresses of members)
http://www.csdl.tamu.edu/FLORA/aspt/aspthome.htm

Aquatic and Wetland Plant Database
http://aquat1.ifas.ufl.edu/database.html

Association of Systematics Collections Standards
gopher://www.keil.ukans.edu:70/11/standards/asc

Authors of Plant Names (standard Brummitt & Powell abbreviations)
http://www.rbgkew.org.uk/web.dbs/authform.html

Balogh Scientific Books WWW Sites in Botany and Gardening
http://www.balogh.com/botany.html

Biodiversity and Biological Collections Web Server
http://muse.bio.cornell.edu/

Biodiversity Worldmap
http://spider.nhm.ac.uk/science/projects/worldmap/

Biological Nomenclature in the 21st Century
http://www.inform.umd.edu/PBIO/nomcl/indx.html

Biota of North America Program (BONAP—to access John Kartesz data)
http://shanana.berkeley.edu/bonap/

Botanical Authors Index
gopher://gopher.mobot.org:70/11/.Author

Botanical Collectors Database
http://herbaria.harvard.edu/Data/Collectors/collectors.html

Botanical Databases at the Smithsonian Institution
http://www.nmnh.si.edu/botany/database.htm

Botanical Library Listing of Herbaria Worldwide
http://www.helsinki.fi/kmus/botmus.html

Botanical Research Institute of Texas (BRIT)
http://www.brit.org

Botanical Society of America
http://www.botany.org

BOTANICAL SOCIETY OF AMERICA BOTANY RELATED WWW SITES
http://www.botany.org/bsa/www-bot.html
BOTANY.COM ENCYCLOPEDIA OF PLANTS (horticultural information)
http://www.botany.com/
BOTANY RESOURCES
http://www.keil.ukans.edu/cgi-bin/hl?botany
CANADIAN POISONOUS PLANTS INFORMATION SYSTEM
http://res.agr.ca/brd/poisonpl/
CENTER FOR CONSERVATION BIOLOGY NETWORK
http://conbio.rice.edu/network
CAREERS IN BOTANY (Botanical Society of America)
http://www.ou.edu/cas/botany-micro/careers/
CARTOGRAPHIC LINKS FOR BOTANISTS
http://www.helsinki.fi/kmus/cartogr.html
CHECKLIST OF THE VASCULAR PLANTS OF TEXAS
http://www.csdl.tamu.edu/FLORA/taes/tracy/coverNF.html
CORNELL UNIVERSITY POISONOUS PLANTS WEBPAGE
http://www.ansci.cornell.edu/plants/plants.html
COUNCIL ON BOTANICAL AND HORTICULTURAL LIBRARIES
http://www.clpgh.org/cmnh/library/cbhl/
DELTA (DESCRIPTIVE LANGUAGE FOR TAXONOMY)
http://biodiversity.uno.edu/delta/delta
DIRECTORIO DE ENLACES RELACIONADOS CON LAS PLANTAS
http://www.arrakis.es/~jmanuel/links.htm
DRAFT BIOCODE (1997): the prospective international rules for the scientific names of organisms
http://www.rom.on.ca/biodiversity/biocode/biocode1997.html
EXPERT CENTER FOR TAXONOMIC IDENTIFICATION (ETI)
http://www.eti.bio.uva.nl/
FAMILIES OF FLOWERING PLANTS (for family information)
http://biodiversity.uno.edu/delta/angio/index.htm
FAMILY NAMES IN CURRENT USE
http://www.inform.umd.edu/PBIO/fam/ncu.html
FLORA EUROPAEA DATABASE
http://www.rbge.org.uk/forms/fe
FLORA OF NORTH AMERICA
http://www.fna.org
FLORA OF TEXAS CONSORTIUM
http://www.csdl.tamu.edu/FLORA/ftc/ftchome.htm
FLORA2K - BIODIVERSITY ON THE INTERNET (to access nomenclatural and family information)
http://www.csdl.tamu.edu/FLORA/kartesz/flora2ka.htm
FLOWERING PLANT GATEWAY (for plant family information)
http://www.isc.tamu.edu/FLORA/newgate/cronang.htm
GENERIC FLORA OF THE SOUTHEASTERN UNITED STATES PROJECT
http://www.flmnh.ufl.edu/natsci/herbarium/genflor/
GEOGRAPHIC NAMES INFORMATION SYSTEM
http://www-nmd.usgs.gov/www/gnis/gnisform.html
GERMPLASM RESOURCES INFORMATION NETWORK (GRIN)
http://www.ars-grin.gov/npgs/tax/
GRAY HERBARIUM CARD INDEX (to find scientific names)
http://herbaria.harvard.edu/Data/Gray/search.html

GRAY HERBARIUM OF HARVARD UNIVERSITY
http://www.herbaria.harvard.edu
ILLUSTRATED TEXAS FLORAS PROJECT
http://artemis.austinc.edu/acad/bio/gdiggs/floras.html
INDEX HERBARIORUM (Database of U.S. Institutions)
http://www.nybg.org/bsci/ih/ih.html
INDEX NOMINUM GENERICORUM
http://nmnhwww.si.edu/ing/
INTEGRATED TAXONOMIC INFORMATION SYSTEM DATABASE QUERY
http://www.itis.usda.gov/itis/itis_query.html
INTERNATIONAL ASSOCIATION FOR PLANT TAXONOMY
http://bgbm3.bgbm.fu-berlin.de/IAPT/default.htm
INTERNATIONAL CODE OF BOTANICAL NOMENCLATURE (Tokyo Code)
http://www.bgbm.fu-berlin.de/iapt/nomenclature/code/tokyo-e/
INTERNATIONAL ORGANIZATION FOR PLANT INFORMATION
http://lorenz.mur.csu.edu.au/iopi/
INTERNET DIRECTORY FOR BOTANY
http://www.uregina.ca/science/biology/liu/bio/idb.shtml
http://herb.biol.uregina.ca/liu/bio/idb.shtml
http://www.ou.edu/cas/botany-micro/idb/
INTERNET DIRECTORY FOR BOTANY - ALPHABETICAL LIST
http://www.uregina.ca/science/biology/liu/bio/botany.shtml
http://herb.biol.uregina.ca/liu/bio/botany.shtml
http://www.ou.edu/cas/botany-micro/idb-alpha/
INTERNET DIRECTORY FOR BOTANY: CHECKLISTS, FLORAS, TAXONOMIC DATABASES, VEGETATION
http://www.helsinki.fi/kmus/botflor.html
INTERNET DIRECTORY FOR BOTANY: SEARCH ENGINES
http://www.helsinki.fi/kmus/botfind.html
INTERNET DIRECTORY FOR BOTANY: SOFTWARE
http://www.helsinki.fi/kmus/botsoft.html
INTERNET DIRECTORY FOR BOTANY: SUBJECT CATEGORY LIST
http://www.helsinki.fi/kmus/botmenu.html
http://www.ou.edu/cas/botany-micro/idb/botmenu.html
INSTITUTE FOR SCIENTIFIC INFORMATION
http://www.isinet.com/
LADY BIRD JOHNSON WILDFLOWER CENTER
http://www.wildflower.org/
LIST OF LINKS TO BOTANICAL GARDENS, ARBORETA, AND RELATED TOPICS
http://www.libertynet.org/~bgmap/links.html
MAJOR WWW AND INTERNET BOTANY ADDRESSES (numerous sites)
http://www.inform.umd.edu/PBIO/pb250/weba.html
MEDICAL AND POISONOUS PLANTS DATABASE
http://www.inform.umd.edu/PBIO/Medicinals/medicinals.html
http://www.inform.umd.edu/EdRes/Colleges/LFSC/life_sciences/plant_biology/Medicinals/medicinals.html
MISSOURI BOTANICAL GARDEN
http://www.mobot.org/welcome.html
MLA (MODERN LANGUAGE ASSOCIATION) (How to cite electronic sources)
http://www.uvm.edu/~ncrane/estyles/mla.html or
http://www.cas.usf.edu/english/walker/mla.html

NAMES IN CURRENT USE FOR EXTANT PLANT GENERA
http://www.bgbm.fu-berlin.de/iapt/ncu/genera/

NATIONAL AGRICULTURAL LIBRARY'S AGRICOLA DATABASE (to search for journal articles, etc.)
www.nal.usda.gov/ag98/ag98.html

NATIVE PLANT CONSERVATION INITIATIVE
http://www.aqd.nps.gov/npci/

NATIVE PLANT SOCIETY OF TEXAS
http://lonestar.texas.net/~jleblanc/npsot_austin.html

NATURE CONSERVANCY
http://www.tnc.org

NATURAL RESOURCES CONSERVATION SERVICE, U.S. DEPARTMENT OF AGRICULTURE, PLANTS PROJECT
http://trident.ftc.nrcs.usda.gov/

NEW YORK BOTANICAL GARDEN
http://www.nybg.org/
http://pathfinder.com/@@x@JHYgUAmjqJ7Waw/vg/Gardens/NYBG/index.html

NEW YORK BOTANICAL GARDEN SPECIMEN CATALOG
http://www.nybg.org/bsci/hcol/hcol.html

NORTH CENTRAL CHAPTER OF THE NATIVE PLANT SOCIETY OF TEXAS
http://www.txnativeplants.org

ORGANIZATION FOR TROPICAL STUDIES (OTS)
http://www.ots.duke.edu/

PHYTOCHEMICAL AND ETHNOBOTANICAL DATABASES
http://www.ars-grin.gov/duke/index.html

PLANT CHROMOSOME NUMBERS DATABASE
gopher://cissus.mobot.org/77/.Chromo/.index/chromo

PLANT-LINK (search engine for plant information)
http://www.plantamerica.com/palink.htm

PLANT TAXONOMISTS ONLINE (to find addresses)
gopher://gopher.unm.edu:70/00/academic/biology/pto/address.test

PLANT TAXONOMY LECTURE NOTES - J.L. REVEAL
http://www.inform.umd.edu/PBIO/pb250/

PLANTS FOR THE FUTURE (to search for plant uses)
http://www.sunsite.unc.edu/pfaf/D_search.html

PLANTS NATIONAL DATABASE (USDA)
http://plants.usda.gov/plants

PLANT TRIVIA TIMELINE
http://www.huntington.org/BotanicalDiv/Timeline.html

POISONOUS PLANT DATABASES
http://www.inform.umd.edu/EdRes/Colleges/LFSC/life_sciences/plant_biology/Medicinals/harmful.html

ROYAL BOTANIC GARDENS, KEW
http://www.rbgkew.org.uk

ROYAL BOTANIC GARDENS, KEW DATABASES
http://www.rbgkew.org.uk/web.dbs/webdbsintro.html

SCOTT'S BOTANICAL LINKS
http://www.ou.edu/cas/botany-micro/bot-linx/

SHINNERS & MAHLER'S ILLUSTRATED FLORA OF NORTH CENTRAL TEXAS
http://artemis.austinc.edu/acad/bio/gdiggs/shinners.html

SIDA, CONTRIBUTIONS TO BOTANY
http://www.brit.org/sida/scb/

SMITHSONIAN INSTITUTION BOTANY DEPARTMENT
http://www.nmnh.si.edu/departments/botany.html
TAXONOMIC RESOURCES AND EXPERTISE DIRECTORY (TRED)
http://www.nbii.gov/tred/
TEXAS A&M UNIVERSITY BIOLOGY DEPARTMENT HERBARIUM
http://csdl.tamu.edu/FLORA/biolherb/tamuhome.htm
TEXAS A&M BIOINFORMATICS WORKING GROUP
http://csdl.tamu.edu/FLORA/tamuherb.htm
TEXAS A&M UNIVERSITY TRACY HERBARIUM
http://www.csdl.tamu.edu/FLORA/taes/tracy/homeNF.html
TEXAS DEPARTMENT OF PARKS AND WILDLIFE
http://www.tpwd.state.tx.us/
TEXAS ENDEMICS CHECKLIST
http://www.csdl.tamu.edu/FLORA/endemics/endemic1.htm
TEXAS NATURAL RESOURCE CONSERVATION COMMISSION
http://www.tnrcc.state.tx.us
TIME LIFE PLANT ENCYCLOPEDIA
http://www.pathfinder.com/@@UW*3PgcAd*SYsqOD/cgi-bin/VG/vg
TROPICOS (Worldwide Nomenclature Database)
http://mobot.mobot.org/Pick/Search/pick.html
TREE OF LIFE
http://phylogeny.arizona.edu/tree/phylogeny.html
UCMP WEB LIFT TO TAXA
http://www.ucmp.berkeley.edu/help/taxaform.html
UNIVERSITY OF MICHIGAN HERBARIUM
http://www.herb.lsa.umich.edu/umherb.htm
UNIVERSITY OF TEXAS HERBARIA
http://www.utexas.edu/ftp/depts/prc/
UNIVERSITY OF WISCONSIN-MADISON, DEPARTMENT OF BOTANY
http://www.wisc.edu/botany
VIRTUAL LIBRARY OF ECOLOGY, BIODIVERSITY AND THE ENVIRONMENT
http://conbio.rice.edu/vl
WHAT ARE ALL THOSE DEAD PLANTS FOR, ANYWAY?
http://ucjeps.berkeley.edu/dead_plants.html
WHAT IS NEW IN BOTANY
http://www.uregina.ca/science/biology/liu/bio/bot-new.html
WILDFLOWER LINKS
http://www.emergence.com/~tnr/wildflower/links.html
WORLD TAXONOMISTS DATABASE
http://www.eti.bio.uva.nl/database/txnmsts/default.shtml
WORLDWATCH INSTITUTE
http://www.worldwatch.org
WWW VIRTUAL LIBRARY: BOTANY
http://www.ou.edu/cas/botany-micro/www-vl/
WWW VIRTUAL LIBRARY: EVOLUTION (Biosciences)
http://golgi.harvard.edu/biopages/evolution.html

TAXONOMY, CLASSIFICATION AND THE DEBATE ABOUT CLADISTICS

Plant taxonomy is the science that deals with the *identification, nomenclature*, and *classification* of plants. The term plant systematics (or systematic botany) is often used synonymously with plant taxonomy (as is done here) but sometimes has the connotation of mainly using recently developed techniques such as chromosomal studies, electron microscopy, or molecular biology to answer questions about plant relationships. From the definition of plant taxonomy it follows that the primary goals of the discipline are to:

- 1) identify and describe all the various kinds of plants;
- 2) develop a uniform, practical, and stable system of naming plants—one that can be used by both plant taxonomists and others needing a way to precisely communicate information about plants [The naming system for plants follows the International Code of Botanical Nomenclature (Greuter et al. 1994)]
- 3) arrange plants with common characteristics into groups that reflect their relationships—in other words, to develop a scheme of classification that is useful. Similar species are thus put into the same genus, similar genera into the same family, etc. (Lawrence 1951; Porter 1967; Radford et al. 1974; Jones & Luchsinger 1986).

Since the time of Darwin, a primary goal of plant taxonomists has been to reflect phylogeny or evolutionary history in the system of plant classification. While this basic premise is agreed on by virtually all botanists, in recent years there has been heated debate between two main schools of taxonomists:

- 1) traditional taxonomists practicing what is sometimes referred to as "Linnaean classification" (Brummitt 1997), a system based on a hierarchy of formal ranks (e.g., family, genus, etc.) and binomial nomenclature (two-part scientific names consisting of a genus name and specific epithet); and
- 2) cladists (whose method of constructing phylogenies is based on the work of the German entomologist Willi Hennig) practicing phylogenetic classification (referred to as cladonomy by Brummitt (1997)). It should be noted that in a clade-based classification system, there are no formal ranks, including the genus, and no binomial nomenclature (de Queiroz & Gauthier 1992; Brummitt 1997; Lidén et al. 1997; Cantino 1998).

In some cases, the classification systems produced by traditional taxonomists and cladists are similar. However, as a result of their different methods, in many instances the classification and nomenclature systems produced are quite different.

Traditional taxonomists, while attempting to have a classification system based on evolutionary relationships, also try to reflect the amount of evolutionary change undergone by groups. They argue that classification is "... more than just branching patterns of evolution" (Stussey 1997). To use an animal example discussed in more detail below, because birds are so different from other vertebrates (e.g., fly, have feathers), they are treated as a different class of animal even though they evolved *from within* the class known as reptiles. Traditional taxonomists also try to incorporate other goals, including practicality and stability, into the classification system (see Brummitt (1997) for a detailed discussion of traditional classification). An example of a classification system produced by a traditional taxonomist can be seen in the work of Cronquist (1981, 1988) whose classification of flowering plants is given (with modifications) in Appendix 1.

The basic goal of cladistics (often referred to as "phylogenetic systematics"), and the only one that is considered important, is that classification should reflect the branching pattern of evolution. Thus one of the central principles of cladistics is that only monophyletic groups (= a common ancestor and all its descendants) should be given taxonomic recognition (Fig. 37). These groupings should be based on shared derived characteristics or character states, which are referred to as synapomorphies. Every organism is a mosaic of *ancestral* characteristics (= pleisomorphies) inherited with little or no change from some remote

POLYPHYLETIC GROUPS

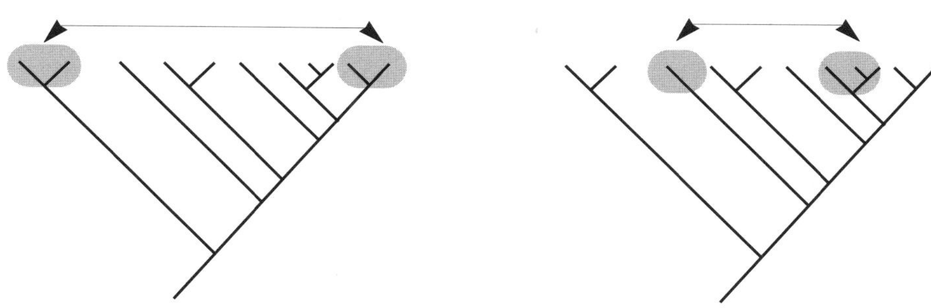

MONOPHYLETIC GROUPS

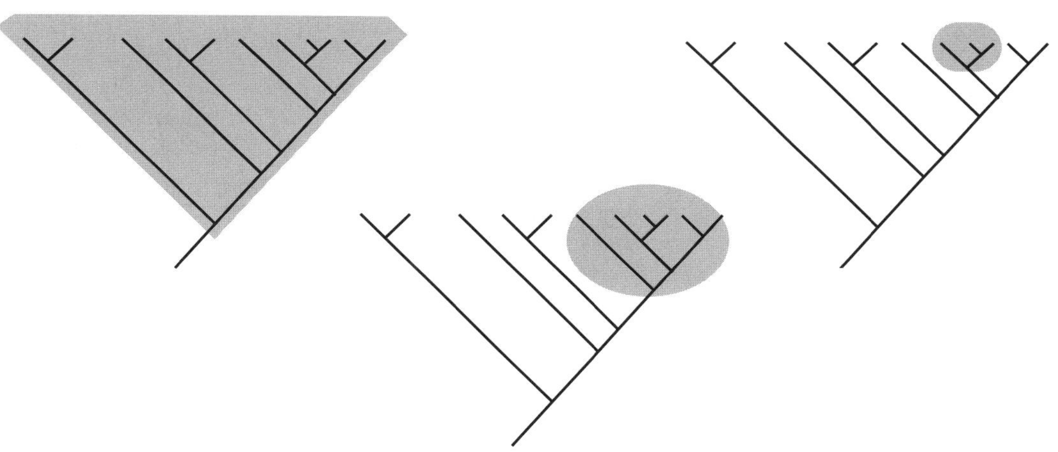

PARAPHYLETIC GROUPS

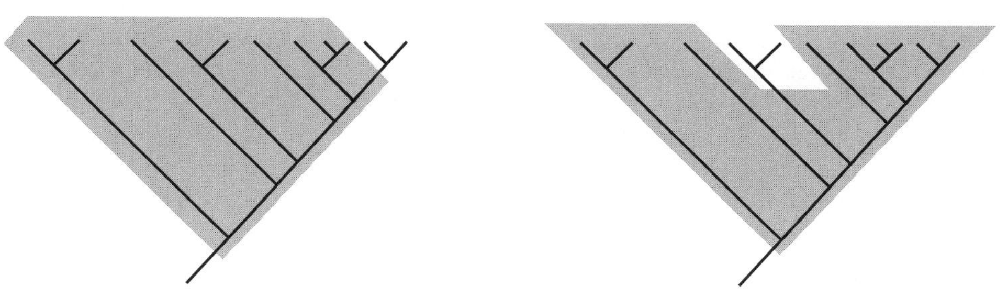

FIG. 37/DIAGRAMMATIC REPRESENTATION OF POLYPHYLETIC, MONOPHYLETIC, AND PARAPHYLETIC GROUPS.

ancestor and *derived* characteristics (apomorphies), which reflect more recent evolutionary change (Futuyma 1986). Apomorphies are thus characteristics that have changed evolutionarily and are different from the characteristics of a distant ancestor. All of the species that *share* an apomorphic characteristic are considered to be derived from a single common ancestor (because they inherited the derived characteristic from that common ancestor). The derived/apomorphic characteristic that they share is referred to as a *synapomorphy*. Mammals are thus considered a monophyletic group because they share a number of synapomorphies (e.g., produce milk, have hair, are warm-blooded) that were inherited from the common ancestor of the mammals. The unique 4+2 arrangement of stamens in the mustard family is another example of a synapomorphy; this derived characteristic occurred in the ancestor of all the mustards and is shared by, and only by, members of that family. According to cladists, only this type of shared derived characteristic provides evidence of phylogenetic relationships and therefore only these characteristics should be used in developing a classification system. They say that shared ancestral characteristics—those characteristics two organisms share because they have been retained from a distant common ancestor—do not accurately reflect recent relationships and should therefore not be used. For example, humans and lizards share the ancestral characteristic of four limbs; snakes, however, have no limbs—legs were lost relatively recently in the evolutionary line leading to snakes. Just because humans and lizards have retained the four limbs found in our common vertebrate ancestor does not mean lizards are more closely related to us than lizards are to snakes. In fact, lizards and snakes are closely related (Fig. 38) and have a number of shared derived characteristics that link them.

According to cladists, polyphyletic groups (containing taxa descended from more than one ancestor) and paraphyletic groups (including the common ancestor and *some, but not all*, of its descendants) should not be recognized (see Fig. 37 for these situations). Further, relationships based on overall similarity (a methodology referred to as phenetics) are not formally recognized—just because two groups appear similar does not necessarily mean they are closely related evolutionarily. For example, the cacti (Cactaceae) and euphorbs (Euphorbiaceae) both have large, desert-adapted, succulent species that are superficially almost indistinguishable but are very distantly related evolutionarily. These similarities are due to convergent evolution, a process by which distantly related, or even unrelated, species evolve similar adaptations in the face of similar selection pressures (such as desert-like conditions).

As indicated above, traditional taxonomists since the time of Darwin have attempted to reflect phylogeny in their systems of classification. However, they have used somewhat different methods than cladists; not only shared derived characteristics, but also shared ancestral characteristics have been utilized. Further, in addition to monophyletic groups, paraphyletic groups often have been recognized if they could be defined phenetically. In fact, our current plant classification system contains numerous examples of paraphyletic groups. The evolutionary tree in Figure 39 is a theoretical example of this situation. Species A, B, C, D, and E are all similar morphologically; species F, however, because of adaptation to some extreme environment (e.g., desert), has become very different morphologically. This phenetic difference of species F is reflected in Figure 39 by its distance from the other species. Traditional taxonomists have in general placed species A, B, C, D and E in one genus, and species F in another. Cladists would argue that this is unacceptable because E and F are more closely related than any of the others (they share the most recent common ancestor); the group A, B, C, D, and E is unacceptable because it is paraphyletic. Either A, B, and C have to be put in one genus and D, E, and F in another, or all six have to be put in the same genus. Traditional taxonomists might counter that these solutions do not reflect the tremendous amount of evolutionary change undergone by species F; they in some cases argue that because F is so different phenetically, it should be recognized as a separate group. An actual example can be seen in the case of the Asclepiadaceae (milkweeds) and Apocynaceae (dogbanes), two families recognized by traditional taxonomists (Fig. 40). The milkweeds, like our theoretical species F, are quite distinctive morphologically and, indeed, are widely recognized as monophyletic. However, when cladistic methods are applied, it becomes readily apparent that the Asclepiadaceae are a monophyletic branch derived from within the Apocynaceae, making the dogbane family (with milkweeds excluded) paraphyletic (like our group A, B, C, D, and E or like the reptiles with the birds removed). From the cladistic standpoint, the two families thus have to be lumped together into a single more inclusive

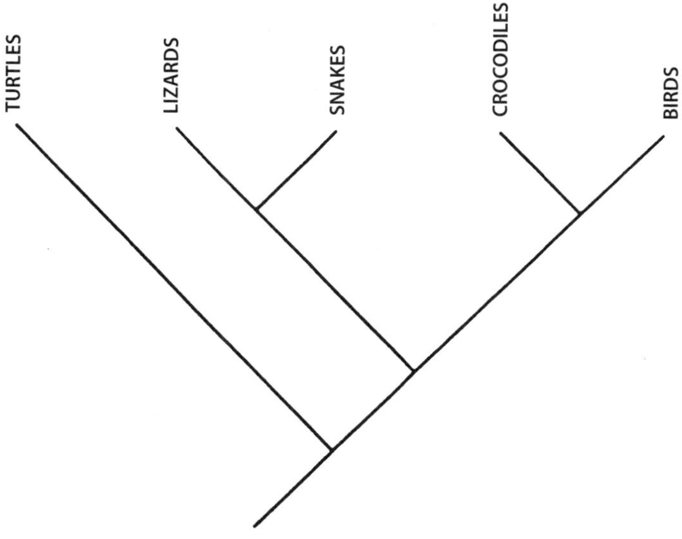

FIG. 38/DENDROGRAM SHOWING RELATIONSHIPS OF SOME VERTEBRATE GROUPS. NOTE THE PARAPHYLETIC NATURE OF THE "REPTILIA."

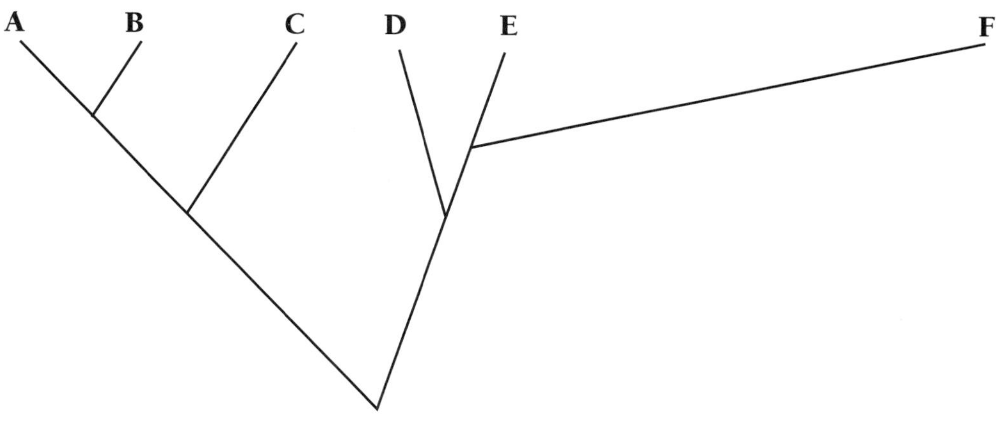

FIG. 39/PHENETIC DISTANCE DIAGRAM; HORIZONTAL DISTANCE BETWEEN SPECIES INDICATES PHENETIC DIFFERENCE. NOTE THAT SPECIES **F**, WHILE MOST CLOSELY RELATED PHYLOGENETICALLY TO SPECIES **E**, IS QUITE DIFFERENT IN TERMS OF PHENETICS.

Apocynaceae (the choice of which name to use is based on the rules of botanical nomenclature) (Judd et al. 1994). To the distress of many traditional taxonomists, the name Apocynaceae is thus used in a very different sense than previously. Traditional taxonomists argue that confusion results from such name changes and that clearly defined and easily recognized groups such as the Asclepiadaceae should be retained. The cladists on the other hand emphasize that the methods used by traditional taxonomists result in groups that do not reflect, and in fact actually distort, our understanding of evolutionary history. Cladists further stress that there are specific objective rules by which their characters are chosen and used. Consequently, they consider the results of their analyses repeatable, and in comparison with traditional taxonomy, less subjective. However, it should be noted that numerous assumptions have to be made in carrying out a cladistic analysis. In particular, because assumptions have to be made as to which character states are derived and which are ancestral, cladistic analyses are subject to various interpretations. Further, in several recent articles (e.g., Brummitt 1997; Sosef 1997) the argument has been made that in a hierarchical system of classification such as that used by plant taxonomists, paraphyletic taxa are inevitable and that a completely cladistic system of classification would be impractical to the point of being nonsensical.

The results of these differing viewpoints are perhaps even more dramatically seen in the case of well known major animal groups. The widely recognized class Reptilia (reptiles) is actually paraphyletic with turtles representing an early branch and birds arising from within the main body of the group (Fig. 38). Even though birds are very different from other members of the group in many ways (e.g., warm-blooded, have feathers, fly, lack teeth), birds and crocodiles are more closely related to one another than crocodiles are to other reptiles; in other words, from an evolutionary standpoint we should no longer recognize a formal group Reptilia. Some traditional taxonomists argue that the amount of evolutionary change, practicality, stability, and tradition in such cases should override phylogeny, while cladists stress the overwhelming importance of accurately reflecting evolutionary history. In any case, at the very least the recognition of the paraphyletic Reptilia de-emphasizes evolutionary history, such as the close relationship between crocodiles and birds.

This controversy over cladistics is currently very heated with clearly articulated positions on both sides. Welzen (1997) for example argued that the outdated "Linnaean system" should be abandoned, and de Queiroz (1997) stated, "The Linnaean hierarchy has become obsolete." De Queiroz (1997) also indicated that the "... next stage in the process of evolutionization [of taxonomy] will extend a central role for the principle of descent into the realm of biological nomenclature." Cantino stated that "Phylogenetic nomenclature is the logical culmination of a revolution that began with Darwin ...." On the other side of the controversy, Brummitt (1997) and Sosef (1997) argued (as indicated above) that paraphyletic taxa are inevitable and that cladistics is unable to cope with the reticulate evolutionary relationships seen in some groups. They further stated (Brummitt & Sosef 1998) that "... attempts to eliminate paraphyletic taxa from Linnaean classification are logically untenable." For a discussion of some of the methodological, conceptual, and philosophical problems associated with cladistics see Stuessy (1997); for a discussion of the importance of consistently applying cladistic methods and thus bringing an evolutionary perspective to biological classification and nomenclatural systems see de Queiroz (1997).

Three things seem clear regarding the controversy:

- 1) This argument over cladistics will not easily or quickly be laid to rest and thus a clear understanding of cladistic methodology and results is important. Further explanation of cladistics can be found in standard works on plant taxonomy such as Zomlefer (1994), Walters and Keil (1996), and Woodland (1997), or in evolutionary biology texts such as Ridley (1996).
- 2) The implementation of cladistic methodology would result in systems of classification and nomenclature radically different from those currently used; all the implications are not yet clear, but levels in the current hierarchy such as family or genus would no longer have meaning. In fact, if cladistic principles are consistently applied, it will be necessary for the binomial system of nomenclature to be abandoned (de Queiroz & Gauthier 1992; Lidén et al. 1997; Cantino 1998). The potential loss of nomenclatural stability is particularly disturbing to many taxonomists. Detailed discussions of some of the implications can be found in recent articles (e.g., Brummitt 1997; Crane & Kendrick 1997; de Queiroz 1997; Kron 1997; Lidén et al. 1997; Nicolson 1997; Sosef 1997; Stuessy 1997; Welzen 1997; Freudenstein 1998; Backlund & Bremer 1998; Brummitt & Sosef 1998; Cantino 1998; Sanders & Judd 1998; Schander 1998a, 1998b; Welzen 1998).
- 3) Our knowledge of phylogenetic relationships, despite the advances of molecular biology, is still incomplete and thus all the necessary information for a completely phylogenetic classification system is not yet available. In fact, many groups are poorly known and for some "... cladistic analysis can yield only the most tentative of hypotheses, subject to drastic change as new relatives are encountered" (Stuessy 1997). There are also methodological problems concerning how the characters used in cladistic analyses are chosen and analyzed (Stuessy 1997). The result can be instability in classification, and more problematically, in nomenclature, if it is linked to a rapidly changing cladistic classification system.

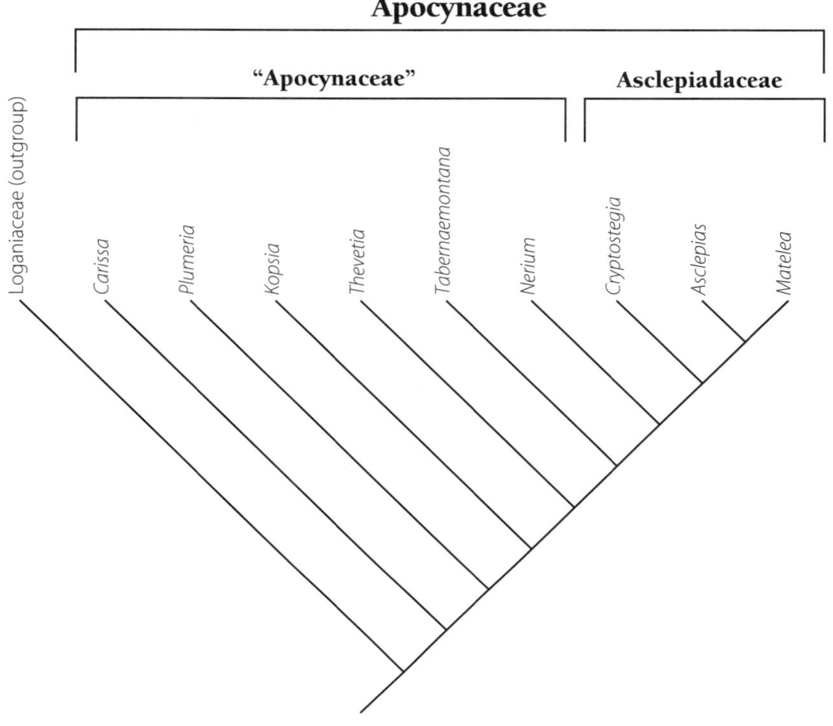

FIG. 40/DENDROGRAM SHOWING RELATIONSHIP OF TWO CLOSELY RELATED FAMILIES, THE "APOCYNACEAE" AND THE ASCLEPIADACEAE (MODIFIED FROM JUDD ET AL. (1994); USED WITH PERMISSION OF THE PRESIDENT AND FELLOWS OF HARVARD COLLEGE AND W.S. JUDD.)

Because of both philosophical and practical implementation problems, Brummitt (1997) pointed out that while the controversy should be debated, it seems unlikely that "Linnaean classification" will soon be abandoned. Brummitt (1997) suggested that both a "Linnaean classification" system and a clade-based phylogeny are desirable because they have different functions. He argued that both be allowed to exist side by side and that the nomenclature of the two should be easily recognized as different (Brummitt 1997). In summarizing his ideas he stated, "...we should not follow traditional practices just because they are traditional, but neither should we adopt new ideas just because they are new. We need to understand the possibilities and appreciate the different objectives and functions of the different options. In the meantime, it seems to me and to many others that the compromise of maintaining Linnaean classification but trying to eliminate paraphyletic taxa is nonsensical and should be abandoned before any more damage is done to existing classifications and nomenclature." Lidén et al. (1997) indicated, "If applied consistently, Phyllis [cladistic methodology] will cause confusion and loss of information content and mnemonic devices, without any substantial scientific or practical advantage.... any attempts to make Phyllis formal would be disastrous. We can find no conclusive, valid arguments against keeping the body of our current system intact." An interesting point was also made by Stuessy (1997) when he said, "...in this urgent climate of seeking to inventory the world's biota (Anonymous 1994), and requesting funds from the rest of society to do so, it would be highly counterproductive to simultaneously recommend whole-scale change of names of organisms for any reason." While strongly supporting a cladistic system, Welzen (1998) also noted that a compromise between the two types of classification is impossible. He also understood that because of practical reasons it is impossible to abandon Linnaean classification "... because too few cladograms are available to replace the existing system with a complete phylogenetic classification. Moreover, quite a few cladograms will not be that trustworthy due to the many homoplasies [result of convergent evolution] that have evolved; they will

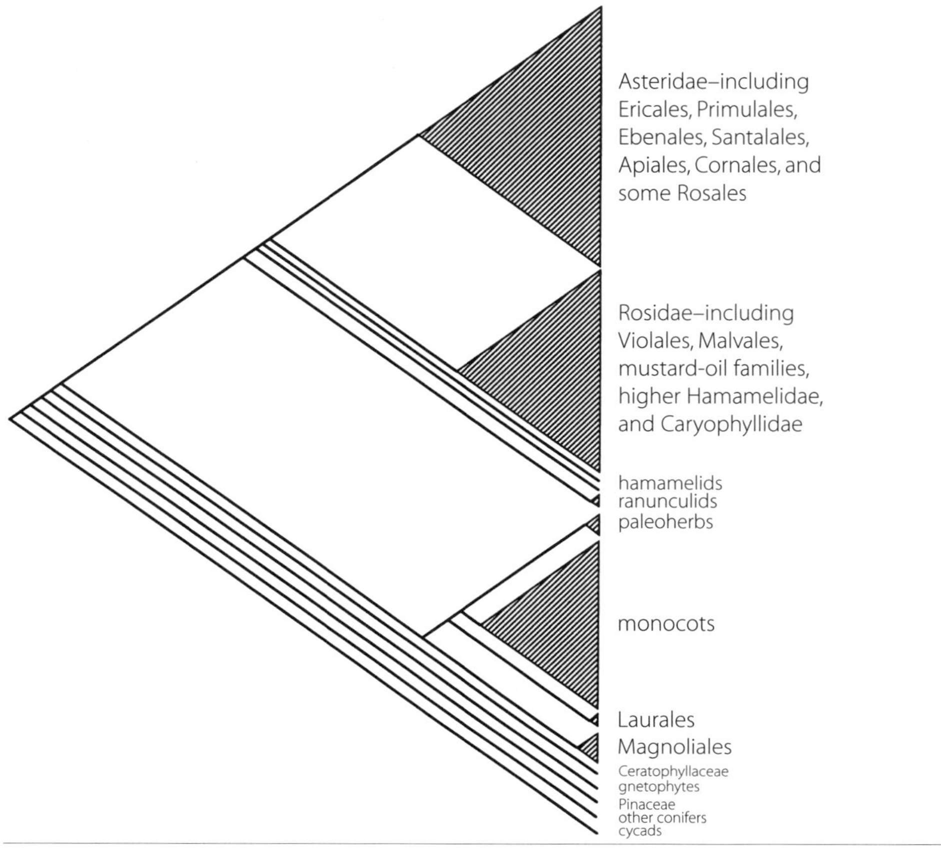

FIG. 41/MAJOR CLADES OF SEED PLANTS BASED ON DNA SEQUENCE DATA (FROM CHASE ET AL. 1993). NOTE THAT THE MONOCOTS ARISE FROM WITHIN THE DICOTS, MAKING THE DICOTS PARAPHYLETIC. USED WITH PERMISSION OF THE MISSOURI BOT. GARD. AND M.W. CHASE.

therefore, provide an unstable classification at best." Welzen (1997) went on to say, "I think, therefore, that the best solution is to choose the second option that Brummitt (1997) provides in his paper, namely, 'retaining Linnaean classification, with paraphyletic taxa, but developing alongside it an independent clade-based dichotomous system with its own separate nomenclature.' "

At present there is no complete, generally accepted, higher level cladistic analysis of the flowering plants; consequently in Appendix 1 we have given family, order, and subclass relationships based largely on the work of Cronquist (1981, 1988). While imperfect from the cladistic standpoint, it is a practical and effective way to organize thinking about plant relationships. Steps are currently being taken toward a new consensus on angiosperm relationships based on both cladistic analyses and much new evidence from molecular systematics; one such scenario is seen in Figure 41. One of the most important implications of this figure is that the monocots appear to be derived from within the dicots, making the dicots paraphyletic and thus according to cladists inappropriate for formal recognition. It might also be noted that the placement of the paleoherb families (such as the Aristolochiaceae, Piperaceae, and Nymphaeaceae) as branches off the line leading to monocots makes sense in terms of characteristics such as their unusual 3-merous perianth and androecium.

It is important for students to gain some understanding of the phylogenetic relationships of the various plant families. As a result we have added notes at the end of many family descriptions (e.g., Asclepiadaceae, Apocynaceae, Brassicaceae, Capparaceae) concerning the implications of cladistics. However, while we believe this understanding is essential, we do not feel it is appropriate or even desirable to undertake a familial rearrangement on cladistic grounds in a largely floristic work such as this; consequently in most cases traditional family boundaries have been maintained.

Categories or Ranks in the Hierarchical System of Plant Classification

The system of classification used by traditional plant taxonomists and reflected in the International Code of Botanical Nomenclature (Greuter et al. 1994) results in the placement of plants into a hierarchical system with the different categories or ranks given such names as class, family, or genus. Every plant species is thus classified in the higher categories. For example, a bluebonnet (*Lupinus texensis*) is in the family Fabaceae (bean family), the subfamily Papilionoideae, the tribe Genisteae, etc. The various categories or ranks are listed below (a number of additional categories can be inserted as needed). Note that not all categories are always used; while every plant species is classified in the categories given below in all capitals (DOMAIN, KINGDOM, CLASS, SUBCLASS, FAMILY, GENUS, and SPECIES), those categories in lower case letters (Subfamily, Tribe, Subspecies, Variety, Form) are often not used.

DOMAIN (sometimes referred to as SUPERKINGDOM; all eukaryotic organisms—those with nuclei in their cells—are in Eukaryota; the other two Domains are Archaea and Eubacteria)
- **KINGDOM** (the Kingdoms include Plantae, Animalia, Fungi, etc.)
  - **DIVISION** (equivalent to Phylum in animal classification; there are nine living divisions of vascular plants; flowering plants are the Division Magnoliophyta)
    - **CLASS** (there are 2 classes of flowering plants: Monocotyledonae & Dicotyledonae)
      - **SUBCLASS** (there are 11 subclasses of flowering plants; see Fig. 36 in Appendix 1)
        - **ORDER** (this level shows relationships between families; it is rarely used by anyone except professional botanists)
          - **FAMILY** (there are ca. 387–685 families of flowering plants depending on the system of classification used; 149 of these are found in North Central Texas)
            - **Subfamily** (this level is important for some families such as the Fabaceae; in other families it is not used)
              - **Tribe** (this level is important for some families such as the Asteraceae; in other families it is not used)
                - **GENUS** (the genus name is the first part of the two-part or binomial scientific name given to each species)
                  - **SPECIES** (the genus name and the specific epithet together make up the scientific name of a species)
                    - **Subspecies** (a subdivision of a species; many species are not divided into subspecies)
                      - **Variety** (many subspecies are not divided into varieties; sometimes varieties are treated in the same sense as subspecies; in other instances the subspecies category is used for grouping varieties within a species)
                        - **Form** (used for minor differences such as flower color)

When studying plants, the most important levels in terms of organizing your thinking are:

- 1) **Division**—Of the nine divisions of vascular plants, is the plant under study a member of Psilophyta (whisk-ferns), Lycophyta (club-mosses), Equisetophyta (Horsetails), Pteridophyta (Ferns), Cycadophyta (Cycads), Ginkgophyta (Ginkgos), Gnetophyta (Joint-firs and relatives), Pinophyta (Conifers), or Magnoliophyta (Flowering plants)?
- 2) **Class** — If a plant is a flowering plant, is it a monocot (in general: herbs, flower parts in 3s, parallel-veined leaves, 1 cotyledon—seed leaf; examples include grasses, lilies, irises, and orchids) or a dicot (in general: herbs to vines, shrubs, or trees, flower parts in 4s or 5s, net-veined leaves, 2 cotyledons; examples include roses, oaks, blueberries, and sunflowers)
- 3) **Subclass** — According to the Cronquist system (Fig. 36 in Appendix 1), within the dicots there are six subclasses and within the monocots there are five subclasses. These groups can be useful in understanding the relationships among families.

- 4) **Family** — Most botanists consider this the most important level in the classification hierarchy in terms of learning about flowering plants. The first thing a botanist tries to do with an unknown plant is to figure out what family it belongs to. As indicated above, there are ca. 387–685 families of flowering plants depending on the system of classification used (Cronquist 1988; Reveal 1993). Cronquist (1988) recognized 387, while Mabberley (1997) recognized 405. According to the International Code of Botanical Nomenclature (Greuter et al. 1994), with the exception of eight families with long-established names, all families are named after one of the genera in the family and all have the ending -aceae. Even for these eight families, use of alternative names ending in "-aceae" is permitted. These are Compositae (Asteraceae), Cruciferae (Brassicaceae), Gramineae (Poaceae), Guttiferae (Clusiaceae), Labiatae (Lamiaceae), Leguminosae (Fabaceae), Palmae (Arecaceae), and Umbelliferae (Apiaceae).
- 5) **Genus** — Many plants are easily recognized to genus (e.g., oaks—*Quercus*; maples—*Acer*, etc.). The genus name and the specific epithet are always underlined, italicized, or set off in some other matter. Note that the term specific epithet is used for the second part of the binomial referring to the species; the name of a species is a combination of the genus name and the specific epithet. Each binomial is followed by the authority or person who named the plant. For example, the common sunflower was first named botanically by Linnaeus (abbreviated L.) and the full citation of the scientific name includes the genus, specific epithet, and authority: *Helianthus annus* L.

The following is an example of the hierarchical system of classification using *Lupinus texensis*, the Texas bluebonnet. Note that subspecies, variety, and form are not given; not all species are divided below the level of species.

Domain Eukaryota
Kingdom Plantae
Division Magnoliophyta
Class Dicotyledonae
Subclass Rosidae
Order Fabales
Family Fabaceae
Subfamily Papilionoideae
Tribe Genisteae
Genus *Lupinus*
Species *texensis*
Subspecies
Variety
Form

Changes in the Scientific Names of Plants

One of the most important goals of plant taxonomy is to develop a uniform, practical, and stable system of naming plants—one that can be used by both taxonomists and others needing a way to communicate information about plants. The internationally accepted system of giving scientific names to plants is set forth in the *International Code of Botanical Nomenclature* (Greuter et al. 1994), often referred to simply as the Code.

Why then do names keep changing? Names of plants are changed for three main reasons:

- There are changes due to *legalistic reasons involving the accepted rules of nomenclature* as outlined in the Code. For example, the oldest validly published name for a species must be used unless a later name is officially conserved (this is referred to as the "rule of priority"). While such changes may be inconvenient, without strict application of nomenclatural rules, scientific names would become as inexact and useless for communication as common names. It should be kept in mind that a particular plant species can have numerous common names in a small geographic area (e.g., a state) and dozens of different common names in different languages and different countries.
- There are changes resulting from *shifts in taxonomic philosophies*, such as those exemplified by "splitters" and "lumpers," or the rejection of paraphyletic groups (for more details see appendix 6 on Taxonomy, Classification, and the Debate about Cladistics).
- Most important, however, are those *changes resulting from an increased understanding of the plant species* themselves. Initial hypotheses on what species exist, and what their diagnostic characteristics are, are sometimes based on a limited number of specimens, little or no experience with the species in the field, and little additional information. These hypotheses are tested whenever more specimens become available for examination, when field work is carried out, or when additional studies are done (including molecular studies, electron microscopy, breeding studies, etc.). Sometimes the initial hypotheses are supported and no name changes are necessary. In other instances the hypotheses need to be modified to reflect the new evidence (e.g., a plant actually belongs in a different genus). This in turn can affect the scientific nomenclature. On-going name changes therefore do not indicate simple equivocation on the part of taxonomists, but rather are an accurate reflection of the dynamic nature of our scientific understanding of the plant kingdom.

In order to minimize the impact of nomenclatural changes on users of this book, we have typically given taxonomic synonyms (particularly commonly used ones) for those species whose names have changed in the recent past. Such synonyms can be found in brackets, [ ], near the end of the species treatments and can be reached using the index.

This write up was modified from one written by B. Ertter (pers. comm.).

Collecting Herbarium Specimens

What Is An Herbarium?

A collection of pressed and dried plant specimens is known by botanists as an herbarium (plural herbaria). The word herbarium, as originally used, referred to a book about medicinal plants. Pitton de Tournefort (early French botanist and physician) around 1700 used the term for a collection of dried plants and his usage was taken up by Linnaeus (Arber 1938; Stearn 1957; Birdson & Forman 1992). Largely through Linnaeus' influence the word herbarium thus replaced such earlier terms as *hortus siccus* (dry garden) or *hortus hyemalis* (winter garden) (Arber 1938; Stearn 1957; Shinners 1958a). Luca Ghini (1490?–1556), a botany professor at the University of Bologna, Italy, is considered to have been the first person to dry plants under pressure, mount them on paper, and thus preserve them as a permanent record (Arber 1938; Birdson & Forman 1992). The usefulness of such specimens was soon apparent and his technique was disseminated over Europe by his pupils (Arber 1938). By the time of Carl Linnaeus (1707–1778), this method was well known and widely used (Stearn 1957). The oldest surviving herbarium is that of Ghini's pupil Gherardo Cibo, who began to collect plants at least as early as 1532 (Arber 1938). Other early herbaria were developed in various countries including England, France, Germany, and Switzerland and, in all, more than twenty 16th century collections survive in different European cities (Arber 1938; Valdés 1993). According to Stearn (1957), "The older herbaria consisted of specimens on sheets bound into [book-like] volumes. Linnaeus never adopted this inelastic and expensive procedure but mounted his specimens on loose sheets stored horizontally which could be easily re-arranged and to which other specimens could be added when necessary. Probably due to Linnaeus' example and teaching this method became general during the second half of the 18th century." Modern herbaria still utilize Linnaeus' basic system of mounting specimens individually on loose sheets. Today about 16,000 specimens that were at one time in Linnaeus' personal herbarium survive in England, Sweden, and France and can still be studied (Stearn 1957). From these beginnings, Holmgren et al. (1990) reported 2,639 herbaria worldwide with an estimated 272,800,926 specimens.

What Are Herbaria Used For?

Herbaria are among the most important tools in studying the plants of a given area, with the reasons for this importance being quite diverse (Benson 1979; Birdson & Forman 1992; Valdés 1993). Specific ways in which herbaria are used include: 1) Herbaria are invaluable *reference collections* used as means of identifying specimens of unknown plants. Even experienced botanists frequently need to refer to herbarium specimens in order to definitively identify a plant in question. In this way botanists are able to identify material for such organizations or individuals as poison centers, medical researchers, ranchers, law enforcement agencies, agricultural extension agents, or gardeners. 2) Herbarium specimens, which have an indefinite life if properly protected (the oldest in existence go back almost 500 years), also provide a valuable *historical record* of where plants occurred in the past in both space and time. A local example can be seen in Julien Reverchon's collections from the late 1800s and early 1900s made in the Dallas area. Because the natural vegetation of Dallas has been almost completely destroyed, without Reverchon's specimens we would have almost no knowledge of the previous richness of that flora. Also, herbarium specimens provide early documentation of the introduction of foreign weeds or the previous geographic limits of native plants (Shinners 1965). These examples demonstrate the importance of herbaria as a special type of museum and as such they are important storehouses of irreplaceable data. In a real sense, herbaria serve as "... a source of primary information about man's explorations and observations of the earth's vegetation. ..." (Radford et al. 1974) and as "... the raw data underpinning our scientific knowledge of what kinds of plants exist, what their diagnostic features are, what range of variation exists within each, and where they occur" (B. Ertter, pers. comm.). To work with the actual specimens collected by Carl Linnaeus, Alexander von Humboldt, Asa Gray, or Charles Darwin not only provides us with valuable data, but also links us in a tangible way with the origins of modern science and our own disciplines. 3) Further, because many plants are available for only a small part of the year and because it is impossible to have live specimens of thousands of species from different regions readily available for study, herbaria provide the only practical way to have material of numerous species to compare and study—they thus are important *research tools*. Without a major herbarium such as BRIT (the official abbreviation of the herbarium of the Botanical Research Institute of Texas), a book such as this would have been impossible to produce. In addition, herbaria serve as a record or repository of the specimens upon which all taxonomic articles or books are usually based (Boom 1996). In this way, other botanists can recheck and judge the validity of the work; these are critical steps in the scientific process. Also, because of the wealth of information contained in herbaria, they are essential research tools not only for taxonomists, but also for such diverse fields as ecology, endangered species research, entomology, environmental science, horticulture, medical botany, mycology, and palynology. 4) Finally, herbaria serve an important *educational purpose*. This ranges from their use by advanced undergraduates or graduate students learning taxonomic botany to grade school students learning about the importance of plants and the natural environment in their lives.

Collecting and Pressing Specimens

One of the most important considerations regarding plant collecting is to *secure appropriate permission*, whether on private or public land. This is critical in order to maintain a working relationship between landowners and botanists. It is particularly important to be sensitive to such landowner concerns as not damaging fences and properly closing gates. On public lands, such as parks and wildlife refuges, there are often strict collecting regulations with legal sanctions for not following these rules.

When selecting plants, collectors need to be sensitive to whether the plant to be collected is rare and whether the population will be adversely affected by having one or several individuals removed. Because populations of many native plants have been dramatically reduced by human activities, this concern is more important now than ever before. A rule of thumb sometimes given is the "1 to 20" rule—for every plant collected, there should be at least 20 others left in the population (Simpson 1997). Collectors should also bear in mind that certain plants (e.g., cacti, orchids) have special legal protection. Once it is ascertained that there is an adequate population for collecting, individuals representative of the range of variation in the population should be chosen. Individuals with herbivore or pathogen damage should not be ignored (Condon & Whalen 1983; McCain & Hennen 1986a, 1986b) and in fact should often be purposefully selected because such specimens are "information-rich" (J. Hennen, pers. comm.)—that is they often contain fungi, gall-inducing insects, or other pathogens and show characteristic plant responses to such organisms. Ideally, the entire plant, including roots or other underground structures, should be collected or, in the case of trees, shrubs, vines, or other large species, ample material representative of the plant should be obtained. Slender plants can be bent or folded to form a V, N, or even W shape on the sheet in order for them to fit (sometimes the point of the V can be stuck through a slit in the bottom of the newspaper to hold the plant in place). For tiny plants, it is appropriate to collect a number of individuals for each specimen needed. A general rule of thumb is that the folded half-sheet of newspaper being used in the pressing process (a full sheet torn in half) should be reasonably well-covered by plant material without excessive overlap or crowding. The best specimens have both flowers and fruits—while this is often not possible, all specimens should have some reproductive structures (either spores, cones, flowers, or fruits). Because most taxonomic keys are based on reproductive characters, sterile specimens are often useless. In fact, many botanists collect extra flowers or fruits to use in identification. Seeds, fruits, or other parts that become easily detached and are in danger of being lost should be put in small envelopes or bags and kept with the specimen. Extremely large structures (e.g., pine cones, large fruits) cannot be pressed and have to be carefully numbered (to match the specimen from which they were detached) and stored separately.

FIELD PRESS,
CUTTING OR DIGGING TOOLS,
AND FIELD BOOK
COLLECTING SPECIMENS
AND RECORDING DATA
IN THE FIELD
JULY 12 HOWARD CO.
SAVAGE ALONG RTE. 1
ON CLAY SOIL
231 PINUS
TREE 3m TALL,
DENSELY BRANCHED
232 LILY
HERB TO 0.5m TALL
FLOWERS WHITE
WITH YELLOW STAMENS
PRESS MATERIALS: FOLDS OF NEWSPAPER
(INTO WHICH SPECIMENS ARE PLACED),
BLOTTERS, AND CORRUGATED CARDBOARDS
PLANT PRESSES – THE PRESS AT THE LEFT
HAS END BOARDS MADE OF WOOD LATTICE;
THE ONE AT RIGHT HAS PLYWOOD END BOARDS
PLANT DRYER WITH LIGHT BULBS
AS HEAT SOURCE

FIG. 42/PLANT PRESSING EQUIPMENT (MODIFIED FROM SMITH 1971).

The highest quality specimens are probably obtained by carrying a lightweight press into the field and doing preliminary pressing there. Such a press is usually referred to as a "field press" and generally consists only of straps, lightweight end boards, two or three cardboards, and newspapers (Fig. 42). Because of the absence of blotters and most cardboards, a field press is much lighter than a regular press and is thus suitable for carrying some distance in the field. Specimens pressed in this way can later be rearranged (leaves flattened, etc.) before being put between blotters and cardboards for final pressing and drying. However, instead of using a field press, practicality sometimes dictates that plastic bags be used to transport the plants back to a car, kitchen table, laboratory, etc., for pressing. We have found that carrying a number of large zip-type bags (which can be reused) inside a plastic trash bag works well. Plants collected in this way and sprinkled with water can often be stored overnight in the plastic bag in a refrigerator with little loss of quality. In the past, botanists temporarily stored specimens in a metal container (called a vasculum) in folds of wet paper.

The pressing process begins by putting the plant in a single half-sheet of newspaper (22" × 14") folded crosswise (the folded half-sheet of newspaper should thus measure 11" × 14"). While this size may seem arbitrary, all subsequent steps in the collecting/herbarium process are tied to this size—these include plant dryers, cardboards and blotters for drying, specially designed herbarium storage cases, certain size shipping boxes, etc. The leaves, flowers, and other structures should be arranged in as natural a manner as possible on the newspaper while at the same time trying to avoid excessive overlap. Folding or bending the plant is often necessary as is the trimming off of excess material. When trimming is done (e.g., excess leaves removed), it should be carried out in a way that makes it clear that material was removed (e.g., a portion of the petiole of a removed leaf should be left). In order to insure the best possible results, delicate structures such as flowers can be given additional padding with small pieces of blotter or pads made of folded pieces of newspaper or paper towels. Another consideration is that leaves should be arranged so that both surfaces can be seen (hair characters of the lower surface are often important); likewise, all parts of the plant, especially reproductive structures, should be accessible for study. At least some flowers should be spread open so that the internal structures will be visible for examination. A number (corresponding to a number in a collecting notebook) is written on the lower corner of the half-sheet of newspaper on the left edge or bottom adjacent to the folds. Permanent, bold red felt-tip markers are very handy for marking newsprint. An absorbent felt or thick paper blotter is then placed on each side of the fold of newspaper containing the specimen and this "sandwich" is placed between two pieces of corrugated cardboard. Additional plants are treated in the same fashion until all have been put in the press. The result is the following sequence: cardboard, blotter, fold of newspaper with plant, blotter, cardboard, blotter, fold of newspaper with plant, etc. (Fig. 42). While it may at first seem a waste of space, specimens of only one species should be placed in each fold of newspaper—this prevents getting "mixed collections" that are often confusing. It is also necessary to treat each species separately because, as discussed below, it is important to accurately record detailed written information about each one. After the plant material has been placed in newspaper between blotters and cardboards, wooden (or other stiff material) end boards are put on each side of the entire stack and straps are used to apply pressure, thus "pressing" the plants (Fig. 42). In this manner, as the plants dry they do not shrivel and high quality specimens can be obtained. The blotters wick moisture away from the plants and the corrugations in the cardboards allow water to easily escape the press. Some botanists use only cardboards (no blotters), with little loss of specimen quality; this is frequently done when the weight of the press is an important consideration or when a source of blotters is not available. In order to speed the drying process (a necessity in humid areas such as many places in the tropics), a heat or forced air source is often necessary. An easy way to heat a press is to leave it in the back seat or trunk of a car. A roof rack is also an excellent place for a plant press. Plant dryers utilizing light bulbs as a heat source are easy and inexpensive to build and are usually used by professional botanists; however, care should be taken to avoid the danger of fire (Fig. 42). Thick materials (e.g., fruits, very thick stems, material of plants such as cacti) may be sliced in order to allow appropriate drying and to prevent unwieldy structures in the press.

Plants should stay in the press until they are dry. The time necessary is quite variable depending on whether a heat source or forced air source is used, the type of plant, the humidity of the ambient air, etc. If upon touching the plant any moisture can be sensed, it needs additional drying. Removing plants too quickly from the press will result in wrinkling or possibly molding. Likewise, leaving plants in a heated press for too long a period will cause damage (e.g., browning or fading of colors). A good rule of thumb is that most plants are dry after one or two days in a heated press while five to ten days is typical for plants in a press without any heat source. Be sure to tighten the press daily to prevent wrinkling. Also, without heat or forced air, it is often good to change blotters daily.

Making a Label

Just as important as the plant specimen itself is a properly done label. All data needed to make the label should be written down in a small notebook or "field book" at the time the plant is collected (e.g., Simpson 1997). Of crucial importance is accurate location data so that the site can be relocated by a stranger in the future. Most important are the state and county; also detailed location information such as landmarks, accurate distances, nearby towns or cities, adjacent streams, rivers, and lakes, or any other data to help relocate the site should be recorded. Other important information includes habitat (e.g., field, forest, weedy roadside, shallow water, soil type, whether the plant is growing in shade or sun), associated plants, collector, collection number, the date the plant was collected, and who was with the collector (these individuals could possibly help in the relocation of the site). The date should be given as 8 May 1996 because 08/05/96 usually means 5 August 1996 in the United States and 8 May 1996 in Europe. It is also very important to record information that will be not be obvious from the specimen (e.g., size or height of plant, in the case of trees diameter-at-breast-height—DBH, appearance of the bark, manner of growth—erect, climbing, prostrate, etc., were the flowers closed at a certain time of the day, pollinators observed, etc.) or that may be lost upon drying (flower color, odor, color of sap, presence of stinging hairs, sticky feel due to glandular hairs). Common names used locally or information on edibility or local uses are also valuable. Latitude and longitude and elevation are particularly valuable for researchers and should be recorded. These can be obtained from standard maps of most areas. However, recent technological advances have made getting such information much easier—inexpensive and accurate Global Positioning System (GPS) units are now readily available and give very accurate information. Specimens with such location data are especially valuable because they can be entered into databases with applications from the local to national levels—in Texas and elsewhere, major database projects are currently underway and GPS locations on specimens are highly recommended. Because all of the other information discussed above is recorded in the field notebook, only the collection number needs to be put on the fold of newspaper in the press. A unique number should be given to each collection that a botanist makes and these numbers should increase sequentially throughout his or her lifetime. Thus, if *Lipscomb 3491* is a collection of *Quercus alba*, there will never be a different *Lipscomb 3491*. If two or more specimens of *Quercus alba* are collected at the same time and same place by the same collector (such specimens are termed "duplicates"), they are given the same number. Such "duplicates" are often distributed to several herbaria so that there is more than one record of a particular collection. Because herbarium specimens are in essence museum collections that need to last hundreds of years, the labels should be printed on acid-free paper with permanent ink. Do not use "white-out" or other correction techniques that will be lost over time. There is no standard label size, but in general a 4" × 4" label should accommodate all necessary information without taking up excessive space on the mounting paper the specimen will ultimately be attached to (such a size will also allow 4 labels to be made from a single 8 1/2" × 11" sheet of paper). The following are two examples of labels containing appropriate information:

Austin College Herbarium, Sherman, Texas
Plants of **TEXAS**

Cnidoscolus texanus (Müll.Arg.) Small — Euphorbiaceae

GRAYSON County: Southwestern corner of county, ca. 4 km south of Tioga, just off (east of) Hwy 377, ca. 200 meters from southern edge of eastern arm of Lake Ray Roberts.

Open, sandy, weedy field with *Cenchrus spinifex* and *Monarda punctata.*

Plants herbaceous, ca. 1/2 to 1 m tall, common. Locally known as bull-nettle. Flowers white, sweet-scented; sap milky; foliage with glass-like hairs which break off in the skin and cause an intense burning sensation.

33° 26' 36.1" N 96° 55' 25.8" W (GPS)

Elevation: ca. 190 m — 24 Sept. 1980

Coll.: Delzie Demaree with Robert Kral and Donna Ware — No.: 65,967

Austin College Herbarium, Sherman, Texas

Cnidoscolus texanus (Müll.Arg.) Small — Euphorbiaceae

Texas, Grayson Co., southwestern corner of county, ca. 4 km south of Tioga, just east of Hwy 377, ca. 200 meters from southern edge of eastern arm of Lake Ray Roberts in open, sandy, weedy field with *Cenchrus spinifex* and *Monarda punctata.* Plants herbaceous, ca. 5–10 dm tall, common. Locally known as bull-nettle. Flowers white, sweet-scented; sap milky; foliage with glass-like hairs that break off in the skin and cause an intense burning sensation.

33° 26' 36.1" N 96° 55' 25.8" W (GPS); ca. 190 m.

24 Sept. 1980

Delzie Demaree 65,967
with Robert Kral and Donna Ware

PUTTING SPECIMENS IN PERMANENT COLLECTIONS

Once a specimen is dried and has an appropriate label, it can be studied or given to an herbarium for permanent storage and use in research and teaching. The proper method of donating specimens is to leave the specimen in the original half-sheet (fold) of newspaper in which it was pressed and insert the label; neither the specimen nor the label should be attached to the newspaper in any way—gluing, stapling, taping, or any other attachment method frequently damages the specimen and sometimes completely destroys its usefulness. Mounting the specimens and labels to museum quality paper for permanent storage and use in the herbarium is done by herbarium personnel properly trained in these techniques—for example, special attachment procedures and long-life glues are used. There are a number of major herbaria in Texas with the largest including those at the University of Texas in Austin, the Botanical Research Institute of Texas (BRIT) in Fort Worth, and Texas A&M University in College Station. Many other schools or organizations have valuable collections; of particular note in North Central Texas are the herbaria at Baylor University in Waco, Fort Worth Museum of Science and History, Fort Worth Nature Center, Howard Payne University in Brownwood, the University of North Texas, and the University of Texas at Arlington. BRIT welcomes the donation of herbarium specimens and botanists there can be contacted at (817) 332-4441 or info@brit.org or Botanical Research Institute of Texas, 509 Pecan Street, Fort Worth, TX 76102-4060. Such specimens will be scientific contributions, will have permanent protection, and will be important resources for the future.

Getting Started Collecting Plants

How do you get started collecting plants? Unfortunately, pressing plants between the pages of books is usually not successful because the plants dry too slowly, loose their color, seldom dry flat, and tend to damage the book. Therefore one of the first steps is to make or buy a plant press (including two end boards, corrugated cardboards, blotters, and two straps). A simple press can be made by cutting two 12" × 18" pieces out of 3/8" or 1/2" plywood and then rounding the corners and sanding the surfaces to avoid injury from splinters. The 12" × 18" size is slightly larger than a folded half-sheet of newspaper and is thus ideal for making the correct size specimens. Other types of end boards can be made out of nearly any reasonably lightweight, sturdy material. Cardboards the same size as the press can be cut from boxes, paper towels can be substituted for blotters, and simple ropes at least 4 feet long can be used as straps. With such a simple system and proper care, excellent specimens can be made. Alternatively, ready-made, convenient presses can be purchased from the sources listed below. Probably the most important parts of the press are the straps—straps that can be easily tightened and thus ensure appropriate pressure increase both the quality of the specimens and the convenience of the process. We thus recommend that a pair of straps be purchased. Because herbaria are museums whose collections are intended to last hundreds of years, the other thing that needs to be purchased is acid-free paper for the labels; this can either be special archival paper or 100% cotton rag bond. Such quality paper will last indefinitely ensuring long-term use of the specimens. Appropriate paper can be obtained from the sources listed below or can often be purchased from or at least ordered through office supply stores. For reasons of clarity, if possible, labels should be typed on a typewriter or printed using a computer (as is now done by most botanists because of speed and practicality). A hand lens (10 power) is another extremely useful tool in working with and identifying plant specimens. Many plant parts are quite small including specialized hairs or scales and moderate magnification is often essential for accurate identification; hand lenses can also be purchased from the sources below.

Herbarium Supply Company
3483 Edison Way
Menlo Park, CA 94025-1813
1-800-348-2338 or
650-366-5492
herbsupp@aol.com

Carolina Biological Supply
2700 York Road
Burlington, NC 27215
1-800-334-5551
FAX 1-800-222-7112

Pacific Papers
15702 119th NE
Bothell, WA 98011
1-800-676-1151
FAX 425-482-0534

Information for this appendix was obtained from Shinners (1958a), Smith (1971), Radford et al. (1974), Benson (1979), Birdson and Forman (1992), MacFarlane (1994), and Simpson (1997). More detailed information about plant collecting techniques can be gained from these sources. Birdson and Forman (1992) in their *Herbarium Handbook*, also provided an extensive treatment of herbarium techniques and management. Stuessy and Sohmer (1996), in a recent edited volume, gave a historical overview of the documentation of plant diversity, an analysis of societal and scientific needs from plant collections, and comprehensive information on collecting, documenting, storing, and preserving plant specimens.

List of Conservation & Environmental Organizations in North Central Texas

Federal

Balcones Canyonlands National Wildlife Refuge
10711 Burnet Road, Suite 201
Austin, TX 78732
512/339-9432
INTERNET: http://www.gorp.com/gorp/resource/us_nwr/tx_balco.htm

Caddo and Lyndon B. Johnson National Grasslands (Fannin and Wise counties)
U. S. Forest Service
P.O. Box 507
1400 N. US 81/287
Decatur, TX 76234
940/627-5475

OR Caddo Work Center (Field Office)
U.S. Forest Service
Rt. 2
Honey Grove, TX 75446
903/378-2103

Fort Hood Center for Cooperative Ecological Research (Bell and Coryell counties)
(Cooperative agreement between Department of Defense and The Nature Conservancy of Texas)
254/287-2885
FAX 254/288-5039
E-MAIL: lsanchez@tnctexas.org
INTERNET: www.tnctexas.org

Hagerman National Wildlife Refuge (Grayson County)
6465 Refuge Road
Sherman, TX 75092-5917
903/786-2826
E-MAIL: r2rw_hg@mail.fws.gov
INTERNET: http://www.gorp.com/gorp/resource/us_nwr/tx_hager.htm

Pat Mayse State Wildlife Management Area (Lamar County)
Pat Mayse Project Office
U.S. Army Corps of Engineers
P.O. Box 129
Powderly, TX 75473
903/732-3020

U.S. Army Corps of Engineers (Environmental Division)
P.O. Box 17300
819 Taylor St.
Fort Worth, TX 76102-0300
817/978-2201

INTERNET: http://www.swf.usace.army.mil
U.S. Environmental Protection Agency (Region 6-EN-XP)
1445 Ross Ave.
Dallas, TX 75202-2733
214/665-2258
INTERNET MAIN SITE: http://www.epa.gov
INTERNET REGION 6 SITE: http://www.epa.gov/earth1r6/index.htm

EPA Office of Planning and Coordination
Contact: Robert D. Lawrence, Chief
EPA Region 6-EN-XP
214/665-2258
INTERNET: http://www.epa.gov/earth1r6/6en/xp/enxp1.htm

EPA Ecosystems Protection Branch
Contact: Richard Hoppers, Chief
EPA Region 6-WQ-E
214/665-7135

EPA Marine and Wetlands Section
Contact: Becky Weber, Chief 214/665-6680
OR Richard Prather, Wetlands Coordinator 214/665-8830
OR Norm Sears, Texas State Lead 214/665-8336

EPA Watershed Management Section
Sharon Parrish, Chief
214/665-7145

STATE

TEXAS AGRICULTURAL EXTENSION SERVICE: (county offices in each county)
Texas Agricultural Extension Service Stephenville Research and Extension Center
Route 2, Box 1
Stephenville, TX 76401
254/968-4144
INTERNET: http://stephenville.tamu.edu

Texas Agricultural Extension Service Dallas Research and Extension Center
17360 Coit Road
Dallas, TX 75252
972/231-5362

Texas Department of Parks and Wildlife
4200 Smith School Road
Austin, TX 78744
1-800/792-1112 - General Information Line
INTERNET MAIN SITE: http://www.tpwd.state.tx.us/
INTERNET LIST OF TEXAS STATE PARKS: http://www.tpwd.state.tx.us/park/parklist.htm

STATE PARKS IN NORTH CENTRAL TEXAS:
Bonham State Park (Fannin Co), RR 1 Box 337, Bonham, TX 75418 903/583-5022
Cedar Hill State Park (Dallas Co.), P.O. Box 2649, Cedar Hill, TX 75104 972/291-3900
Cleburne State Park (Johnson Co.), 5800 Park Road 21, Cleburne, TX 76031 817/645-4215
Cooper Lake State Park (Delta and Hopkins cos.), Rt 1 Box 231-A15, Cooper, TX 75432 903/395-3100

Dinosaur Valley State Park (Somervell Co.), P O Box 396, Glen Rose, TX 76043 254/897-4588
Eisenhower State Park (Grayson Co.) RR 2 Box 50K, Denison TX 75020 903/465-1956, *E-MAIL:* espc@texoma.net
Lake Brownwood State Park (Brown Co.), RR 5, Box 160, Lake Brownwood, TX 76801 915/784-5223
Lake Mineral Wells Park (Parker Co.), RR 4 Box 39 C, Mineral Wells, TX 76067 940/328-1171
Lake Whitney State Park (Hill Co.), Box 1175, Whitney, TX 76692 254/694-3793
Meridian State Park (Bosque Co.), Rt. 2, Box 2465, Meridian, TX 76665 254/435-2536
Mother Neff State Park (Coryell Co.), 1680 Texas 236 Hwy, Moody, TX 76557-3317 254/853-2389
Possum Kingdom State Park (Palo Pinto Co.), P.O. Box 70, Caddo, TX 76429 940/549-1803
Ray Roberts Lake State Park (Cooke and Denton cos.), 100 PW 4137, Pilot Point, TX 76258-8944 940/686-2148
Granger Wildlife Management Area (Williamson Co.) Rt.2, Box 501, Burnet, TX 78611 512/859-2668 *OR* 512/756-2945

Texas Natural Resource Conservation Commission
P. O. Box 13807
Austin, TX 78711
512/239-1000
INTERNET: www.tnrcc.state.tx.us

PUBLIC

Dallas Nature Center
7171 Mountain Creek Parkway
Dallas, TX 75249
(972) 296-1955
FAX (972) 296-0072

Dallas County Park and Open Space Program
411 Elm Street, 3rd Floor
Dallas, TX 75202-3301
214/653-6653
FAX 214/653-651

Harry S. Moss Park (City of Dallas)
Contact: Ruth Andersson May
214/361-7772

Fort Worth Nature Center and Refuge
9601 Fossil Ridge Road
Fort Worth, TX 76135
817/237-1111
FAX 817/237-1168

Parkhill Prairie Preserve
Collin County Parks and Open Space Program
7117 County Road 166
McKinney, TX 75070-7317
972/548-4141

River Legacy Living Science Center
P.O. Box 150392
Arlington, TX 76015
817/860-6752; FAX 817/860-1595
E-MAIL: rlegacy@startext.net
INTERNET: http://www.riverlegacy.com

PRIVATE

Austin College and its Center for Environmental Studies
900 North Grand Avenue
Sherman, TX 75090
903/817-2000 general number
903/817-2342 Science Area Secretary
GENERAL INTERNET: http://www.austinc.edu/
BIOLOGY DEPT. INTERNET: http://www.austinc.edu/academics/MathSci/Biology/index.html
CENTER FOR ENVIRONMENTAL STUDIES INTERNET: http://artemis.austinc.edu/acad/envstud/EnvStudies/index.htm

Botanical Research Institute of Texas
509 Pecan Street
Fort Worth, TX 76102-4060
817/332-4441; FAX 817/332-4112
E-MAIL: info@brit.org
INTERNET: http://www.brit.org

Heard Natural Science Museum and Wildlife Sanctuary
One Nature Place
McKinney, TX 75069-8840
972/562-5566
E-MAIL: heardmuseum@texoma.net
INTERNET: http://www.heardmuseum.org

Lady Bird Johnson Wildflower Center
(previously National Wildflower Research Center)
4801 Lacrosse Ave.
Austin, TX 78739
512/292-4200
E-MAIL: wildflower@wildflower.org
INTERNET: http://www.wildflower.org/

Native Plant Society of Texas (NPSOT)
State Office
Coordinator: Dana Tucker
P.O. Box 891
Georgetown, TX 78627
512/238-0695
FAX 512/238-0703
E-MAIL: dtucker@io.com
INTERNET: http:/lonestar.texas.net/~jleblanc/npsot_austin.html

LOCAL NPSOT CHAPTER CONTACTS IN NORTH CENTRAL TEXAS:

Belton (Tonkawa)—Marie Kline, 134 Woodland Trail, Belton, TX 76513, 817/780-1715
Collin County—Betsy Farris, 4205 Tynes Drive, Garland, TX 75042, 972/494-2241
Dallas—Hannah Larson, 1020 N. Cedar Hill Rd., Cedar Hill, TX 75104, 972/291-7545
Denton—Lou Kraft, 1933 Laurelwood, Denton, TX 76201, 817/387-7725
North Central Texas—Jim Leavy, 4115 Bellaire Dr. S., Fort Worth, TX 76109, 817/923-4189
Waco—Bob Chapman, 441 Lindenwood W., Hewitt, TX 76643, 817/666-7046

Native Prairies Association of Texas
Contact: S. Lee Stone
3503 Lafayette Avenue
Austin, TX 78722-1807
512/327-5437
FAX 512/476-1663
E-MAIL: leeprairie@aol.com

Natural Area Preservation Association
(a land trust; accepts land to preserve in perpetuity)
4144 Cochran Chapel Road
Dallas, TX 75209
214/823-1848
INTERNET: http://www.cep.unt.edu/vol/napa.html

The Nature Conservancy of Texas
State Office
P.O. Box 1440
San Antonio, Texas 78295-1440
210/224-8774
E-MAIL: txfo@tnc.org
INTERNET: http://www.tnc.org

The Nature Conservancy of Texas
North Texas/Blackland Prairies Regional Office
Contact: James Eidson
Clymer Meadow Preserve
County Road 1140
Celeste, TX 75423
903/568-4139

Texas Committee on Natural Resources
5952 Royal Lane, Suite 168
Dallas, TX 75230
214/368-1791
FAX 214/265-1260
E-MAIL: tconr@mindspring.com

Texas Organization for Endangered Species (TOES)
P.O. Box 12773
Austin, TX 78711
INTERNET: http://riceinfo.rice.edu/armadillo/Endanger/AOS/toes.html

Thompson Foundation
2801 Turtle Creek—2W
Dallas, TX 75219
214/522-6142
E-MAIL: blair1189@aol.com

Larval Host Plants of Lepidoptera of North Central Texas

Compiled by JoAnn Karges

I. Butterflies
II. Moths [pg. 1400]

I. LARVAL HOST PLANTS OF NORTH CENTRAL TEXAS BUTTERFLIES

PLANT		BUTTERFLY
CUPRESSACEAE	*Juniperus ashei*	Olive-green hairstreak (*Callophrys gryneus*)
POACEAE	*Festuca* spp.	Sachem (*Atalapodes campestris*)
	Poa pratensis	Least skipper (*Ancyoloxphya numitor*) Fiery skipper (*Hylephila phyleus*) Roadside skipper (*Amblyscirtes vialis*) Whirlabout (*Polites vibex*)
	Chasmanthium latifolium	Bell's roadside skipper (*Amblyscirtes belli*)
	Phragmites australis	Broadwing skipper (*Poanes viator*)
	Tridens spp.	Common wood nymph (*Cercyonis pegala*)
	Nassella spp.	Common wood nymph (Cercyonis pegala)
	Eleusine spp.	Sachem (*Atalapodes campestris*)
	Cynodon dactylon	Julia's skipper (*Nastra lherminier*) Orange skipperling (*Copaeodes aurantiacus*) Southern skipperling (*Copaeodes minimus*) Whirlabout (*Polites vibex*) Sachem (*Atalapodes campestris*) Eufala skipper (*Lerodea eufala*)
	Bouteloua curtipendula	Orange skipperling (*Copaeodes aurantiacus*)
	Bouteloua gracilis	Green skipper (*Hesperia viridis*) Dotted skipper (*Hesperia attalus*)
	Leersia oryzoides	Least skipper (*Ancyoloxphya numitor*)
	Zizaniopsis miliacea	Broadway skipper (*Poanes viator*)
	Digitaria spp.	Sachem (*Atalapodes campestris*) Nysa roadside skipper (*Amblyscirtes nysa*)

PLANT		BUTTERFLY
POACEAE *(continued)*		
	Stenotaphrum secundatum	Clouded skipper (*Lerema accius*) Whirlabout (*Polites vibex*) Sachem (*Atalapodes campestris*) Roadside skipper (*Amblyscirtes vialis*)
	Paspalum setaceum	Whirlabout (*Polites vibex*) Nysa roadside skipper (*Amblyscirtes nysa*)
	Panicum spp.	Least skipper (*Ancyloxypha numitor*) Delaware skipper (*Anatrytrone logan*) Broadwing skipper (*Poanes viator*)
	Dichanthelium spp.	Northern broken dash (*Wallengrenia egremet*)
	Echinochloa crus-galli	Eufala skipper (*Lerodea eufala*)
	Setaria spp.	Common wood nymph (*Cercyonis pegala*) Arogos skipper (*Atrytone aragos*) Delaware skipper (*Atrytone logan*) Dusted skipper (*Atrytonopsis hianna*)
	Schizachyrium scoparium	Red satyr (*Megisto rubricata*) Swarthy skipper (*Nastra lherminier*) Cobweb skipper (*Hesperia metea*) Meske's skipper (*Hesperia meske*) Arogos skipper (*Atrytone arogos*) Dusted skipper (*Atrytonopsis hianna*)
	Sorghum spp.	Eufala skipper (*Lerodea eufala*)
	Zea mays	Eufala skipper (*Lerodea eufala*)

Note: All of the grasses used as larval host plants both by Hesperiinae and by Satyrinae (including *Cercyonis pegala, Megisto rubricata,* and *Megisto cymele,* the most common satyrs in the area) are not recorded.

PLANT		BUTTERFLY
CYPERACEAE	*Carex* spp.	Broadwinged skipper (*Poanes viator*) Dun skipper (*Euphyes vestris*)
AGAVACEAE	*Yucca* spp.	Giant yucca skipper (*Megathymus yuccae*)
CANNACEAE	*Canna* spp.	Brazilian skipper (*Calpodes ethlius*)
SALICACEAE	*Populus deltoides*	Tiger swallowtail (*Papilio glaucus*)
	Salix spp.	Tiger swallowtail (*Papilio glaucus*) Viceroy (*Liminitis archippus*) Red-spotted purple (*Liminitis arthemis astyanax*) Mourning cloak (*Nymphalis antiope*)

PLANT		BUTTERFLY
MYRICACEAE	*Myrica cerifera*	Red-banded hairstreak (*Calycopis cecrops*)
JUGLANDACEAE	*Juglans* spp. *Carya* spp.	Banded hairstreak (*Satyrium calanus*) Striped hairstreak (*Satyrium liparops*)
FAGACEAE	*Quercus* spp.	California sister (*Adelpha bredowi*) Banded hairstreak (*Satyrium calanus*) White M-album (*Parhassius m-album*) Northern hairstreak (*Fixsenia favonius*) Juvenal's duskywing (*Erynnis juvenalis*) Horace's duskywing (*Erynnis horatius*) Sleepy duskywing (*Erynnis brizo*)
ULMACEAE	*Ulmus* spp.	Tiger swallowtail (*Papilio glaucus*) Red-spotted purple (*Limenitis arthemis astyanax*) Mourning cloak (*Nymphalis antiope*) Questionmark (*Polygonia interragationis*)
	Celtis spp.	Hackberry butterfly (*Asterocampa celtis*) Tawny emperor (*Asterocampa clyton*) Questionmark (*Polygonia interragationis*) Snout butterfly (*Libytheana carinenta*)
MORACEAE	*Morus* spp.	Mourning cloak (*Nymphalis antiope*)
URTICACEAE	*Urtica chamaedryoides*	Red admiral (*Vanessa atalanta*)
VISCACEAE (LORANTHACEAE)	*Phoradendron tomentosum*	Great purple hairstreak (*Atlides halesus*)
ARISTOLOCHIACEAE	*Aristolochia tomentosa*	Pipevine swallowtail (*Battus philenor*)
CHENOPODIACEAE	*Chenopodium album*	Hayhurst's scallopwing (*Staphylus hayhurstii*) Common sootywing (*Pholisora catullus*)
AMARANTHACEAE	*Amaranthus* spp.	Common sootywing (*Pholisora catullus*)
BRASSICACEAE	*Arabis* spp., *Brassica* spp., *Draba* spp., *Lesquerella* spp., *Sisymbrium* spp.	The following are on all five of these Brassicaceae genera: Checkered white (*Pontia protidice*) Cabbage white (*Pieris rapae*) Olympia marble (*Euchloë olympia*) Falcate orangetip (*Anthocaris midea*)

PLANT		**BUTTERFLY**
ROSACEAE	*Prunus mexicana, Prunus gracilis*	The following are on both of these *Prunus* spp.: RED-SPOTTED PURPLE (*Limenitis arthemis astyanax*) GRAY HAIRSTREAK (*Strymon melinus*) BANDED HAIRSTREAK (*Satyrium calanus*) CORAL HAIRSTREAK (*Satyrium titus*) [larvae eat fruit] STRIPED HAIRSTREAK (*Satyrium liparops*)
FABACEAE	*Acacia* spp.	OUTIS SKIPPER (*Cogia outis*)
	Mimosa (*Schrankia*) spp., *Desmanthus* spp., *Prosopis* spp.	The following are on all three of these Fabaceae genera: HENRY'S ELFIN (*Incisalia henrici*) REAKIRT'S BLUE (*Hemiargus isola*) CERAUNUS BLUE (*Hemiargus ceraunus*) EASTERN TAILED BLUE (*Everes comyntas*) MARINE BLUE (*Leptotes marina*)
	Cercis canadensis	HENRY'S ELFIN (*Incisalia henrici*)
	Chamaecrista fasciculata	CLOUDLESS SULPHUR (*Phoebis sennae*) LITTLE SULPHUR (*Eurema lisa*)
	Senna roemeriana	SLEEPY ORANGE (*Eurema nicippe*)
	Sophora spp.	HENRY'S ELFIN (*Incisalia henrici*)
	Baptisia spp.	CLOUDED SULPHUR (*Colias philodice*) ORANGE SULPHUR or ALFALFA BUTTERFLY (*Colias eurytheme*) EASTERN TAILED BLUE (*Everes comyntas*) WILD INDIGO DUSKY-WING (*Erynnis baptisiae*)
	Lupinus texensis	CLOUDED SULPHUR (*Colias phiodice*) HENRY'S ELFIN (*Incisalia henrici*) EASTERN TAILED BLUE (*Everes comyntas*) WILD INDIGO DUSKYWING (*Erynnis baptisiae*)
	Medicago spp., *Melilotus* spp., *Trifolium* spp.	The following are on all three of these Fabaceae genera: CLOUDED SULPHUR (*Colias philodice*) ORANGE SULPHUR (*Colias eurytheme*) DOGFACE SULPHUR (*Colias cesonia*)
	Amorpha spp.	DOGFACE SULPHUR (*Colias cesonia*) MARINE BLUE (*Leptotes marina*) SILVER-SPOTTED SKIPPER (*Epargyrus clara*) NORTHERN CLOUDYWING (*Thorybes pylades*)

PLANT		BUTTERFLY
FABACEAE *(continued)*	*Dalea* spp.	Dogface sulphur (*Colias cesonia*) Reakirt's blue (*Hemiargus isola*) Marine blue (*Leptotes marina*)
	Indigofera miniata	Reakirt's blue (*Hemiargus isola*) Funereal duskywing (*Erynnis funeralis*)
	Robinia pseudoacacia	Funereal duskywing (*Erynnis funeralis*)
	Sesbania spp.	Funereal duskywing (*Erynnis funeralis*)
	Wisteria spp.	Silver-spotted skipper (*Epargyrus clarus*) Clouded sulphur (*Colias philodice*) Orange sulphur (*Colias eurytheme*) Southern cloudywing (*Thorybes bathyllus*) Northern cloudywing (*Thorybes pylades*) Wild indigo duskywing (*Erynnis baptisiae*)
	Desmodium spp., *Lespedeza* spp.	The following are on both of these Fabaceae genera: Eastern-tailed blue (*Everes comyntas*) Silver-spotted skipper (*Epargyrus clarus*) Southern cloudywing (*Thorybes bathyllus*) Northern cloudywing (*Thorybes pylades*)
	Vicia spp.	Clouded sulphur (*Colias philodice*) Orange sulphur (*Colias eurytheme*) Northern cloudywing (*Thorybes pylades*) Funereal duskywing (*Erynnis funeralis*)
	Clitoria mariana, *Centrosema virginianum*, *Galactea* spp., *Rhynchosia* spp.	The following are on all four of these Fabaceae taxa: Orange sulphur (*Colias curytheme*) Marine blue (*Leptotes marina*) Eastern tailed blue (*Everes comyntas*) Silver-spotted skipper (*Epargyrus clarus*)
LINACEAE	*Linum* spp.	Variegated fritillary (*Euptoieta claudia*)
RUTACEAE	*Zanthoxylum* spp.	Giant swallowtail (*Papilio cresphontes*)
	Ptelea trifoliata	Giant swallowtail (*Papilio cresphontes*)
EUPHORBIACEAE	*Croton capitatus*	Goatweed leafwing (*Anaea andria*)
	Crotonopsis linearis	Red-banded hairstreak (*Calycopis cecrops*)

PLANT		**BUTTERFLY**
ANACARDIACEAE	*Rhus aromatica*	Red-banded hairstreak (*Calycopis cecrops*)
	Rhus copallinum	Red-banded hairstreak (*Calycopis cecrops*)
SAPINDACEAE	*Sapindus saponaria*	Soapberry hairstreak (*Phaeostrymon alcestis*)
	Ungnadia speciosa	Soapberry hairstreak (*Phaeostrymon alcestis*)
RHAMNACEAE	*Ceanothus americanus*	Red-banded hairstreak (*Calycopis cecrops*) Mottled duskywing (*Erynnis martialis*)
MALVACEAE	*Sphaeralcea* spp., *Malva* spp. *Callirhoe* spp., *Sida* spp., *Hibiscus* spp., *Malvaviscus* spp.	The following three are on all six Malvaceae of these genera: Gray hairstreak (*Strymon melinus*) Common checkered skipper (*Pyrgus communis*) Streaky skipper (*Celotes nessus*)
VIOLACEAE	*Viola* spp.	Variegated fritillary (*Euptoieta claudia*)
PASSIFLORACEAE	*Passiflora affinis* *Passiflora incarnata* *Passiflora lutea* *Passiflora tenuiflora*	Gulf fritillary (*Agraulis vanillae*) on all four Passifloras. Variegated fritillary (*Euptoieta claudia*) on *P. incarnata* and *P. lutea* and suspected on the other two.
APIACEAE	*Bifora* spp.	Black swallowtail (*Papilio polyxenes*)
	Daucus spp.	Black swallowtail (*Papilio polyxenes*)
	Polytaenia nuttallii	Black swallowtail (*Papilio polyxenes*)
EBENACEAE	*Diospyros texana*	Henry's elfin (*Incisalia henrici*) Gray hairstreak (*Strymon melinus*)
OLEACEAE	*Fraxinus* spp.	Tiger swallowtail (*Papilio glaucus*)
APOCYNACEAE	*Apocynum cannabinum*	Monarch (*Danaus plexippus*) Queen (*Danaus gilippus*)
ASCLEPIADACEAE	*Asclepias* spp.	Monarch (*Danaus plexippus*) Queen (*Danaus gilippus*)
	Sarcostemma spp.	Queen (*Danaus gilippus*)
VERBENACEAE	*Lantana* spp.	Gray hairstreak (*Strymon melinus*)
	Lippia (*Phyla*) spp.	Phaon crescent (*Phyciodes phaon*)

PLANT		BUTTERFLY
SCROPHULARIACEAE	*Agalinus* spp., *Linaria texana*, *Castilleja* spp.	The species below is on all three of these Scrophulariaceae taxa: Buckeye (*Junonia coenia*)
ACANTHACEAE	*Ruellia* spp.	Buckeye (*Junonia coenia*) Texas crescent (*Phyciodes texana*)
PLANTAGINACEAE	*Plantago* spp.	Buckeye (*Junonia coenia*)
ASTERACEAE	*Aster* spp.	Pearl crescent (*Phyciodes tharos*)
	Helianthus annuus	Bordered patched (*Chlosyne lacinia*) Gorgone checkerspot (*Chlyosyne gorgone*) Silvery checkerspot (*Chlosyne nycteis*)
	Gnaphalium spp.	Painted lady (*Vanessa cardui*)
	Bidens spp., *Thelesperma* spp., *Helenium* spp., *Palafoxia* spp., *Dysodiopsis tagetoides*	The following species is on all five of these Asteraceae taxa: Dainty sulphur (*Nathalis iole*)
	Artemisia spp.	American painted lady (*Vanessa virginiana*)
	Centaurea spp.	Painted lady (*Vanessa cardui*)
	Cirsium spp.	Painted lady (*Vanessa cardui*)

II. LARVAL HOST PLANTS OF NORTH CENTRAL TEXAS MOTHS

Many moths, like *Automeris io*, are polyphagous, i.e. they may use a variety of leafy plants, often of a number of different families, completing their life-cycle on the plant where the eggs were laid. This list, then, is by no means definitive; it includes some of the largest, most spectacular, and/or most common moths of our region.

PLANT		MOTH
AGAVACEAE	*Yucca* spp.	Yucca moths (*Prodoxidae*), only *Pronuba yuccasella* has a symbiotic relationship with yucca.
SALICACEAE	*Populus* spp., *Salix* spp.	The following are on both *Populus* and *Salix*: Polyphemus (*Antheraea polyphemus*) Cecropia (*Hylaphora cecropia*) Io (*Automeris io*) Big popular sphinx (*Pachysphinx modesta*) One-eyed sphinx (*Smerinthus cerisyi*) Twin-spotted sphinx (*Smerinthus jamaicensis*) Underwings: *Catocala junctura*, *Catocala amatrix*

PLANT		**MOTH**
MYRICACEAE	*Myrica cerifera*	*Catocala muliercula*
JUGLANDACEAE	*Carya illinoinensis*	Luna (*Actias luna*), Underwings (*Catocala* species): *C. dejecta, C. consors, C. epione, C. agrippina, C. insolubilis, C. lacrymosa, C. maestosa, C. neogama, C. sappho, C. ulalume, C. vidua*
	Juglans nigra	Walnut sphinx (*Lathoë juglandis*) Underwings (*Catocala* species): *C. piatrix, C. maestosa, C. neogama, C. vidua*
FAGACEAE	*Quercus* spp.	Imperial (*Eacles imperialis*) Spiny oakworm (*Anisota stigma*) Buckmoth (*Hemileuca maia*) Polyphemus (*Antheraea polyphemus*) Waved sphinx (*Ceratomia undulosa*) Blinded sphinx (*Paonias excaetatus*) Underwings (*Catocala* species): *C. delilah, C. micronympha, C. amica, C. ilia, C. lineela*
ULMACEAE	*Celtis* spp., *Ulmus* spp.	The following are on both *Celtis* and *Ulmus*: Io (*Automeris io*) Waved sphinx (*Ceratomia undulosa*) Blinded sphinx (*Paonias excaetatus*)
MORACEAE	*Maclura pomifera*	Hagen's sphinx (*Ceratomia hageni*)
NYCTAGINACEAE	*Liquidambar styraciflua*	Luna moth (*Actias luna*)
ROSACEAE	*Crataegus* spp.	Waved sphinx (*Ceratomia undulosa*) Underwings (*Catocala* species): *C. alabamae, C. mira, C. texarkana, C. titania*
	Prunus spp.	Great ash sphinx (*Sphinx chersis*) One-eyed sphinx (*Smerinthus cerisyi*) Blinded sphinx (*Paonias excaetatus*) Small-eyed sphinx (*Paonias myops*) Walnut sphinx (*Lathoë juglandis*) White-lined sphinx (*Hyles lineata*) Apple sphinx (*Sphinx gordius*) Underwing (*Catocala ultronia*)
FABACEACE	*Cercis canadensis*	Io moth (*Automeris io*)
	Gleditsia spp.	Honey Locust moth (*Sphingicampa bicolor*) Underwings: (*Catocala undulosa, Catacola minuta*)
ANACARDIACEAE	*Rhus* spp.	Luna (*Actias luna*)

PLANT		MOTH
AQUIFOLIACEAE	*Ilex* spp.	Pawpaw sphinx (*Dolba hyloeus*)
ACERACEAE	*Acer negundo*	Imperial (*Eackes imperialis*) Polyphemus (*Antheraea polyphemus*)
VITACEAE	*Ampelopsis* spp., *Cissus* spp., *Parthenocissus* spp., *Vitis* spp.	The following are on all four of these Vitaceae genera: Hog sphinx (*Darapsa myron*) Vine sphinx (*Eumorpha vitis*) Abbott's sphinx (*Sphecedina abbottii*) Nessus sphinx (*Amphion floridensis*) Gaudy sphinx (*Eumorpha labruscae*) Mournful sphinx (*Enyo lugubris*) Pandora sphinx (*Eumorpha pandorus*) Achemon sphinx (*Eumorpha chemon*) Eight-spotted forester (*Alypia octomaculata*)
TILIACEAE	*Tilia americana* var. *caroliniana*	Waved sphinx (*Ceratomia undulosa*)
PASSIFLORACEAE	*Passiflora* spp.	Plebeian sphinx (*Paratraea plebeja*)
ONAGRACEAE	*Calylophus* spp.	Primrose moth (*Schinia florida*: Noctuidae)
	Gaura spp., *Ludwigia* spp.	The following are on both of these Onagraceae genera: Proud sphinx (*Proserpina gaurae*) White-lined sphinx (*Hyles lineata*) Banded sphinx (*Eumorpha fasciata*)
CORNACEAE	*Cornus* spp.	Cecropia (*Hyalophora cecropia*)
ERICACEAE	*Rhododendron* spp. (cultivated)	Azalea sphinx (*Darapsa pholus*)
EBENACEAE	*Diospyros texana*	Luna moth (*Actias luna*)
OLEACEAE	*Fraxinus* spp., *Ligustrum* spp.	The following are on both of these Oleaceae genera: Waved sphinx (*Ceratomia undulosa*) Ash sphinx (*Manduca Jasminearum*) Great ash sphinx (*Sphinx chersis*) Twin-spotted sphinx (*Sphinx jamaicensis*)
APOCYNACEAE	*Apocynum* spp.	Snowberry clearwing (*Hemaris diffinis*)
ASCLEPIADACEAE	*Asclepias* spp.	Milkweed tussock moth (*Euchaetes egle*: Arctiidae)

PLANT		**MOTH**
SOLANACEAE	*Datura* spp.	PINK-SPOTTED HAWK MOTH (*Agrius cingulata*)
	Solanum spp.	CAROLINA SPHINX (*Manduca sexta*) FIVE-SPOTTED HAWK MOTH (*Manduca quinquemaculata*)
SCROPHULARIACEAE	*Leucophyllum frutescens* (cultivated)	GREAT ASH SPHINX (*Sphinx chersis*)
BIGNONIACEAE	*Catalpa speciosa* *Campsis radicans* *Chilopsis linearis*	The following are on all three of these Bignoniaceae species: CATALPA SPHINX (*Cerestomia catalpe*) RUSTIC SPHINX (*Manduca rustica*)
RUBIACEAE	*Cephalanthus occidentalis*	TERSA SPHINX (*Xylophanes tersa*) HYDRANGEA SPHINX (*Darapsa versicolor*)
CAPRIFOLIACEAE	*Lonicera* spp., *Symphoricarpos* spp., *Viburnum* spp.	The following are on all three of these Caprifoliaceae genera: HUMMINGBIRD CLEARWING (*Hemaris thysbe*) SNOWBERRY CLEARWING (*Hemaris diffinis*)
ASTERACEAE	*Conyza* spp.	LYNX FLOWER MOTH (*Schinia lynx*)
	Ambrosia spp.	RAGWEED FLOWER MOTH (*Schinia rivulosa*)
	Aster spp.	ARCEGUA FLOWER MOTH (*Schinia arcigera*)

LEPIDOPTERA REFERENCES

COVELL, C.V. 1984. A field guide to the moths of eastern North America. Houghton, Mifflin, Boston, MA.

HODGES, R.W. 1971. Sphingoidea, fasc. 21 of the moths of America north of Mexico. Classey, London, England, U.K.

FREEMAN, H.A. 1995. Underwing moths (Noctuidae: Catocala) in my Texas residential list. News of the Lepidopterists' Society 1995:4.

NECK, R. 1996. A field guide to the butterflies of Texas. Gulf Publishing Co., Houston, TX.

NORTH AMERICAN BUTTERFLY ASSOCIATION. 1995. Checkist and English names of North American butterflies. North American Butterfly Association, Morristown, NJ.

OPLER, P.A. 1992. A field guide to eastern butterflies. Houghton, Mifflin, Boston, MA

PYLE, R.M. 1981. The Audubon Society field guide of North American butterflies. Knopf, New York.

SARGENT, T.D. 1976. Legion of night: The underwing moths. Univ. of Massachusetts Press, Amherst.

SCOTT, J.A. 1986. The butterflies of North America. Stanford Univ. Press, Stanford, CA.

TUSKES, P.M., J.P. TUTTLE, and M.M. COLLINS. 1996. The wild silk moths of North America, a natural history of the Saturnidae of the United States and Canada. Cornell Univ. Press, Ithaca, NY.

BOOKS FOR THE STUDY OF TEXAS NATIVE PLANTS

MODIFIED FROM A LIST PROVIDED BY THE NATIVE PLANT SOCIETY OF TEXAS

BOX 891, GEORGETOWN, TX 78627
512/238-0695 FAX: 512/238-0703
E-MAIL: dtucker@io.com

MANY OF THE BOOKS LISTED CAN BE PURCHASED FROM THE NATIVE PLANT SOCIETY OF TEXAS

SOURCES OF GENERAL BOTANICAL INFORMATION

HEYWOOD, V.H., ed. 1993. Flowering plants of the world. Oxford Univ. Press, New York. [concise information about families of flowering plants; beautifully illustrated]

HICKEY, M. and C. KING. 1997. Common families of flowering plants. Cambridge Univ. Press, Cambridge, England, U.K. [information on the more common flowering plant families]

HYAM, R. and R. PANKHURST. 1995. Plants and their names: A concise dictionary. Oxford Univ. Press, New York. [interesting source on where plant names are derived]

KARTESZ, J.T. 1994. A synonymized checklist of the vascular flora of the United States, Canada, and Greenland, 2nd ed. (2 vols.). Timber Press, Portland, OR. [scientific nomenclature of all native and naturalized vascular plants of North America north of Mexico]

MABBERLEY, D.J. 1997. The plant book, a portable dictionary of the higher plants, 2nd ed. Cambridge Univ. Press, Cambridge, England, U.K. [excellent source of plant information worldwide at the family and genus level]

WALTERS, D.R. and D.J. KEIL. 1996. Vascular plant taxonomy, 4th ed. Kendall/Hunt Publishing Co., Dubuque, IA. [a standard text on plant taxonomy/systematics]

WOODLAND, D.W. 1997. Contemporary plant systematics, 2nd ed. Andrews Univ. Press, Berrien Springs, MI. [a standard text on plant taxonomy/systematics, with CD-ROM]

ZOMLEFER, W.B. 1994. Guide to flowering plant families. Univ. of North Carolina Press, Chapel Hill. [concise information about families of flowering plants; beautifully illustrated]

FLORAS AND MANUALS

BURLAGE, H.M. 1973. The wild flowering plants of Highland Lakes country of Texas. Published by the author, Austin, TX.

CORRELL, D.S. 1956. Ferns and fern allies of Texas. Texas Research Foundation, Renner.

______ and H.B. CORRELL. 1972. Aquatic and wetland plants of southwestern United States. U.S. Environmental Protection Agency. U.S. Government Printing Office, Washington, D.C.

CORRELL, D.S. and M.C. Johnson. 1970. Manual of the vascular plants of Texas. Texas Research Foundation, Renner.

CORY, V.L. and H.B. PARKS. 1937. Catalogue of the flora of the state of Texas. Texas Agric. Exp. Sta. Bull. No. 550.

CRONQUIST, A. 1980. Asteraceae. Vascular flora of the southeastern United States 1:1–261. The Univ. of North Carolina Press, Chapel Hill.

FLORA OF NORTH AMERICA EDITORIAL COMMITTEE, eds. 1993a. Flora of North America north of Mexico, Vol. 1, Introduction. Oxford Univ. Press, New York.

______. 1993b. Flora of North America north of Mexico, Vol. 2, Pteridophytes and gymnosperms. Oxford Univ. Press, New York.

______. 1997. Flora of North America north of Mexico, Vol. 3, Magnoliophyta: Magnoliidae and Hamamelidae. Oxford Univ. Press, New York.

GANDHI, K.N. and R.D. THOMAS. 1989. Asteraceae of Louisiana. Sida, Bot. Misc. 4:1–202.

GODFREY, R.K. and J.W. WOOTEN. 1979. Aquatic and wetland plants of southeastern United States, monocotyledons. Vol. 1. Univ. of Georgia Press, Athens.

______ and ______. 1981. Aquatic and wetland plants of the southeastern United States, dicotyledons. Vol. 2. Univ. of Georgia Press, Athens.

GREAT PLAINS FLORA ASSOCIATION. 1977. Atlas of the flora of the Great Plains. Iowa State Univ. Press, Ames.

______. 1986. Flora of the Great Plains. Univ. Press of Kansas, Lawrence.

ISELY, D. 1990. Leguminosae (Fabaceae). Vascular flora of the southeastern United States 3(2):1–258. Univ. of North Carolina Press, Chapel Hill.

JOHNSTON, M.C. 1990. The vascular plants of Texas: A list, up-dating the manual of the vascular plants of Texas, 2nd ed. Published by the author, Austin, TX.

JONES, F.B. 1975. Flora of the Texas Coastal Bend. Welder Wildlife Foundation, Sinton, TX.

______, C.M. ROWELL, JR. and M.C. JOHNSTON. 1961. Flowering plants and ferns of the Texas Coastal Bend counties. Rob and Bessie Welder Wildlife Refuge, Sinton, TX.

LUER, C.A. 1975. The native orchids of the United States and Canada excluding Florida. New York Botanical Garden, Bronx.

LUNDELL, C.L., ed. 1961. Flora of Texas, Vol. III. Texas Research Foundation, Renner, TX.

______. 1966. Flora of Texas, Vol. I. Texas Research Foundation, Renner, TX.

______. 1969. Flora of Texas, Vol. II. Texas Research Foundation, Renner, TX.

MAHLER, W.F. 1970. Keys to the vascular plants of the Black Gap Wildlife Management Area, Brewster County, Texas. Published by the author, Dallas, TX [currently available through BRIT].

______. 1973a. Flora of Taylor County, Texas. Published by the author, Dallas, TX.

______. 1980. The mosses of Texas. Published by the author, Dallas, TX [currently available through BRIT].

______. 1988. Shinners' manual of the North Central Texas flora. Sida, Bot. Misc. 3:1–313 [currently available through BRIT].

MCALISTER, W.H. and M.K. MCALISTER. 1987. Guidebook to the Aransas National Wildlife Refuge. Mince Country Press, Rt. 1, Box 95C, Victoria, TX.

______ and ______. 1993. A naturalist's guide, Matagorda Island. Univ. of Texas Press, Austin.

MCDOUGALL, W.B. and O.E. SPERRY. 1951. Plants of Big Bend National Park. U.S. Government Printing Office, Washington, DC.

METZ, M.C. 1934. A flora of Bexar County Texas. The Catholic Univ. of America, Washington, DC.

METZLER, S. and V. METZLER. 1992. Texas mushrooms. Univ. of Texas Press, Austin.

PETERSON, C.D. and L.E. BROWN. 1983. Vascular flora of the Little Thicket Nature Sanctuary, San Jacinto County, Texas. Outdoor Nature Club, Houston, TX.

REESE, W.O. 1984. Mosses of the Gulf South. Louisiana State Univ. Press, Baton Rouge.

REEVES, R.G. 1972. Flora of Central Texas. Prestige Press, Fort Worth, TX.

______. 1977. Flora of Central Texas, revised ed. Grant Davis Inc., Dallas, TX.

______ and D.C. BAIN. 1947. Flora of South Central Texas. The Exchange Store, Texas A&M Univ., College Station.

RICHARDSON, A. 1990. Plants of southernmost Texas. Gorgas Science Foundation, Inc., Brownsville, TX.

______. 1995. Plants of the Rio Grande Delta. Univ. of Texas Press, Austin.

SHINNERS, L.H. 1958. Spring flora of the Dallas-Fort Worth area Texas. Published by author, Dallas, TX.

STANFORD, J.W. 1976. Keys to the vascular plants of the Texas Edwards Plateau and adjacent areas. Published by the author, Brownwood, TX.

TURNER, B.L. 1959. The legumes of Texas. Univ. of Texas Press, Austin.

WENIGER, D. 1970. Cacti of the southwest. Univ. of Texas Press, Austin.

______. 1984. Cacti of Texas and neighboring states. Univ. of Texas Press, Austin.

GRASSES, RANGE PLANTS

GOULD, F.W. 1975. The grasses of Texas. Texas A&M Univ. Press, College Station.

______. 1978. Common Texas grasses: An illustrated guide. Texas A&M Univ. Press, College Station.

______. and R.B. SHAW. 1983. Grass systematics, 2nd ed. Texas A&M Univ. Press, College Station.

HATCH, S.L. and J. PLUHAR 1993. Texas range plants. Texas A&M Univ. Press, College Station.

HIGNIGHT, K.W., J.K. WIPFF, and S.L. HATCH. 1988. Grasses (Poaceae) of the Texas Cross Timbers and Prairies. Texas Agric. Exp. Sta. Misc. Publ. No. 1657.

PHILLIPS PETROLEUM COMPANY. 1963. Pasture and range plants. Phillips Petroleum Co., Bartlesville, OK.

POWELL, M.A. 1994. Grasses of the Trans-Pecos and adjacent areas. Univ. of Texas Press, Austin.

STUBBENDIECK, J., S.L. HATCH, and C.H. BUTTERFIELD. 1997. North American range plants, 5th ed. Univ. of Nebraska Press, Lincoln.
THARP, B.C. 1952. Texas range grasses. Univ. of Texas Press, Austin.

SHRUBS, TREES

COX, P.W. and P. LESLIE. 1988. Texas trees, a friendly guide. Corona Publishing Co., San Antonio, TX.
ELIAS, T.S. 1980. The complete trees of North America. Van Nostrand Reinhold, NY.
______. 1989. Field guide to North American trees. Grolier Book Clubs, Danbury, CT.
EVERITT, J. H. and D.L. DRAWE. 1993. Trees, shrubs and cacti of South Texas. Texas Tech Univ. Press, Lubbock.
LEONARD, R.I., J.H. EVERITT, and F.W. JUDD. 1991. Woody plants of the Lower Rio Grande Texas. Texas Memorial Museum, Univ. of Texas, Austin.
LYNCH, D. 1981. Native and naturalized woody plants of Austin and the Hill Country. Published by the author, Austin, TX.
MÜLLER, C.H. 1951. The oaks of Texas. Contr. Texas Res. Found. Bot. Stud. 1:21–311. Texas Research Foundation, Renner.
NIXON, E.S. 1985. Trees, shrubs and woody vines of East Texas. Bruce Lyndon Cunningham Productions, Nacogdoches, TX.
POWELL, A.M. 1988. Trees and shrubs of Trans-Pecos Texas including Big Bend and Guadalupe Mountains National Parks. Big Bend Natural History Association, Inc., Big Bend National Park, TX.
______. 1998. Trees and shrubs of the Trans-Pecos and adjacent areas. Univ. of Texas Press, Austin.
SIMPSON, B.J. 1988. A field guide to Texas trees. Texas Monthly Press, Austin.
TEXAS FOREST SERVICE. 1970. Famous trees of Texas. Texas Forest Service, College Station.
______. 1971. Forest trees of Texas. How to know them (B-20). Texas Forest Service, College Station.
VINES, R.A. 1960. Trees, shrubs and woody vines of the Southwest. Univ. of Texas Press, Austin.

WILDFLOWERS

ABBOTT, C. 1979. How to know and grow Texas wildflowers. Green Horizons Press, Kerrville, TX.
AJILVSGI, G. 1990. Butterfly gardening for the South. Taylor Press, Dallas, TX.
______. 1979. Wild flowers of the Big Thicket, East Texas, and western Louisiana. Texas A&M Univ. Press, College Station.
______. 1984. Wildflowers of Texas. Shearer Publishing, Fredericksburg, TX.
ANDREWS, J. 1986. The Texas bluebonnet. Univ. of Texas Press, Austin.
______. 1992. American wildflower florilegium. Univ. of North Texas Press, Denton.
CANNATELLA, M.M. and R.E. ARNOLD. 1985. Plants of the Texas shore. Texas A&M Univ. Press, College Station.
ENQUIST, M. 1987. Wildflowers of the Texas Hill Country. Lone Star Botanical, Austin, TX.
HAM, H. and M. BRUCE. 1984. South Texas wildflowers. Texas A&M Univ., Kingsville.
JOHNSTON, E.G. 1972. Texas wild flowers. Shoal Creek Publishers, Inc., Austin, TX.
KIRKPATRICK, Z.M. 1992. Wildflowers of the western plains. Univ. of Texas Press, Austin.
LOUGHMILLER, C. and L. LOUGHMILLER. 1984. Texas wildflowers. Univ. of Texas Press, Austin.
MATTIZA, D.B. 1993. 100 Texas wildflowers. Southwest Parks and Monuments Assoc., Tucson, AZ.
MEIER, L. and J. REID. 1989. Texas wildflowers. News America and Weldon Owen Publ. Ltd., NY.
NATIONAL WILDFLOWER RESEARCH CENTER. 1992. Wildflower handbook. Voyageur Press, Stillwater, MN.
NIEHAUS, T.F., C.L. RIPPER, and V. SAVAGE. 1984. Southwestern and Texas wildflowers. Easton Press, Norwalk, CT.
O'KENNON, L.E. and R. O'KENNON. 1987. Texas wildflower portraits. Texas Monthly Press, Austin.
RANSON, N.R. and M.J. LAUGHLIN 1989. Wildflowers: Legends, poems and paintings (H.E. Laughlin, ed.). Heard Natural Science Museum, McKinney, TX.
RECHENTHIN, C.A. 1972. Native flowers of Texas. U.S.D.A.-Soil Conservation Service, Temple, TX.
RICKETT, H.W. 1970. Wildflowers of the United States, Texas. McGraw-Hill, NY.
ROSE, F.L. and R.W. STRANDTMANN. 1986. Wildflowers of the Llano Estacado. Taylor Press, Dallas, TX.
SCHULZ, E.D. 1922. 500 wild flowers of San Antonio and vicinity. Published by the author, San Antonio, TX.
______. 1928. Texas wild flowers: A popular account of the common wild flowers of Texas. Laidlaw Brothers, Chicago, IL.
SPELLENBERG, R. 1979. The Audubon Society field guide to North American wildflowers, western region. Alfred A. Knopf, New York.
TVETEN, J.L. and G.A. TVETEN. 1993. Wildflowers of Houston and southeast Texas. Univ. Texas Press, Austin.
WARNOCK, B.H. 1970. Wildflowers of the Big Bend country Texas. Sul Ross State Univ., Alpine, TX.
______. 1977. Wildflowers of the Davis Mountains and the Marathon Basin Texas. Sul Ross State Univ., Alpine, TX.

______. 1974. Wildflowers of the Guadalupe Mountains and the Sand Dune Country Texas. Sul Ross State Univ., Alpine, TX.

WHITEHOUSE, E. 1936. Texas flowers in natural colors. Published by the author, Dallas, TX.

WILLS, M.M. and H.S. IRWIN. 1961. Roadside flowers of Texas. Univ. of Texas Press, Austin.

TEXAS BOTANY, LANDSCAPING, MISCELLANY

AMOS, B.B. and F.R. GEHLBACH. 1988. Edwards Plateau vegetation. Baylor Univ. Press, Waco, TX.

BARLOW, J.C., A.M. POWELL, and B.N. TIMMERMANN. 1983. Invited papers for the 2nd Symposium on Resources of the Chihuahuan Desert Region U.S. and Mexico. Chihuahuan Desert Research Institute, Alpine, TX.

BRITTON, J.C. and B. MORTON. 1989. Shore ecology of the Gulf of Mexico. Univ. of Texas Press, Austin.

COX, P., J. MERRITT, and J. MOLONY. 1998. McMillen's Texas gardening—wildflowers. Gulf Publishing Co., Houston, TX.

CHEATHAM, S. and M.C. JOHNSTON. 1995. The useful wild plants of Texas, the southeastern and southwestern United States, the southern Plains, and northern Mexico. Vol. 1, *Abronia–Arundo*. Useful Wild Plants, Inc., Austin, TX.

FRITZ, E.C. 1986. Realms of beauty, the wilderness areas of East Texas. Univ. of Texas Press, Austin.

GARRETT, J.H. 1993. J. Howard Garrett's organic manual. The Summit Group, Fort Worth, TX.

______. 1993. Howard Garrett's Texas organic gardening book. Gulf Publishing Co., Houston, TX.

______. 1994. Plants of the metroplex III. Univ. of Texas Press, Austin.

______. 1996. Howard Garrett's plants for Texas. Univ. of Texas Press, Austin.

GEISER, S.W. 1945. Horticulture and horticulturists in early Texas. Southern Methodist Univ., Dallas, TX.

______. 1948. Naturalists of the frontier, 2nd ed. Southern Methodist Univ., Dallas, TX.

GOYNE, M.A. 1991. A life among the Texas flora. Texas A&M Univ. Press, College Station.

GUNTER, P.A.Y. 1993. The Big Thicket: An ecological reevaluation. Univ. of North Texas Press, Denton.

HATCH, S.L., K.N. GANDHI, and L.E. BROWN. 1990. Checklist of the vascular plants of Texas. Texas Agric. Exp. Sta. Misc. Publ. No. 1655.

HAYWARD, O.T. and J.C. YELDERMAN. 1991. A field guide to the Blackland Prairie of Texas, from frontier to heartland in one long century. Program for Regional Studies, Baylor Univ., Waco, TX.

HAYWARD, O.T., P.N. DOLLIVER, D.L. AMSBURY, and J.C. YELDERMAN. 1992. A field guide to the Grand Prairie of Texas, land, history, culture. Program for Regional Studies, Baylor Univ., Waco, TX.

Jones, A.G. 1992. *Aster* and *Brachyactis* (Asteraceae) in Oklahoma. Sida, Bot. Misc. 8:1–46.

JONES, S.D., J.K. WIPFF, and P.M. MONTGOMERY. 1997. Vascular plants of Texas: A comprehensive checklist including synonymy, bibliography, and index. Univ. of Texas Press, Austin.

KINDSCHER, K. 1987. Edible wild plants of the prairie. Univ. Press of Kansas, Lawrence.

______. 1992. Medicinal wild plants of the prairie. Univ. of Kansas Press, Lawrence.

MELTZER, S. 1997. Herb gardening in Texas, 3rd ed. Gulf Publishing Co., Houston, TX.

MILLER, G.O. 1991. Landscaping with native plants of Texas and the southwest. Voyageur Press, Stillwater, MN.

NATURAL FIBERS INFORMATION CENTER. 1987. The climates of Texas counties. Monogr. Series No. 2. Office of the State Climatologist, Dept. of Meteorology, College of Geosciences, Texas A&M Univ., College Station.

NIETHAMMER, C.J. 1987. The tumbleweed gourmet - cooking with wild southwestern plants. Univ. of Arizona Press, Tucson.

NOKES, J. 1986. How to grow native plants of Texas and the southwest. Texas Monthly Press, Austin.

PARKS, H.B. 1937. Valuable plants native to Texas. Texas Agric. Exp. Sta. Bull. No. 551.

SHARPLESS, M.R. and J.C. YELDERMAN, eds. 1993. The Texas Blackland Prairie, land, history, and culture. Baylor Univ. Program for Regional Studies, Waco, TX.

SPERRY, N. 1991. Neil Sperry's complete guide to Texas gardening. Taylor Publishing Co., Dallas, TX.

STEITZ, Q. 1987. Grasses, pods, vines, weeds: Decorating with Texas naturals. Univ. of Texas Press, Austin.

TAYLOR, C.E. 1997. Keys to the Asteraceae of Oklahoma. Southeastern Oklahoma State Univ. Herbarium, Durant.

TAYLOR, R.J. and C.E. TAYLOR. 1994. An annotated list of the ferns, fern allies, gymnosperms and flowering plants of Oklahoma, 3rd ed. Southeastern Oklahoma State Univ. Herbarium, Durant, OK.

TOES (Texas Organization for Endangered Species). 1993. Endangered, threatened and watch lists of Texas plants. Publ. No. 9. Texas Organization for Endangered Species, Austin.

TULL, D. 1987. A practical guide to edible and useful plants. Texas Monthly Press, Austin.

TULL, D. and G. O. MILLER. 1991. A field guide to wildflowers, trees and shrubs of Texas. Gulf Publishing Co., Houston, TX.

TYRL, R.J., S.C. BARBER, P. BUCK, J.R. ESTES, P. FOLLEY, L.K. MAGRATH, C.E.S. TAYLOR, and R.A. THOMPSON (Oklahoma Flora Editorial

Committee). 1994 (revised 1 Sep 1997). Key and descriptions for the vascular plant families of Oklahoma. Flora Oklahoma Incorporated, Noble, OK.

WASOWSKI, S. and J. RYAN. 1985. Landscaping with native Texas plants. Texas Monthly Press, Austin.

WASOWSKI, S. and A. WASOWSKI. 1988. Native Texas plants: Landscaping region by region. Texas Monthly Press, Austin.

_____ and _____. 1997. Native Texas gardens. Gulf Publishing Co., Houston, TX.

WATSON, G. 1975. Big Thicket plant ecology. Big Thicket Museum, Saratoga, TX.

WAUER, R. H. 1973. Naturalist's Big Bend. Texas A&M Univ. Press, College Station.

_____ and D. RISKIND. 1974. Transactions of the symposium on the biological resources of the Chihuahuan Desert region U.S. and Mexico. Chihuahuan Desert Research Institute, Alpine, TX.

SUPPORTIVE

BOMAR, G.W. 1983. Texas weather. Univ. of Texas Press, Austin.

CARTER, W.T. 1931. The soils of Texas. Texas Agric. Exp. Sta. Bull. No. 431.

FRAPS, G.S. and J. F. FUDGE. 1937. Chemical composition of soils of Texas. Texas Agric. Exp. Sta. Bull. No. 549.

GODFREY, C.L., G.S. MCKEE and H. OAKES. 1973. General soil map of Texas. Texas Agric. Exp. Sta. Misc. Publ. No. 1034.

JORDAN, T.G., J.L. BEAN, JR., and W.M. HOLMES 1984. Texas: A geography. Westview Press, Boulder, CO.

NORWINE, J., J.R. GIARDINO, G.R. NORTH, and J.B. VALDÉS, eds. 1995. The changing climate of Texas: Predictability and implications for the future. GeoBooks, Texas A&M Univ., College Station.

RENFRO, H.B., D.E. FERAY, and P.B. KING. 1973. Geological highway map of Texas. American Association of Petroleum Geologists, Tulsa, OK.

SELLARDS, E.H., W.S. ADKINS, and F.B. PLUMMER. 1932. The geology of Texas, Vol. I, Stratigraphy. Univ. of Texas Press, Austin.

SHELDON, R.A. 1979. Roadside geology of Texas. Mountain Press, Missoula, MT.

SPEARING, D. 1991. Roadside geology of Texas. Mountain Press, Missourla, MT.

STEPHENS, A.R. and W.M. HOLMES. 1989. Historical atlas of Texas. Univ. of Oklahoma Press, Norman.

A Suggested List of Ornamental Native Plants: Trees, Shrubs, Vines, Grasses, Wildflowers

For Dallas, Texas by Benny Simpson 1928–1996

Range: 50 Mile Radius

Note: Many of the plants listed are not cultivated for sale through retail outlets. Please do not remove them from the wild.
[Nomenclatural changes have been made to match the treatments elsewhere in this volume.]

Trees/ Scientific Name	Common Name
Aesculus arguta	TEXAS BUCKEYE
Carya texana	BLACK HICKORY
Cercis canadensis var. *canadensis*	EASTERN REDBUD
Cercis canadensis var. *texensis*	TEXAS REDBUD
Cornus florida	FLOWERING DOGWOOD
Crataegus reverchonii	REVERCHON'S HAWTHORN
Diospyros virginiana	COMMON PERSIMMON
Forestiera acuminata	SWAMP-PRIVET
Fraxinus americana	WHITE ASH
Fraxinus pennsylvanica	GREEN ASH
Fraxinus texensis	TEXAS ASH
Ilex decidua	DECIDUOUS HOLLY
Juglans nigra	BLACK WALNUT
Maclura pomifera	BOIS D'ARC
Morus microphylla	TEXAS MULBERRY
Morus rubra	RED MULBERRY
Prosopis glandulosa	HONEY MESQUITE, MESQUITE
Prunus mexicana	MEXICAN PLUM
Prunus munsoniana	MUNSON'S PLUM
Ptelea trifoliata	WAFER-ASH
Quercus buckleyi	TEXAS RED OAK
Quercus fusiformis	ESCARPMENT LIVE OAK
Quercus macrocarpa	BUR OAK
Quercus muhlenbergii	CHINKAPIN OAK
Quercus nigra	WATER OAK
Quercus shumardii	SHUMARD'S RED OAK
Quercus sinuata var. *breviloba*	BIGELOW OAK, SCALY-BARK OAK
Rhamnus caroliniana	CAROLINA BUCKTHORN
Rhus copallinum	SHINING SUMAC, WING-RIB SUMAC
Rhus lanceolata	PRAIRIE FLAMELEAF SUMAC
Salix exigua	SANDBAR WILLOW, COYOTE WILLOW
Sapindus drummondii	WESTERN SOAPBERRY
Sophora affinis	EVE'S-NECKLACE
Ulmus americana	AMERICAN ELM
Ulmus crassifolia	CEDAR ELM
Ulmus rubra	SLIPPERY ELM
Zanthoxylum clava-hercules	HERCULES'-CLUB, TICKLETONGUE
Zanthoxylum hirsutum	TICKLETONGUE, PRICKLY-ASH

SHRUBS/

SCIENTIFIC NAME	COMMON NAME
Aloysia gratissima	WHITEBRUSH, COMMON BEEBUSH
Amorpha fruticosa	WILD INDIGO 'DARK LANCE'
Baccharis neglecta	NEW DEAL WEED
Berberis trifoliolata	AGARITO
Callicarpa americana	AMERICAN BEAUTY-BERRY
Ceanothus americanus	NEW JERSEY-TEA
Ceanothus herbaceus	REDROOT
Cephalanthus occidentalis	COMMON BUTTONBUSH
Cornus drummondii	ROUGH-LEAF DOGWOOD
Dalea frutescens	BLACK DALEA
Euonymus atropurpureus	WAHOO
Forestiera pubescens	SPRING-HERALD
Lonicera albiflora	WHITE HONEYSUCKLE
Nolina lindheimeriana	DEVIL'S-SHOESTRING
Prunus angustifolia	CHICKASAW PLUM
Prunus gracilis	OKLAHOMA PLUM
Prunus rivularis	CREEK PLUM
Rhus aromatica	FRAGRANT SUMAC
Rhus glabra	SMOOTH SUMAC
Rosa foliolosa	WHITE PRAIRIE ROSE
Rosa setigera	CLIMBING PRAIRIE ROSE
Sabal minor	DWARF PALMETTO
Sambucus nigra var. *canadensis*	COMMON ELDERBERRY
Sideroxylon lanuginosum subsp. *oblongifolium*	CHITTAMWOOD
Symphoricarpos orbiculatus	CORAL-BERRY
Ungnadia speciosa	MEXICAN-BUCKEYE
Viburnum rufidulum	RUSTY BLACKHAW
Yucca arkansana	ARKANSAS YUCCA
Yucca constricta	BUCKLEY'S YUCCA
Yucca pallida	PALE YUCCA

VINES/

SCIENTIFIC NAME	COMMON NAME
Ampelopsis cordata	HEART-LEAF AMPELOPSIS
Campsis radicans	TRUMPET-CREEPER
Cissus incisa	IVY-TREEBINE, COWITCH
Clematis crispa	CURLY CLEMATIS
Clematis pitcheri	LEATHER-FLOWER
Cocculus carolinus	SNAILSEED
Ibervillea lindheimeri	BALSAM GOURD, LINDHEIMER'S GLOBEBERRY
Lonicera sempervirens	CORAL HONEYSUCKLE
Parthenocissus quinquefolia	VIRGINIA-CREEPER
Passiflora incarnata	PASSION FLOWER
Vitis spp.	GRAPES

GRASSES/

SCIENTIFIC NAME	COMMON NAME
Andropogon gerardii	BIG BLUESTEM
Andropogon glomeratus	BUSHY BLUESTEM
Andropogon ternarius	SPLIT-BEARD BLUESTEM
Andropogon virginicus	BROOMSEDGE BLUESTEM
Bothriochloa laguroides	SILVER BLUESTEM
Bouteloua curtipendula	SIDE-OATS GRAMA
Bouteloua gracilis	BLUE GRAMA
Bouteloua pectinata	TALL GRAMA (HAIRY)
Chasmanthium latifolium	BROADLEAF WOOD-OATS, INLAND SEA OATS
Elymus canadense	CANADIAN WILD RYE
Elymus virginicus	VIRGINIA WILD RYE
Eragrostis trichoides	SAND LOVE GRASS
Eriochola sericea	TEXAS CUP GRASS
Koeleria macrantha	JUNE GRASS, PRAIRIE JUNE GRASS
Muhlenbergia reverchonii	SEEP MUHLY
Panicum virgatum	SWITCH GRASS
Pascopyrum smithii	WESTERN WHEATGRASS
Paspalum floridanum	FLORIDA PASPALUM
Poa arachnifera	TEXAS BLUEGRASS
Schizachyrium scoparium	LITTLE BLUESTEM
Sorghastrum nutans	INDIAN GRASS
Spartina pectinata	PRAIRIE CORD GRASS
Sphenopholis obtustata	PRAIRIE WEDGESCALE
Tridens flavus	PURPLETOP
Tridens strictus	LONG-SPIKE TRIDENS
Tripsacum dactyloides	EASTERN GAMA GRASS

WILDFLOWERS/

SCIENTIFIC NAME	COMMON NAME
Acacia angustissima	FERN ACACIA
Allium drummondii	DRUMMOND'S ONION
Allium stellatum	PRAIRIE ONION
Alophia drummondii	PURPLE PLEAT-LEAF
Amsonia ciliata	NARROW-LEAF SLIMPOD, TEXAS SLIMPOD
Amsonia tabernaemontana	BLUESTAR, WILLOW SLIMPOD
Androstephium caeruleum	BLUE FUNNEL-LILY
Arnoglossum plantagineum	PRAIRIE-PLANTAIN
Aster spp.	PURPLE PRAIRIE ASTER
Baptisia australis	WILD BLUE-INDIGO
Baptisia sphaerocarpa	YELLOW BUSH PEA
Berlandiera texana	GREENEYES
Callirhoe spp.	WINECUPS
Calylophus spp.	HALF-SHRUB SUNDROPS, SUNDROPS
Camassia scilloides	WILD-HYACINTH
Castilleja purpurea (3 var.)	PURPLE PAINTBRUSH
Dalea compacta var. *pubescens*	SHOWY PRAIRIE-CLOVER

WILDFLOWERS/*(CONTINUED)* SCIENTIFIC NAME	COMMON NAME
Dalea multiflora	WHITE PRAIRIE-CLOVER
Datura wrightii	ANGEL-TRUMPET
Delphinium carolinianum var. *carolinianum*	WILD BLUE LARKSPUR
Delphinium carolinianum var. *virescens*	PRAIRIE LARKSPUR
Dodecatheon meadia	COMMON SHOOTING-STAR
Echinacea spp.	CONEFLOWER, PURPLE CONEFLOWER
Engelmannia peristenia	CUT-LEAF DAISY, ENGELMANN'S DAISY
Erigeron philadelphicus	PHILADELPHIA FLEABANE DAISY
Erodium texanum	TEXAS STORK'S-BILL
Eupatorium coelestinum	MISTFLOWER
Gaura spp.	GAURA, BUTTERFLY-WEED
Hedeoma reverchonii	ROCK HEDEOMA
Helianthus grosse-serratus	SAW-TOOTH SUNFLOWER
Helianthus hirsutus	HAIRY SUNFLOWER
Helianthus maximiliani	MAXIMILIAN SUNFLOWER
Helianthus mollis	DOWNY SUNFLOWER
Helianthus pauciflorus	STIFF SUNFLOWER
Helianthus salicifolius	WILLOW-LEAF SUNFLOWER
Helianthus tuberosus	CHOKE SUNFLOWER, JERUSALEM-ARTICHOKE
Hibiscus laevis	HALBERD-LEAF HIBISCUS
Hypericum punctatum	SPOTTED ST. JOHN'S-WORT
Lesquerella engelmannii	ENGELMANN'S BLADDERPOD
Liatris elegans	BLAZING STAR, PINK-SCALE GAYFEATHER
Liatris mucronata	GAYFEATHER, NARROW-LEAF GAYFEATHER
Liatris squarrosa var. *glabrata*	GAYFEATHER, SMOOTH GAYFEATHER
Lippa spp. (PREVIOUSLY *Phyla*)	FROGFRUIT
Lithospermum caroliniense	CAROLINA PUCCOON
Lithospermum incisum	NARROW-LEAF PUCCOON
Lobelia cardinalis	CARDINAL FLOWER
Manfreda virginica	MANFREDA, RATTLE-SNAKE MASTER
Marshallia caespitosa	BARBARA'S-BUTTONS
Melampodium leucanthum	BLACK-FOOT DAISY
Mimosa nuttallii (PREVIOUSLY *Schrankia*)	NUTTALL'S SENSITIVE-BRIAR
Monarda fistulosa	WILD BERGAMOT
Nemastylis geminiflora	PRAIRIE CELESTIAL
Oenothera macrocarpa	MISSOURI EVENING-PRIMROSE
Oenothera speciosa	SHOWY-PRIMROSE
Paronychia virginica	WHITLOW-WORT
Penstemon cobaea	COBAEA PENSTEMON, WILD FOXGLOVE
Penstemon digitalis	SMOOTH BEARDTONGUE
Penstemon laxiflorus	BEARDTONGUE
Phlox pilosa	DOWNY PHLOX
Physostegia spp.	OBEDIENT-PLANT
Phytolacca americana	POKEWEED, POKE SALAT
Pontederia cordata	PICKEREL WEED
Ratibida columnifera	MEXICAN-HAT
Rivina humilis	ROUGEPLANT, PIGEON-BERRY
Rudbeckia maxima	GREAT CONEFLOWER
Salvia azurea	BLUE SAGE

WILDFLOWERS/*(CONTINUED)* **SCIENTIFIC NAME**	**COMMON NAME**
Salvia engelmannii	ENGELMANN'S SAGE
Salvia farinacea	MEALY SAGE
Salvia texana	TEXAS SAGE
Scutellaria resinosa	RESIN-DOT SKULLCAP
Scutellaria wrightii	WRIGHT'S SKULLCAP
Senna marilandica	MARYLAND SENNA
Silphium albiflorum	WHITE ROSINWEED
Silphium laciniatum	COMPASSPLANT
Silphium spp.	ROSINWEED
Solidago spp.	GOLDENROD
Stenosiphon linifolius	STENOSIPHON, FALSE GAURA
Tephrosia virginiana	GOAT'S-RUE
Tetraneuris scaposa	FOUR-NERVE DAISY
Teucrium canadense	GERMANDER
Thelesperma spp.	GREENTHREAD
Vernonia baldwinii	IRONWEED, BALDWIN'S IRONWEED
Vernonia lindheimeri	IRONWEED, WOOLLY IRONWEED

ORNAMENTAL NATIVE PLANTS
FOR DALLAS, TEXAS
RANGE: BEYOND 50 MILE RADIUS

TREES/	
Acacia greggii var. *wrightii*	WRIGHT'S ACACIA, CATCLAW
Acer barbatum	CADDO MAPLE
Acer grandidentatum	BIG-TOOTH MAPLE
Acer leucoderme	CHALK MAPLE
Aesculus pavia var. *flavescens*	PALE BUCKEYE, TEXAS YELLOW BUCKEYE
Aesculus pavia var. *pavia*	RED BUCKEYE
Arbutus xalapensis	TEXAS MADRONE
Cercis canadensis var. *mexicana*	MEXICAN REDBUD
Cercocarpus montanus var. *glabra*	SMOOTHLEAF MOUNTAIN MAHOGANY
Chilopsis linearis	DESERT-WILLOW
Crataegus tracyi	TRACY'S HAWTHORN
Diospyros texana	TEXAS PERSIMMON
Fraxinus cuspidata	FRAGRANT ASH
Ilex vomitoria	YAUPON, YAUPON HOLLY
Juglans microcarpa	LITTLE WALNUT, TEXAS WALNUT
Leucaena retusa	GOLDEN-BALL LEADTREE
Liquidambar styraciflua	SWEETGUM
Magnolia grandiflora	SOUTHERN MAGNOLIA
Pinus cembroides	PINYON, MEXICAN PINYON
Quercus laceyi	LACEY'S OAK
Quercus mohriana	MOHR'S OAK, MOHR'S SHIN OAK
Quercus pungens var. *pungens*	SANDPAPER OAK
Quercus pungens var. *vaseyana*	VASEY'S OAK, VASEY'S SHIN OAK
Taxodium distichum	BALD CYPRESS
Vauquelinia corymbosa var. *angustifolia*	CHISOS ROSEWOOD, SLIM-LEAF VAUQUELINIA

SHRUBS/ SCIENTIFIC NAME	COMMON NAME
Anisacanthus quadrifidus var. *wrightii*	FLAME ACANTHUS, WRIGHT'S ACANTHUS
Atriplex canescens	FOUR-WING SALTBUSH
Atriplex confertifolia	SHADESCALE
Bouchea linifolia	FLAX-LEAF BOUCHEA
Cotinus obovatus	AMERICAN SMOKETREE
Croton alabamensis var. *texensis*	TEXABAMA CROTON
Dalea bicolor var. *argyrea*	SILVER DALEA
Dalea greggii	GREGG'S DALEA
Fallugia paradoxa	APACHE-PLUME
Fendlera spp.	FENDLERBUSH
Forestiera reticulata	NET-LEAF FORESTIERA
Garrya ovata subsp. *lindheimeri*	LINDHEIMER'S SILKTASSEL
Hesperaloe parviflora var. *parviflora*	RED-FLOWER-YUCCA
Lantana urticoides	TEXAS LANTANA
Leucophyllum spp.	SILVERLEAF
Menodora longiflora	SHOWY MENODORA
Myrica cerifera (INCLUDES DWARF FORMS)	SOUTHERN WAX-MYRTLE
Pavonia lasiopetela	WRIGHT'S PAVONIA
Philadelphus spp.	WILD MOCK ORANGE
Rhus microphylla	LITTLE-LEAF SUMAC
Ribes aureum var. *villosum*	BUFFALO CURRANT
Salix exigua	SANDBAR WILLOW, COYOTE WILLOW
Salvia ballotaeflora	SHRUBBY BLUE SAGE
Salvia greggii	AUTUMN SAGE
Salvia regla	MOUNTAIN SAGE
Styrax platanifolia	SYCAMORE-LEAF SNOWBELL
Yucca rupicola	TWIST-LEAF YUCCA

List of Sources for Native Plants

Native Plant Society Of Texas

State Office
Coordinator: Dana Tucker
P.O. Box 891
Georgetown, TX 78627
512/238-0695
FAX 512/238-0703
E-MAIL: dtucker@io.com

Nurseries

Anderson Landscape and Nursery
2222 Pech
Houston, TX 77055
713/984-1342

Barton Springs Nursery
3601 Bee Cave Rd.
Austin, TX 78746
512/328-6655

Bluestem Nursery
4101 Curry Rd.
Arlington, TX 76017
817/478-6202
Contact: John S. Snowden
Native ornamental grasses, catalog

Brazos Rim Farm, Inc.
433 Ridgewood
Ft. Worth, TX 76107
817/740-1184
FAX 817/625-1327
Contact: Pat Needham
Wholesale

Buchanan's Native Plants
611 E. 11th Street
Houston, TX 77008
713/861-5702
Retail

Chaparral Estates Gardens
Rte. 1, Box 425
Killeen, TX 76542
817/526-3973
Contact: Ken and Rita Schoen

Discount Trees of Brenham
2800 N. Park Street
Brenham, TX 77833
409/836-7225
Retail

Dodd Family Tree Nursery
515 West Main
Fredericksburg, TX 78624
830/997-9571
Retail; native plants, organics, special order

Ecotone Gardens
806 Pine/Hwy 69
Kountze, TX 77625
409/246-3070
Contact: Becky Wilder, owner

Garden-Ville of Austin
8648 Old Bee Caves Rd.
Austin, TX 78735
512/288-6113
FAX 512/288-6114
Retail

Gottlieb Gardens
8263 Huber Rd.
Seguin, TX 78155
830/629-9876
Wholesale

Gunsight Mountain Ranch and Nursery
Williams Creek Rd., Box 86
Tarpley, TX 78883
210/562-3225
FAX 562-3266
Wholesale/retail

Hager Nursery
A division of Hager Landscape & Tree, Inc.
1324 Old Martindale Rd.
San Marcos, TX 78666
512/392-1089 or 800/443-TREE
Contact Robert Hager
Retail

Kings Creek Gardens
813 Straus Rd.
Cedar Hill, TX 75104
972/291-7650
FAX 972/293-0920
Contact: Rosa Finsley, owner

Love Creek Nursery
P.O. Box 1401
Medina, TX 78055
210/589-2265
Contact: Ann Landry
Wholesale/retail

Madrone Nursery
2318 Hilliard
San Marcos, TX 78666
512/353-3944
Wholesale/retail

Native American Seed
610 Main Street
Junction, TX 76849
800/728-4043
INTERNET: http://www.seedsource.com

Native Resources Inc.
Rt.1, Box 7J, on FM 971
Georgetown, TX 78626
512/930-3935

Natives of Texas
Spring Canyon Ranch
6520 Medina Hwy.
Kerrville, TX 78028
210/896-2169 or 210/698-3736
Contact: Betty Winningham

Native Texas Nursery
1004 MoPac Circle #101
Austin, TX 78746
512/280-2824
Contact: Henry Chalmers

North Haven Gardens
7700 Northaven Road
Dallas, TX 75230-3297
214/363-5316
Wholesale/retail

The Rustic Wheelbarrow
416 W. Avenue D
San Angelo, TX 76903
915/659-2130

Southwest Landscape and Nursery Company
2220 Sandy Lake Rd.
Carrollton, TX 75006
214/245-4557
Wholesale/retail

Texzen Gardens
4806 Burnet Rd.
Austin, TX 78756
512/454-6471

Weston Gardens in Bloom, Inc.
8101 Anglin Dr.
Ft Worth, TX 76140
817/572-0549

Wichita Valley Landscape
5314 SW Parkway
Wichita Falls, TX 76310
940/696-3082
Contact: Paul or Nila
10% discount to NPSOT members

Wildseed Farms, Inc.
425 Wildflower Hills
P.O. Box 3000
Fredericksburg, TX 78624-3000
800/848-0078
FAX 830/990-8090
INTERNET: www.wildseedfarms.com
Wholesale/retail: seed and live plants

Landscape Professionals

Anderson Landscape & Nursery
2222 Pech
Houston, TX 77055
713/984-1342

Rosa Finsley
Landscape Architect
Kings Creek Landscape
214/653-1160
972/293-0920

Don Gardner
Consulting Arborist
Native Plant Preservation Specialist
13903 Murfin Rd.
Austin, TX 78734
512/263-2586

Landscape Details
324 Cardinal
New Braunfels, TX 78130
830/629-9876
Contact: David E. Will

Place Collaborative, Inc.
8207 Callaghan Rd. #130
San Antonio, TX 78205
210/349-3434
Contact: Larry A. Hicks, ASLA

Dave Shows Associates
17320 Classen Rd.
San Antonio, TX 78247
210/497-3222

Wright Landscape for Texas
2922 High Plains Dr.
Katy, TX 77449
281/578-7304
Contact: LisaGay Wright
Specializing in wildlife habitats

G. Owen Yost, ASLA
Landscape Architect
4516 Coyote Point
Denton, TX 76208
Phone/FAX: 940/383-9655

Nature Centers and Organizations

Heard Natural Science Museum and Wildlife Sanctuary
One Nature Place
McKinney, TX 75069-8840
972/562-5566
E-MAIL: heardmuseum@texoma.net
INTERNET: http://www.heardmuseum.org

Lady Bird Johnson Wildflower Center
4801 LaCross Avenue
Austin, TX 78739
512/292-4100 (info line)
E-MAIL: wildflower@wildflower.org
INTERNET: http://www.wildflower.org/

Riverside Nature Center
150 Lemos St.
Kerrville, TX 78028
830/257-4837 Non-profit organization;
Information and education on the Hill Country

Seeds of Texas Seed Exchange
P.O. Box 9882
College Station, TX 77842
409/693-4485
E-MAIL: jackrowe@compuserve.com
INTERNET: http://csf.Colorado.EDU/perma/stse/
Seeds of native, garden, and landscape plants

These listings were provided by the Native Plant Society of Texas; such listings are published regularly in the *Native Plant Society of Texas News.*

State Botanical Symbols

Information on state botanical symbols can be found in Texas Parks & Wildlife (1995), Tyrl et al. (1994), and Jones et al. (1997).

Texas

Lupinus [Fabaceae], BLUEBONNET. All six *Lupinus* species which occur in the state are the **state flowers of Texas**. *Lupinus subcarnosus* was designated the state flower in 1901 and in 1971 the legislature extended state flower status to the other five *Lupinus* species native in Texas (Andrews 1986).

Carya illinoinensis (Wangenh.) K. Koch [Juglandaceae], PECAN, NOGAL MORADO, NUEZ ENCARCELADA. **State tree of Texas** as designated by the state legislature in 1919 (Jones et al. 1997).

Bouteloua curtipendula (Michx.) Torr. [Poaceae], SIDE-OATS GRAMA. **State grass of Texas** as designated by the 62nd Texas Legislature in 1971 (Jones et al. 1997).

Opuntia [Cactaceae], PRICKLY-PEAR. All members of subgenus *Opuntia* (with flat stems) are considered the **state plant of Texas**, while those of subgenus *Cylindroopuntia* (with cylindrical stems) are not—as designated by the 74th state legislature (Jones et al. 1997).

Citrus paradisi (L.) Macfad. [*C. maxima* × *C. sinensis*] (cultivar "Ruby" (redblush)) [Rutaceae], TEXAS RED GRAPEFRUIT. **State fruit of Texas** as designated in 1993; while this hybrid cultivar was developed in TX, the parents are introduced (Jones et al. 1997).

Capsicum annuum L. var. *annuum* [Solanaceae], JALAPEÑO. **State pepper of Texas** as designated by the 74th state legislature (Jones et al. 1997).

Capsicum annuum L. var. *glabriusculum* (Dunal) Heiser & Pickersgill [Solanaceae], BIRD PEPPER, CHILITEPÍN, CHILIPIQUÍN, CHILE PIQUÍN, BUSH REDPEPPER. **Native pepper of Texas** as designated by Texas House Concurrent Resolution 82 in 1997 (Andrews 1998).

Oklahoma

Gaillardia pulchella Foug. [Asteraceae], FIRE-WHEELS, INDIAN-BLANKET, ROSE-RING GAILLARDIA. **State wildflower of Oklahoma** (Tyrl et al. 1994).

Cercis canadensis L. [Fabaceae], REDBUD, JUDASTREE. **State tree of Oklahoma** (Tyrl et al. 1994).

Sorghastrum nutans (L.) Nash [Poaceae], INDIAN GRASS, YELLOW INDIAN GRASS, INDIAN REED. **State grass of Oklahoma** (S. Barber, pers. comm.).

Phoradendron tomentosum (DC.) Engelm. ex A. Gray [Viscaceae], MISTLETOE, CHRISTMAS MISTLETOE, INJERTO, HAIRY MISTLETOE. **Official floral emblem of Oklahoma** as designated by the Assembly of the Territory of Oklahoma on 11 February 1893 (Tyrl et al. 1994).

SPECIAL RECOGNITION—BENNY J. SIMPSON

IN MEMORIAM

BENNY J. SIMPSON
1928–1996

While a number of individuals have contributed to the native plant movement in Texas (e.g., Carroll Abbott, Rosa Finsley, Lynn Lowrey, Robert Vines, Barton Warnock, Sally Wasowski), only Benny Simpson (Fig. 43) worked his entire career in North Central Texas and devoted much of his professional life to the development of native species for use as landscape plants. From 1954 to his death in 1996, Benny worked at the Texas Research Foundation at Renner which in 1972 became the Texas A&M Research and Extension Center at Dallas.

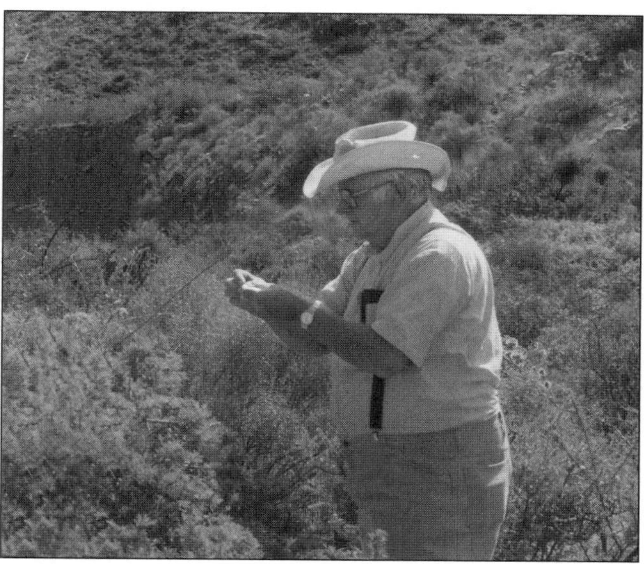

FIG. 43/BENNY SIMPSON (1928-1996).
USED WITH PERMISSION OF THE NATIVE PLANT SOCIETY OF TEXAS.

Simpson is possibly best known as the author of *A Field Guide to Texas Trees* (Simpson 1988) and he published many other scholarly works (for a list of his publications see Davis 1997). However, among botanists and native plant enthusiasts he is correctly best remembered as the "Pioneer of the Native Plant Movement" in Texas (Nokes 1997). Simpson understood that the scarcity of water is one of the biggest challenges facing Texas' future, and that native plants, well-adapted to the state's climate, are an important resource (e.g., Simpson & Hipp 1984; Simpson 1993). Through his research, nine superior selections of native plants were released to the nursery industry including three forms of *Leucophyllum* (Scrophulariaceae), widely known as Texas purple-sage (Nokes 1997; Kiphart 1997). In addition to his other contributions, Simpson was one of the founding members and a former president of the Native Plant Society of Texas and was active in that organization until his death (Nokes 1997; Pickens 1997).

Plants from Simpson's extensive research collection are now at the Dallas Horticultural Center at Fair Park, the Heard Natural Science Museum and Wildlife Sanctuary in McKinney, and the Benny Simpson Outdoor Learning Center at the Texas A&M Research and Extension Center at Dallas.

GLOSSARY

This glossary is modified from those of Shinners (1958a) and Mahler (1988), with additional entries obtained or modified from a variety of sources including Lawrence (1951), Featherly (1954), Correll (1956), Gleason and Cronquist (1963), Radford et al. (1968), Correll and Johnston (1970), Gould (1975), Lewis and Elvin-Lewis (1977), Benson (1979), Schmutz and Hamilton (1979), Fuller and McClintock (1986), Jones and Luchsinger (1986), Schofield (1986), Gandhi and Thomas (1989), Blackwell (1990), Isely (1990), Harris and Harris (1994), Spjut (1994), and Hickey and King (1997).

A

A- A prefix meaning without or not.

ABAXIAL Located on the side away from axis; e.g., lower leaf surface; contrasting with adaxial.

ABERRANT Different from normal or typical condition.

ABORTIVE Not developing or imperfectly developed; barren; defective.

ABSCISSION Act or process of cutting off or shedding; e.g., the shedding or abscising of leaves.

ABSCISSION LAYER Zone at base of petiole or other structure (e.g., pedicel) forming a layer of separation. This layer is important in the drop or shedding of leaves and fruits.

ACAULESCENT Stemless or apparently so; having leaves basal with stems not elongated.

ACCESSORY FRUIT A fruit or assemblage of fruits with fleshy parts derived from organs other than the pistil; e.g., strawberry with fleshy receptacle with achenes (individual fruits) embedded in its surface.

ACCRESCENT Enlarging after anthesis or with age, frequently in reference to the calyx.

ACCUMBENT COTYLEDONS Cotyledons lying face to face with the edges against the radicle.

ACEROSE (= Acicular) Needle-shaped or -like.

ACHENE Small, dry, indehiscent, one-seeded fruit with ovary wall free from seed.

ACHLAMYDEOUS Lacking a perianth.

ACHLOROPHYLLUS Lacking chlorophyll or apparently so; e.g., a number of non-green saprophytes or parasites.

ACICULAR (= Acerose) Needle-shaped or -like.

ACORN Fruit of a *Quercus* species (oak) composed of a nut and its cup or cupule made of fused bracts.

ACRID With sharp and harsh or bitterly pungent taste.

ACROPETAL Developing or maturing in succession from the base toward the apex.

ACTINOMORPHIC (= Regular) Radially symmetrical. The term usually refers to the arrangement of flower parts.

ACUMINATE Having a long, tapering point; longer tapering than acute.

ACUTE Forming a sharp angle of less than 90 degrees; less tapering than acuminate.

AD- A prefix meaning to or toward.

ADAXIAL Located on side towards axis; e.g., upper leaf surface; contrasting with abaxial.

ADHERENT Touching or sticking together, when two organs or parts (typically dissimilar) touch each other but are not grown or fused together.

ADNATE United or fused, when the fusion involves dissimilar structures; e.g., as in fusion of stamens and corolla.

ADPRESSED (= Appressed) Lying flat against a surface.

ADVENTITIOUS Referring to structures or organs that develop in an unusual position; e.g., buds or roots that develop out of their usual place.

ADVENTIVE Not fully naturalized or established; of occasional occurrence.

AERIAL Above ground level.

AESTIVAL Appearing in or pertaining to the summer.

AESTIVATION Arrangement of young flower parts in the bud.

AGAMOSPERMY The production of seeds without fertilization.

AGGREGATE Crowded into a dense cluster or tuft.

AGGREGATE FRUIT A fruit formed by the clustering together of a number of separate pistils from a single flower; e.g., a blackberry is a cluster of druplets.

AGLYCONE The nonsugar component of a glycoside. Glycosides are composed of a sugar plus another compound (the aglycone); many aglycones are toxic.

ALATE Winged.

ALBIDUS White.

ALBUMEN Nutritive material stored within the seed.

ALKALOID Any of a broad class (> 5000 known alkaloids) of bitter, usually basic (alkaline), organic compounds that contain nitrogen and typically have a ring in their structure. They are often physiologically active in animals; many are poisonous; many affect the nervous system; there are a number of general types based on chemical structure including indole, isoquinoline, piperidine, purine, pyrrolidine, quinoline, and tropane alkaloids; well known examples of alkaloids include atropine, caffeine, cocaine, quinine, morphine, nicotine, theobromine, and strychnine.

ALLELOPATHY, ALLELOPATHIC Harmful or detrimental chemical effect by one species upon another; e.g., a plant producing phytotoxic compounds that inhibit the germination or growth of other plants.

ALLERGEN Substance capable of inducing an allergic response.

ALLERGENIC Causing an allergic response or an allergy to become manifest.

ALLERGY Hypersensitivity of the body cells to specific substances as antigens and allergens, resulting in various types of reactions (e.g., anaphylaxis, contact dermatitis, hay fever).

ALLIACEOUS Onion-like.

ALLUVIAL Of or pertaining to alluvium (= organic or inorganic materials, including soils, deposited by running water).

ALTERNATE Bearing one leaf or other structure at a node; having only one attached at a given point; contrasting with opposite or whorled.

ALVEOLATE Honeycombed.

Ament (= Catkin) A flexible often pendulous spike or spike-like raceme of small, inconspicuous, unisexual, apetalous, usually wind-pollinated flowers, the whole falling as one piece; e.g., male inflorescence of oaks or pecan.

Amentiferous Bearing aments.

Amethystine Violet-colored.

Amino acids Compounds containing both an amino group and a carboxyl group. They are the subunits (monomers) that are linked together by peptide bonds to form the polymers known as proteins; some nonprotein amino acids are found free in plants and are sometimes toxic; e.g., in *Lathyrus* (Fabaceae).

Amorphous Without regular or definite form; shapeless.

Amphitropous ovule Ovule that is half inverted so that the point of attachment is near the middle.

Amplexicaul (= Clasping) With base of leaf or other structure (e.g., stipule) wholly or partly surrounding the stem.

Ampliate Enlarged.

Anastomosing Net-like; with veins connecting by cross-veins to form a network.

Anatropous ovule Ovule that is completely inverted, the micropylar end being essentially basal.

Androecium Collective term for the stamens or male structures of a flower.

Androgynophore A stalk bearing both androecium and gynoecium; e.g., in many Passifloraceae.

Androgynous Bearing staminate flowers above (= distal to) the pistillate in the same spike; e.g., in some Cyperaceae.

Androphore A support or column bearing stamens.

Anemophilous, Anemophily Wind-pollinated.

Angiosperm (= Flowering plant) Literally, "vessel seed"; a plant having its seeds enclosed in an ovary (= the proximal part of the carpel or "vessel"); a member of Division Magnoliophyta.

Annual Plant or root system living only one growing season (year); completing the growth cycle within one year.

Annular Arranged in a ring or circle.

Annulate With the appearance of rings; e.g., cross-ribbed or ringed spines of some Cactaceae.

Annulus A group or ring of thick-walled cells, on the sporangia of some ferns, that are involved in spore dehiscence.

Anterior Describing the position of an organ located toward the front in relation to the axis; e.g., in a flower the side away from the axis and toward the subtending bract.

Anther That part of a stamen producing the pollen.

Anther-cell (= Theca) One of the pollen-sacs or locules of an anther.

Antheridium Male sexual organ; structure forming male gametes, typically found in less derived plants (e.g., ferns) but so reduced evolutionarily as to not be present in flowering plants.

Anthesis (a) Time or process of flower expansion or opening; (b) also descriptive of period during which a flower is open and functional.

Anthocarp A structure that includes a fruit united with the perianth or the receptacle; e.g., Nyctaginaceae.

Anthocyanin A red, purplish, or blue water-soluble pigment found in most flowers. The color of these pigments is affected by pH (e.g., in *Hydrangea*); chemically, anthocyanins are phenolic.

Antipetalous Referring to stamens that are of the same number as, and borne in front of (= on the same radius as) the petals or corolla lobes.

Antisepalous Referring to stamens that are of the same number as, and borne in front of (= on the same radius as) the sepals or calyx lobes.

Antrorse Directed toward the summit, upward, or forward; e.g., pubescence directed up the stem, the free end of the hair above or distal to the attached end; contrasting with retrorse.

Antrorsely barbed With barbs (= points) pointing upward toward the summit or apex.

Apetalous Having flowers without petals.

Apetaly The condition of being without petals.

Apeture An opening.

Apex (pl. Apices) The tip or summit.

Aphyllopodic Lacking leaves at the base.

Aphyllous Leafless; e.g., Cuscutaceae.

Apical At the tip or apex; relating to the apex.

Apical bud (= Terminal bud) Bud at the end (= apex) of a stem or branch.

Apiculate Having a small sharp point formed by blade tissue (of a leaf, sepal, or petal) rather than by projection of a rib or vein; with an abrupt tip or projection.

Apocarpous With the carpels separate or free from one another.

Apogamous Forming a sporophyte without the union of gametes.

Apomixis A collective term for reproduction, including vegetative propagation, that does not involve sexual processes; any form of asexual reproduction.

Apopetalous (= Polypetalous) Referring to a corolla consisting of separate petals.

Apophysis Swelling or enlargement of the surface of an organ.

Aposepalous (= Polysepalous) Referring to a calyx consisting of separate sepals.

Appendage Any attached structure that is supplementary or secondary.

Appendiculate With an appendage.

Appressed (= Adpressed) Lying flat against a surface.

Approximate Close together.

Aquatic Living in water.

Arachnoid, Arachnose Cobwebby; cobweb-like, with entangled, slender, loose hairs; thinly pubescent with relatively long, usually appressed and interlaced hairs.

Arborescent Tree-like or becoming tree-like.

Archegonium Female sexual organ; structure forming female gametes, typically found in less derived plants (e.g., ferns) but so reduced evolutionarily as to not be present in flowering plants.

Arcuate Curved or bent like a bow, often used in reference to curving veins.

Arenaceous Sand-like or growing in sand.

Areolate Divided into small angular spaces; marked with areolae.

Areole, Areola (pl. Areolae) (a) Small space marked out on a surface, usually referring to the space bounded by veinlets on the surface of a leaf; (b) the small spine-bearing areas on a cactus stem.

Argillaceous Clayey; growing in clay or clay-like material.

Aril An appendage or covering on a seed, typically involved in dispersal by animals.

Leaf Characters

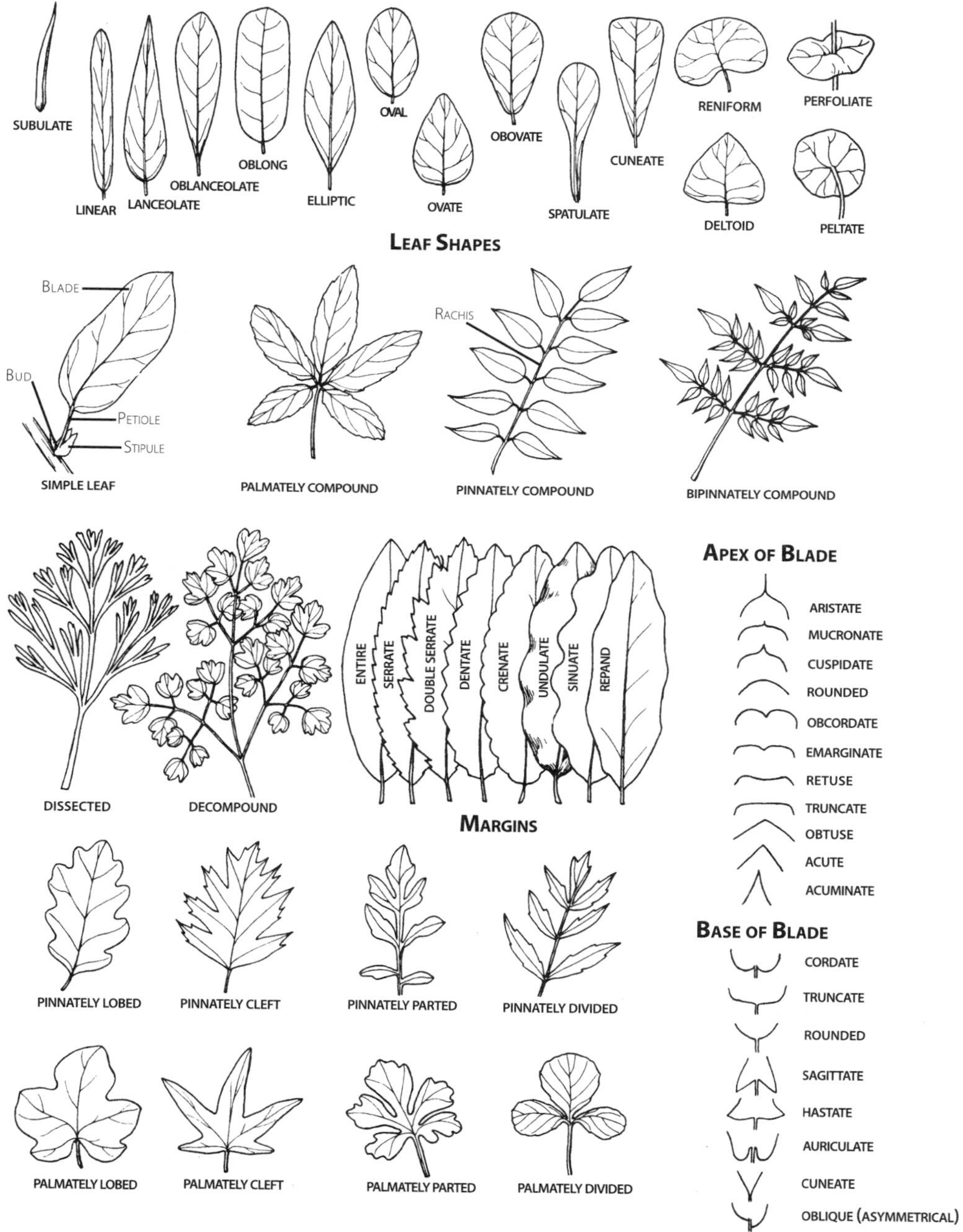
SUBULATE
LINEAR
LANCEOLATE
OBLANCEOLATE
OBLONG
ELLIPTIC
OVAL
OVATE
OBOVATE
SPATULATE
CUNEATE
RENIFORM
DELTOID
PERFOLIATE
PELTATE
LEAF SHAPES
BLADE
BUD
PETIOLE
STIPULE
SIMPLE LEAF
PALMATELY COMPOUND
RACHIS
PINNATELY COMPOUND
BIPINNATELY COMPOUND
DISSECTED
DECOMPOUND
ENTIRE
SERRATE
DOUBLE SERRATE
DENTATE
CRENATE
UNDULATE
SINUATE
REPAND
MARGINS
APEX OF BLADE
ARISTATE
MUCRONATE
CUSPIDATE
ROUNDED
OBCORDATE
EMARGINATE
RETUSE
TRUNCATE
OBTUSE
ACUTE
ACUMINATE
BASE OF BLADE
CORDATE
TRUNCATE
ROUNDED
SAGITTATE
HASTATE
AURICULATE
CUNEATE
OBLIQUE (ASYMMETRICAL)
PINNATELY LOBED
PINNATELY CLEFT
PINNATELY PARTED
PINNATELY DIVIDED
PALMATELY LOBED
PALMATELY CLEFT
PALMATELY PARTED
PALMATELY DIVIDED

ARILLATE With an aril.
ARISTATE Bearing a stiff awn or bristle.
ARISTULATE Diminutive of aristate.
ARMED Possessing sharp projections; e.g., prickles, spines, or thorns.
AROMATIC (a) Generally, having a fragrant odor; (b) chemically, containing or patterned after benzene rings, with or without fragrance.
ARTICLE The individual unit of a constricted or jointed fruit; e.g., in some Fabaceae such as *Desmodium*.
ARTICULATE Jointed; joined.
ARTICULATION A separation place; joint.
ASCENDING, ASCENDENT Rising at an oblique angle.
ASEPALOUS Without sepals.
ASEXUAL Without sex; reproducing without sex.
ASPEROUS Rough to the touch.
ASSURGENT Ascending, rising.
ASTYLOUS Without a style.
ASYMMETRICAL Without symmetry.
ATOMIFEROUS Bearing very fine glands.
ATROCASTANEOUS Very dark chestnut-colored.
ATROPURPUREOUS Dark purple; purple-black.
ATTENUATE Gradually tapering to a very slender tip, the taper more gradual than in acuminate.
ATYPICAL Not typical; deviating from the norm.
AURICLE Earlobe-like lobe or appendage; e.g., at the base of some leaves, sepals, etc.
AURICULATE With an auricle.
AUTOTROPHIC Descriptive of an organism capable of making its own food, usually through photosynthesis; free living, not parasitic or saprophytic; e.g., green plants.
AUTUMNAL Associated with or occurring in the fall of the year.
AWL-SHAPED (= Subulate) Tapering from the base to a slender or stiff point; narrow and sharp-pointed.
AWN Terminal slender bristle or hair-like extension or projection; e.g., in grasses, the prolongation of the midnerve of the glumes or lemmas.
AWN COLUMN In certain grasses, a prominent narrowed beak at the apex of the lemma. The awns arise from this structure.
AXIAL Relating to the axis.
AXIL Angle between two organs; e.g., upper angle formed by a leaf and a stem.
AXILE or **AXILLARY PLACENTATION** Placentation with the ovules attached to the central axis of the ovary.
AXILLARY In an axil; e.g., in the angle between a leaf and a stem.
AXIS (pl. AXES) (a) the central stem from which organs arise; (b) a portion of a plant from which a series of organs arises radially; e.g., the axis of an inflorescence.

B

BACCATE Resembling or having the structure of a berry; berry-like.
BALLISTIC Referring to fruits that are forcibly or elastically dehiscent, whose seeds are thrown catapult-like; e.g., *Phyllanthus* (Euphorbiaceae).
BANNER (= Standard) Adaxial and typically largest petal of a papilionaceous flower.
BARBED With short reflexed points like a multi-pronged fishhook.
BARBELLATE Diminutive of barbed; with short, fine, stiff hairs.
BARK Outer (= external to vascular cambium) protective tissues on the stems or roots of woody plants.
BASAL Located at the base of a plant or of an organ.
BASAL ROSETTE Cluster of leaves on or near the ground.
BASAL STYLE Style projecting from among the lobes of a deeply lobed ovary.
BASIFIXED Attached basally, typically referring to attachment of an anther to a filament; contrasting with either dorsifixed or versatile.
BASILAMINAR At base of blade of leaf or other structure.
BASIPETAL Developing or maturing in succession from the apex toward the base.
BASISCOPIC Directed toward the base.
BASONYM The original epithet assigned to a species (or other taxon of lower rank) by its author.
BEAK A long, prominent, and relatively thickened point; a tapering projection; e.g., projection on a fruit resulting from a persistent style.
BEAKED Ending in a beak.
BEARD A group of long awns or bristle-like trichomes; a zone of pubescence; e.g., on some corollas.
BEARDED Bearing long or stiff hairs, typically in a line or tuft.
BERRY Indehiscent type of fruit with the entire pericarp fleshy and lacking a stone, usually with several to many seeds; e.g., tomato, grape.
BETALAINS Reddish, nitrogen-containing pigments (characteristic of most Caryophyllidae) that derive their name from the genus *Beta* (beets).
BI-, BIS- Latin prefix signifying two, twice, or doubly.
BICOLORED Two-colored.
BICONVEX Convex on both sides.
BIDENTATE Two-toothed.
BIENNIAL Plant or root system living only two years (growing seasons), typically producing only leafy growth the first year, then flowering and dying the second.
BIFID Two-cleft, usually deeply so; with two lobes or segments.
BIFURCATE Two-forked; e.g., some Y-shaped trichomes, stigmas, or styles.
BILABIATE Two-lipped, typically referring to corollas or calyces.
BILATERAL Arranged on two sides; two-sided.
BILATERALLY SYMMETRICAL With only one plane of symmetry; divisible into halves in one plane only.
BILOCULAR Having two cavities.
BINOMIAL The combination of a generic name and a specific epithet given to each species.
BINOMIAL NOMENCLATURE System of nomenclature where each species has a two-part name composed of a generic name and a specific epithet.
BIPARTITE Two-parted; divided into two parts nearly to the base.
BIPINNATE (= Twice-pinnate or 2-pinnate) Descriptive of a leaf with leaflets pinnately arranged on lateral axes that are themselves pinnately arranged on the main axis; with the primary divisions (pinnae) themselves pinnate.

Leaf and Twig Structure and Arrangement

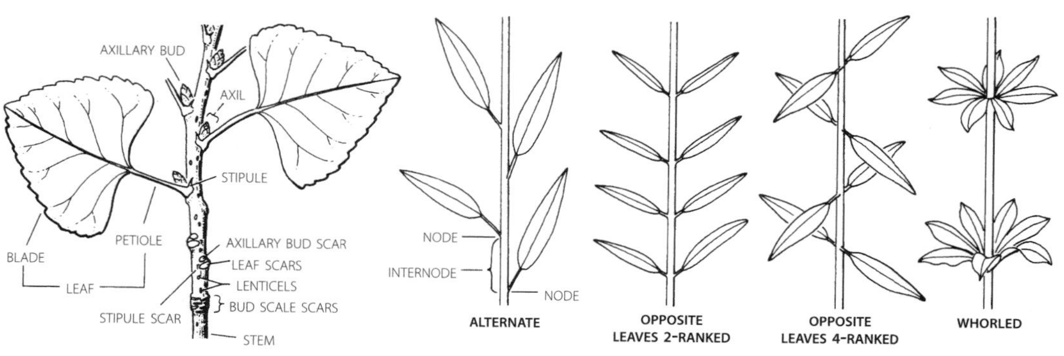

Flower Shape

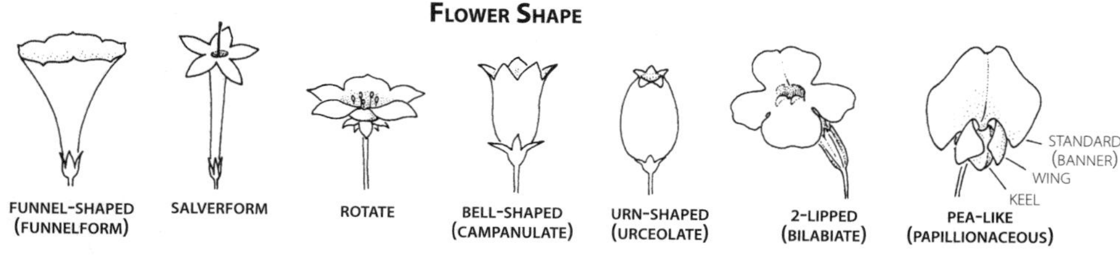

Flower Symmetry

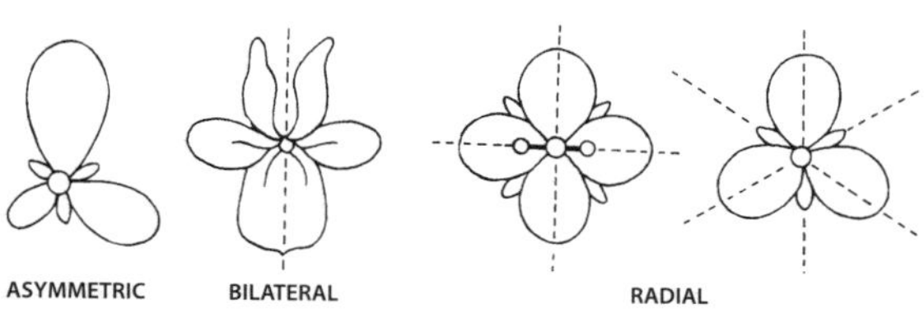

Flower Structure

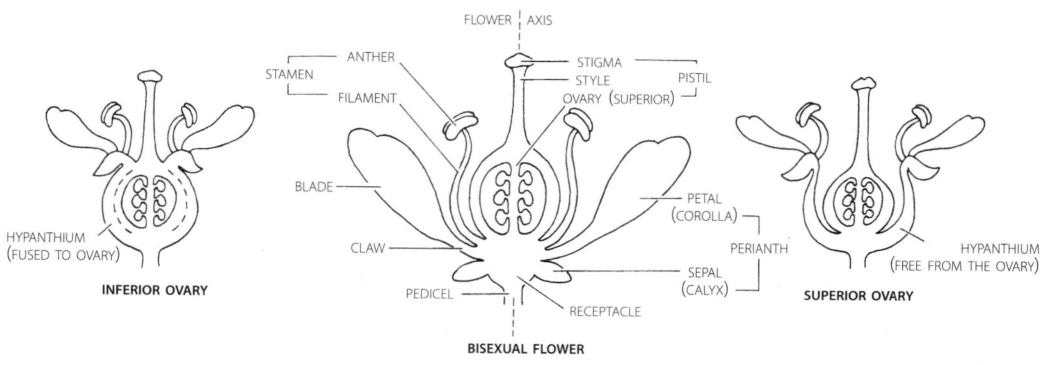

Adapted from Mason (1957); used with permission of Univ. of California Press; ©1957.

Bisected Completely divided into two parts.
Biseriate In two whorls or cycles; e.g., a perianth with both calyx and corolla.
Bisexual flower Type of flower with both stamens and pistil(s) functional within the same flower.
Bivalvate Opening by two valves.
Bladder (a) A thin-walled, inflated structure; (b) a hollow, membranaceous appendage that traps insects; e.g., *Utricularia* (Lentibulariaceae).
Bladdery Thin-walled and inflated; like the bladder of an animal.
Blade Flat, expanded portion, as the main part of a leaf or petal.
Bloom (a) Flower or flowering; (b) coating of white wax or powder, as on plums or grapes.
Bole The main trunk or stem of a tree.
Boss A protrusion.
Brackish Somewhat salty.
Bract A modified reduced leaf typically subtending a flower or cluster of flowers. Bracts can vary from very leaf-like to scale-like or thread-like; in some cases they can be colorful and attract pollinators.
Bracteal Having the form or position of a bract.
Bracteate Having bracts.
Bracteolate Having bracteoles.
Bracteole, Bractlet A usually small bract borne on a secondary axis (e.g., on a pedicel).
Bracteose Having numerous or conspicuous bracts.
Branch A shoot or secondary stem growing from the main stem.
Branchlet The ultimate division of a branch.
Bristle Stiff, strong but slender hair or trichome.
Bristly Bearing bristles.
Broad (= Wide) Distance across a structure (equal to diameter if tubular); sometimes restricted to signify the width or diameter of three-dimensional structures.
Bryophyta Group containing the mosses, liverworts, and hornworts. The Bryophyta is not treated in this flora.
Bud (a) Undeveloped or unopened flower; (b) undeveloped, much-condensed shoots, containing embryonic (meristematic or growing) tissue, usually covered by scales or bracts. Such buds are usually found at the tips of stems or in the axils of leaves.
Bulb Underground structure composed of a short, disc-like stem and one or more buds surrounded by layers of thickened fleshy leaf bases or scales; e.g., an onion.
Bulbil Small bulbs produced in an inflorescence or in leaf axils; e.g., in *Allium* inflorescences.
Bulblet Small bulbs produced alongside a parent bulb; e.g., the numerous underground small bulbs produced by some garlics.
Bulbous, Bulbose Having bulbs or bulb-like structures.
Bullate Describing a surface with rounded elevations resembling blisters or puckers.
Bur, Burr A structure with a rough or prickly envelope or covering; e.g., sandbur.
Bush (= Shrub) A woody perennial usually branching from the base with several main stems.

C

Ca. Latin, circa; abbreviation meaning about, around, approximately.
Caducous Falling off early, quickly, or prematurely; e.g., the sepals in some Papaveraceae.
Caespitose (= Cespitose) Growing in clumps or tufts.
Calcarate With a spur.
Calcareous Containing an excess of available calcium, usually in the form of the compound calcium carbonate; containing limestone or chalk.
Caliche A crust of calcium carbonate formed on stony soils in arid regions.
Callosity (= Callus). A hard protuberance or thickened, raised area.
Callous Having the texture of a callus.
Callus (= Callosity). A hard protuberance or thickened, raised area; e.g., thickened, hardened, basal portion of some lemmas in the Poaceae.
Calycine Resembling or pertaining to a calyx.
Calyculate Calyx-like; e.g., describing bracts that by their size or position are suggestive of a calyx.
Calyptra A lid, cap, cover, or hood; e.g., the lid of certain fruits and moss spore cases.
Calyx (pl. Calyces, Calyxes) Collective term for the sepals; outer series of floral "leaves", often enclosing the other flower parts in bud. The calyx is typically green but can be corolla-like and showy.
Calyx lobe One of the free projecting parts of a synsepalous calyx; also referred to as a calyx tooth.
Calyx tube The basal or tubular portion of a synsepalous calyx, as opposed to the free, distal, calyx lobes.
Cambium The thin layer of delicate, rapidly dividing, meristematic cells that forms wood internally and bark externally; also known as vascular cambium.
Campanulate Bell-shaped; rounded at base with a broad flaring rim.
Campylotropus ovule Ovule curved in its development, so that the morphological apex lies near the base.
Canaliculate Longitudinally channeled or grooved.
Cancellate Latticed.
Cane Stem, specifically (a) floricane, the flowering stem of *Rubus* species (blackberries and dewberries); (b) primocane, first-year leafy stem of the same; (c) persistent woody stems of *Arundinaria gigantea*, giant cane.
Canescent With whitish or grayish-white appearance due to abundance of soft short hairs.
Cap A convex, lid-like, removable covering; e.g., the apical portion of a circumscissile capsule. The term calyptra is used for the cap of some fruits and moss spore cases.
Capillary Hair-like; very slender.
Capitate (a) In heads, head-like, or head-shaped; aggregated into a dense or compact cluster; (b) referring to capitate hairs, like a pin-head on a stalk.
Capitellate Aggregated into a small, dense cluster; diminutive of capitate.

Types of Inflorescences

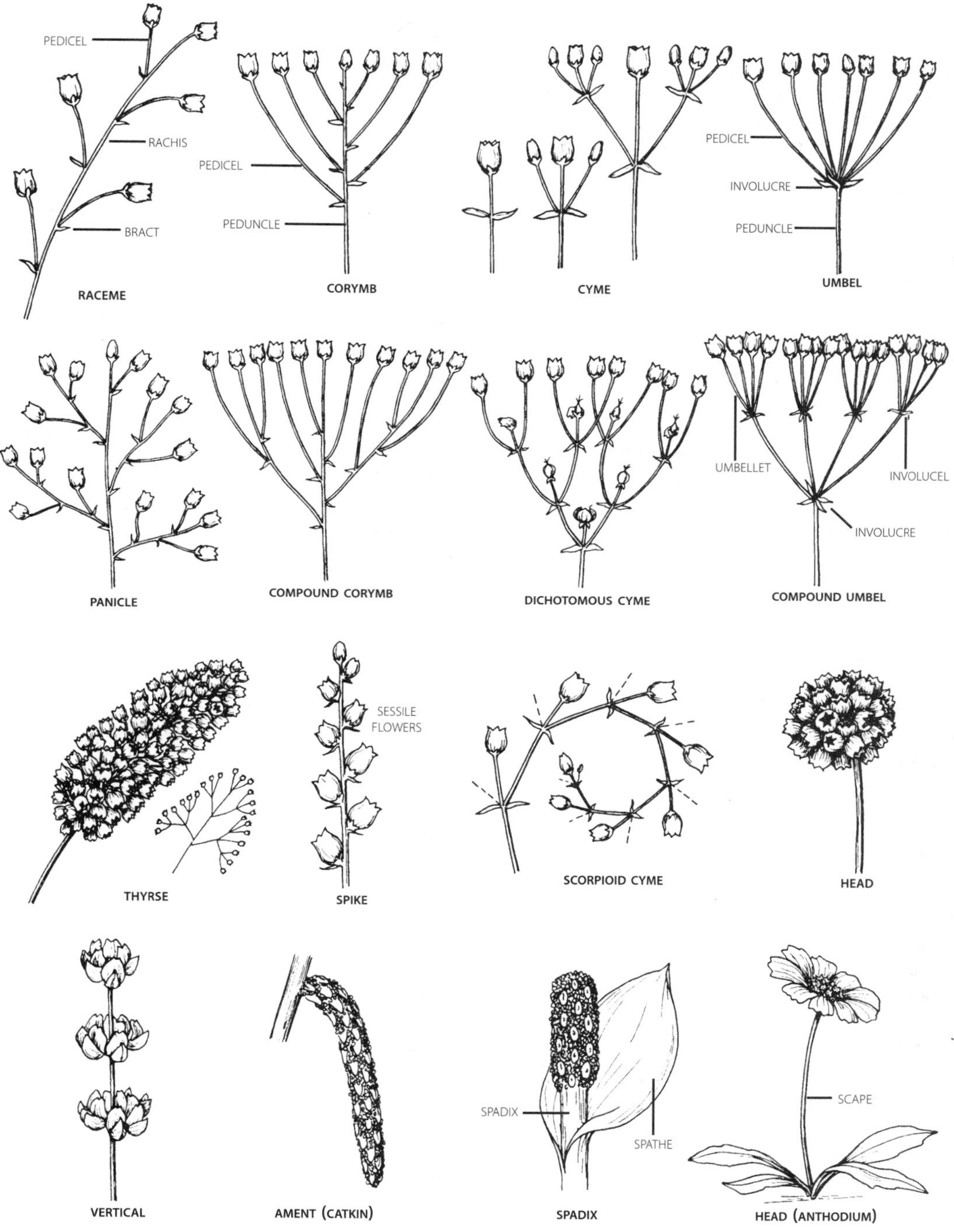

CAPITULUM (= Head) Dense cluster of sessile or nearly sessile flowers. This type of inflorescence is typical of the Asteraceae.

CAPSULAR Having the structure of a capsule.

CAPSULE A dry dehiscent fruit developed from more than one carpel.

CARCINOGEN A substance potentially inducing cancer or malignancy.

CARDIAC GLYCOSIDE A glycoside (two-component molecule) that upon breakdown yields a heart stimulant as the aglycone (= non-sugar component). The aglycones are steroidal in structure and are typically poisonous; e.g., digoxin and digitoxin from *Digitalis*, used as in treating heart trouble.

CARINA (= Keel) The two lower (= abaxial) fused petals of a papilionaceous flower (Fabaceae); (b) prominent longitudinal ridge, shaped like the keel of a boat.

CARINATE Keel-shaped; provided with a ridge or keel extending lengthwise along the middle. If more than one keel is present, the fact may be indicated by a numerical prefix.

CARNIVOROUS Referring to plants that digest animal (primarily insect) tissue to obtain nutrients such as nitrogen.

CARNOSE Fleshy; succulent.

CARPEL A modified leaf bearing ovules; a simple pistil or one unit of a compound pistil; female reproductive structure in flowering plants.

CARPELLATE Possessing carpels. The term is sometimes used with a numerical prefix to indicate the number of carpels.

CARPOPHORE The slender prolongation of the floral axis between the carpels that supports the pendulous fruit segments (= mericarps) in the Apiaceae.

CARTILAGINOUS Tough and hard but not bony; gristly; cartilage-like.

CARUNCLE Enlarged, somewhat spongy, seed appendage.

CARYOPSIS Achene-like, 1-seeded fruit with pericarp adnate to the seed coat; fruit typical of the Poaceae.

CASTANEOUS Chestnut-colored; dark brown.

CATKIN (= Ament) A flexible often pendulous spike or spike-like raceme of small, inconspicuous, unisexual, apetalous, usually wind-pollinated flowers, the whole falling as one piece; e.g., male inflorescence of oaks or pecan.

CAUDATE Having a tail or tail-like appendage.

CAUDEX (pl. CAUDICES) Woody stem base.

CAULESCENT With an evident leafy stem above ground.

CAULIFLOROUS Having flowers borne along the stems or trunks.

CAULINE Growing on or pertaining to the stem.

CELL (a) One of the living units of which a plant is composed; (b) (= locule) cavity or compartment containing the ovules in a carpel or the pollen in an anther.

CENTRIFUGAL Maturation of parts from the center toward the periphery.

CENTRIPETAL Maturation of parts from the periphery toward the center.

CERACEOUS Waxy.

CERIFEROUS Wax-bearing; waxy.

CERNUOUS Nodding or drooping.

CESPITOSE (= Caespitose) Growing in clumps or tufts.

CHAFF Thin, dry, or membranous scales or bracts, often used to refer to receptacular scales or bracts in many Asteraceae; see pale or palea.

CHAFFY Thin, dry, or membranous.

CHALAZA The basal part of the ovule where it attaches to the funiculus.

CHANNELED Deeply grooved.

CHARTACEOUS Stiffly papery.

CHASMOGAMOUS Referring to flowers that open at anthesis; with pollination after opening of flowers; contrasting with cleistogamous.

CHIROPTEROPHILY, CHIROPTEROPHILOUS Bat-pollination.

CHLOROPHYLL The light-capturing pigment giving the green color to plants. Because chlorophyll absorbs less green than other wavelengths of light (and thus reflects and transmits relatively more green), leaves appear green to the human eye.

CHORIPETALOUS Composed of or characterized by separate petals.

CHORISEPALOUS Composed of or characterized by separate sepals.

CHROMOSOMES Thread-like "colored bodies" occurring in the nuclei of cells and containing the genetic material.

CILIATE With a marginal fringe of hairs similar to eye lashes.

CILIOLATE Diminutive of ciliate.

CILIUM (pl. CILIA) Marginal hair or trichome.

CINCINNUS A curl; e.g., a helicoid (= curled or coiled) cyme, as in the Boraginaceae.

CINEREOUS Ash-colored; light-gray.

CIRCINATE Coiled, with the apex innermost, as the young fronds in some ferns.

CIRCUMSCISSILE Dehiscing by a regular transverse line around the fruit or anther, the top coming off like a lid.

CIRRHOUS Tendril-like; e.g., a leaf with a slender coiled apex.

CLADOPHYLL, CLADODE (= Phylloclade) A portion of a stem having the general form and function of a leaf; a flattened photosynthetic stem.

CLAMBERING Vine-like; growing over other plants often without the aid of tendrils or twining stems.

CLASPING (= Amplexicaul) With base of leaf or other structure (e.g., stipule) wholly or partly surrounding the stem.

CLASS The unit, category, or rank in classification made up of one or more orders; ending in -ae or -opsida; sometimes divided into subclasses which in turn are made of orders.

CLATHRATE Latticed; with a series of crossed members.

CLAVATE Club-shaped; becoming gradually enlarged apically.

CLAVELLATE Diminutive of clavate.

CLAW Stalk-like basal portion of some petals or sepals.

CLAWED Having a claw.

CLEFT (a) Cut 1/2 or more the distance from the margin to midrib or from the apex to base; (b) generally, any deep cut.

CLEISTOGAM A small flower that does not open and is necessarily self-pollinating.

CLEISTOGAMOUS, CLEISTOGAMIC Referring to flowers not opening at anthesis and thus self-pollinating; with pollination prior to opening of flowers. Such flowers frequently have reduced or incompletely formed parts (e.g., petals).

CLESITOGAMY The self-pollination of flowers that do not open.

CLONE A group of individuals of the same genotype, usually propagated vegetatively.

Types of Fruits and Root and Stem Variations

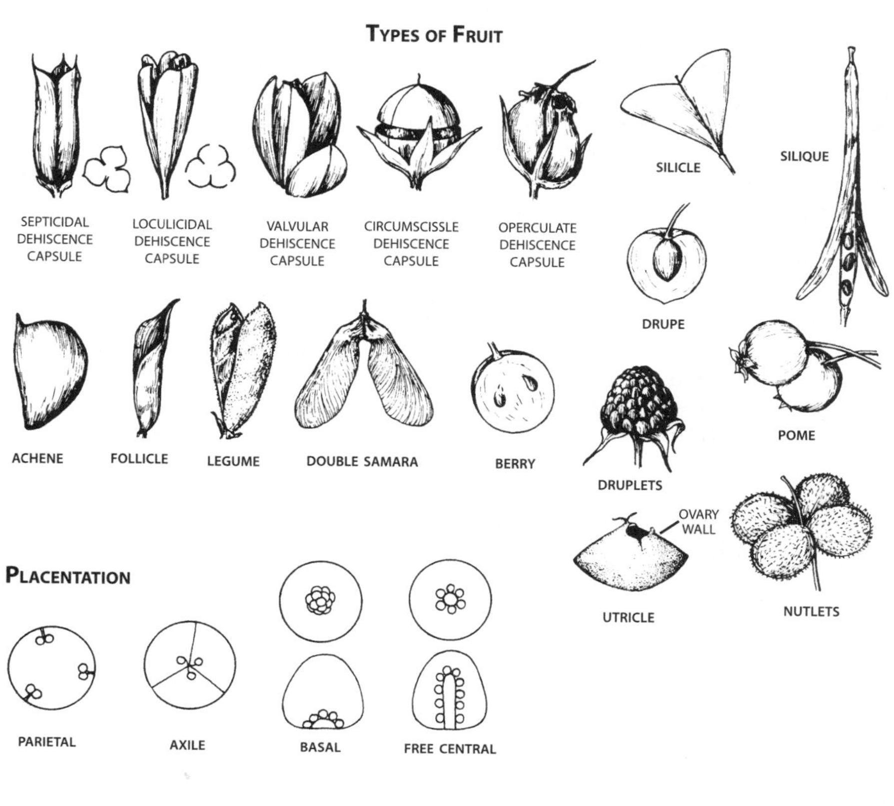
Types of Fruit
SEPTICIDAL DEHISCENCE CAPSULE
LOCULICIDAL DEHISCENCE CAPSULE
VALVULAR DEHISCENCE CAPSULE
CIRCUMSCISSLE DEHISCENCE CAPSULE
OPERCULATE DEHISCENCE CAPSULE
SILICLE
SILIQUE
DRUPE
ACHENE
FOLLICLE
LEGUME
DOUBLE SAMARA
BERRY
DRUPLETS
POME
OVARY WALL
UTRICLE
NUTLETS
Placentation
PARIETAL
AXILE
BASAL
FREE CENTRAL

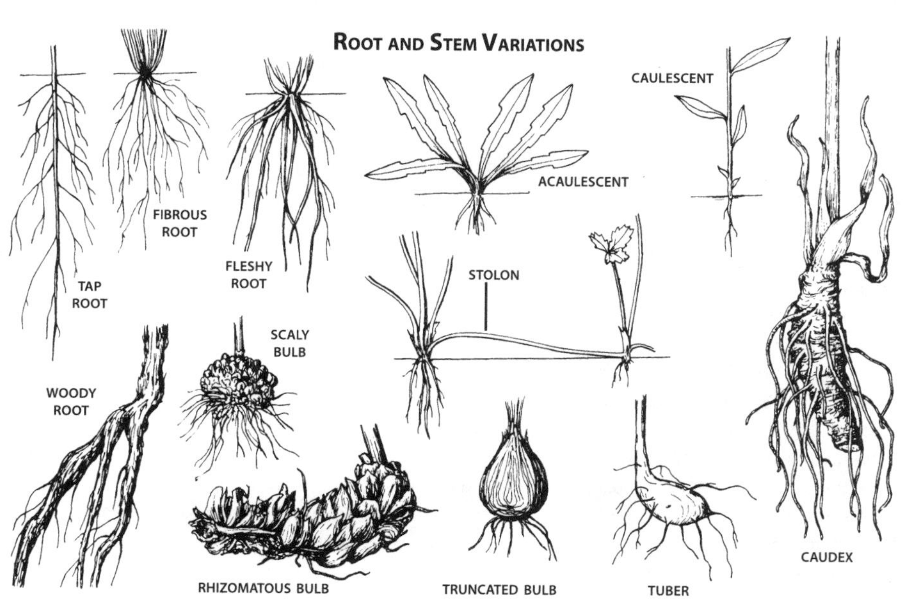
Root and Stem Variations
CAULESCENT
ACAULESCENT
FIBROUS ROOT
FLESHY ROOT
TAP ROOT
STOLON
SCALY BULB
WOODY ROOT
CAUDEX
RHIZOMATOUS BULB
TRUNCATED BULB
TUBER

CLUMP A single plant with two to many, more or less crowded stems arising from a branched rootstock or short rhizome.

CM Centimeter; 10 mm; 1/100 of a meter; ca. 2/5 of an inch.

COALESCENT Referring to organs of one kind that have grown together.

COARCTATE Crowded together.

COB Rachis or central stalk of the pistillate inflorescence of corn.

COCCUS (pl. COCCI) (a) a berry; (b) one of the parts of a lobed or deeply divided fruit with 1-seeded sections.

COCHLEATE Coiled like a snail shell; spiral.

COETANEOUS At the same time or of the same age; e.g., flowers and leaves appearing at the same time.

COHERENT Descriptive of the close association of two similar structures without fusion.

COLLAR The outer side of a grass leaf at the junction of the blade and sheath.

COLLATERAL Located side by side; e.g., ovules located side by side; e.g., accessory buds located on either side of a lateral bud.

COLUMELLA The persistent central axis around which the carpels of some fruits are arranged.

COLONIAL Forming colonies usually by means of underground rhizomes, stolons, etc. The term is commonly used to describe groups of plants with asexual reproduction.

COLONY A stand, group, or population of plants of one species, spreading vegetatively, or from seeds, or both.

-COLPATE A suffix referring to pollen grains having grooves (= colpi).

-COLPORATE A suffix referring to pollen grains having grooves and pores.

COLUMN (a) United style and filaments in Orchidaceae; (b) united filaments in Malvaceae and Asclepiadaceae; (c) basal differentiated portion of the awn(s) in certain grasses.

COLUMNAR Column-shaped.

COMA (a) A tuft of soft hairs or trichomes, as at the apices or bases of some seeds; (b) tuft of structures projecting from something (e.g., tuft of bracts projecting from heads of some *Eryngium* species).

COMATE, COMOSE Resembling or provided with a coma.

COMMISURE The surface where organs are joined; e.g., the face by which one carpel joins another.

COMPLANATE Flattened.

COMPLETE With all of the usual parts; e.g., a flower with all four flower parts: sepals, petals, stamens, and pistils.

COMPLICATE Folded together.

COMPOSITE (a) (= Compound) made up of several distinct parts; (b) common name for species of the Asteraceae.

COMPOUND (= Composite) Made up of several distinct parts.

COMPOUND INFLORESCENCE One having two or more degrees of branching; e.g., a compound umbel is one whose branches bear branchlets rather than ending directly in flowers.

COMPOUND LEAF A leaf that is cut completely to the base or midrib into segments (= leaflets) resembling miniature leaves; a leaf with two or more leaflets.

COMPOUND OVARY Ovary developed from two or more united carpels, as evidenced by the presence of two or more locules, valves, placentae, styles, or stigmas.

COMPOUND PISTIL Pistil composed of two or more united carpels.

COMPRESSED Flattened.

CONCAVE Hollow; with a depression on the surface.

CONCOLOR, CONCOLOROUS Of a uniform color.

CONDUPLICATE Folded together lengthwise.

CONE (= Strobilus) A usually globose or cylindrical structure involved in reproduction and composed of an axis with a spiral, usually dense aggregation of sporophylls, bracts, or scales (these bearing spores, pollen, or seeds).

CONFLUENT Blending of one part into another.

CONGESTED Crowded together.

CONGLOMERATE Densely clustered.

CONICAL, CONIC Cone-shaped.

CONIFEROUS Cone-bearing.

CONJUGATE Jointed in pairs.

CONNATE United or fused, when the fusion involves two or more similar structures; e.g., as in fusion of stamens into a tube.

CONNATE-PERFOLIATE Both connate and perfoliate; e.g., two leaves grown together and completely encircling a stem; e.g., in *Eupatorium perfoliatum*.

CONNECTIVE The tissue connecting the pollen-sacs of an anther. In certain plants the connective is prolonged at its base or apex.

CONNIVENT Converging or nearly or quite in contact, but not fused; e.g., connivent stamens in *Solanum*.

CONSERVED Term applied to a scientific name whose use, even though illegitimate according to nomenclatural rules, is allowed by the International Code of Botanical Nomenclature; e.g., many family names long in use, such as Cactaceae and Caryophyllaceae, have been conserved to prevent confusion.

CONSPECIFIC Of the same species.

CONSTRICTED Tightened or drawn together or narrowed.

CONTACT DERMATITIS Inflammation of the skin due to contact with poisons, irritants, or sensitizers. In some individuals it can be caused by even the slightest contact; caused by a variety of plants including poison-ivy (*Toxicodendron radicans*).

CONTIGUOUS Touching, but not fused.

CONTINUOUS Not interrupted; not articulated; not jointed.

CONTORTED Twisted or distorted.

CONTRACTED Narrowed or shortened; reduced in size.

CONTRARY In an opposite direction or at right angles to.

CONVERGENT Coming together or approaching.

CONVEX Rounded or bulged on the surface.

CONVOLUTE Rolled or twisted together when in an undeveloped stage.

CORALLOID Coral-like.

CORDATE, CORDIFORM (a) Heart-shaped; with a notch at the base and ovate in outline (the words apply specifically to flat surfaces and to solid shapes, respectively); (b) often referring only to the notched base of a structure; e.g., leaf base.

CORIACEOUS With texture like leather; tough; leathery.

CORM Bulb-like usually subterranean stem base, solid instead of with layers of modified leaves as in a true bulb.

CORMOUS Having a corm.

CORNEOUS Horny in texture.

CORNICULATE Bearing a small horn or horns.

CORNUTE Horned or spurred.

Types of Pubescence

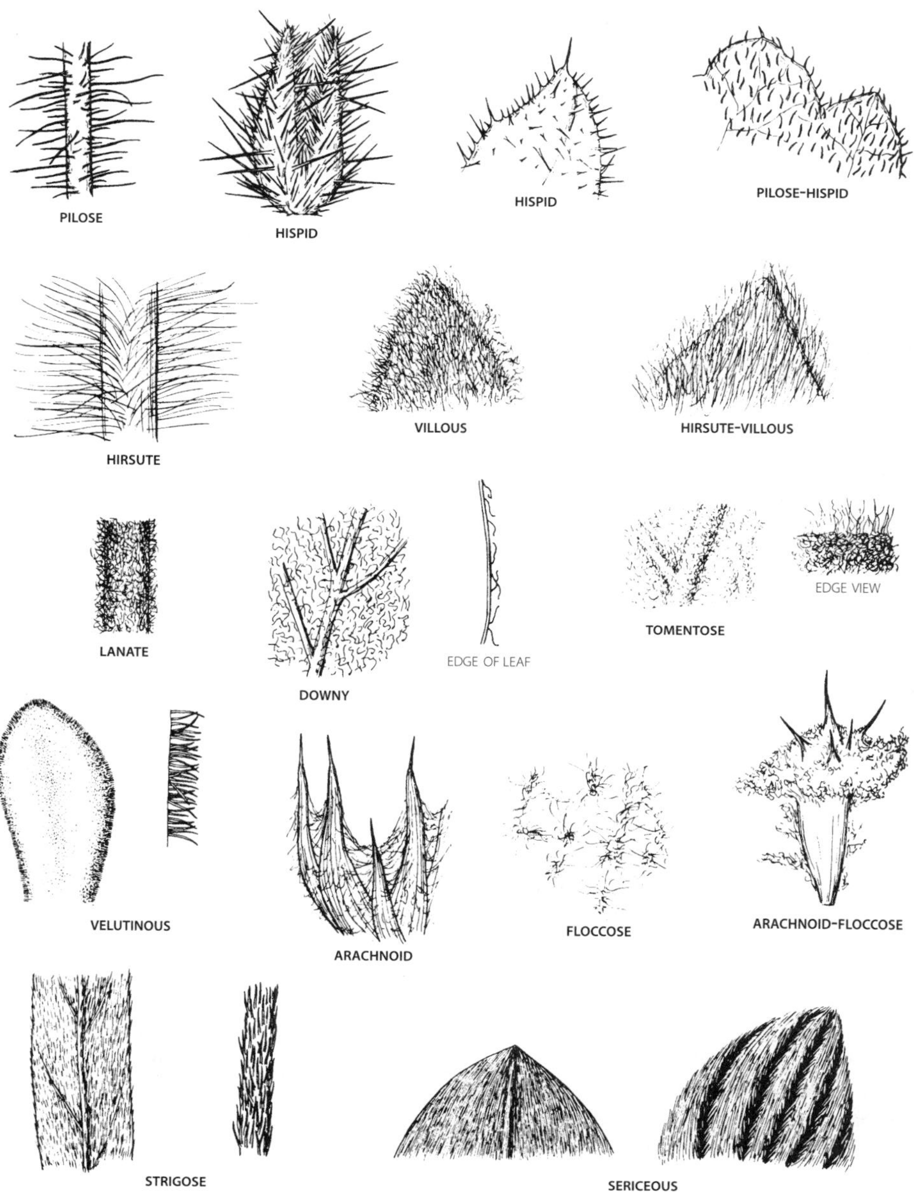

COROLLA Collective term for the petals; the inner series of floral "leaves." Corollas are typically but not always colorful and showy.
COROLLA LOBE A free projecting part of a synsepalous corolla; also referred to as a corolla tooth.
COROLLA TUBE The basal or tubular portion of a synsepalous corolla, as opposed to the free, distal, corolla lobes.
CORONA (= Crown) Projection or outgrowth of a corolla, perianth, or anthers, in the form of a fringe, cup, or tube, sometimes greatly resembling an extra corolla or perianth.
CORONIFORM SCALES Membranous scales broader than long.
CORPUSCULUM In Asclepiadaceae, the gland-like clip connecting the two bands (= filament-like retinacula) attached to the pollinia; part of the translator.
CORRUGATE Having wrinkles or folds.
CORTEX (a) The tissue between the stele and epidermis of a stem; (b) bark or rind.
CORYMB A more or less flat-topped inflorescence (resulting from lower branches being longer than upper) that is indeterminate (i.e., with the outer flowers opening first); inflorescence superficially similar to an umbel but with the branches arising at different points rather than one.
CORYMBIFORM Shaped like a corymb.
CORYMBOSE In corymbs or corymb-like inflorescences (i.e., flat-topped).
COSTA (pl. COSTAE) A rib; the midvein of a leaf.
COSTATE Longitudinally ribbed.
COSTULES Midveins of the pinnules.
COTYLEDON Seed leaf; main leaf or leaves of the embryonic plant.
COUMARIN GLYCOSIDE A glycoside (two-component molecule) that upon breakdown yields coumarin as the aglycone (= non-sugar component). Coumarin can be converted to dicoumarin, a toxic compound which prevents blood-clotting; e.g., in *Melilotus*.
CRATERIFORM In the shape of a saucer; shallow and hemispherical.
CREEPING Growing along the surface of the ground and emitting roots at intervals, usually from the nodes.
CRENATE Scalloped with rounded teeth; shallowly round-toothed or with teeth obtuse.
CRENULATE Diminutive of crenate; with small rounded teeth.
CREST An elevated ridge, process, or appendage on the surface of an organ or structure.
CRESTED Having a crest.
CRISPATE, CRISPED Irregularly curled or twisted.
CROSS-SECTION A slice cut across an object; e.g., a slice of bread.
CROWN (a) An irregular perennial or over-wintering stem or stem-root structure from which new growth arises; (b) (= Corona) projection or outgrowth of a corolla, perianth, or anthers, in the form of a fringe, cup, or tube, sometimes greatly resembling an extra corolla or perianth.
CROZIER A young coiled leaf of some ferns.
CRUCIATE, CRUCIFORM Cross-shaped.
CRUCIFEROUS (a) Cross-bearing; (b) specifically descriptive of cross-like arrangement of petals of members of the Brassicaceae (Cruciferae).
CRUSTACEOUS, CRUSTOSE With a brittle, hard texture.
CRYPTOGAMS An old term for plants that reproduce without flowers or seeds. Cryptogams typically reproduce by spores.
CRYSTALLINE Crystal-like.
CUCULLATE Hood-like.
CULM Stem of Poaceae and Cyperaceae.
CULTIGEN A plant known only in cultivation.
CULTIVAR A variety or race of a cultivated plant; abbreviated cv.
CUNEATE, CUNEIFORM Wedge-shaped; triangular with tapering, straight-sided, narrow base.
CUP, CUPULE The cup-like structure at the base of a fruit; e.g., acorn.
CUPULIFORM, CUPULATE Cup-shaped.
CURVI- A prefix to denote curved or bent.
CUSP A sharp, abrupt, and often rigid point.
CUSPIDATE Bearing a cusp or strong sharp point.
CUT A general term for any dissection of a leaf or petal deeper than a lobe.
CUTICLE The waxy, more or less waterproof coating secreted by the cells of the epidermis. The cuticle prevents water loss.
CYANOGENIC GLYCOSIDE A glycoside (two-component molecule) that upon breakdown yields hydrocyanic (prussic) acid (HCN) as the aglycone (= non-sugar component). Hydrocyanic acid is extremely dangerous, causing cyanide poisoning; e.g., amygdalin in cherry and peach leaves or apple seeds (Rosaceae).
CYATHIFORM Cup-shaped.
CYATHIUM (pl. CYATHIA) (a) Cup-shaped structure producing unisexual flowers; (b) specifically, the units of the inflorescence in *Euphorbia*. In this case the cup contains a single pistillate flower and a number of staminate flowers, each consisting of a single stamen; on the rim of the cup there are glands and these often have a petal-like appendage; the whole structure superficially resembles a single flower.
-CYCLIC A suffix referring to the circles of different parts in a flower, commonly used with a numerical prefix; e.g., a *Verbascum* flower with sepals, a corolla, stamens, and an ovary is four-cyclic; compare with -merous.
CYLINDRICAL, CYLINDRIC Elongate, circular in cross-section; having the form of a cylinder.
CYMBIFORM Boat-like or boat-shaped.
CYME A broad, flattish or convex, determinate inflorescence with the central flowers maturing first.
CYMOSE With the flowers in cymes; having an inflorescence type with the oldest flowers in the center.
CYMULE A small or few-flowered cyme.
CYPSELA Achene derived from an inferior ovary and adnate to the enclosing floral tube; e.g., in Asteraceae.
CYSTOLITH A stone-like mineral concretion, usually of calcium carbonate.

D

DC. De Candolle, name of a distinguished family of Swiss botanists; specifically Augustin Pyramus, who sponsored early botanical exploration in Texas by Berlandier, and named many Texas species; A. DC.: Alphonse, son of the preceding.
DECIDUOUS Falling away; not persistent over a long period of time.
DECLINATE, DECLINED Bent forward or downward.
DECOMPOUND More than once compound.

Types of Hairs and Processes

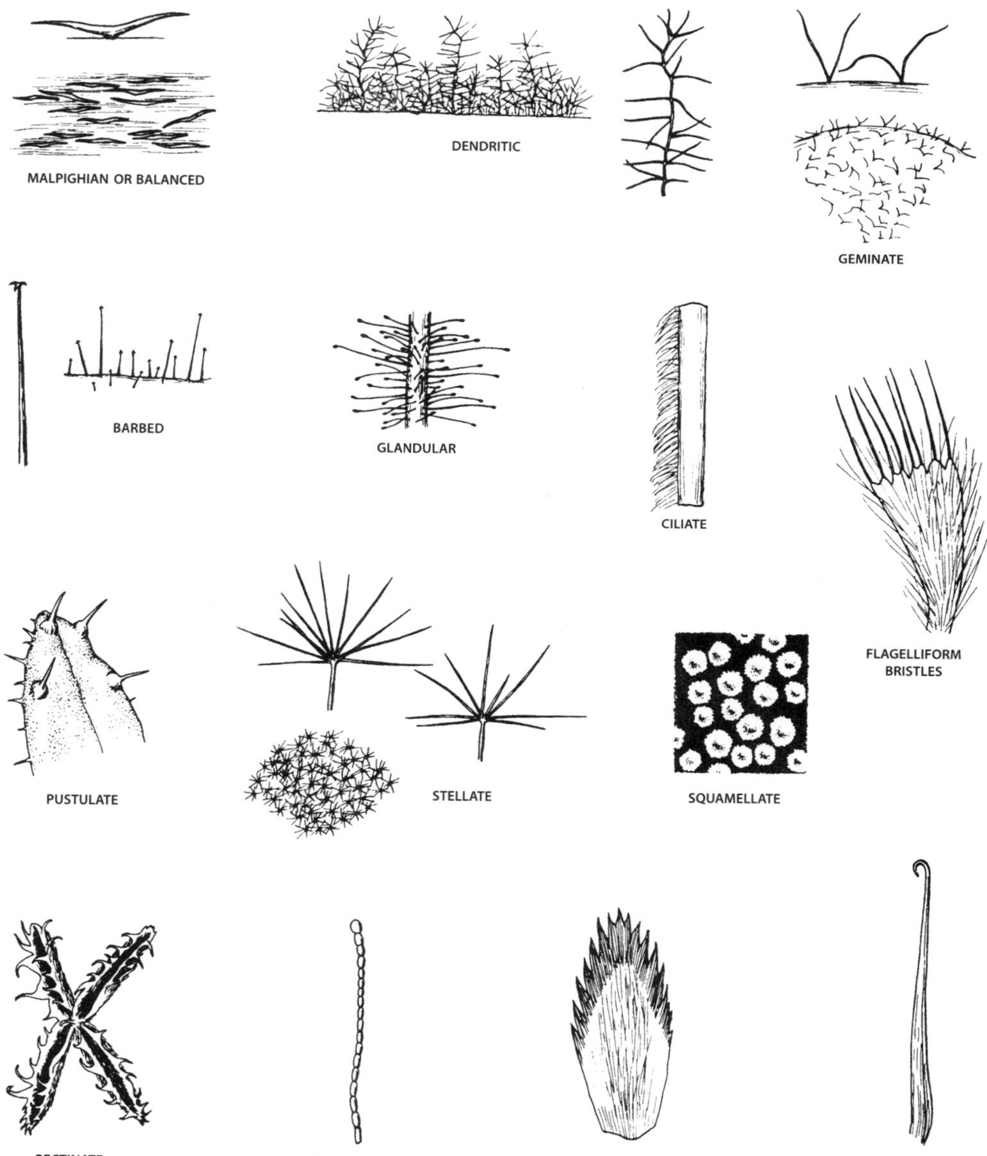

Decumbent Lying flat or reclining with terminal shoots or stem tips ascending.

Decurrent Extending down the stem and united with it, as in the continuation of leaf bases down the stem as wings.

Decurved Curved downward.

Decussate Arranged oppositely in pairs, each successive pair at right angles to the preceding one (resulting in the appearance of four rows), typically referring to arrangement of leaves.

Deflexed Bent downward.

Defoliate To shed or remove leaves.

Dehisce, Dehiscent To open at maturity to discharge the contents; e.g.; fruit releasing seeds or anther releasing pollen.

Dehiscence The process or act of opening.

Deliquescent Softening, dissolving, melting away, or wasting away; e.g., ephemeral petals of *Tradescantia*.

Deltate, Deltoid Shaped like an equilateral triangle, like the Greek letter delta (Δ).

Dendritic Tree-like, as in branching.

Dentate With sharp teeth not directed forward.

Denticulate Minutely dentate.

Depauperate Smaller than the usual natural size; short of the usual development; stunted; impoverished.

Depressed Low as if flattened.

Dermatitis Inflammation of the skin due to exposure to poisons, irritants, or sensitizers. It can be caused by a number of plants including poison ivy.

Descending With a gradual downward tendency.

Determinate Descriptive of an inflorescence whose flowers begin to open first at the top or center, progressing downward or outward; with the number of flowers fixed or limited in number; contrasting with indeterminate.

Di-, Dis- Greek prefix meaning two or double.

Diadelphous With filaments united so as to form two groups of stamens. The most common situation is nine in a group and a tenth separate; e.g., numerous Fabaceae.

Diandrous Having two stamens.

Diaphanous Transparent.

Diaphragm Dividing membrane or partition.

Dichasial With or in the form of a dichasium.

Dichasium A cyme with lateral branches on both sides of the main axis.

Dichotomous Forking regularly into two equal or nearly equal branches.

Dicotyledons (abbreviated Dicots) Flowering plants having two cotyledons, mostly net venation, and flower parts usually in 4s or 5s; one of the two classes of flowering plants which, depending on the system of classification, is known as Dicotyledonae or Magnoliopsida. The dicots are now considered to be a paraphyletic group.

Didymous Developed in or occurring in pairs; twin.

Didynamous Having four stamens, two longer and two shorter.

Diffuse Loosely branching or spreading; of open growth; widely spread.

Digitate Fingered; with a number of structures attached at one point, like fingers on a hand.

Dilated Widened, flattened, broadened, or enlarged.

Dimidiate Halved, as if one half is missing.

Dimorphic, Dimorphism Occurring in two forms.

Dioecious With staminate flowers on one plant (staminate plant) and pistillate flowers on a different plant (pistillate plant); literally, "two houses."

Diploid Possessing two sets of chromosomes in each nucleus; twice the haploid number typical for gametes.

Diplostemonous The stamens in two series, those of the outer series alternating with the petals.

Dipterophily, Dipterophilous Pollination by dipterans (members of the insect order Diptera—flies, gnats, mosquitoes, and their relatives).

Disarticulating Separating; coming apart or falling apart.

Disc, Disk (a) A more or less fleshy or elevated development of the receptacle or of coalesced nectaries or staminodes about the pistil; (b) the central part of an Asteraceae head.

Disc floret (= Disk flower) In Asteraceae, small flower with tubular corolla, in disk (disc) portion of head; contrasting with ray (= ligulate) floret.

Disciform (a) Shaped like a disc; (b) in Asteraceae, in reference to a head with disk florets in center and marginal florets with ligule reduced or lacking.

Disclimax The condition where succession is indefinitely arrested or altered due to factors such as persistent disturbance (e.g., overgrazing).

Discoid (a) Shaped like a disc; (b) in reference to disc (disk) florets of an Asteraceae head; (c) in reference to the head of some Asteraceae with disk florets in center and marginal florets with ligule reduced or lacking; (d) without ligulate flowers.

Discoid head In Asteraceae, a head with only disk (disc) florets.

Discolorous Having the two surfaces of a structure dissimilar in color.

Discrete Separate.

Disjunct (a) Outside the main range of a species; (b) being divided into separate groups; disconnected.

Disk flower (= Disc floret) In Asteraceae, small flower with tubular corolla, in disk (disc) portion of head; contrasting with ray (= ligulate) florets.

Disparate Dissimilar; unequal.

Dissected Divided into numerous narrow or slender segments, the divisions usually deeper than lobes.

Distal Located at or toward the apex of a plant or organ; the terminal portion; the end opposite the attachment; contrasting with proximal.

Distant In reference to similar parts remote from one another; contrasting with approximate.

Distichous In two vertical rows or ranks, not spirally arranged; e.g., leaves occurring in two rows on opposite sides of a stem.

Distinct (= Free) Separate, not united or fused.

Disturbed Referring to a habitat that has been altered or modified but not completely destroyed.

Diurnal Occurring during the daytime.

Divaricate Very widely spreading.

Divergent Spreading, but less broadly than divaricate.

Divided (a) Cut 3/4–completely the distance from margin to midrib or from apex to base; (b) generally, cut deeply.

DIVISION The highest rank, category, or taxon in the plant kingdom; made up of classes; ending in -phyta; equivalent to the rank of phylum in the animal kingdom.

DM Decimeter; 10 cm.

DOCTRINE OF SIGNATURES Ancient belief that a plant structure that resembles a portion of the human body (a sign or signature) gives clues to its use. Some people believed that a creator had placed such signs to indicate the plant's value as a remedy for ailments of those body portions; e.g., *Hepatica*, named for its lobed, liver-shaped leaves, possibly useful for liver problems.

DOLABRIFORM (a) Referring to pubescence where the hairs are attached near the middle or towards (but not at) one end and are thus 2-forked; (b) having the form of an ax or hatchet.

DORMANT Not active.

DORSAL Referring to the back or outer surface of an organ; the side away from the axis; the lower or abaxial surface; contrasting with ventral.

DORSIFIXED Attached by the back or dorsal edge, often in reference to the attachment of an anther to a filament; contrasting with either basifixed or versatile.

DORSIVENTRAL Differentiated into upper and lower surfaces distinct in appearance or structure.

DOUBLE FERTILIZATION Process essentially unique to the angiosperms in which an egg unites with a sperm, forming a zygote, and a second sperm often unites with two nuclei resulting in a triploid endosperm. Double fertilization is also known in the small gymnosperm group the Gnetophyta.

DOUBLE FLOWER (a) One with more than the normal number of petals (anywhere from a few more to many times the usual number); (b) in Asteraceae, double can refer to a head with more than the normal number of ray (= ligulate) florets.

DOUBLY SERRATE With coarse serrations bearing minute teeth on their margins, the teeth angled toward the apex of the structure.

DOWNY Closely covered with short, weak, soft hairs.

DROOPING More or less erect at base but with upper part bending downward.

DRUPACEOUS Pertaining to, or of the nature of a drupe.

DRUPE One-seeded indehiscent fruit with a stony endocarp, the middle part fleshy or juicy, and an outer skin; e.g., plum or cherry.

DRUPELET A small drupe; one drupe from an aggregate fruit composed of many drupes; e.g., a blackberry is an aggregate fruit composed of drupelets.

DUPLEX Double, as in pubescence composed of two kinds of hairs.

E

E East.

E-, EX- Latin prefixes denoting without, that parts are missing.

EBENEOUS Black, ebony-like.

EBRACTEATE Without bracts.

ECHINATE Prickly.

ECILIATE Without cilia.

ECOLOGICAL INDICATOR An organism that is sensitive to pollution or some other environmental problem and can therefore be used as an indicator or gauge of the condition of an ecosystem.

ECOTONE Transition zone between two biological communities; e.g., much of nc TX is an ecotone between the eastern deciduous forest and the central North American grassland.

ECOTYPE Those individuals adapted to only one of the kinds of environment occupied by a widespread species.

EDAPHIC Pertaining to soil conditions.

EDENTATE Without teeth.

EGG A female gamete or sex cell, in flowering plants contained in an ovule.

EGLANDULAR Without glands.

ELAIOSOME An oily appendage on the seeds of some plants. These structures have apparently evolved to attract ants that act as dispersal agents.

ELIMBATE Referring to a corolla without a limb.

ELLIPSOID A solid that is elliptic in outline.

ELLIPTIC Shaped like an ellipse, with widest part at the middle; in the form of a flattened circle usually more than twice as long as wide.

ELONGATE Lengthened; stretched out.

EMARGINATE With a notch in the usually rounded apex.

EMBRYO The new plant in a seed.

EMERSED, EMERGENT Raised above and out of the water.

EMETIC A substance that causes vomiting.

ENATION Outgrowth on the surface; epidermal outgrowth.

ENDEMIC Confined geographically to or native to a single area.

ENDOCARP Innermost layer of pericarp or fruit wall.

ENDOGENOUS Produced deep within another body.

ENDOSPERM The food reserve of many angiosperm seeds.

ENSIFORM Sword-shaped; e.g., leaves of an *Iris*.

ENTIRE With smooth margins; without teeth.

ENTOMOPHILOUS, ENTOMOPHILY Pollinated by insects.

EPAPPOSE Without pappus.

EPHEMERAL Lasting for a brief period; e.g., for only one day.

EPI- Greek prefix meaning upon or on.

EPICALYX A whorl or involucel of sepal-like bracts just below the true sepals; e.g., Malvaceae.

EPICARP (= Exocarp) The outermost layer of the pericarp or fruit wall.

EPICOTYL The portion of the embryo just above the cotyledon(s); the young stem.

EPIDERMAL Relating to the epidermis.

EPIDERMIS The cellular covering of plant tissue below the cuticle.

EPIGEOUS Growing upon or above the ground.

EPIGYNOUS Borne on the ovary; indicating a flower in which the hypanthium or the basal parts of the perianth are adnate to the ovary, the perianth and stamens thus appearing to rise from the summit of an inferior ovary.

EPIGYNY The state of being epigynous.

EPIPETALOUS Descriptive of stamens in which the filaments are adnate to the corolla for all or part of their length; borne upon or arising from the petals or corolla.

EPIPHYTE A plant growing on another plant for physical support only and not parasitic; e.g., many Bromeliaceae including *Tillandsia* species.

EPIPHYTIC Having the character of an epiphyte.

EQUIDISTANT Of equal distance.

Equitant Overlapping in two ranks; folded lengthwise and distichous; e.g., leaves folded around a stem in the manner of the legs of a rider around a horse.

Eradiate Lacking ray (= ligulate) florets; e.g., discoid heads in Asteraceae.

Erect (a) Growing essentially in a vertical position (e.g., whole plant); (b) a structure perpendicular to the object to which it is attached.

Ergot An alkaloid producing fungus, *Claviceps purpurea* (and related species).

Ergotism A sometimes lethal poisoning due to a number of alkaloids (e.g., lysergic acid hydroxyethylamide—LSD-like, egrotamine—vasoconstrictive) produced by ergot fungi, *Claviceps purpurea* (Fr.) Tul., *C. paspali* Stev. & Hall, and related species, which sometimes infect the inflorescences of members of the grass family (e.g., *Secale*—rye). Symptoms resulting from eating ergot-contaminated grain can include hallucinations, psychosis, convulsions, and gangrene of the extremities (due to vasoconstriction); referred to in ages past as Saint Anthony's Fire. The reference to fire resulted from the assumption that the burning sensations and blackened (gangrenous) limbs were retribution for sins. Saint Anthony, supposedly with special powers to protect against fire, infection, and epilepsy, was often prayed to for help by those with the condition. Large scale epidemics of ergotism in Europe prior to 1800, from eating bread made with contaminated grain, resulted in 1,000s of deaths; isolated instances still occur where grain purity is not controlled; ergotism is blamed by some for the hysteria that resulted in the Salem witch trials in 17th century Massachusetts. See treatment of *Secale* (Poaceae) for references.

Erose With ragged margin, as if nibbled or chewed.

Escape A cultivated plant not purposely planted but found growing as though wild.

Escarpment A steep slope.

Essential oils Fragrant (often pleasantly so) substances that are typically rather simple terpenoid compounds (monoterpenes, sesquiterpenes). They are responsible for the fragrance of many flowers and other plant tissues and are currently used in aromatherapy; e.g., in mint family and citrus family.

Estipellate Without stipels.

Estipulate Without stipules.

Etiolate Lengthened and deprived of color by absence of light.

Evanescent Of short duration; quickly lost.

Even-pinnate Descriptive of a compound leaf with terminal leaflet absent; with an even number of leaflets.

Evergreen Remaining green through the winter.

Exalate Without wings.

Exalbuminous In reference to seeds without endosperm.

Excentric Not having the axis placed centrally.

Excurrent Extending beyond the tip or margin.

Excurved Curved outward or away from a central part.

Exfoliate To peel off or come off in scales, flakes, plates, layers, or shreds; e.g., some types of bark.

Exocarp (= Epicarp) The outermost layer of the pericarp or fruit wall.

Exotic Foreign; not native; from another geographic area.

Explanate Spread out flat.

Explosively dehiscent Descriptive of a fruit that suddenly and forcibly dehisces its seeds, with the seeds being thrown away from the plant; e.g., *Impatiens*.

Exserted Projecting out or beyond (e.g., stamens projecting beyond the corolla); contrasting with included.

Exstipulate Lacking stipules.

Extant Still existing; contrast with extinct.

Extinct No longer in existence; descriptive of a species for which living representatives no longer exist. Locally extinct refers to extinction in a given geographic region.

Extrafloral Outside the flower; e.g., extrafloral nectaries.

Extrorse Facing outward, away from the axis, typically used to refer to manner of anther dehiscence.

Exudate Material coming out slowly through small pores or openings.

Eye (a) The marked or contrastingly colored center of a flower; (b) a bud on a tuber; e.g., on potatoes.

F

f. (a) After an author's name: abbreviation of *filius*, the son, or "jr."; (b) abbreviation of forma or form (see next page).

Facet One of a set of small plane surfaces.

Facultative Not necessary or essential; optional; contrasting with obligate.

Falcate Sickle-shaped, with the tip curved to one side.

Fall Outer, spreading, often recurved, and commonly bearded perianth segment in *Iris*.

Family The unit, category, or rank in classification made up of one or more genera; ending in -aceae; sometimes divided into subfamilies, which in turn are made up of genera.

Farinaceous Mealy in texture.

Farinose Covered with a whitish mealy powder or mealiness.

Fasciated With an abnormal widening and flattening of the stem as if several stems had grown together.

Fascicle A condensed or close bundle or cluster.

Fasciculate Congested in bundles or clusters.

Fastigiate In reference to branches, close together and nearly parallel.

Fenestrate Perforated with holes, openings (windows), or translucent areas.

Feral Wild; not cultivated.

Ferrugineous Rust-colored.

Fertile Capable of normal reproductive functions, as a fertile stamen producing pollen, a fertile pistil producing ovules, or a fertile flower normally producing fruit (although it may lack stamens); e.g., used to describe Asteraceae flowers capable of maturing achenes, irrespective of ability to produce pollen.

Fertilization Union of two gametes (e.g., egg and a sperm) to form a zygote.

Fetid With a disagreeable odor.

Fibrillose Having small fibers or appearing finely lined.

Fibrous Resembling or having fibers.

FIBROUS ROOT SYSTEM One with several roots about equal in size and arising from about the same place; contrasting with taproot.

-FID A suffix meaning deeply cut.

FILAMENT (a) The thread-like stalk supporting an anther; (b) a thread or thread-like structure.

FILAMENTOUS, FILAMENTOSE Composed of filaments or threads; thread-like.

FILIFEROUS With coarse marginal threads.

FILIFORM Slender; having the form of a thread; filamentous.

FIMBRIATE Fringed; with narrow or filiform appendages or segments along the margin.

FIMBRILLATE With a minute fringe.

FIRST GLUME Lowermost of the two glumes in a grass spikelet.

FISSURED Cracked or fractured.

FISTULOSE Hollow; lacking pith.

FLABELLATE, FLABELLIFORM Fan-shaped; broadly wedge-shaped.

FLACCID Lax, weak, floppy, not rigid.

FLAGELLIFORM Whip-like.

FLANGE A rim-like structure.

FLAVESCENT Yellowish or becoming yellow.

FLESHY Succulent, juicy, or pulpy.

FLEXUOUS Zigzag; bending or curving alternately in opposite directions.

FLOCCOSE Covered with tufts of soft woolly hairs.

FLOCCULENT Minutely floccose.

FLORA (a) Collective term for the plants of an area; (b) a taxonomic work on the plants of an area.

FLORAL BRACT Reduced leaf subtending a flower in the inflorescence.

FLORAL ENVELOPE The calyx or corolla; the floral "leaves."

FLORAL TUBE, FLORAL CUP Tube or cup formed by union of sepals, petals, and stamen bases. The structure can be either adnate to or free from the ovary; in some cases it is synonymous with hypanthium.

FLORET (a) Small flower in a dense cluster; (b) in reference to individual flowers of Asteraceae and Poaceae; (c) in grasses referring to the lemma and palea together with the enclosed reproductive structures.

FLORICANE Flowering stem of the genus *Rubus* (dewberries and blackberries).

FLORIFEROUS Bearing or producing flowers.

FLOWER An axis bearing stamen(s), pistil(s), or both, and in addition, often floral envelopes (= calyx and corolla); the reproductive structure of an angiosperm.

FLUTED With alternating ridges and grooves.

FOLIACEOUS Leaf-like.

FOLIAGE Collective term for the leaves of a plant.

FOLIATE With leaves.

FOLIOLATE With leaflets.

FOLIOSE Leafy.

FOLLICLE Dry, one-carpellate fruit dehiscing along one suture only.

FORB An herbaceous, non-grass-like plant.

FORKED Dichotomous; divided into two equal or nearly equal branches.

FORMA, FORM (abbreviated f.) A taxon below the rank of variety used to refer to minor variations without distinctive geographic occurrence; e.g., occasional albinos or seasonal growth forms. This category is generally ignored in this book.

FORNICES Internal appendages in the upper throat of a corolla; e.g., in some Boraginaceae.

FOVEA (pl. FOVEAE). A pit or depression.

FOVEATE Pitted.

FOVEOLATE Minutely pitted.

FREE (= Distinct) Separate from one another.

FREE-CENTRAL PLACENTATION Placentation with the seeds attached to a central column and surrounded by a single continuous locule.

FROND The leaf of a fern, often compound or decompound.

FRUCTIFEROUS Producing or bearing fruit.

FRUGIVORE Animal that feeds on fruits.

FRUIT A mature, ripened pistil or ovary. In the case of accessory "fruits" other tissues may be involved.

FRUTICOSE Shrubby or bushy in sense of being woody.

FUGACIOUS Falling or disappearing early, usually in reference to parts of a flower.

FULVOUS Tawny, brownish yellow.

FUNICLE, FUNICULUS The stalk attaching an ovule or seed to the ovary wall or placenta.

FUNNELFORM (= Infundibuliform) Funnel-shaped; gradually widening upwards.

FURROWED With longitudinal channels or grooves.

FUSCOUS Grayish-brown.

FUSED United by normal growth.

FUSIFORM Spindle-shaped; with broadest diameter at middle tapering to each end.

G

GALEA The helmet-like or hood-like upper lip of a bilabiate corolla or calyx, especially one that is strongly concave.

GALEATE Hooded, hood-like, or helmet-like.

GAMETE A sex cell; an egg or sperm.

GAMETOPHYTE The gamete-producing, typically haploid generation alternating with the sporophyte (= spore-producing, typically diploid); the stage in the life-history of a plant that produces male or female cells (= gametes); the dominant generation in mosses and liverworts. In ferns and fern allies it is green and autotrophic, although small; in all flowering plants it is microscopic and develops within the tissues of the sporophyte.

GAMO- A prefix meaning united; e.g., gamopetalous or gamosepalous.

GAMOPETALOUS (= Sympetalous) With petals united, at least basally, forming a tube.

GAMOSEPALOUS (= Synsepalous) With sepals united, at least basally, forming a tube.

GASTROENTERITIS Inflammation of the stomach and intestines. It can be caused by a number of plant materials.

GEMINATE In pairs; twin.

GEMMA An asexual propagule sometimes appearing as, but not homologous with, a vegetative bud.

GENICULATE Bent abruptly, like a knee.

GENUS (pl. GENERA) The unit, category, or rank in classification between family and species; composed of one or more closely related species; sometimes divided into subgenera, which in turn are made up of species.

GIBBOUS Swollen basally on one side.

GLABRATE, GLABRESCENT Becoming hairless with age.

GLABROUS Without hairs.

GLADIATE Sword-shaped.

GLAND A secreting part or appendage, often protruding or wart-like.

GLANDULAR Having or bearing secreting organs, glands, or trichomes.

GLANDULAR-PUBESCENT With gland-tipped, pinhead-like hairs.

GLANDULAR-PUNCTATE With glands recessed in depressions.

GLAUCESCENT Becoming glaucous.

GLAUCOUS With waxy substances forming a whitish or gray-silvery covering or bloom.

GLOBOSE Nearly spherical or rounded.

GLOCHID (pl. GLOCHIDIA) An apically barbed bristle or hair; e.g., in many Cactaceae.

GLOMERATE In a dense cluster or glomerule.

GLOMERULATE Arranged in small dense clusters.

GLOMERULE A dense cluster of two or more structures.

GLUCOSIDE A glycoside with glucose as the sugar.

GLUMACEOUS With greenish bracts or petals similar in appearance to the glumes in grass spikelets.

GLUME One of a pair of bracts at the base of a grass spikelet.

GLUTINOUS Sticky, gluey, or resinous.

GLYCOSIDE Complex, two-component chemical compound that can break down or hydrolyze under certain conditions, yielding a sugar plus another compound (= aglycone) that can be physiologically active including poisonous. Types of glycosides include cardiac, coumarin, cyanogenic, mustard oil, steroidal, and saponic; the term glucoside refers to those in which the sugar molecule is glucose.

GLYCOSINOLATE (= Mustard oil glycoside) A complex molecule that upon breakdown yields a sugar, a sulfate fraction, and isothiocyanates (= mustard oils); e.g., in the Brassicaceae or mustard family.

-GONOUS A suffix meaning angled; e.g., trigonous means three-angled.

GRADUATED Referring to a sequence in shape or size; e.g., leaves becoming narrowed up a stem.

GRAIN The 1-seeded fruit typical of cereal crops; often used synonymously with caryopsis.

GRANULAR, GRANULOSE, GRANULATE Covered with minute, grain-like particles.

GRIT CELL (= Stone cell) A sclerotic or hardened cell, as in the flesh of pears.

GYMNOSPERMS Literally, "naked seed"; a polyphyletic assemblage of plants without flowers, the seeds "naked," (= not enclosed in a special structure), often on the surface of thick or thin, sometimes woody cone scales.

GYNECANDROUS The pistillate flowers above (= distal to) the staminate of the same spike; e.g., in some Cyperaceae.

GYNOBASAL, GYNOBASIC Referring to or having a gynobase.

GYNOBASE An enlargement of the receptacle at the base of the ovary.

GYNODIOECIOUS Basically dioecious, but with some flowers perfect and others pistillate.

GYNOECIUM The pistil or pistils of a flower considered collectively; collective term for the female parts of a flower.

GYNOMONOECIOUS Having female and bisexual flowers on the same plant.

GYNOPHORE Prolonged stipe (= stalk) of a pistil.

GYNOSTEGIUM (a) Sheath or covering of the gynoecium; (b) in Asclepiadaceae, the columnar or disk structure made up of the connate stamens, style, and stigma.

GYPSIFEROUS, GYPSEOUS Containing gypsum (= calcium sulfate).

H

HABIT Style or arrangement of growth; general appearance.

HABITAT Type of locality in which a plant grows; e.g., prairie.

HAIR An epidermal appendage that is usually slender, sometimes branched, not stiff enough to be called a spine, not flattened as a scale; often used synonymously with trichome.

HALBERD-SHAPED (= Hastate). Arrowhead-shaped but with the two basal lobes turned outward.

HALLUCINOGEN A material capable of causing the perception of imaginary sights, sounds, or objects through effects on the nervous systems. Various plant products, including certain alkaloids, are capable of such effects.

HALOPHYTE A plant tolerant of salty or alkaline soils.

HAMATE Hooked.

HAPLOID Having the reduced number of chromosomes typical of gametes; usually with a single set of chromosomes in each nucleus.

HASTATE (= Halberd-shaped). Arrowhead-shaped but with the two basal lobes turned outward.

HAUSTORIUM (pl. HASTORIA) Sucker-like attachment organ of parasitic plants by which they draw their food supply from the host-plant.; e.g., in Cuscutaceae and some Scrophulariaceae.

HAY FEVER Respiratory allergy, frequently due to plant substances or microstructures such as pollen; e.g., *Ambrosia* (ragweed) pollen is a well-known cause of hay fever.

HEAD (= Capitulum) Dense cluster of sessile or nearly sessile flowers; the type of inflorescence typical of the Asteraceae.

HEARTWOOD The innermost and oldest wood, often with materials (e.g., toxins) giving it different characteristics from sapwood (e.g., more durability or resistance to rotting).

-HEDRAL A suffix signifying surface, usually preceded by a number and then indicating the number of sides, as a tetrahedral spore.

HELICOID, HELICAL Coiled or spiraled, usually in reference to inflorescences.

HELIOPHYTE A plant adapted to grow in or tolerate full sun.

HEMI- Greek prefix meaning half.

HEMIPARASITIC (= Semiparasitic) Descriptive of a plant that carries out photosynthesis but obtains some of its food, mineral nutrition, or water needs from another living organism (the host).

Herb A vascular plant lacking a persistent woody stem and typically dying back to the ground each season.

Herbaceous (a) Referring to the aerial shoot of a plant that does not become woody; typically dying back to the ground each year; (b) of a soft texture, as green leaves.

Herbage Collective term for the green or vegetative parts of a plant.

Herbarium (pl. Herbaria) A collection of dried pressed plants prepared for permanent preservation (see Appendix eight for further details).

Herbicidal Having the ability to kill plants.

Hermaphroditic With stamens and pistils in the same flower; bisexual.

Hesperidium A specific type of fruit usually associated with the citrus family; a berry developed from a pistil with numerous carpels, pulpy within, and externally covered with a hard rind; e.g., orange.

Hetero- Greek prefix meaning other, various, or having more than one kind.

Heterocarpous With more than one kind of fruit.

Heterochlamydeous With the perianth differentiated into a calyx and a corolla.

Heterogamous (a) With more than one kind of flower; (b) in Asteraceae, with each head composed of more than one kind of flower.

Heterogeneous Not uniform in kind.

Heterophyllous Having more than one form of leaf.

Heterosporous Having two spore types; e.g., *Selaginella*.

Heterostylous Having styles of different lengths.

Hexagonal Six-angled.

Hexaploid Having six sets of chromosomes.

Hilum Scar or mark on a seed indicating where the seed was attached by a funiculus (= stalk) to the ovary wall or placenta. The hilum is the "eye" of a bean or other large seeds.

Hip The "fruit" of a rose; actually a fleshy hypanthium or floral cup with the true fruits (= achenes) inside.

Hippocrepiform Horseshoe-shaped.

Hirsute With straight moderately stiff hairs.

Hirsutulous Diminutive of hirsute.

Hirtellous Minutely hirsute.

Hispid Resembling hirsute but the hairs stiffer, ± bristly, feeling rough to the touch.

Hispidulose Minutely hispid.

Hoary Covered with a fine, white, whitish, or grayish white pubescence.

Holophyletic A term used to describe a group consisting of a common ancestor and all of its descendants. The term monophyletic is sometimes used in the same sense.

Holotype The one specimen used or designated by the author of a species or other taxon as the nomenclatural type in the original publication. The holotype is the specimen to which the scientific name is permanently attached; it is not necessarily the most typical or representative element of a taxon.

Homo- Greek prefix meaning all alike, very similar, same, or of one sort.

Homochlamydeous With a perianth of tepals undifferentiated into calyx and corolla.

Homogamous (a) With only one kind of flower; (b) with anthers and stigmas maturing simultaneously.

Homogenous Of the same kind or nature; uniform; contrasting with heterogeneous.

Homosporous With spores all of one type.

Hood (a) A segment of the corona in Asclepiadaceae; (b) a hollow arched structure.

Hooded Descriptive of an organ with the lateral margins more or less inrolled and the apex more or less inflexed; helmet-like; shaped like a hood.

Honey-leaf Petal-like perianth part producing nectar; e.g., in some Ranunculaceae.

Horn An exserted tapering appendage resembling a cow's horn; e.g. appendage on the hood in some Asclepiadaceae.

Horny Hard or dense in texture.

Host Organism from which a parasite obtains nourishment.

Humic Consisting of or derived from humus (= organic portion of soil).

Humistrate Laid flat on the soil.

Humus Decomposing organic matter in the soil.

Husk The outer covering of some fruits, typically derived from the perianth or bracts.

Hyaline Thin, membranous, and transparent or translucent.

Hybrid (a) A cross between two unlike parents; (b) specifically, the offspring resulting from a cross between two species.

Hydathode An epidermal structure, usually marginal or terminal, that excretes water.

Hydrophily, Hydrophilous Water-pollination; water-mediated pollination; using water as the mechanism of transferring pollen; e.g., in some Callitrichaceae; see Philbrick (1991).

Hydrophyte A plant typically growing partially or wholly immersed in water; contrasting with mesophyte and xerophyte.

Hygroscopic Susceptible of expanding, shrinking, twisting, or untwisting on the application or removal of water or water vapor.

Hymenopterophily, Hymenopterophilous Pollination by hymenopterans (= members of the insect order Hymenoptera—bees, wasps, and their relatives).

Hypanthium (pl. Hypanthia) Cup-shaped or tubular structure formed by (a) fusion of the sepals, petals, and stamens; or (b) enlargement of the receptacle so that the perianth and androecium are attached above the gynoecium. A hypanthium may be adnate to ovary (resulting in epigyny) or free from ovary (perigyny); formerly referred to as calyx tube.

Hypocotyl Axis of an embryo below the cotyledons.

Hypogeous Below the ground.

Hypogynium A structure below the ovary, as in *Scleria* (Cyperaceae).

Hypogynous Referring to a flower having floral organs attached below the ovary. Hypogynous flowers have superior ovaries.

I

Illegitimate name Name unacceptable as the accepted scientific name because it is not the earliest one given to the plant in question, or published without description, or violating some other specific requirement of the International Code of Botanical Nomenclature.

Imbricate Overlapping like shingles on a roof.

Immaculate Not spotted.

Immersed (= Submerged, Submersed) Growing under water.

Imparipinnate Unequally or odd-pinnate, with a single terminal leaflet.

Imperfect In reference to a flower having either functional stamens or functional pistils, but not both; unisexual.

Incanous Gray or hoary.

Incipient Beginning to be; coming into being.

Incised Cut rather deeply and sharply; intermediate between toothed and lobed.

Included Not exserted; within; not projecting beyond the surrounding organ.

Incomplete Referring to a flower lacking one or more of the flower parts: sepals, petals, stamens, or pistils.

Incomplete veins In *Carex*, referring to veins that do not extend the entire length of the perigynium body.

Incurved Curved inward.

Indehiscent Referring to a fruit that does not open at maturity; contrasting with dehiscent.

Indeterminate Inflorescence whose flowers begin to open first at bottom or outside, progressing upward or inward with the number of flowers not pre-determined at the beginning of flowering; growth of inflorescence not stopped by opening of the first flowers; contrasting with determinate.

Indigenous Native to an area; not introduced.

Indument, Indumentum Surface coating such as hairs, roughening, bloom, or glands.

Induplicate Folded or rolled inward.

Indurate Hardened.

Indusium (pl. Indusia) Epidermal outgrowth covering the sori or sporangia on fern fronds.

Inferior Descriptive of an ovary fused to the hypanthium or to the lower parts of the perianth and therefore appearing to be located below the rest of the flower; an ovary positioned below the point of attachment of the floral organs. This type of ovary is found in epigynous flowers.

Infertile (a) Incapable of normal reproductive functions; (b) specifically used to describe Asteraceae flowers incapable of maturing achenes, irrespective of presence of functional stamens.

Inflated With an internal air space; bladdery.

Inflexed Bent inward.

Inflorescence (a) Term commonly used to refer to the flowering structure of a plant; (b) a flower cluster; (c) arrangement of flowers on the floral axis; (d) manner of bearing flowers.

Infra- Latin prefix meaning below.

Infraspecific Within the species; referring to a unit of classification below the species; e.g., subspecies, variety, form.

Infructescence An inflorescence in the fruiting stage.

Infundibuliform (= Funnelform) Funnel-shaped; gradually widening upwards.

Innocuous Harmless, unarmed, spineless.

Innovation A basal shoot of a perennial grass.

Inrolled Rolled inward.

Insectivorous Consuming insects; referring to plants that digest insect tissue to obtain nutrients such as nitrogen.

Inserted Attached to another part or organ.

Insertion The place or mode of attachment of an organ.

Integument The covering of an organ; e.g., of the ovule.

Inter- Latin prefix meaning between.

Intercalary Medial in position.

Intercostal Located between the ribs or costae.

Interlacunar Between air spaces.

Internode Area of stem or other structure between two nodes.

Interrupted Not continuous or regular.

Interspecific Between different species.

Interstitial Referring to the space intervening between one thing and another.

Intra- Prefix used to denote within.

Introduced Brought from another geographic region; not native.

Introgression, Introgressive hybridization Successive crosses, first between plants of two species, then between the offspring of this cross and plants of one parent species, followed by further interbreeding between mongrels of varying percentage of impurity with purebreds of the parent line. This eventually leads to whole populations of one parent species being contaminated with genes derived from the other.

Introrse Facing inward, toward the axis, typically used to refer to manner of anther dehiscence.

Intruded Projecting inward or forward.

Invaginate To enclose in a sheath.

Involucel Diminutive of involucre; a secondary involucre; e.g., the bracts subtending the secondary umbels in Apiaceae or the whorl of bracts subtending a flower in Malvaceae.

Involucellate With an involucel.

Involucral, Involucrate Pertaining to or having an involucre.

Involucre The whorl of bracts subtending a flower cluster or flower; e.g., involucre of bracts (= phyllaries) subtending a head in Asteraceae or small involucre (= involucel) subtending a flower in many Malvaceae (specifically called an epicalyx).

Involute With margins or edges rolled inward toward the upper side.

Irregular (a) Structures not similar in size or shape; asymmetrical; (b) descriptive of a flower without any plane of symmetry; contrasting to regular and zygomorphic.

-ish Suffix meaning "slightly," often used with color terms.

Isodiametric Of equal dimensions.

Isolateral Equal-sided.

Isothiocyanate (= Mustard oil) Organic compound containing nitrogen and sulfur that has a pungent odor and taste and is irritating to the skin and mucous membranes. Isothiocyanates are toxic and can cause liver and kidney damage as well as other problems; e.g., in the Brassicaceae or mustard family.

Isotype A specimen of the type collection other than the holotype; an extra or duplicate specimen made at the same time and place as the holotype.

J

Jointed With or apparently with nodes or points of articulation; e.g., jointed hairs of *Physalis* (Solanaceae).

Juvenile leaves In plants with more than one leaf type (e.g., leaf dimorphism), the leaves on new growth, often quite different in appearance from adult leaves; e.g., found in *Juniperus*, *Hedera*.

K

Keel (a) (= Carina) The two lower (= abaxial) fused petals of a papilionaceous flower (Fabaceae); (b) prominent longitudinal ridge, shaped like the keel of a boat.

Keeled With a ridge or keel.

Knees Erect woody projections; e.g., found in *Taxodium* (bald-cypress).

L

L. Linnaeus, Swedish naturalist who established the binomial system of nomenclature; L. f., his son.

Labellum (= Lip) In Orchidaceae, the enlarged upper petal that appears to be the lowest petal because of twisting of the pedicel.

Labiate Lipped; differentiated into an upper and a lower portion.

Lacerate Irregularly cleft as if torn.

Laciniate Cut into long, narrow, ± equal divisions or segments.

Lactiferous, Lactescent With milky sap.

Lacuna (pl. Lacunae) A space, hole, cavity, or areole.

Lacunose Perforated, with holes, cavities, or depressions.

Laevigate Smooth as if polished.

Lamella A flat, thin plate.

Lamellate Made up of flat, thin plates.

Lamina The blade or expanded part of an organ.

Laminate In plates or layers.

Lanate, Lanose Woolly; covered with dense, long, entangled hairs resembling wool.

Lanceolate Lance-shaped; several times longer than wide, tapering at both ends, widest about a third above the base.

Lanuginose With a cottony or woolly appearance, the hairs shorter than in lanate.

Latent Dormant.

Lateral Belonging to or borne on the sides.

Lateral bud Bud in a leaf axil; contrasting with terminal or apical bud.

Latex A water insoluble mixture of organic compounds, predominantly hydrocarbons, produced in specialized cells called laticifers and often milky in color. The latex of some species has elastic properties (e.g., *Hevea brasiliensis*, rubber) while that of others contain compounds such as alkaloids (e.g., *Papaver somniferum*, opium poppy).

Laticiferous Latex-bearing.

Latisept With broad partitions in the fruits; e.g., Brassicaceae.

Latrorse Dehiscing laterally and longitudinally; e.g., some anthers.

Lax (a) Spread apart, loose, distant; (b) not rigid.

Leaf The primary photosynthetic organ of most plants, usually composed of a expanded blade and a stalk-like petiole.

Leaflet A single, expanded segment or division of a compound leaf.

Leaf scar The mark or scar left on the stem by the fall of a leaf.

Leaf trace A vascular bundle, one or more in number, extending from the stem into the leaf.

Lectotype A specimen or other material selected by a later worker from the original material studied by the author of the species (or other taxon) to serve as the nomenclatural type when a holotype was not originally designated or was lost or destroyed.

Legume (a) Fruit type with a single carpel typically dehiscent along both sutures (= margins); (b) a member of the Fabaceae; (c) any fruit type within the Fabaceae.

Lemma The outer (= lowermost) of the two bracts enclosing the reproductive structures in the grass floret.

Lenticel A small corky pore or spot on the bark of young twigs found in many trees and shrubs and allowing gas exchange.

Lenticular Two-sided; lens-shaped.

Lepidote Covered with small scales; scurfy.

Lepidopterophily, Lepidopterophilous Pollination by lepidopterans (= members of the insect order Lepidoptera—butterflies and moths).

Lectins Certain plant proteins that cause linking or agglutination between cells. They can be toxic or mitogenic (= capable of stimulating mitosis); they apparently function in the binding of symbiotic nitrogen-fixing bacteria to roots and in protecting against pathenogenic bacteria; often found in members of the Fabaceae.

Liana, Liane A woody climber (e.g., grape vine). Lianas are common in the tropics.

Ligneous Woody.

Lignescent Somewhat woody or becoming woody.

Ligulate Tongue-shaped; strap-shaped.

Ligulate floret (= Ray floret) Flower, with corolla expanded into a ligule, typical of many Asteraceae.

Ligulate head Head having only bisexual flowers with strap-shaped corollas.

Ligule (a) A strap-shaped limb or body; (b) strap-shaped part of ray (= ligulate) corolla in Asteraceae; (c) membranous or hairy appendage on adaxial surface of the leaf at junction of blade and sheath in Poaceae.

Liguliform Strap-shaped.

Limb (a) In a corolla of united petals, the main expanded portion, as distinguished from a basal tube; (b) expanded part of an organ.

Limbate With limb present.

Linear Resembling a line, long and narrow, with margins parallel to one another.

Lingulate Tongue-shaped.

Lip (a) Either of the principal lobes of a bilabiate or strongly zygomorphic corolla or calyx (e.g., Lamiaceae); (b) (= Labellum) the enlarged upper petal in Orchidaceae that appears to be the lowest petal because of twisting of the pedicel.

Lithophyte Plant that grows on rocks but derives its nourishment from the atmosphere and from accumulated humus.

Littoral Of a shore, particularly of the seashore.
Loam Soil consisting of a mixture of sand, clay, silt, and organic matter.
Lobate Having lobes.
Lobe A usually rounded segment or division of a leaf, petal, or other organ.
Lobed Having deep or coarse indentations of the margin, larger than mere teeth (However, there is no sharp distinction between large teeth and small lobes.)
Lobulate Having small lobes.
Locular Having one or more locules.
Locule, Loculus The cavity, compartment, or cell containing the ovules in a carpel or the pollen in an anther.
Loculicidal Descriptive of a capsule dehiscing along the middle of the back of each locule or chamber (= along the midrib of each carpel).
Lodicule One of the two or three minute scales at the base of the ovary in most grasses, thought to be a rudiment of a perianth part. They swell and thus open the lemma and palea, allowing the reproductive parts to be exposed.
Loment, Lomentum An indehiscent fruit separating into one-seeded segments at maturity.
Longitudinal Lengthwise; along the long axis.
Lumper A taxonomist who in general has the tendency to lump segregates into larger groups; contrasting with splitter.
Lunate Crescent-shaped; half-moon-shaped.
Lurid Dirty, dingy.
Lustrous Shining.
Lutescent Becoming yellow.
Lyrate Lyre-shaped; pinnately lobed with the terminal lobe the largest.

M

m Meter; 10 decimeters; 39.37 inches.
Macro- Greek prefix denoting large or long.
Macrospore (= Megaspore) A large spore giving rise to the female gametophyte; the larger of two kinds of spores produced by heterosporous plants; a female spore.
Maculate With a spot or spots.
Malodorous Foul-smelling.
Malpighian Describing hairs lying parallel to a surface and attached by their middle; with two branches and almost no stalk; appearing to be an unbranched hair attached at the middle.
Mammiform Breast-shaped; conical with rounded apex.
Mammillate Having nipple-like structures.
Marbled With irregular streaks or blotches of color.
Marcescent Withering, but the remains persistent; e.g., the corollas of most *Trifolium*.
Margin Edge; the outer portion of a blade or other structure.
Marginal Attached to the edge or pertaining to the edge.
Marginate Distinctly margined; with a distinctly different margin.
Marsh Wet or periodically wet, treeless area.
Medial, Median Central, middle.
Medullary Pertaining to the pith.
Mega- Greek prefix meaning very large.
Megaphyll Leaf with branched veins.
Megasporangium Sporangium in which megaspores are formed.
Megaspore (= Macrospore) A large spore giving rise to the female gametophyte; the larger of two kinds of spores produced by heterosporous plants; a female spore.
Megasporophyll A sporophyll (= spore-bearing leaf) bearing one or more megaspores.
Membranaceous, Membranous Having the nature of a membrane, thin, somewhat flexible, translucent.
Mentum A projection near the base of some flowers in the Orchidaceae.
Mephitic Having an offensive odor.
Mericarp The individual, separated carpels of a schizocarpic fruit; e.g., one of the fruit segments in the Apiaceae or one of the "nutlets" in the Boraginaceae.
Meristem Embryonic or undifferentiated tissue, capable of developing into various organs.
Meristematic Pertaining to or with the nature of a meristem.
-merous Greek suffix used to refer to the number of parts (or multiples of such) in each circle of the floral organs, usually with a numerical prefix (e.g., a 3-merous perianth would mean there are three petals and three sepals or some multiple of three). The term often refers to the perianth only.
Mesa A flat-topped hill with abrupt or steeply sloping side or sides.
Mesocarp The middle layer of the pericarp or fruit wall.
Mesophyte Plant that grows under medium moisture conditions; contrasting with hydrophyte and xerophyte.
Micro- Greek prefix meaning small.
Microphyll A relatively small leaf with a single unbranched vein, typical of the Lycopodiophyta and Equisetophyta.
Microphyllous Having small leaves.
Micropyle A minute opening through the integuments into the ovule through which the pollen-tube usually enters and often distinguished in the mature seed as a slight depression.
Microsporangium Sporangium in which microspores are produced. In angiosperms, the microsporangium is equal to the pollen sac and there are typically four microsporangia per anther.
Microspore A small spore giving rise to the male gametophyte; the smaller of two kinds of spores produced by heterosporous plants; a male spore.
Microsporophyll The sporophyll (= spore-bearing leaf) upon which microspores are produced.
Microtuberculate Minutely tuberculate.
Midrib The central or main rib or vein of a leaf or other similar structure.
Mitriform Shaped like a miter or bishop's hat.
Mixed inflorescence One in which the parts are not consistent in being all determinate or all indeterminate.
mm Millimeter; 1000 microns or 1/1000 of a meter.
Monadelphous With all filaments united into a single tube surrounding the pistil.
Monandrous Having a single stamen.
Moniliform Like a string of beads.
Mono- Greek prefix meaning one or of one.
Monocarpic (a) Descriptive of a plant that flowers only once before dying; (b) having a single carpel.

Monocephalous, Monocephalic Bearing only a single head.

Monochasium A cyme with lateral branching on only one side of the main axis.

Monochlamydeous Having only one set of floral envelopes; having perianth of a single series.

Monocolpate With a single furrow; e.g., on a pollen grain.

Monocotyledons (abbreviated Monocots) Flowering plants having one cotyledon (= seed leaf), mostly parallel venation, and flower parts usually in threes; one of the two classes of flowering plants which, depending on the system of classification, is known as Monocotyledonae or Liliopsida.

Monoecious Plants with staminate flowers and pistillate flowers on the same plant, but lacking perfect flowers.

Monomorphic One form; contrasting with polymorphic.

Monophyletic A term previously used to describe a group of organisms with a common ancestor; more recently it has been used to describe a group consisting of a common ancestor and all of its descendants. Some authorities believe that a different term, holophyletic, should be used for a group consisting of a common ancestor and all of its descendants.

Monopodial With an evident single and continuous axis.

Monotypic Having a single type or representative; e.g., a genus with only one species.

Montane Pertaining to or living in mountains.

Mostly A quantitative term meaning "most of them."

Mucilaginous Slimy; with mucilage.

Mucro A short and small abrupt tip, as with the midrib extending as a short point.

Mucronate With a mucro.

Mucronulate Diminutive of mucronate.

Multi- Latin prefix for many.

Multicipital Literally, "many-headed"; descriptive of a crown of roots or a caudex from which several stems arise.

Multifid Divided into many narrow segments or lobes.

Multiflorous Many-flowered.

Multiple fruit (= Syncarp) A single "fruit" formed by the coalescence of several fruits from separate flowers; e.g., *Morus* (mulberry), *Maclura* (bois d'arc), *Ananas* (pineapple).

Muricate With numerous minute short points; roughened.

Muriculate Diminutive of muricate.

Mustard oil (= Isothiocyanate) Organic compound containing nitrogen and sulfur that has a pungent odor and taste and is irritating to the skin and mucous membranes. Mustard oils are toxic and can cause liver and kidney damage as well as other problems; e.g., in the Brassicaceae or mustard family.

Mustard oil glycoside (= Glycosinolate) A complex molecule that upon breakdown yields a sugar, a sulfate fraction, and an irritating mustard oil or isothiocyanate; e.g., in the mustard family.

Muticous Blunt, lacking a point.

Mycorrhiza (pl. Mychorrhizae) A mutually beneficial, symbiotic association of a fungus and the root of a plant. Mychorrhizal relationships are characteristic of most vascular plants.

Mycorrhizal Pertaining to mycorrhiza.

N

N North.

Naked Lacking various coverings, organs, or appendages, almost always referring to organs or appendages present in other similar plants; e.g., a naked flower lacks perianth.

Napiform. Turnip-shaped; e.g., roots.

Nascent In the act of being formed.

Natant Floating underwater; immersed.

Naturalized Referring to an introduced foreign plant that has become part of the spontaneous, self-perpetuating flora of a region.

Naviculate, Navicular Boat-shaped.

Nectar A sugar-rich solution secreted by plants, typically produced in nectaries. Nectar production has apparently evolved to attract insects or other animals for pollination or other purposes.

Nectar-guide A line or other marking leading insects to the nectary; e.g., contrastingly colored lines on many corollas.

Nectariferous Having or producing nectar.

Nectary A specialized nectar-secreting structure or area; there can be floral nectaries (in the flowers) or extra-floral nectaries (not associated with the flowers).

Needle Linear, often stiff leaf as in *Pinus* (Pinaceae).

Neotype A specimen selected by a later worker to serve as the nomenclatural type of a taxon when all material studied by the original author has been lost or destroyed.

Nerve A simple vein or slender rib of a leaf, bract, or other structure.

Nerved Having nerves.

Net-veined With veinlets branching irregularly and not uniformly angular, forming a net-like pattern.

Neuter Sexless, as a flower that has neither functional stamens nor pistils.

Neutral flower A sterile flower; flower with a perianth but without functional sexual organs.

Nigrescent Turning black.

Nitid Smooth and clear, lustrous, glittering.

Nitrates Nitrogen containing compounds that can be accumulated in plant tissues and cause toxicity in animals if ingested. During digestion, nitrates are converted to nitrites that are toxic due to their ability to render hemoglobin incapable of transporting oxygen.

Nocturnal Occurring at night or lasting for only one night.

Nodal Located at or pertaining to a node.

Node Area of stem or axis at which branches, leaves, bracts, or flower stalks are attached; joint of a stem.

Nodding Hanging down.

Nodose Nodular, knotty, with semispheroid protuberances.

Nodulose Diminutive of nodose; having small, knobby nodes or knots.

Nut Hard-shelled and indehiscent fruit with a single seed.

Nutant Nodding.

Nutlet (a) Diminutive of nut; (b) used to refer to any small, dry, nut-like fruit or seed, thicker-walled than an achene; (c) seed-like sections into which the mature ovary breaks in Boraginaceae, Labiatae, and most Verbenaceae, each section consisting of one seed with extra coat formed from ovary wall or partitions, making it technically a fruit.

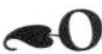

O

OB- Latin prefix indicating the reverse or upside-down, as obcordate, meaning cordate or ovate with wider end at top or away from point of attachment.

OBCONIC, OBCONICAL Inversely cone-shaped, with attachment at the small end.

OBDELTOID Inversely deltoid; triangle-shaped with base pointed.

OBDIPLOSTEMONOUS Describing a flower with two circles of stamens, in which those of the outer circle are borne in front of (= on the same radius as) the petals.

OBLANCEOLATE Lanceolate with broadest part above the middle and tapering toward the base.

OBLATE Nearly spherical but flattened at the poles.

OBLIGATE Necessary or essential; contrasting with facultative.

OBLIQUE Slanting; unequal-sided.

OBLONG Longer than wide with sides nearly parallel.

OBOVATE Egg-shaped with attachment at narrow end; inversely ovate.

OBOVOID Inversely ovoid; a solid that is obovate in outline.

OBPYRIFORM Pear-shaped with attachment at narrow end.

OBSOLESCENT Nearly or becoming obsolete; rudimentary; referring to structures that are not or are only slightly evident.

OBSOLETE Not apparent or evident; rudimentary; vestigial; extinct.

OBTUSE Forming a blunt or rounded angle of more than 90 degrees; not pointed.

OCHRACEOUS Ochre-colored, yellow with a tinge of red.

OCHROLEUCOUS Yellowish-white, buff.

OCREA (pl. OCREAE) Tube formed by fused, sheathing stipules; e.g., in Polygonaceae.

OCREATE With an ocrea.

OCREOLA (pl. OCREOLAE) Secondary sheath; sheath surrounding a fascicle of flowers; e.g., in the Polygonaceae.

ODD-PINNATE Compound leaf with terminal leaflet present; with an odd number of leaflets.

OFFSET A short lateral shoot, arising near the base of a plant, primarily propagative in function and thus can give rise to a new plant.

OLIGO- Greek prefix signifying few.

OLIVACEOUS Olive-green.

ONTOGENY The developmental cycle of an individual organism.

OPAQUE Impervious to light.

OPERCULATE Having an operculum.

OPERCULUM Literally, "a small lid"; term applied to the terminal portion (= lid or cap) of a circumscissile fruit or other organ.

OPPOSITE Arranged two at each node, on opposite sides of the axis.

ORBICULAR, ORBICULATE With round, approximately circular outline.

ORDER The unit, category, or rank in classification made up of one or more families; ending in -ales; sometimes divided into suborders or superfamilies, which in turn are made up of families.

ORIFICE An opening.

ORNITHOPHILY, ORNITHOPHILOUS Pollination by birds.

ORTHO- Greek prefix signifying straight.

ORTHOTROPOUS OVULE An erect ovule, with the micropylar end at the summit.

OVAL Broadly elliptic.

OVARY Basal, ovule-containing portion of the pistil in angiosperms, developing into the fruit.

OVATE Egg-shaped with widest part at the base.

OVOID Solid oval or solid ovate.

OVULATE Bearing ovules.

OVULE An immature or unfertilized seed; the megasporangium of a seed plant; the egg-containing unit of the ovary.

OXALATES Salts of oxalic acid, a carboxylic acid that can occur in plant tissue either as the free acid (e.g., rhubarb leaves) or as salts in the form of calcium oxalate, potassium oxalate, or sodium oxalate. Ingestion of the soluble oxalates by animals results in the formation of calcium oxalate crystals, which when deposited in the kidneys and other organs can cause serious mechanical damage; calcium deficiencies can also result; calcium oxalate also occurs in some plant tissues in the form of raphides or needle-like crystals; e.g., in Araceae.

P

PALATE The raised portion within the throat of a bilaterally symmetrical corolla.

PALE (= Palea) A chaffy scale or bract on the receptacle of many Asteraceae, often subtending the fruit; see chaff.

PALEA (a) In Poaceae, the inner (= uppermost) bract of a floret, often partly enclosed by the lemma; (b) (= pale) the receptacular scale or bract in Asteraceae; see chaff.

PALEACEOUS Chaffy; with small membranaceous scales.

PALLID Pale in color.

PALMATE Attached or radiating from one point, as leaflets in a palmately compound leaf, veins in some leaf blades, or fingers of a hand.

PALMATELY COMPOUND With the leaflets attached at one point at the apex of the petiole, like the fingers all attached to the palm of a hand.

PALMATELY TRIFOLIOLATE Having three leaflets with the terminal leaflet unstalked, sessile.

PALMATIFID Cut so as to appear nearly palmately compound.

PALUDAL Growing in marshes.

PANDURATE, PANDURIFORM Fiddle-shaped; obovate with a contraction on each side.

PANICLE A branched usually elongate (not flat-topped) inflorescence with flowers on the branches of the primary axis; a compound racemose inflorescence. Panicles are indeterminate.

PANICULATE Resembling a panicle.

PANICULIFORM Having the form and appearance but not necessarily the structure of a panicle.

PANNOSE Having the texture or appearance of woolen cloth or felt.

PAPILIONACEOUS, PAPILIONOID Descriptive of a flower having a bilaterally symmetrical corolla composed of a banner (= standard), two lateral wing petals, and a keel of two fused petals; descriptive of flowers characteristic of many Fabaceae in the subfamily Papilionoideae.

PAPILLA (pl. PAPILLAE) Small pimple-like or nipple-like projection.

Papillate, Papilliform, Papillose Shaped like or bearing papillae.

Pappus The crown of bristles, awns, scales, etc. at the apex of the achene of most Asteraceae, thought to be the modified calyx.

Papyraceous Papery.

Parallel-veined With main veins parallel to each other. Such leaves are typical of many monocots.

Paraphyletic A term used to describe a taxonomic group consisting of an ancestral species and some but not all of its descendants. Paraphyletic groups can give an inaccurate view of phylogeny but are often useful in classification; see Appendix 6 for discussion.

Parasite A plant that derives its food, mineral nutrition, and/or water wholly or chiefly from another plant (the host) to which it is attached; contrasting with epiphyte, saprophyte, or autophyte.

Parasitic Deriving food, mineral nutrition, water, or any combination from another living organism (the host).

Parietal Located on the inner side of the exterior wall of the ovary.

Parietal placentation Placentation with ovules attached to the ovary wall instead of the axis.

Paripinnate Referring to a pinnately compound leaf with an even number of leaflets or pinnae.

Parted (a) Cut nearly but not quite the distance from the margin to midrib or from the apex to base; (b) more generally, any deep cut; (c) as a suffix can be combined with a numerical prefix to indicate the number of segments.

Parthenogenetic Developing without fertilization.

-partite A suffix synonymous with -parted.

Patelliform Disk-shaped; circular with narrow rim.

Patent Spreading.

Pauci- Latin prefix denoting few.

Pectinate With narrow closely set segments or units similar to the teeth of a comb.

Pedate Palmately divided with the lateral segments again divided.

Pedicel (a) Stalk supporting a single flower of an inflorescence; (b) in grasses the stalk supporting a single spikelet.

Pedicellate, Pediceled, Pedicelled Borne on a pedicel; contrasting with sessile.

Peduncle Stem or stalk of an inflorescence, flower cluster, or of the only flower in an inflorescence.

Peduncular Pertaining to or on a peduncle.

Pedunculate With a peduncle.

Pellicle A thin skin or filmy covering.

Pellucid Clear, almost transparent.

Peltate Shield-shaped, with stalk attached on the undersurface away from the margin or base (sometimes attached at the middle like the axis of an umbrella).

Pendent, Pendulous Hanging down or suspended.

Penicillate Like a brush, usually referring to a tuft of hairs.

Pentagonal Shaped like a pentagon; five-sided or -angled.

Pentamerous Having the parts in fives.

Pepo A specific type of fruit usually associated with the Cucurbitaceae; a many-seeded fruit with a hard rind; e.g., gourd.

Perennating Surviving from growing season to growing season.

Perennial Root system or plant living at least three growing seasons (years); lasting from year to year.

Perfect flower One with both functional stamens and pistils; a bisexual flower.

Perfoliate Referring to a sessile leaf or bract whose base completely surrounds the stem, the latter seemingly passing through the leaf or with two leaves basally united around the stem.

Perforate Having translucent dots that look like small holes, or pierced through.

Peri- Greek prefix denoting around, enclosing.

Perianth Collective term for the corolla and calyx, often used when corolla and calyx are not well differentiated.

Pericarp Fruit wall; structure enclosing the seeds in angiosperms. The pericarp can be subdivided into endocarp, mesocarp, and exocarp.

Perigynium Urn-shaped, sac-like or pouch-like structure surrounding the pistil in *Carex* (Cyperaceae).

Perigynous Having floral organs united forming a tube (= floral tube) surrounding but not adnate to the pistil, the floral organs thus borne around (versus above or below) the ovary.

Peripheral On or near the margin.

Perisperm The nutritional material of a seed surrounding the embryo and formed outside the embryo sac.

Perisporium, Perispore A thin envelope enclosing a spore.

Pernicious Highly hurtful or harmful.

Persistent Remaining attached; not falling off; contrasting with deciduous.

Personate Referring to a strongly two-lipped corolla with the throat closed by a palate.

Petal One of the next-to-outermost whorl of parts in a complete flower; a segment of the corolla; the inner series of floral "leaves". The petals are collectively referred to as the corolla.

Petal-like (= Petaloid) Having the character or appearance of a petal, usually showy.

Petaloid (= Petal-like) Having the character or appearance of a petal, usually showy.

Petaliferous Bearing petals.

Petiolar Pertaining to or located on the petiole.

Petiolate With a petiole.

Petiole Stalk of a leaf supporting the blade.

Petiolulate Having a petiolule.

Petiolule Stalk supporting a leaflet.

Phanerogam A seed plant or spermatophyte; contrasting with cryptogam.

Phenolic, Phenol An aromatic alcohol; more technically, a substance with an alcohol (= hydroxyl) group attached to an aromatic (= benzene or phenyl) ring structure. Plant phenolics are sometimes toxic; a number of plant resins or resinoids are phenolic in nature; e.g., urushiol from poison ivy, tetrahydrocannabinol (THC) from marijuana; other phenolic compounds include vitamin E and anthocyanins.

Phenology Study of the times at which various events occur in the life of a plant or a flower.

Phloem The food-conducting tissue in a plant.

Photodermatitis Skin inflammation resulting from increased sensitivity to sunlight, often caused by exposure to certain plants or plant-derived materials.

Photosynthesis The process by which plants convert carbon dioxide and water into carbohydrates in the presence of light.

Phyllary (= Involucral bract) One of the bracts subtending a head in the Asteraceae.

Phylloclade (= Cladophyll, Cladode) A portion of stem having the general form and function of a leaf.

Phyllode, Phyllodium An expanded bladeless petiole; a leaf reduced to a flattened petiole. These structures typically serve the photosynthetic function of a blade.

Phyllodial Having the character of a phyllode.

Phyllotaxy Arrangement of the leaves on their axis.

Phylogeny The evolutionary history of a group.

Phytophotodermatitis Plant-induced skin inflammation resulting from increased sensitivity to sunlight.

Phytotoxic Term used to describe materials that are toxic to plants.

Phytotoxin (a) General term for a plant-derived toxin; (b) more typically used to refer to toxic plant proteins or toxalbumins; e.g., ricin from *Ricinus*.

Pilose With long, soft, ± straight hairs; softer than hirsute, not flexuous or curved as in villous.

Pilosulous Diminutive of pilose.

Pinna (pl. Pinnae) A primary division of a compound leaf. A pinna can be simple (and thus equivalent to a leaflet) or compound and divided into leaflets. The term is frequently used in describing fern leaves—here the term is used for any primary division of a leaf (e.g., a leaflet or a leaf division divided into leaflets). In flowering plants the term is generally used to describe a primary division of a decompound leaf (a pinna is thus divided into leaflets).

Pinnate, Pinnatley compound Descriptive of a compound leaf with leaflets distributed along both sides of an elongate axis; feather-like. Bipinnate or 2-pinnate leaves have the leaflets distributed along a secondary axis; tripinnate or 3-pinnate leaves have the leaflets along a tertiary axis, etc.

Pinnately trifoliate Referring to trifoliate leaves in which the terminal leaflet is extended beyond the other two by having a distinct petiolule.

Pinnatifid Pinnately divided into stalkless segments, but the segments not distinct leaflets (i.e., not divided all the way to the midrib).

Pinnatisect Pinnately divided to the midrib.

Pinnipalmate Descriptive of a leaf exhibiting a combination of pinnate and palmate arrangements of the leaflets.

Pinnule A secondary pinna; division of a pinna or a leaflet.

Pisiform Pea-shaped.

Pistil The female reproductive organ of a flower consisting of the ovary, stigma, and style. Pistils can be either simple (of one carpel) or compound (with > one carpel); the pistils are collectively referred to as the gynoecium.

Pistillate With only the female reproductive structures being functional or with female reproductive structures only.

Pistillode, Pistillodium A rudimentary or vestigial pistil.

Pith The soft spongy center of a stem.

Pitted Having little depressions or cavities.

Placenta (pl. Placentae) Place of attachment of the ovule(s) within an ovary.

Placentation Arrangement of placentae and thus ovules in an ovary.

Plait A lengthwise fold or pleat.

Plaited (= Plicate) Folded like a fan.

Plane With a flat even surface.

Plano- A suffix denoting flat.

Pleated Folded.

Plicate (= Plaited) Folded like a fan.

-ploid A suffix used in genetics, prefixed by a term indicating number, to denote the number of sets of chromosomes in the nucleus.

Plumose Feathery; descriptive of a long hair with long, slender, lateral hairs.

Pod A legume, or more generally, a dry dehiscent fruit; sometimes loosely synonymous with capsule.

Pollen, Pollen grain The microscopic spheroidal structures that produce the male cells and borne within the anther of a flowering plant or on the microsporophyll of a gymnosperm; technically microgametophytes develop from microspores.

Pollen sac A microsporangium containing pollen. In angiosperms there are typically four pollen sacs per anther.

Pollen tube The slender tube that develops from a pollen grain, penetrates the tissue of the ovary, and enters the ovule.

Pollinate To transfer pollen from a stamen to a stigma.

Pollination The act or process of pollinating.

Pollinium (pl. Pollinia) A mass of coherent pollen, as developed in Asclepiadaceae and Orchidaceae.

Poly- Greek prefix meaning many.

Polyadelphous With several groups of stamens.

Polygamo-dioecious Descriptive of dioecious plants having some perfect flowers; bearing on one plant flowers partly perfect and partly pistillate, on another plant flowers partly perfect and partly staminate.

Polygamo-monoecious Descriptive of monoecious plants having some perfect flowers; bearing partly perfect, partly unisexual flowers.

Polygamous Bearing unisexual and bisexual flowers on the same plant.

Polymorphic Of various forms; with three or more forms.

Polypetalous (= Apopetalous) With corolla of separate petals; contrasting with gamopetalous.

Polyphyletic A taxonomic group having species derived from more than one common ancestor, the species having been placed in the same group because of similarities due to convergent or parallel evolution. Polyphyletic taxa give an inaccurate view of phylogeny.

Polyploid A plant with three or more basic sets of chromosomes.

Polysepalous (= Aposepalous) Composed of or possessing separate sepals.

Polystichous Arranged in several longitudinal rows.

Pome A fleshy indehiscent fruit developing from a compound inferior ovary embedded in receptacular or floral cup tissue; e.g., apples or pears.

Pore A small aperature or opening; e.g., opening at end of anther in *Solanum* species.

Poricidal Opening by pores.

PORULUS Somewhat porous (= pierced with small round holes).

POSTERIOR Describing the position of an organ located on the side adjacent to the axis. In flowers, the upper lip is posterior, the lower anterior.

PRAEMORSE Short and truncate at the base, as if eaten or chewed off; descriptive of the rhizomes in some species of *Viola* and *Hieracium*.

PRECOCIOUS Developing or appearing very early; denoting flowers that appear in advance of the leaves.

PREHENSILE Clasping or grasping; e.g., tendrils.

PRICKLE A slender, sharp, epidermal outgrowth without vasculature; e.g., the armature of roses.

PRIMOCANE The first year's shoot of woody biennials, typically non-flowering; e.g., *Rubus*.

PRISMATIC Angulate with flat sides; shaped like a prism.

PROCUMBENT Trailing or prostrate, not rooting.

PROLIFEROUS Bearing or developing offshoots or redundant parts; producing numerous new individuals, or parts, of the same kind; bearing offsets, bulbils, or other vegetative progeny.

PROPHYLL, **PROPHYLLUM** (a) In the Poaceae, the first leaf of a lateral shoot or vegetative culm branch, consisting of a sheath without a blade; (b) bracteole subtending an individual flower; e.g., in some species of *Juncus*.

PROPHYLLATE Furnished with prophylls.

PROSTRATE Lying flat.

PROTANDROUS, **PROTERANDROUS** Having the anthers mature before the pistils in the same flower.

PROTHALLUS, **PROTHALLIUM** Gametophyte stage or generation in ferns and fern-allies, bearing the sexual organs.

PROTOGYNOUS Having the stigma receptive to pollen before pollen is shed from the anthers of the same flower.

PROTUBERANCE A protrusion, swelling, bump, or bulge.

PROXIMAL Located nearest to the base or attachment point of a structure; contrasting with distal.

PRUINOSE Having a bloom on the surface; with a waxy or powdery secretion; with a surface coating more pronounced than glaucous.

PSEUDO- Greek prefix meaning false, not genuine.

PSEUDOANTAGONISM Pollination system (e.g., in some Orchidaceae) in which a flower mimics a male insect, thereby attracting a male insect that "fights" with the flower in an effort to protect his territory. In the process of repeating this activity, pollen can be transferred from one plant to another.

PSEUDOBULB The thickened or bulb-like, above ground stems of certain orchids.

PSEUDOCARP A structure made up of the mature ovary combined with some other organ; e.g., the hip of a rose.

PSEUDOCOPULATION Pollination system (e.g., in some Orchidaceae) in which a flower mimics a female insect (e.g., wasp or fly), thereby attracting male insects who copulate with the flowers and in the process of repeating this activity transfer pollen from one plant to another.

PSYCHOACTIVE Mind altering. A number of psychoactive materials are produced by plants.

PTERIDOLOGY The study of ferns and similar plants.

PTERIDOPHYTE A fern or similar plant; any member of the Lycopodiophyta, Psilophyta, Polypodiophyta (formerly Pteridophyta), or Equisetophyta (formerly Sphenophyta).

PUBERULENT Minutely pubescent.

PUBESCENCE The covering or indumentum of hairs on a plant without reference to specific type (e.g., pilose, hirsute).

PUBESCENT (a) General term for covering or indumentum of hairs; (b) sometimes used in a more restricted sense to refer to fine short hairs; downy.

PULP The juicy or fleshy tissue of a fruit.

PULVINATE With a pulvinus; cushion-shaped.

PULVINUS The swollen base of a petiole, as in many Fabaceae. The structure is often involved in leaf movements.

PUNCTAE Dots, depressions, or pits, commonly glandular in nature.

PUNCTATE With translucent or colored dots, depressions, or pits.

PUNCTICULATE Minutely punctate.

PUNGENT (a) Ending in a rigid sharp point or prickle; (b) acrid to the taste or smell.

PURPURASCENT Becoming or turning purple.

PUSTULATE With pustules, blisters, or blister-like swellings.

PUTAMEN The bony endocarp or "pit" of some fruits.

PYRENE The "nutlet," "pit," or stone in a drupe or drupelet; a seed together with the bony endocarp.

PYRIFORM Pear-shaped.

PYXIS (pl. PYXIDES) A capsule with circumscissile dehiscence, the top coming off as a lid; e.g., in *Plantago*.

Q

QUADR- Latin prefix meaning four.

QUADRANGULAR Four-cornered; square.

QUADRATE Nearly square.

QUASI- Prefix meaning "as if."

R

RACEME An unbranched (= simple) inflorescence with pedicelled flowers from an elongate main axis. Racemes are indeterminate.

RACEMIFORM Having the form, but not necessarily the technical character of a raceme.

RACEMOSE In racemes or resembling a raceme.

RACHILLA (a) A small rachis; (b) specifically the axis of a spikelet in the Poaceae.

RACHIS (a) Leaflet-bearing central axis of a pinnately compound leaf; (b) axis of an inflorescence.

RADIALLY SYMMETRICAL Descriptive of a structure that can be cut into halves from any marginal point through the center. When used in reference to flowers, the term is synonymous with actinomorphic or regular.

RADIATE (a) Spreading from a common center; (b) in Asteraceae, describing a head with disk florets in center and a whorl of ray (= ligulate) florets around the periphery.

RADIATE HEAD In Asteraceae, a head with both peripheral ray (= ligulate) florets and central disk florets.

RADICAL LEAVES Leaves arising from a root crown; basal leaves.

RADICLE Embryonic root of a germinating seed.

RAMEAL Pertaining to or located on a branch.

RAMIFICATION The arrangement of branching parts.

-RANKED Suffix, when combined with a numerical prefix, indicating the number of longitudinal rows in which leaves or other structures are arranged along an axis or rachis.

RAPHIDE Needle-shaped crystal in a plant cell, typically of calcium oxalate.

RAPHIDULOUS Resembling or having raphides.

RAY (a) Flat or strap-shaped type of corolla found in many Asteraceae; (b) primary branch of a compound umbel.

RAY FLORET (= Ligulate floret) Flower, with corolla expanded into a ligule, typical of many Asteraceae.

RECEPTACLE (a) (= Torus) Tip of a stem or pedicel, usually more or less enlarged, on which flower parts are attached; (b) in Asteraceae, the structure on which the florets of a head are attached.

RECEPTACULAR Referring to the receptacle.

RECLINED, RECLINATE Bent or turned downward.

RECUMBENT Leaning or reclining.

RECURVED Curved downward or backward.

REDUCED Small but probably derived from larger forerunners.

REFLEXED Abruptly bent downward.

REGULAR (= Actinomorphic or Radially symmetrical) Referring to a structure that can be cut into halves from any marginal point through the center; with the parts in each series alike or uniform. The term usually refers to flowers.

RELIC (a) A long-surviving species whose relatives have become extinct; (b) a plant persisting in a relatively small portion or portions of its former range.

REMOTE Widely or distantly spaced.

RENIFORM Kidney-shaped.

REPAND With a shallowly sinuate, undulating, or slightly wavy margin; less strongly wavy than sinuate.

REPENT Creeping and rooting at the nodes.

REPLICATE Folded backward.

REPLUM The partition between the halves of a fruit; e.g., Brassicaceae.

RESIN, RESINOID A miscellaneous or catchall term for a variety of amorphous, sometimes semisolid, perhaps gummy substances from plants. A number are toxic, sometimes extremely so; some are terpene derivatives while others are phenolic compounds; examples of phenolic resins are tetrahydrocannabinol (THC) from marijuana, urushiol from poison ivy, and hypericin from *Hypericum perforatum*.

RESINOUS, RESINIFEROUS Producing or bearing resin.

RESUPINATE Upside down; inverted; turned 180 degrees; e.g., some members of the Fabaceae and Orchidaceae have resupinate flowers.

RETICULATE Net-veined or with a net-like pattern.

RETINACULUM (pl. RETINACULA) (a) In Asclepiadaceae, the filament-like band connecting a pollinium to the gland-like corpusculum; (b) in Acanthaceae, the curved, hook-like, modified funiculus which retains the seed until maturity.

RETRORSE Bent or turned backward or downward; e.g., hairs pointing down a stem, the free end of the hair below or proximal to the attached end; contrasting with antrorse.

RETRORSELY BARBED With barbs (= points) pointing downward toward base.

RETUSE Shallowly notched at a rounded apex.

REVOLUTE With margins rolled down and inward; e.g., revolute leaves.

RHIZOID A filamentous root-like structure (without the anatomy of a root) on the gametophyte of ferns or other non-seed-producing plants.

RHIZOMATOUS Possessing a rhizome.

RHIZOME Underground stem with nodes and scale leaves.

RHIZOPHORE In certain Pteridophytes, a specialized leafless stem emitting roots.

RHOMBIC Somewhat diamond-shaped; shaped like two adjacent equilateral triangles.

RHOMBOID, RHOMBOIDAL A three-dimensional rhombic figure.

RIB (a) One of the principal longitudinal veins of a leaf or other organ; (b) ridge on a fruit.

RIBBED With prominent ribs or veins.

RINGENT Gaping; open.

ROOT The portion of the main axis (or one of its subdivisions) of a plant usually found below ground and lacking nodes, internodes, or leaves.

ROOTLET (a) A little root; (b) term often applied to the holdfast roots of certain climbing plants.

ROOTSTOCK According to Shinners (1958), a "weasel-word" indicating an elongate crown, rhizome, or rhizome-like structure; an old inaccurate term for rhizome.

ROSEATE Rosy or pinkish.

ROSETTE A cluster or whorl of leaves or other organs closely arranged in a radial pattern.

ROSTELLATE Diminutive of rostrate.

ROSTELLUM (a) A small beak; (b) in Orchidaceae, an extension from the upper edge of the stigma.

ROSTRATE Beaked.

ROSTRUM A beak or any beak-like extension.

ROSULATE In rosettes.

ROSY The rich pink characteristic of the petals of some roses.

ROTATE Wheel-shaped or saucer-shaped as in a sympetalous corolla with an obsolete or very short tube and a flat circular limb.

ROTUND Essentially circular in outline.

RUDERAL Weedy or growing in waste places.

RUDIMENT A structure very imperfectly developed, non-functional, or represented only by a vestige; e.g., rudimentary non-functional florets in some grass spikelets.

RUDIMENTARY Having the character of a rudiment.

RUFESCENT Becoming reddish brown.

RUFOUS Reddish brown.

RUGOSE Wrinkled.

RUGULOSE Diminutive of rugose.

RUMINATE Appearing as through chewed or wadded; roughly wrinkled.

RUNCINATE With margins that are coarsely serrate to sharply incised with the segments pointing toward the base, as in *Taraxacum*.

RUNNER An elongate, slender, prostrate stem taking root at the nodes or tip.

RUPTURING Bursting open along irregular lines.

RUSSET Reddish brown.

s South.

Sac A pouch or bag.

Saccate Sac-like; pouch-like.

Sagittate With the form of an arrowhead; triangular with the basal lobes pointing downward or inward toward the petiole.

Salient Projecting forward.

Saline Of or pertaining to salt.

Salverform Descriptive of a corolla having a narrow, nearly straight basal tube, abruptly expanded at top into a flattened or saucer-shaped limb.

Samara A dry, indehiscent, winged fruit; a fruit that is achene-like but with a flat wing; e.g., in maple, *Acer* (Aceraceae).

Sap The juice of a plant.

Saponaceous Soapy, slippery to the touch.

Saponic or **Saponin glycoside** A glycoside (two-component molecule) that upon breakdown yields a saponin.

Saponin A soap-like molecule that lowers the surface tension of aqueous solutions. Saponins usually have a bitter acrid taste and are commonly irritating to mucous membranes; chemically saponins are either triterpenes or steroids; they can alter the permeability of cell membranes and may react with proteins and are thus potentially toxic; they are highly toxic to cold-blooded animals and have long been used to stun fish.

Saprophyte A non-photosynthetic plant without chlorophyll, deriving its food from dead organic material in the soil by mycorrhizal relationships or otherwise; e.g., *Monotropa* (Ericaceae).

Saprophytic Subsisting on dead organic matter.

Saxicolous Growing on rocks.

Scaberulous Minutely scabrous.

Scabridulous Slightly rough.

Scabrous Rough to the touch due to short stiff hairs or the structure of the epidermis.

Scalariform Ladder-like.

Scale Any small, thin, usually dry, appressed leaf or bract, often only vestigial.

Scalloped Crenate.

Scandent Climbing without aid of tendrils.

Scape A naked (leafless but scales or bracts may be present) flowering stem or stalk arising from the ground.

Scapiform Resembling a scape.

Scapose Bearing or resembling a scape.

Scarious Membranous, dry, papery, translucent or transparent, not green.

Scattered Distributed in an irregular manner; not clustered together.

Schizo- Greek suffix meaning split or divided.

Schizocarp A fruit that splits between carpels into one-seeded portions (= mericarps); e.g., in the Apiaceae or Boraginaceae.

Sciophyte A plant adapted to grow in or tolerate shade.

Scimitar-shaped With the shape of a curved sword or scimitar.

Sclerenchyma An internal tissue composed of hard, thick-walled cells.

Sclerenchymatous Having sclerenchyma.

Sclerotic Hardened; stony in texture.

Scorpioid Uncurling, like a scorpion's tail.

Scrobiculate Marked by minute or shallow depressions.

Scrotiform Pouch-like.

Scrub Vegetation of stunted or densely crowded bushes.

Scurfy With whitish, scaly, blister-like structures or scales on the surface.

Scutellum A shield-like protrusion on the calyx; e.g., in some Lamiaceae.

Secondary compounds Naturally occurring plant materials not essential to the primary (= life-sustaining) metabolism of the plant; examples of categories include alkaloids and glycosides. Many are significant because of their physiological activity when given to animals; they are probably important to plants in defense against herbivores or microbes.

Second glume The uppermost of the two glumes of a spikelet.

Secund Directed to one side; arranged on one side only; unilateral; one-sided.

Seed A fertilized ripened ovule, covered by a seed coat (developed from the integument(s)) and containing the embryo and in flowering plants the endosperm. The seed also contains the remnants of the nucellus (= sporangium) and megagametophyte.

Seed coat (= Testa) Outer covering of a seed, developed from the integument(s).

Seep A moist spot where underground water comes to or near the surface.

Segment One of the parts of a leaf, petal, or calyx that is divided but not truly compound.

Segregate Term used as a noun or adjective to refer to or describe a taxon that is sometimes recognized separately from a more inclusive group; e.g., depending on authority, the segregate, *Tovara* (or the segregate genus, *Tovara*), is either included in, or recognized separately from, the more inclusive genus *Polygonum*.

Seleniferous Containing selenium.

Selenium An element that is concentrated in the tissues of some plants (e.g., poison-vetches in genus *Astragalus*) and can cause a toxic, sometimes fatal response in livestock.

Self-fertile Capable of self-fertilization (= union of gametes from same plant).

Self-incompatible Incapable of self-fertilization.

Semi- Latin prefix meaning half.

Semiparasitic (= Hemiparasitic) Descriptive of a plant that carries out photosynthesis but obtains some of its food, mineral nutrition, or water needs from another living organism (the host).

Senescent Aging or aged.

Sensu lato "In a broad sense"; used to refer to the broad treatment of taxa; e.g., a genus sensu lato is one that has not been split into a number of segregates.

Sensu stricto "In a narrow sense"; used to refer to a restricted or narrow treatment of a taxonomic group; e.g., a genus sensu stricto is viewed in a more restricted sense than previously as the result of segregating or splitting out various taxa.

Sepal A single unit of the calyx; one of the outermost whorl of parts in a complete flower. Sepals typically cover the other flower parts during the bud stage; they are collectively referred to as the calyx.

Sepaloid Sepal-like, usually green and thicker in texture than a petal.

Separate Not joined; of individual units.

Septate With partitions or divisions.

Septicidal Descriptive of a capsule that dehisces along or through the septa or partitions separating its locules or seed chambers; contrasting with loculicidal.

Septifragal Breaking away at the partitions, as the valves of a capsule.

Septum A partition or cross wall within an organ; e.g., the septa of an ovary or of the leaf of a rush (*Juncus*).

Seriate, Serial With series or distinct rows.

Sericeous Silky; with appressed, fine, and straight hairs.

Serotinous Literally, "late"; in *Salix* indicating those species in which the catkins develop later than the leaves; produced or occurring late in the season.

Serrate With pointed teeth sloping forward; saw-toothed.

Serrulate Finely serrate.

Sessile Without a pedicel, petiole, or stalk; inserted directly.

Seta (pl. Setae) A bristle.

Setaceous Bristle-like.

Setiferous Bearing bristles.

Setiform Resembling a bristle.

Setose Bearing bristles.

Setulose Diminutive of setose.

Sheath (a) ± tubular structure surrounding an organ; portion that clasps or encloses; (b) specifically, the basal tubular portion of the leaf in grasses and grass-like plants between the node and the blade.

Shoot (a) A young stem or branch; (b) the ascending axis of a plant.

Shoulder That part of an organ that rather abruptly curves inward.

Shrub A woody perennial usually branching from the base with several main stems.

Sigmoid S-shaped; doubly curved.

Silica Silicon dioxide; a white or colorless, very hard, crystalline mineral substance.

Siliceous Containing or composed of silica (the principal component of glass and sand); e.g., descriptive of certain hairs, with high silica content, that easily break off in the flesh.

Silicle, Silicula A short silique; short and broad capsular fruit of the mustard family. A silicle is not much longer than broad; sometimes defined as < 3 times as long as broad.

Silique, Siliqua Dry, dehiscent, variously shaped, many-seeded, 2-valved capsule with valves splitting from the bottom and leaving a false partition known as a replum; the fruit type typical of the mustard family. The term is sometimes restricted to long and narrowly cylindrical fruits 3 times as long as broad or longer; shorter fruits with the same structure are then referred to as silicles.

Silky Sericeous; with appressed, fine, and straight hairs.

Silvery With a whitish, metallic, more or less shining luster.

Simple (a) Single, of one piece, not compound, as a pistil of one carpel. (b) descriptive of an unbranched stem, inflorescence, or other structure; (c) descriptive of a leaf that is not compound.

Simple leaf Single-bladed leaf, not divided into individual leaflets.

Simple pistil Pistil composed of only one carpel (and with a single style branch).

Sinuate, Sinuous Having the margin wavy with regular strong indentations.

Sinus The space or recess between two lobes, segments, or divisions of a leaf or other expanded organ.

Slough A wet place or deep mud or mire; a sluggish channel.

Smooth Not rough to the touch; without vestiture or other special covering.

Sobol A basal shoot.

Soboliferous Producing basal shoots, clump-forming.

Solitary Borne singly.

Sordid Dirty white.

Sorus (pl. Sori) A cluster or heap of sporangia. The term is used mainly to refer to the sporangial clusters of ferns.

Spadix (pl. Spadices) Fleshy spike with inconspicuous flowers, usually surrounded or partially enclosed by a spathe; e.g., characteristic of Araceae.

Spathaceous, Spathiform Resembling a spathe.

Spathe Enlarged leafy bract surrounding or partially enclosing an inflorescence.

Spatulate Spatula-shaped; rounded above and gradually narrowed to base.

Species Unit of classification below the rank of genus; a group of individuals that are actually or potentially capable of interbreeding in natural populations and are reproductively isolated from other such groups. Generally there are morphological characteristics that distinguish and can thus be used to separate such groups; the definition is complicated by instances of asexual reproduction.

Specific epithet The second half of the scientific name of a species, the scientific name being composed of the genus name and the specific epithet.

Spermatophyte A plant that produces seeds; all gymnosperms and angiosperms.

Spherical Globular; orbicular.

Spicate In spikes or resembling a spike.

Spiciform Having the form of a spike but not necessarily the technical structure.

Spike Indeterminate inflorescence with sessile flowers on a ± elongate floral axis.

Spikelet The basic unit of the grass inflorescence, usually consisting of a short axis (= rachilla) bearing two empty bracts (= glumes) at the basal nodes and one or more florets above. Each floret usually consists of two bracts (the lemma = lower bract and the palea = upper bract) and a flower. The flower usually includes two lodicules (= vestigial perianth segments that become turgid and open the bracts at flowering time), three stamens, and a pistil. The term spikelet is also used to refer to the basic unit of the inflorescence in the Cyperaceae.

Spine A sharp-pointed structure, usually vascularized and thus ± stout or woody, generally modified from part or all of a leaf or stipule. A spine is sometimes distinguished from a thorn, which is a modified branch.

Spinescent Becoming spine-like; ending in a spine; having spines.

Spiniferous, Spinose Having spines.

SPINULOSE With small spines over the surface, margin, or edge.

SPIRAL Describing the arrangement of like organs, such as leaves, at regular angular intervals along an axis.

SPIRAL THICKENING Thickening of the walls of a xylem cell laid down in the form of a spiral.

SPLITTER A taxonomist who in general has the tendency to split or divide larger taxa into a number of segregates; contrasting with lumper.

SPONGIOSE Soft, spongy.

SPORADIC Of irregular occurrence.

SPORANGIOPHORE An appendage holding a sporangium.

SPORANGIUM (pl. SPORANGIA) A spore case or spore sac.

SPORE An asexual, usually one-celled reproductive body; a cell resulting from meiotic cell division in a sporangium representing the first cell of the gametophyte generation.

SPOROCARP A specialized structure containing sporangia.

SPOROPHYLL Spore-bearing leaf.

SPOROPHYTE The spore-producing, typically diploid generation that alternates with the gamatophyte (= gamete-producing, typically haploid); the dominant generation in most plants except mosses and liverworts.

SPREADING Diverging to the side, almost to the horizontal.

SPREADING HAIRS Hairs that are ± erect, not at all appressed.

SPUR (a) Basal, sac-like, hollow projection, short or long and narrow, from a corolla or calyx, and often containing nectar; (b) a short lateral branch with little internode development resulting in closely clustered leaves or flowers.

SQUAMA (pl. SQUAMAE) A scale, usually reduced from and homologous to a leaf.

SQUAMELLA (pl. SQUAMELLAE) A tiny or secondary scale.

SQUAMOSE, SQUAMATE Covered with scales; scaly.

SQUAMULOSE With small scales.

SQUARROSE Spreading rigidly at right angles or more; e.g., with leaves or bracts spreading and bending backward abruptly in the upper part.

STALK The supporting structure of an organ; e.g., petiole, pedicel, peduncle.

STAMEN The male reproductive organ of a flower consisting of a filament (= stalk) and anther (= pollen-bearing structure). The stamens of a flower are collectively referred to as the androecium.

STAMINAL Pertaining to a stamen.

STAMINATE Referring to a flower with only the stamens being functional or with stamens only.

STAMINODE, STAMINODIUM (pl. STAMIDODIA) An abortive, sterile, or nonfunctional stamen with reduced anther or usually none. A staminode can be enlarged or widened, sometimes so much so as to be petal-like.

STANDARD (a) (= Banner) The usually large, upper (= adaxial) petal of a papilionaceous flower (Fabaceae); (b) the inner erect tepals or perianth segments in *Iris* (Iridaceae).

STELE The central vascular cylinder of a plant.

STELLATE Star-shaped or star-like; when used in reference to hairs it means those branched hairs with a central stalk and branch hairs arising at the top of the stalk (like points of light coming out of a star).

STEM A major division of the plant-body in contrast to root and leaf, distinguished from both by certain anatomical features and commonly also by general aspect; the main axis or axes of a plant; the portion of the plant axis bearing nodes, leaves, and buds and usually found above ground.

STERILE Unproductive; nonfruiting; without functional sex organs; without spores.

STERIGMA (pl. STERIGMATA) A very short persistent stipe or stalk.

STERNOTRIBAL Descriptive of flowers in which anthers are positioned to dust pollen on underside of thorax of insects.

STEROID Any of a large group of lipid soluble organic compounds based on a complex framework of four interconnected rings of carbon atoms. A number of plant glycosides have steroids as the aglycone; these aglycones are often toxic.

STEROID GLYCOSIDE A glycoside (two-component molecule) that upon breakdown yields a steroid as the aglycone (= non-sugar component). Cardiac glycosides and some saponic glycosides have steroidal aglycones; these are often poisonous.

STIGMA Portion of pistil (usually terminal and small) modified (roughened, sticky, or branched) to catch or receive pollen; the receptive surface of the pistil.

STIGMATIC Like or pertaining to a stigma.

STIPE (a) In general, a stalk; (b) specifically, the leaf stalk of a fern; (c) specifically, the narrow, stalk-like, basal portion of an ovary or fruit; e.g., in *Cleome*.

STIPEL Appendage at the base of a leaflet, analogous to a stipule at base of a leaf.

STIPELLATE With stipels.

STIPITATE With a stalk or stipe.

STIPULAR Pertaining to or located on a stipule.

STIPULATE With stipules.

STIPULE Paired appendages subtending the petiole of a leaf. Stipules can be conspicuous and persistent, small and rapidly lost, or totally lacking; the evolutionary origin and original function of stipules is unclear. They have become variously modified to serve as spines, additional photosynthetic area, etc.

STOLON A creeping horizontal stem that loops or runs along the surface of the ground and roots at the nodes.

STOLONIFEROUS Producing stolons.

STOMA, STOMATE (pl. STOMATA). A minute opening (= "breathing" pore) between the epidermal cells of a leaf or stem through which gases and water-vapor enter and leave the plant.

STOMATIFEROUS Bearing stomata.

STONE Bony endocarp of a drupe.

STONE CELLS (= Grit cells) Individual cells that have become hardened or sclerotic; e.g., in flesh of a pear.

STRAMINEOUS Straw-colored.

STRIATE With fine longitudinal lines, channels, or ridges.

STRICT Very straight, unbranched, upright.

STRIGILLOSE, STRIGULOSE Diminutive of strigose.

STRIGOSE With sharply bent (at base) but otherwise straight, appressed hairs.

STROBILUS (pl. STROBILI) (= Cone) A usually cone-like, globose or cylindrical structure involved in reproduction and composed of an axis with a spiral, usually dense aggregation of sporophylls, bracts, or scales (these bearing spores, pollen, or seeds).

STROPHIOLATE Having a strophiole.

STROPHIOLE On certain seeds, an appendage to the hilum.

STRUMOSE Descriptive of the surface of an organ bearing cushion-like swellings.

STYLE Portion of pistil between ovary and stigma, often elongated and narrow, frequently branched with the number of branches often indicative of the number of carpels making up the pistil.

STYLOPODIUM Enlarged base of style as in many Apiaceae.

SUB- Latin prefix meaning almost, somewhat, of inferior rank, beneath.

SUBAPICAL Almost at the apex.

SUBBASAL Almost at the base.

SUBCLASS The unit, category, or rank in classification between class and order, composed of one or more orders; e.g., the Liliidae is a subclass of class Monocotyledonae.

SUBCORIACEOUS Somewhat leathery in texture.

SUBEROSE Corky in texture.

SUBFAMILY The unit, category, or rank in classification between family and genus, composed of one or more genera.

SUBGENUS A unit, category, or rank in classification between genus and species, composed of one or more species.

SUBORDER The unit, category, or rank in classification between order and family, composed of one or more families.

SUBMERGED, SUBMERSED (= Immersed) Growing under water.

SUBSESSILE Almost sessile.

SUBSPECIES A unit, category, or rank in classification below the level of species and between species and variety; a geographically distinct variant. The categories of subspecies and variety are not used consistently by taxonomists.

SUBTEND To be present just below; e.g., an involucre of bracts immediately subtends the flowers of some Malvaceae.

SUBTERRANEAN Below ground.

SUBULATE (= Awl-shaped) Tapering from the base to a slender or stiff point; narrow and sharp-pointed.

SUCCULENT Fleshy, thickened.

SUCKER A shoot of subterranean origin. Many tree species have sucker shoots with leaves atypical for the species (e.g., oaks).

SUFFRUTESCENT, SUFFRUTICOSE Woody only at base with the upper parts herbaceous and annual; obscurely shrubby or fruticose.

SULCATE Grooved or furrowed longitudinally.

SULCUS A furrow or groove.

SUPERIOR OVARY One that is separate from the other flower parts. The other flower parts may over-top it but are attached at or under its base; a superior ovary is found in either a hypogynous or a perigynous flower.

SUPINE Laying flat with face upward.

SUPRA-, SUPER- Latin prefix meaning above.

SUPRAMEDIAL Above the middle; when used to refer to the location of fern sori, it means somewhat beyond the middle of the distance between the leaf segment midvein and margin, but not so much so as to be called submarginal.

SUPPRESSED Failing to develop.

SURCULOSE Producing suckers or shoots arising from underground parts.

SURCURRENT Extending upward; said of a pinnule whose base extends upward and forms a wing along the rachis; opposite of decurrent.

SUSPENDED Hanging downward.

SUTURE A line of dehiscence or a longitudinal seam.

SWALE A moist meadowy area lower than the surrounding areas.

SWAMP Wet or periodically wet area with some trees.

SYM-, Greek prefix meaning with or together.

SYMMETRICAL Possessing one or more planes of symmetry; regular in number and size of parts.

SYMPATRIC Growing together with or having the same range as.

SYMPETALOUS (= Gamopetalous) With petals united, at least basally, forming a tube.

SYMPODIAL With the main axis or stem ceasing to elongate but growth being continued by the lateral branches.

SYMPODIUM An apparent main axis formed of successive secondary axes, each of which represents one fork of a dichotomy, the other being much weaker or entirely suppressed.

SYN- Greek prefix meaning united.

SYNCARP (= Multiple fruit) Used to refer to a structure composed of several more or less coalescent fruits from separate flowers; e.g., *Morus* (mulberry), *Maclura* (bois d'arc), *Ananas* (pineapple).

SYNCARPOUS (a) Having carpels united; (b) of or pertaining to a syncarp.

SYNCONIUM The multiple, hollow "fruit" of a fig (*Ficus*), which is actually an enlarged fleshy branch or receptacle enclosing the inflorescence (with flowers borne inside). Much of the tissue of a fig is morphologically derived from the stem.

SYNGENESIOUS With anthers united into a tube surrounding the style. This condition is typical of Asteraceae.

SYNSEPALOUS (= Gamosepalous) With sepals united, at least basally, forming a tube.

SYNONYM A currently unaccepted scientific name for a taxon.

SYNONYMY Referring to the series of names no longer used for a taxon.

SYSTEMATICS Scientific study of the kinds and diversity of living organisms and of the relationships between them. The term is often used synonymously with taxonomy.

T

TANNIN A type of phenol (= aromatic alcohol) that can act as a digestion inhibitor by binding up proteins and thus stopping enzymatic action. They are widespread in plants apparently as a chemical defense against herbivores; used by humans to tan leather; when ingested in quantity they can be toxic to animals.

TAPROOT, TAPROOTED The primary descending root, giving off small laterals but not dividing; the one dominant root markedly larger than the others.

TAWNY Dull brownish-yellow; fulvous.

TAXON (pl. TAXA) (a) General term referring to any unit of classification such as variety, subspecies, species, genus, family, etc.; (b) term used to refer to a specific variety, subspecies, etc.

TAXONOMY The branch of science that deals with classification, identification, and nomenclature.

TEETH (plural of tooth) Marginal projections, protuberances, serrations, or dentations, usually sharply pointed.

Tendril A slender twining or coiling appendage or axis that enables plants to climb; often a modified leaf or stem.

Tepal Part of a perianth, usually of a perianth consisting of only one whorl, or of one not differentiated into sepals and petals; a part of the outermost whorl or whorls of flower parts.

Teratological Distinctly abnormal; malformed.

Terete Rounded or circular in cross-section, cylindrical.

Terminal At the tip or apex; distal.

Terminal bud (= Apical bud) Bud at the end (= apex) of a stem or branch.

Ternate, Ternary In threes.

Terpenes, Terpenoids Common organic compounds in plants that are products of acetate metabolism. Numerous kinds are known resulting from variations in the use of 5-carbon isoprene units in their structures; they are often combined with other substances to form complex molecules; essential oils, saponins, some resins, latex, cartenoid pigments, and steroids are examples of compounds that are terpene derivatives.

Terrestrial Growing in the ground; supported by soil; contrasting with aquatic.

Testa (= Seed coat) The outer covering of a seed; hardened mature integument(s).

Tetra- Greek prefix referring to four.

Tetrad A group of four similar objects; e.g., in Ericaceae, the four pollen grains remaining together.

Tetradynamous With four long stamens and two short stamens.

Tetragonal, Tetragonous Four-angled.

Tetrahedal Four-sided, as a three-sided pyramid and its base.

Tetramerous (= 4-merous) Having flower parts in fours or multiples of four.

Tetraploid With four sets of chromosomes; twice the normal diploid level.

Thalloid Consisting of a thallus; thallus-like.

Thallophytes Algae and fungi.

Thallus (a) A plant body not differentiated into true leaves, stems, or roots; (b) a flat, leaf-like organ.

Theca (= Anther cell) One of the pollen sacs or locules of an anther.

Thorn A sharp-pointed, stiff, woody structure derived from a modified branch.

Throat The area of juncture of limb and tube in a sympetalous corolla.

Thyrse A panicle-like inflorescence consisting of cymules, usually elongate and slender with main axis indeterminate and the lateral axes determinate.

Thyrsoid With the appearance of a thyrse.

Tomentose Covered with short, soft, curly, densely matted or entangled hairs.

Tomentulose Diminutive of tomentose.

Tomentum Densely matted wool.

Tooth (pl. Teeth). Any marginal projection, protuberance, serration, or dentation, usually sharp pointed.

Toothed With minor projections and indentations alternating along the margin.

Topotype A specimen from the original or type locality of that species or other taxon.

Torose Alternately contracted and expanded.

Tortuous Twisted or bent in different directions.

Torulose Cylindrical with slight constrictions at intervals; necklace-like.

Torus (= Receptacle) Tip of a stem or pedicel, usually more or less enlarged, on which flower parts are attached.

Toxalbumin Toxic plant proteins; e.g., ricin from *Ricinus*. The term phytotoxin is also often used in this context.

Trabecula A transverse partition or cross-bar.

Trabecular, Trabeculate Like or with a partition or cross-bar.

Trailing Prostrate, but not rooting.

Translator In Asclepiadaceae, the wishbone-shaped combination of the clip (= gland-like corpusculum) and bands (= filament-like retinacula) connecting a pair of pollinia from adjacent anthers.

Translucent Allowing the passage of light rays, but not transparent.

Transverse Lying or being across or in a cross direction.

Trapezoid A body with four unequal sides.

Tree A woody perennial with usually a solitary trunk or main stem.

Tri- Latin prefix indicating three, or three times.

Triad In threes.

Tribe The unit, category, or rank in classification between subfamily and genus, composed of one or more genera.

Trichome Any hair, hair-like projection, or bristle from the epidermal surface.

Trichotomous Forking into three equal parts.

Tricolpate Three-grooved; e.g., a three-grooved pollen grain.

Trifid Three-cleft.

Trifoliate Having three leaves. The term is often used synonymously with trifoliolate.

Trifoliolate Having three leaflets.

Trifurcate With three prongs; three-forked.

Trigonous, Trigonal Three-angled.

Trilobed, Trilobate Three-lobed.

Trimerous Having the parts in threes.

Trimorphic Of three forms.

Trinerved With three primary nerves; triple-nerved; usually with a midrib and two main lateral nerves arising from the base of the midrib.

Tripartite Three-parted.

Triquetrous Three-angled in cross-section.

Tristichous In three vertical rows.

Trullate Trowel-shaped, widest below the middle.

Trumpet-shaped Describing a gamopetalous corolla or gamosepalous calyx in which the tube gradually widens upward.

Truncate Ending abruptly as if cut off squarely at the end; appearing "chopped off."

Tube The narrow, cylindrical, basal portion of a sympetalous corolla or synsepalous calyx.

Tuber Modified underground stem; stem enlarged and subterranean with nodes, buds, and scale leaves, often serving to store food; e.g. in *Solanum* (potato).

Tubercle (a) A small, tuber-like structure; (b) small rounded protuberance or projection from a surface; (c) the persistent style base in some Cyperaceae.

Tuberculate Covered with tubercles or warty or nipple-like protuberances.

Tuberiferous Bearing tubers.

Tuberous Bearing tubers or resembling a tuber in appearance.

Tubular With the shape of a hollow cylinder.

Tuft; Tufted A cluster or fascicle of trichomes, leaves, or other elongate structures.

Tumid Swollen; inflated.

Tunic A loose, membranous, outer skin or coat.

Tunicate Describing a bulb in which the leaves are arranged in concentric circles; with coats or tunics.

Turbinate Top-shaped; inversely conical.

Turgid Swollen, but solid or full; contrasting with inflated.

Turion A scaly swollen structure or offshoot, often serving to overwinter; e.g., in some *Myriophyllum* species the turions store carbohydrates and serve as propagules.

Twice-pinnately compound (= Bipinnate) Descriptive of a leaf with leaflets pinnately arranged on lateral axes that are themselves pinnately arranged on the main axis; with the primary divisions (= pinnae) themselves pinnate.

Twig A young woody stem; the growth of the current season.

Two-lipped (= Bilabiate) Descriptive of a corolla (or calyx) of united petals (or sepals) cut on the two sides forming an upper and a lower portion.

Type A plant specimen to which the name of a taxon is permanently attached. When any new taxon (e.g., species, variety) is named, the name has to be associated with a particular "type" specimen.

U

Ubiquitous Occurring everywhere.

Umbel Usually flat-topped or convex inflorescence with flower pedicels all attached at the same point, like the rays of an umbrella; inflorescence type typical of Apiaceae (Umbelliferae).

Umbellate Umbel-like, or in an umbel.

Umbellet A secondary or ultimate umbel; one of the smaller umbellate flower clusters in a compound umbel.

Umbelliform Resembling an umbel in appearance.

Umbo A rounded elevation or protuberance at the end of or on the side of a solid organ.

Umbonate With an umbo or projection.

Uncinate With apex hooked; e.g., hairs with a hook at tip.

Uncinulate Minutely uncinate.

Undulate Gently wavy, less pronounced than sinuate.

Unguiculate Narrowed at the base into a claw.

Uni- Latin prefix meaning one.

Unifoliate With only one leaf.

Unifoliolate Referring to a compound leaf reduced to a single leaflet; e.g., in some members of the Fabaceae.

Unilateral One-sided; developed or hanging on one side.

Unilocular Having only one locule or cell.

Uniseriate Arranged in a single horizontal row.

Unisexual Having only stamens or only pistils; of only one sex; having flowers either staminate or pistillate.

United Fused into one unit.

Urceolate Urn-shaped or pitcher-like, ovoid or subcylindrical in shape with narrowed top or opening; e.g., corollas in some Ericaceae.

Utricle (a) A small, bladder-like, one-seeded, usually indehiscent fruit; (b) a small bladder.

V

Valvate Meeting along the margins only and not overlapping; contrasting with imbricate.

Valve A separable part of a capsule; the units or pieces into which a capsule splits or divides in dehiscing.

Variegated Irregularly colored in patches; blotched.

Variety A unit, category, or rank in classification below the level of species, sometimes treated as a subdivision of subspecies; group of plants with minor characters or differences separating them from other similar plants. The terms variety and subspecies are used inconsistently by taxonomists.

Vascular Pertaining to the conducting tissues (xylem and phloem).

Vascular bundle Thread-like fiber of xylem and phloem in a stem or other organ.

Vascular cambium The thin layer of delicate rapidly dividing cells that form wood internally and bark externally; also known as cambium.

Vegetative organ Root, stem, leaf, or other non-reproductive organ of a plant.

Vein Strand or bundle of vascular tissue.

Veinlet A little or ultimate vein.

Velamen A thin sheath or covering; e.g., on orchid roots.

Velamentous With a thin sheath or covering.

Velum Thin flap of tissue.

Velutinous Velvety with numerous erect hairs.

Venation The pattern or arrangement of veins.

Ventral Situated on or pertaining to the adaxial side (= side toward axis) of an organ; typically the upper or inner surface; contrasting with dorsal.

Ventricose (= Gibbous) Swollen or inflated on one side.

Vermiform Worm-like.

Vermillion Scarlet; brillant red.

Vernal Appearing in spring.

Vernation The arrangement of leaves, sepals, or petals in the unopened bud.

Verrucose Covered with wart-like protuberances.

Verruculose Diminutive of verrucose.

Versatile Attached near the middle and often capable of swinging about the attachment point, typically referring to attachment of an anther to a filament; contrasting with either basifixed or dorsifixed.

Verticil (= Whorl) A whorl of three or more members or parts attached at the same node of the supporting axis.

Verticillaster A false whorl composed of pairs of opposite cymes; e.g., in some Lamiaceae.

Verticillate (= Whorled) With three or more leaves or flowers attached at the same node; in a circle or ring.

Vespertine Opening in the evening; e.g., night-blooming *Cereus* (Cactaceae).

VESICLE A small cavity or bladder.

VESSEL Water conducting structure of the xylem, formed from the walls of a series of dead xylem cells stacked end to end.

VESTIGIAL Reduced to a trace, rudiment, or vestige; degenerate; referring to a once more fully developed structure.

VESTURE, VESTITURE Any covering on a surface making it other than glabrous; e.g., hairs, scales.

VEXILLUM The standard or banner in papilionaceous flowers.

VILLOSULOUS Diminutive of villous.

VILLOUS, VILLOSE With long, soft, spreading, or ascending, unmatted hairs; shaggy.

VINE A plant that climbs by tendrils or other means, or that trails or creeps along the ground.

VIRGATE Wand-shaped; slender, straight, and erect.

VISCID, VISCOUS Sticky; with sticky surfaces formed by secretions; glutinous.

VISCIDULOUS Slightly viscid.

VIVIPAROUS Germinating or sprouting from seed or bud while attached to the parent plant.

VOLUBLE Twining.

W

W West.

WAIF A species that is only fleetingly established and probably not permanently naturalized.

WEED (a) A plant growing where it is not wanted; (b) a plant with the genetic endowment to inhabit and thrive in places of continual disturbance, most especially in areas that are repeatedly affected by the activities of humankind.

WHORL, WHORLED (= Verticillate) With three or more leaves or flowers attached at the same node; in a circle or ring.

WIDE (= Broad) Distance across a structure (equal to diameter if tubular).

WING (a) A thin, membranous or flat extension or projection; (b) the two lateral separate petals in some Fabaceae and Polygalaceae.

WINTER ANNUAL An annual plant (the total life cycle taking one year or less) vegetatively persistent through the winter, and flowering and fruiting in the late winter or early spring.

WOOLLY (= Lanate) With long, soft, and more or less matted or entangled hairs; wool-like.

WORT An old word of Anglo-Saxon origin meaning the equivalent of herbaceous plant.

X

XERIC Characterized by or pertaining to conditions of scanty moisture supply; dry.

XERO- Greek prefix signifying dry.

XEROPHYTE A plant that can subsist with a small amount of moisture, such as a desert plant; contrasting with hydrophyte and mesophyte.

XEROPHYTIC Dry-adapted; drought resistant; contrasting with mesophytic and hydrophytic.

XYLEM Water conducting tissue.

Z

ZYGOMORPHIC Referring to a flower or other structure with only one plane of symmetry; divisible into halves in one plane only; bilaterally symmetrical; e.g., with the left half a mirror image of the right; contrasting with both actinomorphic and irregular.

ZYGOTE Cell produced from fertilization or the union of two gametes; a fertilized egg.

LITERATURE CITED

Abu-Asab, M.S. and **P.D. Cantino.** 1989. Pollen morphology of *Trichostema* (Labiatae) and its systematic implications. Syst. Bot. 14:359–369.

_____ and _____. 1993. Phylogenetic implications of pollen morphology in tribe Ajugeae (Labiatae). Syst. Bot. 18:100–122.

Achey, D.M. 1933. A revision of the section *Gymnocaulis* of the genus *Orobanche*. Bull. Torrey Bot. Club 60:441–451.

Adams, R.P. 1972. Chemosystematic and numerical studies of natural populations of *Juniperus pinchotii* Sudworth. Taxon 21:407–427.

_____. 1975. Numerical-chemosystematic studies of infraspecific variation in *Juniperus pinchotii*. Biochem. Syst. & Ecol. 3:71–74.

_____. 1986. Geographical variation in *Juniperus silicola* and *J. virginiana* in the southeastern United States: Multivariate analysis of morphology and terpenoids. Taxon 35:61–75.

_____. 1993. *Juniperus*. In: Flora of North America Editorial Committee, eds. Fl. North Amer. 2:412–420. Oxford Univ. Press, New York and Oxford.

Adams, W.P. 1957. A revision of the genus *Ascyrum* (Hypericaceae). Rhodora 59:73–95.

_____. 1962. Studies in the Guttiferae I. A synopsis of *Hypericum* section *Myriandra*. Contr. Gray Herb. 189:3–51.

_____. 1973. Clusiaceae of the southeastern United States. J. Elisha Mitchell Sci. Soc. 89:62–71.

_____ and **N.K.B. Robson.** 1961. A re-evaluation of the generic status of *Ascyrum* and *Crookea* (Guttiferae). Rhodora 63:10–16.

Aellen, P. and **T. Just.** 1943. Key and synopsis of the American species of the genus *Chenopodium* L. Amer. Midl. Naturalist 30:47–76.

Ahrendt, L.W.A. 1961. *Berberis* and *Mahonia*: A taxonomic revision. J. Linn. Soc., Bot. 57:1–410.

Aiken, S.G. 1981. A conspectus of *Myriophyllum* (Haloragaceae) in North America. Brittonia 33:57–69.

_____ and **L.P. Lefkovitch.** 1993. On the separation of two species within *Festuca* subg. *Obtusae* (Poaceae). Taxon 42:323–337.

Ajilvsgi, G. 1979. Wild flowers of the Big Thicket. Texas A&M Univ. Press, College Station.

_____. 1984. Wildflowers of Texas. Shearer Publishing, Fredericksburg, TX.

Akeroyd, J.R. 1993. *Camelina*. In: T.B. Tutin, N.A. Burges, A.O. Chater, J.R. Edmondson, V.H. Heywood, D.M. Moore, D.H. Valentine, S.M. Walters, and D.A. Webb, eds. Flora Europaea, 2nd ed. 1:380–381.

Albritton, C.C., Jr. 1942. Dinosaur tracks near Comanche, Texas. Field & Lab. 10:161–181.

Alexander, E.J. 1955. *Thelesperma*. N. Amer. Fl. II. 2:65–69.

Allen, C.M. 1992. Grasses of Louisiana, 2nd ed. Cajun Prairie Habitat Preservation Society, Eunice, LA.

Allen, O.N. and **E.K. Allen.** 1981. The Leguminosae: A source book of characteristics, uses, and nodulation. Univ. of Wisconsin Press, Madison.

Allred, B.W. and **H.C. Mitchell.** 1955. Major plant types of Arkansas, Louisiana, Oklahoma and Texas and their relation to climate and soil. Texas J. Sci. 7:7–19.

Allred, K.W. 1981. Cousins to the south: Amphitropical disjunctions in southwestern grasses. Desert Pl. 3:98–106.

_____. 1982. *Paspalum distichum* L. var. *indutum* Shinners (Poaceae). Great Basin Naturalist 42:101–104.

_____. 1984a. Studies in the *Aristida* (Gramineae) of the southeastern United States. I. Spikelet variation in *A. purpurascens*, *A. tenuispica*, and *A. virgata*. Rhodora 86:73–77.

_____. 1984b. Morphologic variation and classification of the North American *Aristida purpurea* complex (Gramineae). Brittonia 36:382–395.

_____. 1985a. Studies in the *Aristida* (Gramineae) of the southeastern United States. II. Morphometric analysis of *A. intermedia* and *A. longespica*. Rhodora 87:137–145.

_____. 1985b. Studies in the *Aristida* (Gramineae) of the southeastern United States. III. Nomenclature and a taxonomic comparison of *A. lanosa* and *A. palustris*. Rhodora 87:147–155.

_____. 1986. Studies in the *Aristida* (Gramineae) of the southeastern United States. IV. Key and conspectus. Rhodora 88:367–387.

_____ and **F.W. Gould.** 1978. Geographic variation in the *Dichanthelium aciculare* complex (Poaceae). Brittonia 30:497–504.

_____ and _____. 1983. Systematics of the *Bothriochloa saccharoides* complex (Poaceae: Andropogoneae). Syst. Bot. 8:168–184.

Al-Shehbaz, I.A. 1984. The tribes of Cruciferae (Brassicaceae) in the southeastern United States. J. Arnold Arbor. 65:343–373.

_____. 1985a. The genera of Thelepodieae (Cruciferae: Brassicaceae) in the southeastern United States. J. Arnold Arbor. 66:95–111.

_____. 1985b. The genera of Brassiceae (Cruciferae: Brassicaceae) in the southeastern United States. J. Arnold Arbor. 66:279–351.

_____. 1986a. New wool-alien Cruciferae (Brassicaceae) in eastern North America. Rhodora 88:347–355.

_____. 1986b. The genera of Lepidieae (Cruciferae; Brassicaceae) in the southeastern United States. J. Arnold Arbor. 67:265–311.

_____. 1987. The genera of Alysseae (Cruciferae: Brassicaceae) in the southeastern United States. J. Arnold Arbor. 68:185–240.

_____. 1988a. The genera of Arabideae (Cruciferae: Brassicaceae) in the southeastern United States. J. Arnold Arbor. 69:85–166.

_____. 1988b. The genera of Anchonieae (Hesperideae) (Cruciferae: Brassicaceae) in the southeastern United States. J. Arnold Arbor. 69:193–212.

_____. 1988c. The genera of Sisymbrieae (Cruciferae: Brassicaceae) in the southeastern United States. J. Arnold Arbor. 69:213–237.

_____. 1991. The genera of Boraginaceae in the southeastern United States. J. Arnold Arbor. Supp. Ser. 1:1–169.

_____ and **B.G. Schubert.** 1989. The Dioscoreaceae in the southeastern United States. J. Arnold Arbor. 70:57–95.

Alston, R.E. and **R.E. Schultes.** 1951. Studies of early specimens and reports of *Ilex vomitoria*. Rhodora 53:274–280.

Alston, R.E. and **B.L. Turner.** 1962. New techniques in analysis of complex natural hybridization. Proc. Natl. Acad. Sci. U.S.A. 48:130–137.

_____ and _____. 1963. Natural hybridization among four species of *Baptisia* (Leguminosae). Amer. J. Bot. 50:159–173.

American Forestry Association. 1996. 1996-97 National register of big trees. Amer. Forests 102:20–47.

Ames, O. 1910. Orchidaceae: Illustrations and studies of the family Orchidaceae, Vol. 4: The genus *Habenaria* in North America. Reprinted 1979 by Earl M. Coleman, Stanfordville, NY.

Ammerman, E. 1944. A monographic study of the genus *Palafoxia* and its immediate allies. Ann. Missouri Bot. Gard. 31:249–277.

Amos, B.B., P.M. Hall, and **J.K. McCoy.** 1998. Phytogeographical analysis of the Texas endemics. Manuscript in preparation.

Amos, B.B. and **C.M. Rowell, Jr.** 1988. Floristic geography of woody and endemic plants. In: B.B. Amos and F.R. Gehlbach, eds. Edwards Plateau vegetation: Plant ecological studies in central Texas. Pp. 25–42. Baylor Univ. Press, Waco, TX.

Anderberg, A.A. 1991. Taxonomy and phylogeny of the tribe Gnaphalieae (Asteraceae). Opera Bot. 104:1–195.

Anderson, D.E. 1961. Taxonomy and distribution of the genus *Phalaris*. Iowa State J. Sci. 36:1–96.

_____. 1974. Taxonomy of the genus *Chloris* (Gramineae). Brigham Young Univ. Sci. Bull., Biol. Ser. 19:1–133.

Anderson, E. 1936. The species problem in *Iris*. Ann. Missouri Bot. Gard. 23:457–509.

_____. 1954. A field survey of chromosome numbers in the species of *Tradescantia* closely related to *Tradescantia virginia*. Ann. Missouri Bot. Gard. 41:305–327.

_____ and **K. Sax.** 1936. A cytological monograph of the American species of *Tradescantia*. Bot. Gaz. (Crawfordsville) 97:433–476.

Anderson, E. and **R.E. Woodson, Jr.** 1935. The species of *Tradescantia* indigenous to the United States. Contr. Arnold Arbor. 9:1–132.

Anderson, L.C. 1970. Studies on *Bigelowia* (Astereae, Compositae). I. Morphology and taxonomy. Sida 3:451–465.

_____. 1977. Studies on *Bigelowia* (Asteraceae). III. Cytotaxonomy and biogeography. Syst. Bot. 2:209–218.

_____ and **J.B. Creech.** 1975. Comparative leaf anatomy of *Solidago* and related Asteraceae. Amer. J. Bot. 62:486–493.

Anderson, R.C. 1990. The historic role of fire in the North American grasslands. In: S.L. Collins and L.L. Wallace, eds. Fire in North American tallgrass prairies. Pp. 8–18. Univ. of Oklahoma Press, Norman.

Anders, B. and **K. Bremer.** 1998. To be or not to be—principles of classification and monotypic families. Taxon 47:391–400.

Andres, T. 1987. *Cucurbita fraterna*, the closest wild relative and progenitor of *C. pepo*. Cucurbit Genetics Cooperative Rep. 10:69–71.

Andrews, E.G. and **Windham, M.D.** 1993. *Pleopeltis*. In: Flora of North America Editorial Committee, eds. Fl. North Amer. 2:324–327. Oxford Univ. Press, New York and Oxford.

Andrews, J. 1984. Peppers, the domesticated capsicums. Univ. of Texas Press, Austin.

_____. 1986. The Texas bluebonnet. Univ. of Texas Press, Austin.

_____. 1995. Peppers, the domesticated capsicums, new edition. Univ. of Texas Press, Austin.

_____. 1998. The pepper lady's pocket pepper primer. Univ. of Texas Press, Austin.

Anonymous. 1994. Systematics agenda 2000: charting the biosphere. Technical Report. Systematics Agenda 2000, New York.

Aplet, G.H., R.D. Laven, M.B. Falkner, and **R.B. Shaw.** 1994. Population and site characteristics of a recently discovered population of *Croton alabamensis* (Euphorbiaceae). Sida 16:37–51.

Appel, D.N. 1995. Epidemiology of oak wilt in Texas. In: D.N. Appel and R.F. Billings, eds. Oak wilt perspectives: The proceedings of the national oak wilt symposium. Pp. 21–28. Information Development, Inc., Houston, TX.

_____ and **R.F. Billings,** eds. 1995. Oak wilt perspectives: The proceedings of the national oak wilt symposium. Information Development, Inc., Houston, TX.

Arber, A. 1938. Herbals, their origin and evolution, 2nd ed. Cambridge Univ. Press, Cambridge, England, U.K.

Arber, E.A.N. and **J. Parkin.** 1908. Studies on the evolution of the angiosperms: The relationship of the angiosperms to the Gnetales. Ann. Bot. 22:489–515.

Argus, G.W. 1986. The genus *Salix* (Salicaceae) in the southeastern United States. Syst. Bot. Monogr. 9:1–170.

_____. 1997. Infrageneric classification of *Salix* (Salicaceae) in the New World. Syst. Bot. Monogr. 52:1–121.

Armada, J. and **A. Barra.** 1992. On *Aloysia* Palau (Verbenaceae). Taxon 41:88–90.

Arnold, R.M. 1981a. Population dynamics and seed dispersal of *Chaenorrhinum minus* on railroad cinder ballast. Amer. Midl. Naturalist 106:80–91.

_____. 1981b. Weeds that ride the rails. Nat. Hist. 90:59–65.

Arnow, L.A. 1994. *Koeleria macrantha* and *K. pyramidata* (Poaceae): Nomenclatural problems and biological distinctions. Syst. Bot. 19:6–20.

Arriagada, J.E. and **N.G. Miller.** 1997. The genera of Anthemideae (Compositae; Asteraceae) in the southeastern United States. Harvard Papers Bot. 2:1–46.

Arridge, R.E. and **P.J. Fonteyn.** 1981. Naturalization of *Colocasia esculenta* (Araceae) in the San Marcos River, Texas. SouthW. Naturalist 26:210–211.

Atwood, D. 1975. A revision of the *Phacelia crenulatae* group (Hydrophyllaceae) for North America. Great Basin Naturalist 35:127–190.

Atwood, J.T. 1986. The size of the Orchidaceae and the systematic distribution of epiphytic orchids. Selbyana 9:171–186.

Austin, D.F. (Daniel). Personal communication. Specialist in Convolvulaceae and Cuscutaceae; Florida Atlantic Univ., Boca Raton.

Austin, D.F. 1975. Bird flowers in the eastern United States. Florida Sci. 38:1–12.

_____. 1976. Varieties of *Ipomoea trichocarpa* (Convolvulaceae). Sida 6:216–220.

_____. 1978. The *Ipomoea batatas* complex—I. Taxonomy. Bull. Torrey Bot. Club. 105:114–129.

_____. 1979a. Plant without beginning or end. Fairchild Trop. Gard. Bull. 34:17–19.

_____. 1979b. Comments on *Cuscuta*—for collectors and curators. Bull. Torrey Bot. Club 106:227–228.

______. 1980. Studies of the Florida Convolvulaceae—III. *Cuscuta*. Florida Sci. 43:294–302.

______. 1986a. Convolvulaceae. In: Great Plains Flora Association. Flora of the Great Plains. Pp. 652–661. Univ. Press of Kansas, Lawrence.

______. 1986b. Cuscutaceae. In: Great Plains Flora Assoc. Flora of the Great Plains. Pp. 661–666. Univ. Press of Kansas, Lawrence.

______. 1988. Taxonomy, evolution and genetic diversity of sweet potatoes and related wild species. In: Exploration, maintenance and utilization of sweet potato genetic resources, Proc. Planning Conf. Pp. 27–60. Centro Internacional de la Papa, Lima, Peru.

______. 1990. Annotated checklist of New Mexican Convolvulaceae. Sida 14:273–286.

______. 1992. Rare Convolvulaceae in the southwestern United States. Ann. Missouri Bot. Gard. 79:8–16.

______. 1998 (in press). Parallel and convergent evolution in the Convolvulaceae. In: P. Mathews, ed. Biodiversity and Taxonomy of Flowering Plants, Calicut Univ., Calicut, India.

______, **G.M. Diggs, Jr.,** and **B.L. Lipscomb.** 1997. *Calystegia* (Convolvulaceae) in Texas. Sida 17:837–840.

Austin, D.F. and **Z. Huáman.** 1996. A synopsis of *Ipomoea* (Convolvulaceae) in the Americas. Taxon 45:3–38.

Averett, J.E. 1972. Two new species of *Chamaesaracha* (Solanaceae) from Texas. Sida 5:48–49.

______. 1973. Biosystematic study of *Chamaesaracha* (Solanaceae). Rhodora 75:325–365.

______. 1979. Biosystematics of the physaloid genera of the Solaneae in North America. In: J.G. Hawkes, R.N. Lester, and A.D. Skelding, eds. The biology and taxonomy of the Solanaceae. Linnean Society Symposium Series No. 7. Academic Press, London, England, U.K.

Avery, A.G., S. Santina, and **J. Rietsema.** 1959. Blakeslee: The genus *Datura*. Ronald Press Co., New York.

Axelrod, D.I. 1950. Evolution of desert vegetation. Publ. Carnegie Inst. Wash. 590:215–306.

______. 1979. Age and origin of Sonoran Desert vegetation. Occas. Pap. Calif. Acad. Sci. 132:1–74.

______. 1983. Biogeography of oaks in the Arcto-Tertiary Province. Ann. Missouri Bot. Gard. 70:629–657.

______. 1985. Rise of the grassland biome, central North America. Bot. Rev. (Lancaster) 51:163–201.

Babcock, E.B. 1947. The genus *Crepis*. I, II. Univ. Calif. Publ. Bot. 21:1–197; 22:199–1030.

______. and **G.L. Stebbins.** 1939. The genus *Youngia*. Publ. Carnegie Inst. Wash. 484:1–106.

Bacon, J.D. (John). Personal communication. Specialist in Hydrophyllaceae; Univ. of Texas, Arlington.

Bacon, J.D. 1974. Chromosome numbers and taxonomic notes in the genus *Nama* (Hydrophyllaceae). Brittonia 26:101–105.

______. 1984. Chromosome numbers and taxonomic notes in the genus *Nama* (Hydrophyllaceae). II. Sida 10:269–275.

Baer, H. 1979. The poisonous Anacardiaceae. In: A.D. Kinghorn, ed. Toxic plants. Pp. 161–170. Columbia Univ. Press, New York.

Baijnath, H. 1975. A study of *Cyperus alternifolius* L., sens. lat. (Cyperaceae). Kew Bull. 30:521–526.

Bailey, L.H. 1922. The standard cyclopedia of horticulture, 6 vols. MacMillan Company, New York.

______. 1925. VI. *Rubus*: Enumeration of the *Eubati* (dewberries and blackberries) native in North America. Gentes Herb. 1:203–300.

______. 1929. The domesticated cucurbitas. Gentes Herb. 2:63–115.

______. 1930a. *Hemerocallis*: The day-lilies. Gentes Herb. 2:143–156.

______. 1930b. Three discussions in Cucurbitaceae. Gentes Herb. 2:175–186.

______. 1932. North American blackberries. Gentes Herb. 2:271–423.

______. 1933. How plants get their names. MacMillan Company. Reprinted 1963 by Dover Publications, Inc., New York.

______. 1934. The species of grapes peculiar to North America. Gentes Herb. 3:151–244.

______. 1938. The garden of pinks. The MacMillan Co., New York.

______. 1941–1945. Species Batorum. The genus *Rubus* in North America I–X. Gentes Herb. 5(in 10 parts):2–918.

______. 1949. Manual of cultivated plants. MacMillan Publishing Co., New York.

______. 1961. Palmaceae [Palmae]. In: C.L. Lundell, ed. Flora of Texas 3:197–199. Texas Research Foundation, Renner.

______ and **E.Z. Bailey.** 1976. Hortus third, a concise dictionary of plants cultivated in the United States and Canada. Revised and expanded by the staff of the Liberty Hyde Bailey Hortorium, Cornell Univ. MacMillan Publishing Co., New York.

Bailey, V. 1905. North American fauna. U.S.D.A. Biol. Surv. No. 25. U.S. Government Printing Office, Washington, D.C.

Bailey, V.L. 1962. Revision. of the genus *Ptelea* (Rutaceae). Brittonia 14:1–45.

Baird, J.R. 1968. A taxonomic revision of the plant family Myricaceae of North America, north of Mexico. Ph.D. dissertation, Univ. of North Carolina, Chapel Hill.

Baird, V.B. 1942. Wild violets of North America. Univ. California Press, Berkeley and Los Angeles.

Baker, A. 1993. Khat—out of the bag: Deployment leads to questions about Somalian drug. Fort Worth Star-Telegram. 12 Jan 1993, Pp. 1,8.

Baker, E.T. 1960. Geology and ground-water resources of Grayson County, Texas. Texas Board of Water Engineers, Bull. 6013.

Baker, H.G. 1986. Yuccas and yucca moths—a historical commentary. Ann. Missouri Bot. Gard. 73:556–564.

Baldon, P.A. (Paul). Personal communication. Park Ranger and naturalist; leads native prairie restoration project at Cedar Hill State Park, Cedar Hill, TX.

Baldwin, J.T. and **B.M. Speese.** 1949. Cytogeography of *Saururus cernuus*. Bull. Torrey Bot. Club 76:213–216.

Ball, C.R. 1961. Salicaceae: *Salix*. In: C.L. Lundell, ed. Flora of Texas 3:369–392. Texas Research Foundation, Renner.

Ball, P.W. 1990. Some aspects of the phytogeography of *Carex*. Canad. J. Bot. 68:1462–1472.

Ballard, H.E., Jr. (Harvey). Personal communication. Specialist in Violaceae; Ohio Univ., Athens.

Ballard, H.E. Jr. 1994. Violets of Michigan. Michigan Bot. 33:131–199.

Ballenger, J.A., E. Dickson, M. Meyer, and **J.J. Doyle.** 1993. DNA sequence data and the phylogeny of the Leguminosae. Amer. J. Bot. 80:130–131.

BALTZER, E.A. 1944. A monographic study of the genus *Palafoxia* and its immediate allies. Ann. Missouri Bot. Gard. 31:249–277.

BANKS, D.J. 1966. Taxonomy of *Paspalum setaceum* (Gramineae). Sida 2:269–284.

BARBER, S.C. (SUSAN). Personal communication. Specialist in *Verbena* (Verbenaceae) and *Pyrrhopappus* (Asteraceae); member of Oklahoma Flora Editorial Committee: Oklahoma City Univ.

BARBER, S.C. 1982. Taxonomic studies in the *Verbena stricta* complex (Verbenaceae). Syst. Bot. 7:433–456.

_____ and **J.R. ESTES.** 1978. Comparative pollination ecology of *Pyrrhopappus geiseri* and *P. carolinianus*. Amer. J. Bot. 65:562–566.

BARBOUR, M.G. and **N.L. CHRISTENSEN.** 1993. Vegetation. In: Flora of North America Editorial Committee, eds. Fl. North Amer. 1:97–131. Oxford Univ. Press, New York and Oxford.

BARKER, W.T. 1986a. Portulacaceae. In: Great Plains Flora Association. Flora of the Great Plains. Pp. 187–190. Univ. Press of Kansas, Lawrence.

_____. 1986b. Sapotaceae. In: Great Plains Flora Association. Flora of the Great Plains. Pp. 341–342. Univ. Press of Kansas, Lawrence.

BARKLEY, F.A. 1937. A monographic study of *Rhus* and its immediate allies in North and Central America including the West Indies. Ann. Missouri Bot. Gard. 24:265–498.

_____. 1961. Anacardiaceae. In: C.L. Lundell, ed. Flora of Texas 3:89–108. Texas Research Foundation, Renner.

BARKLEY, T.M. (THEODORE). Personal communication. Specialist in Asteraceae, especially *Senecio* and *Packera*, and flora of the Great Plains; spent career at Kansas State Univ., Manhattan; currently on Flora of North America Editorial Committee and Research Associate at the Botanical Research Institute of Texas, Fort Worth.

BARKLEY, T.M. 1957. Generic key to the sumac family (Anacardiaceae). Lloydia 20:255–265.

_____. 1962. A revision of *Senecio aureus* Linn. and allied species. Trans. Kansas Acad. Sci. 65:318–408.

_____. 1978. *Senecio*. N. Amer. Fl. II. 10:50–139.

_____. 1981. *Senecio* and *Erechtites* (Compositae) in the North American flora: Supplementary notes. Brittonia 33:523–527.

_____. 1985a. Generic boundaries in the Senecioneae. Taxon 34:17–21.

_____. 1985b. Infrageneric groups in *Senecio*, s.l., and *Cacalia*, s.l. (Asteraceae: Senecioneae) in Mexico and Central America. Brittonia 37:211–218.

_____. 1986. Asteraceae. In: Great Plains Flora Association. Flora of the Great Plains. Pp. 838–1021. Univ. Press of Kansas, Lawrence.

_____. 1988. Variation among aureoid senecios of North America: a geohistorical interpretation. Bot. Rev. 54:82–106.

_____, **B.L. CLARK,** and **A.M. FUNSTON.** 1996. The segregate genera of *Senecio sensu lato* and *Cacalia sensu lato* (Asteraceae: Senecioneae) in Mexico and Central America. In: D.J.N. Hind and H.J. Beentje, eds. Compositae: Systematics. Proceedings of the International Compositae Conference, Kew, 1994. (D.J.N. Hind, Editor-in-Chief), Vol. 1. Pp. 613–620. Royal Botanic Gardens, Kew.

BARKLEY, T.M. and **A. CRONQUIST.** 1978. *Erechtites*. In: N. Amer. Fl. II. 10:139–142.

BARKWORTH, M.E. 1990. *Nassella* (Gramineae: Stipeae): Revised interpretation and nomenclatural changes. Taxon 39:597–614.

_____. 1993. North American Stipeae (Gramineae): Taxonomic changes and other comments. Phytologia 74:1–25.

BARNEBY, R.C. 1952. A revision of the North American species of *Oxytropis*. Proc. Calif. Acad. Sci. 27:177–312.

_____. 1964. Atlas of North American *Astragalus*. Mem. New York Bot. Gard. 13. I:1–596, II:597–1188.

_____. 1977. Daleae imagines. Mem. New York Bot. Gard. 27:1–891.

_____. 1991. Sensitive *Censitae*, a description of the genus *Mimosa* Linnaeus in the New World. Mem. New York Bot. Gard. 65:1–835.

_____ and **E.C. TWISSELMANN.** 1970. Note on *Loeflingia* (Caryophyllaceae). Madroño 20:398–408.

BARNEBY, R.C. and **D. ISELY.** 1986. Reevaluation of *Mimosa biuncifera* and *M. texana* (Leguminosae: Mimosoideae). Brittonia 38:119–122.

BARNEBY, R.C. and **B.A. KRUKOFF.** 1971. Supplementary notes on American Menispermaceae. VIII. A generic survey of the American Triclisieae and Anomospermeae. Mem. New York Bot. Gard. 22:1–90.

BARNES, P.W. 1986. Variation in the big bluestem (*Andropogon gerardii*)-sand bluestem (*Andropogon hallii*) complex along a local dune/meadow gradient in the Nebraska Sandhills. Amer. J. Bot. 73:172–184.

BARNHART, J.H. 1916. Segregation of the genera in Lentibulariaceae. Mem. New York Bot. Gard 6:39–64.

BARRETT, S.C.H. 1988. Evolution of breeding systems in *Eichhornia* (Pontederiaceae): A review. Ann. Missouri Bot. Gard. 75:741–760.

BARRINGER, K. 1997. *Aristolochia*. In: Flora of North America Editorial Committee, eds. Fl. North Amer. 3:45–50. Oxford Univ. Press, New York and Oxford.

_____ and **A.T. WHITTEMORE.** 1997. Aristolochiaceae. In: Flora of North America Editorial Committee, eds. Fl. North Amer. 3:44. Oxford Univ. Press, New York and Oxford.

BARTH, F.G. 1985. Insects and flowers: The biology of a partnership. Translated by M.A. Biederman-Thorson. Princeton Univ. Press, Princeton, NJ.

BARTHLOTT, W. and **D.R. HUNT.** 1993. Cactaceae. In: K. Kubitzki, J.C. Rohwer, and V. Bittrich, eds. The families and genera of vascular plants, Vol. II. Pp. 161–197. Springer-Verlag, Berlin.

BASKIN, J.M., K.M. SNYDER, and **C.C. BASKIN.** 1993. Nomenclatural history and taxonomic status of *Echinacea angustifolia*, *E. pallida*, and *E. tennesseensis* (Asteraceae). Sida 15:597–604.

BASSETT, I.J. 1966. Taxonomy of North American *Plantago* L., section *Micropsyllium* Decne. Canad. J. Bot. 44:467–479.

_____. 1973. The plantains of Canada. Research Branch Canada Dept. Agri. Monogr. 7:1–47.

_____ and **C.W. CROMPTON.** 1968. Pollen morphology and chromosome numbers of the family Plantaginaceae in North America. Canad. J. Bot. 46:349–361.

_____, _____, and **D.W. WOODLAND.** 1974. The family Urticaceae in Canada. Canad. J. Bot. 52:503–516.

BATES, D.M. 1968. Notes on cultivated Malvaceae. 2. *Abelmoschus*. Baileya 16:99–112.

_____, **L.J. DORR,** and **O.J. BLANCHARD, JR.** 1989. Chromosome numbers in *Callirhoe* (Malvaceae). Brittonia 41:143–151.

BATES, D.M., G.J. ANDERSON, and **R.D. LEE.** 1997. Forensic botany: Trichome evidence. J. Forensic Sci. 42:380–386.

BATES, D.M., R.W. ROBINSON, and **C. JEFFREY,** eds. 1990. Biology and utilization of the Cucurbitaceae. Cornell Univ. Press, Ithaca, NY.

BAUM, B.R. 1966. Final research report for the United States Department of Agriculture: Monographic revision of the the genus *Tamarix*. Dept. of Bot. Hebrew Univ., Jerusalem, Israel.

______. 1967. Introduced and naturalized tamarisks in the United States and Canada (Tamaricaceae). Baileya 15:19–25.

______. 1968. Delimitation of the genus *Avena* (Gramineae). Canad. J. Bot. 46:121–132.

______. 1977. Oats: Wild and cultivated, a monograph of the genus *Avena* L. (Poaceae). Monogr. 14:1–463. Biosystematics Research Institute, Canada Department of Agriculture, Research Branch, Ottawa, Ontario, Canada.

______. 1978. The genus *Tamarix*. The Israel Acad. Sci. and Humanities, Jerusalem, Israel.

______. 1991. Proposal to conserve the name *Avena fatua* L. for the common wild oats (Poaceae). Taxon 40:132–134.

______ and **L.G. BAILEY.** 1986. Taxonomy of the North and South American species of *Hordeum* section *Hordeastrum*. Canad. J. Bot. 64:1745–1759.

______ and ______. 1989. Species relationships in the *Hordeum murinum* aggregate viewed from chloroplast DNA restriction fragment patterns. Theor. Appl. Geneti. 78:311–317.

______ and ______. 1990. Key and synopsis of North American *Hordeum* species. Canad. J. Bot. 68:2433–2442.

BAUMGARDT, J.P. 1982. How to identify flowering plant families. Timber Press, Portland, OR.

BAYER, R.J. 1984. Chromosome numbers and taxonomic notes for North American species of *Antennaria* (Asteraceae: Inuleae). Syst. Bot. 9:74–83.

______. 1985. Investigations into the evolutionary history of the polyploid complexes in *Antennaria* (Asteraceae: Inuleae). II. The *A. parlinii* complex. Rhodora 87:321–339.

______, **L. HUFFORD,** and **D.E. SOLTIS.** 1996. Phylogenetic relationships in Sarraceniaceae based on *rbcL* and ITS sequences. Syst. Bot. 21:121–134.

BAYER, R.J. and **G.L. STEBBINS.** 1982. A revised classification of *Antennaria* (Asteraceae: Inuleae) of the eastern United States. Syst. Bot. 7:300–313.

BEAL, E.O. 1956. Taxonomic revision of the genus *Nuphar* Sm. of North America and Europe. J. Elisha Mitchell Sci. Soc. 72:317–346.

______. 1960. The Alismataceae of the Carolinas. J. Elisha Mitchell Sci. Soc. 76:70–79.

______, **J.W. WOOTEN,** and **R.B. KAUL.** 1982. Review of the *Sagittaria engelmanniana* complex (Alismataceae) with environmental correlations. Syst. Bot. 7:417–432.

BEAMAN, J.H. 1957. The systematics and evolution of *Townsendia* (Compositae). Contr. Gray Herb. 183:1–151.

BEARDSLEY, R.L. and **E.T. BROWNE, JR.** 1972. *Lathyrus aphaca* L. new to Tennessee and the southeast. Rhodora 74:155.

BEATTIE, A.J. 1985. The evolutionary ecology of ant-plant mutualism. Cambridge Univ. Press, Cambridge, England, U.K.

______, **D.C. CULVER,** and **R.J. PUDLO.** 1979. Interactions between ants and the diaspores of some common spring flowering herbs in West Virginia. Castanea 44:177–186.

BEATLEY, J.C. 1973. Russian-thistle (*Salsola*) species in western United States. J. Range Managem. 26:225–226.

BEAUCHAMP, R.M. 1980. New names in American acacias. Phytologia 46:5–9.

BEETLE, A.A. 1941. Studies in the genus *Scirpus* L. III. The American species of the section *Lacustres* Clarke. Amer. J. Bot. 28:691–700.

______. 1942. Studies in the genus *Scirpus* L. IV. The section *Bolboschoenus* Palla. Amer. J. Bot. 29:82–88.

______. 1943. Studies in the genus *Scirpus* L. VI. The section *Schoenoplectus* Palla. Amer. J. Bot. 30:395–401.

______. 1947. *Scirpus*. N. Amer. Fl. 18:481–504.

______. 1950. Bulrushes and their multiple uses. Econ. Bot. 4:132–138.

BEHNKE, H.-D. and **T.J. MABRY,** eds. 1994. Caryophyllales: Evolution and systematics. Springer-Verlag, New York.

BELCHER, R.O. 1956. A revision of the genus *Erechtites* (Compositae) with inquiries into *Senecio* and *Arrhenechthites*. Ann. Missouri Bot. Gard. 43:1–85.

BELL, C.R. 1949. A cytotaxonomic study of the Sarraceniaceae of North America. J. Elisha Mitchell Sci. Soc. 65:137–166.

______. 1963. The genus *Eryngium* in the southeastern United States. Castanea 28:73–79.

BELL, N.B. and **L.J. LESTER.** 1980. Morphological and allozyme evidence for *Sabatia formosa* (Gentianaceae) in the section *Campestria*. Amer. J. Bot. 67:327–336.

BELL, P. and **C. WOODCOCK.** 1983. The diversity of green plants. Edward Arnold, London, England, U.K.

BEMIS, W.P., A.M. RHODES, T.W. WHITAKER, and **S.G. CARMER.** 1970. Numerical taxonomy applied to *Cucurbita* relationships. Amer. J. Bot. 57:404–412.

BENHAM, D.M. 1992. Additional taxa in *Astrolepis*. Amer. Fern J. 82:59-62.

______ and **M.D. WINDHAM.** 1992. Generic affinities of the star-scaled cloak ferns. Amer. Fern J. 82:47–58.

______ and ______. 1993. *Astrolepis*. In: Flora of North America Editorial Committee, eds. Fl. North Amer. 2:140–143. Oxford Univ. Press, New York and Oxford.

BENSON, L. 1940. The North American subdivision of *Ranunculus*. Amer. J. Bot. 27:799–807.

______. 1948. A treatise on the North American Ranunculi. Amer. Midl. Naturalist 40:1–261.

______. 1954. Supplement to a treatise on the North American Ranunculi. Amer. Midl. Naturalist 52:328–369.

______. 1969. Cactaceae. In: C.L. Lundell, ed. Flora of Texas 2:221–317. Texas Research Foundation, Renner.

______. 1979. Plant classification, 2nd ed. D.C. Heath and Co., Lexington, MA.

______. 1982. The cacti of the United States and Canada. Stanford Univ. Press, Stanford, CA.

BENZ, B. (BRUCE). Personal communication. Specialist in ethnobotany and the evolution of corn. Has taught at the Univ. of Guadalajara; currently Assistant Professor of Biology at Texas Wesleyan Univ. and Research Associate at Botanical Research Institute of Texas.

Berger, A. 1924. A taxonomic review of currants and gooseberries. New York Agric. Exp. Sta. Bull. 109:1–118.

Berlandier, J.L. 1980. Journey to Mexico during the years 1826–1834. 2 vols. (translated by S.M Ohlendorf et al.; introduction by C.H. Muller). The Texas State Historical Association, Austin.

Bernard, J.M. 1990. Life history and vegetative reproduction in *Carex*. Canad. J. Bot. 68:1441–1448.

Bicknell, E.P. 1895. The genus *Sanicula* in the eastern United States, with descriptions of two new species. Bull. Torrey Bot. Club 22:351–361.

______. 1901. Studies in *Sisyrinchium*–IX. The species of Texas and the southwest. Bull. Torrey Bot. Club 28:570–592.

Biddulph, S.F. 1944. A revision of the genus *Gaillardia*. Res. Stud. State Coll. Wash. 13:195–256.

Bierner, M.W. 1972. Taxonomy of *Helenium* sect. *Tetrodus* and a conspectus of North American *Helenium* (Compositae). Brittonia 24:331–355.

______. 1989. Taxonomy of *Helenium* sect. *Amarum* (Asteraceae). Sida 13:453–459.

______. 1990. Present status of *Amblyolepis* (Asteraceae: Heliantheae). Madroño 37:133–140.

______ and **R.K. Jansen.** 1998. Systematic implications of DNA restriction site variation in *Hymenoxys* and *Tetraneuris* (Asteraceae, Helenieae, Gaillardiinae). Lundellia 1:17–26.

Birdsey, M.R. 1951. The cultivated aroids. The Gillick Press, Berkeley, CA.

Birdson, D. and **L. Forman,** eds. 1992. The herbarium handbook, revised ed. Royal Botanic Gardens, Kew, England, U.K.

Birge, W.I. 1911. The anatomy and some biological aspects of the "Ball Moss," *Tillandsia recurvata* L. Bull. Univ. Texas 194:1–24.

Bittrich, V. 1993. Caryophyllaceae. In: K. Kubitzki, J.C. Rohwer, and V. Bittrich, eds. The families and genera of vascular plants, Vol. II. Pp. 206–236. Springer-Verlag, Berlin.

______ and **U. Kühn.** 1993. Nyctaginaceae. In: K. Kubitzki, J.C. Rohwer, and V. Bittrich, eds. The families and genera of vascular plants, Vol. II. Pp. 473–486. Springer-Verlag, Berlin.

Black, G.A. 1963. Grasses of the genus *Axonopus*. Advancing Frontiers Pl. Sci. 5:1–186.

Black, R.W. 1989. Rangers in Korea. Roster of 14th Co., P. 318. Ivy Books, Random House, New York.

Blackwell, W.H. 1990. Poisonous and medicinal plants. Prentice Hall, Englewood Cliffs, NJ.

______, **M.D. Baechle,** and **G. Williamson.** 1978. Synopsis of *Kochia* (Chenopodiaceae) in North America. Sida 7:248–254.

Blake, S.F. 1916. A revision of the genus *Polygala* in Mexico, Central America, and the West Indies. Contr. Gray Herb. 47:1–122.

______. 1918. A revision of the genus *Viguiera*. Contr. Gray Herb. 54:1–205.

______. 1924. Polygalaceae. N. Amer. Fl. 25:305–379.

Blaney, C. 1995. Who ya gonna call? BioScience 45:746–747.

Blankinship, J.W. 1907. Plantae Lindheimerianae. Part III. Rep. (Annual) Missouri Bot. Gard. 18:123–223.

Block, E. 1985. The chemistry of garlic and onion. Sci. Amer. 252(3):114–119.

Blomquist, H.L. 1955. The genus *Xyris* L. in North Carolina. J. Elisha Mitchell Sci. Soc. 71:35–46.

Bogin, C. 1955. Revision of the genus *Sagittaria* (Alismataceae). Mem. New York Bot. Gard. 9:179–233.

Bogle, A.L. 1969. The genera of Portulacaceae and Basellaceae in the southeastern United States. J. Arnold Arbor. 50:566–598.

______. 1970. The genera of Molluginaceae and Aizoaceae in the southeastern United States. J. Arnold Arbor. 51:431–462.

______. 1974. The genera of Nyctaginaceae in the southeastern United States. J. Arnold Arbor. 55:1–37.

______. 1986. The floral morphology and vascular anatomy of the Hamamelidaceae: Subfamily Liquidambaroideae. Ann. Missouri Bot. Gard. 73:325–347.

Bogler, D.J. and **B.B. Simpson.** 1995. A chloroplast DNA study of the Agavaceae. Syst. Bot. 20:191–205.

______ and ______. 1996. Phylogeny of Agavaceae based on ITS rDNA sequence variation. Amer. J. Bot. 83:1225–1235.

Bogner, J. and **D.H. Nicolson.** 1991. A revised classification of Araceae with dichotomous keys. Willdenowia 21:35–50.

Bogusch, E.R. 1929. A new *Phlox* from Texas. Torreya 29:135–136.

Bohac, J.R., P.D. Dukes, and **D.F. Austin.** 1995. Sweetpotato. In: J. Smartt and N. W. Simmonds, eds. Evolution of Crop Plants, ed. 2. Pp. 57-62. Longman Group, Ltd., London.

Bohs, L. and **R.G. Olmstead.** 1997. Phylogenetic relationships in *Solanum* (Solanaceae) based on *ndh*F sequences. Syst. Bot. 22:5–17.

Boivin, B. 1944. American Thalictra and their Old World allies. Contr. Gray Herb. 152:337–491.

Bold, H.C., C.J. Alexopoulos, and **T. Delevoryas.** 1987. Morphology of plants and fungi. Harper & Row Publishers, New York.

Bolick, M.R. 1985. Cladistic analysis of *Iva*: A case in point. Taxon 34:81–84.

Bolli, R. 1994. Revision of the genus *Sambucus*. Diss. Bot. 223:1–223

Bolton, H.E. 1914. Athanase De Mezieres and the Louisiana-Texas frontier 1768–1780. 2 vols. Arthur H. Clark Co., Cleveland, Ohio.

Bomar, G.W. 1983. Texas weather. Univ. of Texas Press, Austin.

Bonata, M.M., ed. 1995. American women afield: Writings by pioneering women naturalists. Texas A&M Univ. Press, College Station.

Bookman, S.S. 1981. The floral morphology of *Asclepias speciosa* (Asclepiadaceae) in relation to pollination and a clarification in terminology for the genus. Amer. J. Bot. 68:675–679.

Boom, B.M. 1982. Synopsis of *Isoëtes* in the southeastern United States. Castanea 47:38–59.

______. 1996. Societal and scientific information needs from plant collections. In: T.F. Stuessy and S.H. Sohmer, eds, Sampling the green world: Innovative concepts of collection, preservation, and storage of plant diversity. Columbia Univ. Press, New York.

Boothroyd, L.E. 1930. The morphology and anatomy of the inflorescence and flower of the Platanaceae. Amer. J. Bot. 17:678–693.

Bornstein, A.J. 1997. Myricaceae. In: Flora of North America Editorial Committee, eds. Fl. North Amer. 3:429–434. Oxford Univ. Press, New York and Oxford.

Borowski, M., W.C. Holmes, and **J. Singhurst.** 1996. *Phyllostachys aurea* Riv. (Gramineae: Bambuseae) in Texas. Phytologia 80:30–34.

Borror, D.J. 1960. Dictionary of word roots and combining forms. Mayfield Publishing Co., Palo Alto, CA.

Bosland, P.W. 1993. *Capsicum*: A comprehensive bibliography. The Chile Institute, Las Cruces, NM.

Boufford, D.E. 1997a. *Fumaria*. In: Flora of North America Editorial Committee, eds. Fl. North Amer. 3:336–357. Oxford Univ. Press, New York and Oxford.

_____. 1997b. Urticaceae. In: Flora of North America Editorial Committee, eds. Fl. North Amer. 3:400–413. Oxford Univ. Press, New York and Oxford.

_____ and **S.A. Spongberg.** 1983. Eastern Asian-eastern North American phytogeographical relationships–A history from the time of Linnaeus to the twentieth century. Ann. Missouri Bot. Gard. 70:423–439.

Boughton, L.L. 1931. *Croton capitatus* as a poisonous forage plant. Trans. Kansas Acad. Sci. 34:no page number given.

Boulos, L. 1972–74. Révision systématique du genere *Sonchus* L. Bot. Not. 125:287–319; 126:155–196; 127:7–37, 402–451.

_____. 1976. *Sonchus*. In: T.B. Tutin, V.H. Heywood, N.A. Burges, D.M. Moore, D.H. Valentine, S.M. Walters, and D.A. Webb, eds. Flora Europaea 4:327–328.

Bowden, W.M. 1959a. Cytotaxonomy of *Lobelia* L. section *Lobelia*. I. Three diverse species and seven small-flowered species. Canad. J. Genet. Cytol. 1:49–64.

_____. 1959b. The taxonomy and nomenclature of the wheats, barleys and ryes and their wild relatives. Canad. J. Bot. 37:657–684.

_____. 1960a. Cytotaxonomy of *Lobelia* L. section *Lobelia*. II. Four narrow-leaved species and five medium-flowered species. Canad. J. Genet. Cytol. 1:11–27.

_____. 1960b. Cytotaxonomy of *Lobelia* L. Section *Lobelia*. III. *L. siphilitica* L. and *L. cardinalis* L. Canad. J. Genet. Cytol. 2:234–251.

_____. 1961. Interspecific hybridization in *Lobelia* L. Section *Lobelia*. Canad. J. Bot. 39:1679-1693.

_____. 1964a. Cytogenetics of *Lobelia* ×*speciosa* Sweet (*L. cardinalis* L. × *L. siphilitica* L.). Canad. J. Genet. Cytol. 6:121–139.

_____. 1964b. Cytotaxonomy of the species and interspecific hybrids of the genus *Elymus* in Canada and neighboring areas. Canad. J. Bot. 42:547–601.

_____. 1982. The taxonomy of *Lobelia* ×*speciosa* Sweet (*L. cardinalis* L. × *L. siphilitica* L.). Canad. J. Bot. 60:2054–2070.

Box, T.W. and F.W. Gould. 1959. An analysis of the grass vegetation of Texas. SouthW. Naturalist 3:124–129.

Boyd, J.W., D.S. Murray, and **R.J. Tyrl.** 1984. Silverleaf nightshade (*Solanum eleagnifolium*) origin, distribution, and relation to man. Econ. Bot. 38:210–217.

Boyle, W.S. 1945. A cytotaxonomic study of the North American species of *Melica*. Madroño 8:1–26.

Brackett, A. 1923. Contributions from the Gray Herbarium of Harvard University. New Series—No. LXIX. I. Revision of the American species of *Hypoxis*. Rhodora 25:120–147.

Bradfield, H.H. 1957. The petroleum geology of Grayson County, Texas. In: Joint Activities Committee of the Dallas Geological and Geophysical Societies. The geology and geophysics of Cooke and Grayson counties, Texas. Pp. 15–49. The Dallas Geological Society and The Dallas Geophysical Society, Dallas, TX.

Bradley, C.E. and **A.J. Haagen-Smit.** 1949. The essential oil of *Pectis papposa*. Econ. Bot. 3:407–412.

Bradley, T.R. 1975. Hybridization between *Triodanis perfoliata* and *Triodanis biflora* (Campanulaceae). Brittonia 27:110–114.

Bragg, T.B. 1995. The physical environment of Great Plains grasslands. In: A. Joern and K.H. Keller, eds. The Changing Prairie, North American Grasslands. Pp. 49–81. Oxford Univ. Press, New York.

Brainerd, E. 1911. The caulescent violets of the southeastern United States. Bull. Torrey Bot. Club 38:191–198.

_____. 1921. Violets of North America. Vermont Agric. Exp. Sta. Bull. No. 224.

Brandbyge, J. 1993. Polygonaceae. In: K. Kubitzki, J.C. Rohwer, and V. Bittrich, eds. The families and genera of vascular plants, Vol. II. Pp. 531–544. Springer-Verlag, Berlin.

Brandegee, K. 1894. Studies in *Ceanothus*. Proc. Calif. Acad. Sci. 4:173–222.

Brandenburg, D.M., J.R. Estes, and **S.L. Collins.** 1991a. A revision of *Diarrhena* (Poaceae) in the United States. Bull. Torrey Bot. Club 118:128–136.

Brandenburg, D.M., J.R. Estes, and **J.W. Thieret.** 1991b. Hard grass (*Sclerochloa dura*, Poaceae) in the United States. Sida 14:369–376.

Brashier, C.K. 1966. A revision of *Commelina* (Plum.) L. in the U.S.A. Bull. Torrey Bot. Club 93:1–19.

Brawand, H. 1984. Characteristics of the soil of Coit Meadow, a relict area of Blackland Prairie in Collin County, Texas. Wrightia 7:257–265.

Bray, W.L. 1901. Ecological relations of the vegetation of western Texas. Bot. Gaz. (Crawfordsville) 32:99–123, 195–217, 262–291.

Breeden, J.O. 1994. A long ride in Texas: The explorations of John Leonard Riddell. Texas A&M Univ. Press, College Station.

Bremer, K. 1994. Asteraceae cladistics & classification. Timber Press, Portland, OR.

_____, **A.A. Anderberg, P.O. Karis,** and **J. Lundberg.** 1994. Tribe Eupatorieae (Chapter 23). In: Asteraceae cladistics & classification. Pp. 625–680. Timber Press, Portland, OR.

Bretting, P.K. 1983. The taxonomic relationship between *Proboscidea louisianica* and *Proboscidea fragrans* (Martyniaceae). SouthW. Naturalist 28:445–449.

_____ and **S. Nilsson.** 1988. Pollen morphology of the Martyniaceae and its systematic implications. Syst. Bot. 13:51–59.

Brickell, C. and **J.D. Zuk,** eds. 1997. The American Horticultural Society A–Z encyclopedia of garden plants. DK Publishing, New York.

Bridges, E.L. and **S.L. Orzell.** 1987. A new species of *Xyris* (sect. *Xyris*) from the Gulf Coastal Plains. Phytologia 64:56–61.

_____ and _____. 1989. Additions and noteworthy vascular plant collections from Texas and Louisiana, with historical, ecological and geographical notes. Phytologia 66:12–69.

Bridson, G.D.R. and **E.R. Smith.** 1991. B-P-H/S. Botanico-Periodicum-Huntianum/Supplementum. Hunt Institute for Botanical Documentation, Carnegie Mellon Univ., Pittsburg, PA.

Briske, D.D. and **V.J. Anderson.** 1992. Competitive ability of the bunch grass *Schizachyrium scoparium* as affected by grazing history and defoliation. Vegetatio 103:41–49.

Britton, N.L. 1891. The American species of the genus *Anemone* and the genera which have been referred to it. Ann. New York Acad. Sci. 6:215–238.

_____. 1930. Krameriaceae. N. Amer. Fl. 23:195–200.

_____ and **J.N. Rose.** 1905. Crassulaceae. N. Amer. Fl. 22:7–74.

_____ and _____. 1919–1923. The Cactaceae: Descriptions and illustrations of plants of the cactus family, 4 vols. Publ. No. 248. Carnegie Institution of Washington, Washington, D.C.

_____ and _____. 1928. Mimosaceae. N. Amer. Fl. 23:1–194.

_____ and _____. 1930. Caesalpiniaceae. N. Amer. Fl. 23:201–349.

_____ and _____. 1937. The Cactaceae: Descriptions and illustrations of plants of the cactus family, 4 vols. Reprinted from the original Carnegie Institution of Washington printing, Publ. No. 248, 1919–1923. Scott E. Haselton, Abbey San Encino Press, Pasadena, CA.

_____ and _____. 1963. The Cactaceae: descriptions and illustrations of plants of the cactus family. Republication of the 1937 edition in 4 vols. bound as 2 vols. by Dover Publications, Inc., New York.

Brizicky, G.K. 1961a. The genera of Turneraceae and Passifloraceae in the southeastern United States. J. Arnold Arbor. 42:204–218.

_____. 1961b. The genera of Violaceae in the southeastern United States. J. Arnold Arbor. 42:321–333.

_____. 1962a. The genera of Rutaceae in the southeastern United States. J. Arnold Arbor. 43:1–22.

_____. 1962b. Taxonomic and nomenclatural notes on *Zanthoxylum* and *Glycosmis* (Rutaceae). J. Arnold. Arbor. 43:80–93.

_____. 1962c. The genera of Simaroubaceae and Bursuraceae in the southeastern United States. J. Arnold Arbor. 43:173–186.

_____. 1962d. The genera of Anacardiaceae in the southeastern United States. J. Arnold Arbor. 43:359–375.

_____. 1963a. Taxonomic and nomenclatural notes on the genus *Rhus* (Anacardiaceae). J. Arnold Arbor. 44:60–80.

_____. 1963b. The genera of Sapindales in the southeastern United States. J. Arnold Arbor. 44:462–501.

_____. 1964a. The genera of Celastrales in the southeastern United States. J. Arnold Arbor. 45:206–234.

_____. 1964b. The genera of Cistaceae in the southeastern United States. J. Arnold Arbor. 45:346–357.

_____. 1964c. The genera of Rhamnaceae in the southeastern United States. J. Arnold Arbor. 45:439–463.

_____. 1965a. The genera of Vitaceae in the southeastern United States. J. Arnold Arbor. 44:48–67.

_____. 1965b. The genera of Tiliaceae and Elaeocarpaceae in the southeastern United States. J. Arnold Arbor. 46:286–307.

_____. 1966a. The genera of Sterculiaceae in the southeastern United States. J. Arnold Arbor. 47:60–74.

_____. 1966b. On the typification of *Malvastrum*. Taxon 15:311–315.

_____. 1968. *Herissantia*, *Bogenhardia*, and *Gayoides* (Malvaceae). J. Arnold Arbor. 49:278–279.

Brock, J.H. 1994. *Tamarix* spp. (salt cedar), an invasive exotic woody plant in arid and semi-arid riparian habitats of western USA. In: L.C. de Waal, L.E. Child, P.M. Wade, and J.H. Brock, eds. Ecology and management of invasive riverside plants. Pp. 27–44. John Wiley and Sons, Chichester, England, U.K.

Brooke, J. 1848, 1849. From a collection of 22 letters (originals and transcriptions) written from 1848–1860 by Dr. John Brooke, one of the first settlers of Grayson County, Texas. In the collection of the Red River Historical Museum, Sherman, TX.

Brooks, A.B. 1937. *Castanea dentata*. Castanea 2:61–67.

Brooks, R.E. 1977. Poison ivy and poison oak in Kansas. Bull. State. Bio. Sur. Kansas 4:1–38.

_____. 1982. New combinations in *Delphinium* and *Rhus*. Phytologia 52:8.

_____. 1986a. Lamiaceae. In: Great Plains Flora Association. Flora of the Great Plains. Pp. 708–740. Univ. Press of Kansas, Lawrence.

_____. 1986b. Campanulaceae. In: Great Plains Flora Association. Flora of the Great Plains. Pp. 808–816. Univ. Press of Kansas, Lawrence.

_____. 1986c. Caprifoliaceae. In: Great Plains Flora Association. Flora of the Great Plains. Pp. 823–832. Univ. Press of Kansas, Lawrence.

Brown, B.C. 1950. An annotated check list of the reptiles and amphibians of Texas. Baylor Univ. Press, Waco, TX.

Brown, G.D. 1956. Taxonomy of American *Atriplex*. Amer. Midl. Naturalist 55:199–210.

Brown, G.K. and **G.S. Varadarajan.** 1985. Studies in Caryophyllales I: Re-evaluation of classification of Phytolaccaceae s.l. Syst. Bot. 10:49–63.

Brown, L.E. 1984. *Carex rosea* (Cyperaceae), *Trifolium lappaceum* (Fabaceae) and *Aira caryophyllea* (Poaceae) new to Texas. Sida 10:263–264.

_____. 1993. The deletion of *Sporobolus heterolepis* (Poaceae) from the Texas and Louisiana floras, and the addition of *Sporobolus silveanus* to the Oklahoma flora. Phytologia 74:371–381.

_____ and **S.J. Marcus.** 1998. Notes on the flora of Texas with additions and other significant records. Sida 18:315–324.

Brown, L.E. and **C.D. Peterson.** 1984. *Carex rosea* (Cyperaceae), *Trifolium lappaceum* (Fabaceae), and *Aira caryophyllea* (Poaceae) new to Texas. Sida 10:263–264.

Brown, R.W. 1956. Composition of scientific words. Smithsonian Institution Press, Washington, D.C.

Brown, W.V. 1952. The relation of soil moisture to cleistogamy in *Stipa leucotricha*. Bot. Gaz. (Crawfordsville) 113:438–444.

_____ and **G.E. Coe.** 1951. A study of sterility in *Hilaria belangeri* (Steud.) Nash and *Hilaria mutica* (Buckl.) Benth. Amer. J. Bot. 38:823–830.

Browning, J., K.D. Gordon-Gray, and **S.G. Smith.** 1995. Achene structure and taxonomy of North American *Bolboschoenus* (Cyperaceae). Brittonia 47:433–445.

Bruhl, J.J. 1995. Sedge genera of the world: Relationships and a new classification of the Cyperaceae. Austral. Syst. Bot. 8:125–305.

_____, **L. Watson,** and **M.J. Dallwitz.** 1992. Genera of Cyperaceae: Interactive identification and information retrieval. Taxon 41:225–234.

Brummitt, R.K. 1965. New contributions in North American *Calystegia*. Ann. Missouri Bot. Gard. 52:214–216.

______. 1981. Further new names in the genus *Calystegia* (Convolvulaceae). Kew. Bull. 35:327–334.

______. 1988. Report of the committee for Spermatophyta: 35 (proposal 808). Taxon 37:444–450.

______. 1989. Report of the committee for Spermatophyta: 36 (proposal 856). Taxon 38:299–302.

______. 1997. Taxonomy versus cladonomy, a fundamental controversy in biological systematics. Taxon 46:723–734.

______ and **C.E. Powell**, eds. 1992. Authors of plant names: A list of authors of scientific names of plants, with recommended standard forms of their names including abbreviations. Royal Botanic Gardens, Kew.

Brummitt, R.K. and **M.S.M. Sosef.** 1998. Paraphyletic taxa are inherent in Linnaean classification—a reply to Fruedenstein. Taxon 47:411–412.

Brunk, S.K. 1997. The Ruidoso plane crash—the background the the trial verdict. J. Forensic Sci. 42:378–379.

Brunsfeld, S.J., D.E. Soltis, and **P.S. Soltis.** 1992. Evolutionary patterns and processes in *Salix* sect. *Longifoliae*: Evidence from chloroplast DNA. Syst. Bot. 17:239–256.

Brunsfeld, S.J., P.S. Soltis, D.E. Soltis, P.A. Gadek, C.Q. Quinn, D.D. Strenge, and **T.A. Ranker.** 1994. Phylogenetic relationships among the genera of Taxodiaceae and Cupressaceae: Evidence from *rbc*L sequences. Syst. Bot. 19:253–262.

Bryant, V.M. Jr. 1977. A 16,000 year pollen record of vegetational change in central Texas. Palynology 1:143–156.

Bryson, C.T. 1980. A revision of the North American *Carex* section *Laxiflorae* (Cyperaceae). Ph.D. dissertation, Mississippi State Univ., Starksville.

______. 1996. The role of United States Department of Agriculture, Agricultural Research Service in the control of introduced weeds. Castanea 61:261–270.

Buck, R.E., D.W. Taylor, and **A.R. Kruckeberg.** 1993. *Strepanthus*. In: J.C. Hickman, ed. The Jepson manual: Higher plants of California. Pp. 439–444. Univ. of California Press, Berkeley.

Buckingham, G.R. 1996. Biological control of alligatorweed, *Alternanthera philoxeroides*, the world's first aquatic weed success story. Castanea 61:232–243.

Buckley, S.B. 1861 [1862]. Description of new plants from Texas. Proc. Acad. Nat. Sci. Philadelphia 13:448–463.

______. 1870. Remarks on Dr. Asa Gray's notes on Buckley's new plants of Texas. Proc. Acad. Nat. Sci. Philadelphia 22:135–138.

Buddell, G.F., II. and **J.W. Thieret.** 1997. Saururaceae. In: Flora of North America Editorial Committee, eds. Fl. North Amer. 3:37–38. Oxford Univ. Press, New York and Oxford.

Buell, M.F. 1935. *Acorus calamus* in America. Rhodora 37:367–369.

Bullard, F.M. 1931. The geology of Grayson County, Texas. Univ. Texas Bull. No. 3125.

Burch, D. 1966. The application of the Linnaean names of some New World species of *Euphorbia* subgenus *Chamaesyce*. Rhodora 68:155–166.

Burkhalter, L.W. 1965. Gideon Lincecum, 1793–1874: A biography. Univ. of Texas Press, Austin.

Burckhalter, R.E. 1992. The genus *Nyssa* (Cornaceae) in North America: A revision. Sida 15:323–342.

Burkart, A. 1976. A monograph of the genus *Prosopis* (Leguminosae subfam. Mimosoideae). J. Arnold Arbor. 57:219–249, 450–525.

Burkhill, H.M. 1985. The useful plants of west tropical Africa, ed. 2. Vol. 1, Families A–D. Royal Botanic Gardens, Kew, England, U. K.

Burlage, H.M. 1968. Index of plants of Texas with reputed medicinal and poisonous properties. Published by the author, Austin, TX.

Burleson, M.F. 1993. The vanished tallgrass prairie: What we have lost, what we have gained. In: M.R. Sharpless and J.C. Yelderman, eds. The Texas Blackland Prairie, land, history, and culture. Pp. 281–297. Baylor Univ. Program for Regional Studies, Waco, TX.

Burns-Balogh, P. and **V.A. Funk.** 1986. A phylogenetic analysis of the Orchidaceae. Smithsonian Contr. Bot. 61:1–79.

Bush, B.F. 1903. A list of the ferns of Texas. Bull. Torrey Bot. Club 30:343–358.

______. 1904. The Texas tradescantias. Trans. Acad. Sci. St. Louis 14:181–193.

______. 1905. The North American species of *Fuirena*. Rep. (Annual) Missouri Bot. Gard. 16:87–99.

Butler, J.L. and **D.D. Briske.** 1988. Population structure and tiller demography of the bunchgrass *Schizachyrium scoparium* in response to herbivory. Oikos 51:306–312.

Cabrera, A.L. 1949. Sinopsis del género *Soliva* (Compositae). Notas Mus. La Plata, Bot. 14:123–139.

______. 1965. Solanaceae. In: A.L. Cabrera, ed. Flora de la Provincia de Buenos Aires, Parte V, Ericáceas a Calicieráceas. Coleccion Cientifica del Instituto Nacional de Tecnologia Agropecuaria, Buenos Aires, Argentina.

Camp, W.H. 1945. The North American blueberries with notes on other groups of Vacciniaceae. Brittonia 5:203–275.

Campbell, C.S. 1983. Systematics of the *Andropogon virginicus* complex (Gramineae). J. Arnold Arbor. 64:171–254.

______. 1985. The subfamilies and tribes of Gramineae (Poaceae) in the southeastern United States. J. Arnold Arbor. 66:123–199.

______. 1986. Phylogenetic reconstructions and two new varieties in the *Andropogon virginicus* complex (Poaceae: Andropogoneae). Syst. Bot. 11:280–292.

______, **C.W. Greene,** and **T.A. Dickinson.** 1991. Reproductive biology in subfam. Maloideae (Rosaceae). Syst. Bot. 16:333–349.

Campbell, G.R. 1952. The genus *Mysosurus* L. (Ranunculaceae) in North America. Aliso 2:389–403.

Canne, J.M. 1979 [1980]. A light and scanning electron microscope study of seed morphology in *Agalinis* (Scrophulariaceae) and its taxonomic significance. Syst. Bot. 4:281–296.

______. 1980. Seed surface features in *Auerolaria*, *Brachystigma*, *Tomanthera*, and certain South American *Agalinis* (Scrophulariaceae). Syst. Bot. 5:241–252.

______. 1981. Chromosome counts in *Agalinis* and related taxa (Scrophulariaceae). Canad. J. Bot. 59:1111–1116.

Cantino, P.D. 1982. A monograph of the genus *Physostegia* (Labiatae). Contr. Gray Herb. 211:1–105.

_____. 1992. Evidence for a polyphyletic origin of the Labiatae. Ann. Missouri Bot. Gard. 79:361–379.

_____. 1998. Binomials, hyphenated uninomials, and phylogenetic nomenclature. Taxon 47:425–429.

_____ and **R.W. Sanders.** 1986. Subfamilial classification of Labiatae. Syst. Bot. 11:163–185.

Caporael, L.R. 1976. Ergotism: The satan loosed in Salem. Science 192:21–26.

Carlquist, S. 1976. Tribal interrelationships and phylogeny of the Asteraceae. Aliso 8:465–492.

Carmen, J.G. and **D.D. Briske.** 1985. Morphologic and allozymic variation between long-term grazed and non-grazed populations of the bunchgrass *Schizachyrium scoparium* var. *frequens*. Oecologia 66:332–337.

Carolin, R.C. 1993. Portulacaceae. In: K. Kubitzki, J.C. Rohwer, and V. Bittrich, eds. The families and genera of vascular plants, Vol. II. Pp. 544–555. Springer-Verlag, Berlin.

Carr, B.L., J.V. Crisci, and **P.C. Hoch.** 1990. A cladistic analysis of the genus *Gaura* (Onagraceae). Syst. Bot. 15:454–461.

Carr, B.L., D.P. Gregory, P.H. Raven, and **W. Tai.** 1986. Experimental hybridization and chromosomal diversity within *Gaura* sect. *Gaura* (Onagraceae). Syst. Bot. 11:98–111.

_____, _____, _____, and _____. 1988. Experimental hybridization, chromosomal diversity, and phylogeny within *Gaura* sect. *Pterogaura* (Onagraceae). Syst. Bot. 13:324–335.

Carr, W. (William). Personal communication. Texas plant collector; previously with Texas Parks and Wildlife; currently Research Scientist for Botany, The Nature Conservancy, San Antonio, TX.

Carr, W. 1988. *Cyperus difformis* L. (Cyperaceae) new to Texas. Sida 13:255–256.

_____. 1989. Preliminary checklist of vascular plants, Meridian State Recreation Area, Bosque County, Texas. Unpublished September 1989 draft, copy in the library of the Botanical Research Institute of Texas.

_____. 1994. Preliminary checklist of vascular plants, Camp Maxey, Lamar County, Texas. Unpublished October 1994 draft, copy in the library of the Botanical Research Institute of Texas.

_____. 1995. Preliminary checklist of vascular plants, Camp Bowie training site, Brown County, Texas. Unpublished June 1995 draft, copy in the library of the Botanical Research Institute of Texas.

Carter, J.R. 1984. A systematic study of the New World species of section *Umbellati* of *Cyperus*. Ph.D. dissertation, Vanderbilt Univ., Nashville, TN.

_____ and **S.D. Jones.** 1997. Notes on the *Cyperus retroflexus* complex (Cyperaceae) with three nomenclatural proposals. Rhodora 99:319–334.

Carter, J.R. and **R. Kral.** 1990. *Cyerus echinatus* and *Cyperus croceus*, the correct names for North American *Cyperus ovularis* and *Cyperus globulosus* (Cyperaceae). Taxon 39:322–327.

Casado, C.M. 1995. Allelopathic effects of *Lantana camara* (Verbenaceae) on morning glory (*Ipomoea tricolor*). Rhodora 97:264–274.

Casteel, S.W. and **E.M. Bailey.** 1992. Developmental toxicity of an isolate from *Solanum dimidiatum* (potato weed) in Syrian golden hamsters. In: L.F. James, R.F. Keeler, E.M. Bailey, Jr., P.R. Cheeke, and M.P. Hegarty, eds. Poisonous plants: Proceedings of the third international symposium. Pp. 579–583. Iowa State Univ. Press, Ames.

Castetter, E.F., P. Pierce, and **K.H. Schwerin.** 1975. A reassessment of the genus *Escobaria*. Cac. Succ. J. (Los Angeles) 47:60–70.

Catling, P.M. 1981. Taxonomy of autumn flowering *Spiranthes* species of southern Nova Scotia. Canad. J. Bot. 59:1253–1270.

_____. 1982. Breeding systems of northeastern North American *Spiranthes* (Orchidaceae). Canad. J. Bot. 60:3017–3039.

_____. 1983a. Pollination of northeastern North American *Spiranthes* (Orchidaceae). Canad. J. Bot. 61:1080–1093.

_____. 1983b. Morphometrics and ecological isolation in sympatric *Spiranthes* (Orchidaceae) in southwestern Ontario. Canad. J. Bot. 61:2747–2759.

_____ and **I. Dobson.** 1985. The biology of Canadian weeds. 69. *Potamogeton crispus* L. Canad. J. Plant Sci. 65:655–668.

Catling, P.M. and **V.S. Engel.** 1993. Systematics and distribution of *Hexalectris spicata* var. *arizonica* (Orchidaceae). Lindleyana 8:119–125.

Catling, P.M. and **W. Wojtas.** 1986. The waterweeds (*Elodea* and *Egeria*, Hydrocharitaceae) in Canada. Canad. J. Bot. 64:1525–1541.

Caut, B.S. and **J.F. Doebley.** 1997. DNA sequence evidence for the segmental allotetraploid origin of maize. Proc. Natl. Acad. Sci. U.S.A. 94:6809–6814.

Celarier, R.P. 1958. Cytotaxonomic notes on the subsection *Halepensia* of the genus *Sorghum*. Bull. Torrey Bot. Club 85:49–62.

_____ and **J.R. Harlan.** 1958. The cytogeography of the *Bothriochloa ischaemum* complex. Gramineae: I. Taxonomy and geographic distribution. J. Linn. Soc., Bot. 55:755–760.

Chadwell, T.B., S.J. Wagstaff, and **P.D. Cantino.** 1992. Pollen morphology of *Phryma* and some putative relatives. Syst. Bot. 17:210–219.

Chambers, W.T. 1948. Geographic regions of Texas. Texas Geogr. Mag. 12:7–15.

Chandler, J.M., C.-C. Jan, and **B.H. Beard.** 1986. Chromosomal differentiation among the annual *Helianthus* species. Syst. Bot. 11:354–371.

Channell, R.B. 1957. A revisional study of the genus *Marshallia* (Compositae). Contr. Gray Herb. 181:41–132.

_____ and **C.E. Wood, Jr.** 1959. The genera of Primulales of the southeastern United States. J. Arnold Arbor. 40:268–288.

Chapman, A.W. 1860. Flora of the southern United States. Ivison, Phinney and Co., New York.

Chapman, G.C. and **S.B. Jones.** 1978. Biosystematics of the Texanae Vernonias (Vernonieae: Compositae). Sida 7:264–281.

Charrier, A. 1984. Genetic resources of the genus *Abelmoschus* Med. (Okra). International board for plant genetic resources, Rome, Italy.

Chase, A. 1921. The North American species of *Pennisetum*. Contr. U.S. Natl. Herb. 22:209–234.

_____. 1929. The North American species of *Paspalum*. Contr. U.S. Natl. Herb. 28:1–310.

_____. 1938. The carpet grasses. J. Wash. Acad. Sci. 28:178–182.

CHASE, M.W. ET AL. (multiple authors). 1993. Phylogenetics of seed plants: An analysis of nucleotide sequences from the plastid gene *rbc*L. Ann. Missouri Bot. Gard. 80:528–580.

CHATER, A.O. 1964. *Consolida*. In: T.G. Tutin, V.H. Heywood, N.A. Burges, D.H. Valentine, S.M. Walters, and D.A. Webb, eds. 1964. Flora Europaea 1:216–217. Cambridge Univ. Press, Cambridge, England, U.K.

CHEATHAM, S. and **M.C. JOHNSTON.** 1995. The useful wild plants of Texas, the southeastern and southwestern United States, the southern Plains, and northern Mexico. Vol. 1, *Abronia–Arundo*. Useful Wild Plants, Inc., Austin, TX.

CHENG-YIH, W. and **K. KUBITZKI.** 1993. Saururaceae. In: K. Kubitzki, J.C. Rohwer, and V. Bittrich, eds. The families and genera of vascular plants, Vol. II. Pp. 586–588. Springer-Verlag, Berlin.

CHIANG, C.F. 1981. A taxonomic study of the North American species of *Lycium* (Solanaceae). Ph.D. dissertation, Univ. of Texas, Austin.

CHRISPEELS, M.J. and **D. SADAVA.** 1977. Plants, food, and people. W.H. Freeman and Co., San Francisco, CA.

CHURCH, G.L. 1967. Taxonomic and genetic relationships of eastern North American species of *Elymus* with setaceous glumes. Rhodora 69:121–162.

CHURCHILL, S.P. 1986a. Juncaceae. In: Great Plains Flora Association. Flora of the Great Plains. Pp. 1049–1059. Univ. Press of Kansas, Lawrence.

_____. 1986b. Liliaceae. In: Great Plains Flora Association. Flora of the Great Plains. Pp. 1241–1258. Univ. Press of Kansas, Lawrence.

_____. 1986c. Agavaceae. In: Great Plains Flora Association. Flora of the Great Plains. Pp. 1263–1265. Univ. Press of Kansas, Lawrence.

CLARK, C. 1975. Ecogeographic races of *Lesquerella engelmannii* (Cruciferae); distribution, chromosome numbers, and taxonomy. Brittonia 27:263–278.

_____. 1978. Systematic studies of *Eschscholzia* (Papaveraceae). I. The origin and affinities of *E. mexicana*. Syst. Bot. 3:374–385.

_____. 1993. Papaveraceae. In: J.C. Hickman, ed. The Jepson manual: Higher plants of California. Pp. 810–816. Univ. of California Press, Berkeley.

_____. 1997. *Eschscholzia*. In: Flora of North America Editorial Committee, eds. Fl. North Amer. 3:308–312. Oxford Univ. Press, New York and Oxford.

_____ and **J.A. JERNSTEDT.** 1978. Systematic studies of *Eschscholzia* (Papaveraceae). II. Seed coat microsculpturing. Syst. Bot. 3:386–402.

CLARK, H.L. and **J.W. THIERET.** 1968. The duckweeds of Minnesota. Michigan Bot. 7:67–76.

CLARK, L.G. 1990. A new combination in *Chasmanthium* (Poaceae). Ann. Missouri Bot. Gard. 77:601.

_____, **W. ZHANG,** and **J. WENDEL.** 1995. A phylogeny of the grass family (Poaceae) based on *ndhF* sequence data. Syst. Bot. 20:436–460.

CLARK, R.B. 1942. A revision of the genus *Bumelia* in the United States. Ann. Missouri Bot. Gard. 29:155–182.

CLARK, W.D., B.S. GAUT, M.R. DUVALL, and **M.T. CLEGG.** 1993. Phylogenetic relationships of the Bromeliiflorae—Commeliniflorae—Zingiberiflorae complex of monocots based on *rbc*L sequence comparisons. Ann. Missouri Bot. Gard. 80:987–998.

CLARKE, H.D., D.S. SEIGLER, and **J.E. EBINGER.** 1989. *Acacia farnesiana* (Fabaceae: Mimosoideae) and related species from Mexico, the southwestern U.S., and the Caribbean. Syst. Bot. 14:549–564.

CLARY, K. (KAREN). Personal communication. Specialist in *Yucca* (Agavaceae); received training at Univ. of Texas, Austin.

CLAUSEN, J. 1949. Evolutionary patterns in the genus *Crepis*. Evolution 3:185–188.

CLAUSEN, R.T. 1936. Studies in the genus *Najas* in the northern United States. Rhodora 38:333–345.

_____. 1938. A monograph of the Ophioglossaceae. Mem. Torrey Bot. Club. 19:1–177.

_____. 1946. *Selaginella* subgenus *Euselaginella*, in the southeastern United States. Amer. Fern J. 36:65–82.

_____. 1975. *Sedum* of North America, north of the Mexican Plateau. Cornell Univ. Press, Ithaca, NY.

CLAYTON, W.D. 1965. Studies in the Gramineae: VI. Sporoboleae. The *Sporobolus indicus* complex. Kew Bull. 19:287–296.

_____. 1968. The correct name of the common reed. Taxon 17:168–169.

_____. 1970. Studies in the Gramineae: XXI. *Coelorachis* and *Rhytachne*: A study in numerical taxonomy. Kew Bull. 24:309–314.

_____ and **S.A. RENVOIZE.** 1982. Gramineae (Part 3). In: R.M. Polhill, ed. Flora of tropical East Africa. A.A. Balkema on behalf of the East African Governments, Rotterdam, Netherlands.

_____ and _____. 1986. Genera Graminum, grasses of the world. Kew Bull., Addit. Ser. XIII. Royal Botanic Gardens, Her Majesty's Stationary Office, London, England, U.K.

CLEMENT, I.D. 1957. Studies in *Sida* (Malvaceae). Contr. Gray Herb. 180:3–91.

CLEWELL, A.F. 1966. Native North American species of *Lespedeza* (Leguminosae). Rhodora 68:359–408.

_____. 1968. Some comments on American Lespedezas (Leguminosae). Sida 3:206–208.

_____ and **J.W. WOOTEN.** 1971. A revision of *Ageratina* (Compositae: Eupatorieae) from Eastern North America. Brittonia 23:123–143.

CLONTS, J.A. and **S. MCDANIEL.** 1978. *Elephantopus*. N. Amer. Fl. II. 10:196–202.

COBLENTZ, B.E. 1990. Exotic organisms: A dilemma for conservation biology. Conservation Biol. 4:261–265.

CODY, W.J. 1954. A history of *Tillaea aquatica* (Crassulaceae) in Canada and Alaska. Rhodora 56:96–101.

COFFEE, V.J. and **S.B. JONES, JR.** 1980. Biosystematics of *Lysimachia* section *Seleucia* (Primulaceae). Brittonia 32:309–322.

COKER, W.C. 1944. The woody smilaxes of the United States. J. Elisha Mitchell Sci. Soc. 60:27–69.

COLEMAN, J.R. 1966. A taxonomic revision of section *Ximenesia* of the genus *Verbesina* (Compositae). Amer. Midl. Naturalist 76:475–481.

_____. 1968. *Verbesina*, a cytotaxonimic study. Rhodora 70:95–102.

_____. 1977. A summary of experimental hybridization in *Verbesina* (Compositae). Rhodora 79:17–31.

COLLINS, O.B., F.E. SMEINS, and **D.H. RISKIND.** 1975. Plant communities of the Blackland Prairie of Texas. In: M.K. Wali, ed. Prairie: A multiple view. The Univ. of North Dakota Press, Grand Forks.

COLLINS, S.L. and **D.J. GIBSON.** 1990. Effects of fire on community structure in tallgrass and mixed-grass prairie. In: S.L. Collins and L.L. Wallace, eds. Fire in North American tallgrass prairies. Pp. 81–98. Univ. of Oklahoma Press, Norman.

COLLINS, T. 1973. Revision of *Orobanche* (Orobanchaceae). Ph.D. dissertation, Univ. of Wisconsin, Milwaukee.

CONDON, M. and **M.D. WHALEN.** 1983. A plea for collection and preservation of herbivore and pathogen damaged plant materials. Taxon 32:105–107.

CONRAD, H.S. 1905. The waterlilies: A monograph of the genus *Nymphaea*. Publ. Carnegie Inst. Wash. 4:1–279.

CONRAD, J. 1992. A brief history of the bois d'arc tree. Commerce Bois d'Arc Bash, Commerce, TX.

CONSTANCE, L. 1939. The genera of the tribe Hydrophylleae of the Hydrophyllaceae. Madroño 5:28–33.

_____. 1940. The genus *Ellisia*. Rhodora 42:34–39.

_____. 1941. The genus *Nemophila* Nuttall. Univ. Calif. Publ. Bot. 19:341–398.

_____. 1949. A revision of *Phacelia* subgenus *Cosmanthus* (Hydrophyllaceae). Contr. Gray Herb. 168:1–48.

_____. 1950. Some interspecific relationships in *Phacelia* subgenus *Cosmanthus*. Proc. Amer. Acad. Arts 78:135–147.

_____. 1993. Apiaceae. In: J.C. Hickman, ed. The Jepson manual: Higher plants of California. Pp. 136–166. Univ. of California Press, Berkeley.

_____ and **R.H. SHAN.** 1948. The genus *Osmorhiza* (Umbelliferae): A study in geographic affinities. Univ. Calif. Publ. Bot. 23:111–156.

CONTI, E., A. FISCHBACH, and **K.J. SYTSMA.** 1993. Tribal relationships in Onagraceae: Implications from *rbc*L sequence data. Ann. Missouri Bot. Gard. 80:672–685.

CONTRERAS, J., D.F. AUSTIN, F. DE LA PUENTE, and **J. DIAZ.** 1995. Biodiversity of sweetpotato (*Ipomoea batatas*, Convolvulaceae) in southern Mexico. Econ. Bot. 49:286–296.

COOK, C.D.K. and **R. LÜÖND.** 1982. A revision of the genus *Hydrilla* (Hydrocharitaceae). Aquatic Bot. 13:485–504.

COOK, C.D.K. and **K. URMI-KÖNIG.** 1984. A revision of the genus *Egeria* (Hydrocharitaceae). Aquatic Bot. 19:73–96.

COOMBES, A.J. 1985. Dictionary of plant names. Timber Press, Portland, OR.

CORBET, S.A., H. CHAPMAN, and **N. SAVILLE.** 1988. Vibratory pollen collection and flower form: Bumble-bees on *Actinidia*, *Symphytum*, *Borago*, and *Polygonatum*. Functional Ecology 2:147–155.

CORCORAN, SISTER M.L. 1941. A revision of the subgenus *Pycreus* in North and South America. Catholic Univ. Amer., Biol. Ser. 37:1–68.

CORE, E.L. 1936. The American species of *Scleria*. Brittonia 2:1–105.

_____. 1941. The North American species of *Paronychia*. Amer. Midl. Naturalist 26:396–397.

_____. 1966. Cyperaceae: *Scleria*. In: C.L. Lundell, ed. Flora of Texas 1:383–391. Texas Research Foundation, Renner.

_____. 1967. Ethnobotany of the southern Appalachian aborigines. Econ. Bot. 21:199–214.

CORRELL, D.S. 1937. The orchids of North Carolina. J. Elisha Mitchell Sci. Soc. 53:139–173.

_____. 1947. Additions to the orchids of Texas. Wrightia 1:166–182.

_____. 1949. A preliminary survey of the distribution of Texas Pteridophyta. Wrightia 1:247–278.

_____. 1950. Native orchids of North America north of Mexico. Chronica Botanica, Waltham, MA.

_____. 1956. Ferns and fern allies of Texas. Texas Research Foundation, Renner.

_____. 1961a. Orchidaceae. In: C.L. Lundell, ed. Flora of Texas 3:151–196. Texas Research Foundation, Renner.

_____. 1961b. Salicaceae: *Populus*. In: C.L. Lundell, ed. Flora of Texas 3:393–407. Texas Research Foundation, Renner.

_____. 1965. Some additions and contributions to the flora of Texas. Wrightia 3:126–140.

_____. 1966a. Pteridophyta. In: C.L. Lundell, ed. Flora of Texas 1:3–121. Texas Research Foundation, Renner.

_____. 1966b. Gymnospermae. In: C.L. Lundell, ed. Flora of Texas 1:322–368. Texas Research Foundation, Renner.

_____. 1971. Lloyd Herbert Shinners—A portrait. Brittonia 23:101–104.

_____. 1972. The soils and vegetation of a relict area of Blackland Prairie. Part II: An investigation of the vegetational composition of the Stults Meadow. Renner Res. Rep. 1:147–165. Texas Research Foundation, Renner.

_____ and **H.B. CORRELL.** 1972. Aquatic and wetland plants of southwestern United States. U.S. Environmental Protection Agency. U.S. Government Printing Office, Washington, D.C.

CORRELL, D.S. and **M.C. JOHNSTON.** 1970. Manual of the vascular plants of Texas. Texas Research Foundation, Renner.

_____ and _____. 1972. Manual of the vascular plants of Texas: I. Additions and corrections. Amer. Midl. Naturalist 88:490–496.

CORY, V.L. 1940. Six thistles recently introduced into Texas. Madroño 5:200–202.

_____. 1943. The genus *Styrax* in central and western Texas. Madroño 7:110–115.

_____. 1946. The genus *Palafoxia* in Texas. Rhodora 48:84–86.

_____. 1947. Old World plants apparently recently introduced into Texas. Madroño 9:64.

_____. 1948a. Some first records of plant species collected in Texas. Field & Lab. 16:82–89.

_____. 1948b. Curly mesquite grass in Texas and northern Mexico. Wrightia 1:214–217.

_____. 1949. The disappearance of plant species from the range in Texas. Field & Lab. 17:99–115.

_____. 1950. Additional records of plants introduced into Texas. Field & Lab. 18:89–92.

_____ and **H.B. PARKS.** 1937. Catalogue of the flora of the state of Texas. Texas Agric. Exp. Sta. Bull. No. 550.

COUGHENOUR, M.B. 1985. Graminoid response to grazing by large herbivores: Adaptations, exaptations, and interacting processes. Ann. Missouri Bot. Gard. 72:852–863.

COULTER, J.M. 1891–1894. Botany of western Texas. Contr. U.S. Natl. Herb. 2:1–588.

_____ and **W.H. Evans.** 1890. A revision of the North American Cornaceae. Bot. Gaz. (Crawfordsville) 15:30–38, 86–97.

Coulter, J.M. and **J.N. Rose.** 1900. Monograph of the North American Umbelliferae. Contr. U.S. Natl. Herb. 7:1–256.

Covas, G. 1949. Taxonomic observations on the North American species of *Hordeum*. Madroño 10:1–21.

Coville, F.V. 1890. Revision of the United States species of the genus *Fuirena*. Bull. Torrey Bot. Club 17:1–8.

_____ and **N.L. Britton.** 1908. Grossulariaceae. N. Amer. Fl. 22:193–225.

Cox, C. B. and **P.D. Moore.** 1993. Biogeography: An ecological and evolutionary approach, 5th ed. Blackwell Scientific Publications, Oxford, England, U.K.

Cox, P.A. 1988. Hydrophilous pollination. Ann. Rev. Ecol. Syst. 19:261–279.

Cox, P.B. and **L.E. Urbatsch.** 1990. A phylogenetic analysis of the coneflower genera (Asteraceae: Heliantheae). Syst. Bot. 15:394–402.

_____. 1994. A taxonomic revision of *Rudbeckia* sub. *Macrocline* (Asteraceae: Heliantheae: Rudbeckiinae). Castanea 59:300–318.

Cox, P.W. and **P. Leslie.** 1991. Texas trees. Corona Publishing Co., San Antonio, TX.

Cragg, G.M., M.R. Boyd, M.R. Grever, and **S.A. Schepartz.** 1995. Pharmaceutical prospecting and the potential for pharmaceutical crops, natural product drug discovery and development at the United States National Cancer Institute. Ann. Missouri Bot. Gard. 82:47–53.

Craig, R.T. 1945. The *Mammillaria* handbook. Abbey Garden Press, Pasadena, CA.

Crane, P.R. and **P. Kenrick.** 1997. Problems in cladistic classification: Higher-level relationships in land plants. Aliso 15:87–104.

Cranfill, R.B. 1993a. Dennstaedtiaceae. In: Flora of North America Editorial Committee, eds. Fl. North Amer. 2:198–205. Oxford Univ. Press, New York and Oxford.

_____. 1993b. Blechnaceae. In: Flora of North America Editorial Committee, eds. Fl. North Amer. 2:223–227. Oxford Univ. Press, New York and Oxford.

Crawford, D.J. 1975. Systematic relationships in the narrow-leaved species of *Chenopodium* of the western United States. Brittonia 27:279–288.

_____, **J.D. Palmer,** and **M. Kobayashi.** 1992. Chloroplast DNA restriction site variation and the evolution of the annual habit in North American *Coreopsis* (Asteraceae). In: P.S. Soltis, D.E. Soltis, and J.J. Doyle, eds. Molecular Systematics of Plants. Pp. 280–294. Chapman and Hall, New York.

Crawford, D.J. and **E.B. Smith.** 1982. Allozyme divergence between *Coreopsis basalis* and *C. wrightii* (Compositae). Syst. Bot. 7:359–364.

_____ and _____. 1984. Allozyme divergence and intraspecific variation in *Coreopsis grandiflora* (Compositae). Syst. Bot. 9:219–225.

Crawford, D.J. and **H.D. Wilson.** 1986. *Chenopodium*. In: Great Plains Flora Association. Flora of the Great Plains. Pp. 166–173. Univ. Press of Kansas, Lawrence.

Crenshaw, T.C. 1983. Texas blackland heritage. Texian Press, Waco, TX.

Crins, W.J. 1989. The Tamaricaceae in the southeastern United States. J. Arnold Arbor. 70:403–425.

_____. 1990. Phylogenetic considerations below the sectional level in *Carex*. Canad. J. Bot. 68:1433–1440.

_____. 1991. The genera of Paniceae (Gramineae: Panicoideae) in the southeastern United States. J. Arnold Arbor. Supp. Ser. 1:171–312.

Crispeels, M.J. and **D. Sadava.** 1977. Plants, food, and people. W.H. Freeman and Co., San Francisco, CA.

Croat, T. 1972. *Solidago canadensis* complex in the Great Plains. Brittonia 24:317–326.

Crompton, C.W., I.V. Hall, K.I.N. Jensen, and **P.D. Hildebrand.** 1988. The biology of Canadian weeds. 83. *Hypericum perforatum* L. Canad. J. Plant Sci. 68:149–162.

Cronk, Q.C.B. and **J.L. Fuller.** 1995. Plant invaders: The threat to natural ecosystems. Chapman and Hall, New York.

Cronquist, A. 1943. The separation of *Erigeron* from *Conyza*. Bull. Torrey Bot. Club 70:629–632.

_____. 1945. Studies in the Sapotaceae, III. *Dipholis* and *Bumelia*. J. Arnold Arbor. 26:435–471.

_____. 1946. Studies in the Sapotaceae, II Survey of the North American genera. Lloydia 9:241–292.

_____. 1947. A revision of the North American species of *Erigeron* north of Mexico. Brittonia 6:121–302.

_____. 1977. The Compositae revisited. Brittonia 29:137–153.

_____. 1980. Asteraceae. Vascular flora of the southeastern United States 1:1–261. The Univ. of North Carolina Press, Chapel Hill.

_____. 1981. An integrated system of classification of flowering plants. Columbia Univ. Press, New York.

_____. 1985. History of generic concepts in the Compositae. Taxon 34:6–10.

_____. 1988. The evolution and classification of flowering plants, 2nd ed. New York Bot. Garden, Bronx.

_____. 1993. A commentary on the general system of classification of flowering plants. In: Flora of North America Editorial Committee, eds. Fl. North Amer. 1:272–293. Oxford Univ. Press, New York and Oxford.

_____, **A.H. Holmgren, N.H. Holmgren, J.L. Reveal,** and **P.K. Holmgren.** 1977. Intermountain flora: Vascular plants of the intermountain west, U.S.A. Vol. 6. Columbia Univ. Press, New York.

Cronquist, A. and **D.D. Keck.** 1957. A reconstitution of the genus *Machaeranthera*. Brittonia 9:231–239.

Cronquist, A. and **R.F. Thorne.** 1994. Nomenclatural and taxonomic history. In: H.-D. Behnke and T.J. Mabry, eds. Caryophyllales: Evolution and systematics. Pp. 5–25. Springer-Verlag, New York.

Crosswhite, F.S. 1980. Dry country plants of the south Texas Plains. Desert Pl. 2:141–179.

_____. 1982. Corn (*Zea mays*) in relation to its wild relatives. Desert Pl. 3:193–201.

_____. 1983. Selenium and *Castilleja*. Desert Pl. 5:96.

_____. 1984. Crassulacean acid metabolism. Desert Pl. 5:192.

_____ and **C.D. Crosswhite** 1981. Hummingbirds as pollinators of flowers in the red-yellow segment of the color spectrum, with special reference to *Penstemon* and the "open habitat." Desert Pl. 3:156–170.

_____ and _____. 1985. The southwestern pipevine (*Aristolochia watsonii*) in relation to snakeroot oil, swallowtail butterflies, and Ceratopogonid flies. Desert Pl. 6:203–207.

_____ and _____. 1997. Muhly grasses and the Muhlenberg Family, with notes on the Pietist movement and Pietistic ecology. Desert Pl. 13:3–13.

Crosthwaite, J.A. (Andrew). Personal communication. Nature photographer and dentist, Sherman, TX; instructor at Austin College of Tropical Natural History classes in Central America, South America, and Africa.

Crosthwaite, S. (Sally). Personal communication. Botanist trained at Vanderbilt and Columbia, plant collector in north TX, and former instructor at Austin College, Sherman, TX; retired, Grayson Co., TX.

Crow, G.E. 1978. A taxonomic revision of *Sagina* (Caryophyllaceae) in North America. Rhodora 80:1–91.

_____. 1979. The systematic significance of seed morphology in *Sagina* (Caryophyllaceae) under scanning electron microscopy. Brittonia 31:52–63.

Culver, D.C. and **A.J. Beattie.** 1980. The fate of *Viola* seeds dispersed by ants. Amer. J. Bot. 67:710–714.

Cutler, H.C. 1939. Monograph of the North American species of the genus *Ephedra*. Ann. Missouri Bot. Gard. 26:373–428.

_____ and **E. Anderson.** 1941. A preliminary survey of the genus *Tripsacum*. Ann. Missouri Bot. Gard. 28:249–269.

Dahlgren, R.M.T., H.T. Clifford, and **P.F. Yeo.** 1985. The families of the monocotyledons. Springer-Verlag, Berlin, Heidelberg.

Dahling, G.V. 1978. Systematics and evolution of *Garrya*. Contr. Gray Herb. 209:1–104.

Dallas Petroleum Geologists. 1941. Geology of Dallas County Texas. Field & Lab. 10:1–134.

Dallimore, W. and **A.B. Jackson.** 1931. A handbook of Coniferae including Ginkgoaceae. Edward Arnold & Co., London, England, U.K.

Daniel, T.F. 1980. Range extensions of *Carlowrightia* (Acanthaceae) and a key to the species of the United States. SouthW. Naturalist 25:425–426.

_____. 1983. *Carlowrightia* (Acanthaceae). Fl. Neotrop. Monogr. 34:1–116.

_____. 1984. The Acanthaceae of the southwestern United States. Desert Pl. 5:162–179.

_____. 1993. Garryaceae. In: J.C. Hickman, ed. The Jepson manual: Higher plants of California. Pp. 664–666. Univ. of California Press, Berkeley.

Daoud, H.S. and **R.L. Wilbur.** 1965. A revision of the North American species of *Helianthemum* (Cistaceae). Rhodora 67:63–82, 201–216, 255–312.

Darbyshire, S.J. 1993. Realignment of *Festuca* subgenus *Schedonorus* with the genus *Lolium* (Poaceae). Novon 3:239–243.

_____ and **J. Cayouette.** 1989. The biology of Canadian weeds. 92. *Danthonia spicata* (L.) Beauv. in Roem. & Schult. Canad. J. Plant Sci. 69:1217–1233.

Darbyshire, S.J. and **S.I. Warwick.** 1992. Phylogeny of North American *Festuca* (Poaceae) and related genera using chloroplast DNA restriction site variation. Canad. J. Bot. 70:2415–2429.

D'Arcy, W.G., ed. 1986. Solanaceae biology and systematics. Columbia Univ. Press, New York.

_____ and **H. Eshbaugh.** 1973. The name for the common bird pepper. Phytologia 25:350.

Darlington, J. 1934. A monograph of the genus *Mentzelia*. Ann. Missouri Bot. Gard. 21:103–220.

Darwin, S.P. 1976. The subfamilial, tribal and subtribal nomenclature of the Rubiaceae. Taxon 25:595–610.

Daubenmire, R.F. 1974. Plants and environment, a textbook of plant autecology, 3rd ed. John Wiley and Sons, New York.

Davenport, L.J. 1988. A monograph of *Hydrolea* (Hydrophyllaceae). Rhodora 90:169–208.

Davey, J.C. and **W.D. Clayton.** 1978. Some multiple discriminant function studies on *Oplismenus* (Gramineae). Kew Bull. 33:147–157.

Davies, D. 1992. Alliums: The ornamental onions. Timber Press, Portland, OR.

Davies, R.S. 1980. Introgression between *Elymus canadensis* L. and *E. virginicus* L. (Triticeae, Poaceae) in South Central United States. Ph.D. dissertation, Texas A&M Univ., College Station.

Davila, P.D. 1988. Systematic revision of the genus *Sorghastrum* (Poaceae: Androgpogoneae). Ph.D. dissertation, Iowa State Univ., Ames.

Davis, A.S., J. Snowden, and **G. Stanford.** 1995. Switchgrass, *Panicum virgatum* L.: Are there two varieties? An enquiry from the Native Prairies Association of Texas. In: Native Plant Society of Texas. The tallgrass prairies and its many ecosystems. Pp. 88–92. 1995 Symposium Proceedings, Waco.

Davis, J.I. 1995. A phylogenetic structure for the monocotyledons, as inferred from chloroplast DNA restriction site variation, and a comparison of measures of clade support. Syst. Bot. 20:503–527.

Davis, P.H. and **J. Cullen.** 1979. The identification of flowering plant families including a key to those native and cultivated in north temperate regions, 2nd ed. Cambridge Univ. Press, Cambridge, England, U.K.

Davis, T.D. 1997. Benny J. Simpson. Sida 17:855–860.

Davis, W.B. and **D.J. Schmidly.** 1994. The mammals of Texas. Texas Parks & Wildlife Department, Nongame and Urban Program, Austin.

Decker, D.S. 1988. Origin(s), evolution, and systematics of *Cucurbita pepo* (Cucurbitaceae). Econ. Bot. 42:4–15.

_____ and **H.D. Wilson.** 1986. Numerical analysis of seed morphology in *Cucurbita pepo*. Syst. Bot. 11:595–607.

_____ and _____. 1987. Allozyme variation in the *Cucurbita pepo* complex: *C. pepo* var. *ovifera* vs. *C. texana*. Syst. Bot. 12:263–273.

Decker-Walters, D.S. 1990. Evidence for multiple domestication of *Cucurbita pepo*. In: D.M. Bates, R.W. Robinson, and C. Jeffrey, eds. Biology and utilization of the Cucurbitaceae. Pp. 96–101. Cornell Univ. Press, Ithaca, NY.

DeFilipps, R.A. 1976. *Hypochoeris*. In: Tutin, T.G., V.H. Heywood, N.A. Burges, D.M. Moore, D.H. Valentine, S.M. Walters, and D.A. Webb, eds. Flora Europaea 4:308–310.

DeJong, D.C.D. 1965. A systematic study of the genus *Astranthium* (Compositae, Astereae). Publ. Mus. Michigan State Univ., Biol. Ser. 2:429–528.

Delahoussaye, J.A. and **J.W. Thieret.** 1967. *Cyperus* subgenus *Kyllinga* (Cyperaceae) in the Continental United States. Sida 3:128–136.

Delcourt, P.A. and **H.R. Delcourt.** 1993. Paleoclimates, paleovegetation, and paleofloras during the late Quaternary. In: Flora of North America Editorial Committee, eds. Fl. North Amer. 1:71–94. Oxford Univ. Press, New York and Oxford.

DeLisle, D.G. 1963. Taxonomy and distribution of the genus *Cenchrus*. Iowa State J. Sci. 37:259–351.

den Hartog, C. and **F. van der Plas.** 1970. A synopsis of the Lemnaceae. Blumea 18:355–368.

Dennis, W.M. 1976. A biosystematic study of *Clematis* section *Viorna* subsection *Viornae*. Ph.D. dissertation, Univ. of Tennessee, Knoxville.

_____ and **D.H. Webb.** 1981. The distribution of *Pilularia americana* A. Br. (Marsileaceae) in North America, north of Mexico. Sida 9:19–24.

Denton, M.F. 1978. A taxonomic treatment of the Luzulae group of *Cyperus*. Contr. Univ. Michigan Herb. 11:197–271.

DePamphilis, C.W. and **R. Wyatt.** 1989. Hybridization and introgression in buckeyes (*Aesculus*: Hippocastanaceae): A review of the evidence and a hypothesis to explain long-distance gene flow. Syst. Bot. 14:593–611.

de Queiroz, K.D. 1997. The Linnaean hierarchy and the evolutionization of taxonomy, with emphasis on the problem of nomenclature. Aliso 15:125–144.

_____ and **J. Gauthier.** 1992. Phylogenetic taxonomy. Annual Rev. Ecol. Syst. 23:449–480.

Desmaris, Y. 1952. Dynamics of leaf variation in the sugar maples. Brittonia 7:347–387.

Desrochers, A.M., J.F. Bain, and **S.I. Warwick.** 1988. The biology of Canadian weeds. 89. *Carduus nutans* L. and *Carduus acanthoides* L. Canad. J. Plant Sci. 68:1053–1068.

Detling, L.E. 1939. A revision of the North American species of *Descurainia*. Amer. Midl. Naturalist 22:481–520.

de Wet, J.M.J. 1981. Grasses and the culture history of man. Ann. Missouri Bot. Gard. 68:87–104.

_____ and **J.R. Harlan.** 1970. Biosystematics of *Cynodon* L.C. Rich. (Gramineae). Taxon 19:565–569.

Dewey, D.R. 1975. The origin of *Agropyron smithii*. Amer. J. Bot. 62:524–530.

_____. 1982. Genomic and phylogenetic relationships among North American perennial Triticeae. In: J.R. Estes, R.J. Tyrl, and J.N. Brunken, eds. Grasses and grasslands: Systematics and ecology. Pp. 51–88. Univ. of Oklahoma Press, Norman.

_____. 1983. New nomenclatural combinations in the North American perennial Triticeae (Gramineae). Brittonia 35:30–33.

DeWolf, G.P. 1955. A note on the name *Calamintha*. Rhodora 57:73–78.

De-Yuan, H. 1983. The distribution of Scrophulariaceae in the Holarctic with special reference to the floristic relationships between eastern Asia and eastern North America. Ann. Missouri Bot. Gard. 70:701–712.

Diamond, D.D., D.H. Riskind, and **S.L. Orzell.** 1987. A framework for plant community classification and conservation in Texas. Texas J. Sci. 39:203–221.

Diamond, D.D. and **F.E. Smeins.** 1985. Composition, classification, and species response patterns of remnant tallgrass prairie in Texas. Amer. Midl. Naturalist 113:294–309.

_____ and _____. 1993. The native plant communities of the Blackland Prairie. In: M.R. Sharpless and J.C. Yelderman, eds. The Texas Blackland Prairie, land, history, and culture. Pp. 66–81. Baylor Univ. Program for Regional Studies, Waco, TX.

Dick, G. (Gary). Personal communication. Research scientist. U.S. Army Corps of Engineers. Lewisville Aquatic Ecosystem Research Facility, Lewisville, TX.

Dickinson, T.A. and **C.S. Campbell.** 1991a. Evolution in the Maloideae (Rosaceae)—Introduction. Syst. Bot. 16:299–302.

_____ and _____. 1991b. Population structure and reproductive ecology in the Maloideae (Rosaceae). Syst. Bot. 16:350–362.

Dietrich, W. and **W.L. Wagner.** 1987. New taxa of *Oenothera* L. sect. *Oenothera* (Onagraceae). Ann. Missouri Bot. Gard. 74:144-150.

_____ and _____. 1988. Systematics of *Oenothera* section *Oenothera* subsection *Raimannia* and subsection *Nutantigemma* (Onagraceae). Syst. Bot. Monogr. 24:1–91.

_____, _____, and **P.H. Raven.** 1997. Systematics of *Oenothera* section *Oenothera* subsection *Oenothera* (Onagraceae). Syst. Bot. Monogr. 50:1–234.

Diggs, G.M., Jr. 1982. The Campanulaceae (including Lobelioideae) of Virginia. Virginia J. Sci. 33:206-221.

_____, **R.J. O'Kennon,** and **B.L. Lipscomb.** 1997. *Hypochaeris glabra*, a new Asteraceae for Texas. Sida 17:633–634.

Diggs, G.M., Jr., C.E.S. Taylor, and **R.J. Taylor.** 1997. *Chaenorrhinum minus* (Scrophulariaceae) new to Texas. Sida 17:631–632.

Dilcher, D. 1998. The geological history of the SE United States and its influence upon the vegetation. Flora of the Southeast US Symposium, Botanical Research Institute of Texas, 23–25 April 1998. Abstract.

DiMichele, W.A. and **J.E. Skog.** 1992. The Lycopsida: A symposium. Ann. Missouri Bot. Gard. 79:447–449.

Dixon, J.R. 1987. Amphibians and reptiles of Texas. Texas A&M Press, College Station.

Do, L.H., W.C. Holmes, and **J.R. Singhurst.** 1996. New county records for *Bellardia trixago* (Scrophulariaceae) in Texas. Sida 17:295–297.

Doebley, J.F. 1990. Molecular evidence and the evolution of maize. Econ. Bot. 44 (Suppl.):6–27.

_____. 1996. Genetic dissection of the morphological evolution of maize. Aliso 14:297–304.

_____ and **H.H. Iltis.** 1980. Taxonomy of *Zea* I. Subgeneric classification with key to taxa. Amer. J. Bot. 67:982–993.

Domínguez, X.A., R. Franco, J. Verde S., A. Zamudio, and **E.Y. Guevara Z.** 1984. Coumarins from desert rue *Thamnosma texana* (Gray) Torr. Rev. Latinoamer. Quim. 15:138–139.

Donoghue, M.J. 1983. A preliminary analysis of phylogenetic relationships in *Viburnum* (Caprifoliaceae s.l.). Syst. Bot. 8:45–58.

_____, **R.G. Olmstead, J.F. Smith,** and **J.D. Palmer.** 1992. Phylogenetic relationships of Dipsacales based on *rbc*L sequences. Ann. Missouri Bot. Gard. 79:333–345.

Dorado, O. et al. (multiple authors). 1996. The Arbol del Tule (*Taxodium mucronatum* Ten.) is a single genetic individual. Madroño 43:445–452.

Dorn, R.D. 1976. A synopsis of American *Salix*. Canad. J. Bot. 54:2769–2789.

_____. 1988. *Chenopodium simplex*, an older name for *C. gigantospermum* (Chenopodiaceae). Madroño 35:162.

Dorr, L.J. (Laurence). Personal communication. Specialist in botanical biography and bibliography, Ericaceae, and Malvales. Smithsonian Institution, Washington, D.C.

Dorr, L.J. 1990. A revision of the North American genus *Callirhoe* (Malvaceae). Mem. New York Bot. Gard. 56:1–76.

_____. 1997. Botanical libraries and herbaria in North America. 4. The Samuel Botsford Buckley—Rebecca Mann Dean mystery. Taxon 46:661–687.

_____ and **K.C. Nixon.** 1985. Typification of the oak (*Quercus*) taxa described by S.B. Buckley (1809–1884). Taxon 34:211–228.

Downie, S.R. and **J.D. Palmer.** Restriction site mapping of the chloroplast DNA inverted repeat: A molecular phylogeny of the Asteridae. Ann. Missouri Bot. Gard. 79:266–283.

_____ and _____. 1994. A chloroplast DNA phylogeny of the Caryophyllales based on structural and inverted repeat restriction site variation. Syst. Bot. 19:236–252.

Doyle, J.A. 1996. Seed plant phylogeny and the relationships of Gnetales. In: W.E. Friedman (symposium organizer). Biology and evolution of the Gnetales. Int. J. Pl. Sci. 157 (Nov. 1996 suppl.):S3–S39.

_____, **M.J. Donoghue,** and **E.A. Zimmer.** 1994. Integration of morphological and ribosomal RNA data on the origin of angiosperms. Ann. Missouri Bot. Gard. 81:419–450.

Doyle, J.J. 1983. Flavonoid races of *Claytonia virginica* (Portulacaceae). Amer. J. Bot. 70:1085–1091.

_____. 1984. Karyotypic variation of eastern North American *Claytonia* chemical races. Amer. J. Bot. 71:970–978.

Drapalik, D.J. and **R.H. Mohlenbrock.** 1960. A study of *Eleocharis*, series *Ovatae*. Amer. Midl. Naturalist 64:339–341.

Dressler, R.L. 1981. The orchids: Natural history and classification. Harvard Univ. Press, Cambridge, MA.

Duke, J.A. 1973. Utilization of *Papaver*. Econ. Bot. 27:390–400.

_____. 1981. Handbook of legumes of world economic importance. Plenum, New York.

_____. 1985. CRC handbook of medicinal herbs. CRC Press, Inc., Boca Raton, FL.

Duncan, T.A. 1980. A taxonomic study of the *Ranunculus hispidus* Michx. complex in the Western Hemisphere. Univ. Calif. Publ. Bot. 77:1–125.

_____ and **C.S. Keener.** 1991. A classification of the Ranunculaceae with special reference to the Western Hemisphere. Phytologia 70:24–27.

Duncan, W.H. 1959. Leaf variation in *Liquidambar styraciflua* L. Castanea 24:99–111.

_____. 1964. New *Elatine* (Elatinaceae) populations in the southeastern United States. Rhodora 66:48–53.

_____. 1967. Woody vines of the southeastern states. Sida 3:1–76.

_____. 1979. Changes in *Galactia* (Fabaceae) of the southeastern United States. Sida 8:170–180.

_____ and **D.W. DeJong.** 1964. Taxonomy and heterostyly of North American *Gelsemium* (Loganiaceae). Sida 1:346–357.

Duncan, W.H., P.L. Piercy, S.D. Feurt, and **R. Starling.** 1957. Toxicological studies of southeastern plants. II. Compositae. Econ. Bot. 11:75–85.

Dutton, B.E., C.S. Keener, and **B.A. Ford.** 1997. *Anemone*. In: Flora of North America Editorial Committee, eds. Fl. North Amer. 3:139–158. Oxford Univ. Press, New York and Oxford.

Duvall, M.R. and **J.F. Doebley.** 1990. Restriction site variation in the chloroplast genome of *Sorghum* (Poaceae). Syst. Bot. 15:472–480.

Duvall, M.R. et al. (multiple authors). 1993. Phylogenetic hypotheses for the monocotyledons constructed from *rbc*L sequence data. Ann. Missouri Bot. Gard. 80:607–619.

Dyal, S.C. 1938. *Valerianella* of North America. Rhodora 40:185–212.

Dykes, W.R. 1913. The genus *Iris*. Cambridge Univ. Press, Cambridge, England, U.K. Reprinted in 1974 by Dover Publications Inc., New York.

Dyksterhuis, E.J. 1946. The vegetation of the Fort Worth Prairie. Ecol. Monogr. 16:1–29.

_____. 1948. The vegetation of the Western Cross Timbers. Ecol. Monogr. 18:325–376.

Easterly, N.W. 1957. A morphological study of *Ptilimnium*. Brittonia 9:136–145.

Ebinger, J.E. 1962. Validity of the grass species *Digitaria adscendens*. Brittonia 14:248–253.

Eckenwalder, J.E. 1976. Re-evaluation of Cupressaceae and Taxodiaceae: A proposed merger. Madroño 23:237–256.

_____. 1977. North American cottonwoods (*Populus*, Salicaceae) of sections *Abaso* and *Aigeiros*. J. Arnold Arbor. 58:193–208.

_____. 1993. Gymnosperms. In: Flora of North America Editorial Committee, eds. Fl. North Amer. 1:267–271. Oxford Univ. Press, New York and Oxford.

_____ and **S.C.H. Barrett.** 1986. Phylogenetic systematics of Pontederiaceae. Syst. Bot. 11:373–391.

Edwards, A.L. and **R. Wyatt.** 1994. Population genetics of the rare *Asclepias texana* and its widespread sister species, *A. perennis*. Syst. Bot. 19:291–307.

Egger, M. 1994 [1995]. New natural hybrid combinations and comments on interpretation of hybrid populations in *Castilleja* (Scrophulariaceae). Phytologia 77:381–389.

Eggers, D.M. 1969. A revision of *Valerianella* in North America. Ph.D. dissertation, Vanderbilt Univ., Nashville, TN.

Eggers-Ware, D.M. (Donna). Personal communication. Specialist in *Valerianella* (Valerianaceae), floristics of Virginia coastal plain, and endangered species; curator, College of William and Mary, Williamsburg, VA.

Eggers-Ware, D.M. 1983. Genetic fruit polymorphism in North American *Valerianella* (Valerianaceae) and its taxonomic implications. Syst. Bot. 8:33–44.

Egolf, D.R. 1962. A cytological study of the genus *Viburnum*. J. Arnold Arbor. 43:132–172.

Eidson, J.A. (James). Personal communication. North Texas Land Steward, The Nature Conservancy of Texas.

Eiten, G. 1955. The typification of the names "*Oxalis corniculata* L." and "*Oxalis stricta* L." Taxon 4:99–105.

_____. 1963. Taxonomy and regional variation of *Oxalis* section *Corniculatae*. I. Introduction, keys and synopsis of the species. Amer. Midl. Naturalist 69:257–309.

Elias, T.S. 1970. The genera of Ulmaceae in the southeastern United States. J. Arnold Arbor. 51:18–40.

_____. 1971a. The genera of Fagaceae in the southeastern United States. J. Arnold Arbor. 52:159–195.

_____. 1971b. The genera of Myricaceae in the southeastern United States. J. Arnold Arbor. 52:305–318.

_____. 1972. The genera of Juglandaceae in the southeastern United States. J. Arnold Arbor. 53:26–51.

_____. 1974. The genera of Mimosoideae (Leguminosae) in the southeastern United States. J. Arnold Arbor. 55:67–118.

_____. 1977. Threatened and endangered species problems in North America. An overview. In: G.T. Prance and T.S. Elias, eds. Extinction is forever. Pp. 13–16. New York Bot. Garden, Bronx.

_____. 1980. The complete trees of North America. Van Nostrand Reinhold, New York.

_____. 1989. Field guide to North American Trees. Grolier Book Clubs Inc., Danbury, CT.

Elisens, W.J. 1985. Monograph of the Maurandyinae (Scrophulariaceae–Antirrhineae). Syst. Bot. Monogr. 5:1–97.

_____, **R.D. Boyd,** and **A.D. Wolfe.** 1992. Genetic and morphological divergence among varieties of *Aphanostephus skirrhobasis* (Asteraceae-Astereae) and related species with different chromosome numbers. Syst. Bot. 17:380–394.

Ellis, M.D. 1975. Poisonous plants. In: M.D. Ellis, ed. Dangerous plants, snakes, arthropods & marine life of Texas. Pp. 3–120. U.S. Dept. of Health, Education, and Welfare, Public Health Service. U.S. Government Printing Office, Washington, D.C.

Ellstrand, N.C. and **D.A. Levin.** 1980. Evolution of *Oenothera laciniata* (Onagraceae), a permanent translocation heterozygote. Syst. Bot. 5:6–16.

Emboden, W.A. 1974. *Cannabis*—a polytypic genus. Econ. Bot. 28:304–310.

Emery, W.H.P. 1957a. A cyto-taxonomic study of *Setaria macrostachya* (Gramineae) and its relatives in the southwestern United States and Mexico. Bull. Torrey Bot. Club 84:94–105.

_____. 1957b. A study of reproduction in *Setaria macrostachya* and its relatives in the southwestern United States and northern Mexico. Bull. Torrey Bot. Club 84:106–121.

Endress, P.K. 1989. A suprageneric taxonomic classification of the Hamamelidaceae. Taxon 38:371–376.

_____. 1993. Hamamelidaceae. In: K. Kubitzki, J.C. Rohwer, and V. Bittrich, eds. The families and genera of vascular plants, Vol. II. Pp. 322–331. Springer-Verlag, Berlin.

_____. 1994. Evolutionary aspects of the floral structure in *Ceratophyllum*. Pl. Syst. Evol. 8:175–183.

_____ and **V. Bittrich.** 1993. Molluginaceae. In: K. Kubitzki, J.C. Rohwer, and V. Bittrich, eds. The families and genera of vascular plants, Vol. II. Pp. 419–426. Springer-Verlag, Berlin.

Engel, V.S. 1997. Saprophytic orchids of Dallas. North Amer. Native Orchid J. 3:156–167.

Engelmann, G. 1868. A revision of the North American species of the genus *Juncus*, with a description of new or imperfectly known species. Trans. Acad. Sci. St. Louis 2:424–498.

_____ and **A. Gray.** 1845. Plantae Lindheimerianae: An enumeration of F. Lindheimer's collection of Texan plants, with remarks and descriptions of new species, etc. Boston J. Nat. Hist. 5:210–264.

_____ and _____. 1850. Plantae Lindheimerianae, Part II. An account of a collection of plants made by F. Lindheimer in the western part of Texas, in the years 1845–6, and 1847–8, with critical remarks, descriptions of new species, etc. Boston J. Nat. Hist. 6:141–240.

Epling, C. 1934. Preliminary revision of American *Stachys*. Feddes Repert. Spec. Nov. Regni Veg. Beih. 80:1–75.

_____. 1938–1939. A revision of *Salvia*, subgenus *Calosphace*. Repert. Spec. Nov. Regni Veg. Beih. 110:1–380.

_____. 1939. Notes on the Scutellariae of eastern North America. I. Amer. J. Bot. 26:17–24.

_____. 1942. The American species of *Scutellaria*. Univ. Calif. Publ. Bot. 20:1–146.

_____ and **C. Játiva.** 1964. Revisión del género *Satureja* en America del Sur. Brittonia 16:393–416.

_____ and _____. 1966. A descriptive key to the species of *Satureja* indigenous to North America. Brittonia 18:244–248.

Epling, C. and **W.S. Stewart.** 1939. A revision of *Hedeoma* with a review of allied genera. Repert. Spec. Nov. Regni Veg. Beih. 115:1–49.

Erbe, L. 1957. Studies on the crossability of *Lupinus texensis* and *Lupinus subcarnosus*. Madroño 14:17–18.

_____ and **B.L. Turner.** 1962. A biosystematic study of the *Phlox cuspidata-Phlox drummondii* complex. Amer. Midl. Naturalist 67:257–281.

Erdman, K.S. 1965. Taxonomy of the genus *Sphenopholis*. Iowa State J. Sci. 39:289–336.

Erhardt, W. 1992. *Hemerocallis*: Day lilies. Timber Press, Portland, OR.

Erichsen-Brown, C. 1979. Medicinal and other uses of North American plants. Dover Publications, Inc., New York.

Erickson, R.O. 1943. Taxonomy of *Clematis* section *Viorna*. Ann. Missouri Bot. Gard. 30:1–62.

Ernst, W.R. 1962. The genera of Papaveraceae and Fumariaceae in the southeastern United States. J. Arnold Arbor. 43:315–343.

_____. 1963a. The genera of Capparaceae and Moringaceae in the southeastern United States. J. Arnold Arbor. 44:81–95.

_____. 1963b. The genera of Hamamelidaceae and Platanaceae in the southeastern United States. J. Arnold Arbor. 44:193–210.

_____. 1964a. The genera of Berberidaceae, Lardizabalaceae, and Menispermaceae in the southeastern United States. J. Arnold Arbor. 45:1–35.

_____. 1964b. The genus *Eschscholzia* in the south Coast Ranges of California. Madroño 17:281–294.

_____ and **H.J. Thompson.** 1963. The Loasaceae in the southeastern United States. J. Arnold Arbor. 44:138–142.

Ertter, B. (Barbara). Personal communication. Specialist in floristics of the western U.S., Juncaceae, Polygonaceae, and Rosaceae. Univ. of California, Berkeley.

Essig, F.B. 1992. Seedling morphology in *Clematis* (Ranunculaceae) and its taxonomic implications. Sida 15:377–390.

Estes, J.R. and **R.J. Tyrl.** 1982. The generic concept and generic circumscription in the Triticeae: An end paper. In: J.R. Estes, R.J. Tyrl, and J.N. Brunken. Grasses and grasslands: Systematics and ecology. Pp. 145–164. Univ. of Oklahoma Press, Norman.

_____, _____, and **J.N. Brunken.** 1982. Grasses and grasslands: Systematics and ecology. Univ. of Oklahoma Press, Norman.

Eyde, R.H. 1959. The discovery and naming of the genus *Nyssa*. Rhodora 61:209–218.

_____. 1963. Morphological and paleobotanical studies on the Nyssaceae, I. A survey of the modern species and their fruits. J. Arnold Arbor. 44:1–59.

_____. 1966. The Nyssaceae of the southeastern United States. J. Arnold Arbor. 47:117–125.

_____. 1987. The case for keeping *Cornus* in the broad Linnaean sense. Syst. Bot. 12:509–518.

_____ and **E.S. Barghoorn.** 1963. Morphological and paleobotanical studies of the Nyssaceae, II The fossil record. J. Arnold Arbor. 44:328–370.

Eyde, R.H. and **J.A. Teeri.** 1967. Floral anatomy of *Rhexia virginica* (Melastomataceae). Rhodora 69:163–178.

Eyde, R.H. and **C.C. Tseng.** 1971. What is the primitive floral structure of Araliaceae? J. Arnold Arbor. 52:205–239.

Ewan, J. 1945. A synopsis of the North American species of *Delphinium*. Univ. Colorado Stud., Ser. D, Phys. Sci. 2:55–244.

_____. 1951. The genus *Delphinium* in North America: Supplementary notes and distribution records. Bull. Torrey Bot. Club 78:376–381.

Faden, R.B., (Robert). Personal communication. Specialist in Commelinaceae and the flora of tropical Africa. Smithsonian Institution, Washington, D.C.

Faden, R.B. 1985. Commelinaceae. In: R.M.T. Dahlgren, H.T. Clifford, and P.F. Yeo. The families of the monocotyledons. Pp. 381–387. Springer-Verlag, Berlin, Heidelberg.

_____. 1989. *Commelina caroliniana* (Commelinaceae): A misunderstood species in the United States is an old introduction from Asia. Taxon 38:43–53.

_____. 1992. Commelinaceae. Unpublished manuscript of forthcoming Flora of North America treatment.

_____. 1993. The misconstrued and rare species of *Commelina* (Commelinaceae) in the eastern United States. Ann. Missouri Bot. Gard. 80:208–218.

_____ and **D.R. Hunt.** 1991. The classification of the Commelinaceae. Taxon 40:19-31.

Fairey, J.E., III. 1967. The genus *Scleria* in the southeastern United States. Castanea 32:37–55.

Fantz, P.R. 1977. A monograph of the genus *Clitoria* (Leguminosae: Glycineae). Ph.D. dissertation, Univ. of Florida, Gainesville.

_____. 1991. Ethnobotany of *Clitoria* (Leguminosae). Econ. Bot. 45:511–520.

Farr, E.R., J.A. Leussink, and **F.A. Stafleu,** eds. 1979. Index nominum genericorum (plantarum), Vol. 1, Aa–Epochnium. Regnum Veg. 100.

Fassett, N.C. 1944. *Dodecatheon* in eastern North America. Amer. Midl. Naturalist 31:455–486.

_____. 1949. The variations of *Polygonum punctatum*. Brittonia 6:369–393.

_____. 1951. *Callitriche* in the New World. Rhodora 53:137–155, 161–182, 185–194, 209–222.

_____. 1953. A monograph of *Cabomba*. Castanea 18:116–128.

_____. 1955. *Echinodorus* in the American tropics. Rhodora 57:133–156, 174–188, 202–212.

Faust, W.Z. 1972. A biosystematic study of the Interiores species group of the genus *Vernonia* (Compositae). Brittonia 24:363–378.

_____ and **S.B. Jones, Jr.** 1973. The systematic value of trichome complements in a North American group of *Vernonia* (Compositae). Rhodora 75:517–528.

Featherly, H.I. 1931. A new species of *Grama* grass. Bot. Gaz. (Crawfordsville) 91:103–105.

_____. 1954. Taxonomic terminology of the higher plants. Iowa State College Press, Ames.

Ferguson, A.M. 1901. Crotons of the United States. Rep. (Annual) Missouri Bot. Gard. 12:33–74.

Ferguson, D.J. 1989. Revision of the U.S. Members of the *Echinocereus triglochidiatus* group. Cact. Succ. J. (Los Angeles) 61:217–224

Ferguson, I.K. 1965. The genera of the Valerianaceae and Dipsacaceae in the southeastern United States. J. Arnold Arbor. 46:218–231.

_____. 1966a. The genera of Caprifoliaceae in the southeastern United States. J. Arnold Arbor. 47:33–59.

_____. 1966b. Notes on the nomenclature of *Cornus*. J. Arnold Arbor. 47:100–105.

_____. 1966c. The Cornaceae of the southeastern United States. J. Arnold Arbor. 47:106–116.

Fernald, M.L. 1897. A systematic study of the United States and Mexican species of *Pectis*. Proc. Amer. Acad. Arts 33:57–86.

_____. 1932. The linear leaved North American species of *Potamogeton*, section *Axillaries*. Mem. Amer. Acad. Arts 17:1–183.

_____. 1934. *Draba* in temperate northeastern America. Rhodora 36:241–261, 285–305, 314–344, 353–371, 392–404.

_____. 1935. *Geranium carolinianum* and allies of northeastern North America. Rhodora 37:295–301.

_____. 1936. Contributions from the Gray Herbarium of Harvard University—No. CXIII. II. *Pilea* in eastern North America. Rhodora 38:169–170.

_____. 1940a. Some spermatophytes of eastern North America. Rhodora 42:281–302.

_____. 1940b. A synopsis of *Boltonia*. Rhodora 42:482–492.

_____. 1944. The confused publication of *Monarda russeliana*. Rhodora 46:491–493.

_____. 1946a. The North American representatives of *Alisma plantago-aquatica*. Rhodora 48:86–88.

_____. 1946b. Identifications and reidentifications of North American plants. Rhodora 48:137–162, 184–197, 207–216.

_____. 1948. Some spermatophytes of eastern North America. Rhodora 42:281–302.

_____. 1950a. Gray's manual of botany, 8th ed. Reprinted 1987. Dioscorides Press, Portland, OR.

_____. 1950b. *Adiantum capillus-veneris* in the United States.

Rhodora 52:201–208.

_____ and **A.E. Brackett.** 1929. The representatives of *Eleocharis palustris* in North America. Rhodora 31:57–77.

Fernald, M.L. and **B.G. Schubert.** 1949. Some identities in *Breweria*. Rhodora 51:35–42.

Fernandes, R. 1972. *Chaenorrhinum*. In: T.G. Tutin, V.H. Heywood, N.A. Burges, D.M. Moore, D.H. Valentine, S.M. Walters, and D.A. Webb, eds. Flora Europaea 3:224–226. Cambridge Univ. Press, Cambridge, England, U.K.

Finan, J.J. 1948. Maize in the great herbals. Ann. Missouri Bot. Gard. 35:149–191.

Fischer, P.C. 1980. The varieties of *Coryphantha vivipara*. Cact. Succ. J. (Los Angeles) 52:186–191.

Flack, S. and **E. Furlow.** 1996. America's least wanted. Nat. Conservancy Mag. 46(6):17–23.

Flake, R.H., L. Urbatsch, and **B.L. Turner.** 1978. Chemical documentation of allopatric introgression in *Juniperus*. Syst. Bot. 3:129–144.

Flook, J.M. 1973. Guide to the botanical contributions of Lloyd H. Shinners (1918–1971). Sida 5:137–179.

_____. 1974. Eula Whitehouse (1892–1974). Sida 5:354.

Flora of North America Editorial Committee, eds. 1993. Flora of North America north of Mexico, Vol. 2, Pteridophytes and Gymnosperms. Oxford Univ. Press, New York.

_____. 1997. Flora of North America north of Mexico, Vol. 3, Magnoliophyta: Magnoliidae and Hamamelidae. Oxford Univ. Press, New York.

Fonteyn, P.J., M.W. Stone, M.A. Yancy, J.T. Baccus, and **N.M. Nadkarni.** 1988. Determination of community structure by fire. In: B.B. Amos and F.R. Gehlbach, eds. Edwards Plateau vegetation: Plant ecological studies in central Texas. Pp. 79–90. Baylor Univ. Press, Waco, TX.

Forman, L.L. 1966. On the evolution of cupules in the Fagaceae. Kew Bull. 18:385–419.

Forshaw, J.M. 1977. Parrots of the world. T.F.H. Publications, Inc., Neptune, NJ.

Fosberg, F.R. 1936. Varieties of the desert willow, *Chilopsis linearis*. Madroño 3:362–366.

_____. 1958. *Zanthoxylum* L., "*Xanthoxylum* Mill.," and *Thylax* Raf. Taxon 7:94–96.

_____. 1959. Typification of *Zanthoxylum*. Taxon 8:103–105.

Foster, K.E., M.M. Karpiscak, J.G. Taylor, and **N.G. Wright.** 1983. Guayule, jojoba, buffalo gourd and russian thistle: Plant characteristics, products and commercialization potential. Desert Pl. 5:112–117, 126.

Foster, N. and **L.S. Cordell.** 1992. Chilies to chocolate: Food the Americas gave the world. Univ. of Arizona Press, Tucson.

Foster, R.C. 1937. A cyto-taxonomic survey of the North American species of *Iris*. Contr. Gray Herb. 119:3–82.

_____. 1945. Studies in the Iridaceae. III. A revision of the North American species of *Nemastylis* Nutt. Contr. Gray Herb. 155:26–44.

Foster, S. 1991. *Echinacea*: Nature's immune enhancer. Healing Arts Press, Rochester, VT.

Fowler, B.A. and **B.L. Turner.** 1977. Taxonomy of *Selinocarpus* and *Ammocodon* (Nyctaginaceae). Phytologia 37:177–208.

Fox, J.W., C.B. Smith, and **D.O. Lintz.** 1992. Herd bunching at the Waco mammoth site: Preliminary investigations, 1978–1987. In: J.W. Fox, C.B. Smith, and K.T. Wilkins, eds. Proboscidean and paleoindian interactions. Pp. 51–73. Markham Press Fund of Baylor Univ. Press, Waco, TX.

Francaviglia, R.V. (forthcoming). The cast iron forest: A natural and cultural history of the North American Cross Timbers. Univ. of Texas Press, Austin.

Frankel, E. 1989. Distribution of *Pueraria lobata* in and around New York City. Bull. Torrey Bot. Club 116:390–394.

_____. 1991. Poison ivy, poison oak, poison sumac and their relatives pistachios, mangoes, cashews. The Boxwood Press, Pacific Grove, CA.

Fraser, S.V. 1939. *Allium perdulce*: A new *Allium* from Kansas. Trans. Kansas Acad. Sci. 42:123–126.

Freeman, C.C. and **T.M. Barkley.** 1995. A synopsis of the genus *Packera* (Asteraceae: Senecioneae) in Mexico. Sida 16:699–709.

Freeman, D.C., E.D. McArthur, and **K.T. Harper.** 1984. The adaptive significance of sexual lability in plants using *Atriplex canescens* as a principal example. Ann. Missouri Bot. Gard. 71:265–277.

Freudenstein, J.V. 1997. A monograph of *Corallorhiza* (Orchidaceae). Harvard Pap. Bot. 10:5–51.

_____. 1998. Paraphyly, ancestors, and classification—a response to Sosef and Brummitt. Taxon 47:95–104.

Friedland, S. 1941. The American species of *Hemicarpha*. Amer. J. Bot. 28:855–861.

Friedman, W.E. (symposium organizer). 1996. Biology and evolution of the Gnetales. Int. J. Pl. Sci. 157 (Nov. 1966 suppl.):S1–S125.

Friis, I. 1993. Urticaceae. In: K. Kubitzki, J.C. Rohwer, and V. Bittrich, eds. The families and genera of vascular plants, Vol. II. Pp. 612–630. Springer-Verlag, Berlin.

Fryxell, P.A. (Paul). Personal communication. Specialist in Malvaceae; formerly with USDA, currently at Univ. of Texas, Austin.

Fryxell, P.A. 1968. The typification and application of the Linnaean binomials in *Gossypium*. Brittonia 20:378–386.

_____. 1973. New species and other notes in the Malvaceae. Brittonia 25:77–85.

_____. 1974. The North American Malvellas (Malvaceae). SouthW. Naturalist 19:97–103.

_____. 1978. Neotropical segregates from *Sida* L. (Malvaceae). Brittonnia 30:447–462.

_____. 1979a. The natural history of the cotton tribe. Texas A&M Univ. Press, College Station.

_____. 1979b. Una revisión del género *Pavonia* en México. Bol. Soc. Bot. México 38:7–34.

_____. 1980. A revision of the American species of *Hibiscus* sect. *Bombicella*. U.S.D.A. Tech. Bull. 1624:1–53.

_____. 1983. A revision of *Abutilon* sect. *Oligocarpae* (Malvaceae), including a new species from Mexico. Madroño 30:84–92.

_____. 1985. Sidus sidarum–V. The North and Central American species of *Sida*. Sida 11:62–91.

_____. 1988. Malvaceae of Mexico. Syst. Bot. Monogr. 25:1–522.

_____. 1992. A revised taxonomic interpretation of *Gossypium*. Rheedea 2:108–165.

_____. 1997. The American genera of Malvaceae—II. Brittonia 49:204–269.

_____ and **S.R. Hill.** 1977. More on the typification of *Malvastrum* (DC.) A. Gray (Malvaceae). Taxon 26:332–336.

Fuchs, L. 1542. De historia stirpium commentarii insignes, . . . Basil (Isingrin).

Fuller, T.C. and **E. McClintock.** 1986. Poisonous plants of California. Univ. of California Press, Berkeley.

Fulton, M.G., ed. 1941. Diary and letters of Josiah Gregg. Vol. 1., Southwestern enterprises, 1840–1847. Univ. of Oklahoma Press, Norman.

Furlow, J.J. 1979. The systematics of the American species of *Alnus* (Betulaceae). Rhodora 81:1–121, 151–248.

_____. 1987a. The *Carpinus caroliniana* complex in North America. I. A multivariate analysis of geographical variation. Syst. Bot. 12:21–40.

_____. 1987b. The *Carpinus caroliniana* complex in North America. II. Systematics. Syst. Bot. 12:416–434.

_____. 1990. The genera of Betulaceae in the southeastern United States. J. Arnold Arbor. 71:1–67.

_____. 1997. Betulaceae. In: Flora of North America Editorial Committee, eds. Fl. North Amer. 3:507–538. Oxford Univ. Press, New York and Oxford.

Futuyma, D.J. 1986. Evolutionary biology, 2nd ed. Sinauer Associates, Inc., Sunderland, MA.

Gaiser, L.O. 1946. The genus *Liatris*. Rhodora 48:165–183, 216–263, 326, 331–382, 393–412.

Gale, S. 1944. *Rhynchospora* section *Eurhynchospora*, in Canada, the United States and West Indies. Rhodora 46:89–134, 159–197, 207–249, 255–278.

Galloway, L.E. 1975. Systematics of the North American desert species of *Abronia* and *Tripterocalyx* (Nyctaginaceae). Brittonia 27:328–347.

Gandhi, K.N. (Kancheepuram). Personal communication. Specialist in taxonomic nomenclature; formerly of BONAP (Biota of North America Program), currently at Harvard Univ. Herbaria, Cambridge.

Gandhi, K.N. 1989. A nomenclatural note on *Vitis cinera* and *V. berlandieri* (Vitaceae). Sida 13:506–509.

_____ and **B.E. Dutton.** 1993. Palisot de Beauvois, the correct combining author of *Erianthus giganteus* (Poaceae). Taxon 42:855–856.

Gandhi, K.N. and **R.D. Thomas.** 1984. Observations on the floral structure of *Dyssodia tenuiloba* (DC.) B.L. Robinson and *Matricaria matricarioides* (Less.) T.C. Porter (Asteraceae). Phytologia 55:253–255.

_____ and _____. 1989. Asteraceae of Louisiana. Sida, Bot. Misc. 4:1–202.

_____, _____, and **S.L. Hatch.** 1987. Cuscutaceae of Louisiana. Sida 12:361–379.

Garay, L.A. 1980. A generic revision of the Spiranthinae. Bot. Mus. Leafl. 28:277–425.

Garfield, W., R.F. Keeler, D.C. Baker, and **A.E. Stafford.** 1992. Studies on the *Solanum tuberosum* sprout teratogen. In: L.F. James, R.F. Keeler, E.M. Bailey, Jr., P.R. Cheeke, and M.P. Hegarty, eds. Poisonous plants: Proceedings of the third international symposium. Pp. 418–422. Iowa State Univ. Press, Ames.

Garnett, H. (Hugh). Personal communication. Conservationist, plant collector, and Professor of Economics at Austin College, Sherman, TX; restoring an area of native prairie in Montague Co., TX.

Garrett, H. (Howard). Personal communication. Expert on organic gardening, horticulture, and soil building; columnist for the Dallas Morning News and host of the radio talk show, *The Natural Way*, on WBAP, Arlington, TX.

Garrett, H. 1993a. J. Howard Garrett's organic manual. The Summit Group, Fort Worth, TX.

_____. 1993b. Howard Garrett's Texas organic gardening book. Gulf Publishing Co., Houston, TX.

_____. 1994. Plants of the Metroplex III. Univ. of Texas Press, Austin.

_____. 1996. Howard Garrett's plants for Texas. Univ. of Texas Press, Austin.

Gates, F.C. 1945. *Amphiachyris dracunculoides* as a poisonous plant. Trans. Kansas Acad. Sci. 48:87–89.

Gates, R.R. 1917. A revision of the genus *Polygonatum* in North America. Bull. Torrey Bot. Club 44:117–126.

_____. 1958. Taxonomy and genetics of *Oenothera*: Forty years study in the cytology and evolution of the Onagraceae. Uitgeverij Dr. W. Junk, The Hague.

Gaut, B.S. and **J.F. Doebley.** 1997. DNA sequence evidence for the segmental allotetraploid origin of maize. Proc. Natl. Acad. Sci. U.S.A. 94:6809–6814.

Gehlbach, F.R. 1988. Forests and woodlands of the northeastern balcones escarpment. In: B.B. Amos and F.R. Gehlbach, eds. Edwards Plateau vegetation: Plant ecological studies in central Texas. Pp. 57–77. Baylor Univ. Press, Waco, TX.

_____ and **R.C. Gardner.** 1983. Relationships of sugar maples (*Acer saccharum* and *A. grandidentatum*) in Texas and Oklahoma with special reference to relict populations. Texas J. Sci. 35:231–237.

Geiser, S.W. 1945. Horticulture and horticulturists in early Texas. Southern Methodist Univ. Press, Dallas, TX.

_____. 1948. Naturalists of the frontier, 2nd ed. Southern Methodist Univ. Press, Dallas, TX.

George, L.O. 1997. *Podophyllum*. In: Flora of North America Editorial Committee, eds. Fl. North Amer. 3:287–288. Oxford Univ. Press, New York and Oxford.

Giannasi, D.E. 1978. Generic relationships in the Ulmaceae based on flavonoid chemistry. Taxon 27:331–344.

_____, **G. Zurawski, G. Learn,** and **M.T. Clegg.** 1992. Evolutionary relationships of the Caryophyllidae based on comparative *rbcL* sequences. Syst. Bot. 17:1–15.

Gil-ad, N. 1997. Systematics of *Viola* subsection *Boreali-Americanae*. Boissiera 53:1–130.

Gilbert, L.E. 1980. Ecological consequences of a coevolved mutualism between butterflies and plants. In: L.E. Gilbert and P.H. Raven, eds. Coevolution of plants and animals, revised ed. Pp. 210–240. Univ. of Texas Press, Austin.

Gillis, W.T. 1971. The systematics and ecology of poison-ivy and the poison-oaks (*Toxicodendron*, Anacardiaceae). Rhodora 73:72–79, 161–237, 370–443, 465–540.

_____. 1975. Poison-ivy and its kin. Arnoldia 35:93–123.

_____. 1977. *Pluchea* revisited. Taxon 26:587–591.

GINSBURG, R. 1998. Lloyd Herbert Shinners: A biography. Unpublished October 1998 draft, copy in the library of the Botanical Research Institute of Texas.

GINZBARG, S. 1992. A new disjunct variety of *Croton alabamensis* (Euphorbiaceae) from Texas. Sida 15:41–52.

GLEASON, H.A. 1906. A revision of the North American Vernonieae. Bull. New York Bot. Gard. 4:144–243.

_____. 1908. Platanaceae. N. Amer. Fl. 22:227–229.

_____. 1922. Vernonieae. N. Amer. Fl. 33:47–110.

_____. 1923. Evolution and geographical distribution of the genus *Vernonia* in North America. Amer. J. Bot. 10:187–202.

_____. 1947. Notes on some American plants. Phytologia 2:281–291.

_____. 1952. The new Britton and Brown illustrated flora of the northeastern United States and adjacent Canada. New York Bot. Garden, Bronx.

_____ and **A. CRONQUIST.** 1963. Manual of the vascular plants of northeastern United States and adjacent Canada. Van Nostrand Reinhold Company, New York.

_____ and _____. 1991. Manual of the vascular plants of northeastern United States and adjacent Canada, 2nd ed. New York Bot. Garden, Bronx.

GODFREY, R.K. 1952. *Pluchea*, section *Stylimnus*, in North America. J. Elisha Mitchell Sci. Soc. 68:238–271.

_____ and **J.W. WOOTEN.** 1979. Aquatic and wetland plants of southeastern United States, monocotyledons. Vol. 1. Univ. of Georgia Press, Athens.

_____ and _____. 1981. Aquatic and wetland plants of the southeastern United States, dicotyledons. Vol. 2. Univ. of Georgia Press, Athens.

GOETGHEBEUR, P. 1987. A holosystematic approach of the family Cyperaceae. In: W. Greuter, B. Zimmer, and H.-D. Behnke, eds. Abstracts of the 14th International Botanical Congress, Berlin, Jul 24–Aug 1, 1987. Pp. 276. (Abstract). Botanical Museum Berlin-Dahlem, Berlin.

GOLDBERG, A. 1967. The genus *Melochia* L. (Sterculiaceae). Contr. U.S. Natl. Herb. 34:191–357.

GOLDBLATT, P. 1975. Revision of the bulbous Iridaceae of North America. Brittonia 27:373–385.

_____. 1977 [1978]. *Herbertia* (Iridaceae) reinstated as a valid generic name. Ann. Missouri Bot. Gard. 64:378–379.

_____. 1990. Phylogeny and classification of Iridaceae. Ann. Missouri Bot. Gard. 77:607–627.

_____ and **P.K. ENDRESS.** 1977. Cytology and evolution in Hamamelidaceae. J. Arnold Arbor. 58:67–71.

GOLDBLATT, P., J.E. HENRICH, and **R.C. KEATING.** 1989. Seed morphology of *Sisyrinchium* (Iridaceae—Sisyrinchieae) and its allies. Ann. Missouri Bot. Gard. 76:1109–1117.

GOLDBLATT, P. and **M. TAKEI.** 1997. Chromosome cytology of Iridaceae—patterns of variation, determination of ancestral base numbers, and modes of karyotype change. Ann. Missouri Bot. Gard. 84:285–304.

GOLDMAN, D.H. 1995. A new species of *Calopogon* from the midwestern United States. Lindleyana 10:37–42.

GONZÁLEZ-ELIZONDO, M. and **P.M. PETERSON.** 1997. A classification of and key to the supraspecific taxa in *Eleocharis* (Cyperaceae). Taxon 46:433–449.

GOOD, D.A. 1984. A revision of the Mexican and Central American species of *Cerastium* (Caryophyllaceae). Rhodora 86:339–379.

GOODMAN, G.J. and **C.A. LAWSON.** 1992. Two new combinations and a name change from the Long Expedition of 1820. Rhodora 94:381–382.

_____ and _____. 1995. Retracing Major Stephen H. Long's 1820 expedition. Univ. of Oklahoma Press, Norman.

GOODSPEED, T.H. 1954. The genus *Nicotiana*. Chronica Botanica Co., Waltham, MA.

GOODWIN, D. 1983. Pigeons and doves of the world. Cornell Univ. Press, Ithaca, New York.

GORNALL, R.J. 1987. An outline of a revised classification of *Saxifraga* L. J. Linn. Soc., Bot. 95:273–292.

GOULD, F.W. 1942. A systematic treatment of the genus *Camassia* Lind. Amer. Midl. Naturalist 28:712–742.

_____. 1953. A cytotaxonomic study in the genus *Andropogon*. Amer. J. Bot. 40:297–306.

_____. 1957a. New North American Andropogons of subgenus *Amphilophis* and a key to those species occurring in the United States. Madroño 14:18–29.

_____. 1957b. Texas grasses, a preliminary checklist. Texas Agric. Exp. Sta. Misc. Publ. No. 240.

_____. 1959a. Notes on apomixis in sideoats grama. J. Range Managem. 12:25–28.

_____. 1959b. The glume pit of *Andropogon barbinodis*. Brittonia 11:182–187.

_____. 1962. Texas plants—A checklist and ecological summary. Texas Agric. Exp. Sta. Misc. Publ. 585:1–112.

_____. 1963. Cytotaxonomy of *Digitaria sanguinalis* and *D. adscendens*. Brittonia 15:241–244.

_____. 1967. The grass genus *Andropogon* in the United States. Brittonia 19:70–76.

_____. 1968a. Grass systematics. McGraw-Hill Book Company, New York.

_____. 1968b. Chromosome numbers of Texas grasses. Canad. J. Bot. 46:1315–1325.

_____. 1969. Texas plants—A checklist and ecological summary. Texas Agric. Exp. Sta. Misc. Publ. No. 585/revised.

_____. 1973. Grasses of southwestern United States. Reprinted from 1951 ed. Univ. Arizona Press, Tucson.

_____. 1974. Nomenclatural changes in the Poaceae. Brittonia 26:59–60.

_____. 1975a. Texas plants—A checklist and ecological summary. Texas Agric. Exp. Sta. Misc. Publ. No. 585/revised.

_____. 1975b. The grasses of Texas. Texas A&M Univ. Press, College Station.

_____. 1978. Common Texas grasses. Texas A&M Univ. Press, College Station.

_____. 1979. The genus *Bouteloua* (Poaceae). Ann. Missouri Bot. Gard. 66:348–416.

_____, **M.A. Ali,** and **D.E. Fairbrothers.** 1972. A revision of *Echinochloa* in the United States. Amer. Midl. Naturalist 87:36–59.

Gould, F.W. and **C.A. Clark.** 1978. *Dichanthelium* (Poaceae) in the United States and Canada. Ann. Missouri Bot. Gard. 65:1088–1132.

Gould, F.W. and **Z.J. Kapadia.** 1962. Biosystematic studies in the *Bouteloua curtipendula* complex. The aneuploid, rhizomatous *B. curtipendula* of Texas. Amer. J. Bot. 49:887–891.

_____ and _____. 1964. Biosystematic studies in the *Bouteloua curtipendula* complex II. Taxonomy. Brittonia 16:182–207.

Gould, F.W. and **R.B. Shaw.** 1983. Grass systematics, 2nd ed. Texas A&M Univ. Press, College Station.

Gove, P.B., ed. 1993. Webster's third new international dictionary of the English language unabridged. Merriam-Webster Inc., Springfield, MA.

Govoni, D. 1973. The taxonomy of the genus *Lithospermum* L. (Boraginaceae) in the western Great Plains. Ph.D. dissertation, Univ. of Nebraska, Lincoln.

Goyne, M.A. 1991. A life among the Texas flora: Ferdinand Lindheimer's letters to George Engelmann. Texas A&M Univ. Press, College Station.

Graham, A. 1972. Outline of the origin and historical recognition of floristic affinities between Asia and eastern North America. In: A. Graham, ed. Floristics and paleofloristics of Asia and eastern North America. Elsevier Publishing Co., Amsterdam, Netherlands.

_____. 1993a. History of the vegetation: Cretaceous (Maastrichtian)–Tertiary. In: Flora of North America Editorial Committee, eds. Fl. North Amer. 1:57–70. Oxford Univ. Press, New York and Oxford.

_____. 1993b. Historical factors and biological diversity in Mexico. In: T.P. Ramamoorthy, R. Bye, A. Lot, and J. Fa, eds. Biological diversity of Mexico: Origins and distribution. Oxford Univ. Press, New York and Oxford.

Graham, S.A. 1964. The Elaeagnaceae in the southeastern United States. J. Arnold Arbor. 274–278.

_____. 1966. The genera of Araliaceae in the southeastern United States. J. Arnold Arbor. 47:126–136.

_____. 1975. Taxonomy of the Lythraceae in the southeastern United States. Sida 6:80–103.

_____. 1979. The origin of *Ammania* ×*coccinea* Rottboell. Taxon 28:169–178.

_____. 1985. A revision of *Ammania* (Lythraceae) in the Western Hemisphere. J. Arnold Arbor. 66:395–420.

_____. 1986. Lythraceae. In: Great Plains Flora Association. Flora of the Great Plains. Pp. 494–498. Univ. Press of Kansas, Lawrence.

_____ and **C.E. Wood, Jr.** 1965. The genera of Polygonaceae in the southeastern United States. J. Arnold Arbor. 46:91–121.

Grant, A.L. 1924. A monograph of the genus *Mimulus*. Ann. Missouri Bot. Gard. 11:99–388.

Grant, E. and **C. Epling.** 1943. A study of *Pycnanthemum* (Labiatae). Univ. Calif. Publ. Bot. 20:195–240.

Grant, V. 1956. A synopsis of *Ipomopsis*. Aliso 3:351–362.

_____. 1959. Natural history of the *Phlox* family. Martinus Nijhoff, The Hague, Netherlands.

_____ and **K.A. Grant.** 1965. Flower pollination in the *Phlox* family. Columbia Univ. Press, New York.

_____ and _____. 1979a. The pollination spectrum in the southwestern American cactus flora. Pl. Syst. Evol. 133:29–37.

_____ and _____. 1979b. Systematics of the *Opuntia phaeacantha* group in Texas. Bot. Gaz. (Crawfordsville) 140:199–207.

_____ and _____. 1979c. Hybridization and variation in the *Opuntia phaeacantha* group in central Texas. Bot. Gaz. (Crawfordsville) 140:208–215.

_____ and _____. 1982a. Natural pentaploids in the *Opuntia lindheimeri-phaeacantha* group in Texas. Bot. Gaz. (Crawfordsville) 143:117–120.

_____ and _____. 1982b. Systematics of the *Opuntia phaeacantha* group in Texas. Bot. Gaz. (Crawfordsville) 140:199–207.

_____, _____, and **P.D. Hurd.** 1979. Pollination of *Opuntia lindheimeri* and related species. Pl. Syst. Evol. 132:313–320.

Grant, V. and **P.D. Hurd.** 1979. Pollination of the southwestern Opuntias. Pl. Syst. Evol. 133:15–28.

Grant, W.F. 1953. A cytotaxonomic study in the genus *Eupatorium*. Amer. J. Bot. 40:729–742.

Grauke, L.J., J.W. Pratt, W.F. Mahler, and **A.O. Ajayi.** 1986. Proposal to conserve the name of pecan as *Carya illinoensis* (Wang.) K. Koch and reject the orthographic variant *Carya illinoinensis* (Wang.) K. Koch (Juglandaceae). Taxon 35:174–177.

Gray, A. 1846. Analogy between the flora of Japan and that of the United States. Amer. J. Sci. Arts II. 2:135–136 (Reprinted in Stuckey 1978).

_____. 1859. Diagnostic characters of new species of phanerogamous plants collected in Japan by Charles Wright, botanist of the U.S. North Pacific Exploring Expedition. With observations upon the relations of the Japanese flora to that of North America and of other parts of the northern temperate zone. Mem. Amer. Acad. Arts 6:377–453 (Reprinted in Stuckey 1978).

Grayum, M.H. 1987. A summary of evidence and arguments supporting the removal of *Acorus* from the Araceae. Taxon 36:723–729.

_____. 1990. Evolution and phylogeny of the Araceae. Ann. Missouri. Bot. Gard. 77:628–697.

_____. 1992. Comparative external pollen ultrastructure of the Araceae and putatively related taxa. Monogr. Syst. Bot. Missouri Bot. Gard. 43:1–167.

Grear, J.W. 1978. A revision of the New World species of *Rhynchosia* (Leguminosae-Faboideae). Mem. New York Bot. Gard. 31:1–168.

Great Plains Flora Association. 1986. Flora of the Great Plains. Univ. Press of Kansas, Lawrence.

Greene, E.L. 1905. Revision of *Eschscholtzia*. Pittonia 5:205–308.

Green, P.S. 1962. Watercress in the New World. Rhodora 64:32–43.

Greene, G.S., S.G. Patterson, and **E. Warner.** 1996. Ingestion of angel's trumpet: An increasingly common source of toxicity. Southern Medical J. 89:365–369.

Greenman, J.M. 1915–1918. Monograph of the North and Central American species of the genus *Senecio*. Ann.

Missouri Bot. Gard. 2:573–627; 3:85–195; 4:15–37; 5:37–109.

Greenway, J.C., Jr. 1958. Extinct and vanishing birds of the world. Special Publ. No. 13. American Committee for International Wildlife Protection, New York.

Greer, J.K. 1935. Grand Prairie. Tardy Publishing Company, Dallas, TX.

Greer, L.F. 1997. *Thelesperma curvicarpum* (Asteraceae), an achene form in populations of *T. simplicifolium* var. *simplicifolium* and *T. filifolium* var. *filifolium*. SouthW. Naturalist 42:242–244.

Gregg, J. 1844. Commerce of the prairies. Reprinted 1954 (edited by M.L. Moorhead) by Univ. of Oklahoma Press, Norman.

Gregory, M.P. 1956. A phyletic rearrangement of the Aristolochiaceae. Amer. J. Bot. 43:110–122.

Gregory, W.C., A. Krapovickas, and **M.P. Gregory.** 1980. Structure, variation, evolution, and classification in *Arachis*. In: R.J. Summerfield and A.H. Bunting, eds. Advances in legume science. Pp. 469–481. Royal Botanic Gardens, Kew, England, U.K.

Greuter, W., F.R. Barrie, H.M. Burdet, W.G. Chaloner, V. Demoulin, D.L. Hawksworth, P.M. Jørgensen, D.H. Nicolson, P.C. Silva, P. Trehane, and **J. McNeill,** eds. 1994. International code of botanical nomenclature. Regnum Veg. 131.

Greuter, W., R.K. Brummitt, E. Farr, N. Kilian, P.M. Kirk, and **P.S. Silva.** 1993. Names in current use for extant plant genera. Regnum Veg. 129.

Grey-Wilson, C. 1993. Poppies: A guide to the poppy family in the wild and in cultivation. B.T. Batsford Ltd., London, England, U.K.

Griffiths, D. 1912. The grama grasses, *Bouteloua* and related genera. Contr. U.S. Natl. Herb. 14:343–428.

Griffiths, J.F. and **R. Orton.** 1968. Agroclimatic atlas of Texas–Part I. Precipitation probabilities. Texas Agric. Exp. Sta. Misc. Publ. No. 888.

Grimes, J.W. 1990. A revision of the New World species of Psoraleeae (Leguminosae: Papilionoideae). Mem. New York Bot. Gard. 61:1–114.

Grun, P. 1990. The evolution of cultivated potatoes. Econ. Bot. 44(Suppl.):39–55.

Gunn, C.R. 1980. Seeds and fruits of Papaveraceae and Fumariaceae. Seed Sci. & Technol. 8:3–58.

______. 1983. A nomenclator of legume (Fabaceae) genera. U.S.D.A. Tech. Bull. 1680:1–224.

______. 1984. Fruits and seeds of genera in the subfamily Mimosoideae (Fabaceae). U.S.D.A. Tech. Bull. 1681:1–194.

______. 1991. Fruits and seeds of genera in the subfamily Caesalpiniodeae (Fabaceae). U.S.D.A. Tech. Bull. 1755:1–407.

______ and **M. J. Seldin.** 1976. Seeds and fruits of North American Papaveraceae. U.S.D.A. Tech. Bull. 1517:1–96.

Gupta, K.M. 1957: Some American species of *Marsilea* with special reference to their epidermal and soral characters. Madroño 14:113–127.

Gupta, P.K. and **B.R. Baum.** 1989. Stable classification and nomenclature in the Triticeae: Desirability, limitations, and prospects. Euphytica 41:191–197.

______. 1986. Nomenclature and related taxonomic issues in wheats, triticales, and some of their wild relatives. Taxon 35:144–149.

Hackett, M.R. (Marcia). Personal communication. Environmental Resource Specialist; formerly of the Botanical Research Institute of Texas, currently U.S. Army Corps of Engineers, Fort Worth District.

Hagen, S.H. 1941. A revision of the North American species of the genus *Anisacanthus*. Ann. Missouri Bot. Gard. 28:385–408.

Hall, E. 1873. Plantae Texanae: A list of the plants collected in eastern Texas in 1872, and distributed to subscribers. Published by the author, Salem, MA.

Hall, E.R. and **K.R. Kelson.** 1959. The mammals of North America (2 vols.). The Ronald Press Company, New York.

Hall, G.W. 1967. A biosystematic study of the North American complex of the genus *Bidens* (Compositae). Ph.D. dissertation, Indiana Univ., Bloomington.

______, **G.M. Diggs, Jr., D.E. Soltis,** and **P.S. Soltis.** 1990. Genetic uniformity of El Arbol del Tule. Madroño 37:1–5.

Hall, H.M. 1928. The genus *Haplopappus*: A phylogenetic study in the Compositae. Publ. Carnegie Inst. Wash. 389:1–391.

______ and **F.E. Clements.** 1923. The phylogenetic method in taxonomy: The North American species of *Artemisia*, *Chrysothamnus* and *Atriplex*. Publ. Carnegie Inst. Wash. 326:1–355.

Hall, M.T. 1952. Variation and hybridization in *Juniperus*. Ann. Missouri Bot. Gard. 39:1–64.

Hall, T.F. 1940. The biology of *Saururus cernuus* L. Amer. Midl. Naturalist 24:253–260.

______ and **W.T. Pendfound.** 1944. The biology of the American lotus, *Nelumbo lutea* (Willd.) Pers. Amer. Midl. Naturalist 31:744–758.

Haller, K. (Karl). Personal communication. Ornithologist and naturalist; in 1995 named National Volunteer of the Year by the National Wildlife Refuge System for his service at Hagerman National Wildlife Refuge. Austin College, Sherman, TX.

Hallmark, T.C. 1993. The nature and origin of the Blackland soils. In: M.R. Sharpless and J.C. Yelderman, eds. The Texas Blackland Prairie, land, history, and culture. Pp. 41–47. Baylor Univ. Program for Regional Studies, Waco, TX.

Hamby, R.K. and **E. A. Zimmer.** 1992. Ribosomal RNA as a phylogenetic tool in plant systematics. In: P.S. Soltis, D.E. Soltis, and J.J. Doyle, eds. Molecular systematics of plants. Pp. 50–91. Chapman and Hall, New York.

Hamilton, W. 1983. Cretaceous and Cenozoic history of the northern continents. Ann. Missouri Bot. Gard. 70:440–458.

Handell, S.N. 1978. New ant-dispersed species in the genera *Carex*, *Luzula*, and *Claytonia*. Canad. J. Bot. 56:2925–2927.

Hanks, L.T. and **J.K. Small.** 1907. Geraniaceae. N. Amer. Fl. 25:3–24.

Hansen, B.F. and **R.P. Wunderlin.** 1988. Synopsis of *Dichanthelium* (Poaceae) in Florida. Ann. Missouri Bot. Gard. 75:1637–1657.

Hanson, H.C. 1920. Key to the malvaceous plants of Texas. Texas Agric. Exp. Sta. Circ. 22:5–18.

Hara, H. 1969a. The correct author's name of *Citrullus lanatus* (Cucurbitaceae). Taxon 18:346–347.

______. 1969b. Remarkable examples of speciation in Asiatic plants. Amer. J. Bot. 56:732–737.

______. 1975. The identity of *Clematis terniflora* DC. J. Jap. Bot. 50:155–188.

Hardin, J.W. 1957a. A revision of the American Hippocastanaceae. Brittonia 9:145–171.

_____. 1957b. A revision of the American Hippocastanaceae–II. Brittonia 9:173–195.

_____. 1957c. Studies in the Hippocastanaceae. III. A hybrid swarm in the buckeyes. Rhodora 59:45–51.

_____. 1957d. Studies in the Hippocastanaceae. IV. Hybridization in *Aesculus*. Rhodora 59:185–203.

_____. 1971. Studies of the southeastern United States flora. II. The gymnosperms. J. Elisha Mitchell Sci. Soc. 87:43–50.

_____. 1974. Studies of the southeastern United States flora. IV. Oleaceae. Sida 5:274–285.

_____. 1975. Hybridization and introgression in *Quercus alba*. J. Arnold Arbor. 56:336–363.

_____. 1990. Variation patterns and recognition of varieties of *Tilia americana* s.l. Syst. Bot. 15:33–48.

_____ and **J.M. Arena.** 1974. Human poisoning from native and cultivated plants, 2nd ed. Duke Univ. Press, Durham, NC.

Hardin, J.W. and **C.F. Brownie.** 1993. Plants poisonous to livestock and pets in North Carolina. North Carolina Agricultural Research Service, North Carolina State Univ., Raleigh. Bull. No. 414 (revised).

Harley, R.M. and **T. Reynolds,** eds. 1992. Advances in Labiate science. The Royal Botanic Gardens, Kew, England, U.K.

Harms, L.J. 1968. Cytotaxonomic studies in *Eleocharis* subser. *Palustres*: Central United States taxa. Amer. J. Bot. 55:966–974.

_____. 1972. Cytotaxonomy of the *Eleocharis tenuis* complex. Amer. J. Bot. 59:483–487.

Harms, V.L. 1965. Biosystematic studies in the *Heterotheca subaxillaris* complex. Trans. Kansas Acad. Sci. 68:244–257.

_____. 1968. Nomenclatural changes and taxonomic notes on *Heterotheca*, including *Chrysopsis*, in Texas and adjacent states. Wrightia 4:8–20.

_____. 1974. A preliminary conspectus of *Heterotheca* sect. *Chrysopsis*. Castanea 39:155–165.

Harper, R.M. 1936. Is *Acorus calamus* native in the United States? Torreya 36:143–147.

_____. 1945. *Erythronium albidum* in Alabama, and some of its relatives. Castanea 10:1–7.

Harriman, N.A. (Neil). Personal communication. Specialist in floristics, Brassicaceae, and Juncaceae. Univ. of Wisconsin-Oshkosh.

Harriman, N.A. Forthcoming. Proposal to conserve the name *Gymnocladus* (Leguminosae: Caesalpinioideae), with a conserved gender, against *Hyperanthera* (Moringaceae). Taxon (submitted 1998).

Harris, J.G. and **M.W. Harris** 1994. Plant identification terminology: An illustrated glossary. Spring Lake Publishing, Spring Lake, UT.

Harrison, D.E. and **E.O. Beal.** 1964. The Lemnaceae (duckweeds) of North Carolina. J. Elisha Mitchell Sci. Soc. 80:13–18.

Hart, J.A. and **R.A. Price.** 1990. The genera of Cupressaceae (including Taxodiaceae) in the southeastern United States. J. Arnold Arbor. 71:275–322.

Harte, C. 1994. *Oenothera*: Contributions of a plant to biology. Springer-Verlag, New York.

Hartman, L. (Larry). Personal communication. Conservation scientist. Texas Parks and Wildlife Department, Jasper.

Hartman, R.L. 1986. Asclepiadaceae. In: Great Plains Flora Association. Flora of the Great Plains. Pp. 614–637. Univ. Press of Kansas, Lawrence.

_____. 1990. A conspectus of *Machaeranthera* (Asteraceae: Astereae). Phytololgia 68:439–465.

_____, **J.D. Bacon,** and **C.F. Bohnstedt.** 1975. Biosystematics of *Draba cuneifolia* and *D. platycarpa* (Cruciferae) with emphasis on volatile and flavonoid constituents. Brittonia 27:317–327.

Hartmann, H.E.K. 1993. Aizoaceae. In: K. Kubitzki, J.C. Rohwer, and V. Bittrich, eds. The families and genera of vascular plants, Vol. II. Pp. 37–69. Springer-Verlag, Berlin.

Hartzell, H.R., Jr. 1991. Yew tree: A thousand whispers: Biography of a species. Hulogois, Eugene, OR.

_____. 1995. Yew and us: A brief history of the yew tree. In: M. Suffness, ed. Taxol®: Science and applications. Pp. 27–34. CRC Press, Boca Raton, FL.

Harvey, W.D. 1998. Harmful or potentially harmful fish, shellfish, and aquatic plants. In: Texas Administrative Code. Title 31, Part 2 (Texas Parks and Wildlife Department regulations). Chapter 57, subchapter A, section 57.111–112.

Hatch, S.L. (Stephan). Personal communication. Specialist in Poaceae. Tracy Herbarium, Texas A&M Univ., College Station.

Hatch, S.L., K.N. Gandhi, and **L.E. Brown.** 1990. Checklist of the vascular plants of Texas. Texas Agric. Exp. Sta. Misc. Publ. 1655:1–158.

Hauke, R.L. 1993. Equisetaceae. In: Flora of North America Editorial Committee, eds. Fl. North Amer. 2:76–84. Oxford Univ. Press, New York and Oxford.

Hausman, E.H. 1950. A Christmas herb. Horticulture 28:437, 458.

Havard, V. 1885. Report on the flora of western and southern Texas. Proc. U.S. Natl. Mus. 8:449–533.

_____. 1896. Drink plants of the North American Indians. Bull. Torrey Bot. Club 23:33–46.

Hawkes, J.G., R.N. Lester, M. Nee, and **N. Estrada-R.,** eds. 1991. Solanaceae III: Taxonomy, chemistry, evolution. Royal Botanic Gardens, Kew, England, U.K.

Hawkes, J.G., R.N. Lester, and **A.D. Skelding,** eds. 1979. The biology and taxonomy of the Solanaceae. Linnean Society Symposium Series No. 7. Academic Press, London, England, U.K.

Haynes, R.R. (Robert). Personal communication. Specialist in Alismataceae and Potamogetonaceae and flora of Alabama; Univ. of Alabama, Tuscaloosa.

Haynes, R.R. 1968. *Potamogeton* in Louisiana. Proc. Louisiana Acad. Sci. 31:82–90.

_____. 1974. A revision of North American *Potamogeton* subsection *Pusilli* (Potamogetonaceae). Rhodora 76:564–649.

_____. 1977. The Najadaceae in the southeastern United States. J. Arnold Arbor. 58:161–170.

_____. 1978. The Potamogetonaceae in the southeastern United States. J. Arnold Arbor. 59:170–191.

_____. 1979. Revision of North and Central American *Najas* (Najadaceae). Sida 8:34–56.

_____. 1986. Typification of Linnaean species of *Potamogeton* (Potamogetonaceae). Taxon 35:563–573.

_____ and **C.B. Hellquist.** 1996. New combinations in North

American Alismatidae. Novon 6:370–371.

Haynes, R.R. and **L.B. Holm-Nielsen.** 1986. Notes on *Echinodorus* (Alismataceae). Brittonia 38:325–332.

_____. 1987. The Zannichelliaceae in the southeastern United States. J. Arnold Arbor. 68:259–268.

_____. 1994. The Alismataceae. Fl. Neotrop. Monogr. 64:1–112.

Haynes, R.R. and **W.A. Wentz.** 1974. Notes on the genus *Najas* (Najadaceae). Sida 5:259–264.

Hayward, O.T., P.N. Dolliver, D.L. Amsbury, and **J.C. Yelderman.** 1992. A field guide to the Grand Prairie of Texas, land, history, culture. Program for Regional Studies, Baylor Univ., Waco, TX.

Hayward, O.T. and **J.C. Yelderman.** 1991. A field guide to the Blackland Prairie of Texas, from frontier to heartland in one long century. Program for Regional Studies, Baylor Univ., Waco, TX.

Heard, S.B. and **J.C. Semple.** 1988. The *Solidago rigida* complex (Compositae: Astereae): A multivariate morphometric analysis and chromosome numbers. Canad. J. Bot. 66:1800–1807.

Heiser, C.B., Jr. 1951. The sunflower among the North American Indians. Proc. Amer. Phil. Soc. 95:432–448.

_____. 1954. Variation and subspeciation in the common sunflower, *Helianthus annuus*. Amer. Midl. Naturalist 52:287–305.

_____. 1969. Nightshades: The paradoxical plants. W.H. Freeman and Co., San Francisco, CA.

_____. 1978. The totora (*Scirpus californicus*) in Ecuador and Peru. Econ. Bot. 32:222–236.

_____. 1979. The gourd book. Univ. of Oklahoma Press, Norman.

_____. 1987. The fascinating world of the nightshades. Dover Publications, Inc., New York.

_____. 1990a. Seed to civilization, the story of food, new ed. Harvard Univ. Press, Cambridge, MA.

_____. 1990b. New perspectives on the origin and evolution of New World domesticated plants: Summary. Econ. Bot. 44 (Suppl.):111–116.

_____. 1993. Ethnobotany and economic botany. In: Flora of North America Editorial Committee, eds. Fl. North Amer. 1:199–206. Oxford Univ. Press, New York and Oxford.

_____, **S.B. Clevenger,** and **W.C. Martin, Jr.** 1969. The North American sunflowers (*Helianthus*). Mem. Torrey Bot. Club 22:1–218.

Heiser, C.B., Jr. and **B. Pickersgill.** 1975. Names for the bird peppers [*Capsicum* - Solanaceae]. Baileya 19:151–156.

Heiser, C.B., Jr. and **E.E. Shilling.** 1990. The genus *Luffa*: A problem in phytogeography. In: D.M. Bates, R.W. Robinson, and C. Jeffrey, eds. Biology and utilization of the Cucurbitaceae. Pp. 120–133. Cornell Univ. Press, Ithaca, NY.

Heiser, C.B., Jr. and **P.G. Smith.** 1953. The cultivated *Capsicum* peppers. Econ. Bot. 7:214–227.

Heisey, R.M. 1990. Allelopathic and herbicidal effects of extracts from tree of heaven (*Ailanthus altissima*). Amer. J. Bot. 77:662–670.

_____. 1996. Identification of an allelopathic compound from *Ailanthus altissima* (Simaroubaceae) and characterization of its herbicidal activity. Amer. J. Bot. 83:192–200.

Henderson, A., G. Galeano, and **R. Bernal.** 1995. Field guide to the palms of the Americas. Princeton Univ. Press, Princeton, NJ.

Henderson, N.C. 1962. A taxonomic revision of the genus *Lycopus* (Labiatae). Amer. Midl. Naturalist 68:95–138.

Hendrick, A.J. 1957 [1958]. A revision of the genus *Alisma* (Dill.) L. Amer. Midl. Naturalist 58:470–493.

Hennen, J.F. (Joe). Personal communication. Mycologist specializing in rusts; retired from career at Purdue, currently Research Associate at Botanical Research Institute of Texas, Fort Worth.

Hennen, J.F. 1950. The true clovers (*Trifolium*) of Texas. Field & Lab. 18:159–164.

_____. 1951. The sweet clovers (*Melilotus*) of Texas. Field & Lab. 19:87–89.

Henrard, J. T. 1950. Monograph of the genus *Digitaria*. Universitaire Pers Leiden, Leiden, The Netherlands.

Henrickson, J. 1983. A revision of *Samolus ebracteatus* (sensu lato) Primulaceae. SouthW. Naturalist 28:303–314.

_____. 1985. A taxonomic revision of *Chilopsis* (Bignoniaceae). Aliso 11:179–197.

_____. 1986. *Anisacanthus quadrifidus* sensu lato (Acanthaceae). Sida 11:286–299.

_____. 1987. A taxonomic reevaluation of *Gossypianthus* and *Guilleminea* (Amaranthaceae). Sida 12:307-337.

_____. 1989. A new species of *Leucospora* (Scrophulariaceae) from the Chihuahuan Desert of Mexico. Aliso 12:435–439.

_____. 1993. Amaranthaceae. In: J.C. Hickman, ed. The Jepson manual: Higher plants of California. Pp. 130–134. Univ. of California Press, Berkeley.

_____. 1996. Notes on *Spigelia* (Loganiaceae). Sida 17:89–103.

_____ and **T.F. Daniel.** 1979. Three new species of *Carlowrightia* (Acanthaceae) from the Chihuahuan Desert region. Madroño 26:26–36.

Henrickson, J. and **S. Sunberg.** 1986. On the submersion of *Dicraurus* into *Iresine* (Amaranthaceae). Aliso 11:355–364.

Hepper, F.N. 1990. Pharaoh's flowers: The botanical treasures of Tutankhamun. HMSO, London, England, U.K.

_____. 1992. Illustrated encyclopedia of Bible plants. Inter Varsity Press, Leicester, England, U.K.

Hermann, F.J. 1936. Diagnostic characteristics in *Lycopus*. Rhodora 38:373–375.

_____. 1953. A botanical synopsis of the cultivated clovers (*Trifolium*). U.S.D.A. Monogr. 22:1–45.

_____. 1954a. Addenda to North American carices. Amer. Midl. Naturalist 51:265–286.

_____. 1954b. A synopsis of the genus *Arachis*. U.S.D.A. Monogr. 19:1–26.

_____. 1960. *Vicia*: Vetches of the United States–native, naturalized, and cultivated. U.S.D.A. Handb. No. 168.

_____. 1962. A revision of the genus *Glycine* and its immediate allies. U.S.D.A. Techn. Bull. 1268:1–82.

_____. 1970. Manual of the carices of the Rocky Mountains and Colorado Basin. U.S.D.A. Handb. No. 374.

Herndon, A. 1992. Nomenclatural notes on North American *Hypoxis* (Hypoxidaceae). Rhodora 94:43–47.

Hershkovitz, M.A. and **E.A. Zimmer.** 1997. On the evolutionary origins of the cacti. Taxon 46:217–232.

Hess, W.J. (William). Personal communication. Specialist in Rosaceae, Cornaceae, and Tiliaceae; Morton Arboretum, Lisle, IL.

Hess, W.J. 1968. Variation in *Amelanchier* (Rosaceae) and *Centaurium* (Gentianaceae). Proc. Oklahoma Acad. Sci. 47:19–21.

Hewitson, W. 1962. Comparative morphology of the Osmundaceae. Ann. Missouri Bot. Gard. 49:57–93.

Heyn, C.C. and **B. Pazy.** 1989. The annual species of *Adonis* (Ranunculaceae)—A polyploid complex. Pl. Syst. Evol. 168:181–193.

Heywood, J.S. and **D.A. Levin.** 1984. Allozyme variation in *Gaillardia pulchella* and *G. amblyodon* (Compositae): Relation to morphological and chromosomal variation and to geographical isolation. Syst. Bot. 9:448–457.

Heywood, V.H., ed. 1993. Flowering plants of the world. Oxford Univ. Press, New York.

Hibbard, C.W. 1960. An interpretation of Pliocene and Pleistocene climates in North America. Michigan Acad. Sci. Arts Letters, 62nd Ann. Report (for 1959–1960), pp. 5–30.

Hickey, M. and **C. King.** 1997. Common families of flowering plants. Cambridge Univ. Press, Cambridge, England, U.K.

Hickman, J.C., ed. 1993. The Jepson manual: Higher plants of California. Univ. of California Press, Berkeley.

Hicks, L.E. 1937. The Lemnaceae of Indiana. Amer. Midl. Naturalist 18:774–789.

Hiern, W.P. 1873. A monograph of Ebenaceae. Trans. Cambridge Philos. Soc. 12:27–300.

Hignight, K.W., J.K. Wipff, and **S.L. Hatch.** 1988. Grasses (Poaceae) of the Texas Cross Timbers and Prairies. Texas Agric. Exp. Sta. Misc. Publ. No. 1657.

Hill, R.T. 1887. The topography and geology of the Cross Timbers and surrounding regions in northern Texas. Amer. J. Sci. (3rd series) 133:291–303.

_____. 1901. Geography and geology of the Black and Grand prairies, Texas. 21st Annual Report of the U. S. Geological Survey, Part VII–Texas. U. S. Government Printing Office, Washington, D.C.

Hill, S.R. 1980a. A new country record for *Pilularia americana* in Texas. Amer. Fern. J. 70:28.

_____. 1980b. New taxa and combinations in *Malvastrum* A. Gray (Malvaceae: Malveae). Brittonia 32:464–483.

_____. 1982. A monograph of the genus *Malvastrum*. Rhodora 84:1–83, 159–264, 317–409.

Hillard, O.M. and **B.L. Burtt.** 1981. Some generic concepts in Compositae–Gnaphaliinae. J. Linn. Soc., Bot. 82:181–232.

Hiroe, M. 1979. Umbelliferae of World. Ariake Book Company, Tokyo, Japan.

Hitchcock, A.S. 1903. Notes on North American grasses III. New species of *Willkommia*. Bot. Gaz. (Crawfordsville) 35:283–285.

_____. 1905. North American species of *Agrostis*. U.S.D.A. Bull. 68:5–64.

_____. 1920a. The North American species of *Oplismenus*. Contr. U.S. Natl. Herb. 22:123–132.

_____. 1920b. The North American species of *Echinochloa*. Contr. U.S. Natl. Herb. 22:133–153.

_____. 1924. The North American species of *Aristida*. Contr. U.S. Natl. Herb. 22:517–586.

_____. 1925. The North American species of *Stipa*. Contr. U.S. Natl. Herb. 24:215–262.

_____. 1931–1939. Poaceae (in part). N. Amer. Fl. 17:289–638.

_____. 1935. Manual of the grasses of the United States. U.S.D.A. Misc. Publ. No. 200.

_____. 1951. Manual of the grasses of the United States, 2nd ed. revised by A. Chase. Reprinted in 1971 by Dover Publications, Inc., New York.

_____ and **A. Chase.** 1910. The North American species of *Panicum*. Contr. U.S. Natl. Herb. 15:1–396.

Hitchcock, C.L. 1932. A monographic study of the genus *Lycium* of the western hemisphere. Ann. Missouri Bot. Gard. 19:179–374.

_____. 1933a. A taxonomic study of the genus *Nama* I. Amer. J. Bot. 20:415–430, 518–534.

_____. 1933b. A taxonomic study of the genus *Nama* II. Amer. J. Bot. 20:518–534.

_____. 1936. The genus *Lepidium* in the United States. Madroño 3:265–320.

_____. 1945. The Mexican, Central American, and West Indian Lepidia. Madroño 8:118–143.

_____. 1952. A revision of the North American species of *Lathyrus*. Univ. Wash. Publ. Biol. 15:1–104.

_____ and **B. Maguire.** 1947. A revision of the North American species of *Silene*. Univ. Wash. Publ. Biol. 13:1–73.

Hoch, P.C., J.V. Crisci, H. Tobe, and **P.E. Berry.** 1993. A cladistic analysis of the plant family Onagraceae. Syst. Bot. 18:31–47.

Hodgdon, A.R. 1938. A taxonomic study of *Lechea*. Rhodora 40:29–131.

Hoey, M.T. and **C.R. Parks.** 1994. Genetic divergence in *Liquidambar styraciflua*, *L. formosana*, and *L. acalycina* (Hamamelidaceae). Syst. Bot. 19:308–316.

Hoggard, G.D. 1999. *Gaura*. In: Tyrl, R.J., S.C. Barber, P. Buck, J.R. Estes, P. Folley, L.K. Magrath, C.E.S. Taylor, and R.A. Thompson (Oklahoma Flora Editorial Committee). Identification of Oklahoma Plants. Flora Oklahoma Inc., Noble, OK.

Hoggard, R.K. 1999. Plantaginaceae. In: Tyrl, R.J., S.C. Barber, P. Buck, J.R. Estes, P. Folley, L.K. Magrath, C.E.S. Taylor, and R.A. Thompson (Oklahoma Flora Editorial Committee). Identification of Oklahoma Plants. Flora Oklahoma Inc., Noble, OK.

Holley, M.A. 1836. Texas. J. Clarke & Co., Lexington, Kentucky. Reprinted 1985 by the Texas State Historical Association, in cooperation with the Center for Studies in Texas History, Univ. of Texas, Austin.

Holm, L.G., D.L. Plucknett, J.V. Pancho, and **J.P. Herberger.** 1977. The world's worst weeds: Distribution and biology. Published for the East-West Center by the Univ. of Hawaii Press, Honolulu.

Holm, R.W. 1950. The American species of *Sarcostemma* R. Br. (Asclepiadaceae). Ann. Missouri Bot. Gard. 37:477–560.

Holmes, W.C. 1981. *Mikania* (Compositae) of the United States. Sida 9:147–158.

_____. 1990. The genus *Mikania* (Compositae-Eupatorieae) in Mexico. Sida, Bot. Misc. 5:1–45.

Holmes, W.C., T.L. Morgan, J.R. Stevens, and **R.D. Gooch.** 1996. Comments on the distribution of *Botrychium lunarioides* (Ophioglossaceae) in Texas. Phytologia 80:280–283.

Holmes, W.C. and **C.J. Wells.** 1980. The distribution of *Habranthus tubispathus* (L'Her.) Traub in South America and North America—Texas and Louisiana. Sida 8:328–333.

Holmes, W.C. and **D.E. Wivagg.** 1996. Identification and distribution of *Centaurium muhlenbergii* [sic] (Griseb.) Piper and *C. pulchellum* (Sw.) Druce (Gentianaceae) in Louisiana, Mississippi, and Texas. Phytologia 80:23–29.

Holmgren, A.H. and **N.H. Holmgren.** 1977. *Aristida*. In: A. Cronquist, A.H. Holmgren, N.H. Holmgren, J.L. Reveal, and P.K. Holmgren. Intermountain Flora, Vol. 6. Pp. 457–465. Columbia Univ. Press, New York.

Holmgren, N.H. 1986. Scrophulariaceae. In: Great Plains Flora Association. Flora of the Great Plains. Pp. 751–797. Univ. Press of Kansas, Lawrence.

Holmgren, P.K., N.H. Holmgren, and **L.C. Barnett.** 1990. Index herbariorum. Part 1: The herbaria of the world. Regnum Veg. 120.

Hooker, W.J. 1836. *Lupinus subcarnosus* (Plate 3467), *Lupinus texensis* (Plate 3492). Bot. Mag. 63.

Hoot, S.B., J.W. Dadereit, F.R. Blattner, K.B. Jork, A.E. Schwarzbach, and **P.R. Crane.** 1997 [1998]. Data congruence and phylogeny of the Papaveraceae s.l. based on four data sets: *atpB* and *rbcL* sequences, *trnK* restriction sites, and morphological characters. Syst. Bot. 22:575–590.

Hoot, S.B., A.A. Reznicek, and **J. Palmer.** 1994. Phylogenetic relationships in *Anemone* (Ranunculaceae) based on morphology and chloroplast DNA. Syst. Bot. 19:169–200.

Hopkins, M. 1937. *Arabis* in eastern and central North America. Rhodora 39:63–98, 106–148, 155–186.

_____. 1942. *Cercis* in North America. Rhodora 44:193–211.

Hornberger, K.L. 1991. The blue-eyed grasses (*Sisyrinchium*: Iridaceae) of Arkansas. Sida 14:597–604.

Horton, J.H. 1963. A taxonomic revision of *Polygonella* (Polygonaceae). Brittonia 15:177–203.

_____. 1972. Studies of the southeastern United States flora. IV. Polygonaceae. J. Elisha Mitchell Sci. Soc. 88:92–102.

Hotchkiss, N. and **H.L. Dozier.** 1949. Taxonomy and distribution of North American cat-tails. Amer. Midl. Naturalist 41:237–254.

Houghton, J.T., G.J. Jenkins, and **J.J. Ephraums,** eds. 1990. Climate change, the IPCC scientific assessment. Cambridge Univ. Press, Cambridge, England, U.K.

Houghton, J.T., L.G. Miera Filho, B.A. Callander, N. Harris, A. Kattenberg, and **K. Maskell,** eds. 1995. Climate change 1995, the science of climate change. Published for the Intergovernmental Panel on Climate Change by Cambridge Univ. Press, Cambridge, England, U.K.

House, H.D. 1908. The North American species of the genus *Ipomoea*. New York Acad. Sci. 18:181–263.

Howard, T.M. 1990. *Allium texanum* [Amaryllidaceae (Alliaceae)]. A new species from central Texas and adjacent Oklahoma. Herbertia 46:125–127.

Howe, W.T. 1975. The butterflies of North America. Doubleday & Co., Inc., Garden City, NY.

Howell, J.T. 1959. Distribution data on weedy thistles in western North America. Leafl. W. Bot. 9:17–29.

Hsü, J. 1983. Late Cretaceious and Cenozoic vegetation in China, emphasizing their connections with North America. Ann. Missouri Bot. Gard. 70:490–508.

Hu, S.Y. 1954–1956. A monograph of the genus *Philadelphus*. J. Arnold Arbor. 35:275–333; 36:52–109, 326–368; 37:15–90.

Huber, H. 1993. Aristolochiaceae. In: K. Kubitzki, J.C. Rohwer, and V. Bittrich, eds. The families and genera of vascular plants, Vol. II. Pp. 129–137. Springer-Verlag, Berlin.

Hufford, L. 1992. Rosidae and their relationships to other nonmagnoliid dicotyledons: A phylogenetic analysis using morphological and chemical data. Ann. Missouri Bot. Gard. 79:218–248.

Hunt, D.R. 1971. Schumann and Buxbaum reconciled. The Schumann system of *Mammillaria* classification brought provisionally up-to-date. Cact. Succ. J. Gr. Brit. 33:53–72.

_____. 1975. The reunion of *Setcreasea* and *Spathotheca* with *Tradescantia*. American Commelinaceae: I. Kew Bull. 30:443–458.

_____. 1978. Amplification of the genus *Escobaria*. Cact. Succ. J. Gr. Brit. 40:13.

_____. 1980. Sections and series in *Tradescantia*. American Commelinaceae: IX. Kew Bull. 35:437–442.

_____. 1986. *Campelia*, *Rhoeo* and *Zebrina* united with *Tradescantia*. American Commelinaceae: XIII. Kew Bull. 41:401–405.

Hutchinson, J. 1979. The families of flowering plants, 3rd ed. Authorized reprint by Otto Koeltz Science Publishers, Koenigstein, West Germany.

Huxley, A., M. Griffiths, and **M. Levy,** eds. 1992. The new Royal Horticultural Society dictionary of gardening, 4 vols. MacMillan Press Limited, London, and Stockton Press, New York.

Hyam, R. and **R. Pankhurst.** 1995. Plants and their names: A concise dictionary. Oxford Univ. Press, New York.

Hymowitz, T. and **C.A. Newell.** 1981. Taxonomy of the genus *Glycine*, domestication and uses of soybeans. Econ. Bot. 35:272–288.

Ikin, A. 1841. Texas: Its history, topography, agriculture, commerce, and general statistics. Sherwood, Gilbert, and Piper, London, England, U.K. Reprinted in 1964 by Texian Press, Waco, TX.

Iltis, H.H. 1957. Studies in the Capparidaceae. III Evolution and phylogeny of the western North American Cleomoideae. Ann. Missouri Bot. Gard. 44:77–119.

_____. 1958. Studies in the Capparidaceae – IV. *Polanisia* Raf. Brittonia 10:33–58.

_____. 1958 [1959]. Studies in the Capparidaceae – V. Capparidaceae of New Mexico. SouthW. Naturalist 3:133–144.

_____. 1959. Studies in the Capparidaceae – VI. *Cleome* sect. *Physostemon*: Taxonomy, geography, and evolution. Brittonia 11:123–162.

_____. 1966. Studies in the Capparidaceae VIII. *Polanisia dodecandra* (L.) DC.: Further notes on its typification. Rhodora 68:41–47.

_____. 1972. The taxonomy of *Zea mays* (Gramineae). Phytologia 23:248–249.

_____. 1983. From teosinte to maize: The catastrophic sexual transmutation. Science 222:886–894.

_____. 1987. Maize evolution and agricultural origins. In: T.R. Soderstrom, K.W. Hilu, C.S. Campbell, and M.E. Barkworth, eds. Grass systematics and evolution. Pp. 195–213. Smithsonian Institution Press, Washington, D.C.

_____ and **J.F. Doebley.** 1980. Taxonomy of *Zea* (Gramineae). II. Sub-specific categories in the *Zea mays* complex and a generic synopsis. Amer. J. Bot. 67:994–1004.

Ingram, J. 1967. A revisional study of *Argythamnia* subgenus *Argythamnia* (Euphorbiaceae). Gentes Herb. 10:1–38.

_____. 1980. The generic limits of *Argythamnia* (Euphorbiaceae) defined. Gentes Herb. 11:427–436.

Ireland, R.R. 1957. Biosystematics of *Erythronium albidum* and *E. mesochorum*. M.A. Thesis, Univ. Kansas, Manhattan.

Irish, H.C. 1898. Revision of the genus *Capsicum*. Rep. (Annual) Missouri Bot. Gard. 9:53–110.

Irving, R.S. 1976. Chromosome numbers of *Hedeoma* (Labiatae) and related genera. Syst. Bot. 1:46–56.

_____. 1979. Artificial hybridization in *Hedeoma* (Labiatae). Syst. Bot. 4:1–15.

_____. 1980. The systematics of *Hedeoma* (Labiatae). Sida 8:218–295.

Irwin, H.S. and **R.C. Barneby.** 1982. The American Cassiinae. A synoptical revision of Leguminosae tribe Cassieae subtribe Cassiinae in the New World. Mem. New York Bot. Gard. 35:1–918.

Isely, D. 1948. *Lespedeza striata* and *L. stipulacea*. Rhodora 50:21–27.

_____. 1951. The Leguminosae of the north-central United States: I. Loteae and Trifolieae. Iowa State Coll. J. Sci. 25:439–482.

_____. 1954. Keys to sweetclovers (*Melilotus*). Proc. Iowa Acad. Sci. 61:119–131.

_____. 1969. Legumes of the United States: I. Native *Acacia*. Sida 3:365–386.

_____. 1970a. Legumes of the United States: II. *Desmanthus* and *Neptunia*. Iowa State J. Sci. 44:495–511.

_____. 1970b. Legumes of the United States V. *Albizia*, *Lysiloma*, *Leucaena*, *Adenanthera*; and rejected genera of the Mimosoideae. Castanea 35:244–260.

_____. 1971a. Legumes of the United States: III. *Schrankia*. Sida 4:232–245.

_____. 1971b. Legumes of the United States. IV. *Mimosa*. Amer. Midl. Naturalist 85:410–424.

_____. 1972. Legumes of the U.S. VI. *Calliandra*, *Pithecellobium*, and *Prosopis*. Madroño 21:273–298.

_____. 1973. Leguminosae of the United States: I. Subfamily Mimosoideae. Mem. New York Bot. Gard. 25 (1):1–152.

_____. 1975. Leguminosae of the United States: II. Subfamily Caesalpinioideae. Mem. New York Bot. Gard. 25 (2):1–228.

_____. 1981. Leguminosae of the United States. III. Subfamily Papilionoideae: Tribes Sophoreae, Podalyrieae, Loteae. Mem. New York Bot. Gard. 25(3):1–264.

_____. 1982. Leguminosae and *Homo sapiens*. Econ. Bot. 36:46–70.

_____. 1983. The *Desmodium paniculatum* complex revisited. Sida 10:142–158.

_____. 1983–1986. *Astragalus* L. (Leguminosae: Papilionoideae). Iowa State J. Res. 58:3–172; 59:99–209; 60:183–320; 61:157–289.

_____. 1986a. Notes about *Psoralea* sensu auct., *Amorpha*, *Baptisia*, *Sesbania* and *Chamaecrista* (Leguminosae) in the southeastern United States. Sida 11:429–440.

_____. 1986b. Notes on Leguminosae: Papilionoideae of the southeastern United States. Brittonia 38:352–359.

_____. 1990. Leguminosae (Fabaceae). Vascular flora of the southeastern United States 3(2):1–258. Univ. North Carolina Press, Chapel Hill.

_____ and **F.J. Peabody.** 1984. *Robinia* (Leguminosae: Papilionoideae). Castanea 49:187–202.

Isely, D. and **R.M. Polhill.** 1980. Leguminosae: Subfamily Papilionoideae. Taxon 29:105–119.

Jackson, R.C. 1960. A revision of the genus *Iva* L. Univ. Kansas Sci. Bull. 41:793–874.

_____. 1963. Cytotaxonomy of *Helianthus ciliaris* and related species of the southwestern U.S. and Mexico. Brittonia 15:260–271.

Jackson, S.W. 1963. Hybridization among three species of *Ratibida*. Univ. Kansas Sci. Bull. 44:3–27.

Jacobsen, N. 1985. Order Arales. In: R.M.T. Dahlgren, H.T. Clifford, and P.F. Yeo. The families of the monocotyledons. Pp. 278–289. Springer-Verlag, Berlin, Heidelberg.

James, C.W. 1956. A revision of *Rhexia* (Melastomataceae). Brittonia 8:201–230.

James, L.F. and **S.L. Welsh.** 1992. Poisonous plants of North America. In: L.F. James, R.F. Keeler, E.M. Bailey, Jr., P.R. Cheeke, and M.P. Hegarty, eds. Poisonous plants: Proceedings of the third international symposium. Pp. 94–104. Iowa State Univ. Press, Ames.

James, R.L. 1948. Some hummingbird flowers east of the Mississippi. Castanea 12:97–109.

Jansen, R.K. 1981. Systematics of *Spilanthes* (Compositae: Heliantheae). Syst. Bot. 6:231–257.

_____. 1985a. The systematics of *Acmella* (Asteraceae-Heliantheae). Syst. Bot. Monogr. 8:1–115.

_____. 1985b. Systematic significance of chromosome number in *Acmella* (Asteraceae). Amer. J. Bot. 72:1835–1841.

_____, **H.J. Michaels,** and **J.D. Palmer.** 1991. Phylogeny and character evolution in the Asteraceae based on chloroplast DNA restriction site mapping. Syst. Bot. 16:98–115.

Jansen, R.K., H.J. Michaels, R.S. Wallace, K.-J. Kim, S.C. Keeley, L.E. Watson, and **J.D. Palmer.** 1992. Chloroplast DNA variation in the Asteraceae: Phylogenetic and evolutionary implications. In: P.S. Soltis, D.E. Soltis, and J.J. Doyle, eds. Molecular systematics of plants. Pp. 252–279. Chapman and Hall, New York.

Jansen, R.K., E.B. Smith, and **D.J. Crawford.** 1987. A cladistic study of North American *Coreopsis* (Asteraceae: Heliantheae). Pl. Syst. Evol. 157:73–84.

Janzen, D.H. and **P.S. Martin.** 1982. Neotropical anachronisms: The fruits the gomphotheres ate. Science 215:19–27.

Jarman, R. 1968. Eurasian watermilfoil–new menace to Oklahoma waters. Proc. Oklahoma Acad. Sci. 49:171–174.

Jeffrey, C. 1975. Further notes on Cucurbitaceae: IV. Some New-World taxa. Kew Bull. 33:347–380.

_____. 1990. Appendix: An outline classification of the Cucurbitaceae. In: D.M. Bates, R.W. Robinson, and C. Jeffrey, eds. Biology and utilization of the Cucurbitaceae. Pp. 449–463. Cornell Univ. Press, Ithaca, NY.

Jensen, R.J. 1977. Numerical analysis of the scarlet oak complex

(*Quercus* subgen. *Erythrobalanus*) in the eastern United States: Relationships above the species level. Syst. Bot. 2:122–129.

Jensen, S.R. 1992. Systematic implications of the distribution of iridoids and other chemical compounds in the Loganiaceae and other families of the Asteridae. Ann. Missouri Bot. Gard. 79:284–302.

Jensen, U., I. Vogel-Bauer, and **M. Nitschke.** 1994. Leguminlike proteins and the systematics of the Euphorbiaceae. Ann. Missouri Bot. Gard. 81:160–179.

Johnson, D.M. 1986. Systematics of the New World species of *Marsilea* (Marsileaceae). Syst. Bot. Monogr. 11:1–87.

______. 1988. Proposal to conserve *Marsilea* L. (Pteridophyta: Marsileaceae) with *Marsilea quadrifolia* as *typ. conserv.* Taxon 37:483–486.

______. 1993a. Marsileaceae. In: Flora of North America Editorial Committee, eds. Fl. North Amer. 2:331–335. Oxford Univ. Press, New York and Oxford.

______. 1993b. *Onoclea*. In: Flora of North America Editorial Committee, eds. Fl. North Amer. 2:251. Oxford Univ. Press, New York and Oxford.

Johnson, L.A. and **D.E. Soltis.** 1995. Phylogenetic inference in Saxifragaceae sensu stricto and *Gilia* (Polemoniaceae) using matK sequences. Ann. Missouri Bot. Gard. 83:149–175.

Johnston, I.M. 1924. Studies in the Boraginaceae. II. 29. *Lappula*. Contr. Gray Herb. 70:47–51.

______. 1925. Studies in the Boraginaceae. IV. The North American species of *Cryptantha*. Contr. Gray Herb. 74:1–114.

______. 1952. Studies in the Boraginaceae, XXIII. A survey of the genus *Lithospermum*. J. Arnold Arbor. 33:299–366.

______. 1954a. Studies in the Boraginaceae, XXVI. Further reevaluations of the genera of the Lithospermeae. J. Arnold Arbor. 35:1–81.

______. 1954b. Studies in the Boraginaceae, XXVII. Some general observations concerning the Lithospermeae. J. Arnold Arbor. 35:158–166.

______. 1966. Boraginaceae. In: C.L. Lundell, ed. Flora of Texas 1:123–221. Texas Research Foundation, Renner.

Johnston, M.C. 1956. The Texas species of *Dyssodia* (Compositae). Field & Lab. 24:60–69.

______. 1957. Synopsis of the United States species of *Forestiera* (Oleaceae). SouthW. Naturalist 2:140–151.

______. 1959. The Texas species of *Croton* (Euphorbiaceae). SouthW. Naturalist 3:175–203.

______. 1962. Revision of *Condalia* including *Microrhamnus* (Rhamnaceae). Brittonia 14:332–368.

______. 1963a. The geography of the five Texas species of *Dichondra* (Convoluvlaceae). Wrightia 2:252–253.

______. 1963b. The species of *Ziziphus* indigenous to United States and Mexico. Amer. J. Bot. 50:1020–1027.

______. 1964. *Scirpus molestus* M.C. Johnston (Cyperaceae), sp. nov. from Arkansas, Louisiana, and Texas. SouthW. Naturalist 9:310–312.

______. 1966. The Texas species of *Cyperus* subgenus *Kyllinga* (Cyperaceae). SouthW. Naturalist 11:123–124.

______. 1971. Revision of *Colubrina* (Rhamnaceae). Brittonia 23:2–53.

______. 1975. Studies of the *Euphorbia* species of the Chihuahuan Desert region and adjacent areas. Wrightia 5:120–143.

______. 1988. The vascular plants of Texas: A list, up-dating the manual of the vascular plants of Texas. Published by the author, Austin. TX.

______. 1990. The vascular plants of Texas: A list, up-dating the manual of the vascular plants of Texas, 2nd ed. Published by the author, Austin. TX.

______ and **L.A. Johnston.** 1969. Rhamnaceae. In: C.L. Lundell, ed. Flora of Texas 2:357–392. Texas Research Foundation, Renner.

Johnston, M.C. and **B.L. Turner.** 1962. Chromosome numbers of *Dyssodia* (Compositae: Tagetinae) and phyletic interpretations. Rhodora 64:2–15.

Jones, A.G. 1978a. The taxonomy of *Aster* section *Multiflori* (Asteraceae) I. Nomenclatural review and formal presentation of taxa. Rhodora 80:319–357.

______. 1978b. The taxonomy of *Aster* section *Multiflori* (Asteraceae)—II. Biosystematic investigations. Rhodora 80:453–490.

______. 1980. A classification of the New World species of *Aster* (Asteraceae). Brittonia 32:230–239.

______. 1983. Nomenclatural changes in *Aster* (Asteraceae). Bull. Torrey Bot. Club 110:39–42.

______. 1984. Nomenclatural notes on *Aster* (Asteraceae) – II. New combinations and some transfers. Phytologia 55:373–388.

______. 1987. New combinations and status changes in *Aster* (Asteraceae). Phytologia 63:131–133.

______. 1992. *Aster* and *Brachyactis* (Asteraceae) in Oklahoma. Sida, Bot. Misc. 8:1–46.

______ and **M.T. Deonier.** 1965. Interspecific crosses among *Ipomoea lacunosa*, *I. ramoni*, *I. trichoarpa*, and *I. triloba*. Bot. Gaz. (Crawfordsville) 126:226–232.

Jones, A.G. and **D.A. Young.** 1983. Generic concepts of *Aster* (Asteraceae): A comparison of cladistic, phenetic, and cytological approaches. Syst. Bot. 8:71–84.

Jones, F.B. 1975. Flora of the Texas coastal bend. Welder Wildlife Foundation, Sinton, TX.

______. 1977. Flora of the Texas coastal bend, 2nd ed. Welder Wildlife Foundation, Sinton, TX.

______. 1982. Flora of the Texas coastal bend, 3rd ed. Welder Wildlife Foundation, Sinton, TX.

______, **C.M. Rowell, Jr.,** and **M.C. Johnston.** 1961. Flowering plants and ferns of the Texas coastal bend counties. Welder Wildlife Foundation, Sinton, TX.

Jones, G.D. and **S.D. Jones.** 1991. *Sarcostemma clausum*, series *Clausa* (Asclepiadaceae), new to Texas. Phytologia 71:160–162.

Jones, G.N. 1940. A monograph of the genus *Symphoricarpos*. J. Arnold Arbor. 21:201–252.

______. 1946. American species of *Amelanchier*. Illinois Biol. Monogr. 20:1–126.

______. 1968. Taxonomy of American species of Linden (*Tilia*). Illinois Biol. Monogr. 39:1–156.

______ and **F.F. Jones.** 1943. A revision of the perennial species of *Geranium* of the United States and Canada. Rhodora 45:5–26, 32–53.

Jones, J.H. 1986. Evolution of the Fagaceae: The implications of foliar features. Ann. Missouri Bot. Gard. 73:228–275.

Jones, R.L. 1983. A systematic study of *Aster* section *Patentes* (Asteraceae). Sida 10:41–81.

Jones, S.B. 1970. Scanning electron microscopy of pollen as an aid to the systematics of *Vernonia* (Compositae). Bull. Torrey Bot. Club 97:323–335.

_____. 1982. The genera of Vernonieae (Compositae) in the southeastern United States. J. Arnold Arbor. 63:489–507.

_____ and **W.Z. Faust.** 1978. Compositae tribe Vernonieae. N. Amer. Fl. II. 10:180–202.

Jones, S.B. and **A.E. Luchsinger.** 1986. Plant systematics, 2nd ed. McGraw-Hill Book Co., New York.

Jones, S.D. (Stanley). Personal communication. Specialist in *Carex* and other Cyperaceae; botanical consultant at Botanical Research Center Herbarium, Bryan, TX.

Jones, S.D. 1994a. A taxonomic study of the *Carex muhlenbergii* and *C. cephalophora* complexes (Cyperaceae: *Phaestoglochin*). Ph.D. dissertation, Texas A&M Univ., College Station.

_____. 1994b. A new species of *Carex* (Cyperaceae: *Phaestoglochin*) from Oklahoma and Texas; typification of section *Phaestoglochin*, and notes on sections *Bracteosae* and *Phaestoglochin*. Sida 16:341–353.

_____ and **S.L. Hatch.** 1990. Synopsis of *Carex* section Lupulinae (Cyperaceae) in Texas. Sida 14:87–99.

Jones, S.D. and **A.A. Reznicek.** 1991. *Carex bicknellii*, "Bicknell's sedge" (Cyperaceae): New to Texas, with a key to Texas species of section *Ovales*. Phytologia 70:115–118.

_____ and _____. 1997. *Lathyrus aphaca* (Fabaceae), previously unreported for Texas. Phytologia 82:1–2.

Jones, S.D. and **J.K. Wipff.** 1992. *Bulbostylis barbata*, (Cyperaceae), previously unreported for Texas. Phytologia 73:381–383.

_____, _____, and **R. Carter.** 1996. Nomenclatural combinations in *Cyperus* (Cyperaceae). Phytologia 80:288–290.

Jones, S.D., J.K. Wipff., and **P.M. Montgomery.** 1997. Vascular plants of Texas: A comprehensive checklist including synonymy, bibliography, and index. Univ. of Texas Press, Austin.

Joseph, C. and **M. Heimberger.** 1966. Cytotaxonomic studies on New World species of *Anemone* (Section *Eriocephalus*) with tuberous rootstocks. Canad. J. Bot. 44:899–913.

Jordan, T.G. 1973. Pioneer evaluation of vegetation in frontier Texas. SouthW. Hist. Quart. 76:233–254.

Jørgensen, P.M., J.E. Wawesson, and **L.B. Holm-Nielsen.** 1984. A guide to collecting passionflowers. Ann. Missouri Bot. Gard. 71:1172–1174.

Jovet, P. and **R. Willmann.** 1957. Trécul, botaniste français. In: Les botanistes français in Amérique du Nord avant 1850. Pp. 83–106. Paris.

Jubinsky, G. and **L.C. Anderson.** 1996. The invasive potential of chinese tallow-tree (*Sapium sebiferum* Roxb.) in the southeast. Castanea 61:226–231.

Judd, W.S. 1995. *Lyonia*. In: J.L. Luteyn, ed. Ericaceae Part II, the superior-ovaried genera. Fl. Neotrop. Monogr. 66:222–294.

_____ and **K.A. Kron.** 1993. Circumscription of Ericaceae (Ericales) as determined by preliminary cladistic analyses based on morphological, anatomical, and embryological features. Brittonia 45:99–114.

Judd, W.S., R.W. Sanders, and **M.J. Donoghue.** 1994. Angiosperm family pairs: Preliminary phylogenetic analyses. Harvard Pap. Bot. 5:1–51.

Judziewicz, E.J. 1990. A new South American species of *Sacciolepis* (Poaceae: Panicoideae: Paniceae), with a summary of the genus in the New World. Syst. Bot. 15:415–420.

Jurney, D.H. 1987. Presettlement vegetation recorded in land surveyors' notes. In: D.H. Jurney and R.W. Moir. Historic buildings, material culture, and people of the prairie margin. Richland Creek Technical Series Volume V. Pp. 211–227. Archaeology Research Program, Institute for the Study of Earth and Man, Southern Methodist Univ., Dallas, TX.

Kadereit, J.W. 1993. Papaveraceae. In: K. Kubitzki, J.C. Rohwer, and V. Bittrich, eds. The families and genera of vascular plants, Vol. II. Pp. 494–506. Springer-Verlag, Berlin.

Kapadia, Z.J. and **F.W. Gould.** 1964a. Biosystematic studies in the *Bouteloua curtipendula* complex. III. Pollen size as related to chromosome numbers. Amer. J. Bot. 51:166–172.

_____ and _____. 1964b. Biosystematic studies in the *Bouteloua curtipendula* complex IV. Dynamics of variation in *B. curtipendula* var. *caespitosa*. Bull. Torrey Bot. Club 91:465–478.

Kapil, R.N. and **A.K. Bhatnagar.** 1994. The contribution of embryology to the systematics of the Euphorbiaceae. Ann. Missouri Bot. Gard. 81:145–159.

Kapoor, B.M. and **J.R. Beaudry.** 1966. Studies on *Solidago*. VII. The taxonomic status of the taxa *Brachchaeta, Brintonia, Chrysoma, Euthamia, Oligoneuron* and *Petradoria* in relation to *Solidago*. Canad. J. Genet. Cytol. 8:422–433.

Karis, P.O. 1995. Cladistics of the subtribe Ambrosiinae (Asteraceae: Heliantheae). Syst. Bot. 20:40–54.

_____ and **O. Ryding.** 1994. Tribe Helenieae (Chapter 21). In: K. Bremer. Asteraceae cladistics & classification. Pp. 521–558. Timber Press, Portland, OR.

Kartesz, J.T. (John). Personal communication. Specialist in nomenclature of North American plants. Biota of North America Program, Univ. of North Carolina, Chapel Hill.

Kartesz, J.T. 1994. A synonymized checklist of the vascular flora of the United States, Canada, and Greenland, 2nd ed. (2 vols.). Timber Press, Portland, OR.

_____ and **K.N. Gandhi.** 1995. Nomenclatural notes for the North American flora. XIV. Phytologia 78:1–17.

Kato, M. 1993. *Athyrium*. In: Flora of North America Editorial Committee, eds. Fl. North Amer. 2:255–258. Oxford Univ. Press, New York and Oxford.

Kaul, R.B. 1986a. Evolution and reproductive biology of inflorescences in *Lithocarpus*, *Castanopsis*, *Castanea*, and *Quercus* (Fagaceae). Ann. Missouri Bot. Gard. 73:284–296.

_____. 1986b. Polygonaceae. In: Great Plains Flora Assoc. Flora of the Great Plains. Pp. 214–235. Univ. Press of Kansas, Lawrence.

_____. 1986c. Elatinaceae. In: Great Plains Flora Assoc. Flora of the Great Plains. Pp. 236. Univ. Press of Kansas, Lawrence.

_____. 1986d. Loasaceae. In: Great Plains Flora Assoc. Flora of the Great Plains. Pp. 269–273. Univ. Press of Kansas, Lawrence.

_____. 1986e. Zannichelliaceae. In: Great Plains Flora Assoc. Flora of the Great Plains. Pp. 1041–1042. Univ. Press of Kansas, Lawrence.

_____. 1997. Platanaceae. In: Flora of North America Editorial Committee, eds. Fl. North Amer. 3:358–361. Oxford Univ. Press, New York and Oxford.

Kearney, T.H. 1935. The North American species of *Sphaeralcea* subgenus *Eusphaeralcea*. Univ. Calif. Publ. Bot. 19:1–128.

_____. 1951. The American genera of Malvaceae. Amer. Midl. Naturalist 46:93–131.

_____. 1954. A tentative key to the North American species of *Sida*, L. Leafl. W. Bot. 7:138–152.

_____. 1955a. *Malvastrum*, A. Gray—A redefinition of the genus. Leafl. W. Bot. 7:238–241.

_____. 1955b. A tentative key to the North American species of *Abutilon*, Miller. Leafl. W. Bot. 7:241–254.

Kearns, D.M. 1994. The genus *Ibervillea* (Cucurbitaceae): An enumeration of the species and two new combinations. Madroño 41:13–22.

Keeler, R.F. 1979. Toxins and Teratogens of the Solanaceae and Liliaceae. In: A.D. Kinghorn, ed. Toxic plants. Pp. 59–82. Columbia Univ. Press, New York.

Keeley, S.C. and **R.K. Jansen.** 1991. Evidence from chloroplast DNA for the recognition of a new tribe, the Tarchonantheae, and the tribal placement of *Pluchea* (Asteraceae). Syst. Bot. 16:173–181.

Keeley, S.C. and **S.B. Jones, Jr.** 1979. Distribution of pollen types in *Vernonia* (Vernonieae: Compositae). Syst. Bot. 4:195–202.

Keener, C.S. 1975a. Studies in the Ranunculaceae of the southeastern United States. I. *Anemone* L. Castanea 40:36–44.

_____. 1975b. Studies in the Ranunculaceae of the southeastern United States. III. *Clematis* L. Sida 6:33–47.

_____. 1976a. Studies in the Ranunculaceae of the southeastern United States. IV. Genera with zygomorphic flowers. Castanea 41:12–20.

_____. 1976b. Studies in the Ranunculaceae of the southeastern United States. V. *Ranunculus* L. Sida 6:266–283.

_____. 1977. Studies in the Ranunculaceae of the southeastern United States. VI. Miscellaneous genera. Sida 7:1–12.

_____ and **W.M. Dennis.** 1982. The subgeneric classification of *Clematis* (Ranunculaceae) in temperate North America north of Mexico. Taxon 31:37–44.

Keener, C.S. and **B.E. Dutton.** 1994. A new species of *Anemone* (Ranunculaceae) from central Texas. Sida 16:191–202.

Keener, C.S. and **S.B. Hoot.** 1987. *Ranunculus* section *Echinella* (Ranunculaceae) in the southeastern United States. Sida 12:57–68.

Keeney, T.M. and **B.L. Lipscomb.** 1985. Notes on two Texas plants. Sida 11:102–103.

Keil, D.J. 1977. A revision of *Pectis* section *Pectothrix* (Compositae: Tageteae). Rhodora 79:32–78.

Kellogg, E.A., R. Appels, and **R.J. Mason-Gamer.** 1996. When genes tell different stories: The diploid genera of Triticeae (Gramineae). Syst. Bot. 21:321–347.

Kelly, J.P. 1915. Cultivated varieties of *Phlox drummondii*. J. New York Bot. Gard. 16:179–191.

Kendall, G.W. 1845. Narrative of an expedition across the great southwestern prairies from Texas to Santa Fe. 2 vols. David Bogue, London, England, U.K. Readex Microprint Corporation, 1966. [Originally published in New York in 1844 as Narrative of the Texan Santa Fe Expedition].

Kennedy, K. 1992. A systematic study of the genus *Phyla* Lour. (Verbenaceae: Verbenoideae, Lantanae). Ph.D. dissertation, Univ. of Texas, Austin.

Kennedy, W. 1841. The rise, progress, and prospects of the Republic of Texas. London. Reprinted in one volume by Molyneaux Craftsmen, Inc., Fort Worth, Texas, 1925.

Kennemer, G.W. 1987. A quantitative analysis of the vegetation on the Dallas County White Rock Escarpment. Sida, Bot. Misc. 1:1–10.

Kent, D.H. 1967. Index to botanical monographs: A guide to monographs and taxonomic papers relating to phanerogams and vascular cryptogams found growing wild in the British Isles. Published for the Botanical Society of the British Isles by Academic Press, London and New York.

_____. 1997. The correct authority for lesser chickweed, *Stellaria pallida* (Caryophyllaceae). Watsonia 21:364.

Kessler, E. 1987. *Carthamus lanatus* L. (Asteraceae: Cynareae)–a potentially serious plant pest in Oklahoma. Proc. Oklahoma Acad. Sci. 67:39–43.

Kessler, J.W. 1987. A treatment of *Scleria* (Cyperaceae) for North America north of Mexico. Sida 12:391–407.

Kessler, P.J.A. 1993a. Annonaceae. In: K. Kubitzki, J.C. Rohwer, and V. Bittrich, eds. The families and genera of vascular plants, Vol. II. Pp. 93–129. Springer-Verlag, Berlin.

_____. 1993b. Menispermaceae. In: K. Kubitzki, J.C. Rohwer, and V. Bittrich, eds. The families and genera of vascular plants, Vol. II. Pp. 402–418. Springer-Verlag, Berlin.

Keung, W.-M. and **B.L. Vallee.** 1993. Daidzin and daidzein suppress free-choice ethanol intake by Syrian golden hamsters. Proc. Natl. Acad. Sci. U.S.A. 90:10008–10012

Kiger, R.W. 1973. Sectional nomenclature in *Papaver* L. Taxon 22:579–582.

_____. 1975. *Papaver* in North America north of Mexico. Rhodora 77:410–422.

_____. 1985. Revised sectional nomenclature in *Papaver* L. Taxon 34:150–152.

_____. 1997a. Papaveraceae. In: Flora of North America Editorial Committee, eds. Fl. North Amer. 3:300–302. Oxford Univ. Press, New York and Oxford.

_____. 1997b. *Glaucium*. In: Flora of North America Editorial Committee, eds. Fl. North Amer. 3:302–304. Oxford Univ. Press, New York and Oxford.

_____ and **D.F. Murray.** 1997. *Papaver*. In: Flora of North America Editorial Committee, eds. Fl. North Amer. 3:323–333. Oxford Univ. Press, New York and Oxford.

Killip, E.P. 1938. The American species of Passifloraceae. Publ. Field Mus. Nat. Hist., Bot. Ser. 19:1–613.

Kim, K.-J., R.K. Jansen, and **B.L. Turner.** 1992a. Evolutionary implications of intraspecific chloroplast DNA variation in dwarf dandelions (*Krigia*; Asteraceae). Amer. J. Bot. 79:708–715.

Kim, K.-J. and **T.J. Mabry.** 1991. Phylogenetic and evolutionary implications of nuclear ribosomal DNA variation in dwarf dandelions (*Krigia*–Lactuceae–Asteraceae). Pl. Syst. Evol. 177:53–69.

Kim, K.-J. and **B.L. Turner.** 1992. Systematic overview of *Krigia* (Asteraceae-Lactuceae). Brittonia 44:173–198.

______, ______, and **R.K. Jansen.** 1992b. Phylogenetic and evolutionary implications of interspecific chloroplast DNA variation in *Krigia* (Asteraceae-Lactuceae). Syst. Bot. 17:449–469.

Kim, Y.-D. and **R.K. Jansen.** 1996 [1997]. Phylogenetic implications of *rbcL* and ITS sequence variation in the Berberidaceae. Syst. Bot. 21:381–396.

Kindscher, K. 1987. Edible wild plants of the prairie. Univ. Press of Kansas, Lawrence.

______. 1992. Medicinal wild plants of the prairie. Univ. Press of Kansas, Lawrence.

King, B.L. and **S.B. Jones.** 1975. The *Vernonia lindheimeri* complex (Compositae). Brittonia 27:74–86.

King, R.M. and **H. Robinson.** 1970a. Studies in the Eupatorieae (Compositae). XVIII. New combinations in *Fleischmannia*. Phytologia 19:201–207.

______ and ______. 1970b. Studies in the Eupatorieae (Compositae). XIX. New combinations in *Ageratina*. Phytologia 19:208–229.

______ and ______. 1970c. Studies in the Eupatorieae (Compositae). XIII. The genus *Conoclinum*. Phytologia 19:299–300.

______ and ______. 1970d. *Eupatorium*, a composite of arcto-tertiary distribution. Taxon 19:769–774.

______ and ______. 1987. The genera of the Eupatorieae (Asteraceae). Monogr. Syst. Bot. Missouri Bot. Gard. 22:1–581.

Kinghorn, A.D. 1979. Cocarcinogenic irritant Euphorbiaceae. In: A.D. Kinghorn, ed. Toxic plants. Pp. 137–159. Columbia Univ. Press, New York.

Kingsbury, J.M. 1964. Poisonous plants of the United States and Canada. Prentice-Hall, Englewood Cliffs, NJ.

______. 1965. Deadly harvest: A guide to common poisonous plants. Holt, Rinehart and Winston, New York.

Kiphart, T. 1997. Benny Simpson's plant legacy. Sida 17:853–854. Originally published 1997 in Native Pl. Soc. Texas News 15(2):9.

Kirkbride, J.H., Jr. 1993. Biosystematic monograph of the genus *Cucumis* (Cucurbitaceae). Parkway Publishers, Boone, NC.

Kirkpatrick, Z.M. 1992. Wildflowers of the western plains. Univ. of Texas Press, Austin.

______ and **J.K. Williams.** 1998. *Glaucium corniculatum* (Papaveraceae) in Texas. Sida 18: 348–349.

Knobloch, I.W. and **D.M. Britton.** 1963. The chromosome number and possible ancestry of *Pellaea wrightiana*. Amer. J. Bot. 50:52–55.

Kobuski, C.E. 1928. A monograph of the American species of the genus *Dyschoriste*. Ann. Missouri Bot. Gard. 15:9–90.

Koch, S.D. 1974. The *Eragrostis pectinacea–pilosa* complex in North and Central America. Illinois Biol. Monogr. 48:1–75.

______. 1978. Notes on the genus *Eragrostis* (Gramineae) in the southeastern United States. Rhodora 80:390–403.

Kohel, R.J. and **C.F. Lewis.** 1984. Cotton. No. 24 in the series Agronomy. American Society of Agronomy, Inc., Crop Science Society of American, Inc., Soil Science Society of America, Inc., Publishers, Madison, WI.

Kolstad, O.A. 1986. *Fimbristylis*. In: Great Plains Flora Association. Flora of the Great Plains. Pp. 1104–1105. Univ. Press of Kansas, Lawrence.

______. 1986. *Scirpus*. In: Great Plains Flora Assoc. Flora of the Great Plains. Pp. 1107–1112. Univ. Press of Kansas, Lawrence.

Kosnik, M.A., G.M. Diggs, Jr., P.A. Redshaw, and **B.L. Lipscomb.** 1996. Natural hybridization among three sympatric *Baptisia* (Fabaceae) species in North Central Texas. Sida 17:479–500.

Kostermans, A.J.G.H. 1957. The genus *Firmiana* Marsili (Sterculiaceae). Reinwardtia 4:281–310.

______. 1989. Notes on *Firmiana* Marsili (Sterculiaceae). Blumea 34:117–118.

Koyama, T. 1962. The genus *Scirpus* L.: Some North American aphylloid species. Canad. J. Bot. 40:913–937.

______. 1963. The genus *Scirpus* L.: Critical species of the section *Pterolepis*. Canad. J. Bot. 41:1107–1131.

Kral, R. (Robert). Personal communication. Specialist in Cyperaceae, Eriocaulaceae, Melastomataceae, Pinaceae, and Xyridaceae; developed Vanderbilt Univ. herbarium; retired from career at Vanderbilt, currently Research Associate at Botanical Research Institute of Texas, Fort Worth.

Kral, R. 1955. Populations of *Linaria* (Scrophulariaceae) in northeastern Texas. Field & Lab. 23:74–77.

______. 1960a. A revision of *Asimina* and *Deeringothamnus* (Annonaceae). Brittonia 12:233–278.

______. 1960b. The genus *Xyris* in Florida. Rhodora 62:295–319.

______. 1966a. *Xyris* (Xyridaceae) of the continental United States and Canada. Sida 2:177–260.

______. 1966b. Eriocaulaceae of continental North America north of Mexico. Sida 2:285–332.

______. 1971. A treatment of *Abildgaardia*, *Bulbostylis* and *Fimbristylis* (Cyperaceae) for North America. Sida 4:57–227.

______. 1976. A treatment of *Delphinium* for Alabama and Tennessee. Sida 6:243–265.

______. 1978. A synopsis of *Fuirena* (Cyperaceae) for the Americas north of South America. Sida 7:309–354.

______. 1983. The genera of Xyridaceae in the southeastern United States. J. Arnold Arbor. 64:421–429.

______. 1989. The genera of Eriocaulaceae in the southeastern United States. J. Arnold Arbor. 70:131–142.

______. 1993. *Pinus*. In: Flora of North America Editorial Committee, eds. Fl. North Amer. 2:373–398. Oxford Univ. Press, New York and Oxford.

______. 1997. Annonaceae. In: Flora of North America Editorial Committee, eds. Fl. North Amer. 3:11–20. Oxford Univ. Press, New York and Oxford.

______ and **P.E. Bostick.** 1969. The genus *Rhexia* (Melastomataceae). Sida 3:387–440.

Krapovickas, A. and **W.C. Gregory.** 1994. Taxonomía del género *Arachis* (Leguminosae). Bonplandia 8:1–186.

Krochmal, A. 1952. Seeds of weedy *Euphorbia* species and their identification. Weeds 1:243–255.

______, **L. Wilken,** and **M. Chien.** 1972. Lobeline content of four Appalachian lobelias. Lloydia 35:303–304

Kron, K.A. 1996. Phylogenetic relationships of Empetraceae,

Epacridaceae, Ericaceae, Monotropaceae, and Pyrolaceae: Evidence from nuclear ribosomal 18s sequence data. Ann. Bot. 77:293–303.

_____. 1997. Exploring alternative systems of classification. Aliso 15:105–112.

_____ and **M.W. Chase.** 1993. Systematics of the Ericaceae, Empetraceae, Epacridaceae and related taxa based upon *rbc*L sequence data. Ann. Missouri Bot. Gard. 80:735–741.

Kron, K.A. and **W.S. Judd.** 1997 [1998]. Systematics of the *Lyonia* group (Andromedeae, Ericaceae) and the use of species as terminals in higher-level cladistic analyses. Syst. Bot. 22:479–492.

Kruckeberg, A.R. and **R.D. Reeves.** 1995. Nickel accumulation by serpentine species of *Strepanthus* (Brassicaceae): Field and greenhouse studies. Madroño 42:458–469.

Krukoff, B.A. and **R.C. Barneby.** 1974. Conspectus of species of the genus *Erythrina*. Lloydia 37:332–459.

Kruschke, E.P. 1965. Contributions to the taxonomy of *Crataegus*. Milwaukee Public Mus. Publ. Bot. 3:1–273.

Kubitzki, K. 1993a. Betulaceae. In: K. Kubitzki, J.C. Rohwer, and V. Bittrich, eds. The families and genera of vascular plants, Vol. II. Pp. 152–157. Springer-Verlag, Berlin.

_____. 1993b. Cannabaceae. In: K. Kubitzki, J.C. Rohwer, and V. Bittrich, eds. The families and genera of vascular plants, Vol. II. Pp. 204–206. Springer-Verlag, Berlin.

_____. 1993c. Fagaceae. In: K. Kubitzki, J.C. Rohwer, and V. Bittrich, eds. The families and genera of vascular plants, Vol. II. Pp. 301–309. Springer-Verlag, Berlin.

_____. 1993d. Myricaceae. In: K. Kubitzki, J.C. Rohwer, and V. Bittrich, eds. The families and genera of vascular plants, Vol. II. Pp. 453–457. Springer-Verlag, Berlin.

_____. 1993e. Platanaceae. In: K. Kubitzki, J.C. Rohwer, and V. Bittrich, eds. The families and genera of vascular plants, Vol. II. Pp. 521–522. Springer-Verlag, Berlin.

_____, **J.G. Rohwer,** and **V. Bittrich,** eds. 1993. The families and genera of vascular plants II. Flowering plants - dicotyledons. Magnoliid, hamamelid and caryophyllid families. Springer-Verlag, Berlin.

Kuby, J. 1997. Immunology, 3rd ed. W.H. Freeman and Co., New York.

Kuchler, A.W. 1974. A new vegetation map of Kansas. Ecology 55:586–604.

Kühn, U. 1993. Chenopodiaceae. In: K. Kubitzki, J.C. Rohwer, and V. Bittrich, eds. The families and genera of vascular plants, Vol. II. Pp. 253–281. Springer-Verlag, Berlin.

Kuijt, J. 1982. The Viscaceae of the southeastern United States. J. Arnold Arbor. 63:401–410.

Kupicha, F.K. 1976. The infrageneric structure of *Vicia*. Notes Roy. Bot. Gard. Edinburgh 34:287–326.

_____. 1981. Vicieae (Adans.) DC. In: R.M. Polhill and P.H. Raven, eds. Advances in legume systematics 1:377–381.

_____. 1983. The infrageneric structure of *Lathyrus*. Notes Roy. Bot. Gard. Edinburgh 41:209–244.

Lacey, J.B. 1969. Key to violets (*Viola*, Violaceae) of Nacogdoches County, Texas. Sida 3:311–312.

La Duke, J.C. and **J. Doebley.** 1995. A chloroplast DNA based phylogeny of the Malvaceae. Syst. Bot. 20:259–271.

La Duke, J.C. and **D.K. Northington.** 1978. The systematics of *Sphaeralcea coccinea* (Nutt.) Rydb. (Malvaceae). SouthW. Naturalist 23:651–660.

Lammers, T.G. 1992. Circumscription and phylogeny of Campanulales. Ann. Missouri Bot. Gard. 79:388–413.

_____. 1993. Chromosome numbers of Campanulaceae. III. Review and integration of data for subfamily Lobelioideae. Amer. J. Bot. 80:660–675.

LaMotte, C. 1940. *Pilularia* in Texas. Amer. Fern J. 30:99–101.

Lampe, K.E. 1986. Contact dermatitis from Sonoran Desert plants. Desert Pl. 8:32–37.

_____ and **M.A. McCann.** 1985. AMA handbook of poisonous and injurious plants. American Medical Association, Chicago.

Landolt, E. 1975. Morphological differentiation and geographical distribution of the *Lemma gibba–Lemna minor* group. Aquatic Bot. 1:345–363.

_____. 1980. Biosystematic investigation of the family of duckweeds (Lemnaceae) (Vol. 1). Veröff. Geobot. Inst. ETH Stiftung Rübel, Zürich 70:1–247.

_____. 1986. Biosystematic investigations in the family of duckweeds (Lemnaceae) (Vol. 2). Veröff. Geobot. Inst. ETH Stiftung Rübel, Zürich 71:1–566.

Lane, M.A. 1979. Taxonomy of the genus *Amphiachyris*. Syst. Bot. 4:178–189.

_____. 1980. New and re-instated combinations in *Gutierrezia* (Compositae: Astereae). Sida 8:313–314.

_____. 1982. Generic limits of *Xanthocephalum*, *Gutierrezia*, *Amphiachyris*, *Gymnosperma*, *Greenella* and *Thurovia*. Syst. Bot. 7:405–416.

_____. 1985. Taxonomy of *Gutierrezia* (Compositae: Astereae) in North America. Syst. Bot. 10:7–28.

_____ and **R.L. Hartman.** 1996. Reclassification of North American *Haplopappus* (Compositae: Astereae) completed: *Rayjacksonia* gen. nov. Amer. J. Bot. 83:356–370.

Lane, T.M. 1983. Mericarp micromorphology of Great Plains *Scutellaria* (Labiatae). SouthW. Naturalist 28:71–79.

_____. 1986. *Scutellaria*. In: Great Plains Flora Association. Flora of the Great Plains. Pp. 733-737. Univ. Press of Kansas, Lawrence.

Langeland, K.A. 1996. *Hydrilla verticillata* (L.F.) Royle (Hydrocharitaceae), "The perfect aquatic weed." Castanea 61:293–304.

Larisey, M. 1940. A monograph of the genus *Baptisia*. Ann. Missouri Bot. Gard. 27:119–244.

Larsen, E.L. 1927. A revision of the genus *Townsendia*. Ann. Missouri Bot. Gard. 14:1–46.

_____. 1933. *Astranthium* and related genera. Ann. Missouri Bot. Gard. 20:23–44.

_____. 1940. *Lobelia* as a sure cure for venereal disease. Amer. J. of Syphilis, Gonorrhea, and Venereal Diseases 24:13–22.

Larsen, G.E. 1986. Caryophyllaceae. In: Great Plains Flora Association. Flora of the Great Plains. Pp.192–214. Univ. Press of Kansas, Lawrence.

_____ and **W.T. Barker.** 1986. Potamogetonaceae. In: Great Plains Flora Association. Flora of the Great Plains. Pp. 1032–1039. Univ. Press of Kansas, Lawrence.

Larson, S.R. and **J. Doebley.** 1994. Restriction site variation in the chloroplast genome of *Tripsacum* (Poaceae): Phylogeny and rates of sequence evolution. Syst. Bot. 19:21–34.

Lassen, P. 1989. A new delimitation of the genera *Coronilla*, *Hippocrepis*, and *Securigera* (Fabaceae). Willdenowia 19:49–62.

Lassetter, J.S. 1978. Seed characters in some native American vetches. Sida 7:255–263.

_____. 1984. Taxonomy of the *Vicia ludoviciana* complex (Leguminosae). Rhodora 86:475–505.

Laushman, R.H., A. Schnabel, and **J.L. Hamrick.** 1996. Electrophoretic evidence for tetrasomic inheritance in the dioecious tree *Maclura pomifera* (Raf.) Schneid. J. Heredity 87:469–473.

Lawrence, G.H.M. 1951. Taxonomy of vascular plants. MacMillan Publishing Co., New York.

_____, **A.F.G. Buchheim, G.S. Daniels,** and **H. Dolezal,** eds. Botanico-Periodicum-Huntianum. Hunt Botanical Library, Pittsburgh, PA.

Laws, W.D. 1962. The soils and vegetation of a relict area of Blackland Prairie. Part 1: An investigation of the soils of the Stults Meadow, a relict area of Blackland Prairie. Wrightia 2:229–241.

Le Duc, A. 1995. A revision of *Mirabilis* section *Mirabilis* (Nyctaginaceae). Sida 16:613–648.

Lee, D.W. 1975. Population variation and introgression in North American *Typha*. Taxon 24:633–641.

_____ and **D.E. Fairbrothers.** 1969. A serological and disk electrophoretic study of North American *Typha*. Brittonia 21:227–243.

Lee, Y.S. 1981. Serological investigations in *Ambrosia* (Compositae: Ambrosieae) and relatives. Syst. Bot. 113–125.

_____ and **D.B. Dickinson.** 1980. Field observations on hybrids between *Ambrosia bientata* and *A. trifida* (Compositae). Amer. Midl. Naturalist 103:180–184.

Lee, Y-T. 1976. The genus *Gymnocladus* and its tropical affinities. J. Arnold Arbor. 57:91–112.

Legrand, C.D. 1962. Las especies Americanas de *Portulaca*. Anales Mus. Hist. Nat. Montevideo 7:1–147.

Lellinger, D.B. 1985. A field manual of the ferns & fern allies of the United States & Canada. Smithsonian Institution Press, Washington, D.C.

Lelong, M.G. 1984. New combinations for *Panicum* subgenus *Panicum* and subgenus *Dichanthelium* (Poaceae) of the United States. Brittonia 36:262–273.

_____. 1986. A taxonomic treatment of the genus *Panicum* (Poaceae) in Mississippi. Phytologia 61:251–269.

Lemke, D.E. 1991. The genus *Solanum* (Solanaceae) in Texas. Phytologia 71:362–378.

_____ and **E.L. Schneider.** 1988. *Xanthosoma sagittifolium* (Araceae): New to Texas. SouthW. Naturalist 33:498–499.

Lemke, D.E. and **R.D. Worthington.** 1991. *Brassica* and *Rapistrum* (Brassicaceae) in Texas. SouthW. Naturalist 36:194–197.

Leonard, E.C. 1927. The North American species of *Scutellaria*. Contr. U.S. Natl. Herb. 22:703–748.

Les, D.H. 1986a. The phytogeography of *Ceratophyllum demersum* and *C. echinatum* in glaciated North America. Canad. J. Bot. 64:498–509.

_____. 1986b. The evolution of achene morphology in *Ceratophyllum* (Ceratophyllaceae), I. Fruit spine variation and relationships of *C. demersum*, *C. submersum*, and *C. apiculatum*. Syst. Bot. 11:549–558.

_____. 1988a. The origin and affinities of the Ceratophyllaceae. Taxon 37:326–345.

_____. 1988b. The evolution of achene morphology in *Ceratophyllum* (Ceratophyllaceae), II. Fruit variation and systematics of the "spiny-margined" group. Syst. Bot. 13:73–86.

_____. 1988c. The evolution of achene morphology in *Ceratophyllum* (Ceratophyllaceae), III. Relationships of the "facially-spined" group. Syst. Bot. 13:509–518.

_____. 1989. The evolution of achene morphology in *Ceratophyllum* (Ceratophyllaceae), IV. Summary of proposed relationships and evolutionary trends. Syst. Bot. 14:254–262.

_____. 1993. Ceratophyllaceae. In: K. Kubitzki, J.C. Rohwer, and V. Bittrich, eds. The families and genera of vascular plants, Vol. II. Pp. 246–250. Springer-Verlag, Berlin.

_____. 1997. Ceratophyllaceae. In: Flora of North America Editorial Committee, eds. Fl. North Amer. 3:81–88. Oxford Univ. Press, New York and Oxford.

Les, D.H. and **R.R. Haynes.** 1996. *Coleogeton* (Potamogetonaceae), a new genus of pondweeds. Novon 6:389–391.

Lesins, K.A. and **I. Lesins.** 1979. Genus *Medicago* (Leguminosae). W. Junk, The Hague, Netherlands.

Leuenberger, B.E. 1991. Interpretation and typification of *Cactus ficus-indica* L., and *Opuntia ficus-indica* (L.) Miller (Cactaceae). Taxon 40:621–627.

_____. 1993. Interpretation and typification of *Cactus opuntia* L., *Opuntia vulgaris* Mill., and *O. humifusa* (Rafin.) Rafin. (Cactaceae). Taxon 42:419–429.

Leung, A.Y. and **S. Foster.** 1996. Encyclopedia of common natural ingredients used in food, drugs, and cosmetics, 2nd ed. John Wiley and Sons, Inc. New York.

Levin, G.A. (Geoffrey). Personal communication. Specialist in *Acalypha* (Euphorbiaceae). Director of the Center for Biodiversity, Illinois Natural History Survey, Champaign, IL.

Lewis, H. 1945. A revision of the genus *Trichostema*. Brittonia 5:276–303.

_____ and **R. Snow.** 1951. A cytotaxonomic approach to *Eschscholzia*. Madroño 11:141–143.

Lewis, I.M. 1915. The trees of Texas. Univ. Texas Bull. 22:1–169.

Lewis, W.H. 1958 [1959]. Chromosomes of the east Texas *Hedyotis* (Rubiaceae). SouthW. Naturalist 3:204–207.

_____. 1971. Pollen differences between *Stylisma* and *Bonamia* (Convolvulaceae). Brittonia 23:331–334.

_____. 1976. Temporal adaptation correlated with ploidy in *Claytonia virginica*. Syst. Bot. 1:340–347.

_____ and **M.P.F. Elvin-Lewis.** 1977. Medical botany: Plants affecting man's health. John Wiley & Sons, New York.

Lewis, W.H. and **P.R. Fantz.** 1973. Tribal classification of *Triosteum* (Caprifoliaceae). Rhodora 75:120–121.

Lewis, W.H. and **R.L. Oliver.** 1965. Realignment of *Calystegia* and *Convolvulus* (Convolvulaceae). Ann. Missouri Bot. Gard. 52:217–222.

_____ and _____. 1974. Revision of *Richardia* (Rubiaceae). Brittonia 26:271–301.

_____, _____, and **Y. Suda.** 1967. Cytogeography of *Claytonia virginica* and its allies. Ann. Missouri Bot. Gard. 54:153–171.

Lewis, W.H. and **J.C. Semple.** 1977. Geography of *Claytonia virginica* cytotypes. Amer. J. Bot. 64:1078–1082.

Lewis, W.H., Y. Suda, and **B. MacBryde.** 1967. Chromosome numbers of *Claytonia virginica* in the St. Louis, Missouri area. Ann. Missouri Bot. Gard. 54:147–152.

Lewis, W.H., P. Vinay, and **V.E. Zenger.** 1983. Airborne and allergenic pollen of North America. Johns Hopkins Univ. Press, Baltimore, MD.

Li, H.-L. 1952a. The genus *Tovara* (Polygonaceae). Rhodora 54:19–25.

_____. 1952b. Floristic relationships between eastern Asia and eastern North America. Trans. Amer. Philos. Soc. 42:371–409.

Li, S. and **K.T. Adair.** 1994. Species pools in eastern Asia and North America. Sida 16:281–299.

_____ and _____. 1997. Species pools of seed plants in eastern Asia and North America. Arthur Temple College of Forestry, Stephen F. Austin State Univ., Nacogdoches, TX.

Lidén, M. 1986. Synopsis of Fumarioideae (Papaveraceae) with a monograph of the tribe Fumarieae. Opera Bot. 88:1–133.

_____. 1993. Fumariaceae. In: K. Kubitzki, J.C. Rohwer, and V. Bittrich, eds. The families and genera of vascular plants, Vol. II. Pp. 310–318. Springer-Verlag, Berlin.

Liede, S. 1996. *Sarcostemma* (Asclepiadaceae)—a controversial generic circumscription reconsidered: Morphological evidence. Syst. Bot. 21:31–44.

_____. 1997. Subtribes and genera of the tribe Asclepiadaceae (Apocynaceae, Asclepiadoideae) – a synopsis. Taxon 46:233–247.

_____ and **F. Albers.** 1994. Tribal disposition of genera in the Asclepiadaceae. Taxon 43:201–231.

Lim, D. 1998. Microbiology, 2nd. ed. WCB/McGraw-Hill, Boston, MA.

Lincecum, G. 1861. The grasses of Texas. Southern Cultivator 19:33–34, 51–52.

_____. 1862. Agricultural ant of Texas. J. Linn. Soc., Zoo. 6:29–31.

Lincecum, J.B. and E.H. Phillips. 1994. Adventures of a frontier naturalist: The life and times of Dr. Gideon Lincecum. Texas A&M Univ. Press, College Station.

_____, _____, and **P.A. Redshaw.** 1997. Science on the Texas frontier: Observations of Dr. Gideon Lincecum. Texas A&M Univ. Press, College Station.

Lindell, C.M. 1997. Field sampling and chemical analysis. J. Forensic Sci. 42:398–400.

Lidén, M., B. Oxelman, A. Backlund, L. Andersson, B. Bremer, R. Eriksson, R. Moberg, I. Nordal, K. Persson, M. Thulin, and **B. Zimmer.** 1997. Charlie is our darling. Taxon 46:735–738.

Linsley, E.G. and **J.A. Chemsak.** 1984. The Cerambycidae of North America, Part VII, No. 1: Taxonomy and classification of the subfamily Lamiinae, tribes Parmenini through Acanthoderini. Univ. Calif. Publ. Entomology, Univ. of California Press, Berkeley.

Linnaeus, C. 1753. Species Plantarum.... 2 vols. Stockholm.

_____. 1754. Genera Plantarum, ed. 5. Stockholm.

Lipscomb, B.L. 1978a. *Cyperus surinamensis* (Cyperaceae): New to Arkansas, Kansas, and Oklahoma. Sida 7:307.

_____. 1978b. Additions to the Texas flora. Sida 7:392–393.

_____. 1980. *Cyperus difformis* L. (Cyperaceae) in North America. Sida 8:320–327.

_____. 1984. New additions or other noteworthy plants of Texas. Sida 10:326–327.

_____. 1992. Visitors/Clientele [Information on *Maclura pomifera* (bois d'arc) from Fred Tarpley]. Iridos 3(4):1.

_____. 1993. *Pseudotsuga*. In: Flora of North America Editorial Committee, eds. Fl. North Amer. 2:365–366. Oxford Univ. Press, New York and Oxford.

_____ and **G. Ajilvsgi.** 1982. *Bellardia trixago* (L.) All. (Scrophulariaceae) adventive in Texas. Sida 9:370–374.

Lipscomb, B.L. and **G.M. Diggs, Jr.** 1998. The use of animal-dispersed seeds and fruits in forensic botany. Sida 18:335–346.

Lipscomb, B.L. and **E.B. Smith.** 1977. Morphological intergradation of varieties of *Bidens aristosa* (Compositae) in northern Arkansas. Rhodora 79:203–213.

Lira, R., J.L. Villaseñor, and **P.D. Davila.** 1997 [1998]. A cladistic analysis of the subtribe Sicyinae (Cucurbitaceae). Syst. Bot. 22:415–425.

Little, E.L. 1970. Endemic, disjunct and northern trees in the southern Appalachians. In: P.C. Holt and R.A. Paterson, eds. The distributional history of the biota of the southern Appalachians, Part II: Flora. Pp. 249–290. Research Division Monog. 2, Virginia Polytechnic Institute and State Univ., Blacksburg, VA.

_____. 1971. Atlas of United States trees, Vol. 1, Conifers and important hardwoods. U.S.D.A. Forest Serv. Misc. Publ. No. 1146.

_____. 1976. Atlas of United States trees, Vol. 3, Minor western hardwoods. U.S.D.A. Forest Serv. Misc. Publ. No. 1314.

_____. 1976 [1977]. Atlas of United States trees, Vol. 4, Minor eastern hardwoods. U.S.D.A. Forest Serv. Misc. Publ. No. 1342.

_____. 1983. North American trees with relationships in eastern Asia. Ann. Missouri Bot. Gard. 70:605–615.

Lockett, L. 1991. Native Texas palms north of the lower Rio Grande Valley. Principes 35:64–71.

Loconte, H. 1993. Berberidaceae. In: K. Kubitzki, J.C. Rohwer, and V. Bittrich, eds. The families and genera of vascular plants, Vol. II. Pp. 147–152. Springer-Verlag, Berlin.

_____ and **J.R. Estes.** 1989. Phylogenetic systematics of Berberidaceae and Ranunculales (Magnoliidae). Syst. Bot. 14:565–579.

Lonard, R.I. and **F.W. Gould.** 1974. The North American species of *Vulpia* (Gramineae). Madroño 22:217–280.

Long, R.W. 1961. Convergent patterns in *Ruellia caroliniensis* and *R. humilis* (Acanthaceae). Bull. Torrey Bot. Club 88:387–396.

_____. 1966. Artificial interspecific hybridization in *Ruellia* (Acanthaceae). Amer. J. Bot. 53:917–927.

_____. 1970. The genera of Acanthaceae in the southeastern United States. J. Arnold Arbor. 51:257–309.

_____. 1971. Floral polymorphy and amphimictic breeding systems in *Ruellia caroliniensis* (Acanthaceae). Amer. J. Bot. 58:525–531.

_____. 1973. A biosystematic approach to generic delimitation in *Ruellia* (Acanthaceae). Taxon 22:543–555.

_____. 1974. Variation in natural populations of *Ruellia caroliniensis* (Acanthaceae). Bull. Torrey Bot. Club 101:1–6.

_____. 1975. Artificial interspecific hybridization in temperate and tropical species of *Ruellia* (Acanthaceae). Brittonia 27:289–296.

_____ and **O. Lakela.** 1971. A flora of tropical Florida. Univ. of Miami Press, Coral Gables, FL.

Long, R.W. and **L.J. Uttal.** 1962. Some observations on flowering in *Ruellia* (Acanthaceae). Rhodora 64:200–206.

Longley, G. 1996. An introduction to the Edwards Aquifer. Texas Organization for Endangered Species News and Notes, Spring–Fall 1996 issue, Pp. 15–17.

Lourteig, A. 1979. *Oxalidaceae* extra-austroamericanae. II. *Oxalis* L. section *Corniculatae* DC. Phytologia 42:57–198.

Löve, Á. 1961. Some notes on *Myriophyllum spicatum*. Rhodora 63:139–145.

_____. 1980. IOPB chromosome number reports LXVI, reports by Áskell Löve. Taxon 29:166–173.

_____ and **D. Löve.** 1975 [1976]. Nomenclatural notes on Arctic plants. Bot. Not. 128:497–523.

Löve, D. and **P. Dansereau.** 1959. Biosystematic studies on *Xanthium*: Taxonomic appraisal and ecological status. Canad. J. Bot. 37:173–208.

Love, G. 1994. The A–Z of cut flowers. Penguin Books USA, New York.

Lowden, R.M. 1973. Revision of the genus *Pontederia* L. Rhodora 75:426–487.

_____. 1986. Taxonomy of the genus *Najas* L. (Najadaceae) in the Neotropics. Aquatic Bot. 24:147–184.

_____. 1978. Studies on the submerged genus *Ceratophyllum* L. in the Neotropics. Aquatic Bot. 4:127-142.

Lowry, P.P., II and **A.G. Jones.** 1984. Systematics of *Osmorhiza* Raf. (Apiaceae: Apioideae). Ann. Missouri Bot. Gard. 71:1128–1171.

Luckow, M. 1993. Monograph of *Desmanthus* (Leguminosae-Mimosoideae). Syst. Bot. Monogr. 38:1–166.

Luckwill, L.C. 1943. The genus *Lycopersicon*: An historical, biological, and taxonomic survey of the wild and cultivated tomatoes. Aberdeen Univ. Stud. 20:1–44.

Luer, C.A. 1975. The native orchids of the United States and Canada. New York Bot. Gard., Bronx.

Luken, J.O. and **J.W. Thieret.** 1995. Amur honeysuckle (*Lonicera maackii*; Caprifoliaceae): Its ascent, decline, and fall. Sida 16:479–503.

_____ and _____. 1996. Amur honeysuckle, its fall from grace. BioScience 46:18–24.

_____ and _____. 1997. Assessment and management of plant invasions. Springer-Verlag, New York.

_____, _____, and **J.R. Kartesz.** 1993. *Erucastrum gallicum* (Brassicaceae): Invasion and spread in North America. Sida 15:569–582.

Lumpkin, T.A. 1993. Azollaceae. In: Flora of North America Editorial Committee, eds. Fl. North Amer. 2:338–342. Oxford Univ. Press, New York and Oxford.

Lundell, C.L. 1941. Studies of American spermatophytes–I. Contr. Univ. Michigan Herb. 6:1–66.

_____. 1945. New phanerogams from Texas, Mexico and Central America. Wrightia 1:53–61.

_____. 1959. Studies of *Physostegia*–I New species and observations on others. Wrightia 2:4–12.

_____. 1960. Studies of *Physostegia*–II Further notes on the Texas species. Wrightia 2:66–74.

_____, ed. 1961, 1966, 1969. Flora of Texas (3 vols.). Texas Research Foundation, Renner.

_____. 1961. Aquifoliaceae. In: C.L. Lundell, ed. Flora of Texas 3:112–122. Texas Research Foundation, Renner.

_____. 1968. A new species of *Brazoria* (Labiatae) from east Texas. Wrightia 4:29–30.

_____. 1969a. Labiatae: *Brazoria* and *Physostegia*. In: C.L. Lundell, ed. Flora of Texas 2:319–330. Texas Research Foundation, Renner.

_____. 1969b. Celastraceae. In: C.L. Lundell, ed. Flora of Texas 2:339–355. Texas Research Foundation, Renner.

Lusk, S. (Shirley). Personal communication. Wildflower enthusiast and plant collector of Cook Co., TX; has contributed numerous North Central Texas county records to the Botanical Research Institute of Texas, Fort Worth; relative of Julien Reverchon.

Luteyn, J.L., ed. 1995. Ericaceae, Part II, the superior-ovaried genera. Fl. Neotrop. Monogr. 66.

Lynch, Brother D. 1981. Native and naturalized woody plants of Austin and the Hill Country. Published by the author, Austin, TX.

Lyons, A.B. 1900. Plant names scientific and popular. Nelson, Baker and Co., Detroit, MI.

Mabberley, D.J. 1987. The plant book, a portable dictionary of the higher plants. Reprinted with corrections in 1989. Cambridge Univ. Press, Cambridge, England, U.K.

_____. 1997. The plant book, a portable dictionary of the higher plants, 2nd ed. Cambridge Univ. Press, Cambridge, England, U.K.

MacBride, J.F. 1917. A revision of the North American species of *Amsinckia*. Contr. Gray Herb. 49:1–16.

MacDonald, W.L. 1995. Oak wilt: an historical perspective. In: D.N. Appel and R.F. Billings, eds. Oak wilt perspectives: The proceedings of the national oak wilt symposium. Pp. 7–13. Information Development, Inc., Houston, TX.

MacFarlane, R.B.A. 1994. Collecting and preserving plants. Dover Publications, Mineola, NY.

Macior, L.W. 1964. An experimental study of the floral ecology of *Dodecatheon meadia*. Amer. J. Bot. 51:96–108.

_____. 1965. Insect adaptation and behavior in *Asclepias* pollination. Bull. Torrey Bot. Club 92:114–126.

MacKenzie, K.K. 1906. *Onosmodium*. Bull. Torrey Bot. Club 32:495–506.

_____. 1931–1935. *Carex*. N. Amer. Fl. 18:9–478.

_____. 1940. North American Cariaceae, 2 vols. New York Bot. Garden, Bronx.

MacRoberts, D.T. 1977. Notes on *Tradescantia*: *T. diffusa* Bush and *T. pedicillata* Celarier. Phytologia 38:227–228.

_____. 1980. Notes on *Tradescantia* IV (Commelinaceae): The distinction between *T. virginiana* and *T. hirsutiflora*. Phytologia 46:409–416.

Maddox, E. 1986. *Homalocephala texensis* "Texas horse crippler." Cact. Succ. J. (Los Angeles) 58:218–221.

Magrath, L.K. (Lawrence). Personal communication. Specialist in North American Orchidaceae; member of Oklahoma Flora Editorial Committee; Univ. of Science and Arts of Oklahoma, Chickasha.

Magrath, L.K. 1973. The native orchids of the prairies and plains region of North America. Ph.D. dissertation, Univ. of Kansas, Lawrence.

______. 1989. Nomenclatural notes on *Calopogon, Corallorhiza*, and *Cypripedium* (Orchidaceae) in the Great Plains region. Sida 13:371–372.

Maguire, B. 1947. Studies in the Caryophyllaceae–III. A synopsis of the North American species of *Arenaria*, Sect. *Eremogone* Fenzl. Bull. Torrey Bot. Club 74:38–56.

______. 1950. Studies in the Caryophyllaceae–IV. A synopsis of the subfamily Silenoideae. Rhodora 52:233–245.

______. 1951. Studies in the Caryophyllaceae–V. *Arenaria* in America north of Mexico. A conspectus. Amer. Midl. Naturalist 46:493–511.

Mahler, W.F. 1955. The genus *Baccharis* in the southwestern states of Oklahoma, Texas, and New Mexico. M.S. thesis, Oklahoma State Univ., Stillwater.

______. 1971a. Keys to the vascular plants of the Black Gap Wildlife Management Area, Brewster County, Texas. Published by the author, Dallas, TX [currently available through BRIT].

______. 1971b. Lloyd Herbert Shinners 1918–1971. Sida 4:228–231.

______. 1973a. Flora of Taylor County, Texas: A manual of the vascular plants with selected sketches. Published by the author, Dallas, TX.

______. 1973b. By any other name.... Sida 5:180–181.

______. 1979. *Rubus trivialis* Michx. var. *duplaris* (Shinners) Mahler, comb. nov. (Rosaceae). Sida 8:211–212.

______. 1980. The mosses of Texas: A manual of the moss flora with sketches. Published by the author, Dallas, TX [currently available through BRIT].

______. 1984. Shinners' manual of the North Central Texas flora. Southern Methodist Univ. Herbarium, Dallas, TX.

______. 1988. Shinners' manual of the North Central Texas flora. Sida, Bot. Misc. 3:1–313.

______ and **U.T. Waterfall.** 1964. *Baccharis* (Compositae) in Oklahoma, Texas and New Mexico. SouthW. Naturalist 9:189–202.

Manaster, J. 1994. The pecan tree. Univ. of Texas Press, Austin.

Manhart, J.R. 1990. Chemotaxonomy of the genus *Carex* (Cyperaceae). Canad. J. Bot. 68:1457–1461.

Mann, J. 1992. Murder, magic, and medicine. Oxford Univ. Press, New York.

Manning, S.D. 1991. The genera of Pedaliaceae in the southeastern United States. J. Arnold Arbor., Supp. Ser. 1:313–347.

Manos, P.S., K.C. Nixon, and **J.J. Doyle.** 1993. Cladistic analysis of restriction site variation within the chloroplast DNA inverted repeat region of selected Hamamelididae. Syst. Bot. 18:551–562.

Marcks, B.G. 1972. Population studies in North American *Cyperus* section *Laxiglumi* (Cyperaceae). Ph.D. dissertation, Univ. of Wisconsin-Madison.

______. 1974. Preliminary reports on the flora of Wisconsin No. 66. Cyperaceae II—Sedge family II. The genus *Cyperus*—The umbrella sedges. Trans. Wisconsin Acad. Sci. 62:261–284.

Marcy, R.B. 1853. Exploration of the Red River of Louisiana, in the year 1852. Robert Armstrong, Public Printer, Washington, D.C. The journal portion of this work was edited, annotated, and republished by G. Foreman in 1937 as Adventure on the Red River: Report on the exploration of the headwaters of the Red River by Captain Randolph B. Marcy and Captain G.B. McClellan. Univ. of Oklahoma Press, Norman.

______. 1866. Thirty years of army life on the border. Harper and Bros., New York.

Marriott, A.L. 1943. The Cross Timbers as a cultural barrier. Texas Geogr. Mag. 7:14–20.

Marryat, F. 1843. The travels and romantic adventures of Monsieur Violet, 3 vols. Longman, Brown, Green, & Longmans, London, England, U.K.

Marsh, C.D., A.B. Clawson, and **H. Marsh.** 1915. *Zygadenus*, or death camas. U.S.D.A. Bull. 125:1–44.

______, ______, and ______. 1916. Larkspur poisoning of live stock. U.S.D.A. Bull. 365:1–90.

Marsh, V.L. 1952. A taxonomic revision of the genus *Poa* of the United States and southern Canada. Amer. Midl. Naturalist 47:202–250.

Marshall, W.T. and **T.M. Bock.** 1941. Cactaceae. Abbey Garden Press, Pasadena, CA.

Martin, A.C., H.S. Zim, and **A.L. Nelson.** 1951. American Wildlife & Plants. McGraw-Hill Book Co., New York.

Martin, P.S. and **B.E. Harrell.** 1957. The Pleistocene history of temperate biotas in Mexico and eastern United States. Ecology 38:488–479.

Martin, W. (William). Personal communication. Specialist in viticulture; in charge of the T.V. Munson Memorial Vineyard and Viticulture-Enology Center at Grayson County College.

Martin, W.C. and **C.R. Hutchins.** 1981. A flora of New Mexico. 2 vols. J. Cramer, Germany.

Martindale, I.C. 1876. The introduction of foreign plants. Bot. Gaz. (Crawfordsville) 2:55–58.

Mason, H.L. 1957. Flora of the marshes of California. Univ. of California Press, Berkeley.

Massey, J.R. 1975. *Fatoua villosa* (Moraceae): Additional notes on distribution in the southeastern United States. Sida 6:116.

Mathew, B. 1996. A review of *Allium* section *Allium*. Royal Botanic Gardens, Kew, England, U.K.

Mathias, M.E. 1930. Studies in the Umbelliferae. III. A monograph of *Cymopterus* including a critical study of related genera. Ann. Missouri Bot. Gard. 17:213–476.

______. 1965. Distribution patterns of certain Umbelliferae. Ann. Missouri Bot. Gard. 52:387–398.

______ and **L. Constance.** 1941a. *Limnosciadium*, a new genus of Umbelliferae. Amer. J. Bot. 28:162–163.

______ and ______. 1941b. A synopsis of the North American species of *Eryngium*. Amer. Midl. Naturalist 25:361–387.

______ and ______. 1942. A synopsis of the American species of *Cicuta*. Madroño 6:145–151.

______ and ______. 1944–45. Umbelliferae. N. Amer. Fl. 28b:43–295.

______ and ______. 1961. Umbelliferae. In: C.L. Lundell, ed. Flora of Texas 3:263–329. Texas Research Foundation, Renner.

_____ and _____. 1965. A revision of the genus *Bowlesia* Ruiz & Pav. (Umbelliferae-Hydrocotyloideae) and its relatives. Univ. Calif. Publ. Bot. 38:1–73.

_____ and _____. 1970. Umbelliferae. In: D.S. Correll and M.C. Johnston. Manual of the vascular plants of Texas. Pp. 1139–1169. Texas Research Foundation, Renner.

Matossian, M.K. 1989. Poisons of the past: Molds, epidemics and history. Yale Univ. Press, New Haven, CT.

Mat Salleh, K. 1991. *Rafflesia*, magnificent flower of Sabah. Borneo Publishing Co., Kota Kinabalu, Sabah, Malaysia.

Matthews, J.F. 1986. The systematic significance of seed morphology in *Portulaca* (Portulacaceae) under scanning electron microscopy. Syst. Bot. 11:302–308.

_____ and **D.W. Ketron.** 1991. Two new combinations in *Portulaca* (Portulacaceae). Castanea 56:304–305.

_____, _____, and **S.F. Zane.** 1992a. *Portulaca umbraticola* Kunth (Portulacaceae) in the United States. Castanea 57:202–208.

_____, _____, and _____. 1992b. The reevaluation of *Portulaca pilosa* and *P. mundula* (Portulacaceae). Sida 15:71–89.

_____, _____, and _____. 1993. The biology and taxonomy of the *Portulaca oleracea* L. (Portulacaceae) complex in North America. Rhodora 95:166–183.

_____, _____, and _____. 1994. The seed surface morphology and cytology of six species of *Portulaca*. Castanea 59:331–337.

Matthews, J.F. and **P.A. Levins.** 1985a. The genus *Portulaca* in the southeastern U.S. Castanea 50:96–104.

_____ and _____. 1985b. *Portulaca pilosa* L., *P. mundula* I.M. Johnst. and *P. parvula* Gray in the southwest. Sida 11:45–61.

May, R.A. (Ruth). Personal communication. Wildflower artist and naturalist, Dallas, TX; instrumental in conservation activities at Harry S. Moss Park, Dallas; Board Member of the Botanical Research Institute of Texas, Fort Worth.

Mayfield, M. (Mark). Personal communication. Specialist in Eurphobiaceae. Louisiana State Univ., Baton Rouge.

McArthur, E.D. and **B.L. Welch.** 1984. Proceedings – symposium on the biology of *Artemisia* and *Chrysothamnus*. Intermountain Research Station, Ogden, UT.

McCain, J.W. and **J.F. Hennen.** 1982. Is the taxonomy of *Berberis* and *Mahonia* (Berberidaceae) supported by their rust pathogens *Cumminsiella santa* sp. nov. and other *Cumminsiella* species (Uredinales)? Syst. Bot. 7:48–59.

_____ and _____. 1986a. Collection of plant materials damaged by pathogens: An expression of support. Taxon 35:119–121.

_____ and _____. 1986b. "Big fleas have little fleas" (big plants have little plants), even in herbaria. Assoc. Syst. Coll. Newslett. 14:1–4.

McCarley, H. (Howard). Personal communication. Behavioral ecologist and mammalogist retired from a career at Austin College, Sherman, TX; also taught at Univ. of Oklahoma Biological Station.

McCarley, H. 1986. Chapter 13. Ecology. In: L.J. Klosterman, L.S. Swenson, Jr., and S. Rose, eds. 100 years of science and technology in Texas. Pp. 227–242. Rice Univ. Press, Houston, TX.

McClintock, E. 1953. The cultivated species of the genus *Erythrina*. Baileya 1:53–58.

_____. 1993. Haloragaceae. In: J.C. Hickman, ed. The Jepson manual: Higher plants of California. Pp. 860–862. Univ. of California Press, Berkeley.

_____ and **C. Epling.** 1942. A review of the genus *Monarda* (Lamiaceae). Univ. Calif. Publ. Bot. 20:147–194.

_____ and _____. 1946. A revision of *Teucrium* in the New World, with observations on its variation, geographical distribution and history. Brittonia 5:491–510.

McClure, J.W. and **R.E. Alston.** 1966. A chemotaxonomic study of the Lemnaceae. Amer. J. Bot. 53:849–860.

McDaniel, S. 1971. The genus *Sarracenia* (Sarraceniaceae). Bull. Tall Timbers Res. Sta. 9:1–36.

McDermott, L.F. 1910. An illustrated key to the North American species of *Trifolium*. California Press, San Francisco.

McDonald, C.B. 1980. A biosystematic study of the *Polygonum hydropiperoides* (Polygonaceae) complex. Amer. J. Bot. 67:664–670.

McDonnell, M.J., E.W. Stiles, G.P. Cheplick, and **J.A. Armesto.** 1984. Bird-dispersal of *Phytolacca americana* L. and the influence of fruit removal on subsequent fruit development. Amer J. Bot. 71:895–901.

McDougall, W.B. and **O.E. Sperry.** 1951. Plants of Big Bend National Park. U.S. Government Printing Office, Washington, DC.

McGivney, Sister M.V. 1938. A revision of the subgenus *Eucyperus* found in the United States. Catholic Univ. Amer., Biol. Ser. 26:1–74.

_____. 1941a. A revision of the subgenus *Mariscus* found in the United States. Catholic Univ. Amer., Biol. Ser. 33:1–147.

_____. 1941b. A revision of the subgenus *Pycreus* found in the United States. Catholic Univ. Amer., Biol. Ser. 37:1–68.

McGowen, J.H., T.F. Hentz, D.E. Owen, M.K. Pieper, C.A. Shelby, and **V.E. Barnes.** 1991. Geologic atlas of Texas, Sherman sheet. Bureau of Economic Geology, The Univ. of Texas, Austin.

McGregor, R.L. 1968a. A new species and two new varieties of *Echinacea* (Compositae). Trans. Kansas Acad. Sci. 70:366–370.

_____. 1968b. The taxonomy of the genus *Echinacea* (Compositae). Univ. Kansas Sci. Bull. 48:113–142.

_____. 1976. A statistical summary of Kansas vascular plant taxa. Rep. State Biol. Surv. Kansas 1:1–7.

_____. 1984a. A study of *Aesculus glabra* variants in Kansas including presumed hybrids with *Aesculus octandra* and *A. pavia*. Contr. Univ. Kansas Herb. 7:1–6.

_____. 1984b. Studies on the genus *Apocynum* (Apocynaceae) in Kansas. Contr. Univ. Kansas Herb. 9:1–12.

_____. 1984c. *Camelina rumelica*, another weedy mustard established in North America. Phytologia 55:227–228.

_____. 1985a. *Panicum hillmanii* Chase (Poaceae): Validity and distribution. Contr. Univ. Kansas Herb. 13:1–9.

_____. 1985b. Musk thistle in Kansas: Observations from 1940–1985. Contr. Univ. Kansas Herb. 14:1–13.

_____. 1985c. Current status of the genus *Camelina* (Brassicaceae) in the prairies and plains of central North America. Contr. Univ. Kansas Herb. 15:1–13.

_____. 1985d. Studies on the variability of *Lobelia spicata* infraspecific taxa in the prairies and plains of central North America with notes on *Lobelia appendiculata*. Contr. Univ. Kansas Herb. 16:1–10.

_____. 1986. Apiaceae. In: Great Plains Flora Association. Flora of the Great Plains. Pp. 584–604. Univ. Press of Kansas, Lawrence.

_____. 1986. Balsaminaceae. In: Great Plains Flora Association. Flora of the Great Plains. Pp. 582–583. Univ. Press of Kansas, Lawrence.

_____. 1986. Chenopodiaceae. In: Great Plains Flora Association. Flora of the Great Plains. Pp. 160–179. Univ. Press of Kansas, Lawrence.

_____. 1986. Fabaceae. In: Great Plains Flora Association. Flora of the Great Plains. Pp. 416–490. Univ. Press of Kansas, Lawrence.

_____. 1986. Linaceae. In: Great Plains Flora Association. Flora of the Great Plains. Pp. 561–564. Univ. Press of Kansas, Lawrence.

_____. 1986. Onagraceae. In: Great Plains Flora Association. Flora of the Great Plains. Pp. 498–526. Univ. Press of Kansas, Lawrence.

_____. 1986. Polygalaceae. In: Great Plains Flora Association. Flora of the Great Plains. Pp. 564–566. Univ. Press of Kansas, Lawrence.

_____. 1986. Sapindaceae. In: Great Plains Flora Association. Flora of the Great Plains. Pp. 567–568. Univ. Press of Kansas, Lawrence.

_____. 1986. Simaroubaceae. In: Great Plains Flora Association. Flora of the Great Plains. Pp. 575. Univ. Press of Kansas, Lawrence.

_____. 1986. Vitaceae. In: Great Plains Flora Association. Flora of the Great Plains. Pp. 557–561. Univ. Press of Kansas, Lawrence.

_____ and **R.E. Brooks.** 1986. Rubiaceae. In: Great Plains Flora Association. Flora of the Great Plains. Pp. 816–823. Univ. Press of Kansas, Lawrence.

McGregor, R.L., J.L. Gentry, and **R.E. Brooks.** 1986. Solanaceae. In: Great Plains Flora Association. Flora of the Great Plains. Pp. 637–651. Univ. Press of Kansas, Lawrence.

McGuffin, M., C. Hobbs, R. Upton, and **A. Goldberg,** eds. 1997. American Herbal Products Association's botanical safety handbook: Guidelines for the safe use and labeling for herbs in commerce. CRC Press, Boca Raton, FL.

McIntire, M.S., J.R. Guest, and **J.F. Porterfield.** 1992. *Philodendron* poisoning – a delayed death. In: L.F. James, R.F. Keeler, E.M. Bailey, Jr., P.R. Cheeke, and M.P. Hegarty, eds. Poisonous plants: Proceedings of the third international symposium. Pp. 515–520. Iowa State Univ. Press, Ames.

McKelvey, S.D. 1938. Yuccas of the southwestern United States, part 1. Arnold Arboretum of Harvard Univ., Jamaica Plain, MA.

_____. 1947. Yuccas of the southwestern United States, part 2. Arnold Arboretum of Harvard Univ., Jamaica Plain, MA.

_____. 1955. Botanical exploration of the trans-Mississippi west 1790–1850. Arnold Arboretum of Harvard Univ., Jamaica Plain, MA.

_____. 1991. Botanical exploration of the trans-Mississippi west 1790–1850. Reprinted with forward by J. Ewan and Introduction by S.D. Beckham, Oregon State Univ. Press, Corvallis, OR.

McKenzie, P.M., L.E. Urbatsch, and **C. Aulbach-Smith.** 1987. *Eustachys caribaea* (Poaceae), a species new to the United States and a key to *Eustachys* in the United States. Sida 12:227–232.

McKinney, L.E. 1992. A taxonomic revision of the acaulescent blue violets (*Viola*) of North America. Sida, Bot. Misc. 7:1–60.

McLaughlin, S.P. 1982. A revision of the southwestern species of *Amsonia* (Apocynaceae). Ann. Missouri Bot. Gard. 69:336–350.

McLeRoy, S.S. 1993. Black Land, Red River, a pictorial history of Grayson County, Texas. The Donning Company Publishers, Virginia Beach, VA.

McNair, J.B. 1925. The taxonomy of poison ivy with a note on the origin of the generic name. Publ. Field Columbian Mus., Bot. Ser. 4:55–70.

McNeill, J. 1980. The delimitation of *Arenaria* (Caryophyllaceae) and related genera in North America, with 11 new combinations in *Minuartia*. Rhodora 82:495–502.

_____. 1981. Nomenclatural problems in *Polygonum*. Taxon 30:630–641.

McVaugh, R. 1936. Studies in the taxonomy of eastern North American species of *Lobelia*. Rhodora 38:241–263, 273–329, 346–362.

_____. 1940. Campanulaceae (Lobelioideae). In: R.E. Woodson and R.W. Shery. Contributions towards a flora of Panama. Ann. Missouri Bot. Gard. 27:347–353.

_____. 1943. Campanulaceae (Lobelioideae). N. Amer. Fl. 32a:1–134.

_____. 1944. The genus *Cnidoscolus*: Generic limits and intrageneric groups. Bull. Torrey Bot. Club 71:457–474.

_____. 1945. The genus *Triodanis* Rafinesque and its relationships to *Specularia* and *Campanula*. Wrightia 1:13–52.

_____. 1948. Generic status of *Triodanis* and *Specularia*. Rhodora 50:38–49.

_____. 1951. A revision of the North American black cherries (*Prunus serotina* Ehrh., and relatives). Brittonia 7:279–315.

_____. 1961. Campanulaceae. In: C.L. Lundell, ed. Flora of Texas 3:331–366. Texas Research Foundation, Renner.

_____. 1982. The new synantherology vs. *Eupatorium* in Nueva Galicia. Contr. Univ. Michigan Herb. 15:181–190.

_____. 1987. Flora Novo-Galiciana: A descriptive account of the vascular plants of western Mexico, Vol. 5, Leguminosae.

McWilliams, E. (Edward). Personal communication. Horticulturist and specialist in Bromeliaceae; Texas A&M Univ., College Station.

McWilliams, E. 1992. Chronology of the natural range expansion of *Tillandsia recurvata* (Bromeliaceae) in Texas. Sida 15:343–346.

_____. 1995. Phytogeographical implications of long term winter precipitation patterns in central Texas. In: J. Norwine, J.R. Giardino, G.R. North, and J.B. Valdés, eds. The changing climate of Texas: Predictability and implications for the future. Chapter 21. Pp. 294–299. GeoBooks, Texas A&M Univ., College Station.

Meacham, C.A. 1994. Phylogenetic relationships at the basal radiation of angiosperms: Further study by probability of character compatibility. Syst. Bot. 19:506–522.

Mears, J.A. 1967. Revision of *Guilleminea* (*Brayulinea*) including *Gossypianthus* (Amaranthaceae). Sida 3:137–152.

_____. 1975. The taxonomy of *Parthenium* section *Partheniastrum* DC. (Asteraceae-Ambrosiinae). Phytologia 31:463–482.

_____. 1980. The Linnaean species of *Gomphrena* L. (Amaranthaceae). Taxon 29:85–95.

Meeson, B.W. 1977. The pollen morphology of *Dalea* section *Cylipogon* (Psoraleae: Leguminosae). Sida 7:13–21.

Meijer, W. 1993. Rafflesiaceae. In: K. Kubitzki, J.C. Rohwer, and V. Bittrich, eds. The families and genera of vascular plants, Vol. II. Pp. 557–565. Springer-Verlag, Berlin.

Melchert, T.E. 1963. *Thelesperma curvicarpum* T.E. Melchert (Compositae), new species. SouthW. Naturalist 8:179.

Melderis, A. 1972. *Centaurium*. In: T.G. Tutin, V.H. Heywood, N.A. Burges, D.M. Moore, D.H. Valentine, S.M. Walters, and D.A. Webb, eds. Flora Europaea 3:56–59. Cambridge Univ. Press, Cambridge, England, U.K.

Melville, R. 1958. Notes on *Alternanthera*. Kew Bull. 13:171–175.

Menapace, F.J., D.E. Wujek, and **A.A. Reznicek.** 1986. A systematic revision of the genus *Carex* (Cyperaceae) with respect to the section *Lupulinae*. Canad. J. Bot. 64:2785–2788.

Mennema, J. 1989. A taxonomic revision of *Lamium* (Lamiaceae). Leiden Bot. Ser. 11:1–198.

Menzel, M.Y. 1951. The cytotaxonomy and genetics of *Physalis*. Proc. Amer. Phil. Soc. 95:134–185.

Menzies, J.S., C.H. Bridges, and **E.M. Bailey, Jr.** 1979. A neurological disease of cattle associated with *Solanum dimidiatum*. SouthW. Veterin. 32:45–49.

Merrill, W.L. 1977. An investigation of ethnographic and archaeological specimens of mescalbeans (*Sophora secundiflora*) in American museums. Tech. Rep. No. 6. Museum of Anthropology, Univ. of Michigan, Ann Arbor.

Mertens, T.R. and **P.H. Raven.** 1965. Taxonomy of *Polygonum*, section *Polygonum* (*Avicularia*) in North America. Madroño 18:85–92.

Mes, T.H.M, N. Friesen, R.M. Fritsch, M. Klaas, and **K. Bachmann.** 1997 [1998]. Criteria for sampling in *Allium* based on chloroplast DNA PCR-RFLP'S. Syst. Bot. 22:701–712.

Mészáros, S., J. de Laet, and **E. Smets.** 1996. Phylogeny of temperate Gentianaceae: A morphological approach. Syst. Bot. 21:153–168.

Mesfin, T., D.J. Crawford, and **E.B. Smith.** 1995. Pollen morphology of North American *Coreopsis* (Compositae—Heliantheae). Grana 34:21–27.

_____, _____, and _____. 1995b. Comparative capitular morphology and anatomy of *Coreopsis* L. and *Bidens* L. (Compositae), including a review of generic boundaries. Brittonia 47:61–91.

Mestel, R. 1993. Murder trial features tree's genetic fingerprint. New Sci. 138:6.

Metz, Sister M.C. 1934. A flora of Bexar County, Texas. Contr. Biol. Lab. Catholic Univ. Amer. 16:1–214.

Meyer, F.G. 1997. Hamamelidaceae. In: Flora of North America Editorial Committee, eds. Fl. North Amer. 2:362–367. Oxford Univ. Press, New York and Oxford.

Michaels, H.J., K.M. Scott, R.G. Olmstead, T. Szaro, R.K. Jansen, and **J.D. Palmer.** 1993. Interfamilial relationships of the Asteraceae: Insights from *rbc*L sequence variation. Ann. Missouri Bot. Gard. 80:742–751.

Michener, D.C. 1986. Phenotypic instability in *Gleditsia triacanthos* (Fabaceae). Brittonia 38:360–361.

Mickel, J.T. 1979. The fern genus *Cheilanthes* in the continental United States. Phytologia 41:431–437.

_____. 1981. Revision of *Anemia* subgenus *Anemiorrhiza* (Schizaeaceae). Brittonia 33:413–429.

_____. 1993. Anemiaceae. In: Flora of North America Editorial Committee, eds. Fl. North Amer. 2:117–118. Oxford Univ. Press, New York and Oxford.

Mielke, M.E. and M.L. Daughtrey. 1990. How to identify and control dogwood anthracnose. U.S.D.A., Forest Serv., Northeastern Area. Bull. NA-GR-18.

Millán, R. 1941. Revisión de las especies del género *Nierembergia* (Solanaceae). Darwiniana 5:487–547.

Millar, C.I. 1993. Impact of the Eocene on the evolution of *Pinus* L. Ann. Missouri Bot. Gard. 80:471–498.

Miller, D.H. and **F.E. Smeins.** 1988. Vegetation pattern within a remnant San Antonio prairie as influenced by soil and microrelief variation. In: A. Davis and G. Stanford, eds. The prairie: Roots of our culture; foundation of our economy. Unpaged, paper number 0110. Proceedings of the Tenth North American Prairie Conference, Denton, TX. Native Prairie Association of Texas, Dallas.

Miller, G.N. 1955. The genus *Fraxinus*, the ashes, in North America north of Mexico. Cornell Univ. Agric. Exp. Sta. Mem. 335:1–64.

Miller, J.M. 1988. Floral pigments and phylogeny in *Echinocereus* (Cactaceae). Syst. Bot. 13:173–183.

Miller, K.I. and **G.L. Webster.** 1967. A preliminary revision of *Tragia* (Euphorbiaceae) in the United States. Rhodora 69:241–305.

Miller, N.G. 1970. The genera of the Cannabaceae in the southeastern United States. J. Arnold Arbor. 51:185–203.

_____. 1971a. The genera of the Urticaceae in the southeastern United States. J. Arnold Arbor. 52:40–68.

_____. 1971b. The Polygalaceae in the southeastern United States. J. Arnold Arbor. 52:267–284.

_____. 1990. The genera of Meliaceae in the southeastern United States. J. Arnold Arbor. 71:453–486.

Millspaugh, C.F. and **E.E. Sherff.** 1919. Revision of the North American species of *Xanthium*. Publ. Field Columbian Mus., Bot. Ser. 4:9–49.

Miranda, F. and **A.J. Sharp.** 1950. Characteristics of the vegetation in certain temperate regions of eastern Mexico. Ecology 31:313–333.

Mitchell, R.S. 1968. Variation in the *Polygonum amphibium* complex and its taxonomic significance. Univ. Calif. Publ. Bot. 45:1–65.

Mobberley, D.G. 1956. Taxonomy and distribution of the genus *Spartina*. Iowa State Coll. J. Sci. 30:471–574.

Mohlenbrock, R.H. 1957 [1958]. A revision of the genus *Stylosanthes*. Ann. Missouri Bot. Gard. 44:299–355.

_____. 1958. The *Stylosanthes biflora* complex. Bull. Torrey Bot. Club 85:341–346.

_____. 1961. A monograph of the leguminous genus *Zornia*. Webbia 16:1–141.

Moldenke, H.N. 1948. Additional notes on the Eriocaulaceae, Avicenniaceae and Verbenaceae of Texas–I. Wrightia 1:220–246.

_____. 1961a. Eriocaulaceae. In: C.L. Lundell, ed. Flora of Texas 3:3–9. Texas Research Foundation, Renner.

_____. 1961b. Verbenaceae. In: C.L. Lundell, ed. Flora of Texas 3:13–87. Texas Research Foundation, Renner.

_____ and **A.L. Moldenke.** 1952. Plants of the Bible. Chronica Botanica Co., Waltham, MA.

Montgomery, J.A. 1993. The nature and origin of the blackland prairies of Texas. In: M.R. Sharpless and J.C. Yelderman, eds.

The Texas Blackland Prairie, land, history, and culture. Pp. 24–40. Baylor Univ. Program for Regional Studies, Waco, TX.

Moore, D.M., T.G. Tutin, and **S.M. Walters.** 1976. Compositae. In: Tutin, T.G., V.H. Heywood, N.A. Burges, D.M. Moore, D.H. Valentine, S.M. Walters, and D.A. Webb, eds. Flora Europaea 4:103–410. Cambridge Univ. Press, Cambridge, England, U.K.

Moore, H.E., Jr. and **N.W. Uhl.** 1982. Major trends of evolution in palms. Bot. Rev. (Lancaster) 48:2–69.

Moore, L.A. and **M.F. Willson.** 1982. The effect of microhabitat, spatial distribution, and display size on dispersal of *Lindera benzoin* by avian frugivores. Canad. J. Bot. 60:557–560.

Moore, M.O. 1987. A study of selected taxa of *Vitis* (Vitaceae) in the southeastern United States. Rhodora 89:75–91.

_____. 1991. Classification and systematics of eastern North American *Vitis* L. (Vitaceae) north of Mexico. Sida 14:339–367.

Moore, R.J. 1947. Cytotaxonomic studies in the Loganiaceae. I. Chromosome numbers and phylogeny in the Loganiaceae. Amer. J. Bot. 34:527–538.

_____. 1972. Distribution of native and introduced knapweeds in Canada and the United States. Rhodora 74:331–346.

_____ and **C. Frankton.** 1969. Cytotaxonomy of some *Cirsium* species of the eastern United States. Canad. J. Bot. 47:1257–1275.

_____ and _____. 1974. The thistles of Canada. Research Branch, Canada Dept. Agri., Monogr. 10:1–111.

Morden, C.W. and **S.L. Hatch.** 1987. Anatomical study of the *Muhlenbergia repens* complex (Poaceae: Chloridoideae: Eragrostideae). Sida 12:347–359.

_____ and _____. 1989. An analysis of morphological variation in *Muhlenbergia capillaris* (Poaceae) and its allies in the southeastern United States. Sida 13:303–314.

_____ and _____. 1996. Morphological variation and synopsis of the *Muhlenbergia repens* complex (Poaceae). Sida 17:349–365.

Morefield, J.D. 1992. Resurrection and revision of *Hesperevax* (Asteraceae: Inuleae). Syst. Bot. 17:293–310.

Moreno, N.P. and **F. Essig.** 1997. *Clematis* subg. *Clematis*. In: Flora of North America Editorial Committee, eds. Fl. North Amer. 3:159–164. Oxford Univ. Press, New York and Oxford.

Morgan, D.R. 1997 [1998]. Reticulate evolution in *Machaeranthera* (Asteraceae). Syst. Bot. 22:599–615.

_____ and **B.B. Simpson.** 1992. A systematic study of *Machaeranthera* (Asteraceae) and related groups using restriction site analysis of chloroplast DNA. Syst. Bot. 17:511–531.

Morgan, D.R. and **D.E. Soltis.** 1993. Phylogenetic relationships among members of Saxifragaceae s.l. based on *rbc*L sequence data. Ann. Missouri Bot. Gard. 80:631–660.

Morgan, J.J. 1966. A taxonomic study of the genus *Boltonia* (Asteraceae). Ph.D. dissertation, Univ. of North Carolina, Chapel Hill.

Morin, N.R. (Nancy). Personal communication. Specialist in Campanulaceae and Convening Editor of the Flora of North America; formerly of the Missouri Botanical Garden, currently Executive Director of the American Association of Botanical Gardens and Arboreta.

Morong, T. 1893. The Naiadaceae of North America. Mem. Torrey Bot. Club 3:1–65.

Morrone, O. and **F.O. Zuloaga.** 1992. Revisión de las especies sudamericanas nativas e introducidas de los géneros *Brachiaria* y *Urochloa* (Poaceae: Panicoideae: Paniceae). Darwiniana 31:43–109.

Morton, C.M., M.W. Chase, K.A. Kron, and **S.M. Swensen.** 1996 [1997]. A molecular evaluation of the monophyly of the order Ebenales based upon *rbcL* sequence data. Syst. Bot. 21:567–586.

Morton, C.V. 1937. The correct names of the small-flowered mallows. Rhodora 39:98–99.

Morton, J.F. 1977. Major medicinal plants: Botany, culture and uses. Charles C. Thomas, Springfield, IL.

_____. 1982. Plants poisonous to people in Florida and other warm areas, 2nd ed. Southeastern Printing Co., Stuart, FL.

Morton, J.K. 1972. On the occurrence of *Stellaria pallida* in North America. Bull. Torrey Bot. Club 99:95–97.

Mosquin, T. 1970. Chromosome numbers and a proposal for classification in *Sisyrinchium* (Iridaceae). Madroño 20:269–275.

_____. 1971. Biosystematic studies in the North American species of *Linum*, section *Adenolinum* (Linaceae). Canad. J. Bot. 49:1379–1388.

Mosyakin, S.L. 1996. A taxonomic synopsis of the genus *Salsola* (Chenopodiaceae) in North America. Ann. Missouri Bot. Gard. 83:387–395.

Moyer, J.A. and **B.L. Turner.** 1994. Systematic study of Texas populations of *Phacelia patuliflora* (Hydrophyllaceae). Sida 16:245–252.

Muenscher, W.C. 1951. Poisonous plants of the United States. Macmillan Co., New York.

Mühlenbach, V. 1979. Contributions to the synanthropic (adventive) flora of the railroads in St. Louis, Missouri, U.S.A. Ann. Missouri Bot. Gard. 66:1–108.

Mukherjee, S.K. 1958. Revision of the genus *Erianthus* Michx. (Gramineae). Lloydia 21:157–188.

Müller, C.H. 1940. A revision of the genus *Lycopersicon*. U.S.D.A. Misc. Pub. 382:1–28.

_____. 1951. The oaks of Texas. Contr. Texas Res. Found., Bot. Stud. 1:21–311.

Mulligan, G.A. 1980. The genus *Cicuta* in North America. Canad. J. Bot. 58:1755–1767.

_____ and **I.J. Bassett.** 1959. *Achillea millefolium* complex in Canada and portions of the United States. Canad. J. Bot. 37:73–79.

Mulligan, G.A. and **D.B. Munro.** 1989 [1990]. Taxonomy of species of North American *Stachys* (Labiatae) found north of Mexico. Naturaliste Canad. 116:35–51.

_____ and _____. 1990. Poisonous plants of Canada. Agri. Canada, Publ. 1842E. Canadian Government Publishing Centre, Ottawa.

Mullins, M.G., A. Bouquet, and **L.E. Williams.** 1992. Biology of the grapevine. Cambridge Univ. Press, Cambridge, England, U.K.

Munson, T.V. 1883. Forests and forest trees of Texas. Amer. J. Forestry 1:433–451.

_____. 1909. Foundations of American grape culture. Orange Judd Company, New York.

Munz, P.A. 1926. The Antirrhinoideae-Antirrhineae of the New World. Proc. Calif. Acad. Sci. 15:323–397.

______. 1931. The North American species of *Orobanche*, section *Myzorrhiza*. Bull. Torrey Bot. Club 57:611–624.

______. 1938. Studies in Onagraceae XI. A revision of the genus *Gaura*. Bull. Torrey Bot. Club 65:105–122, 211–228.

______. 1944. Studies in Onagraceae-XIII. The American species of *Ludwigia*. Bull. Torrey Bot. Club 71:152–165.

______. 1946. *Aquilegia*: The cultivated and wild columbines. Gentes Herb. 7:1–150.

______. 1961. Onagraceae. In: C.L. Lundell, ed. Flora of Texas 3:208–262. Texas Research Foundation, Renner.

______. 1965. Onagraceae. N. Amer. Fl. II. 5:1–278.

Murray, A.E., Jr. 1970. A monograph of the Aceraceae: A thesis in horticulture. Ph.D. dissertation, Pennsylvania State Univ., University Park.

Musselman, L.J. 1975 [1976]. Parasitism and haustorial structure in *Krameria lanceolata* (Krameriaceae): A preliminary study. Phytomorphology 25:416–422.

______. 1977. Seed germination and seedlings of *Krameria lanceolata* (Krameriaceae). Sida 7:224–225.

Myint, T. 1966. Revision of the genus *Stylisma* (Convolvulaceae). Brittonia 18:97–116.

Nabhan, G.P. 1979. Southwestern Indian sunflowers. Desert Pl. 1:23–26.

Naczi, R.F.C. 1992. Systematics of *Carex* section *Griseae* (Cyperaceae). Ph.D. dissertation, Univ. of Michigan, Ann Arbor.

______ and **C.T. Bryson.** 1990. Noteworthy records of *Carex* (Cyperaceae) from the southeastern United States. Bartonia 56:49–58.

Nash, G.V. 1896. Revision of the genus *Asimina* in North America. Bull. Torrey Bot. Club 23:234–242.

______. 1898. A revision of the genera *Chloris* and *Eustachys* in North America. Bull. Torrey Bot. Club 25:432–450.

______. 1909. Poaceae. N. Amer. Fl. 17:77–98.

______. 1912. Poaceae (cont.). N. Amer. Fl. 17:99–196.

______ and **A.S. Hitchcock.** 1915. Poaceae (cont.). N. Amer. Fl. 17:197–288.

Nauman, C.E. 1993. *Nephrolepis*. In: Flora of North America Editorial Committee, eds. Fl. North Amer. 2:305–308. Oxford Univ. Press, New York and Oxford.

Nee, M. (Michael). Personal communication. Specialist in Solanaceae and Cucurbitaceae; botanist and plant collector (particularly Latin America) of the New York Bot. Garden, Bronx; trained at Univ. of Wisconsin, Madison.

Nee, M. 1986. Solanaceae I. Flora de Veracruz 49:1–191.

______. 1990. The domestication of *Cucurbita* (Cucurbitaceae). Econ. Bot. 44 (Suppl.):56–68.

Nelson, E. 1899. Revision of the western North American phloxes. Published as a part of the ninth report of the Wyoming Agricultural College, Laramie.

Nelson, E.C. 1991. Shamrock: Botany and history of an Irish myth. Boethius Press, Kilkenny, Ireland.

Nelson, J.B. 1980. *Mitreola* vs. *Cynoctonum*, and a new combination. Phytologia 46:338–340.

______. 1981. *Stachys* (Labiatae) in southeastern United States. Sida 9:104–123.

Nesom, G.L. (Guy). Personal communication. Specialist in Asteraceae; formerly of the Univ. of Texas, Austin and the Texas Regional Institute of Environmental Studies, Sam Houston State Univ., Huntsville; currently with the Biota of North America Program of the North Carolina Botanical Garden at the University of North Carolina at Chapel Hill.

Nesom, G.L. 1978. Chromosome numbers in *Erigeron* and *Conyza* (Compositae). Sida 7:375–381.

______. 1988. Synopsis of *Chaetopappa* (Compositae-Astereae) with a new species and the inclusion of *Leucelene*. Phytologia 64:448–456.

______. 1989a. Infrageneric taxonomy of New World *Erigeron* (Compositae: Astereae). Phytologia 67:61–66.

______. 1989b. New species, new sections, and a taxonomic overview of American *Pluchea* (Compositae: Inuleae). Phytologia 67:158–167.

______. 1990a. Taxonomic status of *Gamochaeta* (Asteraceae: Inuleae) and the species of the United States. Phytologia 68:186–198.

______. 1990b. Further definition of *Conyza* (Asteraceae: Astereae). Phytologia 68:229–233.

______. 1990c. Studies in the systematics of Mexican and Texan *Grindelia* (Asteraceae: Astereae). Phytologia 68:303–332.

______. 1990d. Taxonomy of *Solidago petiolaris* (Astereae: Asteraceae) and related Mexican species. Phytologia 69:445–456.

______. 1991a. Union of *Bradburia* with *Chrysopsis* (Asteraceae: Astereae), with a phylogenetic hypothesis for *Chrysopsis*. Phytologia 71:109–121.

______. 1991b. A phylogenetic hypothesis for the goldenasters (Asteraceae: Astereae). Phytologia 71:136–151.

______. 1991c. Morphological definition of the *Gutierrezia* group (Asteraceae: Astereae). Phytologia 71:252–262.

______. 1992a. A new species of *Castilleja* (Scrophulariaceae) from southcentral Texas with comments on other Texas taxa. Phytologia 72:209–230.

______. 1992b. Species rank for the varieties of *Grindelia microcephala* (Asteraceae: Astereae). Phytologia 73:326–329.

______. 1993. Taxonomic infrastructure of *Solidago* and *Oligoneuron* (Asteraceae: Astereae) and observations on their phylogenetic position. Phytologia 75:1–44.

______. 1994a. Subtribal classification of the Astereae (Asteraceae). Phytologia 76:193–274.

______. 1994b. Review of the taxonomy of *Aster* sensu lato (Asteraceae: Astereae), emphasizing the New World species. Phytologia 77:141–297.

______. 1995. Revision of *Chaptalia* (Asteraceae: Mutisieae) from North America and continental Central America. Phytologia 78:153–188.

______. 1997a. Taxonomic adjustments in North American *Aster* sensu latissimo (Asteraceae: Astereae). Phytologia 82:281–288.

______. 1997b. Review: "A revision of *Heterotheca* sect. *Phyllotheca* (Nutt.) Harms (Compositae: Astereae)" by J.C. Semple. Phytologia 83:7–21.

______, **Y. Suh, D.R. Morgan, S.D. Sundberg,** and **B.B. Simpson.** 1991. *Chloracantha*, a new genus of North American

Astereae (Asteraceae). Phytologia 70:371–381.

NESOM, G.L., Y. SUH, and **B.B. SIMPSON.** 1993. *Prinopsis* (Asteraceae: Astereae) united with *Grindelia*. Phytologia 75:341–346.

NESOM, G.L. and **B.L. TURNER.** 1995. Systematics of the *Sedum parvum* group (Crassulaceae) in northeastern Mexico and Texas. Phytologia 79:257–268.

_____ and _____. 1998. Variation in the *Berlandiera pumila* (Asteraceae) complex. Sida 18:493–502.

NEUWINGER, H.D. 1996. African ethnobotany, poisons and drugs, chemistry, pharmacology, toxicology. Chapman & Hall, New York.

NEWSHOLME, C.W. 1992. Willows: The genus *Salix*. B.T. Batsford, Ltd., London, England, U.K.

NEWSOM, V.M. 1929. A revision of the genus *Collinsia* (Scrophulariaceae). Bot. Gaz. (Crawfordsville) 87:260–301.

NICOLSON, D.H. 1997. Hierarchical roots and shoots or *Opera Jehovae Magna!* (Psalms 111:2). Aliso 15:81–86.

NIEFOFF, J. (JERRY). Personal communication. Soil Scientist, Idaho Panhandle National Forest.

NIXON, E.S. 1985. Trees, shrubs, and woody vines of East Texas. B.L. Cunningham Prod., Nacogdoches, TX.

_____, **G.A. SULLIVAN, S.D. JONES, G.D. JONES,** and **J.K. SULLIVAN.** 1990. Species diversity of woody vegetation in the Trinity River Basin, Texas. Castanea 55:97–105.

NIXON, E.S. and **J.R. WARD.** 1982. *Rhynchospora miliacea* and *Scirpus divaricatus* new to Texas. Sida 9:367.

_____, _____, and **B.L. LIPSCOMB.** 1983. Rediscovery of *Lesquerella pallida* (Cruciferae). Sida 10:167–175.

NIXON, K.C. 1984. A biosystematic study of *Quercus* series *Virentes* (the liveoaks) with phylogenetic analysis of Fagales, Fagaceae, and *Quercus*. Ph.D. dissertation, Univ. of Texas, Austin.

_____. 1997a. Fagaceae. In: Flora of North America Editorial Committee, eds. Fl. North Amer. 3:436–437. Oxford Univ. Press, New York and Oxford.

_____. 1997b. *Quercus*. In: Flora of North America Editorial Committee, eds. Fl. North Amer. 3:445–506. Oxford Univ. Press, New York and Oxford.

NOKES, J. 1997. In memoriam, Benny J. Simpson (29 February 1928–27 December 1996). Sida 17:850–852. Originally published 1997 in Native Pl. Soc. Texas News 15(2):1, 8.

NORRIS, F.M. (FIONA). Personal communication. Specialist in Acanthaceae and flora of South Africa; Assistant Director and Head of Public Outreach, Botanical Research Institute of Texas, Fort Worth.

NORTH, G.R. 1995. Global warming. In: J. Norwine, J.R. Giardino, G.R. North, and J.B. Valdés, eds. The changing climate of Texas: Predictability and implications for the future. Chapter 13. Pp. 156–166. GeoBooks, Texas A&M Univ., College Station.

NORTHINGTON, D.K. 1971. Taxonomy of *Pyrrhopappus*, a cytotaxonomic and chemotaxonomic study. Ph.D. dissertation, Univ. of Texas, Austin.

_____. 1974. Systematic studies of the genus *Pyrrhopappus* (Compositae, Cichorieae). Special Publ. No. 6. The Museum, Texas Tech Univ., Lubbock.

NORTON, J.B.S. 1900. A revision of the American species of *Euphorbia* of the section *Tithymalus* occurring north of Mexico. Rep. (Annual) Missouri Bot. Gard. 11:85–144.

NORWINE, J., J.R. GIARDINO, G.R. NORTH, and **J.B. VALDÉS,** eds. 1995. The changing climate of Texas: Predictability and implications for the future. GeoBooks, Texas A&M Univ., College Station.

NOVÁK, J. and **V. PREININGER.** 1987. Chemotaxonomic review of the genus *Papaver*. Preslia 59:1–13.

NOVAK, S.J., D.E. SOLTIS, and **P.S. SOLTIS.** 1991. Ownbey's Tragopogons: 40 years later. Amer. J. Bot. 78:1586–1600.

NOWICKE, J.W. 1969. Palynotaxonomic study of the Phytolaccaceae. Ann. Missouri Bot. Gard. 55:294–364.

_____. 1994. A palynological study of Crotonoideae (Euphorbiaceae). Ann. Missouri Bot. Gard. 81:245–269.

_____. 1996. Pollen morphology, exine structure and the relationships of Basellaceae and Didiereaceae to Portulacaceae. Syst. Bot. 21:187–208.

_____ and **J.J. SKVARLA.** 1981. Pollen morphology and phylogenetic relationships of the Berberidaceae. Smithsonian Contr. Bot. 50:1–83.

OBERHOLSER, H.C. 1974. The bird life of Texas. Univ. of Texas Press, Austin.

OCKENDON, D.J. 1965. A taxonomic study of *Psoralea* subgenus *Pediomelum* (Leguminosae). SouthW. Naturalist 10:81–124.

O'DONELL, C.A. 1959. Las especies americanas de *Ipomoea* L. Sect. *Quamoclit* (Moench) Griseb. Lilloa 29:19–86.

OERTLI, E.H., L.D. ROWE, S.L. LOVERING, G.W. IVIE, and **E.M. BAILEY.** 1983. Phototoxic effect of *Thamnosma texana* (Dutchman's breeches) in sheep. Amer. J. Veterin. Res. 44:1126–1129.

OGDEN, E.C. 1943. The broad-leaved species of *Potamogeton* of North America north of Mexico. Rhodora 45:57–105, 119–163, 171–214.

_____. 1966. Potamogetonaceae. In: C.L. Lundell, ed. Flora of Texas 1:369–382. Texas Research Foundation, Renner.

O'KENNON, R.J. 1991. *Paliurus spina-christi* (Rhamnaceae) new for North America in Texas. Sida 14:606–609.

_____, **G.M. DIGGS, JR.,** and **R.K. HOGGARD.** 1998. *Plantago coronopus* (Plantaginaceae) new to Texas. Sida 18:356–358.

O'KENNON, R.J., G.M. DIGGS, JR., and **B.L. LIPSCOMB.** 1998. *Lactuca saligna* (Asteraceae), a lettuce new for Texas. Sida 18:615–619.

O'KENNON, R.J. and **G. NESOM.** 1988. First report of *Cirsium vulgare* (Asteraceae) in Texas. Sida 13:115.

OLMSTEAD, R.G., B. BREMER, K.M. SCOTT, and **J.D. PALMER.** 1993. A parsimony analysis of the Asteridae sensu lato based on *rbc*L sequences. Ann. Missouri Bot. Gard. 80:700–722.

OLMSTEAD, R.G., H.J. MICHAELS, K.M. SCOTT, and **J.D. PALMER.** 1992. Monophyly of the Asteridae and identification of their major lineages inferred from DNA sequences of *rbc*L. Ann. Missouri Bot. Gard. 79:249–265.

OLMSTEAD, R.G. and **J.D. PALMER.** 1992. A chloroplast DNA phylogeny of the Solanaceae: Subfamilial relationships and character evolution. Ann. Missouri Bot. Gard. 79:346–360.

_____ and _____. 1997. Implications for the phylogeny, classification, and biogeography of *Solanum* from cpDNA restriction site variation. Syst. Bot. 22:19–29.

OLMSTEAD, R.G. and **P.A. REEVES.** 1995. Evidence for the polyphyly of the Scrophulariaceae based on chloroplast *rbc*L and *ndh*F sequences. Ann. Missouri Bot. Gard. 82:176–193.

OLSEN, J. 1979. Taxonomy of the *Verbesina virginica* complex (Asteraceae). Sida 8:128–134.

______. 1985. Synopsis of *Verbesina* section *Ochratinia* (Asteraceae). Pl. Syst. Evol. 149:47–63.

OLSEN, J.D., G.D. MANNERS, and **S.W. PELLETIER.** 1990. Poisonous properties of larkspur (*Delphinium* spp.). Collect. Bot. (Barcelona) 19:141–151.

OLWELL, M. 1982. A population study of the exomorphic variations in *Vicia minutiflora* Dietr. including *V. reverchonii* Wats. (Leguminosae). Sida 9:215–222.

O'NEILL, H.T. 1940. Botany of the Maya area: The sedges of the Yucatan Peninsula. Carnegie Inst. Wash. Publ. No. 19.

ØRGAARD, M. 1991. The genus *Cabomba* (Cabombaceae)—A taxonomic study. Nordic J. Bot. 11:179–203.

ORNDUFF, R. 1966. The origin of dioecism from heterostyly in *Nymphoides* (Menyanthaceae). Evolution 20:309–314.

______. 1970. The systematics and breeding system of *Gelsemium* (Loganiaceae). J. Arnold Arbor. 51:1–17.

OSBORN, J.M. and **E.L. SCHNEIDER.** 1988. Morphological studies of the Nymphaeaceae sensu lato. XVI. The floral biology of *Brasenia schreberi*. Ann. Missouri Bot. Gard. 75:778–794.

OSTERHOUDT, K.C., S.K. LEE, J.M. CALLAHAN, and **F.M. HENRETIG.** 1997. Catnip and the alteration of human consciousness. Veterin. Human Toxicol. 39:373–375.

OSTERHOUT, G.E. 1902. *Hesperaster nudus* (Pursh) Cockerell and its allies. Bull. Torrey Bot. Club 29:173–174.

OTTLEY, A.M. 1944. The American Loti with special consideration of a proposed new section, *Simpeteria*. Brittonia 5:81–123.

OWENS, C. (CHETTA). Personal communication. Scientist. U.S. Army Corps of Engineers. Lewisville Aquatic Ecosystem Research Facility, Lewisville, TX.

OWNBEY, G.B. 1947. Monograph of the North American species of *Corydalis*. Ann. Missouri. Bot. Gard. 34:187–259.

______. 1951. On the cytotaxonomy of the genus *Corydalis*, section *Eucorydalis*. Amer. Midl. Naturalist 45:184–186.

______. 1958. Monograph of the genus *Argemone* for North America and the West Indies. Mem. Torrey Bot. Club 21:1–159.

______. 1997. *Argemone*. In: Flora of North America Editorial Committee, eds. Fl. North Amer. 3:314–322. Oxford Univ. Press, New York and Oxford.

OWNBEY, M. 1950a. Natural hybridization and amphiploidy in the genus *Tragopogon*. Amer. J. Bot. 37:487–499.

______. 1950b. The genus *Allium* in Texas. Res. Stud. State Coll. Wash. 18:181–222.

______ and **H.C. AASE.** 1955. Cytotaxonomic studies in *Allium* I. The *Allium canadense* alliance. Res. Stud. State Coll. Wash., Monogr. Suppl. 1–106.

OWNBEY, R.P. 1944. The liliaceous genus *Polygonatum* in North America. Ann. Missouri Bot. Gard. 31:373–413.

PACKARD, J.M. and **T.L. COOK.** 1995. Effects of climate change on biodiversity and landscape linkages in Texas. In: J. Norwine, J.R. Giardino, G.R. North, and J.B. Valdés, eds. The changing climate of Texas: Predictability and implications for the future. Chapter 23. Pp. 322–336. GeoBooks, Texas A&M Univ., College Station.

PACLT, J. 1998. (1334) Proposal to amend the gender of *Nuphar*, nom. cons. (Nymphaeaceae), to neuter. Taxon 47:167–169.

PAGE, C.N. 1976. The taxonomy and phytogeography of bracken–A review. J. Linn. Soc., Bot. 73:1–34.

PALMER, E.J. 1920. The canyon flora of the Edwards Plateau of Texas. J. Arnold Arbor. 1:233–239.

______. 1925. Synopsis of North American Crataegi. J. Arnold Arbor. 6:5–128.

______. 1946. *Crataegus* in the northeastern and central United States and adjacent Canada. Brittonia 5:471–490.

______. 1948. Hybrid oaks of North America. J. Arnold Arbor. 29:1–48.

______. 1960. Key to *Crataegus* species. In: R.A. Vines. Trees, shrubs, and woody vines of the Southwest. Pp. 329–334. Univ. of Texas Press, Austin.

PAMMEL, L.H. 1911. A manual of poisonous plants. The Torch Press, Cedar Rapids, IA.

PARFITT, B.D. 1997. *Adonis*. In: Flora of North America Editorial Committee, eds. Fl. North Amer. 3:184–187. Oxford Univ. Press, New York and Oxford.

PARIS, C.A. 1993. *Adiantum*. In: Flora of North America Editorial Committee, eds. Fl. North Amer. 2:125–130. Oxford Univ. Press, New York and Oxford.

PARK, K. 1998. Monograph of *Euphorbia* sect. *Tithymalopsis* (Euphorbiaceae). Edinb. J. Bot. 55:161–208.

PARK, M.M. and **D. FESTERLING, JR.** 1997. *Thalictrum*. In: Flora of North America Editorial Committee, eds. Fl. North Amer. 3:258–271. Oxford Univ. Press, New York and Oxford.

PARKER, W.B. 1856. Through unexplored Texas [notes taken during the 1854 Marcy Expedition]. Hayes and Zell, Philadelphia. Reprinted 1990 by the Texas State Historical Association, Austin.

PARKS, H.B. and **V. CORY.** 1936. The fauna and flora of the Big Thicket area. Texas Agric. Exp. Sta., College Station.

PARRY, C.C. 1889. *Ceanothus*, L.: A synoptical list, comprising thirty-three species, with notes and descriptions. Proc. Davenport Acad. Nat. Sci. 5:162–176.

PATON, A. 1990. A global taxonomic investigation of *Scutellaria* (Labiatae). Kew Bull. 45:399–450.

PAYNE, W.W. 1964. A re-evaluation of the genus *Ambrosia* (Compositae). J. Arnold Arbor. 45:401–430.

PAYSON, E.B. 1918. The North American species of *Aquilegia*. Contr. U.S. Natl. Herb. 20:133–157.

______. 1921. A monograph of the genus *Lesquerella*. Ann. Missouri Bot. Gard. 8:103–236.

______. 1922a. Species of *Sisymbrium* native to America north of Mexico. Univ. Wyoming Publ. Bot. 1:1–27.

______. 1922b. A synoptical revision of the genus *Cleomella*. Univ. Wyoming Publ. Bot. 1:29–46.

______. 1926. The genus *Thlaspi* in North America. Univ. Wyoming Publ. Bot. 1:145–163.

PEABODY, F.J. 1984. Revision of the genus *Robinia* (Leguminosae: Papilionoideae). Ph.D. dissertation, Iowa State Univ., Ames.

PEACOCK, H. 1982. Just plain poke. Texas Highways 29:8–11.

PEATTIE, D.C. 1948. A natural history of trees of eastern and central North America. Houghton Mifflin Co., Boston, MA.

PELLMYR, O. and **E.J. AUGENSTEIN.** 1997. Pollination biology of *Hesperaloe parviflora*. SouthW. Naturalist 42:182–187.

PENG, C.I. 1988. The biosystematics of *Ludwigia* sect. *Microcarpium* (Onagraceae). Ann. Missouri Bot. Gard. 75:970–1009.

_____. 1989. The systematics and evolution of *Ludwigia* sect. *Microcarpium* (Onagraceae). Ann. Missouri Bot. Gard. 76:221–302.

Pennell, F.W. 1916. *Commelina* in the United States. Bull. Torrey Bot. Club 43:96–111.

_____. 1920. Scrophulariaceae of the southeastern United States. Contr. New York Bot. Gard. 221:224–291. Reprinted from Proc. Acad. Nat. Sci. Philadelphia, 1919.

_____. 1921. "Veronica" in North and South America. Rhodora 23:1–22, 28–41.

_____. 1928. *Agalinus* and allies in North America–I. Proc. Acad. Nat. Sci. Philadelphia 80:339–449.

_____. 1929. *Agalinus* and allies in North America–II. Proc. Acad. Nat. Sci. Philadelphia 81:111–249.

_____. 1931. "*Polygala verticillata*" in eastern North America. Bartonia 13:7–17.

_____. 1933. *Polygala verticillata* and the problem of typifying Linnean species. Bartonia 15:38–45.

_____. 1935. The Scrophulariaceae of eastern temperate North America. Acad. Nat. Sci. Philadelphia Monogr. 1:i–xv, 1–650.

_____. 1937. "Commelina communis" in the eastern United States. Bartonia 19:19–22.

_____. 1938. What is *Commelina communis*? Proc. Acad. Nat. Sci. Philadelphia 90:31–39.

_____. 1946. Reconsideration of the *Bacopa-Herpestis* problem of the Scrophulariaceae. Proc. Acad. Nat. Sci. Philadelphia 98:83–98.

Pennington, T.D. 1990. Sapotaceae. Fl. Neotrop. Monogr. 52:1–770.

_____. 1991. The genera of Sapotaceae. Royal Botanic Gardens, Kew, England, U.K. and New York Bot. Garden, Bronx.

Perdue, R.E., Jr. 1957. Synopsis of *Rudbeckia* subgenus *Rudbeckia*. Rhodora 59:293–299.

Perry, G. and **J. McNeill.** 1986. The nomenclature of *Eragrostis cilianensis* (Poaceae) and the contribution of Bellardi to Allioni's *Flora pedemontana*. Taxon 35:696–701.

Perry, J.D. 1971. Biosystematic studies in the North American genus *Sabatia* (Gentianaceae). Rhodora 73:309–369.

Perry, L.M. 1933. A revision of the North American species of *Verbena*. Ann. Missouri Bot. Gard. 20:239–362.

_____. 1935. *Evolvulus pilosus* an invalid name. Rhodora 37:63.

_____. 1937. Notes on *Silphium*. Rhodora 39:281–297.

Peterson, K.A., W.J. Elisens, and **J.R. Estes.** 1990. Allozyme variation in *Pyrrhopappus multicaulis* and *P. carolinianus* (Asteraceae): Relation to mating system and purported hybridization. Syst. Bot. 15:534–543.

Peterson, P.M. and **R.D. Webster.** (forthcoming). Gould's grass systematics. Texas A&M Univ. Press, College Station.

_____, _____, and **J. Valdes-Reyna.** 1997. Genera of New World Eragrostideae (Poaceae: Chloridoideae). Smithsonian Institution Press, Washington, D.C.

Peterson, R.E. 1995. Regional climate of Northwest and North Central Texas. In: J. Norwine, J.R. Giardino, G.R. North, and J.B. Valdés, eds. The changing climate of Texas: Predictability and implications for the future. Chapter 10. Pp. 92–121. GeoBooks, Texas A&M Univ., College Station.

Pfeifer, H.W. 1966. Revision of the North and Central American hexandrous species of *Aristolochia* (Aristolochiaceae). Ann. Missouri Bot. Gard. 53:115–196.

Pfeiffer, N.E. 1922. Monograph of the Isoëtaceae. Ann. Missouri Bot. Gard. 9:79–233.

Phelan, R. 1976. Texas wild, the land, plants, and animals of the Lone Star state. Bookthrift, New York.

Philbrick, C.T. 1991. Hydrophily: Phylogenetic and evolutionary considerations. Rhodora 93:36–50.

_____. 1993. Underwater cross-pollination in *Callitriche hermaphroditica* (Callitrichaceae): Evidence from randomly amplified polymorphic DNA markers. Amer. J. Bot. 80:391–394.

_____ and **G.J. Anderson.** 1992. Pollination biology in the Callitrichaceae. Syst. Bot. 17:282–292.

Philbrick, C.T. and **R.K. Jansen.** 1991. Phylogenetic studies of North American *Callitriche* (Callitrichaceae) using Chloroplast DNA restriction fragment analysis. Syst. Bot. 16:478–491.

Philbrick, C.T. and **J.M. Osborn.** 1994. Exine reduction in underwater flowering *Callitriche* (Callitrichaceae): Implications for the evolution of hypohydrophily. Rhodora 96:370–381.

Phipps, J.B. (James). Personal communication. Specialist in Rosaceae, especially Maloideae worldwide and *Crataegus* of North America; Flora of North America Editorial Committee; Univ. of Western Ontario, London.

Phipps, J.B. 1983. Biogeographic, taxonomic, and cladistic relationships between east Asiatic and North American *Crataegus*. Ann. Missouri Bot. Gard. 70:667–700.

_____. 1988. *Crataegus* (Maloideae, Rosaceae) of the southeastern United States, I. Introduction and series *Aestivales*. J. Arnold Arbor. 69:401–431.

_____. 1997. Monograph of northern Mexican *Crataegus* (Rosaceae, subfam. Maloideae). Sida, Bot. Misc. 15:1–94

_____ and **M. Muniyamma.** 1980. A taxonomic revision of *Crataegus* (Rosaceae) in Ontario. Canad. J. Bot. 58:1621–1699.

Phipps, J.B., K.R. Robertson, J.R. Rohrer, and **P.G. Smith.** 1991. Origins and evolution of subfam. Maloideae (Rosaceae). Syst. Bot. 303–332.

Phipps, J.B., K.R. Robertson, P.G. Smith, and **J.R. Rohrer.** 1990. A checklist of the subfamily Maloideae (Rosaceae). Canad. J. Bot. 68:2209–2269.

Pickens, M.A. 1997. Benny Simpson's legacy to NPSOT. Sida 17:852–853. Originally published 1997 in Native Pl. Soc. Texas News 15(2):2.

Piehl, M.A. 1962. The parasitic behavior of *Dasistoma macrophylla*. Rhodora 64:331–336.

_____. 1963. Mode of attachment, haustorium structure, and hosts of *Pedicularis canadensis*. Amer. J. Bot. 50:978–985.

_____. 1965. The natural history and taxonomy of *Comandra* (Santalaceae). Mem. Torrey Bot. Club 22:1–97.

Pielou, E.C. 1988. The world of northern evergreens. Comstock Publishing Associates, a division of Cornell Univ. Press, Ithaca, NY.

Pietropaolo, J. and **P. Pietropaolo.** 1986. Carnivorous plants of the world. Timber Press, Portland, OR.

Pimenov, M.G. and **M.V. Leonov.** 1993. The genera of the Umbelliferae. Royal Botanic Gardens, Kew, England, U.K.

Ping-Shen, H., S. Kurita, Yuzhi-Zhou, and **L. Jin-Zhen.** 1994. Synopsis of the genus *Lycoris* (Amaryllidaceae). Sida 16:301–331.

Pinkava, D.J. 1967. Biosystematic study of *Berlandiera* (Compositae). Brittonia 19:285–298.

Piper, C.V. 1906. North American species of *Festuca*. Contr. U.S. Natl. Herb. 10:1–48.

_____ and **K.S. Bort.** 1915. The early agricultural history of timothy. J. Amer. Soc. Agron. 7:1–14.

Pippen, R.W. 1978. *Cacalia*. N. Amer. Fl. II. 10:151–159.

Platt, S.G. and **C.G. Brantley.** 1997. Canebrakes: An ecological and historical perspective. Castanea 62:8–21

Plunkett, G.M., D.E. Soltis, and **P.S. Soltis.** 1996 [1997]. Evolutionary patterns in Apiaceae: Inferences based on *matK* sequence data. Syst. Bot. 21:477–495.

Pogan, E. 1963. Taxonomical value of *Alisma trivale* Pursh and *Alisma subcordatum* Raf. Canad. J. Bot. 41:1011-1013.

Pohl, R.W. 1951. The genus *Setaria* in Iowa. Iowa State Coll. J. Sci. 25:501–508.

_____. 1959. Morphology and cytology of some hybrids between *Elymus canadensis* and *E. virginicus*. Proc. Iowa Acad. Sci. 66:155–159.

_____. 1962. Notes of *Setaria viridis* and *S. faberi* (Gramineae). Brittonia 14:210–213.

_____. 1969. *Muhlenbergia*, subgenus *Muhlenbergia* (Gramineae) in North America. Amer. Midl. Naturalist 82:512–542.

Polhill, R.M. and **P.H. Raven,** eds. 1981. Advances in legume systematics. Vols. 1, 2. Proc. Intern. Legume Conference, Kew, England, U.K.

Pool, W.C. 1964. Bosque territory: A history of an agrarian community. Chaparral Press, Kyle, Texas.

Poole, M.M. and **D.R. Hunt.** 1980. Pollen morphology and the taxonomy of the Commelinaceae: An exploratory survey. American Commelinaceae: VIII. Kew Bull. 34:639–660.

Porter, C.L. 1967. Taxonomy of flowering plants, 2nd ed. W.H. Freeman and Co., San Francisco, CA.

Porter, D.M. 1969. The genus *Kallstroemia* (Zygophyllaceae). Contr. Gray Herb. 198:41–153.

_____. 1972. The genera of Zygophyllaceae in the southeastern United States. J. Arnold Arbor. 53:531–552.

_____. 1974. Disjunct distributions in the New World Zygophyllaceae. Taxon 23:339–346.

_____. 1976. *Zanthoxylum* (Rutaceae) in North America north of Mexico. Brittonia 28:443–447.

Powell, A.M. 1973. Taxonomy of *Perityle* section *Laphamia* (Compositae-Helenieae-Peritylinae). Sida 5:61–128.

_____. 1988. Trees and shrubs of Trans-Pecos Texas including Big Bend and Guadalupe Mountains National Parks. Big Bend Natural History Association, Inc., Big Bend National Park, TX.

_____. 1994. Grasses of the Trans-Pecos and adjacent areas. Univ. of Texas Press, Austin.

_____. 1998. Trees and shrubs of the Trans-Pecos and adjacent areas. Univ. of Texas Press, Austin.

Price, R.A. 1989. The genera of Pinaceae in the southeastern United States. J. Arnold Arbor. 70:247–305.

_____. 1996. Systematics of the Gnetales: A review of morphological and molecular evidence. In: W.E. Friedman (symposium organizer). Biology and evolution of the Gnetales. Int. J. Pl. Sci. 157 (Nov. 1966 suppl.):S40–S49.

_____ and **J.M. Lowenstein.** 1989. An immunological comparison of the Sciadopityaceae, Taxodiaceae, and Cupressaceae. Syst. Bot. 14:141–149.

Price, R.A. and **J.D. Palmer.** 1993. Phylogenetic relationships of the Geraniaceae and Geraniales from *rbcL* sequence comparisons. Ann. Missouri Bot. Gard. 80:661–671.

Pringle, J.S. (James). Personal communication. Specialist in Gentianaceae and botanical history. Royal Botanical Gardens, Hamilton, Ontario, Canada.

Pringle, J.S. 1997. *Clematis*. In: Flora of North America Editorial Committee, eds. Fl. North Amer. 3:158–159, 164–176. Oxford Univ. Press, New York and Oxford.

Procher, R.D. 1978. *Boerhaavia diffusa* L. (*B. coccinea* Mill.) (Nyctaginaceae) in the Carolinas. Castanea 43:172–174.

Proctor, M., P. Yeo, and **A. Lack.** 1996. The natural history of pollination, Timber Press, Portland, OR.

Pruski, J.F. and **G.L. Nesom.** 1992. *Glandularia* × *hybrida* (Verbenaceae), a new combination for a common horticultural plant. Brittonia 44:494–496.

Puff, C. 1976. The *Galium trifidum* group (Galium sect. *Aparinoides*, Rubiaceae). Canad. J. Bot. 54:1911–1925.

_____. 1977. The *Galium obtusum* group (*Galium* sect. *Aparinoides*, Rubiaceae). Bull. Torrey Bot. Club 104:202–208.

Pulich, W.M. 1988. The birds of North Central Texas. Texas A&M Univ. Press, College Station.

Punt, W. 1962. Pollen morphology of the Euphorbiaceae with special reference to taxonomy. Wentia 7:1–116.

Pyrah, G.L. 1969. Taxonomic and distributional studies in *Leersia* (Gramineae). Iowa State J. Sci. 44:215–270.

Qiu, Y-L., M.W. Chase, D.L. Les, and **C.R. Parks.** 1993. Molecular phylogenetics of the Magnoliidae: Cladistic analyses of nucleotide sequences of the plastid gene *rbcL*. Ann. Missouri Bot. Gard. 80:587–606.

Quayle, J. (Jeff). Personal communication. Plant collector and naturalist of Fort Worth who has contributed numerous North Central Texas county records to the Botanical Research Institute of Texas, Fort Worth.

Quibell, C.F. 1993. Philadelphaceae. In: J.C. Hickman, ed. The Jepson manual: Higher plants of California. Pp. 816–818. Univ. of California Press, Berkeley.

Rabe, E.W. and **M.W. Windham.** 1993. *Cheilanthes*. In: Flora of North America Editorial Committee, eds. Fl. North Amer. 2:152–169. Oxford Univ. Press, New York and Oxford.

Rabeler, R.K. (Richard). Personal communication. Specialist in Caryophyllaceae, especially introduced species; Collections Manager at Univ. of Michigan Herbarium, Ann Arbor.

Rabeler, R.K. 1985. *Petrorhagia* (Caryophyllaceae) of North America. Sida 11:6–44.

_____. 1988. Eurasian introductions to the Michigan flora. IV. Two additional species of Caryophyllaceae in Michigan. Michigan Bot. 27:85–88.

_____. 1992. A new combination in *Minuartia* (Caryophyllaceae). Sida 15:95–96.

_____ and **A.A. Reznicek.** 1997. *Cerastium pumilum* and *Stellaria pallida* (Caryophyllaceae) new to Texas. Sida 17:843–845.

Rabeler, R.K. and **J.W. Thieret.** 1988. Comments on the Caryophyllaceae of the southeastern United States. Sida 13:149–156.

RADFORD, A.E., H.E. AHLES, and **C.R. BELL.** 1968. Manual of the vascular flora of the Carolinas. The Univ. of North Carolina Press, Chapel Hill.

RADFORD, A.E., W.C. DICKISON, J.R. MASSEY, and **C.R. BELL.** 1974. Vascular plant systematics. Harper and Row Publishers, New York.

RAHN, K. 1974. *Plantago* section *Virginica*, a taxonomic revision of a group of American plantains, using experimental, taximetric and classical methods. Dansk Bot. Ark. 30:1–178.

RALPHS, M.H. 1992. Ecology, control, and grazing management of locoweeds in the western U.S. In: L.F. James, R.F. Keeler, E.M. Bailey, Jr., P.R. Cheeke, and M.P. Hegarty, eds. Poisonous plants: Proceedings of the third international symposium. Pp. 528–533. Iowa State Univ. Press, Ames.

RAMAMOORTHY, T.P. and **E.M. ZARDINI.** 1987. The systematics and evolution of *Ludwigia* sect. *Myrtocarpus* sensu lato (Onagraceae). Monogr. Syst. Bot. Missouri Bot. Gard. 19:1–120.

RANDALL, J.B., L. HUFFORD, and **D.E. SOLTIS.** 1996. Phylogenetic relationships in Sarraceniaceae based on *rbc*L and ITS sequences. Syst. Bot. 21:121–134.

RANDALL, J.M. 1997. Defining weeds of natural areas. In: J.O. Luken and J.W. Thieret, eds. Assessment and management of plant invasions. Springer-Verlag, New York.

RANKER, T.A. and **A.F. SCHNABEL.** 1986. Allozymic and morphological evidence for a progenitor-derivative species pair in *Camassia* (Liliaceae). Syst. Bot. 11:433–445.

RASMUSSEN, F.N. 1985. Orchids. In: R.M.T. Dahlgren, H.T. Clifford, and P.F. Yeo. 1985. The families of the monocotyledons. Pp. 249–274. Springer-Verlag, Berlin, Heidelberg.

RAVEN, P.H. 1963. The old world species of *Ludwigia* (including *Jussiaea*), with a synopsis of the genus (Onagraceae). Reinwardtia 6:327–427.

_____. 1964. The generic subdivision of Onagraceae, tribe Onagreae. Brittonia 16:276–288.

_____ and **D.I. AXELROD.** 1974. Angiosperm phylogeny and past continental movements. Ann. Missouri Bot. Gard. 61:539–673.

RAVEN, P.H., W. DEITRICH, and **W. STUBBE.** 1979. An outline of the systematics of *Oenothera* subsect. *Euoenothera* (Onagraceae). Syst. Bot. 4:242–252.

RAVEN, P.H., R.F. EVERT, and **S.E. EICHHORN.** 1986. Biology of plants, 4th ed. Worth Publishers, Inc., New York.

_____, _____, and _____. 1999. Biology of Plants, 6th ed. W.H. Freeman and Co., New York.

RAVEN, P.H. and **D.P. GREGORY.** 1972a. A revision of the genus *Gaura* (Onagraceae). Mem. Torrey Bot. Club 23:1–96.

_____ and _____. 1972b. Observations of meiotic chromosomes in *Gaura* (Onagraceae). Brittonia 24:71–86.

RAY, M.F. 1987. *Soliva* (Asteraceae: Anthemideae) in California. Madroño 34:228–239.

_____. 1995. Systematics of *Lavatera* and *Malva* (Malvaceae, Malveae)—a new perspective. Pl. Syst. Evol. 198:29–53.

RAY, P.M. and **H.F. CHISAKI.** 1957a. Studies on *Amsinckia*. I. A synopsis of the genus, with a study of heterostyly in it. Amer. J. Bot. 44:529–536.

_____ and _____. 1957b. Studies on *Amsinckia*. II. Relationships among the primitive species. Amer. J. Bot. 44:537–544.

_____ and _____. 1957c. Studies on *Amsinckia*. III. Aneuploid diversification in the Muricatae. Amer. J. Bot. 44:545–554.

RAYNAL, J. 1976. Notes cypérologiques: 26. Le genere *Schoenoplectus* II. L'amphicarpie et al sect. *Supini*. Adansonia 16:119–155.

RECHINGER, K.H., JR. 1937. The North American species of *Rumex*. Field Mus. Nat. Hist., Bot. Ser. 17:1–151.

REDMANN, R.E. 1985. Adaptation of grasses to water stress—leaf rolling and stomate distribution. Ann. Missouri Bot. Gard. 72:833–842.

REED, C.F. 1969a. Chenopodiaceae. In: C.L. Lundell, ed. Flora of Texas 2:21–88. Texas Research Foundation, Renner.

_____. 1969b. Amaranthaceae. In: C.L. Lundell, ed. Flora of Texas 2:89–150. Texas Research Foundation, Renner.

_____. 1969c. Nyctaginaceae. In: C.L. Lundell, ed. Flora of Texas 2:151–220. Texas Research Foundation, Renner.

_____. 1977. History and distribution of Eurasian watermilfoil in the United States and Canada. Phytologia 38:417–434.

REESE, G. 1967. Cytologische and taxonomische Untersuchungen an *Zannichellia palustris* L. Biol. Zentralbl. 86(Suppl.):277–306.

REEVES, P.A. and **R.G. OLMSTEAD.** 1993. Polyphyly of the Scrophulariaceae and origin of the aquatic Callitrichaceae inferred from chloroplast DNA sequences. Amer. J. Bot. 80:174.

REEVES, R.G. 1972. Flora of Central Texas. Grant Davis Inc., Dallas, TX.

_____. 1977. Flora of Central Texas, revised ed. Grant Davis Inc., Dallas, TX.

_____ and **D.C. BAIN.** 1947. Flora of South Central Texas. The Exchange Store, Texas A&M Univ., College Station.

REHDER, A. 1903. Synopsis of the genus *Lonicera*. Rep. (Annual) Missouri Bot. Gard. 14:27–232.

_____. 1920. The American and Asiatic species of *Sassafras*. J. Arnold Arbor. 1:242–245.

REINERT, G.W. and **R.K. GODFREY.** 1962. Reappraisal of *Utricularia inflata* and *U. radiata* (Lentibulariaceae). Amer. J. Bot. 49:213–220.

REISEBERG, L.H., S.M. BRECKSTROM-STERNBERG, A. LISTON, and **D.M. ARIAS.** 1991. Phylogenetic and systematic inferences from chloroplast isozyme variation in *Helianthus* sect. *Helianthus* (Asteraceae). Syst. Bot. 16:50–76.

REISFIELD, A.S. 1995. Central Texas plant ecology and oak wilt. In: D.N. Appel and R.F. Billings, eds. Oak wilt perspectives: The proceedings of the national oak wilt symposium. Pp. 133–138. Information Development, Inc., Houston, TX.

REJMÁNEK, M. and **J.M. RANDALL.** 1994. Invasive alien plants in California: 1993 summary and comparison with other areas in North America. Madroño 41:161–177.

RENFRO, H.B., D.E. FERAY, and **P.B. KING.** 1973. Geological highway map of Texas. American Association of Petroleum Geologists, Tulsa, OK.

REVEAL, J.L. 1968. Notes on the Texas Eriogonums. Sida 3:195–205.

_____. 1989. The eriogonoid flora of California (Polygonaceae, Eriogonoideae). Phytologia 66:295–414.

_____. 1990a. The neotypification of *Lemna minuta* Humb., Bonpl. & Dunth, an earlier name for *Lemna minuscula* Herter (Lemnaceae). Taxon 39:328–330.

______. 1990b. Minor new combinations in *Toxicodendron* (Anacardiaceae). Phytologia 69:275.

______. 1993a. A splitter's guide to the higher taxa of the flowering plants (Magnoliophyta) generally arranged to follow the sequence proposed by Thorne (1992) with certain modifications. Phytologia 74:203–263.

______. 1993b. Flowering plant families: An overview. In: Flora of North America Editorial Committee, eds. Fl. North Amer. 1:294–298. Oxford Univ. Press, New York and Oxford.

Reverchon, J. 1879. Flora of Dallas County, Texas. Bot. Gaz. (Crawfordsville) 4:210–211.

______. 1880. Notes on some introduced plants in Dallas County, Texas. Bot. Gaz. (Crawfordsville) 5:10.

______. 1903. The fern flora of Texas. Fern Bull. 11:33–38.

Reznicek, A.A. 1990. Evolution in sedges (*Carex*, Cyperaceae). Canad. J. Bot. 68:1409–1432.

______ and **P.W. Ball.** 1980. The taxonomy of *Carex* section *Stellulatae* in North America north of Mexico. Contr. Univ. Michigan Herb. 14:153–203.

Reznicek, A.A. and **R.S.W. Bobbette.** 1976. The taxonomy of *Potamogeton* subsection *Hybridi* in North America. Rhodora 78:650–673.

Reznicek, A.A. and **R.F.C. Naczi.** 1993. Taxonomic status, ecology, and distribution of *Carex hyalina* (Cyperaceae). Contr. Univ. Michigan Herb. 19:141–147.

Rhodes, A.M., W.P. Bemis, T.W. Whitaker, and **S.G. Carmer.** 1968. A numerical taxonomic study of *Cucurbita*. Brittonia 20:251–266.

Rhodes, D.G. 1997. Menispermaceae. In: Flora of North America Editorial Committee, eds. Fl. North Amer. 3:295–299. Oxford Univ. Press, New York and Oxford.

Richards, E.L. 1968. A monograph of the genus *Ratibida*. Rhodora 70:348–393.

Richardson, C.W. 1993. Disappearing land: Erosion in the Blacklands. In: M.R. Sharpless and J.C. Yelderman, eds. The Texas Blackland Prairie, land, history, and culture. Pp. 262–270. Baylor Univ. Program for Regional Studies, Waco, TX.

Richardson, J.W. 1968. The genus *Euphorbia* of the high plains and prairie plains of Kansas, Nebraska, South and North Dakota. Univ. Kansas Sci. Bull. 48:45–112.

Rick, C.M. and **M. Holle.** 1990. Andean *Lycopersicon esculentum* var. *cerasiforme*: Genetic variation and its evolutionary significance. Econ. Bot. 44 (Suppl.):69–78.

Rickett, H.W. 1945a. Cornaceae. N. Amer. Fl. 28b:299–311.

______. 1945b. Nyssaceae. N. Amer. Fl. 28b:213–216.

Ridley, M. 1996. Evolution. Blackwell Science, Cambridge, MA.

Ridsdale, C.E. 1976. A revision of the tribe Cephalantheae (Rubiaceae). Blumea 23:177–188.

Rieseberg, L.H. 1991. Homoploid reticulate evolution in *Helianthus* (Asteraceae): Evidence from ribosomal genes. Amer. J. Bot. 78:1218–1237.

______ and **G.J. Seiler.** 1990. Molecular evidence and the origin and development of the domesticated sunflower (*Helianthus annus*, Asteraceae). Econ. Bot. 44 (Suppl.):79–91.

Riggins, R. 1977. A biosystematic study of the *Sporobolus asper* complex (Gramineae). Iowa State J. Res. 51:287–321.

______, **S.M. Beckstrom-Sternberg, A. Liston,** and **D.M. Arias.** 1991. Phylogenetic and systematic inferences from chloroplast DNA and isosyme variation in *Helianthus* sect. *Helianthus* (Asteraceae). Syst. Bot. 16:50–76.

Riskind, D.O. and **O.B. Collins.** 1975. The Blackland Prairie of Texas: Conservation of representative climax remnants. In: M.K. Wali. Prairie: A multiple view. Pp. 361–367. The Univ. of North Dakota Press, Grand Forks.

Riskind, D.O. and **D.D. Diamond.** 1988. An introduction to environments and vegetation. In: B.B. Amos and F.R. Gehlbach, eds. Edwards Plateau vegetation: Plant ecological studies in central Texas. Pp. 1–15. Baylor Univ. Press, Waco, TX.

Rivas, M. 1994. Seeking an antidote: Officials hope to learn prevalence of jimson weed abuse, avert new tragedy. Dallas Morning News, 28 August, p. 47–49A.

Robbins, G.T. 1944. North American species of *Androsace*. Amer. Midl. Naturalist 32:137–163.

Roberts, R.P. (Roland). Personal communication. Specialist in Euphorbiaceae, especially *Argythamnia* and *Ditaxis*. Graduate student at Louisiana State Univ., Baton Rouge, LA.

Robertson, K.R. 1966. The genus *Erythronium* (Liliaceae) in Kansas. Ann. Missouri Bot. Gard. 53:197–204.

______. 1971. The Linaceae in the southeastern United States. J. Arnold Arbor. 52:649–655.

______. 1972. The genera of Geraniaceae in the southeastern United States. J. Arnold Arbor. 53:182–201.

______. 1973. The Krameriaceae in the southeastern United States. J. Arnold Arbor. 54:322–327.

______. 1974. The genera of Rosaceae in the southeastern United States. J. Arnold Arbor. 55:303–332, 344–401, 611–662.

______. 1975. The Oxalidaceae in the southeastern United States. J. Arnold Arbor. 56:223–239.

______. 1981. The genera of Amaranthaceae in the southeastern United States. J. Arnold Arbor. 62:267–314.

______ and **Y.-T. Lee.** 1976. The genera of Caesalpinioideae (Leguminosae) in the southeastern United States. J. Arnold Arbor. 57:1–53.

Robertson, K.R., J.B. Phipps, J.R. Rohrer, and **P.G. Smith.** 1991. A synopsis of genera in Maloideae (Rosaceae). Syst. Bot. 16:376–394.

Robinson, B.L. 1917. A monograph of the genus *Brickellia*. Mem. Gray Herb. 1:1–151.

______ and **J.M. Greenman.** 1899. Synopsis of the genus *Verbesina*, with an analytical key to the species. Proc. Amer. Acad. Arts 34:534–566.

Robinson, H. 1974. Studies in the Senecioneae (Asteraceae). VI. the genus *Arnoglossum*. Phytologia 28:294–295.

______. 1977. Studies in the Heliantheae (Asteraceae). VIII. Notes on genus and species limits in the genus *Viguiera*. Phytologia 36:201–215.

______. 1978. Re-establishment of the genus *Smallanthus*. Phytologia 39:47–52.

______ and **J. Cuatrecasas.** 1973. The generic limits of *Pluchea* and *Tessaria* (Inuleae, Asteraceae). Phytologia 27:277–285.

Robinson, H. and **R.M. King.** 1985. Comments on the generic concepts in the Eupatorieae. Taxon 34:11–16.

Robson, N.K.B. 1980. The Linnaean species of *Ascyrum*

(Guttiferae). Taxon 29:267–274.

Rock, H.F.L. 1957. A revision of the vernal species of *Helenium* (Compositae). Rhodora 59:101–116, 128–158, 168–178, 203–216.

Rodman, J.E. 1990. Centrospermae revisited, part 1. Taxon 39:383–393.

_____, **M.K. Oliver, R.R. Nakamura, J.U. McClammer, Jr.,** and **A.H. Bledsoe.** 1984. A taxonomic analysis and revisited classification of Centrospermae. Syst. Bot. 9:297–323.

Roemer, F. 1849. Texas with particular reference to German immigration and the physical appearance of the country. Translated from the German by O. Mueller. Standard Printing Company, San Antonio, TX. Reprinted 1983 by the German-Texan Heritage Society. Texian Press, Waco, TX.

Rogers, C.E., T.E. Thompson, and **G.J. Seiler.** 1982. Sunflower species of the United States. National Sunflower Association, Bismarck, ND.

Rogers, C.M. 1963. Studies in *Linum*: *L. imbricatum* and *L. hudsonioides*. Rhodora 65:51–55.

_____. 1964. Yellow-flowered *Linum* (Linaceae) in Texas. Sida 1:328–336.

_____. 1968. Yellow-flowered species of *Linum* in Central America and western North America. Brittonia 20:107–135.

_____. 1984. Linaceae. N. Amer. Fl. II. 12:1–58.

Rogers, D.G. 1951. A revision of *Stillingia* in the New World. Ann. Missouri Bot. Gard. 38:207–259.

Rogers, G.K. 1983. The genera of Alismataceae in the southeastern United States. J. Arnold Arbor. 62:383–420.

_____. 1984. The Zingiberales (Cannaceae, Marantaceae, and Zingiberaceae) in the southeastern United States. J. Arnold Arbor. 65:5–55.

_____. 1985. The genera of Phytolaccaceae in the southeastern United States. J. Arnold Arbor. 66:1–37.

_____. 1986. The genera of Loganiaceae in the southeastern United States. J. Arnold Arbor. 67:143–185.

_____. 1987. The genera of Cinchonoideae (Rubiaceae) in the southeastern United States. J. Arnold Arbor. 68:137–183.

Rogers, H.J. 1949. The genus *Galactia* in the United States. Ph.D. dissertation, Duke Univ., Durham, NC.

Rohwer, J.G. 1993a. Lauraceae. In: K. Kubitzki, J.C. Rohwer, and V. Bittrich, eds. The families and genera of vascular plants, Vol. II. Pp. 366–391. Springer-Verlag, Berlin.

_____. 1993b. Phytolaccaceae. In: K. Kubitzki, J.C. Rohwer, and V. Bittrich, eds. The families and genera of vascular plants, Vol. II. Pp. 506–515. Springer-Verlag, Berlin.

_____ and **C.C. Berg.** 1993. Moraceae. In: K. Kubitzki, J.C. Rohwer, and V. Bittrich, eds. The families and genera of vascular plants, Vol. II. Pp. 438–453. Springer-Verlag, Berlin.

Rolfsmeier, S.B. 1995. Keys and distributional maps for Nebraska Cyperaceae, Part 1: *Bulbostylis, Cyperus, Dulichium, Eleocharis, Eriophorum, Fimbristylis, Fuirena, Lipocarpha*, and *Scirpus*. Trans. Nebraska Acad. Sci. 22:27–42.

Rollins, R.C. 1941. A monographic study of *Arabis* in western North America. Rhodora 43:289–325, 348–411, 425–481.

_____. 1942. A systematic study of *Iodanthus*. Contr. Dudley Herb. 3:209–239.

_____. 1947. Generic revisions in the Cruciferae: *Sibara*. Contr. Gray Herb. 165:133–143.

_____. 1950. The guayule rubber plant and its relatives. Contr. Gray Herb. 172:1–73.

_____. 1955. The auriculate-leaved species of *Lesquerella* (Cruciferae). Rhodora 57:241–264.

_____. 1956. On the identity of *Lesquerella angustifolia*. Rhodora 58:199–202.

_____. 1981. Weeds of the Cruciferae (Brassicaceae) in North America. J. Arnold Arbor. 62:517–540.

_____. 1993. The Cruciferae of continental North America: Systematics of the mustard family from the Artic to Panama. Stanford Univ. Press, Stanford, CA.

_____ and **R.C. Foster.** 1957. Studies in *Sida* (Malvaceae). A review of the genus and monograph of the sections Malacroideae, Physalodes, Pseudomalvastrum, Incanifolia, Oligandrae, Pseudonapaea, Hookeria, and Stenindа. Contr. Gray Herb. 180:5–87.

Rollins, R.C. and **E.A. Shaw.** 1973. The genus *Lesquerella* (Cruciferae) in North America. Harvard Univ. Press, Cambridge, MA.

Rominger, J.M. 1962. Taxonomy of *Setaria* (Gramineae) in North America. Illinois Biol. Monogr. 29:1–132.

Roose, M.L. and **L.D. Gottlieb.** 1976. Genetic and biochemical consequences of polyploidy in *Tragopogon*. Evolution 30:818–830.

Rosatti, T.J. 1984. The Plantaginaceae in the southeastern United States. J. Arnold Arbor. 65:533–562.

_____. 1986. The genera of Sphenocleaceae and Campanulaceae in the southeastern United States. J. Arnold Arbor. 67:1–64.

_____. 1987. The genera of Pontederiaceae in the southeastern United States. J. Arnold Arbor. 68:35–71.

_____. 1989. The genera of suborder Apocynineae (Apocynaceae and Asclepiadaceae) in the southeastern United States. J. Arnold Arbor. 70:307–401, 443–514.

Rose, F.L. and **R.W. Strandtmann.** 1986. Wildflowers of the Llano Estacado. Taylor Publishing Co., Dallas, TX.

Rose, J.N. 1909. Rafflesiaceae. The North American species of *Pilosytles* (Pp. 262–265). In: Studies of Mexican and Central American Plants—No. 6. Contr. U.S. Natl. Herb. 12:259–302.

Rosendahl, C.O., F.K. Butters, and **O. Lakela.** 1936. A monograph on the genus *Heuchera*. Minnesota Stud. Pl. Sci. 2:1–180.

Rossbach, G.B. 1939. Aquatic Utricularias. Rhodora 41:113–128.

_____. 1958. The genus *Erysimum* in North America north of Mexico—a key to the species and varieties. Madroño 14:261–267.

Rossbach, R.P. 1940. *Spergularia* in North America and South America. Rhodora 42:57–83, 105–143, 158–193, 203–213.

Rothrock, P.E. 1991. The identity of *Carex albolutescens, C. festucacea*, and *C. longii* (Cyperaceae). Rhodora 93:51–66.

Rothwell, N.V. 1959. Aneuploidy in *Claytonia virginica*. Amer. J. Bot. 46:353–360.

_____ and **J.G. Kump.** 1965. Chromosome numbers in populations of *Claytonia virginica* from the New York metropolitan area. Amer. J. Bot. 52:403–407.

Rowell, C.M., Jr. 1972. Lloyd Herbert Shinners, 1918–1971. In Memoriam. Texas J. Sci. 24:266–271.

Roy, G.P. 1968. A systematic study of the *Bouteloua hirsuta-Bouteloua pectinata* complex. Ph.D. dissertation, Texas A&M Univ., College Station.

_____ and **F.W. Gould.** 1971. Biosystematic investigations of *Bouteloua hirsuta* and *B. pectinata*. I. Gross morphology. SouthW. Naturalist 15:377–387.

Rozen, J.G., Jr. and **G.C. Eickwort.** 1997. The entomological evidence. J. Forensic Sci. 42:394–397.

Rudd, V.E. 1972. Leguminosae-Faboideae-Sophoreae. N. Amer. Fl. II 7:1–53.

Ruffin, J. 1974. A taxonomic re-evaluation of the genera *Amphiachyris, Amphipappus, Greenella, Gutierrezia, Gymnosperma, Thurovia*, and *Xanthocephalum* (Compositae). Sida 5:301–333.

Runemark, H. and **W.K. Heneen.** 1968. *Elymus* and *Agropyron*, a problem of generic delimitation. Bot. Not. 121:51–79.

Rupp, R. 1987. Blue corn & square tomatoes: unusual facts about common vegetables. Storey Communications, Inc., Pownal, VT.

Russell, N.H. 1961. Notes: Keys to Louisiana violets (*Viola*-Violaceae). SouthW. Naturalist 6:184–186.

_____. 1965. Violets (*Viola*) of central and eastern United States: An introductory survey. Sida 2:1–113.

Rust, R.W. 1977. Pollination in *Impatiens capensis* and *Impatiens pallida* (Balsaminaceae). Bull. Torrey Bot. Club. 104:361–367.

Rydberg, P.A. 1896. The North American species of *Physalis* and related genera. Mem. Torrey Bot. Club 4:297–372.

_____. 1908. Rosaceae. N. Amer. Fl. 22:239–388.

_____. 1909. Elodeaceae. N. Amer. Fl. 17:67–71.

_____. 1910. Balsaminaceae. N. Amer. Fl. 25:93–96.

_____. 1913. Rosaceae (cont.). N. Amer. Fl. 22:389–480.

_____. 1914–1927. Carduaceae. N. Amer. Fl. 34:1–360.

_____. 1918. Rosaceae (cont.). N. Amer. Fl. 22:481–533.

_____. 1919–1920. Fabaceae: Psoraleae. N. Amer. Fl. 24:1–136.

_____. 1920. Notes on Rosaceae – XII. Bull. Torrey Bot. Club 47:45–66.

_____. 1922. Ambrosiaceae. N. Amer. Fl. 22:3–44.

_____. 1923. Notes on Rosaceae – XIV. Bull. Torrey Bot. Club 50:61–71.

_____. 1923–1929. Fabaceae: Indigofereae, Galegeae. N. Amer. Fl. 24:127–462.

_____. 1932. Portulacaceae. N. Amer. Fl. 21:279–336.

Safford, W.E. 1921. Synopsis of the genus *Datura*. J. Wash. Acad. Sci. 11:173–189.

_____. 1922. *Daturas* of the Old World and New: An account of their narcotic properties and their use in oracular and initiatory ceremonies. Rep. (Annual) Board Regents Smithsonian Inst. (for 1920). Publ. 2644:537–567. Washington, D.C.

Sanchez, L.L. 1997. Vascular plant list of Fort Hood Military Reservation, Bell and Coryell counties, Texas. The Nature Conservancy, Unpublished March 1997 update, copy in the library of the Botanical Research Institute of Texas.

Sanders, R.W. (Roger). Personal communication. Specialist in Arecaceae, Lamiaceae, and Verbenaceae; formerly of Fairchild Tropical Garden, Miami, FL and currently Research Associate at the Botanical Research Institute of Texas, Fort Worth and adjunct assistant professor at Southern Methodist Univ., Dallas, TX.

Sanders, R.W. and **W.S. Judd.** 1998. Incorporating phylogenetic data sets into treatments. Abstract of paper presented at the Flora of the Southeast US Symposium, Botanical Research Institute of Texas, Fort Worth, 23–25 April 1998. Manuscript to be published as part of a symposium proceedings volume in Sida, Bot. Misc.

Sargent, C.S. 1918. Notes on North American trees. III. *Tilia*. Bot. Gaz. (Crawfordsville) 66:421–438, 494–511.

_____. 1922. Manual of the trees of North America. The Riverside Press, Cambridge, MA.

Sarkar, N.M. 1958. Cytotaxonomic studies on *Rumex* section *Axillares*. Canad. J. Bot. 36:947–996.

Sauer, J. 1955. Revision of the dioecious amaranths. Madroño 13:5–46.

_____. 1967. The grain amaranths and their relatives: A revised taxonomic and geographic survey. Ann. Missouri Bot. Gard. 54:103–137.

_____. 1972. Revision of *Stenotaphrum* (Gramineae: Paniceae) with attention to its historical geography. Brittonia 24:202–222.

Saunders, J.H. 1961. The wild species of *Gossypium* and their evolutionary history. Oxford Univ. Press, London.

Schander, C. 1998a. Types, emendations and names—a reply to Lidén et al. Taxon 47:401–406.

_____. 1998b. Mandatory categories and impossible hierarchies—a reply to Sosef. Taxon 47:407–410.

Schery, R.W. 1942. Monograph of *Malvaviscus*. Ann. Missouri Bot. Gard. 29:183–245.

Schilling, E.E. 1981. Systematics of *Solanum* sect. *Solanum* (Solanaceae) in North America. Syst. Bot. 6:172–185.

_____ and **D.M. Spooner.** 1988. Floral flavonoids and the systematics of *Simsia* (Asteraceae: Heliantheae). Syst. Bot. 13:572–575.

Schippers, P., S.J. Ter Borg, and **J.J. Bos.** 1995. A revision of the infraspecific taxonomy of *Cyperus esculentus* (yellow nutsedge) with an experimentally evaluated character set. Syst. Bot. 20:461–481.

Schlessman, M.A. 1984. Systematics of tuberous Lomatiums (Umbelliferae). Syst. Bot. Monogr. 4:1–55.

Schmandt, J. 1995. Effects of climate change on biodiversity and landscape linkages in Texas. In: J. Norwine, J.R. Giardino, G.R. North, and J.B. Valdés, eds. The changing climate of Texas: Predictability and implications for the future. Chapter 17. Pp. 216–240. GeoBooks, Texas A&M Univ., College Station.

Schmid, R. 1997. Some desiderata to make floras and other types of works user (and reviewer) friendly. Taxon 46:179–194.

Schmidly, D.J. 1983. Texas mammals east of the Balcones fault zone. Texas A&M Univ. Press, College Station.

_____, **D.L. Scarbrough,** and **M.A. Horner.** 1993. Wildlife diversity in the Blackland Prairies. In: M.R. Sharpless and J.C. Yelderman, eds. The Texas Blackland Prairie, land, history, and culture. Pp. 82–95. Baylor Univ. Program for Regional Studies, Waco, TX.

Schmutz, E.M. and **L.B. Hamilton.** 1979. Plants that poison. Northland Publishing, Flagstaff, AZ.

Schnack, B. and **G. Covas.** 1944. Nota sobre la validez del género *Glandularia* (Verbenáceas). Darwiniana 6:469–476.

Schnell, D.E. and **D.W. Krider.** 1976. Cluster analysis of the genus *Sarracenia* L. in the southeastern United States. Castanea 41:165–176.

Schneider, E.L. and **S. Carlquist.** 1996. Conductive tissue in *Ceratophyllum demersum* (Ceratophyllaceae). Sida 17:437–443.

Schneider, E.L. and **P.S. Williamson.** 1993. Nymphaeaceae. In: K. Kubitzki, J.C. Rohwer, and V. Bittrich, eds. The families and genera of vascular plants, Vol. II. Pp. 486–493. Springer-Verlag, Berlin.

Schofield, E.K. 1986. Glossary. In: Great Plains Flora Association. Flora of the Great Plains. Pp. 1317–1328. Univ. Press of Kansas, Lawrence.

Scholz, U. 1981. Monographie der gattung *Oplismenus* (Gramineae). Phanerogamarum Monographiae Tomus 13. Pp. 1–213. J. Cramer, Hirschberg, Germany.

Schores, D.M. (Daniel). Personal communication. Sociologist and anthropologist; retired from a career at Austin College, Sherman, TX.

Schubert, B.G. 1950. *Desmodium*: Preliminary studies–III. Rhodora 52:135–155.

______. 1984. Donovan Stewart Correll, 1908–1983. Econ. Bot. 38: 134–136.

Schubert, M.T.R. and **B.-E. van Wyk.** 1995. Two new species of *Centella* (Umbelliferae) with notes on infrageneric taxonomy. Nord. J. Bot. 15:167–171.

Schultes, R.E. and **A.S. Pease.** 1963. Generic names of orchids: Their origin and meaning. Academic Press, New York.

Schulz, E.D. 1922. 500 wild flowers of San Antonio and vicinity. Published by the author, San Antonio, TX.

______. 1928. Texas wild flowers: A popular account of the common wild flowers of Texas. Laidlaw Brothers, Chicago, IL.

______ and **R. Runyon.** 1930. Texas cacti: A popular and scientific account of the cacti native of Texas. Proc. Texas Acad. Sci. 14:1–181.

Schulz, O.E. 1903. Monographie der Gattung *Cardamine*. Bot. Jahrb. Syst. 32:280–623.

Schuyler, A.E. 1966. The taxonomic delineation of *Scirpus lineatus* and *Scirpus pendulus*. Notul. Nat. Acad. Nat. Sci. Philadelphia 390:1–3.

______. 1967. A taxonomic revision of North American leafy species of *Scirpus*. Proc. Acad. Nat. Sci. Philadelphia 119:295–323.

______. 1974. Typification and application of the names *Scirpus americana* Pers., *S. olneyi* Gray, and *S. pungens* Vahl. Rhodora 76:51–52.

Schwarzwalder, R.N., Jr. and **D.L. Dilcher.** 1991. Systematic placement of the Platanaceae in the Hamamelidae. Ann. Missouri Bot. Gard. 78:962–969.

Scora, R.W. 1967. Interspecific relationships in the genus *Monarda* (Labiatae). Univ. Calif. Publ. Bot. 41:1–59.

Scott, J.A. 1986. The butterflies of North America. Stanford Univ. Press, Stanford, CA.

Scott, R.W. 1990. The genera of Cardueae (Compositae; Asteraceae) in the southeastern United States. J. Arnold Arbor. 71:391–451.

Scribner, F.L. 1907. Notes on *Muhlenbergia*. Rhodora 9:17–23.

Seabrook, J.A. and **L.A. Dionne.** 1976. Studies on the genus *Apios*. I. Chromosome number and distribution of *Apios americana* and *A. priceana*. Canad. J. Bot. 54:2567–2572.

Sealy, J.R. 1954. Review of the genus *Hymenocallis*. Kew Bull. 9:201–240.

Seelanan, T., A. Schnabel, and **J.F. Wendel.** 1997. Congruence and consensus in the cotton tribe (Malvaceae). Syst. Bot. 22:259–290.

Seeligmann, P. and **R.E. Alston.** 1967. Complex chemical variation and the taxonomy of *Hymenoxys scaposa* (Compositae). Brittonia 19:205–211.

Seigler, D.S. 1994. Phytochemistry and systematics of the Euphorbiaceae. Ann. Missouri Bot. Gard. 81:380–401.

Sell, P.D. 1976. *Hedypnois*. In: Tutin, T.G., V.H. Heywood, N.A. Burges, D.M. Moore, D.H. Valentine, S.M. Walters, and D.A. Webb, eds. Flora Europaea 4:151–152.

Sellards, E.H., W.S. Adkins, and **F.B. Plummer.** 1932. The geology of Texas, Vol. 1, Stratigraphy. Bureau of Economic Geology, Univ. of Texas, Austin.

Semple, J.C. 1974. The phytogeography and systematics of *Xanthisma texanum* DC. (Asteraceae); proper usage of infraspecific categories. Rhodora 76:1–19.

______. 1977. Cytotaxonomy of *Chrysopsis* and *Heterotheca* (Compositae-Astereae): A new interpretation of phylogeny. Canad. J. Bot. 55:2503–2513.

______. 1981. A revision of the goldenaster genus *Chrysopsis* (Nutt.) Ell. nom. cons. (Compositae-Astereae). Rhodora 83:323–384.

______. 1992. Goldenrods of Ontario: *Solidago* and *Euthamia*. Univ. Waterloo Biol. Ser. 36:1–82.

______. 1993. *Euthamia*. In: J.C. Hickman, ed. The Jepson manual: Higher plants of California. Pp. 266. Univ. of California Press, Berkeley.

______. 1996. A revision of *Heterotheca* sect. *Phyllotheca* (Nutt.) Harms (Compositae: Asteraceae): The prairie and montane goldenasters of North America. Univ. Waterloo Biol. Ser. 37:1–164.

______, **V.C. Blok,** and **P. Heiman.** 1980. Morphological, anatomical, habit, and habitat differences among the goldenaster genera *Chrysopsis*, *Heterotheca*, and *Pityopsis* (Compositae–Astereae). Canad. J. Bot. 58:147–163.

Semple, J.C. and **F.D. Bowers.** 1985. A revision of the goldenaster genus *Pityopsis* Nutt. (Compositae: Asteraceae). Univ. Waterloo Biol. Ser. 29:1–34.

Semple, J.C. and **L. Brouillet.** 1980a. A synopsis of North American asters: The subgenera, sections and subsections of *Aster* and *Lasallea*. Amer. J. Bot. 67:1010–1026.

______ and ______. 1980b. Chromosome numbers and satellite chromosome morphology in *Aster* and *Lasallea*. Amer. J. Bot. 67:1027–1039.

Semple, J.C. and **C.C. Chinnappa.** 1984. Observations on the cytology, morphology, and ecology of *Bradburia hirtella* (Compositae-Astereae). Syst. Bot. 9:95–101.

Semple, J.C. and **J.G. Chmielewski.** 1987. Revision of the *Aster lanceolatus* complex, including *A. simplex* and *A. hesperius* (Compositae: Astereae): A multivariate morphometric study. Canad. J. Bot. 65:1047–1062.

Semple, J.C., S.B. Heard, and **C.S. Xiang.** 1996. The Asters of Ontario (Compositae: Astereae): *Diplactis* Raf., *Oclemena* Greene, *Doellingeria* Nees and *Aster* L. (including *Canadanthus* Nesom, *Symphyotrichum* Nees and *Virgulus* Raf.). Univ. Waterloo Biol. Ser. 38:1–94.

Senn, H.A. 1939. The North American species of *Crotalaria*. Rhodora 41:317–367.

Settle, W.J. and **T.R. Fisher.** 1970. The varieties of *Silphium integrifolium*. Rhodora 72:536–543.

Sexton, C. (Chuck). Personal communication. Wildlife Biologist, Balcones Canyonlands National Wildlife Refuge, U.S. Fish and Wildlife Service, Austin, TX.

Shaffer-Fehre, M. 1991a. The enotegmen tuberculae: An account of little-known structures from the seed coat of the Hydrocharitoideae (Hydrocharitaceae) and *Najas* (Najadaceae). J. Linn. Soc., Bot. 107:169–188.

_____. 1991b. The position of *Najas* within the subclass Alismatidae (Monocotyledones) in the light of new evidence from seed coat structures in the Hydrocharitoideae (Hydrocharitales). J. Linn. Soc., Bot. 107:189–209.

Shan, R.H. and **L. Constance.** 1951. The genus *Sanicula* (Umbelliferae) in the Old World and the New. Univ. Calif. Publ. Bot. 25:1–78.

Sharp, W.M. 1935. A critical study of certain epappose genera of the Heliantheae-Verbesininae of the natural family Compositae. Ann. Missouri Bot. Gard. 22:51–152.

Sharpless, M.R. and **J.C. Yelderman.** 1993. Introduction. In: M.R. Sharpless and J.C. Yelderman, eds. The Texas Blackland Prairie, land, history, and culture. Pp. xv–xvi. Baylor Univ. Program for Regional Studies, Waco, TX.

Shaw, E.A. 1987. Charles Wright on the boundary, 1849–1852, or, Plantae Wrightianae revisited. Meckler Publishing Corp., Westport, CT.

Shaw, R.B. and **R.D. Webster.** 1987. The genus *Eriochloa* (Poaceae: Paniceae) in North and Central America. Sida 12:165–207.

Sherff, E.E. 1936. Revision of the genus *Coreopsis*. Field Mus. Nat. Hist., Bot. Ser. 11:279–475.

_____. 1937. The genus *Bidens*. Publ. Field Mus. Nat. Hist., Bot. Ser. 16:1–709.

_____ and **E.J. Alexander.** 1955. Compositae-Heliantheae-Coreopsidinae. N. Amer. Fl. II. 2:1–190.

Sherman-Broyles, S.L. 1997. *Ulmus*. In: Flora of North America Editorial Committee, eds. Fl. North Amer. 3:369–376. Oxford Univ. Press, New York and Oxford.

_____, **W.T. Barker,** and **L.M. Schulz.** 1997a. Ulmaceae. In: Flora of North America Editorial Committee, eds. Fl. North Amer. 3:368–369. Oxford Univ. Press, New York and Oxford.

_____, _____, and _____. 1997b. *Celtis*. In: Flora of North America Editorial Committee, eds. Fl. North Amer. 3:376–379. Oxford Univ. Press, New York and Oxford.

Sherman-Broyles, S.L., S.B. Broyles, and **J.L. Hamrick.** 1992. Geographic distribution of allozyme variation in *Ulmus crassifolia*. Syst. Bot. 17:33–41.

Shetler, S.G. and **N.R. Morin.** 1986. Seed morphology in North American Campanulaceae. Ann. Missouri Bot. Gard. 73:653–688.

Sheviak, C.J. 1982. Biosystematic study of the *Spiranthes cernua* complex. New York State Mus. Bull. 448:1–73.

Shilling, E.E. 1981. Systematics of *Solanum* sect. *Solanum* (Solanaceae) in North America. Syst. Bot. 6:172–185.

Shinners, L.H. 1946a. Revision of the genus *Chaetopappa*. Wrightia 1:63–81.

_____. 1946b. Revision of the genus *Leucelene* Greene. Wrightia 1:82–89.

_____. 1946c. Revision of the genus *Aphanostephus* DC. Wrightia 95–121.

_____. 1946d. Revision of the genus *Kuhnia* L. Wrightia 1:122–144.

_____. 1947a. Two anomalous new species of *Erigeron* L. from Texas. Wrightia 1:183–186.

_____. 1947b. Revision of the genus *Krigia* Schreber. Wrightia 1:187–206.

_____. 1948a. The vetches and pea vines (*Vicia* and *Lathyrus*) of Texas. Field & Lab. 16:18–29.

_____. 1948b. Geographic limits of some alien weeds in Texas. Texas Geogr. Mag. 12:16–25.

_____. 1949a. Nomenclature of species of dandelion and goatsbeard (*Taraxacum* and *Tragopogon*) introduced into Texas. Field & Lab. 17:13–19.

_____. 1949b. Transfer of Texas species of *Petalostemum* to *Dalea* (Leguminosae). Field & Lab. 17:81–85.

_____. 1949c. The genus *Dalea* (including *Petalostemum*) in North-Central Texas. Field & Lab. 17:85–89.

_____. 1949d. *Arenaria drummondii* Shinners, nom. nov. Field & Lab. 17:89.

_____. 1949e. The Texas species of *Conyza* (Compositae). Field & Lab. 17:142–144.

_____. 1949f. Nomenclature of Texas varieties of *Descurainia pinnata* (Cruciferae). Field & Lab. 17:145.

_____. 1949g. Transfer of Texas species of *Houstonia* to *Hedyotis* (Rubiaeae). Field & Lab. 17:166–169.

_____. 1949h. Early plant collections return to Texas. Texas J. Sci. 1:69–70. [also published in Field & Lab. 17:66–68]

_____. 1950a. The Texas species of *Thelesperma* (Compositae). Field & Lab. 18:17–24.

_____. 1950b. The species of *Matelea* (including *Gonolobus*) in North Central Texas. Field & Lab. 18:73–78.

_____. 1950c. The Texas species of *Cacalia* (Compositae). Field & Lab. 18:79–82.

_____. 1950d. The north Texas species of *Plantago* (Plantaginaceae). Field & Lab. 18:113–119.

_____. 1951a. The Texas species of *Psoralea* (Leguminosae). Field & Lab. 19:14–25.

_____. 1951b. The north Texas species of *Heterotheca*, including *Chrysopsis* (Compositae). Field & Lab. 19:66–71.

_____. 1951c. Notes on Texas Compositae. VII. Field & Lab. 19:74–82.

_____. 1951d. The north Texas species of *Hymenocallis* (Amaryllidaceae). Field & Lab. 19:102–104.

_____. 1951e. The Texas species of *Evax* (Compositae). Field & Lab. 19:125–126.

_____. 1951f. *Agave lata*, a new species from north Texas and Oklahoma. Field & Lab. 19:171–173.

_____. 1951g. The north Texas species of *Mirabilis* (Nyctaginaceae).

Field & Lab. 19:173–182.

_____. 1952. The Texas species of *Palafoxia* (Compositae). Field & Lab. 20:92–102.

_____. 1953a. Synopsis of the United States species of *Lythrum* (Lythraceae). Field & Lab. 21:80–89.

_____. 1953b. Nomenclature of the varieties of *Monarda punctata* L. (Labiatae). Field & Lab. 21:89–92.

_____. 1953c. The bluebonnets (*Lupinus*) of Texas. Field & Lab. 21:149–153.

_____. 1953d. Synopsis of the genus *Brazoria* (Labiatae). Field & Lab. 21:153–154.

_____. 1953e. Notes on Texas Compositae–IX. Field & Lab. 21:155–162.

_____. 1953f. Botanical notes. Field & Lab. 21:164–165.

_____. 1954. Notes on north Texas grasses. Rhodora 56:25–38.

_____. 1955. The Texas species of *Potentilla* (Rosaceae). Field & Lab. 23:19-21.

_____. 1956a. Authorship and nomenclature of burclovers (*Medicago*) found wild in the United States. Rhodora 58:1–13.

_____. 1956b. *Quercus shumardii* Buckley var. *microcarpa* (Torrey) Shinners comb. nov. Field & Lab. 24:37.

_____. 1956c. *Argythamnia humilis* (Engelm. & Gray) Muell.-Arg. var. *laevis* (A. Gray) Shinners, comb. nov. Field & Lab. 24:38.

_____. 1956d. Yellow-flowered *Oxalis* (Oxalidaceae) of eastern Texas and Louisiana. Field & Lab. 24:39–40.

_____. 1956e. *Andropogon ischaemum* L. var. *songaricus* Ruprecht: Technical name for King Ranch bluestem. Field & Lab. 24:101–103.

_____. 1957. Synopsis of the genus *Eustoma* (Gentianaceae). SouthW. Naturalist 2:38–43.

_____. 1958a. Spring flora of the Dallas–Fort Worth area Texas. Published by author, Dallas, TX.

_____. 1958b. *Carthamus lanatus* L. var. *baeticus* (Boissier & Reuter) P. Coutinho (Compositae): Another introduced thistle in Central Texas. SouthW. Naturalist 3:220.

_____. 1961. Nomenclature of the Bignoniaceae of the southern United States. Castanea 26:109–118.

_____. 1962a. Synopsis of United States *Bonamia* including *Breweria* and *Stylisma* (Convolvulaceae). Castanea 27:65–77.

_____. 1962b. Annual Sisyrinchiums (Iridaceae) in the United States. Sida 1:32–42.

_____. 1962c. New names in *Arenaria* (Caryophyllaceae). Sida 1:49–52.

_____. 1962d. *Drosera* (Droseraceae) in the southeastern United States: An interim report. Sida 1:53–59.

_____. 1962e. *Calamintha* (Labiatae) in the southern United States. Sida 1:69–75.

_____. 1963. *Gilia* and *Ipomopsis* (Polemoniaceae) in Texas. Sida 1:171–179.

_____. 1964a. *Calylophus* (*Oenothera* in part: Onagraceae) in Texas. Sida 1:337–345.

_____. 1964b. Texas Asclepiadaceae other than *Asclepias*. Sida 1:358–367.

_____. 1965. *Holosteum umbellatum* (Caryophyllaceae) in the United States: Population and fractionated suicide. Sida 2:119–128.

_____. 1966a. Texas *Polianthes*, including *Manfreda* (*Agave* subgenus *Manfreda*) and *Runyonia* (Agavaceae). Sida 2:333–338.

_____. 1966b. *Hypochoeris microcephala* var. *albiflora* (Compositae) in southeastern Texas: New to North America. Sida 2:393–394.

_____. 1967. Stray notes on Texas *Plantago* (Plantaginaceae). Sida 3:120–122.

_____. 1969. *Petrorhagia prolifera* (*Dianthus prolifer*, *Tunica prolifera*) (Caryophyllaceae) in Arkansas and Texas. Sida 3:345–346.

_____. 1971. *Kuhnia* L. transferred to *Brickellia*. Sida 4:274.

Shuler, E.W. 1934. Collecting fossil elephants at Dallas, Texas. Field & Lab. 3:24–29.

_____. 1935. Dinosaur track mounted in the band stand at Glen Rose, Texas. Field & Lab. 4:9–13.

_____. 1937. Dinosaur tracks at the fourth crossing of the Paluxy River near Glen Rose, Texas. Field & Lab. 5:33–36.

Sieren, D.J. 1981. The taxonomy of the genus *Euthamia*. Rhodora 83:551–579.

Silveus, W.A. 1933. Texas grasses. Published by the author, San Antonio, TX.

_____. 1942. Grasses: *Paspalum* and *Panicum* of the United States. Published by the author, San Antonio, TX.

Simpson, B.B. (Beryl). Personal communication. Expert on Asteraceae, Fabaceae (Caesalpinioideae), and Krameriaceae. Univ. of Texas, Austin.

Simpson, B.B. 1978. Mutisieae. N. Amer. Fl. II. 10:1–13.

_____. 1989. Krameriaceae. Fl. Neotrop. Monogr. 49:1–108.

_____. 1996. Herbaria. In: R. Tyler, D.E. Barnett, R.R. Barkley, P.C. Anderson, and M.F. Odintz. The new handbook of Texas, Vol. 3. Pp. 567. The Texas State Historical Association, Austin.

_____. 1998. A revision of *Pomaria* (Fabaceae) in North America. Lundellia 1:46–71.

_____ and **B. Miao.** 1997. The circumscription of *Hoffmannseggia* (Fabaceae, Caesalpinioideae, Caesalpinieae) and its allies using morphological and cpDNA restriction site data. Pl. Syst. Evol. 205:157–178.

Simpson, B.B., J.L. Neff, and **G. Dieringer.** 1986. Reproductive biology of *Tinantia anomala* (Commelinaceae). Bull. Torrey Bot. Club 113:149–158.

Simpson, B.B. and **M.C. Ogorzaly.** 1986. Economic botany: Plants in our world. McGraw-Hill Book Co., New York.

Simpson, B.B. and **J.J. Skvarla.** 1981. Pollen morphology and ultrastructure of *Krameria* (Krameriaceae): Utility in questions of intrafamilial and interfamilial classification. Amer. J. Bot. 68:277–294.

Simpson, B.J. 1988. A field guide to Texas trees. Texas Monthly Press, Austin.

_____. 1993. The modern urban blacklands: From tallgrass prairie to soccer fields. In: M.R. Sharpless and J.C. Yelderman, eds. The Texas Blackland Prairie, land, history, and culture. Pp. 192–203. Baylor Univ. Program for Regional Studies, Waco, TX.

_____ and **B.W. Hipp.** 1984. Drought tolerant Texas native plants for amenity plantings. In: M.A. Collins, ed., Water for the 21st century: Will it be there? Center for Urban Water Studies, Southern Methodist Univ., Dallas, TX.

Simpson, B.J. and **S.D. Pease.** 1995. The tall grasslands of Texas. In: Native Plant Society of Texas. The tallgrass prairies and its many ecosystems. Pp. 1–10. 1995 Symposium Proceedings, Waco.

SIMPSON, M.G. 1997. Plant collecting and documentation field notebook. San Diego State Univ. Press, San Diego, CA.

SIMS, L.E. and **J. PRICE.** 1985. Nuclear DNA content variation in *Helianthus* (Asteraceae). Amer. J. Bot. 72:1213–1219.

SINK, K.C., ed. 1984a. *Petunia*. Springer-Verlag, New York.

_____. 1984b. Taxonomy. In: K.C. Sink. *Petunia*. Pp. 5–9. Springer-Verlag, New York.

SINNOTT, Q.P. 1985. A revision of *Ribes* L. subg. *Grossularia* (Mill.) Pers. sect. *Grossularia* (Mill.) Nutt. (Grossulariaceae) in North America. Rhodora 87:189–286.

SMALL, E. 1997. Cannabaceae. In: Flora of North America Editorial Committee, eds. Fl. North Amer. 3:381–387. Oxford Univ. Press, New York and Oxford.

_____ and **A. CRONQUIST.** 1976. A practical and natural taxonomy for *Cannabis*. Taxon 25:405–435.

SMALL, E. and **M. JOMPHE.** 1989. A synopsis of the genus *Medicago* (Leguminosae). Canad. J. Bot. 67:3260–3294.

SMALL, E., J.Y. PERRY, and **L.P. LEFKOVITCH.** 1976. A numerical taxonomic analysis of *Cannabis* with special reference to species delimitation. Syst. Bot. 1:67–84.

SMALL, J.K. 1894. New and interesting species of *Polygonum*. Bull. Torrey Bot. Club 21:168–173.

_____. 1895. A monograph of the North American species of the genus *Polygonum*. Mem. Dept. Bot. Columbia Coll. 1:1–183.

_____. 1907a. Oxalidaceae. N. Amer. Fl. 25:25–58.

_____. 1907b. Linaceae. N. Amer. Fl. 25:67–87.

_____. 1909. Alismaceae. N. Amer. Fl. 17:43–62.

_____. 1911. Simaroubaceae. N. Amer. Fl. 25:227–239.

_____. 1914a. Monotropaceae. N. Amer. Fl. 29:11–18.

_____. 1914b. Ericaceae (in part). N. Amer. Fl. 29:33–92.

_____. 1922. The blue-stem – *Sabal minor*. J. New York Bot. Gard. 23:161–168.

_____. 1938. Ferns of the southeastern states. The Science Press, Lancaster, PA.

_____ and **P.A. RYDBERG.** 1905a. Saxifragaceae. N. Amer. Fl. 22:81–158.

_____ and _____. 1905b. Hydrangeaceae. N. Amer. Fl. 22:159–178.

SMART, R.M. (MICHAEL). Personal communication. Aquatic plant ecologist. U.S. Army Corps of Engineers. Lewisville Aquatic Ecosystem Research Facility, Lewisville, TX.

SMEINS, F.E. (FRED). Personal communication. Expert on grassland ecology. Texas A&M Univ., College Station.

SMEINS, F.E. 1984. Origin of the brush problem—a geological and ecological perspective of contemporary distributions. In: K.W. McDaniel, ed. Brush Management Symposium. Texas Tech Univ. Press, Lubbock.

_____. 1988. Grassland (savannah, woodland) regions of Texas—Past and present. In: A. Davis and G. Stanford, eds. The prairie: Roots of our culture; foundation of our economy. Unpaged. Proceedings of the Tenth North American Prairie Conference, Denton, TX. Native Prairie Association of Texas, Dallas.

_____ and **D.D. DIAMOND.** 1986. Grasslands and savannahs of East Central Texas: Ecology, preservation status and management problems. In: D.L. Kulhavy and R.M. Conner, eds. Wilderness and natural areas of the eastern United States: A management challenge. School of Forestry, Stephen F. Austin State Univ., Nacogdoches, TX.

SMITH, A.C. 1944. Araliaceae. N. Amer. Fl. 28b:3–41.

SMITH, A.R. 1971. Systematics of the neotropical species of *Thelpyteris* section *Cyclosorus*. Univ. Calif. Publ. Bot. 59:1–136.

_____. 1993a. Thelypteridaceae. In: Flora of North America Editorial Committee, eds. Fl. North Amer. 2:206–222. Oxford Univ. Press, New York and Oxford.

_____. 1993b. Dryopteridaceae (family description and key to genera). In: Flora of North America Editorial Committee, eds. Fl. North Amer. 2:246–249. Oxford Univ. Press, New York and Oxford.

_____. 1993c. Polypodiaceae (family description and key to genera). In: Flora of North America Editorial Committee, eds. Fl. North Amer. 2:312–313. Oxford Univ. Press, New York and Oxford.

SMITH, A.W. 1963. A gardener's book of plant names: A handbook of the meaning and origins of plant names. Harper & Row, New York.

SMITH, C.B. (CALVIN). Personal communication. Expert on Waco mammoth site; Director of Strecker Museum of Baylor Univ., Waco, TX.

SMITH, C.E., JR. 1971. Preparing herbarium specimens of vascular plants. U.S.D.A. Agric. Info. Bull. 348:1–29.

SMITH, E.B. 1965. Taxonomy of *Haplopappus* section *Isopappus* (Compositae). Rhodora 67:217–238.

_____. 1974. *Coreopsis nuecensis* (Compositae) and a related new species from southern Texas. Brittonia 26:161–171.

_____. 1976. A biosystematic survey of *Coreopsis* in eastern United States and Canada. Sida 6:123–215.

_____. 1981. New combinations in *Croptilon* (Compositae-Astereae). Sida 9:59–63.

_____. 1988. An atlas and annotated list of the vascular plants of Arkansas, 2nd ed. Published by the author, Fayetteville, AR.

_____ and **H.M. PARKER.** 1971. A biosystematic study of *Coreopsis tinctoria* and *C. cardaminefolia* (Compositae). Brittonia 23:161–170.

SMITH, J.G. 1895. North American species of *Sagittaria* and *Lophotocarpus*. Rep. (Annual) Missouri Bot. Gard. 6:27–64.

SMITH, J.M. 1976. A taxonomic study of *Acleisanthes* (Nyctaginaceae). Wrightia 5:261–276.

SMITH, J.P., JR. 1971. Taxonomic revision of the genus *Gymnopogon* (Gramineae). Iowa State J. Sci. 45:319–385.

_____. 1977. Vascular plant families. Mad River Press, Inc., Eureka, CA.

SMITH, L.B. 1938. Bromeliaceae. N. Amer. Fl. 19:61–228.

_____. 1961. Bromeliaceae. In: C.L. Lundell, ed. Flora of Texas 3:200–207. Texas Research Foundation, Renner.

_____ and **C.E. WOOD.** 1975. The genera of Bromeliaceae in the southeastern United States. J. Arnold Arbor. 375–397.

SMITH, R.R. and **D.B. WARD.** 1976. Taxonomy of the genus *Polygala* series *Decurrentes* (Polygalaceae). Sida 6:284–310.

SMITH, S.G. (GALEN). Personal communication. Specialist in Cyperaceae. Univ. of Wisconsin-Whitewater.

SMITH, S.G. 1967. Experimental and natural hybrids in North American *Typha* (Typhaceae). Amer. Midl. Naturalist 78:257–287.

_____. 1995. New combinations in North American *Schoenoplectus*, *Bolboschoenus*, *Isolepis*, and *Trichophorum* (Cyperaceae). Novon 5:97–102.

_____. 1996. *Schoenoplectus*. Unpublished manuscript of forthcoming Flora of North America treatment.

______. 1997a. *Bolboschoenus*. Unpublished manuscript of forthcoming Flora of North America treatment.

______. 1997b. *Isolepis*. Unpublished manuscript of forthcoming Flora of North America treatment.

______ and **G. Yatskievych.** 1996. Notes on the genus *Scirpus* sensu lato in Missouri. Rhodora 98:168–179.

Smythe, D.P. 1852. A journal of the travels of D. Port Smythe, M.D., of Centerville, Texas, from that place to the mouth of the Palo Pinto, on the upper Brazos. Originally published in the Leon Weekly (Centerville in Leon Co., TX) from June 9 to July 14, 1852. Reprinted in Texas Geogr. Mag. 6(2):3–20. 1942.

Snow, D.W. 1981. Tropical frugivorous birds and their food plants: A world survey. Biotropica 13:1–14.

Snow, N. 1996. The phylogenetic utility of lemmatal micromorphology in *Leptochloa* s.l. and related genera in subtribe Eleusininae (Poaceae, Chlorioideae, Eragrostideae). Ann. Missouri Bot. Gard. 83:504–529.

______ and **G. Davidse.** 1993. *Leptochloa mucronata* (Michx.) Kunth is the correct name for *Leptochloa filiformis* (Poaceae). Taxon 42:413–417.

Snowden, J.D. 1936. The cultivated races of *Sorghum*. Adlard & Son, LTD., London, England, U.K.

Soderstrom, T.R. 1967. Taxonomic study of subgenus *Podosemum* and section *Epicampes* of *Muhlenbergia* (Gramineae). Contr. U.S. Natl. Herb. 34:75–189.

______ and **J.H. Beaman.** 1968. The genus *Bromus* (Gramineae) in Mexico and Central America. Publ. Mus. Michigan State Univ., Biol. Ser. 3:465–520.

Soderstrom, T.R., K.W. Hilu, C.S. Campbell, and **M.E. Barkworth,** eds. 1987. Grass systematics and evolution. Smithsonian Institution Press, Washington, D.C.

Sokal, R.R., T.J. Crovello, and **R.S. Unnasch.** 1986. Geographic variation of vegetative characters of *Populus deltoides*. Syst. Bot. 11:419–432.

Solbrig, O.T. 1960. The status of the genera *Amphipappus, Amphiachyris, Greenella, Gutierrezia, Gymnosperma* and *Xanthocephalum*. Rhodora 62:43–54.

______. 1961. Synopsis of the genus *Xanthocephalum* (Compositae). Rhodora 63:151–164.

______. 1963. The tribes of Compositae in the southeastern United States. J. Arnold Arbor. 44:436–461.

______ and **K.S. Bawa.** 1975. Isosyme variation in species of *Prosopis* (Leguminosae). J. Arnold Arbor. 56:398–412.

Solbrig, O.T. and **P.D. Cantino.** 1975. Reproductive adaptations in *Prosopis* (Leguminosae, Mimosoideae). J. Arnold Arbor. 56:185–210.

Solomon, J.D., T.D. Leininger, A.D. Wilson, R.L. Anderson, L.C. Thompson, and **F.I. McCracken.** 1993. Ash pests: A guide to major insects, diseases, air pollution injury, and chemical injury. U.S.D.A., Forest Serv., Southern Forest Exp. Sta. Gen. Tech. Rep. SO-96.

Soltis, D.E., D.R. Morgan, A. Grable, P.S. Soltis, and **R. Kuzoff.** 1993. Molecular systematics of Saxifragaceae sensu stricto. Amer. J. Bot. 80:1056–1081.

Soltis, D.E. and **P.S. Soltis.** 1989. Allopolyploid speciation in *Tragopogon*: Insights from chloroplast DNA. Amer. J. Bot. 1119–1124.

______, ______, **D.L. Nickrent, L.A. Johnson, W.J. Hahn, S.B. Hoot, J.A. Sweere, R.K. Kuzoff, K.A. Kron, M.W. Chase, S.M. Swensen, E.A. Zimmer, S.-M. Chaw, L.J. Gillespie, W.J. Kress,** and **K.J. Sytsma.** 1997. Angiosperm phylogeny inferred from 18S ribosomal DNA sequences. Ann. Missouri Bot. Gard. 84:1–49.

Soltis, P.S., G.M. Plunkett, S.J. Novak, and **D.E. Soltis.** 1995. Genetic variation in *Tragopogon* species: Additional origins of the allotetraploids *T. mirus* and *T. miscellus* (Compositae). Amer. J. Bot. 82:1329–1341.

Soltis, P.S. and **Soltis, D.E.** 1991. Multiple origins of the allotetraploid *Tragopogon mirus* (Compositae): rDNA evidence. Syst. Bot. 16:407–413.

Solymosy, S.L. 1974. *Hydrilla verticillata* (Hydrocharitaceae): New to Louisiana. Sida 5:354.

Sørensen, P.D. 1995. *Arbutus*. In: J.L. Luteyn, ed. Ericaceae Part II, the superior-ovaried genera. Fl. Neotrop. Monogr. 66:194–221.

Sosef, M.S.M. 1997. Hierarchical models, reticulate evolution and the inevitability of paraphyletic taxa. Taxon 46:75–85.

Spearing, D. 1991. Roadside geology of Texas. Mountain Press Publishing Company, Missoula, MT.

Spellenberg, R.W. (Richard). Personal communication. Specialist in Nyctaginaceae and floristics of the southwestern U.S. and northern Mexico; Flora of North America Editorial Committee; New Mexico State Univ., Las Cruces.

Sperlin, C.R. and **V. Bittrich.** 1993. Basellaceae. In: In: K. Kubitzki, J.C. Rohwer, and V. Bittrich, eds. The families and genera of vascular plants, Vol. II. Pp. 143–146. Springer-Verlag, Berlin.

Sperry, B. 1994. A toast to Thomas Volney Munson. Neil Sperry's Gardens 8(8):18–21.

Sperry, N. 1991. Neil Sperry's complete guide to Texas gardening. Taylor Publishing Co., Dallas, TX.

Sperry, O.E., J.W. Dollahite, J. Morrow, and **G.O. Hoffman.** 1955. Texas range plants poisonous to livestock. Texas Agric. Exp. Sta. Bull. No. 796.

Spjut, R.W. 1994. A systematic treatment of fruit types. Mem. New York Bot. Gard. 70:1–181.

Spoerke, D.G., Jr. and **S.C. Smolinske.** 1990. Toxicity of houseplants. CRC Press, Boca Raton, FL.

Spongberg, S.A. 1972. The genera of Saxifragaceae in the southeastern United States. J. Arnold Arbor. 53:409–498.

______. 1975. Lauraceae hardy in temperate North America. J. Arnold Arbor. 56:1–19.

______. 1976. Styracaceae hardy in temperate North America. J. Arnold Arbor. 57:54–73.

______. 1977. Ebenaceae hardy in temperate North America. J. Arnold Arbor. 58:146–160.

______. 1978. The genera of Crassulaceae in the southeastern United States. J. Arnold Arbor. 59:197–248.

Spooner, D.M. 1987. The systematics of *Simsia* (Compositae: Heliantheae). Ph.D. dissertation, Ohio State Univ., Columbus.

Sprague, E.F. 1962. Pollination and evolution in *Pedicularis* (Scrophulariaceae). Aliso 5:181–209.

STACE, H.M. and **L.A. EDYE,** eds. 1984. The biology of agronomy of *Stylosanthes*. Academic Press, San Francisco.

STAFLEU, F.A. and **R.S. COWAN.** 1976–1988. Taxonomic literature, 2nd ed., 7 vols. Bohn, Scheltema & Holkema, Utrecht, Netherlands. Regnum Veg. 94, 98, 105, 110, 112, 115, 116.

STAHLE, D.W. (DAVID). Personal communication. Dendrochronologist at Univ. of Arkansas, Fayetteville.

STAHLE, D.W. 1990. The tree-ring record of false spring in the southcentral USA. Ph.D. dissertation, Arizona State Univ., Tempe.

_____ 1996a. Tree rings and ancient forest history. In: M.B. Davis, ed. Eastern old-growth forests: Prospects for rediscovery and recovery. Chapter 22, Pp. 321–343. Island Press, Washington, D.C.

_____. 1996b. Tree rings and ancient forest relics. Arnoldia 56:2–10.

_____ and **P.L. CHANEY.** 1994. A predictive model for the location of ancient forests. Nat. Areas J. 14:151–158.

STAHLE, D.W. and **M.K. CLEAVELAND.** 1988. Texas drought history reconstructed and analyzed from 1698 to 1980. J. Climate 1:59–74.

_____ and _____. 1992. Reconstruction and analysis of spring rainfall over the southeastern U.S. for the past 1000 years. Bull. Amer. Meterological Soc. 73:1947–1961.

_____ and _____. 1993. Southern oscillation extremes reconstructed from tree rings of the Sierra Madre Occidental and southern Great Plains. J. Climate 6:129–140.

_____ and _____. 1995. Texas paleoclimatic data from daily to millennial time scales. In: J. Norwine, J.R. Giardino, G.R. North, and J.B. Valdés, eds. The changing climate of Texas: Predictability and implications for the future. Chapter 7. Pp. 49–69. GeoBooks, Texas A&M Univ., College Station.

_____, _____, and **J.G. HEHR.** 1988. North Carolina climate changes reconstructed from tree rings: A.D. 372 to 1985. Science 240:1517–1519.

STAHLE, D.W. and **J.G. HEHR.** 1984. Dendroclimatic relationships of post oak across a precipitation gradient in the southcentral United States. Ann. Assoc. Amer. Geogr. 74:561–573.

_____, _____, **G.G. HAWKS, JR., M.K. CLEAVELAND,** and **J.R. BALDWIN.** 1985. Tree-ring chronologies for the southcentral United States. Tree-Ring Laboratory and Office of the State Climatologist, Department of Geography, Univ. of Arkansas, Fayatteville.

STANDLEY, L.A. 1990. Anatomical aspects of the taxonomy of sedges (*Carex*, Cyperaceae). Canad. J. Bot. 68:1449–1456.

STANDLEY, P.C. 1916. Chenopodiaceae. N. Amer. Fl. 21:3–93.

_____. 1917. Amaranthaceae. N. Amer. Fl. 21:95–169.

_____. 1918. Allioniaceae. N. Amer. Fl. 21:171–254.

_____. 1918–1934. Rubiaceae. N. Amer. Fl. 32:1–300.

_____. 1922a. Hamamelidaceae. In: Trees and shrubs of Mexico. Contr. U.S. Natl. Herb. 23:317–319.

_____. 1922b. Leguminosae. In: Trees and shrubs of Mexico. Contr. U.S. Natl. Herb. 23:348–515.

_____. 1923a. Rhamnaceae. In: Trees and shrubs of Mexico. Contr. U.S. Natl. Herb. 23:710–727.

_____. 1923b. Malvaceae. In: Trees and shrubs of Mexico. Contr. U.S. Natl. Herb. 23:746–786.

STANFORD, E.E. 1925. The amphibious group of *Polygonum*, subgenus *Persicaria*. Rhodora 27:109–112, 125–130, 146–152, 156–166.

STANFORD, G. 1995. A chalkland prairie biome? In: Native Plant Society of Texas. The tallgrass prairies and its many ecosystems. Pp. 125–132. 1995 Symposium Proceedings, Waco.

STANFORD, J.W. (JACK). Personal communication. Specialist in plants of the Edwards Plateau and Fabaceae, especially *Mimosa*; Howard Payne Univ., Brownwood, TX.

STANFORD, J.W. 1966. Species separation by pollen morphology of some Texas Mimosoideae. M.S. thesis, Texas Technological College, Lubbock.

_____. 1971. Vascular plants of the three central Texas counties of Brown, Comanche, and Hamilton. Ph.D. Dissertation, Oklahoma State Univ., Stillwater.

_____. 1976. Keys to the vascular plants of the Texas Edwards Plateau and adjacent areas. Published by the author, Brownwood, TX.

_____ and **G.M. DIGGS, JR.** 1998. *Pteris vittata* L. (Pteridaceae), a new fern for Texas. Sida 18:359–360.

STANFORD, N. and **B.L. TURNER.** 1988. The natural distribution and biological status of *Helenium amarum* and *H. badium* (Asteraceae, Heliantheae). Phytologia 65:141–146.

STARR, G. 1995. *Hesperaloe*: Aloes of the west. Desert Pl. 11:3–8.

_____. 1997. A revision of the genus *Hesperaloe* (Agavaceae). Madroño 44:282–296.

STEARN, W.T. 1957. An introduction to the Species Plantarum and cognate botanical works of Carl Linnaeus. Facsimile of first edition of Carl Linnaeus, Species Plantarum. Ray Society, London, England, U.K.

_____. 1983. Botanical Latin: History, grammar, syntax, terminology and vocabulary, 3rd ed. David & Charles Inc., North Pomfret, VT.

_____. 1992. Botanical Latin, 4th ed. Timber Press, Portland, OR.

STEBBINS, G.L. 1937. Critical notes on *Lactuca* and related genera. J. Bot. 75:12–18.

_____. 1949. Speciation, evolutionary trends, and distributional patterns in *Crepis*. Evolution 3:188–193.

_____. 1981. Coevolution of grasses and herbivores. Ann. Missouri Bot. Gard. 68:75–86.

_____. 1982. Major trends of evolution in the Poaceae and their possible significance. In: J.R. Estes, R.J. Tyrl, and J.N. Brunken. Grasses and grasslands: Systematics and ecology. Pp. 3–36. Univ. of Oklahoma Press, Norman.

_____. 1985. Polyploidy, hybridization, and the invasion of new habitats. Ann. Missouri Bot. Gard. 72:824–832.

_____. 1993. Concepts of species and genera. In: Flora of North America Editorial Committee, eds. Fl. North Amer. 1:229–246. Oxford Univ. Press, New York and Oxford.

STEEVES, M.A. and **E.S. BARGHOORN.** 1959. The pollen of *Ephedra*. J. Arnold Arbor. 40:221–259.

STEIGMAN, K.L. and **L. OVENDEN.** 1988. Transplanting tallgrass prairie with a sodcutter. In: A. Davis and G. Stanford, eds. The prairie: Roots of our culture; foundation of our economy. Unpaged, paper number 0901. Proceedings of the Tenth North American Prairie Conference, Denton, TX. Native Prairie Association of Texas, Dallas.

Steila, D. 1993. Soils. In: Flora of North America Editorial Committee, eds. Fl. North Amer. 1:47–54. Oxford Univ. Press, New York and Oxford.

Steinmann, V.W. and **R.S. Felger.** 1997. The Euphorbiaceae of Sonora, Mexico. Aliso 16:1–71.

Stephens, A.R. and W.M. Holmes. 1989. Historical atlas of Texas. Univ. of Oklahoma Press, Norman.

Stephens, H.A. 1980. Poisonous plants of the central United States. Univ. Press of Kansas, Lawrence.

Stermitz, F.R., D.E. Nicodem, C.C. Wei, and **K.D. McMurtrey.** 1969. Alkaloids of *Argemone polyanthemos, A. corymbosa, A. chisosensis, A. sanguinea, A. aurantiaca* and general *Argemone* systematics. Phytochemistry 8:615–620.

Stern, K.R. 1997a. Fumariaceae. In: Flora of North America Editorial Committee, eds. Fl. North Amer. 3:340–341. Oxford Univ. Press, New York and Oxford.

_____. 1997b. *Corydalis*. In: Flora of North America Editorial Committee, eds. Fl. North Amer. 3:348–355. Oxford Univ. Press, New York and Oxford.

Stevens, P.F. 1995. Familial and infrafamilial relationships. In: J.L. Luteyn, ed. Ericaceae, Part II, the superior-ovaried genera. Fl. Neotrop. Monogr. 66:1–12.

Stevenson, D.W. 1993. Ephedraceae. In: Flora of North America Editorial Committee, eds. Fl. North Amer. 2:428–434. Oxford Univ. Press, New York and Oxford.

Stevenson, G.A. 1969. An agronomic and taxonomic review of the genus *Melilotus* Mill. Canad. J. Pl. Sci. 49:1–20.

Steward, K.K., T.K. Van, V. Carter, and **A.H. Pieterse.** 1984. *Hydrilla* invades Washington, D.C. and the Potomac. Amer. J. Bot. 71:162–163.

Steyermark, J.A. 1932. A revision of the genus *Menodora*. Ann. Missouri Bot. Gard. 19:87–176.

_____. 1934. Studies in *Grindelia*. II. A monograph of the North American species of the genus *Grindelia*. Ann. Missouri Bot. Gard. 21:433–608.

_____. 1963. Flora of Missouri. The Iowa State Univ. Press, Ames.

Stiles, E.W. 1984. Fruit for all seasons. Nat. Hist. 93:42–53.

Stirton, C.H., ed. 1987. Advances in legume systematics. Part 3. Royal Botanic Gardens, Kew, England, U.K.

St. John, H. 1916. A revision of the North American species of *Potamogeton* of the section *Coleophylli*. Rhodora 18:121–138.

_____. 1965. Monograph of the genus *Elodea*: Part 4 and summary. Rhodora 67:1–35, 155–180.

Stokes, S.G. 1936. The genus *Eriogonum*. J.H. Neblett Press, San Francisco, CA.

Stone, D.E. 1959. A unique balanced breeding system in the vernal pool mouse-tails. Evolution 13:151–174.

_____. 1993. Juglandaceae. In: K. Kubitzki, J.C. Rohwer, and V. Bittrich, eds. The families and genera of vascular plants, Vol. II. Pp. 348–359. Springer-Verlag, Berlin.

_____. 1997a. Juglandaceae. In: Flora of North America Editorial Committee, eds. Fl. North Amer. 3:416–417. Oxford Univ. Press, New York and Oxford.

_____. 1997b. *Carya*. In: Flora of North America Editorial Committee, eds. Fl. North Amer. 3:417–425. Oxford Univ. Press, New York and Oxford.

Straley, G.B. 1977. Systematics of *Oenothera* sect. *Kneiffia* (Onagraceae). Ann. Missouri Bot. Gard. 64:381–424.

Strausbaugh, P.D. and **E.L. Core.** 1978. Flora of West Virginia, 2nd ed. Seneca Books, Inc., Grantsville, WV.

Straw, R.L. 1966. A redefinition of *Penstemon* (Scrophulariaceae). Brittonia 18:80–95.

Strecker, J.K. 1924. The mammals of McLennan County, Texas. Baylor Bull. 27:3–20.

_____. 1926a. The mammals of McLennan County, Texas (supplementary notes). Contr. Baylor Univ. Mus. 9:1–15.

_____. 1926b. The extension of the range of the nine-banded armadillo. J. Mamm. 7:206–210.

Strickland, S.S. and **J.W. Fox.** 1993. Prehistoric environmental adaptations in the Blackland Prairie. In: M.R. Sharpless and J.C. Yelderman, eds. The Texas Blackland Prairie, land, history, and culture. Pp. 96–121. Baylor Univ. Program for Regional Studies, Waco, TX.

Stritch, L.R. 1984. Nomenclatural contributions to a revision of the genus *Wisteria*. Phytologia 56:183–184.

Strong, M.T. 1993. New combinations in *Schoenoplectus* (Cyperaceae). Novon 3:202–203.

_____. 1994. Taxonomy of *Scirpus*, *Trichophorum*, and *Schoenoplectus* (Cyperaceae) in Virginia. Bartonia 58:29–68.

Strother, J.L. 1966. Chromosome numbers in *Hymenoxys* (Compositae). SouthW. Naturalist 11:223–227.

_____. 1969. Systematics of *Dyssodia* Cavanilles (Compositae: Tageteae). Univ. Calif. Publ. Bot. 48:1–88.

_____. 1986. Renovation of *Dyssodia* (Compositae: Tageteae). Sida 11:371–378.

_____. 1991. Taxonomy of *Complaya*, *Elaphandra*, *Iogeton*, *Jefea*, *Wamalchitamia*, *Wedelia*, *Zexmenia*, and *Zyzyxia* (Compositae-Heliantheae-Ecliptinae). Syst. Bot. Monogr. 33:1–111.

Stubbendieck, J. and **E.C. Conard.** 1989. Common legumes of the Great Plains: An illustrated guide. Univ. of Nebraska Press, Lincoln.

Stuckey, R.L. 1972. Taxonomy and distribution of the genus *Rorippa* (Cruciferae) in North America. Sida 4:279–425.

_____. 1978. Essays on North American plant geography from the nineteenth century. Arno Press, New York.

_____. 1979. Distributional history of *Potamogeton crispus* (curly pondweed) in North America. Bartonia 46:22–42.

_____. 1980. Distributional history of *Lythrum salicaria* (purple loosestrife) in North America. Bartonia 47:3–20.

_____. 1985. Distributional history of *Najas marina* (spiny naiad) in North America. Bartonia 51:2–16.

_____ and **T.M. Barkley.** 1993. Weeds. In: Flora of North America Editorial Committee, eds. Fl. North Amer. 1:193–198. Oxford Univ. Press, New York and Oxford.

Studhalter, R.A. 1931. Mrs. Young's "Familiar Lessons in Botany, with Flora of Texas," a forgotten text of fifty years ago. Texas Tech. Coll. Bull. 7(6):28–52.

Stuessy, T.F. 1971. Systematic relationships in the white-rayed species of *Melampodium* (Compositae). Brittonia 23:177–190.

_____. 1972. Revision of the genus *Melampodium* (Compositae: Heliantheae). Rhodora 74:1–70, 161–219.

_____. 1979. Cladistics of *Melampodium* (Compositae). Taxon 28:179–195.

_____. 1997. Classification: More than just branching patterns of evolution. Aliso 15:113–124.

_____ and **J.V. Crisci.** 1984. Phenetics of *Melampodium* (Compositae, Heliantheae). Madroño 31:8–19.

Stuessy, T.F. and **S.H. Sohmer,** eds. 1996. Sampling the green world: Innovative concepts of collection, preservation, and storage of plant diversity. Columbia Univ. Press, New York.

Subils, R. 1984. Una nueva especie de *Euphorbia* sect. *Poinsettia* (Euphorbiaceae). Kurtziana 17:125–130.

Sudworth, G.B. 1934. Poplars, principle tree willows and walnuts of the Rocky Mountain region. U.S.D.A. Tech. Bull. 420:1–112.

Suffness, M. 1995. Discovery and development of taxol. In: M. Suffness, ed. Taxol®: Science and applications. Pp. 3–25. CRC Press, Boca Raton, FL.

Suh, Y. 1989. Phylogenetic studies of North American Astereae (Asteraceae) based on chloroplast DNA. Ph.D. dissertation, Univ. of Texas, Austin.

_____ and **B.B. Simpson.** 1990. Phylogenetic analysis of chloroplast DNA in North American *Gutierrezia* and related genera (Asteraceae: Astereae). Syst. Bot. 15:660–670.

Sullivan, J.R. 1985. Systematics of the *Physalis viscosa* complex (Solanaceae). Syst. Bot. 10:426–444.

Sullivan, V.I. 1975. Pollen and pollination in the genus *Eupatorium* (Compositae). Canad. J. Bot. 53:582–589.

Summerfield, R.J. and **A.H. Bunting,** eds. 1980. Advances in legume science. Royal Botanic Gardens, Kew, England, U.K.

Sun, V.G. 1946. The evaluation of taxonomic characters of cultivated *Brassica* with a key to species and varieties—I. The characters. Bull. Torrey Bot. Club 73:244–281, 370–377.

Sundberg, S.D. 1991. Infraspecific classification of *Chloracantha spinosa* (Benth.) Nesom (Asteraceae) Astereae. Phytologia 70:382–391.

Sundell, E. 1981. The New World species of *Cynanchum* subgenus *Mellichampia* (Asclepiadaceae). Evol. Monogr. 5:1–63.

Sutherland, D. 1986. Poaceae. In: Great Plains Flora Association. Flora of the Great Plains. Pp. 1113–1235. Univ. Press of Kansas, Lawrence.

Sutton, D.A. 1988. A revision of the tribe Anthirrhineae, Oxford Univ. Press, London, England, U.K.

Svenson, H.K. 1929. Contributions from the Gray Herbarium of Harvard University—No. LXXXVI. Monographic studies in the genus *Eleocharis*. Rhodora 31:121–135, 152–163, 167–191, 199–219, 224–242.

_____. 1939. Monographic studies in the genus *Eleocharis*—V. Rhodora 41:1–19, 43–77, 90–110.

_____. 1944. The New World species of *Azolla*. Amer. Fern J. 34:69–84.

_____. 1953. The *Eleocharis obtusa-ovata* complex. Rhodora 55:1–6.

_____. 1957. Poales: Cyperaceae: Scirpeae (Continuatio) [*Fuirena, Hemicarpha, Eleocharis*]. N. Amer. Fl. 18:505–540.

Swanson, S.D. and **S.H. Sohmer.** 1976. The biology of *Podophyllum peltatum* L. (Berberidaceae), the may apple. II. The transfer of pollen and success of sexual reproduction. Bull. Torrey Bot. Club 103:223–226.

Takahashi, M., J.W. Nowicke, and **G.L. Webster.** 1995. A note on remarkable exines in Acalyphoideae (Euphorbiaceae). Grana 34:282–290.

Takhtajan, A.L. 1997. Diversity and classification of flowering plants. Columbia Univ. Press, New York.

Talalaj, S., D. Talalaj, and **J. Talalaj.** 1991. The strangest plants in the world. Hill of Content Publishing Co., Melbourne, Australia.

Tampion, J. 1977. Dangerous plants. Universe Books, New York.

Tamura, M. 1993. Ranunculaceae. In: K. Kubitzki, J.C. Rohwer, and V. Bittrich, eds. The families and genera of vascular plants, Vol. II. Pp. 563–583. Springer-Verlag, Berlin.

Tateoka, T. 1961. A biosystematic study of *Tridens* (Gramineae). Amer. J. Bot. 48:565–573.

Taylor, C.E.S. (Constance). Personal communication. Specialist in Asteraceae, especially *Solidago* and *Euthamia*, and expert on the flora of Oklahoma; member of Oklahoma Flora Editorial Committee and Oklahoma Native Plant Society; Southeastern Oklahoma State Univ., Durant.

Taylor, C.E.S. *Euthamia gymnospermoides* (Compositae). Ph.D. dissertation, Univ. of Oklahoma, Norman.

_____. 1997. Keys to the Asteraceae of Oklahoma. Southeastern Oklahoma State Univ. Herbarium, Durant.

_____ and **R.J. Taylor.** 1983. New species, new combinations and notes on the goldenrods (*Euthamia* and *Solidago*–Asteraceae). Sida 10:176–183.

_____ and _____. 1984. *Solidago* (Asteraceae) in Oklahoma and Texas. Sida 10:223–251.

Taylor, N. 1909a. Zannichelliaceae. N. Amer. Fl. 17:13–27.

_____. 1909b. Naiadaceae. N. Amer. Fl. 17:33–35.

Taylor, N.P. 1978. Review of the genus *Escobaria*. Cact. Succ. J. Gr. Brit. 40:31–37.

_____. 1985. The genus *Echinocereus*. The Royal Botanic Gardens, Kew in association with Timber Press, Portland, OR.

Taylor, P. 1989. The genus *Utricularia* - a taxonomic monograph. Kew Bull., Addit. Ser. 14:1–724.

Taylor, R.J. and **C.E. Taylor.** 1981. Plants new to Arkansas, Oklahoma and Texas. Sida 9:25–28.

_____ and _____. 1994. An annotated list of the ferns, fern allies, gymnosperms and flowering plants of Oklahoma, 3rd ed. Southeastern Oklahoma State Univ. Herbarium, Durant, OK.

Taylor, T.H. n.d. Rangers lead the way. Roster, P. 153. Turner Publ. Co., Paducah, KY.

Taylor, W.C. 1984. Arkansas ferns and fern allies. Milwaukee Public Museum, Milwaukee, WI.

_____ and **R.J. Hickey.** 1992. Habitat, evolution, and speciation of *Isoëtes*. Ann. Missouri Bot. Gard. 79:613–622.

Taylor, W.C., N.T. Luebke, D.M. Britton, R.J. Hickey, and **D.F. Brunton.** 1993. Isoëtaceae. In: Flora of North America Editorial Committee, eds. Fl. North Amer. 2:64–75. Oxford Univ. Press, New York and Oxford.

Taylor, W.C., R.H. Mohlenbrock, and **J.A. Murray.** 1975. The spores and taxonomy of *Isoëtes butleri* and *I. melanopoda*. Amer. Fern J. 65:33–38.

Tellman, B. 1997. Exotic pest plant introduction in the

American southwest. Desert Pl. 13:3–9.

Terrell, E.E. 1975. Relationships of *Hedyotis fruticosa* L. to *Houstonia* L. and *Oldenlandia* L. Phytologia 31:418–424.

______. 1986a. Nomenclatural notes on *Hedyotis rosea* Rafinesque and a new combination in *Houstonia*. Rhodora 88:389–397.

______. 1986b. Taxonomic and nomenclatural notes on *Houstonia nigricans* (Rubiaceae). Sida 11:471–481.

______. 1990. Synopsis of *Oldenlandia* (Rubiaceae) in the United States. Phytologia 68:125–133.

______. 1991. Overview and annotated list of North American species of *Hedyotis*, *Houstonia*, *Oldenlandia* (Rubiaceae), and related genera. Phytologia 71:212–243.

______. 1996. Revision of *Houstonia* (Rubiaceae-Hedyotideae). Syst. Bot. Monogr. 48:1–118.

______, **W.H. Lewis, H. Robinson,** and **J.W. Nowicke.** 1986. Phylogenetic implications of diverse seed types, chromosome numbers, and pollen morphology in *Houstonia* (Rubiaceae). Amer. J. Bot. 73:103–115.

Teuscher, H. 1978. *Pogonia* and *Nervilia* with a discussion of *Cleistes*, *Isotria* and *Triphora*. Amer. Orchid Soc. Bull. 47:16–23.

Texas Parks & Wildlife. 1995. State symbol quiz. Texas Parks & Wildlife 53(1): 48–53.

Thanikaimoni, G. and **G. Vasanthy.** 1972. Sarraceniaceae: Palynology and systematics. Pollen & Spores 14:143–155.

Tharp, B.C. 1922. *Commelinantia*, a new genus of the Commelinaceae. Bull. Torrey Bot. Club 49:269–275.

______. 1926. Structure of Texas vegetation east of the 98th Meridian. Univ. Texas Bull. 2606:1–99.

______. 1932. *Tradescantia edwardsiana*, nov. sp. Rhodora 34:57–59.

______. 1952. Texas range grasses. Univ. of Texas Press, Austin.

______. 1956. *Commelinantia* (Commelineae): An evaluation of its generic status. Bull. Torrey Bot. Club 83:107–112.

______ and **F.A. Barkley.** 1949. Genus *Ruellia* in Texas. Amer. Midl. Naturalist 42:1–86.

Tharp, B.C. and **M.C. Johnston.** 1961. Recharacterization of *Dichondra* (Convolvulaceae) and a revision of the North American species. Brittonia 13:346–360.

Tharp, B.C. and **C.V. Kielman.** 1962. Mary S. Young's Journal of Botanical Explorations in Trans-Pecos, Texas, August-September, 1914. SouthW. Hist. Quar. 65: 366–393, 512–538.

Thieret, J.W. (John). Personal communication. Specialist in *Cyperus* (Cyperaceae), Poaceae, Scrophulariaceae, and introduced weedy plants; currently associate editor of *Sida, Contributions to Botany*, and on Flora of North America Editorial Committee; retired from a career at Northern Kentucky Univ., Highland Heights.

Thieret, J.W. 1955. The seeds of *Veronica* and allied genera. Lloydia 18:37–45.

______. 1964. *Fatoua villosa* (Moraceae) in Louisiana: New to North America. Sida 1:248.

______. 1966. Seeds of some United States Phytolaccaceae and Aizoaceae. Sida 2:352–360.

______. 1967. Supraspecific classification in the Scrophulariaceae: A review. Sida 3:87–106.

______. 1969a. Orobanchaceae. In: C.L. Lundell, ed. Flora of Texas 2:331–337. Texas Research Foundation, Renner.

______. 1969b. *Trifolium vesiculosum* (Leguminosae) in Mississippi and Louisiana: New to North America. Sida 3:446–447.

______. 1971. The genera of Orobanchaceae in the southeastern United States. J. Arnold Arbor. 52:404–434.

______. 1972. The Phrymaceae in the southeastern United States. J. Arnold Arbor. 53:226–233.

______. 1976. Floral biology of *Proboscidea louisianica* (Martyniaceae). Rhodora 78:169–179.

______. 1977a. The Martyniaceae in the southeastern United States. J. Arnold Arbor. 58:25–39.

______. 1977b. Juvenile leaves in Oklahoma *Marsilea* (Marsileaceae). Sida 7:218.

______. 1980. Louisiana ferns and fern allies. Lafayette Natural History Museum published in conjunction with the Univ. of Southwestern Louisiana, Lafayette.

______. 1993. Pinaceae (family description and key to genera). In: Flora of North America Editorial Committee, eds. Fl. North Amer. 2:352–354. Oxford Univ. Press, New York and Oxford.

______ and **J.O. Luken.** 1996. The Typhaceae in the southeastern United States. Harvard Pap. Bot. 8:27–56.

Thomas, C.C. 1936. The Chinese Jujube (originally issued 1924, revised 1936). U.S.D.A. Bull. 1215:1–14.

Thomas, G.W. 1962. Texas plants – An ecological summary. In: F.W. Gould. Texas plants — A checklist and ecological summary. Texas Agric. Exp. Sta. Misc. Publ. 585:5–14.

Thomas, R.D. 1972. *Botrychium lunarioides*, *Ophioglossum crotalophoroides*, and *Ophioglossum engelmanni* in a Louisiana cemetery. SouthW. Naturalist 16:431–459.

Thomas, W.W. 1984. The systematics of *Rhynchospora* section *Dichromena*. Mem. New York Bot. Gard. 37:1–116.

______. 1992. A synopsis of *Rhynchospora* (Cyperaceae) in Mesoamerica. Brittonia 44:14–44.

Thompson, C.H. 1897. North American Lemnaceae. Rep. (Annual) Missouri Bot. Gard. 9:1–22.

Thompson, C.M. 1993. More from less: Greater demand from fewer acres of productive soils. In: M.R. Sharpless and J.C. Yelderman, eds. The Texas Blackland Prairie, land, history, and culture. Pp. 252–261. Baylor Univ. Program for Regional Studies, Waco, TX.

Thompson, H.J. and **A.M. Powell.** 1981. Loasaceae of the Chihuahuan Desert Region. Phytologia 49:16–32.

Thompson, J.C. and **W.T. Barker.** 1986. Malvaceae. In: Great Plains Flora Association. Flora of the Great Plains. Pp. 240–252. Univ. Press of Kansas, Lawrence.

Thompson, S.W. and **T.G. Lammers.** 1997. Phenetic analysis of morphological variation in the *Lobelia cardinalis* complex (Campanulaceae: Lobelioideae). Syst. Bot. 22:315–331.

Thorne, R.F. 1992. An updated classification of the flowering plants. Aliso 13:365–389.

______. 1993a. Phytogeography. In: Flora of North America Editorial Committee, eds. Fl. North Amer. 1:132–153. Oxford Univ. Press, New York and Oxford.

______. 1993b. Hydrocharitaceae. In: J.C. Hickman, ed. The Jepson manual: Higher plants of California. Pp. 1150–1151. Univ. of California Press, Berkeley.

Tiffney, B.H. 1985. Perspectives on the origin of the floristic similarity between eastern Asia and eastern North America. J. Arnold Arbor. 66:73–94.

Tippo, O. and **W.L. Stern.** 1977. Humanistic botany. W.W. Norton & Co., Inc., New York.

Todzia, C.A. 1993. Ulmaceae. In: K. Kubitzki, J.C. Rohwer, and V. Bittrich, eds. The families and genera of vascular plants, Vol. II. Pp. 603–611. Springer-Verlag, Berlin.

_____. 1998. The Texas plant collections of Mary Sophie Young. Lundellia 1:27–39.

TOES (Texas Organization for Endangered Species). 1992. Endangered, threatened, and watch list of natural communities of Texas. Publ. No. 8. Texas Organization for Endangered Species, Austin.

_____. 1993. Endangered, threatened and watch lists of Texas plants. Publ. No. 9. Texas Organization for Endangered Species, Austin.

Tölkin, H.R. 1977. A revision of the genus *Crassula* in southern Africa. Contr. Bolus Herb. 8:1–595.

Tomb, A.S. 1974. *Hypochoeris* in Texas. Sida 5:287–289.

_____. 1980. Taxonomy of *Lygodesmia* (Asteraceae). Syst. Bot. Monogr. 1:1–51.

Tomlinson, P.B. 1990. The structural biology of palms. Oxford Univ. Press, Oxford, England, U.K.

_____ and **U. Posluszny.** 1976. Generic limits in the Zannichelliaceae (*sensu* Dumortier). Taxon 25:273–279.

Torres, A.M. 1963. Taxonomy of *Zinnia*. Brittonia 15:1–25.

Torrey, J. 1853. Appendix G. Botany. In: R.B. Marcy. Exploration of the Red River of Louisiana, in the year 1852. Pp. 277–304, Plates I–XX. Robert Armstrong, Public Printer, Washington, D.C.

Towner, H.F. 1977. The biosystematics of *Calylophus* (Onagraceae). Ann. Missouri Bot. Gard. 64:48–120.

Townsend, C.C. 1993. Amaranthaceae. In: K. Kubitzki, J.C. Rohwer, and V. Bittrich, eds. The families and genera of vascular plants, Vol. II. Pp. 70–91. Springer-Verlag, Berlin.

Trelease, W. 1902. The Yucceae. Rep. (Annual) Missouri Bot. Gard. 13:27–133.

_____. 1911. The desert group Nolineae. Proc. Amer. Philos. Soc. 50:404–443.

_____. 1924. The American oaks. Mem. Natl. Acad. Sci. U.S.A. 20:1–255.

Tryon, R.M., Jr. 1941. A revision of the genus *Pteridium*. Rhodora 43:1–31, 37–67.

_____. 1955. *Selaginella rupestris* and its allies. Ann. Missouri Bot. Gard. 42:1–99.

_____. 1956. A revision of the American species of *Notholaena*. Contr. Gray Herb. 179:1–106.

_____. 1957. A revision of the fern genus *Pellaea* section *Pellaea*. Ann. Missouri Bot. Gard. 44:125–193.

_____ and **A.F. Tryon.** 1982. Ferns and allied plants with special reference to tropical America. Springer-Verlag, New York.

Tucker, A.O., N.H. Dill, T.D. Pizzolato, and **R.D. Kral.** 1983. Nomenclature, distribution, chromosome numbers, and fruit morphology of *Oxypolis canbyi* and *O. filiformis* (Apiaceae). Syst. Bot. 8:299–304.

Tucker, G.C. 1983. The taxonomy of *Cyperus* (Cyperaceae) in Costa Rica and Panama. Syst. Bot. Monogr. 2:1–85.

_____. 1984. A revision of the genus *Kyllinga* Rottb. (Cyperaceae) in Mexico and Central America. Rhodora 86:507–538.

_____. 1986. The genera of Elatinaceae in the southeastern United States. J. Arnold Arbor. 67:471–483.

_____. 1987. The genera of Cyperaceae in the southeastern United States. J. Arnold Arbor. 68:361–445.

_____. 1988. The genera of Bambusoideae (Gramineae) in the southeastern United States. J. Arnold Arbor. 69:239–273.

_____. 1989. The genera of Commelinaceae in the southeastern United States. J. Arnold Arbor. 70:97–130.

_____. 1990. The genera of Arundinoideae (Gramineae) in the southeastern United States. J. Arnold Arbor. 71:145–177.

_____. 1994. Revision of the Mexican species of *Cyperus* (Cyperaceae). Syst. Bot. Monogr. 43:1–213.

_____. 1996. The genera of Poöideae (Gramineae) in the southeastern United States. Harvard Pap. Bot. 9:11–90.

Tucker, S.C. 1993. Utility of ontogenetic and conventional characters in determining phylogenetic relationships of Saururaceae and Piperaceae (Piperales). Syst. Bot. 18:614–641.

Turner, B.L. 1950a. Vegetative key to Texas *Desmanthus* (Leguminosae) and similar genera. Field & Lab. 18:51–54.

_____. 1950b. Texas species of *Desmanthus* (Leguminosae). Field & Lab. 18:54–65.

_____. 1951. Revision of the United States species of *Neptunia* (Leguminosae). Amer. Midl. Naturalist 46:82–92.

_____. 1955. The *Cassia fasciculata* complex (Leguminosae) in Texas. Field & Lab. 23:87–91.

_____. 1956. A cytotaxonomic study of the genus *Hymenopappus* (Compositae). Rhodora 58:163–186, 208–242, 250–269, 295–308.

_____. 1957. The chromosomal and distributional relationships of *Lupinus texensis* and *L. subcarnosus* (Leguminosae). Madroño 14:13–16.

_____. 1958. Chromosome numbers in the genus *Krameria*: Evidence for familial status. Rhodora 60:101–106.

_____. 1959. The legumes of Texas. Univ. of Texas Press, Austin.

_____. 1983. The Texas species of *Paronychia* (Caryophyllaceae). Phytologia 54:9–23.

_____. 1984. Taxonomy of the genus *Aphanostephus* (Asteraceae-Astereae). Phytologia 56:81–101.

_____. 1988a. New species and combinations in *Wedelia* (Asteraceae-Heliantheae). Phytologia 65:348–358.

_____. 1988b. A new species of, and observations on, the genus *Smallanthus* (Asteraceae-Heliantheae). Phytologia 64:405–407.

_____. 1989. An overview of the *Brickellia* (*Kuhnia*) *eupatorioides* (Asteraceae, Eupatorieae) complex. Phytologia 67:121–131.

_____. 1991a. Texas species of *Ruellia* (Acanthaceae). Phytologia 71:281–299.

_____. 1991b. An overview of the North American species of *Menodora* (Oleaceae). Phytologia 71:340–356.

_____. 1992. New species of *Wedelia* (Asteraceae, Heliantheae) from Mexico and critical assessment of previously described taxa. Phytologia 72:115–126.

_____. 1993a. The Texas species of *Centaurium* (Gentianaceae).

Phytologia 75:259–275.

_____. 1993b. Texas species of *Mirabilis* (Nyctaginaceae). Phytologia 75:432–451.

_____. 1994a. Taxonomic overview of *Gilia*, sect. *Giliastrum* (Polemoniaceae) in Texas and Mexico. Phytologia 76:52–68.

_____. 1994b. A taxonomic overview of *Scutellaria*, section *Resinosa* (Lamiaceae). Phytologia 76:345–382.

_____. 1994c. Taxonomic study of the *Stachys coccinea* (Lamiaceae) complex. Phytologia 76:391–401.

_____. 1994d. Texas species of *Schrankia* (Mimosaceae) transferred to the genus *Mimosa*. Phytologia 76:412–420.

_____. 1994e. Regional variation in the North American elements of *Oxalis corniculata* (Oxalidaceae). Phytologia 77:1–7.

_____. 1994f. Taxonomic treatment of *Monarda* (Lamiaceae) for Texas and Mexico. Phytologia 77:56–79.

_____. 1995a. Synopsis of the genus *Onosmodium* (Boraginaceae). Phytologia 78:39–60.

_____. 1995b. *Paronychia virginica* (Caryophyllaceae), a first report of its occurrence in Mexico. Phytologia 78:446–447.

_____. 1995c [1996]. Taxonomic overview of *Hedyotis nigricans* (Rubiaceae) and closely allied taxa. Phytologia 79:12–21.

_____. 1995d [1996]. *Cerastium texanum* (Caryophyllaceae) does not occur in Texas. Phytologia 79:356–363.

_____. 1998. Texas species of *Glandularia* (Verbenaceae). Lundellia 1:3–16.

_____ and **R. Alston.** 1959. Segregation and recombination of chemical constituents in a hybrid swarm of *Baptisia laevicaulis* × *B. viridis* and their taxonomic implications. Amer. J. Bot. 46:678-686.

Turner, B.L. and **J. Andrews.** 1986. Lectotypification of *Lupinus subcarnosus* and *L. texensis* (Fabaceae). Sida 11:255–257.

Turner, B.L. and **D. Dawson.** 1980. Taxonomy of *Tetragonotheca* (Asteraceae-Heliantheae). Sida 8:296–303.

Turner, B.L. and **L.K. Escobar.** 1991. Documented chromosome numbers 1991: 1. Chromosome numbers in *Hybanthus* (Violaceae). Sida 14:501–503.

Turner, B.L. and **O.S. Fearing.** 1964. A taxonomic study of the genus *Amphicarpaea* (Leguminosae). SouthW. Naturalist 9:207–218.

Turner, B.L. and **R. Hartman.** 1976. Infraspecific categories of *Machaeranthera pinnatifida* (Compositae). Wrightia 5:308–315.

Turner, B.L. and **M.C. Johnston.** 1956. Chromosome numbers and geographic distribution of *Lindheimera*, *Engelmannia*, and *Berlandiera* (Compositae-Heliantheae-Melampodinae). SouthW. Naturalist 1:125–132.

Turner, B.L. and **K.-J. Kim.** 1990. An overview of the genus *Pyrrhopappus* (Asteraceae: Lactuceae) with emphasis on chloroplast DNA restriction site data. Amer. J. Bot. 77:845–850.

Turner, B.L. and **R.M. King.** 1962. A cytotaxonomic survey of *Melampodium* (Compositae-Heliantheae). Amer. J. Bot. 49:263–269.

Turner, B.L. and **M. Mendenhall.** 1993. A revision of *Malvaviscus* (Malvaceae). Ann. Missouri Bot. Gard. 80:439–457.

Turner, B.L. and **M.I. Morris.** 1976. Systematics of *Palafoxia* (Asteraceae: Helenieae). Rhodora 78:567–628.

Turner, B.L. and **M. Whalen.** 1975. Taxonomic study of *Gaillardia pulchella* (Asteraceae-Heliantheae). Wrightia 5:189–192.

Turner, B.L. and **L. Woodruff.** 1993. Annotated distribution of the monotypic genus *Lindheimera* (Asteraceae: Heliantheae). Sida 15:533–537.

Turner, M.W. 1996. Systematic study of the genus *Brazoria* (Lamiaceae), and *Warnockia* (Lamiaceae), a new genus from Texas. Pl. Syst. Evol. 203:65–82.

Turner, N.W. and **A.F. Szczawinski.** 1991. Common poisonous plants and mushrooms of North America. Timber Press, Portland, OR.

Tutin, T.G. 1976. *Soliva* and *Gymnostyles*. In: Tutin, T.G., V.H. Heywood, N.A. Burges, D.M. Moore, D.H. Valentine, S.M. Walters, and D.A. Webb, eds. Flora Europaea 4:178.

_____, **V.H. Heywood, N.A. Burges, D.H. Valentine, S.M. Walters,** and **D.A. Webb,** eds. 1964. Flora Europaea, Vol. 1, Lycopodiaceae to Plantanaceae. Cambridge Univ. Press, Cambridge, England, U.K.

Tutin, T.G., V.H. Heywood, N.A. Burges, D.M. Moore, D.H. Valentine, S.M. Walters, and **D.A. Webb,** eds. 1972. Flora Europaea, Vol. 3, Diapensiaceae to Myoporaceae. Cambridge Univ. Press, Cambridge, England, U.K.

_____, _____, _____, _____, _____, _____, and _____, eds. 1980. Flora Europaea, Vol. 5. Cambridge Univ. Press, Cambridge, England, U.K.

Tveten, J.L. and **G.A. Tveten.** 1993. Wildflowers of Houston and southeast Texas. Univ. Texas Press, Austin.

Tyrl, R.J., S.C. Barber, P. Buck, J.R. Estes, P. Folley, L.K. Magrath, C.E.S. Taylor, and **R.A. Thompson** (Oklahoma Flora Editorial Committee). 1994 (revised 1 Sep 1997). Key and descriptions for the vascular plant families of Oklahoma. Flora Oklahoma Incorporated, Noble, OK.

Tyrl, R.J., J.L. Gentry, Jr., P.G. Risser, and **J.J. Crockett.** 1978. Unpublished status report on *Physostegia micrantha*, copy in the library of the Botanical Research Institute of Texas.

Umber, R.E. 1979. The genus *Glandularia* (Verbenaceae) in North America. Syst. Bot. 4:72–102.

Urbatsch, L.E. 1972. Systematic study of the *Altissimae* and *Giganteae* species groups of the genus *Vernonia* (Compositae). Brittonia 24:229–238.

_____ and **R.K. Jansen.** 1995. Phylogenetic affinities among and within the coneflower genera (Asteraceae, Heliantheae), a chloroplast DNA analysis. Syst. Bot. 20:28–39.

Urich, R.W., D.L. Bowerman, J.A. Levisky, and **J.L. Pflug.** 1982. *Datura stramonium*: A fatal poisoning. J. Forensic Sci. 27:948–954.

Urshel, S. (Susan). Personal communication. Professional gardener and horticulturalist, Fort Worth, TX.

USDA Forest Service (United States Department of Agriculture). 1993. Pacific yew draft environmental impact statement. U.S. Government Printing Office.

Vail, A.M. 1895. A study of the genus *Galactia* in North America. Bull. Torrey Bot. Club 22:500–511.

_____ and **P.A. Rydberg.** 1910. Zygophyllaceae. N. Amer. Fl. 25:103–116.

Valder, P. 1995. Wisterias: A comprehensive guide. Timber Press, Portland, OR.

Valdés, B. 1993. The role of herbaria in scientific research. Webbia 48:163–171.

Valdespino, I.A. 1993. Selaginellaceae. In: Flora of North America Editorial Committee, eds. Fl. North Amer. 2:38–63. Oxford Univ. Press, New York and Oxford.

Valdés-Reyna, J. 1985. A biosystematic study of the genus *Erioneuron* Nash (Poaceae: Eragrostideae). Ph.D. dissertation, Texas A&M Univ., College Station.

_____ and **S.L. Hatch.** 1997. A revision of *Erioneuron* and *Dasyochloa* (Poaceae: Eragrostideae). Sida 17:645–666.

Valentine, D.H. 1962. Variation and evolution in the genus *Viola*. Preslia 34:190–206.

Van den Borre, A. and **L. Watson.** 1994. The infrageneric classification of *Eragrostis* (Poaceae). Taxon 43:383–422.

Vander Kloet, S.P. 1988. The genus *Vaccinium* in North America. Research Branch Agric. Canada Publ. No. 1828.

van der Werff, H. 1997a. Lauraceae. In: Flora of North America Editorial Committee, eds. Fl. North Amer. 3:26–27. Oxford Univ. Press, New York and Oxford.

_____. 1997b. *Sassafras*. In: Flora of North America Editorial Committee, eds. Fl. North Amer. 3:29–30. Oxford Univ. Press, New York and Oxford.

_____ and **H.G. Richter.** 1996. Toward an improved classification of Lauraceae. Ann. Missouri Bot. Gard. 83:409–418.

Van Horn, G.S. 1973. The taxonomic status of *Pentachaeta* and *Chaetopappa* with a revision of *Pentachaeta*. Univ. Calif. Publ. Bot. 65:1–41.

Van Ooststroom, S.J. 1934. A monograph of the genus *Evolvulus*. Meded. Bot. Mus. Herb. Rijks Univ. Utrecht 14:1–267.

Van Vleet, R.L. 1951. Phenotypic variations of *Erigeron strigosus* Muhl. (Compositae) in eastern Texas. Field & Lab. 19:161–163.

Veldkamp, J.F., R. de Koning, and **M.S.M. Sosef.** 1986. Generic delimitation of *Rottboellia* and related genera (Gramineae). Blumea 31:281–307.

Venter, H.J.T. and **R.L. Verhoeven.** 1997. A tribal classification of the Periplocoideae (Apocynaceae). Taxon 46:705–720.

Verdcourt, B. 1976. Rubiaceae (Part 1). In: R.M. Polhill, ed. Fl. Tropical East Africa. Pp. 1–414. Crown Agents for Oversea Governments and Administrations, London, U.K.

Verhoek-Williams, S.E. 1975. A study of the tribe Poliantheae (including *Manfreda*) and revisions of *Manfreda* and *Prochyanthes* (Agavaceae). Unpublished doctoral dissertation, Cornell Univ. Diss. Abstr. Int. 36:356.

Vernon, L. 1965. Biosystematic studies in the *Heterotheca subaxillaris* complex (Compositae: Astereae). Trans. Kansas Acad. Sci. 68:244–257.

Vines, R.A. 1960. Trees, shrubs and woody vines of the southwest. Univ. of Texas Press, Austin.

von Bothmer, R., N. Jacobsen, C. Baden, R.B. Jørgensen, and **I. Linde-Laursen.** 1991. An ecogeographical study of the genus *Hordeum*. Syst. Ecogeogr. Stud. Crop Genepools 7:1–127. International Board for Plant Genetic Resources, Rome, Italy.

Voss, E.G. 1958. Confusion in *Alisma*. Taxon 7:130–133.

Voss, J.W. 1937. A revision of the *Phacelia crenulata* group for North America. Bull. Torrey Bot. Club 64:81–96.

Vuilleumier, B.S. 1969a. The genera of Senecioneae in the southeastern United States. J. Arnold Arb. 50:104–121.

_____. 1969b. The tribe Mutisieae (Compositae) in the southeastern United States. J. Arnold Arbor. 50:620–625.

_____. 1973. The genera of Lactuceae (Compositae) in the southeastern United States. J. Arnold Arbor. 54:42–93.

Wagenitz, G. 1992. The Asteridae: Evolution of a concept and its present status. Ann. Missouri Bot. Gard. 79:209–217.

Wagenknecht, B.L. 1960. Revision of *Heterotheca* section *Heterotheca*. Rhodora 62:61–76, 97–107.

Wagner, D.H. 1993. *Polystichum*. In: Flora of North America Editorial Committee, eds. Fl. North Amer. 2:290–299. Oxford Univ. Press, New York and Oxford.

Wagner, F.H. 1948. The bur clovers (*Medicago*) of Texas. Field & Lab. 16:3–7.

Wagner, W.H., Jr. (Herb). Personal communication. Specialist in pteridophytes and evolution of vascular plants; Flora of North America Editorial Committee; retired from a career at the Univ. of Michigan, Ann Arbor.

Wagner, W.H., Jr. 1954. Reticulate evolution in the Appalachian Aspleniums. Evolution 8:103–118.

_____ and **J.M. Beitel.** 1992. Generic classification of modern North American Lycopodiaceae. Ann. Missouri. Bot. Gard. 79:676–686.

_____ and _____. 1993. Lycopodiaceae. In: Flora of North America Editorial Committee, eds. Fl. North Amer. 2:18–37. Oxford Univ. Press, New York and Oxford.

Wagner, W.H., Jr., R.C. Moran, and **C.R. Werth.** 1993. Aspleniaceae. In: Flora of North America Editorial Committee, eds. Fl. North Amer. 2:228–245. Oxford Univ. Press, New York and Oxford.

Wagner, W.H., Jr. and **A.R. Smith.** 1993. Pteridophytes. In: Flora of North America Editorial Committee, eds. Fl. North Amer. 1:247–266. Oxford Univ. Press, New York and Oxford.

Wagner, W.H., Jr. and **F.S. Wagner.** 1993. Ophioglossaceae. In: Flora of North America Editorial Committee, eds. Fl. North Amer. 2:85–106. Oxford Univ. Press, New York and Oxford.

Wagner, W.L. 1983. New species and combinations in the genus *Oenothera* (Onagraceae). Ann. Missouri Bot. Gard. 70:194–196.

_____. 1986. New taxa in *Oenothera* (Onagraceae). Ann. Missouri Bot. Gard. 73:475–480.

_____, **D.R. Herbst,** and **S.H. Sohmer.** 1990. Manual of the flowering plants of Hawaii, 2 vols. Univ. Hawaii Press and Bishop Museum Press, Honolulu.

Wagnon, H.K. 1952. A revision of the genus *Bromus*, section *Bromopsis*, of North America. Brittonia 7:415–480.

Wagstaff, S.J. and **R.G. Olmstead.** 1997. Phylogeny of Labiatae and Verbenaceae inferred from *rbc*L sequences. Syst. Bot. 22:165–179.

Wahl, H.A. 1954. A preliminary study of the genus *Chenopodium* in North America. Bartonia 27:1–46.

Walker, K.R. (Kevin). Personal communication. Alumnus of Austin College currently in graduate school at Univ. of New

Mexico doing research on the evolutionary ecology of the soapberry bug (*Jadera haematoloma*).

WALLACE, G.D. 1975. Studies of the Monotropoideae (Ericaceae): Taxonomy and distribution. Wasmann J. Biol. 33:1–88.

______. 1993. Ericaceae. In: J.C. Hickman, ed. The Jepson manual: Higher plants of California. Pp. 544–567. Univ. of California Press, Berkeley.

______. 1995. Ericaceae Subfamily Monotropoideae. In: J.L. Luteyn, ed. Ericaceae Part II, the superior-ovaried genera. Fl. Neotrop. Monogr. 66:13–27.

WALTERS, D.R. and **D.J. KEIL.** 1996. Vascular plant taxonomy, 4th ed. Kendall/Hunt Publishing Company, Dubuque, Iowa.

WALTERS, S.M. and **D.A. WEBB.** 1972. *Veronica*. In: T.G. Tutin, V.H. Heywood, N.A. Burges, D.M. Moore, D.H. Valentine, S.M. Walters, and D.A. Webb, eds. Flora Europaea 3:242–251.

WANG, J.K., ed. 1983. Taro: A review of *Colocasia esculenta* and its potentials. Univ. of Hawaii Press, Honolulu.

WARD, D.B. 1977. *Nelumbo lutea*, the correct name for the American lotus. Taxon 26:227–234.

______. 1987. North American collections of *Lepuropetalon spathulatum* (Saxifragaceae). Florida Agric. Exp. Sta. J. Ser. 63:15–35.

______. 1998. *Pueraria montana*: The correct scientific name of the kudzu. Castanea 63:76–77.

______ and **A.K. GHOLSON.** 1987. The hidden abundance of *Lepuropetalon spathulatum* (Saxifragaceae) and its first reported occurrence in Florida. Castanea 52:59–67.

WARNOCK, M.J. 1981. Biosystematics of the *Delphinium carolinianum* complex (Ranunculaceae). Syst. Bot. 6:38–54.

______. 1987a. Synopsis of *Delphinium* (Ranunculaceae) in continental Mexico. Rhodora 89:47–74.

______. 1987b. An index to epithets treated by King and Robinson: Eupatorieae (Asteraceae). Phytologia 62:345–431.

______. 1995. A taxonomic conspectus of North American *Delphinium*. Phytologia 78:73–101.

______. 1997a. *Delphinium*. In: Flora of North America Editorial Committee, eds. Fl. North Amer. 3:196–240. Oxford Univ. Press, New York and Oxford.

______. 1997b. *Consolida*. In: Flora of North America Editorial Committee, eds. Fl. North Amer. 3:240–242. Oxford Univ. Press, New York and Oxford.

WASOWSKI, S. (SALLY). Personal communication. Professional landscape designer and author on gardening and native plants; founding member and past president of the Native Plant Society of Texas; currently living in Santa Fe, NM.

WASSHAUSEN, D.C. 1966. Acanthaceae. In: C.L. Lundell, ed. Flora of Texas 1:223–282. Texas Research Foundation, Renner.

WATERFALL, U.T. 1951. The genus *Callirhoe* (Malvaceae) in Texas. Field & Lab. 19:107–119.

______. 1958. A taxonomic study of the genus *Physalis* in North America north of Mexico. Rhodora 60:107–114, 128–142, 152–173.

______. 1971 [1972]. New taxa, combinations and distribution records for the Oklahoma flora. Rhodora 73:552–555.

______. 1972. Keys to the flora of Oklahoma, 5th ed. Published by the author for sale by the Student Union Bookstore, Oklahoma State Univ., Stillwater.

WATSON, F.D. 1985. The nomenclature of pondcypress and baldcypress. Taxon 34:506–509.

______ and **J.E. ECKENWALDER.** 1993. Cupressaceae (family description and key to genera). In: Flora of North America Editorial Committee, eds. Fl. North Amer. 2:399–401. Oxford Univ. Press, New York and Oxford.

WATSON, L.E. and **J.E. ESTES.** 1990. Biosystematic and phenetic analysis of *Marshallia* (Asteraceae). Syst. Bot. 15:403–414.

WATT, G. 1907. The wild and cultivated cotton plants of the world. A revision of the genus *Gossypium*. Longmans, Green, and Co., New York.

WEATHERBY, C.A. and **S.F. BLAKE.** 1916. *Galium pilosum* and its varieties. Rhodora 18:190–195.

WEAVER, J.E. and **F.E. CLEMENTS.** 1938. Plant ecology, 2nd. ed. McGraw-Hill Book Co., New York.

WEBB, D.A. 1980. *Leucojum*. In: Tutin, T.G., V.H. Heywood, N.A. Burges, D.M. Moore, D.H. Valentine, S.M. Walters, and D.A. Webb, eds. Flora Europaea 5:76–77. Cambridge Univ. Press, Cambridge, England, U.K.

WEBB, R.G. 1970. Reptiles of Oklahoma. Univ. of Oklahoma Press, Norman.

WEBBER, J.M. 1953. Yuccas of the southwest. Agric. Monogr. U.S.D.A. 17:1–97.

WEBSTER, G.L. (GRADY). Personal communication. Specialist in Euphorbiaceae; Flora of North America Editorial Committee; Univ. of California, Davis.

WEBSTER, G.L. 1956. Studies of the Euphorbiaceae, Phyllanthoideae II. The American species of *Phyllanthus* described by Linnaeus. J. Arnold Arbor. 37:1–14.

______. 1967. The genera of Euphorbiaceae in the southeastern United States. J. Arnold Arbor. 48:303–430.

______. 1970. A revision of *Phyllanthus* (Euphorbiaceae) in the continental United States. Brittonia 22:44–76.

______. 1992. Realignments in American *Croton* (Euphorbiaceae). Novon 2:269–273.

______. 1993. A provisional synopsis of the sections of the genus *Croton* (Euphorbiaceae). Taxon 42:793–823.

______. 1994a. Classification of the Euphorbiaceae. Ann. Missouri Bot. Gard. 81:3–32.

______. 1994b. Synopsis of the genera and suprageneric taxa of Euphorbiaceae. Ann. Missouri Bot. Gard. 81:33–144.

______ and **K.I. MILLER.** 1963. The genus *Reverchonia*. Rhodora 65:193–207.

WEBSTER, R.D. 1987. Taxonomy of *Digitaria* section *Digitaria* in North America (Poaceae: Paniceae). Sida 12:209–222.

______. 1988. Genera of the North American Paniceae (Poaceae: Panicoideae). Syst. Bot. 13:576–609.

______. 1992. Character significance and generic similarities in the Paniceae (Poaceae: Panicoideae). Sida 15:185–213.

______. 1993. Nomenclature of *Setaria* (Poaceae: Paniceae). Sida 15:447–489.

______. 1995. Nomenclatural changes in *Setaria* and *Paspalidium* (Poaceae: Paniceae). Sida 16:439–446.

______, **J.H. KIRKBRIDE,** and **J.V. REYNA.** 1989. New World genera of the Paniceae (Poaceae: Panicoideae). Sida 13:393–417.

Webster, R.D. and **R.B. Shaw.** 1995. Taxonomy of the native North American species of *Saccharum* (Poaceae: Andropogoneae). Sida 16:551–580.

Wells, E.F. 1984. A revision of the genus *Heuchera* (Saxifragaceae) in eastern North America. Syst. Bot. Monogr. 3:45–121.

Wells, J.R. 1965. A taxonomic study of *Polymnia* (Compositae). Brittonia 17:144–159.

Welzen, P.C. van. 1997. Paraphyletic groups or what should a classification entail. Taxon 46:99–103.

_____. 1998. Phylogenetic versus Linnaean taxonomy, the continuing story. Taxon 47:413–423.

Wemple, D.K. 1970. Revision of the genus *Petalostemon* (Leguminosae). Iowa State J. Sci. 45:1–102.

_____ and **N.R. Lersten.** 1966. An interpretation of the flower of *Petalostemum* (Leguminosae). Brittonia 18:117–126.

Wen, J. and **T.F. Stuessy.** 1993. The phylogeny and biogeography of *Nyssa* (Cornaceae). Syst. Bot. 18:68–79.

Wendel, J.F. and **V.A. Albert.** 1992. Phylogenetics of the cotton genus (*Gossypium*): Character-state weighted parsimony analysis of chloroplast-DNA restriction site data and its systematic and biogeographic implications. Syst. Bot. 17:115–143.

Wendt, T. 1979. Notes on the genus *Polygala* in the United States and Mexico. J. Arnold Arbor. 60:504–514.

Weniger, D. 1970. Cacti of the southwest. Univ. of Texas Press, Austin.

_____. 1984. Cacti of Texas and neighboring states. Univ. of Texas Press, Austin.

_____. 1996. Catalpa (*Catalpa bignonioides*, Bignoniaceae) and bois d'arc (*Maclura pomifera*, Moraceae) in early Texas records. Sida 17:231–242.

Werner, G., R.K. Brummitt, E. Farr, N. Kilian, P.M. Kirk, and **P.C. Silva.** 1993. Names in current use for extant plant genera. Regnum Veg. 129.

Westbrooks, R.G. and **R.E. Eplee.** 1996. Regulatory exclusion of harmful non-indigenous plants from the United States by USDA APHIS PPQ. Castanea 61:305–312.

Westbrooks, R.G. and **J.W. Preacher.** 1986. Poisonous plants of eastern North America. Univ. of South Carolina Press, Columbia.

Wetherwax, M. 1993. *Linaria*. In: J.C.Hickman, ed. The Jepson manual: Higher plants of California. Pp. 1036–1037. Univ. of California Press, Berkeley.

Whalen, M.D. 1979. Taxonomy of *Solanum* section *Androceras*. Gentes Herb. 11:359–426.

Wheeler, L.C. 1936. Revision of the *Euphorbia polycarpa* group of the southwestern United States and adjacent Mexico; a preliminary treatment. Bull. Torrey Bot. Club 63:397–416.

_____. 1941. *Euphorbia* subgenus *Chamaesyce* in Canada and the United States exclusive of southern Florida. Rhodora 43:97–154, 168–205, 223–286.

_____. 1943. The genera of living Euphorbiaceae. Amer. Midl. Naturalist 30:456–503.

Wherry, E.T. 1955. The genus *Phlox*. Morris Arbor., Monogr. 3. Philadelphia, PA.

_____. 1966. Polemoniaceae. In: C.L. Lundell, ed. Flora of Texas 1:283–321. Texas Research Foundation, Renner.

Whetstone, R.D. 1983. The Sterculiaceae in the flora of the southeastern United States. Sida 10:15–23.

_____ and **T.A. Atkinson.** 1993. Osmundaceae. In: Flora of North America Editorial Committee, eds. Fl. North Amer. 2:107–109. Oxford Univ. Press, New York and Oxford.

_____, _____, and **D.D. Spaulding.** 1997a. Berberidaceae (family description and key to genera). In: Flora of North America Editorial Committee, eds. Fl. North Amer. 3:272–273. Oxford Univ. Press, New York and Oxford.

_____, _____, and _____. 1997b. *Nandina*. In: Flora of North America Editorial Committee, eds. Fl. North Amer. 3:273–274. Oxford Univ. Press, New York and Oxford.

Whitaker, T.W. and **W.P. Bemis.** 1964. Evolution in the genus *Cucurbita*. Evolution 18:553–559.

_____ and _____. 1975. VIII. Origin and evolution of the cultivated *Cucurbita*. Bull. Torrey Bot. Club 102:362–368.

White, H.L., J.R. Branch, W.C. Holmes, and **J.R. Singhurst.** 1998. Comments on the distribution of *Sedum pulchellum* (Crassulaceae) in Texas. Sida 18:622–626.

Whitehead, F.H. and **R.P. Sinha.** 1967. Taxonomy and taximetrics of *Stellaria media* (L.) Vill., *S. neglecta* Weihe and *S. pallida* Dumont.) Piré. New Phytol. 66:769–784.

Whitehouse, E. 1935. Notes on Texas phloxes. Bull. Torrey Bot. Club 62:381–386.

_____. 1936. Texas flowers in natural colors. Published by the author, Dallas, TX.

_____. 1939. A study of the annual *Phlox* species. Ph.D. dissertation, Univ. of Texas, Austin.

_____. 1945. Annual *Phlox* species. Amer. Midl. Naturalist 34:388–401.

_____. 1949. Revision of *Salvia* L., section *Salviastrum* Gray. Field & Lab. 17:151–165.

_____ and **F. McAllister.** 1954. The mosses of Texas. A catalogue with annotations. Bryologist 57:63–146.

Whittemore, A.T. 1997a. *Ranunculus*. In: Flora of North America Editorial Committee, eds. Fl. North Amer. 3:88–135. Oxford Univ. Press, New York and Oxford.

_____. 1997b. *Myosurus*. In: Flora of North America Editorial Committee, eds. Fl. North Amer. 3:135–138. Oxford Univ. Press, New York and Oxford.

_____. 1997c. *Aquilegia*. In: Flora of North America Editorial Committee, eds. Fl. North Amer. 3:249–258. Oxford Univ. Press, New York and Oxford.

_____. 1997d. *Berberis*. In: Flora of North America Editorial Committee, eds. Fl. North Amer. 3:276–286. Oxford Univ. Press, New York and Oxford.

_____ and **B.D. Parfitt.** 1997. Ranunculaceae. In: Flora of North America Editorial Committee, eds. Fl. North Amer. 3:85–87. Oxford Univ. Press, New York and Oxford.

Whittemore, A.T. and **D.E. Stone.** 1997. *Juglans*. In: Flora of North America Editorial Committee, eds. Fl. North Amer. 3:425–428. Oxford Univ. Press, New York and Oxford.

Widrlechner, M.P. 1983. Historical and phenological observations on the spread of *Chaenorrhinum minus* across North America. Canad. J. Bot. 61:179–187.

Wiegand, K.M. 1900. *Juncus tenuis* Willd. and some of its North American allies. Bull. Torrey Bot. Club. 27:511–527.

_____. 1912. *Amelanchier* in eastern North America. Rhodora 14:117–161.

_____. 1925. *Oxalis corniculata* and its relatives in North America. Rhodora 27:113–124, 133–139.

Wiegrefe, S.J., K.J. Sytsma, and **R.P. Guries.** 1994. Phylogeny of elms (*Ulmus*, Ulmaceae): Molecular evidence for a sectional classification. Syst. Bot. 19:590–612.

Wiens, D. 1964. Revision of the acataphyllous species of *Phoradendron*. Brittonia 16:11–54.

Wiersema, J.H. 1987. A monograph of *Nymphaea* subgenus *Hydrocallis* (Nymphaeaceae). Syst. Bot. Monogr. 16:1–112.

_____. 1988. Reproductive biology of *Nymphaea* (Nymphaeaceae). Ann. Missouri Bot. Gard. 75:795–804.

_____. 1997a. Nelumbonaceae. In: Flora of North America Editorial Committee, eds. Fl. North Amer. 3:64–65. Oxford Univ. Press, New York and Oxford.

_____. 1997b. *Nymphaea*. In: Flora of North America Editorial Committee, eds. Fl. North Amer. 3:71–77. Oxford Univ. Press, New York and Oxford.

_____. 1997c. Cabombaceae. In: Flora of North America Editorial Committee, eds. Fl. North Amer. 3:78–80. Oxford Univ. Press, New York and Oxford.

_____ and **C.B. Hellquist.** 1994. Nomenclatural notes in Nymphaeaceae for the North American Flora. Rhodora 96:170–178.

_____ and _____. 1997. Nymphaeaceae. In: Flora of North America Editorial Committee, eds. Fl. North Amer. 3:66–77. Oxford Univ. Press, New York and Oxford.

Wight, W.F. 1915. Native American species of *Prunus*. U.S.D.A. Bull. 179:1–75.

Wijnands, D.O., J.J. Bos, H.J.W. Wijsman, F. Schneider, C.D. Brickell, and **K. Zimmer.** 1986. Proposal to conserve 7436 *Petunia* with *P. nyctaginiflora* as *typ. cons.* (Solanaceae). Taxon 35:748–749.

Wijsman, H.J.W. 1990. On the inter-relationships of certain species of *Petunia* VI. New names for the species of *Calibrachoa* formerly included into *Petunia* (Solanaceae). Acta Bot. Neerl. 39:101–102.

_____ and **J.H. de Jong.** 1985. On the interrelationships of certain species of *Petunia* IV. Hybridization between *P. linearis* and *P. calycina* and nomenclatorial consequences in the *Petunia* group. Acta Bot. Neerl. 34:337–349.

Wilbur, R.L. 1955. A revision of the North American genus *Sabatia* (Gentianaceae). Rhodora 57:1–33, 43–71, 78–104.

_____. 1966. Notes on Rafinesque's species of *Lechea* (Cistaceae). Rhodora 68:192–208.

_____. 1969. Cistaceae. In: C.L. Lundell, ed. Flora of Texas 2:1–17. Texas Research Foundation, Renner.

_____. 1970. Taxonomic and nomenclatural observations on the eastern North American genus *Asimina* (Annonaceae). J. Elisha Mitchell Sci. Soc. 86:88–96.

_____. 1975. A revision of the North American genus *Amorpha* (Leguminoseae–Psoraleae). Rhodora 77:337–409.

_____. 1994. The Myricaceae of the United States and Canada: Genera, subgenera, and series. Sida 16:93–107.

_____ and **H.S. Daoud.** 1961. The genus *Lechea* (Cistaceae) in the southeastern United States. Rhodora 63:103–118.

_____ and _____. 1964. The genus *Helianthemum* (Cistaceae) in the southeastern United States. J. Elisha Mitchell Sci. Soc. 80:38–43.

Wilken, D.H. 1986a. Polemoniaceae. In: Great Plains Flora Association. Flora of the Great Plains. Pp. 666–677. Univ. Press of Kansas, Lawrence.

_____. 1986b. Hydrophyllaceae. In: Great Plains Flora Association. Flora of the Great Plains. Pp. 678–683. Univ. Press of Kansas, Lawrence.

_____. 1993a. Chenopodiaceae. In: J.C. Hickman, ed. The Jepson manual: Higher plants of California. Pp. 500–515. Univ. of California Press, Berkeley.

_____. 1993b. Verbenaceae. In: J.C. Hickman, ed. The Jepson manual: Higher plants of California. Pp. 1085–1089. Univ. of California Press, Berkeley.

_____, **R.R. Halse,** and **R.W. Patterson.** 1993. *Phacelia*. In: J.C. Hickman, ed. The Jepson manual: Higher plants of California. Pp. 691–706. Univ. of California Press, Berkeley.

Williams, J.K. (Justin). Personal communication. Specialist on *Amsonia* (Apocynaceae) and floristics of the Texas panhandle; graduate student at Univ. of Texas, Austin.

Williamson, P.S. and **E.L. Schneider.** 1993a. Cabombaceae. In: K. Kubitzki, J.C. Rohwer, and V. Bittrich, eds. The families and genera of vascular plants, Vol. II. Pp. 157–161. Springer-Verlag, Berlin.

_____ and _____. 1993b. Nelumbonaceae. In: K. Kubitzki, J.C. Rohwer, and V. Bittrich, eds. The families and genera of vascular plants, Vol. II. Pp. 470–473. Springer-Verlag, Berlin.

Wills, M.M. and **H.S. Irwin.** 1961. Roadside flowers of Texas. Univ. of Texas Press, Austin.

Willson, M.F. and **J.N. Thompson.** 1982. Phenology and ecology of color in bird-dispersed fruits, or why some fruits are red when they are "green." Canad. J. Bot. 60:701–713.

Wilson, H.D. 1990. Quinua and relatives (*Chenopodium* sect. *Chenopodium* subsect. *Cellulata*). Econ. Bot. 44 (Suppl.):92–110.

Wilson, J.S. 1964 [1965]. Variation of three taxonomic complexes of the genus *Cornus* in eastern United States. Trans. Kansas Acad. Sci. 67:747–817.

Wilson, K.A. 1960a. The genera of Arales in the southeastern United States. J. Arnold Arbor. 41:47–72.

_____. 1960b. The genera of Hydrophyllaceae and Polemoniaceae in the southeastern United States. J. Arnold Arbor. 41:197–212.

_____. 1960c. The genera of Convolvulaceae in the southeastern United States. J. Arnold Arbor. 41:298–317.

_____ and **C.E. Wood, Jr.** 1959. The genera of Oleaceae in the southeastern United States. J. Arnold Arbor. 40:369–384.

Wilson, P. 1909. Typhaceae. N. Amer. Fl. 17:3–4.

_____. 1911. Rutaceae. N. Amer. Fl. 25:173–224.

_____. 1924. Meliaceae. N. Amer. Fl. 25:263–296.

_____. 1932. Basellaceae. N. Amer. Fl. 21:337–339.

Windham, M.D. 1987a. Chromosomal and electrophoretic studies of the genus *Woodsia* in North America. Amer. J. Bot. 74:715.

_____. 1987b. *Argyrochosma*, a new genus of cheilanthoid ferns. Amer. Fern J. 77:37–41.

_____. 1993a. Pteridaceae (family description and key to genera). In: Flora of North America Editorial Committee, eds. Fl. North Amer. 2:122–124. Oxford Univ. Press, New York and Oxford.

_____. 1993b. *Argyrochosma*. In: Flora of North America Editorial Committee, eds. Fl. North Amer. 2:171–175. Oxford Univ. Press, New York and Oxford.

_____. 1993c. *Pellaea*. In: Flora of North America Editorial Committee, eds. Fl. North Amer. 2:175–186. Oxford Univ. Press, New York and Oxford.

_____. 1993d. *Woodsia*. In: Flora of North America Editorial Committee, eds. Fl. North Amer. 2:270–280. Oxford Univ. Press, New York and Oxford.

_____ and **E.W. Rabe.** 1993. *Cheilanthes*. In: Flora of North America Editorial Committee, eds. Fl. North Amer. 2:152–169. Oxford Univ. Press, New York and Oxford.

Windler, D.R. 1974. A systematic treatment of the native unifoliate crotalarias of North America (Leguminosae). Rhodora 76:151–204.

Winkler, C.H. 1915. The botany of Texas, an account of botanical investigations in Texas and adjoining territory. Univ. Texas Bull. 18:1–27.

Wipff, J.K. (Joseph). Personal communication. Specialist in Poaceae trained at Texas A&M Univ.; Taxonomist and Assistant Plant Breeder, Pure Seed Testing, Inc., Hubbard, OR.

Wipff, J.K. 1996. Nomenclatural combinations in the *Andropogon gerardii* complex (Poaceae: Andropogoneae). Phytologia 80:343–347.

_____ and **S.L. Hatch.** 1994. A systematic study of *Digitaria* sect. *Pennatae* (Poaceae: Paniceae) in the New World. Syst. Bot. 19:613–627.

Wipff, J.K. and **S.D. Jones.** 1994 [1995]. Nomenclatural combinations in Poaceae and Cyperaceae. Phytologia 77:456–464.

_____ and _____. 1995. Nomenclatural combination in Poaceae. Phytologia 78:244–245.

_____ and _____. 1996. A new combination in *Bouteloua* (Poaceae). Sida 17:109–110.

Wipff, J.K., R.I. Lonard, S.D. Jones, and **S.D. Hatch.** 1993. The genus *Urochloa* (Poaceae: Paniceae) in Texas, including one previously unreported species for the state. Sida 15:405–413.

Witherspoon, J.T. 1977. New taxa and combinations in *Eragrostis* (Poaceae). Ann. Missouri Bot. Gard. 64:324–329.

Wofford, B.E. 1981. External seed morphology of *Arenaria* (Caryophyllaceae) of the southeastern United States. Syst. Bot. 6:126–135.

_____. 1997. *Lindera*. In: Flora of North America Editorial Committee, eds. Fl. North Amer. 3:27–29. Oxford Univ. Press, New York and Oxford.

_____ and **R. Kral.** 1993. Checklist of the vascular plants of Tennessee. Sida, Bot. Misc. 10:1–66.

Wolf, S.J. and **J. McNeill.** 1986. Synopsis and achene morphology of *Polygonum* section *Polygonum* (Polygonaceae) in Canada. Rhodora 88:457–479.

Wood, C.E., Jr. 1949. The American barbistyled species of *Tephrosia* (Leguminosae). Rhodora 51:193–231, 233–302, 305–364, 369–384.

_____. 1958. The genera of the woody Ranales in the southeastern United States. J. Arnold Arbor. 39:296–346.

_____. 1959. The genera of the Nymphaeaceae and Ceratophyllaceae in the southeastern United States. J. Arnold Arbor. 40:94–112.

_____. 1960. The genera of Sarraceniaceae and Droseraceae of the southeastern United States. J. Arnold Arbor. 41:152–163.

_____. 1961. The genera of Ericaceae in the southeastern United States. J. Arnold Arbor. 42:10–80.

_____. 1966. On the identity of *Drosera brevifolia*. J. Arnold Arbor. 47:89–99.

_____. 1970. Some floristic relationships between the southern Appalachians and western North America. In: P.C. Holt and R.A. Paterson, eds. The distributional history of the biota of the southern Appalachians, Part II: Flora. Pp. 321–404. Research Division Monog. 2, Virginia Polytechnic Institute and State Univ., Blacksburg, VA.

_____. 1971. The Saururaceae in the southeastern United States. J. Arnold Arbor. 52:479–485.

_____. 1975. The Balsaminaceae in the southeastern United States. J. Arnold Arbor. 56:413–426.

_____. 1983a. The genera of Burmanniaceae in the southeastern United States. J. Arnold Arbor. 64:293–307.

_____. 1983b. The genera of Menyanthaceae in the southeastern United States. J. Arnold Arbor. 64:431–445.

_____ and **P. Adams.** 1976. The genera of Guttiferae (Clusiaceae) in the southeastern United States. J. Arnold Arbor. 57:74–90.

Wood, C.E., Jr. and **R.B. Channell.** 1960. The genera of Ebenales in the southeastern United States. J. Arnold Arbor. 41:1–35.

Wood, C.E., Jr. and **R.E. Weaver, Jr.** 1982. The genera of Gentianaceae in the southeastern United States. J. Arnold Arbor. 63:441–487.

Woodland, D.W. 1982. Biosystematics of the perennial North American taxa of *Urtica*. II. Taxonomy. Syst. Bot. 7:282–290.

_____. 1997. Contemporary plant systematics, 2nd ed. Andrews Univ. Press, Berrien Springs, MI.

_____, **I.J. Bassett,** and **C.W. Crompton.** 1976. The annual species of stinging nettle (*Hesperocnide* and *Urtica*) in North America. Canad. J. Bot. 54:374–383.

Woodruff, L. (Lindsay). Personal communication. Specialist in *Thamnosma* (Rutaceae) trained at Univ. of Texas, Austin; Associate Collections Manager at the Botanical Research Institute of Texas, Fort Worth.

Woodson, R.E., Jr. 1928. Studies in the Apocynaceae. III. A monograph of the genus *Amsonia*. Ann. Missouri Bot. Gard. 15:379–434.

_____. 1929. Studies in the Apocynaceae. III. A new species of *Amsonia* from the south-central states. Ann. Missouri Bot. Gard. 16:407–410.

_____. 1930. Studies in the Apocynaceae. I. A critical study of the Apocynoideae. Ann. Missouri Bot. Gard. 17:1–182.

_____. 1938. Apocynaceae. N. Amer. Fl. 29:103–192.

_____. 1941. The North American Asclepiadaceae. I. Perspective of the genera. Ann. Missouri Bot. Gard. 28:193–244.

_____. 1942. Commentary on the North American genera of Commelinaceae. Ann. Missouri Bot. Gard. 29:141–154.

_____. 1947. Some dynamics of leaf variation in *Asclepias*

tuberosa. Ann. Missouri Bot. Gard. 34:353–432.

______. 1953. Biometric evidence of natural selection in *Asclepias tuberosa*. Proc. Natl. Acad. Sci. U.S.A. 39:74–79.

______. 1954. The North American species of *Asclepias*. Ann. Missouri Bot. Gard. 41:1–211.

______. 1962. Butterflyweed revisited. Evolution 16:168–185.

______, **R.W. Schery,** and **H.J. Kidd.** 1961. Nyctaginaceae. In: Flora of Panama. Ann. Missouri Bot. Gard. 48:51–65.

Woolfe, J.A. 1992. Sweet potato: An untapped food resource. Cambridge Univ. Press, Cambridge, England, U.K.

Wooten, J.W. 1973. Taxonomy of seven species of *Sagittaria* from eastern North America. Brittonia 25:64–74.

______ and **A.F. Clewell.** 1971. *Fleischmannia* and *Conoclinum* (Compositae, Eupatorieae) in eastern North America. Rhodora 73:566–574.

Wu, Z. 1983. On the significance of Pacific intercontinental discontinuity. Ann. Missouri Bot. Gard. 70:577–590.

Wunderlin, R.P. 1997. Moraceae. In: Flora of North America Editorial Committee, eds. Fl. North Amer. 3:388–399. Oxford Univ. Press, New York and Oxford.

Wurdack, J.J. and **R. Kral.** 1982. The genera of Melastomataceae in the southeastern United States. J. Arnold Arbor. 63:429–439.

Wyatt, R. and **L.N. Lodwick.** 1981. Variation and taxonomy of *Aesculus pavia* L. (Hippocastanaceae). Brittonia 33:39–51.

Wynne, F.E. 1944. *Drosera* in eastern North America. Bull. Torrey Bot. Club 71:166–174.

Xiang, Q.-Y., S.J. Brunsfeld, D.E. Soltis, and **P.S. Soltis.** 1996 [1997]. Phylogenetic relationships in *Cornus* based on chloroplast DNA restriction sites: Implications for biogeography and character evolution. Syst. Bot. 21:515–534.

Xiang, Q.-Y., D.E. Soltis, D.R. Morgan, and **P.S. Soltis.** 1993. Phylogenetic relationships of *Cornus* L. sensu lato and putative relatives inferred from *rbc*L sequence data. Ann. Missouri Bot. Gard. 80:723–734.

Xiang, Q.-Y., D.E. Soltis, and **P.S. Soltis.** 1998. Phylogenetic relationships of Cornaceae and close relatives inferred from *matK* and *rbcL* sequences. Amer. J. Bot. 85:285–297.

Yates, H.O. 1966a. Morphology and cytology of *Uniola* (Gramineae). SouthW. Naturalist 11:145–189.

______. 1966b. Revision of grasses traditionally referred to *Uniola* II. *Chasmanthium*. SouthW. Naturalist 11:415–455.

Yatskievych, G. and **R.W. Spellenberg.** 1993. Plant conservation. In: Flora of North America Editorial Committee, eds. Fl. North Amer. 1:207–226. Oxford Univ. Press, New York and Oxford.

Yelderman, J.C. 1993. The water: Nature and distribution in the Blacklands. In: M.R. Sharpless and J.C. Yelderman, eds. The Texas Blackland Prairie, land, history, and culture. Pp. 48–65. Baylor Univ. Program for Regional Studies, Waco, TX.

Ying, T.S. 1983. The floristic relationships of the temperate forest regions of China and the United States. Ann. Missouri Bot. Gard. 70:597–604.

Yoon, C.K. 1993. Botanical witness for the prosecution. Science 260:894–895.

Young, M.J. 1873. Familiar lessons in botany with flora of Texas. A.S. Barnes & Co., New York.

Young, M.S. 1917. A key to the families and genera of the wild plants of Austin Texas. Univ. Texas Bull. 1754:1–71.

______. 1920. The seed plants, ferns, and fern allies of the Austin region. Univ. Texas Bull. 2065:1–98.

Yuncker, T. G. 1932. The genus *Cuscuta*. Mem. Torrey Bot. Club 18:111–331.

______. 1961. Convolvulaceae: *Cuscuta*. In: C.L. Lundell, ed. Flora of Texas 3:123–150. Texas Research Foundation, Renner.

______. 1965. *Cuscuta*. N. Amer. Fl. II. 4:1–51.

Zardini, E. and **P.H. Raven.** 1992. A new section of *Ludwigia* (Onagraceae) with a key to the sections of the genus. Syst. Bot. 17:481–485.

Zhicheng, S. 1992. Research on the pathogenesis of oak leaf poisoning in cattle. In: L.F. James, R.F. Keeler, E.M. Bailey, Jr., P.R. Cheeke, and M.P. Hegarty, eds. Poisonous plants: Proceedings of the third international symposium. Pp. 509–516. Iowa State Univ. Press, Ames.

Ziman, S.N. and **C.S. Keener.** 1989. A geographical analysis of the family Ranunculaceae. Ann. Missouri Bot. Gard. 76:1012–1049.

Zohary, D. and **M. Hopf.** 1994. Domestication of plants in the Old World, 2nd ed. Oxford Univ. Press, New York.

Zohary, M. 1982. Plants of the Bible. Cambridge Univ. Press, Cambridge, England, U.K.

______ and **D. Heller.** 1984. The genus *Trifolium*. Israel Academy of Sciences and Humanities, Jerusalem, Israel.

Zomlefer, W.B. 1994. Guide to flowering plant families. Univ. of North Carolina Press, Chapel Hill.

Zona, S. 1990. A monograph of *Sabal* (Arecaceae: Coryphoideae). Aliso 12:583–666.

______. 1997. The genera of Palmae (Arecaceae) in the southeastern United States. Harvard Pap. Bot. 2:71–107.

Zuloaga, F.O. 1987. Systematics of New World species of *Panicum* (Poaceae: Paniceae). In: T.R. Soderstrom, K.W. Hilu, C.S. Campbell, and M.E. Barkworth, eds. Grass systematics and evolution. Pp. 287–306. Smithsonian Institution Press, Washington, D.C.

______, **R.P. Ellis,** and **O. Morrone.** 1993. A revision of *Panicum* subg. *Dichanthelium* sect. *Dichanthelium* (Poaceae: Panicoideae: Paniceae) in Mesoamerica, the West Indies, and South America. Ann. Missouri Bot. Gard. 80:119–190.

Zuloaga, F.O. and **O. Morrone.** 1996. Revisíon de las especies Americanas de *Panicum* subgénero *Panicum* sección *Panicum* (Poaceae: Panicoideae: Paniceae). Ann. Missouri Bot. Gard. 83:200–280.

INDEX

The accepted scientific names of native or naturalized members of the North Central Texas flora (and other nearby Texas plants discussed in detailed notes) are given in [Roman type]. In addition, accepted generic and family names of plants in the flora are in [**bold**]. Taxonomic synonyms and names of plants casually mentioned are in [*italics*]. Common names are in [SMALL CAPS]. Color photographs are indicated by the symbol ▣.

A

ABELE, 975
Abelia, 507
Abelmoschus, 806
esculentus, 806
Abies, 204
ABRAHAM'S-BALM, 1060
ABROJO, 432
DE FLOR AMARILLO, 1076
Abronia, 835
ameliae, ▣/77, 836
fragrans, 836
speciosa, 836
Abrus precatorius, 617
Abutilon, 806
crispum, 810
fruticosum, 806
incanum, 806
texense, 806
theophrasti, 806
Acacia, 623
angustissima var. hirta, 624
farnesiana, 624
greggii, 624
var. greggii, 625
var. wrightii, 625
hirta, 624
malacophylla, 625
minuta, 625
subsp. *densiflora*, 624
roemeriana, 625
senegal, 624
smallii, 624
wrightii, 625
ACACIA
BASTARD, 692
CATCLAW, 625
FALSE, 692
FERN, 624
PRAIRIE, 624
ROEMER'S, 625
ROSE-, 691
WHITE-BALL, 624
WRIGHT'S, 625
Acalypha, 586
gracilens, 588
var. *delzii*, 588
subsp. *monococca*, 588
var. *monococca*, 588
lindheimeri, 589
monococca, 588
ostryifolia, 588
phleoides, 588
radians, 589
rhomboidea, 589
virginica, 589
var. *rhomboidea*, 589
Acanthaceae, 210
Acanthochiton, 222
wrightii, 224
Acanthus spinosus, 211
ACANTHUS, FLAME-, 212
ACANTHUS FAMILY, 210
Acer, 219
grandidentatum var. sinuosum, 219
negundo, 219
var. negundo, 220
var. texanum, 220
rubrum, 220
saccharinum, 220
saccharum, 219
var. *floridanum*, 219
var. *sinuosum*, 219
Aceraceae, 218
ACHICORIA DULCE, 416
Achillea, 307
lanulosa, 308
millefolium, 308
subsp. *lanulosa*, 308
var. *lanulosa*, 308
var. *occidentalis*, 308
occidentalis, 308
Achyranthes philoxeroides, 222
Acleisanthes, 836
longiflora, 836
Acmella, 308
oppositifolia var. repens, 308
repens, 308
Acnidia, 222
australis, 226
cuspidata, 226
tamariscina, 226
Aconitum, 723, 916
Acoraceae, 1077
ACORN SQUASH, 568
Acorus, 1077
americanus, 1078
calamus, 1077
Acrostichum sinuatum, 195
Adansonia, 804
ADDER'S-TONGUE, 190
BULBOUS, 190
ENGELMANN'S, 190
LIMESTONE, 190
SOUTHERN, 190
ADDER'S-TONGUE FAMILY, 188
ADELIA, TEXAS, 848
Adiantum, 194
capillus-veneris, 194
Adonis, 917
annua, 917
Aegilops, 1235
cylindrica, 1235
squarrosa, 1334
Aesculus, 737
arguta, 738
glabra var. *arguta*, 738
hippocastanum, 737, 738
pavia
var. flavescens, 738
var. pavia, ▣/77, 738
AFRICAN-TULIPTREE, 440
AFRICAN-VIOLET FAMILY, 989
Agalinis, 991
aspera, 993
auriculata, 992
caddoensis, 993
densiflora, 993
edwardsiana, 993
var. *glabra*, 993
fasciculata, 993
gattingeri, 993
heterophylla, 993
homalantha, ▣/77, 993
tenuifolia var. leucanthera, 993
AGALINIS, 991
PRAIRIE, 993
AGARITO, 436
Agavaceae, 1078
Agave, 1078
americana, 1080
lata, 1079, 1080
AGAVE, 1078
VIRGINIA, 1080
AGAVE FAMILY, 1078
Ageratina, 352
altissima, 354
havanensis, 354
Agrimonia, 938
microcarpa, 938

Agrimonia *(cont.)*
parviflora, 938
rostellata, 938
AGRIMONY, 938
MANY-FLOWER, 938
SMALL-FRUIT, 938
WOODLAND, 938
AGRITO, 870
Agropyron smithii, 1304
Agrostemma, 513
githago, 513
Agrostis, 1235
elliottiana, 1236
hyemalis, 1236
perennans, 1236
scabra, 1236
semiverticillata, 1315
AGUEWEED, 354
Ailanthus, 1015
altissima, 1015
moluccana, 1015
Aira, 1236
capillaris, 1236
caryophyllea
var. capillaris, 1236
var. caryophyllea, 1236
elegans, 1236
elegantissima, 1236
AIR-POTATO, 1165
Aizoaceae, 220
Ajuga, 755
reptans, 755
AJUGA, CARPET, 755
ALABAMA
LIP FERN, 196
SUPPLEJACK, 931
ALACRANCILLO, 450
ÁLAMO, 974, 975
BLANCO, 975
ALAMO
COTTONWOOD, 975
-VINE, 558
Albizia, 625
julibrissin, 625
Alcea, 807
rosea, 807
ALDER, 439
COMMON, 439
HAZEL, 439
SMOOTH, 439
Aldrovanda, 576
Aletris, 1193
aurea, ▣/77, 1194
Aleurites, 585
ALEXANDERS, GOLDEN-, 264
ALFALFA, 672, 676
WILD, 690
ALFALFILLA, 676
ALFILARIA, 730
ALFILERILLO, 730
ALFOMBRILLA HEDIONA, 1052
ALGAROBA, 688
ALGERITAS, 436
ALGODÓN, 808
Alisma, 1086
plantago-aquatica var. *parviflorum,* 1087
subcordatum, 1086
Alismataceae, 1086
ALKALI
BULRUSH, 1108
GRASS, 1264
-MALLOW, 815
SIDA, 815
TULE, 1161
Alliaceae, 1191
ALLIGATOR
-BONNET, 845
-WEED, 222
Allium, 1194
acetabulum, 1196
ampeloprasum, 1196
bivalve, 1208
canadense, 1196
var. canadense, 1196
var. ecristatum, 1197
var. fraseri, 1197
var. hyacinthoides, 1197
var. mobilense, 1197
cepa, 1197
drummondii, 1197
fraseri, 1197
hyacinthoides, 1197
mobilense, 1197
perdulce, 1197
porrum, 1196
runyonii, 1197
sativum, 1197
stellatum, 1198
texanum, 1198
ALLSPICE, WILD, 784
ALMOND, 947
Almutaster, 315
Alnus, 439
serrulata, 439
ALOE
AMERICAN-, 1079
FALSE, 1079, 1080
WIDE-LEAF FALSE, 1079
Alopecurus, 1236
carolinianus, 1236
Alophia, 1172
drummondii, ▣/77, 1173
Aloysia, 1049
gratissima, 1049
var. *schulziae*, 1049
ligustrina, 1049
lycioides var. *schulziae*, 1049
ALTA FESCUE, 1276
ALTAMISA, 309
Alternanthera, 221
caracasana, 222
peploides, 222
philoxeroides, 222
Althaea
officinalis, 804
rosea, 807
ALTHAEA, 812
ALUMROOT, 986
AMAPOLA, 876
DE CAMPO, 866, 874
GRANDE, 807
AMARANTH, 222
BERLANDIER'S, 226
COMMON GLOBE-, 228
GLOBE-, 228
GREEN, 226, 227
HAAGE'S GLOBE-, 228
PALMER'S, 226
SANDHILL, 224
SLIM, 226
SOUTHERN, 226
SPLEEN, 226
THORNY, 226
TORREY'S, 224
TROPICAL, 226
TUMBLEWEED, 224
WHITE, 224
AMARANTH FAMILY, 221
Amaranthaceae, 221
Amaranthus, 222
acanthochiton, 224
albus, 224
arenicola, 224
ascendens, 227
australis, 226
berlandieri, 226
blitoides, 226
blitum, 227
caudutus, 222
graecizans, 226
hybridus, 226
lividus, 227
palmeri, 226
polygonoides, 226
retroflexus, 226
rudis, 226
spinosus, 226

Amaranthus *(cont.)*
tamariscinus, 226
torreyi, 226
AMARILLO, 1076
Amaryllidaceae, 1078, 1191
Amaryllis, 1191
AMBER, 547
AMBERIQUE-BEAN, 698
Amblyolepis, 308
setigera, 308
Ambrosia, 309
artemisiifolia, 309
bidentata, 309
confertiflora, 310
cumanensis, 310
psilostachya, 310
trifida var. texana, 310
Amelanchier, 938
arborea, 940
AMELIA'S SAND-VERBENA, 836
AMERICAN
-ALOE, 1079
BASKET-FLOWER, 332
BEAKGRAIN, 1260
BEAUTY-BERRY, 1049
BITTERSWEET, 528
BLUEHEARTS, 996
BROOKLIME, 1012
BUCKWHEAT-VINE, 898
BUGLEWEED, 760
BULRUSH, 1161
BURNWEED, 348
CHESTNUT, 710
COLUMBINE, 918
COWSLIP, 912, 913
CUPSCALE, 1315
DITTANY, 756
ELDERBERRY, 510
ELM, 1039
GERMANDER, 782
HOLLY, 270
HOP-HORNBEAM, 440
HORNBEAM, 439
IPECAC, 946
-IVY, 1068
KNAPWEED, 332
-MANDRAKE, 437
NIGHTSHADE, 1030
PAWPAW, 238
PEPPER-GRASS, 472
PILLWORT, 188
PINESAP, 584
POTATO-BEAN, 628
SCOURING-RUSH, 176
SEEDBOX, 860
SMOKETREE, 231
STAR-JASMINE, 268
SYCAMORE, 888
WATER
-LILY, 845
-WILLOW, 213
WATERWORT, 581
WILD-CELERY, 1170
Ammannia, 799
auriculata, 800
coccinea, 800
subsp. *robusta*, 800
robusta, 800
AMMANNIA
EAR-LEAF, 800
PURPLE, 800
Ammi, 243
majus, 243
visnaga, 243
AMMI, GREATER, 243
Ammoselinum, 243
butleri, 243
popei, 243
Amorpha, 625
canescens, 626
fruticosa, 626
var. *angustifolia*, 626
paniculata, 626
roemeriana, 626
texana, 626
AMORPHA
INDIGO-BUSH, 626
PANICLED, 626
Ampelaster, 315
Ampelopsis, 1066
arborea, 1066
cordata, 1066
AMPELOPSIS, HEART-LEAF, 1066
Amphiachyris, 362
amoena, 364
dracunculoides, 364
Amphicarpaea, 626
bracteata, 626
var. *comosa*, 626
Amsinckia, 445
menziesii, 445
micrantha, 445
Amsonia, 265
ciliata, 265
var. *filifolia*, 265
var. tenuifolia, 265
var. texana, 265
illustris, 266
longiflora var. salpignantha, 265
repens, 266
salpignantha, 266
tabernaemontana, 266
var. *gattingeri*, 266
texana, 265
AMSONIA, TEXAS, 265
AMUR HONEYSUCKLE, 508
Anabaena azollae, 180
Anacardiaceae, 230
Anacardium occidentle, 230
Anacharis densa, 1169
Anagallis, 912
arvensis, 912
minima, 912
ANAGUA, 443
Ananas comosus, 1095
Andrachne, 610
phyllanthoides, 610
Andropogon, 1238
annulatus, 1262
elliottii, 1239
gerardii, 1238
subsp. *chrysocomus*, 1239
subsp. gerardii, /78, 1238
subsp. hallii, 1239
var. *paucipilus*, 1239
glomeratus, 1239
gyrans, 1239
hallii, 1239
paucipilus, 1239
scoparius var. *frequens*, 1316
ternarius, 1239
virginicus, 1239
Androsace, 912
occidentalis, 912
Androstephium, 1198
coeruleum, 1198
Anemia, 179
mexicana, 179
ANEMIA FAMILY, 179
Anemiaceae, 179
Anemone, 917
berlandieri, /78, 918
caroliniana, 918
decapetala var. *heterophylla*, 918
edwardsiana, 918
heterophylla, 918
okennonii, 918
tuberosa var. *texana*, 918
ANEMONE
CAROLINA, 918
O'KENNON'S, 918
TEN-PETAL, 918
TWO-FLOWER, 918
Anemopsis, 984
californica, 984
Anethum, 243
graveolens, 243

- ANGEL
  - -EYES, 968
  - FAVORITA, 1049
  - -TRUMPET, 1020
  - -TRUMPETS, 836
- ANGEL'S
  - -HAIR, 572
  - TRUMPET, 1020
- ANGIOSPERMS, 208
- ANGLE-POD MELOCHIA, 1032
- ANGLEPOD, 283, 284
  - SMOOTH, 282
- ANGLE-STEM WATER-PRIMROSE, 860
- ANISACANTH, 212
- **Anisacanthus,** 212
  - quadrifidus var. wrightii, 212
  - *wrightii*, 212
- ANISACANTHUS, WRIGHT'S, 212
- ANISE, 239
- ANISEROOT, 255, 256
- *Anisostichus capreolata*, 442
- ANNONA FAMILY, 238
- *Annona reticulata*, 238
- **Annonaceae,** 238
- ANNUAL
  - BEARD GRASS, 1314
  - BROOMWEED, 364
  - BULRUSH, 1154
  - FESCUE, 1335
  - FLEABANE, 350
  - GERMANDER, 782
  - HAIR GRASS, 1236
  - LARKSPUR, 922
  - SUNDEW, 578
  - TICKLE GRASS, 1236
  - WATERWORT, 496
  - WILD BUCKWHEAT, 899
- **Anredera,** 434
  - cordifolia, 434
- ANTELOPE-HORNS, 278, 282
  - GREEN, 282
- **Antennaria,** 310
  - *fallax*, 310
  - parlinii subsp. fallax, 310
- **Anthemis,** 310
  - cotula, 310
- ANTHRACNOSE, DOGWOOD, 561
- *Anthurium*, 1092
- **Antigonon,** 898
  - leptopus, 898
- *Antirrhinum*, 988
- APAREJO
  - GRASS, 1290
  - MUHLY, 1290
- **Aphanostephus,** 312
  - pilosus, 312
  - ramosissimus, 312
  - riddellii, 314
  - skirrhobasis, ❁/78, 314
- *Aphora mercurialina*, 604
- **Apiaceae,** 239
- **Apios,** 628
  - americana, ❁/78, 628
- *Apium*
  - *graveolens*, 239
  - *leptophyllum*, 249
- **Apocynaceae,** 264
- **Apocynum,** 266
  - cannabinum, 266
    - var. *glaberrimum*, 266
    - var. *pubescens*, 266
  - *sibiricum*, 266
- *Apogon wrightii*, 380
- APPLE, 936
  - BALSAM-, 569
  - BITTER-, 566
  - COMMON THORN-, 1020
  - CUSTARD-, 238
  - GOLD-, 1030
  - HEDGE-, 831
  - HORSE-, 830, 831
  - INDIAN-, 1020
  - JAPANESE-, 936
  - LOVE-, 1030
  - MAMMEY-, 544
  - MAY-, 437
  - RUST, CEDAR, 940, 203
  - THORN-, 874, 940, 1018
  - VELVET-, 578
- APPLEMINT, 764
- APPRESSED BOG CLUBMOSS, 174
- APRICOT, 947
  - VINE, 878
- **Aquifoliaceae,** 269
- **Aquilegia,** 918
  - canadensis, ❁/78, 918
    - var. *latiuscula*, 920
- **Arabis,** 459
  - canadensis, 459
  - petiolaris, 459
- **Araceae,** 1090
- **Arachis,** 628
  - hypogaea, 628
- Aralia spinosa, 270
- ARALIA FAMILY, 270
- **Araliaceae,** 270
- ARBORVITAE, 202
- **Arbutus,** 582
  - *texana*, 582
  - xalapensis, 582
    - var. *texana*, 582
- ARCE, 219
- *Archilochus*, 434, 442, 498, 510, 555, 658, 738, 799, 815, 891, 922, 998, 1079
- *Arctostaphylos*, 581
- **Arecaceae,** 1094
- **Arenaria,** 514
  - benthamii, 514
  - *drummondii*, 519
  - *muriculata*, 519
  - *patula*, 519
    - var. *robusta*, 519
  - serpyllifolia, 514
  - *stricta* var. *texana*, 519
  - *texana*, 519
- **Argemone,** 873
  - albiflora, 873
    - subsp. texana, 873
    - var. *texana*, 873
  - aurantiaca, 873
  - *intermedia*, 874
    - var. *polyanthemos*, 874
  - mexicana, 874
  - polyanthemos, 874
- ARGENTINE SENNA, 694
- **Argyrochosma,** 194
  - dealbata, 194
  - microphylla, 194
- *Argythamnia*
  - *aphoroides*, 602
  - *humilis*, 602
    - var. *laevis*, 604
  - *mercurialina*, 604
- **Arisaema,** 1092
  - dracontium, ❁/78, 1092
  - triphyllum, 1092
- **Aristida,** 1239
  - basiramea, 1240
  - desmantha, 1240
  - dichotoma, 1242
  - *glauca*, 1243
  - *intermedia*, 1242
  - lanosa, 1242
  - longespica, 1242
    - var. geniculata, 1242
    - var. longespica, 1242
  - *longiseta*, 1243
  - oligantha, 1242
  - purpurascens, 1242
  - purpurea, 1242
    - var. longiseta, 1243
    - var. nealleyi, 1243
    - var. purpurea, 1243
    - var. wrightii, 1243
  - *reverchonii*, 1243
  - *wrightii*, 1243
- ARISTIDA
  - FEW-FLOWER, 1242

ARISTIDA *(cont.)*
LONG-AWNED, 1243
Aristolochia, 272
reticulata, 273
serpentaria, 273
var. *hastata,* 273
tomentosa, 274
Aristolochiaceae, 272
ARIZONA
BLACK WALNUT, 748
COTTONTOP, 1263
LAZY DAISY, 312
WALNUT, 748
ARKANSAS
BUGLEWEED, 762
CALAMINT, 755
CARIC SEDGE, 1118
DOGSHADE, 255
LAZY DAISY, 314
MANNA GRASS, 1277
MEADOW-RUE, 930
YUCCA, 1083
Arnoglossum, 314
ovatum, 314
plantagineum, 314
AROID FAMILY, 1090
AROMATIC ASTER, 319
ARROW
CROTALARIA, 643
-FEATHER THREEAWN, 1242
GRASS, 1242
-LEAF
CLOVER, 706
VIOLET, 1064
-WOOD, 511
INDIAN, 528
ARROWHEAD, 1087
COMMON, 1088
DELTA, 1090
GIANT, 1090
LONG-BARB, 1090
LONG-LOBE, 1090
NIPPLE-BRACT, 1090
SCYTHE-FRUIT, 1088
ARROWHEAD FAMILY, 1086
ARROWROOT, 1210
JAPANESE, 690
WEST INDIAN, 1210
ARROWROOT FAMILY, 1210
Artemisia, 314
absinthium, 314
campestris subsp. caudata, 315
caudata, 315
dracunculus, 314
herba-alba, 314
ludoviciana, 315
subsp. *ludoviciana,* 315
subsp. mexicana, 315
var. *mexicana,* 315
tridentata, 314
Arthuriomyces peckianus, 959
ARTICHOKE, 287
JERUSALEM-, 370
ARTILLERY-PLANT, 1042
Artocarpus altilis, 827
ARUM FAMILY, 1090
Arundinaria, 1243
gigantea, 1243
Arundo, 1243
donax, 1243
Asarina antirrhiniflora, 1003
Asclepiadaceae, 274
Asclepias, 276
amplexicaulis, 277
arenaria, 278
asperula subsp. capricornu, ▣/79, 278
engelmanniana, 278
incarnata, 278
subsp. incarnata, 278
subsp. pulchra, 278
var. *pulchra,* 278
linearis, 278
oenotheroides, 278
perennis, 280
stenophylla, 280
texana, 280
tuberosa, 280
subsp. interior, ▣/79, 280
subsp. *terminalis,* 280
variegata, ▣/79, 280
verticillata, 280
viridiflora, ▣/79, 282
var. *lanceolata,* 282
viridis, 282
Asclepiodora
decumbens, 278
viridis, 282
Ascyrum
hypericoides
var. *multicaule,* 547
var. *oblongifolium,* 547
stans, 547
ASH, 848
GREEN, 849
MOUNTAIN-, 952
PRICKLY-, 973, 974
RED, 849
SOUTHERN PRICKLY-, 974
TEXAS, 849
WHITE, 849
WAFER-, 972
WHITE, 849
ASHE'S JUNIPER, 203
ASH-LEAF MAPLE, 219
ASHY SUNFLOWER, 370
ASIAN-JASMINE, 268
ASIATIC ELM, 1040
Asimina, 238
triloba, 238
Asparagaceae, 1191
Asparagus, 1198
officinalis, 1198
ASPARAGUS, GARDEN, 1198
ASPEN, 974
QUAKING, 975
Aspergillus flavus, 628
Asperula, 961
arvensis, 961
asperula var. *decumbens,* 278
capricornu, 278
Aspleniaceae, 179
Asplenium, 179
nidus, 179
platyneuron, 179
resiliens, 180
Asteliaceae, 1078
Aster, 315
arenosus, 335
azureus, 319
coerulescens, 320
divaricatus, 320
×eulae, 320
drummondii var. texanus, 318
dumosus, 318
var. *cordifolius,* 318
var. *subulifolius,* 318
ericoides, 318
exilis, 320
hesperius, 319
lanceolatus, 318
subsp. hesperius, 318
subsp. lanceolatus, 319
var. *simplex,* 319
lateriflorus, 319
var. *flagellaris,* 319
var. *indutus,* 319
oblongifolius, 319
oolentangiensis, 319
patens, 319
var. gracilis, 320
var. patens, 320
var. patentissimus, 320
praealtus, 320
var. *coerulescens,* 320
var. *texicola,* 320
pratensis, 320
salicifolius, 320
sericeus, 320

Aster *(cont.)*
var. *microphyllus*, 320
simplex, 319
spinosus, 335
subulatus var. ligulatus, 320
texanus, 318
subsp. *texanus*, 318
ASTER, 315
AROMATIC, 319
AZURE, 319
BABY WHITE, 334
BUSHY, 318
CALICO, 319
CALIFORNIA, 319
DEVILWEED-, 335
GOLD-, 372
GOLDEN-, 335, 372
GRASS-LEAF, GOLDEN-, 396
GRAY GOLD-, 372
HEATH, 318
LATE PURPLE, 319
NARROW-LEAF GOLD-, 372
OBLONG-LEAF, 319
PANICLED, 318
RICE-BUTTON, 318
SALTMARSH, 320
SIDE-FLOWER, 319
SILKY, 320
SISKIYOU, 318
SKY
-BLUE, 319
-DROP, 319, 320
SLIM, 320
SOFT GOLDEN-, 336
SPINY-, 335
SPREADING, 319
GOLDEN-, 344
STARVED, 319
TALL, 320
WHITE, 318
TEXAS, 318
WHITE, 334
PRAIRIE, 318
WOODLAND, 319
WILL, 320
WILLOW-LEAF, 320
WREATH, 318
Asteraceae, 287
Astragalus, 628
canadensis, 630
crassicarpus, 630
var. berlandieri, 630
var. crassicarpus, 630
var. trichocalyx, 632
distortus, 632
var. distortus, 632
var. engelmannii, 632
lambertii var. *abbreviatus*, 682
leptocarpus, 632
lindheimeri, 632
lotiflorus, 632
var. *reverchonii*, 634
macilentus, 634
nuttallianus, 634
var. austrinus, 634
var. macilentus, 634
var. nuttallianus, 634
var. pleianthus, 634
var. trichocarpus, 635
plattensis, 635
pleianthus, 634
reflexus, 635
wrightii, 635
Astranthium, 322
integrifolium
subsp. ciliatum, 322
var. *ciliatum*, 322
var. *triflorum*, 322
Astrocasia, 610
Astrolepis, 195
integerrima, 195
sinuata, 195
ATAMOSCO-LILY, 1201
Athyrium, 184
asplenioides, 184
filix-femina subsp. asplenioides, 184
ATLANTIC
CAMASSIA, 1200
PIGEON-WINGS, 642
YAM, 1165
Atriplex, 532
argentea, 532
subsp. argentea, 532
subsp. expansa, 533
canescens, 533
hortensis, 532
AUNT LUCY, 740
Aureolaria, 994
grandiflora, 994
var. cinerea, 994
var. serrata, ▣/79, 994
AUTUMN
BENT GRASS, 1236
CORALROOT, 1215
CROCUS, 1191
SAGE, 776
SQUASH, 569
Avena, 1244
fatua, 1244
var. *sativa*, 1244
sativa, 1244
AVENS, 944
WHITE, 944
Averrahoa carambola, 869
AVOCADO, 784
AWNLESS
BEGGAR-TICKS, 326
BUSH-SUNFLOWER, 406
Axonopus, 1244
affinis, 1244
fissifolius, 1244
AZMIRILLO, 1049
Azolla, 180
caroliniana, 180
AZOLLA, 180
Azollaceae, 180
AZURE
ASTER, 319
SAGE, 774

B

BABY
BLUE-EYES, 742
PONDWEED, 1343
WHITE ASTER, 334
BABY'S-BREATH, 511
BABYLON WEEPING WILLOW, 978
Baccharis, 322
halimifolia, 324
neglecta, 324
salicina, 324
texana, 324
BACCHARIS
EASTERN, 324
PRAIRIE, 324
WILLOW, 324
BACHELOR-BUTTON, 334
Bacopa, 994
monnieri, 994
rotundifolia, 994
BAGPOD, 662
BAHAMA GRASS, 1259
BAHIA
GRASS, 1306
PENSACOLA, 1306
BALD CYPRESS, 204
MEXICAN, 204
MONTEZUMA, 204
BALD-RUSH, SHORT-BEAK, 1160
BALDWIN FLAT SEDGE, 1136
BALDWIN'S IRONWEED, 428
BALL
-MOSS, 1095
-NETTLE, 1028
BALLOONVINE, 980
COMMON, 980
BALMONY, 1007

- BALSAM, 433
  - -APPLE, 569
  - GARDEN, 434
  - GOURD, 569
  - ROSE, 434
- **Balsaminaceae,** 433
- BAMBOO, 1311
  - BLACK, 1311
  - FISHPOLE, 1311
  - GOLDEN, 1311
  - HEAVENLY-, 437
  - SACRED-, 437
  - YELLOW, 1311
- *Bambusa aurea,* 1311
- BANANA WATER-LILY, 845
- BANDAKAI, 806
- BAOBAB, 804
- **Baptisia,** 635
  - alba var. macrophylla, 639
  - australis, 636
    - var. australis, 636
    - var. minor, ▣/80, 636
  - ×bicolor, ▣/80, 636
  - bracteata, 638
    - var. *glabrescens,* 638
    - var. leucophaea, ▣/80, 638
  - ×bushii, ▣/80, 638
  - ×*intermedia,* 638
  - *leucophaea,* 638
    - var. *glabrescens,* 638
  - *minor,* 636
    - var. *aberrans,* 636
  - nuttalliana, 638
  - sphaerocarpa, ▣/80, 638
  - ×*stricta,* 638
  - *texana,* 636
  - *tinctoria,* 635
  - ×variicolor, ▣/80, 638
  - *vespertina,* 636
  - *viridis,* 638
- BARBARA'S-BUTTONS, 387
- BARBAS DE CHIVATO, 921
- BARBERRY, 436
  - EUROPEAN, 1334
- BARBERRY FAMILY, 436
- BARLEY, 1278
  - HARE, 1278
  - LITTLE, 1278
  - MOUSE, 1278
- BARNABY'S STAR-THISTLE, 334
- BARNYARD GRASS, 1264, 1266
- BARREL CACTUS, 486
- BARREN OAK, 716
- BASAL-FRUIT CARIC SEDGE, 1118
- BASELLA FAMILY, 434
- **Basellaceae,** 434
- BASIL, 752
  - BEEBALM, 765
- BASIN
  - BELLFLOWER, 497
  - FLEABANE, 350
  - SNEEZEWEED, 366
- BASKET
  - -FLOWER, 332
    - AMERICAN, 332
  - GRASS, 1290
  - SELAGINELLA, 174
  - WILLOW, 978
- BASSWOOD, 1036
  - CAROLINA, 1036
  - FLORIDA, 1036
  - KENDALL, 1036
- BASTARD
  - ACACIA, 692
  - INDIGO, 626
  - OAK, 718
  - PENNYROYAL, 783
  - SAFFRON, 330
  - TOADFLAX, 979
- BAY LAUREL, 784
- BAYARD LONG'S CARIC SEDGE, 1124
- BAYBERRY, 832
- BAYBERRY FAMILY, 832
- BAYONET-GRASS, 1108
- BAYOU VIOLET, 1064
- BEACH
  - EVENING-PRIMROSE, 854
  - GERARDIA, 993
  - GROUND-CHERRY, 1024
- BEADPOP, 473
- BEAKED
  - CORNSALAD, 1044
  - PANICUM, 1295
  - SIDA, 816
- BEAKGRAIN, 1260
  - AMERICAN, 1260
- BEAK-RUSH, 1156
  - CLUSTER, 1158
  - GLOBE, 1158
  - HARVEY'S, 1158
  - HORNED, 1158
  - PLANK, 1158
  - TALL, 1158
- BEAN, 617
  - AMBERIQUE-, 698
  - AMERICAN POTATO-, 628
  - BROAD, 617, 706
    - -LEAF SNOUT-, 691
  - -CASTOR, 612
  - CHEROKEE-, 658
  - COFFEE-, 696
  - EASTERN CORAL-, 658
  - EGYPTIAN-, 834
  - FIELD-, 706
  - FUZZY-, 697
  - HORSE-, 682, 706
  - LEAST SNOUT-, 691
  - MESCAL-, 697
  - POISON-, 696
  - POTATO-, 628
  - PRECATORY-, 617
  - SACRED-, 834
  - SNOUT, 690
  - SOY-, 662
  - SOYA-, 662
  - TRAILING WILD, 698
  - WILD, 697
- BEAN FAMILY, 617
- BEARBERRY, 270, 581
- BEARD GRASS, 1238, 1244, 1314, 1315
  - ANNUAL, 1314
  - BUSHY, 1239
  - CANE, 1246
  - SILVER, 1246
  - SILVERY, 1239
  - SPLIT-BEARD, 1239
- BEARDED
  - BEGGAR-TICKS, 326
  - DALEA, 651
  - FLAT SEDGE, 1142
  - GRASS-PINK, 1214
  - SKELETON GRASS, 1277
  - SPRANGLETOP, 1281
  - SWALLOW-WORT, 282
- BEARDLESS DARNEL RYE GRASS, 1282
- BEARDTONGUE, 1006, 1007, 1008
  - LARGE, 1008
  - SMOOTH, 1007
- BEAR-GRASS, 1080, 1082
- BEAR'S
  - -BREECHES, 211
  - -FOOT, 408
- *Beaucarnea,* 1080
- BEAUTIFUL FALSE DRAGON'S-HEAD, 770
- BEAUTY-BERRY, 1049
  - AMERICAN, 1049
  - PURPLE, 1049
- BEAVER-POISON, 248
- BEDSTRAW, 962
  - BLUNT-LEAF, 964
  - CATCHWEED, 964
  - DYE, 966
  - FRAGRANT, 966
  - HAIRY, 964
  - SOUTHWEST, 966
  - STIFF MARSH, 966
  - SWEET-SCENTED, 966
  - TEXAS, 966

BEDSTRAW *(cont.)*
    WOODS, 964
BEEBALM, 764
    BASIL, 765
    LEMON, 765
    LINDHEIMER'S, 765
    PLAINS, 765
    PLUMETOOTH, 766
    RUSSELL'S, 766
    SPOTTED, 766
    WESTERN, 766
BEEBLOSSOM, 856, 1049
BEEBRUSH, 658
BEEBUSH, 1049
    COMMON, 1049
BEECH, 710
    BLUE-, 439
    SOUTHERN, 710
    WATER-, 439
BEECH FAMILY, 710
BEEFLOWER, 642
BEEFSTEAK-PLANT, 768
BEEHIVE NIPPLE CACTUS, 489
BEEPLANT, 506
BEET, 533
    SUGAR, 533
BEGGAR'S
    -LICE, 262, 446, 652
    -TICKS, 652
        -SMOOTH, 326
BEGGAR-TICKS, 325, 326
    AWNLESS, 326
    BEARDED, 326
    DEVIL'S, 326
BEJAR
    CLOVER, 704
    MARBLESEED, 454
BELL PEPPER, 1017
BELLADONA, 1015
Bellardia, 994
    trixago, 996
BELLFLOWER FAMILY, 496
BELLTREE, 1034
BELVEDERE SUMMER-CYPRESS, 540
BENJAMIN BUSH'S CARIC SEDGE, 1118
BENT
    -AWN
        FLAT SEDGE, 1140
        PLUME GRASS, 1315
    GRASS, 1235
        AUTUMN, 1236
        ELLIOTT'S, 1236
        SPRING, 1236
        WATER, 1314
        WINTER, 1236
    -POD MILK-VETCH, 632
BEQUILLA, 696
Berberidaceae, 436
Berberis, 436
    aquifolium, 436
    trifoliolata, 436
        var. *glauca,* 437
    vulgaris, 1334
Berchemia, 931
    scandens, 931
BERGAMOT, WILD, 765
Bergia, 580
    texana, 580
BERGIA, TEXAS, 580
BERLANDIER'S
    AMARANTH, 226
    DAISY, 325
    EVENING-PRIMROSE, 854
    FLAX, 792
Berlandiera, 324
    ×betonicifolia, 325
    betonicifolia, 325
    dealbata, 325
    lyrata, 325
    pumila, 325
    texana, 325
BERMUDA
    BLUE-EYED-GRASS, 1176
    GRASS, 1259
    -MULBERRY, 1049
Berula, 244
    erecta, 244
BERULA, STALKY, 244
Beta, 533
    vulgaris, 533
BETONY, 780
    NOSEBURN, 616
    SHADE, 780
    SLENDER-LEAF, 782
    TEXAS, 782
    WOOD-, 1006
Betula, 439
    nigra, 439
    papyrifera, 439
Betulaceae, 438
Bidens, 325
    aristosa, 326
    bipinnata, 326
        var. *biternatoides,* 326
    frondosa, 326
    laevis, 326
    polylepis, 326
Bifora, 244
    americana, 244
BIG
    BLUE LOBELIA, 498
    BLUESTEM, 1238
    -BRACT VERVAIN, 1056
    FLORIDA PASPALUM, 1306
    -FLOWER BLADDERPOD, 474
    -HEAD EVAX, 356
    -LEAF PERIWINKLE, 269
    PASPALUM, 1306
    -POD BONAMIA, 560
    QUAKING GRASS, 1250
    -ROOT
        MORNING-GLORY, 556
        WAVEWING, 249
    -SEED GOOSEFOOT, 538
    -TOOTH MAPLE, 219
        PLATEAU, 219
        UVALDE, 219
    -TOP
        DALEA, 648
        LOVE GRASS, 1270
    -TREE PLUM, 949
Bigelowia, 326
    nuttallii, 328
    virgata, 328
BIGELOWIA, SLENDER, 328
BIGELOW'S OAK, 718
Bignonia, 442
    capreolata, 442
Bignoniaceae, 440
BILBERRY, 584
BINDWEED, 551
    BLACK, 902
    COMMON, 551
    CORN, 902
    FIELD, 551
    GRAY, 552
    HEDGE, 550
    HELIOTROPE, 448
    TEXAS, 552
    TRAILING HEDGE-, 551
BIRCH, 439
    RED, 439
    RIVER, 439
BIRCH FAMILY, 438
BIRD
    -FOOT
        DEER-VETCH, 671
        TREFOIL, 671
    -OF-PARADISE, 639
    PEPPER, 1017
    -WING PASSION-FLOWER, 878
BIRD'S
    -EYE SPEEDWELL, 1014
    -NEST FERN, 179
    RAPE, 460
BIRTHWORT, 272
BIRTHWORT FAMILY, 272

BISHOP
MOCK, 258
PRAIRIE-, 244
BISHOP'S-WEED, 243
MOCK, 258
NUTTALL'S MOCK, 258
THREAD-LEAF MOCK, 258
BITTER
-APPLE, 566
-BLOOM, 728
CRESS, 462, 464
BULB, 464
HAIRY, 462
SAND, 464
-CUCUMBER, 566
MILKWORT, 896
ORANGE, 972
-NUT, 747
HICKORY, 747
PECAN, 747
RUBBERWEED, 375
WATERNUT, 748
BITTERSWEET, 528
AMERICAN, 528
CLIMBING, 528
BITTERSWEET FAMILY, 526
BITTERWEED, 365, 366, 375
POISON, 375
WESTERN, 375
BIZNAGA DE CHILITOS, 489
BLACK
BAMBOO, 1311
BINDWEED, 902
DALEA, 648
-EYED
CLOCKVINE, 218
-SUSAN, 218, 346, 402
-FOOT
DAISY, 388
EUPHORBIA, 592
-FOOTED QUILLWORT, 175
-GUM, 845
HENBANE, 1015
HICKORY, 748
LOCUST, 692
-MOSS, 1096
MEDICK, 674
MUSTARD, 460
OAK, 716, 718
PERSIMMON, 579
POPLAR, 976
SNAKEROOT, 258
-STEM SPLEENWORT, 180
STEM RUST OF WHEAT, 437, 1334
TUPELO, 845
WALNUT, 750
ARIZONA, 748
TEXAS, 748
WILLOW, 978
BLACKBERRY, 958
BLACKCHERRY, 950
ESCARPMENT, 950
WILD, 950
BLACKDRINK, INDIAN, 270
BLACKFOOT, 387
PLAINS, 388
BLACKHAW, 511
RUSTY, 511
SOUTHERN, 511
BLACKJACK, 716
OAK, 716
BLACKLAND THISTLE, 338
BLACKSAMSON, 347
BLACKWEED, 320
BLADDER GROUND-CHERRY, 1026
BLADDERPOD, 472, 662
BIG-FLOWER, 474
DENSE-FLOWER, 473
EAR-LEAF, 473
ENGELMANN'S, 473
FENDLER'S, 473
GORDON'S, 473
LAX, 474
NARROW-LEAF, 473
PEAR-FRUIT, 474
SLENDER, 474
WHITE, 474
BLADDERWORT, 787
CONE-SPUR, 787
HORNED, 787
BLADDERWORT FAMILY, 786
BLADDERY CARIC SEDGE, 1124
BLAEBERRY, 584
BLANKET
-FLOWER, 358
INDIAN-, 358
BLAZINGSTAR, 382
HANDSOME, 384
BLAZINGSTAR FAMILY, 794
Blechnaceae, 180
BLEEDINGHEART, 719
BLESSED MILK-THISTLE, 406
BLIGHT, CHESTNUT, 710
BLISTER BUTTERCUP, 928
BLOOD
MILKWORT, 896
RAGWEED, 310
BLOODBERRY, 882
BLOODLEAF, 228, 230
ROOTSTOCK, 230
BLOODROOT, 872
BLUE
-BEECH, 439
BONESET, 353
-BUGLE, 755
CARDINAL-FLOWER, 498
COMFREY, 446
-CURLS, 743, 783
FORKED, 783
DAISY, 322
-EYED-GRASS, 1174
BERMUDA, 1176
DOTTED, 1178
LEAST, 1178
PALE, 1178
PRAIRIE, 1178
SWORD-LEAF, 1178
WHITE, 1176
-EYED-MARY, 999
-EYES, BABY, 742
FERN, 198
FLAG, 1174
-FLOWER BUFFALO-BUR, 1030
FLAX, 792
FUNNEL-LILY, 1198
GRAMA, 1248
GRASS, 1312
BULBOUS, 1312
CANADA, 1314
CHAPMAN'S, 1314
ELLIOTT'S, 1239
JUNE, 1314
KENTUCKY, 1314
SYLVAN, 1314
TEXAS, 1312
WOODLAND, 1314
HOUND'S-TONGUE, 446
-INDIGO, 636
WILD, 636
-JASMINE, 921
LARKSPUR, 924
-LEAF GRAPE, 1070
MUD-PLANTAIN, 1340
-MUSTARD, 464
PALM, 1094
PANIC, 1295
PRAIRIE VIOLET, 1064
SAGE, 774
-SNAPDRAGON, VINE, 1002
-SAILORS, 336
STEM, 1094
THREEAWN, 1243
VERVAIN, 1056
WATER-LILY, 845
-WEED SUNFLOWER, 369
WILD INDIGO, 636
BLUEBELL, 921
GENTIAN, 727

BLUEBELL FAMILY, 496
BLUEBELLS, 727
  TEXAS, 727
BLUEBERRY, 581, 584
  HAWTHORN, 942
BLUEBERRY FAMILY, 581
BLUEBONNET, 671
  TEXAS, 672
BLUEBOTTLE, 334
BLUEDEVIL, 448
BLUEHEARTS, 996
  AMERICAN, 996
  FLORIDA, 996
BLUEJACK OAK, 716
BLUESTAR, 265
BLUESTEM, 1238, 1244, 1262, 1316
  BIG, 1238
  BRISTLE-JOINT, 1246
  BROOMSEDGE, 1239
  BUSHY, 1239
  CANE, 1246
  ELLIOTT'S, 1239
  FEATHER, 1239
  KING RANCH, 1246
  KLEBERG, 1262
  KR, 1246
  LITTLE, 1316
  PERFORATED, 1246
  PINHOLE, 1246
  SAND, 1239
  SILVER, 1246
  SPLIT-BEARD, 1239
  VIRGINIA, 1239
  WHEAT GRASS, 1304
  YELLOWSEDGE, 1239
BLUET, 966
  LOW, 967
  NODDING, 968
  ROSE, 968
  SLENDER-LEAF, 968
  SMALL, 967
  SOUTHERN, 968
  TINY, 967
BLUETHORN, 936
BLUETS, 584
  FINE-LEAF, 966
  MAT, 967
  PRAIRIE, 966
  SOUTHERN, 967
BLUEVINE, 282
BLUEWEED, 369, 448
  TEXAS, 369
BLUEWOOD, 933
BLUFF OAK, 718
BLUNT
  SPIKE-RUSH, 1147
  -LEAF
    BEDSTRAW, 964
    MILKWEED, 277
  -LOBED WOODSIA, 185
  -SEPAL BRAZORIA, 755
BOATFLOWER, 1098, 1101
BOBSROOT SNAKEROOT, 680
BODKIN MILK-VETCH, 632
Boehmeria, 1042
  cylindrica, 1042
  nivea, 1042
Boerhavia, 836
  coccinea, 838
  diffusa, 838
  erecta, 838
  intermedia, 838
  spicata, 838
BOG
  CLUBMOSS, 173
  -HEMP, 1042
  MARSHCRESS, 478
  -MAT, 1191
  YELLOWCRESS, 478
BOGBEAN FAMILY, 824
BOIS D'ARC, 830, 831
Bolboschoenus, 1108
  maritimus, 1108
    subsp. *paludosus*, 1108
  paludosus, 1108
Boltonia, 328
  asteroides, 328
  diffusa, 328
BOLTONIA, 328
  SMALL-HEAD, 328
  WHITE, 328
Bombax, 804
BONAMIA, 558
  BIG-POD, 560
  HAIRY, 560
  PURPLE, 560
BONESET, 352, 354
  BLUE, 353
  CLIMBING-, 388
  FALL, 354
  FALSE, 328, 329
  PINK, 354
  SHRUBBY, 353
BORAGE, 445
BORAGE FAMILY, 443
Boraginaceae, 443
Borago, 445
  officinalis, 445
BOSTON FERN, 184
  -IVY, 1068
  WILD, 184
Bothriochloa, 1244
  barbinodis, 1246
    var. barbinodis, 1246
    var. perforata, 1246
  ischaemum var. songarica, 1246
  laguroides subsp. torreyana, 1246
  longipaniculata, 1246
  saccharoides, 1246
    var. *longipaniculata*, 1246
    var. *torreyana*, 1246
BO-TREE, 830
Botrychium, 188
  lunaria, 189
  biternatum, 189
  dissectum, 189
    var. *tenuifolium*, 189
  lunarioides, 189
  tenuifolium, 189
  virginianum, 189
BOTTLE GOURD, 569
  WHITE-FLOWER, 569
BOTTLEBRUSH PLANTAIN, 884
BOTTLETREE, CHINESE, 1032
BOTTOM-LAND RED OAK, 714
Bouchetia, 1016
  erecta, 1016
BOUCHETIA, ERECT, 1016
Bougainvillea, 835
BOUNCING-BET, 524
Bouteloua, 1247
  barbata, 1248
  curtipendula, 1248
    var. caespitosa, 1248
    var. curtipendula, 1248
  gracilis, 1248
  hirsuta, 1248
    subsp. *pectinata*, 1248
    var. *pectinata*, 1248
  pectinata, [photo icon]/81, 1248
  rigidiseta, 1248
  trifida, 1248
Bowlesia, 244
  incana, 244
BOWLESIA, HOARY, 244
BOW
  -WILLOW, 443
  -WOOD, 831
BOX
  -ELDER, 219
    TEXAS, 220
  -THORN, 1022
Bracharia, 1334
  ciliatissima, 1335
  fasciculata, 1335
  platyphylla, 1335
  texana, 1335
Brachypodium, 1250

Brachypodium *(cont.)*
distachyon, 1250
BRACKEN FAMILY, 181
BRACKEN FERN, 181
WESTERN, 181
BRACTED
PASSION-FLOWER, 878
PLANTAIN, 884
ZORNIA, 710
BRACTLESS MENTZELIA, 796
Bradburia, 335
hirtella, 336
pilosa, 336
BRADFORD PEAR, 952
BRAKE
BUCKHORN, 192
CLIFF-, 198
PASTURE, 181
PURPLE CLIFF-, 198
WRIGHT'S CLIFF-, 200
BRAKE FAMILY, 193
BRAMBLE, 958
Brasenia, 482
peltata, 482
schreberi, 482
BRASIL, 933
Brassica, 459
campestris, 460
hirta, 479
juncea, 459
kaber, 479
nigra, 460
oleracea, 460
rapa, 460
Brassicaceae, 454
BRAUN'S SCOURING-RUSH, 178
BRAZIL, 933
BRAZILIAN VERVAIN, 1056
Brazoria, 755
scutellarioides, 784
truncata, 755
var. pulcherrima, 755
var. truncata, 755
BRAZORIA
BLUNT-SEPAL, 755
PRAIRIE, 784
BRAZOS
MINT, 755
ROCKCRESS, 459
BREAD WHEAT, 1334
BREADFRUIT, 827, 828
Breweria pickeringii var. *pattersonii,* 560
BRICHO, 694
BRICKELLBUSH, 328
GRAVEL-BAR, 328
Brickellia, 328
cylindracea, 328
eupatorioides, 329
var. corymbulosa, 329
var. eupatorioides, 329
var. texana, 329
BRISTLE
-BRACT PLANTAIN, 886
-CONE PINE, 206
GRASS, 1318
BUR, 1320
GIANT, 1319
GREEN, 1320
HOOKED, 1320
KNOT-ROOT, 1319
REVERCHON'S, 1320
SCHEELE'S, 1320
SOUTHWESTERN, 1320
YELLOW, 1319
GREENBRIER, 1346
-JOINT BLUESTEM, 1246
-LEAF
DYSSODIA, 422
ERYNGO, 252
-THISTLE, SLENDER, 330
BRISTLY
BUTTERCUP, 928
LOCUST, 691
SCALESEED, 262
SENSITIVE-BRIAR, 678
Briza, 1250
maxima, 1250
minor, 1250
BROAD
BEAN, 617, 706
-LEAF
CAT-TAIL, 1348
PLANTAIN, 886
RHYNCHOSIA, 691
SIGNAL GRASS, 1335
SNOUT-BEAN, 691
WOOD-OATS, 1254
-POD DRABA, 468
BROCCOLI, 460
Brodiaea uniflora, 1204
BROME
CHEAT GRASS, 1253
DOWNY, 1253
GRASS, 1250
JAPANESE, 1252
RESCUE, 1252
RYE, 1252
SIXWEEKS GRASS, 1336
SPREADING, 1252
Bromeliaceae, 1095
BROMELIAD FAMILY, 1095
Bromus, 1250
catharticus, 1252
hordeaceus, 1252
japonicus, 1252
molliformis, 1252
mollis, 1252
nottowayanus, 1252
pubescens, 1252
racemosus, 1252, 1253
secalinus, 1252
tectorum, 1253
var. glabratus, 1253
var. tectorum, 1253
unioloides, 1252
willdenowii, 1252
BRONCO GRASS, 1253
BROOK
EVONYMUS, 528
-PIMPERNEL, 1012
BROOKLIME, AMERICAN, 1012
BROOKWEED, 914
LIMEROCK, 914
THIN-LEAF, 914
BROOM
-CORN, 1324
FALSE, 1250
PURPLE FALSE, 1250
NAILWORT, 522
SNAKEWEED, 365
WITCHES', 1038
BROOMCORN MILLET, 1300
BROOMRAPE, 868
CHESTER, 868
LARGE-FLOWER, 868
BROOMRAPE FAMILY, 868
BROOMSEDGE, 1239, 1242
BLUESTEM, 1239
BROOMWEED, 362, 364
ANNUAL, 364
COMMON, 364
PERENNIAL, 365
ROUND-HEAD, 365
TEXAS, 365
BROOMWOOD, 1032
Broussonetia, 828
papyrifera, 828
BROWN
-EYED-SUSAN, 400, 402
LARGEST, 402
LACE CACTUS, 486
-SEED PASPALUM, 1306
-SPINE PRICKLY-PEAR, 492
-TOP SIGNAL GRASS, 1335
BROWNTOP, 1335
Brugmansia, 1020
candida, 1020
BRUNETTA, 1216

Brunnichia, 898
cirrhosa, 898
ovata, 898
BRUSHLAND LANTANA, 1052
BRUSSELS-SPROUTS, 460
Buchloe, 1253
dactyloides, 1253
Buchnera, 996
americana, 996
floridana, 996
BUCKBEAN FAMILY, 824
BUCKBERRY, 584
BUCKBRUSH, 511
BUCKBUSH, 932
BUCKEYE, 737
MEXICAN-, 982
PALE, 738
RED, 738
SCARLET, 738
TEXAS, 738, 982
YELLOW, 738
WESTERN, 738
WHITE, 738
BUCKEYE FAMILY, 737
BUCKHORN, 886
BRAKE, 192
FERN, 192
PLANTAIN, 886
BUCK-HORN PLANTAIN, 884
BUCKLEY'S
HICKORY, 748
SABATIA, 728
CENTAURY, 724
YUCCA, 1083
BUCKTHORN, 884, 933, 934
CAROLINA, 933
LANCE-LEAF, 934
PURGING, 934
WOOLLY-, 982
BUCKTHORN FAMILY, 930
BUCKWHEAT, 897
ANNUAL WILD, 899
CLIMBING-, 902
FALSE, 904
HEART-SEPAL WILD, 899
LONG-LEAF WILD, 899
-VINE, AMERICAN, 898
WILD, 898, 902
Buddleja, 480
davidii, 480
BUDDLEJA FAMILY, 480
Buddlejaceae, 480
BUENA MUJER, 797
BUFFALO
-BUR, 1030
BLUE-FLOWER, 1030
-CLOVER, 632, 672
CURRANT, 732
GOURD, 568
GRASS, 1253
-PLUM, 630
-ROSE, 808
BUFFALOWEED, 310
BUFF-CREST, 1216
BUFFEL GRASS, 1310
BUGLE, 755
BUGLEWEED, 755, 760
AMERICAN, 760
ARKANSAS, 762
STALKED, 762
TAPER-LEAF, 762
VIRGINIA, 762
BUGLOSS, VIPER'S-, 446, 448
Buglossoides, 446
arvensis, 446
BULB
BITTERCRESS, 464
LIP FERN, 195
WOODRUSH, 1186
Bulbostylis, 1109
capillaris, 1109
ciliatifolia, 1109
var. ciliatifolia, 1109
var. coarctata, 1109
BULBOUS
ADDER'S-TONGUE, 190
BLUE GRASS, 1312
BULL
THISTLE, 340
-NETTLE, 596, 1030
TEXAS, 596
BULLACE GRAPE, 1072
BULLBRIER, 1344, 1346
BULLHEAD, 1076
BULRUSH, 1152, 1160, 1162
ALKAI, 1108
AMERICAN, 1161
ANNUAL, 1154
CALIFORNIA, 1161
GIANT, 1161
GREAT, 1161
HARD-STEM, 1161
PALE, 1162
PRAIRIE, 1108
ROCKY MOUNTAIN, 1161
SALTMARSH, 1108
SOFT-STEM, 1161
THREE-SQUARE, 1161
WOOLLY-GRASS, 1162
Bumelia, 982
lanuginosa var. *oblongifolia*, 983
BUMELIA, GUM, 982
BUNCH
-GRASS, 1082
-MOSS, 1095
BUNCHBERRY, 1049, 1053
BUNDLE-FLOWER, 651
ILLINOIS, 652
NET-LEAF, 652
PRAIRIE, 652
SHARP-POD, 652
Bupleurum, 244
lancifolium, 246
rotundifolium, 246
BUR
BLUE-FLOWER BUFFALO-, 1030
BRISTLE GRASS, 1320
BUFFALO-, 1030
-CLOVER, 672, 674
CALIFORNIA, 674
SMALL, 674
SPOTTED, 674
-CUCUMBER, 569, 570
ONE-SEED, 570
WALL, 570
GHERKIN, 566
-MARIGOLD, 325
OAK, 716
PRAIRIE, 750
-SAGE, 310
BURHEAD, 1087
CREEPING, 1087
ERECT, 1087
Burmannia, 1096
biflora, 1096
capitata, 1096
BURMANNIA, CAP, 1096
BURMANNIA FAMILY, 1096
Burmanniaceae, 1096
BURNET, 960
PRAIRIE, 960
BURNING-BUSH, 528
BURNINGBUSH, 970
BURNWEED, 348
AMERICAN, 348
BURSTING-HEART, 528
BURWEED, 415, 432, 672
BUTTON, 416
LAWN, 416
BURY-SEED CHLORIS, 1258
BUSH
-CLOVER, 664, 666
CHINESE, 670
CREEPING, 670
HAIRY, 670
JAPANESE, 665
KOREAN, 664
ROUND-HEAD, 670

BUSH-CLOVER *(cont.)*
SLENDER, 671
STUEVE'S, 670
TALL, 670
TEXAS, 670
TRAILING, 670
GRAPE, 1070
HONEYSUCKLE, 508
MORNING-GLORY, 556
PALMETTO, 1094
-PEA, YELLOW, 638
REDPEPPER, 1017
BUSH'S
HAWTHORN, 942
WILD INDIGO, 638
BUSHY
ASTER, 318
BEARD GRASS, 1239
BLUESTEM, 1239
ERYNGO, 252
HONEYSUCKLE, 508
KNOTWEED, 904
-PONDWEED, 1169
SEEDBOX, 860
WALLFLOWER, 468
BUSY-LIZZIE, 433
BUTLER'S SAND-PARSLEY, 243
BUTTER
DAISY, 426
-AND-EGGS, 1002
BUTTERCUP, 866, 926
BLISTER, 928
BRISTLY, 928
CELERY-LEAF, 928
EARLY, 928
WOODS, 926
KIDNEY-LEAF, 926
LARGE, 928
LITTLE-LEAF, 926
MARSH, 928
PRAIRIE, 928
ROUGE, 928
ROUGH-SEED, 928
SHOWY, 928
STICKTIGHT, 928
TEXAS-, 866
TUFTED, 928
WEAK, 928
BUTTERCUP FAMILY, 916
BUTTERFLY, 239, 273, 274, 276, 479, 877
-BUSH, 480
MILKWEED, 280
ORCHID, 1218
-PEA, 639, 642
-WEED, 280, 855
BUTTERNUT SQUASH, 569
BUTTERPRINT, VELVET-LEAF, 806
BUTTERWEED, 388, 390
BUTTON
BURWEED, 416
-CLOVER, 674
-HEMP, 1042
MEDIC, 674
SNAKE-ROOT, 252
-SNAKEROOT, 382
BUTTONBUSH, 962
COMMON, 962
DODDER, 573
BUTTONPLANT, 970
BUTTONS, BARBARA'S-, 387
HAIRY, 384
BUTTONWEED, 962, 970
LARGE, 962
PRAIRIE, 970
SMOOTH, 970
VIRGINIA, 962
BUTTONWOOD, 888

C

CABBAGE, 460
Cabomba, 482
caroliniana, 482
Cabombaceae, 480
Cacalia plantaginea, 314
CACAO FAMILY, 1031
Cactaceae, 482
CACTUS
BARREL, 486
BEEHIVE NIPPLE, 489
BROWN LACE, 486
CLARET-CUP, 486
DESERT CHRISTMAS, 491
DEVIL'S-HEAD, 486
FINGER, 485
FISHHOOK, 489
HEDGEHOG, 486, 494
JUMPING, 492
LACE, 486
NIPPLE, 486, 489
PENCIL, 491
PINCUSHION, 488, 489
PINEAPPLE, 485
PLAINS NIPPLE, 488
RAT-TAIL, 491
RHINOCEROS, 485
SLENDER-STEM, 491
WHITE LACE, 486
CACTUS FAMILY, 482
CADILLO, 1076, 1254
Caesalpinia, 639
drummondii, 639, 664
gilliesii, 639
jamesii, 639, 688
pulcherrima, 639
Caesalpinioideae, 618
CAFTA, 526
CALABACILLA LOCA, 568
CALABASH
GOURD, 569
-TREE, 440
CALABAZILLA, 568
CALADIUM, 1092
CALAMINT, 755
ARKANSAS, 755
Calamintha, 755
arkansana
Calamus, 1094
CALAMUS, 1077
CALAMUS FAMILY, 1077
Calathea, 1210
Calceolaria, 988
CALERY-PEA, 666
Calibanus, 1080
Calibrachoa, 1017, 1023
parviflora, 1017
CALICO
ASTER, 319
-BUSH, 1053
CALIFORNIA
ASTER, 319
BULRUSH, 1161
BUR-CLOVER, 674
COTTONTOP, 1263
FAN PALM, 1094
FILAREE, 730
LOOSESTRIFE, 802
-POPPY, 874
TULE, 1161
YEW, 201
CALLA FAMILY, 1090
CALLERY PEAR, 952
Calliandra conferta, 625
Callicarpa, 1049
americana, ▣/81, 1049
CALLIOPSIS, 342
Callirhoe, 807
alcaeoides, 807
digitata var. *stipulata,* 808
involucrata, 807
var. involucrata, ▣/81, 807
var. lineariloba, 808
leiocarpa, 808
pedata, 808
Callitrichaceae, 494
Callitriche, 494
heterophylla, 496
nuttallii, 496
palustris, 496

- **Callitriche** *(cont.)*
  - peploides, 496
  - terrestris, 496
  - *verna*, 496
- **Calopogon,** 1214
  - barbatus, 1214
  - oklahomensis, 1214
  - *pulchellus*, 1215
  - tuberosus, 1214
- *Caltha*, 916
- CALTROP, 1076
  - CARIC SEDGE, 1130
  - WARTY, 1076
- CALTROP FAMILY, 1074
- **Calycocarpum,** 824
  - lyonii, 824
- **Calylophus,** 852
  - berlandieri, 854
    - subsp. berlandieri, 854
    - subsp. pinifolius, 854
  - *drummondianus*, 854
    - subsp. *berlandieri*, 854
  - hartwegii
    - subsp. pubescens, 854
    - var. *pubescens*, 854
  - serrulatus, 854
    - var. *spinulosus*, 854
- **Calyptocarpus,** 329
  - vialis, 329
- **Calystegia,** 550
  - macounii, 550
  - sepium
    - subsp. angulata, 551
    - subsp. limnophila, 551
  - silvatica subsp. fraterniflora, 551
- CAMASH, 1198
- CAMASS, 1200
  - DEATH-, 1210
  - EASTERN, 1200
  - -LILY, 1200
  - POISON-, 1210
- **Camassia,** 1198
  - *quamash*, 1198
  - scilloides, ▣/81, 1200
- CAMASSIA, ATLANTIC, 1200
- **Camelina,** 460
  - microcarpa, 460
  - rumelica, 462
- CAMOMILE, STINKING, 310
- CAMOTE DE RATÓN, 663
- **Campanula**
  - *coloradoense*, 501
  - *downingia*, 496
  - reverchonii, 496
- **Campanulaceae,** 496
- CAMPERNELLE JONQUIL, 1208
- CAMPHOR
  - DAISY, 372
  - PLUCHEA, 398
- CAMPHORWEED, 372, 398
- CAMPION, 524
  - STARRY, 524
- **Campsis,** 442
  - radicans, ▣/81, 442
- CANADA
  - BLUE GRASS, 1314
  - CLEARWEED, 1042
  - FLEABANE, 341
  - GARLIC, 1196
  - GOLDENROD, 410
  - MILK-VETCH, 630
  - SANICLE, 260
  - WILD RYE, 1268
- *Canadanthus*, 315
- CANADIAN COLUMBINE, 918
- CANAIGRE, 906
- CANARY GRASS, 1310
  - CAROLINA, 1310
  - SOUTHERN, 1310
  - WILD, 1310
- CAÑATILLA, 208
- CANCERWEED, 776
- CANDELABRA VERVAIN, 1056
- CANDELILLA, 604
- CANDLE
  - -BERRY, 832
  - CHOLLA, 491
- CANE
  - BEARD GRASS, 1246
  - BLUESTEM, 1246
  - GIANT, 1243
  - SOUTHERN, 1243
  - SWITCH, 1243
- CANELA, 398
- **Cannabaceae,** 502
- **Cannabis,** 502
  - sativa, 502
    - subsp. *indica*, 504
- CANTELOPE, 566
- CAÑUELA, 176, 178
- CANYON MUHLY, 1288
- CAP BURMANNIA, 1096
- CAPE
  - PERIWINKLE, 266
  - -WEED, 1054
- CAPER FAMILY, 504
- **Capparaceae,** 504
- Capparidaceae, 504
- *Capparis*, 504
- CAPPED RUSH, 1182
- *Capraria multifida*, 1002
- **Caprifoliaceae,** 507
- **Capsella,** 462
  - bursa-pastoris, 462
  - *rubella*, 462
- **Capsicum,** 1017
  - annuum
    - var. *annuum*, 1018
    - var. *aviculare*, 1017
    - var. glabriusculum, 1017
    - var. *minus*, 1017
- CAPUL NEGRO, 933
- CAPULIN, 950
- CARAMBOLA, 869
- CARAQUITO BLANCO, 1052
- CARAWAY, 239
- **Cardamine,** 462
  - bulbosa, 464
  - *debilis*, 464
  - *flexulosa*, 464
  - hirsuta, 462
  - parviflora var. arenicola, 464
  - pensylvanica, 464
  - *rhomboidea*, 464
- CARDAMINE COREOPSIS, 342
- CARDINAL-FLOWER, 498
  - BLUE, 498
- **Cardiospermum,** 980
  - halicacabum, 980
- CARDO
  - DEL VALLE, 332
  - SANTO, 874
- **Carduus,** 329
  - *macrocephalus*, 330
  - nutans subsp. macrocephalus, 330
  - tenuiflorus, 330
- CARELESSWEED, 226
- **Carex,** 1109
  - abscondita, 1116
  - affinity bicknellii, 1130
  - albicans var. australis, 1116
  - albolutescens, 1118
  - *amphibola*
    - var. *globosa*, 1118
    - var. *turgida*, 1122
  - arkansana, 1118
  - austrina, 1118
  - basiantha, 1118
  - blanda, 1118
  - brevior, 1118
  - *brittoniana*, 1130
  - bulbostylis, 1118
  - bushii, 1118
  - caroliniana, 1118
    - var. *cuspidata*, 1118
  - cephalophora, 1118
    - var. *angustifolia*, 1124
    - var. *leavenworthii*, 1124

Carex *(cont.)*
var. *mesochorea*, 1124
cherokeensis, 1120
complanata, 1120
corrugata, 1120
crinita var. brevicrinis, 1120
crus-corvi, 1120
davisii, 1120
debilis, 1120
edwardsiana, 1120
emoryi, 1104, 1120
festucacea, 1120
fissa, 1122
flaccosperma, 1122
var. *glaucodea*, 1122
frankii, 1122
glaucodea, 1122
granularis, 1122
var. *haleana*, 1122
gravida, 1122
var. *lunelliana*, 1122
grisea, 1122
hyalina, 1122
hyalinolepis, 1122
intumescens, 1124
leavenworthii, 1124
longii, 1124
louisianica, 1124
lunelliana, 1122
lupulina, 1124
lurida, 1124
meadii, 1124
mesochorea, 1124
microdonta, 1124
microrhyncha, 1128
muehlenbergii, 1128
var. *australis*, 1118
var. *austrina*, 1118
var. enervis, 1128
var. muehlenbergii, 1128
oligocarpa, 1120
onusta, 1128
oxylepis, 1128
ozarkana, 1128
perdentata, 1128
physorhyncha, 1116
plana, 1128
planostachys, 1128
reniformis, 1128
retroflexa, 1128
var. *texensis*, 1130
socialis, 1130
striatula, 1130
tetrastachya, 1130
texensis, 1130
triangularis, 1130
tribuloides, 1130
vulpinoidea, 1130
willdenowii, 1118
CARIACO DE SAN JUAN, 1052
CARIC SEDGE, 1109
ARKANSAS, 1118
BASAL-FRUIT, 1118
BAYARD LONG'S, 1124
BENJAMIN BUSH'S, 1118
BLADDERY, 1124
CALTROP, 1130
CAROLINA, 1118
CEDAR, 1128
CHARMING, 1118
CHEROKEE, 1120
COMPANION, 1130
CONSPICUOUSLY-TOOTHED, 1128
CROW-FOOT, 1120
EDWARDS PLATEAU, 1120
EMERSON DAVIS', 1120
FESCUE-LIKE, 1120
FEW-FLOWER, 1122
FINE-LINE, 1130
FLACCID-FRUIT, 1122
FLAT-FRUIT, 1120
FOUR-ANGLED, 1130
FOX-TAIL, 1130
GLOBOSE, 1118
GOTTHILF MUHLENBERG'S, 1128
VEINLESS, 1128
GRANULAR, 1122
GRAY-GREEN-FRUIT, 1122
HEAD-BEARING, 1118
HEAVY-FRUIT, 1122
HIDDEN-FRUIT, 1116
HOP, 1124
HYALINE-SCALE, 1122
INFLATED, 1122
JOSEPH FRANK'S, 1122
KIDNEY-SHAPED, 1128
LOUISIANA, 1124
MEAD'S, 1124
MELINES LEAVENWORTH'S, 1124
MIDLAND, 1124
OZARK, 1128
REFLEXED-FRUIT, 1128
SALLOW, 1124
SAMUEL MEAD'S, 1124
SHARP
-MARGIN, 1122
-SCALE, 1128
SHORT, 1118
-HAIR, 1120
SMALL
-BEAK, 1128
-TOOTH, 1124
SOUTHERN, 1118
BELLOWS-BEAK, 1116
TEXAS, 1130
TRIANGULAR, 1130
WEAK, 1120
WHITISH YELLOW, 1118
WILLIAM EMORY'S, 1120
WRINKLE-FRUIT, 1120
Carlowrightia, 212
texana, 212
CARLOWRIGHTIA, TEXAS, 212
CARNATION, 518
CARNATION FAMILY, 511
CARNAUBA WAX PALM, 1094
CAROLINA
ANEMONE, 918
BASSWOOD, 1036
BUCKTHORN, 933
CANARY GRASS, 1310
CARIC SEDGE, 1118
CLOVER, 704
CRANE'S-BILL, 732
DRABA, 468
DWARF-DANDELION, 380
FALSE DANDELION, 399
FANWORT, 482
FOXTAIL, 1236
GERANIUM, 732
GROMWELL, 452
JESSAMINE, 798
JOINT-TAIL, 1258
LARKSPUR, 924
LAUREL CHERRY, 949
LEAF-FLOWER, 611
MODIOLA, 816
MOONSEED, 824
PUCCOON, 452
ROSE, 956
SNAILSEED, 824
VETCH, 708
VIOLET, 1064
WILLOW, 978
CARPENTER'S-SQUARE, 1010
CARPENTER-WEED, 770
CARPET
AJUGA, 755
GRASS, 1244
COMMON, 1244
CARPETWEED, 1076, 827
GREEN, 827
CARPETWEED FAMILY, 826
Carpinus, 439
caroliniana, 439
CARRION-FLOWER, 274
CARROT, 250
-LEAF LOMATIUM, 255

CARROT *(cont.)*
SOUTHWESTERN, 250
WILD, 250
CARROT FAMILY, 239
Carthamus, 330
lanatus, 330
tinctorius, 330
Carum, 239
Carya, 746
alba, 747
aquatica, 747
cordiformis, 747
illinoinensis, 747
myristiciformis, 748
ovata, 748
pecan, 747
texana, 748
tomentosa, 747
Caryophyllaceae, 511
Caryophyllus, 512
CASCARA SAGRADA, 933
CASHEW, 230
CASHEW FAMILY, 230
CASSAVA, 585
CASSENA, 270
EVERGREEN, 270
Cassia, 617, 692
alata, 694
bicapsularis, 696
chamaecrista, 642
corymbosa, 694
fasciculata, 642
var. *ferrisiae,* 642
var. *puberula,* 642
var. *robusta,* 642
var. *rostrata,* 642
lindheimeriana, 694
marilandica, 694
nictitans, 642
obtusifolia, 694
occidentalis, 694
pumilio, 696
roemeriana, 696
CASSINE, 270
CASSIO-BERRY BUSH, 270
Castanea, 710, 711
dentata, 710
pumila, 718
Castanopsis, 710
Castilleja, 996
indivisa, ▣/82, 996
purpurea, 998
var. citrina, ▣/82, 998
var. lindheimeri, 998
var. purpurea, ▣/82, 998
CASTOR
-BEAN, 612
-OIL-PLANT, 612
CAT
GREENBRIER, 1346
MINT, 768
Catalpa, 442
speciosa, 443
CATALPA
DESERT-, 443
HARDY, 443
NORTHERN, 443
WILLOW-LEAF, 443
CATALPA FAMILY, 440
Catapodium, 1260
rigidum, 1260
CATAWBA-TREE, 443
CATBIRD GRAPE, 1072
CATBRIER, 1344, 1346
CATBRIER FAMILY, 1344
CATCHFLY, 524
GENTIAN, 727
GRASS, 1280
CATCHWEED BEDSTRAW, 964
CATCLAW, 625, 677, 678
ACACIA, 625
SCHRANKIA, 678
SENSITIVE-BRIAR, 678
CATGUT, 698, 700
Catha edulis, 526
Catharanthus, 264, 266
roseus, 266
CATNEP, 768
CATNIP, 768
NOSEBURN, 616
CAT'S
-EAR, 375
-FOOT, 398
CAT-TAIL, 1348
GRASS, 1310
BROAD-LEAF, 1348
COMMON, 1348
NARROW-LEAF, 1348
SOUTHERN, 1348
CAT-TAIL FAMILY, 1348
Cattleya, 1213
CAULIFLOWER, 460
WILD, 374
CAVELÓN, 823
Cayaponia, 565
quinqueloba, 565
CAYENNE
-JASMINE, 266
PEPPER, 1017
Ceanothus, 932
americanus, 932
var. *pitcheri,* 932
herbaceus, 932
var. *pubescens,* 932
CEANOTHUS
FUZZY, 932
INLAND, 932
CEBOLLETA, 1200
CEDAR
APPLE RUST, 203, 940
CARIC SEDGE, 1128
EASTERN RED-, 203
ELM, 1040
MOUNTAIN-, 203
PANIC, 1300
PENCIL-, 203
PLANTAIN, 884
POST-, 203
RED-, 203
ROCK-, 203
ROSETTE GRASS, 1300
SAGE, 776
SALT-, 1035
VIRGINIA RED-, 203
WHITE-, 823
CEDRÓN, 1049
DEL MONTE, 1049
CEIBA, 804
Celastraceae, 526
Celastrus, 528
scandens, 528
CELERY, 239
AMERICAN WILD-, 1170
-LEAF BUTTERCUP, 928
SLIM-LOBE, 249
WATER-, 1170
CELESTIAL, 1174
-LILY, 1174
PRAIRIE, 1174
Celosia, 221
Celtis, 1036
australis, 1038
laevigata, 1038
var. laevigata, 1038
var. reticulata, 1038
var. texana, 1038
reticulata, 1038
tenuifolia, 1038
CELTIS-LEAF GOLDENROD, 414
Cenchrus, 1253
carolinianus, 1254
ciliaris, 1310
echinatus, 1254
incertus, 1254
longispinus, 1254
spinifex, 1254
CENIZA, 988

- **Centaurea,** 332
  - americana, ▣/82, 332
  - cyanus, 334
  - *maculosa*, 332
  - melitensis, 334
  - solstitialis, 334
- CENTAUREA, MALTA, 334
- **Centaurium,** 724
  - beyrichii, ▣/83, 724
    - var. *glanduliferum*, 726
  - calycosum, 724
  - floribundum, 724
  - glanduliferum, 726
  - *muehlenbergii*, 726
  - pulchellum, 726
  - texense, 726
- CENTAURY, 724
  - BUCKLEY'S, 724
  - JUNE, 724
  - LADY BIRD'S, 726
  - ROCK, 724
  - TEXAS, 726
- **Centella,** 246
  - *asiatica*, 246
  - erecta, 246
  - *repanda*, 246
- **Centrosema**
  - virginianum, ▣/83, 639
- *Centunculus minimus*, 912
- CENTURY-PLANT, 1078, 1080
- CENTURY-PLANT FAMILY, 1078
- **Cephalanthus,** 962
  - occidentalis, ▣/82, 962
    - var. *californicus*, 962
    - var. *pubescens*, 962
- **Cerastium,** 514
  - brachypetalum, 516
  - brachypodum, 516
  - fontanum subsp. vulgare, 516
  - glomeratum, 516
  - *holosteoides*, 516
  - nutans, 516
  - pumilum, 516
  - *triviale*, 516
  - *viscosum*, 516
  - *vulgatum*, 516
- *Ceratocystis*
  - *fagacearum*, 711
  - *ulmi*, 1039
- **Ceratophyllaceae,** 529
- **Ceratophyllum,** 529
  - demersum, 529
  - echinatum, 530
  - *muricatum* subsp. *australe*, 530
  - *quinoa*, 530, 533
- **Cercis,** 640
  - canadensis, 640
    - var. canadensis, ▣/83, 640
    - var. texensis, 640
  - *occidentalis*, 640
  - *siliquastrum*, 640
- CEVADILLA, 1210
- **Cevallia,** 794
  - sinuata, ▣/83, 794
- CEVALLIA, STINGING, 794
- CEYLON MAHOGANY, 823
- **Chaenorrhinum,** 998
  - minus, 999
- **Chaerophyllum,** 246
  - tainturieri
    - var. dasycarpum, 246
    - var. tainturieri, 246
- **Chaetopappa,** 334
  - asteroides, 334
  - ericoides, 334
- CHAFF-FLOWER, 221
  - MAT, 222
- CHAFFWEED, 912
- CHAIN
  - FERN, 180, 181
    - NARROW-LEAF, 181
    - VIRGINIA, 181
  - PRICKLY-PEAR, 492
- CHALKHILL WOOLLY-WHITE, 375
- **Chamaecrista,** 640
  - fasciculata, ▣/83, 642
  - nictitans, 642
- *Chamaemelum nobile*, 312
- **Chamaesaracha,** 1018
  - coniodes, 1018
  - coronopus, 1018
  - edwardsiana, 1018
  - sordida, 1018
- **Chamaesyce,** 589
  - albomarginata, 592
  - angusta, 592
  - cordifolia, 594
  - fendleri, 594
  - geyeri, 594
  - glyptosperma, 594
  - hirta, 594
  - hypericifolia, 594
  - lata, 594
  - maculata, 594, 596
  - missurica, 596
    - var. *calcicola*, 596
  - nutans, 596
  - prostrata, 596
  - serpens, 596
  - stictospora, 596
  - *supina*, 596
  - villifera, 596
- CHAMIZA, 533
- CHAMOMILE, 310
  - MAYWEED, 310
- CHAMPIN GRAPE, 1072
- CHAPARRAL, 848
- CHAPMAN CLUBMOSS, 174
- CHAPMAN'S
  - BLUE GRASS, 1314
  - POA, 1314
- CHAPOTE PERSIMMON, 579
- **Chaptalia,** 335
  - *nutans* var. *texana*, 335
  - texana, 335
- CHARD, SWISS, 533
- CHARLOCK, 479
- CHARMING CARIC SEDGE, 1118
- **Chasmanthium,** 1254
  - latifolium, 1254
  - laxum, 1254
    - var. laxum, 1256
    - subsp. *sessiliflorum*, 1256
    - var. sessiliflorum, 1256
  - *sessiliflorum*, 1256
- CHASTE-LAMB-TREE, 1060
- CHASTETREE, 1058
  - COMMON, 1060
  - CUT-LEAF, 1060
  - TRUE, 1060
- CHATTERBOX, 1215
- CHEAT
  - GRASS, 1253
    - BROME, 1253
- CHEESES, 812
- CHEESEWEED, 812
- **Cheilanthes,** 195
  - *aemula*, 196
  - alabamensis, 196
  - *castanea*, 196
  - *dealbata*, 194
  - eatonii, 196
  - feei, 196
  - horridula, 196
  - *integerrima*, 195
  - lanosa, 198
  - lindheimeri, 198
  - *sinuata*, 195
  - tomentosa, 198
- **Chenopodiaceae,** 530
- **Chenopodium,** 533
  - album, 535
    - var. *missouriense*, 538
  - ambrosioides, 535
  - berlandieri, 536
    - var. berlandieri, 536
    - var. boscianum, 536
    - var. sinuatum, 536

Chenopodium *(cont.)*
var. zschackei, 536
botrys, 536
desiccatum var. *leptophylloides,* 538
gigantospermum, 538
leptophyllum, 536
missouriense, 536
murale, 538
pallescens, 538
pratericola, 538
pumilo, 538
simplex, 538
standleyanum, 538
vulvaria, 538
CHERIMOYA, 238
CHEROKEE
-BEAN, 658
CARIC SEDGE, 1120
ROSE, 956
CHERRY, 947
-BARK OAK, 714
BEACH GROUND-, 1024
BLADDER GROUND-, 1026
CAROLINA LAUREL, 949
CLAMMY GROUND-, 1024
COMMON GROUND-, 1024
CUT-LEAF GROUND-, 1024
DOWNY GROUND-, 1026
FIELD GROUND-, 1026
GROUND-, 1023
INDIAN-, 933
LANCE-LEAF GROUND-, 1024
LAUREL, 949
-LAUREL, 947
LOW HAIR GROUND-, 1026
PRAIRIE GROUND-, 1026
PROSTRATE GROUND-, 1018
PURPLE GROUND-, 1026
PURPLE-VEIN GROUND-, 1024
RUM, 950
SOUTHWEST GROUND-, 1024
THICKET GROUND-, 1026
VIRGINIA GROUND-, 1026
CHERVIL, 246
HAIRY-FRUIT, 246
WILD, 246, 249
CHESS, 1250
DOWNY, 1253
JAPANESE, 1252
SOFT, 1252
CHESTER BROOMRAPE, 868
CHESTNUT, 710, 711
AMERICAN, 710
BLIGHT, 710
HORSE-, 737, 738
OAK, 716
SWAMP, 716
WATER-, 1144
CHICALOTE, 874
CHICKASAW PLUM, 949
CHICKENSPIKE, 496, 1031
CHICKENSPIKE FAMILY, 1031
CHICKENTHIEF, 796
CHICK-PEA, 617
CHICKWEED, 514, 525
COMMON, 525
CURTIS' MOUSE-EAR, 516
DWARF MOUSE-EAR, 516
FORKED, 520
GRAY, 516
INDIAN-, 827
LESSER, 526
NODDING, 516
POWDERHORN, 516
SHORT-STALK, 516
TARWORT, 525
WATER-, 494
CHICLE FAMILY, 982
CHICORY, 336
COMMON, 336
CHIGGER-FLOWER, 280
CHIGGERWEED, 280
CHILDING-PINK, 522
CHILE PIQUÍN, 1017
CHILI, 1017
PEPPER, 1015
CHILICOTE, 568
CHILICOYOTE, 568
CHILIPIQUÍN, 1017
CHILITEPÍN, 1017
Chilopsis, 443
linearis, 443
CHINA
-BERRY, 822, 823
-BRIER, 1346
-GRASS, 1042
-TREE, 823
CHINABERRY, WILD, 980
CHINAROOT, 1346
CHINATREE, 979
CHINESE
BOTTLETREE, 1032
BUSH-CLOVER, 670
-DATE, 936
ELM, 1040
-HAT, 910
JUJUBE, 936
-LANTERN, 1026
-LANTERN-OF-THE-PLAINS, 1026
MUSTARD, 459
PARASOL-TREE, 1032
-PARSLEY, 248
PISTACHIO, 230
PRIVET, 850
-PULSEY, 450
TALLOW TREE, 612
TAMARISK, 1035
WISTERIA, 710
CHINGMA, 806
CHINQUAPIN, 710, 711, 718
OAK, 716
WATER-, 834
CHINTÚL, 1136
Chiropetalum, 602
CHISME, 910
CHITTAMWOOD, 231, 982
CHIVES, 1194
Chloracantha, 335
spinosa, 335
Chloris, 1256
argentina, 1276
chloridea, 1258
cucullata, 1256
gayana, 1256
latisquamea, 1256
subdolichostachya, 1256
verticillata, 1256
virgata, 1258
CHLORIS
BURY-SEED, 1258
SHOWY, 1258
CHOCOLATE
DAISY, 325
DEL INDIO, 270
CHOCOLATE FAMILY, 1031
CHOCTAWROOT, 266
CHOLLA, 489
CANDLE, 491
CHRISTMAS, 491
DEVIL, 492
DOG, 492
GREEN-FLOWER, 492
JEFF DAVIS', 492
JUMPING, 492
KLEIN'S, 491
PENCIL, 491
SCHOTT'S, 492
Chondrodendron tomentosum, 823
Chorispora, 464
tenella, 464
CHRISTMAS
CACTUS, DESERT, 491
CHOLLA, 491
FERN, 185
MISTLETOE, 1064
CHRISTMAS MISTLETOE FAMILY, 1064
CHRIST'S-THORN, 934
Chryopsis graminifolia var. *latifolia,* 396

Chrysanthemum, 382
- *leucanthemum*, 382
  - var. *pinnatifidum*, 382

Chrysopsis, 335
- *canescens*, 372
- *hirtella*, 336
- pilosa, 336
- texana, 336
- *villosa* var. *stenophylla*, 372

CHUFA, 1138
CHURCH-MOUSE THREEAWN, 1242
Cicer, 617
Cichorium, 336
- intybus, 336

Cicuta, 246
- maculata, 248
- *mexicana*, 248

CICUTILLA, 393
CIGAR-FLOWER, 799
CIGARTREE, 442, 443
- INDIAN, 442

CILANTRO, 248
CILIATE NUT-RUSH, 1164
CILINDRILLO, 1020
CIMARRÓN, TOBACCO, 1022
Cinchona, 960
CINCO LLAGAS, 880
Cinnamomum, 784
CINNAMON
- OAK, 716
- FERN, 192

CINNAMON FERN FAMILY, 192
CINQUEFOIL, 946
- OLDFIELD, 947
- SULFUR, 947

Cirsium, 338
- altissimum, 338
- engelmannii, 338
- horridulum, ▣/84, 340
  - var. *elliottii*, 340
- *iowense*, 338
- ochrocentrum, 340
- *terrae-nigrae*, 340
- texanum, ▣/84, 340
- undulatum, ▣/84, 340
- vulgare, 340

Cissus, 1066
- *ampelopsis*, 1066
- incisa, 1066
- *trifoliata*, 1066

Cistaceae, 542
Cistus, 542
CITRON, 566
- INDIAN PAINTBRUSH, 998
- PAINTBRUSH, 998

Citrullus, 565
- *citrullus*, 565
- *colocynthis*, 566
- lanatus, 565
  - var. citroides, 566
  - var. lanatus, 565
- *vulgaris*, 565

Citrus, 970
- *maxima*, 970
- *×paradisi*, 970
- *sinensis*, 970
- *trifoliata*, 972

CITRUS FAMILY, 970
Cladium, 1130
- mariscoides, 1132
- mariscus subsp. jamaicense, 1132

CLAMMY GROUND-CHERRY, 1024
CLAMMYWEED, 506
- LARGE, 507

CLAPWEED, 208
CLARET-CUP CACTUS, 486
Clarkia, 852
CLASPING
- CONEFLOWER, 344, 346
- FALSE PIMPERNEL, 1002
- -LEAF CONEFLOWER, 346
- ST. JOHN'S-WORT, 547
- VENUS'-LOOKING-GLASS, 502

CLAVELLINA, 492
Claviceps
- *paspali*, 1304
- *purpurea*, 1304, 1318

Claytonia, 908
- virginica, ▣/84, 908

CLEARWEED, 1042
- CANADA, 1042

CLEAVERS, 962, 964
- GOOSEGRASS, 964

Clematis, 920
- crispa, 921
- *dioscoreifolia*, 921
- drummondii, 921
- *maximowicziana*, 921
- *paniculata*, 921
- pitcheri, 921
- reticulata, 921
- terniflora, 921
- texensis, ▣/84, 922

CLEMATIS
- CURLY, 921
- NET-LEAF, 921
- PITCHER'S, 921
- SCARLET, 922
- SWEET-AUTUMN, 921

Cleome, 506
- hassleriana, 506

Cleomella, 506
- angustifolia, 506

CLEPE, 936
CLIFF
- -BRAKE, 198
  - PURPLE, 198
  - WRIGHT'S, 200
- FERN, 185

CLIMBING
- BITTERSWEET, 528
- -BONESET, 388
- -BUCKWHEAT, 902
- -DOGBANE, 268
- FIG, 830
- HEMPWEED, 388
- -MILKWEED, 283, 284
- PRAIRIE ROSE, 958
- ROSE, 958

Clinopodium, 755
Clitoria, 642
- mariana, 642

CLOAK FERN
- FALSE, 194
- LONG, 195
- POWDERY, 194
- STAR-SCALED, 195
- WAVY, 195

CLOCKVINE, 218
- BLACK-EYED, 218

CLOTBUR, 432
CLOTH-OF-GOLD, 474
CLOVE CURRANT, 732
CLOVER, 700
- ARROW-LEAF, 706
- BEJAR, 704
- BUFFALO-, 632, 672
- BUR-, 672, 674
- BUSH-, 664, 666
- BUTTON-, 674
- CALIFORNIA BUR-, 674
- CAROLINA, 704
- CHINESE BUSH-, 670
- COMANCHE PEAK PRAIRIE-, 651
- CREEPING BUSH-, 670
- CRIMSON, 704
- CROSS-, 964
- DUTCH, 706
- FEW-FLOWER TICK-, 656
- GLANDULAR PRAIRIE-, 650
- HAIRY BUSH-, 670
- HOARY TICK-, 655
- HOP-, 674
- HUBAM-, 676
- ITALIAN, 704
- JAPANESE BUSH-, 665
- KOREAN BUSH-, 664
- LAPPA, 704

- CLOVER *(cont.)*
  - LARGE HOP, 704
  - LEAST HOP, 704
  - LITTLE-LEAF TICK-, 655
  - LONG-BRACT PRAIRIE-, 650
  - LOW HOP, 704
  - MARYLAND TICK-, 655
  - MEXICAN-, 968
  - NUTTALL'S TICK-, 655
  - OLD-FIELD, 702
  - PANICLED TICK-, 656
  - PEANUT, 704
  - PERSIAN, 706
  - PIN-, 730
  - PRAIRIE-, 644, 648, 650, 670
  - PURPLE PRAIRIE-, 650
  - RABBIT-FOOT, 702
  - RED, 706
  - REVERSED, 706
  - RIGID TICK-, 655
  - ROUND-HEAD BUSH-, 670
  - SESSILE-LEAF TICK-, 656
  - SHOWY PRAIRIE-, 648
  - SIMPLE-LEAF TICK-, 656
  - SLENDER BUSH-, 671
  - SLIM-SPIKE PRAIRIE-, 650
  - SMALL
    - BUR-, 674
    - HOP, 704
  - SOUR-, 676
  - SPOTTED BUR-, 674
  - STIFF TICK-, 655
  - STUEVE'S BUSH-, 670
  - SWEET-, 676
  - TALL BUSH-, 670
  - TEXAS BUSH-, 670
  - TICK-, 652
  - TRAILING BUSH-, 670
  - TWEEDY'S TICK-, 656
  - VELVET-LEAF TICK-, 656
  - WATER-, 186
  - WHITE, 706
    - PRAIRIE-, 650
    - SWEET-, 676
  - WRIGHT'S TICK-, 656
  - YELLOW
    - SWEET-, 676
    - SOUR-, 676
- CLOVER FAMILY, WATER-, 185
- CLUBMOSS
  - APPRESSED BOG, 174
  - BOG, 173
  - CHAPMAN, 174
  - CREEPING, 174
  - PROSTRATE BOG, 174
  - SOUTHERN, 174
- CLUBMOSS FAMILY, 173
- **Clusiaceae,** 544
- CLUSTER
  - BEAK-RUSH, 1158
  - DODDER, 573
  - MOUNTAIN MINT, 772
  - SANICLE, 260
  - -STEM NAILWORT, 520
- CLUSTERED POPPY-MALLOW, 807
- CLUSTERVINE, HAIRY, 558
- **Cnidoscolus,** 596
  - texanus, ▣/85, 596
- COAMECATL, 898
- COARSE VERVAIN, 1058
- COAST
  - INDIGO, 664
  - LIVE OAK, 719
  - REDWOOD, 202
- COASTAL
  - PLAIN WILLOW, 978
  - SALT GRASS, 1264
  - WATER-HYSSOP, 994
- COBAEA PENSTEMON, 1007
- **Cocculus,** 824
  - carolinus, 824
- COCKLEBUR, 432, 938
  - COMMON, 432
  - SPINY, 432
- COCK'S
  - -COMB, 1216
  - -FOOT, 1259
- COCKSCOMB, 221
- COCKSPUR HAWTHORN, 942
- COCO
  - -DE-MER, 1094
  - -GRASS, 1142
- COCONUT
  - PALM, 1094
  - DOUBLE-, 1094
- *Cocos nucifera,* 1094
- COCOYAM, 1093
- *Codiaeum,* 598
- **Coelorachis,** 1258
  - cylindrica, 1258
- *Coelostylis lindheimeri,* 798
- *Coffea,* 960
- COFFEE
  - SENNA, 694
  - TREE, KENTUCKY, 663
  - -BEAN, 696
- COFFEE FAMILY, 960
- COFFEEWEED, 694
- COLA
  - DE CABALLO, 178
  - DE MICO, 450
  - DE ZORRILLO, 972
- *Colchicum autumnale,* 1191
- *Coleogeton,* 1343
  - *pectinatus,* 1344
- *Coleosporium ipomoeae,* 554
- COLEUS, 752
- COLICROOT, 1165, 1193
- COLLARDS, 460
- **Collinsia,** 999
  - violacea, 999
- COLLINSIA, VIOLET, 999
- **Colocasia,** 1093
  - esculenta, 1093
- COLOMBIA WOLFFIA, 1191
- COLORADO
  - GRASS, 1335
  - GREENTHREAD, 422
  - RIVER-HEMP, 696
  - VENUS'-LOOKING-GLASS, 501
- COLORIN, 656
- **Colubrina,** 932
  - texensis, 932
- COLUBRINA, TEXAS, 932
- COLUMBINE, 918
  - AMERICAN, 918
  - CANADIAN, 918
  - COMMON, 918
  - WILD, 918
- COMA, 982
- COMANCHE PEAK PRAIRIE-CLOVER, 651
- **Comandra,** 979
  - *pallida,* 979
  - umbellata subsp. pallida, 979
- COMANDRA, WESTERN, 979
- COMB
  - LADY'S-, 260
  - VENUS'-, 260
- COMFREY, 443
  - BLUE, 446
  - WILD, 446
- COMIDA DE VIBORA, 208
- **Commelina,** 1098
  - caroliniana, 1100
  - communis, 1100
  - diffusa, 1100
  - erecta, 1100
    - var. angustifolia, 1100
    - var. deamiana, 1100
    - var. erecta, ▣/85, 1100
  - virginica, 1100
- **Commelinaceae,** 1096
- *Commelinantia anomala,* 1101
- COMMON
  - ALDER, 439
  - ARROWHEAD, 1088
  - BALLOONVINE, 980
  - BEEBUSH, 1049

COMMON *(cont.)*
BINDWEED, 551
BROOMWEED, 364
BUTTONBUSH, 962
CARPET GRASS, 1244
CAT-TAIL, 1348
CHASTETREE, 1060
CHICKWEED, 525
CHICORY, 336
COCKLEBUR, 432
COLUMBINE, 918
CORN-COCKLE, 513
CRAPE-MYRTLE, 800
CURLY-MESQUITE, 1278
DANDELION, 418
DAYFLOWER, 1100
DEVIL'S-CLAW, 880
DODDER, 574
DOGWEED, 422
DUCKMEAT, 1190
ELDERBERRY, 510
EVENING-PRIMROSE, 863
FENNEL, 254
FIG, 830
FLAX, 792
FOUR-O'CLOCK, 840
FOXGLOVE, 989
FROGFRUIT, 1054
FUMITORY, 722
GARDEN PETUNIA, 1023
GLOBE-AMARANTH, 228
GOLDENROD, 410
GOLDSTAR, 1204
GRAPE FERN, 189
GREENBRIER, 1346
GROUND-CHERRY, 1024
GROUNDSEL, 403
HEMICARPHA, 1155
HONEY-LOCUST, 660
HOREHOUND, 762
JUJUBE, 936
LADIES'-TRESSES, 1220
LANTANA, 1053
LEAST DAISY, 334
LESPEDEZA, 665
LOUSEWORT, 1006
MALLOW, 812, 814
MEADOW-BEAUTY, 822
MILLET, 1300
MORNING-GLORY, 556
MOTHERWORT, 760
MOUSE-EAR, 516
MULLEIN, 1010
OATS, 1244
PAWPAW, 238
PEANUT, 628
PEAR, 952
PERILLA, 768
PERIWINKLE, 269
PERSIMMON, 579
PIMPERNEL, 912
PLANTAIN, 886
POOLMAT, 1350
POPPY, 876
PRIVET, 850
PURSLANE, 910
RAGWEED, 309
REED, 1311
ROSE-MALLOW, 810
RUSH, 1182
SAGE, 772
SANDBUR, 1254
SCOURING-RUSH, 176
SELFHEAL, 770
SERVICE-BERRY, 940
SHOOTING-STAR, 913
SIXWEEKS GRASS, 1336
SNEEZEWEED, 366
SOW-THISTLE, 416
SPEEDWELL, 1012
ST. JOHN'S-WORT, 547
SUNFLOWER, 369
THORN-APPLE, 1020
TOAD-FLAX, 1002
TRUMPET-CREEPER, 442
VERVAIN, 1050
VETCH, 707
WATER
-HEMLOCK, 248
-HYACINTH, 1339
-NYMPH, 1170
WATERWORT, 496
WOLFFIA, 1191
WOODSIA, 185
YARROW, 308
ZINNIA, 433
COMPACT DODDER, 573
COMPANION CARIC SEDGE, 1130
COMPASSPLANT, 404
WHITE, 404
Compositae, 287
COMPRESSED SPIKE-RUSH, 1146
Condalia, 932
hookeri, 933
obovata, 933
obtusifolia, 936
CONEFLOWER, 346, 400
CLASPING, 344, 346
-LEAF, 346
GIANT, 402
GREAT, 402
MARSH, 402
PRAIRIE, 400
PURPLE, 346
ROUGH, 402
UPRIGHT PRAIRIE, 400
CONE-SPUR BLADDERWORT, 787
CONFEDERATE
-JASMINE, 268
-VINE, 898
CONIFERS, 201
Conium, 248
maculatum, 248
Conobea multifida, 1000
CONOBEA, NARROW-LEAF, 1000
Conoclinium, 352
coelestinum, 353
Conringia, 464
orientalis, 464
Consolida, 922
ajacis, 922
ambigua, 922
orientalis, 922
CONSPICUOUSLY-TOOTHED CARIC SEDGE, 1128
CONSUMPTION-WEED, 324
Convallariaceae, 1191
Convolvulaceae, 548
Convolvulus, 551
arvensis, 551
equitans, 552
interior, 551
macounii, 551
shumardianus, 558
Conyza, 340
canadensis, 341
var. canadensis, 341
var. glabrata, 341
odorata, 398
ramosissima, 341
CONYZA
HORSE-TAIL, 341
LOW, 341
COON-TAIL, 529
Cooperia, 1200
drummondii, 1200
pedunculata, ▣/85, 1200
COPALTREE, 1015
Copernicia prunifera, 1094
COPPERLEAF, 586
HOP-HORNBEAM, 588
LINDHEIMER'S, 588
RHOMBOID, 589
SLENDER, 588
ONE-SEED, 588
VIRGINIA, 589
COPPER
-LILY, 1201
-MALLOW, 820

CORAL
HONEYSUCKLE, 510
-BEAN, 656, 658
EASTERN, 658
-BERRY, 510
CORALBEAD, 824
CORALBELLS, 986
CORALBERRY, 824
CORALITO, 882
CORALLITA, 898
Corallorrhiza, 1215
arizonica, 1216
odontorhiza, 1215
wisteriana, 1215
CORALROOT, 1215
AUTUMN, 1215
CRESTED-, 1215, 1216
LATE, 1215
SPRING, 1215
WISTER'S, 1215
CORALVINE, 824, 898
Corchorus, 1035
CORD GRASS, 1324
PRAIRIE, 1324
Coreopsis, 341
basalis, 342
var. *wrightii,* 342
cardaminaefolia, 342
grandiflora var. longipes, 342
lanceolata, 342
nuecensis, 342
nuecensoides, 342
tinctoria, ▣/85, 342
wrightii, 342
COREOPSIS, 341
CARDAMINE, 342
GOLDEN-MANE, 342
LANCE, 342
PLAINS, 342
ROCK, 342
CORIANDER, 248
Coriandrum, 248
sativum, 248
CORK
ELM, 1039
OAK, 711
CORN, 1338
BINDWEED, 902
BROOM-, 1324
-COCKLE, 513
COMMON, 513
GUINEA-, 1324
POPPY, 876
SPEEDWELL, 1012
Cornaceae, 560
CORNBIND, 551
DULL-SEED, 902
CORNEL, 560
CORNFLOWER, 334
CORNSALAD, 1043
BEAKED, 1044
HAIRY, 1044
NARROW-CELL, 1044
WOOD'S, 1046
CORNSPERRY, 1012
CORNSTALK PONDWEED, 1343
Cornus, 560
drummondii, 561
florida, 561
priceae, 561
sanguinea, 561
Coronilla, 642
varia, 642
Coronopus, 465
didymus, 465
CORREHUELA DE LAS DOCE, 558
CORREOSA SODA-POP-BUSH, 234
CORSAGE ORCHID, 1213
Cortaderia, 1258
selloana, 1258
Corydalis, 719
aurea
subsp. occidentalis, 720
var. *occidentalis,* 720
crystallina, 720
curvisiliqua, 720
subsp. curvisiliqua, 720
subsp. grandibracteata, 720
var. *grandibracteata,* 720
micrantha, 722
subsp. australis, 722
var. *australis,* 722
subsp. micrantha, 722
subsp. texensis, 722
montana, 720
CORYDALIS
CURVE-POD, 720
GOLDEN, 720
MEALY, 720
SMALL-FLOWER, 722
SOUTHERN, 722
TEXAS, 722
Coryphantha, 485
cornifera var. *echinus,* 485
echinus, 485
missouriensis
var. *caespitosa,* 488
var. *robustior,* 488
sulcata, ▣/85, 485
vivipara var. *radiosa,* 489
Cosmos, 287
COTA, 422
Cotinus, 231
americanus, 231
coggygria, 231
obovatus, 231
Cotoneaster, 936
COTTON, 808
-BATTING CUDWEED, 398
BELT COTTON, 808
DRUMMOND'S SNAKE-, 227
FIELD SNAKE-, 227
-FLOWER, 228
LANCE-LEAF, 228
-WOOLLY, 228
-GUM, 845
MORNING-GLORY, 556
-ROSE, 356
SLENDER SNAKE-, 227
SNAKE-, 227
-THISTLE, 388
UPLAND, 808
WEST INDIAN, 808
WILD, 810
COTTONTOP
ARIZONA, 1263
CALIFORNIA, 1263
COTTONWEED, 227
COTTONWOOD, 974, 975
ALAMO, 975
EASTERN, 975
PLAINS, 976
TEXAS, 976
COVERED-SPIKE DROPSEED, 1327
COW
-CATCHER, 880
-LILY, 844
YELLOW, 844
-TONGUE PRICKLY-PEAR, 491
COWBANE, 248, 256
LEAF-LESS, 256
SPOTTED, 248
WATER DROPWORT, 256
COWBERRY, 584
COWBOY-ROSE, 808
COWBOY'S-TOILET-PAPER, 1010
COWCOCKLE, 526
COWHERB, 526
COWITCH, 1066
VINE, 442
COWPEN DAISY, 426
COWSLIP, AMERICAN, 912, 913
COW'S-TONGUE, 491
COYOTE WILLOW, 978
COYOTILLO, 933
HUMBOLDT'S, 933
CRAB GRASS, 1262
HAIRY, 1264

CRAB GRASS *(cont.)*
LARGE, 1264
SHAGGY, 1264
SLENDER, 1263
SMOOTH, 1264
SOUTHERN, 1263
TEXAS, 1316
CRABWEED, 828
HAIRY, 828
CRAMERIA, 750
CRANBERRY, 581, 584
CRANE'S-BILL, 730, 732
CAROLINA, 732
CRAPE-MYRTLE, 800
COMMON, 800
CRAPE-MYRTLE FAMILY, 799
Crassula, 562
aquatica, 562
drummondii, 562
Crassulaceae, 561
Crataegus, 940
anamesa, 944
antiplasta, 944
berberifolia, 942
brachyacantha, 942
brachyphylla, 942
brazoria, 942
bushii, 942
cherokeensis, 942
crus-galli, 942
dallasiana, 942
engelmannii, 942
glabriuscula, 944
marshallii, 942
mollis, 942
nananixonii, 940
opaca, 940
phaenopyrum, 942
pyracanthoides, 942
reverchonii, 942
spathulata, 942
viridis, 943
CRAZYWEED, 682
LAMBERT'S, 682
CREEK PLUM, 950
CREEK-OATS, 1254
CREEPING
BURHEAD, 1087
BUSH-CLOVER, 670
CLUBMOSS, 174
CUCUMBER, 570
DAYFLOWER, 1100
FIG, 830
LADIES'-SORREL, 870
LESPEDEZA, 670
LOVE GRASS, 1272
ORCHID, 1215
PRIMROSE-WILLOW, 862
RUBBER-PLANT, 830
SEEDBOX, 860
SLIMPOD, 266
SPIKE-RUSH, 1147
SPOTFLOWER, 308
SPURGE, 589
WATER-PRIMROSE, 862
CRENATE-LEAF PHACELIA, 743
CREOSOTE-BUSH, 1074
Crepis, 342
pulchra, 344
Cresentia, 440
CRESPÓN, 800
CRESS, SPRING, 464
CRESTED-CORALROOT, 1215, 1216
CRESTPETAL, LARGE, 507
CRIMSON CLOVER, 704
Cristaltella, 507
erosa, 507
Crocus sativus, 1172
CROCUS, AUTUMN, 1191
CROOK-NECK SQUASH
SUMMER, 568
WINTER, 569
Croptilon, 344
divaricatum, 344
var. *hookerianum*, 344
hookerianum
var. hookerianum, 344
var. validum, 344
CROSS
OAK, 718
-CLOVER, 964
CROSSVINE, 442
Crotalaria, 643
purshii, 643
retusa, 643
sagittalis, 643
spectabilis, 644
CROTALARIA
ARROW, 643
PURSH'S, 643
SHOWY, 644
Croton, 598
alabamensis
var. alabamensis, 599
var. texensis, 599
capitatus, 599
var. capitatus, 599
var. lindheimeri, 599
fruticulosus, 599
glandulosus, 599
var. lindheimeri, 600
var. septentrionalis, 600
lindheimerianus, 600
linearis, 600
michauxii, 600
monanthogynus, 600
texensis, 600
tiglium, 598
CROTON
LINDHEIMER'S, 600
NORTHERN, 600
ONE-SEED, 600
ROUND-, 598
TEXABAMA, 599
TEXAS, 600
THREE-SEED, 600
TROPIC, 599
WOOLLY, 599
Crotonopsis linearis, 600
CROW
-FOOT CARIC SEDGE, 1120
-NEEDLES, 260
-POISON, 1208
CROWFOOT, 1259
GRASS, 1256
DURBAN, 1259
EGYPTIAN, 1259
TEXAS, 1281
CROWN
COREOPSIS, 342
-OF-THORNS, 934
-VETCH, 548, 642
RUST OF OATS, 934
CROWNBEARD, 426
GOLDEN, 426
GRAVELWEED, 427
LINDHEIMER'S, 427
VIRGINIA, 427
WHITE, 427
CROW'S-FOOT, 926
CURSED, 928
Cruciferae, 454
Cryphonectria parasitica, 710
Cryptantha, 446
texana, 446
CRYPTANTHA, TEXAS, 446
Cryptotaenia, 249
canadensis, 249
CUCUMBER, 566
BITTER, 566
BUR-, 569, 570
CREEPING-, 570
-LEAF SUNFLOWER, 369
ONE-SEED, BUR-, 570
WALL BUR-, 570
CUCUMBER FAMILY, 564
Cucumis, 566
anguria, 566

Cucumis *(cont.)*
var. *anguria*, 566
var. *longaculeatus*, 566
melo, 566
sativus, 566
Cucurbita, 568
foetidissima, ⊞/86, 568
maxima, 569
moschata, 569
pepo, 568
subsp. *ovifera* var. *texana*, 568
var. *ovifera*, 568
texana, ⊞/86, 568
Cucurbitaceae, 564
Cudrania, 831
CUDWEED, 359, 398
COTTON-BATTING, 398
FRAGRANT, 398
PURPLE, 359
CULANTRILLO, 194
CULTIVATED
FLAX, 792
OATS, 1244
CULVER'S
-PHYSIC, 1014
-ROOT, 1014
CUMIN, 239
Cuminum, 239
Cumminsiella texana, 437
Cunila, 756
mariana, 756
origanoides, 756
CUP
GRASS, 1274
PRAIRIE, 1274
SILKY, 1274
TEXAS, 1274
-LEAF PENSTEMON, 1008
CUPFLOWER, 1022
TEXAS, 1022
Cuphea, 799
CUPID'S-DELIGHT, 1062
Cupressaceae, 202
Cupressus sempervirens, 202
CUPSCALE, AMERICAN, 1315
CUPSEED, 450, 824
CURARE, 823
CURLTOP SMARTWEED, 902
CURLY
CLEMATIS, 921
-CUP GUMWEED, 362
DOCK, 906
-LEAF DOCK, 906
-MESQUITE, COMMON, 1278
MUCKWEED, 1342
PARSLEY, 256
THREEAWN, 1242
CURRANT, 732
BUFFALO, 732
CLOVE, 732
INDIAN-, 510
-OF-TEXAS, 436
CURRANT FAMILY, 732
CURSED CROW'S-FOOT, 928
CURTIS' MOUSE-EAR CHICKWEED, 516
CURVE-POD CORYDALIS, 720
Cuscuta, 572
campestris, 574
cephalanthi, 573
compacta, 573
coryli, 573
cuspidata, 573
exaltata, 573
glabrior, 574
var. *pubescens*, 574
glandulosa, 574
glomerata, 573
gronovii, 573
var. *latiflora*, 574
indecora, 574
var. indecora, 574
var. longisepala, 574
obtusiflora var. glandulosa, 574
pentagona, 574
var. glabrior, 574
var. pentagona, 574
var. pubescens, 574
Cuscutaceae, 572
CUSP DODDER, 573
CUSPIDATE DODDER, 573
CUSTARD-APPLE, 238
CUSTARD-APPLE FAMILY, 238
CUT
GRASS, 1280, 1338
GIANT, 1338
RICE, 1281
VIRGINIA, 1281
-LEAF
CHASTETREE, 1060
CYCLANTHERA, 569
DAISY, 348
EVENING-PRIMROSE, 864
GERMANDER, 783
GROUND-CHERRY, 1024
IRONPLANT, 387
Cyclamen, 911
CYCLAMEN, 911
Cyclanthera, 569
dissecta, 569
CYCLANTHERA, CUT-LEAF, 569
Cycloloma, 540
atriplicifolium, 540
Cyclospermum, 249
leptophyllum, 249
CYLINDER
FLAT SEDGE, 1136
-FRUIT LUDWIGIA, 860
Cymopterus, 249
macrorhizus, 249
Cynanchum, 282
barbigerum, 282
laeve, 282
racemosum var. unifarium, 282
unifarium, 282
Cynara scolymus, 287
Cynoctonum mitreola, 798
Cynodon, 1259
dactylon, 1259
Cynoglossum, 446
officinale, 446
virginianum, 446
zelyanicum, 446
Cynosciadium, 249
digitatum, 249
pinnatum, 255
Cyperaceae, 1104
Cyperus, 1132
acuminatus, 1136
alternifolius, 1138
subsp. *flabelliformis*, 1138
aristatus, 1144
articulatus, 1136
bipartitus, 1144
brevifolius, 1154
compressus, 1136
croceus, 1136
cyperinus var. *rubricosus*, 1162
difformis, 1136
digitatus, 1144
echinatus, 1136
elegans, 1136
engelmannii, 1140
erythrorhizos, 1136
esculentus, 1138
var. *leptostachyus*, 1138
var. *macrostachyus*, 1138
ferruginescens, 1140
filiculmis, 1138
flabelliformis, 1138
flavescens, 1138
var. *poiformis*, 1138
globulosus, 1136
var. *robustus*, 1142
haspan, 1138
hystricinus, 1138
involucratus, 1138
lanceolatus, 1138
var. *compositus*, 1138

Cyperus *(cont.)*
lupulinus, 1138
niger var. *rivularis,* 1144
ochraceus, 1144
odoratus, 1138
var. *engelmannii,* 1140
var. *squarrosus,* 1140
ovularis, 1136
var. *cylindricus,* 1142
papyrus, 1132
plukenetii, 1140
polystachyos, 1140
var. *texensis,* 1140
pseudovegetus, 1140
reflexus, 1140
var. *fraternus,* 1142
retroflexus, 1142
var. *pumilus,* 1142
retrofractus var. *hystricinus,* 1138
retrorsus, 1142
rivularis, 1144
rotundus, 1142
schweinitzii, 1142
setigerus, 1142
squarrosus, 1142
strigosus, 1144
surinamensis, 1144
tenuifolius, 1154
uniflorus, 1142
var. *pumilus,* 1142
var. *retroflexus,* 1142
virens var. *arenicola,* 1140
Cyphomandra, 1027
CYPRESS, 202
BALD, 204
BELVEDERE SUMMER-, 540
MEXICAN BALD, 204
MOCK, 540
MONTEZUMA BALD, 204
SOUTHERN, 204
STANDING-, 890
SUMMER-, 540
CYPRESS FAMILY, 202
CYPRESSVINE, 558
CYPRESSWEED, 353

D

Dactylis, 1259
glomerata, 1259
Dactyloctenium, 1259
aegyptium, 1259
DAFFODIL, 1207
DAGGER FERN, 185
DAGWOOD, 561
DAISY, 348
FLEABANE, 348
CAMPHOR, 372
ARIZONA LAZY, 312
ARKANSAS LAZY, 314
BERLANDIER'S, 325
BLACK-FOOT, 388
BLUE, 322
BUTTER, 426
CHOCOLATE, 325
COMMON LEAST, 334
COWPEN, 426
CUT-LEAF, 348
DOG, 382
DOLL'S, 328
DOZE, 312
EASTER, 424
ENGELMANN'S, 348
FOUR-NERVE, 420
HAIRY LAZY, 312
HUISACHE, 308
LARGE-FLOWER DOLLS', 328
LAZY, 312
LEAST, 334
LINDHEIMER'S, 386
ROCK, 394
MICHAELMAS, 370
MOON, 382
ORANGE, 430
OX-EYE, 382
PLAINS LAZY, 312
YELLOW, 420
RIDDELL'S LAZY, 314
ROCK, 388, 394
SAW-LEAF, 362
SCRATCH, 344
SLEEPY, 430
SQUARE-BUD, 420
TEXAS SLEEPY, 430
WESTERN, 322
WHITE, 382
YELLOW, 402
DAISY FAMILY, 287
DAKOTA VERVAIN, 1050
Dalea, 644
aurea, ■/86, 646
candida, 648
var. candida, 648
var. oligophylla, 648
compacta var. pubescens, 648
drummondiana, 650
emarginata, 651
enneandra, 648
var. *pumila,* 648
frutescens, 648
var. *laxa,* 648
hallii, 648
helleri, 648
lanata, 648
lasiathera, ■/86, 650
laxiflora, 648
multiflora, 650
nana, 650
var. carnescens, 650
var. *elatior,* 650
var. nana, 650
phleoides
var. microphylla, 650
var. phleoides, 650
pogonathera, 651
purpurea, 650
reverchonii, 651
stanfieldii, 651
tenuis, 651
villosa var. grisea, 651
DALEA
BEARDED, 651
BIG-TOP, 648
BLACK, 648
DWARF, 650
GOLDEN, 646
HALL'S, 648
PURPLE, 650
ROUND-HEAD, 650
SILK-TOP, 646
WOOLLY, 648
DALLAS
FERN, 184
JEWEL FERN™, 184
DALLIS GRASS, 1305
DALMATIAN INSECT-FLOWER, 382
DANDELION, 418
CAROLINA
DWARF-, 380
FALSE, 399
COMMON, 418
DWARF-, 378
FALSE, 399
MANY-STEM FALSE, 399
NATIVE-, 399
PINK-, 394
POTATO-, 380
PURPLE-, 386
RED-SEED, 418
TEXAS, 399
TUBER
DWARF-, 380
FALSE, 399
WEEDY DWARF-, 380
WESTERN DWARF-, 380
Danthonia, 1259
spicata, 1260
DANTHONIA, POVERTY, 1260
DANUBE GRASS, 1311

- DARNEL, 1282
  - RYE GRASS, 1282
    - BEARDLESS, 1282
  - POISON, 1282
- *Darwinia exaltata*, 696
- DASHEEN, 1093
- **Dasistoma,** 999
  - macrophylla, ▣/86, 999
- *Dasylirion*, 1080
- DATE
  - PALM, 1094
  - CHINESE-, 936
  - -PLUMS, 578
- **Datura,** 1018
  - *meteloides*, 1020
  - stramonium, 1020
  - wrightii, ▣/87, 1020
- DATURA, SACRED, 1020
- *Daubentonia*, 696
  - *drummondii*, 696
- **Daucosma,** 249
  - laciniatum, 249
- DAUCOSMA, MEADOW, 249
- **Daucus,** 250
  - carota, 250
    - subsp. *sativus*, 250
  - pusillus, 250
- DAWN REDWOOD, 202
- DAYFLOWER, 1098
  - COMMON, 1100
  - CREEPING, 1100
  - ERECT, 1100
  - FALSE, 1100, 1101
  - NARROW-LEAF, 1100
  - SPREADING, 1100
  - VIRGINIA, 1100
- DAY-LILY, 1201
  - ORANGE, 1202
  - TAWNY, 1202
- DAY-PRIMROSE, 854
  - SQUARE-BUD, 854
- DEAD-NETTLE, 759
  - PURPLE, 759
  - RED, 759
- DEATH-CAMASS, 1210
  - NUTTALL'S, 1210
- DECIDUOUS HOLLY, 270
- DEER
  - PEA VETCH, 707
  - -GRASS, 820
  - -TONGUE, 1296
    - ROSETTE GRASS, 1296
  - -VETCH, 671
    - BIRD-FOOT, 671
    - PURSH'S, 671
- DEER FERN FAMILY, 180
- **Delphinium,** 922
  - *ajacis*, 922
  - *ambiguum*, 922
  - carolinianum, 922
    - subsp. carolinianum, 924
    - subsp. *penardii*, 924
    - subsp. vimineum, 924
    - subsp. virescens, ▣/87, 924
  - *orientale*, 922
  - *vimineum*, 924
  - *virescens*, 924
    - var. *macroceratilis*, 924
    - var. *penardii*, 924
- DELTA ARROWHEAD, 1090
- *Dendranthema*, 382
- *Dendrobium*, 1213
- **Dennstaedtiaceae,** 181
- DENSE-FLOWER BLADDERPOD, 473
- DEPTFORD PINK, 518
- **Descurainia,** 465
  - incana subsp. viscosa, 465
  - pinnata, 465
    - subsp. brachycarpa, 465
    - var. *brachycarpa*, 466
    - subsp. halictorum, 466
    - var. *osmiarum*, 466
    - subsp. pinnata, 466
  - *richardsonii* subsp. *viscosa*, 465
  - sophia, 466
- DESERT
  - -CATALPA, 443
  - CHRISTMAS CACTUS, 491
  - HORSE-PURSLANE, 220
  - -RUE, TEXAS, 973
  - SALT GRASS, 1264
  - SUMAC, 234
  - -THORN, 1020
  - -WILLOW, 443
- **Desmanthus,** 651
  - acuminatus, 652
  - *depressus*, 652
  - illinoensis, 652
  - leptolobus, 652
  - reticulatus, 652
  - velutinus, 652
  - virgatus, 652
    - var. *acuminatus*, 652
    - var. *depressus*, 652
- DESMANTHUS, ILLINOIS, 652
- **Desmazeria,** 1260
  - rigida, 1260
- **Desmodium,** 652
  - canescens, 655
  - ciliare, 655
  - *dichromum*, 656
  - glutinosum, 655
  - marilandicum, 655
  - nuttallii, 655
  - obtusum, 655
  - paniculatum, 656
  - pauciflorum, 656
  - psilophyllum, 656
  - *rigidum*, 656
  - sessilifolium, 656
  - tweedyi, 656
  - viridiflorum, 656
  - *wrightii*, 656
- DEVIL
  - CHOLLA, 492
  - GREENBRIER, 1344, 1346
- DEVIL'S
  - BEGGAR-TICKS, 326
  - -BOUQUET, 842
  - -CLAW, 625, 880
    - COMMON, 880
  - -ELBOW, 848
  - -FIG, 874
  - -GUT, 572
  - -HEAD CACTUS, 486
  - -PINCUSHION, 486
  - -SHOESTRING, 700, 1080
  - -WALKINGSTICK, 270
  - WOLF-APPLE, 1030
- DEVILWEED
  - MEXICAN, 335
  - -ASTER, 335
- DEWBERRY, 958
  - NORTHERN, 959
  - SOUTHERN, 959
- DEWDROP, PURPLE-, 498
- DIAMOND-LEAF FROGFRUIT, 1054
- **Dianthus,** 511, 512, 518
  - ×*allwoodii*, 518
  - armeria, 518
  - barbatus, 518
  - *caryophyllus*, 518
- **Diarrhena,** 1260
  - americana, 1260
    - var. *obovata*, 1262
  - obovata, 1262
- *Dicentra*, 719
- *Dichanthelium*, 1291
  - *aciculare*, 1295
  - *acuminatum*
    - var. *fasciculatum*, 1295
    - var. *implicatum*, 1295
    - var. *lindheimeri*, 1295
    - var. *villosum*, 1295
  - *angustifolium*, 1295
  - *boscii*, 1296
  - *clandestinum*, 1296
  - *commutatum*, 1298

Dichanthelium (cont.)
depauperatum, 1296
dichotomum, 1298
lanuginosum, 1295
var. *lindheimeri*, 1295
var. *villosissimum*, 1295
laxiflorum, 1300
lindheimeri, 1295
linearifolium, 1300
malacophyllum, 1300
oligosanthes, 1300
var. *scribnerianum*, 1300
pedicellatum, 1302
ravenelii, 1302
scoparium, 1302
sphaerocarpon, 1302
villosissimum, 1295
Dichanthium, 1262
annulatum, 1262
Dichondra, 552
carolinensis, 552
recurvata, 552
repens var. *carolinensis*, 552
Dichromena
colorata, 1158
nivea, 1160
reverchonii, 1160
Dicliptera, 212
brachiata, ▣/87, 212
Dicotyledonae, 210
Dictamus albus, 970
Dieffenbachia, 1092
DIFFUSE ERYNGO, 252
Digitalis, 988
purpurea, 989
Digitaria, 1262
adscendens, 1263
californica, 1263
ciliaris, 1263
cognata, 1263, 1296
subsp. cognata, 1263
subsp. pubiflora, 1263
filiformis, 1263
var. *villosa*, 1264
ischaemum, 1264
sanguinalis, 1264
var. *ciliaris*, 1263
villosa, 1264
DILL, 243
WILD, 256
DILLEN'S OXALIS, 870
Diodia, 962
teres, 962
tricocca, 970
virginiana, 962
Dionaea muscipula, 576
Dioscorea, 1164
bulbifera, 1165
quaternata, 1165
villosa, 1165
var. *glabrifolia*, 1165
Dioscoreaceae, 1164
Diospyros, 578
ebenum, 578
kaki, 578
texana, 579
virginiana, 579
Diplotaxis, 466
muralis, 466
Dipsacaceae, 574
Dipteronia, 218
DISC WATER-HYSSOP, 994
Discula, 561
DISPERSAL, FRUIT, 232, 236, 262, 280, 432, 576, 786, 844, 905, 958, 999, 1031, 1066, 1076, 1186, 1201
DISTAFF-THISTLE, 330
Distichlis, 1264
spicata, 1264
var. *stricta*, 1264
Ditaxis, 600
aphoroides, 602
humilis
var. humilis, 602
var. laevis, 602
var. leiosperma, 602
mercurialina, 604
DITAXIS
LOW, 602
SHRUBBY, 602
SMOOTH, 602
TALL, 604
DITCH
-MOSS, 1168
-STONECROP, 562
DITTANY, 756
AMERICAN, 756
DOCK, 905
CURLY, 906
-LEAF, 906
ENGELMANN'S, 906
FIDDLE, 906
PALE, 906
PEACH-LEAF, 906
PURPLE WEN-, 482
SMOOTH, 906
SPATTER-, 844
SWAMP, 906
TANNER'S, 906
DODDER, 572
BUTTONBUSH, 573
CLUSTER, 573
COMMON, 574
COMPACT, 573
CUSP, 573
CUSPIDATE, 573
FIELD, 574
FIVE-ANGLED, 574
GLOMERATE, 573
GRONOVIUS', 574
HAZEL, 573
LARGE-SEED, 574
LONG-SEPAL, 574
PRETTY, 574
RED, 574
SHOWY, 574
TREE, 573
DODDER FAMILY, 572
Dodecatheon, 912
meadia, ▣/87, 913
subsp. *brachycarpum*, 913
subsp. *meadia*, 913
Doellingeria, 315
DOG
CHOLLA, 492
DAISY, 382
-FENNEL, 310
-MUSTARD, 468
-TOOTH-VIOLET, 1200
WHITE, 1201
DOGBANE, 266
CLIMBING-, 268
HAIRY, 266
HEMP, 266
PRAIRIE, 266
SMOOTH, 266
WILLOW, 266
DOGBANE FAMILY, 264
DOGFENNEL, 353
EUPATORIUM, 353
DOGSHADE
ARKANSAS, 255
FINGER, 249
DOGTOOTH NOSEBURN, 617
DOGTOWN GRASS, 1243
DOGWEED, 346, 422
COMMON, 422
MARIGOLD, 346
DOGWOOD, 560
ANTHRACNOSE, 561
EASTERN, 561
FLOWERING, 561
POISON-, 238
ROUGH-LEAF, 561
DOGWOOD FAMILY, 560
DOLLAR WEED, 815
DOLL'S DAISY, 328
LARGE-FLOWER, 328

DON QUIXOTE'S-LANCE, 1084
DOORYARD PLANTAIN, 886
DOTTED
BLUE-EYED-GRASS, 1178
MONARDA, 766
SMARTWEED, 904
WOLFFIA, 1190
DOUBLE-COCONUT, 1094
DOUGLAS FIR, 204
DOVEWEED, 600
DOWNY
BLUE VIOLET, 1064
BROME, 1253
CHESS, 1253
EVENING-PRIMROSE, 864
GAURA, 856
GROUND-CHERRY, 1026
LOBELIA, 498
MILK-PEA, 660
OAKLEECH, 994
PHLOX, 892
SERVICE-BERRY, 940
SUNFLOWER, 370
VIBURNUM, 511
DOWNYHAW, 942
DOZE DAISY, 312
Draba, 466
brachycarpa, 466
cuneifolia, 468
platycarpa, 468
reptans, 468
var. *micrantha,* 468
DRABA
BROAD-POD, 468
CAROLINA, 468
SHORT-POD, 466
WEDGE-LEAF, 468
Dracaena, 1078
Dracaenaceae, 1078
Dracopis, 344
amplexicaulis, ▣/87, 346
DRAGON'S-HEAD
BEAUTIFUL FALSE, 770
FALSE, 679, 768
DRAGONROOT, 1092
DROOPING
MELONETTE, 570
MILK-VETCH, 635
DROPSEED, 1326
COVERED-SPIKE, 1327
GIANT, 1327
MEADOW, 1327
MISSISSIPPI, 1327
OZARK, 1327
POVERTY, 1328
PUFF-SHEATH, 1327
PURPLE-FLOWER, 1327
ROCK-, 1290
SAND, 1327
SILVEUS', 1328
TALL, 1327
WHORLED, 1328
DROPWORT, WATER, 256
Drosera, 576
annua, 578
brevifolia, 578
leucantha, 578
Droseraceae, 576
Drosophyllum, 576
DRUG
FUMITORY, 722
SWEETFLAG, 1077
DRUMMOND'S
FALSE PENNYROYAL, 758
HEDEOMA, 758
LEAF-FLOWER, 611
NAILWORT, 520
ONION, 1197
OXALIS, 870
PHLOX, 892
PINWEED, 544
POST OAK, 716
RUELLIA, 216
RUSH-PEA, 664
SABADILLA, 1210
SANDWORT, 519
SESBANIA, 696
SKULLCAP, 778
SNAKE-COTTON, 227
ST. JOHN'S-WORT, 547
SUNDROPS, 854
WAX-MALLOW, 815
Dryopteridaceae, 182
Dryopteris
normalis, 200
var. *lindheimeri,* 201
Duchesnea, 944
indica, 944
DUCK
-GRASS, 1166
OAK, 718
-POTATO, 1088
DUCKMEAT, 1190
COMMON, 1190
DUCKWEED, 1187
GREATER, 1190
LEAST, 1188
PALE, 1188
VALDIVIANA, 1188
DUCKWEED FAMILY, 1186
DUDLEY'S RUSH, 1182
DULCE, ACHICORIA, 416
Dulichium arundinaceum, 1106
DULL-SEED CORNBIND, 902
DUMBCANE, 1092
DURAND'S WHITE OAK, 718
DURAZNO, 950
DURBAN CROWFOOT GRASS, 1259
DUTCH
CLOVER, 706
ELM DISEASE, 1039
IRIS, 1174
DUTCHMAN'S
-BREECHES, 719, 973
-PIPE
NASH, 273
TEXAS, 273
VIRGINIA, 273
WOOLLY, 274
DWARF
DALEA, 650
-DANDELION, 378
CAROLINA, 380
TUBER, 380
WEEDY, 380
WESTERN, 380
ELM, 1040
FLAMEFLOWER, 911
MALLOW, 814
MEADOW GRASS, 1312
-MORNING-GLORY, 552
MOUSE-EAR CHICKWEED, 516
PALM, 1094
PALMETTO, 1094
PLANTAIN, 886
POST OAK, 716
SENNA, 694
-SNAPDRAGON, 998
SNAPDRAGON, 999
SPIKE-RUSH, 1148
ST. JOHN'S-WORT, 547
SUMAC, 232
WALNUT, 748
DYE BEDSTRAW, 966
Dyschoriste, 212
linearis, ▣/88, 213
DYSCHORISTE, NARROW-LEAF, 213
Dysodiopsis, 346
tagetoides, 346
Dyssodia, 346
pentachaeta, 422
tagetoides, 346
tenuiloba, 424
DYSSODIA, BRISTLE-LEAF, 422

E

EARDROP-VINE, 898
EARED FALSE-FOXGLOVE, 992

EARFLOWER, 498
EAR-LEAF
AMMANNIA, 800
BLADDERPOD, 473
GERARDIA, 992
EARLY
BUTTERCUP, 928
FERN-LEAF LOUSEWORT, 1006
LOUSEWORT, 1006
SCORPION-GRASS, 453
WOODS BUTTERCUP, 926
EARTHNUT, 628
EARTH-SMOKE, 722
EASTER DAISY, 424
EASTERN
BACCHARIS, 324
CAMASS, 1200
CORAL-BEAN, 658
COTTONWOOD, 975
DOGWOOD, 561
GAMMA GRASS, 1332
HOP-HORNBEAM, 440
MANNA GRASS, 1277
PERSIMMON, 579
POISON-OAK, 236
PRICKLY-PEAR, 491
REDBUD, 640
RED-CEDAR, 203
WAHOO, 528
WHORLED MILKWEED, 280
EATON'S LIP FERN, 196
Ebenaceae, 578
EBONY, 578
SPLEENWORT, 179
EBONY FAMILY, 578
Echeveria, 561
Echinacea, 346
angustifolia, 347
atrorubens, 347
pallida, 347
var. *angustifolia*, 347
var. *sanguinea*, 347
var. *strigosa*, 347
purpurea, 347
sanguinea, 347
ECHINACEA, PALE, 347
Echinocactus, 486
setispinus, 494
texensis, ▣/88, 486
Echinocereus, 486
coccineus var. paucispinus, ▣/88, 486
reichenbachii, ▣/88, 486
triglochidiatus var. *paucispinus*, 486
Echinochloa, 1264
colona, 1266
crus-galli, 1266
crus-pavonis var. macera, 1266
muricata, 1266
var. microstachya, 1266
var. muricata, 1266
walteri, 1266
Echinodorus, 1087
berteroi, 1087
var. *lanceolatus*, 1087
cordifolius subsp. fluitans, 1087
fluitans, 1087
rostratus, 1087
Echium, 446
vulgare, 448
Eclipta, 347
alba, 347
prostrata, 347
EDDO, 1093
EDIBLE SCURF-PEA, 684
EDWARDS PLATEAU CARIC SEDGE, 1120
EEL-GRASS, 1170
Egeria, 1168
densa, 1168
EGG-LEAF SKULLCAP, 779
EGGPLANT, 1015
EGGS-AND-BACON, 671
EGLANTINE, 956
EGYPTIAN
-BEAN, 834
CROWFOOT GRASS, 1259
PASPALIDIUM, 1304
Ehretia anacua, 443
Eichhornia, 1339
crassipes, 1339
Elaeagnaceae, 579
Elaeagnus, 579
angustifolia, 580
pungens, 579
ELAEAGNUS, THORNY, 579
Elaeis guineensis, 1094
Elatinaceae, 580
Elatine, 580
americana, 581
brachysperma, 580
triandra, 581
var. *americana*, 581
var. *brachysperma*, 580
ELBOW-BUSH, 848
ELDER
BOX-, 219
MARSH-, 376, 378
POISON-, 238
TEXAS BOX, 220
ELDERBERRY, 510
AMERICAN, 510
COMMON, 510
Eleocharis, 1144
acicularis, 1146
acutisquamata, 1146
arenicola, 1147
baldwinii, 1146
capillacea, 1146
caribaea, 1147
cellulosa, 1146
compressa, 1146
dulcis, 1144
engelmannii, 1146
erythropoda, 1147
flaccida, 1147
flavescens, 1146
geniculata, 1147
lanceolata, 1147
macrostachya, 1147
microcarpa, 1147
montevidensis, 1147
obtusa, 1147
var. *detonsa*, 1146
var. *lanceolata*, 1147
ovata var. *engelmannii*, 1146
palustris, 1147
parvula, 1148
var. *anachaeta*, 1148
quadrangulata, 1148
var. *crassior*, 1148
reverchonii, 1146
xyridiformis, 1147
Elephantopus, 347
carolinianus, 348
ELEPHANTOPUS, LEAFY, 348
ELEPHANT'S
-EAR, 1093, 1094
-FOOT, 347
Eleusine, 1268
coracana, 1268
indica, 1268
ELLIOTT'S
BENT GRASS, 1236, 1239
BLUESTEM, 1239
Ellisia, 740
nyctelea, 740
ELM, 1038
AMERICAN, 1039
ASIATIC, 1040
CEDAR, 1040
CHINESE, 1040
CORK, 1039
DUTCH, DISEASE, 1039
DWARF, 1040
LACE-BARK, 1040
-LEAF GOLDENROD, 415
RED, 1040
SIBERIAN, 1040
SLIPPERY, 1040

ELM *(cont.)*
WAHOO, 1039
WHITE, 1039
WINGED, 1039
ELM FAMILY, 1036
Elodea, 1168
canadensis, 1168, 1169
densa, 1169
Elymus, 1268
canadensis, 1268
var. *brachystachys*, 1268
var. *villosus*, 1268
smithii, 1304
triticoides, 1304
virginicus, 1268
var. *glabriflorus*, 1268
Elytrigia
dasytachya, 1304
smithii, 1304
EMERSON DAVIS' CARIC SEDGE, 1120
Emerus herbacea, 697
EMETIC HOLLY, 270
EMPEROR'S-CANDLESTICKS, 694
ENCINILLA, 599
ENCINO, 719
Endothia parasitica, 710
ENGELMANN'S
ADDER'S-TONGUE, 190
BLADDERPOD, 473
DAISY, 348
DOCK, 906
MILK-VETCH, 632
MILKWEED, 278
PRICKLY-PEAR, 491, 492
SAGE, 776
SPIKE-RUSH, 1146
Engelmannia, 348
peristenia, 348
pinnatifida, 348
ENGLISH
-IVY, 272
PLANTAIN, 886
WALNUT, 744
YEW, 201
Enteropogon chlorideus, 1258
ENTIRE-LEAF PAINTBRUSH, 996
EOLA-WEED, 547
EPAZOTE, 535
Ephedra, 207
antisyphilitica, 208
pedunculata, 208
Ephedraceae, 207
Epidendrum, 1213
Epilobium, 852, 862
Epipactis, 1215
gigantea, 1215
EPIPACTIS, STREAM, 1215
Equisetaceae, 176
Equisetophyta, 176
Equisetum, 176
×*ferrissii*, 178
fluviatile, 176
giganteum, 176
hyemale
subsp. affine, 176
var. *affine*, 178
kansanum, 178
laevigatum, 178
prealtum, 178
Eragrostis, 1268
arida, 1272
barrelieri, 1270
beyrichii, 1272
cilianensis, 1270
curtipedicellata, 1270
curvula, 1270
diffusa, 1272
hirsuta, 1270
hypnoides, 1270
intermedia, 1270
lugens, 1270
oxylepis, 1272
var. *beyrichii*, 1272
pectinacea, 1272
var. miserrima, 1272
var. pectinacea, 1272
pilifera, 1274
pilosa, 1272
reptans, 1272
secundiflora, 1272
subsp. oxylepis, 1272
subsp. *secundiflora*, 1272
sessilispica, 1272
spectabilis, 1272
superba, 1272
tephrosanthos, 1272
trichodes, 1274
var. *pilifera*, 1274
Erechtites, 348
hieraciifolia, 348
var. *intermedia*, 348
ERECT
BOUCHETIA, 1016
BURHEAD, 1087
DAYFLOWER, 1100
SPIDERLING, 838
ERGOT, 1276, 1304, 1318
Erianthus
contortus, 1315
giganteus, 1315
Erica, 581
Ericaceae, 581
Erigeron, 348
annuus, 350
canadensis, 341
var. *glabratus*, 341
divaricatus, 341
geiseri, 350
modestus, 350
philadelphicus, 352
strigosus
var. beyrichii, 352
var. strigosus, 352
tenuis, 352
Eriocaulaceae, 1165
Eriocaulon, 1165
aquaticum, 1166
decangulare, 1166
kornickianum, 1166
septangulare, 1166
texense, 1166
Eriochloa, 1274
contracta, 1274
sericea, 1274
Eriogonum, 898
annuum, 899
longifolium, 899
var. *lindheimeri*, 899
var. *plantagineum*, 899
multiflorum, 899
ovalifolium, 898
Erioneuron, 1274
pilosum, 1274
ERIONEURON, HAIRY, 1274
Erodium, 730
cicutarium, 730
texanum, ■/88, 730
Erucastrum, 468
gallicum, 468
Eryngium, 250
diffusum, 252
hookeri, 252
integrifolium, 252
leavenworthii, ■/89, 252
prostratum, 252
synchaetum, 252
yuccifolium, ■/89, 252
var. *synchaetum*, 252
ERYNGO, 250
BRISTLE-LEAF, 252
BUSHY, 252
DIFFUSE, 252
HOOKER'S, 252
LEAVENWORTH'S, 252
SIMPLE-LEAF, 252
Erysimum, 468
asperum, ■/89, 468
capitatum, 468

Erysimum *(cont.)*
repandum, 468
ERYSIMUM
PLAINS, 468
SPREADING, 468
Erythraea
beyrichii, 724
calycosa, 724
floribunda, 726
pulchella, 726
texensis, 726
Erythrina, 656
herbacea, ▣/89, 658
Erythronium, 1200
albidum, 1201
var. *coloratum,* 1201
var. *mesochoreum,* 1201
mesochoreum, ▣/89, 1201
ESCARPMENT
BLACKCHERRY, 950
LIVE OAK, 716
Eschscholzia, 874
californica
subsp. californica, 874
subsp. mexicana, 874
ESCOBA DE LA VIBORA, 365
Escobaria, 486
missouriensis, 488
var. *caespitosa*, 488
var. robustior, 488
var. similis, ▣/89, 488
vivipara var. radiosa, ▣/90, 488
ESPANTA VAQUEROS, 230
ESPUELA DE CABALLERO, 922
ESTROPAJO, 570
ETERNITY GRASS, 1306
Eucephalus, 315
EULALIA, 1284
Eupatorium, 352
ageratifolium var. *texense,* 354
altissimum, 353
capillifolium, 353
coelestinum, 353
havanense, 353
incarnatum, 354
perfoliatum, 354
rugosum, 354
serotinum, 354
texense, 354
wrightii, 354
EUPATORIUM
DOGFENNEL, 353
LATE, 354
PINK, 354
TALL, 353
Euphorbia, 604
angusta, 594
antisyphilitica, 604
bicolor, 606
chamaesyce, 596
cordifolia, 594
corollata, 606
cyathophora, 606
davidii, 606
dentata, 608
fendleri, 594
geyeri, 594
glyptosperma, 594
hexagona, 608
hirta, 594
hypericifolia, 594
lata, 594
longicruris, 608
maculata, 596
marginata, ▣/90, 608
missurica, 596
nutans, 596
prostrata, 596
pulcherrima, 585, 604
roemeriana, 608
serpens, 596
spathulata, 608
stictospora, 596
supina, 596
tetrapora, 610
villifera, 596
EUPHORBIA
BLACK-FOOT, 592
GEYER'S, 594
HAIRY, 596
HEART-LEAF, 594
HOARY, 594
MAT, 596
PAINTED, 606
PILL-POD, 594
PROSTRATE, 596
RIDGE-SEED, 594
ROEMER'S, 608
SIX-ANGLE, 608
SLIM-SEED, 596
SPOTTED, 594
TROPICAL, 594
WARTY, 608
WEAK, 610
WEDGE-LEAF, 608
WHITE-MARGIN, 592
Euphorbiaceae, 584
EURASIAN WATER-MILFOIL, 734
EUROPEAN
BARBERRY, 1334
SMOKETREE, 231
YEW, 201
Eurybia, 315
Eurytaenia, 254
texana, 254
Eustachys, 1276
retusa, 1276
Eustoma, 727
exaltatum, 727
grandiflorum, 727
russellianum, ▣/90, 727
Eustylis purpurea, 1173
Euthamia, 354, 408
camporum, 356
gymnospermoides, 354
pulverulenta, 356
EUTHAMIA, VISCID, 354
Evax, 356
candida, 356
multicaulis, 356
prolifera, 356
verna, 356
EVAX
BIG-HEAD, 356
MANY-STEM, 356
SILVER, 356
EVENING
TRUMPET-FLOWER, 798
EVENING-PRIMROSE, 852, 862
BEACH, 854
BERLANDIER'S, 854
COMMON, 863
CUT-LEAF, 864
DOWNY, 864
FLOATING, 860
FOUR-POINT, 866
GRAND PRAIRIE, 854
MEXICAN, 866
SHORT-POD, 866
SHOWY, 866
SINUATE-LEAF, 864
SPACH'S, 866
STEMLESS, 866
TRUMPET, 864
VARIABLE, 864
WHITE, 866
WRIGHT'S, 866
YELLOW, 854
EVENING-PRIMROSE FAMILY, 852
EVENINGSTOCK, 474
EVERGREEN
CASSENA, 270
FORESTIERA, 848
HOLLY, 270
HONEYSUCKLE, 510
SUMAC, 234
EVERLASTING, 310, 359, 398
FRAGRANT, 398

EVERLASTING *(cont.)*
-PEA, 666
EVE'S-NECKLACE, 697
Evolvulus, 552
nuttallianus, 552
pilosus, 554
sericeus, 554
EVOLVULUS
HAIRY, 552
NUTTALL'S, 552
SHAGGY, 552
SILKY, 554
Evonymus, 528
americana, 528
atropurpurea, 528
var. atropurpurea, 528
var. cheatumii, 528
Exacum, 723
EYEBANE, 596
Eysenhardtia, 658
angustifolia, 658
texana, 658

F

Faba vulgaris, 617
Fabaceae, 617
Facelis, 356
retusa, 356
Fagaceae, 710
Fagopyrum esculentum, 897
Fagus, 710
FAIRY-SWORDS, 198
FALL
BONESET, 354
PANICUM, 1296
WITCH GRASS, 1263
FALLPOISON, 354
FALSE
ACACIA, 692
ALOE, 1079, 1080
WIDE-LEAF, 1079
BONESET, 328, 329
BROOM, 1250
PURPLE, 1250
BUCKWHEAT, 904
CLOAK FERN, 194
DANDELION, 399
CAROLINA, 399
MANY-STEM, 399
TUBER, 399
DAYFLOWER, 1100, 1101
DRAGON'S-HEAD, 768, 769
BEAUTIFUL, 770
FLAX, 460
SMALL SEED, 460
FOXGLOVE, 994, 1007
EARED, 992
GARLIC, 1208
YELLOW, 1208
GAURA, 866
GROMWELL, 453
HONEYSUCKLE, 218
INDIGO, 626, 635
JALAP, 840
JESSAMINE, 1022
LOOSESTRIFE, 858
MALLOW, 814, 818
RED, 820
THREE-LOBE, 814
WRIGHT'S, 814
MINT, 212
NETTLE, 1042
SMALL-SPIKE, 1042
NIGHTSHADE, 1018
GREEN, 1018
HAIRY, 1018
PLATEAU, 1018
NUT-GRASS, 1144
PENNYROYAL, 756, 783
DRUMMOND'S, 758
REVERCHON'S, 758
ROUGH, 758
PIMPERNEL, 912, 1002
CLASPING, 1002
RAGWEED, 393
SAFFRON, 330
SPOTTED ST. JOHN'S-WORT, 548
FAN-LEAF VERVAIN, 1058
FANWEED, 480
FANWORT, 482
CAROLINA, 482
FARKLE-BERRY, 584
FAROLITOS, 980
Fatoua, 828
villosa, 828
FAWN-LILY, 1200
WHITE, 1201
FEATHER
BLUESTEM, 1239
FINGER GRASS, 1258
-GERANIUM, 536
FEATHERTOP, 1308
FEE'S LIP FERN, 196
FEMALE FERN, 200
FENDLER'S
PENSTEMON, 1007
BLADDERPOD, 473
FENNEL, 254
COMMON, 254
DOG-, 310
HOG-, 256
-LEAF PONDWEED, 1343
FERN, 173, 178
ACACIA, 624
ALABAMA LIP, 196
BLUE, 198
BOSTON, 184
BRACKEN, 181
BUCKHORN, 192
BULB LIP, 195
CHAIN, 180, 181
CHRISTMAS, 185
CINNAMON, 192
CLIFF, 185
COMMON GRAPE, 189
DAGGER, 185
EATON'S LIP, 196
FALSE CLOAK, 194
FEE'S LIP, 196
FEMALE, 200
FLOWERING, 192
GRAPE, 188
HAIRY LIP, 198
HOLLY, 185
LADY, 184
LINDHEIMER'S
LIP, 198
MAIDEN, 200
LIP, 195
LONG CLOAK, 195
LOWLAND LADY, 184
MAIDENHAIR, 194
MOSQUITO, 180
NARROW-LEAF CHAIN, 181
POWDERY CLOAK, 194
PROSTRATE GRAPE, 189
RATTLESNAKE, 189
RESURRECTION, 193
ROUGH LIP, 196
ROYAL, 192
SENSITIVE, 184
SHIELD-SORUS, 193
SLENDER LIP, 196
SMOOTH LIP, 196
SOUTHERN
GRAPE, 189
LADY, 184
SHIELD, 200
SPARSE-LOBED GRAPE, 189
STAR-SCALED CLOAK, 195
SWORD, 184, 185
VENUS'-HAIR, 194
VIRGINIA
CHAIN, 181
GRAPE, 189
WATER, 180
WAVY CLOAK, 195
WESTERN BRACKEN, 181

FERN *(cont.)*
- WIDESPREAD MAIDEN, 200
- WILD BOSTON, 184
- WINTER GRAPE, 189
- WOOLLY LIP, 196, 198

Ferocactus setispinus, 494

FESCUE, 1276
- ALTA, 1276
- ANNUAL, 1335
- GRASS, SIXWEEKS, 1335
- -LIKE CARIC SEDGE, 1120
- NODDING, 1276
- SIXWEEKS, 1336
- TALL, 1276

Festuca, 1276, 1282
- arundinacea, 1276
- *bromoides,* 1336
- *elatior* var. *arundinacea,* 1276
- *myuros,* 1336
- *obtusa,* 1276
- *octoflora,* 1336
- *sciurea,* 1336
- subverticillata, 1276
- *unioloides,* 1252

FETTERBUSH, 581

FEVERBUSH, 784

FEVERFEW, 393

FEVERWEED, 426

FEVERWORT, 511

FEW-FLOWER
- ARISTIDA, 1242
- CARIC SEDGE, 1122
- NUT-RUSH, 1164
- TICK-CLOVER, 656

Ficus, 827, 830
- carica, 830
- *elastica,* 830
- pumila, 830
- *religiosa,* 830
- *repens,* 830

FIDDLE
- DOCK, 906
- -LEAF
  - GREENBRIER, 1346
  - TOBACCO, 1022

FIDDLELEAF, 740
- NAMA, 740

FIDDLENECK, 445
- SMALL-FLOWER, 445

FIELD
- -BEAN, 706
- BINDWEED, 551
- DODDER, 574
- GRASS, 1335
- GROUND-CHERRY, 1026
- -MADDER, 970
- MUSTARD, 479
- PANSY, 1062
- PASPALUM, 1306
- PENNYCRESS, 480
- POPPY, 876
- RAGWEED, 310
- SNAKE-COTTON, 227

FIG, 830
- CLIMBING, 830
- COMMON, 830
- CREEPING, 830
- DEVIL'S-, 874
- INDIAN-, 491
- TREE, 830

FIG FAMILY, 827

FIG-MARIGOLD FAMILY, 220

FIGWORT, 1008
- MARYLAND, 1010

FIGWORT FAMILY, 988

Filago, 356
- *candida,* 356
- *prolifera,* 356
- *verna,* 356

FILAGRANA DE PIÑAR, 1049

FILAREE, 730
- CALIFORNIA, 730
- TEXAS, 730

FILIGRANA DE MAJORCA, 1049

FILLY PANICUM, 1298

Fimbristylis, 1148
- annua, 1148
- autumnalis, 1148
- *baldwiniana,* 1148
- *drummondii,* 1150
- *interior,* 1150
- miliacea, 1150
- puberula, 1150
  - var. interior, 1150
  - var. puberula, 1150
- vahlii, 1150

FIMBRISTYLIS, VAHL'S, 1150

FINE-LEAF
- BLUETS, 966
- GERARDIA, 993

FINE-LINE CARIC SEDGE, 1130

FINGER
- CACTUS, 485
- DOGSHADE, 249
- FLAT SEDGE, 1144
- GRASS, 1262
  - FEATHER, 1258
  - HOODED, 1256
  - SLENDER, 1263
  - WINDMILL, 1256
- LION'S-HEART, 769
- -MILLET, 1268
- ORCHID, 1215, 1281
  - YELLOW, 1218
- POPPY-MALLOW, 808

FIR, 204
- DOUGLAS, 204
- JOINT-, 207, 208
- VINE JOINT-, 208

FIREBUSH MEXICAN, 540

FIRECRACKER-PLANT, 738

FIRE
- -ON-THE-MOUNTAIN, 606
- -THORN, 952
- -WHEELS, 358

FIREWEED, 348, 852
- MEXICAN, 540

Firmiana, 1032
- *platanifolia,* 1032
- simplex, 1032

FISHHOOK CACTUS, 489

FISHPOLE BAMBOO, 1311

FITWEED, 719

FIVE-ANGLED DODDER, 574

FIVE-FINGER, 946

FLACCID-FRUIT CARIC SEDGE, 1122

Flacourtiaceae, 1060

FLAG, 1173
- BLUE, 1174
- YELLOW, 1174

FLAME
- -ACANTHUS, 212
- -LEAF
  - SUMAC, 232
  - PRAIRIE, 234

FLAMEFLOWER, 911
- DWARF, 911
- ORANGE, 911
- PRAIRIE, 911

FLAMETREE, 440

FLANNEL
- -BREECHES, 742
- -LEAF, 1010
- MULLEIN, 1010
- -PLANT, 1010

FLAT
- -FLOWER GERARDIA, 993
- -FRUIT CARIC SEDGE, 1120
- -LEAF RUSH, 1184
- SEDGE, 1132
  - BALDWIN, 1136
  - BEARDED, 1142
  - BENT-AWN, 1140
  - CYLINDER, 1136
  - FRAGRANT, 1138
  - GLOBE, 1136
  - JOINTED, 1136
  - ONE-FLOWER, 1142

FLAT SEDGE *(CONT.)*
POORLAND, 1136
RED-ROOT, 1136
SCHWEINITZ', 1142
SHEATHED, 1138
SHORT-LEAF, 1154
SLENDER, 1138
-LEAF, 1154
STICKY, 1136
TAPER-LEAF, 1136
TROPICAL, 1144
UMBRELLA, 1138
YELLOW, 1138
-SEED-SUNFLOWER, 426
-SPINE STICKSEED, 452
-TOPPED
-GOLDENROD, 354, 412
FLATTENED MAMMILLARIA, 489
FLATWOOD PLUM, 950
FLAX, 788
BERLANDIER'S, 792
BLUE, 792
COMMON, 792
TOAD-, 1002
CULTIVATED, 792
FALSE, 460
FLOWERING, 790
GROOVED, 792
-LEAF STENOSIPHON, 866
LEWIS', 792
MEADOW, 790
NORTON'S, 790
OLDFIELD TOAD-, 1004
PRAIRIE, 792
RIGID, 792
ROCK, 792
SMALLSEED FALSE, 460
STIFF-STEM, 792
SUCKER, 790
TEXAS, 790
TOAD-, 1006
TOAD-, 1002, 1004, 1006
TUFTED, 790
FLAX FAMILY, 787
FLEABANE, 348, 396
ANNUAL, 350
BASIN, 350
CANADA, 341
DAISY, 348
LOW, 341
MARSH-, 396
PHILADELPHIA, 352
PLAINS, 350
PRAIRIE, 352
SLENDER, 352
SPREADING, 341
STINKING-, 398
FLECHA DE AGUA, 1090
Fleischmannia, 352
incarnata, 354
FLESHY-LEAF LUPINE, 672
FLEUR-DE-LIS, 1173
FLIXWEED TANSY-MUSTARD, 466
FLOATING
EVENING-PRIMROSE, 860
-HEART, 826
YELLOW, 826
PRIMROSE-WILLOW, 862
WATER
-PENNYWORT, 254
-PRIMROSE, 862
FLOR DE MIMBRE, 443
FLORIDA
BASSWOOD, 1036
BLUEHEARTS, 996
LETTUCE, 381
LINDEN, 1036
-MOSS, 1096
PASPALUM, 1306
BIG, 1306
FLOWER
BASKET-, 332
BLANKET-, 358
CHAFF-, 221
COTTON-, 228
MAT CHAFF-, 222
-OF-AN-HOUR, 812
FLOWERING
DOGWOOD, 561
FERN, 192
FLAX, 790
SPURGE, 606
-STRAW, 386
-WILLOW, 443
FLOWERING PLANTS, 208
FLUFF GRASS, 1274
FLUTTER-MILL, 864
FLUX-WEED, 783
FLY-AWAY GRASS, 1236
Foeniculum, 254
vulgare, 254
FOETID
GOURD, 568
-MARIGOLD, 346, 422
FOGFRUIT, 1053, 1054
SPATULATE-LEAF, 1054
WEIGHTY, 1054
FOGWEED, 533
FOREST MUHLY, 1290
Forestiera, 846
acuminata, 848
pubescens, 848
var. glabrifolia, 848
var. pubescens, 848
FORESTIERA
EVERGREEN, 848
SMOOTH-LEAF, 848
FORGET-ME-NOT, 453
SOUTHERN, 453
SPRING, 453
FORGET-ME-NOT FAMILY, 443
FORKED
BLUE-CURLS, 783
CHICKWEED, 520
-LEAF WHITE OAK, 714
RUSH, 1182
SCALESEED, 262
FORK-TIP THREEAWN, 1240
Forsythia, 846
FORT THOMPSON GRASS, 1306
FOUNTAIN GRASS, 1308
FOUR
-ANGLED CARIC SEDGE, 1130
-LEAF MANYSEED, 522
-NERVE DAISY, 420
-O'CLOCK, 838, 840
COMMON, 840
GIANT, 839
LINEAR-LEAF, 840
TALL, 839
WHITE, 839
WILD, 840
-POINT EVENING-PRIMROSE, 866
-WING SALTBUSH, 533
FOUR-O'CLOCK FAMILY, 834
FOWL MANNA GRASS, 1277
FOX GRAPE, 1072
FOXBERRY, 584, 1049
FOXGLOVE, 988, 1007
COMMON, 989
FALSE, 994, 1007
MULLEIN, 999
PURPLE, 1007
WILD, 1007
FOXGLOVE FAMILY, 988
FOXTAIL, 1236, 1318
CARIC SEDGE, 1130
CAROLINA, 1236
GRASS, 1320
GIANT, 1319
GREEN, 1320
-MILLET, 1318, 1319
Fragaria, 944
×*ananassa*, 944
chiloensis, 944
virginiana subsp. grayana, 944
FRAGRANT
BEDSTRAW, 966

FRAGRANT *(cont.)*
  CUDWEED, 398
  EVERLASTING, 398
  FLAT SEDGE, 1138
  GAILLARDIA, 359
  GOLDENROD, 412
  MIMOSA, 678
  SPIDER-LILY, 1204
  SUMAC, 232
  WATER-LILY, 845
FRANGIPANI, 264
Frangula, 933
  caroliniana, 933
  purshianus, 933
Fraxinus, 848
  americana, 849
    subsp. *texensis*, 849
    var. *texensis*, 849
  pennsylvanica, 849
    var. *integerrima*, 849
  texensis, 849
FREEMONT'S MONKEY-FLOWER, 1004
FRENCH
  -MULBERRY, 1049
  TAMARISK, 1035
FRENCHWEED, 480
FRESADILLA, TOMATE, 1026
FRESNO, 849
  DE GUAJUCO, 219
FRIJOLITO, 697
FRINGED
  GREENBRIER, 1346
  NUT-RUSH, 1164
  ORCHID, 1218
    YELLOW, 1218
  SIGNAL GRASS, 1334
FRINGE-LEAF RUELLIA, 216
Froelichia, 227
  campestris, 227
  drummondii, 227
  floridana, 227
    var. *campestris,* 227
  gracilis, 227
FROGFRUIT, 1053, 1054
  COMMON, 1054
  DIAMOND-LEAF, 1054
  HOARY, 1054
  LANCE-LEAF, 1054
  NORTHERN, 1054
  SAW-TOOTH, 1054
  TEXAS, 1054
  WEDGE-LEAF, 1054
FROG'S-BIT FAMILY, 1166
FROST GRAPE, 1072
FROSTWEED, 427, 543
FRUIT DISPERSAL, 232, 236, 262, 280, 432, 576, 786, 844, 905, 958, 999, 1031, 1066, 1076, 1186, 1201
FRUITLESS MULBERRY, 832
FRUTILLA BLANCA, 1052
Fuchsia, 852
Fuirena, 1150
  breviseta, 1152
  bushii, 1152
  ciliata, 1152
  hispida, 1152
  simplex, 1152
    var. aristulata, 1152
    var. simplex, 1152
  squarrosa, 1152
    var. *aristulata*, 1152
FULLER'S-HERB, 524
Fumaria, 722
  officinalis, 722
Fumariaceae, 719, 872
FUMEWORT, 719
  SLENDER, 722
FUMITORY, 719, 722
  COMMON, 722
  DRUG, 722
FUMITORY FAMILY, 719
Funastrum, 283
  crispum, ▣/90, 283
  cynanchoides, ▣/90, 283
FUNGI, 232, 437, 438, 554, 561, 581, 582, 628, 688, 710, 711, 732, 917, 932, 934, 950, 959, 1038, 1039, 1092, 1213, 1216, 1276, 1284, 1304, 1318, 1334, 1338
FUNNEL-LILY, 1198
  BLUE, 1198
FUZZY
  -BEAN, 697
  CEANOTHUS, 932
  ROSE, 958

G

Gaillardia, 358
  aestivalis, 358
    var. aestivalis, 358
    var. flavovirens, 358
  fastigiata, 358
  lanceolata, 358
    var. *flavovirens,* 358
  pulchella, ▣/90, 358
  suavis, 359
GAILLARDIA
  FRAGRANT, 359
  LANCE-LEAF, 358
  PRAIRIE, 358
  RAYLESS, 359
  ROSE-RING, 358
Galactia, 658
  canescens, 660
  grayi, 660
  heterophylla, 660
  mississippiensis, 660
  triacanthos var. *inermis,* 662
  volubilis, 660
Galanthus nivalis, 1206
Galium, 962
  aparine, 964
  circaezans, 964
  obtusum, 964
  pilosum, 964
    var. *laevicaule*, 966
    var. *puncticulosum*, 966
  texense, 966
  tinctorium, 966
  triflorum, 966
  vernum, 964
  virgatum, 966
GALLITOS, 1095
GAMMA GRASS, EASTERN, 1332
Gamochaeta, 359
  falcata, 359
  pensylvanica, 359
  purpurea, 359
GAPING PANICUM, 1298
GARCINIA FAMILY, 544
Garcinia mangostana, 544
GARDEN
  ASPARAGUS, 1198
  BALSAM, 434
  HYACINTH, 1202
  IRIS, 1174
  LETTUCE, 381
  PANSY, 1061
  PARSLEY, 256
  SAGE, 772
  VERBENA, 1050
Gardenia, 960
GARLIC, 1194, 1197
  CANADA, 1196
  FALSE, 1208
  WILD, 1196
  YELLOW FALSE, 1208
Garrya, 723
  lindheimeri, 723
  ovata subsp. lindheimeri, 723
Garryaceae, 722
GASPLANT, 970
GATTINGER'S GERARDIA, 993
Gaura, 855
  biennis var. *pitcheri*, 856
  brachycarpa, 856
  coccinea, 856
    var. *glabra*, 856
    var. *parviflora*, 856

Gaura *(cont.)*
drummondii, 856
filiformis, 856
lindheimeri, 856
longiflora, 856
odorata, 856
parviflora, 856
sinuata, 856
suffulta, 856
triangulata, 858
tripetala var. *triangulata*, 858
villosa, 858
GAURA
DOWNY, 856
FALSE, 866
KEARNEY'S, 856
LIZARD-TAIL, 856
MUNZ, 856
PLAINS, 856
ROADSIDE, 856
SCARLET, 856
SCENTED, 856
SMALL-FLOWER, 856
SMOOTH, 856
SWEET, 856
TALL, 856
VELVET-LEAF, 856
WAVY-LEAF, 856
WHITE, 856
WILLOW-HERB, 856
WOOLLY, 858
GAYFEATHER, 382
KANSAS, 384
NARROW-LEAF, 384
PINK-SCALE, 384
ROUGH, 384
SMOOTH, 386
TALL, 384
Gaylussacia, 581
Gelsemium, 797
sempervirens, 798
GENTIAN
BLUEBELL, 727
CATCHFLY, 727
PRAIRIE, 727
ROSE, 728
PURPLE PRAIRIE, 727
ROSE, 727
SHOWY PRAIRIE, 727
SQUARE-STEM ROSE, 728
GENTIAN FAMILY, 723
Gentiana, 723
nigricans, 966
Gentianaceae, 723
GEORGIA
ROCK-ROSE, 543
SUN-ROSE, 543
Geraniaceae, 728
Geranium, 730
carolinianum, 732
var. *texanum*, 732
dissectum, 732
texanum, 732
GERANIUM
CAROLINA, 732
FEATHER-, 536
POPPY-MALLOW, 808
TEXAS, 732
GERANIUM FAMILY, 728
Gerardia, 992
auriculata, 993
densiflora, 993
edwardsiana, 993
gattingeri, 993
heterophylla, 993
homalantha, 993
leucanthera, 993
tenuifolia var. *leucanthera*, 993
GERARDIA, 991
BEACH, 993
EAR-LEAF, 992
FINE-LEAF, 993
FLAT-FLOWER, 993
GATTINGER'S, 993
PLATEAU, 993
ROUGH, 993
SLENDER, 993
STIFF-LEAF, 993
GERMANDER, 782
AMERICAN, 782
ANNUAL, 782
CUT-LEAF, 783
GERMAN-MILLET, 1319
Gesneriaceae, 989
Geum, 944
canadense, 944
var. camporum, 946
var. texanum, 946
GEYER'S EUPHORBIA, 594
GHERKIN
BUR, 566
WEST INDIAN, 566
GIANT
ARROWHEAD, 1090
BRISTLE GRASS, 1319
BULRUSH, 1161
CANE, 1243
CONEFLOWER, 402
CUT GRASS, 1338
DROPSEED, 1327
FOUR-O'CLOCK, 839
FOXTAIL GRASS, 1319
GOLDENROD, 410
GROUNDSEL, 403
HELLEBORINE, 1215
RAGWEED, 310
RAIN-LILY, 1200
REDWOOD, 202
REED, 1243
SNOWFLAKE, 1204
SPIDERWORT, 1102
SPIRAL ORCHID, 1220
TULE, 1161
Gilia, 890
acerosa, 890
incisa, 890
rigidula, 890
subsp. acerosa, 890
subsp. rigidula, 890
rubra, 890
GILIA
PRICK-LEAF, 890
RED, 890
SPLIT-LEAF, 890
Gillenia stipulata, 946
GILLIFLOWER, 474
GILL-OVER-THE-GROUND, 756
GINGELLY, 880
GINSENG, 270
GINSENG FAMILY, 270
GIRASOLE, 370
GLADE-LILY, 864
Gladiolus, 1172
GLAND-TOOTH GUMWEED, 360
GLANDULAR PRAIRIE-CLOVER, 650
Glandularia, 1050
bipinnatifida, ▣/91, 1050
canadensis, 1050
×hybrida, 1050
peruviana, 1052
pumila, 1052
quadrangulata, 1052
Glaucium, 876
corniculatum, 876
GLAUCOUS-LEAF GREENBRIER, 1346
Glechoma, 756
hederacea, 756
Gleditsia, 660
aquatica, 660
triacanthos, 660
GLEN ROSE YUCCA, 1083
Glinus, 827
lotoides, 827
radiatus, 827
GLOBE
-AMARANTH, 228
COMMON, 228
HAAGE'S, 228

GLOBE *(cont.)*
  BEAK-RUSH, 1158
  -CHERRY, NETTED, 1022
  FIMBRISTYLIS, 1150
  FLAT SEDGE, 1136
  -MALLOW, 818
    NARROW-LEAF, 820
    POINT-SEED, 820
    SCARLET, 820
GLOBEBERRY, 569
  LINDHEIMER'S, 569
GLOBEFLOWER, 359, 962
GLOBOSE CARIC SEDGE, 1118
GLOMERATE DODDER, 573
GLOSSY PRIVET, 850
Glottidium, 662
  vesicarium, 662
Glyceria, 1277
  arkansana, 1277
  septentrionalis var. *arkansana,* 1277
  striata, 1277
Glycine, 662
  max, 662
  soya, 662
Gnaphalium
  chilense, 399
  falcatum, 359
  obtusifolium, 398
  pensylvanicum, 359
  peregrinum, 359
  purpureum, 360
  retusum, 358
  stramineum, 399
Gnetophyta, 207
GOATHEAD, 1076
GOAT'S
  -BEARD, 424
  -RUE, 700
GOATWEED, 547
GOBBO, 806
GOBO, 806
GOLD
  -APPLE, 1030
  -ASTER, 372
    GRAY, 372
    NARROW-LEAF, 372
GOLDEN
  -ALEXANDERS, 264
  -ASTER, 335, 372
    GRASS-LEAF, 396
    SOFT, 336
    SPREADING, 344
  BAMBOO, 1311
  CORYDALIS, 720
  CROWNBEARD, 426
  DALEA, 646
  -EYE, 430
    SUNFLOWER, 430
  GROUNDSEL, 390
  HEDGE-HYSSOP, 1000
  -MANE COREOPSIS, 342
  -WAVE, 341
  ZIZIA, 264
GOLDENBELLS, 846
GOLDENGLOW, WILD, 326
GOLDENRAIN-TREE, 979
GOLDENROD, 408
  CANADA, 410
  CELTIS-LEAF, 414
  COMMON, 410
  ELM-LEAF, 415
  FLAT-TOPPED-, 354, 412
  FRAGRANT, 412
  GIANT, 410
  HARSH, 414
  HIGH PLAINS, 415
  LATE, 410
  MISSOURI BASIN, 410
  NOBLE, 414
  OLD-FIELD, 410, 412
  PRAIRIE, 414
  RAYLESS-, 328
  RIGID, 414
  ROUGH, 414
    -LEAF, 414
  SEASIDE, 415
  SHINY, 412
  STIFF, 414
  SWEET, 412
  WILLOW, 410, 412
  WRINKLED, 414
GOLDENWEED, 362
  SLENDER, 344
GOLD-EYE PHLOX, 892
GOLD-POPPY, 874
  MEXICAN, 874
GOLDSTAR, 1204
  COMMON, 1204
GOMBO, 806
Gomphrena, 228
  globosa, 228
  haageana, 228
Gonolobus
  decipiens, 284
  gonocarpus, 284
GOOBER, 628
GOOSEBERRY, 732
GOOSEFOOT, 533
  BIG-SEED, 538
  LIGHT, 538
  MAPLE-LEAF, 538
  NARROW-LEAF, 536
  NETTLE-LEAF, 538
  PIT-SEED, 536
  RIDGED, 538
  SLIM-LEAF, 536
  STANDLEY'S, 538
  THICK-LEAF, 538
GOOSEFOOT FAMILY, 530
GOOSE GRASS, 1268
GOOSEGRASS CLEAVERS, 964
GOOSEWEED, 1031
GORDON'S BLADDERPOD, 473
Gossypianthus, 228
  lanuginosus, 228
    var. lanuginosus, 228
    var. tenuiflorus, 228
Gossypium, 808
  hirsutum, 808
GOTTHILF MUHLENBERG'S
  CARIC SEDGE, 1128
  VEINLESS CARIC SEDE, 1128
GOURD, 568, 569
  BALSAM, 569
  BOTTLE, 569
  BUFFALO, 568
  CALABASH, 569
  FOETID, 568
  STINKING, 568
  TEXAS, 568
  WHITE-FLOWER BOTTLE, 569
GOURD FAMILY, 564
GRAMA
  GRASS, 1247
  BLUE, 1248
  HAIRY, 1248
  RED, 1248
  SIDE-OATS, 1248
  SIXWEEKS, 1248
  TALL, 1248
  TEXAS, 1248
  THREEAWN, 1248
Gramineae, 1222
GRAND PRAIRIE EVENING-PRIMROSE, 854
GRANULAR CARIC SEDGE, 1122
GRAPE, 1065, 1068
  BLUE-LEAF, 1070
  BULLACE, 1072
  BUSH, 1070
  CATBIRD, 1072
  CHAMPIN, 1072
  FERN, 188
    COMMON, 189
    PROSTRATE, 189
    SOUTHERN, 189
    SPARSE-LOBED, 189
    VIRGINIA, 189
    WINTER, 189

GRAPE *(cont.)*
FOX, 1072
FROST, 1072
GRAY BARK, 1070
-HYACINTH, 1206
STARCH, 1206
LONG'S, 1070
MISSOURI, 1072
MUSCADINE, 1072
MUSTANG, 1072
PANHANDLE, 1070
PIGEON, 1070
PINEWOODS, 1070
POSSUM-, 1066
POST OAK, 1070
RACOON-, 1066
RED, 1072
RIVER, 1072
RIVERBANK, 1072
ROUND-LEAF, 1072
SPANISH, 1072
SUMMER, 1070
SWEET, 1070
MOUNTAIN, 1072
VINE, 1068
WINTER, 1072
GRAPE FAMILY, 1065
GRAPEFRUIT, 970
TEXAS RED, 970
GRASS
BROME, CHEAT, 1253
SILKY CUP, 1274
ALKALI, 1264
AMERICAN PEPPER-, 472
ANNUAL
BEARD, 1314
HAIR, 1236
TICKLE, 1236
APAREJO, 1290
ARKANSAS MANNA, 1277
ARROW, 1242
AUTUMN BENT, 1236
BAHAMA, 1259
BAHIA, 1306
BARNYARD, 1264, 1266
BASKET, 1290
BAYONET-, 1108
BEAR-, 1080, 1082
BEARD, 1238, 1244, 1314, 1315
BEARDED SKELETON, 1277
BEARDLESS DARNEL RYE, 1282
BENT, 1235
-AWN PLUME, 1315
BERMUDA, 1259
BIG
QUAKING, 1250
-TOP LOVE, 1270
BLUE, 1312
-EYED-, 1174
BLUESTEM WHEAT, 1304
BRISTLE, 1318
BROAD-LEAF SIGNAL, 1335
BROME, 1250
SIXWEEKS, 1336
BRONCO, 1253
BROWN-TOP SIGNAL, 1335
BUFFALO, 1253
BUFFEL, 1310
BULBOUS BLUE, 1312
BUR BRISTLE, 1320
BUSHY BEARD, 1239
CANADA BLUE, 1314
CANARY, 1310
CANE BEARD, 1246
CAROLINA CANARY, 1310
CARPET, 1244
CATCHFLY, 1280
CAT-TAIL, 1310
CEDAR ROSETTE, 1300
CHAPMAN'S BLUE, 1314
CHEAT, 1253
COASTAL SALT, 1264
COCO-, 1142
COLORADO, 1335
COMMON
CARPET, 1244
SIXWEEKS, 1336
CORD, 1324
CRAB, 1262
CREEPING LOVE, 1272
CROWFOOT, 1256
CUP, 1274
CUT, 1280, 1338
DALLIS, 1305
DANUBE, 1311
DARNEL RYE, 1282
DEER-, 820
-TONGUE ROSETTE, 1296
DESERT SALT, 1264
DOGTOWN, 1243
DUCK-, 1166
DURBAN CROWFOOT, 1259
DWARF MEADOW, 1312
EARLY SCORPION-, 453
EASTERN
GAMMA, 1332
MANNA, 1277
EEL-, 1170
EGYPTIAN CROWFOOT, 1259
ELLIOTT'S
BENT, 1236
BLUE, 1239
ETERNITY, 1306
FALL WITCH, 1263
FALSE NUT-, 1144
FEATHER FINGER, 1258
FIELD, 1335
FINGER, 1262
FLUFF, 1274
FLY-AWAY, 1236
FORT THOMPSON, 1306
FOUNTAIN, 1308
FOWL MANNA, 1277
FOXTAIL, 1320
FRINGED SIGNAL, 1334
GIANT
BRISTLE, 1319
CUT, 1338
FOXTAIL, 1319
GOOSE, 1268
GRAMA, 1247
GREEN
BRISTLE, 1320
FOXTAIL, 1320
-FLOWER PEPPER-, 470
GUMMY LOVE, 1270
HAIR, 1236
HAIRY
CRAB, 1264
SIXWEEKS, 1336
HARD, 1316
HEDGE-HOG, 1254
HOODED
FINGER, 1256
WINDMILL, 1256
HOOKED BRISTLE, 1320
HURRAH, 1335
INDIA LOVE, 1272
INDIAN, 1320
INLAND SALT, 1264
ITALIAN RYE, 1282
JAMAICAN SAW-, 1132
JOHNSON, 1324
JOINT, 1306
JOINTED GOAT, 1235
JUNE, 1280
BLUE, 1314
KENTUCKY BLUE, 1314
KLEIN, 1296
KNOT, 1306
-ROOT BRISTLE, 1319
LARGE CRAB, 1264
-LEAF
GOLDEN-ASTER, 396
LADIES'-TRESSES, 1222
MUD-PLANTAIN, 1340
RUSH, 1184
LINDHEIMER'S ROSETTE, 1295

GRASS *(cont.)*
- LITTLE QUAKING, 1250
- LONG
  - -AWNED HAIR, 1288
  - -BRISTLED INDIAN, 1322
  - -LEAF RUSH, 1327
- LOVE, 1268
- LOW SPEAR, 1312
- MANNA, 1277
- MARSH, 1324
- MEDITERRANEAN LOVE, 1270
- MELIC, 1284
- MESQUITE, 1248
- MONDO-, 1191
- MONKEY-, 1191
- NERVED MANNA, 1277
- NORTHERN NUT-, 1138
- NUT-, 1142
- OPEN-FLOWER ROSETTE, 1300
- ORANGE-, 547
- ORCHARD, 1259
- OZARK, 1282
- PAMPAS, 1258
- PANIC, 1291
- PASPALUM, 1305
- PENSACOLA BAHIA, 1306
- PEPPER-, 470
- PERENNIAL RYE, 1282
- PHILADELPHIA WITCH, 1302
- PINHOLE BEARD, 1246
- -PINK, 1214
  - BEARDED, 1214
- PLAINS LOVE, 1270
- PLUME, 1315
- POVERTY OAT, 1260
- PRAIRIE
  - CORD, 1324
  - CUP, 1274
  - JUNE, 1280
  - LOVE, 1272
  - NEEDLE, 1243
  - SAND, 1332
  - NUT-, 1142
- QUAKING, 1250
- RABBIT'S-FOOT, 1314
- RAT-TAIL SIXWEEKS, 1336
- RED LOVE, 1272
- RESCUE, 1252
- REVERCHON'S BRISTLE, 1320
- RHODES, 1256
- RIB-, 886
- RIBBON-, 886, 1080
- RICE CUT, 1281
- RIPPLE-, 886
- ROUGH RUSH, 1327
- ROUND-SEED ROSETTE, 1302
- RYE, 1282
- SALT, 1264
  - MEADOW, 1281
- SAND LOVE, 1274
- SATIN, 1290
- SAW-, 1130, 1132
- SCHEELE'S BRISTLE, 1320
- SCHRADER'S, 1252
- SCORPION-, 453
- SCRIBNER'S ROSETTE, 1300
- SHAGGY CRAB, 1264
- SHAKING, 1250
- SHORT
  - -SPIKE WINDMILL, 1256
  - -STALKED LOVE, 1270
- SIGNAL, 1334
- SILVER, 1284
- SILVERY BEARD, 1239
- SIXWEEKS, 1335
  - FESCUE, 1335
- SLENDER, 1281
  - CRAB, 1263
  - FINGER, 1263
  - INDIAN, 1322
- SLIM-LEAF ROSETTE, 1300
- SLOUGH, 1324
- SMOOTH
  - CRAB, 1264
  - CREEPING LOVE, 1270
- SMUT, 1327
- SOFT-LEAF ROSETTE, 1300
- SOUR-, 906
- SOUTHERN
  - CANARY, 1310
  - CRAB, 1263
  - POVERTY, 1328
- SOUTHWESTERN BRISTLE, 1320
- SPEAR, 1290
- SPICATE SALT, 1264
- SPIKE, 1264
- SPLIT-BEARD BEARD, 1239
- SPREADING
  - LOVE, 1272
  - WITCH, 1296
- SPRING BENT, 1236
- SQUIRREL SIXWEEKS, 1336
- ST. AUGUSTINE, 1328
- STAGGER-, 1201
- STAR-, 1193
- STARVED ROSETTE, 1296
- STINK, 1270
- STOUT LOVE, 1270
- SUDAN, 1324
- SUGARCANE PLUME, 1315
- SWITCH, 1302
- SWORD-, 1161
- SYLVAN BLUE, 1314
- TALL MARSH, 1324
- TAPE-, 1170
- TEAL LOVE, 1270
- TEXAS
  - BLUE, 1312
  - CRAB, 1316
  - CUP, 1274
  - NEEDLE, 1290
  - SIGNAL, 1335
  - WINTER, 1290
- TICKLE, 1236
- TUMBLE, 1316
  - LOVE, 1272
  - WINDMILL, 1256
- UMBRELLA-, 1158
- VARIABLE ROSETTE, 1298
- VASEY, 1308
- VELVET ROSETTE, 1302
- -VIOLET, 1104, 1174
- VIRGINIA
  - CUT, 1281
  - PEPPER-, 472
- WATER BENT, 1314
- WEDGE, 1324
- WEEPING LOVE, 1270
- WESTERN
  - UMBRELLA-, 1152
  - WHEAT, 1302, 1304
  - WITCH, 1263
- WHIP-, 1164
- WHITE, 1281
  - -HAIRED ROSETTE, 1295
- WILD CANARY, 1310
- WILMANN'S LOVE, 1272
- WINDMILL, 1256
  - FINGER, 1256
- WINTER, 1290
  - BENT, 1236
- WITCH, 1296
- WOOD WITCH, 1302
- WOODLAND BLUE, 1314
- WOOLLY, 1274
  - ROSETTE, 1295
  - TRIPLEAWN, 1242
- WORM-, 798
- WRIGHT'S TRIPLEAWN, 1243
- YARD, 1268
- YELLOW
  - -EYED-, 1349
  - BRISTLE, 1319
  - INDIAN, 1322
  - NUT-, 1138
  - STAR-, 1204

GRASS FAMILY, 1222

GRASSBUR, 1253, 1254

GRASSLAND PRICKLY-PEAR, 492
Gratiola, 1000
flava, 1000
neglecta, 1000
pilosa, 1000
virginiana, 1000
GRAVEL-BAR BRICKELLBUSH, 328
GRAVELWEED CROWNBEARD, 427
GRAY
-BARK GRAPE, 1070
BINDWEED, 552
CHICKWEED, 516
GOLD-ASTER, 372
-GREEN
WOODSORREL, 870
-FRUIT CARIC SEDGE, 1122
POLYPODY, 193
VERVAIN, 1056
GRAY'S LADIES'-TRESSES, 1222
GREAT
BULRUSH, 1161
CONEFLOWER, 402
LOBELIA, 498
PLAINS LADIES'-TRESSES, 1222
PLANTAIN, 886
SCOURING-RUSH, 176
SOLOMON'S-SEAL, 1208
GREATER
AMMI, 243
DUCKWEED, 1190
GREEN
AMARANTH, 226, 227
ANTELOPE-HORNS, 282
ASH, 849
BRISTLE GRASS, 1320
CARPETWEED, 827
-DRAGON, 1092
-EYED LYRE-LEAF, 325
FALSE NIGHTSHADE, 1018
-FLOWER
CHOLLA, 492
MILKWEED, 282
PEPPER-GRASS, 470
FOXTAIL GRASS, 1320
-LILY, 1210
-LIP LADIES'-TRESSES, 1220
MILKWEED, 282
MILKWEEDVINE, 284
PARROT'S-FEATHER, 734
PEPPER, 1015
PIGWEED, 222
SPRANGLETOP, 1281
SPURGE, 608
-VIOLET, 1061
NODDING, 1061
WILD INDIGO, 638
GREENBRIER, 1344
BRISTLE, 1346
CAT, 1346
COMMON, 1346
DEVIL, 1346
FIDDLE-LEAF, 1346
FRINGED, 1346
GLAUCOUS-LEAF, 1346
SAW, 1346
SMALL'S, 1346
GREENBRIER FAMILY, 1344
GREENEYES, 324
LYRE-LEAF, 325
SOFT, 325
GREENHAW, 944
GREENSTRIPE, 224
GREENTHREAD, 421
COLORADO, 422
SLENDER, 422
GREGG'S SAGE, 776
Grindelia, 360
adenodonta, 360
ciliata, 362
lanceolata, 360
microcephala var. *adenodonta,* 360
nuda, 360
papposa, ▣/91, 362
squarrosa, 362
var. *nuda,* 362
Groenlandia, 1340
GROMWELL, 452
CAROLINA, 452
FALSE, 453
NARROW-LEAF, 452
ROUGH, 452
GRONONVIUS'
HAWKWEED, 374
DODDER, 574
GROOVEBUR, 938
MANY-FLOWER, 938
SLENDER, 938
WOODLAND, 938
GROOVED FLAX, 792
GROOVE-STEM INDIAN-PLANTAIN, 314
Grossulariaceae, 732
GROUND
-CHERRY, 1023
BEACH, 1024
BLADDER, 1026
CLAMMY, 1024
COMMON, 1024
CUT-LEAF, 1024
DOWNY, 1026
FIELD, 1026
LANCE-LEAF, 1024
LOW, 1026
LOW HAIR, 1026
PRAIRIE, 1026
PROSTRATE, 1018
PURPLE, 1026
PURPLE-VEIN, 1024
SOUTHWEST, 1024
THICKET, 1026
VIRGINIA, 1026
-IVY, 756
-PLUM, 630, 635
MILK-VETCH, 630
SARACHA, 1018
GROUNDNUT, 628
GROUNDSEL, 388, 403
COMMON, 403
GIANT, 403
GOLDEN, 390
PRAIRIE, 390
ROUND-LEAF, 390
TEXAS, 403
TREE-, 324
GROUNDSELTREE, 322, 324
GUADALUPE PENSTEMON, 1008
Guajacum, 1074
angustifolium, 1074
officinale, 1074
sanctum, 1074
GUAYACÁN, 1074
GUAYULE RUBBER PLANT, 393
Guilleminea, 228
lanuginosa, 228
var. *rigidiflora,* 228
var. *sheldonii,* 228
var. *tenuiflora,* 228
GUINEA-CORN, 1324
GULF
COAST
LARKSPUR, 924
SPIKE-RUSH, 1146
GUMWEED, 360
MUHLY, 1288
VERVAIN, 1058
GUM
BUMELIA, 982
BLACK-, 845
COTTON-, 845
-ELASTIC, 982
SOUR-, 845
GUMBO, 806
GUMDROPBUSH, 936
GUMDROPTREE, 936
GUMMY LOVE GRASS, 1270
GUMPLANT, 360
GUMWEED, 360
CURLY-CUP, 362
GLAND-TOOTH, 360

GUMWEED *(cont.)*
GULF, 360
LITTLE-HEAD, 360
RAYLESS, 360
Gunnera, 733
GUNPOWDER-PLANT, 1042
Gutierrezia, 362
amoena, 364
dracunculoides, 364
microcephala, 364
sarothrae, 365
sphaerocephala, 365
texana, 365
triflora, 362
Guttiferae, 544
Gymnocladus, 663
dioicus, 663
Gymnopogon, 1277
ambiguus, 1277
Gymnosperms, 201
Gymnosporangium, 940
Gymnostachys, 1077
Gymnostyles, 415
stolonifera, 416
GYP PHACELIA, 743
Gypsophila, 511

H

HAAGE'S GLOBE-AMARANTH, 228
Habenaria, 1215
ciliaris, 1218
repens, 1215
HABENARIA, NUTTALL'S, 1215
Habranthus, 1201
texanus, 1201
tubispathus, 1201
HACKBERRY, 1036
NET-LEAF, 1038
SUGAR, 1038
HAYFEVER, 203, 227, 287, 309, 314, 378, 408, 504, 747, 1224, 1259, 1311
HAGBRIER, 1346
HAIR
GRASS, 1236
ANNUAL, 1236
LONG-AWNED, 1288
-SEDGE, 1109
HAIRY
-AWN MUHLY, 1288
BEDSTRAW, 964
BITTERCRESS, 462
BONAMIA, 560
BUSH-CLOVER, 670
BUTTON-SNAKEROOT, 384
CLUSTERVINE, 558
CORNSALAD, 1044
CRAB GRASS, 1264
CRABWEED, 828
DOGBANE, 266
ERIONEURON, 1274
EUPHORBIA, 596
EVOLVULUS, 552
FALSE NIGHTSHADE, 1018
-FLOWER
PASPALUM, 1308
SPIDERWORT, 1102
-FRUIT CHERVIL, 246
GRAMA, 1248
HEDGE-HYSSOP, 1000
HYDROLEA, 740
LAZY DAISY, 312
LEAFCUP, 408
LESPEDEZA, 670
LIP FERN, 198
MISTLETOE, 1064
PEPPERWORT, 186
PHACELIA, 743
PINWEED, 543
-SEED PASPALUM, 1308
SIXWEEKS GRASS, 1336
STICKSEED, 450
SUNFLOWER, 370
TRIDENS, 1274
TUBETONGUE, 218
VERBENA, 1052
VETCH, 708
WEDELIA, 430
HALBERD-LEAF
HIBISCUS, 810
ROSE-MALLOW, 810
Halesia, 1034
HALF-SHRUB SUNDROPS, 854
HALL'S
DALEA, 648
PANIC, 1298
Haloragaceae, 733
Hamamelidaceae, 736
Hamamelis, 736
HAMMERWORT, 1042
HANDSOME BLAZINGSTAR, 384
Haplopappus, 400
annuus, 400
ciliatus, 362
divaricatus, 344
var. *hookerianus*, 344
phyllocephalus subsp. *annuus*, 400
spinulosus, 387
validus, 344
subsp. *torreyi*, 344
HARD
-BARK HICKORY, 747
GRASS, 1316
-STEM BULRUSH, 1161
HARDY CATALPA, 443
HARE BARLEY, 1278
HAREBELL FAMILY, 496
HARE'S-EAR, 464
TREACLE, 464
-MUSTARD, 464
HARSH
GOLDENROD, 414
VERVAIN, 1058
HARVEST-LICE, 938
HARVEY'S BEAK-RUSH, 1158
HAWK'S-BEARD, 342
SHOWY, 344
HAWKWEED, 374
GRONONVIUS', 374
JAPANESE-, 433
HAWTHORN, 940
BLUEBERRY, 942
BUSH'S, 942
COCKSPUR, 942
LITTLE-HIP, 942
PARSLEY, 942
PASTURE, 942
REVERCHON'S, 942
HAZEL ALDER, 439
DODDER, 573
HAZELNUT FAMILY, 438
HEAD-BEARING CARIC SEDGE, 1118
HEALALL, 770
HEARLD-OF-SPRING, 848
HEART
-LEAF
AMPELOPSIS, 1066
EUPHORBIA, 594
NETTLE, 1043
PONDWEED, 1343
SKULLCAP, 778
SPURGE, 594
-SEPAL WILD BUCKWHEAT, 899
-WING SORREL, 906
HEARTS-EASE, 904, 1062
HEARTSEED, 980
HEARTSEEDVINE, 980
HEATH
ASTER, 318
ROSE-, 334
HEATH FAMILY, 581
HEAVENLY-BAMBOO, 437
HEAVY-FRUIT CARIC SEDGE, 1122
Hedeoma, 756
acinoides, 758
drummondii, 758
var. *reverchonii*, 758
var. *serpyllifolium*, 758
hispida, 758

Hedeoma *(cont.)*
reverchonii, 758
var. reverchonii, 758
var. serpyllifolium, 758
serpyllifolium, 758
HEDEOMA
DRUMMOND'S, 758
ROCK, 758
ROUGH, 758
SLENDER, 758
Hedera, 272
helix, 272
HEDGE
-APPLE, 831
-BINDWEED, 550
TRAILING, 551
-HOG GRASS, 1254
-HYSSOP, 1000
GOLDEN, 1000
HAIRY, 1000
SHAGGY, 1000
VIRGINIA, 1000
YELLOW-SEED, 1000
-MUSTARD, 479
-NETTLE, 780
SMOOTH, 782
-PARSLEY, 262
KNOTTED, 262
HEDGEHOG
CACTUS, 486, 494
RED-FLOWER, 486
HEDGEPLANT, 849
Hedyotis, 966
australis, 967
boscii, 968
crassifolia, 968
var. *micrantha*, 967
greenmanii, 967
humifusa, 967
nigricans, 966
var. *filifolia*, 966
rosea, 968
subviscosa, 968
Hedypnois, 365
cretica, 365
HEGARI, 1324
Helenium, 365
amarum, 366
var. amarum, 366
var. badium, 366
autumnale, 366
badium, 366
elegans, 368
microcephalum, 368
setigerum, 309
tenuifolium, 366
Helianthemum, 543
georgianum, ▣/91, 543
rosmarinifolium, 543
Helianthus, 368
annuus, ▣/91, 369
subsp. *lenticularis*, 369
subsp. *texanus*, 369
ciliaris, 369
cucumerifolius, 369
debilis subsp. cucumerifolius, 369
grosseserratus, 369
hirsutus, 370
laetiflorus var. *rigidus*, 370
maximiliani, ▣/91, 370
mollis, 370
pauciflorus, 370
petiolaris, 370
rigidus, 370
salicifolius, 370
simulans, 370
tuberosus, 370
HELIOTROPE, 448
BINDWEED, 448
INDIA, 450
PASTURE, 450
SALT, 450
SEASIDE, 450
VIOLET, 448
Heliotropium, 448
amplexicaule, 448
convolvulaceum, 448
curassavicum, 450
indicum, 450
procumbens, 450
racemosum, 450
tenellum, 450
HELLEBORINE, 1215
GIANT, 1215
HELLER'S MARBLESEED, 454
HELLFETTER, 1344
HELMET-FLOWER, 776
Hemerocallidaceae, 1191
Hemerocallis, 1201
fulva, 1202
lilioasphodelus, 1202
Hemicarpha, 1155
aristulata, 1155
drummondii, 1155
micrantha, 1155, 1156
var. *drummondii*, 1155
var. *minor*, 1156
HEMLOCK, 204, 260
COMMON WATER-, 248
POISON-, 248
SPOTTED WATER-, 248
WATER-, 246
HEMP DOGBANE, 266
HEMP, 502
BOG-, 1042
BUTTON-, 1042
COLORADO RIVER-, 696
INDIAN-, 266
NUTTALL'S WATER-, 226
SOUTHERN WATER-, 226
WATER-, 226
HEMP FAMILY, 502
HEMPTREE, 1060
HEMPVINE, 388
HEMPWEED, 388
CLIMBING, 388
HEN-AND-CHICKENS, 501
HENBIT, 759
HENNA, 799
HERB
SHERARD, 970
HORSE, 329
HERBACEOUS MIMOSA, 680
Herbertia, 1173
caerulea, 1173
lahue subsp. caerulea, ▣/92, 1173
HERCULES'-CLUB, 270, 974
Herissantia, 808
crispa, 810
HERISSANTIA, NET-VEIN, 810
HERONBILL, 730
Hesperaloe, 1079
parviflora, ▣/92, 1079
HESPERALOE, RED, 1079
Heteranthera, 1339
dubia, 1340
limosa, 1340
Heterotheca, 372
canescens, 372
latifolia, 372
pilosa, 336
stenophylla, 372
subaxillaris, 372
var. *latifolia*, 372
Heuchera, 986
americana, 986
var. *brevipetala*, 988
sanguinea, 986
Hevea brasiliensis, 585, 830
Hexalectris, 1215
nitida, ▣/92, 1216
spicata, 1216
var. arizonica, 1216
var. spicata, 1216
warnockii, ▣/92, 1218
HEXALECTRIS, SHINING, 1216
Hibiscus, 810
esculentus, 806

Hibiscus *(cont.)*
laevis, ▣/92, 810
lasiocarpos, 810
militaris, 810
moscheutos, 810
subsp. lasiocarpos, 810
subsp. moscheutos, 810
syriacus, 812
trionum, ▣/92, 812
HIBISCUS, HALBERD-LEAF, 810
HICKORY, 746
BITTER-NUT, 747
BLACK, 748
BUCKLEY'S, 748
HARD-BARK, 747
MOCKERNUT, 747
NUTMEG, 748
OZARK, 748
RED-HEART, 748
SCALY-BARK, 748
SHAG-BARK, 748
SWAMP, 747
TEXAS, 748
UPLAND, 748
WATER, 747
WHITE, 747
-HEART, 747
Hicoria pecan, 747
HIDDEN-FRUIT CARIC SEDGE, 1116
HIEDRA, 236, 1068
Hieracium, 374
gronovii, 374
HIERBA
BUENA MONTES, 1054
DE GOLONDRINA, 596
DE LA PRINCESA, 1049
DE LA VIRGIN MARÍA, 1054
DE POLLO, 1100
DE ZIZOTES, 278
DEL BUEY, 1066
DEL CABALLO, 329
DEL CÁNCER, 802
DEL CORAZON, 651
DEL CRISTO, 1053
DEL PÁJARO, 912
DULCE, 1054
LOCA, 599
MORA, 1031
NEGRA, 1030
NEGRA, 1052
HIGH
MALLOW, 814
PLAINS GOLDENROD, 415
HIGHBELIA, 500
HIGHGROUND OAK, 716
HIGUERA, 830
HIGUERILLA, 612
Hilaria, 1278
belangeri, 1278
HILL COUNTRY WILD MERCURY, 602
HILLMAN'S PANICUM, 1298
HILLSIDE VERVAN, 1056
HILLY SANDWORT, 514
Hippocastanaceae, 737
Hippophae, 579
HOARY
BOWLESIA, 244
EUPHORBIA, 594
FROGFRUIT, 1054
MILK-PEA, 660
-PEA, 698
PLANTAIN, 886
SUN-ROSE, 543
TICK-CLOVER, 655
VERVAIN, 1058
Hoffmannseggia, 663
densiflora, 663
drummondii, 664
glauca, ▣/93, 663
jamesii, 688
HOG
-FENNEL, 256
MILLET, 1300
-PEANUT, 626, 663
SOUTHERN, 626
-PLUM, 932, 950
-POTATO, 663
HOGBRAKE, 309
HOGWORT, 599
HOLLY, 269, 270
AMERICAN, 270
DECIDUOUS, 270
EMETIC, 270
EVERGREEN, 270
FERN, 185
-LEAF WATER-NYMPH, 1170
MEADOW, 270
PRAIRIE, 270
WELK, 270
WHITE, 270
YAUPON, 270
HOLLY FAMILY, 269
HOLLYHOCK, 807
HOLY-THISTLE, 406
Homalocephala texensis, 486
HONESTY, 456
HONEWORT, 249
HONEY
-BALLS, 962
-LOCUST, 660
COMMON, 660
WATER, 660
MESQUITE, 688
HONEYSHUCK, 660
HONEYSUCKLE, 508
AMUR, 508
BUSH, 508
BUSHY, 508
CORAL, 510
EVERGREEN, 510
FALSE, 218
JAPANESE, 508
MAACK'S, 508
TREE, 508
TRUMPET-, 442
TRUMPET, 510
WHITE, 508
WILD, 856
HONEYSUCKLE FAMILY, 507
HOODED
FINGER GRASS, 1256
WINDMILL GRASS, 1256
HOOKED
BRISTLE GRASS, 1320
PEPPERWORT, 186
HOOKER'S
PLANTAIN, 884
ERYNGO, 252
HOP
CARIC SEDGE, 1124
-CLOVER, 674
CLOVER
LEAST, 704
LARGE, 704
LOW, 704
SMALL, 704
-HORNBEAM, 440
AMERICAN, 440
COPPERLEAF, 588
EASTERN, 440
WOOLLY AMERICAN, 440
TREFOIL, 704
HOPS, 502
HOPTREE, 972
WOOLLY, 973
Hordeum, 1278
leporinum, 1278
murinum subsp. leporinum, 1278
pusillum, 1278
vulgare, 1278
subsp. *spontaneum,* 1280
HOREHOUND, 762
COMMON, 762
WATER-, 760, 762
WHITE, 762
HORNBEAM, 439
AMERICAN, 439
HOP-, 440

HORNBEAM *(cont.)*
EASTERN HOP-, 440
HOP-, 440
WOOLLY AMERICAN HOP-, 440
HORNED
BEAK-RUSH, 1158
BLADDERWORT, 787
-PONDWEED, 1350
-POPPY, 876
RED, 876
-RUSH, 1156, 1158
HORNED-PONDWEED FAMILY, 1350
HORNPOD, 798
LAX, 798
HORNWORT, 529
HORNWORT FAMILY, 529
HORRID THISTLE, 340
HORSE
-APPLE, 830, 831
-BEAN, 682, 706
-CHESTNUT, 737
-GENTIAN, 511
YELLOW-FLOWERED, 511
HERB, 329
MINT, 764, 765, 766
LONG-FLOWER, 765
YELLOW, 766
-NETTLE, 1028
WESTERN, 1028
WHITE, 1030
-PURSLANE, 220
DESERT, 220
-TAIL CONYZA, 341
HORSEBRIER, 1346
HORSECRIPPLER, 486
HORSETAIL, 176
KANSAS, 178
SMOOTH, 178
HORSETAIL FAMILY, 176
HORSEWEED, 341
Hosta, 1191
HOUND'S-TONGUE, 446
BLUE, 446
HOUSTON MEADOW-RUE, 930
Houstonia, 966
humifusa, 967
longifolia, 968
micrantha, 967
nigricans, 966
parviflora, 967
pusilla, 967
rosea, 968
subviscosa, 968
HOUSTONIA, LONG-LEAF, 968
Houttuynia, 984
cordata, 984
Hoya, 274
HUBAM-CLOVER, 676
HUBBARD SQUASH, 569
HUCKLE-BERRY, 581, 584
HUISACHE, 624
DAISY, 308
HUISACHILLO, 625
HUMBOLDT'S COYOTILLO, 933
HUMMINGBIRD, 434, 442, 498, 510, 555, 658, 738, 799, 815, 891, 922, 998, 1006, 1008, 1079
HUMMINGBIRD-BUSH, 212
Humulus, 502
HUNGARIAN-MILLET, 1319
HURRAH GRASS, 1335
HYACINTH, 1202
GARDEN, 1202
GRAPE-, 1206
MEADOW-, 1200
WATER-, 1339
WILD-, 1198, 1200
Hyacinthaceae, 1191, 1206
Hyacinthus, 1202
orientalis, 1202
HYALINE-SCALE CARIC SEDGE, 1122
Hybanthus, 1061
linearis, 1061
verticillatus, 1061
HYBRID VERBENA, 1050
HYDRANGEA FAMILY, 738
Hydrangea, 738
Hydrangeaceae, 738
Hydrilla, 1169
verticillata, 1168
Hydrocera, 433
Hydrocharitaceae, 1166
Hydrocotyle, 254
ranunculoides, 254
umbellata, 255
verticillata, 255
var. triradiata, 255
var. verticillata, 255
Hydrolea, 740
ovata, ▣/93, 740
HYDROLEA, HAIRY, 740
Hydrophyllaceae, 739
Hymenatherum tagetoides, 346
Hymenocallis, 1202
caroliniana, 1202
eulae, 1204
liriosme, ▣/93, 1204
Hymenopappus, 374
artemisiifolius, 374
flavescens, 375
scabiosaeus
var. *artemisiifolius*, 374
var. corymbosus, 374
tenuifolius, 375
Hymenoxys, 375, 420
linearifolia, 420
odorata, 375
scaposa, 420
turneri, 421
Hyoscymus, 1015
Hypericaceae, 544
Hypericum, 544
crux-andreae, 547
drummondii, 547
gentianoides, 547
gymnanthum, 547
hypericoides, 547
subsp. hypericoides, 547
subsp. multicaule, 547
mutilum, 547
perforatum, 547
pseudomaculatum, 548
punctatum, 548
sphaerocarpum, 548
walteri, 548
Hypochaeris, 375
brasiliensis var. tweedyi, 376
glabra, 376
microcephala var. albiflora, 376
radicata, 376
tweedyi, 376
Hypochoeris, 375
Hypoxidaceae, 1191
Hypoxis, 1204
hirsuta, 1204
leptocarpa, 1204
rigida, 1204
HYSSOP
COASTAL WATER-, 994
DISC WATER-, 994
GOLDEN HEDGE-, 1000
HAIRY HEDGE-, 1000
HEDGE-, 1000
PROSTRATE WATER-, 1003
SAW-TOOTH WATER-, 1003
SHAGGY HEDGE-, 1000
STALKED WATER-, 1003
VIRGINIA HEDGE-, 1000
WATER-, 994, 1003
YELLOW-SEED HEDGE-, 1000

I

Ibervillea, 569
lindheimeri, ▣/93, 569
ICEPLANT, 220, 427
ICEPLANT FAMILY, 220
Ilex, 269
decidua, 270

Ilex *(cont.)*
opaca, 270
paraguariensis, 269
vomitoria, 270
ILLINOIS
BUNDLE-FLOWER, 652
DESMANTHUS, 652
PONDWEED, 1343
Impatiens, 433
balsamina, 434
biflora, 434
capensis, 434
walleriana, 433
INCA-WHEAT, 222
INDIA
HELIOTROPE, 450
LOVE GRASS, 1272
INDIAN
-APPLE, 1020
ARROW-WOOD, 528
BLACKDRINK, 270
-BLANKET, 358
YELLOW, 358
-CHERRY, 933
-CHICKWEED, 827
CIGARTREE, 442
-CURRANT, 510
-FIG, 491
-FIRE, 774
GRASS, 1320
LONG-BRISTLED, 1322
SLENDER, 1322
YELLOW, 1322
-HEMP, 266
-MALLOW, 806
MOCK STRAWBERRY, 944
MUSTARD, 459
PAINTBRUSH, 996
CITRON, 998
LINDHEIMER'S, 998
LIPPED, 998
PRAIRIE, 998
-PEA, 630
-PHYSIC, 946
-PINK, 799
-PIPE, 584
-PLANTAIN, 314
GROOVE-STEM, 314
LANCE-LEAF, 314
TUBEROUS, 314
-PLUME, 890
-POTATO, 556
REED, 1322
RUBBER-TREE, 830
RUSH-PEA, 663
-SPICE, 1060
-STRAWBERRY, 944
TEA, 422
-TOBACCO, 427, 1010
-TURNIP, 1092
INDIGO, 664
BASTARD, 626
BLUE WILD, 636
BLUE-, 636
-BUSH AMORPHA, 626
BUSH'S WILD, 638
COAST, 664
FALSE, 626, 635
GREEN WILD, 638
NUTTALL'S WILD, 638
PLAINS WILD, 638
TWO-COLOR WILD, 636
VARICOLORED WILD, 638
WESTERN, 664
WILD, 635
BLUE-, 636
Indigofera, 664
miniata, 664
var. leptosepala, ▣/94, 664
var. miniata, 664
suffruticosa, 664
tinctoria, 664
INFLATED CARIC SEDGE, 1122
INJERTO, 1064
INKBERRY, 881
INLAND
CEANOTHUS, 932
MUHLY, 1288
RUSH, 1184
SALT GRASS, 1264
INNOCENCE, 968
INTERMEDIATE LION'S-HEART, 769
Iodanthus, 470
pinnatifidus, 470
Ionactis, 315
IOWA THISTLE, 338
IPECAC, AMERICAN, 946
Ipheion, 1204
uniflorum, 1204
Ipomoea, 554
batatas, 555
carletoni, 558
coccinea, 555
cordatotriloba, 556
var. cordatotriloba, ▣/94, 556
var. torreyana, 556
hederacea, 556
lacunosa, 556
leptophylla, 556
pandurata, ▣/94, 556
purpurea, 556
var. *diversifolia*, 558
quamoclit, 558
shumardiana, 558
sinuata, 558
tamnifolia, 558
trichocarpa, 556
var. *torreyana*, 556
wrightii, 558
Ipomopsis, 890
rubra, ▣/94, 890
Iresine, 228
celosia, 230
diffusa, 230
rhizomatosa, 230
Iridaceae, 1172
Iris, 1173
brevicaulis, 1173
fulva, 1173
germanica, 1174
giganticaerulea, 1173
pallida, 1174
pseudacorus, 1174
xiphium, 1174
IRIS
DUTCH, 1174
GARDEN, 1174
LOUISIANA, 1173
PRAIRIE-, 1174
SPANISH, 1174
IRIS FAMILY, 1172
IRON OAK, 718
IRONPLANT
CUT-LEAF, 387
SPINY, 387
IRONWEED, 427, 1030
BALDWIN'S, 428
MISSOURI, 428
NARROW-LEAF, 428
PLAINS, 428
TALL, 428
TEXAS, 428
WESTERN, 428
WOOLLY, 428
IRONWOOD, 439, 440, 982
Isanthus braciatus, 783
Isoetaceae, 175
Isoetes, 175
melanopoda, 175
Isolepis, 1152
carinata, 1154
koilolepis, 1154
molesta, 1154
ITALIAN
CLOVER, 704
-MILLET, 1319
RYE GRASS, 1282

Iva, 376
angustifolia, 378
annua, 378
IVY, 272
AMERICAN-, 1068
BOSTON-, 1068
ENGLISH-, 272
GROUND-, 756
JAPANESE-, 1068
-LEAF MORNING-GLORY, 556
MARINE-, 1066
POISON-, 234, 236
-TREEBINE, 1066
Ixora, 960

J

JABONCILLO, 980
Jacaranda, 440
JACK
-IN-THE-PULPIT, 1092
OAK, 716
JACKFRUIT, 827
JACOB'S-LADDER, 888
Jacquemontia, 558
tamnifolia, 558
JADE PLANT, 561
JALAP
FALSE, 840
WILD, 437
JALAPEÑO, 1018
JAMAICAN SAW-GRASS, 1132
JAMBERRY, 1023
JAMES' NAILWORT, 520
JAMESTOWN WEED, 1020
JAPANESE
-APPLE, 936
ARROWROOT, 690
BROME, 1252
BUSH-CLOVER, 665
CHESS, 1252
-HAWKWEED, 433
HONEYSUCKLE, 508
-IVY, 1068
LESPEDEZA, 665
PERSIMMON, 578
PRIVET, 850
ROSE, 958
STAR-JASMINE, 268
VARNISHTREE, 1032
WISTERIA, 710
JARA, 324
DULCE, 324
JASMINE, 849
AMERICAN STAR-, 268
ASIAN-, 268
BLUE-, 921
CAYENNE-, 266
CONFEDERATE-, 268
JAPANESE STAR-, 268
ROCK-, 912
WINTER, 849
Jasminum, 849
nudiflorum, 849
Jatropha texana, 598
JEFF DAVIS' CHOLLA, 492
JERSEY-TEA, 932
JERUSALEM
-ARTICHOKE, 370
-OAK, 536
-THORN, 682
JESSAMINE, 849
CAROLINA, 798
FALSE, 1022
YELLOW, 797, 798
JEWELWEED, 433, 434
JIMSONWEED, 1018, 1020
JOCOYOTE, 870
JOHNNY-JUMP-UP, 1062
JOHNSON GRASS, 1324
JOINT GRASS, 1306
JOINTED
FLAT SEDGE, 1136
GOAT GRASS, 1235
RUSH, 1184
JOINT
-FIR, 207, 208
VINE, 208
-TAIL, 1258
CAROLINA, 1258
-VETCH, 625
JOINT-FIR FAMILY, 207
JOINTWEED, 899
SMALL, 899
SOUTHERN, 899
JO-JO WEED, 416
JONQUIL, 1207
JOSEPH FRANK'S CARIC SEDGE, 1122
JUBA'S-BUSH, 230
JUDASTREE, 640
Juglandaceae, 744
Juglans, 748
major, 748
microcarpa, 748
nigra, 750
regia, 744
JUJUBE, 936
CHINESE, 936
COMMON, 936
JUMPING
CACTUS, 492
CHOLLA, 492
JUMPSEED, 905
Juncaceae, 1178
Juncus, 1179
acuminatus, 1182
brachycarpus, 1182
brachyphyllus, 1182
bufonius, 1182
capitatus, 1182
coriaceus, 1182
dichotomus, 1182
diffusissimus, 1182
dudleyi, 1182
effusus var. solutus, 1182
filipendulus, 1184
interior, 1184
marginatus, 1184
var. *setosus,* 1184
nodatus, 1184
polycephalus, 1184
scirpoides, 1184
setosus, 1184
tenuis, 1184
texanus, 1184
torreyi, 1184
validus, 1184
var. fascinatus, 1184
var. validus, 1186
JUNE
-BERRY, 938, 940
BLUE GRASS, 1314
CENTAURY, 724
GRASS, 1280
PRAIRIE, 1280
JUNGLE-RICE, 1266
JUNIPER, 202
ASHE'S, 203
-LEAF, 480
MEXICAN, 203
PINCHOT'S, 203
RED, 203
-BERRY, 203
Juniperus, 202
ashei, 203
communis, 203
pinchotii, 203
virginiana, 203
JUNO'S-TEARS, 1055
JUPITER'S-STAFF, 1010
Jussiaea, 856
repens var. *glabrescens,* 862
Justicia, 213
americana, ▣/94, 213
brandegeeana, 211, 213
lanceolata, 213
ovata var. lanceolata, 213
pilosella, 218

K

KAFIR, 1324
KAKI, 578
Kalanchoe, 561
KALE, ORNAMENTAL, 460
Kallstroemia, 1076
  californica, 1076
  hirsutissima, 1076
  parviflora, 1076
Kalmia, 581
KALO, 1093
KANSAS
  GAYFEATHER, 384
  HORSETAIL, 178
  SCOURING-RUSH, 178
  -THISTLE, 1030
KAOLANG, 1324
KAPOK, 804
KARNES SCHRANKIA, 678
Karwinskia, 933
  humboldtiana, 933
KEARNEY'S
  GAURA, 856
  THREEAWN, 1242
KENDALL BASSWOOD, 1036
KENTUCKY
  BLUE GRASS, 1314
  COFFEE TREE, 663
KHAT, 526
KIDNEY
  -LEAF BUTTERCUP, 926
  -SHAPED CARIC SEDGE, 1128
KIDNEYWOOD, 658
  TEXAS, 658
Kigelia, 440
KINDLING-WEED, 365
KING RANCH BLUESTEM, 1246
KISSES, 856
KLAMATHWEED, 547
KLEBERG BLUESTEM, 1262
KLEIN GRASS, 1296
KLEIN'S CHOLLA, 491
KNAPWEED, 332
  AMERICAN, 332
  SPOTTED, 332
KNOCKAWAY, 443
KNOT
  GRASS, 1306
  -LEAF RUSH, 1182
  -ROOT BRISTLE GRASS, 1319
KNOTTED HEDGE-PARSLEY, 262
KNOTWEED, 899, 902
  BUSHY, 904
  LEAF-FLOWER, 611
  PLEAT-LEAF, 904
  PROSTRATE, 902
  THICKET, 904
KNOTWEED FAMILY, 897
Kochia, 540
  scoparia, 540
Koeleria, 1280
  cristata, 1280
  macrantha, 1280
  pyramidata, 1280
Koelreuteria paniculata, 979
KOHLRABI, 460
Kohlrauschia, 522
KOREAN
  BUSH-CLOVER, 664
  LESPEDEZA, 664
KOYAMA, 1108
KR BLUESTEM, 1246
Krameria, 750
  lanceolata, ⊞/95, 750
  secundiflora, 750
KRAMERIA FAMILY, 750
Krameriaceae, 750
Krigia, 378
  cespitosa, 380
    forma cespitosa, 380
    forma gracilis, 380
  dandelion, 380
  gracilis, 380
  occidentalis, 380
  oppositifolia, 380
  virginica, 380
  wrightii, 380
KUDSU, 690
KUDZUVINE, 690
Kuhnia, 328
  eupatorioides, 329
KUHNIA
  PLAINS, 329
  PRAIRIE, 329
Kummerowia, 664
  stipulacea, 664
  striata, 665
Kyllinga, 1154
  brevifolia, 1154
  pumila, 1154

L

Labiatae, 752
LACE
  CACTUS, 486
    BROWN, 486
    WHITE, 486
  -BARK ELM, 1040
Lactuca, 380
  canadensis, 381
  floridana, 381
  ludoviciana, 381
  saligna, 381
  sativa, 380
  serriola, 381
LADIES'
  -EARDROPS, 898
  -SORREL, CREEPING, 870
  -TOBACCO, 310
  -TRESSES, 1218, 1220
    COMMON, 1220
    GRASS-LEAF, 1222
    GRAY'S, 1222
    GREAT PLAINS, 1222
    GREEN-LIP, 1220
    LITTLE, 1222
    NODDING, 1220
    OCTOBER, 1222
    OVAL, 1222
    SLENDER, 1220
    SPRING, 1222
    TWISTED, 1222
    UPLAND, 1222
LADY
  BIRD'S CENTAURY, 726
  FERN, 184
    LOWLAND, 184
    SOUTHERN, 184
LADY'S
  -COMB, 260
  -EARRINGS, 434
  -FINGER, 806
  -LEG, 582
  -THUMB, 904
Lagenaria, 569
  siceraria, 569
  vulgaris, 570
Lagerstroemia, 800
  indica, 800
LAMBERT'S
  CRAZYWEED, 682
  LOCO, 682
LAMBKILL, 581
LAMB'S
  -LETTUCE, 1043
  -QUARTERS, 533, 535
    NARROW-LEAF, 536
    WORM-SEED, 535
Lamiaceae, 752
Lamium, 759
  amplexicaule, 759
  purpureum, 759
    var. incisum, 759
    var. purpureum, 759
LAMPAZO AMARILLO, 845
LAMPAZOS, 845

LANCE
  COREOPSIS, 342
  -LEAF
    BUCKTHORN, 934
    COTTON-FLOWER, 228
    FROGFRUIT, 1054
    GAILLARDIA, 358
    GROUND-CHERRY, 1024
    INDIAN-PLANTAIN, 314
    LOOSESTRIFE, 802, 913
    RAGWEED, 309
    SAGE, 776
    WATER-WILLOW, 213
  -SPIKE SPIKE-RUSH, 1147
LANGE'S PASPALUM, 1306
Lantana, 1052
  achyranthifolia, 1052
  camara, 1052
    var. *mista*, 1053
  horrida, 1053
    var. *latibracteata*, 1053
  macropoda, 1052
  urticoides, ▣/95, 1053
LANTANA
  BRUSHLAND, 1052
  COMMON, 1053
  LARGE-LEAF, 1052
  TEXAS, 1053
  VEIN-LEAF, 1052
  WEST INDIAN, 1052
LANTEN, 886
LAPPA CLOVER, 704
Lappula, 450
  occidentalis, 450
    var. cupulata, 450
    var. occidentalis, 452
  redowskii, 452
    var. *cupulata*, 452
    var. *occidentalis*, 452
    var. *texana*, 452
  texana, 452
LARGE
  BEARDTONGUE, 1008
  -BRACTED VERVAIN, 1056
  BUTTERCUP, 928
  BUTTONWEED, 962
  CLAMMYWEED, 507
  CRAB GRASS, 1264
  CRESTPETAL, 507
  -FLOWER
    BROOMRAPE, 868
    DOLLS'S DAISY, 328
    NEMOPHILA, 742
    PHLOX, 891
  -FOOT PEPPERWORT, 186
  HOP CLOVER, 704
  -LEAF
    LANTANA, 1052
    PUSSY-TOES, 310
  -SEED DODDER, 574
  -SPIKE SPIKE-RUSH, 1147
  WOODSIA, 185
LARGER WATERWORT, 496
LARGEST BROWN-EYED-SUSAN, 402
LARKSPUR, 922
  ANNUAL, 922
  BLUE, 924
  CAROLINA, 924
  GULF COAST, 924
  PENARD'S, 924
  PINEWOODS, 924
  PLAINS, 924
  PRAIRIE, 924
  ROCKET, 922
  WHITE, 924
  WILD BLUE, 924
Larrea tridentata, 1074
LASATER'S-PRIDE, 836
LATE
  CORALROOT, 1215
  EUPATORIUM, 354
  -FLOWERING THOROUGHWORT, 354
  GOLDENROD, 410
  PURPLE ASTER, 319
Lathyrus, 665
  aphaca, 666
  hirsutus, 666
  latifolius, 666
  pusillus, 666
  venosus, 666
    var. *intonus*, 666
Lauraceae, 784
LAUREL
  CHERRY, 649, 947
    CAROLINA, 949
  MOUNTAIN-, 581, 697
  TEXAS, MOUNTAIN-, 697
LAUREL FAMILY, 784
Laurus nobilis, 784
LAVENDER, 752
  WILD-, 1060
Lavendula, 752
LAWN BURWEED, 416
LAWNFLOWER, PROSTRATE, 329
Lawsonia, 799
LAX
  BLADDERPOD, 474
  HORNPOD, 798
LAZY DAISY, 312
  ARIZONA, 312
  ARKANSAS, 314
  PLAINS, 312
  RIDDELLS', 314
LEADPLANT, 626
LEAF
  -FLOWER, 610
    CAROLINA, 611
    DRUMMOND'S, 611
    KNOTWEED, 611
    PEEWATER, 611
  -LESS COWBANE, 256
  MUSTARD, 459
LEAFCUP, 408
  HAIRY, 408
LEAFLESS ORCHID, 1216
LEAFY
  ELEPHANTOPUS, 348
  PONDWEED, 1343
  ROSE, 956
LEANTREE, 439
LEAST
  BLUE-EYED-GRASS, 1178
  DAISY, 334
    COMMON, 334
  DUCKWEED, 1188
  HOP CLOVER, 704
  SNOUT-BEAN, 691
LEATHER
  -FLOWER, 921
    RED, 922
  RUSH, 1182
LEAVENWORTH'S
  ERYNGO, 252
  VETCH, 707
Lechea, 543
  mucronata, 543
  san-sabeana, 544
  tenuifolia, 544
  villosa, 543
LECHILLO, 439
LEEK, 1194, 1196
  WILD, 1196
Leersia, 1280
  lenticularis, 1280
  oryzoides, 1281
  virginica, 1281
Legousia, 500
  coloradoensis, 501
LEGUME FAMILY, 617
Leguminosae, 617
Lemna, 1187
  aequinoctialis, 1188
  minima, 1188
  minor, 1188
    var. *obscura*, 1188
  minuscula, 1188
  minuta, 1188
  obscura, 1188

Lemna *(cont.)*
perpusilla, 1188
trinervis, 1188
turionifera, 1188
valdiviana, 1188
var. *minima,* 1188
Lemnaceae, 1186
LEMON, 970
BEEBALM, 765
-MINT, 765
PAINTBRUSH, 998
PAINTEDCUP, 998
LENGUA DE VACA, 491
Lens, 617
LENTEJILLA, 472
Lentibulariaceae, 786
LENTIL, 617
LENTISCO, 234
Leonotis, 759
nepetifolia, 759
Leonurus, 759
cardiaca, 760
sibiricus, 760
Lepidium, 470
austrinum, 470
densiflorum, 470
oblongum, 470
ruderale, 472
virginicum var. *medium,* 472
Leptochloa, 1281
dubia, 1281
fascicularis, 1281
filiformis, 1281
mucronata, 1281
uninerva, 1282
Leptoloma, 1262
cognatum, 1263
Leptopus, 610
phyllanthoides, 610
Lepuropetalon, 988
spathulatum, 988
Lespedeza, 666
capitata, 670
cuneata, 670
hirta, 670
intermedia, 671
procumbens, 670
repens, 670
stipulacea, 665
striata, 665
stuevei, 670
texana, 670
violacea, 670
virginica, 671
LESPEDEZA, 664
COMMON, 665
CREEPING, 670
HAIRY, 670
JAPANESE, 665
KOREAN, 664
ROUND-HEAD, 670
SERICEA, 670
SLENDER, 671
TEXAS, 670
TRAILING, 670
VIOLET, 670
Lesquerella, 472
angustifolia, 473
auriculata, 473
densiflora, 473
engelmannii, 473
fendleri, 473
gordonii, 473
gracilis, 474
subsp. gracilis, 474
subsp. nuttallii, 474
var. *repanda,* 474
grandiflora, 474
nuttallii, 474
recurvata, 474
LESSER CHICKWEED, 526
LETTUCE, 380
FLORIDA, 381
GARDEN, 381
LAMB'S-, 1043
PRICKLY, 381
ROCK-, 394
SMALL ROCK-, 394
WATER-, 1093
WESTERN WILD, 381
WILD, 381
WILLOW-LEAF, 381
WOODLAND, 381
Leucanthemum, 382
leucanthemum, 382
vulgare, 382
Leucelene ericoides, 335
Leucojum, 1204
aestivum, 1204
vernum, 1206
Leucophyllum, 988
frutescens, 988
Leucospora, 1000
multifida, 1000
Leucosyris, 335
spinosa, 335
Leucothoe, 581
LEVERWOOD, 440
LEWIS' FLAX, 792
Liatris, 382
acidota, 386
aspera, ■/95, 384
elegans, 384
glabrata, 386
var. *alabamensis,* 386
mucronata, 384
pycnostachya, 384
squarrosa, 384
var. alabamensis, 386
var. glabrata, ■/95, 386
LICE, BEGGAR'S-, 262
LICORICE, WILD, 964
LIGHT
GOOSEFOOT, 538
POPPY-MALLOW, 807
LIGNUM VITAE, 1074
Ligustrum, 849
japonicum, 850
lucidum, 850
quihoui, 850
sinense, 850
vulgare, 850
LIGUSTRUM, WAX-LEAF, 850
LILAC, 846
Liliaceae, 1191
Lilium, 1191
candidum, 1192
LILY, 1191
ATAMOSCO-, 1201
CAMASS-, 1200
CELESTIAL-, 1174
COPPER-, 1201
COW-, 844
DAY-, 1201
FAWY-, 1200
FUNNEL-, 1198
GLADE-, 864
GREEN-, 1210
RAIN-, 1200
SPIDER-, 1202, 1206
SWAMP-, 986
TROUT-, 1201
WATER, 844
YELLOW POND-, 844
LILY FAMILY, 1191
LILYTURF, 1191
LIME, 970, 1036
LIMEROCK
BROOKWEED, 914
MAPLE, 219
LIMESTONE
ADDER'S-TONGUE, 190
RUELLIA, 218
Limnodea, 1282
arkansana, 1282
Limnosciadium, 255
pinnatum, 255
pumilum, 255

Linaceae, 787
Linaria, 1002
canadensis, 1004
var. *texana,* 1006
minor, 999
texana, 1006
vulgaris, 1002
LINAZA, 792
LINDEN, 1036
FLORIDA, 1036
LINDEN FAMILY, 1035
Lindera, 784
benzoin var. pubescens, 786
Lindernia, 1002
anagallidea, 1002
dubia var. anagallidea, 1002
Lindheimera, 386
texana, 386
LINDHEIMER'S
BEEBALM, 765
BLACK WILLOW, 978
COPPERLEAF, 588
CROTON, 600
CROWNBEARD, 427
DAISY, 386
GLOBEBERRY, 569
INDIAN PAINTBRUSH, 998
LIP FERN, 198
MAIDEN FERN, 200
MILK-VETCH, 632
MUHLY, 1288
NAILWORT, 520
NOLINA, 1080
PAINTBRUSH, 998
PANIC, 1295
ROCK DAISY, 394
ROSETTE GRASS, 1295
SENNA, 694
SILKTASSEL, 723
TEPHROSIA, 700
VENUS'-LOOKING-GLASS, 501
LINEAR-LEAF
FOUR-O'CLOCK, 840
LUDWIGIA, 860
LINGBERRY, 584
LINGENBERRY, 584
Linum, 788
alatum, 790
berlandieri, 792
grandiflorum, 790
hudsonioides, 790
imbricatum, 790
lewisii, 792
medium var. texanum, 790
pratense, 790
rigidum, 790
var. berlandieri, ▣/95, 792
var. rigidum, 792
rupestre, 792
schiedeanum, 792
striatum, 792
sulcatum, 792
usitatissimum, 792
LION'S
-EAR, NEP-LEAF, 759
-HEAD, 759
-HEART, 768
FINGER, 769
INTERMEDIATE, 769
-TAIL, 760
LIP FERN, 195
ALABAMA, 196
BULB, 195
EATON'S, 196
FEE'S, 196
HAIRY, 198
LINDHEIMER'S, 198
ROUGH, 196
SLENDER, 196
SMOOTH, 196
WOOLLY, 196, 198
Lipocarpha, 1154
aristulata, 1155
drummondii, 1155
micrantha, 156
LIPPED INDIAN PAINTBRUSH, 998
Lippia, 1053
canescens, 1054
cuneifolia, 1054
graveolens, 1054
lanceolata, 1054
nodiflora, 1054
var. *canescens*, 1054
strigulosa, 1054
LIPPIA
RED BRUSH, 1054
SCENTED, 1054
Liquidambar, 736
orientalis, 736
styraciflua, 737
LIRA DE SAN PEDRO, 727
Liriope muscari, 1191
Litchi, 979
Lithocarpus, 710
Lithops, 220
Lithospermum, 452
arvense, 446
caroliniense, 452
incisum, ▣/96, 452
matamorense, 452
LITTLE
BARLEY, 1278
BLUESTEM, 1316
-CHILIS, 489
EBONY SPLEENWORT, 180
-HEAD
GUMWEED, 360
NUT-RUSH, 1164
-HIP HAWTHORN, 942
LADIES'-TRESSES, 1222
-LEAF
BUTTERCUP, 926
SUMAC, 234
TICK-CLOVER, 655
MALLOW, 812
QUAKING GRASS, 1250
WALNUT, 748
LITTLEPOD, 460
LIVE OAK, 716, 719
COAST, 719
VIRGINIA, 719
LIVING STONES, 220
LIZARD'S-TAIL, 984, 986
GAURA, 856
LIZARD'S-TAIL FAMILY, 983
Loasaceae, 794
Lobelia, 497
appendiculata, 498
cardinalis, ▣/96, 498
subsp. *graminea* var. *phyllostachya*, 498
var. *phyllostachya,* 498
puberula, 498
var. *mineolana*, 498
var. *simulans,* 498
siphilitica var. ludoviciana, ▣/96, 498
spicata, 500
tupa, 497
LOBELIA
BIG BLUE, 498
DOWNY, 498
GREAT, 498
LOUISIANA, 498
PALE-SPIKE, 500
LOBLOLLY PINE, 207
Lobularia, 456
LOCO
LAMBERT'S, 682
POINT, 682
PURPLE, 682
WHITE, 682
LOCOWEED, 628, 682
LOCUST, 691
BLACK, 692
BRISTLY, 691
COMMON HONEY-, 660
HONEY-, 660
MOSSY, 691

LOCUST *(cont.)*
WATER-, 660
WATER HONEY-, 660
Lodoicea maldivica, 1094
Loeflingia, 518
squarrosa, 518
subsp. *texana,* 518
texana, 518
LOEFLINGIA, SPREADING, 518
LOGANIA FAMILY, 797
Loganiaceae, 797
Lolium, 1276, 1282
arundinaceum, 1276
multiflorum, 1282
perenne, 1282
var. *aristatum,* 1282
var. *italicum,* 1282
subsp. multiflorum, 1282
var. *multiflorum,* 1282
subsp. perenne, 1282
temulentum, 1282
var. *leptochaeton,* 1284
Lomatium, 255
daucifolium, 255
foeniculaceum subsp. daucifolium, ▣/96, 255
LOMATIUM, CARROT-LEAF, 255
LOMBARDY POPLAR, 976
LONDON-ROCKET, 479
LONG
-AWNED
ARISTIDA, 1243
HAIR GRASS, 1288
THREEAWN, 1243
-BARB ARROWHEAD, 1090
-BRACT PRAIRIE-CLOVER, 650
-BRISTLED INDIAN GRASS, 1322
CLOAK FERN, 195
-FLOWER HORSE MINT, 765
-LEAF
HOUSTONIA, 968
PONDWEED, 1343
RUSH GRASS, 1327
WILD BUCKWHEAT, 899
WILLOW, 978
-LOBE ARROWHEAD, 1090
-MOSS, 1096
-PEDICEL WILLOW, 978
-SEPAL DODDER, 574
-SPIKE TRIDENS, 1330
-SPINE SANDBUR, 1254
-STYLE SWEETROOT, 256
LONGLEAF
PINE, 207
YELLOW PINE, 207
LONG'S GRAPE, 1070
LONGTAG PINE, 206
Lonicera, 508
albiflora, 508
var. *dumosa,* 508
fragrantissima, 508
japonica, 508
maackii, 508
sempervirens, ▣/96, 510
LOOFAH, 570
LOOSESTRIFE, 913
CALIFORNIA, 802
FALSE, 858
LANCE-LEAF, 802, 913
PURPLE, 800, 802
TRAILING, 913
WINGED, 802
LOOSESTRIFE FAMILY, 799
Lophophora williamsii, 484
LOPSEED, 880, 881
LOPSEED FAMILY, 880
Lorinsera areolata, 181
LOTEBUSH, 936
Lotus, 671
corniculatus, 671
purshianus, 671
unifoliolatus, 671
LOTUS, 834
MILK-VETCH, 632
SACRED, 834
YELLOW, 834
LOTUS FAMILY, 834
LOTUS-LILY FAMILY, 834
LOUISIANA
CARIC SEDGE, 1124
IRIS, 1173
LOBELIA, 498
LOUSEWORT, 1006
COMMON, 1006
EARLY, 1006
FERN-LEAF, 1006
LOVE GRASS, 1268
BIG-TOP, 1270
CREEPING, 1272
GUMMY, 1270
INDIA, 1272
MEDITERRANEAN, 1270
PLAINS, 1270
PURPLE, 1272
RED, 1272
SAND, 1274
SHORT-STALKED, 1270
SMOOTH CREEPING, 1270
SPREADING, 1272
STOUT, 1270
TEAL, 1270
TUMBLE, 1272
WEEPING, 1270
WILMANN'S, 1272
LOVE
-APPLE, 1030
-IN-THE-MIST, 921
LOVELL VIOLET, 1062
LOVEVINE, 572
LOW
BLUET, 967
CONYZA, 341
DITAXIS, 602
FLEABANE, 341
GROUND-CHERRY, 1026
HAIR GROUND-CHERRY, 1026
HOP CLOVER, 704
MENODORA, 850
NUT-RUSH, 1164
PEAVINE, 666
POPPY-MALLOW, 807
RUELLIA, 216
SPEAR GRASS, 1312
WILD MERCURY, 602
WINECUP, 807
LOWLAND LADY FERN, 184
LUCERNE, 676
LUCY, AUNT, 740
Ludwigia, 858
alternifolia, 860
decurrens, 860
glandulosa, 860
leptocarpa, 860
linearis, 860
natans, 862
var. *rotundata,* 862
octovalvis, 860
palustris, 860
peploides, ▣/97, 860
subsp. *glabrescens,* 862
var. *glabrescens,* 862
repens, 862
LUDWIGIA
CYLINDRIC-FRUIT, 860
LINEAR-LEAF, 860
Luffa, 570
aegyptiaca, 570
cylindrica, 570
Lunaria, 456
LUPIN, 671
LUPINE, 671
FLESHY-LEAF, 672
TEXAS, 672
Lupinus, 671
subcarnosus, 672
texensis, ▣/97, 672
Luzula, 1186
bulbosa, 1186

Luzula *(cont.)*
  bulbosa
    var. *bulbosa,* 1186
    var. *echinata,* 1186
  echinata var. mesochorea, 1186
LYCHEE, 979
Lycium, 1020
  barbarum, 1022
  halimifolium, 1022
Lycopersicon, 1027
  esculentum, 1030
Lycopodiaceae, 173
Lycopodiella, 173
  appressa, 174
  prostrata, 174
Lycopodiophyta, 173
Lycopodium
  appressum, 174
  prostratum, 174
Lycopus, 760
  americanus, 760
  rubellus, 762
    var. *arkansanus,* 762
    var. *lanceolatus,* 762
  virginicus, 762
Lycoris, 1206
  radiata, 1206
Lygodesmia, 386
  texana, ⊞/97, 386
Lyonia, 581, 582
  mariana, 582
LYRE-LEAF
  GREENEYES, 325
  SAGE, 776
  GREEN-EYED, 325
Lysimachia, 913
  lanceolata, 913
  radicans, 913
Lythraceae, 799
Lythrum, 800
  alatum var. lanceolatum, 802
  californicum, 802
  lanceolatum, 802
  salicaria, 802

M

MAACK'S HONEYSUCKLE, 508
MACARTNEY ROSE, 956
Machaeranthera, 387
  annua, 400
  pinnatifida, 387
Maclura, 830
  pomifera, 831
MADAGASCAR PERIWINKLE, 266
MADDER, 960
  FIELD-, 970
MADDER FAMILY, 960
MAD-DOG SKULLCAP, 779
MADEIRA-VINE, 434
MADEIRA-VINE FAMILY, 434
MADRONE, 582
  TEXAS, 582
MADROÑO, 582
Magnoliophyta, 208
MAGUEY, 1078
MAHOGANY, 822
  CEYLON, 823
MAHOGANY FAMILY, 822
Mahonia trifoliolata, 437
MAIDEN FERN
  LINDHEIMER'S, 200
  WIDESPREAD, 200
MAIDENBUSH, 610
MAIDENHAIR
  FERN, 194
  SOUTHERN, 194
MAIDENHAIR FERN FAMILY, 193
MAIZE, 1338
MALA MUJER, 596, 1030
MALLOW, 812
  ALKALI-, 815
  CLUSTERED POPPY-, 807
  COMMON, 812, 814
    ROSE-, 810
  COPPER-, 820
  DRUMMOND'S WAX-, 815
  DWARF, 814
  FALSE, 814, 818
  FINGER POPPY-, 808
  GERANIUM POPPY-, 808
  HALBERD-LEAF ROSE-, 810
  HIGH, 814
  INDIAN-, 806
  LIGHT POPPY-, 807
  LITTLE, 812
  LOW POPPY-, 807
  MARSH-, 810
  NARROW-LEAF GLOBE-, 820
  NORTHERN, 814
  PLAINS POPPY-, 807
  POINT-SEED GLOBE-, 820
  POPPY-, 807
  PRICKLY-, 818
  PURPLE-, 808
    POPPY-, 807
  RED FALSE, 820
  ROSE-, 810
  ROUND-LEAF, 814
  RUNNING, 814
  SCARLET
    GLOBE-, 820
    ROSE-, 810
  SLIM-LOBE POPPY-, 808
  SWAMP ROSE-, 810
  TALL POPPY-, 808
  THREE-LOBE FALSE, 814
  VENICE-, 812
  WOOLLY ROSE-, 810
  WRIGHT'S FALSE, 814
MALLOW FAMILY, 804
MALTA
  CENTAUREA, 334
  STAR-THISTLE, 334
Malus, 936
Malva, 812
  neglecta, 812
  parviflora, 812
  rotundifolia, 814
  sylvestris, 814
  wrightii, 814
Malvaceae, 804
Malvastrum, 814
  aurantiacum, 814
  coromandelianum, 814
Malvaviscus, 814
  arboreus var. drummondii, 815
  drummondii, 815
Malvella, 815
  lepidota, 815
  leprosa, 815
  sagittifolia, 816
Mammea americana, 544
MAMMEY-APPLE, 544
Mammillaria, 489
  echinus, 485
  gummifera var. *applanata,* 489
  heyderi, ⊞/97, 489
    var. *applanata,* 489
  similis, 488
  sulcata, 485
  vivipara var. *radiosa,* 489
MAMMILLARIA, FLATTENED, 489
MANCA CABALLO, 486
Mandragora, 1015
MANDRAKE, 437, 1015
  AMERICAN-, 437
Manfreda, 1079
  maculosa, 1080
  virginica, 1079
    subsp. lata, ⊞/98, 1079
    subsp. virginica, 1080
Mangifera indica, 230
MANGLIER, 324
MANGO, 230
MANGOSTEEN, 544
Manihot esculenta, 585
MANIOC, 585
Manisuris cylindrica, 1258

- MANNA GRASS, 1277
  - ARKANSAS, 1277
  - EASTERN, 1277
  - FOWL, 1277
  - NERVED, 1277
- MAN-OF-THE-EARTH, 556
- MANY-FLOWER
  - AGRIMONY, 938
  - GROOVEBUR, 938
- MANYSEED, FOUR-LEAF, 522
- MANY-STEM
  - EVAX, 356
  - FALSE DANDELION, 399
- MANZANILLA SILVESTRE, 342
- MANZANITA, 581
- MAPLE, 219
  - ASH-LEAF, 219
  - LIMEROCK, 219
  - PLATEAU BIG-TOOTH, 219
  - RED, 220
  - SCARLET, 220
  - SILVER, 220
  - SOFT, 220
  - SOUTHERN SUGAR, 219
  - SUGAR, 219
  - UVALDE BIG-TOOTH, 219
- MAPLE FAMILY, 218
- MAPLE-LEAF GOOSEFOOT, 538
- *Maragaranthus,* 1023
- *Maranta,* 1210
  - *arundinacea,* 1210
- **Marantaceae,** 1210
- MARBLESEED, 453
  - BEJAR, 454
  - HELLER'S, 454
  - ROUGH, 454
  - WESTERN, 454
- **Margaranthus,** 1022
  - solanaceus, 1022
- MARGUERITE, 382
- MARIGOLD, 287
  - DOGWEED, 346
  - BUR-, 325
  - FOETID-, 346, 422
  - MARSH-, 916
- MARIJUANA, 502
- MARINE-IVY, 1066
- MARINEVINE, 1066
- *Mariscus mariscoides,* 1132
- MARJORAM, 752
- MARRUBIO, 762
- **Marrubium,** 762
  - vulgare, 762
- MARSH
  - BUTTERCUP, 928
  - CONEFLOWER, 402
  - -ELDER, 376, 378
  - -FLEABANE, 396
  - GRASS, 1324
    - TALL, 1324
  - -MALLOW, 804, 810
  - -MARIGOLD, 916
  - MERMAID-WEED, 736
  - MILKWORT, 896
  - -MILLET, 1338
  - -PURSLANE, 860
  - SANDSPURRY, 525
  - SEEDBOX, 860
- MARSH FERN FAMILY, 200
- **Marshallia,** 387
  - caespitosa, 387
    - var. caespitosa, 387
    - var. signata, ▣/98, 387
- MARSHCRESS, BOG, 478
- **Marsilea,** 186
  - macropoda, 186
  - *mucronata,* 188
  - *tenuifolia,* 186
  - vestita, 186
    - subsp. tenuifolia, 186
    - subsp. vestita, 186
- **Marsileaceae,** 185
- *Martynia louisianica,* 880
- MARVEL-OF-PERU, 840
- MARYLAND
  - FIGWORT, 1010
  - MEADOW-BEAUTY, 822
  - SENNA, 694
  - STONE-MIST, 756
  - TICK-CLOVER, 655
- MASTER, RATTLESNAKE-, 252
- MAT
  - BLUETS, 967
  - CHAFF-FLOWER, 222
  - EUPHORBIA, 596
  - -GRASS, 1054
  - WATERWORT, 496
- MATCHBRUSH, 365
- MATCHWEED, 365
- **Matelea,** 283
  - biflora, ▣/98, 284
  - cynanchoides, 284
  - decipiens, 284
  - edwardsensis, ▣/98, 284
  - gonocarpos, 284
  - reticulata, ▣/98, 284
- MATRIMONY-VINE, 1020
- **Matthiola,** 474
  - *bicornis,* 474
  - longipetala, 474
- MAUCHIA, 336
- *Maurandella antirrhiniflora,* 1003
- MAURANDELLA, SNAPDRAGON, 1002
- **Maurandya,** 1002
  - antirrhiniflora, ▣/99, 1002
- MAXIMILIAN SUNFLOWER, 370
- MAY-APPLE, 437
- MAYHAW, 940
- MAYPOP PASSION-FLOWER, 878
- MAYWEED, 310
  - CHAMOMILE, 310
- **Mazus,** 1003
  - *japonicus,* 1003
  - pumilus, 1003
- MEADOW
  - -BEAUTY, 820
    - COMMON, 822
    - MARYLAND, 822
  - DAUCOSMA, 249
  - DROPSEED, 1327
  - FLAX, 790
  - GRASS
    - DWARF, 1312
    - SALT, 1281
  - HOLLY, 270
  - -HYACINTH, 1200
  - QUILL, 1200
  - -PINK, 728
  - -RUE, 928
    - ARKANSAS, 930
    - HOUSTON, 930
    - PURPLE, 930
  - SPIKE-MOSS, 174
- MEADOW-BEAUTY FAMILY, 820
- MEAD'S CARIC SEDGE, 1124
- MEALY
  - CORYDALIS, 720
  - -CUP SAGE, 776
  - SAGE, 776
- **Mecardonia,** 1003
  - acuminata, 1003
  - procumbens, 1003
  - *vandellioides,* 1003
- MECARDONIA
  - PROSTRATE, 1003
  - PURPLE, 1003
- MEDIC, 672
  - BUTTON, 674
  - SPOTTED, 674
- **Medicago,** 672
  - arabica, 674
  - *hispida,* 676
  - lupulina, 674
  - minima, 674
  - orbicularis, 674
  - polymorpha, 674
    - var. *vulgaris,* 676
  - sativa, 676

MEDICK, 672
BLACK, 674
MEDITERRANEAN LOVE GRASS, 1270
Megapterium oklahomense, 864
MEJORANA, 774, 1052
Melampodium, 387
leucanthum, 388
Melastomataceae, 820
MELASTOME FAMILY, 820
Melia, 822
azedarach, 823
Meliaceae, 822
MELIC, 1284
GRASS, 1284
NARROW, 1284
TALL, 1284
THREE-FLOWER, 1284
TWO-FLOWER, 1284
Melica, 1284
mutica, 1284
nitens, 1284
MELILOT, 676
WHITE, 676
YELLOW, 676
Melilotus, 676
albus, 676
indicus, 676
officinalis, 676
MELINES LEAVENWORTH'S CARIC SEDGE, 1124
Melochia, 1032
pyramidata, 1032
MELOCHIA, ANGLE-POD, 1032
MELON, 566
-LEAF NIGHTSHADE, 1028
PRESERVING, 566
MELONCITO, 570
MELONETTE, 570
DROOPING, 570
Melongena, 1027
Melothria, 570
pendula, 570
Menispermaceae, 823
Menispermum, 824
canadense, 824
Menodora, 850
heterophylla, 850
longiflora, 852
MENODORA, LOW, 850
SHOWY, 852
Mentha, 762
aquatica, 764
arvensis, 764
longifolia, 764
×piperita, 764
×rotundifolia, 764
spicata, 764
suaveolens, 764
Mentzelia, 794
albescens, 796
decapetala, 796
nuda, 796
var. *stricta*, 796
oligosperma, 796
reverchonii, 797
stricta, 796
MENTZELIA
BRACTLESS, 796
TEN-PETAL, 796
WAVY-LEAF, 796
Menyanthaceae, 824
MERCURY
HILL COUNTRY WILD, 602
LOW WILD, 602
SMOOTH WILD, 602
TALL WILD, 604
THREE-SEEDED, 586
WILD, 600
MERMAID-WEED, 736
MARSH, 736
Merremia, 558
dissecta, 558
Mertensia, 443
MESCAL-BEAN, 697
Mesembryanthemum, 220
MESQUITE, 688
GRASS, 1248
COMMON CURLY-, 1278
HONEY, 688
VINE-, 1300
MESQUITEWEED, 663
Metasequoia glyptostroboides, 202
Metastelma barbigerum, 282
MEXICAN
BALD CYPRESS, 204
-BUCKEYE, 982
-CLOVER, 968
PRAIRIE, 970
ROUGH, 968
DEVILWEED, 335
EVENING-PRIMROSE, 866
FERN, 179
FIREBUSH, 540
FIREWEED, 540
GOLD-POPPY, 874
-HAT, 400
JUNIPER, 203
MORNING-GLORY, 556
MULBERRY, 832
PALOVERDE, 682
PERSIMMON, 579
PLUM, 949
-POPPY, 874
-PRIMROSE, 866
SAGEBRUSH, 315
SILKTASSEL, 723
SPRANGLETOP, 1282
-TEA, 207, 535
MICHAELMAS DAISY, 370
MIDLAND CARIC SEDGE, 1124
MIGNONETTE-VINE, 434
Mikania, 388
scandens, 388
MILFOIL, 308
WATER-, 733
MILK
-PEA, 658
DOWNY, 660
HOARY, 660
-PINK, 386
-THISTLE, 406
BLESSED, 406
-VETCH, 628, 630, 632
BENT-POD, 632
BODKIN, 632
CANADA, 630
DROOPING, 635
ENGELMANN'S, 632
GROUND-PLUM, 630
LINDHEIMER'S, 632
LOTUS, 632
NUTTALL'S, 634
OZARK, 632
PLATTE RIVER, 635
SLIM-POD, 632
SMALL-FLOWER, 634
SOUTHWESTERN, 635
TEXAS, 635
WRIGHT'S, 635
MILKVINE, 283, 284
NET-VEIN, 284
TWO-FLOWER, 284
MILKWEED, 276
BLUNT-LEAF, 277
BUTTERFLY, 280
CLIMBING-, 283, 284
EASTERN WHORLED, 280
ENGELMANN'S, 278
GREEN, 282
NARROW-LEAF, 280
ORANGE, 280
SAND, 278
SHORE, 280
SIDE-CLUSTER, 278
SLIM, 278
-LEAF, 280
SWAMP, 278
TEXAS, 280

MILKWEED *(cont.)*
THIN-LEAF, 280
TRAILING, 278
WAND, 282
WHITE-FLOWER, 280
WHORLED, 280
MILKWEED FAMILY, 274
MILKWEEDVINE, GREEN, 284
WAVY-LEAF, 283
MILKWORT, 894
BITTER, 896
BLOOD, 896
MARSH, 896
PINK, 896
RACEMED, 896
ROCK, 896
SHRUBBY, 896
SLENDER, 896
WHITE, 896
WHORLED, 896
MILKWORT FAMILY, 894
MILLET
BROOMCORN, 1300
COMMON, 1300
FINGER-, 1268
FOXTAIL-, 1318, 1319
GERMAN-, 1319
HOG, 1300
HUNGARIAN-, 1319
ITALIAN-, 1319
MARSH-, 1338
RUSSIAN, 1300
SHAMA-, 1266
TEXAS, 1335
MILO, 1324
MIMBRE, 443
Mimosa, 677
aculeaticarpa var. biuncifera, 678
biuncifera, 678
borealis, 678
hystricina, 678
latidens, 678
nuttallii, 678
quadrivalis, 677
var. *hystricina,* 678
var. *latidens,* 678
var. *nuttallii,* 678
var. *platycarpa,* 678
roemeriana, ▣/99, 678
strigillosa, 680
MIMOSA, 625
FRAGRANT, 678
HERBACEOUS, 680
PINK, 678
PRAIRIE-, 652
Mimosoideae, 618
Mimulus, 1003
alatus, 1004
glabratus, 1004
var. *fremontii,* 1004
var. glabratus, 1004
var. jamesii, 1004
jamesii, 1004
MINT, 762
BRAZOS, 755
CAT, 768
CLUSTER MOUNTAIN, 772
FALSE, 212
HORSE, 764, 765, 766
LEMON, 765
LONG-FLOWER HORSE, 765
MOUNTAIN, 772
NARROW-LEAF MOUNTAIN, 772
ROUND-LEAF, 764
SLENDER
MOUNTAIN, 772
-LEAF MOUNTAIN, 772
WHITE
MOUNTAIN, 772
-LEAF MOUNTAIN, 772
YELLOW HORSE, 766
MINT FAMILY, 752
Minuartia, 518
drummondii, 519
michauxii var. texana, 519
muriculata, 519
muscorum, 519
patula, 519
var. *robusta,* 519
Mirabilis, 838
albida, 839
var. *lata,* 839
dumetorum, 840
eutricha, 839
exaltata, 840
gigantea, 839
glabra, 839
jalapa, 840
subsp. *lindheimeri,* 840
latifolia, 840
lindheimeri, 840
linearis, 840
nyctaginea, ▣/99, 840
MIRASOL, 369
MIRTO, 774
Miscanthus, 1284
sinensis, 1284
MISSISSIPPI DROPSEED, 1327
MISSOURI
BASIN GOLDENROD, 410
GRAPE, 1072
IRONWEED, 428
SPURGE, 596
VIOLET, 1062
-PRIMROSE, 864
MISTFLOWER, 353
MISTLETOE, 1064
CHRISTMAS, 1064
HAIRY, 1064
MISTLETOE FAMILY, 1064
MITERWORT, 798
MITHRIDATE-MUSTARD, 480
Mitreola, 798
petiolata, 798
sessilifolia, 798
Mnesithea cylindrica, 1258
MOCK
BISHOP, 258
CYPRESS, 540
ORANGE, 739
PENNYROYAL, 756
STRAWBERRY, 944
MOCK BISHOP'S
-WEED, 258
NUTTALL'S, 258
THREAD-LEAF, 258
MOCKERNUT HICKORY, 747
MOCO DE QUAJOLOTE, 904
Modiola, 816
caroliniana, 816
MODIOLA, CAROLINA, 816
Molluginaceae, 826
Mollugo, 827
verticillata, 827
Monarda, 764
citriodora, ▣/99, 765
clinopodioides, 765
fistulosa
subsp. *fistulosa* var. *mollis,* 765
var. mollis, ▣/99, 765
hirsutissima, 765
lasiodonta, 766
lindheimeri, 765
mollis, 765
pectinata, 765
punctata, 766
subsp. *intermedia,* 766
var. intermedia, ▣/99, 766
var. lasiodonta, 766
subsp. *occidentalis,* 766
var. occidentalis, 766
subsp. *punctata* var. *viridissima,* 766
subsp. *stanfieldii,* 766
var. *stanfieldii,* 766
russeliana, 766
stanfieldii, 766
viridissima, 768

MONARDA, DOTTED, 766
MONDO-GRASS, 1191
MONEYPLANT, 456
MONILLA, 982
MONK'S PEPPER-TREE, 1060
MONKEY
-FLOWER, 1003
FREEMONT'S, 1004
ROUND-LEAF, 1004
SHARP-WING, 1004
-GRASS, 1191
MONKEYNUT, 628
MONKSHOOD, 916
Monocotyledonae, 1077
Monolepis, 540
nuttalliana, 540
MONOLEPIS, NUTTALL'S, 540
Monotropa, 582
hypopithys, 584
latisquama, 584
uniflora, ▣/100, 584
Monstera, 1092
MONTEZUMA BALD CYPRESS, 204
MOON DAISY, 382
MOONFLOWER, 554
MOONPOD, 842
SPREADING, 842
MOONSEED, 824
CAROLINA, 824
RED-BERRIED, 824
MOONSEED FAMILY, 823
MOONWORT, 188
Moraceae, 827
MORADILLA, 1050
MORAL, 832
BLANCO, 831
Morella cerifera, 832
MORMON-TEA, 207
MORMON-TEA FAMILY, 207
MORNING-GLORY, 554
BIG-ROOT, 556
BUSH, 556
COMMON, 556
COTTON, 556
DWARF-, 552
IVY-LEAF, 556
MEXICAN, 556
NARROW-LEAF, 558
PITTED, 556
PURPLE, 556
RED, 555
SCARLET, 555
SHARP-POD, 556
SMALL WHITE, 556
WHITE, 556
WILD, 556
WOOLLY, 556
WRIGHT'S, 558
MORNING-GLORY FAMILY, 548
Morus, 831
alba, 831
microphylla, 832
rubra, 832
var. *tomentosa,* 832
MOSQUITO FERN, 180
MOSS
BALL-, 1095
BLACK-, 1096
BUNCH-, 1095
DITCH-, 1168
FLORIDA-, 1096
LONG-, 1096
MEADOW SPIKE-, 174
PERUVIAN SPIKE-, 175
RIDDELL'S SPIKE-, 175
ROCK-, 564
-ROSE, 908
SPANISH-, 1096
MOSSY
-CUP OAK, 716
LOCUST, 691
-OVERCUP OAK, 716
MOTH MULLEIN, 1010
MOTHER
-IN-LAW'S-TONGUE, 1078
-OF-THOUSANDS, 561
MOTHERWORT, 759, 760
COMMON, 760
SIBERIAN, 760
MOUNTAIN
MINT, 772
CLUSTER, 772
NARROW-LEAF, 772
SLENDER, 772
SLENDER-LEAF, 772
WHITE, 772
WHITE-LEAF, 772
MULBERRY, 832
SAGE, ROCKY, 776
-ASH, 952
-CEDAR, 203
-LAUREL, 581, 697
TEXAS, 697
-PINK, 724
-ROSE, 898
MOUSE
BARLEY, 1278
-EAR, COMMON, 516
MOUSETAIL, 924
TINY, 924
MUCKWEED, CURLY, 1342
MUD
-MIDGET, 1191
-PLANTAIN, 1086, 1339
BLUE, 1340
GRASS-LEAF, 1340
MUGWORT, 314
Muhlenbergia, 1286
brachyphylla, 1288
bushii, 1288
capillaris, 1288
frondosa, 1288
glabrifloris, 1288
×involuta, 1288
lindheimeri, 1288
mexicana, 1288
reverchonii, 1288
schreberi, 1290
sobolifera, 1290
var. *setigera,* 1290
sylvatica, 1290
utilis, 1290
MUHLY, 1286
APAREJO, 1290
CANYON, 1288
FOREST, 1290
GULF, 1288
HAIRY-AWN, 1288
INLAND, 1288
LINDHEIMER'S, 1288
NODDING, 1288
REVERCHON'S, 1288
ROCK, 1290
SCHREBER'S, 1290
SEEP, 1288
SLENDER, 1288
WIRE-STEM, 1288
MULBERRY, 831
BERMUDA-, 1049
FRENCH-, 1049
MEXICAN, 832
MOUNTAIN, 832
PAPER-, 828
RED, 832
RUSSIAN, 831
SILKWORM, 831
SPANISH-, 1049
TEXAS, 832
WHITE, 831
-WEED, 828
MULBERRY FAMILY, 827
MULE-GRAB, 880
MULLEIN, 1010
FOXGLOVE, 999
SEYMERIA, 999
COMMON, 1010
FLANNEL, 1010
MOTH, 1010

MULLEN-LEAF VERVAIN, 1058
MULTI-BLOOM TEPHROSIA, 700
MULTIFLORA ROSE, 958
MUM, 382
MUNSON'S PLUM, 950
MUNZ GAURA, 856
MUSCADINE GRAPE, 1072
Muscari, 1206
neglectum, 1206
racemosus, 1206
MUSCLETREE, 439
MUSKFLOWER, SCARLET, 842
MUSKMELON, 566
MUSKRAT-WEED, 248
MUSK-THISTLE, 330
MUSQUATROOT, 248
MUSTANG GRAPE, 1072
MUSTARD, 478
BLACK, 460
BLUE-, 464
CHINESE, 459
DOG-, 468
FIELD, 479
FLIXWEED TANSY-, 466
HARE'S-EAR-, 464
-HEDGE, 479
INDIAN, 459
LEAF, 459
MITHRIDATE-, 480
PINNATE TANSY-, 466
ROCKET-, 479
-TANSY-, 465, 479
TUMBLE-, 479
WHITE, 478
WILD, 472
MUSTARD FAMILY, 454
Myagrum, 474
perfoliatum, 474
Myosotis, 453
macrosperma, 453
verna, 453
Myosurus, 924
minimus, ▣/100, 924
Myrica, 832
cerifera, 832
Myricaceae, 832
Myriophyllum, 733
aquaticum, 734
brasiliense, 734
exalbescens, 734
heterophyllum, 734
pinnatum, 734
proserpinacoides, 734
sibiricum, 734
spicatum, 734
var. *exalbescens,* 734, 736
verticillatum, 736
MYRTLE
SEA-, 324
SOUTHERN WAX-, 832
WAX-, 832

N

NAIAD, 1169
SOUTHERN, 1170
SPINY, 1170
NAILWORT, 519
BROOM, 522
CLUSTER-STEM, 520
DRUMMOND'S, 520
JAMES', 520
LINDHEIMER'S, 520
PARKS', 522
Najas, 1169
guadalupensis, 1170
marina, 1170
NAKED-INDIAN, 582
NAL, 1311
Nama, 740
hispidum, 740
jamaicense, 740
ovata, 740
NAMA
FIDDLELEAF, 740
ROUGH, 740
Nandina, 437
domestica, 437
NANNY-BERRY, 511
NARANJO CHINO, 831
Narcissus, 1206
×incomparabilis, 1207
jonquilla, 1207
×odorus, 1208
poeticus, 1207
pseudonarcissus, 1207
tazetta, 1207
NARCISSUS
PAPER-WHITE, 1207
POET'S, 1207
POLYANTHUS, 1207
TRUMPET, 1207
NARROW
-CELL CORNSALAD, 1044
-LEAF
BLADDERPOD, 473
CAT-TAIL, 1348
CHAIN FERN, 181
CONOBEA, 1000
DAYFLOWER, 1100
DYSCHORISTE, 213
GAYFEATHER, 384
GLOBE-MALLOW, 820
GOLD-ASTER, 372
GOOSEFOOT, 536
GROMWELL, 452
IRONWEED, 428
LAMB'S-QUARTERS, 536
MILKWEED, 280
MORNING-GLORY, 558
MOUNTAIN MINT, 772
PEPPERWORT, 186
PINWEED, 544
PUCCOON, 452
RHOMBOPOD, 506
RUSHFOIL, 600
SEEDBOX, 860
SNAKEHERB, 213
SUMPWEED, 378
VETCH, 708
WATER-PRIMROSE, 860
WILLOW, 978
WOOD-OATS, 1256
MELIC, 1284
NASH DUTCHMAN'S-PIPE, 273
Nassella, 1290
leucotricha, 1290
Nasturtium officinale, 476
NATIVE-DANDELION, 399
NAVAJO TEA, 422
NECKLACEWEED, 1012, 1014
NECKWEED, 1012, 1014
NECTARINE, 947
NEEDLE
GRASS
PURPLE, 1243
TEXAS, 1290
SPIKE-RUSH, 1146
SHEPHERD'S-, 260
-POD RUSH, 1184
-POINT RUSH, 1184
NEEDLES
CROW-, 260
SPANISH-, 326
Neeragrostis reptans, 1272
NEGUNO, 1060
Nelumbo, 834
lutea, ▣/100, 834
nucifera, 834
NELUMBO, YELLOW, 834
Nelumbonaceae, 834
Nemastylis, 1174
geminiflora, ▣/100, 1174
nuttallii, 1174
purpurea, 1173
Nemophila, 742
phacelioides, ▣/100, 742
NEMOPHILA, LARGE-FLOWER, 742
Neobesseya, 488

NEPAL PRIVET, 850
Nepeta, 768
  cataria, 768
Nephelium, 979
Nephrolepis, 184
  exaltata, 184
NEP-LEAF LION'S-EAR, 759
Neptunia, 680
  lutea, 680
  pubescens, 680
    var. microcarpa, 680
    var. pubescens, 680
NEPTUNIA
  TROPICAL, 680
  YELLOW, 680
Nerium oleander, 264
NERVED MANNA GRASS, 1277
NERVERAY, 418
  PLATEAU, 420
  SAWTOOTH, 420
NET-LEAF
  BUNDLE-FLOWER, 652
  CLEMATIS, 921
  HACKBERRY, 1038
NETTED GLOBE-CHERRY, 1022
NETTLE, 1043
  BALL-, 1028
  BULL-, 596, 1030
  DEAD-, 759
  FALSE, 1042
  HEART-LEAF, 1043
  HEDGE-, 780
  HORSE-, 1028
  -LEAF
    GOOSEFOOT, 538
    NOSEBURN, 616
    VERVAIN, 1058
  PURPLE DEAD-, 759
  RED DEAD-, 759
  SMALL-SPIKE FALSE, 1042
  SMOOTH HEDGE-, 782
  STINGING, 1043
  TEXAS BULL-, 596
  -TREE, 1036
  WHITE HORSE-, 1030
NETTLE FAMILY, 1040
NET-VEIN
  HERISSANTIA, 810
  MILKVINE, 284
NEW
  DEAL WEED, 324
  JERSEY-TEA, 932
NICKELS-AND-DIMES, 254
Nicotiana axillaris, 1023
Nicotiana, 1022
  repanda, 1022
  rustica, 1022
  tabacum, 1022
Nierembergia, 1022
  hippomanica var. coerulea, 1022
NIGHTSHADE, 1027
  AMERICAN, 1030
  FALSE, 1018
  GREEN FALSE, 1018
  HAIRY FALSE, 1018
  MELON-LEAF, 1028
  PLATEAU FALSE, 1018
  SILVER-LEAF, 1030
  TEXAS, 1031
NIGHTSHADE FAMILY, 1015
NIMBLE-WILL, 1290
NIÑARUPÁ, 1049
NINFA ACUÁTICA, 845
NIPPLE
  CACTUS, 486, 489
    BEEHIVE, 489
    PLAINS, 488
  -BRACT ARROWHEAD, 1090
NITS-AND-LICE, 547
NOBLE GOLDENROD, 414
NODDING
  BLUET, 968
  CHICKWEED, 516
  FESCUE, 1276
  GREEN-VIOLET, 1061
  LADIES'-TRESSES, 1220
  MUHLY, 1288
  WILD RYE, 1268
  -THISTLE, 330
NOD-VIOLET, 1061
  WHORLED, 1061
NOGAL, 748
  MORADO, 747
  SILVESTRE, 748
NOGALILLO, 748
NOGALITO, 748
Nolina, 1080
  lindheimeriana, 1080
  texana, 1082
NOLINA, LINDHEIMER'S, 1080
Nolinaceae, 1078, 1080
NONE-SUCH, 674
NOON-FLOWER, 424
NOPAL PRICKLY-PEAR, 490
NORTHERN
  CATALPA, 443
  CROTON, 600
  DEWBERRY, 959
  FROGFRUIT, 1054
  MALLOW, 814
  NUT-GRASS, 1138
NORTON'S FLAX, 790
NOSEBURN, 614
  BETONY, 616
  CATNIP, 616
  DOGTOOTH, 617
  NETTLE-LEAF, 616
  SHORT-SPIKE, 616
Nothofagus, 710
Notholaena
  dealbata, 194
  integerrima, 195
  sinuata, 195
    var. *integerrima,* 195
Nothoscordum, 1208
  bivalve, 1208
NUEZ ENCARCELADA, 747
Nuphar, 844
  advena, 844
  lutea
    subsp. *advena,* 844
    subsp. *macrophylla,* 844
NUT
  -GRASS, 1142
    FALSE, 1144
    NORTHERN, 1138
    PURPLE, 1142
    YELLOW, 1138
  -RUSH, 1164
    CILIATE, 1164
    FEW-FLOWER, 1164
    FRINGED, 1164
    LITTLE-HEAD, 1164
    LOW, 1164
    TALL, 1164
    WHIP, 1164
    WHORLED, 1164
  SEDGE, 1142
NUTMEG HICKORY, 748
Nuttallanthus, 1004
  canadensis, 1004
  texanus, 1006
Nuttallia
  nuda, 796
  stricta, 796
NUTTALL'S
  DEATH-CAMASS, 1210
  EVOLVULUS, 552
  HABENARIA, 1215
  MILK-VETCH, 634
  MOCK BISHOP'S-WEED, 258
  MONOLEPIS, 540
  SENSITIVE-BRIAR, 678
  TICK-CLOVER, 655
  WATER-HEMP, 226
  WATERWORT, 496
  WILD INDIGO, 638
Nyctaginaceae, 834

Nyctaginia, 842
capitata, ■/100, 842
Nymphaea, 844
advena, 844
elegans, 845
mexicana, 845
odorata, ■/101, 845
var. *villosa*, 845
Nymphaeaceae, 842
Nymphoides, 826
peltata, 826
Nyssa, 845
sylvatica, 845
Nyssaceae, 845

O

O'KENNON'S ANEMONE, 918
OAK, 711
BARREN, 716
BASTARD, 718
BIGELOW'S, 718
BLACK, 716, 718
BLACKJACK, 716
BLUEJACK, 716
BLUFF, 718
BOTTOM-LAND RED, 714
BUR, 716
CHERRY-BARK, 714
CHESTNUT, 716
CHINQUAPIN, 716
CINNAMON, 716
COAST LIVE, 719
CORK, 711
CROSS, 718
DRUMMOND'S POST, 716
DUCK, 718
DURAND WHITE, 718
DWARF POST, 716
EASTERN POISON-, 236
ESCARPMENT LIVE, 716
FORKED-LEAF WHITE, 714
HIGHGROUND, 716
IRON, 718
JACK, 716
JERUSALEM-, 536
LIVE, 716, 719
MOSSY
-CUP, 716
-OVERCUP, 716
OVERCUP, 716
PEACH, 718
PIN, 718
PLATEAU LIVE, 716
POISON-, 234, 236
POSSUM, 718
POST, 718
PRAIRIE, 716
PUNK, 718
QUERCITRON, 718
RED, 718
RIDGE WHITE, 714
ROCK, 714, 718
RUNNER, 716
SAND POST, 716
SANDJACK, 716
SCALY-BARK, 718
SCRUB, 718
LIVE, 716
SCRUBBY POST, 716
SHIN, 716, 718
SHUMARD'S, 718
RED, 718
SMOOTH-BARK, 718
SOUTHERN RED, 714
SPANISH, 714
SPOTTED, 714, 718
STAVE, 714
SWAMP
CHESTNUT, 716
POST, 716
RED, 714, 718
SPANISH, 714
WHITE, 716
WILLOW, 718
SWEET-, 832
TAN-BARK, 710
TEXAS RED, 714
THREE-LOBE RED, 714
UPLAND WILLOW, 716
VIRGINIA LIVE, 719
WATER, 718
WHITE, 716
WEST TEXAS LIVE, 716
WHITE, 714, 718
WILLOW, 718
-LEAF, 718
YELLOW, 718
-BARK, 718
TURKEY, 716
WILT, 711
OAK FAMILY, 710
OAKLEECH, 994
DOWNY, 994
OAT GRASS, POVERTY, 1260
OATS, 1244
COMMON, 1244
CREEK-, 1254
CULTIVATED, 1244
POVERTY-, 1260
SEA-1254
WILD, 1244, 1254
WOOD-, 1254
OBEDIENT-PLANT, 768
OBLONG-LEAF ASTER, 319
OCA, 869
Ocimum, 752
Oclemena, 315
OCTOBER LADIES'-TRESSES, 1222
Oenothera, 862
biennis, 863
brachycarpa, 866
var. *typica*, 863
coryi, 863
fulfurriae, 866
grandis, 863
greggii, 854
var. *lampasana*, 854
heterophylla, 864
jamesii, 864
laciniata, 864
linifolia, 864
macrocarpa, 864
subsp. incana, 864
subsp. macrocarpa, ■/101, 864
subsp. oklahomensis, 864
var. *oklahomensis*, 864
mexicana, 866
missouriensis, 864
var. *oklahomensis*, 864
pubescens, 866
rhombipetala, 866
serrulata, 854
subsp. *drummondii*, 854
subsp. *pinifolia*, 854
spachiana, 866
speciosa, ■/101, 866
triloba, 866
OHIO SPIDERWORT, 1104
OIL PALM, 1094
OKLAHOMA PLUM, 949
OKRA, 806
OLD MAN'S-BEARD, 1096
Oldenlandia, 968
boscii, 968
OLD
-FIELD
CLOVER, 702
GOLDENROD, 410, 412
PINE, 207
-MAID, 266
-MAN'S-BEARD, 921
-PLAINSMAN, 374, 375
OLDFIELD
CINQUEFOIL, 947
THREEAWN, 1242
TOAD-FLAX, 1004
Olea europaea, 846
Oleaceae, 846

OLEANDER, 264
OLEANDER FAMILY, 264
OLEASTER, 580
SILVER-BERRY, 579
OLEASTER FAMILY, 579
Oligoneuron, 408
nitidum, 412
rigidum, 414
OLIVE, 846
RUSSIAN-, 580
OLIVE FAMILY, 846
OLMO, 1040
OMBILIGO DE VENUS, 255
Onagraceae, 852
ONE-FLOWER FLAT SEDGE, 1142
ONE-SEED
BUR-CUCUMBER, 570
COPPERLEAF, SLENDER, 588
CROTON, 600
ONION, 1194
DRUMMOND'S, 1197
PINK WILD, 1198
PRAIRIE, 1197, 1198
RUNYON'S, 1197
WILD, 1196, 1198
Onoclea, 184
sensibilis, 184
Onopordum, 388
acanthium, 388
Onosmodium, 453
bejariense, 453
var. bejariense, 454
var. hispidissimum, 454
var. occidentale, 454
helleri, 454
hispidissimum, 454
molle, 454
subsp. *bejariense,* 454
var. *bejariense,* 454
occidentale, 454
OPEN-FLOWER
PANIC, 1300
ROSETTE GRASS, 1300
Ophioglossaceae, 188
Ophioglossum, 190
crotalophoroides, 190
engelmannii, 190
nudicaule, 190
petiolatum, 190
pycnostichum, 190
vulgatum, 190
var. *pycnostichum,* 190
Ophiopogon japonicus, 1191
OPIUM POPPY, 872, 876
Oplismenus, 1290
hirtellus, 1290
subsp. *setarius,* 1291
setarius, 1291
Opuntia, 489
compressa, 491
edwardsii, 492
engelmannii, 490
var. engelmannii, 491
var. lindheimeri, ▣/101, 490
var. linguiformis, 491
ficus-indica, 491
humifusa, 491
kleiniae, 491
leptocaulis, 491
lindheimeri, 491
var. *linguiformis,* 491
macrorhiza, 492
phaeacantha, 492
var. camanchica, 492
var. *discata,* 491
var. major, 492
schottii, 492
tunicata var. davisii, ▣/101, 492
ORACHE, 532
SILVER, 532
ORANGE, 970
BITTER, 972
DAISY, 430
DAY-LILY, 1202
FLAMEFLOWER, 911
-GRASS, 547
MILKWEED, 280
MOCK, 739
OSAGE-, 830, 831
RUST, 959
TRIFOLIATE, 972
ZEXMENIA, 430
Orbexilum, 680
pedunculatum, 680
var. *eglandulosum,* 680
simplex, 682
ORCHARD GRASS, 1259
ORCHID
BUTTERFLY, 1218
CORSAGE, 1213
CREEPING, 1215
FINGER, 1215, 1218
FRINGED, 1218
GIANT SPIRAL, 1220
LEAFLESS, 1216
WATER SPIDER, 1215
YELLOW
FINGER, 1218
FRINGED, 1218
ORCHID FAMILY, 1212
Orchidaceae, 1212
ORCHIS, STREAM-, 1215
OREGANO, 752
CIMARRÓN, 1054
Oreostemma, 315
ORGANILLO, 1049
CIMARRON, 1052
ORIENTAL LACQUER TREE, 236
Origanum, 752
ORNAMENTAL KALE, 460
Ornithogalum, 1208
umbellatum, 1208
Orobanchaceae, 868
Orobanche, 868
fasciculata, 868
var. *subulata,* 868
ludoviciana, 868
subsp. ludoviciana, 868
subsp. multiflora, 868
var. *multiflora,* 868
multiflora, 868
var. *multiflora,* 868
var. *pringlei,* 868
ORPINE, 562
ORPINE FAMILY, 561
ORRIS, 1174
ORTIGUILLA, 1043
Oryza sativa, 1223, 1280
OSAGE-ORANGE, 830, 831
OSIER, 976
WILLOW, 978
Osmorhiza, 255
longistylis, 256
Osmunda, 192
cinnamomea, 192
regalis var. spectabilis, 192
Osmundaceae, 192
Ostrya, 440
virginiana, 440
Ostryopsis, 438
OUR LADY'S
BEDSTRAW, 964
-THISTLE, 406
OVAL LADIES'-TRESSES, 1222
OVATE-LEAF RAGWORT, 390
OVERCUP OAK, 716
MOSSY-, 716
Oxalidaceae, 869
Oxalis, 869
acetosella, 869
articulata subsp. rubra, 870
corniculata, 870
var. *wrightii,* 870
dillenii, 870
var. *filipes,* 870
var. *radicans,* 870
drummondii, 870
rubra, 870

Oxalis *(cont.)*
stricta, 870
tuberosa, 869
violacea, 872
OXALIS
DILLEN'S, 870
DRUMMOND'S, 870
OX-EYE DAISY, 382
Oxypolis, 256
filiformis, 256
rigidor, 256
Oxytropis, 682
lambertii, ▣/101, 682
var. *articulata*, 682
OYSTERPLANT, 426, 1098, 1101
OZARK
CARIC SEDGE, 1128
DROPSEED, 1327
GRASS, 1282
HICKORY, 748
MILK-VETCH, 632
SAVORY, 755
SUNDROPS, 864

P

PACIFIC YEW, 201
Packera, 388
glabella, 390
obovata, 390
plattensis, 390
tampicana, 392
PAGODA TREE, 697
PAIGLE, 926
PAINTBRUSH, 996
CITRON, 998
INDIAN, 998
ENTIRE-LEAF, 996
INDIAN, 996
LEMON, 998
LINDHEIMER'S, 998
INDIAN, 998
LIPPED INDIAN, 998
PRAIRIE, 998
INDIAN, 998
PURPLE, 998
TEXAS, 996
YELLOW, 998
PAINTED
EUPHORBIA, 606
SPURGE, 606
PAINTEDCUP, 996
LEMON, 998
PAINTEDLEAF, 606
Palafoxia, 392
callosa, 392
hookeriana
var. hookeriana, 392
var. minor, 393
reverchonii, 393
rosea, 393
var. macrolepis, 393
var. rosea, 393
sphacelata, 393
PALAFOXIA
RAYED, 393
REVERCHON'S, 393
ROSE, 393
SHOWY, 392
SMALL, 392
PALE
BLUE-EYED-GRASS, 1178
BUCKEYE, 738
BULRUSH, 1162
DOCK, 906
DUCKWEED, 1188
ECHINACEA, 347
-LEAF YUCCA, 1084
-SEED PLANTAIN, 886
SMARTWEED, 902
-SPIKE LOBELIA, 500
SPIKE-RUSH, 1146
VETCH, 708
YUCCA, 1084
Paliurus spina-christi, 934
PALM
BLUE, 1094
CALIFORNIA FAN, 1094
CARNAUBA WAX, 1094
COCONUT, 1094
DATE, 1094
DWARF, 1094
-LEAF SCURF-PEA, 684
OIL, 1094
ROYAL, 1094
SEYCHELLES, 1094
SWAMP, 1094
PALM FAMILY, 1094
PALMA
CHRISTI, 612
DE DÁTILES, 1084
LOCA, 1084
PITA, 1084
Palmae, 1094
PALMER'S
AMARANTH, 226
PIGWEED, 226
PALMETTO
BUSH, 1094
DWARF, 1094
PALO
AMARILLO, 1049
BLANCO, 1038
PALOVERDE, 682
MEXICAN, 682
PAMPAS GRASS, 1258
Panax quinquefolius, 270
PANHANDLE GRAPE, 1070
PANIC
GRASS, 1291
BLUE, 1295
CEDAR, 1300
HALL'S, 1298
LINDHEIMER'S, 1295
OPEN-FLOWER, 1300
RED-TOP, 1302
ROUND-SEED, 1302
SLIM-LEAF, 1300
SOFT-LEAF, 1300
WHITE-HAIRED, 1295
WOOLLY, 1295
PANICLED
AMORPHA, 626
ASTER, 318
TICK-CLOVER, 656
Panicum, 1291
aciculare, 1294
var. aciculare, 1295
var. angustifolium, 1295
acuminatum, 1295
var. acuminatum, 1295
var. *fasciculatum*, 1295
var. lindheimeri, 1295
var. villosum, 1295
agrostoides, 1302
anceps, 1295
var. *rhizomatum*, 1295
angustifolium, 1295
antidotale, 1295
barbulatum, 1298
bergii, 1296
boscii, 1296
brachyanthum, 1296
capillare, 1296
subsp. *hillmanii*, 1300
var. *sylvaticum*, 1302
clandestinum, 1296
coloratum, 1296
commutatum, 1298
depauperatum, 1296
dichotomiflorum, 1296
dichotomum, 1296
diffusum, 1298
divergens, 1298
fasciculatum var. *reticulatum*, 1335
filipes, 1298
hallii, 1298
var. filipes, 1298
var. hallii, 1298

Panicum *(cont.)*
  helleri, 1300
  hians, 1298
  hillmanii, 1298
  lanuginosum, 1295
    var. *lindheimeri,* 1295
  laxiflorum, 1300
  lindheimeri, 1295
  linearifolium, 1300
  lucidum, 1298
  malacophyllum, 1300
  microcarpon, 1298
  miliaceum, 1300
  nitidum, 1298
  obtusum, 1300
  oligosanthes, 1300
    var. oligosanthes, 1300
    var. scribnerianum, 1300
  ovale var. *villosum,* 1295
  ovinum, 1295
  pedicellatum, 1300
  philadelphicum, 1302
  pilcomayense, 1296
  ramisetum, 1320
  ravenelii, 1302
  reverchonii, 1320
  rhizomatum, 1295
  rigidulum, 1302
  roanokense, 1298
  scoparium, 1302
  sphaerocarpon, 1302
  texanum, 1335
  verrucosum, 1296
  villosissimum, 1295
  virgatum, 1302
  yadkinense, 1298
PANICUM
  BEAKED, 1295
  FALL, 1296
  FILLY, 1298
  GAPING, 1298
  HILLMAN'S, 1298
  PIMPLE, 1296
  SPREADING, 1298
  TEXAS, 1335
PANSY, 1060
  FIELD, 1062
  GARDEN, 1061
  WILD, 1062
Papaver, 876
  rhoeas, 876
  somniferum, 872, 876
Papaveraceae, 872
PAPER
  -FLOWER, YELLOW, 420
  -MULBERRY, 828
  -REED, 1132
  -WHITE NARCISSUS, 1207
Papilionoideae, 619
PAPYRUS, 1132
PARÁ RUBBER, 585, 830
PARÁISO, 823
PARASOL-TREE, CHINESE, 1032
Parietaria, 1042
  obtusa, 1042
  pensylvanica, 1042
    var. obtusa, 1042
    var. pensylvanica, 1042
PARILLA, YELLOW, 824
Parkinsonia, 682
  aculeata, 682
PARKS' NAILWORT, 522
Paronychia, 519
  chorizanthoides, 520
  drummondii, 520
    subsp. *parviflora,* 520
  fastigiata, 520
  jamesii, 520
    var. *praelongifolia,* 520
  lindheimeri, 520
  parksii, 522
  virginica, 522
    var. *scoparia,* 522
PARRA, 1068
  SILVESTRE, 1070
PARRALENA, 422
PARROT'S-FEATHER, 733, 734
  GREEN, 734
PARSLEY, 256
  BUTLER'S SAND-, 243
  CHINESE-, 248
  CURLY, 256
  GARDEN, 256
  HAWTHORN, 942
  HEDGE-, 262
  KNOTTED HEDGE-, 262
  PLAINS SAND-, 243
  POISON-, 248
  PRAIRIE-, 256
  SAND-, 243
  TEXAS-, 256
PARSLEY FAMILY, 239
PARSNIP, 239
  PRAIRIE-, 256
  WATER-, 244, 260
Parthenium, 393
  argentatum, 393
  hysterophorus, 393
PARTHENIUM, RAGWEED, 393
Parthenocissus, 1066
  heptaphylla, 1068
  quinquefolia, 1068
  tricuspidata, 1068
PARTRIDGE-PEA, 642
  SENSITIVE, 642
  SHOWY, 642
Pascopyrum, 1302
  smithii, 1302
PASIONARIA, 878
Paspalidium, 1304
  geminatum, 1304
PASPALIDIUM, EGYPTIAN, 1304
Paspalum, 1304
  ciliatifolium, 1308
  dilatatum, 1305
  distichum, 1306
    var. *indutum,* 1306
  floridanum, 1306
    var. *glabratum,* 1306
  laeve, 1306
    var. circulare, 1306
    var. laeve, 1306
  langei, 1306
  notatum, 1306
    var. latiflorum, 1306
    var. saurae, 1306
  platyphyllum, 1335
  plicatulum, 1306
  pubiflorum, 1308
    var. glabrum, 1308
    var. pubiflorum, 1308
  setaceum, 1308
    var. *ciliatifolium,* 1308
    var. *muhlenbergii,* 1308
    var. *stramineum,* 1308
  stramineum, 1308
  urvillei, 1308
PASPALUM
  BIG, 1306
    FLORIDA, 1306
  BROWN-SEED, 1306
  FIELD, 1306
  FLORIDA, 1306
  GRASS, 1305
  HAIRY
    -FLOWER, 1308
    -SEED, 1308
  LANGE'S, 1306
  PLAITED, 1306
  ROUND-SEED, 1306
  RUSTY-SEED, 1306
  SMOOTH, 1306
    -SEED, 1308
  THIN, 1308
  URVILLE'S, 1308
Passiflora, 877
  affinis, ▣/102, 878
  incarnata, ▣/102, 878

Passiflora *(cont.)*
lutea, 878
var. *glabriflora*, 878
tenuiloba, ■/102, 878
Passifloraceae, 877
PASSION-FLOWER, 877
BIRD-WING, 878
BRACTED, 878
MAYPOP, 878
SPREAD-LOBE, 878
YELLOW, 878
PASSION-FLOWER FAMILY, 877
Pastinaca, 239
PASTLE, 1096
PASTURE
BRAKE, 181
HAWTHORN, 942
HELIOTROPE, 450
ROSE, 956
THISTLE, 340
PATA DE LEON, 399
PATH RUSH, 1184
PATTYPAN SQUASH, 568
Pavonia, 816
lasiopetala, ■/102, 816
PAVONIA, WRIGHT'S, 816
PAWPAW, 238
AMERICAN, 238
COMMON, 238
PEA, 617
BUTTERFLY-, 639, 642
CALERY-, 666
CHICK-, 617
DOWNY MILK-, 660
DRUMMOND'S RUSH-, 664
EDIBLE SCURF-, 684
EVERLASTING-, 666
HOARY-, 698
MILK-, 660
INDIAN RUSH-, 663
INDIAN-, 630
MILK-, 658
PALM-LEAF SCURF-, 684
PARTRIDGE-, 642
PERENNIAL SWEET-, 666
ROCK SCURF-, 686
ROSARY-, 617
ROUGH-, 666
ROUND-LEAF SCURF-, 686
RUSH-, 639, 663
SCARLET-, 664
SCURF-, 683, 688
SCURVY, 690
SENSITIVE PARTRIDGE-, 642
SENSITIVE-, 642
SICKLE-POD RUSH-, 663
SINGLETARY-, 666
SLENDER SCURFY, 690
SLIM-LEAF SCURF-, 690
TALL-BREAD SCURF-, 684
-TURKEY-, 634
WESTERN SCARLET-, 664
YELLOW BUSH-, 638
TURNIP-ROOT SCURF-, 684
PEACH, 947, 950
OAK, 718
-LEAF DOCK, 906
PEANUT, 628
CLOVER, 704
COMMON, 628
HOG-, 626, 663
SOUTHERN HOG-, 626
PEAR, 952
BRADFORD, 952
CALLERY, 952
COMMON, 952
-FRUIT BLADDERPOD, 474
PEARLWORT, 522
TRAILING, 524
PEAVINE, 634, 665
LOW, 666
PEBBLE PLANTS, 220
PECAN, 746, 747
BITTER, 747
Pectis, 394
angustifolia var. fastigiata, 394
fastigiata, 394
papposa, 394
texana, 394
Pedaliaceae, 878
Pedicularis, 1006
canadensis, 1006
subsp. *canadensis* var. *dobbsii*, 1006
Pediomelum, 683
cuspidatum, 684
cyphocalyx, 684
digitatum, 684
hypogaeum, 684
var. hypogaeum, 684
var. scaposum, 684
var. subulatum, 684
latestipulatum, 686
var. appressum, 686
var. latestipulatum, 686
linearifolium, 686
reverchonii, 686
rhombifolium, 686
PEEPUL, 830
PEEWATER LEAF-FLOWER, 611
Pellaea, 198
atropurpurea, 198
dealbata, 194
ovata, 200
ternifolia var. *wrightiana*, 200
wrightiana, 200
PELLITORY, 1042
PENNSYLVANIA, 1042
PELOCOTE, 378
PELOTAZO, 806
PENARD'S LARKSPUR, 924
PENCIL
CACTUS, 491
CHOLLA, 491
-CEDAR, 203
-FLOWER, 698
SIDE-BEAK, 698
Pennisetum, 1308
alopecuroides, 1308
americanum, 1308
ciliare, 1310
macrostachyum, 1308
setaceum, 1308
villosum, 1308
PENNSYLVANIA PELLITORY, 1042
PENNSYLVANICA SMARTWEED, 904
PENNYCRESS, 480
FIELD, 480
PENNYROYAL
BASTARD, 783
DRUMMOND'S FALSE, 758
FALSE, 756, 783
MOCK, 756
REVERCHON'S FALSE, 758
ROUGH FALSE, 758
PENNYWORT
FLOATING WATER-, 254
UMBRELLA WATER-, 255
WATER-, 254
WHORLED WATER-, 255
PENSACOLA BAHIA GRASS, 1306
Penstemon, 1006
australis subsp. *laxiflorus*, 1008
bradburii, 1008
cobaea, ■/102, 1007
digitalis, ■/102, 1007
fendleri, 1007
grandiflorus, 1008
guadalupensis, 1008
helleri, 1008
laxiflorus, 1008
murrayanus, 1008
triflorus subsp. integrifolius, 1008
tubaeflorus, 1008
PENSTEMON
COBAEA, 1007
CUP-LEAF, 1008
FENDLER'S, 1007

- PENSTEMON *(cont.)*
  - GUADALUPE, 1008
  - SMOOTH, 1007
  - TUBE, 1008
- *Pentas,* 960
- **Penthorum,** 562
  - sedoides, 562
- PEPPER
  - BELL, 1017
  - BIRD, 1017
  - CAYENNE, 1017
  - CHILI, 1017
  - -GRASS, 470
    - AMERICAN, 472
    - GREEN-FLOWER, 470
    - VIRGINIA, 472
  - POORMAN'S-, 472
  - -TREE, MONK'S, 1060
  - WATER-, 188, 902
  - WILD, 1060
- PEPPER FAMILY, 983
- PEPPERBARK, 973, 974
- PEPPERIDGE, 845
- PEPPERMINT, 764
- PEPPERTREE, 230
- PEPPERVINE, 1066
- PEPPERWEED, 470
  - PRAIRIE, 470
  - SOUTHERN, 470
  - VEINY, 470
- PEPPERWORT, 186
  - HAIRY, 186
  - HOOKED, 186
  - LARGE-FOOT, 186
  - NARROW-LEAF, 186
- PEPPERWORT FAMILY, 185
- PERA, 952
- PERENNIAL
  - BROOMWEED, 365
  - RAGWEED, 310
  - RYE GRASS, 1282
  - SANDY-LAND-SAGE, 766
  - SWEET-PEA, 666
- PERFORATED BLUESTEM, 1246
- PERFUME-BALL, 359
- PERFUMEPLANT, 474
- **Perilla,** 768
  - fructescens, 768
- PERILLA, COMMON, 768
- **Periploca,** 287
  - graeca, 287
- **Perityle,** 394
  - lindheimeri, 394
- PERIWINKLE, 268
  - BIG-LEAF, 269
  - CAPE, 266
  - COMMON, 269
  - MADAGASCAR, 266
  - ROSE, 266
- *Persea americana,* 784
- PERSIAN
  - CLOVER, 706
  - SPEEDWELL, 1014
  - WALNUT, 744
- *Persicaria,* 899
  - *bicornis,* 904
  - *coccinea,* 902
  - *densiflora,* 902
  - *hydropiperoides,* 902
  - *lapathifolia,* 902
  - *pensylvanica,* 904
  - *portoricensis,* 902
  - *punctata,* 904
  - *setacea,* 904
  - *vulgaris,* 904
- PERSIMMON, 578
  - BLACK, 579
  - CHAPOTE, 579
  - COMMON, 579
  - EASTERN, 579
  - MEXICAN, 579
  - TEXAS, 579
- PERSIMMON FAMILY, 578
- PERUVIAN SPIKE-MOSS, 175
- *Petalostemon,* 644
  - *candidus,* 648
    - var. *oligophyllus,* 648
  - *griseus,* 651
  - *multiflorus,* 650
  - *phleoides* var. *microphyllus,* 650
  - *pulcherrimus,* 648
  - *purpureus,* 651
  - *reverchonii,* 651
  - *tenuis,* 651
- **Petrorhagia,** 522
  - dubia, 522
  - *prolifera,* 522
  - *velutina,* 522
- **Petroselinum,** 256
  - crispum, 256
- **Petunia,** 1023
  - ×*atkinsiana,* 1023
  - axillaris, 1023
  - *hybrida,* 1023
  - *integrifolia,* 1017
  - *parviflora,* 1017
  - *violacea,* 1023
- PETUNIA
  - COMMON GARDEN, 1023
  - PRAIRIE-, 216
  - SEASIDE, 1017
  - WILD, 213, 1017
- PEYOTE, 484
- **Phacelia,** 742
  - congesta, 743
  - glabra, 743
  - hirsuta, 743
  - integrifolia, 743
  - patuliflora, 743
    - var. patuliflora, 743
    - var. teucriifolia, 743
  - strictiflora, 743
    - var. connexa, 744
    - var. lundelliana, 744
    - var. robbinsii, 744
    - var. strictiflora, 744
- PHACELIA
  - CRENATE-LEAF, 743
  - GYP, 743
  - HAIRY, 743
  - SAND, 743
  - SMOOTH, 743
  - SPIKE, 743
- **Phalaris,** 1310
  - angusta, 1310
  - canariensis, 1310
  - caroliniana, 1310
- *Phaseolus,* 617, 706
  - *helvula,* 698
  - *leiospermus,* 698
- PHEASANT'S-EYE, 917, 1207
- PHILADELPHIA
  - FLEABANE, 352
  - WITCH GRASS, 1302
- **Philadelphus,** 739
  - pubescens, 739
- *Philodendron,* 1090, 1092
- **Phleum,** 1310
  - pratense, 1311
- **Phlox,** 891
  - *asper,* 892
  - cuspidata, 891
    - var. *grandiflora,* 891
    - var. *humilis,* 891
  - drummondii, 891
    - subsp. drummondi, ▣/103, 892
    - subsp. mcallisterii, 892
    - var. *mcallisterii,* 892
    - subsp. wilcoxiana, ▣/103, 892
    - var. *wilcoxiana,* 892
  - oklahomensis, 892
  - pilosa, 892
    - var. *asper,* 892
    - subsp. latisepala, 892
    - subsp. pilosa, 892
    - subsp. riparia, 892
  - roemeriana, ▣/103, 892

PHLOX
DOWNY, 892
DRUMMOND'S, 892
GOLD-EYE, 892
LARGE-FLOWER, 891
POINTED, 891
PRAIRIE, 892
ROEMER'S, 892
ROUGH, 892
TEXAS, 892
PHLOX FAMILY, 888
Phoenix dactylifera, 1094
PHOENIXTREE, 1032
Phoradendron, 1064
serotinum var. *pubescens*, 1065
tomentosum, 1064
Photinia, 946
serratifolia, 946
serrulata, 946
Phragmites, 1311
australis, 1311
communis, 1311
Phryma, 880
leptostachya, 881
Phrymaceae, 880
Phyla, 1053
canescens, 1054
cuneifolia, 1054
incisa, 1054
lanceolata, 1054
nodiflora
var. *longifolia*, 1054
var. *reptans*, 1054
var. *rosea*, 1054
strigulosa
var. *parviflora*, 1054
var. *sericea*, 1054
Phyllanthus, 610
abnormis, 611
caroliniensis, 611
niruri, 611
polygonoides, 611
tenellus, 611
urinaria, 611
Phyllostachys, 1311
aurea, 1311
nigra, 1311
Physalis, 1023
angulata var. *pendula*, 1024
angulata, 1024
cinerascens, 1024, 1026
heterophylla, 1024
lobata, 1027
longifolia, 1024
var. longifolia, 1024
var. subglabrata, 1026
macrophysa, 1026
mollis, 1026
pendula, 1024
philadelphica, 1023
pubescens, 1026
var. integrifolia, 1026
var. pubescens, 1026
pumila, 1026
turbinata, 1026
virginiana, 1026
forma *macrophysa*, 1026
var. *sonorae*, 1024
var. *subglabrata*, 1026
viscosa
var. *cinerascens*, 1024
subsp. *mollis*, 1026
Physostegia, 768
angustifolia, 769
digitalis, 769
edwardsiana, 769
intermedia, 769
micrantha, 769
praemorsa, 770
pulchella, ▣/103, 770
serotina, 770
virginiana, 770
subsp. praemorsa, 770
subsp. virginiana, 770
Phytelephas macrocarpa, 1094
Phytolacca, 881
americana, 881
Phytolaccaceae, 881
Picea, 204
PICKEREL-WEED, 1340
PICKEREL-WEED FAMILY, 1338
PICKPOCKET, 462
PIEFRUIT, 496, 1031
PIEPLANT, 347, 897
PIGEON
GRAPE, 1070
-BERRY, 882
-WINGS, 642
ATLANTIC, 642
PIGMYWEED, WATER, 562
PIGNUT, 747
WATER, 747
PIGWEED, 533, 535
GREEN, 222
PALMER'S, 226
PROSTRATE, 226
RED-ROOT, 226
ROUGH, 226
SLENDER, 226
SPINY, 226
WINGED-, 540
PIGWEED FAMILY, 221, 530
Pilea, 1042
microphylla, 1042
pumila var. deamii, 1042
Pileolaria
brevipes, 232
patzcuarensis, 232
PILL-POD EUPHORBIA, 594
PILLWORT, 188
AMERICAN, 188
Pilostyles, 914
covillei, 916
thurberi, 648, 914
PILOSTYLES, THURBER'S, 648, 914
Pilularia, 188
americana, 188
PIMPERNEL, 912
BROOK-, 1012
CLASPING FALSE, 1002
COMMON, 912
FALSE, 912, 1002
SCARLET, 912
WATER-, 914
Pimpinella, 239
PIMPLE PANICUM, 1296
PIN OAK, 718
Pinaceae, 204
Pinaropappus, 394
roseus, 394
PINCHOT'S JUNIPER, 203
PIN-CLOVER, 730
PINCUSHION
CACTUS, 488, 489
DEVIL'S-, 486
PINCUSHIONS, 576
PINE, 204
BRISTLE-CONE, 206
LOBLOLLY, 207
LONGLEAF, 207
YELLOW, 207
LONGTAG, 206
OLD-FIELD, 207
PITCH, 206
SHORTLEAF, 206
YELLOW, 206
SLASH, 206
YELLOW SLASH, 207
WHITE, BLISTER RUST, 732
PINE FAMILY, 204
PINE PINWEED, 543
PINEAPPLE, 1095
CACTUS, 485
PINEAPPLE FAMILY, 1095
PINESAP, AMERICAN, 584
PINEWEED, 547
ST. JOHN'S-WORT, 547

PINEWOODS
GRAPE, 1070
LARKSPUR, 924
PINHOLE
BEARD GRASS, 1246
BLUESTEM, 1246
PINK, 518
BONESET, 354
CHILDING-, 522
-DANDELION, 394
DEPTFORD, 518
EUPATORIUM, 354
GRASS-, 1214
INDIAN-, 799
MEADOW-, 728
MILK-, 386
MILKWORT, 896
MIMOSA, 678
MOUNTAIN-, 724
-QUEEN, 506
ROCK-, 911
ROSE-, 728
RUSH-, 386
-SCALE GAYFEATHER, 384
SMARTWEED, 904
TRIDENS, 1330
VERBENA, 1052
VERVAIN, 1052
WILD ONION, 1198
PINK FAMILY, 511
PINKROOT, 798
-QUEEN, 506
PRAIRIE, 798
PINNATE TANSY-MUSTARD, 466
Pinophyta, 201
Pinus, 204
echinata, 206
elliottii, 206
longaeva, 206
palustris, 207
taeda, 207, 554
PINWEED, 543
DRUMMOND'S, 544
HAIRY, 543
NARROW-LEAF, 544
PINE, 543
SAN SABA, 544
Piperaceae, 983
PIPEVINE, 272, 274
PIPEVINE FAMILY, 272
PIPEWORT, 1165, 1166
SEVEN-ANGLE, 1166
SMALL-HEAD, 1166
TEN-ANGLE, 1166
TEXAS, 1166
PIPEWORT FAMILY, 1165
PIQUANTE, 415
PIQUÍN, CHILE, 1017
PISTACHIO, 230
CHINESE, 230
Pistacia
chinensis, 230
texana, 231
vera, 230
Pistia, 1093
stratiotes, 1093
Pisum, 617
PITA, 1084
PITCH PINE, 206
PITCHER
PLANT, 983
SAGE, 774
SANDWORT, 519
CLEMATIS, 921
-FLOWER, 921
PITCHER PLANT FAMILY, 983
PIT-SEED GOOSEFOOT, 536
PITTED MORNING-GLORY, 556
Pityopsis, 396
graminifolia var. latifolia, 396
PLAIN-LEAF PUSSY-TOES, 310
PLAINS
BEEBALM, 765
BLACKFOOT, 388
COREOPSIS, 342
COTTONWOOD, 976
ERYSIMUM, 468
FLEABANE, 350
GAURA, 856
IRONWEED, 428
KUHNIA, 329
LARKSPUR, 924
LAZY DAISY, 312
LOVE GRASS, 1270
NIPPLE CACTUS, 488
POPPY-MALLOW, 807
PRICKLY-PEAR, 492
SAND-PARSLEY, 243
SUNFLOWER, 370
THREEAWN, 1242
TUMBLEWEED, 540
WILD INDIGO, 638
WINECUP, 807
YELLOW DAISY, 420
ZINNIA, 433
PLAINSMAN, OLD-, 374, 375
PLAINTAIN, GRASS-LEAF MUD-, 1340
PLAITED PASPALUM, 1306
PLANETREE, 888
PLANETREE FAMILY, 886
PLANK BEAK-RUSH, 1158
Plantaginaceae, 882
Plantago, 882
afra, 883
aristata, 884
coronopus, 884
elongata, 884
helleri, 884
heterophylla, 884
hookeriana, 884
var. *nuda,* 886
lanceolata, 886
major, 886
patagonica, 886
var. *breviscapa,* 886
var. *gnaphalioides,* 886
var. *spinulosa,* 886
purshii, 886
var. *breviscapa,* 886
var. *spinulosa,* 886
pusilla, 884
rhodosperma, 886
rugelii, 886
virginica, 886
wrightiana, 886
PLANTAGO, SLIM-SPIKE, 884
PLANTAIN, 882
BLUE-MUT-, 1340
BOTTLEBRUSH, 884
BRACTED, 884
BRISTLE-BRACT, 886
BROAD-LEAF, 886
BUCK-HORN, 884, 886
CEDAR, 884
COMMON, 886
DOORYARD, 886
DWARF, 886
ENGLISH, 886
GREAT, 886
GROOVE-STEM INDIAN-, 314
HOARY, 886
HOOKER'S, 884
INDIAN-, 314
LANCE-LEAF INDIAN-, 314
MUD-, 1086, 1339
PALE-SEED, 886
PRAIRIE-, 314
RED-SEED, 886
RUGEL'S, 886
SLENDER, 884
TUBEROUS INDIAN-, 314
WATER-, 1086
WRIGHT'S, 886
PLANTAIN FAMILY, 882
Platanaceae, 886
Platanthera, 1218
ciliaris, ▣/103, 1218

Platanus, 888
 occidentalis, 888
PLATEAU
 BIG-TOOTH MAPLE, 219
 FALSE NIGHTSHADE, 1018
 GERARDIA, 993
 LIVE OAK, 716
 MILKVINE, 284
 NERVERAY, 420
 SPIDERWORT, 1102
 YELLOW BUCKEYE, 738
PLATTE RIVER MILK-VETCH, 635
PLEAT-LEAF, 1174
 KNOTWEED, 904
 PRAIRIE, 1174
 PURPLE, 1173
Plecospermum, 831
Pleopeltis, 193
 polypodioides subsp. michauxiana, 193
PLEURISY-ROOT, 280
Pluchea, 396
 camphorata, 398
 foetida, 398
 odorata, 398
 purpurascens, 398
PLUCHEA
 CAMPHOR, 398
 PURPLE, 398
 STINKING, 398
PLUM, 947
 BIG-TREE, 949
 BUFFALO-, 630
 CHICKASAW, 949
 CREEK, 950
 DATE-, 578
 FLATWOOD, 950
 GROUND-, 630, 635
 HOG-, 932
 HOG, 950
 MEXICAN, 949
 MUNSON'S, 950
 OKLAHOMA, 949
 SAND, 949
 SANDHILL, 949
 SUGAR-, 938
 THICKET, 950
 WILD, 949
 WILDGOOSE, 950
PLUME
 GRASS, 1315
  BENT-AWN, 1315
  SUGARCANE, 1315
 INDIAN-, 890
 TEXAS, 890
PLUMED THISTLE, 338
PLUMELESS-THISTLE, 329
Plumeria rubra, 264
PLUMETOOTH BEEBALM, 766
Poa, 1312
 annua, 1312
 arachnifera, 1312
 bulbosa, 1312
 chapmaniana, 1314
 compressa, 1314
 pratensis, 1314
 sylvestris, 1314
POA, CHAPMAN'S, 1314
Poaceae, 1222
Podophyllum, 437
 peltatum, 437
POET'S NARCISSUS, 1207
Pogonia, 1218
 ophioglossoides, 1218
POGONIA, ROSE, 1218
Poinciana gilliesii, 639
POINCIANA, 639
POINSETTIA, 585, 604
 WILD, 606
POINT
 LOCO, 682
 -SEED GLOBE-MALLOW, 820
POINTED
 PHLOX, 891
 WOLFFIA, 1190
POISON
 BITTERWEED, 375
 DARNEL, 1282
 SUMAC, 238
 BEAVER-, 248
 -BEAN, 696
 -CAMASS, 1210
 -DOGWOOD, 238
 -ELDER, 238
 -HEMLOCK, 248
 -IVY, 234, 236
 -OAK, 234, 236
 -PARSLEY, 248
 -SEGO, 1210
POKE, 881
POKEBERRY, 881
POKEWEED, 881
POKEWEED FAMILY, 881
Polanisia, 506
 dodecandra subsp. trachysperma, 🖼/103, 506
 erosa, 507
 trachysperma, 507
POLECAT-TREE, 933
Polemoniaceae, 888
Polemonium, 888
POLEMONIUM FAMILY, 888
POLEO, 1049
Polianthes, 1079
 lata, 1080
POLLINATION, 227, 273, 274, 287, 309, 314, 332, 408, 434, 437, 438, 442, 445, 446, 494, 498, 502, 510, 529, 553, 656, 658, 692, 700, 830, 862, 913, 1027, 1079, 1082, 1092, 1168, 1169, 1170, 1213, 1223, 1342, 1350
POLLYPRIM, 480
POLYANTHUS NARCISSUS, 1207
POLYCARP, 522
Polycarpon, 522
 tetraphyllum, 522
Polygala, 894
 alba, 896
 cruciata, 896
 incarnata, 896
 lindheimeri, 896
  var. lindheimeri, 896
  var. parviflora, 896
 polygama var. obtusata, 896
 sanguinea, 896
 senega, 894
 tweedyi, 896
 verticillata, 896
  var. *ambigua*, 896
  var. *isocycla*, 896
  var. *sphenostachya*, 896
Polygalaceae, 894
Polygonaceae, 897
Polygonatum, 1208
 biflorum, 1208
Polygonella, 899
 americana, 899
Polygonum, 899
 amphibium, 902
  var. emersum, 902
  var. *stipulaceum*, 902
 aviculare, 902
 bicorne, 904
 coccinum, 902
 convolvulus, 902
 cristatum, 904
 densiflorum, 902
 hydropiperoides, 902
  var. *setaceum*, 904
 lapathifolium, 902
 pensylvanicum, 904
 persicaria, 904
 punctatum, 904
  var. *confertiflorum*, 904
 ramosissimum, 904
 scandens var. cristatum, 904
 setaceum, 904
 tenue, 904

Polygonum *(cont.)*
virginianum, 905
Polymnia, 408
uvedalia, 408
var. *densipilis*, 408
Polypodiaceae, 192
Polypodiophyta, 178
Polypodium polypodioides var.
michauxianum, 193
POLYPODY, GRAY, 193
POLYPODY FAMILY, 192
Polypogon, 1314
monspeliensis, 1314
semiverticillatus, 1315
viridis, 1314
Polypremum, 480
procumbens, 480
Polystichum, 185
acrostichoides, 185
Polytaenia, 256
nuttallii, 256
var. *texana*, 258
texana, 258
Pomaria, 686
jamesii, 688
POMME
BLANCHE, 684
DE PRAIRIE, 630, 684
Poncirus, 972
trifoliata, 972
POND-LILY, 845
YELLOW, 844
POND-NUT, 834
PONDWEED, 1342
BABY, 1343
BUSHY-, 1169
CORNSTALK, 1343
FENNELL-LEAF, 1343
HEART-LEAF, 1343
HORNED-, 1350
ILLINOIS, 1343
LEAFY, 1343
LONG-LEAF, 1343
SAGO, 1343
SHINING, 1343
WATER-THREAD, 1343
PONDWEED FAMILY, 1340
Pontederia, 1340
cordata, 1340
var. *lancifolia*, 1340
Pontederiaceae, 1338
PONY-FOOT, 552
POOLMAT, COMMON, 1350
POOR
-JOE, 962
-MAN'S
-PATCHES, 796
-PEPPER, 472
-ROPE, 798
-WEATHER-GLASS, 912
POORLAND FLAT SEDGE, 1136
POP-BEAN BUSH, 639
POPLAR, 974
BLACK, 976
LOMBARDY, 976
SILVER-LEAF, 975
WHITE, 975
POPOTE, 208
POPPY, 876
CALIFORNIA-, 874
COMMON, 876
CORN, 876
FIELD, 876
GOLD-, 874
HORNED-, 876
-MALLOW, 807
CLUSTERED, 807
FINGER, 808
GERANIUM, 808
LIGHT, 807
LOW, 807
PLAINS, 807
PURPLE, 807
SLIM-LOBE, 808
TALL, 808
MEXICAN-, 874
OPIUM, 872, 876
PRICKLY, 873
RED HORNED-, 876
SHIRLEY, 876
WHITE PRICKLY-, 873
YELLOW PRICKLY-, 874
POPPY FAMILY, 872
Populus, 974
alba, 975
deltoides, 975
subsp. deltoides, 975
subsp. monilifera, 976
var. *occidentalis*, 976
nigra, 976
sargentii, 976
texana, 976
tremuloides, 975
POPWEED, 473
Porlieria angustifolia, 1074
Porteranthus, 946
stipulatus, 946
Portulaca, 908
grandiflora, 908
halimoides, 910
lanceolata, 910
mundula, 910
oleracea, 910
parvula, 910
pilosa, 910
umbraticola subsp. lanceolata, 910
PORTULACA
SHAGGY, 910
SINKER-LEAF, 910
WING-POD, 910
Portulacaceae, 906
POSSESSION-VINE, 551
POSSUM
OAK, 718
-GRAPE, 1066
POSSUMHAW, 270
POST
OAK, 718
GRAPE, 1070
DRUMMOND'S, 716
DWARF, 716
SAND, 716
SCRUBBY, 716
SWAMP, 716
-CEDAR, 203
Potamogeton, 1342
crispus, 1342
diversifolius, 1343
foliosus, 1343
illinoensis, 1343
nodosus, 1343
pectinatus, 1344
pulcher, 1343
pusillus, 1343
Potamogetonaceae, 1340
POTATO, 1027
AIR-, 1165
-BEAN, 628
AMERICAN, 628
-DANDELION, 380
DUCK-, 1088
HOG-, 663
INDIAN-, 556
PRAIRIE-, 684
SWEET-, 555
-VINE, WILD, 556
-WEED, 1028
WILD, 556
POTATO FAMILY, 1015
Potentilla, 946
recta, 947
simplex, 947
POUI, 440
POVERTY
DANTHONIA, 1260
DROPSEED, 1328
GRASS, 1242
SOUTHERN, 1328

POVERTY (*cont.*)
- OAT GRASS, 1260
- -OATS, 1259, 1260
- RUSH, 1184
- -WEED, 540

POWDERED THALIA, 1212
POWDERHORN CHICKWEED, 516
POWDERPUFF, 680
- THISTLE, 332

POWDERY
- CLOAK FERN, 194
- THALIA, 1212

PRAIRIE
- ACACIA, 624
- AGALINIS, 993
- ASTER, WHITE, 318
- BACCHARIS, 324
- -BISHOP, 244
- BLUE-EYED-GRASS, 1178
- BLUETS, 966
- BRAZORIA, 784
- BULRUSH, 1108
- BUNDLE-FLOWER, 652
- BUR, 750
- BURNET, 960
- BUTTERCUP, 928
- BUTTONWEED, 970
- CELESTIAL, 1174
- -CLOVER, 644, 648, 650, 670
  - COMANCHE PEAK, 651
  - GLANDULAR, 650
  - LONG-BRACT, 650
  - PURPLE, 650
  - SHOWY, 648
  - SLIM-SPIKE, 650
  - WHITE, 650
- CONEFLOWER, 400
  - UPRIGHT, 400
- CORD GRASS, 1324
- CUP GRASS, 1274
- DOGBANE, 266
- FLAMEFLOWER, 911
- FLAME-LEAF SUMAC, 234
- FLAX, 792
- FLEABANE, 352
- GAILLARDIA, 358
- GENTIAN, 727
  - PURPLE, 727
  - SHOWY, 727
- GOLDENROD, 414
- GROUND-CHERRY, 1026
- GROUNDSEL, 390
- HOLLY, 270
- INDIAN PAINTBRUSH, 998
- -IRIS, 1174
- JUNE GRASS, 1280
- KUHNIA, 329
- LARKSPUR, 924
- MEXICAN-CLOVER, 970
- -MIMOSA, 652
- OAK, 716
- ONION, 1197, 1198
- PAINTBRUSH, 998
- -PARSLEY, 256
- -PARSNIP, 256
- PEPPERWEED, 470
- -PETUNIA, 216
- PHLOX, 892
- PINKROOT, 798
- -PLANTAIN, 314
- PLEAT-LEAF, 1174
- -POTATO, 684
- RAGWORT, 392
- RAIN-LILY, 1200
- -ROCKET, 468
- ROSE, 958
  - CLIMBING, 958
  - GENTIAN, 728
  - WHITE, 956
- SENNA, 642
- SPIDERWORT, 1102
- SPURGE, 596
- STICKLEAF, 797
- SUMAC, 234
- SUNFLOWER, 370
- -TEA, 600
- THREEAWN, 1242
- TRISETUM, 1332
- VERBENA, 1050
- VERVAIN, 1050
- WEDGESCALE, 1324
- WILD ROSE, 956

PRAYER-PLANT, 1210
PRECATORY-BEAN, 617
PRESERVING MELON, 566
PRETTY DODDER, 574
PRICK-LEAF GILIA, 890
PRICKLEWEED, 652
PRICKLY
- -ASH, 973, 974
  - SOUTHERN, 974
- LETTUCE, 381
- -MALLOW, 818
- -PEAR, 489, 492
  - BROWN-SPINE, 492
  - CHAIN, 492
  - COW-TONGUE, 491
  - EASTERN, 491
  - ENGELMANN'S, 491, 492
  - GRASSLAND, 492
  - NOPAL, 490
  - PLAINS, 492
  - PURPLE-FRUIT, 492
  - TEXAS, 490
  - TUBEROUS-ROOT, 492
- -POPPY, 873
  - WHITE, 873
  - YELLOW, 874
- SIDA, 818
- SOW-THISTLE, 416

PRIDE-OF
- -BARBADOS, 639
- -INDIA, 823
- -TEXAS, 891

PRIMROSE, 911
- ANGLE-STEM WATER-, 860
- COMMON EVENING-, 863
- CREEPING WATER-, 862
- CUT-LEAF EVENING-, 864
- DAY-, 854
- DOWNY EVENING-, 864
- EVENING-, 854, 862
- FLOATING
  - EVENING-, 860
  - WATER-, 862
- FOUR-POINT EVENING-, 866
- MEXICAN EVENING-, 866
- MEXICAN-, 866
- MISSOURI, 864
- NARROW-LEAF WATER-, 860
- SHORT-POD EVENING-, 866
- SHOWY-, 866
  - EVENING-, 866
- SHRUBBY WATER-, 860
- SINUATE-LEAF EVENING-, 864
- SMOOTH WATER-, 860
- SPACH'S EVENING-, 866
- STEMLESS EVENING-, 866
- THREE-LOBED-, 866
- TRUMPET EVENING-, 864
- VARIABLE EVENING-, 864
- WATER-, 858, 860
- WHITE EVENING-, 866
- -WILLOW, 860
  - CREEPING, 862
  - FLOATING, 862
  - UPRIGHT, 860
- WRIGHT'S EVENING-, 866

PRIMROSE FAMILY, 911
Primula, 911
Primulaceae, 911
Prinopsis, 360
- *ciliata*, 362

PRIVET, 849
- CHINESE, 850
- COMMON, 850
- GLOSSY, 850
- JAPANESE, 850

PRIVET *(cont.)*
NEPAL, 850
QUIHOU'S, 850
SWAMP-, 848
WAX-LEAF, 850
Proboscidea, 880
louisianica, [icon]/104, 880
Proserpinaca, 736
palustris var. amblyogona, 736
pectinata, 736
PROSO, 1300
Prosopis, 688
glandulosa, 688
juliflora var. *glandulosa,* 688
PROSTRATE
BOG CLUBMOSS, 174
EUPHORBIA, 596
GRAPE FERN, 189
GROUND-CHERRY, 1018
KNOTWEED, 902
LAWNFLOWER, 329
MECARDONIA, 1003
PIGWEED, 226
VERVAIN, 1056
WATER-HYSSOP, 1003
Prunella, 770
vulgaris
var. *hispida,* 770
subsp. lanceolata, 770
var. *lanceolata,* 770
subsp. *vulgaris,* 770
Prunus, 947
angustifolia, 949
caroliniana, 949
dulcis, 947
gracilis, 949
laurocerasus, 947
mexicana, 949
mitis, 952
munsoniana, 949
persica, 950
rivularis, 950
serotina, 950
subsp. *eximia,* 950
var. eximia, 950
var. serotina, 950
umbellata, 950
var. *tarda,* 952
Pseudognaphalium, 398
obtusifolium, 398
stramineum, 398
Pseudotsuga menziesii, 204
Psilactis, 315
Psilocarya
nitens, 1160
portoricensis, 1160
Psoralea
cuspidata, 684
cyphocalyx, 684
digitata, 684
var. *parvifolia,* 684
hypogaea, 684
var. *scaposa,* 684
latistipulata, 686
var. *appressa,* 686
linearifolia, 686
pedunculata, 680
psoralioides var. *eglandulosa,* 680
reverchonii, 686
rhombifolia, 686
scaposa, 684
var. *breviscapa,* 684
simplex, 682
subulata, 686
var. *minor,* 686
tenuiflora, 690
Psoralidium, 688
linearifolium, 686
tenuiflorum, 690
PSYLLIUM, 883
Ptelea, 972
mollis, 973
persicifolia, 973
tomentosa, 973
trifoliata, 972
subsp. angustifolia var. persicifolia, 973
var. *mollis,* 973
subsp. trifoliata var. mollis, 973
Pteridaceae, 193
Pteridium, 181
aquilinum var. pseudocaudatum, 181
PTERIDOPHYTES, 173
Ptilimnium, 258
capillaceum, 258
costatum, 258
nuttallii, 258
×*texense,* 258
Puccinia
arachidis, 628
coronata, 934
graminis, 437, 1334
podophylli, 438
sorghii, 1338
PUCCOON, 452
CAROLINA, 452
NARROW-LEAF, 452
Pueraria, 690
lobata, 690
montana var. lobata, 690
PUFF-SHEATH DROPSEED, 1327
PULSE FAMILY, 617
PULSEY, CHINESE-, 450
PUMPKINS, 568
PUNCTUREVINE, 1076
PUNCTUREWEED, 1076
PUNK OAK, 718
PURGING BUCKTHORN, 934
PURPLE
AMMANNIA, 800
ASTER, LATE, 319
BEAUTY-BERRY, 1049
BONAMIA, 560
CLIFF-BRAKE, 198
CONEFLOWER, 346
CUDWEED, 359
DALEA, 650
-DANDELION, 386
DEAD-NETTLE, 759
-DEWDROP, 498
FALSE BROOM, 1250
-FLOWER DROPSEED, 1327
FOXGLOVE, 1007
-FRUIT PRICKLY-PEAR, 492
GROUND-CHERRY, 1026
LOCO, 682
LOOSESTRIFE, 800, 802
LOVE GRASS, 1272
-MALLOW, 808
MEADOW-RUE, 930
MECARDONIA, 1003
MORNING-GLORY, 556
NEEDLE GRASS, 1243
NUT-GRASS, 1142
PAINTBRUSH, 998
PLEAT-LEAF, 1173
PLUCHEA, 398
POPPY-MALLOW, 807
PRAIRIE
GENTIAN, 727
-CLOVER, 650
-ROCKET, 470
-SAGE, TEXAS, 988
SAND GRASS, 1332
-SPIKE, TEXAS, 1218
THREEAWN, 1243
-VEIN GROUND-CHERRY, 1024
WEN-DOCK, 482
PURPLETOP, 1330
PURSH'S
CROTALARIA, 643
DEER-VETCH, 671
PURSLANE, 908, 910
COMMON, 910
DESERT HORSE-, 220
HORSE-, 220
MARSH-, 860

- PURSLANE *(cont.)*
  - SEA-, 220
  - SPEEDWELL, 1014
- PURSLANE FAMILY, 906
- PUSLEY, 910
- PUSSY
  - -TOES, 310
    - LARGE-LEAF, 310
    - PLAIN-LEAF, 310
  - WILLOW, 978
- **Pycnanthemum,** 772
  - albescens, 772
  - muticum, 772
  - tenuifolium, 772
- PYGMY
  - -FLOWER VETCH, 707
  - SENNA, 694
- **Pyracantha,** 952
  - koidzumii, 952
- *Pyrethrum cineraiifolium,* 382
- PYRETHRUM, 382
- **Pyrrhopappus,** 399
  - carolinianus, 399
    - var. *georgianus,* 399
  - *geiseri,* 399
  - *georgianus,* 399
  - grandiflorus, 399
  - *multicaulis,* 399
    - var. *geiseri,* 399
  - pauciflorus, 399
- **Pyrus,** 952
  - calleryana, 952
  - communis, 952

Q

- QAT, 526
- QUAILPLANT, 450
- QUAKER-LADIES, 968
- QUAKING
  - ASPEN, 975
  - GRASS, 1250
    - BIG, 1250
    - LITTLE, 1250
- QUAMASH, 1198
- QUARTERVINE, 442
- QUASSIA FAMILY, 1014
- QUEEN
  - -ANNE'S-LACE, 250
  - VICTORIA'S WATER-LILY, 842
- QUEEN'S
  - -DELIGHT, 614
  - -ROOT, 614
  - -WREATH, 898
- QUELITE, 226
  - DE COCHINO, 226
  - ESPINSO, 226
  - MANCHADO, 226
  - MORADO, 226
- QUERCITRON, 718
  - OAK, 718
- **Quercus,** 711
  - alba, 714
  - *breviloba,* 718
  - buckleyi, 714
  - *drummondii,* 716
  - *durandii,* 718
  - falcata, ▣/104, 714
  - fusiformis, 716
  - *gravesii,* 714
  - incana, 716
  - lyrata, 716
  - macrocarpa, 716
  - margarettiae, 716
  - marilandica, 716
  - muehlenbergii, 716
  - nigra, 718
  - phellos, 718
  - *san-sabeana,* 718
  - *schneckii,* 718
  - shumardii, 718
    - var. *microcarpa,* 714
    - var. *schneckii,* 718
    - var. *texana,* 714
  - sinuata, 718
    - var. breviloba, 718
    - var. sinuata, 718
  - stellata, 718
    - var. *margarettiae,* 716
  - *suber,* 711
  - *texana,* 714
  - velutina, 718
  - virginiana, 719
    - var. *fusiformis,* 716
- QUIHOU'S PRIVET, 850
- QUILL, MEADOW, 1200
- QUILLWORT, 173, 175
  - BLACK-FOOTED, 175
- QUILLWORT FAMILY, 175
- **Quincula,** 1026
  - lobata, ▣/104, 1026
- QUINOA, 530, 533
- QUINUA, 533

R

- RABANO, 476
- RABBIT
  - -FACE, 924
  - -FOOT CLOVER, 702
- RABBIT'S
  - TOBACCO, 356
  - -FOOT, 1314
    - GRASS, 1314
- RACEMED MILKWORT, 896
- RACOON-GRAPE, 1066
- RADISH, 476
- *Rafflesia arnoldii,* 914
- RAFFLESIA FAMILY, 914
- **Rafflesiaceae,** 914
- RAGWEED, 309
  - BLOOD, 310
  - COMMON, 309
  - FALSE, 393
  - FIELD, 310
  - GIANT, 310
  - LANCE-LEAF, 309
  - PARTHENIUM, 393
  - PERENNIAL, 310
  - -SAILORS, 336
  - SHORT, 309
  - SOUTHERN, 309
  - VERVAIN, 1050
  - WESTERN, 310
  - WOOLLY-WHITE, 374
- RAGWORT, 388, 403
  - OVATE-LEAF, 390
  - PRAIRIE, 392
- RAIN-LILY, 1200
  - GIANT, 1200
  - PRAIRIE, 1200
  - WHITE, 1200
  - WIDE-LEAF, 1200
- RAM'S-HORN, 880
- RAMIE, 1042
- RANBUTAN, 979
- **Ranunculaceae,** 916
- **Ranunculus,** 926
  - abortivus, 926
  - *carolinianus,* 928
  - fascicularis, 928
    - var. *apricus,* 928
  - hispidus var. nitidus, 928
  - macranthus, 928
  - muricatus, 928
  - parviflorus, 928
  - pusillus, 928
    - var. *angustifolius,* 928
  - sceleratus, 928
- RAPE, 460
  - BIRD'S, 460
- **Raphanus,** 476
  - sativus, 476
- **Rapistrum,** 476
  - rugosum, 476
- RATANY, 750
  - TRAILING, 750
- RATANY FAMILY, 750
- **Ratibida,** 400
  - *columnaris,* 400

Ratibida *(cont.)*
columnifera, [icon]/104, 400
RAT-TAIL
CACTUS, 491
SIXWEEKS GRASS, 1336
RATTAN, 1094
RATTANVINE, 931
RATTLE-BOX, 860
RATTLEBUSH, 696
RATTLEPOD, 643
RATTLESNAKE
FERN, 189
-FLOWER, 784
-MASTER, 252, 1080
-WEED, 250, 755
Rauvolfia serpentina, 264
Ravenelia holwayi, 688
RAYED PALAFOXIA, 393
Rayjacksonia, 387, 400
annua, 400
RAYLESS
GAILLARDIA, 359
-GOLDENROD, 328
GUMWEED, 360
THELESPERMA, 422
RED
ASH, 849
-BERRY
JUNIPER, 203
MOONSEED, 824
BIRCH, 439
BRUSH LIPPIA, 1054
BUCKEYE, 738
-CEDAR, 203
EASTERN, 103
VIRGINIA, 203
CLOVER, 706
DEAD-NETTLE, 759
DODDER, 574
ELM, 1040
FALSE MALLOW, 820
-FLOWER
HEDGEHOG, 486
-YUCCA, 1079
GILIA, 890
GRAMA, 1248
GRAPE, 1072
-HEART HICKORY, 748
HESPERALOE, 1079
HORNED-POPPY, 876
JUNIPER, 203
LEATHER-FLOWER, 922
LOVE GRASS, 1272
MAPLE, 220
MORNING-GLORY, 555
MULBERRY, 832
OAK, 718
BOTTOM-LAND, 714
SHUMARD'S, 718
SOUTHERN, 714
SWAMP, 714, 718
TEXAS, 714
THREE-LOBE, 714
-ROOT
FLAT SEDGE, 1136
PIGWEED, 226
SAVIN, 203
-SEED
DANDELION, 418
PLANTAIN, 886
SPIDER-LILY, 1206
SPRANGLETOP, 1281
THREEAWN, 1243
-TOP PANIC, 1302
REDBUD, 640, 850
EASTERN, 640
TEXAS, 640
REDGUM, 737
REDHAW, 940, 942
REDPEPPER, BUSH, 1017
REDROOT, 226, 932
REDTOP, 1330
REDTWIG-CREEPER, 1068
REDWOOD FAMILY, 202
REED, 1311
COMMON, 1311
GIANT, 1243
INDIAN, 1322
PAPER-, 1132
REFLEXED-FRUIT CARIC SEDGE, 1128
REGAHOSA, 796
Rehnsonia, 708
RESCUE
BROME, 1252
GRASS, 1252
RESEDA DEL CAMPO, 1049
RESIN-DOT SKULLCAP, 780
RESINOUS SKULLCAP, 780
RESURRECTION FERN, 193
RETAMA, 682
REVERCHON'S
BRISTLE GRASS, 1320
FALSE PENNYROYAL, 758
HAWTHORN, 942
MUHLY, 1288
PALAFOXIA, 393
SPIDERWORT, 1104
SPIKE-RUSH, 1146
Reverchonia, 611
arenaria, 612
REVERCHONIA, SAND, 612
REVERSED CLOVER, 706
Rhamnaceae, 930
Rhamnus, 934
caroliniana, 933
cathartica, 934
lanceolata, 934
subsp. glabrata, 934
var. *lanceolata*, 934
purshiana, 933
Rheum rhabarbarum, 897
Rhexia, 820
interior, 822
mariana
var. interior, 822
var. mariana, 822
virginica, 822
RHINOCEROS CACTUS, 485
RHODES GRASS, 1256
Rhododendron, 581
Rhoeo, 1098, 1101
RHOMBOID COPPERLEAF, 589
RHOMBOPOD, 506
NARROW-LEAF, 506
RHUBARB, 897
WILD, 906
Rhus, 232
aromatica
var. *flabelliformis*, 234
var. serotina, 232
copallinum
var. *lanceolata*, 234
var. latifolia, 232
cotinoides, 231
glabra, 234
lanceolata, 234
microphylla, 234
toxicarium, 236
toxicodendron, 236
trilobata, 234
vernix, 238
virens, 234
Rhynchosia, 690
americana, 691
latifolia, 691
minima, 691
var. *diminifolia*, 691
senna
var. *angustifolia*, 691
var. texana, 691
texana, 691
RHYNCHOSIA, BROAD-LEAF, 691
Rhynchosida, 816
physocalyx, [icon]/104, 816
Rhynchospora, 1156
colorata, 1158
corniculata
var. interior, 1158

Rhynchospora *(cont.)*
var. *macrostachya,* 1160
globularis
var. globularis, 1158
var. recognita, 1158
glomerata, 1158
harveyi, 158
macrostachya, 1158
nitens, 1160
nivea, 1160
RIBBON-GRASS, 886, 1080
Ribes, 732
aureum var. villosum, 732
RIB-GRASS, 886
RIBWORT, 882, 886
RICE, 1223, 1280
-BUTTON ASTER, 318
CUT GRASS, 1281
JUNGLE-, 1266
SOUTHERN WILD, 1338
Richardia, 968
scabra, 968
tricocca, 970
RICHWEED, 354, 427, 1042
Ricinus, 612
communis, 612
RIDDELL'S
LAZY DAISY, 314
SELAGINELLA, 175
SPIKE-MOSS, 175
RIDGE
-SEED EUPHORBIA, 594
WHITE OAK, 714
RIDGED GOOSEFOOT, 538
RIGID
GOLDENROD, 414
FLAX, 792
TICK-CLOVER, 655
RING-SEED RUSH, 1184
RINGWING, TUMBLE-, 540
RIPPLE-GRASS, 886
RIVER
BIRCH, 439
GRAPE, 1072
-HEMP, COLORADO, 696
RIVERBANK
GRAPE, 1072
WILLOW, 978
RIVET WHEAT, 1334
Rivina, 882
humilis, 882
ROADSIDE
GAURA, 856
THISTLE, 338
Robinia, 691
hispida, 691
pseudoacacia, 692
ROCK
-CEDAR, 203
CENTAURY, 724
COREOPSIS, 342
DAISY, 388, 394
LINDHEIMER'S, 394
-DROPSEED, 1290
FLAX, 792
HEDEOMA, 758
-JASMINE, 912
WESTERN, 912
-LETTUCE, 394
SMALL, 394
MILKWORT, 896
-MOSS, 564
MUHLY, 1290
OAK, 714, 718
-PINK, 911
-ROSE, 542, 543
GEORGIA, 543
SANDWORT, 519
SCURF-PEA, 686
ROCK-ROSE FAMILY, 542
ROCKCRESS, 459
BRAZOS, 459
ROCKET
LARKSPUR, 922
LONDON-479
-MUSTARD, 479
PRAIRIE-, 468
PURPLE-, 470
ROCKETSALAD, 468
ROCKETWEED, 468
ROCKFOIL, 988
ROCKY MOUNTAIN
BULRUSH, 1161
SAGE, 776
ZINNIA, 433
ROEMER'S
ACACIA, 625
EUPHORBIA, 608
PHLOX, 892
SCHRANKIA, 678
SENSITIVE-BRIAR, 678
SPURGE, 608
ROMAN WORMWOOD, 309
ROMERILLO, 1049
DE MONTE, 1054
ROMPEVIENTOS, 1035
ROOSEVELT-WEED, 324
ROOT, BUTTON SNAKE-, 252
ROOTSTOCK BLOODLEAF, 230
Rorippa, 476
islandica, 478
nasturtium-aquaticum, 476
palustris
subsp. fernaldiana, 478
var. *fernaldiana,* 478
sessiliflora, 478
teres, 478
Rosa, 954
arkansana var. suffulta, 956
bracteata, 956
carolina, 956
eglanteria, 956
foliolosa, 956
ignota, 956
laevigata, 956
micrantha, 958
multiflora, 958
rubiginosa, 956
setigera var. tomentosa, 958
suffulta, 956
Rosaceae, 936
ROSARY-PEA, 617
ROSE, 954
-ACACIA, 691
BALSAM, 434
BLUET, 968
BUFFALO-, 808
CAROLINA, 956
CHEROKEE, 956
CLIMBING, 958
PRAIRIE, 958
COTTON, 356
COWBOY-, 808
FUZZY, 958
GENTIAN, 727
PRAIRIE, 728
SQUARE-STEM, 728
GEORGIA
ROCK-, 543
SUN-, 543
-HEATH, 334
HOARY SUN-, 543
JAPANESE, 958
LEAFY, 956
MACARTNEY, 956
-MALLOW, 810
COMMOM, 810
HALBERD-LEAF, 810
SCARLET, 810
SWAMP, 810
WOOLLY, 810
-MOSS, 908
MOUNTAIN-, 898
MULTIFLORA, 958
-OF-SHARON, 812
PALAFOXIA, 393
PASTURE, 956

ROSE *(cont.)*
PERIWINKLE, 266
-PINK, 728
POGONIA, 1218
PRAIRIE, 958
WILD, 956
-RING GAILLARDIA, 358
ROCK-, 542, 543
ROSIN-, 547
-RUSH, 386
SUN-, 543
SUNSHINE, 956
SWEET BRIAR, 956
VERVAIN, 1050
WHITE PRAIRIE, 956
ROSE FAMILY, 936
ROSEMARY, 752
ROSETTE GRASS
CEDAR, 1300
DEER-TONGUE, 1296
LINDHEIMER'S, 1295
OPEN-FLOWER, 1300
ROUND-SEED, 1302
SCRIBNER'S, 1300
SLIM-LEAF, 1300
SOFT-LEAF, 1300
STARVED, 1296
VARIABLE, 1298
VELVET, 1302
WHITE-HAIRED, 1295
WOOLLY, 1295
ROSIN-ROSE, 547
ROSINWEED, 360, 404
ROUGH-STEM, 406
SIMPSON, 404
SLENDER, 404
WHITE, 404
ROSITA, 724
Rosmarinus, 752
Rotala, 802
ramosior, 802
Rottboellia, 1258
ROUGE BUTTERCUP, 928
ROUGEPLANT, 882
ROUGH
CONEFLOWER, 402
FALSE PENNYROYAL, 758
GAYFEATHER, 384
GERARDIA, 993
GOLDENROD, 414
GROMWELL, 452
HEDEOMA, 758
-LEAF
DOGWOOD, 561
GOLDENROD, 414
LIP FERN, 196
MARBLESEED, 454
MEXICAN-CLOVER, 968
NAMA, 740
-PEA, 666
PHLOX, 892
PIGWEED, 226
RUSH GRASS, 1327
-SEED BUTTERCUP, 928
-STEM ROSINWEED, 406
TRIDENS, 1330
ROUND
-CROTON, 589
-FRUIT ST. JOHN'S-WORT, 548
-HEAD
BROOMWEED, 365
BUSH-CLOVER, 670
DALEA, 650
LESPEDEZA, 670
RUSH, 1186
-LEAF
GRAPE, 1072
GROUNDSEL, 390
MALLOW, 814
MINT, 764
MONKEY-FLOWER, 1004
SCURF-PEA, 686
SEEDBOX, 862
TEPHROSIA, 700
THOROUGHWAX, 246
-SEED
PANIC, 1302
PASPALUM, 1306
ROSETTE GRASS, 1302
ROUNDLEAF, 365
ROYAL
FERN, 192
PALM, 1094
Roystonea regia, 1094
RUBBER, 585, 830
PARÁ, 830
-PLANT, 830
CREEPING, 830
GUAYULE, 393
-TREE, INDIAN, 830
RUBBERWEED, BITTER, 375
Rubiaceae, 960
Rubus, 958
aboriginum, 959
apogaeus, 959
bifrons, 959
duplaris, 960
flagellaris, 959
oklahomus, 959
riograndis, 960
trivialis, 959
var. *duplaris*, 960
uncus, 959
RUDA DEL MONTE, 973
Rudbeckia, 400
amplexicaulis, 346
coryi, 402
fulgida var. palustris, 402
grandiflora var. alismifolia, 402
hirta var. pulcherrima, 402
maxima, 402
triloba, 402
RUE, 970
TEXAS DESERT-, 973
RUE FAMILY, 970
Ruellia, 213
brittoniana, 214
caroliniensis, 216
var. *salicina*, 216
var. *semicalva*, 216
davisiorum, 218
drummondiana, 216
humilis, 216
var. *depauperata*, 216
var. *expansa*, 216
var. *longiflora*, 216
malacosperma, 216
metziae, 216
nudiflora
var. *hispidula*, 216
var. nudiflora, 216
var. runyonii, 216
pedunculata, 216
runyonii, 216
strepens, 218
var. *cleistantha*, 218
RUELLIA
DRUMMOND'S, 216
FRINGE-LEAF, 216
LIMESTONE, 218
LOW, 216
SMALL-FLOWER, 216
SMOOTH, 218
SOFT-SEED, 216
STALKED, 216
VIOLET, 216
RUGEL'S PLANTAIN, 886
RUM CHERRY, 950
Rumex, 905
acetosella, 906
altissimus, 906
crispus, 906
hastatulus, 906
hymenosepalus, 906
pulcher, 906
verticillatus, 906
RUNAWAY-ROBIN, 756
RUNNER OAK, 716

RUNNING MALLOW, 814
RUNYON'S ONION, 1197
RUSH, 1179
  AMERICAN SCOURING-, 176
  BEAK-, 1156
  BRAUN'S SCOURING-, 178
  CAPPED, 1182
  COMMON, 1182
    SCOURING-, 176
  DUDLEY'S, 1182
  FLAT-LEAF, 1184
  FORKED, 1182
  GRASS
    LONG-LEAF, 1327
    ROUGH, 1327
    -LEAF, 1184
  GREAT SCOURING-, 176
  HORNED-, 1156, 1158
  INLAND, 1184
  JOINTED, 1184
  KANSAS SCOURING-, 178
  KNOT-LEAF, 1182
  LEATHERY, 1182
  NEEDLE
    -POD, 1184
    -POINT, 1184
  NUT-, 1164
  PATH, 1184
  -PEA, 663
    DRUMMOND'S, 664
    INDIAN, 663
    SICKLE-POD, 663
  -PINK, 386
  POVERTY, 1184
  RING-SEED, 1184
  ROSE-, 386
  ROUND-HEAD, 1186
  SCOURING-, 176
  SLENDER, 1184
  SLIM-POD, 1182
  SMALL-HEAD, 1182
  SMOOTH SCOURING-, 178
  SOFT, 1182
  SPIKE-, 1144
  SUMMER SCOURING-, 178
  TALL SCOURING-, 176
  TEXAS, 1184
  TOAD, 1182
  TORREY'S, 1184
  TWIG-, 1130, 1132
  TWO-FLOWER, 1184
  WHITE-ROOT, 1182
RUSH FAMILY, 1178
RUSHFOIL, NARROW-LEAF, 600
RUSHWEED, 386
RUSSELL'S BEEBALM, 766
RUSSIAN
  MILLET, 1300
  MULBERRY, 831
  -OLIVE, 580
  -THISTLE, 540
  VETCH, 708
RUST FUNGI
  BLACK STEM RUST OF WHEAT, 437, 1334
  CEDAR APPLE RUST, 940, 203
  CROWN RUST OF OATS, 934
  ORANGE RUST, 959
  WHITE PINE BLISTER RUST, 732
RUSTY BLACKHAW, 511
RUSTY-SEED PASPALUM, 1306
Ruta, 970
Rutaceae, 970
RYE, 1318
  BROME, 1252
  CANADA WILD, 1268
  GRASS, 1282
    BEARDLESS DARNEL, 1282
    DARNEL, 1282
    ITALIAN, 1282
    PERENNIAL, 1282
  NODDING WILD, 1268
  VIRGINIA WILD, 1268
  WILD, 1268

S

SABADILLA, 1210
  DRUMMOND'S, 1210
  TEXAS, 1210
Sabal, 1094
  minor, 1094
Sabatia, 727
  angularis, 728
  campestris, ▣/104, 728
  formosa, 728
SABATIA, BUCKLEY'S, 728
SACAHUISTA, 1082
Saccharum, 1315
  brevibarbe var. contortum, 1315
  contortum, 1315
  giganteum, 1315
  officinarum, 1223
Sacciolepis, 1315
  striata, 1315
SACRED
  -BAMBOO, 437
  -BEAN, 834
  DATURA, 1020
  LOTUS, 834
SAFFLOWER, 330
SAFFRON
  BASTARD, 330
  FALSE, 330
  -THISTLE, 330
SAGE, 314, 772
  AUTUMN, 776
  AZURE, 774
  BLUE, 774
  BUR-, 310
  CEDAR, 776
  COMMON, 772
  ENGELMANN'S, 776
  GARDEN, 772
  GREGG'S, 776
  LANCE-LEAF, 776
  LYRE-LEAF, 776
  MEALY, 776
    -CUP, 776
  PERENNIAL SANDY-LAND-, 766
  PITCHER, 774
  ROCKY MOUNTAIN, 776
  SCARLET, 774
  TEXAS, 774, 776
    PURPLE-, 988
  TROPICAL, 774
  WOOD-, 782
SAGEBRUSH, 314
  MEXICAN, 315
SAGETREE, 1060
SAGEWORT, 314
  THREAD-LEAF, 315
  WESTERN, 315
Sagina, 522
  decumbens, 524
Sagittaria, 1087
  brevirostra, 1088
  calycina, 1090
  falcata, 1088
  graminea var. *platyphylla,* 1090
  lancifolia, 1088
    var. *media,* 1088
  latifolia, 1088
  longiloba, 1090
  montevidensis subsp. calycina, 1090
  papillosa, 1090
  platyphylla, 1090
SAGO, 1343
  PONDWEED, 1343
SAILORS
  BLUE-, 336
  RAGGED-, 336
Salicaceae, 974
Salix, 976
  babylonica, 978
  caroliniana, 978
  exigua, 978
  interior, 978

- **Salix** *(cont.)*
  - nigra, 978
    - var. *lindheimeri*, 979
- SALLOW, 976
  - CARIC SEDGE, 1124
- SALSIFY, 424, 426
  - VEGETABLE-OYSTER, 426
  - WESTERN, 424
- **Salsola,** 540
  - *australis*, 542
  - *kali*
    - subsp. *tenuifolia*, 542
    - subsp. *tragus*, 542
  - *pestifer*, 542
  - tragus, 540
    - subsp. *iberica*, 542
- SALT
  - -CEDAR, 1034, 1035
  - GRASS, 1264
    - COASTAL, 1264
    - DESERT, 1264
    - INLAND, 1264
    - SPICATE, 1264
  - HELIOTROPE, 450
  - MEADOW GRASS, 1281
  - SPRANGLETOP, 1281
- SALT-CEDAR FAMILY, 1034
- SALTBUSH, 324, 532
  - FOUR-WING, 533
  - SILVER-SCALE, 532
  - SPREADING, 533
- SALTMARSH
  - ASTER, 320
  - BULRUSH, 1108
  - -MARSH SANDSPURRY, 525
- **Salvia,** 772
  - azurea var. grandiflora, /105, 774
  - coccinea, 774
  - engelmannii, 776
  - farinacea, /105, 776
  - greggii, 776
  - lyrata, 776
  - *officinalis*, 772
  - reflexa, 776
  - roemeriana, 776
  - texana, 776
- **Sambucus,** 510
  - *canadensis*, 510
    - var. *submollis*, 510
  - nigra, 510
    - var. canadensis, 510
- **Samolus,** 914
  - *cuneatus*, 914
  - ebracteatus subsp. cuneatus, 914
  - *parviflorus*, 914
  - valerandi subsp. parviflorus, 914
- SAMPSON'S SNAKEROOT, 680
- SAMUEL MEAD'S CARIC SEDGE, 1124
- SAN SABA PINWEED, 544
- SAND
  - BITTERCRESS, 464
  - BLUESTEM, 1239
  - DROPSEED, 1327
  - GRASS, PURPLE, 1332
  - -LAND-SAGE, PERENNIAL, 766
  - -LILY, 796
  - LOVE GRASS, 1274
  - MILKWEED, 278
  - -PARSLEY, 243
  - PHACELIA, 743
  - PLUM, 949
  - POST OAK, 716
  - REVERCHONIA, 612
  - -VERBENA, 835, 836
    - AMELIA'S, 836
    - SWEET, 836
- SANDALWOOD FAMILY, 979
- SANDBAR WILLOW, 978
  - TEXAS, 978
- SANDBELL, 740
- SANDBUR, 750, 1253
  - COMMON, 1254
  - LONG-SPINE, 1254
  - SOUTHERN, 1254
- SANDHILL
  - AMARANTH, 224
  - PLUM, 949
- SANDIA, 565
- SANDING VERVAIN, 1056
- SANDJACK OAK, 716
- SANDPAPER VERVAIN, 1058
- SANDROCKET, 466
- SANDSPURRY, 525
  - MARSH, 525
  - SALT-MARSH, 525
- SANDVINE, 282
- SANDWORT, 514, 518
  - DRUMMOND'S, 519
  - HILLY, 514
  - PITCHER, 519
  - ROCK, 519
  - THYME-LEAF, 514
- *Sanguinaria canadensis*, 872
- **Sanguisorba,** 960
  - annua, 960
- SANICLE, 258
  - CANADA, 260
  - CLUSTER, 260
- **Sanicula,** 258
  - canadensis, 260
  - *gregaria*, 260
  - odorata, 260
- *Sansevieria*, 1078
- **Santalaceae,** 979
- SANTA-MARIA, 393
- **Sapindaceae,** 979
- **Sapindus,** 980
  - *drummondii*, 980
  - saponaria var. drummondii, 980
- **Sapium,** 612
  - sebiferum, 612
- SAPODILLA FAMILY, 982
- **Saponaria,** 524
  - officinalis, 524
  - *vaccaria*, 526
- **Sapotaceae,** 982
- SAPOTE FAMILY, 982
- SARACHA, GROUND, 1018
- *Sarcostemma*
  - *crispum*, 283
  - *cynanchoides*, 283
- **Sarracenia,** 983
  - alata, 983
  - *sledgei*, 983
- **Sarraceniaceae,** 983
- SARSAPARILLA
  - WILD, 824, 1346
  - -VINE, 1346
- SARVICE-BERRY, 938
- **Sassafras,** 786
  - albidum, 786
    - var. *molle*, 786
- SASSAFRAS, 786
- SATIN GRASS, 1290
- *Satureja arkansana*, 756
- SAUCE, 976
- **Saururaceae,** 983
- **Saururus,** 984
  - cernuus, 986
- SAUSAGE TREE, 440
- SAÚZ, 978
- *Savia phyllanthoides* var. *reverchonii*, 610
- SAVIN, RED, 203
- SAVORY, 755
  - OZARK, 755
- SAW
  - -GRASS, 1130, 1132
    - JAMAICAN, 1132
  - GREENBRIER, 1346
  - -LEAF DAISY, 362
  - -TOOTH FROGFRUIT, 1054
    - NERVERAY, 420
    - SUNFLOWER, 369
    - WATER-HYSSOP, 1003
- SAWBRIER, 1346
- **Saxifraga,** 988
  - *reevesii*, 988

Saxifraga *(cont.)*
texana, 988
Saxifragaceae, 986
SAXIFRAGE, 988
TEXAS, 988
SAXIFRAGE FAMILY, 986
Scabiosa, 576
atropurpurea, 576
SCABIOUS, SWEET, 576
SCALD, 572
SCALESEED, 260
BRISTLY, 262
FORKED, 262
SPREADING, 262
SCALLOP SQUASH, 568
SCALY-BARK
HICKORY, 748
OAK, 718
Scandix, 260
pecten-veneris, 260
SCARLET
BUCKEYE, 738
CLEMATIS, 922
-CREEPER, 555
GAURA, 856
GLOBE-MALLOW, 820
MAPLE, 220
MORNING-GLORY, 555
MUSKFLOWER, 842
-PEA, 664
WESTERN, 664
PIMPERNEL, 912
ROSE-MALLOW, 810
SAGE, 774
SPIDERLING, 838
SUMAC, 234
SCENTED
GAURA, 856
LIPPIA, 1054
Schedonnardus, 1316
paniculatus, 1316
SCHEELE'S BRISTLE GRASS, 1320
Schefflera, 270
Schinus, 230
Schizachyrium, 1316
scoparium, 1316
var. *frequens,* 1316
Schoenocaulon, 1210
drummondii, 1210
officinale, 1210
texanum, 1210
Schoenoplectus, 1160
acutus, 1161
californicus, 1161
pungens, 1161
saximontanus, 1161
tabernaemontani, 1161
SCHOTT'S CHOLLA, 492
SCHRADER'S GRASS, 1252
Schrankia, 677
hystricina, 678
latidens, 678
microphylla, 678
nuttallii, 678
roemeriana, 678
uncinata, 678
SCHRANKIA
CATCLAW, 678
KARNES, 678
ROEMER'S, 678
SCHREBER'S
MUHLY, 1290
WATERSHED, 482
SCHWEINITZ' FLAT SEDGE, 1142
Scirpus, 1162
acutus, 1161
americanus var. *longispicatus*, 1161
annuus, 1148
atrovirens, 1162
autumnalis, 1148
bergsonii, 1161
californicus, 1161
cyperinus, 1162
georgianus, 1162
koilolepis, 1154
lineatus, 1162
maritimus, 1108
var. *paludosus,* 1108
micranthus, 1156
miliaceus, 1150
molestus, 1154
pendulus, 1162
puberulus, 1150
pungens
var. *longispicatus*, 1161
var. *pungens*, 1161
saximontanus, 1161
supinus var. *saximontanus,* 1161
tabernaemontani, 1162
vahlii, 1150
validus, 1162
Scleria, 1164
ciliata, 1164
var. *elliottii,* 1164
oligantha, 1164
triglomerata, 1164
verticillata, 1164
Sclerochloa, 1316
dura, 1316
Scleropoa rigida, 1260
SCORPION-GRASS, 453
EARLY, 453
SCOTCH-THISTLE, 388
SCOURING-RUSH, 176
AMERICAN, 176
BRAUN'S, 178
COMMON, 176
GREAT, 176
KANSAS, 178
SMOOTH, 178
SUMMER, 178
TALL, 176
SCRAMBLED-EGGS, 719
SCRATCH DAISY, 344
SCRIBNER'S ROSETTE GRASS, 1300
Scrophularia, 1008
marilandica, 1010
Scrophulariaceae, 988
SCRUB
LIVE OAK, 716
OAK, 718
SUMAC, 234
SCRUBBY POST OAK, 716
SCUPPERNONG, 1072
SCURF-PEA, 683, 688
EDIBLE, 684
PALM-LEAF, 684
ROCK, 686
ROUND-LEAF, 686
SLIM-LEAF, 690
TALL-BREAD, 684
TURNIP-ROOT, 684
SCURFY SIDA, 815
SCURVY-PEA, 690
SLENDER, 690
Scutellaria, 776
australis, 780
brevifolia, 780
cardiophylla, 778
drummondii, 778
var. drummondii, 779
var. edwardsiana, 779
integrifolia var. *brevifolia*, 780
lateriflora, 779
leonardii, 780
ovata, 779
subsp. bracteata, 779
subsp. mexicana, 779
parvula, 779
var. australis, 780
var. *leonardii*, 780
var. missouriensis, 780
var. parvula, 780
resinosa, 780
var. *brevifolia*, 780
wrightii, ▣/105, 780
SCYTHE
-FRUIT ARROWHEAD, 1088

SCYTHE *(cont.)*
-LEAF WILLOW, 978
SEA
-COAST SUMPWEED, 378
-MYRTLE, 324
-OATS, 1254
-PURSLANE, 220
SEASIDE
GOLDENROD, 415
HELIOTROPE, 450
PETUNIA, 1017
Secale, 1318
cereale, 1318
montanum, 1318
SEDGE
HAIR-, 1109
NUT, 1142
SNOWY WHITE-TOP, 1160
STARRUSH WHITE-TOP, 1158
THREE-WAY, 1106
TISSUE, 1122
UMBRELLA, 1152
SEDGE FAMILY, 1104
Sedum, 562
nuttallianum, 564
pulchellum, 564
SEEDBOX, 858, 860
AMERICAN, 860
BUSHY, 860
CREEPING, 860
MARSH, 860
NARROW-LEAF, 860
ROUND-LEAF, 862
TORREY'S, 860
SEEDTICKS, 250
SEEP
MARSHALLIA, 387
MUHLY, 1288
-WILLOW, 324
Selaginella, 174
apoda, 174
arenicola subsp. riddellii, 175
peruviana, 175
riddellii, 175
sheldonii, 175
SELAGINELLA
BASKET, 174
RIDDELL'S, 175
Selaginellaceae, 174
SELFHEAL, 770
COMMON, 770
Selinocarpus, 842
diffusus, 842
SENECA-SNAKEROOT, 894
Senecio, 403
ampullaceus, 403
glabellus, 390
greggii, 392
hieraciifolia, 348
imparipinnatus, 392
obovatus, 390
plattensis, 392
tampicanus, 392
vulgaris, 403
SENEGA-ROOT, 894
Senna, 692
alata, 694
corymbosa, 694
lindheimeriana, 694
marilandica, 694
obtusifolia, 694
occidentalis, 694
pendula var. glabrata, 696
pumilio, 694
roemeriana, ▣/105, 696
SENNA, 640
ARGENTINE, 694
COFFEE, 694
DWARF, 694
LINDHEIMER'S, 694
MARYLAND, 694
PRAIRIE, 642
PYGMY, 694
TWO-LEAF, 696
WILD, 694
SENORITA WATER-LILY, 845
SENSITIVE
-BRIAR, 677
BRISTLY, 678
CATCLAW, 678
NUTTALL'S, 678
ROEMER'S, 678
FERN, 184
PARTRIDGE-PEA, 642
-PEA, 642
Sequoia sempervirens, 202
Sequoiadendron, 202
Serenia cespitosa, 380
SERICEA, 670
LESPEDEZA, 670
Sericocarpus, 315
SERVICE-BERRY, 938, 940
COMMON, 940
DOWNY, 940
SESAME, 880
SESAME FAMILY, 878
SESAMUM, 880
Sesamum indicum, 878
SESBANE, 696
Sesbania, 696
vesicaria, 662
drummondii, 696
exaltata, 696
herbacea, 696
macrocarpa, 696
SESBANIA, DRUMMOND'S, 696
SESSILE-LEAF TICK-CLOVER, 656
Setaria, 1304, 1310, 1318
geminata, 1304
geniculata, 1319
glauca, 1320
italica, 1319
leucopila, 1319
lutescens, 1320
magna, 1319
parviflora, 1319
pumila, 1319
ramiseta, 1320
reverchonii, 1320
scheelei, 1320
verticillata, 1320
viridis, 1320
SEVEN-ANGLE PIPEWORT, 1166
SEVENLEAF-CREEPER, 1068
SEYCHELLES PALM, 1094
Seymeria macrophylla, 999
SEYMERIA, MULLEIN, 999
SHAD
-BERRY, 940
-BUSH, 938
SHADE BETONY, 780
SHAG-BARK HICKORY, 748
SHAGGY
CRAB GRASS, 1264
EVOLVULUS, 552
HEDGE-HYSSOP, 1000
PORTULACA, 910
SHAKING GRASS, 1250
SHALLOT, 1194
SHAMA-MILLET, 1266
SHAME-BOY, 678
SHAMROCK, 674, 704, 706, 869
SHARP
-BRACT SUMPWEED, 378
-MARGIN CARIC SEDGE, 1122
-POD
BUNDLE-FLOWER, 652
MORNING-GLORY, 556
-SCALE
CARIC SEDGE, 1128
SPIKE-RUSH, 1146
-WING MONKEY-FLOWER, 1004
SHEATHED FLAT SEDGE, 1138
SHEEP
-SHOWERS, 870
SORREL, 906
SHEEPKILL, 581
SHELLBARK, 748

- SHELL-FLOWER, 1093, 1174
- SHEPHERD'S
  - -WEATHER-GLASS, 912
  - -NEEDLE, 260
  - -PURSE, 462
- SHERARD, 970
  - HERB, 970
- **Sherardia,** 970
  - arvensis, 970
- SHIELD FERN, SOUTHERN, 200
- SHIELD-SORUS FERN, 193
- SHIN OAK, 716, 718
- SHINING
  - HEXALECTRIS, 1216
  - PONDWEED, 1343
  - SUMAC, 232
- SHINY GOLDENROD, 412
- SHIRLEY POPPY, 876
- SHIRLEY'S-NETTLE, 794
- SHOOTING-STAR, 912
  - COMMON, 913
- SHORE MILKWEED, 280
- SHORT
  - -BEAK BALD-RUSH, 1160
  - CARIC SEDGE, 1118
  - -HAIR CARIC SEDGE, 1120
  - -LEAF
    - FLAT SEDGE, 1154
    - SKULLCAP, 780
  - -POD
    - DRABA, 466
    - EVENING-PRIMROSE, 866
  - RAGWEED, 309
  - -SEED WATERWORT, 580
  - -SPIKE
    - NOSEBURN, 616
    - WINDMILL GRASS, 1256
  - -STALK
    - CHICKWEED, 516
    - LOVE GRASS, 1270
- SHORTLEAF
  - PINE, 206
  - YELLOW PINE, 206
- SHOWY
  - BUTTERCUP, 928
  - CHLORIS, 1258
  - CROTALARIA, 644
  - DODDER, 574
  - EVENING-PRIMROSE, 866
  - HAWK'S-BEARD, 344
  - MENODORA, 852
  - PALAFOXIA, 392
  - PARTRIDGE-PEA, 642
  - PRAIRIE
    - GENTIAN, 727
    - -CLOVER, 648
    - -PRIMROSE, 866
- SHRIMP-PLANT, 211, 213
- SHRUB WILLOW, 978
- SHRUBBY
  - BONESET, 353
  - DITAXIS, 602
  - MILKWORT, 896
  - WATER-PRIMROSE, 860
- SHUMARD'S
  - OAK, 718
  - RED OAK, 718
- **Sibara,** 478
  - virginica, 478
- SIBARA, VIRGINIA, 478
- SIBERIAN
  - ELM, 1040
  - MOTHERWORT, 760
- SICKLE-POD, 459, 694
  - RUSH-PEA, 663
- **Sicyos,** 570
  - angulatus, 570
- **Sida,** 818
  - abutifolia, 818
  - *filicaulis,* 818
  - *hastata,* 818
  - *hederacea,* 816
  - *lepidota,* 815
  - *leprosa*
    - var. *depauperata,* 815
    - var. *hederacea,* 816
    - var. *sagittifolia,* 816
  - *physocalyx,* 818
  - *sagittifolia,* 816
  - spinosa, 818
- SIDA
  - ALKALI, 815
  - BEAKED, 816
  - PRICKLY, 818
  - SCURFY, 815
  - SPEAR-LEAF, 816
  - SPREADING, 818
- SIDE
  - -BEAK PENCIL-FLOWER, 698
  - -CLUSTER MILKWEED, 278
  - -FLOWER ASTER, 319
  - -FLOWERING SKULLCAP, 779
  - -OATS GRAMA, 1248
- *Sideranthus annuus,* 400
- **Sideroxylon,** 982
  - lanuginosum
    - subsp. *albicans,* 983
    - subsp. oblongifolium, 982
  - *sessilflorum,* 982
- SIGNAL GRASS, 1334
  - BROAD-LEAF, 1335
  - BROWN-TOP, 1335
  - FRINGED, 1334
  - TEXAS, 1335
- SIKO, 1200
- **Silene,** 524
  - antirrhina, 524
  - stellata, 524
- SILENE, WHORLED, 524
- SILK
  - -GRASS, 396
  - -TOP DALEA, 646
- SILKTASSEL, 723
  - LINDHEIMER'S, 723
  - MEXICAN, 723
- SILKTASSEL FAMILY, 722
- SILKTREE, 625
- SILKVINE, 287
- SILKWEED, 276
- SILKWORM MULBERRY, 831
- SILKY
  - ASTER, 320
  - CUP GRASS, 1274
  - EVOLVULUS, 554
- **Silphium,** 404
  - albiflorum, ▣/105, 404
  - *asperrimum,* 406
  - gracile, 404
  - *integrifolium,* 406
  - laciniatum, ▣/105, 404
  - radula, ▣/105, 406
  - *reverchonii,* 404
  - *simpsonii* var. *wrightii,* 404
- SILVER
  - BEARDGRASS, 1246
  - -BERRY OLEASTER, 579
  - BLUESTEM, 1246
  - EVAX, 356
  - GRASS, 1284
  - -LEAF
    - NIGHTSHADE, 1030
    - POPLAR, 975
  - MAPLE, 220
  - ORACHE, 532
  - -SCALE SALTBUSH, 532
- SILVERBELL, 1034
- SILVERLING, 324
- SILVERPUFF, 335
- SILVERY BEARD GRASS, 1239
- SILVEUS' DROPSEED, 1328
- **Silybum,** 406
  - marianum, 406
- SIMAROUBA FAMILY, 1014
- **Simaroubaceae,** 1014
- SIMPLE-LEAF
  - ERYNGO, 252
  - TICK-CLOVER, 656
- SIMPSON ROSINWEED, 404

- **Simsia,** 406
  - calva, 406
- **Sinapis,** 478
  - alba, 478
  - arvensis, 479
- SINGLE-STEM SNAKEROOT, 682
- SINGLETARY
  - VETCHLING, 666
  - -PEA, 666
- SINKER-LEAF PORTULACA, 910
- SINUATE-LEAF EVENING-PRIMROSE, 864
- **Siphonoglossa,** 218
  - pilosella, 218
- SISKIYOU ASTER, 318
- SISTER VIOLET, 1064
- **Sisymbrium,** 479
  - altissimum, 479
  - irio, 479
  - *nasturtium-aquaticum,* 476
  - officinale, 479
    - var. *leiocarpum,* 479
- **Sisyrinchium,** 1174
  - albidum, 1176
  - angustifolium, 1176
  - *bermudiana,* 1176
  - campestre, 1178
  - chilense, 1178
  - *ensigerum,* 1178
  - langloisii, 1178
  - minus, 1178
  - pruinosum, ▣/106, 1178
- **Sium,** 260
  - *cicutifolium,* 260
  - suave, 260
- SIX-ANGLE EUPHORBIA, 608
- SIXWEEKS
  - FESCUE, 1336
    - GRASS, 1335
  - GRAMA, 1248
  - GRASS, 1335
    - BROME, 1336
    - COMMON, 1336
    - HAIRY, 1336
    - RAT-TAIL, 1336
    - SQUIRREL, 1336
- SKELETON
  - GRASS, BEARDED, 1277
  - -PLANT, 386
    - TEXAS, 386
  - -WEED, 386
- SKEWERWOOD, 561
- SKULLCAP, 776
  - DRUMMOND'S, 778
  - EGG-LEAF, 779
  - HEART-LEAF, 778
  - MAD-DOG, 779
  - RESIN-DOT, 780
  - RESINOUS, 780
  - SHORT-LEAF, 780
  - SIDE-FLOWERING, 779
  - SMALL, 780
  - TUBER, 779
  - VIRGINIAN, 779
  - WRIGHT'S, 780
- SKUNKBUSH, 234, 972
- SKUNKWEED, 600
- SKY
  - -BLUE ASTER, 319
  - -DROP ASTER, 319, 320
- SLASH PINE, 206
  - YELLOW, 207
- SLEEPY DAISY, 430
  - TEXAS, 430
- SLENDER
  - BIGELOWIA, 328
  - BLADDERPOD, 474
  - BRISTLE-THISTLE, 330
  - BUSH-CLOVER, 671
  - COPPERLEAF, 588
  - CRAB GRASS, 1263
  - FIMBRISTYLIS, 1148
  - FINGER GRASS, 1263
  - FLAT SEDGE, 1138
  - FLEABANE, 352
  - FUMEWORT, 722
  - GERARDIA, 993
  - GOLDENWEED, 344
  - GRASS, 1281
  - GREENTHREAD, 422
  - GROOVEBUR, 938
  - HEDEOMA, 758
  - INDIAN GRASS, 1322
  - LADIES'-TRESSES, 1220
  - -LEAF
    - BETONY, 782
    - BLUET, 968
    - FLAT SEDGE, 1154
    - MOUNTAIN MINT, 772
  - LESPEDEZA, 671
  - LIP FERN, 196
  - MILKWORT, 896
  - MOUNTAIN MINT, 772
  - MUHLY, 1288
  - ONE-SEED COPPERLEAF, 588
  - PIGWEED, 226
  - PLANTAIN, 884
  - ROSINWEED, 404
  - RUSH, 1184
  - SCURFY-PEA, 690
  - SNAKE-COTTON, 227
  - -STEM CACTUS, 491
  - THREEAWN, 1242
  - VENUS'-LOOKING-GLASS, 501
  - VERVAIN, 1056
- SLIM
  - AMARANTH, 226
  - ASTER, 320
  - -LEAF
    - GOOSEFOOT, 536
    - MILKWEED, 280
    - PANIC, 1300
    - ROSETTE GRASS, 1300
    - SCURF-PEA, 690
  - -LOBE
    - CELERY, 249
    - POPPY-MALLOW, 808
  - MILKWEED, 278
  - -POD
    - MILK-VETCH, 632
    - RUSH, 1182
  - -SEED EUPHORBIA, 596
  - -SPIKE
    - PLANTAGO, 884
    - PRAIRIE-CLOVER, 650
    - THREEAWN, 1242
  - TRIDENS, 1330
- SLIMPOD, 265
  - CREEPING, 266
  - VENUS'-LOOKING-GLASS, 501
  - TEXAS, 265
  - TRUMPET, 265
  - WILLOW, 266
- SLIPPER-FLOWER, 988
- SLIPPERY ELM, 1040
- SLOUGH GRASS, 1324
- SMALL
  - -BEAK CARIC SEDGE, 1128
  - BLUET, 967
  - BUR-CLOVER, 674
  - -FLOWER
    - CORYDALIS, 722
    - FIDDLENECK, 445
    - GAURA, 856
    - MILK-VETCH, 634
    - RUELLIA, 216
    - VERVAIN, 1050
    - VETCH, 707
    - WATER-PLANTAIN, 1086
  - -FRUIT AGRIMONY, 938
  - -HEAD
    - BOLTONIA, 328
    - PIPEWORT, 1166
    - SNEEZEWEED, 368
    - RUSH, 1182
  - HOP CLOVER, 704
  - JOINTWEED, 899
  - -LEAF SUMAC, 234
  - MEDIC, 674

SMALL *(cont.)*
PALAFOXIA, 392
ROCK-LETTUCE, 394
-SEED
FALSE FLAX, 460
SPIKE-RUSH, 1147
SKULLCAP, 780
-SNAPDRAGON, 999
SNEEZEWEED, 368
-SPIKE FALSE NETTLE, 1042
-TOOTH CARIC SEDGE, 1124
VENUS'-LOOKING-GLASS, 502
WHITE MORNING-GLORY, 556
SMALL'S GREENBRIER, 1346
Smallanthus, 408
uvedalia, 408
SMARTWEED, 899, 904
CURLTOP, 902
DOTTED, 904
PALE, 902
PENNSYLVANICA, 904
PINK, 904
SNOUT, 902
SWAMP, 902
WATER, 902, 904
WILLOW, 902
Smilacaceae, 1344
Smilax, 1344
bona-nox, 1346
glauca, 1346
hispida, 1346
renifolia, 1346
rotundifolia, 1346
smallii, 1346
tamnoides, 1346
SMOKEBUSH, 231
SMOKETREE, 231, 1015
AMERICAN, 231
EUROPEAN, 231
SMOKEWOOD, 231
SMOOTH
ALDER, 439
ANGLEPOD, 282
-BARK OAK, 718
BEARDTONGUE, 1007
BEGGAR'S-TICKS, 326
BUTTONWEED, 970
CRAB GRASS, 1264
CREEPING LOVE GRASS, 1270
DITAXIS, 602
DOCK, 906
DOGBANE, 266
GAURA, 856
GAYFEATHER, 386
HEDGE-NETTLE, 782
HORSETAIL, 178
-LEAF FORESTIERA, 848
LIP FERN, 196
PASPALUM, 1306
PENSTEMON, 1007
PHACELIA, 743
RUELLIA, 218
SCOURING-RUSH, 178
-SEED PASPALUM, 1308
-STICKERS, 416
SUMAC, 234
SWALLOW-WORT, 282
TWISTFLOWER, 479
WATER-PRIMROSE, 860
WILD MERCURY, 602
YELLOW VIOLET, 1062
SMUT GRASS, 1327
SNAILSEED, 824
CAROLINA, 824
SNAKE
-COTTON, 227
DRUMMOND'S, 227
FIELD, 227
SLENDER, 227
-MOUTH, 1218
SNAKEBERRY, 944
SNAKEHERB, 213
NARROW-LEAF, 213
SNAKEROOT, 354, 680
BLACK, 258
BOBSROOT, 680
BUTTON, 252
BUTTON-, 382
HAIRY BUTTON-, 384
SAMPSON'S, 680
SINGLE-STEM, 682
VIRGINIA, 273
WHITE, 354
SNAKEWEED, 362
BROOM, 365
THREAD-LEAF, 364
SNAKEWOOD, 932
TEXAS, 932
SNAPDRAGON, 988
MAURANDELLA, 1002
DWARF-, 998, 999
SMALL-, 999
SPURRED-, 1002
-VINE, 1002
BLUE-, 1002
SNAPDRAGON FAMILY, 988
SNAPWEED, 433
SNEEZEWEED, 365, 368
BASIN, 366
COMMON, 366
SMALL, 368
SMALL-HEAD, 368
TALL, 366
SNOUT
SMARTWEED, 902
-BEAN, 690
BROAD-LEAF, 691
LEAST, 691
SNOWBALL, 836
SNOWBELL, 1034
SYCAMORE-LEAF, 1034
SNOWBERRY, 510
SNOWDROP, 1206
SNOWDROPTREE, 1034
SNOWFLAKE, 1204
GIANT, 1204
SPRING, 1206
SUMMER, 1204
SNOW-ON-THE-
MOUNTAIN, 608
PRAIRIE, 606
SNOWY WHITE-TOP SEDGE, 1160
SOAPBERRY, 980
WESTERN, 980
SOAPBERRY FAMILY, 979
SOAPBUSH, 1074
SOAPWEED, 1082
SOAPWORT, 524
SODA-POP-BUSH, CORREOSA, 234
SOFT
CHESS, 1252
GOLDEN-ASTER, 336
GREENEYES, 325
-LEAF
PANIC, 1300
ROSETTE GRASS, 1300
MAPLE, 220
RUSH, 1182
-SEED RUELLIA, 216
-STEM BULRUSH, 1161
Solanaceae, 1015
Solanum, 1027
americanum, 1030
carolinense, 1028
citrullifolium, ▣/106, 1028
dimidiatum, ▣/106, 1028
elaeagnifolium, 1030
lycopersicum, 1030
nigrum, 1030
perplexum, 1028
ptychanthum, 1030
rostratum, 1030
torreyi, 1028
triquetrum, 1031
tuberosum, 1027
Solenostemon, 752
Solidago, 408
altiplanities, 415

Solidago *(cont.)*
altissima, 410
var. *pluricephala*, 410
aspera, 414
boothii var. *ludoviciana*, 410
canadensis, 410
var. gilvocanescens, 410
var. scabra, 410
celtidifolia, 414
decemflora, 412
delicatula, 415
gigantea, 410
var. *serotina*, 410
gilvocanescens, 410
glaberrima, 410
graminifolia var. *gymnospermoides*, 356
gymnospermoides, 356
ludoviciana, 410
microphylla, 415
missouriensis var. fasciculata, 410
nemoralis, 410
var. *decemflora*, 412
var. longipetiolata, 412
var. nemoralis, 412
nitida, 412
odora, 412
patula var. strictula, 412
petiolaris, 412
var. *angustata*, 412
radula, 414
rigida, 414
subsp. *glabrata*, 414
var. *glabrata*, 414
subsp. *humilis*, 414
var. *laevicaulis*, 414
rigidiuscula, 415
rugosa, 414
subsp. *aspera*, 414
var. aspera, 414
var. *celtidifolia*, 414
var. rugosa, 414
salicina, 412
sempervirens, 415
var. mexicana, 415
var. sempervirens, 415
speciosa var. rigidiuscula, 414
ulmifolia var. microphylla, 415
Soliva, 415
mutisii, 416
pterosperma, 416
sessilis, 416
stolonifera, 416
SOLOMON'S-SEAL, 1208
GREAT, 1208
Sonchus, 416
asper, 416
oleraceus, 416
Sophia viscosa, 465
Sophora, 697
affinis, [photo]/106, 697
japonica, 697
secundiflora, 697
SOPHORA, TEXAS, 697
Sorbus, 952
Sorghastrum, 1320
avenaceum, 1322
elliottii, 1322
nutans, 1322
Sorghum, 1322
almum, 1322
arundinaceum, 1324
bicolor, 1322, 1324
subsp. bicolor, 1324
subsp. drummondii, 1324
×*drummondii*, 1324
halepense, 1322, 1324
sudanese, 1324
SORGHUM, 1324
SORGO, 1324
SORREL, 905, 906
CREEPING LADIES'-, 870
HEART-WING, 906
SHEEP, 906
SOUR
-CLOVER, 676
YELLOW, 676
-GRASS, 906
-GUM, 845
SOUR-GUM FAMILY, 845
SOURBUSH, 1049
SOURSOP, 238
SOUTHERN
ADDER'S-TONGUE, 190
AMARANTH, 226
BEECH, 710
BELLOWS-BEAK CARIC SEDGE, 1116
BLACKHAW, 511
BLUET, 967, 968
CANARY GRASS, 1310
CANE, 1243
CARIC SEDGE, 1118
CAT-TAIL, 1348
CLUBMOSS, 174
CORYDALIS, 722
CRAB GRASS, 1263
CYPRESS, 204
DEWBERRY, 959
FORGET-ME-NOT, 453
GRAPE FERN, 189
HOG-PEANUT, 626
JOINTWEED, 899
LADY FERN, 184
MAIDENHAIR, 194
NAIAD, 1170
PEPPERWEED, 470
POVERTY GRASS, 1328
PRICKLY-ASH, 974
RAGWEED, 309
RED OAK, 714
SANDBUR, 1254
SHIELD FERN, 200
SUGAR MAPLE, 219
THISTLE, 340
WATER-HEMP, 226
WAX-MYRTLE, 832
WILD RICE, 1338
WILLOW, 978
SOUTH-SEA-TEA, 270
SOUTHWEST
BEDSTRAW, 966
GROUND-CHERRY, 1024
SOUTHWESTERN
BRISTLE GRASS, 1320
CARROT, 250
MILK-VETCH, 635
SOWBANE, 538
SOWBRIER, 1346
SOW-THISTLE, 416
COMMON, 416
PRICKLY, 416
SOY-BEAN, 662
SOYA-BEAN, 662
SPACH'S EVENING-PRIMROSE, 866
SPADELEAF, 246
SPANISH
-BAYONET, 1082, 1084
-DAGGER, 1084
GRAPE, 1072
IRIS, 1174
-MOSS, 1096
-MULBERRY, 1049
-NEEDLES, 326
OAK, 714
SWAMP, 714
-TEA, 535
SPARKLE-BERRY, 584
SPARSE-LOBE GRAPE FERN, 189
Spartina, 1324
pectinata, 1324
Spathodea, 440
SPATTER-DOCK, 844
SPATULATE-LEAF FOGFRUIT, 1054
SPEAR
GRASS, 1290
LOW, 1312
-LEAF SIDA, 816
SPEARMINT, 764

- *Specularia*, 500
  - *biflora*, 502
  - *coloradoensis*, 501
  - *holzingeri*, 501
  - *lamprosperma*, 501
  - *leptocarpa*, 501
  - *lindheimeri*, 501
  - *perfoliata*, 502
  - *texana*, 502
- SPEEDWELL, 1010
  - BIRD'S-EYE, 1014
  - COMMON, 1012
  - CORN, 1012
  - PERSIAN, 1014
  - PURSLANE, 1014
  - WATER, 1012
  - WAYSIDE, 1012
  - XALAPA, 1014
- **Spergularia,** 525
  - *marina*, 525
  - platensis, 525
  - salina, 525
- **Spermacoce,** 970
  - glabra, 970
- **Spermolepis,** 260
  - divaricata, 262
  - echinata, 262
  - inermis, 262
- **Sphaeralcea,** 818
  - angustifolia
    - subsp. cuspidata, ▣/106, 820
    - var. *cuspidata*, 820
  - coccinea, 820
  - hastulata, 820
- **Sphenoclea,** 1031
  - zeylanica, 496, 1031
- SPHENOCLEA FAMILY, 1031
- **Sphenocleaceae,** 1031
- **Sphenopholis,** 1324
  - *interrupta*, 1332
  - obtusata, 1324
- SPICATE
  - SALT GRASS, 1264
  - SPIDERLING, 838
- SPICEBUSH, 784, 786, 832
- SPIDER-FLOWER, 506
  - SPINY, 506
- SPIDER-FLOWER FAMILY, 504
- SPIDER-LILY, 1202, 1206
  - FRAGRANT, 1204
  - RED, 1206
  - WESTERN, 1204
  - WHITE, 1202
- SPIDERLING, 836
  - ERECT, 838
  - SCARLET, 838
  - SPICATE, 838
- SPIDERPLANT, 506
- SPIDERWORT, 1101
  - GIANT, 1102
  - HAIRY-FLOWER, 1102
  - OHIO, 1104
  - PLATEAU, 1102
  - PRAIRIE, 1102
  - REVERCHON'S, 1104
  - STEMLESS, 1104
  - THARP'S, 1104
- SPIDERWORT FAMILY, 1096
- **Spigelia,** 798
  - hedyotidea, 798
  - *lindheimeri*, 798
  - marilandica, 799
- SPIKE
  - GRASS, 1264
  - -MOSS, 173, 174
    - MEADOW, 174
    - PERUVIAN, 175
    - RIDDELL'S, 175
  - PHACELIA, 743
  - -RUSH, 1144
    - BLUNT, 1147
    - COMPRESSED, 1146
    - CREEPING, 1147
    - DWARF, 1148
    - ENGELMAN'S, 1146
    - GULFCOAST, 1146
    - LANCE-SPIKE, 1147
    - LARGE-SPIKE, 1147
    - NEEDLE, 1146
    - PALE, 1146
    - REVERCHON'S, 1146
    - SHARP-SCALE, 1146
    - SMALL-SEED, 1147
    - SQUARE-STEM, 1148
- SPIKE-MOSS FAMILY, 174
- *Spilanthes americana* var. *repens*, 308
- SPINACH, 542
- **Spinacia,** 542
  - oleracea, 542
- SPINDLETREE, 528
- SPINY
  - -ASTER, 335
  - COCKLEBUR, 432
  - IRONPLANT, 387
  - NAIAD, 1170
  - PIGWEED, 226
  - SPIDER-FLOWER, 506
  - -STAR, 488
- *Spiraea*, 936
- **Spiranthes,** 1218
  - cernua, 1220
  - *gracilis*, 1220
  - *grayi*, 1222
  - lacera var. gracilis, ▣/107, 1220
  - longilabris, 1220
  - magnicamporum, 1222
  - ovalis, 1222
  - praecox, 1222
  - *reverchonii*, 1222
  - tuberosa, 1222
  - vernalis, 1222
- **Spirodela,** 1190
  - *oligorhiza*, 1190
  - polyrhiza, 1190
  - punctata, 1190
- SPLEEN AMARANTH, 226
- SPLEENWORT, 179
  - BLACK-STEM, 180
  - EBONY, 179
  - LITTLE EBONY, 180
- SPLEENWORT FAMILY, 179
- SPLIT
  - -BEARD
    - GRASS, 1239
    - BLUESTEM, 1239
  - -LEAF GILIA, 890
- SPONGE, VEGETABLE, 570
- SPOONFLOWER, 642
- **Sporobolus,** 1326
  - *asper*, 1327
    - var. *clandestinus*, 1327
  - *clandestinus*, 1327
  - compositus, 1327
    - var. clandestinus, 1327
    - var. compositus, 1327
    - var. drummondii, 1327
    - var. macer, 1327
  - cryptandrus, 1327
  - giganteus, 1327
  - indicus, 1327
  - neglectus, 1327
  - ozarkanus, 1327
  - pyramidatus, 1328
  - silveanus, 1328
  - vaginaeflorus, 1328
    - var. *neglectus*, 1327
- SPOTFLOWER, CREEPING, 308
- SPOTTED
  - BEEBALM, 766
  - BUR-CLOVER, 674
  - COWBANE, 248
  - EUPHORBIA, 594
  - KNAPWEED, 332
  - MEDIC, 674
  - OAK, 714, 718
  - SPURGE, 594
  - ST. JOHN'S-WORT, 548
  - TOUCH-ME-NOT, 434

SPOTTED *(cont.)*
WATER-HEMLOCK, 248
SPRANGLETOP, 1281
BEARDED, 1281
GREEN, 1281
MEXICAN, 1282
RED, 1281
SALT, 1281
SPREADING
ASTER, 319
BROME, 1252
DAYFLOWER, 1100
ERYSIMUM, 468
FLEABANE, 341
GOLDEN-ASTER, 344
LOEFLINGIA, 518
LOVE GRASS, 1272
MOONPOD, 842
PANICUM, 1298
SALTBUSH, 533
SCALESEED, 262
SIDA, 818
WITCH GRASS, 1296
SPREAD-LOBE PASSION-FLOWER, 878
SPREADWING, TEXAS, 254
SPRING
-BEAUTY, 908
VIRGINIA, 908
BENT GRASS, 1236
CORALROOT, 1215
CRESS, 464
FORGET-ME-NOT, 453
-HERALD, 848
LADIES'-TRESSES, 1222
SNOWFLAKE, 1206
-STAR, 1204
SPRUCE, 204
SPURGE, 604
CREEPING, 589
FLOWERING, 606
GREEN, 608
HEART-LEAF, 594
MISSOURI, 596
PAINTED, 606
PRAIRIE, 596
ROEMER'S, 608
SPOTTED, 594
TOOTHED, 608
TRAMP'S, 606
SPURGE FAMILY, 584
SPURRED-SNAPDRAGON, 1002
SPURWORT, 970
SQUARE
-BUD
DAISY, 420
DAY-PRIMROSE, 854
-STEM
ROSE GENTIAN, 728
SPIKE-RUSH, 1148
SQUAREHEAD, 418
SQUASH, 568
ACORN, 568
AUTUMN, 569
BUTTERNUT, 569
HUBBARD, 569
PATTYPAN, 568
SCALLOP, 568
SUMMER
CROOK-NECK, 568
STRAIGHT-NECK, 568
TURBAN, 569
WINTER, 569
CROOK-NECK, 569
SQUASH FAMILY, 564
SQUAW
-BERRY, 1020
-WEED, 403, 427
SQUIRREL SIXWEEKS GRASS, 1336
ST. ANDREW'S-CROSS, 547
ST. AUGUSTINE GRASS, 1328
ST. JOHN'S-WORT, 544, 548
FALSE SPOTTED, 548
CLASPING, 547
COMMON, 547
DRUMMOND'S, 547
DWARF, 547
PINEWEED, 547
ROUND-FRUIT, 548
SPOTTED, 548
ST. JOHN'S-WORT FAMILY, 544
ST. PETER'S-WORT, 547
Stachys, 780
agraria, 782
coccinia, 782
crenata, 780
tenuifolia, 782
STAFFTREE FAMILY, 526
STAGGERBUSH, 582
STAGGER-GRASS, 1201
STAGGERWORT, 366
STALKED
BUGLEWEED, 762
RUELLIA, 216
WATER-HYSSOP, 1003
STALK-LESS YELLOWCRESS, 478
STALKY BERULA, 244
STANDING-CYPRESS, 890
STANDLEY'S GOOSEFOOT, 538
Stapelia, 274
STAR
DAISY, 386
-GRASS, 1193
WATER, 1340
YELLOW, 1194, 1204
-JASMINE, 268
AMERICAN, 268
JAPANESE, 268
-OF-BETHLEHEM, 1208
-RUSH WHITE-TOP SEDGE, 1158
-SCALED CLOAK FERN, 195
SPINY-, 488
TEXAS-, 386
-THISTLE, 332, 334
BARNABY'S, 334
MALTA, 334
YELLOW, 334
-VIOLET, 966, 967
STARCH GRAPE-HYACINTH, 1206
STARFLOWER, 796
STARFRUIT, 869
STARGLORY, 555
STARRY CAMPION, 524
STARVED
ASTER, 319
ROSETTE GRASS, 1296
STARWORT, 525
CHICKWEED, 525
WATER, 494
STAVE OAK, 714
Steinchisma, 1291
hians, 1298
Stellaria, 525
apetala, 526
media, 525
subsp. *pallida*, 526
muscorum, 519
pallida, 526
STEM, BLUE, 1094
STEMLESS
EVENING-PRIMROSE, 866
SPIDERWORT, 1104
TOWNSENDIA, 424
Stemodia multifida, 1000
Stenosiphon, 866
linifolius, 866
STENOSIPHON, FLAX-LEAF, 866
Stenotaphrum, 1328
secundatum, 1328
Sterculiaceae, 1031
STICKERS, 415, 416
SMOOTH-, 416
TRAILING-, 416
STICKERWEED, 415
STICKLEAF, 794, 796
PRAIRIE, 797
STICKLEAF FAMILY, 794
STICKSEED, 450
FLAT-SPINE, 452

STICKSEED *(cont.)*
HAIRY, 450
STICKTIGHT BUTTERCUP, 928
STICKTIGHTS, 326
STICKY FLAT SEDGE, 1136
STIFF
GOLDENROD, 414
-HAIR SUNFLOWER, 370
-LEAF GERARDIA, 993
MARSH BEDSTRAW, 966
-STEM FLAX, 792
SUNFLOWER, 370
TICK-CLOVER, 655
Stillingia, 614
sylvatica, 614
texana, 614
treculeana, 614
STILLINGIA, TEXAS, 614
Stimoryne axillaris, 1023
STINGING
CEVALLIA, 794
NETTLE, 1043
STINK GRASS, 1270
STINKING
CAMOMILE, 310
GOURD, 568
PLUCHEA, 398
WALLROCKET, 466
-FLEABANE, 398
STINKWEED, 310, 396
Stipa leucotricha, 1290
STITCHWORT, 525
STOCK, 474
STONECROP, 562
DITCH-, 562
TEXAS, 564
YELLOW, 564
STONECROP FAMILY, 561
STONE-MIST, 756
MARYLAND, 756
STORAX, 1034
STORAX FAMILY, 1034
STORK'S-BILL, 730
STOUT LOVE GRASS, 1270
STRAIGHT-NECK SQUASH, SUMMER, 568
STRAMONIUM, 1018, 1020
STRANGLER FIG, 827
STRANGLEVINE, 572
STRAW, FLOWERING-, 386
STRAWBERRY, 944
-BUSH, 528
INDIAN-, 944
MOCK, 944
MOCK, 944
VIRGINIA, 944
WILD, 944
YELLOW-, 944
STREAM
EPIPACTIS, 1215
-ORCHIS, 1215
Streptanthus, 479
hyacinthoides, 479
STRETCH-BERRY, 848
STRETCHBERRY, 1346
Striga, 989
Strophanthus, 264
Strophostyles, 697
helvola, 698
helvula, 698
leiosperma, 698
STRYCHNINE FAMILY, 797
Stuckenia, 1343
pectinatus, 1343
STUEVE'S BUSH-CLOVER, 670
Stylisma, 558
aquatica, 560
humistrata, 560
pickeringii var. pattersonii, 560
villosa, 560
Stylosanthes, 698
biflora, 698
var. *hispidissima*, 698
STYPICWEED, 694
Styracaceae, 1034
Styrax, 1034
officinalis, 1034
platanifolius, 1034
STYRAX, SYCAMORE-LEAF, 1034
SUCCORY, 336
SUCCULENT FAMILY, 561
SUCKER FLAX, 790
SUDAN GRASS, 1324
SUGAR
BEET, 533
CANE, 1223
HACKBERRY, 1038
MAPLE, 219
SOUTHERN, 219
-PLUM, 938
SUGARBERRY, 443, 1036, 1038
SUGARCANE, 1315
PLUME GRASS, 1315
SULFUR CINQUEFOIL, 947
SULTAN'S-FLOWER, 433
SUMAC, 232
DESERT, 234
DWARF, 232
EVERGREEN, 234
FLAME-LEAF, 232
FRAGRANT, 232
LITTLE-LEAF, 234
POISON, 238
PRAIRIE, 234
FLAME-LEAF, 234
SCARLET, 234
SCRUB, 234
SHINING, 232
SMALL-LEAF, 234
SMOOTH, 234
TOBACCO, 234
WING-RIB, 232
SUMAC FAMILY, 230
SUMMER
CROOK-NECK SQUASH, 568
-CYPRESS, 540
BELVEDERE, 540
GRAPE, 1070
SCOURING-RUSH, 178
SNOWFLAKE, 1204
STRAIGHT-NECK SQUASH, 568
SUMMERHAW, 942
SUMPWEED, 376
NARROW-LEAF, 378
SEA-COAST, 378
SHARP-BRACT, 378
SUNBONNETS, 335
SUNDEW, 576
ANNUAL, 578
SUNDEW FAMILY, 576
SUNDROPS, 852, 862
DRUMMOND'S, 854
HALF-SHRUB, 854
OZARK, 864
THREAD-LEAF, 864
SUNFLOWER, 368
ASHY, 370
AWNLESS BUSH-, 406
BLUE-WEED, 369
COMMON, 369
CUCUMBER-LEAF, 369
DOWNY, 370
FLAT-SEED-, 426
GOLDEN-EYE, 430
HAIRY, 370
MAXIMILIAN, 370
PLAINS, 370
PRAIRIE, 370
SAW-TOOTH, 369
STIFF, 370
-HAIR, 370
SWAMP-, 366
TICKSEED-, 325, 326
WILLOW-LEAF, 370
SUNFLOWER FAMILY, 287
SUN-ROSE, 543
GEORGIA, 543
HOARY, 543
SUN-ROSE FAMILY, 542

- SUNSHINE ROSE, 956
- SUPPLEJACK, 931
  - ALABAMA, 931
- SWALLOW-WORT, 282
  - BEARDED, 282
  - SMOOTH, 282
- SWAMP
  - CHESTNUT OAK, 716
  - DOCK, 906
  - HICKORY, 747
  - -LILY, 986
  - MILKWEED, 278
  - PALM, 1094
  - POST OAK, 716
  - -PRIVET, 848
  - RED OAK, 714, 718
  - ROSE-MALLOW, 810
  - SMARTWEED, 902
  - SPANISH OAK, 714
  - -SUNFLOWER, 366
  - WHITE OAK, 716
  - WILLOW, 978
    - OAK, 718
- SWEET
  - ACACIA, 624
  - -ALYSSUM, 456
  - -AUTUMN CLEMATIS, 921
  - -BREATH-OF-SPRING, 508
  - BRIAR ROSE, 956
  - CALAMUS, 1078
  - -CICELY, 255
  - -CLOVER, 676
    - WHITE, 676
    - YELLOW, 676
  - GAURA, 856
  - GOLDENROD, 412
  - GRAPE, 1070
  - MOUNTAIN GRAPE, 1072
  - -OAK, 832
  - -PEA, PERENNIAL, 666
  - -POTATO, 555
    - WILD, 556
  - SAND-VERBENA, 836
  - SCABIOUS, 576
  - -SCENTED BEDSTRAW, 966
  - -WILLIAM, 518, 1050
- SWEETBRIER, 956
- SWEETFLAG, 1077
  - DRUG, 1077
- SWEETFLAG FAMILY, 1077
- SWEETGUM, 736, 737
- SWEETHEARTS, 964
- SWEETROOT, LONG-STYLE, 256
- SWEETSOP, 238
- *Swietenia*, 822
- SWINE WARTCRESS, 465
- SWISS CHARD, 533
- SWITCH
  - CANE, 1243
  - GRASS, 1302
- SWORD
  - FERN, 184, 185
  - -GRASS, 1161
  - -LEAF BLUE-EYED-GRASS, 1178
- SWORDS, FAIRY-, 198
- SYCAMORE, 888
  - AMERICAN, 888
  - -LEAF
    - SNOWBELL, 1034
    - STYRAX, 1034
- SYCAMORE FAMILY, 886
- SYLVAN BLUE GRASS, 1314
- **Symphoricarpos,** 510
  - orbiculatus, 510
- *Symphyotrichum,* 315
  - *divaricatum,* 320
  - *drummondii* var. *texanum*, 318
  - *dumosum,* 318
  - *ericoides,* 318
  - *eulae,* 322
  - *lanceolatum,* 319
    - subsp. *hesperium*, 319
  - *lateriflorum,* 319
  - *oblongifolium,* 319
  - *oolentangiense,* 319
  - *patens*
    - var. *gracile,* 320
    - var. *patens,* 320
    - var. *patentissimum,* 320
  - *praealtum,* 320
  - *pratense,* 320
  - *sericeum,* 320
- *Synphytum officinale,* 443
- *Syringa*, 846

T

- *Tabebuia*, 440
- TACOPATE, 272
- *Tagetes*, 287
- TAGUA, 1094
- TALAYOTE, 282
- **Talinum,** 911
  - aurantiacum, ▣/107, 911
  - calycinum, ▣/107, 911
  - parviflorum, 911
- TALL
  - ASTER, 320
  - BEAK-RUSH, 1158
  - -BREAD SCURF-PEA, 684
  - BUSH-CLOVER, 670
  - DITAXIS, 604
  - DROPSEED, 1327
  - EUPATORIUM, 353
  - FESCUE, 1276
  - FOUR-O'CLOCK, 839
  - GAURA, 856
  - GAYFEATHER, 384
  - GRAMA, 1248
  - IRONWEED, 428
  - MARSH GRASS, 1324
  - MELIC, 1284
  - NUT-RUSH, 1164
  - POPPY-MALLOW, 808
  - SCOURING-RUSH, 176
  - SNEEZEWEED, 366
  - THISTLE, 338
  - THOROUGHWORT, 353
  - WHITE ASTER, 318
  - WILD MERCURY, 604
  - WINECUP, 808
- TALLOW
  - -SHRUB, 832
  - TREE
    - CHINESE, 612
    - VEGETABLE, 612
  - -WEED, 884, 886
- **Tamaricaceae,** 1034
- TAMARISK, 1034
  - CHINESE, 1035
  - FRENCH, 1035
- **Tamarix,** 1034
  - chinensis, 1035
  - gallica, 1035
  - ramosissima, 1035
- TAMARIX FAMILY, 1034
- TAN
  - OAK, 710
  - -BARK OAK, 710
- *Tanacetum,* 382
  - *cinerariifolium,* 382
- TANGLEGUT, 572
- TANIA, 1094
- TANNER'S DOCK, 906
- TANNIA, 1094
- TANSY
  - -LEAF YELLOWCRESS, 478
  - -MUSTARD, 465, 479
    - FLIXWEED, 466
    - PINNATE, 466
- TAPE-GRASS, 1170
- TAPE-GRASS FAMILY, 1166
- TAPER-LEAF
  - BUGLEWEED, 762
  - FLAT SEDGE, 1136
- TAPIOCA, 585
- **Taraxacum,** 418
  - *erythrospermum,* 418
  - laevigatum, 418

Taraxacum *(cont.)*
  officinale, 418
TARAY, 978
TARBAY, 1054
TARO, 1093
  WILD, 1093
TARRAGON, 314
TARWEED, 360, 445
TASAJILLO, 491
TAWNY DAY-LILY, 1202
Taxodium, 204
  distichum
    var. mexicanum, 204
    var. distichum, 204
  mucronatum, 204
Taxus
  baccata, 201
  brevifolia, 201
TÉ DE PAÍS, 1054
TEA
  INDIAN, 422
  JERSEY-, 932
  MEXICAN-, 207, 535
  MORMON-, 207
  NAVAJO, 422
  NEW JERSEY-, 932
  PRAIRIE-, 600
  SOUTH-SEA-, 270
  SPANISH-, 535
TEAK, 1048
TEAL LOVE GRASS, 1270
TEASEL FAMILY, 574
Tecoma stans, 440
Tectona grandis, 1048
TEN
  -ANGLE PIPEWORT, 1166
  -PETAL
    ANEMONE, 918
    MENTZELIA, 796
TENPETAL, 525
TEOSINTE, 1338
Tephrosia, 698
  lindheimeri, 700
  onobrychoides, 700
  virginiana, 700
TEPHROSIA
  LINDHEIMER'S, 700
  MULTI-BLOOM, 700
  ROUND-LEAF, 700
  VIRGINIA, 700
TEPOPOTE, 208
TESAJO, 491
Tetragonotheca, 418
  ludoviciana, 420
  texana, 420
Tetraneuris, 420
  linearifolia, 420
  scaposa, 420
  turneri, 421
Teucrium, 782
  canadense, 782
  cubense, 782
    var. cubense, 782
    subsp. *laevigatum*, 783
    var. laevigatum, 782
  laciniatum, ▣/107, 783
TEXABAMA CROTON, 599
TEXAS
  ADELIA, 848
  AMSONIA, 265
  ASH, 849
  ASTER, 318
  -BAYONET, 1084
  BEDSTRAW, 966
  BERGIA, 580
  BETONY, 782
  BINDWEED, 552
  BLACK WALNUT, 748
  BLUE GRASS, 1312
  BLUEBELLS, 727
  BLUEBONNET, 672
  BLUEWEED, 369
  BOX-ELDER, 220
  BROOMWEED, 365
  BUCKEYE, 738
  -BUCKEYE, 982
  BULL-NETTLE, 596
  BUSH-CLOVER, 670
  -BUTTERCUP, 866
  CARIC SEDGE, 1130
  CARLOWRIGHTIA, 212
  CENTAURY, 726
  COLUBRINA, 932
  CORYDALIS, 722
  COTTONWOOD, 976
  CRAB GRASS, 1316
  CROTON, 600
  CROWFOOT, 1281
  CRYPTANTHA, 446
  CUP GRASS, 1274
  CUPFLOWER, 1022
  DANDELION, 399
  DESERT-RUE, 973
  DUTCHMAN'S-PIPE, 273
  FILAREE, 730
  FLAX, 790
  FROGFRUIT, 1054
  GERANIUM, 732
  GOURD, 568
  GRAMA, 1248
  GREENEYES, 325
  GROUNDSEL, 403
  HICKORY, 748
  IRONWEED, 428
  KIDNEYWOOD, 658
  LANTANA, 1053
  LESPEDEZA, 670
  LUPINE, 672
  MADRONE, 582
  -MALLOW, 815
  MILK-VETCH, 635
  MILKWEED, 280
  -MILLET, 1335
  MOUNTAIN-LAUREL, 697
  MULBERRY, 832
  NEEDLE GRASS, 1290
  NIGHTSHADE, 1031
  PAINTBRUSH, 996
  PANICUM, 1335
  -PARSLEY, 256
  PERSIMMON, 579
  PHLOX, 892
  PIPEWORT, 1166
  PLUME, 890
  PRICKLY-PEAR, 490
  PURPLE
    -SAGE, 988
    -SPIKE, 1218
  RED
    GRAPEFRUIT, 970
    OAK, 714
  REDBUD, 640
  RUSH, 1184
  SABADILLA, 1210
  SAGE, 774, 776
  SANDBAR WILLOW, 978
  SAXIFRAGE, 988
  SIGNAL GRASS, 1335
  SKELETON-PLANT, 386
  SLEEPY DAISY, 430
  SLIMPOD, 265
  SNAKEWOOD, 932
  SOPHORA, 697
  SPIDERWORT, 1102
  SPREADWING, 254
  -STAR, 386, 728
    YELLOW, 386
  STILLINGIA, 614
  STONECROP, 564
  THISTLE, 340
  TOAD-FLAX, 1006
  TRIDENS, 1330
  VENUS'-LOOKING-GLASS, 501
  VERVAIN, 1056
  VIRGIN'S-BOWER, 921
  WALNUT, 748
  WHITE ASH, 849
  WILLKOMMIA, 1336

TEXAS *(cont.)*
WINTER GRASS, 1290
YELLOW BUCKEYE, 738
YUCCA, 1084
Thalia, 1212
dealbata, 1212
THALIA, POWDERED, 1212
Thalictrum, 928
arkansanum, 930
dasycarpum, 930
var. *hypoglaucum*, 930
texanum, 930
Thamnosma, 973
texanum, 973
THARP'S SPIDERWORT, 1104
Thelesperma, 421
ambiguum, 421
curvicarpum, 422
filifolium, 421
var. filifolium, 422
var. intermedium, 422
intermedium, 422
megapotamicum, 422
var. *ambiguum*, 421
simplicifolium, 422
trifidum, 422
THELESPERMA
RAYLESS, 422
THREAD-LEAF, 421
Thelocactus, 492
setispinus, 494
Thelypteridaceae, 200
Thelypteris, 200
kunthii, 200
normalis, 200
ovata var. lindheimeri, 200
Theobroma cacao, 1032
THICKET
GROUND-CHERRY, 1026
KNOTWEED, 904
PLUM, 950
THREADVINE, 282
THICK-LEAF GOOSEFOOT, 538
THIMBLE-FLOWER, 400
THIN
PASPALUM, 1308
-LEAF
BROOKWEED, 914
MILKWEED, 280
THISTLE, 329, 338
BARNABY'S STAR-, 334
BLACKLAND, 338
BLESSED MILK-, 406
BULL, 340
COMMON SOW-, 416
COTTON-, 388
DISTAFF-, 330
HOLY-, 406
HORRID, 340
IOWA, 338
KANSAS-, 1030
MALTA STAR-, 334
MILK-, 406
MUSK-, 330
NODDING-, 330
OUR-LADY'S-, 406
PASTURE, 340
PLUMED, 338
PLUMELESS-, 329
POWDERPUFF, 332
PRICKLY SOW-, 416
ROADSIDE, 338
RUSSIAN-, 540
SAFFRON-, 330
SCOTCH-, 388
SLENDER BRISTLE-, 330
SOUTHERN, 340
SOW-, 416
STAR-, 332, 334
TALL, 338
TEXAS, 340
THORNLESS-, 332
TRUE, 338
WAVY-LEAF, 340
YELLOW, 340
-SPINE, 340
STAR-, 334
Thlaspi, 480
arvense, 480
THOMPSON GRASS, FORT, 1306
THORN, 940
-APPLE, 940, 874, 1018
COMMON, 1020
THORNLESS-THISTLE, 332
THORNY
AMARANTH, 226
ELAEAGNUS, 579
THOROUGHWAX, 244
ROUND-LEAF, 246
THOROUGHWORT, 352, 354
LATE-FLOWERING, 354
TALL, 353
THREAD-LEAF
MOCK BISHOP'S- WEED, 258
SAGEWORT, 315
SNAKEWEED, 364
SUNDROPS, 864
THELESPERMA, 421
THREADVINE, THICKET, 282
THREEAWN, 1239
ARROW-FEATHER, 1242
BLUE, 1243
CHURCH-MOUSE, 1242
CURLY, 1242
FORK-TIP, 1240
GRAMA, 1248
KEARNEY'S, 1242
LONG-AWNED, 1243
OLDFIELD, 1242
PLAINS, 1242
PRAIRIE, 1242
PURPLE, 1243
RED, 1243
SLENDER, 1242
SLIM-SPIKE, 1242
WESTERN, 1242
WOOLLY, 1242
-SHEATH, 1242
WRIGHT'S, 1243
THREE
-FLOWER MELIC, 1284
-LOBE
FALSE MALLOW, 814
RED OAK, 714
-PRIMROSE, 866
-SEED CROTON, 600
-SEEDED MERCURY, 586
-SQUARE BULRUSH, 1161
-WAY SEDGE, 1106
Thuja, 202
Thunbergia, 218
alata, 218
THURBER'S PILOSTYLES, 648, 914
Thurovia, 362
Thyella tamnifolia, 558
THYME, 752
-LEAF SANDWORT, 514
Thymophylla, 422
pentachaeta, 422
tenuiloba
var. tenuiloba, 422
var. wrightii, 424
Thymus, 752
TICK-CLOVER, 652
FEW-FLOWER, 656
HOARY, 655
LITTLE-LEAF, 655
MARYLAND, 655
NUTTALL'S, 655
PANICLED, 656
RIGID, 655
SESSILE-LEAF, 656
SIMPLE-LEAF, 656
STIFF, 655
TWEEDY'S, 656
VELVET-LEAF, 656
WRIGHT'S, 656

TICKLE GRASS, 1236
ANNUAL, 1236
TICKLETONGUE, 974
TICKSEED, 341
-SUNFLOWER, 325, 326
Tidestromia, 230
lanuginosa, 230
TIDESTROMIA, WOOLLY, 230
TIEVINE, 556
TIGER-FLOWER, 1172
Tigridia, 1172
Tilia, 1036
americana var. caroliniana, 1036
Tiliaceae, 1035
Tillaea aquatica, 562
Tillandsia, 1095
recurvata, 1095
usneoides, 1096
TIMOTHY, 1310, 1311
Tinantia, 1100
anomala, ▣/107, 1101
TINY
BLUET, 967
MOUSETAIL, 924
-TIM, 422
TIPTONWEED, 547
TISSUE SEDGE, 1122
Tithymalus, 605
TOAD
-FLAX, 1002, 1004, 1006
COMMON, 1002
OLDFIELD, 1004
TEXAS, 1006
RUSH, 1182
TOADFLAX, BASTARD, 979
TOBACCO, 1022
CIMARRÓN, 1022
SUMAC, 234
FIDDLE-LEAF, 1022
INDIAN-, 427, 1010
LADIES'-, 310
-RABBIT'S, 356
WILD, 1022
TOCALOTE, 334, 1020
Tomanthera
auriculata, 993
densiflora, 993
TOMATE FRESADILLA, 1026
TOMATILLO, 1020, 1023
TOMATL, 1030
TOMATO, 1030
TREE, 1027
TOOTHACHETREE, 974
TOOTHCUP, 799, 800, 802
TOOTHED SPURGE, 608
Torilis, 262
arvensis, 262
nodosa, 262
TORREY'S
SEEDBOX, 860
AMARANTH, 224
RUSH, 1184
YUCCA, 1084
TOTORA, 1160
TOUCH-ME-NOT, 433
SPOTTED, 434
TOUCH-ME-NOT FAMILY, 433
Tovara, 899
virginiana, 905
Townsendia, 424
exscapa, 424
TOWNSENDIA, STEMLESS, 424
Toxicodendron, 234
diversilobum var. *pubescens,* 236
pubescens, 236
radicans, 236
subsp. *eximum,* 236
subsp. negundo, 236
subsp. pubens, 236
subsp. *radicans,* 236
subsp. verrucosum, 236
vernicifluum, 236
vernix, 238
Trachelospermum, 268
asiaticum, 268
difforme, 268
jasminoides, 268
Tradescantia, 1101
canaliculata, 1104
edwardsiana, 1102
gigantea, 1102
hirsutiflora, 1102
humilis, 1102
occidentalis, 1102
ohiensis, ▣/108, 1104
reverchonii, 1104
subacaulis, 1104
tharpii, 1104
Tragia, 614
amblyodonta, 617
betonicifolia, 616
brevispica, 616
nepetifolia var. *leptophylla,* 616
ramosa, 616
urticifolia, 616
var. *texana,* 616
Tragopogon, 424
dubius, 424
major, 424
miruus, 424
porrifolius, 426
TRAILING
BUSH-CLOVER, 670
HEDGE-BINDWEED, 551
LESPEDEZA, 670
LOOSESTRIFE, 913
MILKWEED, 278
PEARLWORT, 524
RATANY, 750
-STICKERS, 416
WILD BEAN, 698
TRAMP'S SPURGE, 606
-TROUBLE, 1346
Tranzschelia
cohaesa, 917
discolor, 950
ornata, 917
TREACLE HARE'S-EAR, 464
TREAD-SOFTLY, 596
TRECUL'S YUCCA, 1084
TREE
CATAWBA-, 443
DODDER, 573
-GROUNDSEL, 324
HONEYSUCKLE, 508
-OF-CHASTITY, 1060
-OF-HEAVEN, 1015
PRIVET, 850
TOMATO, 1027
TREFOIL, 671
BIRD-FOOT, 671
HOP, 704
YELLOW, 674
Trepocarpus, 262
aethusae, 262
Triadenum, 548
walteri, 548
Triadica, 612
sebifera, 614
TRIANGULAR CARIC SEDGE, 1130
Trianthema, 220
portulacastrum, 220
Tribulus, 1076
terrestris, 1076
Trichachne californica, 1263
Trichostema, 783
brachiatum, 783
dichotomum, 783
Tridens, 1328
albescens, 1330
congestus, 1330
flavus, 1330
muticus, 1330
var. elongatus, 1330
var. muticus, 1330
pilosus, 1274
strictus, 1330
texanus, 1330

TRIDENS
 HAIRY, 1274
 LONG-SPIKE, 1330
 PINK, 1330
 ROUGH, 1330
 SLIM, 1330
 TEXAS, 1330
 WHITE, 1330
TRIFOLIATE ORANGE, 972
Trifolium, 700
 amphianthum, 704
 arvense, 702
 bejariense, 704
 campestre, 704
 carolinianum, 704
 dubium, 704
 incarnatum, 704
 lappaceum, 704
 polymorphum, 704
 pratense, 706
 repens, 706
 resupinatum, 706
 vesiculosum, 706
Trifurcia lahue subsp. *caerulea,* 1173
TRI-LOBE VIOLET, 1062
Triodanis, 500
 biflora, 502
 coloradoensis, 501
 falcata, 500, 501
 holzingeri, 501
 lamprosperma, 501
 leptocarpa, 501
 perfoliata, 501
  var. biflora, 502
  var. perfoliata, 502
 texana, 502
Triosteum, 511
 angustifolium, 511
Triplasis, 1330
 purpurea, 1332
TRIPLEAWN
 GRASS
  WOOLLY, 1242
  WRIGHT'S, 1243
 WESTERN, 1242
Tripsacum, 1332
 dactyloides, 1332
Trisetum, 1332
 interruptum, 1332
TRISETUM, PRAIRIE, 1332
Tristagma, 1204
Triticum, 1332
 aestivum, 1334
 cylindricum, 1235
 turgidum, 1334
TROMPILLO, 1030
TROPIC CROTON, 599
TROPICAL
 AMARANTH, 226
 EUPHORBIA, 594
 FLAT SEDGE, 1144
 NEPTUNIA, 680
 SAGE, 774
TROUT-LILY, 1201
TRUE
 CHASTETREE, 1060
 LAUREL, 784
 THISTLE, 338
TRUENO DE SETO, 850
TRUMPET, 983
 -CREEPER, 442
  COMMON, 442
 EVENING-PRIMROSE, 864
 -FLOWER, EVENING, 798
 -HONEYSUCKLE, 442, 510
 NARCISSUS, 1207
 SLIMPOD, 265
 YELLOW, 983
TRUMPETVINE FAMILY, 440
TRUMPETS, 836
TUBE PENSTEMON, 1008
TUBER
 DWARF-DANDELION, 380
 FALSE DANDELION, 399
 SKULLCAP, 779
 VERVAIN, 1058
TUBEROUS
 INDIAN-PLANTAIN, 314
 -ROOT PRICKLY-PEAR, 492
TUBETONGUE, 218
 HAIRY, 218
TUFTED
 BUTTERCUP, 928
 FLAX, 790
TULE, 1160, 1161, 1348
 ALKALI, 1161
 CALIFORNIA, 1161
 ESPADILLA, 1348
 GIANT, 1161
TULIP, 1191
Tulipa, 1191
TUMBLE
 GRASS, 1316
 LOVE GRASS, 1272
 -MUSTARD, 479
 -RINGWING, 540
 WINDMILL GRASS, 1256
TUMBLEWEED, 224, 540
 AMARANTH, 224
 PLAINS, 540
TUNG-OIL, 585
TUPELO, 845
 BLACK, 845
TURBAN SQUASH, 569
TURK'S-CAP, 814
TURKEY
 -BERRY, 1049
 OAK, 716
 -PEA, 634
 -TANGLE, 1054
TURKEYFOOT, 1238
TURNIP, 460
 INDIAN-, 1092
 -ROOT SCURF-PEA, 684
TURNSOLE, 448, 450
TURPENTINE-WEED, 365
TURRE HEMBRA, 1054
TWEEDY'S TICK-CLOVER, 656
TWIG-RUSH, 1130, 1132
TWINEVINE, 283
 WAVY-LEAF, 283
 WHITE, 283
TWINING-SNAPDRAGON, VIOLET, 1002
TWINPOD, 850
TWISTED LADIES'-TRESSES, 1222
TWISTFLOWER, 479
 SMOOTH, 479
TWIST-LEAF YUCCA, 1084
TWO
 -COLOR WILD INDIGO, 636
 -FLOWER
  ANEMONE, 918
  MELIC, 1284
  MILKVINE, 284
  RUSH, 1184
 -LEAF SENNA, 696
Typha, 1348
 angustifolia, 1348
 domingensis, 1348
 latifolia, 1348
Typhaceae, 1348

U

Ulmaceae, 1036
Ulmus, 1038
 alata, 1039
 americana, 1039
 crassifolia, 1040
 fulva, 1040
 parvifolia, 1040
 pumila, 1040
 rubra, 1040
Umbelliferae, 239
UMBRELLA
 FLAT SEDGE, 1138
 -GRASS, 1150, 1158
  WESTERN, 1152
  WHITE-TOP, 1158

UMBRELLA *(cont.)*
- -PLANT, 1138
- SEDGE, 1152
- -TREE, 822
- WATER-PENNYWORT, 255

UÑA DE GATO, 625
Ungnadia, 982
- speciosa, ▣/108, 982

UNICORN-PLANT, 880
Uniola
- *latifolia,* 1254
- *laxa,* 1256
- *paniculata,* 1254
- *sessiliflora,* 1256

UPLAND
- COTTON, 808
- HICKORY, 748
- LADIES'-TRESSES, 1222
- WILLOW OAK, 716
- PRAIRIE CONEFLOWER, 400
- PRIMROSE-WILLOW, 860

Urochloa, 1291, 1334
- ciliatissima, 1334
- fasciculata, 1335
- platyphylla, 1335
- texana, 1335

Uromyces ari-triphylli, 1092
Urtica, 1043
- chamaedryoides, 1043
  - var. *runyonii,* 1043
- *villosa,* 828

Urticaceae, 1040
URVILLE'S PASPALUM, 1308
Utricularia, 787
- *biflora,* 787
- cornuta, 787
- gibba, ▣/108, 787

UVA CIMARRONA, 1072
UVALDE BIG-TOOTH MAPLE, 219

V

Vaccaria, 526
- hispanica, 526
- *pyramidata,* 526

Vaccinium, 584
- arboreum, 584
  - var. *glaucescens,* 584
- *macrocarpon,* 584

VAHL'S FIMBRISTYLIS, 1150
VALDIVIANA DUCKWEED, 1188
VALERIAN FAMILY, 1043
Valerianaceae, 1043
Valerianella, 1043
- amarella, 1044
- *florifera,* 1046
- radiata, 1044
  - forma fernaldii, 1046
  - var. *fernaldii,* 1046
  - forma parviflora, 1046
  - forma radiata, 1046
  - var. *radiata,* 1046
- stenocarpa, 1046
  - var. *parviflora,* 1046
- *texana,* 1046
- woodsiana, 1046

Vallisneria, 1170
- americana, 1170
- *spiralis,* 1172

Vanilla planifolia, 1213
VARA DULCE, 658, 1049
VARIABLE
- EVENING-PRIMROSE, 864
- ROSETTE GRASS, 1298

VARICOLORED WILD INDIGO, 638
VARNISHTREE, JAPANESE, 1032
VASEY GRASS, 1308
VEGETABLE
- -IVORY, 1094
- -OYSTER SALSIFY, 426
- -SPONGE, 570
- TALLOW TREE, 612

VEIN-LEAF LANTANA, 1052
VEINY
- PEPPERWEED, 470
- VERVAIN, 1058

VELVET
- -APPLES, 578
- -DOCK, 1010
- -LEAF
  - BUTTERPRINT, 806
  - GAURA, 856
  - TICK-CLOVER, 656
- ROSETTE GRASS, 1302

VENICE-MALLOW, 812
VENUS'
- -COMB, 260
- -FLYTRAP, 576
- -HAIR FERN, 194
- -LOOKING-GLASS, 500
  - CLASPING, 502
  - COLORADO, 501
  - LINDHEIMER'S, 501
  - SLENDER, 501
  - SLIMPOD, 501
  - SMALL, 502
  - TEXAS, 501

Verbascum, 1010
- blattaria, 1010
- thapsus, 1010

Verbena, 1054
- *bipinnatifida,* 1050
- bracteata, 1056
- brasiliensis, 1056
- *canadensis,* 1050
- canescens, 1056
  - var. *roemeriana,* 1056
- *ciliata,* 10501
  - var. *longidentata,* 1050
- halei, ▣/108, 1056
- ×*hybrida,* 1052
- neomexicana, 1056
  - var. hirtella, 1056
  - var. neomexicana, 1058
- *officinalis,* 1055
  - subsp. *halei,* 1056
- ×*oklahomensis,* 1050
- plicata, 1058
- *pumila,* 1052
- rigida, 1058
- scabra, 1058
- stricta, 1058
- urticifolia, 1058
- xutha, 1058

VERBENA, 1050
- AMELIA'S SAND-, 836
- GARDEN, 1050
- HAIRY, 1052
- HYBRID, 1050
- PINK, 1052
- PRAIRIE, 1050
- SAND-, 835, 836
- WESTERN PINK, 1050
- WILD, 1052

VERBENA FAMILY, 1046
Verbenaceae, 1046
Verbesina, 426
- alternifolia, 426
- encelioides, 426
- helianthoides, 427
- lindheimeri, 427
- virginica, 427

VERDOLAGA, 910
- BLANCA, 220
- DE AGUA, 860
- DE PUERCO, 222

VERGONZOSA, 680
Vernonia, 427
- *altissima,* 428
- baldwinii, 428
- gigantea, 428
- ×guadalupensis, 428
- lindheimeri, 428
- marginata, 428
- missurica, 428
- texana, 428
- ×vultrina, 428

Veronica, 1010
agrestis, 1012, 1014
americana, 1012
anagallis-aquatica, 1012
arvensis, 1012
didyma, 1014
peregrina, 1014
subsp. peregrina, 1014
subsp. xalapensis, 1014
var. *xalapensis*, 1014
persica, 1014
polita, 1014
virginica, 1014
xalapensis, 1014
Veronicastrum, 1014
virginicum, 1014
VERVAIN, 1050, 1054, 1055
BIG-BRACT, 1056
BLUE, 1056
BRAZILIAN, 1056
CANDELABRA, 1056
COARSE, 1058
COMMON, 1050
DAKOTA, 1050
FAN-LEAF, 1058
GRAY, 1056
GULF, 1058
HARSH, 1058
HILLSIDE, 1056
HOARY, 1058
LARGE-BRACT, 1056
MULLEN-LEAF, 1058
NETTLE-LEAF, 1058
PINK, 1052
PRAIRIE, 1050
PROSTRATE, 1056
RAGWEED, 1050
ROSE, 1050
SANDING, 1056
SANDPAPER, 1058
SLENDER, 1056
SMALL-FLOWER, 1050
TEXAS, 1056
TUBER, 1058
VEINY, 1058
WESTERN PINK, 1050
WHITE, 1058
WILD, 1050
WOOLLY, 1058
VERVAIN FAMILY, 1046
VETCH, 706
BENT-POD MILK, 632
BIRD-FOOT DEER-, 671
BODKIN MILK-, 632
CANADA MILK-, 630
CAROLINA, 708
COMMON, 707
CROWN-, 642
DEER-, 671
DEER PEA, 707
DROOPING, MILK-, 635
ENGELMANN'S MILK-, 632
GROUND-PLUM MILK-, 630
HAIRY, 708
JOINT-, 625
LEAVENWORTH'S, 707
LINDHEIMER'S MILK-, 632
LOTUS MILK-, 632
MILK, 628, 630, 632
NARROW-LEAF, 708
NUTTALL'S MILK, 634
OZARK MILK-, 632
PALE, 708
PLATTE RIVER MILK-, 635
PURSH'S DEER-, 671
PYGMY-FLOWER, 707
RUSSIAN, 708
SLIM-POD MILK-, 632
SMALL-FLOWER, 707
MILK-, 634
SOUTHWESTERN MILK-, 635
TEXAS MILK-, 635
WINTER, 708
WOOD, 708
WOOLLY-POD, 708
WRIGHT'S MILK-, 635
VETCHLING, 665
SINGLETARY, 666
Viburnum, 511
rufidulum, 511
VIBURNUM, DOWNY, 511
Vicia, 706
angustifolia, 708
caroliniana, 708
dasycarpa, 708
exigua, 707
faba, 617, 706
leavenworthii, 707
var. *occidentalis*, 707
ludoviciana, 707
var. *laxiflora*, 707
subsp. leavenworthii, 707
subsp. ludoviciana, 707
var. *texana*, 707
micrantha, 707
minutiflora, 707
reverchonii, 707
sativa, 707
subsp. nigra, 708
var. *nigra*, 708
subsp. sativa, 708
var. *segetalis*, 708
villosa, 708
var. *glabrescens*, 708
subsp. varia, 708
subsp. villosa, 708
Victoria amazonica, 842
Viguiera, 430
dentata, 430
Vinca, 268
major, 269
minor, 269
rosea, 268
VINE
BLUE-SNAPDRAGON, 1002
JOINT-FIR, 208
-MESQUITE, 1300
Viola, 1061
bicolor, 1062
eriocarpon, 1062
lovelliana, 1062
missouriensis, 1062
palmata, 1062
var. *dilatata*, 1062
var. *triloba*, 1062
papilionacea, 1064
pensylvanica, 1062
pubescens, 1062
var. *eriocarpon*, 1062
rafinesquii, 1062
sagittata, 1064
sororia, 1064
var. *missouriensis*, 1062
triloba, 1062
var. *dilatata*, 1062
villosa, 1064
×*wittrockiana*, 1061
Violaceae, 1060
VIOLET, 1061
ARROW-LEAF, 1064
BAYOU, 1064
BLUE PRAIRIE, 1064
CAROLINA, 1064
COLLINSIA, 999
DOG-TOOTH-, 1200
DOWNY BLUE, 1064
GRASS-, 1104, 1174
GREEN-, 1061
HELIOTROPE, 448
LESPEDEZA, 670
LOVELL, 1062
MISSOURI, 1062
NOD-, 1061
NODDING GREEN-, 1061
RUELLIA, 216
SISTER, 1064
SMOOTH YELLOW, 1062
STAR-, 966, 967

VIOLET *(CONT.)*
- TRI-LOBE, 1062
- TWINING-SNAPDRAGON, 1002
- WHORLED NOD-, 1061
- WOOD, 1062
- WOODSORREL, 872
- YELLOW, 1062

VIOLET FAMILY, 1060
VIPERINA, 710
VIPER'S-BUGLOSS, 446, 448
VIRGIN'S-BOWER, 920
- TEXAS, 921

VIRGINIA
- AGAVE, 1080
- BLUEBELLS, 443
- BLUESTEM, 1239
- BUGLEWEED, 762
- BUTTONWEED, 962
- CHAIN FERN, 181
- COPPERLEAF, 589
- -CREEPER, 1066, 1068
- CROWNBEARD, 427
- CUT GRASS, 1281
- DAYFLOWER, 1100
- DUTCHMAN'S-PIPE, 273
- GRAPE FERN, 189
- GROUND-CHERRY, 1026
- HEDGE-HYSSOP, 1000
- LIVE OAK, 719
- PEPPER-GRASS, 472
- RED-CEDAR, 203
- SIBARA, 478
- SNAKEROOT, 273
- SPRING-BEAUTY, 908
- STRAWBERRY, 944
- TEPHROSIA, 700
- WILD RYE, 1268

VIRGINIAN SKULLCAP, 779
Viscaceae, 1064
VISCID EUTHAMIA, 354
Vitaceae, 1065
Vitex, 1058
- agnus-castus, 1060
  - var. agnus-castus, 1060
  - var. caerulea, 1060
- negundo, 1060
  - var. heterophylla, 1060
  - var. *incisa*, 1060

Vitis, 1068
- acerifolia, 1070
- aestivalis, 1070
  - var. aestivalis, 1070
  - var. *cinerea*, 1070
  - var. lincecumii, 1070
- *berlandieri*, 1072
- *candicans*, 1072
- cinerea, 1070
  - var. *canescens*, 1070
  - var. cinerea, 1070
  - var. helleri, 1072
- ×*champinii*, 1072
- *cordifolia*, 1072
- ×*doaniana*, 1072
- *lincecumii*, 1070
  - var. *glauca*, 1070
  - var. *lactea*, 1070
- monticola, 1072
- mustangensis, 1072
- palmata, 1072
- riparia, 1072
- rotundifolia, 1072
- *rupestris*, 1072
- *vinifera*, 1065, 1068
- vulpina, 1072

Vulpia, 1335
- bromoides, 1336
- *dertonensis*, 1336
- myuros, 1336
- octoflora, 1336
  - var. glauca, 1336
  - var. hirtella, 1336
  - var. octoflora, 1336
  - var. *tenella*, 1336
- sciurea, 1336

W

WAFER-ASH, 972
WAHOO, 528, 972
- ELM, 1039

EASTERN, 528
WAIT-A-
- BIT, 678
- MINUTE, 678

WALL BUR-CUCUMBER, 570
WALLFLOWER, 468
- BUSHY, 468
- WESTERN, 468

WALLROCKET, 466
- STINKING, 466

WALLY, WATER-, 324
WALNUT, 748
- ARIZONA, 748
  - BLACK, 748
- BLACK, 750
- DWARF, 748
- ENGLISH, 744
- LITTLE, 748
- PERSIAN, 744
- TEXAS, 748
  - BLACK, 748

WALNUT FAMILY, 744
WAMPEE, 1340
WAND MILKWEED, 282
WANDERING-JEW, 1098, 1101
WAPATO, 1088
WARD'S WILLOW, 978
Warnockia, 783
- scutellarioides, 784

WARTCRESS, 465
- SWINE, 465

WARTY
- CALTROP, 1076
- EUPHORBIA, 608

Washingtonia filifera, 1094
WASHINGTON-THORN, 942
WATER
- -BEECH, 439
- BENT GRASS, 1314
- -BONNET, 1093
- -CELERY, 1170
- -CHESTNUT, 1144
- -CHICKWEED, 494
- -CHINQUAPIN, 834
- -CLOVER, 186
- CRESS, 476
- -DRAGON, 986
- DROPWORT, 256
  - COWBANE, 256
- -FEATHER, 734
- FERN, 180
- -FRINGE, 826
- -HEMLOCK, 246
  - COMMON, 248
  - SPOTTED, 248
- -HEMP, 226
  - NUTTALL'S, 226
  - SOUTHERN, 226
- HICKORY, 747
- HONEY-LOCUST, 660
- -HOREHOUND, 760, 762
- -HYACINTH, 1339
  - COMMON, 1339
- -HYSSOP, 994, 1003
  - COASTAL, 994
  - DISC, 994
  - PROSTRATE, 1003
  - SAW-TOOTH, 1003
  - STALKED, 1003
- -LETTUCE, 1093
- -LILY, 844
  - AMERICAN, 845
  - BANANA, 845
  - BLUE, 845
  - FRAGRANT, 845
  - QUEEN VICTORIA'S, 842
  - SENORITA, 845
  - WHITE, 845
  - YELLOW, 845

WATER *(cont.)*
- -LOCUST, 660
- -MEAL, 1190
- MELON, 565
- -MILFOIL, 733
  - EURASIAN, 734
- -MILLET, 1338
- -NYMPH, 844, 1169
  - COMMON, 1170
  - HOLLY-LEAF, 1170
- OAK, 718
- -PARSNIP, 244, 260
- -PENNYWORT, 254
  - FLOATING, 254, 255
  - WHORLED, 255
- -PEPPER, 188, 902
- PIGMYWEED, 562
- PIGNUT, 747
- -PIMPERNEL, 914
- -PLANTAIN, 1086
  - SMALL-FLOWER, 1086
- POD, 740
- -PRIMROSE, 858, 860
  - ANGLE-STEM, 860
  - CREEPING, 862
  - FLOATING, 862
  - NARROW-LEAF, 860
  - SHRUBBY, 860
  - SMOOTH, 860
- SHED, SCHREBER'S, 482
- -SHIELD, 482
- SMARTWEED, 902, 904
- SPEEDWELL, 1012
- SPIDER ORCHID, 1215
- STAR-GRASS, 1340
- STARWORT, 494
- -THREAD PONDWEED, 1343
- -WALLY, 324
- WEED, 1168
- WHITE OAK, 716
- -WILLOW, 213, 324
  - AMERICAN, 213
  - LANCE-LEAF, 213

WATER FERN FAMILY, 180
WATER STARWORT FAMILY, 494
WATER WEED FAMILY, 1166
WATER-CLOVER FAMILY, 185
WATER-HYACINTH FAMILY, 1338
WATER-LILY FAMILY, 842
WATER-MILFOIL FAMILY, 733
WATER-PLANTAIN FAMILY, 1086
WATER-SHIELD FAMILY, 480
WATERLEAF FAMILY, 739
WATERNUT, BITTER, 748
WATERWORT, 494, 580
- AMERICAN, 581
- ANNUAL, 496
- COMMON, 496
- LARGER, 496
- MAT, 496
- NUTTALL'S, 496
- SHORT-SEED, 580

WATERWORT FAMILY, 580
WAVEWING, 249
- BIG-ROOT, 249

WAVY
- CLOAK FERN, 195
- -LEAF
  - GAURA, 856
  - MENTZELIA, 796
  - MILKWEEDVINE, 283
  - THISTLE, 340
  - TWINEVINE, 283

WAX
- PLANT, 274
- -LEAF
  - LIGUSTRUM, 850
  - PRIVET, 850
- -MALLOW, DRUMMOND'S, 815
- -MYRTLE, 832
  - SOUTHERN, 832

WAX-MYRTLE FAMILY, 832
WAXBERRY, 832
WAYSIDE SPEEDWELL, 1012
WEAK
- BUTTERCUP, 928
- CARIC SEDGE, 1120
- EUPHORBIA, 610

Wedelia, 430
- *acapulcensis* var. *hispida*, 430
- *hispida*, 430
- texana, 430

WEDELIA, HAIRY, 430
WEDGE
- GRASS, 1324
- -LEAF
  - DRABA, 468
  - EUPHORBIA, 608
  - FROGFRUIT, 1054

WEDGESCALE, 1324
- PRAIRIE, 1324

WEEDY DWARF-DANDELION, 380
WEEPING
- LOVE GRASS, 1270
- WILLOW, 978
  - BABYLON, 978

Weigela, 507
WEIGHTY FOGFRUIT, 1054
WELK HOLLY, 270
WEN-DOCK, PURPLE, 482
WEST INDIAN
- ARROWROOT, 1210
- COTTON, 808
- GHERKIN, 566
- LANTANA, 1052

WEST TEXAS LIVE OAK, 716
WESTERN
- BEEBALM, 766
- BITTERWEED, 375
- BLACK WILLOW, 978
- BRACKEN FERN, 181
- BUCKEYE, 738
- COMANDRA, 979
- DAISY, 322
- DWARF-DANDELION, 380
- HORSE-NETTLE, 1028
- INDIGO, 664
- IRONWEED, 428
- MARBLESEED, 454
- PINK
  - VERBENA, 1050
  - VERVAIN, 1050
- RAGWEED, 310
- ROCK-JASMINE, 912
- SAGEWORT, 315
- SALSIFY, 424
- SCARLET-PEA, 664
- SOAPBERRY, 980
- SPIDER-LILY, 1204
- THREEAWN, 1242
- TRIPLEAWN, 1242
- UMBRELLA-GRASS, 1152
- WALLFLOWER, 468
- WHEAT GRASS, 1302, 1304
- WILD LETTUCE, 381
- WITCH GRASS, 1263
- YARROW, 308

WHEAT, 1235, 1332, 1334
- BLACK STEM RUST OF, 437, 1334
- BREAD, 1334
- GRASS
  - BLUESTEM, 1304
  - WESTERN, 1302, 1304
- INCA, 222
- RIVET, 1334

WHIP
- NUT-RUSH, 1164
- -GRASS, 1164

WHITE
- AMARANTH, 224
- ASH, 849
  - TEXAS, 849
- ASTER, 334
  - TALL, 318
- AVENS, 944
- -BALL ACACIA, 624
- BLADDERPOD, 474
- BLUE-EYED-GRASS, 1176

WHITE *(cont.)*
BOLTONIA, 328
BUCKEYE, 738
-BUTTONS, 1166
-CEDAR, 823
CLOVER, 706
COMPASSPLANT, 404
CROWNBEARD, 427
DAISY, 382
DOG-TOOTH-VIOLET, 1201
ELM, 1039
EVENING-PRIMROSE, 866
FAWN-LILY, 1201
-FLOWER
BOTTLE GOURD, 569
MILKWEED, 280
FOUR-O'CLOCK, 839
GAURA, 856
GRASS, 1281
-HAIRED
PANIC, 1295
ROSETTE GRASS, 1295
-HEART HICKORY, 747
HICKORY, 747
HOLLY, 270
HONEYSUCKLE, 508
HOREHOUND, 762
HORSE-NETTLE, 1030
LACE CACTUS, 486
LARKSPUR, 924
-LEAF MOUNTAIN MINT, 772
LOCO, 682
-MARGIN EUPHORBIA, 592
MELILOT, 676
MILKWORT, 896
MORNING-GLORY, 556
MOUNTAIN MINT, 772
MULBERRY, 831
MUSTARD, 478
OAK, 714, 718
DURAND, 718
FORKED-LEAF, 714
RIDGE, 714
SWAMP, 716
WATER, 716
PINE BLISTER RUST, 732
POPLAR, 975
-ROOT RUSH, 1182
PRAIRIE
ASTER, 318
ROSE, 956
-CLOVER, 650
PRICKLY-POPPY, 873
RAIN-LILY, 1200
ROSINWEED, 404
SNAKEROOT, 354
SPIDER-LILY, 1202
SWEET-CLOVER, 676
-TOP
SEDGE, SNOWY, 1160
SEDGE, STARRUSH, 1158
UMBRELLA-GRASS, 1158
TRIDENS, 1330
TWINEVINE, 283
VERVAIN, 1058
WATER-LILY, 845
WAX TREE, 850
WOODLAND ASTER, 319
WHITEBRUSH, 1049
WHITEMAN'S-FOOT, 886
WHITETOP, 352, 1330
WHITEWEED, 382, 1030
WHITISH YELLOW CARIC SEDGE, 1118
WHITLOW
-GRASS, 466
-WORT, 468, 519
WHORLED
DROPSEED, 1328
MILKWEED, 280
EASTERN, 280
MILKWORT, 896
NOD-VIOLET, 1061
NUT-RUSH, 1164
SILENE, 524
WATER-PENNYWORT, 255
WHORTLE-BERRY, 584
WIDE-LEAF
FALSE ALOE, 1079
RAIN-LILY, 1200
WIDESPREAD MAIDEN FERN, 200
WIDOW'S
-CROSS, 564
-FRILL, 524
-TEARS, 1098, 1100, 1101
WIGTREE, 231
WILD
ALFALFA, 690
ALLSPICE, 784
BEAN, 697
TRAILING, 698
BERGAMOT, 765
BLACKCHERRY, 950
BLUE
LARKSPUR, 924
-INDIGO, 636
BOSTON FERN, 184
BUCKWHEAT, 898, 902
CANARY GRASS, 1310
CARROT, 250
CAULIFLOWER, 374
CELERY, AMERICAN, 1170
CHERVIL, 246, 249
CHINABERRY, 980
COLUMBINE, 918
COMFREY, 446
COTTON, 810
DILL, 256
FOUR-O'CLOCK, 840
FOXGLOVE, 1007
GARLIC, 1196
GOLDENGLOW, 326
HONEYSUCKLE, 855, 856
HYACINTH, 1198, 1200
INDIGO, 635
BLUE, 636
BUSH'S, 638
GREEN, 638
NUTTALL'S, 638
PLAINS, 638
TWO-COLOR, 636
VARICOLORED, 638
JALAP, 437
LAVENDER, 1060
LEEK, 1196
LETTUCE, 381
WESTERN, 381
LICORICE, 964
MERCURY, 600
HILL COUNTRY, 602
LOW, 602
SMOOTH, 602
TALL, 604
MORNING-GLORY, 556
MUSTARD, 472
OATS, 1244, 1254
ONION, 1196, 1198
PANSY, 1062
PEPPER, 1060
PETUNIA, 213, 1017
PLUM, 949
POINSETTIA, 606
POTATO, 556
POTATO-VINE, 556
RHUBARB, 906
RICE, SOUTHERN, 1338
RYE, 1268
CANADA, 1268
NODDING, 1268
VIRGINIA, 1268
SARSAPARILLA, 824, 1346
SENNA, 694
STRAWBERRY, 944
SWEET-POTATO, 556
TARO, 1093
TOBACCO, 1022
VERBENA, 1052
VERVAIN, 1050
YAM, 1165

WILD PETUNIA FAMILY, 210
WILDGOOSE PLUM, 950
WILL ASTER, 320
WILLIAM EMORY'S CARIC SEDGE, 1120
Willkommia, 1336
  texana, 1336
WILLKOMMIA, TEXAS, 1336
WILLOW, 976
  AMERICAN WATER-, 213
  BACCHARIS, 324
  BASKET, 978
  BLACK, 978
  BOW-, 443
  CAROLINA, 978
  COASTAL PLAIN, 978
  COYOTE, 978
  CREEPING PRIMROSE-, 862
  DESERT-, 443
  DOGBANE, 266
  FLOATING PRIMROSE-, 862
  FLOWERING-, 443
  GOLDENROD, 410, 412
  -HERB GAURA, 856
  LANCE-LEAF WATER-, 213
  -LEAF
    ASTER, 320
    CATALPA, 443
    LETTUCE, 381
    OAK, 718
    SUNFLOWER, 370
  LINDHEIMER'S BLACK, 978
  LONG
    -LEAF, 978
    -PEDICEL, 978
  NARROW-LEAF, 978
  OAK, 718
    SWAMP, 718
    UPLAND, 716
  OSIER, 978
  PRIMROSE-, 860
  PUSSY, 978
  RIVERBANK, 978
  SANDBAR, 978
  SCYTHE-LEAF, 978
  SEEP-, 324
  SHRUB, 978
  SLIMPOD, 266
  SMARTWEED, 902
  SOUTHERN, 978
  SWAMP, 978
  TEXAS SANDBAR, 978
  UPRIGHT PRIMROSE-, 860
  WARD'S, 978
  WATER-, 213, 324
  WEEPING, 978
  WESTERN BLACK, 978
WILLOW FAMILY, 974
WILMANN'S LOVE GRASS, 1272
WILT, OAK, 711
WINDFLOWER, 917
WINDMILL
  FINGER GRASS, 1256
  GRASS, 1256
    HOODED, 1256
    SHORT-SPIKE, 1256
    TUMBLE, 1256
WINDOWBOX WOODSORREL, 870
WINECUP, 807, 808
  LOW, 807
  PLAINS, 807
  TALL, 808
WINGED
  ELM, 1039
  LOOSESTRIFE, 802
  -PIGWEED, 540
WING
  -POD PORTULACA, 910
  -RIB SUMAC, 232
WINGSTEM, 426
WINTER
  BENT GRASS, 1236
  CROOK-NECK SQUASH, 569
  GRAPE, 1072
    FERN, 189
  GRASS, 1290
    TEXAS, 1290
  JASMINE, 849
  SQUASH, 569
  VETCH, 708
WINTERBERRY, 270
WIRE-STEM MUHLY, 1288
WIREWEED, 320
Wisteria, 708
  floribunda, 710
  fructescens, 710
  macrostachya, 710
  sinensis, 710
WISTERIA
  CHINESE, 710
  JAPANESE, 710
WISTER'S CORALROOT, 1215
WITCH GRASS, 1296
  PHILADELPHIA, 1302
  SPREADING, 1296
  WOOD, 1302
WITCHES'
  BROOM, 1038
  -SHOELACES, 572
WITCHGRASS
  FALL, 1263
  WESTERN, 1263
WITCH-HAZEL, 736
WITCH-HAZEL FAMILY, 736
WITCHWEED, 989
WITLOOF, 336
WOLFBERRY, 1020
Wolffia, 1190
  brasiliensis, 1190
  columbiana, 1191
  papulifera, 1190
  punctata, 1190
WOLFFIA
  COLOMBIA, 1191
  COMMON, 1191
  DOTTED, 1190
  POINTED, 1190
Wolffiella, 1191
  floridana, 1191
  gladiata, 1191
  lingulata, 1191
WOLFSBANE, 916
WOOD
  ARROW-, 511
  -BETONY, 1006
  -OATS, 1254
    BROAD-LEAF, 1254
    NARROW-LEAF, 1256
  -SAGE, 782
  VETCH, 708
  VIOLET, 1062
  WITCH GRASS, 1302
WOOD FERN FAMILY, 182
WOODBINE, 1066, 1068
WOODLAND
  AGRIMONY, 938
  ASTER, WHITE, 319
  BLUE GRASS, 1314
  GROOVEBUR, 938
  LETTUCE, 381
WOODRUFF, 961
WOODRUSH, 1186
  BULB, 1186
WOODS BEDSTRAW, 964
WOOD'S CORNSALAD, 1046
Woodsia, 185
  obtusa, 185
    subsp. obtusa, 185
    subsp. occidentalis, 185
WOODSIA
  BLUNT-LOBED, 185
  COMMON, 185
  LARGE, 185
WOODSORREL, 869
  GRAY-GREEN, 870
  VIOLET, 872
  WINDOWBOX, 870
WOODSORREL FAMILY, 869

Woodwardia, 180
areolata, 181
virginica, 181
WOOLLY
AMERICAN HOP-HORNBEAM, 440
-BUCKTHORN, 982
COTTON-FLOWER, 228
CROTON, 599
DALEA, 648
DUTCHMAN'S-PIPE, 274
GAURA, 858
-GRASS BULRUSH, 1162
GRASS, 1274
HOPTREE, 973
IRONWEED, 428
LIP FERN, 196, 198
MORNING-GLORY, 556
PANIC, 1295
-POD VETCH, 708
ROSE-MALLOW, 810
ROSETTE GRASS, 1295
-SHEATH THREEAWN, 1242
THREEAWN, 1242
TIDESTROMIA, 230
TRIPLEAWN GRASS, 1242
VERVAIN, 1058
-WHITE, 374, 375
CHALKHILL, 375
RAGWEED, 374
WORM-GRASS, 798
WORM-SEED, 535
LAMB'S-QUARTERS, 535
WORMWOOD, 314
ROMAN, 309
WREATH ASTER, 318
WRIGHT'S
ACACIA, 625
ANISACANTHUS, 212
CLIFF-BRAKE, 200
EVENING-PRIMROSE, 866
FALSE MALLOW, 814
MILK-VETCH, 635
MORNING-GLORY, 558
PAVONIA, 816
PLANTAIN, 886
SKULLCAP, 780
THREEAWN, 1243
TICK-CLOVER, 656
TRIPLEAWN GRASS, 1243
WRINKLE-FRUIT CARIC SEDGE, 1120
WRINKLED GOLDENROD, 414

X

XALAPA SPEEDWELL, 1014
Xanthisma, 430
texanum
subsp. drummondii, 430
var. *drummondii*, 432
Xanthium, 432
italicum, 432
spinosum, 432
strumarium var. canadense, 432
Xanthocephalum, 362
amoenum, 364
var. *intermedium*, 364
dracunculoides, 364
microcephalum, 364
sarothrae, 365
sphaerocephalum, 365
texanum, 365
Xanthosoma, 1093
sagittifolium, 1094
Xanthoxylum, 973
Ximenesia encelioides, 427
Xyridaceae, 1349
Xyris, 1349
ambigua, 1349
baldwiniana, 1349
jupicai, 1349

Y

YAM, 555, 1164
ATLANTIC, 1165
WILD, 1165
YAM FAMILY, 1164
YARD GRASS, 1268
YARROW, 307
COMMON, 308
WESTERN, 308
YAUPON, 270
HOLLY, 270
YAUTIA, 1093
YELLOW
BAMBOO, 1311
BRISTLE GRASS, 1319
BUCKEYE
PLATEAU, 738
TEXAS, 738
BUSH-PEA, 638
COW-LILY, 844
DAISY, 402
PLAINS, 420
EVENING-PRIMROSE, 854
-EYED-GRASS, 1349
FALSE GARLIC, 1208
FINGER ORCHID, 1218
FLAG, 1174
FLAT SEDGE, 1138
FLOATING-HEART, 826
-FLOWERED HORSE-GENTIAN, 511
FRINGED ORCHID, 1218
HORSE MINT, 766
INDIAN
GRASS, 1322
-BLANKET, 358
JESSAMINE, 797, 798
LOTUS, 834
MELILOT, 676
NELUMBO, 834
NEPTUNIA, 680
NUT-GRASS, 1138
OAK, 718
PAINTBRUSH, 998
PAPER-FLOWER, 420
PARILLA, 824
PASSION-FLOWER, 878
PINE
LONGLEAF, 207
SHORTLEAF, 206
POND-LILY, 844
PRICKLY-POPPY, 874
-PUFF, 680
-SEED HEDGE-HYSSOP, 1000
SLASH PINE, 207
SOUR-CLOVER, 676
-SPINE THISTLE, 340
STAR
-GRASS, 1194, 1204
-THISTLE, 334
STONECROP, 564
-STRAWBERRY, 944
SWEET-CLOVER, 676
TEXAS-STAR, 386
THISTLE, 340
TREFOIL, 674
TRUMPET, 983
VIOLET, 1062
WATER-LILY, 845
YELLOW-EYED-GRASS FAMILY, 1349
YELLOWCRESS, 476
BOG, 478
STALK-LESS, 478
TANSY-LEAF, 478
YELLOWSEDGE BLUESTEM, 1239
YELLOWTOP, 390, 392
YELLOWWOOD, 933
YERBA
-DE-LA-RABIA, 836
DE LA RABIA, 589
DE TAGO, 347
DE VIBORA, 365
DEL CHRISTO, 1052
MANSA, 984
MATE, 269
YEW
CALIFORNIA, 201
ENGLISH, 201
EUROPEAN, 201

YEW *(cont.)*
PACIFIC, 201
Youngia, 433
japonica, 433
YOUQUEPEN, 834
YOUTH-AND-OLD-AGE, 433
Yucca, 1082
aloifolia, 1082
arkansana, 1083
baccata var. *macrocarpa,* 1084
constricta, 1083
crassifolia, 1084
filamentosa, 1082
flaccida, 1082, 1084
louisianensis, 1084
macrocarpa, 1084
necopina, ▣/108, 1083
pallida, 1084
parviflora, 1079
rostrata, 1082
rupicola, 1084
thompsoniana, 1082
torreyi, 1084
treculeana, 1084
YUCCA
ARKANSAS, 1083
BUCKLEY'S, 1083
GLEN ROSE, 1083
PALE, 1084
-LEAF, 1084
RED-FLOWERED-, 1079
TEXAS, 1084
TORREY'S, 1084
TRECUL'S, 1084
TWIST-LEAF, 1084
YUCCA FAMILY, 1078
YUCCA MOTH, 1082

Z

ZACATE GUAIMA, 1268
Zannichellia, 1350
palustris, 1350
Zannichelliaceae, 1350
Zanthoxylum, 973
clava-herculis, 974
hirsutum, 974
ZARZAMORA, 959
ZARZAPARRILLA, 1346
Zea, 1338
mays, 1338
subsp. *mexicana,* 1338
mexicana, 1338
Zebrina, 1098, 1101
Zephyranthes, 1200
brazosensis, 1200
drummondii, 1200
herbertiana, 1200
texana, 1201
Zexmenia, 430
hispida, 430
ZEXMENIA, ORANGE, 430
Zigadenus, 1210
nuttallii, 1210
Zinnia, 433
elegans, 433
grandiflora, ▣/108, 433
ZINNIA
COMMON, 433
PLAINS, 433
ROCKY MT., 433
Zizaniopsis, 1338
miliacea, 1338
Zizia, 264
aurea, 264
ZIZIA, GOLDEN, 264
Ziziphus, 934
jujuba, 936
obtusifolia, 936
spina-christi, 934
zizyphus, 936
Zornia, 710
bracteata, 710
ZORNIA, BRACTED, 710
ZUCCHINI, 568
Zygophyllaceae, 1074

Colophon

Three thousand copies were printed using soy based inks via offset lithography.

This book was printed in Texas by *Millet the Printer, Inc.*— craftsmen for four generations.
Book binding was coordinated by *JP & Friends*, Arlington, Texas.
The binding cloth for the cover is *Permalin Iris 845 Evergreen Linen*
with copper foil stamping on front and spine.
Stock for the endpaper binding is acid-free, *100 lb. text Starwhite Vicksburg Archiva Wove,*
manufactured by Fox River® Paper Co., Appleton, Wisconsin.
Text pages of the book are acid-free, *Finch Fine, VHF, Bright White, Basis 50* (626 PPI),
manufactured by Finch, Pruyn & Company, Inc. of Glens Falls, New York.
Color photograph pages and dust jacket are printed on acid-free, *Potlatch McCoy™, Silk, 100 lb. text*,
manufactured by Potlatch Corporation, Cloquet, Minnesota.
Paper consultants were Clampitt Paper Company, Dallas, Texas and
Cathy Blankenbaker of xpedex (Resource Net), DFW Airport, Texas.
Archival permanence was kept in mind throughout the project.
Layout and production of the keys and treatments were by Becky Horn.
Sam Burkett, with the assistance of Amberly Zijewski, scanned the botanical line drawings
as well as managed the layout and production of the illustration pages.
The introduction and treatments are set in 9 pt. *Berkeley Oldstyle Book, Bold*, and *Black* with 12 pt. leading.
The keys, along with the book's captions and the majority of the appendices,
glossary, literature cited, and index, are set in 8 pt. *Myriad* (a multiple master typeface) with 11 pt. leading.
Visual icons (called dingbats) are used for ease of reference throughout the book.
Some of these icons were found in existing commercial typefaces: *Botanical MT* and *Warning Pi*.
Special designs were created to represent other specific "at-a-glance" icons.
These designs were digitized into a typeface which we call *Texicons*.
The majority of the book was electronically produced on the Macintosh platform
with the Adobe Illustrator®, Pagemaker®, Photoshop®, and Quarkxpress™ programs.
The PC platform was used to produce the botanical illustration pages
in the Pagemaker and Photoshop programs.

These specifications are given in the spirit of sharing within the botanical community.
We hope our efforts are well-received and will contribute to furthering an interest
in the "art & science" found throughout the botanical world.

LINNY/DESIGNER, ILLUSTRATOR

LEFT: GEORGE M. DIGGS, JR.
MIDDLE: BARNEY L. LIPSCOMB
RIGHT: ROBERT J. O'KENNON

George M. Diggs, Jr.

George M. Diggs, Jr. was born 4 February 1952 in Charlottesville, Virginia. He attended the College of William and Mary (B.S. 1974, M.A. 1976) and the University of Wisconsin-Madison (Ph.D. 1981). He has been a faculty member in the Biology Department of Austin College, Sherman, Texas, since 1981 and Chairman since 1992. In 1994, he was made a Research Associate at the Botanical Research Institute of Texas. His research interests include the flora of Texas, neotropical floristics, and the taxonomy, numerical systematics, and molecular systematics of the Ericaceae, particularly the Arbuteae. Recent publications include articles on forensic botany and the plants of Texas and systematic treatments of the genera *Arctostaphylos* and *Comarostaphylis* (Ericaceae) for the *Flora Neotropica* series. He has done field work in Africa, Australia, Central and South America, Mexico, Canada, and the U.S. Current research includes work on the *Illustrated Flora of East Texas*, to serve as a companion volume to this book. Diggs frequently takes groups of undergraduates on tropical natural history field courses to Latin America and Africa in an effort to increase knowledge of tropical ecosystems and to raise awareness of current ecological problems including the accelerating destruction of tropical forests. He is active in Austin College's Center for Environmental Studies.

Barney L. Lipscomb

Barney L. Lipscomb was born 24 October 1950 in Temple, Oklahoma. He attended Cameron University, Lawton, Oklahoma (B.S. 1973) and the University of Arkansas, Fayetteville (M.S. 1975). He began his career as herbarium botanist at Southern Methodist University in 1975. Two years later he became the assistant editor of *Sida, Contributions to Botany*. In 1983 he became editor of *Sida*, and in 1987, with Dr. William F. Mahler, founded *Sida, Botanical Miscellany*. Also in 1987, Lipscomb, Mahler, and Andrea McFadden, were instrumental in the establishment of a free-standing research institution, the Botanical Research Institute of Texas (BRIT), located in Fort Worth. Lipscomb has served at BRIT since its inception, was named Assistant Director in 1993, and is also in charge of the library and publications programs as well as serving as editor of the institute's journals. His research specialities include the flora of Texas and the genus *Cyperus* (Cyperaceae). He is the author of numerous scientific papers and is currently working on several research projects including the *Illustrated Flora of East Texas*. He has done field work in various parts of the U.S. as well as Mexico and Central America. In an effort to increase public awareness, Lipscomb frequently gives talks on plants and conservation to preschoolers, garden clubs, plant-oriented societies, high school and college groups, and civic organizations.

Robert J. O'Kennon

Captain Robert (Bob) J. O'Kennon was born 28 January 1942 in Hopewell, Virginia. He received his bachelor's degree in Business from Duke University in 1964. He served in the United States Marine Corps as a fighter pilot where he flew over 300 combat missions in Viet Nam. He has been associated with the Botanical Research Institute of Texas (BRIT) and has served on its Board of Trustees from its inception in 1987, and presently serves as Vice Chairman. O'Kennon has discovered or described more than twenty new plant species. In addition to the current volume, he is presently writing or co-authoring a number of other works including the *Flora and Natural History of Gillespie County, Texas*, the *Illustrated Flora of East Texas,* and a *Field Guide to the Hawthorns of North America*. He is, or has been, on the Boards of Directors of several other organizations involved in landscape ecology and conservation including the Natural Area Preservation Association (NAPA), the Native Prairie Association of Texas (NPAT), the Fort Worth Nature Center, the Dallas Nature Center, and the Useful Wild Plants of Texas Project. He is also a senior Captain with American Airlines where he has piloted commercial jet airliners to Europe, the Caribbean Islands, Central and South America, and throughout North America since 1973.

Authors & Institutions

BRIT

Botanical Research Institute of Texas

The Botanical Research Institute of Texas is a private, nonprofit, international botanical resource center. Its mission is to conserve our natural heritage by deepening our knowledge of the plant world and achieving public understanding of the value plants bring to life. This mission is realized through research and interpretation providing the public with a better understanding of our natural resources which leads to better stewardship of those resources. BRIT holds in trust an herbarium of nearly 860,000 dried plant specimens and a research library containing almost 70,000 items. The core of the collections are the Lloyd H. Shinners' Collection in Systematic Botany, originally at Southern Methodist University, and the Vanderbilt University Herbarium, acquired in 1997. The BRIT library has one of the finest collections of botanical literature in the United States with books dating to 1549. A special part of the collection is the Oliver G. Burk Memorial Library of children's botanical literature—a rare collection of over 2,000 children's books. BRIT is also well known for its scientific press. Two scientific journals, *Sida, Contributions to Botany* and *Sida, Botanical Miscellany*, are published by the institute as well as a quarterly newsletter, *Iridos*. Annually BRIT gives its Award of Excellence in Conservation to an individual whose life and work epitomize the ideals set forth in the BRIT mission. The institute has special programs in research, education, and public outreach. Several of the main activities of the research program include the Florula of Las Orquídeas National Park in Colombia, floristic and ecological work in Papua New Guinea and the Philippines, the Flora of Texas Project, and the Illustrated Texas Floras Project. The education and public outreach programs provide lectures, workshops, and teaching materials to local schools and other educational organizations in order to make botany come alive to students and teachers. Through the new Learning Center many hands-on botanical experiences can be provided for students of all ages.

Austin College

Austin College is a private, residential, coeducational college dedicated primarily to educating undergraduate students in the liberal arts and sciences while also offering select pre-professional programs and a graduate teacher program. The Austin College education emphasizes academic excellence, high achievement, intellectual and personal integrity, and participation in community life. Founded in 1849 by the Presbyterian Church, Austin College is the oldest college in Texas operating under its original charter. The mission of Austin College is to educate individuals in the liberal arts and sciences in order to prepare them for productive and meaningful lives in an increasingly complex world. It does this by providing diverse educational opportunities both on campus and in other countries. Approximately one-half of Austin College students spend some time abroad during their four years at the school, either during the special January Term semester or through the study abroad program. Austin College has a long history of excellence in the sciences, being particularly well known for its pre-medical program. The College also has a tradition of strength in the area of field biology and the environment, and recently the College initiated a Center for Environmental Studies. This Center promotes multidisciplinary inquiry of environmental issues and problems through education, research and outreach programs. These programs are designed to increase scientific knowledge, expand community awareness and foster greater appreciation for the interdependence of humans and other species.

Austin College
1849

VEGETATIONAL AREAS OF NORTH CENTRAL TEXAS

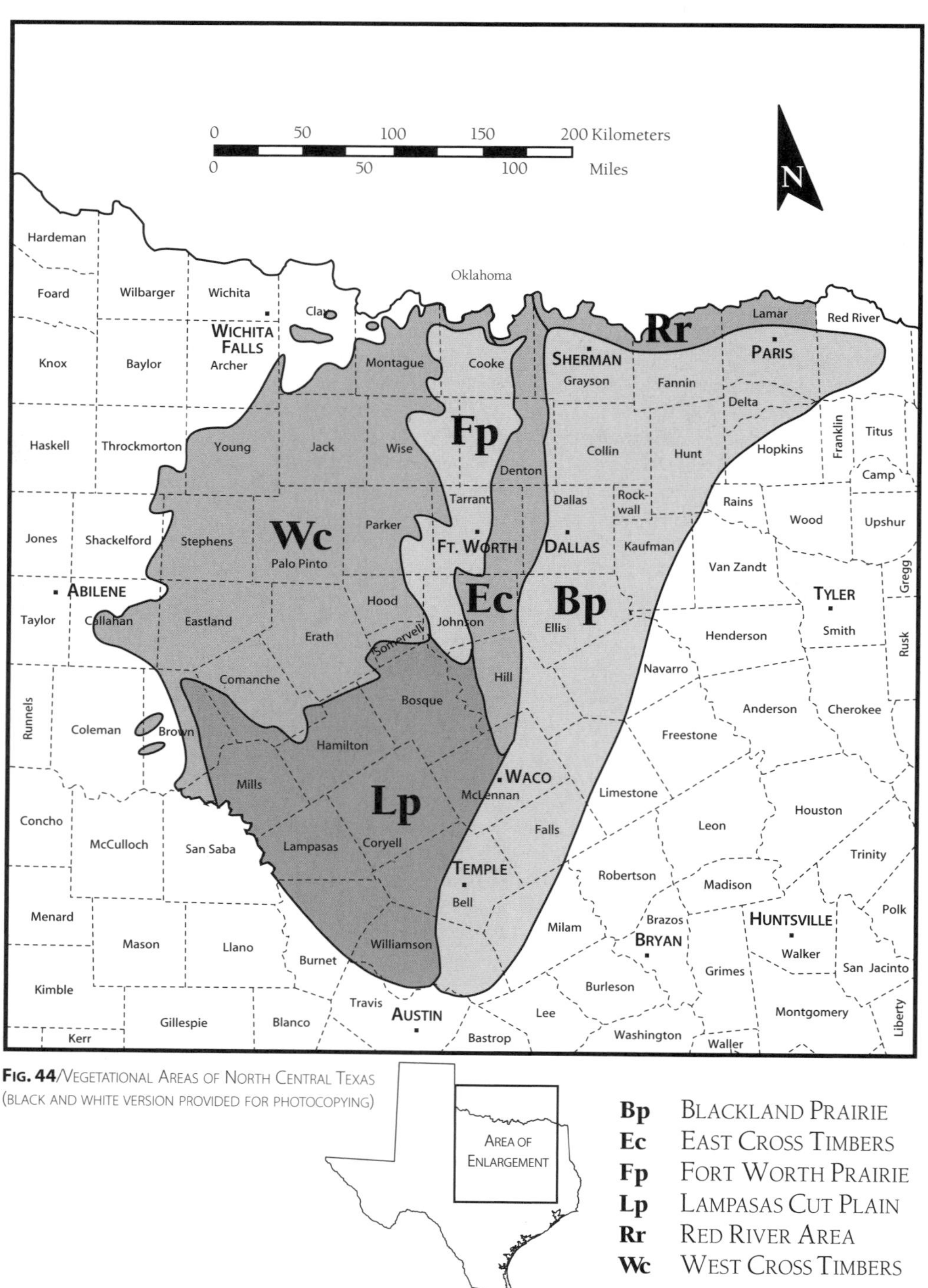

FIG. 44/VEGETATIONAL AREAS OF NORTH CENTRAL TEXAS
(BLACK AND WHITE VERSION PROVIDED FOR PHOTOCOPYING)

DUE	RETURNED	DUE	RETURNED
1.		13.	
2.		14.	
3.		15.	
4.		16.	
5.		17.	
6.		18.	
7.		19.	
8.		20.	
9.		21.	
10.		22.	
11.		23.	
12.		24.	

Alphabetical List of Families

OF VASCULAR PLANTS KNOWN TO OCCUR IN NORTH CENTRAL TEXAS

Acanthaceae 210
Aceraceae 218
Acoraceae 1077
Agavaceae 1078
Aizoaceae 220
Alismataceae 1086
Amaranthaceae 221
Anacardiaceae 230
Anemiaceae 179
Annonaceae 238
Apiaceae 239
Apocynaceae 264
Aquifoliaceae 269
Araceae 1090
Araliaceae 270
Arecaceae 1094
Aristolochiaceae 271
Asclepiadaceae 274
Aspleniaceae 179
Asteraceae 287
Azollaceae 180
Balsaminaceae 433
Basellaceae 434
Berberidaceae 436
Betulaceae 438
Bignoniaceae 440
Blechnaceae 180
Boraginaceae 443
Brassicaceae 454
Bromeliaceae 1095
Buddlejaceae 480
Burmanniaceae 1096
Cabombaceae 480
Cactaceae 482
Callitrichaceae 494
Campanulaceae 496
Cannabaceae 502
Capparaceae 504
Caprifoliaceae 507
Caryophyllaceae 511
Celastraceae 526
Ceratophyllaceae 529
Chenopodiaceae 530
Cistaceae 542
Clusiaceae 544
Commelinaceae 1096
Compositae 287
Convolvulaceae 548
Cornaceae 560
Crassulaceae 561
Cruciferae 454
Cucurbitaceae 564
Cupressaceae 202
Cuscutaceae 572
Cyperaceae 1104
Dennstaedtiaceae 181
Dioscoreaceae 1164
Dipsacaceae 574
Droseraceae 576
Dryopteridaceae 182
Ebenaceae 578
Elaeagnaceae 579
Elatinaceae 580
Ephedraceae 207
Equisetaceae 176
Ericaceae 581
Eriocaulaceae 1165
Euphorbiaceae 584
Fabaceae 617
Fagaceae 710
Fumariaceae 719
Garryaceae 722
Gentianaceae 723
Geraniaceae 728
Gramineae 1222
Grossulariaceae 732
Guttiferae 544
Haloragaceae 733
Hamamelidaceae 736
Hippocastanaceae 737
Hydrangeaceae 738
Hydrocharitaceae 1166
Hydrophyllaceae 739
Iridaceae 1172
Isoetaceae 175
Juglandaceae 744
Juncaceae 1178
Krameriaceae 750
Labiatae 752
Lamiaceae 752
Lauraceae 784
Leguminosae 617
Lemnaceae 1186
Lentibulariaceae 786
Liliaceae 1191
Linaceae 787
Loasaceae 794
Loganiaceae 797
Lycopodiaceae 173
Lythraceae 799
Malvaceae 804
Marantaceae 1210
Marsileaceae 185
Melastomataceae 820
Meliaceae 822
Menispermaceae 823
Menyanthaceae 824
Molluginaceae 826
Moraceae 827
Myricaceae 832
Nelumbonaceae 834
Nyctaginaceae 834
Nymphaeaceae 842
Nyssaceae 845
Oleaceae 846
Onagraceae 852
Ophioglossaceae 188
Orchidaceae 1212
Orobanchaceae 868
Osmundaceae 192
Oxalidaceae 869
Palmae 1094
Papaveraceae 872
Passifloraceae 877
Pedaliaceae 878
Phyrmaceae 880
Phytolaccaceae 881
Pinaceae 204
Plantaginaceae 882
Platanaceae 886
Poaceae 1222
Polemoniaceae 888
Polygalaceae 894
Polygonaceae 897
Polypodiaceae 192
Pontederiaceae 1338
Portulacaceae 906
Potamogetonaceae 1340
Primulaceae 911
Pteridaceae 193
Rafflesiaceae 914
Ranunculaceae 916
Rhamnaceae 930
Rosaceae 936
Rubiaceae 960
Rutaceae 970
Salicaceae 974
Santalaceae 979
Sapindaceae 979
Sapotaceae 982
Sarraceniaceae 983
Saururaceae 983
Saxifragaceae 986
Scrophulariaceae 988
Selaginellaceae 174
Simaroubaceae 1014
Smilacaceae 1344
Solanaceae 1015
Sphenocleaceae 1031
Sterculiaceae 1031
Styracaceae 1032
Tamaricaceae 1034
Thelypteridaceae 200
Tiliaceae 1035
Typhaceae 1348
Ulmaceae 1036
Umbelliferae 239
Urticaceae 1040
Valerianaceae 1043
Verbenaceae 1046
Violaceae 1060
Viscaceae 1064
Vitaceae 1065
Xyridaceae 1349
Zannichelliaceae 1350
Zygophyllaceae 1074